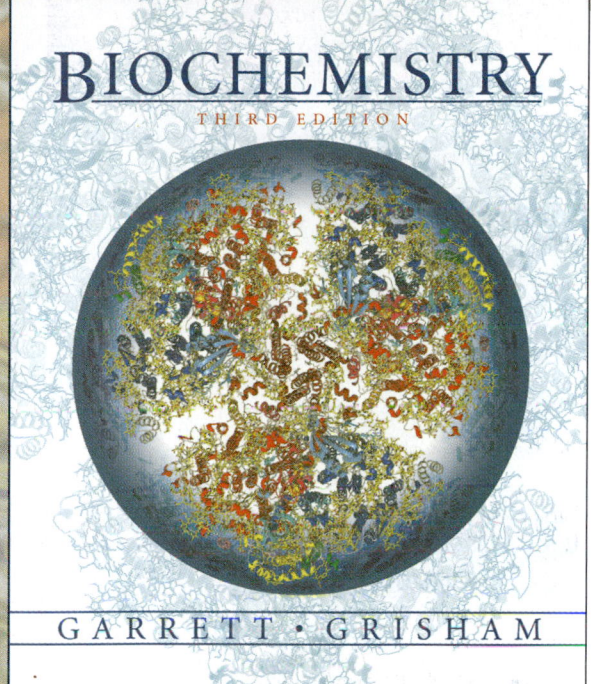

BIOCHEMISTRY
THIRD EDITION

GARRETT · GRISHAM

In biochemistry, the questions can be more revealing than the answers

Engage your students in the fascinating discovery of the molecular nature of life with Reginald H. Garrett and Charles M. Grisham's **Biochemistry, Third Edition.** This beautifully illustrated and clearly written text gives students the most current presentation of biochemistry available.

Now organized around a new inquiry-based framework, **Biochemistry, Third Edition,** emphasizes the importance of asking the essential questions that reveal the principles of biochemistry.

A unique combination of chemical and biological perspectives

Written by a chemist and a biologist, this text presents biochemistry from balanced perspectives. Authors Reginald Garrett, a professor of biology, and Charles Grisham, a professor of chemistry, make biochemistry accessible to a wide audience—one that includes students majoring in biology, chemistry, or premedical programs. The authors' unparalleled blend of chemical insight and medical relevance makes **Biochemistry, Third Edition,** an excellent reference for your course.

Your special Preview of Biochemistry, Third Edition, begins on the next page ➤

THOMSON
—✦—
BROOKS/COLE

Biochemistry ✪ Now™
A powerful online learning system—free access included!
See pages 6 and 7 for details

Preview

1

Essential questions for essential understanding

This third edition offers a unique conceptual and organizing framework that is built around **Essential Questions.** New to this edition, this focused approach guides students through each chapter by using section head questions, supporting concept statements, and summaries—and is enhanced by outstanding text and media integration through **BiochemistryNow™.**

▶ Each chapter opens with an **Essential Question** that highlights the overarching theme of the upcoming material.

> *"I like the use of the Essential Questions and Key Questions at the beginning of each chapter as they focus the students and provide a study guide of learning objectives."*
>
> **Martyn Gunn,**
> Texas A&M University

▶ Also new to this edition are chapter **Summaries.** In keeping with the **Essential Questions** framework, each **Summary** is arranged around the chapter's **Key Questions.** Students can view each **Key Question** and then read a brief summary of its answer.

Carbohydrates and the Glycoconjugates of Cell Surfaces

CHAPTER 7

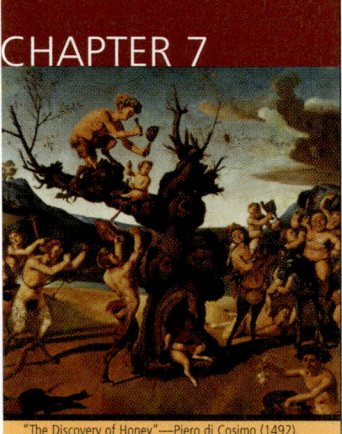

"The Discovery of Honey"—Piero di Cosimo (1492).

Sugar in the gourd and honey in the horn,
I never was so happy since the hour I
was born.
Turkey in the Straw, *stanza 6 (classic American folk tune)*

Essential Question

Carbohydrates are a versatile class of molecules of the formula $(CH_2O)_n$. They are a major form of stored energy in organisms, and they are the metabolic precursors of virtually all other biomolecules. Conjugates of carbohydrates with proteins and lipids perform a variety of functions, including the recognition events that are important in cell growth, transformation, and other processes. *What is the structure, chemistry, and biological function of carbohydrates?*

Carbohydrates are the single most abundant class of organic molecules found in nature. The name *carbohydrate* arises from the basic molecular formula $(CH_2O)_n$, which can be rewritten $(C \cdot H_2O)_n$ to show that these substances are hydrates of carbon, where $n = 3$ or more. Carbohydrates constitute a versatile class of molecules. Energy from the sun captured by green plants, algae, and some bacteria during photosynthesis (see Chapter 21) is stored in the form of carbohydrates. In turn, carbohydrates are the metabolic precursors of virtually all other biomolecules. Breakdown of carbohydrates provides the energy that sustains animal life. In addition, carbohydrates are covalently linked with a variety of other molecules. Carbohydrates linked to lipid molecules, or **glycolipids,** are common components of biological membranes. Proteins that have covalently linked carbohydrates are called **glycoproteins.** These two classes of biomolecules, together called **glycoconjugates,** are important components of cell walls and extracellular structures in plants, animals, and bacteria. In addition to the structural roles such molecules play, they also serve in a variety of processes involving *recognition* between cell types or recognition of cellular structures by other molecules. Recognition events are important in normal cell growth, fertilization, transformation of cells, and other processes.

All of these functions are made possible by the characteristic chemical features of carbohydrates: (1) the existence of at least one and often two or more asymmetric centers, (2) the ability to exist either in linear or ring structures, (3) the capacity to form polymeric structures via *glycosidic* bonds, and (4) the potential to form multiple hydrogen bonds with water or other molecules in their environment.

Key Questions

7.1	How Are Carbohydrates Named?
7.2	What Is the Structure and Chemistry of Monosaccharides?
7.3	What Is the Structure and Chemistry of Oligosaccharides?
7.4	What Is the Structure and Chemistry of Polysaccharides?
7.5	What Are Glycoproteins, and How Do They Function in Cells?
7.6	How Do Proteoglycans Modulate Processes in Cells and Organisms?

◀ **Key Questions** appear at the beginning of each chapter, giving students a preview of what's to come—and what they need to focus on.

◀ The **Key Questions** are repeated as section heads within the chapter.

7.1 How Are Carbohydrates Named?

Carbohydrates are generally classified into three groups: **monosaccharides** (and their derivatives), **oligosaccharides,** and **polysaccharides.** The monosaccharides are also called **simple sugars** and have the formula $(CH_2O)_n$. Monosaccharides cannot be broken down into smaller sugars under mild conditions. Oligosaccharides derive their name from the Greek word *oligo,* meaning "few," and consist of from two to ten simple sugar molecules. Disaccharides are common in nature, and trisaccharides also occur frequently. Four- to six-sugar-unit oligosaccharides are usually bound covalently to other molecules, including glycoproteins. As their name suggests, polysaccharides are polymers of the simple sugars and their derivatives. They may be either linear or branched

Summary

Carbohydrates are a versatile class of molecules of the formula $(CH_2O)_n$. They are a major form of stored energy in organisms, and they are the metabolic precursors of virtually all other biomolecules. Carbohydrates linked to lipids (glycolipids) are components of biological membranes. Carbohydrates linked to proteins (glycoproteins) are important components of cell membranes and function in recognition between cell types and recognition of cells by other molecules. Recognition events are important in cell growth, differentiation, fertilization, tissue formation, transformation of cells, and other processes.

7.1 How Are Carbohydrates Named? Carbohydrates are classified into three groups: monosaccharides, oligosaccharides, and polysaccharides. Monosaccharides cannot be broken down into smaller sugars under mild conditions. Oligosaccharides consist of from two to ten simple sugar molecules. Polysaccharides are polymers of simple sugars and their derivatives and may be branched or linear. Their molecular weights range up to 1 million or more.

7.2 What Is the Structure and Chemistry of Monosaccharides? Monosaccharides consist typically of three to seven carbon atoms and are described as either aldoses or ketoses. Aldoses with at least three carbons and ketoses with at least four carbons contain chiral centers. The prefixes D- and L- are often used to indicate the configuration of the highest numbered asymmetric carbon. The D- and L-forms of a monosaccharide are mirror images of each other, called enantiomers. Pairs of isomers that have opposite configurations at one or more chiral centers, but are not mirror images of each other, are called diastereomers. Sugars that differ in configuration at only one chiral center are epimers. An interesting feature of carbohydrates is their ability to form cyclic structures with formation of an additional asymmetric center. Aldoses and ketoses with five or more carbons can form either furanose or pyranose rings, and the more stable form depends on structural factors. A variety of chemical and enzymatic reactions produce derivatives of simple sugars, such as sugar acids, sugar alcohols, deoxy sugars, sugar esters, amino sugars, acetals, ketals, and glycosides.

7.3 What Is the Structure and Chemistry of Oligosaccharides? The complex array of oligosaccharides in higher organisms is formed from relatively few different monosaccharide units, particularly glucose, fructose, mannose, galactose, ribose, and xylose. Disaccharides consist of two monosaccharide units linked by a glycosidic bond, and each individual unit is termed a residue. The most common disaccharides in nature are sucrose, maltose, and lactose. The anomeric carbons of oligosaccharides may be substituted or unsubsti-

Preview

► The **Key Questions** from the text form the basis of the diagnostic **Pre-Test** questions found at **BiochemistryNow™**, effectively linking text and media like no other program available.

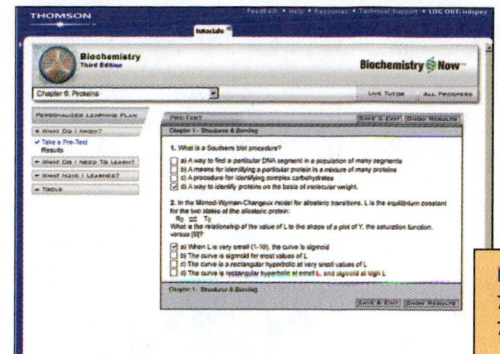

The first Web-based assessment-centered learning tool for the biochemistry course, **BiochemistryNow™** was developed in concert with the text, extending the **Essential Questions** framework. Throughout each chapter, icons with captions alert students to media resources that enhance problem-solving skills and improve conceptual understanding. See page 6 for more information about this seamless learning system.

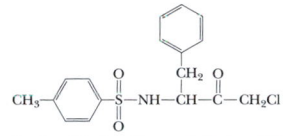

◄ Each chapter concludes with a comprehensive set of **Problems**—the total number of which has been increased by 50% in this edition. With a mix of conceptual and quantitative questions, the **Problems** in each chapter are the perfect complement to the text's **Essential Questions** framework. Also included are select problems to help students prepare for the MCAT exam.

---

472   **Chapter 14**   Mechanisms of Enzyme Action

Problems

1. Tosyl-L-phenylalanine chloromethyl ketone (TPCK) specifically inhibits chymotrypsin by covalently labeling His[57].

Tosyl-L-phenylalanine chloromethyl ketone (TPCK)

a. Propose a mechanism for the inactivation reaction, indicating the structure of the product(s).
b. State why this inhibitor is specific for chymotrypsin.
c. Propose a reagent based on the structure of TPCK that might be an effective inhibitor of trypsin.

2. In this chapter, the experiment in which Craik and Rutter replaced Asp[102] with Asn in trypsin (reducing activity 10,000-fold) was discussed.
a. On the basis of your knowledge of the catalytic triad structure in trypsin, suggest a structure for the "uncatalytic triad" of Asn-His-Ser in this mutant enzyme.
b. Explain why the structure you have proposed explains the reduced activity of the mutant trypsin.
c. See the original journal articles (Sprang, et al., 1987. *Science* **237**:905–909 and Craik, et al., 1987. *Science* **237**:909–913) to see what Craik and Rutter's answer to this question was.

3. Pepstatin (see below) is an extremely potent inhibitor of the monomeric aspartic proteases, with K_i values of less than 1 n*M*.
a. On the basis of the structure of pepstatin, suggest an explanation for the strongly inhibitory properties of this peptide.
b. Would pepstatin be expected to also inhibit the HIV-1 protease? Explain your answer.

4. The k_{cat} for alkaline phosphatase–catalyzed hydrolysis of methylphosphate is approximately 14/sec at pH 8 and 25°C. The rate constant for the uncatalyzed hydrolysis of methylphosphate under the same conditions is approximately 1×10^{-15}/sec. What is the difference in the free energies of activation of these two reactions?

5. Active α-chymotrypsin is produced from chymotrypsinogen, an inactive precursor, as shown in the color figure below. The first intermediate—π-chymotrypsin—displays chymotrypsin activity. Suggest proteolytic enzymes that might carry out these cleavage reactions effectively.

6. Based on the following reaction scheme, derive an expression for k_e/k_u, the ratio of the rate constants for the catalyzed and uncatalyzed reactions, respectively, in terms of the free energies of

Pepstatin

Iva   Val   Val   Sta   Ala   Sta

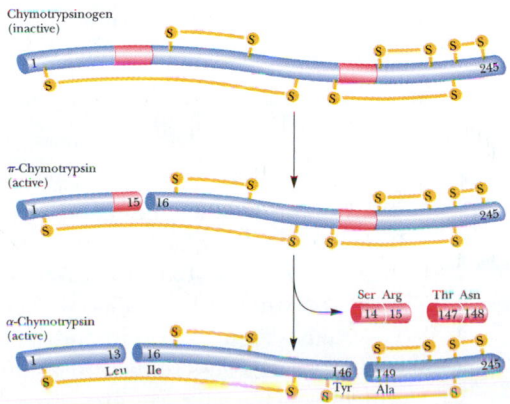

Powerful visuals reveal micro worlds

The text's highly praised art program tells the whole story of biochemistry and provides students with powerful visual aids. What's more, many of the figures found in the text are animated on the **BiochemistryNow**™ Web site.

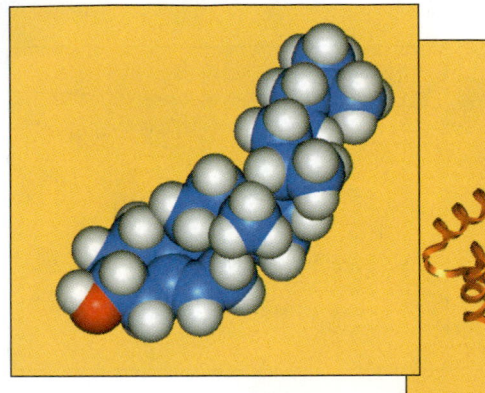

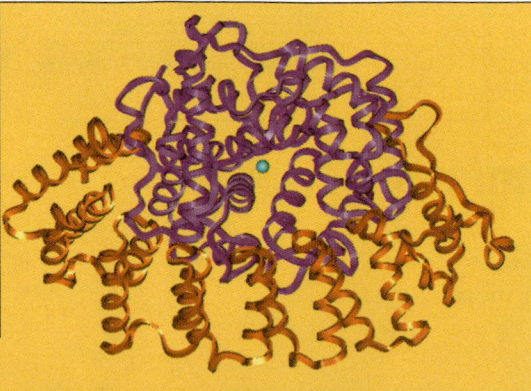

◀ Four-color molecular graphics developed specifically for the book give three dimensions to many complex biomolecules. A gallery of colorful ribbon diagrams and space-filling computer graphics are used throughout the text to reveal the biological structure of many of these biomolecules.

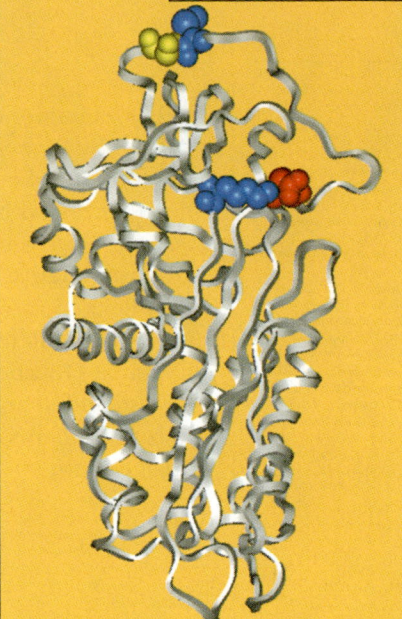

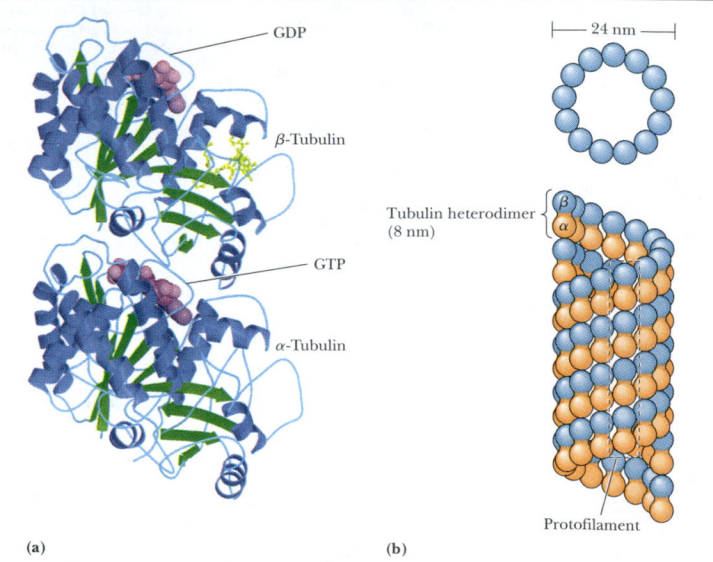

FIGURE 16.2 **(a)** The structure of the tubulin αβ-heterodimer. **(b)** Microtubules may be viewed as consisting of 13 parallel, staggered protofilaments of alternating α-tubulin and β-tubulin subunits. The sequences of the α- and β-subunits of tubulin are homologous, and the αβ-tubulin dimers are quite stable if Ca^{2+} is present. The dimer is dissociated only by strong denaturing agents.

Labels in figure: GDP, β-Tubulin, GTP, α-Tubulin, 24 nm, Tubulin heterodimer (8 nm), β, α, Protofilament, (a), (b)

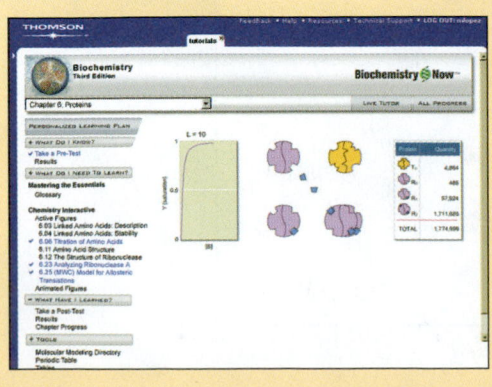

◀ Animations of key concepts are found on **BiochemistryNow**™ as **Active Figures.** Taken straight from the text, these **Active Figures** help students visualize key concepts from the book. The **Active Figures** also include questions so that students can assess their understanding of the concepts. These figures are also available for in-class demonstrations. See page 6 for more information about **BiochemistryNow**™ and the **Active Figures.**

Preview

Keeping pace with the science

Biochemistry, Third Edition, features the most current information available, including the latest on protein folding, proteomics, the Human Genome Project, cellular signaling, and more. A number of updated text boxes draw students into a greater appreciation for the relevance, logic, and historical context of significant biochemical advances.

Critical Developments in Biochemistry

Biochemistry Now™

Green Fluorescent Protein—The "Light Fantastic" from Jellyfish to Gene Expression

Aquorea victoria, a species of jellyfish found in the northwest Pacific Ocean, contains a **green fluorescent protein (GFP)** that works together with another protein, **aequorin**, to provide a defense mechanism for the jellyfish. When the jellyfish is attacked or shaken, aequorin produces a blue light. This light energy is captured by GFP, which then emits a bright green flash that presumably blinds or startles the attacker. Remarkably, the fluorescence of GFP occurs without the assistance of a **prosthetic group**—a "helper molecule" that would mediate GFP's fluorescence. Instead, the light-transducing capability of GFP is the result of a reaction between three amino acids in the protein itself. As shown below, adjacent **serine, tyrosine,** and **glycine** in the sequence of the protein react to form the pigment complex—termed a **chromophore.** No enzymes are required; the reaction is autocatalytic.

Because the light-transducing talents of GFP depend only on the protein itself (upper photo, chromophore highlighted), GFP has quickly become a darling of genetic engineering laboratories. The promoter of any gene whose cellular expression is of interest can be fused to the DNA sequence coding for GFP. Telltale green fluorescence tells the researcher when this fused gene has been expressed (see lower photo and also Chapter 12).

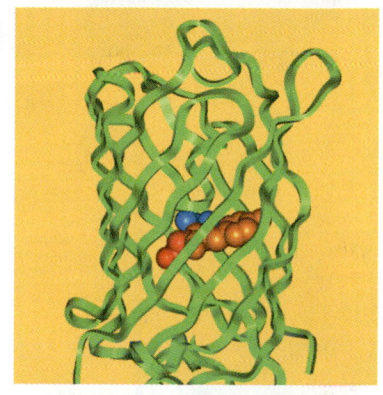

$$\text{Phe-Ser-Tyr-Gly-Val-Gln} \xrightarrow{\ O_2\ }$$
$$\underset{64}{} \qquad \underset{69}{}$$

Boxer, S.G., 1997. Another green revolution. *Nature* **383**:484–485.

◄ **Critical Developments in Biochemistry** boxes pique student interest with information on new advances and technologies. **A Deeper Look** sections explore material more in-depth, such as historical information or special interest topics. The real-world examples found in **Human Biochemistry** boxes focus student attention on topics that are more medical in nature.

▼ This edition includes new part-opening essays, each composed by a prominent scientist, in which the individual explores emerging paradigms or essential questions based on his or her own research.

How Do Cells Coordinate Their Activities?

An Essay by David L. Brautigan, University of Virginia

Despite great diversity among eukaryotic organisms, from single cell microbes like yeast to complex organisms like human beings, the biochemistry is fundamentally the same. This poses a question for continuing research: *What is common, and what is distinctive about divergent organisms?* Part 4 offers us a view of the basis for both unity and diversity.

The relative number of genes, and the conservation of what those genes encode, delivers a message of the fundamental unity among all eukaryotes. Sequence analysis of genomes reveals not a large difference in the number of genes between metazoan organisms, especially those with complex embryology and specialized tissues (see Table 1.5). Many human genes are recognizable as common to worms or fruit flies, implying that they serve the same basic life functions. Indeed, human genes can even replace the corresponding yeast genes, showing that the encoded proteins are functionally equivalent. Consider further that all organisms use the same amino acids, produce proteins and enzymes that fold into the same conserved motifs and three-dimensional conformations, and catalyze the same basic set of chemical reactions. Most of our metabolic machinery is the same (e.g., TCA cycle), yeast to human, from common molecular blueprints in our genomes. Indeed, yeast

> Animal cells have to perform together like a marching band; in contrast, a yeast cell is more like a solo musician.

and other simple organisms can act as models for understanding human diseases and for developing treatments for defects in these conserved processes. On the other hand, although the number of human genes is only about 30,000, we still do not know the function of more than 40% of them! A lot of biochemistry remains to be done.

Diversity among eukaryotic organisms appears on this background of biochemical unity. Worms and fruit flies and humans have many different types of cells. Furthermore, humans have five senses and cognitive abilities not shared by simpler eukaryotes. Without accumulating many new genes, the protein repertoire has been expanded in higher eukaryotes. For example, single gene transcripts are spliced into multiple messages, each encoding a variant protein, and single proteins are cleaved into multiple products, each tailored for a specific biological function. Diversity arises in part from subtle variations.

Specialization of cell function allows for more complex organisms, but the activities of the different cells must be coordinated with one another. Animal cells have to perform together like a marching band; in contrast, a yeast cell is more like a solo musician. To coordinate different biochemical activities, cells communicate with neighboring cells and distant tissues. Over the past 50 years, biochemists have discovered hormones and growth factors that function as chemical messengers between cells. These messengers bind to membrane receptors and trigger a series of intracellular reactions collectively referred to as hormone action, signal transduction, or cell signaling. This has been my area of research specialization. Over the past 20 years, we have deciphered many different linear reaction sequences for signal transduction; however, we are just beginning to learn how these pathways are networked together, governed by feedback loops and multiple inputs, and localized to certain sites within a cell. In my laboratory, we look for sequences conserved between species and among related proteins as keys to the common function, and test distinctive sequences for encoded specificity. Even on a molecular basis, we seek and study unity and diversity.

Also new in this edition:

► A new Chapter 31, "Completing the Protein Life Cycle: Folding, Processing, and Degradation," includes a comprehensive discussion of the structure and function of ribosomes.

► Membranes and membrane transport are now combined into Chapter 9, "Membranes and Membrane Transport," for a more cohesive coverage.

► The order of the "Mechanisms of Enzyme Action" chapter (now Chapter 14) and "Enzyme Regulation" chapter (now Chapter 15) has been switched in the Third Edition so that mechanisms more closely follow kinetics.

► Chapters in Part IV, "Information Transfer," have been combined to make for a tighter presentation and a more concise, 32-chapter book.

Biochemistry Now™   http://chemistry.brookscole.com/ggb3

Help your students take charge of their learning with **BiochemistryNow**™! This powerful online learning companion is the *first* assessment-centered system to help biochemistry students gauge their unique study needs and provide them with a *Personalized Learning Plan* that enhances their chemical problem-solving skills and conceptual understanding. **BiohemistryNow**™ gives your students the individualized resources and responsibility to *manage their concept mastery.* Access to **BiochemistryNow**™ is FREE with every new copy of Garrett and Grisham's **Biochemistry, Third Edition.**

▼ **Totally integrated with this text to enhance and extend the Essential Questions framework!** This dynamic resource and the new edition of the text were developed together, to enhance each other and provide students with a seamless, integrated learning system. As they work through the text, students will see icons

Biochemistry Now™

ACTIVE FIGURE 4.20 The separation of amino acids on a cation exchange column. **Test yourself on the concepts in this figure at http://chemistry.brookscole.com/ggb3**

that direct them to the media-enhanced activities on **BiochemistryNow**™. This precise page-by-page integration enables your students to become a part of the action—experiencing and *doing* biochemistry—not just reading about it!

Easy to use!

When students log on to **BiochemistryNow**™ at **http://chemistry.brookscole.com/ggb3,** they use the free Passcode packaged with their text. The **BiochemistryNow**™ system includes three powerful assessment components:

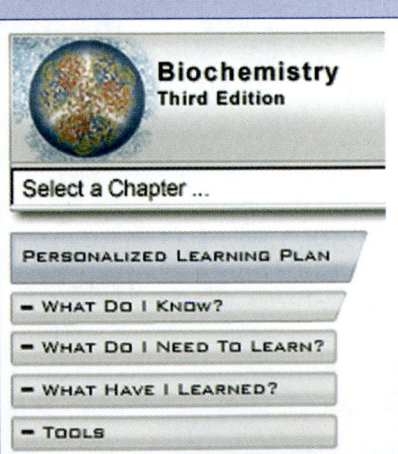

WHAT DO I KNOW? The *Pre-Test,* an initial assessment, is composed of diagnostic quiz questions.

WHAT DO I NEED TO LEARN? Based on the automatically graded *Pre-Test,* students receive a *Personalized Learning Plan* featuring components that are conceptually linked to media resources and to test pages.

WHAT HAVE I LEARNED? After working through a *Personalized Learning Plan,* students complete a *Post-Test* to assess their mastery of core chapter concepts.

WHAT DO I KNOW?

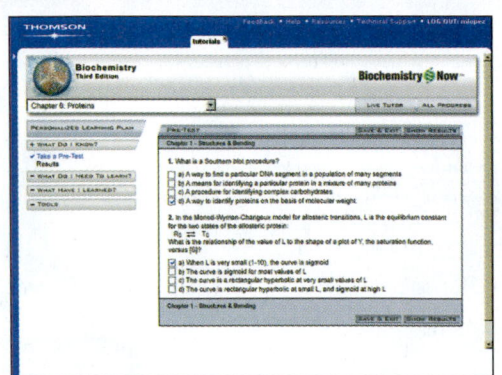

Students create a *Personalized Learning Plan* or review for an exam using the diagnostic *Pre-Test* Web quizzes. Extending the *Essential Questions* framework and based largely on the *Key Questions* found throughout each text chapter, the *Pre-Test* is composed of approximately 20 questions authored by Dr. Martyn Gunn of Texas A&M University. Every incorrect answer is accompanied by specific feedback that directs the student back to the corresponding text section for review or to additional tutorials found within the **BiochemistryNow**™ program.

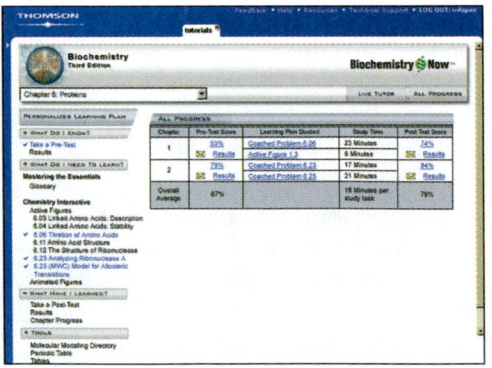

Once they've completed the *Pre-Test,* students are presented with a detailed *Personalized Learning Plan* that includes question-level references outlining the elements they need to review in order to master the chapter's most essential concepts.

WHAT DO I NEED TO LEARN?

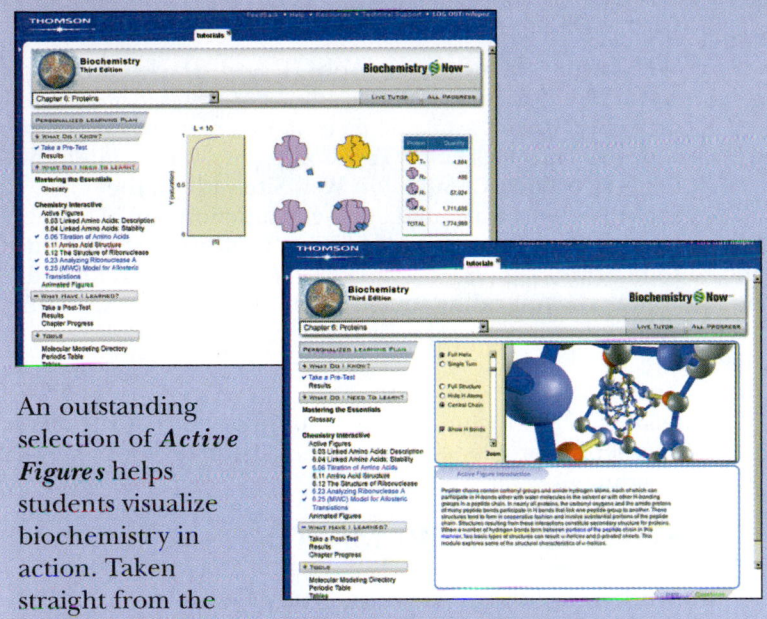

An outstanding selection of *Active Figures* helps students visualize biochemistry in action. Taken straight from the text, these *Active Figures* help students master key concepts from the book—and double as great in-class demonstrations! Each figure is paired with corresponding questions to help students focus on biochemistry at work and to ensure that they understand the concept played out in the animations. Each chapter contains approximately five *Active Figures*.

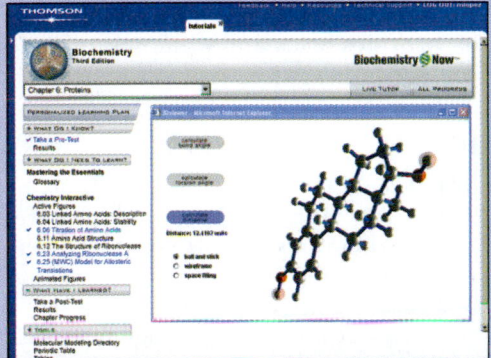

Covering every major area of biochemistry and providing a deeper understanding of challenging concepts, *BiochemistryInteractive* features Java Applets, Chime™ tutorials, and problem-solving simulations. These robust tutorials are composed of biochemical questions that guide the student through each module. Students can also manipulate parameters.

Selected in-text feature boxes—*Critical Developments in Biochemistry, A Deeper Look,* and *Human Biochemistry*—are brought to life in **BiochemistryNow**™ through the application of multimedia. Prompted by an icon, students are encouraged to explore biochemical concepts online.

WHAT HAVE I LEARNED?

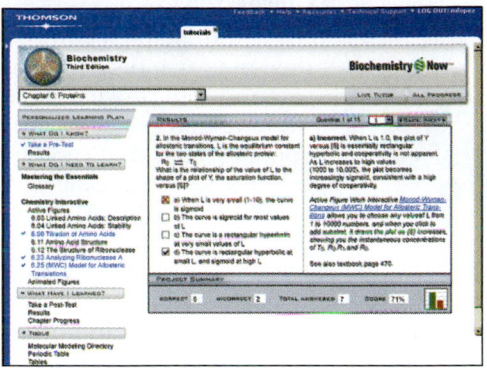

After working through the problems highlighted in their *Personalized Learning Plan,* students move on to a *Chapter Quiz* that will assess their mastery of core concepts and skills. After they've completed the quiz, students receive their results in the form of a percentage. If they need to improve their score, **BiochemistryNow**™ will encourage them to go back through the system, beginning with *What Do I Know?,* and work to build their knowledge and skills to master concepts. Quiz results can be emailed to the instructor.

Explore BiochemistryNow™ at http://chemistry.brookscole.com/ggb3

Multimedia Manager CD-ROM

0-534-49038-7. The simple way to create exciting, multimedia lectures! This easy-to-use, dual-platform digital library and presentation tool provides text art, photos, and tables in a variety of electronic formats that can be exported into other software packages. This enhanced CD-ROM also contains engaging simulations, molecular models, and QuickTime™ movies to supplement your lectures and a lecture outline with integrated media. In addition, you can customize your presentations by importing your personal lecture slides or other material you choose. The **Multimedia Manager** also includes the **Test Bank** and the *Resource Integration Guide.*

Transparency Acetates

0-534-49039-5. This set of full-color acetates includes a selection of the most pedagogically important images from the text.

Resource Integration Guide

Beginning on the next page you'll find a key teaching tool, the *Resource Integration Guide.* The guide provides a grid that links each chapter to corresponding instructional and supplemental resources. At a glance, you'll see which specific **BiochemistryNow**™ assets are appropriate for each key chapter section.

The Brooks/Cole Chemistry Resource Center

http://chemistry.brookscole.com

and **Book-Specific Instructor's Resource Web Site**

http://chemistry.brookscole.com/ggb3

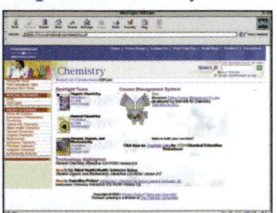

The **Brooks/Cole Chemistry Resource Center** and our password-protected **Book-Specific Instructor's Web Site** give you access to Web links, lecture outlines, a downloadable Solutions Manual, Microsoft® *PowerPoint®* Slides, and a Multimedia Manager demo. Accessible 24 hours a day, this site ensures that substantial resources for your course are never more that a mouse click away!

Student Solutions Manual/Study Guide/ Problems Book

0-534-49035-2. Instructors can use this resource for detailed solutions to all end-of-chapter problems.

ASSESSMENT AND COURSE MANAGEMENT

iLrn Computerized Testing

0-534-49040-9. This dual-platform CD-ROM features approximately 1,000 multiple-choice problems and questions, representing every chapter of the text. The questions are graded in level of difficulty for your convenience, and answers are provided on a separate grading key.

WebTUTOR™ ToolBox

WebTutor™ ToolBox on WebCT and Blackboard

WebCT: 0-534-65667-6 • Blackboard: 0-534-65658-7. Preloaded with content and available free via PIN code when packaged with this text, **WebTutor ToolBox** pairs all the content of this text's rich **Book Companion Web Site** with all the sophisticated course management functionality of a WebCT or Blackboard product. **WebTutor ToolBox** is ready to use as soon as you log on—or, you can customize its preloaded content by uploading images and other resources, adding Web links, or creating your own practice materials.

Test Bank

0-534-49037-9. Includes 25–40 multiple-choice questions per chapter for professors to use as tests, quizzes, or homework assignments.

STUDENT CONCEPT MASTERY

Student Solutions Manual/Study Guide/ Problems Book

0-534-49035-2. This comprehensive combination resource contains chapter summaries, important definitions, illustrations of major metabolic pathways, self-tests, detailed solutions to all end-of-chapter problems, and additional problems with answers.

Student Lecture Notebook

0-534-49036-0. Perfect for note taking during lecture, this convenient booklet provides black and white reproductions of the Transparency Acetates.

Biochemistry ⊜ Now™

http://chemistry.brookscole.com/ggb3

BiochemistryNow™, the first Web-based assessment-centered learning tool for the biochemistry course, was developed in concert with the text, extending the "Essential Questions" framework. Passcode access to **BiochemistryNow**™ is packaged FREE with every new copy of the text. See page 6 of this Preview for a comprehensive look at this powerful resource.

BEYOND THE BOOK

InfoTrac® College Edition

When you adopt **Biochemistry, Third Edition,** you and your students receive four months of free anytime, anywhere access to the reliable resources found on **InfoTrac College Edition.** This fully searchable online library offers 22 years' worth of full-text articles from almost 5,000 scholarly journals and popular publications such as *American Scientist, Chemistry Review, Discover, Science, Scientist,* and more. **NEW**—Students now also gain instant access to critical thinking and paper-writing tools through *InfoWrite.*

Chapter 1: Chemistry Is the Logic of Biological Phenomena

Print Resources	Media and Internet Resources	BiochemistryNow™
Student Solutions Manual/Study Guide/Problems Book Chapter 1 contains Chapter Outline, Objectives, Summaries, Problems and Solutions, Questions for Self-Study w/ Answers, and Additional Problems w/ Abbreviated Answers **Student Lecture Book** Chapter 1 **Test Bank** Chapter 1 **Transparency Package** Chapter 1	**Book Companion Web Site** Online Chapter 1 at **http://chemistry.brookscole.com/ggb3** **Multimedia Manager** Chapter 1 contains image *PowerPoint*® slides, an image file bank, lecture *PowerPoint*® slides, animations, simulations, the Test Bank, and a Resource Integration Guide **WebTutor™ ToolBox** for WebCT and Blackboard Online Chapter 1 **iLrn Computerized Testing** **http://iLrn.thomson.com**	**BiochemistryNow™** Online Chapter 1 at **http://chemistry.brookscole.com/ggb3** **Active Figures** 1.6 Covalent Bond Formation by e^- Pair Sharing 1.9 Biological Macromolecules and Their Building Blocks Have a "Sense" or Directionality 1.10 Biological Macromolecules Are Informational 1.18 Organized Release or Capture of Energy 1.25 The Virus Life Cycle **Animated Figures** 1.5, 1.14, 1.15, 1.17, 1.19, 1.20 **BiochemistryInteractive** Page 17 Immunoglobulin G

Chapter 2: Water: The Medium of Life

Print Resources	Media and Internet Resources	BiochemistryNow™
Student Solutions Manual/Study Guide/Problems Book Chapter 2 contains Chapter Outline, Objectives, Summaries, Problems and Solutions, Questions for Self-Study w/ Answers, and Additional Problems w/ Abbreviated Answers **Student Lecture Book** Chapter 2 **Test Bank** Chapter 2 **Transparency Package** Chapter 2	**Book Companion Web Site** Online Chapter 2 at **http://chemistry.brookscole.com/ggb3** **Multimedia Manager** Chapter 2 contains image *PowerPoint*® slides, an image file bank, lecture *PowerPoint*® slides, animations, simulations, the Test Bank, and a Resource Integration Guide **WebTutor™ ToolBox** for WebCT and Blackboard Online Chapter 2 **iLrn Computerized Testing** **http://iLrn.thomson.com**	**BiochemistryNow™** Online Chapter 2 at **http://chemistry.brookscole.com/ggb3** **Active Figures** 2.1 The Structure of Water 2.3 The Fluid Network of H Bonds Linking Water Molecules in the Liquid State 2.7 Micelle Formation by Amphiphilic Molecules in Aqueous Solution 2.8 The Osmotic Pressure of a 1 Molal (m) Solution is Equal to 22.4 Atmospheres of Pressure 2.9 The Ionization of Water 2.14 Titration Curve for H_3PO_4 **Animated Figures** 2.2, 2.4, 2.5, 2.10, 2.11, 2.12, 2.13

Resource Integration Guide

Chapter 3: Thermodynamics of Biological Systems

Print Resources	Media and Internet Resources	BiochemistryNow™
Student Solutions Manual/Study Guide/Problems Book Chapter 3 contains Chapter Outline, Objectives, Summaries, Problems and Solutions, Questions for Self-Study w/ Answers, and Additional Problems w/ Abbreviated Answers	**Book Companion Web Site** Online Chapter 3 at http://chemistry.brookscole.com/ggb3	**BiochemistryNow™** Online Chapter 3 at http://chemistry.brookscole.com/ggb3
	Multimedia Manager Chapter 3 contains image *PowerPoint®* slides, an image file bank, lecture *PowerPoint®* slides, animations, simulations, the Test Bank, and a Resource Integration Guide	**Active Figures** 3.1 The Characteristics of Isolated, Closed, and Open Systems 3.9 The Triphosphate Chain of ATP 3.10 Electrostatic Repulsion in Reactants 3.12 The Hydrolysis Reactions of Acetyl Phosphate and 1,3-Bisphosphoglycerate 3.18 The Free Energy of Hydrolysis of ATP as a Function of Concentration at $38°C$, pH 7.0
Student Lecture Book Chapter 3	**WebTutor™ ToolBox** for WebCT and Blackboard Online Chapter 3	
Test Bank Chapter 3		**Animated Figures** 3.2, 3.6, 3.7, 3.11, 3.13, 3.14
Transparency Package Chapter 3	**iLrn Computerized Testing** http://iLrn.thomson.com	**BiochemistryInteractive** Page 58 Free Energies
		A Deeper Look Page 55

Chapter 4: Amino Acids

Print Resources	Media and Internet Resources	BiochemistryNow™
Student Solutions Manual/Study Guide/Problems Book Chapter 4 contains Chapter Outline, Objectives, Summaries, Problems and Solutions, Questions for Self-Study w/ Answers, and Additional Problems w/ Abbreviated Answers	**Book Companion Web Site** Online Chapter 4 at http://chemistry.brookscole.com/ggb3	**BiochemistryNow™** Online Chapter 4 at http://chemistry.brookscole.com/ggb3
	Multimedia Manager Chapter 4 contains image *PowerPoint®* slides, an image file bank, lecture *PowerPoint®* slides, animations, simulations, the Test Bank, and a Resource Integration Guide	**Active Figures** 4.8 Titrations of Glumatic Acid and Lysine 4.9 Typical Reactions of the Common Amino Acids 4.20 The Separation of Amino Acids on a Cation Exchange Column
Student Lecture Book Chapter 4	**WebTutor™ ToolBox** for WebCT and Blackboard Online Chapter 4	**Animated Figures** 4.1, 4.2, 4.6, 4.10, 4.11, 4.12, 4.13, 4.14, 4.19
Test Bank Chapter 4		**BiochemistryInteractive** Page 80 Amino Acids Name Game Page 86 Titration Behavior
Transparency Package Chapter 4	**iLrn Computerized Testing** http://iLrn.thomson.com	**Critical Developments in Biochemistry** Pages 89, 92, 94

Chapter 5: Proteins: Their Primary Structure and Biological Functions

Print Resources	Media and Internet Resources	BiochemistryNow™
Student Solutions Manual/Study Guide/Problems Book Chapter 5 contains Chapter Outline, Objectives, Summaries, Problems and Solutions, Questions for Self-Study w/ Answers, and Additional Problems w/ Abbreviated Answers	**Book Companion Web Site** Online Chapter 5 at **http://chemistry.brookscole.com/ggb3**	**BiochemistryNow™** Online Chapter 5 at **http://chemistry.brookscole.com/ggb3**
	Multimedia Manager Chapter 5 contains image *PowerPoint®* slides, an image file bank, lecture *PowerPoint®* slides, animations, simulations, the Test Bank, and a Resource Integration Guide	**Active Figures** 5.3 The Partial Double-Bond Character of the Peptide Bond 5.15 N-Terminal Analysis Using Edman's Reagent, Phenylisothiocyanate 5.19 Disulfide Bridges in Determining the Primary Structure of a Polypeptide
Student Lecture Book Chapter 5	**WebTutor™ ToolBox** for WebCT and Blackboard Online Chapter 5	**Animated Figures** 5.1, 5.2, 5.12, 5.16, 5.17, 5.18, 5.23, 5.34
Test Bank Chapter 5		
Transparency Package Chapter 5	**iLrn Computerized Testing** **http://iLrn.thomson.com**	

Chapter 6: Proteins: Secondary, Tertiary, and Quaternary Structure

Print Resources	Media and Internet Resources	BiochemistryNow™
Student Solutions Manual/Study Guide/Problems Book Chapter 6 contains Chapter Outline, Objectives, Summaries, Problems and Solutions, Questions for Self-Study w/ Answers, and Additional Problems w/ Abbreviated Answers	**Book Companion Web Site** Online Chapter 6 at **http://chemistry.brookscole.com/ggb3**	**BiochemistryNow™** Online Chapter 6 at **http://chemistry.brookscole.com/ggb3**
	Multimedia Manager Chapter 6 contains image *PowerPoint®* slides, an image file bank, lecture *PowerPoint®* slides, animations, simulations, the Test Bank, and a Resource Integration Guide	**Active Figures** 6.3 Steric Crowding about an α-Carbon 6.4 Ramachandran Diagram 6.18 Poly (Gly-Pro-Pro) 6.24 α-Helix Figures
Student Lecture Book Chapter 6	**WebTutor™ ToolBox** for WebCT and Blackboard Online Chapter 6	**Animated Figures** 6.7
Test Bank Chapter 6		**BiochemistryInteractive** Page 159 α-Helix Page 162 β-Sheets Page 165 β-Turns Page 173 Ribonuclease
Transparency Package Chapter 6	**iLrn Computerized Testing** **http://iLrn.thomson.com**	**A Deeper Look** Pages 158, 181
		Critical Developments in Biochemistry Page 185
		Human Biochemisty Page 186

Chapter 7: Carbohydrates and the Glycoconjugates of Cell Surfaces

Print Resources	Media and Internet Resources	BiochemistryNow™
Student Solutions Manual/Study Guide/Problems Book Chapter 7 contains Chapter Outline, Objectives, Summaries, Problems and Solutions, Questions for Self-Study w/ Answers, and Additional Problems w/ Abbreviated Answers	**Book Companion Web Site** Online Chapter 7 at http://chemistry.brookscole.com/ggb3	**BiochemistryNow™** Online Chapter 7 at http://chemistry.brookscole.com/ggb3
Student Lecture Book Chapter 7	**Multimedia Manager** Chapter 7 contains image *PowerPoint®* slides, an image file bank, lecture *PowerPoint®* slides, animations, simulations, the Test Bank, and a Resource Integration Guide	**Active Figures** 7.3 The Structure and Stereochemical Relationships of D-Ketoses with Three to Six Carbons 7.18 The Structures of Several Important Disaccharides 7.19 The Structures of Some Interesting Oligosaccharides
Test Bank Chapter 7	**WebTutor™ ToolBox** for WebCT and Blackboard Online Chapter 7	**Animated Figures** 7.5, 7.6, 7.8, 7.21, 7.23, 7.24, 7.29
Transparency Package Chapter 7	**iLrn Computerized Testing** http://iLrn.thomson.com	**BiochemistryInteractive** Page 205 Simple Sugars **A Deeper Look** Page 213

Chapter 8: Lipids

Print Resources	Media and Internet Resources	BiochemistryNow™
Student Solutions Manual/Study Guide/Problems Book Chapter 8 contains Chapter Outline, Objectives, Summaries, Problems and Solutions, Questions for Self-Study w/ Answers, and Additional Problems w/ Abbreviated Answers	**Book Companion Web Site** Online Chapter 8 at http://chemistry.brookscole.com/ggb3	**BiochemistryNow™** Online Chapter 8 at http://chemistry.brookscole.com/ggb3
Student Lecture Book Chapter 8	**Multimedia Manager** Chapter 8 contains image *PowerPoint®* slides, an image file bank, lecture *PowerPoint®* slides, animations, simulations, the Test Bank, and a Resource Integration Guide	**Active Figures** 8.5 The Absolute Configuration of *sn*-Glycerol-3-Phosphate 8.17 Examples of Classes of Terpenes
Test Bank Chapter 8	**WebTutor™ ToolBox** for WebCT and Blackboard Online Chapter 8	**Animated Figures** 8.1, 8.6
Transparency Package Chapter 8	**iLrn Computerized Testing** http://iLrn.thomson.com	**BiochemistryInteractive** Page 252 Glycerophospholipids **Human Biochemistry** Page 250 **A Deeper Look** Page 254

Chapter 9: Membranes and Membrane Transport

Print Resources	Media and Internet Resources	BiochemistryNow™

Student Solutions Manual/Study Guide/Problems Book
Chapter 9 contains Chapter Outline, Objectives, Summaries, Problems and Solutions, Questions for Self-Study w/ Answers, and Additional Problems w/ Abbreviated Answers

Student Lecture Book
Chapter 9

Test Bank
Chapter 9

Transparency Package
Chapter 9

Book Companion Web Site
Online Chapter 9 at
http://chemistry.brookscole.com/ggb3

Multimedia Manager
Chapter 9 contains image *PowerPoint®* slides, an image file bank, lecture *PowerPoint®* slides, animations, simulations, the Test Bank, and a Resource Integration Guide

WebTutor™ ToolBox
for WebCT and Blackboard
Online Chapter 9

iLrn Computerized Testing
http://iLrn.thomson.com

BiochemistryNow™
Online Chapter 9 at
http://chemistry.brookscole.com/ggb3

Active Figures
9.7 The Frye–Edidin Experiment
9.21 Passive Diffusion of an Uncharged Species Across a Membrane
9.22 Passive Diffusion of a Charged Species Across a Membrane
9.32 The Gastric H^+, K^+-ATPase

Animated Figures
9.8, 9.11, 9.12, 9.28, 9.29, 9.33, 9.46

BiochemistryInteractive
Page 278 Bacteriorhodopsin
Page 281 Maltoporin
Page 298 Hemolysin

Human Biochemistry
Pages 285, 301

Chapter 10: Nucleotides and Nucleic Acids

Print Resources	Media and Internet Resources	BiochemistryNow™

Student Solutions Manual/Study Guide/Problems Book
Chapter 10 contains Chapter Outline, Objectives, Summaries, Problems and Solutions, Questions for Self-Study w/ Answers, and Additional Problems w/ Abbreviated Answers

Student Lecture Book
Chapter 10

Test Bank
Chapter 10

Transparency Package
Chapter 10

Book Companion Web Site
Online Chapter 10 at
http://chemistry.brookscole.com/ggb3

Multimedia Manager
Chapter 10 contains image *PowerPoint®* slides, an image file bank, lecture *PowerPoint®* slides, animations, simulations, the Test Bank, and a Resource Integration Guide

WebTutor™ ToolBox
for WebCT and Blackboard
Online Chapter 10

iLrn Computerized Testing
http://iLrn.thomson.com

BiochemistryNow™
Online Chapter 10 at
http://chemistry.brookscole.com/ggb3

Active Figures
10.24 The Properties of mRNA Molecules in Prokaryotic Versus Eukaryotic Cells During Transcription and Translation
10.29 The Chemical Differences Between DNA and RNA

BiochemistryInteractive
Page 310 Purines and Pyrimidines

Human Biochemistry
Page 314

Chapter 11; Structure of Nucleic Acids

Print Resources	Media and Internet Resources	BiochemistryNow™
Student Solutions Manual/Study Guide/Problems Book Chapter 11 contains Chapter Outline, Objectives, Summaries, Problems and Solutions, Questions for Self-Study w/ Answers, and Additional Problems w/ Abbreviated Answers	**Book Companion Web Site** Online Chapter 11 at **http://chemistry.brookscole.com/ggb3**	**BiochemistryNow™** Online Chapter 11 at **http://chemistry.brookscole.com/ggb3**
Student Lecture Book Chapter 11	**Multimedia Manager** Chapter 11 contains image *PowerPoint®* slides, an image file bank, lecture *PowerPoint®* slides, animations, simulations, the Test Bank, and a Resource Integration Guide	**Active Figures** 11.2 DNA Polymerase Copies ssDNA In Vitro in the Presence of the Four Deoxynucleotide Monomers 11.3 The Chain Termination or Dideoxy Method of DNA Sequencing
Test Bank Chapter 11	**WebTutor™ ToolBox** for WebCT and Blackboard Online Chapter 11	
Transparency Package Chapter 11	**iLrn Computerized Testing** **http://iLrn.thomson.com**	

Chapter 12: Recombinant DNA: Cloning and Creation of Chimeric Genes

Print Resources	Media and Internet Resources	BiochemistryNow™
Student Solutions Manual/Study Guide/Problems Book Chapter 12 contains Chapter Outline, Objectives, Summaries, Problems and Solutions, Questions for Self-Study w/ Answers, and Additional Problems w/ Abbreviated Answers	**Book Companion Web Site** Online Chapter 12 at **http://chemistry.brookscole.com/ggb3**	**BiochemistryNow™** Online Chapter 12 at **http://chemistry.brookscole.com/ggb3**
Student Lecture Book Chapter 12	**Multimedia Manager** Chapter 12 contains image *PowerPoint®* slides, an image file bank, lecture *PowerPoint®* slides, animations, simulations, the Test Bank, and a Resource Integration Guide	**Active Figures** 12.2 Construction of Chimeric Plasmids 12.3 Restriction Endonuclease *Eco*RI Cleaves Double-Stranded DNA 12.7 A Typical Bacterial Transformation Experiment 12.9 Cosmid Vectors for Cloning Large DNA Fragments 12.11 Screening a Genomic Library by Colony Hybridization (or Plaque Hybridization) 12.14 Reverse Transcriptase–Driven Synthesis of cDNA
Student Lecture Book Chapter 12	**WebTutor™ ToolBox** for WebCT and Blackboard Online Chapter 12	**Animated Figures** 12.4, 12.5, 12.6, 12.10, 12.12, 12.13, 12.16, 12.17, 12.18, 12.19, 12.21, 12.22, 12.23, 12.24
Test Bank Chapter 12		**BiochemistryInteractive** Page 376 Plasmids and Genes
Transparency Package Chapter 12	**iLrn Computerized Testing** **http://iLrn.thomson.com**	**Critical Developments in Biochemistry** Pages 388–389
		A Deeper Look Page 395

Chapter 13: Enzymes – Kinetics and Specificity

Print Resources	Media and Internet Resources	BiochemistryNow™
Student Solutions Manual/Study Guide/Problems Book	**Book Companion Web Site** Online Chapter 13 at http://chemistry.brookscole.com/ggb3	**BiochemistryNow™** Online Chapter 13 at http://chemistry.brookscole.com/ggb3

Student Solutions Manual/Study Guide/Problems Book
Chapter 13 contains Chapter Outline, Objectives, Summaries, Problems and Solutions, Questions for Self-Study w/ Answers, and Additional Problems w/ Abbreviated Answers

Student Lecture Book
Chapter 13

Test Bank
Chapter 13

Transparency Package
Chapter 13

Book Companion Web Site
Online Chapter 13 at
http://chemistry.brookscole.com/ggb3

Multimedia Manager
Chapter 13 contains image *PowerPoint®* slides, an image file bank, lecture *PowerPoint®* slides, animations, simulations, the Test Bank, and a Resource Integration Guide

WebTutor™ ToolBox
for WebCT and Blackboard
Online Chapter 13

iLrn Computerized Testing
http://iLrn.thomson.com

BiochemistryNow™
Online Chapter 13 at
http://chemistry.brookscole.com/ggb3

Active Figures
13.9 Lineweaver–Burk Double-Reciprocal Plot
13.13 Lineweaver–Burk Plot of Competitive Inhibition
13.15 Lineweaver–Burk Plot of Pure Noncompetitive Inhibition
13.16 Lineweaver–Burk Plot of Mixed Noncompetitive Inhibition

Animated Figures
13.8, 13.10

BiochemistryInteractive
Page 414 Michaelis–Menten

Chapter 14: Mechanisms of Enzyme Action

Print Resources	Media and Internet Resources	BiochemistryNow™

Student Solutions Manual/Study Guide/Problems Book
Chapter 14 contains Chapter Outline, Objectives, Summaries, Problems and Solutions, Questions for Self-Study w/ Answers, and Additional Problems w/ Abbreviated Answers

Student Lecture Book
Chapter 14

Test Bank
Chapter 14

Transparency Package
Chapter 14

Book Companion Web Site
Online Chapter 14 at
http://chemistry.brookscole.com/ggb3

Multimedia Manager
Chapter 14 contains image *PowerPoint®* slides, an image file bank, lecture *PowerPoint®* slides, animations, simulations, the Test Bank, and a Resource Integration Guide

WebTutor™ ToolBox
for WebCT and Blackboard
Online Chapter 14

iLrn Computerized Testing
http://iLrn.thomson.com

BiochemistryNow™
Online Chapter 14 at
http://chemistry.brookscole.com/ggb3

Active Figures
14.4 Formation of the ES Complex Results in a Loss of Entropy
14.5 Substrates Typically Lose Waters of Hydration in the Formation of the ES Complex
14.24 Diisopropylfluorophosphate (DIFP) Reacts with Active-Site Serine Residues of Serine Proteases, Causing Permanent Inactivation
14.32 HIV-1 Protease Complexed with the Inhibitor Crixivan Made by Merck

BiochemistryInteractive
Page 452 Alcohol Dehydrogenase
Page 455 Chymotrypsin
Page 461 HIV-1 Protease

Chapter 15: Enzyme Regulation

Print Resources	Media and Internet Resources	BiochemistryNow™
Student Solutions Manual/Study Guide/Problems Book Chapter 15 contains Chapter Outline, Objectives, Summaries, Problems and Solutions, Questions for Self-Study w/ Answers, and Additional Problems w/ Abbreviated Answers	**Book Companion Web Site** Online Chapter 15 at http://chemistry.brookscole.com/ggb3	**BiochemistryNow™** Online Chapter 15 at http://chemistry.brookscole.com/ggb3

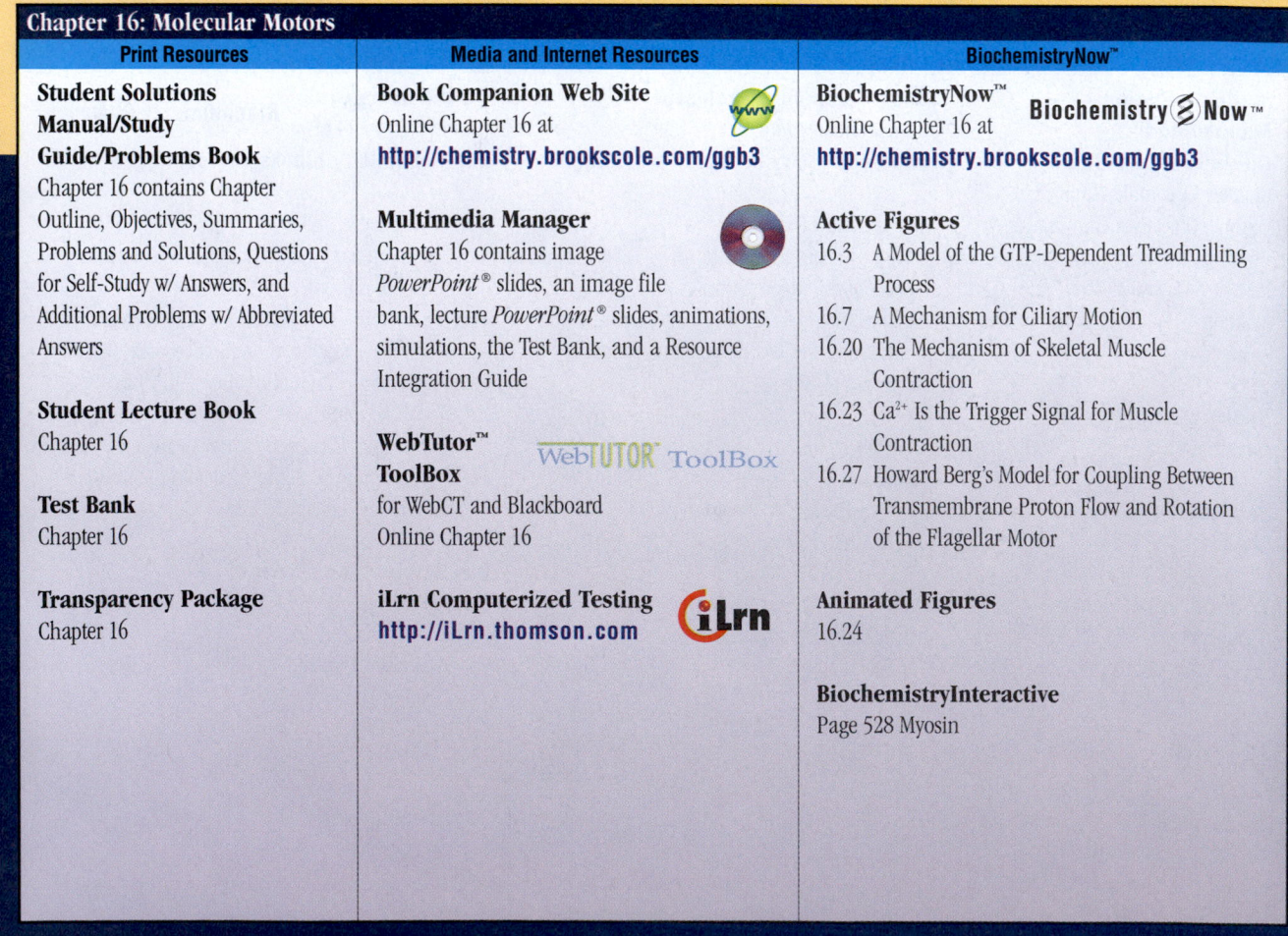

Student Solutions Manual/Study Guide/Problems Book
Chapter 15 contains Chapter Outline, Objectives, Summaries, Problems and Solutions, Questions for Self-Study w/ Answers, and Additional Problems w/ Abbreviated Answers

Student Lecture Book
Chapter 15

Test Bank
Chapter 15

Transparency Package
Chapter 15

Book Companion Web Site
Online Chapter 15 at
http://chemistry.brookscole.com/ggb3

Multimedia Manager
Chapter 15 contains image *PowerPoint*® slides, an image file bank, lecture *PowerPoint*® slides, animations, simulations, the Test Bank, and a Resource Integration Guide

WebTutor™ ToolBox
for WebCT and Blackboard
Online Chapter 15

iLrn Computerized Testing
http://iLrn.thomson.com

BiochemistryNow™
Online Chapter 15 at
http://chemistry.brookscole.com/ggb3

Active Figures
15.5   The Isozymes of Lactate Dehydrogenase (LDH)
15.10 Heterotropic Allosteric Effects: A and I Binding to R and T, Respectively
15.16 The Mechanism of Covalent Modification and Allosteric Regulation of Glycogen Phosphorylase
15.31 Changes in the Position of the Heme Iron Atom upon Oxygenation

Animated Figures
15.3, 15.6, 15.9, 15.30, 15.39

BiochemistryInteractive
Page 486 Glycogen Phosphorylase
Page 491 Hemoglobin/Myoglobin

Chapter 16: Molecular Motors

Print Resources	Media and Internet Resources	BiochemistryNow™

Student Solutions Manual/Study Guide/Problems Book
Chapter 16 contains Chapter Outline, Objectives, Summaries, Problems and Solutions, Questions for Self-Study w/ Answers, and Additional Problems w/ Abbreviated Answers

Student Lecture Book
Chapter 16

Test Bank
Chapter 16

Transparency Package
Chapter 16

Book Companion Web Site
Online Chapter 16 at
http://chemistry.brookscole.com/ggb3

Multimedia Manager
Chapter 16 contains image *PowerPoint*® slides, an image file bank, lecture *PowerPoint*® slides, animations, simulations, the Test Bank, and a Resource Integration Guide

WebTutor™ ToolBox
for WebCT and Blackboard
Online Chapter 16

iLrn Computerized Testing
http://iLrn.thomson.com

BiochemistryNow™
Online Chapter 16 at
http://chemistry.brookscole.com/ggb3

Active Figures
16.3   A Model of the GTP-Dependent Treadmilling Process
16.7   A Mechanism for Ciliary Motion
16.20 The Mechanism of Skeletal Muscle Contraction
16.23 Ca^{2+} Is the Trigger Signal for Muscle Contraction
16.27 Howard Berg's Model for Coupling Between Transmembrane Proton Flow and Rotation of the Flagellar Motor

Animated Figures
16.24

BiochemistryInteractive
Page 528 Myosin

Chapter 17: Metabolism—An Overview

Print Resources	Media and Internet Resources	BiochemistryNow™
Student Solutions Manual/Study Guide/Problems Book Chapter 17 contains Chapter Outline, Objectives, Summaries, Problems and Solutions, Questions for Self-Study w/ Answers, and Additional Problems w/ Abbreviated Answers	**Book Companion Web Site** Online Chapter 17 at **http://chemistry.brookscole.com/ggb3**	**BiochemistryNow™** Online Chapter 17 at **http://chemistry.brookscole.com/ggb3**
	Multimedia Manager Chapter 17 contains image *PowerPoint*® slides, an image file bank, lecture *PowerPoint*® slides, animations, simulations, the Test Bank, and a Resource Integration Guide	**Active Figures** 17.1 A Metabolic Map 17.15 Fractionation of a Cell Extract by Differential Centrifugation 17.18 Thiamine Pyrophosphate 17.27 The Schiff Base 17.32 Biotin is Covalently Linked to a Protein via the ϵ-Amino Group of a Lysine Residue
Student Lecture Book Chapter 17	**WebTutor™ ToolBox** for WebCT and Blackboard Online Chapter 17	**Animated Figures** 17.19, 17.21, 17.24, 17.29
Test Bank Chapter 17		
Transparency Package Chapter 17	**iLrn Computerized Testing** **http://iLrn.thomson.com**	**BiochemistryInteractive** Page 566 Coenzyme-Catalyzed Reactions in Metabolism

Chapter 18: Glycolysis

Print Resources	Media and Internet Resources	BiochemistryNow™
Student Solutions Manual/Study Guide/Problems Book Chapter 18 contains Chapter Outline, Objectives, Summaries, Problems and Solutions, Questions for Self-Study w/ Answers, and Additional Problems w/ Abbreviated Answers	**Book Companion Web Site** Online Chapter 18 at **http://chemistry.brookscole.com/ggb3**	**BiochemistryNow™** Online Chapter 18 at **http://chemistry.brookscole.com/ggb3**
	Multimedia Manager Chapter 18 contains image *PowerPoint*® slides, an image file bank, lecture *PowerPoint*® slides, animations, simulations, the Test Bank, and a Resource Integration Guide	**Active Figures** 18.1 The Glycolytic Pathway 18.6 The Phosphoglucoisomerase Mechanism 18.13 A Mechanism for the Fructose-1,6-Bisphosphate Aldolase Reaction 18.15 A Reaction Mechanism for Triose Phosphate Isomerase 18.18 A Mechanism for the Glyceraldehyde-3-Phosphate Dehydrogenase Reaction
Student Lecture Book Chapter 18	**WebTutor™ ToolBox** for WebCT and Blackboard Online Chapter 18	**Animated Figures** 18.4
Test Bank Chapter 18		
Transparency Package Chapter 18	**iLrn Computerized Testing** **http://iLrn.thomson.com**	**BiochemistryInteractive** Page 585 Phosphofructokinase Reaction
		A Deeper Look Page 590

Chapter 19: The Tricarboxylic Acid Cycle

Print Resources	Media and Internet Resources	BiochemistryNow™
Student Solutions Manual/Study Guide/Problems Book Chapter 19 contains Chapter Outline, Objectives, Summaries, Problems and Solutions, Questions for Self-Study w/ Answers, and Additional Problems w/ Abbreviated Answers	**Book Companion Web Site** Online Chapter 19 at http://chemistry.brookscole.com/ggb3	**BiochemistryNow™** Online Chapter 19 at http://chemistry.brookscole.com/ggb3
	Multimedia Manager Chapter 19 contains image *PowerPoint®* slides, an image file bank, lecture *PowerPoint®* slides, animations, simulations, the Test Bank, and a Resource Integration Guide	**Active Figures** 19.4 The Tricarboxylic Acid Cycle 19.8 The Iron-Sulfur Cluster of Aconitase 19.13 The Mechanism of the Succinyl-CoA Synthetase Reaction 19.21 The Fate of the Carbon Atoms of Acetate in Successive TCA Cycles
Student Lecture Book Chapter 19	**WebTutor™ ToolBox** for WebCT and Blackboard Online Chapter 19	**Animated Figures** 19.10
Test Bank Chapter 19	**iLrn Computerized Testing** http://iLrn.thomson.com	**BiochemistryInteractive** Page 613 Citrate Synthase Page 622 Malate Dehydrogenase
Transparency Package Chapter 19		

Chapter 20: Electron Transport and Oxidative Phosphorylation

Print Resources	Media and Internet Resources	BiochemistryNow™
Student Solutions Manual/Study Guide/Problems Book Chapter 20 contains Chapter Outline, Objectives, Summaries, Problems and Solutions, Questions for Self-Study w/ Answers, and Additional Problems w/ Abbreviated Answers	**Book Companion Web Site** Online Chapter 20 at http://chemistry.brookscole.com/ggb3	**BiochemistryNow™** Online Chapter 20 at http://chemistry.brookscole.com/ggb3
	Multimedia Manager Chapter 20 contains image *PowerPoint®* slides, an image file bank, lecture *PowerPoint®* slides, animations, simulations, the Test Bank, and a Resource Integration Guide	**Active Figures** 20.2 Experimental Apparatus Used to Measure the Standard Reduction Potential of the Indicated Redox Couples 20.6 Proposed Structure and Electron-Transport Pathway for Complex I 20.8 A Probable Scheme for Electron Flow in Complex II 20.12 The Q Cycle in Mitochondria 20.17 The Electron-Transfer Pathway for Cytochrome Oxidase
Student Lecture Book Chapter 20	**WebTutor™ ToolBox** for WebCT and Blackboard Online Chapter 20	
Test Bank Chapter 20	**iLrn Computerized Testing** http://iLrn.thomson.com	**Animated Figures** 20.25, 20.27, 20.28
Transparency Package Chapter 20		**BiochemistryInteractive** Page 646 Cytochrome Oxidase Page 659 ATP Synthase Motor

Resource Integration Guide

Chapter 21: Photosynthesis

Print Resources	Media and Internet Resources	BiochemistryNow™
Student Solutions Manual/Study Guide/Problems Book Chapter 21 contains Chapter Outline, Objectives, Summaries, Problems and Solutions, Questions for Self-Study w/ Answers, and Additional Problems w/ Abbreviated Answers	**Book Companion Web Site** Online Chapter 21 at **http://chemistry.brookscole.com/ggb3**	**BiochemistryNow™** Online Chapter 21 at **http://chemistry.brookscole.com/ggb3**
	Multimedia Manager Chapter 21 contains image *PowerPoint®* slides, an image file bank, lecture *PowerPoint®* slides, animations, simulations, the Test Bank, and a Resource Integration Guide	**Active Figures** 21.12 The *Z* Scheme of Photosynthesis 21.16 The *R. Viridis* Reaction Center 21.20 The Mechanism of Photophosphorylation 21.21 The Pathway of Cyclic Photophosphorylation by PSI 21.24 The Calvin—Benson Cycle of Reactions
Student Lecture Book Chapter 21	**WebTutor™ ToolBox** for WebCT and Blackboard Online Chapter 21	**Animated Figures** 21.4, 21.8, 21.10, 21.26
Test Bank Chapter 21	**iLrn Computerized Testing** **http://iLrn.thomson.com**	**BiochemistryInteractive** Page 681 Photosystem I Page 687 *R. Viridis* Reaction Center Page 695 Rubisco
Transparency Package Chapter 21		**Critical Developments in Biochemistry** Page 691

Chapter 22: Gluconeogenesis, Glycogen Metabolism, and the Pentose Phosphate Pathway

Print Resources	Media and Internet Resources	BiochemistryNow™
Student Solutions Manual/Study Guide/Problems Book Chapter 22 contains Chapter Outline, Objectives, Summaries, Problems and Solutions, Questions for Self-Study w/ Answers, and Additional Problems w/ Abbreviated Answers	**Book Companion Web Site** Online Chapter 22 at **http://chemistry.brookscole.com/ggb3**	**BiochemistryNow™** Online Chapter 22 at **http://chemistry.brookscole.com/ggb3**
	Multimedia Manager Chapter 22 contains image *PowerPoint®* slides, an image file bank, lecture *PowerPoint®* slides, animations, simulations, the Test Bank, and a Resource Integration Guide	**Active Figures** 22.11 The Principal Regulatory Mechanisms in Glycolysis and Gluconeogenesis 22.13 Synthesis and Degradation of Fructose-2,6-Bisphosphate are Catalyzed by the Same Bifunctional Enzyme 22.26 The Pentose Phosphate Pathway
Student Lecture Book Chapter 22	**WebTutor™ ToolBox** for WebCT and Blackboard Online Chapter 22	**Animated Figures** 22.10, 22.18, 22.19, 22.36
Test Bank Chapter 22	**iLrn Computerized Testing** **http://iLrn.thomson.com**	**BiochemistryInteractive** Page 709 Pyruvate Carboxylase Page 725 6-Phosphogluconate Page 729 Transketolase
Transparency Package Chapter 22		

Chapter 23: Fatty Acid Catabolism

Print Resources	Media and Internet Resources	BiochemistryNow™
Student Solutions Manual/Study Guide/Problems Book Chapter 23 contains Chapter Outline, Objectives, Summaries, Problems and Solutions, Questions for Self-Study w/ Answers, and Additional Problems w/ Abbreviated Answers	**Book Companion Web Site** Online Chapter 23 at http://chemistry.brookscole.com/ggb3	**BiochemistryNow™** Online Chapter 23 at http://chemistry.brookscole.com/ggb3
Student Lecture Book Chapter 23	**Multimedia Manager** Chapter 23 contains image *PowerPoint*® slides, an image file bank, lecture *PowerPoint*® slides, animations, simulations, the Test Bank, and a Resource Integration Guide	**Active Figures** 23.10 The β-Oxidation of Saturated Fatty Acids 23.11 The Acyl-CoA Dehydrogenase Reaction 23.19 The Conversion of Propionyl-CoA to Succinyl-CoA
Test Bank Chapter 23	**WebTutor™ ToolBox** for WebCT and Blackboard Online Chapter 23	**Animated Figures** 23.2, 23.8, 23.9
Transparency Package Chapter 23	**iLrn Computerized Testing** http://iLrn.thomson.com	**BiochemistryInteractive** Page 750 Coenzyme A Page 755 Methylmalonate Mutase

Chapter 24: Lipid Biosynthesis

Print Resources	Media and Internet Resources	BiochemistryNow™
Student Solutions Manual/Study Guide/Problems Book Chapter 24 contains Chapter Outline, Objectives, Summaries, Problems and Solutions, Questions for Self-Study w/ Answers, and Additional Problems w/ Abbreviated Answers	**Book Companion Web Site** Online Chapter 24 at http://chemistry.brookscole.com/ggb3	**BiochemistryNow™** Online Chapter 24 at http://chemistry.brookscole.com/ggb3
Student Lecture Book Chapter 24	**Multimedia Manager** Chapter 24 contains image *PowerPoint*® slides, an image file bank, lecture *PowerPoint*® slides, animations, simulations, the Test Bank, and a Resource Integration Guide	**Active Figures** 24.2 The Synthesis of Fatty Acids 24.7 The Pathway of Palmitate Synthesis from Acetyl-CoA and Malonyl-CoA 24.11 The Mechanism of the Fatty Acyl Synthase Reaction in Eukaryotes 24.34 The Conversion of Mevalonate to Squalene 24.35 Cholesterol Is Synthesized from Squalene via Lanosterol
Test Bank Chapter 24	**WebTutor™ ToolBox** for WebCT and Blackboard Online Chapter 24	**Animated Figures** 24.9, 24.14, 24.17, 24.19, 24.32, 24.42
Transparency Package Chapter 24	**iLrn Computerized Testing** http://iLrn.thomson.com	**BiochemistryInteractive** Page 799 Apolipoprotein **A Deeper Look** Pages 790–791

Chapter 25: Nitrogen Acquisition and Amino Acid Metabolism

Print Resources	Media and Internet Resources	BiochemistryNow™
Student Solutions Manual/Study Guide/Problems Book Chapter 25 contains Chapter Outline, Objectives, Summaries, Problems and Solutions, Questions for Self-Study w/ Answers, and Additional Problems w/ Abbreviated Answers	**Book Companion Web Site** Online Chapter 25 at http://chemistry.brookscole.com/ggb3	**BiochemistryNow™** Online Chapter 25 at http://chemistry.brookscole.com/ggb3

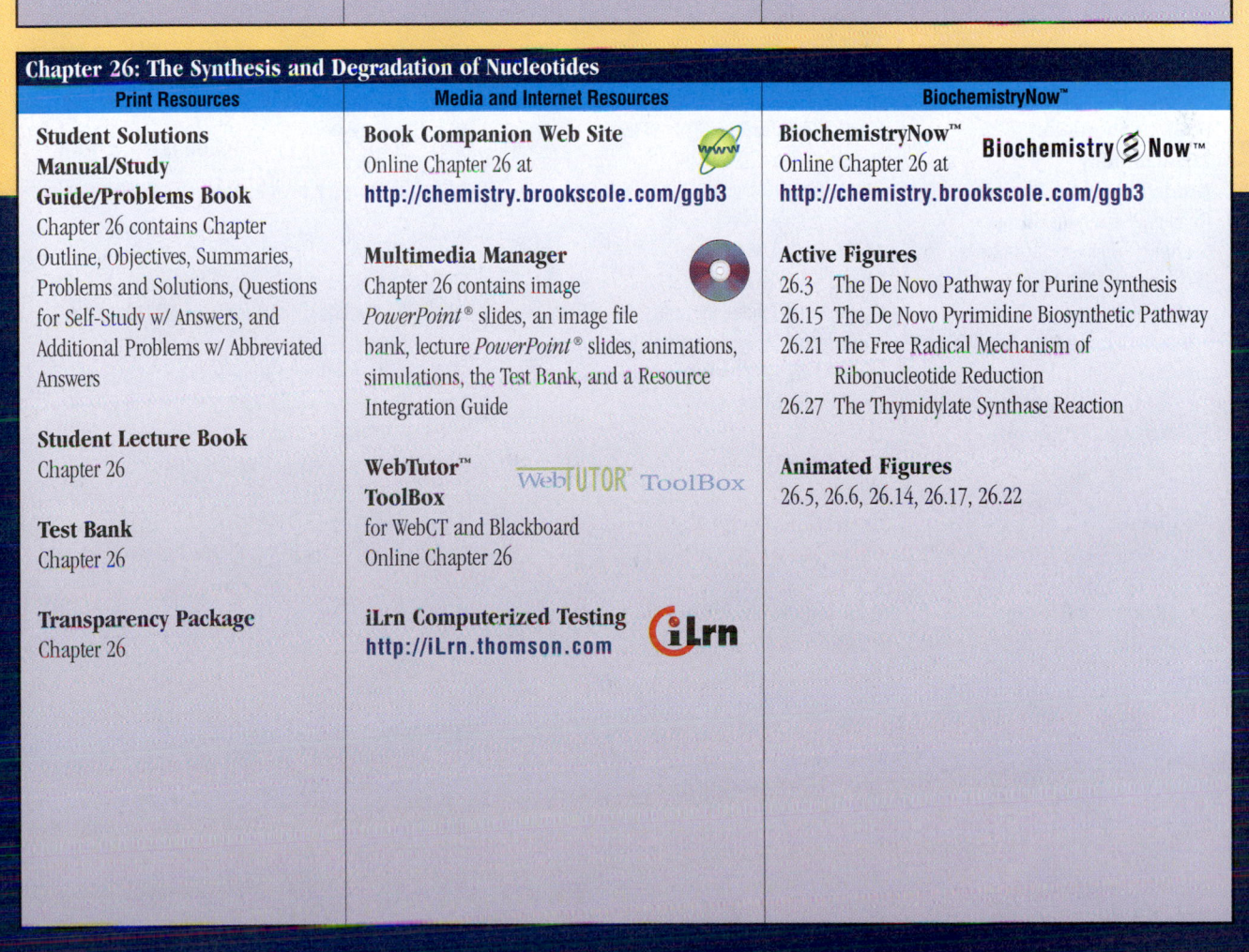

Student Lecture Book
Chapter 25

Multimedia Manager
Chapter 25 contains image *PowerPoint®* slides, an image file bank, lecture *PowerPoint®* slides, animations, simulations, the Test Bank, and a Resource Integration Guide

Active Figures
25.15 The Allosteric Regulation of Glutamine Synthetase Activity by Feedback Inhibition
25.19 Transamination
25.23 The Urea Cycle Series of Reactions
25.41 Metabolic Degradation of the Common Amino Acids

Test Bank
Chapter 25

WebTutor™ ToolBox
for WebCT and Blackboard
Online Chapter 25

Animated Figures
25.22, 25.36, 25.37, 25.40

Transparency Package
Chapter 25

iLrn Computerized Testing
http://iLrn.thomson.com

BiochemistryInteractive
Page 814 Nitrogenase
Page 818 Glutamine Synthetase

A Deeper Look
Page 845

Chapter 26: The Synthesis and Degradation of Nucleotides

Print Resources	Media and Internet Resources	BiochemistryNow™
Student Solutions Manual/Study Guide/Problems Book Chapter 26 contains Chapter Outline, Objectives, Summaries, Problems and Solutions, Questions for Self-Study w/ Answers, and Additional Problems w/ Abbreviated Answers	**Book Companion Web Site** Online Chapter 26 at http://chemistry.brookscole.com/ggb3	**BiochemistryNow™** Online Chapter 26 at http://chemistry.brookscole.com/ggb3

Student Lecture Book
Chapter 26

Multimedia Manager
Chapter 26 contains image *PowerPoint®* slides, an image file bank, lecture *PowerPoint®* slides, animations, simulations, the Test Bank, and a Resource Integration Guide

Active Figures
26.3 The De Novo Pathway for Purine Synthesis
26.15 The De Novo Pyrimidine Biosynthetic Pathway
26.21 The Free Radical Mechanism of Ribonucleotide Reduction
26.27 The Thymidylate Synthase Reaction

Test Bank
Chapter 26

WebTutor™ ToolBox
for WebCT and Blackboard
Online Chapter 26

Animated Figures
26.5, 26.6, 26.14, 26.17, 26.22

Transparency Package
Chapter 26

iLrn Computerized Testing
http://iLrn.thomson.com

Chapter 27: Metabolic Integration and Organ Specialization

Print Resources	Media and Internet Resources	BiochemistryNow™
Student Solutions Manual/Study Guide/Problems Book Chapter 27 contains Chapter Outline, Objectives, Summaries, Problems and Solutions, Questions for Self-Study w/ Answers, and Additional Problems w/ Abbreviated Answers **Student Lecture Book** Chapter 27 **Test Bank** Chapter 27 **Transparency Package** Chapter 27	**Book Companion Web Site** Online Chapter 27 at http://chemistry.brookscole.com/ggb3 **Multimedia Manager** Chapter 27 contains image *PowerPoint*® slides, an image file bank, lecture *PowerPoint*® slides, animations, simulations, the Test Bank, and a Resource Integration Guide **WebTutor™ ToolBox** for WebCT and Blackboard Online Chapter 27 **iLrn Computerized Testing** http://iLrn.thomson.com	**BiochemistryNow™** Online Chapter 27 at http://chemistry.brookscole.com/ggb3 **Active Figures** 27.4 The Oscillation of Energy Charge (E.C.) About a Steady-State Value as a Consequence of the Offsetting Influences of R and U Processes on the Production and Consumption of ATP **Animated Figures** 27.1, 27.5, 27.7 **Human Biochemistry** Pages 888

Chapter 28: DNA Metabolism: Replication, Recombination, and Repair

Print Resources	Media and Internet Resources	BiochemistryNow™
Student Solutions Manual/Study Guide/Problems Book Chapter 28 contains Chapter Outline, Objectives, Summaries, Problems and Solutions, Questions for Self-Study w/ Answers, and Additional Problems w/ Abbreviated Answers **Student Lecture Book** Chapter 28 **Test Bank** Chapter 28 **Transparency Package** Chapter 28	**Book Companion Web Site** Online Chapter 28 at http://chemistry.brookscole.com/ggb3 **Multimedia Manager** Chapter 28 contains image *PowerPoint*® slides, an image file bank, lecture *PowerPoint*® slides, animations, simulations, the Test Bank, and a Resource Integration Guide **WebTutor™ ToolBox** for WebCT and Blackboard Online Chapter 28 **iLrn Computerized Testing** http://iLrn.thomson.com	**BiochemistryNow™** Online Chapter 28 at http://chemistry.brookscole.com/ggb3 **Active Figures** 28.10 General Features of a Replication Fork 28.19 The Holliday Model for Homologous Recombination 28.22 Model for Homologous Recombination as Promoted by RecA Enzyme 28.24 The Typical Transpoon 28.36 Model for V(D)J Recombination **Animated Figures** 28.3, 28.6, 28.11, 28.15, 28.17 **BiochemistryInteractive** Page 906 DNA Polymerase III Page 911 PCNA **Human Biochemistry** Page 916 **A Deeper Look** Page 923

Resource Integration Guide

Chapter 29: Transcription and the Regulation of Gene Expression

Print Resources	Media and Internet Resources	BiochemistryNow™
Student Solutions Manual/Study Guide/Problems Book Chapter 29 contains Chapter Outline, Objectives, Summaries, Problems and Solutions, Questions for Self-Study w/ Answers, and Additional Problems w/ Abbreviated Answers **Student Lecture Book** Chapter 29 **Test Bank** Chapter 29 **Transparency Package** Chapter 29	**Book Companion Web Site** Online Chapter 29 at **http://chemistry.brookscole.com/ggb3** **Multimedia Manager** Chapter 29 contains image *PowerPoint*® slides, an image file bank, lecture *PowerPoint*® slides, animations, simulations, the Test Bank, and a Resource Integration Guide **WebTutor™ ToolBox** WebTUTOR ToolBox for WebCT and Blackboard Online Chapter 29 **iLrn Computerized Testing** **http://iLrn.thomson.com**	**BiochemistryNow™** Biochemistry Now™ Online Chapter 29 at **http://chemistry.brookscole.com/ggb3** **Active Figures** 29.2 Sequence of Events in the Initiation and Elongation Phases of Transcription as it Occurs in Prokaryotes 29.4 Supercoiling Versus Transcription 29.11 The Mode of Action of *lac* Repressor 29.29 A Model for the Transcriptional Regulation of Eukaryotic Genes 29.45 A Unified Theory of Gene Expression **Animated Figures** 29.6, 29.15, 29.16, 29.17, 29.26, 29.27 **BiochemistryInteractive** Page 947 RNA Polymerase II Page 974 Leucine Zipper

Chapter 30: Protein Synthesis

Print Resources	Media and Internet Resources	BiochemistryNow™
Student Solutions Manual/Study Guide/Problems Book Chapter 30 contains Chapter Outline, Objectives, Summaries, Problems and Solutions, Questions for Self-Study w/ Answers, and Additional Problems w/ Abbreviated Answers **Student Lecture Book** Chapter 30 **Test Bank** Chapter 30 **Transparency Package** Chapter 30	**Book Companion Web Site** Online Chapter 30 at **http://chemistry.brookscole.com/ggb3** **Multimedia Manager** Chapter 30 contains image *PowerPoint*® slides, an image file bank, lecture *PowerPoint*® slides, animations, simulations, the Test Bank, and a Resource Integration Guide **WebTutor™ ToolBox** WebTUTOR ToolBox for WebCT and Blackboard Online Chapter 30 **iLrn Computerized Testing** **http://iLrn.thomson.com**	**BiochemistryNow™** Biochemistry Now™ Online Chapter 30 at **http://chemistry.brookscole.com/ggb3** **Active Figures** 30.3 The Aminoacyl-tRNA Synthetase Reaction 30.13 The Basic Steps in Protein Synthesis 30.18 The Sequence of Events in Peptide Chain Initiation 30.19 The Cycle of Events in Peptide Chain Elongation on *E. Coli* Ribosomes 30.23 The Events in Peptide Chain Termination **Animated Figures** 30.9, 30.16, 30.21, 30.28 **A Deeper Look** Page 1010

Chapter 31: Completing the Protein Life Cycle: Folding, Processing, and Degradation

Print Resources	Media and Internet Resources	BiochemistryNow™
Student Solutions Manual/Study Guide/Problems Book Chapter 31 contains Chapter Outline, Objectives, Summaries, Problems and Solutions, Questions for Self-Study w/ Answers, and Additional Problems w/ Abbreviated Answers	**Book Companion Web Site** Online Chapter 31 at http://chemistry.brookscole.com/ggb3 **Multimedia Manager** Chapter 31 contains image *PowerPoint*® slides, an image file bank, lecture *PowerPoint*® slides, animations, simulations, the Test Bank, and a Resource Integration Guide	**BiochemistryNow™** Online Chapter 31 at http://chemistry.brookscole.com/ggb3 **Active Figures** 31.3 Structure and Function of the GroEL–GroES Complex 31.5 Synthesis of a Eukaryotic Secretory Protein and Its Translocation into the Endoplasmic Reticulum 31.11 Diagram of the Ubiquitin-Proteasome Degradation Pathway
Student Lecture Book Chapter 31	**WebTutor™ ToolBox** for WebCT and Blackboard Online Chapter 31	**Animated Figures** 31.1, 31.2, 31.7, 31.8
Test Bank Chapter 31		
Transparency Package Chapter 31	**iiLrn Computerized Testing** http://iLrn.thomson.com	**BiochemistryInteractive** Page 1035 Proteasomes

Chapter 32: The Reception and Transmission of Extracellular Information

Print Resources	Media and Internet Resources	BiochemistryNow™
Student Solutions Manual/Study Guide/Problems Book Chapter 32 contains Chapter Outline, Objectives, Summaries, Problems and Solutions, Questions for Self-Study w/ Answers, and Additional Problems w/ Abbreviated Answers	**Book Companion Web Site** Online Chapter 32 at http://chemistry.brookscole.com/ggb3 **Multimedia Manager** Chapter 32 contains image *PowerPoint*® slides, an image file bank, lecture *PowerPoint*® slides, animations, simulations, the Test Bank, and a Resource Integration Guide	**BiochemistryNow™** Online Chapter 32 at http://chemistry.brookscole.com/ggb3 **Active Figures** 32.4 A Complete Signal Transduction Pathway That Connects a Hormone Receptor with Transcription Events in the Nucleus 32.11 Adenylyl Cyclase Activity is Modulated by the Interplay of Stimulatory (G_s) and Inhibitory (G_i) G Proteins 32.22 IP_3-Mediated Signal Transduction Pathways 32.34 The Propagation of Action Potentials Along an Axon
Student Lecture Book Chapter 32	**WebTutor™ ToolBox** for WebCT and Blackboard Online Chapter 32	**Animated Figures** 32.1, 32.6, 32.7, 32.21, 32.39, 32.42 **BiochemistryInteractive** Page 1047 β_2 Adrenergic Receptor Page 1051 G-Protein Complex Page 1054 Adenylyl Cyclase Page 1064 Protein Kinase C Page 1066 Tyrosine Phosphatase Page 1067 Signaling Domains
Test Bank Chapter 32		**A Deeper Look** Page 1056
Transparency Package Chapter 32	**iLrn Computerized Testing** http://iLrn.thomson.com	**Human Biochemistry** Pages 1058

Resource Integration Guide

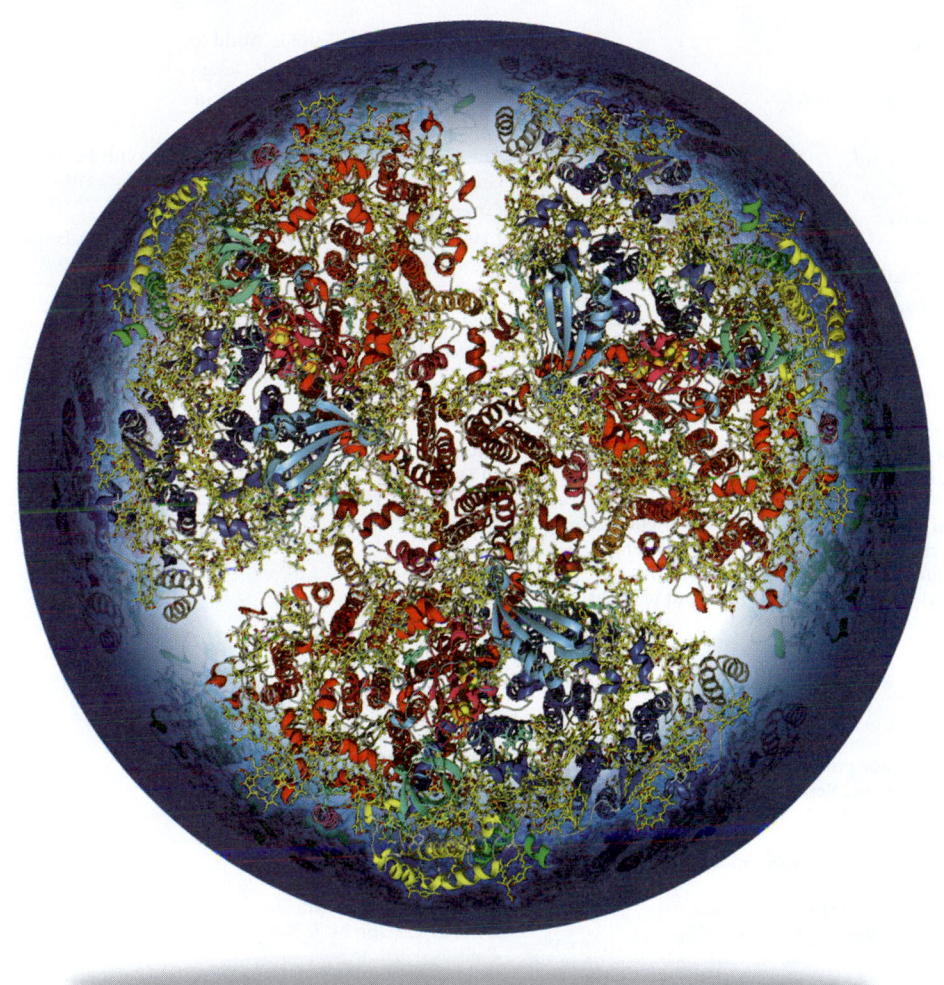

Biochemistry

Third Edition

Reginald H. Garrett
Charles M. Grisham

University of Virginia

THOMSON

BROOKS/COLE

Australia · Canada · Mexico · Singapore · Spain · United Kingdom · United States

THOMSON

BROOKS/COLE

Publisher, Physical Sciences: David Harris
Development Editor: Sandra Kiselica
Development Editor, Media: Peggy Williams
Assistant Editor: Alyssa White
Editorial Assistants: Annie Mac, Jessica Howard
Technology Project Manager: Donna Kelley
Executive Marketing Manager: Julie Conover
Marketing Assistant: Melanie Banfield
Advertising Project Manager: Stacey Purviance
Project Manager, Editorial Production: Lisa Weber
Creative Director: Rob Hugel
Print Buyer: Barbara Britton
Permissions Editor: Kiely Sexton

Production Service: Graphic World Inc.
Text Designer: Patrick Devine Design
Photo Researcher: Rosemary Grisham
Copy Editor: Graphic World Inc.
Illustrators: J/B Woolsey Associates; Dartmouth Publishing, Inc.; Graphic World Inc.; Dr. Michal Sabat; Jane Richardson
Cover Designer: Joan Greenfield Design
Cover Image: © Norbert Krauss, Wolfram Saenger, Horst Tobias Witt, Petra Fromme, Patrick Jordan
Cover Printer: Phoenix Color Corp
Compositor: Graphic World Inc.
Printer: Quebecor World/Versailles

For more information about our products, contact us at:
Thomson Learning Academic Resource Center
1-800-423-0563

For permission to use material from this text, contact us by:
Phone: 1-800-730-2214
Fax: 1-800-730-2215
Web: http://www.thomsonrights.com

Library of Congress Control Number: 2003108540

Student Edition with InfoTrac College Edition: ISBN 0-534-49033-6

Instructor's Edition: ISBN 0-534-49034-4

Thomson Brooks/Cole
10 Davis Drive
Belmont, CA 94002
USA

Asia
Thomson Learning
5 Shenton Way #01-01
UIC Building
Singapore 068808

Australia/New Zealand
Thomson Learning
102 Dodds Street
Southbank, Victoria 3006
Australia

Canada
Nelson
1120 Birchmount Road
Toronto, Ontario M1K 5G4
Canada

Europe/Middle East/Africa
Thomson Learning
High Holborn House
50/51 Bedford Row
London WC1R 4LR
United Kingdom

Latin America
Thomson Learning
Seneca, 53
Colonia Polanco
11560 Mexico D.F.
Mexico

Spain/Portugal
Paraninfo
Calle/Magallanes, 25
28015 Madrid, Spain

About the Cover

"Sun Catcher." The structure of the trimeric Photosystem I from the thermophilic cyanobacterium *Synechococcus elongatus.* This protein complex captures light energy from the sun and converts it into the chemical energy of an oxidation–reduction reaction. Image provided by Norbert Krauss, Petra Fromme, Wolfram Saenger, Horst Tobias Witt, and Patrick Jordan, of the Institute for Crystallography, Free University of Berlin and the Max Volmer Institute for Biophysical Chemistry and Biochemistry at the Technical University Berlin.

Rosemary Jurbala Grisham

Charlie Grisham and Reg Garrett with friends at University of Virginia.

Reginald H. Garrett

Reginald H. Garrett was educated in the Baltimore city public schools and at the Johns Hopkins University, where he received his Ph.D. in biology in 1968. Since that time, he has been at the University of Virginia, where he is currently Professor of Biology. He is the author of previous editions of *Biochemistry,* as well as *Principles of Biochemistry* (Thomson Brooks/Cole), and numerous papers and review articles on the biochemical, genetic, and molecular biological aspects of inorganic nitrogen metabolism. His research interests focused on the pathway of nitrate assimilation in filamentous fungi. His investigations contributed substantially to our understanding of the enzymology, genetics, and regulation of this major pathway of biological nitrogen acquisition. His research has been supported by the National Institutes of Health, the National Science Foundation, and private industry. He is a former Fulbright Scholar at the Universität fur Bodenkultur in Vienna, Austria, and served as Visiting Scholar at the University of Cambridge on two separate occasions. During the second, he was Thomas Jefferson Visiting Fellow in Downing College. Recently, he was Professeur Invité at the Université Paul Sabatier/Toulouse III and the Centre National de la Recherche Scientifique, Institute for Pharmacology and Structural Biology in France. He has taught biochemistry at the University of Virginia for 35 years. He is a member of the American Society for Biochemistry and Molecular Biology.

Charles M. Grisham

Charles M. Grisham was born and raised in Minneapolis, Minnesota, and was educated at Benilde High School. He received his B.S. in chemistry from the Illinois Institute of Technology in 1969 and his Ph.D. in chemistry from the University of Minnesota in 1973. Following a postdoctoral appointment at the Institute for Cancer Research in Philadelphia, he joined the faculty of the University of Virginia, where he is Professor of Chemistry and Chief Technology Officer for the Faculty of Arts and Sciences. He is the author of previous editions of *Biochemistry* and *Principles of Biochemistry* (Thomson Brooks/Cole), as well as of numerous papers and review articles on active transport of sodium, potassium, and calcium in mammalian systems; on protein kinase C; and on the applications of NMR and EPR spectroscopy to the study of biological systems. He has also authored *Interactive Biochemistry CD-ROM and Workbook,* a tutorial CD for students. His work has been supported by the National Institutes of Health, the National Science Foundation, the Muscular Dystrophy Association of America, the Research Corporation, the American Heart Association, and the American Chemical Society. He was a Research Career Development Awardee of the National Institutes of Health, and in 1983 and 1984, he was a Visiting Scientist at the Aarhus University Institute of Physiology Denmark. In 1999, he was Knapp Professor of Chemistry at the University of San Diego. He has taught biochemistry and physical chemistry at the University of Virginia for 29 years. He is a member of the American Society for Biochemistry and Molecular Biology.

Contents in Brief

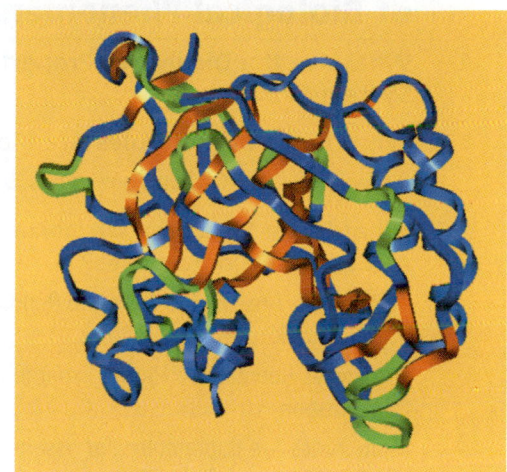

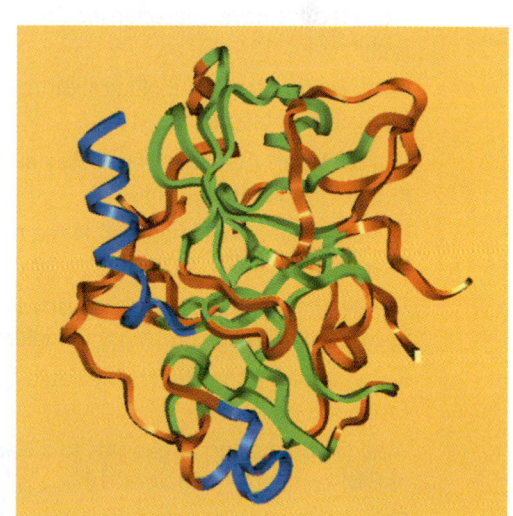

Table of Contents

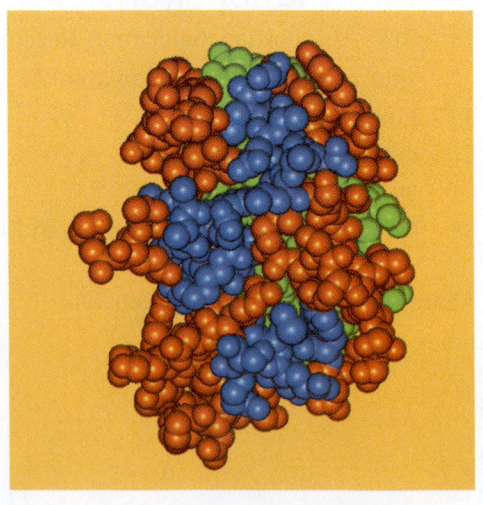

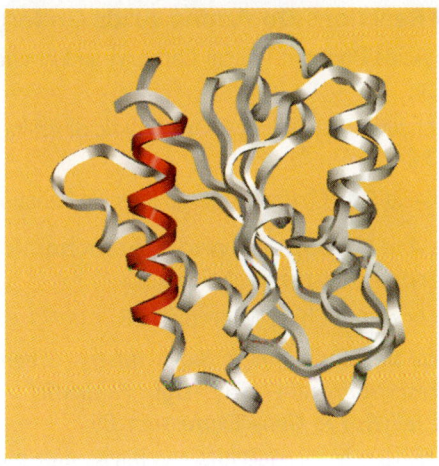

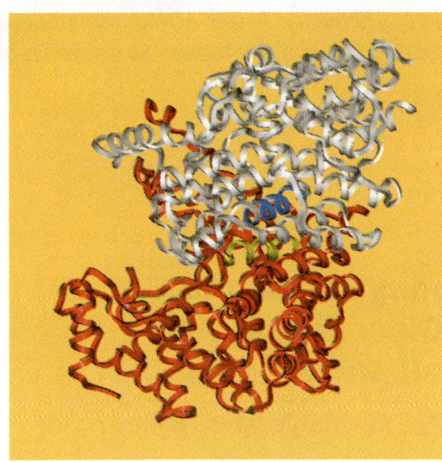

---

7 Carbohydrates and the Glycoconjugates of Cell Surfaces 203

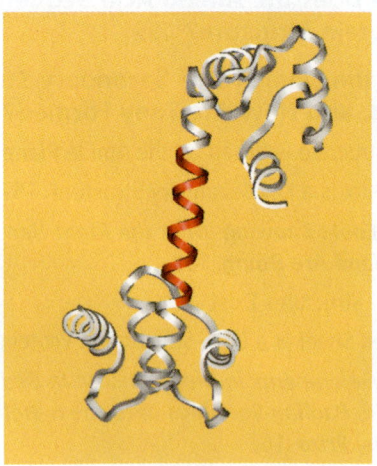

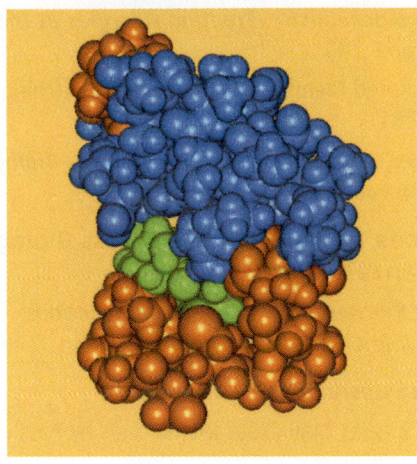

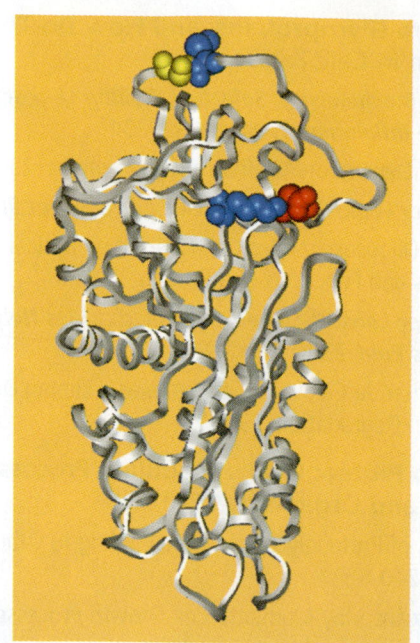

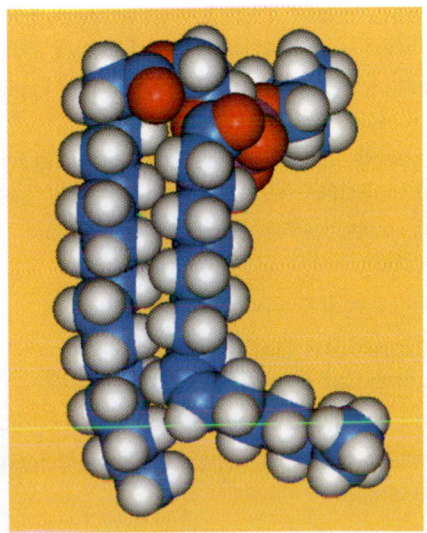

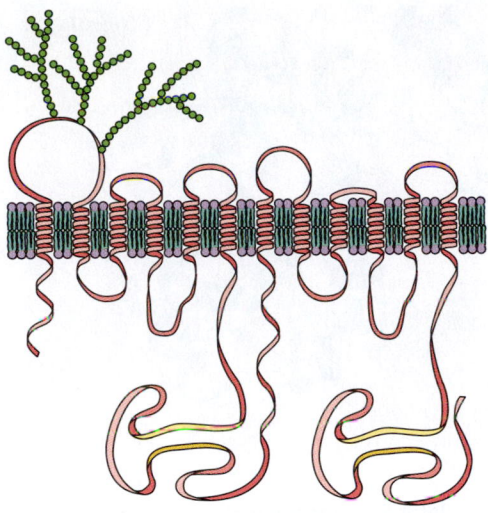

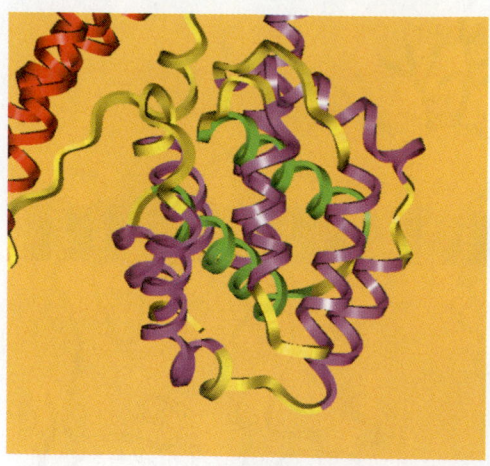

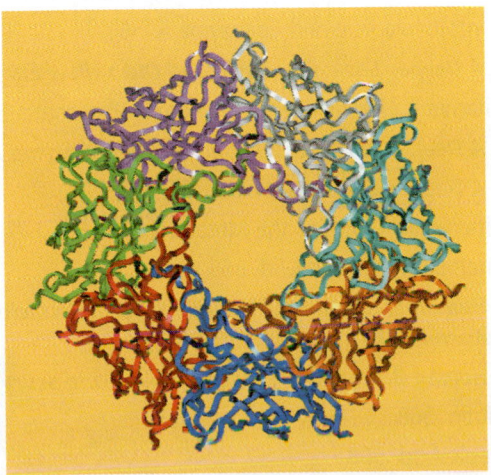

PART III

Metabolism and Its Regulation 536

17 Metabolism—An Overview 538

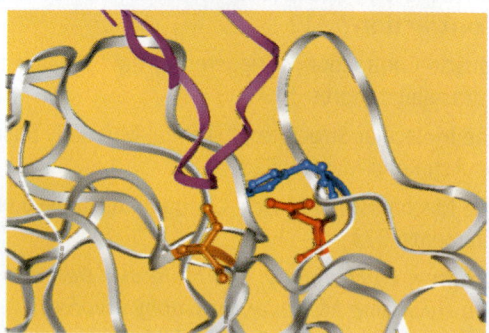

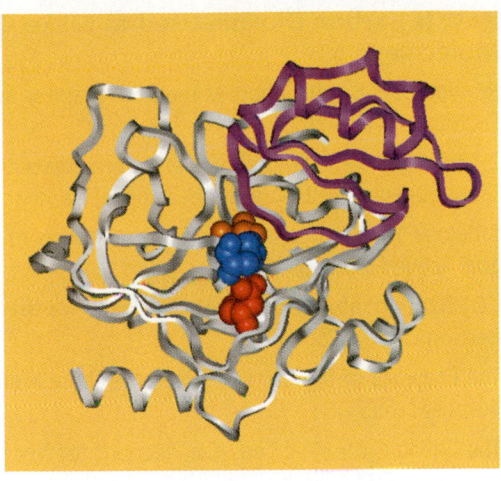

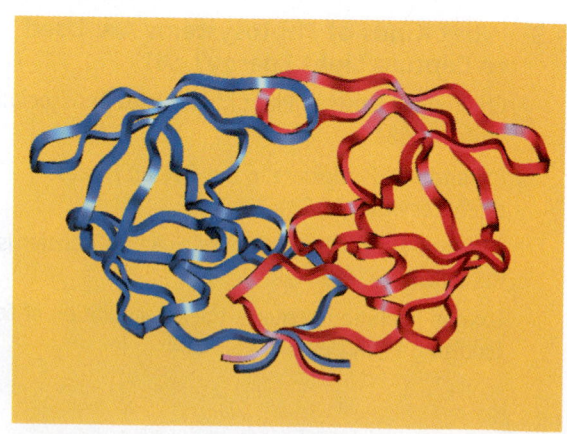

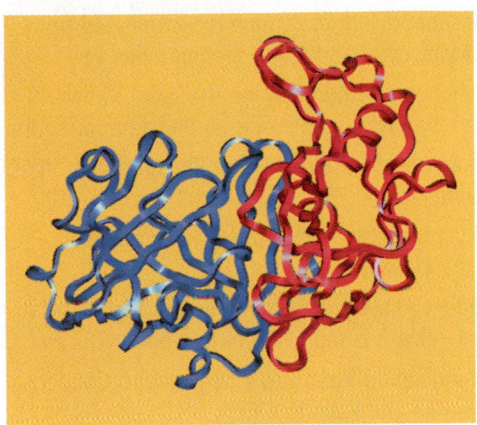

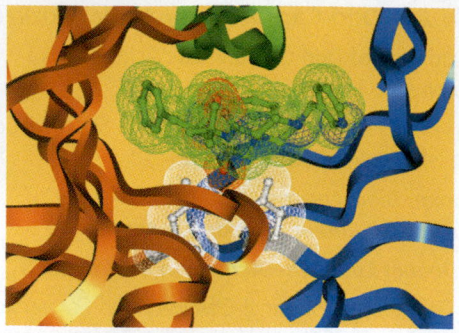

27   Metabolic Integration and Organ Specialization 879

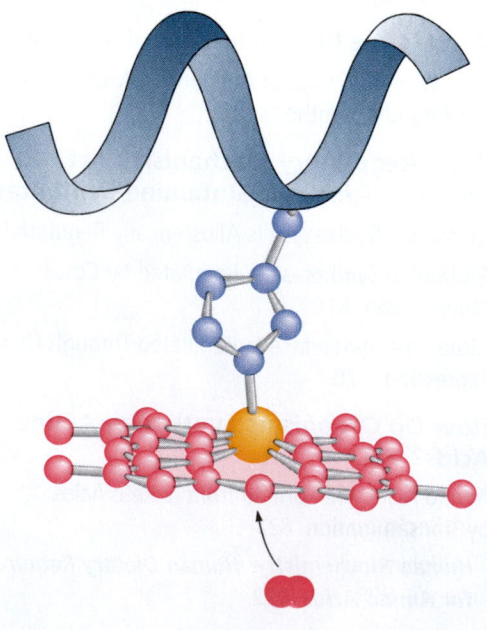

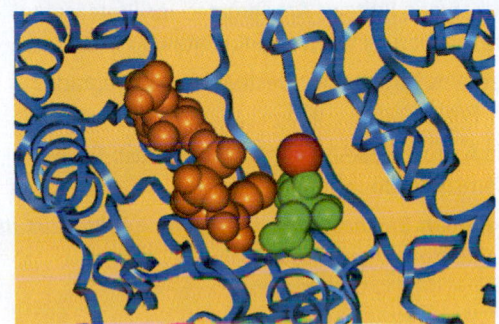

Laboratory Techniques in Biochemistry

All of our knowledge of biochemistry is the outcome of experiments. For the most part, this text presents biochemical knowledge as established fact, but students should never lose sight of the obligatory connection between scientific knowledge and its validation by observation and analysis. The path of discovery by experimental research is often indirect, tortuous, and confounding before the truth is realized. Laboratory techniques lie at the heart of scientific inquiry, and many techniques of biochemistry are presented within these pages to foster a deeper understanding of the biochemical principles and concepts that they reveal.

Recombinant DNA Techniques

Probing the Function of Biomolecules

Techniques Relevant to Clinical Biochemistry

Isolation/Purification of Macromolecules

Analyzing the Physical and Chemical Properties of Biomolecules

BiochemistryNow™ Explore interactive tutorials, animations based on some of these techniques, and test your knowledge on the BiochemistryNow Web site at **http://chemistry.brookscole.com/ggb3**

Asking Questions and Pushing Boundaries

Scientific understanding of the molecular nature of life is growing at an astounding rate. Significantly, society is the prime beneficiary of this increased understanding. Cures for diseases, better public health, remedies for environmental pollution, and the development of cheaper and safer natural products are just a few practical benefits of this knowledge.

In addition, this expansion of information fuels, in the words of Thomas Jefferson, *"the illimitable freedom of the human mind."* Scientists can use the tools of biochemistry and molecular biology to explore all aspects of an organism— from basic questions about its chemical composition; to inquiries into the complexities of its metabolism, its differentiation, and development; to analysis of its evolution and even its behavior. *New procedures based on the results of these explorations lie at the heart of the many modern medical miracles.* Biochemistry is a science whose boundaries now encompass all aspects of biology, from molecules to cells, to organisms, to ecology, and *to all aspects of health care.* Through Essential and Key Questions, we hope that this new edition of *Biochemistry* will encourage students to ask questions of their own and to push the boundaries of their curiosity about science.

Making Connections

As the explication of natural phenomena rests more and more on biochemistry, its inclusion in undergraduate and graduate curricula in biology, chemistry, and the health sciences becomes imperative. The challenge to authors and instructors is a formidable one: how to familiarize students with the essential features of modern biochemistry in an introductory course or textbook. Fortunately, the increased scope of knowledge allows scientists to make generalizations connecting the biochemical properties of living systems with the character of their constituent molecules. As a consequence, these generalizations, validated by repetitive examples, emerge in time as principles of biochemistry, principles that are useful in discerning and describing new relationships between diverse biomolecular functions and in predicting the mechanisms that underlie newly discovered biomolecular processes. Nevertheless, it is increasingly apparent that students must develop skills in inquiry-based learning so that, beyond this first encounter with biochemical principles and concepts, students are equipped to explore science on their own. Much of the design of this new edition is meant to foster the development of such skills.

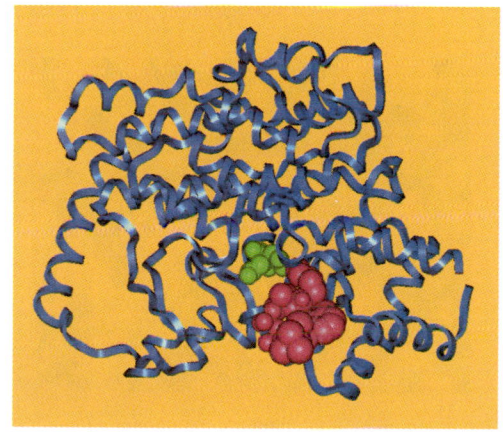

We are both biochemists, but one of us is in a biology department and the other is in a chemistry department. Undoubtedly, we each view biochemistry through the lens of our respective disciplines. We believe, however, that our collaboration on this textbook represents a melding of our perspectives that will provide new dimensions of appreciation and understanding for all students.

Our Audience

This biochemistry textbook is designed to communicate the fundamental principles governing the structure, function, and interactions of biological molecules to students encountering biochemistry for the first time. We aim to bring an appreciation of biochemistry to a broad audience that includes undergraduates majoring in the life sciences, physical sciences, or premedical programs, as well as medical students and graduate students in the various health sciences for whom biochemistry is an important route to understanding human physiology. To make this subject matter more relevant and interesting to all readers, we emphasize, where appropriate, the biochemistry of humans.

Objectives and Building on Previous Editions

We carry forward the clarity of purpose found in previous editions; namely, to illuminate for students the principles governing the structure, function, and interactions of biological molecules. At the same time, this new edition has been revised to reflect tremendous developments in biochemistry. Significantly, emphasis is placed on the interrelationships of ideas so that students can begin to appreciate the overarching questions of biochemistry. We achieve these goals by:

1. Providing a framework that places a chapter in clearer context for students: Questions of a general nature ("Essential Questions") are presented at the beginning of each chapter. These Essential Questions relate the chapter contents to the major ideas of biochemistry.

2. Organizing each chapter by Key Questions: The section headings within chapters are phrased as important questions that serve as organizing principles for a lecture. The subheadings are designed as concept statements that respond to the section headings. Through icons in the margins, in figure legends, and within boxes, students are encouraged to further test their mastery of the Essential and Key Questions and to explore interactive tutorials and animations at the book-specific Web site, BiochemistryNow at **http://chemistry.brookscole.com/ggb3**

3. Repurposing the art program to convey visually the story of biochemistry: More molecular structures are included, and figures that benefit from molecular modeling have been updated.

4. Linking Key Questions to Chapter Summaries: New to this edition are chapter summaries. These summaries recite the key questions posed as section heads and then briefly summarize the important concepts and facts to aid students in organizing the material.

5. Taking advantage of the end-of-chapter Problems: Many more end-of-chapter problems are provided. They serve as meaningful exercises that help students develop problem-solving skills useful in achieving their learning goals. Some problems allow students to become familiar with the quantitative aspects of biochemistry, requiring students to employ calculations to find mathematical answers to relevant structural or functional questions. Other questions address conceptual problems whose answers require application and integration of ideas and concepts introduced in the chapter. Each set of end-of-chapter Problems concludes with MCAT practice questions to aid students in their preparation for standardized examinations such as the MCAT or GRE.

6. Introducing the integrated media package **BiochemistryNow** for students and faculty:

For Students Given that students are very concerned about assessment, we have created the Web site **http://chemistry.brookscole.com/ggb3** for students. This site provides links to resources based on students' responses to typical end-of-chapter/test questions. Students can go to the Web site and work a quiz. If they provide an incorrect answer, they will be directed to the appropriate text reference and/or relevant media tutorial. These include tutorials and animations based on text illustrations. These illustrations are labeled in the text captions as **Active Figures** (see Figure 3.1) and **Animated Figures** (see Figure 3.2). Active Figures have corresponding test questions that quiz students on the concepts of the figures. Animated Figures give life to the art by enabling students to watch the progress of an illustration. This site also includes "Essential Questions" for Biochemistry. These questions are open-ended and may be assigned as student term projects by faculty.

For Faculty Our aim is to provide the best lecture resources in the market. We provide PowerPoint lecture slides and a Multimedia Manager with embedded animations/simulations as well as molecular movies for the classroom.

BiochemistryNow™

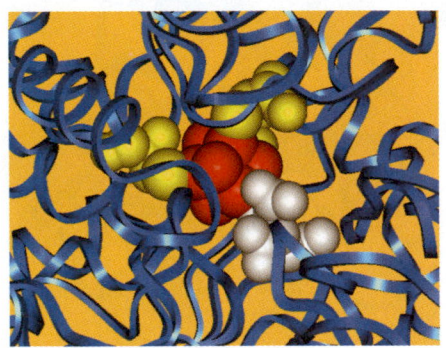

BiochemistryNow™

Organization and Content Changes to This Edition

Part I: Molecular Components of Cells (Chapters 1–12) has been reduced in size, relative to the second edition, from 13 chapters to 12 by bringing various aspects of the carbohydrates of cell surfaces into the carbohydrates chapter and merging previous chapters on membranes and membrane transport into a single chapter. Chapter 3: *Thermodynamics of Biological Systems* provides an early introduction to the central role of thermodynamics in biochemistry. Chapter 7: *Carbohydrates and Glyco-Conjugates* groups together the representative carbohydrates of cells, allowing the range of their structural and functional properties to be treated as a pedagogical unit. And, by combining two previous chapters in one (Chapter 9: *Membranes and Membrane Transport*), we bring together the structure of membranes and one of their primary functions—controlling the movement of materials into and out of the cell—so that students gain a deeper appreciation for the relationship between chemical composition and functional consequences in biological structures.

UPDATED! The power of mass spectrometry in protein identification and amino acid sequencing has been updated and expanded in Chapter 5: *Proteins: Their Primary Structure and Biological Functions.*

UPDATED! Recent advances in our understanding of the protein folding problem are reviewed in Chapter 6: *Proteins: Secondary, Tertiary, and Quaternary Structure.*

NEW! In Chapter 7: *Carbohydrates and Glyco-Conjugates*, the role of boron as an essential element in plant cell wall synthesis is included.

NEW! Chapter 9: *Membranes and Membrane Transport* introduces lipid rafts—recently described aggregates of proteins and lipids giving rise to heterogeneities in the membrane's mosaic of proteins and lipids. The recent scientific excitement deriving from detailed knowledge of the structure of ion channels is featured as well.

NEW! In Chapter 10: *Nucleotides and Nucleic Acids,* material on the newly discovered category of RNAs, the ncRNAs (noncoding RNAs), a class of small, single-stranded RNAs that act through complementary base pairing with their RNA targets is presented.

NEW! In Chapter 11: *Structure of Nucleic Acids,* novel secondary and tertiary structures in RNA, such as pseudoknots, ribose zippers, and coaxial stacking features, are described.

Chapter 12: *Recombinant DNA Technology* covers topic such as cloning, genetic engineering, and PCR, with updates on the emerging sciences of genomics and proteomics that have been spawned by the vast and ever-growing sequence knowledge bases. Proteomics in particular brings a new and exciting global view of metabolism, as reflected in the set of proteins expressed at any moment by a specific cell or cell type.

Part II: Protein Dynamics (Chapters 13–16) presents mechanisms (Chapter 14: *Mechanisms of Enzyme Action*) before regulation (Chapter 15: *Enzyme Regulation*), allowing students to appreciate the catalytic power of enzymes immediately after learning about their kinetic properties (Chapter 13: *Enzyme Kinetics*). Enzymes whose mechanisms are dissected in detail include the serine proteases, the aspartic proteases (including HIV protease), and lysozyme.

NEW! Chapter 14: *Mechanisms of Enzyme Action* highlights the recently revised research of the long-standing classical view of lysozyme as strain-induced destabilization of the substrate followed by enzyme-mediated acid–base catalysis. This research shows that covalent intermediate catalysis plays a prominent role in lysozyme's mechanism of action. Furthermore, emerging appreciation for low-barrier hydrogen bonds in enzymatic catalysis is featured in the aspartic protease mechanism.

NEW! Chapter 16: *Molecular Motors* presents the equation between the chemical energy of ATP and the energy of protein conformational changes. This equation is a unifying concept in biochemistry, applicable to muscle contraction and to oxidative phosphorylation (Chapter 20: *Electron Transport and Oxidative Phosphorylation*).

Part III: Metabolism and Its Regulation (Chapters 17–27) describes the metabolic pathways that orchestrate the synthetic and degradative chemistry of life. The chemical logic of intermediary metabolism is emphasized. Chapter 17: *Metabolism—An Overview* points out the basic similarities in metabolism that unite all forms of life and gives a survey of nutrition and the underlying principles of metabolism, with particular emphasis on the role of vitamins as coenzymes.

The fundamental aspects of catabolic metabolism are described in Chapter 18: Glycolysis, Chapter 19: *The Citric Acid Cycle,* and Chapter 20: *Electron Transport and Oxidative Phosphorylation*. An important highlight in Chapter 20 is the discussion of mitochondrial F_1F_0–ATP synthase as the smallest molecular motor known. ATP synthesis by such integral membrane molecular motors is the principal source of ATP production throughout biology.

UPDATED! Chapter 20 describes how the immediate energy for ATP synthesis is the energy of a protein conformational change (also described in Chapter 16). Conformational energy is delivered to the sites of ATP synthesis in the F_1 part of the ATP synthase by a protein cam that rotates within F_1. Rotation of this cam occurs because it is linked to a proton gradient–driven protein turbine spinning within the plane of membrane.

Chapter 21: *Photosynthesis* describes the photosynthetic processes that capture light energy and use it to carry out the fundamental process of carbohydrate synthesis, upon which virtually all life depends.

UPDATED! A focal point of Chapter 21 is the new information about the molecular structure of photosynthetic reaction centers, those entities that convert the light energy to chemical energy.

Chapters 22–26 complete our coverage of the principal pathways of carbohydrate, lipid, amino acid, purine, and pyrimidine metabolism. Particular emphasis is given to the chemical mechanisms that underlie metabolic reactions and to thermodynamic constraints on metabolism. The regulation of metabolisms is a recurrent theme in these chapters.

Chapter 27: *Metabolic Integration* is unique among textbook chapters in defining the essentially unidirectional nature of metabolic pathways and the stoichiometric role of ATP in driving vital processes that are thermodynamically unfavorable. This chapter also reveals the interlocking logic of metabolic pathways and the metabolic relationships between the various major organs of the human body.

NEW! In Chapter 27, recent advances documenting hormonal controls that govern eating behavior are highlighted in a Human Biochemistry box titled "Are You Hungry?"

Part IV: Information Transfer (Chapters 28–32) addresses the storage and transmission of genetic information in organisms, as well as mechanisms by which organisms interpret and respond to chemical and physical information coming from the environment. The role of DNA molecules as the repository of inheritable information is presented in Chapter 28: *DNA Metabolism,* along with the latest discoveries unraveling the molecular mechanisms underlying the enzymology of DNA replication.

NEW! In Chapter 28, sections on DNA replication and DNA repair treat the biochemistry involved in the maintenance and the replication of genetic information for transmission to daughter cells and accent the exciting new awareness that replication, recombination, and repair are interrelated aspects more appropriately treated together as DNA metabolism.

Chapter 29: *Transcription and the Regulation of Gene Expression* then characterizes the means by which DNA-encoded information is expressed through synthesis of RNA and how expression of this information is regulated.

UPDATED! Highlights of Chapter 29 include recent advances in our understanding of the molecular structure and mechanism of the eukaryotic RNA polymerase II and the DNA-binding transcription factors that modulate its activity.

NEW! In Chapter 29, a unified theory of eukaryotic gene expression is presented, where transcriptional activation, transcription, pre-mRNA processing, nuclear export of mRNA, and translation of mRNA into protein are seen to be parts of a continuous process, with physical and functional connections between the various transcriptional and processing machineries.

NEW! In Chapter 29, detailed emphasis is given to nucleosomes as general repressors of transcription and the prerequisite for chromatin rearrangements in order to activate transcription, along with emphasis on the roles of histone acetylation/deacylation and chromatin remodeling in these processes.

Chapter 30: *Protein Synthesis* discusses the genetic code by which triplets of bases (codons) in mRNA specify particular amino acids in proteins and describes the molecular events that underlie the "second" genetic code—how aminoacyl-tRNA synthetases uniquely recognize their specific tRNA acceptors.

NEW! Chapter 30 presents the structure and function of ribosomes, highlighting new, detailed information on ribosome structure and the interesting realization that 23S rRNA is the peptidyl transferase enzyme responsible for peptide bond formation.

NEW CHAPTER! Chapter 31: *The Protein Life Cycle: Folding, Processing, and Degradation,* a chapter new to this edition, has been added to cover the emerging information on the fate of proteins once they are formed, including their delivery to the cellular sites where they belong. This chapter also reviews the necessity for molecular chaperones in the proper folding of proteins and the emerging importance of proteasome-mediated protein degradation as a means to regulate cellular levels of specific proteins.

Chapter 32: *The Reception and Transmission of Extracellular Information* pulls together an up-to-date perspective on the rapidly changing fields of cellular signaling. It stresses the information transfer aspects involved in the interpretation of environmental information and includes coverage of hormone action, signal transduction cascades, membrane receptors, oncogenes, tumor suppressor genes, sensory transduction and neurotransmission, and the biochemistry of neurological disorders.

NEW! Chapter 32 includes the results of the Human Genome Project, which has revealed 868 protein kinase genes, the so-called kinome. The categorization of these genes is a major step in understanding the evolutionary relationships between these ATP-dependent protein phosphorylating enzymes and a key to understanding the organization of signal transduction pathways.

Key Feature: The Essential Question

The prominent feature of this new edition is the organization of each chapter around an *Essential Question* theme. The term *Essential Question* comes from learning theory. Inquiry-based learning is a powerful way to develop skills for effective comprehension and management of burgeoning scientific information. Inquiry-based learning is a process in which students formulate a hierarchy of questions, seek out information that bears upon or answers the questions, and then build a knowledge base that ultimately reveals insights and understanding about the original question. Skills developed in inquiry-based learning equip students with sound pedagogical techniques for life-long, self-directed learning and with an appreciation for new scientific discoveries. Each chapter in this book is framed around an *Essential Question*. Essential questions are defined as *questions that require decision making or a plan of action*. They force students to become actively engaged in their learning and encourage curiosity and imagination about the subject matter to be learned. Thus, students no longer act merely as passive recipients of information from the instructor. For example, the Essential Question of Chapter 3 asks, "What are the laws and principles of thermodynamics that allow us to describe the flows and interchange of heat, energy, and matter in systems of interest?" The section heads then pose more specific questions, such as, "What Is the Daily Human Requirement for ATP?" (see Section 3.8). The end-of-chapter summary then brings the question and a synopsis of the answer together for the student. In addition, the BiochemistryNow Web site at **http://chemistry.brookscole.com/ggb3** expands on this Essential Question theme by asking students to explore their knowledge of key concepts. It is hoped that the student will then take these questions and formulate more of their own. The desired outcome is knowledge and understanding and acquisition of a critical skill applicable to learning biochemistry.

Biochemistry ⊗ Now™

More Features

- Each part opens with an essay written by a prominent biochemist who addresses an emerging paradigm (or shift in our fundamental thinking) about an aspect of biochemistry. These essays broaden the Essential Question theme of the text.

     Part I, Thomas A. Steitz, Yale University: "How Do Proteins (and Sometimes RNA) Work Together in Large Assemblies to Facilitate Various Processes of the Cell?"

     Part II, Stephen J. Benkovic, The Pennsylvania State University: "How Do Enzymes Work?"

     Part III, Juliet A. Gerrard, University of Canterbury (NZ): "Metabolism: Chemistry of Life or Biology of Molecules?"

     Part IV, David L. Brautigan, University of Virginia School of Medicine: "How Do Cells Coordinate Their Activities?"

- Up-to-date coverage gives students the most current information on biochemistry since the last edition of this text.
- Illustrations are improved by adding steps to drawings and legends to make them easier to follow.
- Many new molecular models are added to give students insight into the structures of biomolecules.
- The number of end-chapter problems is increased by 50%. Chapter Integration problems are marked and incorporate material from other chapters to form connections among topics.
- MCAT practice problems are added at the end of each chapter to help students prepare for this and related exams, such as the GRE.
- Human Biochemistry boxes emphasize the central role of basic biochemistry in medicine and the health sciences. These essays often present clinically important issues such as diet, diabetes, and cardiovascular health.

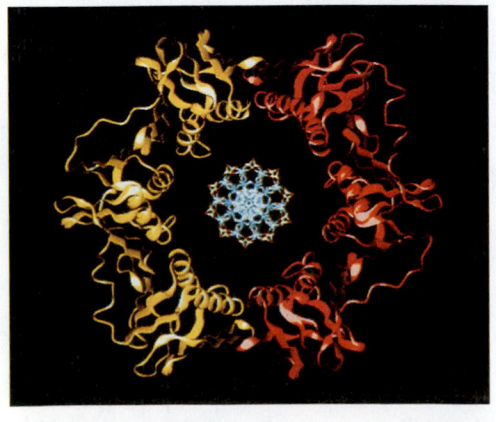

- A Deeper Look boxes expand on the text, highlighting selected topics or experimental observations.
- Critical Developments in Biochemistry boxes emphasize recent and historical advances in the field.
- A critically acclaimed four-color art program complements the text and aids in the students' ability to visualize biochemistry as a three-dimensional science.
- Up-to-date references at the end of each chapter make it easy for students to find additional information about each topic.
- The experimental nature of biochemistry is highlighted, and a list of Laboratory Techniques found in this book can be seen on page xxxiv.
- The Web site at **http://brookscole.com.chemistry/ggb3** that accompanies this book is thoroughly integrated via Web links and annotations in the margins.

Complete Support Package

For Students

The Student Solutions Manual, Study Guide and Problems Book, by David K. Jemiolo (Vassar College) and Steven M. Theg (University of California, Davis) This manual includes summaries of the chapters, detailed solutions to all end-of-chapter problems, a guide to key points of each chapter, important definitions, and illustrations of major metabolic pathways. (0-534-49035-2)

Student Lecture Notebook Perfect for note taking during lecture, this convenient booklet consists of black and white reproductions of the Transparency Acetates. (0-534-49036-0)

BiochemistryNow at **http://chemistry.brookscole.com/ggb3** This is the first Web-based assessment-centered learning tool specifically for biochemistry courses, developed in concert with the text, extending the "Essential Questions" framework. PIN code access to BiochemistryNow is packaged FREE with every new copy of the text.

Biochemistry Now™

InfoTrac® College Edition Four months of access to **InfoTrac College Edition** is automatically packaged FREE with every new copy of this text. This world-class, online university library offers the full text of articles from almost 5000 scholarly and popular publications—updated daily and going back as much as 22 years. With 24-hour access to so many outstanding resources, **InfoTrac College Edition** will help you in *all* of your courses.

For Professors

Instructor materials are available to qualified adopters. Please consult your local Thomson Brooks/Cole sales representative for details. Please visit the *Biochemistry* Web site at **http://chemistry.brookscole.com/ggb3** to see samples of these materials, request a desk copy, locate your sales representative, or purchase a copy online.

Multimedia Manager The simple way to create exciting, multimedia lectures! This easy-to-use, dual-platform digital library and presentation tool provides text art and tables in a variety of electronic formats that can be exported into other software packages. This enhanced CD-ROM also contains engaging simulations, molecular models, and QuickTime™ movies to supplement your lectures and a lecture outline with integrated media. (0-534-49038-7)

Transparency Acetates This set of full-color acetates includes a selection of the most pedagogically important images from the text. (0-534-49039-5)

Printed Test Bank, by Larry Jackson, Montana State University Includes 25 to 40 multiple-choice questions per chapter for professors to use as tests, quizzes, or homework assignments. (0-534-49037-9)

iLrn Testing This dual-platform CD-ROM features approximately 1000 multiple-choice problems and questions, representing every chapter of the text. The questions are graded in level of difficulty for your convenience, and answers are provided on a separate grading key. (0-534-49040-9)

The Brooks/Cole Chemistry Resource Center at http://chemistry.brookscole.com; Book-Specific Instructor's Resource Web site at **http://chemistry.brookscole. com/ggb3** Updated monthly, **The Brooks/Cole Chemistry Resource Center** and our password-protected **Book-Specific Instructor's Web Site** give you access to Web links, lecture outlines, a downloadable *Solutions Manual,* Microsoft® *PowerPoint*® Slides, and a Multimedia Manager demo.

WebTutor™ ToolBox on WebCT and Blackboard Preloaded with content and available free via PIN code when packaged with this text, **WebTutor ToolBox** pairs all the content of this text's rich **Book Companion Web Site** with all the sophisticated course management functionality of a WebCT or Blackboard product. **WebTutor ToolBox** is ready to use as soon as you log on—or, you can customize its preloaded content by uploading images and other resources, adding Web links, or creating your own practice materials. **WebCT (0-534-65667-6)** • **Blackboard (0-534-65658-7)**

Resource Integration Guide This Instructor's Edition includes a key teaching tool, the *Resource Integration Guide.* The guide provides grids that link each chapter to corresponding instructional and supplemental resources. See pages 9–24 of the Preview Section at the beginning of the Instructor's Edition.

Acknowledgments

We are indebted to the many experts in biochemistry and molecular biology who carefully reviewed the third edition manuscript at several stages for their outstanding and invaluable advice on how to construct an effective textbook.

Glenn Cunningham
University of Central Florida

Mark Elliott
Old Dominion University

Eric Fisher
University of Illinois, Springfield

Tim Formosa
University of Utah, School of Medicine

Jon Friesen
Illinois State University

E. M. Gregory
Virginia Polytechnic Institute
and State University

Martyn Gunn
Texas A&M University

Ben Horenstein
University of Florida

Jon Kaguni
Michigan State University

Richard Karpel
University of Maryland,
Baltimore County

Gary Kunkel
Texas A&M University

Robert Marsh
University of Texas, Dallas

Steven Metallo
Georgetown University

Susanne Nonekowski
University of Toledo

Richard Paselk
Humboldt State University

Darrell Peterson
Virginia Commonwealth University

Michael Reddy
University of Wisconsin, Milwaukee

David Schooley
University of Nevada, Reno

Catherine Yang
Rowan University

We particularly thank the four outstanding biochemists who graciously wrote the essays that introduce each part of this book: Thomas A. Steitz, Yale Univer-

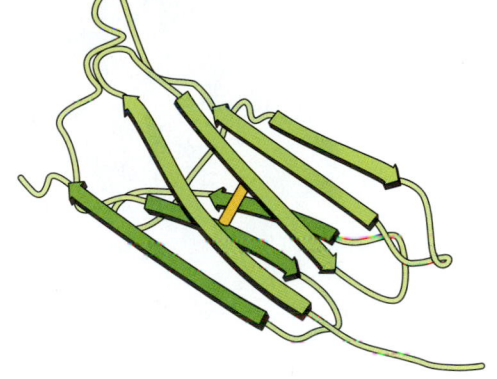

sity; Stephen J. Benkovic, The Pennsylvania State University; Juliet A. Gerrard, University of Canterbury (NZ); and David L. Brautigan, University of Virginia School of Medicine.

We also wish to warmly and gratefully acknowledge many other people who assisted and encouraged us in this endeavor. This book remains a legacy of John Vondeling, who originally recruited us to its authorship. His threats, admonishments, and entreaties, laced with the wisdom he drew from vast experience in the publishing world, were instrumental in urging us to completion of the task. We acknowledge that his presence stills looms large over our book, and we are grateful for it. David Harris, our new publisher, has brought infectious enthusiasm and an unwavering emphasis on student learning as the fundamental purpose of our collective endeavor.

Sandi Kiselica, our Developmental Editor, is a biochemist in her own right. Her fascination with our shared discipline has given her a particular interest in our book and a singular purpose: to keep us focused on the matters at hand, the urgencies of the schedule, and limits of scale in a textbook's dimensions. The dint of her efforts has been a major factor in the fruition of our writing projects. She is truly a colleague in these endeavors. We also applaud the unsung but absolutely indispensable contributions by those whose efforts transformed a rough manuscript into this final product: Lisa Weber, Project Manager, Editorial Production; Rob Hugel, Creative Director; Peggy Williams, Development Editor, Media; Donna Kelley, Technology Project Manager; and Alyssa White, Assistant Editor. If this book has visual appeal and editorial grace, it is due to them. The beautiful illustrations that not only decorate this text but explain its contents are a testament to the creative and tasteful work of Cindy Geiss, Director of Art Services, Graphic World Inc.; to the team at Dartmouth Graphics; and to the legacy of John Woolsey and Patrick Lane at J/B Woolsey Associates. We are thankful to our many colleagues who provided original art and graphic images for this work, particularly Professor Jane Richardson of Duke University, who gave us numerous original line drawings of the protein ribbon structures, and Dr. Michal Sabat, who prepared many of the molecular graphics displayed herein. Lauren Gregg was a big help in compiling thoughts on the key questions. Vera Fleischer, Jeremy Jannotta, Jason Cheatam, and a procession of undergraduates—Catherine Baxter, Megan Doucet, Tiffany Held, Flora Lackner, Edward O'Neil, Maleeha Qazi, Milton Truong, and Justin Watson—are the direct creators of the Flash animations, Java applets, and many of the interactive tutorials on protein structure and function; we are very grateful for their participation. Heidi Creiser Glasgow transformed our text and graphics into electronic format for the e-version of the book. We owe a very special thank-you to Rosemary Jurbala Grisham, devoted spouse of Charles and wonderfully tolerant friend of Reg, who works tirelessly as our cheerleader and our photograph acquisitions specialist; in appreciation for her many contributions spoken and unspoken, we once again dedicate this book to her. Also to be acknowledged with love and pride are Georgia Grant and our children Jeffrey, Randal, and Robert Garrett, and David, Emily, and Andrew Grisham, as well as Clancy, a Golden retriever of epic patience and perspicuity, and Jatszi, Jazmine, and Jasper, three Hungarian Pulis whose unseen eyes view life with an energetic curiosity we all should emulate. With the publication of this third edition of our text, we celebrate and commemorate the role of our mentors in bringing biochemistry to life for us—Alvin Nason, Kenneth R. Schug, William D. McElroy, Ronald E. Barnett, Maurice J. Bessman, Albert S. Mildvan, Ludwig Brand, and Rufus Lumry.

Reginald H. Garrett   Charles M. Grisham
Charlottesville, VA   Ivy, VA

January 2004

Molecular Components of Cells

An Essay by Thomas A. Steitz

The work of a cell is carried out by proteins and RNA macromolecules whose sequences are encoded in the DNA genome. The specific sequences of these molecules dictate their folding into precise three-dimensional structures that are often designed to interact with other macromolecules in order to form a larger complex assembly. It is these three-dimensional structures and the particular conformational mobilities they possess that enable the macromolecules to carry out their assigned tasks. Much of biochemical research involves discovering what task each macromolecule or assembly carries out, measuring how fast it does it, and understanding the chemistry of the process in terms of the three-dimensional structures of the macromolecules.

While many central cellular functions are carried out by individual macromolecules, many are accomplished by large complexes of pro-

How do proteins (and sometimes RNA) work together in large assemblies to facilitate various processes of the cell?

teins or complexes of proteins and RNA that often assemble and disassemble in order to accomplish the processes they promote or regulate. Examples abound in all aspects of cellular metabolism and include the ribosome, which synthesizes proteins; the spliceosome, which removes intervening sequences from messenger RNA; regulatory proteins that act with RNA polymerase to control RNA synthesis; the replisome, which copies genomic DNA; the nuclear pore, which mediates the transport of macromolecules across the nuclear membrane; motility systems such as muscle; and protein and RNA degradation systems, just to name a few. Because these complexes usually undergo large structural changes during their functioning, a complete understanding of the mechanisms by which these assemblies achieve their function requires atomic structures of these assemblies captured at each step in the process that they facilitate. Knowing these structures allows one to create a "movie" that shows how this assembly can carry out the dynamic biological process. This can be accomplished by determining the crystal structures of the assemblies. Particularly in the case of rare and very large assemblies like the nuclear pore, their structures can be approximate from three dimensional cryoelectron microscope images (at 7 to 12 Å resolution) combined with crystal structures of smaller pieces. As is the case with smaller enzymes and other proteins functioning alone, these structural studies need to be integrated and understood within the context of kinetic measurements, mutagenesis, and biochemical studies.

CHAPTER 1

Chemistry Is the Logic of Biological Phenomena

Sperm approaching an egg.

© Dennis Wilson/CORBIS

> *"...everything that living things do can be understood in terms of the jigglings and wigglings of atoms."*
> **Richard P. Feynman** *Lectures on Physics*, Addison-Wesley, 1963

Key Questions

Biochemistry ☰Now™ Test yourself on these Key Questions at BiochemistryNow at **http://chemistry.brookscole.com/ggb3**

Essential Question

Molecules are lifeless. Yet, the properties of living things derive from the properties of molecules. ***Despite the spectacular diversity of life, the elaborate structure of biological molecules, and the complexity of vital mechanisms, are life functions ultimately interpretable in chemical terms?***

Molecules are lifeless. Yet, in appropriate complexity and number, molecules compose living things. These living systems are distinct from the inanimate world because they have certain extraordinary properties. They can grow, move, perform the incredible chemistry of metabolism, respond to stimuli from the environment, and most significantly, replicate themselves with exceptional fidelity. The complex structure and behavior of living organisms veil the basic truth that their molecular constitution can be described and understood. The chemistry of the living cell resembles the chemistry of organic reactions. Indeed, cellular constituents or **biomolecules** must conform to the chemical and physical principles that govern all matter. Despite the spectacular diversity of life, the intricacy of biological structures, and the complexity of vital mechanisms, life functions are ultimately interpretable in chemical terms. *Chemistry is the logic of biological phenomena.*

1.1 What Are the Distinctive Properties of Living Systems?

First, the most obvious quality of **living organisms** is that they are *complicated and highly organized* (Figure 1.1). For example, organisms large enough to be seen with the naked eye are composed of many **cells,** typically of many types. In turn, these cells possess subcellular structures, called **organelles,** which are complex assemblies of very large polymeric molecules, called **macromolecules.**

(a)

Tony Angermayer/Photo Researchers, Inc.

(b)

Thomas C. Boydon/Marie Selby Botanical Gardens

FIGURE 1.1 **(a)** Mandrill *(Mandrillus sphinx)*, a baboon native to West Africa. **(b)** Tropical orchid *(Bulbophyllum blumei)*, New Guinea.

These macromolecules themselves show an exquisite degree of organization in their intricate three-dimensional architecture, even though they are composed of simple sets of chemical building blocks, such as sugars and amino acids. Indeed, the complex three-dimensional structure of a macromolecule, known as its **conformation,** is a consequence of interactions between the monomeric units, according to their individual chemical properties.

Second, *biological structures serve functional purposes.* That is, biological structures play a role in the organism's existence. From parts of organisms, such as limbs and organs, down to the chemical agents of metabolism, such as enzymes and metabolic intermediates, a biological purpose can be given for each component. Indeed, it is this functional characteristic of biological structures that separates the science of biology from studies of the inanimate world such as chemistry, physics, and geology. In biology, it is always meaningful to seek the purpose of observed structures, organizations, or patterns, that is, to ask what functional role they serve within the organism.

Third, *living systems are actively engaged in energy transformations.* Maintenance of the highly organized structure and activity of living systems depends on their ability to extract energy from the environment. The ultimate source of energy is the sun. Solar energy flows from photosynthetic organisms (organisms able to capture light energy by the process of photosynthesis) through food chains to herbivores and ultimately to carnivorous predators at the apex of the food pyramid (Figure 1.2). The biosphere is thus a system through which energy flows. Organisms capture some of this energy, be it from photosynthesis or the metabolism of food, by forming special energized biomolecules, of which **ATP** and **NADPH** are the two most prominent examples (Figure 1.3). (Commonly used abbreviations such as ATP and NADPH are defined on the inside back cover of this book.) ATP and NADPH are energized biomolecules because they represent chemically useful forms of stored energy. We explore the chemical basis of this stored energy in subsequent chapters. For now, suffice it to say that when these molecules react with other molecules in the cell, the energy released can be used to drive unfavorable processes. That is, ATP, NADPH, and related compounds are the power sources that drive the energy-requiring activities of the cell, including biosynthesis, movement, osmotic work against concentration gradients, and in special instances, light emission (bioluminescence). Only upon death does an organism reach equilibrium with its inanimate environment. *The*

Biochemistry ⊗ Now™ This icon, appearing throughout the book, indicates an opportunity to explore interactive tutorials, animations, and test your knowledge for an exam on the BiochemistryNow Web site at http://chemistry.brookscole.com/ggb3

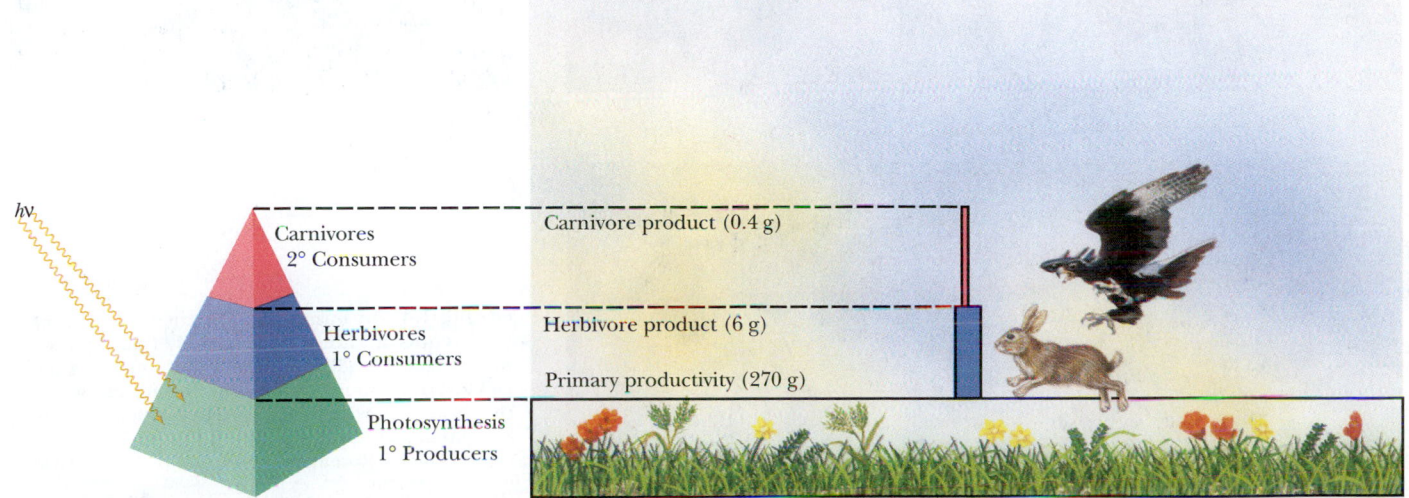

Productivity per square meter of a Tennessee field

FIGURE 1.2 The food pyramid. Photosynthetic organisms at the base capture light energy. Herbivores and carnivores derive their energy ultimately from these primary producers.

FIGURE 1.3 ATP and NADPH, two biochemically important energy-rich compounds.

living state is characterized by the flow of energy through the organism. At the expense of this energy flow, the organism can maintain its intricate order and activity far removed from equilibrium with its surroundings, yet exist in a state of apparent constancy over time. This state of apparent constancy, or so-called **steady state,** is actually a very dynamic condition: Energy and material are consumed by the organism and used to maintain its stability and order. In contrast, inanimate matter, as exemplified by the universe in totality, is moving to a condition of increasing disorder or, in thermodynamic terms, maximum entropy.

Entropy is a thermodynamic term used to designate that amount of energy in a system that is unavailable to do work.

(a)

(b)

(c)

FIGURE 1.4 Organisms resemble their parents. **(a)** The Garrett guys at Three-Chopt Road. Left to right: son Robert, grandson Ricky, Reg Garrett, son Jeff, grandson Jackson (foreground), grandson Reggie, and son Randal. **(b)** Orangutan with infant. **(c)** The Grishams on the Continental Divide, Cottonwood Pass, Colorado. Left to right: Charles, Rosemary, Emily, Andrew, and David.

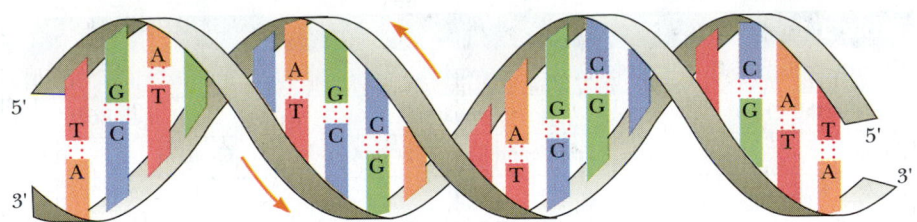

Biochemistry⟋Now™ **ANIMATED FIGURE 1.5** The DNA double helix. Two complementary polynucleotide chains running in opposite directions can pair through hydrogen bonding between their nitrogenous bases. Their complementary nucleotide sequences give rise to structural complementarity. **See this figure animated at http://chemistry.brookscole.com/ggb3**

Fourth, *living systems have a remarkable capacity for self-replication.* Generation after generation, organisms reproduce virtually identical copies of themselves. This self-replication can proceed by a variety of mechanisms, ranging from simple division in bacteria to sexual reproduction in plants and animals; but in every case, it is characterized by an astounding degree of fidelity (Figure 1.4). Indeed, if the accuracy of self-replication were significantly greater, the evolution of organisms would be hampered. This is so because evolution depends upon natural selection operating on individual organisms that vary slightly in their fitness for the environment. The fidelity of self-replication resides ultimately in the chemical nature of the genetic material. This substance consists of polymeric chains of deoxyribonucleic acid, or **DNA,** which are structurally complementary to one another (Figure 1.5). These molecules can generate new copies of themselves in a rigorously executed polymerization process that ensures a faithful reproduction of the original DNA strands. In contrast, the molecules of the inanimate world lack this capacity to replicate. A crude mechanism of replication, or specification of unique chemical structure according to some blueprint, must have existed at life's origin. This primordial system no doubt shared the property of **structural complementarity** (see later section) with the highly evolved patterns of replication prevailing today.

1.2 | What Kinds of Molecules Are Biomolecules?

The elemental composition of living matter differs markedly from the relative abundance of elements in the earth's crust (Table 1.1). Hydrogen, oxygen, carbon, and nitrogen constitute more than 99% of the atoms in the human body, with most of the H and O occurring as H_2O. Oxygen, silicon, aluminum, and iron are the most abundant atoms in the earth's crust, with hydrogen, carbon, and nitrogen being relatively rare (less than 0.2% each). Nitrogen as dinitrogen (N_2) is the predominant gas in the atmosphere, and carbon dioxide (CO_2) is present at a level of 0.05%, a small but critical amount. Oxygen is also abundant in the atmosphere and in the oceans. What property unites H, O, C, and N and renders these atoms so suitable to the chemistry of life? It is their ability to form covalent bonds by electron-pair sharing. Furthermore, H, C, N, and O are among the lightest elements of the periodic table capable of forming such bonds (Figure 1.6). Because the strength of covalent bonds is inversely proportional to the atomic weights of the atoms involved, H, C, N, and O form the strongest covalent bonds. Two other covalent bond–forming elements, phosphorus (as phosphate [$-OPO_3^{2-}$] derivatives) and sulfur, also play important roles in biomolecules.

Biomolecules Are Carbon Compounds

All biomolecules contain carbon. The prevalence of C is due to its unparalleled versatility in forming stable covalent bonds through electron-pair sharing. Carbon can form as many as four such bonds by sharing each of the four electrons

Atoms	e^- pairing	Covalent bond	Bond energy (kJ/mol)
H· + H· ⟶ H:H		H—H	436
·C· + H· ⟶ ·C:H		—C—H	414
·C· + ·C· ⟶ ·C:C·		—C—C—	343
·C· + ·N: ⟶ ·C:N:		—C—N⟨	292
·C· + ·O: ⟶ ·C:O:		—C—O—	351
·C· + ·C· ⟶ C::C		⟩C=C⟨	615
·C· + ·N: ⟶ C::N:		⟩C=N—	615
·C· + ·O: ⟶ C::O:		⟩C=O	686
·O: + ·O: ⟶ ·O:O·		—O—O—	142
·O: + ·O: ⟶ :O::O:		O=O	402
·N: + ·N: ⟶ :N:::N:		N≡N	946
·N: + H· ⟶ :N:H		⟩N—H	393
·O: + H· ⟶ ·O:H		—O—H	460

Biochemistry⟋Now™ **ACTIVE FIGURE 1.6**
Covalent bond formation by e^- pair sharing. **Test yourself on the concepts in this figure at http://chemistry.brookscole.com/ggb3**

Table 1.1					
Composition of the Earth's Crust, Seawater, and the Human Body*					
Earth's Crust		**Seawater**		**Human Body†**	
Element	**%**	**Compound**	**mM**	**Element**	**%**
O	47	Cl^-	548	H	63
Si	28	Na^+	470	O	25.5
Al	7.9	Mg^{2+}	54	C	9.5
Fe	4.5	SO_4^{2-}	28	N	1.4
Ca	3.5	Ca^{2+}	10	Ca	0.31
Na	2.5	K^+	10	P	0.22
K	2.5	HCO_3^-	2.3	Cl	0.08
Mg	2.2	NO_3^-	0.01	K	0.06
Ti	0.46	HPO_4^{2-}	<0.001	S	0.05
H	0.22			Na	0.03
C	0.19			Mg	0.01

*Figures for the earth's crust and the human body are presented as percentages of the total number of atoms; seawater data are in millimoles per liter. Figures for the earth's crust do not include water, whereas figures for the human body do.
†Trace elements found in the human body serving essential biological functions include Mn, Fe, Co, Cu, Zn, Mo, I, Ni, and Se.

in its outer shell with electrons contributed by other atoms. Atoms commonly found in covalent linkage to C are C itself, H, O, and N. Hydrogen can form one such bond by contributing its single electron to the formation of an electron pair. Oxygen, with two unpaired electrons in its outer shell, can participate in two covalent bonds, and nitrogen, which has three unshared electrons, can form three such covalent bonds. Furthermore, C, N, and O can share two electron pairs to form double bonds with one another within biomolecules, a property that enhances their chemical versatility. Carbon and nitrogen can even share three electron pairs to form triple bonds.

Two properties of carbon covalent bonds merit particular attention. One is the ability of carbon to form covalent bonds with itself. The other is the tetrahedral nature of the four covalent bonds when carbon atoms form only single bonds. Together these properties hold the potential for an incredible variety of linear, branched, and cyclic compounds of C. This diversity is multiplied further by the possibilities for including N, O, and H atoms in these compounds (Figure 1.7). We can therefore envision the ability of C to generate complex structures in three dimensions. These structures, by virtue of appropriately included N, O, and H atoms, can display unique chemistries suitable to the living state. Thus, we may ask, is there any pattern or underlying organization that brings order to this astounding potentiality?

1.3 What Is the Structural Organization of Complex Biomolecules?

Examination of the chemical composition of cells reveals a dazzling variety of organic compounds covering a wide range of molecular dimensions (Table 1.2). As this complexity is sorted out and biomolecules are classified according to the similarities of their sizes and chemical properties, an organizational pattern emerges. The biomolecules are built according to a structural hierarchy: Simple molecules are the units for building complex structures.

The molecular constituents of living matter do not reflect randomly the infinite possibilities for combining C, H, O, and N atoms. Instead, only a limited set of the many possibilities is found, and these collections share certain properties essential to the establishment and maintenance of the living state. The

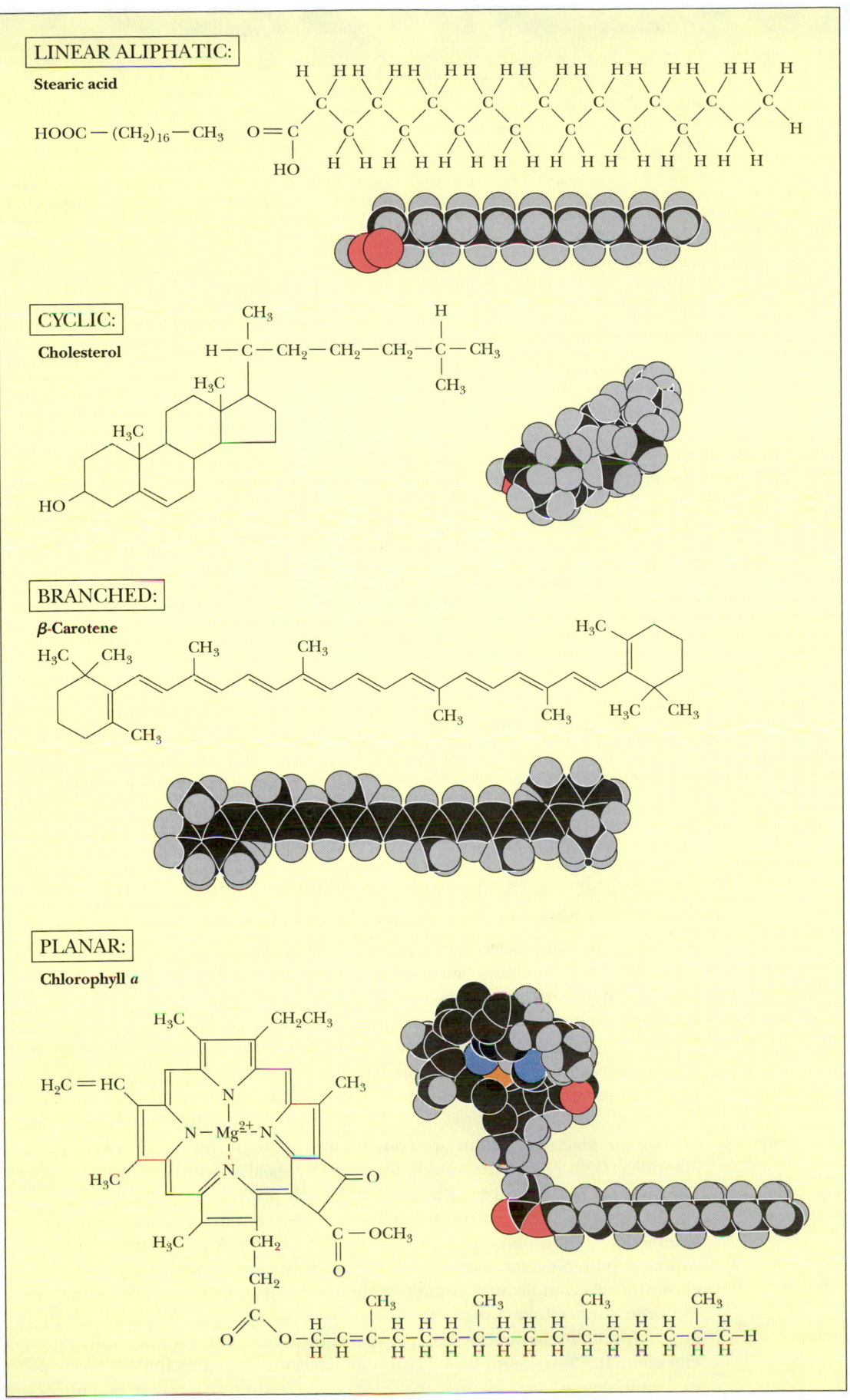

FIGURE 1.7 Examples of the versatility of C—C bonds in building complex structures: linear, cyclic, branched, and planar.

Table 1.2

Biomolecular Dimensions

The dimensions of mass* and length for biomolecules are given typically in daltons and nanometers,[†] respectively. One dalton (D) is the mass of one hydrogen atom, 1.67×10^{-24} g. One nanometer (nm) is 10^{-9} m, or 10 Å (angstroms).

Biomolecule	Length (long dimension, nm)	Mass	
		Daltons	Picograms
Water	0.3	18	
Alanine	0.5	89	
Glucose	0.7	180	
Phospholipid	3.5	750	
Ribonuclease (a small protein)	4	12,600	
Immunoglobulin G (IgG)	14	150,000	
Myosin (a large muscle protein)	160	470,000	
Ribosome (bacteria)	18	2,520,000	
Bacteriophage ϕX174 (a very small bacterial virus)	25	4,700,000	
Pyruvate dehydrogenase complex (a multienzyme complex)	60	7,000,000	
Tobacco mosaic virus (a plant virus)	300	40,000,000	6.68×10^{-5}
Mitochondrion (liver)	1,500		1.5
Escherichia coli cell	2,000		2
Chloroplast (spinach leaf)	8,000		60
Liver cell	20,000		8,000

*Molecular mass is expressed in units of daltons (D) or kilodaltons (kD) in this book; alternatively, the dimensionless term *molecular weight*, symbolized by Mr, and defined as the ratio of the mass of a molecule to 1 dalton of mass, is used.
[†]Prefixes used for powers of 10 are

10^6	mega M	10^{-3}	milli m
10^3	kilo k	10^{-6}	micro μ
10^{-1}	deci d	10^{-9}	nano n
10^{-2}	centi c	10^{-12}	pico p
		10^{-15}	femto f

most prominent aspect of biomolecular organization is that macromolecular structures are constructed from simple molecules according to a hierarchy of increasing structural complexity. What properties do these biomolecules possess that make them so appropriate for the condition of life?

Metabolites Are Used to Form the Building Blocks of Macromolecules

The major precursors for the formation of biomolecules are water, carbon dioxide, and three inorganic nitrogen compounds—ammonium (NH_4^+), nitrate (NO_3^-), and dinitrogen (N_2). Metabolic processes assimilate and transform these inorganic precursors through ever more complex levels of biomolecular order (Figure 1.8). In the first step, precursors are converted to **metabolites,** simple organic compounds that are intermediates in cellular energy transformation and in the biosynthesis of various sets of **building blocks:** amino acids, sugars, nucleotides, fatty acids, and glycerol. Through covalent linkage of these building blocks, the **macromolecules** are constructed: proteins, polysaccharides, polynucleotides (DNA and RNA), and lipids. (Strictly speaking, lipids contain relatively few building blocks and are therefore not really polymeric like other macromolecules; however, lipids are important contributors to higher levels of complexity.) Interactions among macromolecules lead to the next level of structural organization, **supramolecular complexes.** Here, various members of one or more of the classes of macromolecules come together to form specific assemblies that serve important subcellular functions. Examples of these supramolecular assemblies are multifunctional enzyme complexes, ribosomes, chromosomes, and cytoskeletal elements. For example, a eukaryotic ribosome contains four different RNA molecules and at least 70 unique proteins. These supramolecular assemblies are an interesting contrast to their components because their structural integrity is

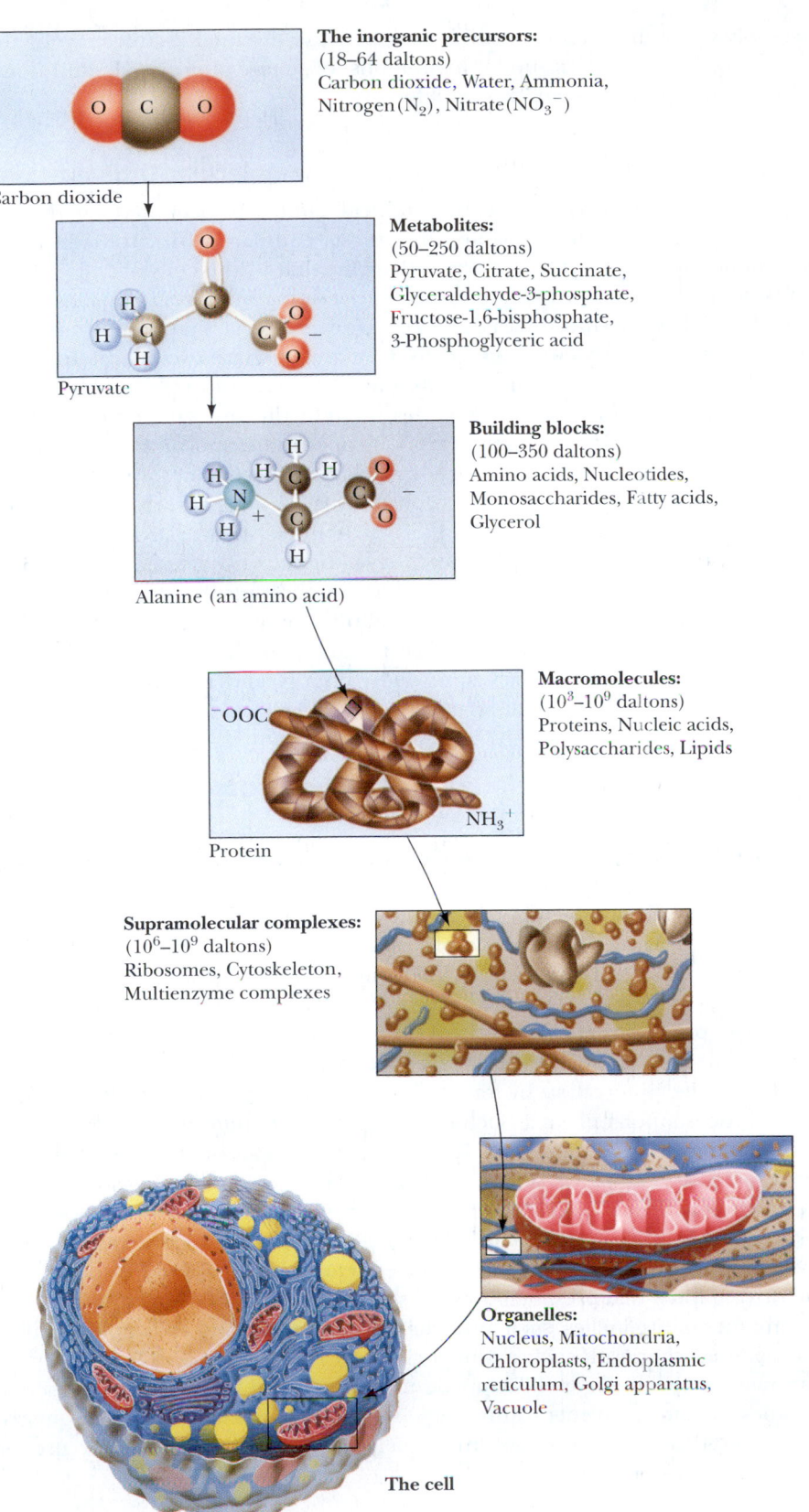

The inorganic precursors:
(18–64 daltons)
Carbon dioxide, Water, Ammonia,
Nitrogen(N_2), Nitrate(NO_3^-)

Carbon dioxide

Metabolites:
(50–250 daltons)
Pyruvate, Citrate, Succinate,
Glyceraldehyde-3-phosphate,
Fructose-1,6-bisphosphate,
3-Phosphoglyceric acid

Pyruvate

Building blocks:
(100–350 daltons)
Amino acids, Nucleotides,
Monosaccharides, Fatty acids,
Glycerol

Alanine (an amino acid)

Macromolecules:
(10^3–10^9 daltons)
Proteins, Nucleic acids,
Polysaccharides, Lipids

Protein

Supramolecular complexes:
(10^6–10^9 daltons)
Ribosomes, Cytoskeleton,
Multienzyme complexes

Organelles:
Nucleus, Mitochondria,
Chloroplasts, Endoplasmic
reticulum, Golgi apparatus,
Vacuole

The cell

FIGURE 1.8 Molecular organization in the cell is a
hierarchy.

maintained by noncovalent forces, not by covalent bonds. These noncovalent
forces include hydrogen bonds, ionic attractions, van der Waals forces, and
hydrophobic interactions between macromolecules. Such forces maintain these
supramolecular assemblies in a highly ordered functional state. Although noncovalent forces are weak (less than 40 kJ/mol), they are numerous in these

assemblies and thus can collectively maintain the essential architecture of the supramolecular complex under conditions of temperature, pH, and ionic strength that are consistent with cell life.

Organelles Represent a Higher Order in Biomolecular Organization

The next higher rung in the hierarchical ladder is occupied by the organelles, entities of considerable dimensions compared with the cell itself. Organelles are found only in **eukaryotic cells,** that is, the cells of "higher" organisms (eukaryotic cells are described in Section 1.5). Several kinds, such as mitochondria and chloroplasts, evolved from bacteria that gained entry to the cytoplasm of early eukaryotic cells. Organelles share two attributes: They are cellular inclusions, usually membrane bounded, and they are dedicated to important cellular tasks. Organelles include the nucleus, mitochondria, chloroplasts, endoplasmic reticulum, Golgi apparatus, and vacuoles, as well as other relatively small cellular inclusions, such as peroxisomes, lysosomes, and chromoplasts. The **nucleus** is the repository of genetic information as contained within the linear sequences of nucleotides in the DNA of chromosomes. **Mitochondria** are the "power plants" of cells by virtue of their ability to carry out the energy-releasing aerobic metabolism of carbohydrates and fatty acids, capturing the energy in metabolically useful forms such as ATP. **Chloroplasts** endow cells with the ability to carry out photosynthesis. They are the biological agents for harvesting light energy and transforming it into metabolically useful chemical forms.

Membranes Are Supramolecular Assemblies That Define the Boundaries of Cells

Membranes define the boundaries of cells and organelles. As such, they are not easily classified as supramolecular assemblies or organelles, although they share the properties of both. Membranes resemble supramolecular complexes in their construction because they are complexes of proteins and lipids maintained by noncovalent forces. **Hydrophobic interactions** are particularly important in maintaining membrane structure. Hydrophobic interactions arise because water molecules prefer to interact with each other rather than with nonpolar substances. The presence of nonpolar molecules lessens the range of opportunities for water–water interaction by forcing the water molecules into ordered arrays around the nonpolar groups. Such ordering can be minimized if the individual nonpolar molecules redistribute from a dispersed state in the water into an aggregated organic phase surrounded by water. The spontaneous assembly of membranes in the aqueous environment where life arose and exists is the natural result of the hydrophobic ("water-fearing") character of their lipids and proteins. Hydrophobic interactions are the creative means of membrane formation and the driving force that presumably established the boundary of the first cell. The membranes of organelles, such as nuclei, mitochondria, and chloroplasts, differ from one another, with each having a characteristic protein and lipid composition tailored to the organelle's function. Furthermore, the creation of discrete volumes or **compartments** within cells is not only an inevitable consequence of the presence of membranes but usually an essential condition for proper organellar function.

The Unit of Life Is the Cell

The cell is characterized as the unit of life, the smallest entity capable of displaying the attributes associated uniquely with the living state: growth, metabolism, stimulus response, and replication. In the previous discussions, we explicitly narrowed the infinity of chemical complexity potentially available to organic life and we previewed an organizational arrangement, moving from simple to complex, that provides interesting insights into the functional and structural plan of the cell. Nevertheless, we find no obvious explanation within these features for the living characteristics of cells. Can we find other themes represented within biomolecules that are explicitly chemical yet anticipate or illuminate the living condition?

1.4 How Do the Properties of Biomolecules Reflect Their Fitness to the Living Condition?

If we consider what attributes of biomolecules render them so fit as components of growing, replicating systems, several biologically relevant themes of structure and organization emerge. Furthermore, as we study biochemistry, we will see that these themes serve as principles of biochemistry. Prominent among them is the *necessity for information and energy in the maintenance of the living state.* Some biomolecules must have the capacity to contain the information, or "recipe," of life. Other biomolecules must have the capacity to translate this information so that the organized structures essential to life are synthesized. Interactions between these structures *are* the processes of life. An orderly mechanism for abstracting energy from the environment must also exist in order to obtain the energy needed to drive these processes. What properties of biomolecules endow them with the potential for such remarkable qualities?

Biological Macromolecules and Their Building Blocks Have a "Sense" or Directionality

The macromolecules of cells are built of units—amino acids in proteins, nucleotides in nucleic acids, and carbohydrates in polysaccharides—that have **structural polarity.** That is, these molecules are not symmetrical, and so they can be thought of as having a "head" and a tail." Polymerization of these units to form macromolecules occurs by head-to-tail linear connections. Because of this, the polymer also has a head and a tail, and hence, the macromolecule has a "sense" or direction to its structure (Figure 1.9).

Biological Macromolecules Are Informational

Because biological macromolecules have a sense to their structure, the sequential order of their component building blocks, when read along the length of the molecule, has the capacity to specify information in the same manner that the letters of the alphabet can form words when arranged in a linear sequence (Figure 1.10). Not all biological macromolecules are rich in information. Polysaccharides are often composed of the same sugar unit repeated over and over, as in cellulose or starch, which are homopolymers of many glucose units. On the other hand, proteins and polynucleotides are typically composed of building blocks arranged in no obvious repetitive way; that is, their sequences are unique, akin to the letters and punctuation that form this descriptive sentence. In these unique sequences lies meaning. Discerning the meaning, however, requires some mechanism for recognition.

Biomolecules Have Characteristic Three-Dimensional Architecture

The structure of any molecule is a unique and specific aspect of its identity. Molecular structure reaches its pinnacle in the intricate complexity of biological macromolecules, particularly the proteins. Although proteins are linear sequences of covalently linked amino acids, the course of the protein chain

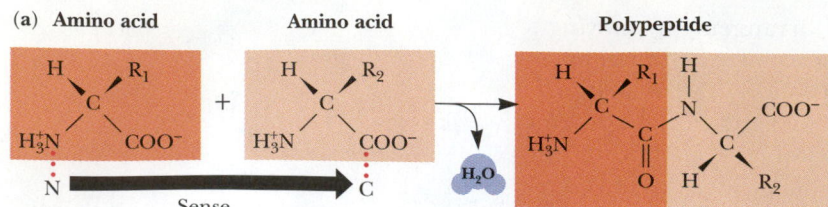

(a) Amino acid   Amino acid   Polypeptide

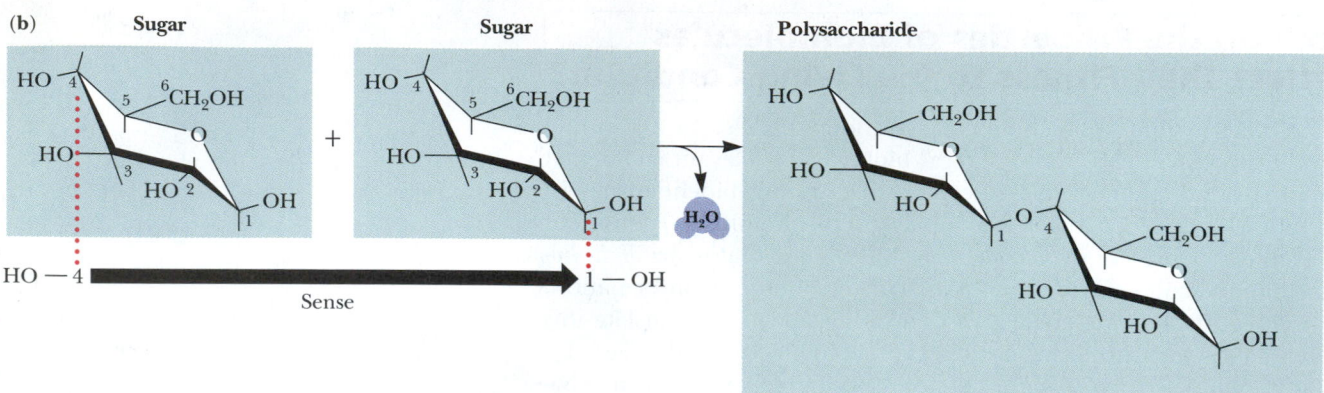

(b)   Sugar   Sugar   Polysaccharide

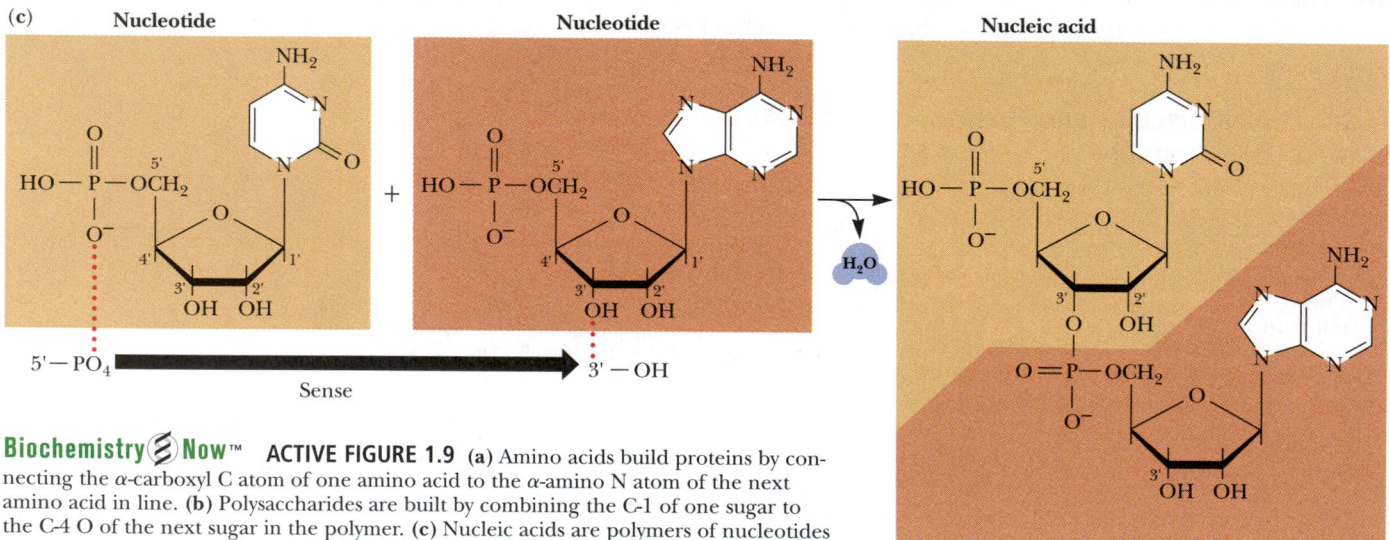

(c)   Nucleotide   Nucleotide   Nucleic acid

Biochemistry⦿Now™   ACTIVE FIGURE 1.9 (a) Amino acids build proteins by connecting the α-carboxyl C atom of one amino acid to the α-amino N atom of the next amino acid in line. **(b)** Polysaccharides are built by combining the C-1 of one sugar to the C-4 O of the next sugar in the polymer. **(c)** Nucleic acids are polymers of nucleotides linked by bonds between the 3′-OH of the ribose ring of one nucleotide to the 5′-PO₄ of its neighboring nucleotide. All three of these polymerization processes involve bond formations accompanied by the elimination of water (dehydration synthesis reactions). **Test yourself on the concepts in this figure at http://chemistry.brookscole.com/ggb3**

A strand of DNA

5′ ┤T┤T┤C┤A┤G┤C┤A┤A┤T┤A┤A┤G┤G┤G┤T┤C┤C┤T┤A┤C┤G┤G┤A┤G├ 3′

A polypeptide segment

Phe — Ser — Asn — Lys — Gly — Pro — Thr — Glu

A polysaccharide chain

Glc — Glc — Glc — Glc — Glc — Glc — Glc — Glc — Glc

Biochemistry⦿Now™   ACTIVE FIGURE 1.10
The sequence of monomeric units in a biological polymer has the potential to contain information if the diversity and order of the units are not overly simple or repetitive. Nucleic acids and proteins are information-rich molecules; polysaccharides are not. **Test yourself on the concepts in this figure at http://chemistry.brookscole.com/ggb3**

can turn, fold, and coil in the three dimensions of space to establish a specific, highly ordered architecture that is an identifying characteristic of the given protein molecule (Figure 1.11).

Weak Forces Maintain Biological Structure and Determine Biomolecular Interactions

Covalent bonds hold atoms together so that molecules are formed. In contrast, **weak chemical forces** or **noncovalent bonds** (hydrogen bonds, van der Waals forces, ionic interactions, and hydrophobic interactions) are intramolecular or intermolecular attractions between atoms. None of these forces, which typically range from 4 to 30 kJ/mol, are strong enough to bind free atoms together (Table 1.3). The average kinetic energy of molecules at 25°C is 2.5 kJ/mol, so the energy of weak forces is only several times greater than the dissociating tendency due to thermal motion of molecules. Thus, these weak forces create interactions that are constantly forming and breaking at physiological temperature, unless by cumulative number they impart stability to the structures generated by their collective action. These weak forces merit further discussion because their attributes profoundly influence the nature of the biological structures they build.

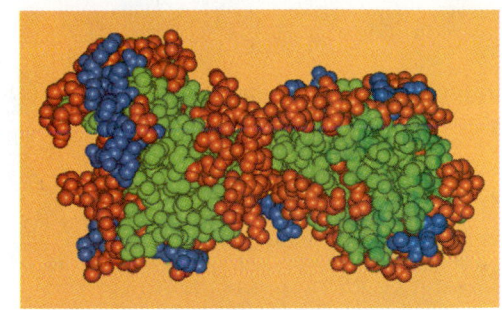

FIGURE 1.11 Three-dimensional space-filling representation of part of a protein molecule, the antigen-binding domain of immunoglobulin G (IgG). IgG is a major type of circulating antibody. Each of the spheres represents an atom in the structure.

Van der Waals Attractive Forces Play an Important Role in Biomolecular Interactions

Van der Waals forces are the result of induced electrical interactions between closely approaching atoms or molecules as their negatively charged electron clouds fluctuate instantaneously in time. These fluctuations allow attractions to occur between the positively charged nuclei and the electrons of nearby atoms. Van der Waals interactions include dipole–dipole interactions, whose interaction energies decrease as $1/r^3$; dipole-induced dipole interactions, which fall off as $1/r^5$; and induced dipole–induced dipole interactions, often called **dispersion** or **London dispersion forces,** which diminish as $1/r^6$. Dispersion forces contribute to the attractive intermolecular forces between all molecules, even those without permanent dipoles, and are thus generally more important than dipole–dipole attractions. Van der Waals attractions operate only over a very limited interatomic distance (0.3 to 0.6 nm) and are an effective bonding interaction at physiological temperatures only when a number of atoms in a molecule can interact with several atoms in a neighboring molecule. For this to occur, the atoms on interacting molecules must pack together neatly. That is,

A *dipole* is any structure with equal and opposite electrical charges separated by a small distance.

Table 1.3			
Weak Chemical Forces and Their Relative Strengths and Distances			
Force	**Strength (kJ/mol)**	**Distance (nm)**	**Description**
Van der Waals interactions	0.4–4.0	0.3–0.6	Strength depends on the relative size of the atoms or molecules and the distance between them. The size factor determines the area of contact between two molecules: The greater the area, the stronger the interaction.
Hydrogen bonds	12–30	0.3	Relative strength is proportional to the polarity of the H bond donor and H bond acceptor. More polar atoms form stronger H bonds.
Ionic interactions	20	0.25	Strength also depends on the relative polarity of the interacting charged species. Some ionic interactions are also H bonds: $—NH_3^+$ . . . $^-OOC—$
Hydrophobic interactions	<40	—	Force is a complex phenomenon determined by the degree to which the structure of water is disordered as discrete hydrophobic molecules or molecular regions coalesce.

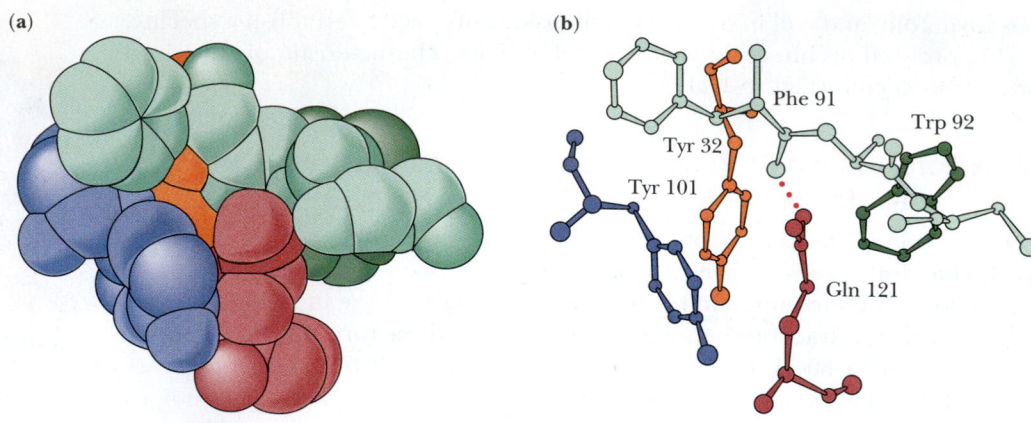

FIGURE 1.12 Van der Waals packing is enhanced in molecules that are structurally complementary. Gln[121] represents a surface protuberance on the protein lysozyme. This protuberance fits nicely within a pocket (formed by Tyr[101], Tyr[32], Phe[91], and Trp[92]) in the antigen-binding domain of an antibody raised against lysozyme. (See also Figure 1.16.) **(a)** A space-filling representation. **(b)** A ball-and-stick model. *(From Amit, A. G., et al., 1986. Three-dimensional structure of an antigen-antibody complex at 2.8 Å resolution. Science* **233:**747–753, *figure 5.)*

their molecular surfaces must possess a degree of structural complementarity (Figure 1.12).

At best, van der Waals interactions are weak and individually contribute 0.4 to 4.0 kJ/mol of stabilization energy. However, the sum of many such interactions within a macromolecule or between macromolecules can be substantial. For example, model studies of heats of sublimation show that each methylene group in a crystalline hydrocarbon accounts for 8 kJ, and each C—H group in a benzene crystal contributes 7 kJ of van der Waals energy per mole. Calculations indicate that the attractive van der Waals energy between the enzyme lysozyme and a sugar substrate that it binds is about 60 kJ/mol.

When two atoms approach each other so closely that their electron clouds interpenetrate, strong repulsion occurs. Such *repulsive* van der Waals forces follow an inverse 12th-power dependence on r ($1/r^{12}$), as shown in Figure 1.13. Between the repulsive and attractive domains lies a low point in the potential curve. This low point defines the distance known as the **van der Waals contact distance,** which is the interatomic distance that results if only van der Waals forces hold two atoms together. The limit of approach of two atoms is determined by the sum of their van der Waals radii (Table 1.4).

Hydrogen Bonds Are Important in Biomolecular Interactions

Hydrogen bonds form between a hydrogen atom covalently bonded to an electronegative atom (such as oxygen or nitrogen) and a second electronegative atom that serves as the hydrogen bond acceptor. Several important biological examples are given in Figure 1.14. Hydrogen bonds, at a strength of 12 to 30 kJ/mol, are stronger than van der Waals forces and have an additional property: H bonds are cylindrically symmetrical and tend to be highly directional, forming straight bonds between donor, hydrogen, and acceptor atoms. Hydrogen bonds are also more specific than van der Waals interactions because they require the presence of complementary hydrogen donor and acceptor groups.

Ionic Interactions   **Ionic interactions** are the result of attractive forces between oppositely charged polar functions, such as negative carboxyl groups and positive amino groups (Figure 1.15). These electrostatic forces average about 20 kJ/mol in aqueous solutions. Typically, the electrical charge is radially distributed, so these interactions may lack the directionality of hydrogen bonds or the precise fit of van der Waals interactions. Nevertheless, because the opposite charges are restricted to sterically defined positions, ionic interactions can impart a high degree of structural specificity.

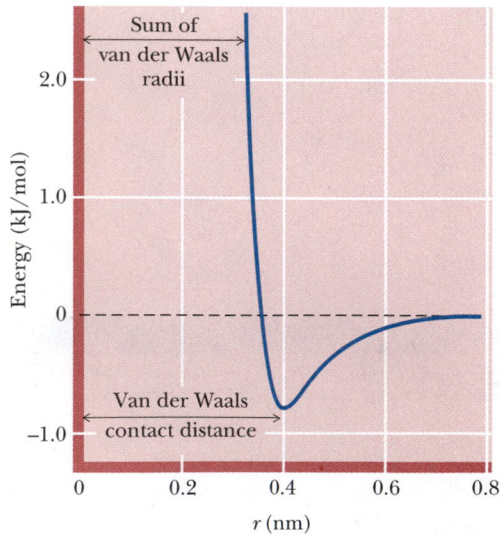

FIGURE 1.13 The van der Waals interaction energy profile as a function of the distance, *r*, between the centers of two atoms. The energy was calculated using the empirical equation $U = B/r^{12} - A/r^{6}$. (Values for the parameters $B = 11.5 \times 10^{-6}$ kJnm[12]/mol and $A = 5.96 \times 10^{-3}$ kJnm[6]/mol for the interaction between two carbon atoms are from Levitt, M., 1974. Energy refinement of hen egg-white lysozyme. *Journal of Molecular Biology* **82:**393–420.)

Table 1.4
Radii of the Common Atoms of Biomolecules

Atom	Van der Waals Radius (nm)	Covalent Radius (nm)	Atom Represented to Scale
H	0.1	0.037	
C	0.17	0.077	
N	0.15	0.070	
O	0.14	0.066	
P	0.19	0.096	
S	0.185	0.104	
Half-thickness of an aromatic ring	0.17	—	

The strength of electrostatic interactions is highly dependent on the nature of the interacting species and the distance, r, between them. Electrostatic interactions may involve **ions** (species possessing discrete charges), **permanent dipoles** (having a permanent separation of positive and negative charge), and **induced dipoles** (having a temporary separation of positive and negative charge induced by the environment). Between two ions, the strength of interaction diminishes as $1/r$. The interaction energy between permanent dipoles falls off as $1/r^3$, whereas the energy between an ion and an induced dipole falls off as $1/r^4$.

Hydrophobic Interactions Hydrophobic interactions result from the strong tendency of water to exclude nonpolar groups or molecules (see Chapter 2). Hydrophobic interactions arise not so much because of any intrinsic affinity of nonpolar substances for one another (although van der Waals forces do promote the weak bonding of nonpolar substances), but because water molecules prefer the stronger interactions that they share with one another, compared to their interaction with nonpolar molecules. Hydrogen-bonding interactions between polar water molecules can be more varied and numerous if nonpolar molecules come together to form a distinct organic phase. This phase separation raises the entropy of water because fewer water molecules are arranged in orderly arrays around individual nonpolar molecules. It is these preferential interactions between water molecules that "exclude" hydrophobic substances from aqueous solution and drive the tendency of nonpolar molecules to cluster together. Thus, nonpolar regions of biological macromolecules are often buried in the molecule's interior to exclude them from the aqueous milieu. The formation of oil droplets as hydrophobic nonpolar lipid molecules coalesce in the presence of water is an approximation of this phenomenon. These tendencies have important consequences in the creation and maintenance of the macromolecular structures and supramolecular assemblies of living cells.

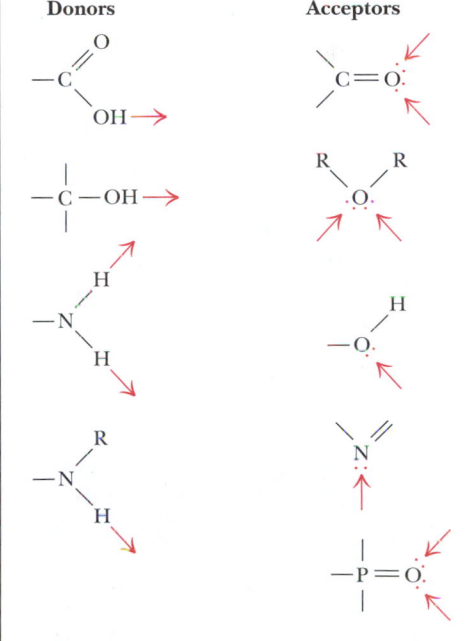

(a) **H bonds**

Bonded atoms	Approximate bond length*
O—H---O	0.27 nm
O—H---O⁻	0.26 nm
O—H---N	0.29 nm
N—H---O	0.30 nm
⁺N—H---O	0.29 nm
N—H---N	0.31 nm

*Lengths given are distances from the atom covalently linked to the H to the atom H bonded to the hydrogen:

O—H---O
|←0.27 nm→|

(b) **Functional groups that are important H-bond donors and acceptors:**

Donors   Acceptors

Biochemistry Now™ ANIMATED FIGURE 1.14
Some of the biologically important H bonds and functional groups that serve as H bond donors and acceptors. **See this figure animated at http://chemistry.brookscole.com/ggb3**

Magnesium ATP

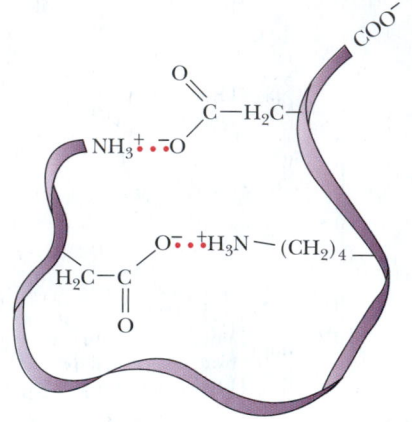

Intramolecular ionic bonds between oppositely charged groups on amino acid residues in a protein

Protein strand

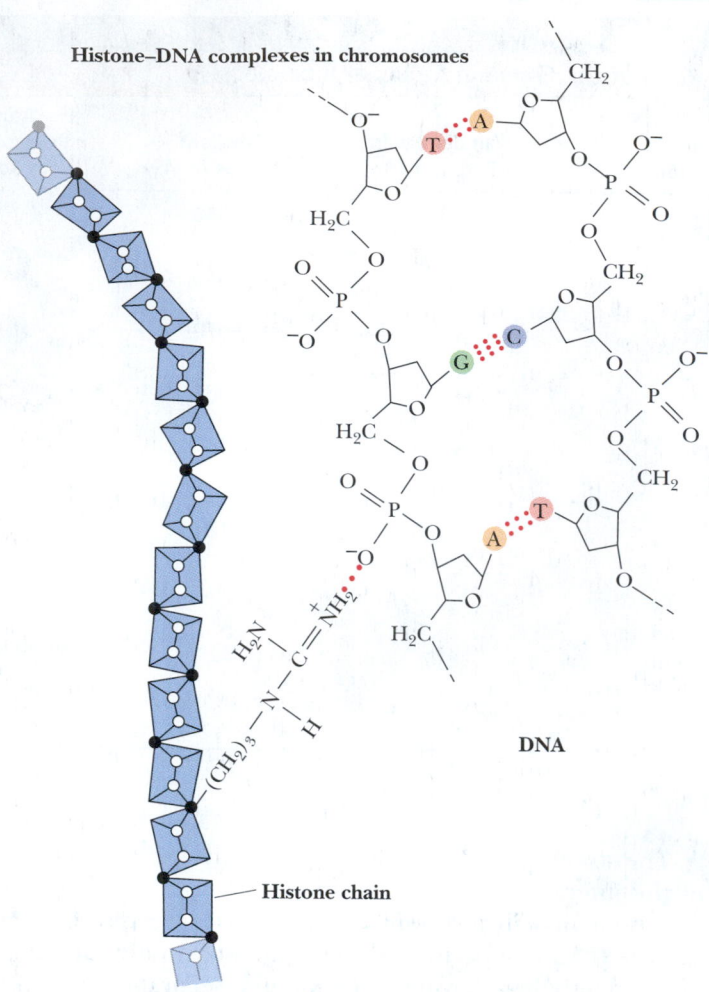

Histone–DNA complexes in chromosomes

DNA

Histone chain

Biochemistry ⒺNow™ **ANIMATED FIGURE 1.15**
Ionic bonds in biological molecules. **See this figure animated at http://chemistry.brookscole.com/ggb3**

The Defining Concept of Biochemistry Is "Molecular Recognition Through Structural Complementarity"

Structural complementarity is the means of recognition in biomolecular interactions. The complicated and highly organized patterns of life depend on the ability of biomolecules to recognize and interact with one another in very specific ways. Such interactions are fundamental to metabolism, growth, replication, and other vital processes. The interaction of one molecule with another, a protein with a metabolite, for example, can be most precise if the structure of one is complementary to the structure of the other, as in two connecting pieces of a puzzle or, in the more popular analogy for macromolecules and their **ligands,** a lock and its key (Figure 1.16). *This principle of structural complementarity is the very essence of biomolecular recognition.* Structural complementarity is the significant clue to understanding the functional properties of biological systems. Biological systems from the macromolecular level to the cellular level operate via specific molecular recognition mechanisms based on structural complementarity: A protein recognizes its specific metabolite, a strand of DNA recognizes its complementary strand, sperm recognize an egg. All these interactions involve structural complementarity between molecules.

Biomolecular Recognition Is Mediated by Weak Chemical Forces

Weak chemical forces underlie the interactions that are the basis of biomolecular recognition. It is important to realize that because these interactions are sufficiently weak, they are readily reversible. Consequently, biomolecular inter-

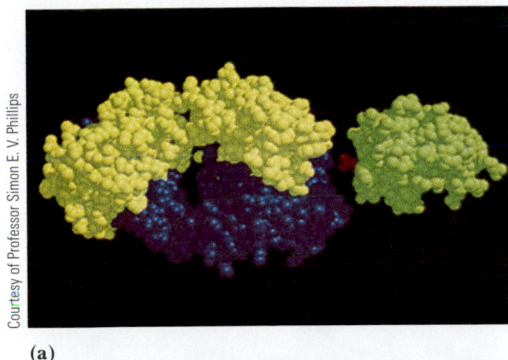

(a)

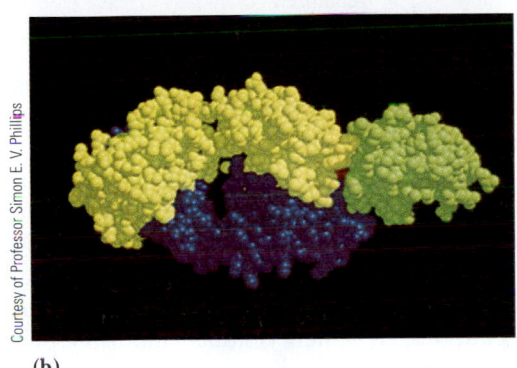

(b)

Puzzle

Lock and key

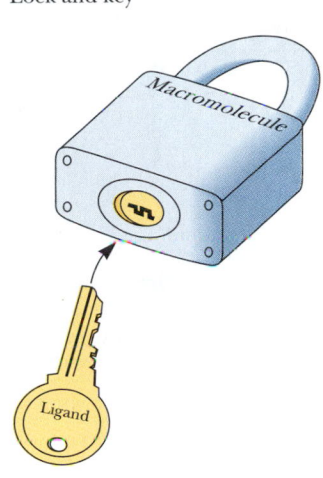

FIGURE 1.16 Structural complementarity: the pieces of a puzzle, the lock and its key, a biological macromolecule and its ligand—an antigen–antibody complex. **(a)** The antigen on the right (green) is a small protein, lysozyme, from hen egg white. The part of the antibody molecule (IgG) shown on the left in blue and yellow includes the antigen-binding domain. **(b)** This domain has a pocket that is structurally complementary to a surface protuberance (Gln^{121}, shown in red between antigen and antigen-binding domain) on the antigen. (See also Figure 1.12.)

actions tend to be transient; rigid, static lattices of biomolecules that might paralyze cellular activities are not formed. Instead, a dynamic interplay occurs between metabolites and macromolecules, hormones and receptors, and all the other participants instrumental to life processes. This interplay is initiated upon specific recognition between complementary molecules and ultimately culminates in unique physiological activities. Biological function is achieved through mechanisms based on structural complementarity and weak chemical interactions.

This principle of structural complementarity extends to higher interactions essential to the establishment of the living condition. For example, the formation of supramolecular complexes occurs because of recognition and interaction between their various macromolecular components, as governed by the weak forces formed between them. If a sufficient number of weak bonds can be formed, as in macromolecules complementary in structure to one another, larger structures assemble spontaneously. The tendency for nonpolar molecules and parts of molecules to come together through hydrophobic interactions also promotes the formation of supramolecular assemblies. Very complex subcellular structures are actually spontaneously formed in an assembly process that is driven by weak forces accumulated through structural complementarity.

Biochemistry ⒺNow™   Go to BiochemistryNow and click BiochemistryInteractive to explore the structure of immunoglobulin G, centering on the role of weak intermolecular forces in controlling structure.

Weak Forces Restrict Organisms to a Narrow Range of Environmental Conditions

Because biomolecular interactions are governed by weak forces, living systems are restricted to a narrow range of physical conditions. Biological macromolecules are functionally active only within a narrow range of environmental conditions, such as temperature, ionic strength, and relative acidity. Extremes of these conditions disrupt the weak forces essential to maintaining the intricate structure of macromolecules. The loss of structural order in these complex macromolecules, so-called **denaturation**, is accompanied by loss of function (Figure 1.17). As a consequence, cells cannot tolerate reactions in which large amounts of energy are

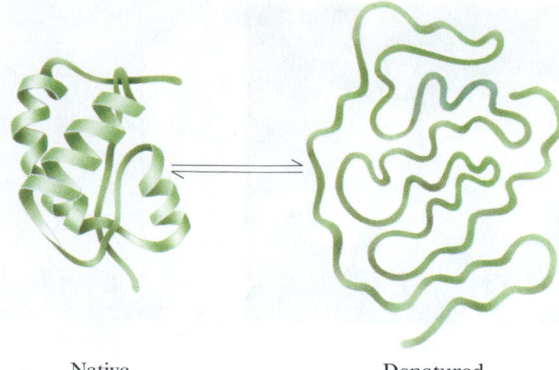

Biochemistry Now™   ANIMATED FIGURE 1.17
Denaturation and renaturation of the intricate structure of a protein. **See this figure animated at http:// chemistry.brookscole.com/ggb3**

Native   Denatured

released, nor can they generate a large energy burst to drive energy-requiring processes. Instead, such transformations take place via sequential series of chemical reactions whose overall effect achieves dramatic energy changes, even though any given reaction in the series proceeds with only modest input or release of energy (Figure 1.18). These sequences of reactions are organized to provide for the release of useful energy to the cell from the breakdown of food or to take such energy and use it to drive the synthesis of biomolecules essential to the living state. Collectively, these reaction sequences constitute cellular **metabolism**—the ordered reaction pathways by which cellular chemistry proceeds and biological energy transformations are accomplished.

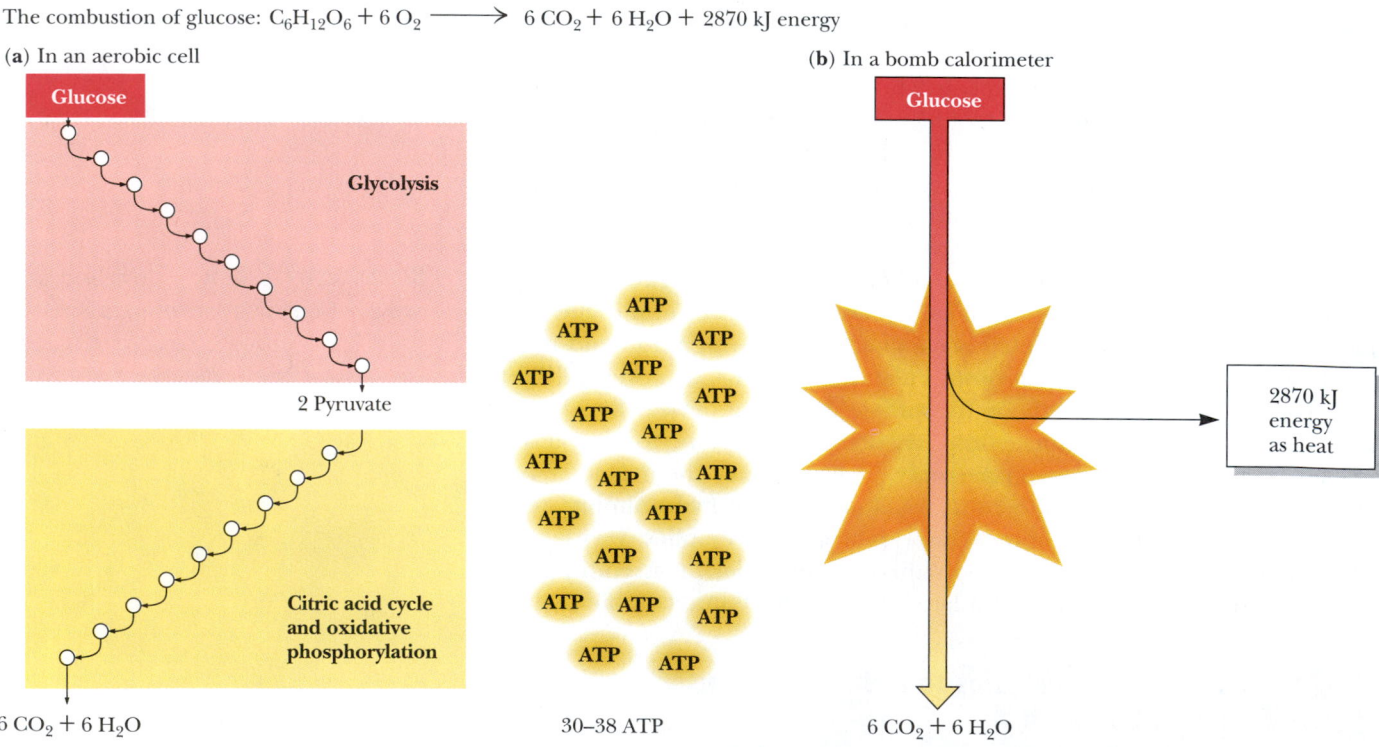

The combustion of glucose: $C_6H_{12}O_6 + 6\,O_2 \longrightarrow 6\,CO_2 + 6\,H_2O + 2870\ kJ$ energy

(**a**) In an aerobic cell

Glucose

Glycolysis

2 Pyruvate

Citric acid cycle and oxidative phosphorylation

$6\,CO_2 + 6\,H_2O$

30–38 ATP

(**b**) In a bomb calorimeter

Glucose

2870 kJ energy as heat

$6\,CO_2 + 6\,H_2O$

Biochemistry Now™   ACTIVE FIGURE 1.18 Metabolism is the organized release or capture of small amounts of energy in processes whose overall change in energy is large. (**a**) For example, the combustion of glucose by cells is a major pathway of energy production, with the energy captured appearing as 30 to 38 equivalents of ATP, the principal energy-rich chemical of cells. The ten reactions of glycolysis, the nine reactions of the citric acid cycle, and the successive linked reactions of oxidative phosphorylation release the energy of glucose in a stepwise fashion and the small "packets" of energy appear in ATP. (**b**) Combustion of glucose in a bomb calorimeter results in an uncontrolled, explosive release of energy in its least useful form, heat. **Test yourself on the concepts in this figure at http://chemistry.brookscole.com/ggb3**

Enzymes Catalyze Metabolic Reactions

The sensitivity of cellular constituents to environmental extremes places another constraint on the reactions of metabolism. The rate at which cellular reactions proceed is a very important factor in maintenance of the living state. However, the common ways chemists accelerate reactions are not available to cells; the temperature cannot be raised, acid or base cannot be added, the pressure cannot be elevated, and concentrations cannot be dramatically increased. Instead, biomolecular catalysts mediate cellular reactions. These catalysts, called **enzymes,** accelerate the reaction rates many orders of magnitude and, by selecting the substances undergoing reaction, determine the specific reaction that takes place. Virtually every metabolic reaction is catalyzed by an enzyme (Figure 1.19).

Metabolic Regulation Is Achieved by Controlling the Activity of Enzymes   Thousands of reactions mediated by an equal number of enzymes are occurring at any given instant within the cell. Metabolism has many branch points, cycles, and interconnections, as a glance at a metabolic pathway map reveals (Figure 1.20). All these reactions, many of which are at apparent cross-purposes in the cell, must be fine-tuned and integrated so that metabolism and life proceed harmoniously. The need for metabolic regulation is obvious. This metabolic regulation is achieved through controls on enzyme activity so that the rates of cellular reactions are appropriate to cellular requirements.

Despite the organized pattern of metabolism and the thousands of enzymes required, cellular reactions nevertheless conform to the same thermodynamic principles that govern any chemical reaction. Enzymes have no influence over energy changes (the thermodynamic component) in their reactions. Enzymes only influence reaction rates. Thus, cells are systems that take in food, release waste, and carry out complex degradative and biosynthetic reactions essential to their survival while operating under conditions of essentially constant temperature and pressure and maintaining a constant internal environment **(homeostasis)** with no outwardly apparent changes. *Cells are open thermodynamic systems exchanging matter and energy with their environment and functioning as highly regulated isothermal chemical engines.*

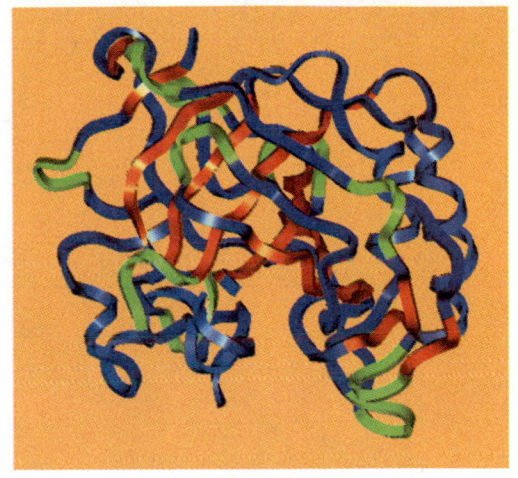

Biochemistry Now™   ANIMATED FIGURE 1.19
Carbonic anhydrase, a representative enzyme, and the reaction that it catalyzes. Dissolved carbon dioxide is slowly hydrated by water to form bicarbonate ion and H^+:

$$CO_2 + H_2O \rightleftharpoons HCO_3^- + H^+$$

At 20°C, the rate constant for this uncatalyzed reaction, k_{uncat}, is 0.03/sec. In the presence of the enzyme carbonic anhydrase, the rate constant for this reaction, k_{cat}, is 10^6/sec. Thus, carbonic anhydrase accelerates the rate of this reaction 3.3×10^7 times. Carbonic anhydrase is a 29-kD protein. **See this figure animated at http://chemistry.brookscole.com/ggb3**

<table><tr><td>**1.5**</td><td>## What Is the Organization and Structure of Cells?</td></tr></table>

All living cells fall into one of two broad categories—**prokaryotic** and **eukaryotic.** The distinction is based on whether the cell has a nucleus. Prokaryotes are single-celled organisms that lack nuclei and other organelles; the word is derived from *pro* meaning "prior to" and *karyot* meaning "nucleus." In conventional biological classification schemes, prokaryotes are grouped together as members of the kingdom Monera, represented by bacteria and cyanobacteria (formerly called blue-green algae). The other four living kingdoms are all eukaryotes—the single-celled Protists, such as amoebae, and all multicellular life forms, including the Fungi, Plant, and Animal kingdoms. Eukaryotic cells have true nuclei and other organelles such as mitochondria, with the prefix *eu* meaning "true."

The Evolution of Early Cells Gave Rise to Eubacteria, Archaea, and Eukaryotes

For a long time, most biologists believed that eukaryotes evolved from the simpler prokaryotes in some linear progression from simple to complex over the course of geological time. However, contemporary evidence favors the view that

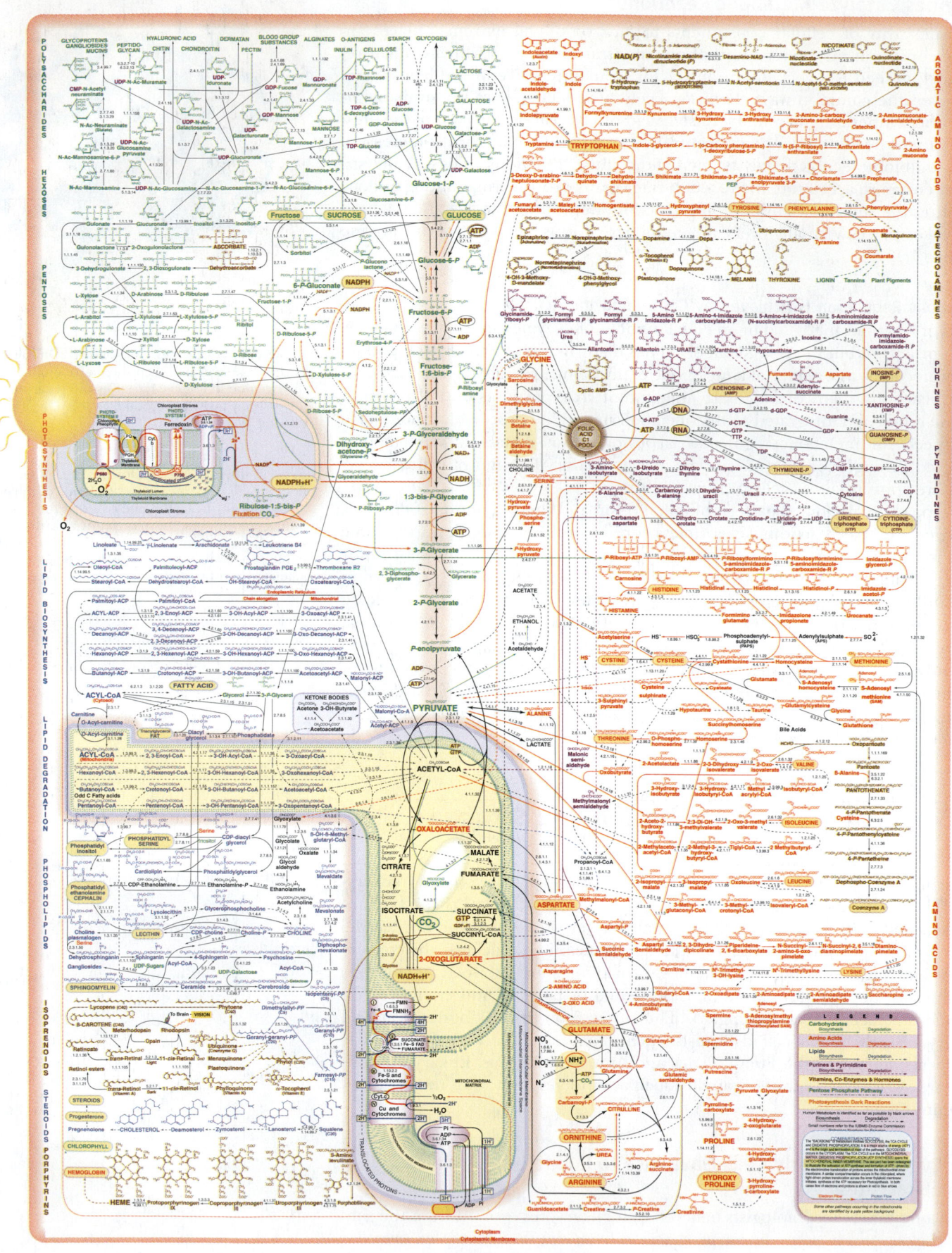

◄ Biochemistry ⓔ Now™ **ANIMATED FIGURE 1.20** Reproduction of a metabolic map. *(Source: Donald Nicholson's Metabolic Map #21. Copyright © International Union of Biochemistry and Molecular Biology. Used with permission.)* **See this figure animated at http://chemistry.brookscole.com/ggb3**

present-day organisms are better grouped into three classes or lineages: eukaryotes and two prokaryotic groups, the **eubacteria** and the **archaea** (formerly designated as **archaebacteria**). All are believed to have evolved approximately 3.5 billion years ago from an ancestral communal gene pool shared among primitive cellular entities. Furthermore, contemporary eukaryotic cells are, in reality, composite cells that harbor various prokaryotic contributions. Thus, the dichotomy between prokaryotic cells and eukaryotic cells, although convenient, is an artificial distinction.

Despite great diversity in form and function, cells and organisms share much biochemistry in common. This commonality and diversity has been substantiated by the results of **whole genome sequencing,** the determination of the complete nucleotide sequence within the DNA of an organism. For example, the genome of the metabolically divergent archaeon *Methanococcus jannaschii* shows 44% similarity to known genes in eubacteria and eukaryotes, yet 56% of its genes are new to science.

How many genes does it take to make a cell or, beyond that, a multicellular organism? Some insight can be drawn from the smallest known genome for an independently replicating organism, that of *Mycoplasma genitalium*, a parasitic eubacterium that causes urogenital tract infection. *M. genitalium* DNA consists of just 580,000 nucleotide pairs, encoding 517 genes (Table 1.5). In contrast, the roughly 3,000,000,000 nucleotide pairs of the human genome encode an estimated 30,000 or so genes.

Gene is a unit of hereditary information, physically defined by a specific sequence of nucleotides in DNA; in molecular terms, a gene is a nucleotide sequence that encodes a protein or RNA product.

Table 1.5

How Many Genes Does It Take To Make An Organism?

Organism	Number of Cells in Adult*	Number of Genes
Mycobacterium genitalium Pathogenic eubacterium	1	517
Methanococcus jannaschii Archaeal methanogen	1	1,800
Escherichia coli K12 Intestinal eubacterium	1	4,400
Saccharomyces cereviseae Baker's yeast (eukaryote)	1	6,000
Caenorhabditis elegans Nematode worm	959	19,000
Drosophila melanogaster Fruit fly	10^4	13,500
Arabidopsis thaliana Flowering plant	10^7	27,000
Fugu rubripes Pufferfish	10^{12}	38,000 (est.)
Homo sapiens Human	10^{14}	30,000 (est.)

The first four of the nine organisms in the table are single-celled microbes; the last six are eukaryotes; the last five are multicellular, four of which are animals; the final two are vertebrates. Although pufferfish and humans have about the same number of genes, the pufferfish genome, at 0.365 billion nucleotide pairs, is only one-eighth the size of the human genome.
*Numbers for *Arabidopsis thaliana*, the pufferfish, and human are "order-of-magnitude" rough estimates.

Prokaryotic Cells Have a Relatively Simple Structural Organization

Among prokaryotes (the simplest cells), most known species are eubacteria and they form a widely spread group. Certain of them are pathogenic to humans. The archaea are remarkable because they can be found in unusual environments where other cells cannot survive. Archaea include the **thermoacidophiles** (heat- and acid-loving bacteria) of hot springs, the **halophiles** (salt-loving bacteria) of salt lakes and ponds, and the **methanogens** (bacteria that generate methane from CO_2 and H_2). Prokaryotes are typically very small, on the order of several microns in length, and are usually surrounded by a rigid **cell wall** that protects the cell and gives it its shape. The characteristic structural organization of a prokaryotic cell is depicted in Figure 1.21.

Prokaryotic cells have only a single membrane, the **plasma membrane** or **cell membrane.** Because they have no other membranes, prokaryotic cells contain no nucleus or organelles. Nevertheless, they possess a distinct **nuclear area** where a single circular chromosome is localized, and some have an internal membranous structure called a **mesosome** that is derived from and continuous with the cell membrane. Reactions of cellular respiration are localized on these membranes. In photosynthetic prokaryotes such as the **cyanobacteria,** flat, sheetlike membranous structures called **lamellae** are formed from cell membrane infoldings. These lamellae are the sites of photosynthetic activity, but in prokaryotes, they are not contained within **plastids,** the organelles of photosynthesis found in higher plant cells. Prokaryotic cells also lack a cytoskeleton; the cell wall maintains their structure. Some bacteria have **flagella,** single, long filaments used for motility. Prokaryotes largely reproduce by asexual division, although sexual exchanges can occur. Table 1.6 lists the major features of prokaryotic cells.

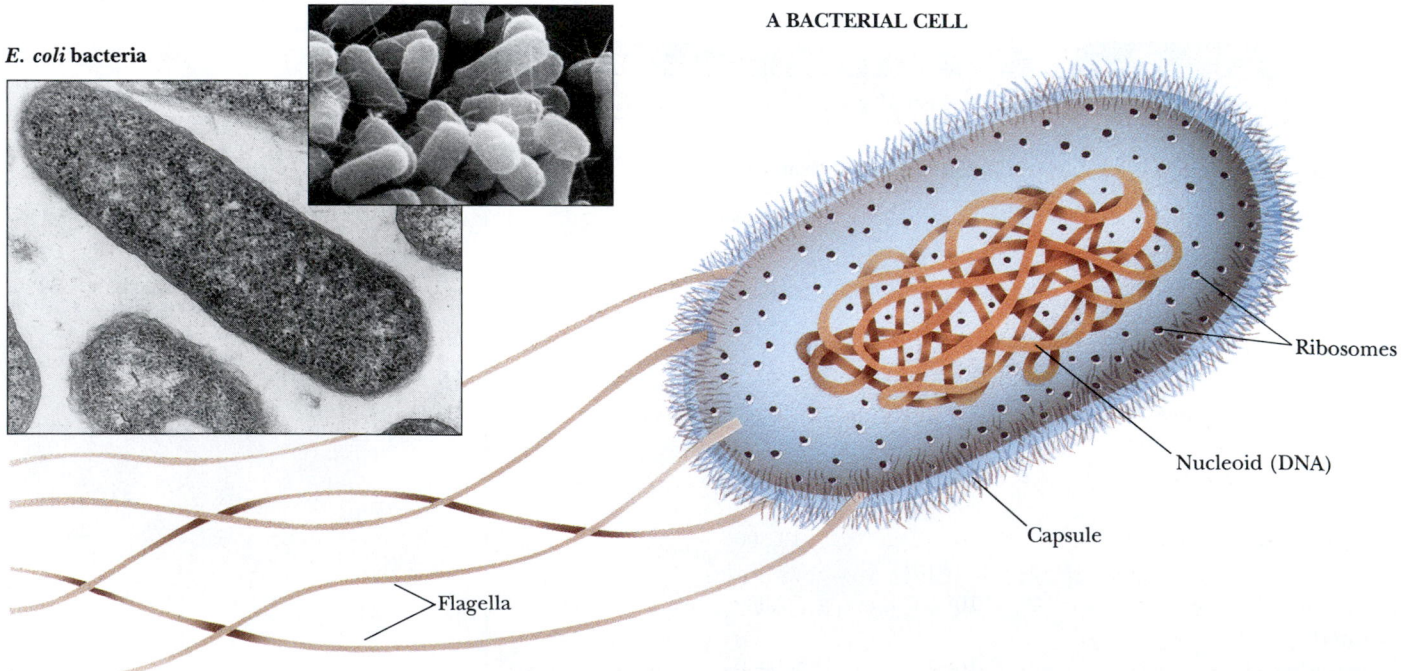

E. coli bacteria

A BACTERIAL CELL

Ribosomes

Nucleoid (DNA)

Capsule

Flagella

FIGURE 1.21 This bacterium is *Escherichia coli,* a member of the coliform group of bacteria that colonize the intestinal tract of humans. *E. coli* cells have rather simple nutritional requirements. They grow and multiply quite well if provided with a simple carbohydrate source of energy (such as glucose), ammonium ions as a source of nitrogen, and a few mineral salts. The simple nutrition of this "lower" organism means that its biosynthetic capacities must be quite advanced. When growing at 37°C on a rich organic medium, *E. coli* cells divide every 20 minutes. Subcellular features include the cell wall, plasma membrane, nuclear region, ribosomes, storage granules, and cytosol (see Table 1.6). *(Photo, Martin Rotker/Phototake, Inc.; inset photo, David M. Phillips/The Population Council/Science Source/Photo Researchers, Inc.)*

Table 1.6
Major Features of Prokaryotic Cells

Structure	Molecular Composition	Function
Cell wall	Peptidoglycan: a rigid framework of polysaccharide crosslinked by short peptide chains. Some bacteria possess a lipopolysaccharide- and protein-rich outer membrane.	Mechanical support, shape, and protection against swelling in hypotonic media. The cell wall is a porous nonselective barrier that allows most small molecules to pass.
Cell membrane	The cell membrane is composed of about 45% lipid and 55% protein. The lipids form a bilayer that is a continuous nonpolar hydrophobic phase in which the proteins are embedded.	The cell membrane is a highly selective permeability barrier that controls the entry of most substances into the cell. Important enzymes in the generation of cellular energy are located in the membrane.
Nuclear area or nucleoid	The genetic material is a single, tightly coiled DNA molecule 2 nm in diameter but more than 1 mm in length (molecular mass of *E. coli* DNA is 3×10^9 daltons; 4.64×10^6 nucleotide pairs).	DNA is the blueprint of the cell, the repository of the cell's genetic information. During cell division, each strand of the double-stranded DNA molecule is replicated to yield two double-helical daughter molecules. Messenger RNA (mRNA) is transcribed from DNA to direct the synthesis of cellular proteins.
Ribosomes	Bacterial cells contain about 15,000 ribosomes. Each is composed of a small (30S) subunit and a large (50S) subunit. The mass of a single ribosome is 2.3×10^6 daltons. It consists of 65% RNA and 35% protein.	Ribosomes are the sites of protein synthesis. The mRNA binds to ribosomes, and the mRNA nucleotide sequence specifies the protein that is synthesized.
Storage granules	Bacteria contain granules that represent storage forms of polymerized metabolites such as sugars or β-hydroxybutyric acid.	When needed as metabolic fuel, the monomeric units of the polymer are liberated and degraded by energy-yielding pathways in the cell.
Cytosol	Despite its amorphous appearance, the cytosol is an organized gelatinous compartment that is 20% protein by weight and rich in the organic molecules that are the intermediates in metabolism.	The cytosol is the site of intermediary metabolism, the interconnecting sets of chemical reactions by which cells generate energy and form the precursors necessary for biosynthesis of macromolecules essential to cell growth and function.

The Structural Organization of Eukaryotic Cells Is More Complex Than That of Prokaryotic Cells

Compared with prokaryotic cells, eukaryotic cells are much greater in size, typically having cell volumes 10^3 to 10^4 times larger. They are also much more complex. These two features require that eukaryotic cells partition their diverse metabolic processes into organized compartments, with each compartment dedicated to a particular function. A system of internal membranes accomplishes this partitioning. A typical animal cell is shown in Figure 1.22 and a typical plant cell in Figure 1.23. Tables 1.7 and 1.8 list the major features of a typical animal cell and a higher plant cell, respectively.

Eukaryotic cells possess a discrete, membrane-bounded **nucleus,** the repository of the cell's genetic material, which is distributed among a few or many **chromosomes.** During cell division, equivalent copies of this genetic material must be passed to both daughter cells through duplication and orderly partitioning of the chromosomes by the process known as **mitosis.** Like prokaryotic cells, eukaryotic cells are surrounded by a plasma membrane. Unlike prokaryotic cells, eukaryotic cells are rich in internal membranes that are differentiated into specialized structures such as the **endoplasmic reticulum (ER)** and the **Golgi apparatus.** Membranes also surround certain organelles (**mitochondria** and **chloroplasts,** for example) and various vesicles, including **vacuoles, lysosomes,** and **peroxisomes.** The common purpose of these membranous partitionings is the creation of cellular compartments that have specific, organized

Rough endoplasmic
reticulum (plant and animal)

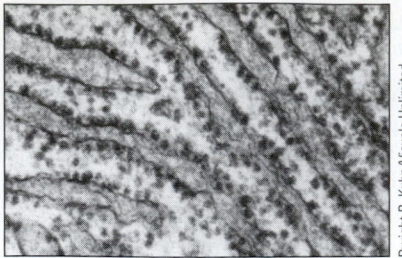

Smooth endoplasmic
reticulum (plant and animal)

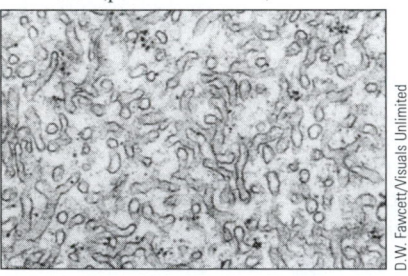

Mitochondrion
(plant and animal)

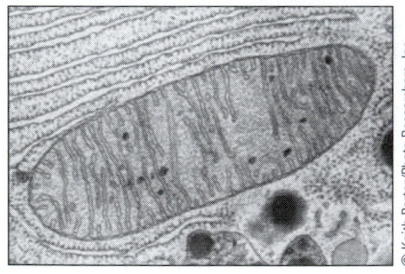

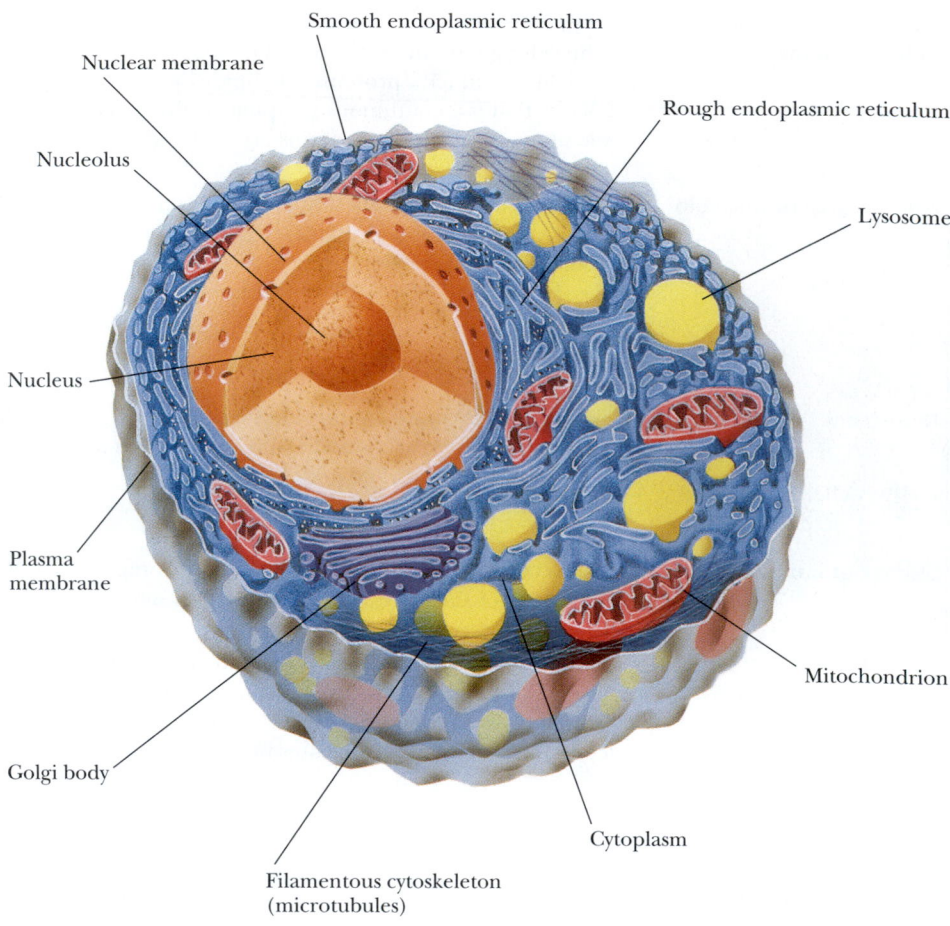

FIGURE 1.22 This figure diagrams a rat liver cell, a typical higher animal cell in which the characteristic features of animal cells—such as a nucleus, nucleolus, mitochondria, Golgi bodies, lysosomes, and endoplasmic reticulum (ER)—are evident. Microtubules and the network of filaments constituting the cytoskeleton are also depicted.

metabolic functions, such as the mitochondrion's role as the principal site of cellular energy production. Eukaryotic cells also have a **cytoskeleton** composed of arrays of filaments that give the cell its shape and its capacity to move. Some eukaryotic cells also have long projections on their surface—cilia or flagella—which provide propulsion.

1.6 | What Are Viruses?

Viruses are supramolecular complexes of nucleic acid, either DNA or RNA, encapsulated in a protein coat and, in some instances, surrounded by a membrane envelope (Figure 1.24). Viruses are acellular, but they act as cellular

Chloroplast (plant cell only)

Dr. Dennis Kunkel/Phototake, NYC

Golgi body (plant and animal)

Dr. Dennis Kunkel/Phototake, NYC

Nucleus (plant and animal)

Biophoto Associates

A PLANT CELL

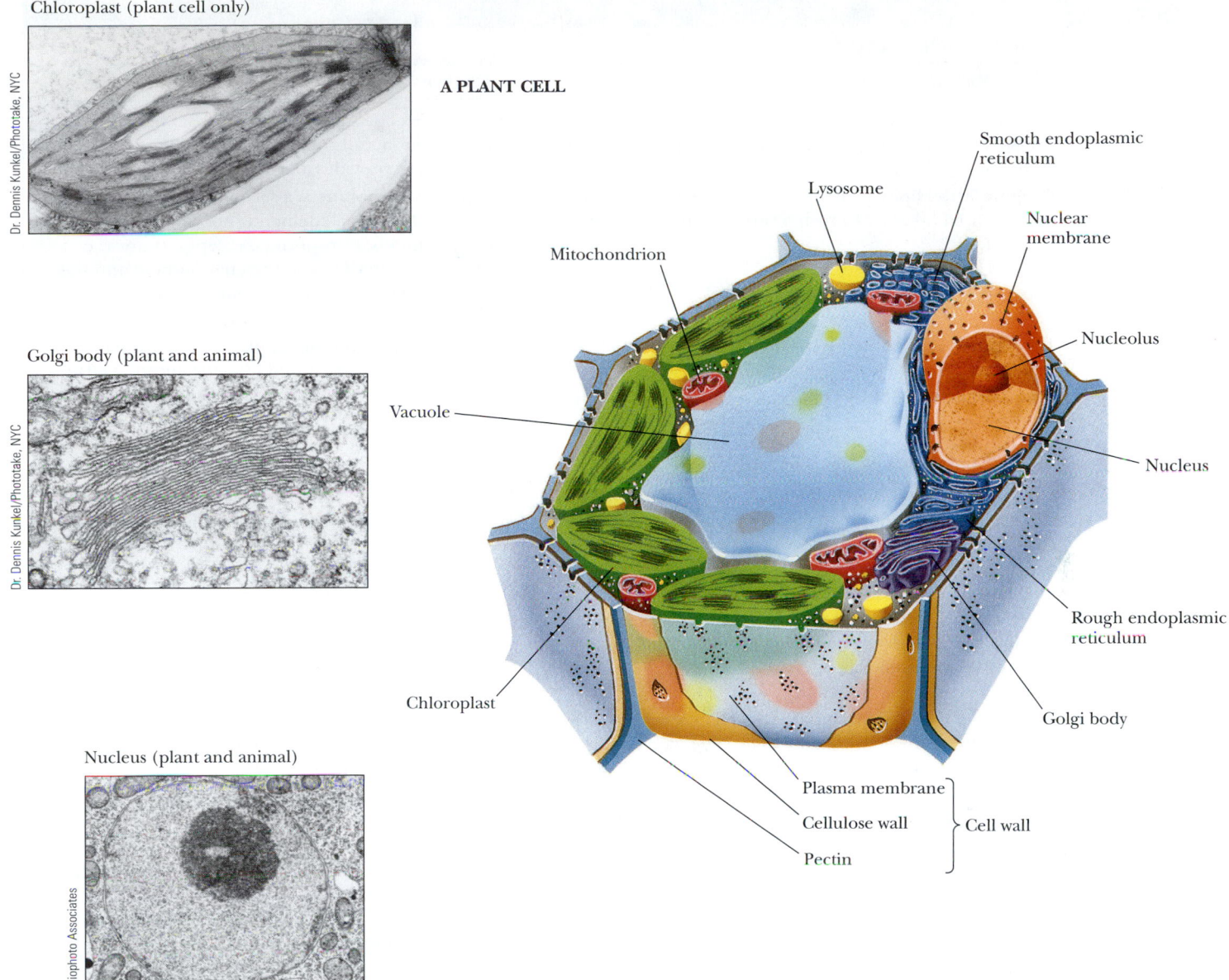

Smooth endoplasmic reticulum

Lysosome

Nuclear membrane

Mitochondrion

Nucleolus

Vacuole

Nucleus

Chloroplast

Rough endoplasmic reticulum

Golgi body

Plasma membrane

Cellulose wall ⎬ Cell wall

Pectin

FIGURE 1.23 This figure diagrams a cell in the leaf of a higher plant. The cell wall, membrane, nucleus, chloroplasts, mitochondria, vacuole, endoplasmic reticulum (ER), and other characteristic features are shown.

parasites in order to reproduce. The bits of nucleic acid in viruses are, in reality, mobile elements of genetic information. The protein coat serves to protect the nucleic acid and allows it to gain entry to the cells that are its specific hosts. Viruses unique for all types of cells are known. Viruses infecting bacteria are called **bacteriophages** ("bacteria eaters"); different viruses infect animal cells and plant cells. Once the nucleic acid of a virus gains access to its specific host, it typically takes over the metabolic machinery of the host cell, diverting it to the production of virus particles. The host metabolic functions are subjugated to the synthesis of viral nucleic acid and proteins. Mature virus particles arise by encapsulating the nucleic acid within a protein coat called the **capsid.** Thus, viruses are supramolecular assemblies that act as parasites of cells (Figure 1.25).

Table 1.7

Major Features of a Typical Animal Cell

Structure	Molecular Composition	Function
Extracellular matrix	The surfaces of animal cells are covered with a flexible and sticky layer of complex carbohydrates, proteins, and lipids.	This complex coating is cell specific, serves in cell–cell recognition and communication, creates cell adhesion, and provides a protective outer layer.
Cell membrane (plasma membrane)	Roughly 50:50 lipid:protein as a 5-nm-thick continuous sheet of lipid bilayer in which a variety of proteins are embedded.	The plasma membrane is a selectively permeable outer boundary of the cell, containing specific systems—pumps, channels, transporters, receptors—for the exchange of materials with the environment and the reception of extracellular information. Important enzymes are also located here.
Nucleus	The nucleus is separated from the cytosol by a double membrane, the nuclear envelope. The DNA is complexed with basic proteins (histones) to form chromatin fibers, the material from which chromosomes are made. A distinct RNA-rich region, the nucleolus, is the site of ribosome assembly.	The nucleus is the repository of genetic information encoded in DNA and organized into chromosomes. During mitosis, the chromosomes are replicated and transmitted to the daughter cells. The genetic information of DNA is transcribed into RNA in the nucleus and passes into the cytosol, where it is translated into protein by ribosomes.
Endoplasmic reticulum (ER) and ribosomes	Flattened sacs, tubes, and sheets of internal membrane extending throughout the cytoplasm of the cell and enclosing a large interconnecting series of volumes called *cisternae*. The ER membrane is continuous with the outer membrane of the nuclear envelope. Portions of the sheet-like areas of the ER are studded with ribosomes, giving rise to *rough ER*. Eukaryotic ribosomes are larger than prokaryotic ribosomes.	The endoplasmic reticulum is a labyrinthine organelle where both membrane proteins and lipids are synthesized. Proteins made by the ribosomes of the rough ER pass through the ER membrane into the cisternae and can be transported via the Golgi to the periphery of the cell. Other ribosomes unassociated with the ER carry on protein synthesis in the cytosol. The nuclear membrane, ER, Golgi, and additional vesicles are all part of a continuous endomembrane system.
Golgi apparatus	An asymmetrical system of flattened membrane-bounded vesicles often stacked into a complex. The face of the complex nearest the ER is the *cis* face; that most distant from the ER is the *trans* face. Numerous small vesicles found peripheral to the *trans* face of the Golgi contain secretory material packaged by the Golgi.	Involved in the packaging and processing of macromolecules for secretion and for delivery to other cellular compartments.
Mitochondria	Mitochondria are organelles surrounded by two membranes that differ markedly in their protein and lipid composition. The inner membrane and its interior volume—the matrix—contain many important enzymes of energy metabolism. Mitochondria are about the size of bacteria, ≈ 1 μm. Cells contain hundreds of mitochondria, which collectively occupy about one-fifth of the cell volume.	Mitochondria are the power plants of eukaryotic cells where carbohydrates, fats, and amino acids are oxidized to CO_2 and H_2O. The energy released is trapped as high-energy phosphate bonds in ATP.
Lysosomes	Lysosomes are vesicles 0.2–0.5 μm in diameter, bounded by a single membrane. They contain hydrolytic enzymes such as proteases and nucleases that act to degrade cell constituents targeted for destruction. They are formed as membrane vesicles budding from the Golgi apparatus.	Lysosomes function in intracellular digestion of materials entering the cell via phagocytosis or pinocytosis. They also function in the controlled degradation of cellular components. Their internal pH is about 5, and the hydrolytic enzymes they contain work best at this pH.
Peroxisomes	Like lysosomes, peroxisomes are 0.2–0.5 μm, single-membrane–bounded vesicles. They contain a variety of oxidative enzymes that use molecular oxygen and generate peroxides. They are also formed from membrane vesicles budding from the smooth ER.	Peroxisomes act to oxidize certain nutrients, such as amino acids. In doing so, they form potentially toxic hydrogen peroxide, H_2O_2, and then decompose it to H_2O and O_2 by way of the peroxide-cleaving enzyme catalase.
Cytoskeleton	The cytoskeleton is composed of a network of protein filaments: actin filaments (or microfilaments), 7 nm in diameter; intermediate filaments, 8–10 nm; and microtubules, 25 nm. These filaments interact in establishing the structure and functions of the cytoskeleton. This interacting network of protein filaments gives structure and organization to the cytoplasm.	The cytoskeleton determines the shape of the cell and gives it its ability to move. It also mediates the internal movements that occur in the cytoplasm, such as the migration of organelles and mitotic movements of chromosomes. The propulsion instruments of cells—cilia and flagella—are constructed of microtubules.

Table 1.8		
Major Features of a Higher Plant Cell: A Photosynthetic Leaf Cell		
Structure	**Molecular Composition**	**Function**
Cell wall	Cellulose fibers embedded in a polysaccharide/protein matrix; it is thick (>0.1 μm), rigid, and porous to small molecules.	Protection against osmotic or mechanical rupture. The walls of neighboring cells interact in cementing the cells together to form the plant. Channels for fluid circulation and for cell–cell communication pass through the walls. The structural material confers form and strength on plant tissue.
Cell membrane	Plant cell membranes are similar in overall structure and organization to animal cell membranes but differ in lipid and protein composition.	The plasma membrane of plant cells is selectively permeable, containing transport systems for the uptake of essential nutrients and inorganic ions. A number of important enzymes are localized here.
Nucleus	The nucleus, nucleolus, and nuclear envelope of plant cells are like those of animal cells.	Chromosomal organization, DNA replication, transcription, ribosome synthesis, and mitosis in plant cells are grossly similar to the analogous features in animals.
Endoplasmic reticulum, Golgi apparatus, ribosomes, lysosomes, peroxisomes, and cytoskeleton	Plant cells also contain all of these characteristic eukaryotic organelles, essentially in the form described for animal cells.	These organelles serve the same purposes in plant cells that they do in animal cells.
Chloroplasts	Chloroplasts have a double-membrane envelope, an inner volume called the **stroma,** and an internal membrane system rich in thylakoid membranes, which enclose a third compartment, the thylakoid **lumen.** Chloroplasts are significantly larger than mitochondria. Other plastids are found in specialized structures such as fruits, flower petals, and roots and have specialized roles.	Chloroplasts are the site of photosynthesis, the reactions by which light energy is converted to metabolically useful chemical energy in the form of ATP. These reactions occur on the thylakoid membranes. The formation of carbohydrate from CO_2 takes place in the stroma. Oxygen is evolved during photosynthesis. Chloroplasts are the primary source of energy in the light.
Mitochondria	Plant cell mitochondria resemble the mitochondria of other eukaryotes in form and function.	Plant mitochondria are the main source of energy generation in photosynthetic cells in the dark and in nonphotosynthetic cells under all conditions.
Vacuole	The vacuole is usually the most obvious compartment in plant cells. It is a very large vesicle enclosed by a single membrane called the **tonoplast.** Vacuoles tend to be smaller in young cells, but in mature cells, they may occupy more than 50% of the cell's volume. Vacuoles occupy the center of the cell, with the cytoplasm being located peripherally around it. They resemble the lysosomes of animal cells.	Vacuoles function in transport and storage of nutrients and cellular waste products. By accumulating water, the vacuole allows the plant cell to grow dramatically in size with no increase in cytoplasmic volume.

Often, viruses cause disintegration of the cells that they have infected, a process referred to as cell **lysis.** It is their cytolytic properties that are the basis of viral disease. In certain circumstances, the viral genetic elements may integrate into the host chromosome and become quiescent. Such a state is termed **lysogeny.** Typically, damage to the host cell activates the replicative capacities of the quiescent viral nucleic acid, leading to viral propagation and release. Some viruses are implicated in transforming cells into a cancerous state, that is, in converting their hosts to an unregulated state of cell division and proliferation. Because all viruses are heavily dependent on their host for the production of viral progeny, viruses must have evolved after cells were established. Presumably, the first viruses were fragments of nucleic acid that developed the ability to replicate independently of the chromosome and then acquired the necessary genes enabling protection, autonomy, and transfer between cells.

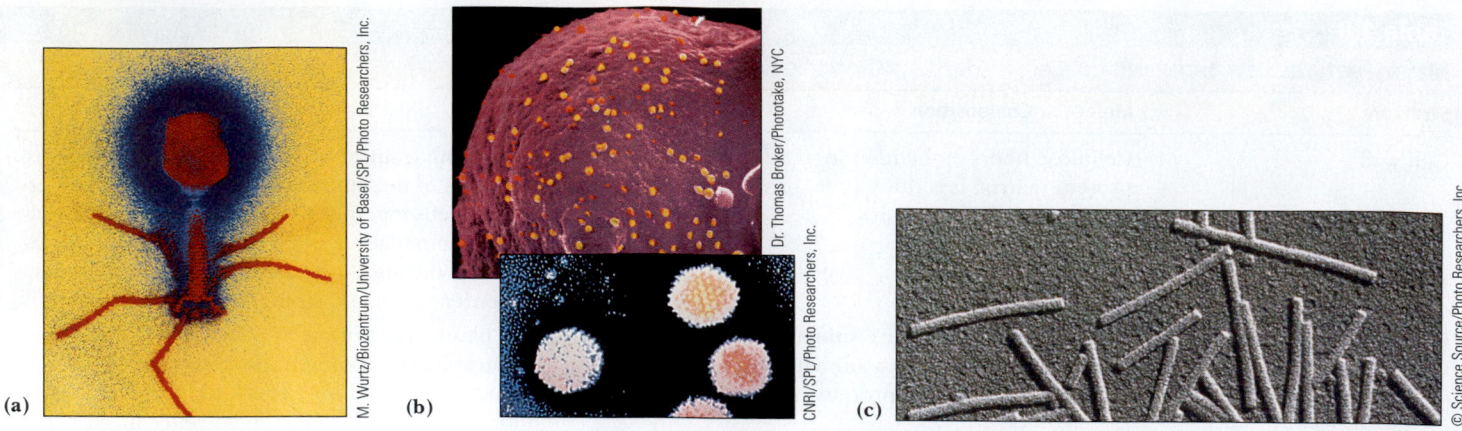

FIGURE 1.24 Viruses are genetic elements enclosed in a protein coat. Viruses are not free-living organisms and can reproduce only within cells. Viruses show an almost absolute specificity for their particular host cells, infecting and multiplying only within those cells. Viruses are known for virtually every kind of cell. Shown here are examples of (**a**) a bacterial virus, bacteriophage T_4; (**b**) an animal virus, adenovirus (inset at greater magnification); and (**c**) a plant virus, tobacco mosaic virus.

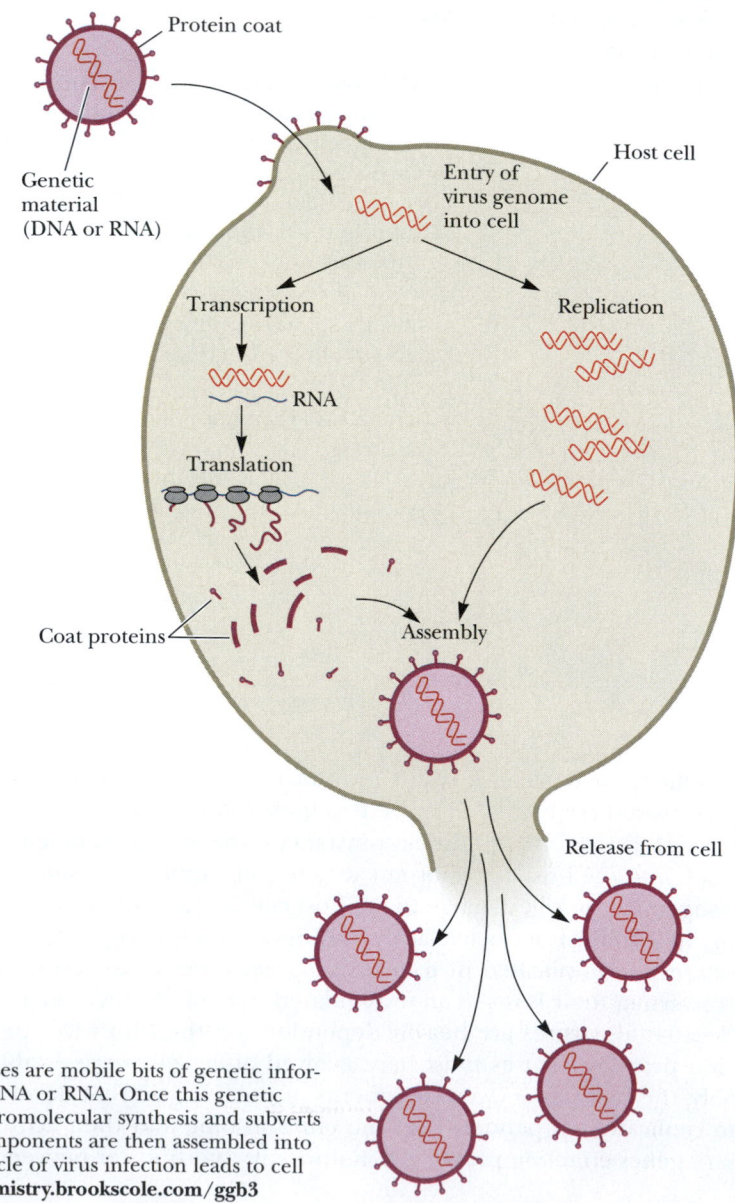

Biochemistry⊜Now™   ACTIVE FIGURE 1.25 The virus life cycle. Viruses are mobile bits of genetic information encapsulated in a protein coat. The genetic material can be either DNA or RNA. Once this genetic material gains entry to its host cell, it takes over the host machinery for macromolecular synthesis and subverts it to the synthesis of viral-specific nucleic acids and proteins. These virus components are then assembled into mature virus particles that are released from the cell. Often, this parasitic cycle of virus infection leads to cell death and disease. **Test yourself on the concepts in this figure at http://chemistry.brookscole.com/ggb3**

Summary

1.1 What Are the Distinctive Properties of Living Systems?

Living systems display an astounding array of activities that collectively constitute growth, metabolism, response to stimuli, and replication. In accord with their functional diversity, living organisms are complicated and highly organized entities composed of many cells. In turn, cells possess subcellular structures known as organelles, which are complex assemblies of very large polymeric molecules, or macromolecules. The monomeric units of macromolecules are common organic molecules (metabolites). Biological structures play a role in the organism's existence. From parts of organisms, such as limbs and organs, down to the chemical agents of metabolism, such as enzymes and metabolic intermediates, a biological purpose can be given for each component. Maintenance of the highly organized structure and activity of living systems requires energy that must be abstracted from the environment. Energy is required to create and maintain structures and to carry out cellular functions. In terms of the capacity of organisms to self-replicate, the fidelity of self-replication resides ultimately in the chemical nature of DNA, the genetic material.

1.2 What Kinds of Molecules Are Biomolecules?

C, H, N, and O are among the lightest elements capable of forming covalent bonds through electron-pair sharing. Because the strength of covalent bonds is inversely proportional to atomic weight, H, C, N, and O form the strongest covalent bonds. Two properties of carbon covalent bonds merit attention: the ability of carbon to form covalent bonds with itself and the tetrahedral nature of the four covalent bonds when carbon atoms form only single bonds. Together these properties hold the potential for an incredible variety of structural forms, whose diversity is multiplied further by including N, O, and H atoms.

1.3 What Is the Structural Organization of Complex Biomolecules?

Biomolecules are built according to a structural hierarchy: Simple molecules are the units for building complex structures. H_2O, CO_2, NH_4^+, NO_3^-, and N_2 are the inorganic precursors for the formation of simple organic compounds from which metabolites are made. These metabolites serve as intermediates in cellular energy transformation and as building blocks (amino acids, sugars, nucleotides, fatty acids, and glycerol) for lipids and for macromolecular synthesis (synthesis of proteins, polysaccharides, DNA, and RNA). The next higher level of structural organization is created when macromolecules come together through noncovalent interactions to form supramolecular complexes, such as multifunctional enzyme complexes, ribosomes, chromosomes, and cytoskeletal elements.

The next higher rung in the hierarchical ladder is occupied by the organelles. Organelles are membrane-bounded cellular inclusions dedicated to important cellular tasks, such as the nucleus, mitochondria, chloroplasts, endoplasmic reticulum, Golgi apparatus, and vacuoles, as well as other relatively small cellular inclusions. At the apex of the biomolecular hierarchy is the cell, the unit of life, the smallest entity displaying those attributes associated uniquely with the living state—growth, metabolism, stimulus response, and replication.

1.4 How Do the Properties of Biomolecules Reflect Their Fitness to the Living Condition?

Some biomolecules carry the information of life; others translate this information so that the organized structures essential to life are formed. Interactions between such structures are the processes of life. Properties of biomolecules that endow them with the potential for creating the living state include the following: Biological macromolecules and their building blocks have directionality, and thus biological macromolecules are informational; in addition, biomolecules have characteristic three-dimensional architectures, providing the means for molecular recognition through structural complementarity. Weak forces (H bonds, van der Waals interactions, ionic attractions, and hydrophobic interactions) mediate the interactions between biological molecules and, as a consequence, restrict organisms to the narrow range of environmental conditions where these forces operate.

1.5 What Is the Organization and Structure of Cells?

All cells share a common ancestor and fall into one of two broad categories—prokaryotic and eukaryotic—depending on whether the cell has a nucleus. Prokaryotes are typically single-celled organisms and have a rather simple cellular organization. In contrast, eukaryotic cells are structurally more complex, having organelles and various subcellular compartments defined by membranes. Other than the Protists, eukaryotes are multicellular.

1.6 What Are Viruses?

Viruses are supramolecular complexes of nucleic acid encapsulated in a protein coat and, in some instances, surrounded by a membrane envelope. Viruses are not alive; they are not even cellular. Instead, they are packaged bits of genetic material that can parasitize cells in order to reproduce. Often, they cause disintegration, or lysis, of the cells they've infected. It is these cytolytic properties that are the basis of viral disease. In certain circumstances, the viral nucleic acid may integrate into the host chromosome and become quiescent, creating a state known as lysogeny. If the host cell is damaged, the replicative capacities of the quiescent viral nucleic acid may be activated, leading to viral propagation and release.

Problems

1. The nutritional requirements of *Escherichia coli* cells are far simpler than those of humans, yet the macromolecules found in bacteria are about as complex as those of animals. Because bacteria can make all their essential biomolecules while subsisting on a simpler diet, do you think bacteria may have more biosynthetic capacity and hence more metabolic complexity than animals? Organize your thoughts on this question, pro and con, into a rational argument.

2. Without consulting the figures in this chapter, sketch the characteristic prokaryotic and eukaryotic cell types and label their pertinent organelle and membrane systems.

3. *Escherichia coli* cells are about 2 μm (microns) long and 0.8 μm in diameter.

   a. How many *E. coli* cells laid end to end would fit across the diameter of a pinhead? (Assume a pinhead diameter of 0.5 mm.)

   b. What is the volume of an *E. coli* cell? (Assume it is a cylinder, with the volume of a cylinder given by $V = \pi r^2 h$, where $\pi = 3.14$.)

   c. What is the surface area of an *E. coli* cell? What is the surface-to-volume ratio of an *E. coli* cell?

   d. Glucose, a major energy-yielding nutrient, is present in bacterial cells at a concentration of about 1 mM. What is the concentration of glucose, expressed as mg/mL? How many glucose molecules are contained in a typical *E. coli* cell? (Recall that Avogadro's number = 6.023×10^{23}.)

   e. A number of regulatory proteins are present in *E. coli* at only one or two molecules per cell. If we assume that an *E. coli* cell contains just one molecule of a particular protein, what is the molar concentration of this protein in the cell? If the molecular weight of this protein is 40 kD, what is its concentration, expressed as mg/mL?

   f. An *E. coli* cell contains about 15,000 ribosomes, which carry out protein synthesis. Assuming ribosomes are spherical and have a diameter of 20 nm (nanometers), what fraction of the *E. coli* cell volume is occupied by ribosomes?

g. The *E. coli* chromosome is a single DNA molecule whose mass is about 3×10^9 daltons. This macromolecule is actually a linear array of nucleotide pairs. The average molecular weight of a nucleotide pair is 660, and each pair imparts 0.34 nm to the length of the DNA molecule. What is the total length of the *E. coli* chromosome? How does this length compare with the overall dimensions of an *E. coli* cell? How many nucleotide pairs does this DNA contain? The average *E. coli* protein is a linear chain of 360 amino acids. If three nucleotide pairs in a gene encode one amino acid in a protein, how many different proteins can the *E. coli* chromosome encode? (The answer to this question is a reasonable approximation of the maximum number of different kinds of proteins that can be expected in bacteria.)

4. Assume that mitochondria are cylinders 1.5 μm in length and 0.6 μm in diameter.

   a. What is the volume of a single mitochondrion?

   b. Oxaloacetate is an intermediate in the citric acid cycle, an important metabolic pathway localized in the mitochondria of eukaryotic cells. The concentration of oxaloacetate in mitochondria is about 0.03 μM. How many molecules of oxaloacetate are in a single mitochondrion?

5. Assume that liver cells are cuboidal in shape, 20 μm on a side.

   a. How many liver cells laid end to end would fit across the diameter of a pinhead? (Assume a pinhead diameter of 0.5 mm.)

   b. What is the volume of a liver cell? (Assume it is a cube.)

   c. What is the surface area of a liver cell? What is the surface-to-volume ratio of a liver cell? How does this compare to the surface-to-volume ratio of an *E. coli* cell (compare this answer with that of problem 3c)? What problems must cells with low surface-to-volume ratios confront that do not occur in cells with high surface-to-volume ratios?

   d. A human liver cell contains two sets of 23 chromosomes, each set being roughly equivalent in information content. The total mass of DNA contained in these 46 enormous DNA molecules is 4×10^{12} daltons. Because each nucleotide pair contributes 660 daltons to the mass of DNA and 0.34 nm to the length of DNA, what is the total number of nucleotide pairs and the complete length of the DNA in a liver cell? How does this length compare with the overall dimensions of a liver cell? The maximal information in each set of liver cell chromosomes should be related to the number of nucleotide pairs in the chromosome set's

DNA. This number can be obtained by dividing the total number of nucleotide pairs just calculated by 2. What is this value? If this information is expressed in proteins that average 400 amino acids in length and three nucleotide pairs encode one amino acid in a protein, how many different kinds of proteins might a liver cell be able to produce? (In reality, liver cells express at most about 30,000 different proteins. Thus, a large discrepancy exists between the theoretical information content of DNA in liver cells and the amount of information actually expressed.)

6. Biomolecules interact with one another through molecular surfaces that are structurally complementary. How can various proteins interact with molecules as different as simple ions, hydrophobic lipids, polar but uncharged carbohydrates, and even nucleic acids?

7. What structural features allow biological polymers to be informational macromolecules? Is it possible for polysaccharides to be informational macromolecules?

8. Why is it important that weak forces, not strong forces, mediate biomolecular recognition?

9. What is the distance between the centers of two carbon atoms (their *limit of approach*) that are interacting through van der Waals forces? What is the distance between the centers of two carbon atoms joined in a covalent bond? (See Table 1.4.)

10. Why does the central role of weak forces in biomolecular interactions restrict living systems to a narrow range of environmental conditions?

11. Describe what is meant by the phrase "cells are steady-state systems."

Preparing for the MCAT Exam

12. Biological molecules often interact via weak forces (H bonds, van der Waals interactions, etc.). What would be the effect of an increase in kinetic energy on such interactions?

13. Proteins and nucleic acids are informational macromolecules. What are the two minimal criteria for a linear informational polymer?

Biochemistry 🧬 **Now**™   Preparing for an exam? Test yourself on key questions at http://chemistry.brookscole.com/ggb3

Further Reading

General Biology Textbooks

Campbell, N. A., and Reece, J. B., 2002. *Biology*, 6th ed. San Francisco: Benjamin/Cummings.

Solomon, E. P., Berg, L. R., and Martin, D. W., 2002. *Biology*, 6th ed. Pacific Grove, CA: Brooks/Cole.

Cell and Molecular Biology Textbooks

Alberts, B., et al., 2002. *Molecular Biology of the Cell*, 4th ed. New York: Garland Press.

Lodish, H., et al., 1999. *Molecular Cell Biology*, 4th ed. New York: W. H. Freeman.

Synder, L., and Champness, W., 2002. *Molecular Genetics of Bacteria*, 2nd ed. Herndon, VA: ASM Press.

Watson, J. D., et al., 1987. *Molecular Biology of the Gene*, 4th ed. Menlo Park, CA: Benjamin/Cummings.

Papers on Cell Structure

Goodsell, D. S., 1991. Inside a living cell. *Trends in Biochemical Sciences* **16:**203–206.

Lloyd, C., ed., 1986. Cell organization. *Trends in Biochemical Sciences* **11:**437–485.

Papers on Genomes

Cho, M. K., et al., 1999. Ethical considerations in synthesizing a minimal genome. *Science* **286:**2087–2090.

Koonin, E. V., et al., 1996. Sequencing and analysis of bacterial genomes. *Current Biology* **6:**404–416.

Papers on Early Cell Evolution

Margulis, L., 1996. Archaeal-eubacterial mergers in the origin of Eukarya: Phylogenetic classification of life. *Proceedings of the National Academy of Science, U.S.A.* **93:**1071–1076.

Pace, N. R., 1996. New perspective on the natural microbial world: Molecular microbial ecology. *ASM News* **62:**463–470.

Service, R. F., 1997. Microbiologists explore life's rich, hidden kingdoms. *Science* **275:**1740–1742.

Wald, G., 1964. The origins of life. *Proceedings of the National Academy of Science, U.S.A.* **52:**595–611.

Woese, C. R., 2002. On the evolution of cells. *Proceedings of the National Academy of Science, U.S.A.* **99:**8742–8747.

A Brief History Life

de Duve, C., 2002. *Life Evolving: Molecules, Mind, and Meaning.* New York: Oxford University Press.

Water: The Medium of Life

CHAPTER 2

Essential Question

Water provided conditions for the origin, evolution, and flourishing of life; water is the medium of life. *What are the properties of water that render it so suited to its role as the medium of life?*

Water is a major chemical component of the earth's surface. It is indispensable to life. Indeed, it is the only liquid that most organisms ever encounter. We are prone to take it for granted because of its ubiquity and bland nature, yet we marvel at its many unusual and fascinating properties. At the center of this fascination is the role of water as the medium of life. Life originated, evolved, and thrives in the seas. Organisms invaded and occupied terrestrial and aerial niches, but none gained true independence from water. Typically, organisms are 70% to 90% water. Indeed, normal metabolic activity can occur only when cells are at least 65% H_2O. This dependency of life on water is not a simple matter, but it can be grasped by considering the unusual chemical and physical properties of H_2O. Subsequent chapters establish that water and its ionization products, hydrogen ions and hydroxide ions, are critical determinants of the structure and function of many biomolecules, including amino acids and proteins, nucleotides and nucleic acids, and even phospholipids and membranes. In yet another essential role, water is an indirect participant—a difference in the concentration of hydrogen ions on opposite sides of a membrane represents an energized condition essential to biological mechanisms of energy transformation. First, let's review the remarkable properties of water.

2.1 | What Are the Properties of Water?

Water Has Unusual Properties

Compared with chemical compounds of similar atomic organization and molecular size, water displays unexpected properties. For example, compare water, the hydride of oxygen, with hydrides of oxygen's nearest neighbors in the periodic table, namely, ammonia (NH_3) and hydrogen fluoride (HF), or with the hydride of its nearest congener, sulfur (H_2S). Water has a substantially higher boiling point, melting point, heat of vaporization, and surface tension. Indeed, all of these physical properties are anomalously high for a substance of this molecular weight that is neither metallic nor ionic. These properties suggest that intermolecular forces of attraction between H_2O molecules are high. Thus, the internal cohesion of this substance is high. Furthermore, water has an unusually high dielectric constant, its maximum density is found in the liquid (not the solid) state, and it has a negative volume of melting (that is, the solid form, ice, occupies more space than does the liquid form, water). It is truly remarkable that so many eccentric properties occur together in this single substance. As chemists, we expect to find an explanation for these apparent eccentricities in the structure of water. The key to its intermolecular attractions must lie in its atomic constitution. Indeed, *the unrivaled ability to form hydrogen bonds is the crucial fact to understanding its properties.*

Hydrogen Bonding in Water Is Key to Its Properties

The two hydrogen atoms of water are linked covalently to oxygen, each sharing an electron pair, to give a nonlinear arrangement (Figure 2.1). This "bent" structure of the H_2O molecule has enormous influence on its properties. If H_2O

Where there's water, there's life.

© Paul Steel/CORBIS

If there is magic on this planet, it is contained in water.
Loren Eisley *(inscribed on the wall of the National Aquarium in Baltimore, Maryland)*

Key Questions

2.1 What Are the Properties of Water?
2.2 What Is pH?
2.3 What Are Buffers, and What Do They Do?
2.4 Does Water Have a Unique Role in the Fitness of the Environment?

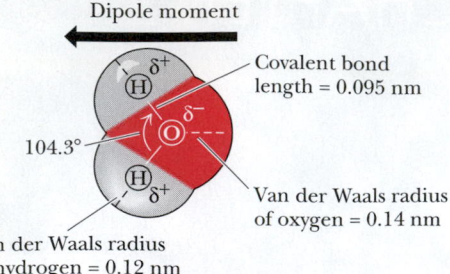

Dipole moment

Covalent bond
length = 0.095 nm

104.3°

Van der Waals radius
of oxygen = 0.14 nm

Van der Waals radius
of hydrogen = 0.12 nm

Biochemistry❀Now™   ACTIVE FIGURE 2.1
The structure of water. Two lobes of negative charge
formed by the lone-pair electrons of the oxygen
atom lie above and below the plane of the diagram.
This electron density contributes substantially to the
large dipole moment and polarizability of the water
molecule. The dipole moment of water corresponds
to the O—H bonds having 33% ionic character.
Note that the H—O—H angle is 104.3°, *not* 109°, the
angular value found in molecules with tetrahedral
symmetry, such as CH_4. Many of the important properties of water derive from this angular value, such
as the decreased density of its crystalline state, ice.
(The dipole moment in this figure points in the
direction from negative to positive, the convention
used by physicists and physical chemists; organic
chemists draw it pointing in the opposite direction.)
**Test yourself on the concepts in this figure at
http://chemistry.brookscole.com/ggb3**

were linear, it would be a nonpolar substance. In the bent configuration, however, the electronegative O atom and the two H atoms form a dipole that renders the molecule distinctly polar. Furthermore, this structure is ideally suited
to H-bond formation. Water can serve as both an H donor and an H acceptor
in H-bond formation. The potential to form four H bonds per water molecule
is the source of the strong intermolecular attractions that endow this substance
with its anomalously high boiling point, melting point, heat of vaporization, and
surface tension. In ordinary ice, the common crystalline form of water, each
H_2O molecule has four nearest neighbors to which it is hydrogen bonded: Each
H atom donates an H bond to the O of a neighbor, and the O atom serves as
an H-bond acceptor from H atoms bound to two different water molecules (Figure 2.2). A local tetrahedral symmetry results.

Hydrogen bonding in water is cooperative. That is, an H-bonded water
molecule serving as an acceptor is a better H-bond donor than an unbonded
molecule (and an H_2O molecule serving as an H-bond donor becomes a better H-bond acceptor). Thus, participation in H bonding by H_2O molecules is
a phenomenon of mutual reinforcement. The H bonds between neighboring
molecules are weak (23 kJ/mol each) relative to the H—O covalent bonds
(420 kJ/mol). As a consequence, the hydrogen atoms are situated asymmetrically between the two oxygen atoms along the O-O axis. There is never any
ambiguity about which O atom the H atom is chemically bound to, nor to
which O it is H bonded.

The Structure of Ice Is Based On H-Bond Formation

In ice, the hydrogen bonds form a space-filling, three-dimensional network.
These bonds are directional and straight; that is, the H atom lies on a direct
line between the two O atoms. This linearity and directionality mean that the
H bonds in ice are strong. In addition, the directional preference of the H
bonds leads to an open lattice structure. For example, if the water molecules
are approximated as rigid spheres centered at the positions of the O atoms in
the lattice, then the observed density of ice is actually only 57% of that expected for a tightly packed arrangement of such spheres. The H bonds in ice

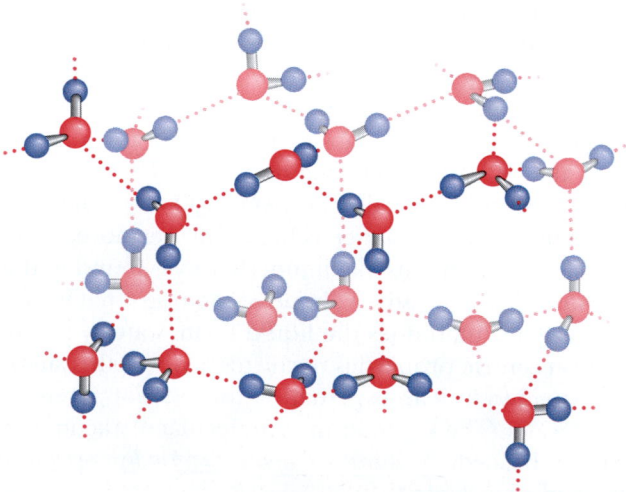

Biochemistry❀Now™   ANIMATED FIGURE 2.2 The structure of normal ice. The hydrogen
bonds in ice form a three-dimensional network. The smallest number of H_2O molecules in any
closed circuit of H-bonded molecules is six, so this structure bears the name hexagonal ice. Covalent
bonds are represented as solid lines, whereas hydrogen bonds are shown as dashed lines. The directional preference of H bonds leads to a rather open lattice structure for crystalline water and, consequently, a low density for the solid state. The distance between neighboring oxygen atoms linked by
a hydrogen bond is 0.274 nm. Since the covalent H—O bond is 0.095 nm, the H—O hydrogen bond
length in ice is 0.18 nm. **See this figure animated at http://chemistry.brookscole.com/ggb3**

hold the water molecules apart. Melting involves breaking some of the H bonds that maintain the crystal structure of ice so that the molecules of water (now liquid) can actually pack closer together. Thus, the density of ice is slightly less than that of water. Ice floats, a property of great importance to aquatic organisms in cold climates.

In liquid water, the rigidity of ice is replaced by fluidity and the crystalline periodicity of ice gives way to spatial homogeneity. The H_2O molecules in liquid water form a random, H-bonded network, with each molecule having an average of 4.4 close neighbors situated within a center-to-center distance of 0.284 nm (2.84 Å). At least half of the hydrogen bonds have nonideal orientations (that is, they are not perfectly straight); consequently, liquid H_2O lacks the regular latticelike structure of ice. The space about an O atom is not defined by the presence of four hydrogens but can be occupied by other water molecules randomly oriented so that the local environment, over time, is essentially uniform. Nevertheless, the heat of melting for ice is but a small fraction (13%) of the heat of sublimation for ice (the energy needed to go from the solid to the vapor state). This fact indicates that the majority of H bonds between H_2O molecules survive the transition from solid to liquid. At 10°C, 2.3 H bonds per H_2O molecule remain and the tetrahedral bond order persists, even though substantial disorder is now present.

Molecular Interactions in Liquid Water Are Based on H Bonds

The present interpretation of water structure is that water molecules are connected by uninterrupted H-bond paths running in every direction, spanning the whole sample. The participation of each water molecule in an average state of H bonding to its neighbors means that each molecule is connected to every other in a fluid network of H bonds. The average lifetime of an H-bonded connection between two H_2O molecules in water is 9.5 psec (picoseconds, where 1 psec = 10^{-12} sec). Thus, about every 10 psec, the average H_2O molecule moves, reorients, and interacts with new neighbors, as illustrated in Figure 2.3.

In summary, pure liquid water consists of H_2O molecules held in a random, three-dimensional network that has a local preference for tetrahedral geometry, yet contains a large number of strained or broken hydrogen bonds. The presence of strain creates a kinetic situation in which H_2O molecules can switch H-bond allegiances; fluidity ensues.

The Solvent Properties of Water Derive from Its Polar Nature

Because of its highly polar nature, water is an excellent solvent for ionic substances such as salts; nonionic but polar substances such as sugars, simple alcohols, and amines; and carbonyl-containing molecules such as aldehydes and ketones. Although the electrostatic attractions between the positive and negative ions in the crystal lattice of a salt are very strong, water readily dissolves salts. For example, sodium chloride is dissolved because dipolar water molecules participate in strong electrostatic interactions with the Na^+ and Cl^- ions, leading to the formation of **hydration shells** surrounding these ions (Figure 2.4). Although hydration shells are stable structures, they are also dynamic. Each water molecule in the inner hydration shell around a Na^+ ion is replaced on average every 2 to 4 nsec (nanoseconds, where 1 nsec = 10^{-9} sec) by another H_2O. Consequently, a water molecule is trapped only several hundred times longer by the electrostatic force field of an ion than it is by the H-bonded network of water. (Recall that the average lifetime of H bonds between water molecules is about 10 psec.)

Water Has a High Dielectric Constant   The attractions between the water molecules interacting with, or **hydrating,** ions are much greater than the tendency of oppositely charged ions to attract one another. Water's ability to surround

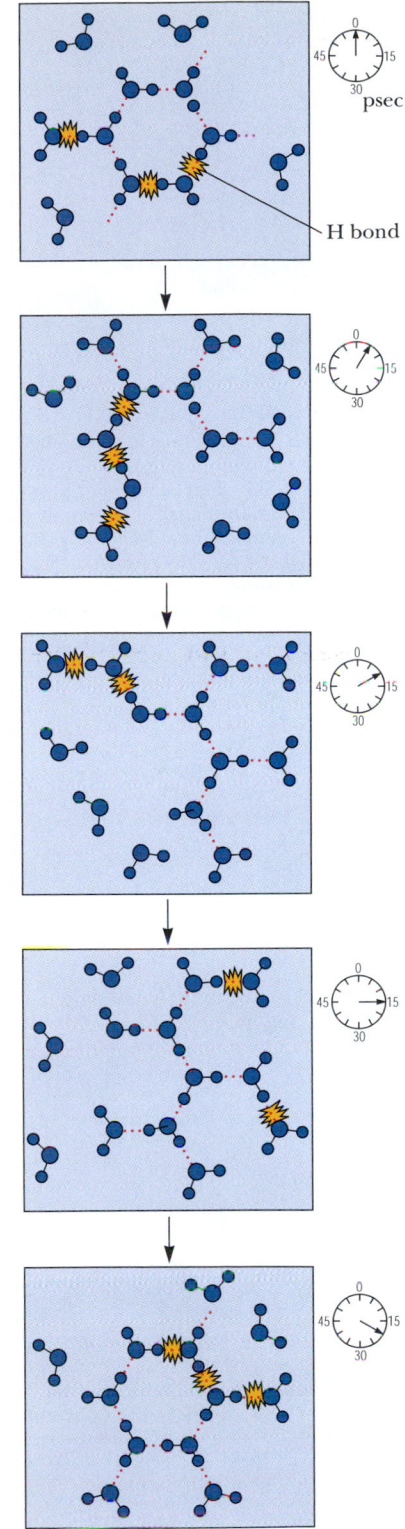

Biochemistry ⧖ **Now**™   **ACTIVE FIGURE 2.3**
The fluid network of H bonds linking water molecules in the liquid state. It is revealing to note that, in 10 psec, a photon of light (which travels at 3×10^8 m/sec) would move a distance of only 0.003 m. **Test yourself on the concepts in this figure at http://chemistry.brookscole.com/ggb3**

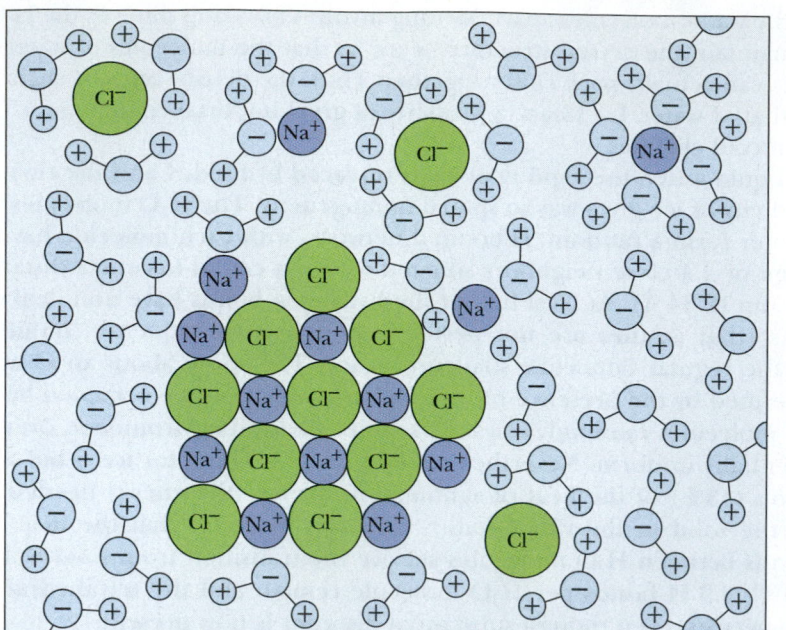

Biochemistry *Now™*   **ANIMATED FIGURE 2.4**
Hydration shells surrounding ions in solution. Water
molecules orient so that the electrical charge on the
ion is sequestered by the water dipole. For positive
ions (cations), the partially negative oxygen atom of
H₂O is toward the ion in solution. Negatively charged
ions (anions) attract the partially positive hydrogen
atoms of water in creating their hydration shells. **See
this figure animated at http://chemistry.brookscole.
com/ggb3**

Keeps Cl⁻ disolved

ions in dipole interactions and diminish their attraction for one another is a
measure of its **dielectric constant,** *D*. Indeed, ionization in solution depends on
the dielectric constant of the solvent; otherwise, the strongly attracted positive
and negative ions would unite to form neutral molecules. The strength of the
dielectric constant is related to the force, *F*, experienced between two ions of
opposite charge separated by a distance, *r*, as given in the relationship

$$F = e_1e_2/Dr^2$$

where e_1 and e_2 are the charges on the two ions. Table 2.1 lists the dielectric constants of some common liquids. Note that the dielectric constant for water is
more than twice that of methanol and more than 40 times that of hexane.

Water Forms H Bonds with Polar Solutes   In the case of nonionic but polar compounds such as sugars, the excellent solvent properties of water stem from its
ability to readily form hydrogen bonds with the polar functional groups on
these compounds, such as hydroxyls, amines, and carbonyls. These polar interactions between solvent and solute are stronger than the intermolecular attractions between solute molecules caused by van der Waals forces and weaker
hydrogen bonding. Thus, the solute molecules readily dissolve in water.

Hydrophobic Interactions   The behavior of water toward nonpolar solutes is
different from the interactions just discussed. Nonpolar solutes (or nonpolar
functional groups on biological macromolecules) do not readily H bond to
H₂O, and as a result, such compounds tend to be only sparingly soluble in
water. The process of dissolving such substances is accompanied by significant
reorganization of the water surrounding the solute so that the response of the
solvent water to such solutes can be equated to "structure making." Because
nonpolar solutes must occupy space, the random H-bonded network of water
must reorganize to accommodate them. At the same time, the water molecules participate in as many H-bonded interactions with one another as the
temperature permits. Consequently, the H-bonded water network rearranges
toward formation of a local cagelike **(clathrate)** structure surrounding each
solute molecule (Figure 2.5). This fixed orientation of water molecules
around a hydrophobic "solute" molecule results in a hydration shell. A major

Table 2.1	
Dielectric Constants* of Some Common Solvents at 25°C	
Solvent	**Dielectric Constant (*D*)**
Formamide	109
Water	78.5
Methyl alcohol	32.6
Ethyl alcohol	24.3
Acetone	20.7
Acetic acid	6.2
Chloroform	5.0
Benzene	2.3
Hexane	1.9

*The dielectric constant is also referred to as *relative permitivity* by
physical chemists.

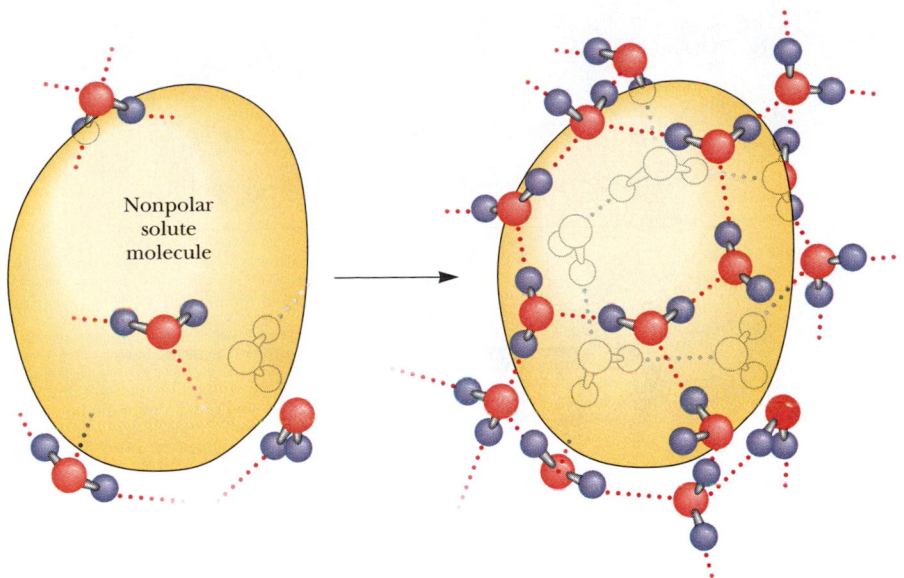

Biochemistry ⊗ **Now**™ **ANIMATED FIGURE 2.5**
Formation of a clathrate structure by water molecules surrounding a hydrophobic solute. **See this figure animated at http://chemistry.brookscole. com/ggb3**

consequence of this rearrangement is that the molecules of H_2O participating in the cage layer have markedly reduced options for orientation in three-dimensional space. Water molecules tend to straddle the nonpolar solute such that two or three tetrahedral directions (H-bonding vectors) are tangential to the space occupied by the inert solute. "Straddling" allows the water molecules to retain their H-bonding possibilities because no H-bond donor or acceptor of the H_2O is directed toward the caged solute. The water molecules forming these clathrates are involved in highly ordered structures. That is, clathrate formation is accompanied by significant ordering of structure or negative entropy.

Under these conditions, nonpolar solute molecules experience a net attraction for one another that is called **hydrophobic interaction.** The basis of this interaction is that when two nonpolar molecules meet, their joint solvation cage involves less surface area and less overall ordering of the water molecules than in their separate cages. The "attraction" between nonpolar solutes is an entropy-driven process due to a net decrease in order among the H_2O molecules. To be specific, hydrophobic interactions between nonpolar molecules are maintained not so much by direct interactions between the inert solutes themselves as by the increase in entropy when the water cages coalesce and reorganize. Because interactions between nonpolar solute molecules and the water surrounding them are of uncertain stoichiometry and do not share the equality of atom-to-atom participation implicit in chemical bonding, the term *hydrophobic interaction* is more correct than the misleading expression *hydrophobic bond.*

Amphiphilic Molecules Compounds containing both strongly polar and strongly nonpolar groups are called **amphiphilic molecules** (from the Greek *amphi* meaning "both" and *philos* meaning "loving"). Such compounds are also referred to as **amphipathic molecules** (from the Greek *pathos* meaning "passion"). Salts of fatty acids are a typical example that has biological relevance. They have a long nonpolar hydrocarbon tail and a strongly polar carboxyl head group, as in the sodium salt of palmitic acid (Figure 2.6). Their behavior in aqueous solution reflects the combination of the contrasting polar and nonpolar nature of these substances. The ionic carboxylate function hydrates readily, whereas the long hydrophobic tail is intrinsically insoluble. Nevertheless, sodium palmitate and other amphiphilic molecules readily disperse in water

The sodium salt of palmitic acid: Sodium palmitate
$(Na^{+-}OOC(CH_2)_{14}CH_3)$

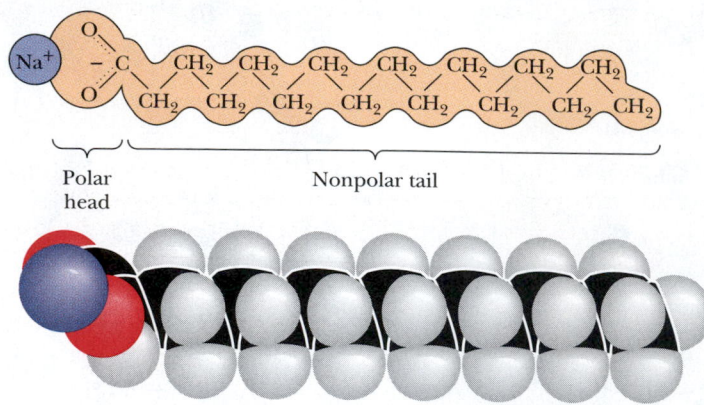

FIGURE 2.6 An amphiphilic molecule: sodium palmitate. Amphiphilic molecules are frequently symbolized by a ball and zigzag line structure, ●〰〰〰, where the ball represents the hydrophilic polar head and the zigzag represents the nonpolar hydrophobic hydrocarbon tail.

because the hydrocarbon tails of these substances are joined together in hydrophobic interactions as their polar carboxylate functions are hydrated in typical hydrophilic fashion. Such clusters of amphipathic molecules are termed **micelles;** Figure 2.7 depicts their structure. Of enormous biological significance is the contrasting solute behavior of the two ends of amphipathic molecules upon introduction into aqueous solutions. The polar ends express their hydrophilicity in ionic interactions with the solvent, whereas their nonpolar counterparts are excluded from the water into a hydrophobic domain constituted from the hydrocarbon tails of many like molecules. It is this behavior that accounts for the formation of membranes, the structures that define the limits and compartments of cells (see Chapter 9).

Influence of Solutes on Water Properties   The presence of dissolved substances disturbs the structure of liquid water, thereby changing its properties. The dynamic H-bonding pattern of water must now accommodate the intruding substance. The net effect is that solutes, regardless of whether they are polar or

Biochemistry《ℰ》Now™   ACTIVE FIGURE 2.7
Micelle formation by amphiphilic molecules in aqueous solution. Negatively charged carboxylate head groups orient to the micelle surface and interact with the polar H_2O molecules via H bonding. The nonpolar hydrocarbon tails cluster in the interior of the spherical micelle, driven by hydrophobic exclusion from the solvent and the formation of favorable van der Waals interactions. Because of their negatively charged surfaces, neighboring micelles repel one another and thereby maintain a relative stability in solution. **Test yourself on the concepts in this figure at http://chemistry.brookscole.com/ggb3**

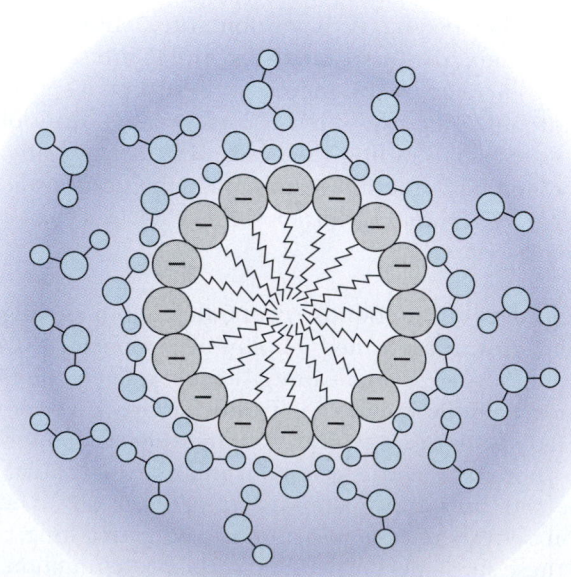

nonpolar, fix nearby water molecules in a more ordered array. Ions, by establishing hydration shells through interactions with the water dipoles, create local order. Hydrophobic substances, for different reasons, make structures within water. To put it another way, by limiting the orientations that neighboring water molecules can assume, solutes give order to the solvent and diminish the dynamic interplay among H_2O molecules that occurs in pure water.

Colligative Properties   This influence of the solute on water is reflected in a set of characteristic changes in behavior termed **colligative properties,** or properties related by a common principle. These alterations in solvent properties are related in that they all depend only on the number of solute particles per unit volume of solvent and not on the chemical nature of the solute. These effects include freezing point depression, boiling point elevation, vapor pressure lowering, and osmotic pressure effects. For example, 1 mol of an ideal solute dissolved in 1000 g of water (a 1 m, or molal, solution) at 1 atm pressure depresses the freezing point by 1.86°C, raises the boiling point by 0.543°C, lowers the vapor pressure in a temperature-dependent manner, and yields a solution whose osmotic pressure relative to pure water is 22.4 atm (at 25°C). In effect, by imposing local order on the water molecules, solutes make it more difficult for water to assume its crystalline lattice (freeze) or escape into the atmosphere (boil or vaporize). Furthermore, when a solution (such as the 1 m solution discussed here) is separated from a volume of pure water by a semipermeable membrane, the solution draws water molecules across this barrier. The water molecules are moving from a region of higher effective concentration (pure H_2O) to a region of lower effective concentration (the solution). This movement of water into the solution dilutes the effects of the solute that is present. The osmotic force exerted by each mole of solute is so strong that it requires the imposition of 22.4 atm of pressure to be negated (Figure 2.8).

Osmotic pressure from high concentrations of dissolved solutes is a serious problem for cells. Bacterial and plant cells have strong, rigid cell walls to contain these pressures. In contrast, animal cells are bathed in extracellular fluids of comparable osmolarity, so no net osmotic gradient exists. Also, to minimize the osmotic pressure created by the contents of their cytosol, cells tend to store substances such as amino acids and sugars in polymeric form. For example, a molecule of glycogen or starch containing 1000 glucose units exerts only 1/1000 the osmotic pressure that 1000 free glucose molecules would.

Water Can Ionize to Form H⁺ and OH⁻

Water shows a small but finite tendency to form ions. This tendency is demonstrated by the electrical conductivity of pure water, a property that clearly establishes the presence of charged species (ions). Water ionizes because the

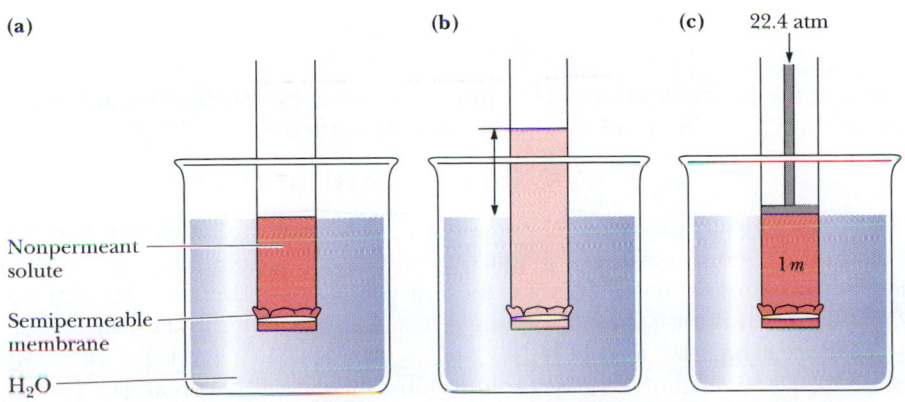

(a)   (b)   (c)   22.4 atm

Nonpermeant solute

Semipermeable membrane

H_2O

1 m

Biochemistry ⊜ Now™   ACTIVE FIGURE 2.8
The osmotic pressure of a 1 molal (m) solution is equal to 22.4 atmospheres of pressure. **(a)** If a nonpermeant solute is separated from pure water by a semipermeable membrane through which H_2O passes freely, **(b)** water molecules enter the solution (osmosis) and the height of the solution column in the tube rises. The pressure necessary to push water back through the membrane at a rate exactly equaled by the water influx is the osmotic pressure of the solution. **(c)** For a 1 m solution, this force is equal to 22.4 atm of pressure. Osmotic pressure is directly proportional to the concentration of the nonpermeant solute. **Test yourself on the concepts in this figure at http://chemistry.brookscole.com/ggb3**

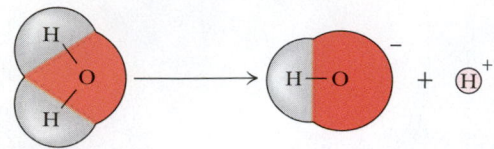

Biochemistry ⊗ Now™   ACTIVE FIGURE 2.9
The ionization of water. **Test yourself on the concepts in this figure at http://chemistry.brookscole.com/ggb3**

larger, strongly electronegative oxygen atom strips the electron from one of its hydrogen atoms, leaving the proton to dissociate (Figure 2.9):

$$H—O—H \longrightarrow H^+ + OH^-$$

Two ions are thus formed: (1) protons or **hydrogen ions,** H^+, and (2) **hydroxyl ions,** OH^-. Free protons are immediately hydrated to form **hydronium ions,** H_3O^+:

$$H^+ + H_2O \longrightarrow H_3O^+$$

Indeed, because most hydrogen atoms in liquid water are hydrogen bonded to a neighboring water molecule, this protonic hydration is an instantaneous process and the ion products of water are H_3O^+ and OH^-:

The amount of H_3O^+ or OH^- in 1 L (liter) of pure water at 25°C is 1×10^{-7} mol; the concentrations are equal because the dissociation is stoichiometric.

Although it is important to keep in mind that the hydronium ion, or hydrated hydrogen ion, represents the true state in solution, the convention is to speak of hydrogen ion concentrations in aqueous solution, even though "naked" protons are virtually nonexistent. Indeed, H_3O^+ itself attracts a hydration shell by H bonding to adjacent water molecules to form an $H_9O_4^+$ species (Figure 2.10) and even more highly hydrated forms. Similarly, the hydroxyl ion, like all other highly charged species, is also hydrated.

K_w, the Ion Product of Water   The dissociation of water into hydrogen ions and hydroxyl ions occurs to the extent that 10^{-7} mol of H^+ and 10^{-7} mol of OH^- are present at equilibrium in 1 L of water at 25°C.

$$H_2O \rightleftharpoons H^+ + OH^-$$

The equilibrium constant for this process is

$$K_{eq} = \frac{[H^+][OH^-]}{[H_2O]}$$

where brackets denote concentrations in moles per liter. Because the concentration of H_2O in 1 L of pure water is equal to the number of grams in a liter divided by the gram molecular weight of H_2O, or 1000/18, the molar concentration of H_2O in pure water is 55.5 M (molar). The decrease in H_2O concentration as a result of ion formation ($[H^+]$, $[OH^-] = 10^{-7} M$) is negligible in comparison; thus its influence on the overall concentration of H_2O can be ignored. Thus,

$$K_{eq} = \frac{(10^{-7})(10^{-7})}{55.5} = 1.8 \times 10^{-16} M$$

Because the concentration of H_2O in pure water is essentially constant, a new constant, K_w, the **ion product of water,** can be written as

$$K_w = 55.5 \, K_{eq} = 10^{-14} M^2 = [H^+][OH^-]$$

This equation has the virtue of revealing the reciprocal relationship between H^+ and OH^- concentrations of aqueous solutions. If a solution is acidic (that is, it has a significant $[H^+]$), then the ion product of water dictates that the OH^- concentration is correspondingly less. For example, if $[H^+]$ is $10^{-2} M$, $[OH^-]$ must be $10^{-12} M$ ($K_w = 10^{-14} M^2 = [10^{-2}][OH^-]$; $[OH^-] = 10^{-12} M$). Similarly, in an alkaline, or basic, solution in which $[OH^-]$ is great, $[H^+]$ is low.

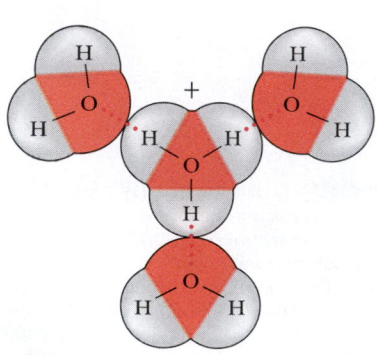

Biochemistry ⊗ Now™   ANIMATED FIGURE 2.10
The hydration of H_3O^+. Solid lines denote covalent bonds; dashed lines represent the H bonds formed between the hydronium ion and its waters of hydration. **See this figure animated at http://chemistry.brookscole.com/ggb3**

2.2 | What Is pH?

To avoid the cumbersome use of negative exponents to express concentrations that range over 14 orders of magnitude, Sørensen, a Danish biochemist, devised the **pH scale** by defining **pH** as *the negative logarithm of the hydrogen ion concentration*[1]:

$$pH = -\log_{10} [H^+]$$

Table 2.2 gives the pH scale. Note again the reciprocal relationship between $[H^+]$ and $[OH^-]$. Also, because the pH scale is based on negative logarithms, low pH values represent the highest H^+ concentrations (and the lowest OH^- concentrations, as K_w specifies). Note also that

$$pK_w = pH + pOH = 14$$

The pH scale is widely used in biological applications because hydrogen ion concentrations in biological fluids are very low, about 10^{-7} *M* or 0.0000001 *M*, a value more easily represented as pH 7. The pH of blood plasma, for example, is 7.4, or 0.00000004 *M* H^+. Certain disease conditions may lower the plasma pH level to 6.8 or less, a situation that may result in death. At pH 6.8, the H^+ concentration is 0.00000016 *M*, four times greater than at pH 7.4.

At pH 7, $[H^+] = [OH^-]$; that is, there is no excess acidity or basicity. The point of **neutrality** is at pH 7, and solutions having a pH of 7 are said to be at **neutral pH.** The pH values of various fluids of biological origin or relevance are given in Table 2.3. Because the pH scale is a logarithmic scale, two solutions whose pH values differ by 1 pH unit have a tenfold difference in $[H^+]$. For example, grapefruit juice at pH 3.2 contains more than 12 times as much H^+ as orange juice at pH 4.3.

[1]To be precise in physical chemical terms, the *activities* of the various components, *not* their molar concentrations, should be used in these equations. The activity *(a)* of a solute component is defined as the product of its molar concentration, *c*, and an *activity coefficient*, γ: *a* = [*c*]γ. Most biochemical work involves dilute solutions, and the use of activities instead of molar concentrations is usually neglected. However, the concentration of certain solutes may be very high in living cells.

Table 2.2				
pH Scale				
The hydrogen ion and hydroxyl ion concentrations are given in moles per liter at 25°C.				
pH	**[H⁺]**		**[OH⁻]**	
0	(10^0)	1.0	0.00000000000001	(10^{-14})
1	(10^{-1})	0.1	0.0000000000001	(10^{-13})
2	(10^{-2})	0.01	0.000000000001	(10^{-12})
3	(10^{-3})	0.001	0.00000000001	(10^{-11})
4	(10^{-4})	0.0001	0.0000000001	(10^{-10})
5	(10^{-5})	0.00001	0.000000001	(10^{-9})
6	(10^{-6})	0.000001	0.00000001	(10^{-8})
7	$\mathbf{(10^{-7})}$	**0.0000001**	**0.0000001**	$\mathbf{(10^{-7})}$
8	(10^{-8})	0.00000001	0.000001	(10^{-6})
9	(10^{-9})	0.000000001	0.00001	(10^{-5})
10	(10^{-10})	0.0000000001	0.0001	(10^{-4})
11	(10^{-11})	0.00000000001	0.001	(10^{-3})
12	(10^{-12})	0.000000000001	0.01	(10^{-2})
13	(10^{-13})	0.0000000000001	0.1	(10^{-1})
14	(10^{-14})	0.00000000000001	1.0	(10^0)

Table 2.3

The pH of Various Common Fluids

Fluid	pH
Household lye	13.6
Bleach	12.6
Household ammonia	11.4
Milk of magnesia	10.3
Baking soda	8.4
Seawater	8.0
Pancreatic fluid	7.8–8.0
Blood plasma	7.4
Intracellular fluids	
Liver	6.9
Muscle	6.1
Saliva	6.6
Urine	5–8
Boric acid	5.0
Beer	4.5
Orange juice	4.3
Grapefruit juice	3.2
Vinegar	2.9
Soft drinks	2.8
Lemon juice	2.3
Gastric juice	1.2–3.0
Battery acid	0.35

Strong Electrolytes Dissociate Completely in Water

Substances that are almost completely dissociated to form ions in solution are called **strong electrolytes.** The term **electrolyte** describes substances capable of generating ions in solution and thereby causing an increase in the electrical conductivity of the solution. Many salts (such as $NaCl$ and K_2SO_4) fit this category, as do strong acids (such as HCl) and strong bases (such as $NaOH$). Recall from general chemistry that acids are proton donors and bases are proton acceptors. In effect, the dissociation of a strong acid such as HCl in water can be treated as a proton transfer reaction between the acid HCl and the base H_2O to give the **conjugate acid** H_3O^+ and the **conjugate base** Cl^-:

$$HCl + H_2O \longrightarrow H_3O^+ + Cl^-$$

The equilibrium constant for this reaction is

$$K = \frac{[H_3O^+][Cl^-]}{[H_2O][HCl]}$$

Customarily, because the term $[H_2O]$ is essentially constant in dilute aqueous solutions, it is incorporated into the equilibrium constant K to give a new term, K_a, the *acid dissociation constant*, where $K_a = K[H_2O]$. Also, the term $[H_3O^+]$ is often replaced by H^+, such that

$$K_a = \frac{[H^+][Cl^-]}{[HCl]}$$

For HCl, the value of K_a is exceedingly large because the concentration of HCl in aqueous solution is vanishingly small. Because this is so, the pH of HCl solutions is readily calculated from the amount of HCl used to make the solution:

$$[H^+] \text{ in solution} = [HCl] \text{ added to solution}$$

Thus, a 1 M solution of HCl has a pH of 0; a 1 mM HCl solution has a pH of 3. Similarly, a 0.1 M $NaOH$ solution has a pH of 13. (Because $[OH^-] = 0.1$ M, $[H^+]$ must be 10^{-13} M.) Viewing the dissociation of strong electrolytes another way, we see that the ions formed show little affinity for one another. For example, in HCl in water, Cl^- has very little affinity for H^+:

$$HCl \longrightarrow H^+ + Cl^-$$

and in $NaOH$ solutions, Na^+ has little affinity for OH^-. The dissociation of these substances in water is effectively complete.

Weak Electrolytes Are Substances That Dissociate Only Slightly in Water

Substances with only a slight tendency to dissociate to form ions in solution are called **weak electrolytes.** Acetic acid, CH_3COOH, is a good example:

$$CH_3COOH + H_2O \rightleftharpoons CH_3COO^- + H_3O^+$$

The acid dissociation constant K_a for acetic acid is 1.74×10^{-5} M:

$$K_a = \frac{[H^+][CH_3COO^-]}{[CH_3COOH]} = 1.74 \times 10^{-5} M$$

K_a is also termed an **ionization constant** because it states the extent to which a substance forms ions in water. The relatively low value of K_a for acetic acid reveals that the un-ionized form, CH_3COOH, predominates over H^+ and CH_3COO^- in aqueous solutions of acetic acid. Viewed another way, CH_3COO^-, the acetate ion, has a high affinity for H^+.

EXAMPLE

What is the pH of a 0.1 M solution of acetic acid? In other words, what is the final pH when 0.1 mol of acetic acid (HAc) is added to water and the volume of the solution is adjusted to equal 1 L?

Answer

The dissociation of HAc in water can be written simply as

$$HAc \rightleftharpoons H^+ + Ac^-$$

where Ac^- represents the acetate ion, CH_3COO^-. In solution, some amount x of HAc dissociates, generating x amount of Ac^- and an equal amount x of H^+. Ionic equilibria characteristically are established very rapidly. At equilibrium, the concentration of HAc + Ac^- must equal 0.1 M. So, [HAc] can be represented as $(0.1 - x)$ M, and [Ac^-] and [H^+] then both equal x molar. From 1.74×10^{-5} $M = ([H^+][Ac^-])/[HAc]$, we get 1.74×10^{-5} $M = x^2/[0.1 - x]$. The solution to quadratic equations of this form ($ax^2 + bx + c = 0$) is $x = -b \pm \sqrt{b^2 - 4ac}/2a$. For $x^2 + (1.74 \times 10^{-5})x - (1.74 \times 10^{-6}) = 0$, $x = 1.319 \times 10^{-3}$ M, so pH = 2.88. (Note that the calculation of x can be simplified here: Because K_a is quite small, $x \ll 0.1$ M. Therefore, K_a is essentially equal to $x^2/0.1$. Thus, $x^2 = 1.74 \times 10^{-6}$ M^2, so $x = 1.32 \times 10^{-3}$ M, and pH = 2.88.)

The Henderson–Hasselbalch Equation Describes the Dissociation of a Weak Acid In the Presence of Its Conjugate Base

Consider the ionization of some weak acid, HA, occurring with an acid dissociation constant, K_a. Then,

$$HA \rightleftharpoons H^+ + A^-$$

and

$$K_a = \frac{[H^+][A^-]}{[HA]}$$

Rearranging this expression in terms of the parameter of interest, [H^+], we have

$$[H^+] = \frac{[K_a][HA]}{[A^-]}$$

Taking the logarithm of both sides gives

$$\log [H^+] = \log K_a + \log_{10} \frac{[HA]}{[A^-]}$$

If we change the signs and define $pK_a = -\log K_a$, we have

$$pH = pK_a - \log_{10} \frac{[HA]}{[A^-]}$$

or

$$\mathbf{pH = pK_a + \log_{10} \frac{[A^-]}{[HA]}}$$

This relationship is known as the **Henderson–Hasselbalch equation.** Thus, the pH of a solution can be calculated, provided K_a and the concentrations of the weak acid HA and its conjugate base A^- are known. Note particularly that

Table 2.4

Acid Dissociation Constants and pK_a Values for Some Weak Electrolytes (at 25°C)

Acid	K_a (M)	pK_a
HCOOH (formic acid)	1.78×10^{-4}	3.75
CH$_3$COOH (acetic acid)	1.74×10^{-5}	4.76
CH$_3$CH$_2$COOH (propionic acid)	1.35×10^{-5}	4.87
CH$_3$CHOHCOOH (lactic acid)	1.38×10^{-4}	3.86
HOOCCH$_2$CH$_2$COOH (succinic acid) pK_1*	6.16×10^{-5}	4.21
HOOCCH$_2$CH$_2$COO$^-$ (succinic acid) pK_2	2.34×10^{-6}	5.63
H$_3$PO$_4$ (phosphoric acid) pK_1	7.08×10^{-3}	2.15
H$_2$PO$_4^-$ (phosphoric acid) pK_2	6.31×10^{-8}	7.20
HPO$_4^{2-}$ (phosphoric acid) pK_3	3.98×10^{-13}	12.40
C$_3$N$_2$H$_5^+$ (imidazole)	1.02×10^{-7}	6.99
C$_6$O$_2$N$_3$H$_{11}^+$ (histidine–imidazole group) pK_R†	9.12×10^{-7}	6.04
H$_2$CO$_3$ (carbonic acid) pK_1	1.70×10^{-4}	3.77
HCO$_3^-$ (bicarbonate) pK_2	5.75×10^{-11}	10.24
(HOCH$_2$)$_3$CNH$_3^+$ (tris-hydroxymethyl aminomethane)	8.32×10^{-9}	8.07
NH$_4^+$ (ammonium)	5.62×10^{-10}	9.25
CH$_3$NH$_3^+$ (methylammonium)	2.46×10^{-11}	10.62

*The pK values listed as pK_1, pK_2, or pK_3 are in actuality pK_a values for the respective dissociations. This simplification in notation is used throughout this book.
†pK_R refers to the imidazole ionization of histidine.
Data from *CRC Handbook of Biochemistry*, The Chemical Rubber Co., 1968.

when [HA] = [A$^-$], pH = pK_a. For example, if equal volumes of 0.1 M HAc and 0.1 M sodium acetate are mixed, then

$$pH = pK_a = 4.76$$

$$pK_a = -\log K_a = -\log_{10}(1.74 \times 10^{-5}) = 4.76$$

(Sodium acetate, the sodium salt of acetic acid, is a strong electrolyte and dissociates completely in water to yield Na$^+$ and Ac$^-$.)

The Henderson–Hasselbalch equation provides a general solution to the quantitative treatment of acid–base equilibria in biological systems. Table 2.4 gives the acid dissociation constants and pK_a values for some weak electrolytes of biochemical interest.

EXAMPLE

What is the pH when 100 mL of 0.1 N NaOH is added to 150 mL of 0.2 M HAc if pK_a for acetic acid = 4.76?

Answer

100 mL 0.1 N NaOH = 0.01 mol OH$^-$, which neutralizes 0.01 mol of HAc, giving an equivalent amount of Ac$^-$:

$$OH^- + HAc \longrightarrow Ac^- + H_2O$$

0.02 mol of the original 0.03 mol of HAc remains essentially undissociated. The final volume is 250 mL.

$$pH = pK_a + \log_{10} \frac{[Ac^-]}{[HAc]} = 4.76 + \log (0.01 \text{ mol})/(0.02 \text{ mol})$$

$$pH = 4.76 - \log_{10} 2 = 4.46$$

If 150 mL of 0.2 M HAc had merely been diluted with 100 mL of water, this would leave 250 mL of a 0.12 M HAc solution. The pH would be given by:

$$K_a = \frac{[H^+][Ac^-]}{[HAc]} = \frac{x^2}{0.12\ M} = 1.74 \times 10^{-5}\ M$$

$$x = 1.44 \times 10^{-3} = [H^+]$$

$$pH = 2.84$$

Titration Curves Illustrate the Progressive Dissociation of a Weak Acid

Titration is the analytical method used to determine the amount of acid in a solution. A measured volume of the acid solution is titrated by slowly adding a solution of base, typically NaOH, of known concentration. As incremental amounts of NaOH are added, the pH of the solution is determined and a plot of the pH of the solution versus the amount of OH$^-$ added yields a **titration curve**. The titration curve for acetic acid is shown in Figure 2.11. In considering the progress of this titration, keep in mind two important equilibria:

1. $HAc \rightleftharpoons H^+ + Ac^-$ $K_a = 1.74 \times 10^{-5}$

2. $H^+ + OH^- \rightleftharpoons H_2O$ $K = \dfrac{[H_2O]}{[K_w]} = 5.55 \times 10^{15}$

As the titration begins, mostly HAc is present, plus some H$^+$ and Ac$^-$ in amounts that can be calculated (see the Example on page 41). Addition of a solution of NaOH allows hydroxide ions to neutralize any H$^+$ present. Note that reaction (2) as written is strongly favored; its apparent equilibrium constant is greater than 10^{15}! As H$^+$ is neutralized, more HAc dissociates to H$^+$ and Ac$^-$. The stoichiometry of the titration is 1:1—for each increment of OH$^-$ added, an equal amount of the weak acid HAc is titrated. As additional NaOH is added, the pH gradually increases as Ac$^-$ accumulates at the expense of diminishing HAc and the neutralization of H$^+$. At the point where half of the HAc has been neutralized (that is, where 0.5 equivalent of OH$^-$ has been added), the concentrations of HAc and Ac$^-$ are equal and pH = pK_a for HAc. Thus, we have an experimental method for determining the pK_a values of weak electrolytes. These pK_a values lie at the midpoint of their respective titration curves. After all of the acid has been neutralized (that is, when one equivalent of base has been added), the pH rises exponentially.

The shapes of the titration curves of weak electrolytes are identical, as Figure 2.12 reveals. Note, however, that the midpoints of the different curves vary in a way that characterizes the particular electrolytes. The pK_a for acetic acid is 4.76, the pK_a for imidazole is 6.99, and that for ammonium is 9.25. These pK_a values are directly related to the dissociation constants of these substances, or, viewed the other way, to the relative affinities of the conjugate bases for protons. NH$_3$ has a high affinity for protons compared to Ac$^-$; NH$_4^+$ is a poor acid compared to HAc.

Phosphoric Acid Has Three Dissociable H$^+$

Figure 2.13 shows the titration curve for phosphoric acid, H$_3$PO$_4$. This substance is a polyprotic acid, meaning it has more than one dissociable proton. Indeed, it has three, and thus three equivalents of OH$^-$ are required to neutralize it, as Figure 2.14 shows. Note that the three dissociable H$^+$ are lost in discrete steps, each dissociation showing a characteristic pK_a. Note that pK_1 occurs at pH = 2.15, and the concentrations of the acid H$_3$PO$_4$ and the

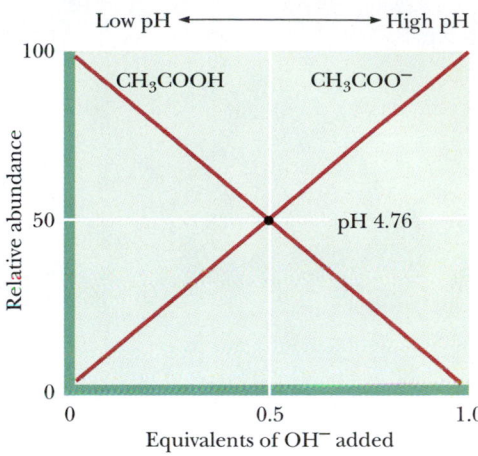

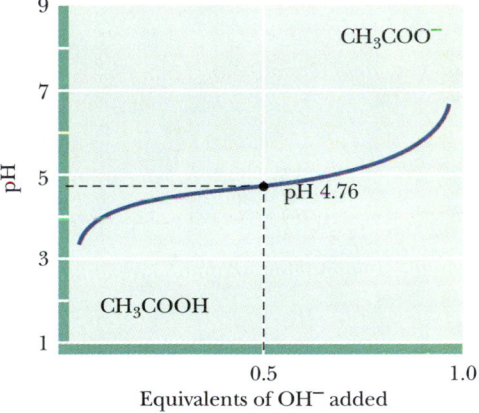

Biochemistry Now™ **ANIMATED FIGURE 2.11**
The titration curve for acetic acid. Note that the titration curve is relatively flat at pH values near the pK_a. In other words, the pH changes relatively little as OH$^-$ is added in this region of the titration curve. **See this figure animated at http://chemistry.brookscole.com/ggb3**

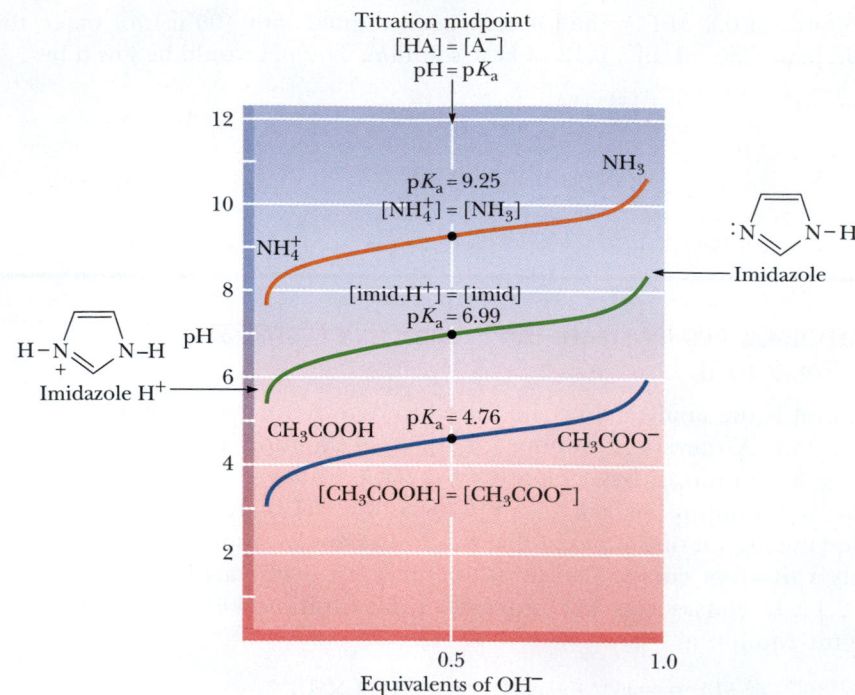

Biochemistry⒮Now™   ANIMATED FIGURE 2.12 The titration curves of several weak electrolytes: acetic acid, imidazole, and ammonium. Note that the shape of these different curves is identical. Only their position along the pH scale is displaced, in accordance with their respective affinities for H⁺ ions, as reflected in their differing pK_a values. **See this figure animated at http:// chemistry.brookscole.com/ggb3**

conjugate base $H_2PO_4^-$ are equal. As the next dissociation is approached, $H_2PO_4^-$ is treated as the acid and HPO_4^{2-} is its conjugate base. Their concentrations are equal at pH 7.20, so pK_2 = 7.20. (Note that at this point, 1.5 equivalents of OH⁻ have been added.) As more OH⁻ is added, the last dissociable hydrogen is titrated, and pK_3 occurs at pH = 12.4, where $[HPO_4^{2-}]$ = $[PO_4^{3-}]$.

The shape of the titration curves for weak electrolytes has a biologically relevant property: In the region of the pK_a, pH remains relatively unaffected as increments of OH⁻ (or H⁺) are added. The weak acid and its conjugate base are acting as a buffer.

Biochemistry⒮Now™   ANIMATED FIGURE 2.13
The titration curve for phosphoric acid. The chemical formulas show the prevailing ionic species present at various pH values. Phosphoric acid (H_3PO_4) has three titratable hydrogens, and therefore three midpoints are seen: at pH 2.15 (pK_1), pH 7.20 (pK_2), and pH 12.4 (pK_3). **See this figure animated at http://chemistry.brookscole.com/ggb3**

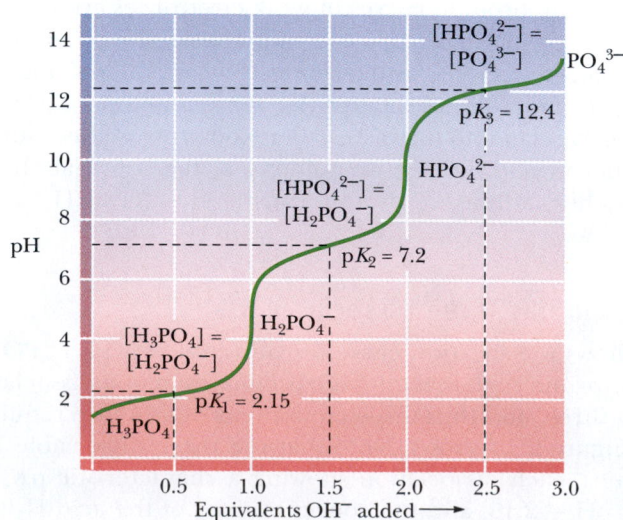

2.3 | **What Are Buffers, and What Do They Do?**

Buffers are solutions that tend to resist changes in their pH as acid or base is added. Typically, a buffer system is composed of a weak acid and its conjugate base. A solution of a weak acid that has a pH nearly equal to its pK_a, by definition, contains an amount of the conjugate base nearly equivalent to the weak acid. Note that in this region, the titration curve is relatively flat (Figure 2.14). Addition of H^+ then has little effect because it is absorbed by the following reaction:

$$H^+ + A^- \longrightarrow HA$$

Similarly, any increase in $[OH^-]$ is offset by the process

$$OH^- + HA \longrightarrow A^- + H_2O$$

Thus, the pH remains relatively constant. The components of a buffer system are chosen such that the pK_a of the weak acid is close to the pH of interest. It is at the pK_a that the buffer system shows its greatest buffering capacity. At pH values more than 1 pH unit from the pK_a, buffer systems become ineffective because the concentration of one of the components is too low to absorb the influx of H^+ or OH^-. The molarity of a buffer is defined as the *sum* of the concentrations of the acid and conjugate base forms.

   Maintenance of pH is vital to all cells. Cellular processes such as metabolism are dependent on the activities of enzymes; in turn, enzyme activity is markedly influenced by pH, as the graphs in Figure 2.15 show. Consequently, changes in pH would be disruptive to metabolism for reasons that become apparent in later chapters. Organisms have a variety of mechanisms to keep the pH of their intracellular and extracellular fluids essentially constant, but the primary protection against harmful pH changes is provided by buffer systems. The buffer systems selected reflect both the need for a pK_a value near pH 7 and the compatibility of the buffer components with the metabolic machinery of cells. Two buffer systems act to maintain intracellular pH essentially constant—the phosphate ($HPO_4^{2-}/H_2PO_4^-$) system and the histidine system. The pH of the extracellular fluid that bathes the cells and tissues of animals is maintained by the bicarbonate/carbonic acid (HCO_3^-/H_2CO_3) system.

The Phosphate Buffer System Is a Major Intracellular Buffering System

The **phosphate system** serves to buffer the intracellular fluid of cells at physiological pH because pK_2 lies near this pH value. The intracellular pH of most cells is maintained in the range between 6.9 and 7.4. Phosphate is an abundant anion in cells, both in inorganic form and as an important functional group on organic molecules that serve as metabolites or macromolecular precursors. In both organic and inorganic forms, its characteristic pK_2 means that the ionic species present at physiological pH are sufficient to donate or accept hydrogen ions to buffer any changes in pH, as the titration curve for H_3PO_4 in Figure 2.14 reveals. For example, if the total cellular concentration of phosphate is 20 mM (millimolar) and the pH is 7.4, the distribution of the major phosphate species is given by

$$pH = pK_2 + \log_{10} \frac{[HPO_4^{2-}]}{[H_2PO_4^-]}$$

$$7.4 = 7.20 + \log_{10} \frac{[HPO_4^{2-}]}{[H_2PO_4^-]}$$

$$\frac{[HPO_4^{2-}]}{[H_2PO_4^-]} = 1.58$$

Thus, if $[HPO_4^{2-}] + [H_2PO_4^-] = 20$ mM, then

$$[HPO_4^{2-}] = 12.25 \text{ m}M \quad \text{and} \quad [H_2PO_4^-] = 7.75 \text{ m}M$$

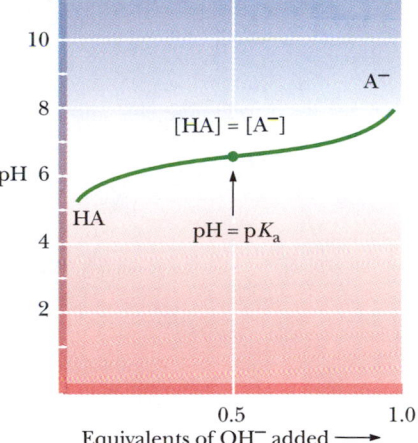

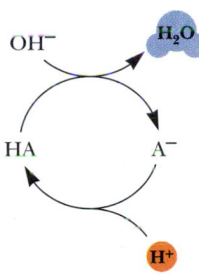

Buffer action:

Biochemistry ⊘ Now™   ACTIVE FIGURE 2.14
A buffer system consists of a weak acid, HA, and its conjugate base, A^-. The pH varies only slightly in the region of the titration curve where $[HA] = [A^-]$. The unshaded box denotes this area of greatest buffering capacity. Buffer action: When HA and A^- are both available in sufficient concentration, the solution can absorb input of either H^+ or OH^-, and pH is maintained essentially constant. **Test yourself on the concepts in this figure at http://chemistry. brookscole.com/ggb3**

(a)

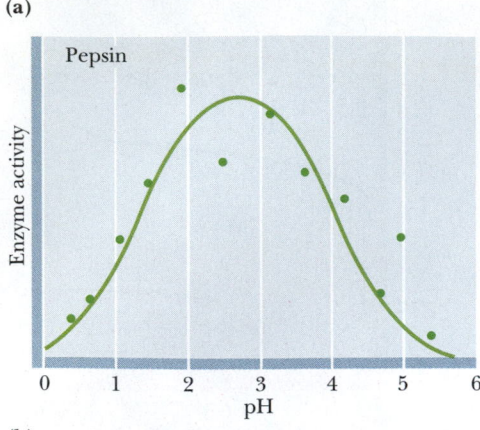

(b)

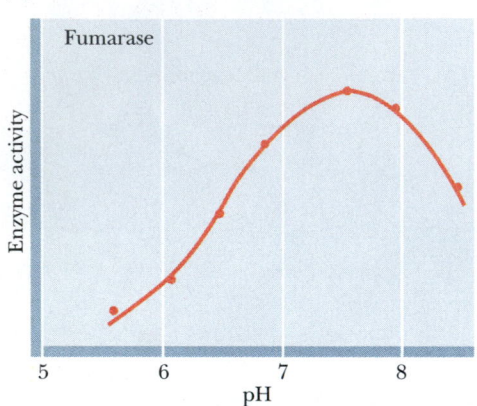

(c)

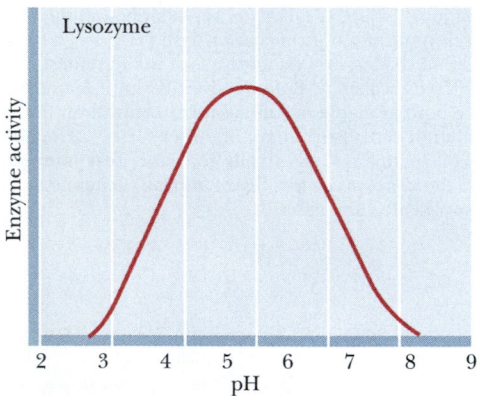

FIGURE 2.15 pH versus enzymatic activity. The activity of enzymes is very sensitive to pH. The pH optimum of an enzyme is one of its most important characteristics. Pepsin is a protein-digesting enzyme active in the gastric fluid. Trypsin is also a proteolytic enzyme, but it acts in the more alkaline milieu of the small intestine. Lysozyme digests the cell walls of bacteria; it is found in tears.

FIGURE 2.16 Anserine (N-β-alanyl-3-methyl-L-histidine) is an important dipeptide buffer in the maintenance of intracellular pH in some tissues. The structure shown is the predominant ionic species at pH 7. pK_1 (COOH) = 2.64; pK_2 (imidazole-N^+H) = 7.04; pK_3 (NH_3^+) = 9.49.

Dissociation of the Histidine–Imidazole Group Also Serves as an Intracellular Buffering System

Histidine is one of the 20 naturally occurring amino acids commonly found in proteins (see Chapter 4). It possesses as part of its structure an imidazole group, a five-membered heterocyclic ring possessing two nitrogen atoms. The pK_a for dissociation of the imidazole hydrogen of histidine is 6.04.

In cells, histidine occurs as the free amino acid, as a constituent of proteins, and as part of dipeptides in combination with other amino acids. Because the concentration of free histidine is low and its imidazole pK_a is more than 1 pH unit removed from prevailing intracellular pH, its role in intracellular buffering is minor. However, protein-bound and dipeptide histidine may be the dominant buffering system in some cells. In combination with other amino acids, as in proteins or dipeptides, the imidazole pK_a may increase substantially. For example, the imidazole pK_a is 7.04 in **anserine,** a dipeptide containing β-alanine and histidine (Figure 2.16). Thus, this pK_a is near physiological pH, and some histidine peptides are well suited for buffering at physiological pH.

"Good" Buffers Are Buffers Useful Within Physiological pH Ranges

Not many common substances have pK_a values in the range from 6 to 8. Consequently, biochemists conducting in vitro experiments were limited in their choice of buffers effective at or near physiological pH. In 1966, N. E. Good devised a set of synthetic buffers to remedy this problem, and over the years the list has expanded so that a "good" selection is available (Figure 2.17). HEPES is an example of a Good buffer (Figure 2.18).

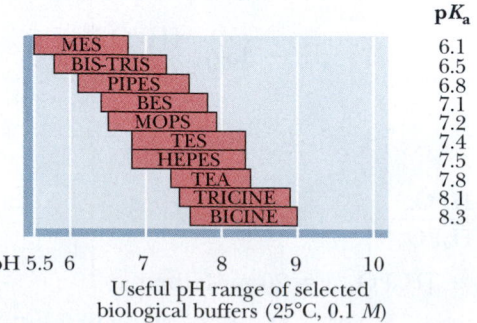

FIGURE 2.17 The pK_a values and pH range of some "Good" buffers.

Human Biochemistry

The Bicarbonate Buffer System of Blood Plasma

The important buffer system of blood plasma is the bicarbonate/carbonic acid couple:

$$H_2CO_3 \rightleftharpoons H^+ + HCO_3^-$$

The relevant pK_a, pK_1 for carbonic acid, has a value far removed from the normal pH of blood plasma (pH 7.4). (The pK_1 for H_2CO_3 at 25°C is 3.77 [Table 2.4], but at 37°C, pK_1 is 3.57.) At pH 7.4, the concentration of H_2CO_3 is a minuscule fraction of the HCO_3^- concentration; thus the plasma appears to be poorly protected against an influx of OH^- ions.

$$pH = 7.4 = 3.57 + \log_{10} \frac{[HCO_3^-]}{[H_2CO_3]}$$

$$\frac{[HCO_3^-]}{[H_2CO_3]} = 6761$$

For example, if $[HCO_3^-] = 24$ mM, then $[H_2CO_3]$ is only 3.55 μM $(3.55 \times 10^{-6}\ M)$, and an equivalent amount of OH^- (its usual concentration in plasma) would swamp the buffer system, causing a dangerous rise in the plasma pH. How, then, can this bicarbonate system function effectively? The bicarbonate buffer system works well because the critical concentration of H_2CO_3 is maintained relatively constant through equilibrium with dissolved CO_2 produced in the tissues and available as a gaseous CO_2 reservoir in the lungs.*

Gaseous CO_2 from the lungs and tissues is dissolved in the blood plasma, symbolized as $CO_2(d)$, and hydrated to form H_2CO_3:

$$CO_2(g) \rightleftharpoons CO_2(d)$$
$$CO_2(d) + H_2O \rightleftharpoons H_2CO_3$$
$$H_2CO_3 \rightleftharpoons H^+ + HCO_3^-$$

Thus, the concentration of H_2CO_3 is itself buffered by the available pools of CO_2. The hydration of CO_2 is actually mediated by an enzyme, *carbonic anhydrase*, which facilitates the equilibrium by rapidly catalyzing the reaction

$$H_2O + CO_2(d) \rightleftharpoons H_2CO_3$$

Under the conditions of temperature and ionic strength prevailing in mammalian body fluids, the equilibrium for this reaction lies far to the left, such that more than 300 CO_2 molecules are present in solution for every molecule of H_2CO_3. Because dissolved CO_2 and H_2CO_3 are in equilibrium, the proper expression for H_2CO_3 availability is $[CO_2(d)] + [H_2CO_3]$, the so-called total carbonic acid pool, consisting primarily of $CO_2(d)$. The overall equilibrium for the bicarbonate buffer system then is

$$CO_2(d) + H_2O \overset{K_h}{\rightleftharpoons} H_2CO_3$$

$$H_2CO_3 \overset{K_a}{\rightleftharpoons} H^+ + HCO_3^-$$

An expression for the ionization of H_2CO_3 under such conditions (that is, in the presence of dissolved CO_2) can be obtained from

K_h, the equilibrium constant for the hydration of CO_2, and from K_a, the first acid dissociation constant for H_2CO_3:

$$K_h = \frac{[H_2CO_3]}{[CO_2(d)]}$$

Thus,

$$[H_2CO_3] = K_h[CO_2(d)]$$

Putting this value for $[H_2CO_3]$ into the expression for the first dissociation of H_2CO_3 gives

$$K_a = \frac{[H^+][HCO_3^-]}{[H_2CO_3]}$$

$$= \frac{[H^+][HCO_3^-]}{K_h[CO_2(d)]}$$

Therefore, the overall equilibrium constant for the ionization of H_2CO_3 in equilibrium with $CO_2(d)$ is given by

$$K_aK_h = \frac{[H^+][HCO_3^-]}{K_h[CO_2(d)]}$$

and K_aK_h, the product of two constants, can be defined as a new equilibrium constant, $K_{overall}$. The value of K_h is 0.003 at 37°C and K_a, the ionization constant for H_2CO_3, is $10^{-3.57} = 0.000269$. Therefore,

$$K_{overall} = (0.000269)(0.003)$$
$$= 8.07 \times 10^{-7}$$
$$pK_{overall} = 6.1$$

which yields the following Henderson–Hasselbalch relationship:

$$pH = pK_{overall} + \log_{10} \frac{[HCO_3^-]}{[CO_2(d)]}$$

Although the prevailing blood pH of 7.4 is more than 1 pH unit away from $pK_{overall}$, the bicarbonate system is still an effective buffer. Note that, at blood pH, the concentration of the acid component of the buffer will be less than 10% of the conjugate base component. One might imagine that this buffer component could be overwhelmed by relatively small amounts of alkali, with consequent disastrous rises in blood pH. However, the acid component is the total carbonic acid pool, that is, $[CO_2(d)] + [H_2CO_3]$, which is stabilized by its equilibrium with $CO_2(g)$. Gaseous CO_2 serves to buffer any losses from the total carbonic acid pool by entering solution as $CO_2(d)$, and blood pH is effectively maintained. Thus, the bicarbonate buffer system is an *open system*. The natural presence of CO_2 gas at a partial pressure of 40 mm Hg in the alveoli of the lungs and the equilibrium

$$CO_2(g) \rightleftharpoons CO_2(d)$$

keep the concentration of $CO_2(d)$ (the principal component of the total carbonic acid pool in blood plasma) in the neighborhood of 1.2 mM. Plasma $[HCO_3^-]$ is about 24 mM under such conditions.

*Well-fed humans exhale about 1 kg of CO_2 daily. Imagine the excretory problem if CO_2 were not a volatile gas.

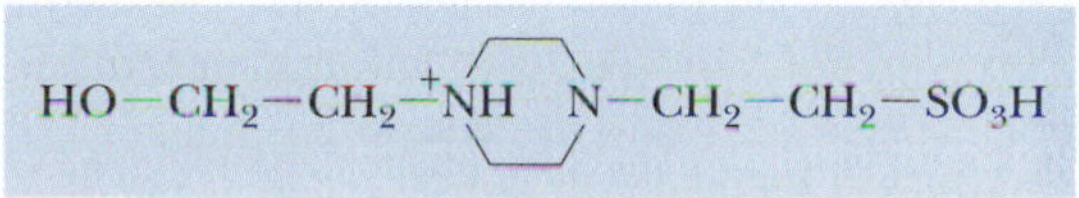

HEPES

◀ **FIGURE 2.18** The structure of HEPES, 4-(2-hydroxy)-1-piperazine ethane sulfonic acid, in its fully protonated form. The pK_a of the sulfonic acid group is about 3; the pK_a of the piperazine-N^+H is 7.55 at 20°C.

Human Biochemistry

Blood pH and Respiration

Hyperventilation, defined as a breathing rate more rapid than necessary for normal CO_2 elimination from the body, can result in an inappropriately low $[CO_2(g)]$ in the blood. Central nervous system disorders such as meningitis, encephalitis, or cerebral hemorrhage, as well as a number of drug- or hormone-induced physiological changes, can lead to hyperventilation. As $[CO_2(g)]$ drops due to excessive exhalation, $[H_2CO_3]$ in the blood plasma falls, followed by a decline in $[H^+]$ and $[HCO_3^-]$ in the blood plasma. Blood pH rises within 20 sec of the onset of hyperventilation, becoming maximal within 15 min. $[H^+]$ can change from its normal value of 40 nM (pH = 7.4) to 18 nM (pH = 7.74). This rise in plasma pH (increase in alkalinity) is termed **respiratory alkalosis.**

Hypoventilation is the opposite of hyperventilation and is characterized by an inability to excrete CO_2 rapidly enough to meet physiological needs. Hypoventilation can be caused by narcotics, sedatives, anesthetics, and depressant drugs; diseases of the lung also lead to hypoventilation. Hypoventilation results in **respiratory acidosis,** as $CO_2(g)$ accumulates, giving rise to H_2CO_3, which dissociates to form H^+ and HCO_3^-.

2.4 | Does Water Have a Unique Role in the Fitness of the Environment?

The remarkable properties of water render it particularly suitable to its unique role in living processes and the environment, and its presence in abundance favors the existence of life. Let's examine water's physical and chemical properties to see the extent to which they provide conditions that are advantageous to organisms.

As a solvent, water is powerful yet innocuous. No other chemically inert solvent compares with water for the substances it can dissolve. Also, it is very important to life that water is a "poor" solvent for nonpolar substances. Thus, through hydrophobic interactions, lipids coalesce, membranes form, boundaries are created delimiting compartments, and the cellular nature of life is established. Because of its very high dielectric constant, water is a medium for ionization. Ions enrich the living environment in that they enhance the variety of chemical species and introduce an important class of chemical reactions. They provide electrical properties to solutions and therefore to organisms. Aqueous solutions are the prime source of ions.

The thermal properties of water are especially relevant to its environmental fitness. It has great power as a buffer resisting thermal (temperature) change. Its heat capacity, or specific heat (4.1840 J/g°C), is remarkably high; it is ten times greater than iron, five times greater than quartz or salt, and twice as great as hexane. Its heat of fusion is 335 J/g. Thus, at 0°C, it takes a loss of 335 J to change the state of 1 g of H_2O from liquid to solid. Its heat of vaporization (2.24 kJ/g) is exceptionally high. These thermal properties mean that it takes substantial changes in heat content to alter the temperature and especially the state of water. Water's thermal properties allow it to buffer the climate through such processes as condensation, evaporation, melting, and freezing. Furthermore, these properties allow effective temperature regulation in living organisms. For example, heat generated within an organism as a result of metabolism can be efficiently eliminated through evaporation or conduction. The thermal conductivity of water is very high compared with that of other liquids. The anomalous expansion of water as it cools to temperatures near its freezing point is a unique attribute of great significance to its natural fitness. As water cools, H bonding increases because the thermal motions of the molecules are lessened. H bonding tends to separate the water molecules (Figure 2.2), thus decreasing the density of water. These changes in density mean that, at temperatures below 4°C, cool water rises and, most important, ice freezes on the surface of bodies of water, forming an insulating layer protecting the liquid water underneath.

Water has the highest surface tension (75 dyne/cm) of all common liquids (except mercury). Together, surface tension and density determine how high a liquid rises in a capillary system. Capillary movement of water plays a prominent role in the life of plants. Last, consider osmosis as it relates to water and, in particular, the bulk movement of water in the direction from a dilute aqueous solution to a more concentrated one across a semipermeable boundary. Such bulk movements determine the shape and form of living things.

Water is truly a crucial determinant of the fitness of the environment. In a very real sense, organisms are aqueous systems in a watery world.

Summary

2.1 What Are the Properties of Water?   Life depends on the unusual chemical and physical properties of H_2O. Its high boiling point, melting point, heat of vaporization, and surface tension indicate that intermolecular forces of attraction between H_2O molecules are high. Hydrogen bonds between adjacent water molecules are the basis of these forces. Liquid water consists of H_2O molecules held in a random, three-dimensional network that has a local preference for tetrahedral geometry, yet contains a large number of strained or broken hydrogen bonds. The presence of strain creates a kinetic situation in which H_2O molecules can switch H-bond allegiances; fluidity ensues. As kinetic energy decreases (the temperature falls), crystalline water (ice) forms.

The solvent properties of water are attributable to the "bent" structure of the water molecule and polar nature of its O—H bonds. Together these attributes yield a liquid that can form hydration shells around salt ions or dissolve polar solutes through H-bond interactions. Hydrophobic interactions in aqueous environments also arise as a consequence of polar interactions between water molecules. The polarity of the O—H bonds means that water also ionizes to a small but finite extent to release H^+ and OH^- ions. K_w, the ion product of water, reveals that the concentration of $[H^+]$ and $[OH^-]$ at 25°C is $10^{-7} M$.

2.2 What Is pH?   pH is defined as $-\log_{10} [H^+]$. pH is an important concept in biochemistry because the structure and function of biological molecules depend strongly on functional groups that ionize, or not, depending on small changes in $[H^+]$ concentration. Weak electrolytes are substances that dissociate incompletely in water. The behavior of weak electrolytes determines the concentration of $[H^+]$ and hence, pH. The Henderson–Hasselbalch equation provides a general solution to the quantitative treatment of acid–base equilibria in biological systems.

2.3 What Are Buffers, and What Do They Do?   Buffers are solutions composed of a weak acid and its conjugate base. Such solutions can resist changes in pH when acid or base is added to the solution.

Maintenance of pH is vital to all cells, and primary protection against harmful pH changes is provided by buffer systems. The buffer systems used by cells reflect a need for a pK_a value near pH 7 and the compatibility of the buffer components with the metabolic apparatus of cells. The phosphate buffer system and the histidine–imidazole system are the two prominent intracellular buffers, whereas the bicarbonate buffer system is the principal extracellular buffering system in animals.

2.4 Does Water Have a Unique Role in the Fitness of the Environment?   Life and water are inextricably related. Water is particularly suited to its unique role in living processes and the environment. As a solvent, water is powerful yet innocuous; no other chemically inert solvent compares with water for the substances it can dissolve. Also, water as a "poor" solvent for nonpolar substances gives rise to hydrophobic interactions, leading lipids to coalesce, membranes to form, and boundaries delimiting compartments to appear.

Water is a medium for ionization. Ions enrich the living environment and introduce an important class of chemical reactions. Ions provide electrical properties to solutions and therefore to organisms.

The thermal properties of water are especially relevant to its environmental fitness. It takes substantial changes in heat content to alter the temperature and especially the state of water. Water's thermal properties allow it to buffer the climate through such processes as condensation, evaporation, melting, and freezing. Furthermore, water's thermal properties allow effective temperature regulation in living organisms.

Osmosis as it relates to water, and in particular, the bulk movement of water in the direction from a dilute aqueous solution to a more concentrated one across semipermeable membranes, determines the shape and form of living things. In large degree, the properties of water define the fitness of the environment. Organisms are aqueous systems in a watery world.

Problems

1. Calculate the pH of the following.
   a. $5 \times 10^{-4} M$ HCl
   b. $7 \times 10^{-5} M$ NaOH
   c. $2 \mu M$ HCl
   d. $3 \times 10^{-2} M$ KOH
   e. 0.04 mM HCl
   f. $6 \times 10^{-9} M$ HCl

2. Calculate the following from the pH values given in Table 2.3.
   a. $[H^+]$ in vinegar
   b. $[H^+]$ in saliva
   c. $[H^+]$ in household ammonia
   d. $[OH^-]$ in milk of magnesia
   e. $[OH^-]$ in beer
   f. $[H^+]$ inside a liver cell

3. The pH of a 0.02 M solution of an acid was measured at 4.6.
   a. What is the $[H^+]$ in this solution?
   b. Calculate the acid dissociation constant K_a and pK_a for this acid.

4. The K_a for formic acid is $1.78 \times 10^{-4} M$.
   a. What is the pH of a 0.1 M solution of formic acid?
   b. 150 mL of 0.1 M NaOH is added to 200 mL of 0.1 M formic acid, and water is added to give a final volume of 1 L. What is the pH of the final solution?

5. Given 0.1 M solutions of acetic acid and sodium acetate, describe the preparation of 1 L of 0.1 M acetate buffer at a pH of 5.4.

6. If the internal pH of a muscle cell is 6.8, what is the $[HPO_4^{2-}]/[H_2PO_4^-]$ ratio in this cell?

7. Given 0.1 M solutions of Na_3PO_4 and H_3PO_4, describe the preparation of 1 L of a phosphate buffer at a pH of 7.5. What are the molar concentrations of the ions in the final buffer solution, including Na^+ and H^+?

8. BICINE is a compound containing a tertiary amino group whose relevant pK_a is 8.3 (Figure 2.17). Given 1 L of 0.05 M BICINE with its tertiary amino group in the unprotonated form, how much 0.1 N HCl must be added to have a BICINE buffer solution of pH 7.5? What is the molarity of BICINE in the final buffer? What is the concentration of the protonated form of BICINE in this final buffer?

9. What are the approximate fractional concentrations of the following phosphate species at pH values of 0, 2, 4, 6, 8, 10, and 12?
   a. H_3PO_4
   b. $H_2PO_4^-$
   c. HPO_4^{2-}
   d. PO_4^{3-}

10. Citric acid, a tricarboxylic acid important in intermediary metabolism, can be symbolized as H_3A. Its dissociation reactions are
$$H_3A \rightleftharpoons H^+ + H_2A^- \qquad pK_1 = 3.13$$
$$H_2A^- \rightleftharpoons H^+ + HA^{2-} \qquad pK_2 = 4.76$$
$$HA^{2-} \rightleftharpoons H^+ + A^{3-} \qquad pK_3 = 6.40$$
If the *total* concentration of the acid *and* its anion forms is 0.02 M, what are the individual concentrations of H_3A, H_2A^-, HA^{2-}, and A^{3-} at pH 5.2?

11. a. If 50 mL of 0.01 M HCl is added to 100 mL of 0.05 M phosphate buffer at pH 7.2, what is the resultant pH? What are the concentrations of $H_2PO_4^-$ and HPO_4^{2-} in the final solution?
    b. If 50 mL of 0.01 M NaOH is added to 100 mL of 0.05 M phosphate buffer at pH 7.2, what is the resultant pH? What are the concentrations of $H_2PO_4^-$ and HPO_4^{2-} in this final solution?

12. At 37°C, if the plasma pH is 7.4 and the plasma concentration of HCO_3^- is 15 mM, what is the plasma concentration of H_2CO_3? What is the plasma concentration of $CO_{2(dissolved)}$? If metabolic activity changes the concentration of $CO_{2(dissolved)}$ to 3 mM and [HCO_3^-] remains at 15 mM, what is the pH of the plasma?

13. Draw the titration curve for anserine (Figure 2.16). The isoelectric point of anserine is the pH where the net charge on the molecule is zero; what is the isoelectric point for anserine? Given a 0.1 M solution of anserine at its isoelectric point and ready access to 0.1 M HCl, 0.1 M NaOH and distilled water, describe the preparation of 1 L of 0.04 M anserine buffer solution, pH 7.2.

14. Given a solution of 0.1 M HEPES in its fully protonated form, and ready access to 0.1 M HCl, 0.1 M NaOH and distilled water, describe the preparation of 1 L of 0.025 M HEPES buffer solution, pH 7.8.

15. A 100-g amount of a solute was dissolved in 1000 g of water. The freezing point of this solution was measured accurately and determined to be −1.12°C. What is the molecular weight of the solute?

Preparing for the MCAT Exam

16. In light of the Human Biochemistry box on page 47, what would be the effect on blood pH if cellular metabolism produced a sudden burst of carbon dioxide?

17. On the basis of Figure 2.12, what will be the pH of the acetate–acetic acid solution when the ratio of [acetate]/[acetic acid] is 10?
   a. 3.76
   b. 4.76
   c. 5.76
   d. 14.76

Biochemistry⩙Now™   Preparing for an exam? Test yourself on key questions at http://chemistry.brookscole.com/ggb3

Further Reading

Properties of Water

Cooke, R., and Kuntz, I. D., 1974. The properties of water in biological systems. *Annual Review of Biophysics and Bioengineering* **3:**95–126.

Franks, F., ed., 1982. *The Biophysics of Water.* New York: John Wiley & Sons.

Stillinger, F. H., 1980. Water revisited. *Science* **209:**451–457.

Properties of Solutions

Cooper, T. G., 1977. *The Tools of Biochemistry,* Chap. 1. New York: John Wiley & Sons.

Segel, I. H., 1976. *Biochemical Calculations,* 2nd ed., Chap. 1. New York: John Wiley & Sons.

Titration Curves

Darvey, I. G., and Ralston, G. B., 1993. Titration curves—misshapen or mislabeled? *Trends in Biochemical Sciences* **18:**69–71.

pH and Buffers

Beynon, R. J., and Easterby, J. S., 1996. *Buffer Solutions: The Basics.* New York: IRL Press: Oxford University Press.

Edsall, J. T., and Wyman, J., 1958. Carbon dioxide and carbonic acid, in *Biophysical Chemistry,* Vol. 1, Chap. 10. New York: Academic Press.

Gillies R. J, and Lynch R. M., 2001. Frontiers in the measurement of cell and tissue pH. *Novartis Foundation Symposium* **240:**7–19.

Kelly, J. A., 2000. Determinants of blood pH in health and disease. *Critical Care* **4:**6–14.

Masoro, E. J., and Siegel, P. D., 1971. *Acid-Base Regulation: Its Physiology and Pathophysiology.* Philadelphia: W.B. Saunders.

Nørby, J. G., 2000. The origin and meaning of the little p in pH. *Trends in Biochemical Sciences* **25:**36–37.

Perrin, D. D., 1982. *Ionization Constants of Inorganic Acids and Bases in Aqueous Solution.* New York: Pergamon Press.

Rose, B. D., 1994. *Clinical Physiology of Acid-Base and Electrolyte Disorders,* 4th ed. New York: McGraw-Hill.

The Fitness of the Environment

Henderson, L. J., 1913. *The Fitness of the Environment.* New York: Macmillan. (Republished 1970. Gloucester, MA: P. Smith.)

Hille, B., 1992. *Ionic Channels of Excitable Membranes,* 2nd ed., Chap. 10. Sunderland, MA: Sinauer Associates.

Thermodynamics of Biological Systems

CHAPTER 3

Essential Question

Living things require energy. Movement, growth, synthesis of biomolecules, and the transport of ions and molecules across membranes all demand energy input. All organisms must acquire energy from their surroundings and must utilize that energy efficiently to carry out life processes. To study such bioenergetic phenomena requires familiarity with **thermodynamics.** Thermodynamics also allows us to determine whether chemical processes and reactions occur spontaneously. The student should appreciate the power and practical value of thermodynamic reasoning and realize that this is well worth the effort needed to understand it. *What are the laws and principles of thermodynamics that allow us to describe the flows and interchanges of heat, energy, and matter in biochemical systems?*

© Nik Wheeler/CORBIS

The sun is the source of energy for virtually all life. We even harvest its energy in the form of electricity using windmills driven by air heated by the sun.

Even the most complicated aspects of thermodynamics are based ultimately on three rather simple and straightforward laws. These laws and their extensions sometimes run counter to our intuition. However, once truly understood, the basic principles of thermodynamics become powerful devices for sorting out complicated chemical and biochemical problems. Once we reach this milestone in our scientific development, thermodynamic thinking becomes an enjoyable and satisfying activity.

Several basic thermodynamic principles are presented in this chapter, including the analysis of heat flow, entropy production, and free energy functions and the relationship between entropy and information. In addition, some ancillary concepts are considered, including the concept of standard states, the effect of pH on standard-state free energies, the effect of concentration on the net free energy change of a reaction, and the importance of coupled processes in living things. The chapter concludes with a discussion of ATP and other energy-rich compounds.

A theory is the more impressive the greater is the simplicity of its premises, the more different are the kinds of things it relates and the more extended is its range of applicability. Therefore, the deep impression which classical thermodynamics made upon me. It is the only physical theory of universal content which I am convinced, that within the framework of applicability of its basic concepts, will never be overthrown.

Albert Einstein

3.1 | What Are the Basic Concepts of Thermodynamics?

In any consideration of thermodynamics, a distinction must be made between the system and the surroundings. The **system** is that portion of the *universe* with which we are concerned. It might be a mixture of chemicals in a test tube, or a single cell, or an entire organism. The **surroundings** include everything else in the universe (Figure 3.1). The nature of the system must also be specified. There are three basic kinds of systems: isolated, closed, and open. An **isolated system** cannot exchange matter or energy with its surroundings. A **closed system** may exchange energy, but not matter, with the surroundings. An **open system** may exchange matter, energy, or both with the surroundings. Living things are typically open systems that exchange matter (nutrients and waste products) and energy (heat from metabolism, for example) with their surroundings.

The First Law: The Total Energy of an Isolated System Is Conserved

It was realized early in the development of thermodynamics that heat could be converted into other forms of energy and moreover that all forms of energy could ultimately be converted to some other form. The **first law of thermodynamics** states that *the total energy of an isolated system is conserved.* Thermodynamicists have

Key Questions

Biochemistry ⊜ Now™ Test yourself on these Key Questions at BiochemistryNow at **http://chemistry.brookscole.com/ggb3**

Isolated system:
No exchange of matter or energy

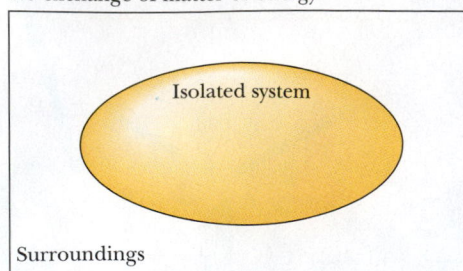

Isolated system

Surroundings

Closed system:
Energy exchange may occur

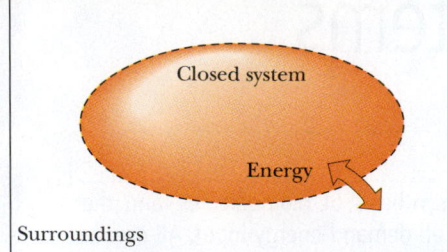

Closed system

Energy

Surroundings

Open system:
Energy exchange and/or matter exchange may occur

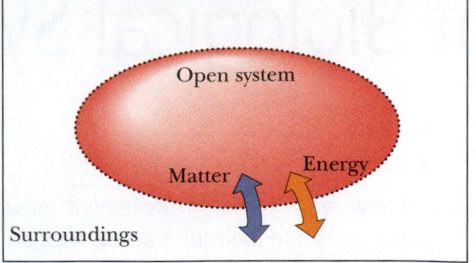

Open system

Matter   Energy

Surroundings

Biochemistry ⓔ Now™   ACTIVE FIGURE 3.1 The characteristics of isolated, closed, and open systems. Isolated systems exchange neither matter nor energy with their surroundings. Closed systems may exchange energy, but not matter, with their surroundings. Open systems may exchange either matter or energy with the surroundings. **Test yourself on the concepts in this figure at http://chemistry.brookscole.com/ggb3**

formulated a mathematical function for keeping track of heat transfers and work expenditures in thermodynamic systems. This function is called the **internal energy,** commonly designated E or U, and it includes all the energies that might be exchanged in physical or chemical processes, including rotational, vibrational, and translational energies of molecules and also the energy stored in covalent and noncovalent bonds. The internal energy depends only on the present state of a system and hence is referred to as a **state function.** The internal energy does not depend on how the system got there and is thus **independent of path.** An extension of this thinking is that we can manipulate the system through any possible pathway of changes, and as long as the system returns to the original state, the internal energy, E, will not have been changed by these manipulations.

The internal energy, E, of any system can change only if energy flows in or out of the system in the form of heat or work. For any process that converts one state (state 1) into another (state 2), the change in internal energy, ΔE, is given as

$$\Delta E = E_2 - E_1 = q + w \tag{3.1}$$

where the quantity q is the *heat absorbed by the system from the surroundings* and w is *the work done on the system by the surroundings.* **Mechanical work** is defined as *movement through some distance caused by the application of a force.* Both movement and force are required for work to have occurred. Examples of work done in biological systems include the flight of insects and birds, the circulation of blood by a pumping heart, the transmission of an impulse along a nerve, and the lifting of a weight by someone who is exercising. On the other hand, if a person strains to lift a heavy weight but fails to move the weight at all, then, in the thermodynamic sense, no work has been done. (The energy expended in the muscles of the would-be weight lifter is given off in the form of heat.) In chemical and biochemical systems, work is often concerned with the pressure and volume of the system under study. The mechanical work done on the system is defined as $w = -P\Delta V$, where P is the pressure and ΔV is the volume change and is equal to $V_2 - V_1$. When work is defined in this way, the sign on the right side of Equation 3.1 is positive. (Sometimes w is defined as work done *by* the system; in this case, the equation is $\Delta E = q - w$.) Work may occur in many forms, such as mechanical, electrical, magnetic, and chemical. ΔE, q, and w must all have the same units. The **calorie,** abbreviated **cal,** and **kilocalorie (kcal)** have been traditional choices of chemists and biochemists, but the SI unit, the **joule,** is now recommended.

Enthalpy Is a More Useful Function for Biological Systems

If the definition of work is limited to mechanical work ($w = -P\Delta V$) and no change in volume occurs, an interesting simplification is possible. In this case, ΔE is merely the *heat exchanged at constant volume.* This is so because if the volume is constant, no mechanical work can be done on or by the system. Then

$\Delta E = q$. Thus ΔE is a very useful quantity in constant volume processes. However, chemical and especially biochemical processes and reactions are much more likely to be carried out at constant pressure. In constant pressure processes, ΔE is not necessarily equal to the heat transferred. For this reason, chemists and biochemists have defined a function that is especially suitable for constant pressure processes. It is called the **enthalpy, H,** and it is defined as

$$H = E + PV \qquad (3.2)$$

The clever nature of this definition is not immediately apparent. However, if the pressure is constant, then we have

$$\Delta H = \Delta E + P\Delta V = q + w + P\Delta V = q - P\Delta V + P\Delta V = q \qquad (3.3)$$

So, ΔE is the heat transferred in a constant volume process, and ΔH is the heat transferred in a constant pressure process.

Often, because biochemical reactions normally occur in liquids or solids rather than in gases, volume changes are typically quite small, and *enthalpy and internal energy are often essentially equal.*

In order to compare the thermodynamic parameters of different reactions, it is convenient to define a *standard state.* For solutes in a solution, the standard state is normally unit activity (often simplified to 1 M concentration). Enthalpy, internal energy, and other thermodynamic quantities are often given or determined for standard-state conditions and are then denoted by a superscript degree sign ("°"), as in $\Delta H°$, $\Delta E°$, and so on.

Enthalpy changes for biochemical processes can be determined experimentally by measuring the heat absorbed (or given off) by the process in a *calorimeter* (Figure 3.2). Alternatively, for any process $A \rightleftharpoons B$ at equilibrium, the standard-state enthalpy change for the process can be determined from the temperature dependence of the equilibrium constant:

$$\Delta H° = -R\frac{d(\ln K_{eq})}{d(1/T)} \qquad (3.4)$$

Here R is the *gas constant,* defined as $R = 8.314$ J/mol · K. A plot of $R(\ln K_{eq})$ versus $1/T$ is called a **van't Hoff plot.** The example below demonstrates how a van't Hoff plot is constructed and how the enthalpy change for a reaction can be determined from the plot itself.

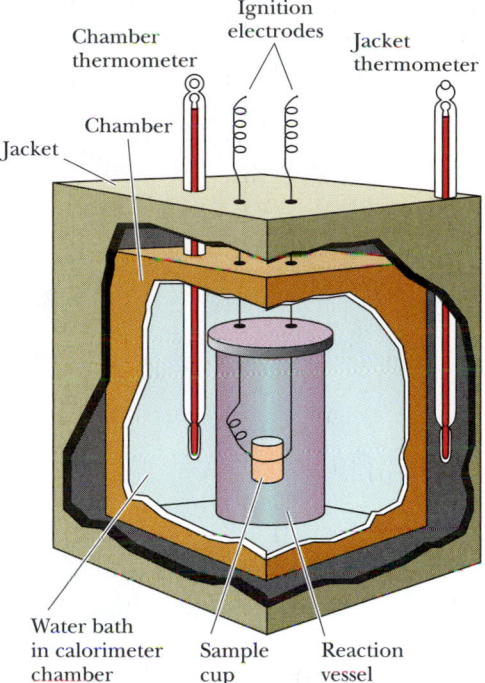

Biochemistry **Now**™ **ANIMATED FIGURE 3.2** Diagram of a calorimeter. The reaction vessel is completely submerged in a water bath. The heat evolved by a reaction is determined by measuring the rise in temperature of the water bath. **See this figure animated at http://chemistry.brookscole. com/ggb3**

EXAMPLE

In a study[1] of the temperature-induced reversible denaturation of the protein chymotrypsinogen,

$$\text{Native state (N)} \rightleftharpoons \text{denatured state (D)}$$
$$K_{eq} = [D]/[N]$$

John F. Brandts measured the equilibrium constants for the denaturation over a range of pH and temperatures. The data for pH 3:

T(K):	324.4	326.1	327.5	329.0	330.7	332.0	333.8
K_{eq}:	0.041	0.12	0.27	0.68	1.9	5.0	21

A plot of $R(\ln K_{eq})$ versus $1/T$ (a van't Hoff plot) is shown in Figure 3.3. $\Delta H°$ for the denaturation process at any temperature is the negative of the slope of the plot at that temperature. As shown, $\Delta H°$ at 54.5°C (327.5 K) is

$$\Delta H° = -[-3.2 - (-17.6)]/[(3.04 - 3.067) \times 10^{-3}] = +533 \text{ kJ/mol}$$

What does this value of $\Delta H°$ mean for the unfolding of the protein? Positive values of $\Delta H°$ would be expected for the breaking of hydrogen bonds as well as for

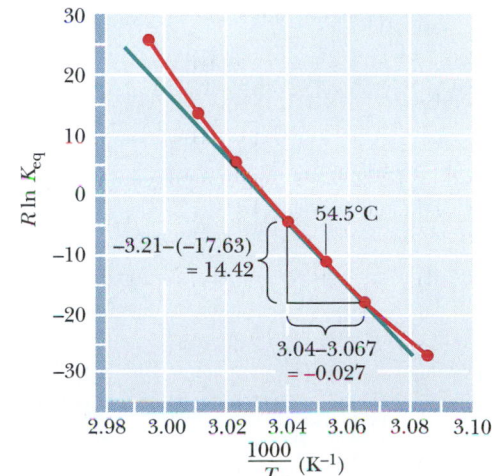

FIGURE 3.3 The enthalpy change, $\Delta H°$, for a reaction can be determined from the slope of a plot of $R \ln K_{eq}$ versus $1/T$. To illustrate the method, the values of the data points on either side of the 327.5 K (54.5°C) data point have been used to calculate $\Delta H°$ at 54.5°C. Regression analysis would normally be preferable. *(Adapted from Brandts, J. F., 1964. The thermodynamics of protein denaturation. I. The denaturation of chymotrypsinogen.* Journal of the American Chemical Society **86:**4291–4301.*)*

[1]Brandts, J. F., 1964. The thermodynamics of protein denaturation. I. The denaturation of chymotrypsinogen. *Journal of the American Chemical Society* **86:**4291–4301.

Table 3.1

Thermodynamic Parameters for Protein Denaturation

Protein (and conditions)	$\Delta H°$ kJ/mol	$\Delta S°$ kJ/mol · K	$\Delta G°$ kJ/mol	ΔC_P kJ/mol · K
Chymotrypsinogen (pH 3, 25°C)	164	0.440	31.0	10.9
β-Lactoglobulin (5 *M* urea, pH 3, 25°C)	−88	−0.300	2.5	9.0
Myoglobin (pH 9, 25°C)	180	0.400	57.0	5.9
Ribonuclease (pH 2.5, 30°C)	240	0.780	3.8	8.4

Adapted from Cantor, C., and Schimmel, P., 1980. *Biophysical Chemistry.* San Francisco: W.H. Freeman; and Tanford, C., 1968. Protein denaturation. *Advances in Protein Chemistry* **23**:121–282.

the exposure of hydrophobic groups from the interior of the native, folded protein during the unfolding process. Such events would raise the energy of the protein–water solution. The magnitude of this enthalpy change (533 kJ/mol) at 54.5°C is large, compared to similar values of $\Delta H°$ for other proteins and for this same protein at 25°C (Table 3.1). If we consider only this positive enthalpy change for the unfolding process, the native, folded state is strongly favored. As we shall see, however, other parameters must be taken into account.

The Second Law: Systems Tend Toward Disorder and Randomness

The **second law of thermodynamics** has been described and expressed in many different ways, including the following:

1. Systems tend to proceed from *ordered* (*low-entropy* or *low-probability*) states to *disordered* (*high-entropy* or *high-probability*) states.
2. The *entropy* of the system plus surroundings is unchanged by *reversible processes;* the entropy of the system plus surroundings increases for *irreversible processes.*
3. All naturally occurring processes proceed toward **equilibrium,** that is, to a state of minimum potential energy.

Several of these statements of the second law invoke the concept of **entropy,** which is a measure of disorder and randomness in the system (or the surroundings). An organized or ordered state is a low-entropy state, whereas a disordered state is a high-entropy state. All else being equal, reactions involving large, positive entropy changes, ΔS, are more likely to occur than reactions for which ΔS is not large and positive.

Entropy can be defined in several quantitative ways. If W is the number of ways to arrange the components of a system without changing the internal energy or enthalpy (that is, the number of energetically equivalent microscopic states at a given temperature, pressure, and amount of material), then the entropy is given by

$$S = k \ln W \tag{3.5}$$

where k is Boltzmann's constant ($k = 1.38 \times 10^{-23}$ J/K). This definition is useful for statistical calculations (in fact, it is a foundation of *statistical thermodynamics*), but a more common form relates entropy to the heat transferred in a process:

$$dS_{\text{reversible}} = \frac{dq}{T} \tag{3.6}$$

A Deeper Look

Biochemistry ⊗ Now™

Entropy, Information, and the Importance of "Negentropy"

When a thermodynamic system undergoes an increase in entropy, it becomes more disordered. On the other hand, a decrease in entropy reflects an increase in order. A more ordered system is more highly organized and possesses a greater information content. To appreciate the implications of decreasing the entropy of a system, consider the random collection of letters in the figure. This disorganized array of letters possesses no inherent information content, and nothing can be learned by its perusal. On the other hand, this particular array of letters can be systematically arranged to construct the first sentence of the Einstein quotation that opened this chapter: "A theory is the more impressive the greater is the simplicity of its premises, the more different are the kinds of things it relates and the more extended is its range of applicability."

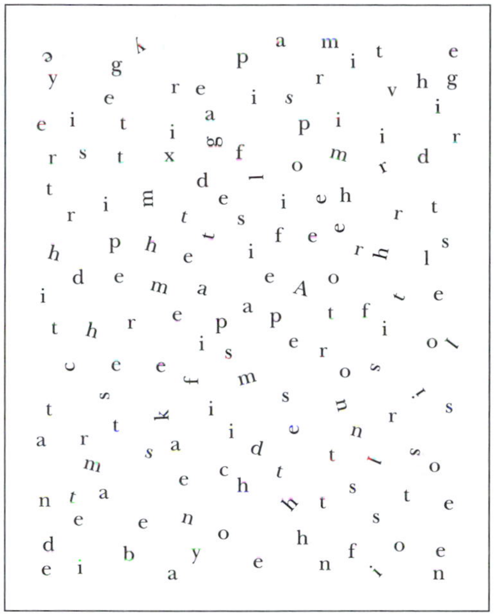

Arranged in this way, this same collection of 151 letters possesses enormous information content—the profound words of a great scientist. Just as it would have required significant effort to rearrange these 151 letters in this way, so large amounts of energy are required to construct and maintain living organisms. Energy input is required to produce information-rich, organized structures such as proteins and nucleic acids. Information content can be thought of as negative entropy. In 1945 Erwin Schrödinger took time out from his studies of quantum mechanics to publish a delightful book titled *What Is Life?* In it, Schrödinger coined the term *negentropy* to describe the negative entropy changes that confer organization and information content to living organisms. Schrödinger pointed out that organisms must "acquire negentropy" to sustain life.

where $dS_{reversible}$ is the entropy change of the system in a reversible[2] process, q is the heat transferred, and T is the temperature at which the heat transfer occurs.

The Third Law: Why Is "Absolute Zero" So Important?

The **third law of thermodynamics** states that the entropy of any crystalline, perfectly ordered substance must approach zero as the temperature approaches 0 K, and at $T = 0$ K *entropy is exactly zero*. Based on this, it is possible to establish a quantitative, absolute entropy scale for any substance as

$$S = \int_0^T C_P d \ln T \qquad (3.7)$$

where C_P is the *heat capacity* at constant pressure. The heat capacity of any substance is the amount of heat 1 mole of it can store as the temperature of that substance is raised by 1 degree. For a constant pressure process, this is described mathematically as

$$C_P = \frac{dH}{dT} \qquad (3.8)$$

If the heat capacity can be evaluated at all temperatures between 0 K and the temperature of interest, an absolute entropy can be calculated. For biological processes, *entropy changes* are more useful than absolute entropies. The entropy change for a process can be calculated if the enthalpy change and *free energy change* are known.

[2]A reversible process is one that can be reversed by an infinitesimal modification of a variable.

Free Energy Provides a Simple Criterion for Equilibrium

An important question for chemists, and particularly for biochemists, is, "Will the reaction proceed in the direction written?" J. Willard Gibbs, one of the founders of thermodynamics, realized that the answer to this question lay in a comparison of the enthalpy change and the entropy change for a reaction at a given temperature. The **Gibbs free energy, G,** is defined as

$$G = H - TS \qquad (3.9)$$

For any process A$\rightleftharpoons$B at constant pressure and temperature, the *free energy change* is given by

$$\Delta G = \Delta H - T\Delta S \qquad (3.10)$$

If ΔG is equal to 0, the process is at *equilibrium* and there is no net flow either in the forward or reverse direction. When $\Delta G = 0$, $\Delta S = \Delta H/T$ and the enthalpic and entropic changes are exactly balanced. Any process with a nonzero ΔG proceeds spontaneously to a final state of lower free energy. If ΔG is negative, the process proceeds spontaneously in the direction written. If ΔG is positive, the reaction or process proceeds spontaneously in the reverse direction. (The sign and value of ΔG do not allow us to determine *how fast* the process will go.) If the process has a negative ΔG, it is said to be **exergonic,** whereas processes with positive ΔG values are **endergonic.**

The Standard-State Free Energy Change   The free energy change, ΔG, for any reaction depends upon the nature of the reactants and products, but it is also affected by the conditions of the reaction, including temperature, pressure, pH, and the concentrations of the reactants and products. As explained earlier, it is useful to define a standard state for such processes. If the free energy change for a reaction is sensitive to solution conditions, what is the particular significance of the standard-state free energy change? To answer this question, consider a reaction between two reactants A and B to produce the products C and D.

$$A + B \rightleftharpoons C + D \qquad (3.11)$$

The free energy change for non–standard-state concentrations is given by

$$\Delta G = \Delta G^{\circ} = RT \ln \frac{[C][D]}{[A][B]} \qquad (3.12)$$

At equilibrium, $\Delta G = 0$ and $[C][D]/[A][B] = K_{eq}$. We then have

$$\Delta G^{\circ} = -RT \ln K_{eq} \qquad (3.13)$$

or, in base 10 logarithms,

$$\Delta G^{\circ} = -2.3RT \log_{10} K_{eq} \qquad (3.14)$$

This can be rearranged to

$$K_{eq} = 10^{-\Delta G^{\circ}/2.3RT} \qquad (3.15)$$

In any of these forms, this relationship allows the standard-state free energy change for any process to be determined if the equilibrium constant is known. More important, it states that the *point of equilibrium for a reaction in solution is a function of the standard-state free energy change for the process.* That is, ΔG° is another way of writing an equilibrium constant.

EXAMPLE

The equilibrium constants determined by Brandts at several temperatures for the denaturation of chymotrypsinogen (see previous Example) can be used to calculate the free energy changes for the denaturation process. For example, the equilibrium constant at 54.5°C is 0.27, so

$$\Delta G° = -(8.314 \text{ J/mol} \cdot \text{K})(327.5 \text{ K}) \ln (0.27)$$
$$\Delta G° = -(2.72 \text{ kJ/mol}) \ln (0.27)$$
$$\Delta G° = 3.56 \text{ kJ/mol}$$

The positive sign of $\Delta G°$ means that the unfolding process is unfavorable; that is, the stable form of the protein at 54.5°C is the folded form. On the other hand, the relatively small magnitude of $\Delta G°$ means that the folded form is only slightly favored. Figure 3.4 shows the dependence of $\Delta G°$ on temperature for the denaturation data at pH 3 (from the data given in the Example on page 53).

Having calculated both $\Delta H°$ and $\Delta G°$ for the denaturation of chymotrypsinogen, we can also calculate $\Delta S°$, using Equation 3.10:

$$\Delta S° = -\frac{(\Delta G - \Delta H°)}{T} \qquad (3.16)$$

At 54.5°C (327.5 K),

$$\Delta S° = -(3560 - 533,000 \text{ J/mol})/327.5 \text{ K}$$
$$\Delta S° = 1620 \text{ J/mol} \cdot \text{K}$$

Figure 3.5 presents the dependence of $\Delta S°$ on temperature for chymotrypsinogen denaturation at pH 3. A positive $\Delta S°$ indicates that the protein solution has become more disordered as the protein unfolds. Comparison of the value of 1.62 kJ/mol · K with the values of $\Delta S°$ in Table 3.1 shows that the present value (for chymotrypsinogen at 54.5°C) is quite large. The physical significance of the thermodynamic parameters for the unfolding of chymotrypsinogen becomes clear in the next section.

3.2 | What Can Thermodynamic Parameters Tell Us About Biochemical Events?

The best answer to this question is that a single parameter (ΔH or ΔS, for example) is not very meaningful. A positive $\Delta H°$ for the unfolding of a protein might reflect either the breaking of hydrogen bonds within the protein or the exposure of hydrophobic groups to water (Figure 3.6). However, *comparison of several thermodynamic parameters can provide meaningful insights about a process.* For example, the transfer of Na$^+$ and Cl$^-$ ions from the gas phase to aqueous solution involves a very large negative $\Delta H°$ (thus a very favorable stabilization of the

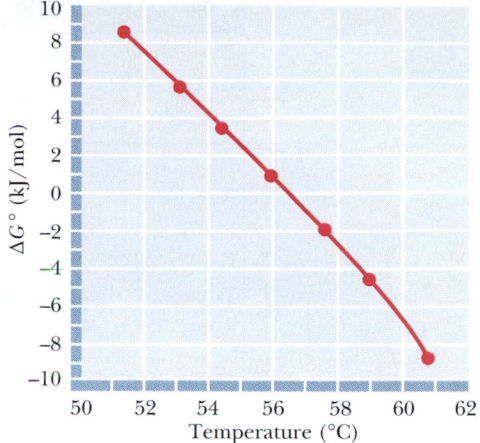

FIGURE 3.4 The dependence of $\Delta G°$ on temperature for the denaturation of chymotrypsinogen. *(Adapted from Brandts, J. F., 1964. The thermodynamics of protein denaturation. I. The denaturation of chymotrypsinogen. Journal of the American Chemical Society 86:4291–4301.)*

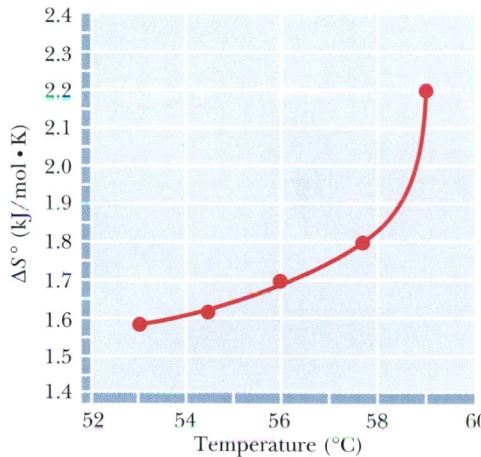

FIGURE 3.5 The dependence of $\Delta S°$ on temperature for the denaturation of chymotrypsinogen. *(Adapted from Brandts, J. F., 1964. The thermodynamics of protein denaturation. I. The denaturation of chymotrypsinogen. Journal of the American Chemical Society 86:4291–4301.)*

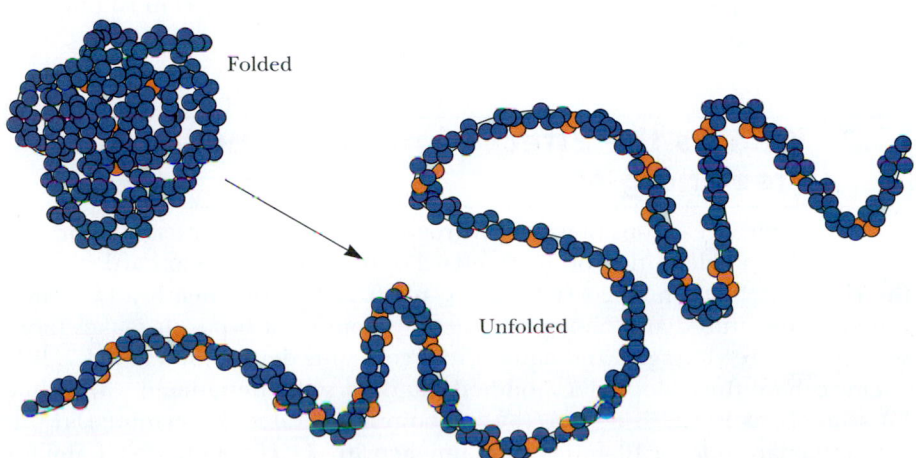

Folded

Unfolded

Biochemistry⚡Now™ ANIMATED FIGURE 3.6
Unfolding of a soluble protein exposes significant numbers of nonpolar groups to water, forcing order on the solvent and resulting in a negative $\Delta S°$ for the unfolding process. Orange spheres represent nonpolar groups; blue spheres are polar and/or charged groups. **See this figure animated at http://chemistry.brookscole.com/ggb3**

Table 3.2

Thermodynamic Parameters for Several Simple Processes*

Process	$\Delta H°$ kJ/mol	$\Delta S°$ kJ/mol · K	$\Delta G°$ kJ/mol	ΔC_P kJ/mol · K
Hydration of ions[†] $Na^+(g) + Cl^-(g) \longrightarrow Na^+(aq) + Cl^-(aq)$	−760.0	−0.185	−705.0	
Dissociation of ions in solution[‡] $H_2O + CH_3COOH \longrightarrow H_3O^+ + CH_3COO^-$	−10.3	−0.126	27.26	−0.143
Transfer of hydrocarbon from pure liquid to water[‡] Toluene (in pure toluene) $\longrightarrow$ toluene (aqueous)	1.72	−0.071	22.7	0.265

*All data collected for 25°C.
[†]Berry, R. S., Rice, S. A., and Ross, J., 1980. *Physical Chemistry.* New York: John Wiley.
[‡]Tanford, C., 1980. *The Hydrophobic Effect.* New York: John Wiley.

ions) and a comparatively small $\Delta S°$ (Table 3.2). The negative entropy term reflects the ordering of water molecules in the hydration shells of the Na^+ and Cl^- ions. The unfavorable $-T\Delta S$ contribution is more than offset by the large heat of hydration, which makes the hydration of ions a very favorable process overall. The negative entropy change for the dissociation of acetic acid in water also reflects the ordering of water molecules in the ion hydration shells. In this case, however, the enthalpy change is much smaller in magnitude. As a result, $\Delta G°$ for dissociation of acetic acid in water is positive, and acetic acid is thus a weak (largely undissociated) acid.

The transfer of a nonpolar hydrocarbon molecule from its pure liquid to water is an appropriate model for the exposure of protein hydrophobic groups to solvent when a protein unfolds. The transfer of toluene from liquid toluene to water involves a negative $\Delta S°$, a positive $\Delta G°$, and a $\Delta H°$ that is small compared to $\Delta G°$ (a pattern similar to that observed for the dissociation of acetic acid). *What distinguishes these two very different processes is the change in heat capacity* (Table 3.2). A positive heat capacity change for a process indicates that the molecules have acquired new ways to move (and thus to store heat energy). A negative ΔC_P means that the process has resulted in less freedom of motion for the molecules involved. ΔC_P is negative for the dissociation of acetic acid and positive for the transfer of toluene to water. The explanation is that polar and nonpolar molecules *both* induce organization of nearby water molecules, *but in different ways.* The water molecules near a nonpolar solute are *organized but labile.* Hydrogen bonds formed by water molecules near nonpolar solutes rearrange more rapidly than the hydrogen bonds of pure water. On the other hand, the hydrogen bonds formed between water molecules near an ion are less labile (rearrange more slowly) than they would be in pure water. This means that ΔC_P should be negative for the dissociation of ions in solution, as observed for acetic acid (Table 3.2).

3.3 | What Is the Effect of pH on Standard-State Free Energies?

For biochemical reactions in which hydrogen ions (H^+) are consumed or produced, the usual definition of the standard state is awkward. Standard state for the H^+ ion is 1 *M,* which corresponds to pH 0. At this pH, nearly all enzymes would be denatured and biological reactions could not occur. It makes more sense to use free energies and equilibrium constants determined at pH 7. Biochemists have thus adopted a modified standard state, designated with prime (′) symbols, as in $\Delta G°′$, $K_{eq}′$, $\Delta H°′$, and so on. For values determined in this way, a standard state of 10^{-7} *M* H^+ and unit activity (1 *M* for solutions, 1 atm for gases and pure solids defined as unit activity) for all other components (in the ionic forms that exist at pH 7) is assumed. The two standard states can be related easily. For a reaction in which H^+ is produced,

$$A \longrightarrow B^- + H^+ \tag{3.17}$$

the relation of the equilibrium constants for the two standard states is

$$K_{eq}' = K_{eq}[H^+] \tag{3.18}$$

and $\Delta G^{\circ\prime}$ is given by

$$\Delta G^{\circ\prime} = \Delta G^\circ + RT \ln [H^+] \tag{3.19}$$

For a reaction in which H^+ is consumed,

$$A^- + H^+ \longrightarrow B \tag{3.20}$$

the equilibrium constants are related by

$$K_{eq}' = \frac{K_{eq}}{[H^+]} \tag{3.21}$$

and $\Delta G^{\circ\prime}$ is given by

$$\Delta G^{\circ\prime} = \Delta G^\circ + RT \ln \left(\frac{1}{[H^+]} \right) = \Delta G^\circ - RT \ln [H^+] \tag{3.22}$$

3.4 | What Is the Effect of Concentration on Net Free Energy Changes?

Equation 3.12 shows that the free energy change for a reaction can be very different from the standard-state value if the concentrations of reactants and products differ significantly from unit activity (1 M for solutions). The effects can often be dramatic. Consider the hydrolysis of phosphocreatine:

$$\text{Phosphocreatine} + H_2O \longrightarrow \text{creatine} + P_i \tag{3.23}$$

This reaction is strongly exergonic, and ΔG° at 37°C is -42.8 kJ/mol. Physiological concentrations of phosphocreatine, creatine, and inorganic phosphate are normally between 1 and 10 mM. Assuming 1 mM concentrations and using Equation 3.12, the ΔG for the hydrolysis of phosphocreatine is

$$\Delta G = -42.8 \text{ kJ/mol} + (8.314 \text{ J/mol} \cdot \text{K})(310 \text{ K}) \ln \left(\frac{[0.001][0.001]}{[0.001]} \right) \tag{3.24}$$

$$\Delta G = -60.5 \text{ kJ/mol} \tag{3.25}$$

At 37°C, the difference between standard-state and 1 mM concentrations for such a reaction is thus approximately -17.7 kJ/mol.

3.5 | Why Are Coupled Processes Important to Living Things?

Many of the reactions necessary to keep cells and organisms alive must run against their **thermodynamic potential,** that is, in the direction of positive ΔG. Among these are the synthesis of adenosine triphosphate (ATP) and other high-energy molecules and the creation of ion gradients in all mammalian cells. These processes are driven in the thermodynamically unfavorable direction via *coupling* with highly favorable processes. Many such *coupled processes* are discussed later in this text. They are crucially important in intermediary metabolism, oxidative phosphorylation, and membrane transport, as we shall see.

Biochemistry Now™ ANIMATED FIGURE 3.7 The pyruvate kinase reaction. **See this figure animated at http://chemistry.brookscole.com/ggb3**

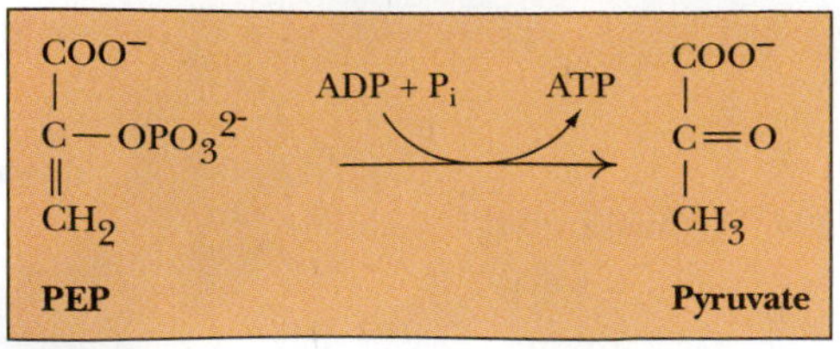

We can predict whether pairs of coupled reactions will proceed spontaneously by simply summing the free energy changes for each reaction. For example, consider the reaction from glycolysis (discussed in Chapter 18) involving the conversion of phospho(enol)pyruvate (PEP) to pyruvate (Figure 3.7). The hydrolysis of PEP is energetically very favorable, and it is used to drive phosphorylation of adenosine diphosphate (ADP) to form ATP, a process that is energetically unfavorable. Using values of ΔG that would be typical for a human erythrocyte:

$$\text{PEP} + H_2O \longrightarrow \text{pyruvate} + P_i \qquad \Delta G = -78 \text{ kJ/mol} \qquad (3.26)$$

$$\text{ADP} + P_i \longrightarrow \text{ATP} + H_2O \qquad \Delta G = +55 \text{ kJ/mol} \qquad (3.27)$$

$$\text{PEP} + \text{ADP} \longrightarrow \text{pyruvate} + \text{ATP} \qquad \text{Total } \Delta G = -23 \text{ kJ/mol} \qquad (3.28)$$

The net reaction catalyzed by this enzyme depends upon coupling between the two reactions shown in Equations 3.26 and 3.27 to produce the net reaction shown in Equation 3.28 with a net negative ΔG. Many other examples of coupled reactions are considered in our discussions of intermediary metabolism (see Part 3). In addition, many of the complex biochemical systems discussed in the later chapters of this text involve reactions and processes with positive ΔG values that are driven forward by coupling to reactions with a negative ΔG.

3.6 What Are the Characteristics of High-Energy Biomolecules?

Virtually all life on earth depends on energy from the sun. Among life forms, there is a hierarchy of energetics: Certain organisms capture solar energy directly, whereas others derive their energy from this group in subsequent processes. Organisms that absorb light energy directly are called **phototrophic organisms.** These organisms store solar energy in the form of various organic molecules. Organisms that feed on these latter molecules, releasing the stored energy in a series of oxidative reactions, are called **chemotrophic organisms.** Despite these differences, both types of organisms share common mechanisms for generating a useful form of chemical energy. Once captured in chemical form, energy can be released in controlled exergonic reactions to drive a variety of life processes (which require energy). A small family of universal biomolecules mediates the flow of energy from exergonic reactions to the energy-requiring processes of life. These molecules are the *reduced coenzymes* and the *high-energy phosphate compounds.* Phosphate compounds are considered high energy if they exhibit large negative free energies of hydrolysis (that is, if $\Delta G°'$ is more negative than -25 kJ/mol).

Table 3.3 lists the most important members of the high-energy phosphate compounds. Such molecules include *phosphoric anhydrides* (ATP, ADP), an *enol phosphate* (PEP), *acyl phosphates* (such as acetyl phosphate), and *guanidino phosphates* (such as creatine phosphate). Also included are thioesters, such as acetyl-CoA, which do not contain phosphorus, but which have a high free energy of hydrolysis. As noted earlier, the exact amount of chemical free energy available from the hydrolysis of such compounds depends on concentration, pH, temperature, and so on, but the $\Delta G°'$ values for hydrolysis of these substances are substantially more negative than those for most other metabolic species. Two important points: First, high-energy phosphate compounds are not long-term energy storage substances. They are transient forms of stored energy, meant to carry energy from point to point, from one enzyme system to another, in the minute-to-minute existence of the cell. (As we shall see in subsequent chapters, other molecules bear the responsibility for long-term storage of energy supplies.) Second, the term *high-energy compound* should not be construed to imply that these molecules are unstable and hydrolyze or decompose unpredictably.

Table 3.3

Free Energies of Hydrolysis of Some High-Energy Compounds*

Compound (and Hydrolysis Product)	$\Delta G^{\circ\prime}$ (kJ/mol)	Structure
Phosphoenolpyruvate (pyruvate + P_i)	−62.2	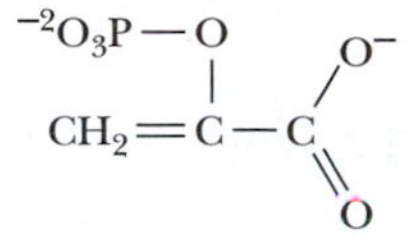
3′,5′-Cyclic adenosine monophosphate (5′-AMP)	−50.4	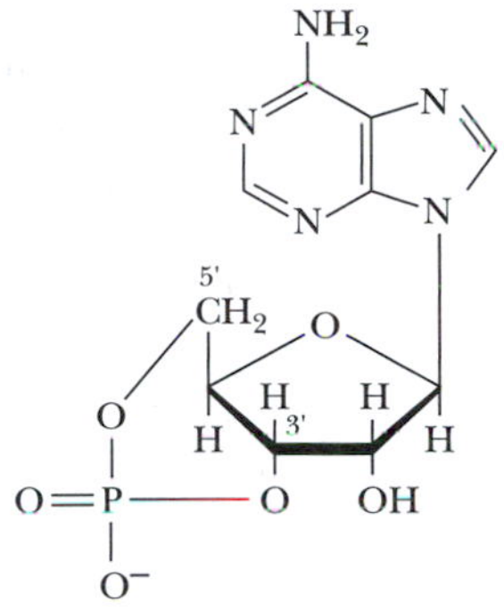
1,3-Bisphosphoglycerate (3-phosphoglycerate + P_i)	−49.6	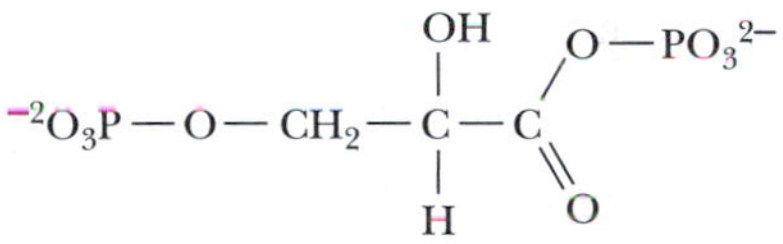
Creatine phosphate (creatine + P_i)	−43.3	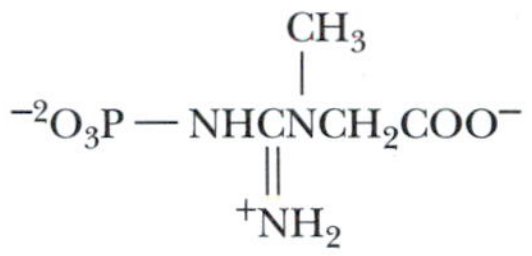
Acetyl phosphate (acetate + P_i)	−43.3	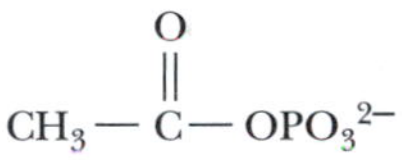
Adenosine-5′-triphosphate (ADP + P_i)	−35.7[†]	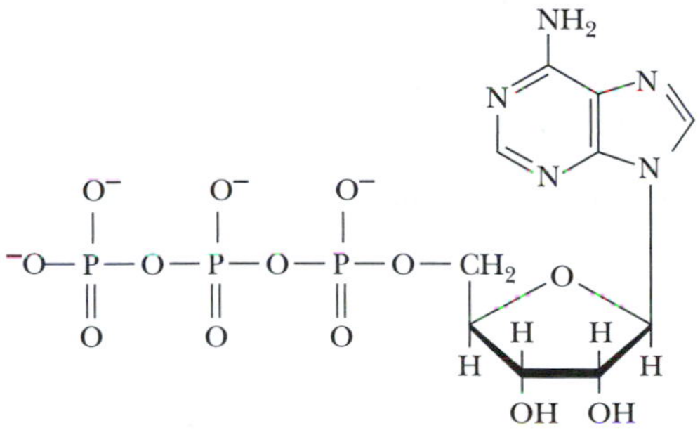
Adenosine-5′-triphosphate (ADP + P_i), excess Mg^{2+}	**−30.5**	
Adenosine-5′-diphosphate (AMP + P_i)	−35.7	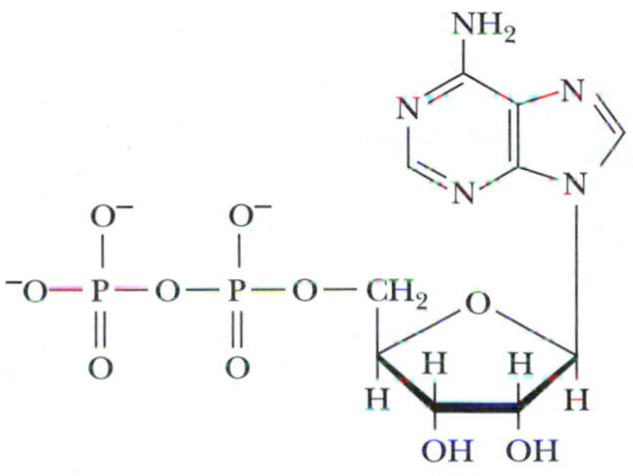

(continued)

Table 3.3

Free Energies of Hydrolysis of Some High-Energy Compounds*—*Cont'd*

Compound (and Hydrolysis Product)	$\Delta G^{\circ\prime}$ (kJ/mol)	Structure
Pyrophosphate ($P_i + P_i$) in 5 mM Mg^{2+}	−33.6	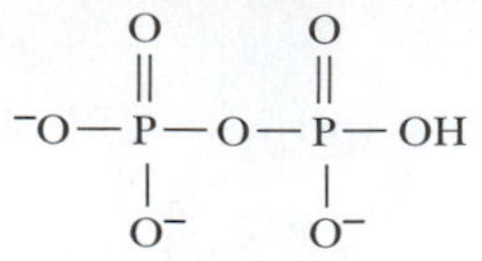
Adenosine-5'-triphosphate (AMP + PP$_i$), excess Mg^{2+}	−32.3	(See ATP structure on previous page)
Uridine diphosphoglucose (UDP + glucose)	−31.9	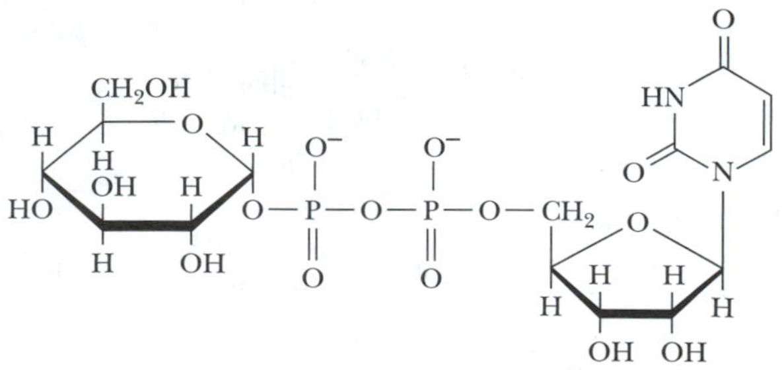
Acetyl-coenzyme A (acetate + CoA)	−31.5	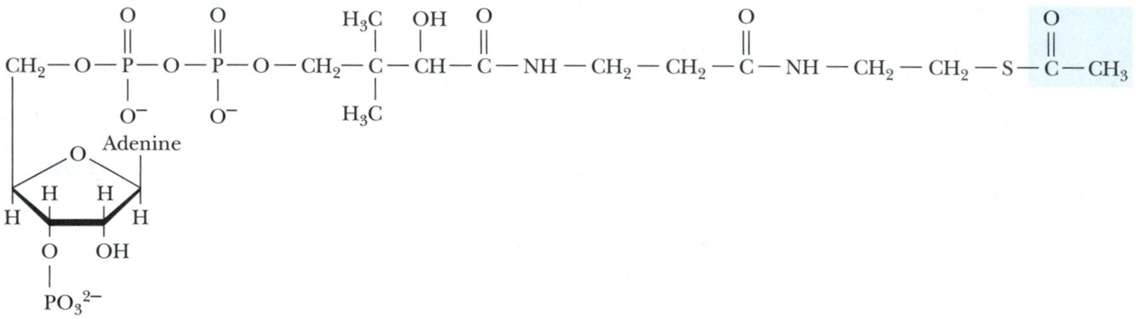
S-adenosylmethionine (methionine + adenosine)	−25.6‡	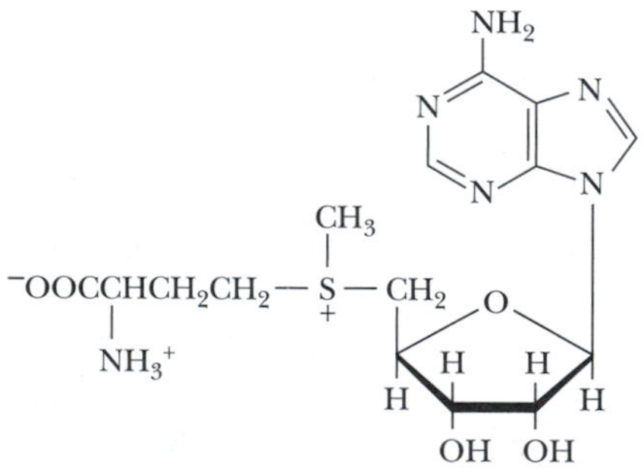

ATP, for example, is quite a stable molecule. A substantial *activation energy* must be delivered to ATP to hydrolyze the terminal, or γ, phosphate group. In fact, as shown in Figure 3.8, the activation energy that must be absorbed by the molecule to break the O—P$_\gamma$ bond is normally 200 to 400 kJ/mol, which is substantially larger than the net 30.5 kJ/mol released in the hydrolysis reaction. Biochemists are much more concerned with the *net release* of 30.5 kJ/mol than with the activation energy for the reaction (because suitable enzymes cope with the latter). The net release of large quantities of free energy distinguishes the high-energy phosphoric anhydrides from their "low-energy" ester cousins, such

Table 3.3

Free Energies of Hydrolysis of Some High-Energy Compounds*—*Cont'd*

Compound (and Hydrolysis Product)	$\Delta G°'$ (kJ/mol)	Structure
Lower-Energy Phosphate Compounds		
Glucose-1-P (glucose + P_i)	−21.0	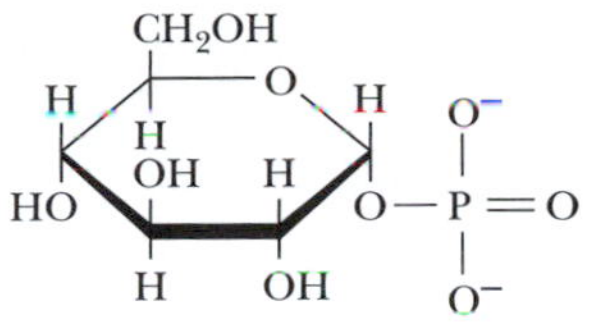
Fructose-1-P (fructose + P_i)	−16.0	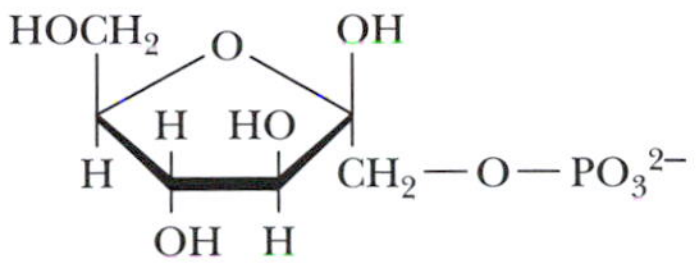
Glucose-6-P (glucose + P_i)	−13.9	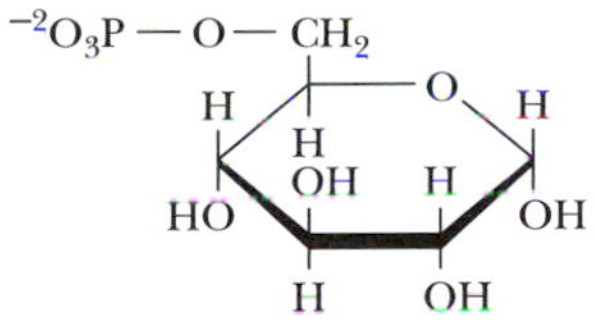
sn-Glycerol-3-P (glycerol + P_i)	−9.2	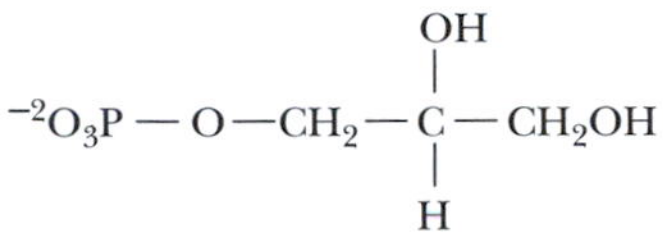
Adenosine-5′-monophosphate (adenosine + P_i)	−9.2	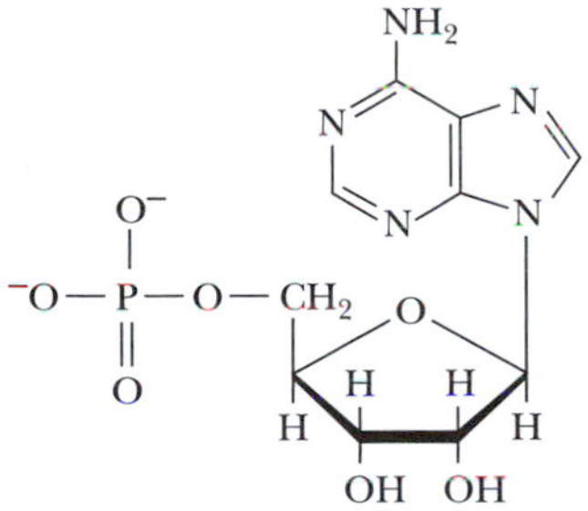

*Adapted primarily from *Handbook of Biochemistry and Molecular Biology*, 1976, 3rd ed. In *Physical and Chemical Data*, G. Fasman, ed., Vol. 1, pp. 296–304. Boca Raton, FL: CRC Press.
†From Gwynn, R. W., and Veech, R. L., 1973. The equilibrium constants of the adenosine triphosphate hydrolysis and the adenosine triphosphate-citrate lyase reactions. *Journal of Biological Chemistry* **248**:6966–6972.
‡From Mudd, H., and Mann, J., 1963. Activation of methionine for transmethylation. *Journal of Biological Chemistry* **238**:2164–2170.

as glycerol-3-phosphate (Table 3.3). The next section provides a quantitative framework for understanding these comparisons.

ATP Is an Intermediate Energy-Shuttle Molecule

One last point about Table 3.3 deserves mention. Given the central importance of ATP as a high-energy phosphate in biology, students are sometimes surprised to find that ATP holds an intermediate place in the rank of high-energy phosphates. PEP, cyclic AMP, 1,3-BPG, phosphocreatine, acetyl phosphate, and pyrophosphate

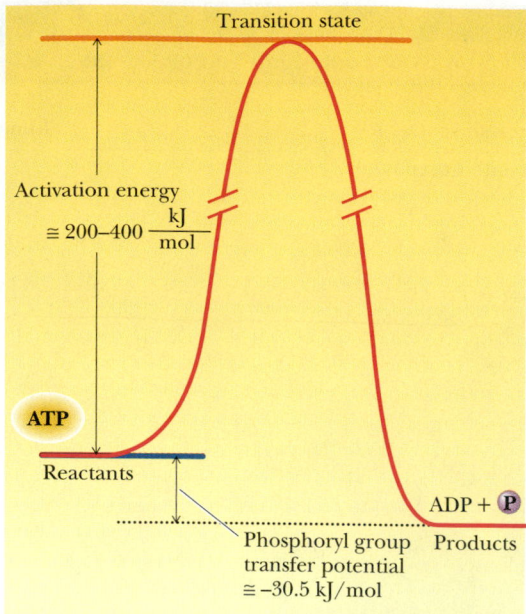

FIGURE 3.8 The activation energies for phosphoryl group transfer reactions (200 to 400 kJ/mol) are substantially larger than the free energy of hydrolysis of ATP (-30.5 kJ/mol).

all exhibit higher values of $\Delta G°'$. This is not a biological anomaly. ATP is uniquely situated between the very-high-energy phosphates synthesized in the breakdown of fuel molecules and the numerous lower-energy acceptor molecules that are phosphorylated in the course of further metabolic reactions. ADP can accept both phosphates and energy from the higher-energy phosphates, and the ATP thus formed can donate both phosphates and energy to the lower-energy molecules of metabolism. The ATP/ADP pair is an intermediately placed acceptor/donor system among high-energy phosphates. In this context, ATP functions as a very versatile but intermediate energy-shuttle device that interacts with many different energy-coupling enzymes of metabolism.

Group Transfer Potentials Quantify the Reactivity of Functional Groups

Many reactions in biochemistry involve the transfer of a functional group from a donor molecule to a specific receptor molecule or to water. The concept of **group transfer potential** explains the tendency for such reactions to occur. Biochemists define the group transfer potential as the free energy change that occurs upon hydrolysis, that is, upon transfer of the particular group to water. This concept and its terminology are preferable to the more qualitative notion of *high-energy bonds*.

The concept of group transfer potential is not particularly novel. Other kinds of transfer (of hydrogen ions and electrons, for example) are commonly characterized in terms of appropriate measures of transfer potential (pK_a and reduction potential, $\mathscr{E}_o$, respectively). As shown in Table 3.4, the notion of group transfer is fully analogous to those of ionization potential and reduction potential. The similarity is anything but coincidental, because all of these are really specific instances of free energy changes. If we write

$$AH \longrightarrow A^- + H^+ \tag{3.29a}$$

we really don't mean that a proton has literally been removed from the acid AH. In the gas phase at least, this would require the input of approximately 1200 kJ/mol! What we really mean is that the proton has been *transferred* to a suitable acceptor molecule, usually water:

$$AH + H_2O \longrightarrow A^- + H_3O^+ \tag{3.29b}$$

A Deeper Look

ATP Changes the K_{eq} by a Factor of 10^8

Consider a process, $A \rightleftharpoons B$. It could be a biochemical reaction, or the transport of an ion against a concentration gradient, or even a mechanical process (such as muscle contraction). Assume that it is a thermodynamically unfavorable reaction. Let's say, for purposes of illustration, that $\Delta G^{\circ\prime} = +13.8$ kJ/mol. From the equation,

$$\Delta G^{\circ\prime} = -RT \ln K_{eq}$$

we have

$$+13,800 = -(8.31 \text{ J/K} \cdot \text{mol})(298 \text{ K}) \ln K_{eq}$$

which yields

$$\ln K_{eq} = -5.57$$

Therefore,

$$K_{eq} = 0.0038 = [B_{eq}]/[A_{eq}]$$

This reaction is clearly unfavorable (as we could have foreseen from its positive $\Delta G^{\circ\prime}$). At equilibrium, there is one molecule of product B for every 263 molecules of reactant A. Not much A was transformed to B.

Now suppose the reaction $A \rightleftharpoons B$ is coupled to ATP hydrolysis, as is often the case in metabolism:

$$A + ATP \rightleftharpoons B + ADP + P_i$$

The thermodynamic properties of this coupled reaction are the same as the sum of the thermodynamic properties of the partial reactions:

$A \rightleftharpoons B$	$\Delta G^{\circ\prime} = +13.8$ kJ/mol
$ATP + H_2O \rightleftharpoons ADP + P_i$	$\Delta G^{\circ\prime} = -30.5$ kJ/mol
$A + ATP + H_2O \rightleftharpoons B + ADP + P_i$	$\Delta G^{\circ\prime} = -16.7$ kJ/mol

That is,

$$\Delta G^{\circ\prime}_{overall} = -16.7 \text{ kJ/mol}$$

So

$$-16,700 = RT \ln K_{eq} = -(8.31)(298) \ln K_{eq}$$
$$\ln K_{eq} = -16,700/-2476 = 6.75$$
$$K_{eq} = 850$$

Using this equilibrium constant, let's now consider the cellular situation in which the concentrations of A and B are brought to equilibrium in the presence of typical prevailing concentrations of ATP, ADP, and P_i.*

*The concentrations of ATP, ADP, and P_i in a normal, healthy bacterial cell growing at 25°C are maintained at roughly 8 mM, 8 mM, and 1 mM, respectively. Therefore, the ratio [ADP][P_i]/[ATP] is about 10^{-3}. Under these conditions, ΔG for ATP hydrolysis is approximately -47.6 kJ/mol.

$$K_{eq} = \frac{[B_{eq}][ADP][P_i]}{[A_{eq}][ATP]}$$

$$850 = \frac{[B_{eq}][8 \times 10^{-3}][10^{-3}]}{[A_{eq}][8 \times 10^{-3}]}$$

$$[B_{eq}]/[A_{eq}] = 850,000$$

Comparison of the $[B_{eq}]/[A_{eq}]$ ratio for the simple $A \rightleftharpoons B$ reaction with the coupling of this reaction to ATP hydrolysis gives

$$\frac{850,000}{0.0038} = 2.2 \times 10^8$$

The equilibrium ratio of B to A is more than 10^8 greater when the reaction is coupled to ATP hydrolysis. A reaction that was clearly unfavorable ($K_{eq} = 0.0038$) has become emphatically spontaneous!

The involvement of ATP has raised the equilibrium ratio of B/A by more than 200 million–fold. It is informative to realize that this multiplication factor does not depend on the nature of the reaction. Recall that we defined $A \rightleftharpoons B$ in the most general terms. Also, the value of this equilibrium constant ratio, some 2.2×10^8, is not at all dependent on the particular reaction chosen or its standard free energy change, $\Delta G^{\circ\prime}$. You can satisfy yourself on this point by choosing some value for $\Delta G^{\circ\prime}$ other than $+13.8$ kJ/mol and repeating these calculations (keeping the concentrations of ATP, ADP, and P_i at 8, 8, and 1 mM, as before).

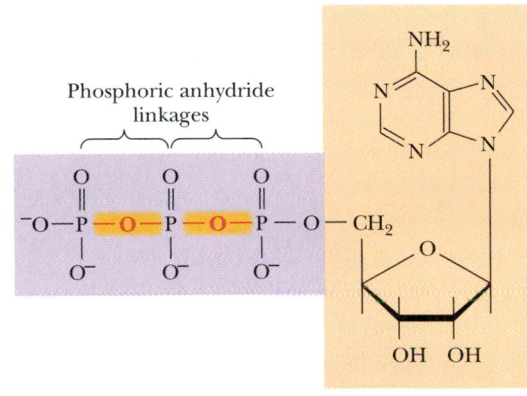

ATP
(adenosine-5'-triphosphate)

The appropriate free energy relationship is of course

$$pK_a = \frac{\Delta G}{2.303 \, RT} \qquad\qquad (3.30)$$

Similarly, in the case of an oxidation-reduction reaction

$$A \longrightarrow A^+ + e^- \qquad\qquad (3.31a)$$

Table 3.4

Types of Transfer Potential

	Proton Transfer Potential (Acidity)	Standard Reduction Potential (Electron Transfer Potential)	Group Transfer Potential (High-Energy Bond)
Simple equation	$AH \rightleftharpoons A^- + H^+$	$A \rightleftharpoons A^+ + e^-$	$A \sim P \rightleftharpoons A + P_i$
Equation including acceptor	$AH + H_2O \rightleftharpoons A^- + H_3O^+$	$A + H^+ \rightleftharpoons A^+ + \frac{1}{2} H_2$	$A \sim PO_4^{2-} + H_2O \rightleftharpoons A-H + HPO_4^{2-}$
Measure of transfer potential	$pK_a = \dfrac{\Delta G^\circ}{2.303\, RT}$	$\Delta \mathscr{E}_o = \dfrac{\Delta G^\circ}{n \mathscr{F}}$	$\ln K_{eq} = \dfrac{-\Delta G^\circ}{RT}$
Free energy change of transfer is given by:	ΔG° per mole of H^+ transferred	ΔG° per mole of e^- transferred	ΔG° per mole of phosphate transferred

Adapted from: Klotz, I. M., 1986. *Introduction to Biomolecular Energetics.* New York: Academic Press.

we don't really mean that A oxidizes independently. What we really mean (and what is much more likely in biochemical systems) is that the electron is transferred to a suitable acceptor:

$$A + H^+ \longrightarrow A^+ + \tfrac{1}{2} H_2 \tag{3.31b}$$

and the relevant free energy relationship is

$$\Delta \mathscr{E}_o = \frac{-\Delta G^\circ}{n \mathscr{F}} \tag{3.32}$$

where n is the number of equivalents of electrons transferred and $\mathscr{F}$ is **Faraday's constant.**

Similarly, the release of free energy that occurs upon the hydrolysis of ATP and other "high-energy phosphates" can be treated quantitatively in terms of *group transfer.* It is common to write for the hydrolysis of ATP

$$ATP + H_2O \longrightarrow ADP + P_i \tag{3.33}$$

The free energy change, which we henceforth call the *group transfer potential,* is given by

$$\Delta G^\circ = -RT \ln K_{eq} \tag{3.34}$$

where K_{eq} is the equilibrium constant for the group transfer, which is normally written as

$$K_{eq} = \frac{[ADP][P]}{[ATP][H_2O]} \tag{3.35}$$

Even this set of equations represents an approximation, because ATP, ADP, and P_i all exist in solutions as a mixture of ionic species. This problem is discussed in a later section. For now, it is enough to note that the free energy changes listed in Table 3.3 are the group transfer potentials observed for transfers to water.

The Hydrolysis of Phosphoric Acid Anhydrides Is Highly Favorable

ATP contains two *pyrophosphoryl* or *phosphoric acid anhydride* linkages, as shown in Figure 3.9. Other common biomolecules possessing phosphoric acid anhydride linkages include ADP, GTP, GDP and the other nucleoside diphosphates and triphosphates, sugar nucleotides such as UDP–glucose, and inorganic pyrophosphate itself. All exhibit large negative free energies of hydrolysis, as shown in Table 3.3. The chemical reasons for the large negative $\Delta G^{\circ\prime}$ values for the hydrolysis reactions include destabilization of the reactant due to bond strain caused by electrostatic repulsion, stabilization of the products by ion-

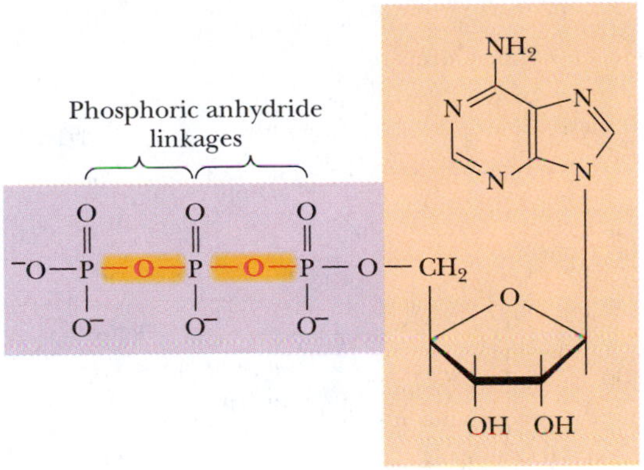

ATP
(adenosine-5'-triphosphate)

Biochemistry ⑧ Now™ ACTIVE FIGURE 3.9
The triphosphate chain of ATP contains two pyrophosphate linkages, both of which release large amounts of energy upon hydrolysis. **Test yourself on the concepts in this figure at http://chemistry. brookscole.com/ggb3.**

ization and resonance, and entropy factors due to hydrolysis and subsequent ionization.

Destabilization Due to Electrostatic Repulsion Electrostatic repulsion in the reactants is best understood by comparing these phosphoric anhydrides with other reactive anhydrides, such as acetic anhydride. As shown in Figure 3.10a, the electronegative carbonyl oxygen atoms withdraw electrons from the C=O bonds, producing partial negative charges on the oxygens and partial positive charges on the carbonyl carbons. Each of these electrophilic carbonyl carbons is further destabilized by the other acetyl group, which is also electron-withdrawing in nature. As a result, acetic anhydride is unstable with respect to the products of hydrolysis.

The situation with phosphoric anhydrides is similar. The phosphorus atoms of the pyrophosphate anion are electron-withdrawing and destabilize PP$_i$ with respect to its hydrolysis products. Furthermore, the reverse reaction, reformation of the anhydride bond from the two anionic products, requires that the electrostatic repulsion between these anions be overcome (see following).

Stabilization of Hydrolysis Products by Ionization and Resonance The pyrophosphate moiety possesses three negative charges at pH values above 7.5 or so (note the pK_a values, Figure 3.10a). The hydrolysis products, two molecules of inorganic phosphate, both carry about two negative charges at pH values above 7.2. The increased ionization of the hydrolysis products helps stabilize the electrophilic phosphorus nuclei.

Resonance stabilization in the products is best illustrated by the reactant anhydrides (Figure 3.10b). The unpaired electrons of the bridging oxygen atom in acetic anhydride (and phosphoric anhydride) cannot participate in resonance structures with both electrophilic centers at once. This **competing resonance** situation is relieved in the product acetate or phosphate molecules.

Entropy Factors Arising from Hydrolysis and Ionization For the phosphoric anhydrides, and for most of the high-energy compounds discussed here, there is an additional "entropic" contribution to the free energy of hydrolysis. Most of the hydrolysis reactions of Table 3.3 result in an increase in the number of molecules in solution. As shown in Figure 3.11, the hydrolysis of ATP (at pH values above 7) creates three species—ADP, inorganic phosphate (P$_i$), and a hydrogen ion—from only two reactants (ATP and H$_2$O). The entropy of the solution increases because the more particles, the more disordered the system.[3] (This

---

[3]Imagine the "disorder" created by hitting a crystal with a hammer and breaking it into many small pieces.

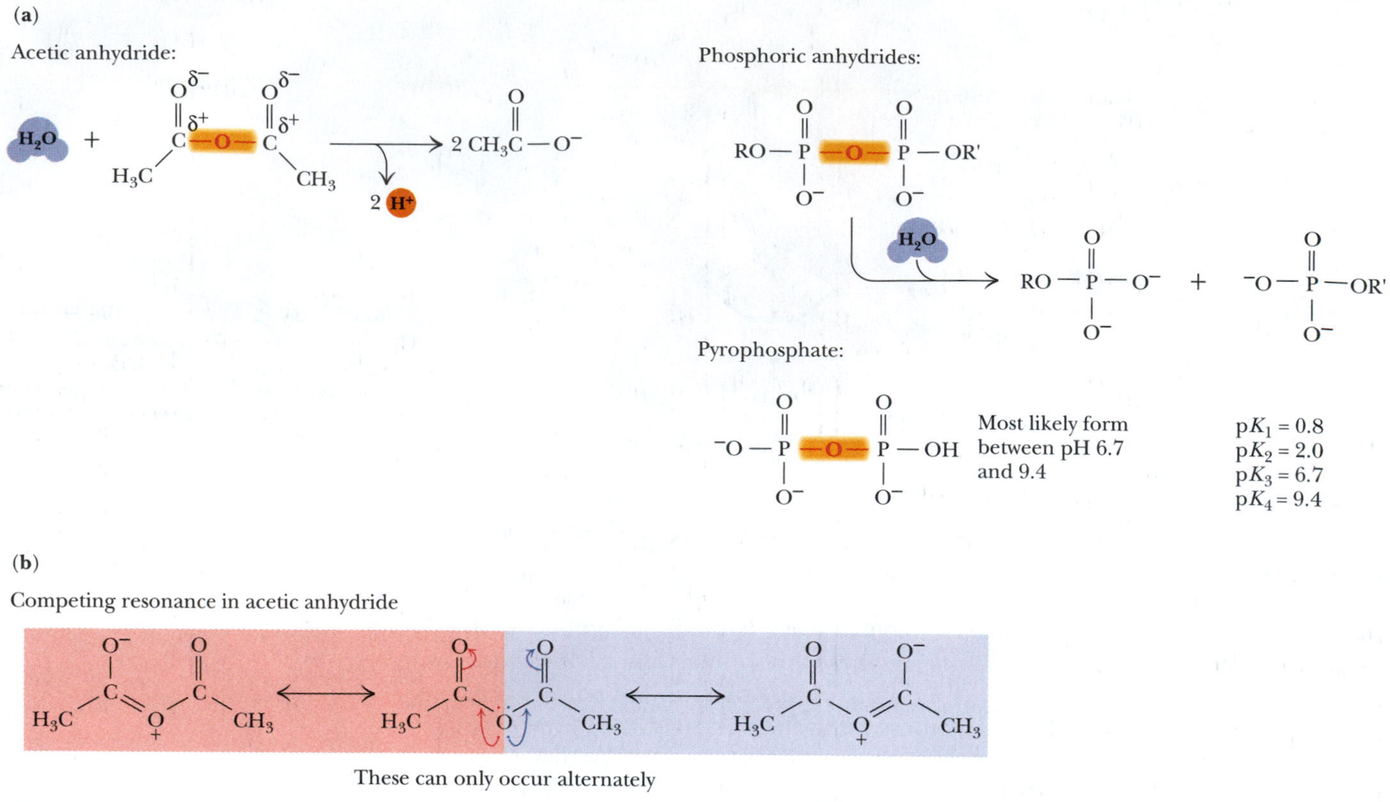

(a)

Acetic anhydride:

Phosphoric anhydrides:

Pyrophosphate:

Most likely form between pH 6.7 and 9.4

$pK_1 = 0.8$
$pK_2 = 2.0$
$pK_3 = 6.7$
$pK_4 = 9.4$

(b)

Competing resonance in acetic anhydride

These can only occur alternately

Simultaneous resonance in the hydrolysis products

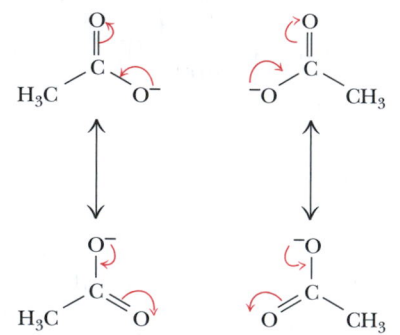

These resonances can occur simultaneously

Biochemistry ⓈNow™ **ACTIVE FIGURE 3.10** **(a)** Electrostatic repulsion between adjacent partial positive charges (on carbon and phosphorus, respectively) is relieved upon hydrolysis of the anhydride bonds of acetic anhydride and phosphoric anhydrides. The predominant form of pyrophosphate at pH values between 6.7 and 9.4 is shown. **(b)** The competing resonances of acetic anhydride and the simultaneous resonance forms of the hydrolysis product, acetate. **Test yourself on the concepts in this figure at http://chemistry.brookscole.com/ggb3**

effect is ionization-dependent because, at low pH, the hydrogen ion created in many of these reactions simply protonates one of the phosphate oxygens, and one fewer "particle" results from the hydrolysis.)

The Hydrolysis $\Delta G^{\circ\prime}$ of ATP and ADP Is Greater Than That of AMP

The concepts of destabilization of reactants and stabilization of products described for pyrophosphate also apply for ATP and other phosphoric anhydrides (Figure 3.11). ATP and ADP are destabilized relative to the hydrolysis products by electrostatic repulsion, competing resonance, and entropy. AMP, on the other hand, is a phosphate ester (not an anhydride) possessing only a single phosphoryl group and is not markedly different from the product inorganic phosphate in terms of electrostatic repulsion and resonance stabilization. Thus, the $\Delta G^{\circ\prime}$ for hydrolysis of AMP is much smaller than the corresponding values for ATP and ADP.

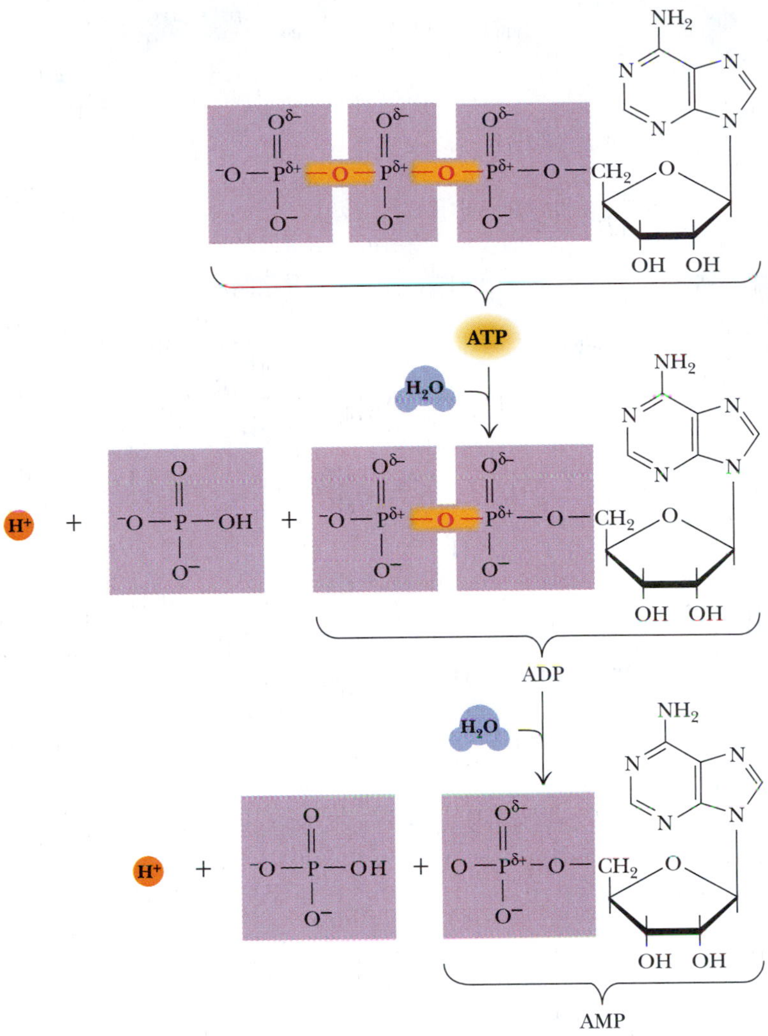

Biochemistry Now™ ANIMATED FIGURE 3.11 Hydrolysis of ATP to ADP (and/or of ADP to AMP) leads to relief of electrostatic repulsion. **See this figure animated at http://chemistry.brookscole. com/ggb3**

Acetyl Phosphate and 1,3-Bisphosphoglycerate Are Phosphoric-Carboxylic Anhydrides

The mixed anhydrides of phosphoric and carboxylic acids, frequently called acyl phosphates, are also energy-rich. Two biologically important acyl phosphates are acetyl phosphate and 1,3-bisphosphoglycerate. Hydrolysis of these species yields acetate and 3-phosphoglycerate, respectively, in addition to inorganic phosphate (Figure 3.12). Once again, the large $\Delta G°'$ values indicate that the reactants are destabilized relative to products. This arises from bond strain, which can be traced to the partial positive charges on the carbonyl carbon and phosphorus atoms of these structures. The energy stored in the mixed anhydride bond (which is required to overcome the charge–charge repulsion) is released upon hydrolysis. Increased resonance possibilities in the products relative to the reactants also contribute to the large negative $\Delta G°'$ values. The value of $\Delta G°'$ depends on the pK_a values of the starting anhydride and the product phosphoric and carboxylic acids, and of course also on the pH of the medium.

Enol Phosphates Are Potent Phosphorylating Agents

The largest value of $\Delta G°'$ in Table 3.3 belongs to *phosphoenolpyruvate* or *PEP,* an example of an enolic phosphate. This molecule is an important intermediate in carbohydrate metabolism, and due to its large negative $\Delta G°'$, it is a potent

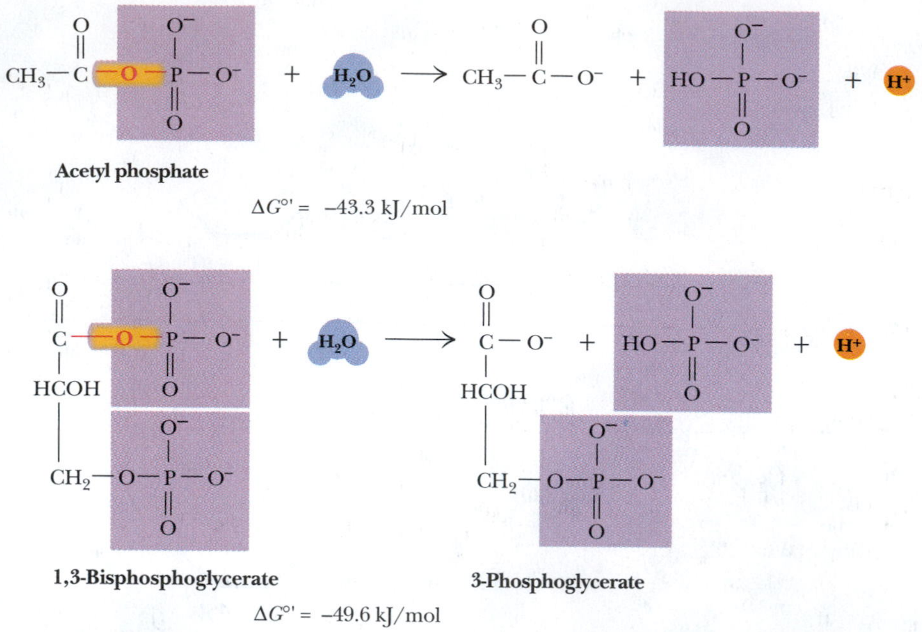

$$CH_3-C \overset{O}{\underset{||}{}} -O-\overset{O^-}{\underset{O}{\overset{|}{\underset{||}{P}}}}-O^- + H_2O \longrightarrow CH_3-\overset{O}{\underset{||}{C}}-O^- + HO-\overset{O^-}{\underset{O}{\overset{|}{\underset{||}{P}}}}-O^- + H^+$$

Acetyl phosphate

$$\Delta G^{\circ\prime} = -43.3 \text{ kJ/mol}$$

1,3-Bisphosphoglycerate   **3-Phosphoglycerate**

$$\Delta G^{\circ\prime} = -49.6 \text{ kJ/mol}$$

Biochemistry ⚡ Now™   ACTIVE FIGURE 3.12 The hydrolysis reactions of acetyl phosphate and 1,3-bisphosphoglycerate. **Test yourself on the concepts in this figure at http://chemistry. brookscole.com/ggb3**

phosphorylating agent. PEP is formed via dehydration of 2-phosphoglycerate by enolase during fermentation and glycolysis. PEP is subsequently transformed into pyruvate upon transfer of its phosphate to ADP by pyruvate kinase (Figure 3.13). The very large negative value of $\Delta G^{\circ\prime}$ for the latter reaction is to a large extent the result of a secondary reaction of the *enol* form of pyruvate. Upon hydrolysis, the unstable enolic form of pyruvate immediately converts to the keto form with a resulting large negative $\Delta G^{\circ\prime}$ (Figure 3.14). Together, the hydrolysis and subsequent *tautomerization* result in an overall $\Delta G^{\circ\prime}$ of -62.2 kJ/mol.

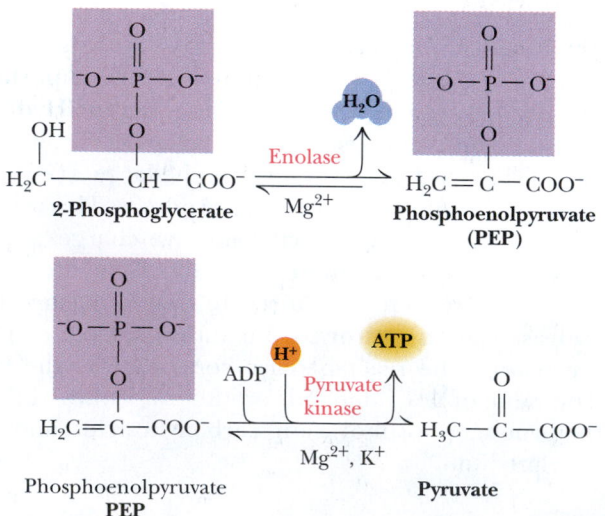

Biochemistry ⚡ Now™   ANIMATED FIGURE 3.13 Phosphoenolpyruvate (PEP) is produced by the enolase reaction (in glycolysis; see Chapter 18) and in turn drives the phosphorylation of ADP to form ATP in the pyruvate kinase reaction. **See this figure animated at http://chemistry. brookscole.com/ggb3**

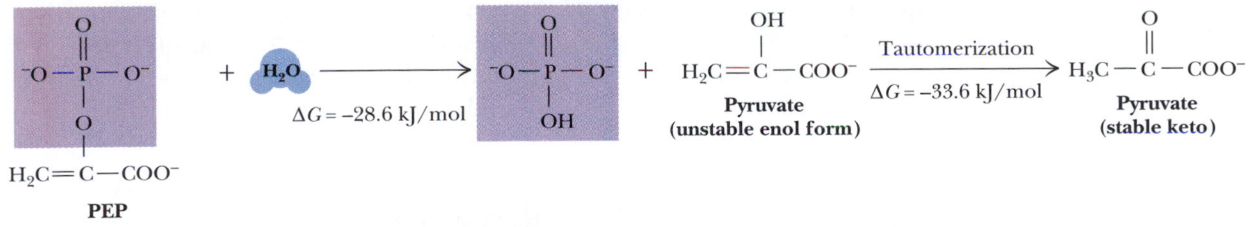

Biochemistry Now™ **ANIMATED FIGURE 3.14** Hydrolysis and the subsequent tautomerization account for the very large $\Delta G^{\circ\prime}$ of PEP. **See this figure animated at http://chemistry. brookscole.com/ggb3**

| 3.7 | **What Are the Complex Equilibria Involved in ATP Hydrolysis?** |

So far, as in Equation 3.33, the hydrolyses of ATP and other high-energy phosphates have been portrayed as simple processes. The situation in a real biological system is far more complex, owing to the operation of several ionic equilibria. First, ATP, ADP, and the other species in Table 3.3 can exist in several different ionization states that must be accounted for in any quantitative analysis. Second, phosphate compounds bind a variety of divalent and monovalent cations with substantial affinity, and the various metal complexes must also be considered in such analyses. Consideration of these special cases makes the quantitative analysis far more realistic. The importance of these multiple equilibria in group transfer reactions is illustrated for the hydrolysis of ATP, but the principles and methods presented are general and can be applied to any similar hydrolysis reaction.

The $\Delta G^{\circ\prime}$ of Hydrolysis for ATP Is pH-Dependent

ATP has five dissociable protons, as indicated in Figure 3.15. Three of the protons on the triphosphate chain dissociate at very low pH. The adenine ring amino group exhibits a pK_a of 4.06, whereas the last proton to dissociate from the triphosphate chain possesses a pK_a of 6.95. At higher pH values, ATP is completely deprotonated. ADP and phosphoric acid also undergo multiple ionizations. These multiple ionizations make the equilibrium constant for ATP hydrolysis more complicated than the simple expression in Equation 3.35. Multiple ionizations must also be taken into account when the pH dependence of ΔG° is considered. The calculations are beyond the scope of this text, but Figure 3.16 shows the variation of ΔG° as a function of pH. The free energy of

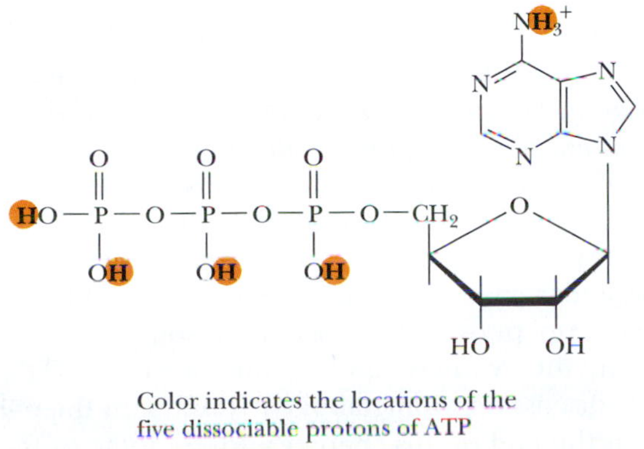

Color indicates the locations of the five dissociable protons of ATP

FIGURE 3.15 Adenosine-5′-triphosphate (ATP).

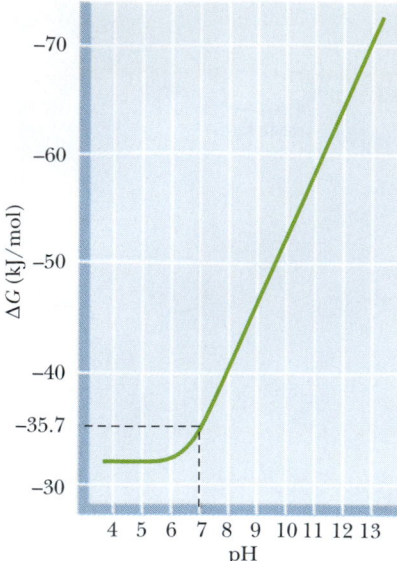

FIGURE 3.16 The pH dependence of the free energy of hydrolysis of ATP. Because pH varies only slightly in biological environments, the effect on ΔG is usually small.

hydrolysis is nearly constant from pH 4 to pH 6. At higher values of pH, $\Delta G°$ varies linearly with pH, becoming more negative by 5.7 kJ/mol for every pH unit of increase at 37°C. Because the pH of most biological tissues and fluids is near neutrality, the effect on $\Delta G°$ is relatively small, but it must be taken into account in certain situations.

Metal Ions Affect the Free Energy of Hydrolysis of ATP

Most biological environments contain substantial amounts of divalent and monovalent metal ions, including Mg^{2+}, Ca^{2+}, Na^+, K^+, and so on. What effect do metal ions have on the equilibrium constant for ATP hydrolysis and the associated free energy change? Figure 3.17 shows the change in $\Delta G°'$ with pMg (that is, $-\log_{10}[Mg^{2+}]$) at pH 7.0 and 38°C. The free energy of hydrolysis of ATP at zero Mg^{2+} is -35.7 kJ/mol, and at 5 mM total Mg^{2+} (the minimum in the plot) the $\Delta G_{obs}°$ is approximately -31 kJ/mol. Thus, in most real biological environments (with pH near 7 and Mg^{2+} concentrations of 5 mM or more) the free energy of hydrolysis of ATP is altered more by metal ions than by protons. A widely used "consensus value" for $\Delta G°'$ of ATP in biological systems is **-30.5 kJ/mol** (Table 3.3). This value, cited in the 1976 *Handbook of Biochemistry and Molecular Biology* (3rd ed., *Physical and Chemical Data*, Vol. 1, pp. 296–304, Boca Raton, FL: CRC Press), was determined in the presence of "excess Mg^{2+}." *This is the value we use for metabolic calculations in the balance of this text.*

Concentration Affects the Free Energy of Hydrolysis of ATP

Through all these calculations of the effect of pH and metal ions on the ATP hydrolysis equilibrium, we have assumed "standard conditions" with respect to concentrations of all species except for protons. The levels of ATP, ADP, and other high-energy metabolites never even begin to approach the standard state of 1 M. In most cells, the concentrations of these species are more typically 1 to 5 mM or even less. Earlier, we described the effect of concentration on equilibrium constants and free energies in the form of Equation 3.12. For the present case, we can rewrite this as

$$\Delta G = \Delta G° + RT \ln \frac{[\Sigma ADP][\Sigma P_i]}{[\Sigma ATP]} \tag{3.36}$$

where the terms in brackets represent the sum (Σ) of the concentrations of all the ionic forms of ATP, ADP, and P_i.

It is clear that changes in the concentrations of these species can have large effects on ΔG. The concentrations of ATP, ADP, and P_i may, of course, vary rather independently in real biological environments, but if, for the sake of some model calculations, we assume that all three concentrations are equal, then the effect of concentration on ΔG is as shown in Figure 3.18. The free energy of hydrolysis of ATP, which is -35.7 kJ/mol at 1 M, becomes -49.4 kJ/mol at 5 mM (that is, the concentration for which p$C = -2.3$ in Figure 3.18). At 1 mM ATP, ADP, and P_i, the free energy change becomes even more negative at -53.6 kJ/mol. *Clearly, the effects of concentration are much greater than the effects of protons or metal ions under physiological conditions.*

Does the "concentration effect" change ATP's position in the energy hierarchy (in Table 3.3)? Not really. All the other high- and low-energy phosphates experience roughly similar changes in concentration under physiological conditions and thus similar changes in their free energies of hydrolysis. The roles of the very-high-energy phosphates (PEP, 1,3-bisphosphoglycerate, and creatine phosphate) in the synthesis and maintenance of ATP in the cell are considered in our discussions of metabolic pathways. In the meantime, several of the problems at the end of this chapter address some of the more interesting cases.

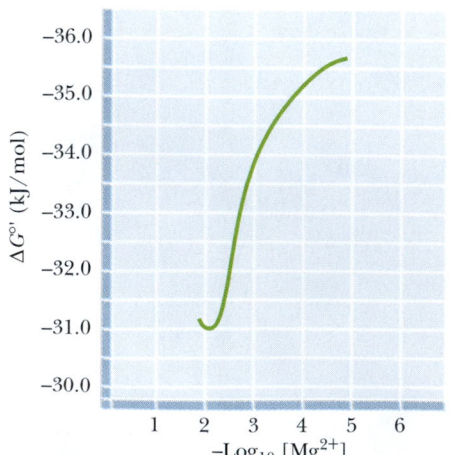

FIGURE 3.17 The free energy of hydrolysis of ATP as a function of total Mg^{2+} ion concentration at 38°C and pH 7.0. (*Adapted from Gwynn, R. W., and Veech, R. L., 1973. The equilibrium constants of the adenosine triphosphate hydrolysis and the adenosine triphosphate-citrate lyase reactions.* Journal of Biological Chemistry **248**:6966–6972.)

3.8 What Is the Daily Human Requirement for ATP?

We can end this discussion of ATP and the other important high-energy compounds in biology by discussing the daily metabolic consumption of ATP by humans. An approximate calculation gives a somewhat surprising and impressive result. Assume that the average adult human consumes approximately 11,700 kJ (2800 kcal, that is, 2800 Calories) per day. Assume also that the metabolic pathways leading to ATP synthesis operate at a thermodynamic efficiency of approximately 50%. Thus, of the 11,700 kJ a person consumes as food, about 5860 kJ end up in the form of synthesized ATP. As indicated earlier, the hydrolysis of 1 mole of ATP yields approximately 50 kJ of free energy under cellular conditions. This means that the body cycles through 5860/50 = 117 moles of ATP each day. The disodium salt of ATP has a molecular weight of 551 g/mol, so an average person hydrolyzes about

$$(117 \text{ moles}) \ \frac{551 \text{ g}}{\text{mole}} = 64,467 \text{ g of ATP per day}$$

The average adult human, with a typical weight of 70 kg or so, thus consumes approximately 65 kg of ATP per day, an amount nearly equal to his or her own body weight! Fortunately, we have a highly efficient recycling system for ATP/ADP utilization. The energy released from food is stored transiently in the form of ATP. Once ATP energy is used and ADP and phosphate are released, our bodies recycle it to ATP through intermediary metabolism so that it may be reused. The typical 70-kg body contains only about 50 grams of ATP/ADP total. Therefore, each ATP molecule in our bodies must be recycled nearly 1300 times each day! Were it not for this fact, at current commercial prices of about $20 per gram, our ATP "habit" would cost more than $1 million per day! In these terms, the ability of biochemistry to sustain the marvelous activity and vigor of organisms gains our respect and fascination.

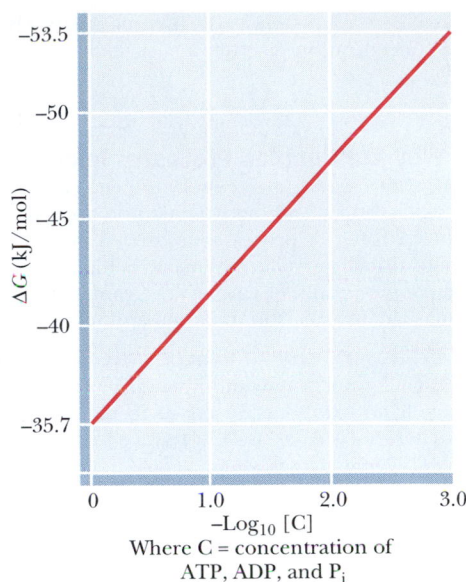

Biochemistry Now™ ACTIVE FIGURE 3.18
The free energy of hydrolysis of ATP as a function of concentration at 38°C, pH 7.0. The plot follows the relationship described in Equation 3.36, with the concentrations [C] of ATP, ADP, and P_i assumed to be equal. **Test yourself on the concepts in this figure at http://chemistry.brookscole.com/ggb3**

Summary

The activities of living things require energy. Movement, growth, synthesis of biomolecules, and the transport of ions and molecules across membranes all demand energy input. All organisms must acquire energy from their surroundings and must utilize that energy efficiently to carry out life processes. To study such bioenergetic phenomena requires familiarity with thermodynamics. Thermodynamics also allows us to determine whether chemical processes and reactions occur spontaneously.

3.1 What Are the Basic Concepts of Thermodynamics?
The system is that portion of the *universe* with which we are concerned. The surroundings include everything else in the universe. An isolated system cannot exchange matter or energy with its surroundings. A closed system may exchange energy, but not matter, with the surroundings. An open system may exchange matter, energy, or both with the surroundings. Living things are typically open systems. The first law of thermodynamics states that the total energy of an isolated system is conserved. Enthalpy, H, is defined as $H = E + PV$. ΔH is equal to the heat transferred in a constant pressure process. For biochemical reactions in liquids, volume changes are typically quite small, and enthalpy and internal energy are often essentially equal. There are several statements of the second law of thermodynamics, including the following: (1) Systems tend to proceed from ordered (low-entropy or low-probability) states to disordered (high-entropy or high-probability) states. (2) The entropy of the system plus surroundings is unchanged by reversible processes; the entropy of the system plus surroundings increases for irreversible processes. (3) All naturally occurring processes proceed toward equilib-

rium, that is, to a state of minimum potential energy. The third law of thermodynamics states that the entropy of any crystalline, perfectly ordered substance must approach zero as the temperature approaches 0 K, and at $T = 0$ K entropy is exactly zero. The Gibbs free energy, G, defined as $G = H - TS$, provides a simple criterion for equilibrium.

3.2 What Can Thermodynamic Parameters Tell Us About Biochemical Events?
A single parameter (ΔH or ΔS, for example) is not very meaningful, but comparison of several thermodynamic parameters can provide meaningful insights about a process. Thermodynamic parameters can be used to predict whether a given reaction will occur as written and to calculate the relative contributions of molecular phenomena (for example, hydrogen bonding or hydrophobic interactions) to an overall process.

3.3 What Is the Effect of pH on Standard-State Free Energies?
For biochemical reactions in which hydrogen ions (H⁺) are consumed or produced, a modified standard state, designated with prime (′) symbols, as in $\Delta G°'$, K_{eq}', $\Delta H°'$, may be employed. For a reaction in which H⁺ is produced, $\Delta G°'$ is given by

$$\Delta G°' = \Delta G° + RT \ln [H^+]$$

3.4 What Is the Effect of Concentration on Net Free Energy Changes?
The free energy change for a reaction can be very different from the standard-state value if the concentrations of reactants and products differ significantly from unit activity (1 M for solutions). For

the reaction A + B $\rightleftharpoons$ C + D, the free energy change for non–standard-state concentrations is given by

$$\Delta G = \Delta G^\circ + RT \ln \frac{[C][D]}{[A][B]}$$

3.5 Why Are Coupled Processes Important to Living Things?

Many of the reactions necessary to keep cells and organisms alive must run against their thermodynamic potential, that is, in the direction of positive ΔG. These processes are driven in the thermodynamically unfavorable direction via coupling with highly favorable processes. Many such coupled processes are crucially important in intermediary metabolism, oxidative phosphorylation, and membrane transport.

3.6 What Are the Characteristics of High-Energy Biomolecules?

A small family of universal biomolecules mediates the flow of energy from exergonic reactions to the energy-requiring processes of life. These molecules are the reduced coenzymes and the high-energy phosphate compounds. High-energy phosphates are not long-term energy storage substances, but rather transient forms of stored energy.

3.7 What Are the Complex Equilibria Involved in ATP Hydrolysis?

ATP, ADP, and similar species can exist in several different ionization states that must be accounted for in any quantitative analysis. Also, phosphate compounds bind a variety of divalent and monovalent cations with substantial affinity, and the various metal complexes must also be considered in such analyses.

3.8 What Is the Daily Human Requirement for ATP?

The average adult human, with a typical weight of 70 kg or so, consumes approximately 2800 calories per day. The energy released from food is stored transiently in the form of ATP. Once ATP energy is used and ADP and phosphate are released, our bodies recycle it to ATP through intermediary metabolism so that it may be reused. The typical 70-kg body contains only about 50 grams of ATP/ADP total. Therefore, each ATP molecule in our bodies must be recycled nearly 1300 times each day.

Problems

1. An enzymatic hydrolysis of fructose-1-P,

   Fructose-1-P + $H_2O \rightleftharpoons$ fructose + P_i

   was allowed to proceed to equilibrium at 25°C. The original concentration of fructose-1-P was 0.2 M, but when the system had reached equilibrium the concentration of fructose-1-P was only 6.52×10^{-5} M. Calculate the equilibrium constant for this reaction and the free energy of hydrolysis of fructose-1-P.

2. The equilibrium constant for some process A $\rightleftharpoons$ B is 0.5 at 20°C and 10 at 30°C. Assuming that ΔH° is independent of temperature, calculate ΔH° for this reaction. Determine ΔG° and ΔS° at 20° and at 30°C. Why is it important in this problem to assume that ΔH° is independent of temperature?

3. The standard-state free energy of hydrolysis for acetyl phosphate is $\Delta G^\circ = -42.3$ kJ/mol.

   Acetyl-P + $H_2O \longrightarrow$ acetate + P_i

   Calculate the free energy change for acetyl phosphate hydrolysis in a solution of 2 mM acetate, 2 mM phosphate, and 3 nM acetyl phosphate.

4. Define a state function. Name three thermodynamic quantities that are state functions and three that are not.

5. ATP hydrolysis at pH 7.0 is accompanied by release of a hydrogen ion to the medium

   $ATP^{4-} + H_2O \rightleftharpoons ADP^{3-} + HPO_4^{2-} + H^+$

   If the $\Delta G^{\circ\prime}$ for this reaction is −30.5 kJ/mol, what is ΔG° (that is, the free energy change for the same reaction with all components, including H^+, at a standard state of 1 M)?

6. For the process A $\rightleftharpoons$ B, K_{eq} (AB) is 0.02 at 37°C. For the process B $\rightleftharpoons$ C, K_{eq} (BC) = 1000 at 37°C.
   a. Determine K_{eq} (AC), the equilibrium constant for the overall process A $\rightleftharpoons$ C, from K_{eq} (AB) and K_{eq} (BC).
   b. Determine standard-state free energy changes for all three processes, and use ΔG°(AC) to determine K_{eq} (AC). Make sure that this value agrees with that determined in part a of this problem.

7. Draw all possible resonance structures for creatine phosphate and discuss their possible effects on resonance stabilization of the molecule.

8. Write the equilibrium constant, K_{eq}, for the hydrolysis of creatine phosphate and calculate a value for K_{eq} at 25°C from the value of $\Delta G^{\circ\prime}$ in Table 3.3.

9. Imagine that creatine phosphate, rather than ATP, is the universal energy carrier molecule in the human body. Repeat the calculation presented in Section 3.8, calculating the weight of creatine phosphate that would need to be consumed each day by a typical adult human if creatine phosphate could not be recycled. If recycling of creatine phosphate were possible, and if the typical adult human body contained 20 grams of creatine phosphate, how many times would each creatine phosphate molecule need to be turned over or recycled each day? Repeat the calculation assuming that glycerol-3-phosphate is the universal energy carrier and that the body contains 20 grams of glycerol-3-phosphate.

10. Calculate the free energy of hydrolysis of ATP in a rat liver cell in which the ATP, ADP, and P_i concentrations are 3.4, 1.3, and 4.8 mM, respectively.

11. Hexokinase catalyzes the phosphorylation of glucose from ATP, yielding glucose-6-P and ADP. Using the values of Table 3.3, calculate the standard-state free energy change and equilibrium constant for the hexokinase reaction.

12. Would you expect the free energy of hydrolysis of acetoacetyl-coenzyme A (see diagram) to be greater than, equal to, or less than that of acetyl-coenzyme A? Provide a chemical rationale for your answer.

$$CH_3 - \overset{\displaystyle O}{\overset{\|}{C}} - CH_2 - \overset{\displaystyle O}{\overset{\|}{C}} - S - CoA$$

13. Consider carbamoyl phosphate, a precursor in the biosynthesis of pyrimidines:

$$\overset{\displaystyle O}{\overset{\|}{\underset{\underset{H_3\overset{+}{N}}{}}{\overset{}{C}}}} \diagup \overset{}{\underset{O - PO_3^{2-}}{}}$$

Based on the discussion of high-energy phosphates in this chapter, would you expect carbamoyl phosphate to possess a high free energy of hydrolysis? Provide a chemical rationale for your answer.

Preparing for the MCAT Exam

14. Consider the data in Figures 3.4 and 3.5. Is the denaturation of chymotrypsinogen spontaneous at 58°C? And what is the temperature at which the native and denatured forms of chymotrypsinogen are in equilibrium?

15. Consider Tables 3.1 and 3.2, as well as the discussion of Table 3.2 in the text, and discuss the meaning of the positive ΔC_P in Table 3.1.

Biochemistry 🌀 Now™ Preparing for an exam? Test yourself on key questions at http://chemistry.brookscole.com/ggb3

Further Reading

General Readings on Thermodynamics

Cantor, C. R., and Schimmel, P. R., 1980. *Biophysical Chemistry*. San Francisco: W.H. Freeman.

Dickerson, R. E., 1969. *Molecular Thermodynamics*. New York: Benjamin Co.

Edsall, J. T., and Gutfreund, H., 1983. *Biothermodynamics: The Study of Biochemical Processes at Equilibrium*. New York: John Wiley.

Edsall, J. T., and Wyman, J., 1958. *Biophysical Chemistry*. New York: Academic Press.

Klotz, I. M., 1967. *Energy Changes in Biochemical Reactions*. New York: Academic Press.

Lehninger, A. L., 1972. *Bioenergetics*, 2nd ed. New York: Benjamin Co.

Morris, J. G., 1968. *A Biologist's Physical Chemistry*. Reading, MA: Addison-Wesley.

Patton, A. R., 1965. *Biochemical Energetics and Kinetics*. Philadelphia: W.B. Saunders.

Chemistry of Adenosine-5′-Triphosphate

Alberty, R. A., 1968. Effect of pH and metal ion concentration on the equilibrium hydrolysis of adenosine triphosphate to adenosine diphosphate. *Journal of Biological Chemistry* **243:**1337–1343.

Alberty, R. A., 1969. Standard Gibbs free energy, enthalpy, and entropy changes as a function of pH and pMg for reactions involving adenosine phosphates. *Journal of Biological Chemistry* **244:**3290–3302.

Alberty, R. A., 2003. *Thermodynamics of Biochemical Reactions*. New York: John Wiley.

Gwynn, R. W., and Veech, R. L., 1973. The equilibrium constants of the adenosine triphosphate hydrolysis and the adenosine triphosphate-citrate lyase reactions. *Journal of Biological Chemistry* **248:**6966–6972.

Special Topics

Brandts, J. F., 1964. The thermodynamics of protein denaturation. I. The denaturation of chymotrypsinogen. *Journal of the American Chemical Society* **86:**4291–4301.

Schrödinger, E., 1945. *What Is Life?* New York: Macmillan.

Segel, I. H., 1976. *Biochemical Calculations*, 2nd ed. New York: John Wiley.

Tanford, C., 1980. *The Hydrophobic Effect*, 2nd ed. New York: John Wiley.

CHAPTER 4 Amino Acids

David W. Grisham

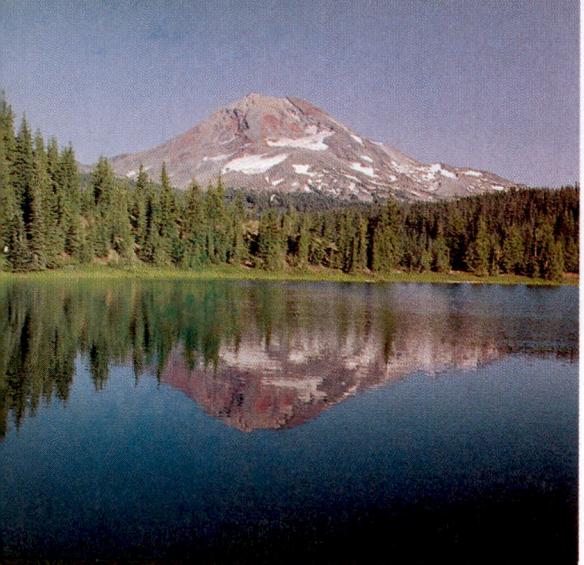

All objects have mirror images. Like many molecules, amino acids exist in mirror-image forms (stereoisomers) that are not superimposable. Only the L-isomers of amino acids commonly occur in nature. (Three Sisters Wilderness, central Oregon. The Middle Sister, reflected in an alpine lake.)

To hold, as 'twere, the mirror up to nature.
William Shakespeare, ***Hamlet***

Key Questions

4.1 What Are the Structures and Properties of Amino Acids, the Building Blocks of Proteins?

4.2 What Are the Acid–Base Properties of Amino Acids?

4.3 What Reactions Do Amino Acids Undergo?

4.4 What Are the Optical and Stereochemical Properties of Amino Acids?

4.5 What Are the Spectroscopic Properties of Amino Acids?

4.6 How Are Amino Acid Mixtures Separated and Analyzed?

BiochemistryNow™ Test yourself on these Key Questions at BiochemistryNow at **http://chemistry.brookscole.com/ggb3**

Essential Question

Proteins are the indispensable agents of biological function, and **amino acids** are the building blocks of proteins. The stunning diversity of the thousands of proteins found in nature arises from the intrinsic properties of only 20 commonly occurring amino acids. These features include (1) the capacity to polymerize, (2) novel acid–base properties, (3) varied structure and chemical functionality in the amino acid side chains, and (4) chirality. This chapter describes each of these properties, laying a foundation for discussions of protein structure (Chapters 5 and 6), enzyme function (Chapters 13–15), and many other subjects in later chapters. ***Why are amino acids uniquely suited to their role as the building blocks of proteins?***

4.1 What Are the Structures and Properties of Amino Acids, the Building Blocks of Proteins?

Typical Amino Acids Contain a Central Tetrahedral Carbon Atom

The structure of a single typical amino acid is shown in Figure 4.1. Central to this structure is the tetrahedral alpha (α) carbon (C_α), which is covalently linked to both the amino group and the carboxyl group. Also bonded to this α-carbon is a hydrogen and a variable side chain. It is the side chain, the so-called R group, that gives each amino acid its identity. The detailed acid–base properties of amino acids are discussed in the following sections. It is sufficient for now to realize that, in neutral solution (pH 7), the carboxyl group exists as —COO^- and the amino group as —NH_3^+. Because the resulting amino acid contains one positive and one negative charge, it is a neutral molecule called a **zwitterion.** Amino acids are also *chiral* molecules. With four different groups attached to it, the α-carbon is said to be *asymmetric.* The two possible configurations for the α-carbon constitute nonidentical mirror-image isomers or enantiomers. Details of amino acid stereochemistry are discussed in Section 4.4.

Amino Acids Can Join via Peptide Bonds

The crucial feature of amino acids that allows them to polymerize to form peptides and proteins is the existence of their two identifying chemical groups: the amino (—NH_3^+) and carboxyl (—COO^-) groups, as shown in Figure 4.2. The amino and carboxyl groups of amino acids can react in a head-to-tail fashion, eliminating a water molecule and forming a covalent amide linkage, which, in the case of peptides and proteins, is typically referred to as a **peptide bond.** The equilibrium for this reaction in aqueous solution favors peptide bond hydroly-

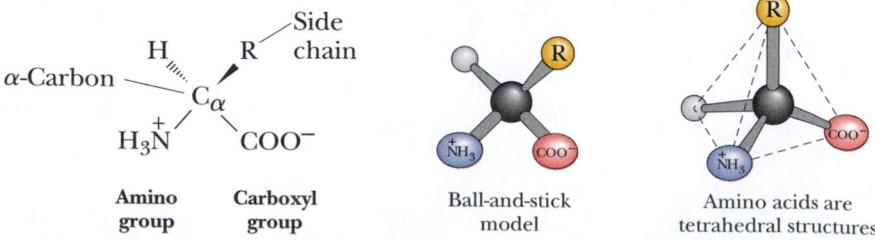

BiochemistryNow™ **ANIMATED FIGURE 4.1** Anatomy of an amino acid. Except for proline and its derivatives, all of the amino acids commonly found in proteins possess this type of structure. **See this figure animated at http://chemistry.brookscole.com/ggb3**

sis. Because peptide bond formation is thermodynamically unfavorable, biological systems as well as peptide chemists in the laboratory must couple peptide bond formation to a thermodynamically favorable reaction.

Iteration of the reaction shown in Figure 4.2 produces **polypeptides** and **proteins.** The remarkable properties of proteins, which we shall discover and come to appreciate in later chapters, all depend in one way or another on the unique properties and chemical diversity of the 20 common amino acids found in proteins.

There Are 20 Common Amino Acids

The structures and abbreviations for the 20 amino acids commonly found in proteins are shown in Figure 4.3. All the amino acids except proline have both free α-amino and free α-carboxyl groups (Figure 4.1). There are several ways to classify the common amino acids. The most useful of these classifications is based on the polarity of the side chains. Thus, the structures shown in Figure 4.3 are grouped into the following categories: (1) nonpolar or hydrophobic amino acids, (2) neutral (uncharged) but polar amino acids, (3) acidic amino acids (which have a net negative charge at pH 7.0), and (4) basic amino acids (which have a net positive charge at neutral pH). In later chapters, the importance of this classification system for predicting protein properties becomes clear. Also shown in Figure 4.3 are the three-letter and one-letter codes used to represent the amino acids. These codes are useful when displaying and comparing the sequences of proteins in shorthand form. (Note that several of the one-letter abbreviations are phonetic in origin: arginine = "Rginine" = R, phenylalanine = "Fenylalanine" = F, aspartic acid = "asparDic" = D.)

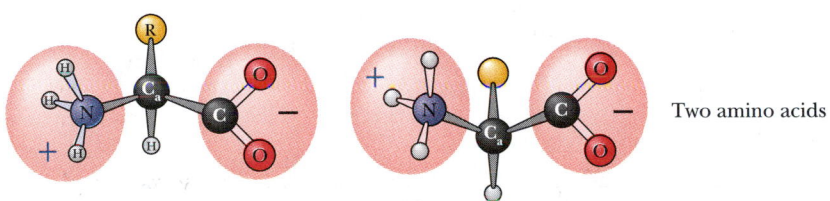

Two amino acids

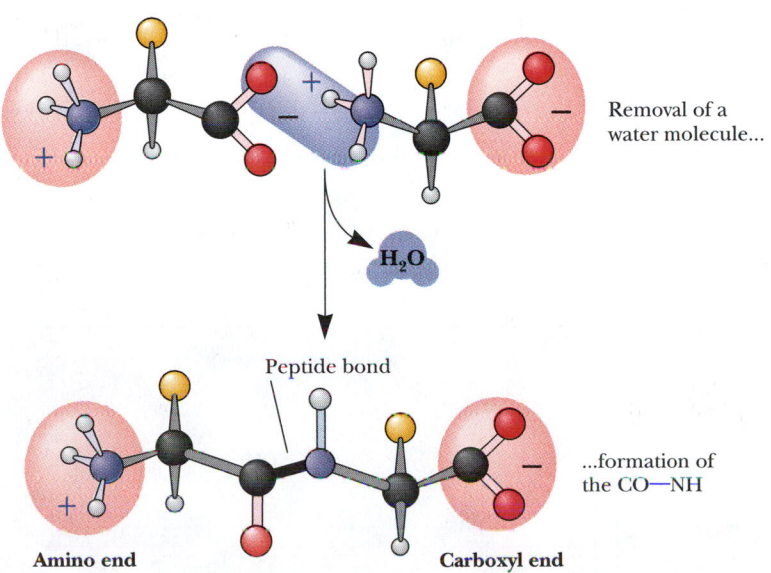

Removal of a water molecule...

H_2O

Peptide bond

...formation of the CO—NH

Amino end   **Carboxyl end**

Biochemistry⚡Now™   **ANIMATED FIGURE 4.2** The α-COOH and α-NH$_3^+$ groups of two amino acids can react with the resulting loss of a water molecule to form a covalent amide bond. *(Illustration. Irving Geis. Rights owned by Howard Hughes Medical Institute. Not to be reproduced without permission.)* **See this figure animated at http://chemistry.brookscole.com/ggb3**

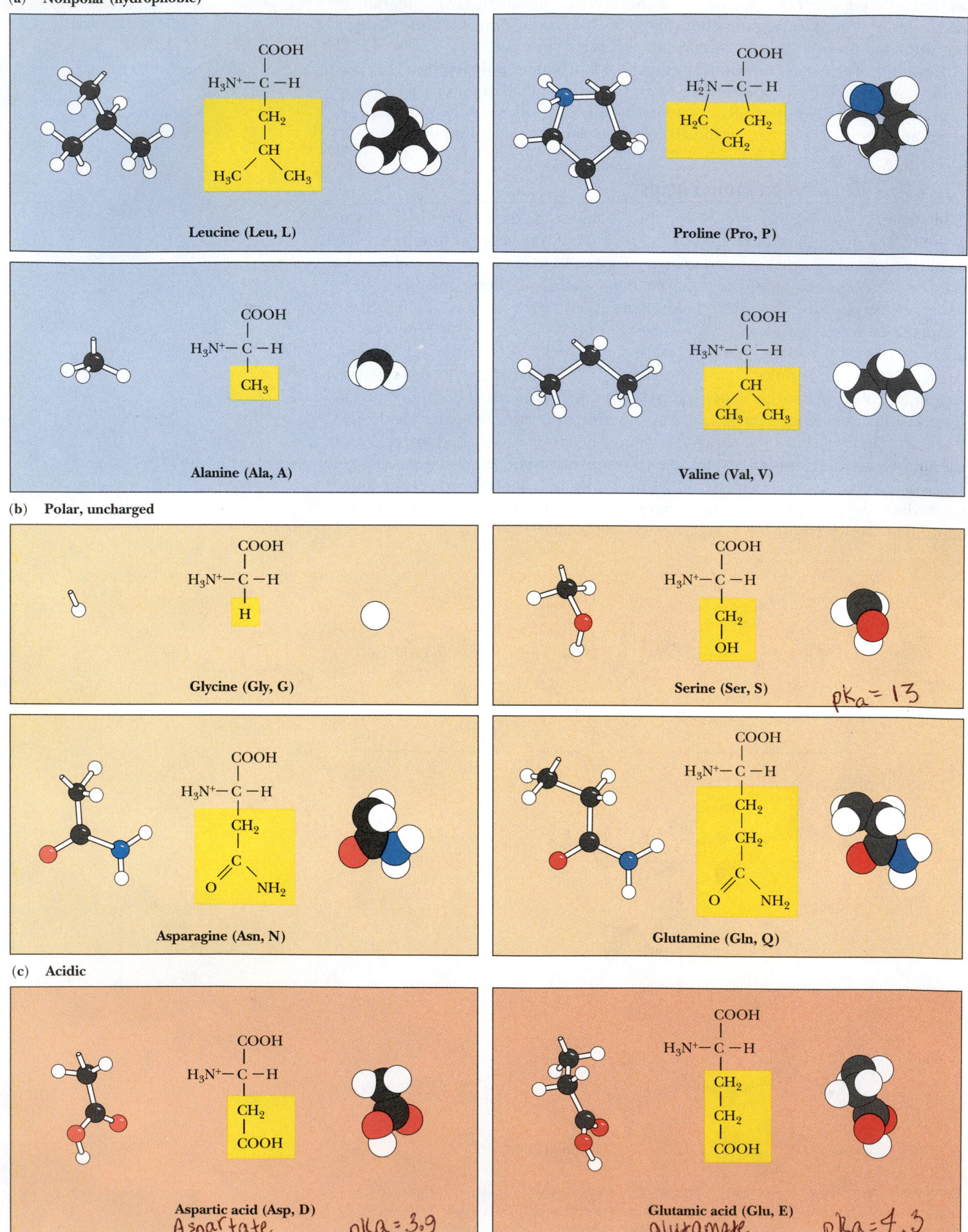

(a) Nonpolar (hydrophobic)

Leucine (Leu, L)

Proline (Pro, P)

Alanine (Ala, A)

Valine (Val, V)

(b) Polar, uncharged

Glycine (Gly, G)

Serine (Ser, S) *pKa = 13*

Asparagine (Asn, N)

Glutamine (Gln, Q)

(c) Acidic

Aspartic acid (Asp, D) *Aspartate   pKa = 3.9*

Glutamic acid (Glu, E) *glutamate   pKa = 4.3*

FIGURE 4.3 The 20 amino acids that are the building blocks of most proteins can be classified as **(a)** nonpolar (hydrophobic); **(b)** polar, neutral; **(c)** acidic; or **(d)** basic. *(Illustration: Irving Geis. Rights owned by Howard Hughes Medical Institute. Not to be produced without permission.)*

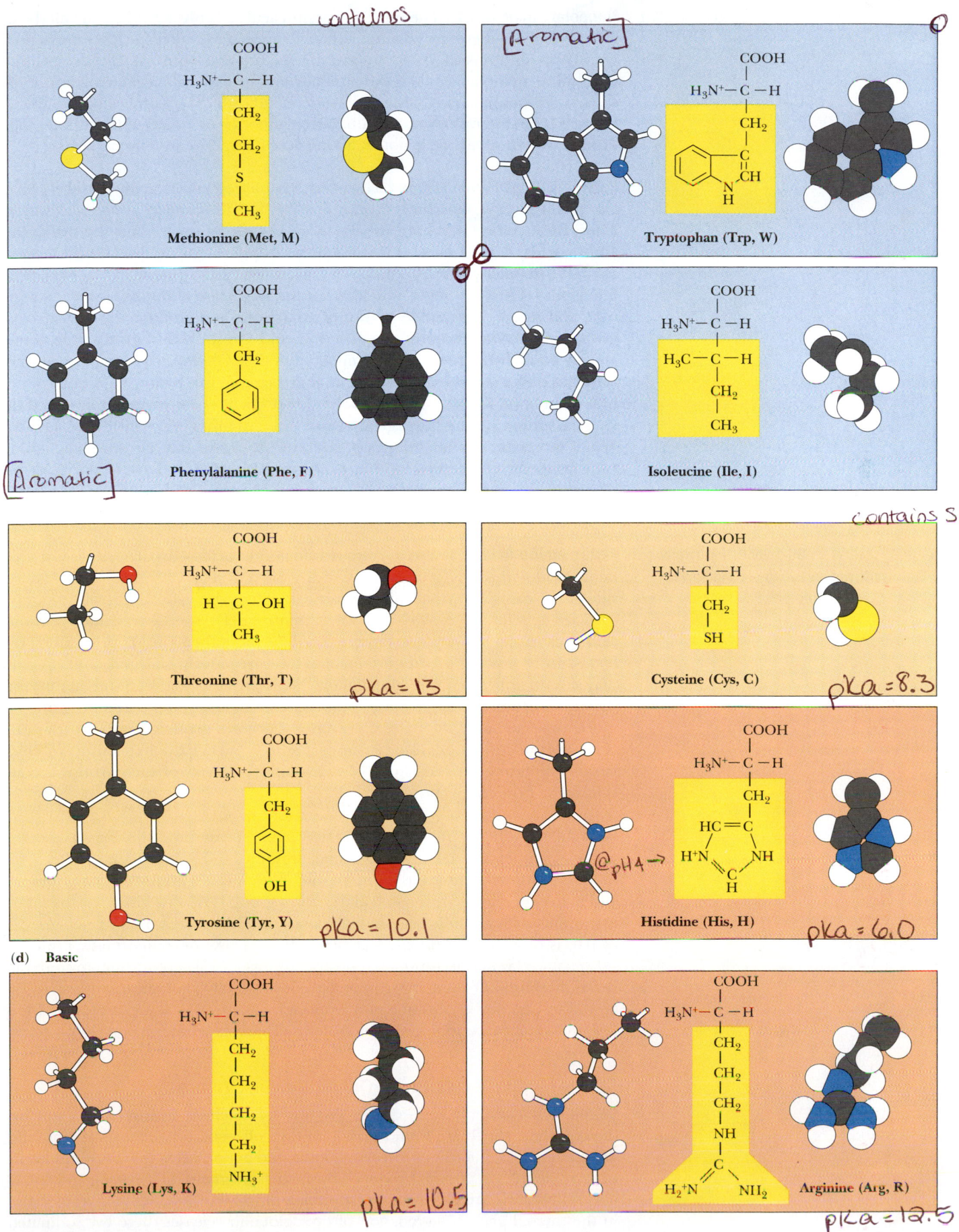

contains

Methionine (Met, M)

COOH
H₃N⁺—C—H
CH₂
CH₂
S
CH₃

[Aromatic]

Tryptophan (Trp, W)

COOH
H₃N⁺—C—H
CH₂
C
CH
NH

[Aromatic]

Phenylalanine (Phe, F)

COOH
H₃N⁺—C—H
CH₂

Isoleucine (Ile, I)

COOH
H₃N⁺—C—H
H₃C—C—H
CH₂
CH₃

Threonine (Thr, T)   pKa = 13

COOH
H₃N⁺—C—H
H—C—OH
CH₃

contains S

Cysteine (Cys, C)   pKa = 8.3

COOH
H₃N⁺—C—H
CH₂
SH

Tyrosine (Tyr, Y)   pKa = 10.1

COOH
H₃N⁺—C—H
CH₂
OH

Histidine (His, H)   pKa = 6.0

COOH
H₃N⁺—C—H
CH₂
HC=C
H⁺N   NH
C
H

@pH4 →

(d)   Basic

Lysine (Lys, K)   pKa = 10.5

COOH
H₃N⁺—C—H
CH₂
CH₂
CH₂
CH₂
NH₃⁺

Arginine (Arg, R)   pKa = 12.5

COOH
H₃N⁺—C—H
CH₂
CH₂
CH₂
NH
C
H₂⁺N   NH₂

FIGURE 4.3 continued

Nonpolar Amino Acids   The nonpolar amino acids (Figure 4.3a) include all those with alkyl chain R groups (alanine, valine, leucine, and isoleucine), as well as proline (with its unusual cyclic structure); methionine (one of the two sulfur-containing amino acids); and two aromatic amino acids, phenylalanine and tryptophan. Tryptophan is sometimes considered a borderline member of this group because it can interact favorably with water via the N—H moiety of the indole ring. Proline, strictly speaking, is not an amino acid but rather an α-imino acid.

Polar, Uncharged Amino Acids   The polar, uncharged amino acids (Figure 4.3b), except for glycine, contain R groups that can form hydrogen bonds with water. Thus, these amino acids are usually more soluble in water than the nonpolar amino acids. Several exceptions should be noted. Tyrosine displays the lowest solubility in water of the 20 common amino acids (0.453 g/L at 25°C). Also, proline is very soluble in water, and alanine and valine are about as soluble as arginine and serine. The amide groups of asparagine and glutamine; the hydroxyl groups of tyrosine, threonine, and serine; and the sulfhydryl group of cysteine are all good hydrogen bond–forming moieties. Glycine, the simplest amino acid, has only a single hydrogen for an R group, and this hydrogen is not a good hydrogen bond former. Glycine's solubility properties are mainly influenced by its polar amino and carboxyl groups, and thus glycine is best considered a member of the polar, uncharged group. It should be noted that tyrosine has significant nonpolar characteristics due to its aromatic ring and could arguably be placed in the nonpolar group (Figure 4.3a). However, with a pK_a of 10.1, tyrosine's phenolic hydroxyl is a charged, polar entity at high pH.

Biochemistry🔊**Now**™   Go to BiochemistryNow and click BiochemistryInteractive to find out how many amino acids you can recognize and name.

Acidic Amino Acids   There are two acidic amino acids—aspartic acid and glutamic acid—whose R groups contain a carboxyl group (Figure 4.3c). These side-chain carboxyl groups are weaker acids than the α-COOH group but are sufficiently acidic to exist as —COO⁻ at neutral pH. Aspartic acid and glutamic acid thus have a net negative charge at pH 7. These forms are appropriately referred to as aspartate and glutamate. These negatively charged amino acids play several important roles in proteins. Many proteins that bind metal ions for structural or functional purposes possess metal-binding sites containing one or more aspartate and glutamate side chains. Carboxyl groups may also act as nucleophiles in certain enzyme reactions and may participate in a variety of electrostatic bonding interactions. The acid–base chemistry of such groups is considered in detail in Section 4.2.

Basic Amino Acids   Three of the common amino acids have side chains with net positive charges at neutral pH: histidine, arginine, and lysine (Figure 4.3d). The ionized group of histidine is an imidazolium, that of arginine is a guanidinium, and lysine contains a protonated alkyl amino group. The side chains of the latter two amino acids are fully protonated at pH 7, but histidine, with a side-chain pK_a of 6.0, is only 10% protonated at pH 7. With a pK_a near neutrality, histidine side chains play important roles as proton donors and acceptors in many enzyme reactions. Histidine-containing peptides are important biological buffers, as discussed in Chapter 2. Arginine and lysine side chains, which are protonated under physiological conditions, participate in electrostatic interactions in proteins.

Several Amino Acids Occur Only Rarely in Proteins

So-called uncommon amino acids (Figure 4.4) include **hydroxylysine** and **hydroxyproline,** which are found mainly in the collagen and gelatin proteins, and **thyroxine** and **3,3′,5-triiodothyronine,** iodinated amino acids that are found only in thyroglobulin, a protein produced by the thyroid gland. (Thyroxine and 3,3′,5-triiodothyronine are produced by iodination of tyrosine residues in thyroglobulin in the thyroid gland. Degradation of thyroglobulin releases these two iodinated amino acids, which act as hormones to regulate growth and development.) Cer-

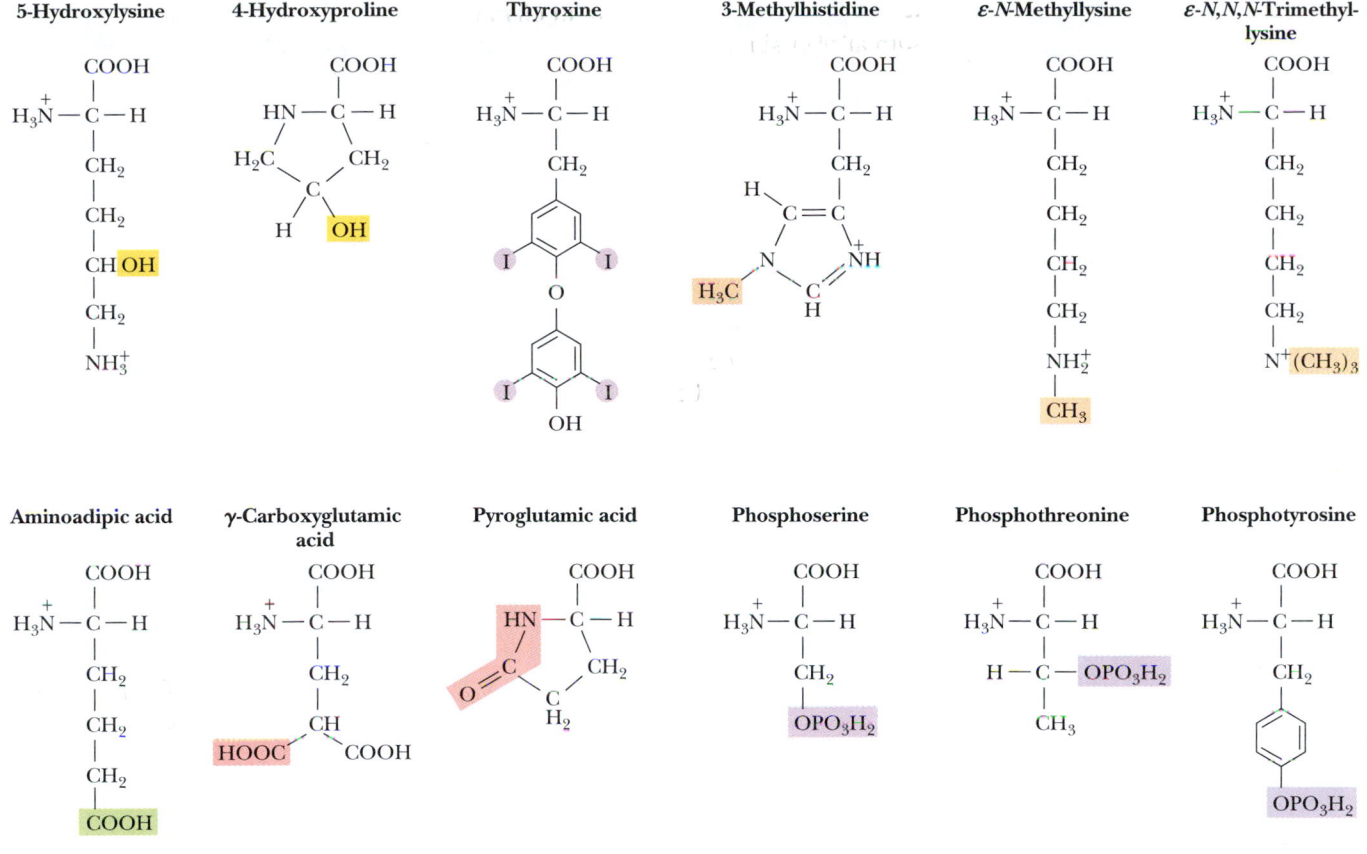

FIGURE 4.4 The structures of several amino acids that are less common but nevertheless found in certain proteins. Hydroxylysine and hydroxyproline are found in connective-tissue proteins, pyroglutamic acid is found in bacteriorhodopsin (a protein in *Halobacterium halobium*), and aminoadipic acid is found in proteins isolated from corn.

tain muscle proteins contain methylated amino acids, including **methylhistidine, ϵ-N-methyllysine,** and **ϵ-N,N,N-trimethyllysine** (Figure 4.4). **γ-Carboxyglutamic acid** is found in several proteins involved in blood clotting, and **pyroglutamic acid** is found in a unique light-driven proton-pumping protein called bacteriorhodopsin, which is discussed elsewhere in this book. Certain proteins involved in cell growth and regulation are reversibly phosphorylated on the —OH groups of serine, threonine, and tyrosine residues. **Aminoadipic acid** is found in proteins isolated from corn. Finally, **N-methylarginine** and **N-acetyllysine** are found in histone proteins associated with chromosomes.

Some Amino Acids Are Not Found in Proteins

Certain amino acids and their derivatives, although not found in proteins, nonetheless are biochemically important. A few of the more notable examples are shown in Figure 4.5. **γ-Aminobutyric acid,** or **GABA,** is produced by the decarboxylation of glutamic acid and is a potent neurotransmitter. **Histamine,**

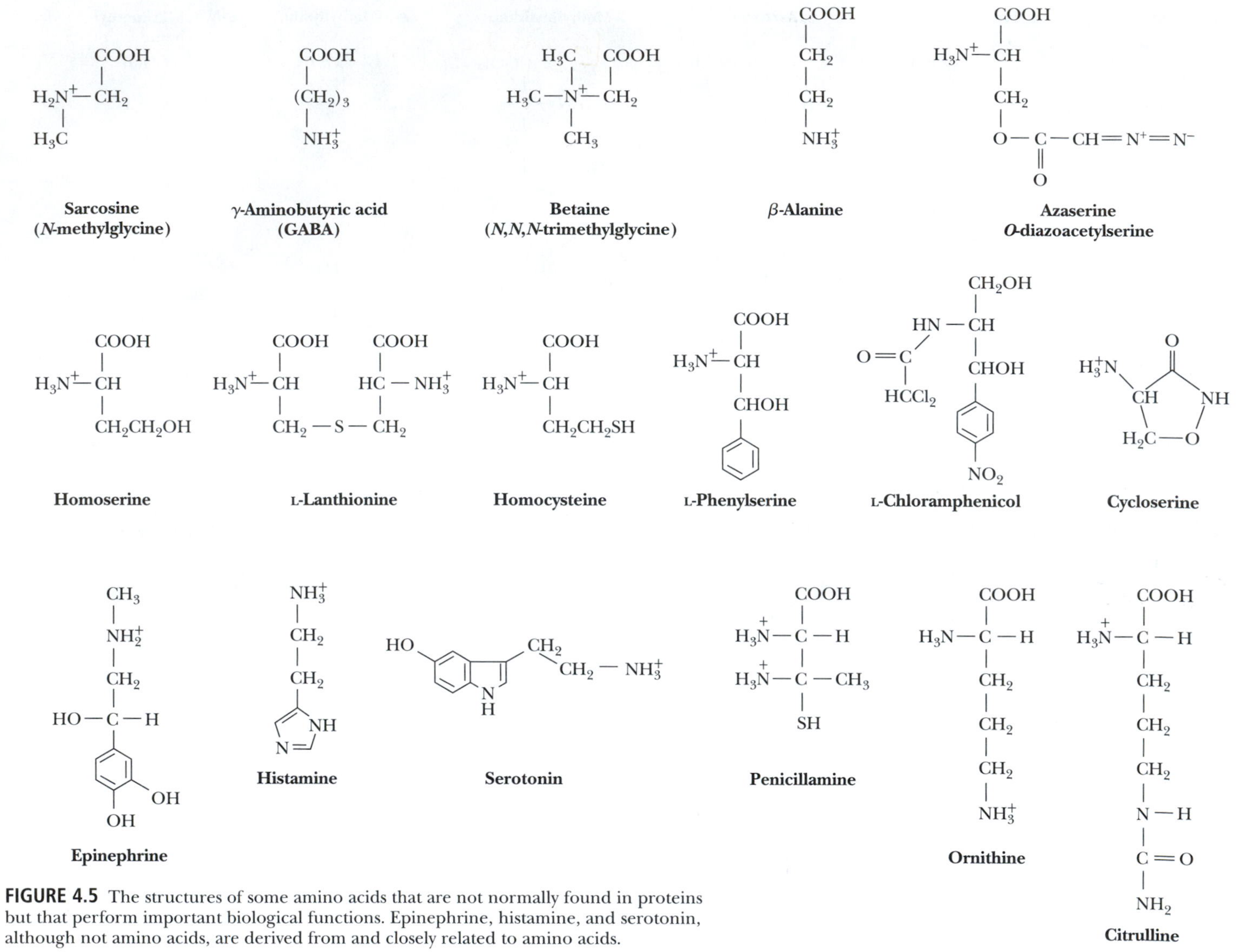

FIGURE 4.5 The structures of some amino acids that are not normally found in proteins but that perform important biological functions. Epinephrine, histamine, and serotonin, although not amino acids, are derived from and closely related to amino acids.

which is synthesized by decarboxylation of histidine, and **serotonin,** which is derived from tryptophan, similarly function as neurotransmitters and regulators. **β-Alanine** is found in nature in the peptides carnosine and anserine and is a component of pantothenic acid (a vitamin), which is a part of coenzyme A. **Epinephrine** (also known as **adrenaline**), derived from tyrosine, is an important hormone. **Penicillamine** is a constituent of the penicillin antibiotics. **Ornithine, betaine, homocysteine,** and **homoserine** are important metabolic intermediates. **Citrulline** is the immediate precursor of arginine.

4.2 | What Are the Acid–Base Properties of Amino Acids?

Amino Acids Are Weak Polyprotic Acids

From a chemical point of view, the common amino acids are all weak polyprotic acids. The ionizable groups are not strongly dissociating ones, and the degree of dissociation thus depends on the pH of the medium. All the amino acids contain at least two dissociable hydrogens.

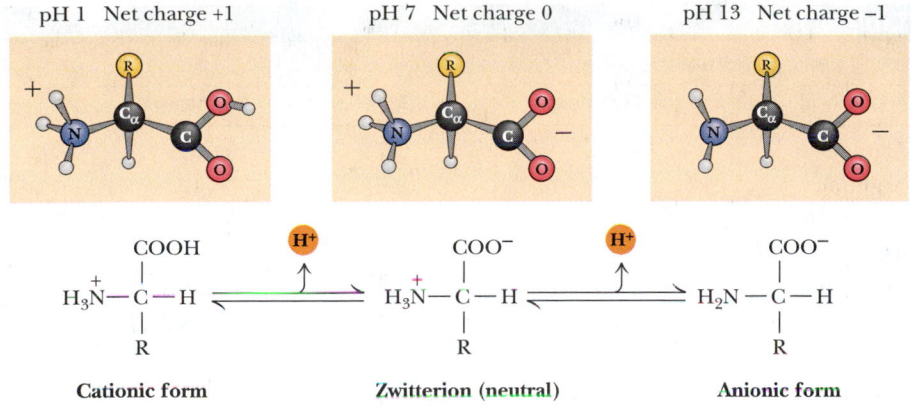

pH 1   Net charge +1   pH 7   Net charge 0   pH 13   Net charge –1

COOH →(H⁺)→ COO⁻ →(H⁺)→ COO⁻

$$\underset{\text{R}}{\overset{\text{COOH}}{\underset{|}{\overset{|}{H_3\overset{+}{N}-C-H}}}} \rightleftharpoons \underset{\text{R}}{\overset{\text{COO}^-}{\underset{|}{\overset{|}{H_3\overset{+}{N}-C-H}}}} \rightleftharpoons \underset{\text{R}}{\overset{\text{COO}^-}{\underset{|}{\overset{|}{H_2N-C-H}}}}$$

Cationic form   **Zwitterion (neutral)**   **Anionic form**

Biochemistry ⓔNow™   ANIMATED FIGURE 4.6 The ionic forms of the amino acids, shown without consideration of any ionizations on the side chain. The cationic form is the low pH form, and the titration of the cationic species with base yields the zwitterion and finally the anionic form. *(Illustration: Irving Geis. Rights owned by Howard Hughes Medical Institute. Not to be reproduced without permission.)* **See this figure animated at http://chemistry.brookscole.com/ggb3**

Consider the acid–base behavior of glycine, the simplest amino acid. At low pH, both the amino and carboxyl groups are protonated and the molecule has a net positive charge. If the counterion in solution is a chloride ion, this form is referred to as **glycine hydrochloride.** If the pH is increased, the carboxyl group is the first to dissociate, yielding the neutral zwitterionic species Gly^0 (Figure 4.6). A further increase in pH eventually results in dissociation of the amino group to yield the negatively charged **glycinate.** If we denote these three forms as Gly^+, Gly^0, and Gly^-, we can write the first dissociation of Gly^+ as

$$Gly^+ + H_2O \rightleftharpoons Gly^0 + H_3O^+$$

and the dissociation constant K_1 as

$$K_1 = \frac{[Gly^0][H_3O^+]}{[Gly^+]}$$

Values for K_1 for the common amino acids are typically 0.4 to 1.0×10^{-2} M, so that typical values of pK_1 center on values of 2.0 to 2.4 (Table 4.1). In a similar manner, we can write the second dissociation reaction as

$$Gly^0 + H_2O \rightleftharpoons Gly^- + H_3O^+$$

and the dissociation constant K_2 as

$$K_2 = \frac{[Gly^-][H_3O^+]}{[Gly^0]}$$

Typical values for pK_2 are in the range of 9.0 to 9.8. At physiological pH, the α-carboxyl group of a simple amino acid (with no ionizable side chains) is completely dissociated, whereas the α-amino group has not really begun its dissociation. The titration curve for such an amino acid is shown in Figure 4.7.

EXAMPLE

What is the pH of a glycine solution in which the α-NH_3^+ group is one-third dissociated?

Answer
The appropriate Henderson–Hasselbalch equation is

$$pH = pK_a + \log_{10} \frac{[Gly^-]}{[Gly^0]}$$

Table 4.1

pK_a Values of Common Amino Acids

Amino Acid	α-COOH pK_a	α-NH$_3^+$ pK_a	R group pK_a
Alanine	2.4	9.7	
Arginine	2.2	9.0	12.5
Asparagine	2.0	8.8	
Aspartic acid	2.1	9.8	3.9
Cysteine	1.7	10.8	8.3
Glutamic acid	2.2	9.7	4.3
Glutamine	2.2	9.1	
Glycine	2.3	9.6	
Histidine	1.8	9.2	6.0
Isoleucine	2.4	9.7	
Leucine	2.4	9.6	
Lysine	2.2	9.0	10.5
Methionine	2.3	9.2	
Phenylalanine	1.8	9.1	
Proline	2.1	10.6	
Serine	2.2	9.2	~13
Threonine	2.6	10.4	~13
Tryptophan	2.4	9.4	
Tyrosine	2.2	9.1	10.1
Valine	2.3	9.6	

If the α-amino group is one-third dissociated, there is 1 part Gly$^-$ for every 2 parts Gly^0. The important pK_a is the pK_a for the amino group. The glycine α-amino group has a pK_a of 9.6. The result is

$$pH = 9.6 + \log_{10}(1/2)$$
$$pH = 9.3$$

Note that the dissociation constants of both the α-carboxyl and α-amino groups are affected by the presence of the other group. The adjacent α-amino group makes the α-COOH group more acidic (that is, it lowers the pK_a), so it gives up a proton more readily than simple alkyl carboxylic acids. Thus, the pK_1 of 2.0 to 2.1 for α-carboxyl groups of amino acids is substantially lower than that of acetic acid (pK_a = 4.76), for example. What is the chemical basis for the low pK_a of the α-COOH group of amino acids? The α-NH$_3^+$ (ammonium) group is strongly electron-withdrawing, and the positive charge of the amino group exerts a strong field effect and stabilizes the carboxylate anion. (The effect of the α-COO$^-$ group on the pK_a of the α-NH$_3^+$ group is the basis for problem 4 at the end of this chapter.)

Side Chains of Amino Acids Undergo Characteristic Ionizations

As we have seen, the side chains of several of the amino acids also contain dissociable groups. Thus, aspartic and glutamic acids contain an additional carboxyl function, and lysine possesses an aliphatic amino function. Histidine contains an ionizable imidazolium proton, and arginine carries a guanidinium function. Typical pK_a values of these groups are shown in Table 4.1.

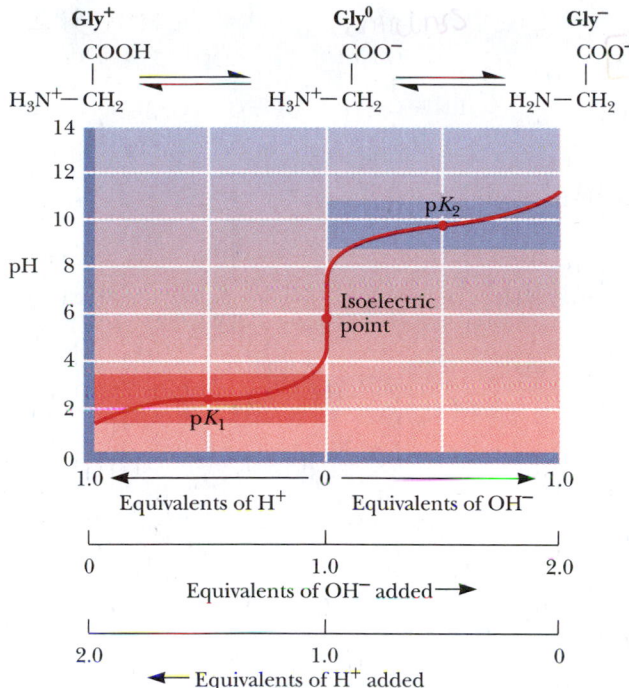

FIGURE 4.7 Titration of glycine, a simple amino acid. The isoelectric point, pI, the pH where glycine has a net charge of 0, can be calculated as $(pK_1 + pK_2)/2$.

The β-carboxyl group of aspartic acid and the γ-carboxyl side chain of glutamic acid exhibit pK_a values intermediate to the α-COOH on one hand and typical aliphatic carboxyl groups on the other hand. In a similar fashion, the ϵ-amino group of lysine exhibits a pK_a that is higher than that of the α-amino group but similar to that for a typical aliphatic amino group. These intermediate side-chain pK_a values reflect the slightly diminished effect of the α-carbon dissociable groups that lie several carbons removed from the side-chain functional groups. Figure 4.8 shows typical titration curves for glutamic acid and lysine, along with the ionic species that predominate at various points in the titration. The only other side-chain groups that exhibit any significant degree of dissociation are the *para*-OH group of tyrosine and the —SH group of cysteine. The pK_a of the cysteine sulfhydryl is 8.32, so it is about 12% dissociated at pH 7. The tyrosine *para*-OH group is a very weakly acidic group, with a pK_a of about 10.1. This group is essentially fully protonated and uncharged at pH 7.

4.3 | What Reactions Do Amino Acids Undergo?

Amino Acids Undergo Typical Carboxyl and Amino Group Reactions

The α-carboxyl and α-amino groups of all amino acids exhibit similar chemical reactivity. The side chains, however, exhibit specific chemical reactivities, depending on the nature of the functional groups. Whereas all of these reactivities are important in the study and analysis of isolated amino acids, it is the characteristic behavior of the side chain that governs the reactivity of amino acids incorporated into proteins. There are three reasons to consider these reactivities. Proteins can be modified in very specific ways by taking advantage of the chemical reactivity of certain amino acid side chains. The detection and quantitation of amino acids and proteins often depend on reactions that are

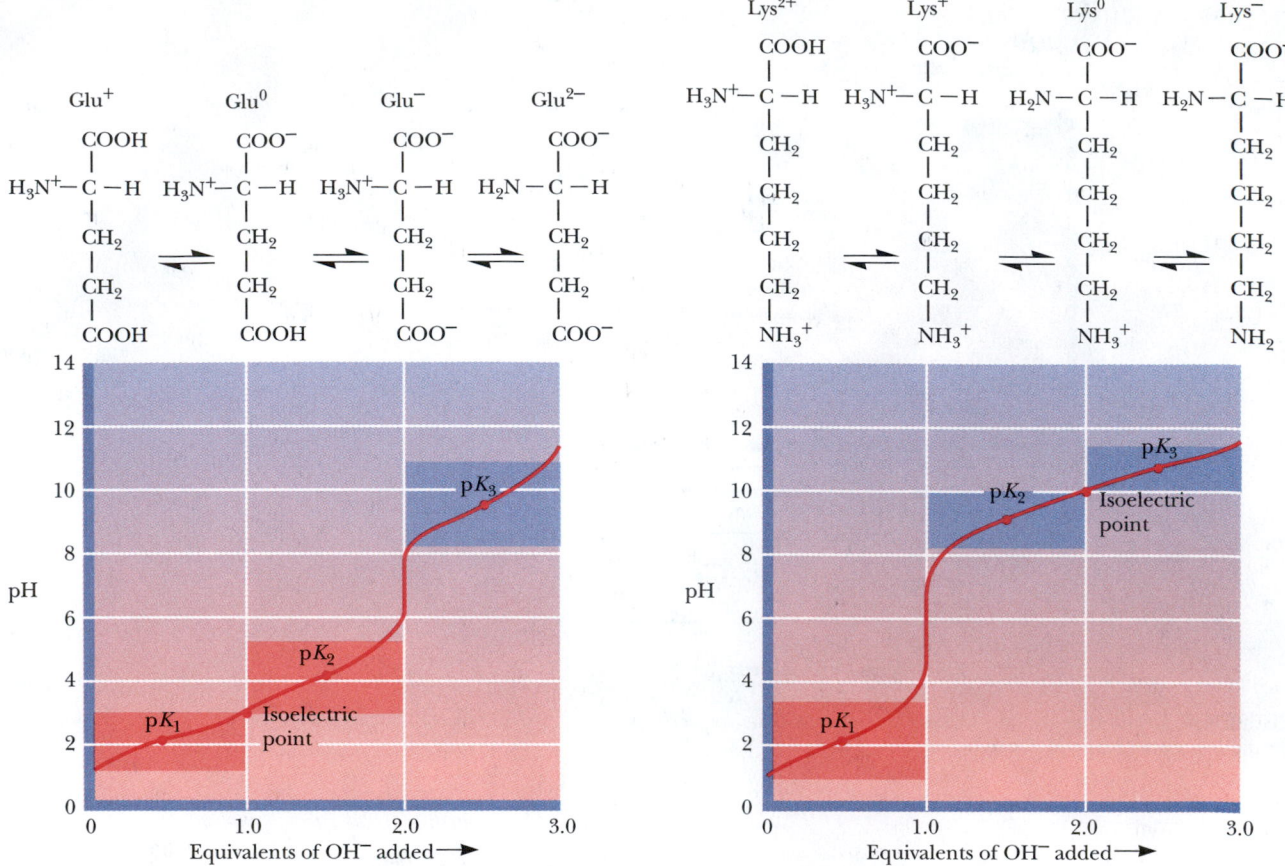

Biochemistry Now™ **ACTIVE FIGURE 4.8** Titrations of glutamic acid and lysine. **Test yourself on the concepts in this figure at http://chemistry.brookscole.com/ggb3**

Biochemistry Now™ Go to BiochemistryNow and click BiochemistryInteractive to explore the titration behavior of amino acids.

specific to one or more amino acids and that result in color, radioactivity, or some other quantity that can be easily measured. Finally and most important, the biological functions of proteins depend on the behavior and reactivity of specific R groups.

The carboxyl groups of amino acids undergo all the simple reactions common to this functional group. Reaction with ammonia and primary amines yields unsubstituted and substituted amides, respectively (Figure 4.9a,b). Esters and acid chlorides are also readily formed. Esterification proceeds in the presence of the appropriate alcohol and a strong acid (Figure 4.9c). Polymerization can occur by repetition of the reaction shown in Figure 4.9d. Free amino groups may react with aldehydes to form Schiff bases (Figure 4.9e) and can be acylated with acid anhydrides and acid halides (Figure 4.9f).

The Ninhydrin Reaction Is Characteristic of Amino Acids

Amino acids can be readily detected and quantified by reaction with ninhydrin. As shown in Figure 4.10, *ninhydrin*, or triketohydrindene hydrate, is a strong oxidizing agent and causes the oxidative deamination of the α-amino function. The products of the reaction are the resulting aldehyde, ammonia, carbon dioxide, and hydrindantin, a reduced derivative of ninhydrin. The ammonia produced in this way can react with the hydrindantin and another molecule of ninhydrin to yield a purple product (Ruhemann's Purple) that can be quantified spectrophotometrically at 570 nm. The appearance of CO_2 can also be moni-

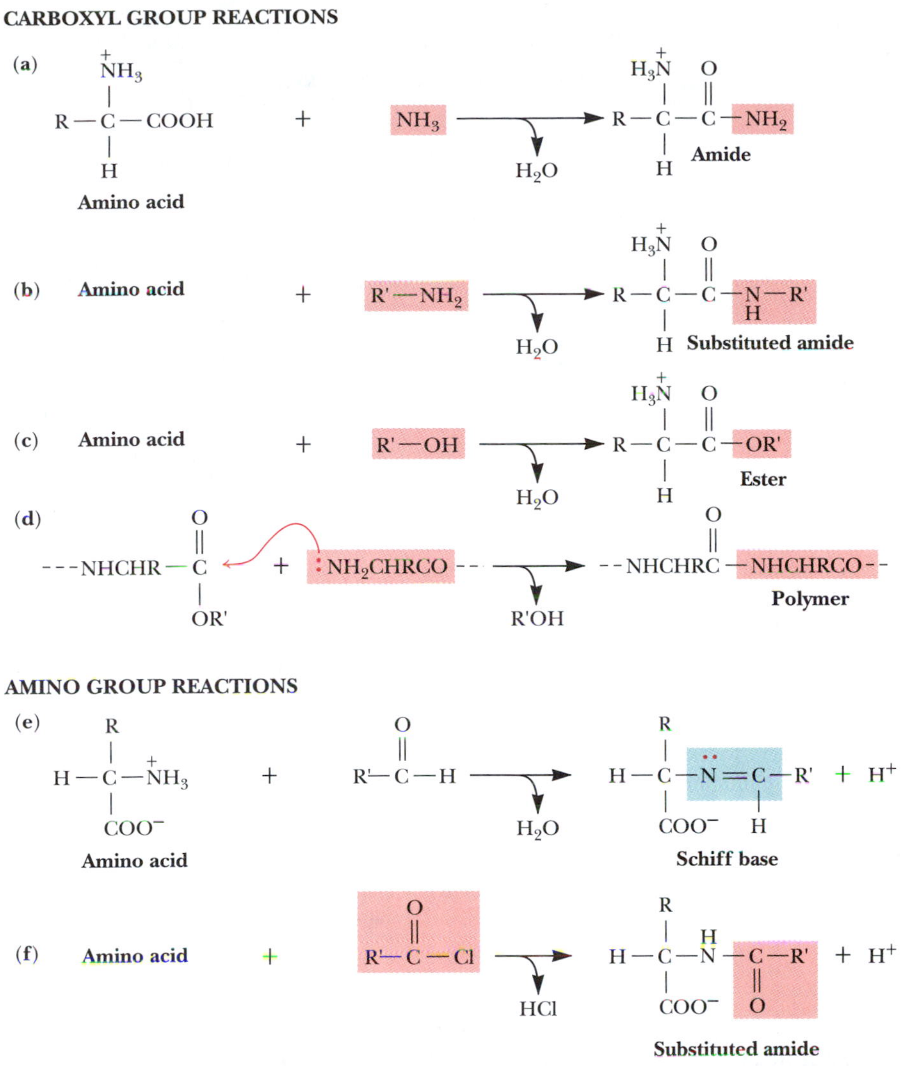

CARBOXYL GROUP REACTIONS

(a)

$$\underset{\text{Amino acid}}{R - \overset{\overset{+}{N}H_3}{\underset{H}{C}} - COOH} + \boxed{NH_3} \xrightarrow[\;H_2O\;]{} R - \overset{\overset{+}{H_3N}}{\underset{H}{C}} - \overset{O}{\overset{\|}{C}} - \boxed{NH_2} \quad \text{Amide}$$

(b)

$$\text{Amino acid} + \boxed{R' - NH_2} \xrightarrow[\;H_2O\;]{} R - \overset{\overset{+}{H_3N}}{\underset{H}{C}} - \overset{O}{\overset{\|}{C}} - \boxed{\underset{H}{N} - R'} \quad \text{Substituted amide}$$

(c)

$$\text{Amino acid} + \boxed{R' - OH} \xrightarrow[\;H_2O\;]{} R - \overset{\overset{+}{H_3N}}{\underset{H}{C}} - \overset{O}{\overset{\|}{C}} - \boxed{OR'} \quad \text{Ester}$$

(d)

$$---NHCHR - \overset{O}{\overset{\|}{\underset{OR'}{C}}} + \boxed{:NH_2CHRCO--} \xrightarrow[\;R'OH\;]{} --NHCHRC - NHCHRCO-- \quad \text{Polymer}$$

AMINO GROUP REACTIONS

(e)

$$\underset{\text{Amino acid}}{H - \overset{R}{\underset{COO^-}{\overset{|}{C}} - \overset{+}{N}H_3}} + \overset{O}{\overset{\|}{R' - C - H}} \xrightarrow[\;H_2O\;]{} H - \overset{R}{\underset{COO^-}{\overset{|}{C}} - \boxed{N = \overset{|}{\underset{H}{C}} - R'}} + H^+ \quad \text{Schiff base}$$

(f)

$$\text{Amino acid} + \boxed{\overset{O}{\overset{\|}{R' - C - Cl}}} \xrightarrow[\;HCl\;]{} H - \overset{R}{\underset{COO^-}{\overset{|}{C}} - \overset{H}{\underset{}{N}} - \boxed{\overset{}{\underset{O}{C}} - R'}} + H^+$$

Substituted amide

Biochemistry Now™ **ACTIVE FIGURE 4.9** Typical reactions of the common amino acids (see text for details). **Test yourself on the concepts in this figure at http://chemistry.brookscole.com/ggb3**

tored. Indeed, CO_2 evolution is diagnostic of the presence of an α-amino acid. α-Imino acids, such as proline and hydroxyproline, give bright yellow ninhydrin products with absorption maxima at 440 nm, allowing these to be distinguished from the α-amino acids. Because amino acids are one of the components of human skin secretions, the ninhydrin reaction was once used extensively by law enforcement and forensic personnel for fingerprint detection. (Fingerprints as old as 15 years can be successfully identified using the ninhydrin reaction.) More sensitive fluorescent reagents are now used routinely for this purpose.

Amino Acid Side Chains Undergo Specific Reactions

A number of reactions of amino acids are noteworthy because they are essential to the degradation, sequencing, and chemical synthesis of peptides and proteins. These reactions are discussed in Chapter 5.

Biochemists have developed an arsenal of reactions that are relatively specific to the side chains of particular amino acids. These reactions can be used

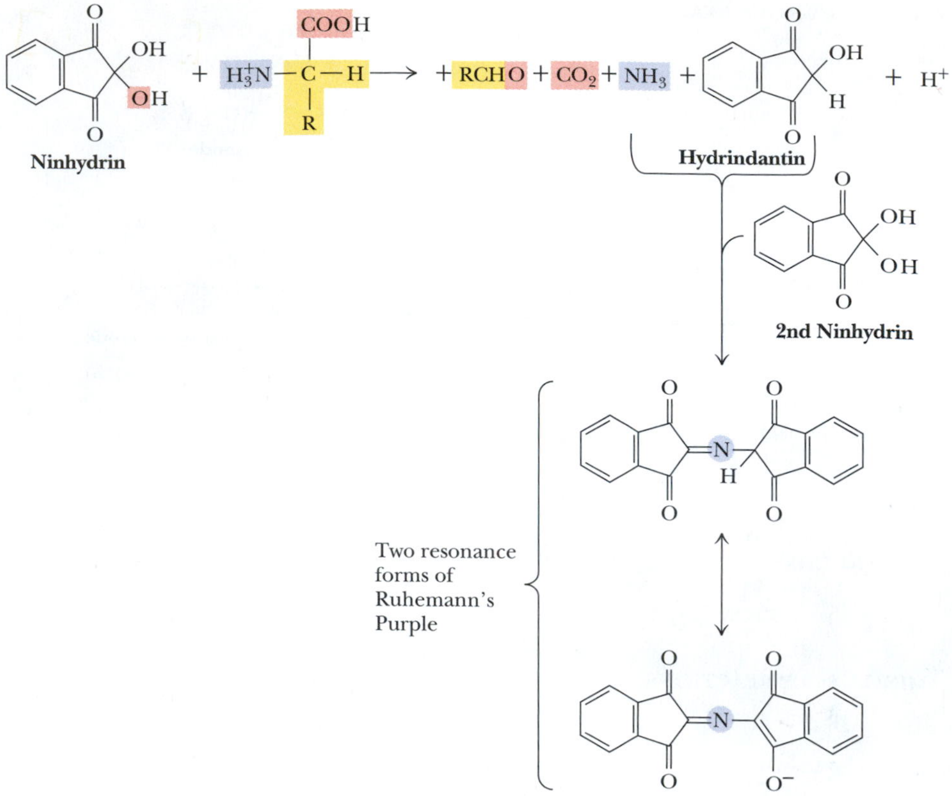

Biochemistry⑤Now™ **ANIMATED FIGURE 4.10** The pathway of the ninhydrin reaction, which produces a colored product called Ruhemann's Purple that absorbs light at 570 nm. Note that the reaction involves and consumes two molecules of ninhydrin. **See this figure animated at** http://chemistry.brookscole.com/ggb3

to identify functional amino acids at the active sites of enzymes or to label proteins with appropriate reagents for further study. Cysteine residues in proteins, for example, react with one another to form disulfide species and also react with a number of reagents, including maleimides (typically *N*-ethylmaleimide), as shown in Figure 4.11. Cysteines also react effectively with iodoacetic acid to yield *S*-carboxymethyl cysteine derivatives. There are numerous other reactions involving specialized reagents specific for particular side-chain functional groups. Figure 4.11 presents a representative list of these reagents and the products that result. It is important to realize that few, if any, of these reactions are truly specific for one functional group; consequently, care must be exercised in their use.

4.4 | What Are the Optical and Stereochemical Properties of Amino Acids?

Amino Acids Are Chiral Molecules

Except for glycine, all of the amino acids isolated from proteins have four different groups attached to the α-carbon atom. In such a case, the α-carbon is said to be **asymmetric** or **chiral** (from the Greek *cheir*, meaning "hand"), and the two possible configurations for the α-carbon constitute nonsuperimposable mirror-image isomers, or **enantiomers** (Figure 4.12). Enantiomeric molecules display a special property called **optical activity**—the ability to rotate the plane of polarization of plane-polarized light. Clockwise rotation of incident light is referred to as **dextrorotatory** behavior, and counterclockwise rotation is called **levorotatory** behavior. The magnitude and direction of the optical rotation

Critical Developments in Biochemistry

Biochemistry ⊘ Now™

Green Fluorescent Protein—The "Light Fantastic" from Jellyfish to Gene Expression

Aquorea victoria, a species of jellyfish found in the northwest Pacific Ocean, contains a **green fluorescent protein (GFP)** that works together with another protein, **aequorin,** to provide a defense mechanism for the jellyfish. When the jellyfish is attacked or shaken, aequorin produces a blue light. This light energy is captured by GFP, which then emits a bright green flash that presumably blinds or startles the attacker. Remarkably, the fluorescence of GFP occurs without the assistance of a **prosthetic group**—a "helper molecule" that would mediate GFP's fluorescence. Instead, the light-transducing capability of GFP is the result of a reaction between three amino acids in the protein itself. As shown below, adjacent **serine, tyrosine,** and **glycine** in the sequence of the protein react to form the pigment complex—termed a **chromophore.** No enzymes are required; the reaction is autocatalytic.

Because the light-transducing talents of GFP depend only on the protein itself (upper photo, chromophore highlighted), GFP has quickly become a darling of genetic engineering laboratories. The promoter of any gene whose cellular expression is of interest can be fused to the DNA sequence coding for GFP. Telltale green fluorescence tells the researcher when this fused gene has been expressed (see lower photo and also Chapter 12).

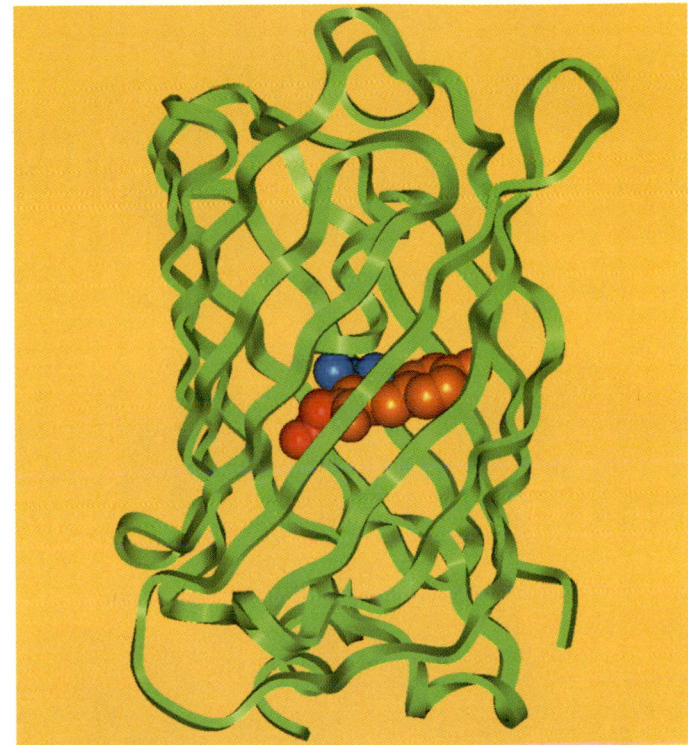

Phe-**Ser-Tyr-Gly**-Val-Gln $\xrightarrow{O_2}$
64   69

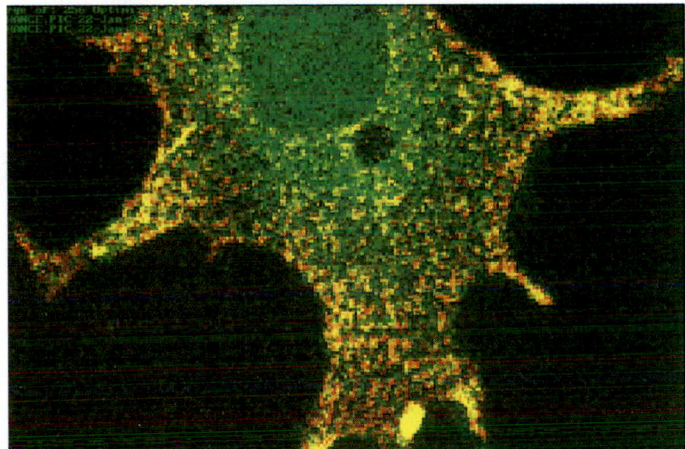

▲ Autocatalytic oxidation of GFP amino acids leads to the chromophore shown on the left. The green fluorescence requires further interactions of the chromophore with other parts of the protein.

Boxer, S.G., 1997. Another green revolution. *Nature* **383**:484–485.

depend on the nature of the amino acid side chain. The temperature, the wavelength of the light used in the measurement, the ionization state of the amino acid, and therefore the pH of the solution can also affect optical rotation behavior. As shown in Table 4.2, some protein-derived amino acids at a given pH are dextrorotatory and others are levorotatory, even though all of them are of the L-configuration. The direction of optical rotation can be specified in the name by using a (+) for dextrorotatory compounds and a (−) for levorotatory compounds, as in L(+)-leucine.

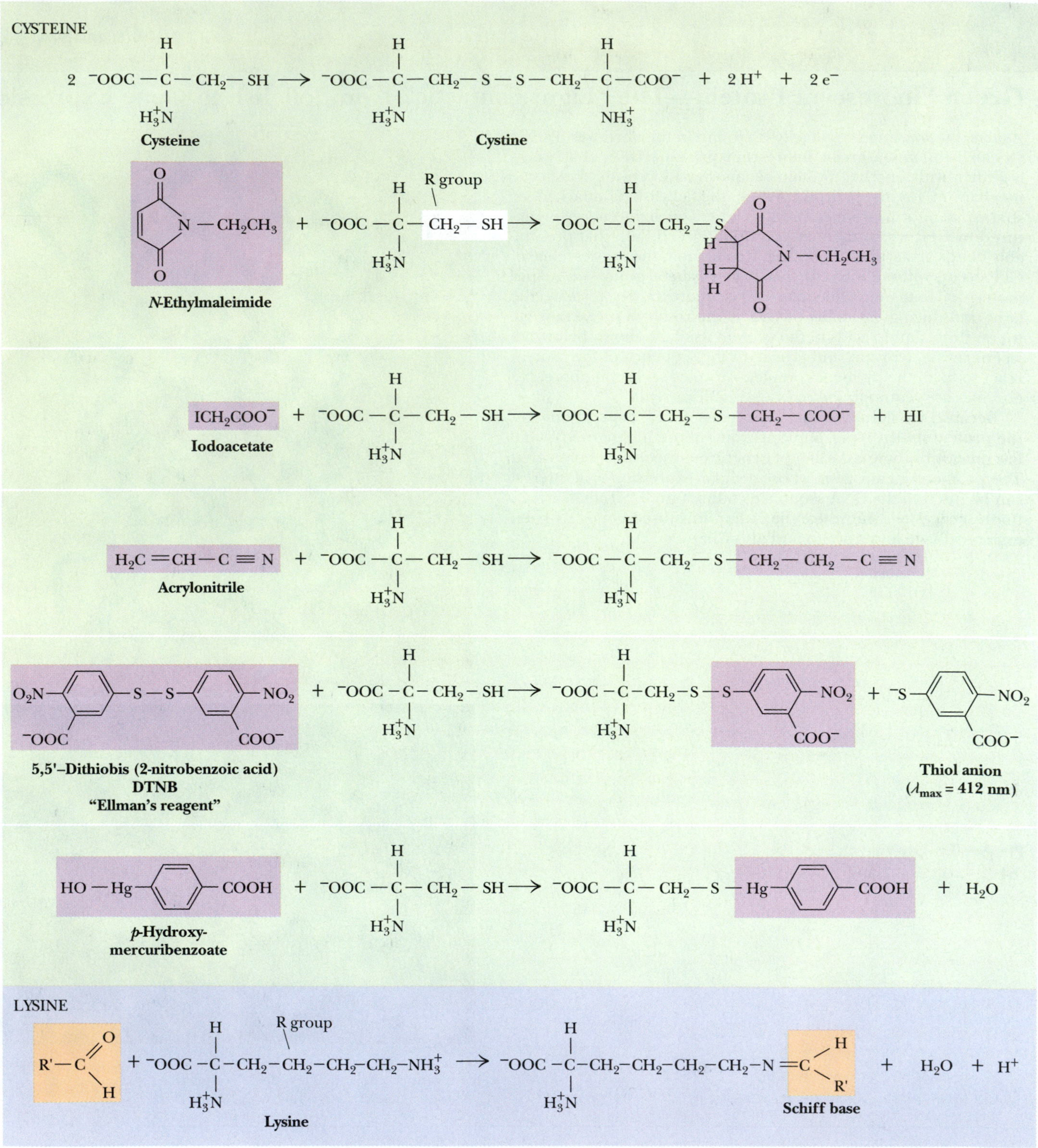

CYSTEINE

$$2 \ ^-OOC-\underset{\underset{H_3^+N}{|}}{\overset{\overset{H}{|}}{C}}-CH_2-SH \longrightarrow \ ^-OOC-\underset{\underset{H_3^+N}{|}}{\overset{\overset{H}{|}}{C}}-CH_2-S-S-CH_2-\underset{\underset{NH_3^+}{|}}{\overset{\overset{H}{|}}{C}}-COO^- + 2H^+ + 2e^-$$

Cysteine   **Cystine**

R group

N-Ethylmaleimide

ICH₂COO⁻
Iodoacetate

H₂C=CH—C≡N
Acrylonitrile

5,5′–Dithiobis (2-nitrobenzoic acid)
DTNB
"Ellman's reagent"

Thiol anion
($\lambda_{max} = 412$ nm)

HO—Hg—⟨⟩—COOH
p-Hydroxy-
mercuribenzoate

LYSINE

Lysine   **Schiff base**

Biochemistry🛇Now™   **ANIMATED FIGURE 4.11** Reactions of amino acid side-chain functional groups. **See this figure animated at http://chemistry.brookscole.com/ggb3**

Chiral Molecules Are Described by the D,L and *R,S* Naming Conventions

The discoveries of optical activity and enantiomeric structures (see Critical Developments in Biochemistry, page 92) made it important to develop suitable nomenclature for chiral molecules. Two systems are in common use today: the so-called D,L system and the (*R,S*) system.

In the **D,L system** of nomenclature, the (+) and (−) isomers of glyceraldehyde are denoted as **D-glyceraldehyde** and **L-glyceraldehyde,** respectively (Figure 4.13). Absolute configurations of all other carbon-based molecules are referenced to D- and L-glyceraldehyde. When sufficient care is taken to avoid racemization of the amino acids during hydrolysis of proteins, it is found that all of the amino acids derived from natural proteins are of the L-configuration. Amino acids of the D-configuration are nonetheless found in nature, especially as components of certain peptide antibiotics, such as valinomycin, gramicidin, and actinomycin D, and in the cell walls of certain microorganisms.

Despite its widespread acceptance, problems exist with the D,L system of nomenclature. For example, this system can be ambiguous for molecules with two or more chiral centers. To address such problems, the (*R,S*) **system** of nomenclature for chiral molecules was proposed in 1956 by Robert Cahn, Sir Christopher Ingold, and Vladimir Prelog. In this more versatile system, priorities are assigned to each of the groups attached to a chiral center on the basis of atomic number, atoms with higher atomic numbers having higher priorities (see the Critical Developments in Biochemistry, page 94).

The newer (*R,S*) system of nomenclature is superior to the older D,L system in one important way: The configuration of molecules with more than one chiral center can be more easily, completely, and unambiguously described with (*R,S*) notation. Several amino acids, including isoleucine, threonine, hydroxyproline, and hydroxylysine, have two chiral centers. In the (*R,S*) system, L-threonine is (2*S*,3*R*)-threonine. A chemical compound with n chiral centers can exist in 2^n-isomeric structures, and the four amino acids just listed can thus each take on four different isomeric configurations. This amounts to two pairs of enantiomers. Isomers that differ in configuration at only one of the asymmetric centers are non–mirror-image isomers, or **diastereomers.** The four stereoisomers of isoleucine are shown in Figure 4.14. The isomer obtained from digests of natural proteins is arbitrarily designated L-isoleucine. In the (*R,S*) system, L-isoleucine is (2*S*,3*S*)-isoleucine. Its diastereomer is referred to as L-alloisoleucine. The D-enantiomeric pair of isomers is named in a similar manner.

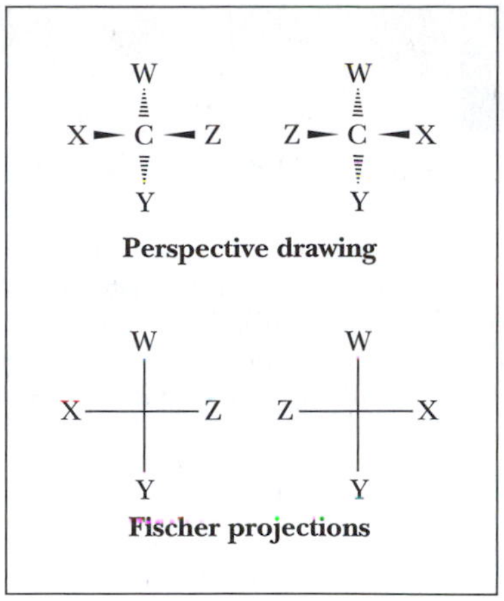

Biochemistry ⧂Now™ ANIMATED FIGURE 4.12
Enantiomeric molecules based on a chiral carbon atom. Enantiomers are nonsuperimposable mirror images of each other. **See this figure animated at http://chemistry.brookscole.com/ggb3**

4.5	**What Are the Spectroscopic Properties of Amino Acids?**

One of the most important and exciting advances in modern biochemistry has been the application of **spectroscopic methods,** which measure the absorption and emission of energy of different frequencies by molecules and atoms. Spectroscopic studies of proteins, nucleic acids, and other biomolecules are providing many new insights into the structure and dynamic processes in these molecules.

Phenylalanine, Tyrosine, and Tryptophan Absorb Ultraviolet Light

Many details of the structure and chemistry of the amino acids have been elucidated or at least confirmed by spectroscopic measurements. None of the amino acids absorbs light in the visible region of the electromagnetic spectrum.

Table 4.2

Specific Rotations for Some Amino Acids

Amino Acid	Specific Rotation $[\alpha]_D^{25}$, Degrees
L-Alanine	+1.8
L-Arginine	+12.5
L-Aspartic acid	+5.0
L-Glutamic acid	+12.0
L-Histidine	−38.5
L-Isoleucine	+12.4
L-Leucine	−11.0
L-Lysine	+13.5
L-Methionine	−10.0
L-Phenylalanine	−34.5
L-Proline	−86.2
L-Serine	−7.5
L-Threonine	−28.5
L-Tryptophan	−33.7
L-Valine	+5.6

Critical Developments in Biochemistry

Biochemistry ⊗ Now™

Discovery of Optically Active Molecules and Determination of Absolute Configuration

The optical activity of quartz and certain other materials was first discovered by Jean-Baptiste Biot in 1815 in France, and in 1848 a young chemist in Paris named Louis Pasteur made a related and remarkable discovery. Pasteur noticed that preparations of optically inactive sodium ammonium tartrate contained two visibly different kinds of crystals that were mirror images of each other. Pasteur carefully separated the two types of crystals, dissolved them each in water, and found that each solution was *optically active*. Even more intriguing, the specific rotations of these two solutions were equal in magnitude and of opposite sign. Because these differences in optical rotation were apparent properties of the dissolved molecules, Pasteur eventually proposed that the molecules themselves were mirror images of each other, just like their respective crystals. Based on this and other related evidence, van't Hoff and LeBel proposed the tetrahedral arrangement of valence bonds to carbon.

In 1888, Emil Fischer decided that it should be possible to determine the *relative* configuration of (+)-glucose, a six-carbon sugar with four asymmetric centers (see figure). Because each of the four C could be either of two configurations, glucose conceivably could exist in any one of 16 possible isomeric structures. It took 3 years to complete the solution of an elaborate chemical and logical puzzle. By 1891, Fischer had reduced his puzzle to a choice between two enantiomeric structures. (Methods for determining *absolute* configuration were not yet available, so Fischer made a simple guess, selecting the structure shown in the figure.) For this remarkable feat, Fischer received the Nobel Prize in Chemistry in 1902.

The absolute choice between Fischer's two enantiomeric possibilities would not be made for a long time. In 1951, J. M. Bijvoet in Utrecht, the Netherlands, used a new X-ray diffraction technique to determine the absolute configuration of (among other things) the sodium rubidium salt of (+)-tartaric acid. Because the tartaric acid configuration could be related to that of glyceraldehyde and because sugar and amino acid configurations could all be related to glyceraldehyde, it became possible to determine the absolute configuration of sugars and the common amino acids. The absolute configuration of tartaric acid determined by Bijvoet turned out to be the configuration that, up to then, had only been assumed. This meant that Emil Fischer's arbitrary guess 60 years earlier had been correct.

It was M. A. Rosanoff, a chemist and instructor at New York University, who first proposed (in 1906) that the isomers of glyceraldehyde be the standards for denoting the stereochemistry of sugars and other molecules. Later, when experiments showed that the configuration of (+)-glyceraldehyde was related to (+)-glucose, (+)-glyceraldehyde was given the designation D. Emil Fischer rejected the **Rosanoff convention,** but it was universally accepted. Ironically, this nomenclature system is often mistakenly referred to as the **Fischer convention.**

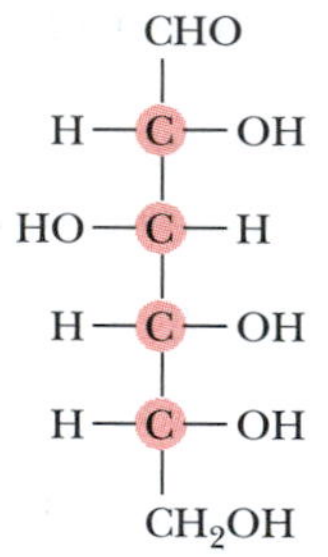

▲ The absolute configuration of (+)-glucose.

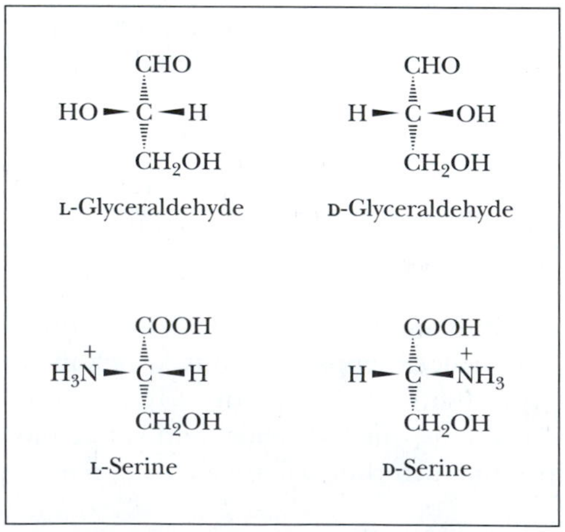

Biochemistry ⊗ Now™ **ANIMATED FIGURE 4.13** The configuration of the common L-amino acids can be related to the configuration of L(−)-glyceraldehyde as shown. These drawings are known as Fischer projections. The horizontal lines of the Fischer projections are meant to indicate bonds coming out of the page from the central carbon, and vertical lines represent bonds extending behind the page from the central carbon atom. **See this figure animated at http://chemistry. brookscole.com/ggb3**

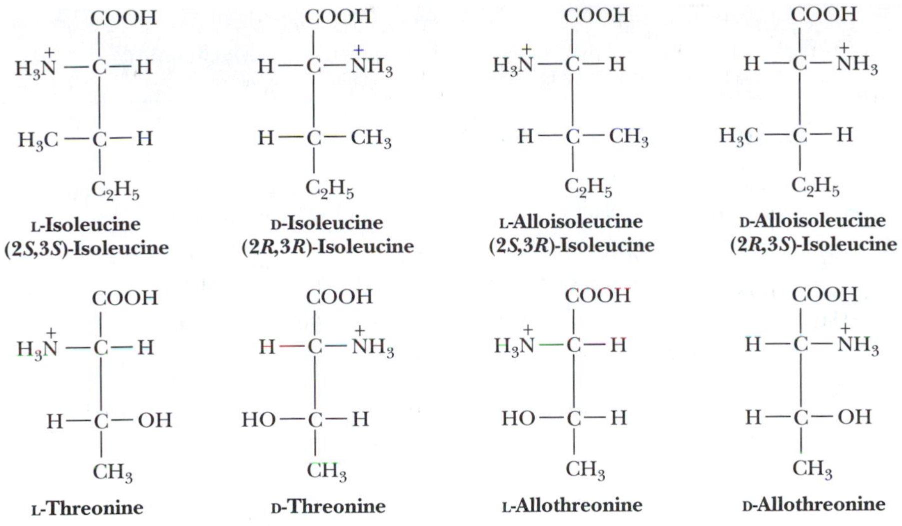

COOH
$H_3\overset{+}{N}$—C—H
H₃C—C—H
C₂H₅
L-Isoleucine
(2S,3S)-Isoleucine

COOH
H—C—$\overset{+}{N}H_3$
H—C—CH₃
C₂H₅
D-Isoleucine
(2R,3R)-Isoleucine

COOH
$H_3\overset{+}{N}$—C—H
H—C—CH₃
C₂H₅
L-Alloisoleucine
(2S,3R)-Isoleucine

COOH
H—C—$\overset{+}{N}H_3$
H₃C—C—H
C₂H₅
D-Alloisoleucine
(2R,3S)-Isoleucine

COOH
$H_3\overset{+}{N}$—C—H
H—C—OH
CH₃
L-Threonine

COOH
H—C—$\overset{+}{N}H_3$
HO—C—H
CH₃
D-Threonine

COOH
$H_3\overset{+}{N}$—C—H
HO—C—H
CH₃
L-Allothreonine

COOH
H—C—$\overset{+}{N}H_3$
H—C—OH
CH₃
D-Allothreonine

Biochemistry⊜Now™   ANIMATED FIGURE 4.14
The stereoisomers of isoleucine and threonine. The structures at the far left are the naturally occurring isomers. **See this figure animated at http://chemistry. brookscole.com/ggb3**

A Deeper Look

The Murchison Meteorite—Discovery of Extraterrestrial Handedness

The predominance of L-amino acids in biological systems is one of life's intriguing features. Prebiotic syntheses of amino acids would be expected to produce equal amounts of L- and D-enantiomers. Some kind of enantiomeric selection process must have intervened to select L-amino acids over their D-counterparts as the constituents of proteins. Was it random chance that chose L- over D-isomers?

Analysis of carbon compounds—even amino acids—from extraterrestrial sources might provide deeper insights into this mystery. John Cronin and Sandra Pizzarello have examined the enantiomeric distribution of unusual amino acids obtained from the Murchison meteorite, which struck the earth on September 28, 1969, near Murchison, Australia. (By selecting unusual amino acids for their studies, Cronin and Pizzarello ensured that they were examining materials that were native to the meteorite and not earth-derived contaminants.) Four α-dialkyl amino acids—α-methylisoleucine, α-methylalloisoleucine, α-methylnorvaline, and isovaline—were found to have an L-enantiomeric excess of 2% to 9%.

This may be the first demonstration that a natural L-enantiomer enrichment occurs in certain cosmological environments. Could these observations be relevant to the emergence of L-enantiomers as the dominant amino acids on the earth? And, if so, could there be life elsewhere in the universe that is based upon the same amino acid handedness?

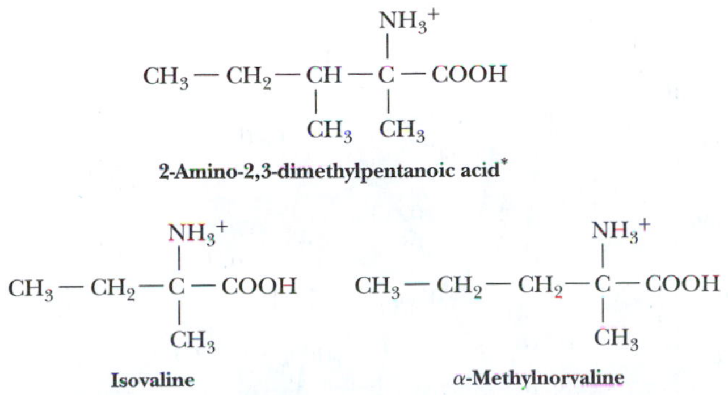

NH_3^+
CH₃—CH₂—CH—C—COOH
CH₃   CH₃
2-Amino-2,3-dimethylpentanoic acid*

NH_3^+
CH₃—CH₂—C—COOH
CH₃
Isovaline

NH_3^+
CH₃—CH₂—CH₂—C—COOH
CH₃
α-Methylnorvaline

◀ Amino acids found in the Murchison meteorite.

*The four stereoisomers of this amino acid include the D- and L-forms of α-methylisoleucine and α-methylalloisoleucine.
Cronin, J. R., and Pizzarello, S., 1997. Enantiomeric excesses in meteoritic amino acids. *Science* **275**:951–955.

Critical Developments in Biochemistry

Biochemistry ⋐ Now™

Rules for Description of Chiral Centers in the (*R,S*) System

Naming a chiral center in the (*R,S*) system is accomplished by viewing the molecule from the chiral center to the atom with the lowest priority. If the other three atoms facing the viewer then decrease in priority in a clockwise direction, the center is said to have the (*R*) configuration (where *R* is from the Latin *rectus*, meaning "right"). If the three atoms in question decrease in priority in a counterclockwise fashion, the chiral center is of the (*S*) configuration (where *S* is from the Latin *sinistrus*, meaning "left"). If two of the atoms coordinated to a chiral center are identical, the atoms bound to these two are considered for pri-

orities. For such purposes, the priorities of certain functional groups found in amino acids and related molecules are in the following order:

$$SH > OH > NH_2 > COOH > CHO > CH_2OH > CH_3$$

From this, it is clear that D-glyceraldehyde is (*R*)-glyceraldehyde and L-alanine is (*S*)-alanine (see figure). Interestingly, the α-carbon configuration of all the L-amino acids *except for cysteine* is (*S*). Cysteine, by virtue of its thiol group, is in fact (*R*)-cysteine.

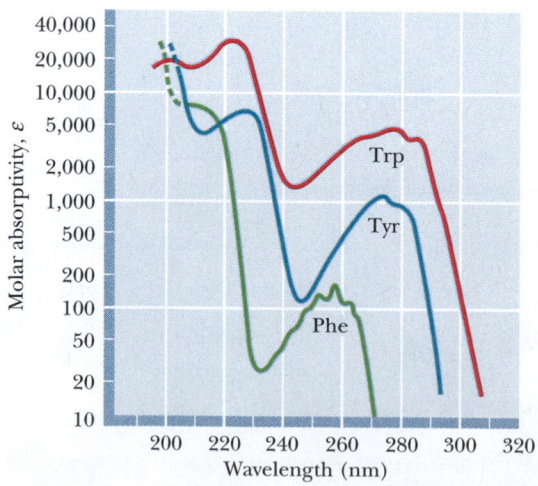

▲ The assignment of (*R*) and (*S*) notation for glyceraldehyde and L-alanine.

Several of the amino acids, however, do absorb **ultraviolet** radiation, and all absorb in the **infrared** region. The absorption of energy by electrons as they rise to higher-energy states occurs in the ultraviolet/visible region of the energy spectrum. Only the aromatic amino acids phenylalanine, tyrosine, and tryptophan exhibit significant ultraviolet absorption above 250 nm, as shown in Figure 4.15. These strong absorptions can be used for spectroscopic determi-

FIGURE 4.15 The ultraviolet absorption spectra of the aromatic amino acids at pH 6. *(From Wetlaufer, D. B., 1962. Ultraviolet spectra of proteins and amino acids. Advances in Protein Chemistry* **17**:303–390.)

nations of protein concentration. The aromatic amino acids also exhibit relatively weak fluorescence, and it has recently been shown that tryptophan can exhibit *phosphorescence*—a relatively long-lived emission of light. These fluorescence and phosphorescence properties are especially useful in the study of protein structure and dynamics (see Chapter 6).

Amino Acids Can Be Characterized by Nuclear Magnetic Resonance

The development in the 1950s of **nuclear magnetic resonance** (NMR), a spectroscopic technique that involves the absorption of radio frequency energy by certain nuclei in the presence of a magnetic field, played an important part in the chemical characterization of amino acids and proteins. Several important principles emerged from these studies. First, the **chemical shift**[1] of amino acid protons depends on their particular chemical environment and thus on the state of ionization of the amino acid. Second, the change in electron density during a titration is transmitted throughout the carbon chain in the aliphatic amino acids and the aliphatic portions of aromatic amino acids, as evidenced by changes in the chemical shifts of relevant protons. Finally, the magnitude of the **coupling constants** between protons on adjacent carbons depends in some cases on the ionization state of the amino acid. This apparently reflects differences in the preferred conformations in different ionization states. Proton NMR spectra of two amino acids are shown in Figure 4.16. Because they are highly sensitive to their environment, the chemical shifts of individual NMR signals can detect the pH-dependent ionizations of amino acids. Figure 4.17 shows the ^{13}C chemical shifts occurring in a titration of lysine. Note that the chemical shifts of the carboxyl C, C_α, and C_β carbons of lysine are sensitive to dissociation of the nearby α-COOH and α-NH$_3^+$ protons (with pK_a values of about 2 and 9, respectively), whereas the C_δ and C_ϵ carbons are sensitive to dissociation of the ϵ-NH$_3^+$ group. Such measurements have been very useful for studies of the ionization behavior of amino acid residues in proteins. More sophisticated NMR measurements at very high magnetic fields are also used to determine the three-dimensional structures of peptides and proteins.

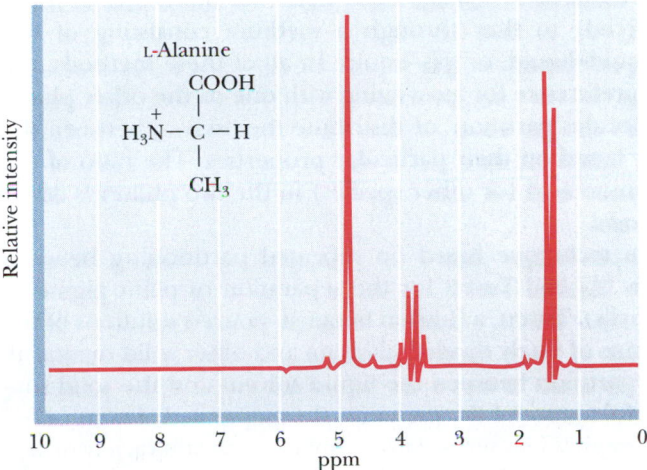
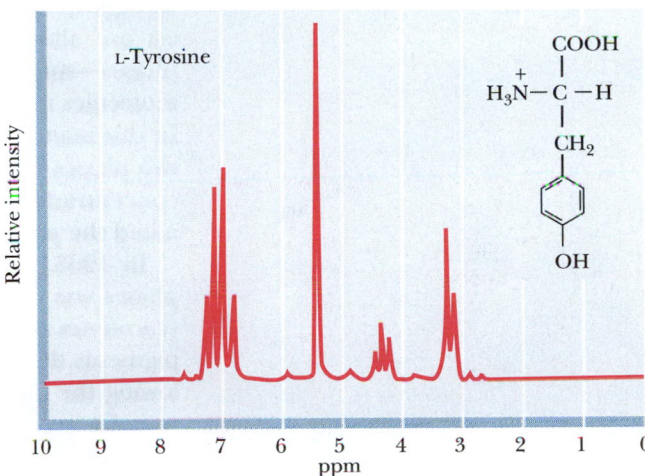

FIGURE 4.16 Proton NMR spectra of several amino acids. Zero on the chemical shift scale is defined by the resonance of tetramethylsilane (TMS). *(Adapted from Aldrich Library of NMR Spectra.)*

[1]The chemical shift for any NMR signal is the difference in resonant frequency between the observed signal and a suitable reference signal. If two nuclei are magnetically coupled, the NMR signals of these nuclei split, and the separation between such split signals, known as the coupling constant, is likewise dependent on the structural relationship between the two nuclei.

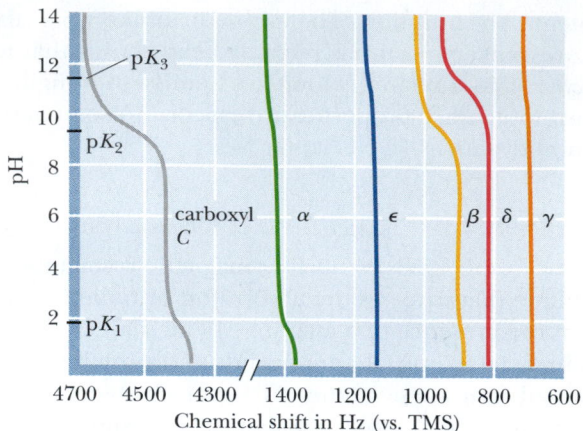

FIGURE 4.17 A plot of chemical shifts versus pH for the carbons of lysine. Changes in chemical shift are most pronounced for atoms near the titrating groups. Note the correspondence between the pK_a values and the particular chemical shift changes. All chemical shifts are defined relative to tetramethylsilane (TMS). *(From Suprenant, H., et al., 1980. Carbon-13 NMR studies of amino acids: Chemical shifts, protonation shifts, microscopic protonation behavior. Journal of Magnetic Resonance **40**:231–243.)*

4.6 | How Are Amino Acid Mixtures Separated and Analyzed?

Amino Acids Can Be Separated by Chromatography

The purification and analysis of individual amino acids from complex mixtures was once a very difficult process. Today, however, the biochemist has a wide variety of methods available for the separation and analysis of amino acids or, for that matter, any of the other biological molecules and macromolecules we encounter. All of these methods take advantage of the relative differences in the physical and chemical characteristics of amino acids, particularly ionization behavior and solubility characteristics. The methods important for amino acids include separations based on **partition** properties (the tendency to associate with one solvent or phase over another) and separations based on **electrical charge.** In all of the partition methods discussed here, the molecules of interest are allowed (or forced) to flow through a medium consisting of two phases—solid–liquid, liquid–liquid, or gas–liquid. In all of these methods, the molecules must show a preference for associating with one or the other phase. In this manner, the molecules partition, or distribute themselves, between the two phases in a manner based on their particular properties. The ratio of the concentrations of the amino acid (or other species) in the two phases is designated the *partition coefficient.*

In 1903, a separation technique based on repeated partitioning between phases was developed by Mikhail Tswett for the separation of plant pigments (carotenes and chlorophylls). Tswett, a Russian botanist, poured solutions of the pigments through columns of finely divided alumina and other solid media, allowing the pigments to partition between the liquid solvent and the solid support. Owing to the colorful nature of the pigments thus separated, Tswett called his technique **chromatography.** This term is now applied to a wide variety of separation methods, regardless of whether the products are colored. The success of all chromatography techniques depends on the repeated microscopic partitioning of a solute mixture between the available phases. The more frequently this partitioning can be made to occur within a given time span or over a given volume, the more efficient is the resulting separation. Chromatographic methods have advanced rapidly in recent years, due in part to the development of sophisticated new solid-phase materials. Methods important for amino acid separations include ion exchange chromatography, gas chromatography (GC), and high-performance liquid chromatography (HPLC).

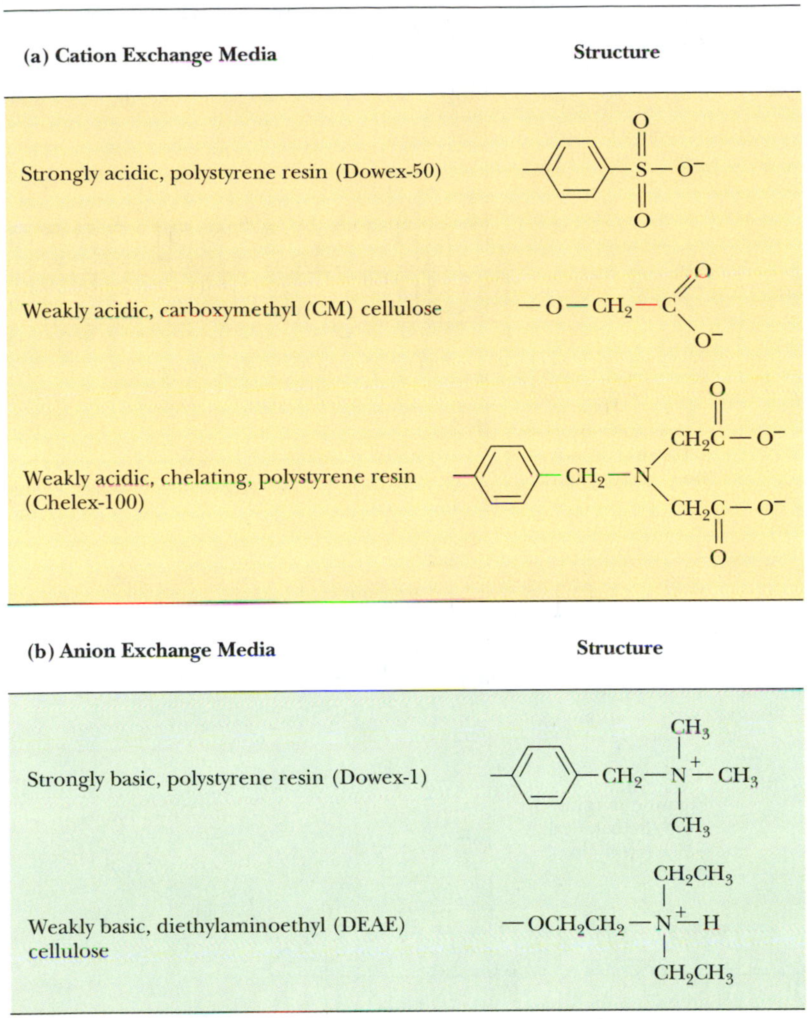

(a) Cation Exchange Media   **Structure**

Strongly acidic, polystyrene resin (Dowex-50)	
Weakly acidic, carboxymethyl (CM) cellulose	
Weakly acidic, chelating, polystyrene resin (Chelex-100)	

(b) Anion Exchange Media   **Structure**

Strongly basic, polystyrene resin (Dowex-1)	
Weakly basic, diethylaminoethyl (DEAE) cellulose	

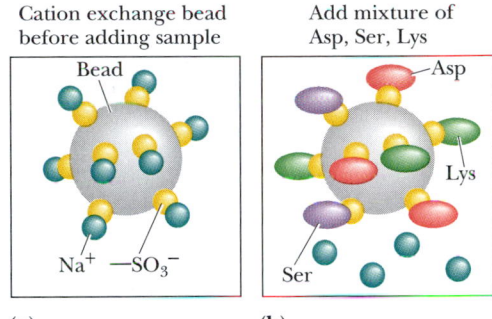

Cation exchange bead before adding sample

Add mixture of Asp, Ser, Lys

(a)   **(b)**

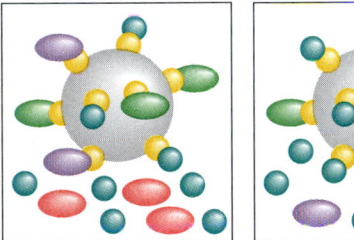
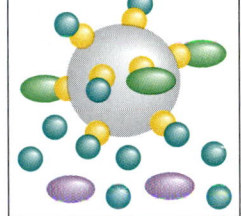

Add Na⁺ (NaCl)   Increase [Na⁺]

(c) Asp, the least positively charged amino acid, is eluted first

(d) Serine is eluted next

Increase [Na⁺]

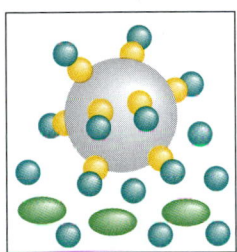

(e) Lysine, the most positively charged amino acid, is eluted last

Ion Exchange Chromatography Separates Amino Acids on the Basis of Charge

The separation of amino acids and other solutes is often achieved by means of **ion exchange chromatography,** in which the molecule of interest is *exchanged* for another ion onto and off of a charged solid support. In a typical procedure, solutes in a liquid phase, usually water, are passed through columns filled with a porous solid phase, usually a bed of synthetic resin particles, containing charged groups. Resins containing positive charges attract negatively charged solutes and are referred to as *anion exchangers*. Solid supports possessing negative charges attract positively charged species and are referred to as *cation exchangers*. Several typical cation and anion exchange resins with different types of charged groups are shown in Figure 4.18. The strength of the acidity or basicity of these groups and their number per unit volume of resin determine the type and strength of binding of an exchanger. Fully ionized acidic groups such as sulfonic acids result in an exchanger with a negative charge, which binds cations very strongly. Weakly acidic or basic groups yield resins whose charge (and binding capacity) depends on the pH of the eluting solvent. The choice of the appropriate resin depends on the strength of binding desired. The bare charges on such solid phases must be counterbalanced by oppositely charged ions in solution ("counterions"). Washing a cation exchange resin, such as Dowex-50, which has strongly acidic phenyl-SO_3^- groups, with a NaCl solution results in the formation of the so-called sodium form of the resin (Figure 4.19). When the mixture whose separation

▲ **Biochemistry ⑤ Now™   ANIMATED FIGURE 4.19**
Operation of a cation exchange column, separating a mixture of Asp, Ser, and Lys. (**a**) The cation exchange resin in the beginning, Na⁺ form. (**b**) A mixture of Asp, Ser, and Lys is added to the column containing the resin. (**c**) A gradient of the eluting salt (for example, NaCl) is added to the column. Asp, the least positively charged amino acid, is eluted first. (**d**) As the salt concentration increases, Ser is eluted. (**e**) As the salt concentration is increased further, Lys, the most positively charged of the three amino acids, is eluted last. See this figure animated at http://chemistry. brookscole.com/ggb3

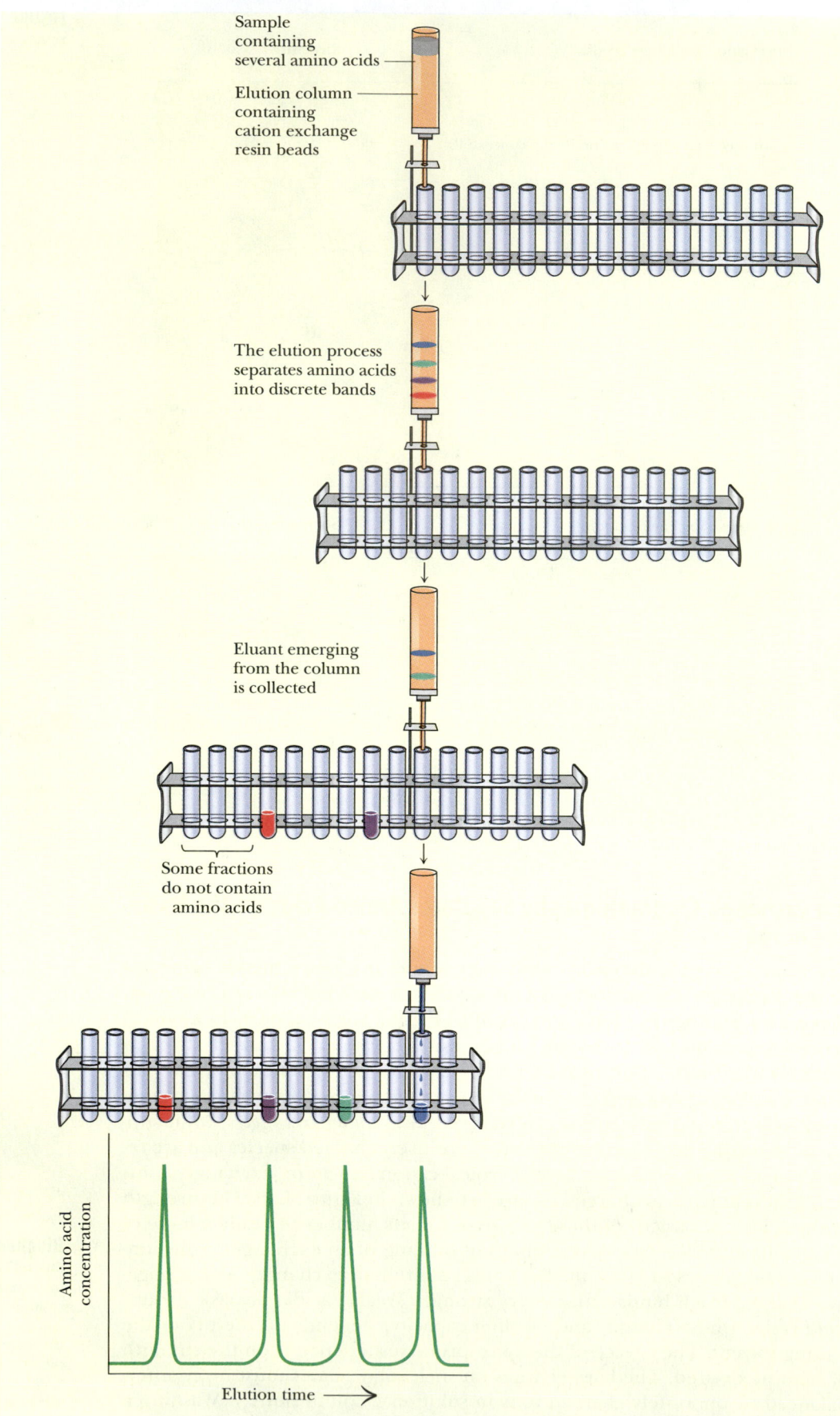

Sample containing several amino acids

Elution column containing cation exchange resin beads

The elution process separates amino acids into discrete bands

Eluant emerging from the column is collected

Some fractions do not contain amino acids

Amino acid concentration

Elution time ⟶

Biochemistry Ⓔ **Now** ™

ACTIVE FIGURE 4.20 The separation of amino acids on a cation exchange column. **Test yourself on the concepts in this figure at http://chemistry.brookscole.com/ggb3**

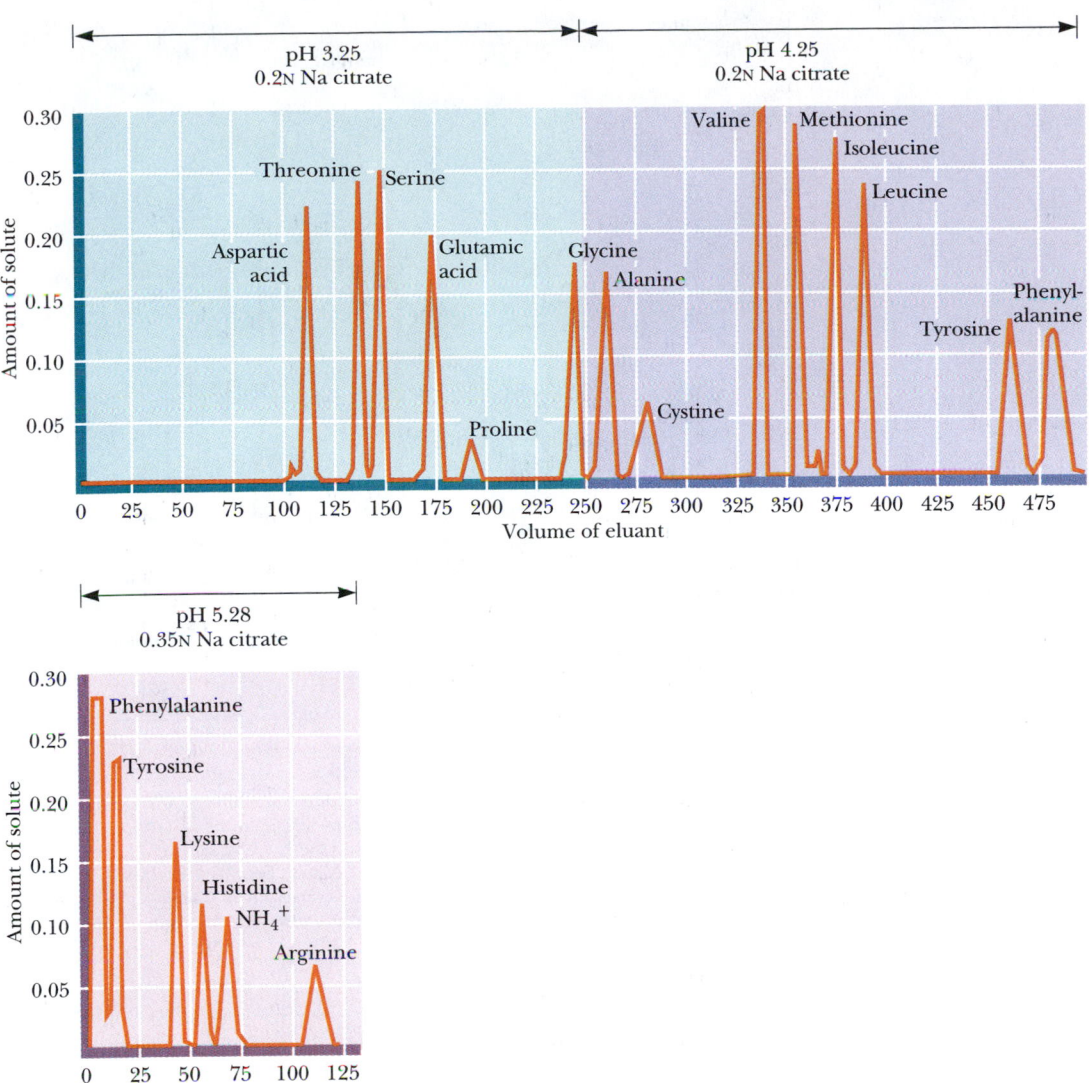

FIGURE 4.21 Chromatographic fractionation of a synthetic mixture of amino acids on ion exchange columns using Amberlite IR-120, a sulfonated polystyrene resin similar to Dowex-50. A second column with different buffer conditions is used to resolve the basic amino acids. *(Adapted from Moore, S., Spackman, D., and Stein, W., 1958. Chromatography of amino acids on sulfonated polystyrene resins. Analytical Chemistry **30**:1185–1190.)*

is desired is added to the column, the positively charged solute molecules displace the Na^+ ions and bind to the resin. A gradient of an appropriate salt is then applied to the column, and the solute molecules are competitively (and sequentially) displaced (eluted) from the column by the rising concentration of cations in the gradient, in an order that is inversely related to their affinities for the column. The separation of a mixture of amino acids on such a column is shown in Figures 4.19 and 4.20. Figure 4.21, taken from a now-classic 1958 paper by Stanford Moore, Darrel Spackman, and William Stein, shows a typical separation of the common amino acids. The events occurring in this separation are essentially those depicted in Figures 4.19 and 4.20. The amino acids are applied to the column at low pH (4.25), under which conditions the acidic amino acids (aspartate and glutamate, among others) are weakly bound and the basic amino acids, such as arginine and lysine, are tightly bound. Sodium citrate solutions, at two different concentrations and three

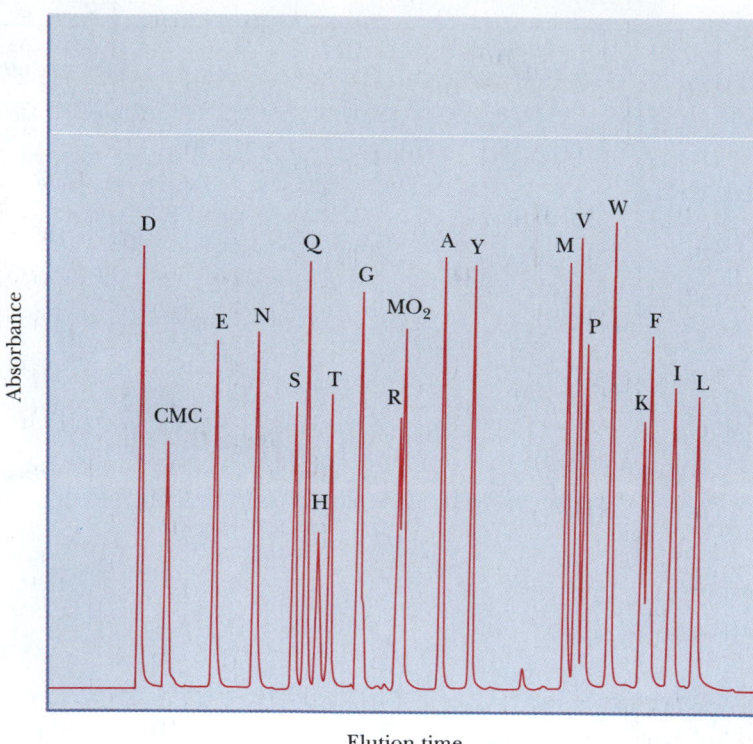

FIGURE 4.22 Gradient separation of common PTH-amino acids, which absorb UV light. Absorbance was monitored at 269 nm. PTH peaks are identified by single-letter notation for amino acid residues and by other abbreviations. D, Asp; CMC, carboxymethyl Cys; E, Glu; N, Asn; S, Ser; Q, Gln; H, His; T, Thr; G, Gly; R, Arg; MO₂, Met sulfoxide; A, Ala; Y, Tyr; M, Met; V, Val; P, Pro; W, Trp; K, Lys; F, Phe; I, Ile; L, Leu. See Figure 5.15 for PTH derivatization. *(Adapted from Persson, B., and Eaker, D., 1990. An optimized procedure for the separation of amino acid phenylthiohydantoins by reversed phase HPLC.* Journal of Biochemical and Biophysical Methods **21:**341-350.)

different values of pH, are used to elute the amino acids gradually from the column.

A typical HPLC chromatogram using precolumn modification of amino acids to form phenylthiohydantoin (PTH) derivatives is shown in Figure 4.22. HPLC is the chromatographic technique of choice for most modern biochemists. The very high resolution, excellent sensitivity, and high speed of this technique usually outweigh the disadvantage of relatively low capacity.

Summary

4.1 What Are the Structures and Properties of Amino Acids, the Building Blocks of Proteins?
The central tetrahedral alpha (α) carbon (C_α) atom of typical amino acids is linked covalently to both the amino group and the carboxyl group. Also bonded to this α-carbon is a hydrogen and a variable side chain. It is the side chain, the so-called R group, that gives each amino acid its identity. In neutral solution (pH 7), the carboxyl group exists as —COO⁻ and the amino group as —NH₃⁺. The amino and carboxyl groups of amino acids can react in a head-to-tail fashion, eliminating a water molecule and forming a covalent amide linkage, which, in the case of peptides and proteins, is typically referred to as a peptide bond. Amino acids are also chiral molecules. With four different groups attached to it, the α-carbon is said to be asymmetric. The two possible configurations for the α-carbon constitute nonidentical mirror-image isomers or enantiomers. The structures of the 20 common amino acids are grouped into the following categories: (1) nonpolar or hydrophobic amino acids, (2) neutral (uncharged) but polar amino acids, (3) acidic amino acids (which have a net negative charge at pH 7.0), and (4) basic amino acids (which have a net positive charge at neutral pH).

4.2 What Are the Acid–Base Properties of Amino Acids?
The common amino acids are all weak polyprotic acids. The ionizable groups are not strongly dissociating ones, and the degree of dissociation thus depends on the pH of the medium. All the amino acids contain at least two dissociable hydrogens. The side chains of several of the amino acids also contain dissociable groups. Thus, aspartic and glutamic acids contain an additional carboxyl function, and lysine possesses an aliphatic amino function. Histidine contains an ionizable imidazolium proton, and arginine carries a guanidinium function.

4.3 What Reactions Do Amino Acids Undergo?

The α-carboxyl and α-amino groups of all amino acids exhibit similar chemical reactivity. The side chains, however, exhibit specific chemical reactivities, depending on the nature of the functional groups. Whereas all of these reactivities are important in the study and analysis of isolated amino acids, it is the characteristic behavior of the side chain that governs the reactivity of amino acids incorporated into proteins. Cysteine residues in proteins, for example, react with one another to form disulfide species, and they also react effectively with iodoacetic acid to yield S-carboxymethyl cysteine derivatives. There are numerous other reactions involving specialized reagents specific for particular side-chain functional groups. It is important to realize that few, if any, of these reactions are truly specific for one functional group; consequently, care must be exercised in their use.

4.4 What Are the Optical and Stereochemical Properties of Amino Acids?

Except for glycine, all of the amino acids isolated from proteins are said to be asymmetric or chiral (from the Greek *cheir*, meaning "hand"), and the two possible configurations for the α-carbon constitute nonsuperimposable mirror-image isomers, or enantiomers. Enantiomeric molecules display a special property called optical activity—the ability to rotate the plane of polarization of plane-polarized light. The magnitude and direction of the optical rotation depend on the nature of the amino acid side chain.

4.5 What Are the Spectroscopic Properties of Amino Acids?

Many details of the structure and chemistry of the amino acids have been elucidated or at least confirmed by spectroscopic measurements. None of the amino acids absorbs light in the visible region of the electromagnetic spectrum. Several of the amino acids, however, do absorb ultraviolet radiation, and all absorb in the infrared region. Proton NMR spectra of amino acids are highly sensitive to their environment, and the chemical shifts of individual NMR signals can detect the pH-dependent ionizations of amino acids.

4.6 How Are Amino Acid Mixtures Separated and Analyzed?

Separation can be achieved on the basis of the relative differences in the physical and chemical characteristics of amino acids, particularly ionization behavior and solubility characteristics. The methods important for amino acids include separations based on partition properties and separations based on electrical charge. The separation of amino acids and other solutes is often achieved by means of ion exchange chromatography, in which the molecule of interest is exchanged for another ion onto and off of a charged solid support. HPLC is the chromatographic technique of choice for most modern biochemists. The very high resolution, excellent sensitivity, and high speed of this technique usually outweigh the disadvantage of relatively low capacity.

Problems

1. Without consulting chapter figures, draw Fischer projection formulas for glycine, aspartate, leucine, isoleucine, methionine, and threonine.

2. Without reference to the text, give the one-letter and three-letter abbreviations for asparagine, arginine, cysteine, lysine, proline, tyrosine, and tryptophan.

3. Write equations for the ionic dissociations of alanine, glutamate, histidine, lysine, and phenylalanine.

4. How is the pK_a of the α-NH_3^+ group affected by the presence on an amino acid of the α-COO^-?

5. (Integrates with Chapter 2.) Draw an appropriate titration curve for aspartic acid, labeling the axes and indicating the equivalence points and the pK_a values.

6. (Integrates with Chapter 2.) Calculate the concentrations of all ionic species in a 0.25 M solution of histidine at pH 2, pH 6.4, and pH 9.3.

7. (Integrates with Chapter 2.) Calculate the pH at which the γ-carboxyl group of glutamic acid is two-thirds dissociated.

8. (Integrates with Chapter 2.) Calculate the pH at which the ϵ-amino group of lysine is 20% dissociated.

9. (Integrates with Chapter 2.) Calculate the pH of a 0.3 M solution of (a) leucine hydrochloride, (b) sodium leucinate, and (c) isoelectric leucine.

10. Quantitative measurements of optical activity are usually expressed in terms of the specific rotation, $[\alpha]_D^{25}$, defined as

$$[\alpha]_D^{25} = \frac{\text{Measured rotation in degrees} \times 100}{(\text{Optical path in dm}) \times (\text{conc. in g/mL})}$$

For any measurement of optical rotation, the wavelength of the light used and the temperature must both be specified. In this case, D refers to the "D line" of sodium at 589 nm and 25 refers to a measurement temperature of 25°C. Calculate the concentration of a solution of L-arginine that rotates the incident light by 0.35° in an optical path length of 1 dm (decimeter).

11. Absolute configurations of the amino acids are referenced to D- and L-glyceraldehyde on the basis of chemical transformations that can convert the molecule of interest to either of these reference isomeric structures. In such reactions, the stereochemical consequences for the asymmetric centers must be understood for each reaction step. Propose a sequence of reactions that would demonstrate that L(−)-serine is stereochemically related to L(−)-glyceraldehyde.

12. Describe the stereochemical aspects of the structure of cystine, the structure that is a disulfide-linked pair of cysteines.

13. Draw a simple mechanism for the reaction of a cysteine sulfhydryl group with iodoacetamide.

Preparing for the MCAT Exam

14. Describe the expected elution pattern for a mixture of aspartate, histidine, isoleucine, valine, and arginine on a column of Dowex-50.

15. Assign (R,S) nomenclature to the threonine isomers of Figure 4.14.

Biochemistry⊗Now™ Preparing for an exam? Test yourself on key questions at http://chemistry.brookscole.com/ggb3

Further Reading

General Amino Acid Chemistry

Barker, R., 1971. *Organic Chemistry of Biological Compounds*, Chap. 4. Englewood Cliffs, NJ: Prentice Hall.

Barrett, G. C., ed., 1985. *Chemistry and Biochemistry of the Amino Acids.* New York: Chapman and Hall.

Greenstein, J. P., and Winitz, M., 1961. *Chemistry of the Amino Acids.* New York: John Wiley & Sons.

Herod, D. W., and Menzel, E. R., 1982. Laser detection of latent fingerprints: Ninhydrin. *Journal of Forensic Science* 27:200–204.

Meister, A., 1965. *Biochemistry of the Amino Acids,* 2nd ed., Vol. 1. New York: Academic Press.

Segel, I. H., 1976. *Biochemical Calculations,* 2nd ed. New York: John Wiley & Sons.

Optical and Stereochemical Properties

Cahn, R. S., 1964. An introduction to the sequence rule. *Journal of Chemical Education* 41:116–125.

Iizuka, E., and Yang, J. T., 1964. Optical rotatory dispersion of L-amino acids in acid solution. *Biochemistry* 3:1519–1524.

Kauffman, G. B., and Priebe, P. M., 1990. The Emil Fischer-William Ramsey friendship. *Journal of Chemical Education* 67:93–101.

Spectroscopic Methods

Bovey, F. A., and Tiers, G. V. D., 1959. Proton N.S.R. spectroscopy. V. Studies of amino acids and peptides in trifluoroacetic acid. *Journal of the American Chemical Society* 81:2870–2878.

Roberts, G. C. K., and Jardetzky, O., 1970. Nuclear magnetic resonance spectroscopy of amino acids, peptides and proteins. *Advances in Protein Chemistry* 24:447–545.

Suprenant, H. L., Sarneski, J. E., Key, R. R., Byrd, J. T., and Reilley, C. N., 1980. Carbon-13 NMR studies of amino acids: Chemical shifts, protonation shifts, microscopic protonation behavior. *Journal of Magnetic Resonance* 40:231–243.

Separation Methods

Heiser, T., 1990. Amino acid chromatography: The "best" technique for student labs. *Journal of Chemical Education* 67:964–966.

Mabbott, G., 1990. Qualitative amino acid analysis of small peptides by GC/MS. *Journal of Chemical Education* 67:441–445.

Moore, S., Spackman, D., and Stein, W. H., 1958. Chromatography of amino acids on sulfonated polystyrene resins. *Analytical Chemistry* 30:1185–1190.

NMR Spectroscopy

de Groot, H. J., 2000. Solid-state NMR spectroscopy applied to membrane proteins. *Current Opinion in Structural Biology* 10:593–600.

Hinds, M. G., and Norton, R. S., 1997. NMR spectroscopy of peptides and proteins. Practical considerations. *Molecular Biotechnology* 7:315–331.

James, T. L., Dötsch, V., and Schmitz, U., eds., 2001. *Nuclear Magnetic Resonance of Biological Macromolecules.* San Diego: Academic Press.

Krishna, N. R., and Berliner, L. J., eds., 2003. *Protein NMR for the Millennium.* New York: Kluwer Academic/Plenum.

Opella, S. J., Nevzorov, A., Mesleb, M. F., and Marassi, F. M., 2002. Structure determination of membrane proteins by NMR spectroscopy. *Biochemistry and Cell Biology* 80:597–604.

Amino Acid Analysis

Prata C., et al., 2001. Recent advances in amino acid analysis by capillary electrophoresis. *Electrophoresis* 22:4129–4138.

Smith, A. J., 1997. Amino acid analysis. *Methods in Enzymology* **289:** 419–426.

Proteins: Their Primary Structure and Biological Functions

CHAPTER 5

Essential Questions

Proteins are polymers composed of hundreds or even thousands of amino acids linked in series by peptide bonds. *What structural forms do these polypeptide chains assume, how can the sequence of amino acids in a protein be determined, and what are the biological roles played by proteins?*

Proteins are a diverse and abundant class of biomolecules, constituting more than 50% of the dry weight of cells. Their diversity and abundance reflect the central role of proteins in virtually all aspects of cell structure and function. An extraordinary diversity of cellular activity is possible only because of the versatility inherent in proteins, each of which is specifically tailored to its biological role. The pattern by which each is tailored resides within the genetic information of cells, encoded in a specific sequence of nucleotide bases in DNA. Each such segment of encoded information defines a gene, and expression of the gene leads to synthesis of the specific protein encoded by it, endowing the cell with the functions unique to that particular protein. Proteins are the agents of biological function; they are also the expressions of genetic information.

Dale Chihuly, Chartreuse Venetian, 1990/Photo by Roger Schreiber

Although helices may appear as decorative motifs in manmade structures, they are a common structural theme in biological macromolecules—proteins, nucleic acids, and even polysaccharides.

…by small and simple things are great things brought to pass.
ALMA 37.6 *The Book of Mormon*

5.1 | What Is the Fundamental Structural Pattern in Proteins?

Chemically, proteins are unbranched polymers of amino acids linked head to tail, from carboxyl group to amino group, through formation of covalent **peptide bonds,** a type of amide linkage (Figure 5.1).

Peptide bond formation results in the release of H_2O. The peptide "backbone" of a protein consists of the repeated sequence $-N-C_\alpha-C_o-$, where the N represents the amide nitrogen, the C_α is the α-carbon atom of an amino acid in the polymer chain, and the final C_o is the carbonyl carbon of the amino acid, which in turn is linked to the amide N of the next amino acid down the line. The geometry of the peptide backbone is shown in Figure 5.2. Note that the carbonyl oxygen and the amide hydrogen are *trans* to each other in this figure. This conformation is favored energetically because it results in less steric hindrance between nonbonded atoms in neighboring amino acids. Because the α-carbon atom of the amino acid is a chiral center (in all amino acids except glycine), the polypeptide chain is inherently asymmetric. Only L-amino acids are found in proteins.

The Peptide Bond Has Partial Double-Bond Character

The peptide linkage is usually portrayed by a single bond between the carbonyl carbon and the amide nitrogen (Figure 5.3a). Therefore, in principle, rotation may occur about any covalent bond in the polypeptide backbone because all three kinds of bonds ($N-C_\alpha$, $C_\alpha-C_o$, and the C_o-N peptide bond) are single bonds. In this representation, the C_o and N atoms of the peptide grouping are both in planar sp^2 hybridization and the C_o and O atoms are

BiochemistryⓈNow™ Test yourself on these Key Questions at BiochemistryNow at **http://chemistry.brookscole.com/ggb3**

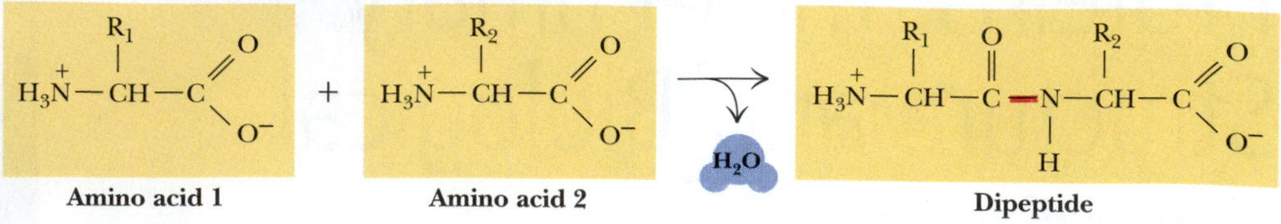

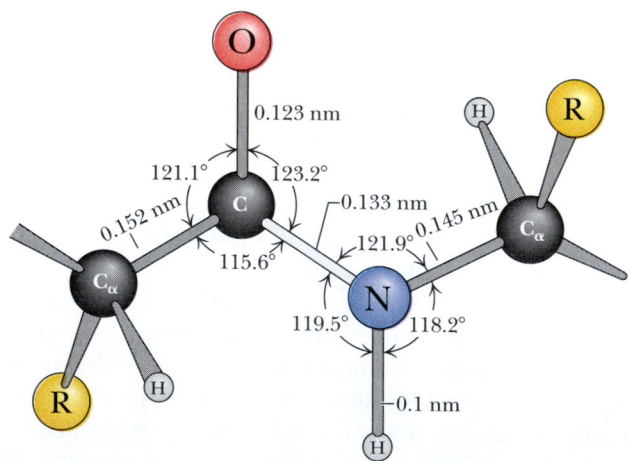

Biochemistry⊗Now™ **ANIMATED FIGURE 5.1** Peptide formation is the creation of an amide bond between the carboxyl group of one amino acid and the amino group of another amino acid. R_1 and R_2 represent the R groups of two different amino acids. **See this figure animated at http://chemistry.brookscole.com/ggb3**

linked by a π bond, leaving the nitrogen with a lone pair of electrons in a $2p$ orbital. However, another resonance form for the peptide bond is feasible in which the C_o and N atoms participate in a π bond, leaving a lone e^- pair on the oxygen (Figure 5.3b). This structure prevents free rotation about the C_o—N peptide bond because it becomes a double bond. The real nature of the peptide bond lies somewhere between these extremes; that is, it has partial double-bond character, as represented by the intermediate form shown in Figure 5.3c.

Peptide bond resonance has several important consequences. First, it restricts free rotation around the peptide bond and leaves the peptide backbone with only two degrees of freedom per amino acid group: rotation around the N—C_α bond and rotation around the C_α—C_o bond.[1] Second, the six atoms composing the peptide bond group tend to be coplanar, forming the so-called **amide plane** of the polypeptide backbone (Figure 5.4). Third, the C_o—N bond length is 0.133 nm, which is shorter than normal C—N bond lengths (for example, the C_α—N bond of 0.145 nm) but longer than typical C=N bonds (0.125 nm). The peptide bond is estimated to have 40% double-bond character.

Biochemistry⊗Now™ **ANIMATED FIGURE 5.2** The peptide bond is shown in its usual *trans* conformation of carbonyl O and amide H. The C_α atoms are the α-carbons of two adjacent amino acids joined in peptide linkage. The dimensions and angles are the average values observed by crystallographic analysis of amino acids and small peptides. The peptide bond is the light gray bond between C and N. (*Adapted from Ramachandran, G. N., et al., 1974. The mean geometry of the peptide unit from crystal structure data. Biochimica Biophysica Acta 359:298–302.*) **See this figure animated at http://chemistry. brookscole.com/ggb3**

[1]The angle of rotation about the N—C_α bond is designated ϕ, phi, whereas the C_α—C_o angle of rotation is designated ψ, psi.

(a)

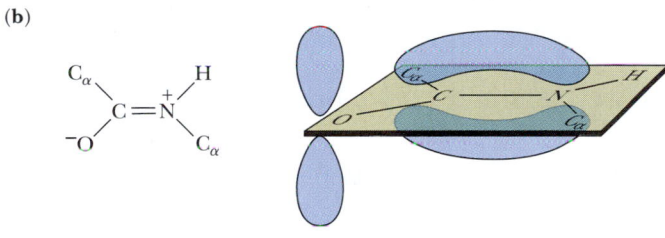

A pure double bond between C
and O would permit free rotation
around the C—N bond.

(b)

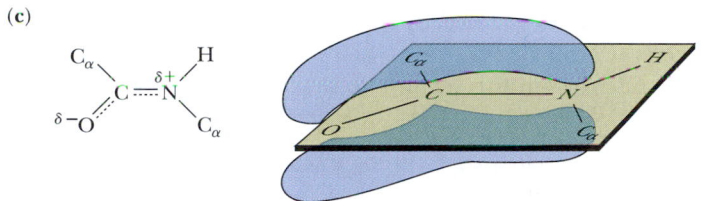

The other extreme would prohibit
C—N bond rotation but would
place too great a charge on
O and N.

(c)

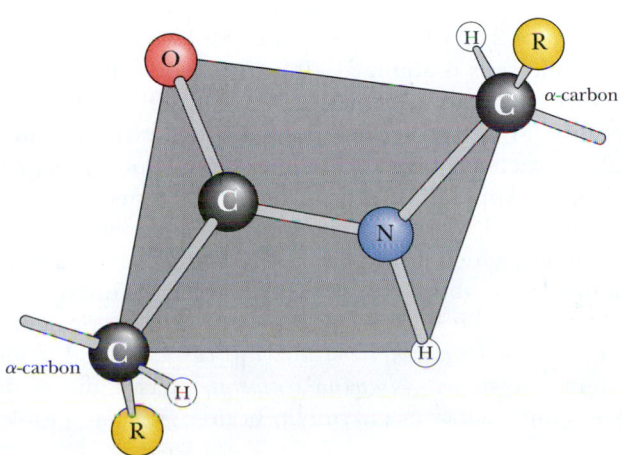

The true electron density is
intermediate. The barrier to
C—N bond rotation of about
88 kJ/mol is enough to
keep the amide group planar.

Biochemistry⊗Now™   ACTIVE FIGURE 5.3 The partial double-bond character of the peptide bond. Resonance interactions among the carbon, oxygen, and nitrogen atoms of the peptide group can be represented by two resonance extremes (**a** and **b**). (**a**) The usual way the peptide atoms are drawn. (**b**) In an equally feasible form, the peptide bond is now a double bond; the amide N bears a positive charge and the carbonyl O has a negative charge. (**c**) The actual peptide bond is best described as a resonance hybrid of the forms in (**a**) and (**b**). Significantly, all of the atoms associated with the peptide group are coplanar, rotation about C_o—N is restricted, and the peptide is distinctly polar. *(Illustration: Irving Geis. Rights owned by Howard Hughes Medical Institute. Not to be reproduced without permission.)* **Test yourself on the concepts in this figure at http://chemistry. brookscole.com/ggb3**

FIGURE 5.4 The coplanar relationship of the atoms in the amide group is highlighted as an imaginary shaded plane lying between two successive α-carbon atoms in the peptide backbone. *(Illustration: Irving Geis. Rights owned by Howard Hughes Medical Institute. Not to be reproduced without permission.)*

The Polypeptide Backbone Is Relatively Polar

Peptide bond resonance also causes the peptide backbone to be relatively polar. As shown in Figure 5.3b, the amide nitrogen is in a protonated or positively charged form, and the carbonyl oxygen is a negatively charged atom in this double-bonded resonance state. In actuality, the hybrid state of the partially double-bonded peptide arrangement gives a net positive charge of 0.28 on the amide N and an equivalent net negative charge of 0.28 on the carbonyl O. The presence of these partial charges means that the peptide bond has a permanent dipole. Nevertheless, the peptide backbone is relatively unreactive chemically, and protons are gained or lost by the peptide groups only at extreme pH conditions.

Peptides Can Be Classified According to How Many Amino Acids They Contain

Peptide is the name assigned to short polymers of amino acids. Peptides are classified according to the number of amino acid units in the chain. Each unit is called an **amino acid residue,** the word *residue* denoting what is left after the release of H_2O when an amino acid forms a peptide link upon joining the peptide chain. **Dipeptides** have two amino acid residues, tripeptides have three, tetrapeptides four, and so on. After about 12 residues, this terminology becomes cumbersome, so peptide chains of more than 12 and less than about 20 amino acid residues are usually referred to as **oligopeptides,** and when the chain exceeds several dozen amino acids in length, the term **polypeptide** is used. The distinctions in this terminology are not precise.

Proteins Are Composed of One or More Polypeptide Chains

The terms *polypeptide* and *protein* are used interchangeably in discussing single polypeptide chains. The term **protein** broadly defines molecules composed of one or more polypeptide chains. Proteins with one polypeptide chain are **monomeric proteins.** Proteins composed of more than one polypeptide chain are **multimeric proteins.** Multimeric proteins may contain only one kind of polypeptide, in which case they are **homomultimeric,** or they may be composed of several different kinds of polypeptide chains, in which instance they are **heteromultimeric.** Greek letters and subscripts are used to denote the polypeptide composition of multimeric proteins. Thus, an α_2-type protein is a dimer of identical polypeptide subunits, or a **homodimer.** Hemoglobin (Table 5.1) consists of four polypeptides of two different kinds; it is an $\alpha_2\beta_2$ heteromultimer.

Polypeptide chains of proteins typically range in length from about 100 amino acids to around 2000, the number found in each of the two polypeptide chains of myosin, the contractile protein of muscle. However, exceptions abound, including human cardiac muscle titin, which has 26,926 amino acid residues and a molecular weight of 2,993,497. The average molecular weight of polypeptide chains in eukaryotic cells is about 31,700, corresponding to about 270 amino acid residues. Table 5.1 is a representative list of proteins according to size. The molecular weights (M_r) of proteins can be estimated by a number of physicochemical methods such as polyacrylamide gel electrophoresis or ultracentrifugation (see Chapter Appendix). Precise determinations of protein molecular masses can be obtained by simple calculations based on knowledge of their amino acid sequence, which is often available in genome databases. No simple generalizations correlate the size of proteins with their functions. For instance, the same function may be fulfilled in different cells by proteins of different molecular weight. The *Escherichia coli* enzyme responsible for glutamine synthesis (a protein known as *glutamine synthetase*) has a molecular weight of 600,000, whereas the analogous enzyme in brain tissue has a molecular weight of 380,000.

Table 5.1

Size of Protein Molecules*

Protein	M_r	Number of Residues per Chain	Subunit Organization
Insulin (bovine)	5,733	21 (A)	$\alpha\beta$
		30 (B)	
Cytochrome c (equine)	12,500	104	α_1
Ribonuclease A (bovine pancreas)	12,640	124	α_1
Lysozyme (egg white)	13,930	129	α_1
Myoglobin (horse)	16,980	153	α_1
Chymotrypsin (bovine pancreas)	22,600	13 (α)	$\alpha\beta\gamma$
		132 (β)	
		97 (γ)	
Hemoglobin (human)	64,500	141 (α)	$\alpha_2\beta_2$
		146 (β)	
Serum albumin (human)	68,500	550	α_1
Hexokinase (yeast)	96,000	200	α_4
γ-Globulin (horse)	149,900	214 (α)	$\alpha_2\beta_2$
		446 (β)	
Glutamate dehydrogenase (liver)	332,694	500	α_6
Myosin (rabbit)	470,000	2,000 (heavy, h)	$h_2\alpha_1\alpha'_2\beta_2$
		190 (α)	
		149 (α')	
		160 (β)	
Ribulose bisphosphate carboxylase (spinach)	560,000	475 (α)	$\alpha_8\beta_8$
		123 (β)	
Glutamine synthetase (E. coli)	600,000	468	α_{12}

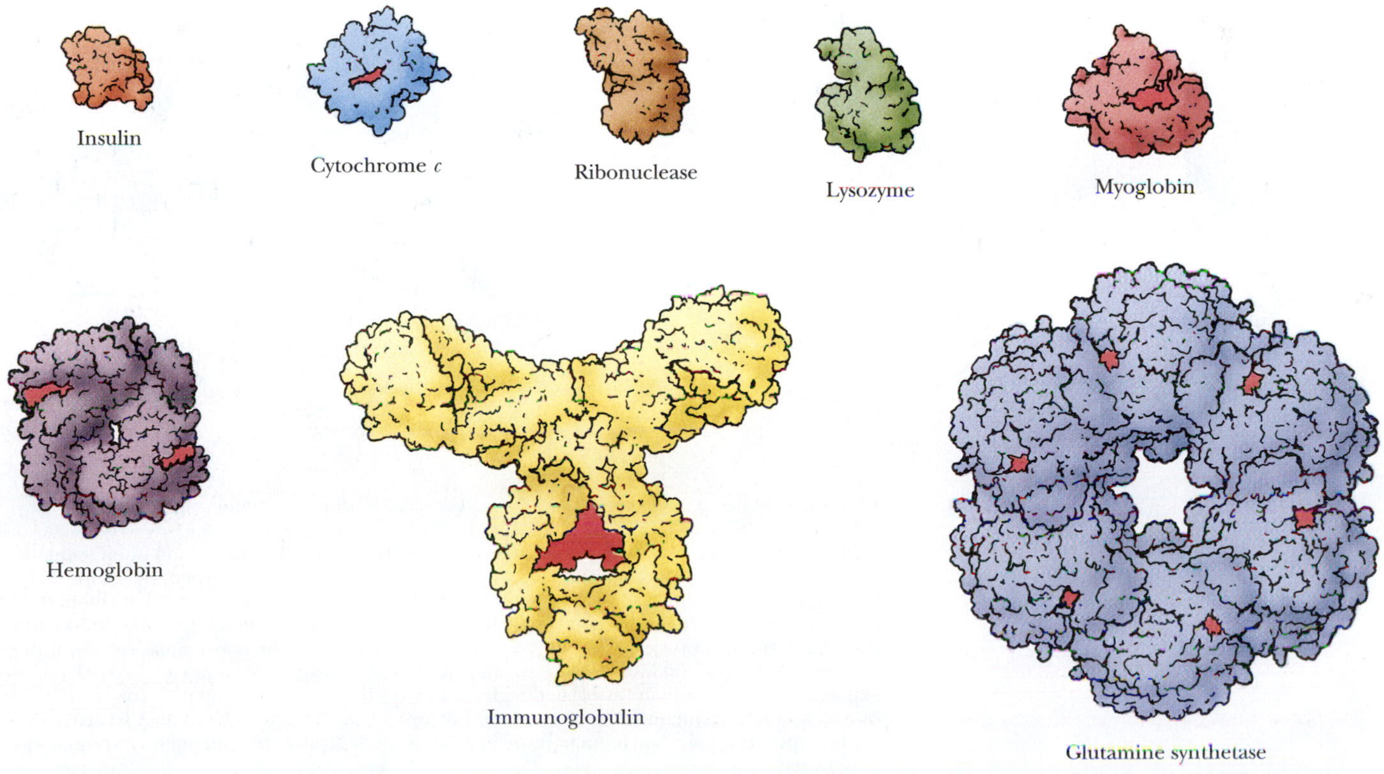

Insulin   Cytochrome c Ribonuclease   Lysozyme   Myoglobin

Hemoglobin   Immunoglobulin   Glutamine synthetase

*Illustrations of selected proteins listed in Table 5.1 are drawn to constant scale.
Adapted from Goodsell, D. S., and Olson, A. J., 1993. Soluble proteins: Size, shape and function. *Trends in Biochemical Sciences* **18**:65–68.

The Chemistry of Peptides and Proteins Is Dictated by the Chemistry of Their Functional Groups

The chemical properties of peptides and proteins are most easily considered in terms of the chemistry of their component functional groups. That is, they possess reactive amino and carboxyl termini, and they display reactions characteristic of the chemistry of the R groups of their component amino acids. These reactions are familiar to us from Chapter 4 and from the study of organic chemistry and need not be repeated here.

5.2	**What Architectural Arrangements Characterize Protein Structure?**

Proteins Fall into Three Basic Classes According to Shape and Solubility

As a first approximation, proteins can be assigned to one of three global classes on the basis of shape and solubility: fibrous, globular, or membrane (Figure 5.5). **Fibrous proteins** tend to have relatively simple, regular linear structures. These proteins often serve structural roles in cells. Typically, they are insoluble in water or in dilute salt solutions. In contrast, **globular proteins** are roughly spherical in shape. The polypeptide chain is compactly folded so that hydrophobic amino acid side chains are in the interior of the molecule and the hydrophilic side chains are on the outside exposed to the solvent, water. Consequently, globular proteins are usually very soluble in aqueous solutions. Most soluble proteins of the cell, such as the cytosolic enzymes, are globular in shape. **Membrane proteins** are found in association with the various membrane systems of cells. For interac-

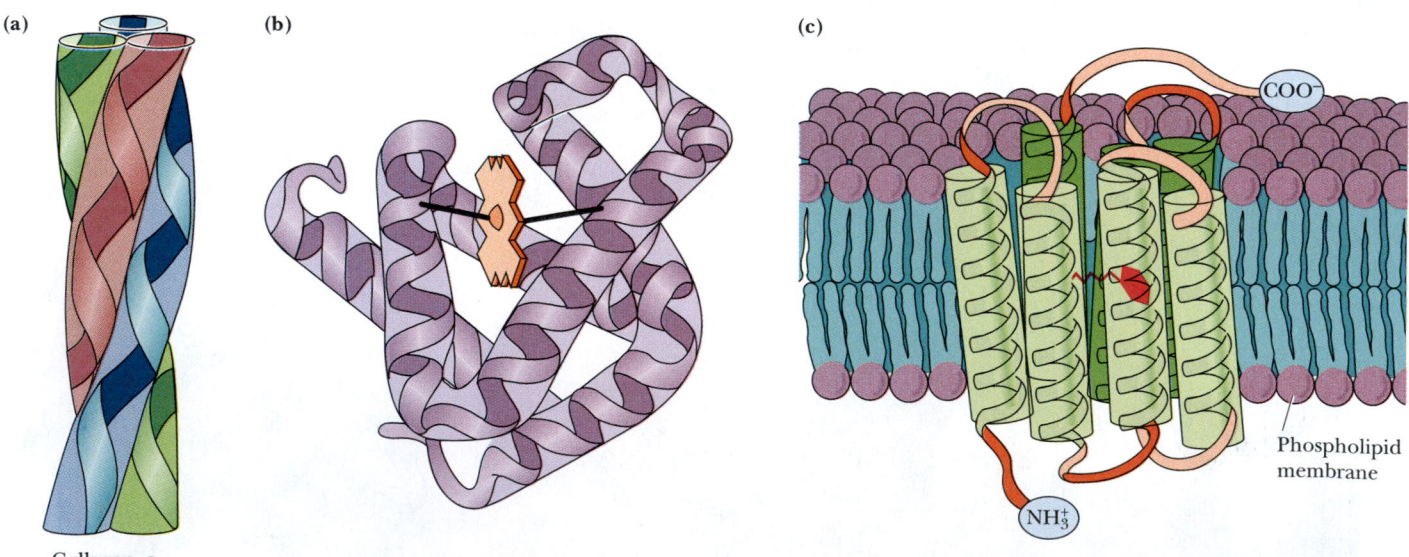

(a)

Collagen, a
fibrous protein

(b)

Myoglobin, a globular protein

(c)

COO⁻

NH₃⁺

Phospholipid
membrane

Bacteriorhodopsin, a membrane protein

FIGURE 5.5 **(a)** Proteins having structural roles in cells are typically fibrous and often water insoluble. Collagen is a good example. Collagen is composed of three polypeptide chains that intertwine. **(b)** Soluble proteins serving metabolic functions can be characterized as compactly folded globular molecules, such as myoglobin. The folding pattern puts hydrophilic amino acid side chains on the outside and buries hydrophobic side chains in the interior, making the protein highly water soluble. **(c)** Membrane proteins fold so that hydrophobic amino acid side chains are exposed in their membrane-associated regions. The portions of membrane proteins extending into or exposed at the aqueous environments are hydrophilic in character, like soluble proteins. Bacteriorhodopsin is a typical membrane protein; it binds the light-absorbing pigment, *cis*-retinal, shown here in red. (*a, b, Illustration: Irving Geis. Rights owned by Howard Hughes Medical Institute. Not to be reproduced without permission.*)

tion with the nonpolar phase within membranes, membrane proteins have hydrophobic amino acid side chains oriented outward. As such, membrane proteins are insoluble in aqueous solutions but can be solubilized in solutions of detergents. Membrane proteins characteristically have fewer hydrophilic amino acids than cytosolic proteins.

Protein Structure Is Described in Terms of Four Levels of Organization

The architecture of protein molecules is quite complex. Nevertheless, this complexity can be resolved by defining various levels of structural organization.

Primary Structure The amino acid sequence is, by definition, the **primary (1°) structure** of a protein, such as that for bovine pancreatic RNase in Figure 5.6, for example.

Secondary Structure Through hydrogen-bonding interactions between adjacent amino acid residues (discussed in detail in Chapter 6), the polypeptide chain can arrange itself into characteristic helical or pleated segments. These segments constitute structural conformities, so-called **regular structures,** which extend along one dimension, like the coils of a spring. Such architectural features of a protein are designated **secondary (2°) structures** (Figure 5.7). Secondary structures are just one of the higher levels of structure that represent the three-dimensional arrangement of the polypeptide in space.

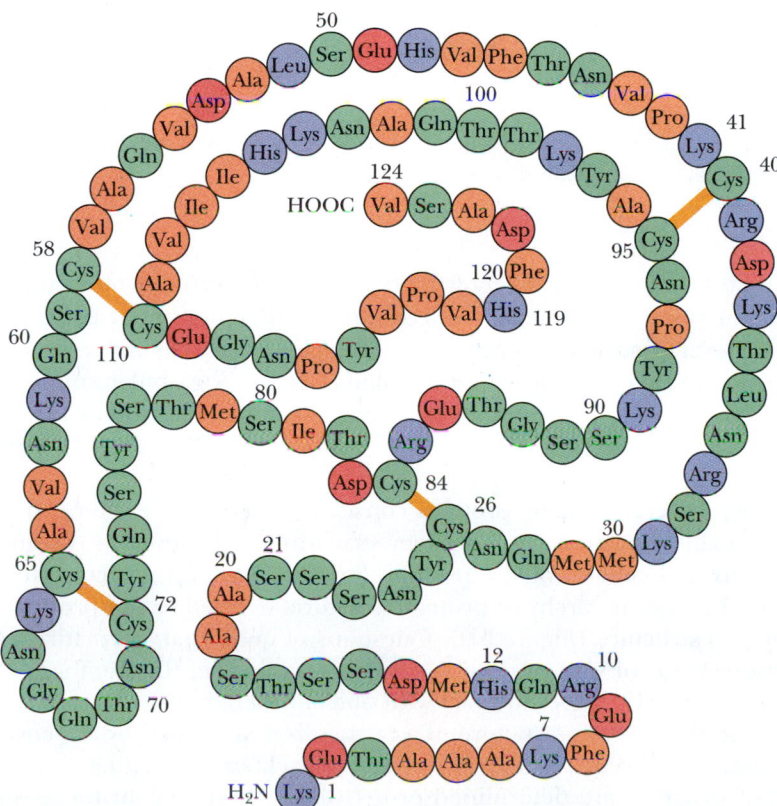

FIGURE 5.6 Bovine pancreatic ribonuclease A contains 124 amino acid residues, none of which are tryptophan. Four intrachain disulfide bridges (S—S) form crosslinks in this polypeptide between Cys[26] and Cys[84], Cys[40] and Cys[95], Cys[58] and Cys[110], and Cys[65] and Cys[72]. These disulfides are depicted by yellow bars.

α-Helix
Only the N—C$_\alpha$—C backbone
is represented. The vertical line
is the helix axis.

β-Strand
The N—C$_\alpha$—C$_O$ backbone as well
as the C$_\beta$ of R groups are represented
here. Note that the amide planes
are perpendicular to the page.

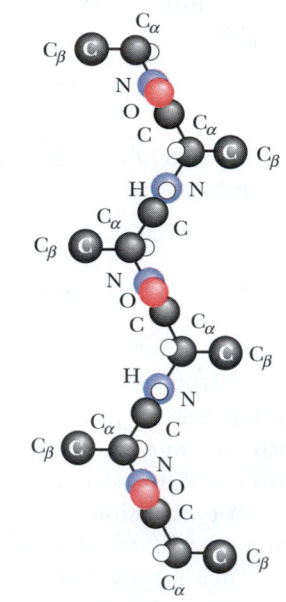

FIGURE 5.7 Two structural motifs that arrange the primary structure of proteins into a higher level of organization predominate in proteins: the α-helix and the β-pleated strand. Atomic representations of these secondary structures are shown here, along with the symbols used by structural chemists to represent them: the flat, helical ribbon for the α-helix and the flat, wide arrow for β-structures. Both of these structures owe their stability to the formation of hydrogen bonds between N—H and O=C functions along the polypeptide backbone (see Chapter 6).

"Shorthand" α-helix

"Shorthand" β-strand

Tertiary Structure When the polypeptide chains of protein molecules bend and fold in order to assume a more compact three-dimensional shape, the **tertiary (3°) level of structure** is generated (Figure 5.8). It is by virtue of their tertiary structure that proteins adopt a globular shape. A globular conformation gives the lowest surface-to-volume ratio, minimizing interaction of the protein with the surrounding environment.

Quaternary Structure Many proteins consist of two or more interacting polypeptide chains of characteristic tertiary structure, each of which is commonly referred to as a **subunit** of the protein. Subunit organization constitutes another level in the hierarchy of protein structure, defined as the protein's **quaternary (4°) structure** (Figure 5.9). Questions of quaternary structure address the various kinds of subunits within a protein molecule, the number of each, and the ways in which they interact with one another.

Whereas the primary structure of a protein is determined by the covalently linked amino acid residues in the polypeptide backbone, secondary and higher orders of structure are determined principally by noncovalent forces such as hydrogen bonds and ionic, van der Waals, and hydrophobic interactions. It is important to emphasize that *all the information necessary for a protein molecule to achieve its intricate architecture is contained within its 1° structure,* that is, within the amino acid sequence of its polypeptide chain(s). Chapter 6 presents a detailed discussion of the 2°, 3°, and 4° structure of protein molecules.

(a)   Chymotrypsin primary structure

H_2N–CGVPAIQPVL$_{10}$SGL[SR]IVNGE$_{20}$EAVPGSWPWQ$_{30}$VSLQDKTGFH$_{40}$GGSLINEN$_{50}$WVVTAAHCGV$_{60}$TTSDVVVAGE$_{70}$FDQGSSSEKI$_{80}$QKLKIA
KVFK$_{90}$NSKYNSLTIN$_{100}$NDITLLKLST$_{110}$AASFSQTVSA$_{120}$VCLPSASDDF$_{130}$AAGTTCVTTG$_{140}$WGLTRY[TN]AN$_{150}$LPSDRLQQASL$_{160}$PLLSNTNCK
K$_{170}$YWGTKIKDAM$_{180}$ICAGASGVSS$_{190}$CMGDSGGPLV$_{200}$CKKNGAWTLV$_{210}$GIVSWGSSTC$_{220}$STSTPGVYAR$_{230}$VTALVNWVQQ$_{240}$TLAAN–COOH

(b)   Chymotrypsin tertiary structure

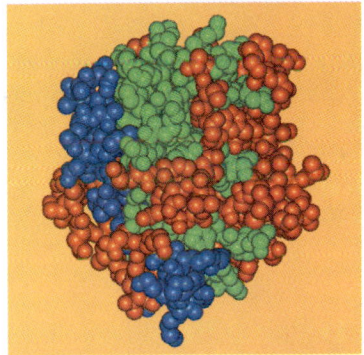

Chymotrypsin space-filling model

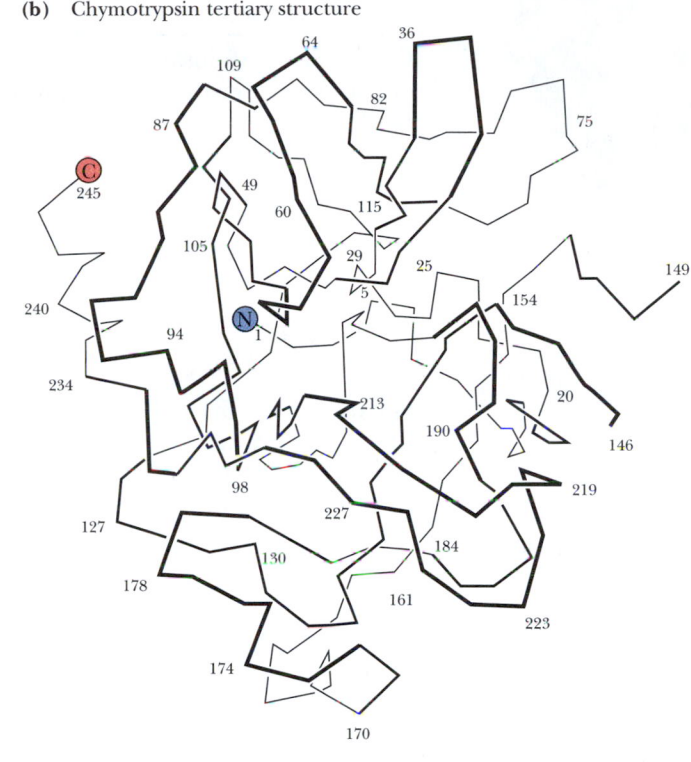

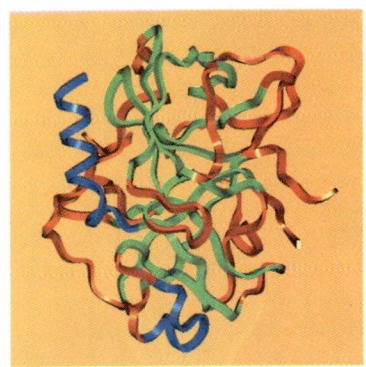

Chymotrypsin ribbon

FIGURE 5.8 Folding of the polypeptide chain into a compact, roughly spherical conformation creates the tertiary level of protein structure. **(a)** The primary structure and **(b)** a representation of the tertiary structure of chymotrypsin, a proteolytic enzyme, are shown here. The tertiary representation in **(b)** shows the course of the chymotrypsin folding pattern by successive numbering of the amino acids in its sequence. (Residues 14 and 15 and 147 and 148 are missing because these residues are removed when chymotrypsin is formed from its larger precursor, chymotrypsinogen.) The ribbon diagram depicts the three-dimensional track of the polypeptide in space.

A Protein's Conformation Can Be Described as Its Overall Three-Dimensional Structure

The overall three-dimensional architecture of a protein is generally referred to as its **conformation.** This term is not to be confused with **configuration,** which denotes the geometric possibilities for a particular set of atoms (Figure 5.10). In going from one configuration to another, covalent bonds must be broken and rearranged. In contrast, the *conformational possibilities* of a molecule are achieved without breaking any covalent bonds. In proteins, rotations about each of the single bonds along the peptide backbone have the potential to alter the course of the polypeptide chain in three-dimensional space. These rotational possibilities create many possible orientations for the protein chain, referred to as its conformational possibilities. Of the great number of theoretical conformations a given protein might adopt, only a very few are favored energetically under physiological conditions. At this time, the rules that direct

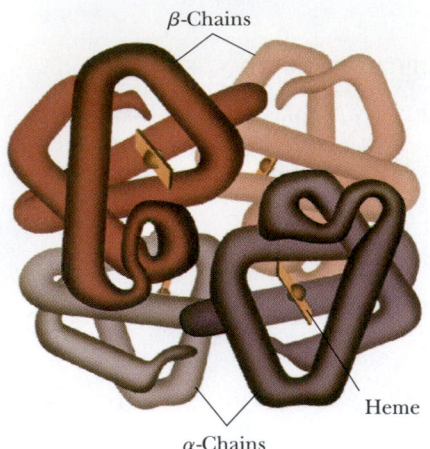

β-Chains

Heme

α-Chains

FIGURE 5.9 Hemoglobin, which consists of two α and two β polypeptide chains, is an example of the quaternary level of protein structure. In this drawing, the β-chains are the two uppermost polypeptides and the two α-chains are the lower half of the molecule. The two closest chains (darkest colored) are the β₂-chain *(upper left)* and the α₁-chain *(lower right)*. The heme groups of the four globin chains are represented by rectangles with spheres (the heme iron atom). Note the symmetry of this macromolecular arrangement. *(Illustration: Irving Geis. Rights owned by Howard Hughes Medical Institute. Not to be reproduced without permission.)*

the folding of protein chains into energetically favorable conformations are still not entirely clear; accordingly, they are the subject of intensive contemporary research.

5.3   How Are Proteins Isolated and Purified from Cells?

Cells contain thousands of different proteins. A major problem for protein chemists is to purify a chosen protein so that they can study its specific properties in the absence of other proteins. Proteins can be separated and purified on the basis of their two prominent physical properties: size and electrical charge. A more direct approach is to use **affinity purification** strategies that take advantage of the biological function or similar specific recognition properties of a protein (see Chapter Appendix).

A Number of Protein Separation Methods Exploit Differences in Size and Charge

Separation methods based on size include size exclusion chromatography, ultrafiltration, and ultracentrifugation (see Chapter Appendix). The ionic properties of peptides and proteins are determined principally by their complement of amino acid side chains. Furthermore, the ionization of these groups is pH-dependent.

A variety of procedures have been designed to exploit the electrical charges on a protein as a means to separate proteins in a mixture. These procedures include ion exchange chromatography (see Chapter 4), electro-

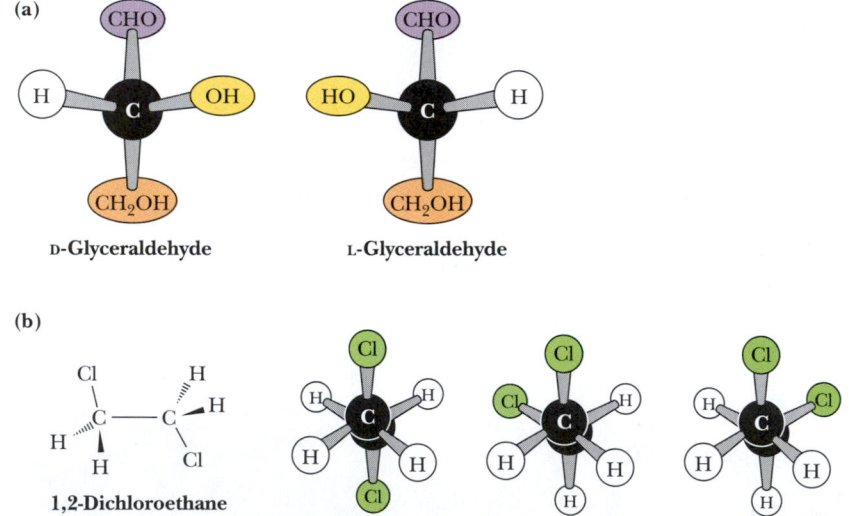

(a)

D-**Glyceraldehyde**   L-**Glyceraldehyde**

(b)

1,2-Dichloroethane

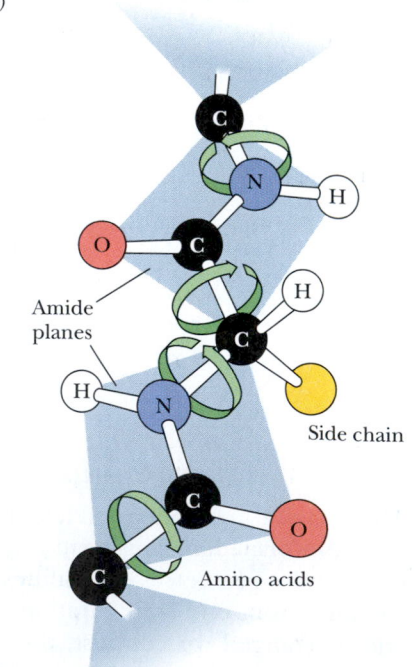

(c)

Amide planes

Side chain

Amino acids

FIGURE 5.10 Configuration and conformation are *not* synonymous. **(a)** Rearrangements between configurational alternatives of a molecule can be achieved only by breaking and remaking bonds, as in the transformation between the D- and L-configurations of glyceraldehyde. No possible rotational reorientation of bonds linking the atoms of D-glyceraldehyde yields geometric identity with L-glyceraldehyde, even though they are mirror images of each other. **(b)** The intrinsic free rotation around single covalent bonds creates a great variety of three-dimensional conformations, even for relatively simple molecules. Consider 1,2-dichloroethane. Viewed end-on in a Newman projection, three principal rotational orientations or conformations predominate. Steric repulsion between eclipsed and partially eclipsed conformations keeps the possibilities at a reasonable number. **(c)** Imagine the conformational possibilities for a protein in which two of every three bonds along its backbone are freely rotating single bonds. *(Illustration: Irving Geis. Rights owned by Howard Hughes Medical Institute. Not to be reproduced without permission.)*

A Deeper Look

Estimation of Protein Concentrations in Solutions of Biological Origin

Biochemists are often interested in knowing the protein concentration in various preparations of biological origin. Such quantitative analysis is not straightforward. Cell extracts are complex mixtures that typically contain protein molecules of many different molecular weights, so the results of protein estimations cannot be expressed on a molar basis. Also, aside from the rather unreactive repeating peptide backbone, little common chemical identity is seen among the many proteins found in cells that might be readily exploited for exact chemical analysis. Most of their chemical properties vary with their amino acid composition, for example, nitrogen or sulfur content or the presence of aromatic, hydroxyl, or other functional groups.

Lowry Procedure

A method that has been the standard of choice for many years is the **Lowry procedure.** This method uses Cu^{2+} ions along with *Folin–Ciocalteau reagent,* a combination of phosphomolybdic and phosphotungstic acid complexes that react with Cu^+. Cu^+ is generated from Cu^{2+} by readily oxidizable protein components, such as cysteine or the phenols and indoles of tyrosine and tryptophan. Although the precise chemistry of the Lowry method remains uncertain, the Cu^+ reaction with the Folin reagent gives intensely colored products measurable spectrophotometrically.

BCA Method

Recently, a reagent that reacts more efficiently with Cu^+ than Folin–Ciocalteau reagent has been developed for protein assays. *Bicinchoninic acid* (BCA) forms a purple complex with Cu^+ in alkaline solution.

Assays Based on Dye Binding

Several other protocols for protein estimation enjoy prevalent usage in biochemical laboratories. The **Bradford assay** is a rapid and reliable technique that uses a dye called *Coomassie Brilliant Blue G-250,* which undergoes a change in its color upon noncovalent binding to proteins. The binding is quantitative and less sensitive to variations in the protein's amino acid composition. The color change is easily measured with a spectrophotometer. A similar, very sensitive method capable of quantifying nanogram amounts of protein is based on the shift in color of colloidal gold upon binding to proteins.

$$Cu^+ + BCA \longrightarrow$$

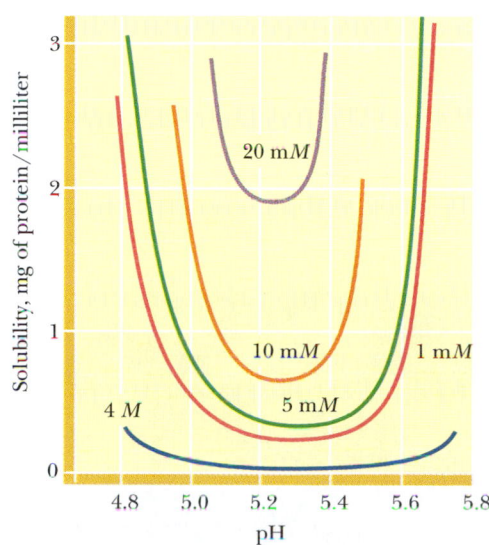

BCA–Cu^+ complex

phoresis (see Chapter Appendix), and solubility. Proteins tend to be least soluble at their **isoelectric point,** the pH value at which the sum of their positive and negative electrical charges is zero. At this pH, electrostatic repulsion between protein molecules is minimal and they are more likely to coalesce and precipitate out of solution. Ionic strength also profoundly influences protein solubility. Most globular proteins tend to become increasingly soluble as the ionic strength is raised. This phenomenon, the salting-in of proteins, is attributed to the diminishment of electrostatic attractions between protein molecules by the presence of abundant salt ions. Such electrostatic interactions between the protein molecules would otherwise lead to precipitation. However, as the salt concentration reaches high levels (greater than 1 *M*), the effect may reverse so that the protein is salted out of solution. In such cases, the numerous salt ions begin to compete with the protein for waters of solvation, and as they win out, the protein becomes insoluble. The solubility properties of a typical protein are shown in Figure 5.11.

Although the side chains of nonpolar amino acids in soluble proteins are usually buried in the interior of the protein away from contact with the aqueous solvent, a portion of them may be exposed at the protein's surface, giving it a partially hydrophobic character. *Hydrophobic interaction chromatography* is a protein purification technique that exploits this hydrophobicity (see Chapter Appendix).

FIGURE 5.11 The solubility of most globular proteins is markedly influenced by pH and ionic strength. This figure shows the solubility of a typical protein as a function of pH and various salt concentrations.

Table 5.2

Example of a Protein Purification Scheme: Purification of the Enzyme Xanthine Dehydrogenase from a Fungus

Fraction	Volume (mL)	Total Protein (mg)	Total Activity*	Specific Activity†	Percent Recovery‡
1. Crude extract	3,800	22,800	2,460	0.108	100
2. Salt precipitate	165	2,800	1,190	0.425	48
3. Ion exchange chromatography	65	100	720	7.2	29
4. Molecular sieve chromatography	40	14.5	555	38.3	23
5. Immunoaffinity chromatography§	6	1.8	275	152	11

*The relative enzymatic activity of each fraction in catalyzing the xanthine dehydrogenase reaction is cited as arbitrarily defined units.

†The specific activity is the total activity of the fraction divided by the total protein in the fraction. This value gives an indication of the increase in purity attained during the course of the purification as the samples become enriched for xanthine dehydrogenase protein.

‡The percent recovery of total activity is a measure of the yield of the desired product, xanthine dehydrogenase.

§The last step in the procedure is an affinity method in which antibodies specific for xanthine dehydrogenase are covalently coupled to a chromatography matrix and packed into a glass tube to make a chromatographic column through which fraction 4 is passed. The enzyme is bound by this immunoaffinity matrix while other proteins pass freely out. The enzyme is then recovered by passing a strong salt solution through the column, which dissociates the enzyme–antibody complex.

Adapted from Lyon, E. S., and Garrett, R. H., 1978. Regulation, purification, and properties of xanthine dehydrogenase in Neurospora crassa. *Journal of Biological Chemistry.* **253**:2604–2614.

A Typical Protein Purification Scheme Uses a Series of Separation Methods

Most purification procedures for a particular protein are developed in an empirical manner, the overriding principle being purification of the protein to a homogeneous state with acceptable yield. Table 5.2 presents a summary of a purification scheme for a selected protein. Note that the **specific activity** of the protein (the enzyme xanthine dehydrogenase) in the immunoaffinity purified fraction (fraction 5) has been increased 152/0.108, or 1407 times the specific activity in the crude extract (fraction 1). Thus, xanthine dehydrogenase in fraction 5 versus fraction 1 is enriched more than 1400-fold by the purification procedure.

5.4　How Is the Amino Acid Analysis of Proteins Performed?

Acid Hydrolysis Liberates the Amino Acids of a Protein

Peptide bonds of proteins are hydrolyzed by either strong acid or strong base. Acid hydrolysis is the method of choice for analysis of the amino acid composition of proteins and polypeptides because it proceeds without racemization and with less destruction of certain amino acids (Ser, Thr, Arg, and Cys). Typically, samples of a protein are hydrolyzed with 6 N HCl at 110°C for 24, 48, and 72 hours in sealed glass vials. Tryptophan is destroyed by acid and must be estimated by other means to determine its contribution to the total amino acid composition. The OH-containing amino acids serine and threonine are slowly destroyed, but the data obtained for the three time points (24, 48, and 72 hours) allow extrapolation to zero time to estimate the original Ser and Thr content (Figure 5.12). In contrast, peptide bonds involving hydrophobic residues such as valine and isoleucine are only slowly hydrolyzed in acid. Another complication arises because the β- and γ-amide linkages in asparagine (Asn) and glutamine (Gln) are acid labile. The amino nitrogen is released as free ammonium, and all of the Asn and Gln residues of the protein are converted to aspartic acid (Asp) and glutamic acid (Glu), respectively. The amount of ammonium released dur-

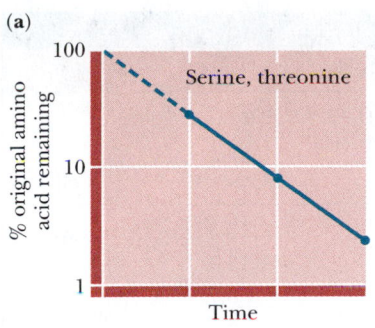

(a)

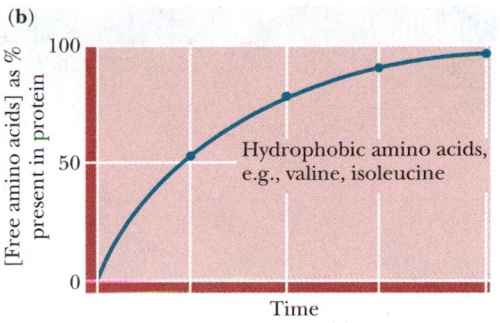

(b)

Biochemistry ⊛ Now™   **ANIMATED FIGURE 5.12**
(a) The hydroxy amino acids serine and threonine are slowly destroyed during the course of protein hydrolysis for amino acid composition analysis. Extrapolation of the data back to time zero allows an accurate estimation of the amount of these amino acids originally present in the protein sample. (b) Peptide bonds involving hydrophobic amino acid residues such as valine and isoleucine resist hydrolysis by HCl. With time, these amino acids are released and their free concentrations approach a limiting value that can be approximated with reliability. **See this figure animated at http://chemistry.brookscole.com/ggb3**

ing acid hydrolysis gives an estimate of the total number of Asn and Gln residues in the original protein, but not the amounts of either. Accordingly, the concentrations of Asp and Glu determined in amino acid analysis are expressed as Asx and Glx, respectively. Because the relative contributions of [Asn + Asp] or [Gln + Glu] cannot be derived from the data, this information must be obtained by alternative means.

Chromatographic Methods Are Used to Separate the Amino Acids

The complex amino acid mixture in the hydrolysate obtained after digestion of a protein in 6 N HCl can be separated into the component amino acids by using either ion exchange chromatography (see Chapter 4) or reversed-phase high-pressure liquid chromatography (HPLC) (see Chapter Appendix). The amount of each amino acid can then be determined. In ion exchange chromatography, the amino acids are separated and then quantified following reaction with ninhydrin (so-called postcolumn derivatization). In HPLC, the amino acids are converted to phenylthiohydantoin (PTH) derivatives via reaction with Edman's reagent (see Figure 5.15) before chromatography (precolumn derivatization). Both of these methods of separation and analysis are fully automated in instruments called **amino acid analyzers.** Analysis of the amino acid composition of a 30-kD protein by these methods requires less than 1 hour and only 6 μg (0.2 nmol) of the protein.

The Amino Acid Compositions of Different Proteins Are Different

Table 5.3 gives the amino acid composition of several selected proteins: ribonuclease A, alcohol dehydrogenase, myoglobin, histone H3, and collagen. Each of the 20 naturally occurring amino acids is usually represented at least once in a polypeptide chain. However, some small proteins may not have a representative of every amino acid. Note that ribonuclease (12.6 kD, 124 amino acid residues) does not contain any tryptophan. Amino acids almost never occur in equimolar ratios in proteins, indicating that proteins are not composed of repeating arrays of amino acids. There are a few exceptions to this rule. Collagen, for example, contains large proportions of glycine and proline, and much of its structure is composed of (Gly-x-Pro) repeating units, where x is any amino acid. Other proteins show unusual abundances of various amino acids. For example, histones are rich in positively charged amino acids such as arginine and lysine. Histones are a class of proteins found associated with the anionic phosphate groups of eukaryotic DNA.

Amino acid analysis itself does not directly give the number of residues of each amino acid in a polypeptide, but it does give amounts from which the percentages or ratios of the various amino acids can be obtained (Table 5.3). If the molecular weight *and* the exact amount of the protein analyzed are known (or the number of amino acid residues per molecule is known), the molar ratios of amino acids in the protein can be calculated. Amino acid analysis provides no information on the

Table 5.3

Amino Acid Composition of Some Selected Proteins

Values expressed are percent representation of each amino acid.

Amino Acid	Proteins*				
	RNase	ADH	Mb	Histone H3	Collagen
Ala	6.9	7.5	9.8	13.3	11.7
Arg	3.7	3.2	1.7	13.3	4.9
Asn	7.6	2.1	2.0	0.7	1.0
Asp	4.1	4.5	5.0	3.0	3.0
Cys	6.7	3.7	0	1.5	0
Gln	6.5	2.1	3.5	5.9	2.6
Glu	4.2	5.6	8.7	5.2	4.5
Gly	3.7	10.2	9.0	5.2	32.7
His	3.7	1.9	7.0	1.5	0.3
Ile	3.1	6.4	5.1	5.2	0.8
Leu	1.7	6.7	11.6	8.9	2.1
Lys	7.7	8.0	13.0	9.6	3.6
Met	3.7	2.4	1.5	1.5	0.7
Phe	2.4	4.8	4.6	3.0	1.2
Pro	4.5	5.3	2.5	4.4	22.5
Ser	12.2	7.0	3.9	3.7	3.8
Thr	6.7	6.4	3.5	7.4	1.5
Trp	0	0.5	1.3	0	0
Tyr	4.0	1.1	1.3	2.2	0.5
Val	7.1	10.4	4.8	4.4	1.7
Acidic	8.4	10.2	13.7	8.1	7.5
Basic	15.0	13.1	21.8	24.4	8.8
Aromatic	6.4	6.4	7.2	5.2	1.7
Hydrophobic	18.0	30.7	27.6	23.0	6.5

*Proteins are as follows:
RNase: Bovine ribonuclease A, an enzyme; 124 amino acid residues. Note that RNase lacks tryptophan.
ADH: Horse liver alcohol dehydrogenase, an enzyme; dimer of identical 374 amino acid polypeptide chains. The amino acid composition of ADH is reasonably representative of the norm for water-soluble proteins.
Mb: Sperm whale myoglobin, an oxygen-binding protein; 153 amino acid residues. Note that Mb lacks cysteine.
Histone H3: Histones are DNA-binding proteins found in chromosomes; 135 amino acid residues. Note the very basic nature of this protein due to its abundance of Arg and Lys residues. It also lacks tryptophan.
Collagen: Collagen is an extracellular structural protein; 1052 amino acid residues. Collagen has an unusual amino acid composition; it is about one-third glycine and is rich in proline. Note that it also lacks Cys and Trp and is deficient in aromatic amino acid residues in general.

order or sequence of amino acid residues in the polypeptide chain. Because the polypeptide chain is unbranched, it has only two ends: an amino-terminal end, or **N-terminal end,** and a carboxyl-terminal end, or **C-terminal end.**

5.5 How Is the Primary Structure of a Protein Determined?

The Sequence of Amino Acids in a Protein Is Distinctive

The unique characteristic of each protein is the distinctive sequence of amino acid residues in its polypeptide chain(s). Indeed, it is the **amino acid sequence** of proteins that is encoded by the nucleotide sequence of DNA. This amino acid sequence, then, is a form of genetic information. By convention, the amino acid sequence is read from the N-terminal end of the polypeptide chain

A Deeper Look

The Virtually Limitless Number of Different Amino Acid Sequences

Given 20 different amino acids, a polypeptide chain of n residues can have any one of 20^n possible sequence arrangements. To portray this, consider the number of tripeptides possible if there were only three different amino acids, A, B, and C (tripeptide $= 3 = n$; $3^n = 3^3 = 27$):

AAA	BBB	CCC
AAB	BBA	CCA
AAC	BBC	CCB
ABA	BAB	CBC
ACA	BCB	CAC
ABC	BAA	CBA
ACB	BCC	CAB
ABB	BAC	CBB
ACC	BCA	CAA

For a polypeptide chain of 100 residues in length, a rather modest size, the number of possible sequences is 20^{100}, or because $20 = 10^{1.3}$, 10^{130} unique possibilities. These numbers are more than astronomical! Because an average protein molecule of 100 residues would have a mass of 13,800 daltons (average molecular mass of an amino acid residue $= 138$), 10^{130} such molecules would have a mass of 1.38×10^{134} daltons. The mass of the observable universe is estimated to be 10^{80} proton masses (about 10^{80} daltons). Thus, the universe lacks enough material to make just one molecule of each possible polypeptide sequence for a protein only 100 residues in length.

through to the C-terminal end. As an example, every molecule of ribonuclease A from bovine pancreas has the same amino acid sequence, beginning with N-terminal lysine at position 1 and ending with C-terminal valine at position 124 (Figure 5.6). Given the possibility of any of the 20 amino acids at each position, the number of unique amino acid sequences is astronomically large. The astounding sequence variation possible within polypeptide chains provides a key insight into the incredible functional diversity of protein molecules in biological systems discussed later in this chapter.

In 1953, Frederick Sanger of Cambridge University in England reported the amino acid sequences of the two polypeptide chains composing the protein insulin (Figure 5.13). Not only was this a remarkable achievement in analytical chemistry, but it helped demystify speculation about the chemical nature of proteins. Sanger's results clearly established that all of the molecules of a given protein have a fixed amino acid composition, a defined amino acid sequence, and therefore an invariant molecular weight. In short, proteins are well defined chemically. Today, the amino acid sequences of hundreds of thousands of proteins are known. Although many sequences have been determined from application of the principles first established by Sanger, most are now deduced from knowledge of the nucleotide sequence of the gene that encodes the protein. In addition, in recent years, the application of mass spectrometry to the sequence analysis of proteins has largely superseded the protocols based on chemical and enzymatic degradation of polypeptides that Sanger pioneered.

Both Chemical and Enzymatic Methodologies Are Used in Protein Sequencing

The chemical strategy for determining the amino acid sequence of a protein involves seven basic steps:

1. If the protein contains more than one polypeptide chain, the chains are separated and purified.
2. Intrachain S—S (disulfide) cross-bridges between cysteine residues in the polypeptide chain are cleaved. (If these disulfides are interchain linkages, then step 2 precedes step 1.)
3. The N-terminal and C-terminal residues are identified.
4. Each polypeptide chain is cleaved into smaller fragments, and the amino acid composition and sequence of each fragment are determined.

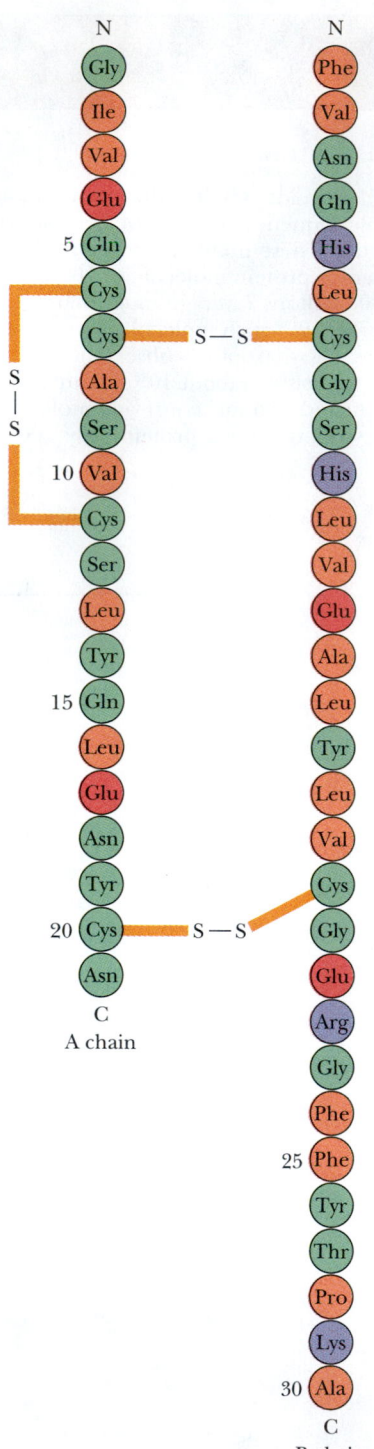

FIGURE 5.13 The hormone insulin consists of two polypeptide chains, A and B, held together by two disulfide cross-bridges (S—S). The A chain has 21 amino acid residues and an intrachain disulfide; the B polypeptide contains 30 amino acids. The sequence shown is for bovine insulin. *(Illustration: Irving Geis. Rights owned by Howard Hughes Medical Institute. Not to be reproduced without permission.)*

5. Step 4 is repeated, using a different cleavage procedure to generate a different and therefore overlapping set of peptide fragments.
6. The overall amino acid sequence of the protein is reconstructed from the sequences in overlapping fragments.
7. The positions of S—S cross-bridges formed between cysteine residues are located.

Each of these steps is discussed in greater detail in the following sections.

Step 1. Separation of Polypeptide Chains

If the protein of interest is a **heteromultimer** (composed of more than one type of polypeptide chain), then the protein must be dissociated into its component polypeptide chains, which then must be separated from one another and sequenced individually. Because subunits in multimeric proteins typically associate through noncovalent interactions, most multimeric proteins can be dissociated by exposure to pH extremes, 8 *M* urea, 6 *M* guanidinium hydrochloride, or high salt concentrations. (All of these treatments disrupt polar interactions such as hydrogen bonds both within the protein molecule and between the protein and the aqueous solvent.) Once dissociated, the individual polypeptides can be isolated from one another on the basis of differences in size and/or charge. Occasionally, heteromultimers are linked together by interchain S—S bridges. In such instances, these crosslinks must be cleaved before dissociation and isolation of the individual chains. The methods described under step 2 are applicable for this purpose.

Step 2. Cleavage of Disulfide Bridges

A number of methods exist for cleaving disulfides (Figure 5.14). An important consideration is to carry out these cleavages so that the original or even new S—S links do not form. Oxidation of a disulfide by performic acid results in the formation of two equivalents of cysteic acid (Figure 5.14a). Because these cysteic acid side chains are ionized SO_3^- groups, electrostatic repulsion (as well as altered chemistry) prevents S—S recombination. Alternatively, sulfhydryl compounds such as 2-mercaptoethanol (Figure 5.14b) or dithiothreitol (DTT) readily reduce S—S bridges to regenerate two cysteine—SH side chains. However, these SH groups recombine to re-form either the original disulfide link or, if other free Cys—SHs are available, new disulfide links. To prevent this, S—S reduction must be followed by treatment with alkylating agents such as iodoacetate or 3-bromopropylamine, which modify the SH groups and block disulfide bridge formation (Figure 5.14a).

Step 3.

A. N-Terminal Analysis The amino acid residing at the N-terminal end of a protein can be identified in a number of ways; one method, **Edman degradation,** has become the procedure of choice. This method is preferable because it allows the sequential identification of a series of residues beginning at the N-terminus (Figure 5.15). In weakly basic solutions, phenylisothiocyanate, or **Edman's reagent** (phenyl—N=C=S), combines with the free amino terminus of a protein (Figure 5.15), which can be excised from the end of the polypeptide chain and recovered as a PTH derivative. Chromatographic methods can be used to identify this PTH derivative. Importantly, in this procedure, the rest of the polypeptide chain remains intact and can be subjected to further rounds of Edman degradation to identify successive amino acid residues in the chain. Often, the carboxyl terminus of the polypeptide under analysis is coupled to an insoluble matrix, allowing the polypeptide to be easily recovered by filtration or centrifugation following each round of Edman

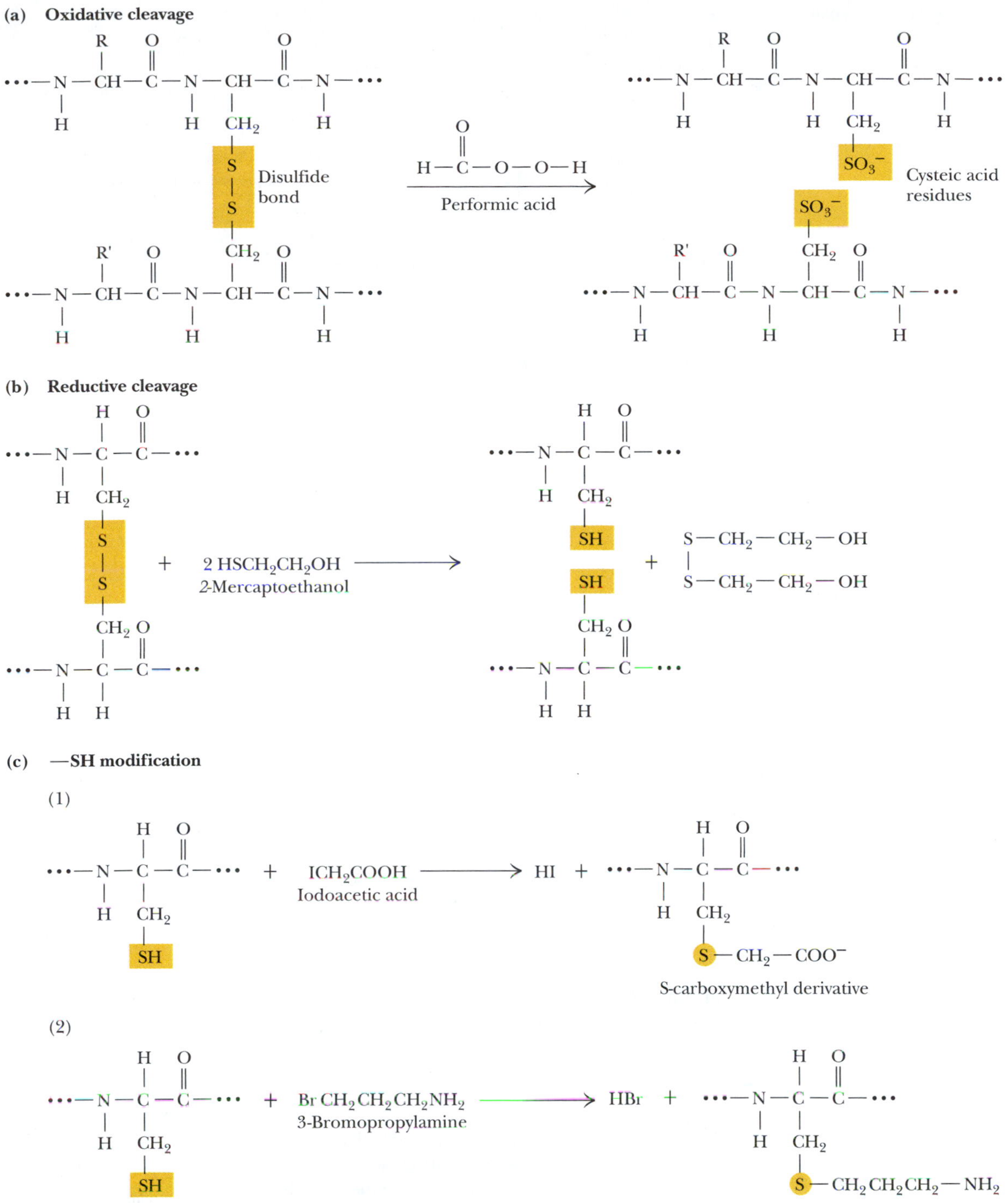

(a) Oxidative cleavage

(b) Reductive cleavage

(c) —SH modification

(1)

S-carboxymethyl derivative

(2)

FIGURE 5.14 Methods for cleavage of disulfide bonds in proteins. **(a)** Oxidative cleavage by reaction with performic acid. **(b)** Reductive cleavage with sulfhydryl compounds. Disulfide bridges can be broken by reduction of the S—S link with sulfhydryl agents such as β-mercaptoethanol or dithiothreitol. Because reaction between the newly reduced —SH groups to reestablish disulfide bonds is a likelihood, S—S reduction must be followed by **(c)** —SH modification: (1) alkylation with iodoacetate (ICH_2COOH) or (2) modification with 3-bromopropylamine (Br—$(CH_2)_3$—NH_2).

reaction. Thus, Edman reaction not only identifies the N-terminal residue of proteins but through successive reaction cycles can reveal further information about sequence. Automated instruments (so-called Edman sequenators) have been designed to carry out repeated rounds of the Edman procedure. In practical terms, as many as 50 cycles of reaction can be accomplished on 50 pmol (about 0.1 μg) of a polypeptide 100 to 200 residues long, revealing the sequential order of the first 50 amino acid residues in the protein. The efficiency with larger proteins is less; a typical 2000–amino acid protein provides only 10 to 20 cycles of reaction.

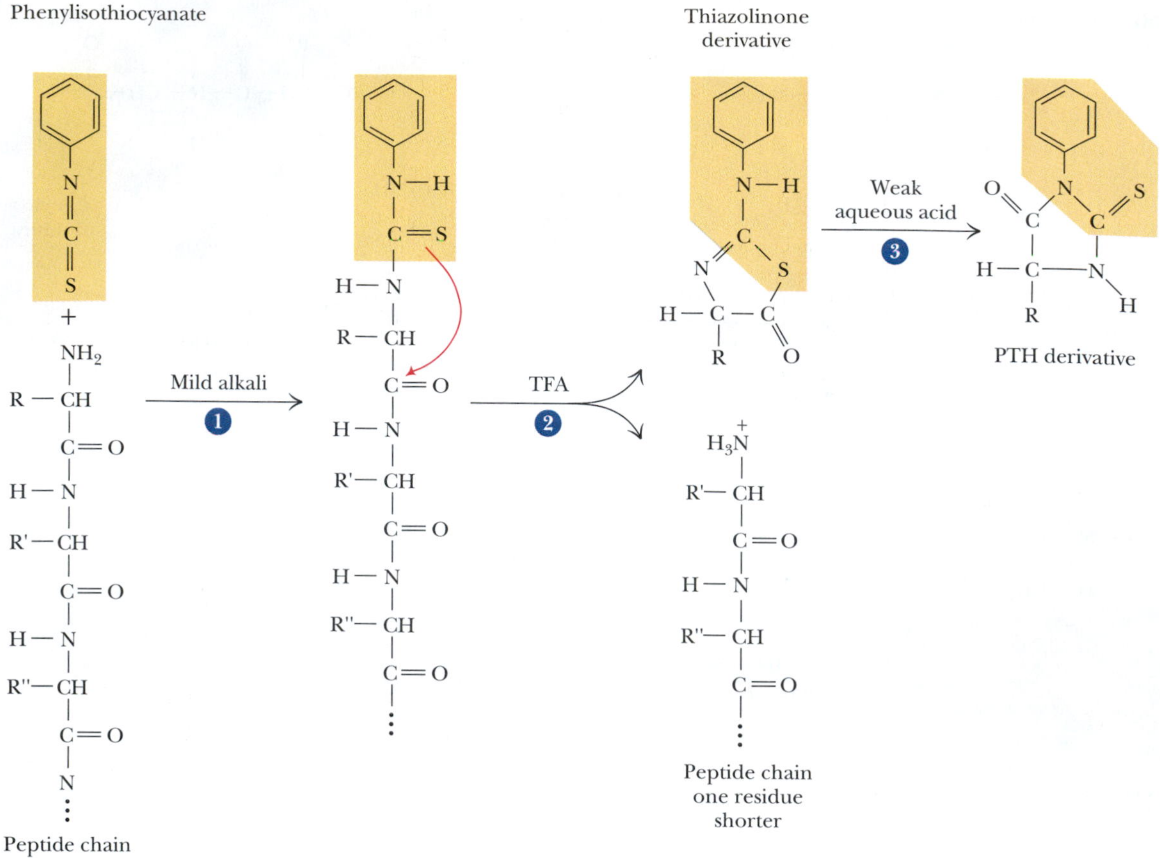

Biochemistry ⊘ **Now**™   **ACTIVE FIGURE 5.15** N-terminal analysis using Edman's reagent, phenylisothiocyanate. (**1**) Phenylisothiocyanate combines with the N-terminus of a peptide under mildly alkaline conditions to form a phenylthiocarbamoyl substitution. (**2**) Upon treatment with TFA (trifluoroacetic acid), this cyclizes to release the N-terminal amino acid residue as a thiazolinone derivative, but the other peptide bonds are not hydrolyzed. (**3**) Organic extraction and treatment with aqueous acid yield the N-terminal amino acid as a phenylthiohydantoin (PTH) derivative. **Test yourself on the concepts in this figure at http://chemistry.brookscole.com/ggb3**

B. C-Terminal Analysis   For the identification of the C-terminal residue of polypeptides, an enzymatic approach is commonly used.

ENZYMATIC ANALYSIS WITH CARBOXYPEPTIDASES.   Carboxypeptidases are enzymes that cleave amino acid residues from the C-termini of polypeptides in a successive fashion. Four carboxypeptidases are in general use: A, B, C, and Y. *Carboxypeptidase A* (from bovine pancreas) works well in hydrolyzing the C-terminal peptide bond of all residues except proline, arginine, and lysine. The analogous enzyme from hog pancreas, *carboxypeptidase B*, is effective only when Arg or Lys are the C-terminal residues. *Carboxypeptidase C* from citrus leaves and *carboxypeptidase Y* from yeast act on any C-terminal residue. Because the nature of the amino acid residue at the end often determines the rate at which it is cleaved and because these enzymes remove residues successively, care must be taken in interpreting results. Carboxypeptidase Y cleavage has been adapted to an automated protocol analogous to that used in Edman sequenators.

Steps 4 and 5. Fragmentation of the Polypeptide Chain

The aim at this step is to produce fragments useful for sequence analysis. The cleavage methods employed are usually enzymatic, but proteins can also be fragmented by specific or nonspecific chemical means (such as partial acid

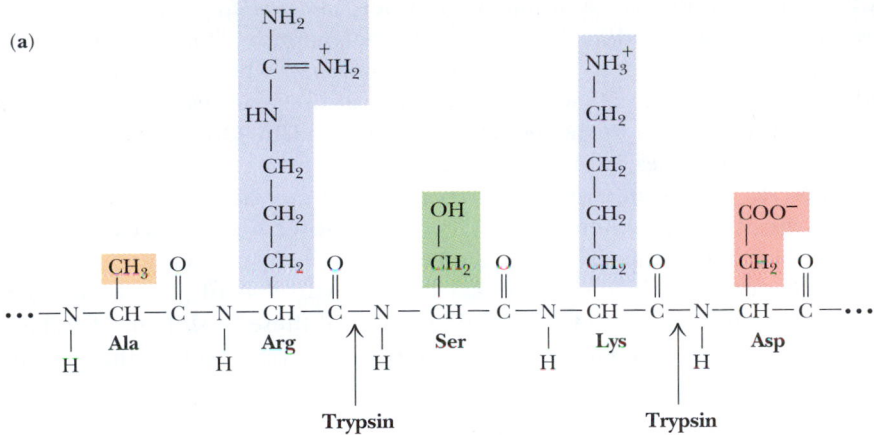

(a)

(b)

hydrolysis). Proteolytic enzymes offer an advantage in that they may hydrolyze only specific peptide bonds, and this specificity immediately gives information about the peptide products. As a first approximation, fragments produced upon cleavage should be small enough to yield their sequences through end-group analysis and Edman degradation, yet not so small that an overabundance of products must be resolved before analysis.

A. Trypsin   The digestive enzyme *trypsin* is the most commonly used reagent for specific proteolysis. Trypsin is specific in hydrolyzing only peptide bonds in which the carbonyl function is contributed by an arginine or a lysine residue. That is, trypsin cleaves on the C-side of Arg or Lys, generating a set of peptide fragments having Arg or Lys at their C-termini. The number of smaller peptides resulting from trypsin action is equal to the total number of Arg and Lys residues in the protein *plus* one—the protein's C-terminal peptide fragment (Figure 5.16).

B. Chymotrypsin   *Chymotrypsin* shows a strong preference for hydrolyzing peptide bonds formed by the carboxyl groups of the aromatic amino acids, phenylalanine, tyrosine, and tryptophan. However, over time, chymotrypsin also hydrolyzes amide bonds involving amino acids other than Phe, Tyr, or Trp. For instance, peptide bonds having leucine-donated carboxyls are also susceptible. Thus, the specificity of chymotrypsin is only relative. Because chymotrypsin produces a very different set of products than trypsin, treatment of separate samples of a protein with these two enzymes generates fragments whose sequences overlap. Resolution of the order of amino acid residues in the fragments yields the amino acid sequence in the original protein.

C. Other Endopeptidases A number of other *endopeptidases* (proteases that cleave peptide bonds within the interior of a polypeptide chain) are also used in sequence investigations. These include *clostripain*, which acts only at Arg residues; *endopeptidase Lys-C*, which cleaves only at Lys residues; and *staphylococcal protease*, which acts at the acidic residues, Asp and Glu. Other, relatively nonspecific endopeptidases are handy for digesting large tryptic or chymotryptic fragments. *Pepsin, papain, subtilisin, thermolysin,* and *elastase* are some examples. Papain is the active ingredient in meat tenderizer, soft contact lens cleaner, and some laundry detergents. The abundance of papain in papaya, and a similar protease (bromelain) in pineapple, causes the hydrolysis of gelatin and prevents the preparation of Jell-O containing either of these fresh fruits. Cooking these fruits thermally denatures their proteolytic enzymes so that they can be used in gelatin desserts.

D. Cyanogen Bromide Several highly specific chemical methods of proteolysis are available, the most widely used being *cyanogen bromide (CNBr)* cleavage. CNBr acts upon methionine residues (Figure 5.17). The nucleophilic sulfur atom of Met reacts with CNBr, yielding a sulfonium ion that undergoes a rapid intramolecular rearrangement to form a cyclic iminolactone. Water readily hydrolyzes this iminolactone, cleaving the polypeptide and generating peptide fragments having C-terminal homoserine lactone residues at the former Met positions.

E. Other Chemical Methods of Fragmentation A number of other chemical methods give specific fragmentation of polypeptides, including cleavage at asparagine–glycine bonds by hydroxylamine (NH_2OH) at pH 9 and selective hydrolysis at aspartyl–prolyl bonds under mildly acidic conditions. Table 5.4 summarizes the various procedures described here for polypeptide cleavage. These methods are only a partial list of the arsenal of reactions available to protein chemists. Cleavage products generated by these procedures must be isolated and individually sequenced to accumulate the information necessary to reconstruct the protein's complete amino acid sequence. Peptide sequencing

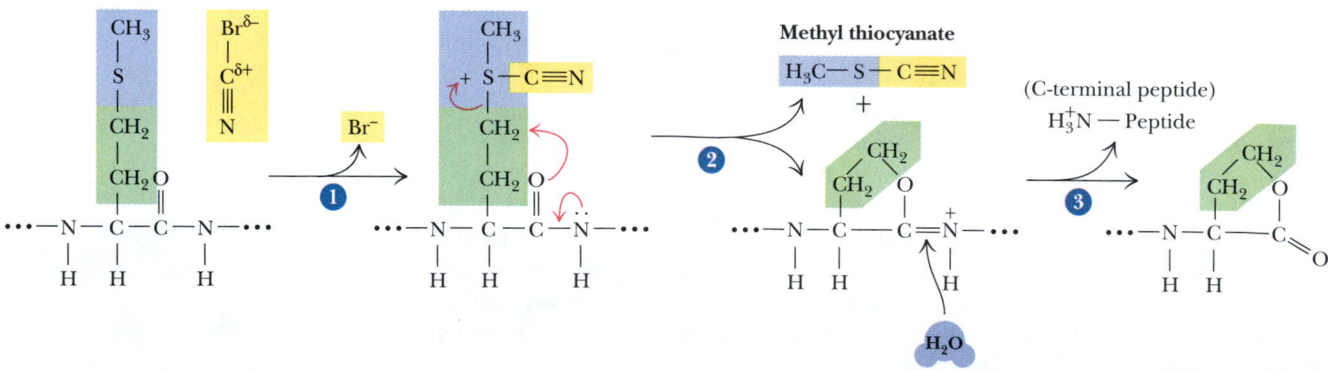

OVERALL REACTION:

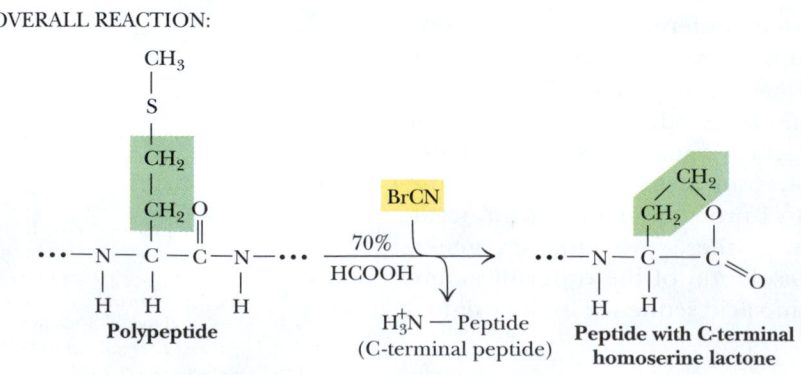

Biochemistry Now™ ANIMATED FIGURE 5.17
Cyanogen bromide (CNBr) is a highly selective reagent for cleavage of peptides only at methionine residues. **(1)** The reaction occurs in 70% formic acid via nucleophilic attack of the Met S atom on the —C≡N carbon atom, with displacement of Br. **(2)** The cyano intermediate undergoes nucleophilic attack by the Met carbonyl oxygen atom on the R group, resulting in formation of the cyclic derivative, which is unstable in aqueous solution. **(3)** Hydrolysis ensues, producing cleavage of the Met peptide bond and release of peptide fragments, with C-terminal homoserine lactone residues where Met residues once were. One peptide does not have a C-terminal homoserine lactone: the original C-terminal end of the polypeptide. **See this figure animated at http://chemistry.brookscole.com/ggb3**

Table 5.4

Specificity of Representative Polypeptide Cleavage Procedures Used in Sequence Analysis

Method	Peptide Bond on Carboxyl (C) or Amino (N) Side of Susceptible Residue	Susceptible Residue(s)
Proteolytic enzymes*		
Trypsin	C	Arg or Lys
Chymotrypsin	C	Phe, Trp, or Tyr; Leu
Clostripain	C	Arg
Staphylococcal protease	C	Asp or Glu
Chemical methods		
Cyanogen bromide	C	Met
NH₂OH	Asn-Gly bonds	
pH 2.5, 40°C	Asp-Pro bonds	

*Some proteolytic enzymes, including trypsin and chymotrypsin, will not cleave peptide bonds where proline is the amino acid contributing the N-atom.

today is most commonly done by Edman degradation of relatively large peptides or by mass spectrometry (see following discussion).

Step 6. Reconstruction of the Overall Amino Acid Sequence

The sequences obtained for the sets of fragments derived from two or more cleavage procedures are now compared, with the objective being to find overlaps that establish continuity of the overall amino acid sequence of the polypeptide chain. The strategy is illustrated by the example shown in Figure 5.18. Peptides generated from specific fragmentation of the polypeptide can be aligned to reveal the overall amino acid sequence. Such comparisons are also

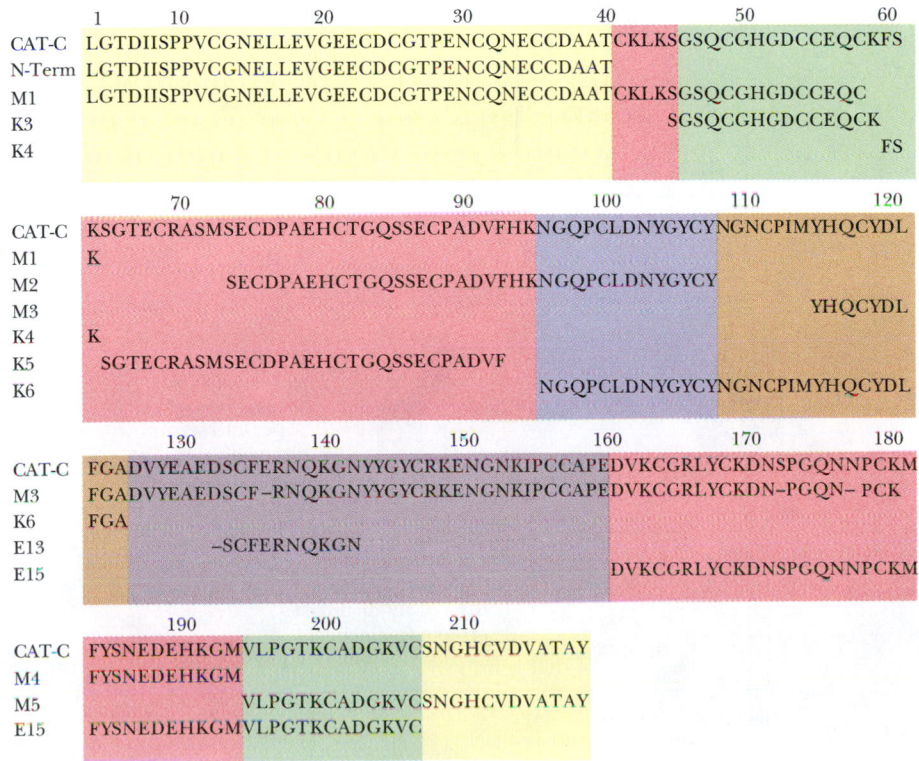

Biochemistry Now™ ANIMATED FIGURE 5.18
Summary of the sequence analysis of catrocollastatin-C, a 23.6-kD protein found in the venom of the western diamondback rattlesnake *Crotalus atrox*. Sequences shown are given in the one-letter amino acid code. The overall amino acid sequence (216 amino acid residues long) for catrocollastatin-C as deduced from the overlapping sequences of peptide fragments is shown on the lines headed **CAT-C**. The other lines report the various sequences used to obtain the overlaps. These sequences were obtained from (a) **N-term:** Edman degradation of the intact protein in an automated Edman sequenator; (b) **M:** proteolytic fragments generated by CNBr cleavage, followed by Edman sequencing of the individual fragments (numbers denote fragments M1 through M5); (c) **K:** proteolytic fragments from endopeptidase Lys-C cleavage, followed by Edman sequencing (only fragments K3 through K6 are shown); (d) **E:** proteolytic fragments from *Staphylococcus* protease digestion of catrocollastatin sequenced in the Edman sequenator (only E13 through E15 are shown). *(Adapted from Shimokawa, K., et al., 1997. Sequence and biological activity of catrocollastatin-C: A disintegrin-like/cysteine-rich two-domain protein from Crotalus atrox venom.* Archives of Biochemistry and Biophysics *343:35–43.)* See this figure animated at http://chemistry.brookscole.com/ggb3

useful in eliminating errors and validating the accuracy of the sequences determined for the individual fragments.

Step 7. Location of Disulfide Cross-Bridges

Strictly speaking, the disulfide bonds formed between cysteine residues in a protein are not a part of its primary structure. Nevertheless, information about their location can be obtained by procedures used in sequencing, provided the disulfides are not broken before cleaving the polypeptide chain. Because these covalent bonds are stable under most conditions used in the cleavage of polypeptides, intact disulfides link the peptide fragments containing their specific cysteinyl residues and thus these linked fragments can be isolated and identified within the protein digest.

An effective way to isolate these fragments is through **diagonal electrophoresis** (Figure 5.19) (the basic technique of *electrophoresis* is described in the

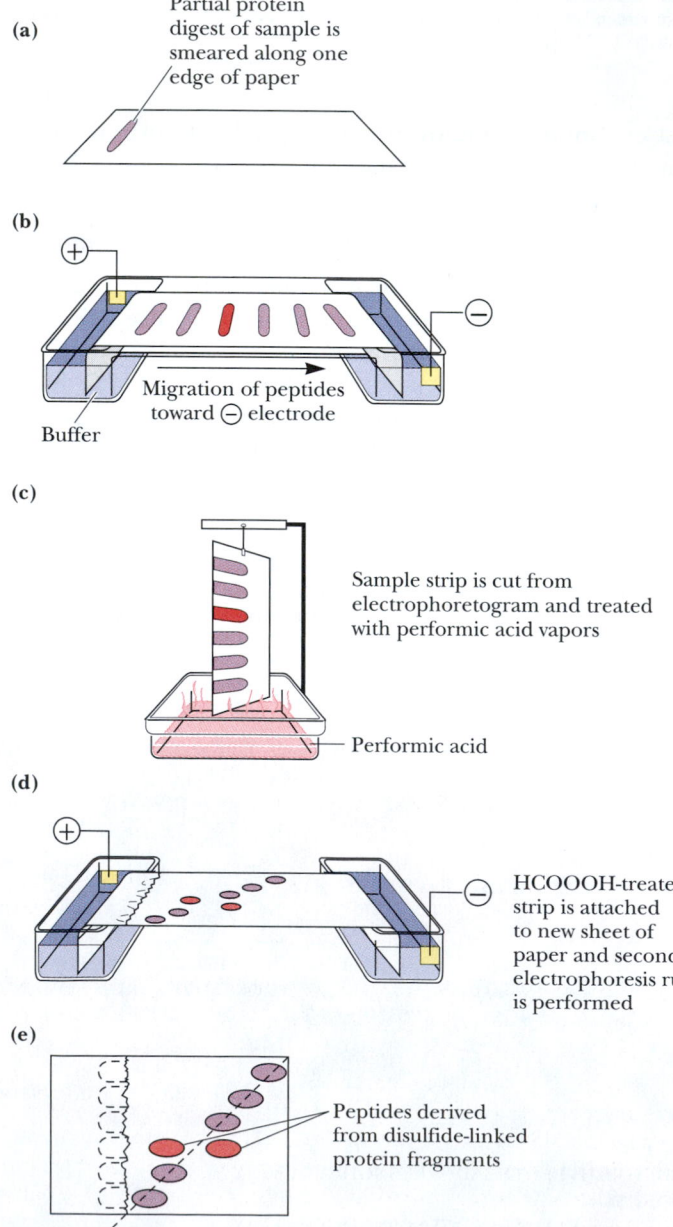

Biochemistry ⑤ Now™   ACTIVE FIGURE 5.19
Disulfide bridges typically are cleaved before determining the primary structure of a polypeptide. Consequently, the positions of disulfide links are not obvious from the sequence data. To determine their location, a sample of the polypeptide with intact S—S bonds can be fragmented and the sites of any disulfides can be elucidated from fragments that remain linked. Diagonal electrophoresis is a technique for identifying such fragments. **(a)** A protein digest in which any disulfide bonds remain intact and link their respective Cys-containing peptides is streaked along the edge of a filter paper and **(b)** subjected to electrophoresis. **(c)** A strip cut from the edge of the paper is then exposed to performic acid fumes to oxidize any disulfide bridges. **(d)** Then the paper strip is attached to a new filter paper so that a second electrophoresis can be run in a direction perpendicular to the first. **(e)** Peptides devoid of disulfides experience no mobility change, and thus their pattern of migration defines a diagonal. Peptides that had disulfides migrate off this diagonal and can be easily identified, isolated, and sequenced to reveal the location of cysteic acid residues formerly involved in disulfide bridges. **Test yourself on the concepts in this figure at** http://chemistry.brookscole.com/ggb3

(a) Partial protein digest of sample is smeared along one edge of paper

(b)
Migration of peptides toward ⊖ electrode
Buffer

(c)
Sample strip is cut from electrophoretogram and treated with performic acid vapors
Performic acid

(d)
HCOOOH-treated strip is attached to new sheet of paper and second electrophoresis run is performed

(e)
Peptides derived from disulfide-linked protein fragments
Diagonal

Chapter Appendix). Peptides that were originally linked by disulfides now migrate as distinct species following disulfide cleavage and are obvious by their location off the diagonal (Figure 5.19e). These cysteic acid–containing peptides are then isolated from the paper and sequenced. From this information, the positions of the disulfides in the protein can be stipulated.

The Amino Acid Sequence of a Protein Can Be Determined by Mass Spectrometry

Mass spectrometers exploit the difference in the mass-to-charge (m/z) ratio of ionized atoms or molecules to separate them from each other. The m/z ratio of a molecule is also a highly characteristic property that can be used to acquire chemical and structural information. Furthermore, molecules can be fragmented in distinctive ways in mass spectrometers, and the fragments that arise also provide quite specific structural information about the molecule. The basic operation of a mass spectrometer is to (1) evaporate and ionize molecules in a vacuum, creating gas-phase ions; (2) separate the ions in space and/or time based on their m/z ratios; and (3) measure the amount of ions with specific m/z ratios. Because proteins (as well as nucleic acids and carbohydrates) decompose upon heating, rather than evaporating, methods to ionize such molecules for mass spectrometry (MS) analysis require innovative approaches. The two most prominent MS modes for protein analysis are summarized in Table 5.5.

Figure 5.20 illustrates the basic features of electrospray mass spectrometry (ES MS). In this technique, the high voltage at the electrode causes proteins to pick up protons from the solvent, such that, on average, individual protein molecules acquire about one positive charge (proton) per kilodalton, leading to the spectrum of m/z ratios for a single protein species (Figure 5.21). Computer

---

Table 5.5

The Two Most Common Methods of Mass Spectrometry for Protein Analysis

Electrospray Ionization (ESI-MS)

A solution of macromolecules is sprayed in the form of fine droplets from a glass capillary under the influence of a strong electrical field. The droplets pick up positive charges as they exit the capillary; evaporation of the solvent leaves multiply charged molecules. The typical 20-kD protein molecule will pick up 10 to 30 positive charges. The MS spectrum of this protein reveals all of the differently charged species as a series of sharp peaks whose consecutive m/z values differ by the charge and mass of a single proton (see Figure 5.21). Note that decreasing m/z values signify increasing number of charges per molecule, z. Tandem mass spectrometers downstream from the ESI source (ESI-MS/MS) can analyze complex protein mixtures (such as tryptic digests of proteins or chromatographically separated proteins emerging from a liquid chromatography column), selecting a single m/z species for collision-induced dissociation and acquisition of amino acid sequence information.

Matrix-Assisted Laser Desorption Ionization-Time of Flight (MALDI-TOF MS)

The protein sample is mixed with a chemical matrix that includes a light-absorbing substance excitable by a laser. A laser pulse is used to excite the chemical matrix, creating a microplasma that transfers the energy to protein molecules in the sample, ionizing them and ejecting them into the gas phase. Among the products are protein molecules that have picked up a single proton. These positively charged species can be selected by the MS for mass analysis. MALDI-TOF MS is very sensitive and very accurate; as little as attomole (10^{-18} moles) quantities of a particular molecule can be detected at accuracies better than 0.001 atomic mass units (0.001 daltons). MALDI-TOF MS is best suited for very accurate mass measurements.

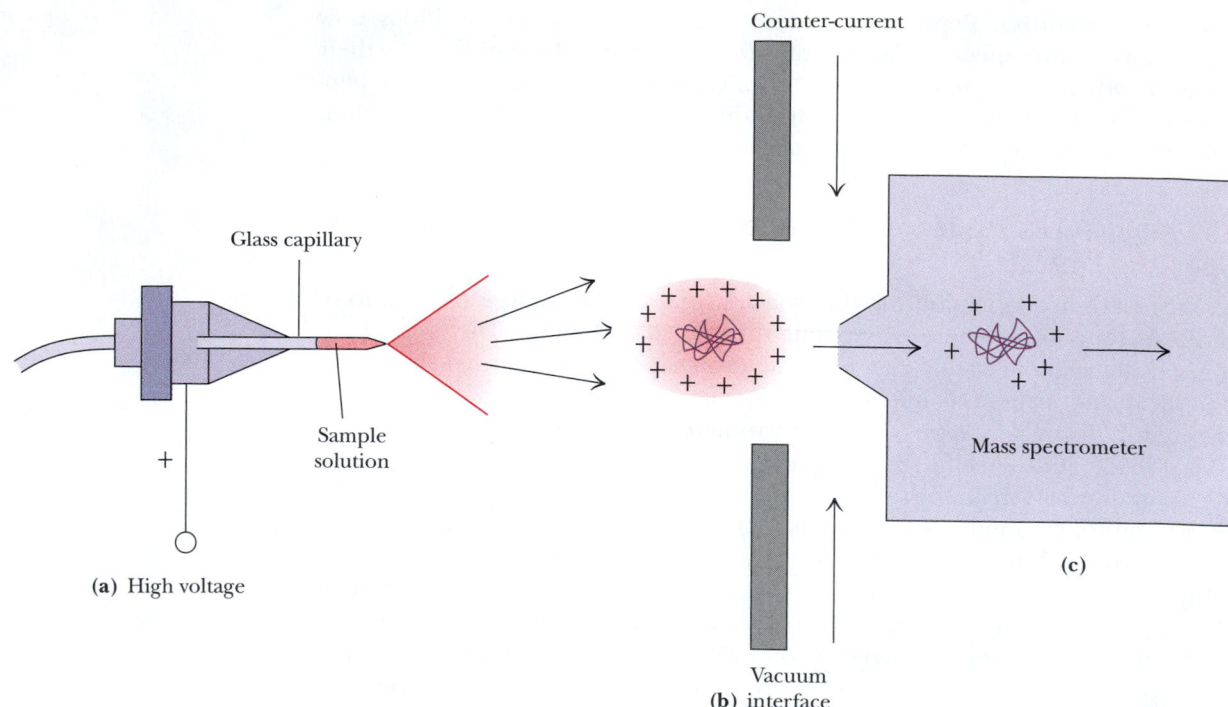

FIGURE 5.20 The three principal steps in electrospray mass spectrometry (ES-MS). **(a)** Small, highly charged droplets are formed by electrostatic dispersion of a protein solution through a glass capillary subjected to a high electric field; **(b)** protein ions are desorbed from the droplets into the gas phase (assisted by evaporation of the droplets in a stream of hot N_2 gas); and **(c)** the protein ions are separated in a mass spectrometer and identified according to their m/z ratios. *(Adapted from Figure 1 in Mann, M., and Wilm, M., 1995. Electrospray mass spectrometry for protein characterization.* Trends in Biochemical Sciences **20**:219–224.)

algorithms can convert these data into a single spectrum that has a peak at the correct protein mass (Figure 5.21, inset).

Sequencing by Tandem Mass Spectrometry *Tandem* MS (or MS/MS) allows sequencing of proteins by hooking two mass spectrometers in tandem. The first mass spectrometer is used as a filter to sort the oligopeptide fragments in a protein digest based on differences in their m/z ratios. Each of these oligopeptides can then be selected by the mass spectrometer for further analysis. A selected ionized oligopeptide is directed toward the second mass spectrometer; on the way, this oligopeptide is fragmented by collision with helium or argon gas molecules (a process called *collision-induced dissociation,* or *c.i.d.*), and the fragments are analyzed by the second mass spectrometer (Figure 5.22). Fragmentation occurs primarily at the peptide bonds linking successive amino acids in the oligopeptide. Thus, the products include a series of fragments that represent a nested set of peptides differing in size by one amino acid residue. The various members of this set of fragments differ in mass by 56 atomic mass units [the mass of the peptide backbone atoms (NH—CH—CO)] *plus* the mass of the R group at each position, which ranges from 1 atomic mass unit (Gly) to 130 (Trp). MS sequencing has the advantages of very high sensitivity, fast sample processing, and the ability to work with mixtures of proteins. Subpicomoles (less than 10^{-12} moles) of peptide can be analyzed with these spectrometers. In practice, tandem MS is limited to rather short sequences (no longer than 15 or so amino acid residues). Nevertheless, capillary HPLC-separated peptide mixtures from trypsin digests of proteins can be directly loaded into the tandem MS spectrometer. Furthermore, separation of a complex mixture of proteins from a whole-cell extract by two-dimensional gel electrophoresis (see Chapter

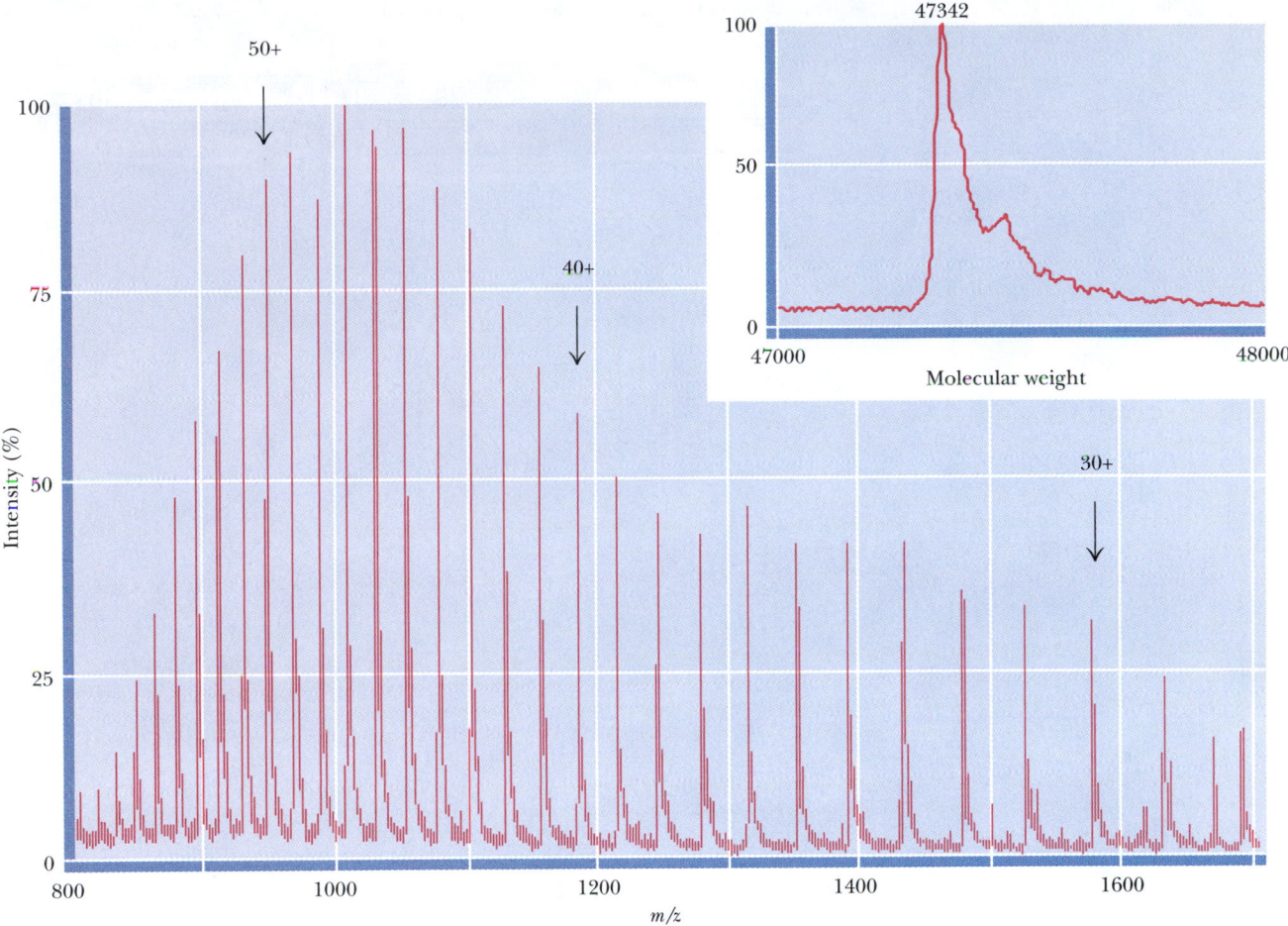

FIGURE 5.21 Electrospray mass spectrum of the protein aerolysin K. The attachment of many protons per protein molecule (from less than 30 to more than 50 here) leads to a series of m/z peaks for this single protein. The equation describing each m/z peak is: $m/z = [M + n(\text{mass of proton})]/n(\text{charge on proton})$, where M = mass of the protein and n = number of positive charges per protein molecule. Thus, if the number of charges per protein molecule is known and m/z is known, M can be calculated. The inset shows a computer analysis of the data from this series of peaks that generates a single peak at the correct molecular mass of the protein. *(Adapted from Figure 2 in Mann, M., and Wilm, M., 1995. Electrospray mass spectrometry for protein characterization. Trends in Biochemical Sciences 20:219–224.)*

Appendix), followed by trypsin digestion of a specific protein spot on the gel and injection of the digest into the HPLC/tandem MS, gives sequence information that can be used to identify specific proteins. Often, by comparing the mass of tryptic peptides from a protein digest with a database of all possible masses for tryptic peptides (based on all known protein and DNA sequences), one can identify a protein of interest without actually sequencing it.

Peptide Mass Fingerprinting *Peptide mass fingerprinting* is used to uniquely identify a protein based on the masses of its proteolytic fragments, usually produced by trypsin digestion. MALDI-TOF MS instruments are ideal for this purpose because they yield highly accurate mass data. The measured masses of the proteolytic fragments can be compared to databases (see following discussion) of peptide masses of known sequence. Such information is easily generated from genomic databases: Nucleotide sequence information can be translated into amino acid sequence information, from which very accurate peptide mass compilations are readily calculated. For example, the SWISS-PROT database lists 1197 proteins with a tryptic fragment of $m/z = 1335.63$ (± 0.2 D), 16 proteins

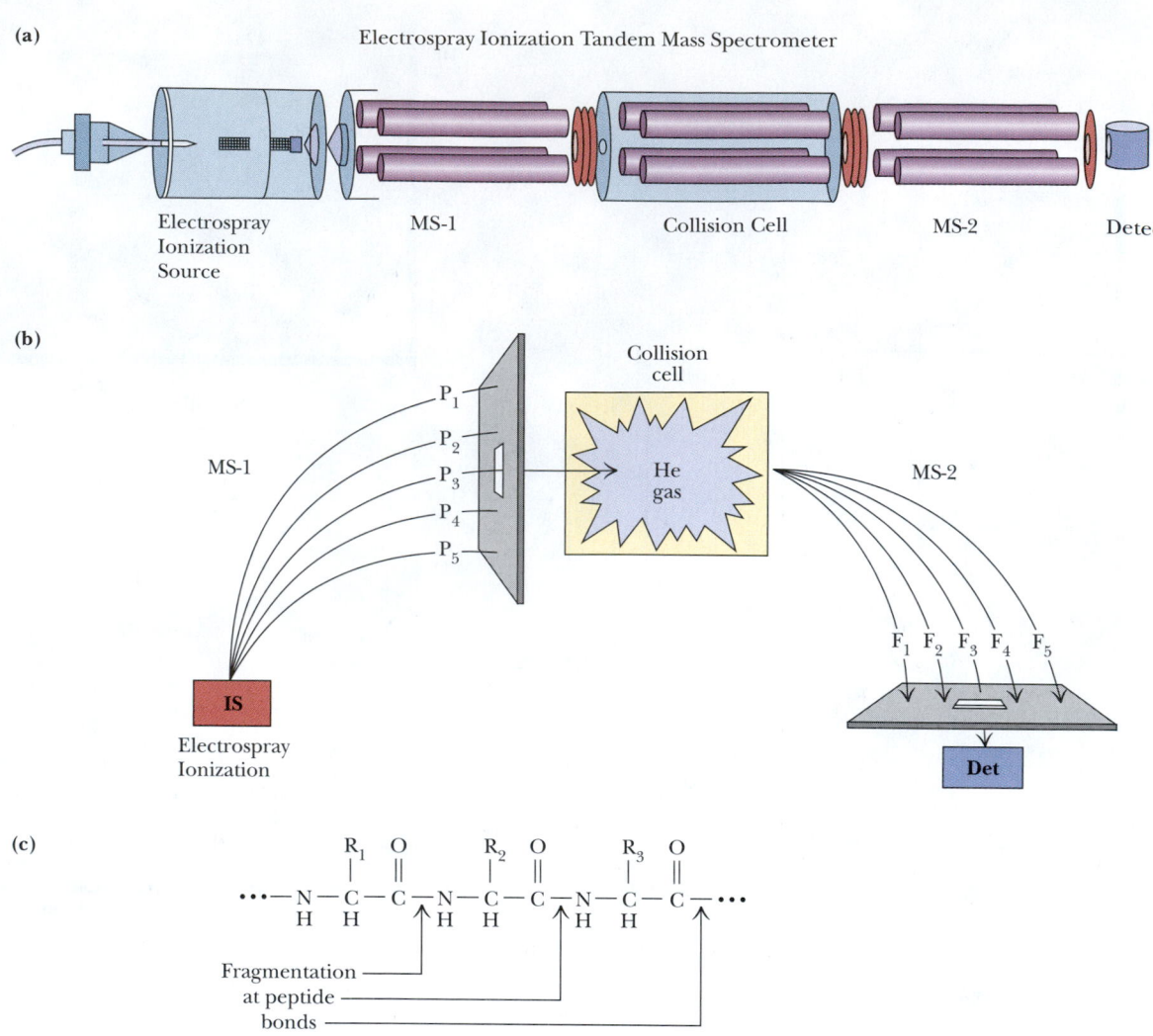

(a)   Electrospray Ionization Tandem Mass Spectrometer

Electrospray
Ionization
Source

MS-1   Collision Cell   MS-2   Detector

FIGURE 5.22 Tandem mass spectrometry. **(a)** Configuration used in tandem MS. **(b)** Schematic description of tandem MS: Tandem MS involves electrospray ionization of a protein digest (IS in this figure), followed by selection of a single peptide ion mass for collision with inert gas molecules (He) and mass analysis of the fragment ions resulting from the collisions. **(c)** Fragmentation usually occurs at peptide bonds, as indicated. *(Adapted from Yates, J. R., 1996. Protein structure analysis by mass spectrometry. Methods in Enzymology **271**:351–376; and Gillece-Castro, B. L., and Stults, J. T., 1996. Peptide characterization by mass spectrometry. Methods in Enzymology **271**:427–447.)*

with tryptic fragments of $m/z = 1335.63$ *and* $m/z = 1405.60$, but only a single protein (human tissue plasminogen activator [tPA]) with tryptic fragments of $m/z = 1335.63$, $m/z = 1405.60$, *and* $m/z = 1272.60$.[2] Although the identities of many proteins revealed by genomic analysis remain unknown, peptide mass fingerprinting can assign a particular protein exclusively to a specific gene in a genomic database.

Sequence Databases Contain the Amino Acid Sequences of a Million Different Proteins

The first protein sequence databases were compiled by protein chemists using chemical sequencing methods. Today, the vast preponderance of protein sequence information has been derived from translating the nucleotide sequences of genes into codons and, thus, amino acid sequences (see Chap-

---

[2]The tPA amino acid sequences corresponding to these masses are $m/z = 1335.63$: HEALSPFYSER; $m/z = 1405.60$: ATCYEDQGISYR; and $m/z = 1272.60$: DSKPWCYVFK.

ter 12). Sequencing the order of nucleotides in cloned genes is a more rapid, efficient, and informative process than determining the amino acid sequences of proteins by chemical methods. Several electronic databases containing continuously updated sequence information are accessible by personal computer. Prominent among these is the SWISS-PROT protein sequence database on the ExPASy (**Ex**pert **P**rotein **A**nalysis **Sy**stem) Molecular Biology server at *http://us.expasy.org* and the PIR (Protein Identification Resource Protein Sequence Database) at *http://pir.georgetown.edu,* as well as protein information from genomic sequences available in databases such as GenBank, accessible via the National Center for Biotechnology Information (NCBI) Web site located at *http://www.ncbi.nlm.nih.gov.* The protein sequence databases contain close to 1 million entries, whereas the genomic databases list tens of millions of nucleotide sequences covering tens of billions of base pairs. The Protein Data Bank (PDB; *http://www.rcsb.org/pdb*) is a protein database that provides three-dimensional structure information on more than 20,000 proteins and nucleic acids.

5.6 Can Polypeptides Be Synthesized in the Laboratory?

Chemical synthesis of peptides and polypeptides of defined sequence can be carried out in the laboratory. Formation of peptide bonds linking amino acids together is not a chemically complex process, but making a specific peptide can be challenging because various functional groups present on side chains of amino acids may also react under the conditions used to form peptide bonds. Furthermore, if correct sequences are to be synthesized, the α-COOH group of residue x must be linked to the α-NH$_2$ group of neighboring residue y in a way that prevents reaction of the amino group of x with the carboxyl group of y. In essence, any functional groups to be protected from reaction must be blocked while the desired coupling reactions proceed. Also, the blocking groups must be removable later under conditions in which the newly formed peptide bonds are stable. An ingenious synthetic strategy to circumvent these technical problems is *orthogonal synthesis*. An orthogonal system is defined as a set of distinctly different blocking groups—one for side-chain protection, another for α-amino protection, and a third for α-carboxyl protection or anchoring to a solid support (see following discussion). Ideally, any of the three classes of protecting groups can be removed in any order and in the presence of the other two, because the reaction chemistries of the three classes are sufficiently different from one another. In peptide synthesis, all reactions must proceed with high yield if peptide recoveries are to be acceptable. Peptide formation between amino and carboxyl groups is not spontaneous under normal conditions (see Chapter 4), so one or the other of these groups must be activated to facilitate the reaction. Despite these difficulties, biologically active peptides and polypeptides have been recreated by synthetic organic chemistry. Milestones include the pioneering synthesis of the nonapeptide posterior pituitary hormones oxytocin and vasopressin by du Vigneaud in 1953 and, in later years, larger proteins such as insulin (21 A-chain and 30 B-chain residues), ribonuclease A (124 residues), and HIV protease (99 residues).

Solid-Phase Methods Are Very Useful in Peptide Synthesis

Bruce Merrifield and his collaborators pioneered a clever solution to the problem of recovering intermediate products in the course of a synthesis. The carboxyl-terminal residues of synthesized peptide chains are covalently anchored to an insoluble resin (polystyrene particles) that can be removed from reaction mixtures simply by filtration. After each new residue is added successively at the free amino-terminus, the elongated product is recovered

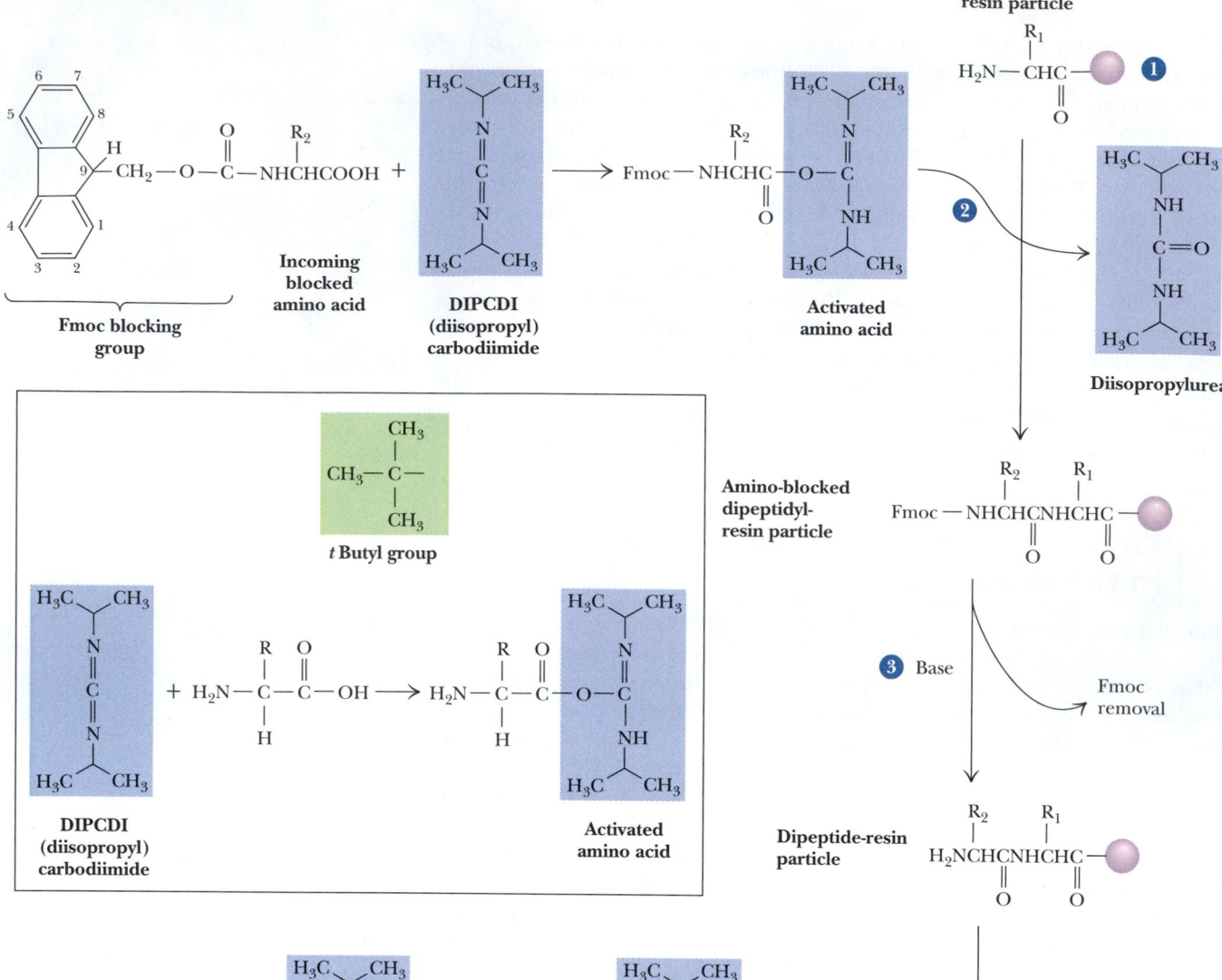

Biochemistry⊜Now™   **ANIMATED FIGURE 5.23** Solid-phase synthesis of a peptide. The 9-fluorenylmethoxycarbonyl (Fmoc) group is an excellent orthogonal blocking group for the α-amino group of amino acids during organic synthesis because it is readily removed under basic conditions that don't affect the linkage between the insoluble resin and the α-carboxyl group of the growing peptide chain. *(inset)* N,N'-diisopropylcarbodiimide (DIPCDI) is one agent of choice for activating carboxyl groups to condense with amino groups to form peptide bonds. **(1)** The carboxyl group of the first amino acid (the carboxyl-terminal amino acid of the peptide to be synthesized) is chemically attached to an insoluble resin particle (the *aminoacyl-resin particle*). **(2)** The second amino acid, with its amino group blocked by a Fmoc group and its carboxyl group activated with DIPCDI, is reacted with the aminoacyl-resin particle to form a peptide linkage, with elimination of DIPCDI as diisopropylurea. **(3)** Then, basic treatment (with piperidine) removes the N-terminal Fmoc blocking group, exposing the N-terminus of the dipeptide for another cycle of amino acid addition **(4).** Any reactive side chains on amino acids are blocked by addition of acid-labile *tertiary* butyl (*t*Bu) groups as an orthogonal protective functions. **(5)** After each step, the peptide product is recovered by collection of the insoluble resin beads by filtration or centrifugation. Following cyclic additions of amino acids, the completed peptide chain is hydrolyzed from linkage to the insoluble resin by treatment with HF; HF also removes any *t*Bu protecting groups from side chains on the peptide. **See this figure animated at http://chemistry.brookscole.com/ggb3**

by filtration and readied for the next synthetic step. Because the growing peptide chain is coupled to an insoluble resin bead, the method is called **solid-phase synthesis.** The procedure is detailed in Figure 5.23. This cyclic process is automated and computer controlled so that the reactions take place in a small cup with reagents being pumped in and removed as programmed.

5.7 | What Is the Nature of Amino Acid Sequences?

Figure 5.24 illustrates the relative frequencies of the amino acids in proteins. Although these data are for all proteins, it is very unusual for a globular protein to have an amino acid composition that deviates substantially from these values. Apparently, these abundances reflect a distribution of amino acid polarities that is optimal for protein stability in an aqueous milieu. Membrane proteins tend to have relatively more hydrophobic and fewer ionic amino acids, a condition consistent with their location. Fibrous proteins may show compositions that are atypical with respect to these norms, indicating an underlying relationship between the composition and the structure of these proteins.

Proteins have unique amino acid sequences, and it is this uniqueness of sequence that ultimately gives each protein its own particular personality. Because the number of possible amino acid sequences in a protein is astronomically large, the probability that two proteins will, by chance, have similar amino acid sequences is negligible. Consequently, sequence similarities between proteins imply evolutionary relatedness.

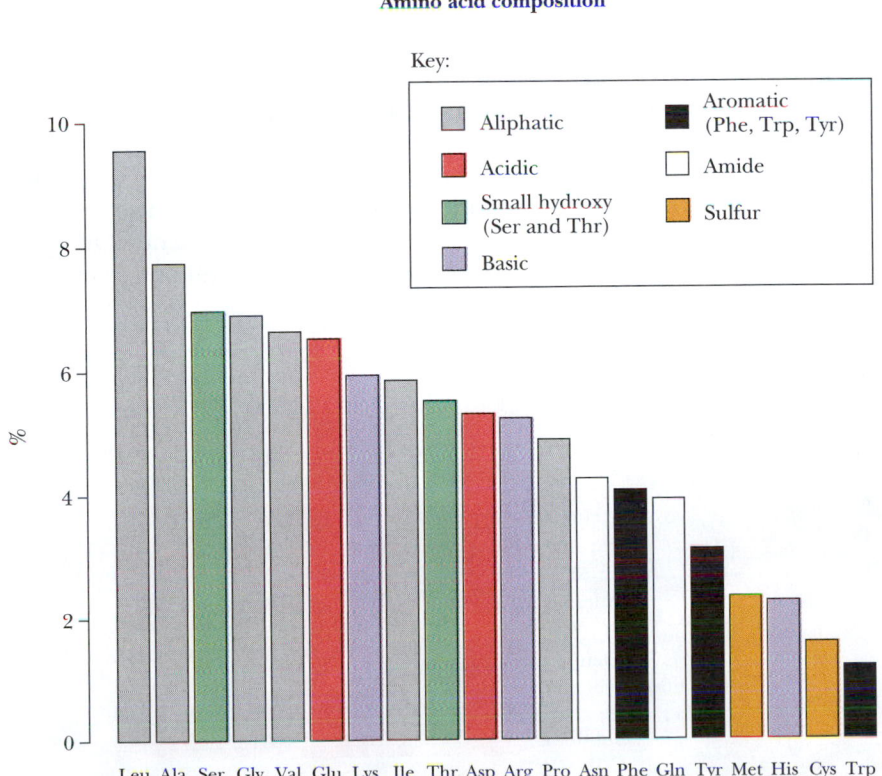

Amino acid composition

Key:

- Aliphatic (gray)
- Acidic (red)
- Small hydroxy (Ser and Thr) (green)
- Basic (lavender)
- Aromatic (Phe, Trp, Tyr) (black)
- Amide (white)
- Sulfur (orange)

(x-axis: Leu Ala Ser Gly Val Glu Lys Ile Thr Asp Arg Pro Asn Phe Gln Tyr Met His Cys Trp)
(y-axis: % from 0 to 10)

FIGURE 5.24 Amino acid composition: Frequencies of the various amino acids in proteins for all the proteins in the SWISS-PROT protein knowledgebase. These data are derived from the amino acid composition of more than 100,000 different proteins (representing more than 40,000,000 amino acid residues). The range is from leucine at 9.55% to tryptophan at 1.18% of all residues.

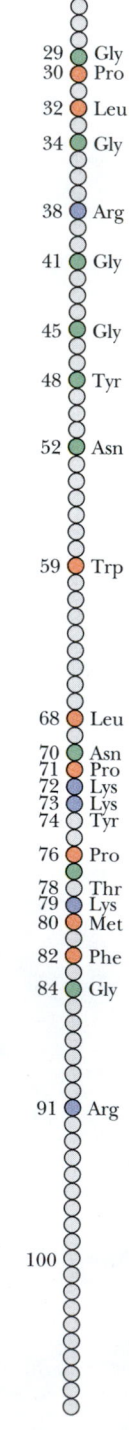

Homologous Proteins from Different Organisms Have Homologous Amino Acid Sequences

Proteins sharing a significant degree of sequence similarity are said to be **homologous.** Proteins that perform the same function in different organisms are also referred to as homologous. For example, the oxygen transport protein hemoglobin serves a similar role and has a similar structure in all vertebrates. The study of the amino acid sequences of homologous proteins from different organisms provides very strong evidence for their evolutionary origin within a common ancestor. Homologous proteins characteristically have polypeptide chains that are nearly identical in length, and their sequences share identity in direct correlation to the relatedness of the species from which they are derived.

Cytochrome c   The electron transport protein **cytochrome c,** found in the mitochondria of all eukaryotic organisms, provides the best-studied example of homology. The polypeptide chain of cytochrome c from most species contains slightly more than 100 amino acids and has a molecular weight of about 12.5 kD. Amino acid sequencing of cytochrome c from more than 40 different species has revealed that there are 28 positions in the polypeptide chain where the same amino acid residues are always found (Figure 5.25). These **invariant residues** serve roles crucial to the biological function of this protein, and thus substitutions of other amino acids at these positions cannot be tolerated.

Furthermore, as shown in Figure 5.26, the number of amino acid differences between two cytochrome c sequences is proportional to the phylogenetic difference between the species from which they are derived. Cytochrome c in humans and in chimpanzees is identical; human and another mammalian (sheep) cytochrome c differ at 10 residues. The human cytochrome c sequence has 14 variant residues from a reptile sequence (rattlesnake), 18 from a fish (carp), 29 from a mollusc (snail), 31 from an insect (moth), and more than 40 from yeast or higher plants (cauliflower).

The Phylogenetic Tree for Cytochrome c   Figure 5.27 displays a **phylogenetic tree** (a diagram illustrating the evolutionary relationships among a group of organisms) constructed from the sequences of cytochrome c. The tips of the branches are occupied by contemporary species whose sequences have been determined. The tree has been deduced by computer analysis of these sequences to find the minimum number of mutational changes connecting the branches. Other computer methods can be used to infer potential ancestral sequences represented

FIGURE 5.25 Cytochrome c is a small protein consisting of a single polypeptide chain of 104 residues in terrestrial vertebrates, 103 or 104 in fishes, 107 in insects, 107 to 109 in fungi and yeasts, and 111 or 112 in green plants. Analysis of the sequence of cytochrome c from more than 40 different species reveals that 28 residues are invariant. These invariant residues are scattered irregularly along the polypeptide chain, except for a cluster between residues 70 and 80. All cytochrome c polypeptide chains have a cysteine residue at position 17, and all but one have another Cys at position 14. These Cys residues serve to link the heme prosthetic group of cytochrome c to the protein, a role explaining their invariable presence.

	Chimpanzee	Sheep	Rattlesnake	Carp	Snail	Moth	Yeast	Cauliflower	Parsnip
Human	0	10	14	18	29	31	44	44	43
Chimpanzee		10	14	18	29	31	44	44	43
Sheep			20	11	24	27	44	46	46
Rattlesnake				26	28	33	47	45	43
Carp					26	26	44	47	46
Garden snail						28	48	51	50
Tobacco hornworm moth							44	44	41
Baker's yeast (iso-1)								47	47
Cauliflower									13

FIGURE 5.26 The number of amino acid differences among the cytochrome c sequences of various organisms can be compared. The numbers bear a direct relationship to the degree of relatedness between the organisms. Each of these species has a cytochrome c of at least 104 residues, so any given pair of species has more than half its residues in common. (Adapted from Creighton, T. E., 1983. Proteins: Structure and Molecular Properties. San Francisco: W. H. Freeman.)

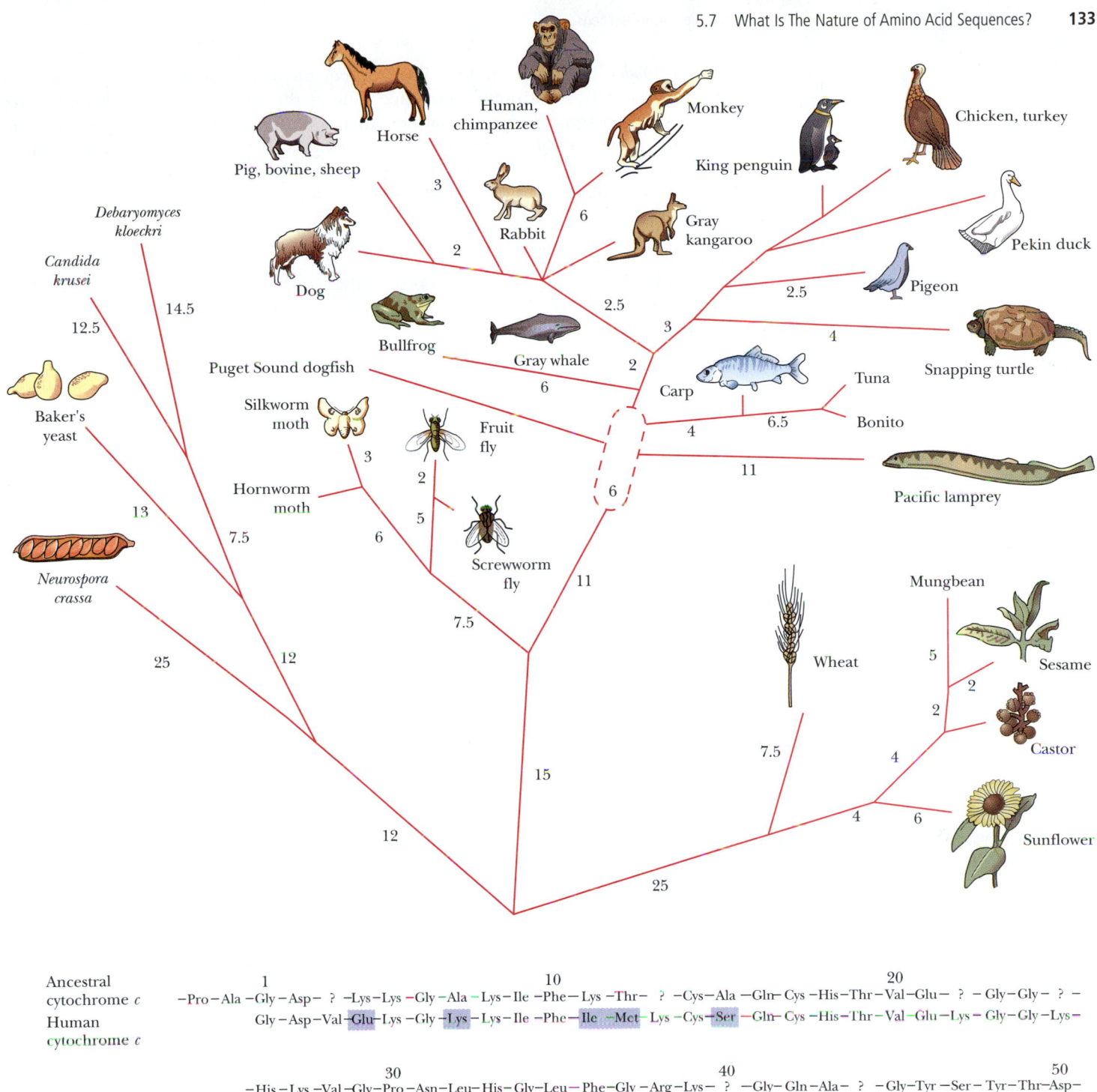

FIGURE 5.27 This phylogenetic tree depicts the evolutionary relationships among organisms as determined by the similarity of their cytochrome *c* amino acid sequences. The numbers along the branches give the amino acid changes between a species and a hypothetical progenitor. Note that extant species are located only at the tips of branches. Below, the sequence of human cytochrome *c* is compared with an inferred ancestral sequence represented by the base of the tree. Uncertainties are denoted by question marks. (*Adapted from Creighton, T. E., 1983. Proteins: Structure and Molecular Properties. San Francisco: W. H. Freeman.*)

Ancestral cytochrome *c*

Human cytochrome *c*

```
                    1                        10                           20
Ancestral      −Pro−Ala −Gly−Asp− ? −Lys−Lys−Gly −Ala−Lys−Ile −Phe−Lys −Thr− ? −Cys−Ala −Gln− Cys −His−Thr−Val−Glu− ? − Gly−Gly− ? −
Human           Gly−Asp−Val−Glu−Lys −Gly −Lys−Lys−Ile −Phe−Ile −Met− Lys−Cys−Ser − Gln− Cys −His−Thr−Val−Glu−Lys − Gly−Gly −Lys−

                         30                           40                              50
               −His−Lys −Val−Gly−Pro−Asn−Leu−His−Gly−Leu− Phe−Gly −Arg−Lys− ? −Gly−Gln−Ala− ? −Gly−Tyr −Ser −Tyr−Thr−Asp−
               −His−Lys −Thr−Gly−Pro−Asn−Leu−His−Gly−Leu− Phe−Gly −Arg−Lys−Thr−Gly−Gln−Ala−Pro −Gly−Tyr −Ser −Tyr−Thr−Ala−

                         60                           70
               −Ala −Asn−Lys−Asn−Lys−Gly− ? − ? −Trp− ? − Glu−Asn−Thr−Leu−Phe−Glu−Tyr−Leu−Glu−Asn−Pro −Lys − Lys−Tyr − Ile −
               −Ala −Asn−Lys−Asn−Lys −Gly − Ile − Ile−Trp−Gly − Glu−Asp−Thr−Leu−Met−Gln− Tyr−Leu−Glu−Asn−Pro−Lys − Lys−Tyr −Pro−

                         80                           90                           100
               −Pro−Gly−Thr−Lys−Met− ? −Phe− ? −Gly−Leu−Lys−Lys − ? − ? −Asp −Arg − Ala−Asp−Leu−Ile −Ala −Tyr−Leu−Lys− ? −
               −Pro−Gly−Thr−Lys−Met−Ile −Phe−Val−Gly−Ile − Lys −Lys −Lys−Glu−Glu −Arg − Ala−Asp−Leu−Ile −Ala −Tyr−Leu−Lys −Lys−

               −Ala−Thr−Ala
               −Ala−Thr−Asn−Glu
```

by *nodes,* or branch points, in the tree. Such analysis ultimately suggests a primordial cytochrome *c* sequence lying at the base of the tree. Evolutionary trees constructed in this manner, that is, solely on the basis of amino acid differences occurring in the primary sequence of one selected protein, show remarkable agreement with phylogenetic relationships derived from more classic approaches and have given rise to the field of *molecular evolution.*

Related Proteins Share a Common Evolutionary Origin

Amino acid sequence analysis reveals that proteins with related functions often show a high degree of sequence similarity. Such findings suggest a common ancestry for these proteins.

Oxygen-Binding Heme Proteins　The oxygen-binding heme protein of muscle, **myoglobin,** consists of a single polypeptide chain of 153 residues. **Hemoglobin,** the oxygen transport protein of erythrocytes, is a tetramer composed of two **α-chains** (141 residues each) and two **β-chains** (146 residues each). These globin polypeptides—myoglobin, α-globin, and β-globin—share a strong degree of sequence homology (Figure 5.28). Human myoglobin and the human α-globin chain show 38 amino acid identities, whereas human α-globin and human

▼ **FIGURE 5.28** Inspection of the amino acid sequences of the globin chains of human hemoglobin and myoglobin reveals a strong degree of homology. The α- and β-globin chains share 64 residues of their approximately 140 residues in common. Myoglobin and the α-globin chain have 38 amino acid sequence identities. This homology is further reflected in these proteins' tertiary structure. *(Illustration: Irving Geis. Rights owned by Howard Hughes Medical Institute. Not to be reproduced without permission.)*

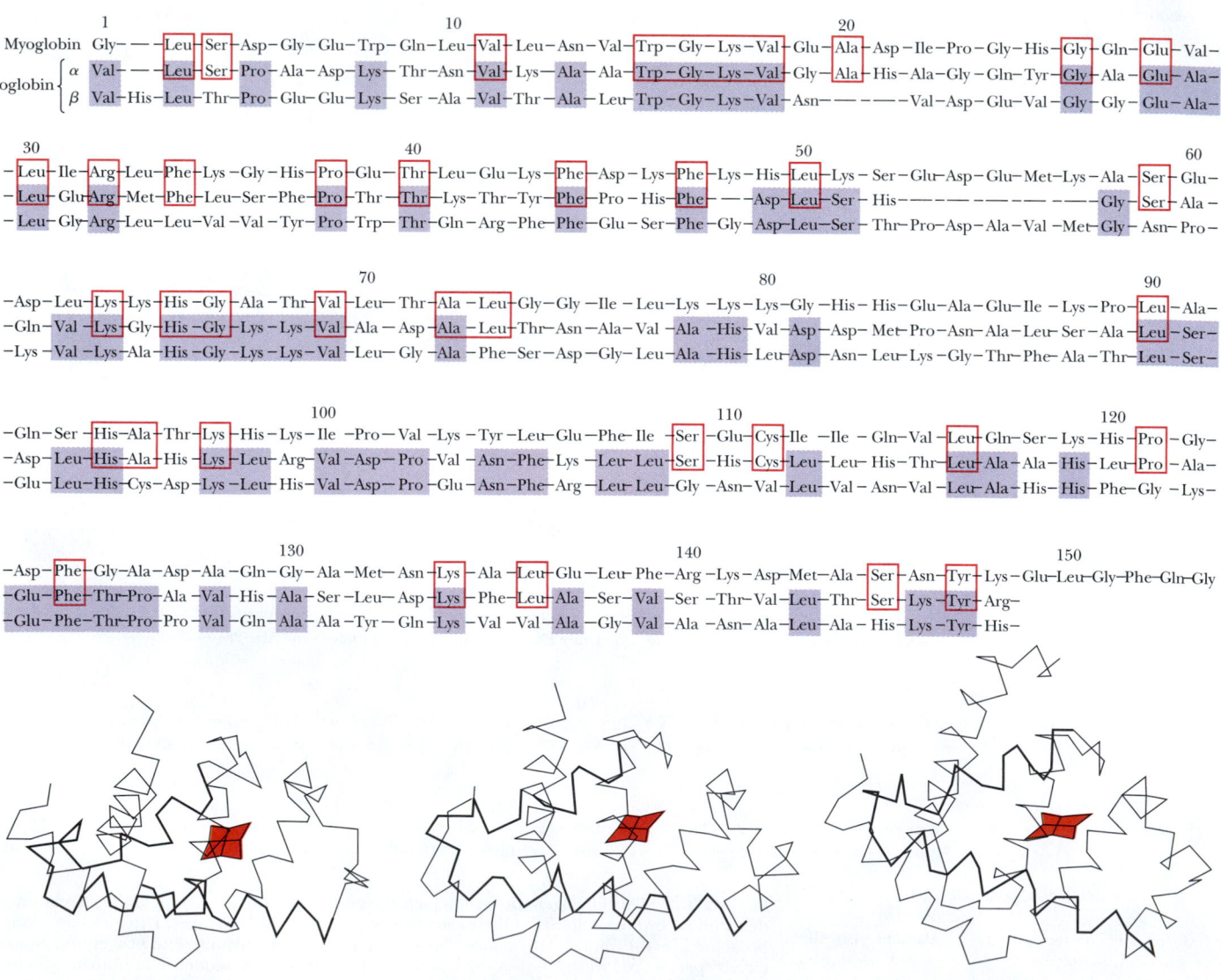

α-chain of horse methemoglobin　　　　β-chain of horse methemoglobin　　　　Sperm whale myoglobin β-globin have 64 residues in common. The relatedness suggests an evolutionary sequence of events in which chance mutations led to amino acid substitutions and divergence in primary structure. The ancestral myoglobin gene diverged first, after duplication of a primordial globin gene had given rise to its progenitor and an ancestral hemoglobin gene (Figure 5.29). Subsequently, the ancestral hemoglobin gene duplicated to generate the progenitors of the present-day α-globin and β-globin genes. The ability to bind O₂ via a heme prosthetic group is retained by all three of these polypeptides.

Serine Proteases   Whereas the globins provide an example of gene duplication giving rise to a set of proteins in which the biological function has been highly conserved, other sets of proteins united by strong sequence homology show more divergent biological functions. **Trypsin, chymotrypsin** (see Section 5.5), and **elastase** are members of a class of proteolytic enzymes called **serine proteases** because of the central role played by specific serine residues in their catalytic activity. **Thrombin,** an essential enzyme in blood clotting, is also a serine protease. These enzymes show sufficient sequence homology to conclude that they arose via duplication of a progenitor serine protease gene, even though their substrate preferences are now quite different.

Apparently Different Proteins May Share a Common Ancestry

A more remarkable example of evolutionary relatedness is inferred from sequence homology between hen egg white **lysozyme** and human milk **α-lactalbumin**, proteins of different biological activity and origin. Lysozyme (129 residues) and α-lactalbumin (123 residues) are identical at 48 positions. Lysozyme hydrolyzes the polysaccharide wall of bacterial cells, whereas α-lactalbumin regulates milk sugar (lactose) synthesis in the mammary gland. Although both proteins act in reactions involving carbohydrates, their functions show little similarity otherwise. Nevertheless, their tertiary structures are strikingly similar (Figure 5.30). It is conceivable that many proteins

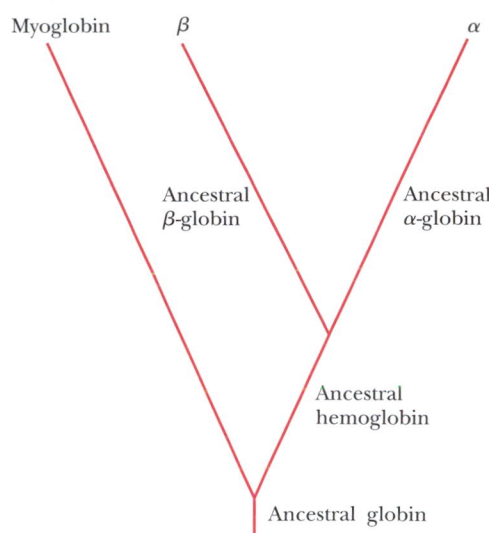

FIGURE 5.29 This evolutionary tree is inferred from the homology between the amino acid sequences of the α-globin, β-globin, and myoglobin chains. Duplication of an ancestral globin gene allowed the divergence of the myoglobin and ancestral hemoglobin genes. Another gene duplication event subsequently gave rise to ancestral α and β forms, as indicated. Gene duplication is an important evolutionary force in creating diversity.

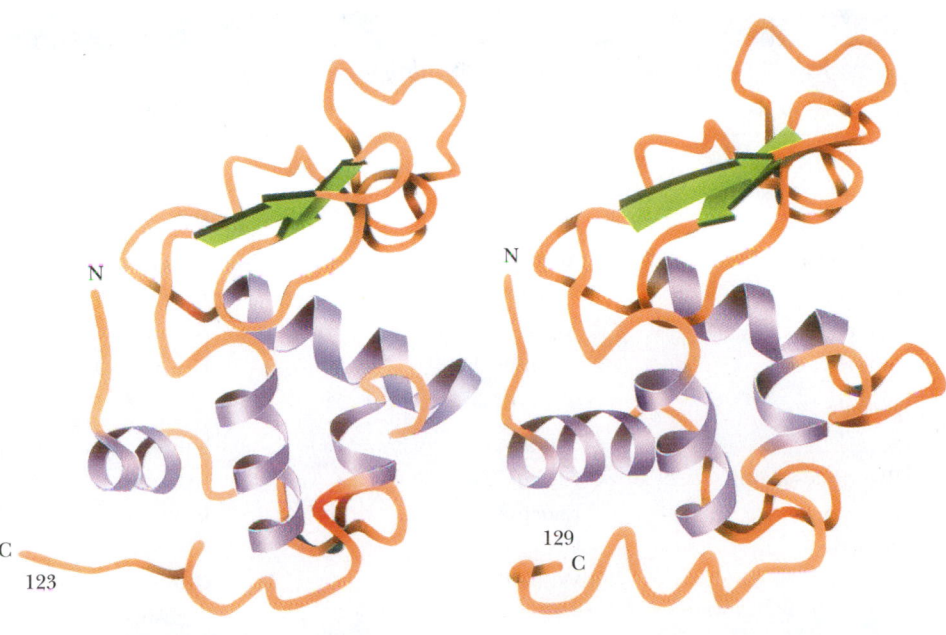

Human milk α-lactalbumin   Hen egg white lysozyme

FIGURE 5.30 The tertiary structures of hen egg white lysozyme and human α-lactalbumin are very similar. *(Adapted from Acharya, K. R., et al., 1990. A critical evaluation of the predicted and X-ray structures of alpha-lactalbumin.* Journal of Protein Chemistry *9:549–563; and Acharya, K. R., et al., 1991. Crystal structure of human alpha-lactalbumin at 1.7 Å resolution.* Journal of Molecular Biology *221:571–581.)*

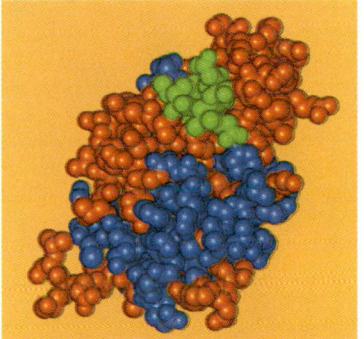

α-Lactalbumin

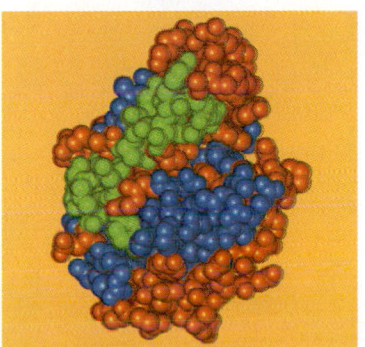

Lysozyme

are related in this way, but time and the course of evolutionary change erased most evidence of their common ancestry. In contrast to this case, the proteins *G-actin* and *hexokinase* share essentially no sequence homology, yet they have strikingly similar three-dimensional structures, even though their biological roles and physical properties are very different. Actin forms a filamentous polymer that is a principal component of the contractile apparatus in muscle; hexokinase is a cytosolic enzyme that catalyzes the first reaction in glucose catabolism.

A Mutant Protein Is a Protein with a Slightly Different Amino Acid Sequence

Given a large population of individuals, a considerable number of sequence variants can be found for a protein. These variants are a consequence of **mutations** in a gene (base substitutions in DNA) that have arisen naturally within the population. Gene mutations lead to mutant forms of the protein in which the amino acid sequence is altered at one or more positions. Many of these mutant forms are "neutral" in that the functional properties of the protein are unaffected by the amino acid substitution. Others may be nonfunctional (if loss of function is not lethal to the individual), and still others may display a range of aberrations between these two extremes. The severity of the effects on function depends on the nature of the amino acid substitution and its role in the protein. These conclusions are exemplified by the more than 300 human hemoglobin variants that have been discovered to date. Some of these are listed in Table 5.6.

A variety of effects on the hemoglobin molecule are seen in these mutants, including alterations in oxygen affinity, heme affinity, stability, solubility, and subunit interactions between the α-globin and β-globin polypeptide chains. Some variants show no apparent changes, whereas others, such as HbS, sickle-cell hemoglobin (see Chapter 15), result in serious illness. This diversity of response indicates that some amino acid changes are relatively unimportant, whereas others drastically alter one or more functions of a protein.

Table 5.6		
Some Pathological Sequence Variants of Human Hemoglobin		
Abnormal Hemoglobin*	**Normal Residue and Position**	**Substitution**
α-chain		
Torino	Phenylalanine 43	Valine
M$_{Boston}$	Histidine 58	Tyrosine
Chesapeake	Arginine 92	Leucine
G$_{Georgia}$	Proline 95	Leucine
Tarrant	Aspartate 126	Asparagine
Suresnes	Arginine 141	Histidine
β-chain		
S	Glutamate 6	Valine
Riverdale–Bronx	Glycine 24	Arginine
Genova	Leucine 28	Proline
Zurich	Histidine 63	Arginine
M$_{Milwaukee}$	Valine 67	Glutamate
M$_{Hyde Park}$	Histidine 92	Tyrosine
Yoshizuka	Asparagine 108	Aspartate
Hiroshima	Histidine 146	Aspartate

*Hemoglobin variants are often given the geographical name of their origin.
Adapted from Dickerson, R. E., and Geis, I., 1983. *Hemoglobin: Structure, Function, Evolution and Pathology*. Menlo Park, CA: Benjamin/Cummings.

5.8 | Do Proteins Have Chemical Groups Other Than Amino Acids?

Many proteins consist of only amino acids and contain no other chemical groups. The enzyme ribonuclease and the contractile protein actin are two such examples. Such proteins are called **simple proteins.** However, many other proteins contain various chemical constituents as an integral part of their structure. These proteins are termed **conjugated proteins.** If the nonprotein part is crucial to the protein's function, it is referred to as a **prosthetic group.** If the nonprotein moiety is not covalently linked to the protein, it can usually be removed by denaturing the protein structure. However, if the conjugate is covalently joined to the protein, it may be necessary to carry out acid hydrolysis of the protein into its component amino acids in order to release it. Conjugated proteins are typically classified according to the chemical nature of their non–amino acid component; a representative selection of them follows. (Note that *chemical composition* [Section 5.8] and *function* [Section 5.9] represent two distinctly different ways of considering the nature of proteins.)

Glycoproteins Are Proteins Containing Carbohydrate Groups

Glycoproteins are proteins that contain carbohydrate. Proteins destined for an extracellular location are characteristically glycoproteins. For example, fibronectin and proteoglycans are important components of the extracellular matrix that surrounds the cells of most tissues in animals. The carbohydrate portions of the proteoglycans may constitute 90% of the mass and the protein only 10%. Immunoglobulin G molecules (less than 2% carbohydrate by weight) are the principal antibody species found circulating free in the blood plasma. Many membrane proteins are glycosylated on their extracellular segments.

Lipoproteins Are Proteins That Are Associated with Lipid Molecules

Blood plasma lipoproteins are prominent examples of the class of proteins conjugated with lipid. The plasma lipoproteins function primarily in the transport of lipids to sites of active membrane synthesis. Lipoprotein complexes may be as much as 75% lipid by weight. Serum levels of *low-density lipoproteins* (LDLs) are often used as a clinical index of susceptibility to vascular disease. Other lipoproteins (such as protein kinase A) are covalently linked to a single acyl group contributed by a fatty acid.

Nucleoproteins Are Proteins Joined with Nucleic Acids

Nucleoprotein conjugates have many roles in the storage and transmission of genetic information. Ribosomes, which possess about 60% RNA by weight, are the sites of protein synthesis. Virus particles and even chromosomes are protein–nucleic acid complexes. And, some enzymes that operate on nucleic acids are nucleoproteins; for example, the human version of telomerase, an enzyme that adds nucleotides at the ends of chromosomes, uses part of its 962-nucleotide RNA prosthetic group as a template for DNA synthesis.

Phosphoproteins Contain Phosphate Groups

Phosphoproteins have phosphate groups esterified to the hydroxyls of serine, threonine, or tyrosine residues. Casein, the major protein of milk, contains many phosphates and serves to bring essential phosphorus to the growing infant. Many key steps in metabolism are regulated between states of activity or inactivity, depending on the presence or absence of phosphate groups on proteins, as we shall see in Chapter 15. Glycogen phosphorylase *a* is one well-studied example.

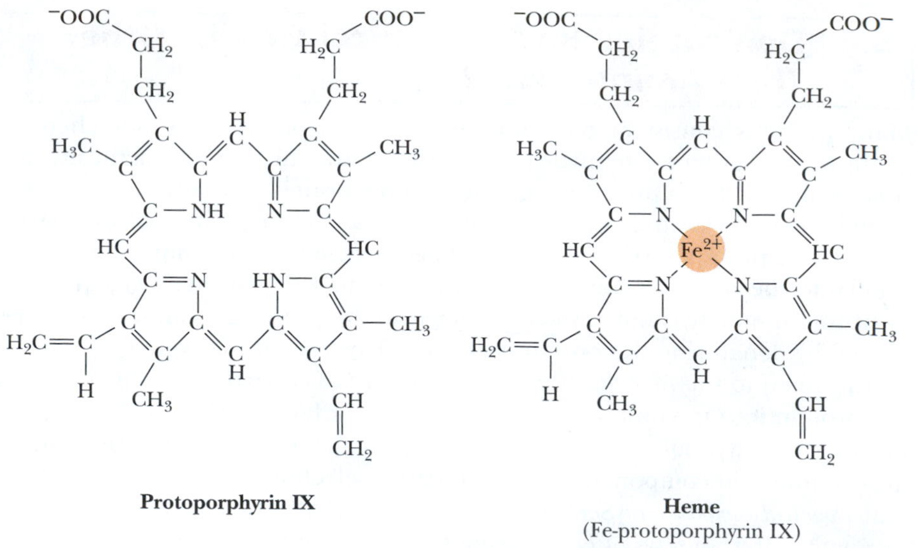

FIGURE 5.31 Heme consists of protoporphyrin IX and an iron atom. Protoporphyrin, a highly conjugated system of double bonds, is composed of four 5-membered heterocyclic rings (pyrroles) fused together to form a tetrapyrrole macrocycle. The specific isomeric arrangement of methyl, vinyl, and propionate side chains shown is protoporphyrin IX. Coordination of an atom of ferrous iron (Fe^{2+}) by the four pyrrole nitrogen atoms yields heme.

Protoporphyrin IX

Heme
(Fe-protoporphyrin IX)

Metalloproteins Are Protein–Metal Complexes

Metalloproteins are either metal storage forms, as in the case of ferritin (35% iron by weight, bearing as many as 4500 Fe atoms), or enzymes in which one or a few metal atoms participate in a catalytically important manner. We encounter many examples throughout this book of the vital metabolic functions served by metalloenzymes.

Hemoproteins Contain Heme

Hemoproteins are actually a subclass of metalloproteins because their prosthetic group is **heme,** the name given to iron protoporphyrin IX (Figure 5.31). Because heme-containing proteins enjoy so many prominent biological functions, they are often placed in a class by themselves. Hemoglobin has 4 hemes, collectively contributing about 4% to its mass.

Flavoproteins Contain Riboflavin

Flavin is an essential substance for the activity of a number of important oxidoreductases. We discuss the chemistry of flavin and its derivatives, FMN and FAD, in Chapter 20.

   Let us now take a brief look at the functional diversity found in proteins, the most interesting of the macromolecules.

5.9 | What Are the Many Biological Functions of Proteins?

Proteins are the agents of biological function. Virtually every cellular activity is dependent on one or more particular proteins. Thus, a convenient way to classify the enormous number of proteins is to group them according to the biological roles they serve. Figure 5.32 summarizes the classification of proteins found in the human **proteome** according to their function. An overview of protein classification by function follows.

Proteome is the complete catalog of proteins encoded by a genome; in cell-specific terms, a proteome is the complete set of proteins found in a particular cell type at a particular time.

Many Proteins Are Enzymes

By far the largest class of proteins is enzymes. Thousands of different enzymes are listed in *Enzyme Nomenclature,* the standard reference volume on enzyme classification, accessible via the International Union of Biochemistry and

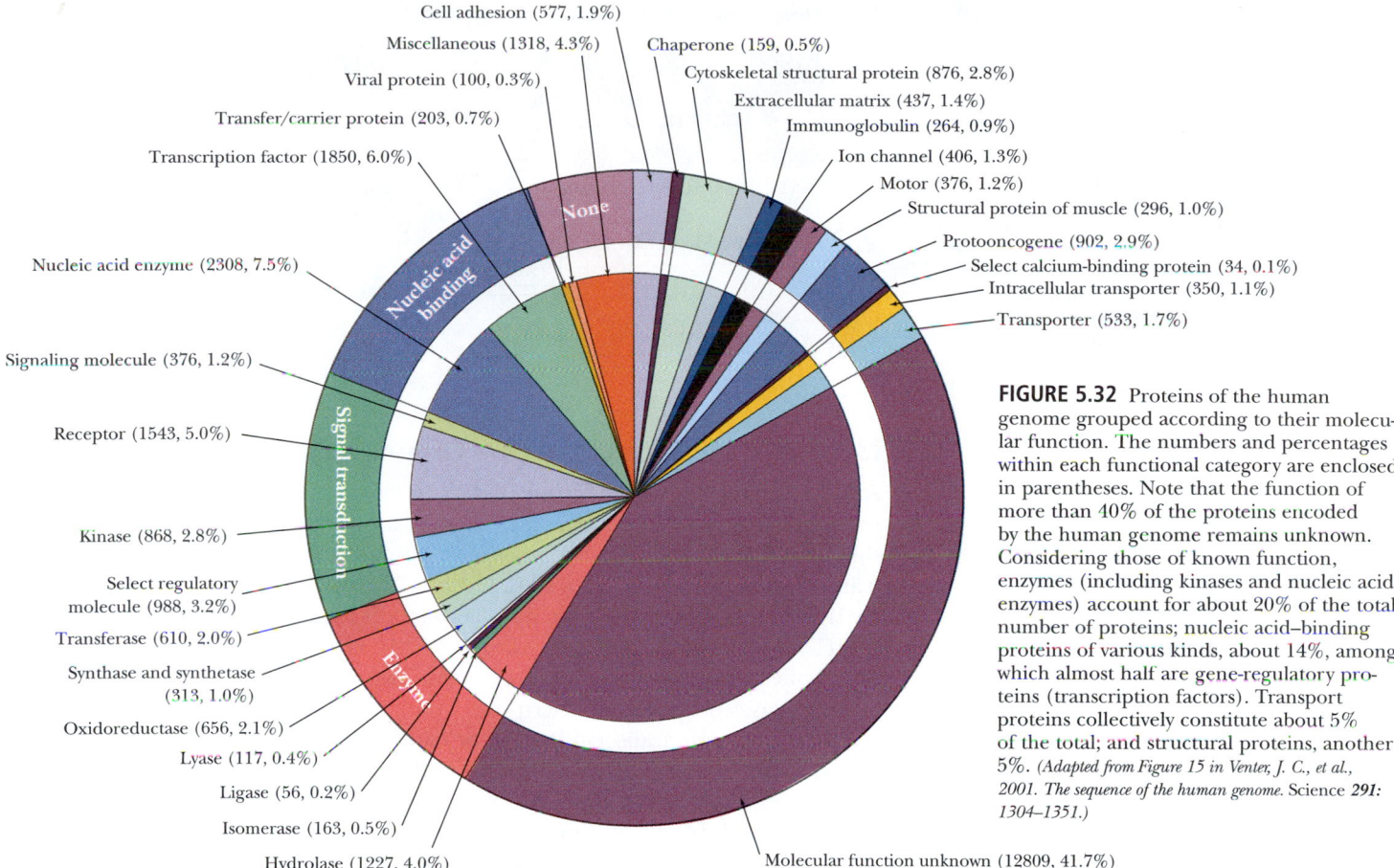

Cell adhesion (577, 1.9%)
Miscellaneous (1318, 4.3%)
Viral protein (100, 0.3%)
Transfer/carrier protein (203, 0.7%)
Transcription factor (1850, 6.0%)
Nucleic acid enzyme (2308, 7.5%)
Signaling molecule (376, 1.2%)
Receptor (1543, 5.0%)
Kinase (868, 2.8%)
Select regulatory molecule (988, 3.2%)
Transferase (610, 2.0%)
Synthase and synthetase (313, 1.0%)
Oxidoreductase (656, 2.1%)
Lyase (117, 0.4%)
Ligase (56, 0.2%)
Isomerase (163, 0.5%)
Hydrolase (1227, 4.0%)

Chaperone (159, 0.5%)
Cytoskeletal structural protein (876, 2.8%)
Extracellular matrix (437, 1.4%)
Immunoglobulin (264, 0.9%)
Ion channel (406, 1.3%)
Motor (376, 1.2%)
Structural protein of muscle (296, 1.0%)
Protooncogene (902, 2.9%)
Select calcium-binding protein (34, 0.1%)
Intracellular transporter (350, 1.1%)
Transporter (533, 1.7%)

None
Nucleic acid binding
Signal transduction
Enzyme

Molecular function unknown (12809, 41.7%)

FIGURE 5.32 Proteins of the human genome grouped according to their molecular function. The numbers and percentages within each functional category are enclosed in parentheses. Note that the function of more than 40% of the proteins encoded by the human genome remains unknown. Considering those of known function, enzymes (including kinases and nucleic acid enzymes) account for about 20% of the total number of proteins; nucleic acid–binding proteins of various kinds, about 14%, among which almost half are gene-regulatory proteins (transcription factors). Transport proteins collectively constitute about 5% of the total; and structural proteins, another 5%. (*Adapted from Figure 15 in Venter, J. C., et al., 2001. The sequence of the human genome.* Science **291:** *1304–1351.*)

Molecular Biology (IUBMB) Web site *http://www.iubmb.org*. **Enzymes** are catalysts that accelerate the rates of biological reactions. Each enzyme is very specific in its function and acts only in a particular metabolic reaction. Virtually every step in metabolism is catalyzed by an enzyme. The catalytic power of enzymes far exceeds that of synthetic catalysts. Enzymes can enhance reaction rates in cells as much as 10^{16} times the uncatalyzed rate. Enzymes are systematically classified according to the nature of the reaction that they catalyze, such as the transfer of a phosphate group (*phosphotransferase*) or an oxidation–reduction (*oxidoreductase*). Although the formal names of enzymes come from the particular reaction within the class that they catalyze, as in ATP : D-fructose-6-phosphate 1-phosphotransferase and alcohol : NAD$^+$ oxidoreductase, enzymes often have common names in addition to their formal names. ATP : D-fructose-6-phosphate 1-phosphotransferase is more commonly known as *phosphofructokinase* (*kinase* is a common name given to ATP-dependent phosphotransferases). Similarly, alcohol : NAD$^+$ oxidoreductase is casually referred to as *alcohol dehydrogenase*. The reactions catalyzed by these two enzymes are shown in Figure 5.33.

Regulatory Proteins Control Metabolism and Gene Expression

A number of proteins do not perform any obvious chemical transformation but nevertheless can regulate the ability of other proteins to carry out their physiological functions. Such proteins are referred to as **regulatory proteins.** Hormones are one class of regulatory proteins. A well-known example is *insulin* (Figure 5.13). Other hormones that are also proteins include pituitary *somatotropin* (21 kD) and *thyrotropin* (28 kD), which stimulates the thyroid gland.

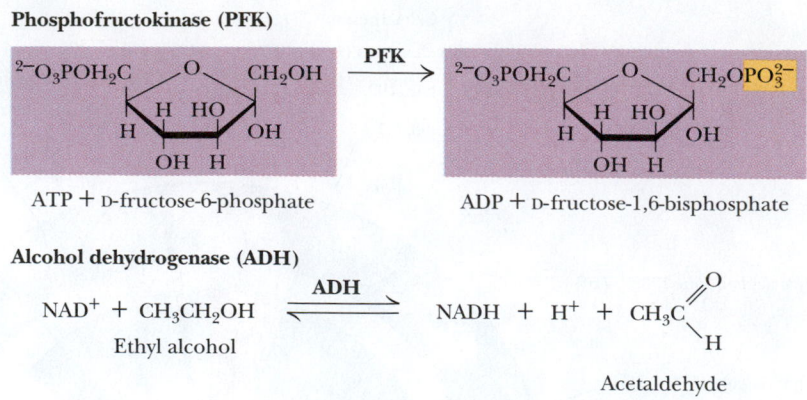

FIGURE 5.33 Enzymes are classified according to the specific biological reaction that they catalyze. Cells contain thousands of different enzymes. Two common examples drawn from carbohydrate metabolism are phosphofructokinase (PFK), or, more precisely, ATP:D-fructose-6-phosphate 1-phosphotransferase, and alcohol dehydrogenase (ADH), or alcohol: NAD$^+$ oxidoreductase, which catalyze the reactions shown here.

Many DNA-Binding Proteins Are Gene-Regulatory Proteins

Another group of regulatory proteins is involved in the regulation of gene expression. These proteins characteristically act by binding to DNA sequences that are adjacent to coding regions of genes, either activating or inhibiting the transcription of genetic information into RNA.

Transcription activators are positively acting control elements. For example, the *E. coli* catabolite gene activator protein (**CAP**) (44 kD), under appropriate metabolic conditions, can bind to specific sites along the *E. coli* chromosome and increase the rate of transcription of adjacent genes. The mammalian AP1 is a heterodimeric transcription factor composed of one polypeptide from the *Jun* family of gene-regulatory proteins and one polypeptide from the *Fos* family of gene-regulatory proteins. Activating expression of the β-globin gene (which encodes the β-subunit of hemoglobin) is one example of AP1's role as a transcription factor. Transcription inhibitors include **repressors,** which, because they block transcription, are considered negative control elements. A prokaryotic representative is *lac repressor* (37 kD), which controls expression of the enzyme system responsible for the metabolism of lactose (milk sugar); a mammalian example is NF1 (*nuclear factor 1,* 60 kD), which inhibits transcription of the β-globin gene. These various DNA-binding regulatory proteins often possess characteristic structural features, such as helix-turn-helix, leucine zipper, and zinc finger motifs (see Chapter 29).

Transport Proteins Carry Substances from One Place to Another

A third class of proteins is the **transport proteins.** Some of these proteins function to transport specific substances from one place to another, as a sort of cargo. This type of transport is exemplified by the transport of oxygen from the lungs to the tissues by *hemoglobin* (Figure 5.34a) or by the transport of fatty acids from adipose tissue to various organs by the blood protein *serum albumin.* A very different type of transport is the movement of metabolites across the permeability barrier imposed by cell membranes, as mediated by specific membrane proteins. These *membrane transport proteins* allow metabolite molecules on one side of a membrane to cross the membrane by creating channels or pores through which the transported molecule can pass. Examples include the transport proteins responsible for the uptake of essential nutrients into the cell, such as glucose or amino acids (Figure 5.34b).

Storage Proteins Serve as Reservoirs of Amino Acids or Other Nutrients

Proteins whose biological function is to provide a reservoir of an essential nutrient are called **storage proteins.** Because proteins are amino acid polymers and because nitrogen is commonly a limiting nutrient for growth, organisms

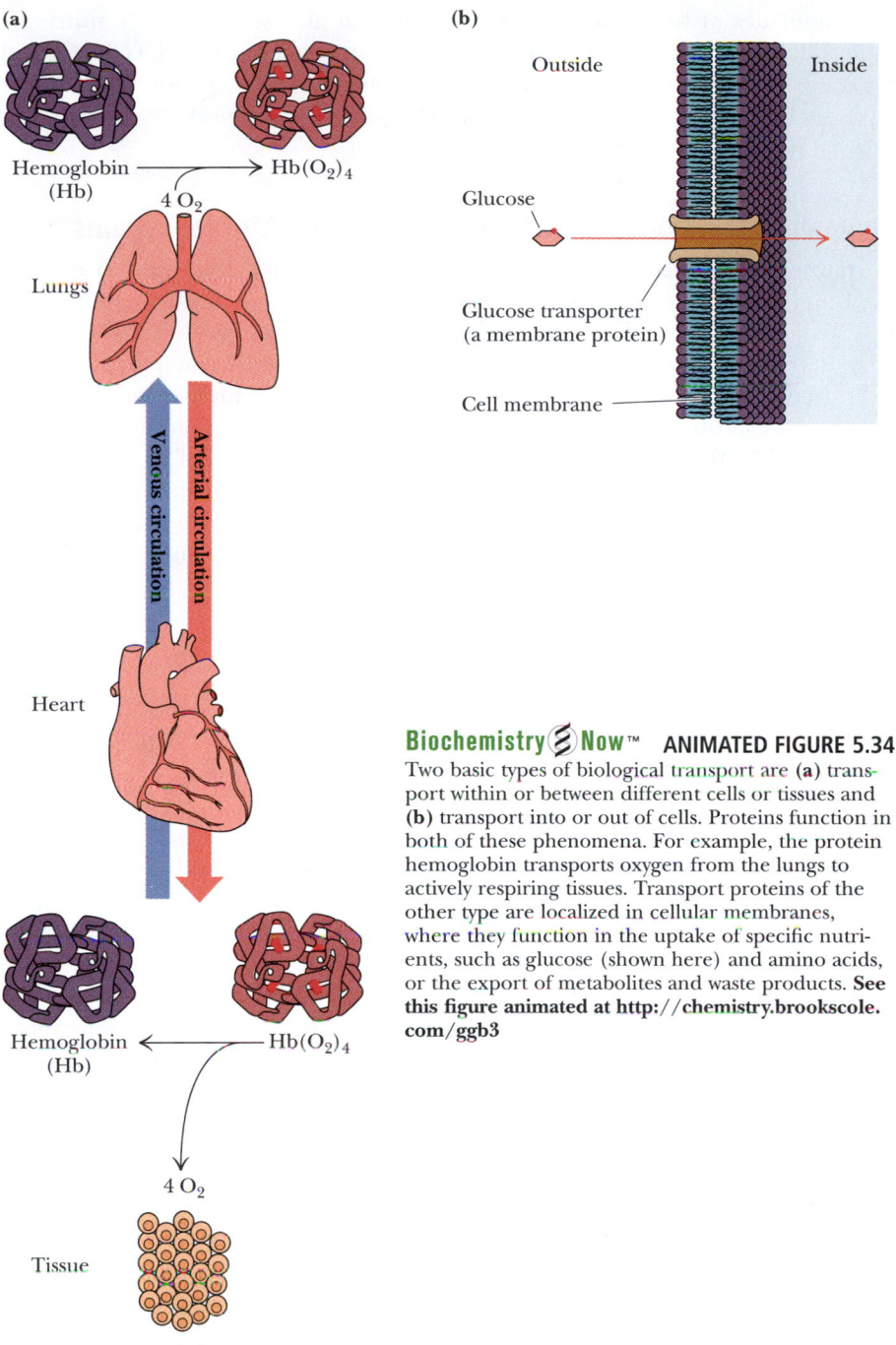

(a)

Hemoglobin (Hb) ⟶ Hb(O$_2$)$_4$

4 O$_2$

Lungs

Venous circulation

Arterial circulation

Heart

Hemoglobin (Hb) ⟵ Hb(O$_2$)$_4$

4 O$_2$

Tissue

(b)

Outside   Inside

Glucose

Glucose transporter (a membrane protein)

Cell membrane

Biochemistry❂Now™ ANIMATED FIGURE 5.34
Two basic types of biological transport are **(a)** transport within or between different cells or tissues and **(b)** transport into or out of cells. Proteins function in both of these phenomena. For example, the protein hemoglobin transports oxygen from the lungs to actively respiring tissues. Transport proteins of the other type are localized in cellular membranes, where they function in the uptake of specific nutrients, such as glucose (shown here) and amino acids, or the export of metabolites and waste products. **See this figure animated at http://chemistry.brookscole. com/ggb3**

have exploited proteins as a means to provide sufficient nitrogen in times of need. For example, *ovalbumin,* the protein of egg white, provides the developing bird embryo with a source of nitrogen during its isolation within the egg. *Casein* is the most abundant protein of milk and thus the major nitrogen source for mammalian infants; it also serves as an important source of phosphate. The seeds of higher plants often contain as much as 60% storage protein to make the germinating seed nitrogen-sufficient during this crucial period of plant development. *Zeins* are a family of low-molecular-weight proteins in the kernels of corn (*Zea mays* or *maize*); peas (the seeds of *Phaseolus vulgaris*) contain a storage protein called *phaseolin.* The use of proteins as a reservoir of nitrogen is more efficient than storing an equivalent amount of amino acids. Not only is the osmotic pressure minimized, but the solvent capacity of the cell is taxed less in solvating one molecule of a polypeptide than in dissolving, for example,

100 molecules of free amino acids. Proteins can also serve to store nutrients other than the more obvious elements composing amino acids (N, C, H, O, and S). As an example, *ferritin,* a iron-binding protein in animals, stores this essential metal so that it is available for the synthesis of important iron-containing proteins such as hemoglobin.

Movement Is Accomplished by Contractile and Motile Proteins

Certain proteins endow cells with unique capabilities for movement. Cell division, muscle contraction, and cell motility represent some of the ways in which cells execute motion. The **contractile** and **motile proteins** underlying these motions share a common property: They are filamentous or polymerize to form filaments. Examples include *actin* and *myosin,* the filamentous proteins forming the contractile systems of cells, and *tubulin,* the major component of microtubules (the filaments involved in the mitotic spindle of cell division as well as in flagella and cilia). Another class of proteins involved in movement includes *dynein* and *kinesin,* so-called **motor proteins** that drive the movement of vesicles, granules, and organelles along microtubules serving as established cytoskeletal "tracks."

Many Proteins Serve a Structural Role

An apparently passive but very important role of proteins is their function in creating and maintaining biological structures. **Structural proteins** provide strength and protection to cells and tissues. Monomeric units of structural proteins typically polymerize to generate long fibers (as in hair) or protective sheets of fibrous arrays, as in cowhide (leather). *α-Keratins* are insoluble fibrous proteins making up hair, horns, and fingernails. *Collagen,* another insoluble fibrous protein, is found in bone, connective tissue, tendons, cartilage, and hide, where it forms inelastic fibrils of great strength. One-third of the total protein in a vertebrate animal is collagen. A structural protein having elastic properties is, appropriately, *elastin,* an important component of ligaments. Because of the way elastin monomers are crosslinked in forming polymers, elastin can stretch in two dimensions. Certain insects make a structurally useful protein known as *fibroin* (a β-keratin), the major constituent of cocoons (silk) and spider webs. An important protective barrier for animal cells is the **extracellular matrix** containing *collagen* and *proteoglycans,* covalent protein–polysaccharide complexes that cushion and lubricate.

Proteins of Signaling Pathways Include Scaffold Proteins (Adapter Proteins)

Some proteins play a recently discovered role in the complex pathways of cellular response to hormones and growth factors. Such pathways are called **signaling pathways.** Signaling pathways have many proteins acting together to convert an extracellular signal into an intracellular response. Among them are hormone receptors and protein kinases that add phosphate groups to other proteins in an ATP-dependent manner. Proteins of signaling pathways can also serve as **scaffold** or **adapter proteins** because they have a modular organization in which specific parts **(modules)** of the protein's structure recognize and bind certain structural elements in other proteins through **protein–protein interactions.** For example, *SH2* modules bind to proteins in which a tyrosine residue has become phosphorylated on its phenolic —OH, and *SH3* modules bind to proteins having a characteristic grouping of proline residues. Others include *PH* modules, which bind to membranes, and *PDZ-*containing proteins, which bind specifically to the C-terminal amino acid of

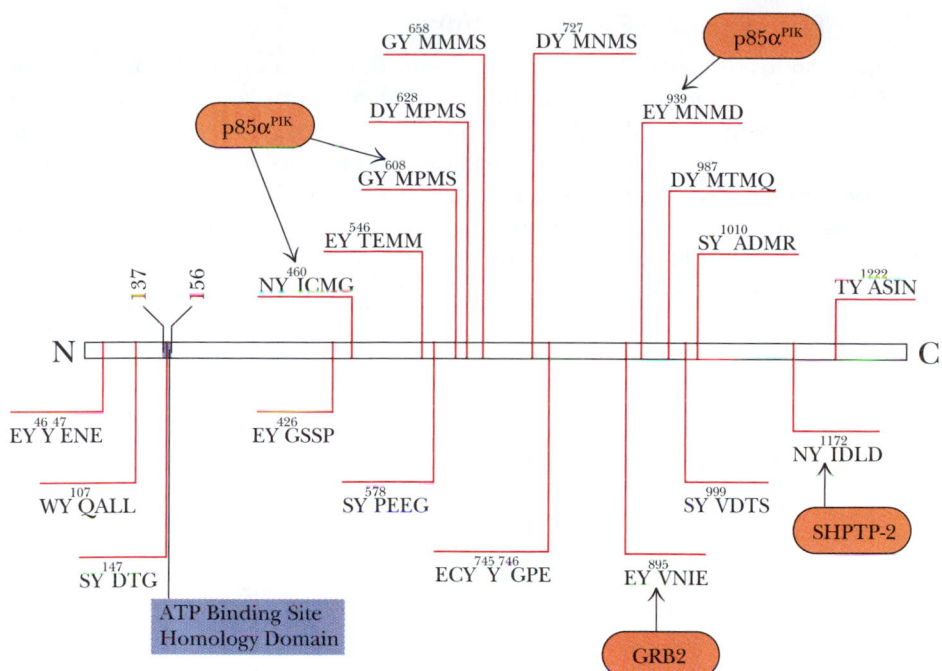

FIGURE 5.35 Diagram of the N → C sequence organization of the adapter protein *insulin receptor substrate-1 (IRS-1)* showing the various amino acid sequences (in one-letter code) that contain tyrosine (Y) residues that are potential sites for phosphorylation. The other adapter proteins that recognize various of these sites are shown as *Grb2, SHPTP-2*, and *p85αPIK*. Insulin binding to the insulin receptor activates the enzymatic activity that phosphorylates these Tyr residues on IRS-1. *(Adapted from White, M. F., and Kahn, C. R., 1994. The insulin signaling system.* Journal of Biological Chemistry *269:1–4.)*

certain proteins. Because scaffold proteins typically possess several of these different kinds of modules, they can act as a scaffold onto which a set of different proteins is assembled into a multiprotein complex. Such assemblages are typically involved in coordinating and communicating the many intracellular responses to hormones or other signaling molecules (Figure 5.35; see also Chapter 32). **Anchoring** (or **targeting**) **proteins** are proteins that bind other proteins, causing them to associate with other structures in the cell. A family of *anchoring proteins,* known as *AKAP* or *A kinase anchoring proteins,* exists in which specific AKAP members bind the regulatory enzyme *protein kinase A* (PKA) to particular subcellular compartments. For example, AKAP100 targets PKA to the endoplasmic reticulum, whereas AKAP79 targets PKA to the plasma membrane.

Other Proteins Have Protective and Exploitive Functions

In contrast to the passive protective nature of some structural proteins, another group can be more aptly classified as **protective** or **exploitive proteins** because of their biologically active role in cell defense, protection, or exploitation. Prominent among the protective proteins are the *immunoglobulins* or *antibodies* produced by the lymphocytes of vertebrates. Antibodies have the remarkable ability to "ignore" molecules that are an intrinsic part of the host organism, yet they can specifically recognize and neutralize "foreign" molecules resulting from the invasion of the organism by bacteria, viruses, or other infectious agents. Another group of protective proteins is the blood-clotting proteins, *thrombin* and *fibrinogen,* which prevent the loss of blood when the circulatory system is damaged. Arctic and Antarctic fishes have *antifreeze proteins* to protect their blood against freezing in the below-zero temperatures of high-latitude seas. In addition, various proteins serve defensive or exploitive roles for organisms, including the lytic and neurotoxic proteins of snake and bee venoms and toxic plant proteins, such as *ricin,* whose apparent purpose is to thwart predation by herbivores. Another class of exploitive proteins includes the toxins produced by bacteria, such as diphtheria toxin and cholera toxin.

A Few Proteins Have Exotic Functions

Some proteins display rather exotic functions that do not quite fit the previous classifications. *Monellin,* a protein found in an African plant, has a very sweet taste and is being considered as an artificial sweetener for human consumption. *Resilin,* a protein with exceptional elastic properties, is found in the hinges of insect wings. Certain marine organisms such as mussels secrete *glue proteins,* allowing them to attach firmly to hard surfaces. It is worth repeating that the great diversity of function in proteins, as reflected in this survey, is attained using just 20 amino acids.

Summary

The primary structure (the amino acid sequence) of a protein is encoded in DNA in the form of a nucleotide sequence. Expression of this genetic information is realized when the polypeptide chain is synthesized and assumes its functional, three-dimensional architecture. Proteins are the agents of biological function.

5.1 What Is the Fundamental Structural Pattern in Proteins?
Proteins are linear polymers joined by peptide bonds. The defining characteristic of a protein is its amino acid sequence. The partially double-bonded character of the peptide bond has profound influences on protein conformation. Proteins are also classified according to the length of their polypeptide chains (how many amino acid residues they contain) and the number and kinds of polypeptide chains (subunit organization).

5.2 What Architectural Arrangements Characterize Protein Structure?
Proteins are generally grouped into three fundamental structural classes—soluble, fibrous, and membrane—based on their shape and solubility. In more detail, protein structure is described in terms of a hierarchy of organization:

Primary (1°) structure—the protein's amino acid sequence

Secondary (2°) structure—regular elements of structure (helices, sheets) within the protein created by hydrogen bonds

Tertiary (3°) structure—the folding of the polypeptide chain in three-dimensional space

Quaternary (4°) structure—the subunit organization of multimeric proteins

The three higher levels of protein structure form and are maintained exclusively through noncovalent interactions.

5.3 How Are Proteins Isolated and Purified from Cells?
Cells contain thousands of different proteins. A protein of choice can be isolated and purified from such complex mixtures by exploiting two prominent physical properties: size and electrical charge. A more direct approach is to employ affinity purification strategies that take advantage of the biological function or similar specific recognition properties of a protein. A typical protein purification strategy will use a series of separation methods to obtain a pure preparation of the desired protein.

5.4 How Is the Amino Acid Analysis of Proteins Performed?
Acid treatment of a protein hydrolyzes all of the peptide bonds, yielding a mixture of amino acids. Chromatographic analysis of this hydrolysate reveals the amino acid composition of the protein. Proteins vary in their amino acid composition, but most proteins contain at least one of each of the 20 common amino acids. To a very rough approximation, proteins contain about 30% charged amino acids and about 30% hydrophobic amino acids (when aromatic amino acids are included in this number), the remaining being polar, uncharged amino acids.

5.5 How Is the Primary Structure of a Protein Determined?
The primary structure (amino acid sequence) of a protein can be determined by a variety of chemical and enzymatic methods. Alternatively, mass spectroscopic methods can also be used. In the chemical and en-zymatic protocols, a pure polypeptide chain whose disulfide linkages have been broken is the starting material. Methods that identify the N-terminal and C-terminal residues of the chain are used to determine which amino acids are at the ends, and then the protein is cleaved into defined sets of smaller fragments using enzymes such as trypsin or chymotrypsin or chemical cleavage by agents such as cyanogen bromide. The sequences of these products can be obtained by Edman degradation. Edman degradation is a powerful method for stepwise release and sequential identification of amino acids from the N-terminus of the polypeptide. The amino acid sequence of the entire protein can be reconstructed once the sequences of overlapping sets of peptide fragments are known. In mass spectrometry, an ionized protein chain is broken into an array of overlapping fragments. Small differences in the masses of the individual amino acids lead to small differences in the masses of the fragments, and the ability of mass spectrometry to measure mass-to-charge ratios very accurately allows computer devolution of the data into an amino acid sequence. The amino acid sequences of about a million different proteins are known. The vast majority of these amino acid sequences were deduced from nucleotide sequences available in genomic databases.

5.6 Can Polypeptides Be Synthesized in the Laboratory?
It is possible, although difficult, to synthesize proteins in the laboratory. The major obstacles involve joining desired amino acids to a growing chain using chemical methods that avoid side reactions and the creation of undesired products, such as the modification of side chains or the addition of more than one residue at a time. Solid-state techniques along with orthogonal protection methods circumvent many of these problems, and polypeptide chains having more than 100 amino acid residues have been artificially created.

5.7 What Is the Nature of Amino Acid Sequences?
Proteins have unique amino acid sequences, and similarity in sequence between proteins implies evolutionary relatedness. Homologous proteins (proteins of similar function) have similar amino acid sequences. These relationships can be used to trace evolutionary histories of proteins and the organisms that contain them, and the study of such relationships has given rise to the field of molecular evolution. Related proteins, such as the oxygen-binding proteins of myoglobin and hemoglobin or the serine proteases, share a common evolutionary origin. Sequence variation within a protein arises from mutations that result in amino acid substitution, and the operation of natural selection on these sequence variants is the basis of evolutionary change. Occasionally, a sequence variant with a novel biological function may appear, upon which selection can operate.

5.8 Do Proteins Have Chemical Groups Other Than Amino Acids?
Although many proteins are composed of just amino acids, other proteins are conjugated with various other chemical components, including carbohydrates, lipids, nucleic acids, metal and other inorganic ions, and a host of novel structures such as heme or flavin. Association with these nonprotein substances dramatically extends the physical and chemical properties that proteins possess, in turn creating a much greater repertoire of functional possibilities.

5.9 What Are the Many Biological Functions of Proteins?

As the agents of biological function, proteins fill essentially every biological role, with the exception of information storage. Catalytic proteins (enzymes) mediate almost every metabolic reaction. Regulatory proteins that bind to specific nucleotide sequences within DNA control gene expression. Hormones are another kind of regulatory protein in that they convey information about the environment and deliver this information to cells when they bind to specific receptors. Transport proteins are engaged in the transport of substances (nutrients, ions, and waste products) across membranes and throughout the body. Structural proteins give form to cells and subcellular structures; contractile and motile proteins endow cells with the ability to change shape or move substances, even the cell itself. Scaffold proteins have as their primary role the recruitment of other proteins into multimeric assemblies that mediate and coordinate the flow of information in cells. The great diversity in function that characterizes biological systems is based on the attributes that proteins possess.

Problems

1. The element molybdenum (atomic weight 95.95) constitutes 0.08% of the weight of nitrate reductase. If the molecular weight of nitrate reductase is 240,000, what is its likely quaternary structure?

2. Amino acid analysis of an oligopeptide 7 residues long gave

   Asp   Leu   Lys   Met   Phe   Tyr

   The following facts were observed:
   a. Trypsin treatment had no apparent effect.
   b. The phenylthiohydantoin released by Edman degradation was

   c. Brief chymotrypsin treatment yielded several products, including a dipeptide and a tetrapeptide. The amino acid composition of the tetrapeptide was Leu, Lys, and Met.
   d. Cyanogen bromide treatment yielded a dipeptide, a tetrapeptide, and free Lys.
   What is the amino acid sequence of this heptapeptide?

3. Amino acid analysis of another heptapeptide gave

   Asp   Glu   Leu   Lys
   Met   Tyr   Trp   NH$_4^+$

   The following facts were observed:
   a. Trypsin had no effect.
   b. The phenylthiohydantoin released by Edman degradation was

   c. Brief chymotrypsin treatment yielded several products, including a dipeptide and a tetrapeptide. The amino acid composition of the tetrapeptide was Glx, Leu, Lys, and Met.
   d. Cyanogen bromide treatment yielded a tetrapeptide that had a net positive charge at pH 7 and a tripeptide that had a zero net charge at pH 7.
   What is the amino acid sequence of this heptapeptide?

4. Amino acid analysis of a decapeptide revealed the presence of the following products:

   NH$_4^+$ Asp   Glu   Tyr   Arg
   Met   Pro   Lys   Ser   Phe

   The following facts were observed:
   a. Neither carboxypeptidase A or B treatment of the decapeptide had any effect.
   b. Trypsin treatment yielded two tetrapeptides and free Lys.

   c. Clostripain treatment yielded a tetrapeptide and a hexapeptide.
   d. Cyanogen bromide treatment yielded an octapeptide and a dipeptide of sequence NP (using the one-letter codes).
   e. Chymotrypsin treatment yielded two tripeptides and a tetrapeptide. The N-terminal chymotryptic peptide had a net charge of −1 at neutral pH and a net charge of −3 at pH 12.
   f. One cycle of Edman degradation gave the PTH derivative

   What is the amino acid sequence of this decapeptide?

5. Analysis of the blood of a catatonic football fan revealed large concentrations of a psychotoxic octapeptide. Amino acid analysis of this octapeptide gave the following results:

   2 Ala   1 Arg   1 Asp   1 Met   2 Tyr   1 Val   1 NH$_4^+$

   The following facts were observed:
   a. Partial acid hydrolysis of the octapeptide yielded a dipeptide of the structure

   b. Chymotrypsin treatment of the octapeptide yielded two tetrapeptides, each containing an alanine residue.
   c. Trypsin treatment of one of the tetrapeptides yielded two dipeptides.
   d. Cyanogen bromide treatment of another sample of the same tetrapeptide yielded a tripeptide and free Tyr.
   e. End-group analysis of the other tetrapeptide gave Asp.
   What is the amino acid sequence of this octapeptide?

6. Amino acid analysis of an octapeptide revealed the following composition:

   2 Arg   1 Gly   1 Met   1 Trp   1 Tyr   1 Phe   1 Lys

   The following facts were observed:
   a. Edman degradation gave

   b. CNBr treatment yielded a pentapeptide and a tripeptide containing phenylalanine.

c. Chymotrypsin treatment yielded a tetrapeptide containing a C-terminal indole amino acid and two dipeptides.

d. Trypsin treatment yielded a tetrapeptide, a dipeptide, and free Lys and Phe.

e. Clostripain yielded a pentapeptide, a dipeptide, and free Phe.

What is the amino acid sequence of this octapeptide?

7. Amino acid analysis of an octapeptide gave the following results:

1 Ala   1 Arg   1 Asp   1 Gly   3 Ile   1 Val   1 NH$_4^+$

The following facts were observed:

a. Trypsin treatment yielded a pentapeptide and a tripeptide.

b. Chemical reduction of the free α-COOH and subsequent acid hydrolysis yielded 2-aminopropanol.

c. Partial acid hydrolysis of the tryptic pentapeptide yielded, among other products, two dipeptides, each of which contained C-terminal isoleucine. One of these dipeptides migrated as an anionic species upon electrophoresis at neutral pH.

d. The tryptic tripeptide was degraded in an Edman sequenator, yielding first **A**, then **B**:

What is an amino acid sequence of the octapeptide? Four sequences are possible, but only one suits the authors. Why?

8. An octapeptide consisting of 2 Gly, 1 Lys, 1 Met, 1 Pro, 1 Arg, 1 Trp, and 1 Tyr was subjected to sequence studies. The following was found:

a. Edman degradation yielded

b. Upon treatment with carboxypeptidases A, B, and C, only carboxypeptidase C had any effect.

c. Trypsin treatment gave two tripeptides and a dipeptide.

d. Chymotrypsin treatment gave two tripeptides and a dipeptide. Acid hydrolysis of the dipeptide yielded only Gly.

e. Cyanogen bromide treatment yielded two tetrapeptides.

f. Clostripain treatment gave a pentapeptide and a tripeptide.

What is the amino acid sequence of this octapeptide?

9. Amino acid analysis of an oligopeptide containing nine residues revealed the presence of the following amino acids:

Arg   Cys   Gly   Leu   Met   Pro   Tyr   Val

The following was found:

a. Carboxypeptidase A treatment yielded no free amino acid.

b. Edman analysis of the intact oligopeptide released

c. Neither trypsin nor chymotrypsin treatment of the nonapeptide released smaller fragments. However, combined trypsin and chymotrypsin treatment liberated free Arg.

d. CNBr treatment of the 8-residue fragment left after combined trypsin and chymotrypsin action yielded a 6-residue fragment containing Cys, Gly, Pro, Tyr, and Val; and a dipeptide.

e. Treatment of the 6-residue fragment with β-mercaptoethanol yielded two tripeptides. Brief Edman analysis of the tripeptide mixture yielded only PTH-Cys. (The sequence of each tripeptide, as read from the N-terminal end, is alphabetical if the one-letter designation for amino acids is used.)

What is the amino acid sequence of this nonapeptide?

10. Describe the synthesis of the dipeptide Lys-Ala by Merrifield's solid-phase chemical method of peptide synthesis. What pitfalls might be encountered if you attempted to add a leucine residue to Lys-Ala to make a tripeptide?

11. Electrospray ionization mass spectrometry (ESI-MS) of the polypeptide chain of myoglobin yielded a series of m/z peaks (similar to those shown in Figure 5.21 for aerolysin K). Two successive peaks had m/z values of 1304.7 and 1413.2, respectively. Calculate the mass of the myoglobin polypeptide chain from these data.

12. Phosphoproteins are formed when a phosphate group is esterified to an —OH group of a Ser, Thr, or Tyr side chain. At typical cellular pH values, this phosphate group bears two negative charges —OPO$_3^{2-}$. Compare this side-chain modification to the 20 side chains of the common amino acids found in proteins and comment on the novel properties that it introduces into side-chain possibilities.

Biochemistry on the Web

13. Peptide mass fingerprinting of tryptic peptides derived from a yeast protein yielded peptides of mass 2164.0, 1702.8, and 1402.7. Go to the PeptIdent peptide identification Web site at *http://us.expasy.org/cgi-bin/peptident.pl* and find the identity of this protein. Check the Peptide Mass Web site at *http://us.expasy.org/tools/peptide-mass.html* to find out its molecular weight and to determine how many tryptic peptides can be obtained from this yeast protein. What is the identity of the human protein having tryptic peptides of masses 2164.0, 1702.8, and 1402.7. What is the molecular weight of this human protein? How many tryptic peptides are found in this protein?

Preparing for the MCAT Exam

14. Proteases such as trypsin and chymotrypsin cleave proteins at different sites, but both use the same reaction mechanism. Based on your knowledge of organic chemistry, suggest a "universal" protease reaction mechanism for hydrolysis of the peptide bond.

15. Table 5.6 presents some of the many known mutations in the genes encoding the α- and β-globin subunits of hemoglobin.

a. Some of these mutations affect subunit interactions between the subunits. In an examination of the tertiary structure of globin chains, where would you expect to find amino acid changes in mutant globins that affect formation of the hemoglobin $\alpha_2\beta_2$ quaternary structure?

b. Other mutations, such as the S form of the β-globin chain, increase the tendency of hemoglobin tetramers to polymerize into very large structures. Where might you expect the amino acid substitutions to be in these mutants?

Biochemistry🅔Now™   Preparing for an exam? Test yourself on key questions at http://chemistry.brookscole.com/ggb3

Further Reading

General References on Protein Structure and Function

Creighton, T. E., 1983. *Proteins: Structure and Molecular Properties.* San Francisco: W. H. Freeman and Co.

Creighton, T. E., ed., 1997. *Protein Function—A Practical Approach,* 2nd ed. Oxford: CRL Press at Oxford University Press.

Fersht, A., 1999. *Structure and Mechanism in Protein Science.* New York: W. H. Freeman and Co.

Goodsell, D. S., and Olson, A. J., 1993. Soluble proteins: Size, shape and function. *Trends in Biochemical Sciences* **18:**65–68.

Lesk, A. M., 2001. *Introduction to Protein Architecture: The Structural Biology of Proteins.* Oxford: Oxford University Press.

Protein Purification

Deutscher, M. P., ed., 1990. *Guide to Protein Purification,* Vol. 182, *Methods in Enzymology.* San Diego: Academic Press.

Amino Acid Sequence Analysis

Dayhoff, M. O., 1972-1978. *The Atlas of Protein Sequence and Structure,* Vols. 1–5. Washington, DC: National Medical Research Foundation.

Heijne, G. von, 1987. *Sequence Analysis in Molecular Biology: Treasure Trove or Trivial Pursuit?* San Diego: Academic Press.

Hill, R. L., 1965. Hydrolysis of proteins. *Advances in Protein Chemistry* **20:**37–107.

Hirs, C. H. W., ed., 1967. *Enzyme Structure,* Vol. XI, *Methods in Enzymology.* New York: Academic Press.

Hirs, C. H. W., and Timasheff, S. E., eds., 1977–1986. *Enzyme Structure,* Parts E–L. New York: Academic Press.

Hsieh, Y. L., et al., 1996. Automated analytical system for the examination of protein primary structure. *Analytical Chemistry* **68:**455–462. An analytical system is described in which a protein is purified by affinity chromatography, digested with trypsin, and its peptides separated by HPLC and analyzed by tandem MS in order to determine its amino acid sequence.

Karger, B. L., and Hancock, W. S., eds. 1996. High Resolution Separation and Analysis of Biological Macromolecules. Part B: Applications. *Methods in Enzymology* **271.** New York: Academic Press. Sections on liquid chromatography, electrophoresis, capillary electrophoresis, mass spectrometry, and interfaces between chromatographic and electrophoretic separations of proteins followed by mass spectrometry of the separated proteins.

Mass Spectrometry

Hernandez, H., and Robinson, C. V., 2001. Dynamic protein complexes: Insights from mass spectrometry. *Journal of Biological Chemistry* **276:**46685–46688. Advances in mass spectrometry open a new view onto the dynamics of protein function, such as protein–protein interactions and the interaction between proteins and their ligands.

Hunt, D. F., et al., 1987. Tandem quadrupole Fourier transform mass spectrometry of oligopeptides and small proteins. *Proceedings of the National Academy of Sciences, U.S.A.* **84:**620–623.

Johnstone, R. A. W., and Rose, M. E., 1996. *Mass Spectrometry for Chemists and Biochemists,* 2nd ed. Cambridge, England: Cambridge University Press.

Karger, B. L., and Hancock, W. S., eds. 1996. High Resolution Separation and Analysis of Biological Macromolecules. Part A: Fundamentals. *Methods in Enzymology* **270.** New York: Academic Press. Separate sections discussing liquid chromatography, columns and instrumentation, electrophoresis, capillary electrophoresis, and mass spectrometry.

Kinter, M., and Sherman, N. E., 2001. *Protein Sequencing and Identification Using Tandem Mass Spectrometry.* Hoboken, NJ: Wiley-Interscience.

Liebler, D. C., 2002. *Introduction to Proteomics.* Towata, NJ: Humana Press. An excellent primer on proteomics, protein purification methods, sequencing of peptides and proteins by mass spectrometry, and identification of proteins in a complex mixture.

Mann, M., and Wilm, M., 1995. Electrospray mass spectrometry for protein characterization. *Trends in Biochemical Sciences* **20:**219–224. A review of the basic application of mass spectrometric methods to the analysis of protein sequence and structure.

Quadroni, M., et al., 1996. Analysis of global responses by protein and peptide fingerprinting of proteins isolated by two-dimensional electrophoresis. Application to sulfate-starvation response of *Escherichia coli. European Journal of Biochemistry* **239:**773–781. This paper describes the use of tandem MS in the analysis of proteins in cell extracts.

Vestling, M. M., 2003. Using mass spectrometry for proteins. *Journal of Chemical Education* **80:**122–124. A report on the 2002 Nobel Prize in Chemistry honoring the scientists who pioneered the application of mass spectrometry to protein analysis.

Solid-Phase Synthesis of Proteins

Aparicio, F., 2000. Orthogonal protecting groups for *N*-amino and C-terminal carboxyl functions in solid-phase peptide synthesis. *Biopolymers* **55:**123–139.

Fields, G. B. ed., 1997. *Solid-Phase Peptide Synthesis,* Vol. 289, *Methods in Enzymology.* San Diego: Academic Press.

Merrifield, B., 1986. Solid phase synthesis. *Science* **232:**341–347.

Wilken, J., and Kent, S. B. H., 1998. Chemical protein synthesis. *Current Opinion in Biotechnology* **9:**412–426.

Protein Techniques[1]

Dialysis and Ultrafiltration

If a solution of protein is separated from a bathing solution by a semipermeable membrane, small molecules and ions can pass through the semipermeable membrane to equilibrate between the protein solution and the bathing solution, called the *dialysis bath* or *dialysate* (Figure 5A.1). This method is useful for removing small molecules from macromolecular solutions or for altering the composition of the protein-containing solution.

Ultrafiltration is an improvement on the dialysis principle. Filters with pore sizes over the range of biomolecular dimensions are used to filter solutions to select for molecules in a particular size range. Because the pore sizes in these filters are microscopic, high pressures are often required to force the solution through the filter. This technique is useful for concentrating dilute solutions of macromolecules. The concentrated protein can then be diluted into the solution of choice.

Size Exclusion Chromatography

Size exclusion chromatography is also known as *gel filtration chromatography* or *molecular sieve chromatography*. In this method, fine, porous beads are packed into a chromatography column. The beads are composed of dextran polymers *(Sephadex)*, agarose *(Sepharose)*, or polyacrylamide *(Sephacryl* or *BioGel P)*. The pore sizes of these beads approximate the dimensions of macromolecules. The total bed volume (Figure 5A.2) of the packed chromatography column, V_t, is equal to the volume outside the porous beads (V_o) plus the volume inside the beads (V_i) plus the

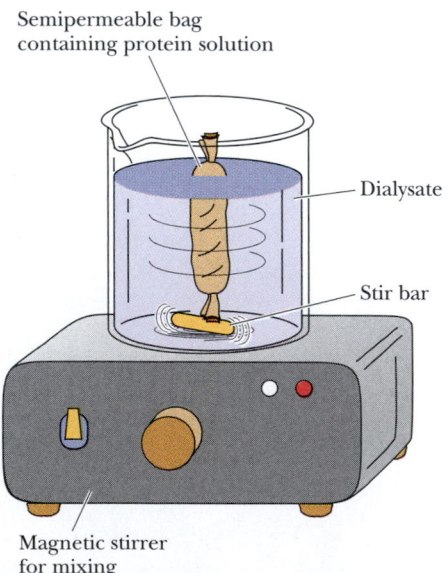

Semipermeable bag
containing protein solution

Dialysate

Stir bar

Magnetic stirrer
for mixing

FIGURE 5A.1 A dialysis experiment. The solution of macromolecules to be dialyzed is placed in a semipermeable membrane bag, and the bag is immersed in a bathing solution. A magnetic stirrer gently mixes the solution to facilitate equilibrium of diffusible solutes between the dialysate and the solution contained in the bag.

[1]Although this appendix is titled *Protein Techniques,* these methods are also applicable to other macromolecules such as nucleic acids.

(a)

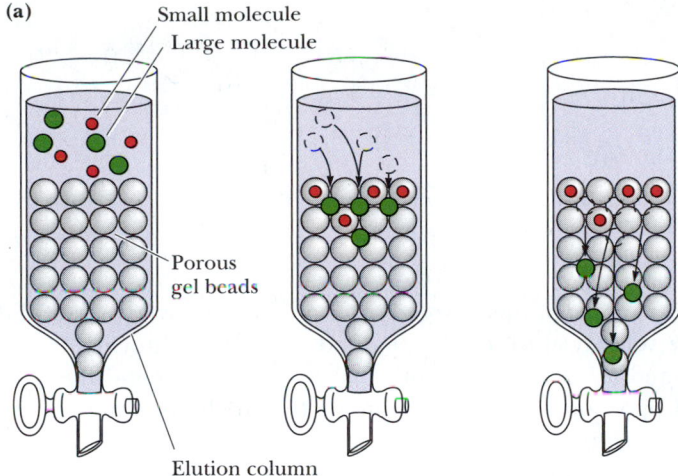

(b)

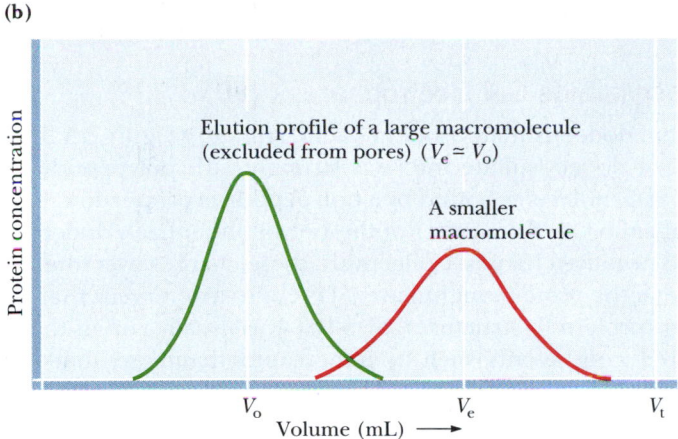

FIGURE 5A.2 **(a)** A gel filtration chromatography column. Larger molecules are excluded from the gel beads and emerge from the column sooner than smaller molecules, whose migration is retarded because they can enter the beads. **(b)** An elution profile.

volume actually occupied by the bead material (V_g): $V_t = V_o + V_i + V_g$. (V_g is typically less than 1% of V_t and can be conveniently ignored in most applications.)

As a solution of molecules is passed through the column, the molecules passively distribute between V_o and V_i, depending on their ability to enter the pores (that is, their size). If a molecule is too large to enter at all, it is totally excluded from V_i and emerges first from the column at an elution volume, V_e, equal to V_o (Figure 5A.1). If a particular molecule can enter the pores in the gel, its distribution is given by the *distribution coefficient,* K_D:

$$K_D = (V_e - V_o)/V_i$$

where V_e is the molecule's characteristic elution volume (Figure 5A.2). The chromatography run is complete when a volume of solvent equal to V_t has passed through the column.

Electrophoresis

Electrophoretic techniques are based on the movement of ions in an electrical field. An ion of charge q experiences a force F given by $F = Eq/d$, where E is the voltage (or *electrical potential*) and d is the distance between the electrodes. In a vacuum, F would cause the molecule to accelerate. In solution, the molecule experiences *frictional drag,* F_f, due to the solvent:

$$F_f = 6\pi r \eta \nu$$

$$Na^+ \ {}^-O-\overset{\overset{\textstyle O}{\|}}{\underset{\underset{\textstyle O^-}{|}}{S}}-O \diagdown CH_2 \diagdown CH_2 \diagdown CH_2 \diagdown CH_2 \diagdown CH_2 \diagdown CH_3$$

$$Na^+$$

FIGURE 5A.3 The structure of sodium dodecylsulfate (SDS).

where r is the radius of the charged molecule, η is the viscosity of the solution, and ν is the velocity at which the charged molecule is moving. So, the velocity of the charged molecule is proportional to its charge q and the voltage E, but inversely proportional to the viscosity of the medium η and d, the distance between the electrodes.

Generally, electrophoresis is carried out *not* in free solution but in a porous support matrix such as polyacrylamide or agarose, which retards the movement of molecules according to their dimensions relative to the size of the pores in the matrix.

SDS-Polyacrylamide Gel Electrophoresis (SDS-PAGE)

SDS is sodium dodecylsulfate (sodium lauryl sulfate) (Figure 5A.3). The hydrophobic tail of dodecylsulfate interacts strongly with polypeptide chains. The number of SDS molecules bound by a polypeptide is proportional to the length (number of amino acid residues) of the polypeptide. Each dodecylsulfate contributes two negative charges. Collectively, these charges overwhelm any intrinsic charge that the protein might have. SDS is also a detergent that disrupts protein folding (protein 3° structure). SDS-PAGE is usually run in the presence of sulfhydryl-reducing agents such as β-mercaptoethanol so that any disulfide links between polypeptide chains are broken. The electrophoretic mobility of proteins upon SDS-PAGE is inversely proportional to the logarithm of the protein's molecular weight (Figure 5A.4). SDS-PAGE is often used to determine the molecular weight of a protein.

Isoelectric Focusing

Isoelectric focusing is an electrophoretic technique for separating proteins according to their *isoelectric points* (pIs). A solution of *ampholytes* (amphoteric electrolytes) is first electrophoresed through a gel, usually contained in a small tube. The migration of these substances in an electric field establishes a pH gradient in the tube. Then a protein mixture is applied to the gel, and electrophoresis is resumed. As the protein molecules move down the gel, they experience the pH gradient and migrate to a position corresponding to their respective pIs. At its pI, a protein has no net charge and thus moves no farther.

Two-Dimensional Gel Electrophoresis

This separation technique uses isoelectric focusing in one dimension and SDS-PAGE in the second dimension to resolve protein mixtures. The proteins in a mixture are first separated according to pI by isoelectric focusing in a polyacrylamide gel in a tube. The gel is then removed and laid along the top of an SDS-PAGE slab, and the proteins are electrophoresed into the SDS polyacrylamide gel, where they are separated according to size (Figure 5A.5). The gel slab can then be stained to reveal the locations of the individual proteins. Using this powerful technique, researchers have the potential to visualize and construct catalogs of virtually *all* the proteins present in particular cell types.

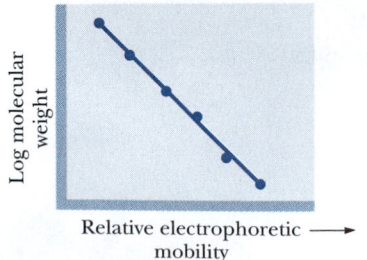

FIGURE 5A.4 A plot of the relative electrophoretic mobility of proteins in SDS-PAGE versus the log of the molecular weights of the individual polypeptides.

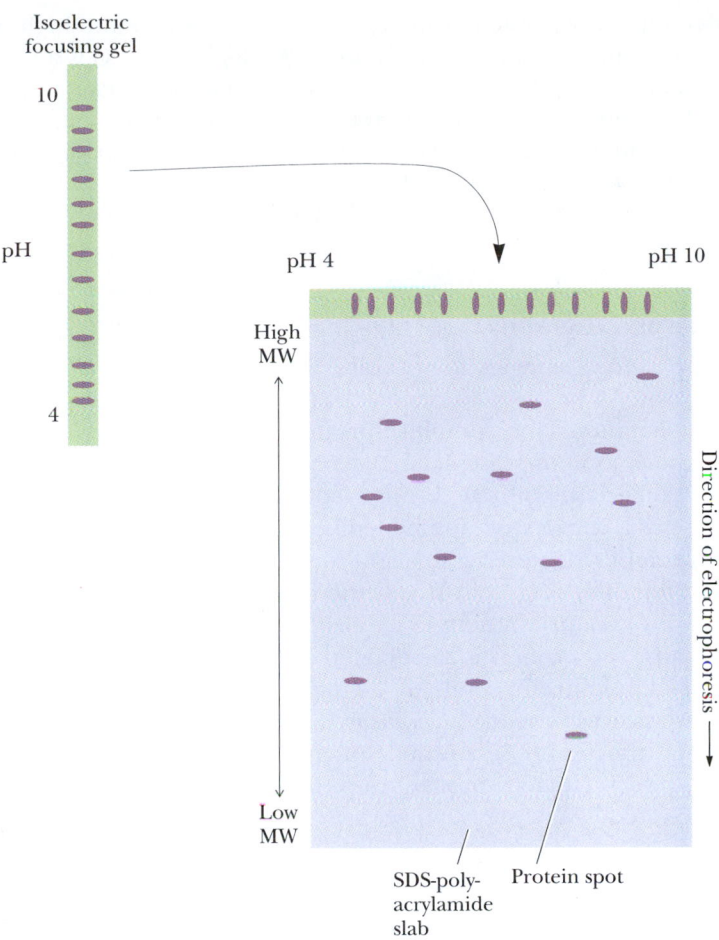

FIGURE 5A.5 A two-dimensional electrophoresis separation. A mixture of macromolecules is first separated according to charge by isoelectric focusing in a tube gel. The gel containing separated molecules is then placed on top of an SDS-PAGE slab, and the molecules are electrophoresed into the SDS-PAGE gel, where they are separated according to size.

The **ExPASy** server *(http://us.expasy.org)* provides access to a two-dimensional polyacrylamide gel electrophoresis database named **SWISS-2DPAGE.** This database contains information on proteins, identified as spots on two-dimensional electrophoresis gels, from many different cell and tissue types.

Hydrophobic Interaction Chromatography

Hydrophobic interaction chromatography (HIC) exploits the hydrophobic nature of proteins in purifying them. Proteins are passed over a chromatographic column packed with a support matrix to which hydrophobic groups are covalently linked. *Phenyl Sepharose,* an agarose support matrix to which phenyl groups are affixed, is a prime example of such material. In the presence of high salt concentrations, proteins bind to the phenyl groups by virtue of hydrophobic interactions. Proteins in a mixture can be differentially eluted from the phenyl groups by lowering the salt concentration or by adding solvents such as polyethylene glycol to the elution fluid.

High-Performance Liquid Chromatography

The principles exploited in *high-performance* (or high-pressure) *liquid chromatography* (HPLC) are the same as those used in the common chromatographic methods such as ion exchange chromatography or size exclusion chromatography.

A protein interacts with a metabolite. The metabolite is thus a ligand that binds specifically to this protein

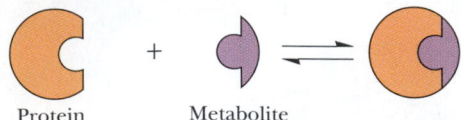

The metabolite can be immobilized by covalently coupling it to an insoluble matrix such as an agarose polymer. Cell extracts containing many individual proteins may be passed through the matrix.

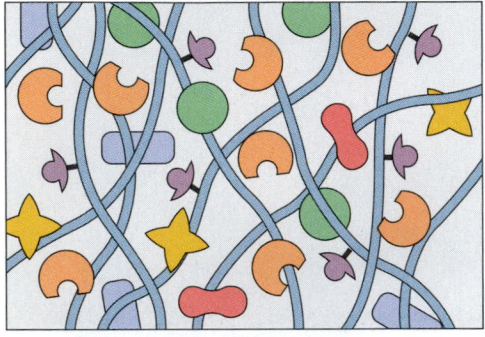

Specific protein binds to ligand. All other unbound material is washed out of the matrix.

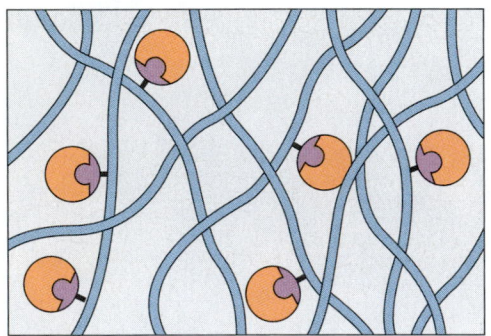

Adding an excess of free metabolite that will compete for the bound protein dissociates the protein from the chromatographic matrix. The protein passes out of the column complexed with free metabolite.

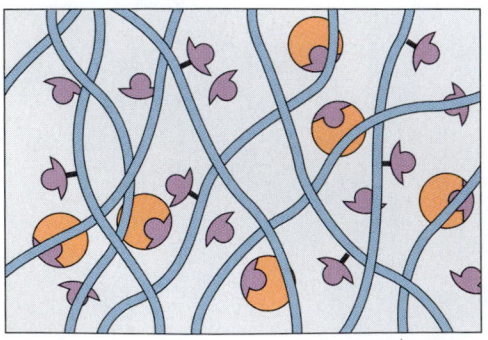

Purifications of proteins as much as 1000-fold or more are routinely achieved in a single affinity chromatographic step like this.

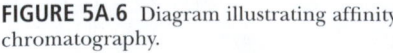

FIGURE 5A.6 Diagram illustrating affinity chromatography.

Very-high-resolution separations can be achieved quickly and with high sensitivity in HPLC using automated instrumentation. *Reverse-phase* HPLC is a widely used chromatographic procedure for the separation of nonpolar solutes. In reverse-phase HPLC, a solution of nonpolar solutes is chromatographed on a column having a nonpolar liquid immobilized on an inert matrix; this nonpolar liquid serves as the *stationary phase*. A more polar liquid that serves as the *mobile phase* is passed over the matrix, and solute molecules are eluted in proportion to their solubility in this more polar liquid.

Affinity Chromatography

Affinity purification strategies for proteins exploit the biological function of the target protein. In most instances, proteins carry out their biological activity through binding or complex formation with specific small biomolecules, or *ligands,* as in the case of an enzyme binding its substrate. If this small molecule can be immobilized through covalent attachment to an insoluble matrix, such as a chromatographic medium like cellulose or polyacrylamide, then the protein of interest, in displaying affinity for its ligand, becomes bound and immobilized itself. It can then be removed from contaminating proteins in the mixture by simple means such as filtration and washing the matrix. Finally, the protein is dissociated or eluted from the matrix by the addition of high concentrations of the free ligand in solution. Figure 5A.6 depicts the protocol for such an *affinity chromatography* scheme. Because this method of purification relies on the biological specificity of the protein of interest, it is a very efficient procedure and proteins can be purified several thousand-fold in a single step.

Ultracentrifugation

Centrifugation methods separate macromolecules on the basis of their characteristic densities. Particles tend to "fall" through a solution if the density of the solution is less than the density of the particle. The velocity of the particle through the medium is proportional to the difference in density between the particle and the solution. The tendency of any particle to move through a solution under centrifugal force is given by the *sedimentation coefficient, S:*

$$S = (\rho_p - \rho_m) V/f$$

where ρ_p is the density of the particle or macromolecule, ρ_m is the density of the medium or solution, V is the volume of the particle, and f is the frictional coefficient, given by

$$f = F_f/v$$

where v is the velocity of the particle and F_f is the frictional drag. Nonspherical molecules have larger frictional coefficients and thus smaller sedimentation coefficients. The smaller the particle and the more its shape deviates from spherical, the more slowly that particle sediments in a centrifuge.

Centrifugation can be used either as a preparative technique for separating and purifying macromolecules and cellular components or as an analytical technique to characterize the hydrodynamic properties of macromolecules such as proteins and nucleic acids.

Proteins: Secondary, Tertiary, and Quaternary Structure

CHAPTER 6

Essential Question

Linus Pauling received the Nobel Prize in Chemistry in 1954. The award cited "his research into the nature of the chemical bond and its application to the elucidation of the structure of complex substances." *How do the forces of chemical bonding determine the formation, stability, and myriad functions of proteins?*

Nearly all biological processes involve the specialized functions of one or more protein molecules. Proteins function to produce other proteins, control all aspects of cellular metabolism, regulate the movement of various molecular and ionic species across membranes, convert and store cellular energy, and carry out many other activities. Essentially all of the information required to initiate, conduct, and regulate each of these functions must be contained in the structure of the protein itself. The previous chapter described the details of primary protein structure. However, proteins do not normally exist as fully extended polypeptide chains but rather as compact, folded structures, and the function of a given protein is rarely, if ever, dependent only on the amino acid sequence. Instead, the ability of a particular protein to carry out its function in nature is normally determined by its overall three-dimensional shape, or *conformation*. This native, folded structure of the protein is dictated by several factors: (1) interactions with solvent molecules (normally water), (2) the pH and ionic composition of the solvent, and most important, (3) the sequence of the protein. The first two of these effects are intuitively reasonable, but the third, the role of the amino acid sequence, may not be. In ways that are just now beginning to be understood, the primary structure facilitates the development of short-range interactions among adjacent parts of the sequence and also long-range interactions among distant parts of the sequence. Although the resulting overall structure of the complete protein molecule may at first look like a disorganized and random arrangement, it is in nearly all cases a delicate and sophisticated balance of numerous forces that combine to determine the protein's unique conformation.

Like the Greek sea god Proteus, who could assume different forms, proteins act through changes in conformation. Proteins (from the Greek *proteios*, meaning "primary") are the primary agents of biological function. (*"Proteus, Old Man of the Sea, Roman period mosaic, from Thessalonika, 1st century A.D. National Archaeological Museum, Athens/Ancient Art and Architecture Collection Ltd./Bridgeman Art Library, London/New York)*

Growing in size and complexity
Living things, masses of atoms, DNA,
*   protein*
Dancing a pattern ever more intricate.
Out of the cradle onto the dry land
Here it is standing
Atoms with consciousness
Matter with curiosity.
Stands at the sea
Wonders at wondering
I
A universe of atoms
An atom in the universe.
Richard P. Feyman (1918–1988)
From "The Value of Science" in Edward Hutchings, Jr., ed. 1958. Frontiers of Science: A Survey. New York: Basic Books.

6.1 What Are the Noncovalent Interactions That Dictate and Stabilize Protein Structure?

Several different kinds of noncovalent interactions are of vital importance in protein structure. Hydrogen bonds, hydrophobic interactions, electrostatic bonds, and van der Waals forces are all noncovalent in nature, yet they are extremely important influences on protein conformation. The stabilization free energies afforded by each of these interactions may be highly dependent on the local environment within the protein, but certain generalizations can still be made.

Hydrogen Bonds Are Formed Whenever Possible

Hydrogen bonds are generally made wherever possible within a given protein structure. In most protein structures that have been examined to date, component atoms of the peptide backbone tend to form hydrogen bonds with one

Key Questions

6.1 What Are the Noncovalent Interactions That Dictate and Stabilize Protein Structure?

6.2 What Role Does the Amino Acid Sequence Play in Protein Structure?

6.3 What Are the Elements of Secondary Structure in Proteins, and How Are They Formed?

6.4 How Do Polypeptides Fold into Three-Dimensional Protein Structures?

6.5 How Do Protein Subunits Interact at the Quaternary Level of Protein Structure?

Biochemistry Now™ Test yourself on these Key Questions at BiochemistryNow at http://chemistry.brookscole.com/ggb3

another. Furthermore, side chains capable of forming H bonds are usually located on the protein surface and form such bonds primarily with the water solvent. Although each hydrogen bond may contribute an average of only about 12 kJ/mol in stabilization energy for the protein structure, the number of H bonds formed in the typical protein is very large. For example, in α-helices, the C=O and N—H groups of every residue participate in H bonds. The importance of H bonds in protein structure cannot be overstated.

Hydrophobic Interactions Drive Protein Folding

Hydrophobic "bonds," or, more accurately, *interactions,* form because nonpolar side chains of amino acids and other nonpolar solutes prefer to cluster in a nonpolar environment rather than to intercalate in a polar solvent such as water. The forming of hydrophobic bonds minimizes the interaction of nonpolar residues with water and is therefore highly favorable. Such clustering is entropically driven. The side chains of the amino acids in the interior or core of the protein structure are almost exclusively hydrophobic. Polar amino acids are almost never found in the interior of a protein, but the protein surface may consist of both polar and nonpolar residues.

Electrostatic Interactions Usually Occur on the Protein Surface

Ionic interactions arise either as electrostatic attractions between opposite charges or repulsions between like charges. Chapter 4 discusses the ionization behavior of amino acids. Amino acid side chains can carry positive charges, as in the case of lysine, arginine, and histidine, or negative charges, as in aspartate and glutamate. In addition, the N-terminal and C-terminal residues of a protein or peptide chain usually exist in ionized states and carry positive or negative charges, respectively. All of these may experience electrostatic interactions in a protein structure. Charged residues are normally located on the protein surface, where they may interact optimally with the water solvent. It is energetically unfavorable for an ionized residue to be located in the hydrophobic core of the protein. Electrostatic interactions between charged groups on a protein surface are often complicated by the presence of salts in the solution. For example, the ability of a positively charged lysine to attract a nearby negative glutamate may be weakened by dissolved NaCl (Figure 6.1). The Na$^+$ and Cl$^-$ ions are highly mobile, compact units of charge, compared to the amino acid side chains, and thus compete effectively for charged sites on the protein. In this manner, electrostatic interactions among amino acid residues on protein surfaces may be damped out by high concentrations of salts. Nevertheless, these interactions are important for protein stability.

Van der Waals Interactions Are Ubiquitous

Both attractive forces and repulsive forces are included in van der Waals interactions. The attractive forces are due primarily to instantaneous dipole-induced dipole interactions that arise because of fluctuations in the electron

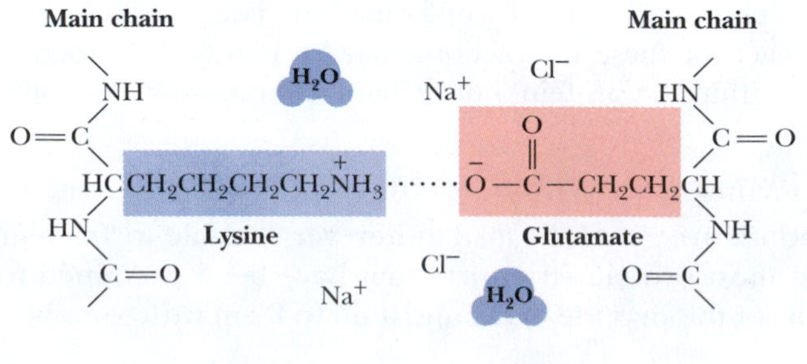

FIGURE 6.1 An electrostatic interaction between the ϵ-amino group of a lysine and the γ-carboxyl group of a glutamate residue.

charge distributions of adjacent nonbonded atoms. Individual van der Waals interactions are weak ones (with stabilization energies of 0.4 to 4.0 kJ/mol), but many such interactions occur in a typical protein, and by sheer force of numbers, they can represent a significant contribution to the stability of a protein. Peter Privalov and George Makhatadze have shown that for pancreatic ribonuclease A, hen egg white lysozyme, horse heart cytochrome c, and sperm whale myoglobin, van der Waals interactions between tightly packed groups in the interior of the protein are a major contribution to protein stability.

<table>
<tr><td>6.2</td><td>

What Role Does the Amino Acid Sequence Play in Protein Structure?
</td></tr>
</table>

It can be inferred from the first section of this chapter that many different forces work together in a delicate balance to determine the overall three-dimensional structure of a protein. These forces operate both within the protein structure itself and between the protein and the water solvent. How, then, does nature dictate the manner of protein folding to generate the three-dimensional structure that optimizes and balances these many forces? *All of the information necessary for folding the peptide chain into its "native" structure is contained in the amino acid sequence of the peptide.* This principle was first appreciated by C. B. Anfinsen and F. White, whose work in the early 1960s dealt with the chemical denaturation and subsequent renaturation of bovine pancreatic ribonuclease. Ribonuclease was first denatured with urea and mercaptoethanol, a treatment that cleaved the four covalent disulfide (S—S) cross-bridges in the protein. Subsequent air oxidation permitted random formation of disulfide cross-bridges, most of which were incorrect. Thus, the air-oxidized material showed little enzymatic activity. However, treatment of these inactive preparations with small amounts of mercaptoethanol allowed a reshuffling of the disulfide bonds and permitted formation of significant amounts of active native enzyme. In such experiments, the only road map for the protein, that is, the only "instructions" it has, are those directed by its primary structure, the linear sequence of its amino acid residues.

Just how proteins recognize and interpret the information that is stored in the amino acid sequence is not yet well understood. It may be assumed that certain loci along the peptide chain act as nucleation points, which initiate folding processes that eventually lead to the correct structures. Regardless of how this process operates, it must take the protein correctly to the final native structure, without getting trapped in a local energy-minimum state that, although stable, may be different from the native state itself. A long-range goal of many researchers in the protein structure field is the prediction of three-dimensional conformation from the amino acid sequence. As the details of secondary and tertiary structure are described in this chapter, the complexity and immensity of such a prediction will be more fully appreciated. This area is one of the greatest uncharted frontiers remaining in molecular biology.

<table>
<tr><td>6.3</td><td>

What Are the Elements of Secondary Structure in Proteins, and How Are They Formed?
</td></tr>
</table>

Any discussion of protein folding and structure must begin with the *peptide bond*, the fundamental structural unit in all proteins. As we saw in Chapter 5, the resonance structures experienced by a peptide bond constrain the oxygen, carbon, nitrogen, and hydrogen atoms of the peptide group, as well as the adjacent α-carbons, to all lie in a plane. The resonance stabilization energy of this

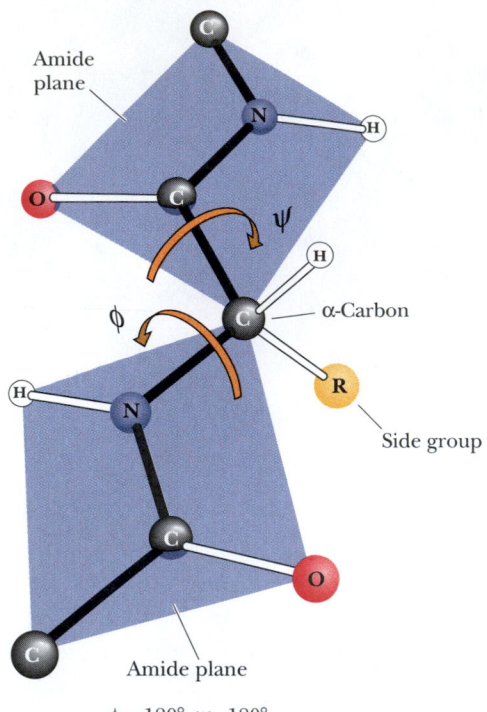

$\phi = 180°, \psi = 180°$

FIGURE 6.2 The amide or peptide bond planes are joined by the tetrahedral bonds of the α-carbon. The rotation parameters are ϕ and ψ. The conformation shown corresponds to $\phi = 180°$ and $\psi = 180°$. Note that positive values of ϕ and ψ correspond to clockwise rotation as viewed from C_α. Starting from 0°, a rotation of 180° in the clockwise direction $(+180°)$ is equivalent to a rotation of 180° in the counterclockwise direction $(-180°)$. *(Illustration: Irving Geis. Rights owned by Howard Hughes Medical Institute. Not to be reproduced without permission.)*

planar structure is approximately 88 kJ/mol, and substantial energy is required to twist the structure about the C—N bond. A twist of θ degrees involves a twist energy of $88 \sin^2\theta$ kJ/mol.

All Protein Structure is Based on the Amide Plane

The planarity of the peptide bond means that there are only two degrees of freedom per residue for the peptide chain. Rotation is allowed about the bond linking the α-carbon and the carbon of the peptide bond and also about the bond linking the nitrogen of the peptide bond and the adjacent α-carbon. As shown in Figure 6.2, each α-carbon is the joining point for two planes defined by peptide bonds. The angle about the C_α—N bond is denoted by the Greek letter ϕ (phi), and that about the C_α—C_o is denoted by ψ (psi). For either of these bond angles, a value of 0° corresponds to an orientation with the amide plane bisecting the H—C_α—R (side-chain) plane and a *cis* conformation of the main chain around the rotating bond in question (Figure 6.3). In any case, the entire path of the peptide backbone in a protein is known if the ϕ and ψ rotation angles are all specified. Some values of ϕ and ψ are not allowed due to steric interference between nonbonded atoms. As shown in Figure 6.4, values of $\phi = 180°$ and $\psi = 0°$ are not allowed because of the forbidden overlap of the N—H hydrogens. Similarly, $\phi = 0°$ and $\psi = 180°$ are forbidden because of unfavorable overlap between the carbonyl oxygens.

G. N. Ramachandran and his co-workers in Madras, India, first showed that it was convenient to plot ϕ values against ψ values to show the distribution of allowed values in a protein or in a family of proteins. A typical **Ramachandran plot** is shown in Figure 6.4. Note the clustering of ϕ and ψ values in a few regions of the plot. Most combinations of ϕ and ψ are sterically forbidden, and the corresponding regions of the Ramachandran plot are sparsely populated. The combinations that are sterically allowed represent the subclasses of structure described in the remainder of this section.

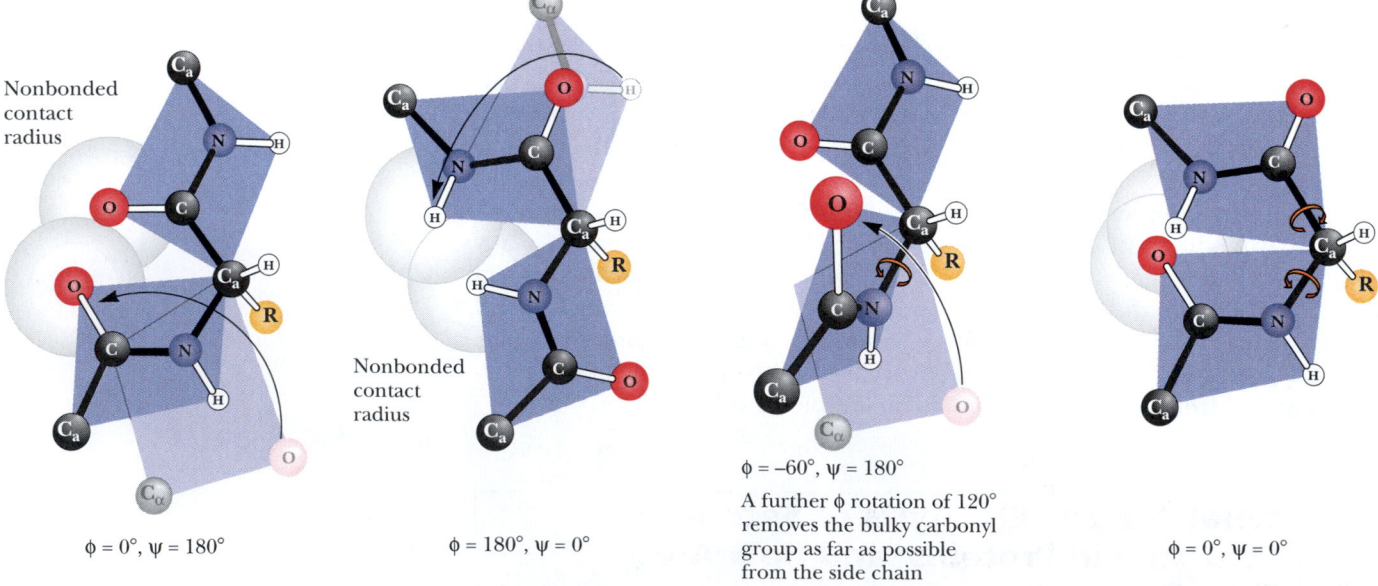

$\phi = 0°, \psi = 180°$ $\phi = 180°, \psi = 0°$ $\phi = -60°, \psi = 180°$

A further ϕ rotation of 120° removes the bulky carbonyl group as far as possible from the side chain

$\phi = 0°, \psi = 0°$

Biochemistry ⊗ Now™   ACTIVE FIGURE 6.3 Many of the possible conformations about an α-carbon between two peptide planes are forbidden because of steric crowding. Several noteworthy examples are shown here.

Note: The formal IUPAC-IUB Commission on Biochemical Nomenclature convention for the definition of the torsion angles ϕ and ψ in a polypeptide chain *(Biochemistry 9:3471–3479, 1970)* is different from that used here, where the C_α atom serves as the point of reference for both rotations, but the result is the same. *(Illustration: Irving Geis. Rights owned by Howard Hughes Medical Institute. Not to be reproduced without permission.)* **Test yourself on the concepts in this figure at http://chemistry.brookscole.com/ggb3**

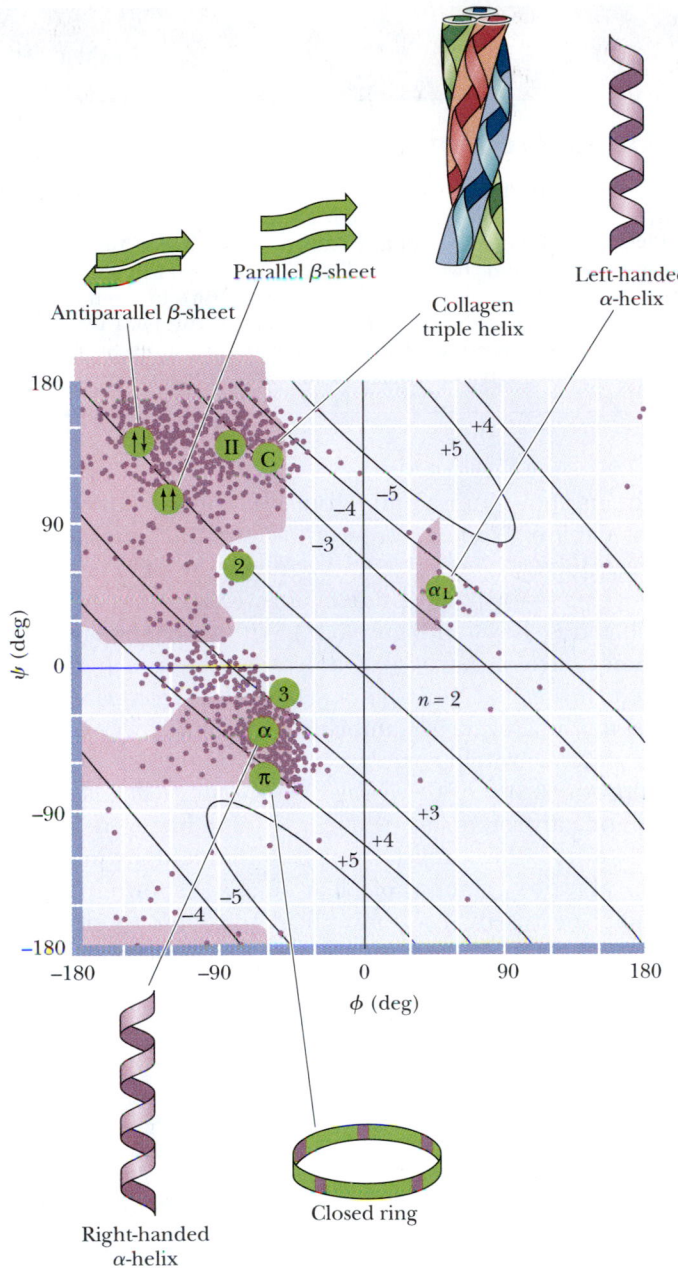

Antiparallel β-sheet

Parallel β-sheet

Collagen triple helix

Left-handed α-helix

Right-handed α-helix

Closed ring

Biochemistry⟨≋⟩Now™ **ACTIVE FIGURE 6.4** A Ramachandran diagram showing the sterically reasonable values of the angles ϕ and ψ. The shaded regions indicate particularly favorable values of these angles. Dots in purple indicate actual angles measured for 1000 residues (excluding glycine, for which a wider range of angles is permitted) in eight proteins. The lines running across the diagram (numbered +5 through 2 and −5 through −3) signify the number of amino acid residues per turn of the helix; "+" means right-handed helices; "−" means left-handed helices. *(After Richardson, J. S., 1981. The anatomy and taxonomy of protein structure. Advances in Protein Chemistry 34:167–339.)* **Test yourself on the concepts in this figure at http://chemistry. brookscole.com/ggb3**

The Alpha-Helix Is a Key Secondary Structure

The discussion of hydrogen bonding in Section 6.1 pointed out that the carbonyl oxygen and amide hydrogen of the peptide bond could participate in H bonds either with water molecules in the solvent or with other H-bonding groups in the peptide chain.

In nearly all proteins, the carbonyl oxygens and the amide protons of many peptide bonds participate in H bonds that link one peptide group to another, as shown in Figure 6.5. These structures tend to form in cooperative fashion and involve substantial portions of the peptide chain. Structures resulting from these interactions constitute **secondary structure** for proteins (see Chapter 5). When a number of hydrogen bonds form between portions of the peptide chain in this manner, two basic types of structures can result: *α-helices* and *β-pleated sheets.*

Evidence for helical structures in proteins was first obtained in the 1930s in studies of fibrous proteins. However, there was little agreement at that time about the exact structure of these helices, primarily because there was also

A Deeper Look

Knowing What the Right Hand and Left Hand Are Doing

Certain conventions related to peptide bond angles and the "hand-edness" of biological structures are useful in any discussion of protein structure. To determine the ϕ and ψ angles between peptide planes, viewers should imagine themselves at the C_α carbon looking outward and should imagine starting from the $\phi = 0°$, $\psi = 0°$ conformation. From this perspective, positive values of ϕ correspond to clockwise rotations about the C_α—N bond of the plane that includes the adjacent N—H group. Similarly, positive values of ψ cor-respond to clockwise rotations about the C_α—C bond of the plane that includes the adjacent C=O group.

Biological structures are often said to exhibit "right-hand" or "left-hand" twists. For all such structures, the sense of the twist can be ascertained by holding the structure in front of you and look-ing along the polymer backbone. If the twist is clockwise as one proceeds outward and through the structure, it is said to be right-handed. If the twist is counterclockwise, it is said to be left-handed.

lack of agreement about interatomic distances and bond angles in peptides. In 1951, Linus Pauling, Robert Corey, and their colleagues at the California Institute of Technology summarized a large volume of crystallographic data in a set of dimensions for polypeptide chains. (A summary of data similar to what they reported is shown in Figure 5.2.) With these data in hand, Pauling, Corey, and their colleagues proposed a new model for a helical structure in proteins, which they called the α-helix. The report from Caltech was of par-ticular interest to Max Perutz in Cambridge, England, a crystallographer who was also interested in protein structure. By taking into account a critical but previously ignored feature of the X-ray data, Perutz realized that the α-helix existed in keratin, a protein from hair, and also in several other proteins. Since then, the α-helix has proved to be a fundamentally important peptide structure. Several representations of the α-helix are shown in Figure 6.6. One turn of the helix represents 3.6 amino acid residues. (A single turn of the α-helix involves 13 atoms from the O to the H of the H bond. For this reason,

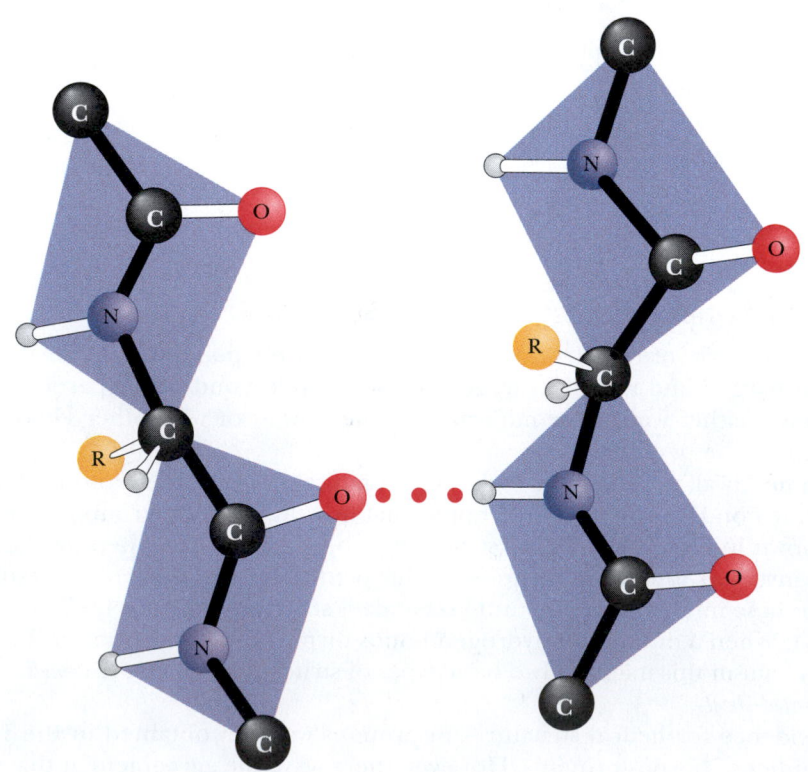

FIGURE 6.5 A hydrogen bond between the amide proton and carbonyl oxygen of adjacent peptide groups.

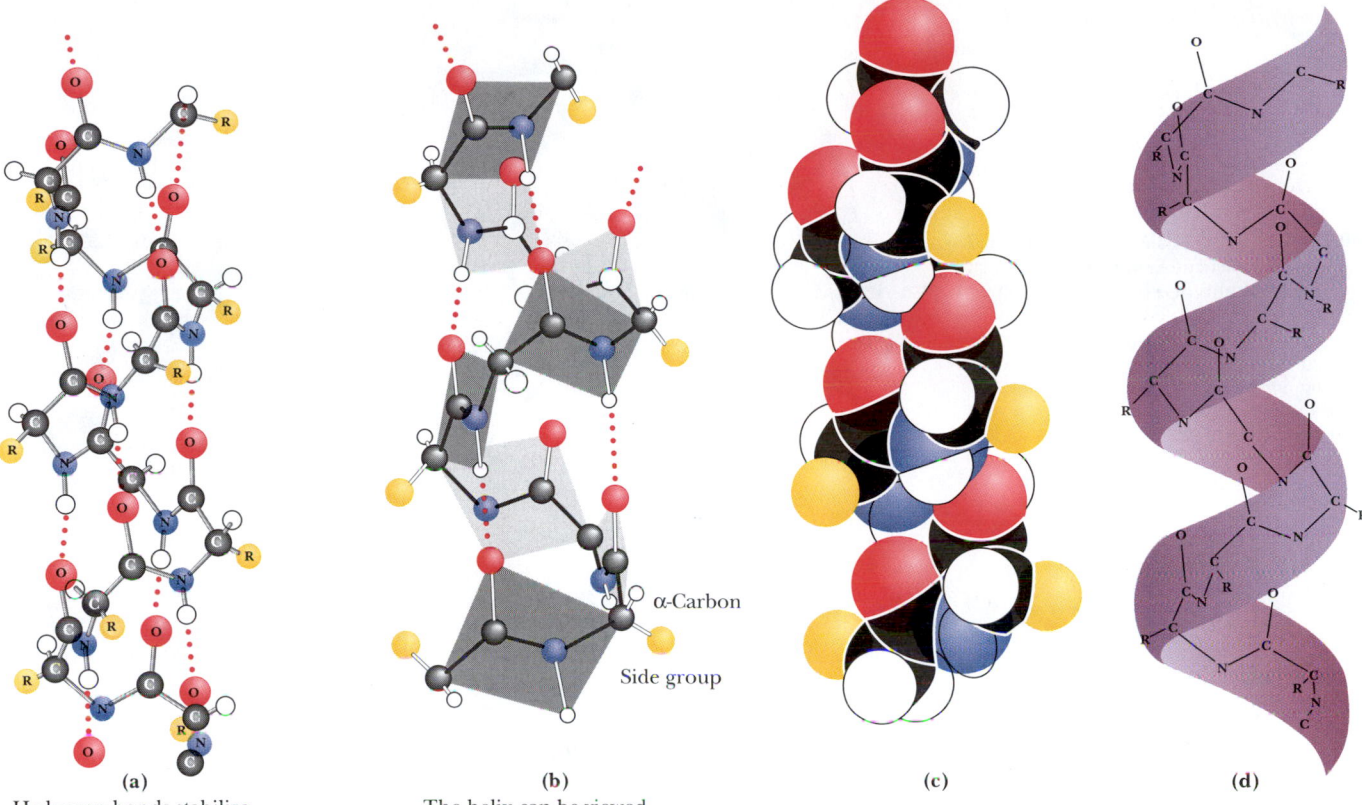

(a)
Hydrogen bonds stabilize the helix structure.

(b)
The helix can be viewed as a stacked array of peptide planes hinged at the α-carbons and approximately parallel to the helix.

α-Carbon

Side group

(c)

(d)

FIGURE 6.6 Four different graphic representations of the α-helix. **(a)** As it originally appeared in Pauling's 1960 *The Nature of the Chemical Bond*. **(b)** Showing the arrangement of peptide planes in the helix. **(c)** A space-filling computer graphic presentation. **(d)** A "ribbon structure" with an inlaid stick figure, showing how the ribbon indicates the path of the polypeptide backbone. *(Illustration: Irving Geis. Rights owned by Howard Hughes Medical Institute. Not to be reproduced without permission.)*

the α-helix is sometimes referred to as the 3.6_{13} helix.) This is in fact the feature that most confused crystallographers before the Pauling and Corey α-helix model. Crystallographers were so accustomed to finding twofold, threefold, sixfold, and similar integral axes in simpler molecules that the notion of a nonintegral number of units per turn was never taken seriously before Pauling and Corey's work.

Each amino acid residue extends **1.5 Å (0.15 nm)** along the helix axis. With **3.6 residues per turn,** this amounts to 3.6×1.5 Å or **5.4 Å (0.54 nm)** of travel along the helix axis per turn. This is referred to as the translation distance or the **pitch** of the helix. If one ignores side chains, the helix is about 6 Å in diameter. The side chains, extending outward from the core structure of the helix, are removed from steric interference with the polypeptide backbone. As can be seen in Figure 6.6, *each peptide carbonyl is hydrogen bonded to the peptide N—H group four residues farther up the chain.* Note that all of the H bonds lie parallel to the helix axis and all of the carbonyl groups are pointing in one direction along the helix axis while the N—H groups are pointing in the opposite direction. Recall that the entire path of the peptide backbone can be known if the ϕ and ψ twist angles are specified for each residue. The α-helix is formed if the values of ϕ are approximately $-60°$ and the values of ψ are in the range of -45 to $-50°$. Figure 6.7 shows the structures of two proteins that contain α-helical segments. The number of residues involved in a given α-helix varies from helix to helix and from protein to protein. On average, there are about 10 residues per helix. Myoglobin, one of the first proteins in which α-helices were observed, has eight stretches of α-helix that form a box to contain the heme prosthetic group (see Figure 5.5).

Biochemistry ⦙Now™   Go to BiochemistryNow and click BiochemistryInteractive to explore the anatomy of the α-helix.

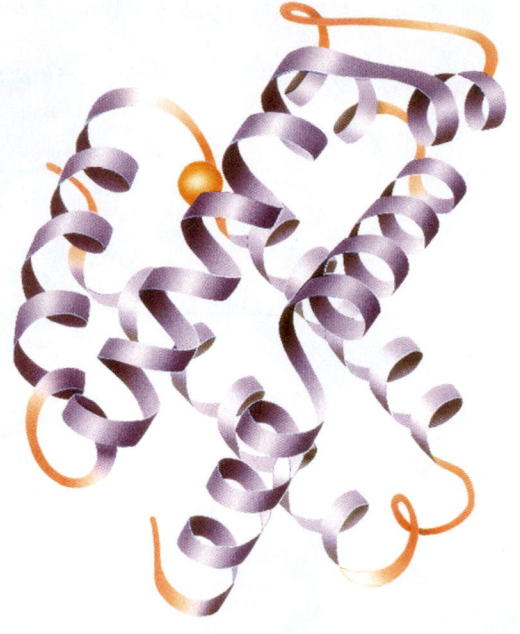

β-Hemoglobin subunit   Myohemerythrin

Biochemistry ⓔ**Now**™ **ANIMATED FIGURE 6.7**
The three-dimensional structures of two proteins that contain substantial amounts of α-helix in their structures. The helices are represented by the regularly coiled sections of the ribbon drawings. Myohemerythrin is the oxygen-carrying protein in certain invertebrates, including *Sipunculids*, a phylum of marine worm. *(Jane Richardson.)* **See this figure animated at http://chemistry.brookscole.com/ggb3**

(a)

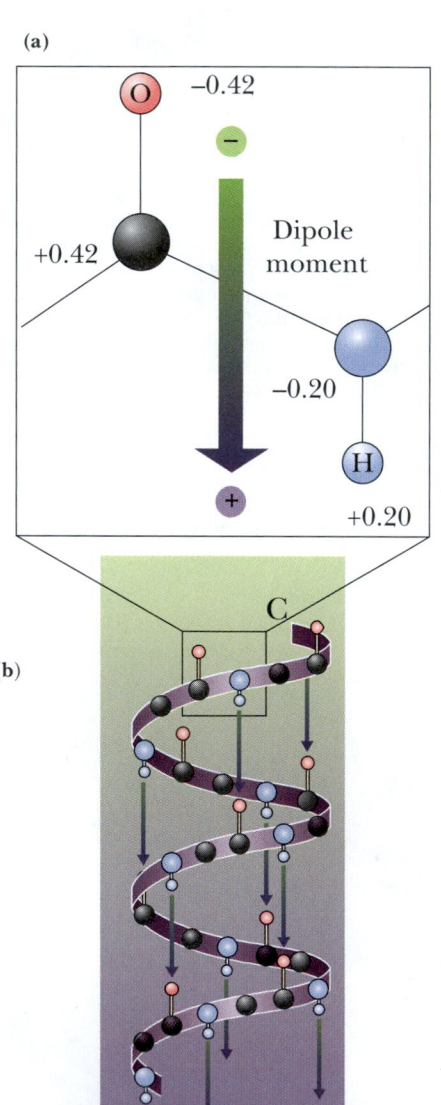

(b)

As shown in Figure 6.6, all of the hydrogen bonds point in the same direction along the α-helix axis. Each peptide bond possesses a dipole moment that arises from the polarities of the N—H and C=O groups, and because these groups are all aligned along the helix axis, the helix itself has a substantial dipole moment, with a partial positive charge at the N-terminus and a partial negative charge at the C-terminus (Figure 6.8). Negatively charged ligands (e.g., phosphates) frequently bind to proteins near the N-terminus of an α-helix. By contrast, positively charged ligands are only rarely found to bind near the C-terminus of an α-helix.

In a typical α-helix of 12 (or n) residues, there are 8 (or $n - 4$) hydrogen bonds. As shown in Figure 6.9, the first 4 amide hydrogens and the last 4 carbonyl oxygens cannot participate in helix H bonds. Also, nonpolar residues situated near the helix termini can be exposed to solvent. Proteins frequently compensate for these problems by **helix capping**—providing H-bond partners for the otherwise bare N—H and C=O groups and folding other parts of the protein to foster hydrophobic contacts with exposed nonpolar residues at the helix termini.

Careful studies of the **polyamino acids,** polymers in which all the amino acids are identical, have shown that certain amino acids tend to occur in α-helices, whereas others are less likely to be found in them. Polyleucine and polyalanine, for example, readily form α-helical structures. In contrast, polyaspartic acid and polyglutamic acid, which are highly negatively charged at pH 7.0, form only random structures because of strong charge repulsion between the R groups along the peptide chain. At pH 1.5 to 2.5, however, where the side chains are protonated and thus uncharged, these latter species spontaneously form α-helical structures. In similar fashion, polylysine is a random coil at pH values below about 11, where repulsion of positive charges prevents helix formation. At pH 12, where polylysine is a neutral peptide chain, it readily forms an α-helix.

◄ **FIGURE 6.8** The arrangement of N—H and C=O groups (each with an individual dipole moment) along the helix axis creates a large net dipole for the helix. Numbers indicate fractional charges on respective atoms.

The tendencies of various amino acids to stabilize or destabilize α-helices are different in typical proteins than in polyamino acids. The occurrence of the common amino acids in helices is summarized in Table 6.1. Notably, proline (and hydroxyproline) act as helix breakers due to their unique structure, which fixes the value of the C_α—N—C bond angle. Helices can be formed from either D- or L-amino acids, but a given helix must be composed entirely of amino acids of one configuration. α-Helices cannot be formed from a mixed copolymer of D- and L-amino acids. An α-helix composed of D-amino acids is left-handed.

Other Helical Structures Exist

There are several other far less common types of helices found in proteins. The most common of these is the 3_{10} helix, which contains 3.0 residues per turn (with 10 atoms in the ring formed by making the hydrogen bond three residues up the chain). It normally extends over shorter stretches of sequence than the α-helix. Other helical structures include the 2_7 ribbon and the π-helix, which has 4.4 residues and 16 atoms per turn and is thus called the 4.4_{16} helix.

The β-Pleated Sheet Is a Core Structure in Proteins

Another type of structure commonly observed in proteins also forms because of local, cooperative formation of hydrogen bonds. That is the pleated sheet, or β-structure, often called the **β-pleated sheet.** This structure was also first postulated by Pauling and Corey in 1951 and has now been observed in many natural proteins. A β-pleated sheet can be visualized by laying thin, pleated strips of paper side by side to make a "pleated sheet" of paper (Figure 6.10). Each

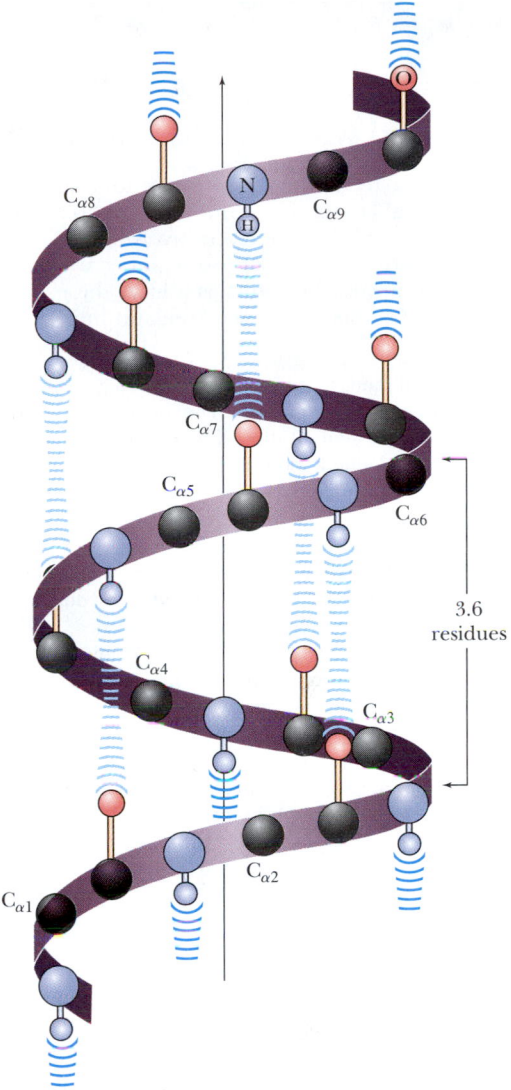

FIGURE 6.9 Four N—H groups at the N-terminal end of an α-helix and four C=O groups at the C-terminal end cannot participate in hydrogen bonding. The formation of H bonds with other nearby donor and acceptor groups is referred to as **helix capping.** Capping may also involve appropriate hydrophobic interactions that accommodate nonpolar side chains at the ends of helical segments.

Table 6.1			
Helix-Forming and Helix-Breaking Behavior of the Amino Acids			
Amino Acid		**Helix Behavior***	
A	Ala	H	(I)
C	Cys	Variable	
D	Asp	Variable	
E	Glu	H	
F	Phe	H	
G	Gly	I	(B)
H	His	H	(I)
I	Ile	H	(C)
K	Lys	Variable	
L	Leu	H	
M	Met	H	
N	Asn	C	(I)
P	Pro	B	
Q	Gln	H	(I)
R	Arg	H	(I)
S	Ser	C	(B)
T	Thr	Variable	
V	Val	Variable	
W	Trp	H	(C)
Y	Tyr	H	(C)

*H = helix former; I = indifferent; B = helix breaker; C = random coil; ( ) = secondary tendency.

Critical Developments in Biochemistry

In Bed with a Cold, Pauling Stumbles onto the α-Helix and a Nobel Prize*

As high technology continues to transform the modern bio-chemical laboratory, it is interesting to reflect on Linus Pauling's discovery of the α-helix. It involved only a piece of paper, a pencil, scissors, and a sick Linus Pauling, who had tired of reading detective novels. The story is told in the excellent book *The Eighth Day of Creation* by Horace Freeland Judson:

> From the spring of 1948 through the spring of 1951...rivalry sputtered and blazed between Pauling's lab and (Sir Lawrence) Bragg's—over protein. The prize was to propose and verify in nature a general three-dimensional structure for the polypeptide chain. Pauling was working up from the simpler structures of components. In January 1948, he went to Oxford as a visiting professor for two terms, to lecture on the chemical bond and on molecular structure and biological specificity. "In Oxford, it was April, I believe, I caught cold. I went to bed, and read detective stories for a day, and got bored, and thought why don't I have a crack at that problem of alpha keratin." Confined, and still fingering the polypeptide chain in his mind, Pauling called for paper, pencil, and straightedge and attempted to reduce the problem to an almost Euclidean purity. "I took a sheet of paper—I still have this sheet of paper—and drew, rather roughly, the way that I thought a polypeptide chain would look if it were spread out into a plane." The repetitious herringbone of the chain he could stretch across the paper as simply as this—

(a)

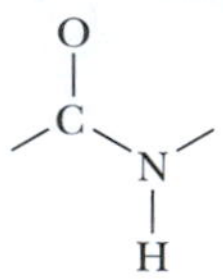

—putting in lengths and bond angles from memory....He knew that the peptide bond, at the carbon-to-nitrogen link, was always rigid:

(b)

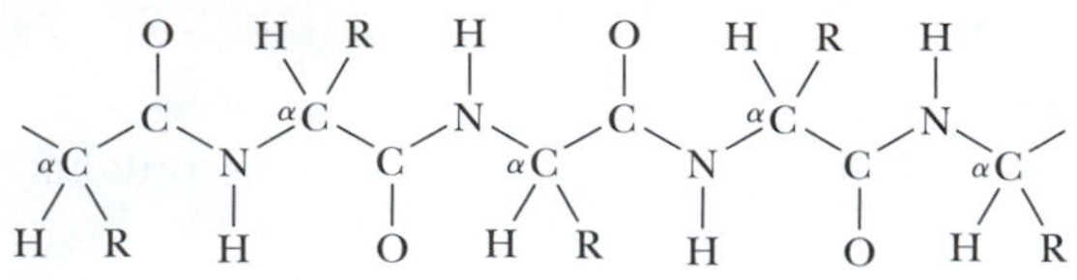

And this meant that the chain could turn corners only at the alpha carbons....."I creased the paper in parallel creases through the alpha carbon atoms, so that I could bend it and make the bonds to the alpha carbons, along the chain, have tetrahedral value. And then I looked to see if I could form hydrogen bonds from one part of the chain to the next." He saw that if he folded the strip like a chain of paper dolls into a helix, and if he got the pitch of the screw right, hydrogen bonds could be shown to form, N—H···O—C, three or four knuckles apart along the backbone, holding the helix in shape. After several tries, changing the angle of the parallel creases in order to adjust the pitch of the helix, he found one where the hydrogen bonds would drop into place, connecting the turns, as straight lines of the right length. He had a model.

*The discovery of the α-helix structure was only one of many achievements that led to Pauling's Nobel Prize in Chemistry in 1954. The official citation for the prize was "for his research into the nature of the chemical bond and its application to the elucidation of the structure of complex substances."

Biochemistry ⊗ Now™   Go to BiochemistryNow and click BiochemistryInteractive to explore β-sheets, one of the principal types of secondary structure in proteins.

strip of paper can then be pictured as a single peptide strand in which the peptide backbone makes a zigzag pattern along the strip, with the α-carbons lying at the folds of the pleats. The pleated sheet can exist in both parallel and antiparallel forms. In the **parallel β-pleated sheet,** adjacent chains run in the same direction (N→C or C→N). In the **antiparallel β-pleated sheet,** adjacent strands run in opposite directions.

Each single strand of the β-sheet structure can be pictured as a twofold helix, that is, a helix with two residues per turn. The arrangement of successive amide planes has a pleated appearance due to the tetrahedral nature of the C_α atom. It is important to note that the hydrogen bonds in this structure are essentially *inter*strand rather than *intra*strand. The peptide backbone in the β-sheet is in its most extended conformation (sometimes called the **ϵ-conformation**). The optimum formation of H bonds in the parallel pleated sheet results in a slightly less extended conformation than in the antiparallel sheet. The H bonds thus formed in the parallel β-sheet are bent significantly. The distance between residues is 0.347 nm for the antiparallel pleated sheet, but only 0.325 nm for the parallel pleated sheet. Figure 6.11 shows examples of both parallel and antiparallel β-pleated sheets. Note that the side chains in the pleated sheet are oriented perpendicular or normal to the plane of the sheet, extending out from the plane on alternating sides.

Parallel β-sheets tend to be more regular than antiparallel β-sheets. The range of ϕ and ψ angles for the peptide bonds in parallel sheets is much smaller

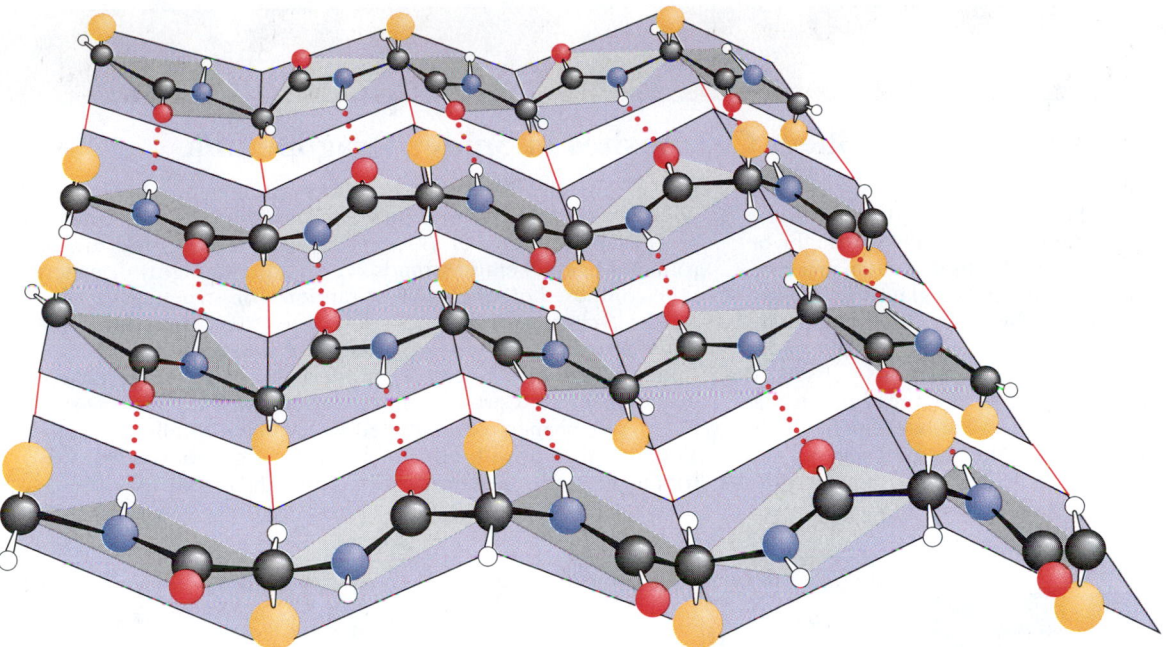

FIGURE 6.10 A "pleated sheet" of paper with an antiparallel β-sheet drawn on it. *(Illustration: Irving Geis. Rights owned by Howard Hughes Medical Institute. Not to be reproduced without permission.)*

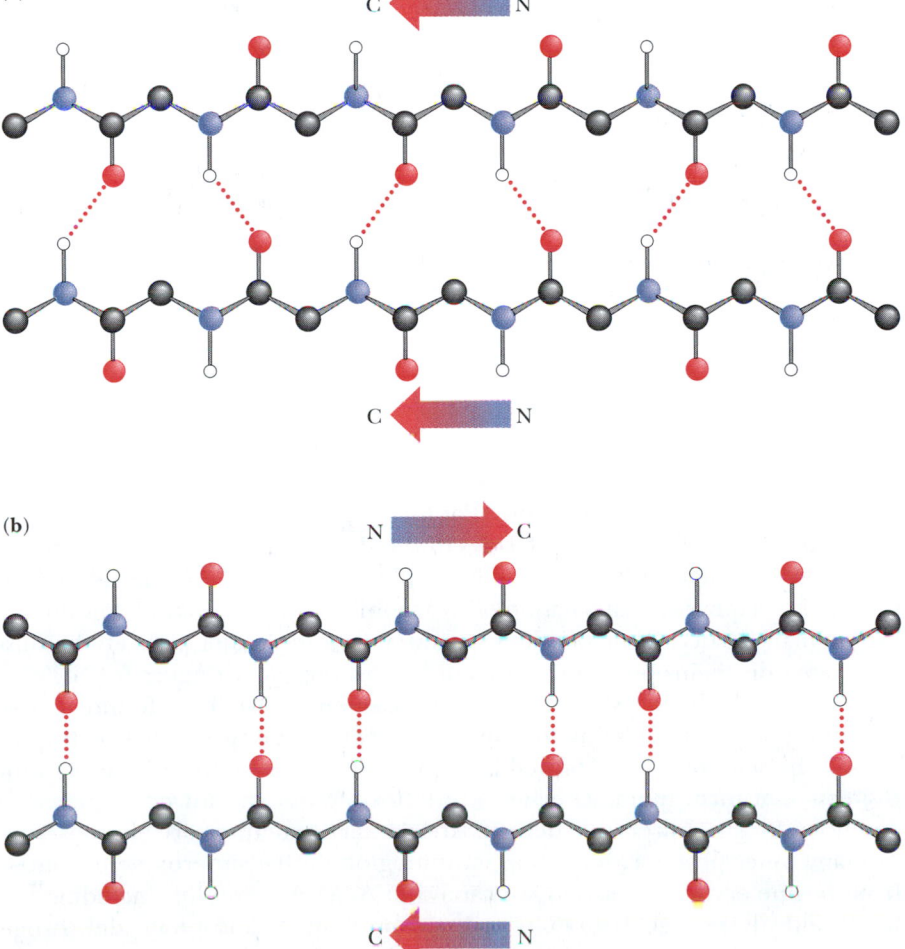

FIGURE 6.11 The arrangement of hydrogen bonds in (a) parallel and (b) antiparallel β-pleated sheets.

A Deeper Look

Charlotte's Web Revisited: Helix—Sheet Composites in Spider Dragline Silk

E. B. White's endearing story *Charlotte's Web* centers around the web-spinning feats of Charlotte the spider. Although the intricate designs of spider webs are eye- (and fly-) catching, it might be argued that the composition of web silk itself is even more remarkable. Spider silk is synthesized in special glands in the spider's abdomen. The silk strands produced by these glands are both strong and elastic. *Dragline silk* (that from which the spider hangs) has a tensile strength of 200,000 psi (pounds per square inch)—stronger than steel and similar to Kevlar, the synthetic material used in bulletproof vests! This same silk fiber is also flexible enough to withstand strong winds and other natural stresses.

This combination of strength and flexibility derives from the *composite nature* of spider silk. As keratin protein is extruded from the spider's glands, it endures shearing forces that break the H bonds stabilizing keratin α-helices. These regions then form microcrystalline arrays of β-sheets. These microcrystals are surrounded by the keratin strands, which adopt a highly disordered state composed of α-helices and random coil structures.

The β-sheet microcrystals contribute strength, and the disordered array of helix and coil make the silk strand flexible. The resulting silk strand resembles modern human-engineered composite materials. Certain tennis racquets, for example, consist of fiberglass polymers impregnated with microcrystalline graphite. The fiberglass provides flexibility, and the graphite crystals contribute strength. Modern high technology, for all its sophistication, is merely imitating nature—and Charlotte's web—after all.

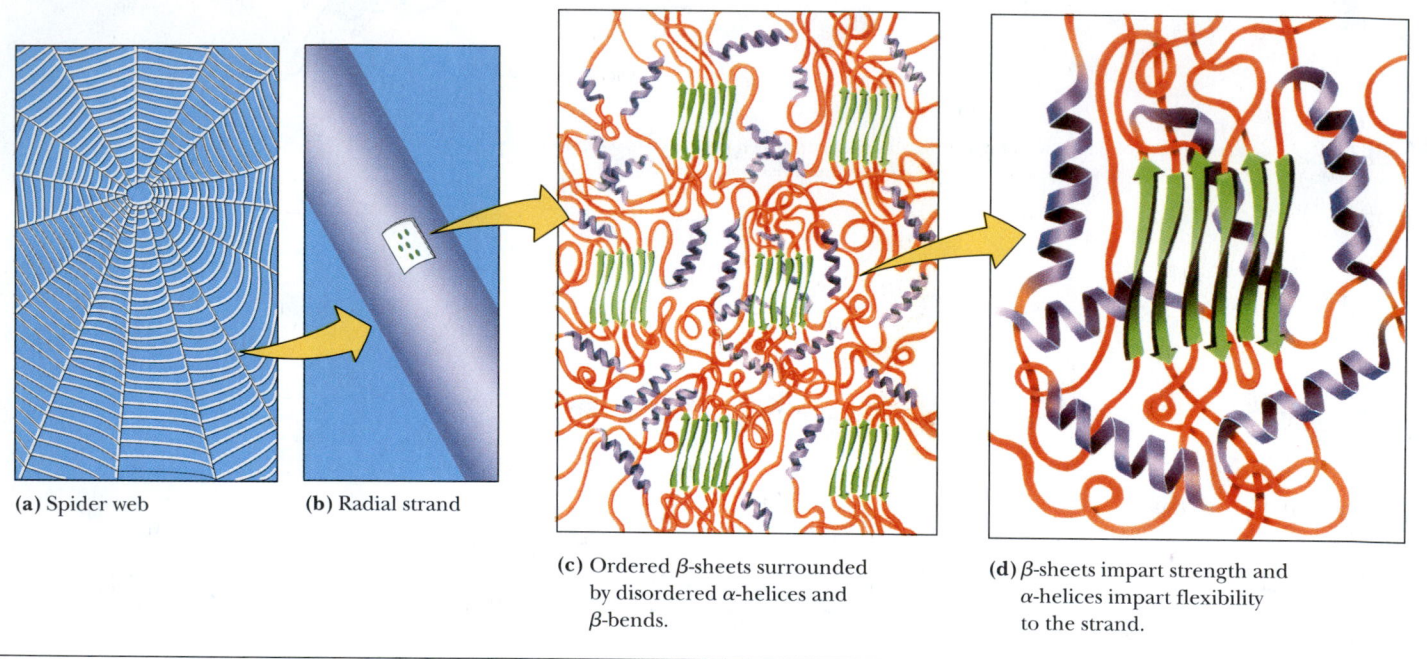

(a) Spider web   **(b)** Radial strand   **(c)** Ordered β-sheets surrounded by disordered α-helices and β-bends.   **(d)** β-sheets impart strength and α-helices impart flexibility to the strand.

than that for antiparallel sheets. Parallel sheets are typically large structures; those composed of less than five strands are rare. Antiparallel sheets, however, may consist of as few as two strands. Parallel sheets characteristically distribute hydrophobic side chains on both sides of the sheet, whereas antiparallel sheets are usually arranged with all their hydrophobic residues on one side of the sheet. This requires an alternation of hydrophilic and hydrophobic residues in the primary structure of peptides involved in antiparallel β-sheets because alternate side chains project to the same side of the sheet (Figure 6.10).

Antiparallel pleated sheets are the fundamental structure found in silk, with the polypeptide chains forming the sheets running parallel to the silk fibers. The silk fibers thus formed have properties consistent with those of the β-sheets that form them. They are quite flexible but cannot be stretched or extended to any appreciable degree. Antiparallel structures are also observed in many other proteins, including immunoglobulin G, superoxide dismutase from bovine erythrocytes, and concanavalin A. Many proteins, including carbonic anhydrase, egg lysozyme, and glyceraldehyde phosphate dehydrogenase, possess both α-helices and β-pleated sheet structures within a single polypeptide chain.

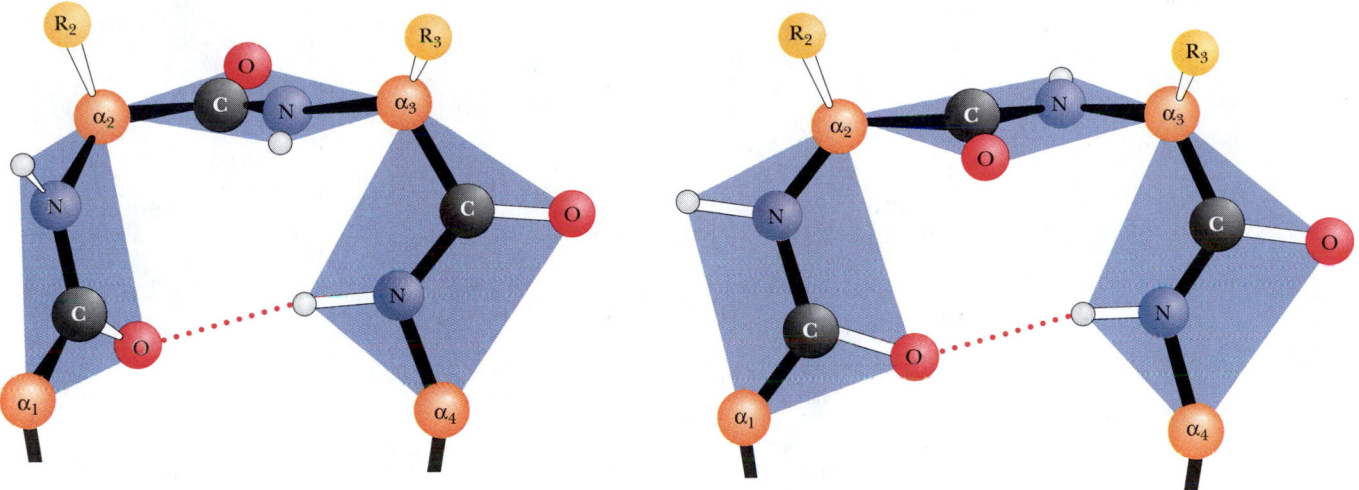

FIGURE 6.12 The structures of two kinds of β-turns (also called tight turns or β-bends). *(Illustration: Irving Geis. Rights owned by Howard Hughes Medical Institute. Not to be reproduced without permission.)*

β-Turns Allow the Protein Strand to Change Direction

Most proteins are globular structures. The polypeptide chain must therefore possess the capacity to bend, turn, and reorient itself to produce the required compact, globular structures. A simple structure observed in many proteins is the **β-turn** (also known as the *tight turn* or *β-bend*), in which the peptide chain forms a tight loop with the carbonyl oxygen of one residue hydrogen bonded with the amide proton of the residue three positions down the chain. This H bond makes the β-turn a relatively stable structure. As shown in Figure 6.12, the β-turn allows the protein to reverse the direction of its peptide chain. This figure shows the two major types of β-turns, but a number of less common types are also found in protein structures. Certain amino acids, such as proline and glycine, occur frequently in β-turn sequences, and the particular conformation of the β-turn sequence depends to some extent on the amino acids composing it. Because it lacks a side chain, glycine is sterically the most adaptable of the amino acids, and it accommodates conveniently to other steric constraints in the β-turn. Proline, however, has a cyclic structure and a fixed ϕ angle, so, to some extent, it forces the formation of a β-turn; in many cases this facilitates the turning of a polypeptide chain upon itself. Such bends promote formation of antiparallel β-pleated sheets.

Biochemistry ⊗ Now™   Go to BiochemistryNow and click BiochemistryInteractive to discover the features of β-turns and how they change the course of a polypeptide strand.

The β-Bulge Is Rare

One final secondary structure, the **β-bulge,** is a small piece of nonrepetitive structure that can occur by itself, although it most often occurs as an irregularity in antiparallel β-structures. A β-bulge can form between two normal β-structure hydrogen bonds and comprises two residues on one strand and one residue on the opposite strand. Figure 6.13 illustrates typical β-bulges. The extra residue on the longer side, which causes additional backbone length, is accommodated partially by creating a bulge in the longer strand and partially by forcing a slight bend in the β-sheet. Bulges thus cause changes in the direction of the polypeptide chain, but to a lesser degree than tight turns do. Many examples of β-bulges are known in protein structures.

The secondary structures we have described here are all found commonly in proteins in nature. In fact, it is hard to find proteins that do not contain one or more of these structures. The energetic (mostly H-bond) stabilization afforded by α-helices, β-pleated sheets, and β-turns is important to proteins, and they seize the opportunity to form such structures wherever possible.

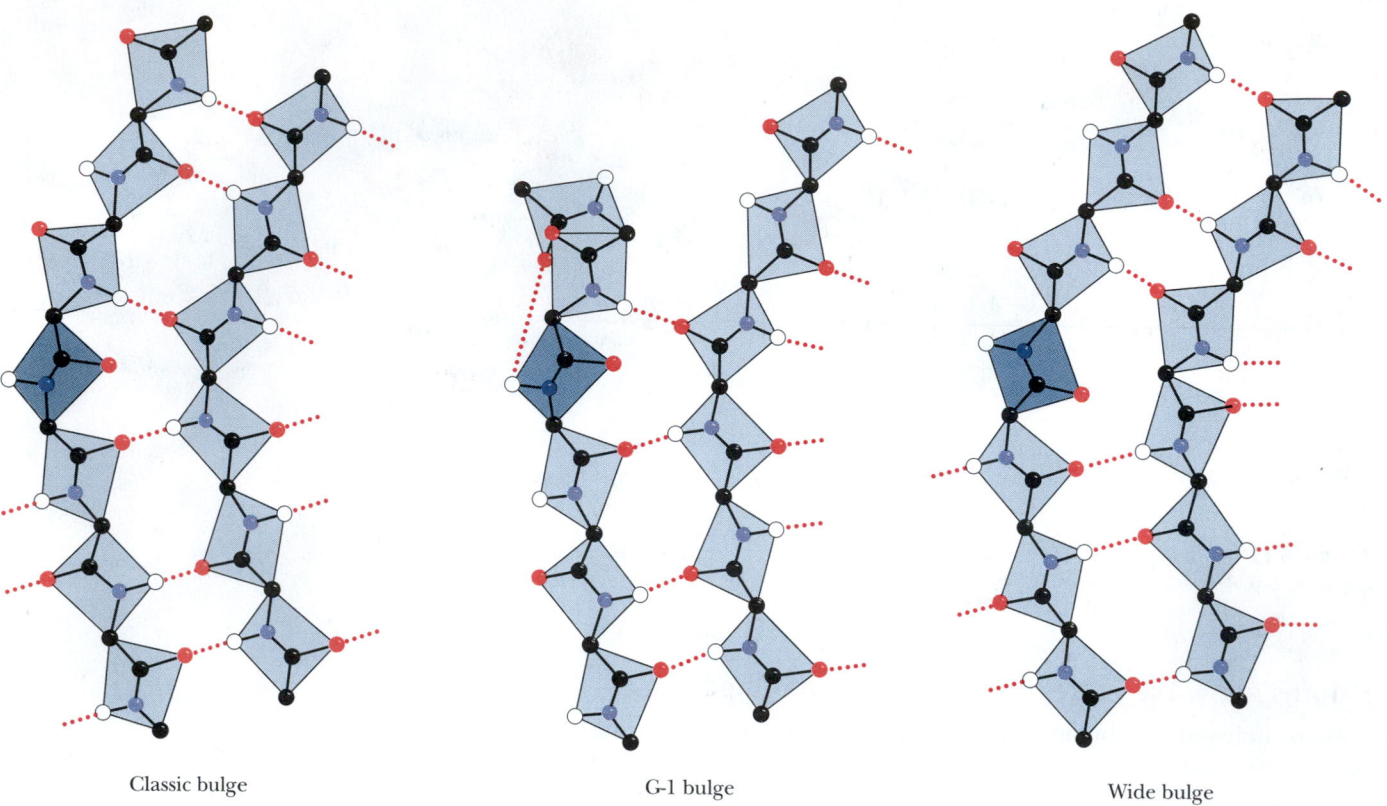

Classic bulge   G-1 bulge   Wide bulge

FIGURE 6.13 Three different kinds of β-bulge structures involving a pair of adjacent polypeptide chains. *(Adapted from Richardson, J. S., 1981. The anatomy and taxonomy of protein structure.* Advances in Protein Chemistry **34**:167–339.)*

6.4	**How Do Polypeptides Fold into Three-Dimensional Protein Structures?**

The folding of a single polypeptide chain in three-dimensional space is referred to as its **tertiary structure.** As discussed in Section 6.2 all of the information needed to fold the protein into its native tertiary structure is contained within the primary structure of the peptide chain itself. With this in mind, it was disappointing to the biochemists of the 1950s when the early protein structures did not reveal the governing principles in any particular detail. It soon became apparent that the proteins knew how they were supposed to fold into tertiary shapes, even if the biochemists did not. Vigorous work in many laboratories has slowly brought important principles to light.

First, secondary structures—helices and sheets—form whenever possible as a consequence of the formation of large numbers of hydrogen bonds. Second, α-helices and β-sheets often associate and pack close together in the protein. No protein is stable as a single-layer structure, for reasons that become apparent later. There are a few common methods for such packing to occur. Third, because the peptide segments between secondary structures in the protein tend to be short and direct, the peptide does not execute complicated twists and knots as it moves from one region of a secondary structure to another. A consequence of these three principles is that protein chains are usually folded so that the secondary structures are arranged in one of a few common patterns. For this reason, there are families of proteins that have similar tertiary structure, with little apparent evolutionary or functional relationship among them. Finally, proteins generally fold so as to form the most stable structures possible. The stability of most proteins arises from (1) the formation of large numbers of intramolecular hydrogen bonds and (2) the reduction in the surface area accessible to solvent that occurs upon folding.

Fibrous Proteins Usually Play a Structure Role

In Chapter 5, we saw that proteins can be grouped into three large classes based on their structure and solubility: *fibrous proteins, globular proteins,* and *membrane proteins.* Fibrous proteins contain polypeptide chains organized approximately parallel along a single axis, producing long fibers or large sheets. Such proteins tend to be mechanically strong and resistant to solubilization in water and dilute salt solutions. Fibrous proteins often play a structural role in nature (see Chapter 5).

α-Keratin As their name suggests, the structure of the α-keratins is dominated by α-helical segments of polypeptide. The amino acid sequence of α-keratin subunits is composed of central α-helix–rich rod domains about 311 to 314 residues in length, flanked by nonhelical N- and C-terminal domains of varying size and composition (Figure 6.14a). The structure of the central rod domain of a typical α-keratin is shown in Figure 6.14b. It consists of four helical strands arranged as twisted pairs of two-stranded **coiled coils.** X-ray diffraction patterns show that these structures resemble α-helices, but with a pitch of 0.51 nm rather than the expected 0.54 nm. This is consistent with a tilt of the helix relative to the long axis of the fiber, as in the two-stranded "rope" in Figure 6.14.

The primary structure of the central rod segments of α-keratin consists of quasi-repeating 7-residue segments of the form (*a-b-c-d-e-f-g*)$_n$. These units are not true repeats, but residues *a* and *d* are usually nonpolar amino acids. In α-helices, with 3.6 residues per turn, these nonpolar residues are arranged in an inclined row or stripe that twists around the helix axis. These nonpolar residues would make the helix highly unstable if they were exposed to solvent, but the association of hydrophobic strips on two α-helices to form the two-stranded rope effectively buries the hydrophobic residues and forms a highly stable structure (Figure 6.14). The helices clearly sacrifice some stability in assuming this twisted conformation, but they gain stabilization energy from the

(a)

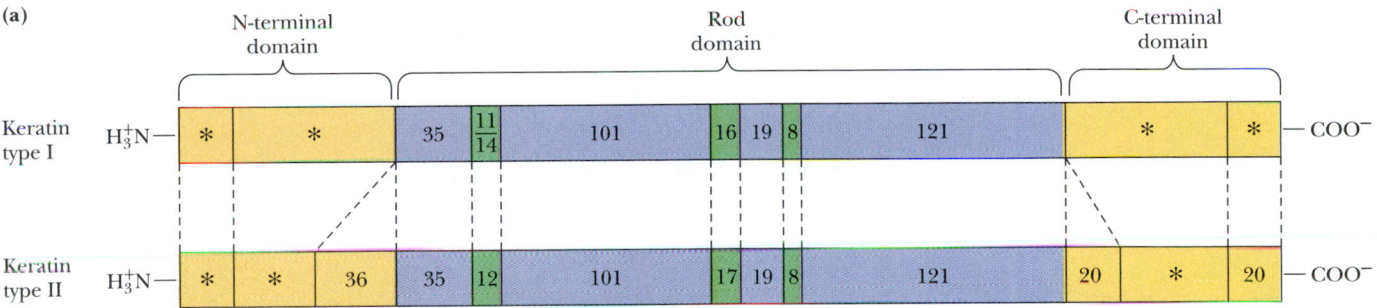

(b)

α-Helix

Coiled coil of two α-helices

Protofilament (pair of coiled coils)

Filament (four right-hand twisted protofibrils)

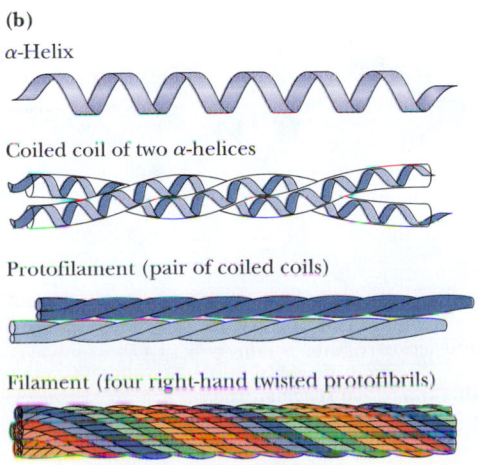

FIGURE 6.14 (a) Both type I and type II α-keratin molecules have sequences consisting of long, central rod domains with terminal cap domains. The numbers of amino acid residues in each domain are indicated. Asterisks denote domains of variable length. **(b)** The rod domains form coiled coils consisting of intertwined right-handed α-helices. These coiled coils then wind around each other in a left-handed twist. Keratin filaments consist of twisted protofibrils (each a bundle of four coiled coils). *(Adapted from Steinert, P., and Parry, D., 1985. Intermediate filaments: Conformity and diversity of expression and structure.* Annual Review of Cell Biology *1:41–65; and Cohlberg, J., 1993. Textbook error: The structure of alpha-keratin.* Trends in Biochemical Sciences *18:360–362.)*

packing of side chains between the helices. In other forms of keratin, covalent disulfide bonds form between cysteine residues of adjacent molecules, making the overall structure rigid, inextensible, and insoluble—important properties for structures such as claws, fingernails, hair, and horns in animals. How and where these disulfides form determines the amount of curling in hair and wool fibers. When a hairstylist creates a permanent wave (simply called a "permanent") in a hair salon, disulfides in the hair are first reduced and cleaved, then reorganized and reoxidized to change the degree of curl or wave. In contrast, a "set" that is created by wetting the hair, setting it with curlers, and then drying it represents merely a rearrangement of the hydrogen bonds between helices and between fibers. (On humid or rainy days, the hydrogen bonds in curled hair may rearrange, and the hair becomes "frizzy.")

Fibroin and β-Keratin: β-Sheet Proteins   The **fibroin** proteins found in silk fibers represent another type of fibrous protein. These are composed of stacked antiparallel β-sheets, as shown in Figure 6.15. In the polypeptide sequence of silk proteins, there are large stretches in which every other residue is a glycine. As previously mentioned, the residues of a β-sheet extend alternately above and below the plane of the sheet. As a result, the glycines all end up on one side of the sheet and the other residues (mainly alanines and serines) compose the opposite surface of the sheet. Pairs of β-sheets can then pack snugly together (glycine surface to glycine surface or alanine–serine surface to alanine—serine surface). The β-keratins found in bird feathers are also made up of stacked β-sheets.

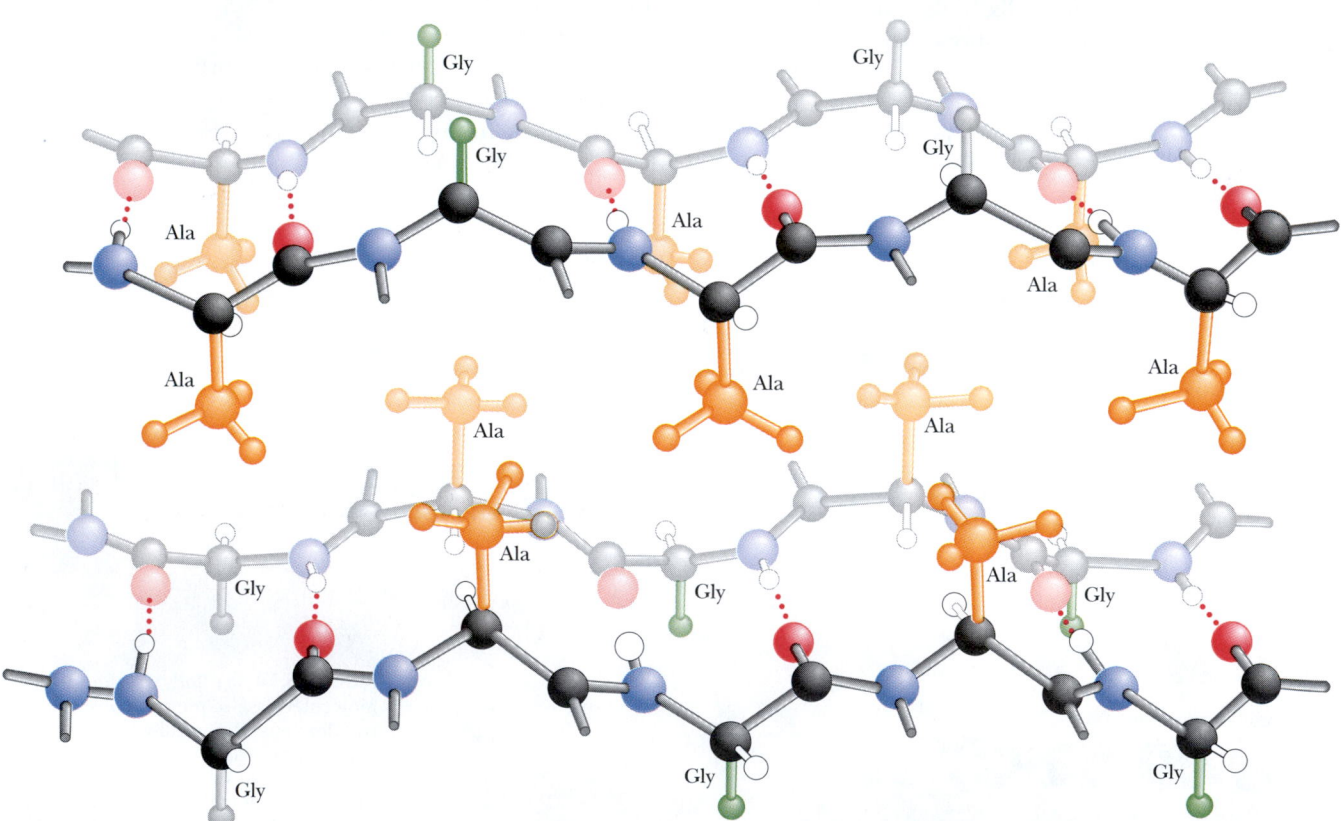

FIGURE 6.15 Silk fibroin consists of a unique stacked array of β-sheets. The primary structure of fibroin molecules consists of long stretches of alternating glycine and alanine or serine residues. When the sheets stack, the more bulky alanine and serine residues on one side of a sheet interdigitate with similar residues on an adjoining sheet. Glycine hydrogens on the alternating faces interdigitate in a similar manner, but with a smaller intersheet spacing. *(Illustration: Irving Geis. Rights owned by Howard Hughes Medical Institute. Not to be reproduced without permission.)*

Collagen: A Triple Helix **Collagen** is a rigid, inextensible fibrous protein that is a principal constituent of connective tissue in animals, including tendons, cartilage, bones, teeth, skin, and blood vessels. The high tensile strength of collagen fibers in these structures makes possible the various animal activities such as running and jumping that put severe stresses on joints and skeleton. Broken bones and tendon and cartilage injuries to knees, elbows, and other joints involve tears or hyperextensions of the collagen matrix in these tissues.

The basic structural unit of collagen is **tropocollagen**, which has a molecular weight of 285,000 and consists of three intertwined polypeptide chains, each about 1000 amino acids in length. Tropocollagen molecules are about 300 nm long and only about 1.4 nm in diameter. Several kinds of collagen have been identified. *Type I collagen*, which is the most common, consists of two identical peptide chains designated $\alpha1(I)$ and one different chain designated $\alpha2(I)$. Type I collagen predominates in bones, tendons, and skin. *Type II collagen*, found in cartilage, *and type III collagen*, found in blood vessels, consist of three identical polypeptide chains.

Collagen has an amino acid composition that is unique and is crucial to its three-dimensional structure and its characteristic physical properties. Nearly one residue out of three is a glycine, and the proline content is also unusually high. Three unusual modified amino acids are also found in collagen: 4-hydroxyproline (Hyp), 3-hydroxyproline, and 5-hydroxylysine (Hyl) (Figure 6.16). Proline and Hyp together compose up to 30% of the residues of collagen. Interestingly, these three amino acids are formed from normal proline and lysine *after* the collagen polypeptides are synthesized. The modifications are effected by two enzymes: *prolyl hydroxylase* and *lysyl hydroxylase*. The prolyl hydroxylase reaction (Figure 6.17) requires molecular oxygen, α-ketoglutarate, and ascorbic acid (vitamin C) and is activated by Fe^{2+}. The hydroxylation of lysine is similar. These processes are referred to as **post-translational modifications** because they occur after genetic information from DNA has been translated into newly formed protein.

Because of their high content of glycine, proline, and hydroxyproline, collagen fibers are incapable of forming traditional structures such as α-helices and β-sheets. Instead, collagen polypeptides intertwine to form a unique **triple helix**, with each of the three strands arranged in a helical fashion (Figure 6.18). Compared to the α-helix, the collagen helix is much more extended, with a rise per residue along the triple helix axis of 2.9 Å (versus 1.5 Å for the α-helix). There are about 3.3 residues per turn of each of these helices. *The triple helix is a structure that forms to accommodate the unique composition and sequence of collagen.* Long stretches of the polypeptide sequence are repeats of a Gly-*x*-*y* motif, where *x* is frequently Pro and *y* is frequently Pro or Hyp. In the triple helix, every third residue faces or contacts the crowded center of the structure. This area is so

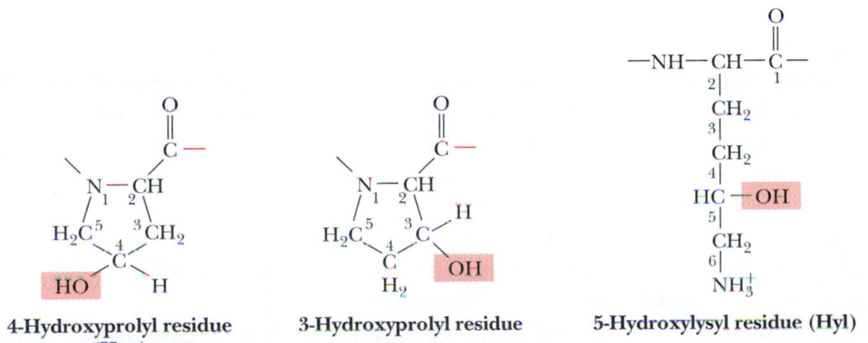

4-Hydroxyprolyl residue (Hyp) **3-Hydroxyprolyl residue** **5-Hydroxylysyl residue (Hyl)**

FIGURE 6.16 The hydroxylated residues typically found in collagen.

Proline + O_2 + **α-Ketoglutarate** + **Ascorbic acid**

Prolyl hydroxylase
Fe^{2+}

Hydroxyproline + CO_2 + **Succinate** + **Dehydroascorbate**

FIGURE 6.17 Hydroxylation of proline residues is catalyzed by prolyl hydroxylase. The reaction requires α-ketoglutarate and ascorbic acid (vitamin C).

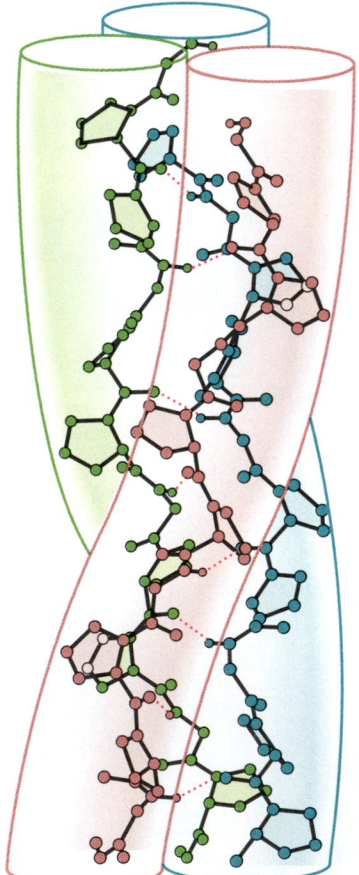

Biochemistry🅔Now™   ACTIVE FIGURE 6.18
Poly(Gly-Pro-Pro), a collagenlike right-handed triple helix composed of three left-handed helical chains. *(Adapted from Miller, M. H., and Scheraga, H. A., 1976. Calculation of the structures of collagen models. Role of interchain interactions in determining the triple-helical coiled-coil conformation. I. Poly(glycyl-prolyl-prolyl). Journal of Polymer Science Symposium 54:171–200.)* **Test yourself on the concepts in this figure at http://chemistry.brookscole.com/ggb3**

crowded that only Gly can fit, and thus every third residue must be a Gly (as observed). Moreover, the triple helix is a *staggered* structure, such that Gly residues from the three strands stack along the center of the triple helix and the Gly from one strand lies adjacent to an *x* residue from the second strand and to a *y* from the third. This allows the N—H of each Gly residue to hydrogen bond with the C=O of the adjacent *x* residue. The triple helix structure is further stabilized and strengthened by the formation of interchain H bonds involving hydroxyproline.

Collagen types I, II, and III form strong, organized **fibrils,** which consist of staggered arrays of tropocollagen molecules (Figure 6.19). The periodic arrangement of triple helices in a head-to-tail fashion results in banded patterns in electron micrographs. The banding pattern typically has a periodicity (repeat distance) of 68 nm. Because collagen triple helices are 300 nm long, 40-nm gaps occur between adjacent collagen molecules in a row along the long axis of the fibrils and the pattern repeats every five rows (5×68 nm = 340 nm). The 40-nm gaps are referred to as *hole regions,* and they are important in at least two ways. First, sugars are found covalently attached to 5-hydroxylysine residues in the hole regions of collagen (Figure 6.20). The occurrence of carbohydrate in the hole region has led to the proposal that it plays a role in organizing fibril assembly. Second, the hole regions may play a role in bone formation. Bone consists of microcrystals of **hydroxyapatite,** $Ca_5(PO_4)_3OH$, embedded in a matrix of collagen fibrils. When new bone tissue forms, the formation of new hydroxyapatite crystals occurs at intervals of 68 nm. The hole regions of collagen fibrils may be the sites of nucleation for the mineralization of bone.

The collagen fibrils are further strengthened and stabilized by the formation of both *intramolecular* (within a tropocollagen molecule) and *intermolecular* (between tropocollagen molecules in the fibril) crosslinks. Intramolecular

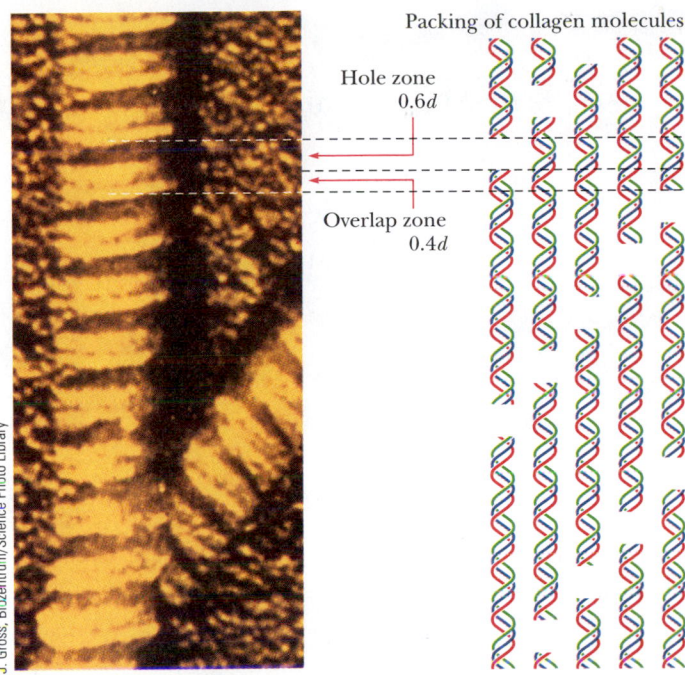

Packing of collagen molecules

Hole zone
0.6*d*

Overlap zone
0.4*d*

FIGURE 6.19 In the electron microscope, collagen fibers exhibit alternating light and dark bands. The dark bands correspond to the 40-nm gaps or "holes" between pairs of aligned collagen triple helices. The repeat distance, *d*, for the light- and dark-banded pattern is 68 nm. The collagen molecule is 300 nm long, which corresponds to 4.41*d*. The molecular repeat pattern of five staggered collagen molecules corresponds to 5*d*.

crosslinks are formed between lysine residues in the (nonhelical) N-terminal region of tropocollagen in a unique pair of reactions shown in Figure 6.21. The enzyme *lysyl oxidase* catalyzes the formation of aldehyde groups at the lysine side chains in a copper-dependent reaction. The aldehyde groups of two such side chains then link covalently in a spontaneous nonenzymatic *aldol condensation*. The intermolecular crosslinking of tropocollagens involves the formation of a unique **hydroxypyridinium** structure from one lysine and two hydroxylysine residues (Figure 6.22). These crosslinks form between the N-terminal region of one tropocollagen and the C-terminal region of an adjacent tropocollagen in the fibril.

Globular Proteins Mediate Cellular Function

Fibrous proteins, although interesting for their structural properties, represent only a small percentage of the proteins found in nature. **Globular proteins,** so named for their approximately spherical shape, are far more numerous.

Helices and Sheets in Globular Proteins   Globular proteins exist in an enormous variety of three-dimensional structures, but nearly all contain substantial amounts of the α-helices and β-sheets that form the basic structures of the simple fibrous proteins. For example, myoglobin, a small, globular, oxygen-carrying protein of muscle (17 kD, 153 amino acid residues), contains eight α-helical segments, each containing 7 to 26 amino acid residues. These are arranged in an apparently irregular (but invariant) fashion (see Figure 5.5). The space between the helices is filled efficiently and tightly with (mostly hydrophobic) amino acid side chains. Most of the polar side chains in myoglobin (and in most other globular proteins) face the outside of the protein structure and interact with solvent water. Myoglobin's structure is unusual because most globular proteins contain

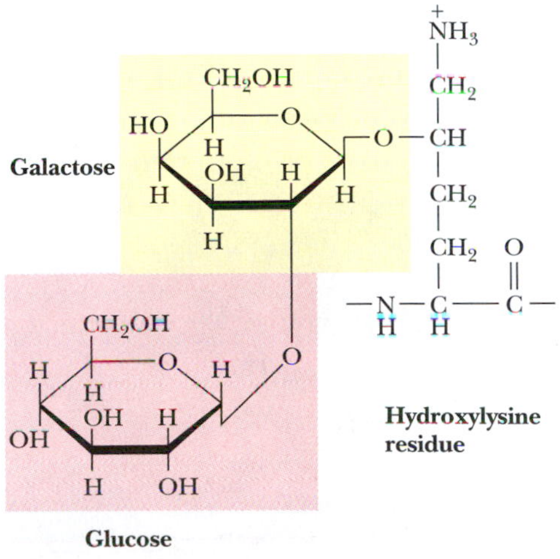

FIGURE 6.20 A disaccharide of galactose and glucose is covalently linked to the 5-hydroxyl group of hydroxylysines in collagen by the combined action of the enzymes galactosyltransferase and glucosyltransferase.

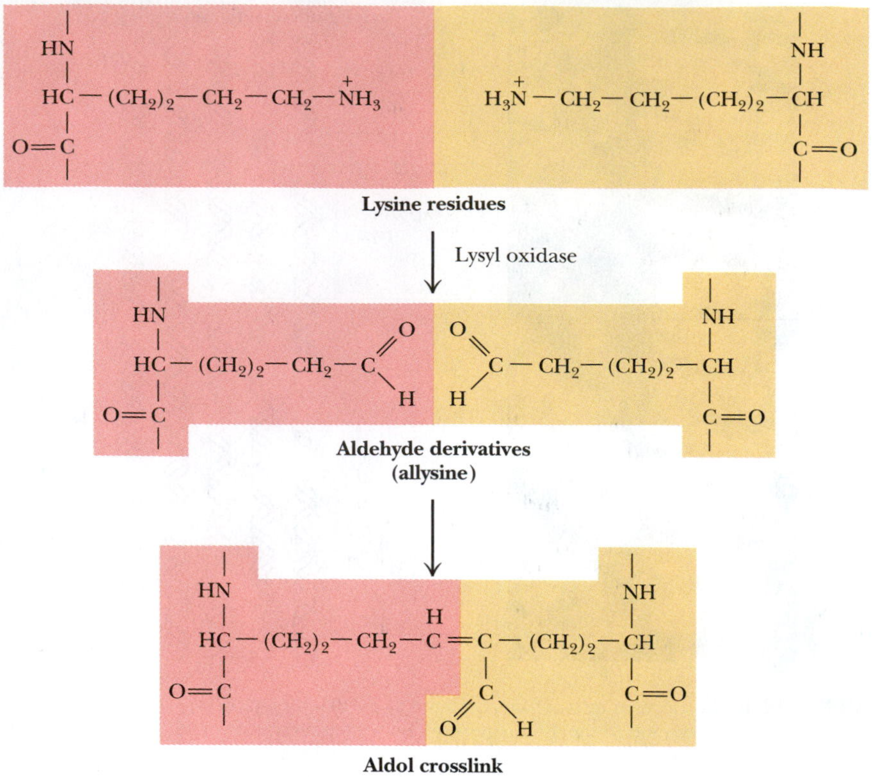

Lysine residues

Lysyl oxidase

Aldehyde derivatives (allysine)

Aldol crosslink

FIGURE 6.21 Collagen fibers are stabilized and strengthened by Lys-Lys crosslinks. Aldehyde moieties formed by lysyl oxidase react in a spontaneous nonenzymatic aldol reaction.

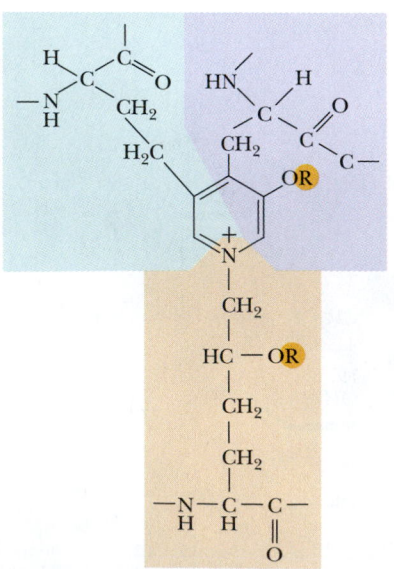

FIGURE 6.22 The hydroxypyridinium structure formed by the crosslinking of a Lys and two hydroxy Lys residues.

a relatively small amount of α-helix. A more typical globular protein (Figure 6.23) is *bovine ribonuclease A,* a small protein (12.6 kD, 124 residues) that contains a few short helices, a broad section of antiparallel β-sheet, a few β-turns, and several peptide segments without defined secondary structure.

Why should the cores of most globular and membrane proteins consist almost entirely of α-helices and β-sheets? The reason is that the highly polar N—H and C=O moieties of the peptide backbone must be neutralized in the hydrophobic core of the protein. The extensively H-bonded nature of α-helices and β-sheets is ideal for this purpose, and these structures effectively stabilize the polar groups of the peptide backbone in the protein core.

In globular protein structures, it is common for one face of an α-helix to be exposed to the water solvent, with the other face toward the hydrophobic interior of the protein. The outward face of such an **amphiphilic helix** consists mainly of polar and charged residues, whereas the inward face contains mostly nonpolar, hydrophobic residues. A good example of such a surface helix is that of residues 153 to 166 of **flavodoxin** from *Anabaena* (Figure 6.24). Note that the **helical wheel presentation** of this helix readily shows that one face contains four hydrophobic residues and that the other is almost entirely polar and charged.

Less commonly, an α-helix can be completely buried in the protein interior or completely exposed to solvent. **Citrate synthase** is a dimeric protein in which α-helical segments form part of the subunit–subunit interface. As shown in Figure 6.24, one of these helices (residues 260 to 270) is highly hydrophobic and contains only two polar residues, as would befit a helix in the protein core. On the other hand, Figure 6.24 also shows the solvent-exposed helix (residues 74 to 87) of **calmodulin,** which consists of 10 charged residues, 2 polar residues, and only 2 nonpolar residues.

Human Biochemistry

Collagen-Related Diseases

Collagen provides an ideal case study of the molecular basis of physiology and disease. For example, the nature and extent of collagen crosslinking depends on the age and function of the tissue. Collagen from young animals is predominantly un-crosslinked and can be extracted in soluble form, whereas collagen from older animals is highly crosslinked and thus insoluble. The loss of flexibility of joints with aging is probably due in part to increased crosslinking of collagen.

Several serious and debilitating diseases involving collagen abnormalities are known. **Lathyrism** occurs in animals due to the regular consumption of seeds of *Lathyrus odoratus,* the sweet pea, and involves weakening and abnormalities in blood vessels, joints, and bones. These conditions are caused by **β-aminopropionitrile** (see figure), which covalently inactivates lysyl oxidase, preventing intramolecular crosslinking of collagen and causing abnormalities in joints, bones, and blood vessels.

Scurvy results from a dietary vitamin C deficiency and involves the inability to form collagen fibrils properly. This is the result of reduced activity of prolyl hydroxylase, which is vitamin C–dependent, as previously noted. Scurvy leads to lesions in the skin and blood vessels, and in its advanced stages, it can lead to grotesque disfiguration and eventual death. Although rare in the modern world, it was a disease well known to sea-faring explorers in earlier times who did not appreciate the importance of fresh fruits and vegetables in the diet.

A number of rare genetic diseases involve collagen abnormalities, including *Marfan's syndrome* and the *Ehlers–Danlos syndromes,* which result in hyperextensible joints and skin. The formation of *atherosclerotic plaques,* which cause arterial blockages in advanced stages, is due in part to the abnormal formation of collagenous structures in blood vessels.

$$N \equiv C - CH_2 - CH_2 - \overset{+}{N}H_3$$

β-Aminopropionitrile

Packing Considerations   The secondary and tertiary structures of myoglobin and ribonuclease A illustrate the importance of packing in tertiary structures. Secondary structures pack closely to one another and also intercalate with (insert between) extended polypeptide chains. If the sum of the van der Waals volumes of a protein's constituent amino acids is divided by the volume occupied by the protein, packing densities of 0.72 to 0.77 are typically obtained. These packing densities are similar to those of solid spheres. This means that even

Biochemistry⑤Now™   Go to BiochemistryNow and click BiochemistryInteractive to examine the secondary and tertiary structure of ribonuclease.

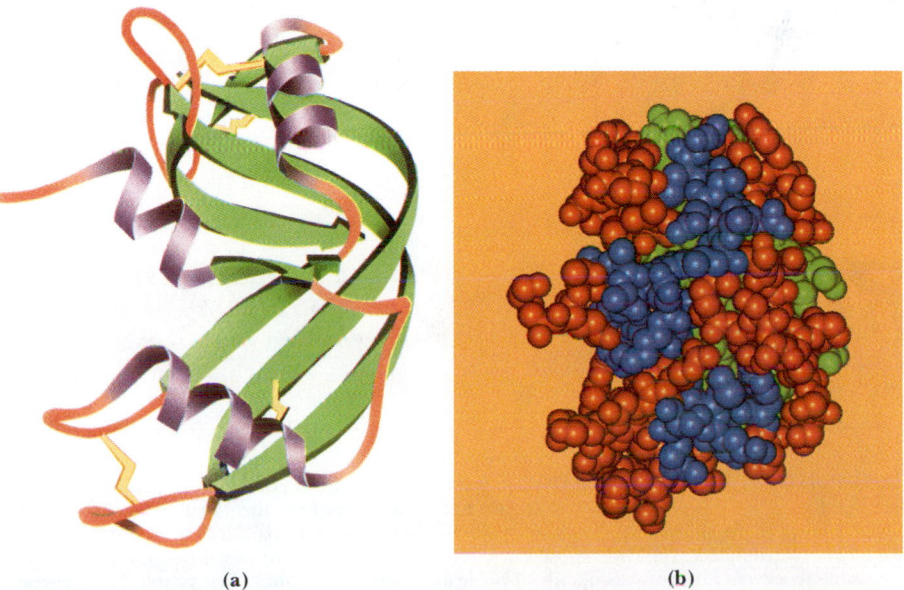

(a)   (b)

FIGURE 6.23 The three-dimensional structure of bovine ribonuclease A, showing the α-helices as ribbons. (*Jane Richardson.*)

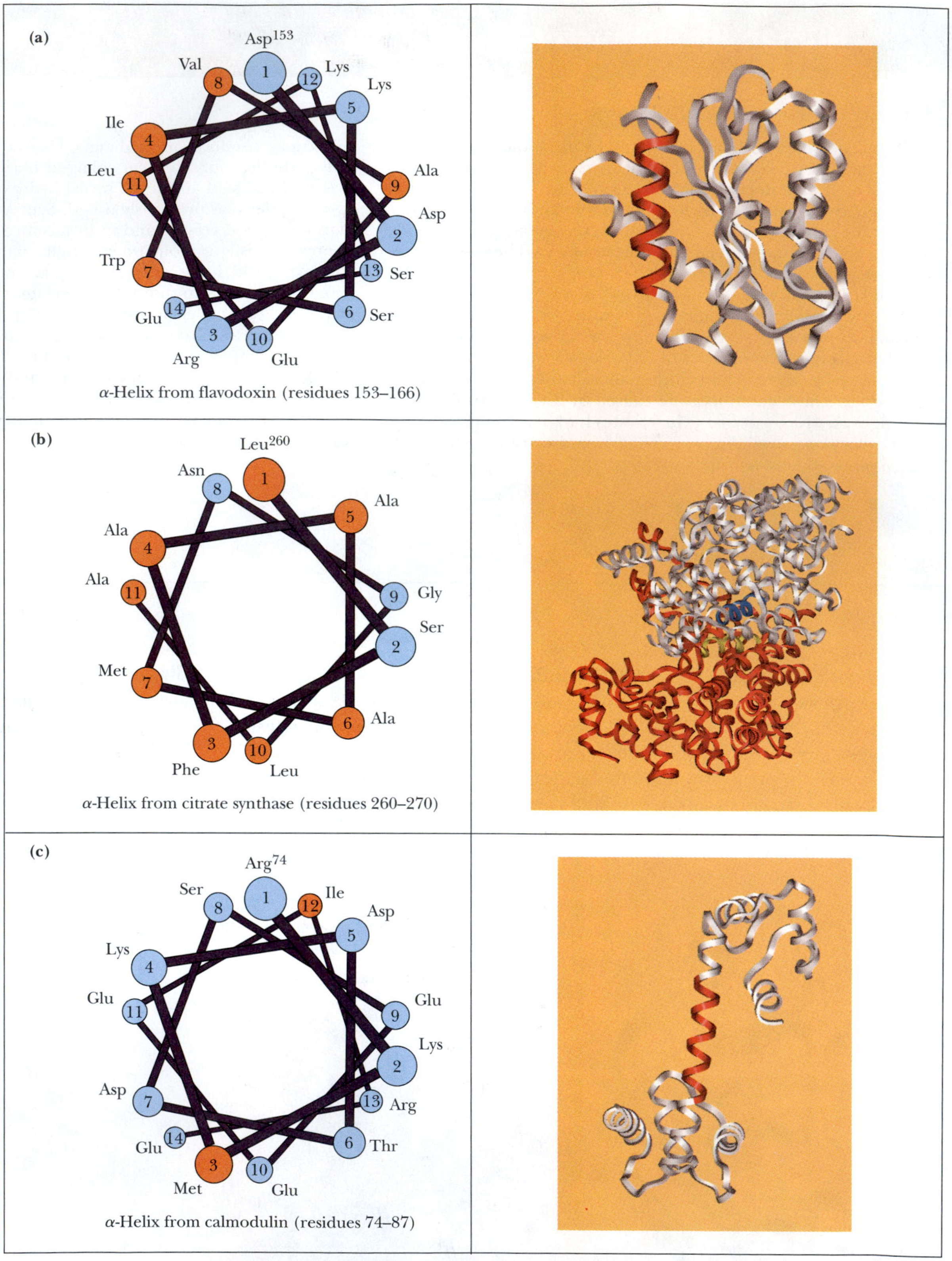

(a)

α-Helix from flavodoxin (residues 153–166)

(b)

α-Helix from citrate synthase (residues 260–270)

(c)

α-Helix from calmodulin (residues 74–87)

Biochemistry ⊜Now™ **ACTIVE FIGURE 6.24** **(a)** The α-helix consisting of residues 153–166 (red) in flavodoxin from *Anabaena* is a surface helix and is amphipathic. **(b)** The two helices (yellow and blue) in the interior of the citrate synthase dimer (residues 260–270 in each monomer) are mostly hydrophobic. **(c)** The exposed helix (residues 74–87—red) of calmodulin is entirely accessible to solvent and consists mainly of polar and charged residues. **Test yourself on the concepts in this figure at http://chemistry.brookscole.com/ggb3**

with close packing, approximately 25% of the total volume of a protein is not occupied by protein atoms. Nearly all of this space is in the form of very small cavities. Cavities the size of water molecules or larger do occasionally occur, but they make up only a small fraction of the total protein volume. It is likely that such cavities provide flexibility for proteins and facilitate conformation changes and a wide range of protein dynamics (discussed later).

Ordered, Nonrepetitive Structures In any protein structure, the segments of the polypeptide chain that cannot be classified as defined secondary structures, such as helices or sheets, have traditionally been referred to as *coil* or *random coil*. Both of these terms are misleading. Most of these segments are neither coiled nor random, in any sense of the words. These structures are every bit as highly organized and stable as the defined secondary structures. They just don't conform to any frequently recurring pattern. These so-called coil structures are strongly influenced by side-chain interactions. Few of these interactions are well understood, but a number of interesting cases have been described. In his early studies of myoglobin structure, John Kendrew found that the —OH group of threonine or serine often forms a hydrogen bond with a backbone NH at the beginning of an α-helix. The same stabilization of an α-helix by a serine is observed in the three-dimensional structure of pancreatic trypsin inhibitor (Figure 6.25). Also in this same structure, an asparagine residue adjacent to a β-strand is found to form H bonds that stabilize the β-structure.

Nonrepetitive but well-defined structures of this type form many important features of enzyme active sites. In some cases, a particular arrangement of "coil" structure providing a specific type of functional site recurs in several functionally related proteins. The peptide loop that binds iron–sulfur clusters in both ferredoxin and high-potential iron protein is one example. Another is the central loop portion of the *E–F hand structure* that binds a calcium ion in several calcium-binding proteins, including calmodulin, carp parvalbumin, troponin C, and the intestinal calcium-binding protein. This loop, shown in Figure 6.26, connects two short α-helices. The calcium ion nestles into the pocket formed by this structure.

Flexible, Disordered Segments In addition to nonrepetitive but well-defined structures, which exist in all proteins, genuinely disordered segments of polypeptide sequence also occur. These sequences either do not show up in electron density maps from X-ray crystallographic studies or give diffuse or ill-defined electron densities. These segments either undergo actual motion in the protein crystals themselves or take on many alternate conformations in different molecules within the crystal. Such behavior is quite common for long, charged side chains on the surface of many proteins. For example, 16 of the 19 lysine side chains in myoglobin have uncertain orientations beyond the δ-carbon, and 5 of these are disordered beyond the β-carbon. Similarly, a majority of the lysine residues are disordered in trypsin, rubredoxin, ribonuclease, and several other proteins. Arginine residues, however, are usually well ordered in protein structures. For the four proteins just mentioned, 70% of the arginine residues are highly ordered, compared to only 26% of the lysines.

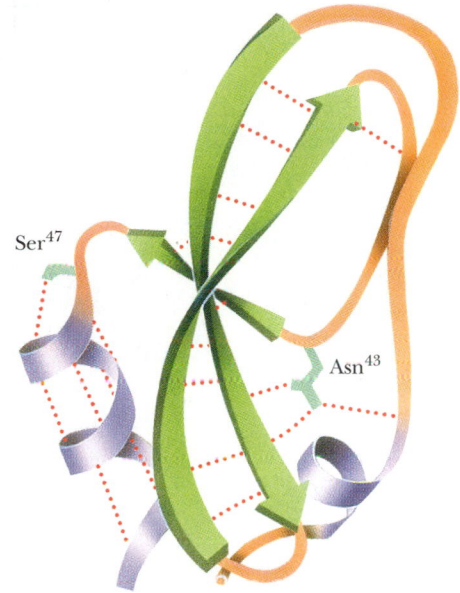

Pancreatic trypsin inhibitor

FIGURE 6.25 The three-dimensional structure of bovine pancreatic trypsin inhibitor. Note the stabilization of the α-helix by a hydrogen bond to Ser⁴⁷ and the stabilization of the β-sheet by Asn⁴³.

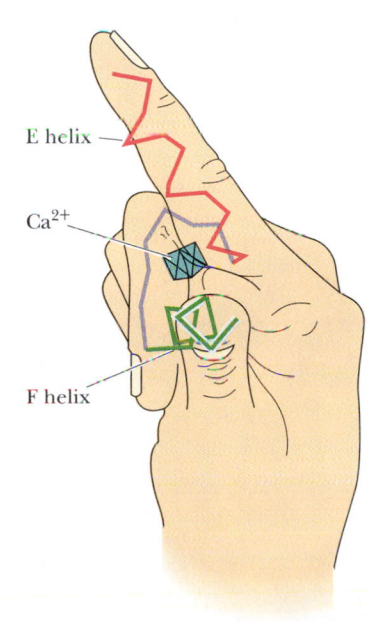

▶ **FIGURE 6.26** A representation of the so-called E–F hand structure, which forms calcium-binding sites in a variety of proteins. The stick drawing shows the peptide backbone of the E–F hand motif. The "E" helix extends along the index finger, a loop traces the approximate arrangement of the curled middle finger, and the "F" helix extends outward along the thumb. A calcium ion (Ca^{2+}) snuggles into the pocket created by the two helices and the loop. Kretsinger and co-workers originally assigned letters alphabetically to the helices in parvalbumin, a protein in carp. The E–F hand derives its name from the letters assigned to the helices at one of the Ca^{2+}-binding sites.

Table 6.2

Motion and Fluctuations in Proteins

Type of Motion	Spatial Displacement (Å)	Characteristic Time (sec)	Source of Energy
Atomic vibrations	0.01–1	10^{-15}–10^{-11}	Kinetic energy
Collective motions	0.01–5 or more	10^{-12}–10^{-3}	Kinetic energy
1. Fast: Tyr ring flips; methyl group rotations			
2. Slow: hinge bending between domains			
Triggered conformation changes	0.5–10 or more	10^{-9}–10^{3}	Interactions with triggering agent

Adapted from Petsko, G. A., and Ringe, D., 1984. Fluctuations in protein structure from X-ray diffraction. *Annual Review of Biophysics and Bioengineering* **13**:331–371.

Motion in Globular Proteins   Although we have distinguished between well-ordered and disordered segments of the polypeptide chain, it is important to realize that even well-ordered side chains in a protein undergo motion, sometimes quite rapid motion. These motions should be viewed as momentary oscillations about a single, highly stable conformation. *Proteins are thus best viewed as dynamic structures.* The allowed motions may be motions of individual atoms, groups of atoms, or even whole sections of the protein. Furthermore, they may arise from either thermal energy or specific, triggered conformational changes in the protein. **Atomic fluctuations** such as vibrations typically are random, are very fast, and usually occur over small distances (less than 0.5 Å), as shown in Table 6.2. These motions arise from the kinetic energy within the protein and are a function of temperature. These very fast motions can be modeled by molecular dynamics calculations and studied by X-ray diffraction.

A class of slower motions, which may extend over larger distances, is **collective motions.** These are movements of groups of atoms covalently linked in such a way that the group moves as a unit. Such groups range in size from a few atoms to hundreds of atoms. Such motions are of two types—(1) those that occur quickly but infrequently, such as tyrosine ring flips, and (2) those that occur slowly, such as *cis–trans* isomerizations of prolines. Whole structural domains within a protein may be involved, as in the case of the flexible antigen-binding domains of immunoglobulins, which move as relatively rigid units to selectively bind separate antigen molecules. These collective motions also arise from thermal energies in the protein and operate on a time scale of 10^{-12} to 10^{-3} sec. These motions can be studied by nuclear magnetic resonance (NMR) and fluorescence spectroscopy.

Conformational changes involve motions of groups of atoms (individual side chains, for example) or even whole sections of proteins. These motions occur on a time scale of 10^{-9} to 10^{3} sec, and the distances covered can be as large as 1 nm. These motions may occur in response to specific stimuli or arise from specific interactions within the protein, such as hydrogen bonding, electrostatic interactions, and ligand binding. More will be said about conformational changes when enzyme catalysis and regulation are discussed (see Chapters 14 and 15).

Forces Driving the Folding of Globular Proteins   As already pointed out, the driving force for protein folding and the resulting formation of a tertiary structure is the formation of the most stable structure possible. Two forces are at work

(a)

(b)

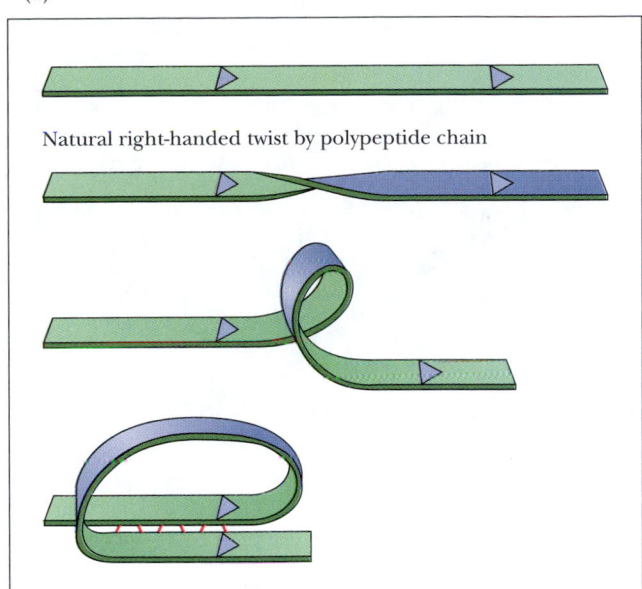

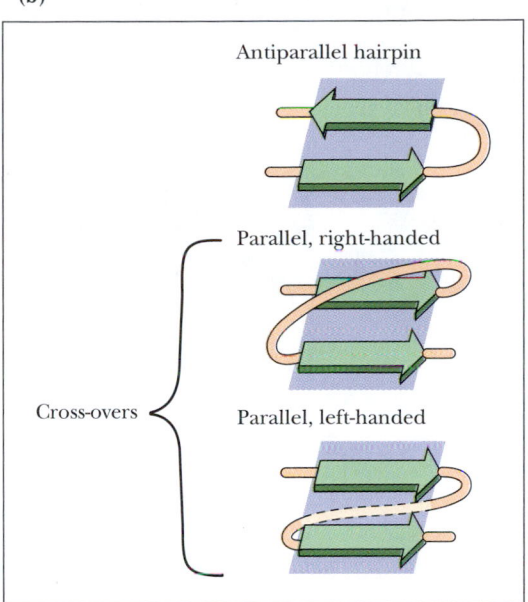

FIGURE 6.27 (a) The natural right-handed twist exhibited by polypeptide chains, and (b) the variety of structures that arise from this twist.

here. The peptide chain must both (1) satisfy the constraints inherent in its own structure and (2) fold so as to "bury" the hydrophobic side chains, minimizing their contact with solvent. The polypeptide itself does not usually form simple straight chains. Even in chain segments where helices and sheets are not formed, an extended peptide chain, being composed of L-amino acids, has a tendency to twist slightly in a right-handed direction. As shown in Figure 6.27, this tendency is apparently the basis for the formation of a variety of tertiary structures having a right-handed sense. Principal among these are the right-handed twists in arrays of β-sheets and right-handed cross-overs in parallel β-sheet arrays. Right-handed twisted β-sheets are found at the center of a number of proteins and provide an extended, highly stable structural core. Phosphoglycerate mutase, adenylate kinase, and carbonic anhydrase, among others, exist as smoothly twisted planes or saddle-shaped structures. Triose phosphate isomerase, soybean trypsin inhibitor, and domain 1 of pyruvate kinase contain right-handed twisted cylinders or barrel structures at their cores.

Connections between β-strands are of two types—hairpins and cross-overs. **Hairpins,** as shown in Figure 6.27, connect adjacent antiparallel β-strands. **Cross-overs** are necessary to connect adjacent (or nearly adjacent) parallel β-strands. Nearly all cross-over structures are right-handed. Isolated left-handed cross-overs have been identified in subtilisin and in phosphoglucoisomerase. In many cross-over structures, the cross-over connection itself contains an α-helical segment. This creates a **βαβ-loop.** As shown in Figure 6.27, the strong tendency in nature to form right-handed cross-overs, the wide occurrence of α-helices in the cross-over connection, and the right-handed twists of β-sheets can all be understood as arising from the tendency of an extended polypeptide chain of L-amino acids to adopt a right-handed twist structure. This is a chiral effect. Proteins composed of D-amino acids would tend to adopt left-handed twist structures.

The second driving force that affects the folding of polypeptide chains is the need to bury the hydrophobic residues of the chain, protecting them from solvent water. From a topological viewpoint, then, all globular proteins must have an "inside" where the hydrophobic core can be arranged and an "outside"

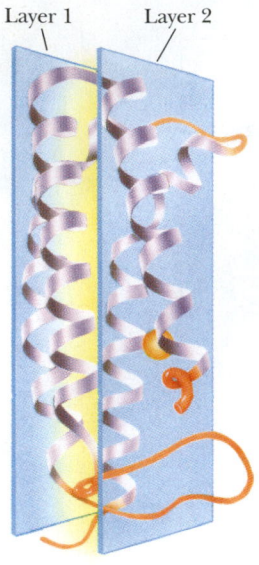

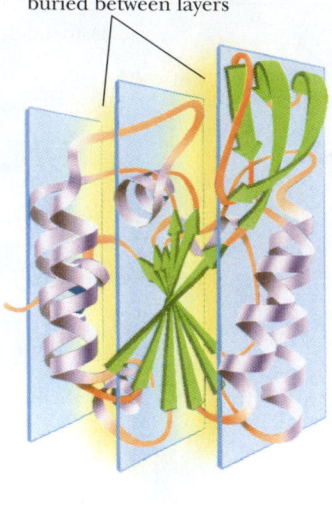

(a) Cytochrome c'

(b) Phosphoglycerate kinase (domain 2)

(c) Phosphorylase (domain 2)

FIGURE 6.28 Examples of protein domains with different numbers of layers of backbone structure. **(a)** Cytochrome c' with two layers of α-helix. **(b)** Domain 2 of phosphoglycerate kinase, composed of a β-sheet layer between two layers of helix, three layers overall. **(c)** An unusual five-layer structure, domain 2 of glycogen phosphorylase, a β-sheet layer sandwiched between four layers of α-helix. **(d)** The concentric "layers" of β-sheet *(inside)* and α-helix *(outside)* in triose phosphate isomerase. Hydrophobic residues are buried between these concentric layers in the same manner as in the planar layers of the other proteins. The hydrophobic layers are shaded yellow. *(Jane Richardson.)*

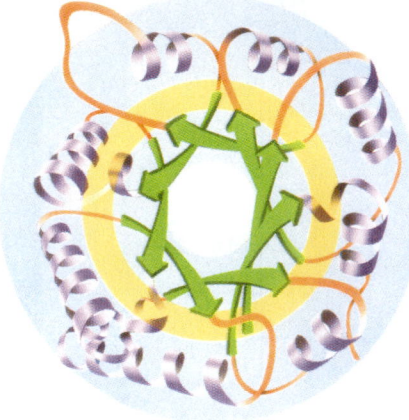

(d) Triose phosphate isomerase

toward which the hydrophilic groups must be directed. The sequestration of hydrophobic residues away from water is the dominant force in the arrangement of secondary structures and nonrepetitive peptide segments to form a given tertiary structure. Globular proteins can be classified mainly on the basis of the particular kind of core or backbone structure they use to accomplish this goal. The term *hydrophobic core,* as used here, refers to a region in which hydrophobic side chains cluster together, away from the solvent. *Backbone* refers to the polypeptide backbone itself, excluding the particular side chains. Globular proteins can be pictured as consisting of "layers" of backbone, with hydrophobic core regions between them. More than half the known globular protein structures have two layers of backbone (separated by one hydrophobic core). Roughly one-third of the known structures are composed of three backbone layers and two hydrophobic cores. There are also a few known four-layer structures and at least one five-layer structure. A few structures are not easily classified in this way, but it is remarkable that most proteins fit into one of these classes. Examples of each are presented in Figure 6.28.

Most Globular Proteins Belong to One of Four Structural Classes

In addition to classification based on layer structure, proteins can be grouped according to the type and arrangement of secondary structure. There are four such broad groups: antiparallel α-helix, parallel or mixed β-sheet, antiparallel β-sheet, and the small metal- and disulfide-rich proteins.

It is important to note that the similarities of tertiary structure within these groups do not necessarily reflect similar or even related functions. Instead, **functional homology** usually depends on structural similarities on a smaller and more intimate scale.

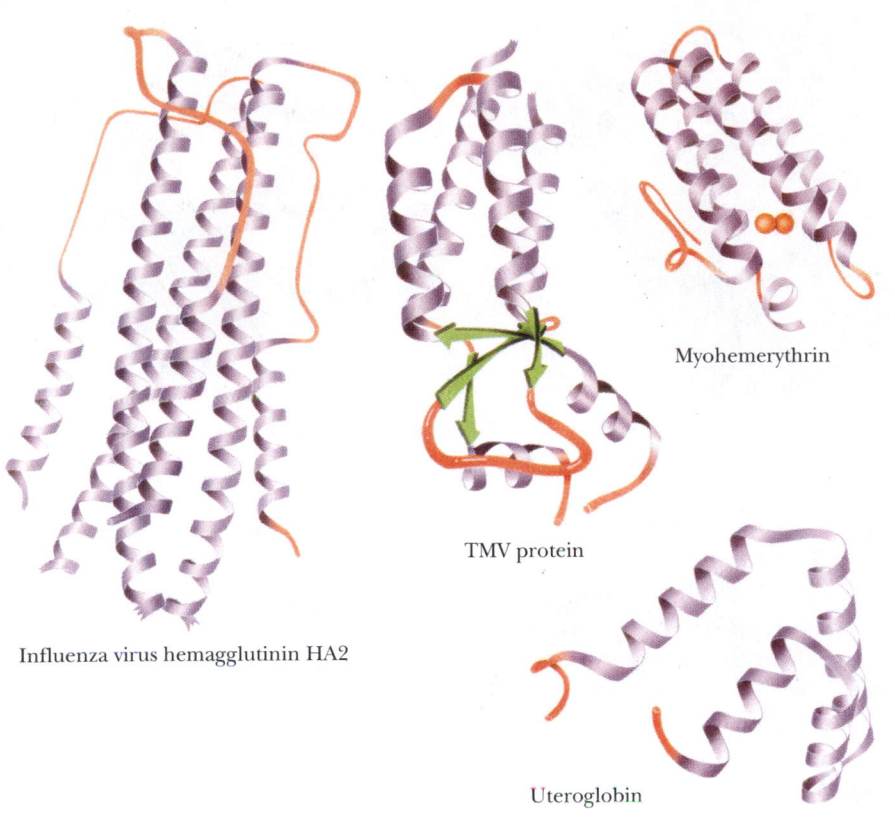

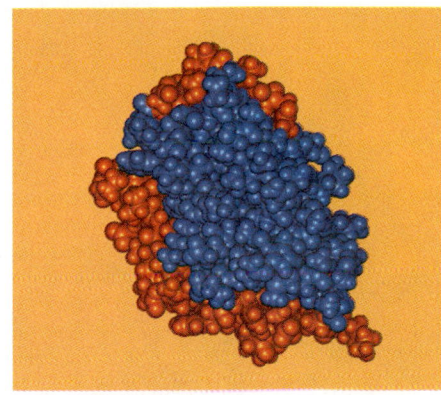

Myohemerythrin

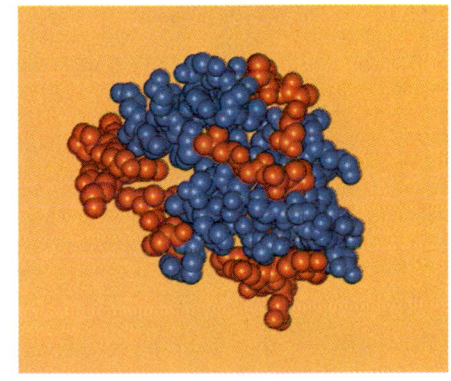

Uteroglobin

FIGURE 6.29 Several examples of antiparallel α-helix proteins. *(Jane Richardson.)*

Antiparallel α-Helix Proteins   **Antiparallel α-helix proteins** are structures heavily dominated by α-helices. The simplest way to pack helices is in an antiparallel manner, and most of the proteins in this class consist of bundles of antiparallel helices. Many of these exhibit a slight (15°) left-handed twist of the helix bundle. Figure 6.29 shows a representative sample of antiparallel α-helix proteins. Many of these are regular, uniform structures, but in a few cases (uteroglobin, for example) one of the helices is tilted away from the bundle. Tobacco mosaic virus protein has small, highly twisted antiparallel β-sheets on one end of the helix bundle with two additional helices on the other side of the sheet. Notice in Figure 6.29 that most of the antiparallel α-helix proteins are made up of four-helix bundles.

The so-called globin proteins are an important group of α-helical proteins. These include hemoglobins and myoglobins from many species. The globin structure can be viewed as two layers of helices, with one of these layers perpendicular to the other and the polypeptide chain moving back and forth between the layers.

Parallel or Mixed β-Sheet Proteins   The second major class of protein structures contains structures based around **parallel** or **mixed β-sheets.** Parallel β-sheet arrays, as previously discussed, distribute hydrophobic side chains on both sides of the sheet. This means that neither side of parallel β-sheets can be exposed to solvent. Parallel β-sheets are thus typically found as core structures in proteins, with little access to solvent.

Another important parallel β-array is the eight-stranded **parallel β-barrel,** exemplified in the structures of triose phosphate isomerase and pyruvate

(a)

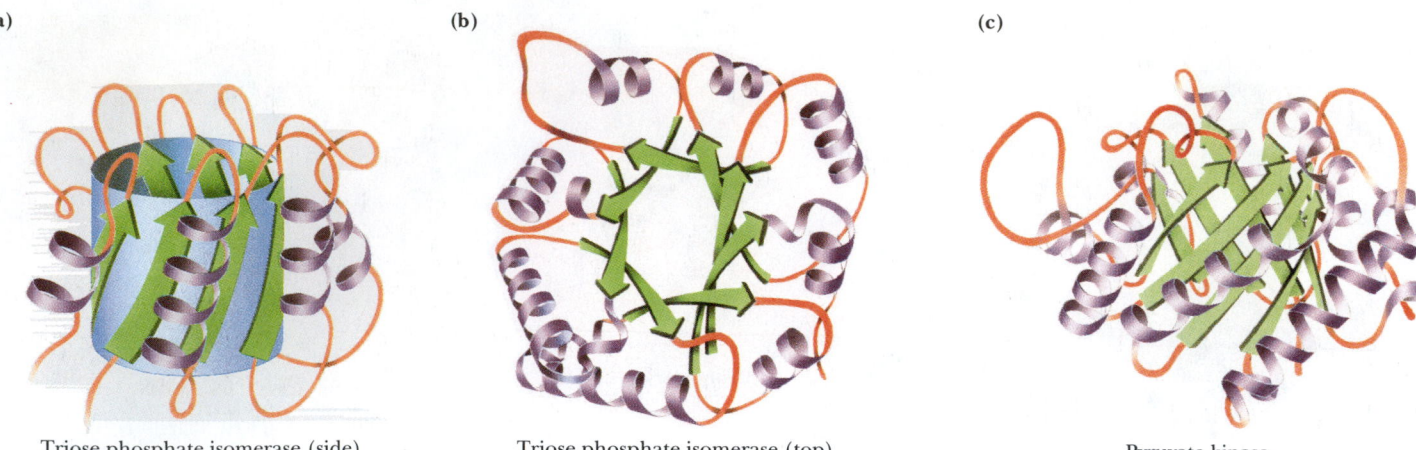

Triose phosphate isomerase (side)

(b)

Triose phosphate isomerase (top)

(c)

Pyruvate kinase

FIGURE 6.30 Parallel β-array proteins—the eight-stranded β-barrels of triose phosphate isomerase (**a,** *side view,* and **b,** *top view*) and (**c**) pyruvate kinase. *(Jane Richardson.)*

kinase (Figure 6.30). Each β-strand in the barrel is flanked by an antiparallel α-helix. The α-helices thus form a larger cylinder of parallel helices concentric with the β-barrel. Both cylinders thus formed have a right-handed twist. Another parallel β-structure consists of an internal twisted wall of parallel or mixed β-sheet protected on both sides by helices or other substructures. This structure is called the **doubly wound parallel β-sheet** because the structure can be imagined to have been wound by strands beginning in the middle and going outward in opposite directions. The essence of this structure is shown in Figure 6.31. Whereas the barrel structures have four layers of backbone

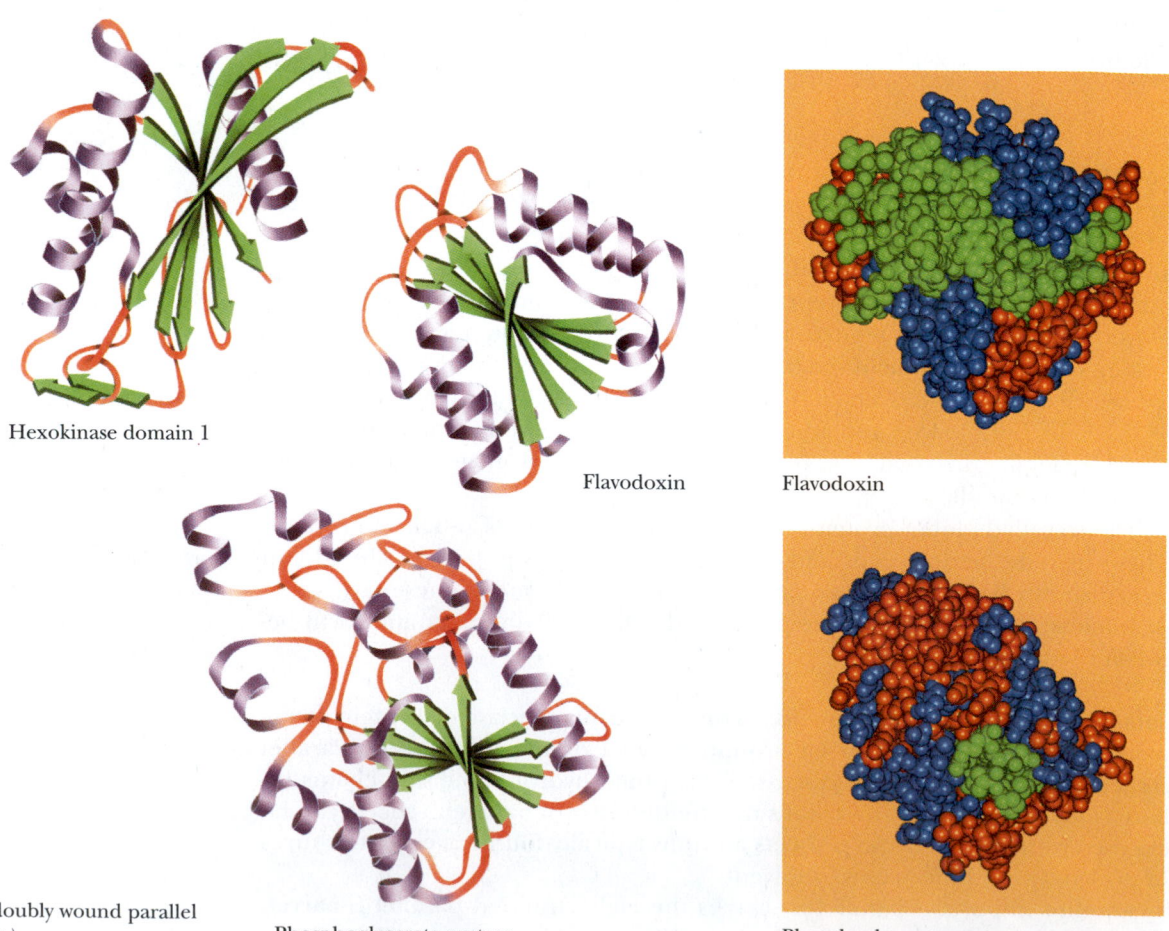

Hexokinase domain 1

Flavodoxin

Flavodoxin

Phosphoglycerate mutase

Phosphoglycerate mutase

FIGURE 6.31 Several typical doubly wound parallel β-sheet proteins. *(Jane Richardson.)*

The Coiled-Coil Motif in Proteins

The **coiled-coil** motif was first identified in 1953 by Linus Pauling, Robert Corey, and Francis Crick as the main structural element of fibrous proteins such as keratin and myosin. Since then, many proteins have been found to contain one or more coiled-coil segments or domains. A coiled coil is a bundle of α-helices that are wound into a superhelix. Two, three, or four helical segments may be found in the bundle, and they may be arranged parallel or antiparallel to one another. Coiled coils are characterized by a distinctive and regular packing of side chains in the core of the bundle. This regular meshing of side chains requires that they occupy equivalent positions turn after turn. This is not possible for undistorted α-helices, which have 3.6 residues per turn. The positions of side chains on their surface shift continuously along the helix surface (see figure). However, giving the right-handed α-helix a left-handed twist reduces the number of residues per turn to 3.5, and because 3.5 × 2 = 7.0, the positions of the side chains repeat after two turns (7 residues). Thus, a **heptad repeat** pattern in the peptide sequence is diagnostic of a coiled-coil structure. The figure shows a sampling of coiled-coil structures (highlighted in color) in various proteins.

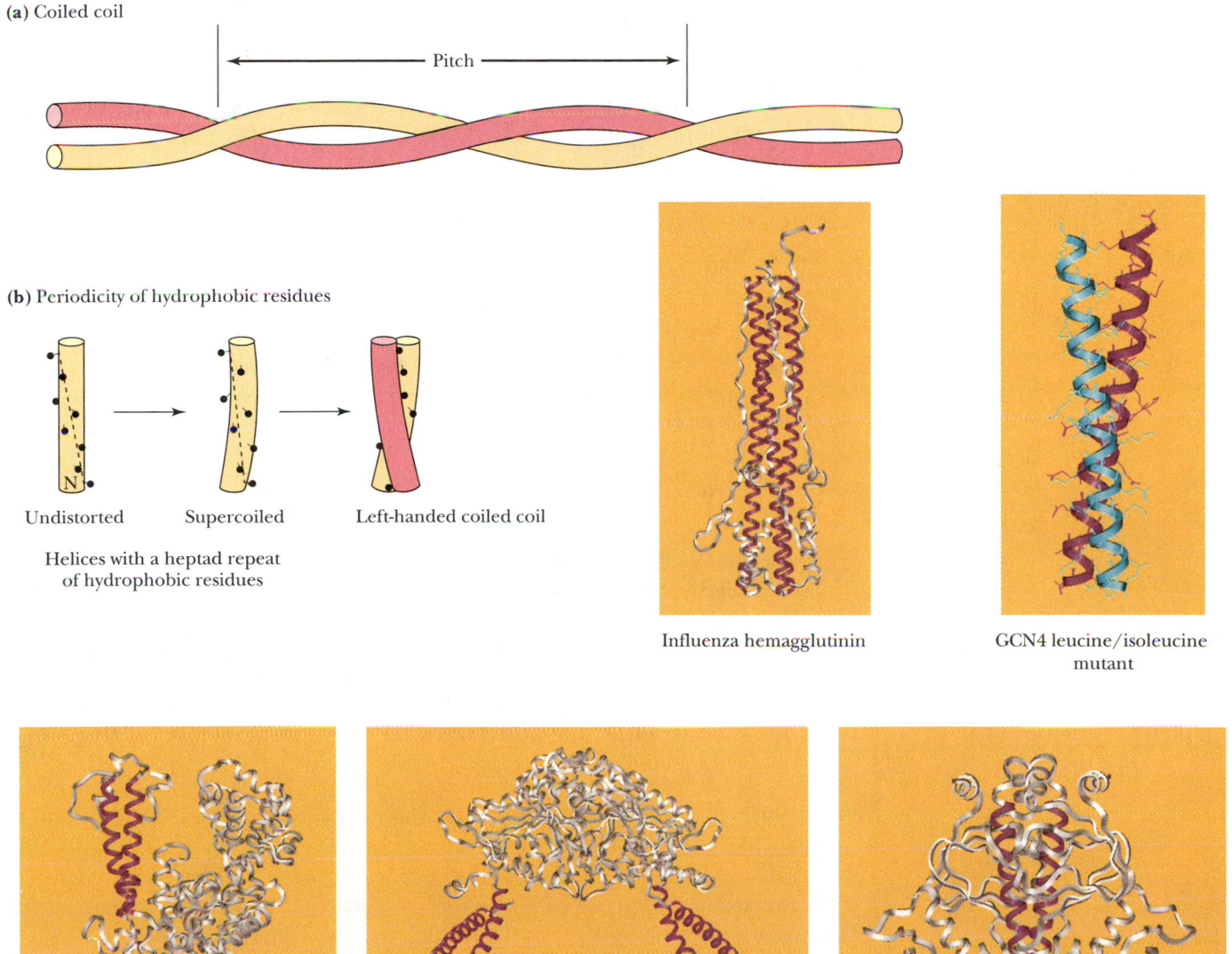

(a) Coiled coil

Pitch

(b) Periodicity of hydrophobic residues

Undistorted   Supercoiled   Left-handed coiled coil

Helices with a heptad repeat of hydrophobic residues

Influenza hemagglutinin

GCN4 leucine/isoleucine mutant

DNA polymerase

Seryl tRNA synthetase

Catabolite activator protein

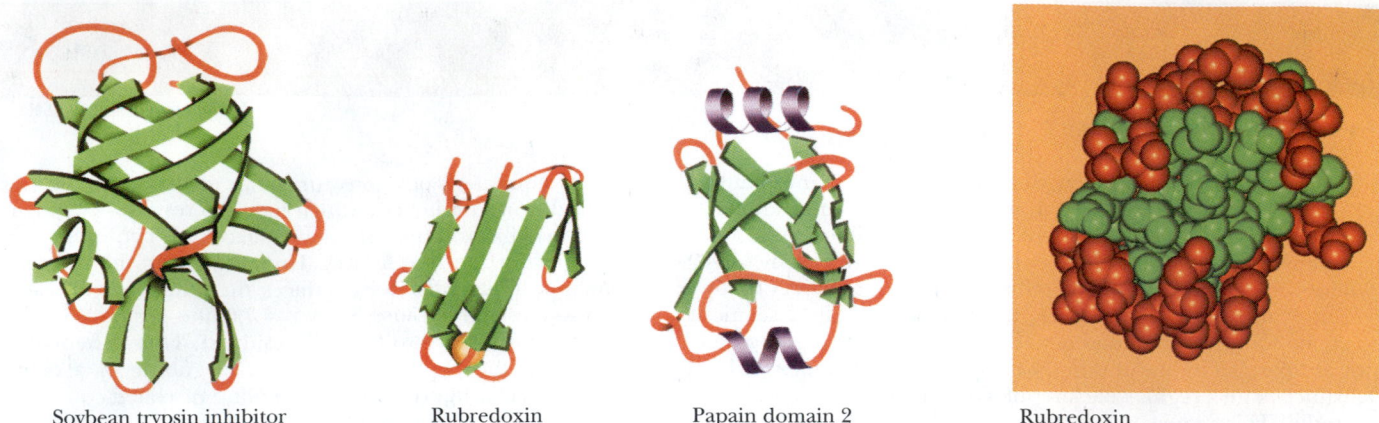

Soybean trypsin inhibitor   Rubredoxin   Papain domain 2   Rubredoxin

FIGURE 6.32 Examples of antiparallel β-sheet structures in proteins. (*Jane Richardson.*)

structure, the doubly wound sheet proteins have three major layers and thus two hydrophobic core regions.

Antiparallel β-Sheet Proteins   Another important class of tertiary protein conformations is the **antiparallel β-sheet** structures. Antiparallel β-sheets, which usually arrange hydrophobic residues on just one side of the sheet, can exist with one side exposed to solvent. The minimal structure for an antiparallel β-sheet protein is thus a two-layered structure, with hydrophobic faces of the two sheets juxtaposed and the opposite faces exposed to solvent. Such domains consist of β-sheets arranged in a cylinder or barrel shape. These structures are usually less symmetric than the singly wound parallel barrels and are not as efficiently hydrogen bonded, but they occur much more frequently in nature. Barrel structures tend to be either all parallel or all antiparallel and usually consist of even numbers of β-strands. Good examples of antiparallel structures include soybean trypsin inhibitor, rubredoxin, and domain 2 of papain (Figure 6.32). Topology diagrams of antiparallel β-sheet barrels reveal that many of them arrange the polypeptide sequence in an interlocking pattern reminiscent of patterns found on ancient Greek vases (Figure 6.33). They are thus described as a *Greek key topology*. Several of these, including concanavalin A and γ-crystallin, contain an extra swirl in the Greek key pattern (see Figure 6.33). Antiparallel arrangements of β-strands can also form sheets as well as barrels. Glyceraldehyde-3-phosphate dehydrogenase, *Streptomyces* subtilisin inhibitor, and glutathione reductase are examples of single-sheet, double-layered topology (Figure 6.34).

Metal- and Disulfide-Rich Proteins   Other than the structural classes just described and a few miscellaneous structures that do not fit nicely into these categories, there is only one other major class of protein tertiary structures—the small metal-rich and disulfide-rich structures. These proteins or fragments of proteins are usually small (fewer than 100 residues), and their conformations are heavily influenced by their high content of either liganded metals or disulfide bonds. The structures of disulfide-rich proteins are unstable if their disulfide bonds are broken. Figure 6.35 shows several representative disulfide-rich proteins, including insulin, phospholipase A_2, and crambin (from the seeds of *Crambe abyssinica*), as well as several metal-rich proteins, including ferredoxin and high-potential iron protein (HiPIP). The structures of some of these proteins bear a striking resemblance to structural classes that have already been discussed. For example, phospholipase A_2 is a distorted α-helix cluster, whereas HiPIP is a distorted β-barrel structure. Others among

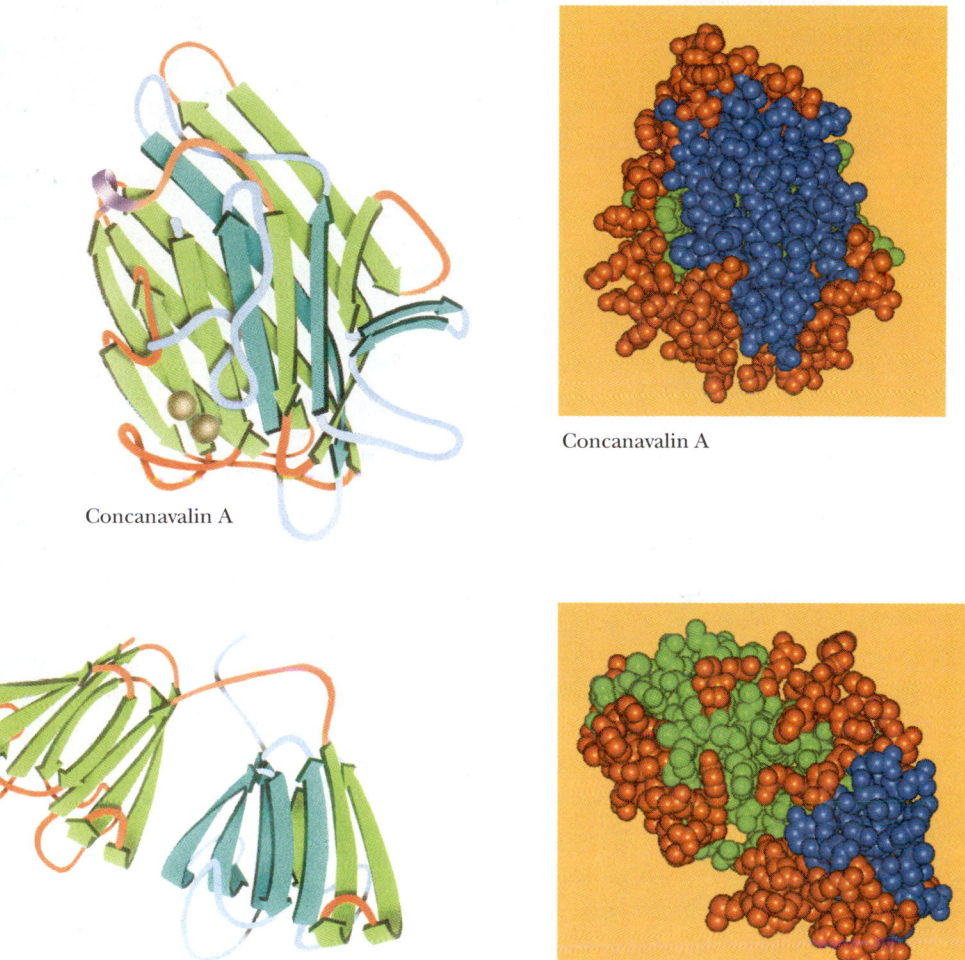

"Greek key" topology

Concanavalin A

Concanavalin A

γ-Crystallin

γ-Crystallin

FIGURE 6.33 Examples of the so-called Greek key antiparallel β-barrel structure in proteins.

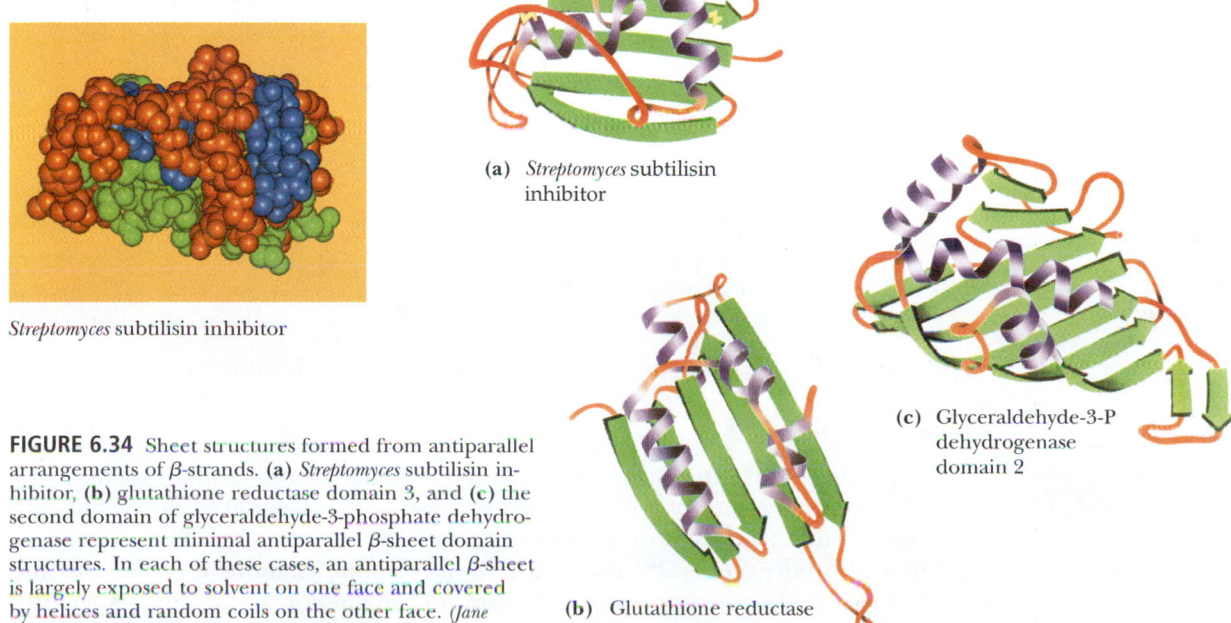

Streptomyces subtilisin inhibitor

(a) *Streptomyces* subtilisin inhibitor

(c) Glyceraldehyde-3-P dehydrogenase domain 2

(b) Glutathione reductase domain 3

FIGURE 6.34 Sheet structures formed from antiparallel arrangements of β-strands. **(a)** *Streptomyces* subtilisin inhibitor, **(b)** glutathione reductase domain 3, and **(c)** the second domain of glyceraldehyde-3-phosphate dehydrogenase represent minimal antiparallel β-sheet domain structures. In each of these cases, an antiparallel β-sheet is largely exposed to solvent on one face and covered by helices and random coils on the other face. *(Jane Richardson.)*

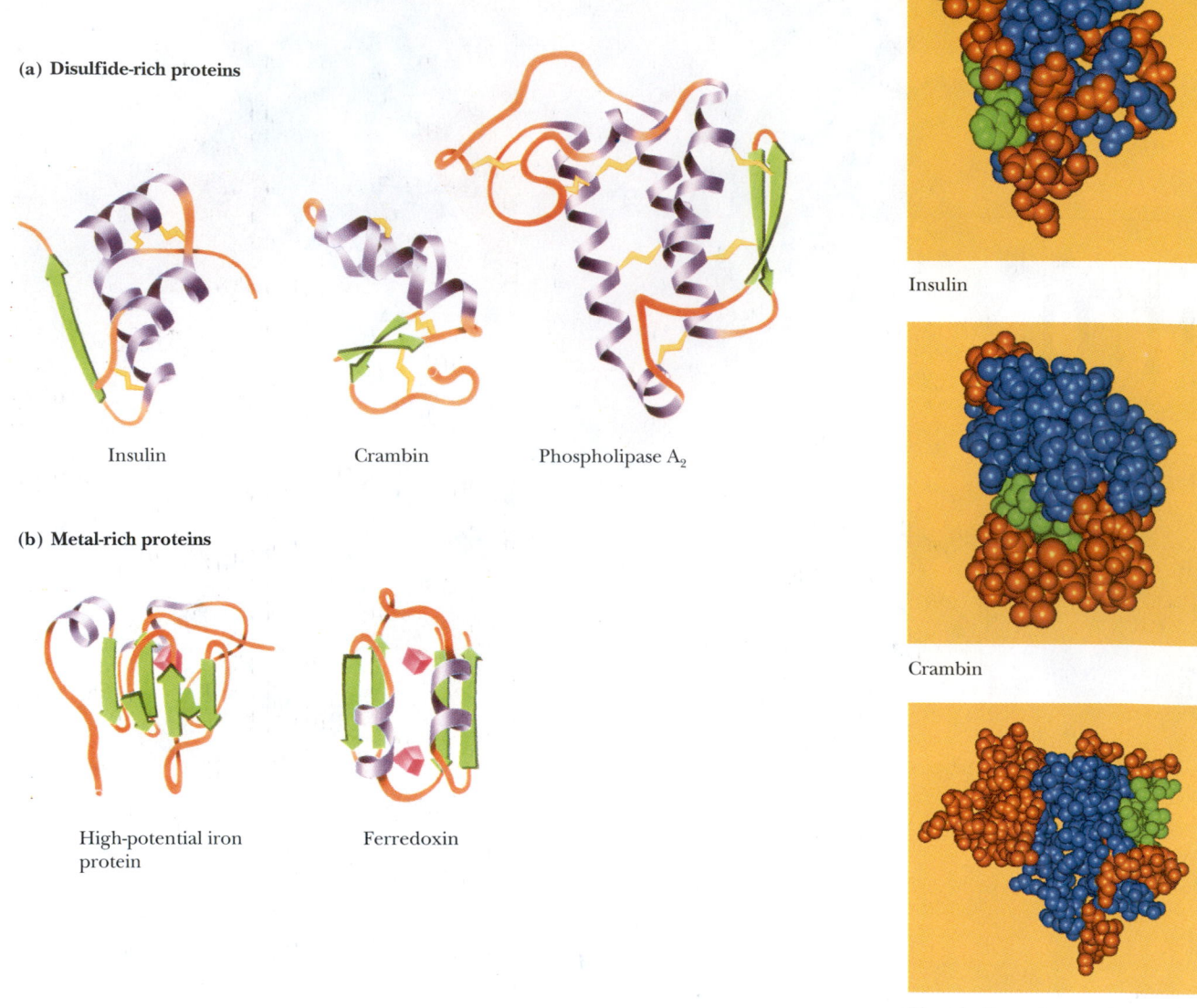

(a) Disulfide-rich proteins

Insulin

Crambin

Phospholipase A$_2$

(b) Metal-rich proteins

High-potential iron protein

Ferredoxin

Insulin

Crambin

Phospholipase A$_2$

FIGURE 6.35 Examples of the **(a)** disulfide-rich and **(b)** metal-rich proteins. *(Jane Richardson.)*

this class (such as insulin and crambin), however, are not easily likened to any of the standard structure classes.

Molecular Chaperones Are Proteins That Help Other Proteins to Fold

The landmark experiments by Christian Anfinsen on the refolding of ribonuclease clearly show that the refolding of a denatured protein in vitro can be a spontaneous process. As noted previously, this refolding is driven by the small Gibbs free energy difference between the unfolded and folded states. It has also been generally assumed that all the information necessary for the correct folding of a polypeptide chain is contained in the primary structure and requires no additional molecular factors. However, the folding of pro-

Thermodynamics of the Folding Process in Globular Proteins

Section 6.1 considered the noncovalent bonding energies that stabilize a protein structure. However, the folding of a protein depends ultimately on the difference in Gibbs free energy (ΔG) between the folded (F) and unfolded (U) states at some temperature T:

$$\Delta G = G_F - G_U = \Delta H - T\Delta S$$
$$= (H_F - H_U) - T(S_F - S_U)$$

In the unfolded state, the peptide chain and its R groups interact with solvent water, and any measurement of the free energy change upon folding must consider contributions to the enthalpy change (ΔH) and the entropy change (ΔS) both for the polypeptide chain and for the solvent:

$$\Delta G_{total} = \Delta H_{chain} + \Delta H_{solvent} - T\Delta S_{chain} - T\Delta S_{solvent}$$

If each of the four terms on the right side of this equation is understood, the thermodynamic basis for protein folding should be clear. A summary of the signs and magnitudes of these quantities for a typical protein is shown in the accompanying figure. The folded protein is a highly ordered structure compared to the unfolded state, so ΔS_{chain} is a negative number and thus $-T\Delta S_{chain}$ is a positive quantity in the equation. The other terms depend on the nature of the particular ensemble of R groups. The nature of ΔH_{chain} depends on both residue–residue interactions and residue–solvent interactions. Nonpolar groups in the folded protein interact mainly with one another via weak van der Waals forces. Interactions between nonpolar groups and water in the unfolded state are stronger because the polar water molecules induce dipoles in the nonpolar groups, producing a significant electrostatic interaction. As a result, ΔH_{chain} is positive for nonpolar groups and favors the unfolded state. $\Delta H_{solvent}$ for nonpolar groups, however, is negative and favors the folded state. This is because folding allows many water molecules to interact (favorably) with one another rather than (less favorably) with the nonpolar side chains. The magnitude of ΔH_{chain} is smaller than that of $\Delta H_{solvent}$, but both these terms are small and usually do not dominate the folding process. However, $\Delta S_{solvent}$ for nonpolar groups is large and positive and strongly favors the folded state. This is be-

cause nonpolar groups force order upon the water solvent in the unfolded state.

For polar side chains, ΔH_{chain} is positive and $\Delta H_{solvent}$ is negative. Because solvent molecules are ordered to some extent around polar groups, $\Delta S_{solvent}$ is small and positive. As shown in the figure, ΔG_{total} for the polar groups of a protein is near zero. Comparison of all the terms considered here makes it clear that *the single largest contribution to the stability of a folded protein is $\Delta S_{solvent}$ for the nonpolar residues.*

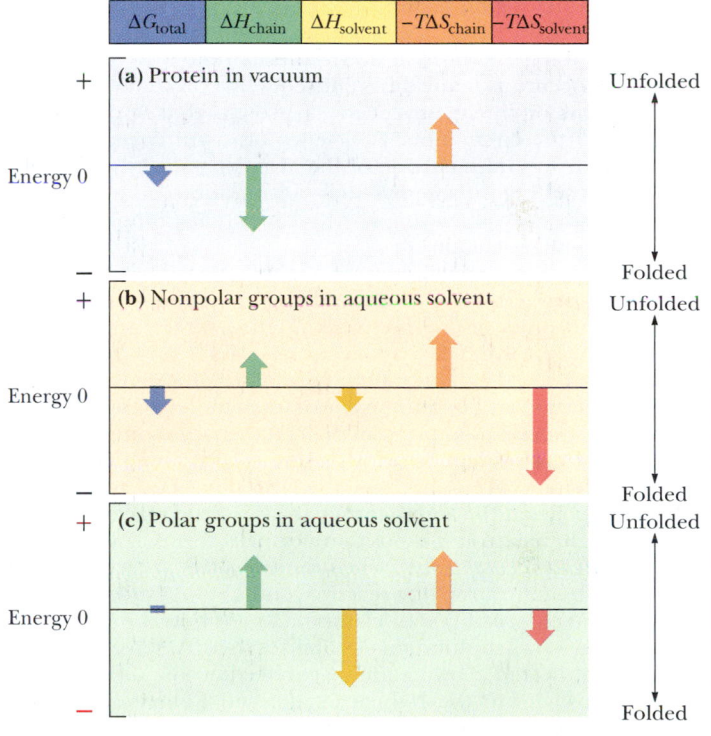

teins in the cell is a different matter. The highly concentrated protein matrix in the cell may adversely affect the folding process by causing aggregation of some unfolded or partially folded proteins. Also, it may be necessary to accelerate slow steps in the folding process or to suppress or reverse incorrect or premature folding. A family of proteins, known as **molecular chaperones,** are essential for the correct folding of certain polypeptide chains in vivo; for their assembly into oligomers; and for preventing inappropriate liaisons with other proteins during their synthesis, folding, and transport. Many of these proteins were first identified as **heat shock proteins,** which are induced in cells by elevated temperature or other stress. The most thoroughly studied proteins are **Hsp70,** a 70-kD heat shock protein, and the so-called **chaperonins,** also known as **Cpn60s** or **Hsp60s,** a class of 60-kD heat shock proteins. A well-characterized **Hsp60** chaperonin is **GroEL,** an *E. coli* protein that has been shown to affect the folding of several proteins. The mechanism of action of chaperones is discussed in Chapter 31.

Human Biochemistry

Biochemistry ◎ Now™

A Mutant Protein That Folds Slowly Can Cause Emphysema and Liver Damage

Lungs enable animals to acquire oxygen from the air and to give off CO_2 produced in respiration. Exchange of oxygen and CO_2 occurs in the alveoli—air sacs surrounded by capillaries that connect the pulmonary veins and arteries. The walls of alveoli consist of the elastic protein **elastin.** Inhalation expands the alveoli, and exhalation compresses them. A pair of human lungs contains 300 million alveoli, and the total area of the alveolar walls in contact with capillaries is about 70 m²—an area about the size of a tennis court! White blood cells naturally secrete **elastase**—a serine protease—which can attack and break down the elastin of the alveolar walls. However, α_1-**antitrypsin**—a 52-kD protein belonging to the **serpin** (*serine protease inhibitor*) family—normally binds to elastase, preventing alveolar damage. The structural gene for α_1-antitrypsin is extremely polymorphic (that is, it occurs as many different sequence variants), and several versions of this gene encode a protein that is poorly secreted into the circulation. Deficiency of α_1-antitrypsin in the blood can lead to destruction of the alveolar walls by white cell elastase, resulting in **emphysema**—a condition in which the alveoli are destroyed, leaving large air sacs that cannot be compressed during exhalation.

α_1-Antitrypsin normally adopts a highly ordered tertiary structure composed of three β-sheets and eight α-helices (see figure). Elastase and other serine proteases interact with a reactive, inhibitory site involving two amino acids—Met[358] and Ser[359]—on the so-called reactive-center loop. Formation of a tight complex between elastase and α_1-antitrypsin renders the elastase inactive. The most common α_1-antitrypsin deficiency involves the so-called **Z-variant** of the protein, in which lysine is substituted for glutamate at position 342 (Glu[342]———→Lys, also described as E342K). Residue 342 lies at the amino-terminal base of the **reactive-center loop,** and glutamate at this position normally forms a crucial salt bridge with Lys[290] on an adjacent strand of sheet A (see figure). In normal α_1-antitrypsin, the reactive-center loop is fully exposed and can interact readily with elastase. However, in the Z-variant, the Glu[342]———→Lys substitution destabilizes sheet A, separating the strands slightly and allowing the reactive-center loop of one molecule to insert into the β-sheet of another. Repetition of this anomalous association of α_1-antitrypsin molecules results in "loop-sheet" polymerization and the formation of large protein aggregates.

Myeong-Hee Yu and co-workers at the Korea Institute of Science and Technology have studied the folding kinetics of normal and Z-variant α_1-antitrypsin and have found that the Z-variant of α_1-antitrypsin folds identically to—but much more slowly than—normal α_1-antitrypsin. Newly synthesized Z-variant protein, incubated for 5 hours at 30°C, eventually adopts a native and active conformation and can associate tightly with elastase. However, incubation of the Z-variant at 37°C results in loop-sheet polymerization and self-aggregation of the protein. These results imply that emphysema arising in individuals carrying the Z-variant of α_1-antitrypsin is due to the slow folding kinetics of the protein rather than the adoption of an altered three-dimensional structure.

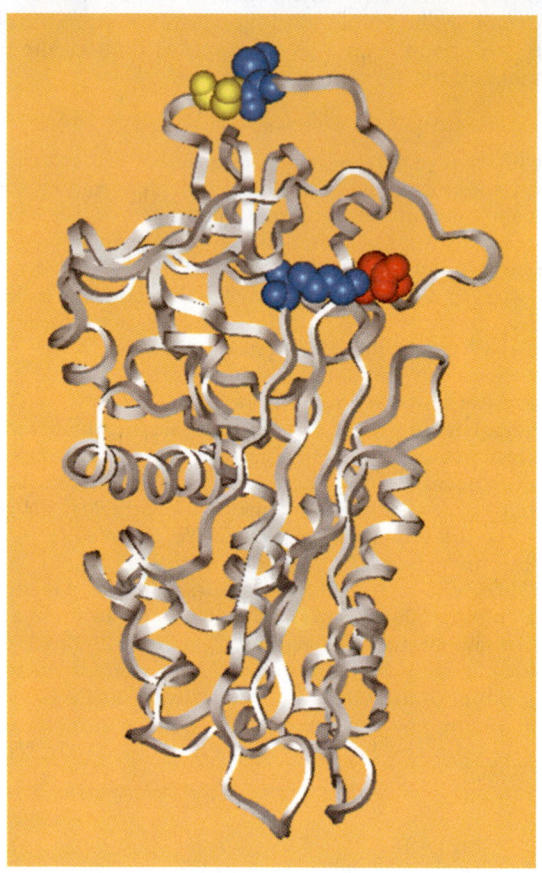

α_1-Antitrypsin. Note Met[358] (blue) and Ser[359] (yellow) at top, as well as Glu[342] (red) and Lys[290] (blue—upper right).

Protein Domains Are Nature's Modular Strategy for Protein Design

On the order of 1 million protein sequences are now known, and it has become obvious that certain protein sequences that give rise to distinct structural domains are used over and over again in modular fashion. These **protein modules** occur in a wide variety of proteins, often being used for different purposes, or they may be used repeatedly in the same protein. Figure 6.36 shows the tertiary structures of five protein modules, and Figure 6.37 presents several proteins that contain versions of these modules. These modules typically contain about 40 to 100 amino acids and often adopt a stable tertiary structure when isolated from their parent protein. One of the best-known examples of a protein module is the **immunoglobulin module,** which has been found not

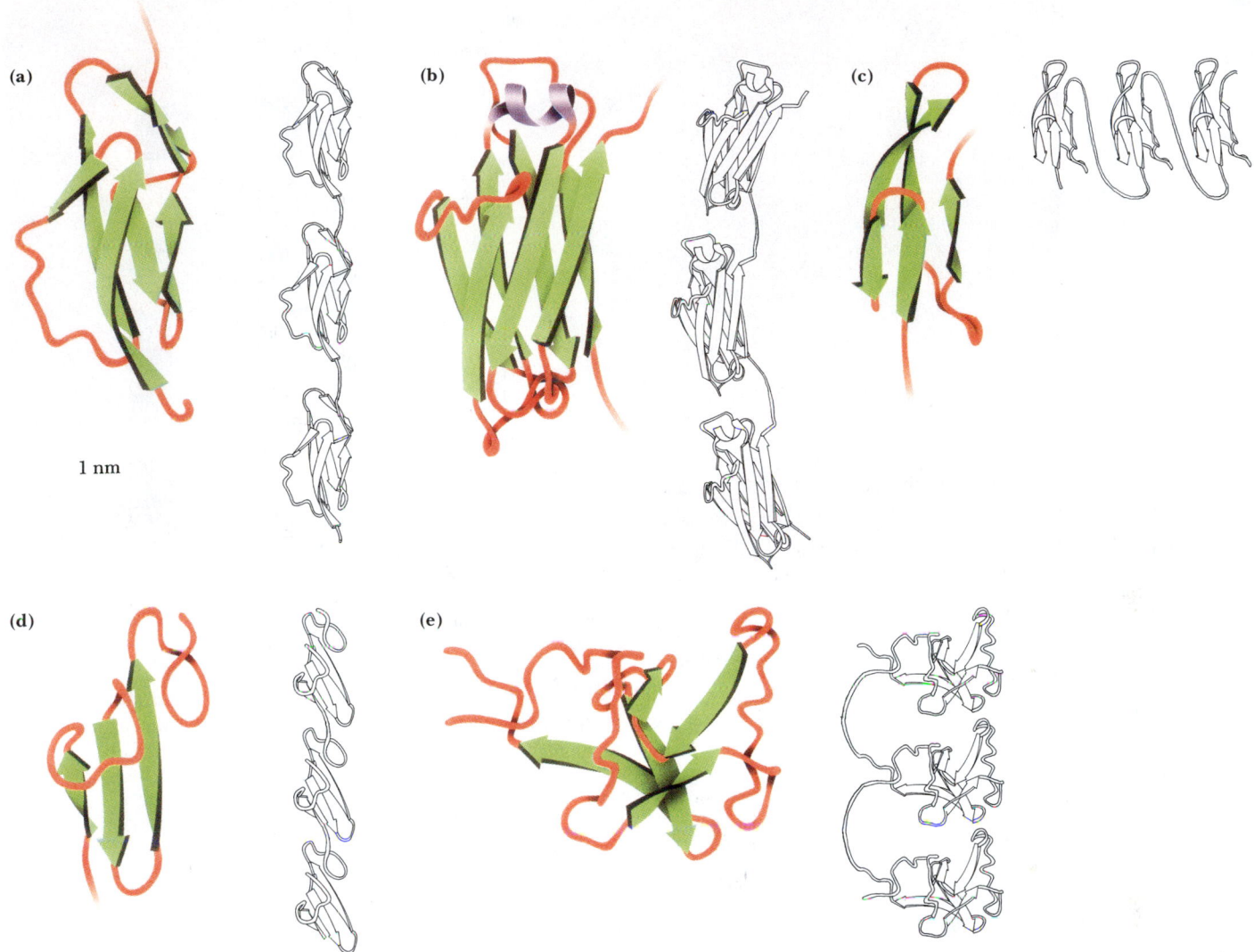

FIGURE 6.36 Ribbon structures of several protein modules used in the construction of complex multimodule proteins. **(a)** The complement control protein module. **(b)** The immunoglobulin module. **(c)** The fibronectin type I module. **(d)** The growth factor module. **(e)** The kringle module. *(Adapted from Baron, M., Norman, D., and Campbell, I., 1991. Protein modules. Trends in Biochemical Sciences* **16:***13-17.)*

only in immunoglobulins but also in a wide variety of cell surface proteins, including cell adhesion molecules and growth factor receptors, and even in *twitchin*, an intracellular protein found in muscle. It is likely that more protein modules will be identified. (The role of protein modules in signal transduction is discussed in Chapter 32.)

How Do Proteins Know How to Fold?

Christian Anfinsen's experiments demonstrated that proteins can fold reversibly. A corollary result of Anfinsen's work is that the native structures of at least some globular proteins are thermodynamically stable states. But the matter of how a given protein achieves such a stable state is a complex one. Cyrus Levinthal pointed out in 1968 that so many conformations are possible for a typical protein that the protein does not have sufficient time to reach its most stable conformational state by sampling all the possible conformations. This argument, termed "Levinthal's paradox," goes as follows: Consider a protein of 100 amino acids. Assume that there are only two conformational possibilities per amino acid, or $2^{100} = 1.27 \times 10^{30}$ possibilities. Allow 10^{-13} sec for the protein to test each conformational possibility in search of the overall energy minimum:

$$(10^{-13} \text{ sec})(1.27 \times 10^{30}) = 1.27 \times 10^{17} \text{ sec} = 4 \times 10^9 \text{ years}$$

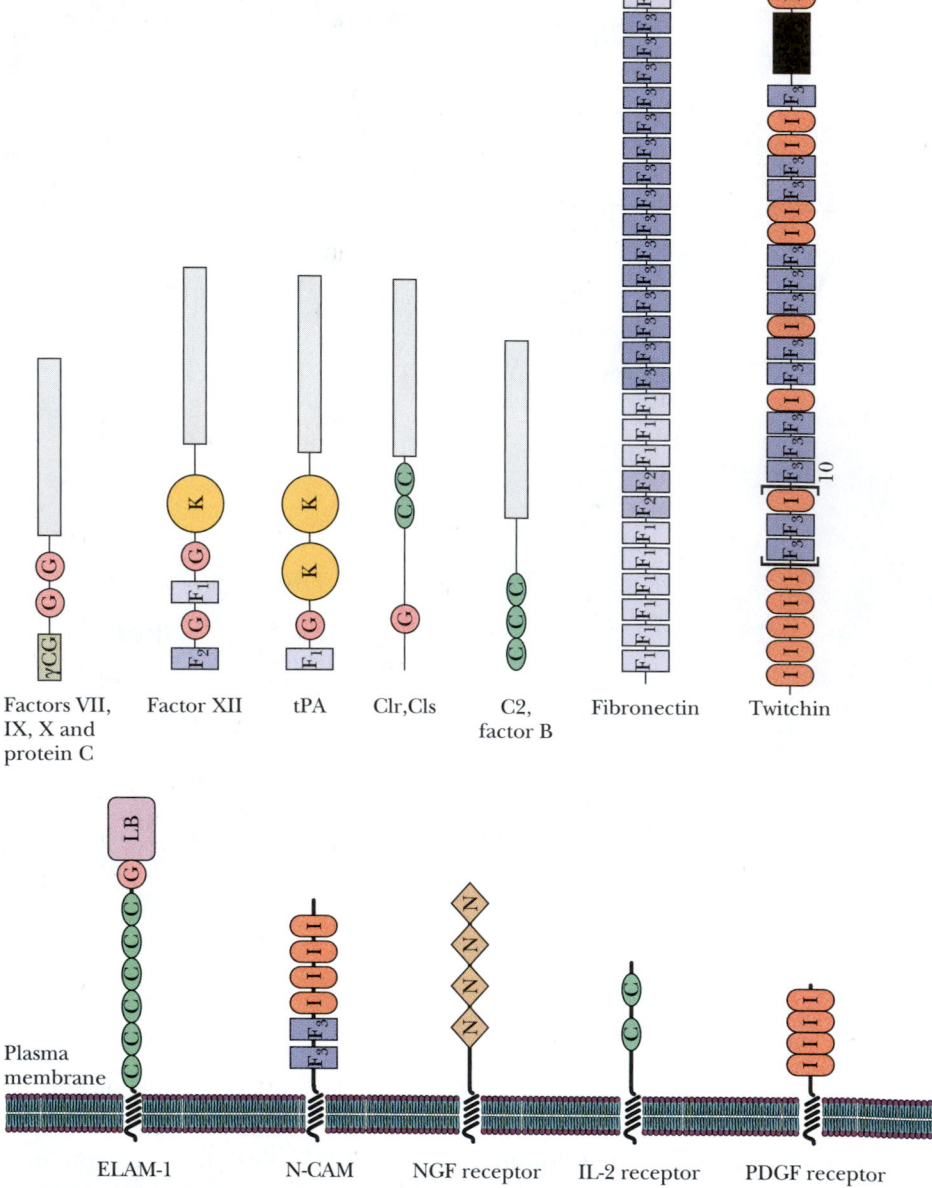

FIGURE 6.37 A sampling of proteins that consist of mosaics of individual protein modules. The modules shown include γCG, a module containing γ-carboxyglutamate residues; G, an epidermal growth factor–like module; K, the "kringle" domain, named for a Danish pastry; C, which is found in complement proteins; F1, F2, and F3, first found in fibronectin; I, the immunoglobulin superfamily domain; N, found in some growth factor receptors; E, a module homologous to the calcium-binding E–F hand domain; and LB, a lectin module found in some cell surface proteins. *(Adapted from Baron, M., Norman, D., and Campbell, I., 1991. Protein modules.* Trends in Biochemical Sciences *16:13–17.)*

Levinthal's paradox led protein chemists to hypothesize that proteins must fold by specific "folding pathways," and many research efforts have been devoted to the search for these pathways.

Implicit in the presumption of folding pathways is the existence of intermediate, partially folded conformational states. The notion of intermediate states on the pathway to a tertiary structure raises the possibility that segments of a protein might independently adopt local and well-defined secondary structures (α-helices and β-sheets). The tendency of a peptide segment to prefer a particular secondary structure depends in turn on its amino acid composition and sequence.

Surveys of the frequency with which various residues appear in helices and sheets show (Figure 6.38) that some residues, such as alanine, glutamate, and methionine, occur much more frequently in α-helices than do others. In contrast, glycine and proline are the least likely residues to be found in an α-helix. Likewise, certain residues, including valine, isoleucine, and the aromatic amino

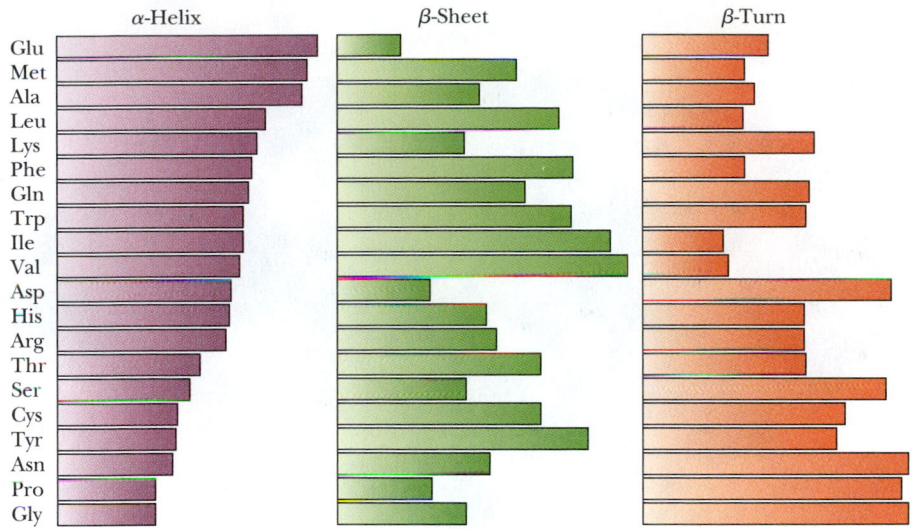

FIGURE 6.38 Relative frequencies of occurrence of amino acid residues in α-helices, β-sheets, and β-turns in proteins of known structure. *(Adapted from Bell, J. E., and Bell, E. T., 1988, Proteins and Enzymes, Englewood Cliffs, NJ: Prentice Hall.)*

acids, are more likely to be found in β-sheets than other residues, and aspartate, glutamate, and proline are much less likely to be found in β-sheets.

Such observations have led to many efforts to predict the occurrence of secondary structure in proteins from knowledge of the peptide sequence. Such **predictive algorithms** consider the composition of short segments of a polypeptide. If these segments are rich in residues that are found frequently in helices or sheets, then that segment is judged likely to adopt the corresponding secondary structure. The predictive algorithm designed by Peter Chou and Gerald Fasman in 1974 attempted to classify the 20 amino acids for their α-helix–forming and β-sheet–forming propensities. By studying the patterns of occurrence of each of these classes in helices and sheets of proteins with known structures, Chou and Fasman formulated a set of rules to predict the occurrence of helices and sheets in sequences of unknown structure. The Chou–Fasman method has been a useful device for some purposes, but it is able to predict the occurrence of helices and sheets in protein structures only about 50% of the time.

Proteins fold and unfold over a vast range of time scales, from microseconds to years. Some proteins fold in a simple two-state manner, with a single energy barrier separating the native (N) and denatured (D) states, whereas others proceed to the folded state through a series of intermediate states (Figure 6.39).

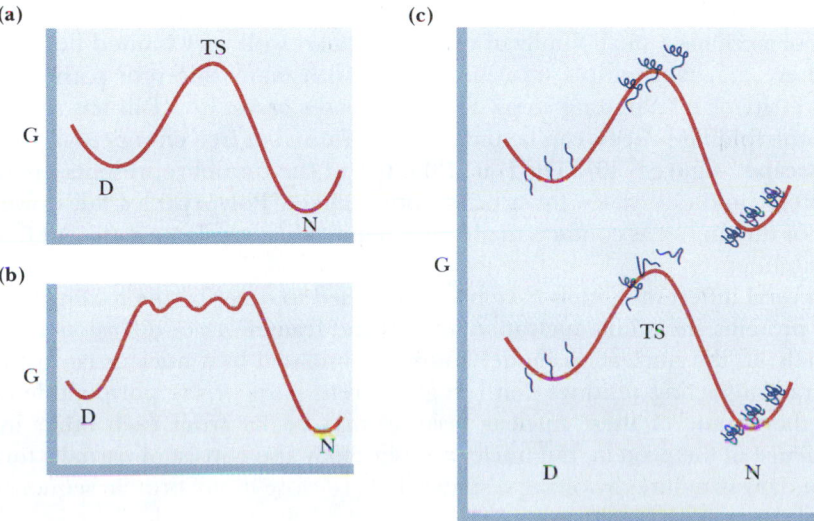

FIGURE 6.39 The transition state model for the folding of globular proteins. **(a)** A single free energy barrier separates the unfolded or denatured (D) state and the folded or native (N) state. **(b)** A model with a single folding pathway with sequential transition states along the folding pathway. **(c)** A model in which there are multiple, similar transition states, and a variety of folding pathways. *(Adapted from Myers, J. K., and Oas, T. G., 2002. Mechanisms of fast protein folding. Annual Review of Biochemistry* **71:**783–815.*)*

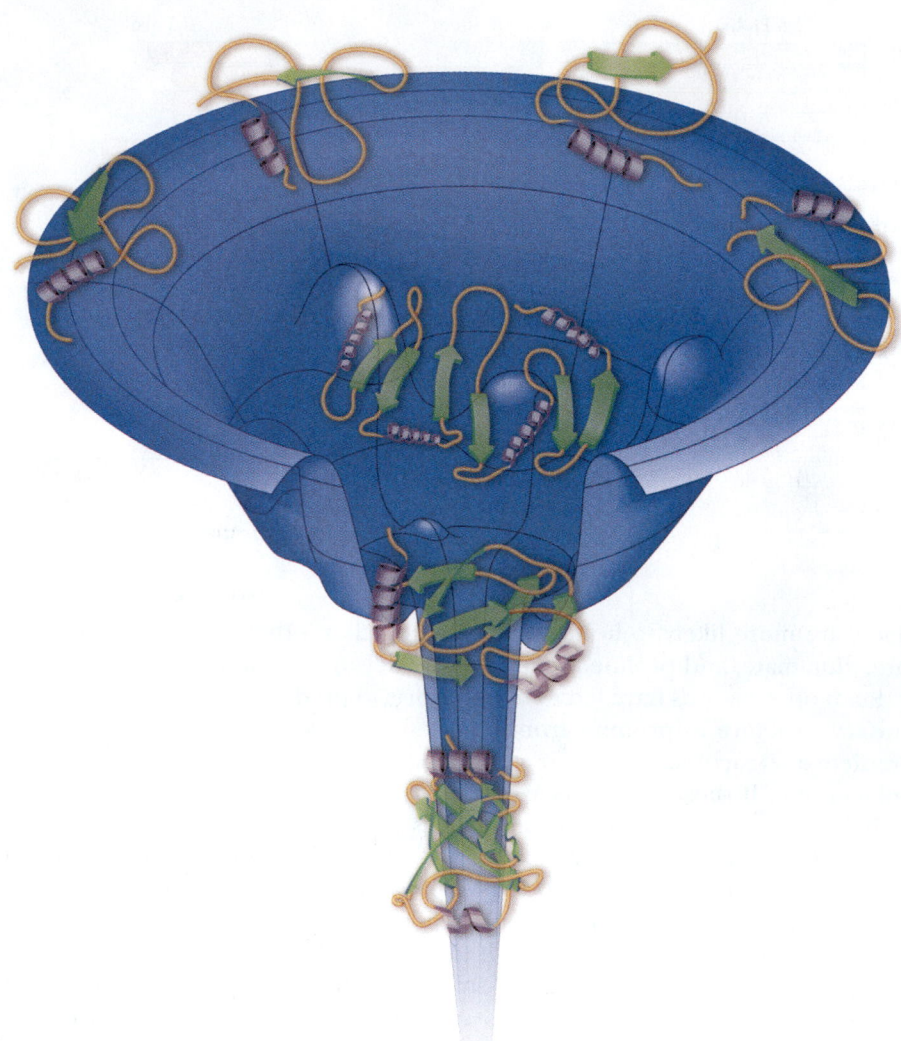

FIGURE 6.40 A model for the steps involved in the folding of globular proteins. The funnel represents a free energy surface or energy landscape for the folding process. The protein folding process is highly cooperative. Rapid and reversible formation of local secondary structures is followed by a slower phase in which establishment of partially folded intermediates leads to the final tertiary structure. Substantial exclusion of water occurs very early in the folding process.

Most single-domain proteins fold in a two-state manner at neutral pH, passing over an energy barrier and through a **transition state (TS).** Even for simple two-state folding behavior, however, there are two extreme possibilities. On one hand, there may be only a single transition state, with only a single conformation or perhaps a small family of transition states with very limited flexibilities, or there may be multiple transition states, with many different pathways and a diversity of rate-limiting steps. For these latter cases, Ken Dill has suggested that the folding process can be pictured as a funnel of free energies—an **energy landscape** (Figure 6.40). The rim at the top of the funnel represents the many possible unfolded states for a polypeptide chain. Polypeptides fall down the wall of the funnel as contacts made between residues nucleate different folding possibilities.

Several different models have been proposed to describe the folding of globular proteins, including **nucleation** models and **framework** or **diffusion–collision** models. In the nucleation model, folding is initiated by a nucleus consisting of several interacting residues that bring different parts of the polypeptide chain together. Some of these nuclear residues may be far from each other in the sequence of the protein, but nucleation sites may also consist of partially formed secondary structures involving residues that are close in the protein sequence. In

(a)   (b)

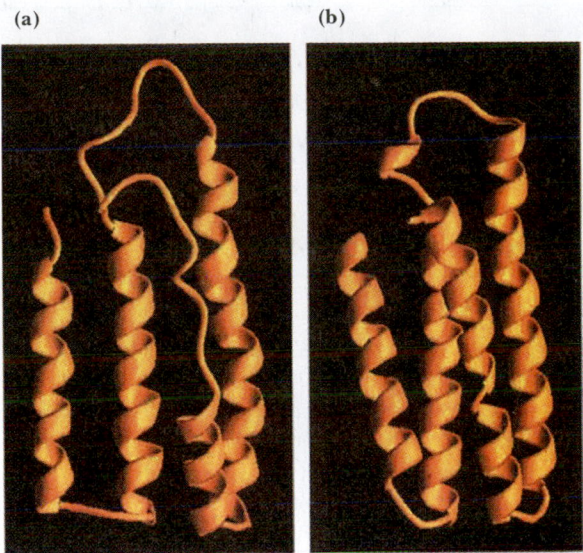

FIGURE 6.41 The structure of the molten globule state **(a)** and the native, folded state **(b)** of cytochrome b562. *(From Alberts, B., Bray, D., Lewis, J., Raff, M., Roberts, K., Watson, J. D., 1994.* Molecular Biology of the Cell, *3rd ed. New York: Garland Press.)*

framework models, relatively stable elements of secondary structure form first, followed by formation of long-range tertiary structure interactions. In diffusion–collision models, the polypeptide chain forms microdomains, which include elements of secondary structure but which also diffuse or wander transiently through a series of nativelike structures. Subsequent collisions between parts of the polypeptide chain enhance the stability of the microdomains and lead to productive folding of the entire protein.

Much of what we know about protein folding has come from studies of protein *unfolding*. Under certain conditions, native folded proteins can be partially denatured to form a **molten globule.** The molten globule state of a protein is a flexible but compact form characterized by significant amounts of secondary structure, virtually no precise tertiary structure, and a loosely packed hydrophobic core (Figure 6.41). These characteristics make the molten globule a close cousin of the initiating structures of the nucleation, framework, and diffusion–collision folding models, and intermediate structures similar to molten globules are postulated to form during the folding of many globular proteins.

Remarkably, it is now becoming clear that many proteins exist and function normally in a partially unfolded state. Such proteins, termed **intrinsically unstructured proteins (IUPs)** or **natively unfolded proteins,** do not possess uniform structural properties but are nonetheless essential for basic cellular functions. These proteins are characterized by an almost complete lack of folded structure and an extended conformation with high intramolecular flexibility. The functions of most IUPs are related to and dependent on their structural disorder (Table 6.3). More than 100 IUPs have been identified.

Intrinsically unstructured proteins contact their targets over a large surface area (Figure 6.42). The p27 protein complexed with cyclin-dependent protein kinase 2 (Cdk2) and cyclin A shows that p27 is in contact with its binding partners across its entire length. It binds in a groove consisting of conserved residues on cyclin A. On Cdk2, it binds to the N-terminal domain and also to the catalytic cleft. One of the most appropriate roles for such long-range interactions is assembly of complexes involved in the transcription of DNA into RNA, where large numbers of proteins must be recruited in macromolecular complexes. Thus the transactivator domain catenin-binding domain (CBD) of tcf3 is bound to several functional domains of β-catenin (Figure 6.42).

Human Biochemistry

Diseases of Protein Folding

A number of human diseases are linked to abnormalities of protein folding. Protein misfolding may cause disease by a variety of mechanisms. For example, misfolding may result in loss of function and the onset of disease. The following table summarizes several other mechanisms and provides an example of each.

Disease	Affected Protein	Mechanism
Alzheimer's disease	β-Amyloid peptide (derived from amyloid precursor protein)	Misfolded β-amyloid peptide accumulates in human neural tissue, forming deposits known as neuritic plaques.
Familial amyloidotic polyneuropathy	Transthyretin	Aggregation of unfolded proteins. Nerves and other organs are damaged by deposits of insoluble protein products.
Cancer	p53	p53 prevents cells with damaged DNA from dividing. One class of p53 mutations leads to misfolding; the misfolded protein is unstable and is destroyed.
Creutzfeldt-Jakob disease (human equivalent of mad cow disease)	Prion	Prion protein with an altered conformation (PrP^{SC}) may seed conformational transitions in normal PrP (PrP^C) molecules.
Hereditary emphysema	α_1-Antitrypsin	Mutated forms of this protein fold slowly, allowing its target, elastase, to destroy lung tissue.
Cystic fibrosis	CFTR (cystic fibrosis transmembrane conductance regulator)	Folding intermediates of mutant CFTR forms don't dissociate freely from chaperones, preventing the CFTR from reaching its destination in the membrane.

Table 6.3

Chou–Fasman Helix and Sheet Propensities (P_α and P_β) of the Amino Acids

Amino Acid	P_α	Helix Classification	P_β	Sheet Classification
A Ala	1.42	H_α	0.83	i_β
C Cys	0.70	i_α	1.19	h_β
D Asp	1.01	I_α	0.54	B_β
E Glu	1.51	H_α	0.37	B_β
F Phe	1.13	h_α	1.38	h_β
G Gly	0.57	B_α	0.75	b_β
H His	1.00	I_α	0.87	h_β
I Ile	1.08	h_α	1.60	H_β
K Lys	1.16	h_α	0.74	b_β
L Leu	1.21	H_α	1.30	h_β
M Met	1.45	H_α	1.05	h_β
N Asn	0.67	b_α	0.89	i_β
P Pro	0.57	B_α	0.55	B_β
Q Gln	1.11	h_α	1.10	h_β
R Arg	0.98	i_α	0.93	i_β
S Ser	0.77	i_α	0.75	b_β
T Thr	0.83	i_α	1.19	h_β
V Val	1.06	h_α	1.70	H_β
W Trp	1.08	h_α	1.37	h_β
Y Tyr	0.69	b_α	1.47	H_β

Source: Chou, P. Y., and Fasman, G. D., 1978. Empirical predictions of protein conformation. *Annual Review of Biochemistry* 47:258.

Human Biochemistry

Structural Genomics

The prodigious advances in genome sequencing in recent years, together with advances in techniques for protein structure determination, have not only provided much new information for biochemists but have also spawned a new field of investigation—**structural genomics,** the large-scale analysis of protein structures and functions based on gene sequences. The scale of this new endeavor is daunting: hundreds of thousands of gene sequences are rapidly being determined, and current estimates suggest that there may be between 1000 and 5000 distinct and stable polypeptide folding patterns in nature. The Protein Data Bank *(www.rcsb.org)* contains the experimental structures of fewer than 700 of these putative chain folds. The feasibility of large-scale, high-throughput structure determination programs is being explored in a variety of pilot studies in Europe, Asia, and North America. These efforts seek to add 20,000 or more new protein structures to our collected knowledge in the near future; from this wealth of new information, it should be possible to predict and determine new structures from sequence information alone. This effort will be vastly more complex and more expensive than the Human Genome Project. It presently costs about $100,000 to determine

the structure of the typical globular protein, and one of the goals of structural genomics is to reduce this number to $20,000 or less. Advances in techniques for protein crystallization, X-ray diffraction, and NMR spectroscopy, the three techniques essential to protein structure determination, will be needed to reach this goal in the near future.

The payoffs anticipated from structural genomics are substantial. Access to large amounts of new three-dimensional structural information should accelerate the development of new families of drugs. The ability to scan databases of chemical entities for activities against drug targets will be enhanced if large numbers of new protein structures are available, especially if complexes of drugs and target proteins can be obtained or predicted. The impact of structural genomics will also extend, however, to **functional genomics**—the study of the functional relationships of genomic content—which will enable the comparison of the composite functions of whole genomes, leading eventually to a complete biochemical and mechanistic understanding of all organisms, including humans.

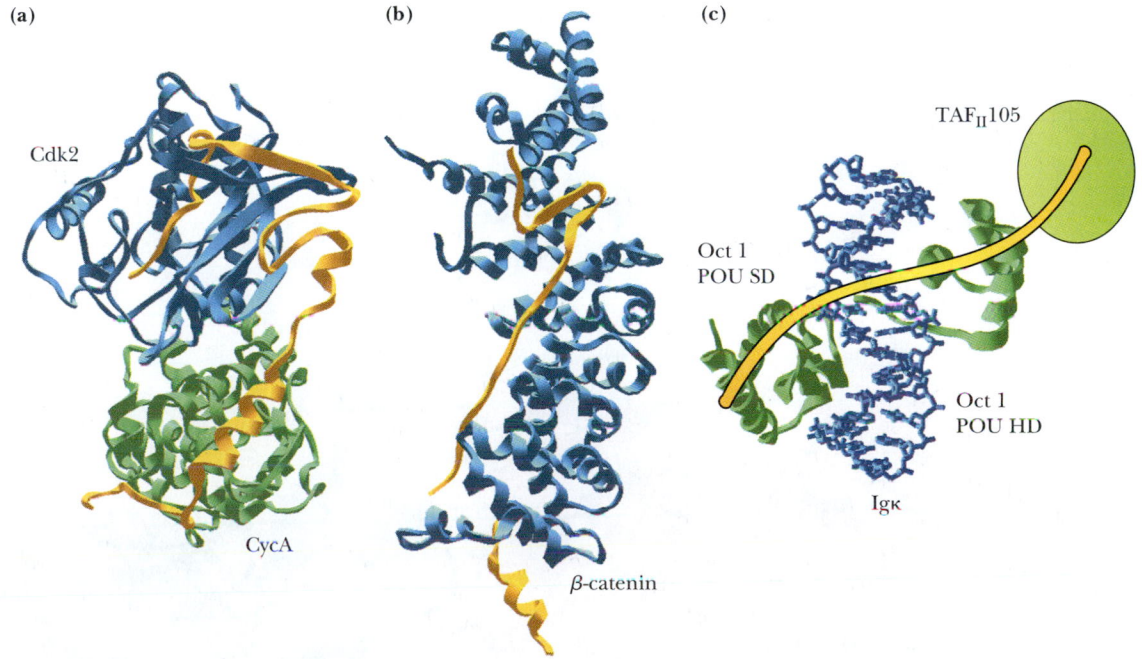

(a)   (b)   (c)

Cdk2

CycA

β-catenin

TAF$_{II}$105

Oct 1
POU SD

Oct 1
POU HD

Igκ

FIGURE 6.42 Intrinsically unstructured proteins (IUPs) contact their target proteins over a large surface area. **(a)** p27^{Kip1} (yellow) complexed with cyclin-dependent kinase 2 (Cdk2, blue) and cyclin A (CycA, green). **(b)** The transactivator domain CBD of Tcf3 (yellow) bound to β-catenin (blue). **Note:** Part of the β-catenin has been removed for a clear view of the CBD. **(c)** Bob 1 transcriptional coactivator (yellow) in contact with its four partners: TAF$_{II}$105 (green oval), the Oct 1 domains POU SD and POU HD (green), and the Igκ promoter (blue). *(From Tompa, P., 2002. Intrinsically unstructured proteins.* Trends in Biochemical Sciences **27:**527–533.)

6.5 | How Do Protein Subunits Interact at the Quaternary Level of Protein Structure?

Many proteins exist in nature as **oligomers,** complexes composed of (often symmetric) noncovalent assemblies of two or more monomer subunits. In fact, subunit association is a common feature of macromolecular organization in biology. Most intracellular enzymes are oligomeric and may be composed either of a single type of monomer subunit (*homomultimers*) or of several different kinds of subunits (*heteromultimers*). The simplest case is a protein composed of identical subunits. Liver alcohol dehydrogenase, shown in Figure 6.43, is such a protein. More complicated proteins may have several different subunits in one, two, or more copies. Hemoglobin, for example, contains two each of two different subunits and is referred to as an $\alpha_2\beta_2$-complex. An interesting counterpoint to these relatively simple cases is made by the proteins that form polymeric structures. Tubulin is an $\alpha\beta$-dimeric protein that polymerizes to form microtubules of the formula $(\alpha\beta)_n$. The way in which separate folded monomeric protein subunits associate to form the oligomeric protein constitutes the **quaternary structure** of that protein. Table 6.4 lists several proteins and their subunit compositions (see also Table 5.1). Clearly, proteins with two to four subunits dominate the list, but many cases of higher numbers exist.

The subunits of an oligomeric protein typically fold into apparently independent globular conformations and then interact with other subunits. The particular surfaces at which protein subunits interact are similar in nature to the interiors of the individual subunits. These interfaces are closely packed and involve both polar and hydrophobic interactions. Interacting surfaces must therefore possess complementary arrangements of polar and hydrophobic groups.

Oligomeric associations of protein subunits can be divided into those between **identical** subunits and those between **nonidentical** subunits. Interactions

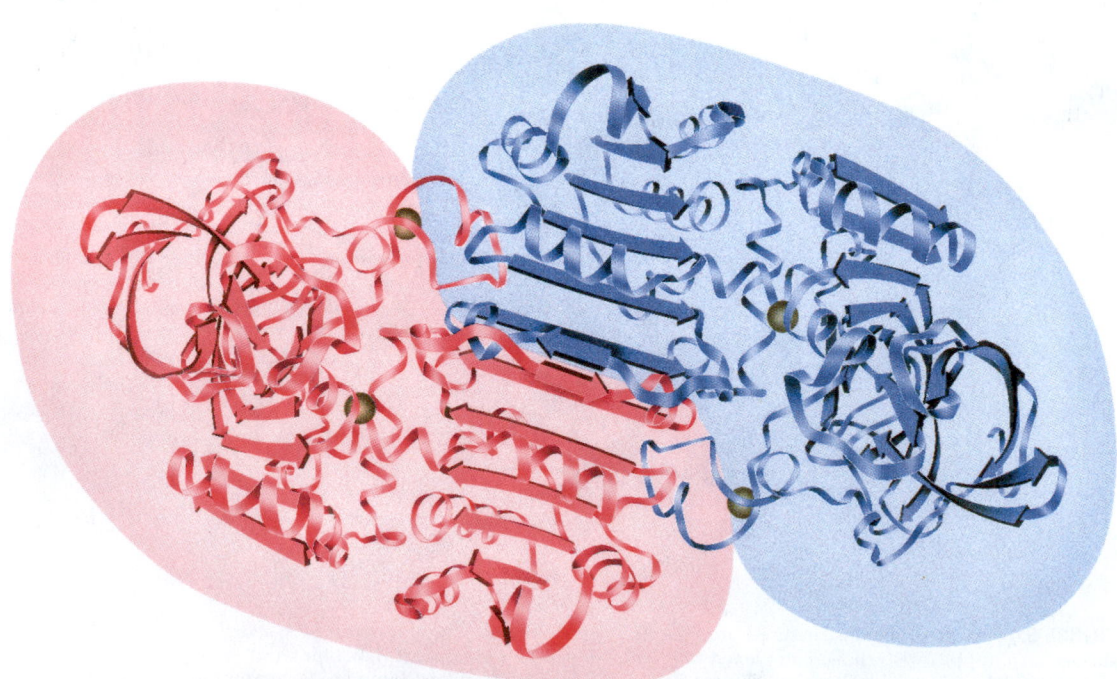

FIGURE 6.43 The quaternary structure of liver alcohol dehydrogenase. Within each subunit is a six-stranded parallel sheet. Between the two subunits is a two-stranded antiparallel sheet. The point in the center is a C_2 symmetry axis. (*Jane Richardson.*)

among identical subunits can be further distinguished as either **isologous** or **heterologous.** In isologous interactions, the interacting surfaces are identical and the resulting structure is necessarily dimeric and closed, with a twofold axis of symmetry (Figure 6.44). If any additional interactions occur to form a trimer or tetramer, these must use different interfaces on the protein's surface. Many proteins, including concanavalin and prealbumin, form tetramers by means of two sets of isologous interactions, one of which is shown in Figure 6.45. Such structures possess three different twofold axes of symmetry. In contrast, heterologous associations among subunits involve nonidentical interfaces. These surfaces must be complementary, but they are generally not symmetric. As shown in Figure 6.45, heterologous interactions are necessarily open-ended. This can give rise either to a closed cyclic structure, if geometric constraints exist, or to large polymeric assemblies. The closed cyclic structures are far more common and include the trimers of aspartate transcarbamoylase catalytic subunits and the tetramers of neuraminidase and hemerythrin.

There Is Symmetry in Quaternary Structures

One useful way to consider quaternary interactions in proteins involves the symmetry of these interactions. Globular protein subunits are always asymmetric objects. All of the polypeptide's α-carbons are asymmetric, and the polypeptide nearly always folds to form a low-symmetry structure. (The long helical arrays formed by some synthetic polypeptides are an exception.) Thus, protein subunits do not have mirror reflection planes, points, or axes of inversion. The only symmetry operation possible for protein subunits is a rotation. The most common symmetries observed for multisubunit proteins are cyclic symmetry and dihedral symmetry. In **cyclic symmetry,** the subunits are arranged around a

Table 6.4	
Aggregation Symmetries of Globular Proteins	
Protein	**Number of Subunits**
Alcohol dehydrogenase	2
Immunoglobulin	4
Malate dehydrogenase	2
Superoxide dismutase	2
Triose phosphate isomerase	2
Glycogen phosphorylase	2
Alkaline phosphatase	2
6-Phosphogluconate dehydrogenase	2
Wheat germ agglutinin	2
Phosphoglucoisomerase	2
Tyrosyl-tRNA synthetase	2
Glutathione reductase	2
Aldolase	3
Bacteriochlorophyll protein	3
TMV protein disc	17
Concanavalin A	4
Glyceraldehyde-3-phosphate dehydrogenase	4
Lactate dehydrogenase	4
Prealbumin	4
Pyruvate kinase	4
Phosphoglycerate mutase	4
Hemoglobin	2 + 2
Insulin	6
Aspartate transcarbamoylase	6 + 6
Glutamine synthetase	12
Apoferritin	24
Coat of tomato bushy stunt virus	180

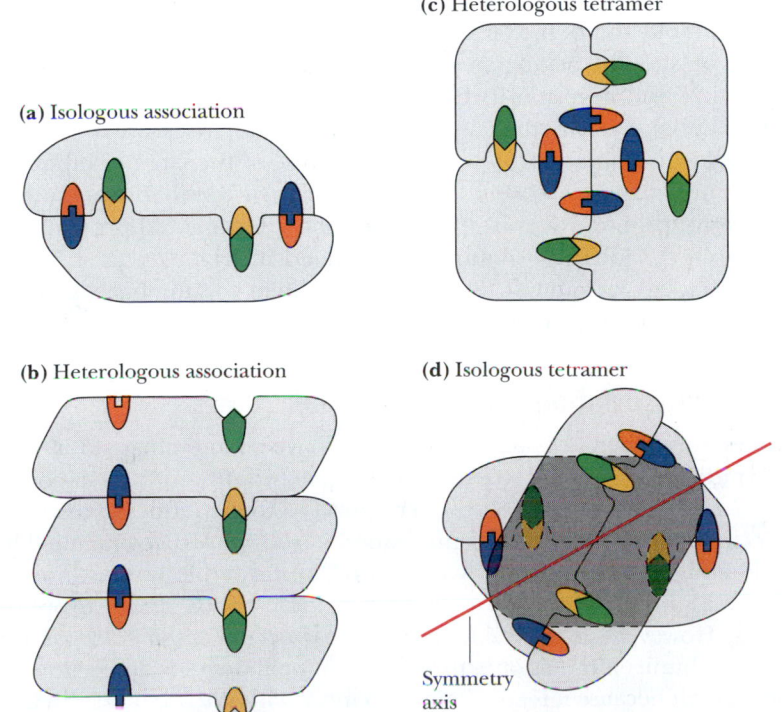

(a) Isologous association

(b) Heterologous association

(c) Heterologous tetramer

(d) Isologous tetramer

Symmetry axis

FIGURE 6.44 Isologous and heterologous associations between protein subunits. **(a)** An isologous interaction between two subunits with a twofold axis of symmetry perpendicular to the plane of the page. **(b)** A heterologous interaction that could lead to the formation of a long polymer. **(c)** A heterologous interaction leading to a closed structure—a tetramer. **(d)** A tetramer formed by two sets of isologous interactions.

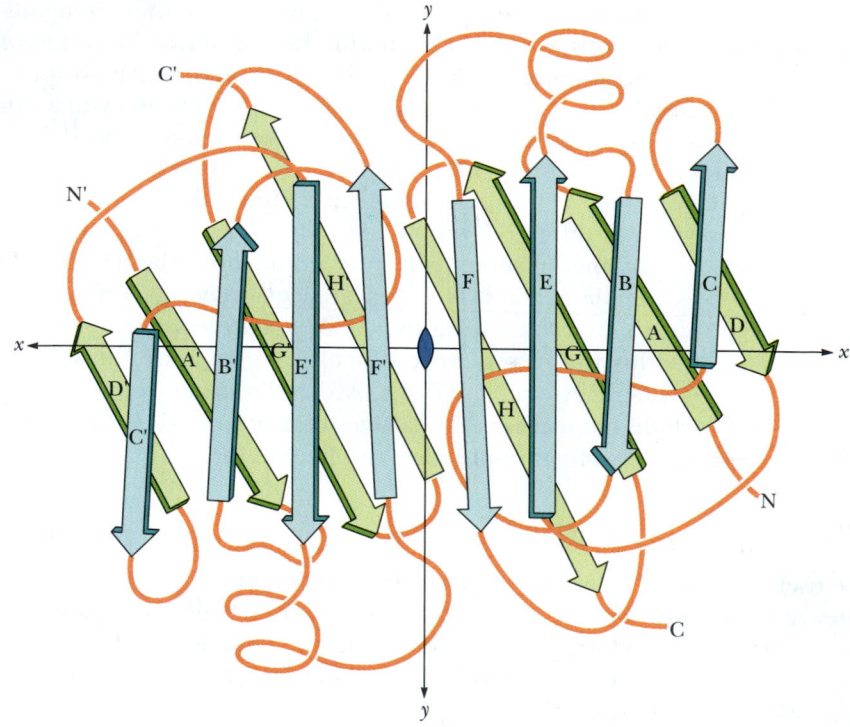

FIGURE 6.45 The polypeptide backbone of the pre-albumin dimer. The monomers associate in a manner that continues the β-sheets. A tetramer is formed by isologous interactions between the side chains extending outward from sheet D'A'G'H'HGAD in both dimers, which pack together nearly at right angles to one another. *(Jane Richardson.)*

single rotation axis, as shown in Figure 6.46. If there are two subunits, the axis is referred to as a *twofold rotation axis*. Rotating the quaternary structure 180° about this axis gives a structure identical to the original one. With three subunits arranged about a threefold rotation axis, a rotation of 120° about that axis gives an identical structure. **Dihedral symmetry** occurs when a structure possesses at least one twofold rotation axis perpendicular to another *n*-fold rotation axis. This type of subunit arrangement (Figure 6.46) occurs in concanavalin A (where *n* = 2) and in insulin (where *n* = 3). Higher symmetry groups, including the tetrahedral, octahedral, and icosahedral symmetries, are much less common among multisubunit proteins, partly because of the large number of asymmetric subunits required to assemble truly symmetric tetrahedra and other high symmetry groups. For example, a truly symmetric tetrahedral protein structure would require 12 identical monomers arranged in triangles, as shown in Figure 6.46. Simple four-subunit tetrahedra of protein monomers, which actually possess dihedral symmetry, are more common in biological systems.

Quaternary Association Is Driven by Weak Forces

The forces that stabilize quaternary structure have been evaluated for a few proteins. Typical dissociation constants for simple two-subunit associations range from 10^{-8} to 10^{-16} M. These values correspond to free energies of association of about 50 to 100 kJ/mol at 37°C. Dimerization of subunits is accompanied by both favorable and unfavorable energy changes. The favorable interactions include van der Waals interactions, hydrogen bonds, ionic bonds, and hydrophobic interactions. However, considerable entropy loss occurs when subunits interact. When two subunits move as one, three translational degrees of freedom are lost for one subunit because it is constrained to move with the other one. In addition, many peptide residues at the subunit interface, which were previously free to move on the protein surface, now have their movements restricted by the subunit association. This unfavorable energy of association is in the range of 80 to 120 kJ/mol for temperatures of 25° to 37°C. Thus, to achieve stability, the dimerization of two subunits must involve approximately 130 to 220 kJ/mol of favor-

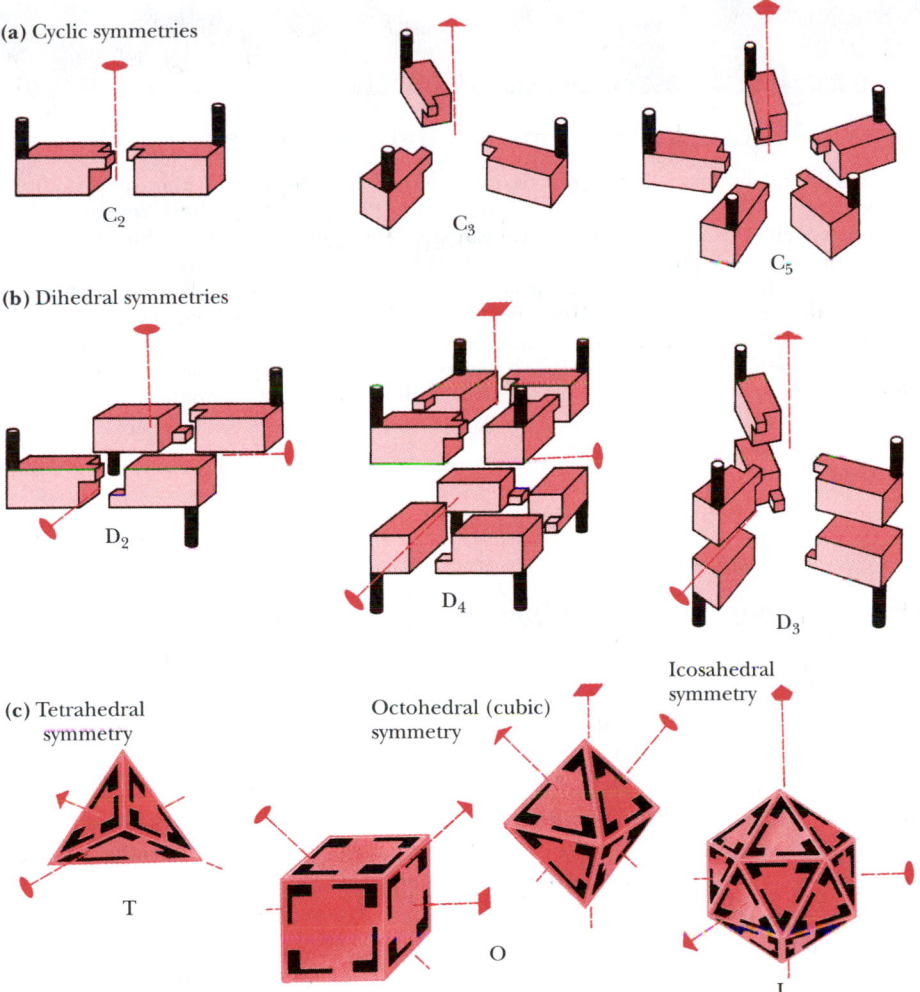

FIGURE 6.46 Several possible symmetric arrays of identical protein subunits, including (**a**) cyclic symmetry; (**b**) dihedral symmetry; and (**c**) cubic symmetry, including examples of tetrahedral (T), octahedral (O), and icosahedral (I) symmetry. *(Illustration: Irving Geis. Rights owned by Howard Hughes Medical Institute. Not to be reproduced without permission.)*

able interactions.[1] Van der Waals interactions at protein interfaces are numerous, often running to several hundred for a typical monomer–monomer association. This would account for about 150 to 200 kJ/mol of favorable free energy of association. However, when solvent is removed from the protein surface to form the subunit–subunit contacts, nearly as many van der Waals associations are lost as are made. One subunit is simply trading water molecules for peptide residues in the other subunit. As a result, the energy of subunit association due to van der Waals interactions actually contributes little to the stability of the dimer. Hydrophobic interactions, however, are generally very favorable. For many proteins, the subunit association process effectively buries as much as 20 nm² of surface area previously exposed to solvent, resulting in as much as 100 to 200 kJ/mol of favorable hydrophobic interactions. Together with whatever polar interactions occur at the protein–protein interface, this is sufficient to account for the observed stabilization that occurs when two protein subunits associate.

An additional and important factor contributing to the stability of subunit associations for some proteins is the formation of disulfide bonds between different subunits. All antibodies are $\alpha_2\beta_2$-tetramers composed of two heavy chains (53 to 75 kD) and two relatively light chains (23 kD). In addition to *intrasubunit* disulfide bonds (four per heavy chain, two per light chain), two *intersubunit* disulfide bridges hold the two heavy chains together and a disulfide bridge links each of the two light chains to a heavy chain (Figure 6.47).

[1]For example, 130 kJ/mol of favorable interaction minus 80 kJ/mol of unfavorable interaction equals a net free energy of association of 50 kJ/mol.

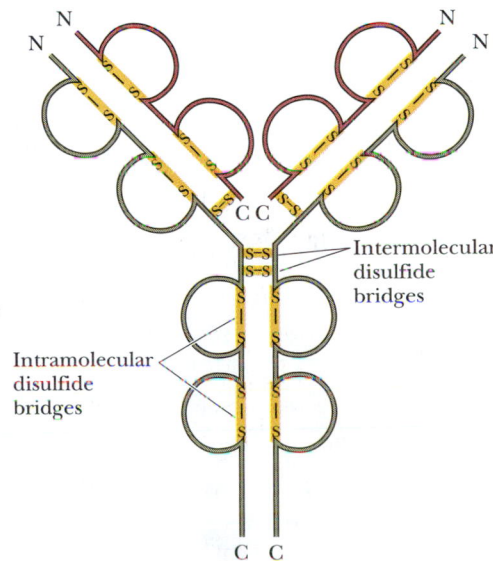

FIGURE 6.47 Schematic drawing of an immunoglobulin molecule showing the intramolecular and intermolecular disulfide bridges. (A space-filling model of the antigen-binding domain of an IgG molecule is shown in Figure 1.11.)

A Deeper Look

Immunoglobulins—All the Features of Protein Structure Brought Together

The immunoglobulin structure in Figure 6.47 represents the confluence of all the details of protein structure that have been thus far discussed. As for all proteins, the primary structure determines other aspects of structure. There are numerous elements of secondary structure, including β-sheets and tight turns. The tertiary structure consists of 12 distinct domains, and the protein adopts a heterotetrameric quaternary structure. To make matters more interesting, both intrasubunit and intersubunit disulfide linkages act to stabilize the discrete domains and to stabilize the tetramer itself.

One more level of sophistication awaits. As discussed in Chapter 28, the amino acid sequences of both light and heavy immunoglobulin chains are not constant! Instead, the primary structure of these chains is highly variable in the N-terminal regions (first 108 residues). Heterogeneity of the amino acid sequence leads to variations in the conformation of these variable regions. This variation accounts for antibody diversity and the ability of antibodies to recognize and bind a virtually limitless range of antigens. This full potential of antibody:antigen recognition enables organisms to mount immunological responses to almost any antigen that might challenge the organism.

Proteins Form a Variety of Quaternary Structures

When a protein is composed of only one kind of polypeptide chain, the manner in which the subunits interact and the arrangement of the subunits to produce the quaternary structure are usually simple matters. Sometimes, however, the same protein derived from several different species can exhibit different modes of quaternary interactions. Hemerythrin, the oxygen-carrying protein in certain species of marine invertebrates, is composed of a compact arrangement of four antiparallel α-helices. It is capable of forming dimers, trimers, tetramers, octamers, and even higher aggregates (Figure 6.48).

When two or more distinct peptide chains are involved, the nature of their interactions can be quite complicated. Multimeric proteins with more than one kind of subunit often display different affinities between different pairs of subunits. Whereas strongly denaturing solvents may dissociate the protein entirely into monomers, more subtle denaturing conditions may dissociate the oligomeric structure in a carefully controlled stepwise manner. Hemoglobin is a good example. Strong denaturants dissociate hemoglobin into α- and β-monomers. Using mild denaturing conditions, however, it is possible to dissociate hemoglobin almost completely into αβ-dimers, with few or no

(a)

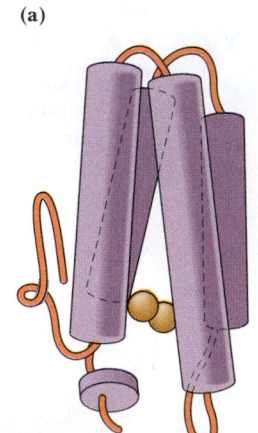

(b)

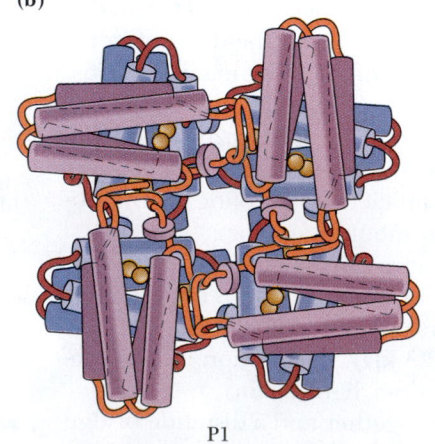

(c)

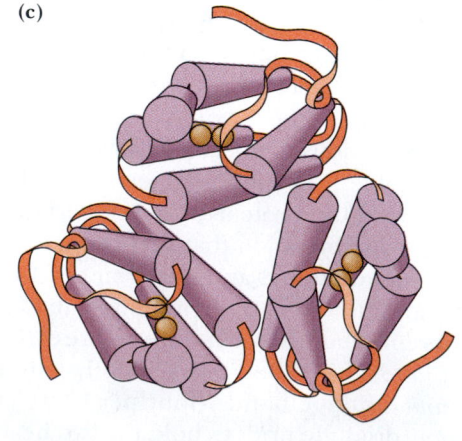

P1

FIGURE 6.48 The oligomeric states of hemerythrin from various marine worms. **(a)** The hemerythrin in *Thermiste zostericola* crystallized as a monomer; **(b)** the octameric hemerythrin crystallized from *Phascolopsis gouldii;* **(c)** the trimeric hemerythrin crystallized from *Siphonosoma* collected in mangrove swamps in Fiji.

free monomers occurring. In this sense, hemoglobin behaves functionally like a two-subunit protein, with each "subunit" composed of an $\alpha\beta$-dimer.

Open Quaternary Structures Can Polymerize

All of the quaternary structures we have considered to this point have been **closed** structures, with a limited capacity to associate. Many proteins in nature associate to form **open** heterologous structures, which can polymerize more or less indefinitely, creating structures that are both esthetically attractive and functionally important to the cells or tissue in which they exist. One such protein is **tubulin,** the $\alpha\beta$-dimeric protein that polymerizes into long, tubular structures that are the structural basis of cilia, flagella, and the cytoskeletal matrix. The microtubule thus formed (Figure 6.49) may be viewed as consisting of 13 parallel filaments arising from end-to-end aggregation of the tubulin dimers. Human immunodeficiency virus, HIV, the causative agent of AIDS (also discussed in Chapter 14), is enveloped by a spherical shell composed of hundreds of coat protein subunits, a large-scale quaternary association.

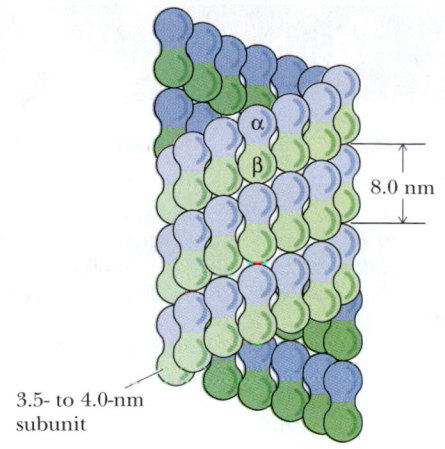

FIGURE 6.49 The structure of a typical microtubule, showing the arrangement of the α- and β-monomers of the tubulin dimer.

There Are Structural and Functional Advantages to Quaternary Association

There are several important reasons for protein subunits to associate in oligomeric structures.

Stability   One general benefit of subunit association is a favorable reduction of the protein's surface-to-volume ratio. The surface-to-volume ratio becomes smaller as the radius of any particle or object becomes larger. (This is because surface area is a function of the radius squared and volume is a function of the radius cubed.) Because interactions within the protein usually tend to stabilize the protein energetically and because the interaction of the protein surface with solvent water is often energetically unfavorable, decreased surface-to-volume ratios usually result in more stable proteins. Subunit association may also serve to shield hydrophobic residues from solvent water. Subunits that recognize either themselves or other subunits avoid any errors arising in genetic translation by binding mutant forms of the subunits less tightly.

Genetic Economy and Efficiency   Oligomeric association of protein monomers is genetically economical for an organism. Less DNA is required to code for a monomer that assembles into a homomultimer than for a large polypeptide of the same molecular mass. Another way to look at this is to realize that virtually all of the information that determines oligomer assembly and subunit–subunit interaction is contained in the genetic material needed to code for the monomer. For example, HIV protease, an enzyme that is a dimer of identical subunits, performs a catalytic function similar to homologous cellular enzymes that are single polypeptide chains of twice the molecular mass (see Chapter 14).

Bringing Catalytic Sites Together   Many enzymes (see Chapters 13 to 15) derive at least some of their catalytic power from oligomeric associations of monomer subunits. This can happen in several ways. The monomer may not constitute a complete enzyme active site. Formation of the oligomer may bring all the necessary catalytic groups together to form an active enzyme. For example, the active sites of bacterial glutamine synthetase are formed from pairs of adjacent subunits. The dissociated monomers are inactive.

Oligomeric enzymes may also carry out different but related reactions on different subunits. Thus, tryptophan synthase is a tetramer consisting of pairs of different subunits, $\alpha_2\beta_2$. Purified α-subunits catalyze the following reaction:

$$\text{Indoleglycerol phosphate} \rightleftharpoons \text{indole} + \text{glyceraldehyde-3-phosphate}$$

Human Biochemistry

Faster-Acting Insulin: Genetic Engineering Solves a Quaternary Structure Problem

Insulin is a peptide hormone secreted by the pancreas that regulates glucose metabolism in the body. Insufficient production of insulin or failure of insulin to stimulate target sites in liver, muscle, and adipose tissue leads to the serious metabolic disorder known as *diabetes mellitus*. Diabetes afflicts millions of people worldwide. Diabetic individuals typically exhibit high levels of glucose in the blood, but insulin injection therapy allows these individuals to maintain normal levels of blood glucose.

Insulin is composed of two peptide chains covalently linked by disulfide bonds (see Figures 5.13 and 6.35). This "monomer" of insulin is the active form that binds to receptors in target cells. However, in solution, insulin spontaneously forms dimers, which themselves aggregate to form hexamers. The surface of the insulin molecule that self-associates to form hexamers is also the surface that binds to insulin receptors in target cells. Thus, hexamers of insulin are inactive.

Insulin released from the pancreas is monomeric and acts rapidly at target tissues. However, when insulin is administered (by injection) to a diabetic patient, the insulin hexamers dissociate slowly and the patient's blood glucose levels typically drop slowly (over several hours).

In 1988, G. Dodson showed that insulin could be genetically engineered to prefer the monomeric (active) state. Dodson and his colleagues used recombinant DNA technology (discussed in Chapter 12) to produce insulin with an aspartate residue replacing a proline at the contact interface between adjacent subunits. The negative charge on the Asp side chain creates electrostatic repulsion between subunits and increases the dissociation constant for the hexamer⇌monomer equilibrium. Injection of this mutant insulin into test animals produced more rapid decreases in blood glucose than did ordinary insulin. This mutant insulin, marketed by the Danish pharmaceutical company Novo as NovoLog in the United States and as NovoRapid in Europe, may eventually replace ordinary insulin in the treatment of diabetes. NovoLog has a faster rate of absorption, a faster onset of action, and a shorter duration of action than regular human insulin. It is particularly suited for mealtime dosing to control postprandial glycemia, the rise in blood sugar following consumption of food. Regular human insulin acts more slowly, so patients must usually administer it 30 minutes before eating.

and the β-subunits catalyze this reaction:

$$\text{Indole} + \text{L-serine} \rightleftharpoons \text{L-tryptophan}$$

Indole, the product of the α-reaction and the reactant for the β-reaction, is passed directly from the α-subunit to the β-subunit and cannot be detected as a free intermediate.

Cooperativity   There is another, more important reason for monomer subunits to associate into oligomeric complexes. Most oligomeric enzymes regulate catalytic activity by means of subunit interactions, which may give rise to cooperative phenomena. Multisubunit proteins typically possess multiple binding sites for a given ligand. If the binding of ligand at one site changes the affinity of the protein for ligand at the other binding sites, the binding is said to be **cooperative.** Increases in affinity at subsequent sites represent positive cooperativity, whereas decreases in affinity correspond to negative cooperativity. The points of contact between protein subunits provide a mechanism for communication between the subunits. This in turn provides a way in which the binding of ligand to one subunit can influence the binding behavior at the other subunits. Such cooperative behavior, discussed in greater depth in Chapter 15, is the underlying mechanism for regulation of many biological processes.

Summary

6.1 What Are the Noncovalent Interactions That Dictate and Stabilize Protein Structure?

Several different kinds of noncovalent interactions are of vital importance in protein structure. Hydrogen bonds, hydrophobic interactions, electrostatic bonds, and van der Waals forces are all noncovalent in nature yet are extremely important influences on protein conformations. The stabilization free energies afforded by each of these interactions are highly dependent on the local environment within the protein.

Hydrogen bonds are generally made wherever possible within a given protein structure. Hydrophobic interactions form because nonpolar side chains of amino acids and other nonpolar solutes prefer to cluster in a nonpolar environment rather than to intercalate in a polar solvent such as water. Electrostatic interactions include the attraction between opposite charges and the repulsion of like charges in the protein. Van der Waals interactions involve instantaneous dipoles and induced dipoles that arise because of fluctuations in the electron charge distributions of adjacent nonbonded atoms.

6.2 What Role Does the Amino Acid Sequence Play in Protein Structure?

All of the information necessary for folding the peptide chain into its "native" structure is contained in the amino acid sequence of the peptide. Just how proteins recognize and interpret the information that is stored in the polypeptide sequence is not yet well understood. It may be assumed that certain loci along the peptide chain act as nucleation points, which initiate folding processes that eventually lead to the correct structures. Regardless of how this process operates, it must take the protein correctly to the final native structure, without getting trapped in a local energy-minimum state, which, although stable, may be different from the native state itself.

6.3 What Are the Elements of Secondary Structure in Proteins, and How Are They Formed?

Secondary structure in proteins forms so as to maximize hydrogen bonding and maintain the planar nature of the peptide bond. Secondary structures include α-helices, β-sheets, and tight turns.

6.4 How Do Polypeptides Fold into Three-Dimensional Protein Structures?

First, secondary structures—helices and sheets—form whenever possible as a consequence of the formation of large numbers of hydrogen bonds. Second, α-helices and β-sheets often associate and pack close together in the protein. There are a few common methods for such packing to occur. Third, because the peptide segments between secondary structures in the protein tend to be short and direct, the peptide does not execute complicated twists and knots as it moves from one region of a secondary structure to another. A consequence of these three principles is that protein chains are usually folded so that the secondary structures are arranged in one of a few common patterns. For this reason, there are families of proteins that have similar tertiary structure, with little apparent evolutionary or functional relationship among them. Finally, proteins generally fold so as to form the most stable structures possible. The stability of most proteins arises from (1) the formation of large numbers of intramolecular hydrogen bonds and (2) the reduction in the surface area accessible to solvent that occurs upon folding.

6.5 How Do Protein Subunits Interact at the Quaternary Level of Protein Structure?

The subunits of an oligomeric protein typically fold into apparently independent globular conformations and then interact with other subunits. The particular surfaces at which protein subunits interact are similar in nature to the interiors of the individual subunits. These interfaces are closely packed and involve both polar and hydrophobic interactions. Interacting surfaces must therefore possess complementary arrangements of polar and hydrophobic groups.

Problems

1. The central rod domain of a keratin protein is approximately 312 residues in length. What is the length (in Å) of the keratin rod domain? If this same peptide segment were a true α-helix, how long would it be? If the same segment were a β-sheet, what would its length be?

2. A teenager can grow 4 inches in a year during a "growth spurt." Assuming that the increase in height is due to vertical growth of collagen fibers (in bone), calculate the number of collagen helix turns synthesized per minute.

3. Discuss the potential contributions to hydrophobic and van der Waals interactions and ionic and hydrogen bonds for the side chains of Asp, Leu, Tyr, and His in a protein.

4. Figure 6.38 shows that Pro is the amino acid least commonly found in α-helices but most commonly found in β-turns. Discuss the reasons for this behavior.

5. For flavodoxin in Figure 6.31, identify the right-handed cross-overs and the left-handed cross-overs in the parallel β-sheet.

6. Choose any three regions in the Ramachandran plot and discuss the likelihood of observing that combination of ϕ and ψ in a peptide or protein. Defend your answer using suitable molecular models of a peptide.

7. A new protein of unknown structure has been purified. Gel filtration chromatography reveals that the native protein has a molecular weight of 240,000. Chromatography in the presence of 6 M guanidine hydrochloride yields only a peak for a protein of M_r 60,000. Chromatography in the presence of 6 M guanidine hydrochloride and 10 mM β-mercaptoethanol yields peaks for proteins of M_r 34,000 and 26,000. Explain what can be determined about the structure of this protein from these data.

8. Two polypeptides, A and B, have similar tertiary structures, but A normally exists as a monomer, whereas B exists as a tetramer, B$_4$. What differences might be expected in the amino acid composition of A versus B?

9. The hemagglutinin protein in influenza virus contains a remarkably long α-helix, with 53 residues.
   a. How long is this α-helix (in nm)?
   b. How many turns does this helix have?
   c. Each residue in an α-helix is involved in two H bonds. How many H bonds are present in this helix?

10. It is often observed that Gly residues are conserved in proteins to a greater degree than other amino acids. From what you have learned in this chapter, suggest a reason for this observation.

11. Which amino acids would be capable of forming H bonds with a lysine residue in a protein?

12. Poly-L-glutamate adopts an α-helical structure at low pH but becomes a random coil above pH 5. Explain this behavior.

13. Imagine that the dimensions of the alpha helix were such that there were exactly 3.5 amino acids per turn, instead of 3.6. What would be the consequences for coiled-coil structures?

Preparing for the MCAT Exam

14. Consider the following peptide sequences:

    EANQIDEMLYNVQCSLTTLEDTVPW
    LGVHLDITVPLSWTWTLYVKL
    QQNWGGLVVILTLVWFLM
    CNMKHGDSQCDERTYP
    YTREQSDGHIPKMNCDS
    AGPFGPDGPTIGPK

    Which of the preceding sequences would be likely to be found in each of the following:
    a. A parallel β-sheet
    b. An antiparallel β-sheet
    c. A tropocollagen molecule
    d. The helical portions of a protein found in your hair

15. To fully appreciate the elements of secondary structure in proteins, it is useful to have a practical sense of their structures. On a piece of paper, draw a simple but large zigzag pattern to represent a β-strand. Then fill in the structure, drawing the locations of the atoms of the chain on this zigzag pattern. Then draw a simple, large coil on a piece of paper to represent an α-helix. Then fill in the structure, drawing the backbone atoms in the correction locations along the coil and indicating the locations of the R groups in your drawing.

Biochemistry ❂Now™ Preparing for an exam? Test yourself on key questions at http://chemistry.brookscole.com/ggb3

Further Reading

General

Branden, C., and Tooze, J., 1991. *Introduction to Protein Structure*. New York: Garland Publishing.

Chothia, C., 1984. Principles that determine the structure of proteins. *Annual Review of Biochemistry* **53:**537–572.

Dickerson, R. E., and Geis, I., 1969. *The Structure and Action of Proteins*. New York: Harper and Row.

Hardie, D. G., and Coggins, J. R., eds., 1986. *Multidomain Proteins: Structure and Evolution*. New York: Elsevier.

Harper, E., and Rose, G. D., 1993. Helix stop signals in proteins and peptides: The capping box. *Biochemistry* **32:**7605–7609.

Judson, H. F., 1979. *The Eighth Day of Creation*. New York: Simon and Schuster.

Klotz, I. M., 1996. Equilibrium constants and free energies in unfolding of proteins in urea solutions. *Proceedings of the National Academy of Sciences* **93:**14411–14415.

Lupas, A., 1996. Coiled coils: New structures and new functions. *Trends in Biochemical Sciences* **21:**375–382.

Richardson, J. S., 1981. The anatomy and taxonomy of protein structure. *Advances in Protein Chemistry* **34:**167–339.

Richardson, J. S., and Richardson, D. C., 1988. Amino acid preferences for specific locations at the ends of α-helices. *Science* **240:**1648–1652.

Schulze, A. J., Huber, R., Bode, W., and Engh, R. A., 1994. Structural aspects of serpin inhibition. *FEBS Letters* **344:**117–124.

Smith, T., 2000. Structural Genomics—special supplement. *Nature Structural Biology* Volume **7.** This entire supplemental issue is devoted to structural genomics and contains a trove of information about this burgeoning field.

Tompa, P., 2002. Intrinsically unstructured proteins. *Trends in Biochemical Sciences* **27:**527–533.

Uversky, V.N., 2002. Natively unfolded proteins: A point where biology waits for physics. *Protein Science* **11:**739–756.

Webster, D. M., 2000. *Protein Structure Prediction—Methods and Protocols*. New Jersey: Humana Press.

Protein Folding

Aurora, R., Creamer, T., Srinivasan, R., and Rose, G. D., 1997. Local interactions in protein folding: Lessons from the α-helix. *The Journal of Biological Chemistry* **272:**1413–1416.

Baker, D., 2000. A surprising simplicity to protein folding. *Nature* **405:** 39-42.

Creighton, T. E., 1997. How important is the molten globule for correct protein folding? *Trends in Biochemical Sciences* **22:**6–11.

Deber, C. M., and Therien, A. G., 2002. Putting the β-breaks on membrane protein misfolding. *Nature Structural Biology* **9:**318–319.

Dill, K. A., and Chan, H. S., 1997. From Levinthal to pathways to funnels. *Nature Structural Biology* **4:**10–19.

Dinner, A. R., Sali, A., Smith, L. J., Dobson, C. M., and Karplus, M., 2001. Understanding protein folding via free-energy surfaces from theory and experiment. *Trends in Biochemical Sciences* **25:**331–339.

Mirny, L., and Shakhnovich, E., 2001. Protein folding theory: From lattice to all-atom models. *Annual Review of Biophysics and Biolmolecular Structure* **30:**361–396.

Murphy, K. P., 2001. *Protein Structure, Stability, and Folding*. New Jersey: Humana Press.

Myers, J. K., and Oas, T. G., 2002. Mechanisms of fast protein folding. *Annual Review of Biochemistry* **71:**783–815.

Privalov, P. L., and Makhatadze, G. I., 1993. Contributions of hydration to protein folding thermodynamics. II. The entropy and Gibbs energy of hydration. *Journal of Molecular Biology* **232:**660–679.

Radford, S. E., 2000. Protein folding: Progress made and promises ahead. *Trends in Biochemical Sciences* **25:**611–618.

Raschke, T. M., and Marqusee, S., 1997. The kinetic folding intermediate of ribonuclease H resembles the acid molten globule and partially unfolded molecules detected under native conditions. *Nature Structural Biology* **4:**298–304.

Srinivasan, R., and Rose, G. D., 1995. LINUS: A hierarchic procedure to predict the fold of a protein. *Proteins: Structure, Function and Genetics* **22:**81–99.

Secondary Structure

Salemme, F. R., 1983. Structural properties of protein β-sheets. *Progress in Biophysics and Molecular Biology* **42:**95–133.

Xiong, H., Buckwalter, B., Shieh, H-M, and Hecht, M. H., 1995. Periodicity of polar and nonpolar amino acids is the major determinant of secondary structure in self-assembling oligomeric peptides. *Proceedings of the National Academy of Sciences* **92:**6349–6353.

Structural Studies

Petsko, G. A., and Ringe, D., 1984. Fluctuations in protein structure from X-ray diffraction. *Annual Review of Biophysics and Bioengineering* **13:**331–371.

Torchia, D. A., 1984. Solid state NMR studies of protein internal dynamics. *Annual Review of Biophysics and Bioengineering* **13:**125–144.

Wand, A. J., 2001. Dynamic activation of protein function: A view emerging from NMR spectroscopy. *Nature Structural Biology* **8:**926–931.

Wagner, G., Hyberts, S., and Havel, T., 1992. NMR structure determination in solution: A critique and comparison with X-ray crystallography. *Annual Review of Biophysics and Biomolecular Structure* **21:**167–242.

Diseases of Protein Folding

Bucchiantini, M., et al., 2002. Inherent toxicity of aggregates implies a common mechanism for protein misfolding diseases. *Nature* **416:**507–511.

Sifers, R. M., 1995. Defective protein folding as a cause of disease. *Nature Structural Biology* **2:**355–367.

Stein, P. E., and Carrell, R. W., 1995. What do dysfunctional serpins tell us about molecular mobility and disease? *Nature Structural Biology* **2:**96–113.

Thomas, P. J., Qu, B-H., and Pedersen, P. L., 1995. Defective protein folding as a basis of human disease. *Trends in Biochemical Sciences* **20:**456–459.

Carbohydrates and the Glycoconjugates of Cell Surfaces

CHAPTER 7

"The Discovery of Honey"—Piero di Cosimo (1492).

*Sugar in the gourd and honey in the horn,
I never was so happy since the hour I
was born.*
Turkey in the Straw, *stanza 6 (classic American folk tune)*

Essential Question

Carbohydrates are a versatile class of molecules of the formula $(CH_2O)_n$. They are a major form of stored energy in organisms, and they are the metabolic precursors of virtually all other biomolecules. Conjugates of carbohydrates with proteins and lipids perform a variety of functions, including the recognition events that are important in cell growth, transformation, and other processes. ***What is the structure, chemistry, and biological function of carbohydrates?***

Carbohydrates are the single most abundant class of organic molecules found in nature. The name *carbohydrate* arises from the basic molecular formula $(CH_2O)_n$, which can be rewritten $(C \cdot H_2O)_n$ to show that these substances are hydrates of carbon, where $n = 3$ or more. Carbohydrates constitute a versatile class of molecules. Energy from the sun captured by green plants, algae, and some bacteria during photosynthesis (see Chapter 21) is stored in the form of carbohydrates. In turn, carbohydrates are the metabolic precursors of virtually all other biomolecules. Breakdown of carbohydrates provides the energy that sustains animal life. In addition, carbohydrates are covalently linked with a variety of other molecules. Carbohydrates linked to lipid molecules, or **glycolipids,** are common components of biological membranes. Proteins that have covalently linked carbohydrates are called **glycoproteins.** These two classes of biomolecules, together called **glycoconjugates,** are important components of cell walls and extracellular structures in plants, animals, and bacteria. In addition to the structural roles such molecules play, they also serve in a variety of processes involving *recognition* between cell types or recognition of cellular structures by other molecules. Recognition events are important in normal cell growth, fertilization, transformation of cells, and other processes.

All of these functions are made possible by the characteristic chemical features of carbohydrates: (1) the existence of at least one and often two or more asymmetric centers, (2) the ability to exist either in linear or ring structures, (3) the capacity to form polymeric structures via *glycosidic* bonds, and (4) the potential to form multiple hydrogen bonds with water or other molecules in their environment.

Key Questions

7.1   How Are Carbohydrates Named?
7.2   What Is the Structure and Chemistry of Monosaccharides?
7.3   What Is the Structure and Chemistry of Oligosaccharides?
7.4   What Is the Structure and Chemistry of Polysaccharides?
7.5   What Are Glycoproteins, and How Do They Function in Cells?
7.6   How Do Proteoglycans Modulate Processes in Cells and Organisms?

7.1 | How Are Carbohydrates Named?

Carbohydrates are generally classified into three groups: **monosaccharides** (and their derivatives), **oligosaccharides,** and **polysaccharides.** The monosaccharides are also called **simple sugars** and have the formula $(CH_2O)_n$. Monosaccharides cannot be broken down into smaller sugars under mild conditions. Oligosaccharides derive their name from the Greek word *oligo*, meaning "few," and consist of from two to ten simple sugar molecules. Disaccharides are common in nature, and trisaccharides also occur frequently. Four- to six-sugar-unit oligosaccharides are usually bound covalently to other molecules, including glycoproteins. As their name suggests, polysaccharides are polymers of the simple sugars and their derivatives. They may be either linear or branched

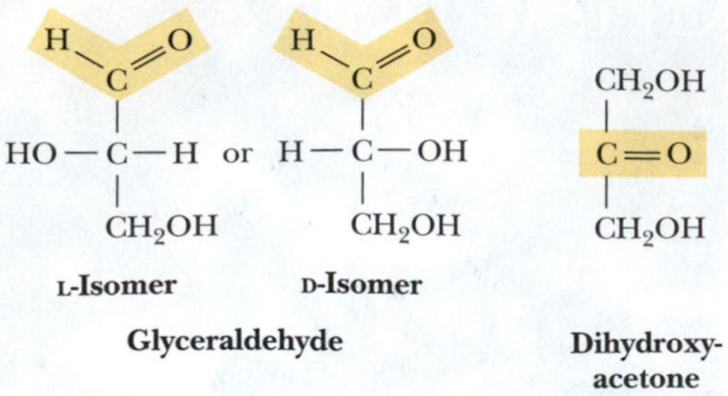

L-Isomer   D-Isomer

Glyceraldehyde

Dihydroxy-
acetone

FIGURE 7.1 Structure of a simple aldose (glyceraldehyde) and a simple ketose (dihydroxyacetone).

polymers and may contain hundreds or even thousands of monosaccharide units. Their molecular weights range up to 1 million or more.

7.2 | What Is the Structure and Chemistry of Monosaccharides?

Monosaccharides Are Classified as Aldoses and Ketoses

Monosaccharides consist typically of three to seven carbon atoms and are described either as **aldoses** or **ketoses**, depending on whether the molecule contains an aldehyde function or a ketone group. The simplest aldose is glyceraldehyde, and the simplest ketose is dihydroxyacetone (Figure 7.1). These two simple sugars are termed **trioses** because they each contain three carbon atoms. The structures and names of a family of aldoses and ketoses with three, four, five, and six carbons are shown in Figures 7.2 and 7.3. *Hexoses* are the most abundant sugars in nature. Nevertheless, sugars from all these classes are important in metabolism.

Monosaccharides, either aldoses or ketoses, are often given more detailed generic names to describe both the important functional groups and the total number of carbon atoms. Thus, one can refer to *aldotetroses* and *ketotetroses*, *aldopentoses* and *ketopentoses*, *aldohexoses* and *ketohexoses*, and so on. Sometimes the ketone-containing monosaccharides are named simply by inserting the letters -ul- into the simple generic terms, such as *tetruloses, pentuloses, hexuloses, heptuloses,* and so on. The simplest monosaccharides are water soluble, and most taste sweet.

Stereochemistry Is a Prominent Feature of Monosaccharides

Aldoses with at least three carbons and ketoses with at least four carbons contain **chiral centers** (see Chapter 4). The nomenclature for such molecules must specify the **configuration** about each asymmetric center, and drawings of these molecules must be based on a system that clearly specifies these configurations. As noted in Chapter 4, the **Fischer projection** system is used almost universally for this purpose today. The structures shown in Figures 7.2 and 7.3 are Fischer projections. For monosaccharides with two or more asymmetric carbons, the prefix D or L refers to the configuration of the highest numbered asymmetric carbon (the asymmetric carbon farthest from the carbonyl carbon). A monosaccharide is designated D if the hydroxyl group on the highest numbered asymmetric carbon is drawn to the right in a Fischer projection, as in D-glyceraldehyde (Figure 7.1). Note that the designation D or L merely relates the configuration of a given molecule to that of glyceraldehyde and does *not* specify the sign of rotation of plane-polarized light. If the sign of optical rotation is to be specified in the name, the Fischer convention of D or L designations may be used along with a + (plus) or − (minus) sign. Thus, D-glucose (Figure 7.2) may also be called D(+)-glucose because it is dextrorotatory, whereas D-fructose (Figure 7.3), which is levorotatory, can also be named D(−)-fructose.

All of the structures shown in Figures 7.2 and 7.3 are D-configurations, and the D-forms of monosaccharides predominate in nature, just as L-amino acids do. These preferences, established in apparently random choices early in evolution, persist uniformly in nature because of the stereospecificity of the enzymes that synthesize and metabolize these small molecules. L-Monosaccharides do exist in nature, serving a few relatively specialized roles. L-Galactose is a constituent of certain polysaccharides, and L-arabinose is a constituent of bacterial cell walls.

According to convention, the D- and L-forms of a monosaccharide are *mirror images* of each other, as shown in Figure 7.4 for fructose. Stereoisomers that are

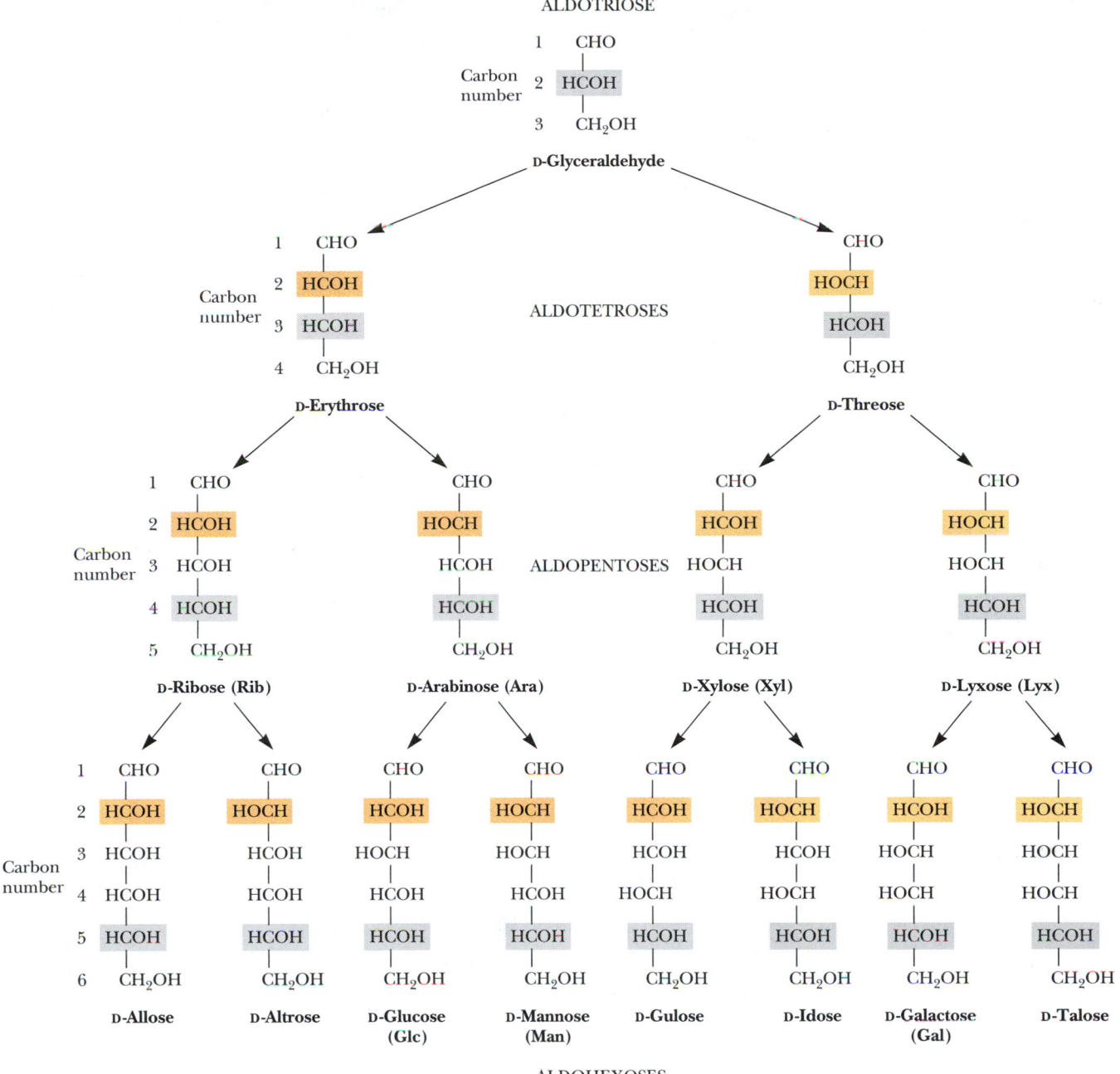

FIGURE 7.2 The structure and stereochemical relationships of D-aldoses with three to six carbons. The configuration in each case is determined by the highest numbered asymmetric carbon (shown in gray). In each row, the "new" asymmetric carbon is shown in yellow.

mirror images of each other are called **enantiomers,** or sometimes *enantiomeric pairs*. For molecules that possess two or more chiral centers, more than two stereoisomers can exist. Pairs of isomers that have opposite configurations at one or more of the chiral centers but that are not mirror images of each other are called **diastereomers** or *diastereomeric pairs*. Any two structures in a given row in Figures 7.2 and 7.3 are diastereomeric pairs. Two sugars that differ in configuration at only one chiral center are described as **epimers.** For example, D-mannose and D-talose are epimers and D-glucose and D-mannose are epimers, whereas D-glucose and D-talose are *not* epimers but merely diastereomers.

Biochemistry ⊜ Now™   Go to BiochemistryNow and click BiochemistryInteractive to learn how to identify the structures of simple sugars.

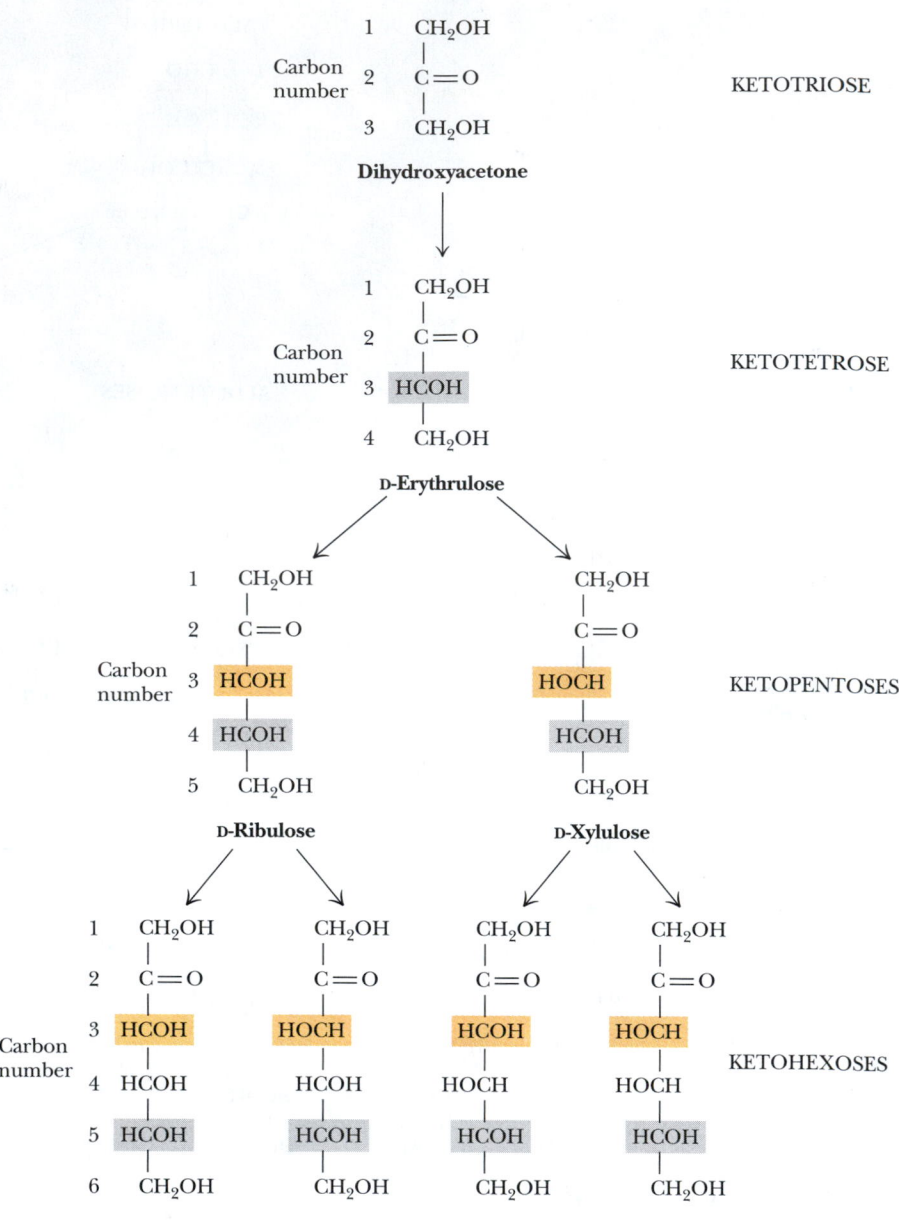

The structure and stereochemical relationships of
D-ketoses with three to six carbons. The configuration in each case is determined by the highest numbered asymmetric carbon (shown in gray). In each
row, the "new" asymmetric carbon is shown in yellow.

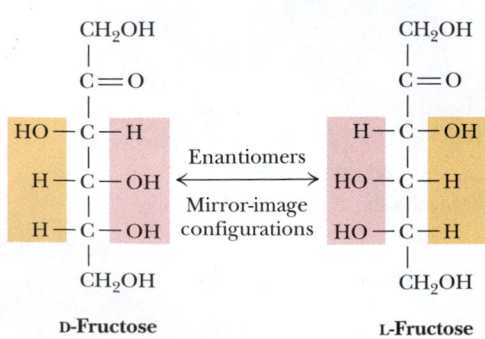

FIGURE 7.4 D-Fructose and L-fructose, an enantiomeric pair. Note that changing the configuration
only at C_5 would change D-fructose to L-sorbose.

Monosaccharides Exist in Cyclic and Anomeric Forms

Although Fischer projections are useful for presenting the structures of particular monosaccharides and their stereoisomers, they ignore one of the most
interesting facets of sugar structure—*the ability to form cyclic structures with formation of an additional asymmetric center.* Alcohols react readily with aldehydes to
form **hemiacetals** (Figure 7.5). The British carbohydrate chemist Sir Norman
Haworth showed that the linear form of glucose (and other aldohexoses) could
undergo a similar *intramolecular* reaction to form a *cyclic hemiacetal.* The resulting six-membered, oxygen-containing ring is similar to *pyran* and is designated
a **pyranose.** The reaction is catalyzed by acid (H^+) or base (OH^-) and is readily reversible.

In a similar manner, ketones can react with alcohols to form **hemiketals.** The
analogous intramolecular reaction of a ketose sugar such as fructose yields a
cyclic hemiketal (Figure 7.6). The five-membered ring thus formed is reminiscent
of *furan* and is referred to as a **furanose.** The cyclic pyranose and furanose
forms are the preferred structures for monosaccharides in aqueous solution. At

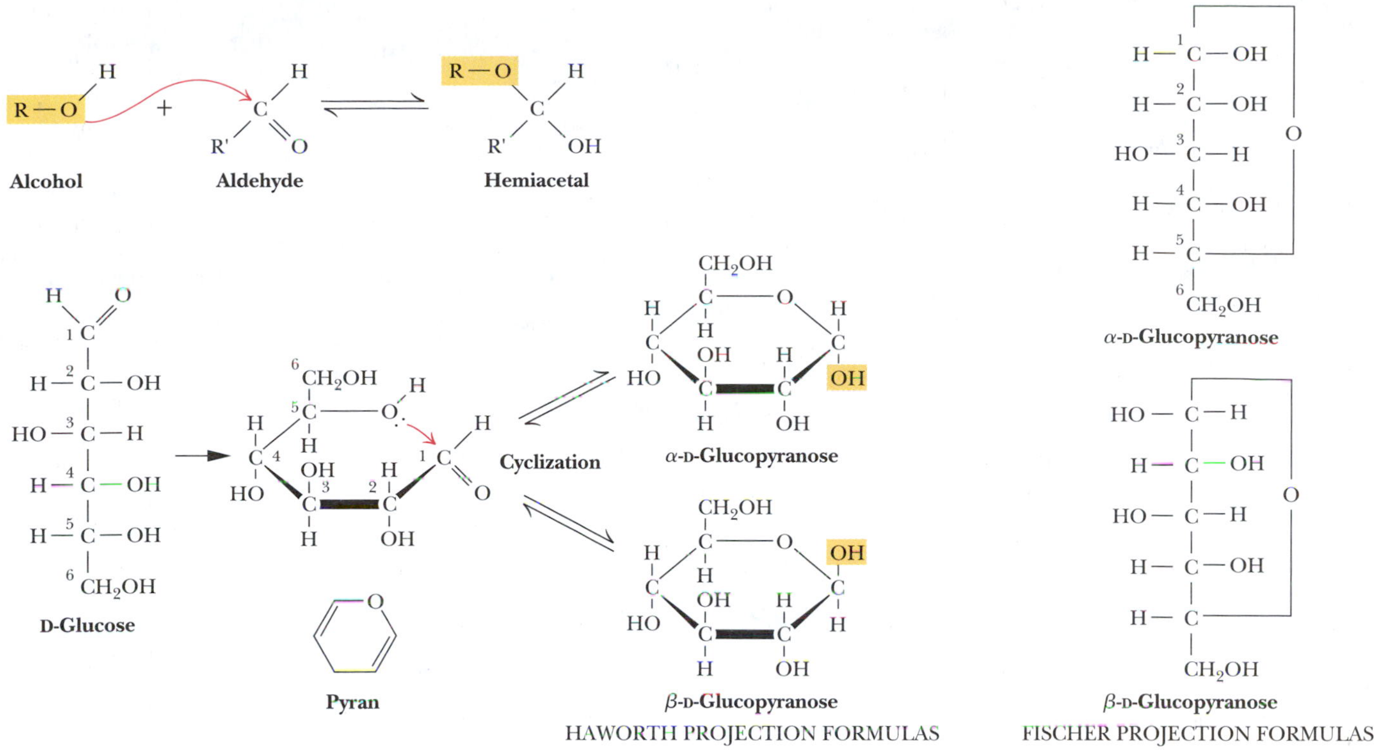

Alcohol   Aldehyde   Hemiacetal

D-Glucose

Pyran

Cyclization

α-D-Glucopyranose

β-D-Glucopyranose

HAWORTH PROJECTION FORMULAS

α-D-Glucopyranose

β-D-Glucopyranose

FISCHER PROJECTION FORMULAS

Biochemistry Now™ ANIMATED FIGURE 7.5 The linear form of D-glucose undergoes an intramolecular reaction to form a cyclic hemiacetal. **See this figure animated at http://chemistry.brookscole.com/ggb3**

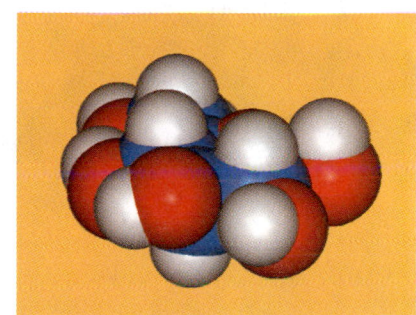

β-D-Glucopyranose

equilibrium, the linear aldehyde or ketone structure is only a minor component of the mixture (generally much less than 1%).

When hemiacetals and hemiketals are formed, the carbon atom that carried the carbonyl function becomes an asymmetric carbon atom. Isomers of monosaccharides that differ only in their configuration about that carbon atom are called **anomers,** designated as α or β, as shown in Figure 7.5, and the carbonyl carbon is thus called the **anomeric carbon.** When the hydroxyl group at the anomeric carbon is on the *same side* of a Fischer projection as the oxygen atom at the highest numbered asymmetric carbon, the configuration at the anomeric carbon is α, as in α-D-glucose. When the anomeric hydroxyl is on the *opposite side* of the Fischer projection, the configuration is β, as in β-D-glucopyranose (Figure 7.5).

The addition of this asymmetric center upon hemiacetal and hemiketal formation alters the optical rotation properties of monosaccharides, and the original assignment of the α and β notations arose from studies of these properties. Early carbohydrate chemists frequently observed that the optical rotation of glucose (and other sugar) solutions could change with time, a process called **mutarotation.** This indicated that a structural change was occurring. It was eventually found that α-D-glucose has a specific optical rotation, $[\alpha]_D^{20}$, of 112.2°, and that β-D-glucose has a specific optical rotation of 18.7°. Mutarotation involves interconversion of α- and β-forms of the monosaccharide with intermediate formation of the linear aldehyde or ketone, as shown in Figures 7.5 and 7.6.

Haworth Projections Are a Convenient Device for Drawing Sugars

Another of Haworth's lasting contributions to the field of carbohydrate chemistry was his proposal to represent pyranose and furanose structures as hexagonal and pentagonal rings lying perpendicular to the plane of the

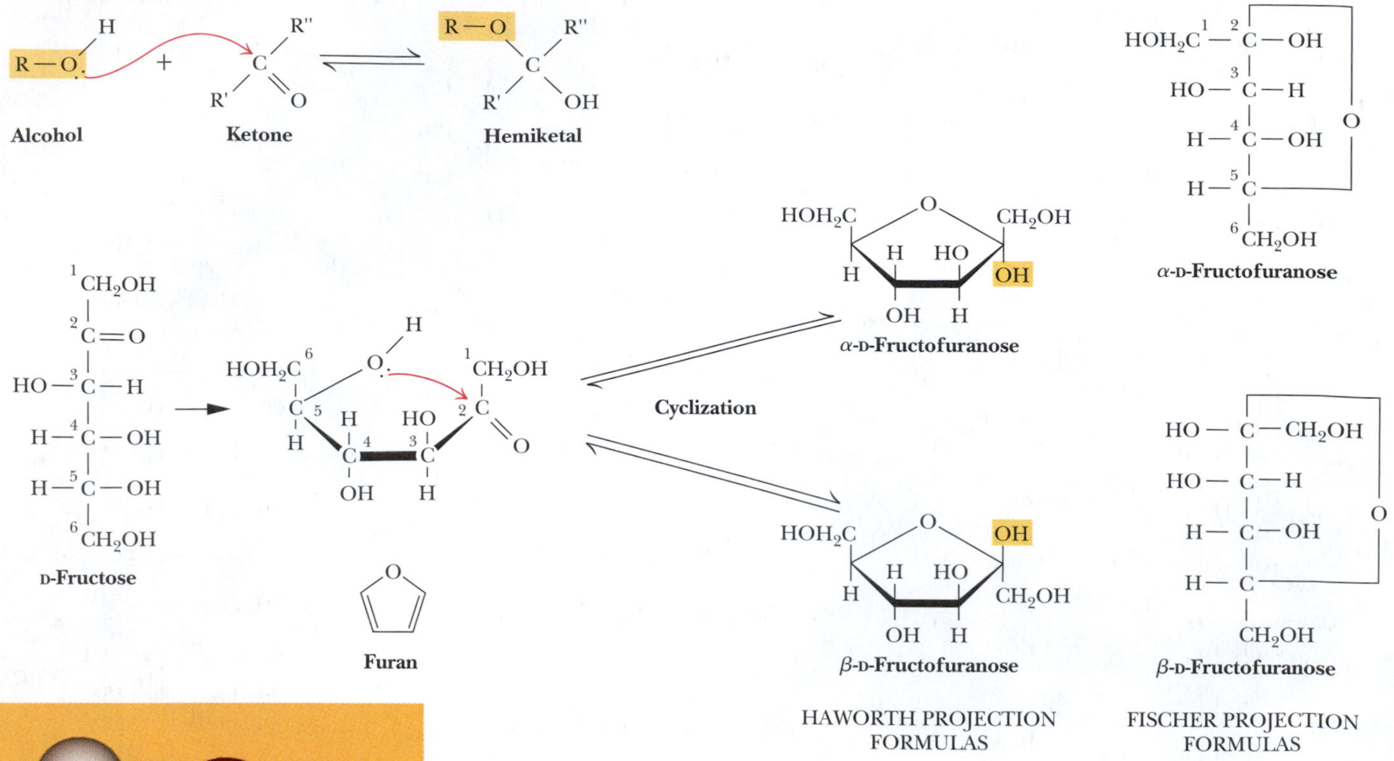

Alcohol Ketone Hemiketal

D-Fructose Furan

Cyclization

α-D-Fructofuranose

β-D-Fructofuranose

α-D-Fructofuranose

β-D-Fructofuranose

HAWORTH PROJECTION FORMULAS

FISCHER PROJECTION FORMULAS

Biochemistry Now™ ANIMATED FIGURE 7.6 The linear form of D-fructose undergoes an intramolecular reaction to form a cyclic hemiketal. **See this figure animated at http://chemistry. brookscole.com/ggb3**

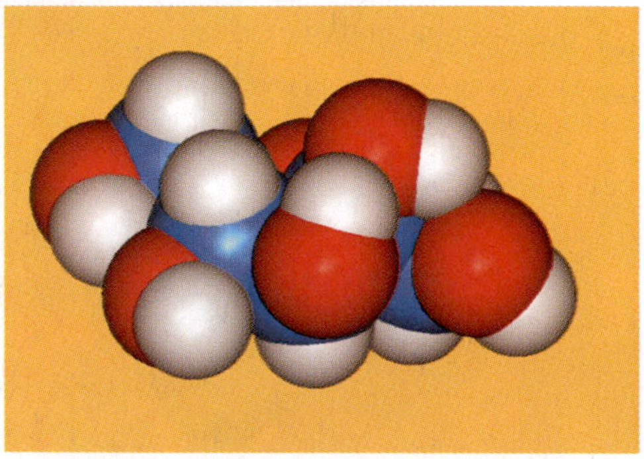

β-D-Fructofuranose

paper, with thickened lines indicating the side of the ring closest to the reader. Such **Haworth projections,** which are now widely used to represent saccharide structures (Figures 7.5 and 7.6), show substituent groups extending either above or below the ring. Substituents drawn to the left in a Fischer projection are drawn above the ring in the corresponding Haworth projection. Substituents drawn to the right in a Fischer projection are below the ring in a Haworth projection. Exceptions to these rules occur in the formation of furanose forms of pentoses and the formation of furanose or pyranose forms of hexoses. In these cases, the structure must be redrawn with a rotation about the carbon whose hydroxyl group is involved in the formation of the cyclic form (Figures 7.7 and 7.8) in order to orient the appropriate hydroxyl group for ring formation. This is merely for illustrative purposes and involves no change in configuration of the saccharide molecule.

The rules previously mentioned for assignment of α- and β-configurations can be readily applied to Haworth projection formulas. For the D-sugars, the anomeric hydroxyl group is below the ring in the α-anomer and above the ring in the β-anomer. For L-sugars, the opposite relationship holds.

As Figures 7.7 and 7.8 imply, in most monosaccharides there are two or more hydroxyl groups that can react with an aldehyde or ketone at the other end of the molecule to form a hemiacetal or hemiketal. Consider the possibilities for glucose, as shown in Figure 7.7. If the C-4 hydroxyl group reacts with the aldehyde of glucose, a five-membered ring is formed, whereas if the C-5 hydroxyl reacts, a six-membered ring is formed. The C-6 hydroxyl does not react effectively because a seven-membered ring is too strained to form a stable hemiacetal. The same is true for the C-2 and C-3 hydroxyls, and thus five- and six-membered rings are by far the most likely to be formed from six-membered monosaccharides. D-Ribose, with five carbons, readily forms either

FIGURE 7.7 D-Glucose can cyclize in two ways, forming either furanose or pyranose structures.

five-membered rings (α- or β-D-ribofuranose) or six-membered rings (α- or β-D-ribopyranose) (Figure 7.8). In general, aldoses and ketoses with five or more carbons can form *either* furanose or pyranose rings, and the more stable form depends on structural factors. The nature of the substituent groups on the carbonyl and hydroxyl groups and the configuration about the asymmetric carbon will determine whether a given monosaccharide prefers the pyranose or furanose structure. In general, the pyranose form is favored over the furanose ring for aldohexose sugars, although, as we shall see, furanose structures are more stable for ketohexoses.

Although Haworth projections are convenient for displaying monosaccharide structures, they do not accurately portray the conformations of pyranose and furanose rings. Given C—C—C tetrahedral bond angles of 109° and C—O—C angles of 111°, neither pyranose nor furanose rings can adopt true planar structures. Instead, they take on puckered conformations, and in the case of pyranose rings, the two favored structures are the **chair conformation** and the **boat conformation,** shown in Figure 7.9. Note that the ring substituents in these structures can be **equatorial,** which means approximately coplanar with the ring, or **axial,** that is, parallel to an axis drawn through the ring as shown. Two general rules dictate the conformation to be adopted by a

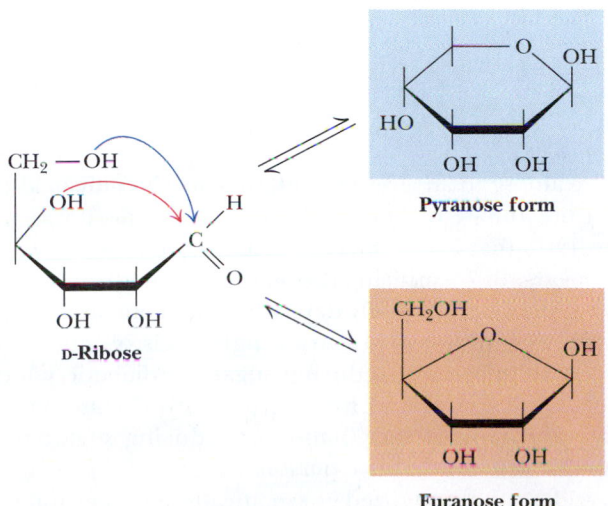

Biochemistry ⓔNow™   ANIMATED FIGURE 7.8
D-Ribose and other five-carbon saccharides can form either furanose or pyranose structures. **See this figure animated at http://chemistry.brookscole. com/ggb3**

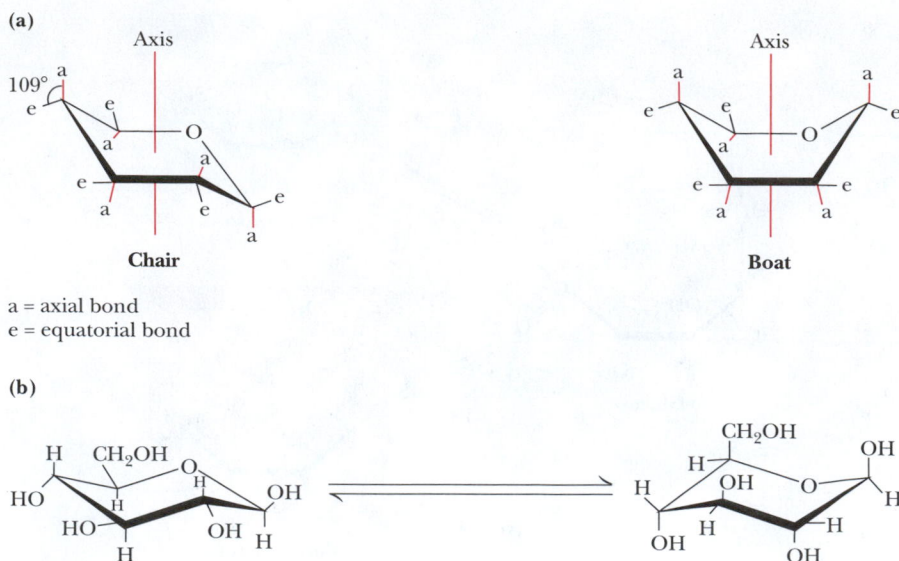

FIGURE 7.9 **(a)** Chair and boat conformations of a pyranose sugar. **(b)** Two possible chair conformations of β-D-glucose.

given saccharide unit. First, bulky substituent groups on such rings are more stable when they occupy equatorial positions rather than axial positions, and second, chair conformations are slightly more stable than boat conformations. For a typical pyranose, such as β-D-glucose, there are two possible chair conformations (Figure 7.9). Of all the D-aldohexoses, β-D-glucose is the only one that can adopt a conformation with all its bulky groups in an equatorial position. With this advantage of stability, it may come as no surprise that β-D-glucose is the most widely occurring organic group in nature and the central hexose in carbohydrate metabolism.

Monosaccharides Can Be Converted to Several Derivative Forms

A variety of chemical and enzymatic reactions produce **derivatives** of the simple sugars. These modifications produce a diverse array of saccharide derivatives. Some of the most common derivations are discussed here.

Sugar Acids Sugars with free anomeric carbon atoms are reasonably good reducing agents and will reduce hydrogen peroxide, ferricyanide, certain metals (Cu^{2+} and Ag^+), and other oxidizing agents. Such reactions convert the sugar to a **sugar acid.** For example, addition of alkaline $CuSO_4$ (called *Fehling's solution*) to an aldose sugar produces a red cuprous oxide (Cu_2O) precipitate:

$$\underset{\textbf{Aldehyde}}{R\overset{\displaystyle O}{\overset{\displaystyle \|}{C}}-H} + 2\ Cu^{2+} + 5\ OH^- \longrightarrow \underset{\textbf{Carboxylate}}{R\overset{\displaystyle O}{\overset{\displaystyle \|}{C}}-O^-} + Cu_2O\downarrow + 3\ H_2O$$

and converts the aldose to an **aldonic acid,** such **as gluconic acid** (Figure 7.10). Formation of a precipitate of red Cu_2O constitutes a positive test for an aldehyde. Carbohydrates that can reduce oxidizing agents in this way are referred to as **reducing sugars.** By quantifying the amount of oxidizing agent reduced by a sugar solution, one can accurately determine the concentration of the sugar. *Diabetes mellitus* is a condition that causes high levels of glucose in urine and blood, and frequent analysis of reducing sugars in diabetic patients is an important part of the diagnosis and treatment of this disease. Over-the-counter kits for the easy and rapid determination of reducing sugars have made this procedure a simple one for diabetic persons.

Monosaccharides can be oxidized enzymatically at C-6, yielding **uronic acids,** such as D-glucuronic and L-iduronic acids (Figure 7.10). L-Iduronic acid is sim-

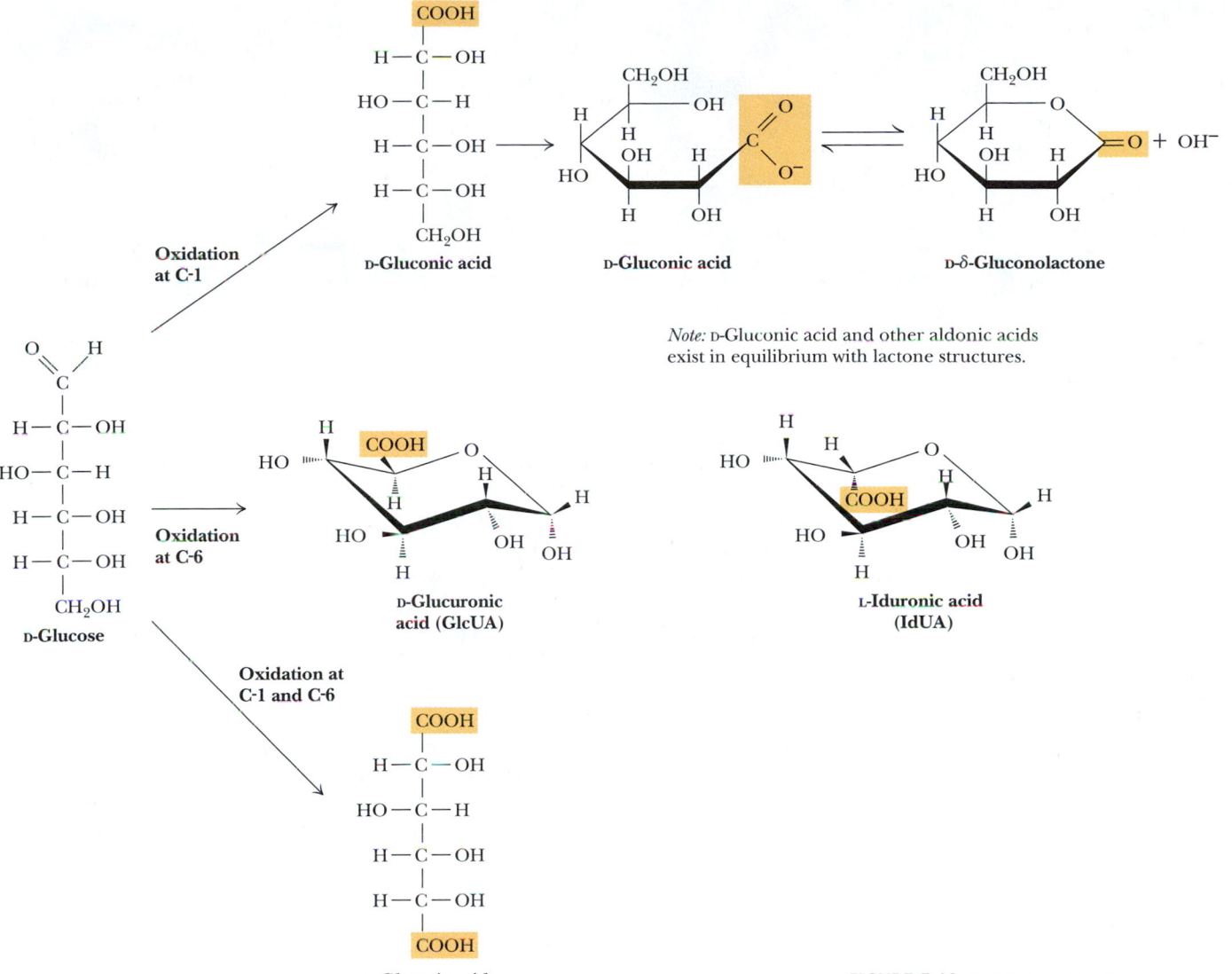

Note: D-Gluconic acid and other aldonic acids exist in equilibrium with lactone structures.

FIGURE 7.10 Oxidation of D-glucose to sugar acids.

ilar to D-glucuronic acid, except it has an opposite configuration at C-5. Oxidation at both C-1 and C-6 produces **aldaric acids,** such as D-glucaric acid.

Sugar Alcohols   **Sugar alcohols,** another class of sugar derivative, can be prepared by the mild reduction (with $NaBH_4$ or similar agents) of the carbonyl groups of aldoses and ketoses. Sugar alcohols, or **alditols,** are designated by the addition of *-itol* to the name of the parent sugar (Figure 7.11). The alditols are linear molecules that cannot cyclize in the manner of aldoses. Nonetheless, alditols are characteristically sweet tasting, and **sorbitol, mannitol,** and **xylitol** are widely used to sweeten sugarless gum and mints. Sorbitol buildup in the eyes of diabetic persons is implicated in cataract formation. **Glycerol** and *myo-inositol,* a cyclic alcohol, are components of lipids (see Chapter 8). There are nine different stereoisomers of inositol; the one shown in Figure 7.11 was first isolated from heart muscle and thus has the prefix *myo-* for muscle. **Ribitol** is a constituent of flavin coenzymes (see Chapter 17).

Deoxy Sugars   The **deoxy sugars** are monosaccharides with one or more hydroxyl groups replaced by hydrogens. 2-Deoxy-D-ribose (Figure 7.12), whose systematic name is 2-deoxy-D-erythropentose, is a constituent of DNA in all living things (see Chapter 10). Deoxy sugars also occur frequently in glycoproteins

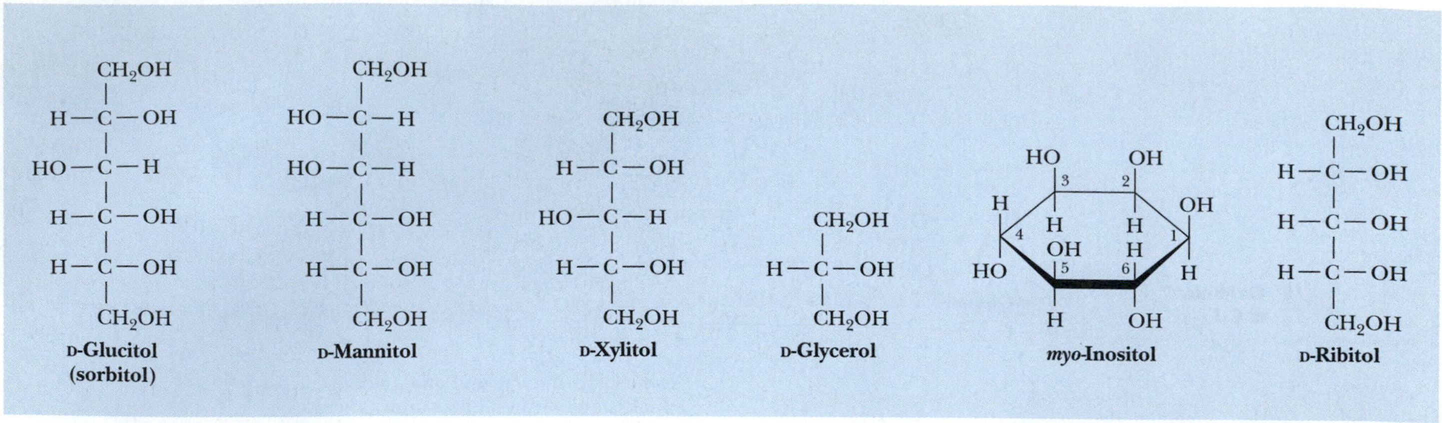

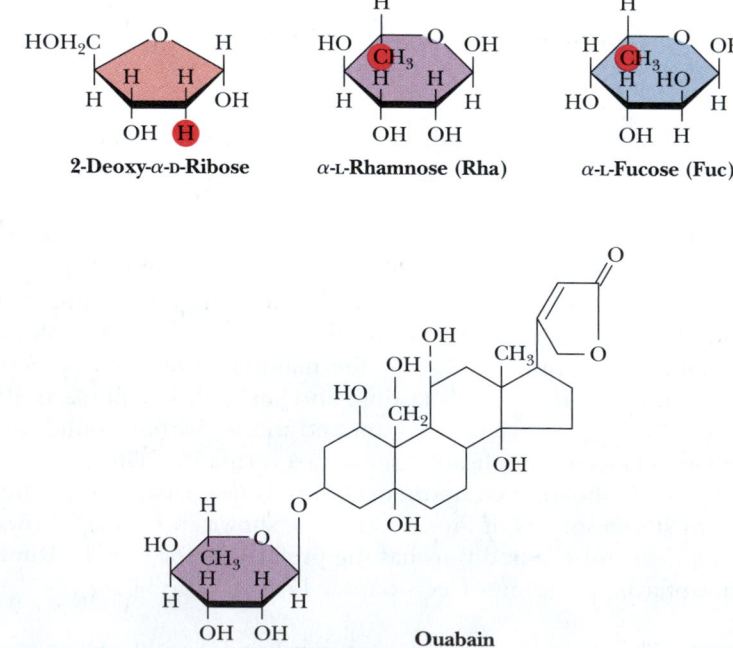

CH2OH
H—C—OH
HO—C—H
H—C—OH
H—C—OH
CH2OH
D-Glucitol (sorbitol)
D-Mannitol
D-Xylitol
D-Glycerol
myo-Inositol
D-Ribitol

FIGURE 7.11 Structures of some sugar alcohols.

and polysaccharides. L-Fucose and L-rhamnose, both 6-deoxy sugars, are components of some cell walls, and rhamnose is a component of **ouabain,** a highly toxic *cardiac glycoside* found in the bark and root of the ouabaio tree. Ouabain is used by the East African Somalis as an arrow poison. The sugar moiety is not the toxic part of the molecule (see Chapter 9).

Sugar Esters **Phosphate esters** of glucose, fructose, and other monosaccharides are important metabolic intermediates, and the ribose moiety of nucleotides such as ATP and GTP is phosphorylated at the 5′-position (Figure 7.13).

Amino Sugars **Amino sugars,** including D-glucosamine and D-galactosamine (Figure 7.14), contain an amino group (instead of a hydroxyl group) at the C-2 position. They are found in many oligosaccharides and polysaccharides, including *chitin,* a polysaccharide in the exoskeletons of crustaceans and insects.

2-Deoxy-α-D-Ribose
α-L-Rhamnose (Rha)
α-L-Fucose (Fuc)
Ouabain

FIGURE 7.12 Several deoxy sugars and ouabain, which contains α-L-rhamnose (Rha). Hydrogen atoms highlighted in red are "deoxy" positions.

α-D-Glucose-1-phosphate

α-D-Fructose-1,6-bisphosphate

Adenosine-5'-triphosphate

FIGURE 7.13 Several sugar esters important in metabolism.

A Deeper Look

Biochemistry ◉ Now™

Honey—An Ancestral Carbohydrate Treat

Honey, the first sweet known to humankind, is the only sweetening agent that can be stored and used exactly as produced in nature. Bees process the nectar of flowers so that their final product is able to survive long-term storage at ambient temperature. Used as a ceremonial material and medicinal agent in earliest times, honey was not regarded as a food until the Greeks and Romans. Only in modern times have cane and beet sugar surpassed honey as the most frequently used sweetener. What is the chemical nature of this magical, viscous substance?

The bees' processing of honey consists of (1) reducing the water content of the nectar (30% to 60%) to the self-preserving range of 15% to 19%, (2) hydrolyzing the significant amount of sucrose in nectar to glucose and fructose by the action of the enzyme invertase, and (3) producing small amounts of gluconic acid from glucose by the action of the enzyme **glucose oxidase.** Most of the sugar in the final product is glucose and fructose, and the final product is supersaturated with respect to these monosaccharides. Honey actually consists of an emulsion of microscopic glucose hydrate and fructose hydrate crystals in a thick syrup. Sucrose accounts for only about 1% of the sugar in the final product, with fructose at about 38% and glucose at 31% by weight.

The accompanying figure shows a ^{13}C nuclear magnetic resonance spectrum of honey from a mixture of wildflowers in southeastern Pennsylvania. Interestingly, five major hexose species contribute to this spectrum. Although most textbooks show fructose exclusively in its furanose form, the predominant form of fructose (67% of total fructose) is β-D-fructopyranose, with the β- and α-fructofuranose forms accounting for 27% and 6% of the fructose, respectively. In polysaccharides, fructose invariably prefers the furanose form, but free fructose (and crystalline fructose) is predominantly β-fructopyranose.

Sources: White, J. W., 1978. Honey. *Advances in Food Research* **24**:287–374; and Prince, R. C., Gunson, D. E., Leigh, J. S., and McDonald, G. G., 1982. The predominant form of fructose is a pyranose, not a furanose ring. *Trends in Biochemical Sciences* **7**:239–240.

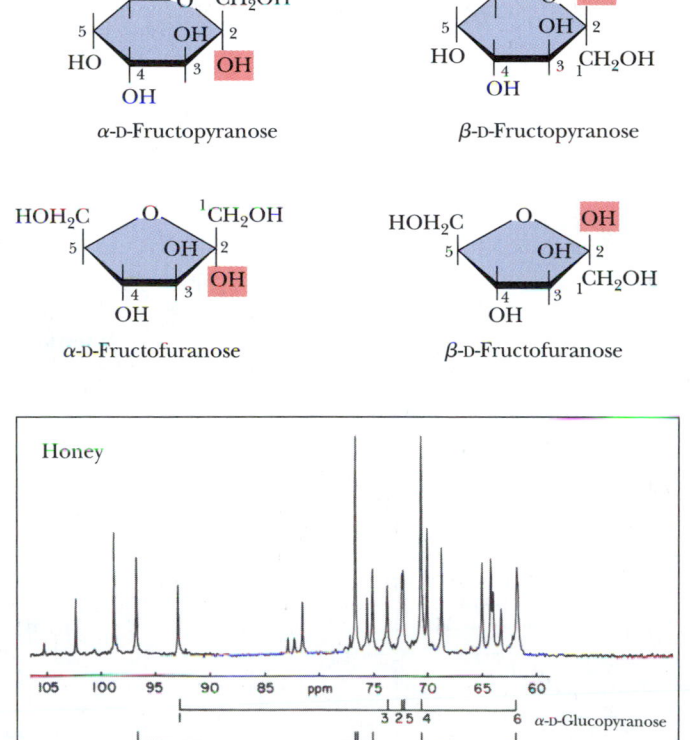

α-D-Fructopyranose

β-D-Fructopyranose

α-D-Fructofuranose

β-D-Fructofuranose

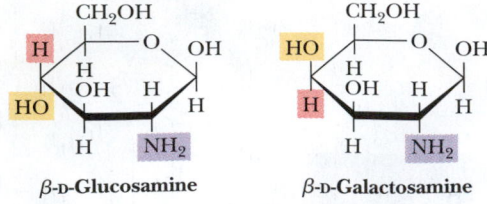

FIGURE 7.14 Structures of D-glucosamine and D-galactosamine.

Muramic acid and **neuraminic acid,** which are components of the polysaccharides of cell membranes of higher organisms and also bacterial cell walls, are glucosamines linked to three-carbon acids at the C-1 or C-3 positions. In muramic acid (thus named as an *amine* isolated from bacterial cell wall polysaccharides; *murus* is Latin for "wall"), the hydroxyl group of a lactic acid moiety makes an ether linkage to the C-3 of glucosamine. Neuraminic acid (an *amine* isolated from *neural* tissue) forms a C—C bond between the C-1 of *N*-acetylmannosamine and the C-3 of pyruvic acid (Figure 7.15). The *N*-acetyl and *N*-glycolyl derivatives of neuraminic acid are collectively known as **sialic acids** and are distributed widely in bacteria and animal systems.

Acetals, Ketals, and Glycosides Hemiacetals and hemiketals can react with alcohols in the presence of acid to form **acetals** and **ketals,** as shown in Figure 7.16. This reaction is another example of a *dehydration synthesis* and is similar in this respect to the reactions undergone by amino acids to form peptides and nucleotides to form nucleic acids. The pyranose and furanose forms of monosaccharides react with alcohols in this way to form **glycosides** with retention of the α- or β-configuration at the C-1 carbon. The new bond between the anomeric carbon atom and the oxygen atom of the alcohol is called a

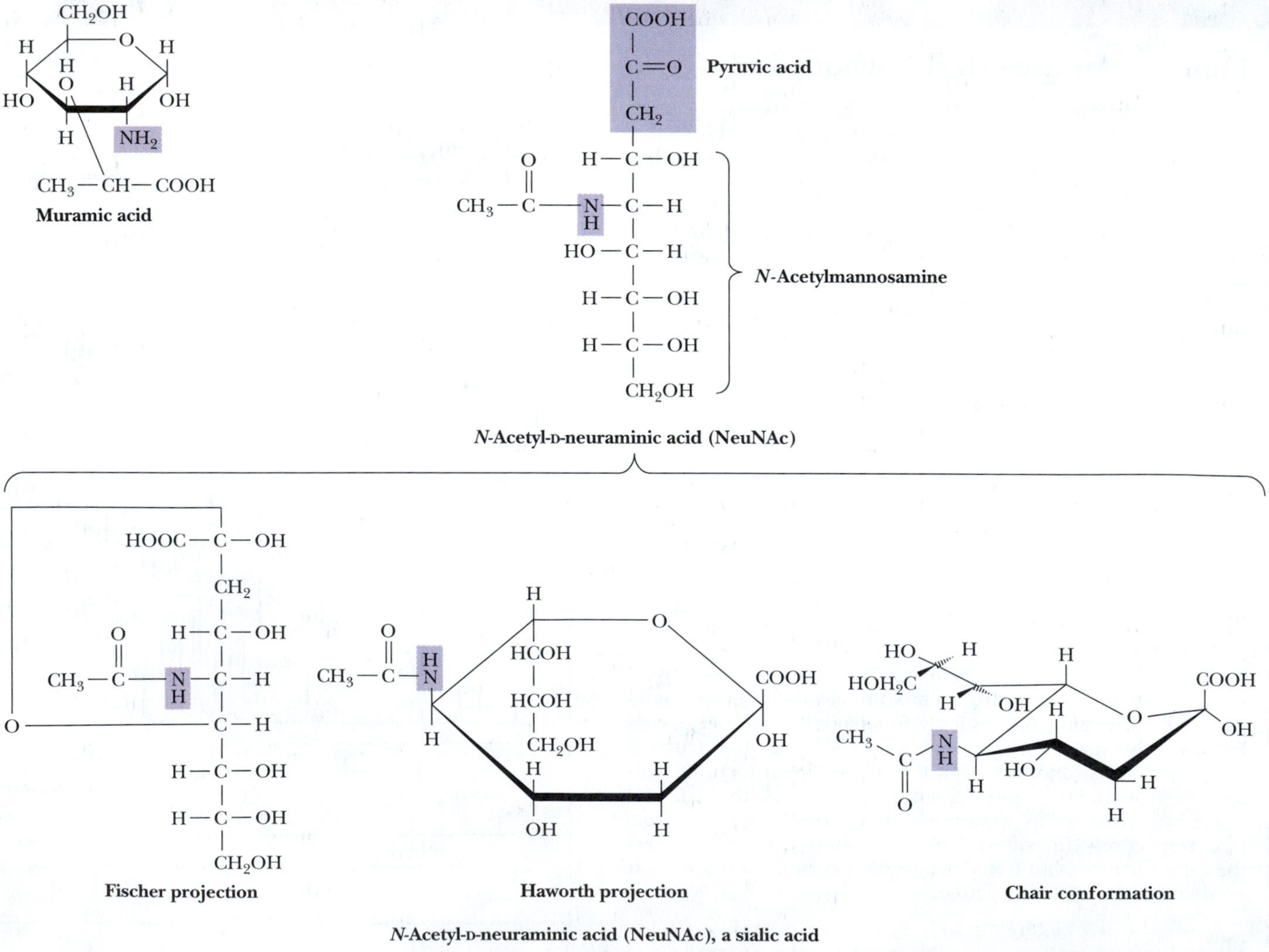

Muramic acid

N-Acetyl-D-neuraminic acid (NeuNAc)

Fischer projection

Haworth projection

Chair conformation

N-Acetyl-D-neuraminic acid (NeuNAc), a sialic acid

FIGURE 7.15 Structures of muramic acid and neuraminic acid and several depictions of sialic acid.

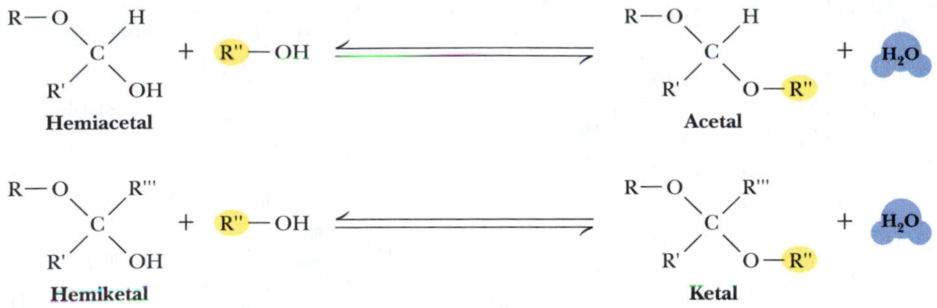

Hemiacetal + R″—OH ⇌ **Acetal** + H_2O

Hemiketal + R″—OH ⇌ **Ketal** + H_2O

FIGURE 7.16 Acetals and ketals can be formed from hemiacetals and hemiketals, respectively.

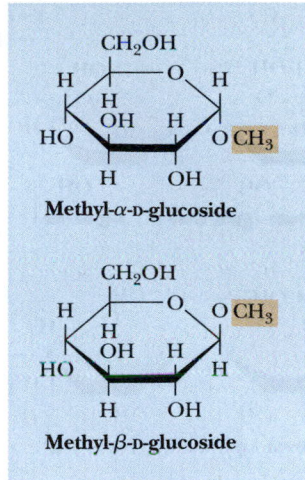

Methyl-α-D-glucoside

Methyl-β-D-glucoside

FIGURE 7.17 The anomeric forms of methyl-D-glucoside.

glycosidic bond. Glycosides are named according to the parent monosaccharide. For example, *methyl-β-D-glucoside* (Figure 7.17) can be considered a derivative of β-D-glucose.

7.3 What Is the Structure and Chemistry of Oligosaccharides?

Given the relative complexity of oligosaccharides and polysaccharides in higher organisms, it is perhaps surprising that these molecules are formed from relatively few different monosaccharide units. (In this respect, the oligosaccharides and polysaccharides are similar to proteins; both form complicated structures based on a small number of different building blocks.) Monosaccharide units include the hexoses glucose, fructose, mannose, and galactose and the pentoses ribose and xylose.

Disaccharides Are the Simplest Oligosaccharides

The simplest oligosaccharides are the **disaccharides,** which consist of two monosaccharide units linked by a glycosidic bond. As in proteins and nucleic acids, each individual unit in an oligosaccharide is termed a *residue*. The disaccharides shown in Figure 7.18 are all commonly found in nature, with sucrose, maltose, and lactose being the most common. Each is a mixed acetal, with one hydroxyl group provided intramolecularly and one hydroxyl from the other monosaccharide. Except for sucrose, each of these structures possesses one free unsubstituted anomeric carbon atom, and thus each of these disaccharides is a reducing sugar. The end of the molecule containing the free anomeric carbon is called the **reducing end,** and the other end is called the **nonreducing end.** In the case of sucrose, both of the anomeric carbon atoms are substituted, that is, neither has a free —OH group. The substituted anomeric carbons cannot be converted to the aldehyde configuration and thus cannot participate in the oxidation–reduction reactions characteristic of reducing sugars. Thus, sucrose is *not* a reducing sugar.

Maltose, isomaltose, and cellobiose are all **homodisaccharides** because they each contain only one kind of monosaccharide, namely, glucose. **Maltose** is produced from starch (a polymer of α-D-glucose produced by plants) by the action of amylase enzymes and is a component of malt, a substance obtained by allowing grain (particularly barley) to soften in water and germinate. The enzyme **diastase,** produced during the germination process, catalyzes the hydrolysis of starch to maltose. Maltose is used in beverages (malted milk, for example), and because it is fermented readily by yeast, it is important in the brewing of beer. In both maltose and cellobiose, the glucose units are **1→4 linked,** meaning that the C-1 of one glucose is linked by a glycosidic bond to

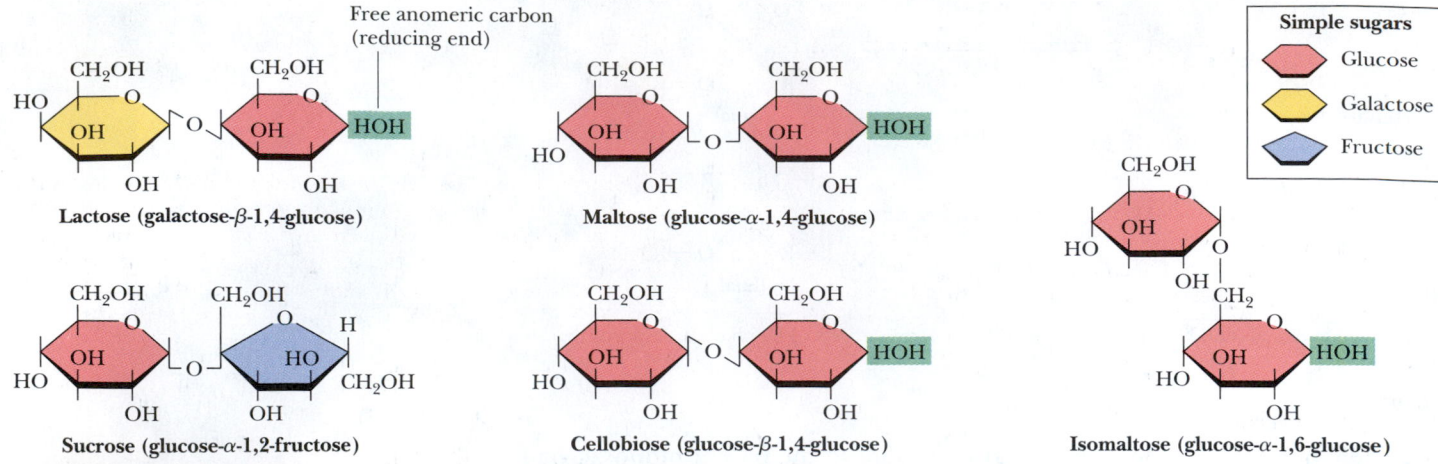

Free anomeric carbon
(reducing end)

Lactose (galactose-β-1,4-glucose)

Maltose (glucose-α-1,4-glucose)

Simple sugars

- Glucose
- Galactose
- Fructose

Sucrose (glucose-α-1,2-fructose)

Cellobiose (glucose-β-1,4-glucose)

Isomaltose (glucose-α-1,6-glucose)

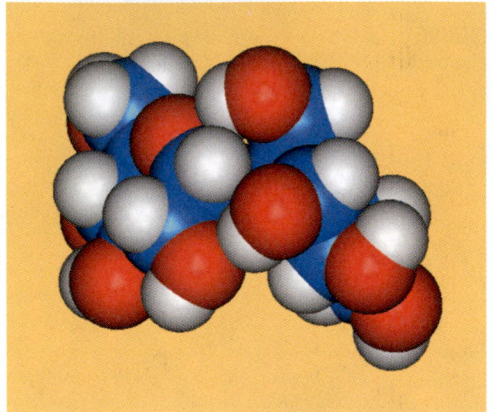

Sucrose

Biochemistry ⊜ Now™   ACTIVE FIGURE 7.18 The structures of several important disaccharides. Note that the notation —HOH means that the configuration can be either α or β. If the —OH group is above the ring, the configuration is termed β. The configuration is α if the —OH group is below the ring. Also note that sucrose has no free anomeric carbon atoms. **Test yourself on the concepts in this figure at http://chemistry.brookscole.com/ggb3**

the C-4 oxygen of the other glucose. The only difference between them is in the configuration at the glycosidic bond. Maltose exists in the α-configuration, whereas cellobiose is a β-configuration. **Isomaltose** is obtained in the hydrolysis of some polysaccharides (such as dextran), and **cellobiose** is obtained from the acid hydrolysis of cellulose. Isomaltose also consists of two glucose units in a glycosidic bond, but in this case, C-1 of one glucose is linked to C-6 of the other, and the configuration is α.

The complete structures of these disaccharides can be specified in shorthand notation by using abbreviations for each monosaccharide, α or β, to denote configuration, and appropriate numbers to indicate the nature of the linkage. Thus, cellobiose is Glcβ1–4Glc, whereas isomaltose is Glcα1–6Glc. Often the glycosidic linkage is written with an arrow so that cellobiose and isomaltose would be Glcβ1→4Glc and Glcα1→6Glc, respectively. Because the linkage carbon on the first sugar is always C-1, a newer trend is to drop the 1– or 1→ and describe these simply as Glcβ4Glc and Glcα6Glc, respectively. More complete names can also be used, however; for example, maltose would be O-α-D-glucopyranosyl-(1→4)-D-glucopyranose. Cellobiose, because of its β-glycosidic linkage, is formally O-β-D-glucopyranosyl-(1→4)-D-glucopyranose.

β-D-Lactose (O-β-D-galactopyranosyl-(1→4)-D-glucopyranose) (Figure 7.18) is the principal carbohydrate in milk and is of critical nutritional importance to mammals in the early stages of their lives. It is formed from D-galactose and D-glucose via a β(1→4) link, and because it has a free anomeric carbon, it is capable of mutarotation and is a reducing sugar. It is an interesting quirk of nature that lactose cannot be absorbed directly into the bloodstream. It must first be broken down into galactose and glucose by **lactase,** an intestinal enzyme that exists in young, nursing mammals but is not produced in significant quantities in the mature mammal. Most humans, with the exception of certain groups in Africa and northern Europe, produce only low levels of lactase. For most individuals, this is not a problem, but some cannot tolerate lactose and experience intestinal pain and diarrhea upon consumption of milk.

Sucrose, in contrast, is a disaccharide of almost universal appeal and tolerance. Produced by many higher plants and commonly known as *table sugar,* it is one of the products of photosynthesis and is composed of fructose and glucose.

A Deeper Look

Trehalose—A Natural Protectant for Bugs

Insects use an open circulatory system to circulate **hemolymph** (insect blood). The "blood sugar" is not glucose but rather **trehalose,** an unusual, nonreducing disaccharide (see figure). Trehalose is found typically in organisms that are naturally subject to temperature variations and other environmental stresses—bacterial spores, fungi, yeast, and many insects. (Interestingly, honeybees do not have trehalose in their hemolymph, perhaps because they practice a colonial, rather than solitary, lifestyle. Bee colonies maintain a rather constant temperature of 18°C, protecting the residents from large temperature changes.)

What might explain this correlation between trehalose utilization and environmentally stressful lifestyles? Konrad Bloch* suggests that trehalose may act as a natural cryoprotectant. Freezing and thawing of biological tissues frequently causes irreversible

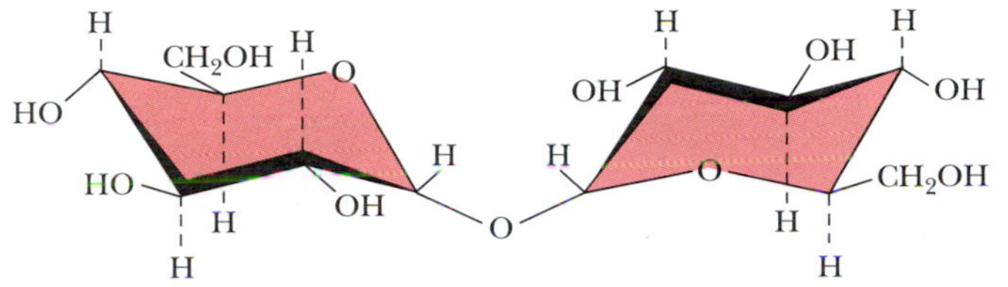

*Bloch, K., 1994. *Blondes in Venetian Paintings, the Nine-Banded Armadillo, and Other Essays in Biochemistry.* New Haven: Yale University Press.
†Attfield, P. A., 1987. Trehalose accumulates in *Saccharomyces cerevisiae* during exposure to agents that induce heat shock responses. *FEBS Letters* **225**:259.

structural changes, destroying biological activity. High concentrations of polyhydroxy compounds, such as sucrose and glycerol, can protect biological materials from such damage. Trehalose is particularly well suited for this purpose and has been shown to be superior to other polyhydroxy compounds, especially at low concentrations. Support for this novel idea comes from studies by P. A. Attfield,† which show that trehalose levels in the yeast *Saccharomyces cerevisiae* increase significantly during exposure to high salt and high growth temperatures—the same conditions that elicit the production of heat shock proteins!

Sucrose has a specific optical rotation, $[\alpha]_D^{20}$, of $+66.5°$, but an equimolar mixture of its component monosaccharides has a net negative rotation ($[\alpha]_D^{20}$ of glucose is $+52.5°$ and of fructose is $-92°$). Sucrose is hydrolyzed by the enzyme **invertase,** so named for the inversion of optical rotation accompanying this reaction. Sucrose is also easily hydrolyzed by dilute acid, apparently because the fructose in sucrose is in the relatively unstable furanose form. Although sucrose and maltose are important to the human diet, they are not taken up directly in the body. In a manner similar to lactose, they are first hydrolyzed by **sucrase** and **maltase,** respectively, in the human intestine.

A Variety of Higher Oligosaccharides Occur in Nature

In addition to the simple disaccharides, many other oligosaccharides are found in both prokaryotic and eukaryotic organisms, either as naturally occurring substances or as hydrolysis products of natural materials. Figure 7.19 lists a number of simple oligosaccharides, along with descriptions of their origins and interesting features. Several are constituents of the sweet nectars or saps exuded or extracted from plants and trees. One particularly interesting and useful group of oligosaccharides is the **cycloamyloses.** These oligosaccharides are cyclic structures, and in solution they form molecular "pockets" of various diameters. These pockets are surrounded by the chiral carbons of the saccharides themselves and are able to form stereospecific inclusion complexes with chiral molecules that can fit into the pockets. Thus, mixtures of stereoisomers of small organic molecules can be separated into pure isomers on columns of **cycloheptaamylose,** for example.

Stachyose is typical of the oligosaccharide components found in substantial quantities in beans, peas, bran, and whole grains. These oligosaccharides are not affected by digestive enzymes, but *are* metabolized readily by bacteria in the intestines. This is the source of the flatulence that often accompanies the consumption of such foods. Commercial products are now available that assist in the digestion of the gas-producing components of these foods. These products contain an enzyme that hydrolyzes the culprit oligosaccharides before they become available to intestinal microorganisms.

Melezitose (a constituent of honey)

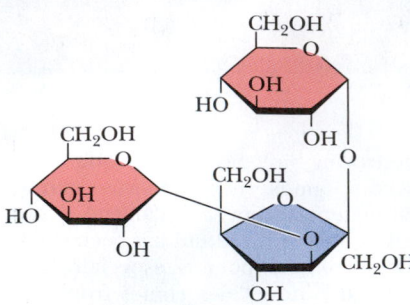

Amygdalin (occurs in seeds of *Rosaceae*; glycoside of bitter almonds, in kernels of cherries, peaches, apricots)

Laetrile (claimed to be an anticancer agent, but there is no scientific evidence for this)

Stachyose (a constituent of many plants: white jasmine, yellow lupine, soybeans, lentils, etc.; causes flatulence because humans cannot digest it)

Cycloheptaamylose (a breakdown product of starch; useful in chromatographic separations)

Dextrantriose (a constituent of saké and honeydew)

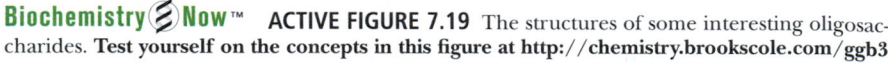

Biochemistry ⒺNow™   **ACTIVE FIGURE 7.19** The structures of some interesting oligosaccharides. **Test yourself on the concepts in this figure at http://chemistry.brookscole.com/ggb3**

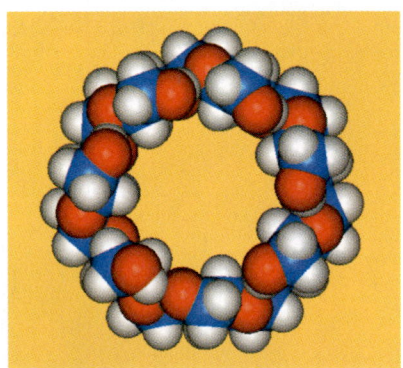

Cycloheptaamylose

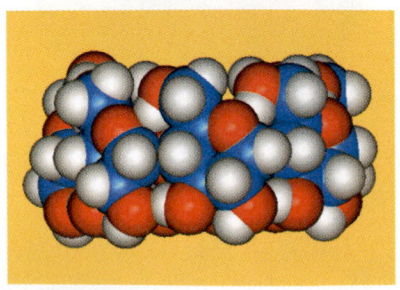

Cycloheptaamylose (side view)

Another notable glycoside is **amygdalin,** which occurs in bitter almonds and in the kernels or pits of cherries, peaches, and apricots. Hydrolysis of this substance and subsequent oxidation yield **laetrile,** which has been claimed by some to have anticancer properties. There is no scientific evidence for these claims, and the U.S. Food and Drug Administration has never approved laetrile for use in the United States.

Oligosaccharides also occur widely as components (via glycosidic bonds) of *antibiotics* derived from various sources. Figure 7.20 shows the structures of a few representative carbohydrate-containing antibiotics. Some of these antibiotics also show antitumor activity. One of the most important of this type is **bleomycin A₂,** which is used clinically against certain tumors.

7.4 | What Is the Structure and Chemistry of Polysaccharides?

Nomenclature for Polysaccharides Is Based on Their Composition and Structure

By far the majority of carbohydrate material in nature occurs in the form of polysaccharides. By our definition, polysaccharides include not only those substances composed only of glycosidically linked sugar residues but also molecules

Bleomycin A₂ (an antitumor agent used clinically against specific tumors)

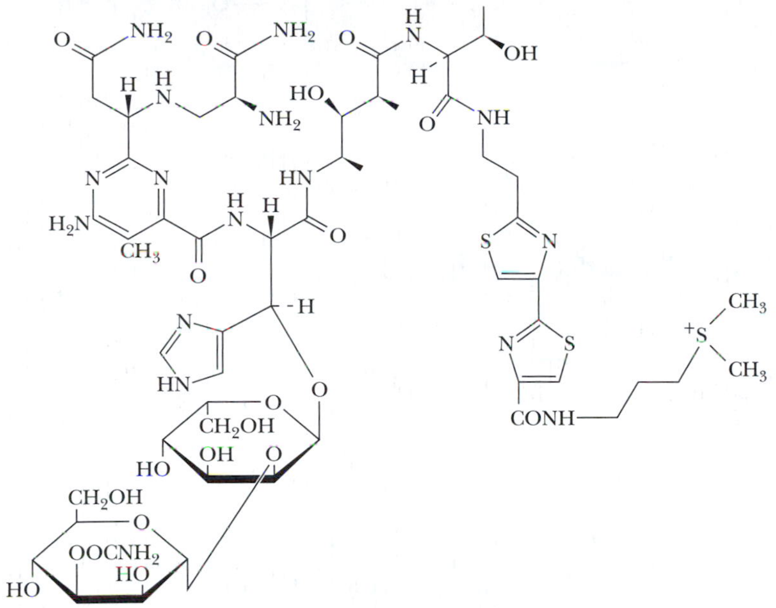

Streptomycin (a broad-spectrum antibiotic)

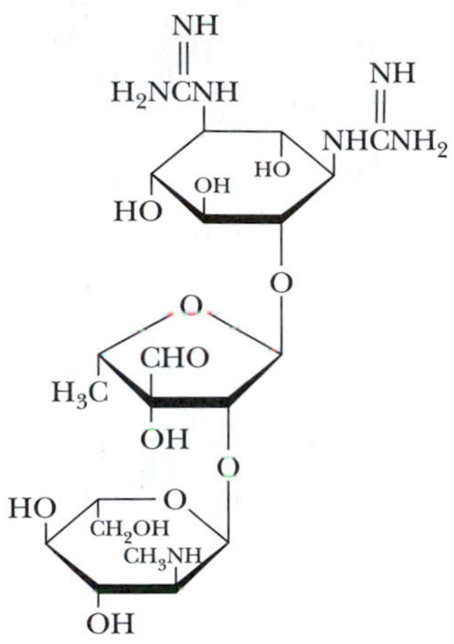

FIGURE 7.20 Some antibiotics are oligosaccharides or contain oligosaccharide groups.

that contain polymeric saccharide structures linked via covalent bonds to amino acids, peptides, proteins, lipids, and other structures.

Polysaccharides, also called **glycans,** consist of monosaccharides and their derivatives. If a polysaccharide contains only one kind of monosaccharide molecule, it is a **homopolysaccharide,** or **homoglycan,** whereas those containing more than one kind of monosaccharide are **heteropolysaccharides.** The most common constituent of polysaccharides is D-glucose, but D-fructose, D-galactose, L-galactose, D-mannose, L-arabinose, and D-xylose are also common. Common monosaccharide derivatives in polysaccharides include the amino sugars (D-glucosamine and D-galactosamine), their derivatives (N-acetylneuraminic acid and N-acetylmuramic acid), and simple sugar acids (glucuronic and iduronic acids). Homopolysaccharides are often named for the sugar unit they contain, so glucose homopolysaccharides are called **glucans,** and mannose homopolysaccharides are **mannans.** Other homopolysaccharide names are just as obvious: *galacturonans, arabinans,* and so on. Homopolysaccharides of uniform linkage type are often named by including notation to denote ring size and linkage type. Thus, cellulose is a *(1→4)-β-D-glucopyranan.* Polysaccharides differ not only in the nature of their component monosaccharides but also in the length of their chains and in the amount of chain branching that occurs. Although a given sugar residue has only one anomeric carbon and thus can form only one glycosidic linkage with hydroxyl groups on other molecules, each sugar residue carries several hydroxyls, one or more of which may be an acceptor of glycosyl substituents (Figure 7.21). This ability to form branched structures distinguishes polysaccharides from proteins and nucleic acids, which occur only as linear polymers.

Polysaccharides Serve Energy Storage, Structure, and Protection Functions

The functions of many individual polysaccharides cannot be assigned uniquely, and some of their functions may not yet be appreciated. Traditionally, biochemistry textbooks have listed the functions of polysaccharides as storage materials, structural components, or protective substances. Thus, *starch, glycogen,* and other storage polysaccharides, as readily metabolizable

CH_2OH CH_2OH CH_2OH CH_2OH CH_2OH

Amylose

CH_2OH CH_2OH CH_2OH

CH_2OH CH_2OH CH_2 CH_2OH CH_2OH

Amylopectin

Biochemistry ⊛ Now™   **ANIMATED FIGURE 7.21** Amylose and amylopectin are the two forms of starch. Note that the linear linkages are $\alpha(1\rightarrow4)$ but the branches in amylopectin are $\alpha(1\rightarrow6)$. Branches in polysaccharides can involve any of the hydroxyl groups on the monosaccharide components. Amylopectin is a highly branched structure, with branches occurring every 12 to 30 residues. **See this figure animated at http://chemistry.brookscole.com/ggb3**

food, provide energy reserves for cells. *Chitin* and *cellulose* provide strong support for the skeletons of arthropods and green plants, respectively. Mucopolysaccharides, such as the *hyaluronic acids*, form protective coats on animal cells. In each of these cases, the relevant polysaccharide is either a homopolymer or a polymer of small repeating units. Recent research indicates, however, that oligosaccharides and polysaccharides with varied structures may also be involved in much more sophisticated tasks in cells, including a variety of cellular recognition and intercellular communication events, as discussed later.

Polysaccharides Provide Stores of Energy

Storage polysaccharides are an important carbohydrate form in plants and animals. It seems likely that organisms store carbohydrates in the form of polysaccharides rather than as monosaccharides to lower the osmotic pressure of the sugar reserves. Because osmotic pressures depend only on *numbers of molecules,* the osmotic pressure is greatly reduced by formation of a few polysaccharide molecules out of thousands (or even millions) of monosaccharide units.

Starch   By far the most common storage polysaccharide in plants is **starch,** which exists in two forms: **α-amylose** and **amylopectin,** the structures of which are shown in Figure 7.21. Most forms of starch in nature are 10% to 30% α-amylose and 70% to 90% amylopectin. Typical cornstarch produced in the United States is about 25% α-amylose and 75% amylopectin. α-Amylose is composed of linear chains of D-glucose in $\alpha(1\rightarrow4)$ linkages. The chains are of varying length, having molecular weights from several thousand to half a million. As can be seen from the structure in Figure 7.21, the chain has a reducing end and a nonreducing end. Although poorly soluble in water, α-amylose forms micelles in which the polysaccharide chain adopts a helical conformation (Figure 7.22). Iodine reacts with α-amylose to give a characteristic blue color, which arises from the insertion of iodine into the middle of the hydrophobic amylose helix.

In contrast to α-amylose, amylopectin, the other component of typical starches, is a highly branched chain of glucose units (Figure 7.21). Branches occur in these chains every 12 to 30 residues. The average branch length is be-

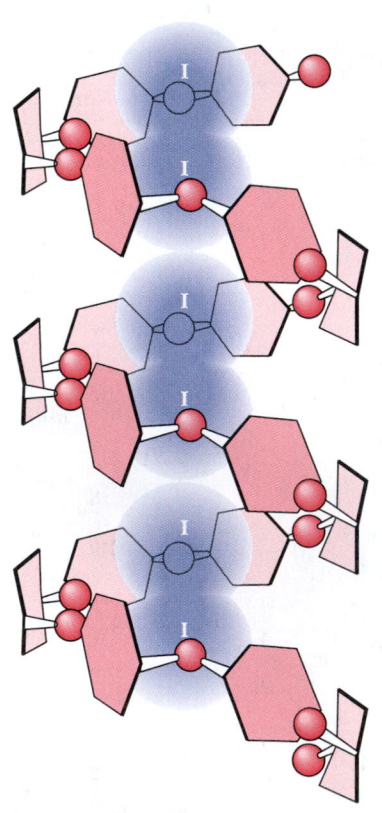

FIGURE 7.22 Suspensions of amylose in water adopt a helical conformation. Iodine (I_2) can insert into the middle of the amylose helix to give a blue color that is characteristic and diagnostic for starch.

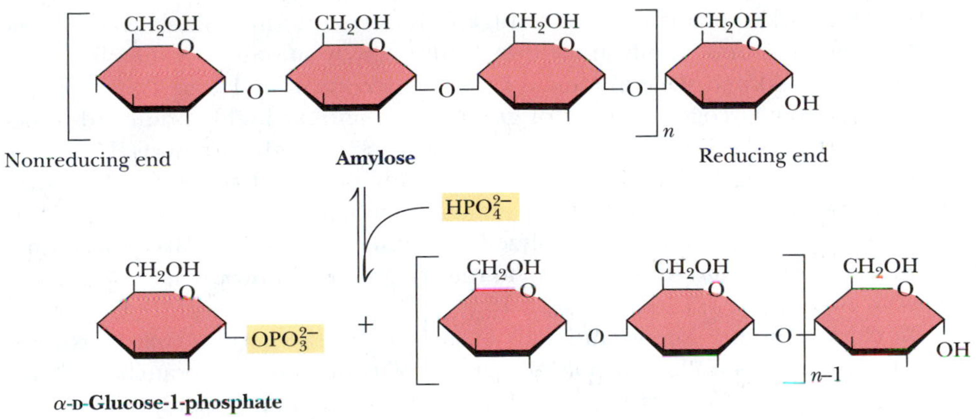

Biochemistry Now™ **ANIMATED FIGURE 7.23** The starch phosphorylase reaction cleaves glucose residues from amylose, producing α-D-glucose-1-phosphate. **See this figure animated at** http://chemistry.brookscole.com/ggb3

tween 24 and 30 residues, and molecular weights of amylopectin molecules can range up to 100 million. The linear linkages in amylopectin are $\alpha(1\rightarrow4)$, whereas the branch linkages are $\alpha(1\rightarrow6)$. As is the case for α-amylose, amylopectin forms micellar suspensions in water; iodine reacts with such suspensions to produce a red-violet color.

Starch is stored in plant cells in the form of granules in the stroma of plastids (plant cell organelles) of two types: **chloroplasts,** in which photosynthesis takes place, and **amyloplasts,** plastids that are specialized starch accumulation bodies. When starch is to be mobilized and used by the plant that stored it, it must be broken down into its component monosaccharides. Starch is split into its monosaccharide elements by stepwise phosphorolytic cleavage of glucose units, a reaction catalyzed by **starch phosphorylase** (Figure 7.23). This is formally an $\alpha(1\rightarrow4)$-glucan phosphorylase reaction, and at each step, the products are one molecule of glucose-1-phosphate and a starch molecule with one less glucose unit. In α-amylose, this process continues all along the chain until the end is reached. However, the $\alpha(1\rightarrow6)$ branch points of amylopectin are not susceptible to cleavage by phosphorylase, and thorough digestion of amylopectin by phosphorylase leaves a *limit dextrin,* which must be attacked by an $\alpha(1\rightarrow6)$-glucosidase to cleave the $1\rightarrow6$ branch points and allow complete hydrolysis of the remaining $1\rightarrow4$ linkages. Glucose-1-phosphate units are thus delivered to the plant cell, suitable for further processing in glycolytic pathways (see Chapter 18).

In animals, digestion and use of plant starches begin in the mouth with **salivary α-amylase** ($\alpha(1\rightarrow4)$-glucan 4-glucanohydrolase), the major enzyme secreted by the salivary glands. Although the capability of making and secreting salivary α-amylases is widespread in the animal world, some animals (such as cats, dogs, birds, and horses) do not secrete them. Salivary α-amylase is an **endoamylase** that splits $\alpha(1\rightarrow4)$ glycosidic linkages only within the chain. Raw starch is not very susceptible to salivary endoamylase. However, when suspensions of starch granules are heated, the granules swell, taking up water and causing the polymers to become more accessible to enzymes. Thus, cooked starch is more digestible. In the stomach, salivary α-amylase is inactivated by the lower pH, but pancreatic secretions also contain α-amylase. β-Amylase, an enzyme absent in animals but prevalent in plants and microorganisms, cleaves disaccharide (maltose) units from the termini of starch chains and is an **exoamylase.** Neither α-amylase nor β-amylase, however, can cleave the $\alpha(1\rightarrow6)$ branch points of amylopectin, and once again, $\alpha(1\rightarrow6)$-glucosidase is required to cleave at the branch points and allow complete hydrolysis of starch amylopectin.

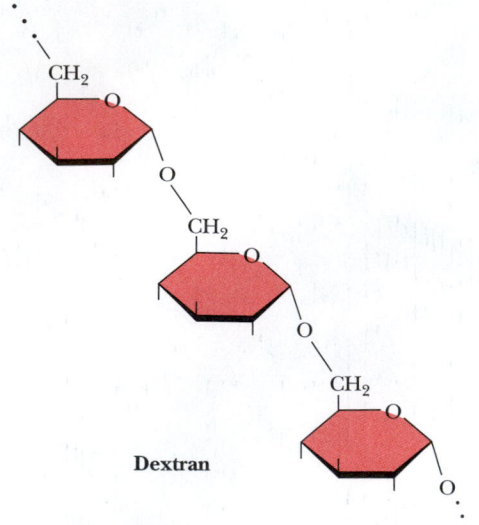

Dextran

Biochemistry⊗Now™   ANIMATED FIGURE 7.24
Dextran is a branched polymer of D-glucose units.
The main chain linkage is α(1→6), but 1→2, 1→3,
or 1→4 branches can occur. **See this figure animated
at http://chemistry.brookscole.com/ggb3**

Glycogen   The major form of storage polysaccharide in animals is **glycogen.** Glycogen is found mainly in the liver (where it may amount to as much as 10% of liver mass) and skeletal muscle (where it accounts for 1% to 2% of muscle mass). Liver glycogen consists of granules containing highly branched molecules, with α(1→6) branches occurring every 8 to 12 glucose units. Like amylopectin, glycogen yields a red-violet color with iodine. Glycogen can be hydrolyzed by both α- and β-amylases, yielding glucose and maltose, respectively, as products and can also be hydrolyzed by **glycogen phosphorylase,** an enzyme present in liver and muscle tissue, to release glucose-1-phosphate.

Dextran   Another important family of storage polysaccharides is the **dextrans,** which are α(1→6)-linked polysaccharides of D-glucose with branched chains found in yeast and bacteria (Figure 7.24). Because the main polymer chain is α(1→6) linked, the repeating unit is *isomaltose,* Glcα1→6Glc. The branch points may be 1→2, 1→3, or 1→4 in various species. The degree of branching and the average chain length between branches depend on the species and strain of the organism. Bacteria growing on the surfaces of teeth produce extracellular accumulations of dextrans, an important component of *dental plaque.* Bacterial dextrans are often used in research laboratories as the support medium for column chromatography of macromolecules. Dextran chains crosslinked with epichlorohydrin yield the structure shown in Figure 7.25. These preparations (known by various trade names, such as Sephadex and Bio-Gel) are extremely hydrophilic and swell to form highly hydrated gels in water.

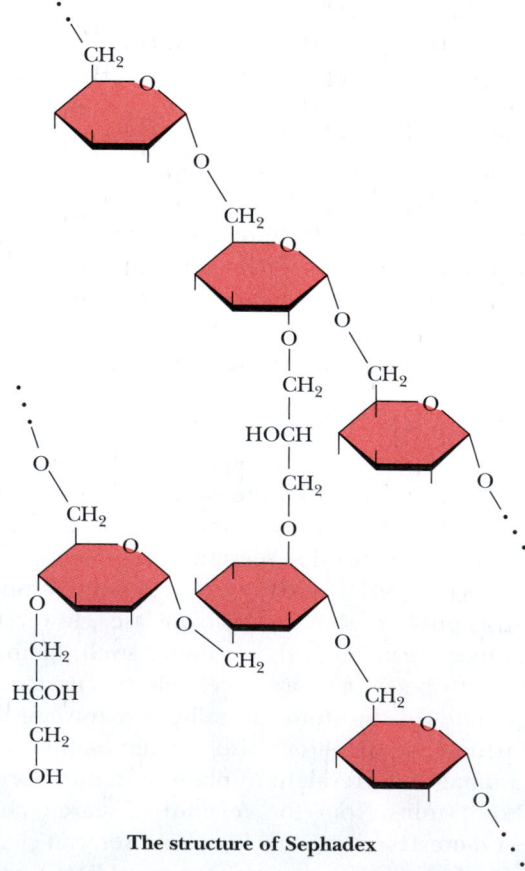

The structure of Sephadex

FIGURE 7.25 Sephadex gels are formed from dextran chains crosslinked with epichlorohydrin. The degree of crosslinking determines the chromatographic properties of Sephadex gels. Sephacryl gels are formed by crosslinking of dextran polymers with *N,N'*-methylene bisacrylamide.

Depending on the degree of crosslinking and the size of the gel particle, these materials form gels containing from 50% to 98% water. Dextran can also be crosslinked with other agents, forming gels with slightly different properties.

Polysaccharides Provide Physical Structure and Strength to Organisms

Cellulose   The **structural polysaccharides** have properties that are dramatically different from those of the storage polysaccharides, even though the compositions of these two classes are similar. The structural polysaccharide **cellulose** is the most abundant natural polymer found in the world. Found in the cell walls of nearly all plants, cellulose is one of the principal components providing physical structure and strength. The wood and bark of trees are insoluble, highly organized structures formed from cellulose and also from *lignin* (see Figure 25.35). It is awe-inspiring to look at a large tree and realize the amount of weight supported by polymeric structures derived from sugars and organic alcohols. Cellulose also has its delicate side, however. **Cotton,** whose woven fibers make some of our most comfortable clothing fabrics, is almost pure cellulose. Derivatives of cellulose have found wide use in our society. **Cellulose acetates** are produced by the action of acetic anhydride on cellulose in the presence of sulfuric acid and can be spun into a variety of fabrics with particular properties. Referred to simply as *acetates,* they have a silky appearance, a luxuriously soft feel, and a deep luster and are used in dresses, lingerie, linings, and blouses.

Cellulose is a linear homopolymer of D-glucose units, just as in α-amylose. The structural difference, which completely alters the properties of the polymer, is that in cellulose the glucose units are linked by $\beta(1\rightarrow4)$-glycosidic bonds, whereas in α-amylose the linkage is $\alpha(1\rightarrow4)$. The conformational difference between these two structures is shown in Figure 7.26. The $\alpha(1\rightarrow4)$-linkage sites of amylose are naturally bent, conferring a gradual turn to the polymer chain, which results in the helical conformation already described (see Figure 7.22). The most stable conformation about the $\beta(1\rightarrow4)$ linkage involves alternating 180° flips of the glucose units along the chain so that the chain adopts a fully extended conformation, referred to as an **extended ribbon.** Juxtaposition of several such chains permits efficient interchain hydrogen bonding, the basis of much of the strength of cellulose.

The structure of one form of cellulose, determined by X-ray and electron diffraction data, is shown in Figure 7.27. The flattened sheets of the chains lie side by side and are joined by hydrogen bonds. These sheets are laid on top of one another in a way that staggers the chains, just as bricks are staggered to give strength and stability to a wall. Cellulose is extremely resistant to hydrolysis, whether by acid or by the digestive tract amylases described earlier. As a result, most animals (including humans) cannot digest cellulose

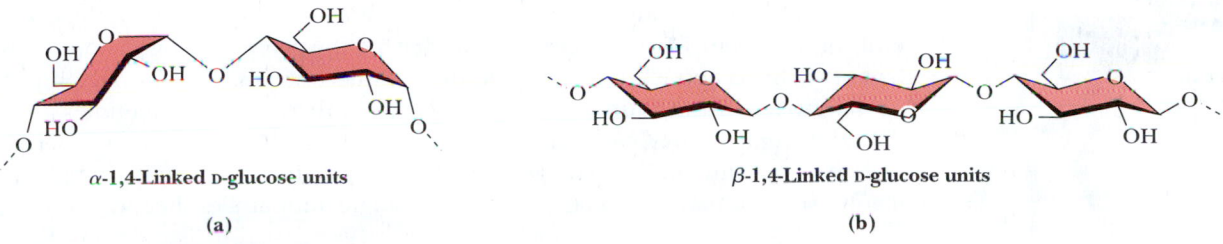

α-1,4-Linked D-glucose units

(a)

β-1,4-Linked D-glucose units

(b)

FIGURE 7.26 (a) Amylose, composed exclusively of the relatively bent $\alpha(1\rightarrow4)$ linkages, prefers to adopt a helical conformation, whereas **(b)** cellulose, with $\beta(1\rightarrow4)$-glycosidic linkages, can adopt a fully extended conformation with alternating 180° flips of the glucose units. The hydrogen bonding inherent in such extended structures is responsible for the great strength of tree trunks and other cellulose-based materials.

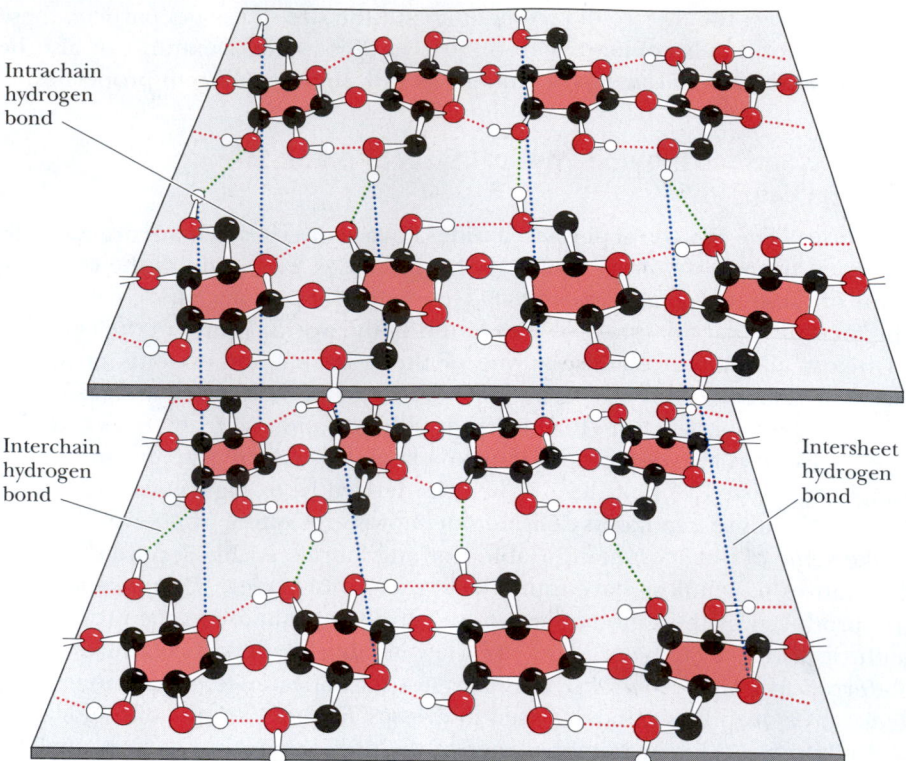

► **FIGURE 7.27** The structure of cellulose, showing the hydrogen bonds (blue) between the sheets, which strengthen the structure. Intrachain hydrogen bonds are in red, and interchain hydrogen bonds are in green. *(Illustration: Irving Geis. Rights owned by Howard Hughes Medical Institute. Not to be reproduced without permission.)*

Labels on figure: Intrachain hydrogen bond; Interchain hydrogen bond; Intersheet hydrogen bond

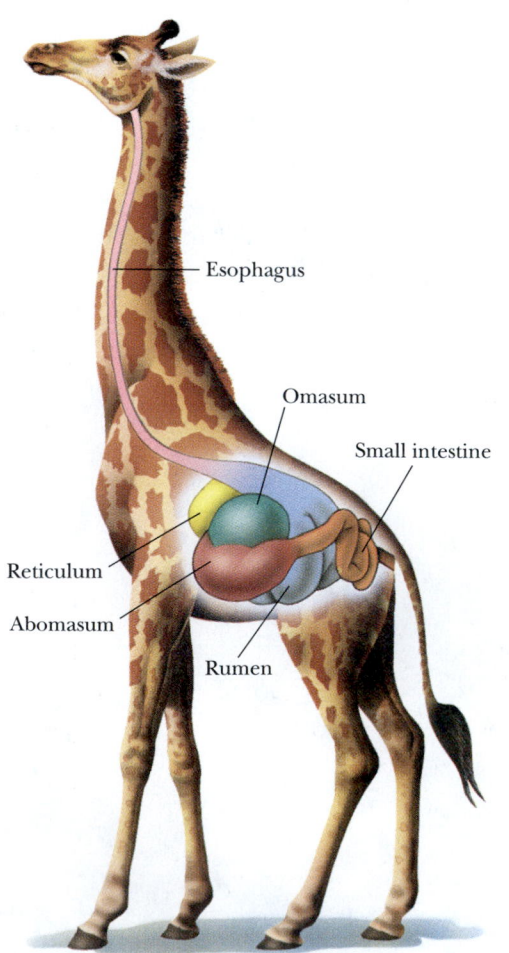

Labels: Esophagus; Omasum; Small intestine; Reticulum; Abomasum; Rumen

to any significant degree. Ruminant animals, such as cattle, deer, giraffes, and camels, are an exception because bacteria that live in the rumen (Figure 7.28) secrete the enzyme **cellulase**, a β-glucosidase effective in the hydrolysis of cellulose. The resulting glucose is then metabolized in a fermentation process to the benefit of the host animal. Termites and shipworms (*Teredo navalis*) similarly digest cellulose because their digestive tracts also contain bacteria that secrete cellulase.

Chitin   A polysaccharide that is similar to cellulose, both in its biological function and its primary, secondary, and tertiary structure, is **chitin.** Chitin is present in the cell walls of fungi and is the fundamental material in the exoskeletons of crustaceans, insects, and spiders. The structure of chitin, an extended ribbon, is identical to that of cellulose, except that the —OH group on each C-2 is replaced by —NHCOCH$_3$, so the repeating units are *N-acetyl-D-glucosamines* in β(1→4) linkage. Like cellulose (Figure 7.27), the chains of chitin form extended ribbons (Figure 7.29) and pack side by side in a crystalline, strongly hydrogen-bonded form. One significant difference between cellulose and chitin is whether the chains are arranged in **parallel** (all the reducing ends together at one end of a packed bundle and all the nonreducing ends together at the other end) or **antiparallel** (each sheet of chains having the chains arranged oppositely from the sheets above and below). Natural cellulose seems to occur only in parallel arrangements. Chitin, however, can occur in three forms, sometimes all in the same organism. α-*Chitin* is an all-parallel arrangement of the chains, whereas β-*chitin* is an antiparallel arrangement. In δ-*chitin,* the structure is thought to involve pairs of parallel sheets separated by single antiparallel sheets.

◄ **FIGURE 7.28** Giraffes, cattle, deer, and camels are ruminant animals that are able to metabolize cellulose, thanks to bacterial cellulase in the rumen, a large first compartment in the stomach of a ruminant.

A Deeper Look

A Complex Polysaccharide in Red Wine—The Strange Story of Rhamnogalacturonan II

For many years, cotton and grape growers and other farmers have known that boron is an essential trace element for their crops. Until recently, however, the role or roles of boron in sustaining plant growth were unknown. Recent reports show that at least one role for boron in plants is that of crosslinking an unusual polysaccharide called rhamnogalacturonan II (RGII). RGII is a low-molecular-weight (5 to 10 kDa) polysaccharide, but it is thought to be the most complex polysaccharide on earth, comprised as it is of 11 different sugar monomers. It can be released from plant cell walls by treatment with a galacturonase, and it is also present in red wine. Part of the structure of RGII is shown in the accompanying figure. The nature of the borate ester crosslinks (also indicated in the figure) was elucidated by Malcolm O'Neill and his colleagues, who used a combination of chemical methods and boron-11 NMR.

Why is rhamnogalacturonan II essential for the structure and growth of plant walls? Plant walls are extremely sophisticated composite materials, composed of networks of protein, polysaccharides, and phenolic compounds. Cellulose microfibrils as strong as steel provide a load-bearing framework for the plant. These microfibrils are tiny wires made of crystalline arrays of α-1,4-linked chains of glucose residues, which are extruded from hexameric spinnerets in the plasma membrane of the plant cell, surrounding the growing plant cell like hoops around a barrel. These microfibrils thus constrain the directions of cell expansion and determine the shapes of the plant cells and the plant as well. The separation of the barrel hoops is controlled by hemicelluloses, such as xyloglucans, which form H-bonded crosslinks with the cellulose microfibrils. The hemicellulose network is embedded in a hydrated gel inside the plant wall. This gel consists of complex galacturonic acid–rich polysaccharides, including RGII—it provides a dynamic operating environment for cell wall processes.

It is interesting to note that the tiny spinnerets of plant cells are nature's version of the viscose process, developed in 1910, for the production of rayon fibers. In this process, viscose—literally a *visc*ous solution of cellul*ose*—is forced through a spinneret (a device resembling a shower head with many tiny holes). Each hole produces a fine filament of viscose. The fibers precipitate in an acid bath and are stretched to form interchain H bonds that give the filaments the properties essential for use as textile fibers.

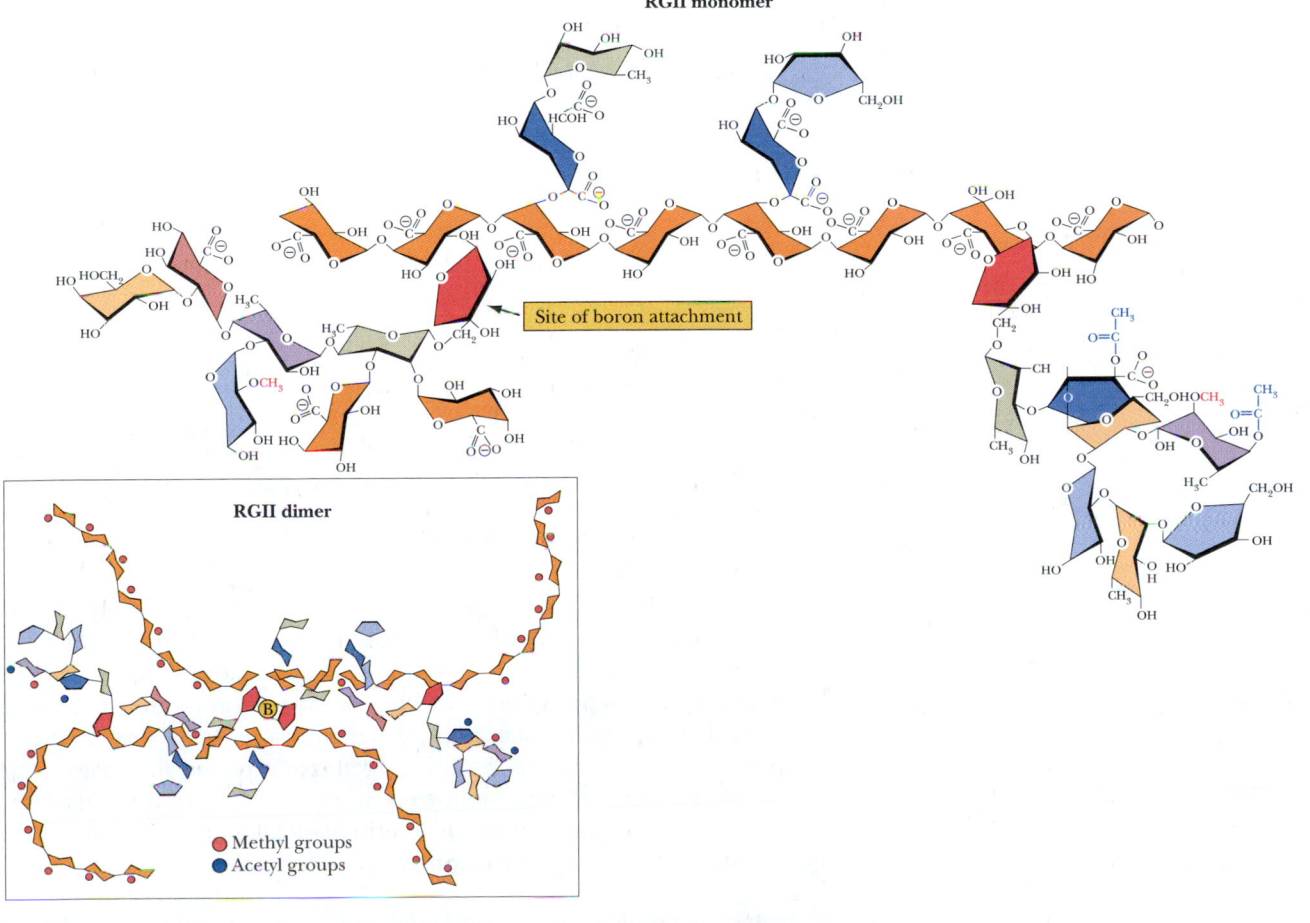

Source: Hofte, H., 2001. A baroque residue in red wine. *Science* **294**:795–797.

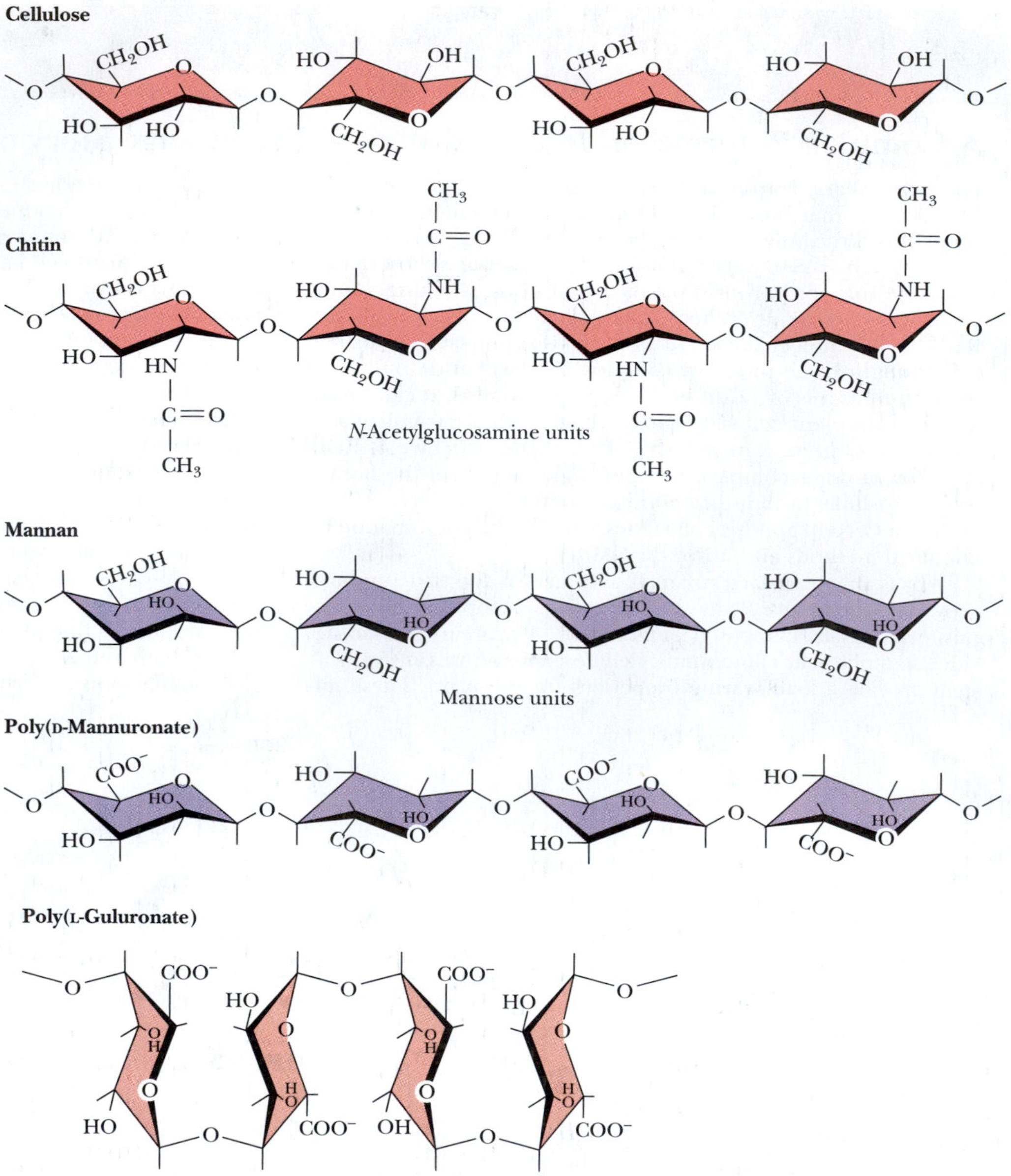

Cellulose

Chitin

N-Acetylglucosamine units

Mannan

Mannose units

Poly(D-Mannuronate)

Poly(L-Guluronate)

Biochemistry 🅔 Now™ **ANIMATED FIGURE 7.29** Like cellulose, chitin, mannan, and poly(D-mannuronate) form extended ribbons and pack together efficiently, taking advantage of multiple hydrogen bonds. **See this figure animated at http://chemistry.brookscole.com/ggb3**

Chitin is the earth's second most abundant carbohydrate polymer (after cellulose), and its ready availability and abundance offer opportunities for industrial and commercial applications. Chitin-based coatings can extend the shelf life of fruits, and a chitin derivative that binds to iron atoms in meat has been found to slow the reactions that cause rancidity and flavor loss. Without such a coating, the iron in meats activates oxygen from the air, forming reactive free radicals that attack and oxidize polyunsaturated lipids, causing most of the flavor loss associated with rancidity. Chitin-based coatings coordinate the iron atoms, preventing their interaction with oxygen.

Alginates A family of novel extended ribbon structures that bind metal ions, particularly calcium, in their structure are the **alginate** polysaccharides of marine brown algae (*Phaeophyceae*). These include **poly(β-D-mannuronate)** and **poly(α-L-guluronate),** which are (1→4)-linked chains formed from β-D-mannuronic acid and α-L-guluronic acid, respectively. Both of these homopolymers are found

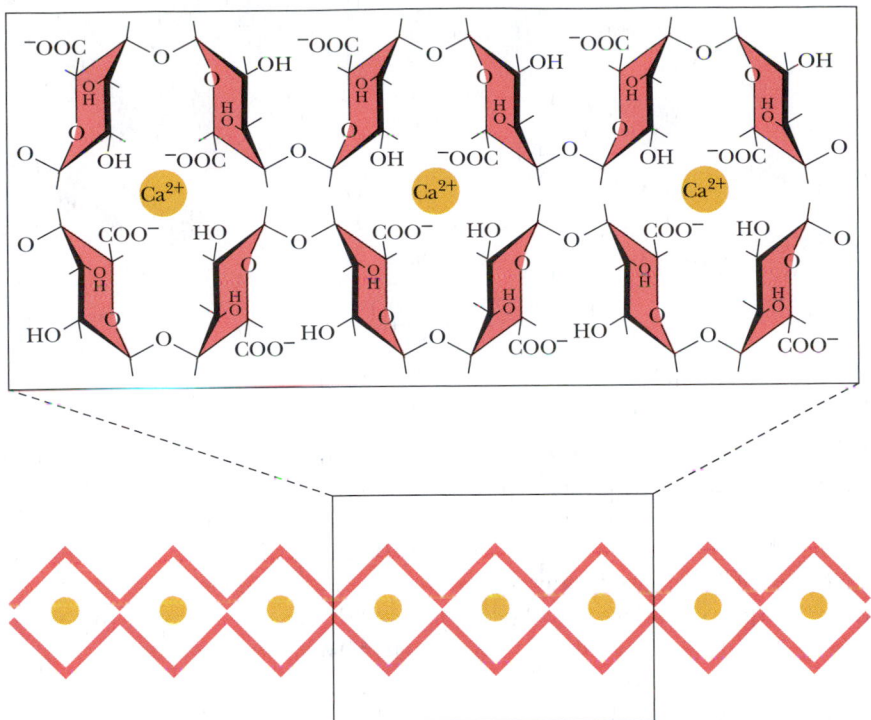

FIGURE 7.30 Poly(α-L-guluronate) strands dimerize in the presence of Ca^{2+}, forming a structure known as an "egg carton."

together in most marine alginates, although to widely differing extents, and mixed chains containing both monomer units are also found. As shown in Figure 7.29, the conformation of poly(β-D-mannuronate) is similar to that of cellulose. In the solid state, the free form of the polymer exists in celluloselike form. However, complexes of the polymer with cations (such as lithium, sodium, potassium, and calcium) adopt a threefold helix structure, presumably to accommodate the bound cations. For poly(α-L-guluronate) (Figure 7.29), the axial–axial configuration of the glycosidic linkage leads to a distinctly buckled ribbon with limited flexibility. Cooperative interactions between such buckled ribbons can be strong only if the interstices are filled effectively with water molecules or metal ions. Figure 7.30 shows a molecular model of a Ca^{2+}-induced dimer of poly(α-L-guluronate).

Agarose   An important polysaccharide mixture isolated from marine red algae *(Rhodophyceae)* is **agar,** which consists of two components: **agarose** and **agaropectin.** Agarose (Figure 7.31) is a chain of alternating D-galactose and 3,6-anhydro-L-galactose, with side chains of 6-methyl-D-galactose. Agaropectin is similar, but in addition, it contains sulfate ester side chains and D-glucuronic acid. The three-dimensional structure of agarose is a double helix with a threefold screw axis, as shown in Figure 7.31. The central cavity is large enough to accommodate water molecules. Agarose and agaropectin readily form gels containing large amounts (up to 99.5%) of water. Agarose can be processed to remove most of the charged groups, yielding a material (trade name Sepharose) useful for purification of macromolecules in gel exclusion chromatography. Pairs of chains form double helices that subsequently aggregate in bundles to form a stable gel, as shown in Figure 7.32.

Glycosaminoglycans   A class of polysaccharides known as **glycosaminoglycans** is involved in a variety of extracellular (and sometimes intracellular) functions. Glycosaminoglycans consist of linear chains of repeating disaccharides in which

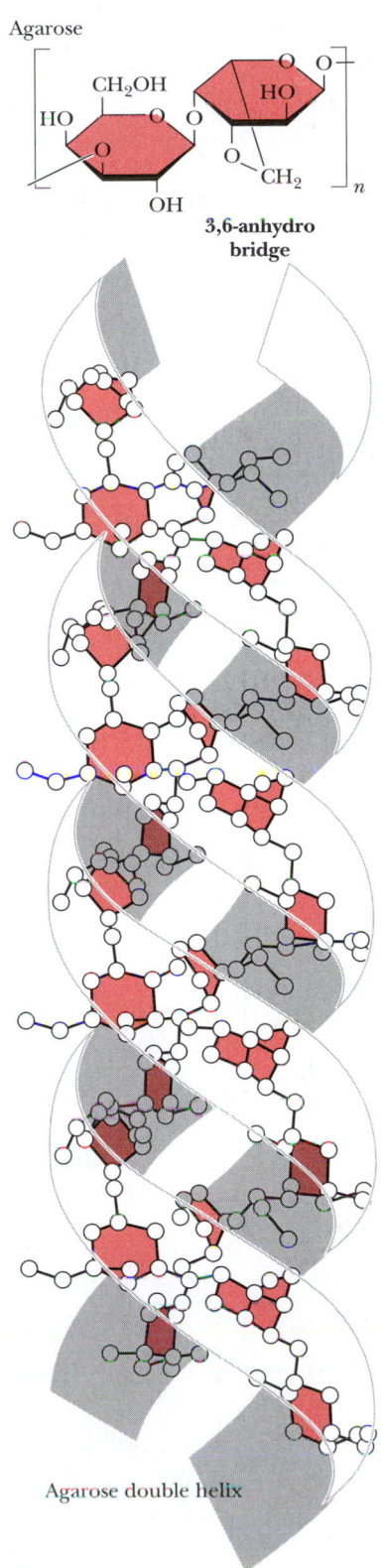

Agarose

3,6-anhydro bridge

Agarose double helix

FIGURE 7.31 The favored conformation of agarose in water is a double helix with a threefold screw axis.

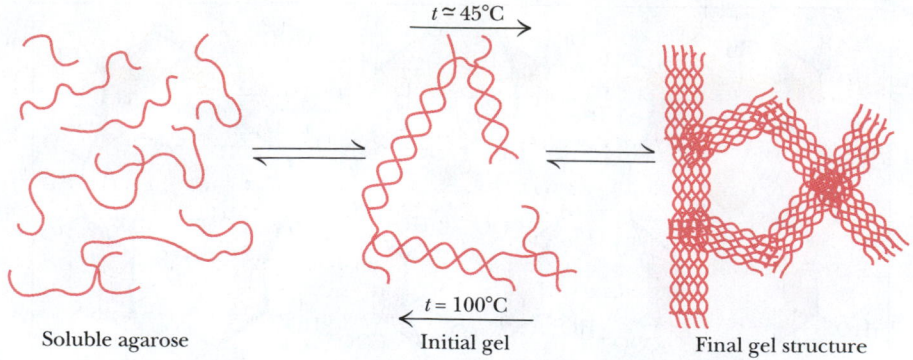

FIGURE 7.32 The ability of agarose to assemble in complex bundles to form gels in aqueous solution makes it useful in numerous chromatographic procedures, including gel exclusion chromatography and electrophoresis. Cells grown in culture can be embedded in stable agarose gel "threads" so that their metabolic and physiological properties can be studied.

Soluble agarose   Initial gel   Final gel structure

one of the monosaccharide units is an amino sugar and one (or both) of the monosaccharide units contains at least one negatively charged sulfate or carboxylate group. The repeating disaccharide structures found commonly in glycosaminoglycans are shown in Figure 7.33. **Heparin,** with the highest net negative charge of the disaccharides shown, is a natural anticoagulant substance. It binds strongly to *antithrombin III* (a protein involved in terminating the clotting process) and inhibits blood clotting. **Hyaluronate** molecules may consist of as many as 25,000 disaccharide units, with molecular weights of up to 10^7. Hyaluronates are important components of the vitreous humor in the eye and of *synovial fluid,* the lubricant fluid of joints in the body. The **chondroitins** and **keratan sulfate** are found in tendons, cartilage, and other connective tissue,

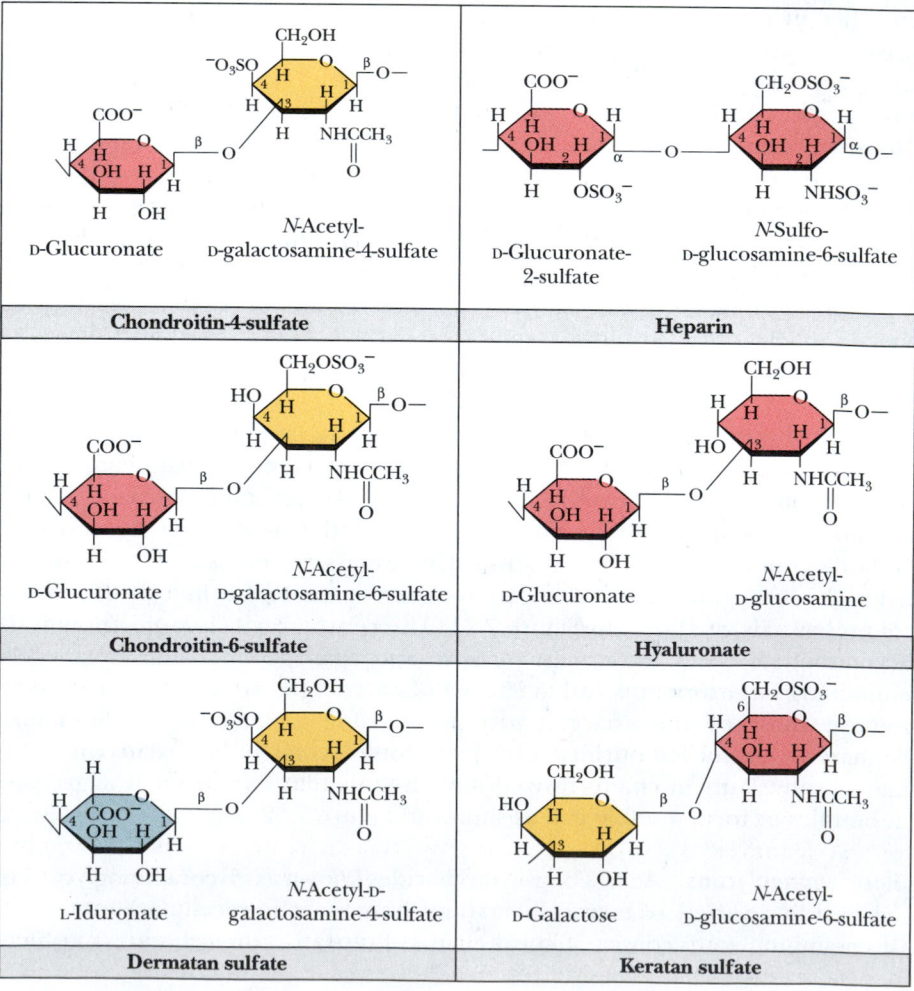

FIGURE 7.33 Glycosaminoglycans are formed from repeating disaccharide arrays. Glycosaminoglycans are components of the proteoglycans.

A Deeper Look

Billiard Balls, Exploding Teeth, and Dynamite—The Colorful History of Cellulose

Although humans cannot digest it and most people's acquaintance with cellulose is limited to comfortable cotton clothing, cellulose has enjoyed a colorful and varied history of utilization. In 1838, Théophile Pelouze in France found that paper or cotton could be made explosive if dipped in concentrated nitric acid. Christian Schönbein, a professor of chemistry at the University of Basel, prepared "nitrocotton" in 1845 by dipping cotton in a mixture of nitric and sulfuric acids and then washing the material to remove excess acid. In 1860, Major E. Schultze of the Prussian Army used the same material, now called **guncotton,** as a propellant replacement for gunpowder, and its preparation in brass cartridges quickly made it popular for this purpose. The only problem was that it was too explosive and could detonate unpredictably in factories where it was produced. The entire town of Faversham, England, was destroyed in such an accident. In 1868, Alfred Nobel mixed guncotton with ether and alcohol, thus preparing **nitrocellulose,** and in turn mixed this with nitroglycerin and sawdust to produce **dynamite.** Nobel's income from dynamite and also from his profitable development of the

Russian oil fields in Baku eventually formed the endowment for the Nobel Prizes.

In 1869, concerned over the precipitous decline (from hunting) of the elephant population in Africa, the billiard ball manufacturers Phelan and Collander offered a prize of $10,000 for production of a substitute for ivory. Brothers Isaiah and John Hyatt in Albany, New York, produced a substitute for ivory by mixing guncotton with camphor, then heating and squeezing it to produce **celluloid.** This product found immediate uses well beyond billiard balls. It was easy to shape, strong, and resilient, and it exhibited a high tensile strength. Celluloid was eventually used to make dolls, combs, musical instruments, fountain pens, piano keys, and a variety of other products. The Hyatt brothers eventually formed the Albany Dental Company to make false teeth from celluloid. Because camphor was used in their production, the company advertised that their teeth smelled "clean," but as reported in the *New York Times* in 1875, the teeth also occasionally exploded!

Portions adapted from Burke, J., 1996. *The Pinball Effect: How Renaissance Water Gardens Made the Carburetor Possible and Other Journeys Through Knowledge.* New York: Little, Brown, & Company.

whereas **dermatan sulfate,** as its name implies, is a component of the extracellular matrix of skin. Glycosaminoglycans are fundamental constituents of *proteoglycans* (discussed later).

Polysaccharides Provide Strength and Rigidity to Bacterial Cell Walls

Some of nature's most interesting polysaccharide structures are found in *bacterial cell walls.* Given the strength and rigidity provided by polysaccharide structures, it is not surprising that bacteria use such structures to provide protection for their cellular contents. Bacteria normally exhibit high internal osmotic pressures and frequently encounter variable, often hypotonic exterior conditions. The rigid cell walls synthesized by bacteria maintain cell shape and size and prevent swelling or shrinkage that would inevitably accompany variations in solution osmotic strength.

Peptidoglycan Is the Polysaccharide of Bacterial Cell Walls

Bacteria are conveniently classified as either **Gram-positive** or **Gram-negative** depending on their response to the so-called Gram stain. Despite substantial differences in the various structures surrounding these two types of cells, nearly all bacterial cell walls have a strong, protective peptide–polysaccharide layer called **peptidoglycan.** Gram-positive bacteria have a thick (approximately 25 nm) cell wall consisting of multiple layers of peptidoglycan. This thick cell wall surrounds the bacterial plasma membrane. Gram-negative bacteria, in contrast, have a much thinner (2 to 3 nm) cell wall consisting of a single layer of peptidoglycan sandwiched between the inner and outer lipid bilayer membranes. In either case, peptidoglycan, sometimes called **murein** (from the Latin *murus,* meaning "wall"), is a continuous crosslinked structure—in essence, a single molecule—built around the cell. The structure is shown in Figure 7.34. The backbone is a $\beta(1\rightarrow4)$-linked polymer of

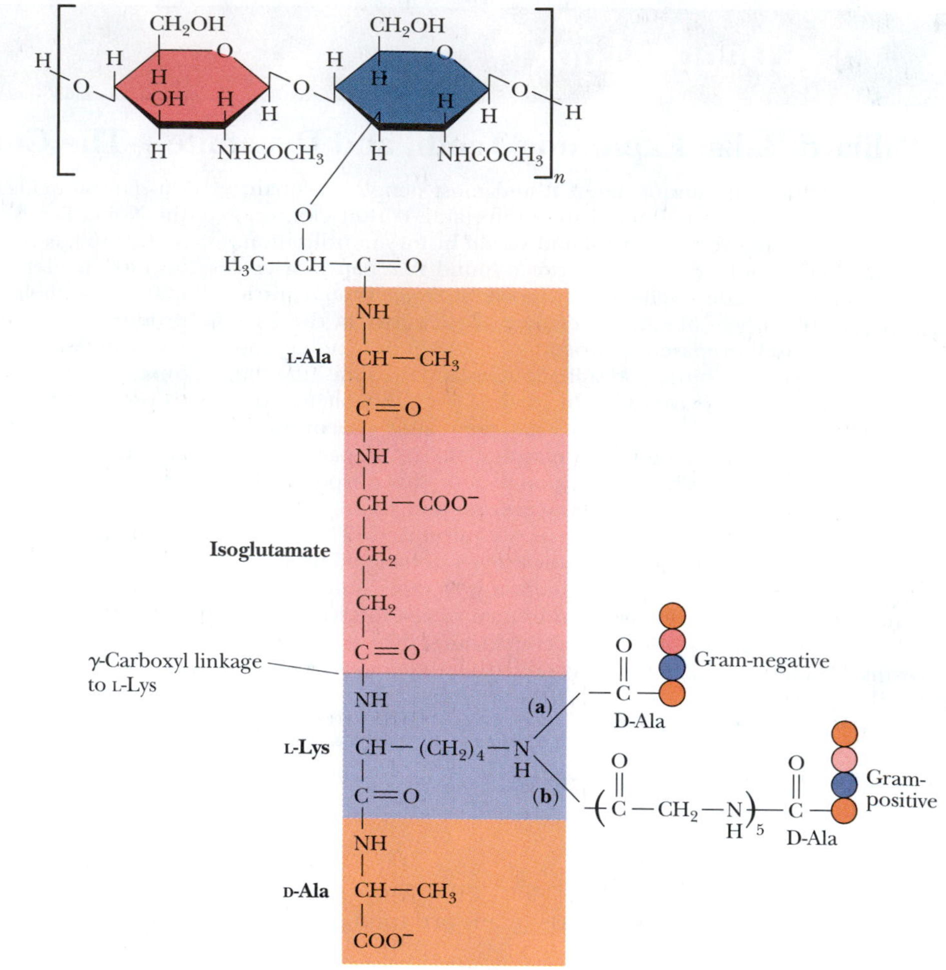

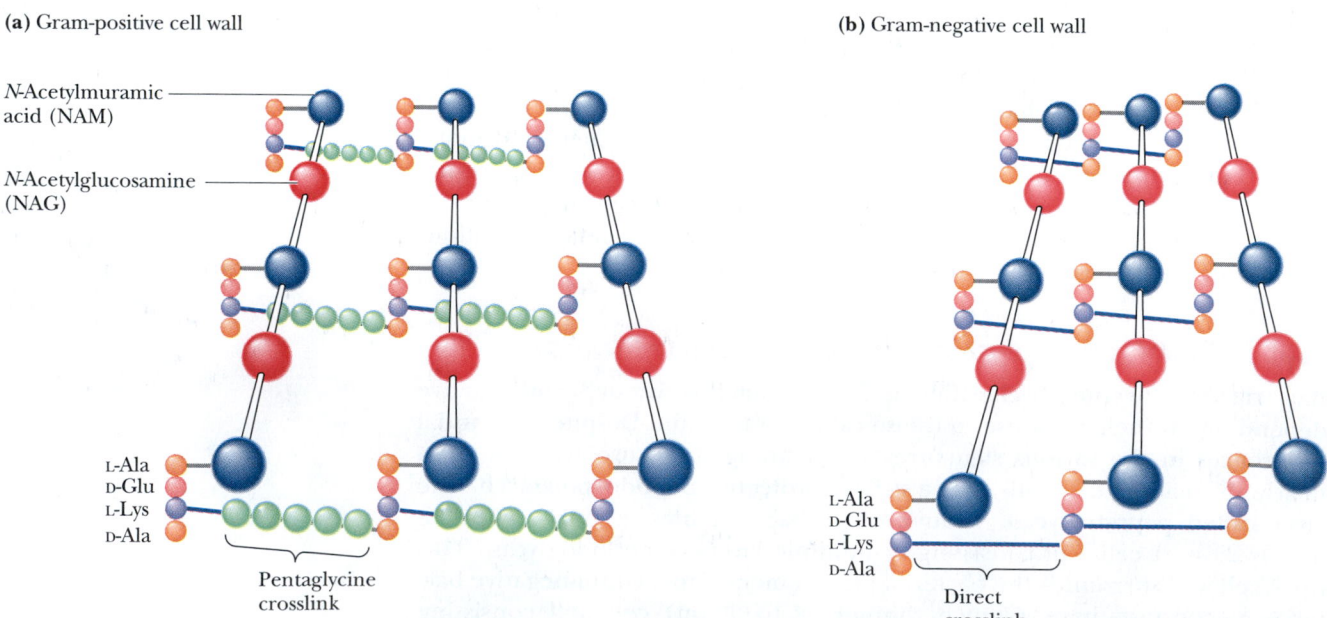

FIGURE 7.34 The structure of peptidoglycan. The tetrapeptides linking adjacent backbone chains contain an unusual γ-carboxyl linkage.

(a) Gram-positive cell wall

(b) Gram-negative cell wall

N-Acetylmuramic acid (NAM)

N-Acetylglucosamine (NAG)

L-Ala
D-Glu
L-Lys
D-Ala

Pentaglycine crosslink

L-Ala
D-Glu
L-Lys
D-Ala

Direct crosslink

FIGURE 7.35 **(a)** The crosslink in Gram-positive cell walls is a pentaglycine bridge. **(b)** In Gram-negative cell walls, the linkage between the tetrapeptides of adjacent carbohydrate chains in peptidoglycan involves a direct amide bond between the lysine side chain of one tetrapeptide and D-alanine of the other.

(a)   Gram-positive bacteria

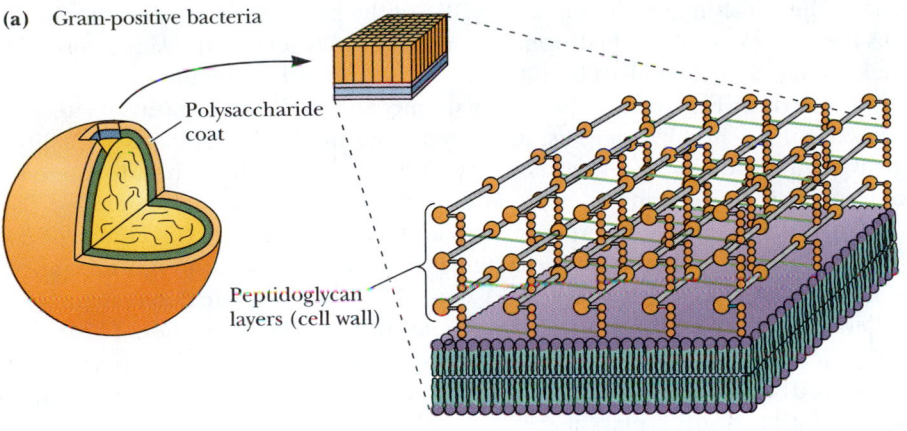

Polysaccharide coat

Peptidoglycan layers (cell wall)

(b)   Gram-negative bacteria

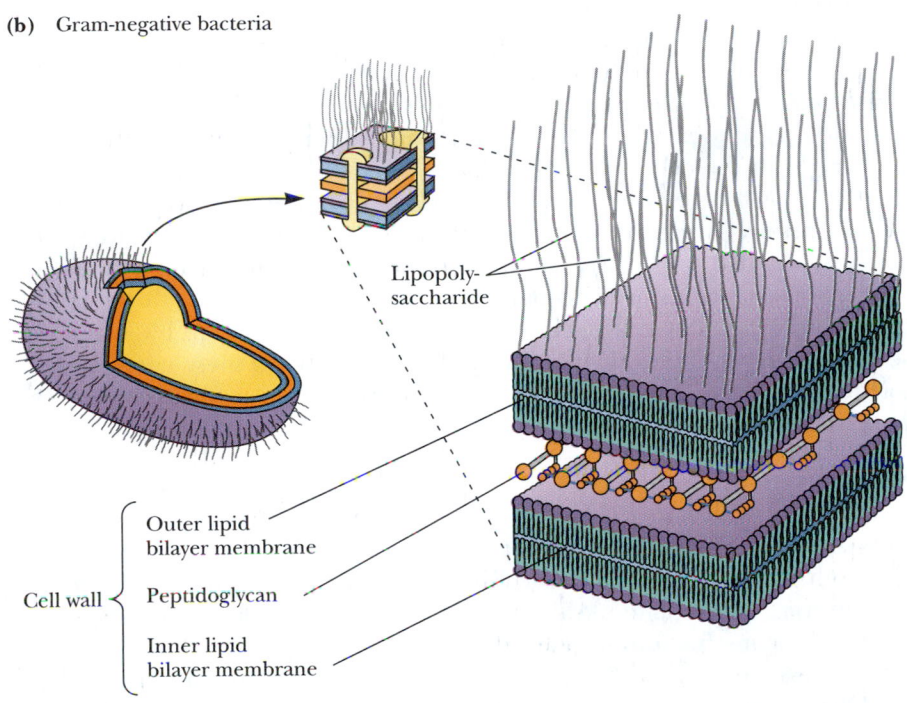

Lipopoly-saccharide

Cell wall
- Outer lipid bilayer membrane
- Peptidoglycan
- Inner lipid bilayer membrane

FIGURE 7.36 The structures of the cell wall and membrane(s) in Gram-positive and Gram-negative bacteria. The Gram-positive cell wall is thicker than that in Gram-negative bacteria, compensating for the absence of a second (outer) bilayer membrane.

alternating *N*-acetylglucosamine and *N*-acetylmuramic acid units. This part of the structure is similar to that of chitin, but it is joined to a tetrapeptide, usually L-Ala · D-Glu · L-Lys · D-Ala, in which the L-lysine is linked to the γ-COOH of D-glutamate. The peptide is linked to the *N*-acetylmuramic acid units via its D-lactate moiety. The ϵ-amino group of lysine in this peptide is linked to the —COOH of D-alanine of an adjacent tetrapeptide. In Gram-negative cell walls, the lysine ϵ-amino group forms a *direct amide bond* with this D-alanine carboxyl (Figure 7.35). In Gram-positive cell walls, a **pentaglycine chain** bridges the lysine ϵ-amino group and the D-Ala carboxyl group.

Cell Walls of Gram-Negative Bacteria   In Gram-negative bacteria, the peptidoglycan wall is the rigid framework around which is built an elaborate membrane structure (Figure 7.36). The peptidoglycan layer encloses the *periplasmic space* and is attached to the outer membrane via a group of **hydrophobic proteins.** These proteins, each having 57 amino acid residues, are attached through amide linkages from the side chains of C-terminal lysines of the proteins to diaminopimelic acid groups on the peptidoglycan. Diaminopimelic acid replaces

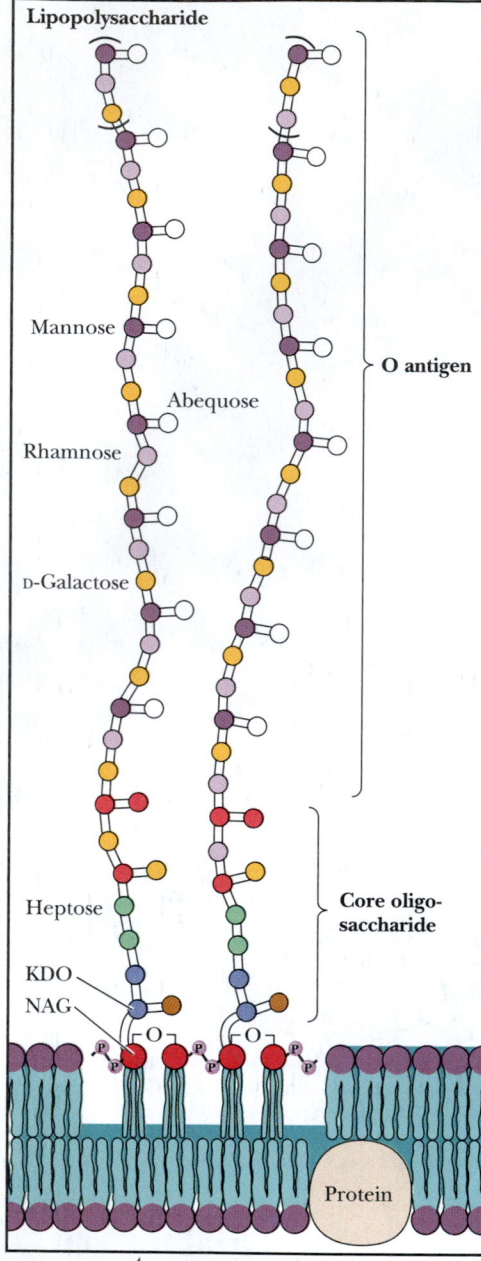

Lipopolysaccharide

Mannose

O antigen

Abequose

Rhamnose

D-Galactose

Heptose

Core oligo-
saccharide

KDO
NAG

Protein

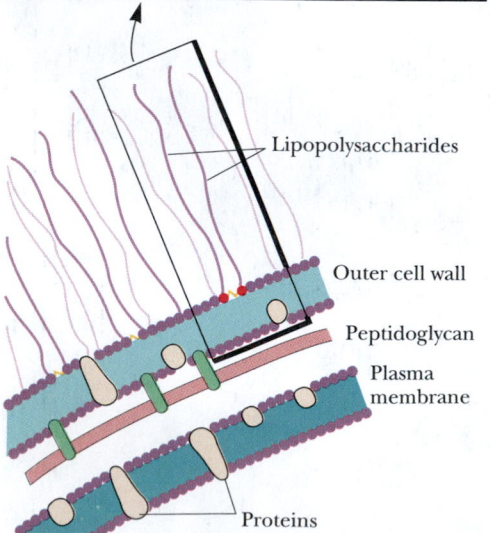

Lipopolysaccharides

Outer cell wall

Peptidoglycan

Plasma
membrane

Proteins

one of the D-alanine residues in about 10% of the peptides of the peptidoglycan. On the other end of the hydrophobic protein, the N-terminal residue, a serine, makes a covalent bond to a lipid that is part of the outer membrane.

As shown in Figure 7.37, the outer membrane of Gram-negative bacteria is coated with a highly complex **lipopolysaccharide,** which consists of a lipid group (anchored in the outer membrane) joined to a polysaccharide made up of long chains with many different and characteristic repeating structures (Figure 7.37). These many different unique units determine the antigenicity of the bacteria; that is, animal immune systems recognize them as foreign substances and raise antibodies against them. As a group, these **antigenic determinants** are called the **O antigens,** and there are thousands of different ones. The *Salmonella* bacteria alone have well over a thousand known O antigens that have been organized into 17 different groups. The great variation in these O antigen structures apparently plays a role in the recognition of one type of cell by another and in evasion of the host immune system.

Cell Walls of Gram-Positive Bacteria   In Gram-positive bacteria, the cell exterior is less complex than for Gram-negative cells. Having no outer membrane, Gram-positive cells compensate with a thicker wall. Covalently attached to the peptidoglycan layer are **teichoic acids,** which often account for 50% of the dry weight of the cell wall (Figure 7.38). The teichoic acids are polymers of *ribitol phosphate* or *glycerol phosphate* linked by phosphodiester bonds. In these heteropolysaccharides, the free hydroxyl groups of the ribitol or glycerol are often substituted by glycosidically linked monosaccharides (often glucose or *N*-acetylglucosamine) or disaccharides. D-Alanine is sometimes found in ester linkage to the saccharides. Teichoic acids are not confined to the cell wall itself, and they may be present in the inner membranes of these bacteria. Many teichoic acids are antigenic, and they also serve as the receptors for bacteriophages in some cases.

Animals Display a Variety of Cell Surface Polysaccharides

Compared to bacterial cells, which are identical within a given cell type (except for O antigen variations), animal cells display a wondrous diversity of structure, constitution, and function. Although each animal cell contains, in its genetic material, the instructions to replicate the entire organism, each differentiated animal cell carefully controls its composition and behavior within the organism. A great part of each cell's uniqueness begins at the cell surface. This surface uniqueness is critical to each animal cell because cells spend their entire life span in intimate contact with other cells and must therefore communicate with one another. That cells are able to pass information among themselves is evidenced by numerous experiments. For example, heart *myocytes,* when grown in culture (in glass dishes), establish *synchrony* when they make contact, so that they "beat" or contract in unison. If they are removed from the culture and separated, they lose their synchronous behavior, but if allowed to reestablish cell-to-cell contact, they spontaneously restore their synchronous contractions. Kidney cells grown in culture with liver cells seek out and make contact with other kidney cells and avoid contact with liver cells. Cells grown in culture grow freely until they make contact with one another, at which point growth stops, a phenomenon well known as **contact inhibition.** One important characteristic of cancerous cells is the loss of contact inhibition.

As these and many other related phenomena show, it is clear that molecular structures on one cell are recognizing and responding to molecules on the

◀ **FIGURE 7.37** Lipopolysaccharide (LPS) coats the outer membrane of Gram-negative bacteria. The lipid portion of the LPS is embedded in the outer membrane and is linked to a complex polysaccharide.

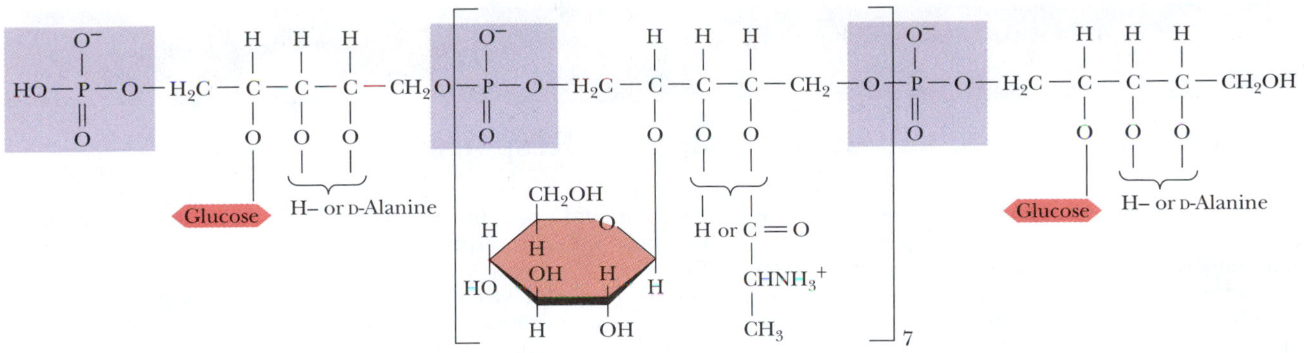

Ribitol teichoic acid from *Bacillus subtilis*

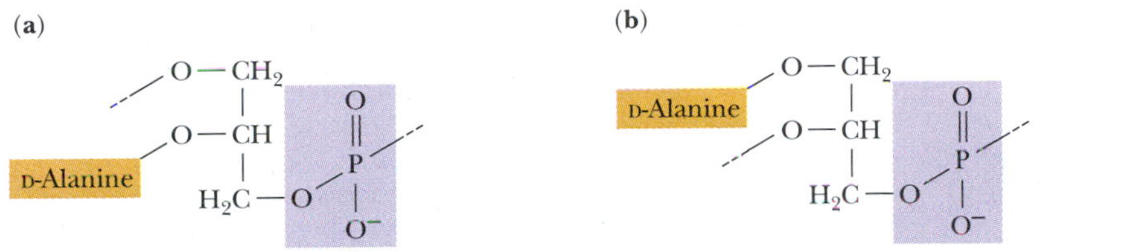

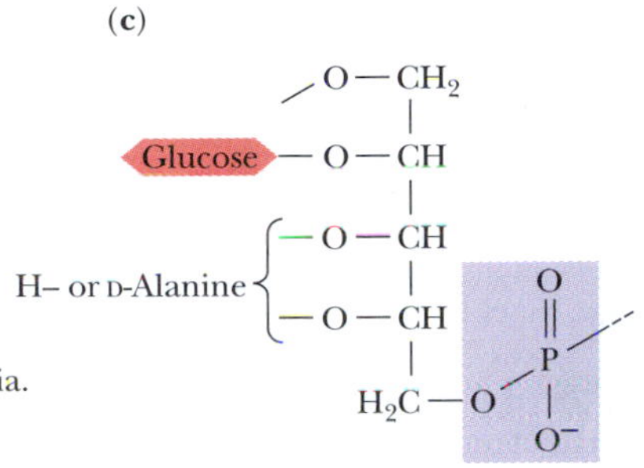

(a)

(b)

(c)

FIGURE 7.38 Teichoic acids are covalently linked to the peptidoglycan of Gram-positive bacteria. These polymers of (a, b) glycerol phosphate or (c) ribitol phosphate are linked by phosphodiester bonds.

adjacent cell or to molecules in the **extracellular matrix,** the complex "soup" of connective proteins and other molecules that exists outside of and among cells. Many of these interactions involve *glycoproteins* on the cell surface and *proteoglycans* in the extracellular matrix. The "information" held in these special carbohydrate-containing molecules is not encoded directly in the genes (as with proteins) but is determined instead by expression of the appropriate enzymes that assemble carbohydrate units in a characteristic way on these molecules. Also, by virtue of the several hydroxyl linkages that can be formed with each carbohydrate monomer, these structures are arguably more information-rich than proteins and nucleic acids, which can form only linear polymers. A few of these glycoproteins and their unique properties are described in the following sections.

7.5 | What Are Glycoproteins, and How Do They Function in Cells?

Many proteins found in nature are glycoproteins because they contain covalently linked oligosaccharide and polysaccharide groups. The list of known glycoproteins includes structural proteins, enzymes, membrane receptors, transport proteins, and immunoglobulins, among others. In most cases, the precise function of the bound carbohydrate moiety is not understood.

Carbohydrate groups may be linked to polypeptide chains via the hydroxyl groups of serine, threonine, or hydroxylysine residues (in **O-linked saccharides**) (Figure 7.39a) or via the amide nitrogen of an asparagine residue (in **N-linked saccharides**) (Figure 7.39b). The carbohydrate residue linked to the protein in O-linked saccharides is usually an *N*-acetylgalactosamine, but mannose, galactose, and xylose residues linked to protein hydroxyls are also found (Figure 7.39a). Oligosaccharides O-linked to glycophorin (see Figure 9.14) involve *N*-acetylgalactosamine linkages and are rich in sialic acid residues. N-linked saccharides always have a unique

Human Biochemistry

Selectins, Rolling Leukocytes, and the Inflammatory Response

Human bodies are constantly exposed to a plethora of bacteria, viruses, and other inflammatory substances. To combat these infectious and toxic agents, the body has developed a carefully regulated inflammatory response system. Part of that response is the orderly migration of leukocytes to sites of inflammation. Leukocytes literally roll along the vascular wall and into the tissue site of inflammation. This rolling movement is mediated by reversible adhesive interactions between the leukocytes and the vascular surface.

These interactions involve adhesion proteins called selectins, which are found both on the rolling leukocytes and on the endothelial cells of the vascular walls. Selectins have a characteristic domain structure, consisting of an N-terminal extracellular lectin domain, a single epidermal growth factor (EGR) domain, a series of two to nine short consensus repeat (SCR) domains, a single transmembrane segment, and a short cytoplasmic domain. Lectin domains, first characterized in plants, bind carbohydrates with high affinity and specificity. Selectins of three types are known—E-selectins, L-selectins, and P-selectins. L-selectin is found on the surfaces of leukocytes, including neutrophils and lymphocytes, and binds to carbohydrate ligands on endothelial cells. The pres-

ence of L-selectin is a necessary component of leukocyte rolling. P-selectin and E-selectin are located on the vascular endothelium and bind with carbohydrate ligands on leukocytes. Typical neutrophil cells possess 10,000 to 20,000 P-selectin–binding sites. Selectins are expressed on the surfaces of their respective cells by exposure to inflammatory signal molecules, such as histamine, hydrogen peroxide, and bacterial endotoxins. P-selectins, for example, are stored in intracellular granules and are transported to the cell membrane within seconds to minutes of exposure to a triggering agent.

Substantial evidence supports the hypothesis that selectin–carbohydrate ligand interactions modulate the rolling of leukocytes along the vascular wall. Studies with L-selectin–deficient and P-selectin–deficient leukocytes show that L-selectins mediate weaker adherence of the leukocyte to the vascular wall and promote faster rolling along the wall. Conversely, P-selectins promote stronger adherence and slower rolling. Thus, leukocyte rolling velocity in the inflammatory response could be modulated by variable exposure of P-selectins and L-selectins at the surfaces of endothelial cells and leukocytes, respectively.

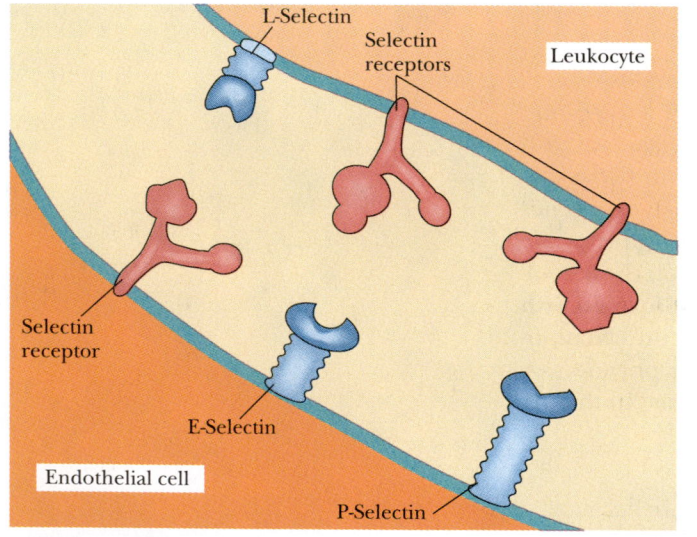

▲ A diagram showing the interactions of selectins with their receptors.

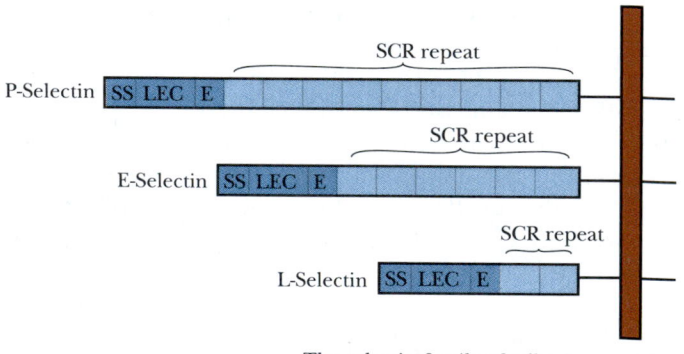

The selectin family of adhesion proteins.

core structure composed of two *N*-acetylglucosamine residues linked to a branched mannose triad (Figure 7.39b, c). Many other sugar units may be linked to each of the mannose residues of this branched core.

O-linked saccharides are often found in cell surface glycoproteins and in **mucins,** the large glycoproteins that coat and protect mucous membranes in the respiratory and gastrointestinal tracts in the body. Certain viral glycoproteins also contain O-linked sugars. O-linked saccharides in glycoproteins are often found clustered in richly glycosylated domains of the polypeptide chain. Physical studies on mucins show that they adopt rigid, extended structures. An individual mucin molecule ($M_r = 10^7$) may extend over a distance of 150 to 200 nm in solution. Inherent steric interactions between the sugar residues and the protein residues in these cluster regions cause the peptide core to fold

(a) O-linked saccharides

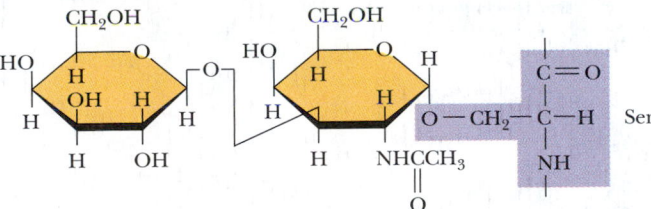

β-Galactosyl-1,3-α-N-acetylgalactosyl-serine

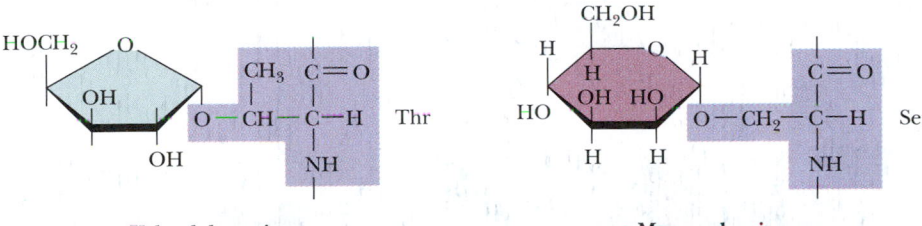

α-Xylosyl-threonine   α-Mannosyl-serine

FIGURE 7.39 The carbohydrate moieties of glycoproteins may be linked to the protein via **(a)** serine or threonine residues (in the O-linked saccharides) or **(b)** asparagine residues (in the N-linked saccharides). **(c)** N-linked glycoproteins are of three types: high mannose, complex, and hybrid, the latter of which combines structures found in the high mannose and complex saccharides.

(b) Core oligosaccharides in N-linked glycoproteins

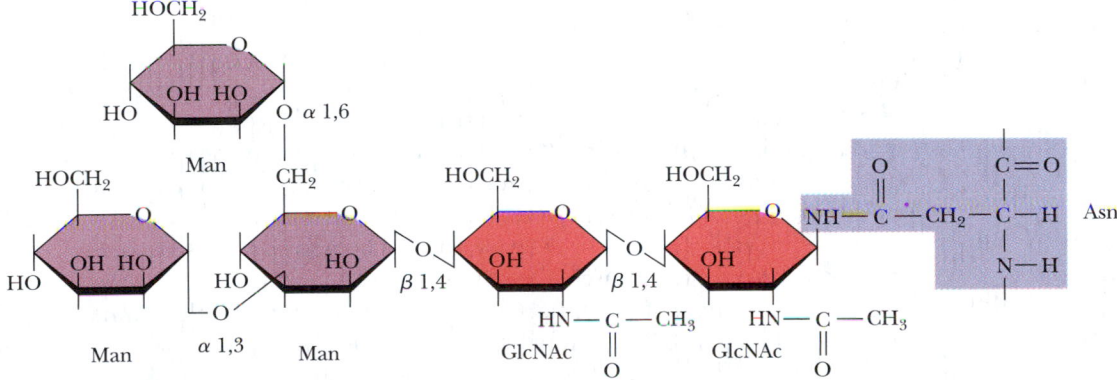

(c) N-linked glycoproteins

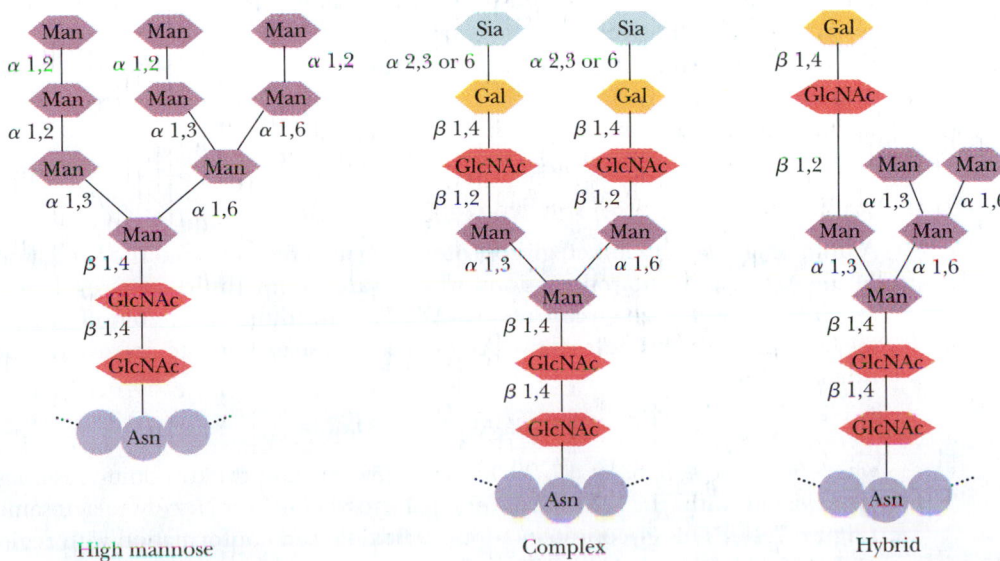

High mannose   Complex   Hybrid

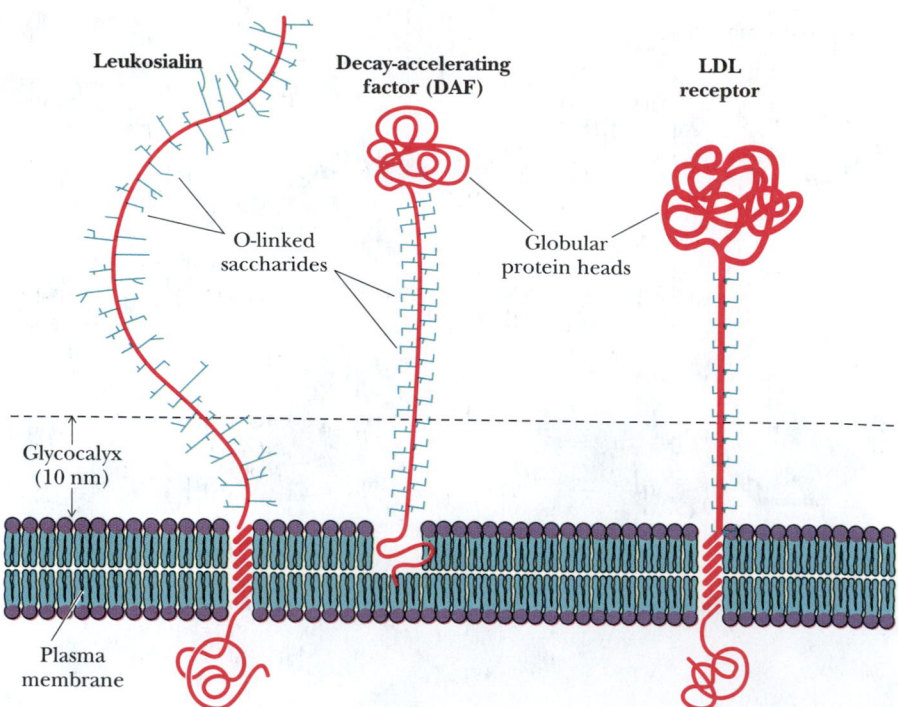

Leukosialin

Decay-accelerating factor (DAF)

LDL receptor

O-linked saccharides

Globular protein heads

Glycocalyx (10 nm)

Plasma membrane

FIGURE 7.40 The O-linked saccharides of glycoproteins appear in many cases to adopt extended conformations that serve to extend the functional domains of these proteins above the membrane surface. (*Adapted from Jentoft, N., 1990. Why are proteins O-glycosylated?* Trends in Biochemical Sciences *15:291–294.*)

into an extended and relatively rigid conformation. This interesting effect may be related to the function of O-linked saccharides in glycoproteins. It allows aggregates of mucin molecules to form extensive, intertwined networks, even at low concentrations. These viscous networks protect the mucosal surface of the respiratory and gastrointestinal tracts from harmful environmental agents.

There appear to be two structural motifs for membrane glycoproteins containing O-linked saccharides. Certain glycoproteins, such as **leukosialin,** are O-glycosylated throughout much or most of their extracellular domain (Figure 7.40). Leukosialin, like mucin, adopts a highly extended conformation, allowing it to project great distances above the membrane surface, perhaps protecting the cell from unwanted interactions with macromolecules or other cells. The second structural motif is exemplified by the **low-density lipoprotein (LDL) receptor** and by **decay-accelerating factor (DAF).** These proteins contain a highly O-glycosylated stem region that separates the transmembrane domain from the globular, functional extracellular domain. The O-glycosylated stem serves to raise the functional domain of the protein far enough above the membrane surface to make it accessible to the extracellular macromolecules with which it interacts.

Polar Fish Depend on Antifreeze Glycoproteins

A unique family of O-linked glycoproteins permits fish to live in the icy seawater of the Arctic and Antarctic regions, where water temperature may reach as low as $-1.9°C$. **Antifreeze glycoproteins (AFGPs)** are found in the blood of nearly all Antarctic fish and at least five Arctic fish. These glycoproteins have the peptide structure

$$[\text{Ala-Ala-Thr}]_n\text{-Ala-Ala}$$

where n can be 4, 5, 6, 12, 17, 28, 35, 45, or 50. Each of the threonine residues is glycosylated with the disaccharide β-galactosyl-$(1\rightarrow3)$-α-N-acetylgalactosamine (Figure 7.41). This glycoprotein adopts a **flexible rod** conformation with regions of threefold left-handed helix. The evidence suggests that antifreeze glycoproteins

A Deeper Look

Drug Research Finds a Sweet Spot

A variety of diseases are being successfully treated with sugar-based therapies. As this table shows, several carbohydrate-based drugs are either on the market or at various stages of clinical trials. Some of these drugs are enzymes, whereas others are glycoconjugates.

Drug	Description	Manufacturer
Cerzyme (imiglucerase)	This enzyme degrades glycolipids, compensating for an enzyme deficiency that causes Gaucher's disease.	Genzyme Cambridge, MA
Vancocin (vancomycin)	A very potent glycopeptide antibiotic that is typically used against antibiotic-resistant infections. It inhibits synthesis of peptidoglycan in the bacterial cell wall.	Eli Lilly Indianapolis, IN
Vevesca (OGT 918)	A sugar analog that inhibits synthesis of the glycolipid that accumulates in Gaucher's disease.	Oxford GlycoSciences Abingdon, UK
GMK	A vaccine containing ganglioside GM2; it triggers an immune response against cancer cells carrying GM2.	Progenics Pharmaceuticals Tarrytown, NY
Staphvax	A vaccine that is a protein with a linked bacterial sugar; it is intended to treat *Staphylococcus* infection.	NABI Pharmaceuticals Boca Raton, FL
Bimosiamose (TBC1269)	A sugar analog that inhibits selectin-based inflammation in blood vessels.	Texas Biotechnology Houston, TX
GCS-100	A sugar that blocks action of a sugar-binding protein on tumors.	GlycoGenesys Boston
GD0039 (swainsonine)	A sugar analog that inhibits synthesis of carbohydrates essential to tumor metastasis.	GlycoDesign Toronto, Canada
PI-88	A sugar that inhibits growth factor–dependent angiogenesis and enzymes that promote metastasis.	Progen Darra, Australia
UT231B	A sugar analog that prevents hepatitis C viral infections.	United Technologies Silver Spring, MD

Adapted from Maeder, T., 2002. Sweet Medicines. *Scientific American* **287**:40–47.

Additional Reference: Alper, J., 2001. Searching for Medicine's Sweet Spot. *Science* **291**:2338–2343.

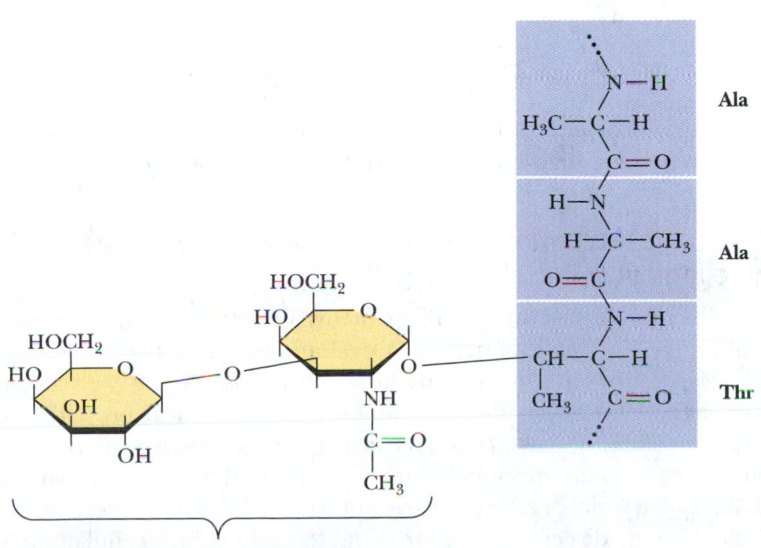

β-Galactosyl-1,3-α-N-acetylgalactosamine

Repeating unit of antifreeze glycoproteins

FIGURE 7.41 The structure of the repeating unit of antifreeze glycoproteins, a disaccharide consisting of β-galactosyl-(1→3)-α-N-acetylgalactosamine in glycosidic linkage to a threonine residue.

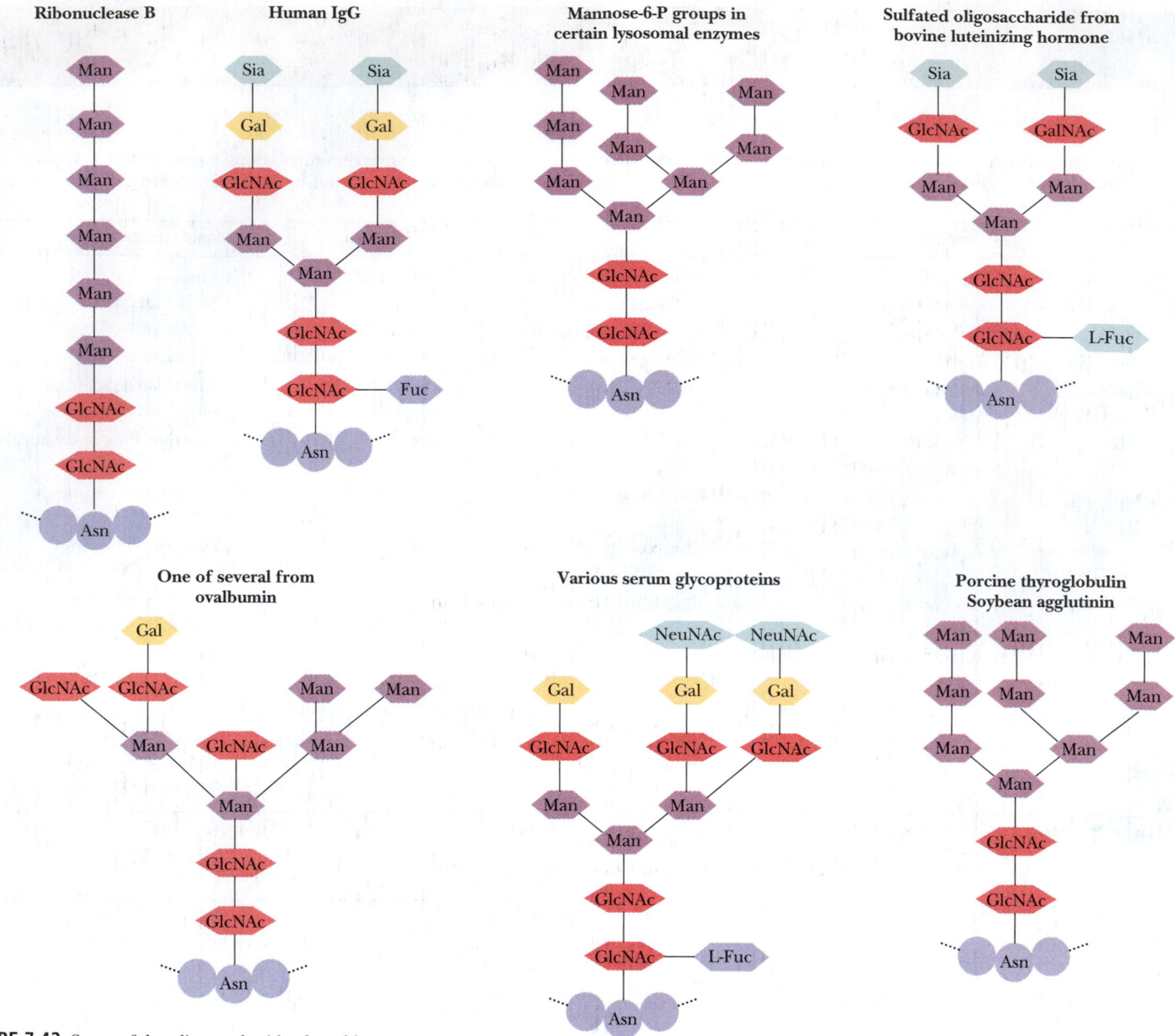

FIGURE 7.42 Some of the oligosaccharides found in N-linked glycoproteins.

may inhibit the formation of ice in the fish by binding specifically to the growth sites of ice crystals, inhibiting further growth of the crystals.

N-Linked Oligosaccharides Can Affect the Physical Properties and Functions of a Protein

N-linked oligosaccharides are found in many different proteins, including immunoglobulins G and M, ribonuclease B, ovalbumin, and peptide hormones (Figure 7.42). Many different functions are known or suspected for *N*-glycosylation of proteins. Glycosylation can affect the physical and chemical properties of proteins, altering solubility, mass, and electrical charge. Carbohydrate moieties have been shown to stabilize protein conformations and protect proteins against proteolysis. Eukaryotic organisms use post-translational additions of *N*-linked oligosaccharides to direct selected proteins to various intracellular organelles. Recent evidence indicates that N-linked oligosaccharides promote the proper folding of newly synthesized polypeptides in the endoplasmic reticulum (see A Deeper Look on page 239).

A Deeper Look

N-Linked Oligosaccharides Help Proteins Fold

The most important effect of N-linked oligosaccharides in eukaryotic organisms may be their contribution to the correct folding of certain globular proteins. This adaptation of saccharide function allows cells to produce and secrete larger and more complex proteins at high levels. Inhibition of glycosylation leads to production of misfolded, aggregated proteins that lack function. Certain proteins are highly dependent on glycosylation, whereas others are much less so, and certain glycosylation sites are more important for protein folding than are others.

Studies with model peptides show that oligosaccharides can alter the conformational preferences near the glycosylation sites. In addition, the presence of polar saccharides may serve to orient that portion of a peptide toward the surface of protein domains. However, it has also been found that saccharides are not typically essential for maintaining the overall folded structure after a glycoprotein has reached its native, folded structure.

Source: Helenius, A., and Aebi, M., 2001. Intracellular functions of N-linked glycans. *Science* **291**:2364–2369.

Oligosaccharide Cleavage Can Serve as a Timing Device for Protein Degradation

The slow cleavage of monosaccharide residues from N-linked glycoproteins circulating in the blood targets these proteins for degradation by the organism. The liver contains specific receptor proteins that recognize and bind glycoproteins that are ready to be degraded and recycled. Newly synthesized serum glycoproteins contain N-linked **triantennary** (three-chain) oligosaccharides having structures similar to those in Figure 7.43, in which sialic acid residues cap galactose residues. As these glycoproteins circulate, enzymes on the blood vessel walls cleave off the sialic acid groups, exposing the galactose residues. In the liver, the **asialoglycoprotein receptor** binds the exposed galactose residues of

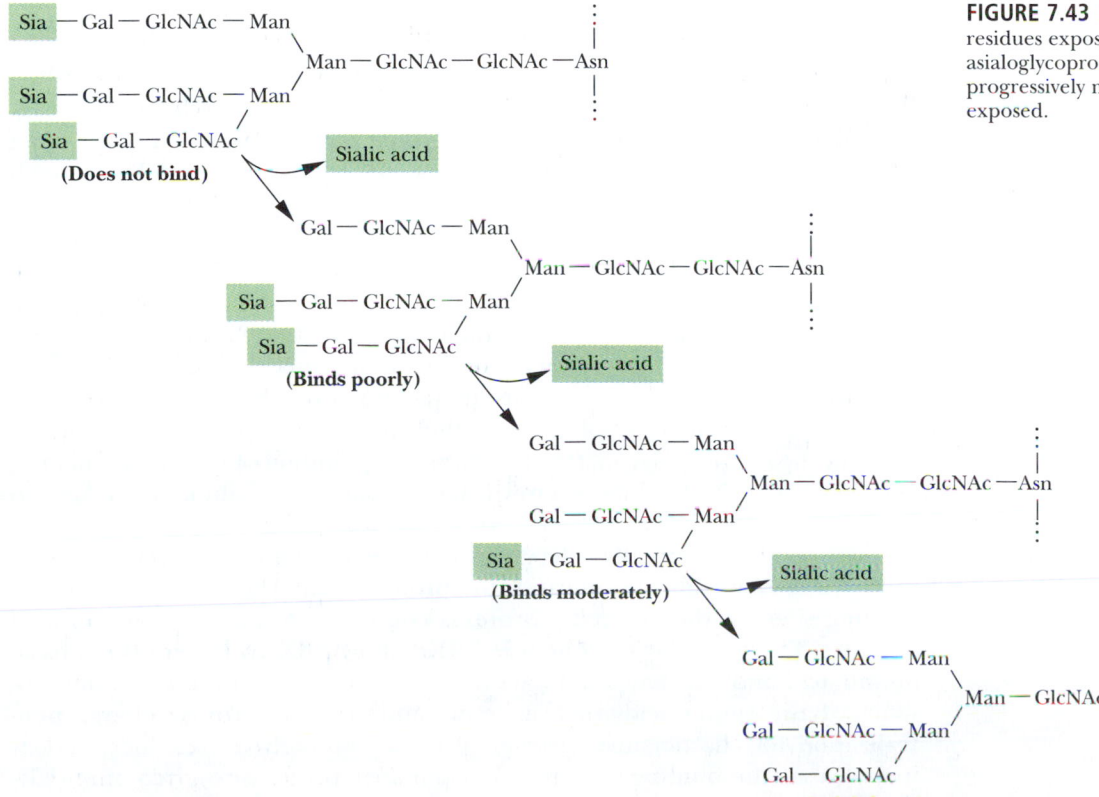

FIGURE 7.43 Progressive cleavage of sialic acid residues exposes galactose residues. Binding to the asialoglycoprotein receptor in the liver becomes progressively more likely as more Gal residues are exposed.

these glycoproteins with very high affinity ($K_D = 10^{-9}$ to 10^{-8} M). The complex of receptor and glycoprotein is then taken into the cell by **endocytosis,** and the glycoprotein is degraded in cellular lysosomes. Highest affinity binding of glycoprotein to the asialoglycoprotein receptor requires three free galactose residues. Oligosaccharides with only one or two exposed galactose residues bind less tightly. This is an elegant way for the body to keep track of how long glycoproteins have been in circulation. Over a period of time—anywhere from a few hours to weeks—the sialic acid groups are cleaved one by one. The longer the glycoprotein circulates and the more sialic acid residues are removed, the more galactose residues become exposed so that the glycoprotein is eventually bound to the liver receptor.

7.6 How Do Proteoglycans Modulate Processes in Cells and Organisms?

Proteoglycans are a family of glycoproteins whose carbohydrate moieties are predominantly **glycosaminoglycans.** The structures of only a few proteoglycans are known, and even these few display considerable diversity (Figure 7.44). Those known range in size from serglycin, having 104 amino acid residues (10.2 kD), to **versican,** having 2409 residues (265 kD). Each of these proteoglycans contains one or two types of covalently linked glycosaminoglycans. In the known proteoglycans, the glycosaminoglycan units are O-linked to serine residues of Ser-Gly dipeptide sequences. Serglycin is named for a unique central domain of 49 amino acids composed of alternating serine and glycine residues. The **cartilage matrix proteoglycan** contains 117 Ser-Gly pairs to which chondroitin sulfates attach. **Decorin,** a small proteoglycan secreted by fibroblasts and found in the extracellular matrix of connective tissues, contains only three Ser-Gly pairs, only one of which is normally glycosylated. In addition to glycosaminoglycan units, proteoglycans may also contain other N-linked and O-linked oligosaccharide groups.

Functions of Proteoglycans Involve Binding to Other Proteins

Proteoglycans may be *soluble* and located in the extracellular matrix, as is the case for serglycin, versican, and the cartilage matrix proteoglycan, or they may be *integral transmembrane proteins,* such as **syndecan.** Both types of proteoglycan appear to function by interacting with a variety of other molecules through their glycosaminoglycan components and through specific receptor domains in the polypeptide itself. For example, syndecan (from the Greek *syndein,* meaning "to bind together") is a transmembrane proteoglycan that associates intracellularly with the actin cytoskeleton (see Chapter 16). Outside the cell, it interacts with **fibronectin,** an extracellular protein that binds to several cell surface proteins and to components of the extracellular matrix. The ability of syndecan to participate in multiple interactions with these target molecules allows them to act as a sort of "glue" in the extracellular space, linking components of the extracellular matrix, facilitating the binding of cells to the matrix, and mediating the binding of growth factors and other soluble molecules to the matrix and to cell surfaces (Figure 7.45).

Many of the functions of proteoglycans involve the binding of specific proteins to the glycosaminoglycan groups of the proteoglycan. The glycosaminoglycan-binding sites on these specific proteins contain multiple basic amino acid residues. The amino acid sequences BBXB and BBBXXB (where B is a basic amino acid and X is any amino acid) recur repeatedly in these binding domains. Basic amino acids such as lysine and arginine provide charge neutralization for the negative charges of glycosaminoglycan residues, and in many cases, the binding of extracellular matrix proteins to glycosaminogly-

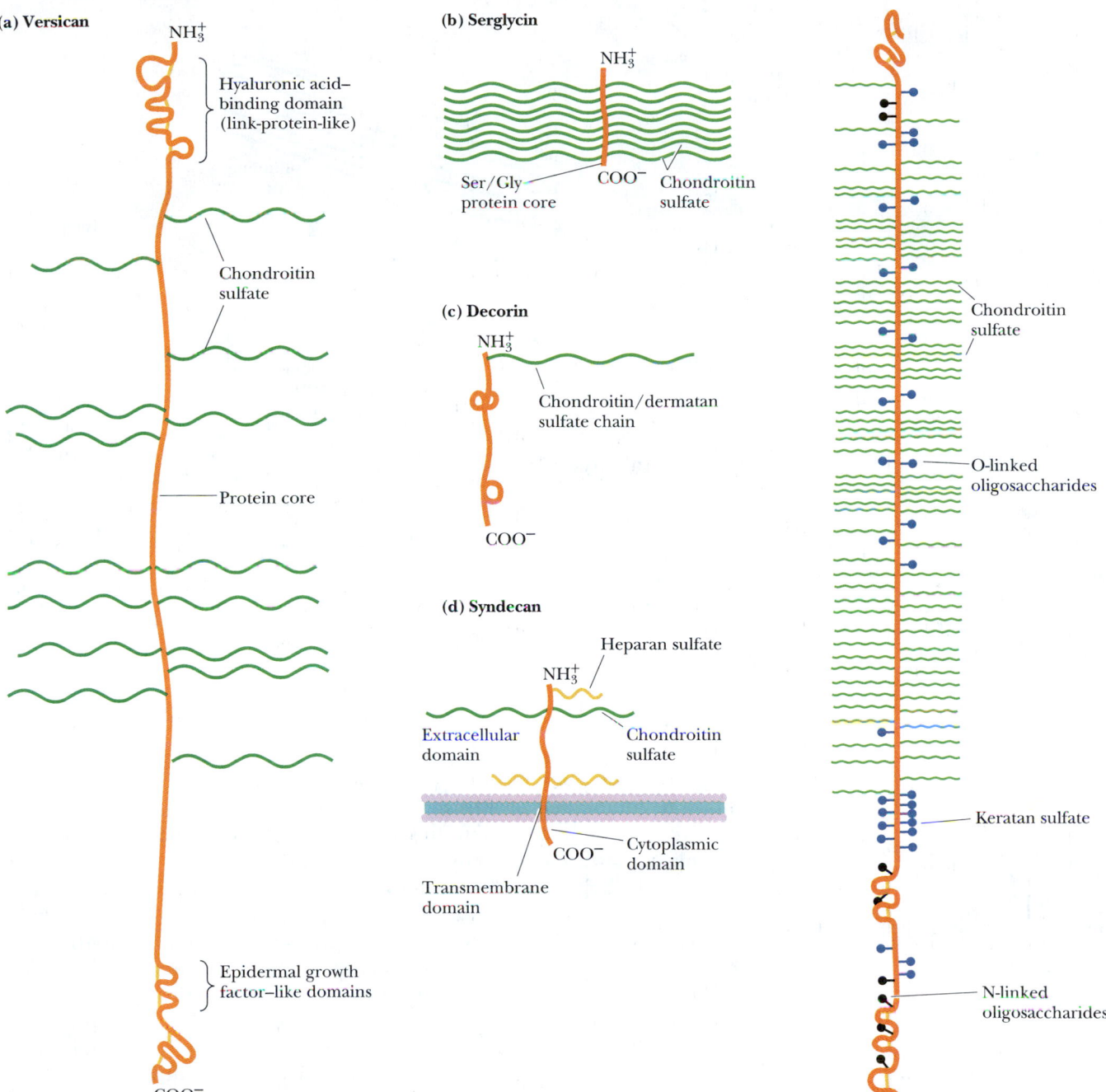

(a) Versican

Hyaluronic acid–binding domain (link-protein-like)

Chondroitin sulfate

Protein core

Epidermal growth factor–like domains

(b) Serglycin

Ser/Gly protein core

Chondroitin sulfate

(c) Decorin

Chondroitin/dermatan sulfate chain

(d) Syndecan

Heparan sulfate

Extracellular domain

Chondroitin sulfate

Cytoplasmic domain

Transmembrane domain

(e) Rat cartilage proteoglycan

Chondroitin sulfate

O-linked oligosaccharides

Keratan sulfate

N-linked oligosaccharides

FIGURE 7.44 The known proteoglycans include a variety of structures. The carbohydrate groups of proteoglycans are predominantly glycosaminoglycans O-linked to serine residues. Proteoglycans include both soluble proteins and integral transmembrane proteins.

cans is primarily charge-dependent. For example, more highly sulfated glycosaminoglycans bind more tightly to fibronectin. However, certain protein–glycosaminoglycan interactions require a specific carbohydrate sequence. A particular pentasaccharide sequence in heparin, for example, binds tightly to antithrombin III (Figure 7.46), accounting for the anticoagulant properties of heparin. Other glycosaminoglycans interact much more weakly.

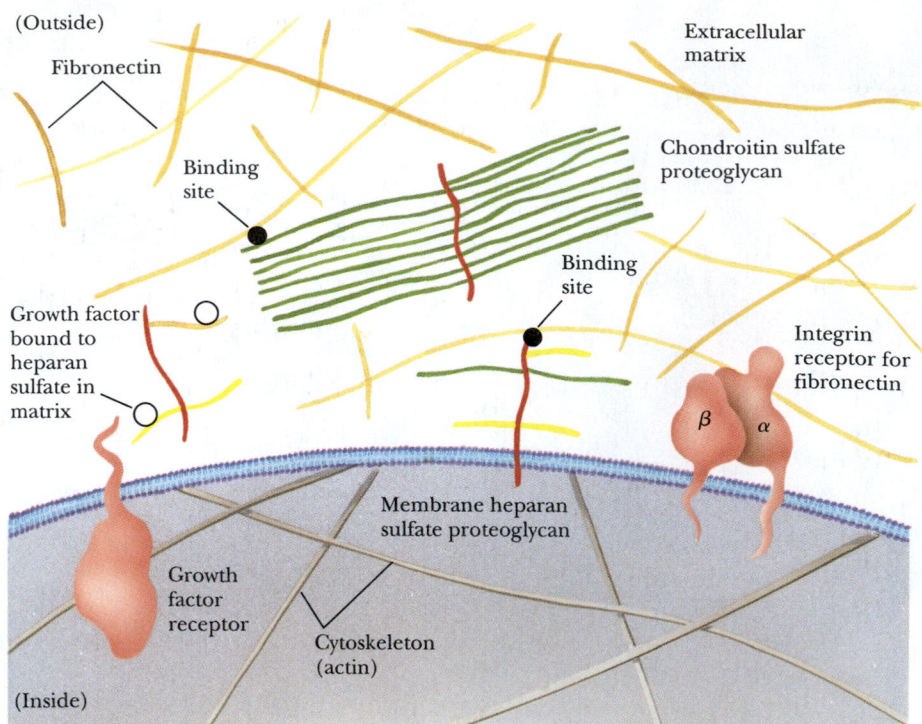

FIGURE 7.45 Proteoglycans serve a variety of functions on the cytoplasmic and extracellular surfaces of the plasma membrane. Many of these functions appear to involve the binding of specific proteins to the glycosaminoglycan groups.

Proteoglycans May Modulate Cell Growth Processes

Several lines of evidence raise the possibility of modulation or regulation of cell growth processes by proteoglycans. First, heparin and heparan sulfate are known to inhibit cell proliferation in a process involving internalization of the glycosaminoglycan moiety and its migration to the cell nucleus. Second, **fibroblast growth factor** binds tightly to heparin and other glycosaminoglycans, and the heparin–growth factor complex protects the growth factor from degradative enzymes, thus enhancing its activity. There is evidence that binding of fibroblast growth factors by proteoglycans and glycosaminoglycans in the extracellular matrix creates a reservoir of growth factors for cells to use. Third, **transforming growth factor β** has been shown to stimulate the synthesis and secretion of proteoglycans in certain cells. Fourth, several proteoglycan core proteins, including versican and **lymphocyte homing receptor,** have domains similar in sequence to those of **epidermal growth factor** and **complement regulatory factor.** These growth factor domains may interact specifically with growth factor receptors in the cell membrane in processes that are not yet understood.

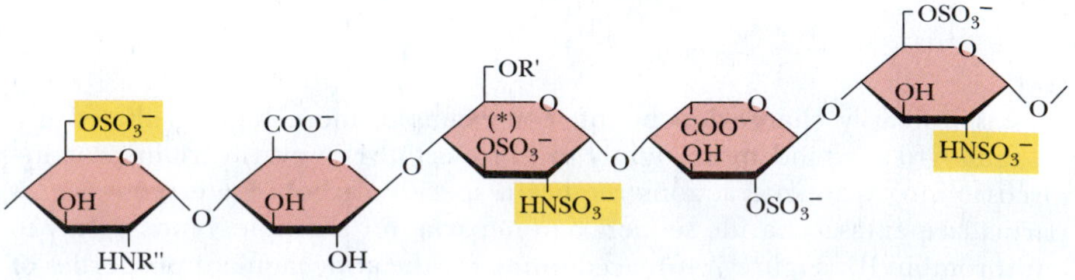

FIGURE 7.46 A portion of the structure of heparin, a carbohydrate having anticoagulant properties. It is used by blood banks to prevent the clotting of blood during donation and storage and also by physicians to prevent the formation of life-threatening blood clots in patients recovering from serious injury or surgery. This sulfated pentasaccharide sequence in heparin binds with high affinity to antithrombin III, accounting for this anticoagulant activity. The 3-O-sulfate marked by an asterisk is essential for high-affinity binding of heparin to antithrombin III.

Proteoglycans Make Cartilage Flexible and Resilient

Cartilage matrix proteoglycan is responsible for the flexibility and resilience of cartilage tissue in the body. In cartilage, long filaments of hyaluronic acid are studded or coated with proteoglycan molecules, as shown in Figure 7.47. The hyaluronate chains can be as long as 4 μm and can coordinate 100 or more proteoglycan units. Cartilage proteoglycan possesses a **hyaluronic**

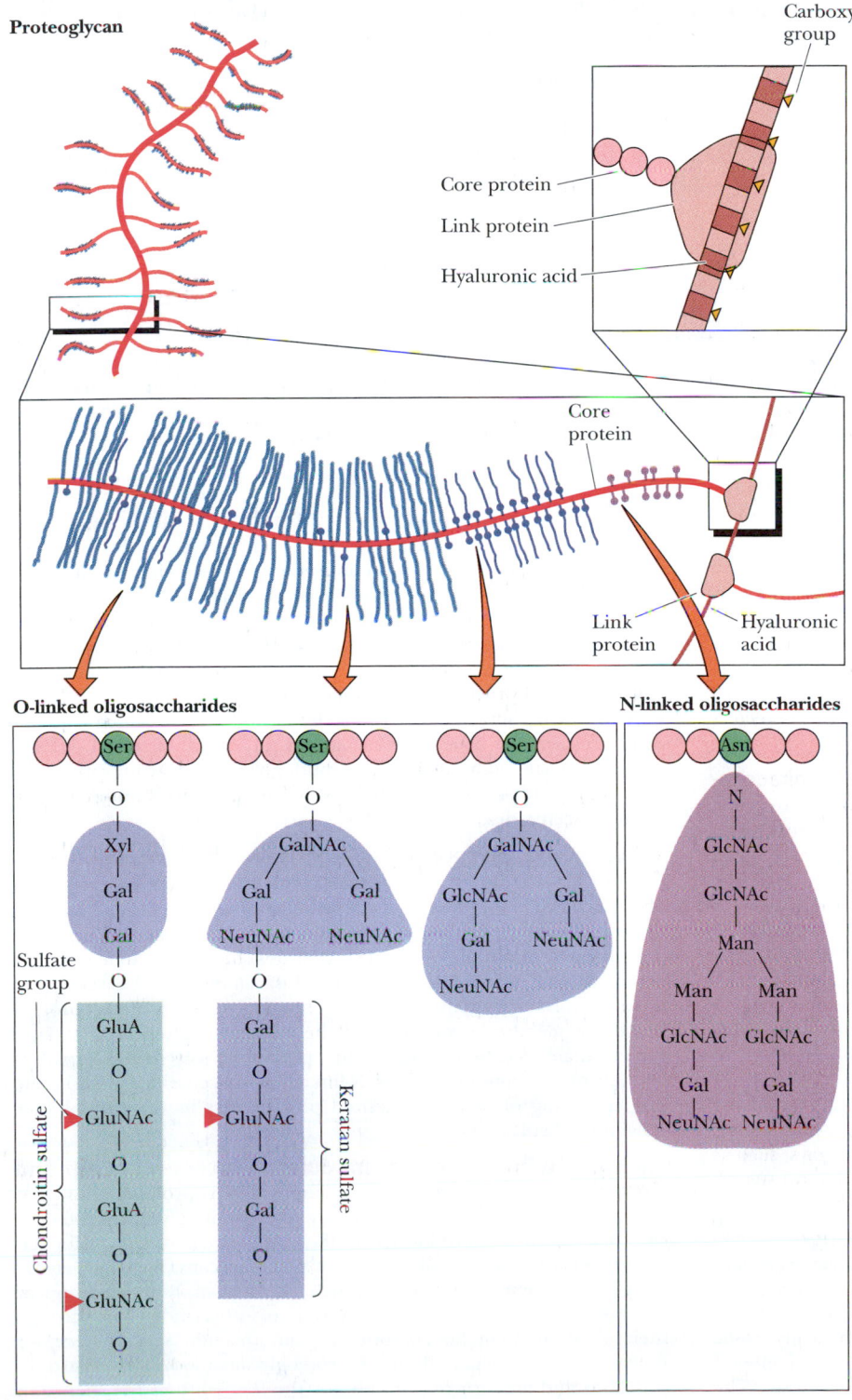

FIGURE 7.47 Hyaluronate (see Figure 7.33) forms the backbone of proteoglycan structures, such as those found in cartilage. The proteoglycan subunits consist of a core protein containing numerous O-linked and N-linked glycosaminoglycans. In cartilage, these highly hydrated proteoglycan structures are enmeshed in a network of collagen fibers. Release (and subsequent reabsorption) of water by these structures during compression accounts for the shock-absorbing qualities of cartilaginous tissue.

acid–binding domain on the NH_2-terminal portion of the polypeptide, which binds to hyaluronate with the assistance of a **link protein.** The proteoglycan–hyaluronate aggregates can have molecular weights of 2 million or more.

The proteoglycan–hyaluronate aggregates are highly hydrated by virtue of strong interactions between water molecules and the polyanionic complex. When cartilage is compressed (such as when joints absorb the impact of walking or running), water is briefly squeezed out of the cartilage tissue and then reabsorbed when the stress is diminished. This reversible hydration gives cartilage its flexible, shock-absorbing qualities and cushions the joints during physical activities that might otherwise injure the involved tissues.

Summary

Carbohydrates are a versatile class of molecules of the formula $(CH_2O)_n$. They are a major form of stored energy in organisms, and they are the metabolic precursors of virtually all other biomolecules. Carbohydrates linked to lipids (glycolipids) are components of biological membranes. Carbohydrates linked to proteins (glycoproteins) are important components of cell membranes and function in recognition between cell types and recognition of cells by other molecules. Recognition events are important in cell growth, differentiation, fertilization, tissue formation, transformation of cells, and other processes.

7.1 How Are Carbohydrates Named? Carbohydrates are classified into three groups: monosaccharides, oligosaccharides, and polysaccharides. Monosaccharides cannot be broken down into smaller sugars under mild conditions. Oligosaccharides consist of from two to ten simple sugar molecules. Polysaccharides are polymers of simple sugars and their derivatives and may be branched or linear. Their molecular weights range up to 1 million or more.

7.2 What Is the Structure and Chemistry of Monosaccharides? Monosaccharides consist typically of three to seven carbon atoms and are described as either aldoses or ketoses. Aldoses with at least three carbons and ketoses with at least four carbons contain chiral centers. The prefixes D- and L- are often used to indicate the configuration of the highest numbered asymmetric carbon. The D- and L-forms of a monosaccharide are mirror images of each other, called enantiomers. Pairs of isomers that have opposite configurations at one or more chiral centers, but are not mirror images of each other, are called diastereomers. Sugars that differ in configuration at only one chiral center are epimers. An interesting feature of carbohydrates is their ability to form cyclic structures with formation of an additional asymmetric center. Aldoses and ketoses with five or more carbons can form either furanose or pyranose rings, and the more stable form depends on structural factors. A variety of chemical and enzymatic reactions produce derivatives of simple sugars, such as sugar acids, sugar alcohols, deoxy sugars, sugar esters, amino sugars, acetals, ketals, and glycosides.

7.3 What Is the Structure and Chemistry of Oligosaccharides? The complex array of oligosaccharides in higher organisms is formed from relatively few different monosaccharide units, particularly glucose, fructose, mannose, galactose, ribose, and xylose. Disaccharides consist of two monosaccharide units linked by a glycosidic bond, and each individual unit is termed a residue. The most common disaccharides in nature are sucrose, maltose, and lactose. The anomeric carbons of oligosaccharides may be substituted or unsubstituted. Disaccharides with a free, unsubstituted anomeric carbon can reduce oxidizing agents and thus are termed reducing sugars. More complex oligosaccharides include the cycloamyloses; stachyose (found in beans, peas, bran, and whole grains); and amygdalin, a constituent of bitter almonds and the kernals and pits of cherries, peaches, and apricots.

7.4 What Is the Structure and Chemistry of Polysaccharides? Polysaccharides are formed from monosaccharides and their derivatives. If a polysaccharide consists of only one kind of monosaccharide, it is a homopolysaccharide, whereas those with more than one kind of monosaccharide are heteropolysaccharides. Polysaccharides may function as energy storage materials, structural components of organisms, or protective substances. Starch and glycogen are readily metabolizable and provide energy reserves for cells. Chitin and cellulose provide strong support for the skeletons of arthropods and green plants, respectively. Mucopolysaccharides such as hyaluronic acid form protective coats on animal cells. Peptidoglycan, the strong protective macromolecule of bacterial cell walls, is one of nature's more interesting polysaccharides.

7.5 What Are Glycoproteins, and How Do They Function in Cells? Glycoproteins are proteins that contain covalently linked oligosaccharides and polysaccharides. Carbohydrate groups may be linked to proteins via the hydroxyl groups of serine, threonine, or hydroxylysine residues (in O-linked saccharides) or via the amide nitrogen of an asparagine residue (in N-linked saccharides). O-Glycosylated stems of certain proteins raise the functional domain of the protein above the membrane surface and the associated glycocalyx, making these domains accessible to interacting proteins. *N*-Glycosylation confers a variety of functions to proteins. N-linked oligosaccharides promote the proper folding of newly synthesized polypeptides in the endoplasmic reticulum of eukaryotic cells.

7.6 How Do Proteoglycans Modulate Processes in Cells and Organisms? Proteoglycans are a family of glycoproteins whose carbohydrate moieties are predominantly glycosaminoglycans. Proteoglycans may be soluble and located in the extracellular matrix, as for serglycin, versican, and cartilage matrix proteoglycans, or they may be integral transmembrane proteins, such as syndecan. Both types appear to function by interacting with a variety of other molecules through their glycosaminoglycan components and through specific receptor domains in the polypeptide itself. Proteoglycans modulate cell growth processes and are also responsible for the flexibility and resilience of cartilage tissue in the body.

Problems

1. Draw Haworth structures for the two possible isomers of D-altrose (Figure 7.2) and D-psicose (Figure 7.3).

*2. (Integrates with Chapters 4 and 5.) Consider the peptide DGNILSR, where N has a covalently linked galactose and S has a covalently linked glucose. Draw the structure of this glycopeptide, and also draw titration curves for the glycopeptide and for the free peptide that would result from hydrolysis of the two sugar residues.

3. Give the systematic name for stachyose (Figure 7.19).

4. (Integrates with Chapters 5 and 6.) Human hemoglobin can react with sugars in the blood (usually glucose) to form covalent adducts. The α-amino groups of N-terminal valine in the Hb β-subunits react with the C-1 (aldehyde) carbons of monosaccharides to form aldimine adducts, which rearrange to form very stable ketoamine products. Quantitation of this "glycated hemoglobin" is important clinically, especially for diabetic individuals. Suggest at least three methods by which glycated Hb could be separated from normal Hb and quantitated.

5. Trehalose, a disaccharide produced in fungi, has the following structure:

   a. What is the systematic name for this disaccharide?
   b. Is trehalose a reducing sugar? Explain.

6. Draw a Fischer projection structure for L-sorbose (D-sorbose is shown in Figure 7.3).

7. α-D-Glucose has a specific rotation, $[\alpha]_D^{20}$, of $+112.2°$, whereas β-D-glucose has a specific rotation of $+18.7°$. What is the composition of a mixture of α-D- and β-D-glucose, which has a specific rotation of $83.0°$?

8. Use the information in the Critical Developments in Biochemistry box titled "Rules for Description of Chiral Centers in the (R,S) System" (Chapter 4) to name D-galactose using (R,S) nomenclature. Do the same for L-altrose.

*9. A 0.2-g sample of amylopectin was analyzed to determine the fraction of the total glucose residues that are branch points in the structure. The sample was exhaustively methylated and then digested, yielding 50 μmol of 2,3-dimethylglucose and 0.4 μmol of 1,2,3,6-tetramethylglucose.

   a. What fraction of the total residues are branch points?
   b. How many reducing ends does this amylopectin have?

10. (Integrates with Chapters 5, 6, and 9.) Consider the sequence of glycophorin (see Figure 9.14), and imagine subjecting glycophorin, and also a sample of glycophorin treated to remove all sugars, to treatment with trypsin and chymotrypsin. Would the presence of sugars in the native glycophorin make any difference to the results?

11. (Integrates with Chapters 4, 5, and 23.) The caloric content of protein and carbohydrate are quite similar, at approximately 16 to 17 kJ/g, whereas that of fat is much higher, at 38 kJ/g. Discuss the chemical basis for the similarity of the values for carbohydrate and for protein.

12. Write a reasonable chemical mechanism for the starch phosphorylase reaction (Figure 7.23).

*13. The commercial product Beano contains an enzyme that hydrolyzes stachyose and related oligosaccharides. Write a chemical mechanism for this reaction, name the enzyme, and explain why this product prevents intestinal gas.

14. Laetrile treatment is offered in some countries as a cancer therapy. This procedure is dangerous, and there is no valid clinical evidence of its efficacy. Suggest at least one reason that laetrile treatment could be dangerous for human patients.

15. Treatment with chondroitin and glucosamine is offered as one popular remedy for arthritis pain. Suggest an argument for the efficacy of this treatment, and then comment on its validity, based on what you know of polysaccharide chemistry.

Preparing for the MCAT Exam

16. Heparin has a characteristic pattern of hydroxy and anionic functions. What amino acid side chains on antithrombin III might be the basis for the strong interactions between this protein and the anticoagulant heparin?

17. What properties of hyaluronate, chondroitin sulfate, and keratan sulfate make them ideal components of cartilage?

Biochemistry⊗Now™   Preparing for an exam? Test yourself on key questions at http://chemistry.brookscole.com/ggb3

Further Reading

Carbohydrate Structure and Chemistry

Collins, P. M., 1987. *Carbohydrates.* London: Chapman and Hall.

Davison, E. A., 1967. *Carbohydrate Chemistry.* New York: Holt, Rinehart and Winston

Pigman, W., and Horton, D., 1972. *The Carbohydrates.* New York: Academic Press.

Sharon, N., 1980. Carbohydrates. *Scientific American* **243:**90–102.

Polysaccharides

Aspinall, G. O., 1982. *The Polysaccharides,* Vols. 1 and 2. New York: Academic Press.

Höfte, H., 2001. A baroque residue in red wine. *Science* **294:**795–797.

McNeil, M., Darvill, A. G., Fry, S. C., and Albersheim, P., 1984. Structure and function of the primary cell walls of plants. *Annual Review of Biochemistry* **53:**625–664.

O'Neill, M. A., Eberhard, S., Albersheim, P., and Darvill, A. G., 2002. Requirements of borate cross-linking of cell wall rhamnogalacturonan II for *Arabidopsis* growth. *Science* **294:**846–849.

Glycoproteins

Feeney, R. E., Burcham, T. S., and Yeh, Y., 1986. Antifreeze glycoproteins from polar fish blood. *Annual Review of Biophysical Chemistry* **15:**59–78.

Helenius, A., and Aebi, M., 2001. Intracellular functions of N-linked glycans. *Science* **291**:2364–2369.

Jentoft, N., 1990. Why are proteins O-glycosylated? *Trends in Biochemical Sciences* **155**:291–294.

Sharon, N., 1984. Glycoproteins. *Trends in Biochemical Sciences* **9**:198–202.

Proteoglycans

Day, A. J., and Prestwich, G. D., 2002. Hyaluronan-binding proteins: Tying up the giant. *Journal of Biological Chemistry* **277**:4585–4588.

Kjellen, L., and Lindahl, U., 1991. Proteoglycans: Structures and interactions. *Annual Review of Biochemistry* **60**:443–475.

Lennarz, W. J., 1980. *The Biochemistry of Glycoproteins and Proteoglycans.* New York: Plenum Press.

Ruoslahti, E., 1989. Proteoglycans in cell regulation. *Journal of Biological Chemistry* **264**:13369–13372.

Glycobiology

Bertozzi, C. R., and Kiessling, L. L., 2001. Chemical glycobiology. *Science* **291**:2357–2363.

Lodish, H. F., 1991. Recognition of complex oligosaccharides by the multisubunit asialoglycoprotein receptor. *Trends in Biochemical Sciences* **16**:374–377.

Maeder, T., 2002. Sweet medicines. *Scientific American* **287**:40–47.

Rademacher, T. W., Parekh, R. B., and Dwek, R. A., 1988. Glycobiology. *Annual Review of Biochemistry* **57**:785–838.

Lipids

CHAPTER 8

Essential Question

Lipids are a class of biological molecules defined by low solubility in water and high solubility in nonpolar solvents. As molecules that are largely hydrocarbon in nature, lipids represent highly reduced forms of carbon and, upon oxidation in metabolism, yield large amounts of energy. Lipids are thus the molecules of choice for metabolic energy storage. *What is the structure, chemistry, and biological function of lipids?*

The lipids found in biological systems are either **hydrophobic** (containing only nonpolar groups) or **amphipathic** (possessing both polar and nonpolar groups). The hydrophobic nature of lipid molecules allows membranes to act as effective barriers to more polar molecules. In this chapter, we discuss the chemical and physical properties of the various classes of lipid molecules. The following chapter considers membranes, whose properties depend intimately on their lipid constituents.

© Brandon D. Cole/CORBIS

"The mighty whales which swim in a sea of water, and have a sea of oil swimming in them." Herman Melville, "Extracts." *Moby Dick*. New York: Penguin Books, 1972. (*Humpback whale* [Megaptera novaeangliae] *breaching, Cape Cod, MA*)

A feast of fat things, a feast of wines on the lees.
Isiah 25:6

8.1 | What Is the Structure and Chemistry of Fatty Acids?

A **fatty acid** is composed of a long hydrocarbon chain ("tail") and a terminal carboxyl group (or "head"). The carboxyl group is normally ionized under physiological conditions. Fatty acids occur in large amounts in biological systems but only rarely in the free, uncomplexed state. They typically are esterified to glycerol or other backbone structures. Most of the fatty acids found in nature have an even number of carbon atoms (usually 14 to 24). Certain marine organisms, however, contain substantial amounts of fatty acids with odd numbers of carbon atoms. Fatty acids are either **saturated** (all carbon–carbon bonds are single bonds) or **unsaturated** (with one or more double bonds in the hydrocarbon chain). If a fatty acid has a single double bond, it is said to be **monounsaturated,** and if it has more than one, **polyunsaturated.** Fatty acids can be named or described in at least three ways, as listed in Table 8.1. For example, a fatty acid composed of an 18-carbon chain with no double bonds can be called by its systematic name (**octadecanoic acid),** its common name (stearic acid), or its shorthand notation, in which the number of carbons is followed by a colon and the number of double bonds in the molecule (18:0 for stearic acid). The structures of several fatty acids are given in Figure 8.1. **Stearic acid** (18:0) and **palmitic acid** (16:0) are the most common saturated fatty acids in nature.

Free rotation around each of the carbon–carbon bonds makes saturated fatty acids extremely flexible molecules. Owing to steric constraints, however, the fully extended conformation (Figure 8.1) is the most stable for saturated fatty acids. Nonetheless, the degree of stabilization is slight, and (as will be seen) saturated fatty acid chains adopt a variety of conformations.

Unsaturated fatty acids are slightly more abundant in nature than saturated fatty acids, especially in higher plants. The most common unsaturated fatty acid is **oleic acid,** or 18:1(9), with the number in parentheses indicating that the double bond is between carbons 9 and 10. The number of double bonds in an unsaturated fatty acid typically varies from one to four, but in the fatty acids found in most bacteria, this number rarely exceeds one.

The double bonds found in fatty acids are nearly always in the *cis* configuration. As shown in Figure 8.1, this causes a bend or "kink" in the fatty acid chain. This bend has very important consequences for the structure of biological

Key Questions

Biochemistry⊗**Now**™ Test yourself on these Key Questions at BiochemistryNow at **http://chemistry.brookscole.com/ggb3**

Table 8.1

Common Biological Fatty Acids

Number of Carbons	Common Name	Systematic Name	Symbol	Structure
Saturated fatty acids				
12	Lauric acid	Dodecanoic acid	12:0	$CH_3(CH_2)_{10}COOH$
14	Myristic acid	Tetradecanoic acid	14:0	$CH_3(CH_2)_{12}COOH$
16	Palmitic acid	Hexadecanoic acid	16:0	$CH_3(CH_2)_{14}COOH$
18	Stearic acid	Octadecanoic acid	18:0	$CH_3(CH_2)_{16}COOH$
20	Arachidic acid	Eicosanoic acid	20:0	$CH_3(CH_2)_{18}COOH$
22	Behenic acid	Docosanoic acid	22:0	$CH_3(CH_2)_{20}COOH$
24	Lignoceric acid	Tetracosanoic acid	24:0	$CH_3(CH_2)_{22}COOH$
Unsaturated fatty acids (all double bonds are cis)				
16	Palmitoleic acid	9-Hexadecenoic acid	16:1	$CH_3(CH_2)_5CH=CH(CH_2)_7COOH$
18	Oleic acid	9-Octadecenoic acid	18:1	$CH_3(CH_2)_7CH=CH(CH_2)_7COOH$
18	Linoleic acid	9,12-Octadecadienoic acid	18:2	$CH_3(CH_2)_4(CH=CHCH_2)_2(CH_2)_6COOH$
18	α-Linolenic acid	9,12,15-Octadecatrienoic acid	18:3	$CH_3CH_2(CH=CHCH_2)_3(CH_2)_6COOH$
18	γ-Linolenic acid	6,9,12-Octadecatrienoic acid	18:3	$CH_3(CH_2)_4(CH=CHCH_2)_3(CH_2)_3COOH$
20	Arachidonic acid	5,8,11,14-Eicosatetraenoic acid	20:4	$CH_3(CH_2)_4(CH=CHCH_2)_4(CH_2)_2COOH$
24	Nervonic acid	15-Tetracosenoic acid	24:1	$CH_3(CH_2)_7CH=CH(CH_2)_{13}COOH$

membranes. Saturated fatty acid chains can pack closely together to form ordered, rigid arrays under certain conditions, but unsaturated fatty acids prevent such close packing and produce flexible, fluid aggregates.

Some fatty acids are not synthesized by mammals and yet are necessary for normal growth and life. These essential fatty acids include **linoleic** and **γ-linolenic acids.** These must be obtained by mammals in their diet (specifically from plant sources). **Arachidonic acid,** which is not found in plants, can be synthesized by mammals only from linoleic acid. At least one function of the essential fatty acids is to serve as a precursor for the synthesis of **eicosanoids,** such as prostaglandins, a class of compounds that exert hormonelike effects in many physiological processes (discussed in Chapter 24).

In addition to unsaturated fatty acids, several other modified fatty acids are found in nature. Microorganisms, for example, often contain branched-chain fatty acids, such as **tuberculostearic acid** (Figure 8.2). When these fatty acids are incorporated in membranes, the methyl group constitutes a local structural perturbation in a manner similar to the double bonds in unsaturated fatty acids (see Chapter 9). Some bacteria also synthesize fatty acids containing cyclic structures such as cyclopropane, cyclopropene, and even cyclopentane rings.

8.2 | What Is the Structure and Chemistry of Triacylglycerols?

A significant number of the fatty acids in plants and animals exist in the form of **triacylglycerols** (also called **triglycerides**). Triacylglycerols are a major energy reserve and the principal neutral derivatives of glycerol found in animals. These molecules consist of a glycerol esterified with three fatty acids (Figure 8.3). If all three fatty acid groups are the same, the molecule is called a simple triacylglycerol. Examples include **tristearoylglycerol** (common name tristearin) and **trioleoylglycerol** (triolein). Mixed triacylglycerols contain two or three different fatty acids. Triacylglycerols in animals are found primarily in the adipose tissue

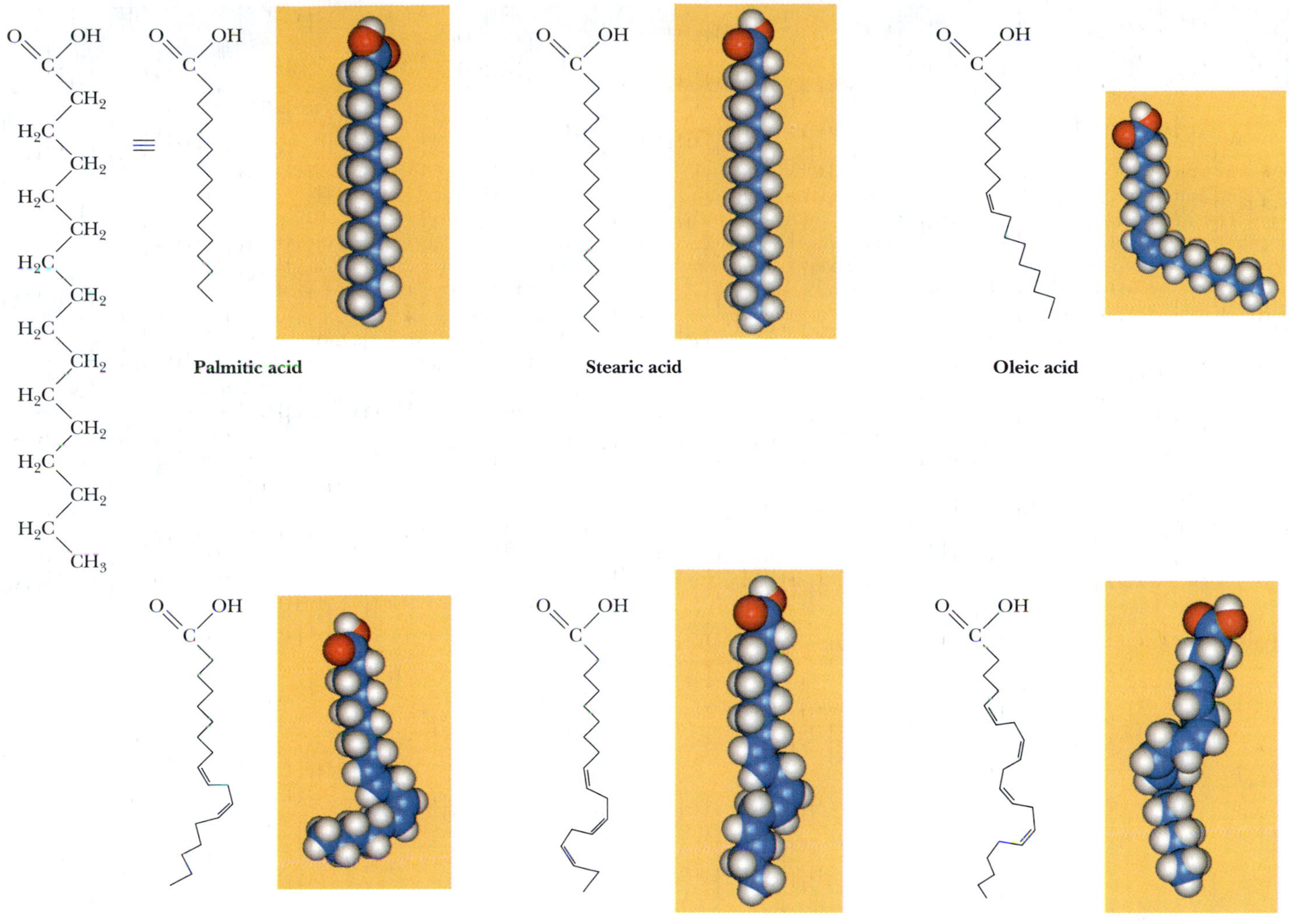

Palmitic acid

Stearic acid

Oleic acid

Linoleic acid

α-Linolenic acid

Arachidonic acid

Biochemistry ⋐Now™   ANIMATED FIGURE 8.1 The structures of some typical fatty acids. Note that most natural fatty acids contain an even number of carbon atoms and that the double bonds are nearly always *cis* and rarely conjugated. **See this figure animated at http://chemistry. brookscole.com/ggb3**

(body fat), which serves as a depot or storage site for lipids. Monoacylglycerols and diacylglycerols also exist, but they are far less common than the triacylglycerols. Most natural plant and animal fat is composed of mixtures of simple and mixed triacylglycerols.

Acylglycerols can be hydrolyzed by heating with acid or base or by treatment with lipases. Hydrolysis with alkali is called **saponification** and yields salts of free fatty acids and glycerol. This is how our ancestors made **soap** (a metal salt of an acid derived from fat). One method used potassium hydroxide (potash) leached from wood ashes to hydrolyze animal fat (mostly triacylglycerols). (The tendency of such soaps to be precipitated by Mg^{2+} and Ca^{2+} ions in hard water makes them less useful than modern detergents.) When the fatty acids esterified at the first and third carbons of glycerol are different, the second carbon is asymmetric. The various acylglycerols are normally soluble in benzene, chloroform, ether, and hot ethanol. Although triacylglycerols are insoluble in water, monoacylglycerols and diacylglycerols readily form organized structures in water (see Chapter 9), owing to the polarity of their free hydroxyl groups.

Triacylglycerols are rich in highly reduced carbons and thus yield large amounts of energy in the oxidative reactions of metabolism. Complete oxidation of 1 g of triacylglycerols yields about 38 kJ of energy, whereas proteins and

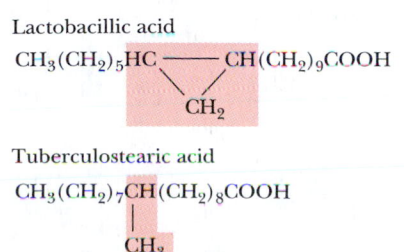

Lactobacillic acid

$$CH_3(CH_2)_5HC \underset{\underset{CH_2}{\diagdown \diagup}}{\text{————}} CH(CH_2)_9COOH$$

Tuberculostearic acid

$$CH_3(CH_2)_7\underset{\underset{CH_3}{|}}{CH}(CH_2)_8COOH$$

FIGURE 8.2 Structures of two unusual fatty acids: lactobacillic acid, a fatty acid containing a cyclopropane ring, and tuberculostearic acid, a branched-chain fatty acid.

Human Biochemistry

Fatty Acids in Food: Saturated Versus Unsaturated

Fats consumed in the modern human diet vary widely in their fatty acid compositions. The following table provides a brief summary. The incidence of cardiovascular disease is correlated with diets high in saturated fatty acids. By contrast, a diet that is relatively higher in unsaturated fatty acids (especially polyunsaturated fatty acids) may reduce the risk of heart attacks and strokes. Corn oil, abundant in the United States and high in (polyunsaturated) linoleic acid, is an attractive dietary choice. *Margarine* made from corn, safflower, or sunflower oils is much lower in saturated fatty acids than is butter, which is made from milk fat. However, margarine may present its own health risks. Its fatty acids contain *trans* double bonds (introduced by the hydrogenation process), which also contribute to cardiovascular disease. (Margarine was invented by a French chemist, H. Mège Mouriès, who won a prize from Napoleon III in 1869 for developing a substitute for butter.)

Although vegetable oils usually contain a higher proportion of unsaturated fatty acids than do animal oils and fats, several plant oils are actually high in saturated fats. Palm oil is low in polyunsaturated fatty acids and particularly high in (saturated) palmitic acid (whence the name *palmitic*). Coconut oil is particularly high in lauric and myristic acids (both saturated) and contains very few unsaturated fatty acids.

Some of the fatty acids found in the diets of developed nations (often 1 to 10 g of daily fatty acid intake) are *trans* fatty acids—fatty acids with one or more double bonds in the *trans* configuration. Some of these derive from dairy fat and ruminant meats, but the bulk are provided by partially hydrogenated vegetable or fish oils. Numerous studies have shown that *trans* fatty acids raise plasma low-density lipoprotein (LDL) cholesterol levels when exchanged for *cis*-unsaturated fatty acids in the diet and may also lower high-density lipoprotein (HDL) cholesterol levels and raise triglyceride levels. The effects of *trans* fatty acids on LDL, HDL, and cholesterol levels are similar to those of saturated fatty acids, and diets aimed at reducing the risk of coronary heart disease should be low in both *trans* and saturated fatty acids.

Fatty Acid Compositions of Some Dietary Lipids*

Source	Lauric and Myristic	Palmitic	Stearic	Oleic	Linoleic
Beef	5	24–32	20–25	37–43	2–3
Milk		25	12	33	3
Coconut	74	10	2	7	
Corn		8–12	3–4	19–49	34–62
Olive		9	2	84	4
Palm		39	4	40	8
Safflower		6	3	13	78
Soybean		9	6	20	52
Sunflower		6	1	21	66

Data from *Merck Index,* 10th ed. Rahway, NJ: Merck and Co.; and Wilson, E. D., et al., 1979, *Principles of Nutrition,* 4th ed. New York: Wiley.
*Values are percentages of total fatty acids.

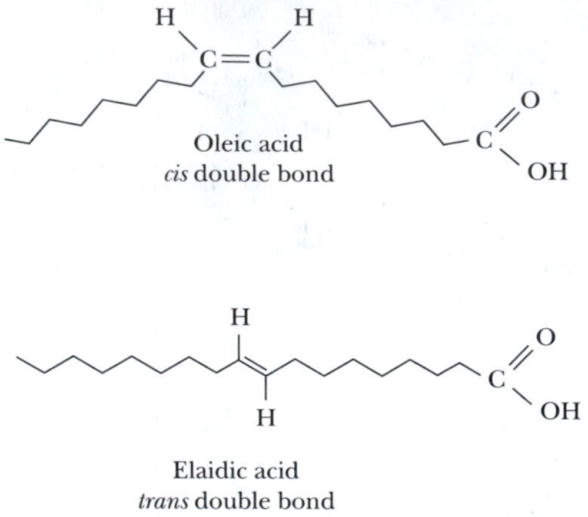

Oleic acid
cis double bond

Elaidic acid
trans double bond

▲ Structure of *cis* and *trans* monounsaturated C_{18} fatty acids.

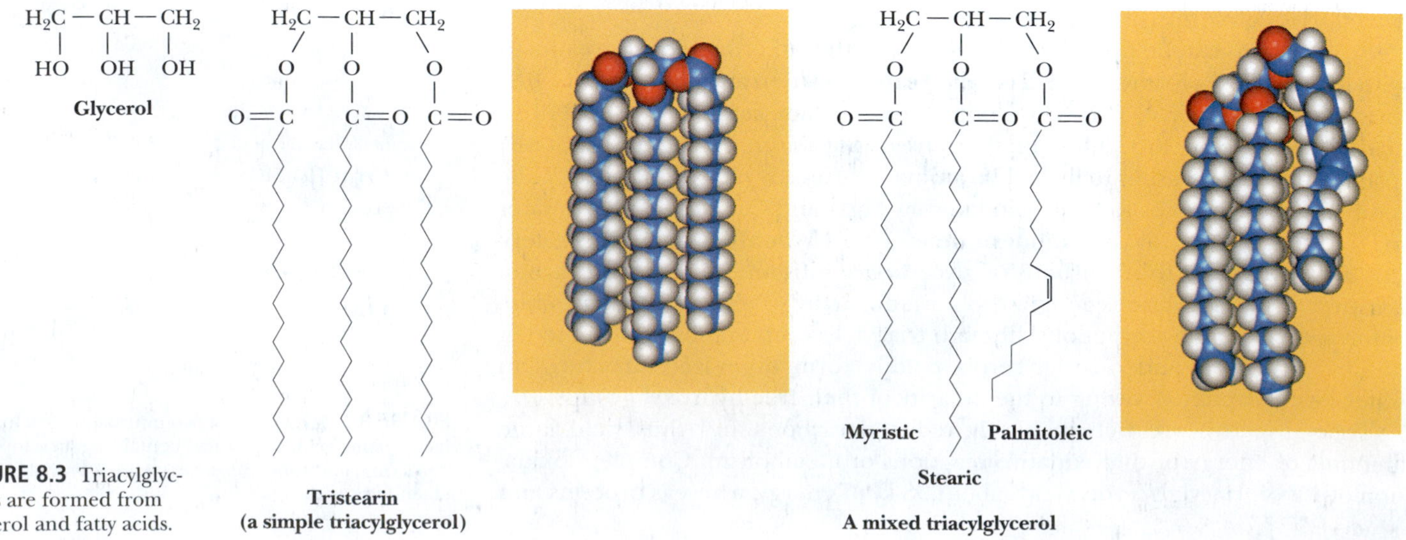

Glycerol

Tristearin
(a simple triacylglycerol)

Myristic   Palmitoleic

Stearic

A mixed triacylglycerol

FIGURE 8.3 Triacylglycerols are formed from glycerol and fatty acids.

A Deeper Look

Polar Bears Prefer Nonpolar Food

The polar bear is magnificently adapted to thrive in its harsh Arctic environment. Research by Malcolm Ramsay (at the University of Saskatchewan in Canada) and others has shown that polar bears eat only during a few weeks out of the year and then fast for periods of 8 months or more, consuming no food or water during that time. Eating mainly in the winter, the adult polar bear feeds almost exclusively on seal blubber (largely composed of triacylglycerols), thus building up its own triacylglycerol reserves. Through the Arctic summer, the polar bear maintains normal physical activity, roaming over long distances, but relies entirely on its body fat for sustenance, burning as much as 1 to 1.5 kg of fat per day. It neither urinates nor defecates for extended periods. All the water needed to sustain life is provided from the metabolism of triacylglycerides (because oxidation of fatty acids yields carbon dioxide and water).

Ironically, the word *Arctic* comes from the ancient Greeks, who understood that the northernmost part of the earth lay under the stars of the constellation Ursa Major, the Great Bear. Although unaware of the polar bear, they called this region *Arktikós*, which means "the country of the great bear."

© Kennan Ward/CORBIS

carbohydrates yield only about 17 kJ/g. Also, their hydrophobic nature allows them to aggregate in highly anhydrous forms, whereas polysaccharides and proteins are highly hydrated. For these reasons, triacylglycerols are the molecules of choice for energy storage in animals. Body fat (mainly triacylglycerols) also provides good insulation. Whales and Arctic mammals rely on body fat for both insulation and energy reserves.

8.3 | What Is the Structure and Chemistry of Glycerophospholipids?

A 1,2-diacylglycerol that has a phosphate group esterified at carbon atom 3 of the glycerol backbone is a **glycerophospholipid,** also known as a phosphoglyceride or a glycerol phosphatide (Figure 8.4). These lipids form one of the largest and most important classes of natural lipids. They are essential components of cell membranes and are found in small concentrations in other parts of the cell. It should be noted that all glycerophospholipids are members of the broader class of lipids known as **phospholipids.**

The numbering and nomenclature of glycerophospholipids present a dilemma in that the number 2 carbon of the glycerol backbone of a phospholipid

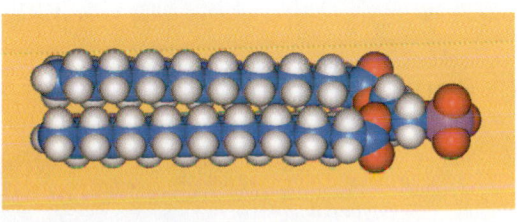

FIGURE 8.4 Phosphatidic acid, the parent compound for glycerophospholipids.

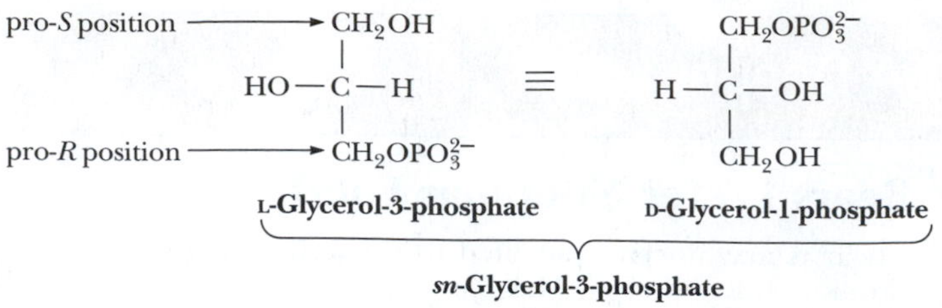

Biochemistry Now™ ACTIVE FIGURE 8.5 The absolute configuration of *sn*-glycerol-3-phosphate. The pro-*R* and pro-*S* positions of the parent glycerol are also indicated. **Test yourself on the concepts in this figure at http://chemistry.brookscole.com/ggb3**

is asymmetric. It is possible to name these molecules either as D- or L-isomers. Thus, glycerol phosphate itself can be referred to either as D-glycerol-1-phosphate or as L-glycerol-3-phosphate (Figure 8.5). Instead of naming the glycerol phosphatides in this way, biochemists have adopted the *stereospecific numbering* or *sn-* system. In this system, the *pro-S* position of a prochiral atom is denoted as the *1-position,* the prochiral atom as the *2-position,* and so on. When this scheme is used, the prefix *sn-* precedes the molecule name (glycerol phosphate in this case) and distinguishes this nomenclature from other approaches. In this way, the glycerol phosphate in natural phosphoglycerides is named *sn*-glycerol-3-phosphate.

Biochemistry Now™ Go to BiochemistryNow and click BiochemistryInteractive to learn the structures and names of the glycerophospholipids.

Glycerophospholipids Are the Most Common Phospholipids

Phosphatidic acid, the parent compound for the glycerol-based phospholipids (Figure 8.4), consists of *sn*-glycerol-3-phosphate, with fatty acids esterified at the 1- and 2-positions. Phosphatidic acid is found in small amounts in most natural systems and is an important intermediate in the biosynthesis of the more common glycerophospholipids (Figure 8.6). In these compounds, a variety of polar groups are esterified to the phosphoric acid moiety of the molecule. The phosphate, together with such esterified entities, is referred to as a "head" group. Phosphatides with choline or ethanolamine are referred to as phosphatidylcholine (known commonly as **lecithin**) or **phosphatidylethanolamine,** respectively. These phosphatides are two of the most common

A Deeper Look

Prochirality

If a tetrahedral center in a molecule has two identical substituents, it is referred to as **prochiral** because if either of the like substituents is converted to a different group, the tetrahedral center then becomes chiral. Consider glycerol: The central carbon of glycerol is prochiral because replacing either of the —CH_2OH groups would make the central carbon chiral. Nomenclature for prochiral centers is based on the (*R,S*) system (see Chapter 4). To name the otherwise identical substituents of a prochiral center, imagine increasing slightly the priority of one of them (by substituting a deuterium for a hydrogen, for example) as shown: The resulting molecule has an (*S*)-configuration about the (now chiral) central carbon atom. The group that contains the deuterium is thus referred to as the *pro-S* group. As a useful exercise, you should confirm that labeling the other CH_2OH group with a deuterium produces the (*R*)-configuration at the central carbon so that this latter CH_2OH group is the *pro-R* substituent.

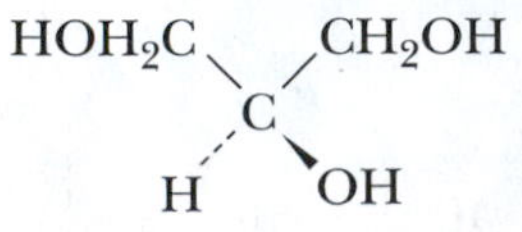

Glycerol

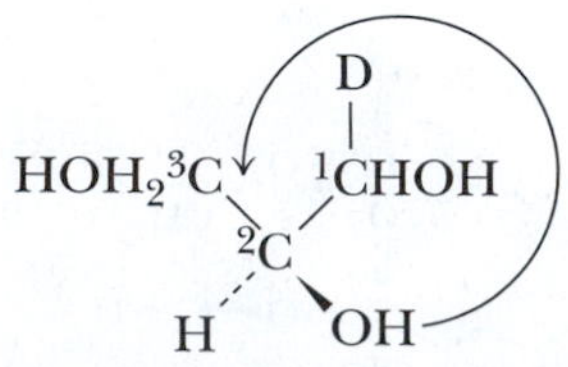

1-d, 2(*S*)-Glycerol
(*S*-configuration at C-2)

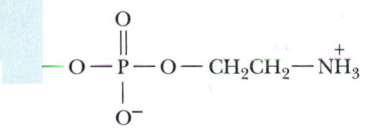

Phosphatidylcholine

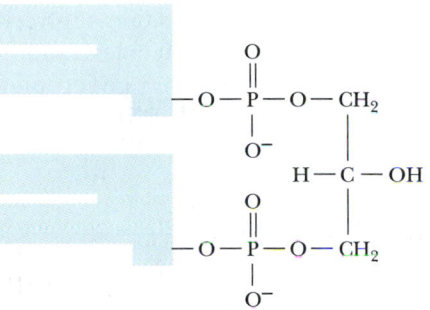

GLYCEROLIPIDS WITH OTHER HEAD GROUPS:

Phosphatidylethanolamine

Phosphatidylserine

Diphosphatidylglycerol (Cardiolipin)

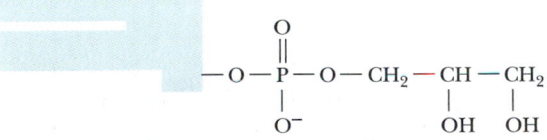

Phosphatidylglycerol

Phosphatidylinositol

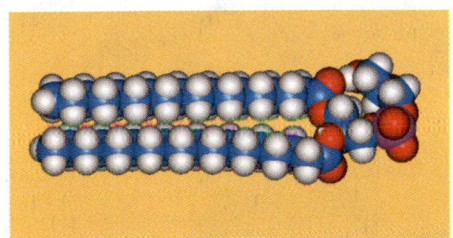

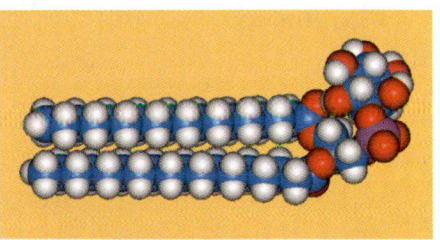

Biochemistry Now™   ANIMATED

FIGURE 8.6 Structures of several glycerophospholipids and space-filling models of phosphatidylcholine, phosphatidylglycerol, and phosphatidylinositol. **See this figure animated at http:// chemistry.brookscole.com/ggb3**

constituents of biological membranes. Other common head groups found in phosphatides include glycerol, serine, and inositol (Figure 8.6). Another kind of glycerol phosphatide found in many tissues is **diphosphatidylglycerol.** First observed in heart tissue, it is also called **cardiolipin.** In cardiolipin, a phosphatidylglycerol is esterified through the C-1 hydroxyl group of the glycerol moiety of the head group to the phosphoryl group of another phosphatidic acid molecule.

Phosphatides exist in many different varieties, depending on the fatty acids esterified to the glycerol group. As we shall see, the nature of the fatty acids can greatly affect the chemical and physical properties of the phosphatides and the

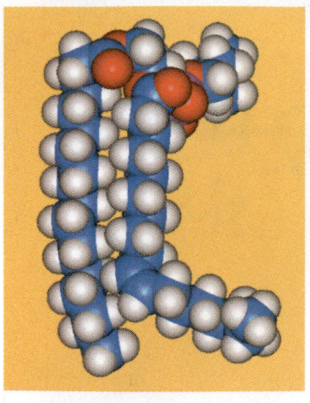

FIGURE 8.7 A space-filling model of 1-stearoyl-2-oleoyl-phosphatidylcholine.

membranes that contain them. In most cases, glycerol phosphatides have a saturated fatty acid at position 1 and an unsaturated fatty acid at position 2 of the glycerol. Thus, **1-stearoyl-2-oleoyl-phosphatidylcholine** (Figure 8.7) is a common constituent in natural membranes, but **1-linoleoyl-2-palmitoylphosphatidylcholine** is not.

Both structural and functional strategies govern the natural design of the many different kinds of glycerophospholipid head groups and fatty acids. The structural roles of these different glycerophospholipid classes are described in Chapter 9. Certain phospholipids, including phosphatidylinositol and phosphatidylcholine, participate in complex cellular signaling events. These roles, appreciated only in recent years, are described in Chapter 32.

Ether Glycerophospholipids Include PAF and Plasmalogens

Ether glycerophospholipids possess an ether linkage instead of an acyl group at the C-1 position of glycerol (Figure 8.8). One of the most versatile biochemical signal molecules found in mammals is **platelet-activating factor,** or **PAF,** a unique ether glycerophospholipid (Figure 8.9). The alkyl group at C-1 of PAF is typically a 16-carbon chain, but the acyl group at C-2 is a 2-carbon acetate

---

A Deeper Look

Glycerophospholipid Degradation: One of the Effects of Snake Venom

The venoms of poisonous snakes contain (among other things) a class of enzymes known as **phospholipases,** enzymes that cause the breakdown of phospholipids. For example, the venoms of the eastern diamondback rattlesnake (*Crotalus adamanteus*) and the Indian cobra (*Naja naja*) both contain phospholipase A₂, which catalyzes the hydrolysis of fatty acids at the C-2 position of glycerophospholipids. The phospholipid breakdown product

of this reaction, *lysolecithin,* acts as a detergent and dissolves the membranes of red blood cells, causing them to rupture. Indian cobras kill several thousand people each year.

▲ Western diamondback rattlesnake.

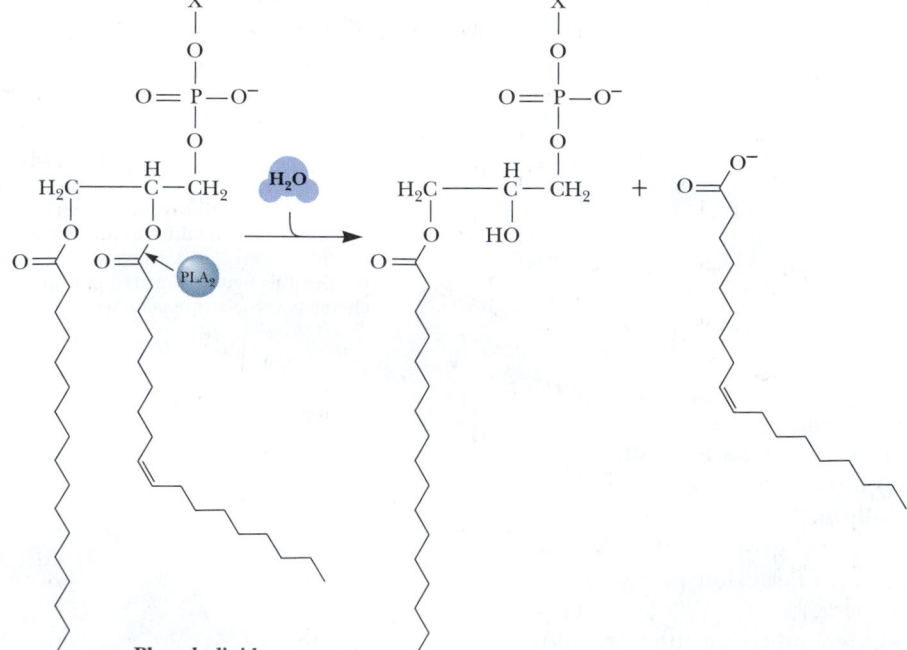

Phospholipid

▲ Indian cobra.

Human Biochemistry

Platelet-Activating Factor: A Potent Glyceroether Mediator

Platelet-activating factor (PAF) was first identified by its ability (at low levels) to cause platelet aggregation and dilation of blood vessels, but it is now known to be a potent mediator in inflammation, allergic responses, and shock. PAF effects are observed at tissue concentrations as low as $10^{-12}M$. PAF causes a dramatic inflammation of air passages and induces asthmalike symptoms in laboratory animals. **Toxic shock syndrome** occurs when fragments of destroyed bacteria act as toxins and induce the synthesis of PAF. PAF causes a drop in blood pressure and a reduced volume of blood pumped by the heart, which leads to shock and, in severe cases, death.

Beneficial effects have also been attributed to PAF. In reproduction, PAF secreted by the fertilized egg is instrumental in the implantation of the egg in the uterine wall. PAF is produced in significant quantities in the lungs of the fetus late in pregnancy and may stimulate the production of fetal lung surfactant, a protein–lipid complex that prevents collapse of the lungs in a newborn infant.

---

unit. By virtue of this acetate group, PAF is much more water soluble than other lipids, allowing PAF to function as a soluble messenger in signal transduction.

Plasmalogens are ether glycerophospholipids in which the alkyl moiety is cis-α,β-unsaturated (Figure 8.10). Common plasmalogen head groups include choline, ethanolamine, and serine. These lipids are referred to as phosphatidal choline, phosphatidal ethanolamine, and phosphatidal serine.

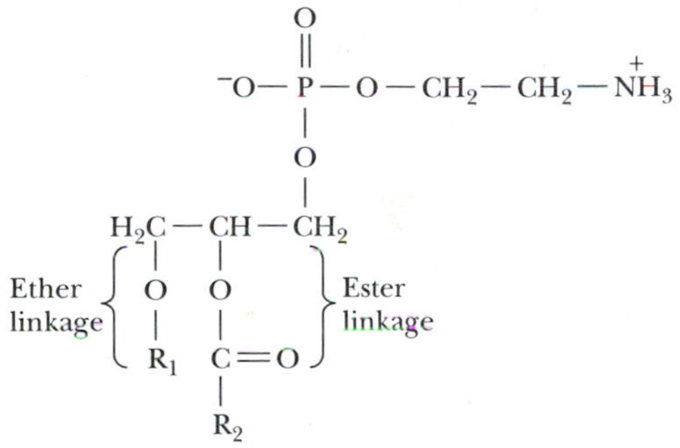

8.4 | What Are Sphingolipids, and How Are They Important for Higher Animals?

Sphingolipids represent another class of lipids frequently found in biological membranes. An 18-carbon amino alcohol, **sphingosine** (Figure 8.11), forms the backbone of these lipids rather than glycerol. Typically, a fatty acid is joined to a sphingosine via an amide linkage to form a **ceramide. Sphingomyelins** represent a phosphorus-containing subclass of sphingolipids and are especially

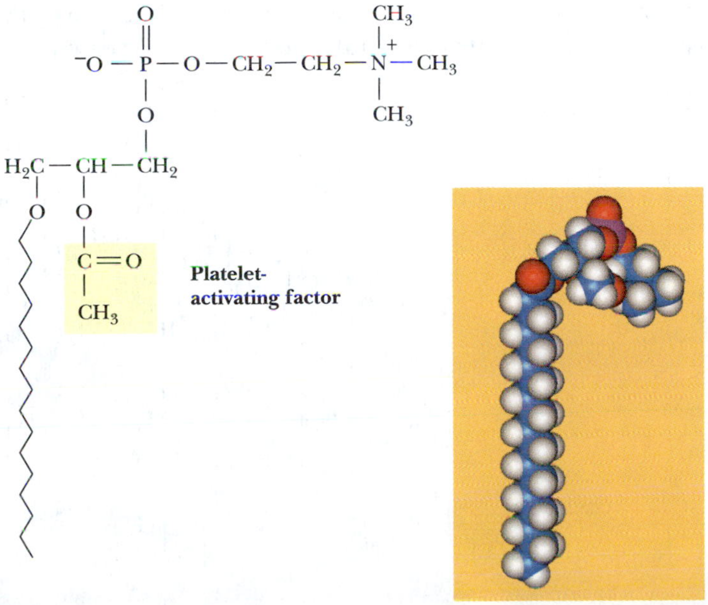

FIGURE 8.8 A 1-alkyl 2-acyl-phosphatidylethanolamine (an ether glycerophospholipid).

FIGURE 8.9 The structure of 1-alkyl 2-acetyl-phosphatidylcholine, also known as platelet-activating factor or PAF.

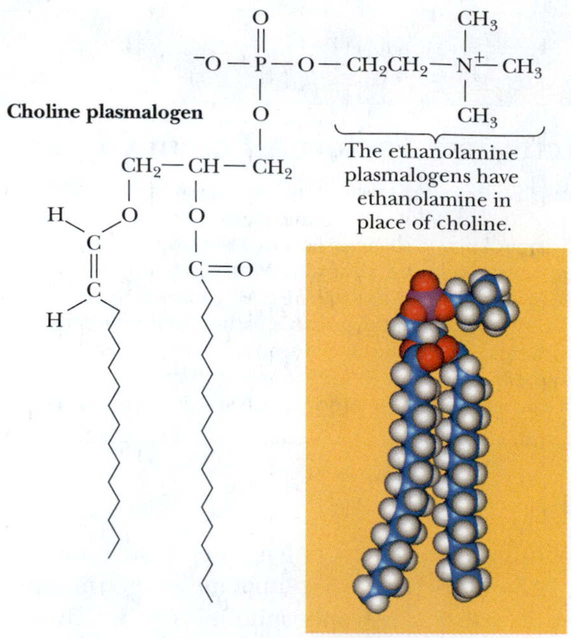

FIGURE 8.10 The structure and a space-filling model of a choline plasmalogen.

important in the nervous tissue of higher animals. A **sphingomyelin** is formed by the esterification of a phosphorylcholine or a phosphorylethanolamine to the 1-hydroxy group of a ceramide (Figure 8.12).

There is another class of ceramide-based lipids that, like the sphingomyelins, are important components of muscle and nerve membranes in animals. These are the **glycosphingolipids,** and they consist of a ceramide with one or more sugar residues in a β-glycosidic linkage at the 1-hydroxyl moiety. The neutral glycosphingolipids contain only neutral (uncharged) sugar residues. When a single glucose or galactose is bound in this manner, the molecule is a **cerebroside** (Figure 8.13). Another class of lipids is formed when a sulfate is esterified at the 3-position of the galactose to make a **sulfatide. Gangliosides** (Figure 8.14) are more complex glycosphingolipids that consist of a ceramide backbone with three or more sugars esterified, one of these being a **sialic acid** such as *N*-**acetylneuraminic acid.** These latter compounds are

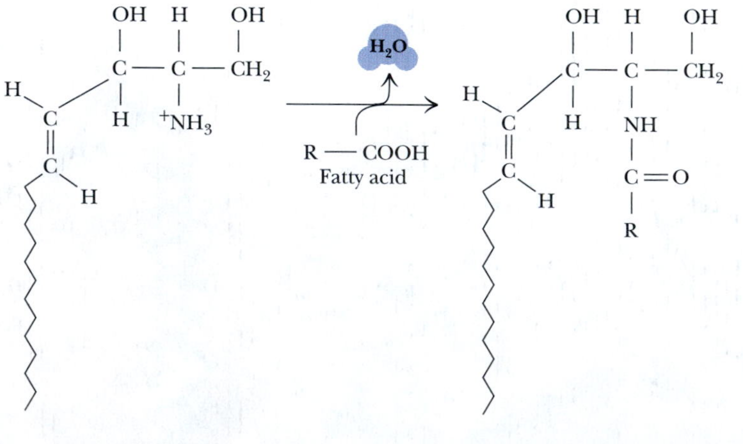

FIGURE 8.11 Formation of an amide linkage between a fatty acid and sphingosine produces a ceramide.

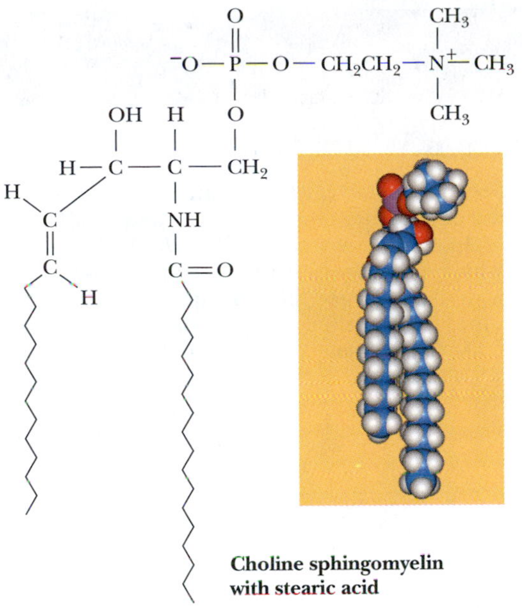

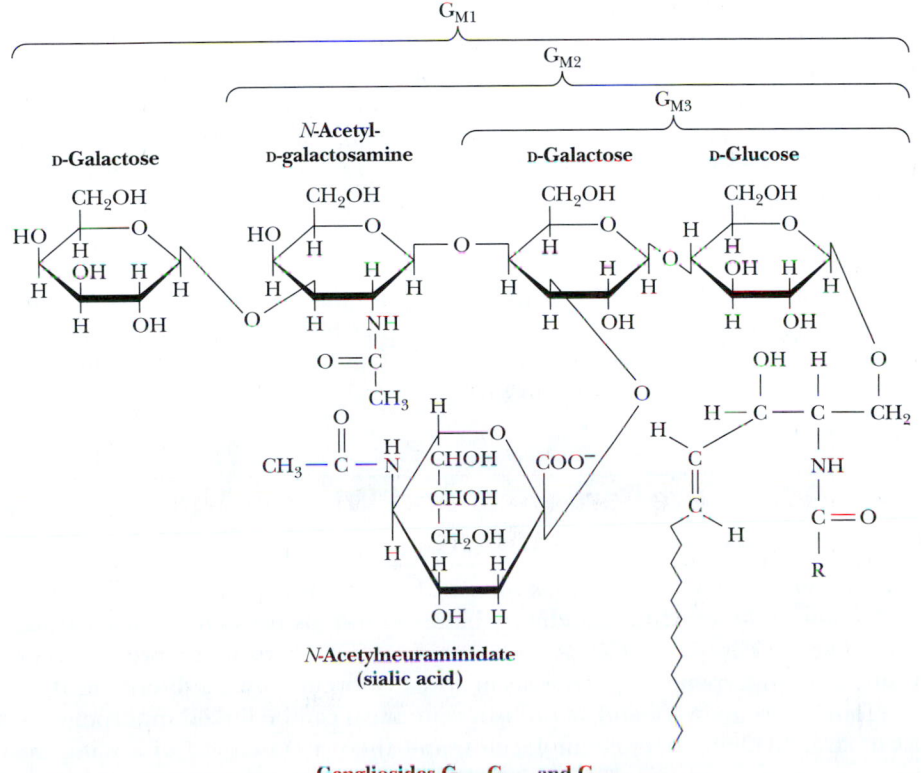

**Choline sphingomyelin
with stearic acid**

FIGURE 8.12 A structure and a space-filling model of a choline sphingomyelin formed from stearic acid.

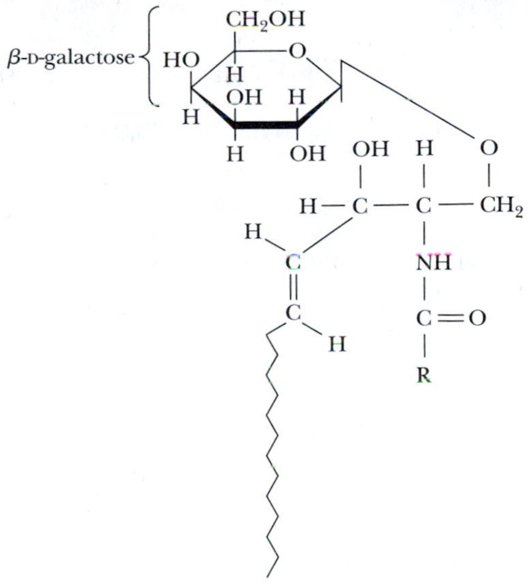

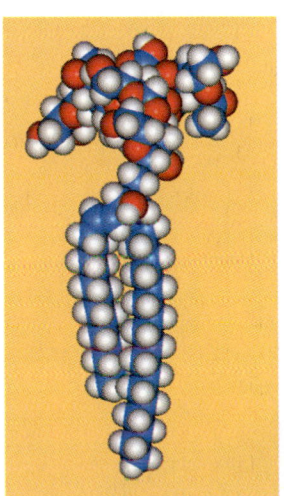

A cerebroside

FIGURE 8.13 The structure of a cerebroside. Note the sphingosine backbone.

referred to as acidic glycosphingolipids, and they have a net negative charge at neutral pH.

The glycosphingolipids have a number of important cellular functions, despite the fact that they are present only in small amounts in most membranes. Glycosphingolipids at cell surfaces appear to determine, at least in part, certain elements of tissue and organ specificity. Cell–cell recognition and tissue

Gangliosides G$_{M1}$, G$_{M2}$, and G$_{M3}$

FIGURE 8.14 The structures of several important gangliosides. Also shown is a space-filling model of ganglioside G$_{M1}$.

A Deeper Look

Moby Dick and Spermaceti: A Valuable Wax from Whale Oil

When oil from the head of the sperm whale is cooled, **spermaceti,** a translucent wax with a white, pearly luster, crystallizes from the mixture. Spermaceti, which makes up 11% of whale oil, is composed mainly of the wax **cetyl palmitate:**

$$CH_3(CH_2)_{14}-COO-(CH_2)_{15}CH_3$$

as well as smaller amounts of cetyl alcohol:

$$HO-(CH_2)_{15}CH_3$$

*Melville, H., 1984. *Moby Dick*. London: Octopus Books, p. 205. (Adapted from Waddell, T. G., and Sanderlin, R. R., 1986. Chemistry in *Moby Dick*. Journal of Chemical Education **63**:1019–1020.)

Spermaceti and cetyl palmitate have been widely used in the making of cosmetics, fragrant soaps, and candles.

In the literary classic *Moby Dick*, Herman Melville describes Ishmael's impressions of spermaceti, when he muses that the waxes "discharged all their opulence, like fully ripe grapes their wine; as I snuffed that uncontaminated aroma—literally and truly, like the smell of spring violets."*

immunity appear to depend on specific glycosphingolipids. Gangliosides are present in nerve endings and appear to be important in nerve impulse transmission. A number of genetically transmitted diseases involve the accumulation of specific glycosphingolipids due to an absence of the enzymes needed for their degradation. Such is the case for ganglioside G_{M2} in the brains of Tay-Sachs disease victims, a rare but fatal disease characterized by a red spot on the retina, gradual blindness, and loss of weight, especially in infants and children.

8.5 | What Are Waxes, and How Are They Used?

Waxes are esters of long-chain alcohols with long-chain fatty acids. The resulting molecule can be viewed (in analogy to the glycerolipids) as having a weakly polar head group (the ester moiety itself) and a long, nonpolar tail (the hydrocarbon chains) (Figure 8.15). Fatty acids found in waxes are usually saturated. The alcohols found in waxes may be saturated or unsaturated and may include sterols, such as cholesterol (see later section). Waxes are water insoluble due to the weakly polar nature of the ester group. As a result, this class of molecules confers water-repellant character to animal skin, to the leaves of certain plants, and to bird feathers. The glossy surface of a polished apple results from a wax coating. **Carnauba wax,** obtained from the fronds of a species of palm tree in Brazil, is a particularly hard wax used for high gloss finishes, such as in automobile wax, boat wax, floor wax, and shoe polish. **Lanolin,** a component of wool wax, is used as a base for pharmaceutical and cosmetic products because it is rapidly assimilated by human skin.

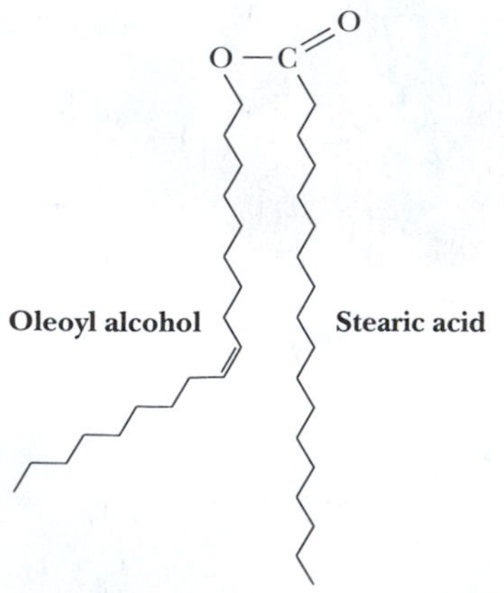

FIGURE 8.15 An example of a wax. Oleoyl alcohol is esterified to stearic acid in this case.

8.6 | What Are Terpenes, and What Is Their Relevance to Biological Systems?

The **terpenes** are a class of lipids formed from combinations of two or more molecules of 2-methyl-1,3-butadiene, better known as **isoprene** (a five-carbon unit that is abbreviated C_5). A **monoterpene** (C_{10}) consists of two isoprene units, a **sesquiterpene** (C_{15}) consists of three isoprene units, a **diterpene** (C_{20}) has four isoprene units, and so on. Isoprene units can be linked in terpenes to form straight-chain or cyclic molecules, and the usual method of linking isoprene units is head to tail (Figure 8.16). Monoterpenes occur in all higher

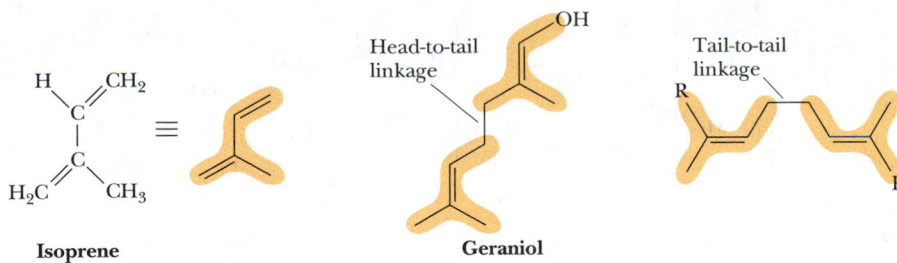

FIGURE 8.16 The structure of isoprene (2-methyl-1,3-butadiene) and the structure of head-to-tail and tail-to-tail linkages. Isoprene itself can be formed by distillation of natural rubber, a linear head-to-tail polymer of isoprene units.

plants, whereas sesquiterpenes and diterpenes are less widely known. Several examples of these classes of terpenes are shown in Figure 8.17. The **triterpenes** are C_{30} terpenes and include **squalene** and **lanosterol,** two of the precursors of cholesterol and other steroids (discussed later). **Tetraterpenes** (C_{40}) are less common but include the carotenoids, a class of colorful photosynthetic pigments. β-Carotene is the precursor of vitamin A, whereas lycopene, similar to β-carotene but lacking the cyclopentene rings, is a pigment found in tomatoes.

Long-chain polyisoprenoid molecules with a terminal alcohol moiety are called **polyprenols.** The **dolichols,** one class of polyprenols (Figure 8.18), consist of 16 to 22 isoprene units and, in the form of dolichyl phosphates, function

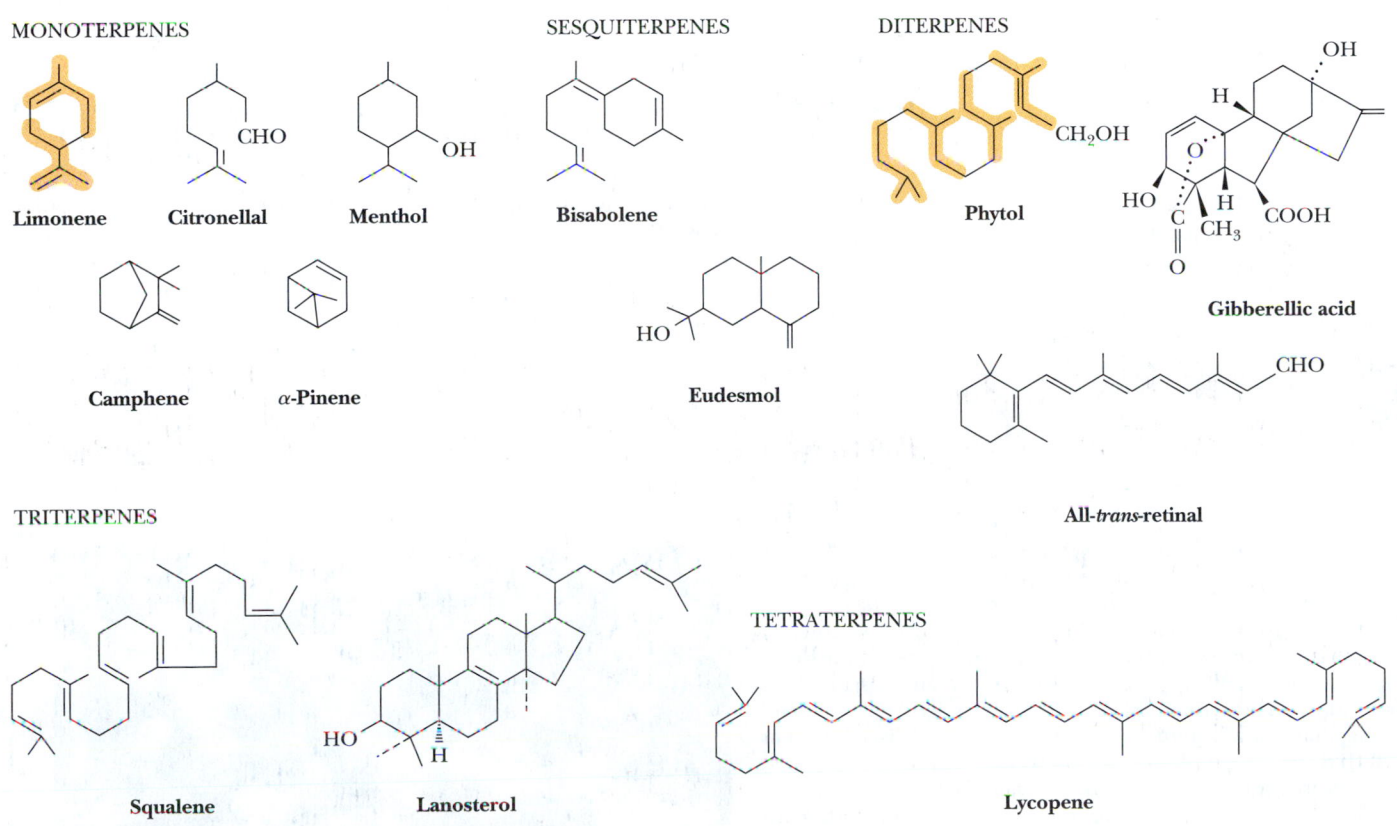

Biochemistry⑤Now™ ACTIVE FIGURE 8.17 Many monoterpenes are readily recognized by their characteristic flavors or odors (limonene in lemons; citronellal in roses, geraniums, and some perfumes; pinene in turpentine; and menthol from peppermint, used in cough drops and nasal inhalers). The diterpenes, which are C_{20} terpenes, include retinal (the essential light-absorbing pigment in rhodopsin, the photoreceptor protein of the eye), phytol (a constituent of chlorophyll), and the gibberellins (potent plant hormones). The triterpene lanosterol is a constituent of wool fat. Lycopene is a carotenoid found in ripe fruit, especially tomatoes. **Test yourself on the concepts in this figure at http://chemistry.brookscole.com/ggb3**

Dolichol phosphate

Coenzyme Q (Ubiquinone, UQ)

Vitamin E (α-tocopherol)

Vitamin K$_1$ (phylloquinone)

Undecaprenyl alcohol (bactoprenol)

Vitamin K$_2$ (menaquinone)

FIGURE 8.18 Dolichol phosphate is an initiation point for the synthesis of carbohydrate polymers in animals. The analogous alcohol in bacterial systems, *undecaprenol*, also known as *bactoprenol*, consists of 11 isoprene units. Undecaprenyl phosphate delivers sugars from the cytoplasm for the synthesis of cell wall components such as peptidoglycans, lipopolysaccharides, and glycoproteins. Polyprenyl compounds also serve as the side chains of vitamin K, the ubiquinones, plastoquinones, and tocopherols (such as vitamin E).

A Deeper Look

Why Do Plants Emit Isoprene?

The Blue Ridge Mountains of Virginia are so named for the misty blue vapor or haze that hangs over them through much of the summer season. This haze is composed in part of isoprene that is produced and emitted by the plants and trees of the mountains. Global emission of isoprene from vegetation is estimated at 3×10^{14} g/yr. Plants frequently emit as much as 15% of the carbon fixed in photosynthesis as isoprene, and Thomas Sharkey, a botanist at the University of Wisconsin, has shown that the kudzu plant can emit as much as 67% of its fixed carbon as isoprene as the result of water stress. Why should plants and trees emit large amounts of isoprene and other hydrocarbons? Sharkey has shown that an isoprene atmosphere or "blanket" can protect leaves from irreversible damage induced by high (summerlike) temperatures. He hypothesizes that isoprene in the air around plants dissolves into leaf-cell membranes, altering the lipid bilayer and/or lipid–protein and protein–protein interactions within the membrane to increase thermal tolerance.

▲ Blue Ridge Mountains.

Randy Wells/Getty Images

Human Biochemistry

Coumadin or Warfarin—Agent of Life or Death

The isoprene-derived molecule whose structure is shown here is known alternately as **Coumadin** and **warfarin.** By the former name, it is a widely prescribed anticoagulant. By the latter name, it is a component of rodent poisons. How can the same chemical species be used for such disparate purposes? The key to both uses lies in its ability to act as an antagonist of vitamin K in the body.

Vitamin K stimulates the carboxylation of glutamate residues on certain proteins, including some proteins in the blood-clotting cascade (including **prothrombin, factor VII, factor IX,** and **factor X,** which undergo a Ca^{2+}-dependent conformational change in the course of their biological activity, as well as **protein C** and **protein S,** two regulatory proteins in coagulation). Carboxylation of these coagulation factors is catalyzed by a carboxylase that requires the reduced form of vitamin K (vitamin KH_2), molecular oxygen, and carbon dioxide. KH_2 is oxidized to vitamin K epoxide, which is recycled to KH_2 by the enzymes **vitamin K epoxide reductase** (1) and **vitamin K reductase** (2, 3). Coumadin/warfarin exerts its anticoagulant effect by inhibiting vitamin K epoxide reductase and possibly also vitamin K reductase. This inhibition depletes vitamin KH_2 and reduces the activity of the carboxylase.

Coumadin/warfarin, given at a typical dosage of 4 to 5 mg/day, prevents the deleterious formation in the bloodstream of small blood clots and thus reduces the risk of heart attacks and strokes for individuals whose arteries contain sclerotic plaques. Taken in much larger doses, as for example in rodent poisons, Coumadin/warfarin can cause massive hemorrhages and death.

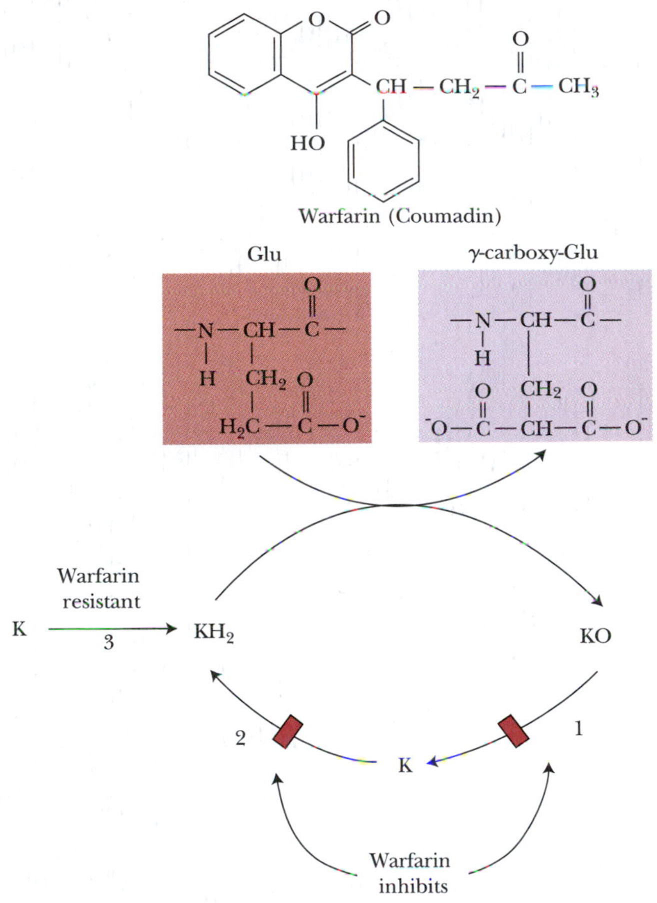

Warfarin (Coumadin)

to carry carbohydrate units in the biosynthesis of glycoproteins in animals. Polyprenyl groups serve to anchor certain proteins to biological membranes (discussed in Chapter 9).

8.7 | What Are Steroids, and What Are Their Cellular Functions?

Cholesterol

A large and important class of terpene-based lipids is the **steroids.** This molecular family, whose members effect an amazing array of cellular functions, is based on a common structural motif of three 6-membered rings and one 5-membered ring all fused together. **Cholesterol** (Figure 8.19) is the most common steroid in animals and the precursor for all other animal steroids. The numbering system for cholesterol applies to all such molecules. Many steroids contain methyl groups at positions 10 and 13 and an 8- to 10-carbon alkyl side chain at position 17. The polyprenyl nature of this compound is particularly evident in the side chain. Many steroids contain an oxygen at C-3, either a hydroxyl group in sterols or a carbonyl group in other steroids.

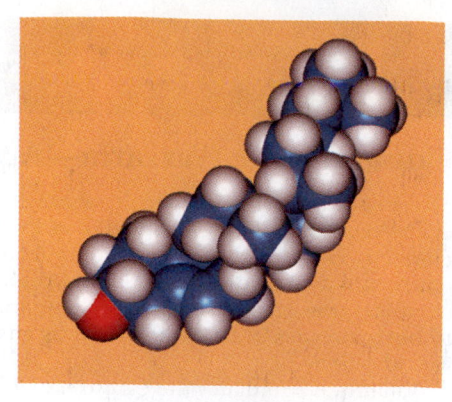

FIGURE 8.19 The structure of cholesterol, shown with steroid ring designations and carbon numbering.

Cholesterol

Note also that the carbons at positions 10 and 13 and the alkyl group at position 17 are nearly always oriented on the same side of the steroid nucleus, the β-orientation. Alkyl groups that extend from the other side of the steroid backbone are in an α-orientation.

Cholesterol is a principal component of animal cell plasma membranes, and much smaller amounts of cholesterol are found in the membranes of intracellular organelles. The relatively rigid fused ring system of cholesterol and the weakly polar alcohol group at the C-3 position have important consequences for the properties of plasma membranes. Cholesterol is also a component of lipoprotein complexes in the blood, and it is one of the constituents of plaques that form on arterial walls in atherosclerosis.

Steroid Hormones Are Derived from Cholesterol

Steroids derived from cholesterol in animals include five families of hormones (the androgens, estrogens, progestins, glucocorticoids, and mineralocorticoids) and bile acids (Figure 8.20). **Androgens** such as **testosterone** and **estrogens** such

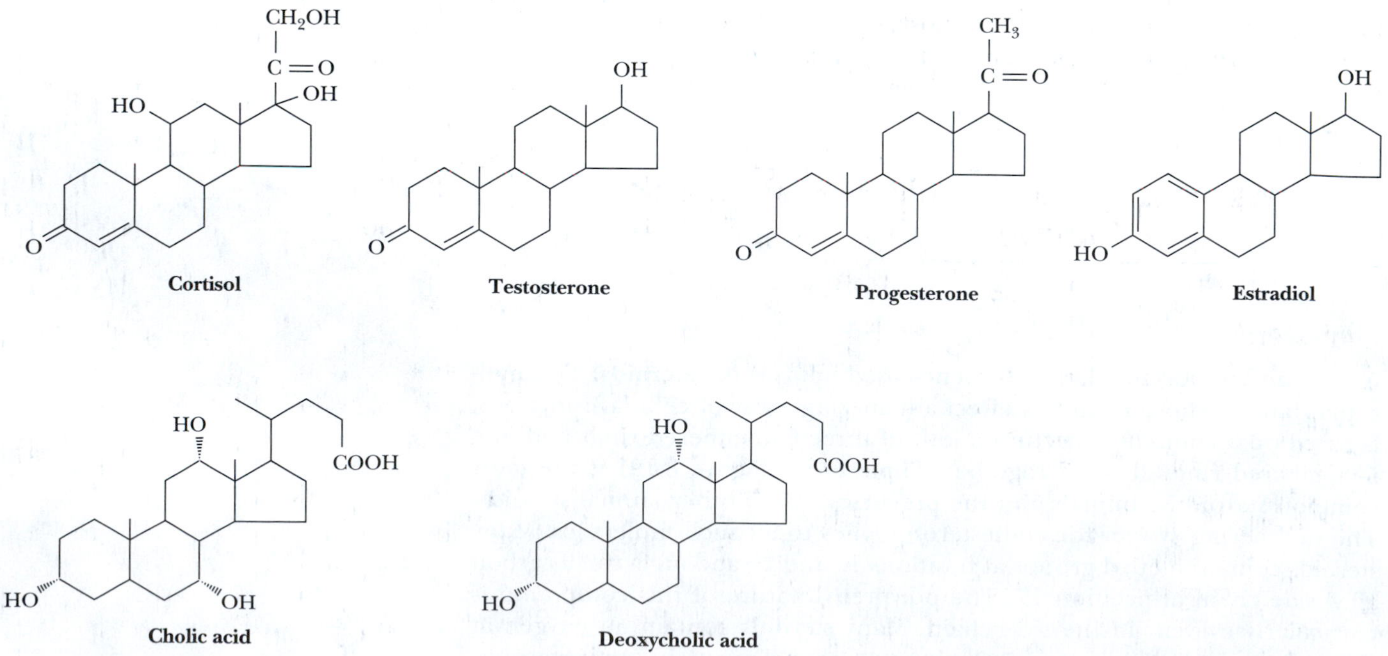

Cortisol

Testosterone

Progesterone

Estradiol

Cholic acid

Deoxycholic acid

FIGURE 8.20 The structures of several important sterols derived from cholesterol.

Human Biochemistry

Plant Sterols—Natural Cholesterol Fighters

Dietary guidelines for optimal health call for reducing the intake of cholesterol. One strategy for doing so involves the plant sterols, including sitosterol, stigmasterol, stigmastanol, and campesterol, shown in the accompanying figure. Despite their structural similarity to cholesterol, minor isomeric differences and/or the presence of methyl and ethyl groups in the side chains of these substances result in their poor absorption by intestinal mucosal cells. Interestingly, although plant sterols are not effectively absorbed by the body, they nonetheless are highly effective in blocking the absorption of cholesterol itself by intestinal cells.

The practical development of plant sterol drugs as cholesterol-lowering agents will depend both on structural features of the sterols themselves and on the form of the administered agent. For example, the unsaturated sterol sitosterol is poorly absorbed in the human intestine, whereas sitostanol, the saturated analog, is almost totally unabsorbable. In addition, there is evidence that plant sterols administered in a soluble, micellar form (see page 268 for a description of micelles) are more effective in blocking cholesterol absorption than plant sterols administered in a solid, crystalline form.

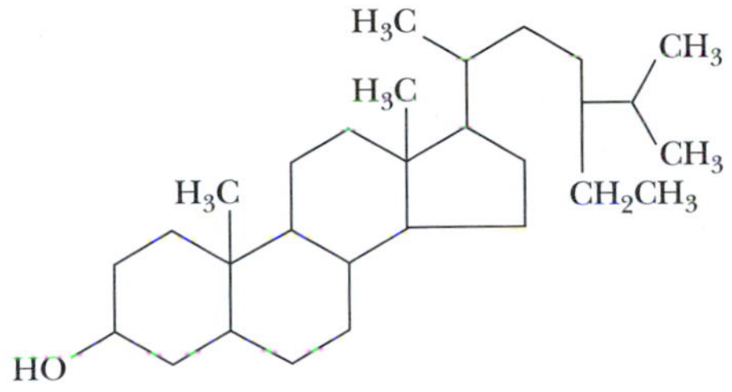

Stigmastanol

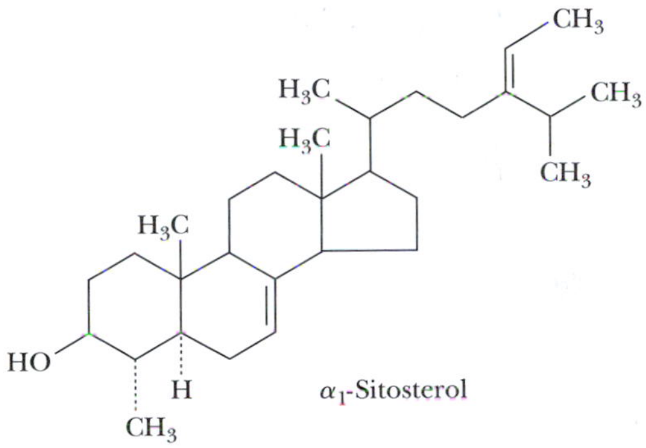

α_1-Sitosterol

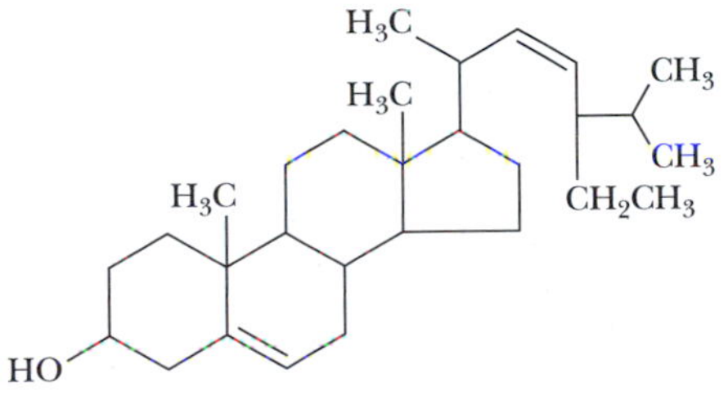

Stigmasterol

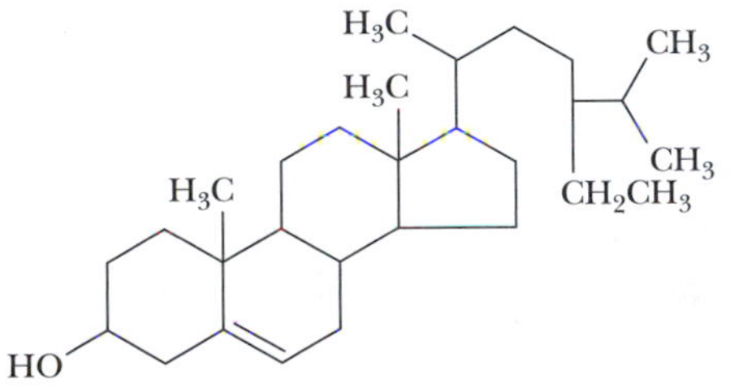

β-Sitosterol

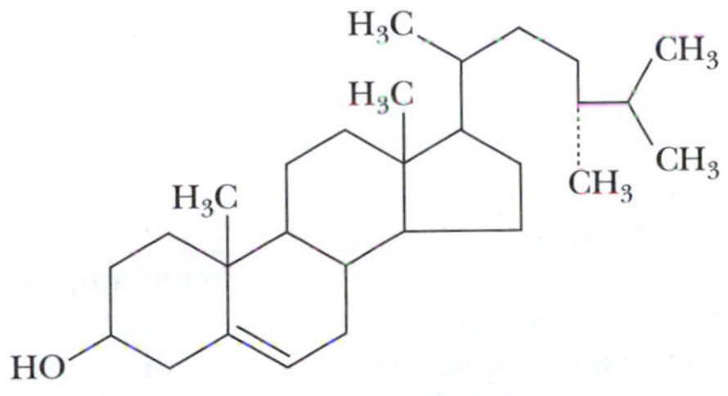

Campesterol

as **estradiol** mediate the development of sexual characteristics and sexual function in animals. The **progestins** such as **progesterone** participate in control of the menstrual cycle and pregnancy. **Glucocorticoids** (**cortisol,** for example) participate in the control of carbohydrate, protein, and lipid metabolism, whereas the **mineralocorticoids** regulate salt (Na^+, K^+, and Cl^-) balances in tissues. The **bile acids** (including **cholic** and **deoxycholic acid**) are detergent molecules secreted in bile from the gallbladder that assist in the absorption of dietary lipids in the intestine.

Human Biochemistry

17β-Hydroxysteroid Dehydrogenase 3 Deficiency

Testosterone, the principal male sex steroid hormone, is synthesized in five steps from cholesterol, as shown in the following figure. In the last step, five isozymes catalyze the 17β-hydroxysteroid dehydrogenase reactions that interconvert 4-androstenedione and testosterone. Defects in the synthesis or action of testosterone can impair the development of the male phenotype during embryogenesis and cause the disorders of human sexuality termed male pseudohermaphroditism. Specifically, mutations in isozyme 3 of the 17β-hydroxysteroid dehydrogenase in the fetal testes impair the formation of testosterone and give rise to genetic males with female external genitalia and blind-ending vaginas. Such individuals are typically raised as females but virilize at puberty, due to an increase in serum testosterone, and develop male hair growth patterns. Fourteen different mutations of 17β-hydroxysteroid dehydrogenase 3 have been identified in 17 affected families in the United States, the Middle East, Brazil, and Western Europe. These families account for about 45% of the patients with this disorder reported in scientific literature.

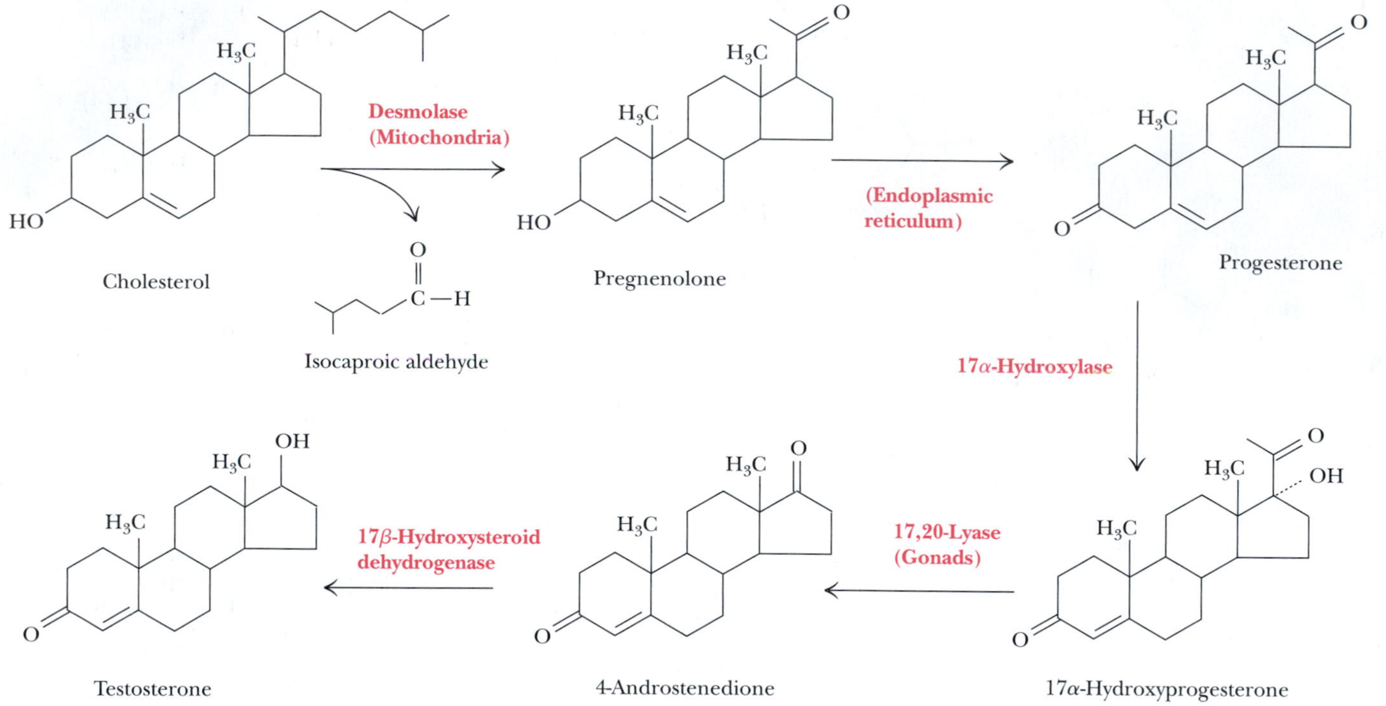

Cholesterol
Isocaproic aldehyde
Desmolase (Mitochondria)
Pregnenolone
(Endoplasmic reticulum)
Progesterone
17α-Hydroxylase
17α-Hydroxyprogesterone
17,20-Lyase (Gonads)
4-Androstenedione
17β-Hydroxysteroid dehydrogenase
Testosterone

Summary

Lipids are a class of biological molecules defined by low solubility in water and high solubility in nonpolar solvents. As molecules that are largely hydrocarbon in nature, lipids represent highly reduced forms of carbon and, upon oxidation in metabolism, yield large amounts of energy. Lipids are thus the molecules of choice for metabolic energy storage. The lipids found in biological systems are either hydrophobic (containing only nonpolar groups) or amphipathic (containing both polar and nonpolar groups). The hydrophobic nature of lipid molecules allows membranes to act as effective barriers to more polar molecules.

8.1 What Is the Structure and Chemistry of Fatty Acids? A fatty acid is composed of a long hydrocarbon chain ("tail") and a terminal carboxyl group (or "head"). The carboxyl group is normally ionized under physiological conditions. Fatty acids occur in large amounts in biological systems but only rarely in the free, uncomplexed state. They typically are esterified to glycerol or other backbone structures.

8.2 What Is the Structure and Chemistry of Triacylglycerols? A significant number of the fatty acids in plants and animals exist in the form of triacylglycerols (also called triglycerides). Triacylglycerols are a major energy reserve and the principal neutral derivatives of glycerol found in animals. These molecules consist of a glycerol esterified with three fatty acids. Triacylglycerols in animals are found primarily in the adipose tissue (body fat), which serves as a depot or storage site for lipids. Monoacylglycerols and diacylglycerols also exist, but they are far less common than the triacylglycerols.

8.3 What Is the Structure and Chemistry of Glycerophospholipids? A 1,2-diacylglycerol that has a phosphate group esterified at carbon atom 3 of the glycerol backbone is a glycerophospholipid, also known as a phosphoglyceride or a glycerol phosphatide. These lipids form one of the largest and most important classes of natural lipids. They are essential components of cell membranes and are found in small concentrations in other parts of the cell. All glycerophospholipids are members of the broader class of lipids known as phospholipids.

8.4 What Are Sphingolipids, and How Are They Important for Higher Animals?

Sphingolipids represent another class of lipids in biological membranes. An 18-carbon amino alcohol, sphingosine, forms the backbone of these lipids rather than glycerol. Typically, a fatty acid is joined to a sphingosine via an amide linkage to form a ceramide. Sphingomyelins are a phosphorus-containing subclass of sphingolipids especially important in the nervous tissue of higher animals. A sphingomyelin is formed by the esterification of a phosphorylcholine or a phosphorylethanolamine to the 1-hydroxy group of a ceramide. Glycosphingolipids are another class of ceramide-based lipids that, like the sphingomyelins, are important components of muscle and nerve membranes in animals. Glycosphingolipids consist of a ceramide with one or more sugar residues in a β-glycosidic linkage at the 1-hydroxyl moiety.

8.5 What Are Waxes, and How Are They Used?

Waxes are esters of long-chain alcohols with long-chain fatty acids. The resulting molecule can be viewed (in analogy to the glycerolipids) as having a weakly polar head group (the ester moiety itself) and a long, nonpolar tail (the hydrocarbon chains). Fatty acids found in waxes are usually saturated. The alcohols found in waxes may be saturated or unsaturated and may include sterols, such as cholesterol. Waxes are water insoluble due to the weakly polar nature of the ester group.

8.6 What Are Terpenes, and What Is Their Relevance to Biological Systems?

The terpenes are a class of lipids formed from combinations of two or more molecules of 2-methyl-1,3-butadiene, better known as isoprene (a five-carbon unit abbreviated C_5). A monoterpene (C_{10}) consists of two isoprene units, a sesquiterpene (C_{15}) consists of three isoprene units, a diterpene (C_{20}) has four isoprene units, and so on. Isoprene units can be linked in terpenes to form straight-chain or cyclic molecules, and the usual method of linking isoprene units is head to tail. Monoterpenes occur in all higher plants, whereas sesquiterpenes and diterpenes are less widely known.

8.7 What Are Steroids, and What Are Their Cellular Functions?

A large and important class of terpene-based lipids is the steroids. This molecular family, whose members effect an amazing array of cellular functions, is based on a common structural motif of three 6-membered rings and one 5-membered ring all fused together. Cholesterol is the most common steroid in animals and the precursor for all other animal steroids. The numbering system for cholesterol applies to all such molecules. The polyprenyl nature of this compound is particularly evident in the side chain. Many steroids contain an oxygen at C-3, either a hydroxyl group in sterols or a carbonyl group in other steroids. The methyl groups at positions 10 and 13 and the alkyl group at position 17 are nearly always oriented on the same side of the steroid nucleus, the β-orientation. Alkyl groups that extend from the other side of the steroid backbone are in an α-orientation. Cholesterol is a principal component of animal cell plasma membranes. Steroids derived from cholesterol in animals include five families of hormones (the androgens, estrogens, progestins, glucocorticoids, and mineralocorticoids) and bile acids.

Problems

1. Draw the structures of (a) all the possible triacylglycerols that can be formed from glycerol with stearic and arachidonic acid and (b) all the phosphatidylserine isomers that can be formed from palmitic and linolenic acids.

2. Describe in your own words the structural features of
   a. a ceramide and how it differs from a cerebroside.
   b. a phosphatidylethanolamine and how it differs from a phosphatidylcholine.
   c. an ether glycerophospholipid and how it differs from a plasmalogen.
   d. a ganglioside and how it differs from a cerebroside.
   e. testosterone and how it differs from estradiol.

3. From your memory of the structures, name
   a. the glycerophospholipids that carry a net positive charge.
   b. the glycerophospholipids that carry a net negative charge.
   c. the glycerophospholipids that have zero net charge.

4. Compare and contrast two individuals, one whose diet consists largely of meats containing high levels of cholesterol and the other whose diet is rich in plant sterols. Are their risks of cardiovascular disease likely to be similar or different? Explain your reasoning.

5. James G. Watt, Secretary of the Interior (1981–1983) in Ronald Reagan's first term, provoked substantial controversy by stating publicly that trees cause significant amounts of air pollution. Based on your reading of this chapter, evaluate Watt's remarks.

6. In a departure from his usual and highly popular westerns, author Louis L'Amour wrote a novel in 1987, *Best of the Breed* (Bantam Press), in which a military pilot of Native American ancestry is shot down over the former Soviet Union and is forced to use the survival skills of his ancestral culture to escape his enemies. On the rare occasions when he is able to trap and kill an animal for food, he selectively eats the fat, not the meat. Based on your reading of this chapter, what is his reasoning for doing so?

7. Consult a grocery store near you and look for a product in the dairy cooler called Benecol. Examine the package and suggest what the special ingredient is in this product that is credited with blockage of cholesterol uptake in the body. What is the structure of this ingredient, and how does it function?

8. If you are still at the grocery store working on problem 7, stop by the rodent poison section and examine a container of warfarin or a related product. From what you can glean from the packaging, how much warfarin would a typical dog (40 lbs) have to consume to risk hemorrhages and/or death?

9. Refer to Figure 8.17 and draw each of the structures shown and try to identify the isoprene units in each of the molecules. (Note that there may be more than one correct answer for some of these molecules, unless you have the time and facilities to carry out ^{14}C labeling studies with suitable organisms.)

10. (Integrates with Chapter 3.) As noted in the Deeper Look box on polar bears, a polar bear may burn as much as 1.5 kg of fat resources per day. What weight of seal blubber would you have to ingest if you were to obtain all your calories from this energy source?

11. Just in case you are *still* at the grocery store working on problems 7 and 8, stop by the cookie shelves and choose your three favorite cookies from the shelves. Estimate how many calories of fat, and how many other calories from other sources, are contained in 100 g of each of these cookies. Survey the ingredients listed on each package, and describe the contents of the package in terms of (a) saturated fat, (b) cholesterol, and (c) *trans* fatty acids. (Note that food makers are required to list ingredients in order of decreasing amounts in each package.)

12. Describe all of the structural differences between cholesterol and stigmasterol.

13. Describe in your own words the functions of androgens, glucocorticoids, and mineralocorticoids.

14. Look through your refrigerator, your medicine cabinet, and your cleaning solutions shelf or cabinet, and find at least three commercial products that contain fragrant monoterpenes. Identify each one by its scent and then draw its structure.

15. Gibberellic acid is described in Figure 8.17 as a plant hormone. Look it up on the Internet or in an encyclopedia and describe at least one of its functions in your own words.

Preparing for the MCAT Exam

16. Make a list of the advantages polar bears enjoy from their nonpolar diet. Why wouldn't juvenile polar bears thrive on an exclusively nonpolar diet?

17. Snake venom phospholipase A_2 causes death by generating membrane-soluble anionic fragments from glycerophospholipids. Predict the fatal effects of such molecules on membrane proteins and lipids.

Biochemistry ⊘ Now™ Preparing for an exam? Test yourself on key questions at http://chemistry.brookscole.com/ggb3

Further Readings

General

Robertson, R. N., 1983. *The Lively Membranes.* Cambridge: Cambridge University Press.

Seachrist, L., 1996. A fragrance for cancer treatment and prevention. *The Journal of NIH Research* **8**:43.

Vance, D. E., and Vance, J. E. (eds.), 1985. *Biochemistry of Lipids and Membranes.* Menlo Park, CA.: Benjamin/Cummings.

Sterols

Anderson, S., Russell, D. W., and Wilson, J. D., 1996. 17β-Hydroxysteroid dehydrogenase 3 deficiency. *Trends in Endocrinology and Metabolism* **7**:121–126.

DeLuca, H. F., and Schneos, H. K., 1983. Vitamin D: recent advances. *Annual Review of Biochemistry* **52**:411–439.

Denke, M. A., 1995. Lack of efficacy of low-dose sitostanol therapy as an adjunct to a cholesterol-lowering diet in men with moderate hypercholesterolemia. *American Journal of Clinical Nutrition* **61**:392–396.

Vanhanen, H. T., Blomqvist, S., Ehnholm, C., et al., 1993. Serum cholesterol, cholesterol precursors, and plant sterols in hypercholesterolemic subjects with different apoE phenotypes during dietary sitostanol ester treatment. *Journal of Lipid Research* **34**:1535–1544.

Isoprenes and Prenyl Derivatives

Dowd, P., Ham, S.-W., Naganathan, S., and Hershline, R., 1995. The mechanism of action of vitamin K. *Annual Review of Nutrition* **15**:419–440.

Hirsh, J., Dalen, J. E., Deykin, D., Poller, L., and Bussey, H., 1995. Oral anticoagulants: Mechanism of action, clinical effectiveness, and optimal therapeutic range. *Chest* **108**:231S–246S.

Sharkey, T. D., 1995. Why plants emit isoprene. *Nature* **374**:769.

Sharkey, T. D., 1996. Emission of low molecular-mass hydrocarbons from plants. *Trends in Plant Science* **1**:78–82.

Eicosanoids

Chakrin, L. W., and Bailey, D. M., 1984. *The Leukotrienes—Chemistry and Biology.* Orlando: Academic Press.

Keuhl, F. A., and Egan, R. W., 1980. Prostaglandins, arachidonic acid and inflammation. *Science* **210**:978–984.

Sphingolipids

Hakamori, S., 1986. Glycosphingolipids. *Scientific American* **254**:44–53.

Trans Fatty Acids

Katan, M. B., Zock, P. L., and Mensink, R. P., 1995. *Trans* fatty acids and their effects on lipoproteins in humans. *Annual Review of Biochemistry* **15**:473–493.

Membranes and Membrane Transport

Membranes are thin films that surround cells, ephemeral yet stable, like the soap film surrounding bubbles.

Essential Question

Membranes serve a number of essential cellular functions. They constitute the boundaries of cells and intracellular organelles, and they provide a surface where many important biological reactions and processes occur. Membranes have proteins that mediate and regulate the transport of metabolites, macromolecules, and ions. Hormones and many other biological signal molecules and regulatory agents exert their effects via interactions with membranes. Photosynthesis, electron transport, oxidative phosphorylation, muscle contraction, and electrical activity all depend on membranes and membrane proteins. For example, 30 percent of the genes of *Mycoplasma genitalium* are thought to encode membrane proteins. *What are the properties and characteristics of biological membranes that account for their broad influence on cellular processes and transport?*

Biological membranes are uniquely organized arrays of lipids and proteins (either of which may be decorated with carbohydrate groups). The lipids found in biological systems are often **amphipathic,** signifying that they possess both polar and nonpolar groups. The hydrophobic nature of lipid molecules allows membranes to act as effective barriers to polar molecules. The polar moieties of amphipathic lipids typically lie at the surface of membranes, where they interact with water. Proteins interact with the lipids of membranes in a variety of ways. Some proteins associate with membranes via electrostatic interactions with polar groups on the membrane surface, whereas other proteins are embedded to various extents in the hydrophobic core of the membrane. Other proteins are *anchored* to membranes via covalently bound lipid molecules that associate strongly with the hydrophobic membrane core.

This chapter discusses the composition, structure, and dynamic processes of biological membranes.

It takes a membrane to make sense out of disorder in biology.
Lewis Thomas, *The World's Biggest Membrane,* The Lives of a Cell (1974)

Key Questions

9.1 What Are the Chemical and Physical Properties of Membranes?

9.2 What Is the Structure and Chemistry of Membrane Proteins?

9.3 How Does Transport Occur Across Biological Membranes?

9.4 What Is Passive Diffusion?

9.5 How Does Facilitated Diffusion Occur?

9.6 How Does Energy Input Drive Active Transport Processes?

9.7 How Are Certain Transport Processes Driven by Light Energy?

9.8 How Are Amino Acid and Sugar Transport Driven by Ion Gradients?

9.9 How Are Specialized Membrane Pores Formed by Toxins?

9.10 What Is the Structure and Function of Ionophore Antibiotics?

9.1 What Are the Chemical and Physical Properties of Membranes?

Cells make use of many different types of membranes. All cells have a cytoplasmic membrane, or *plasma membrane,* that functions (in part) to separate the cytoplasm from the surroundings. In the early days of biochemistry, the plasma membrane was not accorded many functions other than this one of partition. We now know that the plasma membrane is also responsible for (1) the exclusion of certain toxic ions and molecules from the cell, (2) the accumulation of cell nutrients, and (3) energy transduction. It functions in (4) cell locomotion, (5) reproduction, (6) signal transduction processes, and (7) interactions with molecules or other cells in the vicinity.

Even the plasma membranes of prokaryotic cells (bacteria) are complex (Figure 9.1). With no intracellular organelles to divide and organize the work, bacteria carry out processes either at the plasma membrane or in the cytoplasm itself. Eukaryotic cells, however, contain numerous intracellular organelles that perform specialized tasks. Nucleic acid biosynthesis is handled in the nucleus; mitochondria are the site of electron transport, oxidative phosphorylation, fatty acid oxidation, and the tricarboxylic acid cycle; and secretion of proteins and other substances is handled by the endoplasmic reticulum (ER) and the

Biochemistry ⚡ Now™ Test yourself on these Key Questions at BiochemistryNow at **http://chemistry.brookscole.com/ggb3**

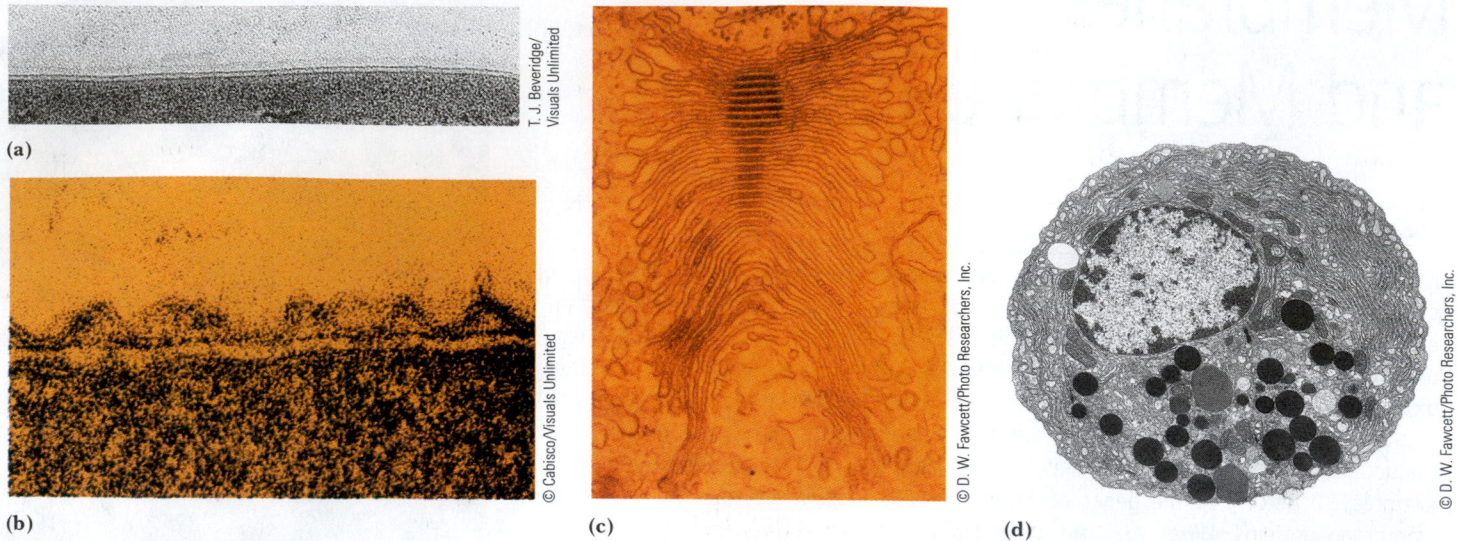

(a)

T. J. Beveridge/Visuals Unlimited

(b)

© Cabisco/Visuals Unlimited

(c)

© D. W. Fawcett/Photo Researchers, Inc.

(d)

© D. W. Fawcett/Photo Researchers, Inc.

FIGURE 9.1 Electron micrographs of several different membrane structures: **(a)** Plasma membrane of *Menoidium,* a protozoan; **(b)** Gram-negative envelope of *Aquaspirillum serpens;* **(c)** Golgi apparatus. **(d)** Many membrane structures are evident in pancreatic acinar cells.

Golgi apparatus. This partitioning of labor is not the only contribution of the membranes in these cells. Many of the processes occurring in these organelles (or in the prokaryotic cell) actively involve membranes. Thus, some of the enzymes involved in nucleic acid metabolism are membrane associated. The electron transfer chain and its associated system for ATP synthesis are embedded in the mitochondrial membrane. Many enzymes responsible for aspects of lipid biosynthesis are located in the ER membrane.

Lipids Form Ordered Structures Spontaneously in Water

Monolayers and Micelles   Amphipathic lipids spontaneously form a variety of structures when added to aqueous solution. All these structures form in ways that minimize contact between the hydrophobic lipid chains and the aqueous milieu. For example, when small amounts of a fatty acid are added to an aqueous solution, a monolayer is formed at the air–water interface, with the polar head groups in contact with the water surface and the hydrophobic tails in contact with the air (Figure 9.2). Few lipid molecules are found as monomers in solution.

Further addition of fatty acid eventually results in the formation of micelles. **Micelles** formed from an amphipathic lipid in water position the hydrophobic tails in the center of the lipid aggregation with the polar head groups facing outward. Amphipathic molecules that form micelles are characterized by a

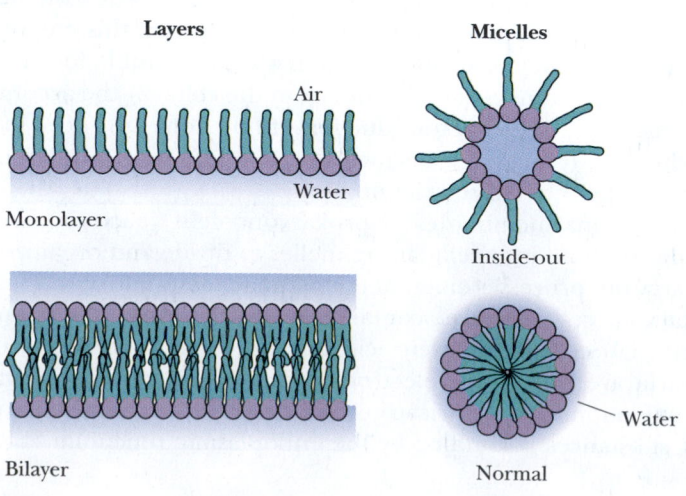

FIGURE 9.2 Several spontaneously formed lipid structures.

Structure	M_r	CMC	Micelle M_r
Triton X-100 CH$_3$—C(CH$_3$)—CH$_2$—C(CH$_3$)$_2$—⬡—(OCH$_2$CH$_2$)$_{10}$—OH	625	0.24 mM	90–95,000
Octyl glucoside CH$_2$OH ... O—(CH$_2$)$_7$—CH$_3$	292	25 mM	
C$_{12}$E$_8$ (Dodecyl octaoxyethylene ether) C$_{12}$H$_{25}$—(OCH$_2$CH$_2$)$_8$—OH	538	0.071 mM	

FIGURE 9.3 The structures of some common detergents and their physical properties. Micelles formed by detergents can be quite large. Triton X-100, for example, typically forms micelles with a total molecular mass of 90 to 95 kD. This corresponds to approximately 150 molecules of Triton X-100 per micelle.

unique **critical micelle concentration,** or **CMC.** Below the CMC, individual lipid molecules predominate. Nearly all the lipid added above the CMC, however, spontaneously forms micelles. Micelles are the preferred form of aggregation in water for detergents and soaps. Some typical CMC values are listed in Figure 9.3.

Lipid Bilayers **Lipid bilayers** consist of back-to-back arrangements of monolayers (Figure 9.2). Phospholipids prefer to form bilayer structures in aqueous solution because their pairs of fatty acyl chains do not pack well in the interior of a micelle. Phospholipid bilayers form rapidly and spontaneously when phospholipids are added to water, and they are stable structures in aqueous solution. As opposed to micelles, which are small, self-limiting structures of a few hundred molecules, bilayers may form spontaneously over large areas (10^8 nm^2 or more). Because exposure of the edges of the bilayer to solvent is highly unfavorable, extensive bilayers normally wrap around themselves and form closed vesicles (Figure 9.4). The nature and integrity of these vesicle structures are very much dependent on the lipid composition. Phospholipids can form either *unilamellar vesicles* (with a single lipid bilayer), known as *liposomes*, or *multilamellar vesicles*. These latter structures are reminiscent of the layered structure of onions. Multilamellar vesicles were discovered by Sir Alex Bangham and are sometimes referred to as "Bangosomes" in his honor.

Liposomes are highly stable structures that can be subjected to manipulations such as gel filtration chromatography and dialysis. With such methods, it is possible to prepare liposomes having different inside and outside solution compositions. Liposomes can be used as drug and enzyme delivery systems in therapeutic applications. For example, liposomes can be used to introduce contrast agents into the body for diagnostic imaging procedures, including *computed tomography* (CT) and *magnetic resonance imaging* (MRI) (Figure 9.5). Liposomes can fuse with cells, mixing their contents with the intracellular medium. If methods can be developed to target liposomes to selected cell populations, it may be possible to deliver drugs, therapeutic enzymes, and contrast agents to particular kinds of cells (such as cancer cells).

That vesicles and liposomes form at all is a consequence of the amphipathic nature of the phospholipid molecule. Ionic interactions between the polar

Bilayer

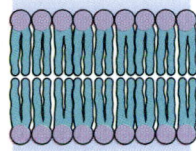

(a)

Unilamellar vesicle

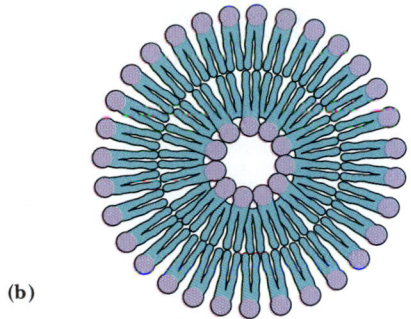

(b)

Multilamellar vesicle

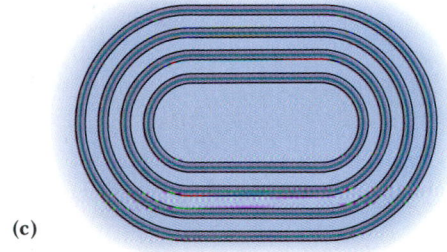

(c)

David Phillips/Visuals Unlimited

(d)

FIGURE 9.4 Drawings of **(a)** a bilayer, **(b)** a unilamellar vesicle, **(c)** a multilamellar vesicle, and **(d)** an electron micrograph of a multilamellar Golgi structure (×94,000).

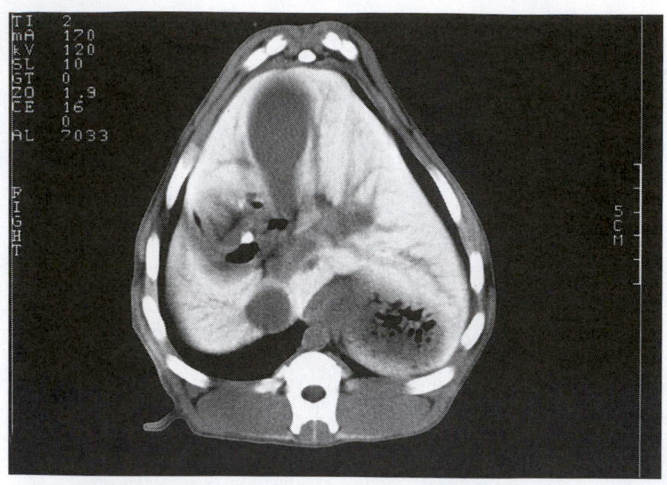

FIGURE 9.5 A computed tomography (CT) image of the upper abdomen of a dog following administration of liposome-encapsulated iodine, a contrast agent that improves the light/dark contrast of objects in the image. The spine is the bright white object at the bottom, and the other bright objects on the periphery are ribs. The liver (white) occupies most of the abdominal space. The gallbladder (bulbous object at the center top) and blood vessels appear dark in the image. The liposomal iodine contrast agent has been taken up by Kuppfer cells, which are distributed throughout the liver, except in tumors. The dark object in the lower right is a large tumor. None of these anatomical features would be visible in a CT image in the absence of the liposomal iodine contrast agent.

Courtesy of Walter Perkins, The Liposome Co., Inc., Princeton, NJ, and Brigham and Women's Hospital, Boston, MA

head groups and water are maximized, whereas hydrophobic interactions (see Chapter 2) facilitate the association of hydrocarbon chains in the interior of the bilayer. The formation of vesicles results in a favorable increase in the entropy of the solution, because the water molecules are not required to order themselves around the lipid chains. It is important to consider for a moment the physical properties of the bilayer membrane, which is the basis of vesicles and also of natural membranes. Bilayers have a polar surface and a nonpolar core. This hydrophobic core provides a substantial barrier to ions and other polar entities. The rates of movement of such species across membranes are thus quite slow. However, this same core also provides a favorable environment for nonpolar molecules and hydrophobic proteins. We will encounter numerous cases of hydrophobic molecules that interact with membranes and regulate biological functions in some way by binding to or embedding themselves in membranes.

The Fluid Mosaic Model Describes Membrane Dynamics

In 1972, S. J. Singer and G. L. Nicolson proposed the **fluid mosaic model** for membrane structure, which suggested that membranes are dynamic structures composed of proteins and phospholipids. In this model, the phospholipid bilayer is a *fluid* matrix, in essence, a two-dimensional solvent for proteins. Both lipids and proteins are capable of rotational and lateral movement.

Singer and Nicolson also pointed out that proteins can be associated with the surface of this bilayer or embedded in the bilayer to varying degrees (Figure 9.6). They defined two classes of membrane proteins. The first, called **peripheral proteins** (or **extrinsic proteins**), includes those that do not penetrate the bilayer to any significant degree and are associated with the membrane by virtue of ionic interactions and hydrogen bonds between the membrane surface and the surface of the protein. Peripheral proteins can be dissociated from the membrane by treatment with salt solutions or by changes in pH (treatments that disrupt hydrogen bonds and ionic interactions). **Integral proteins** (or **intrinsic proteins**), in contrast, possess hydrophobic surfaces that can readily penetrate the lipid bilayer itself, as well as surfaces that prefer contact with the aqueous medium. These proteins can either insert into the membrane or extend all the way across the membrane and expose themselves to the aqueous solvent on both sides. Singer and Nicolson also suggested that a portion of the bilayer lipid interacts in specific ways with integral membrane proteins and that these interactions might be important for the function of certain membrane proteins. Because of these intimate associations with membrane lipid, integral proteins can be removed from the membrane only by agents capable of breaking up the hydrophobic interactions within the lipid bilayer itself (such as

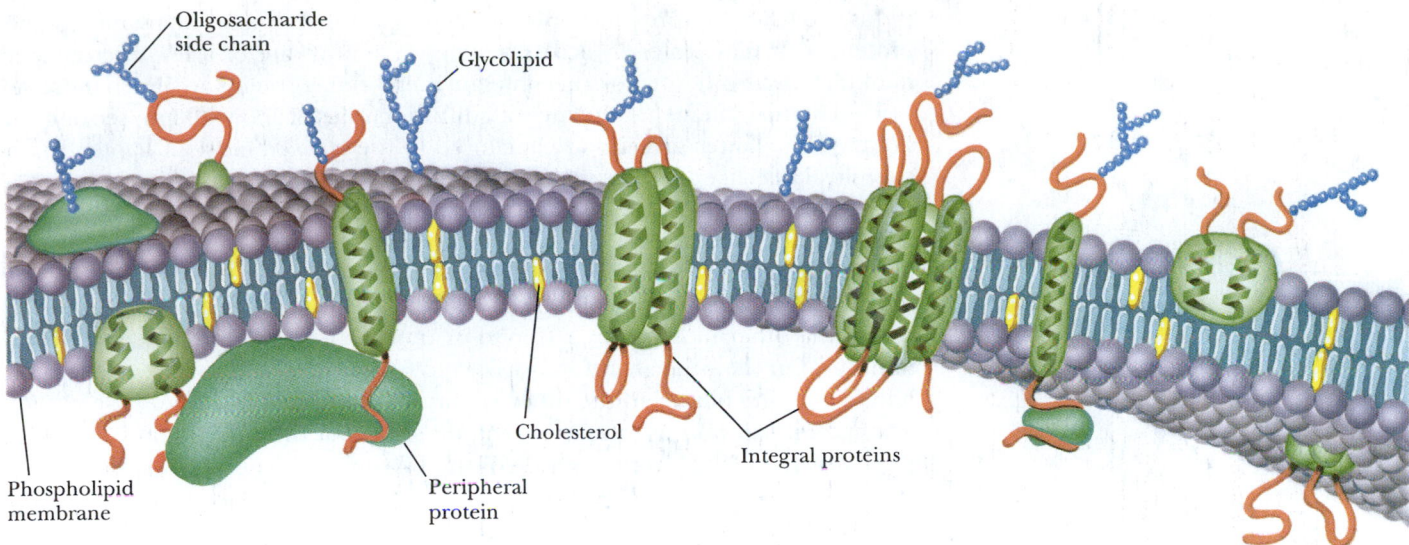

Oligosaccharide
side chain

Glycolipid

Cholesterol

Integral proteins

Phospholipid
membrane

Peripheral
protein

FIGURE 9.6 The fluid mosaic model of membrane structure proposed by S. J. Singer and G. L. Nicolson. In this model, the lipids and proteins are assumed to be mobile; they can move rapidly and laterally in the plane of the membrane. Transverse motion may also occur, but it is much slower.

detergents and organic solvents). The fluid mosaic model has become the paradigm for modern studies of membrane structure and function.

Membrane Bilayer Thickness The Singer–Nicolson model suggested a value of approximately 5 nm for membrane thickness, the same thickness as a lipid bilayer itself. Low-angle X-ray diffraction studies in the early 1970s showed that many natural membranes were approximately 5 nm in thickness and that the interiors of these membranes were low in electron density. This is consistent with the arrangement of bilayers having the hydrocarbon tails (low in electron density) in the interior of the membrane. The outside edges of these same membranes exhibit high electron density, which is consistent with the arrangement of the polar lipid head groups on the outside surfaces of the membrane.

Hydrocarbon Chain Orientation in the Bilayer An important aspect of membrane structure is the orientation or ordering of lipid molecules in the bilayer. In the bilayers sketched in Figures 9.2 and 9.4, the long axes of the lipid molecules are portrayed as being perpendicular (or normal) to the plane of the bilayer. In fact, the hydrocarbon tails of phospholipids may tilt and bend and adopt a variety of orientations. Typically, the portions of a lipid chain near the membrane surface lie most nearly perpendicular to the membrane plane, and lipid chain ordering decreases toward the end of the chain (toward the middle of the bilayer).

Membrane Bilayer Mobility The idea that lipids and proteins could move rapidly in biological membranes was a relatively new one when the fluid mosaic model was proposed. Many of the experiments designed to test this hypothesis involved the use of specially designed probe molecules. The first experiment demonstrating protein lateral movement in the membrane was described by L. Frye and M. Edidin in 1970. In this experiment, human cells and mouse cells were allowed to fuse together. Frye and Edidin used fluorescent antibodies to determine whether integral membrane proteins from the two cell types could move and intermingle in the newly formed, fused cells. The antibodies specific for human cell proteins were labeled with rhodamine, a red fluorescent marker, and the antibodies specific for mouse cell proteins were labeled with fluorescein, a green fluorescent marker. When both types of antibodies were added to newly fused cells, the binding pattern indicated that integral membrane proteins from the two cell types had moved laterally and were dispersed throughout the surface of the fused cell (Figure 9.7). This clearly demonstrated that integral membrane proteins possess significant lateral mobility.

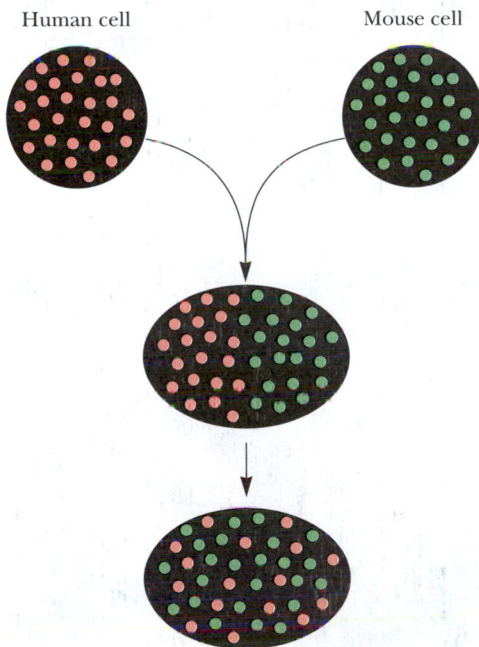

Human cell

Mouse cell

Biochemistry ⋑Now™ ACTIVE FIGURE 9.7
The Frye–Edidin experiment. Human cells with membrane antigens for red fluorescent antibodies were mixed and fused with mouse cells having membrane antigens for green fluorescent antibodies. Treatment of the resulting composite cells with red- and green-fluorescent–labeled antibodies revealed a rapid mixing of the membrane antigens in the composite membrane. This experiment demonstrated the lateral mobility of membrane proteins. **Test yourself on the concepts in this figure at http://chemistry. brookscole.com/ggb3**

Just how fast can proteins move in a biological membrane? Many membrane proteins can move laterally across a membrane at a rate of a few microns per minute. On the other hand, some integral membrane proteins are much more restricted in their lateral movement, with diffusion rates of about 10 nm/sec or even slower. These latter proteins are anchored to the *cytoskeleton* (see Chapter 16), a complex latticelike structure that maintains the cell's shape and assists in the controlled movement of various substances through the cell.

Lipids also undergo rapid lateral motion in membranes. A typical phospholipid can diffuse laterally in a membrane at a linear rate of several microns per second. At that rate, a phospholipid could travel from one end of a bacterial cell to the other in less than a second or traverse a typical animal cell in a few minutes. On the other hand, *transverse* movement of lipids (or proteins) from one face of the bilayer to the other is much slower (and much less likely). For example, it can take as long as several days for half the phospholipids in a bilayer vesicle to "flip" from one side of the bilayer to the other.

Membranes Are Asymmetric Structures

Biological membranes are **asymmetric** structures. There are several kinds of asymmetry to consider. Both the lipids and the proteins of membranes exhibit lateral and transverse asymmetries. **Lateral asymmetry** arises when lipids or proteins of particular types cluster in the plane of the membrane.

Lipids Can Form Clusters in the Membrane   Lipids in model systems are often found in asymmetric clusters (Figure 9.8). Such behavior is referred to as a **phase separation,** which arises either spontaneously or as the result of some extraneous influence. Phase separations can be induced in model membranes by divalent cations, which interact with negatively charged moieties on the surface of the bilayer. For example, Ca^{2+} induces phase separations in membranes formed from phosphatidylserine (PS) and phosphatidylethanolamine (PE) or from PS, PE, and phosphatidylcholine. Ca^{2+} added to these membranes forms complexes with the negatively charged serine carboxyls, causing the PS to cluster and separate from the other lipids. Such metal-induced lipid phase separations have been shown to regulate the activity of membrane-bound enzymes.

There are other ways in which the lateral organization (and asymmetry) of lipids in biological membranes can be altered. For example, cholesterol can intercalate between the phospholipid fatty acid chains, its polar hydroxyl group associated with the polar head groups. In this manner, patches of cholesterol and phospholipids can form in an otherwise homogeneous sea of pure phospholipid. This lateral asymmetry can in turn affect the function of membrane proteins and enzymes. The lateral distribution of lipids in a membrane can also be affected by proteins in the membrane. Certain integral membrane proteins prefer associations with specific lipids. Proteins may select unsaturated lipid chains over saturated chains or may prefer a specific head group over others.

Proteins Aggregate in Membranes   Membrane proteins in many cases are randomly distributed through the plane of the membrane. This was one of the corollaries of the fluid mosaic model of Singer and Nicolson and has been experimentally verified using electron microscopy. Electron micrographs show that integral membrane proteins are often randomly distributed in the membrane, with no apparent long-range order.

However, membrane proteins can also be distributed in nonrandom ways across the surface of a membrane. This can occur for several reasons. Some proteins must interact intimately with certain other proteins, forming multisubunit complexes that perform specific functions in the membrane. A few integral membrane proteins are known to *self-associate* in the membrane, forming large multimeric clusters. **Bacteriorhodopsin,** a light-driven proton pump pro-

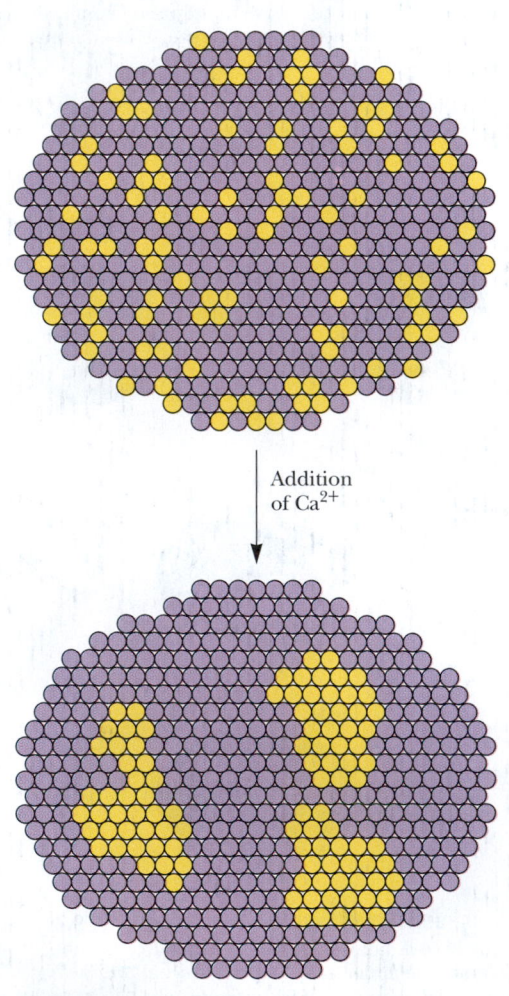

Addition of Ca^{2+}

Biochemistry ⊜ Now™ ANIMATED FIGURE 9.8
An illustration of the concept of lateral phase separations in a membrane. Phase separations of phosphatidylserine (gold circles) can be induced by divalent cations such as Ca^{2+}. **See this figure animated at http://chemistry.brookscole.com/ggb3**

tein, forms such clusters, known as "purple patches," in the membranes of *Halobacterium halobium* (Figure 9.9). The bacteriorhodopsin protein in these purple patches forms highly ordered, two-dimensional crystals.

The Two Sides of a Membrane Bilayer Are Different   Membrane proteins and lipids are oriented specifically in the **transverse** direction (from one side of the membrane to the other). This can be appreciated when one considers that many properties of a membrane depend on its two-sided nature. Properties that are a consequence of membrane "sidedness" include membrane transport, which is driven in one direction only; the effects of hormones at the outsides of cells; and the immunological reactions that occur between cells (necessarily involving only the outside surfaces of the cells). One would surmise that the proteins involved in these and other interactions must be arranged asymmetrically in the membrane.

Protein Transverse Asymmetry   Protein transverse asymmetries have been characterized using chemical, enzymatic, and immunological labeling methods. Working with **glycophorin,** the major glycoprotein in the erythrocyte membrane (discussed in Section 9.2), Mark Bretscher was the first to demonstrate the asymmetric arrangement of an integral membrane protein. Treatment of whole erythrocytes with trypsin released the carbohydrate groups of glycophorin (in the form of several small glycopeptides). Because trypsin is much too large to penetrate the erythrocyte membrane, the N-terminus of glycophorin, which contains the carbohydrate moieties, must be exposed to the outside surface of the membrane. Bretscher showed that [35S]-formylmethionylsulfone methyl phosphate could label the C-terminus of glycophorin with 35S in erythrocyte membrane fragments but not in intact erythrocytes. This clearly demonstrated that the C-terminus of glycophorin is uniformly exposed to the interior surface of the erythrocyte membrane. Since that time, many integral membrane proteins have been shown to be oriented uniformly in their respective membranes.

Lipid Transverse Asymmetry   Phospholipids are also distributed asymmetrically across many membranes. In the erythrocyte, phosphatidylcholine (PC) comprises about 30% of the total phospholipid in the membrane. Of this amount, 76% is found in the outer monolayer and 24% is found in the inner monolayer. Since this early observation, the lipids of many membranes have been found to be asymmetrically distributed between the inner and outer monolayers. Figure 9.10 shows the asymmetric distribution of phospholipids observed in the human erythrocyte membrane. Asymmetric lipid distributions are important to cells in several ways. The carbohydrate groups of glycolipids (and of glycoproteins) always face the outside surface of plasma membranes, where they participate in cell recognition phenomena. Asymmetric lipid distributions may also be important to various integral membrane proteins, which may prefer particular lipid classes in the inner and outer monolayers. The total charge on the inner and outer surfaces of

FIGURE 9.9 The purple patches of *Halobacterium halobium.*

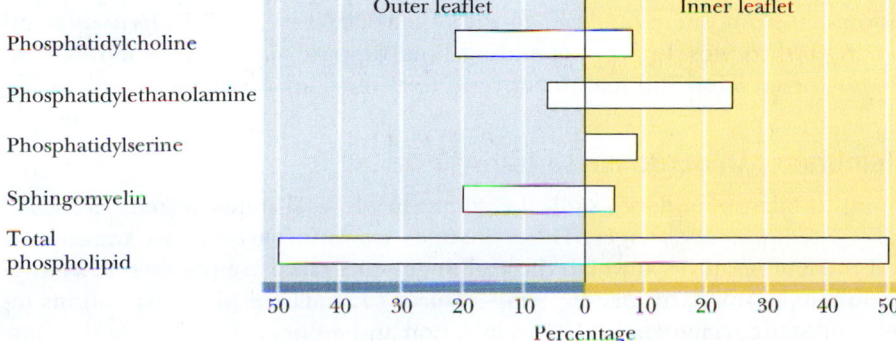

Phosphatidylcholine

Phosphatidylethanolamine

Phosphatidylserine

Sphingomyelin

Total phospholipid

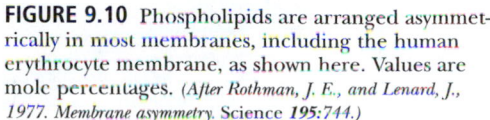

FIGURE 9.10 Phospholipids are arranged asymmetrically in most membranes, including the human erythrocyte membrane, as shown here. Values are mole percentages. *(After Rothman, J. E., and Lenard, J., 1977. Membrane asymmetry.* Science **195**:744.)

Critical Developments in Biochemistry

Rafting Down the Cellular River: How the Cell Sorts and Signals

From earliest times, wooden rafts have been used to move cargo on rivers and streams. In eukaryotic cells, clusters of lipids—called "rafts" for obvious reasons—have a similar function for the trafficking of lipids, proteins, and even chemical signals. First observed in epithelial cells—cells with anatomically distinct plasma membrane domains*—it now appears likely that lipid rafts may be ubiquitous regulators of molecular interactions in membranes. The best-characterized rafts are clusters of sphingolipid and cholesterol that form in the endoplasmic reticulum or Golgi apparatus and eventually are moved to the outer leaflet of the plasma membrane. These sphingolipid–cholesterol clusters can selectively incorporate or exclude proteins and thereby govern protein–protein and protein–lipid interactions. For example, in most epithelial cell lines, glycosyl phosphatidylinositol–anchored proteins are raft associated and are eventually delivered selectively to the apical surface of the cell. Rafts of specific lipids also serve as platforms for the triggering of signaling cascades. Rafts also facilitate the concentration of particular lipids and proteins during the docking and fusion of pairs of eukaryotic cells. In all these functions, it appears that raft association is necessary but not sufficient to accomplish either protein or lipid trafficking, cell fusion, or cell signaling, since for any function there appear to be multiple layers of specificity and organizing forces.

*All cavities and compartments in animals—for example, blood vessels—are lined with a tightly packed single layer of cells called epithelial cells. The surface of the epithelial cell that faces the cavity is referred to as the apical face. The rest of the epithelial cell membrane is referred to as the basolateral face of the membrane.

a membrane depends on the distribution of lipids. The resulting charge differences affect the membrane potential, which in turn is known to modulate the activity of certain ion channels and other membrane proteins.

How are transverse lipid asymmetries created and maintained in cell membranes? From a thermodynamic perspective, these asymmetries could occur only by virtue of asymmetric syntheses of the bilayer itself or by energy-dependent asymmetric transport mechanisms. Without at least one of these, lipids of all kinds would eventually distribute equally between the two monolayers of a membrane. In eukaryotic cells, phospholipids, glycolipids, and cholesterol are synthesized by enzymes located in (or on the surface of) the ER and the Golgi system (discussed in Chapter 24). Most, if not all, of these biosynthetic processes are asymmetrically arranged across the membranes of the ER and Golgi. There is also a separate and continuous flow of phospholipids, glycolipids, and cholesterol from the ER and Golgi to other membranes in the cell, including the plasma membrane. This flow is mediated by specific **lipid transfer proteins.** Most eukaryotic cells appear to contain such proteins.

Flippases: Proteins That Flip Lipids Across the Membrane   Proteins that can "flip" phospholipids from one side of a bilayer to the other have also been identified in several tissues (Figure 9.11). Called **flippases,** these proteins reduce the half-time for phospholipid movement across a membrane from 10 days or more to a few minutes or less. Some of these systems may operate passively, with no required input of energy, but passive transport alone cannot establish or maintain asymmetric transverse lipid distributions. However, rapid phospholipid movement from one monolayer to the other occurs in an *ATP-dependent* manner in erythrocytes. Energy-dependent lipid flippase activity may be responsible for the creation and maintenance of transverse lipid asymmetries.

Membranes Undergo Phase Transitions

Lipids in bilayers undergo radical changes in physical state over characteristic narrow temperature ranges. These changes are in fact true **phase transitions,** and the temperatures at which these changes take place are referred to as **transition temperatures** or **melting temperatures** (T_m). These phase transitions involve substantial changes in the organization and motion of the fatty acyl chains

Flippase protein

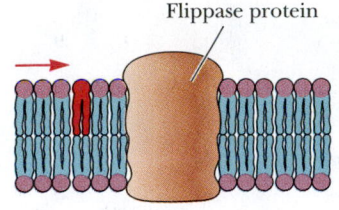 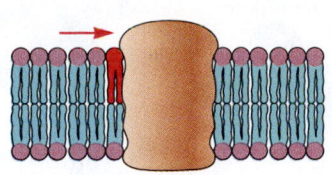 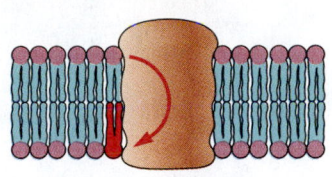 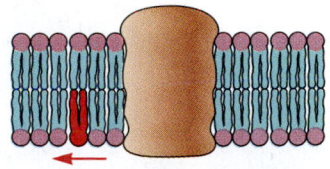

1 Lipid molecule diffuses to flippase protein

2 Flippase flips lipid to opposite side of bilayer

3 Lipid diffuses away from flippase

Biochemistry Now™ ANIMATED FIGURE 9.11
Phospholipids can be "flipped" across a bilayer membrane by the action of flippase proteins. When, by normal diffusion through the bilayer, the lipid encounters a flippase, it can be moved quickly to the other face of the bilayer. **See this figure animated at http://chemistry.brookscole.com/ggb3**

within the bilayer. The bilayer below the phase transition exists in a closely packed gel state, with the fatty acyl chains relatively immobilized in a tightly packed array (Figure 9.12). In this state, the anti conformation is adopted by all the carbon–carbon bonds in the lipid chains. This leaves the lipid chains in their fully extended conformation. As a result, the surface area per lipid is minimal and the bilayer thickness is maximal. Above the transition temperature, a liquid crystalline state exists in which the mobility of fatty acyl chains is intermediate between solid and liquid alkane. In this more fluid, liquid crystalline state, the carbon–carbon bonds of the lipid chains more readily adopt gauche conformations (Figure 9.13). As a result, the surface area per lipid increases and the bilayer thickness decreases by 10% to 15%.

The sharpness of the transition in pure lipid preparations shows that the phase change is a cooperative behavior. This is to say that the behavior of one or a few molecules affects the behavior of many other molecules in the vicinity. The sharpness of the transition then reflects the number of molecules that are acting in concert. Sharp transitions involve large numbers of molecules all "melting" together.

Phase transitions have been characterized in a number of different pure and mixed lipid systems. Table 9.1 shows a comparison of the transition temperatures observed for several different phosphatidylcholines with different fatty acyl chain compositions. General characteristics of bilayer phase transitions include the following:

1. The transitions are always endothermic; heat is absorbed as the temperature increases through the transition (Figure 9.13).
2. Particular phospholipids display characteristic transition temperatures (T_m). As shown in Table 9.1, T_m increases with chain length, decreases with unsaturation, and depends on the nature of the polar head group.
3. For pure phospholipid bilayers, the transition occurs over a narrow temperature range. The phase transition for dimyristoyl lecithin has a peak width of about 0.2°C.
4. Native biological membranes also display characteristic phase transitions, but these are broad and strongly dependent on the lipid and protein composition of the membrane.
5. With certain lipid bilayers, a change of physical state referred to as a *pretransition* occurs 5° to 15°C below the phase transition itself. These pretransitions involve a tilting of the hydrocarbon chains.

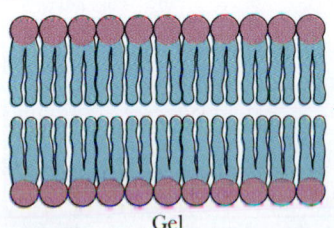 Heat →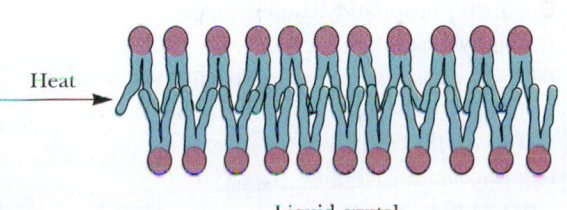

Gel   Liquid crystal

Biochemistry Now™ ANIMATED FIGURE 9.12
An illustration of the gel-to-liquid crystalline phase transition, which occurs when a membrane is warmed through the transition temperature, T_m. Notice that the surface area must increase and the thickness must decrease as the membrane goes through a phase transition. The mobility of the lipid chains increases dramatically. **See this figure animated at http://chemistry.brookscole.com/ggb3**

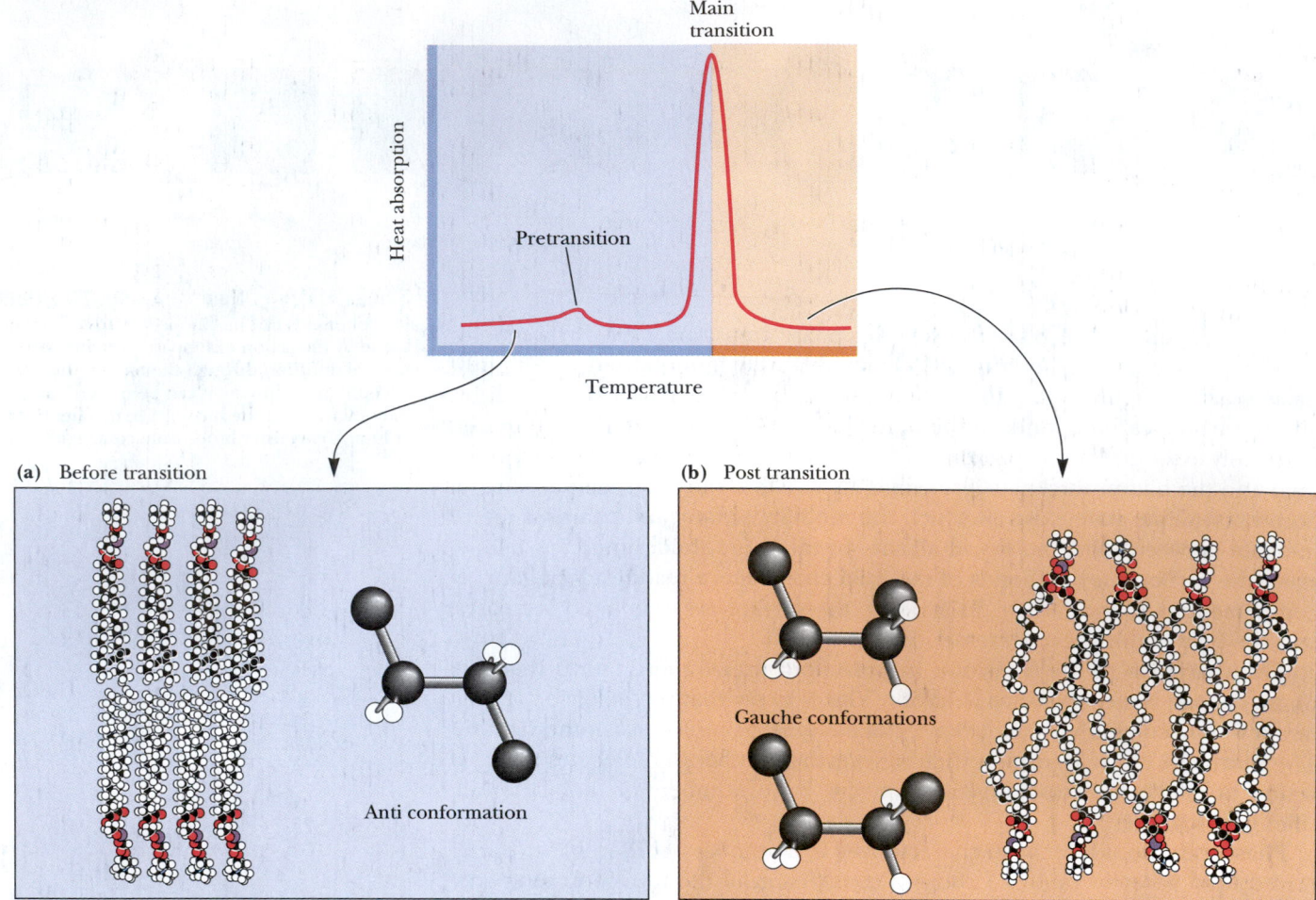

FIGURE 9.13 Membrane lipid phase transitions can be detected and characterized by measuring the rate of absorption of heat by a membrane sample in a calorimeter (see Chapter 3 for a detailed discussion of calorimetry). Pure, homogeneous bilayers (containing only a single lipid component) give sharp calorimetric peaks. Egg PC contains a variety of fatty acid chains and thus yields a broad calorimetric peak. Below the phase transition, lipid chains primarily adopt the anti conformation. Above the phase transition, lipid chains have absorbed a substantial amount of heat. This is reflected in the adoption of higher-energy conformations, including the gauche conformations shown.

6. A volume change is usually associated with phase transitions in lipid bilayers.

7. Bilayer phase transitions are sensitive to the presence of solutes that interact with lipids, including multivalent cations, lipid-soluble agents, peptides, and proteins.

Table 9.1

Phase Transition Temperatures for Phospholipids in Water

Phospholipid	Transition Temperature (T_m), °C
Dipalmitoyl phosphatidic acid (Di 16:0 PA)	67
Dipalmitoyl phosphatidylethanolamine (Di 16:0 PE)	63.8
Dipalmitoyl phosphatidylcholine (Di 16:0 PC)	41.4
Dipalmitoyl phosphatidylglycerol (Di 16:0 PG)	41.0
Dilauroyl phosphatidylcholine (Di 14:0 PC)	23.6
Distearoyl phosphatidylcholine (Di 18:0 PC)	58
Dioleoyl phosphatidylcholine (Di 18:1 PC)	−22
1-Stearoyl-2-oleoyl-phosphatidylcholine (1-18:0, 2-18:1 PC)	3
Egg phosphatidylcholine (Egg PC)	−15

Adapted from Jain, M., and Wagner, R. C., 1980. *Introduction to Biological Membranes.* New York: John Wiley and Sons; and Martonosi, A., ed., 1982. *Membranes and Transport,* Vol. 1. New York: Plenum Press.

Cells adjust the lipid composition of their membranes to maintain proper fluidity as environmental conditions change.

| 9.2 | **What Is the Structure and Chemistry of Membrane Proteins?** |

The lipid bilayer constitutes the fundamental structural unit of all biological membranes. Proteins, in contrast, carry out essentially all of the active functions of membranes, including transport activities, receptor functions, and other related processes. As suggested by Singer and Nicolson, most membrane proteins can be classified as peripheral or integral. The peripheral proteins are globular proteins that interact with the membrane mainly through electrostatic and hydrogen-bonding interactions with integral proteins. Although peripheral proteins are not discussed further here, many proteins of this class are described in the context of other discussions throughout this textbook. Integral proteins are those that are strongly associated with the lipid bilayer, with a portion of the protein embedded in, or extending all the way across, the lipid bilayer. Another class of proteins not anticipated by Singer and Nicolson, the **lipid-anchored proteins,** is important in a variety of functions in different cells and tissues. These proteins associate with membranes by means of a variety of covalently linked lipid anchors.

Integral Membrane Proteins Are Firmly Anchored in the Membrane

Despite the diversity of integral membrane proteins, most fall into two general classes. One of these includes proteins attached or anchored to the membrane by only a small hydrophobic segment, such that most of the protein extends out into the water solvent on one or both sides of the membrane. The other class includes those proteins that are more or less globular in shape and more totally embedded in the membrane, exposing only a small surface to the water solvent outside the membrane. In general, those structures of integral membrane protein within the nonpolar core of the lipid bilayer are dominated by α-helices or β-sheets because these secondary structures neutralize the highly polar N—H and C=O functions of the peptide backbone through H-bond formation.

A Protein with a Single Transmembrane Segment In the case of the proteins that are anchored by a small hydrophobic polypeptide segment, that segment typically takes the form of a single α-helix. One of the best examples of a membrane protein with such an α-helical structure is **glycophorin.** Most of glycophorin's mass is oriented on the outside surface of the cell, exposed to the aqueous milieu (Figure 9.14). A variety of hydrophilic oligosaccharide units are attached to this extracellular domain. These oligosaccharide groups constitute the ABO and MN blood group antigenic specificities of the red cell. This extracellular portion of the protein also serves as the receptor for the influenza virus. Glycophorin has a total molecular weight of about 31,000 and is approximately 40% protein and 60% carbohydrate. The glycophorin primary structure consists of a segment of 19 hydrophobic amino acid residues with a short hydrophilic sequence on one end and a longer hydrophilic sequence on the other end. The 19-residue sequence is just the right length to span the cell membrane if it is coiled in the shape of an α-helix. The large hydrophilic sequence includes the amino terminal residue of the polypeptide chain.

Bacteriorhodopsin: A Seven-Transmembrane Segment Protein Membrane proteins that take on a more globular shape, instead of the single TMS structure previously described, are often involved with transport activities and other functions requiring a substantial portion of the peptide to be embedded in the membrane. These proteins may consist of numerous hydrophobic α-helical

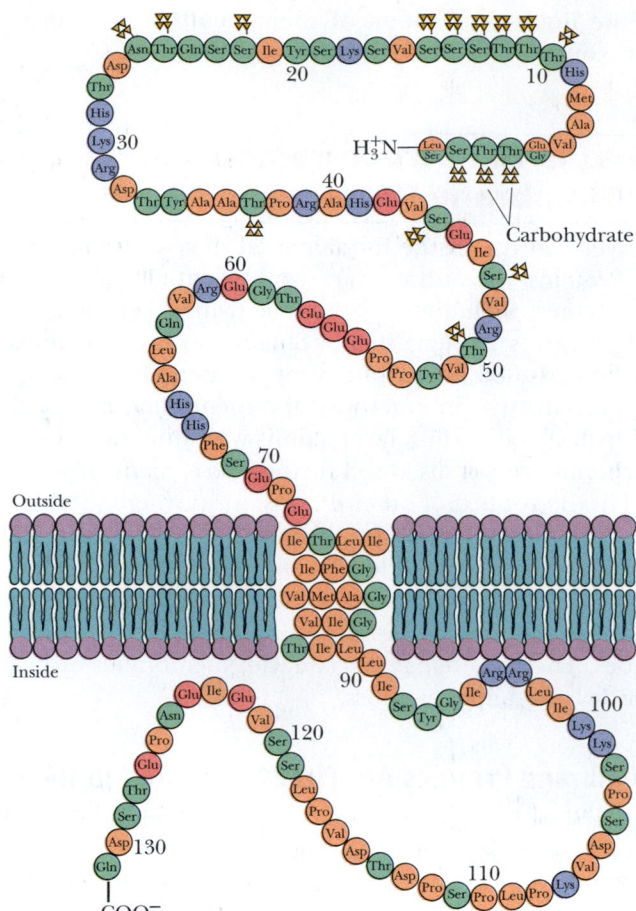

FIGURE 9.14 Glycophorin A spans the membrane of the human erythrocyte via a single α-helical transmembrane segment. The C-terminus of the peptide, whose sequence is shown here, faces the cytosol of the erythrocyte; the N-terminal domain is extracellular. Points of attachment of carbohydrate groups are indicated.

segments joined by hinge regions so that the protein winds in a zigzag pattern back and forth across the membrane. A well-characterized example of such a protein is **bacteriorhodopsin,** which clusters in purple patches in the membrane of the bacterium *Halobacterium halobium*. The name *Halobacterium* refers to the fact that this bacterium thrives in solutions having high concentrations of sodium chloride, such as the salt beds of San Francisco Bay. *Halobacterium* carries out a light-driven proton transport by means of bacteriorhodopsin, named in reference to its spectral similarities to rhodopsin in the rod outer segments of the mammalian retina. When this organism is deprived of oxygen for oxidative metabolism, it switches to the capture of energy from sunlight, using this energy to pump protons out of the cell. The proton gradient generated by such light-driven proton pumping represents potential energy, which is exploited elsewhere in the membrane to synthesize ATP.

Bacteriorhodopsin clusters in hexagonal arrays (Figure 9.15) in the purple membrane patches of *Halobacterium,* and it was this orderly, repeating arrangement of proteins in the membrane that enabled Nigel Unwin and Richard Henderson in 1975 to determine the bacteriorhodopsin structure. The polypeptide chain crosses the membrane seven times, in seven α-helical segments, with very little of the protein exposed to the aqueous milieu. The bacteriorhodopsin structure has become a model of globular membrane protein structure. Many other integral membrane proteins contain numerous hydrophobic sequences that, like those of bacteriorhodopsin, could form α-helical transmembrane segments. For example, the amino acid sequence of the sodium–potassium transport ATPase contains ten hydrophobic segments of length sufficient to span the plasma membrane. By analogy with bacteriorhodopsin, one would expect that these segments form a globular

A Deeper Look

Single TMS Proteins

In addition to glycophorin, numerous other membrane proteins are attached to the membrane by means of a single hydrophobic α-helix, with hydrophilic segments extending either into the cytoplasm or the extracellular space. These proteins often function as receptors for extracellular signaling molecules or as recognition sites that allow the immune system to recognize and distinguish the cells of the host organism from invading foreign cells or viruses. The proteins that represent the *major transplantation antigens H2* in mice and *human leukocyte associated (HLA) proteins* in humans are members of this class. Other such proteins include the *surface immunoglobulin receptors* on B lymphocytes and the *spike proteins* of many membrane viruses. The function of many of these proteins depends primarily on their extracellular domain, and thus the segment facing the intracellular surface is often a shorter one.

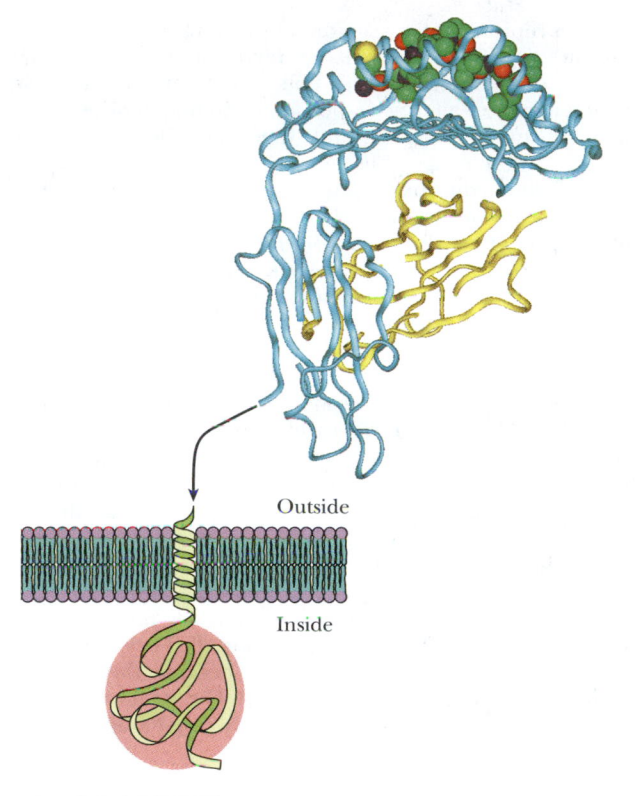

▶ The major histocompatibility antigen HLA-A2 is a membrane-associated protein with a single transmembrane helical segment. The extracellular domain of this protein (in blue and yellow) is shown here complexed to a decapeptide (as a space-filling model) from calreticulin.

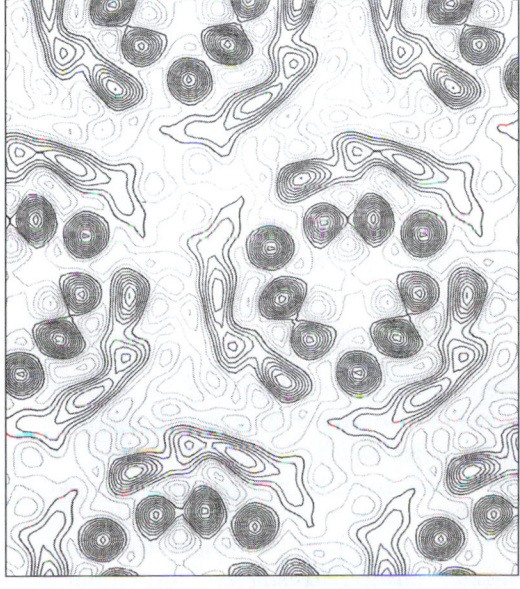
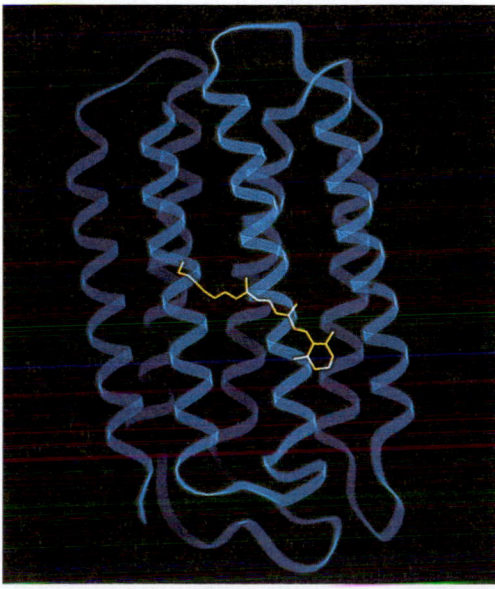

FIGURE 9.15 An electron density profile illustrating the three centers of threefold symmetry in arrays of bacteriorhodopsin in the purple membrane of *Halobacterium halobium*, together with a computer-generated model showing the seven α-helical transmembrane segments in bacteriorhodopsin. *(Electron density map from Stoecknius, W., 1980. Purple membrane of Halobacteria: A new light-energy converter. Accounts of Chemical Research **13:**337–344. Model on right from Henderson, R., 1990. Model for the structure of bacteriorhodopsin based on high-resolution electron cryo-microscopy. Journal of Molecular Biology **213:**899–929.)*

Human Biochemistry

Treating Allergies at the Cell Membrane

Allergies represent overreactions of the immune system caused by exposure to foreign substances referred to as allergens. The inhalation of allergens, such as pollen, pet dander, and dust, can cause a variety of allergic responses, including itchy eyes, a runny nose, shortness of breath, and wheezing. Allergies can also be caused by food, drugs, dyes, and other chemicals.

The visible symptoms of such an allergic response are caused by the release of histamine (see accompanying figure) by mast cells, a type of cell found in loose connective tissue. Histamine dilates blood vessels, increases the permeability of capillaries (allowing antibodies to pass from the capillaries to surrounding tissue), and constricts bronchial air passages. Histamine acts by binding to specialized membrane proteins called histamine H_1 receptors. These integral membrane proteins possess seven transmembrane α-helical segments, with an extracellular amino terminus and a cytoplasmic carboxy terminus. When histamine binds to the extracellular domain of an H_1 receptor, the intracellular domain undergoes a conformation change that stimulates a GTP-binding protein, which in turn activates the allergic response in the affected cell.

A variety of highly effective antihistamine drugs are available for the treatment of allergy symptoms. These drugs share the property of binding tightly to histamine H_1 receptors, without eliciting the same effects as histamine itself. They are referred to as histamine H_1 receptor antagonists because they prevent the binding of histamine to the receptors. The structures of Allegra (made by Aventis, Inc.), Claritin (by Schering-Plough Corp.), and Zyrtec (by Pfizer) are all shown at right.

Histamine

Allegra
(Aventis, Inc.)

Zyrtec
(Pfizer)

Claritin
(Schering)

▶ The structures of histamine and three antihistamine drugs.

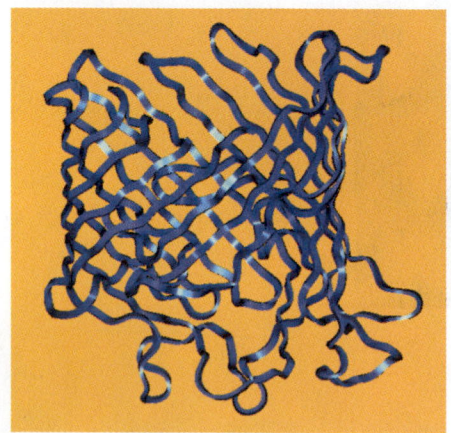

FIGURE 9.16 The three-dimensional structure of maltoporin from *E. coli.*

hydrophobic core that anchors the ATPase in the membrane. The helical segments may also account for the transport properties of the enzyme itself.

Porins—A β-Sheet Motif for Membrane Proteins The β-sheet is another structural motif that provides extensive hydrogen bonding for transmembrane peptide segments. **Porin** proteins found in the outer membranes (OMs) of Gram-negative bacteria such as *Escherichia coli,* and also in the outer mitochondrial membranes of eukaryotic cells, span their respective membranes with large β-sheets. A good example is **maltoporin,** also known as **LamB protein** or **lambda receptor,** which participates in the entry of maltose and maltodextrins into *E. coli.* Maltoporin is active as a trimer. The 421-residue monomer is an aesthetically pleasing 18-strand β-barrel (Figure 9.16). The β-strands are connected to their nearest neighbors either by long loops or by β-turns (Figure 9.17). The long loops are found at the end of the barrel that is exposed to the cell exterior, whereas the turns are located on the intracellular face of the barrel. Three of the loops fold into the center of the barrel.

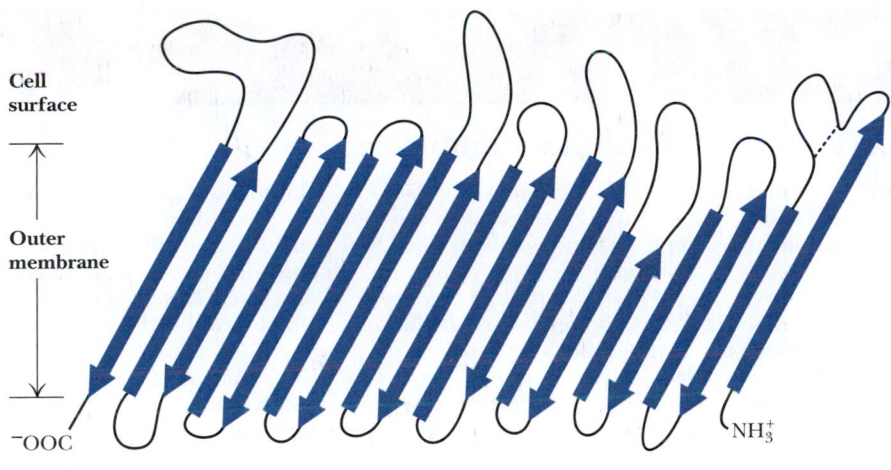

Cell
surface

Outer
membrane

⁻OOC

Periplasmic space

NH₃⁺

FIGURE 9.17 The arrangement of the peptide chain in maltoporin from *E. coli*.

The amino acid compositions and sequences of the β-strands in porin proteins are novel. Polar and nonpolar residues alternate along the β-strands, with polar residues facing the central pore or cavity of the barrel and nonpolar residues facing out from the barrel, where they can interact with the hydrophobic lipid milieu of the membrane. The smallest diameter of the porin channel is about 5 Å. Thus, a maltodextrin polymer (composed of two or more glucose units) must pass through the porin in an extended conformation (like a spaghetti strand).

Porins and the other OM proteins of Gram-negative bacteria are a prominent class of membrane proteins that have chosen the β-strand over the α-helix. Why might this be? Among other reasons, there is an advantage of genetic economy in the use of β-strands to traverse the membrane instead of α-helices. An α-helix requires 21 to 25 amino acid residues to span a typical biological membrane; a β-strand can cross the same membrane with 9 to 11 residues. Therefore, a given amount of genetic information could encode a larger number of membrane-spanning segments using a β-strand motif instead of α-helical arrays. Furthermore, β-strands can present alternating hydrophobic and hydrophilic R groups along their length, with hydrophobic R groups facing the lipid bilayer and hydrophilic R groups facing the water-filled channel (see Figure 9.17).

Biochemistry Now™   Go to BiochemistryNow and click BiochemistryInteractive to discover how a β-sheet motif for membrane proteins is exploited by maltoporin.

Lipid-Anchored Membrane Proteins Are Switching Devices

Certain proteins are found to be covalently linked to lipid molecules. For many of these proteins, covalent attachment of lipid is required for association with a membrane. The lipid moieties can insert into the membrane bilayer, effectively **anchoring** their linked proteins to the membrane. Some proteins with covalently linked lipid normally behave as soluble proteins; others are integral membrane proteins and remain membrane associated even when the lipid is removed. Covalently bound lipid in these latter proteins can play a role distinct from membrane anchoring. In many cases, attachment to the membrane via the lipid anchor serves to modulate the activity of the protein.

Another interesting facet of lipid anchors is that they are transient. Lipid anchors can be reversibly attached to and detached from proteins. This provides a "switching device" for altering the affinity of a protein for the membrane. Reversible lipid anchoring is one factor in the control of **signal transduction pathways** in eukaryotic cells (see Chapter 32).

Four different types of lipid-anchoring motifs have been found to date. These are **amide-linked myristoyl** anchors, **thioester-linked fatty acyl** anchors, **thioether-linked prenyl** anchors, and **amide-linked glycosyl phosphatidylinositol** anchors. Each of these anchoring motifs is used by a variety of membrane

A Deeper Look

Exterminator Proteins—Biological Pest Control at the Membrane

Control of biological pests, including mosquitoes, houseflies, gnats, and tree-consuming predators like the eastern tent caterpillar, is frequently achieved through the use of microbial membrane proteins. For example, several varieties of *Bacillus thuringiensis* produce proteins that bind to cell membranes in the digestive systems of insects that consume them, creating transmembrane ion channels. Leakage of Na^+, K^+, and H^+ ions through these membranes in the insect gut destroys crucial ion gradients and interferes with digestion of food. Insects that ingest these toxins eventually die of starvation. *B. thuringiensis* toxins account for more than 90% of sales of biological pest control agents.

B. thuringiensis is a common Gram-positive, spore-forming soil bacterium that produces inclusion bodies, microcrystalline clusters of many different proteins. These crystalline proteins, called δ-endotoxins, are the ion channel toxins that are sold commercially for pest control. Most such endotoxins are protoxins, which are inactive until cleaved to smaller, active proteins by proteases in the gut of a susceptible insect. One such crystalline protoxin,

lethal to mosquitoes, is a 27-kD protein, which is cleaved to form the active 25-kD toxin in the mosquito. This toxin has no effect on membranes at neutral pH, but at pH 9.5 (the pH of the mosquito gut) the toxin forms cation channels in the gut membranes.

This 25-kD protein is not toxic to tent caterpillars, but a larger, 130-kD protein in the *B. thuringiensis* inclusion bodies is cleaved by a caterpillar gut protease to produce a 55-kD toxin that is active in the caterpillar. Remarkably, the strain of *B. thuringiensis* known as azawai produces a protoxin with dual specificity: In the caterpillar gut, this 130-kD protein is cleaved to form a 55-kD toxin active in the caterpillar. However, when the same 130-kD protoxin is consumed by mosquitoes or houseflies, it is cleaved to form a 53-kD protein (15 amino acid residues shorter than the caterpillar toxin) that is toxic to these latter organisms. Understanding the molecular basis of the toxicity and specificity of these proteins and the means by which they interact with membranes to form lethal ion channels is a fascinating biochemical challenge with far-reaching commercial implications.

proteins, but each nonetheless exhibits a characteristic pattern of structural requirements.

Amide-Linked Myristoyl Anchors   Myristic acid may be linked via an amide bond to the α-amino group of the N-terminal glycine residue of selected proteins (Figure 9.18). The reaction is referred to as **N-myristoylation** and is catalyzed by *myristoyl–CoA:protein N-myristoyltransferase,* known simply as **NMT.** N-Myristoyl–anchored proteins include the catalytic subunit of *cAMP-dependent protein kinase,* the *pp60^src tyrosine kinase,* the phosphatase known as *calcineurin B,* the α-subunit of *G proteins* (involved in GTP-dependent transmembrane signaling events), and the *gag proteins* of certain retroviruses (including the HIV-1 virus that causes AIDS).

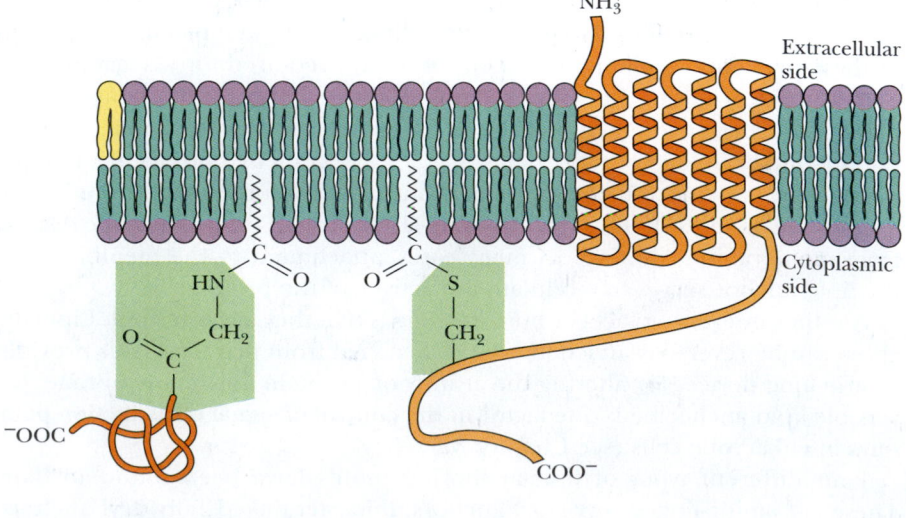

FIGURE 9.18 Certain proteins are anchored to biological membranes by lipid anchors. Particularly common are the **(a)** *N*-myristoyl– and **(b)** *S*-palmitoyl–anchoring motifs shown here. *N*-Myristoylation always occurs at an *N*-terminal glycine residue, whereas thioester linkages occur at cysteine residues within the polypeptide chain. G-protein–coupled receptors, with seven transmembrane segments, may contain one (and sometimes two) palmitoyl anchors in thioester linkage to cysteine residues in the C-terminal segment of the protein.

(a) *N*-Myristoylation   **(b)** *S*-Palmitoylation

Thioester-Linked Fatty Acyl Anchors   A variety of cellular and viral proteins contain fatty acids covalently bound via ester linkages to the side chains of cysteine and sometimes to serine or threonine residues within a polypeptide chain (Figure 9.18). This type of fatty acyl chain linkage has a broader fatty acid specificity than *N*-myristoylation. Myristate, palmitate, stearate, and oleate can all be esterified in this way, with the C_{16} and C_{18} chain lengths being most commonly found. Proteins anchored to membranes via fatty acyl thioesters include *G-protein–coupled receptors,* the *surface glycoproteins* of several viruses, and the *transferrin receptor* protein.

Thioether-Linked Prenyl Anchors   As noted in Chapter 8, polyprenyl (or simply prenyl) groups are long-chain polyisoprenoid groups derived from isoprene units. Prenylation of proteins destined for membrane anchoring can involve either **farnesyl** or **geranylgeranyl** groups (Figure 9.19). The addition of a prenyl group typically occurs at the cysteine residue of a carboxy-terminal CAAX sequence of the target protein, where C is cysteine, A is any aliphatic residue, and X can be any amino acid. As shown in Figure 9.19, the result is a thioether-linked farnesyl or geranylgeranyl group. Once the prenylation reaction has occurred, a specific protease cleaves the three carboxy-terminal residues, and the carboxyl group of the now terminal Cys is methylated to produce an ester. All of these modifications appear to be important for subsequent activity of the prenyl-anchored protein. Proteins anchored to membranes via prenyl groups include *yeast mating factors,* the *p21^{ras} protein* (the protein product of the *ras*

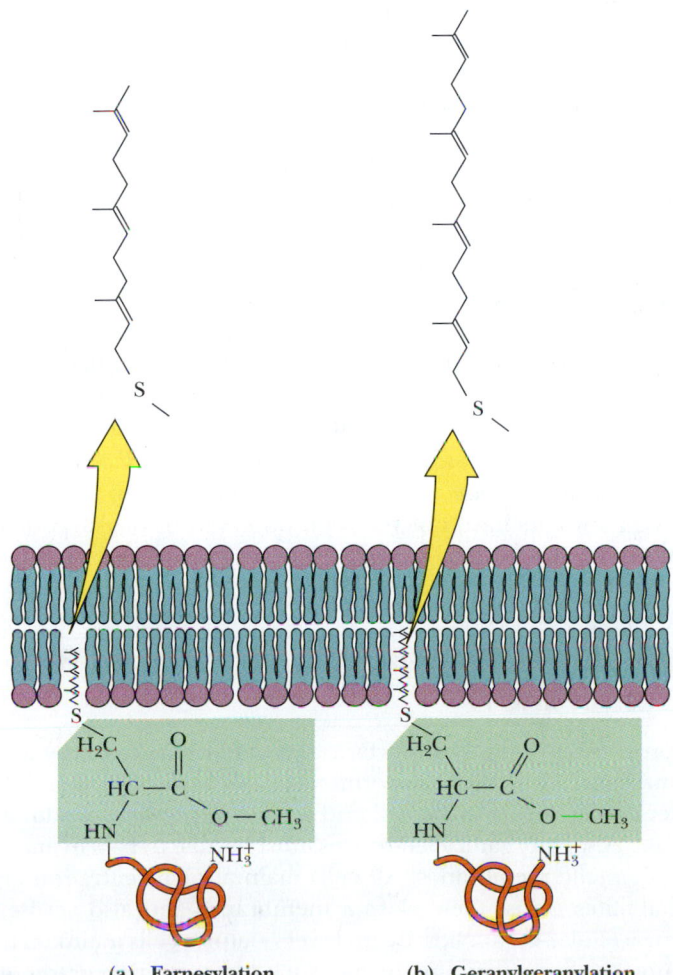

(a) **Farnesylation**   (b) **Geranylgeranylation**

FIGURE 9.19 Proteins containing the C-terminal sequence CAAX can undergo prenylation reactions that place thioether-linked **(a)** farnesyl or **(b)** geranylgeranyl groups at the cysteine side chain. Prenylation is accompanied by removal of the AAX peptide and methylation of the carboxyl group of the cysteine residue, which has become the C-terminal residue.

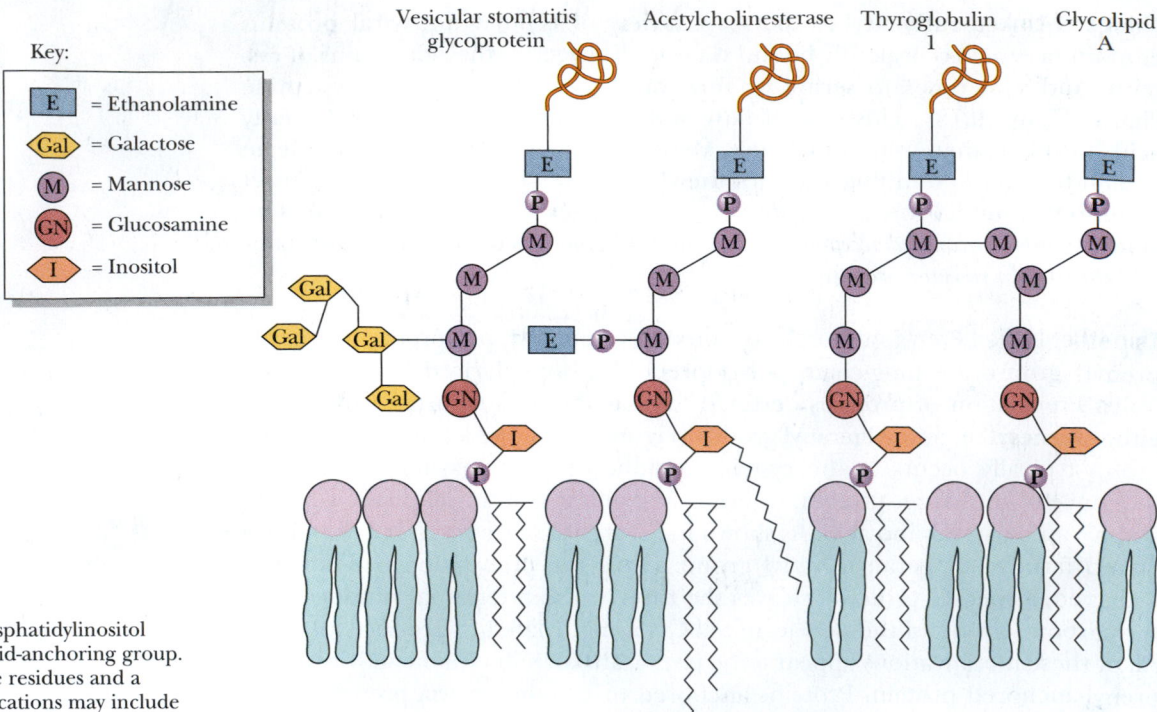

FIGURE 9.20 The glycosyl phosphatidylinositol (GPI) moiety is an elaborate lipid-anchoring group. Note the core of three mannose residues and a glucosamine. Additional modifications may include fatty acids at the inositol and glycerol —OH groups.

oncogene; see Chapter 32), and the *nuclear lamins,* structural components of the lamina of the inner nuclear membrane.

Glycosyl Phosphatidylinositol Anchors   Glycosyl phosphatidylinositol, or **GPI,** groups are structurally more elaborate membrane anchors than fatty acyl or prenyl groups. GPI groups modify the carboxy-terminal amino acid of a target protein via an ethanolamine residue linked to an oligosaccharide, which is linked in turn to the inositol moiety of a phosphatidylinositol (Figure 9.20). The oligosaccharide typically consists of a conserved tetrasaccharide core of three mannose residues and a glucosamine, which can be altered by modifications of the mannose residues or addition of galactosyl side chains of various sizes, extra phosphoethanolamines, or additional *N*-acetylgalactose or mannosyl residues (Figure 9.20). The inositol moiety can also be modified by an additional fatty acid, and a variety of fatty acyl groups are found linked to the glycerol group. GPI groups anchor a wide variety of *surface antigens, adhesion molecules,* and *cell surface hydrolases* to plasma membranes in various eukaryotic organisms. GPI anchors have not yet been observed in prokaryotic organisms or plants.

| 9.3 | **How Does Transport Occur Across Biological Membranes?** |

Transport processes are vitally important to all life forms, because all cells must exchange materials with their environment. Cells obviously must have ways to bring nutrient molecules into the cell and ways to send waste products and toxic substances out. Also, inorganic electrolytes must be able to pass in and out of cells and across organelle membranes. All cells maintain **concentration gradients** of various metabolites across their plasma membranes and also across the membranes of intracellular organelles. By their very nature, cells maintain a very large amount of potential energy in the form of such concentration gradients. Sodium

Human Biochemistry

Biochemistry Now™

Prenylation Reactions as Possible Chemotherapy Targets

The protein called p21[ras], or simply Ras, is a small GTP-binding protein involved in cell signaling pathways that regulate growth and cell division. Mutant forms of Ras cause uncontrolled cell growth, and Ras mutations are involved in one-third of all human cancers. Because the signaling activity of Ras is dependent on prenylation, the prenylation reaction itself, as well as the proteolysis of the -AAX motif and the methylation of the prenylated Cys residue, have been considered targets for development of new chemotherapy strategies.

Farnesyl transferase from rat cells is a heterodimer consisting of a 48-kD α-subunit and a 46-kD β-subunit. In the structure shown here, helices 2 to 15 of the α-subunit are folded into seven short, coiled coils that together form a crescent-shaped envelope partially surrounding the β-subunit. Twelve helices of the β-subunit form a novel barrel motif that creates the active site of the enzyme. Farnesyl transferase inhibitors, one of which is shown here, are potent suppressors of tumor growth in mice, but their value in humans has not been established.

Mutations that inhibit prenyl transferases cause defective growth or death of cells, raising questions about the usefulness of prenyl transferase inhibitors in chemotherapy. However, Victor Boyartchuk and his colleagues at the University of California, Berkeley, and Acacia Biosciences have shown that the protease that cleaves the -AAX motif from Ras following the prenylation reaction may be a better chemotherapeutic target. They have identified two genes for the prenyl protein protease in the yeast *Saccharomyces cerevisiae* and have shown that deletion of these genes results in loss of proteolytic processing of prenylated proteins, including Ras. Interestingly, normal yeast cells are unaffected by this gene deletion. However, in yeast cells that carry mutant forms of Ras and that display aberrant growth behaviors, deletion of the protease gene restores normal growth patterns. If these remarkable results translate from yeast to human tumor cells, inhibitors of CAAX proteases may be more valuable chemotherapeutic agents than prenyl transferase inhibitors.

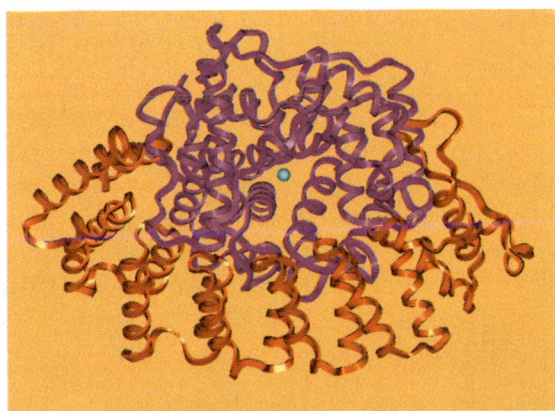

▲ The structure of the farnesyl transferase heterodimer. A novel barrel structure is formed from 12 helical segments in the β-subunit (purple). The α-subunit (yellow) consists largely of seven successive pairs of α-helices that form a series of right-handed antiparallel coiled coils running along the bottom of the structure. These "helical hairpins" are arranged in a double-layered, right-handed superhelix resulting in a crescent-shaped subunit that envelopes part of the subunit.

2(S)-{(S)-[2(R)-amino-3-mercapto]propylamino-3(S)-methyl}pentyloxy-3-phenylpropionyl-methioninesulfone methyl ester

▲ This substance, also known as I-739,749, is a farnesyl transferase inhibitor that is a potent tumor growth suppressor.

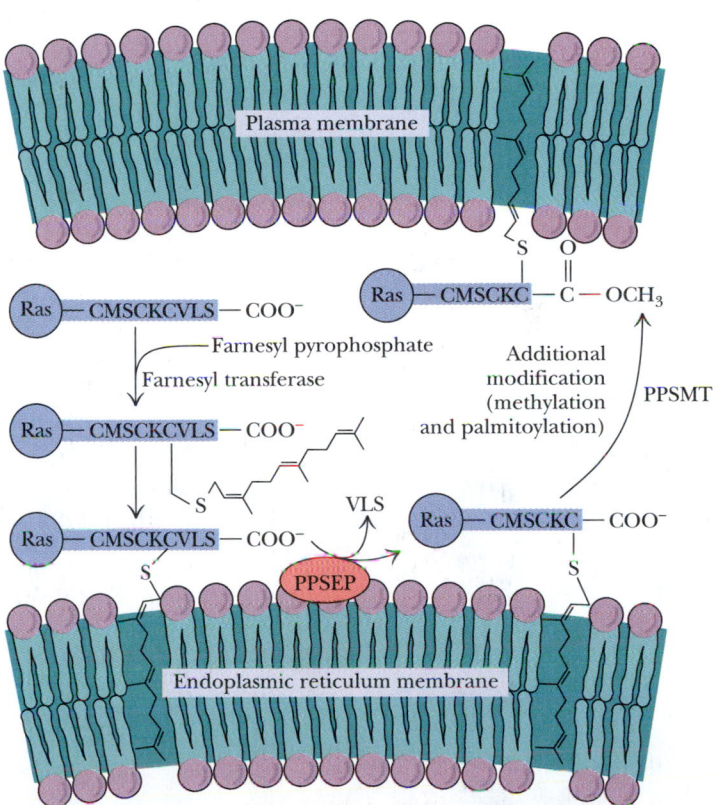

▲ The farnesylation and subsequent processing of the Ras protein. Following farnesylation by the FTase, the carboxy-terminal VLS peptide is removed by a prenyl protein-specific endoprotease (PPSEP) in the ER; then a prenylprotein-specific methyltransferase (PPSMT) donates a methyl group from S-adenosylmethionine (SAM) to the carboxy-terminal S-farnesylated cysteine. Finally, palmitates are added to cysteine residues near the C-terminus of the protein.

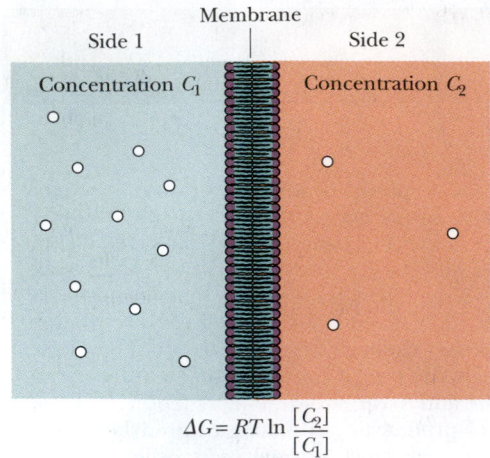

$$\Delta G = RT \ln \frac{[C_2]}{[C_1]}$$

Biochemistry ⊜ Now™ ACTIVE FIGURE 9.21
Passive diffusion of an uncharged species across a membrane depends only on the concentrations (C_1 and C_2) on the two sides of the membrane. **Test yourself on the concepts in this figure at http://chemistry.brookscole.com/ggb3**

and potassium ion gradients across the plasma membrane mediate the transmission of nerve impulses and the normal functions of the brain, heart, kidneys, and liver, among other organs. Storage and release of calcium from cellular compartments controls muscle contraction, as well as the response of many cells to hormonal signals. High acid concentrations in the stomach are required for the digestion of food. Extremely high hydrogen ion gradients are maintained across the plasma membranes of the mucosal cells lining the stomach in order to maintain high acid levels in the stomach yet protect the cells that constitute the stomach walls from the deleterious effects of such acid.

We shall consider the molecules and mechanisms that mediate these transport activities. In nearly every case, the molecule or ion transported is water soluble, yet moves across the hydrophobic, impermeable lipid membrane at a rate high enough to serve the metabolic and physiological needs of the cell. This perplexing problem is solved in each case by a specific transport protein. The transported species either diffuses through a channel-forming protein or is carried by a carrier protein. Transport proteins are all classed as **integral membrane proteins.**

From a thermodynamic and kinetic perspective, there are only three types of membrane transport processes: *passive diffusion, facilitated diffusion,* and *active transport.* To be thoroughly appreciated, membrane transport phenomena must be considered in terms of thermodynamics. Some of the important kinetic considerations also will be discussed.

9.4 | What Is Passive Diffusion?

Passive diffusion is the simplest transport process. In passive diffusion, the transported species moves across the membrane in the thermodynamically favored direction without the help of any specific transport system/molecule. For an uncharged molecule, passive diffusion is an entropic process, in which movement of molecules across the membrane proceeds until the concentration of the substance on both sides of the membrane is the same. For an uncharged molecule, the free energy difference between side 1 and side 2 of a membrane (Figure 9.21) is given by

$$\Delta G = G_2 - G_1 = RT \ln \frac{[C_2]}{[C_1]} \tag{9.1}$$

The difference in concentrations, $[C_2] - [C_1]$, is termed the *concentration gradient,* and ΔG here is the *chemical potential difference.*

Charged Species May Cross Membranes by Passive Diffusion

For a charged species, the situation is slightly more complicated. In this case, the movement of a molecule across a membrane depends on its **electrochemical potential.** This is given by

$$\Delta G = G_2 - G_1 = RT \ln \frac{[C_2]}{[C_1]} + Z\mathcal{F}\Delta\psi \tag{9.2}$$

Biochemistry ⊜ Now™ ACTIVE FIGURE 9.22
The passive diffusion of a charged species across a membrane depends on the concentration and also on the charge of the particle, Z, and the electrical potential difference across the membrane, $\Delta\psi$. **Test yourself on the concepts in this figure at http://chemistry.brookscole.com/ggb3**

where Z is the **charge** on the transported species, $\mathcal{F}$ is **Faraday's constant** (the charge on 1 mole of electrons = 96,485 coulombs/mol = 96,485 joules/volt/mol, since 1 volt = 1 joule/coulomb), and $\Delta\psi$ is the electric potential difference (that is, voltage difference) across the membrane. The second term in the expression thus accounts for the movement of a charge across a potential difference. Note that the effect of this second term on ΔG depends on the magnitude and the sign of both Z and $\Delta\psi$. For example, as shown in Figure 9.22, if side 2 has a higher potential than side 1 (so that $\Delta\psi$ is positive), for a negatively charged ion the term $Z\mathcal{F}\Delta\psi$ makes a negative contribution to ΔG.

Below Figure 9.22:
$$\Psi_2 - \Psi_1 = \Delta\Psi > 0$$
$$Z = -1$$
$$Z\mathcal{F}\Delta\Psi < 0$$

In other words, the negative charge is spontaneously attracted to the more positive potential—and ΔG is negative. In any case, if the sum of the two terms on the right side of Equation 9.2 is a negative number, transport of the ion in question from side 1 to side 2 would occur spontaneously. The driving force for passive transport is the ΔG term for the transported species itself.

9.5 | How Does Facilitated Diffusion Occur?

The transport of many substances across simple lipid bilayer membranes via passive diffusion is far too slow to sustain life processes. On the other hand, the transport rates for many ions and small molecules across actual biological membranes are much higher than anticipated from passive diffusion alone. This difference is due to specific proteins in the cell membranes that **facilitate** transport of these species across the membrane. Proteins capable of effecting **facilitated diffusion** of a variety of solutes are present in essentially all natural membranes. Such proteins have two features in common: (1) They facilitate net movement of solutes only in the thermodynamically favored direction (that is, $\Delta G < 0$), and (2) they display a measurable affinity and specificity for the transported solute. Consequently, facilitated diffusion rates display **saturation behavior** similar to that observed with substrate binding by enzymes (see Chapter 13). Such behavior provides a simple means for distinguishing between passive diffusion and facilitated diffusion experimentally. The dependence of transport rate on solute concentration takes the form of a rectangular hyperbola (Figure 9.23), so the transport rate approaches a limiting value, V_{max}, at very high solute concentration. Figure 9.23 also shows the graphical behavior exhibited by simple passive diffusion. Because passive diffusion does not involve formation of a specific solute:protein complex, the plot of rate versus concentration is linear, not hyperbolic.

Glucose Transport in Erythrocytes Occurs by Facilitated Diffusion

Many transport processes in a variety of cells occur by facilitated diffusion. Table 9.2 lists just a few of these. The **glucose transporter** of erythrocytes illustrates many of the important features of facilitated transport systems. Although glucose transport operates variously by passive diffusion, facilitated diffusion, or active transport mechanisms, depending on the particular cell, the **glucose**

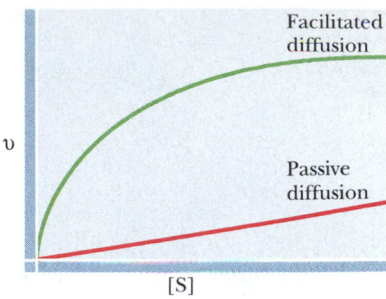

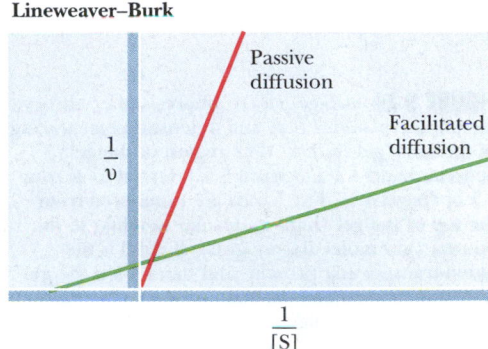

Lineweaver–Burk

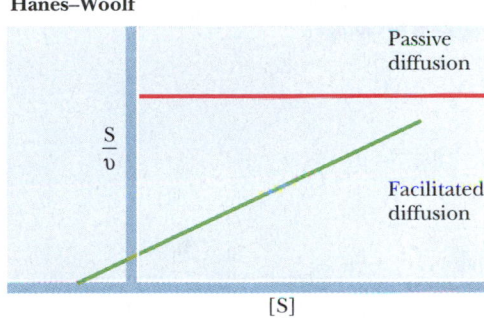

Hanes–Woolf

FIGURE 9.23 Passive diffusion and facilitated diffusion may be distinguished graphically. The plots for facilitated diffusion are similar to plots of enzyme-catalyzed processes (see Chapter 13), and they display saturation behavior.

Table 9.2			
Facilitated Transport Systems			
Permeant	**Cell Type**	K_m **(mM)**	V_{max} **(mM/min)**
D-Glucose	Erythrocyte	4–10	100–500
Chloride	Erythrocyte	25–30	
cAMP	Erythrocyte	0.0047	0.028
Phosphate	Erythrocyte	80	2.8
D-Glucose	Adipocytes	20	
D-Glucose	Yeast	5	
Sugars and amino acids	Tumor cells	0.5–4	2–6
D-Glucose	Rat liver	30	
D-Glucose	*Neurospora crassa*	8.3	46
Choline	Synaptosomes	0.083	
L-Valine	*Arthrobotrys conoides*	0.15–0.75	

Adapted from Jain, M., and Wagner, R., 1980. *Introduction to Biological Membranes.* New York: Wiley.

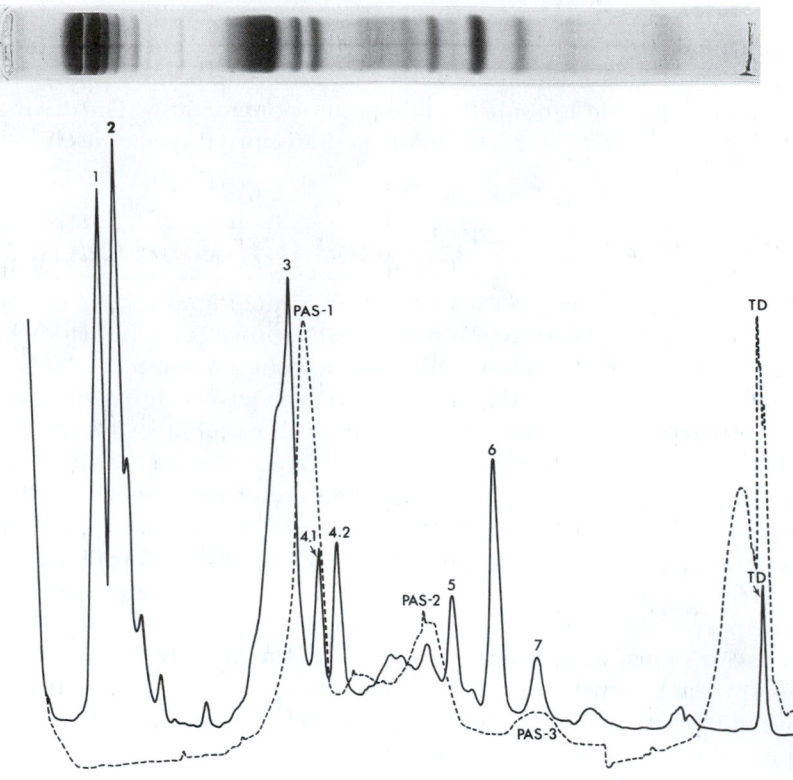

FIGURE 9.24 SDS-gel electrophoresis of erythrocyte membrane proteins *(top)* and a densitometer tracing of the same gel *(bottom)*. The region of the gel between band 4.2 and band 5 is referred to as zone 4.5 or "band 4.5." The bands are numbered from the top of the gel (high molecular weights) to the bottom (low molecular weights). Band 3 is the anion-transporting protein, and band 4.5 is the glucose transporter. The dashed line shows the staining of the gel by periodic acid–Schiff's reagent (PAS), which stains carbohydrates. Three "PAS bands" (PAS-1, PAS-2, PAS-3) indicate the positions of glycoproteins in the gel. *(Photo courtesy of Theodore Steck, University of Chicago.)*

transport system of erythrocytes (red blood cells) operates exclusively by facilitated diffusion. The erythrocyte glucose transporter has a molecular mass of approximately 55 kD and is found on SDS polyacrylamide electrophoresis gels (Figure 9.24) as **band 4.5.** Typical erythrocytes contain around 500,000 copies of this protein. The active form of this transport protein in the erythrocyte membrane is a trimer. Hydropathy analysis of the amino acid sequence of the erythrocyte glucose transporter has provided a model for the structure of the protein (Figure 9.25). In this model, the protein spans the membrane 12 times, with both the N- and C-termini located on the cytoplasmic side. Transmembrane segments 7, 8, and 11 comprise a hydrophilic transmembrane channel, with segments 9 and 10 forming a relatively hydrophobic pocket adjacent to the glucose-binding site. Cytochalasin B, a fungal metabolite (Figure 9.26), is a

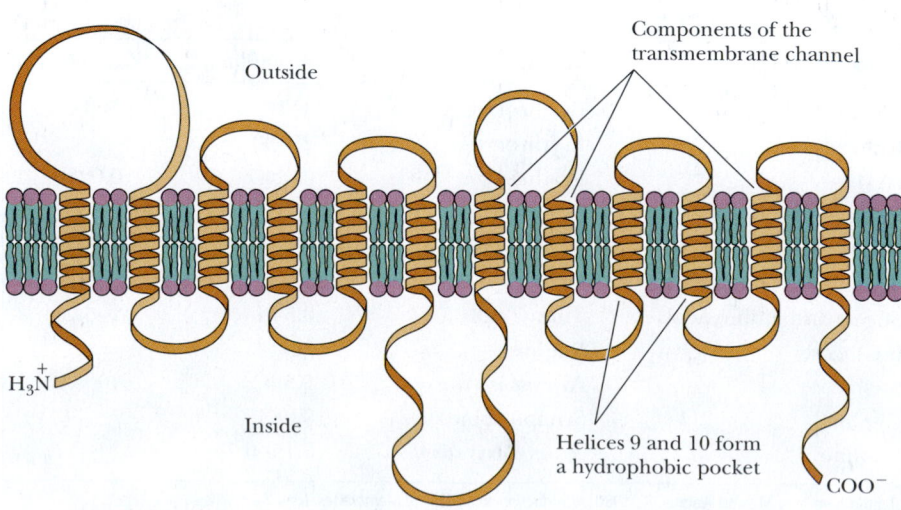

FIGURE 9.25 A model for the arrangement of the glucose transport protein in the erythrocyte membrane. Hydropathy analysis is consistent with 12 transmembrane α-helical segments.

competitive inhibitor of glucose transport. The reduced ability of insulin to stimulate glucose transport in diabetic patients is due to reduced expression of certain glucose transport proteins.

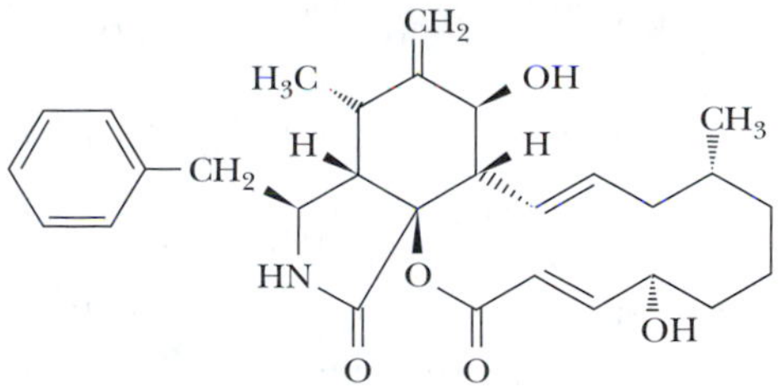

The Anion Transporter of Erythrocytes Also Operates by Facilitated Diffusion

The **anion transport system** is another facilitated diffusion system of the erythrocyte membrane. Chloride and bicarbonate (HCO_3^-) ions are exchanged across the red cell membrane by a 95-kD transmembrane protein. This protein is abundant in the red cell membrane and is represented by **band 3** on SDS electrophoresis gels (Figure 9.24). The gene for the human erythrocyte anion transporter has been sequenced, and hydropathy analysis has yielded a model for the arrangement of the protein in the red cell membrane (Figure 9.27). The model has 14 transmembrane segments, and the sequence includes three regions: a hydrophilic, cytoplasmic domain (residues 1 through 403) that interacts with numerous cytoplasmic and membrane proteins; a hydrophobic domain (residues 404 through 882) that comprises the anion transporting channel; and an acidic, C-terminal domain (residues 883 through 911). This transport system facilitates a one-for-one exchange of chloride and bicarbonate, so the net transport process is electrically neutral. The net direction of anion flow through this protein depends on the sum of the chloride and bicarbonate concentration gradients. Typically, carbon dioxide is collected by red cells in respiring tissues (by means of $Cl^- \rightleftharpoons HCO_3^-$ exchange) and is then carried in the blood to the lungs, where bicarbonate diffuses out of the erythrocytes in exchange for Cl^- ions.

FIGURE 9.26 The structure of cytochalasin B.

9.6 | How Does Energy Input Drive Active Transport Processes?

Passive and facilitated diffusion systems are relatively simple, in the sense that the transported species flow downhill energetically, that is, from high concentration to low concentration. However, other transport processes in biological systems must be *driven* in an energetic sense. In these cases, the transported species move from low concentration to high concentration, and thus the transport requires *energy input*. As such, it is considered an **active transport system.** The most common energy input is **ATP hydrolysis,** with hydrolysis being *tightly coupled* to the transport event. Other energy sources also drive active transport processes, including *light energy* and the *energy stored in ion gradients*. The original ion gradient

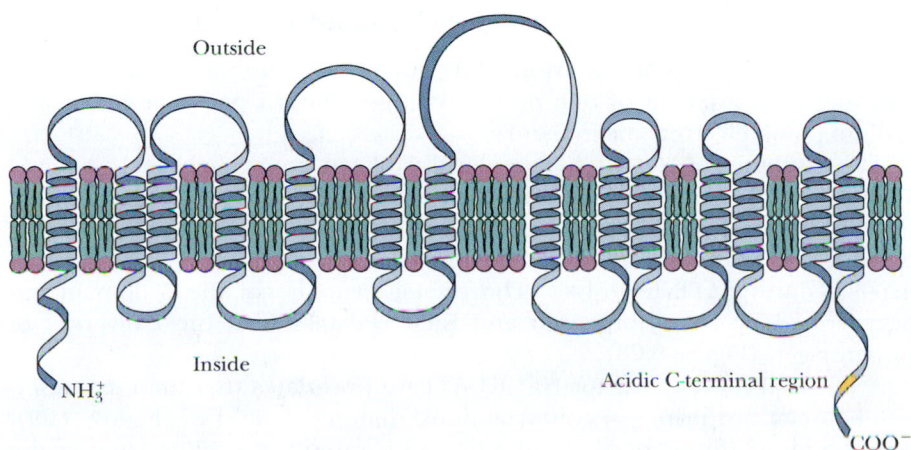

FIGURE 9.27 A model for the arrangement of the anion transport protein in the membrane, based on hydropathy analysis.

is said to arise from a **primary active transport** process, and the transport that depends on the ion gradient for its energy input is referred to as a **secondary active transport** process (see later discussion of amino acid and sugar transport). When transport results in a net movement of electric charge across the membrane, it is referred to as an **electrogenic transport** process. If no net movement of charge occurs during transport, the process is electrically neutral.

All Active Transport Systems Are Energy-Coupling Devices

Hydrolysis of ATP is essentially a chemical process, whereas movement of species across a membrane is a mechanical process (that is, movement). An active transport process that depends on ATP hydrolysis thus couples chemical free energy to mechanical (translational) free energy. The bacteriorhodopsin protein in *Halobacterium halobium* couples light energy and mechanical energy. Oxidative phosphorylation (see Chapter 20) involves coupling between electron transport, proton translocation, and the capture of chemical energy in the form of ATP synthesis. Similarly, the overall process of photosynthesis (see Chapter 21) amounts to a coupling between captured light energy, proton translocation, and chemical energy stored in ATP.

Many Active Transport Processes are Driven by ATP

Monovalent Cation Transport: Na$^+$,K$^+$-ATPase   All animal cells actively extrude Na$^+$ ions and accumulate K$^+$ ions. These two transport processes are driven by **Na$^+$,K$^+$-ATPase**, also known as the **sodium pump,** an integral protein of the plasma membrane. Most animal cells maintain cytosolic concentrations of Na$^+$ and K$^+$ of 10 mM and 100 mM, respectively. The extracellular milieu typically contains about 100 to 140 mM Na$^+$ and 5 to 10 mM K$^+$. Potassium is required within the cell to activate a variety of processes, whereas high intracellular sodium concentrations are inhibitory. The transmembrane gradients of Na$^+$ and K$^+$ and the attendant gradients of Cl$^-$ and other ions provide the means by which neurons communicate. They also serve to regulate cellular volume and shape. Animal cells also depend upon these Na$^+$ and K$^+$ gradients to drive transport processes involving amino acids, sugars, nucleotides, and other substances. In fact, maintenance of these Na$^+$ and K$^+$ gradients consumes large amounts of energy in animal cells—20% to 40% of total metabolic energy in many cases and up to 70% in neural tissue.

The Na$^+$- and K$^+$-dependent ATPase comprises two subunits: an α-subunit of 1016 residues (120 kD) and a 35-kD β-subunit. The sodium pump actively pumps three Na$^+$ ions out of the cell and two K$^+$ ions into the cell per ATP hydrolyzed:

$$\text{ATP}^{4-} + \text{H}_2\text{O} + 3\,\text{Na}^+(\text{inside}) + 2\,\text{K}^+(\text{outside}) \longrightarrow \text{ADP}^{3-} + \text{H}_2\text{PO}_4^- +$$
$$3\,\text{Na}^+(\text{outside}) + 2\,\text{K}^+(\text{inside}) \quad (9.3)$$

ATP hydrolysis occurs on the cytoplasmic side of the membrane (Figure 9.28), and the net movement of one positive charge outward per cycle makes the sodium pump electrogenic in nature.

The α-subunit of Na$^+$,K$^+$-ATPase consists of ten transmembrane α-helices, with three cytoplasmic domains, denoted A, P, and N. A large cytoplasmic loop between transmembrane helices 4 and 5 forms the P (phosphorylation) and N (nucleotide-binding) domains. The enzyme is covalently phosphorylated at Asp-369 during ATP hydrolysis. The crystal structure of the N-domain has been solved by Peter Jørgensen and Kjell Håkansson at the University of Copenhagen (Figure 9.28).

A minimal mechanism for Na$^+$,K$^+$-ATPase postulates that the enzyme cycles between two principal conformations, denoted E$_1$ and E$_2$ (Figure 9.29). E$_1$ has a high affinity for Na$^+$ and ATP and is rapidly phosphorylated in the

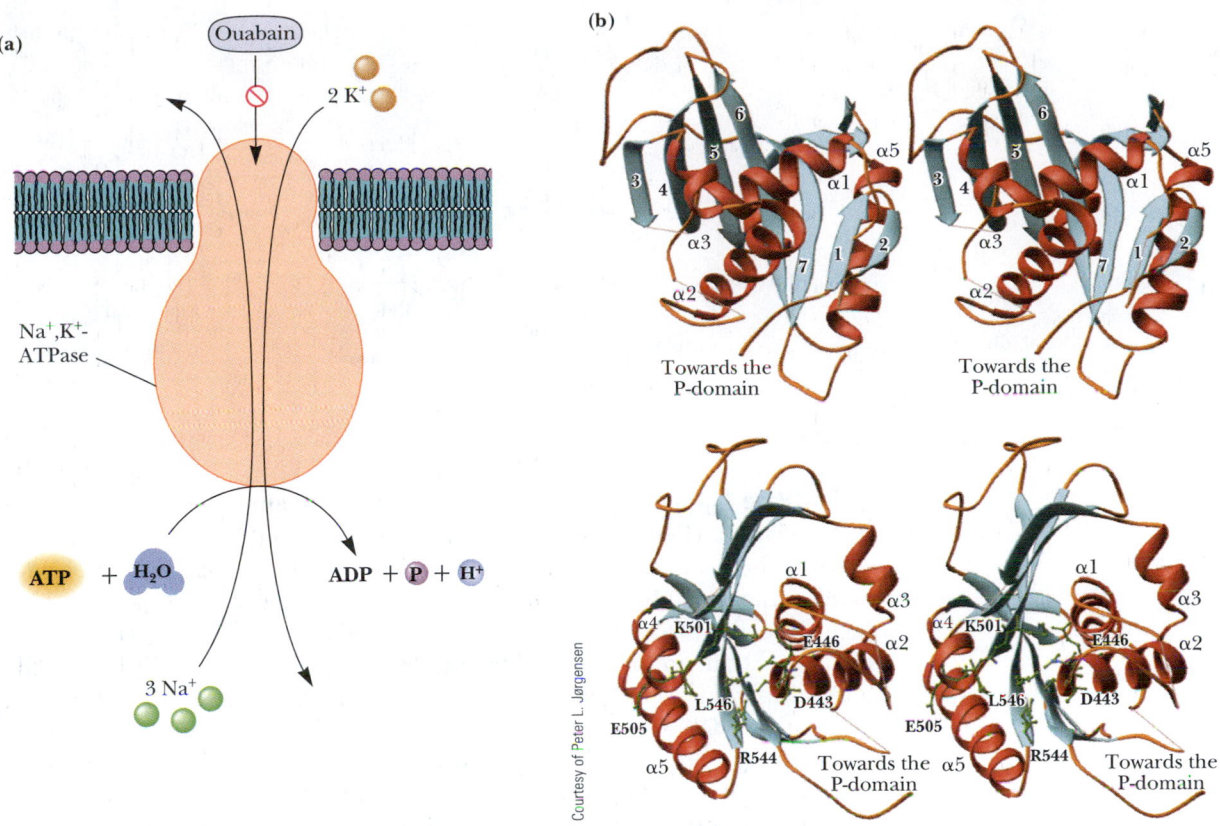

(a)

(b)

Courtesy of Peter L. Jørgensen

Biochemistry ⊗ Now™   ANIMATED FIGURE 9.28   (a) A schematic diagram of the Na⁺,K⁺-ATPase in mammalian plasma membrane. ATP hydrolysis occurs on the cytoplasmic side of the membrane, Na⁺ ions are transported out of the cell, and K⁺ ions are transported in. The transport stoichiometry is 3 Na⁺ out and 2 K⁺ in per ATP hydrolyzed. The specific inhibitor ouabain (Figure 7.12) and other cardiac glycosides inhibit Na⁺,K⁺-ATPase by binding on the extracellular surface of the pump protein. **(b)** Stereo views of the Na⁺,K⁺-ATPase. *(From Håkansson, K. O., 2003. The crystallographic structure of Na,K-ATPase N-domain of 2.6 Å resolution.* J Mol Biol **332:**1175–1182.) **See this figure animated at http://chemistry.brookscole.com/ggb3**

presence of Mg^{2+} to form E_1-P, a state that *contains three occluded Na^+ ions* (occluded in the sense that they are tightly bound and not easily dissociated from the enzyme in this conformation). A conformation change yields E_2-P, a form of the enzyme with relatively low affinity for Na^+ but a high affinity for K^+. This state presumably releases 3 Na^+ ions and binds 2 K^+ ions on the outside of the cell. Dephosphorylation leaves E_2K_2, a form of the enzyme with *two occluded K^+ ions.* A conformation change, which appears to be accelerated by the binding of ATP (with a relatively low affinity), releases the bound K^+ inside the cell and returns the enzyme to the E_1 state. Enzyme forms with occluded cations represent states of the enzyme with cations bound in the transport channel. The alternation between high and low affinities for Na^+, K^+, and ATP serves to tightly couple the hydrolysis of ATP and ion binding and transport.

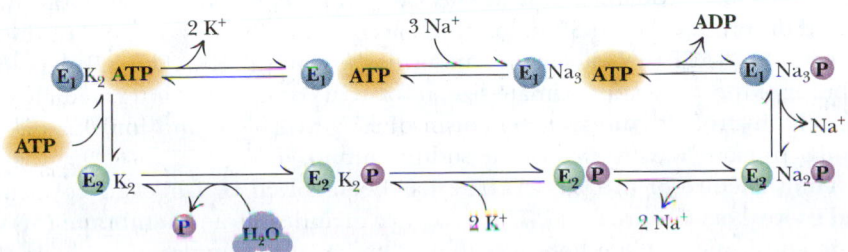

Biochemistry ⊗ Now™   ANIMATED FIGURE 9.29
A mechanism for Na⁺,K⁺-ATPase. The model assumes two principal conformations, E_1 and E_2. Binding of Na⁺ ions to E_1 is followed by phosphorylation and release of ADP. Na⁺ ions are transported and released, and K⁺ ions are bound before dephosphorylation of the enzyme. Transport and release of K⁺ ions complete the cycle. **See this figure animated at http://chemistry.brookscole.com/ggb3**

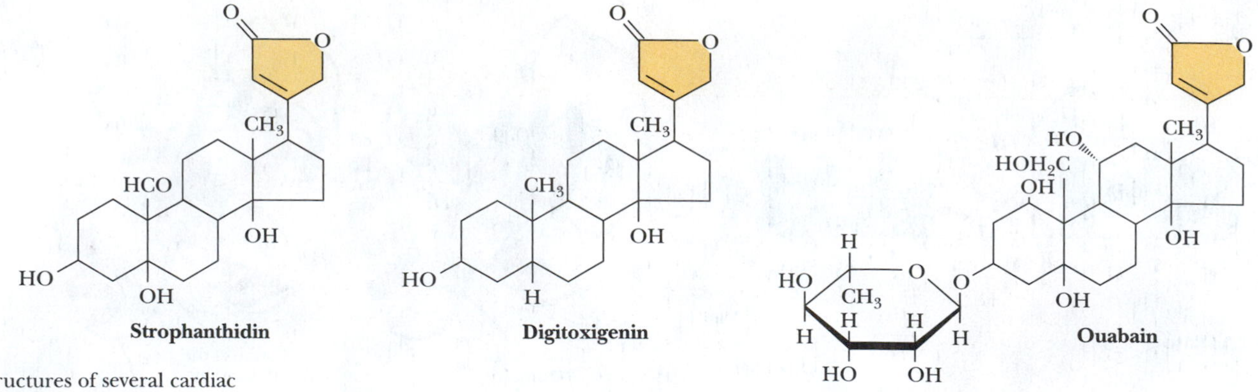

Strophanthidin   **Digitoxigenin**   **Ouabain**

FIGURE 9.30 The structures of several cardiac glycosides. The lactone rings are yellow.

Na⁺,K⁺-ATPase Is Inhibited by Cardiac Glycosides   Plant and animal steroids such as *ouabain* (Figure 9.30) specifically inhibit Na^+,K^+-ATPase and ion transport. These substances are traditionally referred to as **cardiac glycosides** or **cardiotonic steroids,** both names derived from the potent effects of these molecules on the heart. These molecules all possess a *cis*-configuration of the C-D ring junction, an unsaturated lactone ring (five- or six-membered) in the β-configuration at C-17, and a β-OH at C-14. There may be one or more sugar residues at C-3. The sugars are not required for inhibition, but do contribute to water solubility of the molecule. Cardiac glycosides bind exclusively to the extracellular surface of Na^+,K^+-ATPase when it is in the E_2-P state, forming a very stable E_2-P(cardiac glycoside) complex.

Medical researchers studying high blood pressure have consistently found that people with hypertension have high blood levels of an endogenous Na^+,K^+-ATPase inhibitor. In such patients, inhibition of the sodium pump in the cells lining the blood vessel wall results in accumulation of sodium and calcium in these cells and the narrowing of the vessels to create hypertension. An 8-year study aimed at the isolation and identification of the agent responsible for these effects by researchers at the University of Maryland Medical School and the Upjohn Laboratories in Michigan recently yielded a surprising result. Mass spectrometric analysis of compounds isolated from many hundreds of gallons of blood plasma has revealed that the hypertensive agent is ouabain itself or a closely related molecule!

Calcium Transport: Ca²⁺-ATPase   Calcium, an ion acting as a cellular signal in virtually all cells (see Chapter 32), plays a special role in muscles. It is the signal that stimulates muscles to contract (see Chapter 16). In the resting state, the levels of Ca^{2+} near the muscle fibers are very low (approximately 0.1 μM), and nearly all of the calcium ion in muscles is sequestered inside a complex network of vesicles called the **sarcoplasmic reticulum,** or **SR** (see Figure 16.11). Nerve impulses induce the SR membrane to quickly release large amounts of Ca^{2+}, with cytosolic levels rising to approximately 10 μM. At these levels, Ca^{2+} stimulates contraction. Relaxation of the muscle requires that cytosolic Ca^{2+} levels be reduced to their resting values. This is accomplished by an ATP-driven Ca^{2+} transport protein known as the **Ca²⁺-ATPase.** This enzyme is the most abundant protein in the SR membrane, accounting for 70% to 80% of the SR protein. Ca^{2+}-ATPase bears many similarities to the Na^+,K^+-ATPase. It has an α-subunit of the same approximate size, it forms a covalent E-P intermediate during ATP hydrolysis, and its mechanism of ATP hydrolysis and ion transport is similar in many ways to that of the sodium pump.

The structure of the Ca^{2+}-ATPase has been solved by Chikashi Toyoshima and co-workers (Figure 9.31). The structure includes a transmembrane (M) domain consisting of ten α-helical segments and a large cytoplasmic domain that

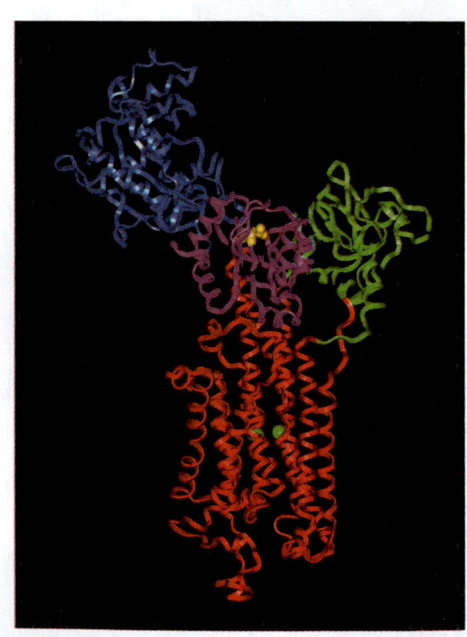

FIGURE 9.31 The structure of Ca²⁺-ATPase. The transmembrane (M) domain is shown in red, the nucleotide (N) domain is shown in blue, the phosphorylation (P) domain is purple, and the actuator (A) domain is green. The phosphorylation site, Asp-351, is yellow. Two Ca²⁺ ions in the transmembrane site are green.

A Deeper Look

Cardiac Glycosides: Potent Drugs from Ancient Times

The cardiac glycosides have a long and colorful history. Many species of plants producing these agents grow in tropical regions and have been used by natives in South America and Africa to prepare poisoned arrows used in fighting and hunting. Zulus in South Africa, for example, have used spears tipped with cardiac glycoside poisons. The sea onion, found commonly in southern Europe and northern Africa, was used by the Romans and the Egyptians as a cardiac stimulant, diuretic, and expectorant. The Chinese have long used a medicine made from the skins of certain toads for similar purposes. Cardiac glycosides are also found in several species of domestic plants, including the foxglove, lily of the valley, oleander (figure part a), and milkweed plants. Monarch butterflies (figure part b) acquire these compounds by feeding on milkweed and then storing the cardiac glycosides in their exoskeletons. Cardiac glycosides deter predation of monarch butterflies by birds, which learn by experience not to feed on monarchs. Viceroy butterflies (figure part c) mimic monarchs in overall appearance. Although viceroys contain no cardiac glycosides and are edible, they are avoided by birds that mistake them for monarchs.

In 1785, the physician and botanist William Withering described the medicinal uses for agents derived from the foxglove plant. In modern times, **digitalis** (a preparation of dried leaves prepared from the foxglove, *Digitalis purpurea*) and other purified cardiotonic steroids have been used to increase the contractile force of heart muscle, to slow the rate of beating, and to restore normal function in hearts undergoing fibrillation (a condition in which heart valves do not open and close rhythmically but rather remain partially open, fluttering in an irregular and ineffective way). Inhibition of the cardiac sodium pump increases the intracellular Na^+ concentration, leading to stimulation of the Na^+-Ca^{2+} exchanger, which extrudes sodium in exchange for inward movement of calcium. Increased intracellular Ca^{2+} stimulates muscle contraction. Careful use of digitalis drugs has substantial therapeutic benefit for patients with heart problems.

(a) Oleander

(b) Monarch butterfly

(c) Viceroy butterfly

(a) Cardiac glycoside inhibitors of Na^+,K^+-ATPase are produced by many plants, including foxglove, lily of the valley, milkweed, and oleander (shown here). (b) The monarch butterfly, which concentrates cardiac glycosides in its exoskeleton, is shunned by predatory birds. (c) Predators also avoid the viceroy, even though it contains no cardiac glycosides, because it is similar in appearance to the monarch.

itself consists of a nucleotide-binding (N) domain, a phosphorylation (P) domain, and an actuator (A) domain. In this structure, two Ca^{2+} ions are buried deep in the membrane-spanning portion of the enzyme (Figure 9.31).

The Gastric H$^+$,K$^+$-ATPase   Production of protons is a fundamental activity of cellular metabolism, and proton production plays a special role in the stomach. The highly acidic environment of the stomach is essential for the digestion of food in all animals. The pH of the stomach fluid is normally 0.8 to 1. The pH of the parietal cells of the gastric mucosa in mammals is approximately 7.4. This represents a **pH gradient** across the mucosal cell membrane of 6.6, the largest known transmembrane gradient in eukaryotic cells. This enormous gradient must be maintained constantly so that food can be digested in the stomach without damage to the cells and organs adjacent to the stomach. The gradient of H^+ is maintained by an **H$^+$,K$^+$-ATPase**, which uses the energy of hydrolysis of ATP to pump H^+ out of the mucosal cells and into the stomach interior in exchange for K^+ ions. This transport is electrically neutral, and the K^+ that is transported into the mucosal cell is subsequently pumped back out of the cell

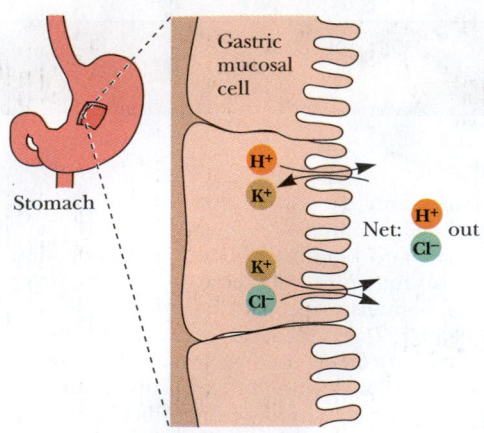

Biochemistry ✏ Now™ **ACTIVE FIGURE 9.32**
The H^+,K^+-ATPase of gastric mucosal cells mediates proton transport into the stomach. Potassium ions are recycled by means of an associated K^+/Cl^- co-transport system. The action of these two pumps results in net transport of H^+ and Cl^- into the stomach. **Test yourself on the concepts in this figure at http://chemistry.brookscole.com/ggb3**

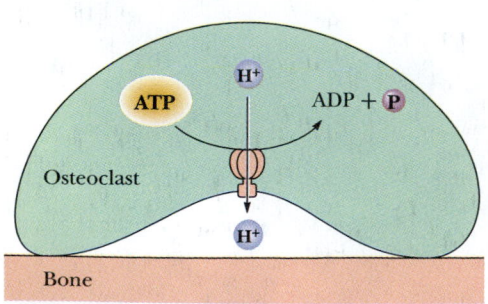

Biochemistry ✏ Now™ **ANIMATED FIGURE 9.33**
Proton pumps cluster on the ruffled border of osteoclast cells and function to pump protons into the space between the cell membrane and the bone surface. High proton concentration in this space dissolves the mineral matrix of the bone. **See this figure animated at http://chemistry.brookscole.com/ggb3**

together with Cl^- in a second electroneutral process (Figure 9.32). Thus, the net transport effected by these two systems is the movement of HCl into the interior of the stomach. (Only a small amount of K^+ is needed, because it is recycled.) The H^+,K^+-ATPase bears many similarities to the plasma membrane Na^+,K^+-ATPase and the SR Ca^{2+}-ATPase described earlier. It has a similar molecular weight, it forms an E-P intermediate, and many parts of its peptide sequence are homologous with the Na^+,K^+-ATPase and Ca^{2+}-ATPase.

Bone Remodeling by Osteoclast Proton Pumps Other proton-translocating ATPases exist in eukaryotic and prokaryotic systems. **Vacuolar ATPases** (V-type ATPases) are found in vacuoles, lysosomes, endosomes, Golgi, chromaffin granules, and coated vesicles. Various H^+-transporting ATPases occur in yeast and bacteria as well. H^+-transporting ATPases found in **osteoclasts** (multinucleate cells that break down bone during normal bone remodeling) provide a source of circulating calcium for soft tissues such as nerves and muscles. About 5% of bone mass in the human body undergoes remodeling at any given time. Once growth is complete, the body balances formation of new bone tissue by cells called **osteoblasts** with resorption of existing bone matrix by osteoclasts. Osteoclasts possess proton pumps—which are in fact V-type ATPases—on the portion of the plasma membrane that attaches to the bone. This region of the osteoclast membrane is called the ruffled border. The osteoclast attaches to the bone in the manner of a cup turned upside down on a saucer (Figure 9.33), leaving an extracellular space between the bone surface and the cell. The H^+-ATPases in the ruffled border pump protons into this space, creating an acidic solution that dissolves the bone mineral matrix. Bone mineral consists mainly of poorly crystalline hydroxyapatite $[Ca_{10}(PO_4)_6(OH)_2]$ with some carbonate (HCO_3^-) replacing OH^- or PO_4^{3-} in the crystal lattice. Transport of protons out of the osteoclasts lowers the pH of the extracellular space near the bone to about 4, solubilizing the hydroxyapatite.

ATPases That Transport Peptides and Drugs Species other than protons and inorganic ions are also transported across certain membranes by specialized ATPases. Yeast (*Saccharomyces cerevisiae*) has one such system. Yeasts exist in two haploid mating types, designated **a** and **α.** Each mating type produces a mating factor (**a-factor** or **α-factor,** respectively) and responds to the mating factor of the opposite type. The α-factor is a peptide that is inserted into the ER during translation on the ribosome, thanks to the presence of a signal sequence. α-Factor is glycosylated in the ER and then secreted from the cell. On the other hand, the a-factor is a 12–amino acid peptide made from a short precursor with no signal sequence or glycosylation site. Export of this peptide from the cell is carried out by a 1290-residue protein, which consists of two identical halves joined together—a tandem duplication. Each half contains six putative transmembrane segments arranged in pairs and a conserved hydrophilic cytoplasmic domain containing a consensus sequence for an ATP-binding site (Figure 9.34). This protein uses the energy of ATP hydrolysis to export a-factor from the cell. In yeast cells that produce mutant forms of the a-factor ATPase, a-factor is not excreted and accumulates to high levels inside the cell.

Proteins very similar to the yeast a-factor transporter have been identified in a variety of prokaryotic and eukaryotic cells, and one of these appears to be responsible for the acquisition of **drug resistance** in many human malignancies. Clinical treatment of human cancer often involves chemotherapy, the treatment with one or more drugs that selectively inhibit the growth and proliferation of tumorous tissue. However, the efficacy of a given chemotherapeutic drug often decreases with time as a result of an acquired resistance. Even worse, the acquired resistance to a single drug usually results in a simultaneous resistance to a wide spectrum of drugs with little structural or even functional similarity to the original drug, a phenomenon referred to as **multidrug resistance,** or **MDR.** This perplexing problem has been traced to the induced expression of a 170-kD plasma membrane glycoprotein known as the **P-glycoprotein** or the **MDR ATPase.** Like the yeast a-factor transporter, MDR ATPase is a tandem repeat, each half consisting of a hydrophobic sequence with six transmembrane segments followed by a hydrophilic, cytoplasmic sequence containing a consensus ATP-binding site (Figure 9.34). The protein uses the energy of ATP hydrolysis to actively transport a wide variety of drugs (Figure 9.35) out of the cell. Ironically, it is probably part of a sophisticated protection system for the cell and the organism. Organic molecules of various types and structures that might diffuse across the plasma membrane are apparently recognized by this protein and actively extruded from the cell. Despite the cancer-fighting nature of chemotherapeutic agents, the MDR ATPase recognizes these agents as cellular intruders and rapidly removes them. It is not yet understood how this large protein can recognize, bind, and transport such a broad group of diverse molecules, but it is known that the yeast a-factor ATPase and the MDR ATPase are just two members of a **superfamily** of transport proteins, many of whose functions are not yet understood.

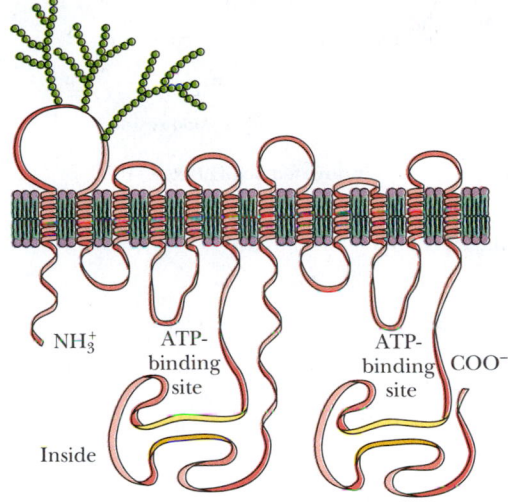

Outside

NH_3^+ ATP-binding site   ATP-binding site COO^-

Inside

FIGURE 9.34 A model for the structure of the a-factor transport protein in the yeast plasma membrane. Gene duplication has yielded a protein with two identical halves, each half containing six transmembrane helical segments and an ATP-binding site. Like the yeast a-factor transporter, the multidrug transporter is postulated to have 12 transmembrane α-helices and 2 ATP-binding sites.

9.7 | How Are Certain Transport Processes Driven by Light Energy?

As noted previously, certain biological transport processes are driven by light energy rather than by ATP. Two well-characterized systems are **bacteriorhodopsin,** the light-driven H^+-pump, and **halorhodopsin,** the light-driven Cl^- pump, of *Halobacterium halobium,* an archaebacterium that thrives in high-salt media. *H. halobium* grows optimally at an NaCl concentration of 4.3 *M.* It was extensively characterized by Walther Stoeckenius, who found it growing prolifically in the salt pools near San Francisco Bay, where salt is commercially extracted from seawater. *H. halobium* carries out normal respiration if oxygen and metabolic energy sources are plentiful. However, when these substrates are lacking, *H. halobium* survives by using bacteriorhodopsin and halorhodopsin to capture light energy. In oxygen- and nutrient-deficient conditions, **purple patches** appear on

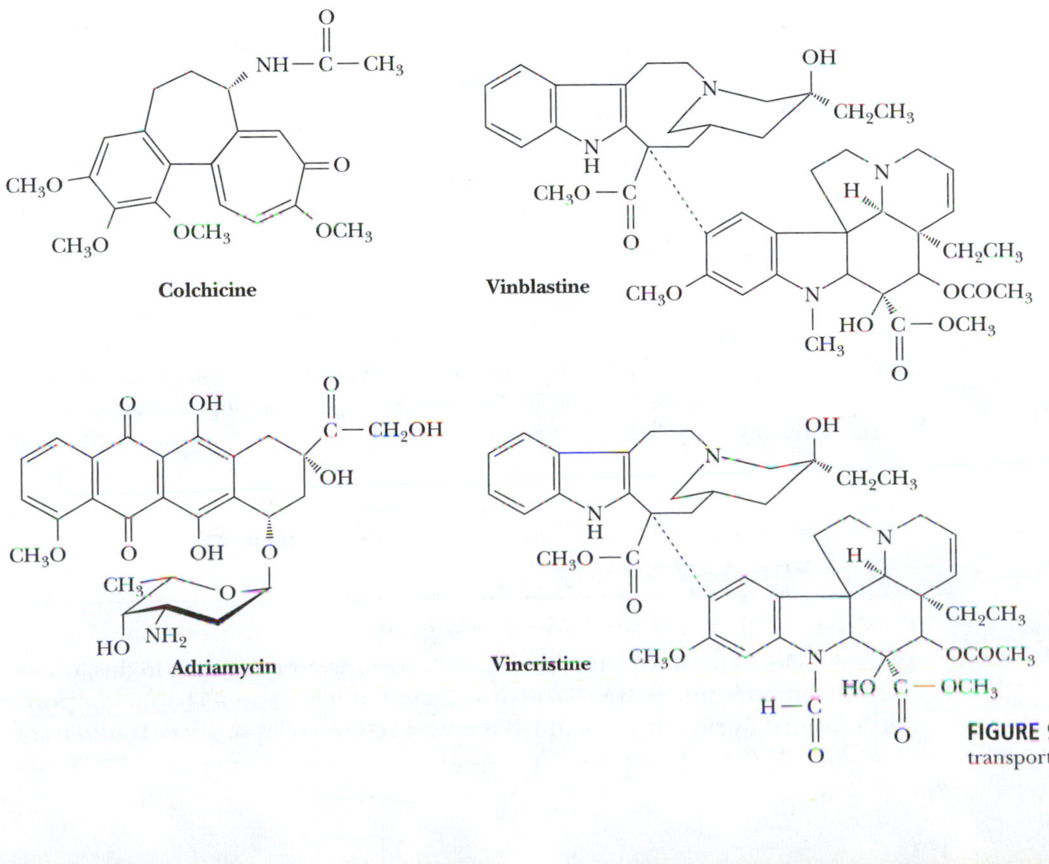

Colchicine

Vinblastine

Adriamycin

Vincristine

FIGURE 9.35 Some of the cytotoxic drugs that are transported by the MDR ATPase.

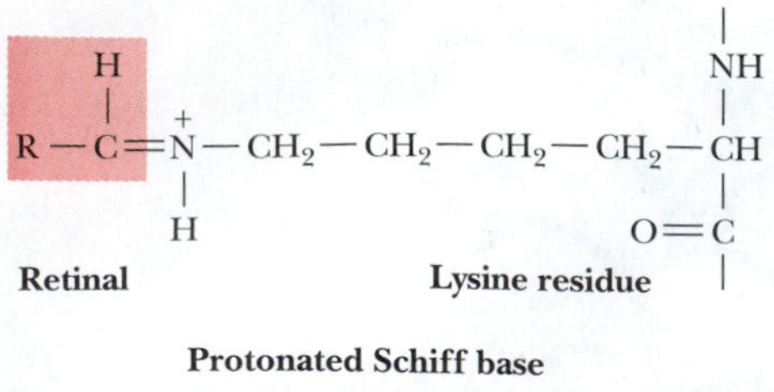

FIGURE 9.36 The Schiff base linkage between the retinal chromophore and Lys^{216}.

the surface of *H. halobium* (Figure 9.9). These purple patches of membrane are 75% protein, the only protein being **bacteriorhodopsin (bR).** The purple color arises from a retinal molecule that is covalently bound in a Schiff base linkage with an ϵ-NH$_2$ group of Lys^{216} on each bacteriorhodopsin protein (Figure 9.36). Bacteriorhodopsin is a 26-kD transmembrane protein that packs so densely in the membrane that it naturally forms a two-dimensional crystal in the plane of the membrane. The structure of bR has been elucidated by image enhancement analysis of electron microscopic data, which reveals seven transmembrane helical protein segments. The retinal moiety lies parallel to the membrane plane, about 1 nm below the membrane's outer surface (Figure 9.15).

Bacteriorhodopsin Effects Light-Driven Proton Transport

The mechanism of the light-driven transport of protons by bacteriorhodopsin is complex, but a partial model has emerged (Figure 9.37). A series of intermediate states, named for the wavelengths (in nm) of their absorption spectra, has been identified. Absorption of a photon of light by the bR$_{568}$ form (in which the Schiff base at Lys^{216} is protonated) converts the retinal from the all-*trans* configuration to the 13-*cis* isomer. Passage through several different intermediate states results in outward transport of 2 H$^+$ ions per photon absorbed, and the return of the bound retinal to the all-*trans* configuration. It appears that the transported protons are in fact protons from the protonated Schiff base. The proton gradient thus established represents chemical energy that can be used by *H. halobium* to drive ATP synthesis and the movement of molecules across the cell membrane (see Chapter 20).

9.8 | How Are Amino Acid and Sugar Transport Driven by Ion Gradients?

Na$^+$ and H$^+$ Drive Secondary Active Transport

The gradients of H$^+$, Na$^+$, and other cations and anions established by ATPases and other energy sources can be used for **secondary active transport** of various substrates. The best-understood systems use Na$^+$ or H$^+$ gradients to transport amino acids and sugars in certain cells. Many of these systems operate as **symports,** with the ion and the transported amino acid or sugar moving in the same direction (that is, into the cell). In **antiport** processes, the ion and the other transported species move in opposite directions. (For example, the anion transporter of erythrocytes is an antiport.) **Proton symport** proteins are used by *E. coli* and other bacteria to accumulate lactose, arabinose, ribose, and a variety of amino acids. *E. coli* also possesses Na$^+$-symport systems for melibiose, as well as for glutamate and other amino acids.

Table 9.3 lists several systems that transport amino acids into mammalian cells. The accumulation of neutral amino acids in the liver by System A represents an important metabolic process. Thus, plasma membrane transport of alanine is the rate-limiting step in hepatic alanine metabolism. This system is normally expressed at low levels in the liver, but substrate deprivation and hormonal activation both stimulate System A expression.

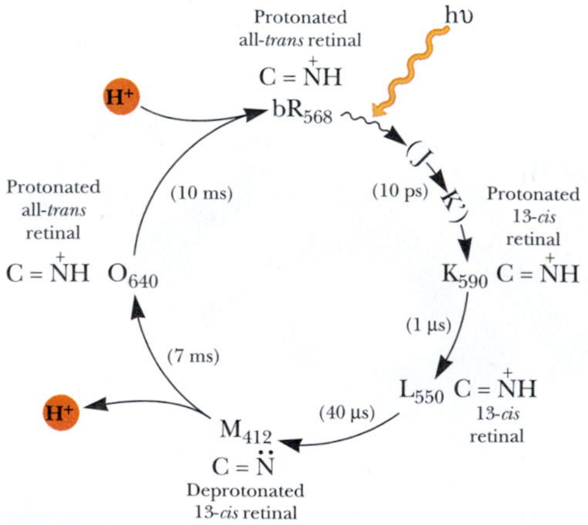

FIGURE 9.37 The reaction cycle of bacteriorhodopsin. The intermediate states are indicated by letters, with subscripts to indicate the absorption maxima of the states. Also indicated for each state is the configuration of the retinal chromophore (all-*trans* or 13-*cis*) and the protonation state of the Schiff base (C=N: or C=N$^+$H).

9.9 | How Are Specialized Membrane Pores Formed by Toxins?

Pore-Forming Toxins Collapse Ion Gradients

Many organisms produce lethal molecules known as **pore-forming toxins,** which insert themselves in a host cell's plasma membrane to form a channel or pore. Pores formed by such toxins can kill the host cell by collapsing ion gradients or

Table 9.3

Some Mammalian Amino Acid Transport Systems

System Designation	Ion Dependence	Amino Acids Transported	Cellular Source
A	Na$^+$	Neutral amino acids	
ASC	Na$^+$	Neutral amino acids	
L	Na$^+$-independent	Branched-chain and aromatic amino acids	Ehrlich ascites cells Chinese hamster ovary cells Hepatocytes
N	Na$^+$	Nitrogen-containing side chains (Gln, Asn, His, etc.)	
y$^+$	Na$^+$-independent	Cationic amino acids	
x$_{AG}^-$	Na$^+$	Aspartate and glutamate	Hepatocytes
P	Na$^+$	Proline	Chinese hamster ovary cells

Adapted from Collarini, E. J., and Oxender, D. L., 1987. Mechanisms of transport of amino acids across membranes. *Annual Review of Nutrition* **7**:75–90.

by facilitating the entry of toxic agents into the cell. Produced by a variety of organisms and directed toward a similarly diverse range of target cells, these toxins nonetheless share certain features in common. The structures of these remarkable toxins have provided valuable insights into the mechanisms of their membrane insertion and also into the architecture of membrane proteins.

Colicins are pore-forming proteins, produced by certain strains of *E. coli*, that kill or inhibit the growth of competing bacteria, even other strains of *E. coli* (a process known as *allelopathy*). Channel-forming colicins are released as soluble monomers. Upon encountering a host cell, the colicin molecule traverses the bacterial outer membrane and periplasm, then inserts itself into the inner (plasma) membrane. The channel thus formed is monomeric and a single colicin molecule can kill a host cell. The structure of colicin Ia, a 626-residue protein, is shown in Figure 9.38. It consists of three domains, termed the **T (translocation) domain,** the **R (receptor-binding) domain,** and the **C (channel-forming) domain.** The T domain mediates translocation across the outer membrane, the R domain binds to an outer-membrane receptor, and the C-domain creates a voltage-gated channel across the inner membrane. The T, R, and C domains are separated by long (160 Å) α-helical segments. The peptide is folded at the

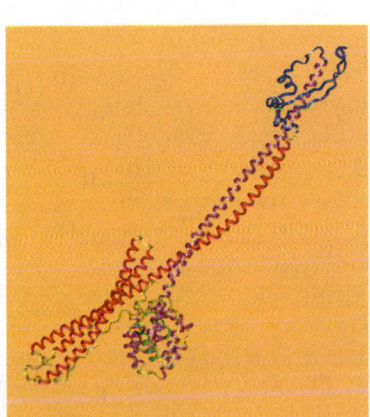
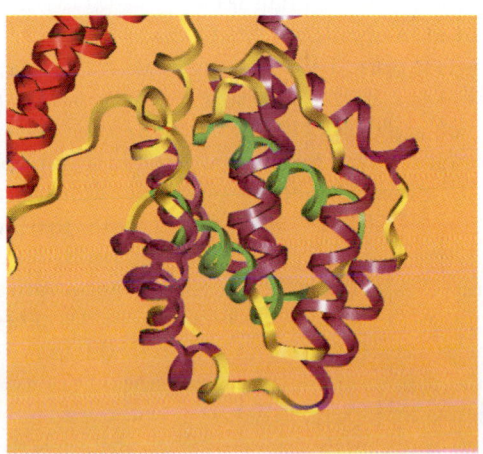

FIGURE 9.38 The structure of colicin Ia. Colicin Ia, with a total length of 210 Å, spans the periplasmic space of a Gram-negative bacterium host, with the R (receptor-binding) domain (blue) anchored to proteins in the outer membrane and the C domain (violet) forming a channel in the inner membrane. The T (translocation) domain is shown in red. The image on the right shows details of the C domain, including helices 8 and 9 (green), which are highly hydrophobic.

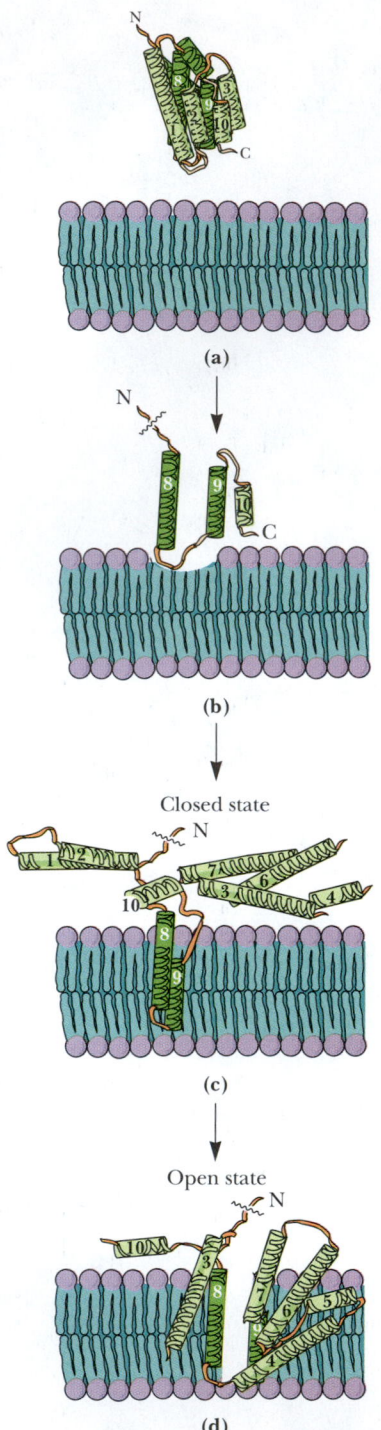

FIGURE 9.39 The umbrella model of membrane channel protein insertion. Hydrophobic helices insert directly into the core of the membrane, with amphipathic helices arrayed on the surface like an open umbrella. A trigger signal (low pH or a voltage gradient) draws some of the amphipathic helices into and across the membrane, causing the pore to open.

Biochemistry ⒺNow™   Go to BiochemistryNow and click BiochemistryInteractive to examine a pore-forming membrane protein toxin.

R domain, so the C and T domains are juxtaposed and the two long helices form an underwound antiparallel coiled coil. The protein is unusually elongated—210 Å from end to end—with the T and C domains at one end and R at the other. This unusual design permits colicin Ia to span the periplasmic space (which has an average width of 150 Å) and insert in the inner membrane.

The nature of the channel-forming domain provides clues to the process of channel formation in the inner membrane. The C domain consists of a 10-helix bundle, with helices 8 and 9 forming an unusually hydrophobic hairpin structure. The other eight helices are amphipathic and serve to stabilize hydrophobic helices 8 and 9 in solution. When this domain inserts in the inner membrane, helices 8 and 9 inject themselves into the hydrophobic membrane core, leaving the other helices behind on the membrane surface (Figure 9.39). Application of a transmembrane potential (voltage) then triggers the amphipathic helices to insert into the membrane, with their hydrophobic faces facing the hydrophobic bilayer and their polar faces forming the channel surface. This model is hypothetical, but it is supported by studies showing that channel opening involves dramatic structural changes and that helices 2 to 5 move across the membrane during channel opening.

Interestingly, certain other pore-forming toxins possess helix-bundle motifs that may participate in channel formation, in a manner similar to that proposed for colicin Ia. For example, the **δ-endotoxin** produced by *Bacillus thuringiensis* is toxic to Coleoptera insects (beetles) and is composed of three domains, including a seven-helix bundle, a three-sheet domain, and a β-sandwich. In the seven-helix bundle, helix 5 is highly hydrophobic and the other six helices are amphipathic. In solution (Figure 9.40), the six amphipathic helices surround helix 5, with their nonpolar faces apposed to helix 5 and their polar faces directed to the solvent. Membrane insertion and channel formation may involve initial insertion of helix 5, as in Figure 9.40, followed by insertion of the amphipathic helices, so that their nonpolar faces contact the bilayer lipids and their polar faces line the channel.

There are a number of other toxins for which the helical channel model is inappropriate. These include α-hemolysin from *Staphylococcus aureus*, **aerolysin** from *Aeromonas hydrophila*, and the **anthrax toxin protective antigen** from *Bacillus anthracis*. The membrane-spanning domains of these proteins do not possess long stretches of hydrophobic residues that could form α-helical transmembrane segments. They do, however, contain substantial peptide segments of alternating hydrophobic and polar residues. Like the porins, such segments can adopt β-strand structures, such that one side of the β-strand is hydrophobic and the other side is polar. Oligomeric association of several such segments can produce a β-barrel motif, with the inside of the barrel lined with polar residues and the outside of the barrel coated with hydrophobic residues, a motif that can be accommodated readily in a bilayer membrane, creating a polar transmembrane channel.

α-Hemolysin, a 33.2-kD monomer protein, forms a mushroom-shaped, heptameric pore, 100 Å in length, with a diameter that ranges from 14 to 46 Å (Figure 9.41). In this structure, each monomer contributes two β-strands 65 Å long, which are connected by a hairpin turn. The interior of the 14-stranded β-barrel structure is hydrophilic, and the hydrophobic outer surface of the barrel is 28 Å wide. Pores formed by α-hemolysin in human erythrocytes, platelets, and lymphocytes allow rapid Ca^{2+} influx into these cells with toxic consequences.

Aeromonas hydrophila is a bacterium that causes diarrheal diseases and deep wound infections. These complications arise due to pore formation in sensitive cells by the protein toxin aerolysin. Proteolytic processing of the 52-kD precursor **proaerolysin** (Figure 9.42) produces the toxic form of the protein, aerolysin. Like α-hemolysin, aerolysin monomers associate to form a heptameric transmembrane pore. Michael Parker and co-workers have proposed that each monomer in this aggregate contributes three β-strands to the β-barrel pore. Each of these β-strands (residues 277 to 287, 290 to 302, and 410 to 422) con-

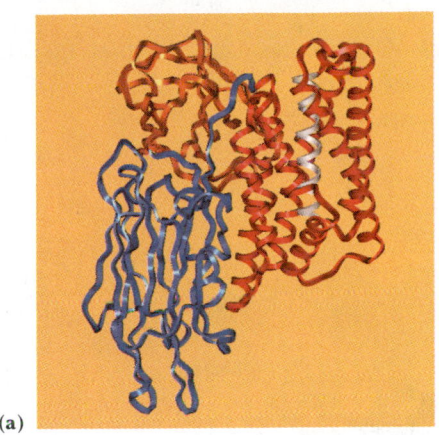
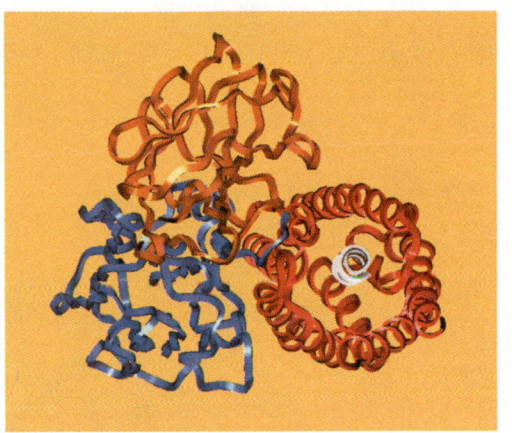

(a)

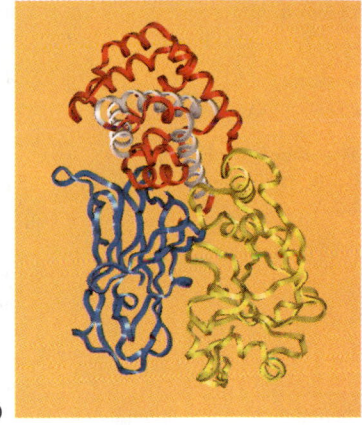

(b)

FIGURE 9.40 The structures of (**a**) δ-endotoxin (two views) from *Bacillus thuringiensis* and (**b**) diphtheria toxin from *Corynebacterium diphtheriae.* Each of these toxins possesses a bundle of α-helices, which is presumed to form the transmembrane channel when the toxin is inserted across the host membrane. In δ-endotoxin, helix 5 (white) is surrounded by 6 helices (red) in a 7-helix bundle. In diphtheria toxin, three hydrophobic helices (white) lie at the center of the transmembrane domain (red).

sists of alternating hydrophobic and polar residues, so the pore once again places polar residues toward the water-filled channel and nonpolar residues facing the lipid bilayer.

Whether crossing the membrane with aggregates of amphipathic α-helices or β-barrels, these pore-forming toxins represent nature's accommodation to a structural challenge facing all protein-based transmembrane channels: the need to provide hydrogen-bonding partners for the polypeptide backbone N—H and C=O groups in an environment (the bilayer interior) that lacks hydrogen-bond donors or acceptors. The solution to this problem is found, of course, in the extensive hydrogen-bonding possibilities of α-helices and β-sheets.

Amphipathic Helices Form Transmembrane Ion Channels

Recently, a variety of natural peptides that form transmembrane channels have been identified and characterized. Melittin (Figure 9.43) is a bee venom toxin peptide of 26 residues. The cecropins are peptides induced in *Hyalophora cecropia* (Figure 9.44) and other related silkworms when challenged by bacterial infections. These peptides are thought to form α-helical aggregates in membranes, creating an ion channel in the center of the aggregate. The unifying feature of these helices is their **amphipathic** character, with polar residues clustered on one face of the helix and nonpolar residues elsewhere. In the membrane, the polar residues face the ion channel, leaving the nonpolar residues elsewhere on the helix to interact with the hydrophobic interior of the lipid bilayer.

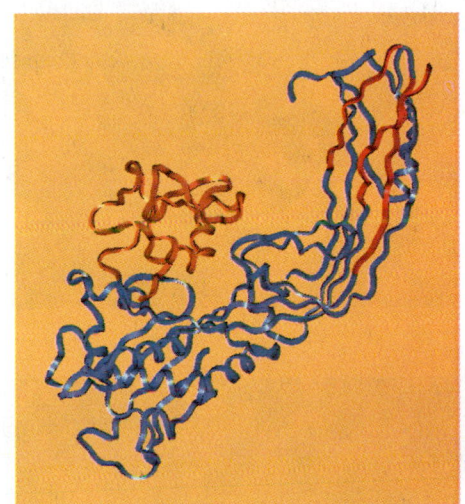

FIGURE 9.42 The structure of proaerolysin, produced by *Aeromonas hydrophila.* Proteolysis of this precursor yields the active form, aerolysin, which is responsible for the pathogenic effects of the bacterium in deep wound infections and diarrheal diseases. Like hemolysin, aerolysin monomers associate to form heptameric membrane pores. The three β-strands that contribute to the formation of the heptameric pore are shown in red. The N-terminal domain (residues 1–80, yellow) is a small lobe that protrudes from the rest of the protein.

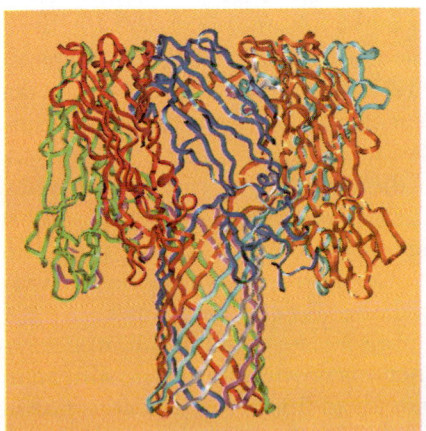
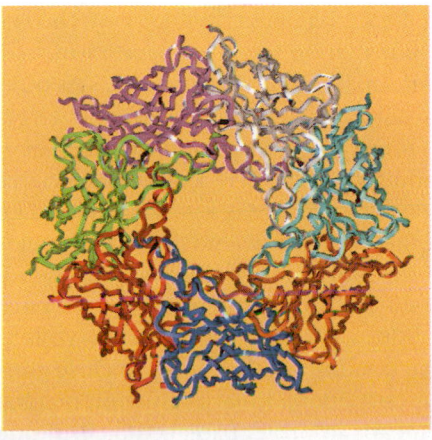

FIGURE 9.41 The structure of the heptameric channel formed by α-hemolysin. Each of the seven subunits contributes a β-sheet hairpin to the transmembrane channel.

Cecropin A:

Lys-Trp-Lys-Leu-Phe-Lys-Lys-Ile-Glu-Lys-Val-Gly-Gln-Asn-Ile-Arg-Asp-Gly-Ile-Ile-Lys-Ala-Gly-Pro-Ala-Val-Ala-Val-Val-Gly-Gln-Ala-Thr-Gln-Ile-Ala-Lys-NH$_2$

Melittin:

Gly-Ile-Gly-Ala-Val-Leu-Lys-Val-Leu-Thr-Thr-Gly-Leu-Pro-Ala-Leu-Ile-Ser-Trp-Ile-Lys-Arg-Lys-Arg-Gln-Gln-NH$_2$

Magainin 2 amide:

Gly-Ile-Gly-Lys-Phe-Leu-His-Ser-Ala-Lys-Lys-Phe-Gly-Lys-Ala-Phe-Val-Gly-Glu-Ile-Met-Asn-Ser-NH$_2$

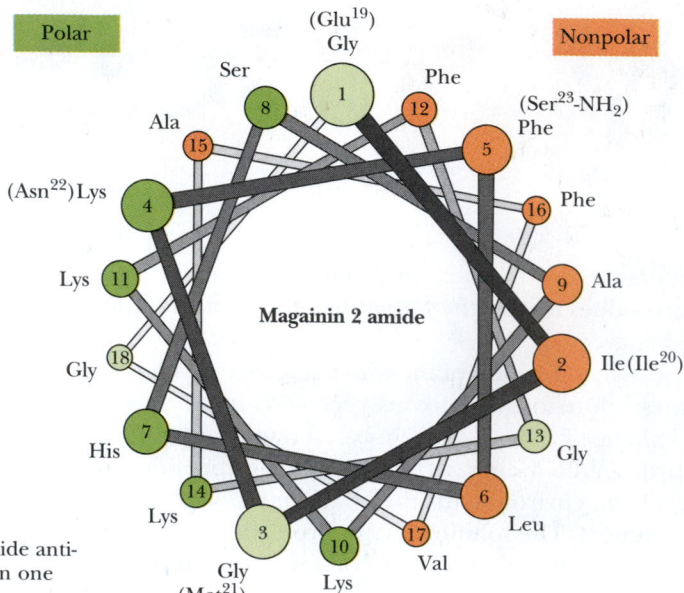

FIGURE 9.43 The amino acid sequences of several amphipathic peptide antibiotics. α-Helices formed from these peptides cluster polar residues on one face of the helix, with nonpolar residues at other positions.

FIGURE 9.44 **(a)** Adult and **(b)** caterpillar stages of the cecropia moth, *Hyalophora cecropia*.

Gap Junctions Connect Cells in Mammalian Cell Membranes

When cells lie adjacent to each other in animal tissues, they are often connected by **gap junction** structures, which permit the passive flow of small molecules from one cell to the other. Such junctions essentially connect the cells metabolically, providing a means of chemical transfer and communication. In certain tissues, such as heart muscle that is not innervated, gap junctions permit very large numbers of cells to act synchronously. Gap junctions also provide a means for transport of nutrients to cells disconnected from the circulatory system, such as the lens cells of the eye.

Gap junctions are formed from hexameric arrays of a single 32-kD protein. Each subunit of the array is cylindrical, with a length of 7.5 nm and a diameter of 2.5 nm. The subunits of the hexameric array are normally tilted with respect to the sixfold axis running down the center of the hexamer (Figure 9.45). In this conformation, a central pore having a diameter of about 1.8 to 2.0 nm is created, and small molecules (up to masses of 1 to 1.2 kD) can pass through unimpeded. Proteins, nucleic acids, and other large structures cannot. A complete gap junction is formed from two such hexameric arrays, one from each cell. A twisting, sliding movement of the subunits narrows the channel and closes the gap junction. This closure is a cooperative process, and a localized conformation change at the cytoplasmic end assists in the closing of the channels. Because the closing of the gap junction does not appear to involve massive conformational changes in the individual subunits, the free energy change for closure is small.

Although gap junctions allow cells to communicate metabolically under normal conditions, the ability to close gap junctions provides the tissue with an important intercellular regulation mechanism. In addition, gap junctions provide a means to protect adjacent cells if one or more cells are damaged or stressed. To these ends, gap junctions are sensitive to membrane potentials, hormonal signals, pH changes, and intracellular calcium levels. Dramatic changes in pH

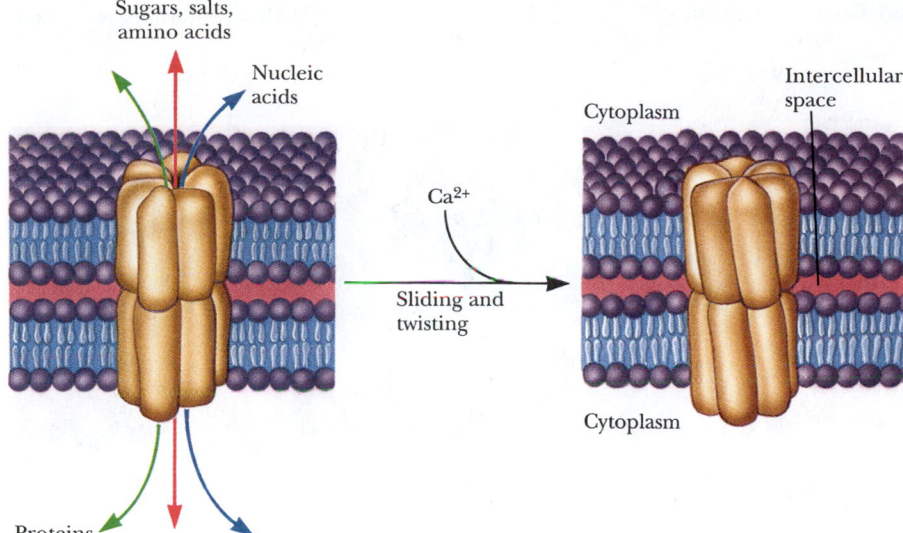

FIGURE 9.45 Gap junctions consist of hexameric arrays of cylindrical protein subunits in the plasma membrane. The subunit cylinders are tilted with respect to the axis running through the center of the gap junction. A gap junction between cells is formed when two hexameric arrays of subunits in separate cells contact each other and form a pore through which cellular contents may pass. Gap junctions close by means of a twisting, sliding motion in which the subunits decrease their tilt with respect to the central axis. Closure of the gap junction is Ca^{2+}-dependent.

or Ca^{2+} concentration in a cell may be a sign of cellular damage or death. To protect neighboring cells from the propagation of such effects, gap junctions close in response to decreased pH or prolonged increases in intracellular Ca^{2+}. Under normal conditions of intracellular Ca^{2+} levels ($<10^{-7}$ M), gap junctions are open and intercellular communication is maintained. When calcium levels rise to 10^{-5} M or higher, the junctions, sensing danger, rapidly close.

9.10 What Is the Structure and Function of Ionophore Antibiotics?

All of the protein-based transport systems examined thus far are relatively large. Nevertheless, several small molecule toxins produced by microorganisms can also facilitate ion transport across membranes. Due to their relative simplicity,

---

Human Biochemistry

Biochemistry ⦚ Now™

Melittin—How to Sting Like a Bee

The stings of many stinging insects, like wasps, hornets, and bumblebees, cause a pain that, although mild at first, increases in intensity over 2 to 30 minutes, with a following period of swelling that may last for several days. The sting of the honeybee (*Apis mellifera*), on the other hand, elicits a sharp, stabbing pain within 10 seconds. This pain may last for several minutes and is followed by several hours of swelling and itching. The immediate, intense pain is caused by **melittin,** a 26-residue peptide that constitutes about half of the 50 μg (dry weight) of material injected during the "sting" (in a total volume of only 0.5 μL). How does this simple peptide cause the intense pain that accompanies a bee sting?

The pain appears to arise from the formation of melittin pores in the membranes of **nociceptors,** free nerve endings that detect harmful ("noxious"—thus the name) stimuli of violent mechanical stress, high temperatures, and irritant chemicals. The creation of pores by melittin depends on the nociceptor membrane potential. Melittin in water solution is tetrameric. However, melittin interacting with membranes in the absence of a membrane potential is monomeric and shows no evidence of oligomer formation.

When an electrical potential (voltage) is applied across the membrane, melittin tetramers form and the membrane becomes permeable to anions such as chloride. Nociceptor membranes maintain a resting potential of 70 mV (negative inside). When melittin binds to the nociceptor membrane, the flow of chloride ions out of the cell diminishes the transmembrane potential, stimulating the nerve and triggering a pain response and also inducing melittin tetramers to dissociate. When the membrane potential is reestablished, melittin tetramers reform and the cycle is repeated over and over, causing a prolonged and painful stimulation of the nociceptors. The pain of the sting eventually lessens, perhaps due to the molecules of melittin diffusing apart, so that tetramers can no longer form.

Although the honeybee's sting is unpleasant, this tiny creature is crucial to the world's agricultural economy. Honeybees produce more than $100 million worth of honey each year, and, more important, the pollination of numerous plants by honeybees is responsible for the production of $20 billion worth of crops in the United States alone.

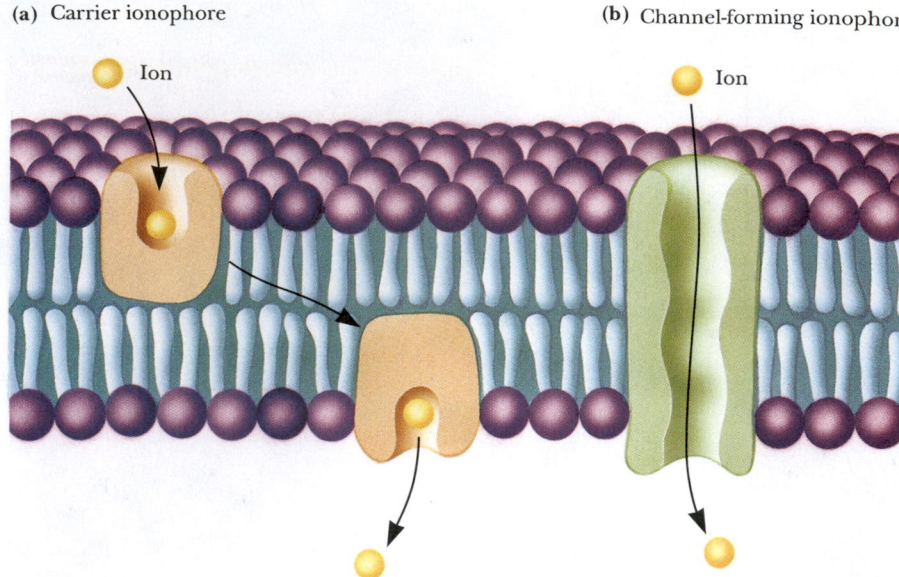

(a) Carrier ionophore

Ion

(b) Channel-forming ionophore

Ion

Biochemistry⊛Now™ **ANIMATED FIGURE 9.46**
Schematic drawings of mobile carrier and channel ionophores. Carrier ionophores must move from one side of the membrane to the other, acquiring the transported species on one side and releasing it on the other side. Channel ionophores span the entire membrane. **See this figure animated at** http://chemistry.brookscole.com/ggb3

these molecules, the **ionophore antibiotics,** represent paradigms of the **mobile carrier** and **pore** or **channel** models for membrane transport. Mobile carriers are molecules that form complexes with particular ions and diffuse freely across a lipid membrane (Figure 9.46). Pores or channels, on the other hand, adopt a fixed orientation in a membrane, creating a hole that permits the transmembrane movement of ions. These pores or channels may be formed from monomeric or (more often) multimeric structures in the membrane.

Carriers and channels may be distinguished on the basis of their temperature dependence. Channels are comparatively insensitive to membrane phase transitions and show only a slight dependence of transport rate on temperature. Mobile carriers, on the other hand, function efficiently above a membrane phase transition but only poorly below it. Consequently, mobile carrier systems often show dramatic increases in transport rate as the system is heated through its phase transition. Figure 9.47 displays the structures of several of these interesting molecules. As might be anticipated from the variety of structures represented here, these molecules associate with membranes and facilitate transport by different means.

Valinomycin Is a Mobile Carrier Ionophore

Valinomycin (isolated from *Streptomyces fulvissimus*) is a cyclic structure containing 12 units made from 4 different residues. Two are amino acids (L-valine and D-valine); the other two residues, L-lactate and D-hydroxyisovalerate, contribute ester linkages. Valinomycin is a **depsipeptide,** that is, a molecule with both peptide and ester bonds. (Considering the 12 units in the structure, valinomycin is called a **dodecadepsipeptide.**) Valinomycin consists of the 4-unit sequence (D-valine, L-lactate, L-valine, D-hydroxyisovaleric acid), repeated three times to form the cyclic structure in Figure 9.47. The structures of uncomplexed valinomycin and the K^+-valinomycin complex have been studied by X-ray crystallography (Figure 9.48). The structure places K^+ at the center of the valinomycin ring, coordinated with the carbonyl oxygens of the 6 valines. The polar groups of the valinomycin structure are positioned toward the center of the ring, whereas the nonpolar groups (the methyl and isopropyl side chains) are directed outward from the ring. The hydrophobic exterior of valinomycin interacts favorably with low dielectric solvents and with the hydrophobic interiors of lipid bilayers. Moreover, the central carbonyl groups completely surround the K^+ ion, shielding it from contact with nonpolar solvents or the hydrophobic membrane interior. As a result, the

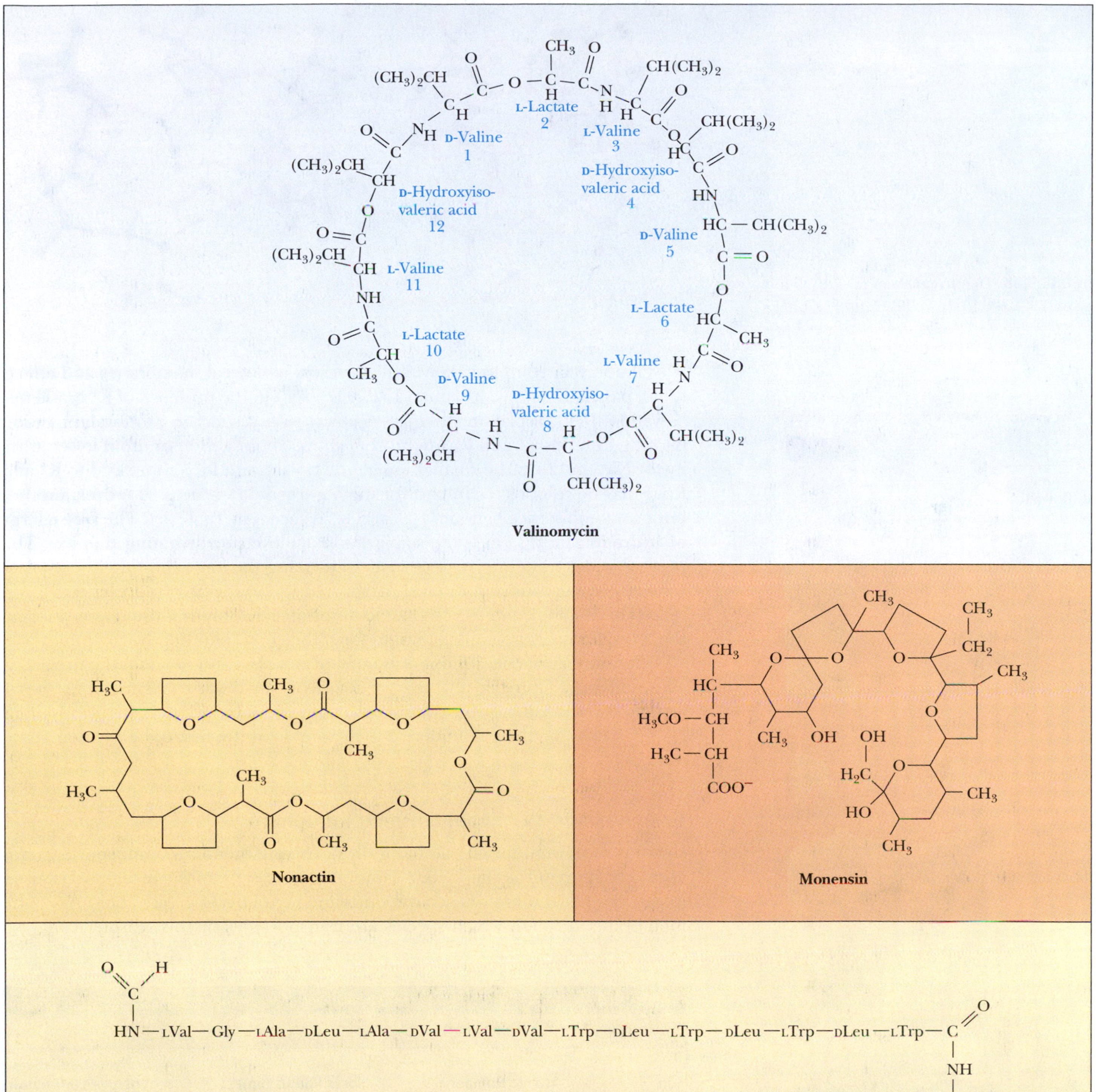

FIGURE 9.47 Structures of several ionophore antibiotics. Valinomycin consists of three repeats of a four-unit sequence. Because it contains both peptide and ester bonds, it is referred to as a depsipeptide.

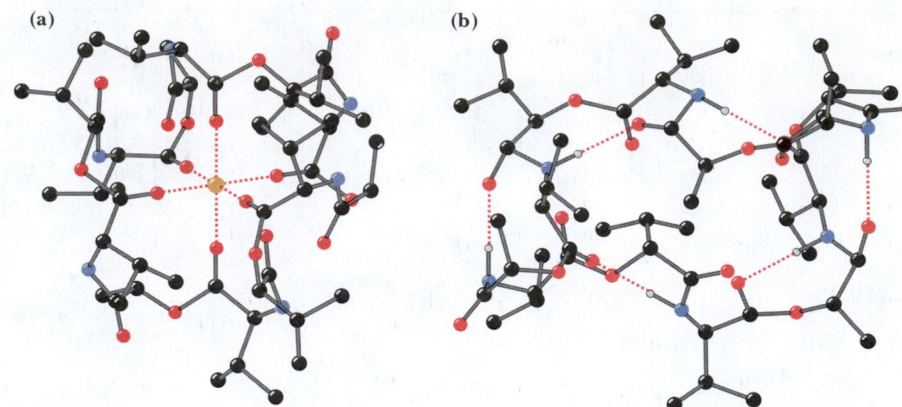

(a) **(b)**

FIGURE 9.48 The structures of **(a)** the valinomycin–K⁺ complex and **(b)** uncomplexed valinomycin.

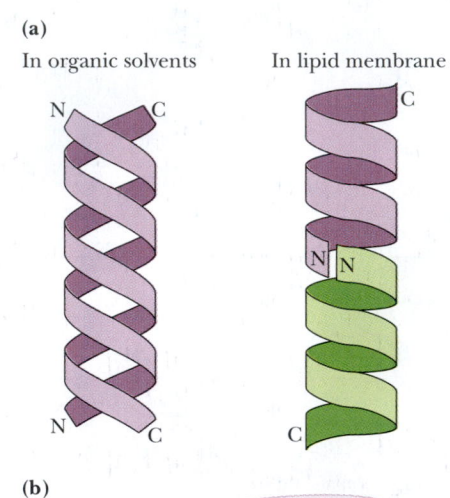

(a)

In organic solvents   In lipid membrane

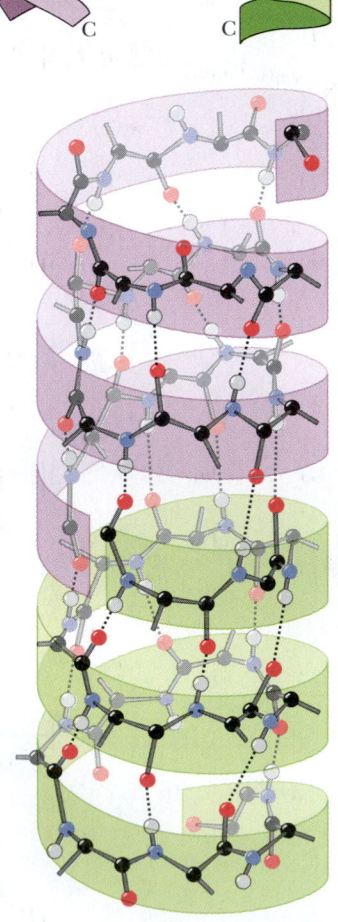

(b)

K⁺–valinomycin complex freely diffuses across biological membranes and effects rapid, passive K⁺ transport (up to 10,000 K⁺/sec) in the presence of K⁺ gradients.

Valinomycin displays a striking selectivity with respect to monovalent cation binding. It binds K⁺ and Rb⁺ tightly but shows about a thousandfold lower affinity for Na⁺ and Li⁺. The smaller ionic radii of Na⁺ and Li⁺ (compared to K⁺ and Rb⁺) may be responsible in part for the observed differences. However, another important difference between Na⁺ and K⁺ is shown in Table 9.4. The **free energy of hydration** for an ion is the stabilization achieved by hydrating that ion. The process of dehydration, a prerequisite to forming the ion–valinomycin complex, requires energy input. As shown in Table 9.4, considerably more energy is required to desolvate an Na⁺ ion than to desolvate a K⁺ ion. It is thus easier to form the K⁺–valinomycin complex than to form the corresponding Na⁺ complex.

Other mobile carrier ionophores include *monensin* and *nonactin* (Figure 9.47). The unifying feature in all these structures is an inward orientation of polar groups (to coordinate the central ion) and outward orientation of nonpolar residues (making these complexes freely soluble in the hydrophobic membrane interior).

Gramicidin Is a Channel-Forming Ionophore

In contrast to valinomycin, all naturally occurring membrane transport systems appear to function as channels, not mobile carriers. All of the proteins discussed in this chapter use multiple transmembrane segments to create channels in the membrane, through which species are transported. For this reason, it may be

Table 9.4			
Properties of Alkali Cations			
Ion	**Atomic Number**	**Ionic Radius (nm)**	**Hydration Free Energy, ΔG (kJ/mol)**
Li⁺	3	0.06	−410
Na⁺	11	0.095	−300
K⁺	19	0.133	−230
Rb⁺	37	0.148	−210
Cs⁺	55	0.169	−200

◄ **FIGURE 9.49** **(a)** Gramicidin forms a double helix in organic solvents; a helical dimer is the preferred structure in lipid bilayers. The structure is a head-to-head, left-handed helix, with the carboxy-termini of the two monomers at the ends of the structure. **(b)** The hydrogen-bonding pattern resembles that of a parallel β-sheet.

more relevant to consider the **pore** or **channel** ionophores. **Gramicidin** from *Bacillus brevis* (Figure 9.49) is a linear peptide of 15 residues and is a prototypical channel ionophore. Gramicidin contains alternating L- and D-residues, a formyl group at the N-terminus, and an ethanolamine at the C-terminus. The predominance of hydrophobic residues in the gramicidin structure facilitates its incorporation into lipid bilayers and membranes. Once incorporated in lipid bilayers, it permits the rapid diffusion of many different cations. Gramicidin possesses considerably less ionic specificity than does valinomycin but permits higher transport rates. A single gramicidin channel can transport as many as 10 million K^+ ions per second. Protons and all alkali cations can diffuse through gramicidin channels, but divalent cations such as Ca^{2+} block the channel.

Gramicidin forms two different helical structures. A double helical structure predominates in organic solvents (Figure 9.49), whereas a helical dimer is formed in lipid membranes. (An α-helix cannot be formed by gramicidin, because it has both D- and L-amino acid residues.) The helical dimer is a head-to-head or amino terminus-to-amino terminus (N-to-N) dimer oriented perpendicular to the membrane surface, with the formyl groups at the bilayer center and the ethanolamine moieties at the membrane surface. The helix is unusual, with 6.3 residues per turn and a central hole approximately 0.4 nm in diameter. The hydrogen-bonding pattern in this structure, in which N—H groups alternately point up and down the axis of the helix to hydrogen bond with carbonyl groups, is reminiscent of a β-sheet. For this reason, this structure has often been referred to as a β-helix.

Summary

Membranes constitute the boundaries of cells and intracellular organelles, and they provide a surface where many important biological reactions and processes occur. Membranes have proteins that mediate and regulate the transport of metabolites, macromolecules, and ions.

9.1 What Are the Chemical and Physical Properties of Membranes?
Amphipathic lipids spontaneously form a variety of structures when added to aqueous solution, including micelles and lipid bilayers. The fluid mosaic model for membrane structure suggests that membranes are dynamic structures composed of proteins and phospholipids. In this model, the phospholipid bilayer is a *fluid* matrix, in essence, a two-dimensional solvent for proteins. Both lipids and proteins are capable of rotational and lateral movement. Biological membranes exhibit both lateral and transverse asymmetries of lipid and protein distribution. Lipid bilayers typically undergo gel-to-liquid crystalline phase transitions, with the transition temperature being dependent upon bilayer composition.

9.2 What Is the Structure and Chemistry of Membrane Proteins?
Peripheral proteins interact with the membrane mainly through electrostatic and hydrogen-bonding interactions with integral proteins. Integral proteins are those that are strongly associated with the lipid bilayer, with a portion of the protein embedded in, or extending all the way across, the lipid bilayer. Another class of proteins not anticipated by Singer and Nicolson, the lipid-anchored proteins, associate with membranes by means of a variety of covalently linked lipid anchors.

9.3 How Does Transport Occur Across Biological Membranes?
In most biological transport processes, the molecule or ion transported is water soluble, yet moves across the hydrophobic, impermeable lipid membrane at a rate high enough to serve the metabolic and physiological needs of the cell. Most of these processes occur with the assistance of specific transport protein. The transported species either diffuses through a channel-forming protein or is carried by a carrier protein. Transport proteins are all classed as integral membrane proteins. From a thermodynamic and kinetic perspective, there are only three types of membrane transport processes: *passive diffusion, facilitated diffusion,* and *active transport.*

9.4 What Is Passive Diffusion?
In passive diffusion, the transported species moves across the membrane in the thermodynamically favored direction without the help of any specific transport system/molecule. For an uncharged molecule, passive diffusion is an entropic process, in which movement of molecules across the membrane proceeds until the concentration of the substance on both sides of the membrane is the same. The passive transport of charged species depends on their electrochemical potentials.

9.5 How Does Facilitated Diffusion Occur?
Certain metabolites and ions move across biological membrane more readily than can be explained by passive diffusion alone. In all such cases, a protein that binds the transported species is said to facilitate its transport. Facilitated diffusion rates display saturation behavior similar to that observed with substrate binding by enzymes.

9.6 How Does Energy Input Drive Active Transport Processes?
Active transport involves the movement of a given species against its thermodynamic potential. Such systems require energy input and are referred to as active transport systems. Active transport may be driven by the energy of ATP hydrolysis, by light energy, or by the potential stored in ion gradients. The original ion gradient arises from a primary active transport process, and the transport that depends on the ion gradient for its energy input is referred to as a secondary active transport process. When transport results in a net movement of electric charge across the membrane, it is referred to as an electrogenic transport process. If no net movement of charge occurs during transport, the process is electrically neutral. The Na^+,K^+-ATPase of animal plasma membranes, the Ca^{2+}-ATPase of muscle sarcoplasmic reticulum, the gastric ATPase, the osteoclast proton pump, and the multidrug transporter all use the free energy of hydrolysis of ATP to drive transport processes.

9.7 How Are Certain Transport Processes Driven by Light Energy?
Light energy drives a series of conformation changes in the transmembrane protein bacteriorhodopsin that drive proton transport. The transport involves the *cis–trans* isomerization of retinal in Schiff base linkage to the protein via a lysine residue.

9.8 How Are Amino Acid and Sugar Transport Driven by Ion Gradients?
The gradients of H^+, Na^+, and other cations and anions established by ATPases and other energy sources can be used for secondary active transport of various substrates. Many of these systems operate as symports, with the ion and the transported amino acid or sugar moving in the same direction (that is, into the cell). In antiport processes, the ion and the other transported species move in opposite directions.

9.9 How Are Specialized Membrane Pores Formed by Toxins?
Many specialized membrane pores are formed from hydrophobic and amphipathic α-helices. Such helical structures may either insert into the membrane spontaneously or be driven by association of a larger protein structure with a bilayer membrane. The various pore-forming toxins typically insert a pair of hydrophobic α-helices into a membrane bilayer, followed by amphipathic helices that create the membrane pore. Other pore-forming toxins assemble into multimeric β-barrels made from β-strands contributed by individual monomers.

9.10 What Is the Structure and Function of Ionophore Antibiotics?
The ionophore antibiotics represent paradigms of the mobile carrier and pore or channel models for membrane transport. Mobile carriers are molecules that form complexes with particular ions and diffuse freely across a lipid membrane. Pores or channels, on the other hand, adopt a fixed orientation in a membrane, creating a hole that permits the transmembrane movement of ions. These pores or channels may be formed from monomeric or (more often) multimeric structures in the membrane. Carriers and channels may be distinguished on the basis of the temperature dependence of ion transport.

Problems

1. In problem 1 (b) in Chapter 8 (page 265), you were asked to draw all the possible phosphatidylserine isomers that can be formed from palmitic and linolenic acids. Which of the PS isomers are not likely to be found in biological membranes?

2. The purple patches of the *Halobacterium halobium* membrane, which contain the protein bacteriorhodopsin, are approximately 75% protein and 25% lipid. If the protein molecular weight is 26,000 and an average phospholipid has a molecular weight of 800, calculate the phospholipid-to-protein mole ratio.

3. Sucrose gradients for separation of membrane proteins must be able to separate proteins and protein–lipid complexes having a wide range of densities, typically 1.00 to 1.35 g/mL.
   a. Consult reference books (such as the *CRC Handbook of Biochemistry*) and plot the density of sucrose solutions versus percent sucrose by weight (g sucrose per 100 g solution), and versus percent by volume (g sucrose per 100 mL solution). Why is one plot linear and the other plot curved?
   b. What would be a suitable range of sucrose concentrations for separation of three membrane-derived protein–lipid complexes with densities of 1.03, 1.07, and 1.08 g/mL?

4. Phospholipid lateral motion in membranes is characterized by a diffusion coefficient of about 1×10^{-8} cm^2/sec. The distance traveled in two dimensions (across the membrane) in a given time is $r = (4Dt)^{1/2}$, where r is the distance traveled in centimeters, D is the diffusion coefficient, and t is the time during which diffusion occurs. Calculate the distance traveled by a phospholipid across a bilayer in 10 msec (milliseconds).

5. Protein lateral motion is much slower than that of lipids because proteins are larger than lipids. Also, some membrane proteins can diffuse freely through the membrane, whereas others are bound or anchored to other protein structures in the membrane. The diffusion constant for the membrane protein fibronectin is approximately 0.7×10^{-12} cm^2/sec, whereas that for rhodopsin is about 3×10^{-9} cm^2/sec.
   a. Calculate the distance traversed by each of these proteins in 10 msec.
   b. What could you surmise about the interactions of these proteins with other membrane components?

6. Discuss the effects on the lipid phase transition of pure dimyristoyl phosphatidylcholine vesicles of added (a) divalent cations, (b) cholesterol, (c) distearoyl phosphatidylserine, (d) dioleoyl phosphatidylcholine, and (e) integral membrane proteins.

7. Calculate the free energy difference at 25°C due to a galactose gradient across a membrane, if the concentration on side 1 is 2 mM and the concentration on side 2 is 10 mM.

8. Consider a phospholipid vesicle containing 10 mM Na^+ ions. The vesicle is bathed in a solution that contains 52 mM Na^+ ions, and the electrical potential difference across the vesicle membrane $\Delta\psi = \psi_{outside} - \psi_{inside} = -30$ mV. What is the electrochemical potential at 25°C for Na^+ ions?

9. Transport of histidine across a cell membrane was measured at several histidine concentrations:

[Histidine], μM	Transport, μmol/min
2.5	42.5
7	119
16	272
31	527
72	1220

Does this transport operate by passive diffusion or by facilitated diffusion?

10. (Integrates with Chapter 3.) Fructose is present outside a cell at 1 μM concentration. An active transport system in the plasma membrane transports fructose into this cell, using the free energy of ATP hydrolysis to drive fructose uptake. What is the highest intracellular concentration of fructose that this transport system can generate? Assume that one fructose is transported per ATP hydrolyzed; that ATP is hydrolyzed on the intracellular surface of the membrane; and that the concentrations of ATP, ADP, and P$_i$ are 3 mM, 1 mM, and 0.5 mM, respectively. T = 298 K. (*Hint:* Refer to Chapter 3 to recall the effects of concentration on free energy of ATP hydrolysis.)

11. In this chapter, we have examined coupled transport systems that rely on ATP hydrolysis, on primary gradients of Na^+ or H^+, and on phosphotransferase systems. Suppose you have just discovered an unusual strain of bacteria that transports rhamnose across its plasma membrane. Suggest experiments that would test whether it was linked to any of these other transport systems.

12. Which of the following peptides would be the most likely to acquire an N-terminal myristoyl lipid anchor?
    a. VLIHGLEQN
    b. THISISIT
    c. RIGHTHERE
    d. MEMEME
    e. GETREAL

13. Which of the following peptides would be the most likely to acquire a prenyl anchor?
    a. RIGHTCALL
    b. PICKME
    c. ICANTICANT
    d. AINTMEPICKA
    e. none of the above

Preparing for the MCAT Exam

14. Singer and Nicolson's fluid mosaic model of membrane structure presumed all of the following statements to be true EXCEPT:
    a. The phospholipid bilayer is a fluid matrix.
    b. Proteins can be anchored to the membrane by covalently linked lipid chains.
    c. Proteins can move laterally across a membrane.
    d. Membranes should be about 5 nm thick.
    e. Transverse motion of lipid molecules can occur occasionally.

15. The rate of K^+ transport across bilayer membranes reconstituted from dipalmitoylphosphatidylcholine (DPPC) and monensin is approximately the same as that observed across membranes reconstituted from DPPC and cecropin a at 35°C. Based on your reading of sections 9.8 and 9.9 of this chapter, would you expect the transport rates across these two membranes to also be similar at 50°C? Explain.

Biochemistry ⋐ Now™   Preparing for an exam? Test yourself on key questions at http://chemistry.brookscole.com/ggb3

Further Reading

Membrane Structure

Balasubramanian, K., and Schroit, A. J., 2003. Aminophospholipid asymmetry: A matter of life and death. *Annual Review of Physiology* **65:**701–734.

Bretscher, M., 1985. The molecules of the cell membrane. *Scientific American* **253:**100–108.

Dawidowicz, E. A., 1987. Dynamics of membrane lipid metabolism and turnover. *Annual Review of Biochemistry* **56:**43–61.

De Weer, P., 2000. A century of thinking about cell membranes. *Annual Review of Physiology* **62:**919–926.

Dowhan, W., 1997. Molecular basis for membrane phospholipid diversity: Why are there so many lipids? *Annual Review of Biochemistry* **66:**199–232.

Edidin, M., 2003. The state of lipid rafts: From model membranes to cells. *Annual Review of Biophysics and Biomolecular Structure* **32:**257–283.

Frye, C. D., and Edidin, M., 1970. The rapid intermixing of cell surface antigens after formation of mouse-human heterokaryons. *Journal of Cell Science* **7:**319–335.

Jain, M. K., 1988. *Introduction to Biological Membranes,* 2nd ed. New York: John Wiley & Sons.

Op den Kamp, J. A. F., 1979. Lipid asymmetry in membranes. *Annual Review of Biochemistry* **48:**47–71.

Robertson, R. N., 1983. *The Lively Membranes.* Cambridge: Cambridge University Press.

Singer, S. J., and Nicolson, G. L., 1972. The fluid mosaic model of the structure of cell membranes. *Science* **175:**720–731.

Tanford, C., 1980. *The Hydrophobic Effect: Formation of Micelles and Biological Membranes,* 2nd ed. New York: Wiley-Interscience.

Membrane Transport

Blair, H. C., et al., 1989. Osteoclastic bone resorption by a polarized vacuolar proton pump. *Science* **245:**855–857.

Christensen, B., et al., 1988. Channel-forming properties of cecropins and related model compounds incorporated into planar lipid membranes. *Proceedings of the National Academy of Sciences, U.S.A.* **85:**5072–5076.

Collarini, E. J., and Oxender, D., 1987. Mechanisms of transport of amino acids across membrane. *Annual Review of Nutrition* **7:**75–90.

Featherstone, C., 1990. An ATP-driven pump for secretion of yeast mating factor. *Trends in Biochemical Sciences* **15:**169–170.

Gould, G. W., and Bell, G. I., 1990. Facilitative glucose transporters: An expanding family. *Trends in Biochemical Sciences* **15:**18–23.

Ikonen, E., 2001. Roles of lipid rafts in membrane transport. *Current Opinions in Cell Biology* **13:**470–477.

Inesi, G., Sumbilla, C., and Kirtley, M., 1990. Relationships of molecular structure and function in Ca^{2+}-transport ATPase. *Physiological Reviews* **70:**749–759.

Jap, B., and Walian, P. J., 1990. Biophysics of the structure and function of porins. *Quarterly Reviews of Biophysics* **23:**367–403.

Jay, D., and Cantley, L., 1986. Structural aspects of the red cell anion exchange protein. *Annual Review of Biochemistry* **55:**511–538.

Jørgensen, P. L., Hakansson, K. O., and Karlish, S. J. D., 2003. Structure and mechanism of Na,K-ATPase: Functional sites and their interactions. *Annual Review of Physiology* **65:**817–849.

Juranka, P. F., Zastawny, R. L., and Ling, V., 1989. P-Glycoprotein: Multidrug-resistance and a superfamily of membrane-associated transport proteins. *The FASEB Journal* **3:**2583–2592.

Kaplan, J. H., 2002. Biochemistry of Na,K-ATPase. *Annual Review of Biochemistry* **71:**511–535.

Meadow, N. D., Fox, D. K., and Roseman, S., 1990. The bacterial phosphoenolpyruvate:glycose phosphotransferase system. *Annual Review of Biochemistry* **59:**497–542.

Oesterhelt, D., and Tittor, J., 1989. Two pumps, one principle: Light-driven ion transport in *Halobacteria. Trends in Biochemical Sciences* **14:**57–61.

Palmgren, M. G., 2001. Plant plasma membrane H^+-ATPases: Powerhouses for nutrient uptake. *Annual Review of Plant Physiology and Plant Molecular Biology* **52:**817–845.

Spencer, R. H., and Rees, D. C., 2002. The alpha-helix and the organization and gating of ion channels. *Annual Review of Biophysics and Biomolecular Structure* **31:**207–233.

Spudich, J. L., and Bogomolni, R. A., 1988. Sensory rhodopsins of *Halobacteria. Annual Review of Biophysics and Biophysical Chemistry* **17:**193–215.

Tanner, W., and Caspari, T., 1996. Membrane transport carriers. *Annual Review of Plant Physiology and Plant Molecular Biology* **47:**595–626.

Wallace, B. A., 1990. Gramicidin channels and pores. *Annual Review of Biophysics and Biophysical Chemistry* **19:**127–157.

Walmsley, A. R., 1988. The dynamics of the glucose transporter. *Trends in Biochemical Sciences* **13:**226–231.

Wheeler, T. J., and Hinkle, P., 1985. The glucose transporter of mammalian cells. *Annual Review of Physiology* **47:**503–517.

Wirtz, K. W. A., 1991. Phospholipid transfer proteins. *Annual Review of Biochemistry* **60:**73–99.

Structure of Membrane Proteins

Cartailler, J-P., and Luecke, H., 2003. X-ray crystallographic analysis of lipid-protein interactions in the bacteriorhodopsin purple membrane. *Annual Review of Biophysics and Biomolecular Structure* **32:**285–310.

Doering, T. L., Masterson, W. J., Hart, G. W., and Englund, P. T., 1990. Biosynthesis of glycosyl phosphatidylinositol membrane anchors. *Journal of Biological Chemistry* **265:**611–614.

Fasman, G. D., and Gilbert, W. A., 1990. The prediction of transmembrane protein sequences and their conformation: An evaluation. *Trends in Biochemical Sciences* **15:**89–92.

Gelb, M. H., 1997. Protein prenylation, et cetera: Signal transduction in two dimensions. *Science* **275:**1750–1751.

Glomset, J. A., Gelb, M. H., and Farnsworth, C. C., 1990. Prenyl proteins in eukaryotic cells: A new type of membrane anchor. *Trends in Biochemical Sciences* **15:**139–142.

Gordon, J. I., Duronio, R. J., Rudnick, D. A., Adams, S. P., and Gokel, G. W., 1991. Protein *N*-myristoylation. *Journal of Biological Chemistry* **266:**8647–8650.

Jennings, M. L., 1989. Topography of membrane proteins. *Annual Review of Biochemistry* **58:**999–1027.

Jentoft, N., 1990. Why are proteins O-glycosylated? *Trends in Biochemical Sciences* **15:**291–294.

Jump, D. B., 2002. The biochemistry of n-3 polyunsaturated fatty acids. *Journal of Biological Chemistry* **277:**8755–8758.

Koblan, K. S., Kohl, N. E., Omer, C. A., et al., 1996. Farnesyltransferase inhibitors: A new class of cancer chemotherapeutics. *Biochemical Society Transactions* **24:**688–692.

Park, H-W., Boduluri, S. R., Moomaw, J. F., et al., 1997. Crystal structure of protein farnesyltransferase at 2.25 Angstrom resolution. *Science* **275:**1800–1804.

Sefton, B., and Buss, J. E., 1987. The covalent modification of eukaryotic proteins with lipid. *Journal of Cell Biology* **104:**1449–1453.

Van Meer, G., and Lisman, Q., 2002. Sphingolipid transport: Rafts and translocators. *Journal of Biological Chemistry* **277:**25855–25858.

Wirtz, K. W. A., 1991. Phospholipid transfer proteins. *Annual Review of Biochemistry* **60:**73–99.

Nucleotides and Nucleic Acids

Francis Crick *(right)* and James Watson *(left)* point out features of their model for the structure of DNA.

Essential Question

Nucleotides and nucleic acids are compounds containing nitrogen bases (aromatic cyclic structures possessing nitrogen atoms) as part of their structure. Nucleotides are essential to cellular metabolism, and nucleic acids are the molecules of genetic information storage and expression. *What are the structures of the nucleotides? How are nucleotides joined together to form nucleic acids? How is information stored in nucleic acids? What are the biological functions of nucleotides and nucleic acids?*

Nucleotides and **nucleic acids** are biological molecules that possess heterocyclic nitrogenous bases as principal components of their structure. The biochemical roles of nucleotides are numerous; they participate as essential intermediates in virtually all aspects of cellular metabolism. Serving an even more central biological purpose are the nucleic acids, the elements of heredity and the agents of genetic information transfer. Just as proteins are linear polymers of amino acids, nucleic acids are linear polymers of nucleotides. Like the letters in this sentence, the orderly sequence of nucleotide residues in a nucleic acid can encode information. The two basic kinds of nucleic acids are **deoxyribonucleic acid (DNA)** and **ribonucleic acid (RNA).** Complete hydrolysis of nucleic acids liberates nitrogenous bases, a five-carbon sugar, and phosphoric acid in equal amounts. The five-carbon sugar in DNA is 2-deoxyribose; in RNA, it is ribose. (See Chapter 7 for a detailed discussion of sugars and other carbohydrates.) DNA is the repository of genetic information in cells, whereas RNA serves in the expression of this information through the processes of **transcription** and **translation** (Figure 10.1). An interesting exception to this rule is that some viruses have their genetic information stored as RNA.

This chapter describes the chemistry of nucleotides and the major classes of nucleic acids. Chapter 11 presents methods for determination of nucleic acid primary structure (nucleic acid sequencing) and describes the higher orders of nucleic acid structure. Chapter 12 introduces the *molecular biology of recombinant DNA:* the construction and uses of novel DNA molecules assembled by combining segments from other DNA molecules.

We have discovered the secret of life!
Proclamation by **Francis H. C. Crick** to patrons of the Eagle, a pub in Cambridge, England (1953)

Key Questions

10.1 What Is the Structure and Chemistry of Nitrogenous Bases?

10.2 What Are Nucleosides?

10.3 What Is the Structure and Chemistry of Nucleotides?

10.4 What Are Nucleic Acids?

10.5 What Are the Different Classes of Nucleic Acids?

10.6 Are Nucleic Acids Susceptible to Hydrolysis?

10.1 What Is the Structure and Chemistry of Nitrogenous Bases?

The bases of nucleotides and nucleic acids are derivatives of either **pyrimidine** or **purine.** Pyrimidines are six-membered heterocyclic aromatic rings containing two nitrogen atoms (Figure 10.2a). The atoms are numbered in a clockwise fashion, as shown in Figure 10.2. The purine ring system consists of two rings of atoms: one resembling the pyrimidine ring and another resembling the imidazole ring (Figure 10.2b). The nine atoms in this fused ring system are numbered according to the convention shown.

The pyrimidine ring system is planar, whereas the purine system deviates somewhat from planarity in having a slight pucker between its imidazole and pyrimidine portions. Both are relatively insoluble in water, as might be expected from their pronounced aromatic character.

Biochemistry Now™ Test yourself on these Key Questions at BiochemistryNow at **http://chemistry.brookscole.com/ggb3**

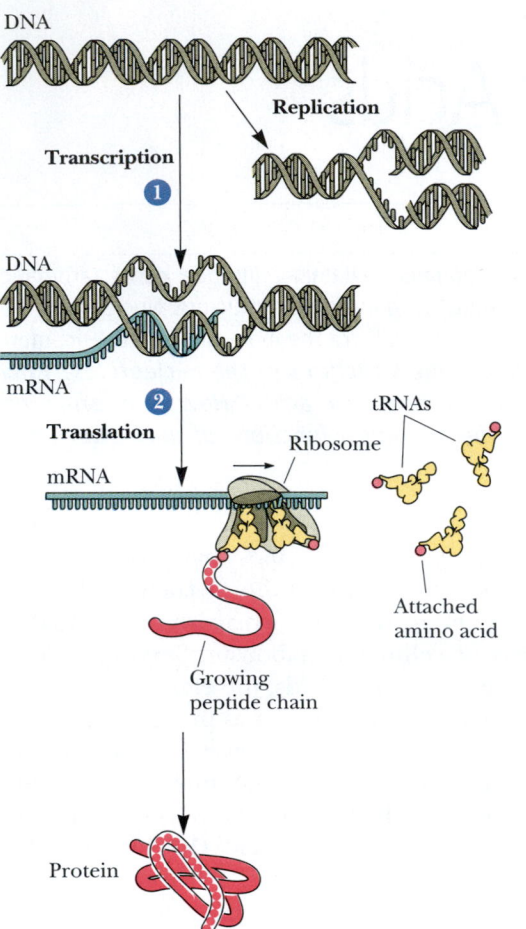

Replication
DNA replication yields two DNA molecules identical to the original one, ensuring transmission of genetic information to daughter cells with exceptional fidelity.

Transcription
The sequence of bases in DNA is recorded as a sequence of complementary bases in a single-stranded mRNA molecule.

Translation
Three-base codons on the mRNA corresponding to specific amino acids direct the sequence of building a protein. These codons are recognized by tRNAs (transfer RNAs) carrying the appropriate amino acids. Ribosomes are the "machinery" for protein synthesis.

FIGURE 10.1 The fundamental process of information transfer in cells. **(1)** Information encoded in the nucleotide sequence of DNA is transcribed through synthesis of an RNA molecule whose sequence is dictated by the DNA sequence. **(2)** As the sequence of this RNA is read (as groups of three consecutive nucleotides) by the protein synthesis machinery, it is translated into the sequence of amino acids in a protein. This information transfer system is encapsulated in the dogma: DNA → RNA → protein.

Biochemistry⊛Now™   Go to BiochemistryNow and click BiochemistryInteractive to learn the structures of the common purines and pyrimidines.

Three Pyrimidines and Two Purines Are Commonly Found in Cells

The common naturally occurring pyrimidines are **cytosine, uracil,** and **thymine** (5-methyluracil) (Figure 10.3). Cytosine and thymine are the pyrimidines typically found in DNA, whereas cytosine and uracil are common in RNA. To view this generality another way, the uracil component of DNA occurs as the 5-methyl variety thymine. Various pyrimidine derivatives, such as dihydrouracil, are present as minor constituents in certain RNA molecules.

Adenine (6-amino purine) and **guanine** (2-amino-6-oxy purine), the two common purines, are found in both DNA and RNA (Figure 10.4). Other naturally occurring purine derivatives include **hypoxanthine, xanthine,** and **uric acid** (Figure 10.5). Hypoxanthine and xanthine are found only rarely as constituents of nucleic acids. Uric acid, the most oxidized state for a purine derivative, is never found in nucleic acids.

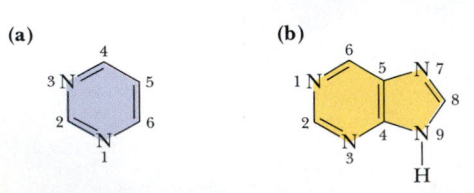

(a) The pyrimidine ring

(b) The purine ring system

FIGURE 10.2 (a) The pyrimidine ring system; by convention, atoms are numbered as indicated. **(b)** The purine ring system, atoms numbered as shown.

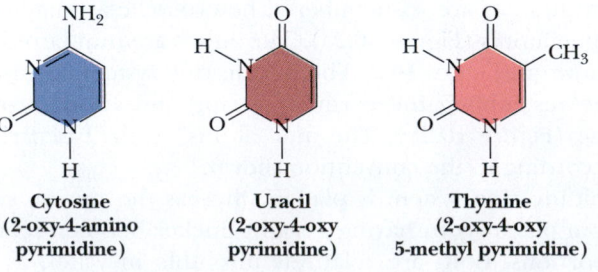

Cytosine
(2-oxy-4-amino pyrimidine)

Uracil
(2-oxy-4-oxy pyrimidine)

Thymine
(2-oxy-4-oxy 5-methyl pyrimidine)

FIGURE 10.3 The common pyrimidine bases—cytosine, uracil, and thymine—in the tautomeric forms predominant at pH 7.

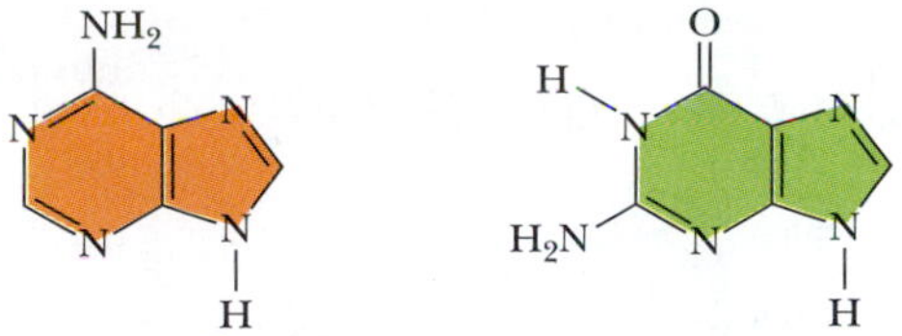

Adenine
(6-amino purine)

Guanine
(2-amino-6-oxy purine)

FIGURE 10.4 The common purine bases—adenine and guanine—in the tautomeric forms predominant at pH 7.

The Properties of Pyrimidines and Purines Can Be Traced to Their Electron-Rich Nature

The aromaticity of the pyrimidine and purine ring systems and the electron-rich nature of their —OH and —NH_2 substituents endow them with the capacity to undergo **keto–enol tautomeric shifts.** That is, pyrimidines and purines exist as tautomeric pairs, as shown in Figure 10.6 for uracil. The keto tautomer is called a **lactam,** whereas the enol form is a **lactim.** The lactam form vastly predominates at neutral pH. In other words, pK_a values for ring nitrogen atoms 1 and 3 in uracil are greater than 8 (the pK_a value for N-3 is 9.5) (Table 10.1). In contrast, as might be expected from the form of cytosine that predominates at pH 7, the pK_a value for N-3 in this pyrimidine is 4.5. Similarly, keto–enol tautomeric forms can be represented for purines, as given for guanine in Figure 10.7.[1] Here, the pK_a value is 9.4 for N-1 and less than 5 for N-3. These pK_a values specify whether hydrogen atoms are associated with the various ring nitrogens at neutral pH. As such, they are important in determining whether these nitrogens serve as H-bond donors or acceptors. Hydrogen bonding between purine and pyrimidine bases is fundamental to the biological functions of nucleic acids, as in the formation of the double-helix structure of DNA (see Section 10.5). The important functional groups participating in H-bond formation are the amino groups of cytosine, adenine, and guanine; the ring nitrogens at position 3 of pyrimidines and position 1 of purines; and the strongly electronegative oxygen atoms attached at position 4 of uracil and thymine, position 2 of cytosine, and position 6 of guanine (see Figure 10.20).

Another property of pyrimidines and purines is their strong absorbance of ultraviolet (UV) light, which is also a consequence of the aromaticity of their heterocyclic ring structures. Figure 10.8 shows characteristic absorption spectra of several of the common bases of nucleic acids—adenine, uracil, cytosine, and guanine—in their nucleotide forms: AMP, UMP, CMP, and GMP (see Section 10.3). This property is particularly useful in quantitative and qualitative analysis of nucleotides and nucleic acids.

[1]The 2-, 4-, and 6-pyrimidine and purine amino groups can undergo tautomerism as well, changing from amino to imino functions.

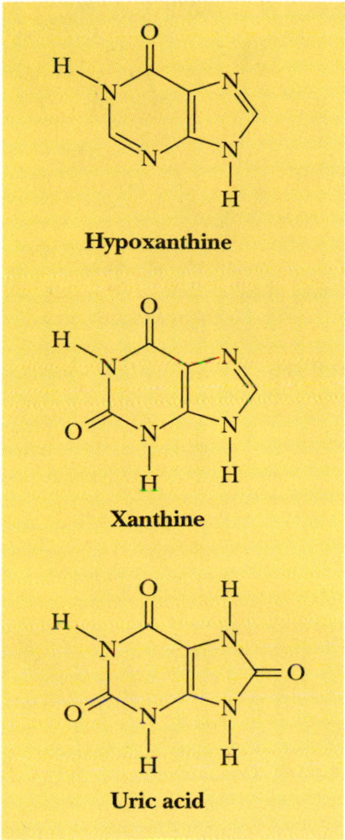

Hypoxanthine

Xanthine

Uric acid

FIGURE 10.5 Other naturally occurring purine derivatives—hypoxanthine, xanthine, and uric acid.

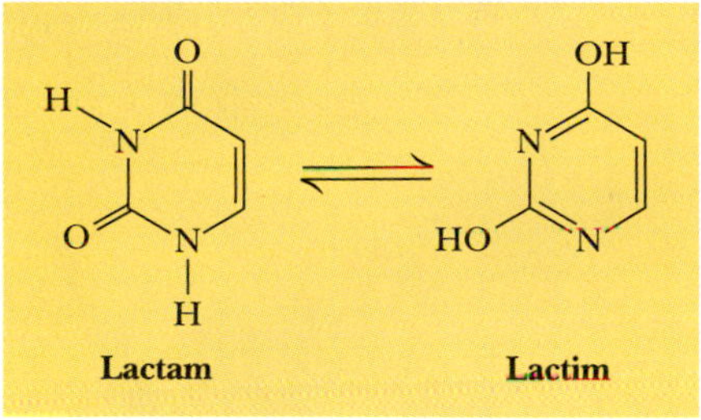

Lactam

Lactim

FIGURE 10.6 The keto–enol tautomerization of uracil.

Table 10.1			
Proton Dissociation Constants (pK_a Values) for Nucleotides			
Nucleotide	pK_a Base-N	pK_1 Phosphate	pK_2 Phosphate
5'-AMP	3.8 (N-1)	0.9	6.1
5'-GMP	9.4 (N-1)	0.7	6.1
	2.4 (N-7)		
5'-CMP	4.5 (N-3)	0.8	6.3
5'-UMP	9.5 (N-3)	1.0	6.4

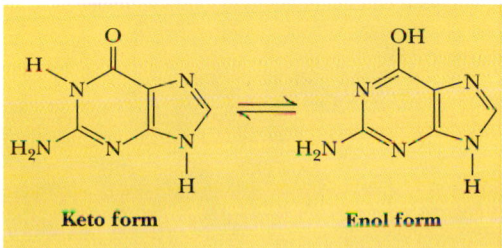

Keto form

Enol form

FIGURE 10.7 The tautomerization of the purine guanine.

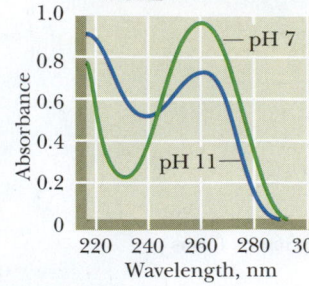
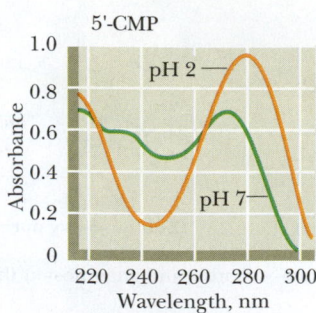
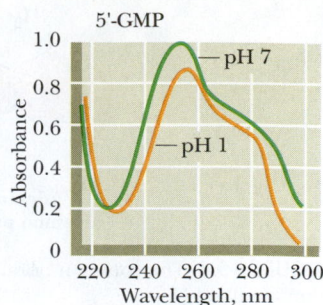

FIGURE 10.8 The UV absorption spectra of the common ribonucleotides.

10.2 | What Are Nucleosides?

Nucleosides are compounds formed when a base is linked to a sugar. The sugars of nucleosides are **pentoses** (five-carbon sugars, see Chapter 7). **Ribonucleosides** contain the pentose D-ribose, whereas 2-deoxy-D-ribose is found in **deoxyribonucleosides.** In both instances, the pentose is in the five-membered ring form known as furanose: D-ribofuranose for ribonucleosides and 2-deoxy-D-ribofuranose for deoxyribonucleosides (Figure 10.9). In nucleosides, these ribofuranose atoms are numbered as 1′, 2′, 3′, and so on to distinguish them from the ring atoms of the nitrogenous bases. (As we shall see, the seemingly minor difference of a hydroxyl group at the 2′-position has far-reaching effects on the secondary structures available to RNA and DNA, as well as their relative susceptibilities to chemical and enzymatic hydrolysis.)

In nucleosides, the base is linked to the sugar via a **glycosidic bond** (Figure 10.10). Glycosidic bonds by definition involve the carbonyl carbon atom of the sugar, which in cyclic structures is joined to the ring O atom. As discussed in Chapter 7, such carbon atoms are called **anomeric.** In nucleosides, the bond is an *N*-glycoside because it connects the anomeric C-1′ to N-1 of a pyrimidine or to N-9 of a purine. Recall that glycosidic bonds can be either α or β, depending on their orientation relative to the anomeric C atom. Glycosidic bonds in nucleosides (and nucleotides, see following discussion) are always of the β-configuration, as represented in Figure 10.10. Nucleosides are named by adding the ending *-idine* to the root name of a pyrimidine or *-osine* to the root name of a purine. The common nucleosides are thus **cytidine, uridine, thymidine, adenosine,** and **guanosine.** Structures of the common ribonucleosides are shown in Figure 10.11. The nucleoside formed by hypoxanthine and ribose is **inosine.**

Nucleosides Usually Adopt an Anti Conformation About the Glycosidic Bond

In nucleosides, rotation of the base about the glycosidic bond is sterically hindered, principally by the hydrogen atom on the C-2′ carbon of the furanose. (This hindrance is most easily seen and appreciated by manipulating accurate

FIGURE 10.9 Furanose structures—ribose and deoxyribose.

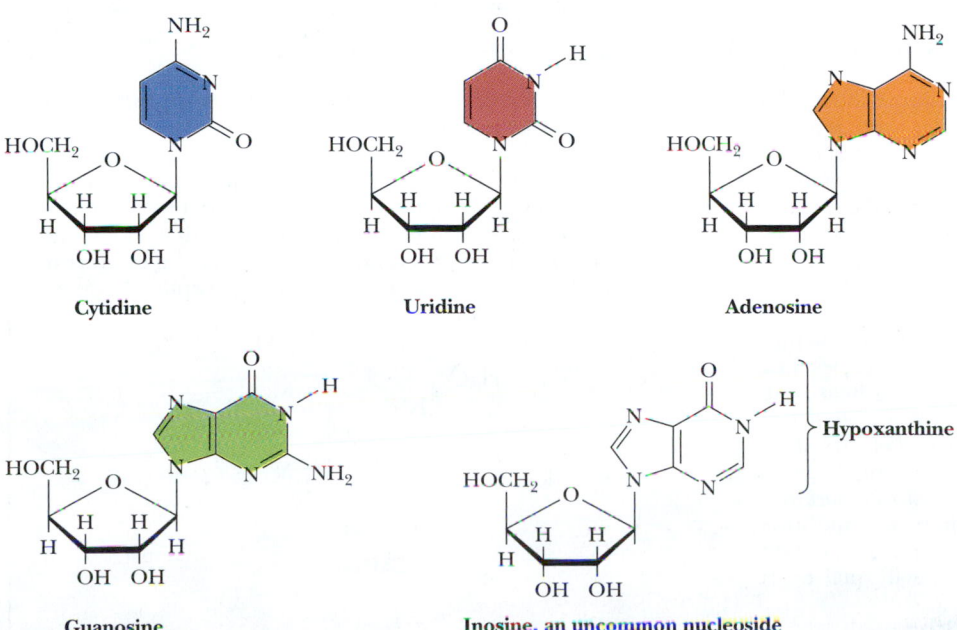

FIGURE 10.10 β-Glycosidic bonds link nitrogenous bases and sugars to form nucleosides.

molecular models of these structures.) Consequently, nucleosides (and nucleotides, see next section) exist in either of two conformations, designated *syn* and *anti* (Figure 10.12). For pyrimidines in the syn conformation, the oxygen substituent at position C-2 would lie in a sterically hindered position immediately above the furanose ring; in the anti conformation, this steric interference is avoided. Consequently, pyrimidine nucleosides adopt the anti conformation. Purine nucleosides can assume either the syn or anti conformation, although the anti conformation is favored. In either conformation, the roughly planar furanose and base rings are not coplanar but lie at approximately right angles to one another.

Nucleosides Are More Water Soluble Than Free Bases

Nucleosides are much more water soluble than the free bases because of the hydrophilicity of the sugar moiety. Like glycosides (see Chapter 7), nucleosides are relatively stable in alkali. Pyrimidine nucleosides are also resistant to acid hydrolysis, but purine nucleosides are easily hydrolyzed in acid to yield the free base and pentose.

FIGURE 10.11 The common ribonucleosides—cytidine, uridine, adenosine, and guanosine. Also, inosine drawn in anti conformation.

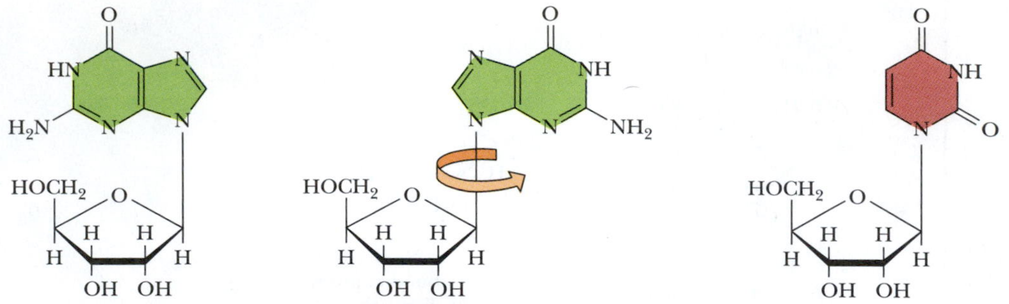

FIGURE 10.12 Rotation around the glycosidic bond is sterically hindered; syn versus anti conformations in nucleosides are shown.

syn **Guanosine**   *anti* **Guanosine**   *anti* **Uridine**

10.3 What Is the Structure and Chemistry of Nucleotides?

A **nucleotide** results when phosphoric acid is esterified to a sugar —OH group of a nucleoside. The nucleoside ribose ring has three —OH groups available for esterification, at C-2′, C-3′, and C-5′ (although 2′-deoxyribose has only two). The vast majority of monomeric nucleotides in the cell are **ribonucleotides** having 5′-phosphate groups. Figure 10.13 shows the structures of the common four *ribonucleotides,* whose formal names are **adenosine 5′-monophosphate, guanosine 5′-monophosphate, cytidine 5′-monophosphate,** and **uridine 5′-monophosphate.** These compounds are more often referred to by their abbreviations: **5′-AMP, 5′-GMP, 5′-CMP,** and **5′-UMP,** or even more simply as **AMP, GMP, CMP,** and **UMP.** Nucleoside 3′-phosphates and nucleoside 2′-phosphates (3′-NMP and 2′-NMP, where N is a generic designation for "nucleoside") are uncommon, except as products of nucleic acid hydrolysis. Because the pK_a value for the first dissociation of a proton from the phosphoric acid moiety is 1.0 or less (Table 10.1), the nucleotides have acidic properties. This acidity is implicit in the other names by which these substances are known—**adenylic acid, guanylic acid, cytidylic acid,**

Human Biochemistry

Biochemistry ⑤ Now™

Adenosine: A Nucleoside with Physiological Activity

For the most part, nucleosides have no biological role other than to serve as component parts of nucleotides. Adenosine is a rare exception. In mammals, adenosine functions as an **autocoid,** or "local hormone," and as a neuromodulator. This nucleoside circulates in the bloodstream, acting locally on specific cells to influence such diverse physiological phenomena as blood vessel dilation, smooth muscle contraction, neuronal discharge, neurotransmitter release, and metabolism of fat. For example, when muscles work hard, they release adenosine, causing the surrounding blood vessels to dilate, which in turn increases the flow of blood and its delivery of O_2 and nutrients to the muscles. In a different autocoid role, adenosine acts in regulating heartbeat. The natural rhythm of the heart is controlled by a pacemaker, the sinoatrial node, which cyclically sends a wave of electrical excitation to the heart muscles. By blocking the flow of electrical current, adenosine slows the heart rate. *Supraventricular tachycardia* is a heart condition characterized by a rapid heartbeat. Intravenous injection of adenosine causes a momentary interruption of the rapid cycle of contraction and restores a normal heart

rate. Adenosine is licensed and marketed as *Adenocard* to treat supraventricular tachycardia.

In addition, adenosine is implicated in sleep regulation. During periods of extended wakefulness, extracellular adenosine levels rise as a result of metabolic activity in the brain, and this increase promotes sleepiness. During sleep, adenosine levels fall. Caffeine promotes wakefulness by blocking the interaction of extracellular adenosine with its neuronal receptors.*

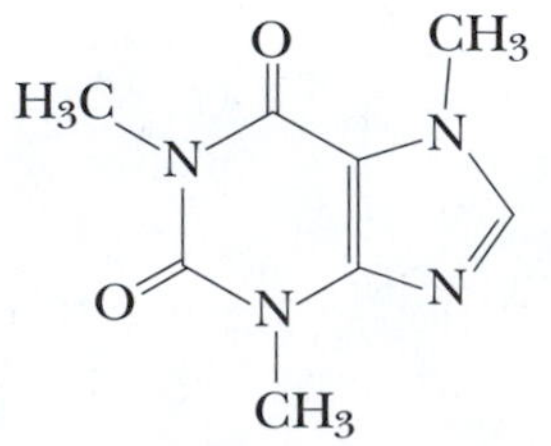

Caffeine

---

*Porrka-Heiskanen, T., et al., 1997. Adenosine: A mediator of the sleep-inducing effects of prolonged wakefulness. *Science* **276**:1265–1268; and Vaugeois, J-M., 2002. Positive feedback from caffeine. *Nature* **418**:734–726.

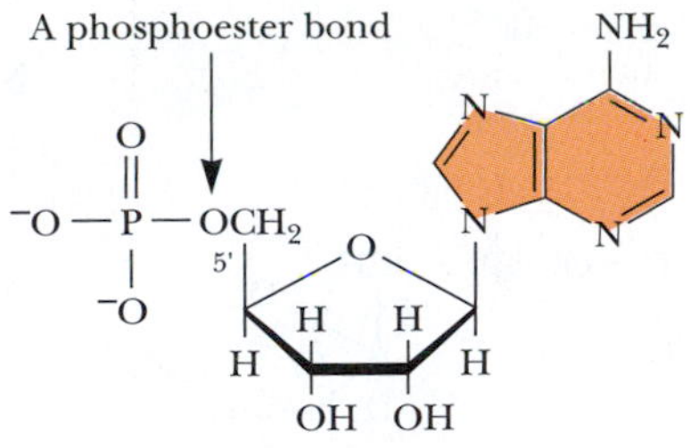

Adenosine 5'-monophosphate
(or AMP or adenylic acid)

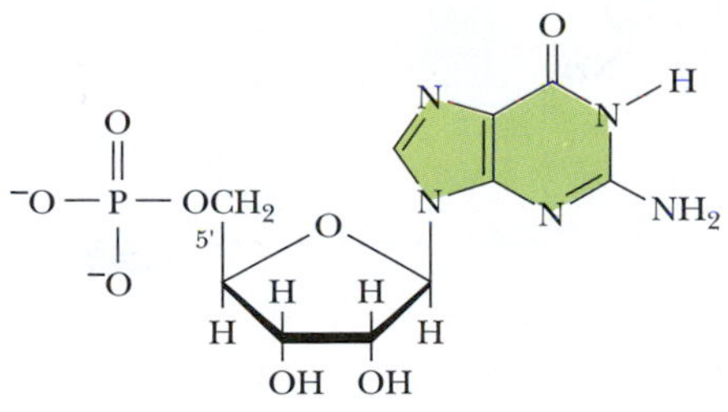

A phosphoester bond

Guanosine 5'-monophosphate
(or GMP or guanylic acid)

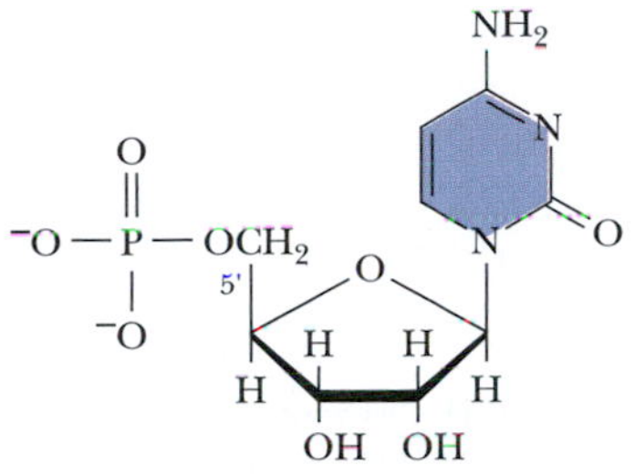

Cytidine 5'-monophosphate
(or CMP or cytidylic acid)

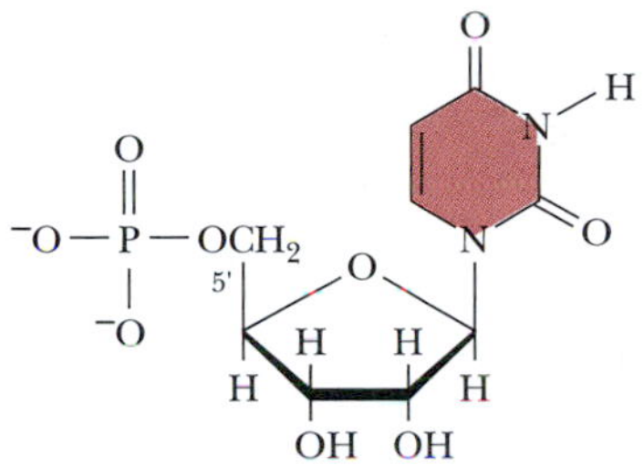

Uridine 5'-monophosphate
(or UMP or uridylic acid)

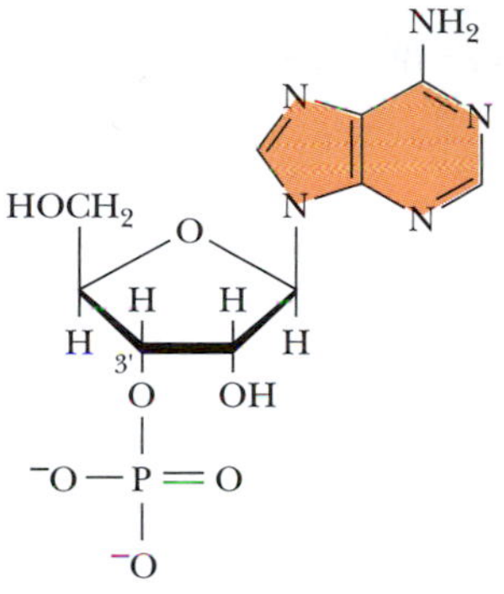

A nucleoside 3'-monophosphate
3'-AMP

FIGURE 10.13 Structures of the four common ribonucleotides—AMP, GMP, CMP, and UMP—together with their two sets of full names, for example, adenosine 5'-monophosphate and adenylic acid. Also shown is the nucleoside 3'-AMP.

and **uridylic acid.** The pK_a value for the second dissociation, pK_2, is about 6.0, so at neutral pH or above, the net charge on a nucleoside monophosphate is -2. Nucleic acids, which are polymers of nucleoside monophosphates, derive their name from the acidity of these phosphate groups.

Cyclic Nucleotides Are Cyclic Phosphodiesters

Nucleoside monophosphates in which the phosphoric acid is esterified to *two* of the available ribose hydroxyl groups (Figure 10.14) are found in all cells. Forming two such ester linkages with one phosphate results in a cyclic phosphodiester structure. **3',5'-cyclic AMP,** often abbreviated **cAMP,** and its guanine analog **3',5'-cyclic GMP,** or **cGMP,** are important regulators of cellular metabolism (see Parts 3 and 4).

Nucleoside Diphosphates and Triphosphates Are Nucleotides with Two or Three Phosphate Groups

Additional phosphate groups can be linked to the phosphoryl group of a nucleotide through the formation of phosphoric anhydride linkages, as shown in Figure 10.15. Addition of a second phosphate to AMP creates **adenosine 5'-diphosphate,** or **ADP,** and adding a third yields **adenosine 5'-triphosphate,** or **ATP.** The respective phosphate groups are designated by the Greek letters α, β, and γ, starting with the α-phosphate as the one linked directly to the pentose. The abbreviations **GTP, CTP,** and **UTP** represent the other corresponding nucleoside 5'-triphosphates. Like the nucleoside 5'-monophosphates, the nucleoside 5'-diphosphates and 5'-triphosphates all occur in the free state in the cell, as do their deoxyribonucleoside phosphate counterparts, represented as dAMP, dADP, and dATP; dGMP, dGDP, and dGTP; dCMP, dCDP, and dCTP; dUMP, dUDP, and dUTP; and dTMP, dTDP, and dTTP.

NDPs and NTPs Are Polyprotic Acids

Nucleoside 5'-diphosphates (NDPs) and **nucleoside 5'-triphosphates (NTPs)** are relatively strong *polyprotic acids* in that they dissociate three and four protons, respectively, from their phosphoric acid groups. The resulting phosphate

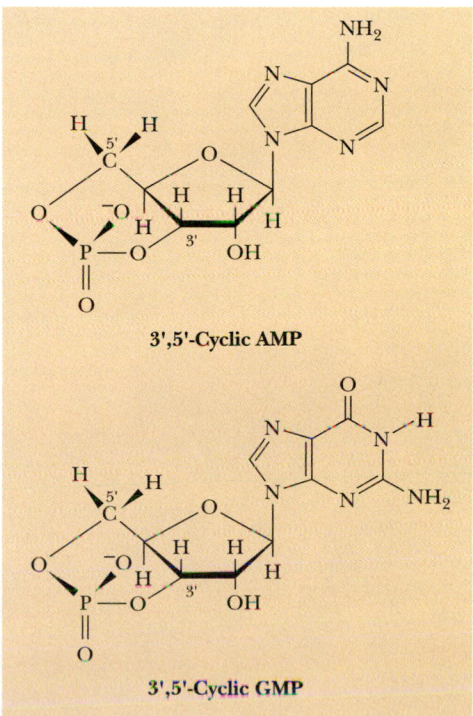

3',5'-Cyclic AMP

3',5'-Cyclic GMP

FIGURE 10.14 Structures of the cyclic nucleotides cAMP and cGMP.

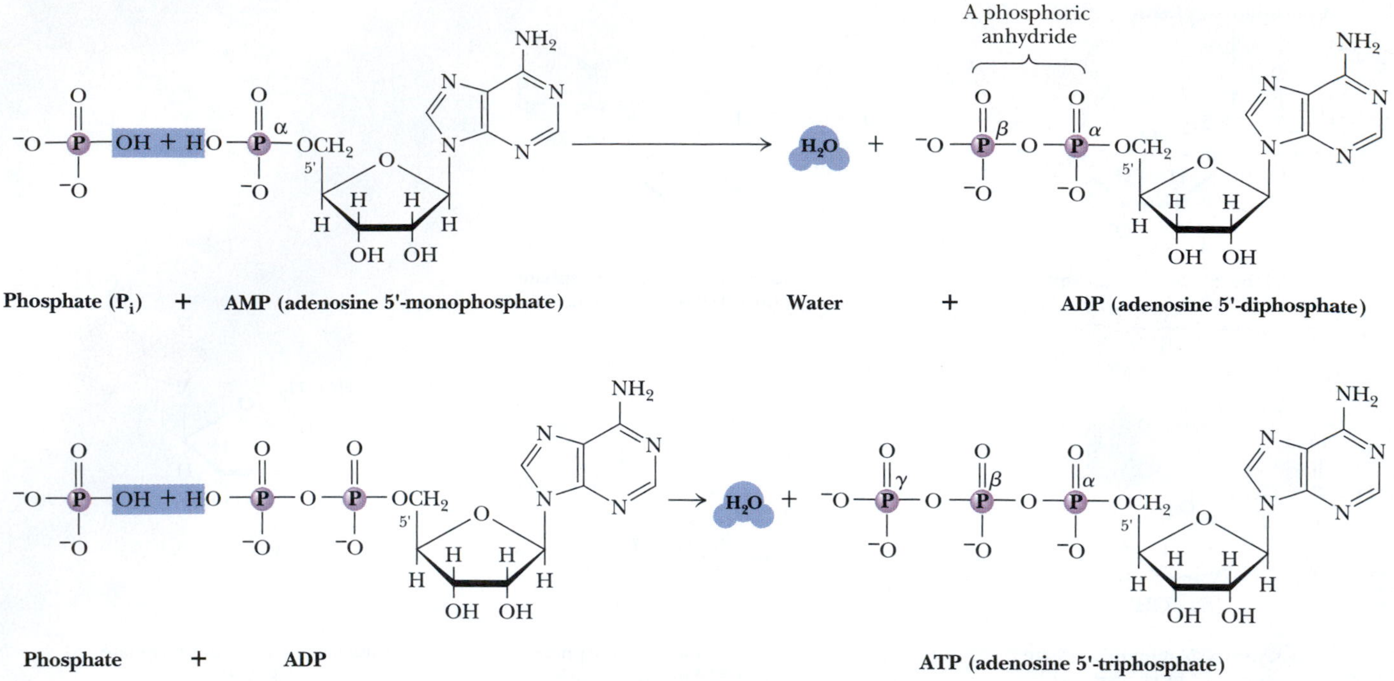

FIGURE 10.15 Formation of ADP and ATP by the successive addition of phosphate groups via phosphoric anhydride linkages. Note the removal of equivalents of H_2O in these dehydration synthesis reactions.

anions on NDPs and NTPs form stable complexes with divalent cations such as Mg^{2+} and Ca^{2+}. Because Mg^{2+} is present at high concentrations (as much as 40 mM) intracellularly, NDPs and NTPs occur primarily as Mg^{2+} complexes in the cell. The phosphoric anhydride linkages in NDPs and NTPs are readily hydrolyzed by acid, liberating inorganic phosphate (often symbolized as P_i) and the corresponding NMP. A diagnostic test for NDPs and NTPs is quantitative liberation of P_i upon treatment with 1 N HCl at 100°C for 7 minutes.

Nucleoside 5′-Triphosphates Are Carriers of Chemical Energy

Nucleoside 5′-triphosphates are indispensable agents in metabolism because the phosphoric anhydride bonds they possess are a prime source of chemical energy to do biological work. ATP has been termed the energy currency of the cell (see Chapter 3). GTP is the major energy source for protein synthesis (see Chapter 30), CTP is an essential metabolite in phospholipid synthesis (see Chapter 24), and UTP forms activated intermediates with sugars that go on to serve as substrates in the biosynthesis of complex carbohydrates and polysaccharides (see Chapter 22). The evolution of metabolism has led to the dedication of one of these four NTPs to each of these major branches of metabolism. To complete the picture, the four NTPs and their dNTP counterparts are the substrates for the synthesis of the remaining great class of biomolecules—the nucleic acids.

The Bases of Nucleotides Serve as "Information Symbols"

Are the bases of nucleotides directly involved in the biochemistry of metabolism? Not really. Virtually all of the biochemical reactions of nucleotides involve either *phosphate* or *pyrophosphate group transfer:* the release of a phosphoryl group from an NTP to give an NDP, the release of a pyrophosphoryl group to give an NMP unit, or the acceptance of a phosphoryl group by an NMP or an NDP to give an NDP or an NTP (Figure 10.16). Interestingly, the pentose and the base are *not* directly involved in this chemistry. However, as noted, a "division of labor" directs ATP to serve as the primary nucleotide in central pathways of energy metabolism, while GTP, for example, is used to drive protein synthesis. Thus, the various nucleotides are channeled in appropriate metabolic

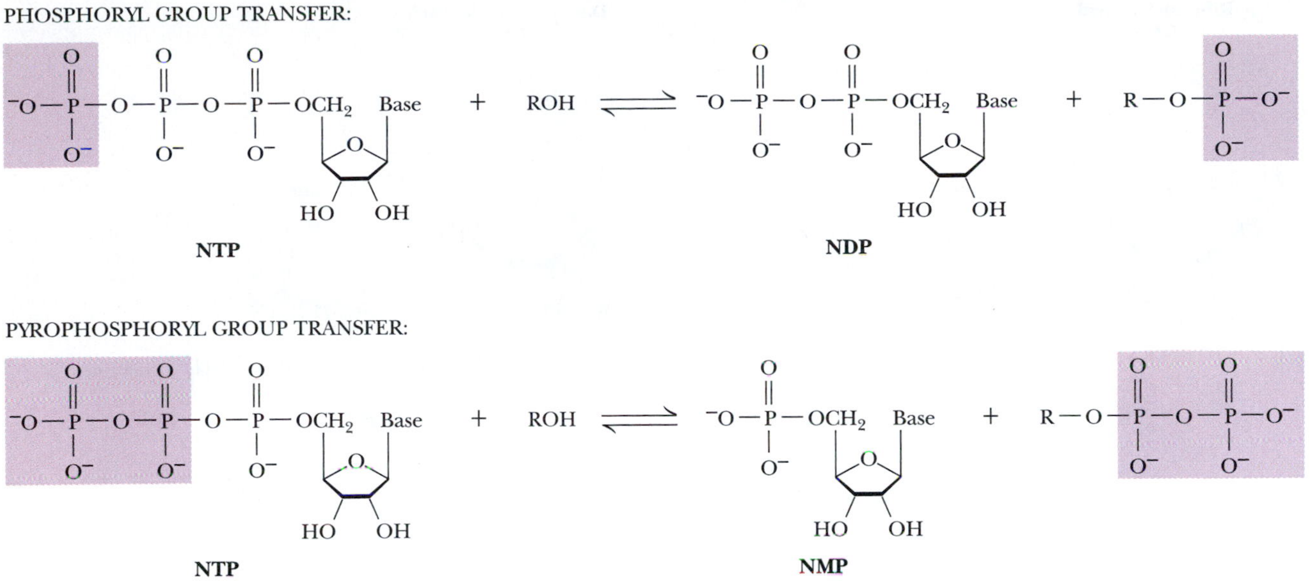

PHOSPHORYL GROUP TRANSFER:

PYROPHOSPHORYL GROUP TRANSFER:

FIGURE 10.16 Phosphoryl and pyrophosphoryl group transfer, the major biochemical reactions of nucleotides.

directions through specific recognition of the base of the nucleotide. That is, the bases of nucleotides serve as *information symbols,* never participating directly in the covalent bond chemistry that goes on. This role as information symbols extends to nucleotide polymers, the nucleic acids, where the bases serve as the information symbols for the code of genetic information.

10.4	**What Are Nucleic Acids?**

Nucleic acids are **polynucleotides:** linear polymers of nucleotides linked 3′ to 5′ by **phosphodiester bridges** (Figure 10.17). They are formed as 5′-nucleoside monophosphates are successively added to the 3′-OH group of the preceding nucleotide, a process that gives the polymer a directional sense. Polymers of ribonucleotides are named **ribonucleic acid,** or **RNA.** Deoxyribonucleotide polymers are called **deoxyribonucleic acid,** or **DNA.** Because C-1′ and C-4′ in deoxyribonucleotides are involved in furanose ring formation and because there is no 2′-OH, only the 3′- and 5′-hydroxyl groups are available for internucleotide phosphodiester bonds. In the case of DNA, a polynucleotide chain may contain hundreds of millions of nucleotide units. *The convention in all notations of nucleic acid structure is to read the polynucleotide chain from the 5′-end of the polymer to the 3′-end.* Note that this reading direction actually passes through each phosphodiester from 3′ to 5′ (Figure 10.18). A repetitious uniformity exists in the covalent backbone of polynucleotides.

The Base Sequence of a Nucleic Acid Is Its Distinctive Characteristic

The only significant variation that commonly occurs in the chemical structure of nucleic acids is the nature of the base at each nucleotide position. These bases are not part of the sugar–phosphate backbone but instead serve as distinctive side chains, much like the R groups of amino acids along a polypeptide backbone. They give the polymer its unique identity. A simple notation for nucleic acid structures is merely to list the order of bases in the polynucleotide using single capital letters—A, G, C, and U (or T). Occasionally, a lowercase "p" is written between each successive base to indicate the phosphodiester bridge, as in GpApCpGpUpA. A "p" preceding the sequence indicates that the nucleic acid carries a PO₄ on its 5′-end, as in pGpApCpGpUpA; a "p" terminating the sequence connotes the presence of a phosphate on the 3′-OH end, as in GpApCpGpUpApAp.

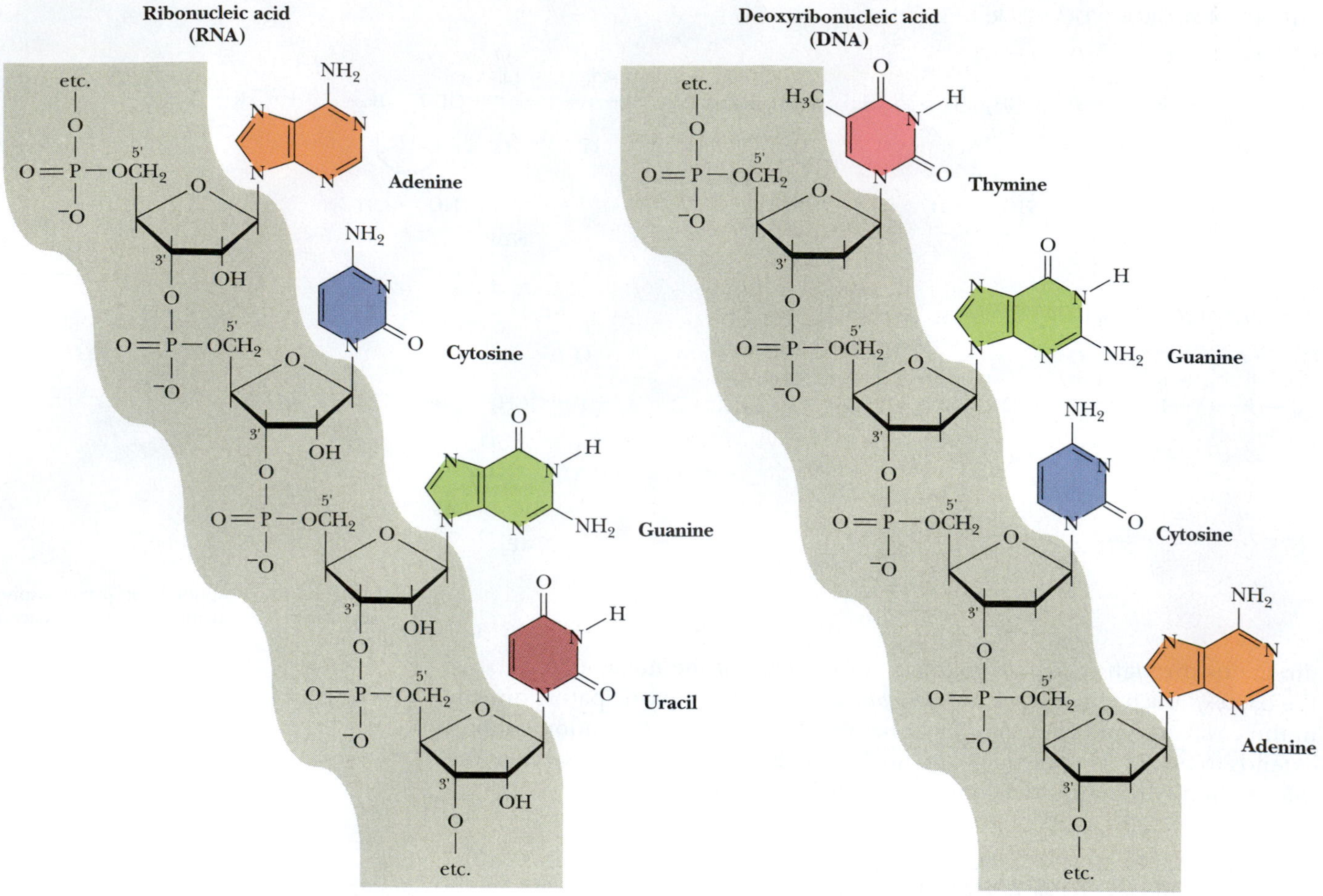

**Ribonucleic acid
(RNA)**

Adenine

Cytosine

Guanine

Uracil

**Deoxyribonucleic acid
(DNA)**

Thymine

Guanine

Cytosine

Adenine

FIGURE 10.17 3′, 5′-phosphodiester bridges link nucleotides together to form polynucleotide chains.

A more common method of representing nucleotide sequences is to omit the "p" and write only the order of bases, such as GACGUA. This notation assumes the presence of the phosphodiesters joining adjacent nucleotides. The presence of 3′- or 5′-phosphate at the termini may be specified, as in GACGUAp for a 3′-PO₄ terminus. To distinguish between RNA and DNA sequences, DNA sequences may be preceded by a lowercase "d" to denote deoxy, as in d-GACGTA. From a simple string of letters such as this, any biochemistry student should be able to draw the unique chemical structure for a pentanucleotide, even though it may contain more than 200 atoms.

10.5	**What Are the Different Classes of Nucleic Acids?**

The two major classes of nucleic acids are DNA and RNA. DNA has only one biological role, but it is the more central one. The information to make all the functional macromolecules of the cell (even DNA itself) is preserved in DNA and accessed through transcription of the information into RNA copies. Coincident

FIGURE 10.18 Shorthand notations for polynucleotide structures: By convention, the "sense," or direction, of polynucleotide chains is defined as 5′→3′. That is, the sugar–phosphate backbone is read running from 5′ to 3′ along the atoms of one furanose and thence across the phosphodiester bridge to the 5′-carbon in the furanose of the next nucleotide in line. In a convenient shorthand notation, this backbone can be diagrammed as a series of vertical lines (representing the furanoses) and slashes (representing the phosphodiester links), as shown. Each diagonal slash runs from the middle of a furanose line to the bottom of an adjacent one to indicate the 3′- (middle) to 5′- (bottom) carbons of neighboring furanoses joined by the phosphodiester bridge. The base attached to each furanose is indicated above it in one-letter designation: A, C, G, or U (or T).

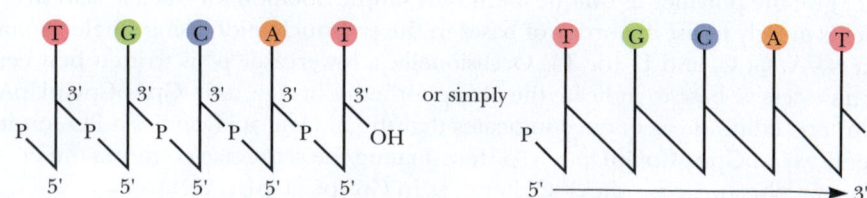

with its singular purpose, there is only a single DNA molecule (or "chromosome") in simple life forms such as viruses or bacteria. Such DNA molecules must be quite large in order to embrace enough information for making the macromolecules necessary to maintain a living cell. The *Escherichia coli* chromosome has a molecular mass of 2.9×10^9 D and contains more than 9 million nucleotides. Eukaryotic cells have many chromosomes, and DNA is found principally in two copies in the diploid chromosomes of the nucleus, but it also occurs in mitochondria and in chloroplasts, where it encodes some of the proteins and RNAs unique to these organelles.

In contrast, RNA occurs in multiple copies and various forms (Table 10.2). Cells typically contain about eight times as much RNA as DNA. RNA has a number of important biological functions, its central one being information transfer from DNA to protein. RNA molecules playing this role are categorized into several major types: **messenger RNA, ribosomal RNA,** and **transfer RNA.** Eukaryotic cells contain an additional type: **small nuclear RNA (snRNA).** Beyond its role in information transfer, RNA participates in a number of metabolic functions, including (1) processing and modification of tRNA, rRNA, and mRNA; (2) regulation of gene expression; and (3) several maintenance or "housekeeping" functions, such as preservation of telomeres.

With these basic definitions in mind, let's now briefly consider the chemical and structural nature of DNA and the various RNAs. Chapter 11 elaborates on methods to determine the primary structure of nucleic acids by sequencing methods and discusses the secondary and tertiary structures of DNA and RNA. Part 4, Information Transfer, includes a detailed treatment of the dynamic role of nucleic acids in the molecular biology of the cell.

Telomeres are specialized nucleotide sequences at the ends of chromosomes.

The Fundamental Structure of DNA Is a Double Helix

The DNA isolated from different cells and viruses characteristically consists of two polynucleotide strands wound together to form a long, slender, helical molecule, the **DNA double helix.** The strands run in opposite directions; that is, they are *antiparallel.* The two strands are held together in the double helical structure through *interchain hydrogen bonds* (Figure 10.19). These H bonds pair the bases of nucleotides in one chain to complementary bases in the other, a phenomenon called **base pairing.**

Erwin Chargaff's Analysis of the Base Composition of Different DNAs Provided a Key Clue to DNA Structure   A clue to the chemical basis of base pairing in DNA came from the analysis of the base composition of various DNAs by Erwin Chargaff in the late 1940s. His data showed that the four bases commonly found in DNA (A, C, G, and T) do not occur in equimolar amounts and that the relative amounts of each vary from species to species (Table 10.3). Nevertheless, Chargaff noted that certain pairs of bases, namely, adenine and thymine, and guanine and cytosine, are always found in a 1:1 ratio and that the number of pyrimidine residues always

Table 10.2				
Principle Kinds of RNA Found in an *E. coli* Cell				
Type	Sedimentation Coefficient	Molecular Weight	Number of Nucleotide Residues	Percentage of Total Cell RNA
mRNA	6–25	25,000–1,000,000	75–3,000	~2
tRNA	~4	23,000–30,000	73–94	16
rRNA	5	35,000	120 ⎫	
	16	550,000	1,542 ⎬	82
	23	1,100,000	2,904 ⎭	

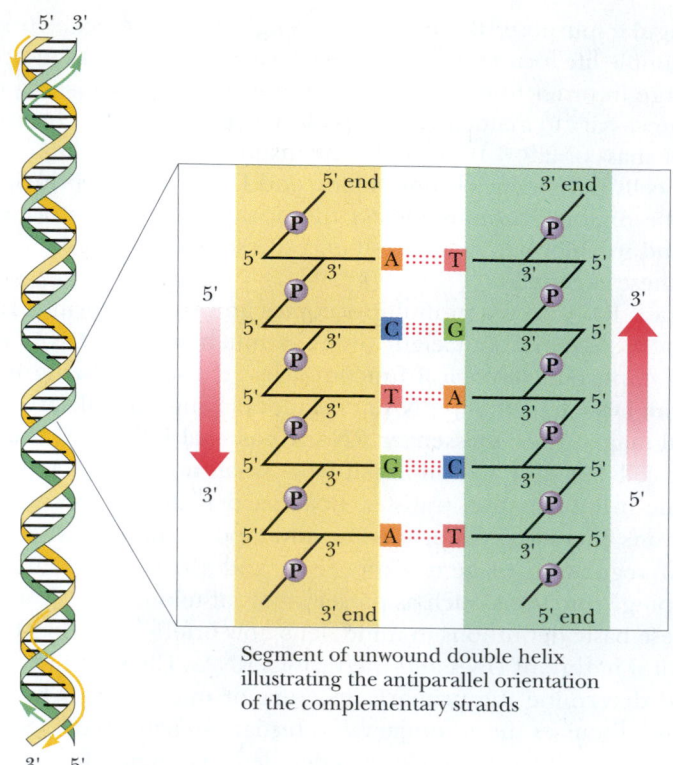

Segment of unwound double helix
illustrating the antiparallel orientation
of the complementary strands

FIGURE 10.19 The antiparallel nature of the DNA double helix.

equals the number of purine residues. These findings are known as *Chargaff's rules:* **[A] = [T]; [C] = [G]; [pyrimidines] = [purines].**

Watson and Crick's Postulate of the DNA Double Helix Became the Icon of DNA Structure   James Watson and Francis Crick, working in the Cavendish Laboratory at Cambridge University in 1953, took advantage of Chargaff's results and the data obtained by Rosalind Franklin and Maurice Wilkins in X-ray diffraction studies on the structure of DNA to conclude that DNA was a *complementary double helix*. Two strands of deoxyribonucleic acid (sometimes referred to as the *Watson strand* and the *Crick strand*) are held together by the bonding interac-

Table 10.3					
Molar Ratios Leading to the Formulation of Chargaff's Rules					
Source	**Adenine to Guanine**	**Thymine to Cytosine**	**Adenine to Thymine**	**Guanine to Cytosine**	**Purines to Pyrimidines**
Ox	1.29	1.43	1.04	1.00	1.1
Human	1.56	1.75	1.00	1.00	1.0
Hen	1.45	1.29	1.06	0.91	0.99
Salmon	1.43	1.43	1.02	1.02	1.02
Wheat	1.22	1.18	1.00	0.97	0.99
Yeast	1.67	1.92	1.03	1.20	1.0
Haemophilus influenzae	1.74	1.54	1.07	0.91	1.0
E. coli K-12	1.05	0.95	1.09	0.99	1.0
Avian tubercle bacillus	0.4	0.4	1.09	1.08	1.1
Serratia marcescens	0.7	0.7	0.95	0.86	0.9
Bacillus schatz	0.7	0.6	1.12	0.89	1.0

Source: After Chargaff, E., 1951. Structure and function of nucleic acids as cell constituents. *Federation Proceedings* **10**:654–659.

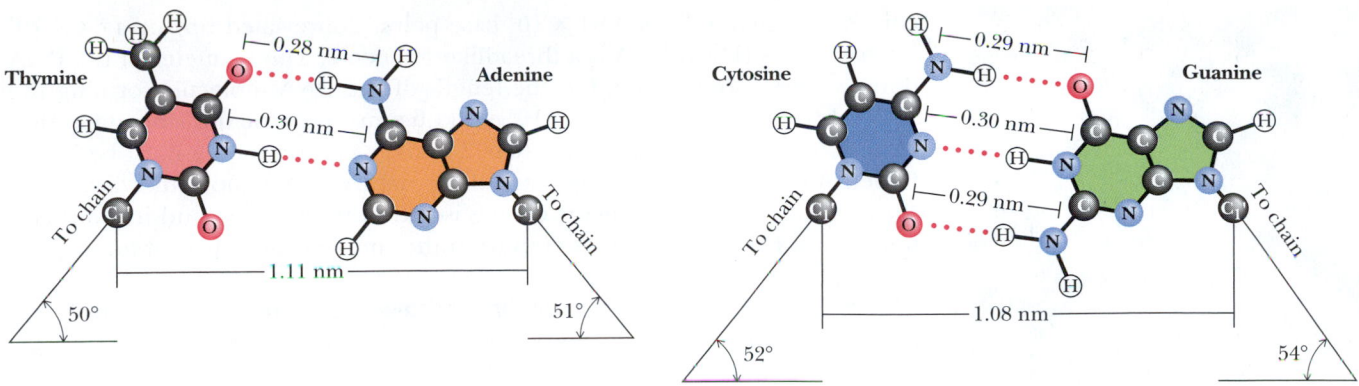

FIGURE 10.20 The Watson–Crick base pairs A:T and G:C.

tions between unique base pairs, always consisting of a purine in one strand and a pyrimidine in the other. Base pairing is very specific: If the purine is adenine, the pyrimidine must be thymine. Similarly, guanine pairs only with cytosine (Figure 10.20). Thus, if an A occurs in one strand of the helix, T must occupy the complementary position in the opposing strand. Likewise, a G in one dictates a C in the other. Because exceptions to this exclusive pairing of A only with T and G only with C are rare, these pairs are taken as the standard or accepted law, and the A:T and G:C base pairs are often referred to as canonical. As Watson recognized from testing various combinations of bases using structurally accurate models, the A:T pair and the G:C pair form spatially equivalent units (Figure 10.20). The backbone-to-backbone distance of an A:T pair is 1.11 nm, virtually identical to the 1.08 nm chain separation in G:C base pairs.

The DNA molecule not only conforms to Chargaff's rules but also has a profound property relating to heredity: *The sequence of bases in one strand has a complementary relationship to the sequence of bases in the other strand.* That is, the information contained in the sequence of one strand is conserved in the sequence of the other. Therefore, separation of the two strands and faithful replication of each, through a process in which base pairing specifies the nucleotide sequence in the newly synthesized strand, leads to two progeny molecules identical in every respect to the parental double helix (Figure 10.21). Elucidation of the double helical structure of DNA represented one of the most significant events in the history of science. This discovery more than any other marked the beginning of molecular biology. Indeed, upon solving the structure of DNA, Crick proclaimed in The Eagle, a pub just across from the Cavendish lab, "We have discovered the secret of life!"

The Information in DNA Is Encoded in Digital Form In this digital age, we are accustomed to electronic information encoded in the form of extremely long arrays of just two digits: ones (1s) and zeros (0s). DNA uses four digits to encode biological information: A, C, G, and T. A significant feature of the DNA double helix is that virtually any base sequence (encoded information) is possible: Other than the base-pairing rules, no structural constraints operate to limit the potential sequence of bases in DNA.

DNA contains two kinds of information:

1. The base sequences of genes that encode the amino acid sequences of proteins and the nucleotide sequences of functional RNA molecules such as rRNA and tRNA (see following discussion)
2. The gene regulatory networks that control the expression of protein-encoding (and functional RNA-encoding) genes (see Chapter 29)

DNA Is in the Form of Enormously Long, Threadlike Molecules Because of the double helical nature of DNA molecules, their size can be represented in terms of the numbers of nucleotide base pairs they contain. For example, the *E. coli*

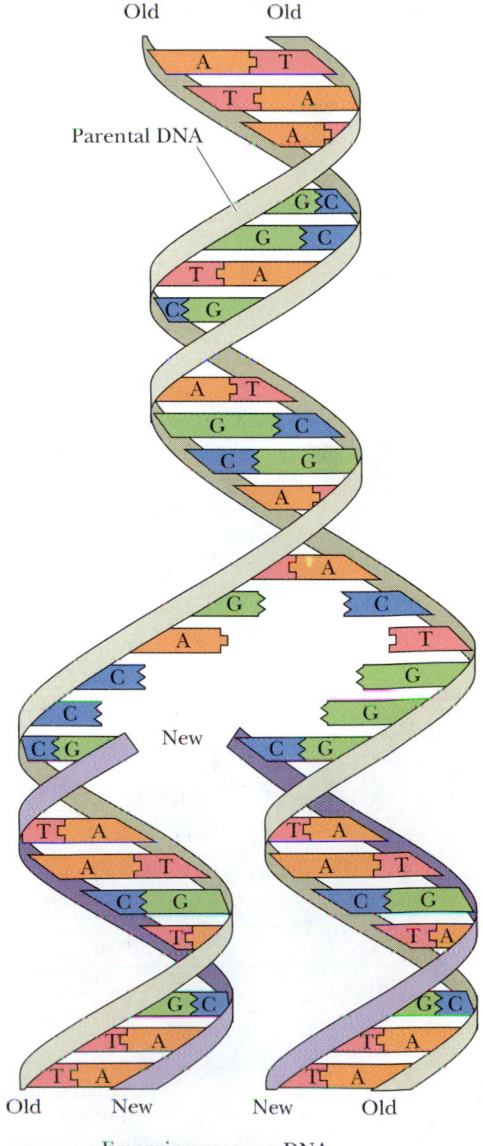

FIGURE 10.21 Replication of DNA gives identical progeny molecules because base pairing is the mechanism determining the nucleotide sequence synthesized within each of the new strands during replication.

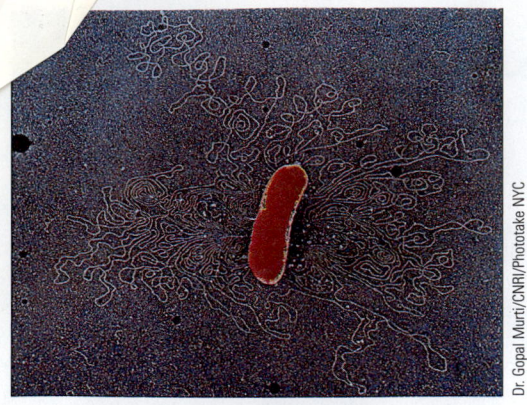

Dr. Gopal Murti/CNRI/Phototake NYC

FIGURE 10.22 If the cell walls of bacteria such as *Escherichia coli* are partially digested and the cells are then osmotically shocked by dilution with water, the contents of the cells are extruded to the exterior. In electron micrographs, the most obvious extruded component is the bacterial chromosome, shown here surrounding the cell.

Histone "core" octamer
(here shown in cross section)

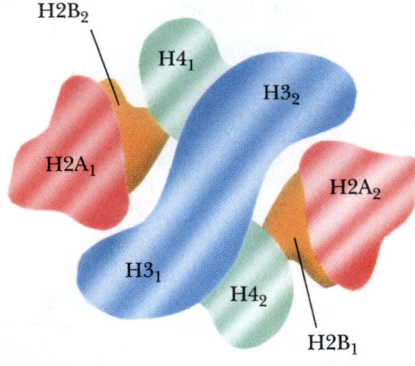

Nucleosome

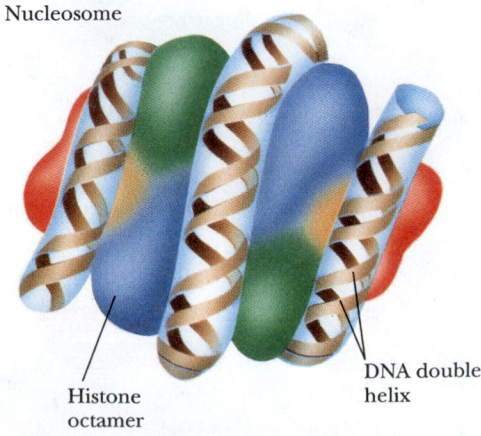

FIGURE 10.23 A diagram of the histone octamer. Nucleosomes consist of two turns of DNA super-coiled about a histone "core" octamer.

chromosome consists of 4.64×10^6 base pairs (abbreviated bp) or 4.64×10^3 kilobase pairs (kbp). DNA is a threadlike molecule. The diameter of the DNA double helix is only 2 nm, but the length of the DNA molecule forming the *E. coli* chromosome is over 1.6×10^6 nm (1.6 mm). Because the long dimension of an *E. coli* cell is only 2000 nm (0.002 mm), its chromosome must be highly folded. Because of their long, threadlike nature, DNA molecules are easily sheared into shorter fragments during isolation procedures, and it is difficult to obtain intact chromosomes even from the simple cells of prokaryotes.

DNA in Cells Occurs in the Form of Chromosomes DNA occurs in various forms in different cells. The single chromosome of prokaryotic cells (Figure 10.22) is typically a circular DNA molecule. Proteins are associated with prokaryotic DNA, but unlike eukaryotic chromosomes, prokaryotic chromosomes are not uniformly organized into ordered nucleoprotein arrays. The DNA molecules of eukaryotic cells, each of which defines a chromosome, are linear and richly adorned with proteins. A class of arginine- and lysine-rich basic proteins called **histones** interact ionically with the anionic phosphate groups in the DNA backbone to form **nucleosomes,** structures in which the DNA double helix is wound around a protein "core" composed of pairs of four different histone polypeptides (Figure 10.23; see also Section 11.5 in Chapter 11). Chromosomes also contain a varying mixture of other proteins, so-called **nonhistone chromosomal proteins,** many of which are involved in regulating which genes in DNA are transcribed at any given moment. The amount of DNA in a diploid mammalian cell is typically more than 1000 times that found in an *E. coli* cell. Some higher plant cells contain more than 50,000 times as much.

Various Forms of RNA Serve Different Roles in Cells

Messenger RNA Carries the Sequence Information for Synthesis of a Protein
Messenger RNA (**mRNA**) serves to carry the information or "message" that is encoded in genes to the sites of protein synthesis in the cell, where this information is translated into a polypeptide sequence. Because mRNA molecules are transcribed copies of the protein-coding genetic units that comprise most of DNA, mRNA is said to be "the DNA-like RNA."

Messenger RNA is synthesized during transcription, an enzymatic process in which an RNA copy is made of the sequence of bases along one strand of DNA. This mRNA then directs the synthesis of a polypeptide chain as the information that is contained within its nucleotide sequence is translated into an amino acid sequence by the protein-synthesizing machinery of the ribosomes. Ribosomal RNA and tRNA molecules are also synthesized by transcription of DNA sequences, but unlike mRNA molecules, these RNAs are not subsequently translated to form proteins. In prokaryotes, a single mRNA may contain the information for the synthesis of several polypeptide chains within its nucleotide sequence (Figure 10.24). In contrast, eukaryotic mRNAs encode only one polypeptide but are more complex in that they are synthesized in the nucleus in the form of much larger precursor molecules called **heterogeneous nuclear RNA,** or **hnRNA.** hnRNA molecules contain stretches of nucleotide sequence that have no protein-coding capacity. These noncoding regions are called **intervening sequences** or **introns** because they intervene between coding regions, which are called **exons.** Introns interrupt the continuity of the information specifying the amino acid sequence of a protein and must be spliced out before the message can be translated. In addition, eukaryotic hnRNA and mRNA molecules have a run of 100 to 200 adenylic acid residues attached at their 3′-ends, so-called **poly(A) tails.** This polyadenylylation occurs after transcription has been completed and is essential for efficient translation and stability of the mRNA. The properties of mRNA molecules as they move through transcription and translation in prokaryotic versus eukaryotic cells are summarized in Figure 10.24.

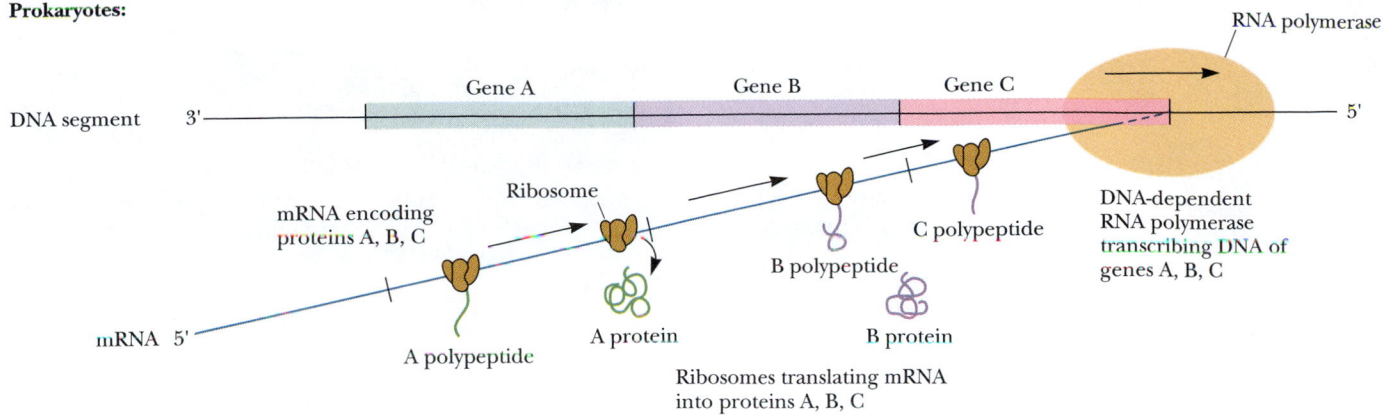

Prokaryotes:

RNA polymerase

DNA segment 3'— Gene A | Gene B | Gene C —5'

DNA-dependent RNA polymerase transcribing DNA of genes A, B, C

mRNA encoding proteins A, B, C

Ribosome

C polypeptide

B polypeptide

mRNA 5'

A polypeptide

A protein

B protein

Ribosomes translating mRNA into proteins A, B, C

Eukaryotes:

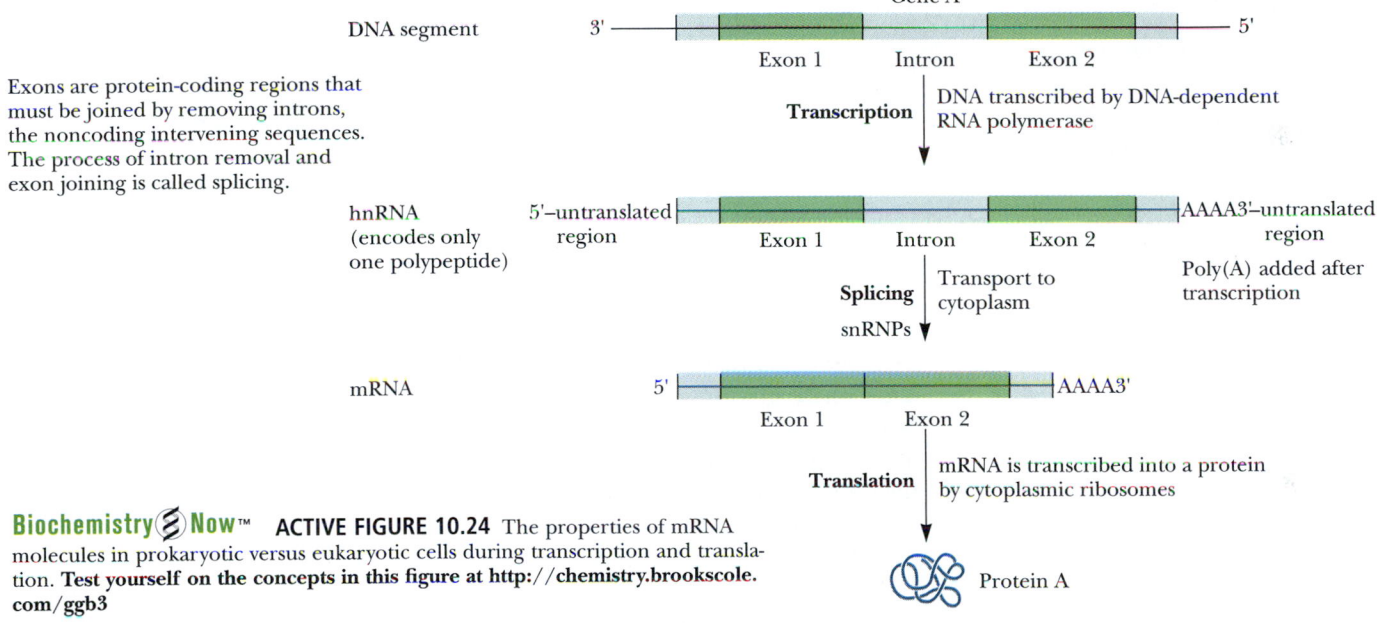

Exons are protein-coding regions that must be joined by removing introns, the noncoding intervening sequences. The process of intron removal and exon joining is called splicing.

Gene A

DNA segment 3'———5'

Exon 1 Intron Exon 2

Transcription — DNA transcribed by DNA-dependent RNA polymerase

hnRNA (encodes only one polypeptide)

5'–untranslated region — Exon 1 Intron Exon 2 — AAAA3'–untranslated region

Poly(A) added after transcription

Splicing — Transport to cytoplasm

snRNPs

mRNA 5' — Exon 1 Exon 2 — AAAA3'

Translation — mRNA is transcribed into a protein by cytoplasmic ribosomes

Protein A

Biochemistry Now™ ACTIVE FIGURE 10.24 The properties of mRNA molecules in prokaryotic versus eukaryotic cells during transcription and translation. **Test yourself on the concepts in this figure at http://chemistry.brookscole. com/ggb3**

Ribosomal RNA Provides the Structural and Functional Foundation for Ribosomes

Ribosomes, the supramolecular assemblies where protein synthesis occurs, are about 65% RNA of the ribosomal RNA type. Ribosomal RNA **(rRNA)** molecules fold into characteristic secondary structures as a consequence of intramolecular base-pairing interactions (marginal figure). The different species of rRNA are generally referred to according to their **sedimentation coefficients**[2] (see the Appendix to Chapter 5), which are a rough measure of their relative size (Table 10.2 and Figure 10.25).

Ribosomes are composed of two subunits of different sizes that dissociate from each other if the Mg^{2+} concentration is below 10^{-3} M. Each subunit is a supramolecular assembly of proteins and RNA and has a total mass of 10^6 D or more. *E. coli* ribosomal subunits have sedimentation coefficients of 30S (the small subunit) and 50S (the large subunit). Eukaryotic ribosomes are somewhat larger than prokaryotic ribosomes, consisting of 40S and 60S subunits. The properties of ribosomes and their rRNAs are summarized in Figure 10.25.

[2]Sedimentation coefficients are a measure of the velocity with which a particle sediments in a centrifugal force field. Sedimentation coefficients are typically expressed in **Svedbergs** (symbolized S), named to honor The Svedberg, developer of the ultracentrifuge. One S equals 10^{-13} sec.

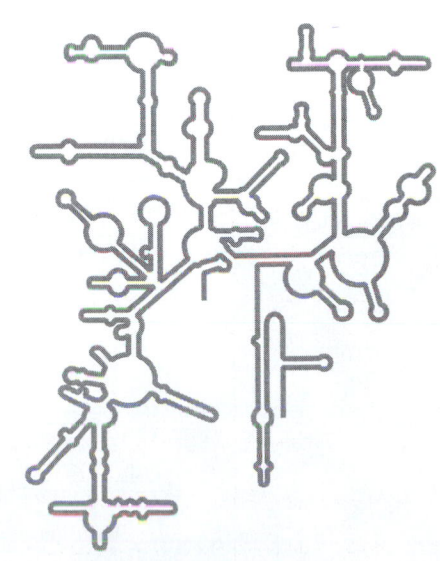

Ribosomal RNA has a complex secondary structure due to many intrastrand hydrogen bonds.

PROKARYOTIC RIBOSOMES
(*E. coli*)

EUKARYOTIC RIBOSOMES
(Rat)

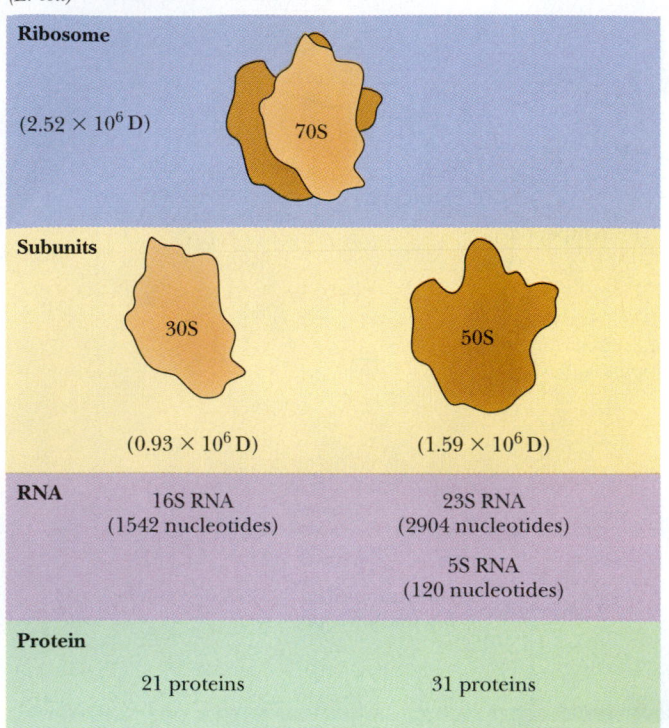

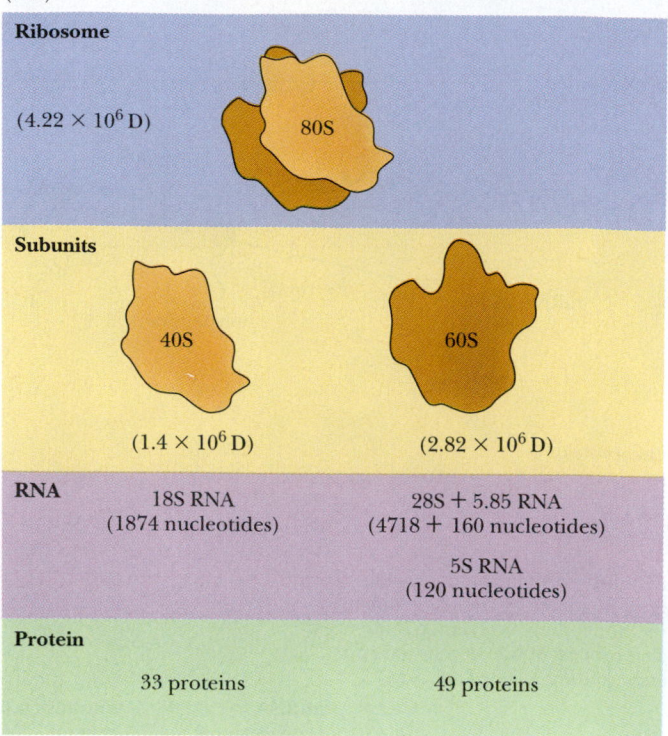

	PROKARYOTIC RIBOSOMES (*E. coli*)		EUKARYOTIC RIBOSOMES (Rat)	
Ribosome	$(2.52 \times 10^6 \text{ D})$ 70S		$(4.22 \times 10^6 \text{ D})$ 80S	
Subunits	30S $(0.93 \times 10^6 \text{ D})$	50S $(1.59 \times 10^6 \text{ D})$	40S $(1.4 \times 10^6 \text{ D})$	60S $(2.82 \times 10^6 \text{ D})$
RNA	16S RNA (1542 nucleotides)	23S RNA (2904 nucleotides); 5S RNA (120 nucleotides)	18S RNA (1874 nucleotides)	28S + 5.85 RNA (4718 + 160 nucleotides); 5S RNA (120 nucleotides)
Protein	21 proteins	31 proteins	33 proteins	49 proteins

FIGURE 10.25 The organization and composition of prokaryotic and eukaryotic ribosomes.

The 30S subunit of *E. coli* contains a single RNA chain of 1542 nucleotides. This small subunit rRNA itself has a sedimentation coefficient of 16S. The large *E. coli* subunit has two rRNA molecules: a 23S (2904 nucleotides) and a 5S (120 nucleotides). The ribosomes of a representative eukaryote, the rat, have rRNA molecules of 18S (1874 nucleotides) and 28S (4718 bases), 5.8S

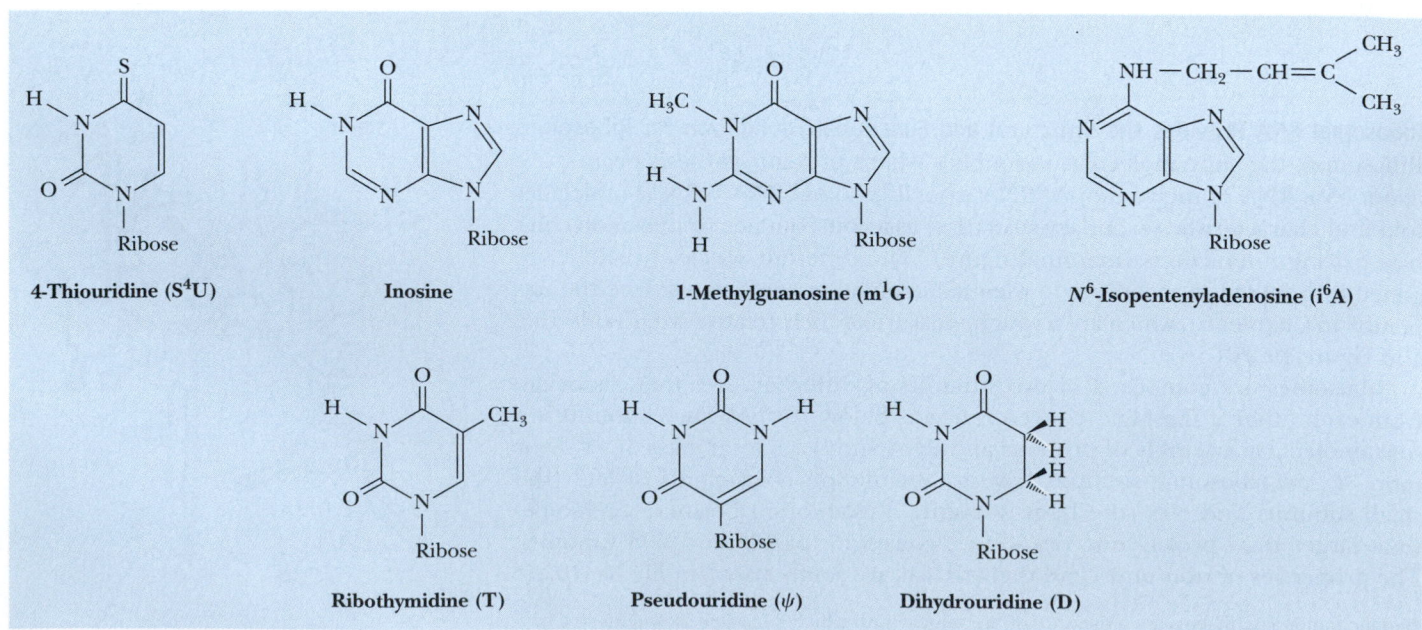

4-Thiouridine (S⁴U)   **Inosine**   **1-Methylguanosine (m¹G)**   **N⁶-Isopentenyladenosine (i⁶A)**

Ribothymidine (T)   **Pseudouridine (ψ)**   **Dihydrouridine (D)**

FIGURE 10.26 Unusual bases of RNA—pseudouridine, ribothymidylic acid, and various methylated bases.

(160 bases), and 5S (120 bases). The 18S rRNA is in the 40S subunit, and the latter three are all part of the 60S subunit.

Ribosomal RNAs characteristically contain a number of specially modified nucleotides, including **pseudouridine** residues, **ribothymidylic acid,** and **methylated bases** (Figure 10.26). The central role of ribosomes in the biosynthesis of proteins is treated in detail in Chapter 30. Here we briefly note the significant point that genetic information in the nucleotide sequence of an mRNA is translated into the amino acid sequence of a polypeptide chain by ribosomes.

Transfer RNAs Carry Amino Acids to Ribosomes for Use in Protein Synthesis

Transfer RNAs (**tRNAs**) serve as the carrier of amino acids for protein synthesis (see Chapter 30). tRNA molecules also fold into a characteristic secondary structure (marginal figure). tRNAs are small RNA molecules (size range, 23 to 30 kD), containing 73 to 94 residues, a substantial number of which are methylated or otherwise unusually modified. Each of the 20 amino acids in proteins has at least one unique tRNA species dedicated to chauffeuring its delivery to ribosomes for insertion into growing polypeptide chains, and some amino acids are served by several tRNAs. In eukaryotes, there are even discrete sets of tRNA molecules for each site of protein synthesis—the cytoplasm, the mitochondrion, and in plant cells, the chloroplast. All tRNA molecules possess a 3'-terminal nucleotide sequence that reads -CCA, and the amino acid is carried to the ribosome attached as an acyl ester to the free 3'-OH of the terminal A residue. These **aminoacyl-tRNAs** are the substrates of protein synthesis, the amino acid being transferred to the carboxyl end of a growing polypeptide. The peptide bond–forming reaction is a catalytic process intrinsic to ribosomes.

Small Nuclear RNAs Mediate the Splicing of Eukaryotic Gene Transcripts (hnRNA) into mRNA

Small nuclear RNAs, or snRNAs, are a class of RNA molecules found in eukaryotic cells, principally in the nucleus. They are neither tRNA nor small rRNA molecules, although they are similar in size to these species. They contain from 100 to about 200 nucleotides, some of which, like tRNA and rRNA, are methylated or otherwise modified. No snRNA exists as naked RNA. Instead, snRNA is found in stable complexes with specific proteins forming **small nuclear ribonucleoprotein particles,** or **snRNPs,** which are about 10S in size. Their occurrence in eukaryotes, their location in the nucleus, and their relative abundance (1% to 10% of the number of ribosomes) are significant clues to their biological purpose: snRNPs are important in the processing of eukaryotic gene transcripts (hnRNA) into mature messenger RNA for export from the nucleus to the cytoplasm (Figure 10.24).

Small RNAs Serve a Number of Roles, Including Post-Transcriptional Gene Silencing

Recently, a new class of RNA molecules even smaller than tRNAs has been recognized, the **small RNAs,** so-called because they are only 21 to 28 nucleotides long. (Some have referred to this new class as the **noncoding RNAs** [or **ncRNAs**].) Small RNAs are involved in a number of novel biological functions. These small RNAs can target DNA or RNA through complementary base pairing. Base pairing of the small RNA with particular nucleotide sequences in the target is called *direct readout.*

Small RNAs are classified into a number of subclasses on the basis of their function. The recently discovered phenomenon of **RNA interference (RNAi)** is mediated by one subclass: the **small interfering RNAs (siRNAs).** siRNAs disrupt gene expression by blocking specific protein production, even though the mRNA encoding the protein has been synthesized. The 21- to 23-nucleotide-long siRNAs act by base pairing with complementary sequences within a particular mRNA to form regions of double-stranded RNA (dsRNA). These dsRNA regions are then specifically degraded, eliminating the mRNA from the cell. Thus, RNAi is a mechanism to silence the expression of

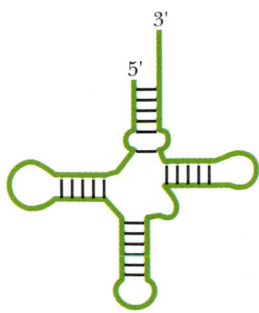

Transfer RNA also has a complex secondary structure due to many intrastrand hydrogen bonds.

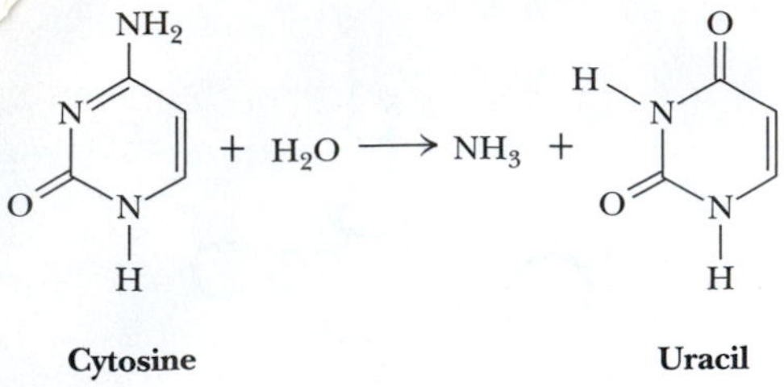

FIGURE 10.27 Deamination of cytosine forms uracil.

specific genes, even after they have been transcribed, a phenomenon referred to as **post-transcriptional gene silencing.** RNAi is also implicated in modifying the structure of chromatin and causing large-scale influences in gene expression. Another subclass, the **micro RNAs (miRNAs)** (also known as **small temporal RNAs [stRNAs]**), control developmental timing by base pairing with and preventing the translation of certain mRNAs, thus blocking protein production. However, unlike siRNAs, stRNAs (22 nucleotides long) do not cause mRNA degradation. A third subclass is the **small nucleolar RNAs (snoRNAs).** snoRNAs (60 to 300 nucleotides long) are catalysts that accomplish some of the chemical modifications found in tRNA, rRNA, and even DNA (see Figure 10.26, for example). *Small RNAs* in bacteria (known by the acronym **sRNAs**) play an important role altering gene expression in response to stressful environmental situations.

The Chemical Differences Between DNA and RNA Have Biological Significance

Two fundamental chemical differences distinguish DNA from RNA:

1. DNA contains 2-deoxyribose instead of ribose.
2. DNA contains thymine instead of uracil.

What are the consequences of these differences, and do they hold any significance in common? An argument can be made that, because of these differences, DNA is chemically more stable than RNA. The greater stability of DNA over RNA is consistent with the respective roles these macromolecules have assumed in heredity and information transfer.

Consider first why DNA contains thymine instead of uracil. The key observation is that *cytosine deaminates to form uracil* at a finite rate in vivo (Figure 10.27). Because C in one DNA strand pairs with G in the other strand, whereas U would pair with A, conversion of a C to a U could potentially result in a heritable change of a CG pair to a UA pair. Such a change in nucleotide sequence would constitute a *mutation* in the DNA. To prevent this C deamination from leading to permanent changes in nucleotide sequence, a cellular repair mechanism "proofreads" DNA, and when a U arising from C deamination is encountered, it is treated as inappropriate and is replaced by a C. If DNA normally contained U rather than T, this repair system could not readily distinguish U formed by C deamination from U correctly paired with A. However, the U in DNA is "5-methyl-U" or, as it is conventionally known, thymine (Figure 10.28). That is, the 5-methyl group on T labels it as if to say "this U belongs; do not replace it."

The other chemical difference between RNA and DNA is that the ribose 2′-OH group on each nucleotide in RNA is absent in DNA. Consequently, the ubiquitous 3′-O of polynucleotide backbones lacks a vicinal hydroxyl neighbor in DNA. This difference leads to a greater resistance of DNA to alkaline hydrolysis, examined in detail in the following section. To view it another way, RNA is less stable than DNA because its vicinal 2′-OH group makes the 3′-phosphodiester bond susceptible to nucleophilic cleavage (Figure 10.29). For just this reason, it is selectively advantageous for the heritable form of genetic information to be DNA rather than RNA.

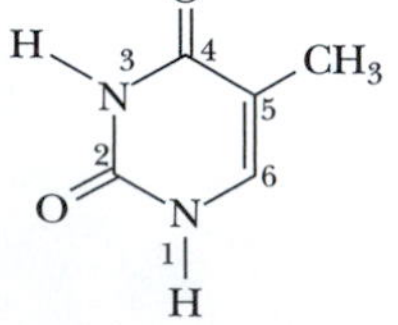

FIGURE 10.28 The 5-methyl group on thymine labels it as a special kind of uracil.

10.6 | Are Nucleic Acids Susceptible to Hydrolysis?

Most reactions of nucleic acid hydrolysis break bonds in the polynucleotide backbone. Such reactions are important because they can be used to manipulate these polymeric molecules. For example, hydrolysis of polynucleotides generates smaller fragments whose nucleotide sequence can be more easily determined.

(a) RNA:

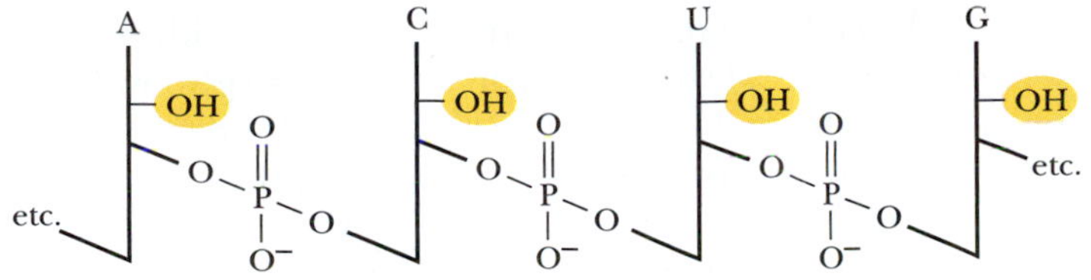

A nucleophile such as OH⁻ can abstract the H of the 2'–OH, generating 2'–O⁻ which attacks the δ^+P of the phosphodiester bridge:

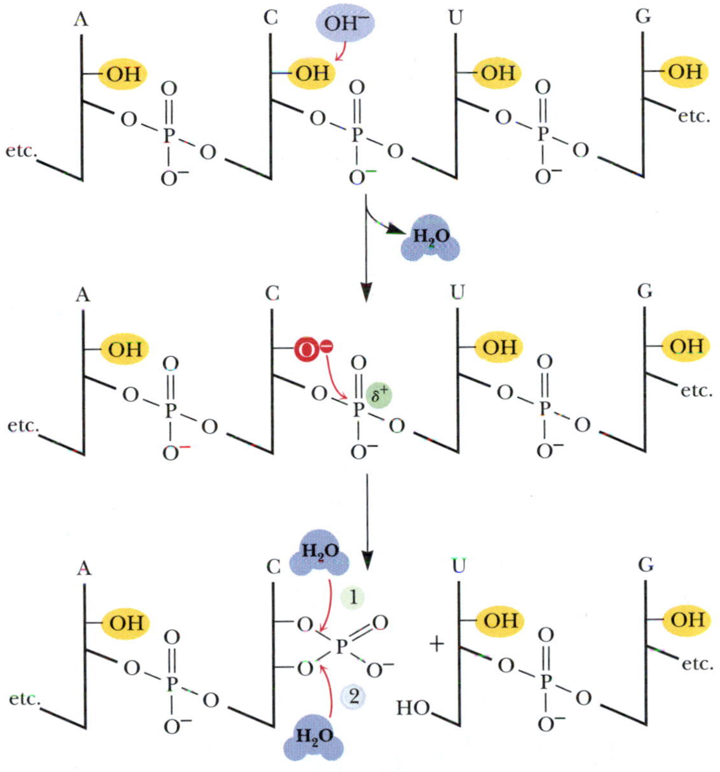

Sugar–PO₄ backbone cleaved

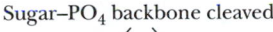

or

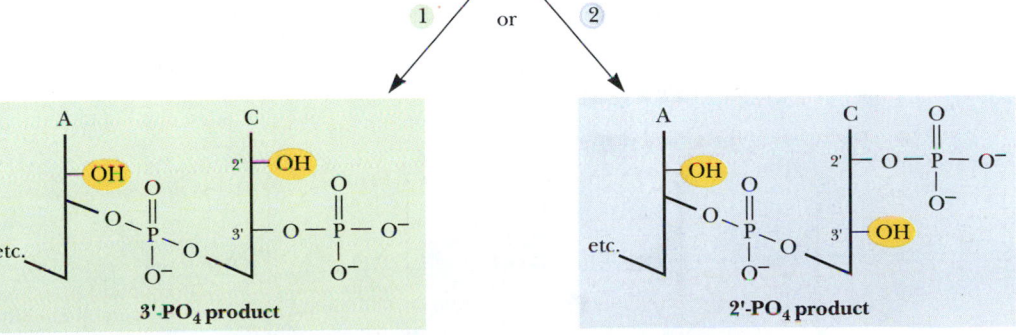

3'-PO₄ product

2'-PO₄ product

Complete hydrolysis of RNA by alkali yields a random mixture of 2'-NMPs and 3'-NMPs.

(b) DNA: no 2'-OH; resistant to OH⁻:

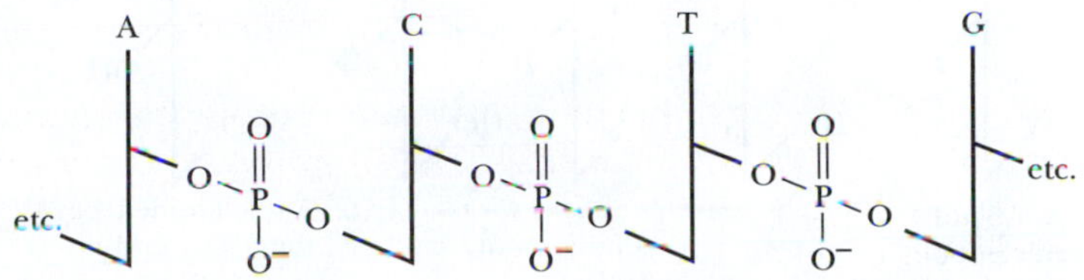

Biochemistry ⬩Now™ ACTIVE FIGURE 10.29
(a) The vicinal —OH groups of RNA are susceptible to nucleophilic attack leading to hydrolysis of the phosphodiester bond and fracture of the polynucleotide chain. Alkaline hydrolysis of RNA results in the formation of a mixture of 2'- and 3'-nucleoside monophosphates. **(b)** DNA lacks a 2'-OH vicinal to its 3'-O-phosphodiester backbone. **Test yourself on the concepts in this figure at http://chemistry. brookscole.com/ggb3**

RNA Is Susceptible to Hydrolysis by Base, But DNA Is Not

RNA is relatively resistant to the effects of dilute acid, but gentle treatment of DNA with 1 mM HCl leads to hydrolysis of purine glycosidic bonds and the loss of purine bases from the DNA. The glycosidic bonds between pyrimidine bases and 2′-deoxyribose are not affected, and in this case, the polynucleotide's sugar–phosphate backbone remains intact. The purine-free polynucleotide product is called **apurinic acid.**

DNA is not susceptible to alkaline hydrolysis. On the other hand, RNA is alkali labile and is readily hydrolyzed by dilute sodium hydroxide. Cleavage is random in RNA, and the ultimate products are a mixture of nucleoside 2′- and 3′-monophosphates. These products provide a clue to the reaction mechanism (Figure 10.29). Abstraction of the 2′-OH hydrogen by hydroxyl anion leaves a 2′-O$^-$ that carries out a nucleophilic attack on the δ^+ phosphorus atom of the phosphate moiety, resulting in cleavage of the 5′-phosphodiester bond and formation of a cyclic 2′,3′-phosphate. This cyclic 2′,3′-phosphodiester is unstable and decomposes randomly to either a 2′- or 3′-phosphate ester. DNA has no 2′-OH; therefore, DNA is alkali stable.

The Enzymes That Hydrolyze Nucleic Acids Are Phosphodiesterases

Enzymes that hydrolyze nucleic acids are called **nucleases.** Virtually all cells contain various nucleases that serve important housekeeping roles in the normal course of nucleic acid metabolism. Organs that provide digestive fluids, such as the pancreas, are rich in nucleases and secrete substantial amounts to hydrolyze ingested nucleic acids. Fungi and snake venom are often good sources of nucleases. As a class, nucleases are **phosphodiesterases** because the reaction that they catalyze is the cleavage of phosphodiester bonds by H_2O. Because each internal phosphate in a polynucleotide backbone is involved in two phosphoester linkages, cleavage can potentially occur on either side of the phosphorus (Figure 10.30). Convention labels the 3′-side as a and the 5′-side

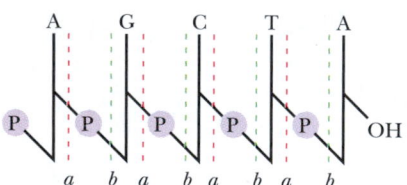

Convention: The 3′-side of each phosphodiester is termed a; the 5′-side is termed b.

(a) Hydrolysis of the a bond yields 5′-PO$_4$ products:

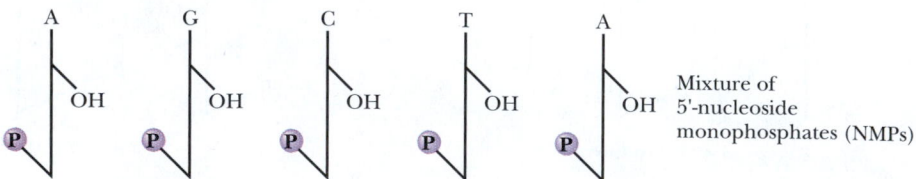

Mixture of 5′-nucleoside monophosphates (NMPs)

(b) Hydrolysis of the b bond yields 3′-PO$_4$ products:

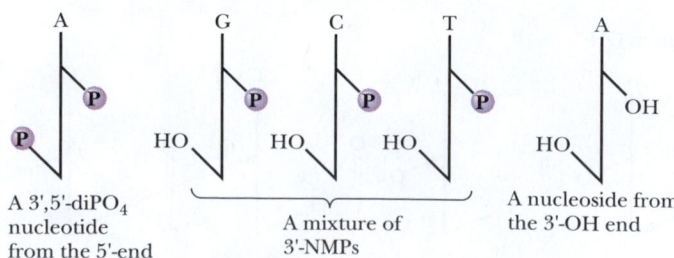

A 3′,5′-diPO$_4$ nucleotide from the 5′-end

A mixture of 3′-NMPs

A nucleoside from the 3′-OH end

FIGURE 10.30 Cleavage in polynucleotide chains. **(a)** Cleavage on the a side leaves the phosphate attached to the 5′-position of the adjacent nucleotide, while **(b)** b-side hydrolysis yields 3′-phosphate products. Enzymes or reactions that hydrolyze nucleic acids are characterized as acting at either a or b.

as *b*. Cleavage on the *a* side leaves the phosphate attached to the 5′-position of the adjacent nucleotide, whereas *b*-side hydrolysis yields 3′-phosphate products. Enzymes or reactions that hydrolyze nucleic acids are characterized as acting at either *a* or *b*. A second convention denotes whether the nucleic acid chain was cleaved at some internal location, *endo*, or whether a terminal nucleotide residue was hydrolytically removed, *exo*. Note that exo *a* cleavage characteristically occurs at the 3′-end of the polymer, whereas exo *b* cleavage involves attack at the 5′-terminus (Figure 10.31).

Nucleases Differ in Their Specificity for Different Forms of Nucleic Acid

Like most enzymes (see Chapter 13), nucleases exhibit selectivity or *specificity* for the nature of the substance on which they act. That is, some nucleases act only on DNA **(DNases),** whereas others are specific for RNA (the **RNases**). Still others are nonspecific and are referred to simply as **nucleases,** as in *nuclease S1* (Table 10.4). Nucleases may also show specificity for only single-stranded nucleic acids or may act only on double helices. Single-stranded nucleic acids are abbreviated by an *ss* prefix, as in ssRNA; the prefix *ds* denotes double-stranded. Nucleases may also display a decided preference for acting only at certain bases in a polynucleotide (Figure 10.32), or as we shall see for *restriction endonucleases*, some nucleases will act only at a particular nucleotide sequence four to eight nucleotides (or more) in length. Table 10.4 lists the various permutations in specificity displayed by these nucleases and gives prominent examples of each. To the molecular biologist, nucleases are the surgical tools for the dissection and manipulation of nucleic acids in the laboratory.

Exonucleases degrade nucleic acids by sequentially removing nucleotides from their ends. Two in common use are *snake venom phosphodiesterase* and *bovine spleen phosphodiesterase* (Figure 10.31). Because they act on either DNA or RNA, they are referred to by the generic name *phosphodiesterase*. These two enzymes have complementary specificities. Snake venom phosphodiesterase acts by *a* cleavage and starts at the free 3′-OH end of a polynucleotide chain, liberating nucleoside 5′-monophosphates. In contrast, the bovine spleen enzyme starts at the 5′-end of a nucleic acid, cleaving *b* and releasing 3′-NMPs.

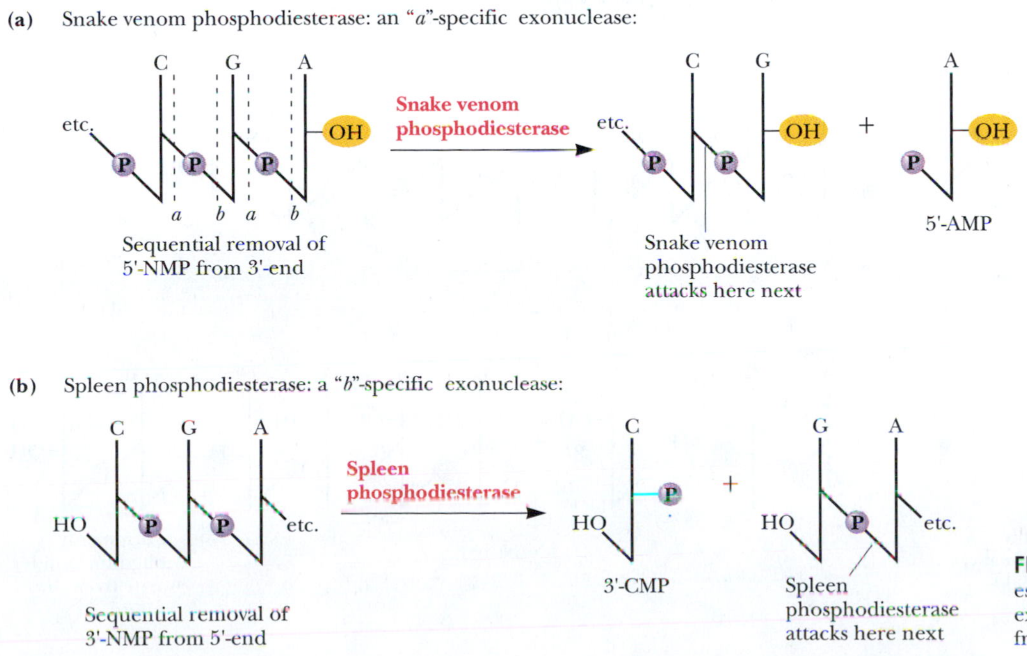

(a)   Snake venom phosphodiesterase: an "*a*"-specific exonuclease:

Sequential removal of 5′-NMP from 3′-end

Snake venom phosphodiesterase attacks here next

5′-AMP

(b)   Spleen phosphodiesterase: a "*b*"-specific exonuclease:

Sequential removal of 3′-NMP from 5′-end

3′-CMP

Spleen phosphodiesterase attacks here next

FIGURE 10.31 **(a)** Snake venom phosphodiesterase and **(b)** spleen phosphodiesterase are exonucleases that degrade polynucleotides from opposite ends.

Table 10.4

Specificity of Various Nucleases

Enzyme	DNA, RNA, or Both	a or b	Specificity
Exonucleases			
Snake venom phosphodiesterase	Both	a	Starts at 3'-end, 5'-NMP products
Spleen phosphodiesterase	Both	b	Starts at 5'-end, 3'-NMP products
Endonucleases			
RNase A (pancreas)	RNA	b	Where 3'-PO$_4$ is to pyrimidine; oligos with pyrimidine 3'-PO$_4$ ends
Bacillus subtilis RNase	RNA	b	Where 3'-PO$_4$ is to purine; oligos with purine 3'-PO$_4$ ends
RNase T$_1$	RNA	b	Where 3'-PO$_4$ is to guanine
RNase T$_2$	RNA	b	Where 3'-PO$_4$ is to adenine
DNase I (pancreas)	DNA	a	Preferably between Py and Pu; nicks dsDNA, creating 3'-OH ends
DNase II (spleen, thymus, *Staphylococcus aureus*)	DNA	b	Oligo products
Nuclease S1	Both	a	Cleaves single-stranded but not double-stranded nucleic acids

Restriction Enzymes Are Nucleases That Cleave Double-Stranded DNA Molecules

Restriction endonucleases are enzymes, isolated chiefly from bacteria, that have the ability to cleave double-stranded DNA. The term *restriction* comes from the capacity of prokaryotes to defend against or "restrict" the possibility of takeover by foreign DNA that might gain entry into their cells. Prokaryotes degrade foreign DNA by using their unique restriction enzymes to chop it into relatively large but noninfective fragments. Restriction enzymes are classified into three types: I, II, or III. Types I and III require ATP to hydrolyze DNA and can

Pancreatic RNase is an enzyme specific for *b* cleavage where a pyrimidine base lies to the 3'-side of the phosphodiester; it acts endo. The products are oligonucleotides with pyrimidine–3'-PO$_4$ ends:

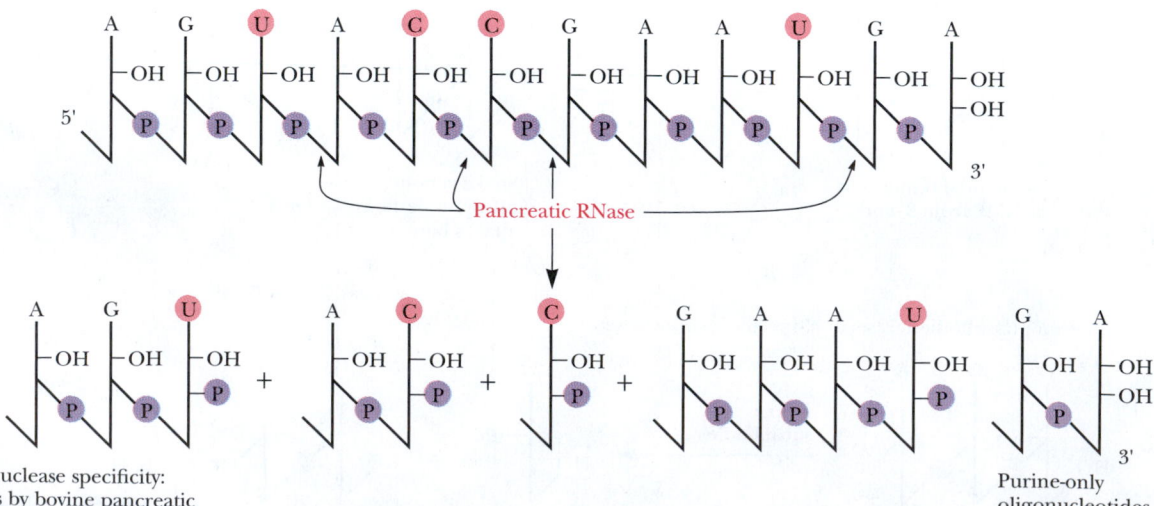

FIGURE 10.32 An example of nuclease specificity: The specificity of RNA hydrolysis by bovine pancreatic RNase. This RNase cleaves *b* at 3'-pyrimidines, yielding oligonucleotides with pyrimidine 3'-PO$_4$ ends.

Purine-only oligonucleotides arise from the 3'-end

A Deeper Look

Peptide Nucleic Acids (PNAs) Are Synthetic Mimics of DNA and RNA

Synthetic chemists have invented analogs of DNA (and RNA) in which the sugar–phosphate backbone is replaced by a peptide backbone, creating a polymer appropriately termed a *peptide nucleic acid,* or *PNA.* The PNA peptide backbone was designed so that the space between successive bases was the same as in natural DNA (see accompanying figure). PNA consists of repeating units of *N*-(2-aminoethyl)-glycine residues linked by peptide bonds; the bases are attached to this backbone through methylene carbonyl linkages. This chemistry provides six bonds along the backbone between bases and three bonds between the backbone and each base, just like natural DNA. PNA oligomers interact with DNA (and RNA) through specific base-pairing interactions, just as would be expected for a pair of complementary oligonucleotides. PNAs are resistant to nucleases and also are poor substrates for proteases. PNAs thus show great promise as specific diagnostic probes for unique DNA or RNA nucleotide sequences. PNAs also have potential application as antisense drugs (see problem 7 in the end-of-chapter problems).

Buchardt, O., et al., 1993. Peptide nucleic acids and their potential applications in biotechnology. *Trends in Biotechnology* **11**:384–386.

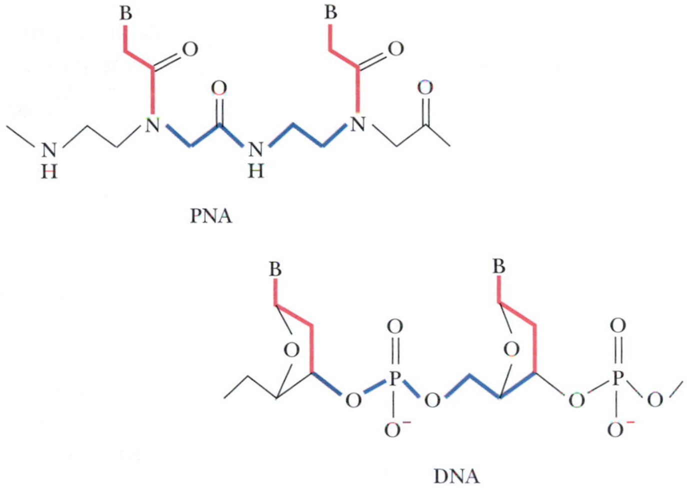

▲ Note the repeating six bonds (in blue) between base attachments and the three-bond linker between base (B) and backbone.

also catalyze chemical modification of DNA through addition of methyl groups to specific bases. Type I restriction endonucleases cleave DNA randomly, whereas type III recognize specific nucleotide sequences within dsDNA and cut the DNA at or near these sites.

Type II Restriction Endonucleases Are Useful for Manipulating DNA in the Lab

Type II restriction enzymes have received widespread application in the cloning and sequencing of DNA molecules. Their hydrolytic activity is not ATP-dependent, and they do not modify DNA by methylation or other means. Most important, they cut DNA within or near particular nucleotide sequences that they specifically recognize. These recognition sequences are typically four or six nucleotides in length and have a twofold axis of symmetry. For example, *E. coli* has a restriction enzyme, *Eco*RI, that recognizes the hexanucleotide sequence GAATTC:

$$5'\text{——N—N—N—N—G—A—A—T—T—C—N—N—N—N——}3'$$
$$3'\text{——N—N—N—N—C—T—T—A—A—G—N—N—N—N——}5'$$

Note the twofold symmetry: the sequence read $5' \rightarrow 3'$ is the same in both strands.

When *Eco*RI encounters this sequence in dsDNA, it causes a staggered, double-stranded break by hydrolyzing each chain between the G and A residues:

$$5'\text{——N—N—N—N—G \quad A—A—T—T—C—N—N—N—N——}3'$$
$$3'\text{——N—N—N—N—C—T—T—A—A \quad G—N—N—N—N——}5'$$

Staggered cleavage results in fragments with protruding single-stranded 5′-ends:

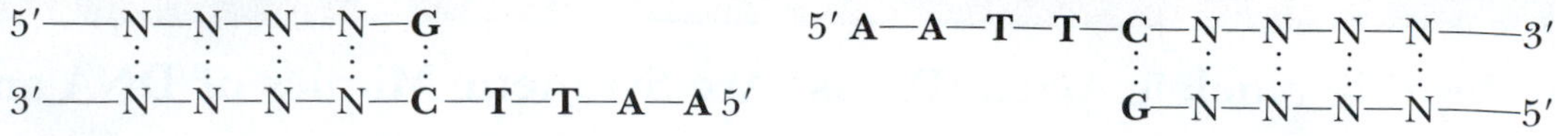

$$5'——N—N—N—N—G \qquad\qquad 5'\,A—A—T—T—C—N—N—N—N——3'$$
$$3'——N—N—N—N—C—T—T—A\,5' \qquad\qquad G—N—N—N—N——5'$$

Because the protruding termini of *Eco*RI fragments have complementary base sequences, they can form base pairs with one another.

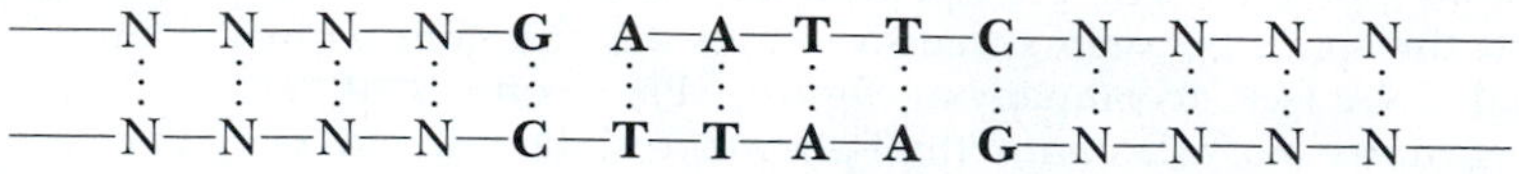

$$——N—N—N—N—G\ A—A—T—T—C—N—N—N—N——$$
$$——N—N—N—N—C—T—T—A\ G—N—N—N—N——$$

Therefore, DNA restriction fragments having such "sticky" ends can be joined together to create new combinations of DNA sequence. If fragments derived from DNA molecules of different origin are combined, novel recombinant forms of DNA are created.

*Eco*RI leaves staggered 5′-termini. Other restriction enzymes, such as *Pst*I, which recognizes the sequence 5′-CTGCAG-3′ and cleaves between A and G, produce cohesive staggered 3′-ends. Still others, such as *Bal*I, act at the center of the twofold symmetry axis of their recognition site and generate blunt ends that are noncohesive. *Bal*I recognizes 5′-TGGCCA-3′ and cuts between G and C.

Table 10.5 lists many of the commonly used restriction endonucleases and their recognition sites. Because these sites all have twofold symmetry, only the sequence on one strand needs to be designated. Different restriction enzymes sometimes recognize and cleave within identical target sequences. Such enzymes are called **isoschizomers,** meaning that they cut at the same site; for example, *Mbo*I and *Sau*3A are isoschizomers.

Restriction Fragment Size Assuming random distribution and equimolar proportions for the four nucleotides in DNA, a particular tetranucleotide sequence should occur every 4^4 nucleotides, or every 256 bases. Therefore, the fragments generated by a restriction enzyme that acts at a four-nucleotide sequence should average about 250 bp in length. "Six-cutters," enzymes such as *Eco*RI or *Bam*HI, will find their unique hexanucleotide sequences on the average once in every 4096 (4^6) bp of length. Because the genetic code is a triplet code with three bases of DNA specifying one amino acid in a polypeptide sequence, and because polypeptides typically contain at most 1000 amino acid residues, the fragments generated by six-cutters are approximately the size of prokaryotic genes. This property makes these enzymes useful in the construction and cloning of genetically useful recombinant DNA molecules. For the isolation of even larger nucleotide sequences, such as those of genes encoding large polypeptides (or those of eukaryotic genes that are disrupted by large introns), partial or limited digestion of DNA by restriction enzymes can be employed. However, restriction endonucleases that cut only at specific nucleotide sequences 8 or even 13 nucleotides in length are also available, such as *Not*I and *Sfi*I.

Restriction Endonucleases Can Be Used to Map the Structure of a DNA Fragment

The application of these sequence-specific nucleases to problems in molecular biology is considered in detail in Chapter 12, but one prominent application is described here. Because restriction endonucleases cut dsDNA at unique sites to generate large fragments, they provide a means for mapping DNA molecules that are many kilobase pairs in length. Restriction digestion of a DNA molecule is in many ways analogous to proteolytic digestion of a protein by an

Table 10.5

Restriction Endonucleases

About 1000 restriction enzymes have been characterized. They are named by italicized three-letter codes; the first is a capital letter denoting the genus of the organism of origin, and the next two letters are an abbreviation of the particular species. Because prokaryotes often contain more than one restriction enzyme, the various representatives are assigned letter and number codes as they are identified. Thus, *Eco*RI is the initial restriction endonuclease isolated from *Escherichia coli,* strain R. With one exception (*Nci*I), all known type II restriction endonucleases generate fragments with 5'-PO$_4$ and 3'-OH ends.

Enzyme	Common Isoschizomers	Recognition Sequence	Compatible Cohesive Ends
*Alu*I		AG↓CT	Blunt
*Apy*I	*Atu*I, *Eco*RII	CC↓G(^A_T)GG	
*Asu*II		TT↓CGAA	*Cla*I, *Hpa*II, *Taq*I
*Ava*I		G↓PyCGPuG	*Sal*I, *Xho*I, *Xma*I
*Avr*II		C↓CTAGG	
*Bal*I		TGG↓CCA	Blunt
*Bam*HI		G↓GATCC	*Bcl*I, *Bgl*II, *Mbo*I, *Sau*3A, *Xho*II
*Bcl*I		T↓GATCA	*Bam*HI, *Bgl*II, *Mbo*I, *Sau*3A, *Xho*II
*Bgl*II		A↓GATCT	*Bam*HI, *Bcl*I, *Mbo*I, *Sau*3A, *Xho*II
*Bst*EII		G↓GTNACC	
*Bst*XI		CCANNNNN↓NTGG	
*Cla*I		AT↓CGAT	*Acc*I, *Acy*I, *Asu*II, *Hpa*II, *Taq*I
*Dde*I		C↓TNAG	
*Eco*RI		G↓AATTC	
*Eco*RII	*Atu*I, *Apy*I	↓CC(^A_T)GG	
*Fnu*DII	*Tha*I	CG↓CG	Blunt
*Hae*I		(^A_T)GG↓CC(^T_A)	Blunt
*Hae*II		PuGCGC↓Py	
*Hae*III		GG↓CC	Blunt
*Hinc*II		GTPy↓PuAC	Blunt
*Hind*III		A↓AGCTT	
*Hpa*I		GTT↓AAC	Blunt
*Hpa*II		C↓CGG	*Acc*I, *Acy*I, *Asu*II, *Cla*I, *Taq*I
*Kpn*I		GGTAC↓C	
*Mbo*I	*Sau*3A	↓GATC	*Bam*HI, *Bcl*I, *Bgl*II, *Xho*II
*Msp*I		C↓CGG	
*Mst*I		TGC↓GCA	Blunt
*Not*I		GC↓GGCCGC	
*Pst*I		CTGCA↓G	
*Sac*I	*Sst*I	GAGCT↓C	
*Sal*I		G↓TCGAC	*Ava*I, *Xho*I
*Sau*3A		↓GATC	*Bam*HI, *Bcl*I, *Bgl*II, *Mbo*I, *Xho*II
*Sfi*I		GGCCNNNN↓NGGCC	
*Sma*I	*Xma*I	CCC↓GGG	Blunt
*Sph*I		GCATG↓C	
*Sst*I	*Sac*I	GAGCT↓C	
*Taq*I		T↓CGA	*Acc*I, *Acy*I, *Asu*II, *Cla*I, *Hpa*II
*Xba*I		T↓CTAGA	
*Xho*I		C↓TCGAG	*Ava*I, *Sal*I
*Xho*II		(^A_G)↓GATC(^T_C)	*Bam*HI, *Bcl*I, *Bgl*II, *Mbo*I, *Sau*3A
*Xma*I	*Sma*I	C↓CCGGG	*Ava*I

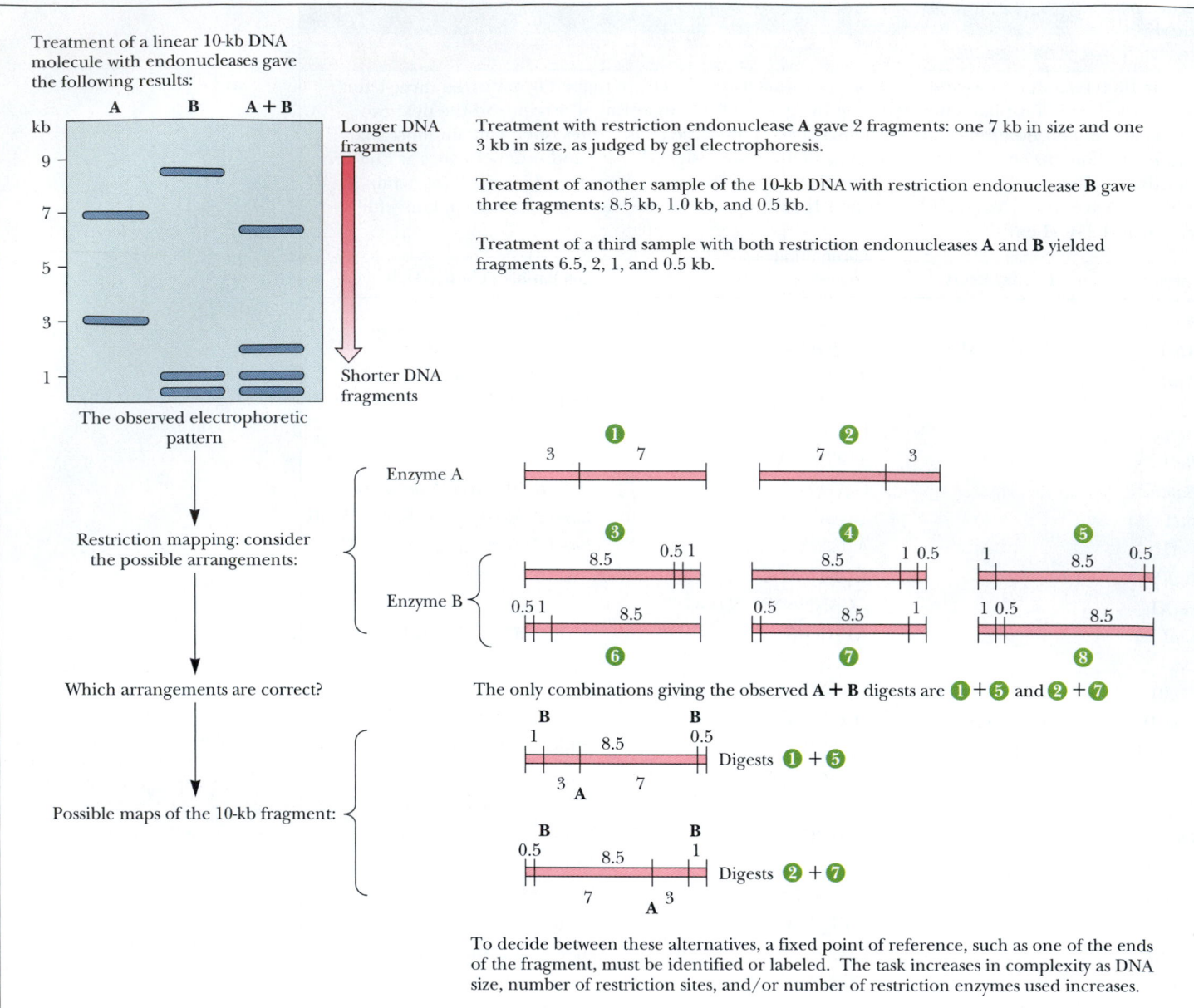

Treatment of a linear 10-kb DNA molecule with endonucleases gave the following results:

The observed electrophoretic pattern

Treatment with restriction endonuclease **A** gave 2 fragments: one 7 kb in size and one 3 kb in size, as judged by gel electrophoresis.

Treatment of another sample of the 10-kb DNA with restriction endonuclease **B** gave three fragments: 8.5 kb, 1.0 kb, and 0.5 kb.

Treatment of a third sample with both restriction endonucleases **A** and **B** yielded fragments 6.5, 2, 1, and 0.5 kb.

Restriction mapping: consider the possible arrangements:

Which arrangements are correct?

The only combinations giving the observed **A + B** digests are ❶+❺ and ❷+❼

Possible maps of the 10-kb fragment:

To decide between these alternatives, a fixed point of reference, such as one of the ends of the fragment, must be identified or labeled. The task increases in complexity as DNA size, number of restriction sites, and/or number of restriction enzymes used increases.

FIGURE 10.33 Restriction mapping of a DNA molecule as determined by an analysis of the electrophoretic pattern obtained for different restriction endonuclease digests. (Keep in mind that a dsDNA molecule has a unique nucleotide sequence and therefore a definite polarity; thus, fragments from one end are distinctly different from fragments derived from the other end.)

enzyme such as trypsin (see Chapter 5): The restriction endonuclease acts only at its specific sites so that a discrete set of nucleic acid fragments is generated. This action is analogous to trypsin cleavage only at Arg and Lys residues to yield a particular set of tryptic peptides from a given protein. The restriction fragments represent a unique collection of different-sized DNA pieces. Fortunately, this complex mixture can be resolved by *electrophoresis* (see the Appendix to Chapter 5). Electrophoresis of DNA molecules on gels of restricted pore size (as formed in agarose or polyacrylamide media) separates them according to size, the largest being retarded in their migration through the gel pores while the smallest move relatively unhindered. Figure 10.33 shows a hypothetical electrophoretogram obtained for a DNA molecule treated with two different restriction nucleases, alone and in combination. Just as cleavage of a protein with different proteases to generate overlapping fragments allows an ordering of the peptides, restriction fragments can be ordered or "mapped" according to their sizes, as deduced from the patterns depicted in Figure 10.33.

Summary

Nucleotides and nucleic acids possess heterocyclic nitrogenous bases as principal components of their structure. Nucleotides participate as essential intermediates in virtually all aspects of cellular metabolism. Nucleic acids are the substances of heredity (DNA) and the agents of genetic information transfer (RNA).

10.1 What Is the Structure and Chemistry of Nitrogenous Bases? The bases of nucleotides and nucleic acids are derivatives of either pyrimidine (cytosine, uracil, and thymine) or purine (adenine and guanine). The aromaticity of the pyrimidine and purine ring systems and the electron-rich nature of their —OH and —NH$_2$ substituents allow them to undergo keto–enol tautomeric shifts and endow them with the capacity to absorb UV light.

10.2 What Are Nucleosides? Nucleosides are formed when a base is linked to a sugar. The usual sugars of nucleosides are pentoses; ribonucleosides contain the pentose D-ribose, whereas 2-deoxy-D-ribose is found in deoxyribonucleosides. Nucleosides are more water soluble than free bases.

10.3 What Is the Structure and Chemistry of Nucleotides? A nucleotide results when phosphoric acid is esterified to a sugar —OH group of a nucleoside. Successive phosphate groups can be linked to the phosphoryl group of a nucleotide through phosphoric anhydride linkages. Nucleoside 5′-triphosphates, as carriers of chemical energy, are indispensable agents in metabolism because phosphoric anhydride bonds are a prime source of chemical energy to do biological work. Virtually all of the biochemical reactions of nucleotides involve either *phosphate* or *pyrophosphate group transfer*. The bases of nucleotides serve as "information symbols."

10.4 What Are Nucleic Acids? Nucleic acids are polynucleotides: linear polymers of nucleotides linked 3′ to 5′ by phosphodiester bridges. The only significant variation in the chemical structure of nucleic acids is the particular base at each nucleotide position. These bases are not part of the sugar–phosphate backbone but instead serve as distinctive side chains.

10.5 What Are the Different Classes of Nucleic Acids? The two major classes of nucleic acids are DNA and RNA. Two fundamental chemical differences distinguish DNA from RNA: The nucleotides in DNA contain 2-deoxyribose instead of ribose as their sugar component, and DNA contains the base thymine instead of uracil. These differences confer important biological properties on DNA.

DNA consists of two antiparallel polynucleotide strands wound together to form a long, slender, double helix. The strands are held together through specific base pairing of A with T and C with G. The information in DNA is encoded in digital form in terms of the sequence of bases along each strand. Because base pairing is specific, the information in the two strands is complementary. DNA molecules may contain tens or even hundreds of millions of base pairs. In eukaryotic cells, DNA is complexed with histone proteins to form a nucleoprotein complex known as chromatin.

RNA occurs in multiple forms in cells. Messenger RNA (mRNA) molecules are direct copies of the base sequences of protein-coding genes. Ribosomal RNA (rRNA) molecules provide the structural and functional foundations for ribosomes, the agents for translating mRNAs into proteins. In protein synthesis, the amino acids are delivered to the ribosomes in the form of aminoacyl-tRNA (transfer RNA) derivatives. Small nuclear RNAs (snRNAs) are characteristic of eukaryotic cells and are necessary for processing the RNA transcripts of protein-coding genes into mature mRNA molecules. Small RNAs are a recently discovered class of RNA molecules. A prominent role of small RNAs is post-transcriptional gene silencing, particularly in the phenomenon of RNA interference (RNAi).

10.6 Are Nucleic Acids Susceptible to Hydrolysis? Like all biological polymers, nucleic acids are susceptible to hydrolysis, particularly hydrolysis of the phosphoester bonds in the polynucleotide backbone. RNA is susceptible to hydrolysis by base: DNA is not. Nucleases are hydrolytic enzymes that cleave the phosphoester linkages in the sugar–phosphate backbone of nucleic acids. Nucleases abound in nature, with varying specificity for RNA or DNA, single- or double-stranded nucleic acids, endo versus exo action, and 3′- versus 5′-cleavage of phosphodiesters. Restriction endonucleases of the type II class are sequence-specific endonucleases useful in mapping the structure of DNA molecules.

Problems

1. From the pK_a values for nucleotides presented in Table 10.1, draw the principal ionic species of 5′-GMP occurring at pH 2.

2. Draw the chemical structure of pACG.

3. Chargaff's results (Table 10.3) yielded a molar ratio of 1.29 for A to G in ox DNA, 1.43 for T to C, 1.04 for A to T, and 1.00 for G to C. Given these values, what are the approximate mole fractions of A, C, G, and T in ox DNA?

4. Results on the human genome published in *Science* (*Science* **291**:1304–1350 [2001]) indicate that the haploid human genome consists of 2.91 gigabase pairs (2.91 × 10^9 base pairs) and that 27% of the bases in human DNA are A. Calculate the number of A, T, G, and C residues in a typical human cell.

5. Adhering to the convention of writing nucleotide sequences in the 5′→3′ direction, what is the nucleotide sequence of the DNA strand that is complementary to d-ATCGCAACTGTCACTA?

6. Messenger RNAs are synthesized by RNA polymerases that read along a DNA template strand in the 3′→5′ direction, polymerizing ribonucleotides in the 5′→3′ direction (see Figure 10.24). Give the nucleotide sequence (5′→3′) of the DNA template strand from which the following mRNA segment was transcribed: 5′-UAGUGACAGUUGCGAU-3′.

7. The DNA strand that is complementary to the template strand copied by RNA polymerase during transcription has a nucleotide sequence identical to that of the RNA being synthesized (except T residues are found in the DNA strand at sites where U residues occur in the RNA). An RNA transcribed from this nontemplate DNA strand would be complementary to the mRNA synthesized by RNA polymerase. Such an RNA is called antisense RNA because its base sequence is complementary to the "sense" mRNA. A promising strategy to thwart the deleterious effects of genes activated in disease states (such as cancer) is to generate antisense RNAs in affected cells. These antisense RNAs would form double-stranded hybrids with mRNAs transcribed from the activated genes and prevent their translation into protein. Suppose transcription of a cancer-activated gene yielded an mRNA whose sequence included the segment 5′-UACGGUCUAAGCUGA. What is the corresponding nucleotide sequence (5′→3′) of the template strand in a DNA duplex that might be introduced into these cells so that an antisense RNA could be transcribed from it?

8. A 10-kb DNA fragment digested with restriction endonuclease *Eco*RI yielded fragments 4 kb and 6 kb in size. When digested with *Bam*HI, fragments 1, 3.5, and 5.5 kb were generated. Concomitant digestion with both *Eco*RI and *Bam*HI yielded fragments 0.5, 1, 3,

and 5.5 kb in size. Give a possible restriction map for the original fragment.

9. Based on the information in Table 10.5, describe two different 20-base nucleotide sequences that have restriction sites for *Bam*H1, *Pst*I, *Sal*I, and *Sma*I. Give the sequences of the *Sma*I cleavage products of each.

10. (Integrates with Chapter 3.) The synthesis of RNA can be summarized by the reaction:

$$n \text{ NTP} \longrightarrow (\text{NMP})_n + n \text{ PP}_i$$

What is the $\Delta G^{\circ\prime}_{\text{overall}}$ for synthesis of an RNA molecule 100 nucleotides in length, assuming that the $\Delta G^{\circ\prime}$ for transfer of an NMP from an NTP to the 3′-O of polynucleotide chain is the same as the $\Delta G^{\circ\prime}$ for transfer of an NMP from an NTP to H_2O? (Use data given in Table 3.3.)

11. Gene expression is controlled through the interaction of proteins with specific nucleotide sequences in double-stranded DNA.
    a. List the kinds of noncovalent interactions that might take place between a protein and DNA.
    b. How do you suppose a particular protein might specifically interact with a particular nucleotide sequence in DNA? That is, how might proteins recognize specific base sequences within the double helix?

12. Restriction endonucleases also recognize specific base sequences and then act to cleave the double-stranded DNA at a defined site. Speculate on the mechanisms by which this sequence recognition and cleavage reaction might occur by listing a set of requirements for the process to take place.

13. A carbohydrate group is an integral part of a nucleoside.
    a. What advantage does the carbohydrate provide?
    Polynucleotides are formed through formation of a sugar–phosphate backbone.

    b. Why might ribose be preferable for this backbone instead of glucose?
    c. Why might 2-deoxyribose be preferable to ribose in some situations?

14. Phosphate groups are also integral parts of nucleotides, with the second and third phosphates of a nucleotide linked through phosphoric anhydride bonds, an important distinction in terms of the metabolic role of nucleotides.
    a. What property does a phosphate group have that a nucleoside lacks?
    b. How are phosphoric anhydride bonds useful in metabolism?
    c. How are phosphate anhydride bonds an advantage to the energetics of polynucleotide synthesis?

Preparing for the MCAT Exam

15. The bases of nucleotides and polynucleotides are "information symbols." Their central role in providing information content to DNA and RNA is clear. What advantages might bases as "information symbols" bring to the roles of nucleotides in metabolism?

16. Structural complementarity is the key to molecular recognition, a lesson learned in Chapter 1. The principle of structural complementarity is relevant to answering problems 5, 6, 7, 11, 12, and 15. The quintessential example of structural complementarity in all of biology is the DNA double helix. What features of the DNA double helix exemplify structural complementarity?

Biochemistry ⋐ Now™ Preparing for an exam? Test yourself on key questions at http://chemistry.brookscole.com/ggb3

Further Reading

Nucleic Acid Biochemistry and Molecular Biology

Adams, R. L. P., Knowler, J. T., and Leader, D. P., 1992. *The Biochemistry of the Nucleic Acids*, 11th ed. New York: Chapman and Hall (Methuen and Co., distrib.).

Watson, J. D., Hopkins, N. H., Roberts, J. W., Steitz, J. A., and Weiner, A. M., 1987. *The Molecular Biology of the Gene*, Vol. I, *General Principles*, 4th ed. Menlo Park, CA: Benjamin/Cummings. Still a classic.

The History of Discovery of the DNA Double Helix

Judson, H. F., 1979. *The Eighth Day of Creation*. New York: Simon and Schuster.

DNA as Information

Hood, L., and Galas, D., 2003. The digital code of DNA. *Nature* **421:**444–448.

The Catalytic Properties of RNA and Its Role in Early Evolution

Caprara, M. G., and Nilsen, T. W., 2000. RNA: Versatility in form and function. *Nature Structural Biology* **7:**831–833.

Gray, M. W., and Cedergren, R., eds., 1993. The new age of RNA. *The FASEB Journal* **7:**4–239. A collection of articles emphasizing the new appreciation for RNA in protein synthesis, in evolution, and as a catalyst.

RNAi (RNA Interference): A Newly Discovered Role for RNA—Post-Transcriptional Control of Gene Expression

Hannon, G. J., 2002. RNA interference. *Nature* **418:**244–251. A review of RNAi, a widely conserved biological response to the intracellular presence of double-stranded RNA. RNAi provides an experimental method for manipulating gene expression as well as a mechanism to investigate specific gene function at the whole genome level.

Tuschi, T., 2003. RNA sets the standard. *Nature* **421:**220–221. Overview of the use of RNA interference to inactivate all the genes in a model organism (*Caenorhabditis elegans*) as a means of identifying gene function.

Nucleases and DNA Manipulation

Mishra, N. C., 2002. *Nucleases: Molecular Biology and Applications*. Hoboken, NJ: Wiley-Interscience.

Sambrook, J., and Russell, D., 2000. *Molecular Cloning: A Laboratory Manual*, 3rd ed. Cold Spring Harbor, NY: Cold Spring Harbor Laboratory.

Structure of Nucleic Acids

CHAPTER 11

Essential Question

The nucleotide sequence—the primary structure—of DNA not only is the determinant of its higher-order structure but is also the physical representation of genetic information in organisms. RNA sequences, as copies of specific DNA segments, determine both the higher-order structure and the function of RNA molecules in information transfer processes. ***What is the higher-order structure of DNA and RNA, and what methodologies have allowed scientists to probe these structures and the functions that derive from them?***

Chapter 10 presented the structure and chemistry of nucleotides and how these units are joined via phosphodiester bonds to form nucleic acids, the biological polymers for information storage and transmission. In this chapter, we investigate biochemical methods that reveal this information by determining the sequential order of nucleotides in a polynucleotide, the so-called primary structure of nucleic acids. Then, we consider the higher orders of structure in the nucleic acids: the secondary and tertiary levels. Although the focus here is primarily on the structural and chemical properties of these macromolecules, it is fruitful to keep in mind the biological roles of these remarkable substances. The sequence of nucleotides in nucleic acids is the embodiment of genetic information (see Part 4). We can anticipate that the cellular mechanisms for accessing this information, as well as reproducing it with high fidelity, will be illuminated by knowledge of the chemical and structural qualities of these polymers.

Reginald H. Garrett

What do you suppose those masons, who created this double helix adorning the cathedral in Orvieto, Italy, some 500 years ago, might have thought about the DNA double helix and heredity?

The Structure of DNA: "A melody for the eye of the intellect, with not a note wasted."
Horace Freeland Judson, *The Eighth Day of Creation*

Key Questions

Biochemistry Now™ Test yourself on these Key Questions at BiochemistryNow at **http://chemistry.brookscole.com/ggb3**

11.1 How Do Scientists Determine the Primary Structure of Nucleic Acids?

As recently as 1975, determining the primary structure of nucleic acids (the nucleotide sequence) was a more formidable problem than amino acid sequencing of proteins, simply because nucleic acids contain only 4 unique monomeric units, whereas proteins have 20. With only 4, there are *apparently* fewer specific sites for selective cleavage, distinctive sequences are more difficult to recognize, and the likelihood of ambiguity is greater. The much greater number of monomeric units in most polynucleotides as compared to polypeptides is a further difficulty. Two important breakthroughs reversed this situation so that now sequencing nucleic acids is substantially easier than sequencing polypeptides. One was the discovery of *restriction endonucleases* that cleave DNA at specific oligonucleotide sites, generating unique fragments of manageable size (see Chapter 10). The second is the power of *polyacrylamide gel electrophoresis* separation methods to resolve nucleic acid fragments that differ from one another in length by just one nucleotide.

The Nucleotide Sequence of DNA Can Be Determined from the Electrophoretic Migration of a Defined Set of Polynucleotide Fragments

Two basic protocols for nucleic acid sequencing were developed, both of which depended on the resolving power of polyacrylamide gel electrophoresis: (1) the **chain termination** or **dideoxy method** of F. Sanger that relies on enzymatic replication of the DNA to be sequenced and (2) a **base-specific chemical cleavage method** developed by A. M. Maxam and W. Gilbert that exploits chemical methods for cleaving the sugar–phosphate backbone of a DNA strand at the location

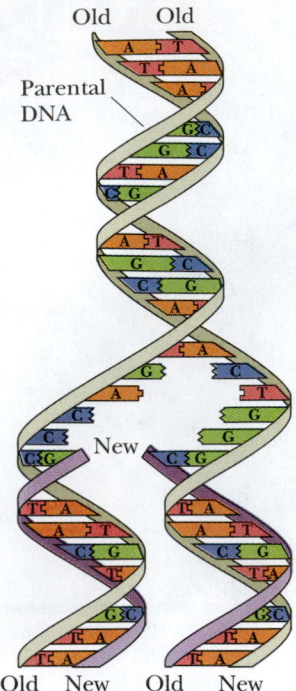

FIGURE 11.1 DNA replication yields two daughter DNA duplexes identical to the parental DNA molecule. Each original strand of the double helix serves as a template, and the sequence of nucleotides in each of these strands is copied to form a new complementary strand by the enzyme DNA polymerase. By this process, biosynthesis yields two daughter DNA duplexes from the parental double helix.

of particular bases. Over time, the Sanger method has proved to be the method of choice, and thus is the only one discussed here. These methods are carried out on nanogram amounts of DNA, requiring very sensitive analytical techniques that can detect the DNA chains following electrophoretic separation on polyacrylamide gels. Typically, the DNA molecules are labeled with radioactive ^{32}P,[1] and following electrophoresis, the pattern of their separation is visualized by **autoradiography.** A piece of X-ray film is placed over the gel, and the radioactive disintegrations emanating from ^{32}P decay create a pattern on the film that is an accurate image of the resolved oligonucleotides. Recently, sensitive biochemical and chemiluminescent methods have begun to supersede the use of radioisotopes as tracers in these experiments.

Sanger's Chain Termination or Dideoxy Method Uses DNA Replication to Generate a Defined Set of Polynucleotide Fragments

To appreciate the rationale of the chain termination or dideoxy method, we first must briefly examine the biochemistry of DNA replication. DNA is a double helical molecule. In the course of its replication, the sequence of nucleotides in one strand is copied in a complementary fashion to form a new second strand by the enzyme **DNA polymerase.** Each original strand of the double helix serves as **template** for the biosynthesis that yields two daughter DNA duplexes from the parental double helix (Figure 11.1). DNA polymerase carries out this reaction in vitro in the presence of the four deoxynucleotide monomers and copies single-stranded DNA, provided a double-stranded region of DNA is artificially generated by adding a **primer.** This primer is merely an oligonucleotide capable of forming a short stretch of dsDNA by base pairing with the ssDNA (Figure 11.2). The primer must have a free 3′-OH end from which the new polynucleotide chain can grow as the first residue is added in the initial step of the polymerization process. DNA polymerases synthesize new strands by adding successive nucleotides in the 5′→3′ direction.

The Chain Termination Protocol In the chain termination method of DNA sequencing, a DNA fragment of unknown sequence serves as template in a polymerization reaction using some type of DNA polymerase, usually a genetically engineered version of DNA polymerase that lacks all traces of exonuclease activity that might otherwise degrade the DNA. (DNA polymerases have an intrinsic exonuclease activity that allows proofreading and correction of the DNA strand being synthesized; see Chapter 28.) The primer requirement is met by an appropriate oligonucleotide (this method is also known as the **primed synthesis method** for this reason). Four parallel reactions are run; all four contain the four deoxynucleoside triphosphates dATP, dGTP, dCTP, and dTTP, which are the substrates for DNA polymerase (Figure 11.3).

[1]Because its longer half-life and lower energy make it more convenient to handle, ^{35}S is replacing ^{32}P as the radioactive tracer of choice in sequencing by the Sanger method. ^{35}S-α-labeled deoxynucleotide analogs provide the source for incorporating radioactivity into DNA.

Biochemistry Now™ ACTIVE FIGURE 11.2
DNA polymerase copies ssDNA in vitro in the presence of the four deoxynucleotide monomers, provided a double-stranded region of DNA is artificially generated by adding a primer, an oligonucleotide capable of forming a short stretch of dsDNA by base pairing with the ssDNA. The primer must have a free 3′-OH end from which the new polynucleotide chain can grow as the first residue is added in the initial step of the polymerization process. **Test yourself on the concepts in this figure at http://chemistry. brookscole.com/ggb3**

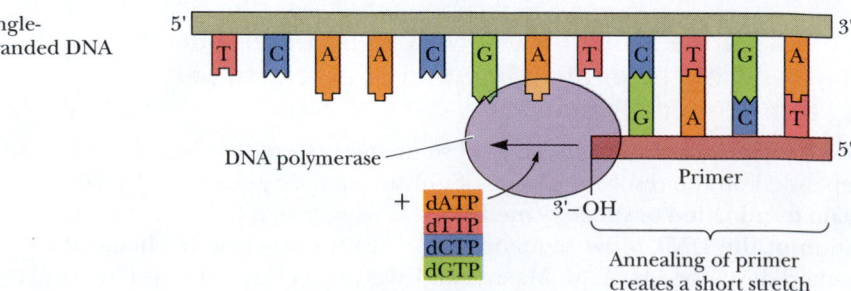

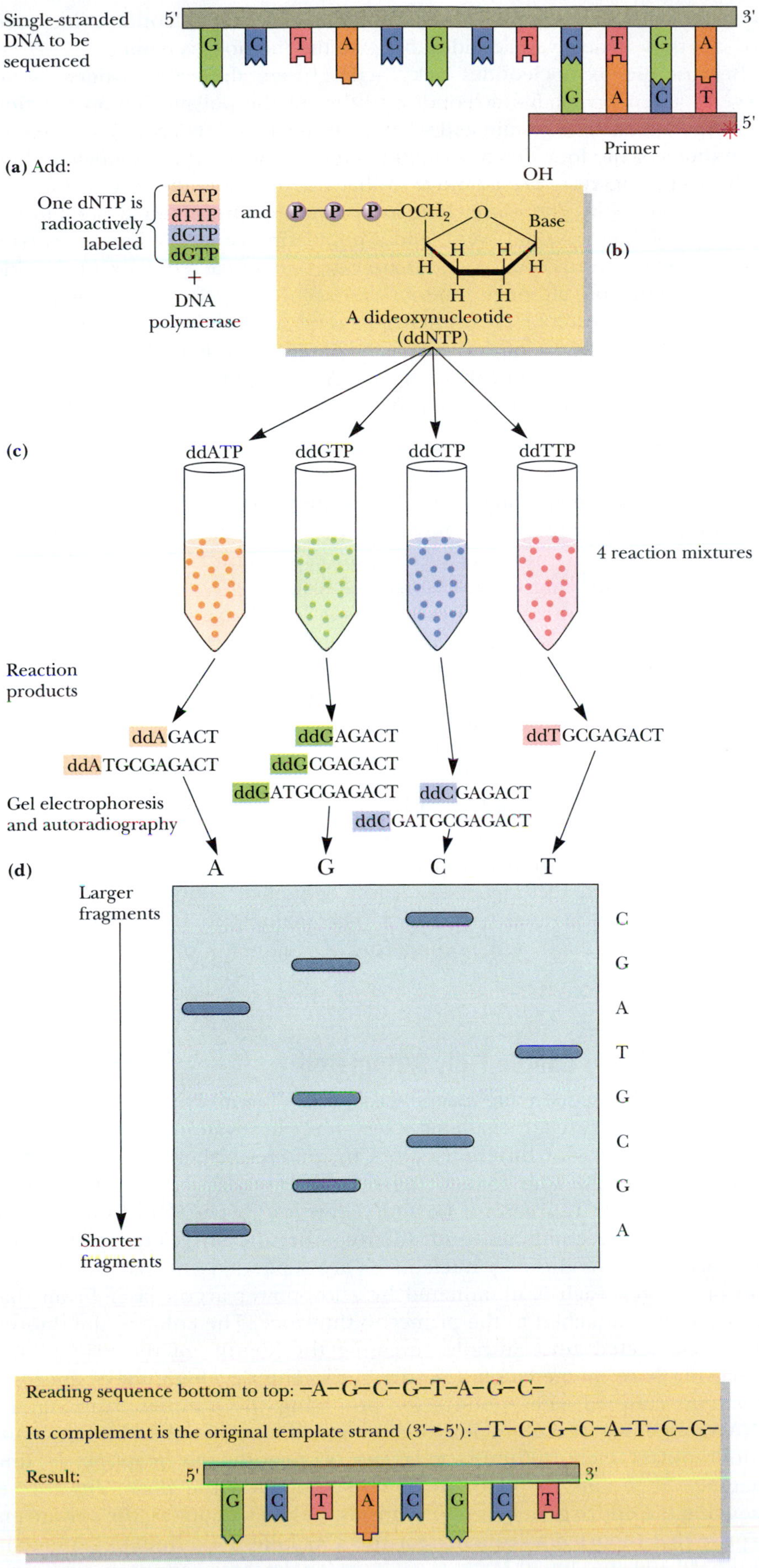

Single-stranded DNA to be sequenced

5' G C T A C G C T C T G A 3'

G A C T 5'

Primer

(a) Add:

One dNTP is radioactively labeled
{ dATP
dTTP
dCTP
dGTP }
and

+
DNA polymerase

P—P—P—OCH₂   O   Base

OH

H H H H
H H

A dideoxynucleotide (ddNTP)

(b)

(c)   ddATP   ddGTP   ddCTP   ddTTP

4 reaction mixtures

Reaction products

ddAGACT   ddGAGACT   ddTGCGAGACT
ddATGCGAGACT   ddGCGAGACT
               ddGATGCGAGACT   ddCGAGACT
Gel electrophoresis and autoradiography   ddCGATGCGAGACT

(d)   A   G   C   T

Larger fragments

C
G
A
T
G
C
G
A

Shorter fragments

Biochemistry ⊗ Now™   **ACTIVE FIGURE 11.3**
The chain termination or dideoxy method of DNA sequencing. **(a)** DNA polymerase reaction. **(b)** Structure of dideoxynucleotide. **(c)** Four reaction mixtures with nucleoside triphosphates plus one dideoxynucleoside triphosphate. **(d)** Electrophoretogram. Note that the nucleotide sequence as read from the bottom to the top of the gel is the order of nucleotide addition carried out by DNA polymerase. **Test yourself on the concepts in this figure at http://chemistry.brookscole.com/ggb3**

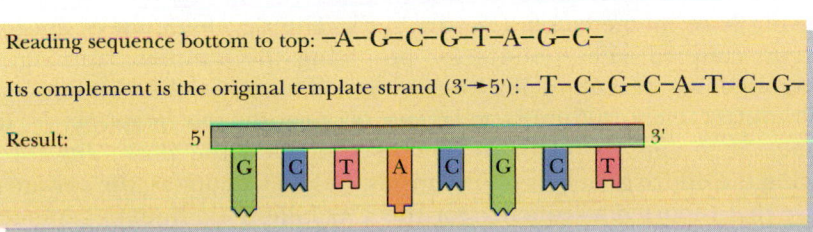

Reading sequence bottom to top: –A–G–C–G–T–A–G–C–

Its complement is the original template strand (3'→5'): –T–C–G–C–A–T–C–G–

Result: 5'   3'

G C T A C G C T

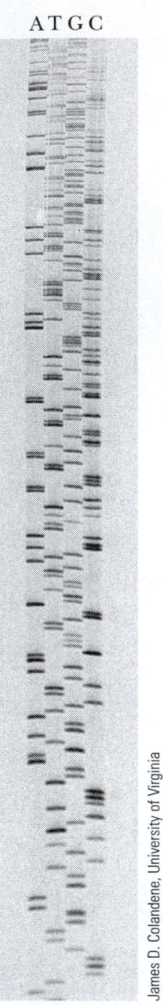

A T G C

James D. Colandene, University of Virginia

FIGURE 11.4 A photograph of the autoradiogram from an actual sequencing gel. A portion of the DNA sequence of *nit-6,* the *Neurospora* gene encoding the enzyme nitrite reductase.

In each of the four reactions, a different 2',3'-**di**deoxynucleotide is included, and it is these dideoxynucleotides that give the method its name.

Because dideoxynucleotides lack 3'-OH groups, these nucleotides cannot serve as acceptors for 5'-nucleotide addition in the polymerization reaction, and thus the chain is terminated where they become incorporated. The concentrations of the four deoxynucleotides and the single dideoxynucleotide in each reaction mixture are adjusted so that the dideoxynucleotide is incorporated infrequently. Therefore, base-specific premature chain termination is only a random, occasional event, and a population of new strands of varying lengths is synthesized. Four reactions are run, one for each dideoxynucleotide, so that termination, although random, can occur everywhere in the sequence. In each mixture, each newly synthesized strand has a dideoxynucleotide at its 3'-end, and its presence at that position demonstrates that a base of that particular kind was specified by the template. A radioactively labeled dNTP is included in each reaction mixture to provide a tracer for the products of the polymerization process.

Reading Dideoxy Sequencing Gels   The sequencing products are visualized by autoradiography (or similar means) following their separation according to size by polyacrylamide gel electrophoresis (Figure 11.3). Because the smallest fragments migrate fastest upon electrophoresis and because fragments differing by only single nucleotides in length are readily resolved, the autoradiogram of the gel can be read from bottom to top, noting which lane has the next largest band at each step. Thus, the gel in Figure 11.3 is read AGCGTAGC (5'→3'). Because of the way DNA polymerase acts, this observed sequence is complementary to the corresponding unknown template sequence. Knowing this, the template sequence now can be written GCTACGCT (5'→3').

With such simple technology, it is possible to read the order of as many as 400 bases from the autoradiogram of a sequencing gel (Figure 11.4). The actual enzymatic reactions, electrophoresis, and autoradiography are routine, and a skilled technician can easily sequence about several kbp per week using these manual techniques. The major effort in DNA sequencing is in the isolation and preparation of fragments of interest, such as cloned genes.

DNA Sequencing Can Be Fully Automated

Automated DNA sequencing machines capable of identifying about 10^5 bases per day are commercially available. One clever innovation has been the use of fluorescent dyes of different colors to uniquely label the primer DNA introduced into the four sequencing reactions; for example, red for the A reaction, blue for T, green for G, and yellow for C. Then, all four reaction mixtures can be combined and run together in one lane of the electrophoretic gel. As the oligonucleotides are separated and pass to the bottom of the gel, each is illuminated by a low-power argon laser beam that causes the dye attached to the primer to fluoresce. The color of the fluorescence is detected automatically, revealing the identity of the primer, and hence the base, immediately (Figure 11.5). The development of such automation, coupled with robotics for preparing the samples, running the DNA sequencing reactions, loading the chain-terminated DNA fragments onto capillary electrophoresis gels, performing the electrophoresis, and imaging the results for computer analysis, opened the possibility for sequencing the entire genomes of organisms. Celera Genomics, the private enterprise that reported a sequence for the 2.91 billion–bp human genome in 2001 used 300 automated DNA sequencers/analyzers to sequence more than 1 billion bases every month.

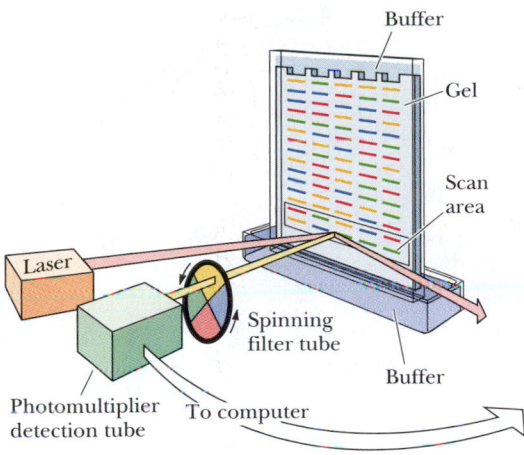

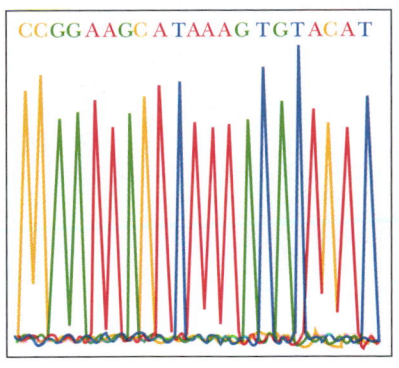

CCGGAAGCATAAAGTGTACAT

FIGURE 11.5 Schematic diagram of the methodology used in fluorescent labeling and automated sequencing of DNA. Four reactions are set up, one for each base, and the primer in each is end-labeled with one of four different fluorescent dyes; the dyes serve to color-code the base-specific sequencing protocol (a unique dye is used in each dideoxynucleotide reaction). The four reaction mixtures are then combined and run in one lane. Thus, each lane in the gel represents a different sequencing experiment. As the differently sized fragments pass down the gel, a laser beam excites the dye in the scan area. The emitted energy passes through a rotating color filter and is detected by a fluorometer. The color of the emitted light identifies the final base in the fragment. (*Applied Biosystems, Inc., Foster City, CA.*)

| 11.2 | What Sorts of Secondary Structures Can Double-Stranded DNA Molecules Adopt? |

Double-stranded DNA molecules assume one of three secondary structures, termed A, B, and Z. In a moment, we will address the "ABZs of DNA secondary structure"; first we must consider some general features of DNA double helices. Fundamentally, double-stranded DNA is a regular two-chain structure with hydrogen bonds formed between opposing bases on the two chains (see Chapter 10). Such H bonding is possible only when the two chains are antiparallel. The polar sugar–phosphate backbones of the two chains are on the outside. The bases are stacked on the inside of the structure; these heterocyclic bases, as a consequence of their π-electron clouds, are hydrophobic on their flat sides. One purely hypothetical conformational possibility for a two-stranded arrangement would be a ladderlike structure (Figure 11.6) in which the base pairs are fixed at 0.6 nm apart because this is the distance between adjacent sugars in the DNA backbone. Because H_2O molecules would be accessible to the spaces between the hydrophobic surfaces of the bases, this conformation is energetically unfavorable. This ladderlike structure converts to a helix when given a simple right-handed twist. Helical twisting brings the base-pair rungs of the ladder closer together, stacking them 0.34 nm apart, without affecting the sugar–sugar distance of 0.6 nm. Because this helix repeats itself approximately every 10 bp, its **pitch** is 3.4 nm. This is the major conformation of DNA in solution, and it is called **B-DNA.**

Watson–Crick Base Pairs Have Virtually Identical Dimensions

As indicated in Chapter 10, the base pairing in DNA is very specific: The purine adenine pairs with the pyrimidine thymine; the purine guanine pairs with the pyrimidine cytosine. Furthermore, the A:T pair and G:C pair have virtually identical dimensions (Figure 11.7). Watson and Crick realized that units of such structural equivalence could serve as spatially invariant substructures to build a polymer whose exterior dimensions would be uniform along its length, regardless of the sequence of bases.

The DNA Double Helix Is a Stable Structure

Several factors account for the stability of the double helical structure of DNA.

H BONDS Although it has long been emphasized that the two strands of DNA are held together by H bonds formed between the complementary purines and pyrimidines, two in an A:T pair and three in a G:C pair (Figure 11.7), the H bonds between base pairs impart little net stability to the double-stranded

(a) Ladder

Base-pair spacing   0.6 nm

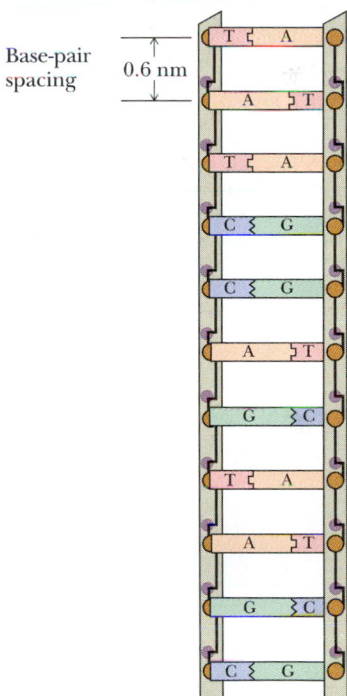

(b) Helix

Base-pair spacing   0.34 nm

Pitch length 3.4 nm

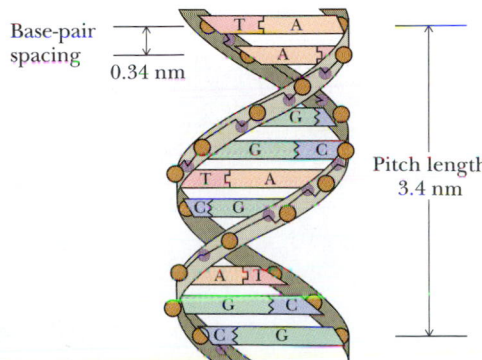

FIGURE 11.6 **(a)** Double-stranded DNA as an imaginary ladderlike structure. **(b)** A simple right-handed twist converts the ladder to a helix.

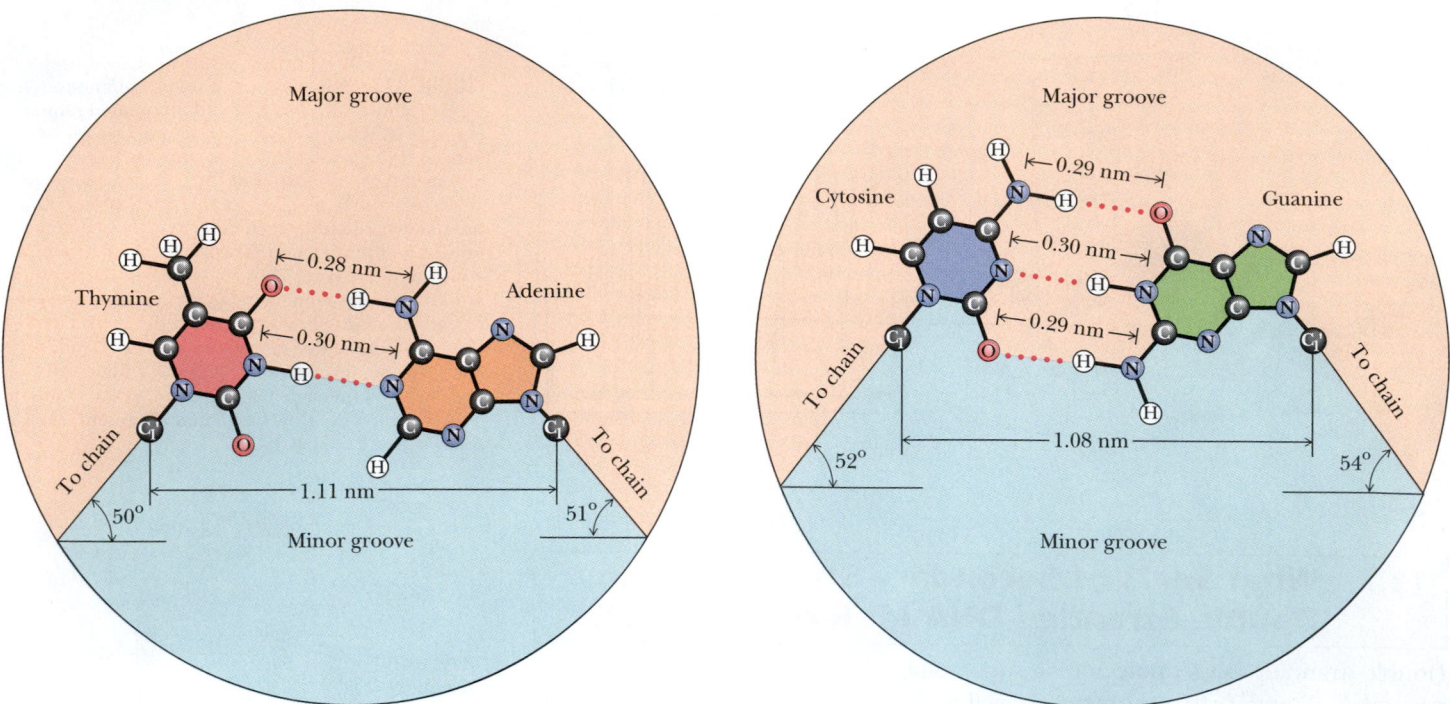

FIGURE 11.7 Watson–Crick A:T and G:C base pairs. All H bonds in both base pairs are straight, with each H atom pointing directly at its acceptor N or O atom. Linear H bonds are the strongest. The mandatory binding of larger purines with smaller pyrimidines leads to base pairs that have virtually identical dimensions, allowing the two sugar–phosphate backbones to adopt identical helical conformations.

structure compared to the separated strands in solution. When the two strands of the double helix are separated, the H bonds between base pairs are replaced by H bonds between individual bases and surrounding water molecules. Polar atoms in the sugar–phosphate backbone do form external H bonds with surrounding water molecules, but these form with separated strands as well.

ELECTROSTATIC INTERACTIONS   The negatively charged phosphate groups are all situated on the exterior surface of the helix in such a way that repulsive effects on one another are minimized. In fact, the phosphate ions are electrostatically shielded from one another because divalent cations, particularly Mg^{2+}, bind strongly to the anionic phosphates.

VAN DER WAALS AND HYDROPHOBIC INTERACTIONS   The core of the helix consists of the base pairs, and these base pairs stack together through π, π electronic interactions (a form of van der Waals interaction) and hydrophobic forces. These base-pair stacking interactions range from -16 to -51 kJ/mol (expressed as the energy of interaction between adjacent base pairs), contributing significantly to the overall stabilizing energy.

   A stereochemical consequence of the way A:T and G:C base pairs form is that the sugars of the respective nucleotides have opposite orientations, and thus the sugar–phosphate backbones of the two chains run in opposite or "antiparallel" directions. Furthermore, the two glycosidic bonds holding the bases in each base pair are not directly across the helix from each other, defining a common diameter (Figure 11.8). Consequently, the sugar–phosphate backbones of the helix are not equally spaced along the helix axis and the grooves between them are not the same size. Instead, the intertwined chains create a **major groove** and a **minor groove** (Figure 11.8). The edges of the base pairs have a specific relationship to these grooves. The "top" edges of the base pairs ("top" as defined by placing the glycosidic bond at the bottom, as in Figure 11.7) are exposed along the interior surface or "floor" of the major groove; the base-pair edges nearest to the glycosidic bond form the interior surface of the minor groove. Some proteins that bind to DNA can actually recognize specific nucleotide sequences by "reading" the pattern of H-bonding possibilities presented by the edges of the bases in these grooves. Such DNA–protein interactions provide one step toward under-

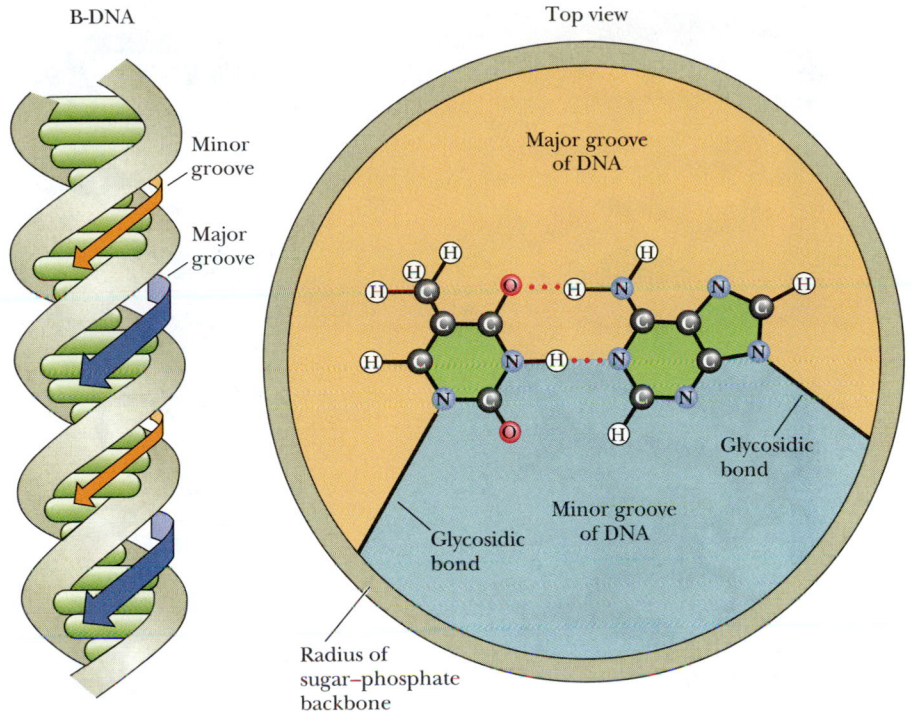

FIGURE 11.8 The bases in a base pair are not directly across the helix axis from one another along some diameter but rather are slightly displaced. This displacement, and the relative orientation of the glycosidic bonds linking the bases to the sugar–phosphate backbone, leads to differently sized grooves in the cylindrical column created by the double helix, the major groove, and the minor groove, each coursing along its length.

standing how cells regulate the expression of genetic information encoded in DNA (see Chapter 29).

Double Helical Structures Can Adopt a Number of Stable Conformations

In solution, DNA ordinarily assumes the structure we have been discussing: B-DNA. However, nucleic acids also occur naturally in other double helical forms. The base-pairing arrangement remains the same, but the sugar–phosphate groupings that constitute the backbone are inherently flexible and can adopt different conformations. One conformational variation is **propeller twist** (Figure 11.9). Propeller twist allows greater overlap between successive bases along a strand of DNA and diminishes the area of contact between bases and solvent water.

A-Form DNA Is an Alternative Form of Right-Handed DNA

An alternative form of the right-handed double helix is **A-DNA**. A-DNA molecules differ from B-DNA molecules in a number of ways. The pitch, or distance required to complete one helical turn, is different. In B-DNA, it is 3.4 nm, whereas in A-DNA it is 2.46 nm. One turn in A-DNA requires 11 bp to complete. Depending on local sequence, 10 to 10.6 bp define one helical turn in B-form DNA. In A-DNA, the base pairs are no longer nearly perpendicular to the helix axis but instead are tilted 19° with respect to this axis. Successive base pairs occur every 0.23 nm along the axis, as opposed to 0.332 nm in B-DNA. The B-form of DNA is thus longer and thinner than the short, squat A-form, which has its base pairs displaced around, rather than centered on, the helix axis. Figure 11.10 shows the relevant structural characteristics of the A- and B-forms of DNA. (Z-DNA, another form of DNA to be discussed shortly, is also depicted in Figure 11.10.) A comparison of the structural properties of A-, B-, and Z-DNA is summarized in Table 11.1.

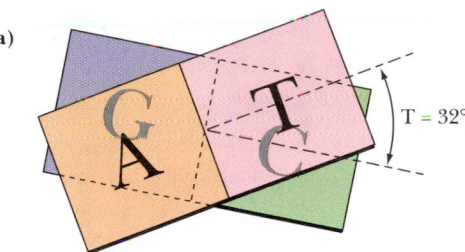

(a)

Two base pairs with 32° of right-handed helical twist: the *minor-groove edges are drawn with heavy shading.*

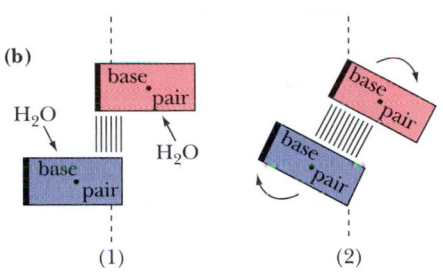

(b)

Propeller twist, as in (2), allows greater overlap of bases within the same strand and reduces the area of contact between the bases and water.

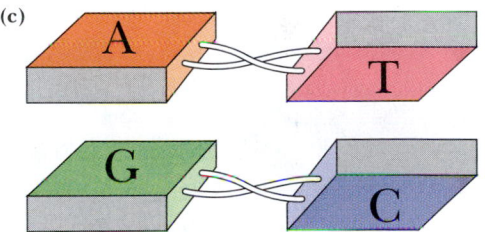

(c)

Propeller-twisted base pairs. Note how the hydrogen bonds between bases are distorted by this motion, yet remain intact. The minor-groove edges of the bases are shaded.

FIGURE 11.9 Helical twist and propeller twist in DNA. **(a)** Successive base pairs in B-DNA show a rotation with respect to each other (so-called helical twist) of 36° or so, as viewed down the cylindrical axis of the DNA. **(b)** Rotation in a different dimension—**propeller twist**—allows the hydrophobic surfaces of bases to overlap better. The view here is edge-on to two successive bases in one DNA strand (as if the two bases on the right-hand strand of DNA in (a) were viewed from the right-hand margin of the page; dots represent end-on views down the glycosidic bonds). Clockwise rotation (as shown here) has a positive sign. **(c)** The two bases on the left-hand strand of DNA in (a) also show positive propeller twist [a clockwise rotation of the two bases in (a) as viewed from the left-hand margin of the paper]. *(Adapted from Figure 3.4 in Callandine, C. R., and Drew, H. R., 1992. Understanding DNA: The Molecule and How It Works. London: Academic Press.)*

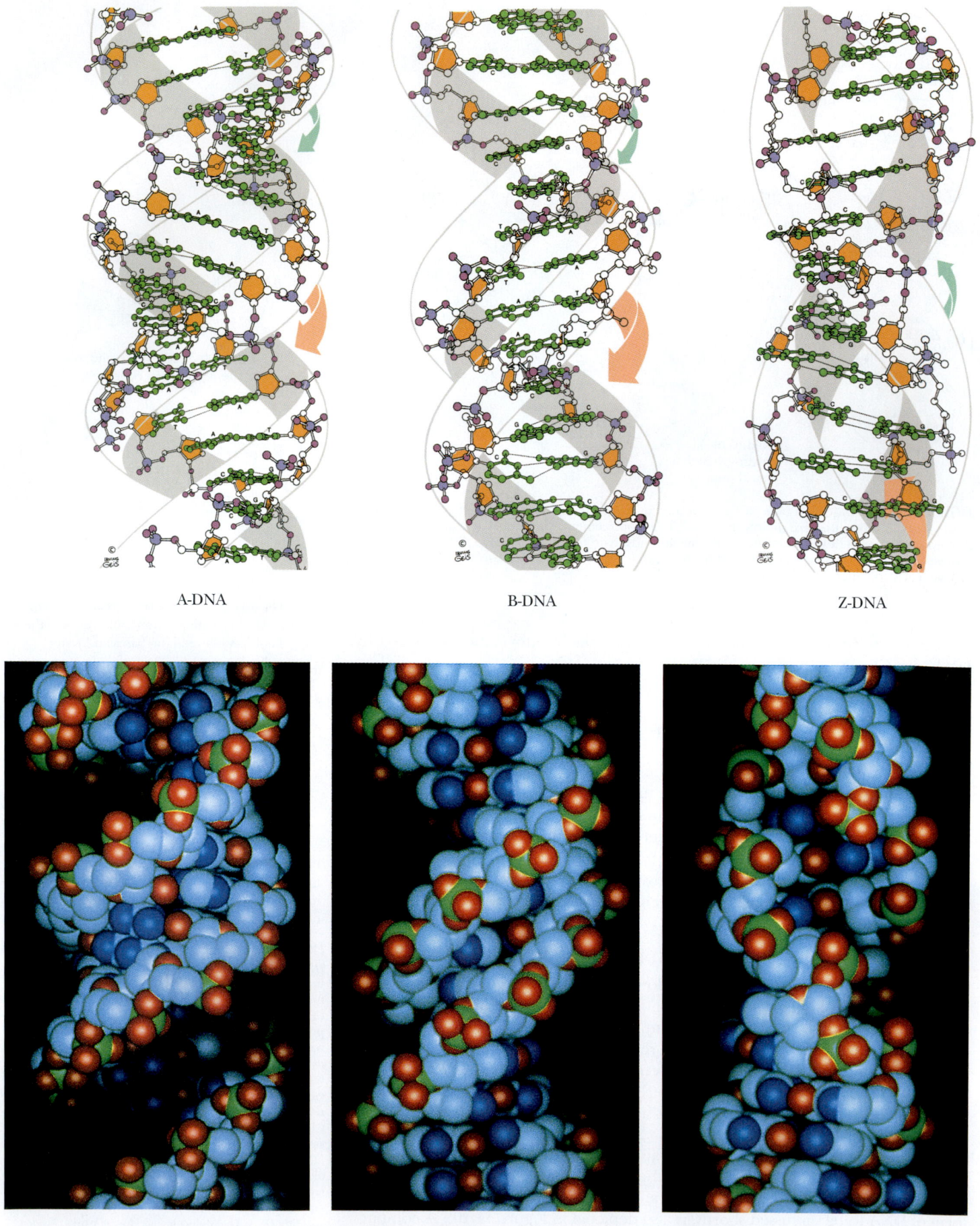

A-DNA   B-DNA   Z-DNA

FIGURE 11.10 DNA forms.

continued

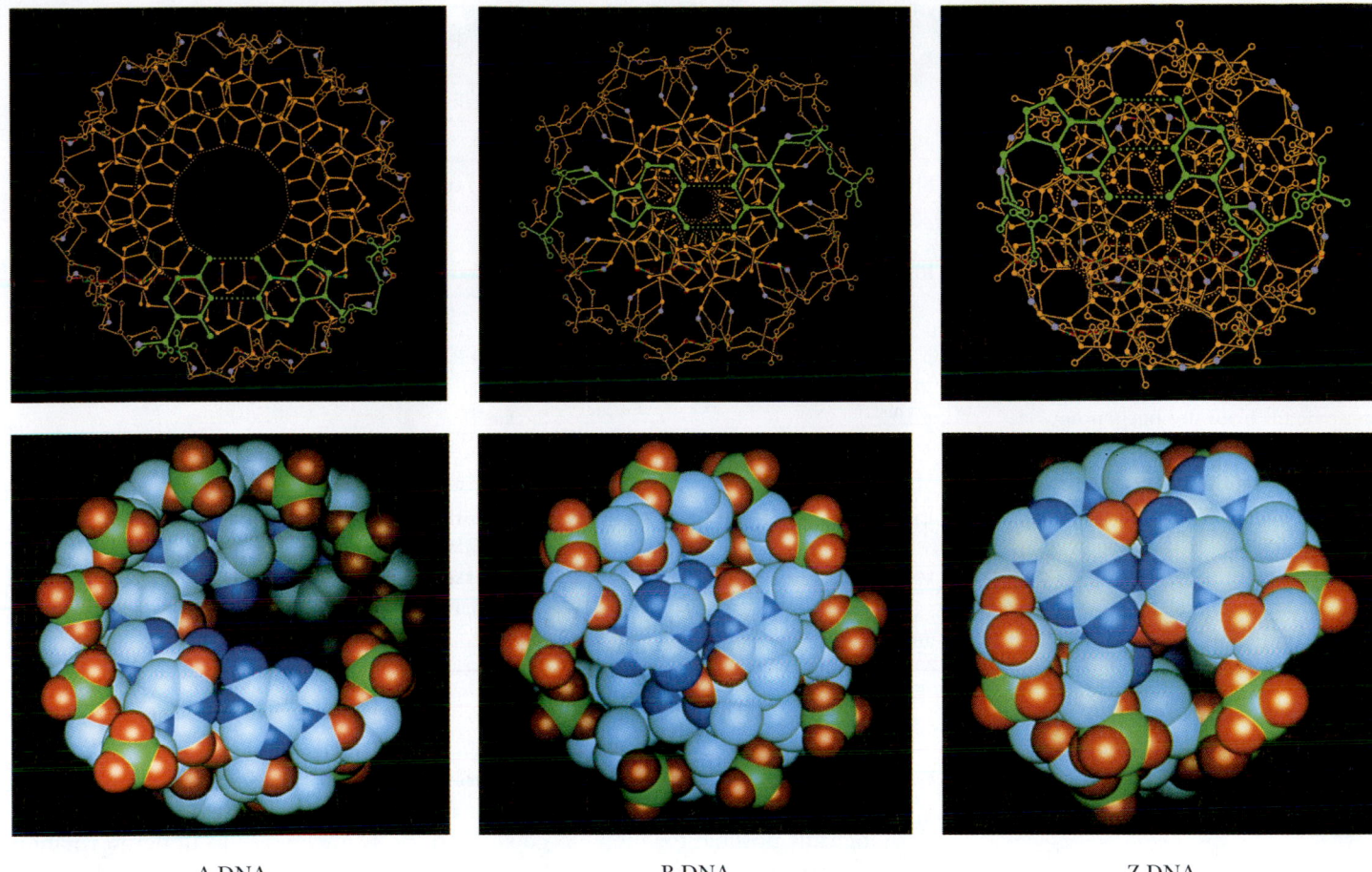

A-DNA   B-DNA   Z-DNA

FIGURE 11.10 (*here and on the facing page*) Comparison of the A-, B-, and Z-forms of the DNA double helix. The distance required to complete one helical turn is shorter in A-DNA than it is in B-DNA. The alternating pyrimidine–purine sequence of Z-DNA is the key to the "left-handedness" of this helix. (*Illustrations: Irving Geis; computer images: Robert Stodola, Fox Chase Cancer Research Center and Irving Geis.*)

Although relatively dehydrated DNA fibers can be shown to adopt the A-conformation under physiological conditions, it is unclear whether DNA ever assumes this form in vivo. However, double helical DNA:RNA hybrids probably have an A-like conformation. The 2′-OH in RNA sterically prevents double helical regions of RNA chains from adopting the B-form helical arrangement. Importantly, double-stranded regions in RNA chains often assume an A-like conformation, with their bases strongly tilted with respect to the helix axis.

Z-DNA Is a Conformational Variation in the Form of a Left-Handed Double Helix

Z-DNA was first recognized by Alexander Rich and his colleagues at MIT in X-ray analysis of the synthetic deoxynucleotide dCpGpCpGpCpG, which crystallized into an antiparallel double helix of unexpected conformation. The alternating pyrimidine–purine (Py–Pu) sequence of this oligonucleotide is the key to its unusual properties. The *N*-glycosyl bonds of G residues in this alternating copolymer are rotated 180° with respect to their conformation in B-DNA, so now the purine ring is in the syn rather than the anti conformation (Figure 11.11). The C residues remain in the anti form. Because the G ring is "flipped," the C ring must also flip to maintain normal Watson–Crick base pairing. However, pyrimidine nucleosides do not readily adopt the syn conformation because it creates steric interference between the pyrimidine C-2 oxy substituent and atoms of the

Table 11.1

Comparison of the Structural Properties of A-, B-, and Z-DNA

	Double Helix Type		
	A	**B**	**Z**
Overall proportions	Short and broad	Longer and thinner	Elongated and slim
Rise per base pair	2.3 Å	3.32 Å ± 0.19 Å	3.8 Å
Helix packing diameter	25.5 Å	23.7 Å	18.4 Å
Helix rotation sense	Right-handed	Right-handed	Left-handed
Base pairs per helix repeat	1	1	2
Base pairs per turn of helix	~11	~10	12
Mean rotation per base pair	33.6°	35.9° ± 4.2°	−60°/2
Pitch per turn of helix	24.6 Å	33.2 Å	45.6 Å
Base-pair tilt from the perpendicular	+19°	−1.2° ± 4.1°	−9°
Base-pair mean propeller twist	+18°	+16° ± 7°	~0°
Helix axis location	Major groove	Through base pairs	Minor groove
Major groove proportions	Extremely narrow but very deep	Wide and with intermediate depth	Flattened out on helix surface
Minor groove proportions	Very broad but shallow	Narrow and with intermediate depth	Extremely narrow but very deep
Glycosyl bond conformation	anti	anti	anti at C, syn at G

Adapted from Dickerson, R. L., et al., 1982. *Cold Spring Harbor Symposium on Quantitative Biology* **47**:14.

pentose. Because the cytosine ring does not rotate relative to the pentose, the whole C nucleoside (base and sugar) must flip 180° (Figure 11.12). It is topologically possible for the G to go syn and the C nucleoside to undergo rotation by 180° without breaking and re-forming the G:C hydrogen bonds. In other words, the B-to-Z structural transition can take place without disrupting the bonding relationships among the atoms involved.

Because alternate nucleotides assume different conformations, the repeating unit on a given strand in the Z-helix is the dinucleotide. That is, for any number of bases, *n*, along one strand, *n* − 1 dinucleotides must be considered. For example, a GpCpGpC subset of sequence along one strand is composed of *three* successive dinucleotide units: GpC, CpG, and GpC. (In B-DNA, the nucleotide conformations are essentially uniform and the repeating unit is the mononucleotide.) It follows that the CpG sequence is distinct conformationally from the GpC sequence along the alternating copolymer chains in the Z-double helix. The conformational alterations going from B to Z realign the sugar–phosphate backbone along a zigzag course that has a left-handed orientation (Figure 11.10), thus the designation Z-DNA. Note that in any GpCpGp subset, the sugar–phosphates of GpC form the horizontal "zig" while the CpG backbone segment forms the

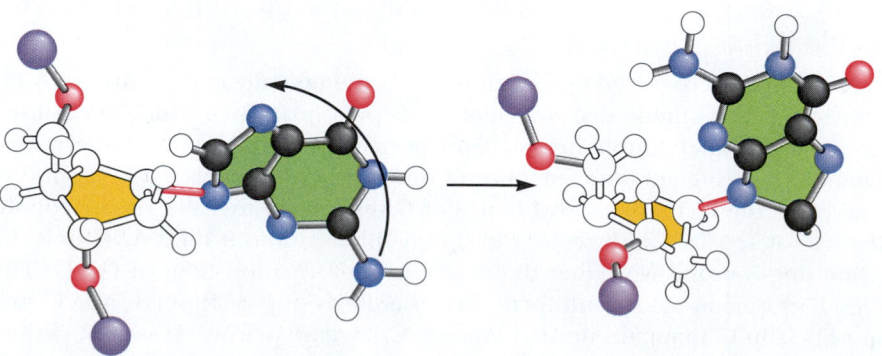

FIGURE 11.11 Comparison of the deoxyguanosine conformation in B- and Z-DNA. In B-DNA, the conformation about the C1′–N9 glycosyl bond is always anti **(left)**. In contrast, in the left-handed Z-DNA structure, this bond rotates (as shown) to adopt the syn conformation.

Deoxyguanosine in B-DNA (anti position)   Deoxyguanosine in Z-DNA (syn position)

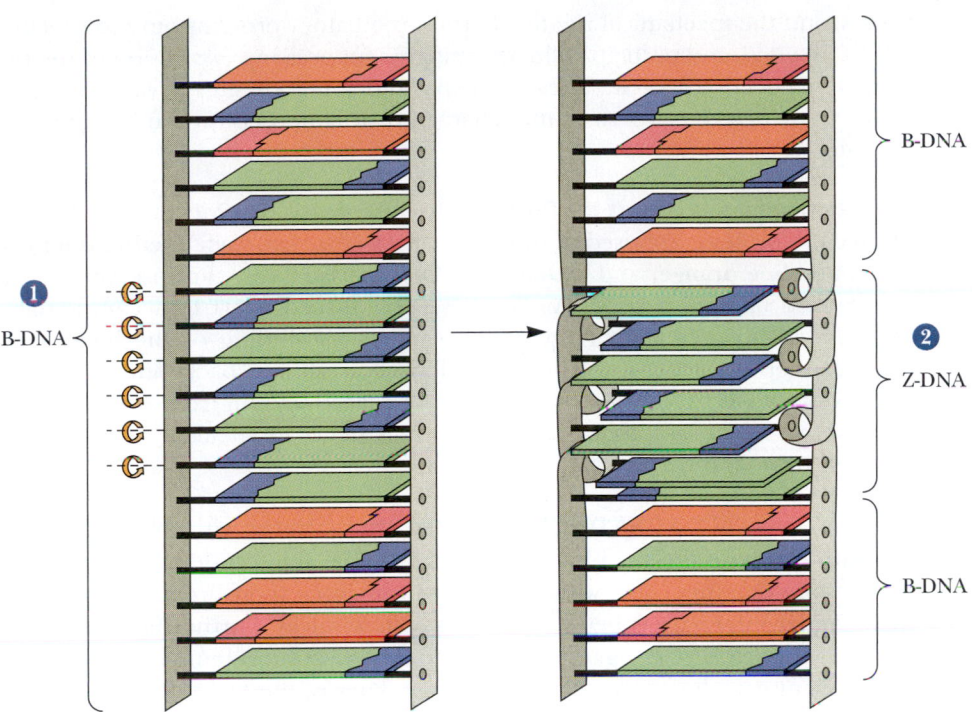

FIGURE 11.12 The change in topological relationships of base pairs from B- to Z-DNA. A six-base-pair segment of B-DNA (**1**) is converted to Z-DNA (**2**) through rotation of the base pairs, as indicated by the curved arrows. The purine rings (green) of the deoxyguanosine nucleosides rotate via an anti to syn change in the conformation of the guanine–deoxyribose glycosidic bond; the pyrimidine rings (blue) are rotated by flipping the entire deoxycytidine nucleoside (base *and* deoxyribose). As a consequence of these conformational changes, the base pairs in the Z-DNA region no longer share π, π stacking interactions with adjacent B-DNA regions.

vertical "zag." The mean rotation angle circumscribed around the helix axis is $-15°$ for a CpG step and $-45°$ for a GpC step (giving $-60°$ for the dinucleotide repeat). The minus sign denotes a left-handed or counterclockwise rotation about the helix axis. Z-DNA is more elongated and slimmer than B-DNA.

Cytosine Methylation and Z-DNA The Z-form can arise in sequences that are not strictly alternating Py–Pu. For example, the hexanucleotide ^{m5}CGAT^{m5}CG, a Py-Pu-Pu-Py-Py-Pu sequence containing two 5-methylcytosines (^{m5}C), crystallizes as Z-DNA. Indeed, the in vivo methylation of C at the 5-position is believed to favor a B-to-Z switch because, in B-DNA, these hydrophobic methyl groups would protrude into the aqueous environment of the major groove and destabilize its structure. In Z-DNA, the same methyl groups can form a stabilizing hydrophobic patch. It is likely that the Z-conformation naturally occurs in specific regions of cellular DNA, which otherwise is predominantly in the B-form. Furthermore, because methylation is implicated in gene regulation, the occurrence of Z-DNA may affect the expression of genetic information (see Part 4).

The Double Helix Is a Very Dynamic Structure

The long-range structure of B-DNA in solution is not a rigid, linear rod. Instead, DNA behaves as a dynamic, flexible molecule. Localized thermal fluctuations temporarily distort and deform DNA structure over short regions. Base and backbone ensembles of atoms undergo elastic motions on a time scale of nanoseconds. To some extent, these effects represent changes in rotational angles of the bonds comprising the polynucleotide backbone. These changes are also influenced by sequence-dependent variations in base-pair stacking. The consequence is that the helix bends gently. When these variations are summed over the great length of a DNA molecule, the net result of these bending motions is that at any given time, the double helix assumes a roughly spherical shape, as might be expected for a long, semirigid rod undergoing apparently random coiling. It is also worth noting that, on close scrutiny, the surface of the double helix is *not* that of a totally featureless, smooth, regular "barber pole" structure. Different base sequences impart their own special signatures to the molecule by subtle influences on such factors as the groove width, the angle between the helix axis and base planes, and the mechanical rigidity. Certain regulatory proteins bind to specific DNA sequences and participate in activating or suppressing expression of the information encoded therein. These proteins bind at unique sites by virtue of their ability to recognize novel structural characteristics imposed on the DNA by the local nucleotide sequence.

Intercalating Agents Distort the Double Helix Aromatic macrocycles, flat hydrophobic molecules composed of fused, heterocyclic rings, such as **ethidium bromide, acridine orange,** and **actinomycin D** (Figure 11.13), can insert between the stacked base pairs of DNA. The bases are forced apart to accommodate these so-called **intercalating agents,** causing an unwinding of the helix to a more ladderlike structure. The deoxyribose–phosphate backbone is almost fully extended as successive base pairs are displaced 0.7 nm from one another, and the rotational angle about the helix axis between adjacent base pairs is reduced from 36° to 10°.

Dynamic Nature of the DNA Double Helix in Solution Intercalating substances insert with ease into the double helix, indicating that the van der Waals stacking interactions that they share with the bases sandwiching them are more favorable than similar interactions between the bases themselves. Furthermore, the fact that these agents slip in suggests that the double helix must temporarily unwind and present gaps for these agents to occupy. That is, the DNA double helix in solution must be represented by a set of metastable alternatives to the standard B-conformation. These alternatives constitute a flickering repertoire of dynamic structures.

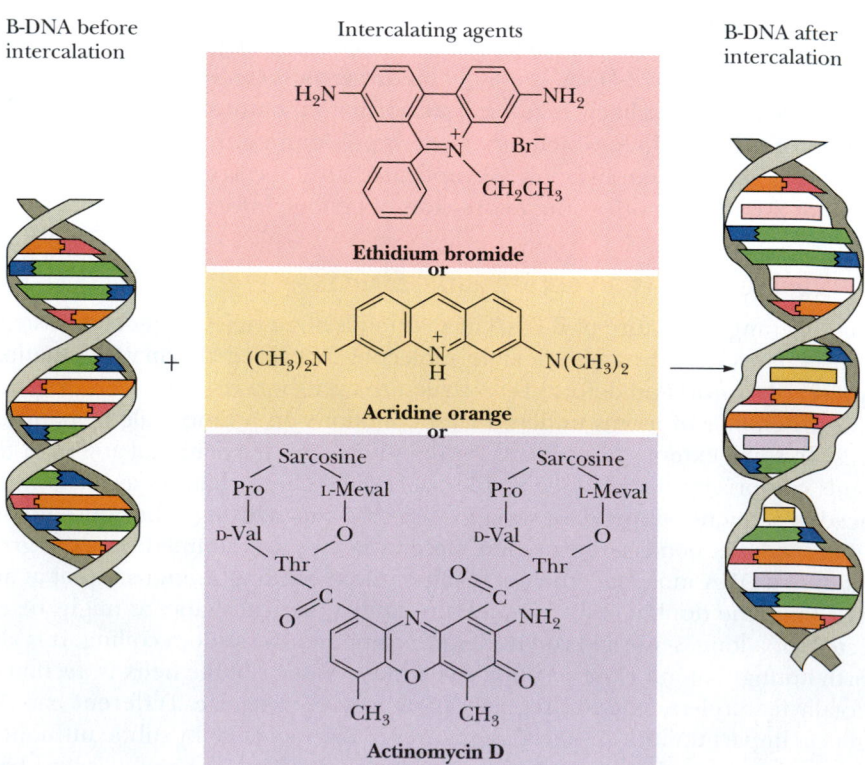

FIGURE 11.13 The structures of ethidium bromide, acridine orange, and actinomycin D, three intercalating agents, and their effects on DNA structure.

Can the Secondary Structure of DNA
Be Denatured and Renatured?

Thermal Denaturation of DNA Can Be Observed by Changes in UV Absorbance

When duplex DNA molecules are subjected to conditions of pH, temperature, or ionic strength that disrupt base-pairing interactions, the strands are no longer held together. That is, the double helix is **denatured,** and the strands separate as individual random coils. If temperature is the denaturing agent, the double helix is said to *melt*. The course of this dissociation can be followed spectrophotometrically because the relative absorbance of the DNA solution at 260 nm increases as much as 40% as the bases unstack. This absorbance increase, or **hyperchromic shift,** is due to the fact that the aromatic bases in DNA interact via their π-electron clouds when stacked together in the double helix. Because the UV absorbance of the bases is a consequence of π electron transitions, and because the potential for these transitions is diminished when the bases stack, the bases in duplex DNA absorb less 260-nm radiation than expected for their numbers. Unstacking alleviates this suppression of UV absorbance. The rise in absorbance coincides with strand separation, and the midpoint of the absorbance increase is termed the **melting temperature,** T_m (Figure 11.14). DNAs differ in their T_m values because they differ in relative G + C content. The higher the G + C content of a DNA, the higher its melting temperature because G:C pairs have higher base stacking energies than A:T pairs. The dependence of T_m on the G + C content is depicted in Figure 11.15. Also note that T_m is dependent on the ionic strength of the solution; the lower the ionic strength, the lower the melting temperature. At 0.2 M Na$^+$, $T_m = 69.3 + 0.41(\% \text{ G} + \text{C})$. Ions suppress the electrostatic repulsion between the negatively charged phosphate groups in the complementary strands of the helix, thereby stabilizing it. (DNA in pure water melts even at room temperature.) At high concentrations of ions, T_m is raised and the transition between helix and coil is sharp.

pH Extremes or Strong H-Bonding Solutes Also Denature DNA Duplexes

At pH values greater than 10, extensive deprotonation of the bases occurs, destroying their base-pairing potential and denaturing the DNA duplex. Similarly, extensive protonation of the bases below pH 2.3 disrupts base pairing. Alkali is

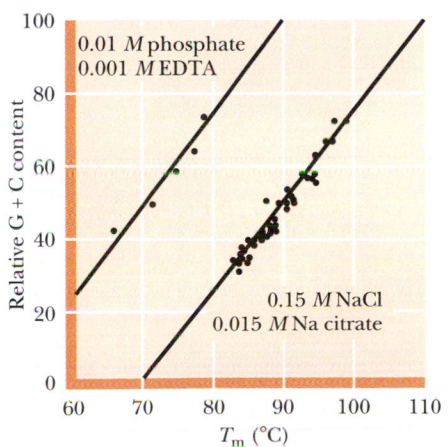

FIGURE 11.15 The dependence of melting temperature on relative (G + C) content in DNA. Note that T_m increases as ionic strength is raised at constant pH (pH 7); 0.01 M phosphate + 0.001 M EDTA versus 0.15 M NaCl/0.015 M Na citrate. In 0.15 M NaCl/0.015 M Na citrate, duplex DNA consisting of 100% A:T pairs melts at less than 70°C, whereas DNA of 100% G:C has a T_m greater than 110°C. *(From Marmur, J., and Doty, P., 1962. Determination of the base composition of deoxyribonucleic acid from its thermal denaturation temperature.* Journal of Molecular Biology **5:**120.)

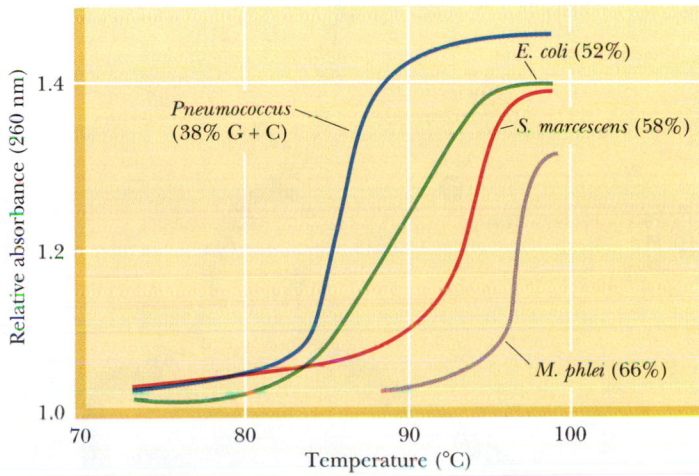

FIGURE 11.14 Heat denaturation of DNA from various sources, so-called melting curves. The midpoint of the melting curve is defined as the melting temperature, T_m. *(From Marmur, J., 1959. Heterogenity in deoxyribonucleic acids.* Nature **183:**1427–1429.)

the preferred denaturant because, unlike acid, it does not hydrolyze the glycosidic linkages in the sugar–phosphate backbone. Small solutes that readily form H bonds are also DNA denaturants at temperatures below T_m if present in sufficiently high concentrations to compete effectively with the H bonding between the base pairs. Examples include formamide and urea.

Single-Stranded DNA Can Renature to Form DNA Duplexes

Denatured DNA will **renature** to re-form the duplex structure if the denaturing conditions are removed (that is, if the solution is cooled, the pH is returned to neutrality, or the denaturants are diluted out). Renaturation requires reassociation of the DNA strands into a double helix, a process termed **reannealing.** For this to occur, the strands must realign themselves so that their complementary bases are once again in register and the helix can be zipped up (Figure 11.16). Renaturation is dependent on both DNA concentration and time. Many of the realignments are imperfect, and thus the strands must dissociate again to allow for proper pairings to be formed. The process occurs more quickly if the temperature is warm enough to promote diffusion of the large DNA molecules but not so warm as to cause melting.

The Rate of DNA Renaturation Is an Index of DNA Sequence Complexity

The renaturation rate of DNA is an excellent indicator of the sequence complexity of DNA. For example, bacteriophage T_4 DNA contains about 2×10^5 nucleotide pairs, whereas *Escherichia coli* DNA possesses 4.64×10^6. *E. coli* DNA is considerably more complex in that it encodes more information. Expressed another way, for any given amount of DNA (in grams), the sequences represented in an *E. coli* sample are more heterogeneous, that is, more dissimilar from one another, than those in an equal weight of phage T_4 DNA. Therefore, it will take the *E. coli* DNA strands longer to find their complementary partners and reanneal. This situation can be analyzed quantitatively.

If c is the concentration of single-stranded DNA at time t, then the second-order rate equation for two complementary strands coming together is given by the rate of decrease in c:

$$-dc/dt = k_2 c^2$$

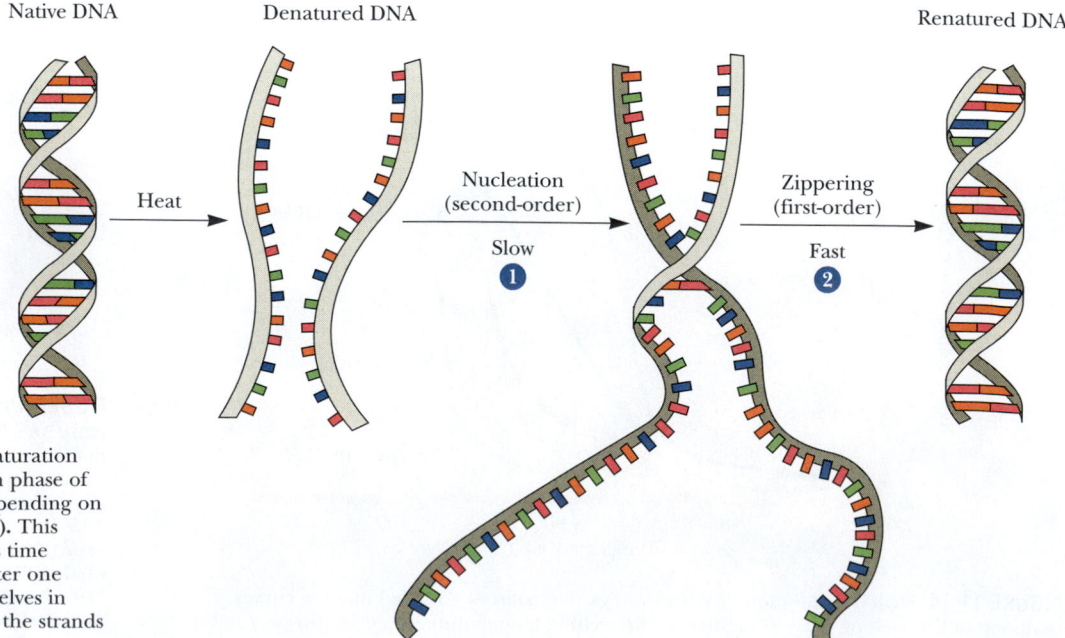

FIGURE 11.16 Steps in the thermal denaturation and renaturation of DNA. The nucleation phase of the reaction is a second-order process depending on sequence alignment of the two strands (**1**). This process takes place slowly because it takes time for complementary sequences to encounter one another in solution and then align themselves in register. Once the sequences are aligned, the strands zipper up quickly (**2**).

Native DNA   Denatured DNA   Renatured DNA

Heat

Nucleation (second-order)   Slow   **1**

Zippering (first-order)   Fast   **2**

where k_2 is the second-order rate constant. Starting with a concentration, c_0, of completely denatured DNA at $t = 0$, the amount of single-stranded DNA remaining at some time t is

$$c/c_0 = 1/(1 + k_2 c_0 t)$$

where the units of c are mol of nucleotide per L and t is in seconds. The time for half of the DNA to renature (when $c/c_0 = 0.5$) is defined as $t = t_{1/2}$. Then,

$$0.5 = 1/(1 + k_2 c_0 t_{1/2}) \quad \text{and thus} \quad 1 + k_2 c_0 t_{1/2} = 2$$

yielding

$$c_0 t_{1/2} = 1/k_2$$

A graph of the fraction of single-stranded DNA reannealed (c/c_0) as a function of $c_0 t$ on a semilogarithmic plot is referred to as a $c_0 t$ **curve** ($c_0 t$ is pronounced "cot") (Figure 11.17). The rate of reassociation can be followed spectrophotometrically by the UV absorbance decrease as duplex DNA is formed. Note that relatively more complex DNAs take longer to renature, as reflected by their greater $c_0 t_{1/2}$ values. Poly A and poly U (Figure 11.17) are minimally complex in sequence and anneal rapidly to form a double-stranded A:U polynucleotide. *Mouse satellite DNA* is a highly repetitive subfraction of mouse DNA. Its lack of sequence heterogeneity is seen in its low $c_0 t_{1/2}$ value. MS-2 is a small bacteriophage whose genetic material is RNA. Calf thymus DNA is the mammalian representative in Figure 11.17.

Nucleic Acid Hybridization: Different DNA Strands of Similar Sequence Can Form Hybrid Duplexes

If DNA from two different species are mixed, denatured, and allowed to cool slowly so that reannealing can occur, artificial **hybrid duplexes** may form, provided the DNA from one species is similar in nucleotide sequence to the DNA of the other. The degree of hybridization is a measure of the sequence similarity or *relatedness* between the two species. Depending on the conditions of the experiment, about 25% of the DNA from a human forms hybrids with mouse DNA, implying that some of the nucleotide sequences (genes) in humans are very similar to those in mice (Figure 11.18). Mixed RNA:DNA hybrids can be created in vitro if single-stranded DNA is allowed to anneal with RNA copies of itself, such as those formed when genes are transcribed into mRNA molecules.

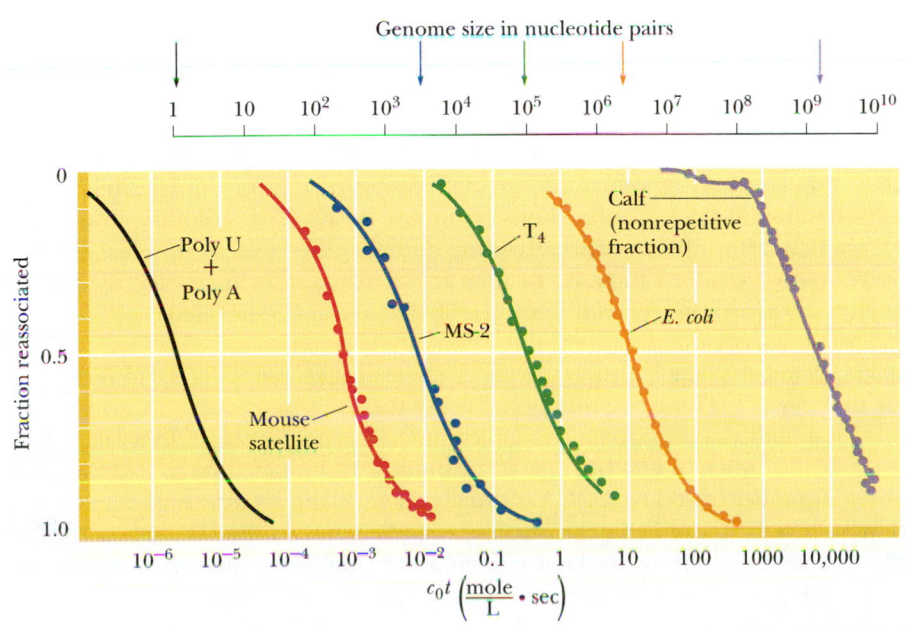

FIGURE 11.17 These $c_0 t$ curves show the rates of reassociation of denatured DNA from various sources and illustrate how the rate of reassociation is inversely proportional to genome complexity. The DNA sources are as follows: poly A + poly U, a synthetic DNA duplex of poly A and poly U polynucleotide chains; mouse satellite DNA, a fraction of mouse DNA in which the same sequence is repeated many thousands of times; MS-2 dsRNA, the double-stranded form of RNA found during replication of MS-2, a simple bacteriophage; T₄ DNA, the DNA of a more complex bacteriophage; *E. coli* DNA, bacterial DNA; calf DNA (nonrepetitive fraction), mammalian DNA (calf) from which the highly repetitive DNA fraction (satellite DNA) has been removed. Arrows indicate the genome size (in bp) of the various DNAs. *(From Britten, R. J., and Kohne, D. E., 1968. Repeated sequences in DNA. Hundreds of thousands of copies of DNA sequences have been incorporated into genomes of higher organisms. Science **161**:529–540.)*

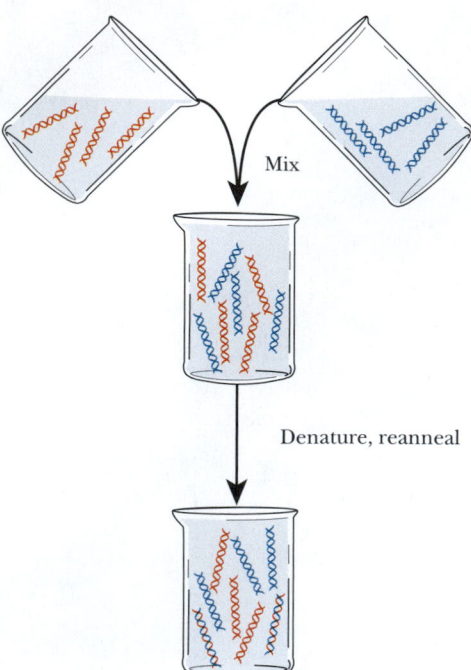

FIGURE 11.18 Solutions of human DNA (red) and mouse DNA (blue) are mixed and denatured, and the single strands are allowed to reanneal. About 25% of the human DNA strands form hybrid duplexes (one red and one blue strand) with mouse DNA.

Nucleic acid hybridization is a commonly employed procedure in molecular biology. First, it can reveal evolutionary relationships. Second, it gives researchers the power to identify specific genes selectively against a vast background of irrelevant genetic material. An appropriately labeled oligonucleotide or polynucleotide, referred to as a **probe,** is constructed so that its sequence is complementary to a target gene. The probe specifically base pairs with the target gene, allowing identification and subsequent isolation of the gene. Also, the quantitative expression of genes (in terms of the amount of mRNA synthesized) can be assayed by hybridization experiments.

The Buoyant Density of DNA Is an Index of Its G:C Content

Not only the melting temperature of DNA but also its density in solution is dependent on relative G:C content. G:C-rich DNA has a significantly higher density than A:T-rich DNA. Furthermore, a linear relationship exists between the buoyant densities of DNA from different sources and their G:C content (Figure 11.19). The density of DNA, ρ (in g/mL), as a function of its G:C content is given by the equation $\rho = 1.660 + 0.098(GC)$, where (GC) is the mole fraction of (G + C) in the DNA. Because of its relatively high density, DNA can be purified from cellular material by a form of density gradient centrifugation known as *isopycnic centrifugation* (see Chapter Appendix).

11.4 | What Is the Tertiary Structure of DNA?

The conformations of DNA discussed thus far are variations sharing a common secondary structural theme, the double helix, in which the DNA is assumed to be in a regular, linear form. DNA can also adopt regular structures of higher complexity in several ways. For example, many DNA molecules are circular. Most, but not all, bacterial chromosomes are covalently closed, circular DNA duplexes, as are most plasmid DNAs. **Plasmids** are naturally occurring, self-replicating, extrachromosomal DNA molecules found in bacteria; plasmids carry genes specifying novel metabolic capacities advantageous to the host bacterium. Various animal virus DNAs are circular as well.

Supercoils Are One Kind of DNA Tertiary Structure

In duplex DNA, the two strands are wound about each other once every 10 bp, that is, once every turn of the helix. Double-stranded circular DNA (or linear DNA duplexes whose ends are not free to rotate) form **supercoils** if the strands are underwound (*negatively supercoiled*) or overwound (*positively supercoiled*) (Figure 11.20). Underwound duplex DNA has fewer than the normal number of turns, whereas overwound DNA has more. DNA supercoiling is analogous to twisting or untwisting a two-stranded rope so that it is torsionally stressed. Negative supercoiling introduces a torsional stress that favors unwinding of the right-handed B-DNA double helix, whereas positive supercoiling overwinds such a helix. Both forms of supercoiling compact the DNA so that it sediments faster upon ultracentrifugation or migrates more rapidly in an electrophoretic gel in comparison to **relaxed DNA** (DNA that is not supercoiled).

Linking Number   The basic parameter characterizing supercoiled DNA is the **linking number** (L). This is the number of times the two strands are intertwined, and provided both strands remain covalently intact, L cannot change. In a relaxed circular DNA duplex of 400 bp, L is 40 (assuming 10 bp per turn in B-DNA). The linking number for relaxed DNA is usually taken as the reference parameter and is written as L_0. L can be equated to the **twist** (T) and **writhe** (W) of the duplex, where twist is the number of helical turns and writhe is the number of supercoils:

$$L = T + W$$

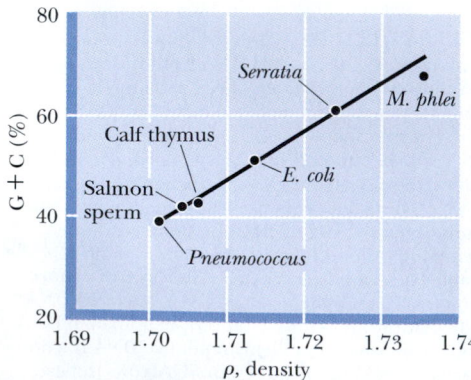

FIGURE 11.19 The relationship of the densities (in g/mL) of DNAs from various sources and their G:C content. *(From Doty, P., 1961. Harvey Lectures 55:103.)*

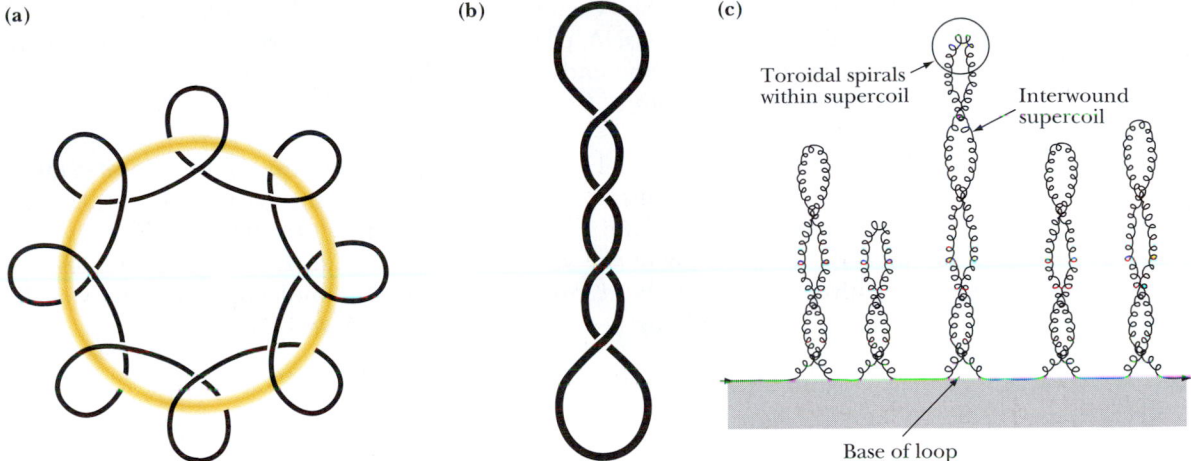

FIGURE 11.20 Toroidal and interwound varieties of DNA supercoiling. **(a)** The DNA is coiled in a spiral fashion about an imaginary toroid. **(b)** The DNA interwinds and wraps about itself. **(c)** Supercoils in long, linear DNA arranged into loops whose ends are restrained—a model for chromosomal DNA. *(Adapted from Figures 6.1 and 6.2 in Callandine, C. R., and Drew, H. R., 1992.* Understanding DNA: The Molecule and How It Works. *London: Academic Press.)*

Figure 11.21 shows the values of T and W for a simple striped circular tube in various supercoiled forms. In any closed, circular DNA duplex that is relaxed, $W = 0$. A relaxed circular DNA of 400 bp has 40 helical turns, $T = L = 40$. This linking number can be changed only by breaking one or both strands of the DNA, winding them tighter or looser, and rejoining the ends. Enzymes capable of carrying out such reactions are called **topoisomerases** because they change the topological state of DNA. Topoisomerase falls into two basic classes: I and II. Topoisomerases of the I type cut one strand of a DNA double helix, pass the

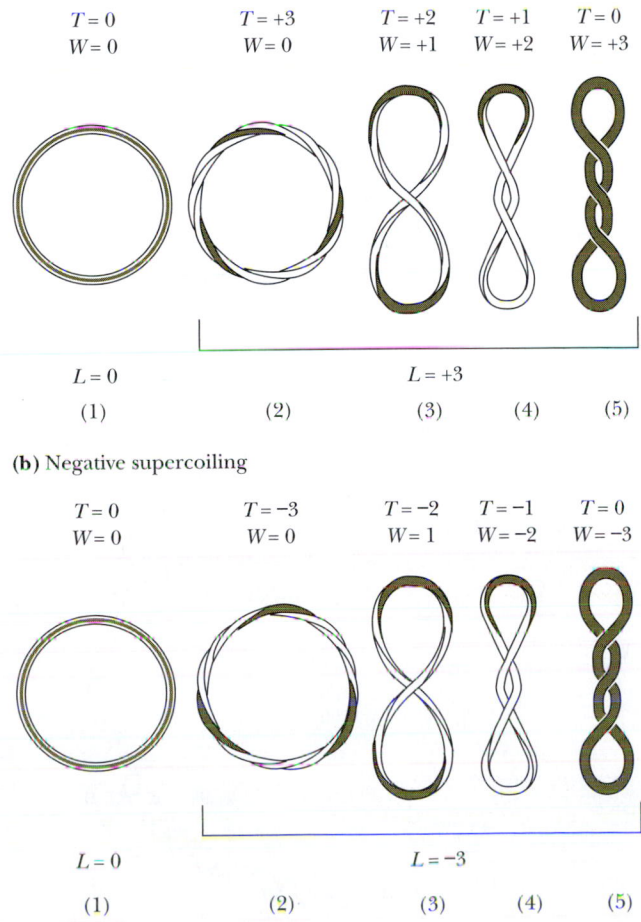

FIGURE 11.21 Supercoil topology for a simple circular tube with a single stripe along it. *(Adapted from Figures 6.5 and 6.6 in Callandine, C. R., and Drew, H. R., 1992.* Understanding DNA: The Molecule and How It Works. *London: Academic Press.)*

other strand through, and then rejoin the cut ends. Topoisomerase II enzymes cut both strands of a dsDNA, pass a region of the DNA duplex between the cut ends, and then rejoin the ends (Figure 11.22). Topoisomerases are important players in DNA replication (see Chapter 28).

DNA Gyrase   The bacterial enzyme **DNA gyrase** is a topoisomerase that introduces negative supercoils into DNA in the manner shown in Figure 11.22. Suppose DNA gyrase puts four negative supercoils into the 400-bp circular duplex, then $W = -4$, T remains the same, and $L = 36$ (Figure 11.23). In actuality, the negative supercoils cause a torsional stress on the molecule, so T tends to decrease; that is, the helix becomes a bit unwound, so base pairs are separated.

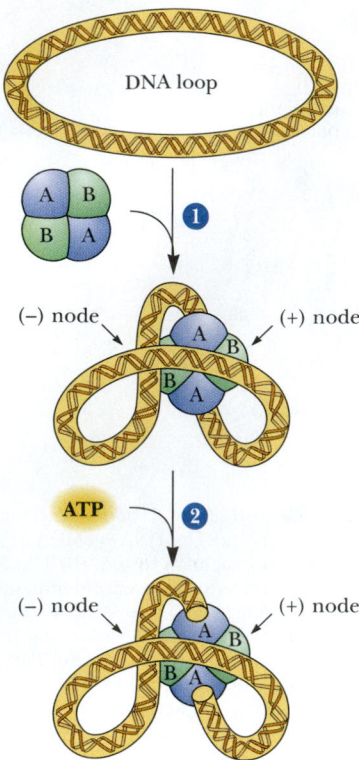

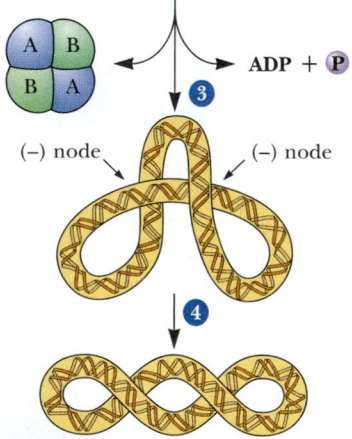

DNA is cut and a conformational change allows the DNA to pass through. Gyrase religates the DNA and then releases it.

FIGURE 11.22 A simple model for the action of bacterial DNA gyrase (topoisomerase II). The A-subunits cut the DNA duplex (**1**) and then hold onto the cut ends (**2**). Conformational changes occur in the enzyme, which allow an intact region of the DNA duplex to pass between the cut ends and into an internal cavity of the protein. The cut ends are then re-ligated (**3**), and the covalently complete DNA duplex is released from the enzyme. The circular DNA now contains two negative supercoils as a consequence of DNA gyrase action (**4**).

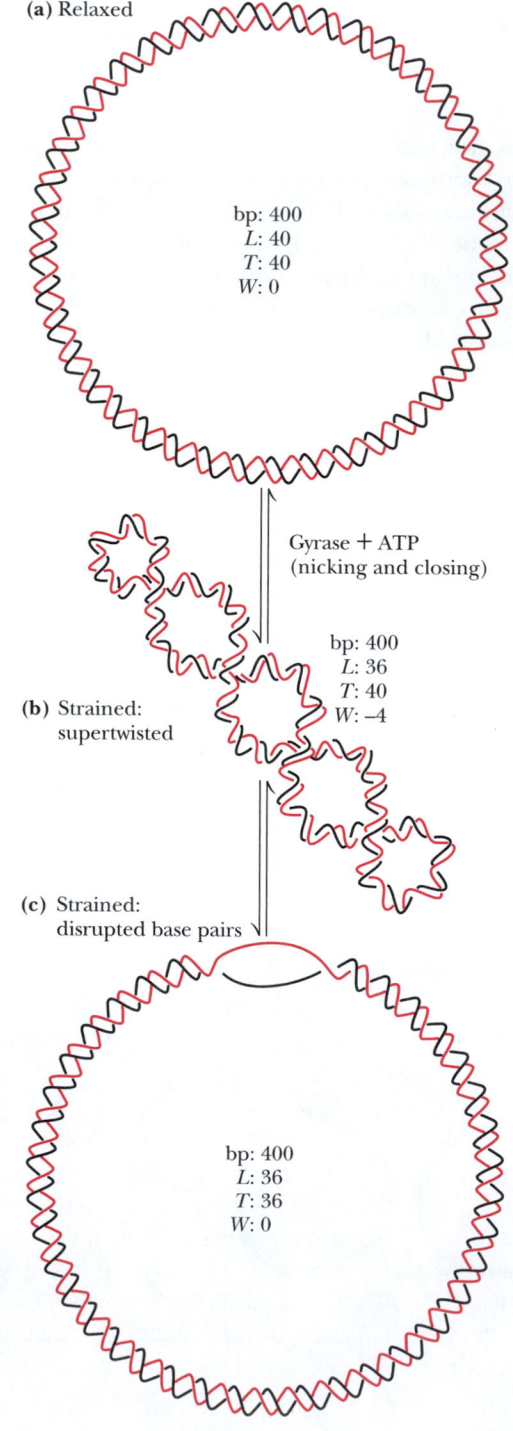

(a) Relaxed

bp: 400
L: 40
T: 40
W: 0

Gyrase + ATP
(nicking and closing)

bp: 400
L: 36
T: 40
W: −4

(b) Strained:
supertwisted

(c) Strained:
disrupted base pairs

bp: 400
L: 36
T: 36
W: 0

FIGURE 11.23 A 400-bp circular DNA molecule in different topological states: **(a)** relaxed, **(b)** negative supercoils distributed over the entire length, and **(c)** negative supercoils creating a localized single-stranded region. Negative supercoiling has the potential to cause localized unwinding of the DNA double helix so that single-stranded regions (or bubbles) are created.

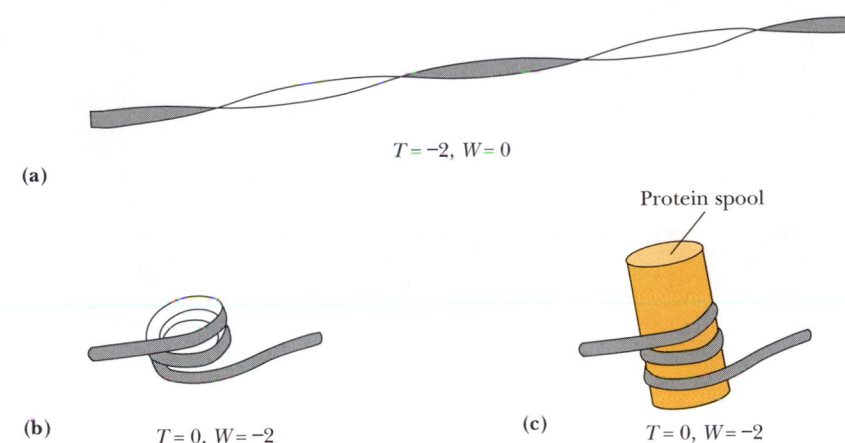

(a)

$T = -2, W = 0$

(b)

$T = 0, W = -2$

Protein spool

(c)

$T = 0, W = -2$

FIGURE 11.24 Supercoiled DNA in a toroidal form wraps readily around protein "spools." A twisted segment of linear DNA with two negative supercoils **(a)** can collapse into a toroidal conformation if its ends are brought closer together **(b)**. Wrapping the DNA toroid around a protein "spool" stabilizes this conformation of supercoiled DNA **(c)**. *(Adapted from Figure 6.6 in Callandine, C. R., and Drew, H. R., 1992. Understanding DNA: The Molecule and How It Works. London: Academic Press.)*

The extreme would be that T would decrease by 4 and the supercoiling would be removed ($T = 36$, $L = 36$, and $W = 0$). Usually the real situation is a compromise in which the negative value of W is reduced, T decreases slightly, and these changes are distributed over the length of the circular duplex so that no localized unwinding of the helix ensues. Although the parameters T and W are conceptually useful, neither can be measured experimentally at present.

Superhelix Density   The difference between the linking number of a DNA and the linking number of its relaxed form is ΔL: $\Delta L = (L - L_0)$. In our example with four negative supercoils, $\Delta L = -4$. The **superhelix density** or **specific linking difference** is defined as $\Delta L / L_0$ and is sometimes termed *sigma*, σ. For our example, $\sigma = -4/40$, or -0.1. As a ratio, σ is a measure of supercoiling that is independent of length. Its sign reflects whether the supercoiling tends to unwind *(negative σ)* or overwind *(positive σ)* the helix. In other words, the superhelix density states the number of supercoils per 10 bp, which also is the same as the number of supercoils per B-DNA repeat. Circular DNA isolated from natural sources is always found in the underwound, negatively supercoiled state.

Toroidal Supercoiled DNA   Negatively supercoiled DNA can arrange into a toroidal state (Figure 11.24). The toroidal state of negatively supercoiled DNA is stabilized by wrapping around proteins that serve as spools for the DNA "ribbon." This toroidal conformation of DNA is found in protein: DNA interactions that are the basis of phenomena as diverse as chromosome structure (see Figure 11.28) and gene expression.

Cruciforms Can Contribute to DNA Tertiary Structure

Palindromes are words, phrases, or sentences that are the same when read backward or forward, such as "radar," "sex at noon taxes," "Madam, I'm Adam," and "a man, a plan, a canal, Panama." DNA sequences that are **inverted repeats**, or palindromes, have the potential to form a tertiary structure known as a **cruciform** (literally meaning "cross-shaped") if the normal interstrand base pairing is replaced by intrastrand pairing (Figure 11.25). In effect, each DNA strand folds back on itself in a hairpin structure to align the palindrome in base-pairing register. Such cruciforms are never as stable as normal DNA duplexes because an unpaired segment must exist in the loop region. However, negative supercoiling causes a localized disruption of hydrogen bonding between base pairs in DNA and may promote formation of cruciform loops. Cruciform structures have a twofold rotational symmetry about their centers and potentially create distinctive recognition sites for specific DNA-binding proteins.

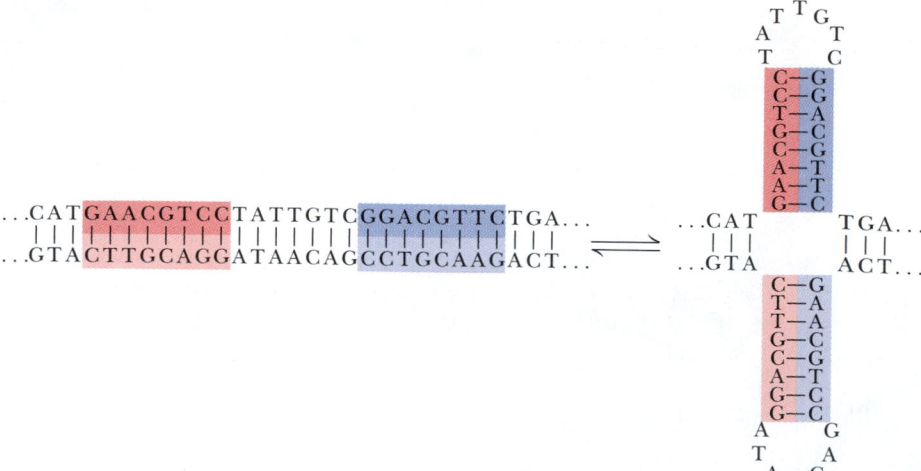

FIGURE 11.25 The formation of a cruciform structure from a palindromic sequence within DNA. The self-complementary inverted repeats can rearrange to form hydrogen-bonded cruciform loops.

| 11.5 | # What Is the Structure of Eukaryotic Chromosomes? |

A typical human cell is 20 μm in diameter. Its genetic material consists of 23 pairs of dsDNA molecules in the form of **chromosomes,** the average length of which is 3×10^9 bp/23 or 1.3×10^8 nucleotide pairs. At 0.34 nm/bp in B-DNA, this represents a DNA molecule 5 cm long. Together, these 46 dsDNA molecules amount to more than 2 m of DNA that must be packaged into a nucleus perhaps 5 μm in diameter! Clearly, the DNA must be condensed by a factor of more than 10^5. This packing problem is solved by neatly wrapping the DNA around protein spools called **nucleosomes,** and the string of nucleosomes is then coiled to form a helical filament. Next, this filament is arranged in loops associated with the **nuclear matrix,** a skeleton or scaffold of proteins providing a structural framework within the nucleus (see following discussion).

Nucleosomes Are the Fundamental Structural Unit in Chromatin

The DNA in a eukaryotic cell nucleus during the interphase between cell divisions exists as a nucleoprotein complex called **chromatin.** The proteins of chromatin fall into two classes: **histones** and **nonhistone chromosomal proteins.** Histones are abundant and play an important role in chromatin structure. In contrast, the nonhistone class is defined by a great variety of different proteins, all of which are involved in genetic regulation; typically, there are only a few molecules of each per cell. Five distinct histones are known: **H1, H2A, H2B, H3,** and **H4** (Table 11.2). All five are relatively small, positively charged, arginine- or lysine-rich proteins that interact via ionic bonds with the negatively charged phosphate groups on the polynucleotide backbone. Pairs of histones H2A, H2B, H3, and H4 aggregate

Table 11.2			
Properties of Histones			
Histone	Ratio of Lysine to Arginine	M_r	Copies per Nucleosome
H1	59/3	21,200	1 (not in bead)
H2A	13/13	14,100	2 (in bead)
H2B	20/8	13,900	2 (in bead)
H3	13/17	15,100	2 (in bead)
H4	11/14	11,400	2 (in bead)

to form octameric core structures, and the DNA helix is wound about these core octamers, creating **nucleosomes.**

If chromatin is swelled suddenly in water and prepared for viewing in the electron microscope, the nucleosomes are evident as "beads on a string," dsDNA being the string (Figure 11.26). The structure of the histone octamer core wrapped with DNA has been solved by T. J. Richmond and collaborators (Figure 11.27). The core octamer has surface landmarks that guide the course of the DNA; 146 bp of B-DNA in a flat, left-handed superhelical conformation make 1.65 turns around the histone core (Figure 11.27), which itself is a protein superhelix consisting of a spiral array of the four histone dimers. Histone H1, a three-domain protein, serves to seal the ends of the DNA turns to the nucleosome core and to organize the additional 40 to 60 bp of DNA that link consecutive nucleosomes. The N-terminal tails of histones H3 and H4 are accessible on the surface of the nucleosome. Lysine and serine residues in these tails can be covalently modified in myriad ways (lysines may be acetylated, methylated, or ubiquitinated; serines may be phosphorylated). These modifications play an important role in chromatin dynamics and gene expression (see Chapter 29).

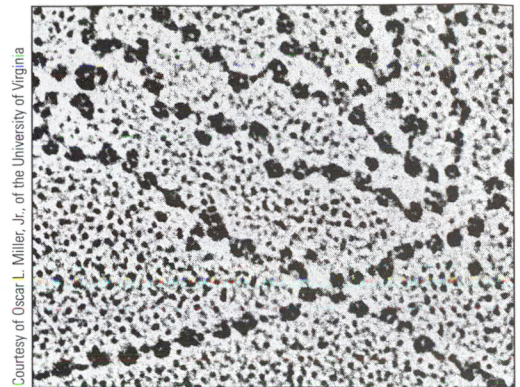

FIGURE 11.26 Electron micrograph of *Drosophila melanogaster* chromatin after swelling reveals the presence of nucleosomes as "beads on a string."

Higher-Order Structural Organization of Chromatin Gives Rise to Chromosomes

A higher order of chromatin structure is created when the array of nucleosomes, in their characteristic beads-on-a-string motif, is wound in the fashion of a *solenoid* having six nucleosomes per turn (Figure 11.28). The resulting 30-nm filament contains about 1200 bp in each of its solenoid turns. Interactions between the respective H1 components of successive nucleosomes stabilize the 30-nm filament. This 30-nm filament then forms long DNA loops of variable length, each containing on average between 60,000 and 150,000 bp. Electron microscopic analysis of human chromosome 4 suggests that 18 such loops are then arranged radially about the circumference of a single turn to form a **miniband unit** of the chromosome. According to this model, approximately 10^6 of these minibands are arranged along a central axis in each of the chromatids of human chromosome 4

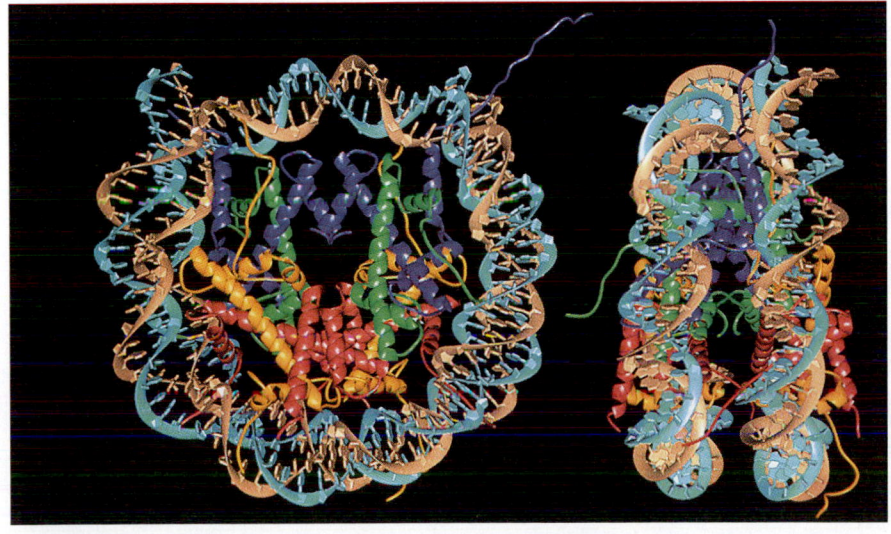
(a)

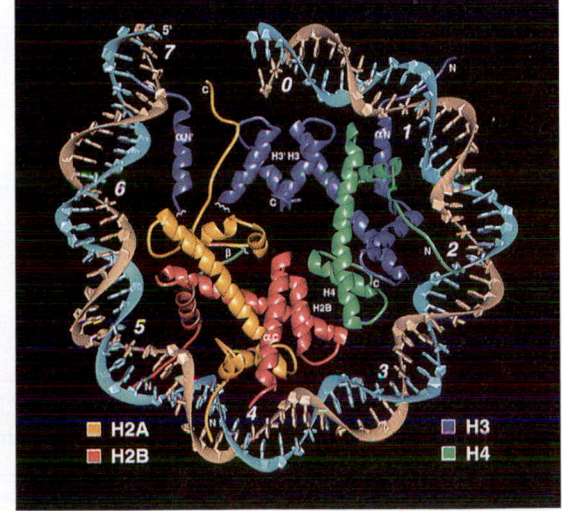
(b)

FIGURE 11.27 (a) Deduced structure of the nucleosome core particle wrapped with 1.65 turns of DNA (146 bp). The DNA is shown as a ribbon. (*left*) View down the axis of the nucleosome; (*right*) view perpendicular to the axis. (b) One-half of the nucleosome core particle with 73 bp of DNA, as viewed down the nucleosome axis. Note that the DNA does not wrap in a uniform circle about the histone core but instead follows a course consisting of a series of somewhat straight segments separated by bends. (*Adapted from Luger, K., et al., 1997. Crystal structure of the nucleosome core particle at 2.8 Å resolution. Nature* **389:**251–260. *Photos courtesy of T. J. Richmond, ETH-Hönggerberg, Zurich, Switzerland.*)

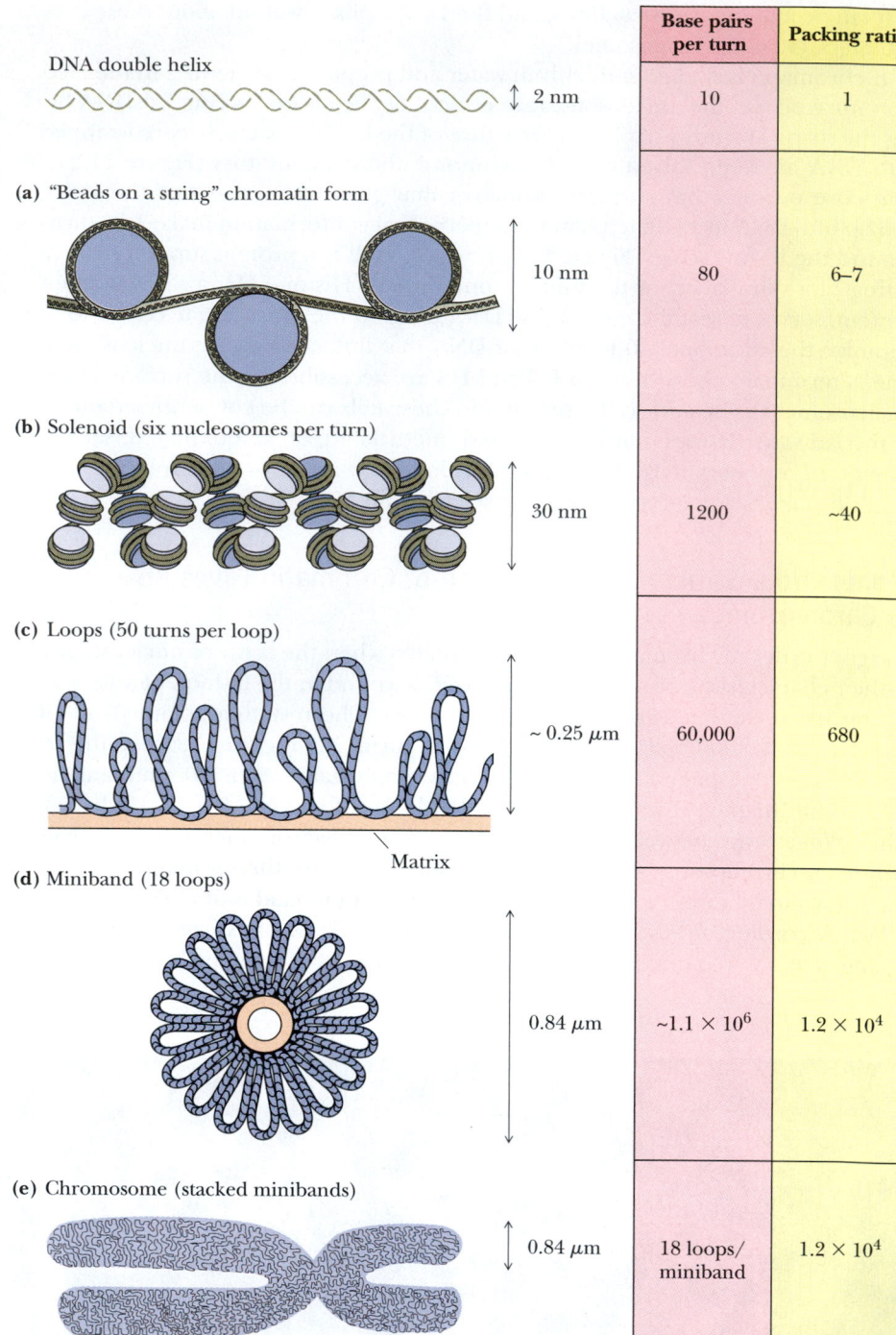

	Base pairs per turn	Packing ratio
DNA double helix ‖ 2 nm	10	1
(a) "Beads on a string" chromatin form — 10 nm	80	6–7
(b) Solenoid (six nucleosomes per turn) — 30 nm	1200	~40
(c) Loops (50 turns per loop) — ~ 0.25 μm	60,000	680
(d) Miniband (18 loops) — 0.84 μm	~1.1 × 10^6	1.2 × 10^4
(e) Chromosome (stacked minibands) — 0.84 μm	18 loops/ miniband	1.2 × 10^4

Matrix

FIGURE 11.28 A model for chromosome structure, human chromosome 4. **(a)** The 2-nm DNA helix is wound twice around histone octamers to form 10-nm nucleosomes, each of which contains 160 bp (80 per turn). **(b)** These nucleosomes are then wound in solenoid fashion with six nucleosomes per turn to form a 30-nm filament. **(c)** In this model, the 30-nm filament forms long DNA loops, each containing about 60,000 bp, which are attached at their base to the nuclear matrix **(d)**. Eighteen of these loops are then wound radially around the circumference of a single turn to form a miniband unit of a chromosome **(e)**. Approximately 10^6 of these minibands occur in each chromatid of human chromosome 4 at mitosis.

that form at mitosis (Figure 11.28). Despite intensive study, much about the higher-order structure of chromosomes remains to be discovered.

11.6	**Can Nucleic Acids Be Chemically Synthesized?**

Laboratory synthesis of oligonucleotide chains of defined sequence presents some of the same problems encountered in chemical synthesis of polypeptides (see Chapter 5). First, functional groups on the monomeric units (in this case,

Human Biochemistry

Telomeres and Tumors

Eukaryotic chromosomes are linear. The ends of chromosomes have specialized structures known as **telomeres.** The telomeres of virtually all eukaryotic chromosomes consist of short, tandemly repeated nucleotide sequences at the ends of the chromosomal DNA. For example, the telomeres of human germline (sperm and egg) cells contain between 1000 and 1700 copies of the hexameric repeat TTAGGG (see accompanying figure). Telomeres contribute to the maintenance of chromosomal integrity by protecting against DNA degradation or rearrangement. Telomeres are added to the ends of chromosomal DNA by an RNA-containing enzyme known as **telomerase** (see Chapter 28); telomerase is an unusual DNA polymerase that was discovered in 1985 by Elizabeth Blackburn and Carol Greider of the University of California, San Francisco. However, most normal somatic cells lack telomerase. Consequently, upon every cycle of cell division when the cell replicates its DNA, about 50-nucleotide portions are lost from the end of each telomere. Thus, over time, the telomeres of somatic cells in animals become shorter and shorter, eventually leading to chromosome instability and cell death. This phenomenon has led some scientists to espouse a "telomere theory of aging" that implicates telomere shortening as the principal factor in cell, tissue, and even organism aging. Interestingly, cancer cells appear "immortal" because they continue to reproduce indefinitely. A survey of 20 different tumor types by Geron Corporation of Menlo Park, California, revealed that all contained telomerase activity.

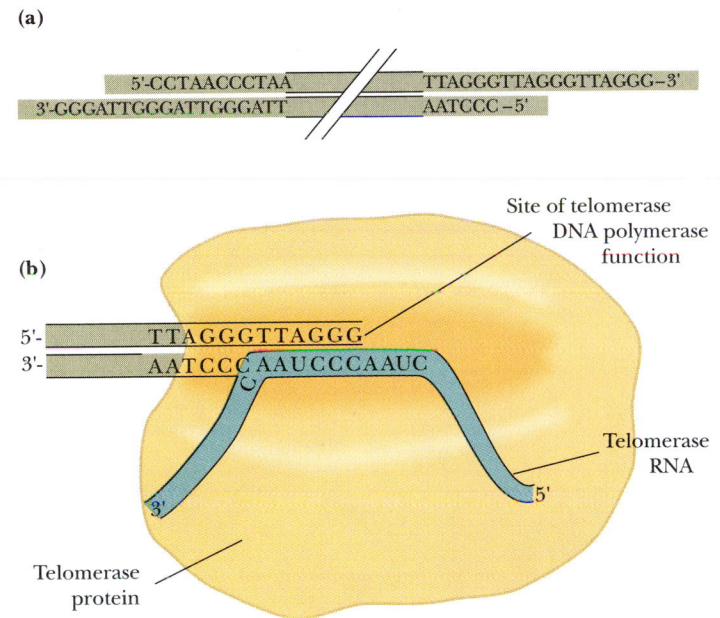

(a)

(b)

◄ **(a)** Telomeres on human chromosomes consist of the hexanucleotide sequence TTAGGG repeated between 1000 and 1700 times. These TTAGGG tandem repeats are attached to the 3′-ends of the DNA strands and are paired with the complementary sequence 3′-AATCCC-5′ on the other DNA strand. Thus, a G-rich region is created at the 3′-end of each DNA strand, and a C-rich region is created at the 5′-end of each DNA strand. Typically, at each end of the chromosome, the G-rich strand protrudes 12 to 16 nucleotides beyond its complementary C-rich strand. **(b)** Like other telomerases, human telomerase is a ribonucleoprotein. The ribonucleic acid of human telomerase is an RNA molecule 962 nucleotides long. This RNA serves as the template for the DNA polymerase activity of telomerase. Nucleotides 46 to 56 of this RNA are CUAA**CCCUAA**C and provide the template function for the telomerase-catalyzed addition of TTAGGG units to the 3′-end of a DNA strand.

bases) are reactive under conditions of polymerization and therefore must be protected by blocking agents. Second, to generate the desired sequence, a phosphodiester bridge must be formed between the 3′-O of one nucleotide (B) and the 5′-O of the preceding one (A) in a way that precludes the unwanted bridging of the 3′-O of A with the 5′-O of B. Finally, recoveries at each step must be high so that overall yields in the multistep process are acceptable. As in peptide synthesis (see Chapter 5), orthogonal *solid-phase methods* are used to overcome some of these problems. Commercially available automated instruments, called **DNA synthesizers** or "gene machines," are capable of carrying out the synthesis of oligonucleotides of 150 bases or more.

Phosphoramidite Chemistry Is Used to Form Oligonucleotides from Nucleotides

Phosphoramidite chemistry is currently the accepted method of oligonucleotide synthesis. The general strategy involves the sequential addition of nucleotide units as *nucleoside phosphoramidite* derivatives to a nucleoside covalently attached to the insoluble resin. Excess reagents, starting materials, and side products are removed after each step by filtration. After the desired oligonucleotide has been formed, it is freed of all blocking groups, hydrolyzed from the resin, and purified by gel electrophoresis. The four-step cycle is shown in

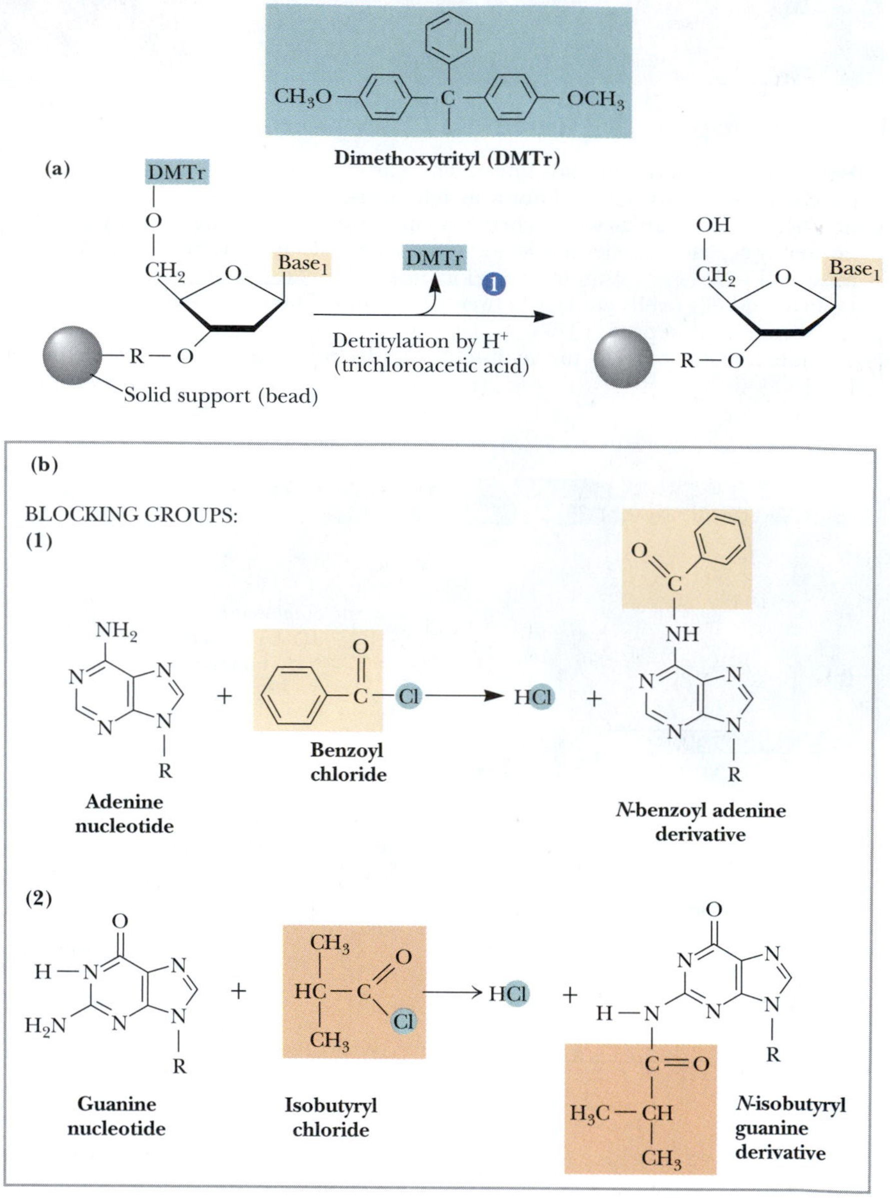

FIGURE 11.29 Solid-phase oligonucleotide synthesis. The four-step cycle starts with the first base in nucleoside form (N-1) attached by its 3'-OH group to an insoluble, inert resin or matrix, typically either controlled pore glass (CPG) or silica beads. Its 5'-OH is blocked with a dimethoxytrityl (DMTr) group **(a)**. If the base has reactive —NH$_2$ functions, as in A, G, or C, then N-benzoyl or N-isobutyryl derivatives are used to prevent their reaction **(b)**. In step 1, the DMTr protecting group is removed by trichloroacetic acid treatment. Step 2 is the coupling step: The second base (N-2) is added in the form of a nucleoside phosphoramidite derivative whose 5'-OH bears a DMTr blocking group so it cannot polymerize with itself **(c)**. *continued*

Figure 11.29. Chemical synthesis takes place in the 3'→5' direction (the reverse of the biological polymerization direction).

Genes Can Be Chemically Synthesized

Table 11.3 lists some of the genes that have been chemically synthesized. Because protein-coding genes are characteristically much larger than the 150-bp practical limit on oligonucleotide synthesis, their synthesis involves joining a series of oligonucleotides to assemble the overall sequence. A prime example of such synthesis is the gene for rhodopsin (see A Deeper Look box on page 363).

(c)

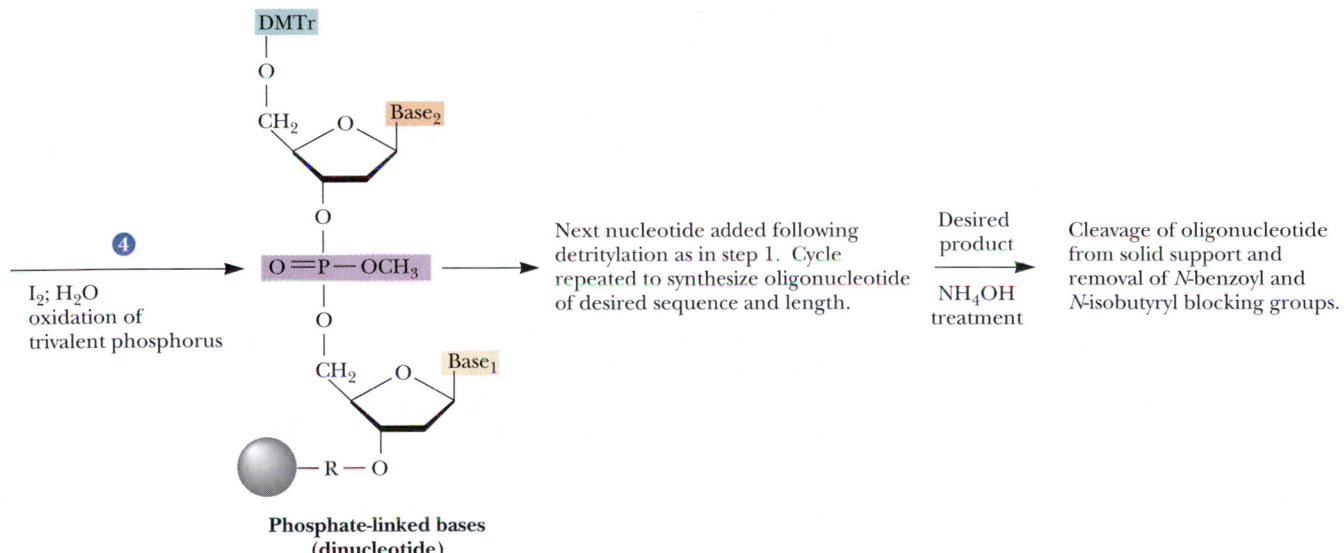

Phosphoramidite derivative of nucleotides 2

Phosphite-linked bases (dinucleotide)

Phosphate-linked bases (dinucleotide)

④
I_2; H_2O
oxidation of trivalent phosphorus

Next nucleotide added following detritylation as in step 1. Cycle repeated to synthesize oligonucleotide of desired sequence and length.

Desired product

NH_4OH treatment

Cleavage of oligonucleotide from solid support and removal of N-benzoyl and N-isobutyryl blocking groups.

FIGURE 11.29 continued The presence of a weak acid, such as tetrazole, activates the phosphoramidite, and it rapidly reacts with the free 5'-OH of N-1, forming a dinucleotide linked by a phosphite group. Chemical synthesis thus takes place in the 3'→5' direction. Unreacted free 5'-OHs of N-1 (usually only 2%–6% of the total) are blocked from further participation in the polymerization process by acetylation with acetic anhydride in step 3, referred to as *capping*. The phosphite linkage between N-1 and N-2 is highly reactive, and in step 4, it is oxidized by aqueous iodine (I_2) to form the desired more stable phosphate group. This completes the cycle. Subsequent cycles add successive residues to the resin-immobilized chain. When the chain is complete, it is cleaved from the support with NH_4OH, which also removes the N-benzoyl and N-isobutyryl protecting groups from the amino functions on the A, G, and C residues.

Table 11.3

Some Chemically Synthesized Genes

Gene	Size (bp)
tRNA	126
α-Interferon	542
Secretin	81
γ-Interferon	453
Rhodopsin	1057
Proenkephalin	77
Connective tissue activating peptide III	280
Lysozyme	385
Tissue plasminogen activator	1610
c-Ha-ras	576
RNase T1	324
Cytochrome b_5	330
Bovine intestinal Ca-binding protein	298
Hirudin	226
RNase A	375

11.7 What Is the Secondary and Tertiary Structure of RNA?

RNA molecules (see Chapter 10) are typically single-stranded. The course of a single-stranded RNA in three-dimensional space conceivably would have six degrees of freedom per nucleotide, represented by rotation about each of the six single bonds along the sugar–phosphate backbone per nucleotide unit. (Rotation about the β-glycosidic bond creates a seventh degree of freedom in terms of the total conformational possibilities at each nucleotide.) Compare this situation with DNA, whose separated strands would obviously enjoy the same degrees of freedom. However, the double-stranded nature of DNA imposes great constraint on its conformational possibilities. Compared to dsDNA, an RNA molecule has a much greater number of conformational possibilities. Intramolecular interactions and other stabilizing influences limit these possibilities, but the higher-order structure of RNA remains an area for fruitful scientific discovery.

Although single-stranded, RNA molecules are often rich in double-stranded regions that form when complementary sequences within the chain come together and join via **intrastrand base pairing.** These interactions create hairpin **stem-loop structures,** in which the base-paired regions form the stem and the unpaired regions between base pairs are the loop (see Figures 11.30 and 11.35). Paired regions of RNA cannot form B-DNA type double helices because the RNA 2′-OH groups are a steric hindrance to this conformation. Instead, these paired regions adopt a conformation similar to the A-form of DNA, having about 11 bp per turn, with the bases strongly tilted from the plane perpendicular to the helix axis (see Figure 11.10). A-form double helices are the most prominent secondary structural elements in RNA. Both tRNA and rRNA have large amounts of A-form double helix. In addition, a number of defined structural motifs recur within the loops of stem-loop structures, such as **U-turns** (a loop motif of consensus sequence UNRN, where N is any nucleotide and R is a purine) and **tetraloops** (another class of four-nucleotide loops found at the termini of stem-loop structures). Stems of stem-loop structures may also have **bulges** (or **internal loops**) where the RNA strand is forced into a short single-stranded loop because one or more bases along one strand in an RNA double helix finds no base-pairing partners. Regions where several stem-loop structures meet are termed **junctions.** Stems, loops, bulges, and junctions are the four basic secondary structural elements in RNA. These secondary structural patterns were revealed by studies on tRNA and rRNA, but secondary structure exists in mRNA species as well, although its nature is unique to the specific mRNA. Some mRNA secondary structures form ligand-binding sites, and the binding of ligand can influence whether the mRNA is completely transcribed and translated, or whether one or the other of these processes is aborted. (The functions of tRNA, rRNA, and mRNA are discussed in detail in Part 4.)

The single-stranded loops in RNA stem-loops create base-pairing opportunities between distant, complementary, single-stranded loop regions. These interactions, mostly based on Watson–Crick base pairing, lead to tertiary structure in RNA. Other tertiary structural motifs arise from **coaxial stacking, pseudoknot formation,** and **ribose zippers.** In coaxial stacking, the blunt, non-loop ends of stem-loops situated next to one another in the RNA sequence stack upon each other to create an uninterrupted stack of base pairs. Pseudoknots occur when bases in the loops of stem-loop structures form a short double helix by base pairing with nearby single-stranded regions in the RNA. Ribose zippers are found when two antiparallel, single-stranded regions of RNA align as an H-bonded network forms between the 2′-OH groups of the respective strands, the O at the 2′-OH position of one strand serving as the H-bond acceptor while the H on the 2′-OH of the other strand is the H-bond donor.

A Deeper Look

Total Synthesis of the Rhodopsin Gene

The strategy used in the total synthesis of the gene for bovine rhodopsin is shown in the accompanying figure. This gene, which is 1057 base pairs long, encodes the 348–amino acid photorecep-tor protein of the vertebrate retina; rhodopsin is the protein that allows us to detect light and enjoy vision.

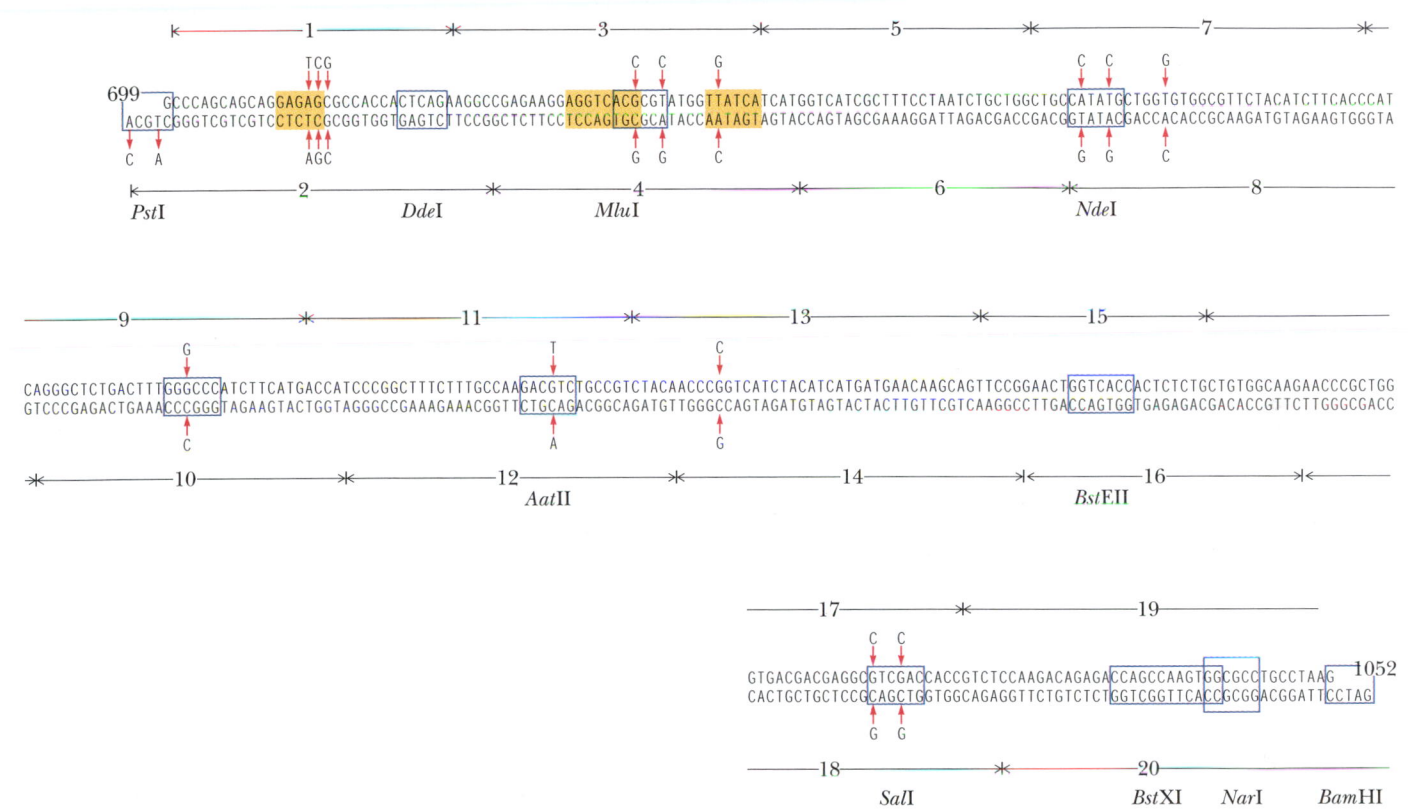

▲ Total synthesis of the bovine rhodopsin gene was achieved by joining 72 synthetic oligonucleotides, 36 representing one DNA strand and 36 the other, complementary strand. These oligonucleotides were overlapping. Once synthesized, the various oligonucleotides, each 15 to 40 nucleotides long, were assembled by annealing and enzymatic ligation into three large fragments, representing nucleotides −5 to 338 (−5 meaning 5 nucleo-tides before the start of the region encoding the rhodopsin amino acid sequence), 335 to 702, and 699 to 1052. Finally, the total gene was created by joining these fragments. This figure shows only one fragment (fragment PB, comprising nucleotides 699 through 1052), assembled from 20 com-plementary oligonucleotides whose ends overlap. Odd-numbered oligonucleotides (1, 3, 5, . . .) compose the 5′→3′ strand; even-numbered ones (2, 4, 6, . . .) represent the 3′→5′ strand. (Vertical arrows indicate nucleotides that were changed from the native gene sequence. Restriction sites are shown boxed in blue lines; those removed from the gene through nucleotide substitutions are shown as yellow shaded boxes.) Note the single-stranded overhangs at either end of the 3′→5′ strand. The sequences at these overhangs correspond to restriction endonuclease sites (*Pst*I and *Bam*H1), which facilitate subsequent manipulation of the fragment in gene assembly and cloning. Theoretically, no gene is beyond the scope of these methods, a fact that opens the door to an incredibly exciting range of possibilities for investigating structure–function relationships in the organization and expression of hereditary material.

Transfer RNA Adopts Higher-Order Structure Through Intrastrand Base Pairing

In tRNA molecules, which contain 73 to 94 nucleotides in a single chain, a ma-jority of the bases are hydrogen bonded to one another. Figure 11.30 shows the structure that typifies tRNAs. *Hairpin turns* bring complementary stretches of bases in the chain into contact so that double helical regions form, creating stem-loop secondary structures. Because of the arrangement of the comple-mentary stretches along the chain, the overall pattern of base pairing can be represented as a *cloverleaf*. Each cloverleaf consists of four base-paired segments—three loops and the stem where the 3′- and 5′-ends of the molecule meet. These

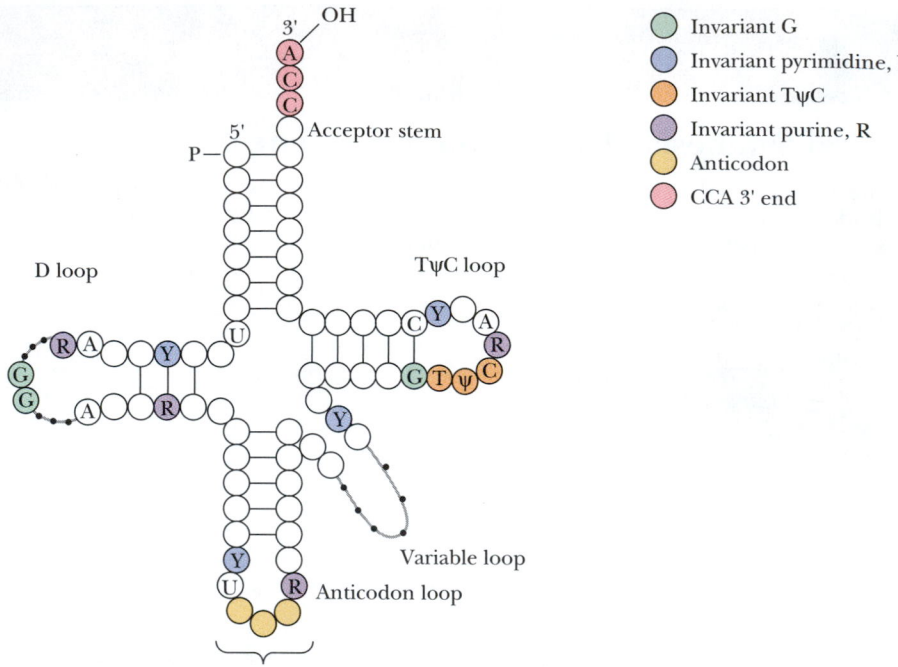

FIGURE 11.30 A general diagram for the structure of tRNA. The positions of invariant bases as well as bases that seldom vary are shown in color. The numbering system is based on yeast tRNA^Phe. R = purine; Y = pyrimidine. Dotted lines denote sites in the D loop and variable loop regions where varying numbers of nucleotides are found in different tRNAs.

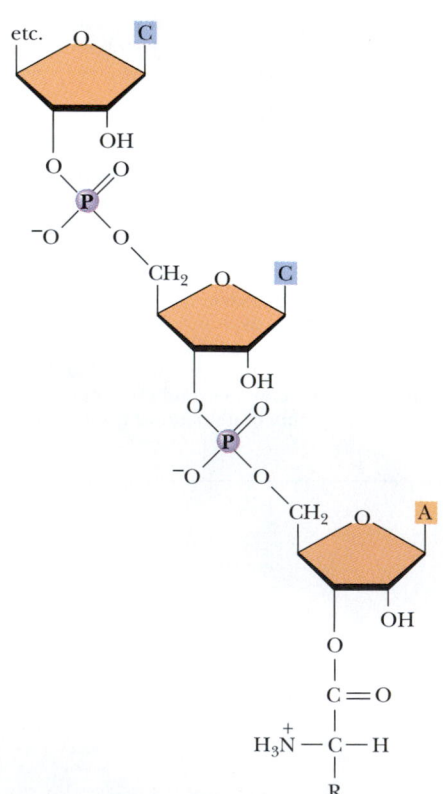

FIGURE 11.31 Amino acids are linked to the 3'-OH end of tRNA molecules by an ester bond formed between the carboxyl group of the amino acid and the 3'-OH of the terminal ribose of the tRNA.

four segments are designated the **acceptor stem**, the **D loop**, the **anticodon loop**, and the **TψC loop** (the latter two are U-turn motifs).

tRNA Secondary Structure The *acceptor stem* is where the amino acid is linked to form the aminoacyl-tRNA derivative, which serves as the amino acid–donating species in protein synthesis; this is the physiological role of tRNA. The amino acid adds to the 3'-OH of the 3'-terminal A nucleotide (Figure 11.31). The 3'-end of tRNA is invariantly CCA-3'-OH. This CCA sequence plus a fourth nucleotide extends beyond the double helical portion of the acceptor stem. The *D loop* is so named because this tRNA loop often contains dihydrouridine, or D, residues. In addition to dihydrouridine, tRNAs characteristically contain a number of unusual bases, including inosine, thiouridine, pseudouridine, and hypermethylated purines (see Figure 10.26). The *anticodon stem-loop* consists of a double helical segment and seven unpaired bases, three of which are the **anticodon**—a three-nucleotide unit that recognizes and base pairs with a particular mRNA **codon,** a complementary three-base unit in mRNA providing the genetic information that specifies an amino acid. Anticodon base pairing to the codon on mRNA allows a particular RNA species to deliver its amino acid to the protein-synthesizing apparatus. It represents the key event in translating the information in the nucleic acid sequence into the amino acid sequence of a protein. Codon:anticodon pairing ensures that the appropriate amino acid is inserted at the right place in the amino acid sequence of the protein being synthesized. Continuing along the tRNA sequence in the 5'→3' direction beyond the anticodon stem-loop lies a loop that varies from tRNA to tRNA in the number of residues that it has, the so-called **extra or variable loop.** The last loop in the tRNA, reading 5'→3', is the loop found in the **TψC stem-loop.** It contains seven unpaired bases, including the sequence TψC, where ψ is the symbol for **pseudouridine.** Ribosomes bind tRNAs through recognition of this TψC loop. Almost all of the invariant residues common to tRNAs lie within the non–hydrogen-bonded regions of the cloverleaf structure (Figure 11.32).

tRNA Tertiary Structure Tertiary structure in tRNA arises from base-pairing interactions between bases in the D loop with bases in the variable and TψC loops, as shown for yeast phenylalanine tRNA in Figure 11.33. Note that these

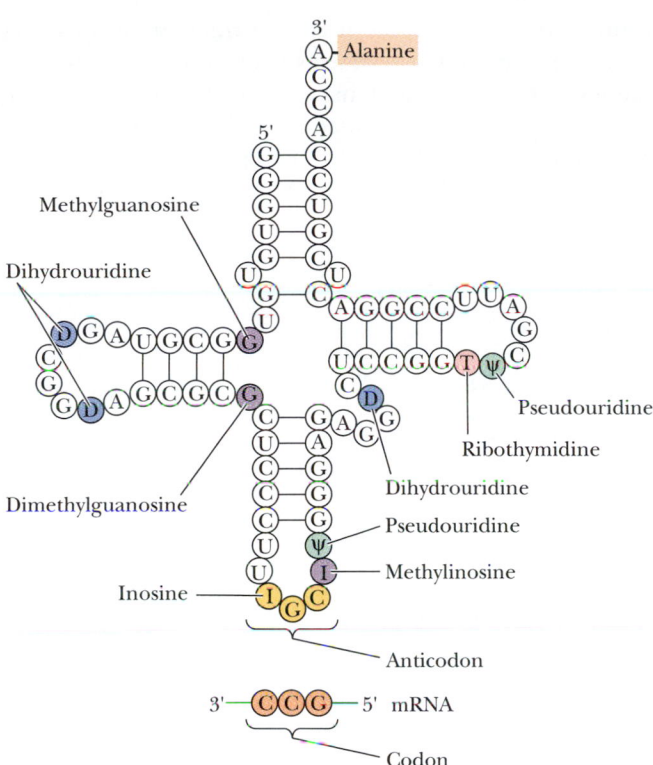

FIGURE 11.32 The complete nucleotide sequence and cloverleaf structure of yeast alanine tRNA.

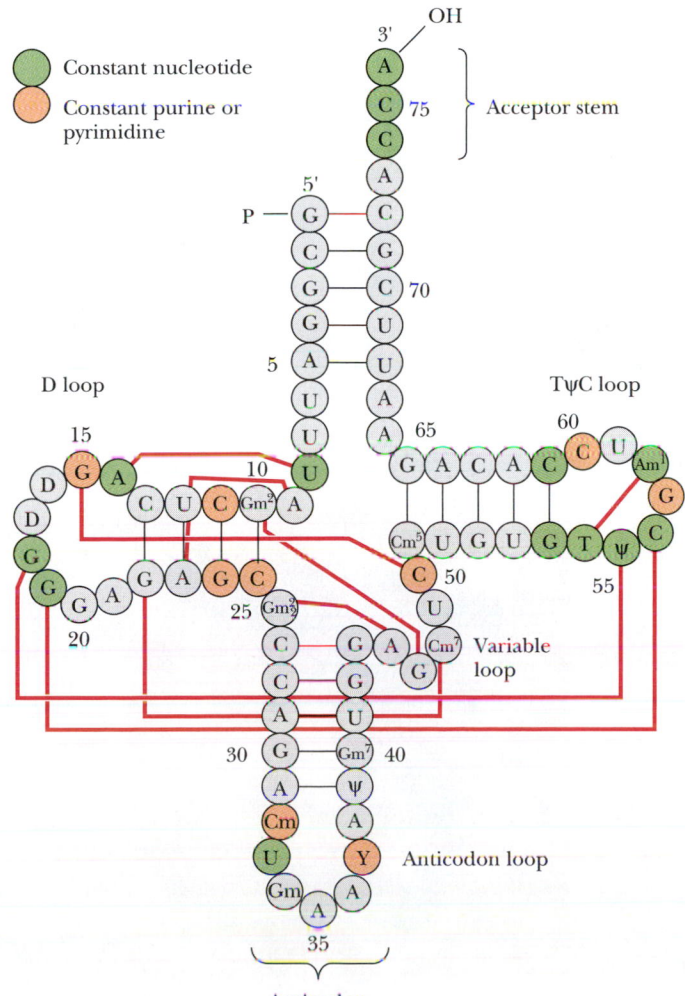

FIGURE 11.33 Tertiary interactions in yeast phenylalanine tRNA. The molecule is presented in the conventional cloverleaf secondary structure generated by intrastrand hydrogen bonding. Solid lines connect bases that are hydrogen bonded when this cloverleaf pattern is folded into the characteristic tRNA tertiary structure (see also Figure 11.34).

base-pairing interactions involve the invariant nucleotides of tRNAs, thus emphasizing the importance of the tertiary structure they create to the function of tRNAs in general. These interactions fold the D and TψC arms together and bend the cloverleaf into the stable L-shaped tertiary form (Figure 11.34). Many of these base-pairing interactions involve base pairs that are not canonical A:T or G:C pairings (Figure 11.34). The amino acid acceptor stem is at one end of

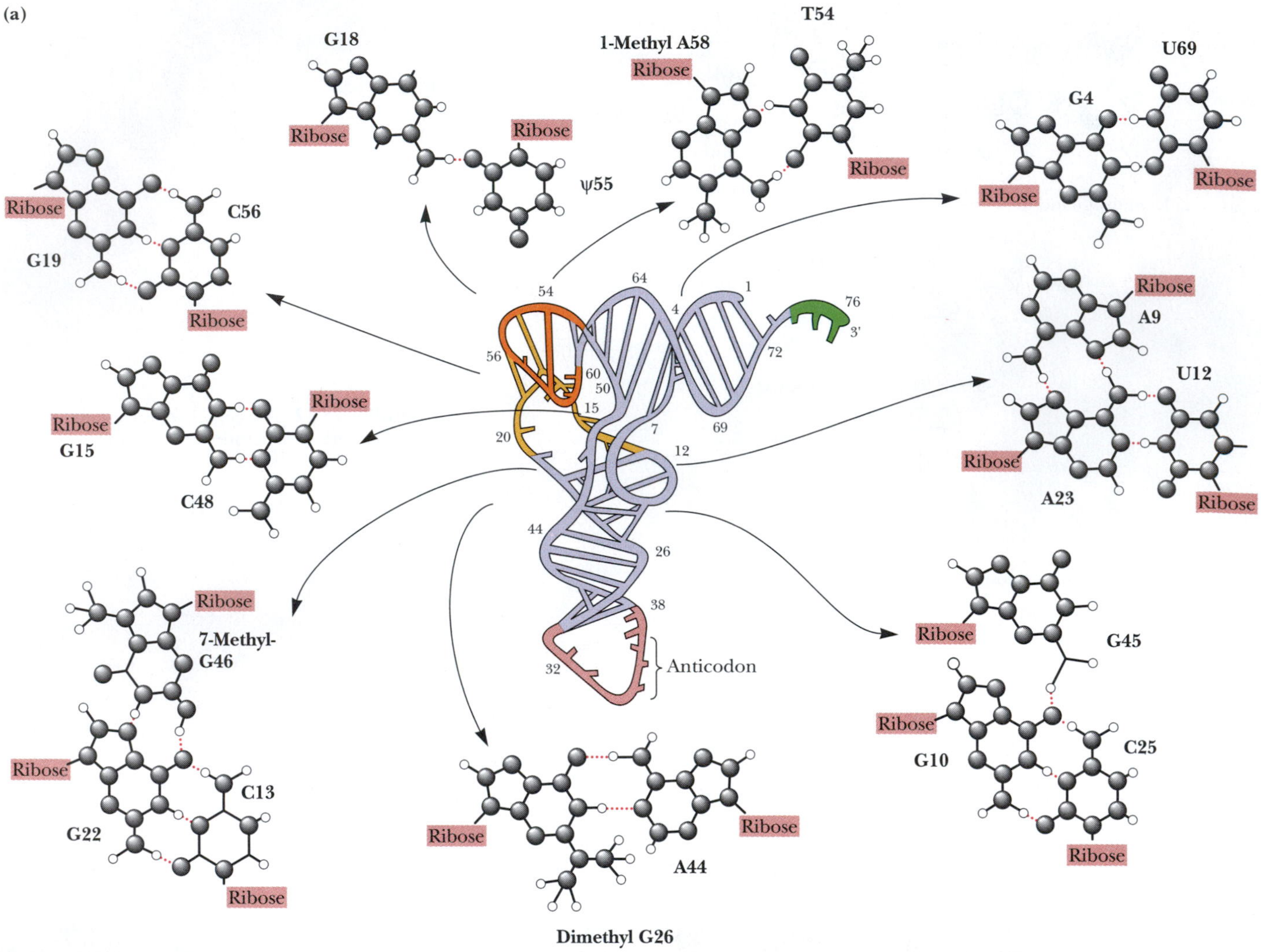

(a)

G18

C56

G19

G15

C48

Ribose

7-Methyl-G46

C13

G22

Dimethyl G26

A44

1-Methyl A58

ψ55

T54

U69

G4

A9

U12

A23

G45

G10

C25

FIGURE 11.34 **(a)** The three-dimensional structure of yeast phenylalanine tRNA as deduced from X-ray diffraction studies of its crystals. The tertiary folding is illustrated in the center of the diagram with the ribose–phosphate backbone presented as a continuous ribbon; H bonds are indicated by crossbars. Unpaired bases are shown as short, unconnected rods. The anticodon loop is at the bottom and the -CCA 3′-OH acceptor end is at the top right. The various types of noncanonical hydrogen-bonding interactions observed between bases surround the central molecule. Three of these structures show examples of unusual H-bonded interactions involving three bases; these interactions aid in establishing tRNA tertiary structure. **(b)** A space-filling model of the molecule. *(After Kim, S. H., in Schimmel, P., Söll, D., and Abelson, J. N., eds., 1979.* Transfer RNA: Structure, Properties, and Recognition. *New York: Cold Spring Harbor Laboratory.)*

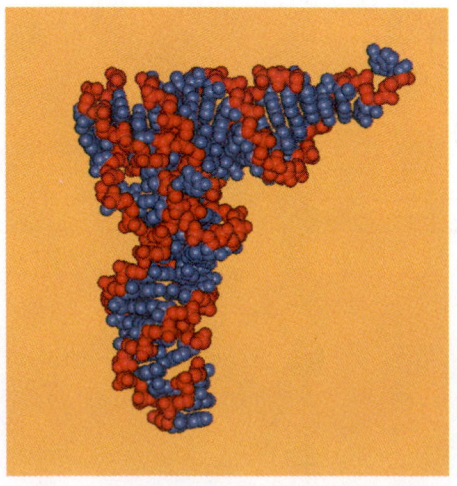

(b)

the L, separated by 7 nm or so from the anticodon at the opposite end of the L. The D and TψC loops form the corner of the L. In the L-conformation, the bases are oriented to maximize hydrophobic stacking interactions between their flat faces. Such stacking is a second major factor contributing to L-form stabilization.

Ribosomal RNA Also Adopts Higher-Order Structure Through Intrastrand Base Pairing

rRNA Secondary Structure   A large degree of *intrastrand sequence complementarity* is found in all ribosomal RNA strands, and all assume a highly folded pattern that allows base pairing between these complementary segments, giving rise to multiple stem-loop structures. Furthermore, the loop regions of stem-loops contain the characteristic structural motifs, such as U-turns, tetraloops, and bulges. Figure 11.35 shows the secondary structure assigned to the *E. coli* 16S rRNA. This structure is based on computer alignment of the nucleotide sequence into optimal H-bonding segments. The reliability of these alignments is then tested through a comparative analysis of whether identical secondary structures can be predicted from nucleotide sequences of 16S-like rRNAs from other species. If so, then such structures are apparently conserved. The approach is based on the thesis that because ribosomal RNA

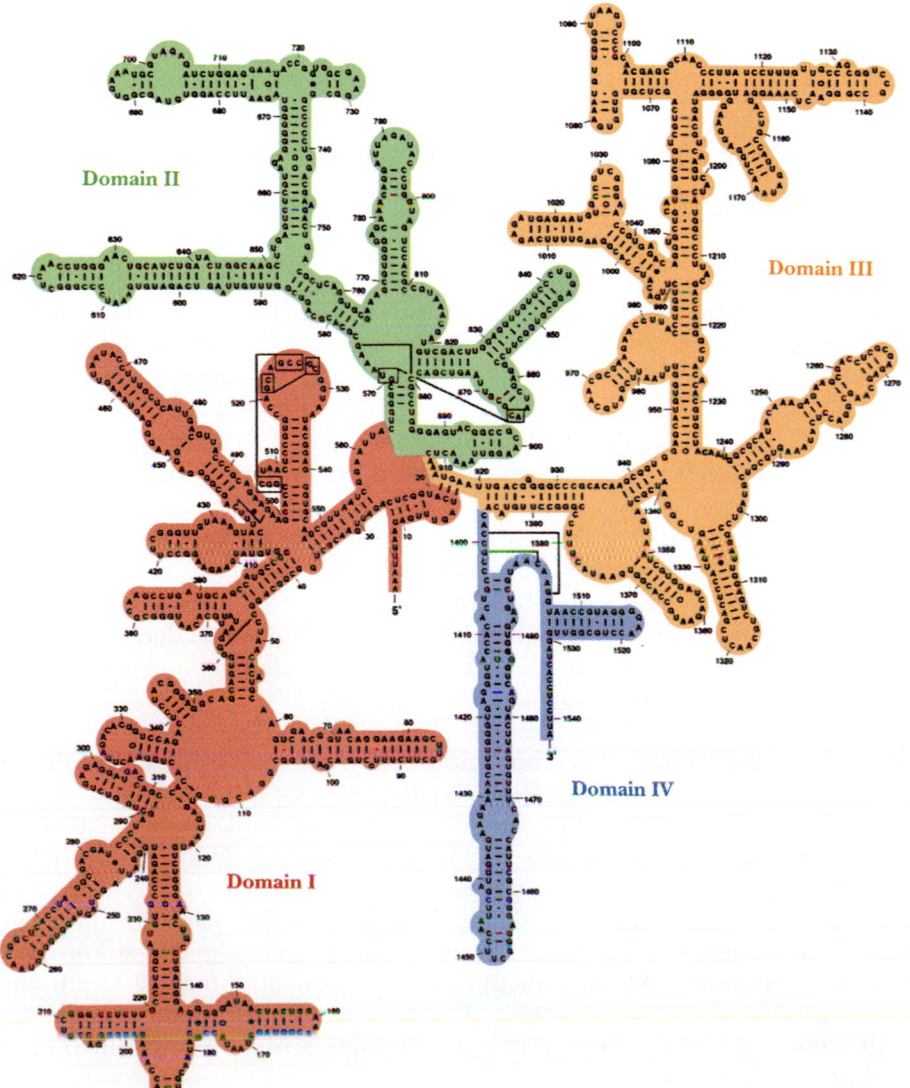

FIGURE 11.35 The proposed secondary structure for *E. coli* 16S rRNA, based on comparative sequence analysis in which the folding pattern is assumed to be conserved across different species. The molecule can be subdivided into four domains—**I, II, III,** and **IV**—on the basis of contiguous stretches of the chain that are closed by long-range base-pairing interactions. **I,** the 5′-domain, includes nucleotides 27 through 556. **II,** the central domain, runs from nucleotide 564 to 912. Two domains comprise the 3′-end of the molecule. **III,** the major one, comprises nucleotides 923 to 1391. **IV,** the 3′-terminal domain, covers residues 1392 to 1541.

(a) *E. coli* (a eubacterium)   **(b)** *H. volcanii* (an archaebacterium)   **(c)** *S. cerevisiae* (yeast, a lower eukaryote)

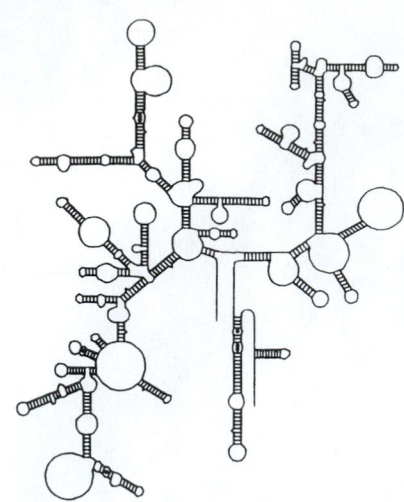
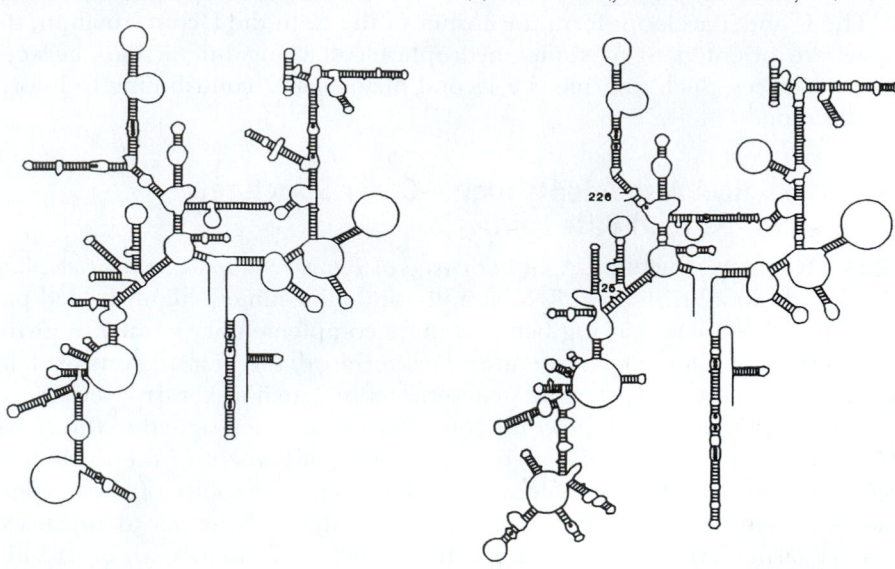

FIGURE 11.36 Phylogenetic comparison of secondary structures of 16S-like rRNAs from **(a)** a eubacterium *(E. coli)*, **(b)** an archaebacterium *(H. volcanii)*, and **(c)** a eukaryote *(S. cerevisiae, a yeast)*.

species (regardless of source) serve common roles in protein synthesis, it may be anticipated that they share structural features. As usual with RNAs, the single-stranded regions of rRNA create the possibility of base-pairing opportunities with distant, complementary, single-stranded regions. Such interactions are the driving force for tertiary structure formation in RNAs. Furthermore, such tertiary interactions can on occasion alter the base-pairing arrangements in adjacent base-paired regions.

Comparison of rRNAs from Various Species   If a phylogenetic comparison is made of the 16S-like rRNAs from an archaebacterium *(Halobacterium volcanii)*, a eubacterium *(E. coli)*, and a eukaryote (the yeast *Saccharomyces cerevisiae)*, a striking similarity in secondary structure emerges (Figure 11.36). Remarkably, these secondary structures are similar despite the fact that the nucleotide sequences of these rRNAs themselves exhibit a low degree of similarity. Apparently, evolution is acting at the level of rRNA secondary structure, not rRNA nucleotide sequence. Similar conserved folding patterns are seen for the 23S-like and 5S-like rRNAs that reside in the large ribosomal subunits of various species. An insightful conclusion may be drawn regarding the persistence of such strong secondary structure conservation despite the millennia that have passed since these organisms diverged: *All ribosomes are constructed to a common design, and all function in a similar manner.*

rRNA Tertiary Structure   Recently, the overall three-dimensional, or tertiary, structure of rRNAs has been revealed through X-ray crystallography and cryoelectron microscopy of ribosomes (see Chapter 30). These detailed images of ribosome structure also disclose the tertiary structure of the rRNAs (Figure 11.37), as well as the quaternary interactions that must occur when ribosomal proteins combine with rRNAs and when the ensuing ribonucleoprotein complexes, the small and large subunits, come together to form the complete ribosome. An assortment of tertiary structural features are found in the rRNAs, including coaxial stacks, pseudoknots, and ribose zippers. We will consider of the role of rRNA in ribosome structure and function in Chapter 30.

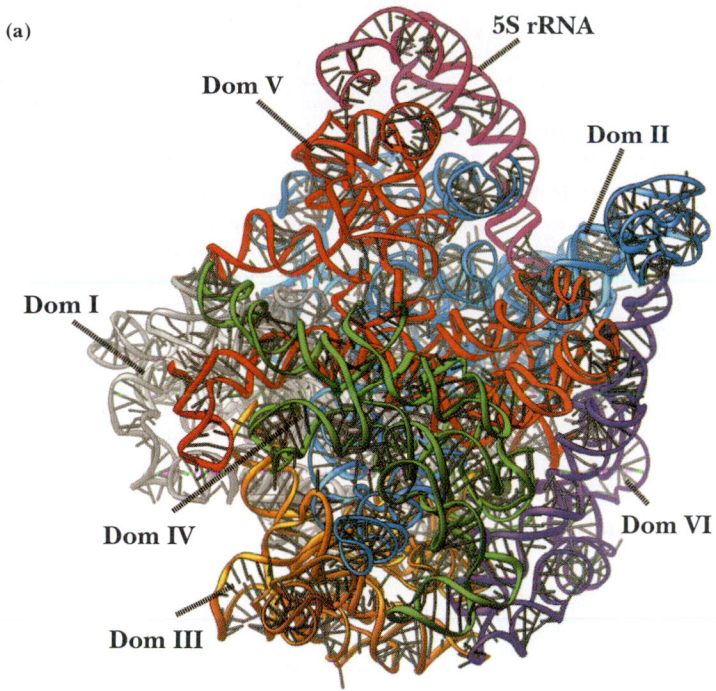

(a)

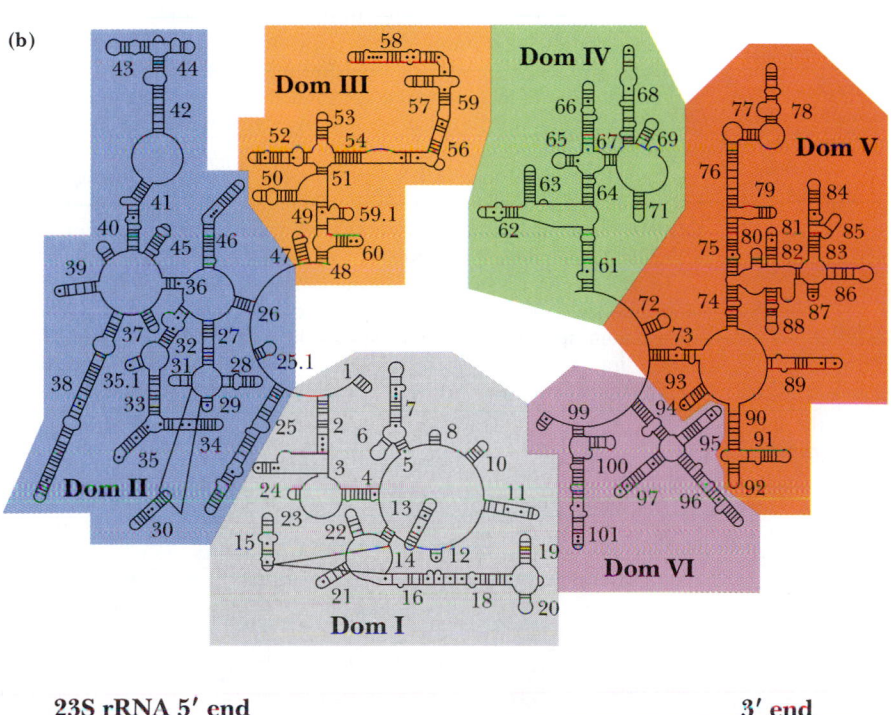

(b)

23S rRNA 5′ end   **3′ end**

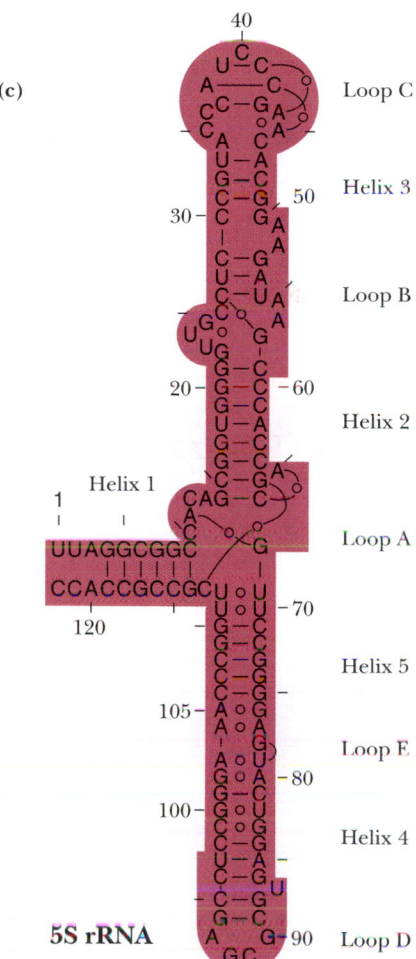

(c)

5S rRNA

FIGURE 11.37 The secondary and tertiary structures of rRNAs in the 50S ribosomal subunit from the archaeon *Haloarcula marismortui*. **(a)** Tertiary structure of the rRNAs within the 50S subunit. The 5S rRNA lies atop the 23S rRNA, as indicated. Domains are color-coded according to the schematic in **(b)**. No ribosomal proteins are shown in this illustration, only the 5S and 23S rRNAs. Note that the overall anatomy of the 50S ribosomal subunit (shown diagrammatically in Figure 10.25) is essentially the same as that of the rRNA molecules within this subunit, despite the fact that these rRNAs account for only 65% of the mass of this particle. **(b)** A schematic diagram of the secondary structure of 23S rRNA. **(c)** Secondary structure of the 5S rRNA in the 50S ribosomal subunit. (*Adapted from Figure 4 in Ban, N., et al., 2000. The complete atomic structure of the large ribosomal subunit at 2.4 Å resolution.* Science **289**:905–920.)

Summary

11.1 How Do Scientists Determine the Primary Structure of Nucleic Acids?
In Sanger's chain termination method of DNA sequencing, a DNA fragment serves as template in a polymerization reaction using DNA polymerase and a primer. Four parallel reactions are run; in each, a different $2',3'$-**di**deoxynucleotide is included (ddATP, ddGTP, ddCTP, or ddTTP). Because dideoxynucleotides lack $3'$-OH groups, they cannot serve as acceptors for $5'$-nucleotide addition, and thus the chain is terminated. This protocol generates a nested set of DNA fragments, with the terminal nucleotide's identity revealed by the dideoxynucleotide that was incorporated. The DNA sequence is read from the gel pattern, and the template sequence is written out from this information.

11.2 What Sorts of Secondary Structures Can Double-Stranded DNA Molecules Adopt?
DNA typically occurs as a double helical molecule, with the two DNA strands running antiparallel to one another, bases inside, sugar–phosphate backbone outside. The double helical arrangement dramatically curtails the conformational possibilities otherwise available to single-stranded DNA. DNA double helices can be in a number of stable conformations, with the three predominant forms termed A-, B-, and Z-DNA. B-DNA, has about 10.5 base pairs per turn, each contributing about 0.332 nm to the length of the double helix. The base pairs in B-DNA are nearly perpendicular to the helix axis. In A-DNA, the pitch is 2.46 nm, with 11 bp per turn. A-DNA has its base pairs displaced around, rather than centered on, the helix axis. Z-DNA has four distinctions: It is left-handed, it is G:C-rich, the repeating unit on a given strand is the dinucleotide, and the sugar–phosphate backbone follows a zigzag course.

11.3 Can the Secondary Structure of DNA Be Denatured and Renatured?
When duplex DNA is subjected to conditions that disrupt base-pairing interactions, the double helix is denatured and the two DNA strands separate as individual random coils. Denatured DNA will renature to re-form a duplex structure if the denaturing conditions are removed. The rate of DNA renaturation is an index of DNA sequence complexity.

If DNA from two different species are mixed, denatured, and allowed to anneal, artificial hybrid duplexes may form, provided the DNA from one species is similar in nucleotide sequence to the DNA of the other. Nucleic acid hybridization can reveal evolutionary relationships, and it can be exploited to identify specific DNA sequences.

11.4 What Is the Tertiary Structure of DNA?
Supercoils are one kind of DNA tertiary structure. In relaxed, B-form DNA, the two strands wind about each other once every 10 bp or so (once every turn of the helix). DNA duplexes form supercoils if the strands are underwound (*negatively supercoiled*) or overwound (*positively supercoiled*). The basic parameter characterizing supercoiled DNA is the linking number, L. L can be equated to the twist (T) and writhe (W), where twist is the number of helical turns and writhe is the number of supercoils: $L = T + W$. L can be changed only if one or both strands of the DNA are broken, the strands are wound tighter or looser, and their ends are rejoined. DNA gyrase is a topoisomerase that introduces negative supercoils into bacterial DNA. DNA sequences that are inverted repeats can form a cruciform if the normal interstrand base pairing is replaced by intrastrand pairing.

11.5 What Is the Structure of Eukaryotic Chromosomes?
The DNA in a eukaryotic cell exists as chromatin, a nucleoprotein complex mostly composed of DNA wrapped around a protein core consisting of eight histone polypeptide chains—two copies each of histones H2A, H2B, H3, and H4. This DNA:histone core structure is termed a nucleosome, the fundamental structural unit of chromosomes. A higher order of chromatin structure is created when the array of nucleosomes is wound into a solenoid with six nucleosomes per turn, creating a 30-nm filament. This 30-nm filament then is formed into long DNA loops, and loops are arranged radially about the circumference of a single turn to form a miniband unit of a chromosome.

11.6 Can Nucleic Acids Be Chemically Synthesized?
Laboratory synthesis of oligonucleotide chains of defined sequence is accomplished through orthogonal solid-phase methods based on phosphoramidite chemistry. Chemical synthesis takes place in the $3' \rightarrow 5'$ direction (the reverse of the biological polymerization direction). Commercially available automated instruments called DNA synthesizers can synthesize oligonucleotide chains with 150 bases or more.

11.7 What Is the Secondary and Tertiary Structure of RNA?
Compared to double-stranded DNA, single-stranded RNA has many more conformational possibilities, but intramolecular interactions and other stabilizing influences limit these possibilities. RNA molecules have many double-stranded regions formed via intrastrand hydrogen bonding. Such double-stranded regions give rise to hairpin stem-loop structures. A number of defined structural motifs recur within the loops of stem-loop structures, such as U-turns and tetraloops.

Single-stranded loops in RNA stem-loops create base-pairing opportunities between distant, complementary, single-stranded loop regions. Other tertiary structural motifs arise from coaxial stacking, pseudoknot formation, and ribose zippers.

In tRNAs, the formation of stem-loops leads to a cloverleaf pattern of secondary structure formed from four base-paired segments: the acceptor stem, the D loop, the anticodon loop, and the TψC loop. Base-pairing interactions between bases in the D and TψC loops give rise to tertiary structure by bending the cloverleaf into the stable L-shaped form.

Substantial intrastrand sequence complementarity also is found in ribosomal RNA molecules, leading to a highly folded pattern based on base pairing between complementary segments. The complete three-dimensional structure of rRNAs has revealed an assortment of the tertiary structural features common to RNAs, including coaxial stacks, pseudoknots, and ribose zippers.

Problems

1. The oligonucleotide d-AGATGCCTGACT was subjected to sequencing by Sanger's dideoxy method, and the products were analyzed by electrophoresis on a polyacrylamide gel. Draw a diagram of the gel-banding patterns that were obtained.

2. The result of sequence determination by the Sanger dideoxy chain termination method is displayed at right. What is the sequence of the original oligonucleotide?

3. X-ray diffraction studies indicate the existence of a novel double-stranded DNA helical conformation in which ΔZ (the rise per base pair) = 0.32 nm and P (the pitch) = 3.36 nm. What are the other parameters of this novel helix: (a) the number of base pairs per turn, (b) $\Delta\phi$ (the mean rotation per base pair), and (c) c (the true repeat)?

4. A 41.5-nm-long duplex DNA molecule in the B-conformation adopts the A-conformation upon dehydration. How long is it now? What is its approximate number of base pairs?

5. If 80% of the base pairs in a duplex DNA molecule (12.5 kbp) are in the B-conformation and 20% are in the Z-conformation, what is the length of the molecule?

6. A "relaxed," circular, double-stranded DNA molecule (1600 bp) is in a solution where conditions favor 10 bp per turn. What is the value of L_0 for this DNA molecule? Suppose DNA gyrase introduces 12 negative supercoils into this molecule. What are the values of L, W, and T now? What is the superhelical density, σ?

7. Suppose one double helical turn of a superhelical DNA molecule changes conformation from B- to Z-form. What are the changes in L, W, and T? Why do you suppose the transition of DNA from B- to Z-form is favored by negative supercoiling?

8. Assume that there is one nucleosome for every 200 bp of eukaryotic DNA. How many nucleosomes are there in a diploid human cell? Nucleosomes can be approximated as disks 11 nm in diameter and 6 nm long. If all the DNA molecules in a diploid human cell are in the B-conformation, what is the sum of their lengths? If this DNA is now arrayed on nucleosomes in the beads-on-a-string motif, what would be the approximate total height of the nucleosome column if these disks were stacked atop one another?

9. The characteristic secondary structures of tRNA and rRNA molecules are achieved through intrastrand hydrogen bonding. Even for the small tRNAs, remote regions of the nucleotide sequence interact via H bonding when the molecule adopts the cloverleaf pattern. Using Figure 11.30 as a guide, draw the primary structure of a tRNA and label the positions of its various self-complementary regions.

10. Using the data in Table 10.3, arrange the DNAs from the following sources in order of increasing T_m: human, salmon, wheat, yeast, E. coli.

11. The DNAs from mice and rats have (G + C) contents of 44% and 40%, respectively. Calculate the T_ms for these DNAs in 0.2 M NaCl. If samples of these DNAs were inadvertently mixed, how might they be separated from one another? Describe the procedure and the results. (Hint: See Chapter Appendix to this chapter.)

12. Calculate the density (ρ) of avian tubercle bacillus DNA from the data presented in Table 10.3 and the equation $\rho = 1.660 + 0.098(GC)$, where (GC) is the mole fraction of (G + C) in DNA.

A	C	G	T

---------			

13. (Integrates with Chapter 10.) Pseudouridine (ψ) is an invariant base in the TψC loop of tRNA; ψ is also found in strategic places in rRNA. (Figure 10.26 shows the structure of pseudouridine.) Draw the structure of the base pair that ψ might form with G.

14. The plasmid pBR322 is a closed circular dsDNA containing 4363 base pairs. What is the length in nm of this DNA (that is, what is its circumference if it were laid out as a perfect circle)? The E. coli K12 chromosome is a closed circular dsDNA of about 4,639,000 base pairs. What would be the circumference of a perfect circle formed from this chromosome? What is the diameter of a dsDNA molecule? Calculate the ratio of the length of the circular plasmid pBR322 to the diameter of the DNA of which it's made. Do the same for the E. coli chromosome.

Preparing for the MCAT Exam

15. (Integrates with Chapter 10.) Erwin Chargaff did not have any DNA samples from thermoacidophilic bacteria such as those that thrive in the geothermal springs of Yellowstone National Park. (Such bacteria had not been isolated by 1951 when Chargaff reported his results.) If he had obtained such a sample, what do you think its relative G:C content might have been? Why?

*16. Think about the structure of DNA in its most common B-form double helical conformation and then list its most important structural features (deciding what is "important" from the biological role of DNA as the material of heredity). Arrange your answer with the most significant features first.

Biochemistry ⑤Now™ Preparing for an exam? Test yourself on key questions at http://chemistry.brookscole.com/ggb3

Further Reading

General References

Adams, R. L. P., Knowler, J. T., and Leader, D. P., 1992. *The Biochemistry of the Nucleic Acids,* 11th ed. London: Chapman and Hall.

Kornberg, A., and Baker, T. A., 1991. *DNA Replication,* 2nd ed. New York: W. H. Freeman.

Watson, J. D., Hopkins, N. H., Roberts, J. W., Steitz, J. A., and Weiner, A. M., 1987. *The Molecular Biology of the Gene,* Vol. I, *General Principles,* 4th ed. Menlo Park, CA: Benjamin/Cummings.

DNA Sequencing

Meldrum, D. 2000. Automation for genomics, Part One: Preparation for sequencing. *Genome Research* **10:**1081–1092.

Meldrum, D., 2000. Automation for genomics, Part Two: Sequencers, microarrays, and future trends. *Genome Research* **10:**1288–1303.

Wu, R., 1993. Development of enzyme-based methods for DNA sequence analysis and their application in genome projects. *Methods in Enzymology* **67:**431–468.

Higher-Order DNA Structure

Bates, A. D., and Maxwell, A., 1993. *DNA Topology.* New York: IRL Press at Oxford University Press.

Callandine, C. R., and Drew, H. R., 1992. *Understanding DNA: The Molecule and How It Works.* London: Academic Press.

Rich, A., 2003. The double helix: A tale of two puckers. *Nature Structural Biology* **10:**247–249.

Rich, A., Nordheim, A., and Wang, A. H-J., 1984. The chemistry and biology of left-handed Z-DNA. *Annual Review of Biochemistry* **53:**791–846.

Watson, J. D., ed., 1983. *Structures of DNA. Cold Spring Harbor Symposia on Quantitative Biology,* Volume XLVII. New York: Cold Spring Harbor Laboratory.

Nucleosomes

Arents, G., et al., 1991. The nucleosome core histone octamer at 3.1 Å resolution: A tripartite protein assembly and a left-hand superhelix. *Proceedings of the National Academy of Sciences U.S.A.* **88:**10148–10152.

Luger, C., et al., 1997. Crystal structure of the nucleosome core particle at 2.8 Å resolution. *Nature* **389:**251–260.

Rhodes, D., 1997. The nucleosome core all wrapped up. *Nature* **389:**231–233.

Wang, B-C., et al., 1994. The octameric histone core of the nucleosome. *Journal of Molecular Biology* **236:**179–188.

Chromosome Structure

Pienta, K. J., and Coffey, D. S., 1984. A structural analysis of the role of the nuclear matrix and DNA loops in the organization of the nucleus and chromosomes. In Cook, P. R., and Laskey, R. A., eds., Higher order structure in the nucleus. *Journal of Cell Science Supplement* **1:**123–135.

Sumner, A. T., 2003. *Chromosomes: Organization and Function.* Malden, MA: Blackwell Science.

Telomeres

Axelrod, N., 1996. Of telomeres and tumors. *Nature Medicine* **2:**158–159.

Feng, J., Funk, W. D., Wang, S-S., Weinrich, S. L., et al., 1995. The RNA component of human telomerase. *Science* **269:**1236–1241.

Chemical Synthesis of Genes

Ferretti, L., Karnik, S. S., Khorana, H. G., Nassal, M., and Oprian, D. D., 1986. Total synthesis of a gene for bovine rhodopsin. *Proceedings of the National Academy of Sciences U.S.A.* **83:**599–603.

Gray, M. W., and Cedergren, R., eds., 1993. The new age of RNA. *The FASEB Journal* **7:**4–239. A collection of articles emphasizing the new appreciation for RNA in protein synthesis, in evolution, and as a catalyst.

Higher-Order RNA Structure

Ban, N, et al., 2000. The complete atomic structure of the large ribosomal subunit at 2.4 Å resolution. *Science* **289:**905–920.

Cannone, J. J., Subashchandran, S., Schnare, M. N., et al., 2003. The comparative RNA web (CRW) site: An online database of comparative sequence and structure information for ribosomal, intron, and other RNAs. URL: *http://www.rna.icmb.utexas.edu/*

Moore, P. B., 1999. Structural motifs in RNA. *Annual Review of Biochemistry* **67:**287–300.

Tinoco, I., Jr., and Bustamente, C., 1999. How RNA folds. *Journal of Molecular Biology* **293:**271–281.

Isopycnic Centrifugation and Buoyant Density of DNA

Density gradient ultracentrifugation is a variant of the basic technique of ultracentrifugation (discussed in the Appendix to Chapter 5). Density gradient centrifugation can be used to isolate DNA. The densities of DNAs are about the same as those of concentrated solutions of cesium chloride, CsCl (1.6 to 1.8 g/mL). Centrifugation of CsCl solutions at very high rotational speeds, where the centrifugal force becomes 10^5 times stronger than the force of gravity, causes the formation of a density gradient within the solution. This gradient is the result of a balance that is established between the sedimentation of the salt ions toward the bottom of the tube and their diffusion upward toward regions of lower concentration. If DNA is present in the centrifuged CsCl solution, it moves to a position of equilibrium in the gradient equivalent to its buoyant density (Figure A11.1). For this reason, this technique is also called **isopycnic centrifugation.**

Cesium chloride centrifugation is an excellent means of removing RNA and proteins in the purification of DNA. The density of DNA is typically slightly greater than 1.7 g/cm^3, whereas the density of RNA is more than 1.8 g/cm^3. Proteins have densities less than 1.3 g/cm^3. In CsCl solutions of appropriate density, the DNA bands near the center of the tube, RNA pellets to the bottom, and the proteins float near the top. Single-stranded DNA is denser than double helical DNA. The irregular structure of randomly coiled ssDNA allows the atoms to pack together through van der Waals interactions. These interactions compact the molecule into a smaller volume than that occupied by a hydrogen-bonded double helix.

The net movement of solute particles in an ultracentrifuge is the result of two processes: diffusion (from regions of higher concentration to regions of lower concentration) and sedimentation due to centrifugal force (in the direction away from the axis of rotation). In general, diffusion rates for molecules are inversely proportional to their molecular weight—larger molecules diffuse more slowly than smaller ones. On the other hand, sedimentation rates increase with increasing molecular weight. A macromolecular species that has reached its position of equilibrium in isopycnic centrifugation has formed a concentrated band of material.

Essentially three effects are influencing the movement of the molecules in creating this concentration zone: (1) diffusion away to regions of lower concentration, (2) sedimentation of molecules situated at positions of slightly lower solution density in the density gradient, and (3) flotation (buoyancy or "reverse sedimentation") of molecules that have reached positions of slightly greater solution density in the gradient. The consequence of the physics of these effects is that, at equilibrium, *the width of the concentration band established by the macromolecular species is inversely proportional to the square root of its molecular weight.* That is, a population of large molecules will form a concentration band that is narrower than the band formed by a population of small molecules. For example, the bandwidth formed by dsDNA will be less than the bandwidth formed by the same DNA when dissociated into ssDNA.

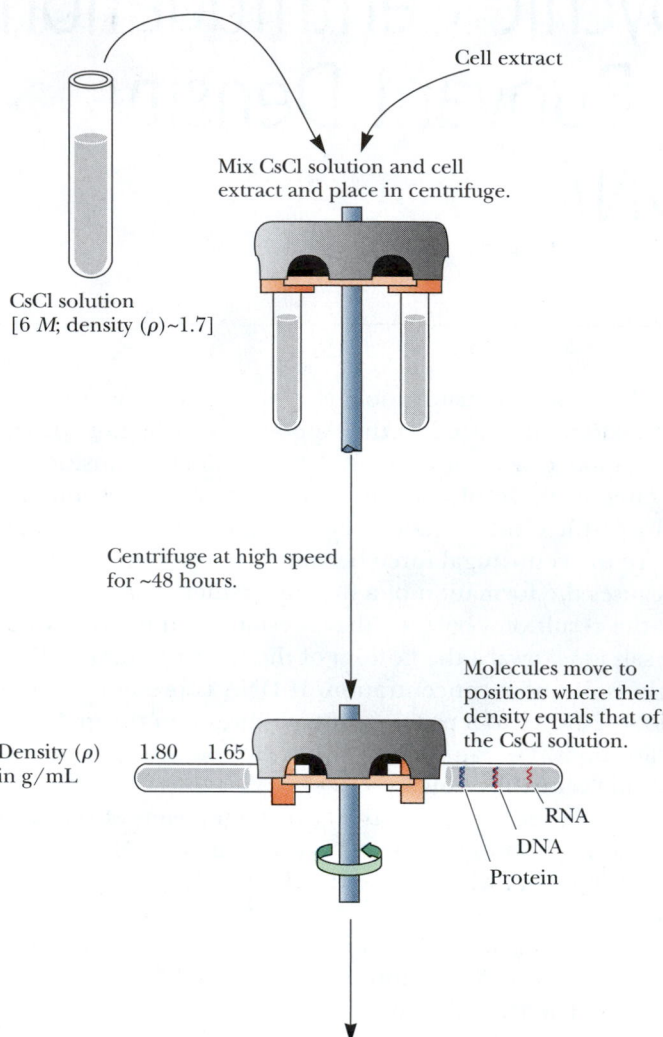

Cell extract

Mix CsCl solution and cell
extract and place in centrifuge.

CsCl solution
[6 *M*; density (ρ)~1.7]

Centrifuge at high speed
for ~48 hours.

Molecules move to
positions where their
density equals that of
the CsCl solution.

Density (ρ)   1.80   1.65
in g/mL

RNA
DNA
Protein

Proteins and nucleic acids absorb UV light.
The positions of these molecules within the
centrifuge can be determined by ultraviolet optics.

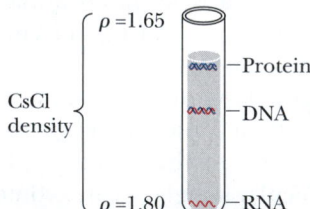

ρ =1.65

CsCl
density

ρ =1.80

Protein

DNA

RNA

FIGURE A11.1 Density gradient centrifugation is a
common method of separating macromolecules, particularly nucleic acids, in solution. A cell extract is
mixed with a solution of CsCl to a final density of
about 1.7 g/cm³ and centrifuged at high speed
(40,000 rpm, giving relative centrifugal forces of
about 200,000 *g*). The biological macromolecules in
the extract will move to equilibrium positions in the
CsCl gradient that reflect their buoyant densities.

Recombinant DNA: Cloning and Creation of Chimeric Genes

CHAPTER 12

Essential Question

Emerging techniques to manipulate nucleic acids in the laboratory allowed scientists to combine DNA segments derived from different sources. Such manmade products are called recombinant DNA molecules, and the use of such molecules to alter the genetics of organisms is termed genetic engineering. *What are the methods that scientists use to create recombinant DNA molecules; can scientists create genes from recombinant DNA molecules; and can scientists modify the heredity of an organism using recombinant DNA?*

Scala/Art Resource, NY

The Chimera of Arezzo, of Etruscan origin and probably from the fifth century B.C., was found near Arezzo, Italy, in 1553. Chimeric animals existed only in the imagination of the ancients. But the ability to create chimeric DNA molecules is a very real technology that has opened up a whole new field of scientific investigation.

In the early 1970s, technologies for the laboratory manipulation of nucleic acids emerged. In turn, these technologies led to the construction of DNA molecules composed of nucleotide sequences taken from different sources. The products of these innovations, **recombinant DNA molecules**,[1] opened exciting new avenues of investigation in molecular biology and genetics, and a new field was born—**recombinant DNA technology. Genetic engineering** is the application of this technology to the manipulation of genes. These advances were made possible by methods for **amplification** of any particular DNA segment, regardless of source, within bacterial host cells. Or, in the language of recombinant DNA technology, the **cloning** of virtually any DNA sequence became feasible.

...how many vain chimeras have you created?...Go and take your place with the seekers after gold.
Leonardo da Vinci, *The Notebooks* (1508–1518), Volume II, Chapter 25

12.1 | What Does It Mean: "To Clone"?

In classical biology, a *clone* is a population of identical organisms derived from a single parental organism. For example, the members of a colony of bacterial cells that arise from a single cell on a petri plate are a clone. Molecular biology has borrowed the term to mean a collection of molecules or cells all identical to an original molecule or cell. So, if the original cell on the petri plate harbored a recombinant DNA molecule in the form of a plasmid, the plasmids within the millions of cells in a bacterial colony represent a clone of the original DNA molecule, and these molecules can be isolated and studied. Furthermore, if the cloned DNA molecule is a gene (or part of a gene)—that is, it encodes a functional product—a new avenue to isolating and studying this product has opened. Recombinant DNA methodology offers exciting new vistas in biochemistry.

Plasmids Are Very Useful in Cloning Genes

Plasmids are naturally occurring, circular, extrachromosomal DNA molecules (see Chapter 11). Natural strains of the common colon bacterium *Escherichia coli* isolated from various sources harbor diverse plasmids. Often these plasmids carry genes specifying novel metabolic activities that are advantageous to the

Key Questions

12.1 What Does It Mean: "To Clone"?
12.2 What Is a DNA Library?
12.3 What Is the Polymerase Chain Reaction (PCR)?
12.4 Is It Possible to Make Directed Changes in the Heredity of an Organism?

[1]The advent of molecular biology, like that of most scientific disciplines, has generated a jargon all its own. Learning new fields often requires gaining familiarity with a new vocabulary. We will soon see that many words—*vector, amplification,* and *insert* are but a few examples—have been bent into new meanings to describe the marvels of this new biology.

host bacterium. These activities range from catabolism of unusual organic substances to metabolic functions that endow the host cells with resistance to antibiotics, heavy metals, or bacteriophages. Plasmids that are able to perpetuate themselves in *E. coli,* the bacterium favored by bacterial geneticists and molecular biologists, have become the darlings of recombinant DNA technology. Because restriction endonuclease digestion of plasmids can generate fragments with overlapping or "sticky" ends, artificial plasmids can be constructed by ligating different fragments together. Such artificial plasmids were among the earliest recombinant DNA molecules. These recombinant molecules can be autonomously replicated, and hence propagated, in suitable bacterial host cells, provided they still possess a site signaling where DNA replication can begin (a so-called **origin of replication** or *ori* sequence).

Plasmids as Cloning Vectors   The idea arose that "foreign" DNA sequences could be inserted into artificial plasmids and that these foreign sequences would be carried into *E. coli* and propagated as part of the plasmid. That is, these plasmids could serve as **cloning vectors** to carry genes. (The word *vector* is used here in the sense of "a vehicle or carrier.") Plasmids useful as cloning vectors possess three common features: **a replicator, a selectable marker,** and **a cloning site** (Figure 12.1). A *replicator* is an origin of replication, or *ori*. The *selectable marker* is typically a gene conferring resistance to an antibiotic. Only cells containing the cloning vector will grow in the presence of the antibiotic. Therefore, growth on antibiotic-containing media "selects for" plasmid-containing cells. Typically, the *cloning site* is a sequence of nucleotides representing one or more restriction endonuclease cleavage sites. Cloning sites are located where the insertion of foreign DNA neither disrupts the plasmid's ability to replicate nor inactivates essential markers.

Virtually Any DNA Sequence Can Be Cloned   Nuclease cleavage at a restriction site opens, or *linearizes,* the circular plasmid so that a foreign DNA fragment can be inserted. The ends of this linearized plasmid are joined to the ends of the fragment so that the circle is closed again, creating a recombinant plasmid (Figure 12.2).

Biochemistry ⒺNow™   Go to BiochemistryNow and click BiochemistryInteractive to explore the restriction sites of plasmids and genes.

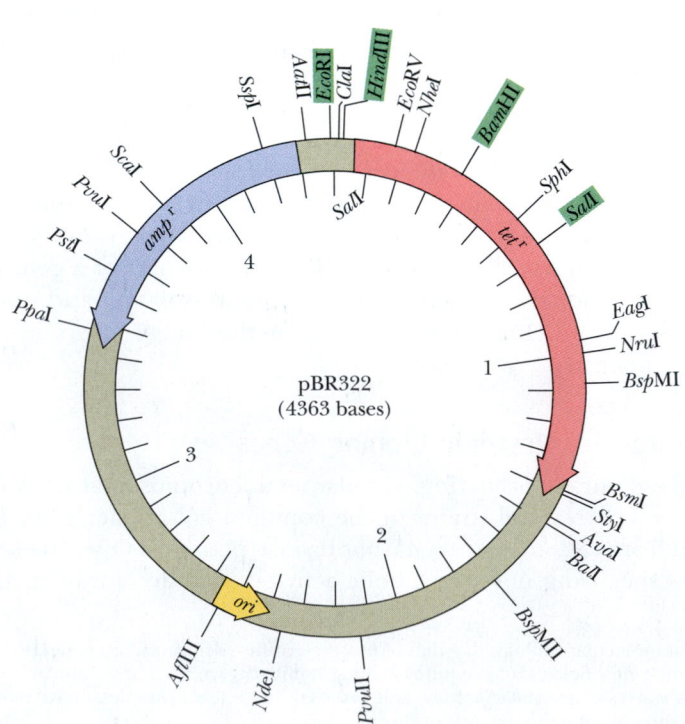

FIGURE 12.1   One of the first widely used cloning vectors, the plasmid pBR322. This 4363-bp plasmid contains an origin of replication *(ori)* and genes encoding resistance to the drugs ampicillin *(amp^r)* and tetracycline *(tet^r)*. The locations of restriction endonuclease cleavage sites are indicated.

Recombinant plasmids are hybrid DNA molecules consisting of plasmid DNA sequences plus inserted DNA elements (called *inserts*). Such hybrid molecules are also called **chimeric constructs** or **chimeric plasmids.** (The term *chimera* is borrowed from mythology and refers to a beast composed of the body and head of a lion, the heads of a goat and a snake, and the wings of a bat.) The presence of foreign DNA sequences does not adversely affect replication of the plasmid, so chimeric plasmids can be propagated in bacteria just like the original plasmid. Bacteria often harbor several hundred copies of common cloning vectors per cell. Hence, large amounts of a cloned DNA sequence can be recovered from bacterial cultures. The enormous power of recombinant DNA technology stems in part from the fact that *virtually any DNA sequence can be selectively cloned and amplified in this manner.* DNA sequences that are difficult to clone include inverted repeats, origins of replication, centromeres, and telomeres. The only practical limitation is the size of the foreign DNA segment: Most plasmids with inserts larger than about 10 kbp are not replicated efficiently.

Bacterial cells may contain one or many copies of a particular plasmid, depending on the nature of the plasmid replicator. That is, plasmids are classified as *high copy number* or *low copy number.* The copy number of most genetically engineered plasmids is high (30–40), but some are lower.

Construction of Chimeric Plasmids   Creation of chimeric plasmids requires joining the ends of the foreign DNA insert to the ends of a linearized plasmid (Figure 12.2). This ligation is facilitated if the ends of the plasmid and the insert have complementary, single-stranded overhangs. Then these ends can base-pair with one another, annealing the two molecules together. One way to generate such ends is to cleave the DNA with restriction enzymes that make staggered cuts; many such restriction endonucleases are available (see Table 10.5). For example, if the sequence to be inserted is an *Eco*RI fragment and the plasmid is cut with *Eco*RI, the single-stranded sticky ends of the two DNAs can anneal (Figure 12.3). The interruptions in the sugar–phosphate backbone of DNA can then be sealed with DNA ligase to yield a covalently closed, circular chimeric plasmid. DNA ligase is an enzyme that covalently links adjacent $3'$-OH and $5'$-PO$_4$ groups. An inconvenience of this strategy is that *any* pair of *Eco*RI sticky ends can anneal with each other. So, plasmid molecules can reanneal with themselves, as can the foreign DNA restriction fragments. These DNAs can be eliminated by selection schemes designed to identify only those bacteria containing chimeric plasmids.

Blunt-end ligation is an alternative method for joining different DNAs. This method depends on the ability of **phage T4 DNA ligase** to covalently join the ends of any two DNA molecules (even those lacking $3'$- or $5'$-overhangs) (Figure 12.4). Some restriction endonucleases cut DNA so that blunt ends are formed (see Table 10.5). Because there is no control over which pair of DNAs are blunt-end ligated by T4 DNA ligase, strategies to identify the desired products must be applied.

A great number of variations on these basic themes have emerged. For example, short synthetic DNA duplexes whose nucleotide sequence consists of little more than a restriction site can be blunt-end ligated onto any DNA. These short DNAs are known as **linkers.** Cleavage of the ligated DNA with the restriction enzyme then leaves tailor-made sticky ends useful in cloning reactions (Figure 12.5). Similarly, many vectors contain a **polylinker** cloning site, a short region of DNA sequence bearing numerous restriction sites.

Promoters and Directional Cloning   Note that the strategies discussed thus far create hybrids in which the orientation of the DNA insert within the chimera is random. Sometimes it is desirable to insert the DNA in a particular orientation. For example, an experimenter might wish to insert a particular DNA (a gene) in a vector so that its gene product is synthesized. To do this, the DNA must be

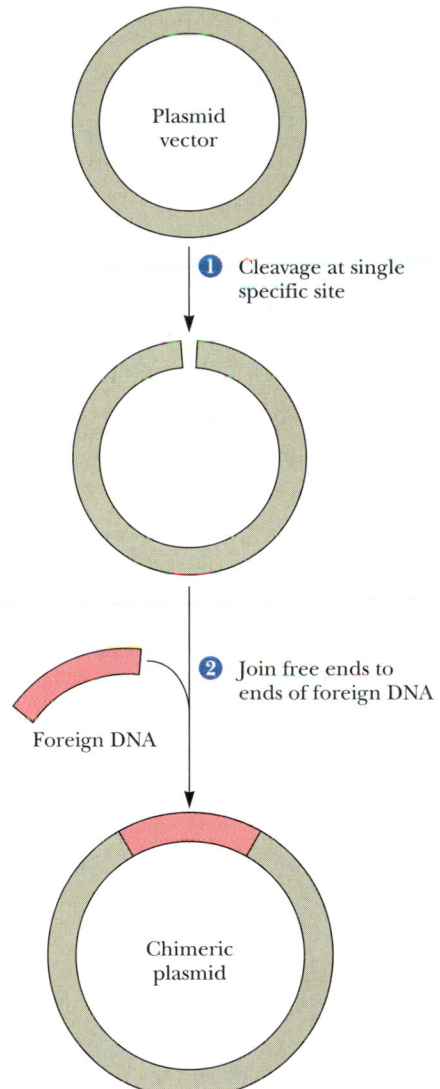

Biochemistry ⊘ Now™   ACTIVE FIGURE 12.2
(1) Foreign DNA sequences can be inserted into plasmid vectors by opening the circular plasmid with a restriction endonuclease. **(2)** The ends of the linearized plasmid DNA are then joined with the ends of a foreign sequence, reclosing the circle to create a chimeric plasmid. **Test yourself on the concepts in this figure at http://chemistry.brookscole. com/ggb3**

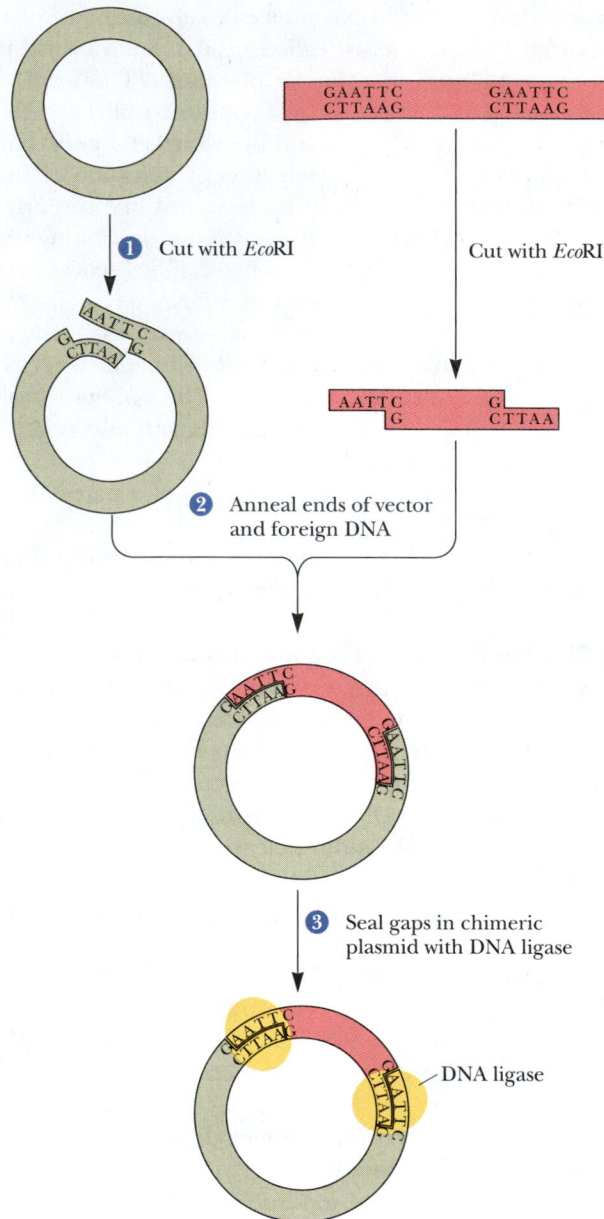

Biochemistry☰**Now**™ **ACTIVE FIGURE 12.3** Restriction endonuclease *Eco*RI cleaves double-stranded DNA. The recognition site for *Eco*RI is the hexameric sequence GAATTC:

$$5' \ldots \text{NpNpNpNp}\textbf{GpApApTpTpC}\text{pNpNpNpNp} \ldots 3'$$
$$3' \ldots \text{NpNpNpNp}\textbf{CpTpTpApApG}\text{pNpNpNpNp} \ldots 5'$$

Cleavage occurs at the G residue on each strand, so the DNA is cut in a staggered fashion, leaving 5′-overhanging single-stranded ends (sticky ends):

$$5' \ldots \text{NpNpNpNp}\textbf{G} \qquad \text{pApApTpTpCpNpNpNpNp} \ldots 3'$$
$$3' \ldots \text{NpNpNpNp}\textbf{CpTpTpApAp} \qquad \text{GpNpNpNpNp} \ldots 5'$$

An *Eco*RI restriction fragment of foreign DNA can be inserted into a plasmid having an *Eco*RI cloning site by (**1**) cutting the plasmid at this site with *Eco*RI, (**2**) annealing the linearized plasmid with the *Eco*RI foreign DNA fragment, and (**3**) sealing the nicks with DNA ligase. **Test yourself on the concepts in this figure at http://chemistry.brookscole.com/ggb3**

placed downstream from a **promoter.** A promoter is a nucleotide sequence lying upstream of a gene. The promoter controls expression of the gene. RNA polymerase molecules bind specifically at promoters and initiate transcription of adjacent genes, copying template DNA into RNA products. One way to insert DNA so that it will be properly oriented with respect to the promoter is to create DNA molecules whose ends have different overhangs. Ligation of such

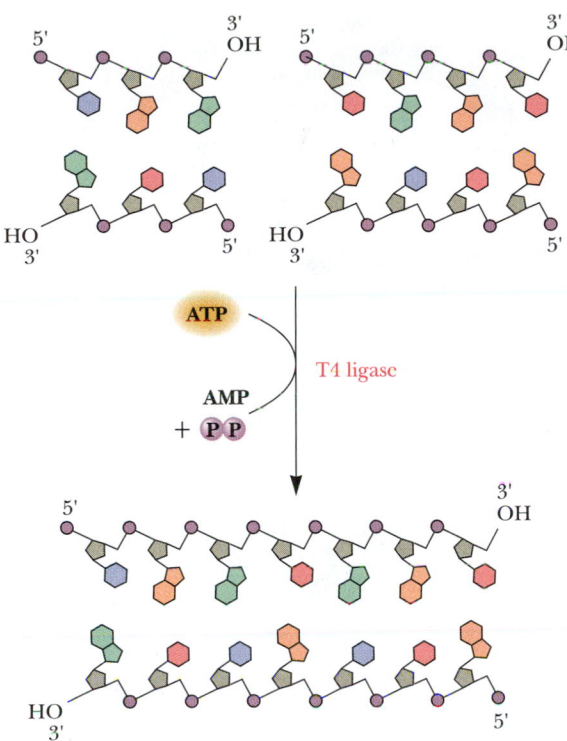

Biochemistry ⑤ Now™   **ANIMATED FIGURE 12.4**
Blunt-end ligation using phage T4 DNA ligase, which
catalyzes the ATP-dependent ligation of DNA molecules. AMP and PP$_i$ are by-products. **See this figure
animated at http://chemistry.brookscole.com/ggb3**

molecules into the plasmid vector can only take place in one orientation to give
directional cloning (Figure 12.6).

Biologically Functional Chimeric Plasmids   The first biologically functional
chimeric DNA molecules constructed in vitro were assembled from parts of different plasmids in 1973 by Stanley Cohen, Annie Chang, Herbert Boyer, and
Robert Helling. These plasmids were used to **transform** recipient *E. coli* cells
(*transformation* means the uptake and replication of exogenous DNA by a recipient cell). To facilitate transformation, the bacterial cells were rendered

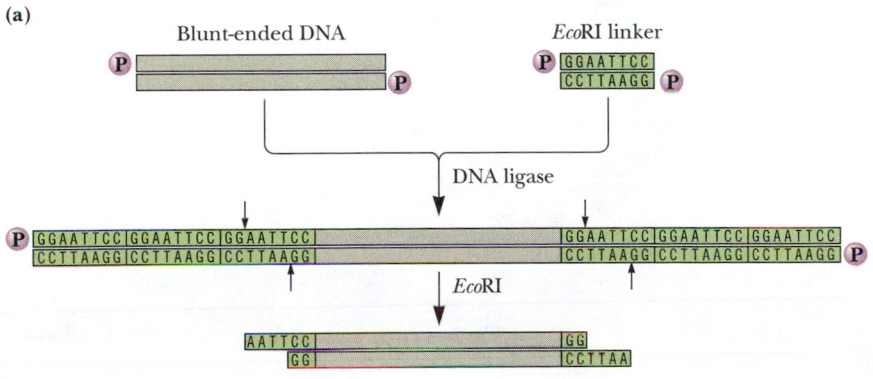

(b) A vector cloning site containing multiple restriction sites,
a so-called polylinker.

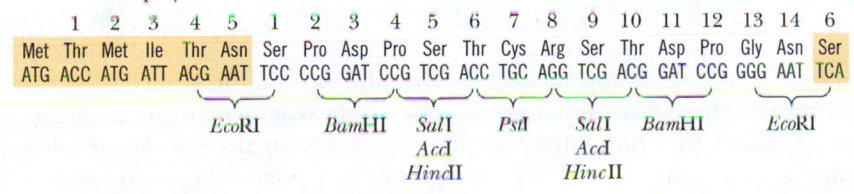

Biochemistry ⑤ Now™   **ANIMATED FIGURE 12.5**
(a) The use of linkers to create tailor-made ends on
cloning fragments. Synthetic oligonucleotide duplexes
whose sequences represent *Eco*RI restriction sites are
blunt-end ligated to a DNA molecule using T4 DNA
ligase. Note that the ligation reaction can add multiple linkers on each end of the blunt-ended DNA.
*Eco*RI digestion removes all but the terminal one,
leaving the desired 5'-overhangs. **(b)** Cloning
vectors often have polylinkers consisting of a multiple array
of restriction sites at their cloning sites, so restriction
fragments generated by a variety of endonucleases
can be incorporated into the vector. Note that the
polylinker is engineered not only to have multiple
restriction sites but also to have an uninterrupted
sequence of codons, so this region of the vector has
the potential for translation into protein. The
sequence shown is the cloning site for the vectors
M13mp7 and pUC7; the colored amino acid residues
are contiguous with the coding sequence of the *lacZ*
gene carried by this vector (see Figure 12.19). *(a, Adapted
from Figure 3.16.3; b, adapted from Figure 1.14.2, in Ausubel, F. M.,
et al., 1987, Current Protocols in Molecular Biology. New York:
John Wiley & Sons.)* **See this figure animated at http://
chemistry.brookscole.com/ggb3**

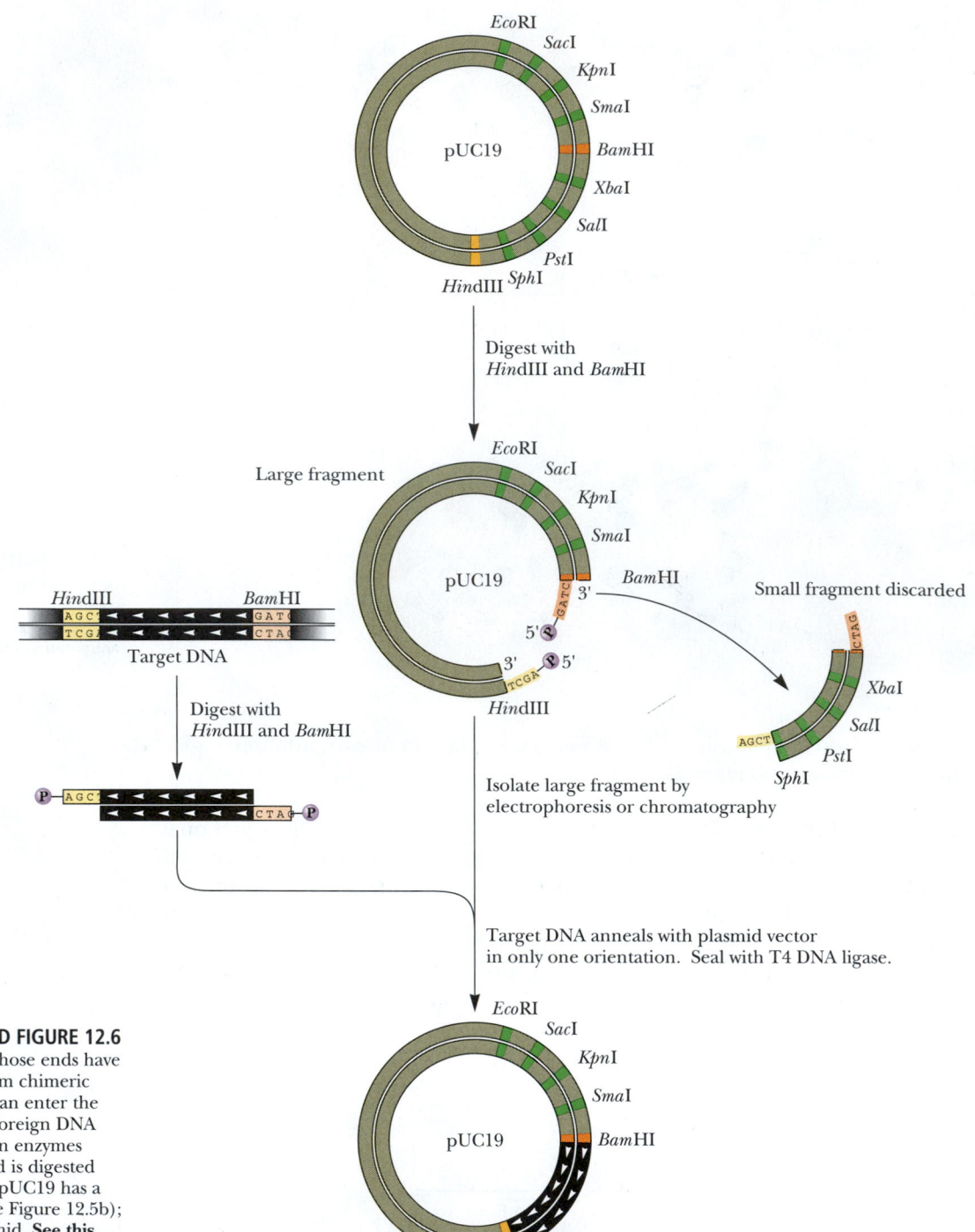

Biochemistry✺Now™ ANIMATED FIGURE 12.6
Directional cloning. DNA molecules whose ends have different overhangs can be used to form chimeric constructs in which the foreign DNA can enter the plasmid in only one orientation. The foreign DNA is digested with two different restriction enzymes (*Hind*III and *Bam*HI), and the plasmid is digested with the same two enzymes. Note that pUC19 has a polylinker or universal cloning site (see Figure 12.5b); pUC stands for universal cloning plasmid. **See this figure animated at http://chemistry.brookscole. com/ggb3**

somewhat permeable to DNA by Ca^{2+} treatment and a brief 42°C heat shock. Although less than 0.1% of the Ca^{2+}-treated bacteria became competent for transformation, transformed bacteria could be selected by their resistance to certain antibiotics (Figure 12.7). Consequently, the chimeric plasmids must have been biologically functional in at least two aspects: They replicated stably within their hosts, and they expressed the drug resistance markers they carried.

In general, plasmids used as cloning vectors are engineered to be small (2.5 kbp to about 10 kbp in size) so that the size of the insert DNA can be maximized. These plasmids have only a single origin of replication, so the time necessary for complete replication depends on the size of the plasmid. Under selective pressure in a growing culture of bacteria, overly large plas-

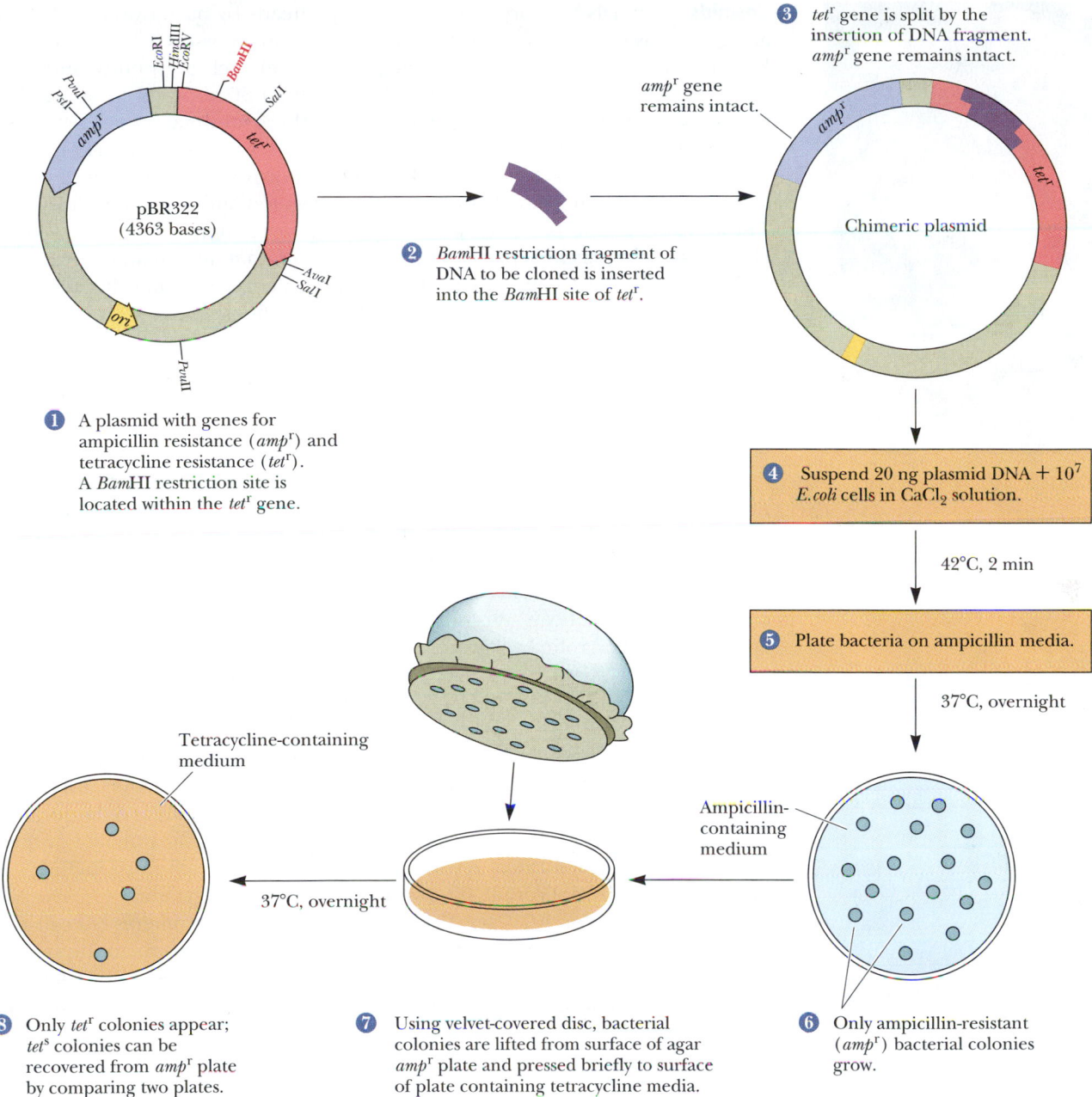

❸ *tet*^{r} gene is split by the insertion of DNA fragment. *amp*^{r} gene remains intact.

amp^r gene remains intact.

Chimeric plasmid

❷ *Bam*HI restriction fragment of DNA to be cloned is inserted into the *Bam*HI site of *tet*^{r}.

❶ A plasmid with genes for ampicillin resistance (*amp*^{r}) and tetracycline resistance (*tet*^{r}). A *Bam*HI restriction site is located within the *tet*^{r} gene.

pBR322 (4363 bases)

❹ Suspend 20 ng plasmid DNA + 10^7 *E. coli* cells in CaCl$_2$ solution.

42°C, 2 min

❺ Plate bacteria on ampicillin media.

37°C, overnight

Tetracycline-containing medium

Ampicillin-containing medium

37°C, overnight

❽ Only *tet*^{r} colonies appear; *tet*^{s} colonies can be recovered from *amp*^{r} plate by comparing two plates.

❼ Using velvet-covered disc, bacterial colonies are lifted from surface of agar *amp*^{r} plate and pressed briefly to surface of plate containing tetracycline media.

❻ Only ampicillin-resistant (*amp*^{r}) bacterial colonies grow.

Biochemistry ⊘ Now™   ACTIVE FIGURE 12.7
A typical bacterial transformation experiment. Here the plasmid pBR322 is the cloning vector. (1) Cleavage of pBR322 with restriction enzyme *Bam*HI, followed by (2) annealing and ligation of inserts generated by *Bam*HI cleavage of some foreign DNA, (3) creates a chimeric plasmid. (4) The chimeric plasmid is then used to transform Ca^{2+}-treated heat-shocked *E. coli* cells, and the bacterial sample is plated on a petri plate. (5) Following incubation of the petri plate overnight at 37°C, (6) colonies of *amp*^{r} bacteria are evident. (7) Replica plating of these bacteria on plates of tetracycline-containing media (8) reveals which colonies are *tet*^{r} and which are tetracycline sensitive (*tet*^{s}). Only the *tet*^{s} colonies possess plasmids with foreign DNA inserts. **Test yourself on the concepts in this figure at http://chemistry. brookscole.com/ggb3**

mids are prone to delete any nonessential "genes," such as any foreign inserts. Such deletion would thwart the purpose of most cloning experiments. The useful upper limit on cloned inserts in plasmids is about 10 kbp. Many eukaryotic genes exceed this size.

Bacteriophage λ Can Be Used as a Cloning Vector

The genome of bacteriophage λ (lambda) (Figure 12.8) is a 48.5-kbp linear DNA molecule that is packaged into the head of the bacteriophage. The middle one-third of this genome is not essential to phage infection, so λ phage DNA has been engineered to accommodate the insertion of foreign DNA molecules up to 16 kbp into this region for cloning purposes. In vitro packaging systems are then used to package the chimeric DNA into phage heads which, when assembled with phage tails, form infective phage particles. Bacteria infected with recombinant phage produce large numbers of phage progeny before they lyse, and large amounts of recombinant DNA can be easily purified from the lysate.

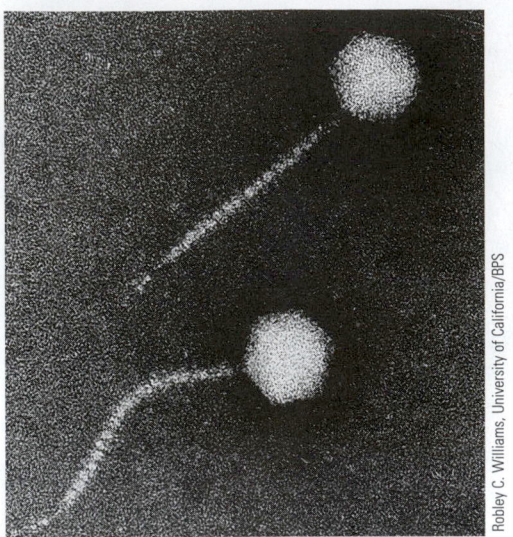

FIGURE 12.8 Electron micrograph of bacterio-phage λ.

Robley C. Williams, University of California/BPS

Cosmids   The DNA incorporated into phage heads by bacteriophage λ packaging systems must satisfy only a few criteria. It must possess a 14-bp sequence known as *cos* (which stands for *co*hesive end *s*ite) at each of its ends, and these *cos* sequences must be separated by no fewer than 36 kbp and no more than 51 kbp of DNA. Essentially, any DNA satisfying these minimal requirements will be packaged and assembled into an infective phage particle. Other cloning features, such as an *ori*, selectable markers, and a polylinker, are joined to the *cos* sequence so that the cloned DNA can be propagated and selected in host cells. These features have been achieved by placing *cos* sequences on either side of cloning sites in plasmids to create **cosmid vectors** that are capable of carrying DNA inserts about 40 kbp in size (Figure 12.9). Because cosmids lack essential phage genes, they reproduce in host bacteria as plasmids.

Shuttle Vectors Are Plasmids That Can Propagate in Two Different Organisms

Shuttle vectors are plasmids capable of propagating and transferring ("shuttling") genes between two different organisms, one of which is typically a prokaryote (*E. coli*) and the other a eukaryote (for example, yeast). Shuttle vectors must have unique origins of replication for each cell type as well as different markers for selection of transformed host cells harboring the vector (Figure 12.10). Shuttle vectors have the advantage that eukaryotic genes can be cloned in bacterial hosts, yet the expression of these genes can be analyzed in appropriate eukaryotic backgrounds.

Artificial Chromosomes Can Be Created from Recombinant DNA

DNA molecules 2 megabase pairs in length have been successfully propagated in yeast by creating **yeast artificial chromosomes** or **YACs.** Furthermore, such YACs have been transferred into transgenic mice for the analysis of large genes or multigenic DNA sequences in vivo, that is, within the living animal. For these large DNAs to be replicated in the yeast cell, YAC constructs must include not only an origin of replication (known in yeast terminology as an *autonomously replicating sequence* or *ARS*) but also a centromere and telomeres. Recall that centromeres provide the site for attachment of the chromosome to the spindle during mitosis and meiosis, and telomeres are nucleotide sequences defining the ends of chromosomes. Telomeres are essential for proper replication of the chromosome.

12.2 | What Is a DNA Library?

A DNA library is a set of cloned fragments that collectively represent the genes of a specific organism. Particular genes can be isolated from DNA libraries, much as books can be obtained from conventional libraries. The secret is knowing where and how to look.

Genomic Libraries Are Prepared from the Total DNA in an Organism

Any particular gene constitutes only a small part of an organism's genome. For example, if the organism is a mammal whose entire genome encompasses some 10^6 kbp and the gene is 10 kbp, then the gene represents only 0.001% of the total nuclear DNA. It is impractical to attempt to recover such rare sequences directly from isolated nuclear DNA because of the overwhelming amount of extraneous DNA sequences. Instead, a **genomic library** is prepared by isolating total DNA from the organism, digesting it into fragments of suitable size, and cloning the fragments into an appropriate vector. This approach is called *shotgun cloning*

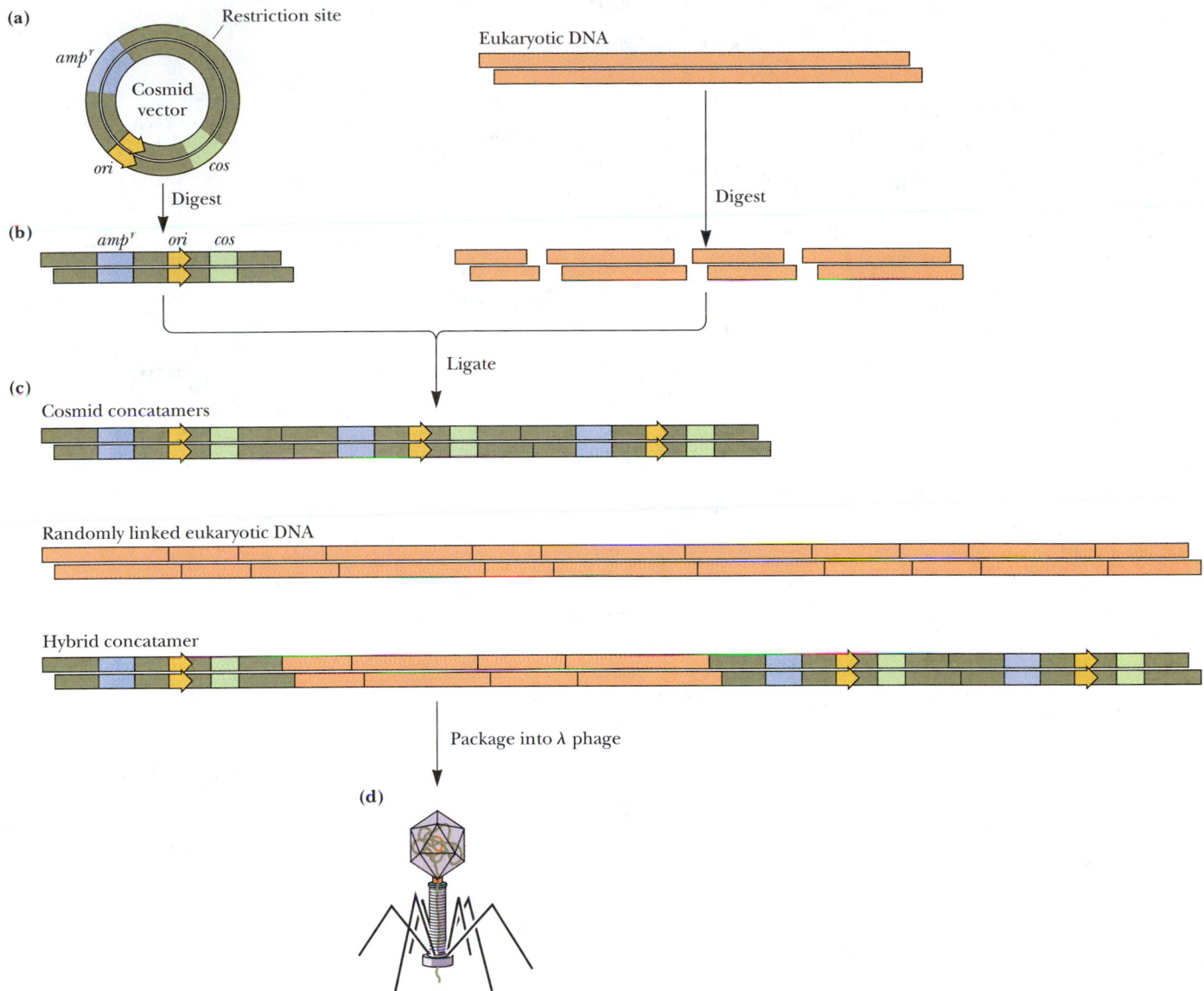

Biochemistry ⊗ Now™ **ACTIVE FIGURE 12.9** Cosmid vectors for cloning large DNA fragments. **(a)** Cosmid vectors are plasmids that carry a selectable marker such as *amp*ʳ, an origin of replication *(ori)*, a polylinker suitable for insertion of foreign DNA, and **(b)** a *cos* sequence. Both the plasmid and the foreign DNA to be cloned are cut with a restriction enzyme, and the two DNAs are then ligated together. **(c)** The ligation reaction leads to the formation of hybrid concatamers, molecules in which plasmid sequences and foreign DNAs are linked in series in no particular order. The bacteriophage λ packaging extract contains the restriction enzyme that recognizes *cos* sequences and cleaves at these sites. **(d)** DNA molecules of the proper size (36 to 51 kbp) are packaged into phage heads, forming infective phage particles. **(e)** The *cos* sequence is

$$\downarrow$$
5′-TACG**GGGCGGCGACCT**CGCG-3′
3′-ATGC**CCCGCCGCTGGA**GCGC-5′
$$\uparrow$$

Endonuclease cleavage at the sites indicated by arrows leaves 12-bp cohesive ends. *(a–d, Adapted from Figure 1.10.7 in Ausubel, F. M., et al., eds., 1987. Current Protocols in Molecular Biology. New York: John Wiley & Sons; e, from Figure 4 in Murialdo, H., 1991. Bacteriophage lambda DNA maturation and packaging. Annual Review of Biochemistry 60:136.)* **Test yourself on the concepts in this figure at http://chemistry.brookscole.com/ggb3**

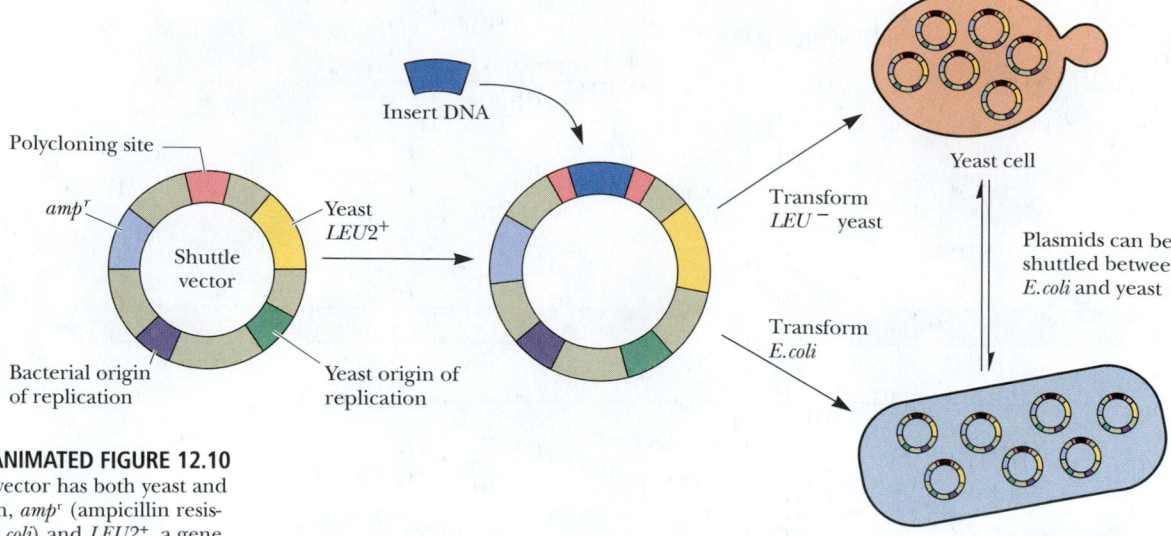

Biochemistry ⏣ Now™ **ANIMATED FIGURE 12.10**
A typical shuttle vector. This vector has both yeast and
bacterial origins of replication, *amp*ᵣ (ampicillin resistance gene for selection in *E. coli*) and *LEU2+*, a gene
in the yeast pathway for leucine biosynthesis. The recipient yeast cells are *LEU2⁻* (defective in this gene)
and thus require leucine for growth. *LEU2⁻* yeast
cells transformed with this shuttle vector can be selected on medium lacking any leucine supplement.
*(Adapted from Figure 19-5 in Watson, J. D., et al., 1987. The
Molecular Biology of the Gene. Menlo Park, CA: Benjamin/
Cummings.)* **See this figure animated at http://
chemistry.brookscole.com/ggb3**

because the strategy has no way of targeting a particular gene but instead seeks
to clone all the genes of the organism at one time. The intent is that at least one
recombinant clone will contain at least part of the gene of interest. Usually, the
isolated DNA is only partially digested by the chosen restriction endonuclease so
that not every restriction site is cleaved in every DNA molecule. Then, even if the
gene of interest contains a susceptible restriction site, some intact genes might
still be found in the digest. Genomic libraries have been prepared from hundreds of different species.

Many clones must be created to be confident that the genomic library contains the gene of interest. The probability, *P*, that some number of clones, *N*,
contains a particular fragment representing a fraction, *f*, of the genome is

$$P = 1 - (1 - f)^N$$

Thus,

$$N = \frac{\ln (1 - P)}{\ln (1 - f)}$$

For example, if the library consists of 10-kbp fragments of the *E. coli* genome
(4640 kbp total), more than 2000 individual clones must be screened to have a
99% probability ($P = 0.99$) of finding a particular fragment. Since $f = 10/4640 =
0.0022$ and $P = 0.99$, $N = 2093$. For a 99% probability of finding a particular se-
quence within the 3×10^6 kbp human genome, N would equal almost 1.4 million
if the cloned fragments averaged 10 kbp in size. The need for cloning vectors capable of carrying very large DNA inserts becomes obvious from these numbers.

Libraries Can Be Screened for the Presence of Specific Genes

A common method of screening plasmid-based genomic libraries is to carry
out a **colony hybridization experiment.** (The protocol is similar for phage-
based libraries except that bacteriophage plaques, not bacterial colonies, are
screened.) In a typical experiment, host bacteria containing either a plasmid-
based or bacteriophage-based library are plated out on a petri dish and allowed
to grow overnight to form colonies (or in the case of phage libraries, plaques)
(Figure 12.11). A replica of the bacterial colonies (or plaques) is then obtained
by overlaying the plate with a flexible, absorbent disc. The disc is removed,
treated with alkali to dissociate bound DNA duplexes into single-stranded

Critical Developments in Biochemistry

Combinatorial Libraries

Specific recognition and binding of other molecules is a defining characteristic of any protein or nucleic acid. Often, target ligands of a particular protein are unknown, or in other instances, a unique ligand for a known protein may be sought in the hope of blocking the activity of the protein or otherwise perturbing its function. Or, the hybridization of nucleic acids with each other according to base-pairing rules, as an act of specific recognition, can be exploited to isolate or identify pairing partners. **Combinatorial libraries** are the products of emerging strategies to facilitate the identification and characterization of macromolecules (proteins, DNA, RNA) that interact with small-molecule ligands or with other macromolecules. Unlike genomic libraries, combinatorial libraries consist of synthetic oligomers. Arrays of synthetic oligonucleotides printed as tiny dots on miniature solid supports are known as **DNA chips.** (See the section titled "DNA Microarrays (Gene Chips) Are Arrays of Different Oligonucleotides Immobilized on a Chip.")

Specifically, combinatorial libraries contain very large numbers of chemically synthesized molecules (such as peptides or oligonucleotides) with randomized sequences or structures. Such libraries are designed and constructed with the hope that one molecule among a vast number will be recognized as a ligand by the protein (or nucleic acid) of interest. If so, perhaps that molecule will be useful in a pharmaceutical application. For instance, the synthetic oligomer may serve as a drug to treat a disease involving the protein to which it binds.

An example of this strategy is the preparation of a **synthetic combinatorial library** of hexapeptides. The maximum number of sequence combinations for hexapeptides is 20^6, or 64,000,000. One approach to simplify preparation and screening possibilities for such a library is to specify the first two amino acids in the hexapeptide while the next four are randomly chosen. In this approach, 400 libraries (20^2) are synthesized, each of which is unique in terms of the amino acids at positions 1 and 2 but random at the other four positions (as in AAXXXX, ACXXXX, ADXXXX, etc.), so each of the 400 libraries contains 20^4, or 160,000, different sequence combinations. Screening these libraries with the protein of interest reveals which of the 400 libraries contains a ligand with high affinity. Then, this library is expanded systematically by specifying the first three amino acids (knowing from the chosen 1-of-400 libraries which amino acids are best as the first two); only 20 synthetic libraries (each containing 20^3, or 8000, hexapeptides) are made here (one for each third-position possibility, the remaining three positions being randomized). Selection for ligand binding, again with the protein of interest, reveals the best of these 20, and this particular library is then varied systematically at the fourth position, creating 20 more libraries (each containing 20^2, or 400, hexapeptides). This cycle of synthesis, screening, and selection is repeated until all six positions in the hexapeptide are optimized to create the best ligand for the protein. A variation on this basic strategy using synthetic oligonucleotides rather than peptides identified a unique 15-mer (sequence GGTTGGTGTGGTTGG) with high affinity ($K_D = 2.7$ nM) toward thrombin, a serine protease in the blood coagulation pathway. Thrombin is a major target for the pharmacological prevention of clot formation in coronary thrombosis.

From Cortese, R., 1996. *Combinatorial Libraries: Synthesis, Screening and Application Potential.* Berlin: Walter de Gruyter.

DNA, dried, and placed in a sealed bag with labeled probe (see the Critical Developments in Biochemistry box on page 388). If the probe DNA is duplex DNA, it must be denatured by heating at 70°C. The probe and target DNA complementary sequences must be in a single-stranded form if they are to hybridize with one another. Any DNA sequences complementary to probe DNA will be revealed by autoradiography of the absorbent disc. Bacterial colonies (phage plaques) containing clones bearing target DNA are identified on the film and can be recovered from the master plate.

Probes for Southern Hybridization Can Be Prepared in a Variety of Ways

Clearly, specific probes are essential reagents if the goal is to identify a particular gene against a background of innumerable DNA sequences. Usually, the probes that are used to screen libraries are nucleotide sequences that are complementary to some part of the target gene. Making useful probes requires some information about the gene's nucleotide sequence. Sometimes such information is available. Alternatively, if the amino acid sequence of the protein encoded by the gene is known, it is possible to work backward through the genetic code to the DNA sequence (Figure 12.12). Because the genetic code is *degenerate* (that is, several codons may specify the same amino acid; see Chapter 30), probes designed by this approach are usually **degenerate oligonucleotides** about

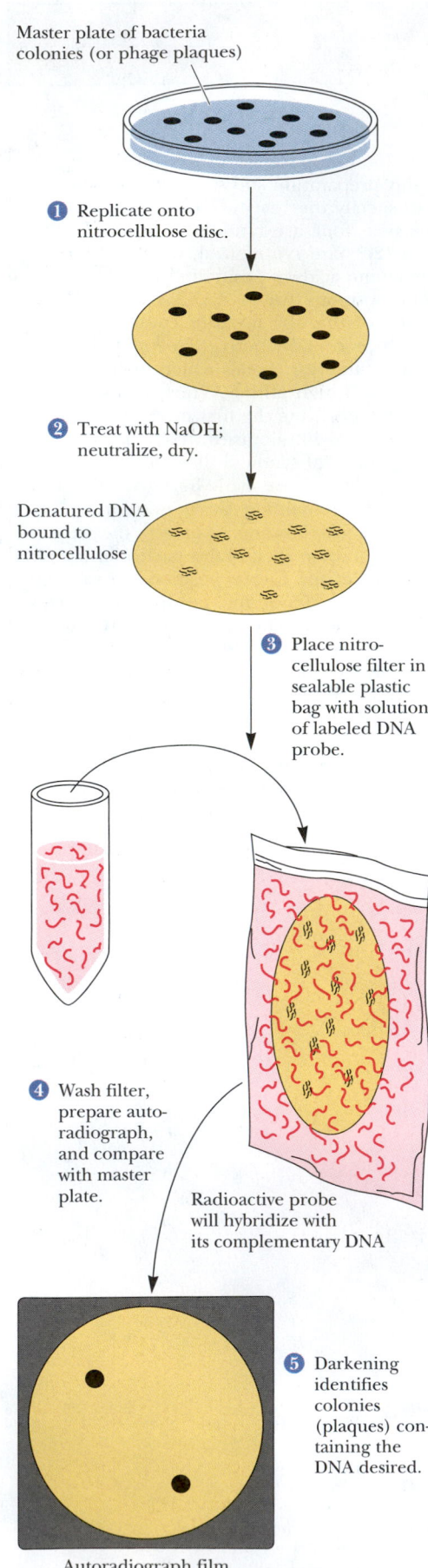

Master plate of bacteria colonies (or phage plaques)

1 Replicate onto nitrocellulose disc.

2 Treat with NaOH; neutralize, dry.

Denatured DNA bound to nitrocellulose

3 Place nitro-cellulose filter in sealable plastic bag with solution of labeled DNA probe.

4 Wash filter, prepare auto-radiograph, and compare with master plate.

Radioactive probe will hybridize with its complementary DNA

5 Darkening identifies colonies (plaques) containing the DNA desired.

Autoradiograph film

17 to 50 residues long (such oligonucleotides are so-called 17- to 50-mers). The oligonucleotides are synthesized so that different bases are incorporated at sites where degeneracies occur in the codons. The final preparation thus consists of a mixture of equal-length oligonucleotides whose sequences vary to accommodate the degeneracies. Presumably, one oligonucleotide sequence in the mixture will hybridize with the target gene. These oligonucleotide probes are at least 17-mers because shorter degenerate oligonucleotides might hybridize with sequences unrelated to the target sequence.

A piece of DNA from the corresponding gene in a related organism can also be used as a probe in screening a library for a particular gene. Such probes are termed **heterologous probes** because they are not derived from the homologous (same) organism.

Problems arise if a complete eukaryotic gene is the cloning target; eukaryotic genes can be tens or even hundreds of kilobase pairs in size. Genes this size are fragmented in most cloning procedures. Thus, the DNA identified by the probe may represent a clone that carries only part of the desired gene. However, most cloning strategies are based on a partial digestion of the genomic DNA, a technique that generates an overlapping set of genomic fragments. This being so, DNA segments from the ends of the identified clone can now be used to probe the library for clones carrying DNA sequences that flanked the original isolate in the genome. Repeating this process ultimately yields the complete gene among a subset of overlapping clones.

cDNA Libraries Are DNA Libraries Prepared from mRNA

cDNAs are DNA molecules copied from mRNA templates. cDNA libraries are constructed by synthesizing cDNA from purified cellular mRNA. These libraries present an alternative strategy for gene isolation, especially eukaryotic genes. Because most eukaryotic mRNAs carry 3′-poly(A) tails, mRNA can be selectively isolated from preparations of total cellular RNA by oligo(dT)-cellulose chromatography (Figure 12.13). DNA copies of the purified mRNAs are synthesized by first annealing short oligo(dT) chains to the poly(A) tails. These oligo(dT) chains serve as primers for reverse transcriptase–driven synthesis of DNA (Figure 12.14). [Random oligonucleotides can also be used as primers, with the advantages being less dependency on poly(A) tracts and increased likelihood of creating clones representing the 5′-ends of mRNAs.] **Reverse transcriptase** is an enzyme that synthesizes a DNA strand, copying RNA as the template. DNA polymerase is then used to copy the DNA strand and form a double-stranded (duplex DNA) molecule. Linkers are then added to the DNA duplexes rendered from the mRNA templates, and the cDNA is

Biochemistry ⑀ Now™   ACTIVE FIGURE 12.11 Screening a genomic library by colony hybridization (or plaque hybridization). Host bacteria transformed with a plasmid-based genomic library or infected with a bacteriophage-based genomic library are plated on a petri plate and incubated overnight to allow bacterial colonies (or phage plaques) to form. A replica of the bacterial colonies (or plaques) is then obtained by overlaying the plate with a nitrocellulose disc (1). Nitrocellulose strongly binds nucleic acids; single-stranded nucleic acids are bound more tightly than double-stranded nucleic acids. (Nylon membranes with similar nucleic acid- and protein-binding properties are also used.) Once the nitrocellulose disc has taken up an impression of the bacterial colonies (or plaques), it is removed and the petri plate is set aside and saved. The disc is treated with 2 *M* NaOH, neutralized, and dried (2). NaOH both lyses any bacteria (or phage particles) and dissociates the DNA strands. When the disc is dried, the DNA strands become immobilized on the filter. The dried disc is placed in a sealable plastic bag, and a solution containing heat-denatured (single-stranded), labeled probe is added (3). The bag is incubated to allow annealing of the probe DNA to any target DNA sequences that might be present on the nitrocellulose. The filter is then washed, dried, and placed on a piece of X-ray film to obtain an autoradiogram (4). The position of any spots on the X-ray film reveals where the labeled probe has hybridized with target DNA (5). The location of these spots can be used to recover the genomic clone from the bacteria (or plaques) on the original petri plate. **Test yourself on the concepts in this figure at http://chemistry.brookscole.com/ggb3**

cloned into a suitable vector. Once a cDNA derived from a particular gene has been identified, the cDNA becomes an effective probe for screening genomic libraries for isolation of the gene itself.

Because different cell types in eukaryotic organisms express selected subsets of genes, RNA preparations from cells or tissues in which genes of interest are selectively transcribed are enriched for the desired mRNAs. cDNA libraries prepared from such mRNA are representative of the pattern and extent of gene expression that uniquely define particular kinds of differentiated cells. cDNA libraries of many normal and diseased human cell types are commercially available, including cDNA libraries of many tumor cells. Comparison of normal and abnormal cDNA libraries, in conjunction with two-dimensional gel electrophoretic analysis (see Appendix to Chapter 5) of the proteins produced in normal and abnormal cells, is a promising new strategy in clinical medicine to understand disease mechanisms.

Expressed Sequence Tags When a cDNA library is prepared from the mRNAs synthesized in a particular cell type under certain conditions, these cDNAs represent the nucleotide sequences (genes) that have been expressed in this cell type under these conditions. **Expressed sequence tags (ESTs)** are relatively short (~200 nucleotides or so) sequences obtained by determining a portion of the nucleotide sequence for each insert in randomly selected cDNAs. An EST represents part of a gene that is being expressed. Probes derived from ESTs can be labeled, radioactively or otherwise, and used in hybridization experiments to identify which genes in a genomic library are being expressed in the cell. For example, labeled ESTs can be hybridized to a *gene chip* (see following discussion).

Known amino acid sequence:

Phe	Met	Glu	Trp	His	Lys	Asn

Possible mRNA sequence:

UUU	AUG	GAA	UGG	CAU	AGG	AAU
UUC		GAG		CAC	AAA	AAC

① Nitrocellulose filter replica of bacterial colonies carrying different DNA fragments

② Synthesize 32 possible DNA oligonucleotides and end label with radioactive ^{32}P

③ Incubate nitrocellulose filter with probe solution in plastic bag

④ Hybridization of the correct oligonucleotide to the DNA

⑤ Detection by autoradiography

Autoradiograph film

Biochemistry Now™ ANIMATED FIGURE 12.12
Cloning genes using oligonucleotide probes designed from a known amino acid sequence. A radioactively labeled set of DNA (degenerate) oligonucleotides representing all possible mRNA coding sequences is synthesized. (In this case, there are 2^5, or 32.) The complete mixture is used to probe the genomic library by colony hybridization (see Figure 12.11). *(From Figure 19.18 in Watson, J. D., et al., 1987. Molecular Biology of the Gene, Vol. 1, 4th ed. Copyright © 1987 by James D. Watson. Reprinted by permission of Addison Wesley Longman Publishers, Inc.)* **See this figure animated at http://chemistry. brookscole.com/ggb3**

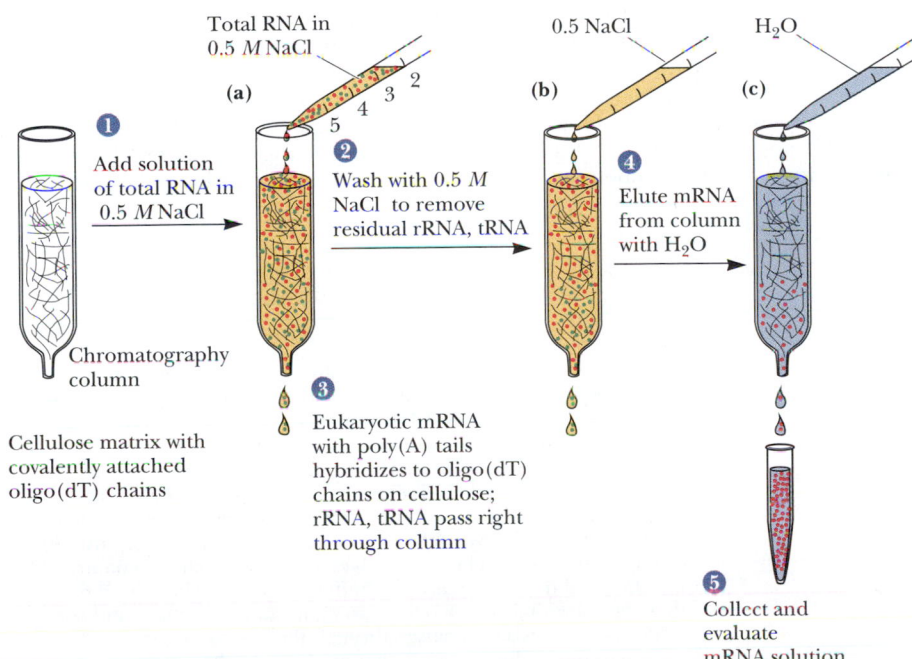

① Add solution of total RNA in 0.5 *M* NaCl

Chromatography column

Cellulose matrix with covalently attached oligo(dT) chains

(a) Total RNA in 0.5 *M* NaCl

② Wash with 0.5 *M* NaCl to remove residual rRNA, tRNA

③ Eukaryotic mRNA with poly(A) tails hybridizes to oligo(dT) chains on cellulose; rRNA, tRNA pass right through column

(b) 0.5 NaCl

④ Elute mRNA from column with H₂O

(c) H₂O

⑤ Collect and evaluate mRNA solution

Biochemistry Now™ ANIMATED FIGURE 12.13 Isolation of eukaryotic mRNA via oligo(dT)-cellulose chromatography. **(a)** In the presence of 0.5 *M* NaCl, the poly(A) tails of eukaryotic mRNA anneal with short oligo(dT) chains covalently attached to an insoluble chromatographic matrix such as cellulose. Other RNAs, such as rRNA (green), pass right through the chromatography column. **(b)** The column is washed with more 0.5 *M* NaCl to remove residual contaminants. **(c)** Then the poly(A) mRNA is recovered by washing the column with water because the base pairs formed between the poly(A) tails of the mRNA and the oligo(dT) chains are unstable in solutions of low ionic strength. **See this figure animated at http://chemistry.brookscole. com/ggb3**

Critical Developments in Biochemistry

Identifying Specific DNA Sequences by Southern Blotting (Southern Hybridization)

Any given DNA fragment is unique solely by virtue of its specific nucleotide sequence. The only practical way to find one particular DNA segment among a vast population of different DNA fragments (such as you might find in genomic DNA preparations) is to exploit its sequence specificity to identify it. In 1975, E. M. Southern invented a technique capable of doing just that.

Electrophoresis

Southern first fractionated a population of DNA fragments according to size by gel electrophoresis (see step 2 in figure). The electrophoretic mobility of a nucleic acid is inversely proportional to its molecular mass. Polyacrylamide gels are suitable for separation of nucleic acids of 25 to 2000 bp. Agarose gels are better if the DNA fragments range up to 10 times this size. Most preparations of genomic DNA show a broad spectrum of sizes, from less than 1 kbp to more than 20 kbp. Typically, no discrete-size fragments are evident following electrophoresis, just a "smear" of DNA throughout the gel.

Blotting

Once the fragments have been separated by electrophoresis (step 3), the gel is soaked in a solution of NaOH. Alkali denatures duplex DNA, converting it to single-stranded DNA. After the pH of the gel is adjusted to neutrality with buffer, a sheet of absorbent material such as nitrocellulose or nylon soaked in a concentrated salt solution is then placed over the gel, and salt solution is drawn through the gel in a direction perpendicular to the direction of electrophoresis (step 4). The salt solution is pulled through the gel in one of three ways: capillary action *(blotting)*, suction *(vacuum blotting)*, or electrophoresis *(electroblotting)*. The movement of salt solution through the gel carries the DNA to the absorbent sheet. Nitrocellulose binds single-stranded DNA molecules very tightly, effectively immobilizing them in place on the sheet.* Note that the distribution pattern of the electrophoretically separated DNA is maintained when the single-stranded DNA molecules bind to the nitrocellulose sheet (step 5 in figure). Next, the nitrocellulose is dried by baking in a vacuum oven;† baking tightly fixes the single-stranded DNAs to the nitrocellulose. Next, in the *prehybridization step,* the nitrocellulose sheet is incubated with a solution containing protein (serum albumin, for example) and/or a detergent such as sodium dodecylsulfate. The protein and detergent molecules saturate any remaining binding sites for DNA on the nitrocellulose. Thus, no more DNA can bind nonspecifically to the nitrocellulose sheet.

Hybridization

To detect a particular DNA within the electrophoretic smear of countless DNA fragments, the prehybridized nitrocellulose sheet is incubated in a sealed plastic bag with a solution of specific probe molecules (step 6 in figure). A **probe** is usually a single-stranded DNA of defined sequence that is distinctively labeled, either with a radioactive isotope (such as ^{32}P) or some other easily detectable tag. The nucleotide sequence of the probe is designed to be complementary to the sought-for or *target* DNA fragment. The single-stranded probe DNA **anneals** with the single-stranded target DNA bound to the nitrocellulose through specific base pairing to form a DNA duplex. This annealing, or **hybridization** as it is usually called, labels the target DNA, revealing its position on the nitrocellulose. For example, if the probe is ^{32}P-labeled, its location can be detected by autoradiographic exposure of a piece of X-ray film laid over the nitrocellulose sheet (step 7 in figure).

Southern's procedure has been extended to the identification of specific RNA and protein molecules. In a play on Southern's name, the identification of particular RNAs following separation by gel electrophoresis, blotting, and probe hybridization is called **Northern blotting.** The analogous technique for identifying protein molecules is termed **Western blotting.** In Western blotting, the probe of choice is usually an antibody specific for the target protein.

▶ The Southern blotting technique involves the transfer of electrophoretically separated DNA fragments to a nitrocellulose sheet and subsequent detection of specific DNA sequences. A preparation of DNA fragments [typically a restriction digest, (**1**)] is separated according to size by gel electrophoresis (**2**). The separation pattern can be visualized by soaking the gel in ethidium bromide to stain the DNA and then illuminating the gel with UV light (**3**). Ethidium bromide molecules intercalated between the hydrophobic bases of DNA are fluorescent under UV light. The gel is soaked in strong alkali to denature the DNA and then neutralized in buffer. Next, the gel is placed on a sheet of nitrocellulose (or DNA-binding nylon membrane), and concentrated salt solution is passed through the gel (**4**) to carry the DNA fragments out of the gel where they are bound tightly to the nitrocellulose (**5**). Incubation of the nitrocellulose sheet with a solution of labeled, single-stranded probe DNA (**6**) allows the probe to hybridize with target DNA sequences complementary to it. The location of these target sequences is then revealed by an appropriate means of detection, such as autoradiography (**7**).

---

*The underlying cause of DNA binding to nitrocellulose is unclear, but probably involves a combination of hydrogen bonding, hydrophobic interactions, and salt bridges.
†Vacuum drying is essential because nitrocellulose reacts violently with O_2 if heated. For this reason, nylon-based membranes are preferable to nitrocellulose membranes.

Biochemistry ⊜ Now™

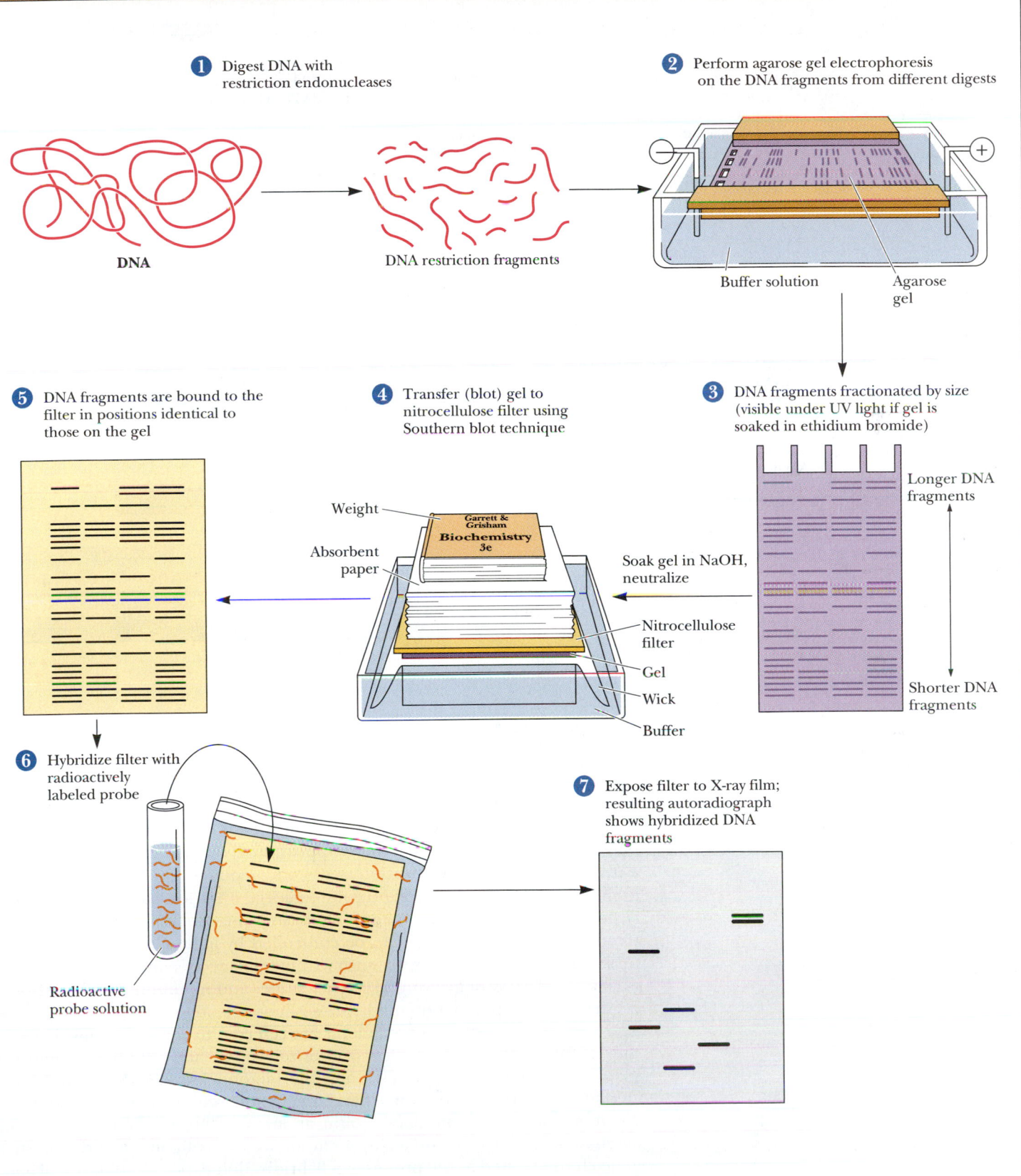

1 Digest DNA with restriction endonucleases

DNA

DNA restriction fragments

2 Perform agarose gel electrophoresis on the DNA fragments from different digests

Buffer solution

Agarose gel

5 DNA fragments are bound to the filter in positions identical to those on the gel

4 Transfer (blot) gel to nitrocellulose filter using Southern blot technique

Weight

Garrett & Grisham
Biochemistry
3e

Absorbent paper

Nitrocellulose filter

Gel

Wick

Buffer

3 DNA fragments fractionated by size (visible under UV light if gel is soaked in ethidium bromide)

Longer DNA fragments

Soak gel in NaOH, neutralize

Shorter DNA fragments

6 Hybridize filter with radioactively labeled probe

Radioactive probe solution

7 Expose filter to X-ray film; resulting autoradiograph shows hybridized DNA fragments

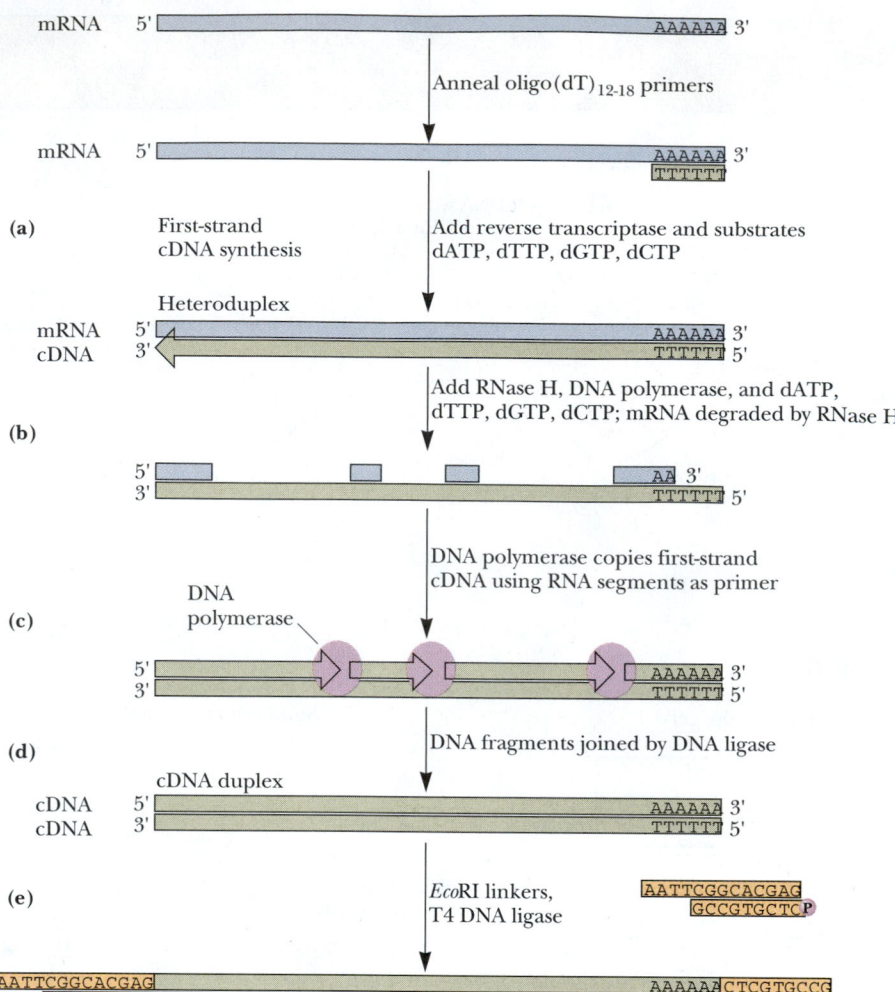

Biochemistry⑤Now™ **ACTIVE FIGURE 12.14**
Reverse transcriptase–driven synthesis of cDNA from oligo(dT) primers annealed to the poly(A) tails of purified eukaryotic mRNA. **(a)** Oligo(dT) chains serve as primers for synthesis of a DNA copy of the mRNA by reverse transcriptase. Following completion of first-strand cDNA synthesis by reverse transcriptase, RNase H and DNA polymerase are added **(b)**. RNase H specifically digests RNA strands in DNA:RNA hybrid duplexes. DNA polymerase copies the first-strand cDNA, using as primers the residual RNA segments after RNase H has created nicks and gaps **(c)**. DNA polymerase has a 5′→3′ exonuclease activity that removes the residual RNA as it fills in with DNA. The nicks remaining in the second-strand DNA are sealed by DNA ligase **(d)**, yielding duplex cDNA. *Eco*RI adapters with 5′-overhangs are then ligated onto the cDNA duplexes **(e)** using phage T4 DNA ligase to create *Eco*RI-ended cDNA for insertion into a cloning vector. **Test yourself on the concepts in this figure at http://chemistry.brookscole.com/ggb3**

*Eco*RI-ended cDNA duplexes for cloning

DNA Microarrays *(Gene Chips)* Are Arrays of Different Oligonucleotides Immobilized on a Chip

Robotic methods can be used to synthesize combinatorial libraries of DNA oligonucleotides directly on a solid support, such that the completed library is a two-dimensional array of different oligonucleotides (see the Critical Developments in Biochemistry box on combinatorial libraries, page 385). Synthesis is performed by phosphoramidite chemistry (Figure 11.29) adapted into a photochemical process that can be controlled by light. Computer-controlled masking of the light allows chemistry to take place at some spots in the two-dimensional array of growing oligonucleotides and not at others, so each spot on the array is a population of identical oligonucleotides of unique sequence. The final products of such procedures are referred to as "gene chips" because the oligonucleotide sequences synthesized upon the chip represent the sequences of chosen genes. Typically, the oligonucleotides are up to 25 nucleotides long (there are more than 10^{15} possible sequence arrangements for 25-mers made from four bases), and as many as 40,000 different oligonucleotides can be arrayed on a chip 1 cm square. The oligonucleotides on such gene chips are used as the probes in a hybridization experiment to reveal gene expression patterns. Figure 12.15 show one design for gene chip analysis of gene expression.

Human Biochemistry

The Human Genome Project

The Human Genome Project is a collaborative international, government- and private-sponsored effort to map and sequence the entire human genome, some 3 billion base pairs distributed among the two sex chromosomes (**X** and **Y**) and 22 **autosomes** (chromosomes that are not sex chromosomes). A primary goal was to identify and map at least 3000 genetic **markers** (genes or other recognizable loci on the DNA), which were evenly distributed throughout the chromosomes at roughly 100-kb intervals. At the same time, determination of the entire nucleotide sequence of the human genome was undertaken. A working draft of the human genome was completed in June 2000 and published in February 2001. An ancillary part of the project has focused on sequencing the genomes of other species (such as yeast, *Drosophila melanogaster* [the fruit fly], mice, and *Arabidopsis thaliana* [a plant]) to reveal comparative aspects of genetic and sequence organization (Table 12.1). Information about whole genome sequences of organisms has created a new branch of science called **bioinformatics:** the study of the nature and organization of biological information. Bioinformatics includes such approaches as **functional genomics** and **proteomics.** *Functional genomics* addresses global issues of gene expression, such as looking at *all* the genes that are activated during major metabolic shifts (as from growth under aerobic to growth under anaerobic conditions) or during embryogenesis and development of organisms. **Transcriptome** is the word used in functional genomics to define the entire set of genes expressed (as mRNAs transcribed from DNA) in a particular cell or tissue under defined conditions. Functional genomics also provides new insights into evolutionary relationships between organisms. *Proteomics* is the study of all the proteins expressed by a certain cell or tissue under specified conditions. Typically, this set of proteins is revealed by running two-dimensional polyacrylamide gel electrophoresis on a cellular extract or by coupling protein separation techniques to mass spectrometric analysis.

The Human Genome Project is also vital to medicine. Many human diseases have been traced to genetic defects whose position within the human genome has been identified. As of 2003, the Human Gene Mutation Database (HGMD) listed more than 32,000 mutations in more than 1300 nuclear genes associated with human disease. Among these are

cystic fibrosis gene

the *breast cancer* genes, BRCA1 and BRCA2

Duchenne muscular dystrophy gene* (at 2.4 megabases, one of the largest known genes in any organism)

Huntington's disease gene

neurofibromatosis gene

neuroblastoma gene (a form of brain cancer)

amyotrophic lateral sclerosis gene (Lou Gehrig's disease)

melanocortin-4 receptor gene (obesity and binge eating)

fragile X-linked mental retardation gene*

as well as genes associated with the development of diabetes, a variety of other cancers, and affective disorders such as *schizophrenia* and *bipolar affective disorder* (manic depression).

Table 12.1

Completed Genome Nucleotide Sequences[1]

Genome	Genome Size[2]	Year Completed
Bacteriophage φX174	0.0054	1977
Bacteriophage λ	0.048	1982
Marchantia[3] chloroplast genome	0.187	1986
Vaccinia virus	0.192	1990
Cytomegalovirus (CMV)	0.229	1991
Marchantia[3] mitochondrial genome	0.187	1992
Variola (smallpox) virus	0.186	1993
Haemophilus influenzae[4] (Gram-negative bacterium)	1.830	1995
Mycobacterium genitalium (mycobacterium)	0.58	1995
Escherichia coli (Gram-negative bacterium)	4.64	1996
Saccharomyces cerevisiae (yeast)	12.1	1996
Methanococcus jannaschii (archaeon)	1.66	1998
Arabidopsis thaliana (green plant)	115	2000
Caenorhabditis elegans (simple animal: nematode worm)	88	1998
Drosophila melanogaster (fruit fly)	117	2000
Homo sapiens (human)	3,038	2001

[1]Data available from the National Center for Biotechnology Information at the National Library of Medicine. Website: *http://www.ncbi.nlm.nih.gov/*
[2]Genome size is given as millions of base pairs (mb).
[3]*Marchantia* is a bryophyte (a nonvascular green plant).
[4]The first complete sequence for the genome of a free-living organism.

*X-chromosome–linked gene. As of 2003, more than 260 disease-related genes have been mapped to the X chromosome (source: the *GeneCards* website at the Weizmann Institute of Science, Israel.)

Expression Vectors Are Engineered So That the RNA or Protein Products of Cloned Genes Can Be Expressed

Expression vectors are engineered so that any cloned insert can be transcribed into RNA, and, in many instances, even translated into protein. cDNA expression libraries can be constructed in specially designed vectors derived from either plasmids or bacteriophage λ. Proteins encoded by the various cDNA

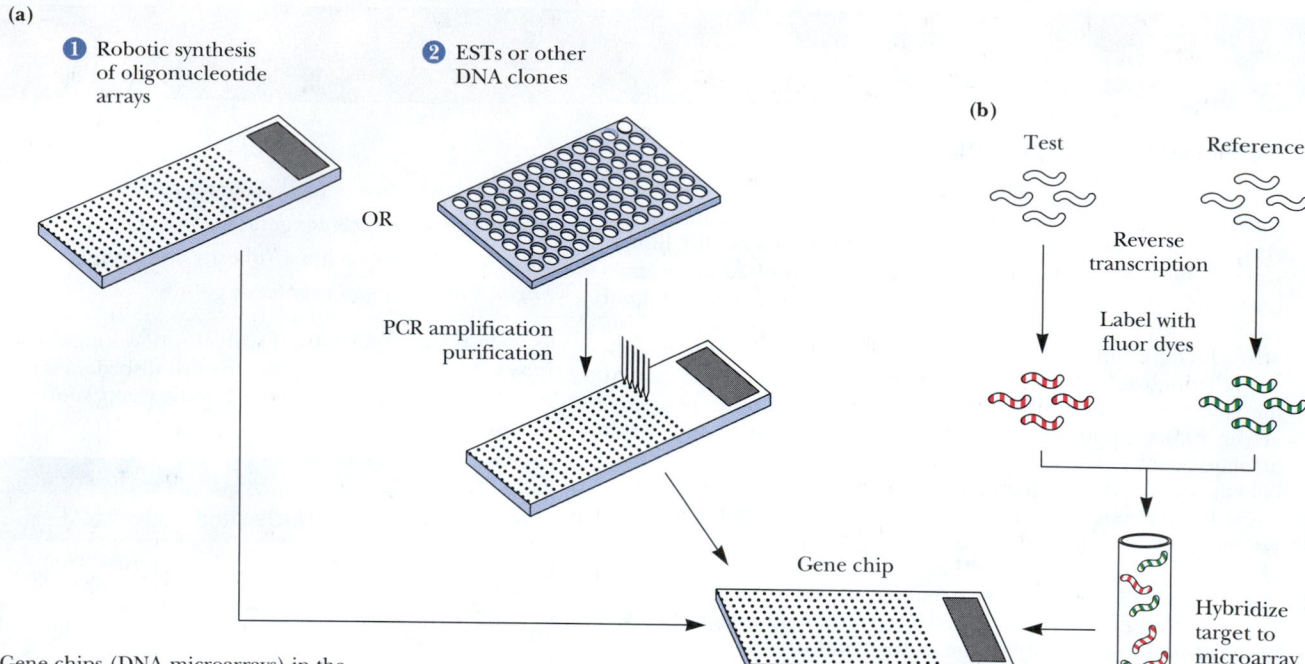

(a)

① Robotic synthesis of oligonucleotide arrays

② ESTs or other DNA clones

OR

PCR amplification purification

(b)

Test   Reference

Reverse transcription

Label with fluor dyes

Gene chip

Hybridize target to microarray

Excitation

Laser 1   Laser 2

Emission

Computer analysis

FIGURE 12.15 Gene chips (DNA microarrays) in the analysis of gene expression. Here is one of many analytical possibilities based on DNA microarray technology: **(1)** Gene segments (for example, ESTs) are isolated and amplified by PCR (see Figure 12.21), and the PCR products are robotically printed onto coated glass microscope slides to create a gene chip. The gene chip usually is considered the "probe" in a "target:probe" screening experiment. **(2)** Target preparation: Total RNA from two sets of cell treatments (control and test treatment) are isolated, and cDNA is produced from the two batches of RNA via reverse transcriptase. During cDNA production, one sample (for example, the control) is labeled through use of a Cy3-linked dUTP derivative and the other (for example, the test treatment) is labeled via a Cy5-linked UTP derivative. (Cy3 and Cy5 are two of a family of highly fluorescent cyanine dyes; Cy3 fluoresces at 563 nm and Cy5 fluoresces at 662 nm, so the wavelength of fluorescence allows discrimination between Cy3- versus Cy5-labeled compounds.) The two batches of cyanine-labeled cDNA are pooled and hybridized to the gene chip. Laser excitation of the hybridized gene chip with light of appropriate wavelength allows collection of data indicating the intensities of fluorescence, and hence the degree of hybridization of the two different probes with the gene chips. Because the location of genes on the gene chip is known, which genes are expressed (or not) and the degree to which they are expressed is revealed by the fluorescent patterns. *(Adapted from Figure 1 in Duggan, D. J., et al., 1999. Expression profiling using cDNA microarrays.* Nature Genetics **21** *supplement:10–14.)*

clones within such expression libraries can be synthesized in the host cells, and if suitable assays are available to identify a particular protein, its corresponding cDNA clone can be identified and isolated. Expression vectors designed for RNA expression or protein expression, or both, are available.

RNA Expression A vector for in vitro expression of DNA inserts as RNA transcripts can be constructed by putting a highly efficient promoter adjacent to a versatile cloning site. Figure 12.16 depicts such an expression vector. Linearized recombinant vector DNA is transcribed in vitro using SP6 RNA polymerase. Large amounts of RNA product can be obtained in this manner; if radioactive ribonucleotides are used as substrates, labeled RNA molecules useful as probes are made.

Protein Expression   Because cDNAs are DNA copies of mRNAs, cDNAs are un-interrupted copies of the exons of expressed genes. Because cDNAs lack introns, it is feasible to express these cDNA versions of eukaryotic genes in prokaryotic hosts that cannot process the complex primary transcripts of eukaryotic genes. To express a eukaryotic protein in *E. coli*, the eukaryotic cDNA must be cloned in an *expression vector* that contains regulatory signals for both transcription and translation. Accordingly, a *promoter* where RNA polymerase initiates transcription as well as a *ribosome-binding site* to facilitate translation are engineered into the vector just upstream from the restriction site for inserting foreign DNA. The AUG initiation codon that specifies the first amino acid in the protein (the *translation start site*) is contributed by the insert (Figure 12.17).

Strong promoters have been constructed that drive the synthesis of foreign proteins to levels equal to 30% or more of total *E. coli* cellular protein. An example is the hybrid promoter, p_{tac}, which was created by fusing part of the promoter for the *E. coli* genes encoding the enzymes of lactose metabolism (the *lac* promoter) with part of the promoter for the genes encoding the enzymes of tryptophan biosynthesis (the *trp* promoter) (Figure 12.18). In cells carrying p_{tac} expression vectors, the p_{tac} promoter is not induced to drive transcription of the foreign insert until the cells are exposed to *inducers* that lead to its activation. Analogs of lactose (a β-galactoside) such as *isopropyl-β-thiogalactoside*, or *IPTG*, are excellent inducers of p_{tac}. Thus, expression of the foreign protein is easily controlled. (See Chapter 29 for detailed discussions of inducible gene expression.) The bacterial production of valuable eukaryotic proteins represents one of the most important uses of recombinant DNA technology. For example, human insulin for the clinical treatment of diabetes is now produced in bacteria.

Analogous systems for expression of foreign genes in eukaryotic cells include vectors carrying promoter elements derived from mammalian viruses, such as *simian virus 40 (SV40)*, the *Epstein–Barr virus*, and the human *cytomegalovirus (CMV)*. A system for high-level expression of foreign genes uses insect cells infected with the *baculovirus* expression vector. **Baculoviruses** infect *lepidopteran* insects (butterflies and moths). In engineered baculovirus vectors, the foreign gene is cloned downstream of the promoter for **polyhedrin,** a major viral-encoded structural protein, and the recombinant vector is incorporated into insect cells grown in culture. Expression from the polyhedrin promoter can lead to accumulation of the foreign gene product to levels as high as 500 mg/L.

Screening cDNA Expression Libraries with Antibodies

Antibodies that specifically cross-react with a particular protein of interest are often available. If so, these antibodies can be used to screen a cDNA expression library to identify and isolate cDNA clones encoding the protein. The cDNA library is introduced into host bacteria, which are plated out and grown overnight, as in the colony hybridization scheme previously described. DNA-binding nylon membranes are placed on the plates to obtain a replica of the bacterial colonies. The nylon membrane is then incubated under conditions that induce protein synthesis from the cloned cDNA inserts, and the cells are treated to release the synthesized protein. The synthesized protein binds tightly to the nylon membrane, which can then be incubated with the specific antibody. Binding of the antibody to its target protein product reveals the position of any cDNA clones expressing the protein, and these clones can be recovered from the original plate. Like other libraries, expression libraries can be screened with oligonucleotide probes, too.

Fusion Protein Expression

Some expression vectors carry cDNA inserts cloned directly into the coding sequence of a vector-borne protein-coding gene (Figure 12.19). Translation of the recombinant sequence leads to synthesis of a *hybrid protein* or *fusion protein*. The N-terminal region of the fused protein represents amino acid sequences encoded in the vector, whereas the remainder of

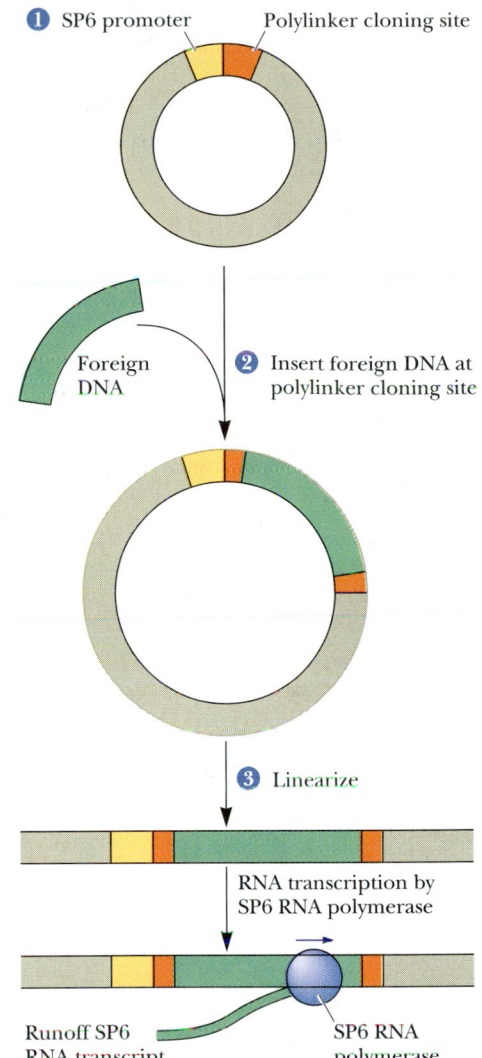

① SP6 promoter   Polylinker cloning site

Foreign DNA   ② Insert foreign DNA at polylinker cloning site

③ Linearize

RNA transcription by SP6 RNA polymerase

Runoff SP6 RNA transcript   SP6 RNA polymerase

Biochemistry ⊜**Now**™   **ANIMATED FIGURE 12.16**
Expression vectors carrying the promoter recognized by the RNA polymerase of bacteriophage SP6 are useful for making RNA transcripts in vitro. SP6 RNA polymerase works efficiently in vitro and recognizes its specific promoter with high specificity. (**1**) These vectors typically have a polylinker adjacent to the SP6 promoter. (**2**) Successive rounds of transcription initiated by SP6 RNA polymerase at its promoter lead to the production of multiple RNA copies of any DNA inserted at the polylinker. (**3**) Before transcription is initiated, the circular expression vector is linearized by a single cleavage at or near the end of the insert so that transcription terminates at a fixed point. **See this figure animated at http://chemistry.brookscole. com/ggb3**

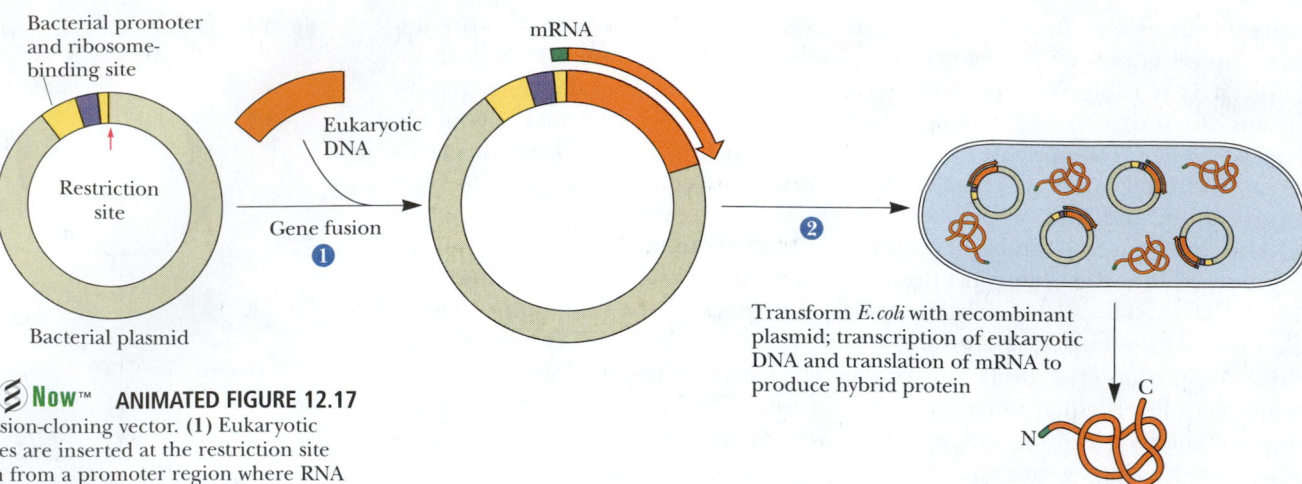

Biochemistry Now™ ANIMATED FIGURE 12.17
A typical expression-cloning vector. **(1)** Eukaryotic coding sequences are inserted at the restriction site just downstream from a promoter region where RNA polymerase binds and initiates transcription. **(2)** Transcription proceeds through a region encoding a bacterial ribosome-binding site and into the cloned insert. The presence of the bacterial ribosome-binding site in the RNA transcript ensures that the RNA can be translated into protein by the ribosomes of the host bacteria. (*From Figure 19.19 from* Molecular Biology of the Gene, *Vol. 1, 4th ed. Copyright © 1987 by James D. Watson. Reprinted by permission of Addison Wesley Longman Publishers, Inc.*) **See this figure animated at http://chemistry. brookscole.com/ggb3**

the protein is encoded by the foreign insert. Keep in mind that the triplet codon sequence within the cloned insert must be in phase with codons contributed by the vector sequences to make the right protein. The N-terminal protein sequence contributed by the vector can be chosen to suit purposes. Furthermore, adding an N-terminal signal sequence that targets the hybrid protein for secretion from the cell simplifies recovery of the fusion protein. A variety of gene fusion systems have been developed to facilitate isolation of a specific protein encoded by a cloned insert. The isolation procedures are based on affinity chromatography purification of the fusion protein through exploitation of the unique ligand-binding properties of the vector-encoded protein (Table 12.2).

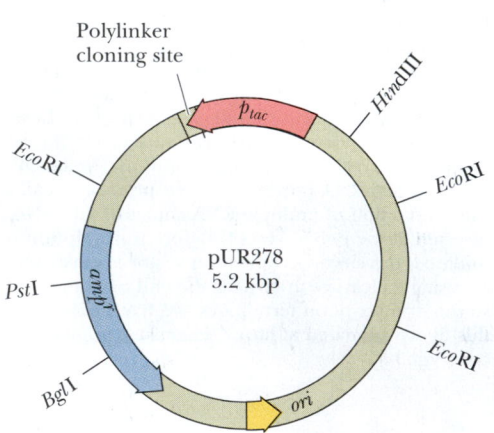

Biochemistry Now™ ANIMATED FIGURE 12.18
A p_{tac} protein expression vector contains the hybrid promoter p_{tac} derived from fusion of the *lac* and *trp* promoters. Expression from p_{tac} is more than 10 times greater than expression from either the *lac* or *trp* promoter alone. Isopropyl-β-D-thiogalactoside, or IPTG, induces expression from p_{tac} as well as *lac*. **See this figure animated at http://chemistry.brookscole. com/ggb3**

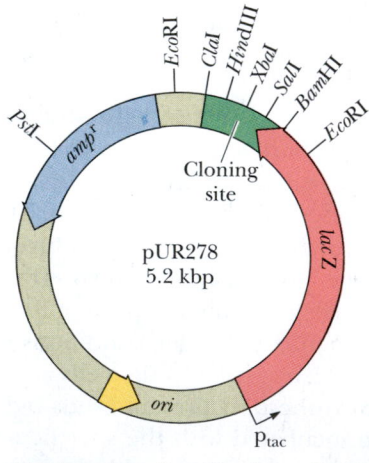

Codon:	Cys	Gln	Lys	Gly	Asp	Pro	Ser	Thr	Leu	Glu	Ser	Leu	Ser	Met
Cloning site:	TGT	CAA	AAA	GGG	GAT	CCG	TCG	ACT	CTA	GAA	AGC	TTA	TCG	ATG

BamHI	SalI	XbaI	HindIII	ClaI

Biochemistry Now™ ANIMATED FIGURE 12.19 A typical expression vector for the synthesis of a hybrid protein. The cloning site is located at the end of the coding region for the protein β-galactosidase. Insertion of foreign DNAs at this site fuses the foreign sequence to the β-galactosidase coding region (the *lacZ* gene). IPTG induces the transcription of the *lacZ* gene from its promoter p_{lac}, causing expression of the fusion protein. (*Adapted from Figure 1.5.4, Ausubel, F. M., et al., 1987.* Current Protocols in Molecular Biology. *New York: John Wiley & Sons.*) **See this figure animated at http://chemistry.brookscole.com/ggb3**

Table 12.2
Gene Fusion Systems for Isolation of Cloned Fusion Proteins

Gene Product	Origin	Molecular Mass (kD)	Secreted?*	Affinity Ligand
β-Galactosidase	*Escherichia coli*	116	No	*p*-Aminophenyl-β-D-thiogalactoside (APTG)
Protein A	*Staphylococcus aureus*	31	Yes	Immunoglobulin G (IgG)
Chloramphenicol acetyltransferase (CAT)	*E. coli*	24	Yes	Chloramphenicol
Streptavidin	*Streptomyces*	13	Yes	Biotin
Glutathione-S-transferase (GST)	*E. coli*	26	No	Glutathione
Maltose-binding protein (MBP)	*E. coli*	40	Yes	Starch

*This indicates whether combined secretion–fusion gene systems have led to secretion of the protein product from the cells, which simplifies its isolation and purification.
Adapted from Uhlen, M., and Moks, T., 1990. Gene fusions for purpose of expression: An introduction. *Methods in Enzymology* **185**:129–143.

A Deeper Look

Biochemistry Now™

The Two-Hybrid System to Identify Proteins Involved in Specific Protein–Protein Interactions

Specific interactions between proteins (so-called protein–protein interactions) lie at the heart of many essential biological processes. Stanley Fields, Cheng-Ting Chien, and their collaborators have invented a method to identify specific protein–protein interactions in vivo through expression of a reporter gene whose transcription is dependent on a functional transcriptional activator, the *GAL4* protein. The *GAL4* protein consists of two domains: a DNA-binding (or **DB**) domain and a transcriptional activation (or **TA**) domain. Even if expressed as separate proteins, these two domains will still work, provided they can be brought together. The method depends on two separate plasmids encoding two hybrid proteins, one consisting of the *GAL4* DB domain fused to protein X and the other consisting of the *GAL4* TA domain fused to protein Y (part a of accompanying figure). If proteins X and Y interact in a specific protein–protein interaction, the *GAL4* DB and TA domains are brought together so that transcription of a reporter gene driven by the *GAL4* promoter can take place (part b of figure). Protein X, fused to the *GAL4*-DNA–binding domain (DB), serves as the "bait" to fish for the protein Y "target" and its fused *GAL4* TA domain. This method can be used to screen cells for protein "targets" that interact specifically with a particular "bait" protein. To do so, cDNAs encoding proteins from the cells of interest are inserted into the TA-containing plasmid to create fusions of the cDNA coding sequences with the *GAL4* TA domain coding sequences, so a fusion protein library is expressed. Identification of a target of the "bait" protein by this method also yields directly a cDNA version of the gene encoding the "target" protein.

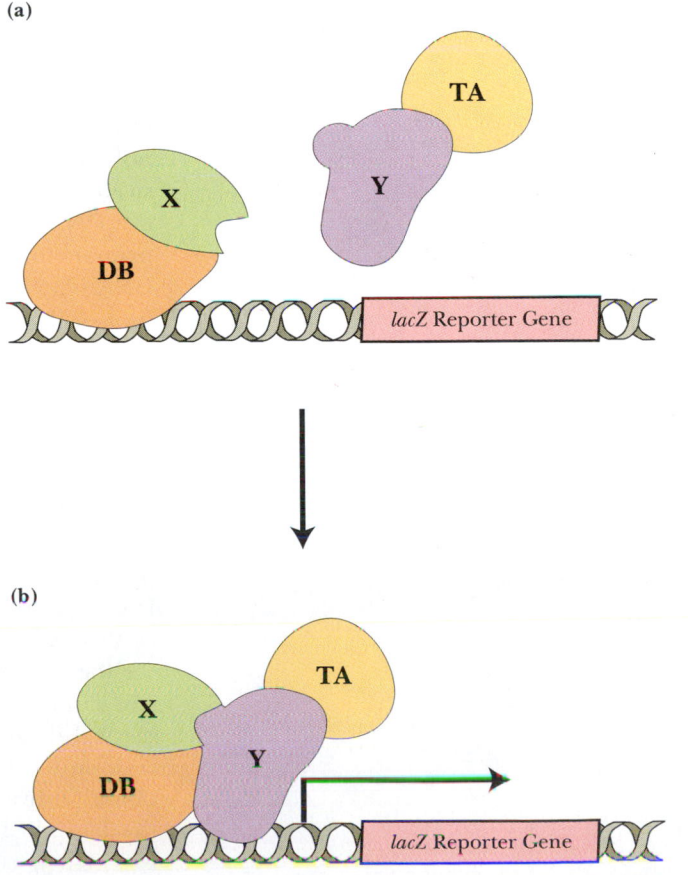

(a)

(b)

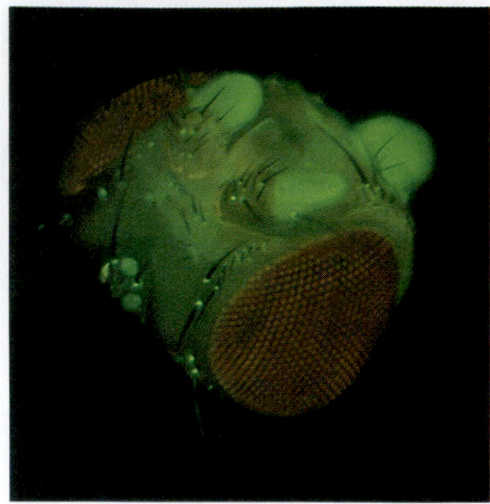

FIGURE 12.20 Green fluorescent protein (GFP) as a reporter gene. The promoter from the *per* gene was placed upstream of the GFP gene in a plasmid and transformed into *Drosophila* (fruit flies). The *per* gene encodes a protein involved in establishing the circadian (daily) rhythmic activity of fruit flies. The fluorescence shown here in an isolated fly head follows a 24-hour rhythmic pattern and occurs to a lesser extent throughout the entire fly, indicating that *per* gene expression can occur in cells throughout the animal. Such uniformity suggests that individual cells have their own independent clocks. *(Image courtesy of Jeffrey D. Plautz and Steve A. Kay, Scripps Research Institute, La Jolla, California. See also Plautz, J. D., et al., 1997. Independent photoreceptive circadian clocks throughout Drosophila. Science **278**:1632–1635.)*

Reporter Gene Constructs Are Chimeric DNA Molecules Composed of Gene Regulatory Sequences Positioned Next to an Easily Expressible Gene Product

Potential regulatory regions of genes (such as promoters) can be investigated by placing these regulatory sequences into plasmids upstream of a gene, called a **reporter gene,** whose expression is easy to measure. Such chimeric plasmids are then introduced into cells of choice (including eukaryotic cells) to assess the potential function of the nucleotide sequence in regulation because expression of the reporter gene serves as a report on the effectiveness of the regulatory element. A number of different genes have been used as reporter genes. A reporter gene with many inherent advantages is that encoding the **green fluorescent protein** (or **GFP**), described in Chapter 4. Unlike the protein expressed by other reporter gene systems, GFP does not require any substrate to measure its activity, nor is it dependent on any cofactor or prosthetic group. Detection of GFP requires only irradiation with near-UV or blue light (400-nm light is optimal), and the green fluorescence (light of 500 nm) that results is easily observed with the naked eye, although it can also be measured precisely with a fluorometer. Figure 12.20 demonstrates the use of GFP as a reporter gene.

12.3 | What Is the Polymerase Chain Reaction (PCR)?

Polymerase chain reaction, or **PCR,** is a technique for dramatically amplifying the amount of a specific DNA segment. A preparation of denatured DNA containing the segment of interest serves as template for DNA polymerase, and two specific oligonucleotides serve as primers for DNA synthesis (as in Figure 12.21). These primers, designed to be complementary to the two 3′-ends of the specific DNA segment to be amplified, are added in excess amounts of 1000 times or greater (Figure 12.21). They prime the DNA polymerase–catalyzed synthesis of the two complementary strands of the desired segment, effectively doubling its concentration in the solution. Then the DNA is heated to dissociate the DNA duplexes and then cooled so that primers bind to both the newly formed and the old strands. Another cycle of DNA synthesis follows. The protocol has been automated through the invention of **thermal cyclers** that alternately heat the reaction mixture to 95°C to dissociate the DNA, followed by cooling, annealing of primers, and another round of DNA synthesis. The isolation of heat-stable DNA polymerases from thermophilic bacteria (such as the *Taq* DNA polymerase from *Thermus aquaticus*) has made it unnecessary to add fresh enzyme for each round of synthesis. Because the amount of target DNA theoretically doubles each round, 25 rounds would increase its concentration about 33 million times. In practice, the increase is actually more like a million times, which is more than ample for gene isolation. Thus, starting with a tiny amount of total genomic DNA, a particular sequence can be produced in quantity in a few hours.

PCR amplification is an effective cloning strategy if sequence information for the design of appropriate primers is available. Because DNA from a single cell can be used as a template, the technique has enormous potential for the clinical diagnosis of infectious diseases and genetic abnormalities. With PCR techniques, DNA from a single hair or sperm can be analyzed to identify particular individuals in criminal cases without ambiguity. **RT-PCR,** a variation on the basic PCR method, is useful when the nucleic acid to be amplified is an RNA (such as mRNA). Reverse transcriptase (RT) is used to synthesize a cDNA strand complementary to the RNA, and this cDNA serves as the template for further cycles of PCR.

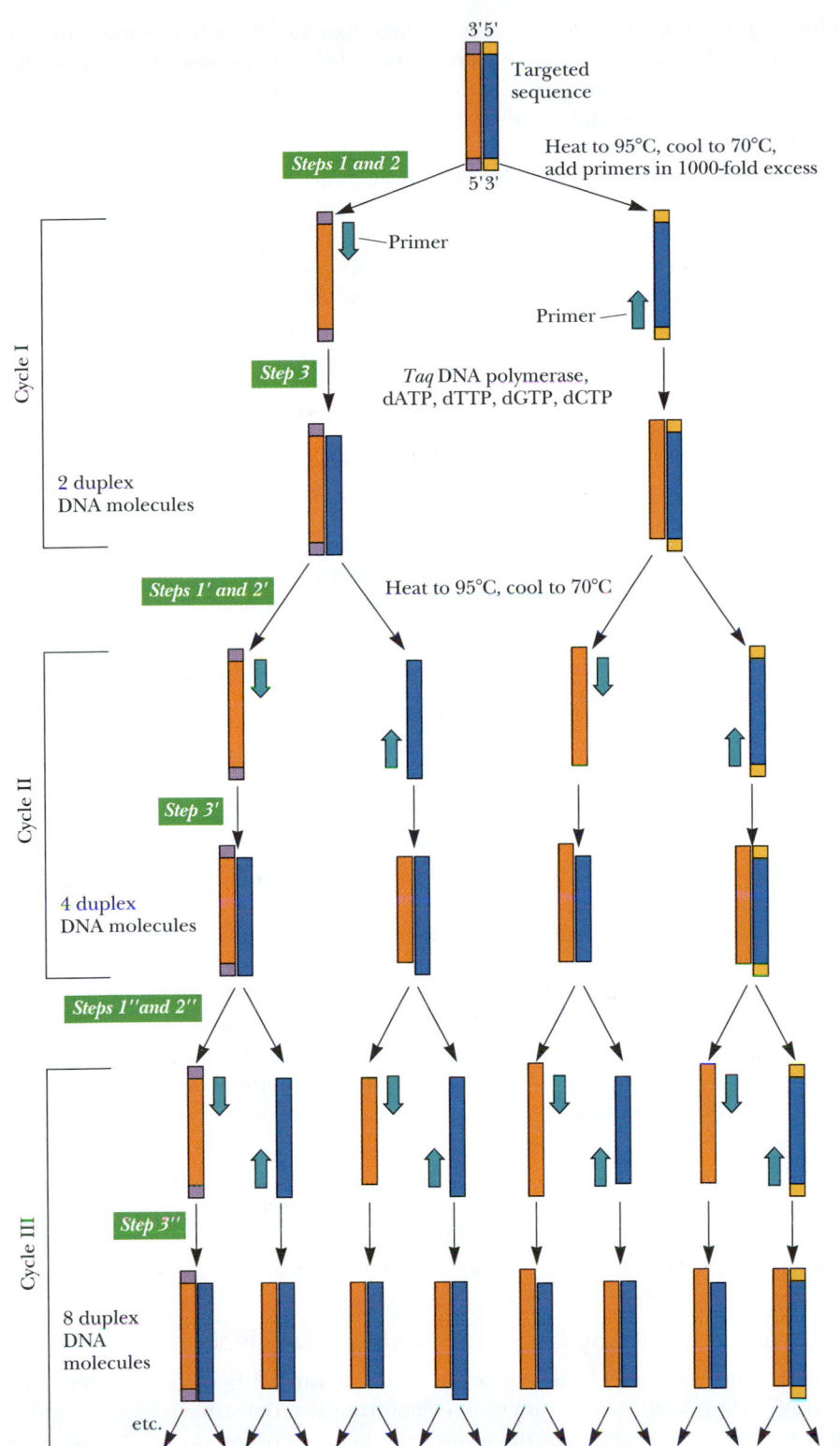

Biochemistry⊜**Now**™   **ANIMATED FIGURE 12.21**
Polymerase chain reaction (PCR). Oligonucleotides complementary to a given DNA sequence prime the synthesis of only that sequence. Heat-stable *Taq* DNA polymerase survives many cycles of heating. Theoretically, the amount of the specific primed sequence is doubled in each cycle. **See this figure animated at http://chemistry.brookscole.com/ggb3**

In Vitro Mutagenesis

The advent of recombinant DNA technology has made it possible to clone genes, manipulate them in vitro, and express them in a variety of cell types under various conditions. The function of any protein is ultimately dependent on its amino acid sequence, which in turn can be traced to the nucleotide sequence of its gene. The introduction of purposeful changes in the nucleotide sequence of a

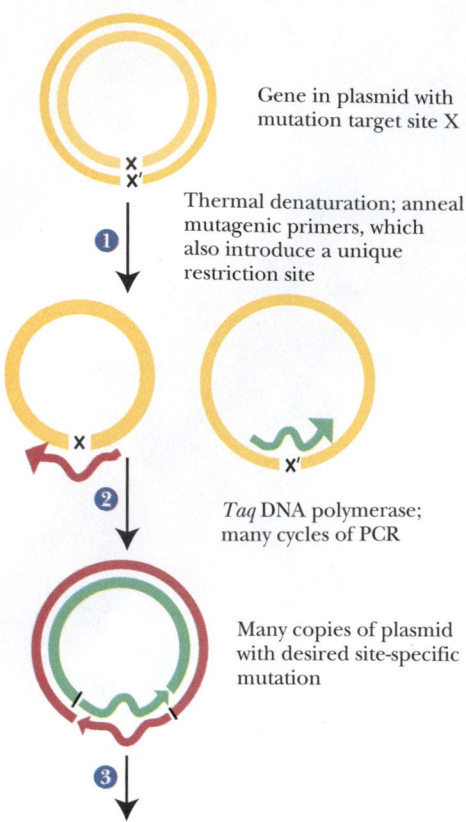

Gene in plasmid with mutation target site X

1 Thermal denaturation; anneal mutagenic primers, which also introduce a unique restriction site

2 *Taq* DNA polymerase; many cycles of PCR

Many copies of plasmid with desired site-specific mutation

3 Transform *E. coli* cells; screen single colonies for plasmids with unique restriction site (≡ mutant gene)

Biochemistry⊜Now™ ANIMATED FIGURE 12.22
One method of PCR-based site-directed mutagenesis. **(1)** Template DNA strands are separated by increased temperature, and the single strands are amplified by PCR using mutagenic primers (represented as bent arrows) whose sequences introduce a single base substitution at site X (and its complementary base X′; thus, the desired amino acid change in the protein encoded by the gene). Ideally, the mutagenic primers also introduce a unique restriction site into the plasmid that was not present before. **(2)** Following many cycles of PCR, the DNA product can be used to transform *E. coli* cells. Single colonies of the transformed cells can be picked. **(3)** The plasmid DNA within each colony can be isolated and screened for the presence of the mutation by screening for the presence of the unique restriction site by restriction endonuclease cleavage. For example, the nucleotide sequence GGATCT within a gene codes for amino acid residues Gly-Ser. Using mutagenic primers of nucleotide sequence AGATCT (and its complement AGATCT) changes the amino acid sequence from Gly-Ser to Arg-Ser and creates a *Bgl*II restriction site (see Table 10.5). Gene expression of the isolated mutant plasmid in *E. coli* allows recovery and analysis of the mutant protein. **See this figure animated at http://chemistry.brookscole.com/ggb3**

cloned gene represents an ideal way to make specific structural changes in a protein. The effects of these changes on the protein's function can then be studied. Such changes constitute *mutations* introduced in vitro into the gene. In vitro **mutagenesis** makes it possible to alter the nucleotide sequence of a cloned gene systematically, as opposed to the chance occurrence of mutations in natural genes.

One efficient technique for in vitro mutagenesis is **PCR-based mutagenesis.** Mutant primers are added to a PCR reaction in which the gene (or segment of a gene) is undergoing amplification. The *mutant primers* are primers whose sequence has been specifically altered to introduce a directed change at a particular place in the nucleotide sequence of the gene being amplified (Figure 12.22). Mutant versions of the gene can then be cloned and expressed to determine any effects of the mutation on the function of the gene product.

12.4 Is It Possible to Make Directed Changes in the Heredity of an Organism?

Recombinant DNA technology is a powerful tool for the genetic modification of organisms. The strategies and methodologies described in this chapter are but an overview of the repertoire of experimental approaches that have been devised by molecular biologists in order to manipulate DNA and the information inherent in it. The enormous success of recombinant DNA technology means that the molecular biologist's task in searching genomes for genes is now akin to that of a lexicographer compiling a dictionary, a dictionary in which the "letters" (the nucleotide sequences), spell out not words but rather genes and what they mean. Molecular biologists have no index or alphabetical arrangement to serve as a guide through the vast volume of information in a genome; nevertheless, this information and its organization is rapidly being disclosed by the imaginative efforts and diligence of these scientists and their growing arsenal of analytical schemes.

Recombinant DNA technology now verges on the ability to engineer at will the heredity (or genetic makeup) of organisms for desired ends. The commercial production of therapeutic biomolecules in microbial cultures is already established (for example, the production of human insulin in quantity in *E. coli* cells). Agricultural crops with desired attributes, such as enhanced resistance to herbicides or elevated vitamin levels, are in cultivation. The rat growth hormone gene has been cloned and transferred into mouse embryos, creating *transgenic mice* that at adulthood are twice normal size (see Chapter 28). Already, transgenic versions of domestic animals such as pigs, sheep, and even fish have been developed for human benefit. Perhaps most important, in a number of instances, clinical trials have been approved for **gene replacement therapy** (or, more simply, *gene therapy*) to correct particular human genetic disorders.

Human Gene Therapy Can Repair Genetic Deficiencies

Human gene therapy seeks to repair the damage caused by a genetic deficiency through introduction of a functional version of the defective gene. To achieve this end, a cloned variant of the gene must be incorporated into the organism in such a manner that it is expressed only at the proper time *and* only in appropriate cell types. At this time, these conditions impose serious technical and clinical difficulties. Many gene therapies have received approval from the National Institutes of Health for trials in human patients, including the introduction of gene constructs into patients. Among these are constructs designed to cure ADA⁻ SCID (**s**evere **c**ombined **i**mmuno**d**eficiency due to adenosine deaminase [ADA] deficiency), neuroblastoma, or cystic fibrosis or to treat cancer through expression of the *E1A* and *p53* tumor suppressor genes.

A basic strategy in human gene therapy involves incorporation of a functional gene into target cells. The gene is typically in the form of an **expression cassette**

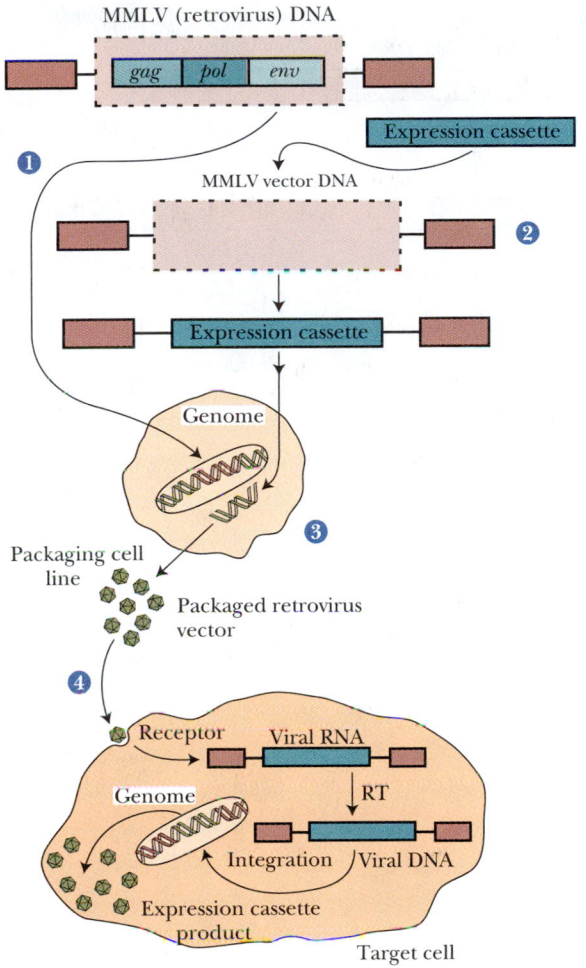

Biochemistry ⊜ **Now**™   **ANIMATED FIGURE 12.23**
Retrovirus-mediated gene delivery ex vivo. Retroviruses are RNA viruses that replicate their RNA
genome by first making a DNA intermediate. The
Maloney murine leukemia virus (MMLV) is the retrovirus used in human gene therapy. Deletion of the
essential genes *gag*, *pol*, and *env* from MMLV makes it
replication deficient (so it can't reproduce) (**1**) and
creates a space for insertion of an expression cassette
(**2**). The modified MMLV acts as a vector for the expression cassette; although replication defective, it is
still infectious. Infection of a packaging cell line that
carries intact *gag*, *pol*, and *env* genes allows the modified MMLV to reproduce (**3**), and the packaged recombinant viruses can be collected and used to infect a
patient (**4**). In the cytosol of the patient's cells, a DNA
copy of the viral RNA is synthesized by viral reverse
transcriptase, which accompanies the viral RNA into
the cells. This DNA is then randomly integrated into
the host cell genome, where its expression leads to production of the expression cassette product. *(Adapted from
Figure 1 in Crystal, R. G., 1995. Transfer of genes to humans: Early
lessons and obstacles to success. Science 270:404.)* **See this figure
animated at http://chemistry.brookscole.com/ggb3**

Human Biochemistry

The Biochemical Defects in Cystic Fibrosis and ADA⁻ SCID

The gene defective in cystic fibrosis codes for CFTR (cystic fibrosis transmembrane conductance regulator), a membrane protein
that pumps Cl^- out of cells. If this Cl^- pump is defective, Cl^- ions
remain in cells, which then take up water from the surrounding
mucus by osmosis. The mucus thickens and accumulates in various organs, including the lungs, where its presence favors infections such as pneumonia. Left untreated, children with cystic
fibrosis seldom survive past the age of 5 years.

ADA⁻ SCID (adenosine deaminase–defective severe combined
immunodeficiency) is a fatal genetic disorder caused by defects in
the gene that encodes ADA. The consequence of ADA deficiency
is accumulation of adenosine and 2′-deoxyadenosine, substances
toxic to lymphocytes, important cells in the immune response.
2′-Deoxyadenosine is particularly toxic because its presence leads
to accumulation of its nucleotide form, dATP, an essential substrate in DNA synthesis. Elevated levels of dATP actually block
DNA replication and cell division by inhibiting synthesis of the
other deoxynucleoside 5′-triphosphates (see Chapter 26). Accumulation of dATP also leads to selective depletion of cellular ATP,
robbing cells of energy. Children with ADA⁻ SCID fail to develop
normal immune responses and are susceptible to fatal infections,
unless kept in protective isolation.

▲ David, the Boy in the Bubble. David was born with SCID and lived all
12 years of his life inside a sterile plastic "bubble" to protect him from
germs common in the environment. He died in 1984 following an
unsuccessful bone marrow transplant.

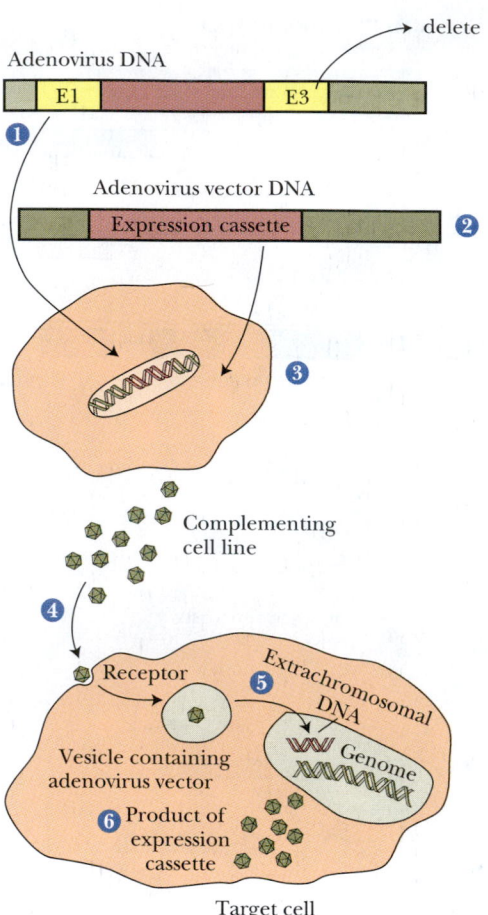

Biochemistry Now™ ANIMATED FIGURE 12.24
Adenovirus-mediated gene delivery in vivo. Adenoviruses are DNA viruses. The adenovirus genome
(36 kb) is divided into *early genes* (E1 through E4) and
late genes (L1 to L5) **(1)**. Adenovirus vectors are generated by deleting gene E1 (and sometimes E3 if more
space for an expression cassette is needed) **(2)**; deletion of E1 renders the adenovirus incapable of replication unless introduced into a complementing cell
line carrying the E1 gene **(3)**. Adenovirus progeny
from the complementing cell line can be used to infect a patient. In the patient, the adenovirus vector
with its expression cassette enters the cells via specific
receptors **(4)**. Its linear dsDNA ultimately gains access
to the cell nucleus **(5)**, where it functions extrachro-
mosomally and expresses the product of the expression cassette **(6)**. *(Adapted from Figure 2 in Crystal, R. G.,
1995. Transfer of genes to humans: Early lessons and obstacles to
success.* Science *270:404.)* **See this figure animated at
http://chemistry.brookscole.com/ggb3**

consisting of a cDNA version of the gene downstream from a promoter that will
drive expression of the gene. A vector carrying such an expression cassette is introduced into target cells, either ex vivo via gene transfer into cultured cells in
the laboratory and administration of the modified cells to the patient or in vivo
via direct incorporation of the gene into the cells of the patient. Because retroviruses can transfer their genetic information directly into the genome of host
cells, retroviruses provide one route to permanent modification of host cells ex
vivo. A replication-deficient version of *Maloney murine leukemia virus* can serve as
a vector for expression cassettes up to 9 kb. Figure 12.23 describes a strategy for
retrovirus vector-mediated gene delivery. In this strategy, it is hoped that the expression cassette will become stably integrated into the DNA of the patient's own
cells and expressed to produce the desired gene product. In 2000, scientists at
the Pasteur Institute in Paris used such an ex vivo approach to successfully treat
infants with X-linked SCID. The gene encoding the γc cytokine receptor subunit
gene was defective in these infants, and gene therapy was used to deliver a functional γc cytokine receptor subunit gene to stem cells harvested from the infants. Transformed stem cells were reintroduced into the patients, who were
then able to produce functional lymphocytes and lead normal lives. This
achievement represents the first successful outcome in human gene therapy.

Adenovirus vectors, which can carry expression cassettes up to 7.5 kb, are a
possible in vivo approach to human gene therapy (Figure 12.24). Recombinant, replication-deficient adenoviruses enter target cells via specific receptors
on the target cell surface; the transferred genetic information is expressed directly from the adenovirus recombinant DNA and is never incorporated into
the host cell genome. Although many problems remain to be solved, human
gene therapy as a clinical strategy is feasible.

Summary

12.1 What Does It Mean: "To Clone"? A clone is a collection of molecules or cells all identical to an original molecule or cell. Plasmids (naturally occurring, circular, extrachromosomal DNA molecules) are very useful in cloning genes. Artificial plasmids can be created by ligating different DNA fragments together. In this manner, "foreign" DNA sequences can be inserted into artificial plasmids, carried into *E. coli*, and propagated as part of the plasmid. Recombinant plasmids are hybrid DNA molecules consisting of plasmid DNA sequences plus inserted DNA elements. A great number of cloning strategies have emerged to make recombinant plasmids for different purposes.

12.2 What Is a DNA Library? A DNA library is a set of cloned fragments representing all the genes of an organism. Particular genes can be isolated from DNA libraries, even though a particular gene constitutes only a small part of an organism's genome. Genomic libraries have been prepared from hundreds of different species. Libraries can be screened for the presence of specific genes. A common method of screening plasmid-based genomic libraries is colony hybridization. Making useful probes requires some information about the gene's nucleotide sequence (or the amino acid sequence of a protein whose gene is sought). DNA from the corresponding gene in a related organism can also be used as a probe in screening a library for a particular gene.

cDNA libraries are DNA libraries prepared from mRNA. Because different cell types in eukaryotic organisms express selected subsets of genes, cDNA libraries prepared from such mRNA are representative of the pattern and extent of gene expression that uniquely define particular kinds of differentiated cells.

Expressed sequence tags (ESTs) are relatively short (~200 nucleotides or so) sequences derived from determining a portion of the nucleotide sequence for each insert in randomly selected cDNAs. ESTs can be used to identify which genes in a genomic library are being expressed in the cell. For example, labeled ESTs can be hybridized to DNA microarrays *(gene chips)*. DNA microarrays are arrays of different oligonucleotides immobilized on a solid support, or *chip*. The oligonucleotides on the chip represent a two-dimensional array of different oligonucleotides. Such gene chips are used to reveal gene expression patterns.

Expression vectors are engineered so that any cloned insert can be transcribed into RNA and, in many instances, translated into protein.

Strong promoters have been constructed that drive the synthesis of foreign proteins to levels equal to 30% or more of total *E. coli* cellular protein. cDNA expression libraries can also be screened with antibodies to identify and isolate cDNA clones encoding a particular protein.

Reporter gene constructs are chimeric DNA molecules composed of gene regulatory sequences positioned next to an easily expressible gene product, such as green fluorescent protein. Reporter gene constructs introduced into cells of choice (including eukaryotic cells) can reveal the function of nucleotide sequences involved in regulation.

12.3 What Is the Polymerase Chain Reaction (PCR)? PCR is a technique for dramatically amplifying the amount of a specific DNA segment. Denatured DNA containing the segment of interest serves as template for DNA polymerase, and two specific oligonucleotides serve as primers for DNA synthesis. The protocol has been automated through the invention of thermal cyclers that alternately heat the reaction mixture to 95°C to dissociate the DNA, followed by cooling, annealing of primers, and another round of DNA synthesis. Because DNA from a single cell can be used as a template, the technique has enormous potential for the clinical diagnosis of infectious diseases and genetic abnormalities.

Recombinant DNA technology makes it possible to clone genes, manipulate them in vitro, and express them in a variety of cell types under various conditions. The introduction of changes in the nucleotide sequence of a cloned gene represents an ideal way to make specific structural changes in a protein; such changes constitute mutations introduced in vitro into the gene. One efficient technique for in vitro mutagenesis is PCR-based mutagenesis.

12.4 Is It Possible to Make Directed Changes in the Heredity of an Organism? Recombinant DNA technology now verges on the ability to engineer at will the heredity (or genetic makeup) of organisms for desired ends. In a number of instances, clinical trials have been approved for gene replacement therapy (or, more simply, *gene therapy*) to correct particular human genetic disorders. Human gene therapy seeks to repair the damage caused by a genetic deficiency through the introduction of a functional version of the defective gene. In 2000, scientists at the Pasteur Institute in Paris used an ex vivo approach to successfully treat infants with X-linked SCID.

Problems

1. A DNA fragment isolated from an *Eco*RI digest of genomic DNA was combined with a plasmid vector linearized by *Eco*RI digestion so that sticky ends could anneal. Phage T4 DNA ligase was then added to the mixture. List all possible products of the ligation reaction.

2. The nucleotide sequence of a polylinker in a particular plasmid vector is

   -GAATTCCCGGGGATCCTCTAGAGTCGACCTGCAGGCATGC-

   This polylinker contains restriction sites for *Bam*HI, *Eco*RI, *Pst*I, *Sal*I, *Sma*I, *Sph*I, and *Xba*I. Indicate the location of each restriction site in this sequence. (See Table 10.5 of restriction enzymes for their cleavage sites.)

3. A vector has a polylinker containing restriction sites in the following order: *Hind*III, *Sac*I, *Xho*I, *Bgl*II, *Xba*I, and *Cla*I.
   a. Give a possible nucleotide sequence for the polylinker.
   b. The vector is digested with *Hind*III and *Cla*I. A DNA segment contains a *Hind*III restriction site fragment 650 bases upstream from a *Cla*I site. This DNA fragment is digested with *Hind*III and *Cla*I, and the resulting *Hind*III–*Cla*I fragment is directionally cloned into the *Hind*III–*Cla*I-digested vector. Give the nucleotide sequence at each end of the vector and the insert and show that the insert can be cloned into the vector in only one orientation.

4. Yeast (*Saccharomyces cerevisiae*) has a genome size of 1.21×10^7 bp. If a genomic library of yeast DNA was constructed in a bacteriophage λ vector capable of carrying 16-kbp inserts, how many individual clones would have to be screened to have a 99% probability of finding a particular fragment?

5. The South American lungfish has a genome size of 1.02×10^{11} bp. If a genomic library of lungfish DNA was constructed in a cosmid vector capable of carrying inserts averaging 45 kbp in size, how many individual clones would have to be screened to have a 99% probability of finding a particular DNA fragment?

6. Given the following short DNA duplex of sequence ($5' \rightarrow 3'$)

   ATGCCGTAGTCGATCATTACGATAGCATAGCACAGGGATCCA-

   CATGCACACACATGACATAGGACAGATAGCAT

   what oligonucleotide primers (17-mers) would be required for PCR amplification of this duplex?

7. Figure 12.5b shows a polylinker that falls within the β-galactosidase coding region of the *lacZ* gene. This polylinker serves as a cloning site in a fusion protein expression vector where the closed insert is expressed as a β-galactosidase fusion protein. Assume the vector polylinker was cleaved with *Bam*HI and then ligated with an insert whose sequence reads

GATCCATTTATCCACCGGAGAGCTGGTATCCCCAAAAGACG-

GCC . . .

What is the amino acid sequence of the fusion protein? Where is the junction between β-galactosidase and the sequence encoded by the insert? (Consult the genetic code table on the inside front cover to decipher the amino acid sequence.)

8. The amino acid sequence across a region of interest in a protein is

Asn-Ser-Gly-Met-His-Pro-Gly-Lys-Leu-Ala-Ser-Trp-Phe-Val-Gly-Asn-Ser

The nucleotide sequence encoding this region begins and ends with an *Eco*RI site, making it easy to clone out the sequence and amplify it by the polymerase chain reaction (PCR). Give the nucleotide sequence of this region. Suppose you wished to change the middle Ser residue to a Cys to study the effects of this change on the protein's activity. What would be the sequence of the mutant oligonucleotide you would use for PCR amplification?

9. Combinatorial chemistry can be used to synthesize polymers such as oligopeptides or oligonucleotides. The number of sequence possibilities for a polymer is given by x^y, where x is the number of different monomer types (for example, 20 different amino acids in a protein or 4 different nucleotides in a nucleic acid) and y is the number of monomers in the oligomers.
   a. Calculate the number of sequence possibilities for RNA oligomers 15 nucleotides long.
   b. Calculate the number of amino acid sequence possibilities for pentapeptides.

10. Imagine that you are interested in a protein that interacts with proteins of the cytoskeleton in human epithelial cells. Describe an experimental protocol based on the yeast two-hybrid system that would allow you to identify proteins that might interact with your protein of interest.

11. Describe an experimental protocol for the preparation of two cDNA libraries, one from anaerobically grown yeast cells and the second from aerobically grown yeast cells.

12. Describe an experimental protocol based on DNA microarrays (gene chips) that would allow you to compare gene expression in anaerobically grown yeast versus aerobically grown yeast.

13. You have an antibody against yeast hexokinase A (hexokinase is the first enzyme in the glycolytic pathway). Describe an experimental protocol using the cDNA libraries prepared in problem 11 that would allow you to identify and isolate the cDNA for hexokinase. Consulting Chapter 5 for protein analysis protocols, describe an experimental protocol to verify that the protein you have identified is hexokinase A.

14. In your experiment in problem 12, you discover a gene that is strongly expressed in anaerobically grown yeast but turned off in aerobically grown yeast. You name this gene *nox* (for "no oxygen"). You have the "bright idea" that you can engineer a yeast strain that senses O_2 levels if you can isolate the *nox* promoter. Describe how you might make a reporter gene construct using the *nox* promoter and how the yeast strain bearing this reporter gene construct might be an effective oxygen sensor.

Biochemistry on the Web
15. Search the National Center for Biotechnology Information (NCBI) website at *http://www.ncbi.nlm.nih.gov/Sitemap/index.html* to discover the number of organisms whose genome sequences have been completed. Explore the rich depository of sequence information available here by selecting one organism from the list and browsing through the contents available.

Preparing for the MCAT Exam
16. Figure 12.1 shows restriction endonuclease sites for the plasmid pBR322. You want to clone a DNA fragment and select for it in transformed bacteria by using resistance to tetracycline and sensitivity to ampicillin as a way of identifying the recombinant plasmid. What restriction endonucleases might be useful for this purpose?

17. Suppose in the Figure 12.12 known acid sequence, tryptophan was replaced by cysteine. How would that affect the possible mRNA sequence? (Consult the inside front cover of this textbook for amino acid codons.) How many nucleotide changes are necessary in replacing Trp with Cys in this coding sequence? What is the total number of possible oligonucleotide sequences for the mRNA if Cys replaces Trp?

Biochemistry ⊜ Now™ Preparing for an exam? Test yourself on key questions at http://chemistry.brookscole.com/ggb3

Further Reading

Cloning Manuals and Procedures

Ausubel, F. M., et al., eds., 1999. *Short Protocols in Molecular Biology*, 4th ed. New York: John Wiley & Sons. A popular cloning manual.

Berger, S. L., and Kimmel, A. R., eds., 1987. *Guide to Molecular Cloning Techniques. Methods in Enzymology*, Volume 152. New York: Academic Press.

Cohen, S. N., Chang, A. C. Y., Boyer, H. W., and Helling, R. B., 1973. Construction of biologically functional bacterial plasmids in vitro. *Proceedings of the National Academy of Sciences U.S.A.* **70**:3240–3244. The classic paper on the construction of chimeric plasmids.

Peterson, K. R., et al., 1997. Production of transgenic mice with yeast artificial chromosomes. *Trends in Genetics* **13**:61–66.

Sambrook, J., 2001. *Molecular Cloning: A Laboratory Manual*, 3rd ed. Long Island: Cold Spring Harbor Laboratory Press.

Expression and Screening of DNA Libraries

Glorioso, J. C., and Schmidt, M. C., eds., 1999. Expression of recombinant genes in eukaryotic cells. *Methods in Enzymology* **306**:1–403.

Hillier, L., et al., 1996. Generation and analysis of 280,000 human expressed sequence tags. *Genome Research* **6**:807–828.

Southern, E. M., 1975. Detection of specific sequences among DNA fragments separated by gel electrophoresis. *Journal of Molecular Biology* **98**:503–517. The classic paper on the identification of specific DNA sequences through hybridization with unique probes.

Thorner, J., and Emr, S., eds., 2000. Applications of chimeric genes and hybrid proteins. *Methods in Enzymology* **328**:1–690.

Weissman, S., ed., 1999. cDNA preparation and display. *Methods in Enzymology* **303**:1–575.

Young, R. A., and Davis, R. W., 1983. Efficient isolation of genes using antibody probes. *Proceedings of the National Academy of Sciences U.S.A.*

80:1194–1198. Using antibodies to screen protein expression libraries to isolate the structural gene for a specific protein.

Combinatorial Libraries and Microarrays

Botwell, D., 2003. *DNA Microarrays: A Molecular Cloning Manual.* Long Island, New York: Cold Spring Harbor Laboratory Press. Techniques used in preparing microarrays and using them in genomic analysis and bioinformatics.

Duggan, D. J., et al., 1999. Expression profiling using cDNA microarrays. *Nature Genetics* **21:**10–14. This is one of a number of articles published in a special supplement of *Nature Genetics* **21** devoted to the use of DNA microarrays to study global gene expression.

Geysen, H. M., et al., 2003. Combinatorial compound libraries for drug discovery: An ongoing challenge. *Nature Reviews Drug Discovery* **2:**222–230.

MacBeath, G., and Schreiber, S. L., 2000. Printing proteins as microarrays for high-throughput function determination. *Science* **289:**1760–1763. This paper describes robotic construction of protein arrays (functionally active proteins immobilized on a solid support) to study protein function.

Southern, E. M., 1996. DNA chips: Analysing sequence by hybridization to oligonucleotides on a large scale. *Trends in Genetics* **12:**110–115.

Genomes

Collins, F., and the International Human Genome Consortium, 2001. Initial sequencing and analysis of the human genome. *Nature* **409:**860–921.

Ewing, B., and Green, P., 2002. Analysis of expressed sequence tags indicates 35,000 human genes. *Nature Genetics* **25:**232–234.

Lander, E., Page, D., and Lifton, R., eds., 2000–2002. *Annual Review of Genomics and Human Genetics,* Vols. 1–3. Palo Alto, CA: Annual Reviews, Inc. A review series on genomics and human diseases.

Venter, J. C., et al., 2001. The sequence of the human genome. *Science* **291:**1304–1351.

The Two-Hybrid System

Chien, C-T., et al., 1991. The two-hybrid system: A method to identify and clone genes for proteins that interact with a protein of interest. *Proceedings of the National Academy of Sciences U.S.A.* **88:**9578–9582.

Golemis, E. A., 2002. *Protein-Protein Interactions: A Molecular Cloning Manual.* Long Island, New York: Cold Spring Harbor Laboratory Press.

Uetz, P, et al., 2000. A comprehensive analysis of protein-protein interactions in *Saccharomyces cerevisiae*. *Nature* **403:**623–627.

Reporter Gene Constructs

Chalfie, M., et al., 1994. Green fluorescent protein as a marker for gene expression. *Science* **263:**802–805.

Polymerase Chain Reaction (PCR)

Saiki, R. K., Gelfand, D. H., Stoeffel, B., et al., 1988. Primer-directed amplification of DNA with a thermostable DNA polymerase. *Science* **239:**487–491. Discussion of the polymerase chain reaction procedure.

Timmer, W. C., and Villalobos, J. M., 1993. The polymerase chain reaction. *The Journal of Chemical Education* **70:**273–280.

Gene Therapy

Cavazzana-Calvo, M., et al., 2000. Gene therapy of human severe combined immunodeficiency (SCID)-X1 disease. *Science* **288:**669–672.

Crystal, R. G., 1995. Transfer of genes to humans: Early lessons and obstacles to success. *Science* **270:**404–410.

Lyon, J., and Gorner, P., 1995. *Altered Fates. Gene Therapy and the Retooling of Human Life.* New York: Norton.

Morgan, R. A., and Anderson, W. F., 1993. Human gene therapy. *Annual Review of Biochemistry* **62:**191–217.

PART II

Protein Dynamics

How Do Enzymes Work?

An Essay by Stephen J. Benkovic, The Pennsylvania State University

How do enzymes achieve accelerated rates for difficult chemical transformations and exquisite specificity toward substrates distinguished only by their stereochemistry?

The early historical hypothesis of a "lock and key" model where the binding of a substrate (key) to an active site of an enzyme (lock) forced a conformation of the substrate that was activated for chemical reaction has been largely replaced by the concept of enzyme-transition state complementarity, where specific binding of the reaction's transition state leads to a catalytic process. The comparison of enzyme-catalyzed and noncatalytic rates has provided an estimate of the degree of enzymatic transition-state stabilization achieved through binding. Enzymes are capable of enhancing the rates of a chemical transformation by 10^{15}- to 10^{17}-fold, requiring an astonishing affinity for the transition state of 10^{-15} to 10^{-17} M.

"**Thus, we can think of a highly, flexible protein able to give rise to conformations that facilitate the chemical transformation of the substrate by favoring orientations and conformations that provide a framework for optimal catalytic activity.**"

How is this achieved? X-ray crystallographic structures of enzymes bound to various transition state analogs show the precise, optimal positioning of active site residues necessary for the acid/base/nucleophilic catalysis required to accelerate the chemical transformation. In order to form such enzyme substrate complexes and to obtain these precise alignments, transient reorganizations of the active site or more distal regions of the protein are required to occur along the reaction coordinate. There are many cases where after substrate binding, the active site of an enzyme is closed by a loop of peptide acting as a lid only to have the loop open when the product is released. The folded enzyme thus can provide a preorganized polar environment that is already partially oriented in the initial Michaelis complex to stabilize the transition state as well as to sequester the substrate into conformations more favorable for reaction.

Recent theoretical and experimental studies indicate that thermally averaged, equilibrium motions exist within the protein framework and can occur on the time scale of substrate turnover, that is, the chemical transformation catalyzed by the enzyme. In particular, the introduction of specific amino acid changes by site-specific mutagenesis can produce striking effects on the rates of the enzyme-catalyzed reaction even though the changes are far from the active site. Thus, we can think of a highly flexible protein able to give rise to conformations that facilitate the chemical transformation of the substrate by favoring orientations and conformations that provide a framework for optimal catalytic activity. The conservation of key amino acids throughout an enzyme from many species over the course of evolution hints at the operation of such a network. More evidence bearing on this viewpoint will come from the application of both developing theoretical and experimental methods.

Enzymes—Kinetics and Specificity

The space shuttle must accelerate from zero velocity to a velocity of more than 25,000 miles per hour in order to escape earth's gravity.

Essential Question

At any moment, thousands of chemical reactions are taking place in any living cell. Virtually all of these reactions would proceed at rates that could not sustain life were it not for enzymes. **What are enzymes, and what do they do?**

Living organisms seethe with metabolic activity. Thousands of chemical reactions are proceeding very rapidly at any given instant within all living cells. Virtually all of these transformations are mediated by **enzymes**—proteins (and occasionally RNA) specialized to catalyze metabolic reactions. The substances transformed in these reactions are often organic compounds that show little tendency for reaction outside the cell. An excellent example is glucose, a sugar that can be stored indefinitely on the shelf with no deterioration. Most cells quickly oxidize glucose, producing carbon dioxide and water and releasing lots of energy:

$$C_6H_{12}O_6 + 6\ O_2 \longrightarrow 6\ CO_2 + 6\ H_2O + 2870\ kJ\ of\ energy$$

(-2870 kJ/mol is the standard free energy change [$\Delta G°'$] for the oxidation of glucose; see Chapter 3.) In chemical terms, 2870 kJ is a large amount of energy, and glucose can be viewed as an energy-rich compound even though at ambient temperature it is not readily reactive with oxygen outside of cells. Stated another way, glucose represents **thermodynamic potentiality:** Its reaction with oxygen is strongly exergonic, but it doesn't occur under just normal conditions. On the other hand, enzymes can catalyze such thermodynamically favorable reactions, causing them to proceed at extraordinarily rapid rates (Figure 13.1). In glucose oxidation and countless other instances, enzymes provide cells with the ability to exert *kinetic control over thermodynamic potentiality*. That is, living systems use enzymes to accelerate and control the rates of vitally important biochemical reactions.

Enzymes Are the Agents of Metabolic Function

Acting in sequence, enzymes form metabolic pathways by which nutrient molecules are degraded, energy is released and converted into metabolically useful forms, and precursors are generated and transformed to create the literally thousands of distinctive biomolecules found in any living cell (Figure 13.2). Situated at key junctions of metabolic pathways are specialized **regulatory enzymes** capable of sensing the momentary metabolic needs of the cell and adjusting their catalytic rates accordingly. The responses of these enzymes ensure the harmonious integration of the diverse and often divergent metabolic activities of cells so that the living state is promoted and preserved.

13.1 | What Characteristic Features Define Enzymes?

Enzymes are remarkably versatile biochemical catalysts that have in common three distinctive features: **catalytic power, specificity,** and **regulation.**

> *There is more to life than increasing its speed.*
> **Mahatma Gandhi** (1869–1948)

Key Questions

13.1 What Characteristic Features Define Enzymes?

13.2 Can the Rate of an Enzyme-Catalyzed Reaction Be Defined in a Mathematical Way?

13.3 What Equations Define the Kinetics of Enzyme-Catalyzed Reactions?

13.4 What Can Be Learned from the Inhibition of Enzyme Activity?

13.5 What Is the Kinetic Behavior of Enzymes Catalyzing Bimolecular Reactions?

13.6 Are All Enzymes Proteins?

13.7 How Can Enzymes Be So Specific?

Biochemistry⊜Now™ Test yourself on these Key Questions at BiochemistryNow at **http://chemistry.brookscole.com/ggb3**

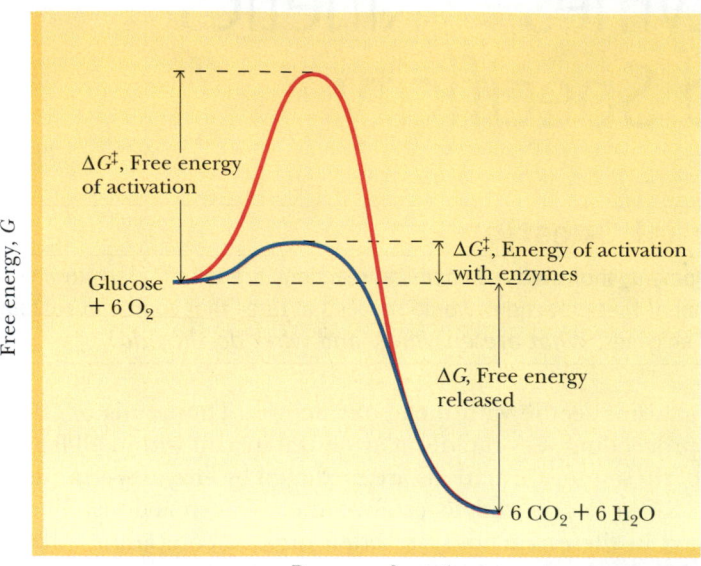

FIGURE 13.1 Reaction profile showing large $\Delta G^{\ddagger}$ for glucose oxidation, free energy change of $-2{,}870$ kJ/mol; catalysts lower $\Delta G^{\ddagger}$, thereby accelerating rate.

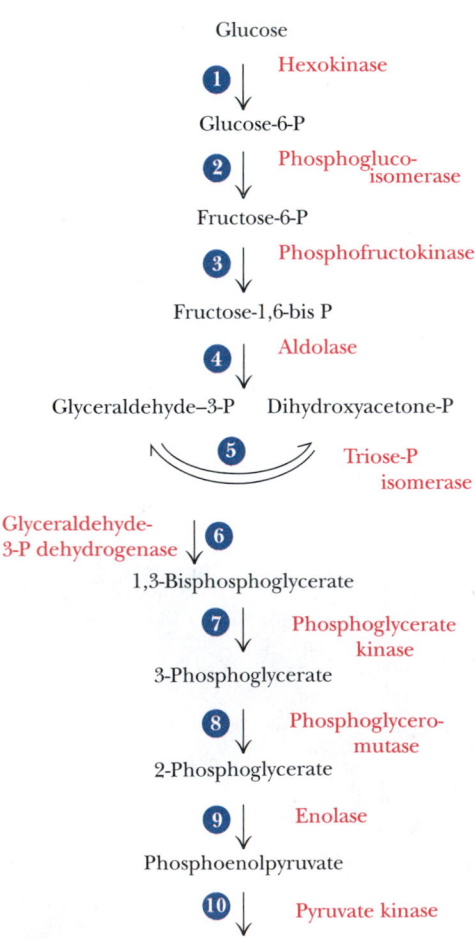

FIGURE 13.2 The breakdown of glucose by *glycolysis* provides a prime example of a metabolic pathway. Ten enzymes mediate the reactions of glycolysis. Enzyme 4, *fructose 1,6-bisphosphate aldolase*, catalyzes the C—C bond-breaking reaction in this pathway.

Catalytic Power Is Defined as the Ratio of the Enzyme-Catalyzed Rate of a Reaction to the Uncatalyzed Rate

Enzymes display enormous catalytic power, accelerating reaction rates as much as 10^{16} over uncatalyzed levels, which is far greater than any synthetic catalysts can achieve, and enzymes accomplish these astounding feats in dilute aqueous solutions under mild conditions of temperature and pH. For example, the enzyme jack bean *urease* catalyzes the hydrolysis of urea:

$$H_2N-\overset{\overset{\displaystyle O}{\|}}{C}-NH_2 + 2\,H_2O + H^+ \longrightarrow 2\,NH_4^+ + HCO_3^-$$

At 20°C, the rate constant for the enzyme-catalyzed reaction is 3×10^4/sec; the rate constant for the uncatalyzed hydrolysis of urea is 3×10^{-10}/sec. Thus, 10^{14} is the ratio of the catalyzed rate to the uncatalyzed rate of reaction. Such a ratio is defined as the relative **catalytic power** of an enzyme, so the catalytic power of urease is 10^{14}.

Specificity Is the Term Used to Define the Selectivity of Enzymes for the Reactants They Act Upon

A given enzyme is very selective, both in the substances with which it interacts and in the reaction that it catalyzes. The substances upon which an enzyme acts are traditionally called **substrates.** In an enzyme-catalyzed reaction, none of the substrate is diverted into nonproductive side reactions, so no wasteful by-products are produced. It follows then that the products formed by a given enzyme are also very specific. This situation can be contrasted with your own experiences in the organic chemistry laboratory, where yields of 50% or even 30% are viewed as substantial accomplishments (Figure 13.3). The selective qualities of an enzyme are collectively recognized as its **specificity.** Intimate interaction between an enzyme and its substrates occurs through molecular recognition based on structural complementarity; such mutual recognition is the basis of specificity. The specific site on the enzyme where substrate binds and catalysis occurs is called the **active site.**

Regulation of Enzyme Activity Ensures That the Rate of Metabolic Reactions Is Appropriate to Cellular Requirements

Regulation of enzyme activity is essential to the integration and regulation of metabolism. Enzyme regulation is achieved in a variety of ways, ranging from controls over the amount of enzyme protein produced by the cell to more rapid, reversible interactions of the enzyme with metabolic inhibitors and activators. Chapter 15 is devoted to discussions of this topic. Because most enzymes are proteins, we can anticipate that the functional attributes of enzymes are due to the remarkable versatility found in protein structures.

Enzyme Nomenclature Provides a Systematic Way of Naming Metabolic Reactions

Traditionally, enzymes often were named by adding the suffix *-ase* to the name of the substrate upon which they acted, as in *urease* for the urea-hydrolyzing enzyme or *phosphatase* for enzymes hydrolyzing phosphoryl groups from organic phosphate compounds. Other enzymes acquired names bearing little resemblance to their activity, such as the peroxide-decomposing enzyme *catalase* or the proteolytic enzymes *(proteases)* of the digestive tract, *trypsin* and *pepsin*. Because of the confusion that arose from these trivial designations, an International Commission on Enzymes was established in 1956 to create a systematic basis for enzyme nomenclature. Although common names for many enzymes remain in use, all enzymes now are classified and formally named according to the reaction they catalyze. Six classes of reactions are recognized (Table 13.1). Within each class are subclasses, and under each subclass are sub-subclasses within which individual enzymes are listed. Classes, subclasses, sub-subclasses, and individual entries are each numbered so that a series of four numbers serves to specify a particular enzyme. A systematic name, descriptive of the reaction, is also assigned to each entry. To illustrate, consider the enzyme that catalyzes this reaction:

$$ATP + \text{D-glucose} \longrightarrow ADP + \text{D-glucose-6-phosphate}$$

A phosphate group is transferred from ATP to the C-6-OH group of glucose, so the enzyme is a *transferase* (Class 2, Table 13.1). Subclass 7 of transferases is *enzymes transferring phosphorus-containing groups,* and sub-subclass 1 covers those *phosphotransferases with an alcohol group as an acceptor.* Entry 2 in this sub-subclass is **ATP:D-glucose-6-phosphotransferase,** and its classification number is **2.7.1.2.** In use, this number is written preceded by the letters **E.C.,** denoting the Enzyme Commission. For example, entry 1 in the same sub-subclass is E.C.2.7.1.1, ATP:D-hexose-6-phosphotransferase, an ATP-dependent enzyme that transfers a phosphate to the 6-OH of hexoses (that is, it is nonspecific regarding its hexose acceptor). These designations can be cumbersome, so in everyday usage, trivial names are commonly used. The glucose-specific enzyme E.C.2.7.1.2 is called *glucokinase,* and the nonspecific E.C.2.7.1.1 is known as *hexokinase. Kinase* is a trivial term for enzymes that are ATP-dependent phosphotransferases.

Coenzymes and Cofactors Are Nonprotein Components Essential to Enzyme Activity

Many enzymes carry out their catalytic function relying solely on their protein structure. Many others require nonprotein components, called **cofactors** (Table 13.2). Cofactors may be metal ions or organic molecules referred to as **coenzymes.** Coenzymes and cofactors provide proteins with chemically versatile functions not found in amino acid side chains. Cofactors, because they are structurally less complex than proteins, tend to be stable to heat (incubation in a boiling water bath). Typically, proteins are denatured under such conditions. Many coenzymes are vitamins or contain vitamins as part of their structure. Usually coenzymes are actively involved in the catalytic reaction of the enzyme, often serving as intermediate carriers of functional groups in the conversion of substrates to products. In most cases, a coenzyme is firmly associated with its enzyme, perhaps even by covalent bonds, and it is difficult to separate the two. Such tightly bound coenzymes are referred to as **prosthetic groups** of the enzyme. The catalytically active complex of protein and prosthetic group is called the **holoenzyme.** The protein without the prosthetic group is called the **apoenzyme;** it is catalytically inactive.

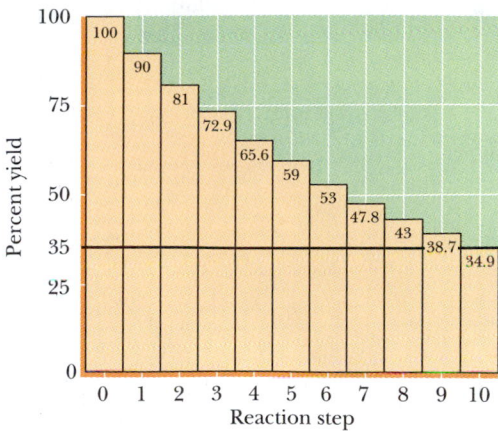

FIGURE 13.3 A 90% yield over 10 steps, for example, in a metabolic pathway, gives an overall yield of 35%. Therefore, yields in biological reactions *must be substantially greater;* otherwise, unwanted by-products would accumulate to unacceptable levels.

Table 13.1

Systematic Classification of Enzymes According to the Enzyme Commission

E.C. Number	Systematic Name and Subclasses
1	*Oxidoreductases* (oxidation–reduction reactions)
1.1	Acting on CH—OH group of donors
1.1.1	With NAD or NADP as acceptor
1.1.3	With O_2 as acceptor
1.2	Acting on the $\diagdown C{=}O$ group of donors
1.2.3	With O_2 as acceptor
1.3	Acting on the CH—CH group of donors
1.3.1	With NAD or NADP as acceptor
2	*Transferases* (transfer of functional groups)
2.1	Transferring C-1 groups
2.1.1	Methyltransferases
2.1.2	Hydroxymethyltransferases and formyltransferases
2.1.3	Carboxyltransferases and carbamoyltransferases
2.2	Transferring aldehydic or ketonic residues
2.3	Acyltransferases
2.4	Glycosyltransferases
2.6	Transferring N-containing groups
2.6.1	Aminotransferases
2.7	Transferring P-containing groups
2.7.1	With an alcohol group as acceptor
3	*Hydrolases* (hydrolysis reactions)
3.1	Cleaving ester linkage
3.1.1	Carboxylic ester hydrolases
3.1.3	Phosphoric monoester hydrolases
3.1.4	Phosphoric diester hydrolases
4	*Lyases* (addition to double bonds)
4.1	C=C lyases
4.1.1	Carboxy lyases
4.1.2	Aldehyde lyases
4.2	C=O lyases
4.2.1	Hydrolases
4.3	C=N lyases
4.3.1	Ammonia-lyases
5	*Isomerases* (isomerization reactions)
5.1	Racemases and epimerases
5.1.3	Acting on carbohydrates
5.2	Cis-trans isomerases
6	*Ligases* (formation of bonds with ATP cleavage)
6.1	Forming C—O bonds
6.1.1	Amino acid–RNA ligases
6.2	Forming C—S bonds
6.3	Forming C—N bonds
6.4	Forming C—C bonds
6.4.1	Carboxylases

13.2 | Can the Rate of an Enzyme-Catalyzed Reaction Be Defined in a Mathematical Way?

Kinetics is the branch of science concerned with the rates of chemical reactions. The study of **enzyme kinetics** addresses the biological roles of enzymatic catalysts and how they accomplish their remarkable feats. In enzyme kinetics, we seek to determine the maximum reaction velocity that the enzyme can attain and its

Table 13.2

Enzyme Cofactors: Some Metal Ions and Coenzymes and the Enzymes with Which They Are Associated

Metal Ions and Some Enzymes That Require Them		Coenzymes Serving as Transient Carriers of Specific Atoms or Functional Groups		
Metal Ion	Enzyme	Coenzyme	Entity Transferred	Representative Enzymes Using Coenzymes
Fe^{2+} or Fe^{3+}	Cytochrome oxidase	Thiamine pyrophosphate (TPP)	Aldehydes	Pyruvate dehydrogenase
	Catalase	Flavin adenine dinucleotide (FAD)	Hydrogen atoms	Succinate dehydrogenase
	Peroxidase	Nicotinamide adenine dinucleotide (NAD)	Hydride ion (H^-)	Alcohol dehydrogenase
Cu^{2+}	Cytochrome oxidase			
Zn^{2+}	DNA polymerase	Coenzyme A (CoA)	Acyl groups	Acetyl-CoA carboxylase
	Carbonic anhydrase	Pyridoxal phosphate (PLP)	Amino groups	Aspartate aminotransferase
	Alcohol dehydrogenase	5′-Deoxyadenosylcobalamin (vitamin B_{12})	H atoms and alkyl groups	Methylmalonyl-CoA mutase
Mg^{2+}	Hexokinase			
	Glucose-6-phosphatase	Biotin (biocytin)	CO_2	Propionyl-CoA carboxylase
Mn^{2+}	Arginase	Tetrahydrofolate (THF)	Other one-carbon groups	Thymidylate synthase
K^+	Pyruvate kinase (also requires Mg^{2+})			
Ni^{2+}	Urease			
Mo	Nitrate reductase			
Se	Glutathione peroxidase			

binding affinities for substrates and inhibitors. Coupled with studies on the structure and chemistry of the enzyme, analysis of the enzymatic rate under different reaction conditions yields insights regarding the enzyme's mechanism of catalytic action. Such information is essential to an overall understanding of metabolism.

Significantly, this information can be exploited to control and manipulate the course of metabolic events. The science of pharmacology relies on such a strategy. **Pharmaceuticals, or drugs,** are often special inhibitors specifically targeted at a particular enzyme in order to overcome infection or to alleviate illness. A detailed knowledge of the enzyme's kinetics is indispensable to rational drug design and successful pharmacological intervention.

Chemical Kinetics Provides a Foundation for Exploring Enzyme Kinetics

Before beginning a quantitative treatment of enzyme kinetics, it will be fruitful to review briefly some basic principles of chemical kinetics. **Chemical kinetics** is the study of the rates of chemical reactions. Consider a reaction of overall stoichiometry:

$$A \longrightarrow P$$

Although we treat this reaction as a simple, one-step conversion of A to P, it more likely occurs through a sequence of elementary reactions, each of which is a simple molecular process, as in

$$A \longrightarrow I \longrightarrow J \longrightarrow P$$

where I and J represent intermediates in the reaction. Precise description of all of the elementary reactions in a process is necessary to define the overall reaction mechanism for $A \rightarrow P$.

Let us assume that $A \rightarrow P$ *is* an elementary reaction and that it is spontaneous and essentially irreversible. Irreversibility is easily assumed if the rate of P conversion to A is very slow *or* the concentration of P (expressed as [P]) is negligible under the conditions chosen. The **velocity,** v, or **rate,** of the reaction A→P is the amount of P formed or the amount of A consumed per unit time, t. That is,

$$v = \frac{d[P]}{dt} \quad \text{or} \quad v = \frac{-d[A]}{dt} \tag{13.1}$$

The mathematical relationship between reaction rate and concentration of reactant(s) is the **rate law.** For this simple case, the rate law is

$$v = \frac{-d[A]}{dt} = k[A] \tag{13.2}$$

From this expression, it is obvious that the rate is proportional to the concentration of A, and k is the proportionality constant, or **rate constant.** k has the units of $(\text{time})^{-1}$, usually $\sec^{-1}$. v is a function of [A] to the first power, or in the terminology of kinetics, v is first-order with respect to A. For an elementary reaction, the **order** for any reactant is given by its exponent in the rate equation. The number of molecules that must simultaneously interact is defined as the **molecularity** of the reaction. Thus, the simple elementary reaction of A→P is a **first-order reaction.** Figure 13.4 portrays the course of a first-order reaction as a function of time. The rate of decay of a radioactive isotope, like ^{14}C or ^{32}P, is a first-order reaction, as is an intramolecular rearrangement, such as A→P. Both are **unimolecular reactions** (the molecularity equals 1).

Bimolecular Reactions Are Reactions Involving Two Reactant Molecules

Consider the more complex reaction, where two molecules must react to yield products:

$$A + B \longrightarrow P + Q$$

Assuming this reaction is an elementary reaction, its molecularity is 2; that is, it is a **bimolecular reaction.** The velocity of this reaction can be determined from the rate of disappearance of either A or B, or the rate of appearance of P or Q:

$$v = \frac{-d[A]}{dt} = \frac{-d[B]}{dt} = \frac{d[P]}{dt} = \frac{d[Q]}{dt} \tag{13.3}$$

The rate law is

$$v = k[A][B] \tag{13.4}$$

Since A and B must collide in order to react, the rate of their reaction will be proportional to the concentrations of both A and B. Because it is proportional to the product of two concentration terms, the reaction is **second-order** overall, first-order with respect to A and first-order with respect to B. (Were the

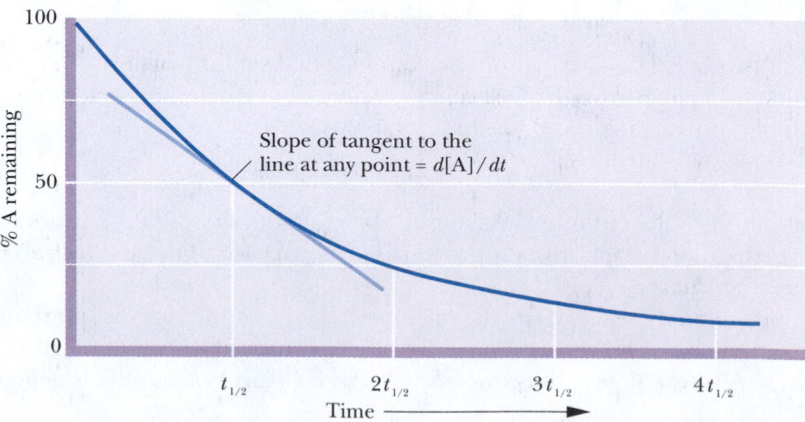

FIGURE 13.4 Plot of the course of a first-order reaction. The half-time, $t_{1/2}$, is the time for one-half of the starting amount of A to disappear.

elementary reaction $2A \rightarrow P + Q$, the rate law would be $v = k[A]^2$, second-order overall and second-order with respect to A.) Second-order rate constants have the units of (concentration)$^{-1}$(time)$^{-1}$, as in M^{-1} sec^{-1}.

Molecularities greater than 2 are rarely found (and greater than 3, never). (The likelihood of simultaneous collision of three molecules is very, very small.) When the overall stoichiometry of a reaction is greater than two (for example, as in $A + B + C \rightarrow$ or $2A + B \rightarrow$), the reaction almost always proceeds via unimolecular or bimolecular elementary steps, and the overall rate obeys a simple first- or second-order rate law.

At this point, it may be useful to remind ourselves of an important caveat that is the first principle of kinetics: *Kinetics cannot prove a hypothetical mechanism.* Kinetic experiments can only rule out various alternative hypotheses because they don't fit the data. However, through thoughtful kinetic studies, a process of elimination of alternative hypotheses leads ever closer to the reality.

Catalysts Lower the Free Energy of Activation for a Reaction

In a first-order chemical reaction, the conversion of A to P occurs because, at any given instant, a fraction of the A molecules has the energy necessary to achieve a reactive condition known as the **transition state.** In this state, the probability is very high that the particular rearrangement accompanying the A→P transition will occur. This transition state sits at the apex of the energy profile in the energy diagram describing the energetic relationship between A and P (Figure 13.5). The average free energy of A molecules defines the initial state, and the average free energy of P molecules is the final state along the reaction coordinate. The rate of any chemical reaction is proportional to the concentration of reactant molecules (A in this case) having this transition-state energy. Obviously, the higher this energy is above the average energy, the smaller the fraction of molecules that will have this energy and the slower the reaction will proceed. The height of this energy barrier is called the **free energy of activation, $\Delta G^{\ddagger}$.** Specifically, $\Delta G^{\ddagger}$ is the energy required to raise the average energy of 1 mole of reactant (at a given temperature) to the transition-state energy. The relationship between activation energy and the rate constant of the reaction, k, is given by the **Arrhenius equation:**

$$k = Ae^{-\Delta G^{\ddagger}/RT} \tag{13.5}$$

where A is a constant for a particular reaction (not to be confused with the reactant species, A, that we're discussing). Another way of writing this is

FIGURE 13.5 Energy diagram for a chemical reaction (A→P) and the effects of **(a)** raising the temperature from T_1 to T_2 or **(b)** adding a catalyst. Raising the temperature raises the average energy of A molecules, which increases the population of A molecules having energies equal to the activation energy for the reaction, thereby increasing the reaction rate. In contrast, the average free energy of A molecules remains the same in uncatalyzed versus catalyzed reactions (conducted at the same temperature). The effect of the catalyst is to lower the free energy of activation for the reaction.

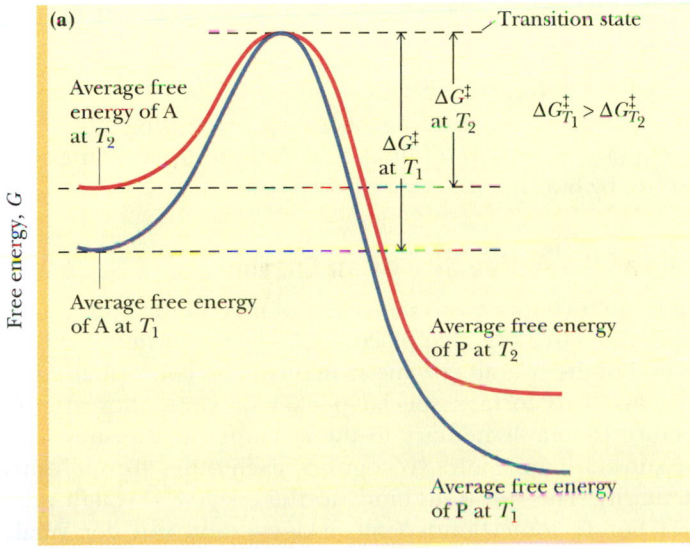

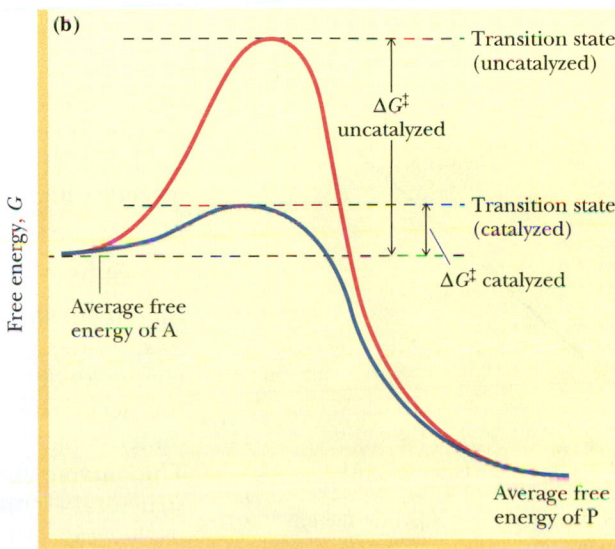

$1/k = (1/A)e^{\Delta G^{\ddagger}/RT}$. That is, k is inversely proportional to $e^{\Delta G^{\ddagger}/RT}$. Therefore, if the energy of activation decreases, the reaction rate increases.

Decreasing $\Delta G^{\ddagger}$ Increases Reaction Rate

We are familiar with two general ways that rates of chemical reactions may be accelerated. First, the temperature can be raised. This will increase the kinetic energy of reactant molecules, and more reactant molecules will possess the energy to reach the transition state (Figure 13.5a). In effect, increasing the average energy of reactant molecules makes the energy difference between the average energy and the transition-state energy smaller. (Also note that the equation $k = Ae^{-\Delta G^{\ddagger}/RT}$ demonstrates that k increases as T increases.) The rates of many chemical reactions are doubled by a 10°C rise in temperature. Second, the rates of chemical reactions can also be accelerated by catalysts. Catalysts work by lowering the energy of activation rather than by raising the average energy of the reactants (Figure 13.5b). Catalysts accomplish this remarkable feat by combining transiently with the reactants in a way that promotes their entry into the reactive, transition-state condition. Two aspects of catalysts are worth noting: (1) They are regenerated after each reaction cycle (A→P), and therefore can be used over and over again; and (2) catalysts have *no* effect on the overall free energy change in the reaction, the free energy difference between A and P (Figure 13.5b).

13.3 | What Equations Define the Kinetics of Enzyme-Catalyzed Reactions?

Examination of the change in reaction velocity as the reactant concentration is varied is one of the primary measurements in kinetic analysis. Returning to A→P, a plot of the reaction rate as a function of the concentration of A yields a straight line whose slope is k (Figure 13.6). The more A that is available, the greater the rate of the reaction, v. Similar analyses of enzyme-catalyzed reactions involving only a single substrate yield remarkably different results (Figure 13.7). At low concentrations of the substrate S, v is proportional to [S], as expected for a first-order reaction. However, v does not increase proportionally as [S] increases, but instead begins to level off. At high [S], v becomes virtually independent of [S] and approaches a maximal limit. The value of v at this limit is written V_{max}. Because rate is no longer dependent on [S] at these high concentrations, the enzyme-catalyzed reaction is now obeying **zero-order kinetics;** that is, the rate is independent of the reactant (substrate) concentration. This behavior is a **saturation effect:** When v shows no increase even though [S] is increased, the system is saturated with substrate. Such plots are called **substrate saturation curves.** The physical interpretation is that every enzyme molecule in the reaction mixture has its substrate-binding site occupied by S. Indeed, such curves were the initial clue that an enzyme interacts directly with its substrate by binding it.

The Substrate Binds at the Active Site of an Enzyme

An enzyme molecule is often (but not always) orders of magnitude larger than its substrate. In any case, its **active site,** that place on the enzyme where S binds, comprises only a portion of the overall enzyme structure. The conformation of the active site is structured to form a special pocket or cleft whose three-dimensional architecture is complementary to the structure of the substrate. The enzyme and the substrate molecules "recognize" each other through this structural complementarity. The substrate binds to the enzyme through relatively weak forces—H bonds, ionic bonds (salt bridges), and van der Waals interactions between sterically complementary clusters of atoms.

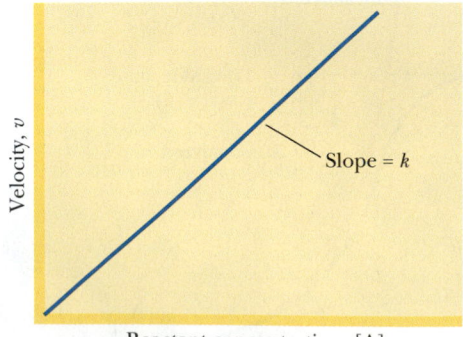

Velocity, v

Slope = k

Reactant concentration, [A]

FIGURE 13.6 A plot of v versus [A] for the unimolecular chemical reaction, A→P, yields a straight line having a slope equal to k.

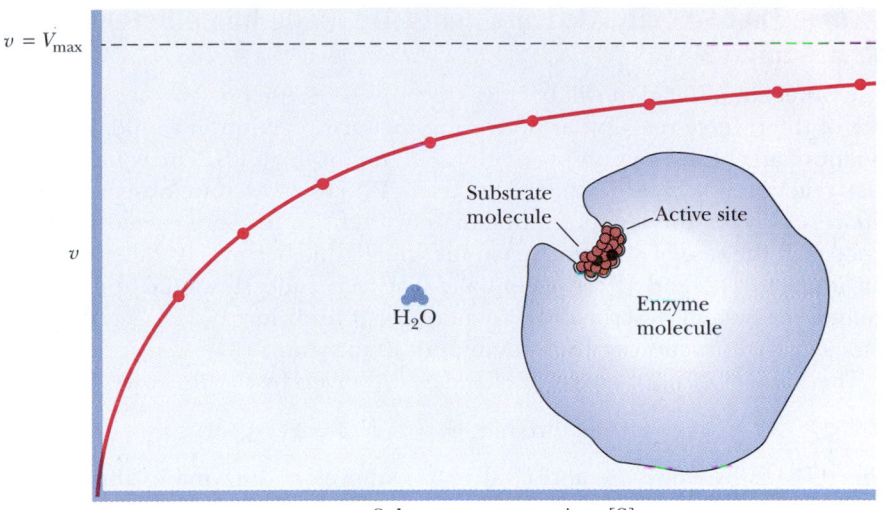

FIGURE 13.7 Substrate saturation curve for an enzyme-catalyzed reaction. The amount of enzyme is constant, and the velocity of the reaction is determined at various substrate concentrations. The reaction rate, v, as a function of [S] is described by a rectangular hyperbola. At very high [S], v approaches V_{max} ($v \approx V_{max}$ at very high [S]). That is, the velocity is limited only by conditions (temperature, pH, ionic strength) and by the amount of enzyme present; v becomes independent of [S]. Such a condition is termed *zero-order kinetics*. Under zero-order conditions, velocity is directly dependent on [enzyme]. The H_2O molecule provides a rough guide to scale. The substrate is bound at the active site of the enzyme.

The Michaelis–Menten Equation Is the Fundamental Equation of Enzyme Kinetics

Lenore Michaelis and Maud L. Menten proposed a general theory of enzyme action in 1913 consistent with observed enzyme kinetics. Their theory was based on the assumption that the enzyme, E, and its substrate, S, associate reversibly to form an enzyme–substrate complex, ES:

$$E + S \underset{k_{-1}}{\overset{k_1}{\rightleftharpoons}} ES \tag{13.6}$$

This association/dissociation is assumed to be a rapid equilibrium, and K_s is the *enzyme: substrate dissociation constant*. At equilibrium,

$$k_{-1}[ES] = k_1[E][S] \tag{13.7}$$

and

$$K_s = \frac{[E][S]}{[ES]} = \frac{k_{-1}}{k_1} \tag{13.8}$$

Product, P, is formed in a second step when ES breaks down to yield E + P.

$$E + S \underset{k_{-1}}{\overset{k_1}{\rightleftharpoons}} ES \overset{k_2}{\longrightarrow} E + P \tag{13.9}$$

E is then free to interact with another molecule of S.

Assume That [ES] Remains Constant During an Enzymatic Reaction

The interpretations of Michaelis and Menten were refined and extended in 1925 by Briggs and Haldane, who assumed the concentration of the enzyme–substrate complex ES quickly reaches a constant value in such a dynamic system. That is, ES is formed as rapidly from E + S as it disappears by its two possible fates: dissociation to regenerate E + S and reaction to form E + P. This assumption is termed the **steady-state assumption** and is expressed as

$$\frac{d[ES]}{dt} = 0 \tag{13.10}$$

That is, the change in concentration of ES with time, t, is 0. Figure 13.8 illustrates the time course for formation of the ES complex and establishment of the steady-state condition.

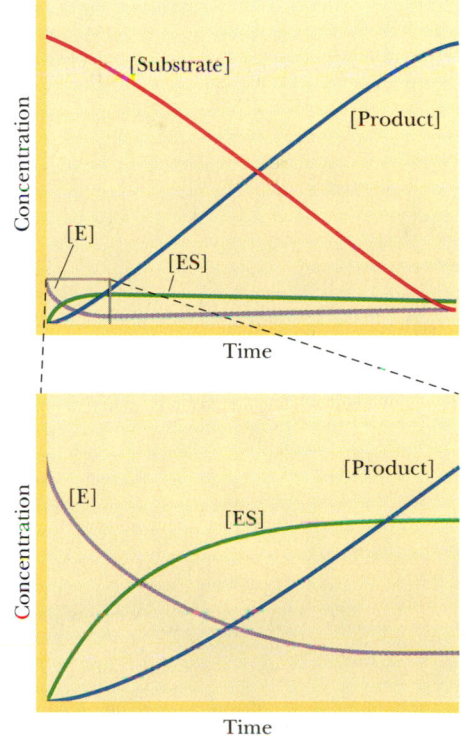

Biochemistry Now™ ANIMATED FIGURE 13.8
Time course for the consumption of substrate, the formation of product, and the establishment of a steady-state level of the enzyme-substrate [ES] complex for a typical enzyme obeying the Michaelis–Menten, Briggs–Haldane models for enzyme kinetics. The early stage of the time course is shown in greater magnification in the bottom graph. **See this figure animated at http://chemistry.brookscole.com/ggb3**

Assume That Velocity Measurements Are Made Immediately After Adding S

One other simplification will be advantageous. Because enzymes accelerate the rate of the reverse reaction as well as the forward reaction, it would be helpful to ignore any back reaction by which E + P might form ES. The velocity of this back reaction would be given by $v = k_{-2}[E][P]$. However, if we observe only the *initial velocity* for the reaction immediately after E and S are mixed in the absence of P, the rate of any back reaction is negligible because its rate will be proportional to [P] and [P] is essentially 0. Given such simplification, we now analyze the system described by Equation 13.9 in order to describe the initial velocity v as a function of [S] and amount of enzyme.

The total amount of enzyme is fixed and is given by the formula

$$\text{Total enzyme, } [E_T] = [E] + [ES] \tag{13.11}$$

where [E] is free enzyme and [ES] is the amount of enzyme in the enzyme–substrate complex. From Equation 13.9, the rate of [ES] formation is

$$v_f = k_1([E_T] - [ES])[S]$$

where

$$[E_T] - [ES] = [E] \tag{13.12}$$

From Equation 13.9, the rate of [ES] disappearance is

$$v_d = k_{-1}[ES] + k_2[ES] = (k_{-1} + k_2)[ES] \tag{13.13}$$

At steady state, $d[ES]/dt = 0$, and therefore, $v_f = v_d$. So,

$$K_1([E_T] - [ES])[S] = (k_{-1} + k_2)[ES] \tag{13.14}$$

Rearranging gives

$$\frac{([E_T] - [ES])[S]}{[ES]} = \frac{(k_{-1} + k_2)}{k_1} \tag{13.15}$$

The Michaelis Constant, K_m, Is Defined as $(k_{-1} + k_2)/k_1$

The ratio of constants $(k_{-1} + k_2)/k_1$ is itself a constant and is defined as the **Michaelis constant, K_m**

$$K_m = \frac{(k_{-1} + k_2)}{k_1} \tag{13.16}$$

Note from Equation 13.15 that K_m is given by the ratio of two concentrations $(([E_T] - [ES])$ and [S]) to one ([ES]), so K_m has the units of *molarity*. (Also, because the units of k_{-1} and k_2 are in time^{-1} and the units of k_1 are M^{-1}time^{-1}, it becomes obvious that the units of K_m are M.) From Equation 13.15, we can write

$$\frac{([E_T] - [ES])[S]}{[ES]} = K_m \tag{13.17}$$

which rearranges to

$$[ES] = \frac{[E_T][S]}{K_m + [S]} \tag{13.18}$$

Now, the most important parameter in the kinetics of any reaction is the **rate of product formation.** This rate is given by

$$v = \frac{d[P]}{dt} \tag{13.19}$$

and for this reaction

$$v = k_2[ES] \tag{13.20}$$

Substituting the expression for [ES] from Equation 13.18 into Equation 13.20 gives

$$v = \frac{k_2[E_T][S]}{K_m + [S]} \tag{13.21}$$

The product $k_2[E_T]$ has special meaning. When [S] is high enough to saturate all of the enzyme, the velocity of the reaction, v, is maximal. At saturation, the amount of [ES] complex is equal to the total enzyme concentration, E_T, its maximum possible value. From Equation 13.20, the initial velocity v then equals $k_2[E_T] = V_{max}$. Written symbolically, when $[S] \gg [E_T]$ (and K_m), $[E_T] = [ES]$ and $v = V_{max}$. Therefore,

$$V_{max} = k_2[E_T] \tag{13.22}$$

Substituting this relationship into the expression for v gives the **Michaelis–Menten equation:**

$$v = \frac{V_{max}[S]}{K_m + [S]} \tag{13.23}$$

This equation says that the initial rate of an enzyme-catalyzed reaction, v, is determined by two constants, K_m and V_{max}, and the initial concentration of substrate.

When $[S] = K_m$, $v = V_{max}/2$

We can provide an operational definition for the constant K_m by rearranging Equation 13.23 to give

$$K_m = [S]\left(\frac{V_{max}}{v} - 1\right) \tag{13.24}$$

Then, at $v = V_{max}/2$, $K_m = [S]$. That is, K_m is defined by the substrate concentration that gives a velocity equal to one-half the maximal velocity. Table 13.3 gives the K_m values of some enzymes for their substrates.

Plots of v Versus [S] Illustrate the Relationships Between V_{max}, K_m, and Reaction Order

The Michaelis–Menten equation (Equation 13.23) describes a curve known from analytical geometry as a *rectangular hyperbola*.[1] In such curves, as [S] is increased, v approaches the limiting value, V_{max}, in an asymptotic fashion. V_{max} can be approximated experimentally from a substrate saturation curve (Figure 13.7), and K_m can be derived from $V_{max}/2$, so the two constants of the Michaelis–Menten equation can be obtained from plots of v versus [S]. Note, however, that actual estimation of V_{max}, and consequently K_m, is only approximate from such graphs. That is, according to Equation 13.23, to get $v = 0.99$ V_{max}, [S] must equal 99 K_m, a concentration that may be difficult to achieve in practice.

From Equation 13.23, when $[S] \gg K_m$, then $v = V_{max}$. That is, v is no longer dependent on [S], so the reaction is obeying zero-order kinetics. Also, when $[S] < K_m$, then $v \approx (V_{max}/K_m)[S]$. That is, the rate, v, approximately follows a first-order rate equation, $v = k'[A]$, where $k' = V_{max}/K_m$.

K_m and V_{max}, once known explicitly, define the rate of the enzyme-catalyzed reaction, *provided:*

1. The reaction involves only one substrate, *or* if the reaction is multisubstrate, the concentration of only one substrate is varied while the concentrations of all other substrates are held constant.

[1]A proof that the Michaelis–Menten equation describes a rectangular hyperbola is given by Naqui, A., 1986. Where are the asymptotes of Michaelis–Menten? *Trends in Biochemical Sciences* **11**:64–65.

Table 13.3

K_m Values for Some Enzymes

Enzyme	Substrate	K_m (mM)
Carbonic anhydrase	CO_2	12
Chymotrypsin	N-Benzoyltyrosinamide	2.5
	Acetyl-L-tryptophanamide	5
	N-Formyltyrosinamide	12
	N-Acetyltyrosinamide	32
	Glycyltyrosinamide	122
Hexokinase	Glucose	0.15
	Fructose	1.5
β-Galactosidase	Lactose	4
Glutamate dehydrogenase	NH_4^+	57
	Glutamate	0.12
	α-Ketoglutarate	2
	NAD^+	0.025
	NADH	0.018
Aspartate aminotransferase	Aspartate	0.9
	α-Ketoglutarate	0.1
	Oxaloacetate	0.04
	Glutamate	4
Threonine deaminase	Threonine	5
Arginyl-tRNA synthetase	Arginine	0.003
	tRNA^{Arg}	0.0004
	ATP	0.3
Pyruvate carboxylase	HCO_3^-	1.0
	Pyruvate	0.4
	ATP	0.06
Penicillinase	Benzylpenicillin	0.05
Lysozyme	Hexa-N-acetylglucosamine	0.006

2. The reaction ES⟶E + P is irreversible, *or* the experiment is limited to observing only initial velocities where [P] = 0.
3. $[S]_0 > [E_T]$ and $[E_T]$ is held constant.
4. All other variables that might influence the rate of the reaction (temperature, pH, ionic strength, and so on) are constant.

Turnover Number Defines the Activity of One Enzyme Molecule

The **turnover number** of an enzyme, k_{cat}, is a measure of its maximal catalytic activity. k_{cat} is defined as the number of substrate molecules converted into product per enzyme molecule per unit time when the enzyme is saturated with substrate. The turnover number is also referred to as the **molecular activity** of the enzyme. For the simple Michaelis–Menten reaction (13.9) under conditions of initial velocity measurements, $k_2 = k_{cat}$. Provided the concentration of enzyme, $[E_T]$, in the reaction mixture is known, k_{cat} can be determined from V_{max}. At saturating [S], $v = V_{max} = k_2 [E_T]$. Thus,

$$k_2 = \frac{V_{max}}{[E_T]} = k_{cat} \qquad (13.25)$$

The term k_{cat} represents the kinetic efficiency of the enzyme. Table 13.4 lists turnover numbers for some representative enzymes. Catalase has the highest turnover number known; each molecule of this enzyme can degrade 40 million molecules of H_2O_2 in 1 second! At the other end of the scale, lysozyme requires 2 seconds to cleave a glycosidic bond in its glycan substrate.

Table 13.4

Values of k_{cat} (Turnover Number) for Some Enzymes

Enzyme	k_{cat} (sec^{-1})
Catalase	40,000,000
Carbonic anhydrase	1,000,000
Acetylcholinesterase	14,000
Penicillinase	2,000
Lactate dehydrogenase	1,000
Chymotrypsin	100
DNA polymerase I	15
Lysozyme	0.5

The Ratio, k_{cat}/K_m, Defines the Catalytic Efficiency of an Enzyme

Under physiological conditions, [S] is seldom saturating and k_{cat} itself is not particularly informative. That is, the in vivo ratio of $[S]/K_m$ usually falls in the range of 0.01 to 1.0, so active sites often are not filled with substrate. Nevertheless, we can derive a meaningful index of the efficiency of Michaelis–Menten-type enzymes under these conditions by using the following equations. As presented in Equation 13.23, if

$$v = \frac{V_{max}[S]}{K_m + [S]}$$

and $V_{max} = k_{cat}[E_T]$, then

$$v = \frac{k_{cat}[E_T][S]}{K_m + [S]} \tag{13.26}$$

When $[S] \ll K_m$, the concentration of free enzyme, [E], is approximately equal to $[E_T]$, so

$$v = \left(\frac{k_{cat}}{K_m}\right)[E][S] \tag{13.27}$$

That is, k_{cat}/K_m is an apparent second-order rate constant for the reaction of E and S to form product. Because K_m is inversely proportional to the affinity of the enzyme for its substrate and k_{cat} is directly proportional to the kinetic efficiency of the enzyme, k_{cat}/K_m provides an index of the catalytic efficiency of an enzyme operating at substrate concentrations substantially below saturation amounts.

An interesting point emerges if we restrict ourselves to the simple case where $k_{cat} = k_2$. Then

$$\frac{k_{cat}}{K_m} = \frac{k_1 k_2}{k_{-1} + k_2} \tag{13.28}$$

But k_1 must always be greater than or equal to $k_1 k_2/(k_{-1} + k_2)$. That is, the reaction can go no faster than the rate at which E and S come together. Thus, k_1 sets the upper limit for k_{cat}/K_m. In other words, *the catalytic efficiency of an enzyme cannot exceed the diffusion-controlled rate of combination of E and S to form ES.* In H_2O, the rate constant for such diffusion is approximately $10^9/M \cdot sec$ for small substrates (for example, glyceraldehydes-3-P) and an order of magnitude smaller ($\approx 10^8/M \cdot sec$) for substrates the size of nucleotides. Those enzymes that are most efficient in their catalysis have k_{cat}/K_m ratios approaching this value. Their catalytic velocity is limited only by the rate at which they encounter S; enzymes this efficient have achieved so-called catalytic perfection. All E and S encounters lead to reaction because such "catalytically perfect" enzymes can channel S to the active site, regardless of where S hits E. Table 13.5 lists the kinetic parameters of several enzymes in this category. Note that k_{cat} and K_m both show a substantial range of variation in this table, even though their ratio falls around $10^8/M \cdot sec$.

Enzyme Units Are Used to Define the Activity of an Enzyme

In many situations, the actual molar amount of the enzyme is not known. However, its amount can be expressed in terms of the activity observed. The International Commission on Enzymes defines **One International Unit** of enzyme *as the amount that catalyzes the formation of 1 micromole of product in 1 minute.* (Because enzymes are very sensitive to factors such as pH, temperature, and ionic strength, the conditions of assay must be specified.) Another definition for units of enzyme activity is the **katal.** One katal is *that amount of enzyme catalyzing the conversion of 1 mole of substrate to product in 1 second.* Thus, 1 katal equals 6×10^7 international units. In the process of purifying enzymes from their cellular sources, many extraneous proteins may be present. Then, units of enzyme

Table 13.5

Enzymes Whose k_{cat}/K_m Approaches the Diffusion-Controlled Rate of Association with Substrate

Enzyme	Substrate	k_{cat} (sec^{-1})	K_m (M)	k_{cat}/K_m (M^{-1} sec^{-1})
Acetylcholinesterase	Acetylcholine	1.4×10^4	9×10^{-5}	1.6×10^8
Carbonic	CO_2	1×10^6	0.012	8.3×10^7
anhydrase	HCO_3^-	4×10^5	0.026	1.5×10^7
Catalase	H_2O_2	4×10^7	1.1	4×10^7
Crotonase	Crotonyl-CoA	5.7×10^3	2×10^{-5}	2.8×10^8
Fumarase	Fumarate	800	5×10^{-6}	1.6×10^8
	Malate	900	2.5×10^{-5}	3.6×10^7
Triosephosphate isomerase	Glyceraldehyde-3-phosphate*	4.3×10^3	1.8×10^{-5}	2.4×10^8
β-Lactamase	Benzylpenicillin	2×10^3	2×10^{-5}	1×10^8

*K_m for glyceraldehyde-3-phosphate is calculated on the basis that only 3.8% of the substrate in solution is unhydrated and therefore reactive with the enzyme.
Adapted from Fersht, A., 1985. *Enzyme Structure and Mechanism*, 2nd ed. New York: W. H. Freeman.

activity are expressed as enzyme units per mg protein, a term known as **specific activity.** As extraneous proteins are removed in the purification process, the specific activity of the enzyme preparation increases (see Table 5.2).

Linear Plots Can Be Derived from the Michaelis–Menten Equation

Because of the hyperbolic shape of v versus [S] plots, V_{max} can be determined only from an extrapolation of the asymptotic approach of v to some limiting value as [S] increases indefinitely (Figure 13.7); and K_m is derived from that value of [S] giving $v = V_{max}/2$. However, several rearrangements of the Michaelis–Menten equation transform it into a straight-line equation. The best known of these is the **Lineweaver–Burk double-reciprocal plot:**

Taking the reciprocal of both sides of the Michaelis–Menten equation, Equation 13.23, yields the equality

$$\frac{1}{v} = \left(\frac{K_m}{V_{max}} \right)\left(\frac{1}{[S]} \right) + \frac{1}{V_{max}} \tag{13.29}$$

This conforms to $y = mx + b$ (the equation for a straight line), where $y = 1/v$; m, the slope, is K_m/V_{max}; $x = 1/[S]$; and $b = 1/V_{max}$. Plotting $1/v$ versus $1/[S]$ gives a straight line whose x-intercept is $-1/K_m$, whose y-intercept is $1/V_{max}$, and whose slope is K_m/V_{max} (Figure 13.9).

The **Hanes–Woolf plot** is another rearrangement of the Michaelis–Menten equation that yields a straight line:

Multiplying both sides of Equation 13.29 by [S] gives

$$\frac{[S]}{v} = [S] \left(\frac{K_m}{V_{max}} \right)\left(\frac{1}{[S]} \right) + \frac{[S]}{V_{max}} = \frac{K_m}{V_{max}} + \frac{[S]}{V_{max}} \tag{13.30}$$

and

$$\frac{[S]}{v} = \left(\frac{1}{V_{max}} \right)[S] + \frac{K_m}{V_{max}} \tag{13.31}$$

Graphing $[S]/v$ versus [S] yields a straight line where the slope is $1/V_{max}$, the y-intercept is K_m/V_{max}, and the x-intercept is $-K_m$, as shown in Figure 13.10. The Hanes–Woolf plot has the advantage of not overemphasizing the data obtained at low [S], a fault inherent in the Lineweaver–Burk plot. The common advantage of these plots is that they allow both K_m and V_{max} to be accurately

A Deeper Look

An Example of the Effect of Amino Acid Substitutions on K_m and k_{cat}: Wild-Type and Mutant Forms of Human Sulfite Oxidase

Mammalian sulfite oxidase is the last enzyme in the pathway for degradation of sulfur-containing amino acids. Sulfite oxidase (SO) catalyzes the oxidation of sulfite (SO_3^{2-}) to sulfate (SO_4^{2-}), using the heme-containing protein, cytochrome c, as electron acceptor:

$$SO_3^{2-} + 2 \text{ cytochrome } c_{oxidized} + H_2O \rightleftharpoons$$
$$SO_4^{2-} + 2 \text{ cytochrome } c_{reduced} + 2 H^+$$

Isolated sulfite oxidase deficiency is a rare and often fatal genetic disorder in humans. The disease is characterized by severe neurological abnormalities, revealed as convulsions shortly after birth. R. M. Garrett and K. V. Rajagopalan at Duke University Medical Center have isolated the human cDNA for sulfite oxidase from the cells of normal *(wild-type)* and SO-deficient individuals. Expression of these SO cDNAs in transformed *Escherichia coli* cells allowed the isolation and kinetic analysis of wild-type and mutant forms of SO, including one (designated R160Q) in which the Arg at position 160 in the polypeptide chain is replaced by Gln. A genetically engineered version of SO (designated R160K) in which Lys replaces Arg[160] was also studied.

Kinetic Constants for Wild-Type and Mutant Sulfite Oxidase			
Enzyme	$K_m^{sulfite}$ (μM)	k_{cat} (sec^{-1})	k_{cat}/K_m (10^6 M^{-1} sec^{-1})
Wild-type	17	18	1.1
R160Q	1900	3	0.0016
R160K	360	5.5	0.015

Replacing R[160] in sulfite oxidase by Q increases K_m, decreases k_{cat}, and markedly diminishes the catalytic efficiency (k_{cat}/K_m) of the enzyme. The R160K mutant enzyme has properties intermediate between wild-type and the R160Q mutant form. The substrate, SO_3^{2-}, is strongly anionic, and R[160] is one of several Arg residues situated within the SO substrate-binding site. Positively charged side chains in the substrate-binding site facilitate SO_3^{2-} binding and catalysis, with Arg being optimal in this role.

---

estimated by extrapolation of straight lines rather than asymptotes. Computer fitting of v versus [S] data to the Michaelis–Menten equation is more commonly done than graphical plotting.

Nonlinear Lineweaver–Burk or Hanes–Woolf Plots Are a Property of Regulatory Enzymes

If the kinetics of the reaction disobey the Michaelis–Menten equation, the violation is revealed by a departure from linearity in these straight-line graphs. We shall see in the next chapter that such deviations from linearity are characteristic of the kinetics of regulatory enzymes known as **allosteric enzymes.** Such regulatory enzymes are very important in the overall control of metabolic pathways.

$$\frac{1}{v} = \frac{K_m}{V_{max}}\left(\frac{1}{[S]}\right) + \frac{1}{V_{max}}$$

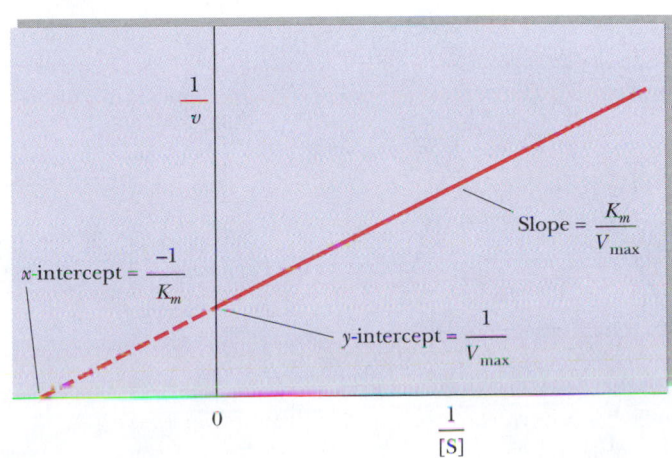

Biochemistry Now™   ACTIVE FIGURE 13.9
The Lineweaver–Burk double-reciprocal plot, depicting extrapolations that allow the determination of the *x*- and *y*-intercepts and slope. **Test yourself on the concepts in this figure at http://chemistry. brookscole.com/ggb3**

$$\frac{[S]}{v} = \left(\frac{1}{V_{max}}\right)[S] + \frac{K_m}{V_{max}}$$

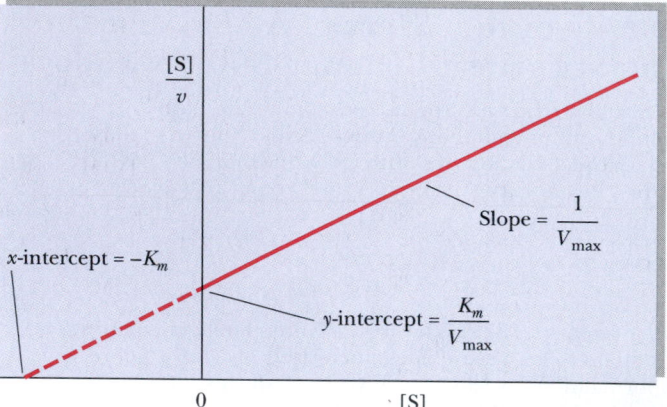

Biochemistry Now™ **ANIMATED FIGURE 13.10**
A Hanes–Woolf plot of [S]/v versus [S], another straight-line rearrangement of the Michaelis–Menten equation. **See this figure animated at http://chemistry.brookscole.com/ggb3**

Enzymatic Activity Is Strongly Influenced by pH

Enzyme–substrate recognition and the catalytic events that ensue are greatly dependent on pH. An enzyme possesses an array of ionizable side chains and prosthetic groups that not only determine its secondary and tertiary structure but may also be intimately involved in its active site. Furthermore, the substrate itself often has ionizing groups, and one or another of the ionic forms may preferentially interact with the enzyme. Enzymes in general are active only over a limited pH range, and most have a particular pH at which their catalytic activity is optimal. These effects of pH may be due to effects on K_m or V_{max} or both. Figure 13.11 illustrates the relative activity of four enzymes as a function of pH. Although the pH optimum of an enzyme often reflects the pH of its normal environment, the optimum may not be precisely the same. This difference suggests that the pH-activity response of an enzyme may be a factor in the intracellular regulation of its activity.

The Response of Enzymatic Activity to Temperature Is Complex

Like most chemical reactions, the rates of enzyme-catalyzed reactions generally increase with increasing temperature. However, at temperatures above 50° to 60°C, enzymes typically show a decline in activity (Figure 13.12). Two effects are operating here: (1) the characteristic increase in reaction rate with temperature and (2) thermal denaturation of protein structure at higher temperatures. Most

FIGURE 13.11 The pH activity profiles of four different enzymes. *Trypsin*, an intestinal protease, has a slightly alkaline pH optimum, whereas *pepsin*, a gastric protease, acts in the acidic confines of the stomach and has a pH optimum near 2. *Papain*, a protease found in papaya, is relatively insensitive to pHs between 4 and 8. *Cholinesterase* activity is pH sensitive below pH 7 but not between pH 7 and 10. The cholinesterase pH activity profile suggests that an ionizable group with a pK' near 6 is essential to its activity. Might it be a histidine residue within the active site?

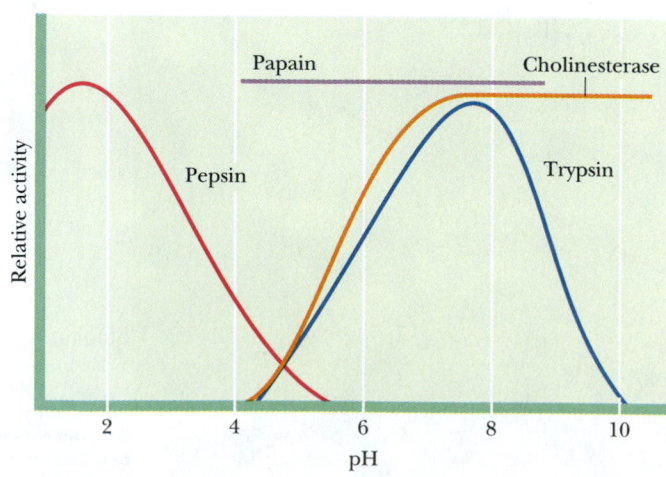

Optimum pH of Some Enzymes	
Enzyme	Optimum pH
Pepsin	1.5
Catalase	7.6
Trypsin	7.7
Fumarase	7.8
Ribonuclease	7.8
Arginase	9.7

enzymatic reactions double in rate for every 10°C rise in temperature (that is, $Q_{10} = 2$, where Q_{10} is defined as *the ratio of activities at two temperatures 10° apart*) as long as the enzyme is stable and fully active. Some enzymes, those catalyzing reactions having very high activation energies, show proportionally greater Q_{10} values. The increasing rate with increasing temperature is ultimately offset by the instability of higher orders of protein structure at elevated temperatures, where the enzyme is inactivated. Not all enzymes are quite so thermally labile. For example, the enzymes of thermophilic bacteria (*thermophilic* = "heat-loving") found in geothermal springs retain full activity at temperatures in excess of 85°C.

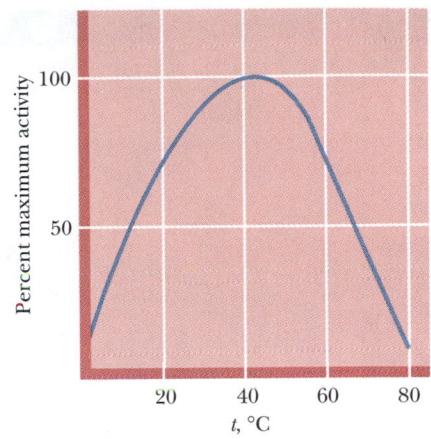

FIGURE 13.12 The effect of temperature on enzyme activity. The relative activity of an enzymatic reaction as a function of temperature. The decrease in the activity above 50°C is due to thermal denaturation.

13.4 | What Can Be Learned from the Inhibition of Enzyme Activity?

If the velocity of an enzymatic reaction is decreased or **inhibited**, the kinetics of the reaction obviously have been perturbed. Systematic perturbations are a basic tool of experimental scientists; much can be learned about the normal workings of any system by inducing changes in it and then observing the effects of the change. The study of enzyme inhibition has contributed significantly to our understanding of enzymes.

Enzymes May Be Inhibited Reversibly or Irreversibly

Enzyme inhibitors are classified in several ways. The inhibitor may interact either reversibly or irreversibly with the enzyme. **Reversible inhibitors** interact with the enzyme through noncovalent association/dissociation reactions. In contrast, **irreversible inhibitors** usually cause stable, covalent alterations in the enzyme. That is, the consequence of irreversible inhibition is a decrease in the concentration of active enzyme. The kinetics observed are consistent with this interpretation, as we shall see later.

Reversible Inhibitors May Bind at the Active Site or at Some Other Site

Reversible inhibitors fall into two major categories: competitive and noncompetitive (although other more unusual and rare categories are known). **Competitive inhibitors** are characterized by the fact that the substrate and inhibitor compete for the same binding site on the enzyme, the so-called **active site** or **substrate-binding site.** Thus, increasing the concentration of S favors the likelihood of S binding to the enzyme instead of the inhibitor, I. That is, high [S] can overcome the effects of I. The effects of the other major type, noncompetitive inhibition, cannot be overcome by increasing [S]. The two types can be distinguished by the particular patterns obtained when the kinetic data are analyzed in linear plots, such as double-reciprocal (Lineweaver–Burk) plots. A general formulation for common inhibitor interactions in our simple enzyme kinetic model would include

$$E + I \rightleftharpoons EI \quad \text{and/or} \quad I + ES \rightleftharpoons IES \qquad (13.32)$$

That is, we consider here reversible combinations of the inhibitor with E and/or ES.

Competitive Inhibition   Consider the following system:

$$E + S \underset{k_{-1}}{\overset{k_1}{\rightleftharpoons}} ES \overset{k_2}{\longrightarrow} E + P \qquad E + I \underset{k_{-3}}{\overset{k_3}{\rightleftharpoons}} EI \qquad (13.33)$$

where an inhibitor, I, binds *reversibly* to the enzyme at the same site as S. S-binding and I-binding are mutually exclusive, *competitive* processes. Formation of the ternary complex, IES, where both S and I are bound, is physically impossible. This

Table 13.6

The Effect of Various Types of Inhibitors on the Michaelis–Menten Rate Equation and on Apparent K_m and Apparent V_{max}

Inhibition Type	Rate Equation	Apparent K_m	Apparent V_{max}
None	$v = V_{max}[S]/(K_m + [S])$	K_m	V_{max}
Competitive	$v = V_{max}[S]/([S] + K_m(1 + [I]/K_I))$	$K_m(1 + [I]/K_I)$	V_{max}
Noncompetitive	$v = (V_{max}[S]/(1 + [I]/K_I))/(K_m + [S])$	K_m	$V_{max}/(1 + [I]/K_I)$
Mixed	$v = V_{max}[S]/((1 + [I]/K_I)K_m + (1 + [I]/K_I'[S]))$	$K_m(1 + [I]/K_I)/(1 + [I]/K_I')$	$V_{max}/(1 + [I]/K_I')$
Uncompetitive	$v = V_{max}[S]/(K_m + [S](1 + [I]/K_I'))$	$K_m/(1 + [I]/K_I')$	$V_{max}/(1 + [I]/K_I')$

K_I is defined as the enzyme:inhibitor dissociation constant $K_I = [E][I]/[EI]$; K_I' is defined as the enzyme substrate complex:inhibitor dissociation constant $K_I' = [ES][I]/[ESI]$.

condition leads us to anticipate that S and I must share a high degree of structural similarity because they bind at the same site on the enzyme. Also notice that, in our model, EI does not react to give rise to E + P. That is, I is not changed by interaction with E. The rate of the product-forming reaction is $v = k_2[ES]$.

It is revealing to compare the equation for the uninhibited case, Equation 13.23 (the Michaelis–Menten equation) with Equation 13.43 for the rate of the enzymatic reaction in the presence of a fixed concentration of the competitive inhibitor, [I]

$$v = \frac{V_{max}[S]}{K_m + [S]}$$

$$v = \frac{V_{max}[S]}{[S] + K_m\left(1 + \dfrac{[I]}{K_I}\right)}$$

(see also Table 13.6). The K_m term in the denominator in the inhibited case is increased by the factor $(1 + [I]/K_I)$; thus, v is less in the presence of the inhibitor, as expected. Clearly, in the absence of I, the two equations are identical. Figure 13.13 shows a Lineweaver–Burk plot of competitive inhibition. Several features of competitive inhibition are evident. First, at a given [I], v decreases ($1/v$ increases). When [S] becomes infinite, $v = V_{max}$ and is unaffected by I because all of the enzyme is in the ES form. Note that the value of the $-x$-intercept decreases as [I] increases. This $-x$-intercept is often termed the *apparent* K_m (or K_{mapp}) because it is the K_m apparent under these conditions. The diagnostic criterion for competitive inhibition is that V_{max} is unaffected by I; that is, all lines share a common y-intercept. This criterion is also the best experimental indication of binding at the same site by two substances. Competitive inhibitors resemble S structurally.

Biochemistry Now™ ACTIVE FIGURE 13.13
Lineweaver–Burk plot of competitive inhibition, showing lines for no I, [I], and 2[I]. Note that when [S] is infinitely large ($1/[S] \approx 0$), V_{max} is the same, whether I is present or not. In the presence of I, the negative x-intercept $= \dfrac{-1}{K_m\left(1 + \dfrac{[I]}{K_I}\right)}$. **Test yourself** on the concepts in this figure at http://chemistry.brookscole.com/ggb3

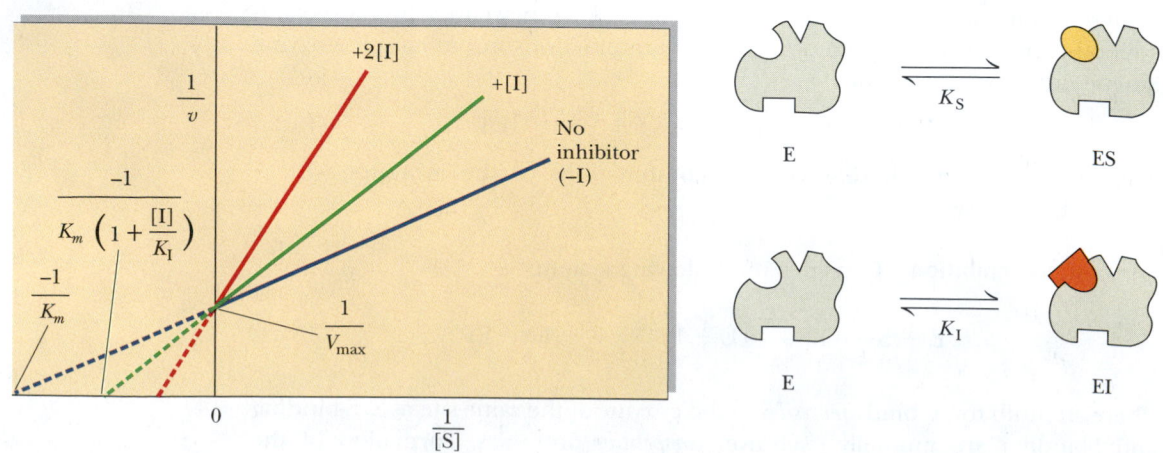

A Deeper Look

The Equations of Competitive Inhibition

Given the relationships between E, S, and I described previously and recalling the steady-state assumption that $d[ES]/dt = 0$, from Equations (13.14) and (13.16) we can write

$$ES = \frac{k_1[E][S]}{(k_2 + k_{-1})} = \frac{[E][S]}{K_m} \tag{13.34}$$

Assuming that $E + I \rightleftharpoons EI$ reaches rapid equilibrium, the rate of EI formation, $v_f' = k_3[E][I]$, and the rate of disappearance of EI, $v_d' = k_{-3}[EI]$, are equal. So,

$$k_3[E][I] = k_{-3}[EI] \tag{13.35}$$

Therefore,

$$[EI] = \frac{k_3}{k_{-3}}[E][I] \tag{13.36}$$

If we define K_I as k_{-3}/k_3, an enzyme-inhibitor dissociation constant, then

$$[EI] = \frac{[E][I]}{K_I} \tag{13.37}$$

knowing $[E_T] = [E] + [ES] + [EI]$. Then

$$[E_T] = [E] + \frac{[E][S]}{K_m} + \frac{[E][I]}{K_I} \tag{13.38}$$

Solving for [E] gives

$$[E] = \frac{K_I K_m[E_T]}{(K_I K_m + K_I[S] + K_m[I])} \tag{13.39}$$

Because the rate of product formation is given by $v = k_2[ES]$, from Equation 13.34 we have

$$v = \frac{k_2[E][S]}{K_m} \tag{13.40}$$

So,

$$v = \frac{(k_2 K_I[E_T][S])}{(K_I K_m + K_I[S] + K_m[I])} \tag{13.41}$$

Because $V_{max} = k_2[E_T]$,

$$v = \frac{V_{max}[S]}{K_m + [S] + \dfrac{K_m[I]}{K_I}} \tag{13.42}$$

or

$$v = \frac{V_{max}[S]}{[S] + K_m\left(1 + \dfrac{[I]}{K_I}\right)} \tag{13.43}$$

Succinate Dehydrogenase—A Classic Example of Competitive Inhibition The enzyme *succinate dehydrogenase (SDH)* is competitively inhibited by malonate. Figure 13.14 shows the structures of succinate and malonate. The structural similarity between them is obvious and is the basis of malonate's ability to mimic succinate and bind at the active site of SDH. However, unlike succinate, which is oxidized by SDH to form fumarate, malonate cannot lose two hydrogens; consequently, it is unreactive.

Noncompetitive Inhibition Noncompetitive inhibitors interact with both E and ES (or with S and ES, but this is a rare and specialized case). Obviously, then, the inhibitor is not binding to the same site as S, and the inhibition cannot be overcome by raising [S]. There are two types of noncompetitive inhibition: pure and mixed.

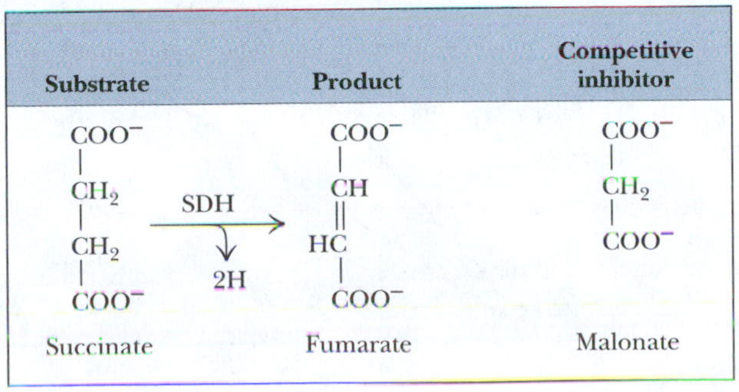

Substrate		Product	Competitive inhibitor
COO⁻		COO⁻	COO⁻
	SDH		
CH₂		CH	CH₂
CH₂		HC	COO⁻
COO⁻	2H	COO⁻	
Succinate		Fumarate	Malonate

FIGURE 13.14 Structures of succinate, the substrate of succinate dehydrogenase (SDH), and malonate, the competitive inhibitor. Fumarate (the product of SDH action on succinate) is also shown.

Pure Noncompetitive Inhibition In this situation, the binding of I by E has no effect on the binding of S by E. That is, S and I bind at different sites on E, and binding of I does not affect binding of S. Consider the system

$$E + I \overset{K_I}{\rightleftharpoons} EI \qquad ES + I \overset{K_I'}{\rightleftharpoons} IES \tag{13.44}$$

Pure noncompetitive inhibition occurs if $K_I = K_I'$. This situation is relatively uncommon; the Lineweaver–Burk plot for such an instance is given in Figure 13.15. Note that K_m is unchanged by I (the x-intercept remains the same, with or without I). Note also that the apparent V_{max} decreases. A similar pattern is seen if the amount of enzyme in the experiment is decreased. Thus, it is as if I lowered [E].

Mixed Noncompetitive Inhibition In this situation, the binding of I by E influences the binding of S by E. Either the binding sites for I and S are near one another or conformational changes in E caused by I affect S binding. In this case, K_I and K_I', as defined previously, are not equal. Both the apparent K_m and the apparent V_{max} are altered by the presence of I, and K_m/V_{max} is not constant (Figure 13.16). This inhibitory pattern is commonly encountered. A reasonable explanation is that the inhibitor is binding at a site distinct from the active site yet is influencing the binding of S at the active site. Presumably, these effects are transmitted via alterations in the protein's conformation. Table 13.6 includes the rate equations and apparent K_m and V_{max} values for both types of noncompetitive inhibition.

Uncompetitive Inhibition Completing the set of inhibitory possibilities is uncompetitive inhibition. Unlike competitive inhibition (where I combines only with E) or noncompetitive inhibition (where I combines with E and ES), in uncompetitive inhibition, I combines only with ES.

$$ES + I \overset{K_I'}{\rightleftharpoons} IES \tag{13.45}$$

The pattern obtained in Lineweaver–Burk plots is a set of parallel lines (Figure 13.17). A clinically important example is the action of lithium in alleviating manic depression; Li^+ ions are uncompetitive inhibitors of *myo*-inositol monophosphatase. Some pesticides are also uncompetitive inhibitors, such as Roundup, an uncompetitive inhibitor of 3-enolpyruvylshikimate-5-P synthase, an enzyme essential to aromatic amino acid biosynthesis (see Chapter 25).

Enzymes Also Can Be Inhibited in an Irreversible Manner

If the inhibitor combines irreversibly with the enzyme—for example, by covalent attachment—the kinetic pattern seen is like that of noncompetitive inhibition, because the net effect is a loss of active enzyme. Usually, this type of

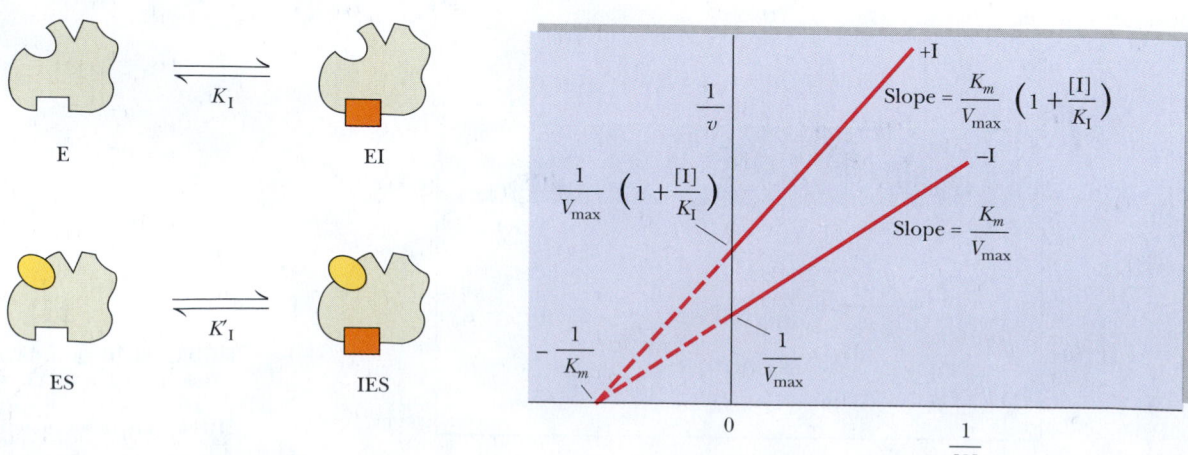

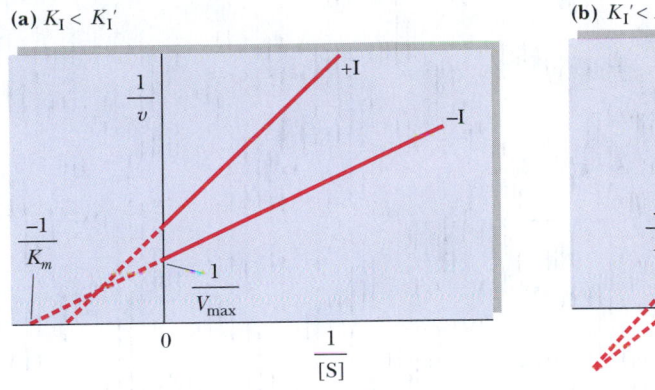

(a) $K_I < K_I'$

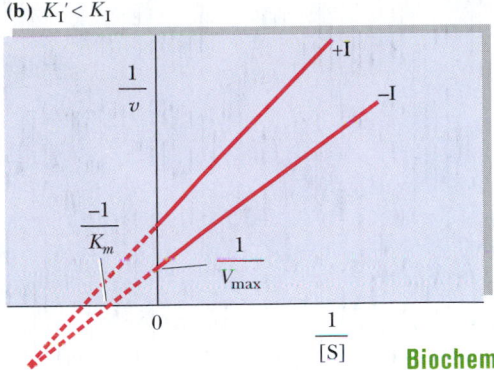

(b) $K_I' < K_I$

Biochemistry ⦿ Now™ **ACTIVE FIGURE 13.16**
Lineweaver–Burk plot of mixed noncompetitive inhibition. Note that both intercepts and the slope change in the presence of I. (a) When K_I is less than K_I'; (b) when K_I is greater than K_I'. **Test yourself on the concepts in this figure at** http://chemistry.brookscole.com/ggb3

inhibition can be distinguished from the noncompetitive, reversible inhibition case because the reaction of I with E (and/or ES) is not instantaneous. Instead, there is a *time-dependent decrease in enzymatic activity* as E + I→EI proceeds, and the rate of this inactivation can be followed. Also, unlike reversible inhibitions, dilution or dialysis of the enzyme:inhibitor solution does not dissociate the EI complex and restore enzyme activity.

Suicide Substrates—Mechanism-Based Enzyme Inactivators Suicide sub-strates are inhibitory substrate analogs designed so that, via normal catalytic action of the enzyme, a very reactive group is generated. This reactive group then forms a covalent bond with a nearby functional group within the active site of the enzyme, thereby causing irreversible inhibition. Suicide sub-strates, also called *Trojan horse substrates*, are a type of **affinity label.** As sub-strate analogs, they bind with specificity and high affinity to the enzyme active site; in their reactive form, they become covalently bound to the en-zyme. This covalent link effectively labels a particular functional group within the active site, identifying the group as a key player in the enzyme's catalytic cycle.

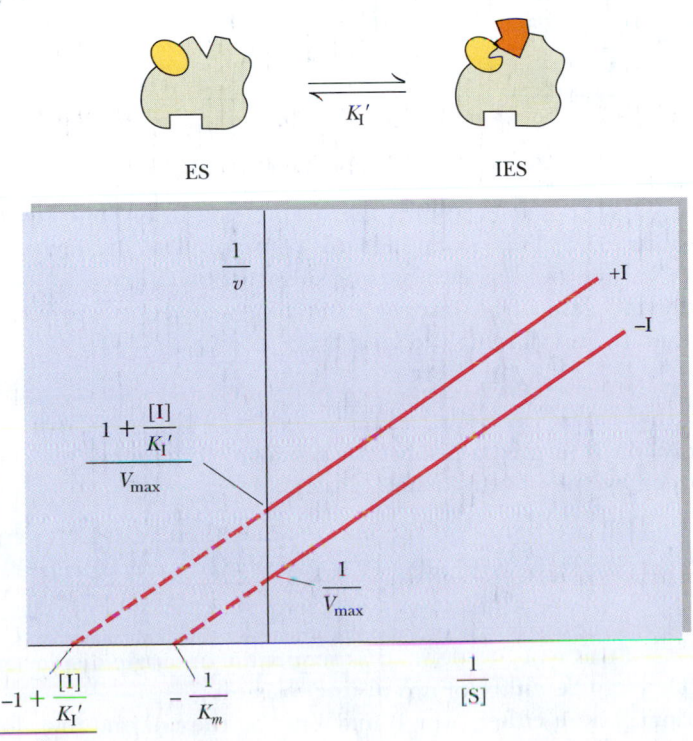

FIGURE 13.17 Lineweaver–Burk plot of uncompeti-tive inhibition. Note that both intercepts change but the slope (K_m/V_{max}) remains constant in the pres-ence of I. Uncompetitive inhibition occurs when I combines only with ES. Because IES does not lead to product formation, the observed rate constant for product formation, k_2, is uniquely affected. In sim-ple Michaelis–Menten kinetics, k_2 is the only rate constant that is part of both V_{max} and K_m [$V_{max} = k_2 E_T$ and $K_m = (k_{-1} + k_2)/k_1$].

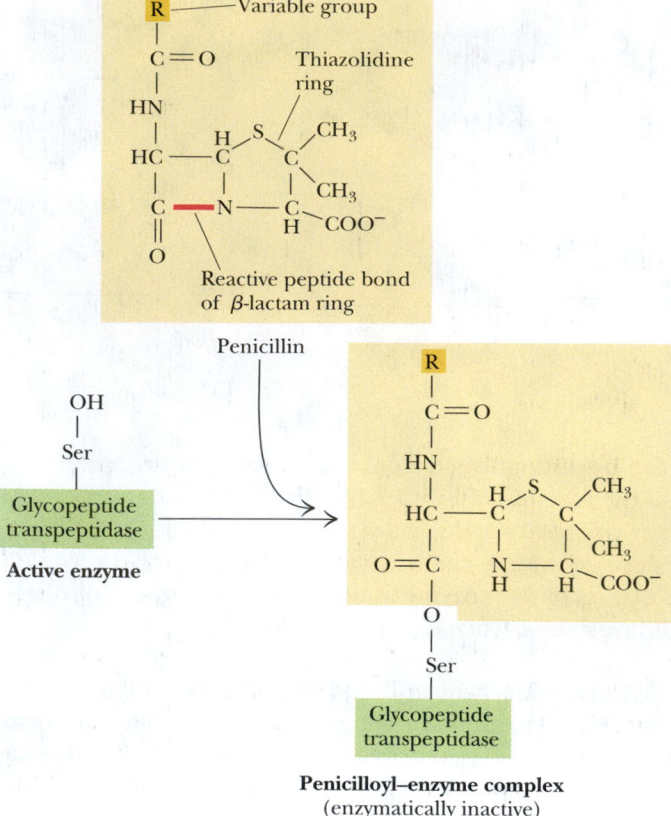

FIGURE 13.18 Penicillin is an irreversible inhibitor of the enzyme *glycoprotein peptidase,* which catalyzes an essential step in bacterial cell wall synthesis. Penicillin consists of a thiazolidine ring fused to a β-lactam ring to which a variable group R is attached. A reactive peptide bond in the β-lactam ring covalently attaches to a serine residue in the active site of the glycopeptide transpeptidase. (The conformation of penicillin around its reactive peptide bond resembles the transition state of the normal glycoprotein peptidase substrate.) The penicilloyl–enzyme complex is catalytically inactive. The bond between the enzyme and penicillin is indefinitely stable; that is, penicillin binding is irreversible.

Penicillin—A Suicide Substrate   Several drugs in current medical use are mechanism-based enzyme inactivators. For example, the antibiotic **penicillin** exerts its effects by covalently reacting with an essential serine residue in the active site of *glycoprotein peptidase,* an enzyme that acts to crosslink the peptidoglycan chains during synthesis of bacterial cell walls (Figure 13.18). Once cell wall synthesis is blocked, the bacterial cells are very susceptible to rupture by osmotic lysis and bacterial growth is halted.

13.5 What Is the Kinetic Behavior of Enzymes Catalyzing Bimolecular Reactions?

Thus far, we have considered only the simple case of enzymes that act upon a single substrate, S. This situation is not common. Usually, enzymes catalyze reactions in which two (or even more) substrates take part.

Consider the case of an enzyme catalyzing a reaction involving two substrates, A and B, and yielding the products P and Q:

$$A + B \underset{}{\overset{enzyme}{\rightleftharpoons}} P + Q \tag{13.46}$$

Such a reaction is termed a **bisubstrate reaction.** In general, bisubstrate reactions proceed by one of two possible routes:

1. Both A and B are bound to the enzyme and then reaction occurs to give P + Q:

$$E + A + B \longrightarrow AEB \longrightarrow PEQ \longrightarrow E + P + Q \tag{13.47}$$

Reactions of this type are defined as **sequential** or **single-displacement reactions.** They can be either of two distinct classes:

a. **random,** where either A or B may bind to the enzyme first, followed by the other substrate, or

Human Biochemistry

Viagra—An Unexpected Outcome in a Program of Drug Design

Prior to the accumulation of detailed biochemical information on metabolism, enzymes, and receptors, drugs were fortuitous discoveries made by observant scientists; the discovery of penicillin as a bacteria-killing substance by Fleming is an example. Today, **drug design** is the rational application of scientific knowledge and principles to the development of pharmacologically active agents. A particular target for therapeutic intervention is identified (such as an enzyme or receptor involved in illness), and chemical analogs of its substrate or ligand are synthesized in hopes of finding an inhibitor (or activator) that will serve as a drug to treat the illness. Sometimes the outcome is unanticipated, as the story of **Viagra** (sildenafil citrate) reveals.

When the smooth muscle cells of blood vessels relax, blood flow increases and blood pressure drops. Such relaxation is the result of decreases in intracellular $[Ca^{2+}]$ triggered by increases in intracellular [cGMP] (which in turn is triggered by nitric oxide, NO; see Chapter 32). Cyclic GMP (cGMP) is hydrolyzed by *phosphodiesterases* to form 5'-GMP, and the muscles contract again. Scientists at Pfizer reasoned that, if phosphodiesterase inhibitors could be found, they might be useful drugs to treat *angina* (chest pain due to inadequate blood flow to heart muscle) or *hypertension* (high blood pressure). The phosphodiesterase (PDE) prevalent in vascular muscle is PDE 5, one of at least nine different substypes of PDE in human cells. The search was on for substances that inhibit PDE 5, but not the other prominent PDE types, and Viagra was found. Disappointingly, Viagra showed no significant benefits for angina or hypertension, but some men in clinical trials reported penile erection. Apparently, Viagra led to an increase in [cGMP] in penile vascular tissue, allowing vascular muscle relaxation, improved blood flow, and erection. A drug was born.

In a more focused way, detailed structural data on enzymes, receptors, and the ligands that bind to them has led to **rational drug design,** in which *computer modeling of enzyme and ligand interactions* replaces much of the initial chemical synthesis and clinical prescreening of potential therapeutic agents, saving much time and effort in drug development.

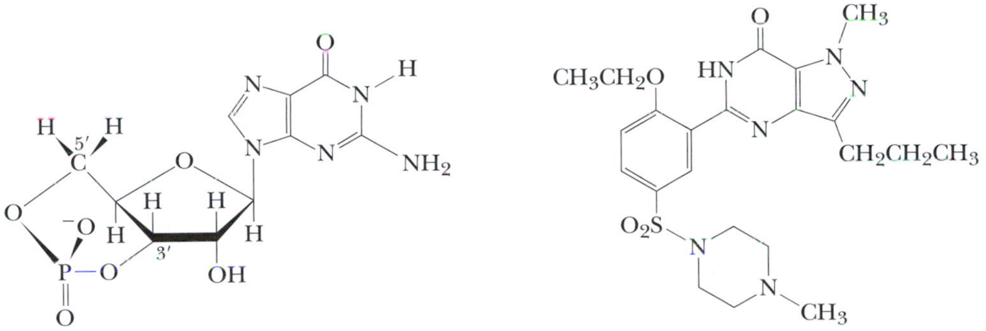

◀ Note the structural similarity between cGMP *(left)* and Viagra *(right)*.

b. **ordered,** where A, designated the *leading substrate*, must bind to E first before B can be bound.

Both classes of single-displacement reactions are characterized by lines that intersect to the left of the $1/v$ axis in Lineweaver–Burk double-reciprocal plots (Figure 13.19).

Double-reciprocal form of the rate equation:
$$\frac{1}{v} = \frac{1}{V_{max}} \left(K_m^A + \frac{K_S^A \, K_m^B}{[B]} \right) \left(\frac{1}{[A]} + \frac{1}{V_{max}} \left(1 + \frac{K_m^B}{[B]} \right) \right)$$

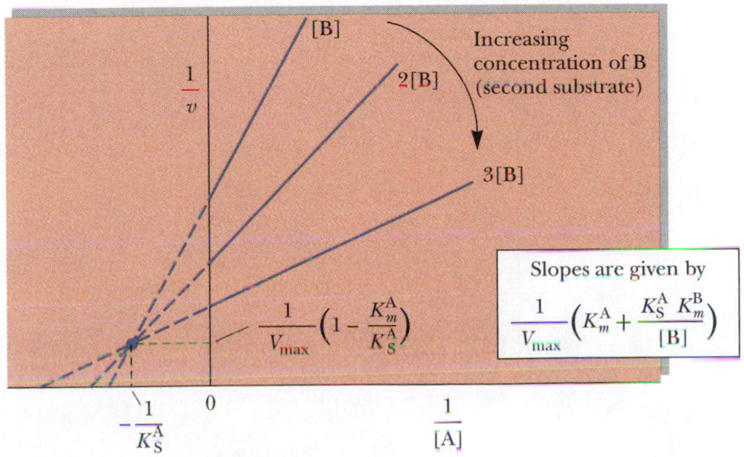

Slopes are given by
$$\frac{1}{V_{max}} \left(K_m^A + \frac{K_S^A \, K_m^B}{[B]} \right)$$

$$-\frac{1}{V_{max}} \left(1 - \frac{K_m^A}{K_S^A} \right)$$

FIGURE 13.19 Single-displacement bisubstrate mechanism. Double-reciprocal plots of the rates observed with different fixed concentrations of one substrate (B here) are graphed versus a series of concentrations of A. Note that, in these Lineweaver–Burk plots for single-displacement bisubstrate mechanisms, the lines intersect to the left of the $1/v$ axis.

2. The other general possibility is that one substrate, A, binds to the enzyme and reacts with it to yield a chemically modified form of the enzyme (E′) plus the product, P. The second substrate, B, then reacts with E′, regenerating E and forming the other product, Q.

$$E + A \longrightarrow EA \longrightarrow E'P \searrow E' \nearrow E'B \longrightarrow EQ \longrightarrow E + Q$$

$$P \qquad B \qquad \qquad (13.48)$$

Reactions that fit this model are called **ping-pong** or **double-displacement reactions.** Two distinctive features of this mechanism are the obligatory formation of a modified enzyme intermediate, E′, and the pattern of parallel lines obtained in double-reciprocal plots (see Figure 13.22).

The Conversion of AEB to PEQ Is the Rate-Limiting Step in Random, Single-Displacement Reactions

In this type of sequential reaction, all possible binary enzyme–substrate complexes (AE, EB, PE, EQ) are formed rapidly and reversibly when the enzyme is added to a reaction mixture containing A, B, P, and Q:

$$A + E \rightleftharpoons AE \searrow \qquad \nearrow EP \rightleftharpoons P + E$$
$$\qquad \qquad AEB \rightleftharpoons PEQ$$
$$E + B \rightleftharpoons EB \nearrow \qquad \searrow QE \rightleftharpoons E + Q \quad (13.49)$$

The rate-limiting step is the reaction AEB→PEQ. It doesn't matter whether A or B binds first to E, or whether Q or P is released first from QEP. Sometimes, reactions that follow this random order of addition of substrates to E can be distinguished mechanistically from reactions obeying an ordered, single-displacement mechanism, *if* A has *no* influence on the binding constant for B (and vice versa); that is, the mechanism is purely random. Then, the lines in a Lineweaver–Burk plot intersect at the $1/[A]$ axis (Figure 13.20).

Creatine Kinase Acts by a Random, Single-Displacement Mechanism   An example of a random, single-displacement mechanism is seen in the enzyme creatine kinase, a phosphoryl transfer enzyme that uses ATP as a phosphoryl donor to form creatine phosphate (CrP) from creatine (Cr). Creatine-P is an important reservoir of phosphate-bond energy in muscle cells (Figure 13.21).

FIGURE 13.20 Random, single-displacement bisubstrate mechanism where A does not affect B binding, and vice versa. Note that the lines intersect at the $1/[A]$ axis. (If [B] were varied in an experiment with several fixed concentrations of A, the lines would intersect at the $1/[B]$ axis in a $1/v$ versus $1/[B]$ plot.)

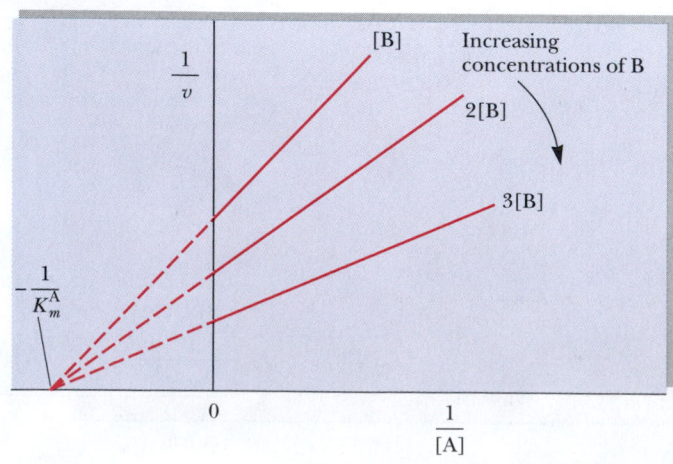

$$\text{ATP} + \text{E} \rightleftharpoons \text{ATP:E}$$

$$\text{ATP:E:Cr} \rightleftharpoons \text{ADP:E:CrP}$$

$$\text{E} + \text{Cr} \rightleftharpoons \text{E:Cr}$$

$$\text{ADP:E} \rightleftharpoons \text{ADP} + \text{E}$$

$$\text{E:CrP} \rightleftharpoons \text{E} + \text{CrP}$$

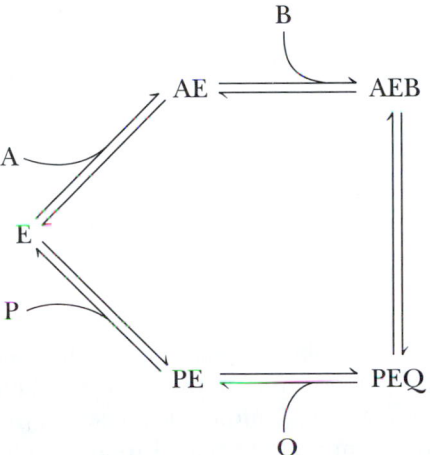

Creatine

Creatine-P

FIGURE 13.21 The structures of creatine and creatine phosphate, guanidinium compounds that are important in muscle energy metabolism.

The overall direction of the reaction will be determined by the relative concentrations of ATP, ADP, Cr, and CrP and the equilibrium constant for the reaction. The enzyme can be considered to have two sites for substrate (or product) binding: an adenine nucleotide site, where ATP or ADP binds, and a creatine site, where Cr or CrP is bound. In such a mechanism, ATP and ADP compete for binding at their unique site while Cr and CrP compete at the specific Cr-, CrP-binding site. Note that no modified enzyme form (E′), such as an E-PO$_4$ intermediate, appears here. The reaction is characterized by rapid and reversible binary ES complex formation, followed by addition of the remaining substrate, and the rate-determining reaction taking place within the ternary complex.

In an Ordered, Single-Displacement Reaction, the Leading Substrate Must Bind First

In this case, the **leading substrate,** A (also called the **obligatory** or **compulsory substrate**), must bind first. Then the second substrate, B, binds. Strictly speaking, B cannot bind to free enzyme in the absence of A. Reaction between A and B occurs in the ternary complex and is usually followed by an ordered release of the products of the reaction, P and Q. In the following schemes, P is the product of A and is released last. One representation, suggested by W. W. Cleland, follows:

$$
\begin{array}{ccccccc}
\text{A} & \text{B} & & \text{Q} & & \text{P} & \\
\downarrow & \downarrow & & \uparrow & & \uparrow & \\
\text{E} \underline{\quad} \text{AE} \underline{\quad} \text{AEB} & \rightleftharpoons & \text{PEQ} \underline{\quad} & \text{PE} \underline{\quad} & \text{E} & (13.50)
\end{array}
$$

Another way of portraying this mechanism is as follows:

Note that A and P are competitive for binding to the free enzyme, E, but not A and B (or P and B).

NAD$^+$-Dependent Dehydrogenases Show Ordered Single-Displacement Mechanisms

Nicotinamide adenine dinucleotide (NAD$^+$)-dependent dehydrogenases are enzymes that typically behave according to the kinetic pattern just described. A general reaction of these dehydrogenases is

$$\text{NAD}^+ + \text{BH}_2 \rightleftharpoons \text{NADH} + \text{H}^+ + \text{B}$$

The leading substrate (A) is nicotinamide adenine dinucleotide (NAD$^+$), and NAD$^+$ and NADH (product P) compete for a common site on E. A specific example is offered by alcohol dehydrogenase (ADH):

$$\text{NAD}^+ + \text{CH}_3\text{CH}_2\text{OH} \rightleftharpoons \text{NADH} + \text{H}^+ + \text{CH}_3\text{CHO}$$

(A)   ethanol   (P)   acetaldehyde
                    (B)   (Q)

We can verify that this ordered mechanism is not random by demonstrating that no B (ethanol) is bound to E in the absence of A (NAD$^+$).

Double-Displacement (Ping-Pong) Reactions Proceed Via Formation of a Covalently Modified Enzyme Intermediate

Double-displacement reactions are characterized by a pattern of parallel lines when $1/v$ is plotted as a function of $1/[A]$ at different concentrations of B, the second substrate (Figure 13.22). Reactions conforming to this kinetic pattern are characterized by the fact that the product of the enzyme's reaction with A (called P in the following schemes) is released *prior* to reaction of the enzyme with the second substrate, B. As a result of this process, the enzyme, E, is converted to a modified form, E′, which then reacts with B to give the second product, Q, and regenerate the unmodified enzyme form, E:

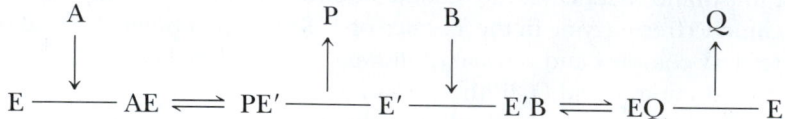

or

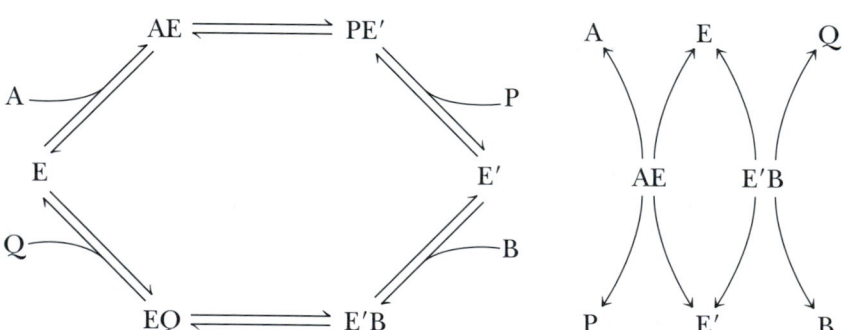

Note that these schemes predict that A and Q compete for the free enzyme form, E, while B and P compete for the modified enzyme form, E′. A and Q do not bind to E′, nor do B and P combine with E.

Aminotransferases Show Double-Displacement Catalytic Mechanisms   One class of enzymes that follow a ping-pong–type mechanism are *aminotransferases* (previously known as transaminases). These enzymes catalyze the transfer of an amino group from an amino acid to an α-keto acid. The products are a new amino acid and the keto acid corresponding to the carbon skeleton of the amino donor:

$$\text{amino acid}_1 + \text{keto acid}_2 \longrightarrow \text{keto acid}_1 + \text{amino acid}_2$$

A specific example would be *glutamate : aspartate aminotransferase*. Figure 13.23 depicts the scheme for this mechanism. Note that glutamate and aspartate are competitive for E and that oxaloacetate and α-ketoglutarate compete for E′. In glutamate : aspartate aminotransferase, an enzyme-bound coenzyme, *pyridoxal phosphate* (a vitamin B_6 derivative), serves as the amino group acceptor/donor

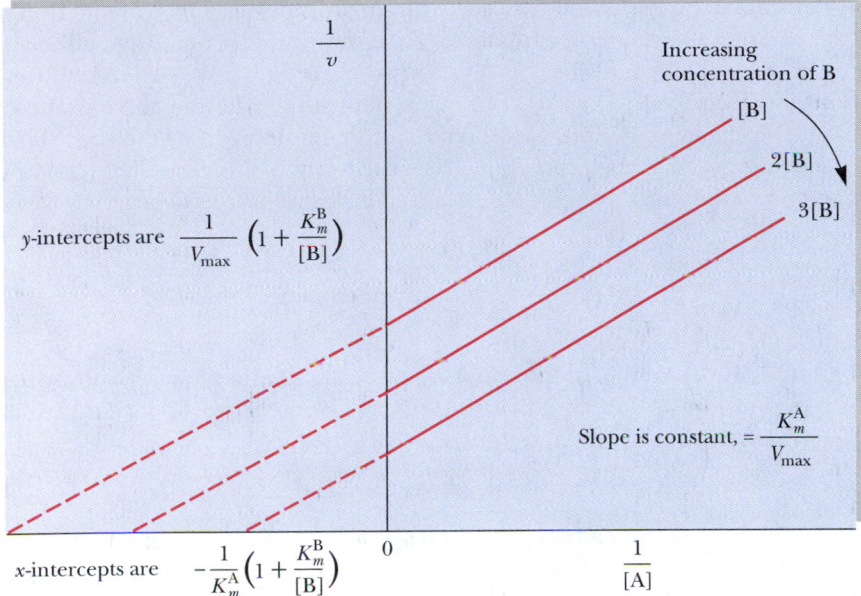

Double-reciprocal form of the rate equation:
$$\frac{1}{v} = \frac{K_m^A}{V_{max}}\left(\frac{1}{[A]}\right) + \left(1 + \frac{K_m^B}{[B]}\right)\left(\frac{1}{V_{max}}\right)$$

y-intercepts are $\dfrac{1}{V_{max}}\left(1 + \dfrac{K_m^B}{[B]}\right)$

Slope is constant, $= \dfrac{K_m^A}{V_{max}}$

x-intercepts are $-\dfrac{1}{K_m^A}\left(1 + \dfrac{K_m^B}{[B]}\right)$

FIGURE 13.22 Double-displacement (ping-pong) bisubstrate mechanisms are characterized by Lineweaver–Burk plots of parallel lines when double-reciprocal plots of the rates observed with different fixed concentrations of the second substrate, B, are graphed versus a series of concentrations of A.

in the enzymatic reaction. The unmodified enzyme form, E, has the coenzyme in the aldehydic pyridoxal form, whereas the modified enzyme form, E′, is actually pyridoxamine phosphate (Figure 13.23). Not all enzymes displaying ping-pong–type mechanisms require coenzymes as carriers for the chemical substituent transferred in the reaction.

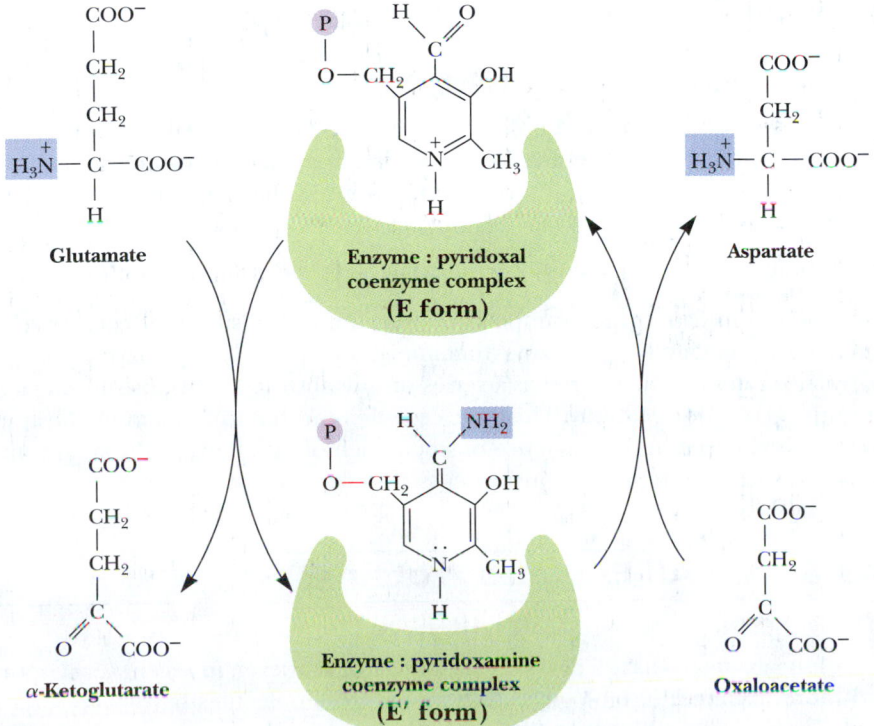

FIGURE 13.23 *Glutamate : aspartate aminotransferase,* an enzyme conforming to a double-displacement bisubstrate mechanism. Glutamate : aspartate aminotransferase is a pyridoxal phosphate–dependent enzyme. The pyridoxal serves as the —NH₂ acceptor from glutamate to form pyridoxamine. Pyridoxamine is then the amino donor to oxaloacetate to form aspartate and regenerate the pyridoxal coenzyme form. (The pyridoxamine : enzyme is the E′ form.)

Exchange Reactions Are One Way to Diagnose Bisubstrate Mechanisms

Kineticists rely on a number of diagnostic tests for the assignment of a reaction mechanism to a specific enzyme. One is the graphic analysis of the kinetic patterns observed. It is usually easy to distinguish between single- and double-displacement reactions in this manner, and examining competitive effects between substrates aids in assigning reactions to random versus ordered patterns of S binding. A second diagnostic test is to determine whether the enzyme catalyzes an **exchange reaction.** Consider as an example the two enzymes *sucrose phosphorylase* and *maltose phosphorylase.* Both catalyze the phosphorolysis of a disaccharide and both yield glucose-1-phosphate and a free hexose:

$$\text{Sucrose} + \text{P}_i \rightleftharpoons \text{glucose-1-phosphate} + \text{fructose}$$

$$\text{Maltose} + \text{P}_i \rightleftharpoons \text{glucose-1-phosphate} + \text{glucose}$$

Interestingly, in the absence of sucrose and fructose, sucrose phosphorylase will catalyze the exchange of inorganic phosphate, P_i, into glucose-1-phosphate. This reaction can be followed by using $^{32}\text{P}_i$ as a radioactive tracer and observing the incorporation of ^{32}P into glucose-1-phosphate:

$$^{32}\text{P}_i + \text{G-1-P} \rightleftharpoons \text{P}_i + \text{G-1-}^{32}\text{P}$$

Maltose phosphorylase cannot carry out a similar reaction. The ^{32}P exchange reaction of sucrose phosphorylase is accounted for by a double-displacement mechanism where E′ is E-glucose:

$$\text{Sucrose} + \text{E} \rightleftharpoons \text{E-glucose} + \text{fructose}$$

$$\text{E-glucose} + \text{P}_i \rightleftharpoons \text{E} + \text{glucose-1-phosphate}$$

Thus, in the presence of just $^{32}\text{P}_i$ and glucose-1-phosphate, sucrose phosphorylase still catalyzes the second reaction and radioactive P_i is incorporated into glucose-1-phosphate over time.

Maltose phosphorylase proceeds via a single-displacement reaction that necessarily requires the formation of a ternary maltose : E : P_i (or glucose : E : glucose-1-phosphate) complex for any reaction to occur. Exchange reactions are a characteristic of enzymes that obey double-displacement mechanisms at some point in their catalysis.

Multisubstrate Reactions Can Also Occur in Cells

Thus far, we have considered enzyme-catalyzed reactions involving one or two substrates. How are the kinetics described in those cases in which more than two substrates participate in the reaction? An example might be the glycolytic enzyme *glyceraldehyde-3-phosphate dehydrogenase* (see Chapter 18):

$$\text{NAD}^+ + \text{glyceraldehyde-3-P} + \text{P}_i \rightleftharpoons \text{NADH} + \text{H}^+ + \text{1,3-bisphosphoglycerate}$$

Many other multisubstrate examples abound in metabolism. In effect, these situations are managed by realizing that the interaction of the enzyme with its many substrates can be treated as a series of unisubstrate or bisubstrate steps in a multistep reaction pathway. Thus, the complex mechanism of a multisubstrate reaction is resolved into a sequence of steps, each of which obeys the single- and double-displacement patterns just discussed.

13.6 | Are All Enzymes Proteins?

RNA Molecules That Are Catalytic Have Been Termed "Ribozymes"

It was long assumed that all enzymes are proteins. However, in recent years, more and more instances of biological catalysis by RNA molecules have been discovered. These catalytic RNAs, or **ribozymes,** satisfy several enzymatic criteria: They are substrate specific, they enhance the reaction rate, and they emerge from the reaction unchanged. For example, RNase P, an enzyme responsible for the formation of mature tRNA molecules from tRNA precursors, requires an RNA component as well as a protein subunit for its activity in the cell. In vitro, the protein alone is incapable of catalyzing the maturation reaction, but the RNA component by itself can carry out the reaction under appropriate conditions. In another case, in the ciliated protozoan *Tetrahymena*, formation of mature ribosomal RNA from a pre-rRNA precursor involves the removal of an internal RNA segment and the joining of the two ends in a process known as **splicing.** The excision of this intervening internal sequence of RNA and ligation of the ends is, remarkably, catalyzed by the intervening sequence of RNA itself, in the presence of Mg^{2+} and a free molecule of guanosine nucleoside or nucleotide (Figure 13.24). In vivo, the intervening sequence RNA probably acts only in splicing itself out; in vitro, however, it can act many times, turning over like a true enzyme.

Protein-Free 50S Ribosomal Subunits Catalyze Peptide Bond Formation In Vitro

Perhaps the most significant case of catalysis by RNA occurs in protein synthesis. Harry F. Noller and his colleagues demonstrated that the **peptidyl transferase**

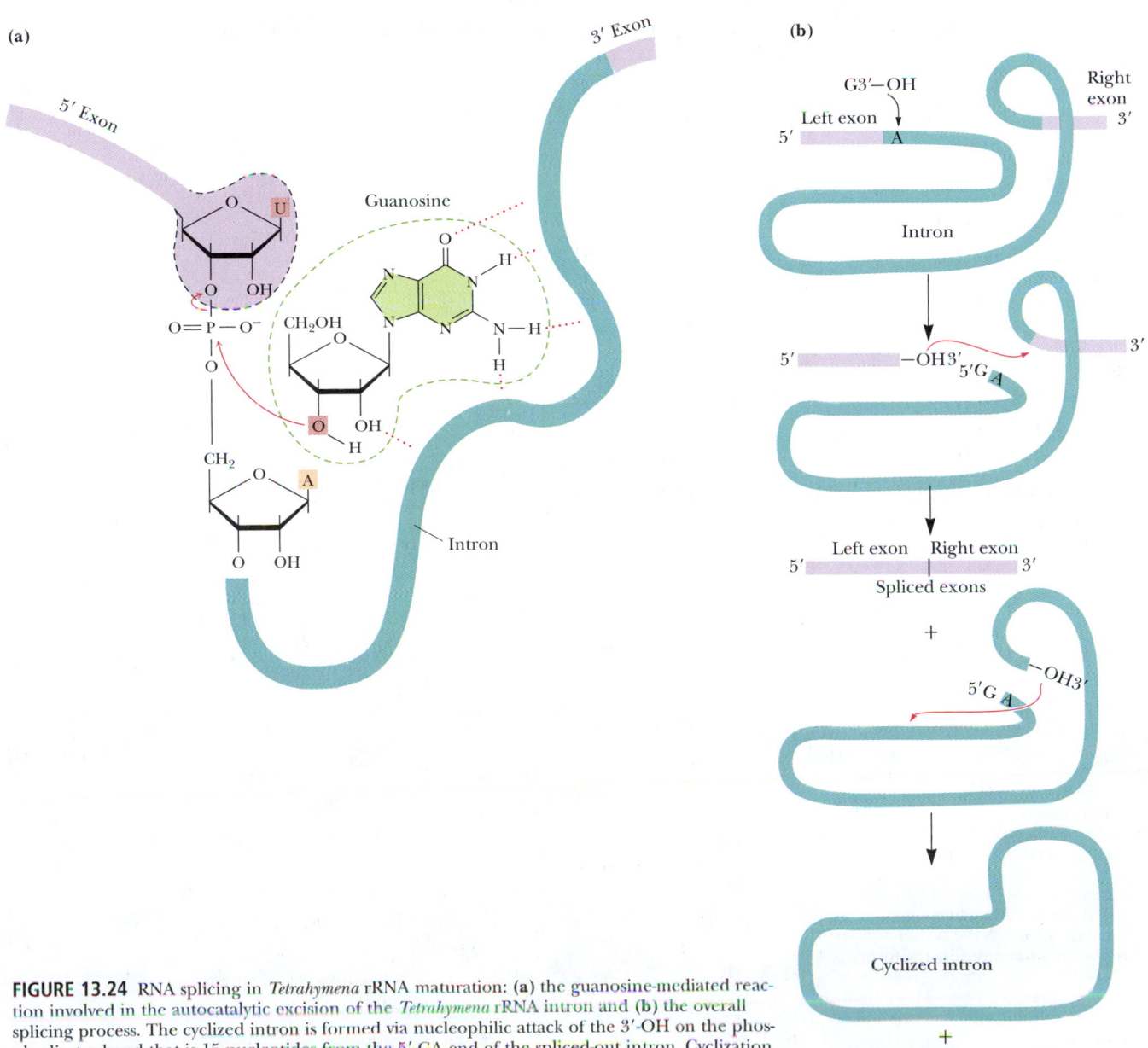

FIGURE 13.24 RNA splicing in *Tetrahymena* rRNA maturation: (a) the guanosine-mediated reaction involved in the autocatalytic excision of the *Tetrahymena* rRNA intron and (b) the overall splicing process. The cyclized intron is formed via nucleophilic attack of the 3'-OH on the phosphodiester bond that is 15 nucleotides from the 5'-GA end of the spliced-out intron. Cyclization frees a linear 15-mer with a 5'-GA end.

reaction, which is the reaction of peptide bond formation during protein synthesis (Figure 13.25), can be catalyzed by 50S ribosomal subunits (see Chapter 10) from which virtually all of the protein has been removed. These experiments imply that just the 23S rRNA by itself is capable of catalyzing peptide bond formation. This observation has been substantiated by studies with intact ribosomes that show that the adenine ring of nucleotide 2451 in the 2904-nucleotide-long 23S rRNA is a nucleophile that initiates the reaction leading to peptide bond formation (see Chapter 30). Also, the laboratory of Thomas R. Cech has created a synthetic 196-nucleotide-long ribozyme capable of performing the peptidyl transferase reaction.

Several features of these "RNA enzymes," or ribozymes, led to the realization that their biological efficiency does not challenge that achieved by proteins. First, RNA enzymes often do not fulfill the criterion of catalysis in vivo because they act only once in intramolecular events such as self-splicing. Second, the catalytic rates achieved by RNA enzymes in vivo and in vitro are significantly enhanced by the participation of protein subunits. Furthermore, nucleic acids lack the range of hydrophobic and electrostatic interactions that proteins exploit in substrate recognition and catalysis. Ribozymes rely mostly on H bonds to achieve catalysis. Nevertheless, the fact that RNA can catalyze certain reactions is experimental support for the idea that a primordial world dominated

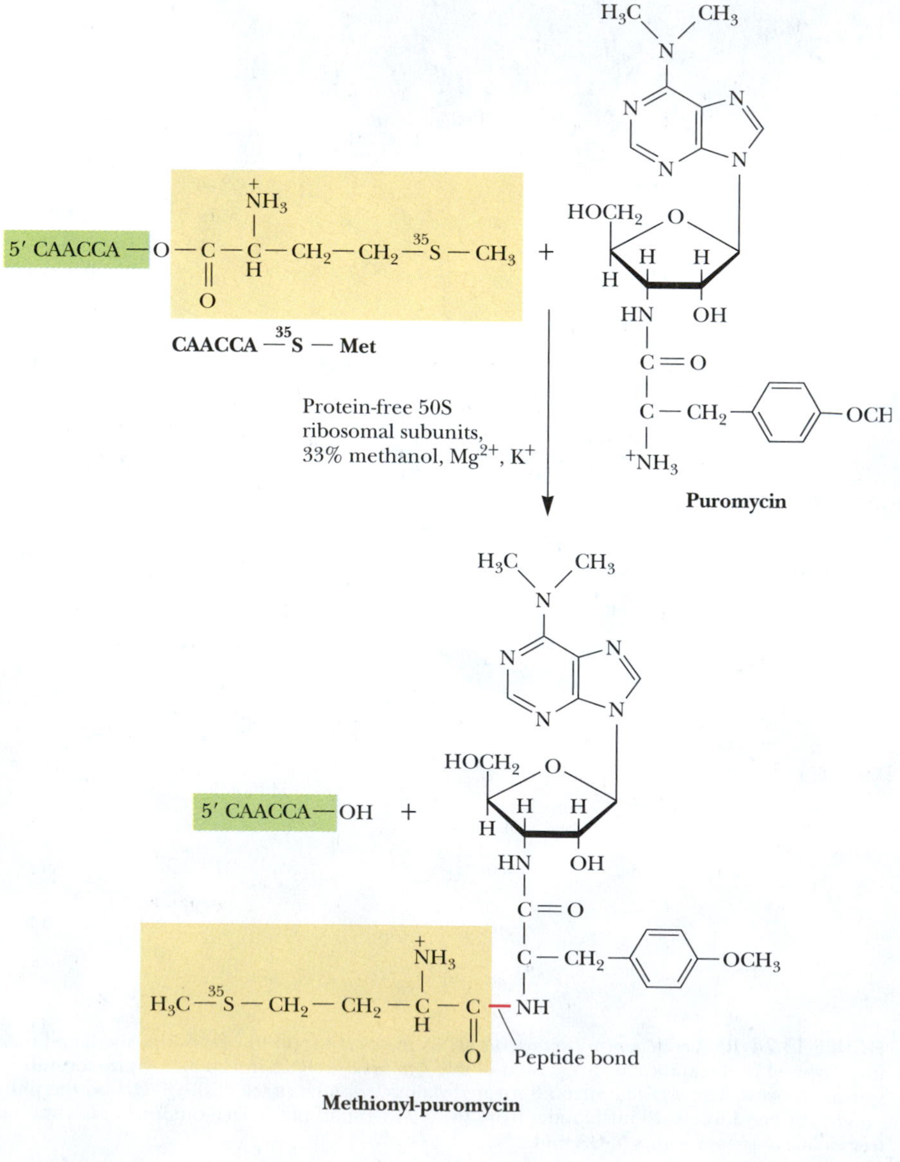

FIGURE 13.25 Protein-free 50S ribosomal subunits have peptidyl transferase activity. Peptidyl transferase is the name of the enzymatic function that catalyzes peptide bond formation. The presence of this activity in protein-free 50S ribosomal subunits was demonstrated using a model assay for peptide bond formation in which an aminoacyl-tRNA analog (a short RNA oligonucleotide of sequence CAACCA carrying ^{35}S-labeled methionine attached at its 3'-OH end) served as the peptidyl donor and puromycin (another aminoacyl-tRNA analog) served as the peptidyl acceptor. Activity was measured by monitoring the formation of ^{35}S-labeled methionyl-puromycin. *(Adapted from Noller, H. F., Hoffarth, V., and Zimniak, L., 1992. Unusual resistance of peptidyl transferase to protein-extraction procedures. Science 256:1416–1419.)*

by RNA molecules existed before the evolution of DNA and proteins. Sidney Altman and Thomas R. Cech shared the 1989 Nobel Prize in Chemistry for their discovery of the catalytic properties of RNA.

Antibody Molecules Can Have Catalytic Activity

Antibodies are *immunoglobulins,* which, of course, are proteins. Catalytic antibodies are antibodies with catalytic activity (catalytic antibodies are also called **abzymes,** a word created by combining "Ab," the abbreviation for antibody, with "enzyme.") Like other antibodies, catalytic antibodies are elicited in an organism in response to immunological challenge by a foreign molecule called an **antigen** (see Chapter 28 for discussions on the molecular basis of immunology). In this case, however, the antigen is purposefully engineered to be *an analog of the transition-state intermediate in a reaction.* The rationale is that a protein specific for binding the transition-state intermediate of a reaction will promote entry of the normal reactant into the reactive, transition-state conformation. Thus, a catalytic antibody facilitates, or catalyzes, a reaction by forcing the conformation of its substrate in the direction of its transition state. (A prominent explanation for the remarkable catalytic power of conventional enzymes is their great affinity for the transition-state intermediates in the reactions they catalyze; see Chapter 14.)

One strategy has been to prepare ester analogs by substituting a phosphorus atom for the carbon in the ester group (Figure 13.26). The phospho compound mimics the natural transition state of ester hydrolysis, and antibodies elicited against these analogs act like enzymes in accelerating the rate of ester hydrolysis as much as 1000-fold. Abzymes have been developed for a number of other classes of reactions, including C—C bond formation via aldol condensation (the reverse of the aldolase reaction [see Figure 13.2, reaction 4, and Chapter 18]) and the pyridoxal 5′-P–dependent aminotransferase reaction shown in Figure 13.23. In this latter instance, N^α-(5′-phosphopyridoxyl)-lysine (Figure 13.27a) coupled to a carrier protein served as the antigen. An antibody raised against this antigen catalyzed the conversion of D-alanine and pyridoxal 5′-P to pyruvate and pyridoxamine 5′-P (Figure 13.27b). This biotechnology offers the real possibility of creating **"designer enzymes,"** specially tailored enzymes designed to carry out specific catalytic processes.

In an interesting twist, it was recently discovered that all antibodies have an intrinsic hydrogen peroxide–generating (and under appropriate circumstances, ozone-generating) enzymatic activity that can kill bacteria, regardless of the antigen specificity of the antibody.

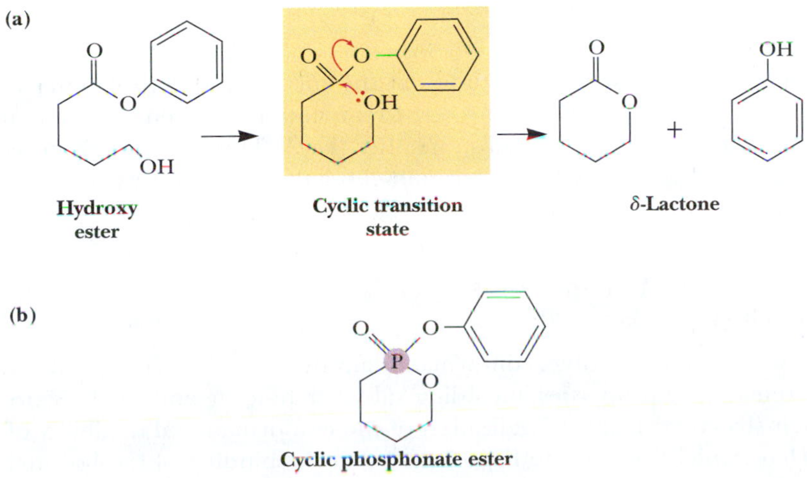

(a)

Hydroxy ester → Cyclic transition state → δ-Lactone +

(b)

Cyclic phosphonate ester

FIGURE 13.26 Catalytic antibodies are designed to specifically bind the transition-state intermediate in a chemical reaction. **(a)** The intramolecular hydrolysis of a hydroxy ester to yield as products a δ-lactone and the alcohol phenol. Note the cyclic transition state. **(b)** The cyclic phosphonate ester analog of the cyclic transition state. Antibodies raised against this phosphonate ester act as *enzymes:* They are catalysts that markedly accelerate the rate of ester hydrolysis.

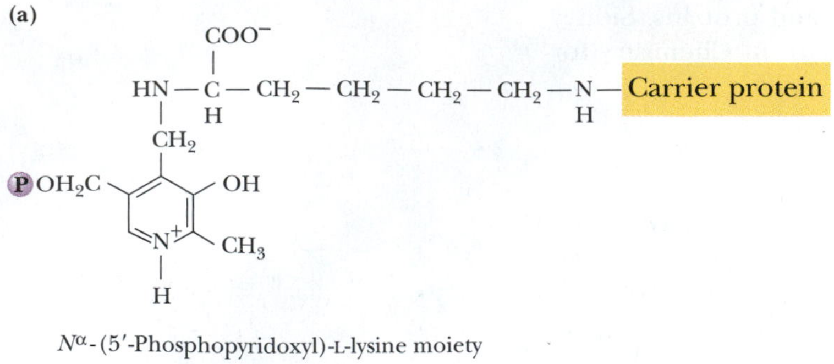

(a)

N^{α}-(5'-Phosphopyridoxyl)-L-lysine moiety

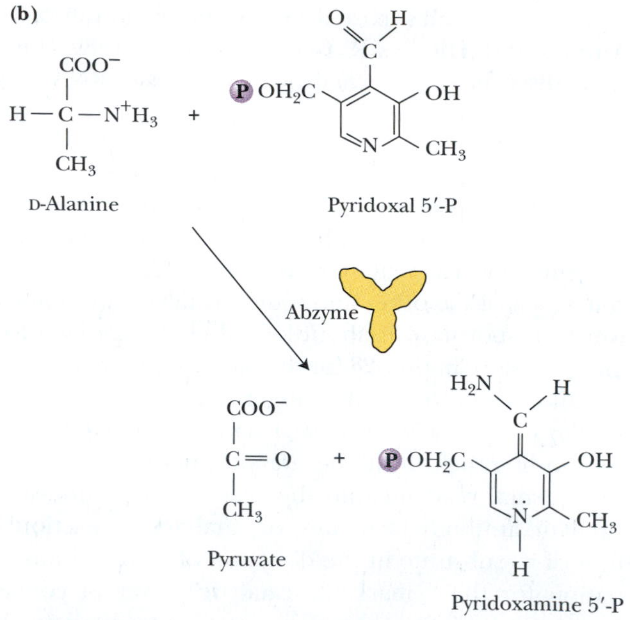

(b)

D-Alanine

Pyridoxal 5'-P

Abzyme

Pyruvate

Pyridoxamine 5'-P

FIGURE 13.27 (a) Antigen used to create an abzyme with aminotransferase activity. (b) Aminotransferase reaction catalyzed by the abzyme.

13.7 | How Can Enzymes Be So Specific?

The extraordinary ability of an enzyme to catalyze only one particular reaction is a quality known as **specificity.** Specificity means an enzyme acts only on a specific substance, its substrate, invariably transforming it into a specific product. That is, an enzyme binds only certain compounds, and then, only a specific reaction ensues. Some enzymes show absolute specificity, catalyzing the transformation of only one specific substrate to yield a unique product. Other enzymes carry out a particular reaction but act on a class of compounds. For example, *hexokinase* (ATP:hexose-6-phosphotransferase) will carry out the ATP-dependent phosphorylation of a number of hexoses at the 6-position, including glucose. Specificity studies on enzymes entail an examination of the rates of the enzymatic reaction obtained with various **structural analogs** of the substrate. By determining which functional and structural groups within the substrate affect binding or catalysis, enzymologists can map the properties of the active site, analyzing questions such as: Can the active site accommodate sterically bulky groups? Are ionic interactions between E and S important? Are H bonds formed?

The "Lock and Key" Hypothesis Was the First Explanation for Specificity

Pioneering enzyme specificity studies at the turn of the 20th century by the great organic chemist Emil Fischer led to the notion of an enzyme resembling a **"lock"** and its particular substrate the **"key."** This analogy captures the essence of the specificity that exists between an enzyme and its substrate, but enzymes are not rigid templates like locks.

The "Induced Fit" Hypothesis Provides a More Accurate Description of Specificity

Enzymes are highly flexible, conformationally dynamic molecules, and many of their remarkable properties, including substrate binding and catalysis, are due to their structural pliancy. Realization of the conformational flexibility of proteins led Daniel Koshland to hypothesize that the binding of a substrate by an enzyme is an interactive process. That is, the shape of the enzyme's active site is actually modified upon binding S, in a process of dynamic recognition between enzyme and substrate aptly called **induced fit.** In essence, substrate binding alters the conformation of the protein, so that the protein and the substrate "fit" each other more precisely. The process is truly interactive in that the conformation of the substrate also changes as it adapts to the conformation of the enzyme.

This idea also helps explain some of the mystery surrounding the enormous catalytic power of enzymes: In enzyme catalysis, precise orientation of catalytic residues comprising the active site is necessary for the reaction to occur; substrate binding induces this precise orientation by the changes it causes in the protein's conformation.

"Induced Fit" Favors Formation of the Transition-State Intermediate

The catalytically active enzyme:substrate complex is an interactive structure in which the enzyme causes the substrate to adopt a form that mimics the transition-state intermediate of the reaction. Thus, a poor substrate would be one that was less effective in directing the formation of an optimally active enzyme:transition-state intermediate conformation. This active conformation of the enzyme molecule is thought to be relatively unstable in the absence of substrate, and free enzyme thus reverts to a conformationally different state.

Specificity and Reactivity

Consider, for example, why hexokinase catalyzes the ATP-dependent phosphorylation of hexoses but not smaller phosphoryl-group acceptors such as glycerol, ethanol, or even water. Surely these smaller compounds are not sterically forbidden from approaching the active site of hexokinase (Figure 13.28). Indeed, water should penetrate the active site easily and serve as a highly effective phosphoryl-group acceptor. Accordingly, hexokinase should display high ATPase activity. It does not. Only the binding of hexoses induces hexokinase to assume its fully active conformation.

In Chapter 14, we explore in greater detail the factors that contribute to the remarkable catalytic power of enzymes and examine specific examples of enzyme reaction mechanisms.

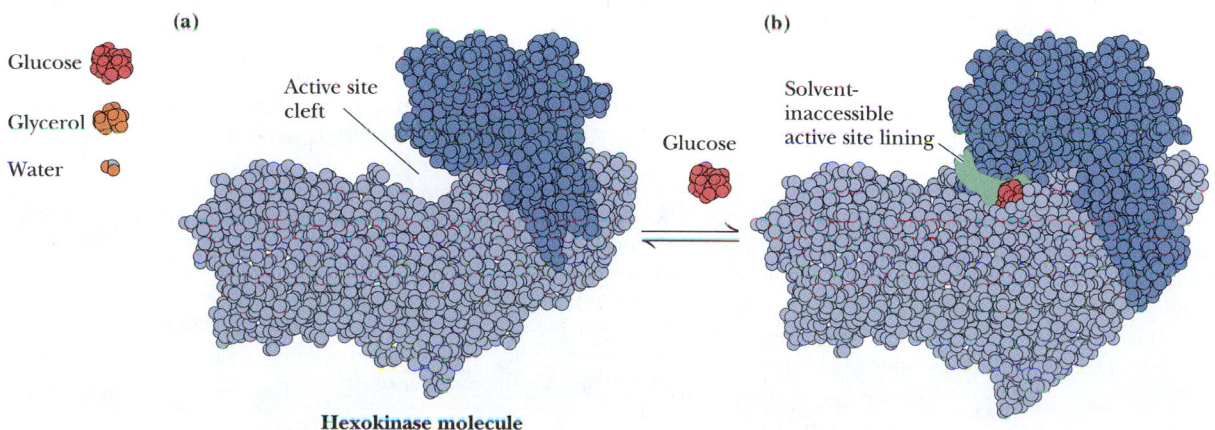

FIGURE 13.28 A drawing, roughly to scale, of H_2O, glycerol, glucose, and an idealized hexokinase molecule. Note the two domains of structure in hexokinase **(a)**, between which the active site is located. Binding of glucose induces a conformational change in hexokinase. The two domains close together, creating the catalytic site **(b).** The shaded area in **(b)** represents solvent-inaccessible surface area in the active site cleft that results when the enzyme binds substrate.

Summary

Living systems use enzymes to accelerate and control the rates of vitally important biochemical reactions. Enzymes provide kinetic control over thermodynamic potentiality: Reactions occur in a timeframe suitable to the metabolic requirements of cells. Enzymes are the agents of metabolic function.

13.1 What Characteristic Features Define Enzymes? Enzymes can be characterized in terms of three prominent features: catalytic power, specificity, and regulation. The site on the enzyme where substrate binds and catalysis occurs is called the active site. Regulation of enzyme activity is essential to the integration and regulation of metabolism.

13.2 Can the Rate of an Enzyme-Catalyzed Reaction Be Defined in a Mathematical Way? Enzyme kinetics seeks to determine the maximum reaction velocity that the enzyme can attain, its binding affinities for substrates and inhibitors, and the mechanism by which it accomplishes its catalysis. The kinetics of simple chemical reactions provides a foundation for exploring enzyme kinetics. Enzymes, like other catalysts, act by lowering the free energy of activation for a reaction.

13.3 What Equations Define the Kinetics of Enzyme-Catalyzed Reactions? A plot of the velocity of an enzyme-catalyzed reaction v versus the concentration of the substrate S is called a substrate saturation curve. The Michaelis–Menten equation is derived by assuming that E combines with S to form ES and then ES reacts to give E + P. Rapid, reversible combination of E and S and ES breakdown to yield P reach a steady-state condition where [ES] is essentially constant. The Michaelis–Menten equation says that the initial rate of an enzyme reaction, v, is determined by two constants, K_m and V_{max}, and the initial concentration of substrate. The turnover number of an enzyme, k_{cat}, is a measure of its maximal catalytic activity (the number of substrate molecules converted into product per enzyme molecule per unit time when the enzyme is saturated with substrate). However, the ratio k_{cat}/K_m defines the catalytic efficiency of an enzyme. This ratio, k_{cat}/K_m, cannot exceed the diffusion-controlled rate of combination of E and S to form ES.

Several rearrangements of the Michaelis–Menten equation transform it into a straight-line equation, a better form for experimental determination of the constants K_m and V_{max} and for detection of regulatory properties of enzymes.

13.4 What Can Be Learned from the Inhibition of Enzyme Activity? Inhibition studies on enzymes have contributed significantly to our understanding of enzymes. Inhibitors may interact either reversibly or irreversibly with an enzyme. Reversible inhibitors bind to the enzyme through noncovalent association/dissociation reactions. Irreversible inhibitors typically form stable, covalent bonds with the enzyme. Reversible inhibitors may bind at the active site of the enzyme (competitive inhibition) or at some other site on the enzyme (noncompetitive inhibition). Uncompetitive inhibitors bind only to the ES complex.

13.5 What Is the Kinetic Behavior of Enzymes Catalyzing Bimolecular Reactions? Usually, enzymes catalyze reactions in which two (or even more) substrates take part, so the reaction is bimolecular. Several possibilities arise. In single-displacement reactions, both substrates, A and B, are bound before reaction occurs. In double-displacement (or ping-pong) reactions, one substrate (A) is bound and reaction occurs to yield product P and a modified enzyme form, E′. The second substrate (B) then binds to E′ and reaction occurs to yield product Q and E, the unmodified form of enzyme. Graphical methods can be used to distinguish these possibilities. Exchange reactions are another way to diagnose bisubstrate mechanisms.

13.6 Are All Enzymes Proteins? Not all enzymes are proteins. Catalytic RNA molecules ("ribozymes") play important cellular roles in RNA processing and protein synthesis, among other things. Catalytic RNAs give support to the idea that a primordial world dominated by RNA molecules existed before the evolution of DNA and proteins.

Antibodies that have catalytic activity ("abzymes") can be elicited in an organism in response to immunological challenge with an analog of the transition-state intermediate for a reaction. Such antibodies are catalytic because they bind the transition-state intermediate of a reaction and promote entry of the normal substrate into the reactive, transition-state conformation.

13.7 How Can Enzymes Be So Specific? Early enzyme specificity studies by Emil Fischer led to the hypothesis that an enzyme resembles a "lock" and its particular substrate the "key." However, enzymes are not rigid templates like locks. Koshland noted that the conformation of an enzyme is dynamic and hypothesized that the interaction of E with S is also dynamic. The enzyme's active site is actually modified upon binding S, in a process of dynamic recognition between enzyme and substrate called induced fit. Hexokinase provides a good illustration of the relationship between substrate binding, induced fit, and catalysis.

Problems

1. According to the Michaelis–Menten equation, what is the v/V_{max} ratio when [S] = 4 K_m?

2. If V_{max} = 100 μmol/mL sec and K_m = 2 mM, what is the velocity of the reaction when [S] = 20 mM?

3. For a Michaelis–Menten reaction, $k_1 = 7 \times 10^7$/M · sec, $k_{-1} = 1 \times 10^3$/sec, and $k_2 = 2 \times 10^4$/sec. What are the values of K_S and K_m? Does substrate binding approach equilibrium, or does it behave more like a steady-state system?

4. The following kinetic data were obtained for an enzyme in the absence of any inhibitor (1), and in the presence of two different inhibitors (2) and (3) at 5 mM concentration. Assume [E$_T$] is the same in each experiment.

[S] (mM)	(1) v (μmol/ mL sec)	(2) v (μmol/ mL sec)	(3) v (μmol/ mL sec)
1	12	4.3	5.5
2	20	8	9
3	29	14	13
8	35	21	16
12	40	26	18

  a. Determine V_{max} and K_m for the enzyme.
  b. Determine the type of inhibition and the K_I for each inhibitor.

5. Using Figure 13.7 as a model, draw curves that would be obtained in v versus [S] plots when
  a. twice as much enzyme is used.
  b. half as much enzyme is used.
  c. a competitive inhibitor is added.
  d. a pure noncompetitive inhibitor is added.
  e. an uncompetitive inhibitor is added.
  For each example, indicate how V_{max} and K_m change.

6. The general rate equation for an ordered, single-displacement reaction where A is the leading substrate is

$$v = \frac{V_{max}[A][B]}{(K_S^A K_m^B + K_m^A[B] + K_m^B[A] + [A][B])}$$

Write the Lineweaver–Burk (double-reciprocal) equivalent of this equation and from it calculate algebraic expressions for the following:
  a. The slope
  b. The y-intercepts

c. The horizontal and vertical coordinates of the point of intersection when $1/v$ is plotted versus $1/[B]$ at various *fixed* concentrations of A

7. The following graphical patterns obtained from kinetic experiments have several possible interpretations depending on the nature of the experiment and the variables being plotted. Give at least two possibilities for each.

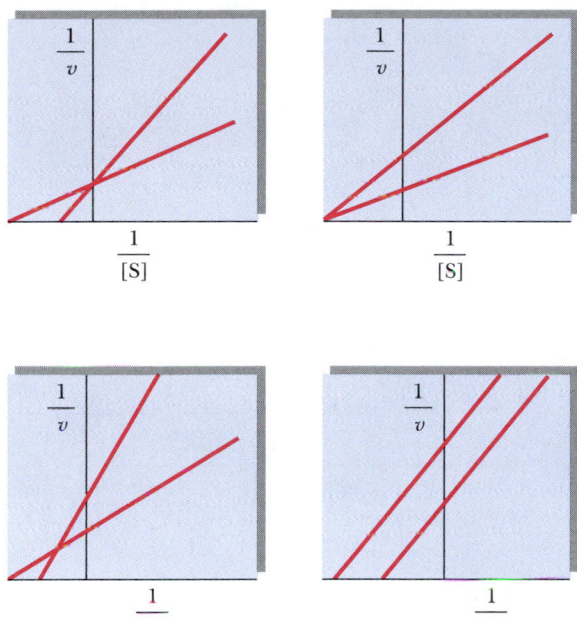

8. Liver alcohol dehydrogenase (ADH) is relatively nonspecific and will oxidize ethanol or other alcohols, including methanol. Methanol oxidation yields formaldehyde, which is quite toxic, causing, among other things, blindness. Mistaking it for the cheap wine he usually prefers, my dog Clancy ingested about 50 mL of windshield washer fluid (a solution 50% in methanol). Knowing that methanol would be excreted eventually by Clancy's kidneys if its oxidation could be blocked, and realizing that, in terms of methanol oxidation by ADH, ethanol would act as a competitive inhibitor, I decided to offer Clancy some wine. How much of Clancy's favorite vintage (12% ethanol) must he consume in order to lower the activity of his ADH on methanol to 5% of its normal value if the K_m values of canine ADH for ethanol and methanol are 1 millimolar and 10 millimolar, respectively? (The K_I for ethanol in its role as competitive inhibitor of methanol oxidation by ADH is the same as its K_m.) Both the methanol and ethanol will quickly distribute throughout Clancy's body fluids, which amount to about 15 L. Assume the densities of 50% methanol and the wine are both 0.9 g/mL.

9. Measurement of the rate constants for a simple enzymatic reaction obeying Michaelis–Menten kinetics gave the following results:
$k_1 = 2 \times 10^8 \ M^{-1} \ sec^{-1}$
$k_{-1} = 1 \times 10^3 \ sec^{-1}$
$k_2 = 5 \times 10^3 \ sec^{-1}$
a. What is K_s, the dissociation constant for the enzyme–substrate complex?
b. What is K_m, the Michaelis constant for this enzyme?
c. What is k_{cat} (the turnover number) for this enzyme?
d. What is the catalytic efficiency (k_{cat}/K_m) for this enzyme?
e. Does this enzyme approach "kinetic perfection"? (That is, does k_{cat}/K_m approach the diffusion-controlled rate of enzyme association with substrate?)
f. If a kinetic measurement was made using 2 nanomoles of enzyme per mL and saturating amounts of substrate, what would V_{max} equal?
g. Again, using 2 nanomoles of enzyme per mL of reaction mixture, what concentration of substrate would give $v = 0.75 \ V_{max}$?

h. If a kinetic measurement was made using 4 nanomoles of enzyme per mL and saturating amounts of substrate, what would V_{max} equal? What would K_m equal under these conditions?

10. Triose phosphate isomerase catalyzes the conversion of glyceraldehyde-3-phosphate to dihydroxyacetone phosphate.

$$\text{Glyceraldehyde-3-P} \rightleftharpoons \text{dihydroxyacetone-P}$$

The K_m of this enzyme for its substrate glyceraldehyde-3-phosphate is $1.8 \times 10^{-5} \ M$. When [glyceraldehydes-3-phosphate] = 30 μM, the rate of the reaction, v, was 82.5 $\mu mol \ mL^{-1} \ sec^{-1}$.
a. What is V_{max} for this enzyme?
b. Assuming 3 nanomoles per mL of enzyme was used in this experiment ([E_{total}] = 3 nanomol/mL), what is k_{cat} for this enzyme?
c. What is the catalytic efficiency (k_{cat}/K_m) for triose phosphate isomerase?
d. Does the value of k_{cat}/K_m reveal whether triose phosphate isomerase approaches "catalytic perfection"?
e. What determines the ultimate speed limit of an enzyme-catalyzed reaction? That is, what is it that imposes the physical limit on kinetic perfection?

11. The citric acid cycle enzyme *fumarase* catalyzes the conversion of fumarate to form malate.

$$\text{Fumarate} + H_2O \rightleftharpoons \text{malate}$$

The turnover number, k_{cat}, for fumarase is 800/sec. The K_m of fumarase for its substrate fumarate is 5 μM.
a. In an experiment using 2 nanomole/L of fumarase, what is V_{max}?
b. The cellular concentration of fumarate is 47.5 μM. What is v when [fumarate] = 47.5 μM?
c. What is the catalytic efficiency of fumarase?
d. Does fumarase approach "catalytic perfection"?

12. Carbonic anhydrase catalyzes the hydration of CO_2:

$$CO_2 + H_2O \rightleftharpoons H_2CO_3$$

The K_m of carbonic anhydrase for CO_2 is 12 mM. Carbonic anhydrase gave an initial velocity $v_o = 4.5 \ \mu mol \ H_2CO_3 \ formed/mL \cdot sec$ when [CO_2] = 36 mM.
a. What is V_{max} for this enzyme?
b. Assuming 5 pmol/mL (5×10^{-12} moles/mL) of enzyme were used in this experiment, what is k_{cat} for this enzyme?
c. What is the catalytic efficiency of this enzyme?
d. Does carbonic anhydrase approach "catalytic perfection"?

13. Acetylcholinesterase catalyzes the hydrolysis of the neurotransmitter acetylcholine:

$$\text{Acetylcholine} + H_2O \longrightarrow \text{acetate} + \text{choline}$$

The K_m of acetylcholinesterase for its substrate acetylcholine is $9 \times 10^{-5} \ M$. In a reaction mixture containing 5 nanomoles/mL of acetylcholinesterase and 150 μM acetylcholine, a velocity $v_o = 40 \ \mu mol/mL \cdot sec$ was observed for the acetylcholinesterase reaction.
a. Calculate V_{max} for this amount of enzyme.
b. Calculate k_{cat} for acetylcholinesterase.
c. Calculate the catalytic efficiency (k_{cat}/K_m) for acetylcholinesterase.
d. Does acetylcholinesterase approach "catalytic perfection"?

14. The enzyme catalase catalyzes the decomposition of hydrogen peroxide:

$$2 \ H_2O_2 \rightleftharpoons 2 \ H_2O + O_2$$

The turnover number (k_{cat}) for catalase is 40,000,000 sec^{-1}. The K_m of catalase for its substrate H_2O_2 is 0.11 M.
a. In an experiment using 3 nanomole/L of catalase, what is V_{max}?
b. What is v when [H_2O_2] = 0.75 M?
c. What is the catalytic efficiency of fumarase?
d. Does catalase approach "catalytic perfection"?

15. Equation 13.9 presents the simple Michaelis–Menten situation where the reaction is considered to be irreversible ([P] is negligible). Many enzymatic reactions are reversible, and P does accumulate.
    a. Derive an equation for v, the rate of the enzyme-catalyzed reaction S→P in terms of a modified Michaelis–Menten model that incorporates the reverse reaction that will occur in the presence of product, P.
    b. Solve this modified Michaelis–Menten equation for the special situation when $v = 0$ (that is, S⇌P is at equilibrium, or in other words, $K_{eq} = [P]/[S]$).

    (J. B. S. Haldane first described this reversible Michaelis–Menten modification, and his expression for K_{eq} in terms of the modified M–M equation is known as the Haldane relationship.)

Preparing for the MCAT Exam

16. Enzyme A follows simple Michaelis–Menten kinetics.
    a. The K_m of enzyme A for its substrate S is $K_m^S = 1$ mM. Enzyme A also acts on substrate T and its $K_m^T = 10$ mM. Is S or T the preferred substrate for enzyme A?

    b. The rate constant k_2 with substrate S is 2×10^4 sec^{-1}; with substrate T, $k_2 = 4 \times 10^5$ sec^{-1}. Does enzyme A use substrate S or substrate T with greater catalytic efficiency?

17. Use Figure 13.12 to answer the following questions.
    a. Is the enzyme whose temperature versus activity profile is shown in Figure 13.12 likely to be from an animal or a plant? Why?
    b. What do you think the temperature versus activity profile for an enzyme from a thermophilic bacterium growing in a 80°F pool of water would resemble?

Biochemistry ⑤ Now™   Preparing for an exam? Test yourself on key questions at http://chemistry.brookscole.com/ggb3

Further Reading

Enzymes in General

Bell, J. E., and Bell, E. T., 1988. *Proteins and Enzymes.* Englewood Cliffs, NJ: Prentice Hall. This text describes the structural and functional characteristics of proteins and enzymes.

Creighton, T. E., 1997. *Protein Structure: A Practical Approach* and *Protein Function: A Practical Approach.* Oxford: Oxford University Press.

Fersht, A., 1999. *Structure and Mechanism in Protein Science.* New York: Freeman & Co. A guide to protein structure, chemical catalysis, enzyme kinetics, enzyme regulation, protein engineering, and protein folding.

Catalytic Power

Miller, B. G., and Wolfenden, R., 2002. Catalytic proficiency: The unusual case of OMP decarboxylase. *Annual Review of Biochemistry* **71:**847–885.

General Reviews of Enzyme Kinetics

Cleland, W. W., 1990. Steady-state kinetics. In *The Enzymes,* 3rd ed. Sigman, D. S., and Boyer, P. D., eds. Volume XIX, pp. 99–158. See also, *The Enzymes,* 3rd ed. Boyer, P. D., ed., Volume II, pp. 1–65, 1970.

Cornish-Bowden, A., 1994. *Fundamentals of Enzyme Kinetics.* Cambridge: Cambridge University Press.

Smith, W. G., 1992. In vivo kinetics and the reversible Michaelis–Menten model. *Journal of Chemical Education* **12:**981–984.

Graphical and Statistical Analysis of Kinetic Data

Cleland, W. W., 1979. Statistical analysis of enzyme kinetic data. *Methods in Enzymology* **82:**103–138.

Naqui, A., 1986. Where are the asymptotes of Michaelis–Menten? *Trends in Biochemical Sciences* **11:**64–65.

Rudolph, F. B., and Fromm, H. J., 1979. Plotting methods for analyzing enzyme rate data. *Methods in Enzymology* **63:**138–159. A review of the various rearrangements of the Michaelis–Menten equation that yield straight-line plots.

Segel, I. H., 1976. *Biochemical Calculations,* 2nd ed. New York: John Wiley & Sons. An excellent guide to solving problems in enzyme kinetics.

Effect of Active Site Amino Acid Substitutions on k_{cat}/K_m

Garrett, R. M., et al., 1998. Human sulfite oxidase R160Q: Identification of the mutation in a sulfite oxidase-deficient patient and expression and characterization of the mutant enzyme. *Proceedings of the National Academy of Sciences U.S.A.* **95:**6394–6398.

Garrett, R. M., and Rajagopalan, K. V., 1996. Site-directed mutagenesis of recombinant sulfite oxidase. *Journal of Biological Chemistry* **271:**7387–7391.

Enzymes and Rational Drug Design

Cornish-Bowden, A., and Eisenthal, R., 1998. Prospects for antiparasitic drugs: The case of *Trypanosoma brucei,* the causative agent of African sleeping sickness. *Journal of Biological Chemistry* **273:**5500–5505. An analysis of why drug design strategies have had only limited success.

Kling, J., 1998. From hypertension to angina to Viagra. *Modern Drug Discovery* **1:**31–38. The story of the serendipitous discovery of Viagra in a search for agents to treat angina and high blood pressure.

Enzyme Inhibition

Cleland, W. W., 1979. Substrate inhibition. *Methods in Enzymology* **63:**500–513.

Pollack, S. J., et al., 1994. Mechanism of inositol monophosphatase, the putative target of lithium therapy. *Proceedings of the National Academy of Sciences U.S.A.* **91:**5766–5770.

Silverman, R. B., 1988. *Mechanism-Based Enzyme Inactivation: Chemistry and Enzymology,* Vols. I and II. Boca Raton, FL: CRC Press.

Catalytic RNA

Altman, S., 2000. The road to RNase P. *Nature Structural Biology* **7:**827–828.

Cech, T. R., and Bass, B. L., 1986. Biological catalysis by RNA. *Annual Review of Biochemistry* **55:**599–629. A review of the early evidence that RNA can act like an enzyme.

Doherty, E. A., and Doudna, J. A., 2000. Ribozyme structures and mechanisms. *Annual Review of Biochemistry* **69:**597–615.

Frank, D. N., Pace, N. R., 1998. Ribonuclease P: Unity and diversity in a tRNA processing ribozyme. *Annual Review of Biochemistry* **67:**153–180.

Narlikar, G. J., and Herschlag, D., 1997. Mechanistic aspects of enzymatic catalysis: Comparison of RNA and protein enzymes. *Annual Review of Biochemistry* **66:**19–59. A comparison of RNA and protein enzymes that addresses fundamental principles in catalysis and macromolecular structure.

Nissen, P., et al., 2000. The structural basis of ribosome activity in peptide bond synthesis. *Science* **289:**920–930. Peptide bond formation by the ribosome: the ribosome is a ribozyme.

Schimmel, P., and Kelley, S. O., 2000. Exiting an RNA world. *Nature Structural Biology* **7**:5–7. Review of the in vitro creation of an RNA capable of catalyzing the formation of an aminoacyl-tRNA. Such a ribozyme would be necessary to bridge the evolutionary gap between a primordial RNA world and the contemporary world of proteins.

Watson, J. D., ed., 1987. Evolution of catalytic function. *Cold Spring Harbor Symposium on Quantitative Biology* **52**:1–955. Publications from a symposium on the nature and evolution of catalytic biomolecules (proteins and RNA) prompted by the discovery that RNA could act catalytically.

Wilson, D. S., and Szostak, J. W., 1999. In vitro selection of functional nucleic acids. *Annual Review of Biochemistry* **68**:611–647. Screening libraries of random nucleotide sequences for catalytic RNAs.

Catalytic Antibodies

Hilvert, D., 2000. Critical analysis of antibody catalysis. *Annual Review of Biochemistry* **69**:751–793. A review of catalytic antibodies that were elicited with rationally designed transition-state analogs.

Janda, K. D., 1997. Chemical selection for catalysis in combinatorial antibody libraries. *Science* **275**:945.

Wagner, J., Lerner, R. A., and Barbas, C. F., III, 1995. Efficient adolase catalytic antibodies that use the enamine mechanism of natural enzymes. *Science* **270**:1797–1800.

Wentworth, P., Jr., et al., 2002. Evidence for antibody-catalyzed ozone formation in bacterial killing and inflammation. *Science* **298**:2195–2199.

Specificity

Jencks, W. P., 1975. Binding energy, specificity, and enzymic catalysis: the Circe effect. *Advances in Enzymology* **43**:219–410. Enzyme specificity stems from the favorable binding energy between the active site and the substrate and unfavorable binding or exclusion of nonsubstrate molecules.

CHAPTER 14

Mechanisms of Enzyme Action

Like the workings of machines, the details of enzyme mechanisms are at once complex and simple.

No single thing abides but all things flow.
Fragment to fragment clings and thus they
grow
Until we know them by name.
Then by degrees they change and are no
more the things we know.
Lucretius (CA. 94 B.C.–50 B.C.)

Key Questions

Biochemistry ⊛ Now™ Test yourself on these Key Questions at BiochemistryNow at http://chemistry.brookscole.com/ggb3

Essential Question

Although the catalytic properties of enzymes may seem almost magical, it is simply chemistry—the breaking and making of bonds—that gives enzymes their prowess. This chapter will explore the unique features of this chemistry. The mechanisms of hundreds of enzymes have been studied in at least some detail. In this chapter, it will be possible to examine only a few of these. *What are the universal chemical principles that influence the mechanisms of these and other enzymes, and how may we understand the many other cases, in light of the knowledge gained from these examples?*

14.1 What Role Does Transition-State Stabilization Play in Enzyme Catalysis?

In all chemical reactions, the reacting atoms or molecules pass through a state that is intermediate in structure between the reactant(s) and the product(s). Consider the transfer of a proton from a water molecule to a chloride anion:

$$H-O-H + Cl^- \rightleftharpoons H-O^{\delta-}\cdots H\cdots Cl^{\delta-} \rightleftharpoons HO^- + H-Cl$$

Reactants   **Transition state**   **Products**

In the middle structure, the proton undergoing transfer is shared equally by the hydroxyl and chloride anions. This structure represents, as nearly as possible, the transition between the reactants and products, and it is known as the **transition state.**[1]

Chemical reactions in which a substrate (S) is converted to a product (P) can be pictured as involving a transition state (which we henceforth denote as $X^{\ddagger}$), a species intermediate in structure between S and P (Figure 14.1). As seen in Chapter 13, the catalytic role of an enzyme is to reduce the energy barrier between substrate and transition state. This is accomplished through the formation of an **enzyme–substrate complex** (ES). This complex is converted to product by passing through a transition state, $EX^{\ddagger}$ (Figure 14.1). As shown, the energy of $EX^{\ddagger}$ is clearly lower than that of $X^{\ddagger}$. One might be tempted to conclude that this decrease in energy explains the rate enhancement achieved by the enzyme, but there is more to the story.

The energy barrier for the uncatalyzed reaction (Figure 14.1) is of course the difference in energies of the S and $X^{\ddagger}$ states. Similarly, the energy barrier to be surmounted in the enzyme-catalyzed reaction, assuming that E is saturated with S, is the energy difference between ES and $EX^{\ddagger}$. *Reaction rate acceleration by an enzyme means simply that the energy barrier between ES and $EX^{\ddagger}$ is less than the energy barrier between S and $X^{\ddagger}$.* In terms of the free energies of activation, $\Delta G_e^{\ddagger} < \Delta G_u^{\ddagger}$.

There are important consequences for this statement. The enzyme must stabilize the transition-state complex, $EX^{\ddagger}$, more than it stabilizes the substrate complex, ES. Put another way, enzymes are "designed" by nature to bind the transition-state structure more tightly than the substrate (or the product). The dissociation constant for the enzyme–substrate complex is

$$K_S = \frac{[E][S]}{[ES]} \tag{14.1}$$

[1]It is important to distinguish **transition states** from **intermediates.** A transition state is envisioned as an extreme distortion of a bond, and thus the lifetime of a typical transition state is viewed as being on the order of the lifetime of a bond vibration, typically 10^{-13} sec. Intermediates, on the other hand, are longer lived, with lifetimes in the range of 10^{-13} to 10^{-3} sec.

David W. Grisham

(a)

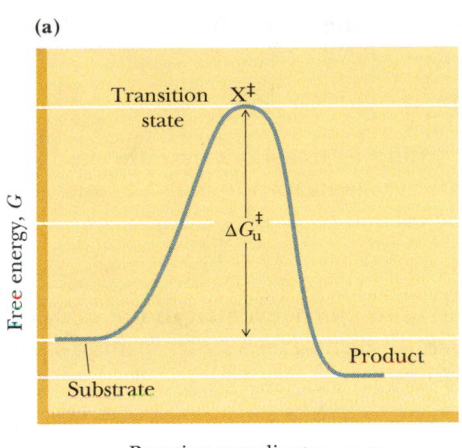

(b)

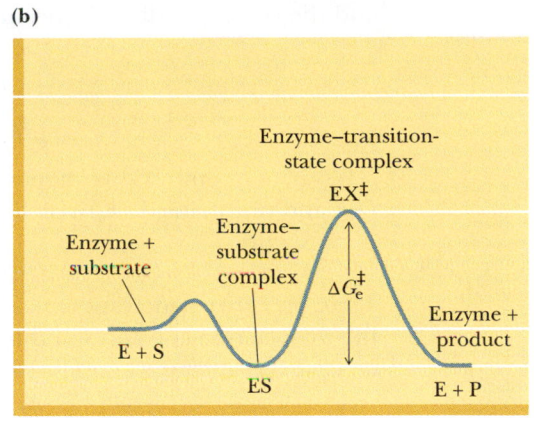

FIGURE 14.1 Enzymes catalyze reactions by lowering the activation energy. Here the free energy of activation for **(a)** the uncatalyzed reaction, $\Delta G_u^{\ddagger}$, is larger than that for **(b)** the enzyme-catalyzed reaction, $\Delta G_e^{\ddagger}$.

A Deeper Look

What Is the Rate Enhancement of an Enzyme?

Enigmas abound in the world of enzyme catalysis. One surrounds the discussion of how the rate enhancement by an enzyme can be best expressed. Notice that the uncatalyzed conversion of a substrate S to a product P is usually a simple first-order process, described by a first-order rate constant k_u:

$$v_u = k_u[S]$$

On the other hand, for an enzyme that obeys Michaelis–Menten kinetics, the reaction is viewed as being first-order in S at low S and zero-order in S at high S. (See Chapter 13, where this distinction is discussed.)

$$v_e = \frac{k_{cat}[E_T][S]}{K_m + [S]}$$

If the "rate enhancement" effected by the enzyme is defined as

$$\text{rate enhancement} = v_e/v_u$$

then we can write:

$$\text{rate enhancement} = \frac{k_{cat}}{k_u}\left(\frac{[E_T]}{K_m + [S]}\right)$$

Depending on the relative sizes of K_m and [S], there are two possible results:

Case 1: When [S] is large compared to K_m, the enzyme is saturated with S and the kinetics are zero-order in S.

$$\text{rate enhancement} = \frac{k_{cat}}{k_u}\left(\frac{[E_T]}{[S]}\right)$$

where $[E_T]/[S]$ is the fraction of the total S that is in the ES complex. Note here that defining the rate enhancement in terms of k_{cat}/k_u is equivalent to comparing the quantities $\Delta G_e^{\ddagger}$ and $\Delta G_u^{\ddagger}$ in the figure to the right.

Case 2: When [S] is small compared to K_m, not all the enzyme molecules have S bound, and the kinetics are first-order in S.

$$\text{rate enhancement} = \frac{k_{cat}}{k_u}\left(\frac{[E_T]}{K_m}\right)$$

Here, defining the rate enhancement in terms of $\frac{k_{cat}}{k_u K_m}$ is equivalent to comparing the quantities $\Delta G_e^{\ddagger}$, and $\Delta G_u^{\ddagger}$ in the accompanying figure. Moreover, to the extent that K_m is approximated by K_S (see Equation 14.1), this rate enhancement can be rewritten as

$$\text{rate enhancement} = \frac{[E_T]}{K_T}$$

where K_T is the dissociation constant for the EX‡ complex (see Equation 14.2).

Viewed in this way, the best definition of "rate enhancement" depends upon the relationship between enzyme and substrate concentrations and the enzyme's kinetic parameters.

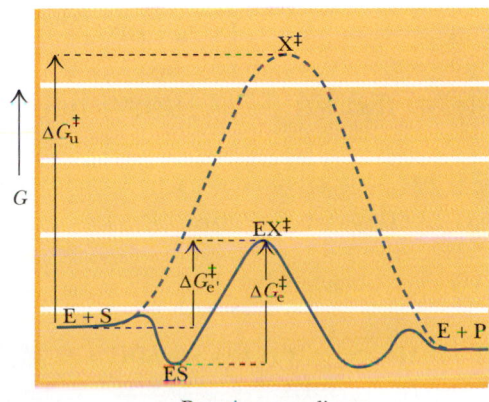

and the corresponding dissociation constant for the transition-state complex is

$$K_T = \frac{[E][X^\ddagger]}{[EX^\ddagger]} \qquad (14.2)$$

Enzyme catalysis requires that $K_T < K_S$. According to **transition-state theory** (see references at end of this chapter), the rate constants for the enzyme-catalyzed (k_e) and uncatalyzed (k_u) reactions can be related to K_S and K_T by:

$$k_e/k_u > K_S/K_T \qquad (14.3)$$

Thus, the enzymatic rate enhancement is approximately equal to the ratio of the dissociation constants of the enzyme–substrate and enzyme–transition-state complexes, at least when E is saturated with S.

14.2 What Are the Magnitudes of Enzyme-Induced Rate Accelerations?

Enzymes are powerful catalysts. Enzyme-catalyzed reactions are typically 10^7 to 10^{14} times faster than their uncatalyzed counterparts (Table 14.1). (There is even a report of a rate acceleration of $>10^{16}$ for the alkaline phosphatase–catalyzed hydrolysis of methylphosphate!)

These large rate accelerations correspond to substantial changes in the free energy of activation for the reaction in question. The urease reaction, for example,

$$H_2N-\overset{\overset{\displaystyle O}{\|}}{C}-NH_2 + 2\,H_2O + H^+ \longrightarrow 2\,NH_4^+ + HCO_3^-$$

shows an energy of activation some 84 kJ/mol smaller than that for the corresponding uncatalyzed reaction. To fully understand any enzyme reaction, it is important to account for the rate acceleration in terms of the structure of the enzyme and its mechanism of action. There are a limited number of catalytic mechanisms or factors that contribute to the remarkable performance of enzymes. These include the following:

1. Entropy loss in ES formation
2. Destabilization of ES due to strain, desolvation, or electrostatic effects

Table 14.1

A Comparison of Enzyme-Catalyzed Reactions and Their Uncatalyzed Counterparts

Reaction	Enzyme	Uncatalyzed Rate, v_u (sec^{-1})	Catalyzed Rate, v_e (sec^{-1})	v_e/v_u
$CH_3-O-PO_3^{2-} + H_2O \longrightarrow CH_3OH + HPO_4^{2-}$	Alkaline phosphatase	1×10^{-15}	14	1.4×10^{16}
$H_2N-\overset{\overset{O}{\|}}{C}-NH_2 + 2\,H_2O + H^+ \longrightarrow 2\,NH_4^+ + HCO_3^-$	Urease	3×10^{-10}	3×10^4	1×10^{14}
$R-\overset{\overset{O}{\|}}{C}-O-CH_2CH_3 + H_2O \longrightarrow RCOOH + HOCH_2CH_3$	Chymotrypsin	1×10^{-10}	1×10^2	1×10^{12}
Glycogen + P_i $\longrightarrow$ Glycogen + Glucose-1-P (n) $\qquad\qquad (n-1)$	Glycogen phosphorylase	$<5 \times 10^{-15}$	1.6×10^{-3}	$>3.2 \times 10^{11}$
Glucose + ATP $\longrightarrow$ Glucose-6-P + ADP	Hexokinase	$<1 \times 10^{-13}$	1.3×10^{-3}	$>1.3 \times 10^{10}$
$CH_3CH_2OH + NAD^+ \longrightarrow CH_3\overset{\overset{O}{\|}}{C}H + NADH + H^+$	Alcohol dehydrogenase	$<6 \times 10^{-12}$	2.7×10^{-5}	$>4.5 \times 10^6$
$CO_2 + H_2O \longrightarrow HCO_3^- + H^+$	Carbonic anhydrase	10^{-2}	10^5	1×10^7
Creatine + ATP $\longrightarrow$ Cr-P + ADP	Creatine kinase	$<3 \times 10^{-9}$	4×10^{-5}	$>1.33 \times 10^4$

Adapted from Koshland, D., 1956. Molecular geometry in enzyme action. *Journal of Cellular Comparative Physiology*, Supp. 1, **47**:217.

3. Covalent catalysis
4. General acid or base catalysis
5. Metal ion catalysis
6. Proximity and orientation

Any or all of these mechanisms may contribute to the net rate acceleration of an enzyme-catalyzed reaction relative to the uncatalyzed reaction. A thorough understanding of any enzyme would require that the net acceleration be accounted for in terms of contributions from one or (usually) more of these mechanisms. Each of these will be discussed in detail in this chapter, but first it is important to appreciate how the formation of the enzyme–substrate complex makes all these mechanisms possible.

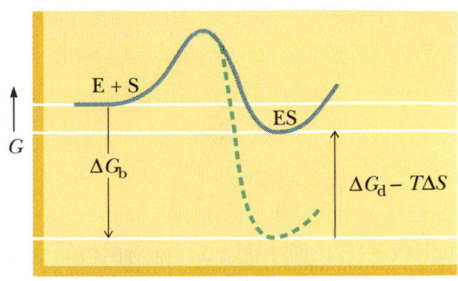

Reaction coordinate

FIGURE 14.2 The intrinsic binding energy of the enzyme–substrate (ES) complex (ΔG_b) is compensated to some extent by entropy loss due to the binding of E and S ($T\Delta S$) and by destabilization of ES (ΔG_d) by strain, distortion, desolvation, and similar effects. If ΔG_b were not compensated by $T\Delta S$ and ΔG_d, the formation of ES would follow the dashed line.

| 14.3 | **Why Is the Binding Energy of ES Crucial to Catalysis?** |

How is it that $X^{\ddagger}$ is stabilized more than S at the enzyme active site? To understand this, we must dissect and analyze the formation of the enzyme–substrate complex, ES. There are a number of important contributions to the free energy difference between the uncomplexed enzyme and substrate (E + S) and the ES complex (Figure 14.2). The favorable interactions between the substrate and amino acid residues on the enzyme account for the **intrinsic binding energy, ΔG_b.** The intrinsic binding energy ensures the favorable formation of the ES complex, but if uncompensated, it makes the activation energy for the enzyme-catalyzed reaction unnecessarily large and wastes some of the catalytic power of the enzyme.

Compare the two cases in Figure 14.3. Because the enzymatic reaction rate is determined by the difference in energies between ES and EX‡, the smaller this difference, the faster the enzyme-catalyzed reaction. Tight binding of the substrate deepens the energy well of the ES complex and actually lowers the rate of the reaction.

| 14.4 | **What Roles Do Entropy Loss and Destabilization of the ES Complex Play?** |

The message of Figure 14.3 is that raising the energy of ES will increase the enzyme-catalyzed reaction rate. This is accomplished in two ways: (1) **loss of entropy** due to the binding of S to E and (2) **destabilization of ES** by strain, distortion, desolvation, or other similar effects. The entropy loss arises from the

(a)

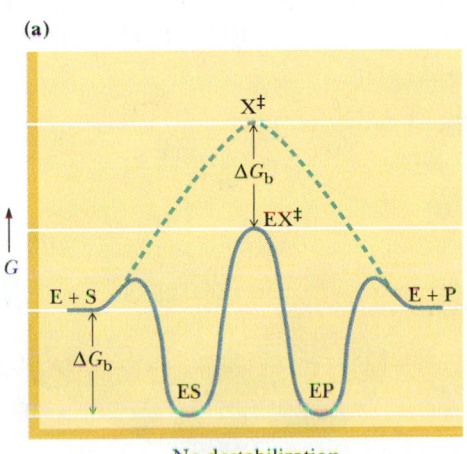

No destabilization, thus no catalysis

(b)

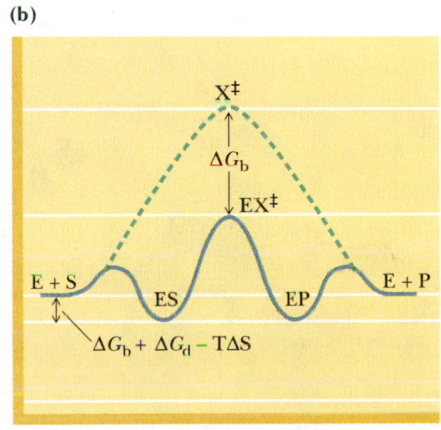

Destabilization of ES facilitates catalysis

FIGURE 14.3 **(a)** Catalysis does not occur if the ES complex and the transition state for the reaction are stabilized to equal extents. **(b)** Catalysis *will* occur if the transition state is stabilized to a greater extent than the ES complex *(right)*. Entropy loss and destabilization of the ES complex ΔG_d ensure that this will be the case.

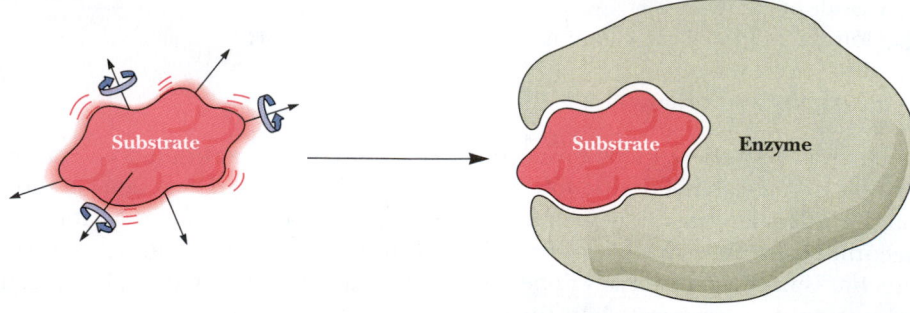

Formation of the ES complex results in a loss of
entropy. Prior to binding, E and S are free to undergo
translational and rotational motion. By comparison,
the ES complex is a more highly ordered, low-entropy
complex. **Test yourself on the concepts in this figure
at http://chemistry.brookscole.com/ggb3**

Substrate (and enzyme) are free
to undergo translational motion.
A disordered, high-entropy situation

The highly ordered, low-entropy complex

fact that the ES complex (Figure 14.4) is a highly organized (low-entropy) entity
compared to E + S in solution (a disordered, high-entropy situation). The entry
of the substrate into the active site brings all the reacting groups and coordinating
residues of the enzyme together with the substrate in just the proper position for
reaction, with a net loss of entropy. The substrate and enzyme both possess **translational entropy,** the freedom to move in three dimensions, as well as **rotational
entropy,** the freedom to rotate or tumble about any axis through the molecule.
Both types of entropy are lost to some extent when two molecules (E and S) interact to form one molecule (the ES complex). Because ΔS is negative for this
process, the term $-T\Delta S$ is a positive quantity, and *the intrinsic binding energy of ES
is compensated to some extent by the entropy loss that attends the formation of the complex.*

Destabilization of the ES complex can involve **structural strain, desolvation,**
or **electrostatic effects.** Destabilization by strain or distortion is usually just a
consequence of the fact (noted previously) that *the enzyme is designed to bind the
transition state more strongly than the substrate.* When the substrate binds, the imperfect nature of the "fit" results in distortion or strain in the substrate, the
enzyme, or both. This means that the amino acid residues that make up the
active site are oriented to coordinate the transition-state structure precisely but
will interact with the substrate or product less effectively.

Destabilization may also involve desolvation of charged groups on the substrate upon binding in the active site. Charged groups are highly stabilized in
water. For example, the transfer of Na^+ and Cl^- from the gas phase to aqueous
solution is characterized by an **enthalpy of solvation, ΔH_{solv},** of -775 kJ/mol.
(Energy is given off and the ions become more stable.) When a substrate with
charged groups moves from water into an enzyme active site (Figure 14.5), the
charged groups are often desolvated to some extent, becoming less stable and
therefore more reactive.

When a substrate enters the active site, charged groups may be forced to
interact (unfavorably) with charges of like sign, resulting in **electrostatic**

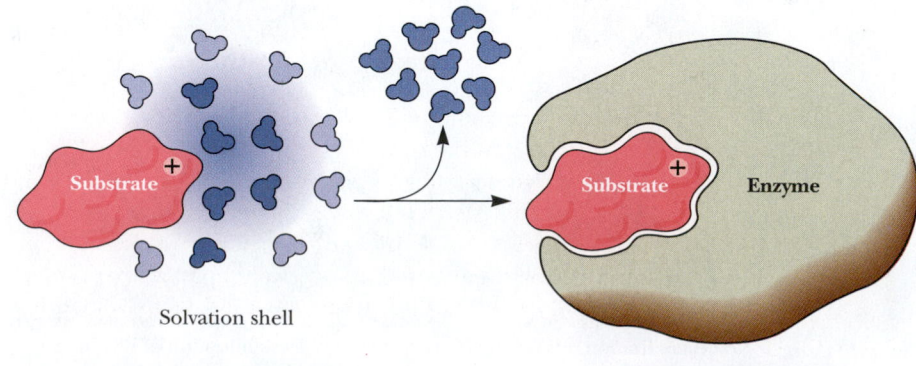

Substrates typically lose waters of hydration in the
formation of the ES complex. Desolvation raises the
energy of the ES complex, making it more reactive.
**Test yourself on the concepts in this figure at
http://chemistry.brookscole.com/ggb3**

Solvation shell

Desolvated ES complex

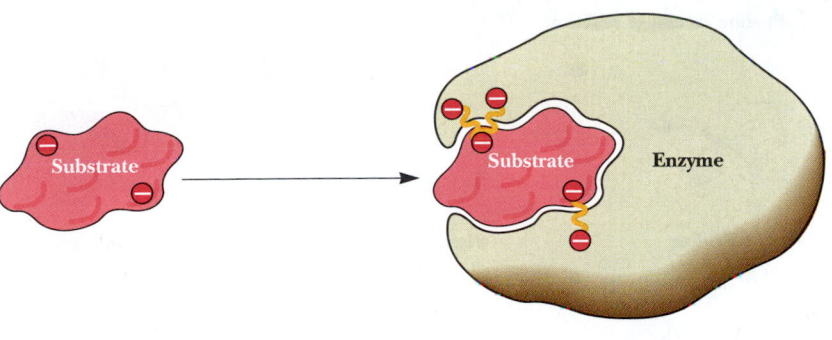

Electrostatic destabilization
in ES complex

destabilization (Figure 14.6). The reaction pathway acts in part to remove this stress. If the charge on the substrate is diminished or lost in the course of reaction, electrostatic destabilization can result in rate acceleration.

Whether by strain, desolvation, or electrostatic effects, destabilization raises the energy of the ES complex, and this increase is summed in the term ΔG_d, the free energy of destabilization. As noted in Figure 14.2, the net energy difference between E + S and the ES complex is the sum of the intrinsic binding energy, ΔG_b; the entropy loss on binding, $-T\Delta S$; and the distortion energy, ΔG_d. ES is destabilized (raised in energy) by the amount $\Delta G_d - T\Delta S$. The transition state is subject to no such destabilization, and the difference between the energies of $X^\ddagger$ and $EX^\ddagger$ is essentially ΔG_b, the full intrinsic binding energy.

14.5 How Tightly Do Transition-State Analogs Bind to the Active Site?

Although not apparent at first, there are other important implications of Equation 14.3. It is important to consider the magnitudes of K_S and K_T. The ratio k_e/k_u may even exceed 10^{16}, as noted previously. Given a typical ratio of 10^{12} and a typical K_S of 10^{-3} M, the value of K_T should be 10^{-15} M! This is the dissociation constant for the transition-state complex from the enzyme, and this very low value corresponds to very tight binding of the transition state by the enzyme.

It is unlikely that such tight binding in an enzyme transition state will ever be measured experimentally, however, because the transition state itself is a "moving target." It exists only for about 10^{-14} to 10^{-13} sec, less than the time required for a bond vibration. The nature of the elusive transition state can be explored, on the other hand, using **transition-state analogs,** stable molecules that are chemically and structurally similar to the transition state. Such molecules should bind more strongly than a substrate and more strongly than competitive inhibitors that bear no significant similarity to the transition state. Hundreds of examples of such behavior have been reported. For example, Robert Abeles studied a series of inhibitors of **proline racemase** (Figure 14.7) and found that *pyrrole-2-carboxylate* bound to the enzyme 160 times more tightly than L-proline, the normal substrate. This analog binds so tightly because it is planar and is similar in structure to the planar transition state for the racemization of proline. Two other examples of transition-state analogs are shown in Figure 14.8. *Phosphoglycolohydroxamate* binds 40,000 times more tightly to yeast aldolase than the substrate dihydroxyacetone phosphate. Even more remarkable, the 1,6-hydrate of *purine ribonucleoside* has been estimated to bind to adenosine deaminase with a K_I of 3×10^{-13} M!

It should be noted that transition-state analogs are only approximations of the transition state itself and will never bind as tightly as would be expected for the true transition state. These analogs are, after all, stable molecules and cannot be expected to resemble a true transition state too closely.

Proline racemase reaction

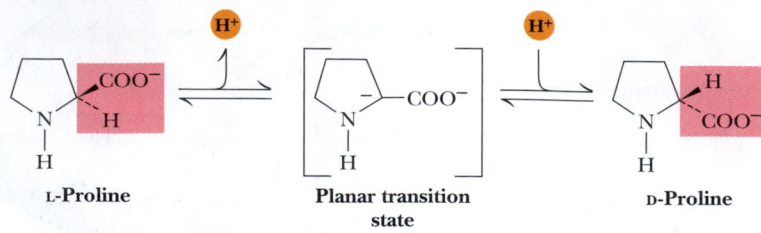

FIGURE 14.7 The proline racemase reaction. Pyrrole-2-carboxylate and Δ-1-pyrroline-2-carboxylate mimic the planar transition state of the reaction.

(a) Yeast aldolase reaction

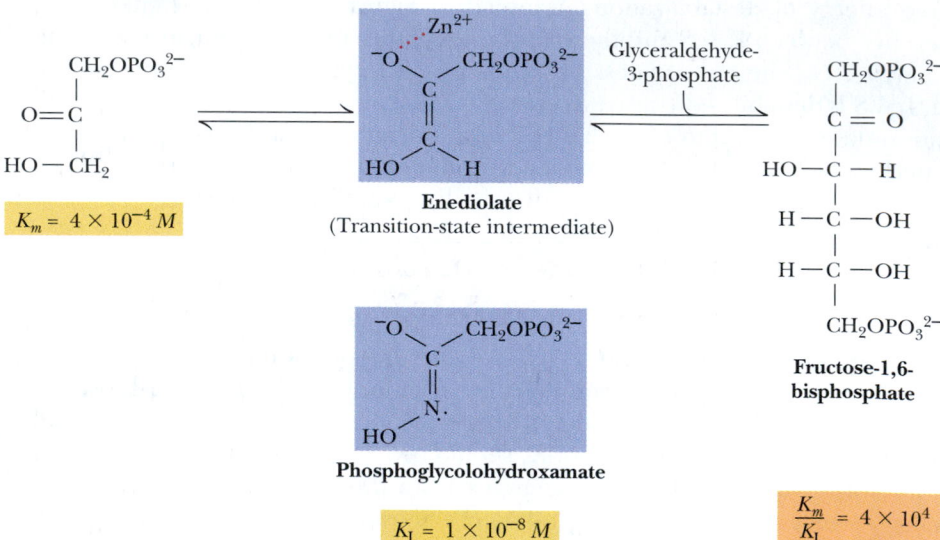

(b) Calf intestinal adenosine deaminase reaction

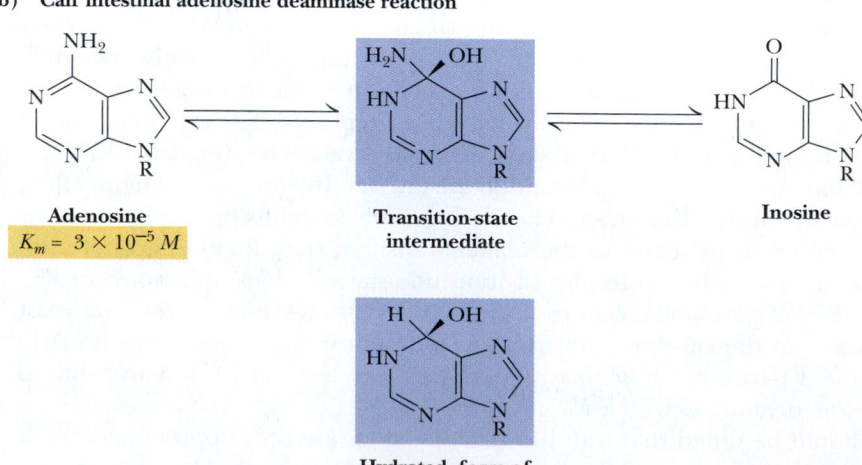

FIGURE 14.8 **(a)** Phosphoglycolohydroxamate is an analog of the enediolate transition state of the yeast aldolase reaction. **(b)** Purine riboside, a potent inhibitor of the calf intestinal adenosine deaminase reaction, binds to adenosine deaminase as the 1,6-hydrate. The hydrated form of purine riboside is an analog of the proposed transition state for the reaction.

14.6 | What Are the Mechanisms of Catalysis?

Covalent Catalysis

Some enzyme reactions derive much of their rate acceleration from the formation of **covalent bonds** between enzyme and substrate. Consider the reaction:

$$BX + Y \longrightarrow BY + X$$

and an enzymatic version of this reaction involving formation of a **covalent intermediate:**

$$BX + Enz \longrightarrow E:B + X + Y \longrightarrow Enz + BY$$

If the enzyme-catalyzed reaction is to be faster than the uncatalyzed case, the acceptor group on the enzyme must be a better attacking group than Y and a better leaving group than X. Note that most enzymes that carry out covalent catalysis have ping-pong kinetic mechanisms.

The side chains of amino acids in proteins offer a variety of **nucleophilic** centers for catalysis, including amines, carboxylates, aryl and alkyl hydroxyls, imidazoles, and thiol groups. These groups readily attack electrophilic centers of substrates, forming covalently bonded enzyme–substrate intermediates. Typical electrophilic centers in substrates include phosphoryl groups, acyl groups, and glycosyl groups (Figure 14.9). The covalent intermediates thus formed can be attacked in a subsequent step by a water molecule or a second substrate, giving the desired product. **Covalent electrophilic catalysis** is also observed, but it usually involves coenzyme adducts that generate electrophilic centers. Well over 100 enzymes are now known to form covalent intermediates during catalysis. Table 14.2 lists some typical examples, including that of glyceraldehyde-3-phosphate dehydrogenase, which catalyzes the reaction:

Glyceraldehyde-3-P + NAD$^+$ + P$_i \longrightarrow$

1,3-bisphosphoglycerate + NADH + H$^+$

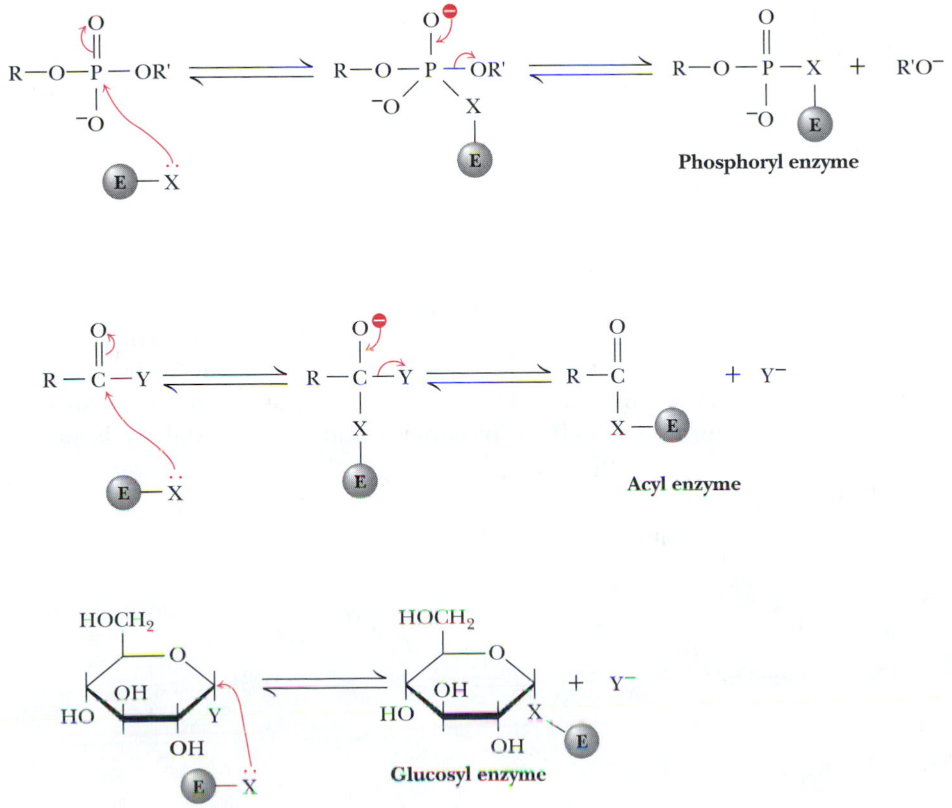

FIGURE 14.9 Examples of covalent bond formation between enzyme and substrate. In each case, a nucleophilic center (X:) on an enzyme attacks an electrophilic center on a substrate.

Table 14.2

Enzymes That Form Covalent Intermediates

Enzymes	Reacting Group	Covalent Intermediate
1. Chymotrypsin Elastase Esterases Subtilisin Thrombin Trypsin	>CH–CH_2–OH (Ser)	>CH–CH_2–O–C(=O)–R (Acyl-Ser)
2. Glyceraldehyde-3-phosphate dehydrogenase Papain	>CH–CH_2–SH (Cys)	>CH–CH_2–S–C(=O)–R (Acyl-Cys)
3. Alkaline phosphatase Phosphoglucomutase	>CH–CH_2–OH (Ser)	>CH–CH_2–O–PO_3^{2-} (Phosphoserine)
4. Phosphoglycerate mutase Succinyl-CoA synthetase	–CH_2–(imidazole HN, N) (His)	–CH_2–(imidazole, N–P(=O)(O^-)–O^-) (Phosphohistidine)
5. Aldolase Decarboxylases Pyridoxal phosphate–dependent enzymes	R–NH_3^+ (Amino)	R–N=C< (Schiff base)

As shown in Figure 14.10, this reaction mechanism involves nucleophilic attack by —SH on the substrate glyceraldehyde-3-P to form a covalent acylcysteine (or hemithioacetal) intermediate. Hydride transfer to NAD^+ generates a thioester intermediate. Nucleophilic attack by phosphate yields the desired mixed carboxylic–phosphoric anhydride product, 1,3-bisphosphoglycerate. Several examples of covalent catalysis will be discussed in detail in later chapters.

General Acid–Base Catalysis

Nearly all enzyme reactions involve some degree of acid or base catalysis. There are two types of acid–base catalysis: (1) **specific acid–base catalysis,** in which H^+ or OH^- accelerates the reaction, and (2) **general acid–base catalysis,** in which an acid or base other than H^+ or OH^- accelerates the reaction. For ordinary solution reactions, these two cases can be distinguished on the basis of simple experiments. As shown in Figure 14.11, in specific acid or base catalysis, the buffer concentration has no effect. In general acid or base catalysis, however,

FIGURE 14.10 Formation of a covalent intermediate in the glyceraldehyde-3-phosphate dehydrogenase reaction. Nucleophilic attack by a cysteine —SH group forms a covalent acylcysteine intermediate. Following hydride transfer to NAD^+, nucleophilic attack by phosphate yields the product, 1,3-bisphosphoglycerate.

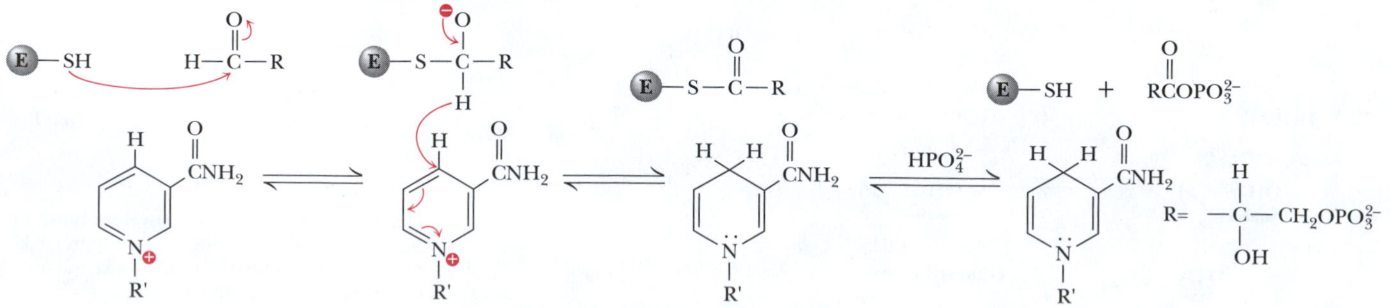

the buffer may donate or accept a proton in the transition state and thus affect the rate. *By definition, general acid–base catalysis is catalysis in which a proton is transferred in the transition state.* Consider the hydrolysis of *p*-nitrophenylacetate with imidazole acting as a general base (Figure 14.12). Proton transfer apparently stabilizes the transition state here. The water has been made more nucleophilic without generation of a high concentration of OH^- or without the formation of unstable, high-energy species. General acid or general base catalysis may increase reaction rates 10- to 100-fold. In an enzyme, ionizable groups on the protein provide the H^+ transferred in the transition state. Clearly, an ionizable group will be most effective as a H^+ transferring agent at or near its pK_a. Because the pK_a of the histidine side chain is near 7, histidine is often the most effective general acid or base. Descriptions of several cases of general acid–base catalysis in typical enzymes follow.

Low-Barrier Hydrogen Bonds

As previously noted, the typical strength of a hydrogen bond is 10 to 30 kJ/mol. For an O—H—O hydrogen bond, the O—O separation is typically 0.28 nm and the interaction is a relatively weak electrostatic interaction. The hydrogen is firmly linked to one of the oxygens at a distance of approximately 0.1 nm, and the distance to the other oxygen is thus about 0.18 nm, which corresponds to a bond order of about 0.07. Not all hydrogen bonds are weak, however. As the distance between heteroatoms becomes smaller, the overall bond becomes stronger, the hydrogen becomes centered, and the bond order approaches 0.5 for both O—H interactions (Figure 14.13). These interactions are more nearly covalent in nature, and the stabilization energy is much higher. Notably, the barrier that the hydrogen atom must surmount to exchange oxygens becomes lower as the O—O separation decreases (Figure 14.13). When the barrier to hydrogen exchange has dropped to the point that it is at or below the zero-point energy level of hydrogen, the interaction is referred to as a **low-barrier hydrogen bond (LBHB)**. The hydrogen is now free to move anywhere between the two oxygens (or, more generally, two heteroatoms). The stabilization energy of LBHBs may approach 100 kJ/mol in the gas phase and 60 kJ/mol or more in solution. LBHBs require matched pK_as for the two electronegative atoms that share the hydrogen. As the two pK_a values diverge, the stabilization energy of the LBHB is decreased. Widely divergent pK_a values thus correspond to ordinary, weak hydrogen bonds.

How may low-barrier hydrogen bonds affect enzyme catalysis? A weak hydrogen bond in an enzyme ground state may become an LBHB in a transient intermediate, or even in the transition state for the reaction. In such a case, the energy released in forming the LBHB is used to help the reaction which forms it, lowering the activation barrier for the reaction. Alternatively, the purpose of the LBHB may be to redistribute electron density in the reactive intermediate, achieving rate acceleration by facilitation of "hydrogen tunneling." Enzyme

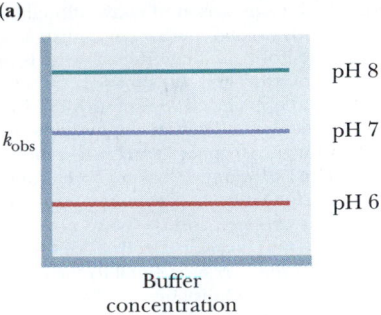
(a)

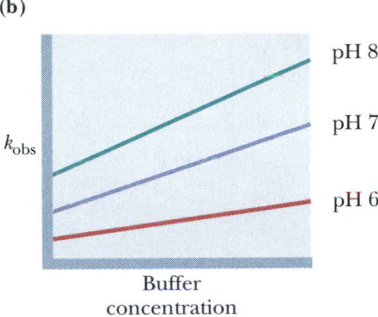
(b)

FIGURE 14.11 Specific and general acid–base catalysis of simple reactions in solution may be distinguished by determining the dependence of observed reaction rate constants (k_{obs}) on pH and buffer concentration. **(a)** In specific acid–base catalysis, H^+ or OH^- concentration affects the reaction rate, k_{obs} is pH-dependent, but buffers (which accept or donate H^+/OH^-) have no effect. **(b)** In general acid–base catalysis, in which an ionizable buffer may donate or accept a proton in the transition state, k_{obs} is dependent on buffer concentration.

Bond order refers to the number of electron pairs in a bond. (For a single bond, the bond order = 1.)

FIGURE 14.12 Catalysis of *p*-nitrophenylacetate hydrolysis by imidazole—an example of general base catalysis. Proton transfer to imidazole in the transition state facilitates hydroxyl attack on the substrate carbonyl carbon.

Reaction

$$CH_3\overset{\displaystyle O}{\overset{\|}{C}}-O-\bigcirc-NO_2 \; + \; H_2O \; \rightleftharpoons \; CH_3\overset{\displaystyle O}{\overset{\|}{C}}-O^- \; + \; HO-\bigcirc-NO_2 \; + \; H^+$$

Mechanism

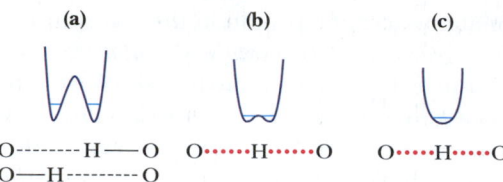

FIGURE 14.13 Comparison of conventional (weak) hydrogen bonds **(a)** and low-barrier hydrogen bonds **(b and c)**. The horizontal line in each case is the zero-point energy of hydrogen. **(a)** shows an O—H—O hydrogen bond of length 0.28 nm, with the hydrogen attached to one or the other of the oxygens. The bond order for the stronger O—H interaction is approximately 1.0, and the weaker O—H interaction is 0.07. As the O-O distance decreases, the hydrogen bond becomes stronger, and the bond order of the weakest interaction increases. In **(b)**, the O-O distance is 0.25 nm, and the barrier is equal to the zero-point energy. In **(c)**, the O-O distance is 0.23 to 0.24 nm, and the bond order of each O—H interaction is 0.5.

mechanisms that will be examined later in this chapter (the serine proteases and aspartic proteases) appear to depend upon one or the other of these effects.

Metal Ion Catalysis

Many enzymes require metal ions for maximal activity. If the enzyme binds the metal very tightly or requires the metal ion to maintain its stable, native state, it is referred to as a **metalloenzyme.** Enzymes that bind metal ions more weakly, perhaps only during the catalytic cycle, are referred to as **metal activated.** One role for metals in metal-activated enzymes and metalloenzymes is to act as electrophilic catalysts, stabilizing the increased electron density or negative charge that can develop during reactions. Among the enzymes that function in this manner (Figure 14.14) is liver alcohol dehydrogenase. Another potential function of metal ions is to provide a powerful nucleophile at neutral pH. Coordination to a metal ion can increase the acidity of a nucleophile with an ionizable proton:

$$M^{2+} + NucH \rightleftharpoons M^{2+} (NucH) \rightleftharpoons M^{2+} (Nuc^-) + H^+$$

The reactivity of the coordinated, deprotonated nucleophile is typically intermediate between that of the un-ionized and ionized forms of free nucleophile. Carboxypeptidase (see Chapter 5) contains an active site Zn^{2+}, which facilitates deprotonation of a water molecule in this manner.

Proximity

Chemical reactions go faster when the reactants are in proximity, that is, near each other. In solution or in the gas phase, this means that increasing the concentrations of reacting molecules, which raises the number of collisions, causes higher rates of reaction. Enzymes, which have specific binding sites for particular reacting molecules, essentially take the reactants out of dilute solution and hold them close to each other. This proximity of reactants is said to raise the "effective" concentration over that of the substrates in solution and leads to an increased reaction rate. In order to measure proximity effects in enzyme reactions, enzymologists have turned to model studies comparing intermolecular reaction rates with corresponding or similar intramolecular reaction rates. A typical case is the imidazole-catalyzed hydrolysis of *p*-nitrophenylacetate (Figure 14.15a). Under certain conditions the rate constant for this bimolecular reaction is 35 M^{-1} min^{-1}. By comparison, the first-order rate constant for the analogous but intramolecular reaction shown in Figure 14.15b is 839 min^{-1}. The ratio of these two rate constants

$$(839 \ min^{-1})/(35 \ M^{-1} \ min^{-1}) = 23.97 \ M$$

has the units of concentration and can be thought of as an effective concentration of imidazole in the intramolecular reaction. Put another way, a concentration of imidazole of 23.9 M would be required in the intermolecular reaction to make it proceed as fast as the intramolecular reaction.

There is more to this story, however. Enzymes not only bring substrates and catalytic groups close together, they orient them in a manner suitable for catalysis as well. Comparison of the rates of reaction of the molecules shown in Figure 14.16 makes it clear that the bulky methyl groups force an orientation on the alkyl carboxylate and the aromatic hydroxyl groups that makes

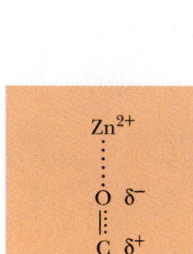

FIGURE 14.14 Liver alcohol dehydrogenase catalyzes the transfer of a hydride ion ($H:^-$) from NADH to acetaldehyde (CH_3CHO), forming ethanol (CH_3CH_2OH). An active-site zinc ion stabilizes negative charge development on the oxygen atom of acetaldehyde, leading to an induced partial positive charge on the carbonyl C atom. Transfer of the negatively charged hydride ion to this carbon forms ethanol.

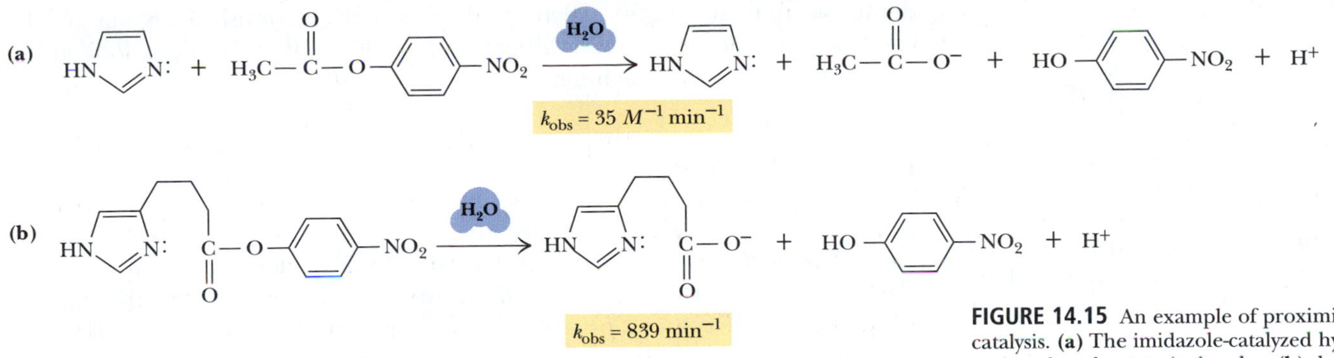

FIGURE 14.15 An example of proximity effects in catalysis. **(a)** The imidazole-catalyzed hydrolysis of *p*-nitrophenylacetate is slow, but **(b)** the corresponding intramolecular reaction is 24-fold faster (assuming [imidazole] = 1 *M* in [a]).

them approximately 250 billion times more likely to react. Enzymes function similarly by placing catalytically functional groups (from the protein side chains or from another substrate) in the proper position for reaction.

Clearly, proximity and orientation play a role in enzyme catalysis, but there is a problem with each of the aforementioned comparisons. In both cases, it is impossible to separate true proximity and orientation effects from the effects of entropy loss when molecules are brought together (described the Section 14.4). The actual rate accelerations afforded by proximity and orientation effects in Figures 14.15 and 14.16, respectively, are much smaller than the values given in these figures. Simple theories based on probability and nearest-neighbor models, for example, predict that proximity effects may actually provide rate increases of only fivefold to tenfold. For any real case of enzymatic catalysis, it is nonetheless important to remember that proximity and orientation effects are significant.

14.7 | What Can Be Learned from Typical Enzyme Mechanisms?

The balance of this chapter will be devoted to several classic and representative enzyme mechanisms. These particular cases are well understood, because the three-dimensional structures of the enzymes and the bound substrates are known at atomic resolution and because great efforts have been devoted to kinetic and mechanistic studies. They are important because they represent reaction types that appear again and again in living systems and because they

Reaction	Rate const. ($M^{-1}\ sec^{-1}$)	Ratio
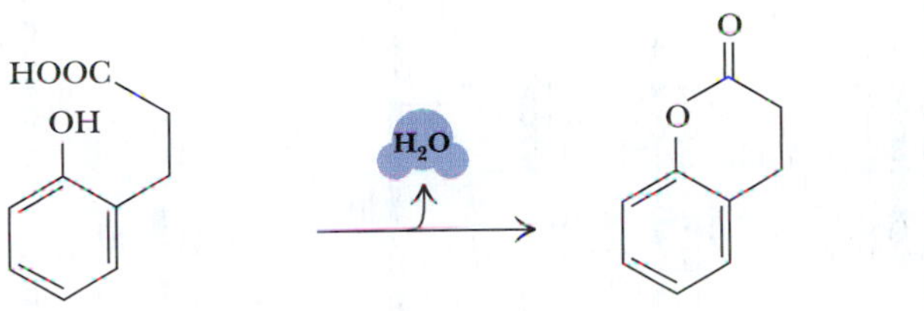	5.9×10^{-6}	
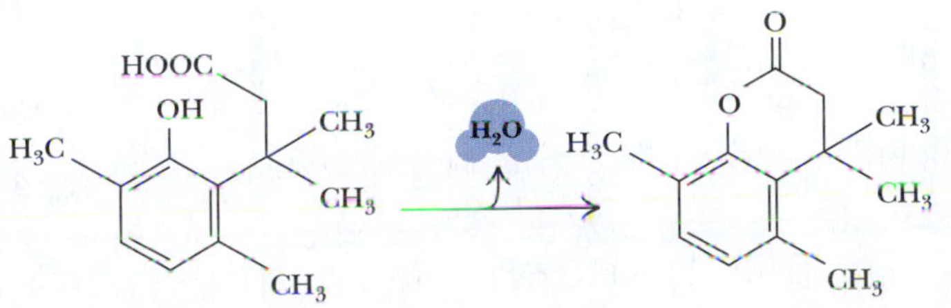	1.5×10^{6}	2.5×10^{11}

FIGURE 14.16 Orientation effects in intramolecular reactions can be dramatic. Steric crowding by methyl groups provides a rate acceleration of 2.5×10^{11} for the lower reaction compared to the upper reaction. *(Adapted from Milstien, S., and Cohen, L. A., 1972. Stereopopulation control I. Rate enhancements in the lactonization of o-hydroxyhydrocinnamic acid.* Journal of the American Chemical Society *94:9158–9165.)*

demonstrate many of the catalytic principles cited previously. Enzymes are the catalytic machines that sustain life, and what follows is an intimate look at the inner workings of the machinery.

Serine Proteases

Serine proteases are a class of proteolytic enzymes whose catalytic mechanism is based on an active-site serine residue. Serine proteases are one of the best-characterized families of enzymes. This family includes *trypsin, chymotrypsin, elastase, thrombin, subtilisin, plasmin, tissue plasminogen activator,* and other related enzymes. The first three of these are digestive enzymes and are synthesized in the pancreas and secreted into the digestive tract as inactive **proenzymes,** or **zymogens.** Within the digestive tract, the zymogen is converted into the active enzyme form by cleaving off a portion of the peptide chain. Thrombin is a crucial enzyme in the blood-clotting cascade, subtilisin is a bacterial protease, and plasmin breaks down the fibrin polymers of blood clots. Tissue plasminogen activator (TPA) specifically cleaves the proenzyme *plasminogen,* yielding plasmin. Owing to its ability to stimulate breakdown of blood clots, TPA can minimize the harmful consequences of a heart attack, if administered to a patient within 30 minutes of onset. Finally, although not itself a protease, *acetylcholinesterase is a serine esterase* and is related mechanistically to the serine proteases. It degrades the neurotransmitter acetylcholine in the synaptic cleft between neurons.

The Digestive Serine Proteases

Trypsin, chymotrypsin, and elastase all carry out the same reaction—the cleavage of a peptide chain—and although their structures and mechanisms are quite similar, they display very different specificities. Trypsin cleaves peptides on the car-

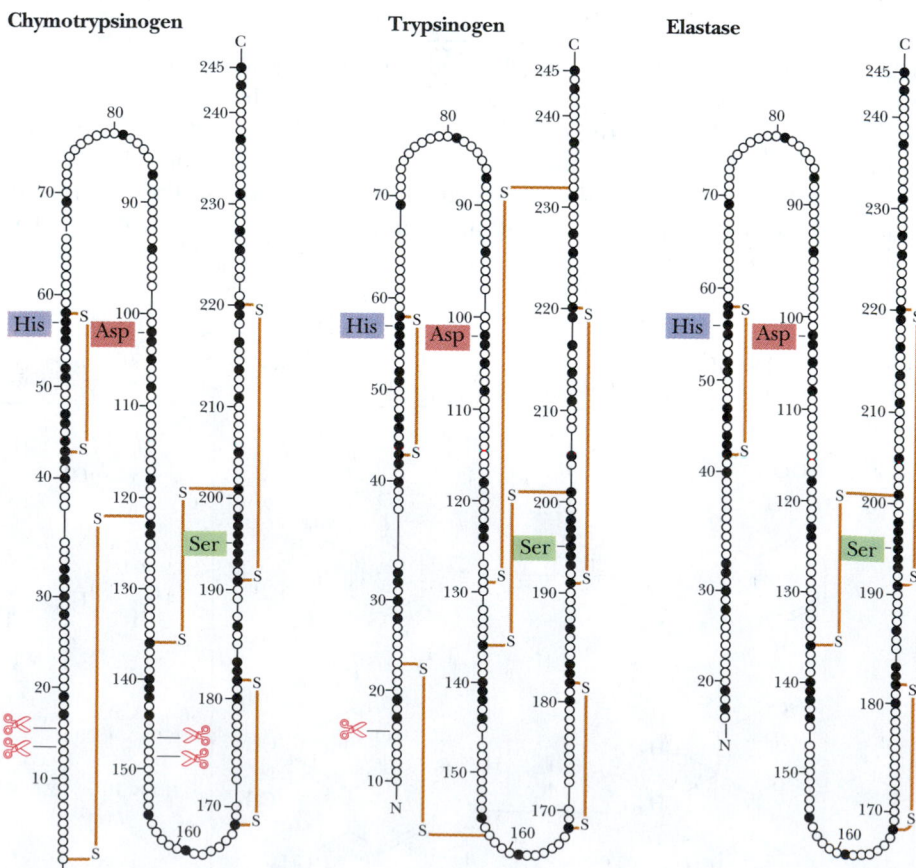

FIGURE 14.17 Comparison of the amino acid sequences of chymotrypsinogen, trypsinogen, and elastase. Each circle represents one amino acid. Numbering is based on the sequence of chymotrypsinogen. Filled circles indicate residues that are identical in all three proteins. Disulfide bonds are indicated in yellow. The positions of the three catalytically important active-site residues (His[57], Asp[102], and Ser[195]) are indicated.

bonyl side of the basic amino acids, arginine or lysine (see Table 5.6). Chymotrypsin prefers to cleave on the carbonyl side of aromatic residues, such as phenylalanine and tyrosine. Elastase is not as specific as the other two; it mainly cleaves peptides on the carbonyl side of small, neutral residues. These three enzymes all possess molecular weights in the range of 25,000, and all have similar sequences (Figure 14.17) and three-dimensional structures. The structure of chymotrypsin is typical (Figure 14.18). The molecule is ellipsoidal in shape and contains an α-helix at the C-terminal end (residues 230 to 245) and several β-sheet domains. Most of the aromatic and hydrophobic residues are buried in the interior of the protein, and most of the charged or hydrophilic residues are on the surface. Three polar residues—His[57], Asp[102], and Ser[195]—form what is known as a **catalytic triad** at the active site (Figure 14.19). These three residues are conserved in trypsin and elastase as well. The active site is actually a depression on the surface of the enzyme, with a small pocket that the enzyme uses to identify the residue for which it is specific (Figure 14.20). Chymotrypsin, for example, has a pocket surrounded by hydrophobic residues and large enough to accommodate an aromatic side chain. The pocket in trypsin has a negative charge (Asp[189]) at its bottom, facilitating the binding of positively charged arginine and lysine residues. Elastase, on the other hand, has a shallow pocket with bulky threonine and valine residues at the opening. Only small, nonbulky residues can be accommodated in its pocket. The backbone of the peptide substrate is hydrogen bonded in antiparallel fashion to residues 215 to 219 and bent so that the peptide bond to be cleaved is bound close to His[57] and Ser[195].

The Chymotrypsin Mechanism in Detail: Kinetics

Much of what is known about the chymotrypsin mechanism is based on studies of the hydrolysis of artificial substrates—simple organic esters, such as *p*-nitrophenylacetate, and methyl esters of amino acid analogs, such as formylphenylalanine methyl ester and acetylphenylalanine methyl ester (Figure 14.21). *p*-Nitrophenylacetate is an especially useful model substrate, because the nitrophenolate product is easily observed, owing to its strong absorbance at 400 nm. When large amounts of chymotrypsin are used in kinetic studies with this substrate, a **rapid initial burst** of *p*-nitrophenolate is observed (in an amount approximately equal to the enzyme concentration), followed by a much slower, linear rate of nitrophenolate release (Figure 14.22). Observation of a burst, followed by slower, steady-state product release, is strong evidence for a multistep mechanism, with a fast first step and a slower second step.

In the chymotrypsin mechanism, the nitrophenylacetate combines with the enzyme to form an ES complex. This is followed by a rapid second step in

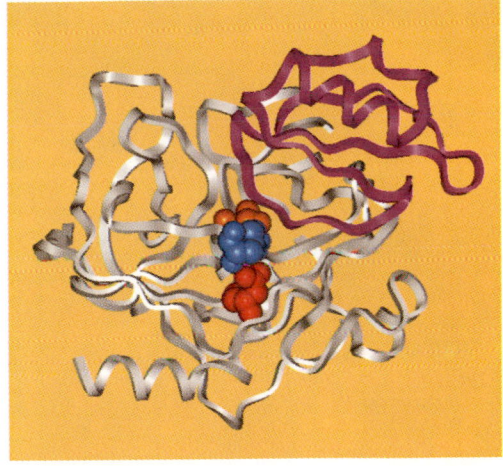

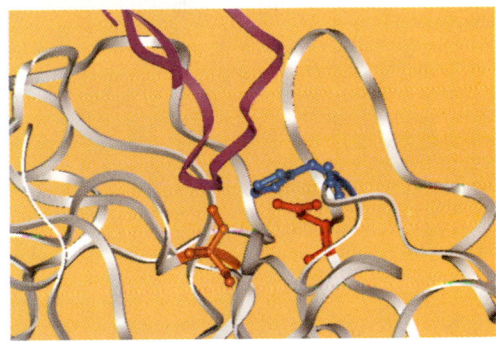

FIGURE 14.18 Structure of chymotrypsin (white) in a complex with eglin C (blue ribbon structure), a target protein. The residues of the catalytic triad (His[57], Asp[102], and Ser[195]) are highlighted. His[57] (blue) is flanked above by Asp[102] (red) and on the right by Ser[195] (yellow). The catalytic site is filled by a peptide segment of eglin. Note how close Ser[195] is to the peptide that would be cleaved in a chymotrypsin reaction.

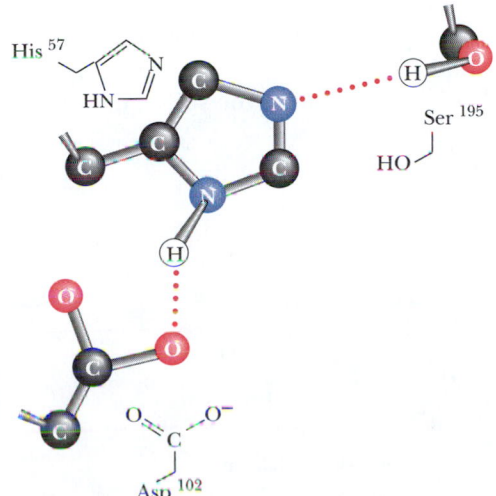

FIGURE 14.19 The catalytic triad of chymotrypsin.

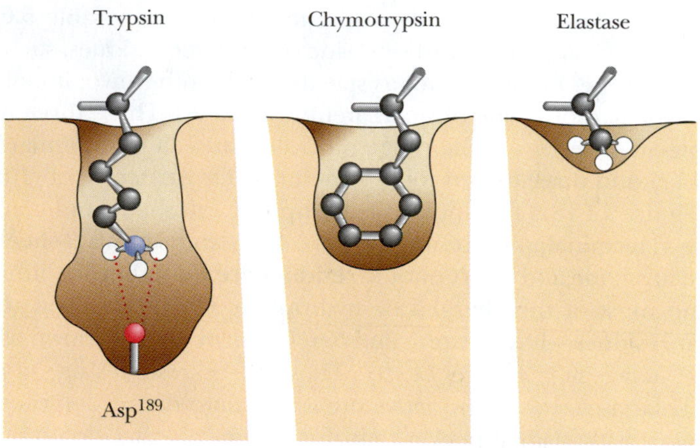

FIGURE 14.20 The substrate-binding pockets of trypsin, chymotrypsin, and elastase. *(Illustration: Irving Geis. Rights owned by Howard Hughes Medical Institute. Not to be reproduced without permission.)*

which an **acyl-enzyme intermediate** is formed, with the acetyl group covalently bound to the very reactive Ser[195]. The nitrophenyl moiety is released as nitrophenolate (Figure 14.23), accounting for the burst of nitrophenolate product. Attack of a water molecule on the acyl-enzyme intermediate yields acetate as the second product in a subsequent, slower step. The enzyme is now free to bind another molecule of *p*-nitrophenylacetate, and the *p*-nitrophenolate product produced at this point corresponds to the slower, steady-state formation of product in the upper right portion of Figure 14.22. In this mechanism, the release of acetate is the **rate-limiting step** and accounts for the observation of **burst kinetics**—the pattern shown in Figure 14.22.

Serine proteases like chymotrypsin are susceptible to inhibition by **organic fluorophosphates,** such as *diisopropylfluorophosphate* (*DIFP*, Figure 14.24). DIFP reacts rapidly with active-site serine residues, such as Ser[195] of chymotrypsin and the other serine proteases (but not with any of the other serines in these proteins), to form a DIP–enzyme. This covalent enzyme–inhibitor complex is extremely stable, and chymotrypsin is thus permanently inactivated by DIFP.

The Serine Protease Mechanism in Detail: Events at the Active Site

A likely mechanism for peptide hydrolysis is shown in Figure 14.25. As the backbone of the substrate peptide binds adjacent to the catalytic triad, the specific side chain fits into its pocket. Asp[102] of the catalytic triad positions His[57] and immobilizes it through a hydrogen bond as shown. In the first step of the reaction, His[57] acts as a general base to withdraw a proton from Ser[195], facilitating nucleophilic attack by Ser[195] on the carbonyl carbon of the peptide bond to be cleaved. This is probably a *concerted step*, because proton transfer prior to Ser[195] attack on the acyl carbon would leave a relatively unstable negative charge on the serine oxygen. In the next step, donation of a proton from His[57] to the peptide's amide nitrogen creates a protonated amine on the covalent, tetrahedral intermediate, facilitating the subsequent bond breaking and disso-

FIGURE 14.21 Artificial substrates used in studies of the mechanism of chymotrypsin.

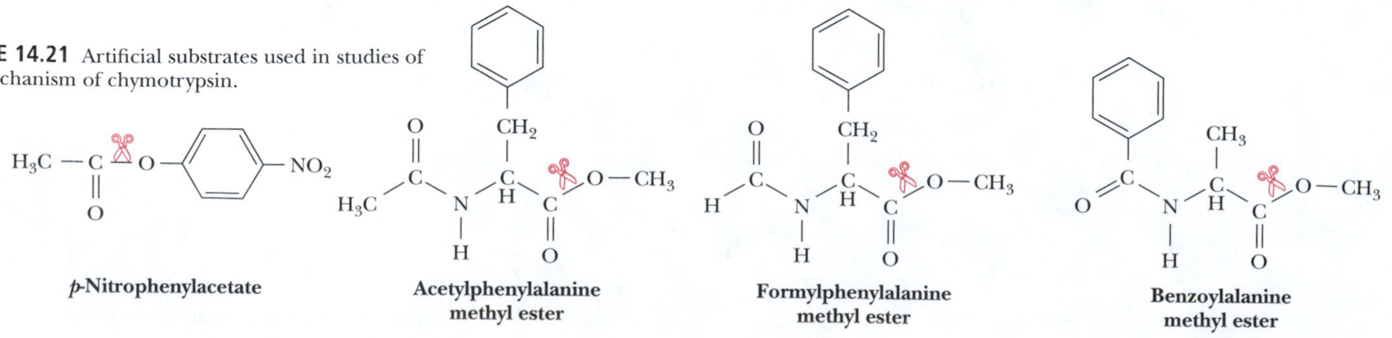

p-Nitrophenylacetate

Acetylphenylalanine methyl ester

Formylphenylalanine methyl ester

Benzoylalanine methyl ester

ciation of the amine product. The negative charge on the peptide oxygen is unstable; the tetrahedral intermediate is short lived and rapidly breaks down to expel the amine product. The acyl-enzyme intermediate that results is reasonably stable; it can even be isolated using substrate analogs for which further reaction cannot occur. With normal peptide substrates, however, subsequent nucleophilic attack at the carbonyl carbon by water generates another transient tetrahedral intermediate (Figure 14.25). His[57] acts as a general base in this step, accepting a proton from the attacking water molecule. The subsequent collapse of the tetrahedral intermediate is assisted by proton donation from His[57] to the serine oxygen in a concerted manner. Deprotonation of the carboxyl group and its departure from the active site complete the reaction as shown.

Until recently, the catalytic role of Asp[102] in trypsin and the other serine proteases had been surmised on the basis of its proximity to His[57] in structures obtained from X-ray diffraction studies, but it had never been demonstrated with certainty in physical or chemical studies. As can be seen in Figure 14.18, Asp[102] is buried at the active site and is normally inaccessible to chemical modifying reagents. In 1987, however, Charles Craik, William Rutter, and their colleagues used site-directed mutagenesis (see Chapter 12) to prepare a mutant trypsin with an asparagine in place of Asp[102]. This mutant trypsin possessed a hydrolytic activity with ester substrates only 1/10,000 that of native trypsin, demonstrating that Asp[102] is indeed essential for catalysis and that its ability to immobilize and orient His[57] by formation of a hydrogen bond is crucial to the function of the catalytic triad.

The serine protease mechanism relies in part on a low-barrier hydrogen bond. In the free enzyme, the pK_as of Asp[102] and His[57] are very different, and the H bond between them is a weak one. However, donation of the proton of Ser[195] to His[57] lowers the pK_a of the protonated imidazole ring so it becomes a close match to that of Asp[102], and the H bond between them becomes an LBHB. The energy released in the formation of this LBHB is used to facilitate the formation of the subsequent tetrahedral intermediate (Figure 14.25).

The Aspartic Proteases

Mammals, fungi, and higher plants produce a family of proteolytic enzymes known as **aspartic proteases.** These enzymes are active at acidic (or sometimes neutral) pH, and each possesses two aspartic acid residues at the active site. Aspartic proteases carry out a variety of functions (Table 14.3), including digestion (*pepsin* and *chymosin*), lysosomal protein degradation (*cathepsin D and E*), and regulation of blood pressure (*renin* is an aspartic protease involved in the production of *angiotensin*, a hormone that stimulates smooth muscle contraction and reduces excretion of salts and fluid). The aspartic proteases display a variety of substrate specificities, but normally they are most active in the cleavage of peptide bonds between two hydrophobic amino acid residues. The preferred substrates of pepsin, for example, contain aromatic residues on both sides of the peptide bond to be cleaved.

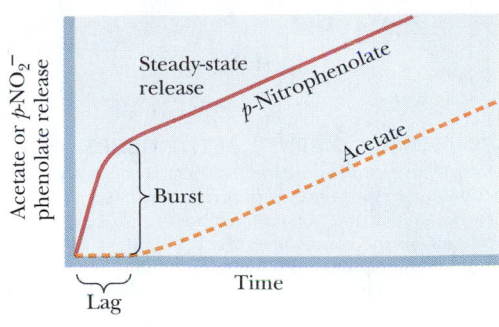

FIGURE 14.22 Burst kinetics observed in the chymotrypsin reaction. A burst of nitrophenolate production is followed by a slower, steady-state release. After an initial lag period, acetate release is also observed. This kinetic pattern is consistent with rapid formation of an acyl-enzyme intermediate (and the burst of nitrophenolate). The slower, steady-state release of products corresponds to rate-limiting breakdown of the acyl-enzyme intermediate.

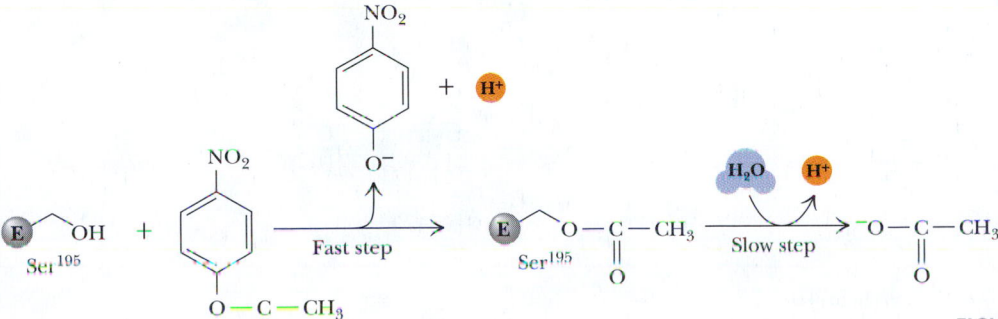

FIGURE 14.23 Rapid formation of the acyl-enzyme intermediate is followed by slower product release.

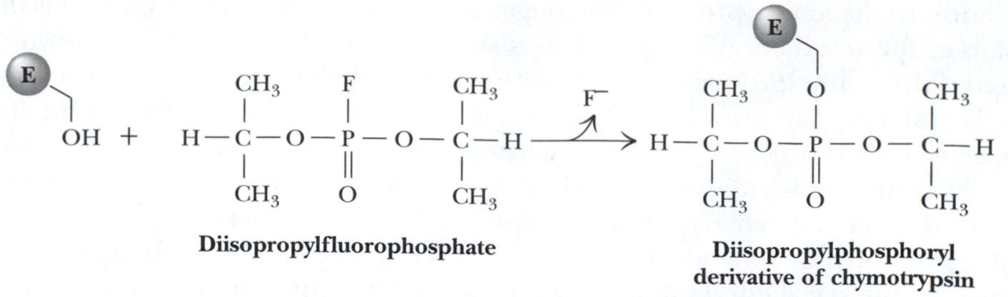

Biochemistry ⊜ Now™   **ACTIVE FIGURE 14.24**
Diisopropylfluorophosphate (DIFP) reacts with
active-site serine residues of serine proteases (and
esterases), causing permanent inactivation. **Test
yourself on the concepts in this figure at**
http://chemistry.brookscole.com/ggb3

Diisopropylfluorophosphate

**Diisopropylphosphoryl
derivative of chymotrypsin**

Most aspartic proteases are composed of 323 to 340 amino acid residues,
with molecular weights near 35,000. Aspartic protease polypeptides consist of
two homologous domains that fold to produce a tertiary structure composed
of two similar lobes, with approximate twofold symmetry (Figure 14.26). Each of
these lobes or domains consists of two β-sheets and two short α-helices. The two
domains are bridged and connected by a six-stranded, antiparallel β-sheet. The
active site is a deep and extended cleft, formed by the two juxtaposed domains
and large enough to accommodate about seven amino acid residues. The two

A Deeper Look

Transition-State Stabilization in the Serine Proteases

X-ray crystallographic studies of serine protease complexes with
transition-state analogs have shown how chymotrypsin stabilizes
the **tetrahedral oxyanion transition states** [structures (c) and
(g) in Figure 14.25] of the protease reaction. The amide nitrogens of Ser^{195} and Gly^{193} form an "oxyanion hole" in which the
substrate carbonyl oxygen is hydrogen bonded to the amide
N—H groups.

Formation of the tetrahedral transition state increases the interaction of the carbonyl oxygen with the amide N—H groups
in two ways. Conversion of the carbonyl double bond to the
longer tetrahedral single bond brings the oxygen atom closer to
the amide hydrogens. Also, the hydrogen bonds between the
charged oxygen and the amide hydrogens are significantly
stronger than the hydrogen bonds with the uncharged carbonyl
oxygen.

Transition-state stabilization in chymotrypsin also involves the
side chains of the substrate. The side chain of the departing
amine product forms stronger interactions with the enzyme
upon formation of the tetrahedral intermediate. When the tetrahedral intermediate breaks down (Figure 14.25d and e), steric
repulsion between the product amine group and the carbonyl
group of the acyl-enzyme intermediate leads to departure of the
amine product.

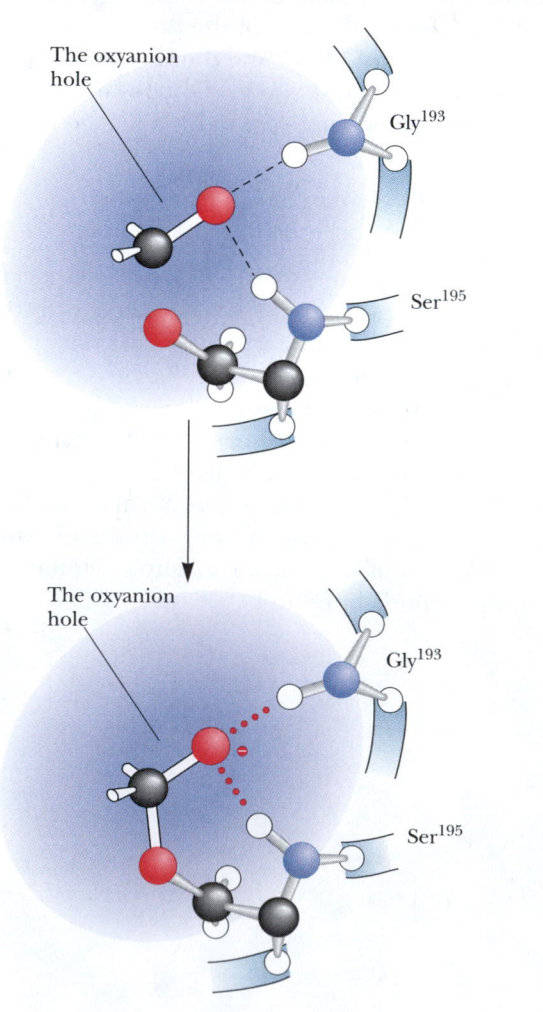

▶ The "oxyanion hole" of chymotrypsin stabilizes the tetrahedral oxyanion intermediate of the mechanism in Figure 14.25.

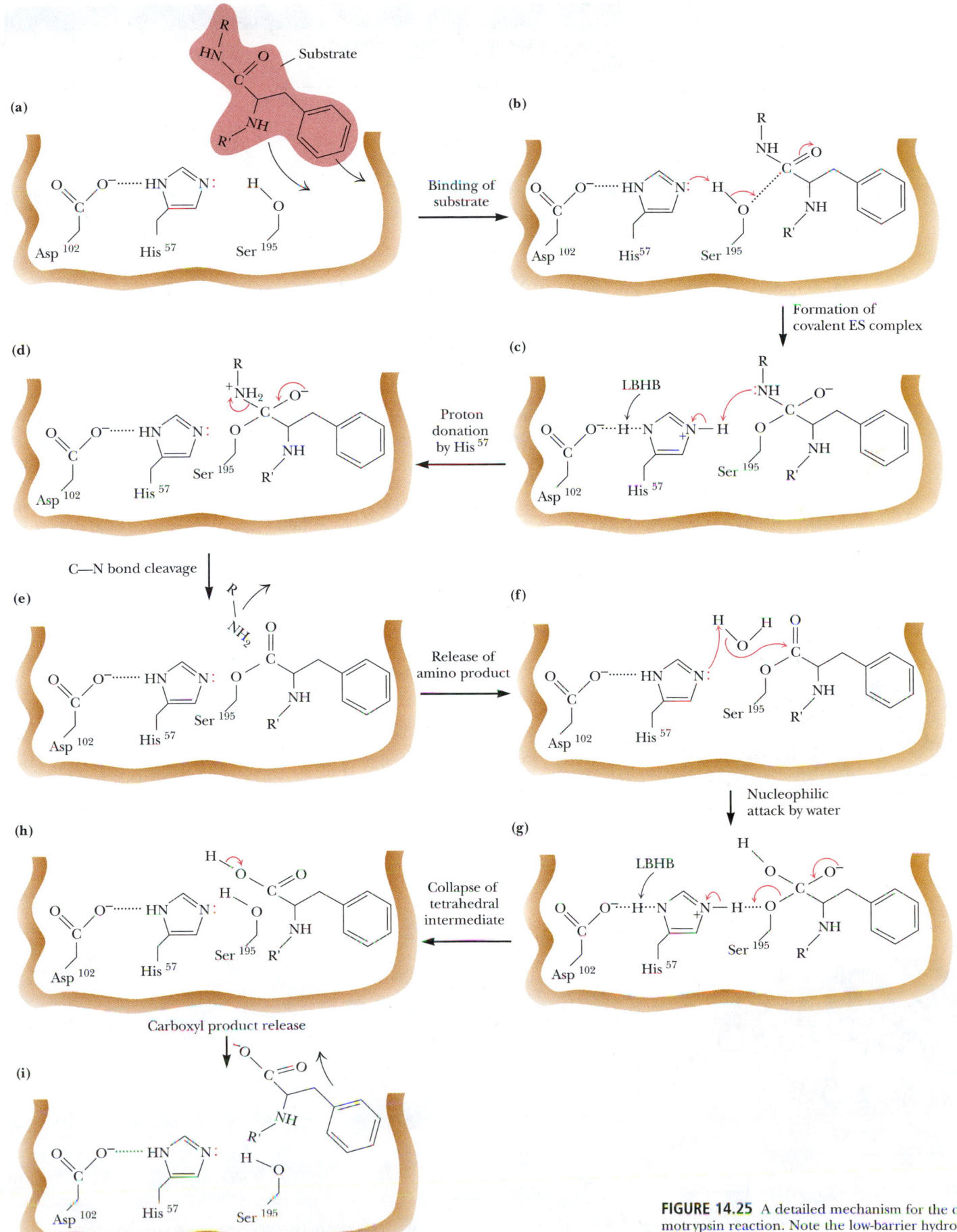

FIGURE 14.25 A detailed mechanism for the chymotrypsin reaction. Note the low-barrier hydrogen bond (LBHB) in (c) and (g).

Table 14.3

Some Representative Aspartic Proteases

Name	Source	Function
Pepsin[*]	Stomach	Digestion of dietary protein
Chymosin[†]	Stomach	Digestion of dietary protein
Cathepsin D	Spleen, liver, and many other animal tissues	Lysosomal digestion of proteins
Renin[‡]	Kidney	Conversion of angiotensinogen to angiotensin I; regulation of blood pressure
HIV-protease[§]	AIDS virus	Processing of AIDS virus proteins

[*]The second enzyme to be crystallized (by John Northrop in 1930). Even more than urease before it, pepsin study by Northrop established that enzyme activity comes from proteins.
[†]Also known as rennin, it is the major pepsinlike enzyme in gastric juice of fetal and newborn animals. It has been used for thousands of years, in a gastric extract called rennet, in the making of cheese.
[‡]A drop in blood pressure causes release of renin from the kidneys, which converts more angiotensinogen to angiotensin.
[§]A dimer of identical monomers, homologous to pepsin.

catalytic aspartate residues, residues 32 and 215 in porcine pepsin, for example, are located deep in the center of the active site cleft. The N-terminal domain forms a "flap" that extends over the active site, which may help to immobilize the substrate in the active site.

On the basis, in part, of comparisons with chymotrypsin, trypsin, and the other serine proteases, it was hypothesized that aspartic proteases might function by formation of covalent enzyme–substrate intermediates involving the active-site aspartate residues. Two possibilities were proposed: an acyl-enzyme intermediate involving an acid anhydride bond and an amino-enzyme intermediate involving an amide (peptide) bond (Figure 14.27). All attempts to trap or isolate a covalent intermediate failed, and a mechanism (see following section) favoring noncovalent enzyme–substrate intermediates and general acid–general base catalysis is now favored for aspartic proteases.

The Mechanism of Action of Aspartic Proteases

A crucial datum supporting the general acid–general base model is the pH dependence of protease activity (Figure 14.28). For many years, enzymologists hypothesized that the aspartate carboxyl groups functioned alternately as general acid and general base. This model requires that one of the aspartate carboxyls be

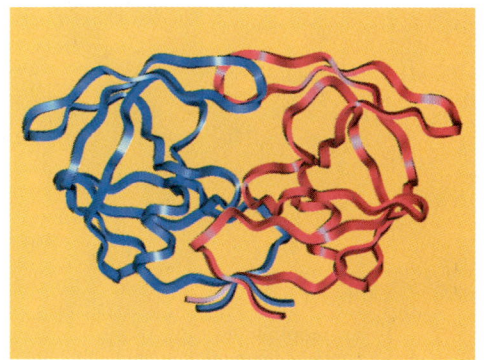

(a)

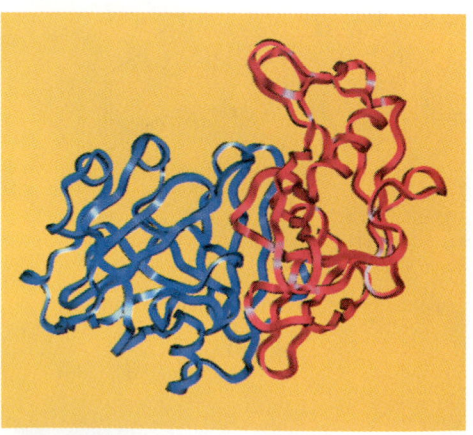

(b)

FIGURE 14.26 Structures of (a) HIV-1 protease, a dimer, and (b) pepsin, a monomer. Pepsin's N-terminal half is shown in red; C-terminal half is shown in blue.

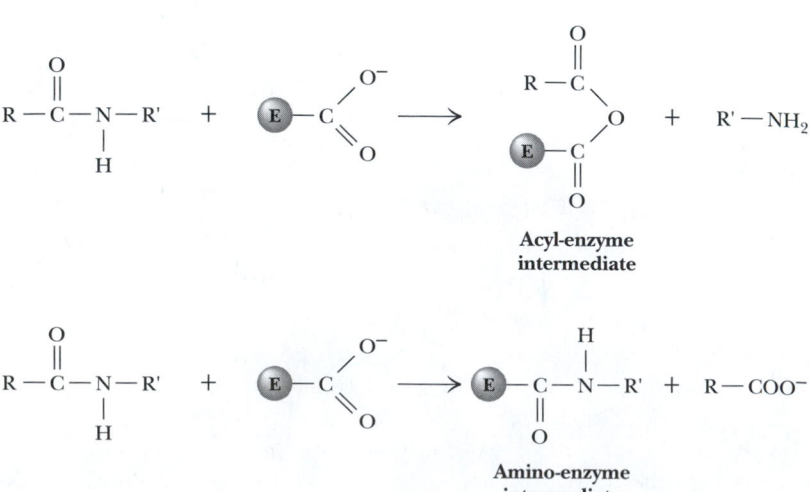

Acyl-enzyme intermediate

Amino-enzyme intermediate

FIGURE 14.27 Acyl-enzyme and amino-enzyme intermediates originally proposed for aspartic proteases were modeled after the acyl-enzyme intermediate of the serine proteases.

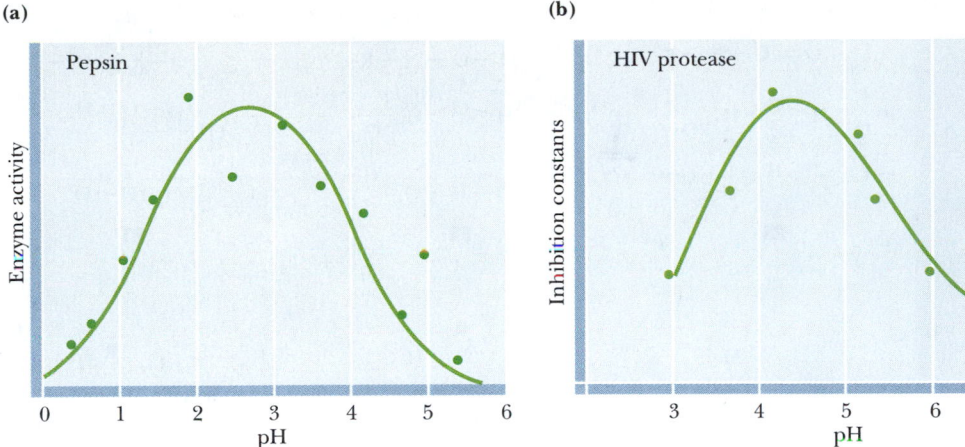

(a) Pepsin

(b) HIV protease

FIGURE 14.28 pH-rate profiles for **(a)** pepsin and **(b)** HIV protease. *(Adapted from Denburg, J., et al., 1968. The effect of pH on the rates of hydrolysis of three acylated dipeptides by pepsin.* Journal of the American Chemical Society *90:479–486; and Hyland, J., et al., 1991. Human immuno-deficiency virus-1 protease. 2. Use of pH rate studies and solvent kinetic isotope effects to elucidate details of chemical mechanism.* Biochemistry *30:8454–8463.)*

protonated and one be deprotonated when substrate binds. (This made sense, because X-ray diffraction data on aspartic proteases had shown that the active-site structure in the vicinity of the two aspartates is highly symmetric.) However, Stefano Piana and Paolo Carloni reported in 2000 that molecular dynamics simulations of aspartic proteases were consistent with a low-barrier hydrogen bond involving the two active-site aspartates. This led to a new mechanism for the aspartic proteases (Figure 14.29) that begins with Piana and Carloni's model of the LBHB structure of the free enzyme (state E). In this model, the LBHB holds the twin aspartate carboxyls in a coplanar conformation, with the catalytic water molecule on the opposite side of a ten-atom cyclic structure.

Following substrate binding, a counterclockwise flow of electrons moves two protons clockwise and creates a tetrahedral intermediate bound to a diprotonated enzyme form (FT). Then a clockwise movement of electrons moves two protons counterclockwise and generates the zwitterion intermediate bound to a monoprotonated enzyme form (ET'). Collapse of the zwitterion cleaves the C—N bond of the substrate. Dissociation of one product leaves the enzyme in the diprotonated FQ form. Finally, deprotonation and rehydration lead to regeneration of the ten-atom cyclic structure, E.

What is the purpose of the low-barrier hydrogen bond in the aspartic protease mechanism? It may be to disperse electron density in the ten-atom cyclic structure, accomplishing rate acceleration by means of "hydrogen tunneling" (Figure 14.30). The barrier between the ES and ET' states of Figure 14.29 is imagined to be large, and the state FT may not exist as a discrete intermediate but rather may exist transiently to facilitate conversion of ES and ET'.

The AIDS Virus HIV-1 Protease Is an Aspartic Protease

Recent research on acquired immunodeficiency syndrome (AIDS) and its causative viral agent, the human immunodeficiency virus (HIV-1), has brought a new aspartic protease to light. **HIV-1 protease** cleaves the polyprotein products of the HIV-1 genome, producing several proteins necessary for viral growth and cellular infection. HIV-1 protease cleaves several different peptide linkages in the HIV-1 polyproteins, including those shown in Figure 14.31. For example, the protease cleaves between the Tyr and Pro residues of the sequence Ser-Gln-Asn-Tyr-Pro-Ile-Val, which joins the p17 and p24 HIV-1 proteins.

The HIV-1 protease is a remarkable viral imitation of mammalian aspartic proteases: It is a **dimer of identical subunits** that mimics the two-lobed monomeric structure of pepsin and other aspartic proteases. The HIV-1 protease subunits are 99-residue polypeptides that are homologous with the individual domains of the monomeric proteases. Structures determined by X-ray diffraction studies reveal that the active site of HIV-1 protease is formed at the interface of the homodimer and consists of two aspartate residues, designated Asp^{25} and $Asp^{25'}$,

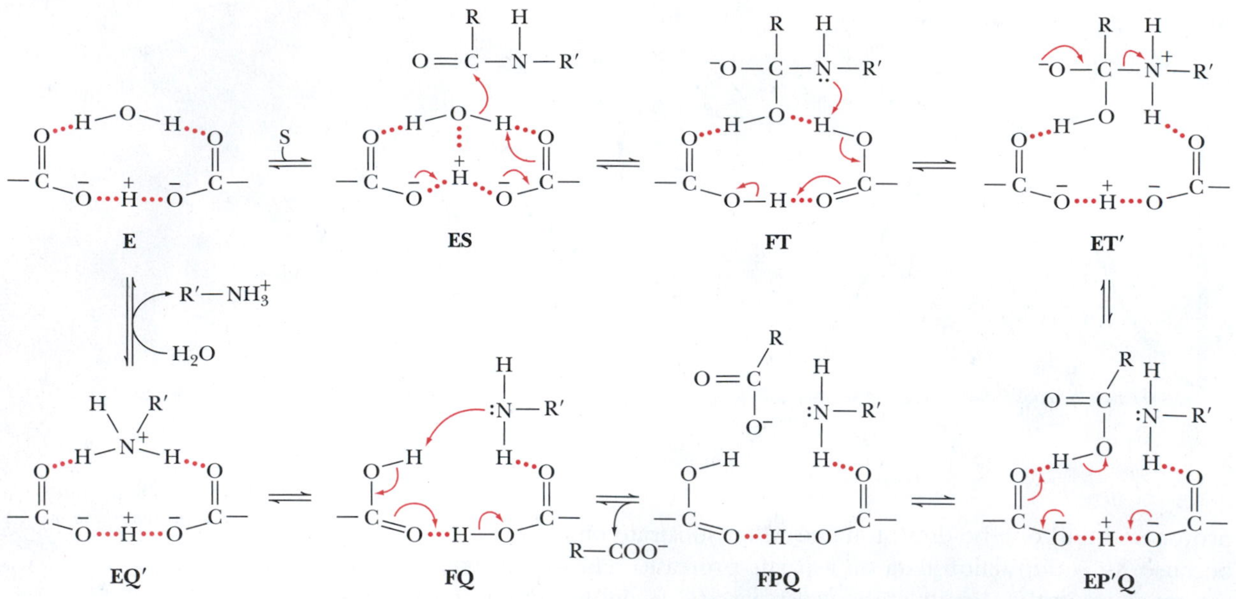

FIGURE 14.29 A mechanism for the aspartic proteases. The letter titles describe the states as follows: E represents the enzyme form with a low-barrier hydrogen bond between the catalytic aspartates, F represents the enzyme form with one aspartate protonated and the other sharing in a conventional hydrogen bond, S represents bound substrate, T represents a tetrahedral amide hydrate intermediate, P represents bound carboxyl product, and Q represents bound amine product. This mechanism is based in part on a mechanism proposed by Dexter Northrop, a distant relative of John Northrop, who had first crystallized pepsin in 1930. *(Northrop, D. B., 2001. Follow the protons: a low-barrier hydrogen bond unifies the mechanisms of the aspartic proteases.* Accounts of Chemical Research **34**:790–797.) The mechanism is also based on data of Thomas Meek. *(Meek, T. D., Catalytic mechanisms of the aspartic proteinases. In Sinnott, M., ed,* Comprehensive Biological Catalysis: A Mechanistic Reference, *San Diego: Academic Press, 1998.)*

one contributed by each subunit (Figure 14.32). In the homodimer, the active site is covered by two identical "flaps," one from each subunit, in contrast to the monomeric aspartic proteases, which possess only a single active-site flap. Enzyme kinetic measurements by Thomas Meek and his collaborators at SmithKline Beecham Pharmaceuticals have shown that the mechanism of HIV-1 protease is very similar to those of other aspartic proteases.

Lysozyme

Lysozyme is an enzyme that hydrolyzes polysaccharide chains. It ruptures certain bacterial cells by cleaving the polysaccharide chains that make up their cell wall. Lysozyme is found in many body fluids, but the most thoroughly studied form is from hen egg whites. The Russian scientist P. Laschtchenko first described the bacteriolytic properties of hen egg white lysozyme in 1909. In 1922,

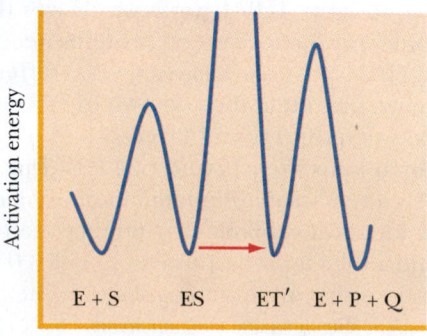

FIGURE 14.30 Energy level diagram showing ground-state hydrogen tunneling (arrow), with consequent rate acceleration.

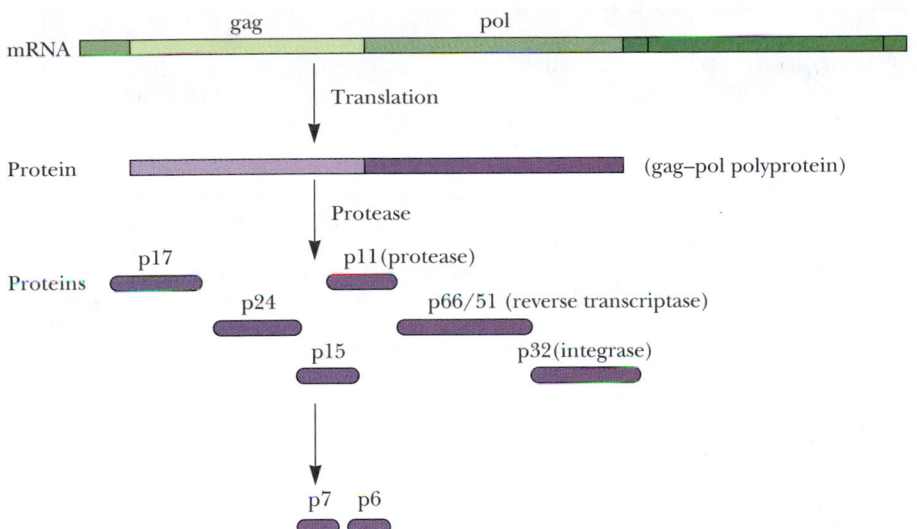

FIGURE 14.31 HIV mRNA provides the genetic information for synthesis of a polyprotein. Proteolytic cleavage of this polyprotein by HIV protease produces the individual proteins required for viral growth and cellular infection.

Alexander Fleming, the London bacteriologist who later discovered penicillin, gave the name *lysozyme* to the agent in mucus and tears that destroyed certain bacteria, because it was an en*zyme* that caused bacterial *lysis*.

As seen in Chapter 7, bacterial cells are surrounded by a rigid, strong wall of peptidoglycan, a copolymer of two sugar units, *N*-acetylmuramic acid (NAM) and *N*-acetylglucosamine (NAG). Both of these sugars are *N*-acetylated analogs of glucosamine, and in bacterial cell wall polysaccharides, they are joined in $\beta(1\rightarrow4)$ glycosidic linkages (Figure 14.31). Lysozyme hydrolyzes the glycosidic bond between C-1 of NAM and C-4 of NAG, as shown in Figure 14.33, but does not act on the $\beta(1\rightarrow4)$ linkages between NAG and NAM.

Lysozyme is a small globular protein composed of 129 amino acids (14 kD) in a single polypeptide chain. It has eight cysteine residues linked in four disulfide bonds. The structure of this very stable protein was determined by X-ray crystallographic methods in 1965 by David Phillips (Figure 14.34). Although X-ray structures had previously been reported for proteins (hemoglobin and myoglobin), lysozyme was the first enzyme structure to be solved by crystallographic (or any other) methods. Although the location of the active site was not obvious from the X-ray structure of the protein alone, X-ray studies of lysozyme-inhibitor complexes soon revealed the location and nature of the active site. Since it is an enzyme, lysozyme cannot form stable ES complexes for structural studies, because the substrate is rapidly transformed into products. On the other hand, several substrate analogs have proved to be good competitive inhibitors of lysozyme that can form complexes with the enzyme stable enough to be characterized by X-ray crystallography and other physical techniques. One of the best is a trimer of *N*-acetylglucosamine, (NAG)$_3$ (Figure 14.35), which is hydrolyzed

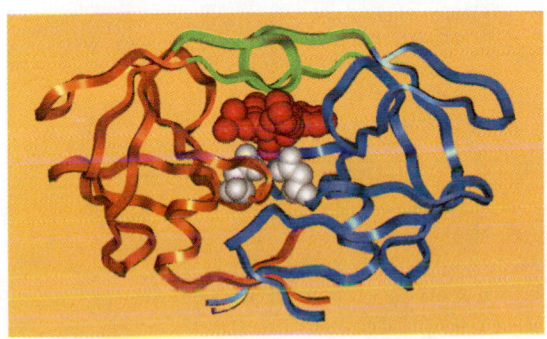 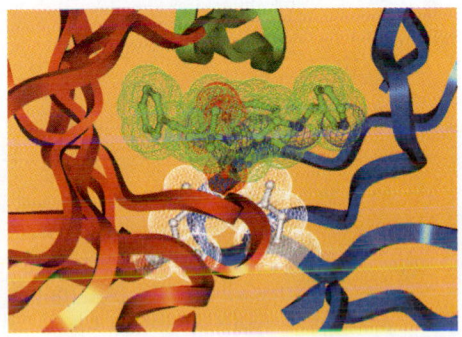

Biochemistry 🔵**Now**™ **ACTIVE FIGURE 14.32** (*left*) HIV-1 protease complexed with the inhibitor Crixivan (red) made by Merck. The flaps (residues 46–55 from each subunit) covering the active site are shown in green, and the active-site aspartate residues involved in catalysis are shown in white. (*right*) The close-up of the active site shows the interaction of Crixivan with the carboxyl groups of the essential aspartate residues. **Test yourself on the concepts in this figure at http://chemistry.brookscole. com/ggb3**

Human Biochemistry

Protease Inhibitors Give Life to AIDS Patients

Infection with HIV was once considered a death sentence, but the emergence of a new family of drugs called protease inhibitors has made it possible for some AIDS patients to improve their overall health and extend their lives. These drugs are all specific inhibitors of the HIV protease. By inhibiting the protease, they prevent the development of new virus particles in the cells of infected patients. Clinical testing has shown that a combination of drugs—including a protease inhibitor together with a reverse transcriptase inhibitor like AZT—can reduce the human immunodeficiency virus (HIV) to undetectable levels in about 40% to 50% of infected individuals. Patients who respond successfully to this combination therapy have experienced dramatic improvement in their overall health and a substantially lengthened life span.

Four of the protease inhibitors approved for use in humans by the U.S. Food and Drug Administration are shown below: Crixivan by Merck, Invirase by Hoffman-LaRoche, Norvir by Abbott, and Viracept by Agouron. These drugs were all developed from a "structure-based" design strategy; that is, the drug molecules were designed to bind tightly to the active site of the HIV-1 protease. The backbone OH-group in all these substances inserts between the two active-site carboxyl groups of the protease.

In the development of an effective drug, it is not sufficient merely to show that a candidate compound can cause the desired biochemical effect. It must also be demonstrated that the drug can be effectively delivered in sufficient quantities to the desired site(s) of action in the organism and that the drug does not cause undesirable side effects. The HIV-1 protease inhibitors shown here fulfill all of these criteria. Other drug candidates have been found that are even better inhibitors of HIV-1 protease in cell cultures, but many of these fail the test of bioavailability—the ability of a drug to be delivered to the desired site(s) of action in the organism.

Candidate protease inhibitor drugs must be relatively specific for the HIV-1 protease. Many other aspartic proteases exist in the human body and are essential to a variety of body functions, including digestion of food and processing of hormones. An ideal drug thus must strongly inhibit the HIV-1 protease, must be delivered effectively to the lymphocytes where the protease must be blocked, and should not adversely affect the activities of the essential human aspartic proteases.

A final but important consideration is viral mutation. Certain mutant HIV strains are resistant to one or more of the protease inhibitors, and even for patients who respond initially to protease inhibitors it is possible that mutant viral forms may eventually arise and thrive in the infected individual. The search for new and more effective protease inhibitors is ongoing.

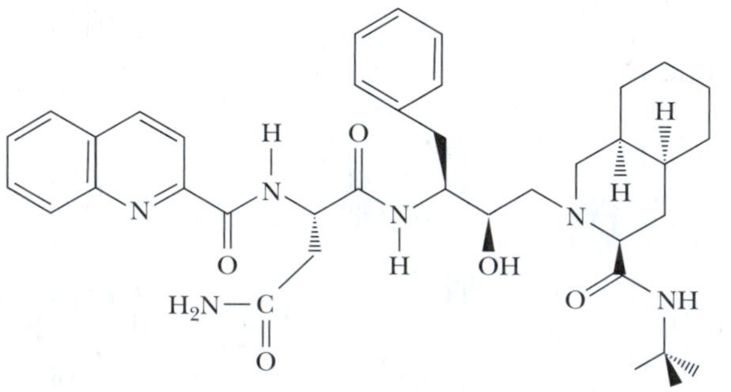

Invirase (saquinavir)

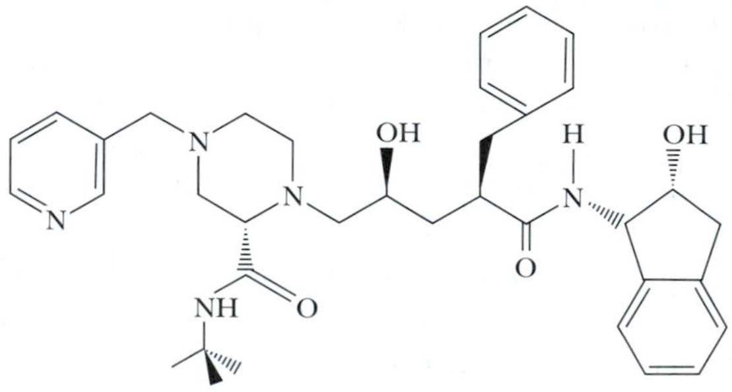

Crixivan (indinavir)

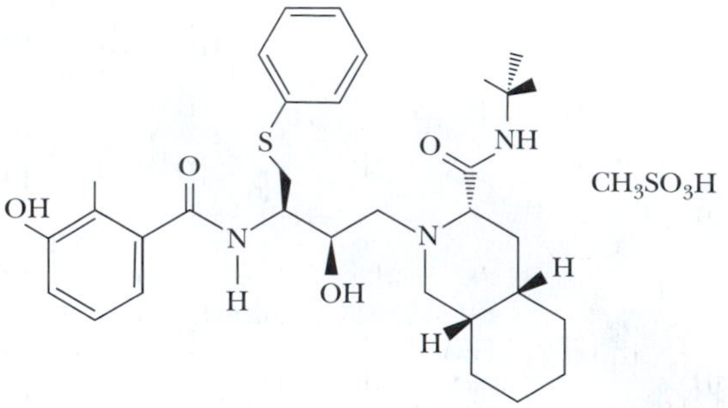

Viracept (nelfinavir mesylate)

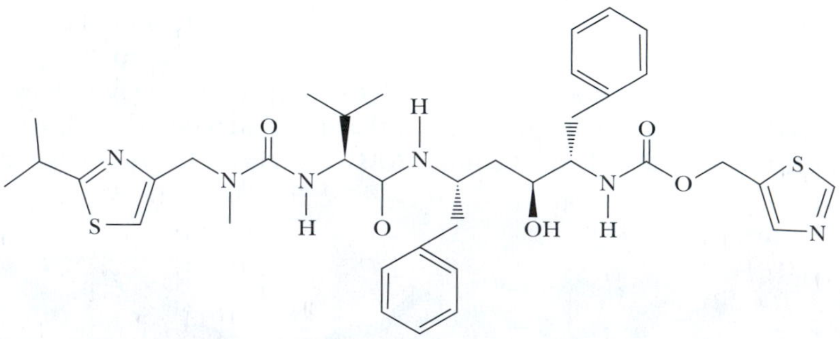

Norvir (ritonavir)

by lysozyme at a rate only 1/60,000 that of the native substrate (Table 14.4). (NAG)$_3$ binds at the enzyme active site by forming five hydrogen bonds with residues located in one-half of a depression or crevice that spans the surface of the enzyme (Figure 14.36). The few hydrophobic residues that exist on the surface of lysozyme are located in this depression, and they may participate in

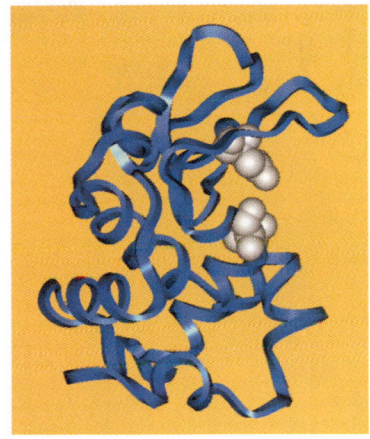

FIGURE 14.34 The structure of lysozyme. Glu^{35} and Asp^{52} are shown in white.

FIGURE 14.33 The lysozyme reaction.

hydrophobic and van der Waals interactions with $(NAG)_3$, as well as the normal substrate. The absence of charged groups on $(NAG)_3$ precludes the involvement of electrostatic interactions with the enzyme. Comparisons of the X-ray structures of the native lysozyme and the lysozyme–$(NAG)_3$ complex reveal that several amino acid residues at the active site move slightly upon inhibitor binding, including Trp^{62}, which moves about 0.75 Å to form a hydrogen bond with a hydroxymethyl group (Figure 14.37).

Model Studies Reveal a Strain-Induced Destabilization of a Bound Substrate on Lysozyme

One of the premises of lysozyme models is that the native substrate would occupy the rest of the crevice or depression running across the surface of the enzyme, because there is room to fit three more sugar residues into the crevice and because the hexamer $(NAG)_6$ is in fact a good substrate for lysozyme (Table 14.4). The model-building studies refer to the six sugar residue-binding subsites in the crevice with the letters A through F, with A, B, and C representing the part of the crevice occupied by the $(NAG)_3$ inhibitor (Figure 14.37). Modeling studies clearly show that NAG residues fit nicely into subsites A, B, C, E, and F of the crevice but that fitting a residue of the $(NAG)_6$ hexamer into site D requires a

Table 14.4	
Hydrolysis Rate Constants for Model Oligosaccharides with Lysozyme	
Oligosaccharide	**Rate Constant, k_{cat} (s^{-1})**
$(NAG\text{-}NAM)_3$	0.5
$(NAG)_6$	0.25
$(NAG)_5$	0.033
$(NAG)_4$	7×10^{-5}
$(NAG)_3$	8×10^{-6}
$(NAG)_2$	2.5×10^{-8}

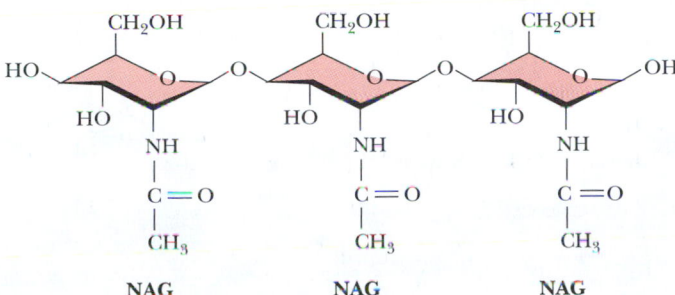

FIGURE 14.35 $(NAG)_3$, a substrate analog, forms stable complexes with lysozyme.

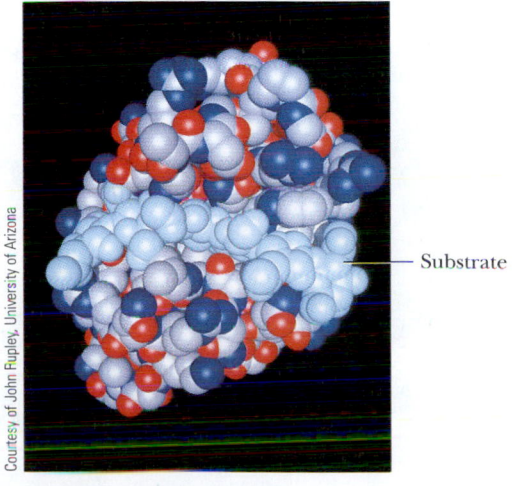

Courtesy of John Rupley, University of Arizona

FIGURE 14.36 The lysozyme-enzyme–substrate complex.

FIGURE 14.37 Enzyme–substrate interactions at the six sugar residue-binding subsites of the lysozyme active site. *(Illustration: Irving Geis. Rights owned by Howard Hughes Medical Institute. Not to be reproduced without permission.)*

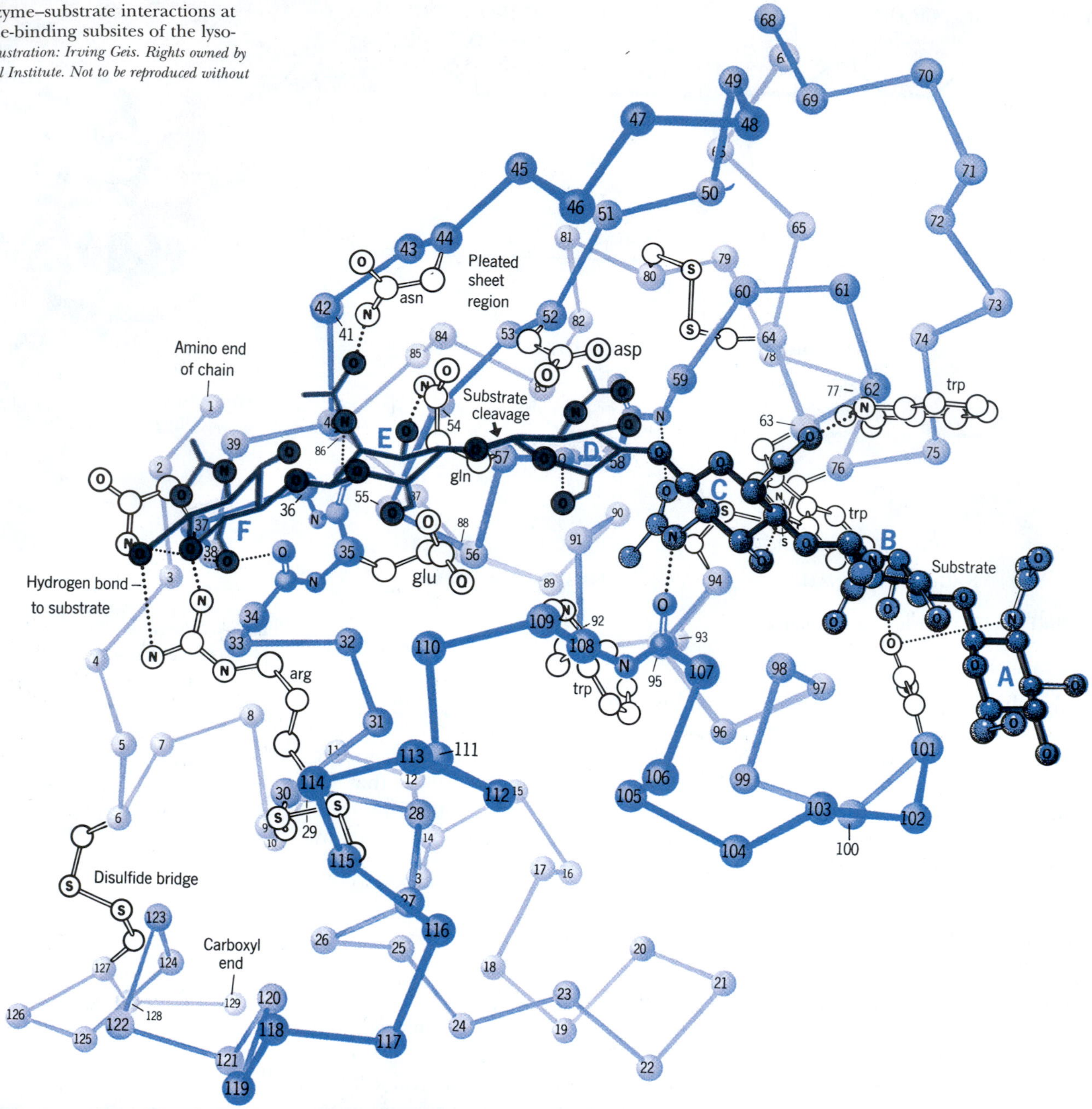

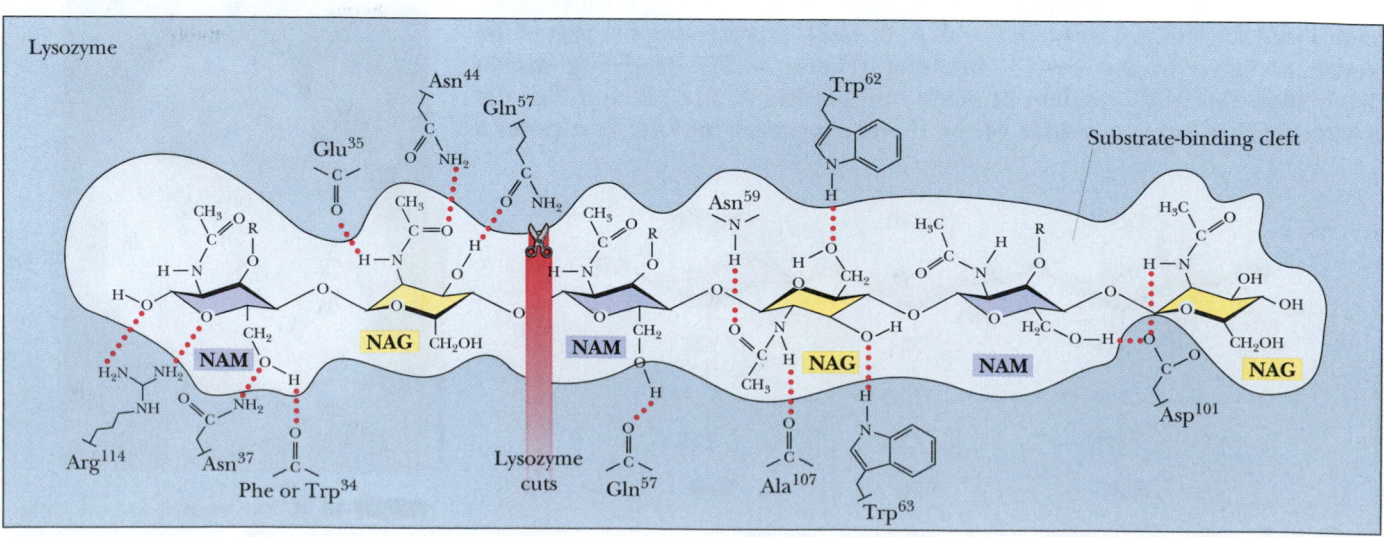

substantial distortion of the sugar (out of its preferred chair conformation) to prevent steric crowding and overlap between atoms C-6 and O-6 of the sugar at the D site and Ile[98] of the enzyme. This distorted sugar residue is adjacent to the glycosidic bond to be cleaved (between sites D and E), and the inference is made that this distortion or strain brings the substrate closer to the transition state for hydrolysis. This is a good example of strain-induced destabilization of an otherwise favorably binding substrate (see Section 14.4). Thus, the overall binding interaction of the rest of the sugar substrate would be favorable ($\Delta G < 0$), but distortion of the ring at the D site uses some of this binding energy to raise the substrate closer to the transition state for hydrolysis, *an example of stabilization of a transition state (relative to the simple enzyme–substrate complex).* As noted in Section 14.4, distortion is one of the molecular mechanisms that can lead to such transition-state stabilization.

The Lysozyme Mechanism—A Classic Choice, and Recent Evidence

There are two mechanisms that would be consistent with the early X-ray structures of lysozyme and its model substrate complexes, and these two reactions represent a classic choice for the student enzymologist. In order to choose between these two, consider the following evidence: Studies using ^{18}O-enriched water showed that the C_1—O bond is cleaved on the substrate between the D and E sites. Hydrolysis under these conditions incorporates ^{18}O into the C_1 position of the sugar at the D site, not into the oxygen at C_4 at the E site (Figure 14.38). Model building studies place the cleaved bond approximately between protein residues Glu[35] and Asp[52]. Glu[35] is in a nonpolar or hydrophobic region of the protein, whereas Asp[52] is located in a much more polar environment. Glu[35] is protonated, but Asp[52] is ionized (Figure 14.39).

In the lysozyme mechanism that was accepted for many years (Figure 14.39a), Glu[35] may act as a general acid, donating a proton to the oxygen atom of the glycosidic bond and accelerating the reaction. Asp[52], on the other hand, stabilizes the carbocation (also called a carbonium ion or an oxocarbenium ion) generated at the D site upon bond cleavage. Formation of the carbocation ion may also be enhanced by the strain on the ring at the D site. Following bond cleavage, the product formed at the E site diffuses away, and the carbocation intermediate can then react with H_2O from the solution. Glu[35] can now act as a general base, accepting a proton from the attacking water. The tetramer of NAG thus formed at sites A through D can now be dissociated from the enzyme. If this were indeed the true mechanism for lysozyme, the rate acceleration afforded by lysozyme would be due to (1) general acid catalysis by Glu[35]; (2) distortion of the sugar ring at the D site, which may stabilize the carbonium ion *(and the transition state);* and (3) electrostatic stabilization of the carbocation by nearby Asp[52].

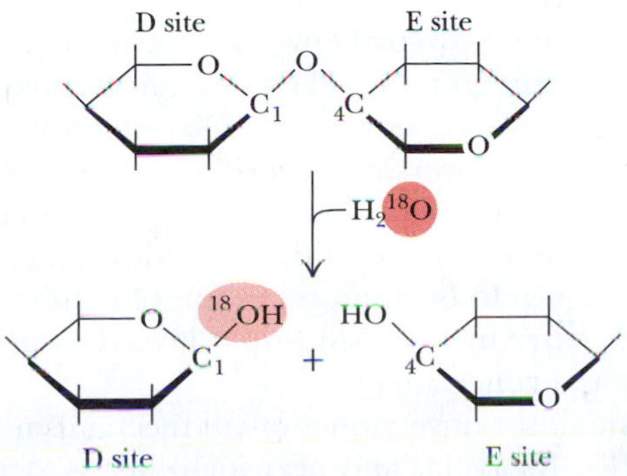

FIGURE 14.38 The C_1—O bond, not the O—C_4 bond, is cleaved in the lysozyme reaction. ^{18}O from $H_2{}^{18}O$ is thus incorporated at the C_1 position.

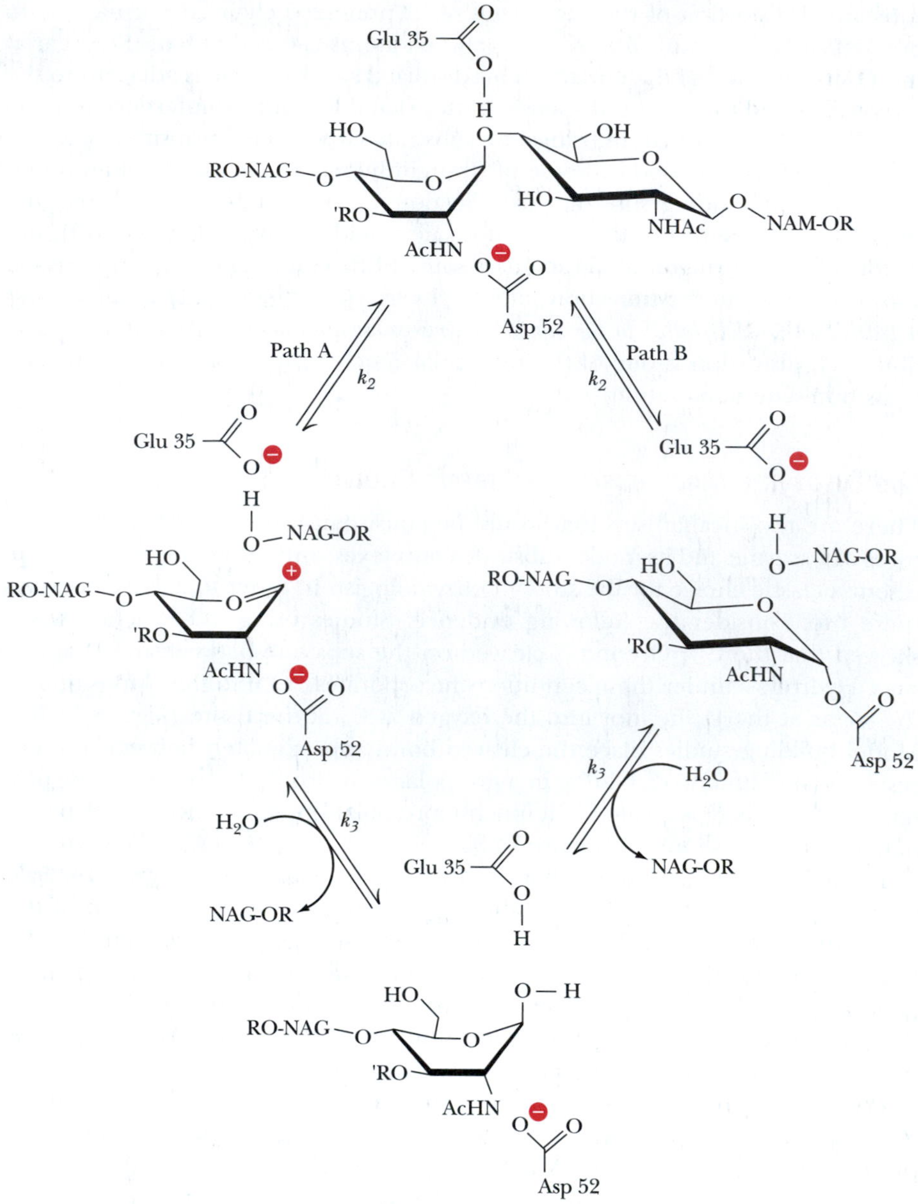

FIGURE 14.39 Two possible mechanisms for the lysozyme reaction. In Path A, the intermediate is a noncovalent oxocarbenium ion (carbocation). Path B depends upon a covalent intermediate involving Asp[52] and the C-1 oxygen of the cleaved glycosidic bond.

The other possible mechanism for lysozyme (Figure 14.39b) involves an initial nucleophilic attack by the carboxylate anion of Asp[52], in an associative S_N2 reaction, to form a covalent glycosyl-enzyme intermediate, a step that would occur with inversion of configuration. The enzyme carboxylate would then be displaced from the glycosyl-enzyme intermediate by water in a second step, which would also occur with inversion of configuration. The second step would involve Glu[35] as a general base, withdrawing a proton from a water molecule in the active site to produce hydroxide in the transition state. The result of these two inversion reactions would be a net retention of configuration at the glycosyl C-1 position. This latter mechanism would involve Asp[52] in covalent catalysis, with Glu[35] acting as a general base.

For many years, the dissociative, noncovalent mechanism (Figure 14.39a) was the favored choice for lysozyme, and also for other enzymatic reactions that cleave glycosidic linkages with net retention of configuration. However, recent experiments by Stephen Withers and his colleagues provide convincing evidence

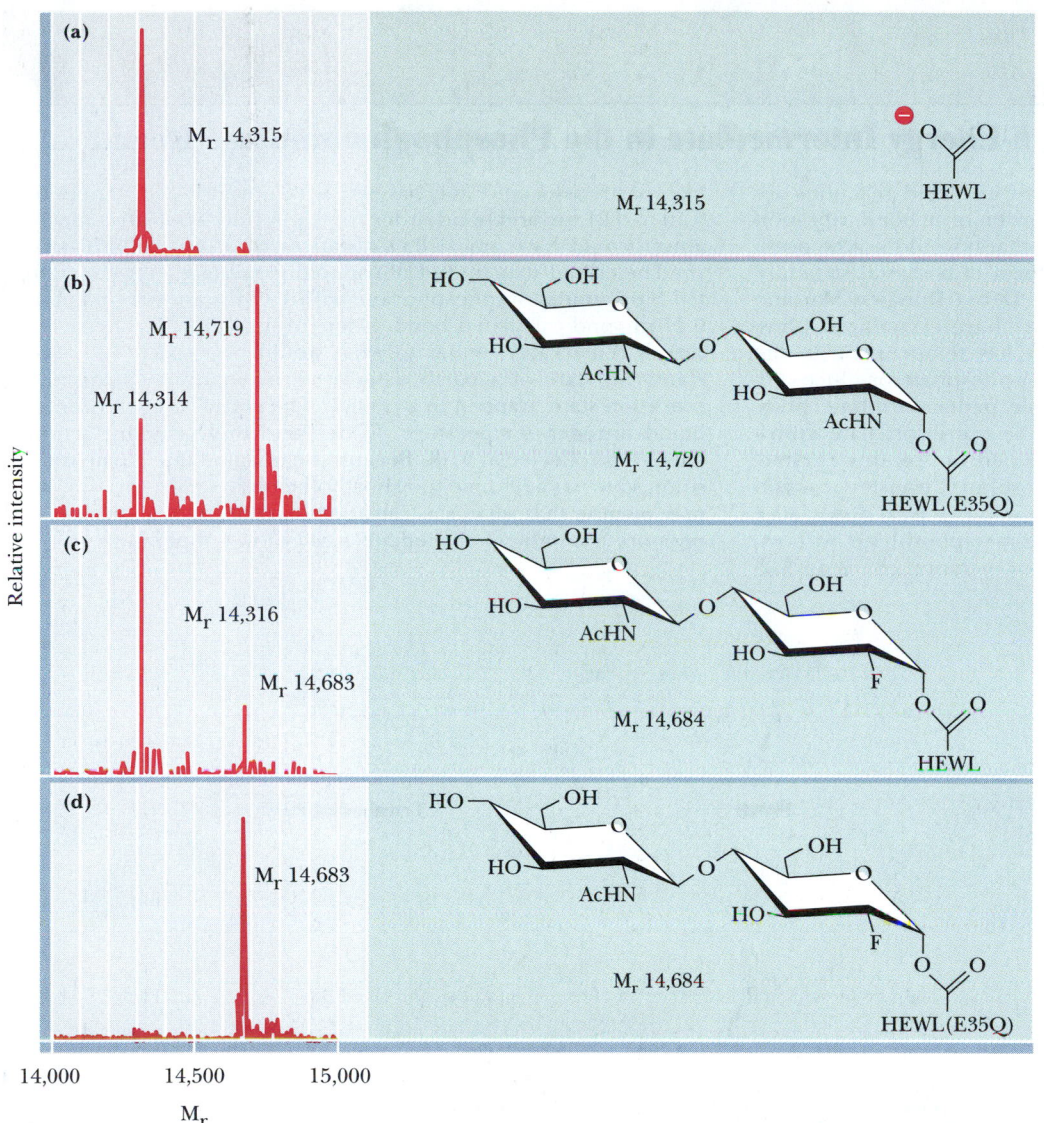

FIGURE 14.40 Mass spectra of lysozyme complexes. (a) Wild-type hen egg white lysozyme (HEWL). (b) Mutant lysozyme with Glu[35] replaced with glutamine [HEWL(E35Q)], incubated with chitobiosyl fluoride (NAG2F). (c) Wild-type HEWL, incubated with 2-acetamido-2-deoxy-β-D-glucopyranosyl-(1→4)-2-deoxy-2-fluoro-β-D-glucopyranosyl fluoride (NAG2FGlcF). (d) HEWL(E35Q), incubated with NAG2FGlcF. Structures of the species corresponding to each peak observed in the mass spectra are shown to the right, with their expected relative molecular mass. *(From Vocadlo, D. J., et al., 2000. Catalysis by hen egg-white lysozyme proceeds via a covalent intermediate. Nature **412**:835–838.)*

for a covalent Asp[52]–substrate intermediate and the pair of associative S_N2 mechanisms shown in Figure 14.39b. The challenge for Withers and his colleagues was to find conditions in which the rate of formation of the covalent intermediate would be faster than its rate of breakdown. They used a mutant lysozyme (in which Glu[35] is replaced by Gln) and several substrate analogs to prepare covalent enzyme–substrate complexes that could be observed unambiguously by electrospray ionization mass spectrometry (Figure 14.40). Then they succeeded in crystallizing the mutant lysozyme in a covalent complex with a difluorinated substrate analog (Figure 14.41). The X-ray diffraction structure clearly shows the covalent bond between Asp[52] and the C-1 carbon of the sugar in the D position in the active site. The structures in Figure 14.41 also show that carbon C-1 is located above the sugar ring in the noncovalent enzyme–substrate complex. Formation of the covalent complex involves an electrophilic migration of the C-1 carbon from above the ring plane to below the ring plane, where it approaches to within 1.6 to 1.8 Å of the Asp[52] oxygen—close enough to form a covalent bond

Critical Developments in Biochemistry

Caught in the Act! A High-Energy Intermediate in the Phosphoglucomutase Reaction

Because the transition states of enzyme-catalyzed reactions are imagined to have lifetimes on the order of a bond vibration (10^{-13} sec), it has long been assumed that it would not be possible to see a transition state in the form of a crystal structure solved by X-ray diffraction. However, Debra Dunaway-Mariano and Karen Allen and their colleagues have crystallized phosphorylated β-phosphoglucomutase at low temperature in the presence of Mg^{2+} and either glucose-1-phosphate or glucose-6-phosphate and have observed a stable pentacoordinate phosphorane that looks very much like the transition state anticipated for the phosphoryl transfer carried out by this enzyme. The most likely mechanisms for a phosphoryl transfer reaction are shown in the accompanying figure: (a) is a dissociative mechanism involving an intermediate metaphosphate, with expected apical P-O distances of 0.33 nm or more. (b) is an S_N2-

like, partly associative mechanism, with apical P-O distances of 0.19 to 0.21 nm and bond orders of 0.5. A fully-associative mechanism would have apical P-O distances of 0.166 to 0.176 nm. (c) The crystal structure of phosphoglucomutase shows a trigonal bipyramidal oxyphosphorane with P-O distances of 0.2 and 0.21 nm and calculated bond orders of 0.24 to 0.45. The structure is remarkably similar to what would be expected for the transition state of a partly associative mechanism. Is this the transition state, trapped in a crystal? The crystals were frozen at liquid nitrogen temperature (77 K), and the X-ray diffraction data were collected at 93 K. Because we imagine that a true transition state has a lifetime too short to be observed in this way, we may surmise that what is a transition state at physiological temperature is a stable intermediate at very low temperature.

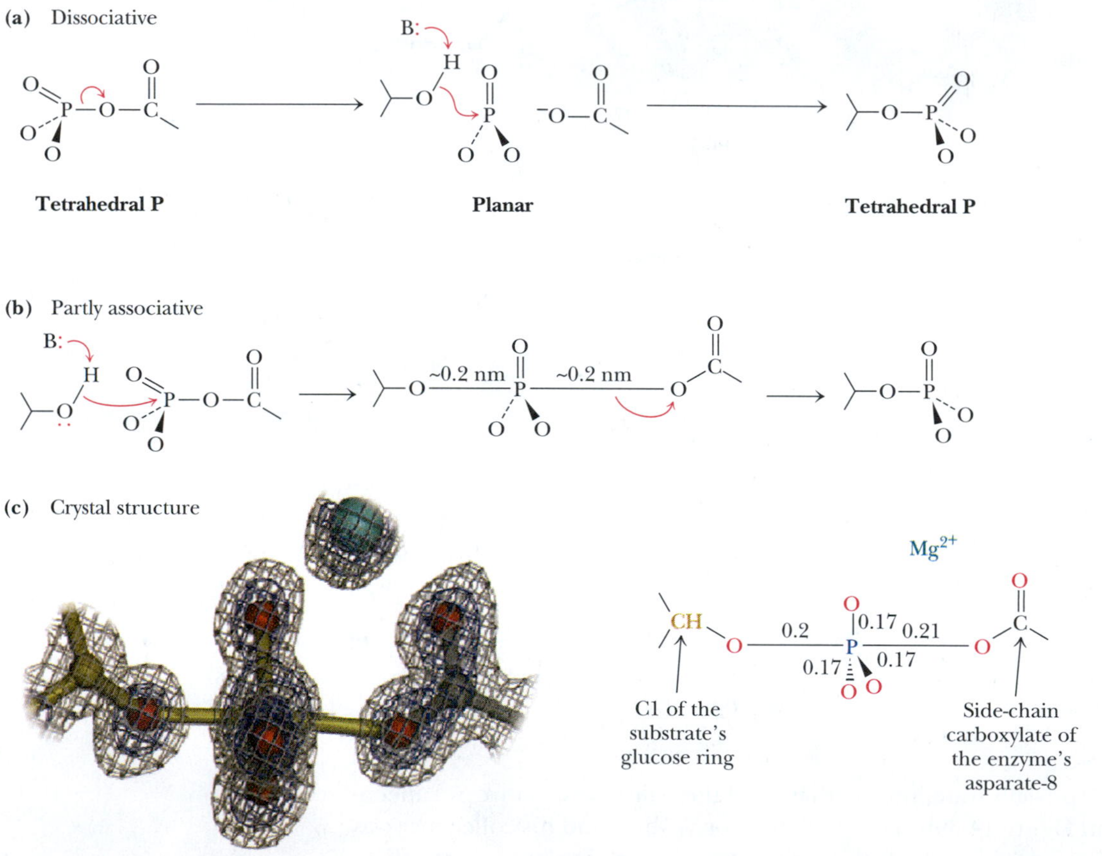

(a) Dissociative

Tetrahedral P   Planar   Tetrahedral P

(b) Partly associative

(c) Crystal structure

C1 of the substrate's glucose ring   Side-chain carboxylate of the enzyme's asparate-8

with this residue. The mass spectrometry and X-ray diffraction data provide unequivocal evidence that the mechanism of hen egg white lysozyme involves a covalent intermediate, as portrayed in Figure 14.39b.

The overall k_{cat} for lysozyme is about 0.5/sec, which is quite slow (Table 13.4) compared with that for other enzymes. On the other hand, the destruction of a bacterial cell wall may require hydrolysis of only a few polysaccharide chains. The high osmotic pressure of the cell ensures that cell rupture will follow rapidly. Thus, lysozyme can accomplish cell lysis without a particularly high k_{cat}.

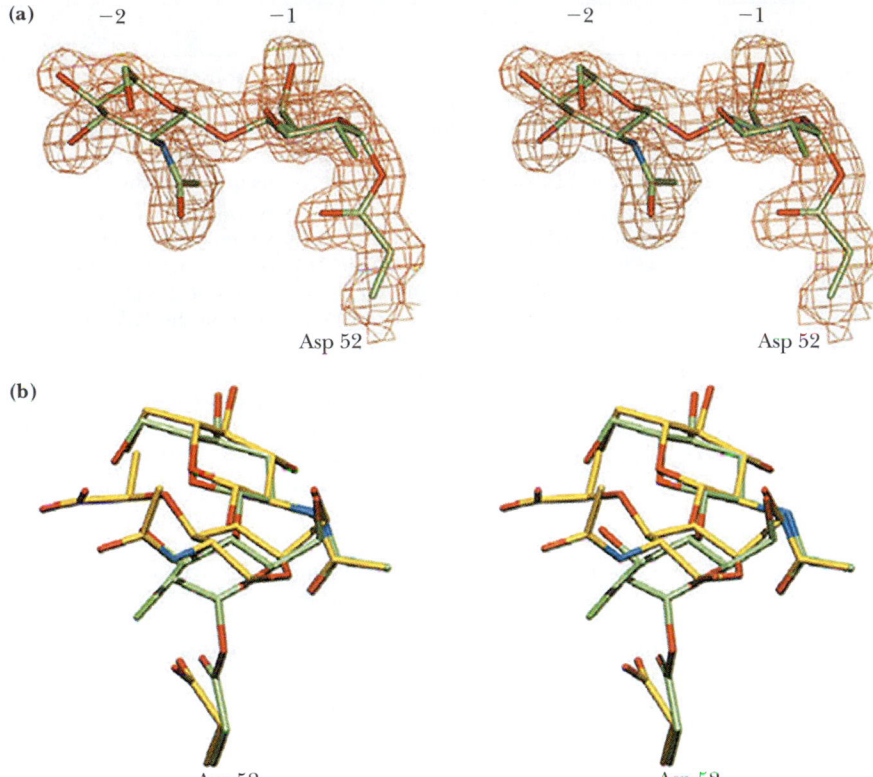

(a)

−2   −1   −2   −1

Asp 52   Asp 52

(b)

Asp 52   Asp 52

FIGURE 14.41 Stereo view of the covalent NAG2FG1cF intermediate of the lysozyme reaction. *(From Vocadlo, D. J., et al., 2000. Catalysis by hen egg-white lysozyme proceeds via a covalent intermediate.* Nature *412:835–838.)*

Summary

It is simply chemistry—the breaking and making of bonds—that gives enzymes their prowess. This chapter explores the unique features of this chemistry. The mechanisms of hundreds of enzymes have been studied in at least some detail.

14.1 What Role Does Transition-State Stabilization Play in Enzyme Catalysis?
The energy barrier for the uncatalyzed reaction is the difference in energies of the S and $X^{\ddagger}$ states. Similarly, the energy barrier to be surmounted in the enzyme-catalyzed reaction, assuming that E is saturated with S, is the energy difference between ES and $EX^{\ddagger}$. Reaction rate acceleration by an enzyme means simply that the energy barrier between ES and $EX^{\ddagger}$ is less than the energy barrier between S and $X^{\ddagger}$. In terms of the free energies of activation, $\Delta G_e^{\ddagger} < \Delta G_u^{\ddagger}$.

14.2 What Are the Magnitudes of Enzyme-Induced Rate Accelerations?
Enzymes are powerful catalysts. Enzyme-catalyzed reactions are typically 10^7 to 10^{14} times faster than their uncatalyzed counterparts and may exceed 10^{16}.

14.3 Why Is the Binding Energy of ES Crucial to Catalysis?
The favorable interactions between the substrate and amino acid residues on the enzyme account for the intrinsic binding energy, ΔG_b. The intrinsic binding energy ensures the favorable formation of the ES complex, but if uncompensated, it makes the activation energy for the enzyme-catalyzed reaction unnecessarily large and wastes some of the catalytic power of the enzyme. Because the enzymatic reaction rate is determined by the difference in energies between ES and $EX^{\ddagger}$, the smaller this difference, the faster the enzyme-catalyzed reaction. Tight binding of the substrate deepens the energy well of the ES complex and actually lowers the rate of the reaction.

14.4 What Roles Do Entropy Loss and Destabilization of the ES Complex Play?
Entropy is lost when two molecules (E and S) interact to form one molecule (the ES complex). Because ΔS is negative for this process, the term $-T\Delta S$ is a positive quantity, and the intrinsic binding energy of ES is compensated to some extent by the entropy loss that attends the formation of the complex. Destabilization of the ES complex can involve structural strain, desolvation, or electrostatic effects. Destabilization by strain or distortion is usually just a consequence of the fact that the enzyme is designed to bind the transition state more strongly than the substrate.

14.5 How Tightly Do Transition-State Analogs Bind to the Active Site?
Given a ratio k_e/k_u of 10^{12} and a typical K_S of 10^{-3} M, the value of K_T should be 10^{-15} M. This is the dissociation constant for the transition-state complex from the enzyme, and this very low value corresponds to very tight binding of the transition state by the enzyme. It is unlikely that such tight binding in an enzyme transition state will ever be measured experimentally, however, because the transition state itself is a "moving target."

14.6 What Are the Mechanisms of Catalysis?
Enzyme reaction mechanisms involve covalent bond formation, general acid–base catalysis, low-barrier hydrogen bonds, metal ion effects, and proximity of reactants. Most enzymes display involvement of two of these or more in any given reaction.

14.7 What Can Be Learned from Typical Enzyme Mechanisms?
The enzymes examined in this chapter—serine proteases, aspartic proteases, and lysozyme—all embody two or more of the rate enhancement contributions.

Problems

1. Tosyl-L-phenylalanine chloromethyl ketone (TPCK) specifically inhibits chymotrypsin by covalently labeling His[57].

Tosyl-L-phenylalanine chloromethyl ketone (TPCK)

   a. Propose a mechanism for the inactivation reaction, indicating the structure of the product(s).
   b. State why this inhibitor is specific for chymotrypsin.
   c. Propose a reagent based on the structure of TPCK that might be an effective inhibitor of trypsin.

2. In this chapter, the experiment in which Craik and Rutter replaced Asp[102] with Asn in trypsin (reducing activity 10,000-fold) was discussed.
   a. On the basis of your knowledge of the catalytic triad structure in trypsin, suggest a structure for the "uncatalytic triad" of Asn-His-Ser in this mutant enzyme.

   b. Explain why the structure you have proposed explains the reduced activity of the mutant trypsin.
   c. See the original journal articles (Sprang, et al., 1987. *Science* **237**:905–909 and Craik, et al., 1987. *Science* **237**:909–913) to see what Craik and Rutter's answer to this question was.

3. Pepstatin (see below) is an extremely potent inhibitor of the monomeric aspartic proteases, with K_I values of less than 1 nM.
   a. On the basis of the structure of pepstatin, suggest an explanation for the strongly inhibitory properties of this peptide.
   b. Would pepstatin be expected to also inhibit the HIV-1 protease? Explain your answer.

4. The k_{cat} for alkaline phosphatase–catalyzed hydrolysis of methylphosphate is approximately 14/sec at pH 8 and 25°C. The rate constant for the uncatalyzed hydrolysis of methylphosphate under the same conditions is approximately 1×10^{-15}/sec. What is the difference in the free energies of activation of these two reactions?

5. Active α-chymotrypsin is produced from chymotrypsinogen, an inactive precursor, as shown in the color figure below. The first intermediate—π-chymotrypsin—displays chymotrypsin activity. Suggest proteolytic enzymes that might carry out these cleavage reactions effectively.

6. Based on the following reaction scheme, derive an expression for k_e/k_u, the ratio of the rate constants for the catalyzed and uncatalyzed reactions, respectively, in terms of the free energies of

Pepstatin

Iva   Val   Val   Sta   Ala   Sta

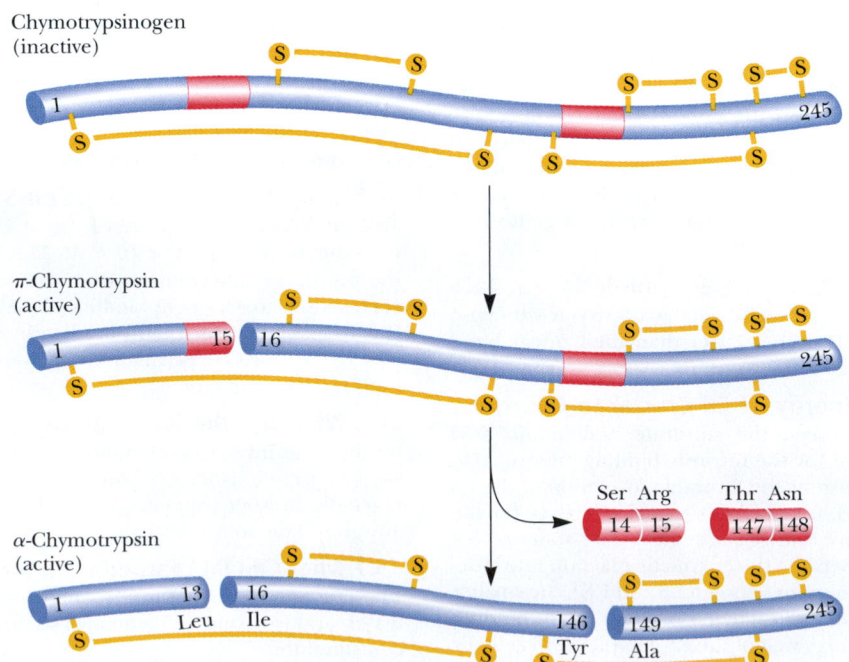

activation for the catalyzed $(\Delta G_e^{\ddagger})$ and the uncatalyzed $(\Delta G_u^{\ddagger})$ reactions.

$$S \xrightleftharpoons{K_u} X^{\ddagger} \xrightarrow{k_u'} P$$
$$E \xrightleftharpoons{K_s} \qquad \qquad \rightarrow E$$
$$ES \xrightleftharpoons{K_e} EX^{\ddagger} \xrightarrow{k_e'} EP$$

7. Consult a classic paper by William Lipscomb (1982. *Accounts of Chemical Research* **15**:232–238), and on the basis of this article write a simple mechanism for the enzyme carboxypeptidase A.

8. Consider the figure in the Deeper Look box on page 443. If the energy of the ES complex is 10 kJ/mol lower than the energy of E + S, the value of $\Delta G_e^{\ddagger}$ is 20 kJ/mol, and the value of $\Delta G_u^{\ddagger}$ is 90 kJ/mol. What is the rate enhancement achieved by an enzyme in this case?

Preparing for the MCAT Exam

The following graphs show the temperature and pH dependencies of four enzymes, A, B, X, and Y. Problems 9 through 15 refer to these graphs.

(a)

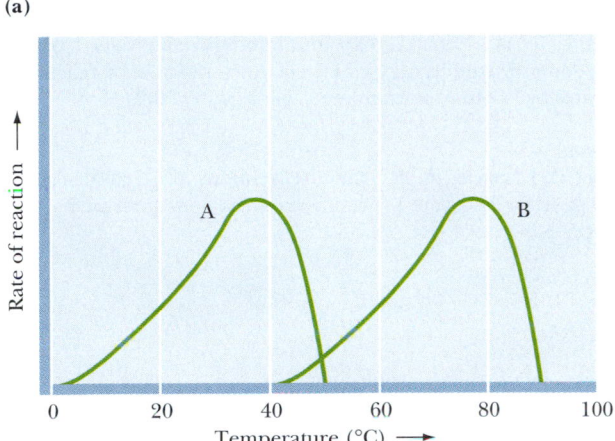

(b)

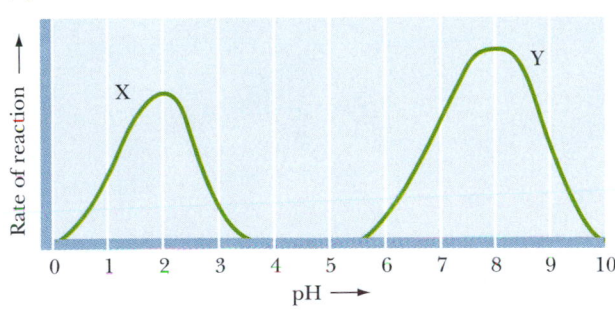

9. Enzymes X and Y in the figure are both protein-digesting enzymes found in humans. Where would they most likely be at work?
   a. X is found in the mouth, Y in the intestine.
   b. X in the small intestine, Y in the mouth.
   c. X in the stomach, Y in the small intestine.
   d. X in the small intestine, Y in the stomach.

10. Which statement is true concerning enzymes X and Y?
    a. They could not possibly be at work in the same part of the body at the same time.
    b. They have different temperature ranges at which they work best.
    c. At a pH of 4.5, enzyme X works slower than enzyme Y.
    d. At their appropriate pH ranges, both enzymes work equally fast.

11. What conclusion may be drawn concerning enzymes A and B?
    a. Neither enzyme is likely to be a human enzyme.
    b. Enzyme A is more likely to be a human enzyme.
    c. Enzyme B is more likely to be a human enzyme.
    d. Both enzymes are likely to be human enzymes.

12. At which temperatures might enzymes A and B both work?
    a. Above 40°C
    b. Below 50°C
    c. Above 50°C and below 40°C
    d. Between 40° and 50°C

13. An enzyme–substrate complex can form when the substrate(s) bind(s) to the active site of the enzyme. Which environmental condition might alter the conformation of an enzyme in the figure to the extent that its substrate is unable to bind?
    a. Enzyme A at 40°C
    b. Enzyme B at pH 2
    c. Enzyme X at pH 4
    d. Enzyme Y at 37°C

14. At 35°C, the rate of the reaction catalyzed by enzyme A begins to level off. Which hypothesis best explains this observation?
    a. The temperature is too far below optimum.
    b. The enzyme has become saturated with substrate.
    c. Both A and B.
    d. Neither A nor B.

15. In which of the following environmental conditions would digestive enzyme Y be unable to bring its substrate(s) to the transition state?
    a. At any temperature below optimum
    b. At any pH where the rate of reaction is not maximum
    c. At any pH lower than 5.5
    d. At any temperature higher than 37°C

Biochemistry⊗Now™ Preparing for an exam? Test yourself on key questions at http://chemistry.brookscole.com/ggb3

Further Reading

General

Eigen, M., 1964. Proton transfer, acid-base catalysis, and enzymatic hydrolysis. *Angewandte Chemie, Int. Ed.* **3**:1–72.

Gerlt, J. A., Kreevoy, M. M., Cleland, W. W., and Frey, P. A., 1997. Understanding enzymic catalysis: The importance of short, strong hydrogen bonds. *Chemistry and Biology* **4**:259–267.

Jencks, W., 1997. From chemistry to biochemistry to catalysis to movement. *Annual Review of Biochemistry* **66**:1–18.

Northrop, D. B., 1998. On the meaning of K_m and V/K in enzyme kinetics. *Journal of Chemical Education* **75**:1153–1157.

Radzicka, A., and Wolfenden, R., 1995. A proficient enzyme. *Science* **267**:90–93.

Simopoulos, T. T., and Jencks, W. P., 1994. Alkaline phosphatase is an almost perfect enzyme. *Biochemistry* **33**:10375–10380.

Smithrud, D. B., and Benkovic, S. J., 1997. The state of antibody catalysis. *Current Opinions in Biotechnology* **8**:459–466.

Walsh, C., 1979. *Enzymatic Reaction Mechanisms.* San Francisco: W. H. Freeman.

Transition-State Stabilization and Transition-State Analogs

Knowles, J. 2003. Seeing is believing. *Science* **299:**2002–2003.

Kraut, J., 1988. How do enzymes work? *Science* **242:**533–540.

Kreevoy, M., and Truhlar, D. G., 1986. Transition-state theory. Chapter 1 in *Investigations of Rates and Mechanisms of Reactions,* Bernasconi, C. F., ed. Vol. 6, Part 1. New York: John Wiley & Sons.

Lahiri, S. D., Zhang, G., Dunaway-Mariano, D., and Allen, K. N., 2003. The pentacovalent phosphorus intermediate of a phosphoryl transfer reaction. *Science* **299:**2067–2071.

Miller, B. G., and Wolfenden, R., 2002. Catalytic proficiency: The unusual case of OMP decarboxylase. *Annual Review of Biochemistry* **71:**847–885.

Radzicka, A., and Wolfenden, R., 1995. Transition state and multisubstrate analog inhibitors. *Methods in Enzymology* **249:**284–312.

Richards, N. G. J., 2000. Reaction mechanism. Part (iii) Bioorganic enzyme-catalyzed reactions. *Annual Report on the Progress of Chemistry, Sect. B* **96:**347–397.

Wolfenden, R., and Kati, W. M., 1991. Testing the limits of protein-ligand binding discrimination with transition-state analogue inhibitors. *Accounts of Chemical Research* **24:**209–215.

Serine Proteases

Cassidy, C. S., Lin, J., and Frey, P. A., 1997. A new concept for the mechanism of action of chymotrypsin: The role of the low-barrier hydrogen bond. *Biochemistry* **36:**4576–4584.

Craik, C. S., et al., 1987. The catalytic role of the active site aspartic acid in serine proteases. *Science* **237:**909–919.

Renatus, M., Engh, R. A., Stubbs, M. T., et al., 1997. Lysine-156 promotes the anomalous proenzyme activity of tPA: X-ray crystal structure of single-chain human tPA. *EMBO Journal* **16:**4797–4805.

Sprang, S., et al., 1987. The three-dimensional structure of Asn102 mutant of trypsin: Role of Asp102 in serine protease catalysis. *Science* **237:**905–909.

Aspartic Proteases

Northrop, D. B., 2001. Follow the protons: A low-barrier hydrogen bond unifies the mechanisms of the aspartic proteases. *Accounts of Chemical Research* **34:**790–797.

Toulokhonova, L., Metzler, W. J., Witmer, M. R., Copeland, R. A., and Marcinkeviciene, J., 2003. Kinetic studies on β-site amyloid precursor protein-cleaving enzyme (BACE). *Journal of Biological Chemistry* **278:**4582–4589.

HIV-1 Protease

Beaulieu, P. L., Wernic, D., Abraham, A., et al., 1997. Potent HIV protease inhibitors containing a novel (hydroxyethyl)amide isostere. *Journal of Medicinal Chemistry* **40:**2164–2176.

Carr, A., and Cooper, D. A., 1996. HIV protease inhibitors. *AIDS* **10:**S151–S157.

Chen, Z., Li, Y., Chen, E., et al., 1994. Crystal structure at 1.9-Å resolution of human immunodeficiency virus (HIV) II protease complexed with L-735,524, an orally bioavailable inhibitor of the HIV proteases. *Journal of Biological Chemistry* **269:**26344–26348.

Hyland, L., Tomaszek, T., and Meek, T., 1991. Human immunodeficiency virus-1 protease 2: Use of pH rate studies and solvent isotope effects to elucidate details of chemical mechanism. *Biochemistry* **30:**8454–8463.

Hyland, L., et al., 1991. Human immunodeficiency virus-1 protease 1: Initial velocity studies and kinetic characterization of reaction intermediates by ^{18}O isotope exchange. *Biochemistry* **30:**8441–8453.

Lysozyme

Vocadlo, D. J., Davies, G. J., Laine, R, Withers, S. G., 2001. Catalysis by hen egg-while lysozyme proceeds via a covalent intermediate. *Nature* **412:**835–838.

Enzyme Regulation

CHAPTER 15

Essential Question

Enzymes catalyze essentially all of the thousands of metabolic reactions taking place in cells. Many of these reactions are at cross-purposes: Some enzymes catalyze the breakdown of substances, whereas others catalyze synthesis of the same substances; many metabolic intermediates have more than one fate; and energy is released in some reactions and consumed in others. At key positions within the metabolic pathways, regulatory enzymes sense the momentary needs of the cell and adjust their catalytic activity accordingly. Regulation of these enzymes ensures the harmonious integration of the diverse and often divergent reactions of metabolism. *What are the properties of regulatory enzymes? How do regulatory enzymes sense the momentary needs of cells? What molecular mechanisms are used to regulate enzyme activity?*

15.1 | What Factors Influence Enzymatic Activity?

The activity displayed by enzymes is affected by a variety of factors, some of which are essential to the harmony of metabolism. Two of the more obvious ways to regulate the amount of activity at a given time are (1) to increase or decrease the number of enzyme molecules and (2) to increase or decrease the activity of each enzyme molecule. Although these ways are obvious, the cellular mechanisms that underlie them are complex and varied, as we shall see. A general overview of factors influencing enzyme activity includes the following considerations.

The Availability of Substrates and Cofactors Usually Determines How Fast the Reaction Goes

The availability of substrates and cofactors typically determines the enzymatic reaction rate. In general, enzymes have evolved such that their K_m values approximate the prevailing in vivo concentration of their substrates. (It is also true that the concentration of some enzymes in cells is within an order of magnitude or so of the concentrations of their substrates.)

As Product Accumulates, the Apparent Rate of the Enzymatic Reaction Will Decrease

The enzymatic rate, $v = d[P]/dt$, "slows down" as product accumulates and equilibrium is approached. The apparent decrease in rate is due to the conversion of P to S by the reverse reaction as [P] rises. Once $[P]/[S] = K_{eq}$, no further reaction is apparent. K_{eq} defines thermodynamic equilibrium. Enzymes have no influence on the thermodynamics of a reaction. Also, product inhibition can be a kinetically valid phenomenon: Some enzymes are actually inhibited by the products of their action.

Genetic Regulation of Enzyme Synthesis and Decay Determines the Amount of Enzyme Present at Any Moment

The amounts of enzyme synthesized by a cell are determined by transcription regulation (see Chapter 29). If the gene encoding a particular enzyme protein is turned on or off, changes in the amount of enzyme activity soon follow. **Induction,** which is the activation of enzyme synthesis, and **repression,** which is the shutdown of enzyme synthesis, are important mechanisms for the regulation of metabolism. By controlling the amount of an enzyme that is present at any moment, cells can either activate or terminate various metabolic routes. Genetic controls over enzyme levels have a response time ranging from minutes in rapidly

© Christie's Images/CORBIS

Metabolic regulation is achieved through an exquisitely balanced interplay among enzymes and small molecules, a process symbolized by the delicate balance of forces in this mobile by Alexander Calder.

Allostery is a key chemical process that makes possible intracellular and intercellular regulation: "…the molecular interactions which ensure the transmission and interpretation of (regulatory) signals rest upon (allosteric) proteins endowed with discriminatory stereospecific recognition properties."

Jacques Monod in *Chance and Necessity*

Key Questions

15.1 What Factors Influence Enzymatic Activity?

15.2 What Are the General Features of Allosteric Regulation?

15.3 Can a Simple Equilibrium Model Explain Allosteric Kinetics?

15.4 Is the Activity of Some Enzymes Controlled by Both Allosteric Regulation and Covalent Modification?

Special Focus: Is There an Example in Nature That Exemplifies the Relationship Between Quaternary Structure and the Emergence of Allosteric Properties? Hemoglobin and Myoglobin—Paradigms of Protein Structure and Function

Biochemistry ⊛ Now™ Test yourself on these Key Questions at BiochemistryNow at http://chemistry.brookscole.com/ggb3

dividing bacteria to hours (or longer) in higher eukaryotes. Once synthesized, the enzyme may also be degraded, either through normal turnover of the protein or through specific decay mechanisms that target the enzyme for destruction. These mechanisms are discussed in detail in Chapter 31.

Enzyme Activity Can Be Regulated Allosterically

Enzymatic activity can also be activated or inhibited through noncovalent interaction of the enzyme with small molecules (metabolites) other than the substrate. This form of control is termed **allosteric regulation,** because the activator or inhibitor binds to the enzyme at a site *other* than (*allo* means "other") the active site. Furthermore, such allosteric regulators, or **effector molecules,** are often quite different sterically from the substrate. Because this form of regulation results simply from reversible binding of regulatory ligands to the enzyme, the cellular response time can be virtually instantaneous.

Enzyme Activity Can Be Regulated Through Covalent Modification

Enzymes can be regulated by **covalent modification,** the reversible covalent attachment of a chemical group. Thus, a fully active enzyme can be converted into an inactive form simply by the covalent attachment of a functional group. For

A Deeper Look

Protein Kinases: Target Recognition and Intrasteric Control

Protein kinases are converter enzymes that catalyze the ATP-dependent phosphorylation of serine, threonine, or tyrosine hydroxyl groups in target proteins (see accompanying table). Phosphorylation introduces a bulky group bearing two negative charges, causing conformational changes that alter the target protein's function. (Unlike a phosphoryl group, no amino acid side chain can provide *two* negative charges.) Protein kinases represent a protein superfamily whose members are widely diverse in terms of size, subunit structure, and subcellular localization. Nevertheless, all share a common catalytic mechanism based on a conserved catalytic core/ kinase domain of approximately 260 amino acid residues (see accompanying figure). Protein kinases are classified as Ser/Thr and/ or Tyr specific. They also differ in terms of the target proteins that they recognize and phosphorylate; target selection depends on the presence of an amino acid sequence within the target protein that is recognized by the kinase. For example, cAMP-dependent protein kinase **(PKA)** phosphorylates proteins having Ser or Thr residues within an R(R/K)X(S*/T*) target consensus sequence (* denotes the residue that becomes phosphorylated). That is, PKA phosphorylates Ser or Thr residues that occur in an Arg-(Arg or Lys)-(any amino acid)-(Ser or Thr) sequence segment (see table).

Targeting of protein kinases to particular consensus sequence elements within proteins creates a means to regulate these kinases by **intrasteric control.** Intrasteric control occurs when a regulatory subunit (or protein domain) has a **pseudosubstrate sequence** that mimics the target sequence but lacks a OH-bearing side chain at the right place. For example, the cAMP-binding regulatory subunits of PKA (R subunits in Figure 15.6) possess the pseudosubstrate sequence RRGA*I, and this sequence binds to the active site of PKA catalytic subunits, blocking their activity. This pseudosubstrate sequence has an alanine residue where serine occurs in the PKA target sequence; Ala is sterically similar to serine but lacks a phosphorylatable OH group. When these PKA regulatory subunits bind cAMP, they undergo a conformational change and dissociate from the catalytic (C) subunits, and the active site of PKA is free to

bind and phosphorylate its targets. In other protein kinases, the pseudosubstrate sequence involved in intrasteric control and the kinase domain are part of the same polypeptide chain. In these cases, binding of an allosteric effector (like cAMP) induces a conformational change in the protein that releases the pseudosubstrate sequence from the active site of the kinase domain.

The abundance of many protein kinases in cells is an indication of the great importance of protein phosphorylation in cellular regulation. Exactly 113 protein kinase genes have been recognized in yeast, and 868 putative protein kinase genes have been identified in the human genome. **Tyrosine kinases** (protein kinases that phosphorylate Tyr residues) occur only in multicellular organisms (yeast has no tyrosine kinases). Tyrosine kinases are components of signaling pathways involved in cell–cell communication (see Chapter 32).

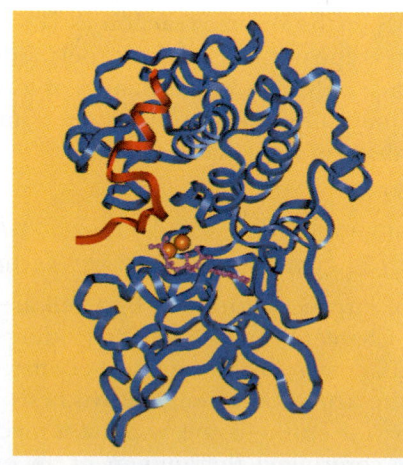

▲ Cyclic AMP-dependent protein kinase is shown complexed with a pseudosubstrate peptide (red). This complex also includes ATP (yellow) and two Mn²⁺ ions (violet) bound at the active site.

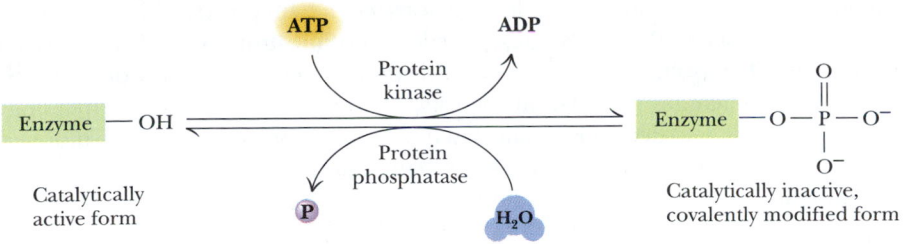

FIGURE 15.1 Enzymes regulated by covalent modification are called **interconvertible enzymes.** The enzymes (*protein kinase* and *protein phosphatase* in the example shown here) catalyzing the conversion of the interconvertible enzyme between its two forms are called **converter enzymes.** In this example, the free enzyme form is catalytically active, whereas the phosphoryl-enzyme form represents an inactive state. The —OH on the interconvertible enzyme represents an —OH group on a specific amino acid side chain in the protein (for example, a particular Ser residue) capable of accepting the phosphoryl group.

example, **protein kinases** are enzymes that act in covalent modification by attaching a phosphoryl moiety to target proteins (Figure 15.1). Alternatively, some enzymes exist in an inactive state unless specifically converted into the active form through covalent addition of a functional group. Covalent modification reactions are catalyzed by special **converter enzymes,** which are themselves subject to metabolic regulation. (Protein kinases are one class of converter enzymes.) Although covalent modification represents a stable alteration of the enzyme, a different converter enzyme operates to remove the modification, so when the conditions that favored modification of the enzyme are no longer present, the process can be reversed, restoring the enzyme to its unmodified state. Many examples of covalent

Classification of Protein Kinases		
Protein Kinase Class	**Target Sequence***	**Activators**
I. Ser/Thr protein kinases		
A. Cyclic nucleotide–dependent		
cAMP-dependent (PKA)	—R(R/K)X(S*/T*)—	cAMP
cGMP-dependent	—(R/K)KKX(S*/T*)—	cGMP
B. Ca^{2+}-calmodulin (CaM)–dependent		
Phosphorylase kinase (PhK)	—KRKQIS*VRGL—	phosphorylation by PKA
Myosin light-chain kinase (MLCK)	—KKRPQRATS*NV—	Ca^{2+}-CaM
C. Protein kinase C (PKC)		Ca^{2+}, diacylglycerol
D. Mitogen-activated protein kinases	—PXX(S*/T*)P—	phosphorylation
(MAP kinases)		by MAPK kinase
E. G-protein–coupled receptors		
β-Adrenergic receptor kinase (BARK)		
Rhodopsin kinase		
II. Ser/Thr/Tyr protein kinases		
MAP kinase kinase (MAPK kinase)	—TEY—	phosphorylation by *Raf* (a protein kinase)
III. Tyr protein kinases		
A. Cytosolic tyrosine kinases (*src, fgr, abl,* etc.)		
B. Receptor tyrosine kinases (RTKs)		
Plasma membrane receptors for hormones such as *epidermal growth factor* (EGF) or *platelet-derived growth factor* (PDGF)		

*X denotes any amino acid.

modification at important metabolic junctions will be encountered in our discussions of metabolic pathways. Because covalent modification events are catalyzed by enzymes, they occur very quickly, with response times of seconds or even less for significant changes in metabolic activity. The 1992 Nobel Prize in Physiology or Medicine was awarded to Edmond Fischer and Edwin Krebs for their pioneering studies of reversible protein phosphorylation as an important means of cellular regulation via covalent modification.

Regulation of Enzyme Activity Also Can Be Accomplished in Other Ways

Enzyme regulation is an important matter to cells, and evolution has provided a variety of additional options, including zymogens, isozymes, and modulator proteins. We will discuss these options first and then return to the major topics of this chapter—enzyme regulation through allosteric mechanisms and covalent modification.

Zymogens Are Inactive Precursors of Enzymes

Most proteins become fully active as their synthesis is completed and they spontaneously fold into their native, three-dimensional conformations. Some proteins, however, are synthesized as inactive precursors, called **zymogens** or **proenzymes,** that acquire full activity only upon specific proteolytic cleavage of one or several of their peptide bonds. Unlike allosteric regulation or covalent modification, zymogen activation by specific proteolysis is an irreversible process. Activation of enzymes and other physiologically important proteins by specific proteolysis is a strategy frequently exploited by biological systems to switch on processes at the appropriate time and place, as the following examples illustrate.

INSULIN. Some protein hormones are synthesized in the form of inactive precursor molecules, from which the active hormone is derived by proteolysis. For instance, **insulin,** an important metabolic regulator, is generated by proteolytic excision of a specific peptide from **proinsulin** (Figure 15.2).

PROTEOLYTIC ENZYMES OF THE DIGESTIVE TRACT. Enzymes of the digestive tract that serve to hydrolyze dietary proteins are synthesized in the stomach and pancreas as zymogens (Table 15.1). Only upon proteolytic activation are these enzymes able to form a catalytically active substrate-binding site. The activation of chymotrypsinogen is an interesting example (Figure 15.3). **Chymotrypsinogen** is a 245-residue polypeptide chain crosslinked by five disulfide bonds. Chymotrypsinogen is converted to an enzymatically active form called π-chymotrypsin when trypsin cleaves the peptide bond joining Arg^{15} and Ile^{16}. The enzymatically active π-chymotrypsin acts upon other π-chymotrypsin molecules, excising two dipeptides: Ser^{14}-Arg^{15} and Thr^{147}-Asn^{148}. The end product of this processing pathway is the mature protease **α-chymotrypsin,** in which the three peptide chains, A (residues 1 through 13), B (residues 16 through 146), and C (residues

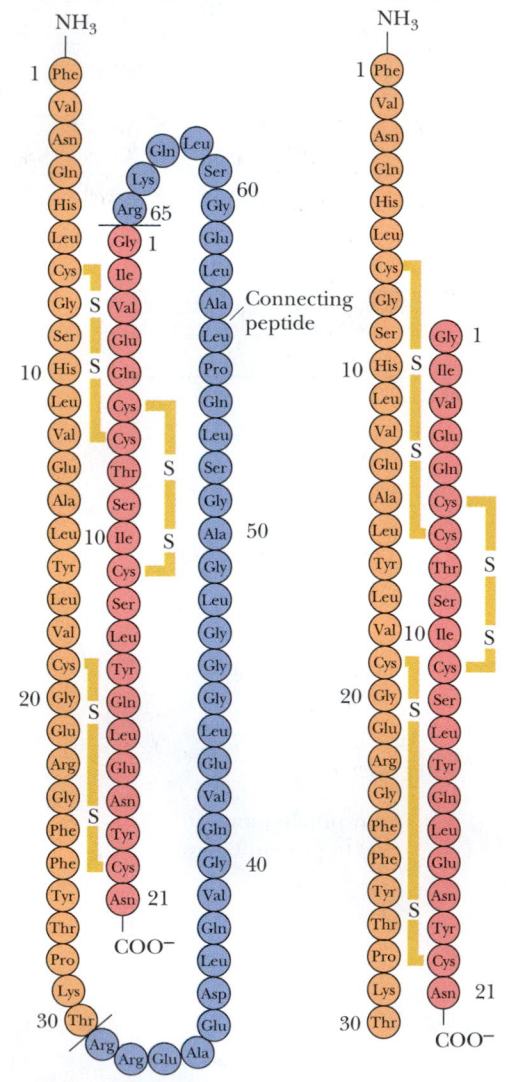

FIGURE 15.2 Proinsulin is an 86-residue precursor to insulin (the sequence shown here is human proinsulin). Proteolytic removal of residues 31 to 65 yields insulin. Residues 1 through 30 (the B chain) remain linked to residues 66 through 87 (the A chain) by a pair of interchain disulfide bridges.

Table 15.1		
Pancreatic and Gastric Zymogens		
Origin	**Zymogen**	**Active Protease**
Pancreas	Trypsinogen	Trypsin
Pancreas	Chymotrypsinogen	Chymotrypsin
Pancreas	Procarboxypeptidase	Carboxypeptidase
Pancreas	Proelastase	Elastase
Stomach	Pepsinogen	Pepsin

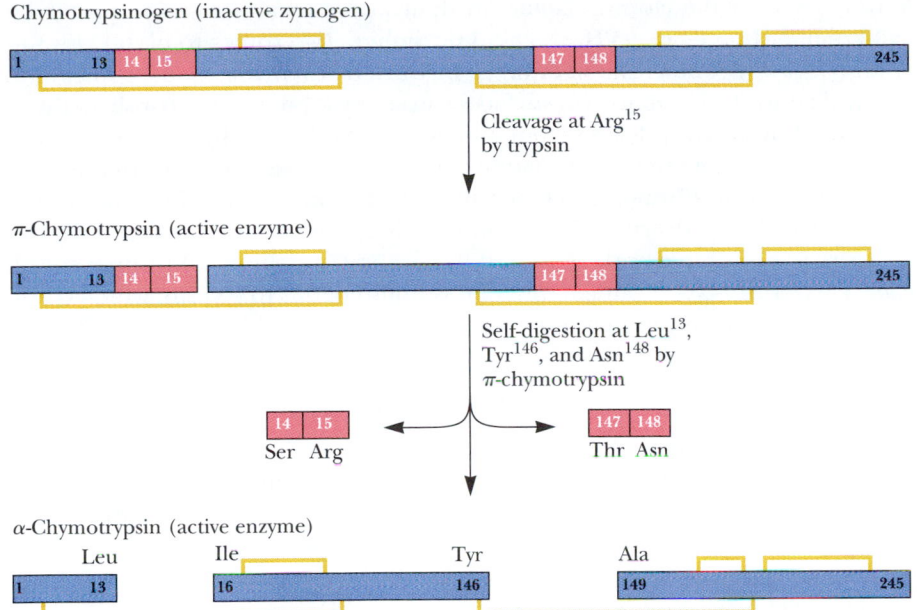

Chymotrypsinogen (inactive zymogen)

Cleavage at Arg^{15}
by trypsin

π-Chymotrypsin (active enzyme)

Self-digestion at Leu^{13},
Tyr^{146}, and Asn^{148} by
π-chymotrypsin

14 15
Ser Arg

147 148
Thr Asn

α-Chymotrypsin (active enzyme)

Biochemistry ⑤ **Now**™   **ANIMATED FIGURE 15.3**
The proteolytic activation of chymotrypsinogen. **See this figure animated at http://chemistry.brookscole.com/ggb3**

149 through 245), remain together because they are linked by two disulfide bonds, one from A to B and one from B to C. It is interesting to note that the transformation of inactive chymotrypsinogen to active π-chymotrypsin requires the cleavage of just one particular peptide bond.

BLOOD CLOTTING. The formation of blood clots is the result of a series of zymogen activations (Figure 15.4). The amplification achieved by this cascade of enzymatic activations allows blood clotting to occur rapidly in response to

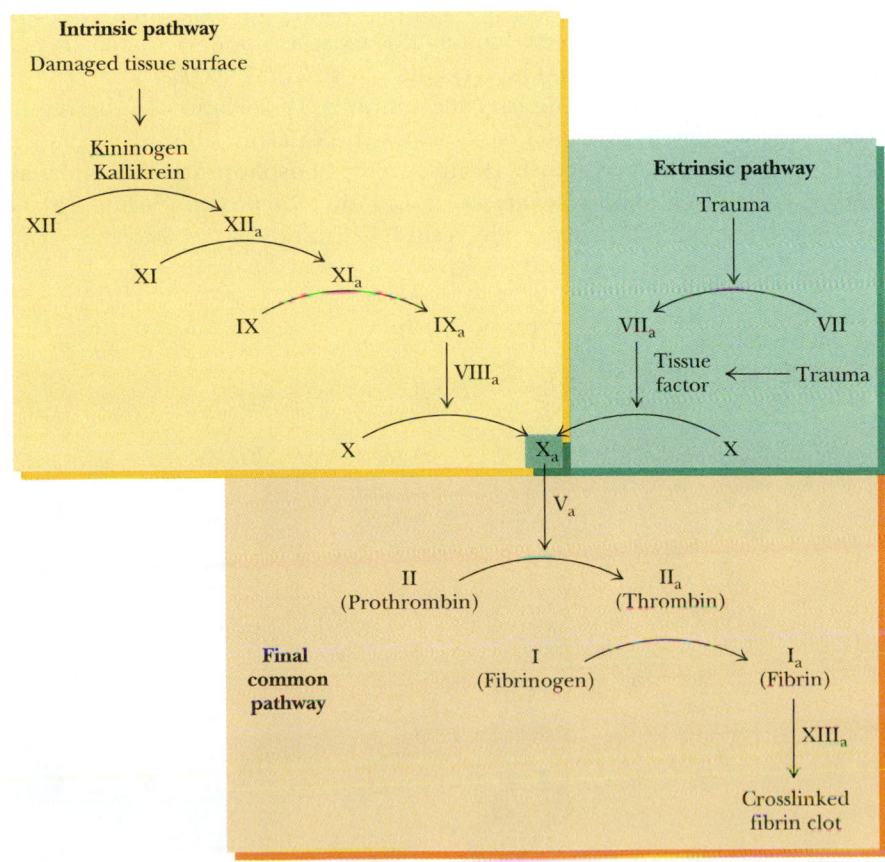

FIGURE 15.4 The cascade of activation steps leading to blood clotting. The intrinsic and extrinsic pathways converge at factor X, and the final common pathway involves the activation of thrombin and its conversion of fibrinogen into fibrin, which aggregates into ordered filamentous arrays that become crosslinked to form the clot.

injury. Seven of the clotting factors in their active form are serine proteases: **kallikrein, XII$_a$, XI$_a$, IX$_a$, VII$_a$, X$_a$,** and **thrombin.** Two routes to blood clot formation exist. The **intrinsic pathway** is instigated when the blood comes into physical contact with abnormal surfaces caused by injury; the **extrinsic pathway** is initiated by factors released from injured tissues. The pathways merge at factor X and culminate in clot formation. Thrombin excises peptides rich in negative charge from **fibrinogen,** converting it to **fibrin,** a molecule with a different surface charge distribution. Fibrin readily aggregates into ordered fibrous arrays that are subsequently stabilized by covalent crosslinks. Thrombin specifically cleaves Arg-Gly peptide bonds and is homologous to trypsin, which is also a serine protease (recall that trypsin acts only at Arg and Lys residues).

Isozymes Are Enzymes with Slightly Different Subunits

A number of enzymes exist in more than one quaternary form, differing in their relative proportions of structurally equivalent but catalytically distinct polypeptide subunits. A classic example is mammalian **lactate dehydrogenase (LDH),** which exists as five different isozymes, depending on the tetrameric association of two different subunits, A and B: A$_4$, A$_3$B, A$_2$B$_2$, AB$_3$, and B$_4$ (Figure 15.5). The kinetic properties of the various LDH isozymes differ in terms of their relative affinities for the various substrates and their sensitivity to inhibition by product. Different tissues express different isozyme forms, as appropriate to their particular metabolic needs. By regulating the relative amounts of A and B subunits they synthesize, the cells of various tissues control which isozymic forms are likely to assemble and thus which kinetic parameters prevail.

Modulator Proteins Regulate Enzymes Through Reversible Binding

Modulator proteins are yet another way that cells mediate metabolic activity. **Modulator proteins** are proteins that bind to enzymes and, by binding, influence the activity of the enzyme. For example, some enzymes, such as **cAMP-dependent protein kinase** (see Chapter 23), exist as dimers of catalytic subunits and regulatory subunits. These regulatory subunits are *modulator proteins* that suppress the activity of the catalytic subunits. Dissociation of the regulatory subunits (modulator proteins) activates the catalytic subunits; reassociation once again suppresses activity (Figure 15.6). **Phosphoprotein phosphatase inhibitor-1 (PPI-1)** is another example of a modulator protein. When PPI-1 is phosphorylated on one of its serine residues, it binds to *phosphoprotein phos-*

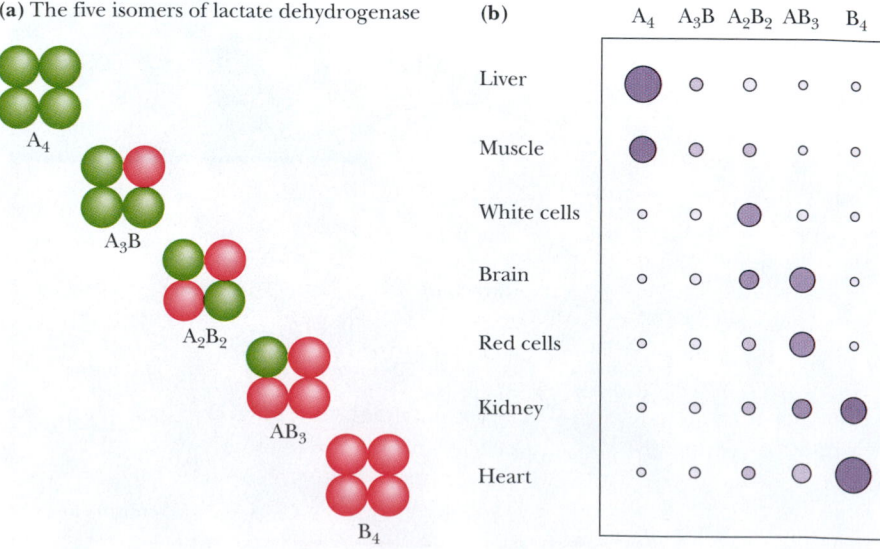

Biochemistry ⊜ Now™ ACTIVE FIGURE 15.5
The isozymes of lactate dehydrogenase (LDH). Active muscle tissue becomes anaerobic and produces pyruvate from glucose via glycolysis (see Chapter 18). It needs LDH to regenerate NAD$^+$ from NADH so that glycolysis can continue. The lactate produced is released into the blood. The muscle LDH isozyme (A$_4$) works best in the NAD$^+$-regenerating direction. Heart tissue is aerobic and uses lactate as a fuel, converting it to pyruvate via LDH and using the pyruvate to fuel the citric acid cycle to obtain energy. The heart LDH isozyme (B$_4$) is inhibited by excess pyruvate so that the fuel won't be wasted. **Test yourself on the concepts in this figure at http://chemistry. brookscole.com/ggb3**

(a) The five isomers of lactate dehydrogenase

A$_4$
A$_3$B
A$_2$B$_2$
AB$_3$
B$_4$

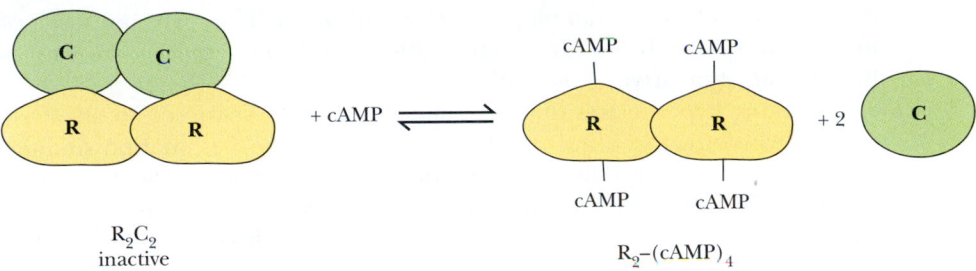

R_2C_2
inactive

$R_2-(cAMP)_4$

Biochemistry ⟨ Now™ ANIMATED
FIGURE 15.6 Cyclic AMP–dependent protein kinase (also known as *PKA*) is a 150- to 170-kD R_2C_2 tetramer in mammalian cells. The two R (regulatory) subunits bind cAMP ($K_D = 3 \times 10^{-8}$ *M*); cAMP binding releases the R subunits from the C (catalytic) subunits. C subunits are enzymatically active as monomers. **See this figure animated at http://chemistry. brookscole.com/ggb3**

phatase (Figure 15.1), inhibiting its phosphatase activity. The result is an increased phosphorylation of the interconvertible enzyme targeted by the protein kinase/phosphoprotein phosphatase cycle (Figure 15.1). We will meet other important representatives of this class as the processes of metabolism unfold in subsequent chapters. For now, let us focus our attention on the fascinating kinetics of allosteric enzymes.

15.2 What Are the General Features of Allosteric Regulation?

Allosteric regulation acts to modulate enzymes situated at key steps in metabolic pathways. Consider as an illustration the following pathway, where A is the precursor for formation of an end product, F, in a sequence of five enzyme-catalyzed reactions:

$$A \xrightarrow{\text{enz 1}} B \xrightarrow{\text{enz 2}} C \xrightarrow{\text{enz 3}} D \xrightarrow{\text{enz 4}} E \xrightarrow{\text{enz 5}} F$$

In this scheme, F symbolizes an essential metabolite, such as an amino acid or a nucleotide. In such systems, F, the essential end product, inhibits *enzyme 1*, the *first step* in the pathway. Therefore, when sufficient F is synthesized, it blocks further synthesis of itself. This phenomenon is called **feedback inhibition** or **feedback regulation**.

Regulatory Enzymes Have Certain Exceptional Properties

Enzymes such as enzyme 1, which are subject to feedback regulation, represent a distinct class of enzymes, the **regulatory enzymes.** As a class, these enzymes have certain exceptional properties:

1. Their kinetics do not obey the Michaelis–Menten equation. Their *v* versus [S] plots yield **sigmoid-** or **S-shaped** curves rather than rectangular hyperbolas (Figure 15.7). Such curves suggest a second-order (or higher) relationship between *v* and [S]; that is, *v* is proportional to [S]^n, where $n > 1$. A qualitative description of the mechanism responsible for the S-shaped curves is that binding of one S to a protein molecule makes it easier for additional substrate molecules to bind to the same protein molecule. In the jargon of allostery, substrate binding is **cooperative.**

2. Inhibition of a regulatory enzyme by a feedback inhibitor does not conform to any normal inhibition pattern, and the feedback inhibitor F bears little structural similarity to A, the substrate for the regulatory enzyme. F apparently acts at a binding site distinct from the substrate-binding site. The term *allosteric* is apt, because F is sterically dissimilar and, moreover, acts at a site other than the site for S. Its effect is called **allosteric inhibition.**

3. Regulatory or allosteric enzymes like enzyme 1 are, in some instances, regulated by activation. That is, whereas some effector molecules such as F exert negative effects on enzyme activity, other effectors show stimulatory, or positive, influences on activity.

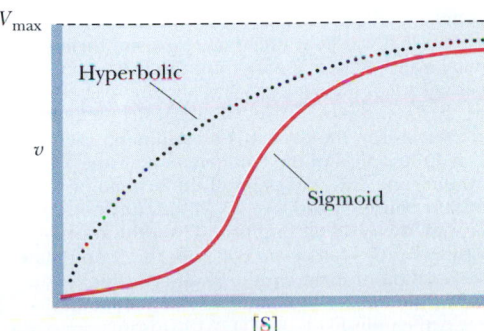

FIGURE 15.7 Sigmoid *v* versus [S] plot. The dotted line represents the hyperbolic plot characteristic of normal Michaelis–Menten-type enzyme kinetics.

4. Allosteric enzymes have an oligomeric organization. They are composed of more than one polypeptide chain (subunit) and have more than one S-binding site per enzyme molecule.

5. The working hypothesis is that, by some means, interaction of an allosteric enzyme with effectors alters the distribution of conformational possibilities or subunit interactions available to the enzyme. That is, the regulatory effects exerted on the enzyme's activity are achieved by conformational changes occurring in the protein when effector metabolites bind.

In addition to enzymes, noncatalytic proteins may exhibit many of these properties; hemoglobin is the classic example. The allosteric properties of hemoglobin are the subject of a Special Focus at the end of this chapter.

15.3 | Can a Simple Equilibrium Model Explain Allosteric Kinetics?

Monod, Wyman, and Changeux Proposed the Symmetry Model for Allosteric Regulation

In 1965, Jacques Monod, Jeffries Wyman, and Jean-Pierre Changeux proposed a theoretical model of allosteric transitions based on the observation that allosteric proteins are oligomers. They suggested that allosteric proteins can exist in (at least) two conformational states, designated **R,** signifying "relaxed," and **T,** or "taut," and that, in each protein molecule, all of the subunits have the same conformation (either R or T). That is, molecular symmetry is conserved. Molecules of mixed conformation (having subunits of both R and T states) are not allowed by this model.

In the absence of ligand, the two states of the allosteric protein are in equilibrium:

$$R_0 \rightleftharpoons T_0$$

(Note that the subscript "0" signifies "in the absence of ligand.") The equilibrium constant is termed L: $L = T_0/R_0$. L is assumed to be large; that is, the amount of the protein in the T conformational state is much greater than the amount in the R conformation. Let us suppose that $L = 10^4$.

The affinities of the two states for substrate, S, are characterized by the respective dissociation constants, K_R and K_T. The model supposes that $K_T \gg K_R$. That is, the affinity of R_0 for S is much greater than the affinity of T_0 for S. Let us choose the extreme where $K_R/K_T = 0$ (that is, K_T is infinitely greater than K_R). In effect, we are picking conditions in which S binds only to R. (If K_T is infinite, T does not bind S.)

Given these parameters, consider what happens when S is added to a solution of the allosteric protein at conformational equilibrium (Figure 15.8). Although the relative $[R_0]$ concentration is small, S will bind "only" to R_0, forming R_1. This depletes the concentration of R_0, perturbing the T_0/R_0 equilibrium. To restore equilibrium, molecules in the T_0 conformation undergo a transition to R_0. This shift renders more R_0 available to bind S, yielding R_1, diminishing $[R_0]$, perturbing the T_0/R_0 equilibrium, and so on. Thus, these linked equilibria (Figure 15.8) are such that S-binding by the R_0 state of the allosteric protein perturbs the T_0/R_0 equilibrium with the result that S-binding drives the conformational transition, $T_0 \rightarrow R_0$.

In just this simple system, *cooperativity* is achieved because each subunit has a binding site for S, and thus, *each protein molecule has more than one binding site for S.* Therefore, the increase in the population of R conformers gives a progressive increase in the number of sites available for S. The extent of cooperativity depends on the relative T_0/R_0 ratio and the relative affinities of R and T for S. If L is large (that is, the equilibrium lies strongly in favor of T_0) and if $K_T \gg K_R$, as in the example we have chosen, cooperativity is great (Figure 15.9). Ligands

(a) A dimeric protein can exist in either of two conformational states at equilibrium.

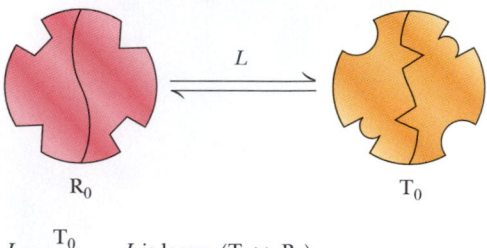

$$L = \frac{T_0}{R_0} \qquad L \text{ is large. } (T_0 \gg R_0)$$

(b) Substrate binding shifts equilibrium in favor of R.

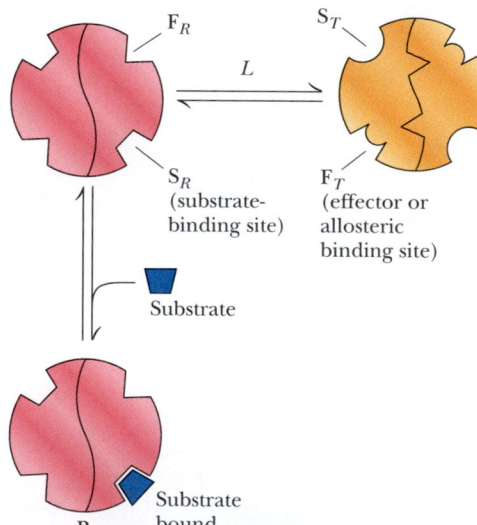

F_R S_T L

S_R F_T
(substrate-binding site)   (effector or allosteric binding site)

Substrate

Substrate bound

FIGURE 15.8 Monod–Wyman–Changeux (MWC) model for allosteric transitions. Consider a dimeric protein that can exist in either of two conformational states, R or T. Each subunit in the dimer has a binding site for substrate S and an allosteric effector site, F. The promoters are symmetrically related to one another in the protein, and symmetry is conserved regardless of the conformational state of the protein. The different states of the protein, with or without bound ligand, are linked to one another through the various equilibria. Thus, the relative population of protein molecules in the R or T state is a function of these equilibria and the concentration of the various ligands, substrate (S), and effectors (which bind at F_R or F_T). As [S] is increased, the T/R equilibrium shifts in favor of an increased proportion of R conformers in the total population (that is, more protein molecules in the R conformational state).

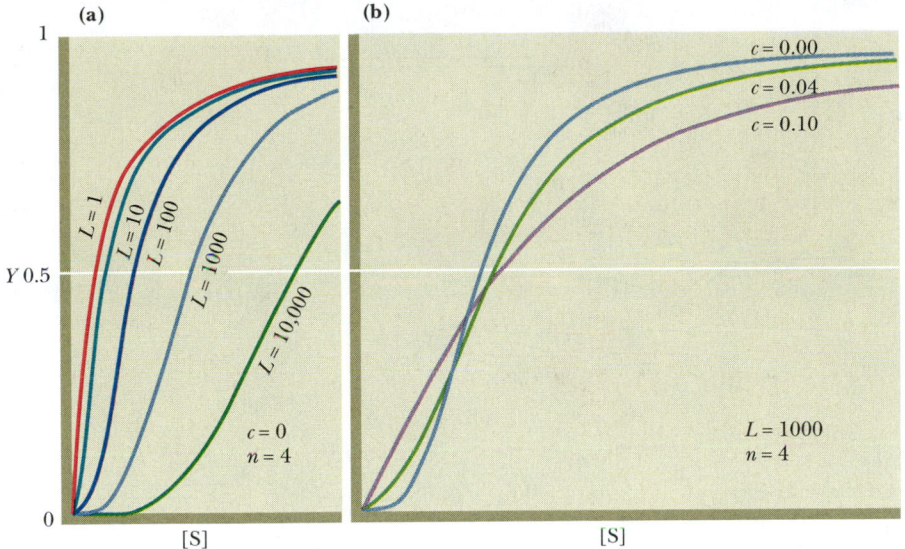

Biochemistry Now™   ANIMATED FIGURE 15.9
The Monod–Wyman–Changeux model. Graphs of allosteric effects for a tetramer ($n = 4$) in terms of Y,
the saturation function, versus [S]. Y is defined as
[ligand-binding sites that are occupied by ligand]/
[total ligand-binding sites]. **(a)** A plot of Y as a function of [S], at various L values. **(b)** Y as a function of
[S], at different c, where $c = K_R/K_T$. (When $c = 0$,
K_T is infinite.) *(Adapted from Monod, J., Wyman, J., and
Changeux, J-P., 1965. On the nature of allosteric transitions: A
plausible model.* Journal of Molecular Biology *12:92.)* **See
this figure animated at http://chemistry.brookscole.
com/ggb3**

such as S here that bind in a cooperative manner, so that binding of one equivalent enhances the binding of additional equivalents of S to the same protein
molecule, are termed **positive homotropic effectors.** (The prefix *homo* indicates
that the ligand influences the binding of like molecules.)

Heterotropic Effectors Influence the Binding of Other Ligands

This simple system also provides an explanation for the more complex substrate-
binding responses to positive and negative effectors. Effectors that influence the
binding of something other than themselves are termed **heterotropic effectors.**
For example, effectors that promote S binding are termed **positive heterotropic
effectors** or **allosteric activators.** Effectors that diminish S binding are **negative
heterotropic effectors** or **allosteric inhibitors.** Feedback inhibitors fit this class.
Consider a protein composed of two subunits, each of which has two binding
sites: one for the substrate, S, and one to which allosteric effectors bind, the
allosteric site. Assume that S binds preferentially ("only") to the R conformer; further assume that the *positive heterotropic effector,* A, binds to the allosteric site only
when the protein is in the R conformation and the *negative allosteric effector,* I,
binds at the allosteric site only if the protein is in the T conformation. Thus, with
respect to binding at the allosteric site, A and I are competitive with each other.

Positive Effectors Increase the Number of Binding Sites for a Ligand

If A binds to R_0, forming the new species $R_{1(A)}$, the relative concentration of
R_0 is decreased and the T_0/R_0 equilibrium is perturbed (Figure 15.10). As a
consequence, a relative $T_0 \rightarrow R_0$ shift occurs in order to restore equilibrium.
The net effect is an increase in the number of R conformers in the presence
of A, meaning that more binding sites for S are available. For this reason, A
leads to a decrease in the cooperativity of the substrate saturation curve, as
seen by a shift of this curve to the left (Figure 15.10). Effectively, the presence
of A lowers the apparent value of L.

Negative Effectors Decrease the Number of Binding Sites Available to a Ligand

The converse situation applies in the presence of I, which binds "only" to T.
I binding will lead to an increase in the population of T conformers, at the
expense of R_0 (Figure 15.10). The decline in [R_0] means that it is less likely for
S (or A) to bind. Consequently, the presence of I increases the cooperativity

A dimeric protein that can exist in
either of two states: R_0 or T_0.
This protein can bind three ligands:

1) Substrate (S) : A positive homotropic
   effector that binds
   only to R at site S

2) Activator (A) : A positive heterotropic
   effector that binds
   only to R at site F

3) Inhibitor (I) : A negative heterotropic
   effector that binds
   only to T at site F

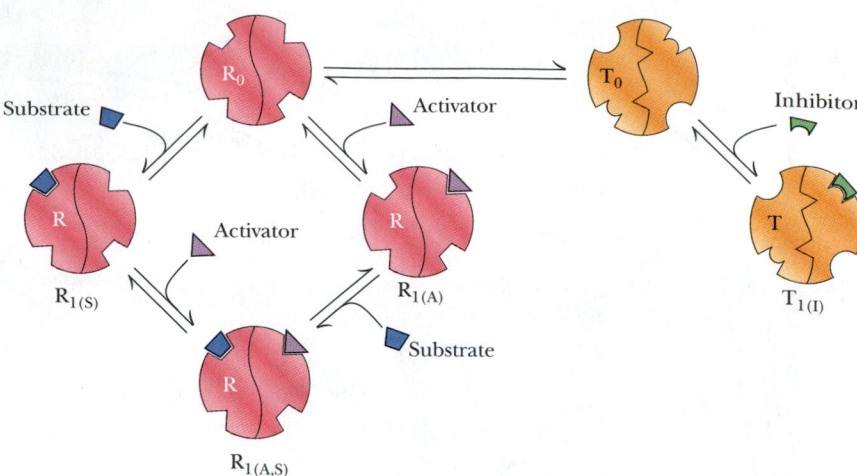

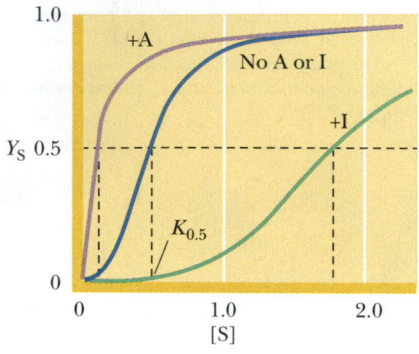

Effects of A:
$A + R_0 \longrightarrow R_{1(A)}$
Increase in number of
R-conformers shifts $R_0 \rightleftharpoons T_0$
so that $T_0 \longrightarrow R_0$

(1) More binding sites for S made
available.

(2) Decrease in cooperativity of
substrate saturation curve. Effector A
lowers the apparent value of L.

Effects of I:
$I + T_0 \longrightarrow T_{1(I)}$
Increase in number of
T-conformers (decrease in R_0 as $R_0 \longrightarrow T_0$
to restore equilibrium)

Thus, I inhibits association of S and A
with R by lowering R_0 level. I increases
cooperativity of substrate saturation curve.
I raises the apparent value of L.

Biochemistry Now™   ACTIVE FIGURE 15.10 Heterotropic allosteric effects: A and I binding to R and T, respectively. The linked equilibria lead to changes in the relative amounts of
R and T and, therefore, shifts in the substrate saturation curve. This behavior, depicted by the
graph, defines an allosteric "K system." The parameters of such a system are that (1) S and A
(or I) have different affinities for R and T and (2) A (or I) modifies the apparent $K_{0.5}$ for S by
shifting the relative R versus T population. **Test yourself on the concepts in this figure at
http://chemistry.brookscole.com/ggb3**

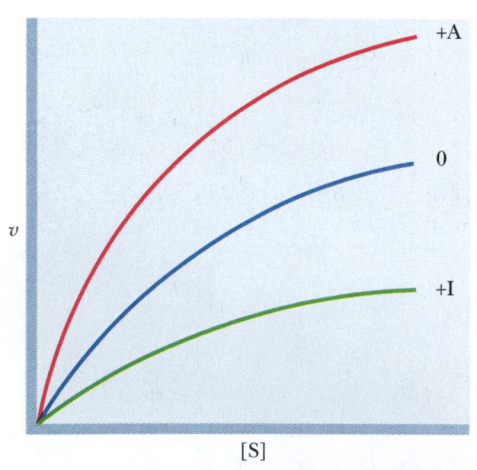

FIGURE 15.11 v versus [S] curves for an allosteric
"V system." The V system fits the model of Monod,
Wyman, and Changeux, given the following conditions: (1) R and T have the *same* affinity for the substrate, S. (2) The effectors A and I have different
affinities for R and T and thus can shift the relative
T/R distribution. (That is, A and I change the
apparent value of L.) Assume as before that A binds
"only" to the R state and I binds "only" to the T state.
(3) R and T differ in their catalytic ability. Assume
that R is the enzymatically active form, whereas T is
inactive. Because A perturbs the T/R equilibrium in
favor of more R, A increases the apparent V_{max}. I
favors transition to the inactive T state.

(that is, the sigmoidicity) of the substrate saturation curve, as evidenced by
the shift of this curve to the right (Figure 15.10). The presence of I raises the
apparent value of L.

K Systems and V Systems Are Two Different Forms of the MWC Model

The allosteric model just presented is called a **K system** because the concentration of substrate giving half-maximal velocity, defined as $K_{0.5}$, changes in
response to effectors (Figure 15.10). Note that V_{max} is constant in this system.

An allosteric situation in which $K_{0.5}$ is constant but the apparent V_{max} changes
in response to effectors is termed a **V system.** In a V system, all v versus S plots
are hyperbolic rather than sigmoid (Figure 15.11). The positive heterotropic
effector A activates by raising V_{max}, whereas I, the negative heterotropic effector,
decreases it. Note that neither A nor I affects $K_{0.5}$. This situation arises if R and
T have the *same* affinity for the substrate, S, but differ in their catalytic ability and
their affinities for A and I. A and I thus can shift the relative T/R distribution.
Acetyl-coenzyme A carboxylase, the enzyme catalyzing the committed step in
the fatty acid biosynthetic pathway, behaves as a V system in response to its allosteric activator, citrate (see Chapter 24).

K Systems and V Systems Fill Different Biological Roles

The K and V systems have design features that mean they work best under different physiological situations. "K system" enzymes are adapted to conditions in
which the prevailing substrate concentration is rate limiting, as when [S] in vivo

A Deeper Look

Cooperativity and Conformational Changes: The Sequential Allosteric Model of Koshland, Nemethy, and Filmer

Daniel Koshland has championed the idea that proteins are inherently flexible molecules whose conformations may be altered when ligands bind. This notion serves as the fundamental tenet of the "induced-fit hypothesis" discussed in Chapter 13. Given that ligand binding can cause conformational changes in a protein, Koshland and his associates postulated that the induced conformational change when a ligand binds to one subunit of a multimeric protein could be transmitted via subunit contacts to the other subunits, causing their conformations to change. As a consequence of changing conformation, the other subunits might have greater (or for that matter, lesser) affinity for the ligand (or for other ligands). That is, the binding of one molecule of ligand to one subunit could result in conformational transitions in the protein that make it easier or harder for other ligand molecules to bind to the other subunits. Depending on the nature of such coupled conformational changes, virtually any sort of allosteric interaction is possible.

Because ligand binding and conformational transitions are distinct steps in a sequential pathway, the Koshland, Nemethy, Filmer (or KNF) model is dubbed the **sequential model** for allosteric transitions. The accompanying figure depicts the essential features of this model in a hypothetical dimeric protein. Binding of the ligand S induces a conformational change in the subunit to which it binds. Note that there is no requirement for conservation of symmetry here; the two subunits can assume different conformations (represented here as a square and a circle). If the subunit interactions are tightly coupled, then binding of S to one subunit could cause the other subunit(s) to assume a conformation having more, or less, affinity for S (or some other ligand). The underlying mechanism rests on the fact that the ligand-induced conformational change in one subunit can transmit its effects to neighboring subunits by changing the interactions and alignments of amino acid residues at the interface between the subunits. Depending on the relative ligand affinity of the conformation adopted by the neighboring subunit, the overall effect on further ligand binding may be positive, negative, or neutral (see accompanying graph). Note that in negative cooperativity, the response (binding) at $[S] > K_{0.5}$ is less than that seen for the "no cooperativity" (or Michaelis–Menten) situation. Negative cooperativity is not possible in the MWC model. Thus, the KNF model is more general than the MWC model in covering all allosteric possibilities—positive, negative, or no cooperativity. Approximately half of all known allosteric enzymes display negative cooperativity.

(a) Binding of S induces a conformational change.

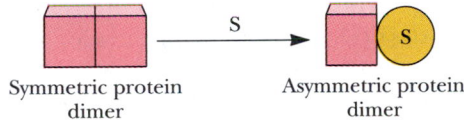

Symmetric protein   Asymmetric protein
dimer   dimer

(b)

Transmitted conformational change

If the relative affinities of the various conformations for S are:

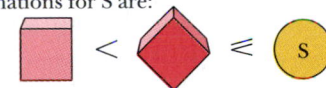

positive homotropic effects ensue.

If the relative affinities of the various conformations for S are:

negative homotropic effects are seen.

▲ The Koshland–Nemethy–Filmer sequential model for allosteric behavior. **(a)** S binding can, by induced fit, cause a conformational change in the subunit to which it binds. **(b)** If subunit interactions are tightly coupled, binding of S to one subunit may cause the other subunit to assume a conformation having a greater (positive homotropic) or lesser (negative homotropic) affinity for S. That is, the ligand-induced conformational change in one subunit can affect the adjoining subunit. Such effects could be transmitted between neighboring peptide domains by changing alignments of non-bonded amino acid residues.

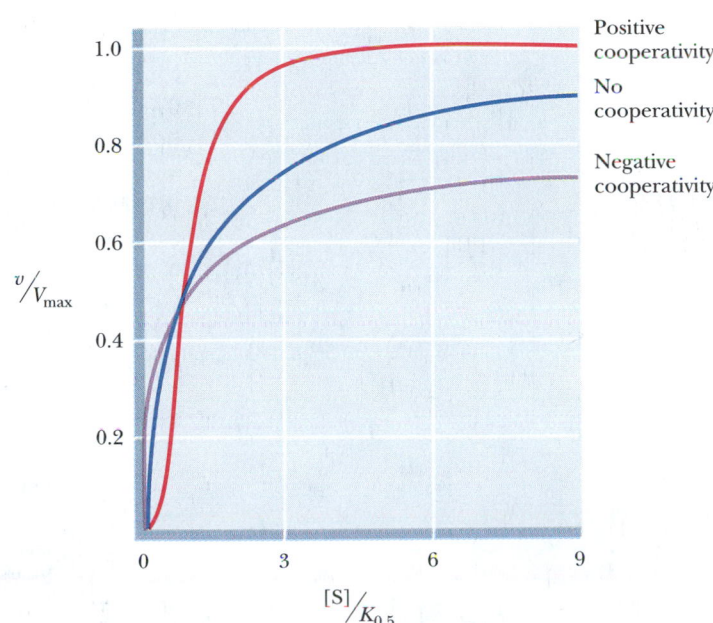

▲ Theoretical curves for the binding of a ligand to a protein having four identical subunits, each with one binding site for the ligand. The fraction of maximal binding is plotted as a function of $[S]/K_{0.5}$.

$\approx K_{0.5}$. On the other hand, when the physiological conditions are such that $[S]$ is usually saturating for the regulatory enzyme of interest, an enzyme must be of the "V system" type in order to have an effective regulatory response.

15.4	**Is the Activity of Some Enzymes Controlled by Both Allosteric Regulation and Covalent Modification?**

Glycogen phosphorylase, the enzyme that catalyzes the release of glucose units from glycogen, serves as an excellent example of the many enzymes regulated both by allosteric controls and by covalent modification.

The Glycogen Phosphorylase Reaction Converts Glycogen into Readily Usable Fuel in the Form of Glucose-1-Phosphate

The cleavage of glucose units from the nonreducing ends of glycogen molecules is catalyzed by **glycogen phosphorylase,** an allosteric enzyme. The enzymatic reaction involves phosphorolysis of the bond between C-1 of the departing glucose unit and the glycosidic oxygen, to yield *glucose-1-phosphate* and a glycogen molecule that is shortened by one residue (Figure 15.12). (Because the reaction involves attack by phosphate instead of H_2O, it is referred to as a **phosphorolysis** rather than a hydrolysis.) Phosphorolysis produces a phosphorylated sugar product, glucose-1-P, which is converted to the glycolytic substrate, glucose-6-P, by *phosphoglucomutase* (Figure 15.13). In muscle, glucose-6-P proceeds into glycolysis, providing needed energy for muscle contraction. In the liver, hydrolysis of glucose-6-P yields glucose, which is exported to other tissues via the circulatory system.

Glycogen Phosphorylase Is a Homodimer

Muscle glycogen phosphorylase is a dimer of two identical subunits (842 residues, 97.44 kD). Each subunit contains a pyridoxal phosphate cofactor, covalently linked as a Schiff base to Lys[680]. Each subunit contains an active site (at the center of the subunit) and an allosteric effector site near the subunit interface (Figure 15.14). In addition, a regulatory phosphorylation site is located at Ser[14] on each subunit. A glycogen-binding site on each subunit facilitates prior association of glycogen phosphorylase with its substrate and also exerts regulatory control on the enzymatic reaction.

Each subunit contributes a tower helix (residues 262 to 278) to the subunit–subunit contact interface in glycogen phosphorylase. In the phosphorylase

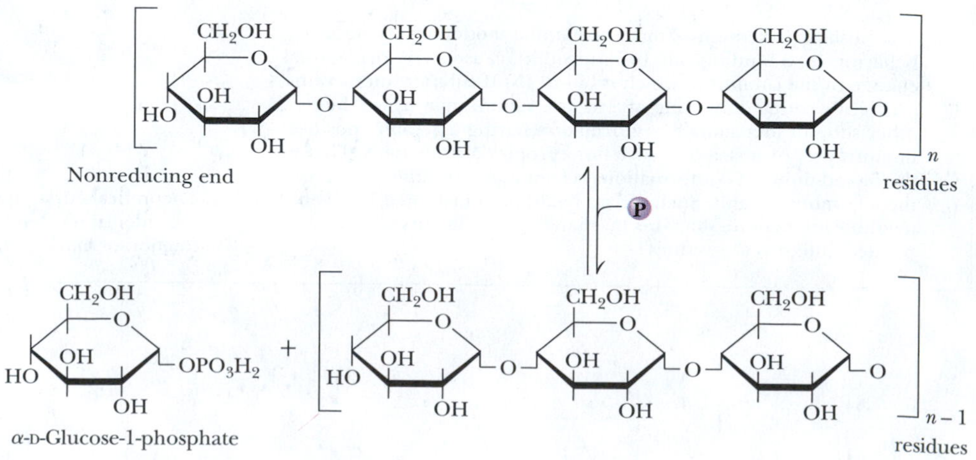

FIGURE 15.12 The glycogen phosphorylase reaction.

α-D-Glucose-1-phosphate

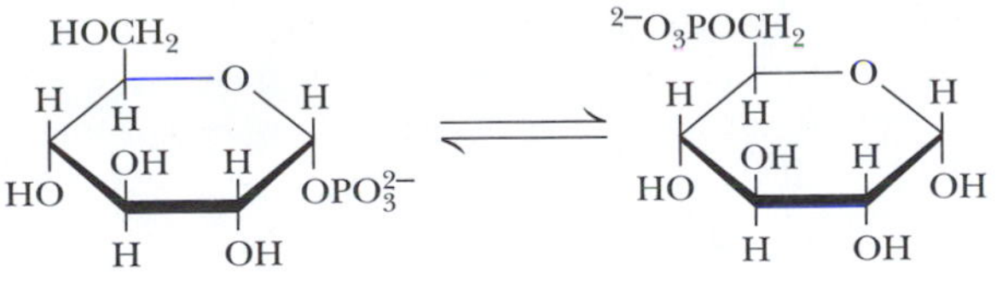

Glucose-1-phosphate   Glucose-6-phosphate

FIGURE 15.13 The phosphoglucomutase reaction.

dimer, the tower helices extend from their respective subunits and pack against each other in an antiparallel manner.

Glycogen Phosphorylase Activity Is Regulated Allosterically

Muscle Glycogen Phosphorylase Shows Cooperativity in Substrate Binding   The binding of the substrate *inorganic phosphate* (P_i) to muscle glycogen phosphorylase is highly cooperative (Figure 15.15a), which allows the enzyme activity to increase markedly over a rather narrow range of substrate concentration. P_i is a *positive homotropic effector* with regard to its interaction with glycogen phosphorylase.

ATP and Glucose-6-P Are Allosteric Inhibitors of Glycogen Phosphorylase   ATP can be viewed as the "end product" of glycogen phosphorylase action, in that the glucose-1-P liberated by glycogen phosphorylase is degraded in muscle via metabolic pathways whose purpose is energy (ATP) production. Glucose-1-P is readily converted into glucose-6-P to feed such pathways. (In the liver, glucose-1-P from glycogen is converted to glucose and released into the bloodstream to raise blood glucose levels.) Thus, feedback inhibition of glycogen phosphorylase by ATP and glucose-6-P provides a very effective way to regulate glycogen breakdown. Both ATP and glucose-6-P act by decreasing the affinity of glycogen phosphorylase for its substrate P_i (Figure 15.15b). Because the binding of ATP or glucose-6-P has a

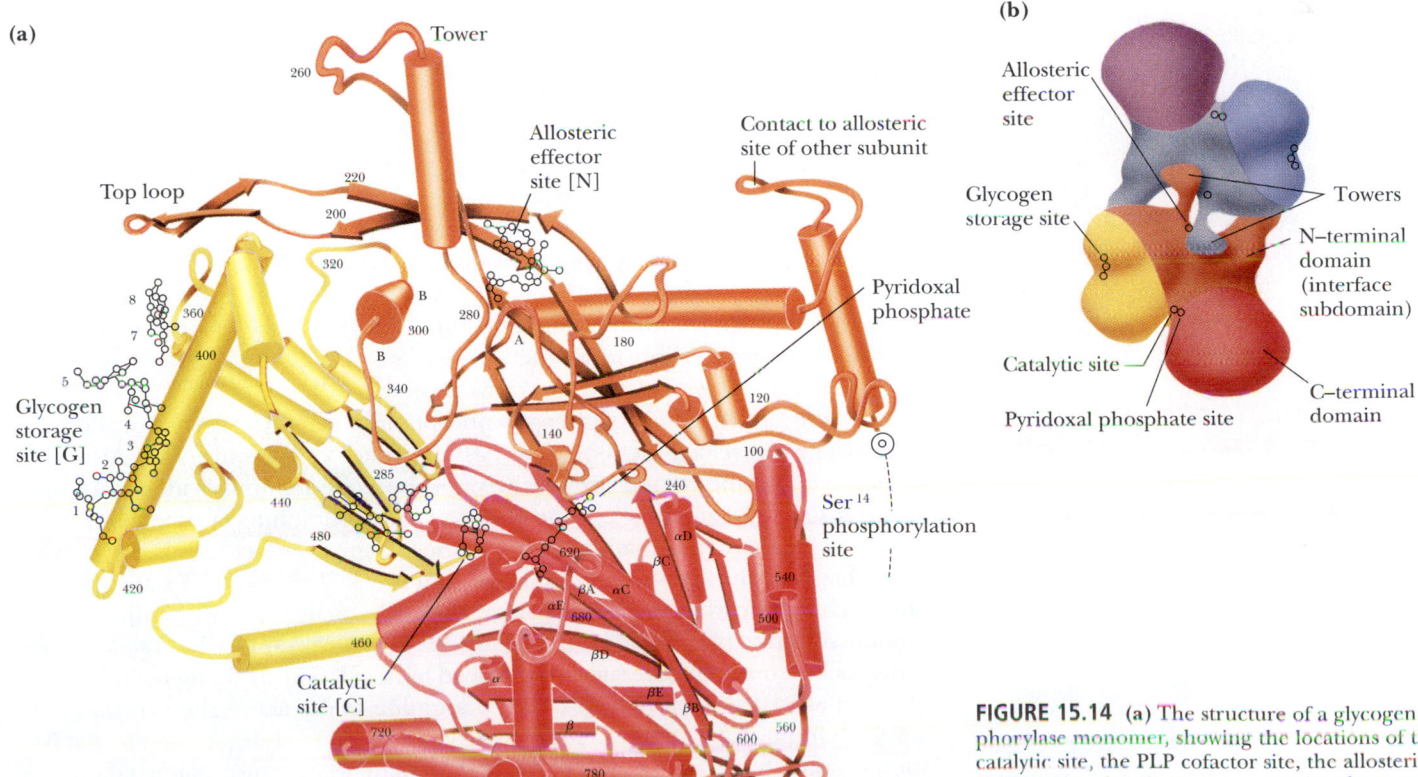

FIGURE 15.14 (a) The structure of a glycogen phosphorylase monomer, showing the locations of the catalytic site, the PLP cofactor site, the allosteric effector site, the glycogen storage site, the tower helix (residues 262 through 278), and the subunit interface. **(b)** Glycogen phosphorylase dimer.

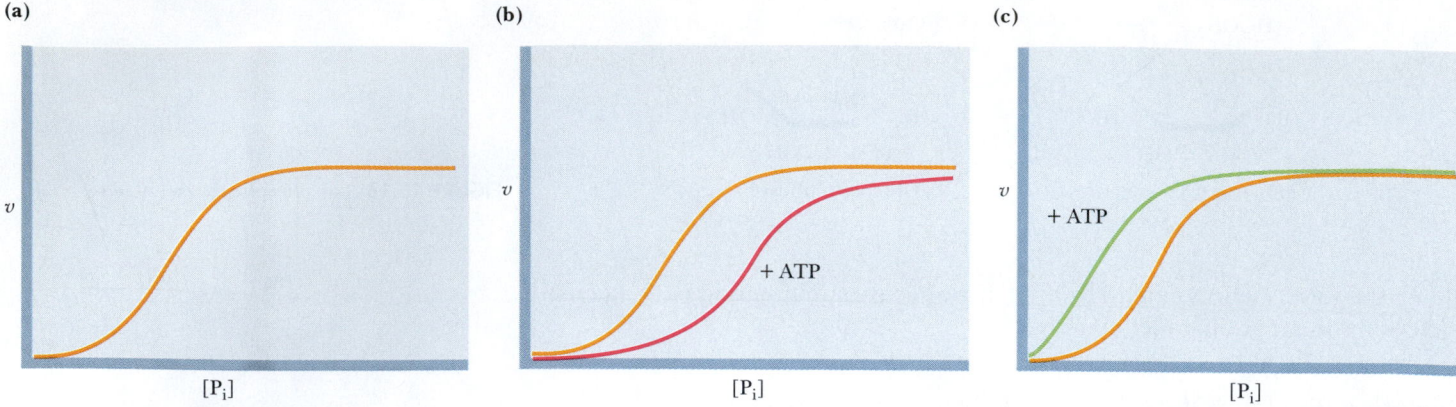

(a)   (b)   (c)

FIGURE 15.15 v versus S curves for glycogen phosphorylase. **(a)** The sigmoid response of glycogen phosphorylase to the concentration of the substrate phosphate (P_i) shows strong positive cooperativity. **(b)** ATP is a feedback inhibitor that affects the affinity of glycogen phosphorylase for its substrates but does not affect V_{max}. (Glucose-6-P shows similar effects on glycogen phosphorylase.) **(c)** AMP is a positive heterotropic effector for glycogen phosphorylase. It binds at the same site as ATP. AMP and ATP are competitive. Like ATP, AMP affects the affinity of glycogen phosphorylase for its substrates but does not affect V_{max}.

negative effect on substrate binding, these substances act as *negative heterotropic effectors*. Note in Figure 15.15b that the substrate saturation curve is displaced to the right in the presence of ATP or glucose-6-P, and a higher substrate concentration is needed to achieve half-maximal velocity ($V_{max}/2$). When concentrations of ATP or glucose-6-P accumulate to high levels, glycogen phosphorylase is inhibited; when [ATP] and [glucose-6-P] are low, the activity of glycogen phosphorylase is regulated by availability of its substrate, P_i.

AMP Is an Allosteric Activator of Glycogen Phosphorylase AMP also provides a regulatory signal to glycogen phosphorylase. It binds to the same site as ATP, but it stimulates glycogen phosphorylase rather than inhibiting it (Figure 15.15c). AMP acts as a *positive heterotropic effector,* meaning that it enhances the binding of substrate to glycogen phosphorylase. Significant levels of AMP indicate that the energy status of the cell is low and that more energy (ATP) should be produced. Reciprocal changes in the cellular concentrations of ATP and AMP and their competition for binding to the same site (the *allosteric site*) on glycogen phosphorylase, with opposite effects, allow these two nucleotides to exert *rapid and reversible control* over glycogen phosphorylase activity. Such reciprocal regulation ensures that the production of energy (ATP) is commensurate with cellular needs.

To summarize, muscle glycogen phosphorylase is allosterically activated by AMP and inhibited by ATP and glucose-6-P; caffeine can also act as an allosteric inhibitor (Figure 15.16). When ATP and glucose-6-P are abundant, glycogen breakdown is inhibited. When cellular energy reserves are low (i.e., high [AMP] and low [ATP] and [G-6-P]), glycogen catabolism is stimulated.

Glycogen phosphorylase conforms to the Monod–Wyman–Changeux model of allosteric transitions, with the active form of the enzyme designated the **R state** and the inactive form denoted as the **T state** (Figure 15.16). Thus, AMP promotes the conversion to the active R state, whereas ATP, glucose-6-P, and caffeine favor conversion to the inactive T state.

X-ray diffraction studies of glycogen phosphorylase in the presence of allosteric effectors have revealed the molecular basis for the $T \rightleftharpoons R$ conversion. Although the structure of the central core of the phosphorylase subunits is identical in the T and R states, a significant change occurs at the subunit interface between the T and R states. This conformation change at the subunit interface is linked to a structural change at the active site that is important for catalysis. In the T state, the negatively charged carboxyl group of Asp[283] faces the active site, so binding of the anionic substrate phosphate is unfavorable. In the conversion to the R state, Asp[283] is displaced from the active site and replaced by Arg[569]. The exchange of negatively charged aspartate for positively charged arginine at the active site provides a favorable binding site for phosphate. These allosteric controls serve as a mechanism for adjusting the activity of glycogen phosphorylase to meet normal metabolic demands. However, in crisis situations in which abundant energy (ATP) is needed immediately, these controls can be overridden by covalent modification of glycogen

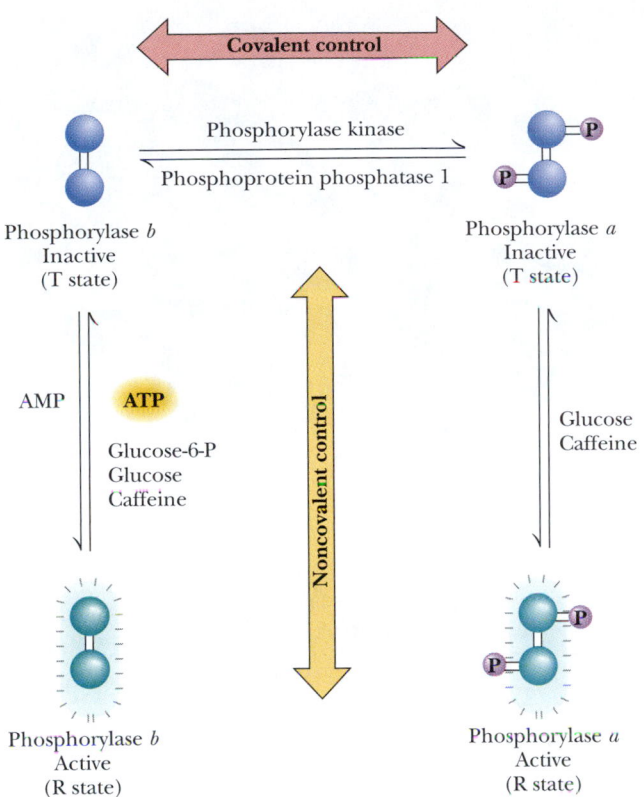

Biochemistry⟨𝒮⟩Now™   ACTIVE FIGURE 15.16
The mechanism of covalent modification and allosteric regulation of glycogen phosphorylase. The T states are blue, and the R states blue-green. **Test yourself on the concepts in this figure at http://chemistry.brookscole.com/ggb3**

phosphorylase. Covalent modification through phosphorylation of Ser[14] in glycogen phosphorylase converts the enzyme from a less active, allosterically regulated form (the *b* form) to a more active, allosterically unresponsive form (the *a* form).

Covalent Modification of Glycogen Phosphorylase Trumps Allosteric Regulation

As early as 1938, it was known that glycogen phosphorylase existed in two forms: the less active **phosphorylase *b*** and the more active **phosphorylase *a***. In 1956, Edwin Krebs and Edmond Fischer reported that a "converting enzyme" could convert phosphorylase *b* to phosphorylase *a*. Three years later, Krebs and Fischer demonstrated that the conversion of phosphorylase *b* to phosphorylase *a* involved covalent phosphorylation, as shown in Figure 15.16.

Phosphorylation of Ser[14] causes a dramatic conformation change in phosphorylase. Upon phosphorylation, the amino-terminal end of the protein (including residues 10 through 22) swings through an arc of 120°, moving into the subunit interface (Figure 15.17). This conformation change moves Ser[14] by more than 3.6 nm. The phosphorylated or *a* form of glycogen phosphorylase is much less sensitive to allosteric regulation that the *b* form. Thus, covalent modification of glycogen phosphorylase converts this enzyme from an allosterically regulated form into a persistently active form. Covalent modification overrides the allosteric regulation.

Dephosphorylation of glycogen phosphorylase is carried out by **phosphoprotein phosphatase 1.** The action of phosphoprotein phosphatase 1 inactivates glycogen phosphorylase.

Enzyme Cascades Regulate Glycogen Phosphorylase Covalent Modification

The phosphorylation reaction that activates glycogen phosphorylase is mediated by an **enzyme cascade** (Figure 15.18). The first part of the cascade leads to hormonal stimulation (described in the next section) of **adenylyl cyclase,** a

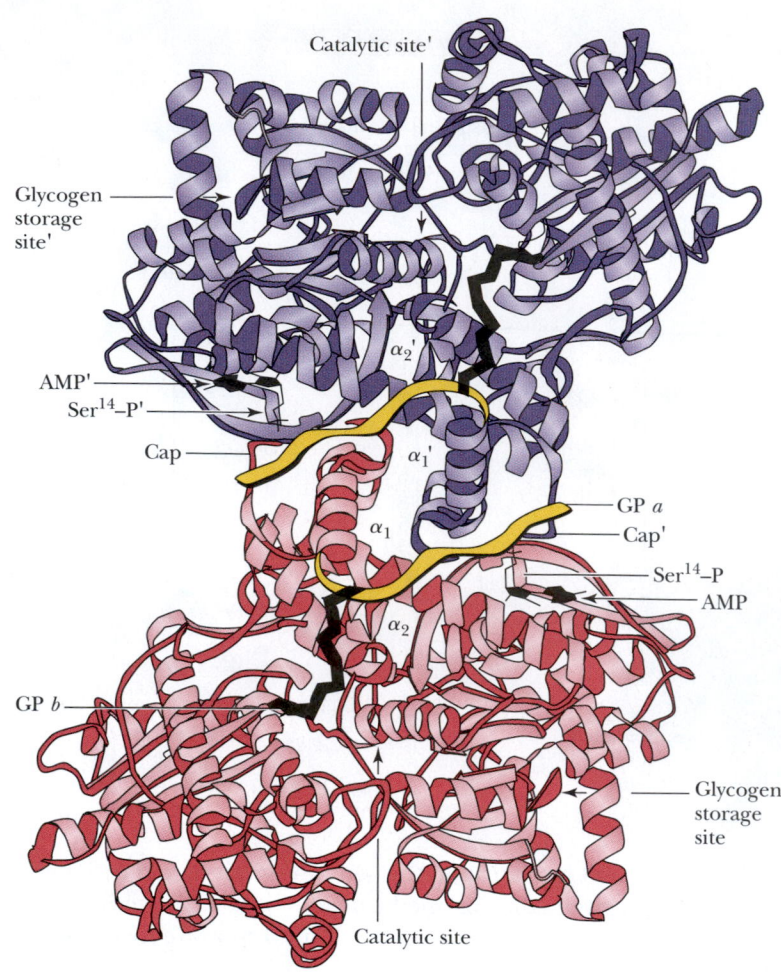

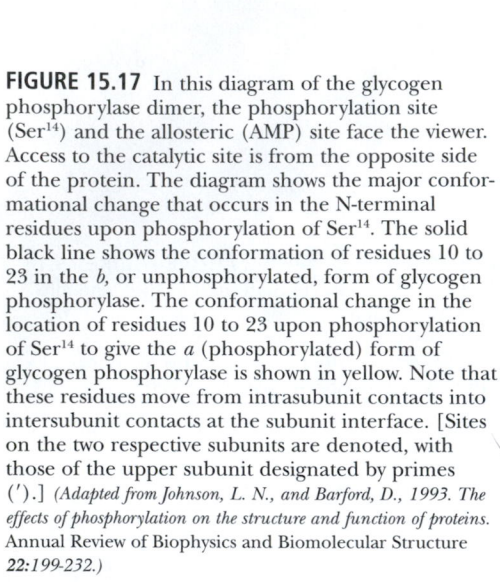

FIGURE 15.17 In this diagram of the glycogen phosphorylase dimer, the phosphorylation site (Ser^{14}) and the allosteric (AMP) site face the viewer. Access to the catalytic site is from the opposite side of the protein. The diagram shows the major conformational change that occurs in the N-terminal residues upon phosphorylation of Ser^{14}. The solid black line shows the conformation of residues 10 to 23 in the *b*, or unphosphorylated, form of glycogen phosphorylase. The conformational change in the location of residues 10 to 23 upon phosphorylation of Ser^{14} to give the *a* (phosphorylated) form of glycogen phosphorylase is shown in yellow. Note that these residues move from intrasubunit contacts into intersubunit contacts at the subunit interface. [Sites on the two respective subunits are denoted, with those of the upper subunit designated by primes (′).] *(Adapted from Johnson, L. N., and Barford, D., 1993. The effects of phosphorylation on the structure and function of proteins. Annual Review of Biophysics and Biomolecular Structure 22:199-232.)*

membrane-bound enzyme that converts ATP to *adenosine-3′,5′-cyclic monophosphate,* denoted as *cyclic AMP* or simply *cAMP* (Figure 15.19). This regulatory molecule is found in all eukaryotic cells and acts as an intracellular messenger molecule, controlling a wide variety of processes. Cyclic AMP is known as a **second messenger** because it is the intracellular agent of a hormone (the "first messenger"). (The myriad cellular roles of cyclic AMP are described in detail in Chapter 32.)

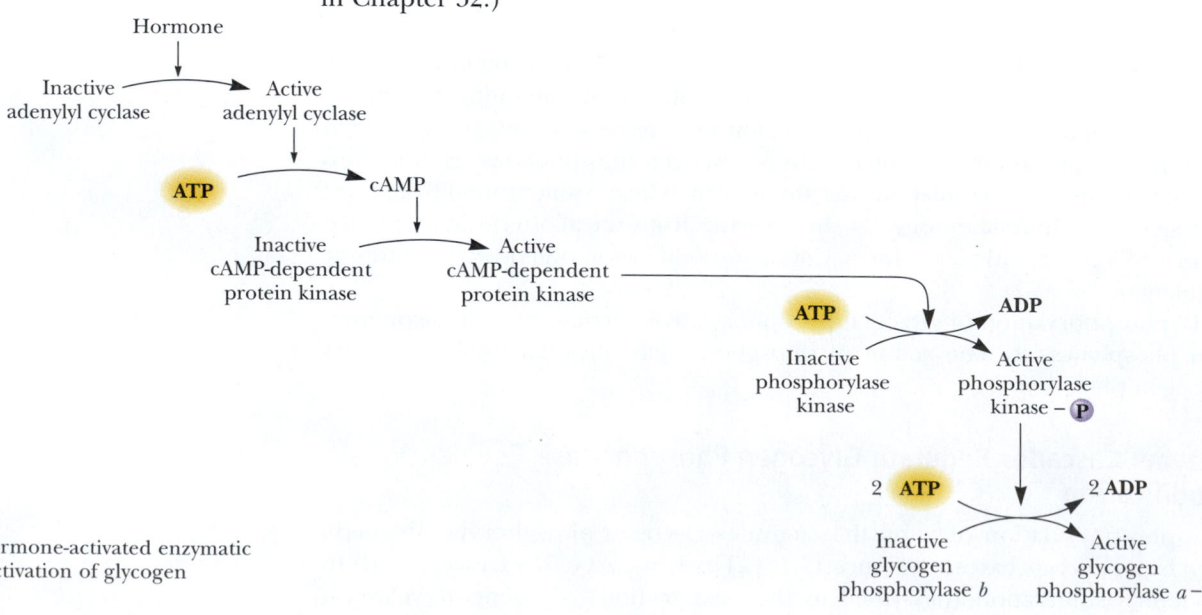

FIGURE 15.18 The hormone-activated enzymatic cascade that leads to activation of glycogen phosphorylase.

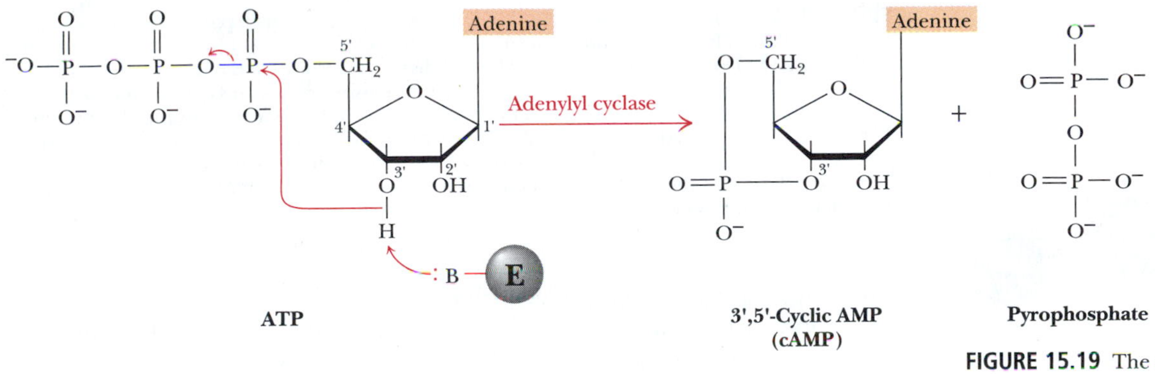

ATP   **3',5'-Cyclic AMP (cAMP)**   **Pyrophosphate**

FIGURE 15.19 The adenylyl cyclase reaction yields 3',5'-cyclic AMP and pyrophosphate. The reaction is driven forward by subsequent hydrolysis of pyrophosphate by the enzyme inorganic pyrophosphatase.

The hormonal stimulation of adenylyl cyclase is effected by a transmembrane signaling pathway consisting of three components, all membrane associated. Binding of hormone to the external surface of a hormone receptor causes a conformational change in this transmembrane protein, which in turn stimulates a **GTP-binding protein** (abbreviated **G protein**). G proteins are heterotrimeric proteins consisting of α- (45–47 kD), β- (35 kD), and γ- (7–9 kD) subunits. The α-subunit binds GDP or GTP and has an intrinsic, slow GTPase activity. In the inactive state, the $G_{\alpha\beta\gamma}$ complex has GDP at the nucleotide site. When a G protein is stimulated by a hormone-receptor complex, GDP dissociates and GTP binds to G_α, causing it to dissociate from $G_{\beta\gamma}$ and to associate with adenylyl cyclase (Figure 15.20). *Binding of G_α (GTP) activates adenylyl cyclase to form cAMP from ATP.* However, the intrinsic GTPase activity of G_α eventually hydrolyzes GTP to GDP, leading to dissociation of G_α (GDP) from adenylyl cyclase and reassociation with $G_{\beta\gamma}$ to form the inactive $G_{\alpha\beta\gamma}$ complex. This cascade amplifies the hormonal signal because a single hormone-receptor complex can activate many G proteins before the hormone dissociates from the receptor, and because the G_α-activated adenylyl cyclase can synthesize many cAMP molecules before bound GTP is hydrolyzed by G_α. More than 100 different G-protein–coupled receptors and at least 21 distinct G_α proteins are known (see Chapter 32).

Cyclic AMP is an essential activator of *cAMP-dependent protein kinase (PKA)*. This enzyme is normally inactive because its two catalytic subunits (C) are strongly associated with a pair of regulatory subunits (R), which serve to block activity. Binding of cyclic AMP to the regulatory subunits induces a conformation change that causes the dissociation of the C monomers from the R dimer (Figure 15.6). The free C subunits are active and can phosphorylate other proteins. One of the many proteins phosphorylated by PKA is *phosphorylase kinase* (Figure 15.18). Phosphorylase kinase is inactive in the unphosphorylated state and active in the phosphorylated form. As its name implies, phosphorylase kinase functions to phosphorylate (and activate) glycogen phosphorylase. Thus, stimulation of adenylyl cyclase leads to activation of glycogen breakdown.

Special Focus: Is There an Example in Nature That Exemplifies the Relationship Between Quaternary Structure and the Emergence of Allosteric Properties? Hemoglobin and Myoglobin— Paradigms of Protein Structure and Function

Ancient life forms evolved in the absence of oxygen and were capable only of anaerobic metabolism. As the earth's atmosphere changed over time, so too did living things. Indeed, the production of O_2 by photosynthesis was a major factor in altering the atmosphere. Evolution to an oxygen-based metabolism was highly beneficial. Aerobic metabolism of sugars, for example, yields far more energy than corresponding anaerobic processes. Two important oxygen-binding

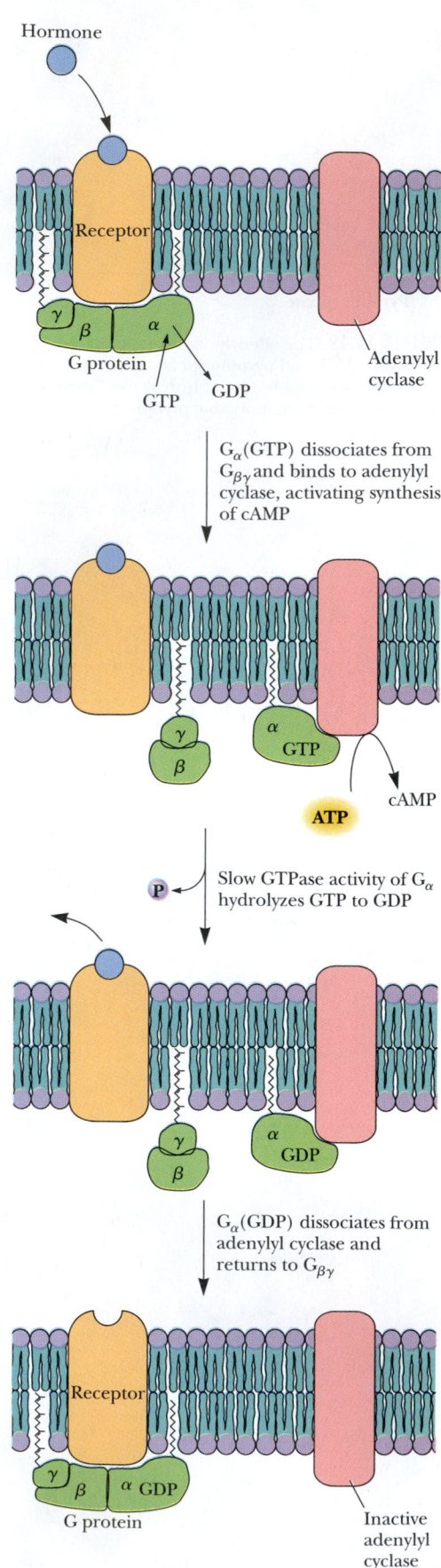

Hormone

Receptor

G protein

γ β α

GTP   GDP

Adenylyl cyclase

G_α(GTP) dissociates from $G_{\beta\gamma}$ and binds to adenylyl cyclase, activating synthesis of cAMP

γ
β

α
GTP

cAMP

ATP

Slow GTPase activity of G_α hydrolyzes GTP to GDP

P

γ
β

α
GDP

G_α(GDP) dissociates from adenylyl cyclase and returns to $G_{\beta\gamma}$

Receptor

γ β α GDP

G protein

Inactive adenylyl cyclase

◀ **FIGURE 15.20** Hormone binding to its receptor creates a hormone:receptor complex that catalyzes GDP–GTP exchange on the α-subunit of the heterotrimer G protein ($G_{\alpha\beta\gamma}$), replacing GDP with GTP. The G_α-subunit with GTP bound dissociates from the βγ-subunits and binds to adenylyl cyclase. Adenylyl cyclase becomes active upon association with G_α:GTP and catalyzes the formation of cAMP from ATP. With time, the intrinsic GTPase activity of the G_α-subunit hydrolyzes the bound GTP, forming GDP; this leads to dissociation of G_α:GDP from adenylyl cyclase, reassociation of G_α with the βγ-subunits, and cessation of adenylyl cyclase activity. Adenylyl cyclase and the hormone receptor are integral plasma membrane proteins; G_α and $G_{\beta\gamma}$ are membrane-anchored proteins.

proteins appeared in the course of evolution so that aerobic metabolic processes were no longer limited by the solubility of O_2 in water. These proteins are represented in animals as **hemoglobin (Hb)** in blood and **myoglobin (Mb)** in muscle. Because hemoglobin and myoglobin are two of the most-studied proteins in nature, they have become paradigms of protein structure and function. Moreover, hemoglobin is a model for protein quaternary structure and allosteric function. The binding of O_2 by hemoglobin, and its modulation by effectors such as protons, CO_2, and 2,3-bisphosphoglycerate, depend on interactions between subunits in the Hb tetramer. Subunit–subunit interactions in Hb reveal much about the functional significance of quaternary associations and allosteric regulation.

The Comparative Biochemistry of Myoglobin and Hemoglobin Reveals Insights into Allostery

A comparison of the properties of hemoglobin and myoglobin offers insights into allosteric phenomena, even though these proteins are *not* enzymes. Hemoglobin displays sigmoid-shaped O_2-binding curves (Figure 15.21). The unusual shape of these curves was once a great enigma in biochemistry. Such curves closely resemble allosteric enzyme:substrate saturation graphs (see Figure 15.7). In contrast, myoglobin's interaction with oxygen obeys classical Michaelis–Menten-type substrate saturation behavior.

Before examining myoglobin and hemoglobin in detail, let us first encapsulate the lesson: Myoglobin is a compact globular protein composed of a single polypeptide chain 153 amino acids in length; its molecular mass is 17.2 kD (Figure 15.22). It contains **heme,** a porphyrin ring system complexing an iron ion, as its prosthetic group (see Figure 5.15). Oxygen binds to Mb via its heme. Hemoglobin (Hb) is also a compact globular protein, but Hb is a tetramer. It consists of four polypeptide chains, each of which is very similar structurally to the

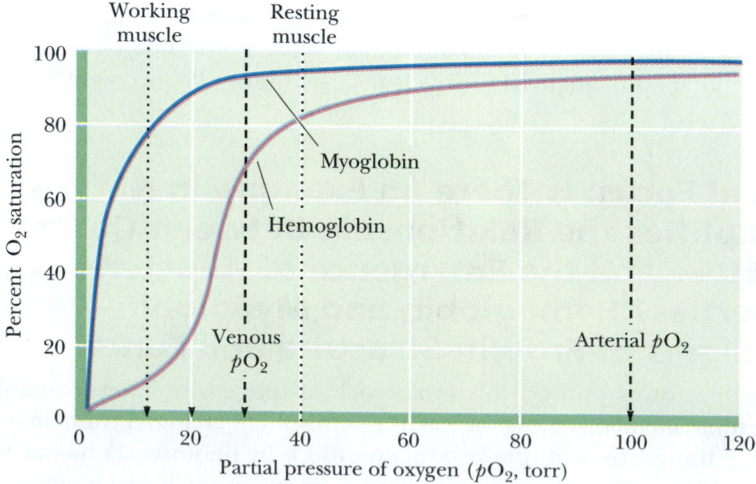

FIGURE 15.21 O_2-binding curves for hemoglobin and myoglobin.

myoglobin polypeptide chain, and each bears a heme group. Thus, a hemoglobin molecule can bind four O_2 molecules. In adult human Hb, there are two identical chains of 141 amino acids, the α-chains, and two identical β-chains, each of 146 residues. The human Hb molecule is an $\alpha_2\beta_2$-type tetramer of molecular mass 64.45 kD. The tetrameric nature of Hb is crucial to its biological function: *When a molecule of O_2 binds to a heme in Hb, the heme Fe ion is drawn into the plane of the porphyrin ring. This slight movement sets off a chain of conformational events that are transmitted to adjacent subunits, dramatically enhancing the affinity of their heme groups for O_2.* That is, the binding of O_2 to one heme of Hb makes it easier for the Hb molecule to bind additional equivalents of O_2. Hemoglobin is a marvelously constructed molecular machine. Let us dissect its mechanism, beginning with its monomeric counterpart, the myoglobin molecule.

Myoglobin Is an Oxygen-Storage Protein

Myoglobin is the oxygen-storage protein of muscle. The muscles of diving mammals such as seals and whales are especially rich in this protein, which serves as a store for O_2 during the animal's prolonged periods underwater. Myoglobin is abundant in skeletal and cardiac muscle of nondiving animals as well. Myoglobin is the cause of the characteristic red color of muscle.

The Mb Polypeptide Cradles the Heme Group

The myoglobin polypeptide chain is folded to form a cradle ($4.4 \times 4.4 \times 2.5$ nm) that nestles the heme prosthetic group (Figure 15.23). O_2 binding depends on the heme's oxidation state. The iron ion in the heme of myoglobin is in the +2 oxidation state, that is, the ferrous form. This is the form that binds O_2. Oxidation of the *ferrous* form to the +3 *ferric* form yields **metmyoglobin,** which will not bind O_2. It is interesting to note that free heme in solution will readily interact with O_2 also, but the oxygen quickly oxidizes the iron atom to the ferric state. Fe^{3+} : protoporphyrin IX is referred to as **hematin.** Thus, the polypeptide of myoglobin may be viewed as serving three critical functions: It cradles the heme group, it protects the heme iron atom from oxidation, and it provides a pocket into which the O_2 can fit.

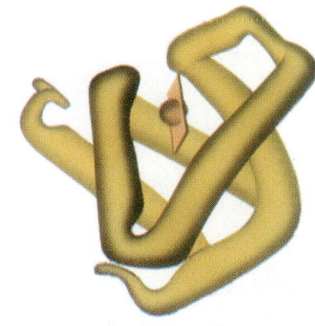

Myoglobin (Mb)

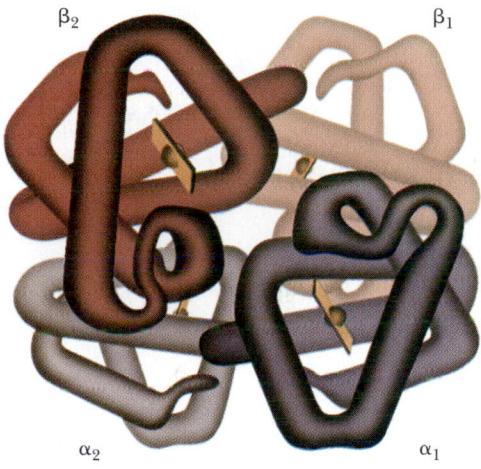

Hemoglobin (Hb)

FIGURE 15.22 The myoglobin and hemoglobin molecules. *Myoglobin* (sperm whale): one polypeptide chain of 153 amino acid residues (mass = 17.2 kD) has one heme (mass = 652 D) and binds one O_2. *Hemoglobin* (human): four polypeptide chains, two of 141 amino acid residues (α) and two of 146 residues (β); mass = 64.45 kD. Each polypeptide has a heme; the Hb tetramer binds four O_2. (*Illustration: Irving Geis. Rights owned by Howard Hughes Medical Institute. Not to be reproduced without permission.*)

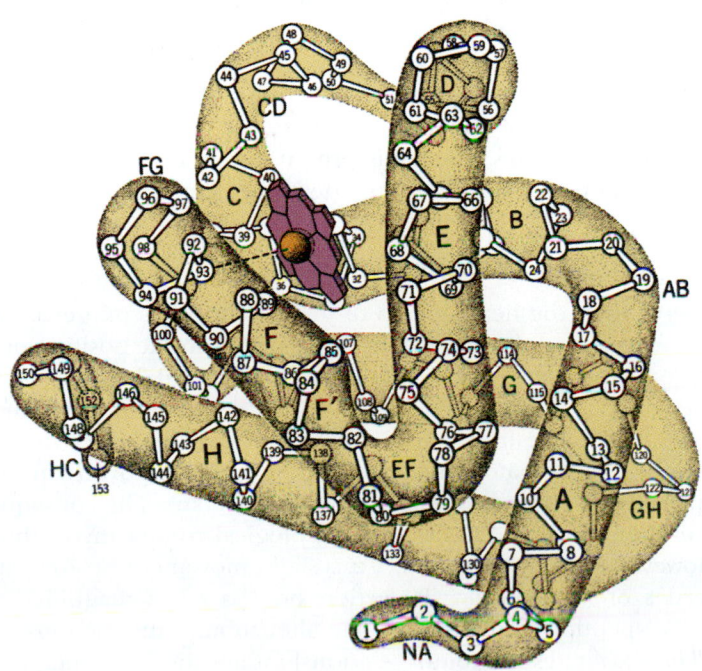

FIGURE 15.23 Detailed structure of the myoglobin molecule. The myoglobin polypeptide chain consists of eight helical segments, designated by the letters A through H, counting from the N-terminus. These helices, ranging in length from 7 to 26 residues, are linked by short, unordered regions that are named for the helices they connect, as in the AB region or the EF region. The individual amino acids in the polypeptide are indicated according to their position within the various segments, as in His F8, the eighth residue in helix F, or Phe CD1, the first amino acid in the interhelical CD region. Occasionally, amino acids are specified in the conventional way, that is, by the relative position in the chain, as in Gly[153]. The heme group is cradled within the folded polypeptide chain. (*Illustration: Irving Geis. Rights owned by Howard Hughes Medical Institute. Not to be reproduced without permission.*)

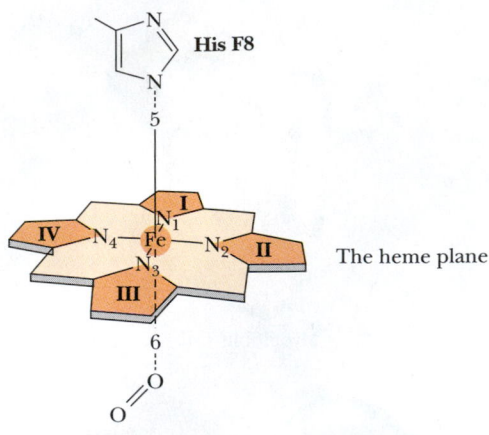

FIGURE 15.24 The six liganding positions of an iron ion. Four ligands lie in the same plane; the remaining two are, respectively, above and below this plane. In myoglobin, His F8 is the fifth ligand; in oxymyoglobin, O_2 becomes the sixth.

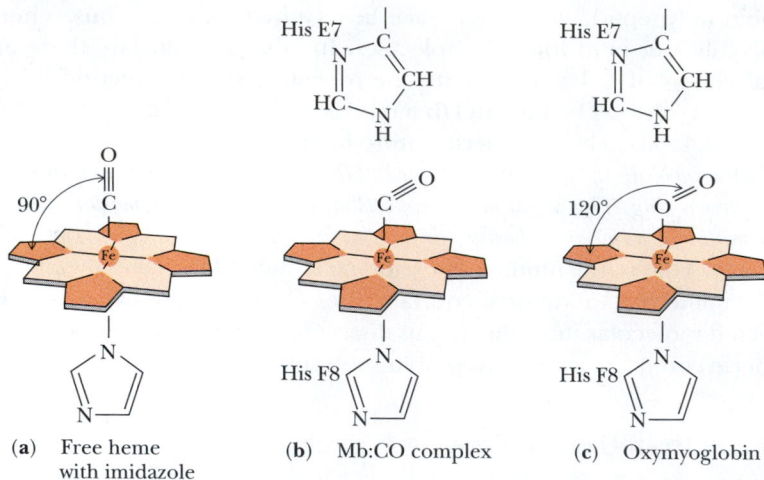

(a) Free heme with imidazole (b) Mb:CO complex (c) Oxymyoglobin

FIGURE 15.25 Oxygen and carbon monoxide binding to the heme group of myoglobin.

O_2 Binds to the Mb Heme Group

Iron ions, whether ferrous or ferric, prefer to interact with six ligands, four of which share a common plane. The fifth and sixth ligands lie above and below this plane (see Figure 15.24). In heme, four of the ligands are provided by the nitrogen atoms of the four pyrroles. A fifth ligand is donated by the imidazole side chain of amino acid residue His F8. When myoglobin binds O_2 to become **oxymyoglobin,** the O_2 molecule adds to the heme iron ion as the sixth ligand (Figure 15.24). O_2 adds end on to the heme iron, but it is not oriented perpendicular to the plane of the heme. Rather, it is tilted about 60° with respect to the perpendicular. In **deoxymyoglobin,** the sixth ligand position is vacant, and in metmyoglobin, a water molecule fills the O_2 site and becomes the sixth ligand for the ferric atom. On the oxygen-binding side of the heme lies another histidine residue, His E7. Although its imidazole function lies too far away to interact with the Fe atom, it is close enough to contact the O_2. Therefore, the O_2-binding site is a sterically hindered region. Biologically important properties stem from this hindrance. For example, the affinity of *free heme* in solution for carbon monoxide (CO) is 25,000 times greater than its affinity for O_2. But CO only binds 250 times more tightly than O_2 to the heme of myoglobin, because His E7 forces the CO molecule to tilt away from a preferred perpendicular alignment with the plane of the heme (Figure 15.25). This diminished affinity of myoglobin for CO guards against the possibility that traces of CO produced during metabolism might occupy all of the heme sites, effectively preventing O_2 from binding. Nevertheless, CO is a potent poison and can cause death by asphyxiation.

O_2 Binding Alters Mb Conformation

What happens when the heme group of myoglobin binds oxygen? X-ray crystallography has revealed that a crucial change occurs in the position of the iron atom relative to the plane of the heme. In deoxymyoglobin, the ferrous ion has but five ligands, and it lies 0.055 nm above the plane of the heme, in the direction of His F8. The iron:porphyrin complex is therefore dome-shaped. When O_2 binds, the iron atom is pulled back toward the porphyrin plane and is now displaced from it by only 0.026 nm (Figure 15.26). The consequences of this small motion are trivial as far as the biological role of myoglobin is concerned. However, as we shall soon see, this slight movement profoundly affects the properties of hemoglobin. Its action on His F8 is magnified through changes in polypeptide conformation that alter subunit interactions in the Hb tetramer. These changes in subunit relationships are the fundamental cause of the allosteric properties of hemoglobin.

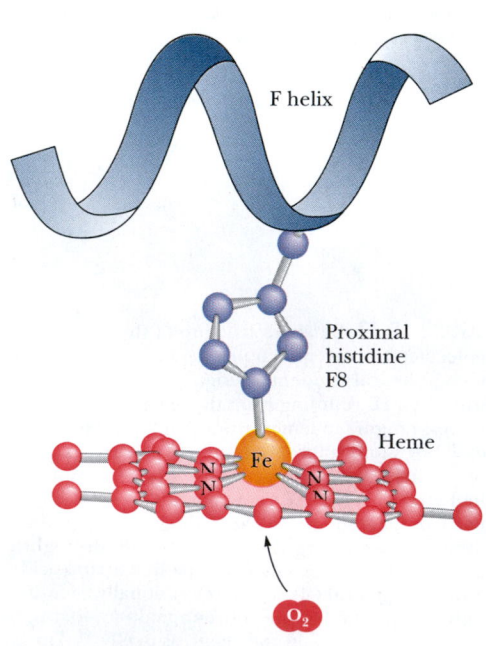

FIGURE 15.26 The displacement of the Fe ion of the heme of deoxymyoglobin from the plane of the porphyrin ring system by the pull of His F8. In oxymyoglobin, the bound O_2 counteracts this effect.

Cooperative Binding of Oxygen by Hemoglobin Has Important Physiological Significance

The oxygen-binding equations for myoglobin and hemoglobin are described in detail in the Chapter Appendix. The relative oxygen affinities of hemoglobin and myoglobin reflect their respective physiological roles (see Figure 15.21). Myoglobin, as an oxygen storage protein, has a greater affinity for O_2 than hemoglobin at all oxygen pressures. Hemoglobin, as the oxygen carrier, becomes saturated with O_2 in the lungs, where the partial pressure of O_2 (pO_2) is about 100 torr.[1] In the capillaries of tissues, pO_2 is typically 40 torr, and oxygen is released from Hb. In muscle, some of it can be bound by myoglobin, to be stored for use in times of severe oxygen deprivation, such as during strenuous exercise.

Hemoglobin Has an $\alpha_2\beta_2$ Tetrameric Structure

As noted, hemoglobin is an $\alpha_2\beta_2$ tetramer. Each of the four subunits has a conformation virtually identical to that of myoglobin. Two different types of subunits, α and β, are necessary to achieve cooperative O_2-binding by Hb. The β-chain at 146 amino acid residues is shorter than the myoglobin chain (153 residues), mainly because its final helical segment (the H helix) is shorter. The α-chain (141 residues) also has a shortened H helix and lacks the D helix as well (Figure 15.27). Max Perutz, who devoted his career to elucidating the atomic structure of Hb, noted very early in his studies that the molecule was highly symmetric. The actual arrangement of the four subunits with respect to one another is shown in Figure 15.28 for horse methemoglobin. All vertebrate hemoglobins show a three-dimensional structure essentially the same as this. The subunits pack in a tetrahedral array, creating a roughly spherical molecule $6.4 \times 5.5 \times 5.0$ nm. The four heme groups, nestled within the easily recognizable cleft formed between the E and F helices of each polypeptide, are exposed at the surface of the molecule. The heme groups are quite far apart; 2.5 nm separates the closest iron ions, those of hemes α_1 and β_2, and those of hemes α_2 and β_1. The subunit interactions are mostly between dissimilar chains: Each of the α-chains is in contact with both β-chains, but there are few α–α or β–β interactions.

Oxygenation Markedly Alters the Quaternary Structure of Hb

Crystals of deoxyhemoglobin shatter when exposed to O_2. Furthermore, X-ray crystallographic analysis reveals that oxyhemoglobin and deoxyhemoglobin differ markedly in quaternary structure. In particular, specific $\alpha\beta$-subunit interactions

[1]The **torr** is a unit of pressure named for Torricelli, inventor of the barometer. One torr corresponds to 1 mm Hg (1/760th of an atmosphere).

FIGURE 15.27 Conformational drawings of the α- and β-chains of Hb and the myoglobin chain. *(Illustration: Irving Geis. Rights owned by Howard Hughes Medical Institute. Not to be reproduced without permission.)*

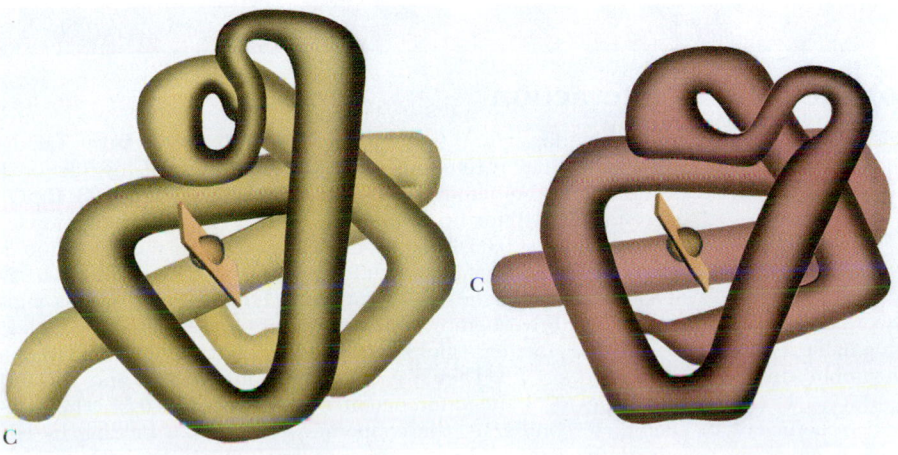

Myoglobin (Mb) α-Globin (Hbα)

β-Globin (Hbβ)

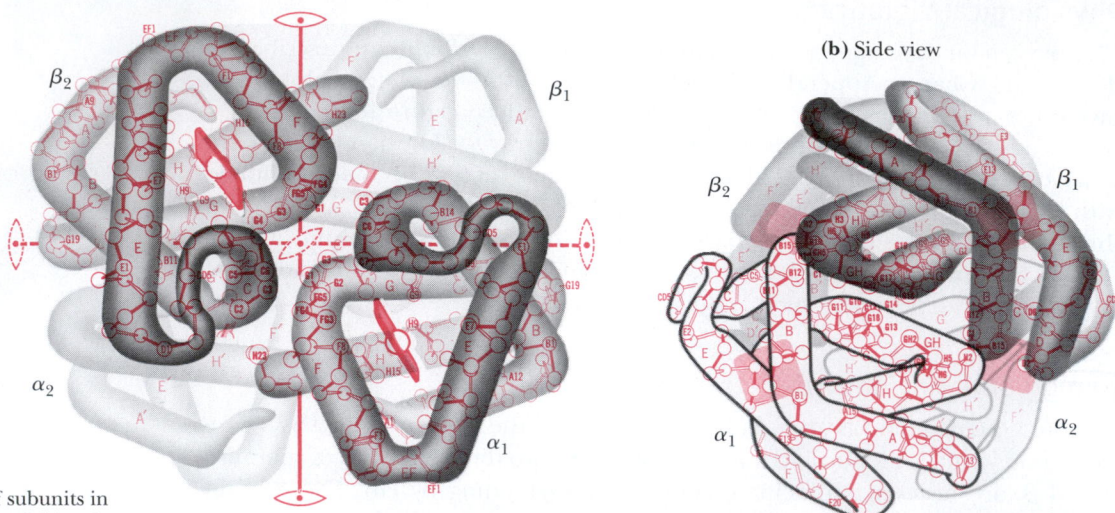

(a) Front view

(b) Side view

FIGURE 15.28 The arrangement of subunits in horse methemoglobin, the first hemoglobin whose structure was determined by X-ray diffraction. The iron atoms on metHb are in the oxidized, ferric (Fe^{3+}) state. *(Illustration: Irving Geis. Rights owned by Howard Hughes Medical Institute. Not to be reproduced without permission.)*

change. The $\alpha\beta$ contacts are of two kinds. The $\alpha_1\beta_1$ and $\alpha_2\beta_2$ contacts involve helices B, G, and H and the GH corner. These contacts are extensive and important to subunit packing; they remain unchanged when hemoglobin goes from its deoxy to its oxy form. The $\alpha_1\beta_2$ and $\alpha_2\beta_1$ contacts are called **sliding contacts.** They principally involve helices C and G and the FG corner (Figure 15.29). When hemoglobin undergoes a conformational change as a result of ligand binding to the heme, these contacts are altered (Figure 15.30). Hemoglobin, as a conformationally dynamic molecule, consists of two dimeric halves, an $\alpha_1\beta_1$-subunit pair and an $\alpha_2\beta_2$-subunit pair. Each $\alpha\beta$-dimer moves as a rigid body, and the two halves of the molecule slide past each other upon oxygenation of the heme. The two halves rotate some 15° about an imaginary pivot passing through the $\alpha\beta$-subunits; some atoms at the interface between $\alpha\beta$-dimers are relocated by as much as 0.6 nm.

Movement of the Heme Iron by Less Than 0.04 nm Induces the Conformational Change in Hemoglobin

In deoxyhemoglobin, histidine F8 is liganded to the heme iron ion, but steric constraints force the Fe^{2+}:His-N bond to be tilted about 8° from the perpendicular to the plane of the heme. Steric repulsion between histidine F8 and the

A Deeper Look

The Physiological Significance of the Hb:O_2 Interaction

We can determine quantitatively the physiological significance of the sigmoid nature of the hemoglobin oxygen-binding curve, or, in other words, the biological importance of cooperativity. The equation

$$\frac{Y}{1-Y} = \frac{[pO_2]^n}{P_{50}}$$

describes the relationship between pO_2, the affinity of hemoglobin for O_2 (defined as P_{50}, the partial pressure of O_2 giving half-maximal saturation of Hb with O_2), and the fraction of hemoglobin with O_2 bound, Y, versus the fraction of Hb with no O_2 bound, $(1 - Y)$ (see Appendix Equation A15.16). The coefficient n is the Hill coefficient, an index of the cooperativity (sigmoidicity) of the hemoglobin

oxygen-binding curve (see Chapter Appendix for details). Taking pO_2 in the lungs as 100 torr, P_{50} as 26 torr, and n as 2.8, the fractional saturation of the hemoglobin heme groups with O_2, is 0.98. If pO_2 were to fall to 10 torr within the capillaries of an exercising muscle, Y would drop to 0.06. The oxygen delivered under these conditions would be proportional to the difference, $Y_{lungs} - Y_{muscle}$, which is 0.92. That is, virtually all the oxygen carried by Hb would be released. Suppose instead that hemoglobin binding of O_2 were not cooperative; in that case, the hemoglobin oxygen-binding curve would be hyperbolic, and $n = 1.0$. Then Y in the lungs would be 0.79 and Y in the capillaries, 0.28; the difference in Y values would be 0.51. Thus, under these conditions, the cooperativity of oxygen binding by Hb means that 0.92/0.51 or 1.8 times as much O_2 can be delivered.

nitrogen atoms of the porphyrin ring system, combined with electrostatic repulsions between the electrons of Fe^{2+} and the porphyrin π-electrons, forces the iron atom to lie out of the porphyrin plane by about 0.06 nm. Changes in electronic and steric factors upon heme oxygenation allow the Fe^{2+} atom to move about 0.039 nm closer to the plane of the porphyrin, so now it is displaced only 0.021 nm above the plane. It is as if the O_2 were drawing the heme Fe^{2+} into the porphyrin plane (Figure 15.31). This modest displacement of 0.039 nm seems a trivial distance, but its biological consequences are far reaching. As the iron atom moves, it drags histidine F8 along with it, causing helix F, the EF corner, and the FG corner to follow. These shifts are transmitted to the subunit interfaces, where they trigger conformational readjustments that lead to the rupture of interchain salt links.

The Oxy and Deoxy Forms of Hemoglobin Represent Two Different Conformational States

Hemoglobin resists oxygenation (see Figure 15.21) because the deoxy form is stabilized by specific hydrogen bonds and salt bridges (ion-pair bonds). All of these interactions are broken in oxyhemoglobin, as the molecule stabilizes into a new conformation. A crucial H bond in this transition involves a particular tyrosine residue. Both α- and β-subunits have Tyr as the penultimate C-terminal residue (Tyr α140 = Tyr HC2; Tyr β145 = Tyr HC2, respectively[2]). The phenolic —OH groups of these Tyr residues form intrachain H bonds to the peptide C=O function contributed by Val FG5 in deoxyhemoglobin. (Val FG5 is α93 and β98, respectively.) The shift in helix F upon oxygenation leads to rupture of this Tyr HC2:Val FG5 hydrogen bond. Furthermore, eight salt bridges linking the polypeptide chains are broken as hemoglobin goes from the deoxy to the oxy form (Figure 15.32). Six of these salt links are between different subunits. Four of these six involve either carboxyl-terminal or amino-terminal amino acids in the chains; two are between the amino termini and the carboxyl termini of the α-chains, and two join the carboxyl termini of the β-chains to the ϵ-NH_3^+ groups of the two Lys α140 residues. The other two interchain electrostatic bonds link Arg and Asp residues in the two α-chains. In addition, ionic interactions between Asp β94 and His β146 form an intrachain salt bridge in each β-subunit. In deoxyhemoglobin, with all of these interactions intact, the C-termini of the four subunits are restrained, and this conformational state is termed **T,** the **tense** or **taut form.** In oxyhemoglobin, these C-termini have almost complete freedom of rotation, and the molecule is now in its **R, or relaxed, form.**

[2]C here designates the C-terminus; the H helix is C-terminal in these polypeptides. "C2" symbolizes the next-to-last residue.

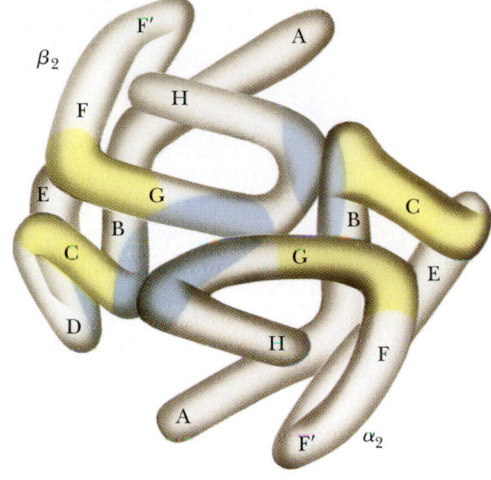

FIGURE 15.29 Side view of one of the two $\alpha\beta$-dimers in Hb, with packing contacts indicated in blue. The sliding contacts made with the other dimer are shown in yellow. The changes in these sliding contacts are shown in Figure 15.30. *(Illustration: Irving Geis. Rights owned by Howard Hughes Medical Institute. Not to be reproduced without permission.)*

Biochemistry Now™   ANIMATED FIGURE 15.30
Subunit motion in hemoglobin when the molecule goes from the **(a)** deoxy to the **(b)** oxy form. *(Illustration: Irving Geis. Rights owned by Howard Hughes Medical Institute. Not to be reproduced without permission.)* **See this figure animated at http://chemistry.brookscole.com/ggb3**

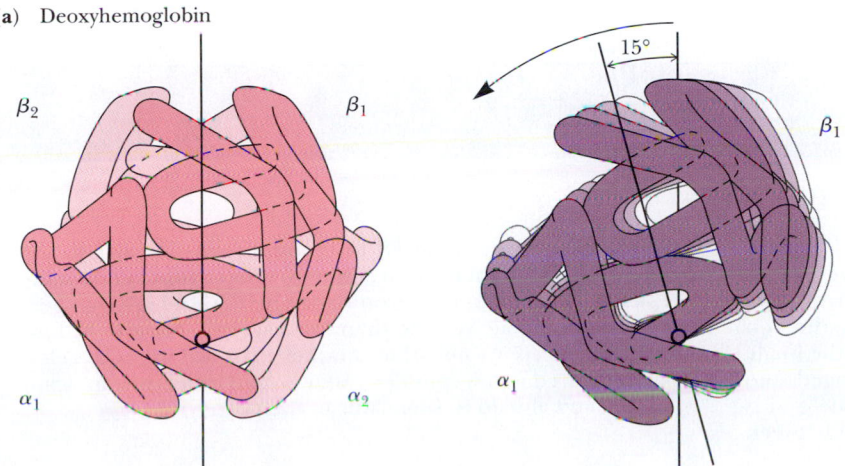

(a) Deoxyhemoglobin

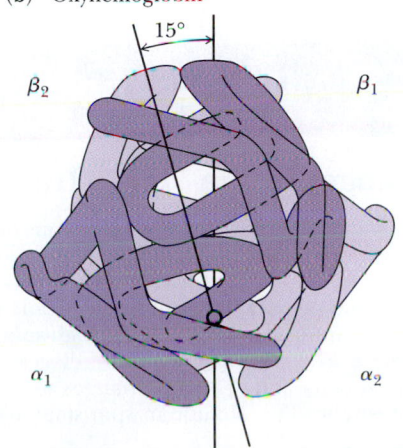

(b) Oxyhemoglobin

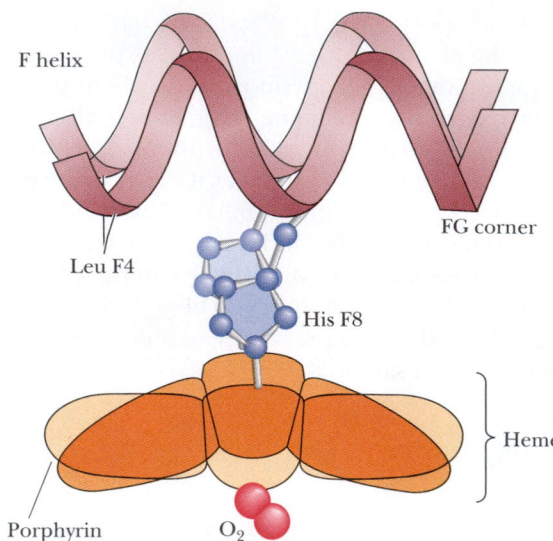

Biochemistry Now™   ACTIVE FIGURE 15.31
Changes in the position of the heme iron atom
upon oxygenation lead to conformational changes
in the hemoglobin molecule.

The Allosteric Behavior of Hemoglobin Has Both Symmetry (MWC) Model and Sequential (KNF) Model Components

Oxygen is accessible only to the heme groups of the α-chains when hemoglobin is in the T conformational state. Max Perutz has pointed out that the heme environment of β-chains in the T state is virtually inaccessible because of steric hindrance by amino acid residues in the E helix. This hindrance disappears when the hemoglobin molecule undergoes transition to the R conformational state. Binding of O_2 to the β-chains is thus dependent on a T-to-R conformational shift, and this shift is triggered by the subtle changes that occur when O_2 binds to the α-chain heme groups. Together these observations lead to a model that is partially MWC and partially KNF (see A Deeper Look Box, page 485): O_2 binding to one α-subunit and then the other leads to sequential changes in conformation, followed by a switch in quaternary structure at the Hb:$2O_2$ state from T to R. Thus, the real behavior of this protein is an amalgam of the two prominent theoretical models for allosteric behavior.

H⁺ Promotes the Dissociation of Oxygen from Hemoglobin

Protons, carbon dioxide, and chloride ions, as well as the metabolite 2,3-bisphosphoglycerate (or BPG), all affect the binding of O_2 by hemoglobin. Their effects have interesting ramifications, which we shall see as we discuss them in turn. Deoxyhemoglobin has a higher affinity for protons than oxyhemoglobin.

A Deeper Look

Changes in the Heme Iron upon O_2 Binding

In deoxyhemoglobin, the six d electrons of the heme Fe^{2+} exist as four unpaired electrons and one electron pair, and five ligands can be accommodated: the four N-atoms of the porphyrin ring system and histidine F8. In this electronic configuration, the iron atom is paramagnetic and in the **high-spin state.** When the heme binds O_2 as a sixth ligand, these electrons are rearranged into three e^- pairs and the iron changes to the **low-spin state** and is diamagnetic. This change in spin state allows the bond between the Fe^{2+} ion and histidine F8 to become perpendicular to the heme plane and to shorten. In addition, interactions between the porphyrin N atoms and the iron strengthen. Also, high-spin Fe^{2+} has a greater atomic volume than low-spin Fe^{2+} because its four unpaired e^- occupy four orbitals rather than two when the electrons are paired in low-spin Fe^{2+}. So, low-spin iron is less sterically hindered and able to move nearer to the porphyrin plane.

(a)

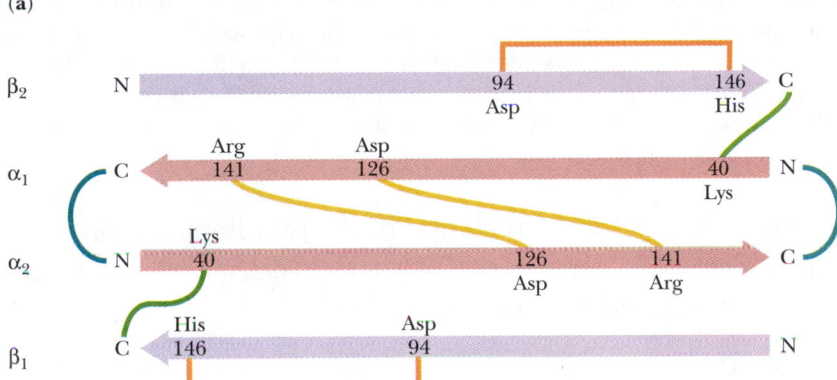

(b)

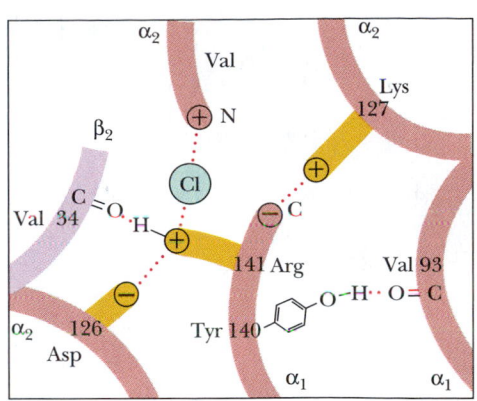

FIGURE 15.32 Salt bridges between different subunits in hemoglobin. These noncovalent, electrostatic interactions are disrupted upon oxygenation. Arg α141 and His β146 are the C-termini of the α- and β-polypeptide chains. **(a)** The various intrachain and interchain salt links formed among the α- and β-chains of deoxyhemoglobin. **(b)** A focus on those salt bridges and hydrogen bonds involving interactions between N-terminal and C-terminal residues in the α-chains. Note the Cl$^-$ ion, which bridges ionic interactions between the N-terminus of α_2 and the R group of Arg α141. **(c)** A focus on the salt bridges and hydrogen bonds in which the residues located at the C-termini of β-chains are involved. All of these links are abolished in the deoxy to oxy transition. *(Illustration: Irving Geis. Rights owned by Howard Hughes Medical Institute. Not to be reproduced without permission.)*

(c)

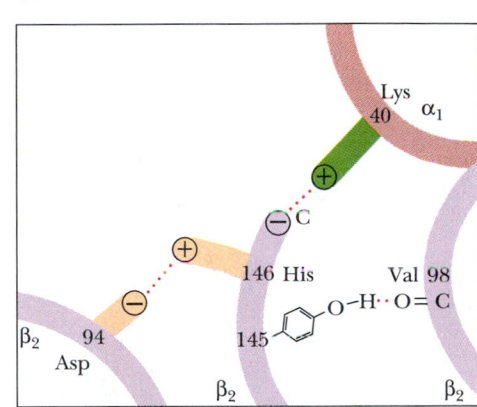

Thus, as the pH decreases, dissociation of O_2 from hemoglobin is enhanced. In simple symbolism, ignoring the stoichiometry of O_2 or H^+ involved:

$$HbO_2 + H^+ \rightleftharpoons HbH^+ + O_2$$

Expressed another way, H^+ is an antagonist of oxygen binding by Hb, and the saturation curve of Hb for O_2 is displaced to the right as acidity increases (Figure 15.33). This phenomenon is called the **Bohr effect**, after its discoverer, the Danish physiologist Christian Bohr (the father of Niels Bohr, the atomic physicist). The effect has important physiological significance because actively metabolizing tissues produce acid, promoting O_2 release where it is most needed. About two protons are taken up by deoxyhemoglobin. The N-termini of the two α-chains and the His β146 residues have been implicated as the major players in the Bohr effect. (The pK_a of a free amino terminus in a protein

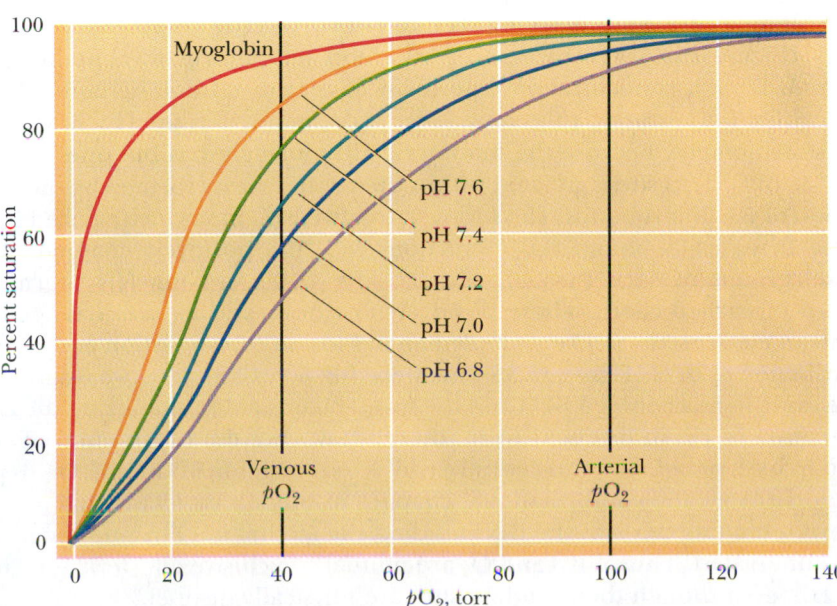

FIGURE 15.33 The oxygen saturation curves for myoglobin and for hemoglobin at five different pH values: 7.6, 7.4, 7.2, 7.0, and 6.8.

is about 8.0, but the pK_a of a protein histidine imidazole is around 6.5.) Neighboring carboxylate groups of Asp β94 residues help stabilize the protonated state of the His β146 imidazoles that occur in deoxyhemoglobin. However, when Hb binds O_2, changes in the conformation of β-chains upon Hb oxygenation move the negative Asp function away, and dissociation of the imidazole protons is favored.

CO_2 Also Promotes the Dissociation of O_2 from Hemoglobin

Carbon dioxide has an effect on O_2 binding by Hb that is similar to that of H^+, partly because it produces H^+ when it dissolves in the blood:

$$CO_2 + H_2O \xmrightarrow{\text{carbonic anhydrase}} \underset{\text{carbonic acid}}{H_2CO_3} \rightleftharpoons H^+ + \underset{\text{bicarbonate}}{HCO_3^-}$$

The enzyme *carbonic anhydrase* promotes the hydration of CO_2. Many of the protons formed upon ionization of carbonic acid are picked up by Hb as O_2 dissociates. The bicarbonate ions are transported with the blood back to the lungs. When Hb becomes oxygenated again in the lungs, H^+ is released and reacts with HCO_3^- to re-form H_2CO_3, from which CO_2 is liberated. The CO_2 is then exhaled as a gas.

In addition, some CO_2 is directly transported by hemoglobin in the form of *carbamate* ($-NHCOO^-$). Free α-amino groups of Hb react with CO_2 reversibly:

$$R-NH_2 + CO_2 \rightleftharpoons R-NH-COO^- + H^+$$

This reaction is driven to the right in tissues by the high CO_2 concentration; the equilibrium shifts the other way in the lungs where $[CO_2]$ is low. Thus, carbamylation of the N-termini converts them to anionic functions, which then form salt links with the cationic side chains of Arg α141 that stabilize the deoxy or T state of hemoglobin.

In addition to CO_2, Cl^- and BPG also bind better to deoxyhemoglobin than to oxyhemoglobin, causing a shift in equilibrium in favor of O_2 release. These various effects are demonstrated by the shift in the oxygen saturation curves for Hb in the presence of one or more of these substances (Figure 15.34). Note that the O_2-binding curve for Hb + BPG + CO_2 fits that of whole blood very well.

2,3-Bisphosphoglycerate Is an Important Allosteric Effector for Hemoglobin

The binding of 2,3-bisphosphoglycerate (BPG) to Hb promotes the release of O_2 (Figure 15.34). Erythrocytes (red blood cells) normally contain about 4.5 mM BPG, a concentration equivalent to that of tetrameric hemoglobin molecules. Interestingly, this equivalence is maintained in the Hb:BPG binding stoichiometry because the tetrameric Hb molecule has but one binding site for BPG. This site is situated within the central cavity formed by the association of the four subunits. The strongly negative BPG molecule (Figure 15.35) is electrostatically bound via interactions with the positively charged functional groups of each Lys β82, His β2, His β143, and the NH_3^+-terminal group of each β-chain. These positively charged residues are arranged to form an electrostatic pocket complementary to the conformation and charge distribution of BPG (Figure 15.36). In effect, BPG crosslinks the two β-subunits. The ionic bonds between BPG and the two β-chains aid in stabilizing the conformation of Hb in its deoxy form, thereby favoring the dissociation of oxygen. In oxyhemoglobin, this central cavity is too small for BPG to fit. Or, to put it another way, the conformational changes in the Hb molecule that accompany O_2 binding perturb the BPG-binding site so that BPG can no longer be accommodated. Thus, BPG and O_2 are mutually exclusive allosteric effectors for Hb, even though their binding sites are physically distinct.

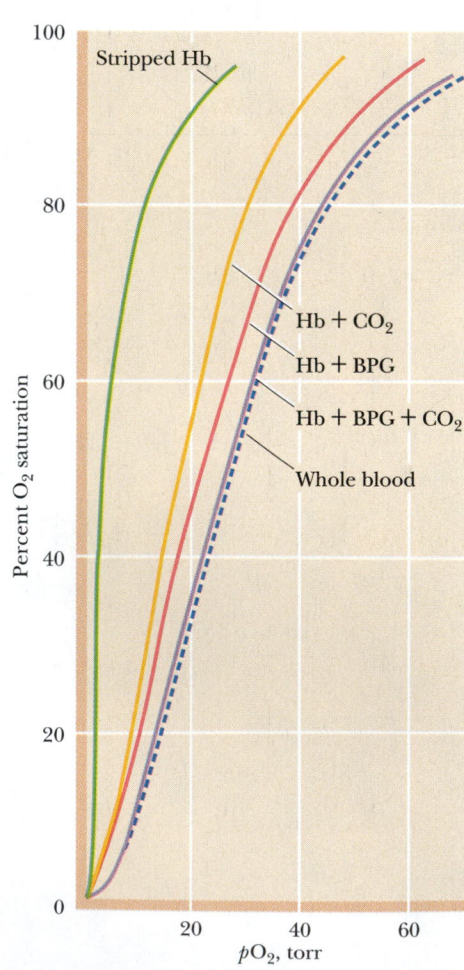

FIGURE 15.34 Oxygen-binding curves of blood and of hemoglobin in the absence and presence of CO_2 and BPG. From left to right: stripped Hb, Hb + CO_2, Hb + BPG, Hb + BPG + CO_2, and whole blood.

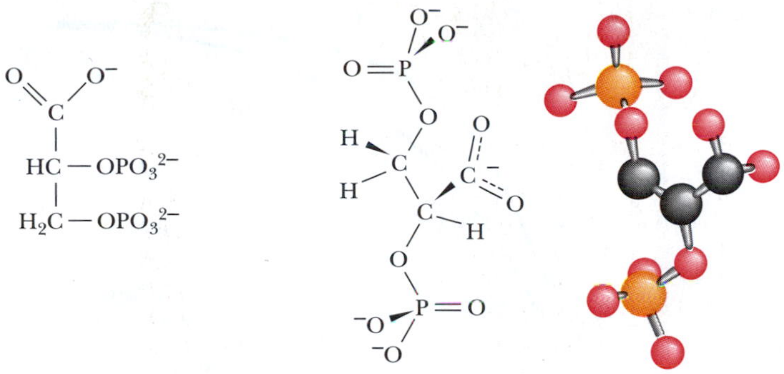

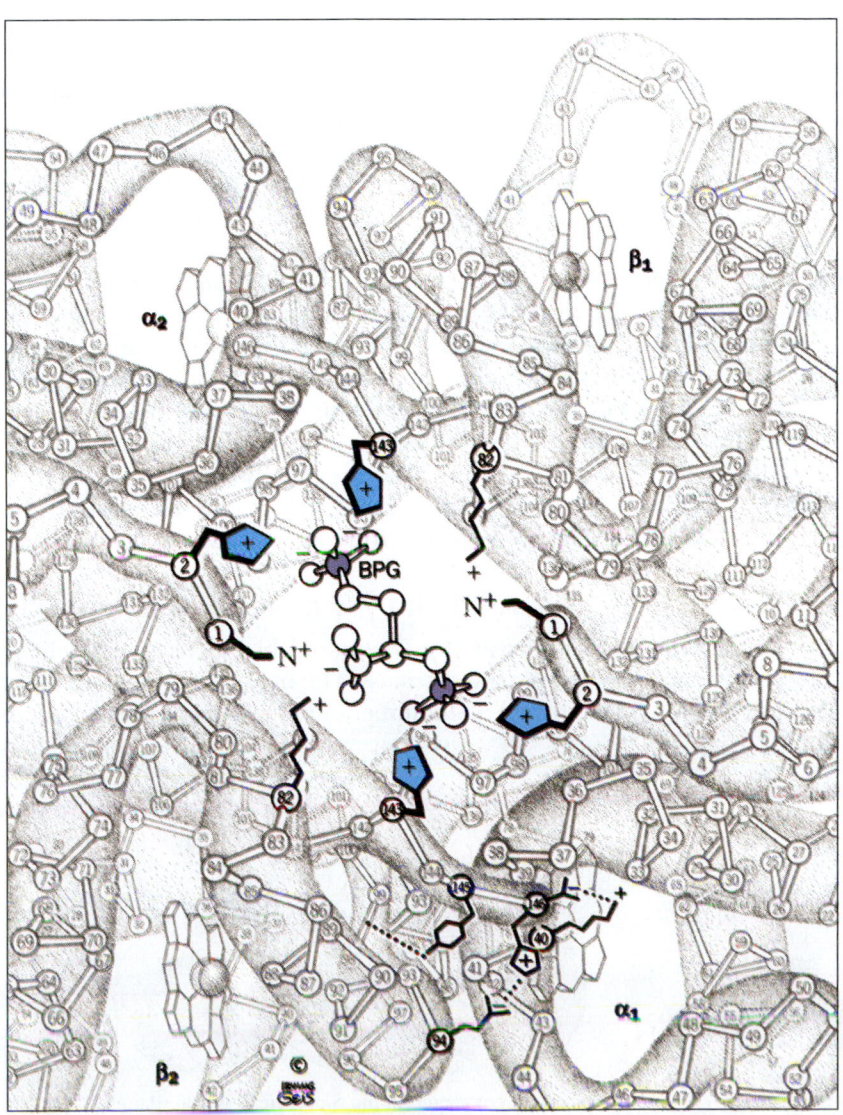

FIGURE 15.35 The structure, in ionic form, of BPG or 2,3-bisphosphoglycerate, an important allosteric effector for hemoglobin.

BPG Binding to Hb Has Important Physiological Significance

The importance of the BPG effect is evident in Figure 15.34. Hemoglobin stripped of BPG is virtually saturated with O_2 at a pO_2 of only 20 torr, and it cannot release its oxygen within tissues, where the pO_2 is typically 40 torr. BPG shifts the oxygen saturation curve of Hb to the right, making the Hb an O_2 delivery system eminently suited to the needs of the organism. BPG serves this vital function in humans, most primates, and a number of other mammals. However, the hemoglobins of cattle, sheep, goats, deer, and other animals have an intrinsically lower affinity for O_2, and these Hbs are relatively unaffected by BPG. In fish,

FIGURE 15.36 The ionic binding of BPG to the two β-subunits of Hb. *(Illustration: Irving Geis. Rights owned by Howard Hughes Medical Institute. Not to be reproduced without permission.)*

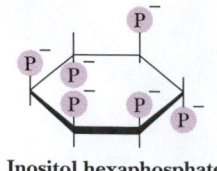

Inositol pentaphosphate
(IPP)   **Inositol hexaphosphate**
                        (IHP)

FIGURE 15.37 The structures of inositol pentaphosphate and inositol hexaphosphate, the functional
analogs of BPG in birds and reptiles.

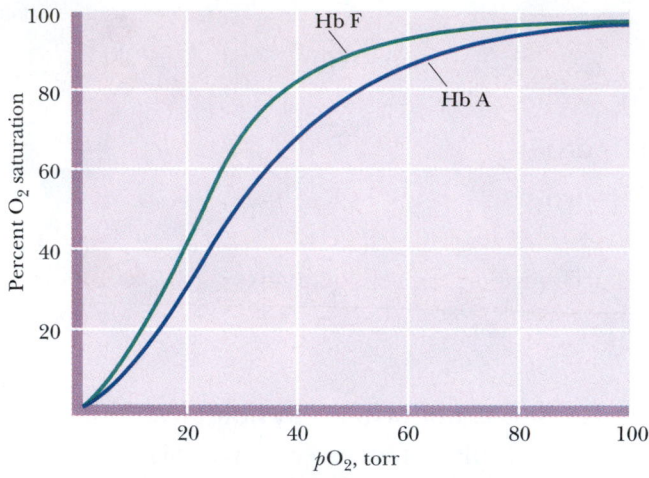

FIGURE 15.38 Comparison of the oxygen saturation curves of Hb A and Hb F under similar
conditions of pH and [BPG].

whose erythrocytes contain mitochondria, the regulatory role of BPG is filled by
ATP or GTP. In reptiles and birds, a different organophosphate serves, namely
inositol pentaphosphate (IPP) or inositol hexaphosphate (IHP) (Figure 15.37).

Fetal Hemoglobin Has a Higher Affinity for O_2 Because It Has a Lower Affinity for BPG

The fetus depends on its mother for an adequate supply of oxygen, but its circulatory system is entirely independent. Gas exchange takes place across the pla-
centa. Ideally then, fetal Hb should be able to absorb O_2 better than maternal Hb
so that an effective transfer of oxygen can occur. Fetal Hb differs from adult Hb
in that the β-chains are replaced by very similar, but not identical, 146-residue subunits called γ-chains (gamma chains). Fetal Hb is thus $\alpha_2\gamma_2$. Recall that BPG functions through its interaction with the β-chains. BPG binds less effectively with the
γ-chains of fetal Hb (also called Hb F). (Fetal γ-chains have Ser instead of His at
position 143 and thus lack two of the positive charges in the central BPG-binding
cavity.) Figure 15.38 compares the relative affinities of adult Hb (also known as
Hb A) and Hb F for O_2 under similar conditions of pH and [BPG]. Note that
Hb F binds O_2 at pO_2 values where most of the oxygen has dissociated from Hb A.
Much of the difference can be attributed to the diminished capacity of Hb F to
bind BPG (compare Figures 15.34 and 15.38); Hb F thus has an intrinsically
greater affinity for O_2, and oxygen transfer from mother to fetus is ensured.

Sickle-Cell Anemia Is Characterized by Abnormal Red Blood Cells

In 1904, a Chicago physician treated a 20-year-old black college student com-
plaining of headache, weakness, and dizziness. The blood of this patient revealed
serious anemia—only half the normal number of red cells were present. Many of

Biochemistry🅔Now™   ANIMATED FIGURE 15.39
The polymerization of Hb S via the interactions between the hydrophobic Val side chains at position $\beta6$
and the hydrophobic pockets in the EF corners of
β-chains in neighboring Hb molecules. The protruding "block" on Oxy S represents the Val hydrophobic
protrusion. The complementary hydrophobic pocket
in the EF corner of the β-chains is represented by a
square-shaped indentation. (This indentation is probably present in Hb A also.) Only the β_2 Val protrusions
and the β_1 EF pockets are shown. (The β_1 Val protrusions and the β_2 EF pockets are not involved, although
they are present.) **See this figure animated at http://
chemistry.brookscole.com/ggb3**

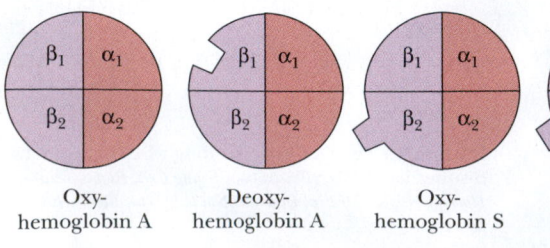

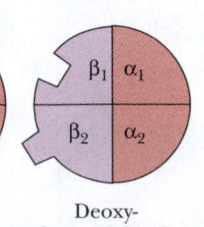

Oxy-
hemoglobin A   Deoxy-
                     hemoglobin A   Oxy-
                                          hemoglobin S   Deoxy-
                                                               hemoglobin S

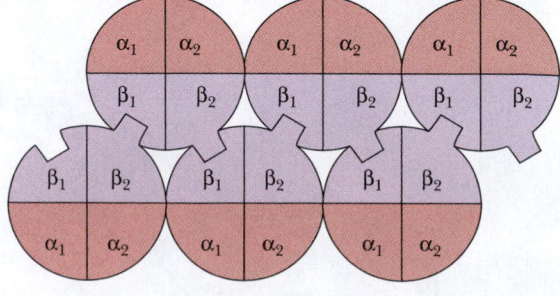

Deoxyhemoglobin S polymerizes into filaments

these cells were abnormally shaped; in fact, instead of the characteristic disc shape, these erythrocytes were elongated and crescentlike in form, a feature that eventually gave name to the disease **sickle-cell anemia.** These sickle cells pass less freely through the capillaries, impairing circulation and causing tissue damage. Furthermore, these cells are more fragile and rupture more easily than normal red cells, leading to anemia.

Sickle-Cell Anemia Is a Molecular Disease

A single amino acid substitution in the β-chains of Hb causes sickle-cell anemia. Replacement of the glutamate residue at position 6 in the β-chain by a valine residue marks the only chemical difference between Hb A and sickle-cell hemoglobin, Hb S. The amino acid residues at position β6 lie at the surface of the hemoglobin molecule. In Hb A, the ionic R groups of the Glu residues fit this environment. In contrast, the aliphatic side chains of the Val residues in Hb S create hydrophobic protrusions where none existed before. To the detriment of individuals who carry this trait, a hydrophobic pocket forms in the EF corner of each β-chain of Hb when it is in the deoxy state, and this pocket nicely accommodates the Val side chain of a neighboring Hb S molecule (Figure 15.39). This interaction leads to the aggregation of Hb S molecules into long, chainlike polymeric structures. The obvious consequence is that deoxyHb S is less soluble than

Human Biochemistry

Hemoglobin and Nitric Oxide

Nitric oxide (NO $\cdot$) is a simple gaseous molecule whose many remarkable physiological functions are still being discovered. For example, NO $\cdot$ is known to act as a neurotransmitter and as a second messenger in signal transduction (see Chapter 32). Furthermore, **endothelial relaxing factor** (ERF, also known as endothelium-derived relaxing factor, or EDRF), an elusive hormonelike agent that acts to relax the musculature of the walls **(endothelium)** of blood vessels and lower blood pressure, has been identified as NO $\cdot$. It has long been known that NO $\cdot$ is a high-affinity ligand for Hb, binding to its heme-Fe^{2+} atom with an affinity 10,000 times greater than that of O$_2$. An enigma thus arises: Why isn't NO $\cdot$ instantaneously bound by Hb within human erythrocytes and prevented from exerting its vasodilation properties?

The reason that Hb doesn't block the action of NO $\cdot$ is due to a unique interaction between Cys 93β of Hb and NO $\cdot$ discovered by Li Jia, Celia and Joseph Bonaventura, and Johnathan Stamler at Duke University. Nitric oxide reacts with the sulfhydryl group of Cys 93β, forming an S-nitroso derivative:

$$-CH_2-S-N=O$$

This S-nitroso group is in equilibrium with other S-nitroso compounds formed by reaction of NO $\cdot$ with small-molecule thiols such as free cysteine or glutathione (an isoglutamylcystcinylglycine tripeptide):

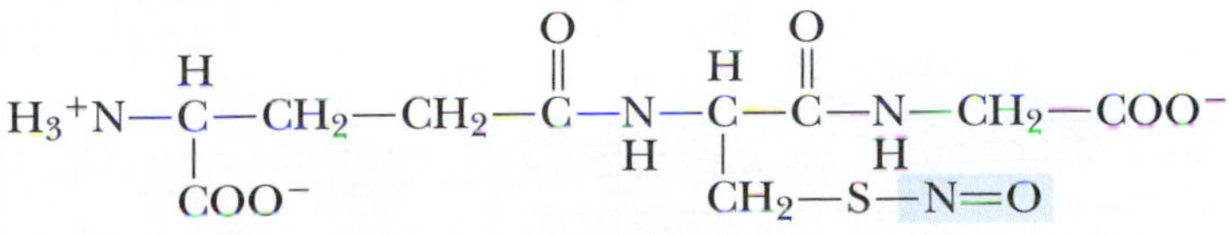

$$H_3{}^+N-\underset{\underset{COO^-}{|}}{C}-CH_2-CH_2-\overset{\overset{O}{\|}}{C}-\underset{H}{N}-\underset{\underset{CH_2-S-N=O}{|}}{\underset{H}{C}}-\overset{\overset{O}{\|}}{C}-\underset{H}{N}-CH_2-COO^-$$

S-nitrosoglutathione

These small-molecule thiols serve to transfer NO $\cdot$ from erythrocytes to endothelial receptors, where it acts to relax vascular tension. NO $\cdot$ itself is a reactive free-radical compound whose biological half-life is very short (1–5 sec). S-nitrosoglutathione has a half-life of several hours.

The reactions between Hb and NO $\cdot$ are complex. NO $\cdot$ forms a ligand with the heme-Fe^{2+} that is quite stable in the absence of O$_2$. However, in the presence of O$_2$, NO $\cdot$ is oxidized to NO$_3{}^-$ and the heme-Fe^{2+} of Hb is oxidized to Fe^{3+}, forming methemoglobin. Fortunately, the interaction of Hb with NO $\cdot$ is controlled by the allosteric transition between R-state Hb (oxyHb) and T-state Hb (deoxyHb). Cys 93β is more exposed and reactive in R-state Hb than in T-state Hb, and binding of NO $\cdot$ to Cys 93β precludes reaction of NO $\cdot$ with heme iron. Upon release of O$_2$ from Hb in tissues, Hb shifts conformation from R state to T state, and binding of NO $\cdot$ at Cys 93β is no longer favored. Consequently, NO $\cdot$ is released from Cys 93β and transferred to small-molecule thiols for delivery to endothelial receptors, causing capillary vasodilation. This mechanism also explains the puzzling observation that free Hb produced by recombinant DNA methodology for use as a whole-blood substitute causes a transient rise of 10 to 12 mm Hg in diastolic blood pressure in experimental clinical trials. (Conventional whole-blood transfusion has no such effect.) It is now apparent that the "synthetic" Hb, which has no bound NO $\cdot$, is binding NO $\cdot$ in the blood and preventing its vasoregulatory function.

In the course of hemoglobin evolution, the only invariant amino acid residues in globin chains are His F8 (the obligatory heme ligand) and a Phe residue acting to wedge the heme into its pocket. However, in mammals and birds, Cys 93β is also invariant, no doubt due to its vital role in NO $\cdot$ delivery.

Adapted from Jia, L., et al., 1996. S-Nitrosohaemoglobin: A dynamic activity of blood involved in vascular control. *Nature* **380**:221–226.

deoxyHb A. The concentration of hemoglobin in red blood cells is high (about 150 mg/mL), so even in normal circumstances it is on the verge of crystallization. The formation of insoluble deoxyHb S fibers distorts the red cell into the elongated sickle shape characteristic of the disease.[3]

[3]In certain regions of Africa, the sickle-cell trait is found in 20% of the people. Why does such a deleterious heritable condition persist in the population? For reasons as yet unknown, individuals with this trait are less susceptible to the most virulent form of malaria. The geographic distribution of malaria and the sickle-cell trait are positively correlated.

Summary

15.1 What Factors Influence Enzymatic Activity?
The two prominent ways to regulate enzyme activity are (1) to increase or decrease the number of enzyme molecules or (2) to increase or decrease the intrinsic activity of each enzyme molecule. Changes in enzyme amounts are typically regulated via gene expression and protein degradation. Changes in the intrinsic activity of enzyme molecules are achieved principally by allosteric regulation or covalent modification.

15.2 What Are the General Features of Allosteric Regulation?
Allosteric enzymes show a sigmoid response of velocity, v, to increasing [S], indicating that binding of S to the enzyme is cooperative. Allosteric enzymes often are susceptible to feedback inhibition. Allosteric enzymes may also respond to allosteric activation. Allosteric activators signal a need for the end product of the pathway in which the allosteric enzyme functions. As a general rule, allosteric enzymes are oligomeric, with each monomer possessing a substrate-binding site and an allosteric site where effectors bind. Interaction of one subunit of an allosteric enzyme with its substrate (or its effectors) is communicated to the other subunits of the enzyme through intersubunit interactions. These interactions can lead to conformational transitions that make it easier (or harder) for additional equivalents of ligand (S, A, or I) to bind to the enzyme.

15.3 Can a Simple Equilibrium Model Explain Allosteric Kinetics?
Monod, Wyman, and Changeux postulated that the subunits of allosteric enzymes can exist in two conformational states (R and T), that all subunits in any enzyme molecule are in the same conformational state *(symmetry)*, that equilibrium strongly favors the T conformational state, and that S binds preferentially ("only") to the R state. Sigmoid binding curves result, provided that $[T_0] \gg [R_0]$ in the absence of S and that S binds "only" to R ($K_R/K_T < 0.1$). Positive or negative effectors influence the relative T/R equilibrium by binding preferentially to T (negative effectors) or R (positive effectors), and the substrate saturation curve is shifted to the right (negative effectors) or left (positive effectors). These features define the MWC K system, so-called because $K_{0.5}$ changes and V_{max} is constant. Another MWC possibility is the V system, so-called because V_{max} changes and $K_{0.5}$ is constant. The V system arises if S binds equally well to R and T. Thus, no cooperativity is seen in S binding. However, if only R is catalytically active and a positive effector binds preferentially to R, the R state is favored and V_{max} increases. If a negative effector binds preferentially to T, the T state will be favored and V_{max} will decrease.

In an alternative allosteric model suggested by Koshland, Nemethy, and Filmer (the KNF model), S binding leads to conformational changes in the enzyme. The altered conformation of the enzyme may display higher affinity for the substrate (positive cooperativity) or lower affinity for the substrate or other ligand (negative cooperativity). Negative cooperativity is not possible within the MWC model.

15.4 Is the Activity of Some Enzymes Controlled by Both Allosteric Regulation and Covalent Modification?
Some enzymes are subject to both allosteric regulation and regulation by covalent modification. A prime example is glycogen phosphorylase. Glycogen phosphorylase exists in two forms, *a* and *b*, which differ only in whether or not Ser^{14}-OH is phosphorylated *(a)* or not *(b)*. Glycogen phosphorylase *b* shows positive cooperativity in binding its substrate, phosphate. In addition, glycogen phosphorylase *b* is allosterically activated by the positive effector AMP. In contrast, ATP and glucose-6-P are negative effectors for glycogen phosphorylase *b*. Covalent modification of glycogen phosphorylase *b* by phosphorylase kinase converts it from a less active, allosterically regulated form to the more active *a* form that is less responsive to allosteric regulation. Glycogen phosphorylase is both activated and freed from allosteric control by covalent modification.

Special Focus: Is There an Example in Nature That Exemplifies the Relationship Between Quaternary Structure and the Emergence of Allosteric Properties? Hemoglobin and Myoglobin—Paradigms of Protein Structure and Function
Myoglobin and hemoglobin have illuminated our understanding of protein structure and function. Myoglobin is monomeric, whereas hemoglobin has a quaternary structure. Myoglobin functions as an oxygen-storage protein in muscle; Hb is an O_2-transport protein. When Mb binds O_2, its heme iron atom is drawn within the plane of the heme, slightly shifting the position of the F helix of the protein. Hemoglobin shows cooperative binding of O_2 and allosteric regulation by H^+, CO_2, and 2,3-bisphosphoglycerate. The allosteric properties of Hb can be traced to the movement of the F helix upon O_2 binding to Hb heme groups and the effects of F-helix movement on interactions between the protein's subunits that alter the intrinsic affinity of the other subunits for O_2. The allosteric transitions in Hb partially conform to the MWC model in that a concerted conformational change from a T-state, low-affinity conformation to an R-state, high-affinity form takes place after 2 O_2 are bound (by the 2 Hb α-subunits). However, Hb also behaves somewhat according to the KNF model of allostery in that oxygen binding leads to sequential changes in the conformation and O_2 affinity of hemoglobin subunits. Sickle-cell anemia is a molecular disease traceable to a tendency for Hb S to polymerize as a consequence of having a βE6V amino acid substitution that creates a "sticky" hydrophobic patch on the Hb surface.

Problems

1. List six general ways in which enzyme activity is controlled.

2. Why do you suppose proteolytic enzymes are often synthesized as inactive zymogens?

3. (Integrates with Chapter 13.) First draw both Lineweaver–Burk plots and Hanes–Woolf plots for the following: a Monod–Wyman–Changeux allosteric K enzyme system, showing separate curves for the kinetic response in (a) the absence of any effectors, (b) the presence of allosteric activator A, and (c) the presence of allosteric inhibitor I. Then draw a similar set of curves for a Monod–Wyman–Changeux allosteric V enzyme system.

4. In the Monod–Wyman–Changeux model for allosteric regulation, what values of L and relative affinities of R and T for A will lead activator A to exhibit positive homotropic effects? (That is, under what conditions will the binding of A enhance further A binding, in the same manner that S binding shows positive cooperativity?) What values of L and relative affinities of R and T for I will lead inhibitor I to exhibit positive homotropic effects? (That is, under what conditions will the binding of I promote further I binding?)

*5. The KNF model for allosteric transitions includes the possibility of negative cooperativity. Draw Lineweaver–Burk and Hanes–Woolf plots for the case of negative cooperativity in substrate binding. (As a point of reference, include a line showing the classic Michaelis–Menten response of v to [S].)

6. The equation $\dfrac{Y}{(1 - Y)} = \left(\dfrac{pO_2}{P_{50}}\right)^n$ allows the calculation of Y (the fractional saturation of hemoglobin with O_2), given P_{50} and n (see box on page 496). Let $P_{50} = 26$ torr and $n = 2.8$. Calculate Y in the lungs, where $pO_2 = 100$ torr, and Y in the capillaries, where $pO_2 = 40$ torr. What is the efficiency of O_2 delivery under these conditions (expressed as $Y_{lungs} - Y_{capillaries}$)? Repeat the calculations, but for $n = 1$. Compare the values for $Y_{lungs} - Y_{capillaries}$ for $n = 2.8$ versus $Y_{lungs} - Y_{capillaries}$ for $n = 1$ to determine the effect of cooperative O_2 binding on oxygen delivery by hemoglobin.

7. The cAMP formed by adenylyl cyclase (Figure 15.19) does not persist because 5′-phosphodiesterase activity prevalent in cells hydrolyzes cAMP to give 5′-AMP. Caffeine inhibits 5′-phosphodiesterase activity. Describe the effects on glycogen phosphorylase activity that arise as a consequence of drinking lots of caffeinated coffee.

8. If no precautions are taken, blood that has been stored for some time becomes depleted in 2,3-BPG. What happens if such blood is used in a transfusion?

9. Enzymes have evolved such that their K_m values (or $K_{0.5}$ values) for substrate(s) are roughly equal to the in vivo concentration(s) of the substrate(s). Assume that glycogen phosphorylase is assayed at $[P_i] \approx K_{0.5}$ in the absence and presence of AMP or ATP. Estimate from Figure 15.15 the relative glycogen phosphorylase activity when (a) neither AMP or ATP is present, (b) AMP is present, and (c) ATP is present. (Hint: Use a ruler to get relative values for the velocity v at the appropriate midpoints of the saturation curves.)

10. Cholera toxin is an enzyme that covalently modifies the G_α-subunit of G proteins. (Cholera toxin catalyzes the transfer of ADP-ribose from NAD^+ to an arginine residue in G_α, an ADP-ribosylation reaction.) Covalent modification of G_α inactivates its GTPase activity. Predict the consequences of cholera toxin on cellular cAMP and glycogen levels.

*11. Allosteric enzymes that sit at branch points leading to several essential products sometimes display negative cooperativity for feedback inhibition (allosteric inhibition) by one of the products. What might be the advantage of negative cooperativity instead of positive cooperativity in feedback inhibitor binding by such enzymes?

12. Consult the table in the A Deeper Look box on page 477.
   a. Suggest a consensus amino acid sequence within phosphorylase kinase that makes it a target of protein kinase A (the cAMP-dependent protein kinase).
   b. Suggest an effective amino acid sequence for a regulatory domain pseudosubstrate sequence that would exert intrasteric control on phosphorylase kinase by blocking its active site.

13. What are the relative advantages (and disadvantages) of allosteric regulation versus covalent modification?

*14. You land a post as scientific investigator with a pharmaceutical company that would like to develop drugs to treat people with sickle-cell anemia. They want ideas from you! What molecular properties of Hb S might you suggest as potential targets of drug therapy?

*15. Under appropriate conditions, nitric oxide (NO ·) combines with Cys 93β in hemoglobin and influences its interaction with O_2. Is this interaction an example of allosteric regulation or covalent modification?

Preparing for the MCAT Exam
16. On the basis of the graphs shown in Figures 15.9 and 15.10:
   a. If a tetrameric enzyme with an allosteric L value of 1000 had a c value of 0.2, would this enzyme show cooperative binding of S?
   b. Suppose this tetrameric enzyme bound an allosteric inhibitor I. Would its cooperativity with regard to S binding increase, decrease, or stay the same?

17. Figure 15.18 traces the activation of glycogen phosphorylase from hormone to phosphorylation of the b form of glycogen phosphorylase to the a form. These effects are reversible when hormone disappears. Suggest reactions by which such reversibility is achieved.

Biochemistry Now™   Preparing for an exam? Test yourself on key questions at http://chemistry.brookscole.com/ggb3

Further Reading

General References
Fersht, A., 1999. *Structure and Mechanism in Protein Science: A Guide to Enzyme Catalysis and Protein Folding.* New York: W. H. Freeman. An advanced textbook on protein structure and function, including principles of enzyme regulation.

Protein Kinases
Manning, G., et al., 2002. The protein kinase complement of the human genome. *Science* **298:**1912–1934. A catalog of the protein kinase genes identified within the human genome. About 2% of all eukaryotic genes encode protein kinases.

Allosteric Regulation
Helmstaedt, K., Krappman, S., and Braus, G. H., 2001. Allosteric regulation of catalytic activity: *Escherichia coli* aspartate transcarbamoylase versus yeast chorismate mutase. *Microbiology and Molecular Biology Reviews* **65:**404–421. The authors present evidence to show that the MWC two-state model is oversimplified, as Monod, Wyman, and Changeux themselves originally stipulated.

Koshland, D. E., Jr., and Hamadani, K., 2002. Proteomics and models for enzyme cooperativity. *Journal of Biological Chemistry* **277:**46841–46844. An overview of both the MWC and the KNF models for allostery and a discussion of the relative merits of these models. The fact that the number of allosteric enzymes showing negative cooperativity is about the same as the number showing positive cooperativity is an important focus of this review.

Koshland, D. E., Jr., Nemethy, G., and Filmer, D., 1966. Comparison of experimental binding data and theoretical models in proteins containing subunits. *Biochemistry* **5:**365–385. The KNF model.

Monod, J., Wyman, J., and Changeux, J-P., 1965. On the nature of allosteric transitions: A plausible model. *Journal of Molecular Biology* **12:**88–118. The classic paper that provided the first theoretical analysis of allosteric regulation.

Schachman, H. K., 1990. Can a simple model account for the allosteric transition of aspartate transcarbamoylase? *Journal of Biological Chemistry* **263:**18583–18586. Tests of the postulates of the allosteric models through experiments on aspartate transcarbamoylase.

Glycogen Phosphorylase

Johnson, L. N., and Barford, D., 1993. The effects of phosphorylation on the structure and function of proteins. *Annual Review of Biophysics and Biomolecular Structure* **22:**199–232. A review of protein phosphorylation and its role in regulation of enzymatic activity, with particular emphasis on glycogen phosphorylase.

Johnson, L. N., and Barford, D., 1994. Electrostatic effects in the control of glycogen phosphorylase by phosphorylation. *Protein Science* **3:**1726–1730. Discussion of the phosphate group's ability to deliver two negative charges to a protein, a property that no amino acid side chain can provide.

Lin, K., et al., 1996. Comparison of the activation triggers in yeast and muscle glycogen phosphorylase. *Science* **273:**1539–1541. Despite structural and regulatory differences between yeast and muscle glycogen phosphorylases, both are activated through changes in their intersubunit interface.

Lin, K., et al., 1997. Distinct phosphorylation signals converge at the catalytic center in glycogen phosphorylases. *Structure* **5:**1511–1523.

Rath, V. L., et al., 1996. The evolution of an allosteric site in phosphorylase. *Structure* **4:**463–473.

Hemoglobin

Ackers, G. K., 1998. Deciphering the molecular code of hemoglobin allostery. *Advances in Protein Chemistry* **51:**185–253.

Dickerson, R. E., and Geis, I., 1983. *Hemoglobin: Structure, Function, Evolution and Pathology.* Menlo Park, CA: Benjamin/Cummings.

Gill, S. J., et al., 1988. New twists on an old story: Hemoglobin. *Trends in Biochemical Sciences* **13:**465–467.

Weiss, J. N., 1997. The Hill equation revisited: Uses and abuses. *The FASEB Journal* **11:**835–841.

The Oxygen-Binding Curves of Myoglobin and Hemoglobin

Myoglobin

The reversible binding of oxygen to myoglobin,

$$MbO_2 \rightleftharpoons Mb + O_2$$

can be characterized by the equilibrium dissociation constant, K.

$$K = \frac{[Mb][O_2]}{[MbO_2]} \tag{A15.1}$$

If Y is defined as the **fractional saturation** of myoglobin with O_2, that is, the fraction of myoglobin molecules having an oxygen molecule bound, then

$$Y = \frac{[MbO_2]}{[MbO_2] + [Mb]} \tag{A15.2}$$

The value of Y ranges from 0 (no myoglobin molecules carry an O_2) to 1.0 (all myoglobin molecules have an O_2 molecule bound). Substituting from Equation A15.1, $([Mb][O_2])/K$ for $[MbO_2]$ gives

$$Y = \frac{\left(\dfrac{[Mb][O_2]}{K}\right)}{\left(\dfrac{[Mb][O_2]}{K} + [Mb]\right)} = \frac{\left(\dfrac{[O_2]}{K}\right)}{\left(\dfrac{[O_2]}{K} + 1\right)} = \frac{[O_2]}{[O_2] + K} \tag{A15.3}$$

and, if the concentration of O_2 is expressed in terms of the partial pressure (in torr) of oxygen gas in equilibrium with the solution of interest, then

$$Y = \frac{pO_2}{pO_2 + K} \tag{A15.4}$$

(In this form, K has the units of torr.) The relationship defined by Equation A15.4 plots as a hyperbola. That is, the MbO_2 saturation curve resembles an enzyme:substrate saturation curve. For myoglobin, a partial pressure of 1 torr for pO_2 is sufficient for half-saturation (Figure A15.1). We can define P_{50} as the partial pressure of O_2 at which 50% of the myoglobin molecules have a molecule of O_2 bound (that is, $Y = 0.5$), then

$$0.5 = \frac{pO_2}{pO_2 + P_{50}} \tag{A15.5}$$

(Note from Equation A15.1 that when $[MbO_2] = [Mb]$, $K = [O_2]$, which is the same as saying when $Y = 0.5$, $K = P_{50}$.) The general equation for O_2 binding to Mb becomes

$$Y = \frac{pO_2}{pO_2 + P_{50}} \tag{A15.6}$$

The ratio of the fractional saturation of myoglobin, Y, to free myoglobin, $1 - Y$, depends on pO_2 and K according to the equation

$$\frac{Y}{1 - Y} = \frac{pO_2}{K} \tag{A15.7}$$

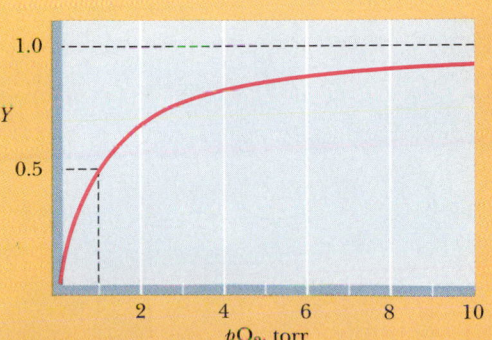

FIGURE A15.1 Oxygen saturation curve for myoglobin in the form of Y versus pO_2 showing P_{50} is at a pO_2 of 1 torr.

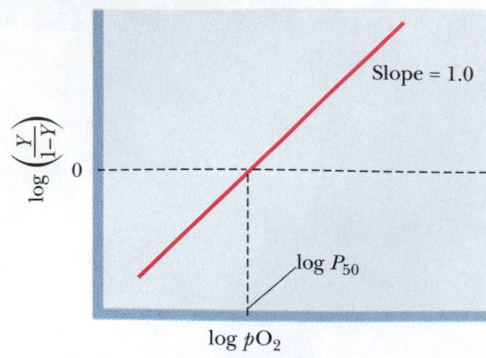

FIGURE A15.2 Hill plot for the binding of O_2 to myoglobin. The slope of the line is the **Hill coefficient.** For Mb, the Hill coefficient is 1.0. At log $[Y/(1 - Y)] = 0$, log $pO_2 = $ log P_{50}.

Taking the logarithm yields

$$\log\left(\frac{Y}{1 - Y}\right) = \log pO_2 - \log K \qquad \text{(A15.8)}$$

A graph of log $[Y/(1 - Y)]$ versus log pO_2 is known as a **Hill plot** (in honor of Archibald Hill, a pioneer in the study of O_2 binding by hemoglobin). A Hill plot for myoglobin (Figure A15.2) gives a straight line. At half-saturation, defined as $Y = 0.5$, $Y/(1 - Y) = 1$, and log $[Y/(1 - Y)] = 0$. At this value of log $[Y/(1 - Y)]$, the value for $pO_2 = K = P_{50}$. The slope of the Hill plot at the point where log $[Y/(1 - Y)] = 0$, the midpoint of binding, is known as the **Hill coefficient.** The Hill coefficient for myoglobin is 1.0. A Hill coefficient of 1.0 means that O_2 molecules bind independently of one another to myoglobin, a conclusion entirely logical because each Mb molecule can bind only one O_2.

Hemoglobin

New properties emerge when four heme-containing polypeptides come together to form a tetramer. The O_2-binding curve of hemoglobin is sigmoid rather than hyperbolic (see Figure 15.21), and Equation A15.4 does not describe such curves. Of course, each hemoglobin molecule has four hemes and can bind up to four oxygen molecules. Suppose for the moment the O_2 binding to hemoglobin is an "all-or-none" phenomenon, where Hb exists either free of O_2 or with four O_2 molecules bound. This supposition represents the extreme case for cooperative binding of a ligand by a protein with multiple binding sites. In effect, it says that if one ligand binds to the protein molecule, then all other sites are immediately occupied by ligand. Or, to say it another way for the case in hand, suppose that four O_2 molecules bind to Hb simultaneously:

$$\text{Hb} + 4\,O_2 \rightleftharpoons \text{Hb}(O_2)_4$$

Then the dissociation constant, K, would be

$$K = \frac{[\text{Hb}][O_2]^4}{[\text{Hb}(O_2)_4]} \qquad \text{(A15.9)}$$

By analogy with Equation A15.4, the equation for fractional saturation of Hb is given by

$$Y = \frac{[pO_2]^4}{[pO_2]^4 + K} \qquad \text{(A15.10)}$$

A plot of Y versus pO_2 according to Equation A15.10 is presented in Figure A15.3. This curve has the characteristic sigmoid shape seen for O_2 binding by Hb. Half-saturation is set to be a pO_2 of 26 torr. Note that when pO_2 is low, the fractional saturation, Y, changes very little as pO_2 increases. The interpretation is that Hb has little affinity for O_2 at these low partial pressures of O_2. However, as pO_2 reaches some threshold value and the first O_2 is bound, Y, the fractional saturation, increases rapidly. Note that the slope of the curve is steepest in the region where $Y = 0.5$. The sigmoid character of this curve is diagnostic of the fact that the binding of O_2 to one site on Hb strongly enhances binding of additional O_2 molecules to the remaining vacant sites on the same Hb molecule, a phenomenon aptly termed **cooperativity.** (If each O_2 bound independently, exerting no influence on the affinity of Hb for more O_2 binding, this plot would be hyperbolic.)

The experimentally observed oxygen-binding curve for Hb does not fit the graph given in Figure A15.3 exactly. If we generalize Equation A15.10 by replacing the exponent 4 with n, we can write the equation as

$$Y = \frac{[pO_2]^n}{[pO_2]^n + K} \qquad \text{(A15.11)}$$

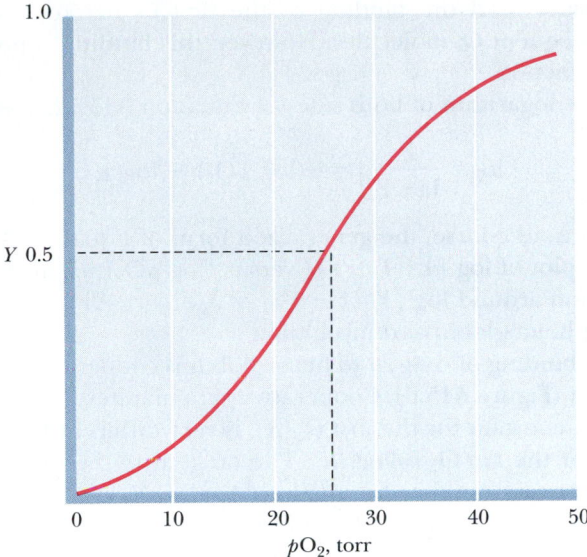

FIGURE A15.3 Oxygen saturation curve for Hb in the form of Y versus pO_2, assuming $n = 4$, and $P_{50} = 26$ torr. The graph has the characteristic experimentally observed sigmoid shape.

Rearranging yields

$$\frac{Y}{1-Y} = \frac{[pO_2]^n}{K} \tag{A15.12}$$

This equation states that the ratio of oxygenated heme groups (Y) to O_2-free heme ($1 - Y$) is equal to the nth power of the pO_2 divided by the apparent dissociation constant, K.

Archibald Hill demonstrated in 1913, well before any knowledge about the molecular organization of Hb existed, that the O_2-binding behavior of Hb could be described by Equation A15.12. If a value of 2.8 is taken for n, Equation A15.12 fits the experimentally observed O_2-binding curve for Hb very well (Figure A15.4). If the binding of O_2 to Hb were an all-or-none phenomenon, n would equal 4, as discussed previously. If the O_2-binding sites on Hb were completely noninteracting, that is, if the binding of one O_2 to Hb had no influence on the binding of additional O_2 molecules to the same Hb, n would equal 1. Figure A15.4 compares these extremes. Obviously, the real situation falls between the extremes of $n = 1$ or 4. The qualitative answer is that O_2 binding by Hb is

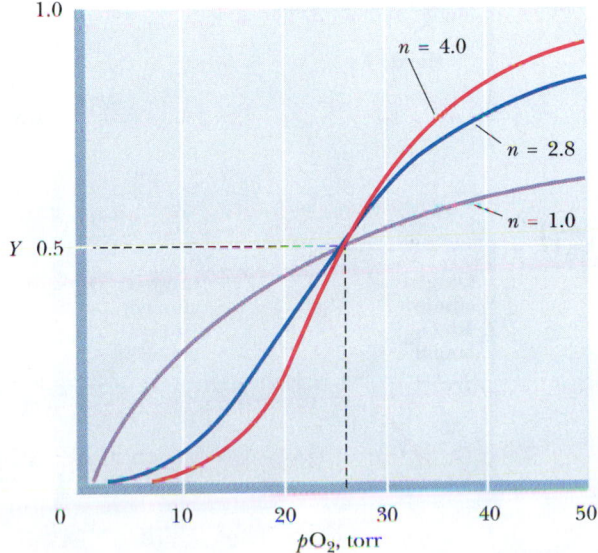

FIGURE A15.4 A comparison of the experimentally observed O_2 curve for Hb yielding a value for n of 2.8, the hypothetical curve if $n = 4$, and the curve if $n = 1$ (noninteracting O_2-binding sites).

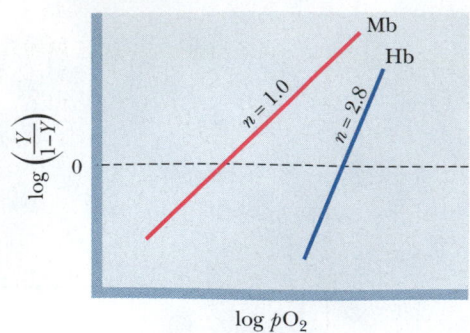

FIGURE A15.5 Hill plot (log $[Y/(1-Y)]$ versus log pO_2) for Mb and Hb, showing that at log $[Y/(1-Y)] = 0$, that is, $Y = (1-Y)$, the slope for Mb is 1.0 and for Hb is 2.8. The plot for Hb only approximates a straight line.

highly cooperative, and the binding of the first O_2 markedly enhances the binding of subsequent O_2 molecules. However, this binding is not quite an all-or-none phenomenon.

If we take the logarithm of both sides of Equation A15.12:

$$\log\left(\frac{Y}{1-Y}\right) = n(\log pO_2) - \log K \tag{A15.13}$$

this expression is, of course, the generalized form of Equation A15.8, the *Hill equation*, and a plot of log $[Y/(1-Y)]$ versus (log pO_2) *approximates* a straight line in the region around log $[Y/(1-Y)] = 0$. Figure A15.5 represents a *Hill plot* comparing hemoglobin and myoglobin.

Because the binding of oxygen to hemoglobin is cooperative, the Hill plot is actually sigmoid (Figure A15.6). Cooperativity is a manifestation of the fact that the dissociation constant for the *first* O_2, K_1, is very different from the dissociation constant for the *last* O_2 *bound*, K_4. The tangent to the lower asymptote of the Hill plot, when extrapolated to the log $[Y/(1-Y)] = 0$ axis, gives the dissociation constant, K_1, for the binding of the first O_2 by Hb. Note that the value of K_1 is quite large ($>10^2$ torr), indicating a low affinity of Hb for this first O_2 [or conversely, a ready dissociation of the Hb $(O_2)_1$ complex]. By a similar process, the tangent to the upper asymptote gives K_4, the dissociation constant for the last O_2 to bind. K_4 has a value of less than 1 torr. The K_1/K_4 ratio exceeds 100, meaning the affinity of Hb for binding the fourth O_2 is more than 100 times greater than for binding the first oxygen.

The value P_{50} has been defined for myoglobin as the pO_2 that gives 50% saturation of the oxygen-binding protein with oxygen. Noting that at 50% saturation, $Y = (1-Y)$, then we have from Equation A15.13.

$$0 = n(\log pO_2) - \log K = n(\log P_{50}) - \log K \tag{A15.14}$$

$$\log K = n(\log P_{50}) \ or \ K = (P_{50})^n \tag{A15.15}$$

That is, the situations for myoglobin and hemoglobin differ; therefore, P_{50} and K cannot be equated for Hb because of its multiple, interacting, O_2-binding sites. The relationship between pO_2 and P_{50} for hemoglobin, by use of Equation A15.12, becomes

$$\frac{Y}{1-Y} = \left(\frac{pO_2}{P_{50}}\right)^n \tag{A15.16}$$

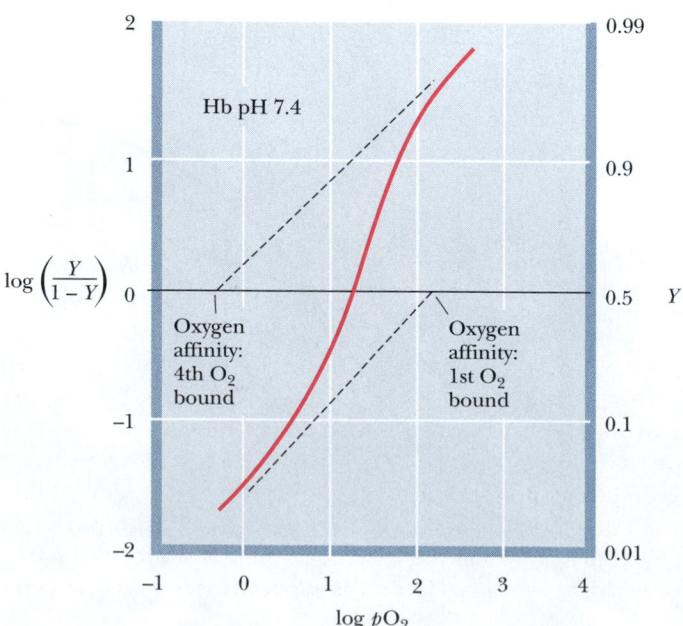

FIGURE A15.6 Hill plot of Hb showing its nonlinear nature and the fact that its asymptotes can be extrapolated to yield the dissociation constants, K_1 and K_4, for the first and fourth oxygens.

Molecular Motors

CHAPTER 16

Essential Question

Movement is an intrinsic property associated with all living things. Within cells, molecules undergo coordinated and organized movements, and cells themselves may move across a surface. At the tissue level, **muscle contraction** allows higher organisms to carry out and control crucial internal functions, such as peristalsis in the gut and the beating of the heart. Muscle contraction also enables the organism to perform organized and sophisticated movements, such as walking, running, flying, and swimming. *How can biological macromolecules, carrying out conformational changes on the microscopic, molecular level, achieve these feats of movement that span the molecular and macroscopic worlds?*

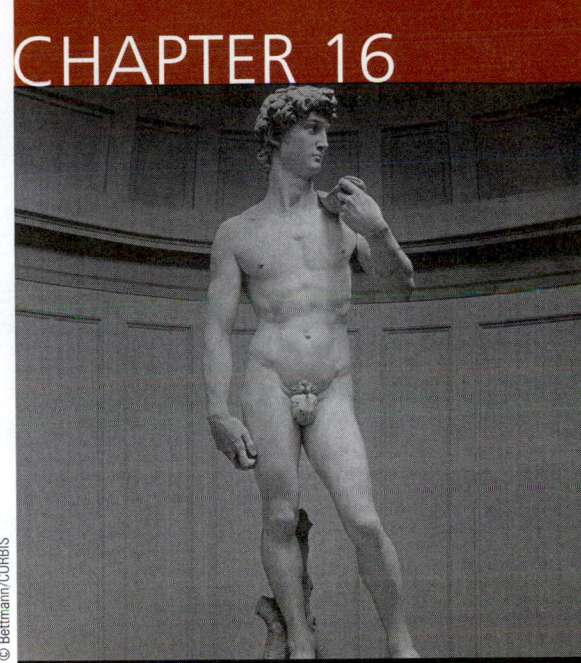

Michelangelo's *David* epitomizes the musculature of the human form.

Nature does nothing needlessly.
Aristotle, *Politics,* book 1, chapter 2

16.1 What Is a Molecular Motor?

Motor proteins, also known as **molecular motors,** use chemical energy (ATP) to orchestrate movements, transforming ATP energy into the mechanical energy of motion. In all cases, ATP hydrolysis is presumed to drive and control protein conformational changes that result in sliding or walking movements of one molecule relative to another. To carry out directed movements, molecular motors must be able to associate and dissociate reversibly with a polymeric protein array, a surface or substructure in the cell. ATP hydrolysis drives the process by which the motor protein ratchets along the protein array or surface. As fundamental and straightforward as all this sounds, elucidation of these basically simple processes has been extremely challenging for biochemists, involving the application of many sophisticated chemical and physical methods in many different laboratories. This chapter describes the structures and chemical functions of molecular motor proteins and some of the experiments by which we have come to understand them.

Molecular motors may be **linear** or **rotating.** Linear motors crawl or creep along a polymer lattice, whereas rotating motors consist of a rotating element (the "rotor") and a stationary element (the "stator"), in a fashion much like a simple electrical motor. The linear motors we will discuss include **kinesins** and **dyneins** (which crawl along microtubules), **myosin** (which slides along actin filaments in muscle), and **DNA helicases** (which move along a DNA lattice, unwinding duplex DNA to form single-stranded DNA). Rotating motors include the flagellar motor complex, described in this chapter, and the ATP synthase, which will be described in Chapter 20.

Key Questions

- **16.1** What Is a Molecular Motor?
- **16.2** What Are the Molecular Motors That Orchestrate the Mechanochemistry of Microtubules?
- **16.3** How Do Molecular Motors Unwind DNA?
- **16.4** What Is the Molecular Mechanism of Muscle Contraction?
- **16.5** How Do Bacterial Flagella Use a Proton Gradient to Drive Rotation?

16.2 What Are the Molecular Motors That Orchestrate the Mechanochemistry of Microtubules?

One of the simplest self-assembling structures found in biological systems is the microtubule, one of the fundamental components of the eukaryotic cytoskeleton and the primary structural element of cilia and flagella (Figure 16.1). **Microtubules** are hollow, cylindrical structures, approximately 30 nm in diameter, formed from **tubulin,** a dimeric protein composed of two similar 55-kD subunits known as α-*tubulin* and β-*tubulin*. Eva Nogales, Sharon Wolf, and Kenneth Downing have determined the structure of the bovine tubulin αβ-dimer to 3.7 Å resolution (Figure 16.2a). Tubulin dimers polymerize as shown in Figure 16.2b to form microtubules, which are essentially helical structures, with 13 tubulin monomer "residues" per turn. Microtubules grown in vitro are dynamic structures

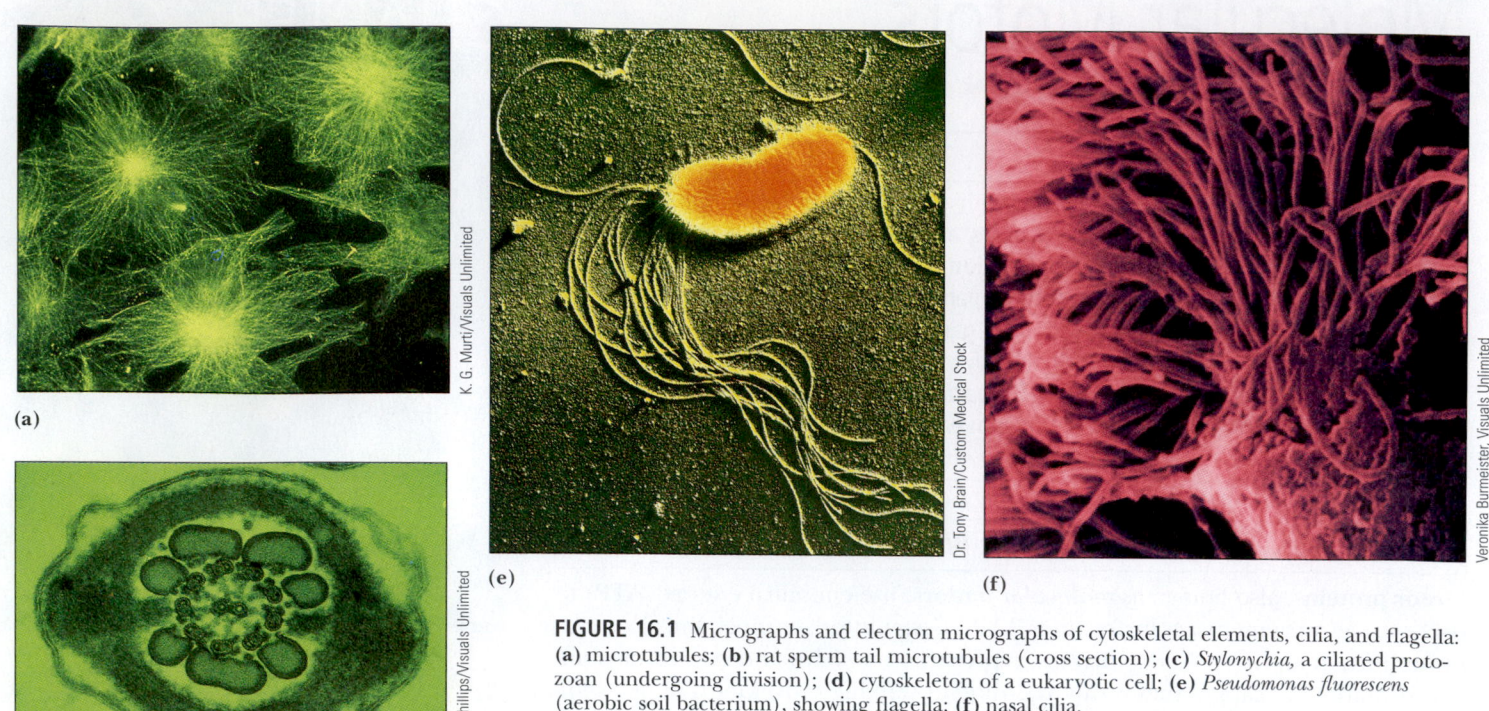

K. G. Murti/Visuals Unlimited

David Phillips/Visuals Unlimited

Eric Grave/Phototake

Fawcett and Heuser/Photo Researchers, Inc.

Dr. Tony Brain/Custom Medical Stock

Veronika Burmeister, Visuals Unlimited

FIGURE 16.1 Micrographs and electron micrographs of cytoskeletal elements, cilia, and flagella: **(a)** microtubules; **(b)** rat sperm tail microtubules (cross section); **(c)** *Stylonychia*, a ciliated protozoan (undergoing division); **(d)** cytoskeleton of a eukaryotic cell; **(e)** *Pseudomonas fluorescens* (aerobic soil bacterium), showing flagella; **(f)** nasal cilia.

that are constantly being assembled and disassembled. Because all tubulin dimers in a microtubule are oriented similarly, microtubules are polar structures. The end of the microtubule at which growth occurs is the **plus end,** and the other is the **minus end.** Microtubules in vitro carry out a GTP-dependent process called **treadmilling,** in which tubulin dimers are added to the plus end at about the same rate at which dimers are removed from the minus end (Figure 16.3).

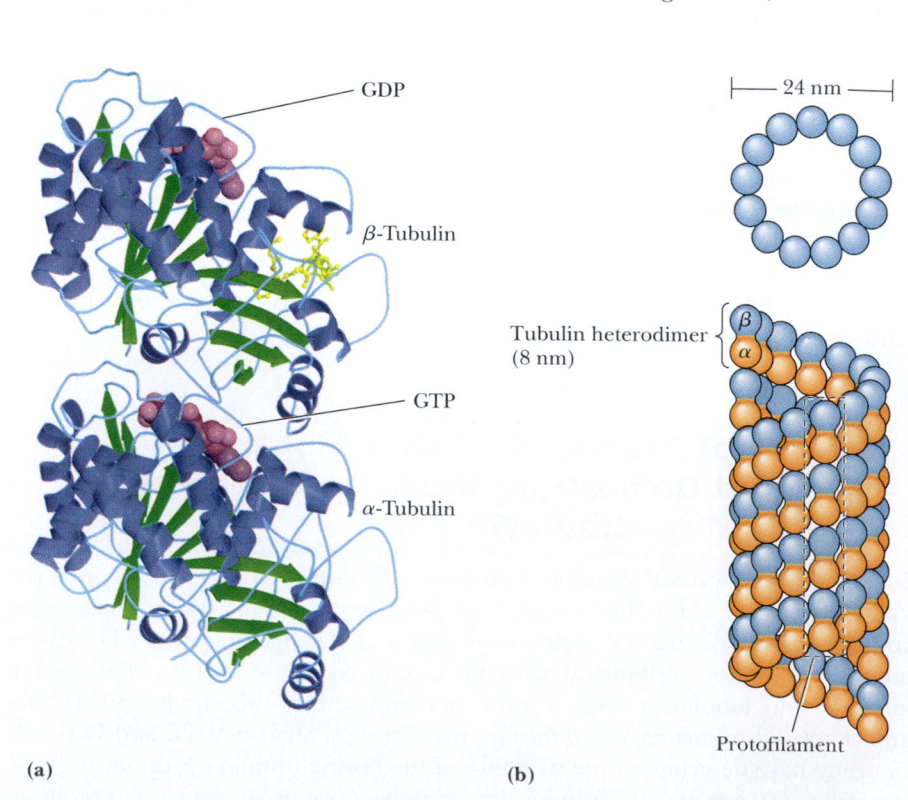

FIGURE 16.2 **(a)** The structure of the tubulin αβ-heterodimer. **(b)** Microtubules may be viewed as consisting of 13 parallel, staggered protofilaments of alternating α-tubulin and β-tubulin subunits. The sequences of the α- and β-subunits of tubulin are homologous, and the αβ-tubulin dimers are quite stable if Ca^{2+} is present. The dimer is dissociated only by strong denaturing agents.

Microtubules Are Constituents of the Cytoskeleton

Although composed only of 55-kD tubulin subunits, microtubules can grow sufficiently large to span a eukaryotic cell or to form large structures such as cilia and flagella. Inside cells, networks of microtubules play many functions, including formation of the mitotic spindle that segregates chromosomes during cell division, the movement of organelles and various vesicular structures through the cell, and the variation and maintenance of cell shape. Microtubules are, in fact, a significant part of the **cytoskeleton**, a sort of intracellular scaffold formed of microtubules, intermediate filaments, and microfilaments (Figure 16.4). In most cells, microtubules are oriented with their minus ends toward the centrosome and their plus ends toward the cell periphery. This consistent orientation is important for mechanisms of intracellular transport.

Microtubules Are the Fundamental Structural Units of Cilia and Flagella

As already noted, microtubules are also the fundamental building blocks of cilia and flagella. **Cilia** are short, cylindrical, hairlike projections on the surfaces of the cells of many animals and lower plants. The beating motion of cilia functions either to move cells from place to place or to facilitate the movement of extracellular fluid over the cell surface. Flagella are much longer structures found singly or a few at a time on certain cells (such as sperm cells). They propel cells through fluids. Cilia and flagella share a common design (Figure 16.5). The **axoneme** is a complex bundle of microtubule fibers that includes two central, separated microtubules surrounded by nine pairs of joined microtubules. The axoneme is surrounded by a plasma membrane that is continuous with the plasma membrane of the cell. Removal of the plasma membrane by detergent and subsequent treatment of the exposed axonemes with high concentrations of salt releases the **dynein** molecules (Figure 16.6), which form the dynein arms.

Ciliary Motion Involves Bending of Microtubule Bundles

The motion of cilia results from the ATP-driven sliding or walking of dyneins along one microtubule while they remain firmly attached to an adjacent microtubule. The flexible stems of the dyneins remain permanently attached to

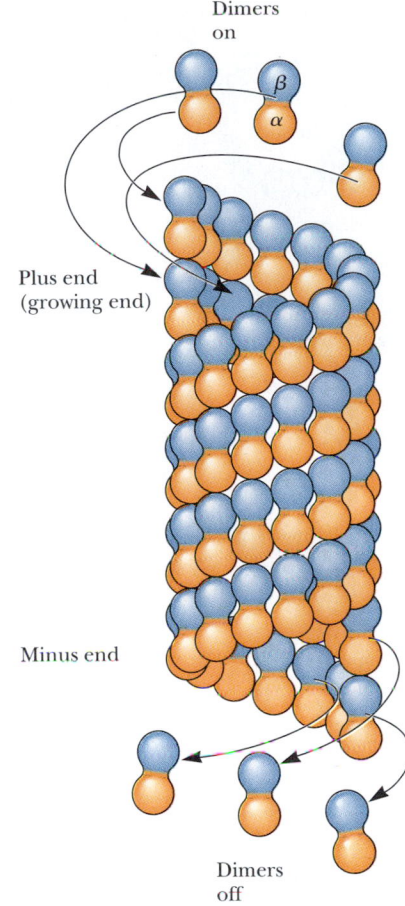

Biochemistry Now™ **ACTIVE FIGURE 16.3**
A model of the GTP-dependent treadmilling process. Both α- and β-tubulin possess two different binding sites for GTP. The polymerization of tubulin to form microtubules is driven by GTP hydrolysis in a process that is only beginning to be understood in detail. **Test yourself on the concepts in this figure at http://chemistry.brookscole.com/ggb3**

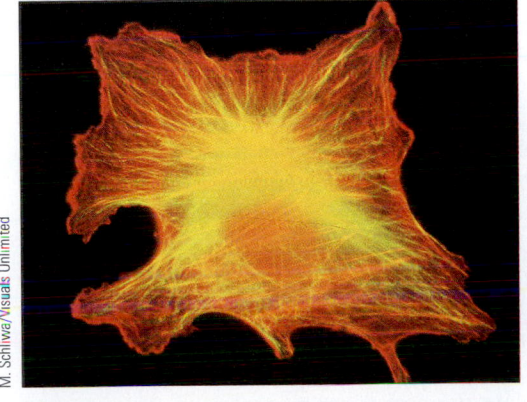

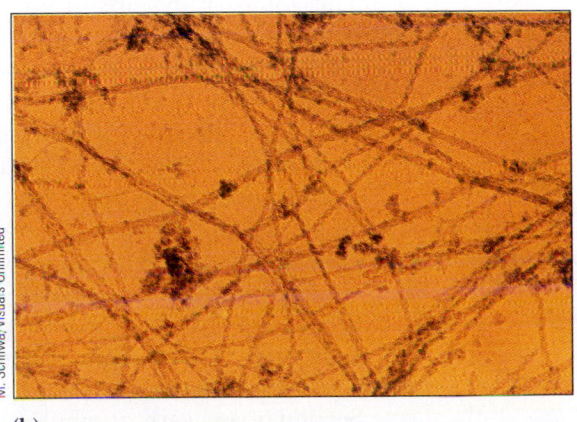

(a) **(b)**

FIGURE 16.4 Intermediate filaments have diameters of approximately 7 to 12 nm, whereas microfilaments, which are made from actin, have diameters of approximately 7 nm. The intermediate filaments appear to play only a structural role (maintaining cell shape), but the microfilaments and microtubules play more dynamic roles. Microfilaments are involved in cell motility, whereas microtubules act as long filamentous tracks, along which cellular components may be rapidly transported by specific mechanisms. **(a)** Cytoskeleton, double-labeled with actin in red and tubulin in green. **(b)** Cytoskeletal elements in a eukaryotic cell, including microtubules (thickest strands), intermediate filaments, and actin microfilaments (smallest strands).

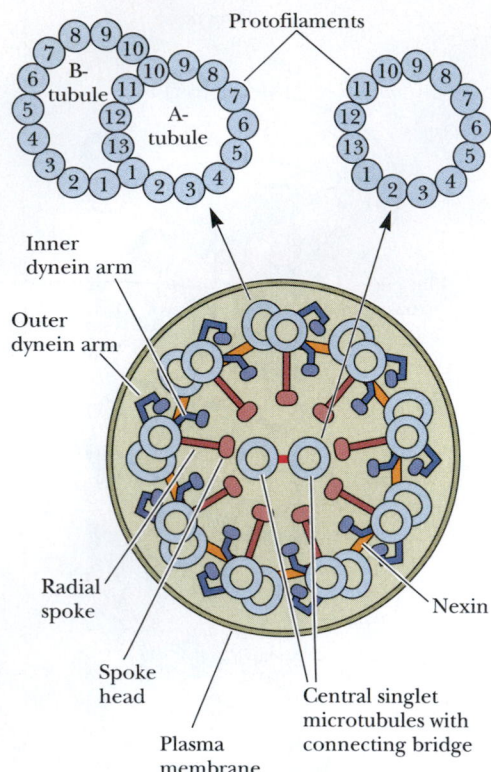

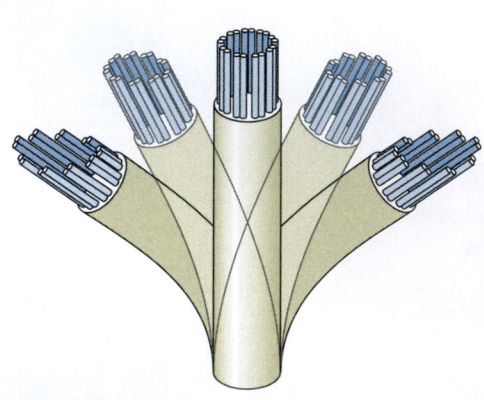

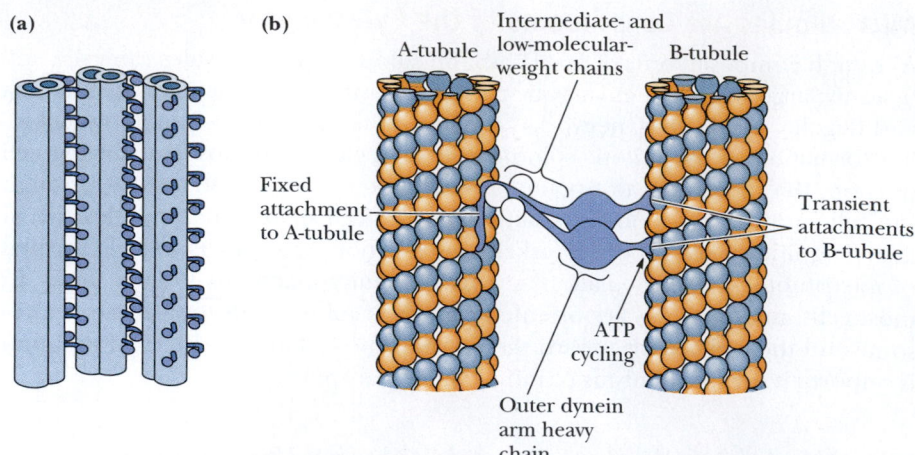

FIGURE 16.6 **(a)** Diagram showing dynein interactions between adjacent microtubule pairs. **(b)** Detailed views of dynein crosslinks between the A-tubule of one microtubule pair and the B-tubule of a neighboring pair. (The B-tubule of the first pair and the A-tubule of the neighboring pair are omitted for clarity.) Isolated axonemal dyneins, which possess ATPase activity, consist of two or three "heavy chains" with molecular masses of 400 to 500 kD, referred to as α and β (and γ when present), as well as several chains with intermediate (40–120 kD) and low (15–25 kD) molecular masses. Each outer-arm heavy chain consists of a globular domain with a flexible stem on one end and a shorter projection extending at an angle with respect to the flexible stem. In a dynein arm, the flexible stems of several heavy chains are joined in a common base, where the intermediate- and low-molecular-weight proteins are located.

FIGURE 16.5 The structure of an axoneme. Note the manner in which two microtubules are joined in the nine outer pairs. The smaller-diameter tubule of each pair, which is a true cylinder, is called the A-tubule and is joined to the center sheath of the axoneme by a spoke structure. Each outer pair of tubules is joined to adjacent pairs by a nexin bridge. The A-tubule of each outer pair possesses an outer dynein arm and an inner dynein arm. The larger-diameter tubule is known as the B-tubule.

A-tubules (Figure 16.6). However, the projections on the globular heads form transient attachments to adjacent B-tubules. Binding of ATP to the dynein heavy chain causes dissociation of the projections from the B-tubules. These projections then reattach to the B-tubules at a position closer to the minus end. Repetition of this process causes the sliding of A-tubules relative to B-tubules. The crosslinked structure of the axoneme dictates that this sliding motion will occur in an asymmetric fashion, resulting in a bending motion of the axoneme, as shown in Figure 16.7.

Microtubules Also Mediate the Intracellular Motion of Organelles and Vesicles

The ability of dyneins to effect **mechanochemical coupling**—that is, motion coupled with a chemical reaction—is also vitally important inside eukaryotic cells, which, as already noted, contain microtubule networks as part of the cytoskeleton. The mechanisms of intracellular, microtubule-based transport of organelles and vesicles were first elucidated in studies of **axons,** the long projections of neurons that extend great distances away from the body of the cell. In these cells, it was found that subcellular organelles and vesicles could travel at surprisingly fast rates—as great as 2 to 5 μm/sec—in either direction. Unraveling the molecular mechanism for this rapid transport turned out to be a challenging biochemical problem. The early evidence that these movements occur by association with specialized proteins on the microtubules was met with some resistance, for two reasons. First, the notion that a network of microtubules could mediate transport was novel and, like all novel ideas, difficult to accept. Second, many early attempts to isolate dyneins from neural tissue were unsuccessful, and the dyneinlike proteins that were first isolated from cytosolic fractions were thought to represent contaminations from axoneme structures. However, things changed dramatically in 1985 with a report by Michael Sheetz and his co-workers of a new ATP-driven, force-generating protein, different from myosin and dynein, which they called *kinesin.* Then, in 1987, Richard McIntosh and Mary Porter described the isolation of *cytosolic dynein* proteins from *Caenorhabditis elegans,* a nematode worm that never makes motile axo-

Biochemistry💿Now™ **ACTIVE FIGURE 16.7** A mechanism for ciliary motion. The sliding motion of dyneins along one microtubule while attached to an adjacent microtubule results in a bending motion of the axoneme. **Test yourself on the concepts in this figure at http://chemistry.brookscole.com/ggb3**

Human Biochemistry

Effectors of Microtubule Polymerization as Therapeutic Agents

Microtubules in eukaryotic cells are important for the maintenance and modulation of cell shape and the disposition of intracellular elements during the growth cycle and mitosis. It may thus come as no surprise that the inhibition of microtubule polymerization can block many normal cellular processes. The alkaloid **colchicine** (see accompanying figure), a constituent of the swollen, underground stems of the autumn crocus (*Colchicum autumnale*) and meadow saffron, inhibits the polymerization of tubulin into microtubules. This effect blocks the mitotic cycle of plants and animals. Colchicine also inhibits cell motility and intracellular transport of vesicles and organelles (which in turn blocks secretory processes of cells). Colchicine has been used for hundreds of years to alleviate some of the acute pain of gout and rheumatism. In gout, white cell lysosomes surround and engulf small crystals of uric acid. The subsequent rupture of the lysosomes and the attendant lysis of the white cells initiate an inflammatory response that causes intense pain. The mechanism of pain alleviation by colchicine is not known for certain, but appears to involve inhibition of white cell movement in tissues. Interestingly, colchicine's ability to inhibit mitosis has given it an important role in the commercial development of new varieties of agricultural and ornamental plants. When mitosis is blocked by colchicine, the treated cells may be left with an extra set of chromosomes. Plants with extra sets of chromosomes are typically larger and more vigorous than normal plants. Flowers developed in this way may grow with double the normal number of petals, and fruits may produce much larger amounts of sugar.

Another class of alkaloids, the **vinca alkaloids** from *Vinca rosea*, the Madagascar periwinkle, can also bind to tubulin and inhibit microtubule polymerization. **Vinblastine** and **vincristine** are used as potent agents for cancer chemotherapy because of their ability to inhibit the growth of fast-growing tumor cells. For reasons that are not well understood, colchicine is not an effective chemotherapeutic agent, although it appears to act similarly to the vinca alkaloids in inhibiting tubulin polymerization.

The antitumor drug **taxol** was originally isolated from the bark of *Taxus brevifolia*, the Pacific yew tree. Like vinblastine and colchicine, taxol inhibits cell replication by acting on microtubules. Unlike these other antimitotic drugs, however, taxol stimulates microtubule polymerization and stabilizes microtubules. The remarkable success of taxol in the treatment of breast and ovarian cancers stimulated research efforts to synthesize taxol directly and to identify new antimitotic agents that, like taxol, stimulate microtubule polymerization.

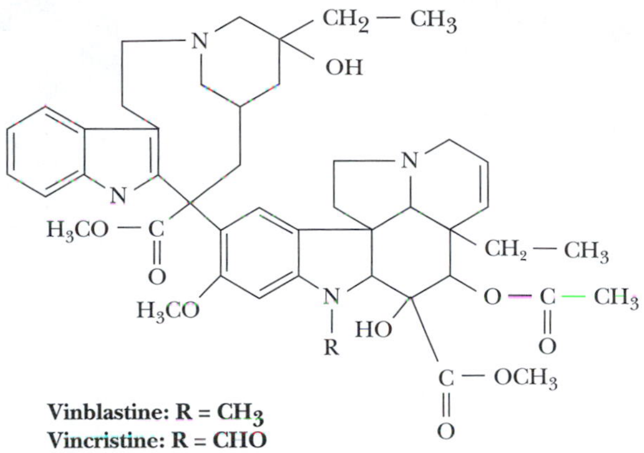

Vinblastine: R = CH₃
Vincristine: R = CHO

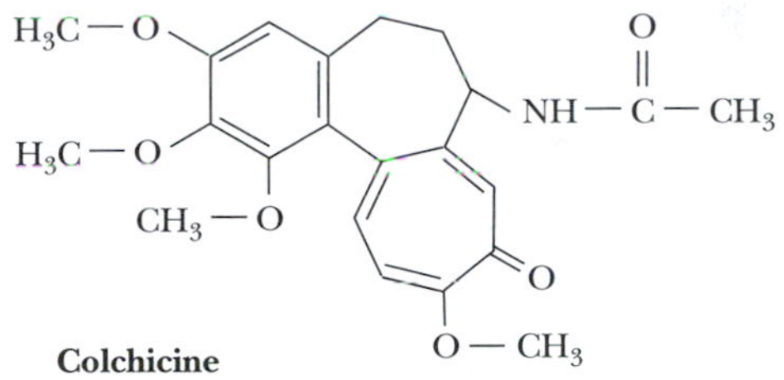

Colchicine

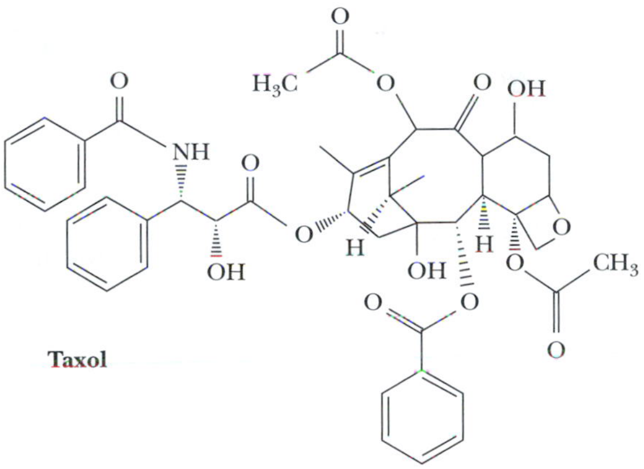

Taxol

▲ The structures of vinblastine, vincristine, colchicine, and taxol.

nemes at any stage of its life cycle. Kinesins have now been found in many eukaryotic cell types, and similar cytosolic dyneins have been found in fruit flies, amoebae, and slime molds; in vertebrate brain and testes; and in HeLa cells (a human tumor cell line).

Dyneins Move Organelles in a Plus-to-Minus Direction; Kinesins, in a Minus-to-Plus Direction—Mostly

The cytosolic dyneins bear many similarities to axonemal dynein. The protein isolated from *C. elegans* includes a "heavy chain" with a molecular mass of approximately 400 kD, as well as smaller peptides with molecular mass ranging from 53 to 74 kD. The protein possesses a microtubule-activated ATPase activity,

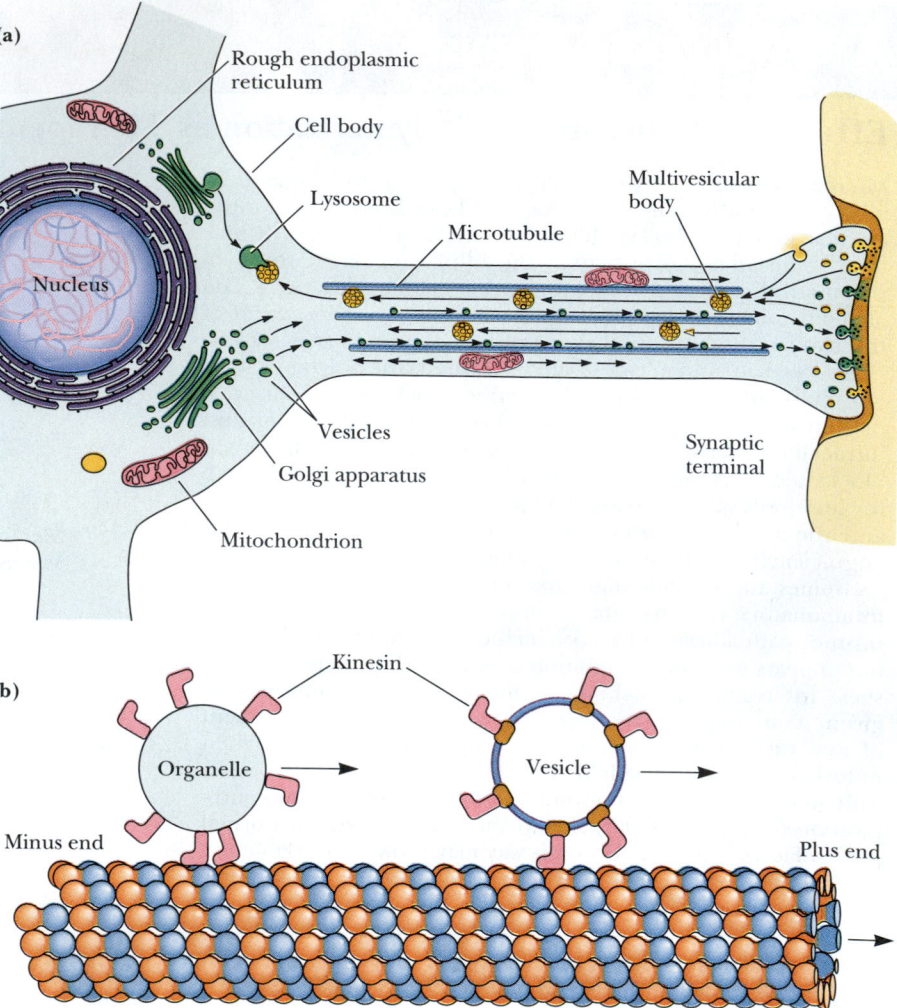

(a)
Rough endoplasmic reticulum
Cell body
Lysosome
Multivesicular body
Microtubule
Nucleus
Synaptic terminal
Vesicles
Golgi apparatus
Mitochondrion

(b)
Kinesin
Organelle
Vesicle
Minus end
Plus end

FIGURE 16.8 **(a)** Rapid axonal transport along microtubules permits the exchange of material between the synaptic terminal and the body of the nerve cell. **(b)** Vesicles, multivesicular bodies, and mitochondria are carried through the axon by this mechanism. *(Adapted from a drawing by Ronald Vale.)*

and when anchored to a glass surface in vitro, these proteins, in the presence of ATP, can bind microtubules and move them through the solution. In the cell, cytosolic dyneins specifically move organelles and vesicles from the plus end of a microtubule to the minus end. Thus, as shown in Figure 16.8, dyneins move vesicles and organelles from the cell periphery toward the centrosome (or, in an axon, from the synaptic termini toward the cell body). Most **kinesins,** on the other hand, assist the movement of organelles and vesicles from the minus end to the plus end of microtubules, resulting in outward movement of organelles and vesicles. Kinesin is similar to cytosolic dyneins but smaller in size (360 kD) and contains subunits of 110 kD and 65 to 70 kD. Its length is 100 nm. Like dyneins, kinesins possess ATPase activity in their globular heads, and it is the free energy of ATP hydrolysis that drives the movement of vesicles along the microtubules. A few kinesins are known to move in a plus-to-minus direction.

The N-terminal "head" domain of the kinesin heavy chain (38 kD, approximately 340 residues) contains the ATP- and microtubule-binding sites and is the domain responsible for movement. Electron microscopy and image analysis of tubulin–kinesin complexes reveals (Figure 16.9) that the kinesin head domain is compact and primarily contacts a single tubulin subunit on a microtubule surface, inducing a conformational change in the tubulin subunit. Optical trapping experiments (see page 530) demonstrate that kinesin heads move in 8-nm (80-Å) steps along the long axis of a microtubule. Kenneth Johnson and his co-workers have shown that the ability of a single kinesin tetramer to move unidirectionally for long distances on a microtubule depends upon cooperative interactions between the two mechanochemical head domains of the protein.

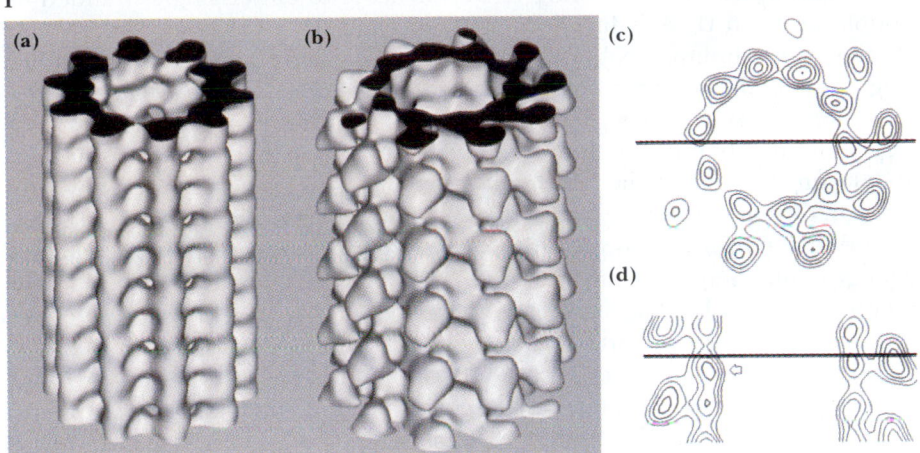

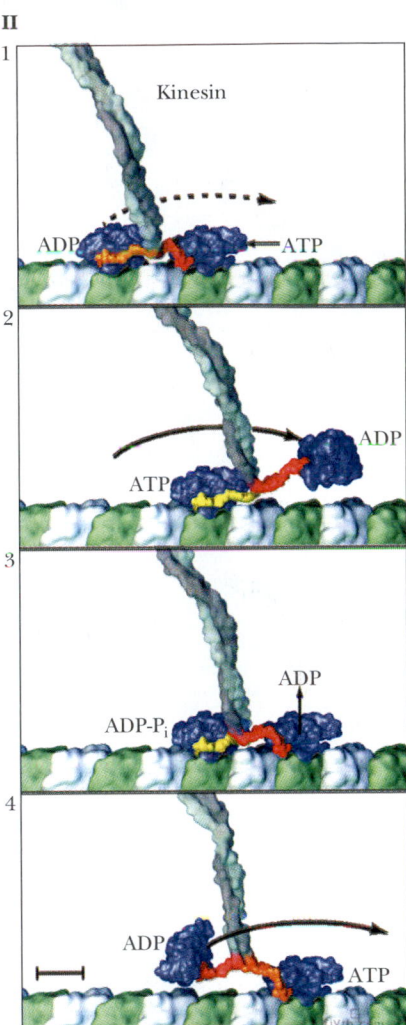

FIGURE 16.9 (I) The structure of the tubulin–kinesin complex, as revealed by image analysis of cryoelectron microscopy data. (a) The computed three-dimensional map of a microtubule, (b) the kinesin globular head domain–microtubule complex, (c) a contour plot of a horizontal section of the kinesin–microtubule complex, and (d) a contour plot of a vertical section of the same complex. *(Taken from Kikkawa et al., 1995. Nature **376**:274–277. Photo courtesy of Nobutaka Hirokawa.)* (II) A model for the motility cycle of kinesin. The two heads of the kinesin dimer work together to move processively along a microtubule. **Frame 1:** Each kinesin head is bound to the tubulin surface. The heads are connected to the coiled coil by "neck linker" segments (orange and red). **Frame 2:** Conformation changes in the neck linkers flip the trailing head by 160°, over and beyond the leading head and toward the next tubulin binding site. **Frame 3:** The new leading head binds to a new site on the tubulin surface (with ADP dissociation), completing an 80 Å movement of the coiled coil and the kinesin's cargo. During this time, the trailing head hydrolyzes ATP to ADP and P_i. **Frame 4:** ATP binds to the leading head, and P_i dissociates from the trailing head, completing the cycle. *(Adapted from Vale, R., and Milligan, R., 2000. The way things move: Looking under the hood of molecular motor proteins. Science **288**:88–95.)*

16.3 How Do Molecular Motors Unwind DNA?

DNA normally exists as a double-stranded duplex, but when DNA is to be replicated or repaired, the strands of the double helix must be unwound and separated to form single-stranded DNA intermediates. This separation is carried out by molecular motors known as **DNA helicases** that move along the length of the DNA lattice, sequentially destabilizing the interactions between complementary base pairs. The movement along the lattice and the separation of the DNA strands are coupled to the hydrolysis of nucleoside 5′-triphosphates. An important property shared by all helicases is the ability to move along the DNA lattice for long distances without dissociating. This is termed **processive movement**, and helicases are said to have a high processivity. For example, the *E. coli* BCD helicase, which is involved in recombination processes, can unwind 33,000 base pairs before it dissociates from the DNA lattice. Processive movement is essential for helicases involved in DNA replication, where millions of base pairs must be replicated rapidly.

Helicases have evolved at least two structural and functional strategies for achieving high processivity. Certain hexameric helicases form ringlike structures that completely encircle at least one of the strands of a DNA duplex. Other helicases, notably **Rep helicase** from *E. coli*, are homodimeric and move processively along the DNA helix by means of a "hand-over-hand" movement that is remarkably similar to that of kinesin's movement along microtubules. A key feature of hand-over-hand movement of a dimeric motor protein along a polymer is that at least one of the motor subunits must be bound to the polymer at any moment.

Negative Cooperativity Facilitates Hand-Over-Hand Movement

How does hand-over-hand movement of a motor protein along a polymer occur? Clues have come from the structures of Rep helicase and its complexes with DNA. The Rep helicase from *E. coli* is a 76-kD protein that is monomeric in the absence of DNA. Binding of Rep helicase to either single-stranded or double-stranded DNA induces dimerization, and the Rep dimer is the active species in unwinding DNA. Each subunit of the Rep dimer can bind either single-stranded (ss) or double-stranded (ds) DNA. However, *the binding of Rep dimer subunits to DNA is negatively cooperative* (see Chapter 15). Once the first Rep subunit is bound, the affinity of DNA for the second subunit is at least 10,000 times weaker than that for the first! This negative cooperativity provides an obvious advantage for hand-over-hand walking. When one "hand" has bound the polymer substrate, the other "hand" releases. A conformation change could then move the unbound "hand" one step farther along the polymer where it can bind again.

But what would provide the energy for such a conformation change? ATP hydrolysis is the driving force for Rep helicase movement along DNA, and the negative cooperativity of Rep binding to DNA is regulated by nucleotide binding. In the absence of nucleotide, a Rep dimer is favored, in which only one subunit is bound to ssDNA. In Figure 16.10a, this state is represented as P_2S [a Rep dimer (P_2) bound to ssDNA (S)]. Timothy Lohman and his colleagues at Washington University in St. Louis have shown that binding of ATP analogs induces formation of a complex of the Rep dimer with both ssDNA and dsDNA, one to each Rep subunit (shown as P_2SD in Figure 16.10a). In their model, unwinding of the dsDNA and ATP hydrolysis occur at this point, leaving a P_2S_2 state in which both Rep subunits are bound to ssDNA. Dissociation of ADP and P_i leave the P_2S state again (Figure 16.10a).

Work by Lohman and his colleagues has shown that coupling of ATP hydrolysis and hand-over-hand movement of Rep over the DNA involves the existence of the Rep dimer in an asymmetric state. A crystal structure of the Rep dimer in complex with ssDNA and ADP shows that the two Rep monomers are in different conformations (Figure 16.10b). The two conformations differ by a 130° rotation about a hinge region between two subdomains within the monomer subunit. The hand-over-hand walking of the Rep dimer along the DNA surface may involve alternation of each subunit between these two conformations, with coordination of the movements by nucleotide binding and hydrolysis.

(a)

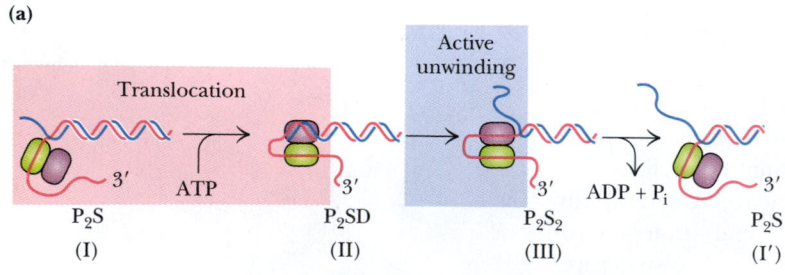

(b)

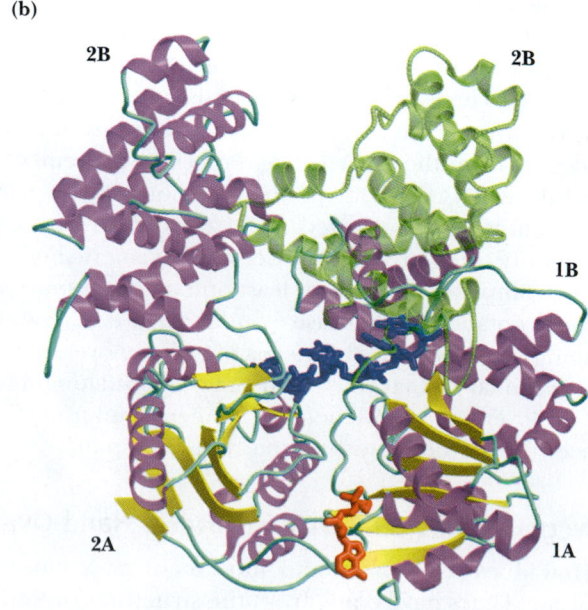

FIGURE 16.10 (a) A hand-over-hand model for movement along (and unwinding of) DNA by *E. coli* Rep helicase. The P_2S state consists of a Rep dimer bound to ssDNA. The P_2SD state involves one Rep monomer bound to ssDNA and the other bound to dsDNA. The P_2S_2 state has ssDNA bound to each Rep monomer. ATP binding and hydrolysis control the interconversion of these states and walking along the DNA substrate. **(b)** Crystal structure of the *E. coli* Rep helicase dimer. *(With permission from Korolev, S., Hsieh, J., Gauss, G., Lohman, T. L., and Waksman, G., 1997. Major domain swiveling revealed by the crystal structures of complexes of* E. coli *Rep helicase bound to single-stranded DNA and ADP. Cell* **90:**635–647.)

16.4 What Is the Molecular Mechanism of Muscle Contraction?

Muscle Contraction Is Triggered by Ca²⁺ Release from Intracellular Stores

The cells of skeletal muscle are long and multinucleate and are referred to as **muscle fibers.** At the microscopic level, skeletal muscle and cardiac muscle display alternating light and dark bands and for this reason are often referred to as **striated** muscles. Skeletal muscles in higher animals consist of 100-μm-diameter **fiber bundles,** some as long as the muscle itself. Each of these muscle fibers contains hundreds of **myofibrils** (Figure 16.11), each of which spans the length of the fiber and is about 1 to 2 μm in diameter. Myofibrils are linear arrays of cylindrical **sarcomeres,** the basic structural units of muscle contraction. The sarcomeres are surrounded on each end by a membrane system that is actually an elaborate extension of the muscle fiber plasma membrane or **sarcolemma.** These extensions of the sarcolemma, which are called **transverse tubules** or **t-tubules,** enable the sarcolemmal membrane to contact the ends of each myofibril in the muscle fiber (Figure 16.11). This topological feature is crucial to the initiation of contractions.

Between the t-tubules, the sarcomere is covered with a specialized endoplasmic reticulum called the **sarcoplasmic reticulum,** or **SR.** The SR contains high concentrations of Ca²⁺, and the release of Ca²⁺ from the SR and its interactions within the sarcomeres trigger muscle contraction, as we will see. Each SR structure consists of two domains. **Longitudinal tubules** run the length of the sarcomere and are capped on either end by the **terminal cisternae** (Figure 16.11). The structure at the end of each sarcomere, which consists of a t-tubule and two apposed terminal cisternae, is called a **triad,** and the intervening gaps of approximately 15 nm are called **triad junctions.** The junctional face of each terminal cisterna is joined to its respective t-tubule by a **foot structure.**

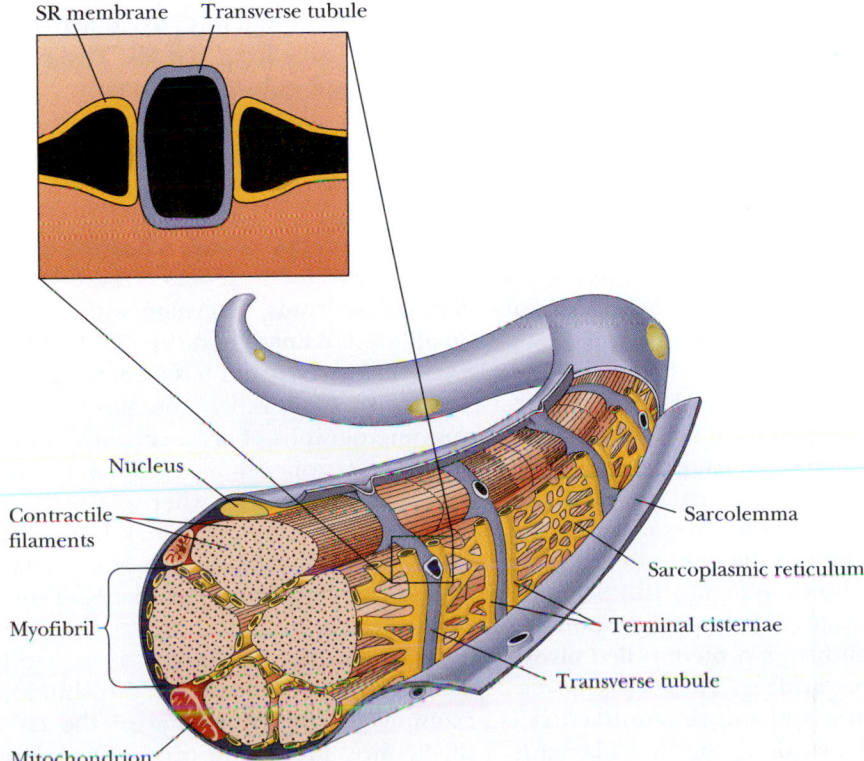

FIGURE 16.11 The structure of a skeletal muscle cell, showing the manner in which t-tubules enable the sarcolemmal membrane to contact the ends of each myofibril in the muscle fiber. The foot structure is shown in the box.

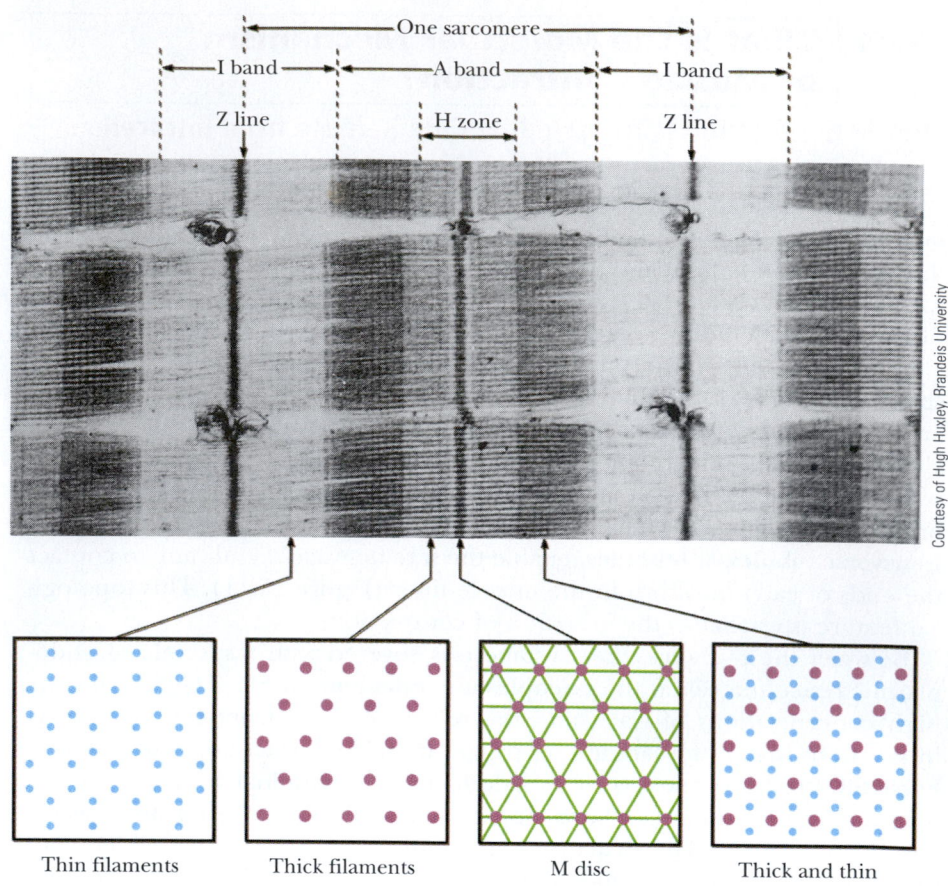

Thin filaments Thick filaments M disc Thick and thin filaments

FIGURE 16.12 Electron micrograph of a skeletal muscle myofibril (in longitudinal section). The length of one sarcomere is indicated, as are the A and I bands, the H zone, the M disc, and the Z lines. Cross sections from the H zone show a hexagonal array of thick filaments, whereas the I band cross section shows a hexagonal array of thin filaments.

Skeletal muscle contractions are initiated by nerve stimuli that act directly on the muscle. Nerve impulses produce an electrochemical signal (see Chapter 32) called an action potential that spreads over the sarcolemmal membrane and into the fiber along the t-tubule network. This signal is passed across the triad junction and induces the release of Ca^{2+} ions from the SR. These Ca^{2+} ions bind to proteins within the muscle fibers and induce contraction.

The Molecular Structure of Skeletal Muscle Is Based on Actin and Myosin

Examination of myofibrils in the electron microscope reveals a banded or striated structure. The bands are traditionally identified by letters (Figure 16.12). Regions of high electron density, denoted **A bands,** alternate with regions of low electron density, the **I bands.** Small, dark **Z lines** lie in the middle of the I bands, marking the ends of the sarcomere. Each A band has a central region of slightly lower electron density called the **H zone,** which contains a central **M disc** (also called an **M line**). Electron micrographs of cross sections of each of these regions reveal molecular details. The H zone shows a regular, hexagonally arranged array of **thick filaments** (15 nm diameter), whereas the I band shows a regular, hexagonal array of **thin filaments** (7 nm diameter). In the dark regions at the ends of each A band, the thin and thick filaments interdigitate, as shown in Figure 16.12. The thin filaments are composed primarily of three proteins called **actin, troponin,** and **tropomyosin.** The thick filaments consist mainly of a protein called **myosin.** The thin and thick filaments are joined by **cross-bridges.** These cross-bridges are actually extensions of the myosin molecules, and muscle contraction is accomplished by the sliding of the cross-bridges along the thin filaments, a mechanical movement driven by the free energy of ATP hydrolysis.

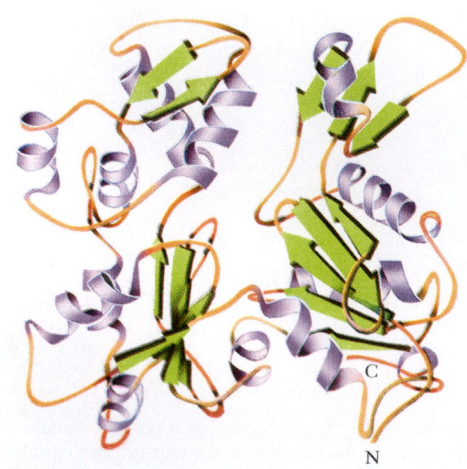

FIGURE 16.13 The three-dimensional structure of an actin monomer from skeletal muscle. This view shows the two domains (left and right) of actin.

The Composition and Structure of Thin Filaments   Actin, the principal component of thin filaments, can be isolated in two forms. Under conditions of low ionic strength, actin exists as a 42-kD globular protein, denoted **G-actin.** G-actin consists of two principal lobes or domains (Figure 16.13). Under physiological conditions (higher ionic strength), G-actin polymerizes to form a fibrous form of actin, called **F-actin.** As shown in Figure 16.14, F-actin is a right-handed helical structure, with a helix pitch of about 72 nm per turn. The F-actin helix is the core of the thin filament, to which **tropomyosin** and the **troponin complex** also add. Tropomyosin is a dimer of homologous but nonidentical 33-kD subunits. These two subunits form long α-helices that intertwine, creating 38- to 40-nm-long coiled coils, which join in head-to-tail fashion to form long rods. These rods bind to the F-actin polymer and lie almost parallel to the long axis of the F-actin helix (Figure 16.15a–c). Each tropomyosin heterodimer contacts approximately seven actin subunits. The troponin complex consists of three different proteins: **troponin T,** or **TnT** (37 kD); **troponin I,** or **TnI** (24 kD); and **troponin C,** or **TnC** (18 kD). TnT binds to tropomyosin, specifically at the head-to-tail junction. Troponin I binds both to tropomyosin and to actin. Troponin C is a Ca^{2+}-binding protein that binds to TnI. TnC shows 70% homology with the important Ca^{2+} signaling protein, calmodulin (see Chapter 32). The release of Ca^{2+} from the SR, which signals a contraction, raises the cytosolic Ca^{2+} concentration high enough to saturate the Ca^{2+} sites on TnC. Ca^{2+} binding induces a conformational change in the amino-terminal domain of TnC, which in turn causes a rearrangement of the troponin complex and tropomyosin with respect to the actin fiber.

The Composition and Structure of Thick Filaments   Myosin, the principal component of muscle thick filaments, is a large protein consisting of six polypeptides, with an aggregate molecular weight of approximately 540 kD. As shown

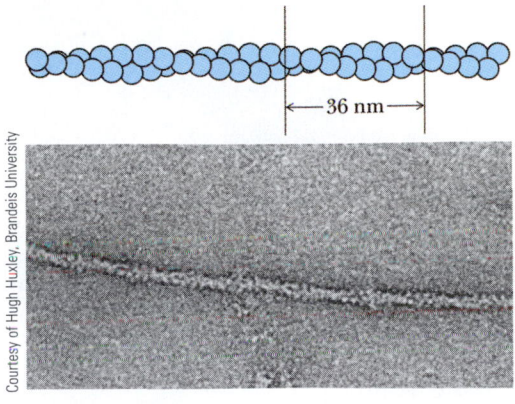

FIGURE 16.14 The helical arrangement of actin monomers in F-actin. The F-actin helix has a pitch of 72 nm and a repeat distance of 36 nm.

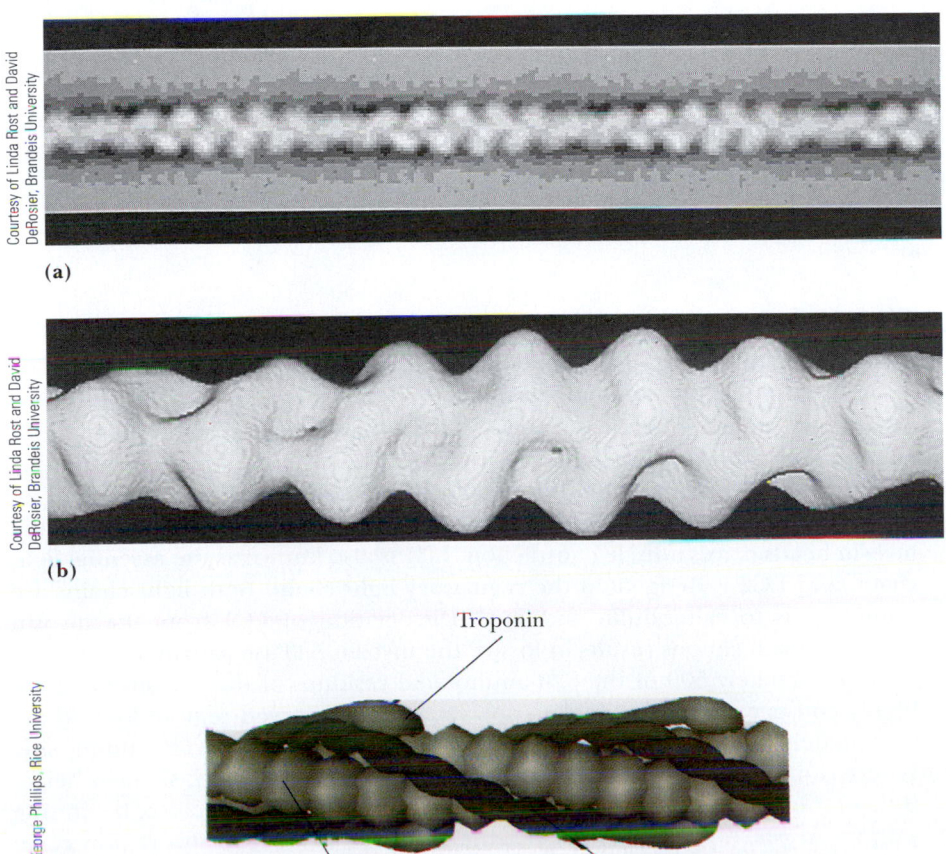

(a)

(b)

Troponin

Tropomyosin

Actin

(c)

FIGURE 16.15 **(a)** An electron micrograph of a thin filament, **(b)** a corresponding image reconstruction, and **(c)** a schematic drawing based on the images in (a) and (b). The tropomyosin coiled coil winds around the actin helix, each tropomyosin dimer interacting with seven consecutive actin monomers. Troponin T binds to tropomyosin at the head-to-tail junction.

(a)

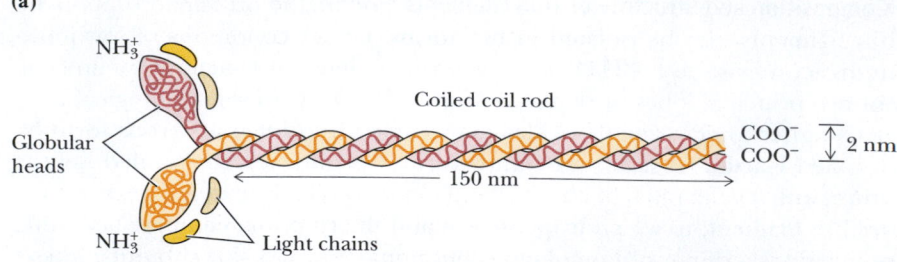

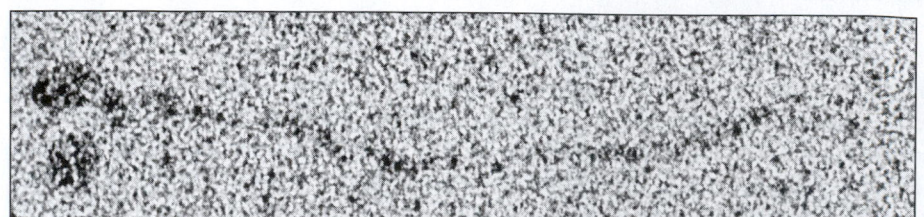

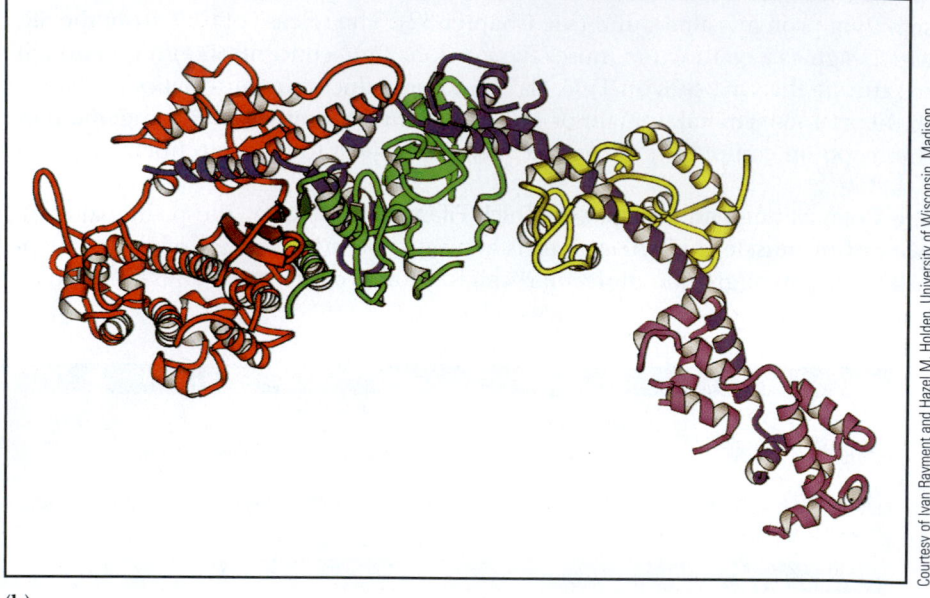

Courtesy of Henry Slayter, Harvard Medical School

Courtesy of Ivan Rayment and Hazel M. Holden, University of Wisconsin, Madison

FIGURE 16.16 **(a)** An electron micrograph of a myosin molecule and a corresponding schematic drawing. The tail is a coiled coil of intertwined α-helices extending from the two globular heads. One of each of the myosin light-chain proteins, LC1 and LC2, is bound to each of the globular heads. **(b)** A ribbon diagram shows the structure of the S1 myosin head (green, red, and blue segments) and its associated essential (yellow) and regulatory (violet) light chains.

(b)

in Figure 16.16, the six peptides include two 230-kD **heavy chains,** as well as two pairs of different 20-kD **light chains,** denoted **LC1** and **LC2.** The heavy chains consist of globular amino-terminal **myosin heads,** joined to long α-helical carboxy-terminal segments, the **tails.** These tails are intertwined to form a left-handed coiled coil approximately 2 nm in diameter and 130 to 150 nm long. Each of the heads in this dimeric structure is associated with an LC1 and an LC2. The myosin heads exhibit **ATPase activity,** and hydrolysis of ATP by the myosin heads drives muscle contraction. LC1 is also known as the **essential light chain,** and LC2 is designated the **regulatory light chain.** Both light chains are homologous to calmodulin and TnC. Dissociation of LC1 from the myosin heads by alkali cations results in loss of the myosin ATPase activity.

Approximately 500 of the 820 amino acid residues of the myosin head are highly conserved between various species. One conserved region, located approximately at residues 170 to 214, constitutes part of the ATP-binding site. Whereas many ATP-binding proteins and enzymes employ a β-sheet–α-helix–β-sheet motif, this region of myosin forms a related α-β-α structure, beginning with an Arg at (approximately) residue 192. The β-sheet in this region of all myosins includes the amino acid sequence

Gly-Glu-Ser-Gly-Ala-Gly-Lys-Thr

The Gly-X-X-Gly-X-Gly sequence in this segment is found in many ATP- and nucleotide-binding enzymes. The Lys of this segment is thought to interact with the α-phosphate of bound ATP.

Repeating Structural Elements Are the Secret of Myosin's Coiled Coils   Myosin tails show less homology than the head regions, but several key features of the tail sequence are responsible for the α-helical coiled coils formed by myosin tails. Several orders of repeating structure are found in all myosin tails, including 7-residue, 28-residue, and 196-residue repeating units. Large stretches of the tail domain are composed of 7-residue repeating segments. The first and fourth residues of these 7-residue units are generally small, hydrophobic amino acids, whereas the second, third, and sixth are likely to be charged residues. The consequence of this arrangement is shown in Figure 16.17. Seven residues form two turns of an α-helix, and in the coiled coil structure of the myosin tails, the first and fourth residues face the interior contact region of the coiled coil. Residues b, c, and f (2, 3, and 6) of the 7-residue repeat face the periphery, where charged residues can interact with the water solvent. Groups of four 7-residue units with distinct patterns of alternating side-chain charge form 28-residue repeats that establish alternating regions of positive and negative charge on the surface of the myosin coiled coil. These alternating charged regions interact with similar regions in the tails of adjacent myosin molecules to assist in stabilizing the thick filament.

At a still higher level of organization, groups of seven of these 28-residue units—a total of 196 residues—also form a repeating pattern, and this large-scale repeating motif contributes to the packing of the myosin molecules in the thick filament. The myosin molecules in thick filaments are offset (Figure 16.18) by approximately 14 nm, a distance that corresponds to 98 residues of a coiled coil, or exactly half the length of the 196-residue repeat. Thus, several layers of repeating structure play specific roles in the formation and stabilization of the myosin coiled coil and the thick filament formed from them.

The Mechanism of Muscle Contraction Is Based on Sliding Filaments

When muscle fibers contract, the thick myosin filaments slide or walk along the thin actin filaments. The basic elements of the **sliding filament model** were first described in 1954 by two different research groups: Hugh Huxley and his colleague Jean Hanson, and the physiologist Andrew Huxley and his colleague Ralph Niedergerke. Several key discoveries paved the way for this model. Electron microscopic studies of muscle revealed that sarcomeres decreased in length during contraction and that this decrease was due to decreases in the width of both the I band and the H zone (Figure 16.19). At the same time, the width of the A band (which is the length of the thick filaments) and the distance from the Z discs to the nearby H zone (that is, the length of the thin filaments) did not change. These observations made it clear that the lengths of both the thin and thick filaments were constant during contraction. This conclusion was consistent with a sliding filament model.

The Sliding Filament Model   The shortening of a sarcomere (Figure 16.19) involves sliding motions in opposing directions at the two ends of a myosin thick filament. Net sliding motions in a specific direction occur because the thin and thick

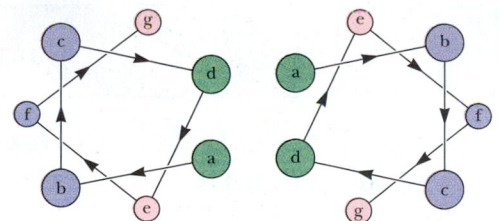

FIGURE 16.17 An axial view of the two-stranded, α-helical coiled coil of a myosin tail. Hydrophobic residues a and d of the 7-residue repeat sequence align to form a hydrophobic core. Residues b, c, and f face the outer surface of the coiled coil and are typically ionic.

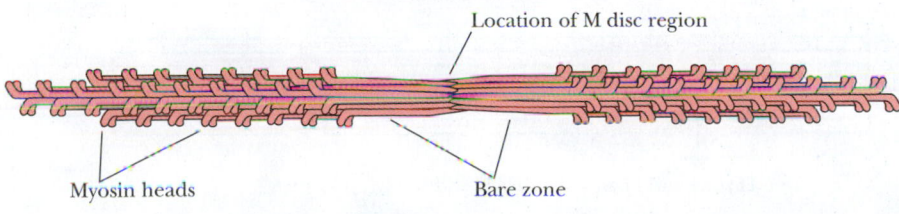

Location of M disc region

Myosin heads   Bare zone

FIGURE 16.18 The packing of myosin molecules in a thick filament. Adjoining molecules are offset by approximately 14 nm, a distance corresponding to 98 residues of the coiled coil.

Human Biochemistry

The Molecular Defect in Duchenne Muscular Dystrophy Involves an Actin-Anchoring Protein

Discovery of an actinin/spectrin-like protein provided insights into the molecular basis for at least one form of muscular dystrophy. Duchenne muscular dystrophy is a degenerative and fatal disorder of muscle affecting approximately 1 in 3500 boys. Victims of Duchenne dystrophy show early abnormalities in walking and running. By the age of 5, the victim cannot run and has difficulty standing, and by early adolescence, walking is difficult or impossible. The loss of muscle function progresses upward in the body, affecting next the arms and the diaphragm. Respiratory problems or infections usually result in death by the age of 30. Louis Kunkel and his co-workers identified the Duchenne muscular dystrophy gene in 1986. This gene produces a protein called **dystrophin**, which is highly homologous to α-actinin and spectrin. A defect in dystrophin is responsible for the muscle degeneration of Duchenne dystrophy.

Dystrophin is located on the cytoplasmic face of the muscle plasma membrane, linked to the plasma membrane via an integral membrane glycoprotein. Dystrophin has a high molecular mass (427 kD) but constitutes less than 0.01% of the total muscle protein. It folds into four principal domains (see accompanying figure, part a), including an N-terminal domain similar to the actin-binding domains of actinin and spectrin, a long repeat domain, a cysteine-rich domain, and a C-terminal domain that is unique to dystrophin. The repeat domain consists of 24 repeat units of approximately 109 residues each. "Spacer sequences" high in proline content, which do not align with the repeat consensus sequence, occur at the beginning and end of the repeat domain. Spacer segments are found between repeat elements 3 and 4 and 19 and 20. The high proline content of the spacers suggests that they may represent hinge domains. The spacer/hinge segments are sensitive to proteolytic enzymes, indicating that they may represent more exposed regions of the polypeptide.

Dystrophin itself appears to be part of an elaborate protein–glycoprotein complex that bridges the inner cytoskeleton (actin fil-

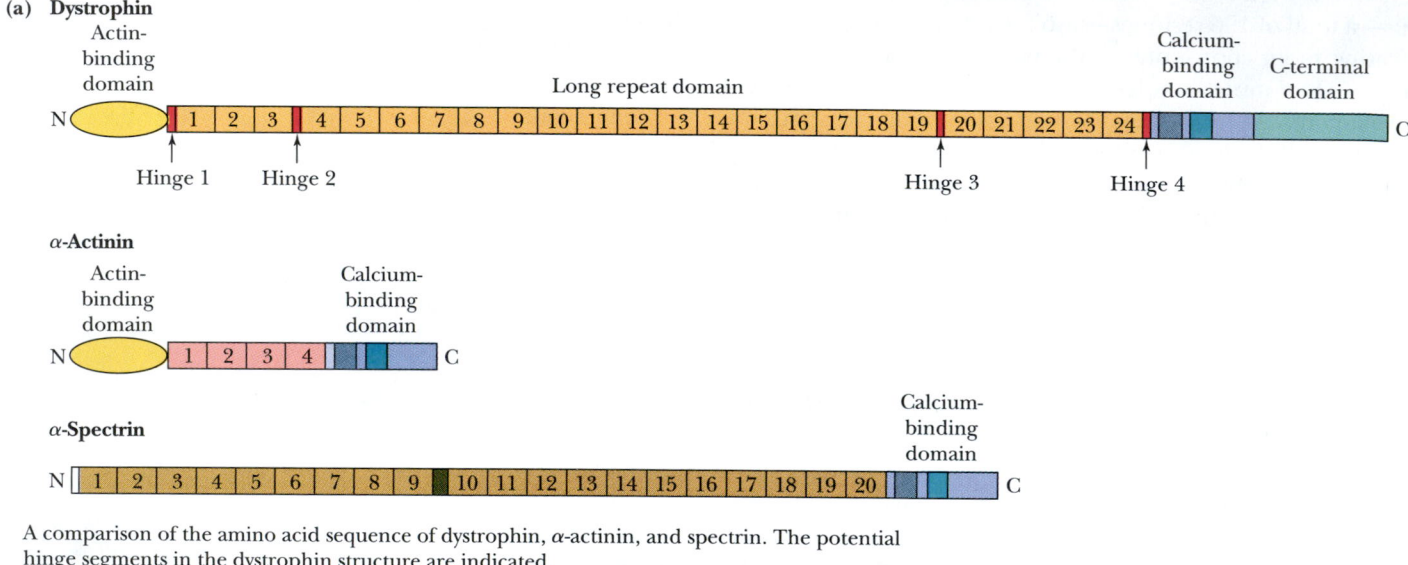

A comparison of the amino acid sequence of dystrophin, α-actinin, and spectrin. The potential hinge segments in the dystrophin structure are indicated.

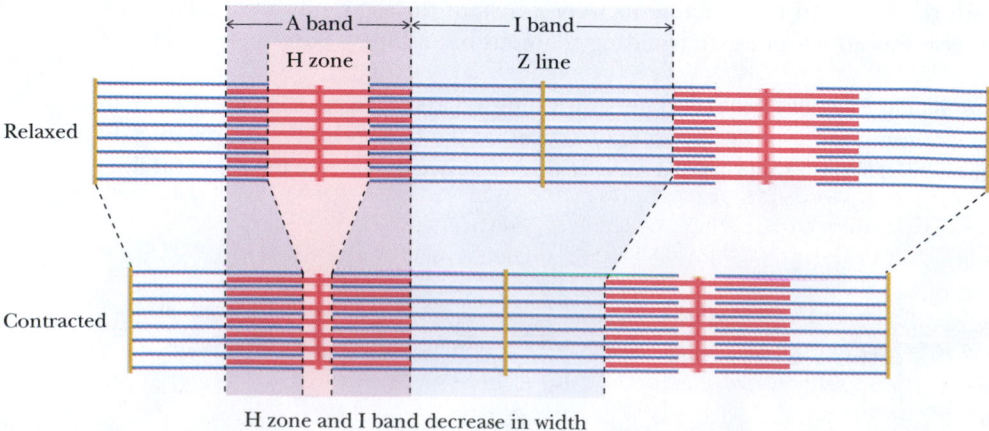

FIGURE 16.19 The sliding filament model of skeletal muscle contraction. The decrease in sarcomere length is due to decreases in the width of the I band and H zone, with no change in the width of the A band. These observations mean that the lengths of both the thick and thin filaments do not change during contraction. Rather, the thick and thin filaments slide along one another.

H zone and I band decrease in width

aments) and the extracellular matrix (via a matrix protein called laminin) (see figure). It is now clear that defects in one or more of the proteins in this complex are responsible for many of the other forms of muscular dystrophy. The glycoprotein complex is composed of two subcomplexes, the dystroglycan complex and the sarcoglycan complex. The dystroglycan complex consists of α-dystroglycan, an extracellular protein that binds to merosin, a laminin subunit and component of the extracellular matrix, and

β-dystroglycan, a transmembrane protein that binds the C-terminal domain of dystrophin inside the cell (see figure). The sarcoglycan complex is composed of α-, β-, and γ-sarcoglycans, all of which are transmembrane glycoproteins. Alterations of the sarcoglycan proteins are linked to limb-girdle muscular dystrophy and autosomal recessive muscular dystrophy. Mutations in the gene for merosin, which binds to α-dystroglycan, are linked to severe congenital muscular dystrophy, yet another form of the disease.

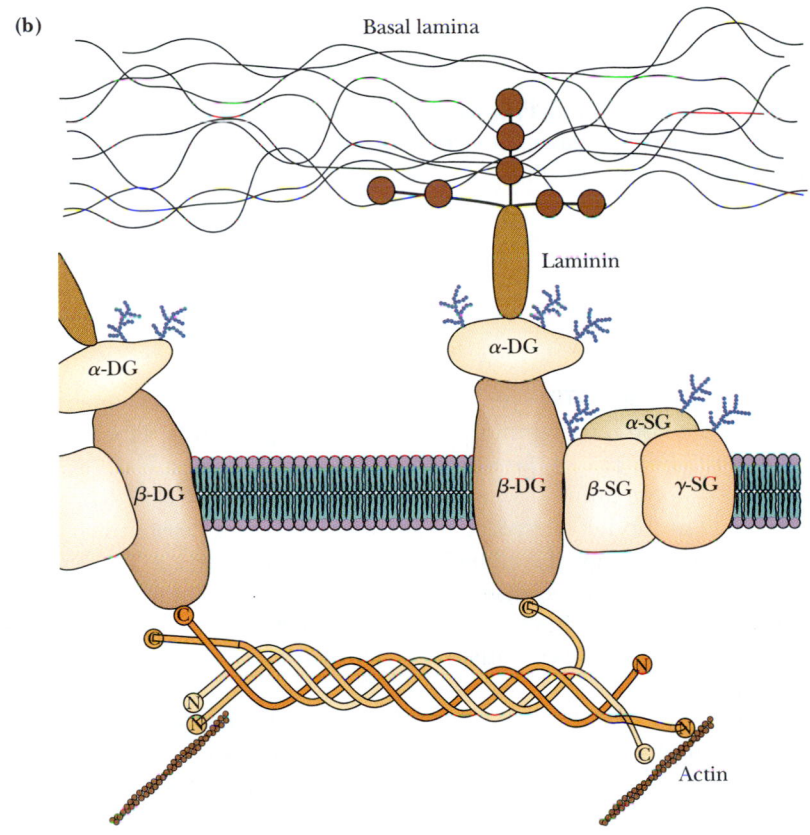

(b)

Basal lamina

Laminin

α-DG

β-DG

α-DG

β-DG

α-SG

β-SG

γ-SG

Actin

◄ A model for the actin–dystrophin–glycoprotein complex in skeletal muscle. Dystrophin is postulated to form tetramers of antiparallel monomers that bind actin at their N-termini and a family of dystrophin-associated glycoproteins at their C-termini. This dystrophin-anchored complex may function to stabilize the sarcolemmal membrane during contraction–relaxation cycles, link the contractile force generated in the cell (fiber) with the extracellular environment, or maintain local organization of key proteins in the membrane. The dystrophin-associated membrane proteins (dystroglycans and sarcoglycans) range from 25 to 154 kD. *(Adapted from Ahn, A. H., and Kunkel, L. M., 1993.* Nature Genetics *3:283–291; and Worton, R., 1995.* Science *270:755–756.)*

filaments both have **directional character.** The organization of the thin and thick filaments in the sarcomere takes particular advantage of this directional character. Actin filaments always extend outward from the Z lines in a uniform manner. Thus, between any two Z lines, the two sets of actin filaments point in opposing directions. The myosin thick filaments, on the other hand, also assemble in a directional manner. The polarity of myosin thick filaments reverses at the M disc. The nature of this reversal is not well understood but presumably involves structural constraints provided by proteins in the M disc, such as the M protein and myomesin described earlier. The reversal of polarity at the M disc means that actin filaments on either side of the M disc are pulled toward the M disc during contraction by the sliding of the myosin heads, causing net shortening of the sarcomere.

Albert Szent-Györgyi's Discovery of the Effects of Actin on Myosin   The molecular events of contraction are powered by the ATPase activity of myosin. Much of our present understanding of this reaction and its dependence on actin can be traced to several key discoveries by Albert Szent-Györgyi at the University of Szeged in Hungary in the early 1940s. Szent-Györgyi showed that solution viscosity is dramatically increased when solutions of myosin and actin are mixed. Increased viscosity is a manifestation of the formation of an **actomyosin complex.**

Szent-Györgyi further showed that the viscosity of an actomyosin solution was lowered by the addition of ATP, indicating that ATP decreases myosin's affinity for actin. Kinetic studies demonstrated that myosin ATPase activity was increased substantially by actin. (For this reason, Szent-Györgyi gave the name **actin** to the thin filament protein.) The ATPase turnover number of pure myosin is 0.05/sec. In the presence of actin, however, the turnover number increases to about 10/sec, a number more like that of intact muscle fibers.

The specific effect of actin on myosin ATPase becomes apparent if the product release steps of the reaction are carefully compared. In the absence of actin, the addition of ATP to myosin produces a rapid release of H^+, one of the products of the ATPase reaction:

$$ATP^{4-} + H_2O \longrightarrow ADP^{3-} + P_i^{2-} + H^+$$

However, release of ADP and P_i from myosin is much slower. Actin activates myosin ATPase activity by stimulating the release of P_i and then ADP. Product release is followed by the binding of a new ATP to the actomyosin complex, which causes actomyosin to dissociate into free actin and myosin. The cycle of ATP hydrolysis then repeats, as shown in Figure 16.20a. The crucial point of this model is that ATP hydrolysis and the association and dissociation of actin and myosin are coupled. It is this coupling that enables ATP hydrolysis to power muscle contraction.

The Coupling Mechanism: ATP Hydrolysis Drives Conformation Changes in the Myosin Heads   The only remaining piece of the puzzle is this: How does the close coupling of actin-myosin binding and ATP hydrolysis result in the shortening of myofibrils? Put another way, how are the model for ATP hydrolysis and the sliding filament model related? The answer to this puzzle is shown in Figure 16.20b. The free energy of ATP hydrolysis is translated into a conformation change in the myosin head, so dissociation of myosin and actin, hydrolysis of ATP, and rebinding of myosin and actin occur with stepwise movement of the myosin S1 head along the actin filament. The conformation change in the myosin head is driven by the hydrolysis of ATP.

As shown in the cycle in Figure 16.20a, the myosin heads—with the hydrolysis products ADP and P_i bound—are mainly dissociated from the actin filaments in resting muscle. When the signal to contract is presented (see following discussion), the myosin heads move out from the thick filaments to bind to actin on the thin filaments (Step 1). Binding to actin stimulates the release of phosphate, and this is followed by the crucial conformational change by the S1 myosin heads—the so-called **power** stroke—and ADP dissociation. In this step (Step 2), the thick filaments move along the thin filaments as the myosin heads relax to a lower-energy conformation. In the power stroke, the myosin heads tilt by approximately 45° and the conformational energy of the myosin heads is lowered by about 29 kJ/mol. This moves the thick filament approximately 10 nm along the thin filament (Step 3). Subsequent binding (Step 4) and hydrolysis (Step 5) of ATP cause dissociation of the heads from the thin filaments and also cause the myosin heads to shift back to their high-energy conformation with the heads' long axis nearly perpendicular to the long axis of the thick filaments. The heads may then begin another cycle by binding to actin filaments. This cycle is repeated at rates up to 5/sec in a typical skeletal muscle contraction. The conformational changes occurring in this cycle are the secret of the energy coupling that allows ATP binding and hydrolysis to drive muscle contraction.

The conformation change in the power stroke has been studied in two ways: (1) Cryoelectron microscopy together with computerized image analysis has

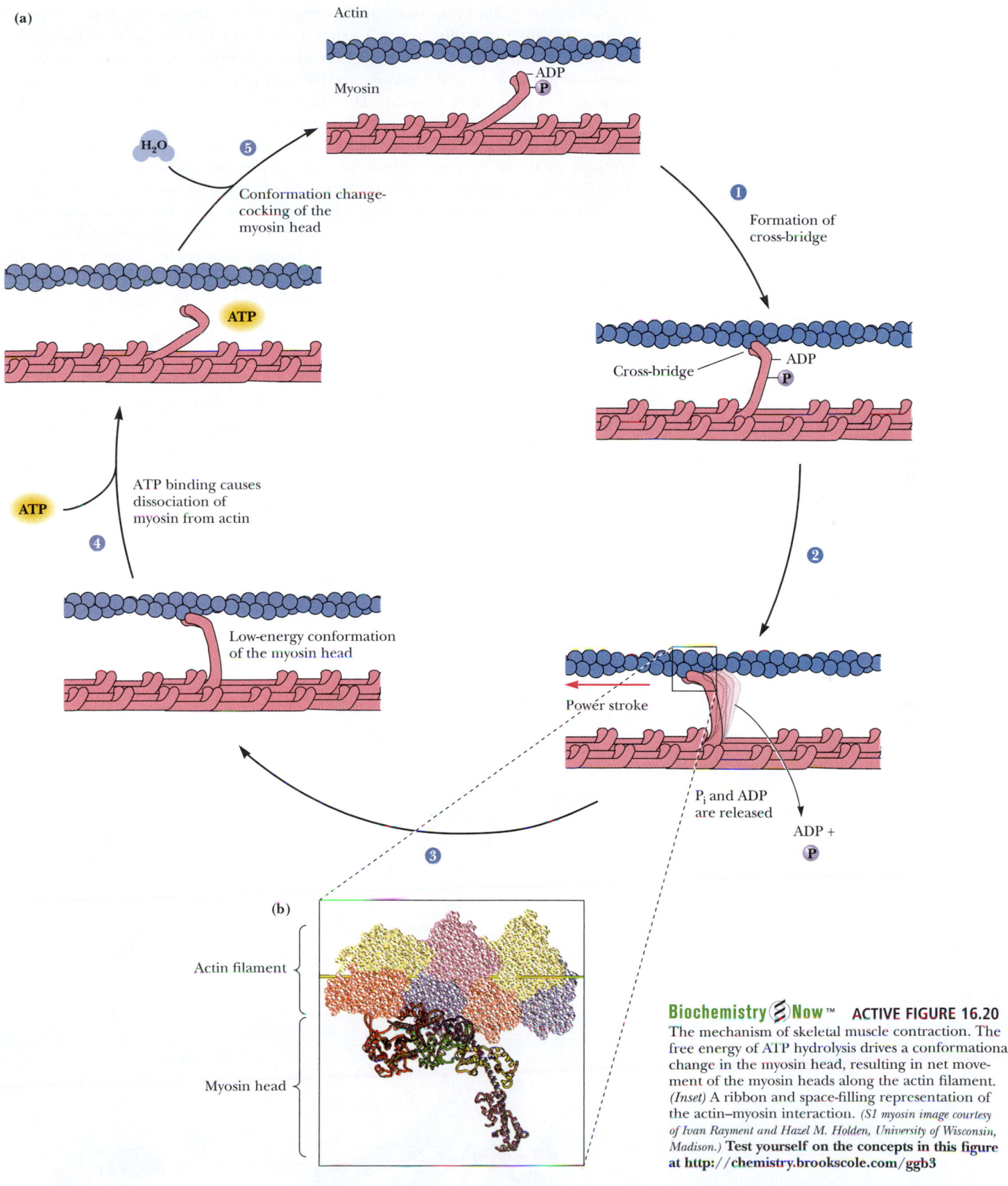

(a)

Actin

Myosin

ADP
P

H_2O 5   Conformation change-cocking of the myosin head

1   Formation of cross-bridge

Cross-bridge   ADP
P

ATP

ATP binding causes dissociation of myosin from actin

4

2

Low-energy conformation of the myosin head

ATP

Power stroke

3

P_i and ADP are released

ADP +
P

(b)

Actin filament

Myosin head

Biochemistry ⒺNow™   ACTIVE FIGURE 16.20
The mechanism of skeletal muscle contraction. The free energy of ATP hydrolysis drives a conformational change in the myosin head, resulting in net movement of the myosin heads along the actin filament. *(Inset)* A ribbon and space-filling representation of the actin–myosin interaction. *(S1 myosin image courtesy of Ivan Rayment and Hazel M. Holden, University of Wisconsin, Madison.)* **Test yourself on the concepts in this figure at http://chemistry.brookscole.com/ggb3**

yielded low-resolution images of S1-decorated actin in the presence and absence of MgADP (corresponding approximately to the states before and after the power stroke), and (2) feedback-enhanced laser optical trapping experiments have measured the movements and forces exerted during single turnovers of single myosin molecules along an actin filament. The images of myosin, when compared with the X-ray crystal structure of myosin S1, show that the long α-helix of S1 that binds the light chains (ELC and RLC) may behave as a lever arm and that this arm swings through an arc of 23° upon release of ADP. (A glycine residue at position 770 in the S1 myosin head lies at the N-terminal end of this helix/lever arm and may act as a hinge.) *This results in a 3.5-nm (35-Å) movement of the last myosin heavy-chain residue of the X-ray structure in a direction nearly parallel to the actin filament.* These two imaging "snapshots" of the myosin S1 conformation may represent only part of the working power stroke of the contraction cycle, and the total movement of a myosin head with respect to the apposed actin filament may thus be more than 3.5 nm.

The Initial Events of Myosin and Kinesin Action Are Similar

The ATP hydrolysis cycle must be linked to a conformational change cycle for motors to produce directed motion. How is ATP hydrolysis coupled to the conformation change cycle? For both myosin and kinesin, a part of the protein must act as a "γ-phosphate sensor" to detect the presence or absence of the γ-P of ATP in the active site. In both myosin and kinesin, this sensor consists of two loops of the protein, termed "switch I" and "switch II," which form H bonds with the γ-P and which orient a water molecule and crucial protein residues involved in ATP hydrolysis. Small movements of the γ-P sensor are communicated to distant parts of the protein by a long "relay helix" at the amino-terminus of switch II. *The relay helix moves back and forth like a piston to link tiny movements in switch II to larger movements of the protein* (Figure 16.21).

Biochemistry ⊜ Now™ Go to BiochemistryNow and click BiochemistryInteractive to learn more about the structure and function of myosin.

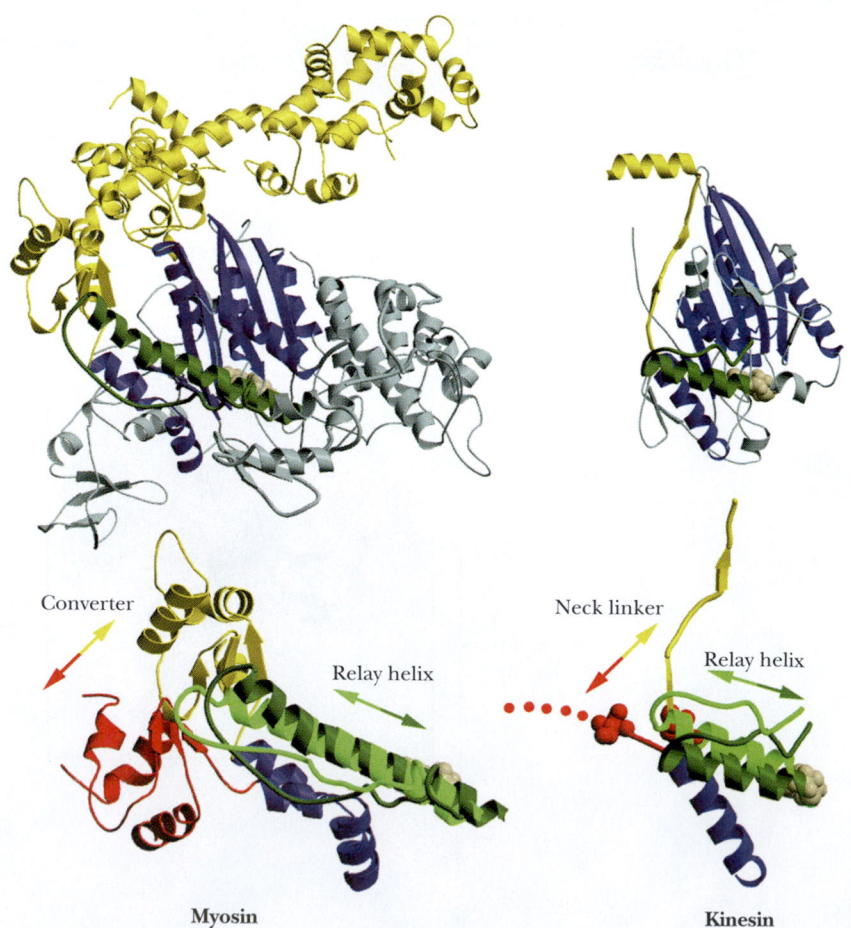

FIGURE 16.21 Ribbon structures of the myosin and kinesin motor domains and the conformational changes triggered by the γ-P sensor and the relay helix. The upper panels represent the motor domains of myosin and kinesin, respectively, in the ATP- or ADP-P$_i$–like state. Similar structural elements in the catalytic cores of the two domains are shown in blue, the relay helices are dark green, and the mechanical elements (neck linker for kinesin, lever arm domains for myosin) are yellow. The nucleotide is shown as a white space-filling model. The similarity of the conformation changes caused by the relay helix in going from the ATP/ADP-P$_i$–bound state to the ADP-bound or nucleotide-free state is shown in the lower panels. In both cases, the mechanical elements of the protein shift their positions in response to relay helix motion. Note that the direction of mechanical element motion is nearly perpendicular to the relay helix motion. *(Adapted from Vale, R. D., and Milligan, R. A., 2000. The way things move: Looking under the hood of molecular motor proteins. Science **288**:88–95.)*

Converter

Relay helix

Myosin

Neck linker

Relay helix

Kinesin

The Conformation Change That Leads to Movement Is Different in Myosins, Kinesins, and Dyneins

The linkage of conformation change at the active site to structural changes in the rest of the motor protein is different in myosins, kinesins, and dyneins. In skeletal muscle myosin, the long α-helix (stabilized by light chains) acts as a lever arm that swings through an angle of up to 70° (Figure 16.22a). In most kinesins, the amplification of movement depends on a short, flexible segment of about ten amino acids. Mobility of this flexible segment drives kinesin movement (Figure 16.22b).

Dynein motors are different. The motor domain of dyneins comprises a ring of six protein modules, members of a large family of proteins known as AAA ATPases. ATP-dependent conformational changes in the ring of AAA-modules are transmitted to a stalk that has the microtubule-binding site on its tip. Swings of this stalk, driven by ATP hydrolysis, lead to a 15-nm movement of the tip (Figure 16.22c).

Calcium Channels and Pumps Control the Muscle Contraction–Relaxation Cycle

The trigger for all muscle contraction is an increase in Ca^{2+} concentration in the vicinity of the muscle fibers of skeletal muscle or the myocytes of cardiac and smooth muscle. In all these cases, this increase in Ca^{2+} is due to the flow of

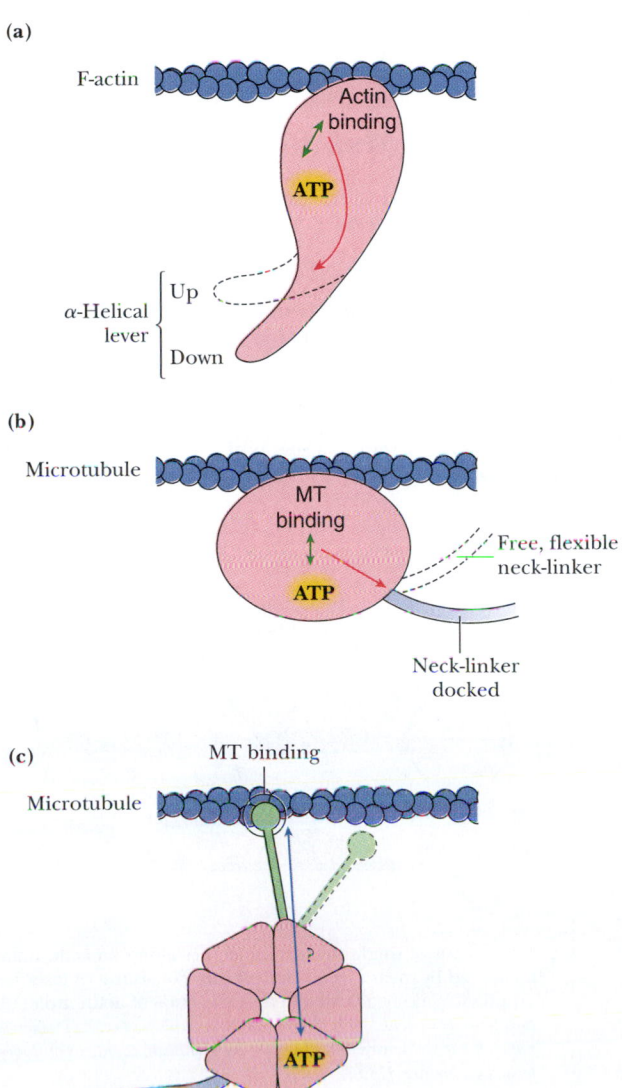

(a) F-actin · Actin binding · ATP · α-Helical lever · Up · Down

(b) Microtubule · MT binding · ATP · Free, flexible neck-linker · Neck-linker docked

(c) MT binding · Microtubule · ? · ATP

FIGURE 16.22 Models for the intramolecular communication and conformational changes that lead to movement within the motor domains of myosin, kinesin, and dynein. In both myosin **(a)** and kinesin **(b)**, ATP hydrolysis causes a conformational change near the ATP-binding site that is communicated to the track-binding site (green arrow). The information is then relayed (red arrow) via homologous structural elements to a mechanical amplifier. (In myosin, the amplifier is the long helix stabilized by light chains; in kinesin, it is a flexible peptide segment, the "neck-linker," that connects the motor domain with the neck helix.) **(c)** The mechanism of intramolecular communication in dynein is not well understood, but a conformational change at the ATP-binding site must be communicated to the stalk that contains the microtubule (MT)-binding site, inducing an angular swinging of the stalk.

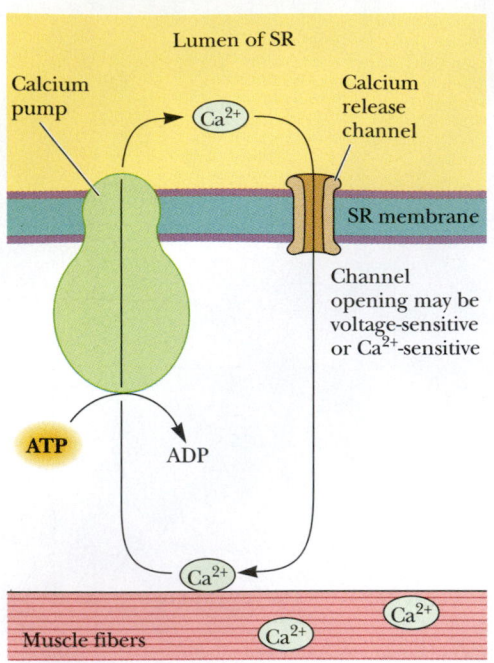

Lumen of SR

Calcium pump

Ca²⁺

Calcium release channel

SR membrane

Channel opening may be voltage-sensitive or Ca²⁺-sensitive

ATP

ADP

Ca²⁺

Muscle fibers

Ca²⁺   Ca²⁺   Ca²⁺

Biochemistry ⊜ Now™   ACTIVE FIGURE 16.23 Ca²⁺ is the trigger signal for muscle contraction. Release of Ca²⁺ through voltage- or Ca²⁺-sensitive channels activates contraction. Ca²⁺ pumps induce relaxation by reducing the concentration of Ca²⁺ available to the muscle fibers. **Test yourself on the concepts in this figure at http://chemistry.brookscole.com/ggb3**

Ca²⁺ through **calcium channels** (Figure 16.23). A muscle contraction ends when the Ca²⁺ concentration is reduced by specific calcium pumps (such as the SR Ca²⁺-ATPase; see Chapter 9).

Muscle Contraction Is Regulated by Ca²⁺

The importance of Ca²⁺ ion as the triggering signal for muscle contraction was described earlier. Ca²⁺ is the intermediary signal that allows striated muscle to respond to motor nerve impulses (Figure 16.23). The Ca²⁺ signal is correctly interpreted by muscle only when tropomyosin and the troponins are present. Specifically, actomyosin prepared from pure preparations of actin and myosin (thus containing no tropomyosin and troponins) was observed to contract when ATP was added, even in the absence of Ca²⁺. However, actomyosin prepared directly from whole muscle would contract in the presence of ATP only

---

Critical Developments in Biochemistry

Molecular "Tweezers" of Light Take the Measure of a Muscle Fiber's Force

The optical trapping experiment involves the attachment of myosin molecules to silica beads that are immobilized on a microscope coverslip (see accompanying figure). Actin filaments are then prepared such that a polystyrene bead is attached to each end of the filament. These beads can be "caught" and held in place in solution by a pair of "optical traps"—two high-intensity infrared laser beams, one focused on the polystyrene bead at one end of the actin filament and the other focused on the bead at the other end of the actin filament. The force acting on each bead in such a trap is proportional to the position of the bead in the "trap," so displacement and forces acting on the bead (and thus on the actin filament) can both be measured. When the "trapped" actin filament is brought close to the myosin-coated silica bead, one or a few myosin molecules may interact with sites on the actin and ATP-induced interactions of individual myosin molecules with the trapped actin filament can be measured and quantitated. Such optical trapping experiments have shown that *a single cycle or turnover of a single myosin molecule along an actin filament involves an average movement of 4 to 11 nm (40–110 Å) and generates an average force of 1.7 to 4 × 10⁻¹² newton (1.7–4 piconewtons [pN]).*

The magnitudes of the movements observed in the optical trapping experiments are consistent with the movements predicted by the cryoelectron microscopy imaging data. Can the movements and forces detected in a single contraction cycle by optical trapping also be related to the energy available from hydrolysis of a single ATP molecule? The energy required for a contraction cycle is defined by the "work" accomplished by contraction, and work *(w)* is defined as force *(F)* times distance *(d)*:

$$w = F \cdot d$$

For a movement of 4 nm against a force of 1.7 pN, we have

$$w = (1.7 \text{ pN}) \cdot (4 \text{ nm}) = 0.68 \times 10^{-20} \text{ J}$$

For a movement of 11 nm against a force of 4 pN, the energy requirement is larger:

$$w = (4 \text{ pN}) \cdot (11 \text{ nm}) = 4.4 \times 10^{-20} \text{ J}$$

If the cellular free energy of hydrolysis of ATP is taken as −50 kJ/mol, the free energy available from the hydrolysis of a single ATP molecule is

$$\Delta G = (-50 \text{ kJ/mol})/(6.02 \times 10^{23} \text{ molecules/mol}) = 8.3 \times 10^{-20} \text{ J}$$

Thus, the free energy of hydrolysis of a single ATP molecule is sufficient to drive the observed movements against the forces that have been measured.

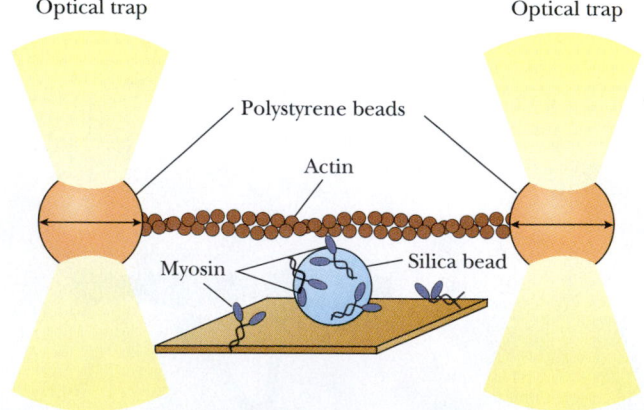

Optical trap   Optical trap

Polystyrene beads

Actin

Myosin   Silica bead

▲ Movements of single myosin molecules along an actin filament can be measured by means of an optical trap consisting of laser beams focused on polystyrene beads attached to the ends of actin molecules. *(Adapted from Finer, J. T., et al., 1994. Single myosin molecule mechanics: Piconewton forces and nanometre steps. Nature **368**:113–119. See also Block, S. M., 1995. Macromolecular physiology. Nature **378**:132–133.)*

when Ca^{2+} was added. Clearly the muscle extracts contained a factor that conferred normal Ca^{2+} sensitivity to actomyosin. The factor turned out to be the tropomyosin–troponin complex.

Actin thin filaments consist of actin, tropomyosin, and the troponins in a 7:1:1 ratio (Figure 16.15). Each tropomyosin molecule spans seven actin molecules, lying along the thin filament groove, between pairs of actin monomers. As shown in a cross-section view in Figure 16.24, in the absence of Ca^{2+}, troponin I is thought to interact directly with actin to prevent the interaction of actin with myosin S1 heads. Troponin I and troponin T interact with tropomyosin to keep tropomyosin away from the groove between adjacent actin monomers. However, the binding of Ca^{2+} ions to troponin C appears to increase the binding of troponin C to troponin I, simultaneously decreasing the interaction of troponin I with actin. As a result, tropomyosin slides deeper into the actin–thin filament groove, exposing myosin-binding sites on actin and initiating the muscle contraction cycle (Figure 16.24). Because the troponin complexes can interact only with every seventh actin in the thin filament, the conformational changes that expose myosin-binding sites on actin may well be cooperative. Binding of an S1 head to an actin may displace tropomyosin and the troponin complex from myosin-binding sites on adjacent actin subunits.

The Interaction of Ca^{2+} with Troponin C There are four Ca^{2+}-binding sites on troponin C—two high-affinity sites on the carboxy-terminal end of the molecule, labeled III and IV in Figure 16.25, and two low-affinity sites on the amino-terminal end, labeled I and II. Ca^{2+} binding to sites III and IV is sufficiently strong ($K_D = 0.1\ \mu M$) that these sites are presumed to be filled under resting conditions. Sites I and II, however, where the K_D is approximately $10\ \mu M$, are empty in resting muscle. The rise of Ca^{2+} levels when contraction is signaled leads to the filling of sites I and II, causing a conformation change in the amino-terminal domain of TnC. This conformational change apparently facilitates a more intimate binding of TnI to TnC that involves the C helix, and also possibly the E helix of TnC. The increased interaction between TnI and TnC results in a decreased interaction between TnI and actin.

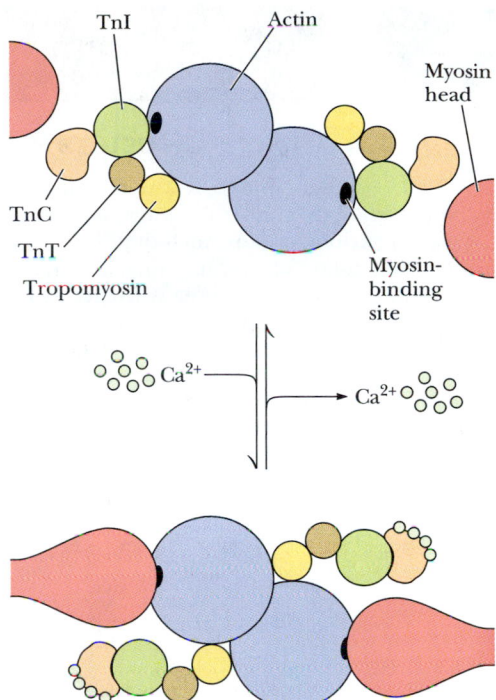

Biochemistry Now™ **ANIMATED FIGURE 16.24**
A drawing of the thick and thin filaments of skeletal muscle in cross section showing the changes that are postulated to occur when Ca^{2+} binds to troponin C. **See this figure animated at http://chemistry. brookscole.com/ggb3**

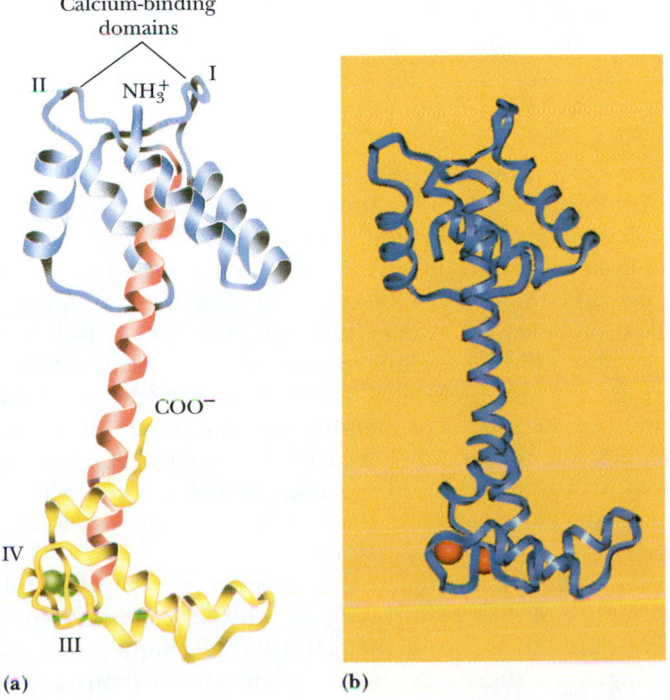

FIGURE 16.25 Two slightly different views of the structure of troponin C: **(a)** a ribbon diagram and **(b)** a molecular graphic. Note the long α-helical domain connecting the N-terminal and C-terminal lobes of the molecule.

Human Biochemistry

Smooth Muscle Effectors Are Useful Drugs

Not all vertebrate muscle is skeletal muscle. Vertebrate organisms employ smooth muscle for long, slow, and involuntary contractions in various organs, including large blood vessels, intestinal walls, the gums of the mouth, and in the female, the uterus. Smooth muscle contraction is triggered by Ca^{2+}-activated phosphorylation of myosin by myosin light-chain kinase (MLCK). The action of epinephrine and related agents forms the basis of therapeutic control of smooth muscle contraction. Breathing disorders, including asthma and various allergies, can result from excessive contraction of bronchial smooth muscle tissue. Treatment with epinephrine, whether by tablets or aerosol inhalation, inhibits MLCK and relaxes bronchial muscle tissue. More specific **bronchodilators,** such as **albuterol** (see accompanying figure), act more selectively on the lungs and avoid the undesirable side effects of epinephrine on the heart. Albuterol is also used to prevent premature labor in pregnant women because of its relaxing effect on uterine smooth muscle. Conversely, **oxytocin,** known also as **Pitocin,** stimulates contraction of uterine smooth muscle. This natural secretion of the pituitary gland is often administered to induce labor.

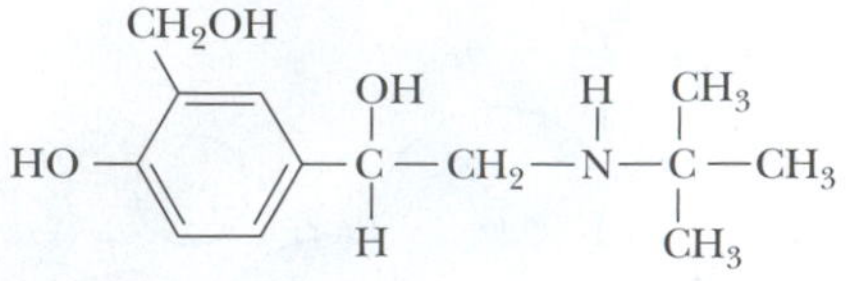

Albuterol

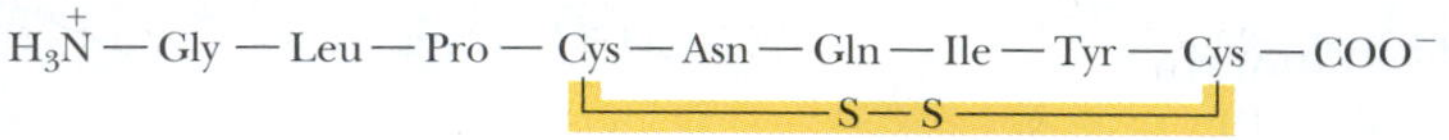

Oxytocin (Pitocin)

$$H_3\overset{+}{N} - Gly - Leu - Pro - Cys - Asn - Gln - Ile - Tyr - Cys - COO^-$$

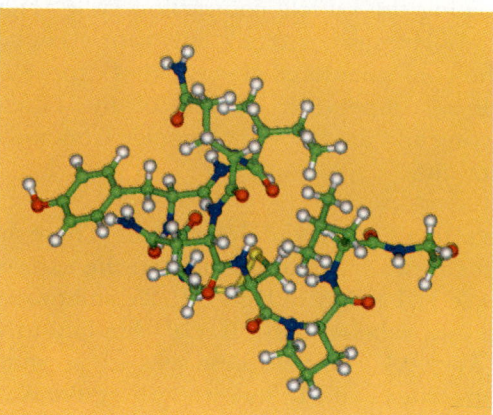

▲ The structure of oxytocin.

| 16.5 | ### How Do Bacterial Flagella Use a Proton Gradient to Drive Rotation? |

Bacterial cells swim and move by rotating their flagella. The flagella of *E. coli* are helical filaments about 10,000 nm (10 μm) in length and 15 nm in diameter. The direction of rotation of these filaments affects the movements of the cell. When the half-dozen filaments on the surface of the bacterial cell rotate in a counterclockwise direction, they twist and bundle together and rotate in a concerted fashion, propelling the cell through the medium. (On the other hand, clockwise-rotating flagella cannot bundle together, and under such conditions the cell merely tumbles and moves erratically.)

The rotations of bacterial flagellar filaments are the result of the rotation of motor protein complexes in the bacterial plasma membrane. The flagellar motor consists of at least two rings (including the M ring and the S ring) with diameters of about 25 nm assembled around and connected rigidly to a rod attached in turn to the helical filament (Figure 16.26). The rings are surrounded by a circular array of membrane proteins. In all, at least 40 genes appear to code for proteins involved in this magnificent assembly. One of these, the motB protein, lies on the edge of the M ring, where it interacts with the motA protein, located in the membrane protein array and facing the M ring.

In contrast to the many other motor proteins described in this chapter, a proton gradient, not ATP hydrolysis, drives the flagellar motor. The concentration of protons, [H^+], outside the cell is typically higher than that inside the cell. Thus, there is a thermodynamic tendency for protons to move into the

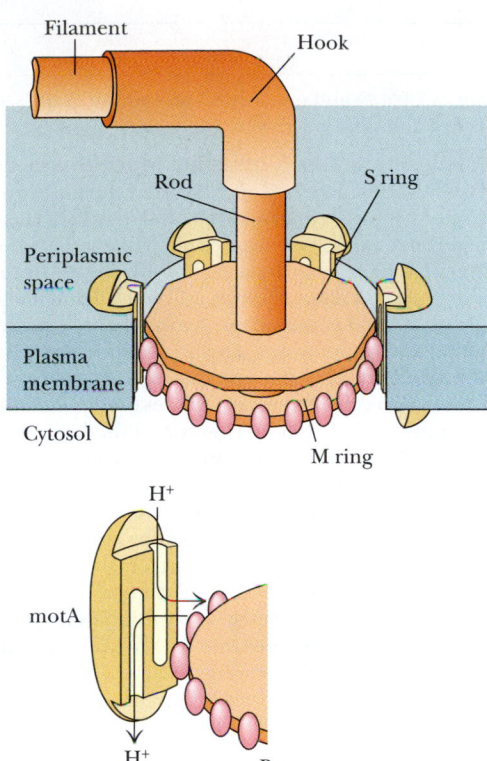

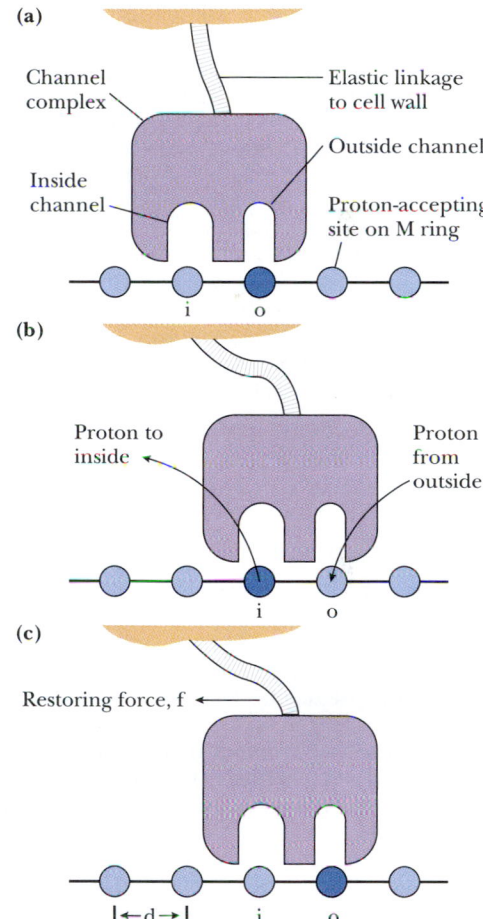

FIGURE 16.26 A model of the flagellar motor assembly of *Escherichia coli*. The M ring carries an array of about 100 motB proteins at its periphery. These juxtapose with motA proteins in the protein complex that surrounds the ring assembly. Motion of protons through the motA–motB complexes drives the rotation of the rings and the associated rod and helical filament.

Biochemistry Now™   ACTIVE FIGURE 16.27
Howard Berg's model for coupling between transmembrane proton flow and rotation of the flagellar motor. A proton moves through an outside channel to bind to an exchange site on the M ring. When the channel protein slides one step around the ring, the proton is released and flows through an inside channel and into the cell while another proton flows into the outside channel to bind to an adjacent exchange site. When the motA channel protein returns to its original position under an elastic restoring force, the associated motB protein moves with it, causing a counterclockwise rotation of the ring, rod, and helical filament. *(Adapted from Meister, M., Caplan, S. R., and Berg, H. C., 1989. Dynamics of a tightly coupled mechanism for flagellar rotation. Biophysical Journal 55:905–914.)* **Test yourself on the concepts in this figure at http:// chemistry.brookscole.com/ggb3**

cell. The motA and motB proteins together form a proton-shuttling device that is coupled to motion of the motor discs. Proton movement into the cell through this protein complex or "channel" drives the rotation of the flagellar motor. A model for this coupling has been proposed by Howard Berg and his co-workers (Figure 16.27). In this model, the motB proteins possess proton exchanging sites—for example, carboxyl groups on aspartate or glutamate residues or imidazole moieties on histidine residues. The motA proteins, on the other hand, possess a pair of "half-channels," with one half-channel facing the inside of the cell and the other facing the outside. In Berg's model, the outside edges of the motA channel protein cannot move past a proton-exchanging site on motB when that site has a proton bound, and the center of the channel protein cannot move past an exchange site when that site is empty. As shown in Figure 16.27, these constraints lead to coupling between proton translocation and rotation of the flagellar filament. For example, imagine that a proton has entered the outside channel of motA and is bound to an exchange site on motB (Figure 16.27a). An oscillation by motA, linked elastically to the cell wall, can then position the inside channel over the proton at the exchange site (Figure 16.27b), whereupon the proton can travel through the inside channel and into the cell while another proton travels up the outside channel to bind to an adjacent exchange site. The restoring force acting on the channel protein then pulls the motA–motB complex to the left as shown (Figure 16.27c), leading to counterclockwise rotation of the disc, rod, and helical filament. The flagellar motor is driven entirely by the proton gradient. Thus, a reversal of the proton gradient (which would occur, for example, if the external medium became alkaline) would drive the flagellar filaments in a clockwise direction. Extending this picture of a single motA–motB complex to the whole motor disc array, one can imagine the torrent of protons that pass through the motor assembly to drive flagellar rotation at a typical speed of 100 rotations per second. Berg estimates that the M ring carries 100 motB proton-exchange sites, and various models predict that 800 to 1200 protons must flow through the complex during a single rotation of the flagellar filament!

Summary

16.1 What Is a Molecular Motor?
Motor proteins, also known as molecular motors, use chemical energy (ATP) to orchestrate different movements, transforming ATP energy into the mechanical energy of motion. In all cases, ATP hydrolysis is presumed to drive and control protein conformational changes that result in sliding or walking movements of one molecule relative to another. To carry out directed movements, molecular motors must be able to associate and dissociate reversibly with a polymeric protein array, a surface, or substructure in the cell. ATP hydrolysis drives the process by which the motor protein ratchets along the protein array or surface. Molecular motors may be linear or rotating. Linear motors crawl or creep along a polymer lattice, whereas rotating motors consist of a rotating element (the "rotor") and a stationary element (the "stator"), in a fashion much like a simple electrical motor.

16.2 What Are the Molecular Motors That Orchestrate the Mechanochemistry of Microtubules?
Microtubules are hollow, cylindrical structures, approximately 30 nm in diameter, formed from tubulin, a dimeric protein composed of two similar 55-kD subunits known as α-tubulin and β-tubulin. Tubulin dimers polymerize to form microtubules, which are essentially helical structures, with 13 tubulin monomer "residues" per turn. Microtubules are, in fact, a significant part of the cytoskeleton, a sort of intracellular scaffold formed of microtubules, intermediate filaments, and microfilaments. In most cells, microtubules are oriented with their minus ends toward the centrosome and their plus ends toward the cell periphery. This consistent orientation is important for mechanisms of intracellular transport. Microtubules are also the fundamental building blocks of cilia and flagella. The motion of cilia results from the ATP-driven sliding or walking of dyneins along one microtubule while they remain firmly attached to an adjacent microtubule. Microtubules also mediate the intracellular motion of organelles and vesicles.

16.3 How Do Molecular Motors Unwind DNA?
When DNA is to be replicated or repaired, the strands of the double helix must be unwound and separated to form single-stranded DNA intermediates. This separation is carried out by molecular motors known as DNA helicases that move along the length of the DNA lattice, sequentially destabilizing the hydrogen bonds between complementary base pairs. The movement along the lattice and the separation of the DNA strands are coupled to the hydrolysis of nucleoside 5'-triphosphates. The *E. coli* BCD helicase, which is involved in recombination processes, can unwind 33,000 base pairs before it dissociates from the DNA lattice. Processive movement is essential for helicases involved in DNA replication, where millions of base pairs must be replicated rapidly. Certain hexameric helicases form ringlike structures that completely encircle at least one of the strands of a DNA duplex. Other helicases, notably Rep helicase from *E. coli*, are homodimeric and move processively along the DNA helix by

means of a "hand-over-hand" movement that is remarkably similar to that of kinesin's movement along microtubules.

16.4 What Is the Molecular Mechanism of Muscle Contraction?
Examination of myofibrils in the electron microscope reveals a banded or striated structure. The so-called H zone shows a regular, hexagonally arranged array of thick filaments, whereas the I band shows a regular, hexagonal array of thin filaments. In the dark regions at the ends of each A band, the thin and thick filaments interdigitate. The thin filaments are composed primarily of three proteins called actin, troponin, and tropomyosin. The thick filaments consist mainly of a protein called myosin. The thin and thick filaments are joined by cross-bridges. These cross-bridges are actually extensions of the myosin molecules, and muscle contraction is accomplished by the sliding of the cross-bridges along the thin filaments, a mechanical movement driven by the free energy of ATP hydrolysis.

Myosin, the principal component of muscle thick filaments, is a large protein consisting of six polypeptides, including light chains and heavy chains. The heavy chains consist of globular amino-terminal myosin heads, joined to long α-helical carboxy-terminal segments, the tails. These tails are intertwined to form a left-handed coiled coil approximately 2 nm in diameter and 130 to 150 nm long. The myosin heads exhibit ATPase activity, and hydrolysis of ATP by the myosin heads drives muscle contraction.

The free energy of ATP hydrolysis is translated into a conformation change in the myosin head, so dissociation of myosin and actin, hydrolysis of ATP, and rebinding of myosin and actin occur with stepwise movement of the myosin S1 head along the actin filament. The conformation change in the myosin head is driven by the hydrolysis of ATP.

16.5 How Do Bacterial Flagella Use a Proton Gradient to Drive Rotation?
Bacterial cells swim and move by rotating their flagella. The direction of rotation of these filaments affects the movements of the cell. When the half-dozen filaments on the surface of the bacterial cell rotate in a counterclockwise direction, they twist and bundle together and rotate in a concerted fashion, propelling the cell through the medium. The rotations of bacterial flagellar filaments are the result of rotation of motor protein complexes in the bacterial plasma membrane. The flagellar motor consists of at least two rings (including the M ring and the S ring). The rings are surrounded by a circular array of membrane proteins. In all, at least 40 genes appear to code for proteins involved in this magnificent assembly. One of these, the motB protein, lies on the edge of the M ring, where it interacts with the motA protein, located in the membrane protein array and facing the M ring. In contrast to the many other motor proteins described in this chapter, a proton gradient, not ATP hydrolysis, drives the flagellar motor.

Problems

1. The cheetah is generally regarded as nature's fastest mammal, but another amazing athlete in the animal kingdom (and almost as fast as the cheetah) is the pronghorn antelope, which roams the plains of Wyoming. Whereas the cheetah can maintain its top speed of 70 mph for only a few seconds, the pronghorn antelope can run at 60 mph for about an hour! (It is thought to have evolved to do so in order to elude now-extinct ancestral cheetahs that lived in North America.) What differences would you expect in the muscle structure and anatomy of pronghorn antelopes that could account for their remarkable speed and endurance?

2. An ATP analog, β,γ-methylene-ATP, in which a —CH₂— group replaces the oxygen atom between the β- and γ-phosphorus atoms, is a potent inhibitor of muscle contraction. At which step in the contraction cycle would you expect β,γ-methylene-ATP to block contraction?

3. ATP stores in muscle are augmented or supplemented by stores of phosphocreatine. During periods of contraction, phosphocreatine

is hydrolyzed to drive the synthesis of needed ATP in the creatine kinase reaction:

$$\text{Phosphocreatine} + \text{ADP} \longrightarrow \text{creatine} + \text{ATP}$$

Muscle cells contain two different isozymes of creatine kinase, one in the mitochondria and one in the sarcoplasm. Explain.

4. *Rigor* is a muscle condition in which muscle fibers, depleted of ATP and phosphocreatine, develop a state of extreme rigidity and cannot be easily extended. (In death, this state is called *rigor mortis*, the rigor of death.) From what you have learned about muscle contraction, explain the state of rigor in molecular terms.

5. Skeletal muscle can generate approximately 3 to 4 kg of tension or force per square centimeter of cross-sectional area. This number is roughly the same for all mammals. Because many human muscles have large cross-sectional areas, the force that these muscles can (and must) generate is prodigious. The gluteus maximus (on which

you are probably sitting as you read this) can generate a tension of 1200 kg! Estimate the cross-sectional area of all of the muscles in your body and the total force that your skeletal muscles could generate if they all contracted at once.

6. Calculate a diameter for a tubulin monomer, assuming that the monomer MW is 55,000, that the monomer is spherical, and that the density of the protein monomer is 1.3 g/ml. How does the number that you calculate compare to the dimension portrayed in Figure 16.2?

7. Use the number you obtained in problem 6 to calculate how many tubulin monomers would be found in a microtubule that stretched across the length of a liver cell. (See Table 1.2 for the diameter of a liver cell.)

8. The giant axon of the squid may be up to 4 inches in length. Use the value cited in this chapter for the rate of movement of vesicles and organelles across axons to determine the time required for a vesicle to traverse the length of this axon.

9. As noted in this chapter, the myosin molecules in thick filaments of muscle are offset by approximately 14 nm. To how many residues of a coiled coil structure does this correspond?

10. (Integrates with Chapter 9.) Use the equations of Chapter 9 to determine the free energy difference represented by a Ca^{2+} gradient across the sarcoplasmic reticulum membrane if the luminal (inside) concentration of Ca^{2+} is 1 mM and the concentration of Ca^{2+} in the solution bathing the muscle fibers is 1 μM.

11. (Integrates with Chapter 3.) Use the equations of Chapter 3 to determine the free energy of hydrolysis of ATP by the sarcoplasmic reticulum Ca-ATPase if the concentration of ATP is 3 mM, the concentration of ADP is 1 mM, and the concentration of P_i is 2 mM.

12. Under the conditions described in problems 10 and 11, what is the maximum number of Ca^{2+} ions that could be transported per ATP hydrolyzed by the Ca-ATPase?

Preparing for the MCAT Exam

13. Consult Figure 16.9 and use the data in problem 8 to determine how many steps a kinesin motor must take to traverse the length of the squid giant axon.

14. When athletes overexert themselves on hot days, they often suffer immobility from painful muscle cramps. Which of the following is a reasonable hypothesis to explain such cramps?
    a. Muscle cells do not have enough ATP for normal muscle relaxation.
    b. Excessive sweating has affected the salt balance within the muscles.
    c. Prolonged contractions have temporarily interrupted blood flow to parts of the muscle.
    d. All of the above.

15. Duchenne muscular dystrophy is a sex-linked recessive disorder associated with severe deterioration of muscle tissue. The gene for the disease:
    a. is inherited by males from their mothers.
    b. should be more common in females than in males.
    c. both a and b.
    d. neither a nor b.

Biochemistry⑤Now™   Preparing for an exam? Test yourself on key questions at http://chemistry.brookscole.com/ggb3

Further Reading

Tubulin

Hoenger, A., Sablin, E., Vale, R., et al., 1995. Three-dimensional structure of a tubulin-motor-protein complex. *Nature* **376:**271–274.

Howard, J., and Hyman, A. A., 2003. Dynamics and mechanics of the microtubule plus end. *Nature* **422:**753–758.

Nogales, E., Wolf, S., and Downing, K. H., 1998. Structure of the $\alpha\beta$ tubulin dimer by electron crystallography. *Nature* **391:**199–203.

Muscle Contraction

Ahn, A. H., and Kunkel, L. M., 1993. The structural and functional diversity of dystrophin. *Nature Genetics* **3:**283–291.

Allen, B., and Walsh, M., 1994. The biochemical basis of the regulation of smooth-muscle contraction. *Trends in Biochemical Sciences* **19:**362–368.

Cooke, R., 1995. The actomyosin engine. *The FASEB Journal* **9:**636–642.

Fisher, A., Smith, C., Thoden, J., et al., 1995. Structural studies of myosin:nucleotide complexes: A revised model for the molecular basis of muscle contraction. *Biophysical Journal* **68:**19S–26S.

Goldman, Y. E., 1998. Wag the tail: Structural dynamics of actomyosin. *Cell* **93:**1–4.

Labeit, S., and Kolmerer, B., 1995. Titins: Giant proteins in charge of muscle ultrastructure and elasticity. *Science* **270:**293–296.

Molloy, J., Burns, J., Kendrick-Jones, J., et al., 1995. Movement and force produced by a single myosin head. *Nature* **378:**209–213.

Rayment, I., 1996. Kinesin and myosin: Molecular motors with similar engines. *Structure* **4:**501–504.

Rayment, I., and Holden, H., 1994. The three-dimensional structure of a molecular motor. *Trends in Biochemical Sciences* **19:**129–134.

Wagenknecht, T., et al., 1989. Three-dimensional architecture of the calcium channel/foot structure of sarcoplasmic reticulum. *Nature* **338:**167–170.

Whittaker, M., Wilson-Kubalek, E., Smith, J. E., et al., 1995. A 35 Å movement of smooth muscle myosin on ADP release. *Nature* **378:**748–753.

Worton, R., 1995. Muscular dystrophies: Diseases of the dystrophin–glycoprotein complex. *Science* **270:**755–756.

Kinesins, Dyneins, and Organelle Transport

Burgess, S. A., Walker, M. L., Sakakibara, H., Knight, P. J., and Oiwa, K., 2003. Dynein structure and power stroke. *Nature* **421:**715–718.

Coppin, C. M., Finer, J. T., Spudich, J. A., and Vale, R. D., 1996. Detection of sub-8-nm movements of kinesin by high-resolution optical-trap microscopy. *Proceedings of the National Academy of Sciences* **93:**1913–1917.

Deacon, S. W., Serpinskaya, A. S., Vaughan, P. S., Fanarraga, M. L., Vernos, I., Vaughan, K. T., and Gelfand, V. I., 2003. Dynactin is required for bidirectional organelle transport. *Journal of Cell Biology* **160:**297–301.

Dell, K. R., 2003. Dynactin polices two-way organelle traffic. *Journal of Cell Biology* **160:**291–293.

Hirose, K., Lockhart, A., Cross, R., and Amos, L., 1995. Nucleotide-dependent angular change in kinesin motor domain bound to tubulin. *Nature* **376:**277–279.

Naber, N., Minehardt, T. J., Rice, S., et al., 2003. Cloning of the nucleotide pocket of kinesin-family motors upon binding to microtubules. *Science* **300:**798–801.

Schliwa, M., and Woehlke, G., 2003. Molecular motors. *Nature* **422:**759–765.

Vale, R. D., 2003. The molecular motor toolbox for intracellular transport. *Cell* **112:**467–480.

Vale, R. D., and Milligan, R. A., 2000. The way things move: Looking under the hood of molecular motor proteins. *Science* **288:**88–95.

Rotating Motors

DeRosier, D. J., 1998. The turn of the screw: The bacterial flagellar motor. *Cell* **93:**17–20.

Kinosita, K., Jr., Yasuda, R., Noji, H., et al., 1998. F1-ATPase: A rotary motor made of a single molecule. *Cell* **93:**21–24.

PART III

Molecular Components of Cells

Metabolism: Chemistry of Life or Biology of Molecules?

An Essay by Juliet A. Gerrard, University of Canterbury

Since its inception, biochemistry has been framed as the "chemistry of life." In Part III of this text, metabolism—the chemistry that takes place in cells—is discussed in detail. Thousands of molecules have been isolated and studied, giving us insights into the structure of cellular metabolites and macromolecular machinery, the mechanisms of enzyme-catalyzed reactions, and the constraints that chemistry imposes on the workings of biology.

As the information describing individual cellular components has amassed, the value of the set of data as a whole has become limited by its own complexity. The original goal—to relate biochemical function, as observed in the laboratory, to biological function, as emerges in the cell—often becomes more elusive as new pieces of the chemical puzzle come to light. (Imagine listening to each individual sound from the latest Red Hot Chili Peppers tracks, in no particular order, and being asked to recreate

"In short, it is time to re-frame biochemistry as the "biology of molecules" and find new ways to look at metabolism."

their new album.) In the postgenomic era, we must assemble all these individual pieces of biochemical information and understand them in the context of living cells. It is time to focus on how the chemistry of life is organized in space and time, how individual components interact, and what new characteristics emerge from these interactions, which are not features of any of the isolated units. In short, it is time to reframe biochemistry as the "biology of molecules" and find new ways to look at metabolism.

This is no simple task, but finding new ways to think about metabolism may give us new insights into how cells work. One approach, which draws on the thinking of systems biology, is to take a holistic view of the cell as a "community" of molecules. The properties of the cell may be more dependent on the qualities of the community than on the individual characteristics of the molecular components themselves.

There are many ways to think about such a community of molecules, not only the traditional one that focuses on metabolites and emphasizes the organic chemistry of the cell, with the enzymes relegated to sit above the conversion arrows. We could instead focus on the enzymes, with metabolites as signals between interacting protein components. This "protein-centric" alternative affords an opportunity to simplify our view of metabolism, as illustrated in the accompanying figure. It is an alluring thought that this "new view" may give insights into the physical organization of cells—for example, it may help us locate multienzyme complexes, or metabolons.

As you read the chapters that follow, remember that in vivo every isolated piece of biochemistry must take place in a highly organized fashion. Many components of cells may be well understood, but our knowledge of their interactions remains in its infancy. New ways of thinking will be required to gain a true appreciation of biology at its smallest scale.

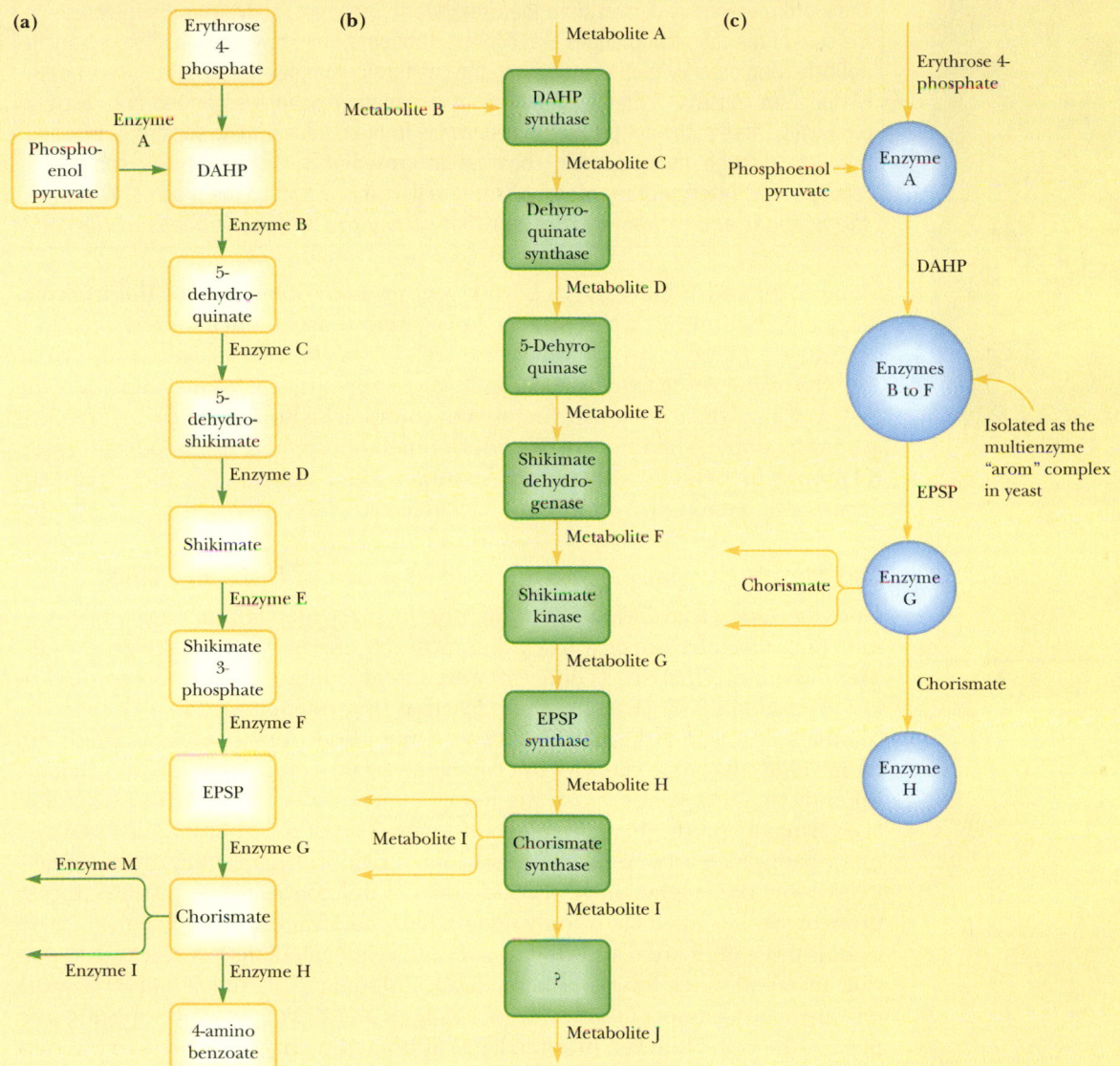

(a)

Erythrose 4-phosphate

Phospho-enol pyruvate → **Enzyme A** → DAHP

DAHP → **Enzyme B** → 5-dehydro-quinate

5-dehydro-quinate → **Enzyme C** → 5-dehydro-shikimate

5-dehydro-shikimate → **Enzyme D** → Shikimate

Shikimate → **Enzyme E** → Shikimate 3-phosphate

Shikimate 3-phosphate → **Enzyme F** → EPSP

EPSP → **Enzyme G** → Chorismate

Enzyme M ←

Enzyme I ←

Chorismate → **Enzyme H** → 4-amino benzoate

(b)

Metabolite A

Metabolite B → **DAHP synthase**

Metabolite C

Dehyro-quinate synthase

Metabolite D

5-Dehyro-quinase

Metabolite E

Shikimate dehydro-genase

Metabolite F

Shikimate kinase

Metabolite G

EPSP synthase

Metabolite H

Metabolite I → **Chorismate synthase**

Metabolite I

?

Metabolite J

(c)

Erythrose 4-phosphate

Phosphoenol pyruvate → **Enzyme A**

DAHP

Enzymes B to F ← Isolated as the multienzyme "arom" complex in yeast

EPSP

Chorismate → **Enzyme G**

Chorismate

Enzyme H

▲ Turning biochemistry inside out. Three alternate perspectives on part of the biosynthesis of the aromatic amino acids in *Escherichia coli*. **(a)** The traditional metabolite-centric view, with emphasis on the metabolites (see Chapter 25). **(b)** An "inside out" view, in which the enzymes and metabolites have been transposed to reveal a related version of the metabolic map with a new emphasis: Here the metabolites are acting as "signals" in a cellular network of proteins. **(c)** A simplified version of (b), highlighting a predicted cellular compartment or multienzyme complex. "Redundant" metabolic steps have been eliminated, condensing the map to show only key metabolic signals and multienzyme nodes. The result is a dramatic simplification of the network, containing only essential signaling information. In this case, the postulated compartment or multienzyme complex has been found to exist in yeast: The arom complex is a pentafunctional enzyme complex, performing the functions of all five enzymes, b–f. (For more on the protein-centric view of metabolism, see Gerrard, J. A., Sparrow, A. D., and Wells, J. A., 2001. Metabolic databases—what next? *Trends in Biochemical Sciences* 26:137.)

CHAPTER 17

Metabolism—An Overview

© Gray Hardel/CORBIS

Anise swallowtail butterfly *(Papilio zelicans)* with its pupal case. Metamorphosis of butterflies is a dramatic example of metabolic change.

All is flux, nothing stays still. Nothing endures but change.
Heraclitus (c. 540–c. 480 B.C.)

Key Questions

17.1 Are There Similarities of Metabolism Between Organisms?

17.2 How Do Anabolic and Catabolic Processes Form the Core of Metabolic Pathways?

17.3 What Experiments Can Be Used to Elucidate Metabolic Pathways?

17.4 What Food Substances Form the Basis of Human Nutrition?

Special Focus: Vitamins

Biochemistry ⊜ Now™ Test yourself on these Key Questions at BiochemistryNow at **http://chemistry.brookscole.com/ggb3**

Essential Question

The word *metabolism* derives from the Greek word for "change." **Metabolism** represents the sum of the chemical changes that convert **nutrients,** the "raw materials" necessary to nourish living organisms, into energy and the chemically complex finished products of cells. Metabolism consists of literally hundreds of enzymatic reactions organized into discrete pathways. These pathways proceed in a stepwise fashion, transforming substrates into end products through many specific chemical **intermediates.** Metabolism is sometimes referred to as **intermediary metabolism** to reflect this aspect of the process. *What are the anabolic and catabolic processes that satisfy the metabolic needs of the cell?*

Metabolic maps (Figure 17.1) portray the principal reactions of the intermediary metabolism of carbohydrates, lipids, amino acids, nucleotides, and their derivatives. These maps are very complex at first glance and seem to be virtually impossible to learn easily. Despite their appearance, these maps become easy to follow once the major metabolic routes are known and their functions are understood. The underlying order of metabolism and the important interrelationships between the various pathways then appear as simple patterns against the seemingly complicated background.

The Metabolic Map Can Be Viewed as a Set of Dots and Lines

One interesting transformation of the intermediary metabolism map is to represent each intermediate as a black dot and each enzyme as a line (Figure 17.2). Then, the more than 1000 different enzymes and substrates are represented by just two symbols. This chart has about 520 dots (intermediates). Table 17.1 lists the numbers of dots that have one or two or more lines (enzymes) associated with them. Thus, this table classifies intermediates by the number of enzymes that act upon them. A dot connected to just a single line must be either a nutrient, a storage form, an end product, or an excretory product of metabolism. Also, because many pathways tend to proceed in only one direction (that is, they are essentially irreversible under physiological conditions), a dot connected to just two lines is probably an intermediate in only one pathway and has only one fate in metabolism. If three lines are connected to a dot, that intermediate has at least two possible metabolic fates; four lines, three fates; and so on. Note that about 80% of the intermediates connect to only one or two lines and thus have only a specific purpose in the cell. However, intermediates at branch points are subject to a variety of fates. In such instances, the pathway followed is an important regulatory choice. Indeed, whether any substrate is routed down a particular metabolic pathway is the consequence of a regulatory decision made in response to the cell's (or organism's) momentary requirements for energy or nutrition. The regulation of metabolism is an interesting and important subject to which we will return often.

17.1 | Are There Similarities of Metabolism Between Organisms?

One of the great unifying principles of modern biology is that organisms show marked similarity in their major pathways of metabolism. Given the almost unlimited possibilities within organic chemistry, this generality would appear

▶ **Biochemistry ⊜ Now™ ACTIVE FIGURE 17.1** A metabolic map, indicating the reactions of intermediary metabolism and the enzymes that catalyze them. More than 500 different chemical intermediates, or metabolites, and a greater number of enzymes are represented here. *(Source: Donald Nicholson's Metabolic Map #21. Copyright © International Union of Biochemistry and Molecular Biology. Used with permission.)* **Test yourself on the concepts in this figure at http://chemistry.brookscole.com/ggb3**

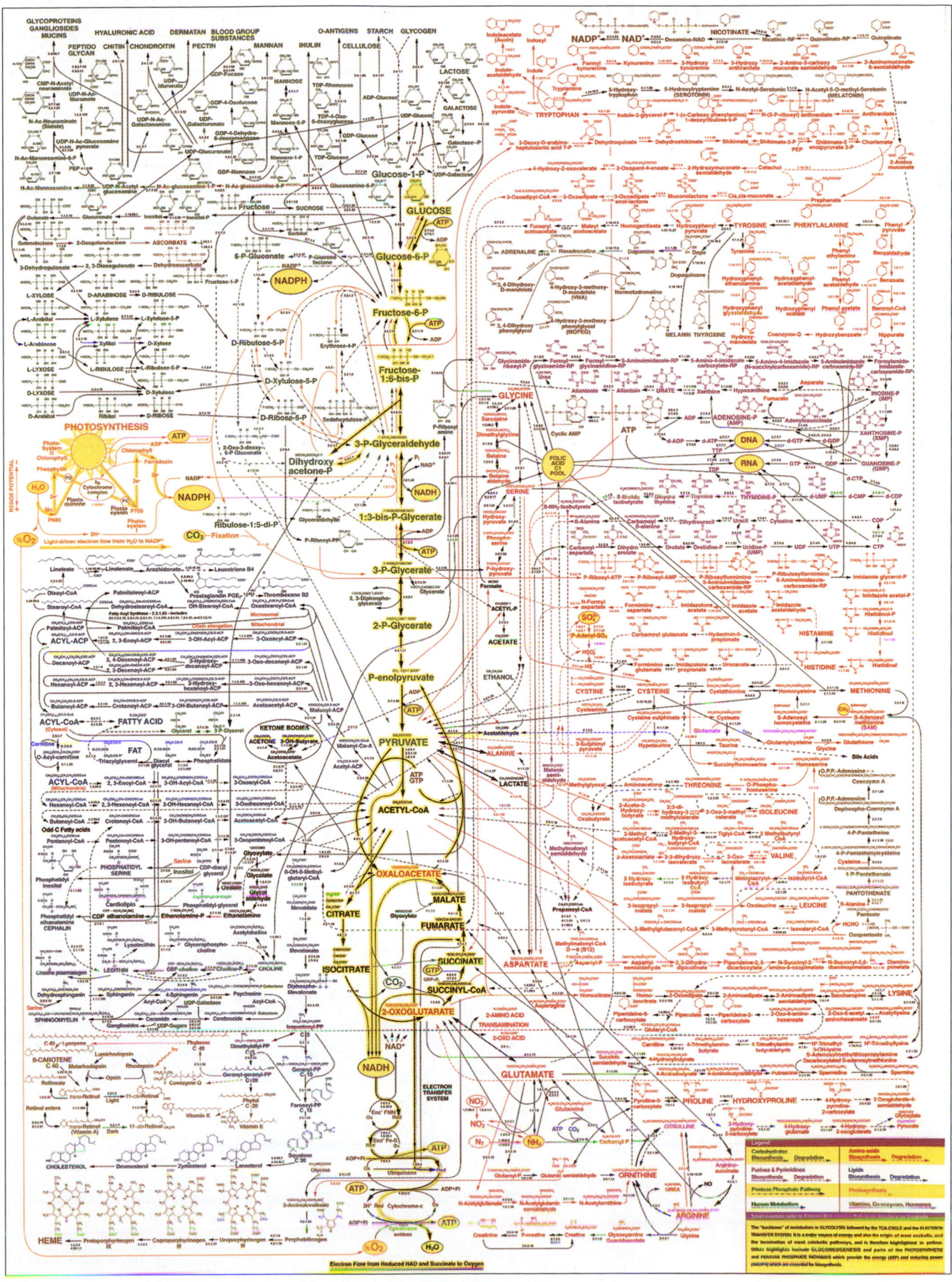

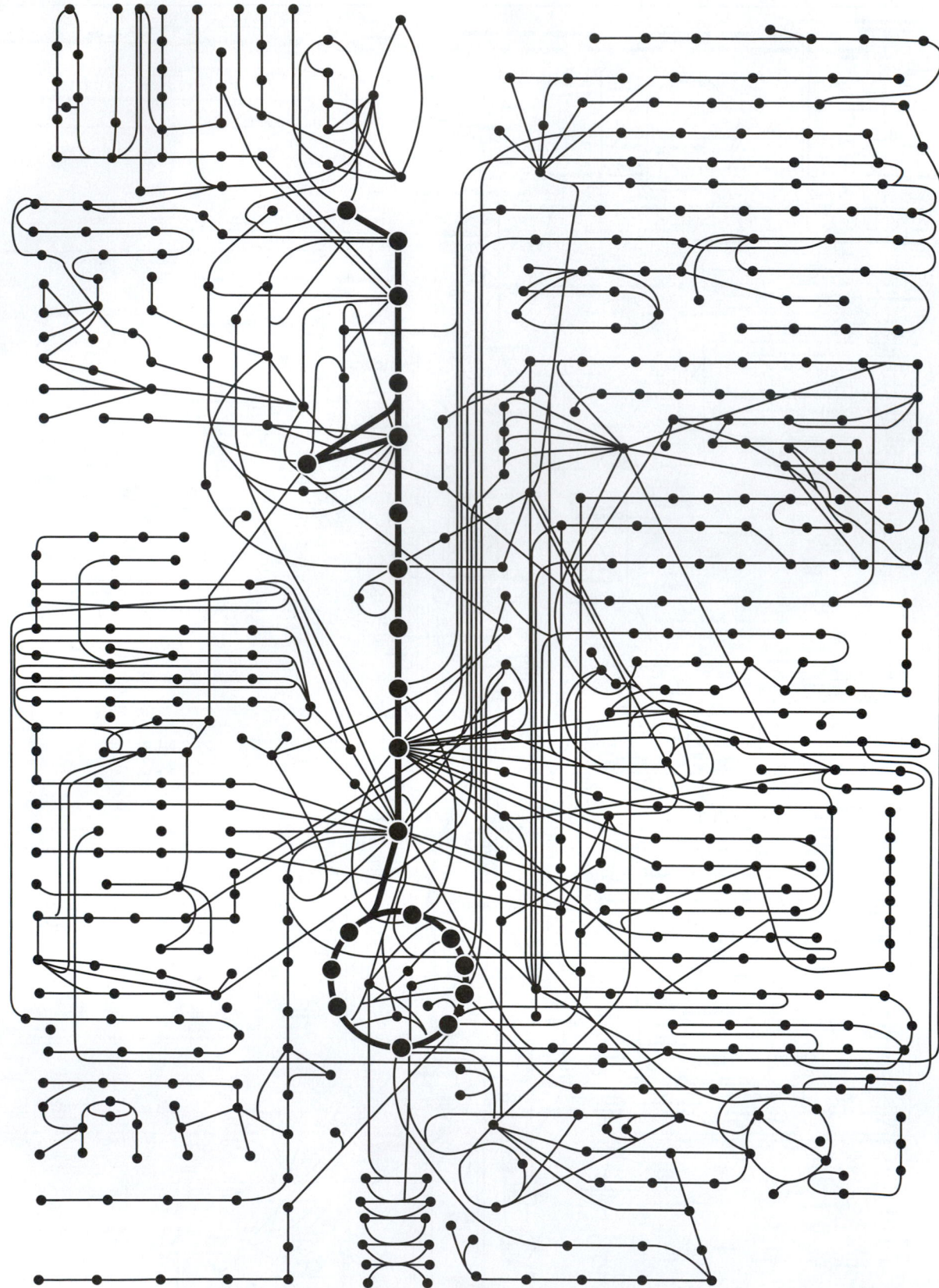

FIGURE 17.2 The metabolic map as a set of dots and lines. The heavy dots and lines trace the central energy-releasing pathways known as glycolysis and the citric acid cycle. *(Adapted from Alberts, B., et al., 1989. Molecular Biology of the Cell, 2nd ed. New York: Garland Publishing Co.)*

most unlikely. Yet it's true, and it provides strong evidence that all life has descended from a common ancestral form. All forms of nutrition and almost all metabolic pathways evolved in early prokaryotes prior to the appearance of eukaryotes 1 billion years ago. For example, **glycolysis,** the metabolic pathway by which energy is released from glucose and captured in the form of ATP under anaerobic conditions, is common to almost every cell. It is believed to be the most ancient of metabolic pathways, having arisen prior to the appearance of oxygen in abundance in the atmosphere. All organisms, even those that can synthesize their own glucose, are capable of glucose degradation and ATP synthesis via glycolysis. Other prominent pathways are also virtually ubiquitous among organisms.

Living Things Exhibit Metabolic Diversity

Although most cells have the same basic set of central metabolic pathways, different cells (and, by extension, different organisms) are characterized by the alternative pathways they might express. These pathways offer a wide diversity of metabolic possibilities. For instance, organisms are often classified according to the major metabolic pathways they exploit to obtain carbon or energy. Classification based on carbon requirements defines two major groups: autotrophs and heterotrophs. **Autotrophs** are organisms that can use just carbon dioxide as their sole source of carbon. **Heterotrophs** require an organic form of carbon, such as glucose, in order to synthesize other essential carbon compounds.

Classification based on energy sources also gives two groups: phototrophs and chemotrophs. **Phototrophs** are *photosynthetic organisms,* which use light as a source of energy. **Chemotrophs** use organic compounds such as glucose or, in some instances, oxidizable inorganic substances such as Fe^{2+}, NO_2^-, NH_4^+, or elemental sulfur as sole sources of energy. Typically, the energy is extracted through oxidation–reduction reactions. Based on these characteristics, every organism falls into one of four categories (Table 17.2).

Metabolic Diversity Among the Five Kingdoms   Prokaryotes (the kingdom Monera—bacteria) show a greater metabolic diversity than all the four eukaryotic kingdoms (Protoctista [previously called Protozoa], Fungi, Plants, and Animals) put together. Prokaryotes are variously chemoheterotrophic, photoautotrophic, photoheterotrophic, or chemoautotrophic. No protoctista are chemoautotrophs; fungi and animals are exclusively chemoheterotrophs; plants are characteristically photoautotrophs, although some are heterotrophic in their mode of carbon acquisition.

Table 17.1
Number of Dots (Intermediates) in the Metabolic Map of Figure 17.2, and the Number of Lines Associated with Them

Lines	Dots
1 or 2	410
3	71
4	20
5	11
6 or more	8

Table 17.2				
Metabolic Classification of Organisms According to Their Carbon and Energy Requirements				
Classification	**Carbon Source**	**Energy Source**	**Electron Donors**	**Examples**
Photoautotrophs	CO_2	Light	H_2O, H_2S, S, other inorganic compounds	Green plants, algae, cyanobacteria, photosynthetic bacteria
Photoheterotrophs	Organic compounds	Light	Organic compounds	Nonsulfur purple bacteria
Chemoautotrophs	CO_2	Oxidation–reduction reactions	Inorganic compounds: H_2, H_2S, NH_4^+, NO_2^-, Fe^{2+}, Mn^{2+}	Nitrifying bacteria; hydrogen, sulfur, and iron bacteria
Chemoheterotrophs	Organic compounds	Oxidation–reduction reactions	Organic compounds (e.g., glucose)	All animals, most microorganisms, nonphotosynthetic plant tissue such as roots, photosynthetic cells in the dark

A Deeper Look

Calcium Carbonate—A Biological Sink for CO_2

A major biological sink for CO_2 that is often overlooked is the calcium carbonate shells of corals, molluscs, and crustacea. These invertebrate animals deposit $CaCO_3$ in the form of protective exoskeletons. In some invertebrates, such as the *scleractinians* (hard corals) of tropical seas, photosynthetic dinoflagellates (kingdom Protoctista) known as *zooxanthellae* live within the ani-

mal cells as **endosymbionts.** These phototrophic cells use light to drive the resynthesis of organic molecules from CO_2 released (as bicarbonate ion) by the animal's metabolic activity. In the presence of Ca^{2+}, the photosynthetic CO_2 fixation "pulls" the deposition of $CaCO_3$, as summarized in the following coupled reactions:

$$Ca^{2+} + 2\ HCO_3^- \rightleftharpoons CaCO_{3(s)}\downarrow + H_2CO_3$$
$$H_2CO_3 \rightleftharpoons H_2O + CO_2$$
$$H_2O + CO_2 \longrightarrow \text{carbohydrate} + O_2$$

Oxygen Is Essential to Life for Aerobes

A further metabolic distinction among organisms is whether or not they can use oxygen as an electron acceptor in energy-producing pathways. Those that can are called **aerobes** or *aerobic organisms;* others, termed **anaerobes,** can subsist without O_2. Organisms for which O_2 is obligatory for life are called **obligate aerobes;** humans are an example. Some species, the so-called **facultative anaerobes,** can adapt to anaerobic conditions by substituting other electron acceptors for O_2 in their energy-producing pathways; *Escherichia coli* is an example. Yet others cannot use oxygen at all and are even poisoned by it; these are the **obligate anaerobes.** *Clostridium botulinum,* the bacterium that produces botulin toxin, is representative.

The Flow of Energy in the Biosphere and the Carbon and Oxygen Cycles Are Intimately Related

The primary source of energy for life is the sun. Photoautotrophs utilize light energy to drive the synthesis of organic molecules, such as carbohydrates, from atmospheric CO_2 and water (Figure 17.3). Heterotrophic cells then use these organic products of photosynthetic cells both as fuels and as building blocks, or precursors, for the biosynthesis of their own unique complement of biomolecules. Ultimately, CO_2 is the end product of heterotrophic carbon metabolism, and CO_2 is returned to the atmosphere for reuse by the photoautotrophs. In effect, solar energy is converted to the chemical energy of organic molecules by photoautotrophs, and heterotrophs recover this energy by metabolizing the organic substances. The flow of energy in the biosphere is thus conveyed within the carbon cycle, and the impetus driving the cycle is light energy.

17.2 | How Do Anabolic and Catabolic Processes Form the Core of Metabolic Pathways?

Metabolism serves two fundamentally different purposes: the generation of energy to drive vital functions and the synthesis of biological molecules. To achieve these ends, metabolism consists largely of two contrasting processes: catabolism and anabolism. *Catabolic pathways are characteristically energy yielding, whereas anabolic pathways are energy requiring.* **Catabolism** involves the oxidative degradation of complex nutrient molecules (carbohydrates, lipids, and proteins) obtained either from the environment or from cellular reserves. The breakdown of these molecules by catabolism leads to the formation of simpler molecules such as lactic acid, ethanol, carbon dioxide, urea, or ammonia. Catabolic reactions are usually exergonic, and often the chemical energy

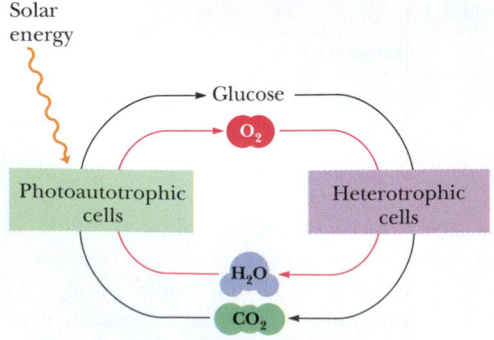

FIGURE 17.3 The flow of energy in the biosphere is coupled primarily to the carbon and oxygen cycles.

released is captured in the form of ATP (see Chapter 3). Because catabolism is oxidative for the most part, part of the chemical energy may be conserved as energy-rich electrons transferred to the coenzymes NAD^+ and $NADP^+$. These two reduced coenzymes have very different metabolic roles: NAD^+ reduction is part of catabolism; NADPH oxidation is an important aspect of anabolism. The energy released upon oxidation of NADH is coupled to the phosphorylation of ADP in aerobic cells, and so NADH oxidation back to NAD^+ serves to generate more ATP; in contrast, NADPH is the source of the reducing power needed to drive reductive biosynthetic reactions.

Thermodynamic considerations demand that the energy necessary for biosynthesis of any substance exceed the energy available from its catabolism. Otherwise, organisms could achieve the status of perpetual motion machines: A few molecules of substrate whose catabolism yielded more ATP than required for its resynthesis would allow the cell to cycle this substance and harvest an endless supply of energy.

Anabolism Is Biosynthesis

Anabolism is a synthetic process in which the varied and complex biomolecules (proteins, nucleic acids, polysaccharides, and lipids) are assembled from simpler precursors. Such biosynthesis involves the formation of new covalent bonds, and an input of chemical energy is necessary to drive such endergonic processes. The ATP generated by catabolism provides this energy. Furthermore, NADPH is an excellent donor of high-energy electrons for the reductive reactions of anabolism. Despite their divergent roles, anabolism and catabolism are interrelated in that the products of one provide the substrates of the other (Figure 17.4). Many metabolic intermediates are shared between the two processes, and the precursors needed by anabolic pathways are found among the products of catabolism.

Anabolism and Catabolism Are Not Mutually Exclusive

Interestingly, anabolism and catabolism occur simultaneously in the cell. The conflicting demands of concomitant catabolism and anabolism are managed by cells in two ways. First, the cell maintains tight and separate regulation of both catabolism and anabolism, so metabolic needs are served in an immediate and orderly fashion. Second, competing metabolic pathways are often localized within different cellular compartments. Isolating opposing activities

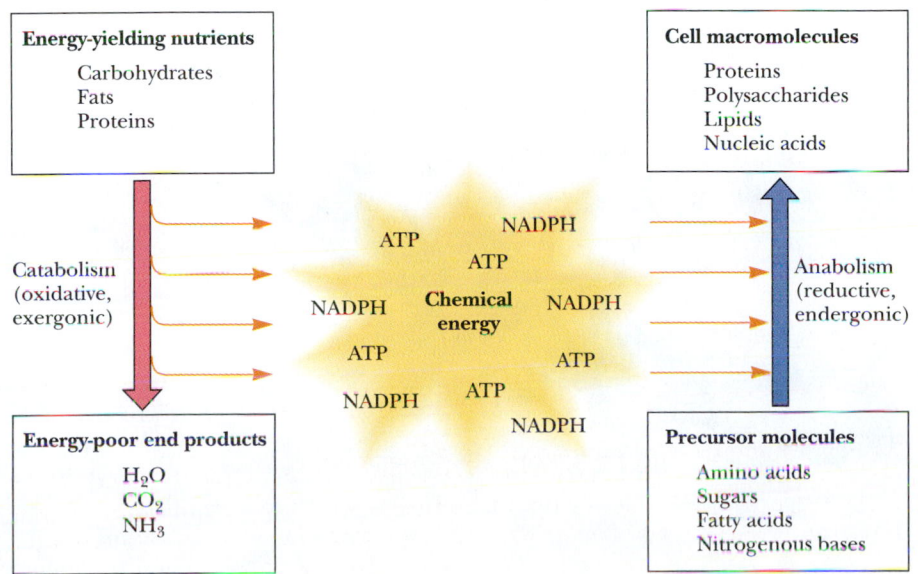

FIGURE 17.4 Energy relationships between the pathways of catabolism and anabolism. Oxidative, exergonic pathways of catabolism release free energy and reducing power that are captured in the form of ATP and NADPH, respectively. Anabolic processes are endergonic, consuming chemical energy in the form of ATP and using NADPH as a source of high-energy electrons for reductive purposes.

within distinct compartments, such as separate organelles, avoids interference between them. For example, the enzymes responsible for catabolism of fatty acids, the *fatty acid oxidation pathway,* are localized within mitochondria. In contrast, *fatty acid biosynthesis* takes place in the cytosol. In subsequent chapters, we shall see that the particular molecular interactions responsible for the regulation of metabolism become important for an understanding and appreciation of metabolic biochemistry.

Enzymes Are Organized into Metabolic Pathways

The individual metabolic pathways of anabolism and catabolism consist of sequential enzymatic steps (Figure 17.5). Several types of organization are possible. The enzymes of some multienzyme systems may exist as physically separate, soluble entities, with diffusing intermediates (Figure 17.5a). In other instances, the enzymes of a pathway are collected to form a discrete *multienzyme complex,* and the substrate is sequentially modified as it is passed along from enzyme to enzyme (Figure 17.5b). This type of organization has the advantage that intermediates are not lost or diluted by diffusion. In a third pattern of organization, the enzymes common to a pathway reside together as a *membrane-bound system* (Figure 17.5c). In this case, the enzyme participants (and perhaps the substrates as well) must diffuse in just the two dimensions of the membrane to interact with their neighbors.

As research reveals the ultrastructural organization of the cell in ever greater detail, more and more of the so-called soluble enzyme systems are found to be physically united into functional complexes. Thus, in many (perhaps all) metabolic pathways, the consecutively acting enzymes are associated into stable multienzyme complexes that are sometimes referred to as **metabolons,** a word meaning "units of metabolism."

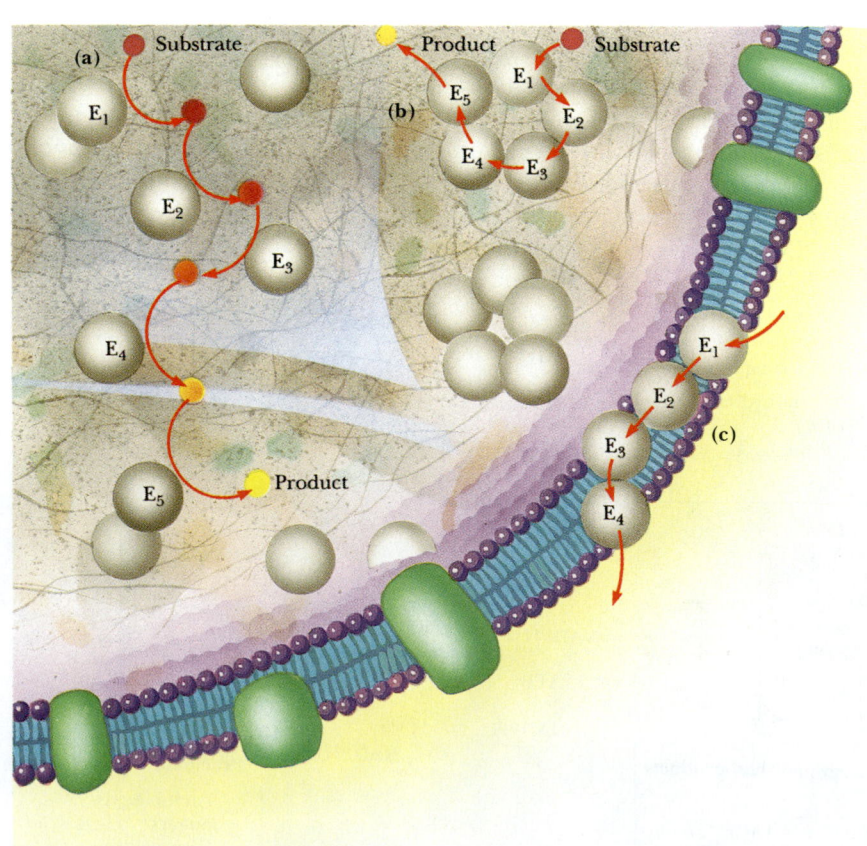

FIGURE 17.5 Schematic representation of types of multienzyme systems carrying out a metabolic pathway: **(a)** Physically separate, soluble enzymes with diffusing intermediates. **(b)** A multienzyme complex. Substrate enters the complex and becomes covalently bound and then sequentially modified by enzymes E_1 to E_5 before product is released. No intermediates are free to diffuse away. **(c)** A membrane-bound multienzyme system.

The Pathways of Catabolism Converge to a Few End Products

If we survey the catabolism of the principal energy-yielding nutrients (carbohydrates, lipids, and proteins) in a typical heterotrophic cell, we see that the degradation of these substances involves a succession of enzymatic reactions. In the presence of oxygen *(aerobic catabolism)*, these molecules are degraded ultimately to carbon dioxide, water, and ammonia. Aerobic catabolism consists of three distinct stages. In **stage 1,** the nutrient macromolecules are broken down into their respective building blocks. Despite the great diversity of macromolecules, these building blocks represent a rather limited number of products. Proteins yield up their 20 component amino acids, polysaccharides give rise to carbohydrate units that are convertible to glucose, and lipids are broken down into glycerol and fatty acids (Figure 17.6).

In **stage 2,** the collection of product building blocks generated in stage 1 is further degraded to yield an even more limited set of simpler metabolic intermediates. The deamination of amino acids leaves α-keto acid carbon skeletons. Several of these α-keto acids are citric acid cycle intermediates and are fed directly into stage 3 catabolism via this cycle. Others are converted either to the three-carbon α-keto acid *pyruvate* or to the acetyl groups of *acetyl-coenzyme* A (acetyl-CoA). Glucose and the glycerol from lipids also generate pyruvate, whereas the fatty acids are broken into two-carbon units that appear as *acetyl-CoA*. Because pyruvate also gives rise to acetyl-CoA, we see that the degradation of macromolecular nutrients converges to a common end product, acetyl-CoA (Figure 17.6).

The combustion of the acetyl groups of acetyl-CoA by the citric acid cycle and *oxidative phosphorylation* to produce CO_2 and H_2O represents **stage 3** of catabolism. The end products of the citric acid cycle, CO_2 and H_2O, are the ultimate waste products of aerobic catabolism. As we shall see in Chapter 19, the oxidation of acetyl-CoA during stage 3 metabolism generates most of the energy produced by the cell.

Anabolic Pathways Diverge, Synthesizing an Astounding Variety of Biomolecules from a Limited Set of Building Blocks

A rather limited collection of simple precursor molecules is sufficient to provide for the biosynthesis of virtually any cellular constituent, be it protein, nucleic acid, lipid, or polysaccharide. All of these substances are constructed from appropriate building blocks via the pathways of anabolism. In turn, the building blocks (amino acids, nucleotides, sugars, and fatty acids) can be generated from metabolites in the cell. For example, amino acids can be formed by amination of the corresponding α-keto acid carbon skeletons, and pyruvate can be converted to hexoses for polysaccharide biosynthesis.

Amphibolic Intermediates Play Dual Roles

Certain of the central pathways of intermediary metabolism, such as the citric acid cycle, and many metabolites of other pathways have dual purposes—they serve in both catabolism and anabolism. This dual nature is reflected in the designation of such pathways as **amphibolic** rather than solely catabolic or anabolic. In any event, in contrast to catabolism—which converges to the common intermediate, acetyl-CoA—the pathways of anabolism diverge from a small group of simple metabolic intermediates to yield a spectacular variety of cellular constituents.

Amphi is from the Greek for "on both sides."

Corresponding Pathways of Catabolism and Anabolism Differ in Important Ways

The anabolic pathway for synthesis of a given end product usually does not precisely match the pathway used for catabolism of the same substance. Some of the intermediates may be common to steps in both pathways, but different enzymatic reactions and unique metabolites characterize other steps. A good example of

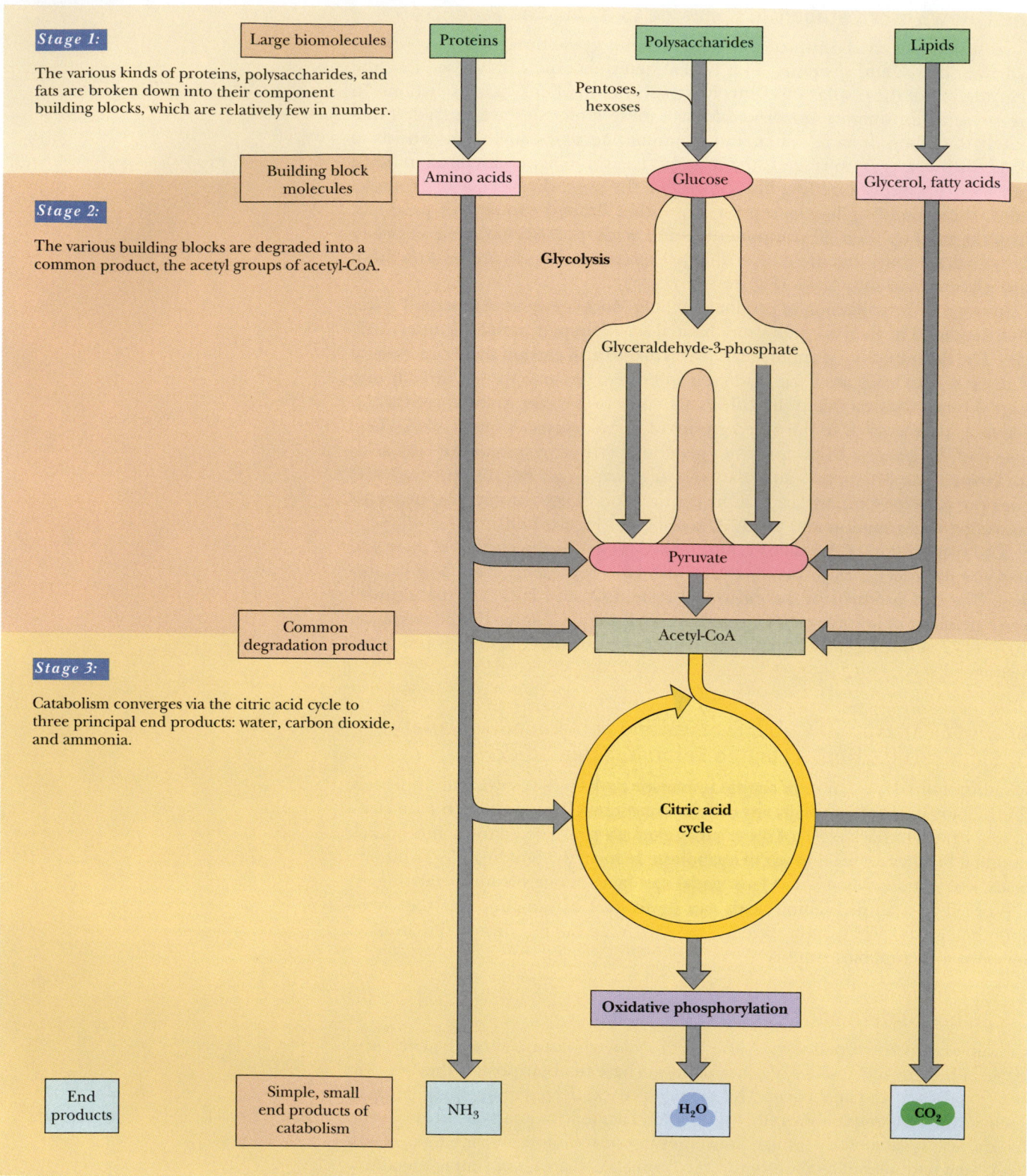

FIGURE 17.6 The three stages of catabolism.
Stage 1: Proteins, polysaccharides, and lipids are
broken down into their component building blocks,
which are relatively few in number. **Stage 2:** The
various building blocks are degraded into the
common product, the acetyl groups of acetyl-CoA.
Stage 3: Catabolism converges to three principal
end products: water, carbon dioxide, and ammonia.

these differences is found in a comparison of the catabolism of glucose to pyruvic acid by the pathway of glycolysis and the biosynthesis of glucose from pyruvate
by the pathway called *gluconeogenesis*. The glycolytic pathway from glucose to
pyruvate consists of ten enzymes. Although it may seem efficient for glucose synthesis from pyruvate to proceed by a reversal of all ten steps, gluconeogenesis
uses only seven of the glycolytic enzymes in reverse, replacing the remaining

Labels within Figure 17.6:

Stage 1:
The various kinds of proteins, polysaccharides, and fats are broken down into their component building blocks, which are relatively few in number.

Large biomolecules

Building block molecules

Stage 2:
The various building blocks are degraded into a common product, the acetyl groups of acetyl-CoA.

Common degradation product

Stage 3:
Catabolism converges via the citric acid cycle to three principal end products: water, carbon dioxide, and ammonia.

End products

Simple, small end products of catabolism

Proteins   Polysaccharides   Lipids

Pentoses, hexoses

Amino acids   Glucose   Glycerol, fatty acids

Glycolysis

Glyceraldehyde-3-phosphate

Pyruvate

Acetyl-CoA

Citric acid cycle

Oxidative phosphorylation

NH_3 H_2O CO_2

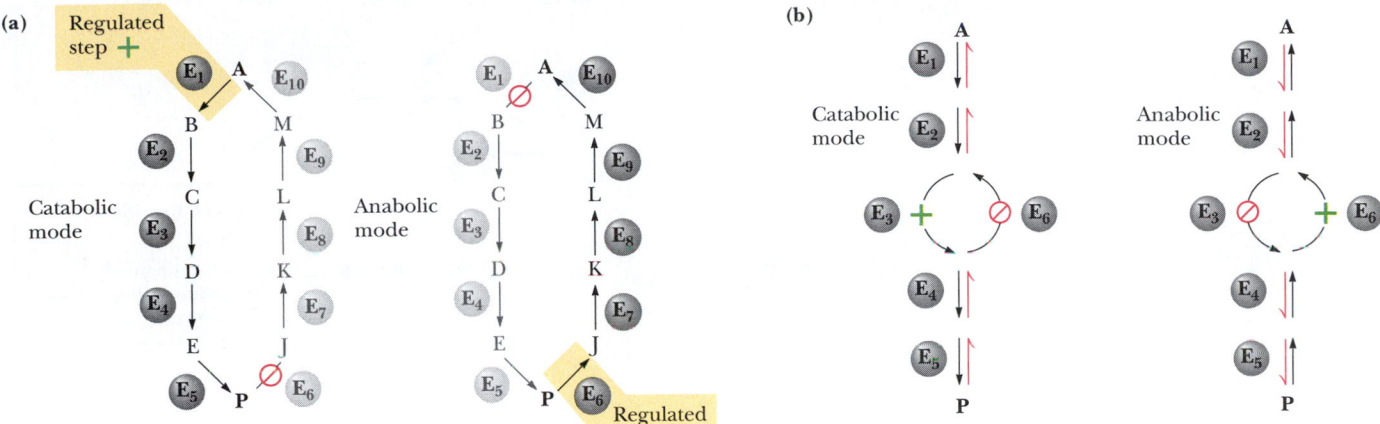

Activation of one mode is accompanied by reciprocal inhibition of the other mode.

FIGURE 17.7 Parallel pathways of catabolism and anabolism must differ in at least one metabolic step in order that they can be regulated independently. Shown here are two possible arrangements of opposing catabolic and anabolic sequences between A and P. **(a)** The parallel sequences proceed via independent routes. **(b)** Only one reaction has two different enzymes, a catabolic one (E_3) and its anabolic counterpart (E_6). These provide sites for regulation.

three with four enzymes specific to glucose biosynthesis. In similar fashion, the pathway responsible for degrading proteins to amino acids differs from the protein synthesis system, and the oxidative degradation of fatty acids to two-carbon acetyl-CoA groups does not follow the same reaction path as the biosynthesis of fatty acids from acetyl-CoA.

Metabolic Regulation Requires Different Pathways for Oppositely Directed Metabolic Sequences A second reason for different pathways serving in opposite metabolic directions is that such pathways must be independently regulated. If catabolism and anabolism passed along the same set of metabolic tracks, equilibrium considerations would dictate that slowing the traffic in one direction by inhibiting a particular enzymatic reaction would necessarily slow traffic in the opposite direction. Independent regulation of anabolism and catabolism can be accomplished only if these two contrasting processes move along different routes or, in the case of shared pathways, the rate-limiting steps serving as the points of regulation are catalyzed by enzymes that are unique to each opposing sequence (Figure 17.7).

ATP Serves in a Cellular Energy Cycle

We saw in Chapter 3 that ATP is the energy currency of cells. In phototrophs, ATP is one of the two energy-rich primary products resulting from the transformation of light energy into chemical energy. (The other is NADPH; see the following discussion.) In heterotrophs, the pathways of catabolism have as their major purpose the release of free energy that can be captured in the form of energy-rich phosphoric anhydride bonds in ATP. In turn, ATP provides the energy that drives the manifold activities of all living cells—the synthesis of complex biomolecules, the osmotic work involved in transporting substances into cells, the work of cell motility, the work of muscle contraction. These diverse activities are all powered by energy released in the hydrolysis of ATP to ADP and P_i. Thus, there is an energy cycle in cells where ATP serves as the vessel carrying energy from photosynthesis or catabolism to the energy-requiring processes unique to living cells (Figure 17.8).

NAD$^+$ Collects Electrons Released in Catabolism

The substrates of catabolism—proteins, carbohydrates, and lipids—are good sources of chemical energy because the carbon atoms in these molecules are in a relatively reduced state (Figure 17.9). In the oxidative reactions of catabolism,

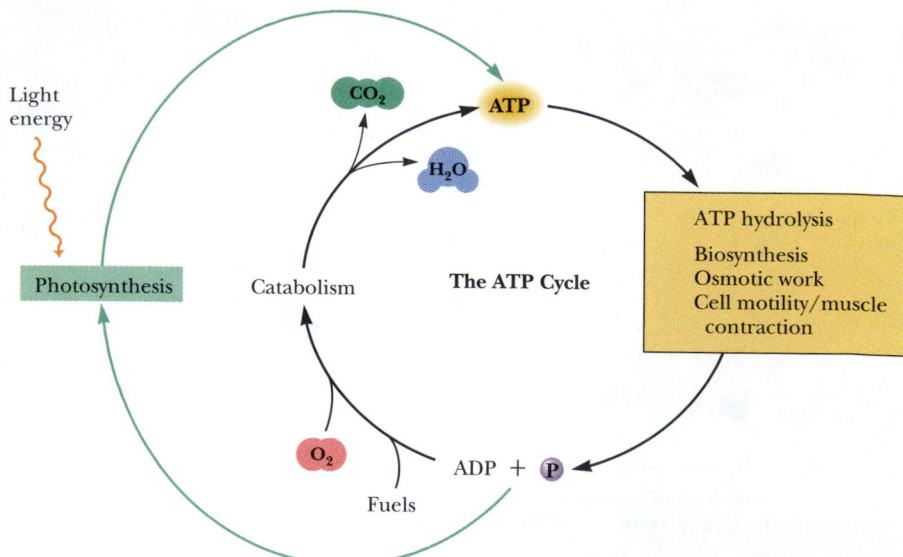

FIGURE 17.8 The ATP cycle in cells. ATP is formed via photosynthesis in phototrophic cells or catabolism in heterotrophic cells. Energy-requiring cellular activities are powered by ATP hydrolysis, liberating ADP and P_i.

reducing equivalents are released from these substrates, often in the form of **hydride ions** (a proton coupled with two electrons, $H:^-$). These hydride ions are transferred in enzymatic **dehydrogenase** reactions from the substrates to NAD^+ molecules, reducing them to NADH. A second proton accompanies these reactions, appearing in the overall equation as H^+ (Figure 17.10). In turn, NADH is oxidized back to NAD^+ when it transfers its reducing equivalents to electron acceptor systems that are part of the metabolic apparatus of the mitochondria. The ultimate oxidizing agent (e^- acceptor) is O_2, becoming reduced to H_2O.

Oxidation reactions are exergonic, and the energy released is coupled with the formation of ATP in a process called **oxidative phosphorylation.** The NAD^+–NADH system can be viewed as a *shuttle* that carries the electrons released from catabolic substrates to the mitochondria, where they are transferred to O_2, the ultimate electron acceptor in catabolism. In the process, the free energy released is trapped in ATP. The NADH cycle is an important player in the transformation of the chemical energy of carbon compounds into the chemical energy of phosphoric anhydride bonds. Such transformations of energy from one form to another are referred to as **energy transduction.** Oxidative phosphorylation is one cellular mechanism for energy transduction. Chapter 20 is devoted to electron transport reactions and oxidative phosphorylation.

NADPH Provides the Reducing Power for Anabolic Processes

Whereas catabolism is fundamentally an oxidative process, anabolism is, by its contrasting nature, reductive. The biosynthesis of the complex constituents of the cell begins at the level of intermediates derived from the degradative pathways of catabolism; or, less commonly, biosynthesis begins with oxidized substances available in the inanimate environment, such as carbon dioxide. When the hydrocarbon chains of fatty acids are assembled from acetyl-CoA units, activated hydrogens are needed to reduce the carbonyl (C=O) carbon of acetyl-CoA into a —CH₂— at every other position along the chain. When glucose is synthesized from CO_2 during photosynthesis in plants, reducing power is

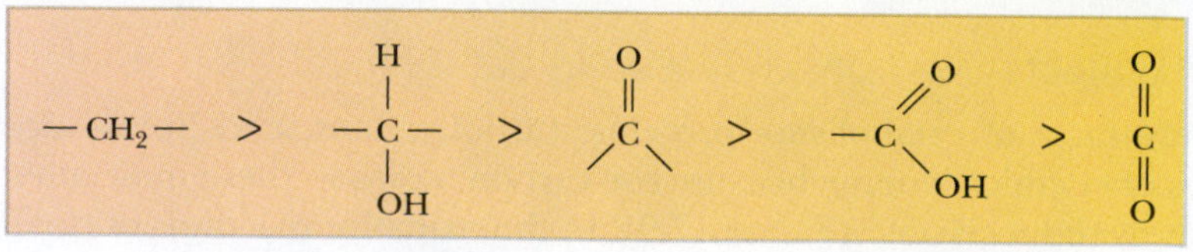

FIGURE 17.9 Comparison of the state of reduction of carbon atoms in biomolecules: —CH₂— (fats) > —CHOH— (carbohydrates) > C=O (carbonyls) > —COOH (carboxyls) > CO_2 (carbon dioxide, the final product of catabolism).

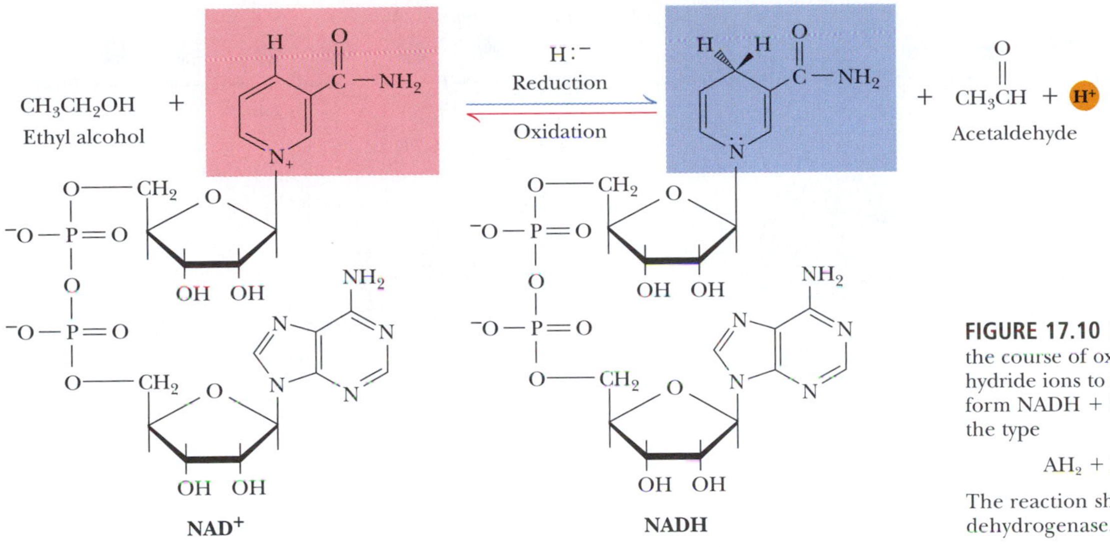

$$CH_3CH_2OH \quad + \qquad\qquad\qquad\qquad \xrightleftharpoons[\text{Oxidation}]{\substack{H:^- \\ \text{Reduction}}} \qquad\qquad\qquad + \quad CH_3CH + H^+$$

Ethyl alcohol

Acetaldehyde

NAD$^+$

NADH

FIGURE 17.10 Hydrogen and electrons released in the course of oxidative catabolism are transferred as hydride ions to the pyridine nucleotide, NAD$^+$, to form NADH + H$^+$ in dehydrogenase reactions of the type

$$AH_2 + NAD^+ \longrightarrow A + NADH + H^+$$

The reaction shown is catalyzed by alcohol dehydrogenase.

required. These reducing equivalents are provided by NADPH, the usual source of high-energy hydrogens for reductive biosynthesis. NADPH is generated when NADP$^+$ is reduced with electrons in the form of hydride ions. In heterotrophic organisms, these electrons are removed from fuel molecules by NADP$^+$-specific dehydrogenases. In these organisms, NADPH can be viewed as the carrier of electrons from catabolic reactions to anabolic reactions (Figure 17.11). In photosynthetic organisms, the energy of light is used to pull electrons from water and transfer them to NADP$^+$; O$_2$ is a by-product of this process.

17.3 What Experiments Can Be Used to Elucidate Metabolic Pathways?

Armed with the knowledge that metabolism is organized into pathways of successive reactions, we can appreciate by hindsight the techniques employed by early biochemists to reveal their sequence. A major intellectual advance took place at the end of the 19th century when Eduard Buchner showed that the fermentation of glucose to yield ethanol and carbon dioxide can occur in extracts of broken yeast cells. Until this discovery, many thought that metabolism was a vital property, unique to intact cells; even the eminent microbiologist Louis Pasteur, who contributed so much to our understanding of fermentation, was a *vitalist*, one of those who believed that the processes of living substance transcend the laws of chemistry and physics. After Buchner's revelation, biochemists searched for intermediates in the transformation of glucose and soon learned that inorganic phosphate was essential to glucose breakdown. This observation gradually led to the discovery of a variety of phosphorylated organic compounds that serve as intermediates along the fermentative pathway.

An important tool for elucidating the steps in the pathway was the use of *metabolic inhibitors*. Adding an enzyme inhibitor to a cell-free extract caused an accumulation of intermediates in the pathway prior to the point of inhibition (Figure 17.12). Each inhibitor was specific for a particular site in the sequence of metabolic events. As the arsenal of inhibitors was expanded, the individual steps in metabolism were revealed.

Mutations Create Specific Metabolic Blocks

Genetics provides an approach to the identification of intermediate steps in metabolism that is somewhat analogous to inhibition. Mutation in a gene encoding an enzyme often results in an inability to synthesize the enzyme in an

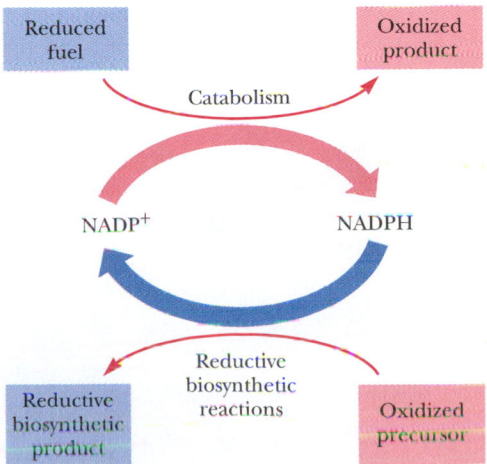

FIGURE 17.11 Transfer of reducing equivalents from catabolism to anabolism via the NADPH cycle.

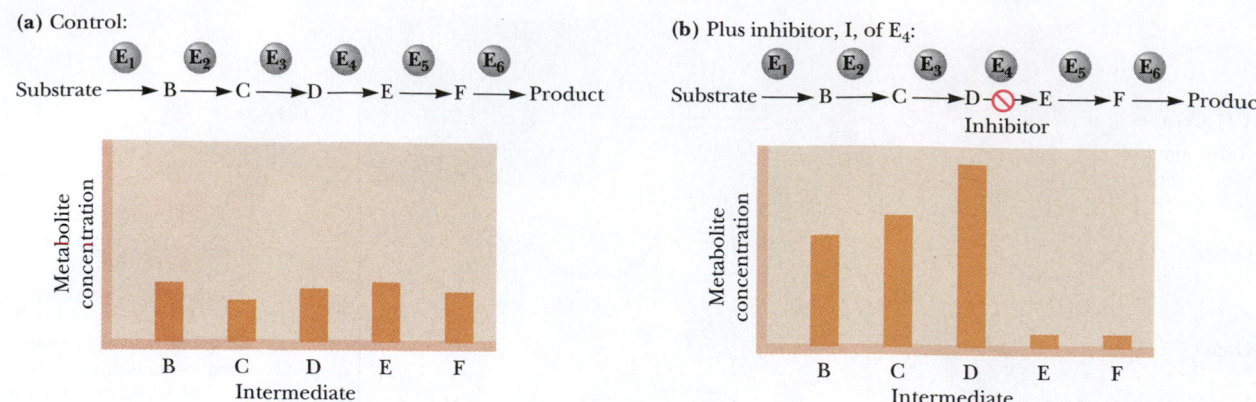

FIGURE 17.12 The use of inhibitors to reveal the sequence of reactions in a metabolic pathway. **(a) Control:** Under normal conditions, the steady-state concentrations of a series of intermediates will be determined by the relative activities of the enzymes in the pathway. **(b) Plus inhibitor:** In the presence of an inhibitor (in this case, an inhibitor of *enzyme 4*), intermediates upstream of the metabolic block (B, C, and D) accumulate, revealing themselves as intermediates in the pathway. The concentration of intermediates lying downstream (E and F) will fall.

active form. Such a defect leads to a block in the metabolic pathway at the point where the enzyme acts, and the enzyme's substrate accumulates. Such genetic disorders are lethal if the end product of the pathway is essential or if the accumulated intermediates have toxic effects. In microorganisms, however, it is often possible to manipulate the growth medium so that essential end products are provided. Then the biochemical consequences of the mutation can be investigated. Studies on mutations in genes of the filamentous fungus *Neurospora crassa* led G. W. Beadle and E. L. Tatum to hypothesize in 1941 that genes are units of heredity that encode enzymes (a principle referred to as the "one gene–one enzyme" hypothesis).

Isotopic Tracers Can Be Used as Metabolic Probes

Another widely used approach to the elucidation of metabolic sequences is to "feed" cells a substrate or metabolic intermediate labeled with a particular isotopic form of an element that can be traced. Two sorts of isotopes are useful in this regard: radioactive isotopes, such as ^{14}C, and stable "heavy" isotopes, such as ^{18}O or ^{15}N (Table 17.3). Because the chemical behavior of isotopically

Table 17.3

Properties of Radioactive and Stable "Heavy" Isotopes Used as Tracers in Metabolic Studies

Isotope	Type	Radiation Type	Half-Life	Relative Abundance*
^2H	Stable			0.0154%
^3H	Radioactive	β^-	12.1 years	
^{13}C	Stable			1.1%
^{14}C	Radioactive	β^-	5700 years	
^{15}N	Stable			0.365%
^{18}O	Stable			0.204%
^{24}Na	Radioactive	β^-, γ	15 hours	
^{32}P	Radioactive	β^-	14.3 days	
^{35}S	Radioactive	β^-	87.1 days	
^{36}Cl	Radioactive	β^-	310,000 years	
^{42}K	Radioactive	β^-	12.5 hours	
^{45}Ca	Radioactive	β^-	152 days	
^{59}Fe	Radioactive	β^-, γ	45 days	
^{131}I	Radioactive	β^-, γ	8 days	

*The relative natural abundance of a stable isotope is important because, in tracer studies, the amount of stable isotope is typically expressed in terms of atoms percent excess over the natural abundance of the isotope.

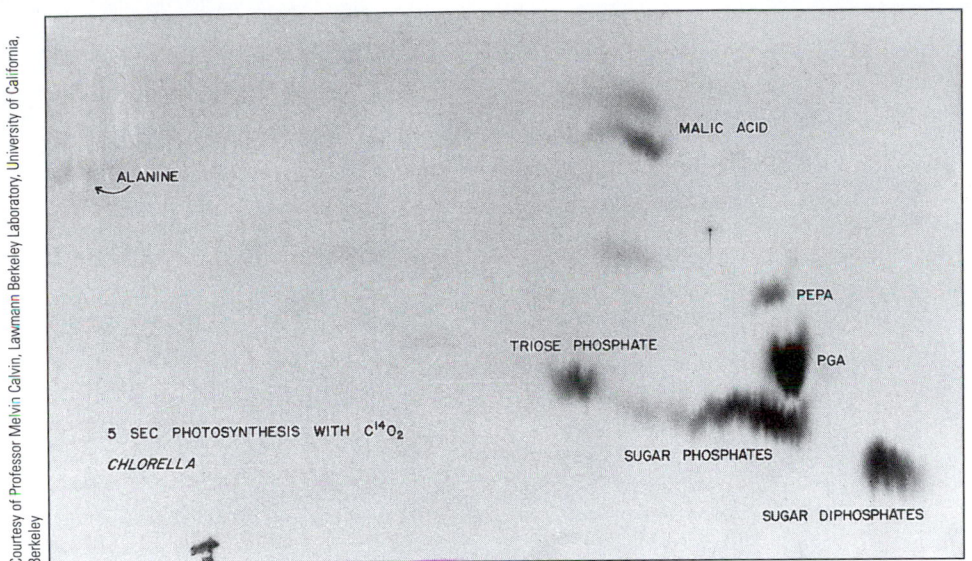

FIGURE 17.13 One of the earliest experiments using a radioactive isotope as a metabolic tracer. Cells of *Chlorella* (a green alga) synthesizing carbohydrate from carbon dioxide were exposed briefly (5 sec) to ^{14}C-labeled CO_2. The products of CO_2 incorporation were then quickly isolated from the cells, separated by two-dimensional paper chromatography, and observed via autoradiographic exposure of the chromatogram. Such experiments identified radioactive 3-phosphoglycerate (PGA) as the primary product of CO_2 fixation. The 3-phosphoglycerate was labeled in the 1-position (in its carboxyl group). Radioactive compounds arising from the conversion of 3-phosphoglycerate to other metabolic intermediates included phosphoenolpyruvate (PEP), malic acid, triose phosphate, alanine, and sugar phosphates and diphosphates.

labeled compounds is rarely distinguishable from that of their unlabeled counterparts, isotopes provide reliable "tags" for observing metabolic changes. The metabolic fate of a radioactively labeled substance can be traced by determining the presence and position of the radioactive atoms in intermediates derived from the labeled compound (Figure 17.13).

Heavy Isotopes Heavy isotopes endow the compounds in which they appear with slightly greater masses than their unlabeled counterparts. These compounds can be separated and quantitated by mass spectrometry (or density gradient centrifugation, if they are macromolecules). For example, ^{18}O was used in separate experiments as a tracer of the fate of the oxygen atoms in water and carbon dioxide to determine whether the atmospheric oxygen produced in photosynthesis arose from H_2O, CO_2, or both:

$$CO_2 + H_2O \longrightarrow (CH_2O) + O_2$$

If ^{18}O-labeled CO_2 was presented to a green plant carrying out photosynthesis, none of the ^{18}O was found in O_2. Curiously, it was recovered as $H_2{}^{18}$O. In contrast, when plants fixing CO_2 were equilibrated with $H_2{}^{18}$O, $^{18}O_2$ was evolved. These latter labeling experiments established that photosynthesis is best described by the equation

$$C^{16}O_2 + 2\ H_2{}^{18}O \longrightarrow (CH_2{}^{16}O) + {}^{18}O_2 + H_2{}^{16}O$$

That is, in the process of photosynthesis, the two oxygen atoms in O_2 come from two H_2O molecules. One O is lost from CO_2 and appears in H_2O, and the other O of CO_2 is retained in the carbohydrate product. Two of the four H atoms are accounted for in (CH_2O), and two reduce the O lost from CO_2 to H_2O.

NMR Spectroscopy Is a Noninvasive Metabolic Probe

A technology analogous to isotopic tracers is provided by **nuclear magnetic resonance (NMR) spectroscopy.** The atomic nuclei of certain isotopes, such as the naturally occurring isotope of phosphorus, ^{31}P, have *magnetic moments*. The resonance frequency of a magnetic moment is influenced by the local chemical environment. That is, the NMR signal of the nucleus is influenced in an identifiable way by the chemical nature of its neighboring atoms in the compound. In many ways, these nuclei are ideal tracers because their signals contain a great deal of structural information about the environment around the atom and

(a)

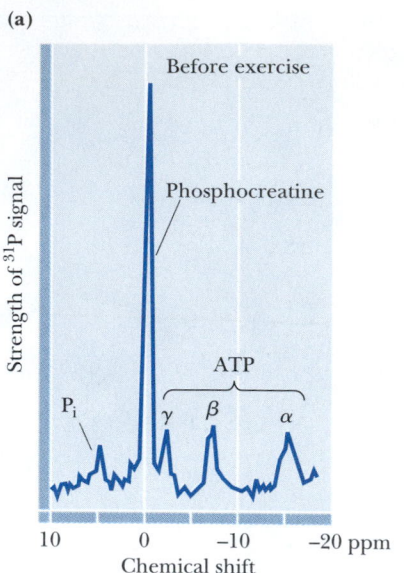

(b)

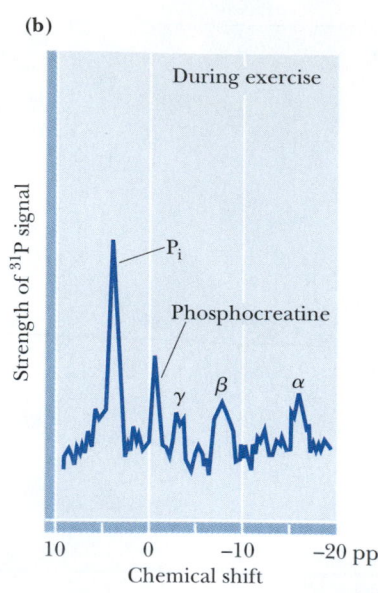

FIGURE 17.14 With NMR spectroscopy, one can observe the metabolism of a living subject in real time. These NMR spectra show the changes in ATP, creatine-P (phosphocreatine), and P_i levels in the forearm muscle of a human subjected to 19 minutes of exercise. Note that the three P atoms of ATP (α, β, and γ) have different chemical shifts, reflecting their different chemical environments.

thus the nature of the compound containing the atom. Transformations of substrates and metabolic intermediates labeled with magnetic nuclei can be traced by following changes in NMR spectra. Furthermore, NMR spectroscopy is a noninvasive procedure. Whole-body NMR spectrometers are being used today in hospitals to directly observe the metabolism (and clinical condition) of living subjects (Figure 17.14). NMR promises to be a revolutionary tool for clinical diagnosis and for the investigation of metabolism in situ (literally "in site," meaning, in this case, "where and as it happens").

Metabolic Pathways Are Compartmentalized Within Cells

Although the interior of a prokaryotic cell is not subdivided into compartments by internal membranes, the cell still shows some segregation of metabolism. For example, certain metabolic pathways, such as phospholipid synthesis and oxidative phosphorylation, are localized in the plasma membrane. Protein biosynthesis is carried out on ribosomes.

In contrast, eukaryotic cells are extensively compartmentalized by an endomembrane system. Each of these cells has a true nucleus bounded by a double membrane called the *nuclear envelope*. The nuclear envelope is continuous with the endomembrane system, which is composed of differentiated regions: the endoplasmic reticulum; the Golgi complex; various membrane-bounded vesicles such as lysosomes, vacuoles, and microbodies; and, ultimately, the plasma membrane itself. Eukaryotic cells also possess mitochondria and, if they are photosynthetic, chloroplasts. Disruption of the cell membrane and fractionation of the cell contents into the component organelles have allowed an analysis of their respective functions (Figure 17.15). Each compartment is dedicated to specialized metabolic functions, and the enzymes appropriate to these specialized functions are confined together within the organelle. In many instances, the enzymes of a metabolic sequence occur together within the organellar membrane. Thus, *the flow of metabolic intermediates in the cell is spatially as well as chemically segregated.* For example, the ten enzymes of glycolysis are found in the cytosol, but pyruvate, the product of glycolysis, is fed into the mitochondria. These organelles contain the citric acid cycle enzymes, which oxidize pyruvate to CO_2. The great amount of energy released in the process is captured by the oxidative phosphorylation system of mitochondrial membranes and used to drive the formation of ATP (Figure 17.16).

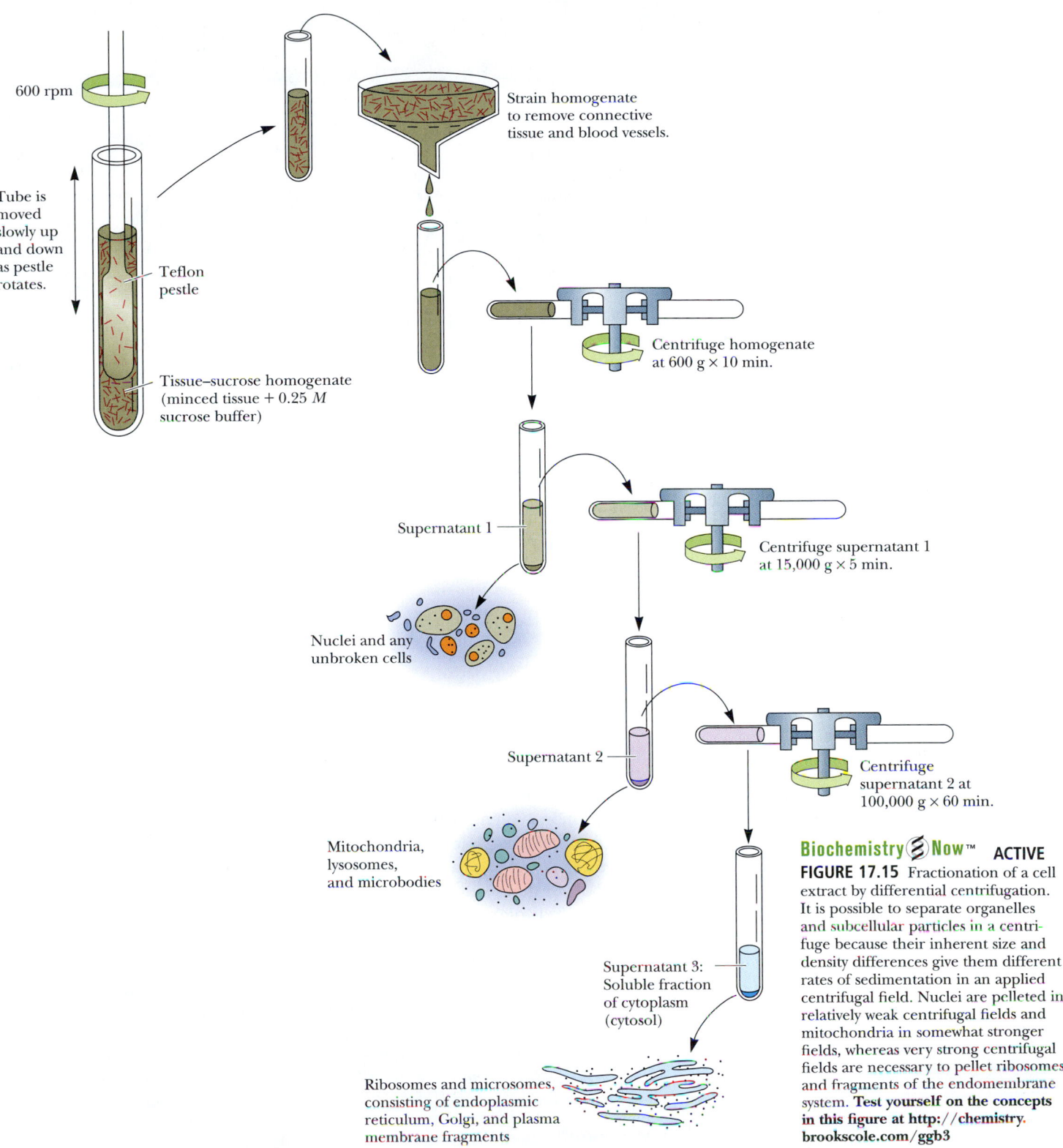

600 rpm

Tube is moved slowly up and down as pestle rotates.

Teflon pestle

Tissue–sucrose homogenate (minced tissue + 0.25 M sucrose buffer)

Strain homogenate to remove connective tissue and blood vessels.

Centrifuge homogenate at 600 g × 10 min.

Supernatant 1

Centrifuge supernatant 1 at 15,000 g × 5 min.

Nuclei and any unbroken cells

Supernatant 2

Centrifuge supernatant 2 at 100,000 g × 60 min.

Mitochondria, lysosomes, and microbodies

Supernatant 3: Soluble fraction of cytoplasm (cytosol)

Ribosomes and microsomes, consisting of endoplasmic reticulum, Golgi, and plasma membrane fragments

Biochemistry Now™ ACTIVE
FIGURE 17.15 Fractionation of a cell extract by differential centrifugation. It is possible to separate organelles and subcellular particles in a centrifuge because their inherent size and density differences give them different rates of sedimentation in an applied centrifugal field. Nuclei are pelleted in relatively weak centrifugal fields and mitochondria in somewhat stronger fields, whereas very strong centrifugal fields are necessary to pellet ribosomes and fragments of the endomembrane system. **Test yourself on the concepts in this figure at http://chemistry. brookscole.com/ggb3**

17.4 What Food Substances Form the Basis of Human Nutrition?

The use of foods by organisms is termed **nutrition.** The ability of an organism to use a particular food material depends upon its chemical composition and upon the metabolic pathways available to the organism. In addition to essential

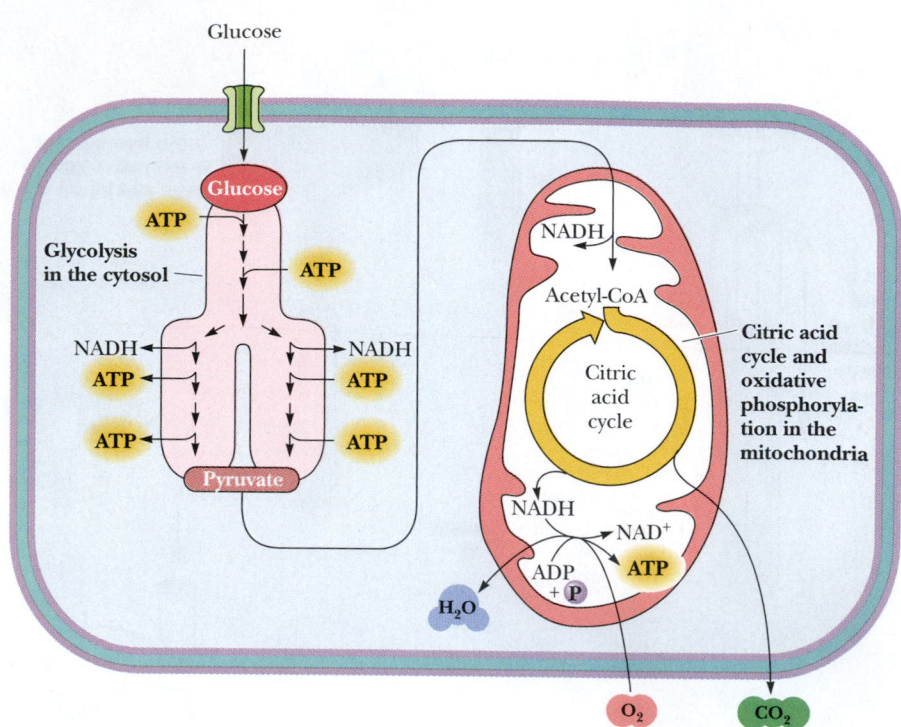

FIGURE 17.16 Compartmentalization of glycolysis, the citric acid cycle, and oxidative phosphorylation.

fiber, food includes the macronutrients—protein, carbohydrate, and lipid—and the micronutrients—including vitamins and minerals.

Humans Require Protein

Humans must consume protein in order to make new proteins. Dietary protein is a rich source of nitrogen, and certain amino acids—the so-called **essential amino acids**—cannot be synthesized by humans (and other animals) and can be obtained only in the diet. The average adult in the United States consumes far more protein than required for synthesis of essential proteins. Excess dietary protein is then merely a source of metabolic energy. Some of the amino acids (termed **glucogenic**) can be converted into glucose, whereas others, the **ketogenic** amino acids, can be converted to fatty acids and/or keto acids. If fat and carbohydrate are already adequate for the energy needs of the individual, then both kinds of amino acids will be converted to triacylglycerol and stored in adipose tissue.

An individual's protein undergoes a constant process of degradation and resynthesis. Together with dietary protein, this recycled protein material participates in a nitrogen equilibrium, or **nitrogen balance.** A positive nitrogen balance occurs whenever there is a net increase in the organism's protein content, such as during periods of growth. A negative nitrogen balance exists when dietary intake of nitrogen is insufficient to meet the demands for new protein synthesis.

Carbohydrates Provide Metabolic Energy

The principal purpose of carbohydrate in the diet is production of metabolic energy. Simple sugars are metabolized in the glycolytic pathway (see Chapter 18). Complex carbohydrates are degraded into simple sugars, which then can enter the glycolytic pathway. Carbohydrates are also essential components of nucleotides, nucleic acids, glycoproteins, and glycolipids. Human metabolism can adapt to a wide range of dietary carbohydrate levels, but the brain requires glucose for fuel. When dietary carbohydrate consumption exceeds the energy needs of the individual, excess carbohydrate is converted to triacylglycerols and glycogen for long-term energy storage. On the other hand, when dietary

A Deeper Look

A Popular Fad Diet—Low Carbohydrates, High Protein, High Fat

Possibly the most serious nutrition problem in the United States is excessive food consumption, and many people have experimented with fad diets in the hope of losing excess weight. One of the most popular of the fad diets has been the high-protein, high-fat (low-carbohydrate) diet. The premise for such diets is tantalizing: Because the tricarboxylic acid (TCA) cycle (see Chapter 20) is the primary site of fat metabolism and because glucose is usually needed to replenish intermediates in the TCA cycle, if carbohydrates are restricted in the diet, dietary fat should merely be converted to ketone bodies and excreted. This so-called diet appears to work at first because a low-carbohydrate diet results in an initial water (and weight) loss. This occurs because glycogen

reserves are depleted by the diet and because about 3 grams of water of hydration are lost for every gram of glycogen.

However, the rationale for this diet is problematic for several reasons. First, ketone body excretion by the human body usually does not exceed 20 grams (400 kJ) per day. Second, amino acids can function effectively to replenish TCA cycle intermediates, making the reduced carbohydrate regimen irrelevant. Third, the typical fare in a high-protein, high-fat, low-carbohydrate diet is expensive but not very tasty, and it is thus difficult to maintain. Finally, a diet high in saturated and *trans* fatty acids is a high risk factor for atherosclerosis and coronary artery disease.

carbohydrate intake is low, **ketone bodies** are formed from acetate units to provide metabolic fuel for the brain and other organs.

Lipids Are Essential, But in Moderation

Fatty acids and triacylglycerols can be used as fuel by many tissues in the human body, and phospholipids are essential components of all biological membranes. Even though the human body can tolerate a wide range of fat intake levels, there are disadvantages in either extreme. Excess dietary fat is stored as triacylglycerols in adipose tissue, but high levels of dietary fat can also increase the risk of atherosclerosis and heart disease. Moreover, high dietary fat levels are also correlated with increased risk for colon, breast, and prostate cancers. When dietary fat consumption is low, there is a risk of **essential fatty acid** deficiencies. As will be seen in Chapter 24, the human body cannot synthesize linoleic and linolenic acids, so these must be acquired in the diet. In addition, arachidonic acid can by synthesized in humans only from linoleic acid, so it too is classified as essential. The essential fatty acids are key components of biological membranes, and arachidonic acid is the precursor to prostaglandins, which mediate a variety of processes in the body.

Fiber May Be Soluble or Insoluble

The components of food materials that cannot be broken down by human digestive enzymes are referred to as **dietary fiber.** There are several kinds of dietary fiber, each with its own chemical and biological properties. **Cellulose** and **hemicellulose** are insoluble fiber materials that stimulate regular function of the colon. They may play a role in reducing the risk of colon cancer. **Lignins,** another class of insoluble fibers, absorb organic molecules in the digestive system. Lignins bind cholesterol and clear it from the digestive system, reducing the risk of heart disease. Pectins and gums are water-soluble fiber materials that form viscous gel-like suspensions in the digestive system, slowing the rate of absorption of many nutrients, including carbohydrates, and lowering serum cholesterol in many cases. The insoluble fibers are prevalent in vegetable grains. Water-soluble fiber is a component of fruits, legumes, and oats.

Special Focus: Vitamins

Vitamins are essential nutrients that are required in the diet, usually in trace amounts, because they cannot be synthesized by the organism itself. The requirement for any given vitamin depends on the organism. Not all "vitamins"

Table 17.4

Vitamins and Coenzymes

Vitamin	Coenzyme Form
Water-Soluble	
Thiamine (vitamin B_1)	Thiamine pyrophosphate
Niacin (nicotinic acid)	Nicotinamide adenine dinucleotide (NAD^+)
	Nicotinamide adenine dinucleotide phosphate ($NADP^+$)
Riboflavin (vitamin B_2)	Flavin adenine dinucleotide (FAD)
	Flavin mononucleotide (FMN)
Pantothenic acid	Coenzyme A
Pyridoxal, pyridoxine, pyridoxamine (vitamin B_6)	Pyridoxal phosphate
Cobalamin (vitamin B_{12})	5′-Deoxyadenosylcobalamin
	Methylcobalamin
Biotin	Biotin–lysine complexes (biocytin)
Lipoic acid	Lipoyl–lysine complexes (lipoamide)
Folic acid	Tetrahydrofolate
Fat-Soluble	
Retinol (vitamin A)	
Retinal (vitamin A)	
Retinoic acid (vitamin A)	
Ergocalciferol (vitamin D_2)	
Cholecalciferol (vitamin D_3)	
α-Tocopherol (vitamin E)	
Vitamin K	

are required by all organisms. Vitamins required in the human diet are listed in Table 17.4. These important substances are traditionally distinguished as being either water soluble or fat soluble. Except for vitamin C (ascorbic acid), the water-soluble vitamins are all components or precursors of important biological substances known as **coenzymes.** These are low-molecular-weight molecules that bring unique chemical functionality to certain enzyme reactions. Coenzymes may also act as *carriers* of specific functional groups, such as methyl groups and acyl groups. The side chains of the common amino acids provide only a limited range of chemical reactivities and carrier properties. Coenzymes, acting in concert with appropriate enzymes, provide a broader range of catalytic properties for the reactions of metabolism. Coenzymes are typically modified by these reactions and are then converted back to their original forms by other enzymes, so small amounts of these substances can be used repeatedly. The coenzymes derived from the water-soluble vitamins are listed in Table 17.4. Each will be discussed in this chapter. The fat-soluble vitamins are not directly related to coenzymes, but they play essential roles in a variety of critical biological processes, including vision, maintenance of bone structure, and blood coagulation. The mechanisms of action of fat-soluble vitamins are not as well understood as their water-soluble counterparts, but modern research efforts are gradually closing this gap.

Vitamin B_1: Thiamine and Thiamine Pyrophosphate

As shown in Figure 17.17, thiamine is composed of a substituted *thiazole* ring joined to a substituted pyrimidine by a methylene bridge. It is the precursor of **thiamine pyrophosphate (TPP),** a coenzyme involved in reactions of carbohy-

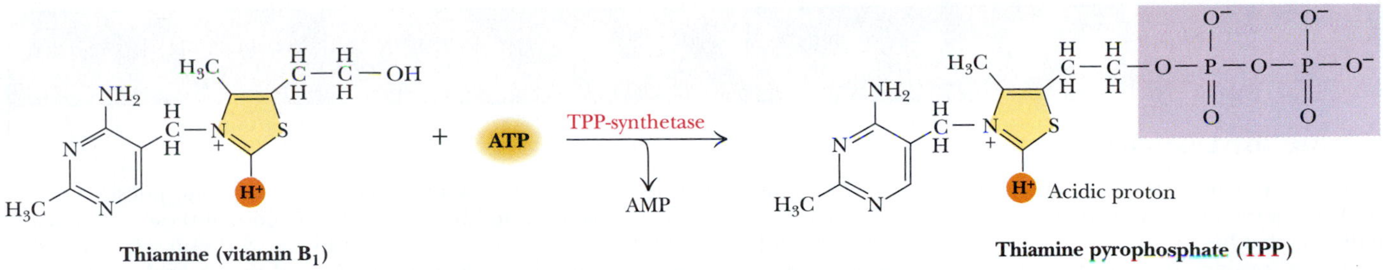

Thiamine (vitamin B₁)

Thiamine pyrophosphate (TPP)

FIGURE 17.17 Thiamine pyrophosphate (TPP), the active form of vitamin B_1, is formed by the action of TPP-synthetase.

drate metabolism in which bonds to carbonyl carbons (aldehydes or ketones) are made or broken. In particular, the *decarboxylations of α-keto acids and the formation and cleavage of α-hydroxyketones* depend on thiamine pyrophosphate. The first of these is illustrated in Figure 17.18a by the decarboxylation of pyruvate by **yeast pyruvate decarboxylase** to yield carbon dioxide and acetaldehyde. An example of the formation and cleavage of α-hydroxyketones is presented in Figure 17.18b, the condensation of two molecules of pyruvate in the **acetolactate synthase** reaction. Another example is provided by a reaction from the pentose phosphate pathway (see Chapters 21 and 22) called the **transketolase** reaction. This latter reaction is referred to as an α-ketol transfer for obvious reasons.

Some Vitamins Contain Adenine Nucleotides

Several classes of vitamins are related to, or are precursors of, coenzymes that contain adenine nucleotides as part of their structure. These coenzymes include the flavin dinucleotides, the pyridine dinucleotides, and coenzyme A. The adenine nucleotide portion of these coenzymes does not participate actively in the reactions of these coenzymes; rather, it enables the proper enzymes to recognize the coenzyme. Specifically, the adenine nucleotide greatly increases both the *affinity* and the *specificity* of the coenzyme for its site on the enzyme, owing to its numerous sites for hydrogen bonding, and also the hydrophobic and ionic bonding possibilities it brings to the coenzyme structure.

Nicotinic Acid and the Nicotinamide Coenzymes

Nicotinamide is an essential part of two important coenzymes: **nicotinamide adenine dinucleotide (NAD⁺)** and **nicotinamide adenine dinucleotide phosphate (NADP⁺)** (Figure 17.19). The reduced forms of these coenzymes are NADH and NADPH. *The nicotinamide coenzymes (also known as pyridine nucleotides) are electron carriers.* They play vital roles in a variety of enzyme-catalyzed oxidation–reduction reactions. (NAD^+ is an electron acceptor in oxidative [catabolic] pathways, and NADPH is an electron donor in reductive [biosynthetic] pathways.) These reactions involve direct transfer of hydride anion either to $NAD(P)^+$ or from $NAD(P)H$. The enzymes that facilitate such transfers are thus known as **dehydrogenases.** The hydride anion contains two electrons, and thus NAD^+ and $NADP^+$ act exclusively as **two-electron carriers.** The C-4 position of the pyridine ring,

Biochemistry ⓔNow™ ACTIVE FIGURE 17.18
Thiamine pyrophosphate participates in (a) the decarboxylation of α-keto acids and (b) the formation and cleavage of α-hydroxyketones. **Test yourself on the concepts in this figure at http://chemistry. brookscole.com/ggb3**

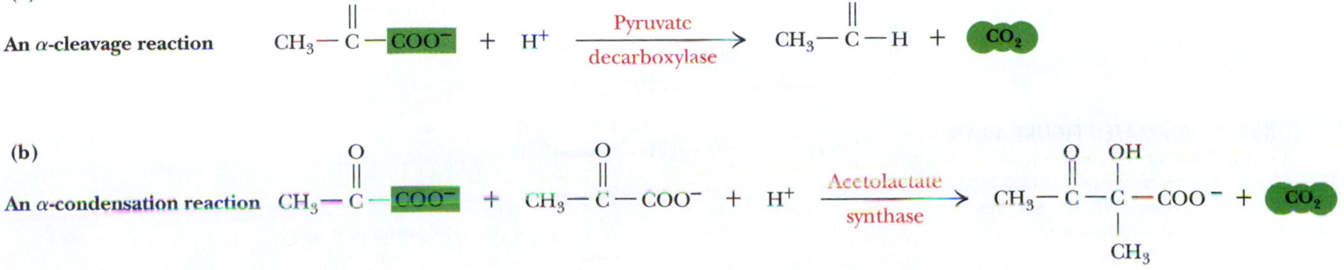

(a)

An α-cleavage reaction

(b)

An α-condensation reaction

Human Biochemistry

Thiamine and Beriberi

Thiamine, whose structure is shown in Figure 17.17, is known as vitamin B₁ and is essential for the prevention of **beriberi,** a nervous system disease that has occurred in the Far East for centuries and has resulted in considerable sickness and death in these countries. (As recently as 1958, it was the fourth leading cause of death in the Philippine Islands.) It was shown in 1882 by the director-general of the medical department of the Japanese navy that beriberi could be prevented by dietary modifications. Ten years later, Christiaan Eijkman, a Dutch medical scientist working in Java, began research that eventually showed that thiamine was the "anti-

beriberi" substance. He found that chickens fed only polished rice exhibited paralysis and head retractions and that these symptoms could be reversed if the rice polishings (the outer layers and embryo of the rice kernel) were also fed to the birds. In 1911, Casimir Funk prepared a crystalline material from rice bran that cured beriberi in birds. He named it **beriberi vitamine,** because he viewed it as a "vital amine," and thus he is credited with coining the word *vitamin.* The American biochemist R. R. Williams and his research group were the first to establish the structure of thiamine (in 1935) and a route for its synthesis.

which can either accept or donate hydride ion, is the reactive center of both NAD and NADP. The quaternary nitrogen of the nicotinamide ring functions as an electron sink to facilitate hydride transfer to NAD⁺, as shown in Figure 17.20. The adenine portion of the molecule is not directly involved in redox processes.

Examination of the structures of NADH and NADPH reveals that the 4-position of the nicotinamide ring is **pro-chiral,** meaning that although this carbon is not chiral, it *would* be if either of its hydrogens were replaced by something else. As shown in Figure 17.20, the hydrogen "projecting" out of the page toward you is the "pro-*R*" hydrogen because, if a deuterium were substituted at this position, the molecule would have the *R*-configuration. Substitution of the other hydrogen would yield an *S*-configuration. An interesting aspect of the enzymes that require nicotinamide coenzymes is that they are **stereospecific** and withdraw hydrogen from either the pro-*R* or the pro-*S* position selectively. This stereospecificity arises from the fact that enzymes (and the active sites of enzymes) are inherently asymmetric structures. These same enzymes are stereospecific with respect to the substrates as well.

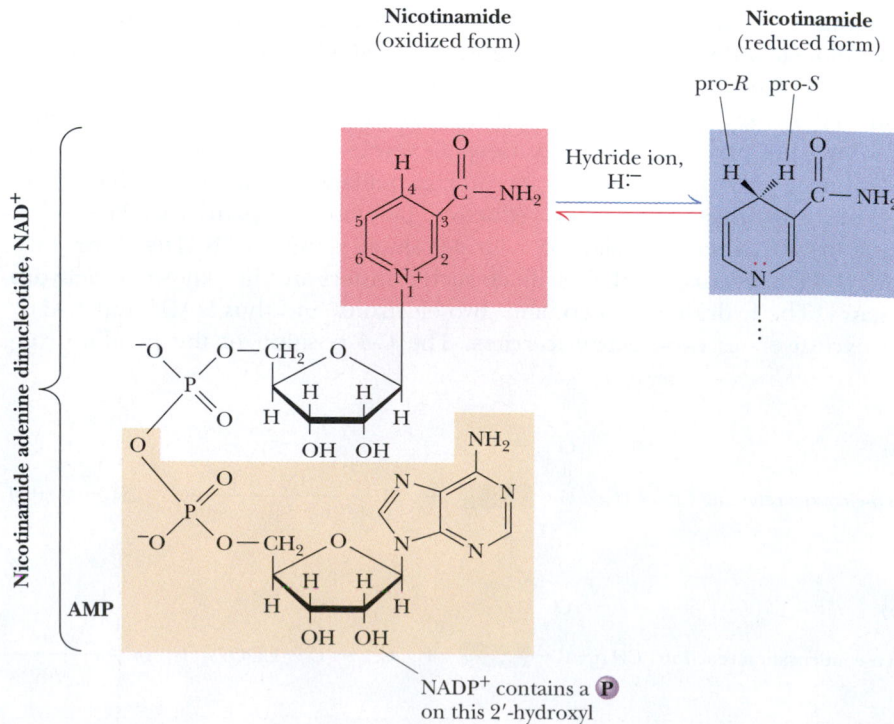

Biochemistry⊗Now™   ANIMATED FIGURE 17.19
The structures and redox states of the nicotinamide coenzymes. Hydride ion (H:⁻, a proton with two electrons) transfers to NAD⁺ to produce NADH. **See this** figure animated at **http://chemistry.brookscole. com/ggb3**

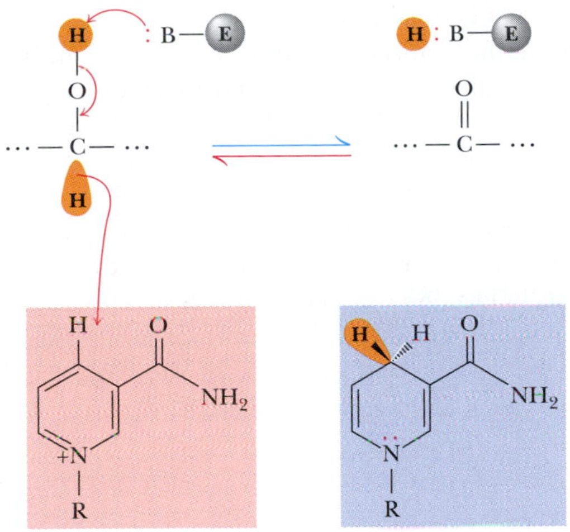

Oxidized coenzyme (NAD$^+$ or NADP$^+$)

Reduced coenzyme (NADH or NADPH)

FIGURE 17.20 NAD$^+$ and NADP$^+$ participate exclusively in two-electron transfer reactions. For example, alcohols can be oxidized to ketones or aldehydes via hydride transfer to NAD(P)$^+$.

The NAD- and NADP-dependent dehydrogenases catalyze at least six different types of reactions: simple hydride transfer, deamination of an amino acid to form an α-keto acid, oxidation of β-hydroxy acids followed by decarboxylation of the β-keto acid intermediate, oxidation of aldehydes, reduction of isolated double bonds, and oxidation of carbon–nitrogen bonds (as with dihydrofolate reductase).

Riboflavin and the Flavin Coenzymes

Riboflavin, or **vitamin B$_2$,** is a constituent and precursor of both **riboflavin 5′-phosphate,** also known as **flavin mononucleotide (FMN),** and **flavin adenine dinucleotide (FAD).** The name *riboflavin* is a synthesis of the names for the molecule's component parts, **ribitol** and **flavin.** The structures of riboflavin, FMN, and FAD are shown in Figure 17.21. The *isoalloxazine ring* is the core

Human Biochemistry

Niacin and Pellagra

Pellagra, a disease characterized by dermatitis, diarrhea, and dementia, has been known for centuries. It was once prevalent in the southern part of the United States and is still a common problem in some parts of Spain, Italy, and Romania. Pellagra was once thought to be an infectious disease, but Joseph Goldberger showed early in the 20th century that it could be cured by dietary actions. Soon thereafter, it was found that brewer's yeast would prevent pellagra in humans. Studies of a similar disease in dogs, called **blacktongue,** eventually led to the identification of **nicotinic acid** as the relevant dietary factor. Elvehjem and his colleagues at the University of Wisconsin in 1937 isolated **nicotinamide** from liver and showed that it and nicotinic acid could prevent and cure blacktongue in dogs. That same year, nicotinamide and nicotinic acid were both shown to cure pellagra in humans. Interestingly, many animals, including humans, can synthesize nicotinic acid from tryptophan and other precursors, and nicotinic acid is thus not absolutely essential in the diet. However, if dietary intake of tryptophan is low, nicotinic acid is required for optimal health. Nicotinic acid is structurally related to **nicotine,** a highly toxic tobacco alkaloid. To avoid confusion of nicotinic acid and nicotinamide with nicotine itself, **niacin** was adopted as a common name for nicotinic acid. Cowgill, at Yale University, suggested the name from the letters of three words—*ni*cotinic, *ac*id, and vitam*in*.

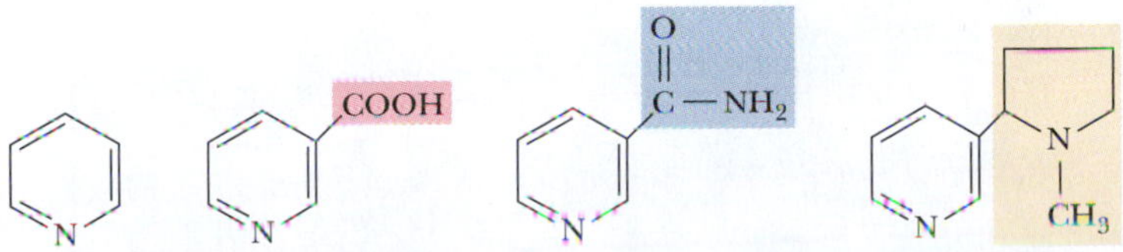

Pyridine Nicotinic acid Nicotinamide Nicotine

The structures of pyridine, nicotinic acid, nicotinamide, and nicotine.

structure of the various flavins. Because the ribityl group is not a true pentose sugar (it is a sugar alcohol) and is not joined to riboflavin in a glycosidic bond, the molecule is not truly a "nucleotide" and the terms *flavin mononucleotide* and *dinucleotide* are incorrect. Nonetheless, these designations are so deeply ingrained in common biochemical usage that the erroneous nomenclature persists. The flavins have a characteristic bright yellow color and take their name from the Latin *flavus* for "yellow." As shown in Figure 17.22, the oxidized form of the isoalloxazine structure absorbs light around 450 nm (in the visible region) and also at 350 to 380 nm. The color is lost, however, when the ring is reduced or "bleached." Similarly, the enzymes that bind flavins, known as **flavoenzymes,** can be yellow, red, or green in their oxidized states. Nevertheless, these enzymes also lose their color on reduction of the bound flavin group.

Flavin coenzymes can exist in any of three different redox states. Fully oxidized flavin is converted to a **semiquinone** by a one-electron transfer, as shown in Figure 17.22. At physiological pH, the semiquinone is a neutral radical, blue in color, with a λ_{max} of 570 nm. The semiquinone possesses a pK_a of about 8.4. When it loses a proton at higher pH values, it becomes a radical anion, displaying a red color with a λ_{max} of 490 nm. The semiquinone radical is particularly stable, owing to extensive delocalization of the unpaired electron across the π-electron system of the isoalloxazine. A second one-electron transfer converts the semiquinone to the completely reduced dihydroflavin as shown in Figure 17.22.

Access to three different redox states allows flavin coenzymes to participate in *one-electron transfer* and *two-electron transfer reactions*. Partly because of this, flavoproteins catalyze many different reactions in biological systems and work together with many different electron acceptors and donors. These include two-electron acceptor/donors, such as NAD^+ and $NADP^+$; one- or two-electron carriers, such as quinones; and a variety of one-electron acceptor/donors, such as cytochrome proteins. Many of the components of the respiratory electron transport chain are one-electron acceptor/donors (see Chapter 20).

Biochemistry⊗**Now™** **ANIMATED FIGURE 17.21**
The structures of riboflavin, flavin mononucleotide (FMN), and flavin adenine dinucleotide (FAD). Flavin coenzymes bind tightly to the enzymes that use them, with typical dissociation constants in the range of 10^{-8} to 10^{-11} M, so only very low levels of free flavin coenzymes occur in most cells. Even in organisms that rely on the nicotinamide coenzymes (NADH and NADPH) for many of their oxidation–reduction cycles, the flavin coenzymes fill essential roles. Flavins are stronger oxidizing agents than NAD^+ and $NADP^+$. They can be reduced by both one-electron and two-electron pathways and can be reoxidized easily by molecular oxygen. Enzymes that use flavins to carry out their reactions—*flavoenzymes*—are involved in many kinds of oxidation–reduction reactions. **See this figure animated at http://chemistry. brookscole.com/ggb3**

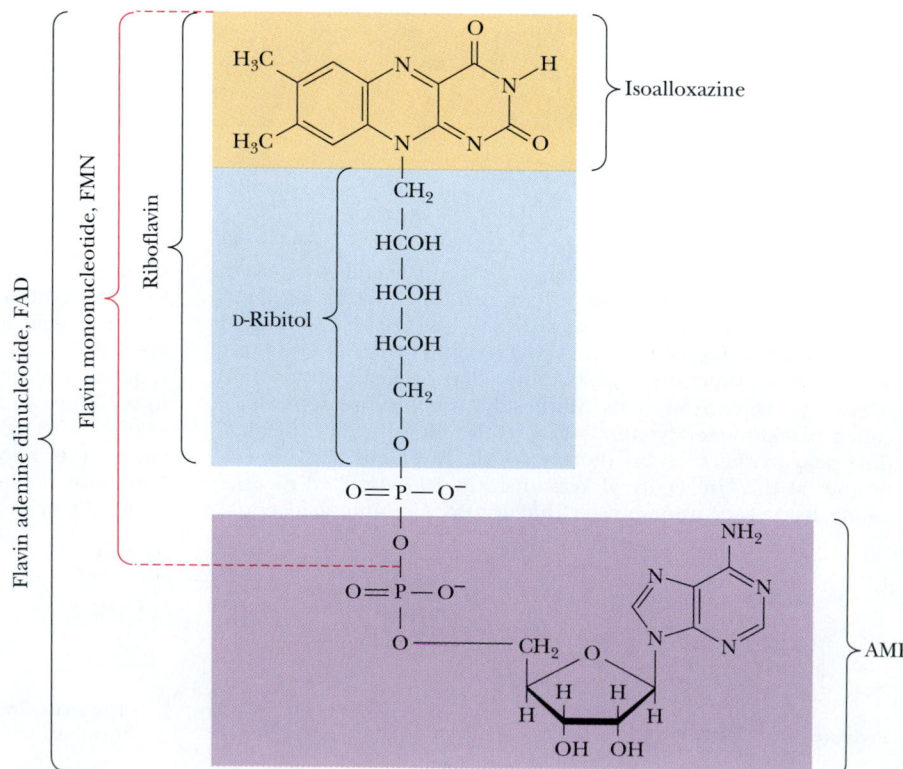

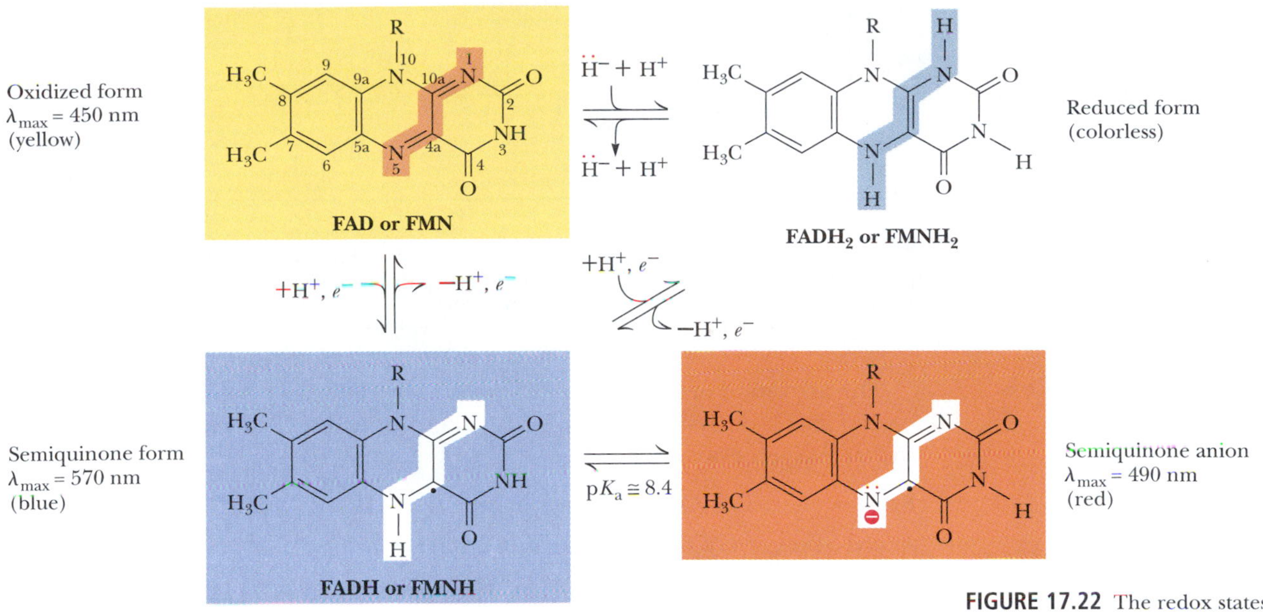

FIGURE 17.22 The redox states of FAD and FMN. The boxes correspond to the colors of each of these forms. The atoms primarily involved in electron transfer are indicated by red shading in the oxidized form, white in the semiquinone form, and blue in the reduced form.

Pantothenic Acid and Coenzyme A

Pantothenic acid, sometimes called vitamin B_3, is a vitamin that makes up one part of a complex coenzyme called **coenzyme A (CoA)** (Figure 17.23). Pantothenic acid is also a constituent of **acyl carrier proteins.** Coenzyme A consists of 3′,5′-adenosine bisphosphate joined to 4-phosphopantetheine in a phosphoric anhydride linkage. Phosphopantetheine in turn consists of three parts: β-mercaptoethylamine linked to β-alanine, which makes an amide bond with a branched-chain dihydroxy acid. As was the case for the nicotinamide and flavin coenzymes, the adenine nucleotide moiety of CoA acts as a recognition site, increasing the affinity and specificity of CoA binding to its enzymes.

The two main functions of coenzyme A are

1. *activation of acyl groups for transfer by nucleophilic attack* and
2. *activation of the α-hydrogen of the acyl group for abstraction as a proton.*

Both of these functions are mediated by the reactive sulfhydryl group on CoA, which forms **thioester** linkages with acyl groups.

A Deeper Look

Riboflavin and Old Yellow Enzyme

Riboflavin was first isolated from whey in 1879 by Blyth, and the structure was determined by Kuhn and co-workers in 1933. For the structure determination, this group isolated 30 mg of pure riboflavin from the whites of about 10,000 eggs. The discovery of the actions of riboflavin in biological systems arose from the work of Otto Warburg in Germany and Hugo Theorell in Sweden, both of whom identified yellow substances bound to a yeast enzyme involved in the oxidation of pyridine nucleotides. Theorell showed that riboflavin 5′-phosphate was the source of the yellow color in this *old yellow enzyme*. By 1938, Warburg had identified FAD, the second common form of riboflavin, as the coenzyme in D-amino acid oxidase, another yellow protein. Riboflavin deficiencies are not at all common. Humans require only about 2 mg per day, and the vitamin is prevalent in many foods. This vitamin is extremely light sensitive, and it is degraded in foods (milk, for example) left in the sun.

The milling and refining of wheat, rice, and other grains causes a loss of riboflavin and other vitamins. In order to correct and prevent dietary deficiencies, the Committee on Food and Nutrition of the National Research Council began in the 1940s to recommend enrichment of cereal grains sold in the United States. Thiamine, riboflavin, niacin, and iron were the first nutrients originally recommended for enrichment. As a result of these actions, generations of American children have become accustomed to reading (on their cereal boxes and bread wrappers) that their foods contain certain percentages of the "U.S. Recommended Daily Allowance" of various vitamins and nutrients.

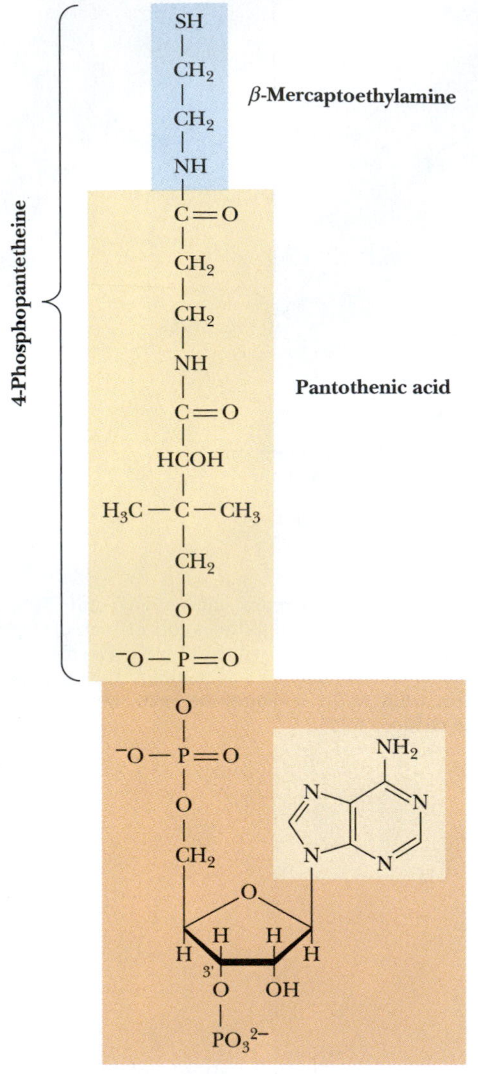

FIGURE 17.23 The structure of coenzyme A. Acyl groups form thioester linkages with the —SH group of the β-mercaptoethylamine moiety.

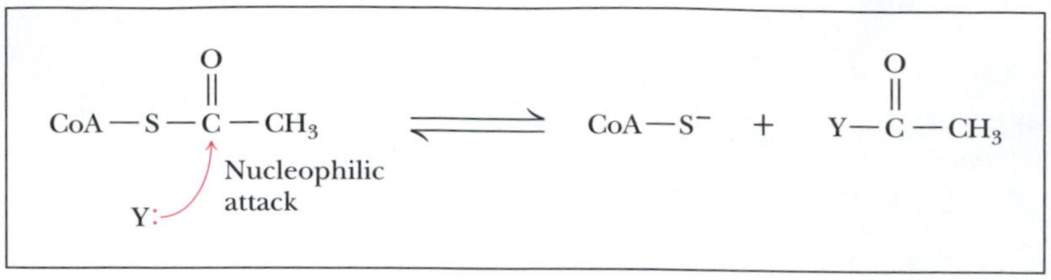

Biochemistry⊛Now™ **ANIMATED FIGURE 17.24** Acyl transfer from acyl-CoA to a nucleophile is more favorable than transfer of an acyl group from an oxygen ester. **See this figure animated at http://chemistry.brookscole.com/ggb3**

The activation of acyl groups for transfer by CoA can be appreciated by comparing the hydrolysis of the thioester bond of acetyl-CoA with hydrolysis of a simple oxygen ester:

$$\text{Ethyl acetate} + H_2O \longrightarrow \text{acetate} + \text{ethanol} + H^+ \qquad \Delta G^{\circ\prime} = -20.0 \text{ kJ/mol}$$

$$\text{Acetyl-SCoA} + H_2O \longrightarrow \text{acetate} + \text{CoA-SH} + H^+ \qquad \Delta G^{\circ\prime} = -31.5 \text{ kJ/mol}$$

Hydrolysis of the thioester is more favorable than that of oxygen esters, presumably because the carbon–sulfur bond has less double-bond character than the corresponding carbon–oxygen bond. This means that transfer of the acetyl group from acetyl-CoA to a given nucleophile (Figure 17.24) will be more spontaneous than transfer of an acetyl group from an oxygen ester. For this reason, acetyl-CoA is said to have a high group-transfer potential.

The 4-phosphopantetheine group of CoA is also utilized (for essentially the same purposes) in **acyl carrier proteins (ACPs)** involved in fatty acid biosynthesis (see Chapter 24). In acyl carrier proteins, the 4-phosphopantetheine is covalently linked to a serine hydroxyl group. Pantothenic acid is an essential factor for the metabolism of fat, protein, and carbohydrates in the tricarboxylic acid cycle and other pathways. In view of its universal importance in metabolism, it is surprising that pantothenic acid deficiencies are not a more serious problem in humans, but this vitamin is abundant in almost all foods, so deficiencies are rarely observed.

Vitamin B₆: Pyridoxine and Pyridoxal Phosphate

The biologically active form of vitamin B₆ is **pyridoxal-5-phosphate (PLP)**, a coenzyme that exists under physiological conditions in two tautomeric forms (Figure 17.25). PLP participates in the catalysis of a wide variety of reactions involving amino acids, including transaminations, α- and β-decarboxylations, β- and γ-eliminations, racemizations, and aldol reactions (Figure 17.26). Note that these reactions include cleavage of any of the bonds to the amino acid

A Deeper Look

Fritz Lipmann and Coenzyme A

Pantothenic acid is found in extracts from nearly all plants, bacteria, and animals, and the name derives from the Greek *pantos,* meaning "everywhere." It is required in the diet of all vertebrates, but some microorganisms produce it in the rumens of animals such as cattle and sheep. This vitamin is widely distributed in foods common to the human diet, and deficiencies are observed only in cases of severe malnutrition. The eminent German-born biochemist Fritz Lipmann was the first to show that a coenzyme was required to facilitate biological acetylation reactions. (The

"A" in coenzyme A in fact stands for *acetylation.*) In studies of acetylation of sulfanilic acid (a reaction chosen because of a favorable colorimetric assay) by liver extracts, Lipmann found that a heat-stable cofactor was required. Eventually Lipmann isolated and purified the required cofactor—coenzyme A—from both liver and yeast. For his pioneering work in elucidating the role of this important coenzyme, Fritz Lipmann received the Nobel Prize in Physiology or Medicine in 1953.

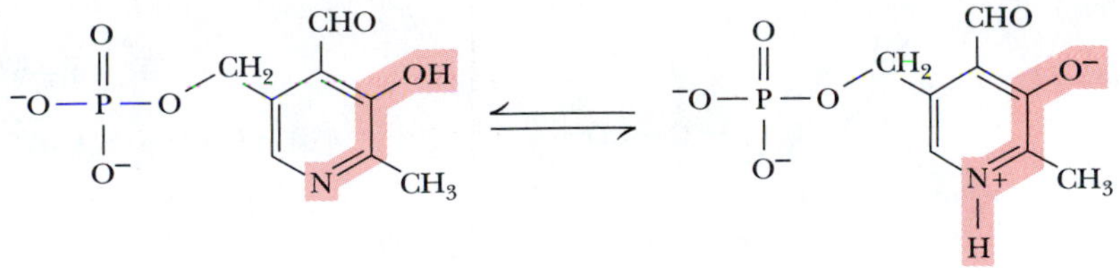

FIGURE 17.25 The tautomeric forms of pyridoxal-5-phosphate (PLP).

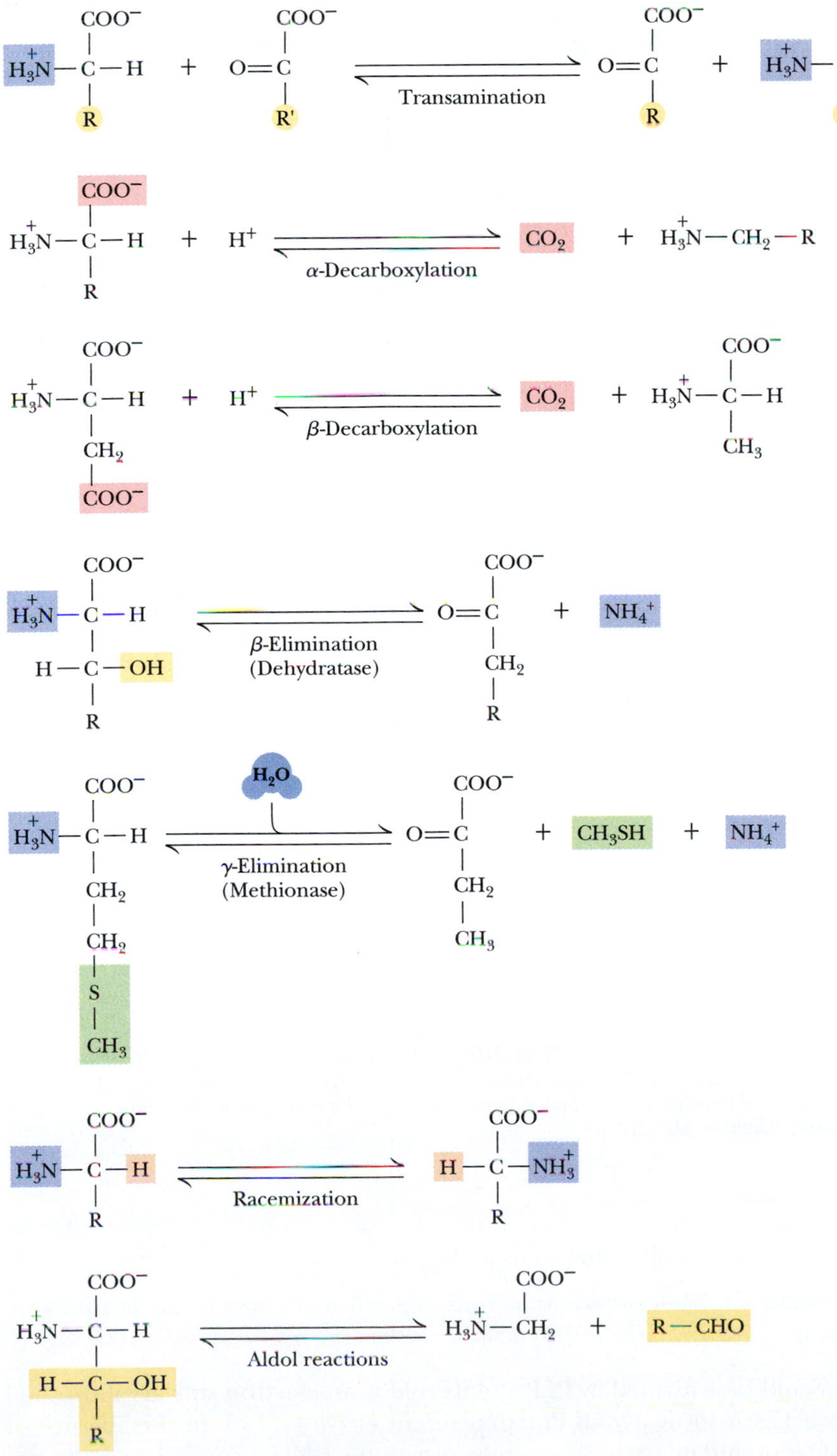

FIGURE 17.26 The seven classes of reactions catalyzed by pyridoxal-5-phosphate.

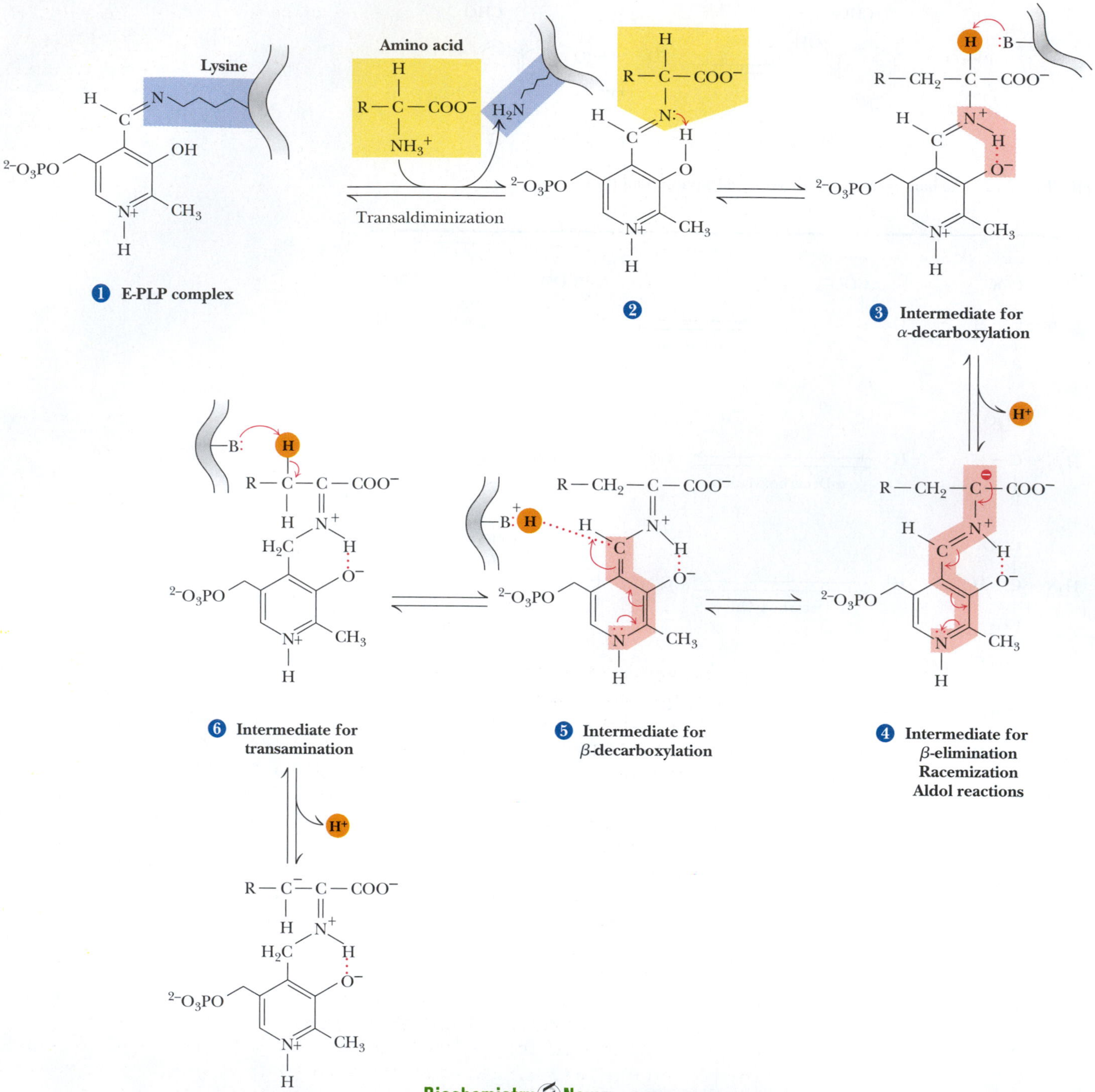

BiochemistryⒺNow™ **ACTIVE FIGURE 17.27** Pyridoxal-5-phosphate forms stable Schiff base adducts with amino acids and acts as an effective electron sink to stabilize a variety of reaction intermediates. **Test yourself on the concepts in this figure at** **http://chemistry.brookscole.com/ggb3**

alpha carbon, as well as several bonds in the side chain. The remarkably versatile chemistry of PLP is due to its ability to

1. *form stable Schiff base (aldimine) adducts with α-amino groups of amino acids* and
2. *act as an effective electron sink to stabilize reaction intermediates.*

The Schiff base formed by PLP and its role as an electron sink are illustrated in Figure 17.27. In nearly all PLP-dependent enzymes, PLP in the absence of substrate is bound in a Schiff base linkage with the ε-NH₂ group of an active site lysine. Rearrangement to a Schiff base with the arriving substrate is a **transal-**

A Deeper Look

Vitamin B$_6$

Goldberger and Lillie in 1926 found that rats fed certain nutritionally deficient diets developed **dermatitis acrodynia,** a skin disorder characterized by edema and lesions of the ears, paws, nose, and tail. Szent-Györgyi later found that a factor he had isolated prevented these skin lesions in the rat. He proposed the name **vitamin B$_6$** for his factor. **Pyridoxine,** a form of this vitamin found in plants (and the form of B$_6$ sold commercially), was isolated in 1938 by three research groups working independently. **Pyridoxal** and **pyridoxamine,** the forms that predominate in animals, were identified in 1945. A metabolic role for pyridoxal was postulated by Esmond Snell, who had shown that when pyridoxal was heated with glutamate (in the absence of any enzymes), the amino group of glutamate was transferred to pyridoxal, forming pyridoxamine. Snell postulated (correctly) that pyridoxal might be a component of a coenzyme needed for transamination reactions in which the α-amino group of an amino acid is transferred to the α-carbon of an α-keto acid.

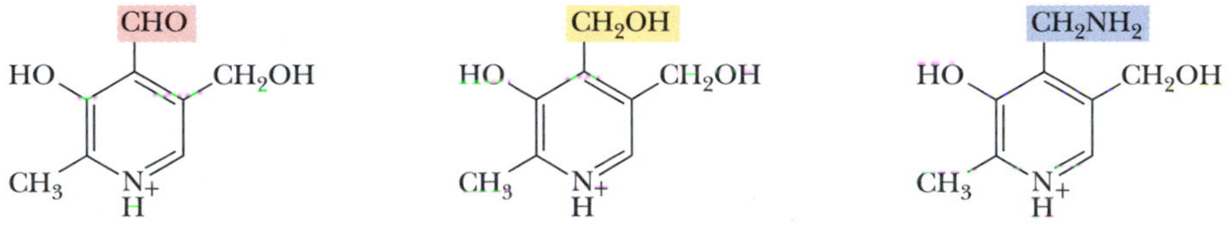

The structures of pyridoxal, pyridoxine, and pyridoxamine.

diminization reaction. One key to PLP chemistry is the protonation of the Schiff base, which is stabilized by H bonding to the ring oxygen, increasing the acidity of the C$_\alpha$ proton [as shown in (3) of Figure 17.27]. The carbanion formed by loss of the C$_\alpha$ proton is stabilized by electron delocalization into the pyridinium ring, with the positively charged ring nitrogen acting as an electron sink. Another important intermediate is formed by protonation of the aldehyde carbon of PLP. As shown, this produces a new substrate–PLP Schiff base, which plays a role in transamination reactions and increases the acidity of the proton at C$_\beta$, a feature important in γ-elimination reactions.

The versatile chemistry of pyridoxal phosphate offers a rich learning experience for the student of mechanistic chemistry. William Jencks, in his classic text, *Catalysis in Chemistry and Enzymology,* writes:

> It has been said that God created an organism especially adapted to help the biologist find an answer to every question about the physiology of living systems; if this is so it must be concluded that pyridoxal phosphate was created to provide satisfaction and enlightenment to those enzymologists and chemists who enjoy pushing electrons, for no other coenzyme is involved in such a wide variety of reactions, in both enzyme and model systems, which can be reasonably interpreted in terms of the chemical properties of the coenzyme. Most of these reactions are made possible by a common structural feature. That is, electron withdrawal toward the cationic nitrogen atom of the imine and into the electron sink of the pyridoxal ring from the α carbon atom of the attached amino acid activates all three of the substituents on this carbon atom for reactions which require electron withdrawal from this atom.[1]

Vitamin B$_{12}$ Contains the Metal Cobalt

Vitamin B$_{12}$, or **cyanocobalamin,** is converted in the body into two coenzymes. The predominant coenzyme form is **5′-deoxyadenosylcobalamin** (Figure 17.28), but smaller amounts of **methylcobalamin** also exist in liver, for example. The crystal structure of 5′-deoxyadenosylcobalamin was determined by X-ray diffraction in 1961 by Dorothy Hodgkin and co-workers in England. The structure consists of a *corrin ring* with a *cobalt ion* in the center. The corrin ring, with four *pyrrole groups,* is similar to the heme porphyrin ring, except that two of the pyrrole rings

[1]Jencks, William P., 1969. *Catalysis in Chemistry and Enzymology.* New York: McGraw-Hill.

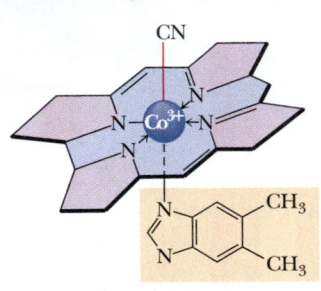

FIGURE 17.28 The structure of cyanocobalamin *(top)* and simplified structures showing several coenzyme forms of vitamin B$_{12}$. The Co—C bond of 5′-deoxyadenosylcobalamin is predominantly covalent (note the short bond length of 0.205 nm) but with some ionic character. Note that the convention of writing the cobalt atom as Co^{3+} attributes the electrons of the Co—C and Co—N bonds to carbon and nitrogen, respectively.

Dimethylbenzimidazole (DMBz)

Cyanocobalamin

Cyanocobalamin
Vitamin B$_{12}$

5′-Deoxyadenosylcobalamin

Methylcobalamin

Hydroxocobalamin
Vitamin B$_{12b}$

Coenzyme Forms

are linked directly. Methylene bridges form the other pyrrole–pyrrole linkages, as for porphyrin. The cobalt is coordinated to the four (planar) pyrrole nitrogens. One of the axial cobalt ligands is a nitrogen of the dimethylbenzimidazole group. The other axial cobalt ligand may be —CN, —CH$_3$, —OH, or the 5′-carbon of a 5′-deoxyadenosyl group, depending on the form of the coenzyme. The most striking feature of 5′-deoxyadenosylcobalamin is the cobalt–carbon bond distance of 0.205 nm. *This bond is predominantly covalent*, and the structure is actually an **alkyl cobalt.** Such alkyl cobalts were thought to be highly unstable until Hodgkin's pioneering X-ray study. The Co–carbon–carbon bond angle of 130° indicates partial ionic character.

The B$_{12}$ coenzymes participate in three types of reactions (Figure 17.29):

1. *Intramolecular rearrangements*
2. *Reductions of ribonucleotides to deoxyribonucleotides (in certain bacteria)*
3. *Methyl group transfers*

The first two of these are mediated by 5′-deoxyadenosylcobalamin, whereas methyl transfers are effected by methylcobalamin. The mechanism of ribonucleotide reductase is discussed in Chapter 26. Methyl group transfers that employ *tetrahydrofolate* as a coenzyme are described later in this chapter.

Vitamin C: Ascorbic Acid

L-Ascorbic acid, better known as **vitamin C,** has the simplest chemical structure of all the vitamins (Figure 17.30). It is widely distributed in the animal and plant kingdoms, and only a few vertebrates—humans and other primates,

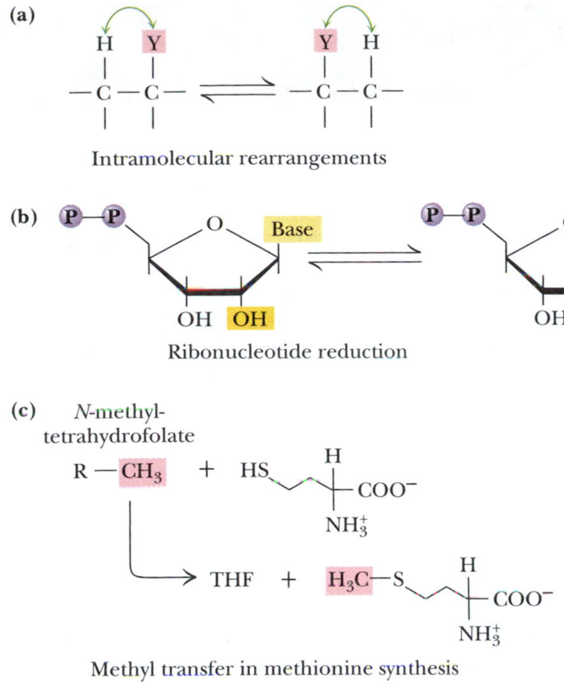

(a)

Intramolecular rearrangements

(b)

Ribonucleotide reduction

(c) *N*-methyl-
tetrahydrofolate

Methyl transfer in methionine synthesis

Biochemistry ⓢ Now™   ANIMATED FIGURE 17.29 Vitamin B$_{12}$ functions as a coenzyme in intramolecular rearrangements, reduction of ribonucleotides, and methyl group transfers. **See this figure animated at http://chemistry.brookscole.com/ggb3**

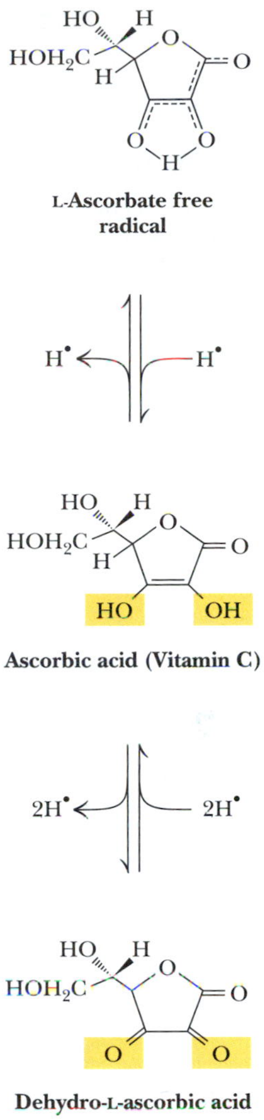

L-Ascorbate free
radical

Ascorbic acid (Vitamin C)

Dehydro-L-ascorbic acid

FIGURE 17.30 The physiological effects of ascorbic acid (vitamin C) are the result of its action as a reducing agent. A two-electron oxidation of ascorbic acid yields dehydroascorbic acid.

guinea pigs, fruit-eating bats, certain birds, and some fish (rainbow trout, carp, and Coho salmon, for example)—are unable to synthesize it. In all these organisms, the inability to synthesize ascorbic acid stems from a lack of a liver enzyme, L-gulono-γ-lactone oxidase.

Ascorbic acid is a reasonably strong reducing agent. The biochemical and physiological functions of ascorbic acid most likely derive from its reducing properties—it functions as an electron carrier. Loss of one electron due to interactions with oxygen or metal ions leads to **semidehydro-L-ascorbate,** a reactive free radical (Figure 17.30) that can be reduced back to L-ascorbic acid by various enzymes in animals and plants. A characteristic reaction of ascorbic acid is its oxidation to *dehydro-L-ascorbic acid*. Ascorbic acid and dehydroascorbic acid form an effective redox system.

Human Biochemistry

Vitamin B$_{12}$ and Pernicious Anemia

The most potent known vitamin (that is, the one needed in the smallest amounts) was the last to be discovered. Vitamin B$_{12}$ is best known as the vitamin that prevents **pernicious anemia.** Minot and Murphy in 1926 demonstrated that such anemia could be prevented by eating large quantities of liver, but the active agent was not identified for many years. In 1948, Rickes and co-workers (in the United States) and Smith (in England) both reported the first successful isolation of vitamin B$_{12}$. West showed that injections of the vitamin induced dramatic beneficial responses in pernicious anemia patients. Eventually, two different crystalline preparations of the vitamin were distinguished. The first appeared to be true cyanocobalamin. The second showed the same biological activity as a cyanocobalamin but had a different spectrum and was named **vitamin B$_{12b}$** and also **hydroxocobalamin.** It was eventually found

that the cyanide group in cyanocobalamin originated from the charcoal used in the purification process!

Vitamin B$_{12}$ is not synthesized by animals or by plants. Only a few species of bacteria synthesize this complex substance. Carnivorous animals easily acquire sufficient amounts of B$_{12}$ from meat in their diet, but herbivorous creatures typically depend on intestinal bacteria to synthesize B$_{12}$ for them. This is sometimes not sufficient, and certain animals, including rabbits, occasionally eat their feces in order to accumulate the necessary quantities of B$_{12}$.

The nutritional requirement for vitamin B$_{12}$ is low. Adult humans require only about 3 μg per day, an amount easily acquired with normal eating habits. However, because plants do not synthesize vitamin B$_{12}$, pernicious anemia symptoms are sometimes observed in strict vegetarians.

Human Biochemistry

Ascorbic Acid and Scurvy

Ascorbic acid is effective in the treatment and prevention of **scurvy**, a potentially fatal disorder characterized by anemia; alteration of protein metabolism; and weakening of collagenous structures in bone, cartilage, teeth, and connective tissues (see Chapter 6). Western world diets are now routinely so rich in vitamin C that it is easy to forget that scurvy affected many people in ancient Egypt, Greece, and Rome and that, in the Middle Ages, it was endemic in northern Europe in winter when fresh fruits and vegetables were scarce. Ascorbic acid is a vitamin that has routinely altered the course of history, ending ocean voyages and military campaigns when food supplies became depleted of vitamin C and fatal outbreaks of scurvy occurred.

The isolation of ascorbic acid was first reported by Albert Szent-Györgyi (who called it *hexuronic acid*) in 1928. The structure was determined by Hirst and Haworth in 1933, and simultaneously, Reichstein reported its synthesis. Haworth and Szent-Györgyi, who together suggested that the name be changed to L-ascorbic acid to describe its **antiscorbutic** (antiscurvy) activity, were awarded the Nobel Prize in 1937 for their studies of vitamin C.

FIGURE 17.31 The structure of biotin.

In addition to its role in preventing scurvy (see Human Biochemistry box above and also Chapter 6), ascorbic acid also plays important roles in the brain and nervous system. It also mobilizes iron in the body, prevents anemia, ameliorates allergic responses, and stimulates the immune system.

Biotin

Biotin (Figure 17.31) acts as a **mobile carboxyl group carrier** in a variety of enzymatic carboxylation reactions. In each of these, biotin is bound covalently to the enzyme as a prosthetic group via the ε-amino group of a lysine residue on the protein (Figure 17.32). The biotin–lysine function is referred to as a **biocytin** residue. The result is that *the biotin ring system is tethered to the protein by a long, flexible chain.* The ten atoms in this chain separate the biotin ring and the lysine α-carbon by approximately 1.5 nm. This chain allows biotin to acquire carboxyl groups at one subsite of the enzyme active site and deliver them to a substrate acceptor at another subsite.

Most biotin-dependent carboxylations (Table 17.5) use *bicarbonate* as the carboxylating agent and transfer the carboxyl group to a *substrate carbanion.* Bicarbonate is plentiful in biological fluids, but it is a poor electrophile at carbon and must be "activated" for attack by the substrate carbanion.

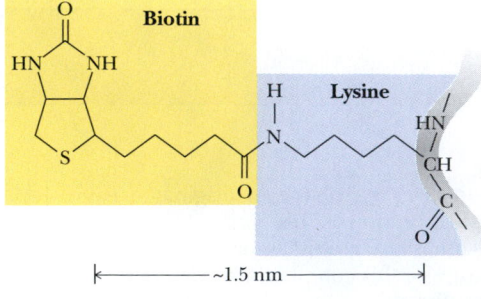

The biotin–lysine (biocytin) complex

Biochemistry ⊘ Now™ ACTIVE FIGURE 17.32
Biotin is covalently linked to a protein via the ε-amino group of a lysine residue. The biotin ring is thus tethered to the protein by a ten-atom chain. It functions by carrying carboxyl groups between distant sites on biotin-dependent enzymes. **Test yourself on the concepts in this figure at http://chemistry.brookscole. com/ggb3**

Lipoic Acid

Lipoic acid exists as a mixture of two structures: a closed-ring disulfide form and an open-chain reduced form (Figure 17.33). Oxidation–reduction cycles interconvert these two species. As is the case for biotin, lipoic acid does not often occur free in nature but rather is covalently attached in amide linkage with lysine residues on enzymes. The enzyme that catalyzes the formation of the *lipoamide* linkage requires ATP and produces lipoamide-enzyme conjugates, AMP, and pyrophosphate as products of the reaction.

Lipoic acid is an **acyl group carrier.** It is found in *pyruvate dehydrogenase* and *α-ketoglutarate dehydrogenase,* two multienzyme complexes involved in carbohydrate metabolism (Figure 17.34). *Lipoic acid functions to couple acyl-group transfer and electron transfer during oxidation and decarboxylation of α-keto acids.*

The special properties of lipoic acid arise from the ring strain experienced by oxidized lipoic acid. The closed ring form is about 20 kJ/mol higher in energy than the open-chain form, and this results in a strong negative reduction potential of about −0.30 V. The oxidized form readily oxidizes cyanides to isothiocyanates and sulfhydryl groups to mixed disulfides.

A Deeper Look

Biotin

Early in the 1900s, it was observed that certain strains of yeast required a material called **bios** for growth. Bios was eventually found to contain four different substances: myoinositol, β-alanine, pantothenic acid, and a compound later shown to be biotin. Kögl and Tönnis first isolated biotin from egg yolk in 1936. Boas, in 1927, and Szent-Györgyi, in 1931, found substances in liver that were capable of curing and preventing the dermatitis, loss of hair, and paralysis that occurred in rats fed large amounts of raw egg whites (a condition known as *egg white injury*). Boas called the factor "protective factor X" and Szent-Györgyi named the substance *vitamin H* (from the German *haut,* meaning "skin"), but both were soon shown to be identical to biotin. It is now known that egg white contains a basic protein called **avidin,** which has an extremely high affinity for biotin ($K_D = 10^{-15}\ M$). The sequestering of biotin by avidin is the cause of the egg white injury condition.

The structure of biotin was determined in the early 1940s by Kögl in Europe and by du Vigneaud and co-workers in the United States. Interestingly, the biotin molecule contains three asymmetric carbon atoms, and biotin could thus exist as eight different stereoisomers. Only one of these shows biological activity.

Table 17.5

Principal Biotin-Dependent Carboxylations

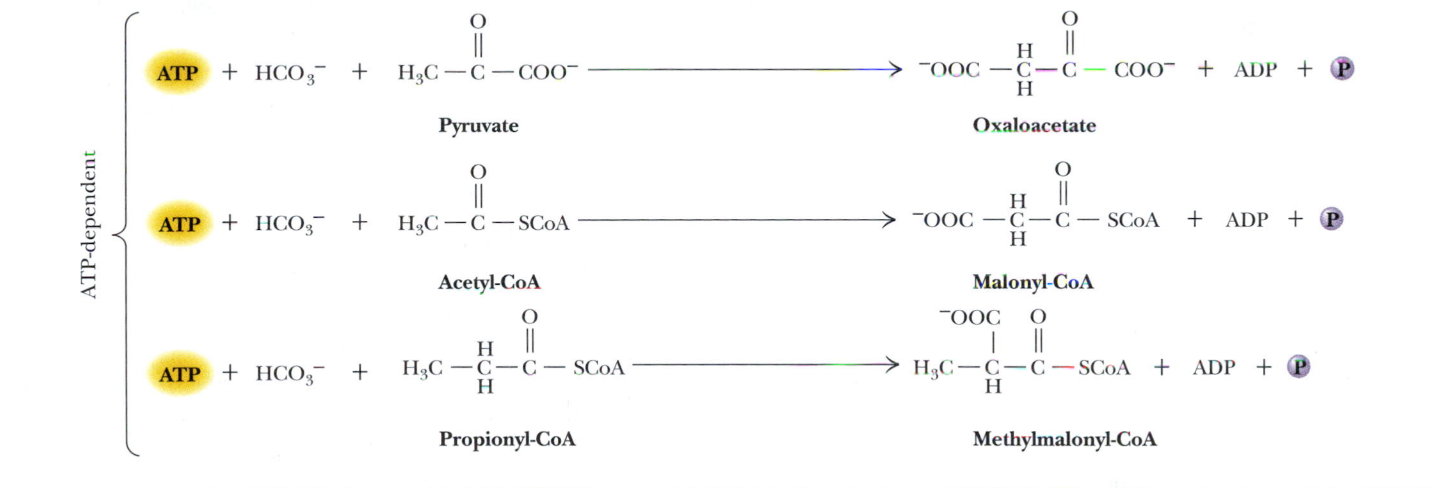

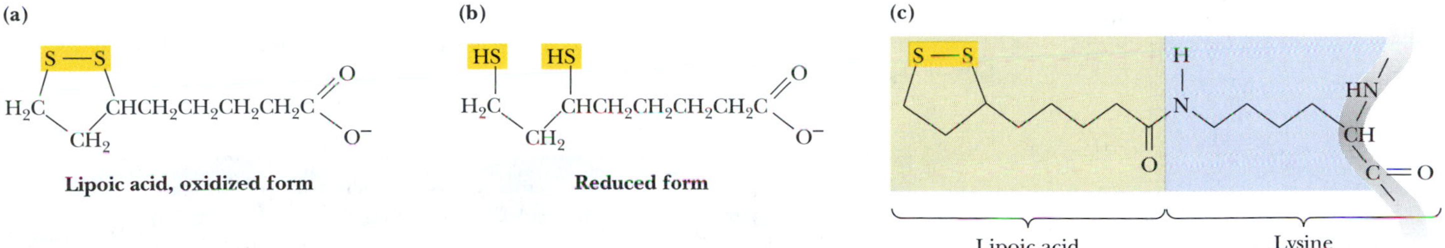

FIGURE 17.33 The oxidized and reduced forms of lipoic acid and the structure of the lipoic acid–lysine conjugate.

$$\text{Pyruvate} + \text{CoA} + \text{NAD}^+ \xrightarrow[\text{Pyruvate dehydrogenase}]{} \text{Acetyl-CoA} + CO_2 + \text{NADH} + H^+$$

$$\alpha\text{-Ketoglutarate} + \text{CoA} + \text{NAD}^+ \xrightarrow[\alpha\text{-Ketoglutarate dehydrogenase}]{} \text{Succinyl-CoA} + CO_2 + \text{NADH} + H^+$$

FIGURE 17.34 The enzyme reactions catalyzed by lipoic acid.

A Deeper Look

Lipoic Acid

Lipoic acid (6,8-dithiooctanoic acid) was isolated and characterized in 1951 in studies that showed that it was required for the growth of certain bacteria and protozoa. This accomplishment was one of the most impressive feats of isolation in the early history of biochemistry. Eli Lilly and Co., in cooperation with Lester J. Reed at the University of Texas and I. C. Gunsalus at the University of Illinois, isolated just 30 mg of lipoic acid from approximately 10 tons of liver! No evidence exists of a dietary lipoic acid requirement by humans; strictly speaking, it is not considered a vitamin. Nevertheless, it is an essential component of several enzymes of intermediary metabolism and is present in body tissues in small amounts.

Folic Acid

Folic acid derivatives (folates) are acceptors and donors of one-carbon units for all oxidation levels of carbon except that of CO_2 (where biotin is the relevant carrier). The active coenzyme form of folic acid is **tetrahydrofolate (THF).** THF is formed via two successive reductions of folate by *dihydrofolate reductase* (Figure 17.35). One-carbon units in three different oxidation states may be bound to tetrahydrofolate at the N^5 and/or N^{10} nitrogens (Table 17.6). These one-carbon units may exist at the oxidation levels of methanol, formaldehyde, or formate (carbon atom oxidation states of -2, 0, and 2, respectively). The biosynthetic pathways for methionine and homocysteine (Chapter 25), purines (Chapter 26), and the pyrimidine thymine (Chapter 26) rely on the incorporation of one-carbon units from THF derivatives.

The Vitamin A Group Includes Retinol, Retinal, and Retinoic Acid

Vitamin A or **retinol** (Figure 17.36) often occurs in the form of esters, called **retinyl esters.** The aldehyde form is called **retinal** or **retinaldehyde.** Like all the fat-soluble vitamins, retinol is an *isoprenoid* molecule and is biosynthesized from isoprene building blocks (see Chapter 8). Retinol can be absorbed in the diet from animal sources or synthesized from β-carotene from plant sources. The absorption by the body of fat-soluble vitamins proceeds by mechanisms different from those of the water-soluble vitamins. Once ingested, preformed vitamin A or β-carotene and its analogs are released from proteins by the action of proteolytic enzymes in the stomach and small intestine. The free carotenoids and retinyl esters aggregate in fatty globules that enter the duodenum. The detergent actions of bile salts break these globules down into small aggregates that can be digested by pancreatic lipase, cholesteryl ester hydrolase, retinyl ester

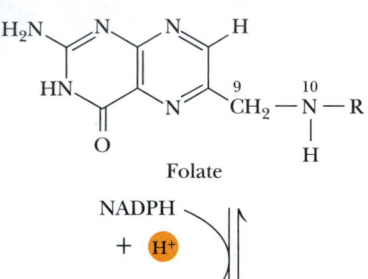
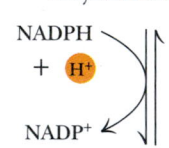
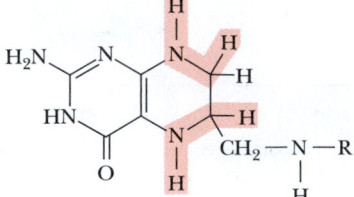

FIGURE 17.35 Formation of THF from folic acid by the dihydrofolate reductase reaction. The R group on these folate molecules includes the one to seven (or more) glutamate units that folates characteristically contain. All of these glutamates are bound in γ-carboxyl amide linkages (as in the folic acid structure shown in the A Deeper Look box on page 571). The one-carbon units carried by THF are bound at N^5, or at N^{10}, or as a single carbon attached to both N^5 and N^{10}.

Table 17.6			
Oxidation States of Carbon in One-Carbon Units Carried by Tetrahydrofolate			
Oxidation Number*	**Oxidation Level**	**One-Carbon Form†**	**Tetrahydrofolate Form**
-2	Methanol (most reduced)	$-CH_3$	N^5-Methyl-THF
0	Formaldehyde	$-CH_2-$	N^5,N^{10}-Methylene-THF
2	Formate (most oxidized)	$-CH=O$	N^5-Formyl-THF
		$-CH=O$	N^{10}-Formyl-THF
		$-CH=NH$	N^5-Formimino-THF
		$-CH=$	N^5,N^{10}-Methenyl-THF

*Calculated by assigning valence bond electrons to the more electronegative atom and then counting the charge on the quasi ion. A carbon assigned four valence electrons would have an oxidation number of 0. The carbon in N^5-methyl-THF is assigned six electrons from the three C—H bonds and thus has an oxidation number of -2.
†Note: All vacant bonds in the structures shown are to atoms more electronegative than C.

A Deeper Look

Folic Acid, Pterins, and Insect Wings

Folic acid is a member of the vitamin B complex found in green plants, fresh fruit, yeast, and liver. Folic acid takes its name from *folium*, Latin for "leaf." Pterin compounds are named from the Greek word *pté´ryj* for "wing" because these substances were first identified in insect wings. Two pterins are familiar to any child who has seen (and chased) the common yellow sulfur butterfly

and its white counterpart, the cabbage butterfly. *Xanthopterin* and *leucopterin* are the respective pigments in these butterflies' wings. Mammalian organisms cannot synthesize pterins; they derive folates from their diet or from microorganisms active in the intestines.

Folic acid

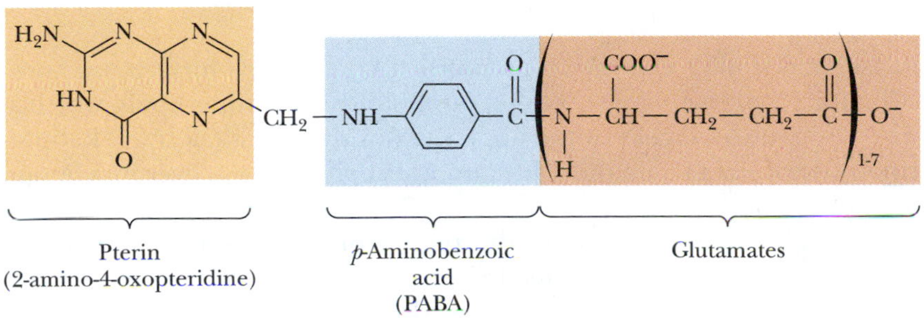

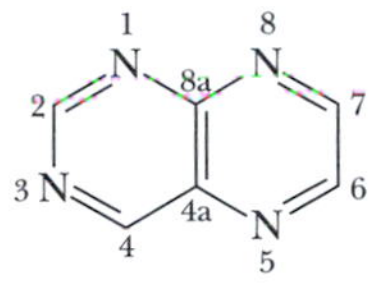

Pterin
(2-amino-4-oxopteridine)

p-Aminobenzoic acid (PABA)

Glutamates

Xanthopterin (yellow)   Leucopterin (white)

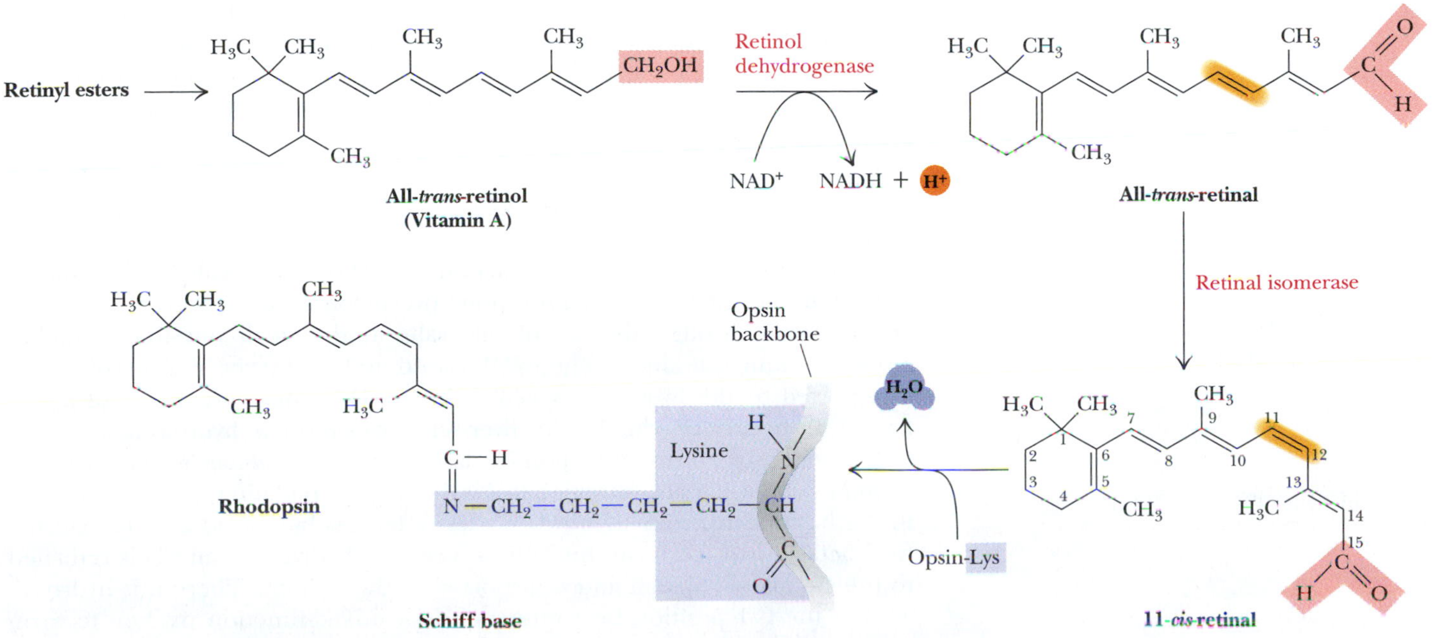

Pteridine

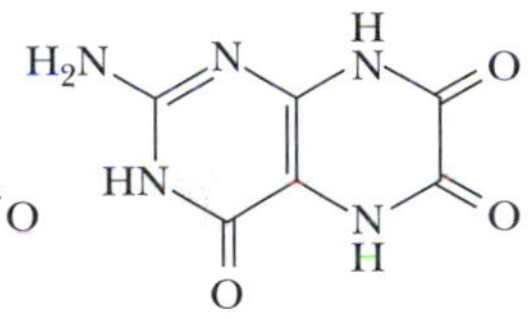

Pterin: 2-amino-4-oxopteridine

FIGURE 17.36 The incorporation of retinal into the light-sensitive protein rhodopsin involves several steps. All-*trans*-retinol is oxidized by retinol dehydrogenase and then isomerized to 11-*cis*-retinal, which forms a Schiff base linkage with opsin to form light-sensitive rhodopsin.

Human Biochemistry

β-Carotene and Vision

Night blindness was probably the first disorder to be ascribed to a nutritional deficiency. The ancient Egyptians left records as early as 1500 B.C. of recommendations that the juice squeezed from cooked liver could cure night blindness if applied topically, and the method may have been known much earlier. Frederick Gowland Hopkins, working in England in the early 1900s, found that alcoholic extracts of milk contained a growth-stimulating factor. Marguerite Davis and Elmer McCollum at Wisconsin showed that egg yolk and butter contain a similar growth-stimulating lipid, which, in 1915, they called "fat-soluble A." Moore in England showed that β-carotene, the plant pigment, could be converted to the colorless form of the liver-derived vitamin. In 1935, George Wald of Harvard showed that *retinene* found in visual pigments of the eye was identical with *retinaldehyde,* a derivative of vitamin A.

hydrolase, and similar enzymes. The product compounds form *mixed micelles* (see Chapter 8) containing the retinol, carotenoids, and other lipids, which are absorbed into mucosal cells in the upper half of the intestinal tract. Retinol is esterified (usually with palmitic acid) and transported to the liver in a lipoprotein complex.

The retinol that is delivered to the retinas of the eyes in this manner is accumulated by **rod** and **cone cells.** In the rods (which are the better characterized of the two cell types), retinol is oxidized by a specific **retinol dehydrogenase** to become all-*trans* retinal and then converted to 11-*cis*-retinal by **retinal isomerase** (Figure 17.36). The aldehyde group of retinal forms a Schiff base with a lysine on **opsin,** to form light-sensitive **rhodopsin.**

Retinoic acid is essential for proper cell division and differentiation, the immune response, and embryonic development.

Vitamin D Is Essential for Proper Calcium Metabolism

The two most prominent members of the **vitamin D** family are **ergocalciferol** (known as vitamin D_2) and **cholecalciferol** (vitamin D_3). Cholecalciferol is produced in the skin of animals by the action of ultraviolet light (sunlight, for example) on its precursor molecule, 7-dehydrocholesterol (Figure 17.37). The absorption of light energy induces a photoisomerization via an excited singlet state, which results in breakage of the 9,10 carbon bond and formation of **previtamin D_3.** The next step is a spontaneous isomerization to yield vitamin D_3, cholecalciferol. Ergocalciferol, which differs from cholecalciferol only in the side-chain structure, is similarly produced by the action of sunlight on the plant sterol **ergosterol.** (Ergosterol is so named because it was first isolated from ergot, a rye fungus.) Because humans can produce vitamin D_3 from 7-dehydrocholesterol by the action of sunlight on the skin, "vitamin D" is not strictly speaking a vitamin at all.

On the basis of its mechanism of action in the body, cholecalciferol should be called a **prohormone,** that is, a hormone precursor. Dietary forms of vitamin D are absorbed through the aid of bile salts in the small intestine. Whether absorbed in the intestine or photosynthesized in the skin, cholecalciferol is then transported to the liver by a specific **vitamin D–binding protein (DBP),** also known as **transcalciferin.** In the liver, cholecalciferol is hydroxylated at the C-25 position by a mixed-function oxidase to form *25-hydroxyvitamin D* (that is, *25-hydroxycholecalciferol*). Although this is the major circulating form of vitamin D in the body, 25-hydroxyvitamin D possesses far less biological activity than the final active form. To form this latter species, 25-hydroxyvitamin D is returned to the circulatory system and transported to the kidneys. There it is hydroxylated at the C-1 position by a mitochondrial mixed-function oxidase to form *1,25-dihydroxyvitamin D_3* (that is, *1,25-dihydroxycholecalciferol*), the active form of vitamin D. 1,25-Dihydroxycholecalciferol is then transported to target tissues, where it acts like a hormone to regulate calcium and phosphate metabolism.

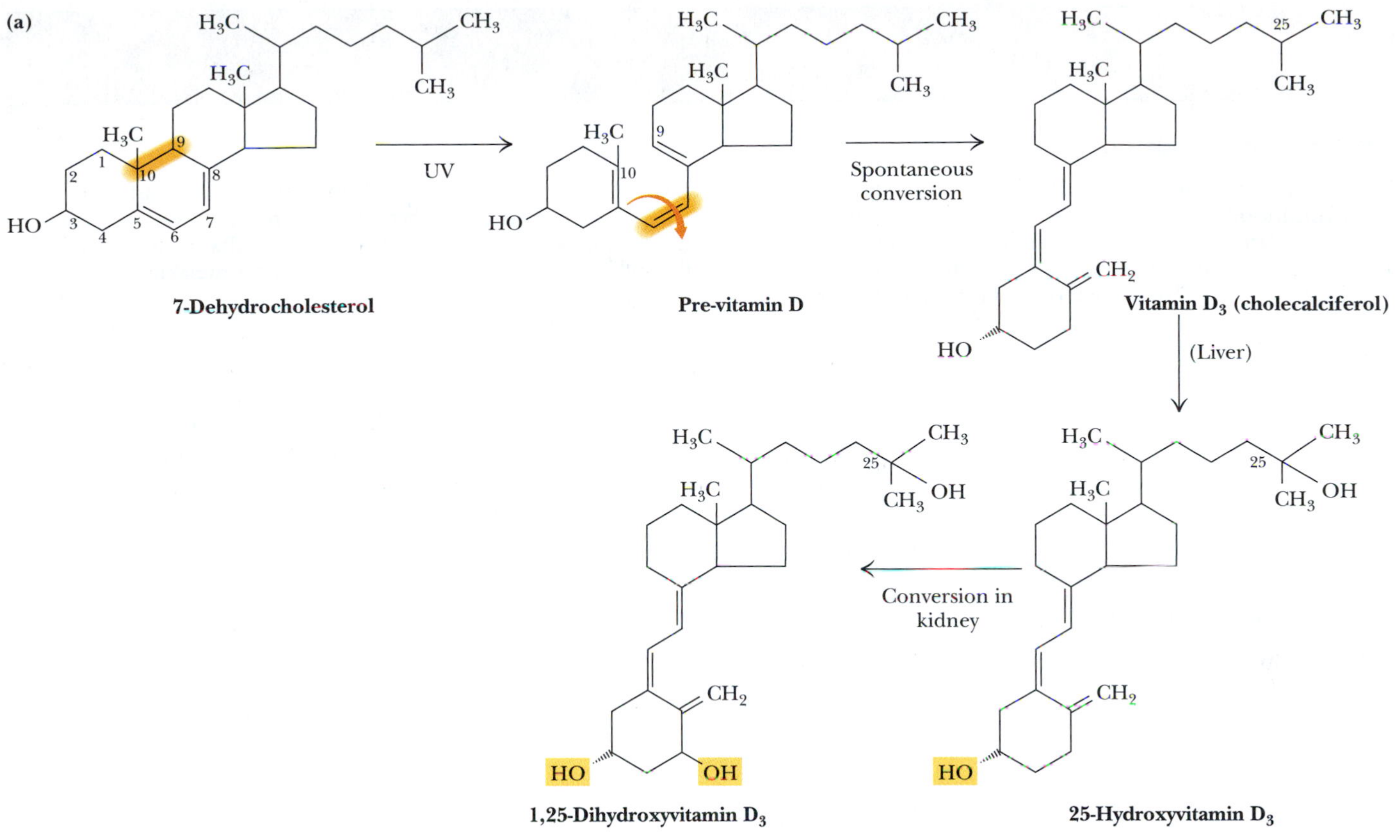

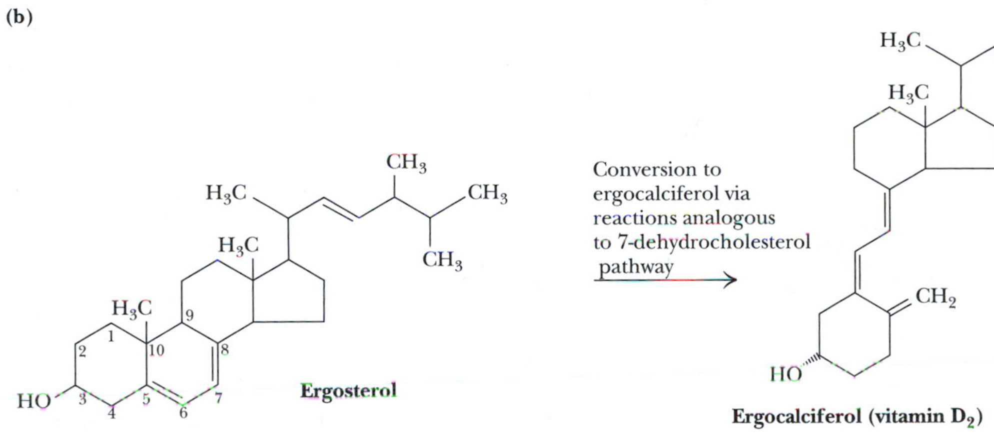

(a)

7-Dehydrocholesterol

Pre-vitamin D

Vitamin D₃ (cholecalciferol)

1,25-Dihydroxyvitamin D₃

25-Hydroxyvitamin D₃

(b)

Ergosterol

Ergocalciferol (vitamin D₂)

FIGURE 17.37 **(a)** Vitamin D₃ (cholecalciferol) is produced in the skin by the action of sunlight on 7-dehydrocholesterol. The successive action of mixed-function oxidases in the liver and kidney produces 1,25-dihydroxyvitamin D₃, the active form of vitamin D. **(b)** Ergocalciferol is produced in analogous fashion from ergosterol.

1,25-Dihydroxyvitamin D₃, together with two peptide hormones, *calcitonin* and *parathyroid hormone* (PTH), functions to regulate calcium homeostasis and plays a role in phosphorus homeostasis. As described elsewhere in this text, calcium is important for many processes, including muscle contraction, nerve impulse transmission, blood clotting, and membrane structure. Phosphorus, of course, is of critical importance to DNA, RNA, lipids, and many metabolic processes. Phosphorylation of proteins is an important regulatory signal for many biological processes. Phosphorus and calcium are also critically important for the formation of bones. Any disturbance of normal serum phosphorus and calcium levels will result in alterations of bone structure, as in rickets. The mechanism of calcium homeostasis involves precise coordination of calcium (1) absorption in the intestine, (2) deposition in the bones, and (3) excretion by the kidneys. If a decrease in serum calcium occurs, vitamin D is converted to its active form, which acts in the intestine to increase calcium absorption. PTH

Human Biochemistry

Vitamin D and Rickets

Vitamin D is a family of closely related molecules that prevent **rickets,** a childhood disease characterized by inadequate intestinal absorption and kidney reabsorption of calcium and phosphate. These inadequacies eventually lead to the demineralization of bones. The symptoms of rickets include bowlegs, knock-knees, curvature of the spine, and pelvic and thoracic deformities, the results of normal mechanical stresses on demineralized bones. Vitamin D deficiency in adults leads to a weakening of bones and cartilage, known as **osteomalacia.**

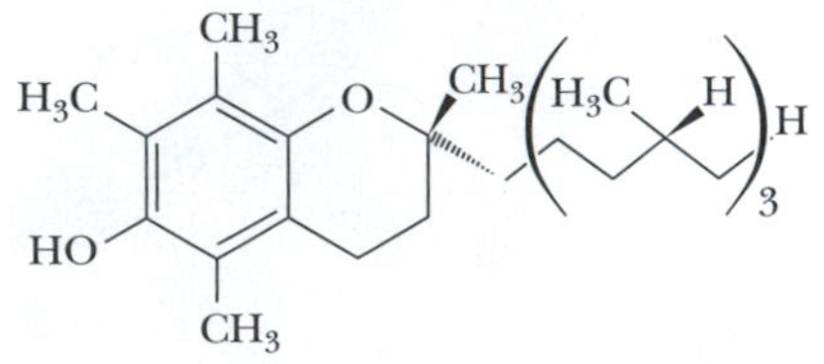

Vitamin E (α-tocopherol)

FIGURE 17.38 The structure of vitamin E (α-tocopherol).

and vitamin D act on bones to release calcium into the blood, and PTH acts on the kidney to cause increased calcium reabsorption. If serum calcium levels get too high, calcitonin induces calcium excretion from the kidneys and inhibits calcium mobilization from bone while inhibiting vitamin D metabolism and PTH secretion.

Vitamin E Is an Antioxidant

The structure of **vitamin E** in its most active form, **α-tocopherol,** is shown in Figure 17.38. α-Tocopherol is a potent antioxidant, and its function in animals and humans is often ascribed to this property. On the other hand, the molecular details of its function are almost entirely unknown. One possible role for vitamin E may relate to the protection of unsaturated fatty acids in membranes because these fatty acids are particularly susceptible to oxidation. When human plasma levels of α-tocopherol are low, red blood cells are increasingly subject to oxidative hemolysis. Infants, especially premature infants, are deficient in vitamin E. When low-birth-weight infants are exposed to high oxygen levels for the purpose of alleviating respiratory distress, the risk of oxygen-induced retina damage can be reduced with vitamin E administration. The mechanisms of action of vitamin E remain obscure.

Vitamin K Is Essential for Carboxylation of Protein Glutamate Residues

The function of vitamin K (Figure 17.39) in the activation of blood clotting was not elucidated until the early 1970s, when it was found that animals and humans treated with coumadin-type anticoagulants contained an inactive form of prothrombin (an essential protein in the coagulation cascade). It was soon shown that a post-translational modification of prothrombin is essential to its function. In this modification, ten glutamic acid residues on the amino terminal end of prothrombin are carboxylated to form γ-carboxyglutamyl residues. These residues are effective in the coordination of calcium, which is required for the coag-

A Deeper Look

Vitamin E

In a study of the effect of nutrition on reproduction in the rat in the 1920s, Herbert Evans and Katherine Bishop found that rats failed to reproduce on a diet of rancid lard, unless lettuce or whole wheat was added to the diet. The essential factor was traced to a vitamin in the wheat germ oil. Named *vitamin E* by Evans (using the next available letter following on the discovery of vitamin D), the factor was purified by Emerson, who named it *tocopherol,* from the Greek *tokos,* for "childbirth," and *pherein,* for "to bring forth." Vitamin E is now recognized as a generic term for a family of substances, all of them similar in structure to the most active form, α-tocopherol.

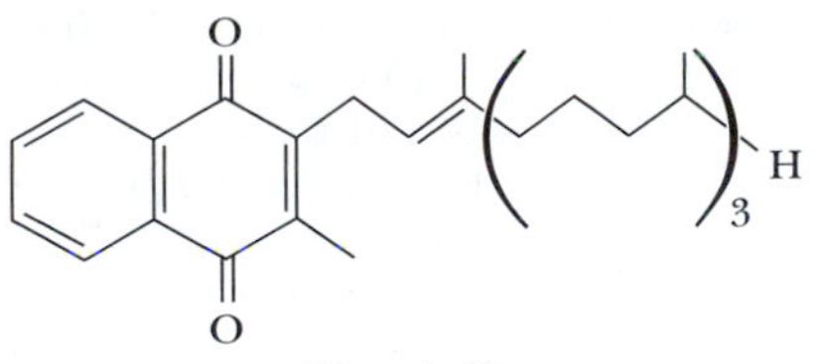

Vitamin K₁
(phylloquinone)

Vitamin K₂
(menaquinone series)

FIGURE 17.39 The structures of the K vitamins.

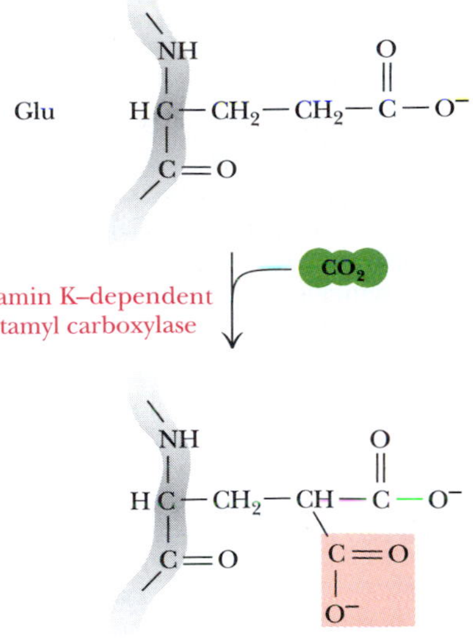

γ-**Carboxyglutamic acid
in a protein**

FIGURE 17.40 The glutamyl carboxylase reaction is vitamin K–dependent. This enzyme activity is essential for the formation of γ-carboxyglutamyl residues in a variety of proteins, including several proteins of the blood-clotting cascade (Figure 15.5). These latter carboxylations account for the vitamin K dependence of coagulation.

ulation process. The enzyme responsible for this modification, a liver microsomal glutamyl carboxylase, requires vitamin K for its activity (Figure 17.40). Not only prothrombin (called "factor II" in the clotting pathway) but also clotting factors VII, IX, and X and several plasma proteins—proteins C, M, S, and Z—contain γ-carboxyglutamyl residues in a manner similar to prothrombin. Other examples of γ-carboxyglutamyl residues in proteins are known.

Human Biochemistry

Vitamin K and Blood Clotting

In studies in Denmark in the 1920s, Henrik Dam noticed that chicks fed a diet extracted with nonpolar solvents developed hemorrhages. Moreover, blood taken from such animals clotted slowly. Further studies by Dam led him to conclude in 1935 that the antihemorrhage factor was a new fat-soluble vitamin, which he called *vitamin K* (from *koagulering*, the Danish word for "coagulation").

Dam, along with Karrar of Zurich, isolated the pure vitamin from alfalfa as a yellow oil. Another form, which was crystalline at room temperature, was soon isolated from fishmeal. These two compounds were named *vitamins K₁* and *K₂*. Vitamin K₂ can actually occur as a family of structures with different chain lengths at the C-3 position.

Summary

Metabolism represents the sum of the chemical changes that convert nutrients, the "raw materials" necessary to nourish living organisms, into energy and the chemically complex finished products of cells. Metabolism consists of literally hundreds of enzymatic reactions organized into discrete pathways.

17.1 Are There Similarities of Metabolism Between Organisms?
One of the great unifying principles of modern biology is that organisms show marked similarity in their major pathways of metabolism. Given the almost unlimited possibilities within organic chemistry, this generality would appear most unlikely. Yet it's true, and it provides strong evidence that all life has descended from a common ancestral form. All forms of nutrition and almost all metabolic pathways evolved in early prokaryotes prior to the appearance of eukaryotes 1 billion years ago. All organisms, even those that can synthesize

their own glucose, are capable of glucose degradation and ATP synthesis via glycolysis. Other prominent pathways are also virtually ubiquitous among organisms.

17.2 How Do Anabolic and Catabolic Processes Form the Core of Metabolic Pathways?
Catabolism involves the oxidative degradation of complex nutrient molecules (carbohydrates, lipids, and proteins) obtained either from the environment or from cellular reserves. The breakdown of these molecules by catabolism leads to the formation of simpler molecules such as lactic acid, ethanol, carbon dioxide, urea, or ammonia. Catabolic reactions are usually exergonic, and often the chemical energy released is captured in the form of ATP. *Anabolism* is a synthetic process in which the varied and complex biomolecules (proteins, nucleic acids, polysaccharides, and lipids) are assembled from simpler precursors. Such biosynthesis involves the formation of new covalent

bonds, and an input of chemical energy is necessary to drive such endergonic processes. The ATP generated by catabolism provides this energy. Furthermore, NADPH is an excellent donor of high-energy electrons for the reductive reactions of anabolism.

17.3 What Experiments Can Be Used to Elucidate Metabolic Pathways?
An important tool for elucidating the steps in the pathway is the use of *metabolic inhibitors*. Adding an enzyme inhibitor to a cell-free extract causes an accumulation of intermediates in the pathway prior to the point of inhibition. Each inhibitor is specific for a particular site in the sequence of metabolic events. Genetics provides an approach to the identification of intermediate steps in metabolism that is somewhat analogous to inhibition. Mutation in a gene encoding an enzyme often results in an inability to synthesize the enzyme in an active form. Such a defect leads to a block in the metabolic pathway at the point where the enzyme acts, and the enzyme's substrate accumulates. Such genetic disorders are lethal if the end product of the pathway is essential or if the accumulated intermediates have toxic effects. In microorganisms, however, it is often possible to manipulate the growth medium so that essential end products are provided. Then the biochemical consequences of the mutation can be investigated.

17.4 What Food Substances Form the Basis of Human Nutrition?
In addition to essential fiber, the food that human beings require includes the macronutrients—protein, carbohydrate, and lipid—and the micronutrients—including vitamins and minerals.

Problems

1. If 3×10^{14} kg of CO_2 are cycled through the biosphere annually, how many human equivalents (70-kg persons composed of 18% carbon by weight) could be produced each year from this amount of CO_2?

2. Define the differences in carbon and energy metabolism between *photoautotrophs* and *photoheterotrophs* and between *chemoautotrophs* and *chemoheterotrophs*.

3. Name three principal inorganic sources of oxygen atoms that are commonly available in the inanimate environment and readily accessible to the biosphere.

4. What are the features that generally distinguish pathways of catabolism from pathways of anabolism?

5. Name the three principal modes of enzyme organization in metabolic pathways.

6. (Integrates with Chapter 1.) Why do metabolic pathways have so many different steps?

7. Why is the pathway for the biosynthesis of a biomolecule at least partially different from the pathway for its catabolism? Why is the pathway for the biosynthesis of a biomolecule inherently more complex than the pathway for its degradation?

8. (Integrates with Chapters 1 and 3.) What are the metabolic roles of ATP, NAD⁺, and NADPH?

9. (Integrates with Chapter 15.) Metabolic regulation is achieved via regulating enzyme activity in three prominent ways: allosteric regulation, covalent modification, and enzyme synthesis and degradation. Which of these three modes of regulation is likely to be the quickest; which the slowest? For each of these general enzyme regulatory mechanisms, cite conditions in which cells might employ that mode in preference to either of the other two.

10. What are the advantages of compartmentalizing particular metabolic pathways within specific organelles?

11. Maple syrup urine disease (MSUD) is an autosomal recessive genetic disease characterized by progressive neurological dysfunction and a sweet, burnt-sugar or maple-syrup smell in the urine. Affected individuals carry high levels of branched-chain amino acids (leucine, isoleucine, and valine) and their respective branched-chain α-keto acids in cells and body fluids. The genetic defect has been traced to the mitochondrial branched-chain α-keto acid dehydrogenase (BCKD). Affected individuals exhibit mutations in their BCKD, but these mutant enzymes exhibit normal levels of activity. Nonetheless, treatment of MSUD patients with substantial doses of thiamine can alleviate the symptoms of the disease. Suggest an explanation for the symptoms described and for the role of thiamine in ameliorating the symptoms of MSUD.

12. (Integrates with Chapter 14.) Write a simple enzyme mechanism for liver alcohol dehydrogenase, which interconverts ethanol and acetaldehyde and which uses NADH as a coenzyme.

13. (Integrates with Chapter 14.) Write a simple enzyme mechanism for tyrosine racemase, an enzyme that interconverts L-tyrosine and D-tyrosine, and which uses pyridoxal phosphate as a coenzyme.

14. Write a simple enzyme mechanism for the carboxylation of pyruvate using biotin as a coenzyme.

Preparing for the MCAT Exam

15. Consult Table 17.3, and consider the information presented for ^{32}P and ^{35}S. Write reactions for the decay events for these two isotopes, indicating clearly the products of the decays, and calculate what percentage of each would remain from a sample that contained both and decayed for 100 days.

16. Which statement is most likely to be true concerning obligate anaerobes?
    a. These organisms can use oxygen if it is present in their environment.
    b. These organisms cannot use oxygen as their final electron acceptor.
    c. These organisms carry out fermentation for at least 50% of their ATP production.
    d. Most of these organisms are vegetative fungi.

17. Foods rich in fiber are basically plant materials high in cellulose, a cell wall polysaccharide that we cannot digest. The nutritional benefits provided by such foods result from
    a. other nutrients present that can be digested and absorbed.
    b. macromolecules (like cellulose) that are absorbed without digestion and then catabolized inside the cells.
    c. microbes that are the normal symbionts of plant tissues.
    d. All of the above.

Biochemistry❸Now™ Preparing for an exam? Test yourself on key questions at http://chemistry.brookscole.com/ggb3

Further Reading

Metabolism

Atkinson, D. E., 1977. *Cellular Energy Metabolism and Its Regulation.* New York: Academic Press. A monograph on energy metabolism that is filled with novel insights regarding the ability of cells to generate energy in a carefully regulated fashion while contending with the thermodynamic realities of life.

Cooper, T. G., 1977. *The Tools of Biochemistry.* New York: Wiley-Interscience. Chapter 3, "Radiochemistry," discusses techniques for using radioisotopes in biochemistry.

Reed, L., 1974. Multienzyme complexes. *Accounts of Chemical Research* **7:**40–46.

Srere, P. A., 1987. Complexes of sequential metabolic enzymes. *Annual Review of Biochemistry* **56:**89–124. A review of how enzymes in some metabolic pathways are organized into complexes.

Vitamins

Boyer, P. D., 1970. *The Enzymes,* 3rd ed. New York: Academic Press. A good reference source for the mechanisms of action of vitamins and coenzymes.

Boyer, P. D., 1970. *The Enzymes,* Vol. 6. New York: Academic Press. See discussion of carboxylation and decarboxylation involving TPP, PLP, lipoic acid, and biotin; B_{12}-dependent mutases.

Boyer, P. D., 1972. *The Enzymes,* Vol. 7. New York: Academic Press. See especially elimination reactions involving PLP.

Boyer, P. D., 1974. *The Enzymes,* Vol. 10. New York: Academic Press. See discussion of pyridine nucleotide–dependent enzymes.

Boyer, P. D., 1976. *The Enzymes,* Vol. 13. New York: Academic Press. See discussion of flavin-dependent enzymes.

DeLuca, H., and Schnoes, H., 1983. Vitamin D: Recent advances. *Annual Review of Biochemistry* **52:**411–439.

Jencks, W. P., 1969. *Catalysis in Chemistry and Enzymology.* New York: McGraw-Hill.

Knowles, J. R., 1989. The mechanism of biotin-dependent enzymes. *Annual Review of Biochemistry* **58:**195–221.

Page, M. I., and Williams, A., eds., 1987. *Enzyme Mechanisms.* London: Royal Society of London.

Walsh, C. T., 1979. *Enzymatic Reaction Mechanisms.* San Francisco: W. H. Freeman.

CHAPTER 18 Glycolysis

Louis Pasteur's scientific investigations into fermentation of grape sugar were pioneering studies of glycolysis.

Living organisms, like machines, conform to the law of conservation of energy, and must pay for all their activities in the currency of catabolism.

Ernest Baldwin, *Dynamic Aspects of Biochemistry* (1952)

Key Questions

Essential Question

Nearly every living cell carries out a catabolic process known as **glycolysis**—the stepwise degradation of glucose (and other simple sugars). Glycolysis is a paradigm of metabolic pathways. Carried out in the cytosol of cells, it is basically an anaerobic process; its principal steps occur with no requirement for oxygen. Living things first appeared in an environment lacking O_2, and glycolysis was an early and important pathway for extracting energy from nutrient molecules. It played a central role in anaerobic metabolic processes during the first 2 billion years of biological evolution on earth. Modern organisms still employ glycolysis to provide precursor molecules for aerobic catabolic pathways (such as the tricarboxylic acid cycle) and as a short-term energy source when oxygen is limiting. *What is the chemical basis and logic for this central pathway of metabolism: how does glycolysis work?*

18.1   What Are the Essential Features of Glycolysis?

An overview of the glycolytic pathway is presented in Figure 18.1. Most of the details of this pathway (the first metabolic pathway to be elucidated) were worked out in the first half of the 20th century by the German biochemists Otto Warburg, G. Embden, and O. Meyerhof. In fact, the sequence of reactions in Figure 18.1 is often referred to as the **Embden–Meyerhof pathway.**

Glycolysis consists of two phases. In the first phase, a series of five reactions, glucose is broken down to two molecules of glyceraldehyde-3-phosphate. In the second phase, five subsequent reactions convert these two molecules of glyceraldehyde-3-phosphate into two molecules of pyruvate. Phase 1 consumes two molecules of ATP (Figure 18.2). The later stages of glycolysis result in the production of four molecules of ATP. The net is $4 - 2 = 2$ molecules of ATP produced per molecule of glucose.

Rates and Regulation of Glycolytic Reactions Vary Among Species

Microorganisms, plants, and animals (including humans) carry out the ten reactions of glycolysis in more or less similar fashion, although the rates of the individual reactions and the means by which they are regulated differ from species to species. The most significant difference among species, however, is the way in which the product pyruvate is utilized. The three possible paths for pyruvate are shown in Figure 18.1. In aerobic organisms, including humans, pyruvate is oxidized (with loss of the carboxyl group as CO_2), and the remaining two-carbon unit becomes the acetyl group of acetyl-coenzyme A. This acetyl group is metabolized by the tricarboxylic acid cycle (and fully oxidized) to yield CO_2. The electrons removed in this oxidation process are subsequently passed through the mitochondrial electron transport system and used to generate molecules of ATP by oxidative phosphorylation, thus capturing most of the metabolic energy available in the original glucose molecule.

18.2   Why Are Coupled Reactions Important in Glycolysis?

The process of glycolysis converts some, but not all, of the metabolic energy of the glucose molecule into ATP. The free energy change for the conversion of glucose to two molecules of lactate (the anaerobic route shown in Figure 18.1) is -183.6 kJ/mol:

$$C_6H_{12}O_6 \rightarrow 2\ H_3C\text{—}CHOH\text{—}COO^- + 2\ H^+ \tag{18.1}$$
$$\Delta G^{\circ\prime} = -183.6\ \text{kJ/mol}$$

This process occurs with no net oxidation or reduction. Although several individual steps in the pathway involve oxidation or reduction, these steps compensate each other exactly. Thus, the conversion of a molecule of glucose to two molecules of lactate involves simply a rearrangement of bonds, with no net loss or gain of electrons. The energy made available through this rearrangement into a more stable (lower-energy) form is a relatively small part of the total energy obtainable from glucose.

The production of two molecules of ATP in glycolysis is an energy-requiring process:

$$2\ ADP + 2\ P_i \longrightarrow 2\ ATP + 2\ H_2O \qquad (18.2)$$
$$\Delta G^{\circ\prime} = 2 \times 30.5\ kJ/mol = 61.0\ kJ/mol$$

Glycolysis couples these two reactions:

$$Glucose + 2\ ADP + 2\ P_i \rightarrow 2\ lactate + 2\ ATP + 2\ H^+ + 2\ H_2O \quad (18.3)$$
$$\Delta G^{\circ\prime} = -183.6 + 61 = -122.6\ kJ/mol$$

Thus, under standard-state conditions, $(61/183.6) \times 100\%$, or 33%, of the free energy released is preserved in the form of ATP in these reactions. However, as we discussed in Chapter 3, the various solution conditions, such as pH, concentration, ionic strength, and presence of metal ions, can substantially alter the free energy change for such reactions. Under actual cellular conditions, the free energy change for the synthesis of ATP (Equation 18.2) is much larger, and approximately 50% of the available free energy is converted into ATP. Clearly, then, more than enough free energy is available in the conversion of glucose into lactate to drive the synthesis of two molecules of ATP.

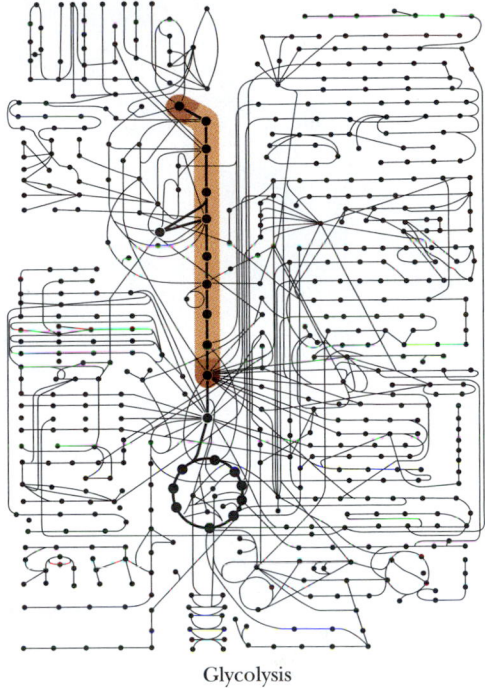

Glycolysis

18.3	**What Are the Chemical Principles and Features of the First Phase of Glycolysis?**

One way to synthesize ATP using the metabolic free energy contained in the glucose molecule would be to convert glucose into one (or more) of the high-energy phosphates in Table 3.3 that have standard-state free energies of hydrolysis more negative than that of ATP. Those molecules in Table 3.3 that can be synthesized easily from glucose are phosphoenolpyruvate, 1,3-bisphosphoglycerate, and acetyl phosphate. In fact, in the first stage of glycolysis, glucose is converted into two molecules of glyceraldehyde-3-phosphate. Energy released from this high-energy molecule in the second phase of glycolysis is then used to synthesize ATP.

Reaction 1: Glucose Is Phosphorylated by Hexokinase or Glucokinase—The First Priming Reaction

The initial reaction of the glycolysis pathway involves phosphorylation of glucose at carbon atom 6 by either hexokinase or glucokinase. The formation of such a phosphoester is thermodynamically unfavorable and requires energy input to operate in the forward direction (see Chapter 3). The energy comes from ATP, a requirement that at first seems counterproductive. Glycolysis is designed to *make* ATP, not consume it. However, the hexokinase, glucokinase reaction (Figure 18.2) is one of two **priming reactions** in the cycle. Just as old-fashioned, hand-operated water pumps (Figure 18.3) have to be primed with a small amount of water to deliver more water to the thirsty pumper, the glycolysis pathway requires two priming ATP molecules to start the sequence of reactions and delivers four molecules of ATP in the end.

The complete reaction for the first step in glycolysis is

$$\alpha\text{-D-Glucose} + ATP^{4-} \longrightarrow \alpha\text{-D-glucose-6-phosphate}^{2-} + ADP^{3-} + H^+ \qquad (18.4)$$
$$\Delta G^{\circ\prime} = -16.7\ kJ/mol$$

The hydrolysis of ATP makes 30.5 kJ/mol available in this reaction, and the phosphorylation of glucose "costs" 13.8 kJ/mol (Table 18.1). Thus, the reaction

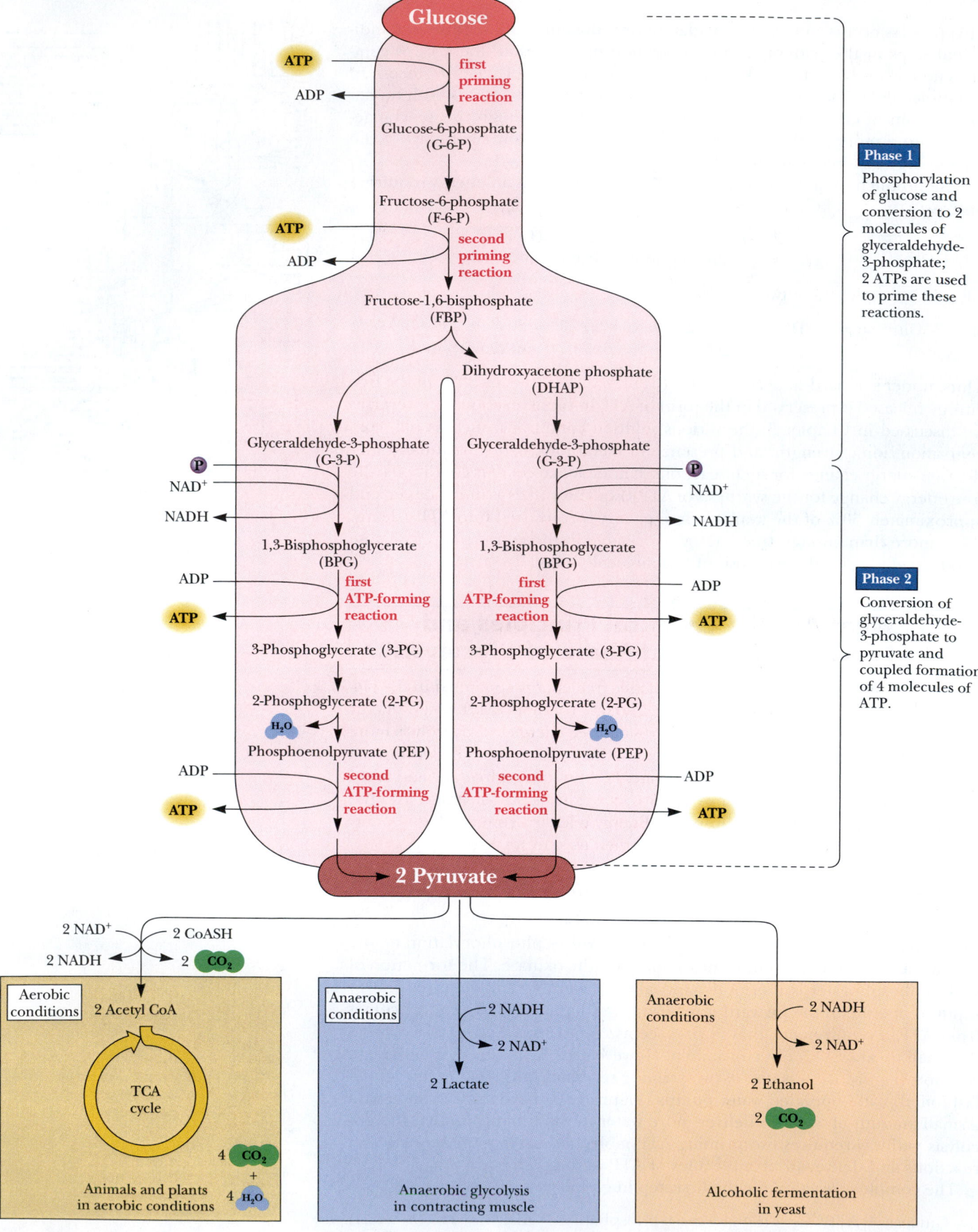

Biochemistry Now™ ACTIVE FIGURE 18.1 The glycolytic pathway. **Test yourself on the concepts in this figure at http://chemistry.brookscole.com/ggb3**

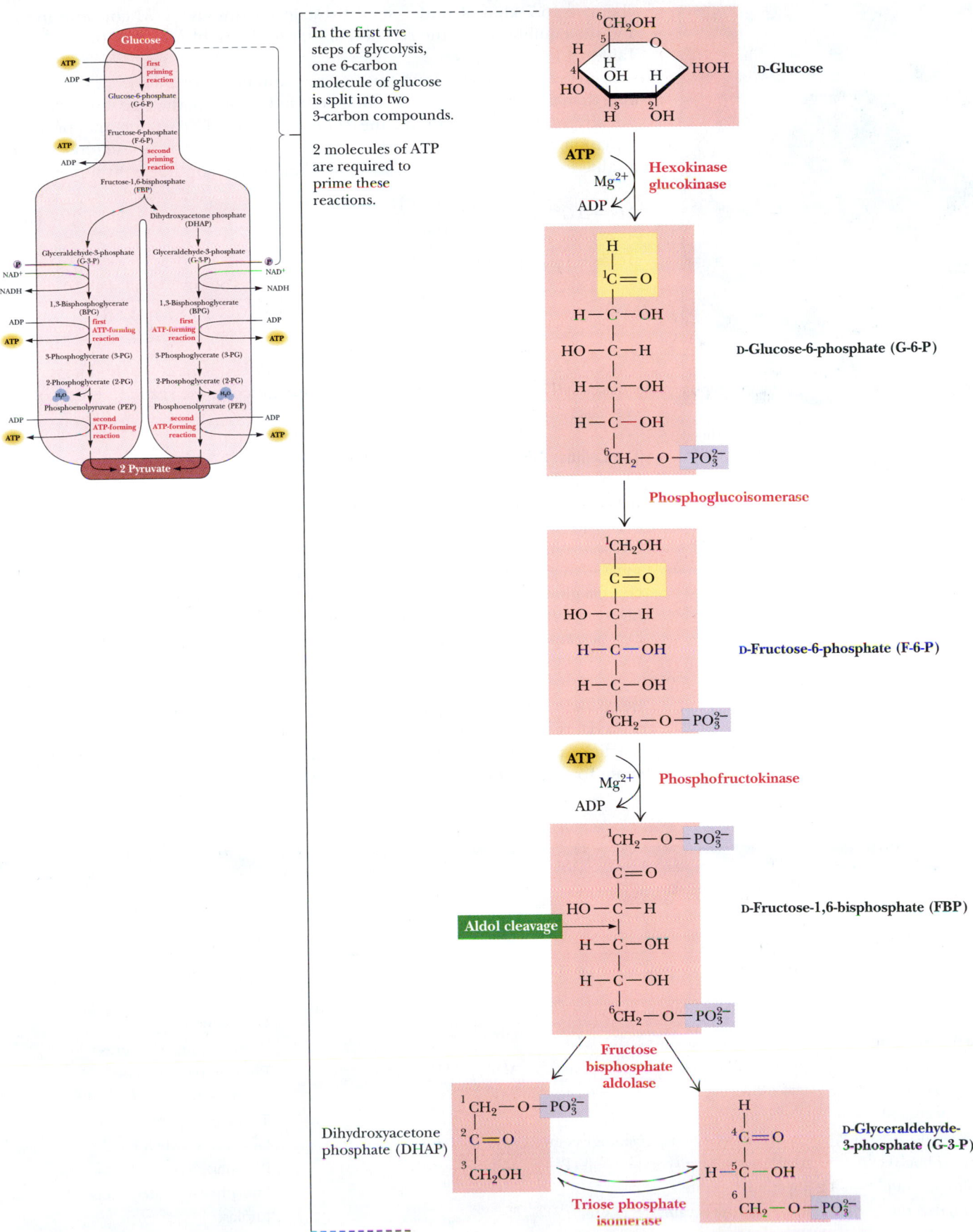

FIGURE 18.2 In the first phase of glycolysis, five reactions convert a molecule of glucose to two molecules of glyceraldehyde-3-phosphate.

Michelle Sassi/The Stock Market

FIGURE 18.3 Just as a water pump must be "primed" with water to get more water out, the glycolytic pathway is primed with ATP in steps 1 and 3 in order to achieve net production of ATP in the second phase of the pathway.

liberates 16.7 kJ/mol under standard-state conditions (1 M concentrations), and the equilibrium of the reaction lies far to the right ($K_{eq} = 850$ at 25°C; see Table 18.1).

Under cellular conditions, this first reaction of glycolysis is even more favorable than at standard state. As pointed out in Chapter 3, the free energy change for any reaction depends on the concentrations of reactants and products. Equation 3.12 in Chapter 3 and the data in Table 18.2 can be used to calculate a value for ΔG for the hexokinase, glucokinase reaction in erythrocytes:

$$\Delta G = \Delta G°' + RT \ln\left(\frac{[\text{G-6-P}][\text{ADP}]}{[\text{Glu}][\text{ATP}]}\right) \qquad (18.5)$$

$$\Delta G = -16.7 \text{ kJ/mol} +$$

$$(8.314 \text{ J/mol} \cdot \text{K})(310 \text{ K}) \ln\left(\frac{(8.3 \times 10^{-5} \, M)(1.4 \times 10^{-4} \, M)}{(5.0 \times 10^{-3} \, M)(1.85 \times 10^{-3} \, M)}\right)$$

$$\Delta G = -33.9 \text{ kJ/mol}$$

Thus, ΔG is even more favorable under cellular conditions than at standard state. As we will see later in this chapter, the hexokinase, glucokinase reaction is one of several that drive glycolysis forward.

The Cellular Advantages of Phosphorylating Glucose   The incorporation of a phosphate into glucose in this energetically favorable reaction is important for several reasons. First, phosphorylation keeps the substrate in the cell. Glucose is a neutral molecule and could diffuse across the cell membrane, but phosphorylation confers a negative charge on glucose and the plasma membrane is essentially impermeable to glucose-6-phosphate (Figure 18.4). Moreover, rapid conversion of glucose to glucose-6-phosphate keeps the *intracellular* concentration of glucose low, favoring diffusion of glucose *into* the cell. In addition, because regulatory control can be imposed only on reactions not at equilibrium, the favorable thermodynamics of this first reaction makes it an important site for regulation.

Hexokinase   In most animal, plant, and microbial cells, the enzyme that phosphorylates glucose is **hexokinase**. Magnesium ion (Mg^{2+}) is required for this reaction, as for the other kinase enzymes in the glycolytic pathway. The true substrate for the hexokinase reaction is $MgATP^{2-}$. The apparent K_m for glucose of

Table 18.1

Reactions and Thermodynamics of Glycolysis

Reaction	Enzyme
α-D-Glucose + ATP^{4-} ⇌ glucose-6-phosphate^{2-} + ADP^{3-} + H$^+$	Hexokinase
	Hexokinase
	Glucokinase
Glucose-6-phosphate^{2-} ⇌ fructose-6-phosphate^{2-}	Phosphoglucoisomerase
Fructose-6-phosphate^{2-} + ATP^{4-} ⇌ fructose-1,6-bisphosphate^{4-} + ADP^{3-} + H$^+$	Phosphofructokinase
Fructose-1,6-bisphosphate^{4-} ⇌ dihydroxyacetone-P^{2-} + glyceraldehyde-3-P^{2-}	Fructose bisphosphate aldolase
Dihydroxyacetone-P^{2-} ⇌ glyceraldehyde-3-P^{2-}	Triose phosphate isomerase
Glyceraldehyde-3-P^{2-} + P$_i^{2-}$ + NAD$^+$ ⇌ 1,3-bisphosphoglycerate^{4-} + NADH + H$^+$	Glyceraldehyde-3-P dehydrogenase
1,3-Bisphosphoglycerate^{4-} + ADP^{3-} ⇌ 3-P-glycerate^{3-} + ATP^{4-}	Phosphoglycerate kinase
3-Phosphoglycerate^{3-} ⇌ 2-phosphoglycerate^{3-}	Phosphoglycerate mutase
2-Phosphoglycerate^{3-} ⇌ phosphoenolpyruvate^{3-} + H$_2$O	Enolase
Phosphoenolpyruvate^{3-} + ADP^{3-} + H$^+$ ⇌ pyruvate$^-$ + ATP^{4-}	Pyruvate kinase
Pyruvate$^-$ + NADH + H$^+$ ⇌ lactate$^-$ + NAD$^+$	Lactate dehydrogenase

the animal skeletal muscle enzyme is approximately 0.1 mM, and the enzyme thus operates efficiently at normal blood glucose levels of 4 mM or so. Different body tissues possess different isozymes of hexokinase, each exhibiting somewhat different kinetic properties. The animal enzyme is allosterically inhibited by the product, glucose-6-phosphate. High levels of glucose-6-phosphate inhibit hexokinase activity until consumption by glycolysis lowers its concentration. The hexokinase reaction is one of three points in the glycolysis pathway that are *regulated*. As the generic name implies, hexokinase can phosphorylate a variety of hexose sugars, including glucose, mannose, and fructose.

Glucokinase Liver contains an enzyme called **glucokinase,** which also carries out the reaction in Figure 18.4 but is highly specific for D-glucose, has a much higher K_m for glucose (approximately 10 mM), and is not product inhibited. With such a high K_m for glucose, glucokinase becomes important metabolically only when liver glucose levels are high (for example, when the individual has consumed large amounts of sugar). When glucose levels are low, hexokinase is primarily responsible for phosphorylating glucose. However, when glucose levels are high, glucose is converted by glucokinase to glucose-6-phosphate and is eventually stored in the liver as glycogen. Glucokinase is an *inducible* enzyme—the amount present in the liver is controlled by *insulin* (secreted by the pancreas). (Patients with **diabetes mellitus** produce insufficient insulin. They have low levels of glucokinase, cannot tolerate high levels of blood glucose, and produce little liver glycogen.) Because glucose-6-phosphate is common to several metabolic pathways (Figure 18.5), it occupies a branch point in glucose metabolism.

Reaction 2: Phosphoglucoisomerase Catalyzes the Isomerization of Glucose-6-Phosphate

The second step in glycolysis is a common type of metabolic reaction: the isomerization of a sugar. In this particular case, the carbonyl oxygen of glucose-6-phosphate is shifted from C-1 to C-2. This amounts to isomerization of an aldose (glucose-6-phosphate) to a ketose—fructose-6-phosphate (Figure 18.6). The reaction is necessary for two reasons. First, the next step in glycolysis is phosphorylation at C-1, and the hemiacetal —OH of glucose, would be more difficult to phosphorylate than a simple primary hydroxyl. Second, the isomerization to

Source	Subunit Molecular Weight (M_r)	Oligomeric Composition	$\Delta G^{\circ\prime}$ (kJ/mol)	K_{eq} at 25°C	ΔG (kJ/mol)
Mammals	100,000	Monomer	−16.7	850	−33.9*
Yeast	55,000	Dimer			
Mammalian liver	50,000	Monomer			
Human	65,000	Dimer	+1.67	0.51	−2.92
Rabbit muscle	78,000	Tetramer	−14.2	310	−18.8
Rabbit muscle	40,000	Tetramer	+23.9	6.43×10^{-5}	−0.23
Chicken muscle	27,000	Dimer	+7.56	0.0472	+2.41
Rabbit muscle	37,000	Tetramer	+6.30	0.0786	−1.29
Rabbit muscle	64,000	Monomer	−18.9	2060	+0.1
Rabbit muscle	27,000	Dimer	+4.4	0.169	+0.83
Rabbit muscle	41,000	Dimer	+1.8	0.483	+1.1
Rabbit muscle	57,000	Tetramer	−31.7	3.63×10^5	−23.0
Rabbit muscle	35,000	Tetramer	−25.2	2.63×10^4	−14.8

*ΔG values calculated for 310K (37°C) using the data in Table 18.2 for metabolite concentrations in erythrocytes. $\Delta G^{\circ\prime}$ values are assumed to be the same at 25° and 37°C.

Table 18.2

Steady-State Concentrations of Glycolytic Metabolites in Erythrocytes

Metabolite	mM
Glucose	5.0
Glucose-6-phosphate	0.083
Fructose-6-phosphate	0.014
Fructose-1,6-bisphosphate	0.031
Dihydroxyacetone phosphate	0.14
Glyceraldehyde-3-phosphate	0.019
1,3-Bisphosphoglycerate	0.001
2,3-Bisphosphoglycerate	4.0
3-Phosphoglycerate	0.12
2-Phosphoglycerate	0.030
Phosphoenolpyruvate	0.023
Pyruvate	0.051
Lactate	2.9
ATP	1.85
ADP	0.14
P_i	1.0

Adapted from Minakami, S., and Yoshikawa, H., 1965. Thermodynamic considerations on erythrocyte glycolysis. *Biochemical and Biophysical Research Communications* **18**:345.

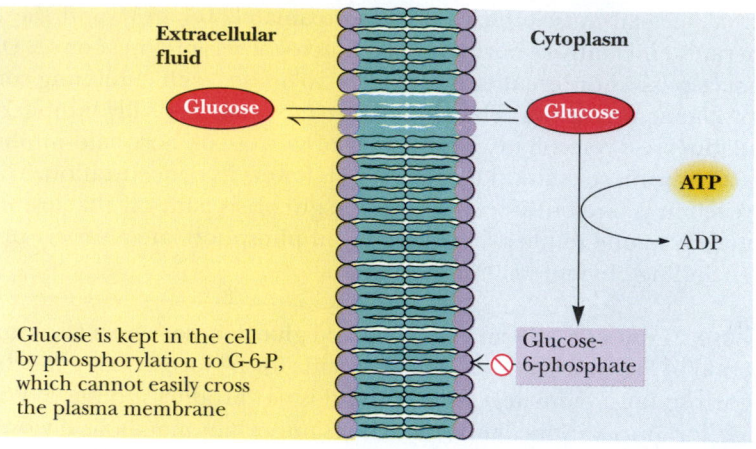

Biochemistry ⊛ Now™   ANIMATED FIGURE 18.4 Phosphorylation of glucose to glucose-6-phosphate by ATP creates a charged molecule that cannot easily cross the plasma membrane. **See this figure animated at http://chemistry.brookscole.com/ggb3**

fructose (with a carbonyl group at position 2 in the linear form) activates carbon C-3 for cleavage in the fourth step of glycolysis. The enzyme responsible for this isomerization is **phosphoglucoisomerase,** also known as **glucose phosphate isomerase.** In humans, the enzyme requires Mg²⁺ for activity and is highly specific for glucose-6-phosphate. The $\Delta G°'$ is 1.67 kJ/mol, and the value of ΔG under cellular conditions (Table 18.1) is −2.92 kJ/mol. This small value means that the reaction operates near equilibrium in the cell and is readily reversible. Phosphoglucoisomerase proceeds through an *enediol* intermediate, as shown in Figure 18.6. Although the predominant forms of glucose-6-phosphate and fructose-6-phosphate in solution are the ring forms (Figure 18.6), the isomerase interconverts the open-chain form of G-6-P with the open-chain form of F-6-P. The first reaction catalyzed by the isomerase is the opening of the pyranose ring (Figure 18.6, Step A). In the next step, the C-2 proton is removed from the substrate by a basic residue on the enzyme, facilitating formation of the enediol intermediate (Figure 18.6, Step B). This process then operates somewhat in reverse (Figure 18.6, Step C), creating a carbonyl group at C-2 to complete the formation of fructose-6-phosphate. The furanose form of the product is formed in the usual manner by attack of the C-5 hydroxyl on the carbonyl group, as shown.

Reaction 3: ATP Drives a Second Phosphorylation by Phosphofructokinase—The Second Priming Reaction

The action of phosphoglucoisomerase, "moving" the carbonyl group from C-1 to C-2, creates a new primary alcohol function at C-1 (see Figure 18.6). The next step in the glycolytic pathway is the phosphorylation of this group by **phosphofructokinase.** Once again, the substrate that provides the phosphoryl group is

FIGURE 18.5 Glucose-6-phosphate is the branch point for several metabolic pathways.

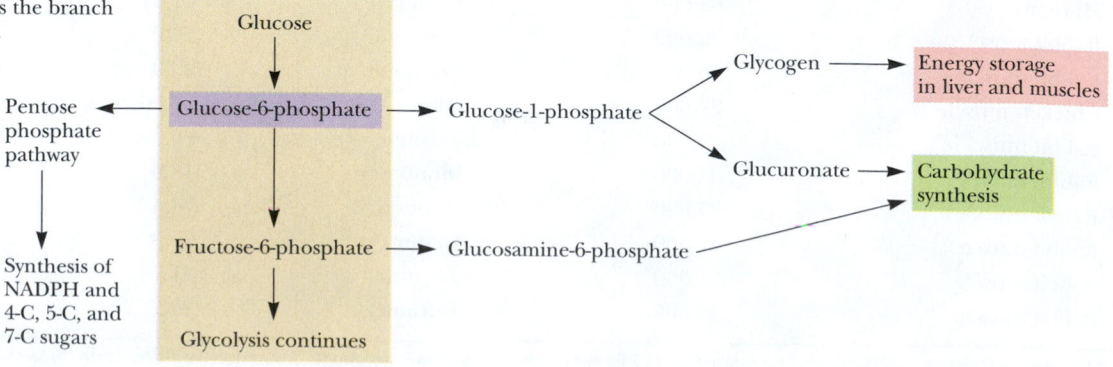

Biochemistry≡Now™ **ACTIVE FIGURE 18.6** The phosphoglucoisomerase mechanism involves opening of the pyranose ring (Step A), proton abstraction leading to enediol formation (Step B), and proton addition to the double bond, followed by ring closure (Step C). **Test yourself on the concepts in this figure at http://chemistry.brookscole.com/ggb3**

ATP. Like the hexokinase, glucokinase reaction, the phosphorylation of fructose-6-phosphate is a priming reaction and is endergonic:

$$\text{Fructose-6-P} + P_i \longrightarrow \text{fructose-1,6-bisphosphate} \qquad (18.6)$$
$$\Delta G^{\circ\prime} = 16.3 \text{ kJ/mol}$$

When coupled (by phosphofructokinase) with the hydrolysis of ATP, the overall reaction (Figure 18.7) is strongly exergonic:

$$\text{Fructose-6-P} + \text{ATP} \longrightarrow \text{fructose-1,6-bisphosphate} + \text{ADP} \qquad (18.7)$$
$$\Delta G^{\circ\prime} = -14.2 \text{ kJ/mol}$$
$$\Delta G \text{ (in erythrocytes)} = -18.8 \text{ kJ/mol}$$

At pH 7 and 37°C, the phosphofructokinase reaction equilibrium lies far to the right. Just as the hexokinase reaction commits the cell to taking up glucose, *the phosphofructokinase reaction commits the cell to metabolizing glucose* rather than converting it to another sugar or storing it. Similarly, just as the large free energy change of the hexokinase reaction makes it a likely candidate for regulation, so the phosphofructokinase reaction is an important site of regulation—indeed, the most important site in the glycolytic pathway.

Regulation of Phosphofructokinase Phosphofructokinase is the "valve" controlling the rate of glycolysis. In addition to its role as a substrate, ATP is also an allosteric inhibitor of this enzyme. Thus, phosphofructokinase has two distinct binding sites for ATP; a high-affinity substrate site and a low-affinity regulatory

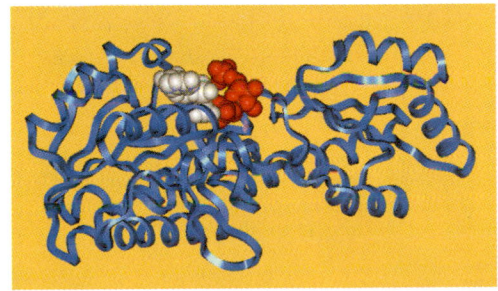

Phosphofructokinase with ADP shown in white and fructose-6-P in red.

Biochemistry≡Now™ Go to BiochemistryNow and click BiochemistryInteractive to learn more about the regulation of phosphofructokinase.

Fructose-6-phosphate $+$ **ATP** $\xrightarrow[\text{Phosphofructokinase}\ (\text{PFK})]{\text{Mg}^{2+}}$ **Fructose-1,6-bisphosphate** $+$ **ADP**

$$\Delta G^{\circ\prime} = -14.2 \text{ kJ/mol}$$
$$\Delta G_{\text{erythrocyte}} = -18.8 \text{ kJ/mol}$$

FIGURE 18.7 The phosphofructokinase reaction.

A Deeper Look

Phosphoglucoisomerase—A Moonlighting Protein

When someone has a day job but also works at night (that is, under the moon) at a second job, they are said to be "moonlighting." Similarly, a number of proteins have been found to have two or more different functions, and Constance Jeffery at Brandeis University has dubbed these "moonlighting proteins." Phosphoglucoisomerase catalyzes the second step of glycolysis but also moonlights as a nerve growth factor outside animal cells. In fact, outside the cell, this protein is known as neuroleukin (NL), autocrine motility factor (AMF), and differentiation and maturation mediator (DMM). Neuroleukin is secreted by (immune system) T cells and promotes the survival of certain spinal neurons and sensory nerves. AMF is secreted by

tumor cells and stimulates cancer cell migration. DMM causes certain leukemia cells to differentiate.

How phosphoglucoisomerase is secreted by the cell for its moonlighting functions is unknown, but there is evidence that the organism itself is surprised by this secretion. Diane Mathis and Christophe Benoist at the University of Strasbourg have shown that, in mice with disorders similar to rheumatoid arthritis, the immune system recognizes extracellular phosphoglucoisomerase as an antigen—that is, a protein that is "nonself." That a protein can be vital to metabolism inside the cell and also function as a growth factor and occasionally act as an antigen outside the cell is indeed remarkable.

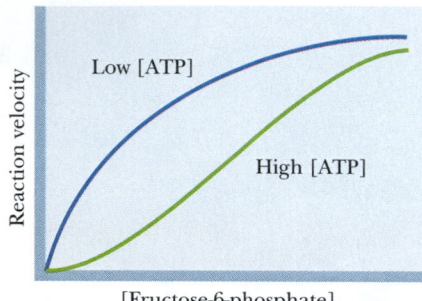

FIGURE 18.8 At high [ATP], phosphofructokinase (PFK) behaves cooperatively and the plot of enzyme activity versus [fructose-6-phosphate] is sigmoid. High [ATP] thus inhibits PFK, decreasing the enzyme's affinity for fructose-6-phosphate.

site. In the presence of high ATP concentrations, phosphofructokinase behaves cooperatively, plots of enzyme activity versus fructose-6-phosphate are sigmoid, and the K_m for fructose-6-phosphate is increased (Figure 18.8). Thus, when ATP levels are sufficiently high in the cytosol, glycolysis "turns off." Under most cellular conditions, however, the ATP concentration does not vary over a large range. The ATP concentration in muscle during vigorous exercise, for example, is only about 10% lower than that during the resting state. The rate of glycolysis, however, varies much more. A large range of glycolytic rates cannot be directly accounted for by only a 10% change in ATP levels.

AMP reverses the inhibition due to ATP, and AMP levels in cells *can* rise dramatically when ATP levels decrease, due to the action of the enzyme *adenylate kinase,* which catalyzes the reaction

$$ADP + ADP \rightleftharpoons ATP + AMP$$

with the equilibrium constant:

$$K_{eq} = \frac{[ATP][AMP]}{[ADP]^2} = 0.44 \tag{18.8}$$

Adenylate kinase rapidly interconverts ADP, ATP, and AMP to maintain this equilibrium. ADP levels in cells are typically 10% of ATP levels, and AMP levels are often less than 1% of the ATP concentration. Under such conditions, a small net change in ATP concentration due to ATP hydrolysis results in a much larger relative increase in the AMP levels because of adenylate kinase activity.

EXAMPLE

Calculate the change in concentration in AMP that would occur if 8% of the ATP in an erythrocyte (red blood cell) were suddenly hydrolyzed to ADP. In erythrocytes (Table 18.2), the concentration of ATP is typically 1850 μM, the concentration of ADP is 145 μM, and the concentration of AMP is 5 μM. The total adenine nucleotide concentration is 2000 μM.

Answer

The problem can be solved using the equilibrium expression for the adenylate kinase reaction:

$$K_{eq} = 0.44 = \frac{[ATP][AMP]}{[ADP]^2}$$

If 8% of the ATP is hydrolyzed to ADP, then [ATP] becomes $1850(0.92) = 1702\ \mu M$, and [AMP] + [ADP] becomes $2000 - 1702 = 298\ \mu M$, and [AMP] may be calculated from the adenylate kinase equilibrium:

$$0.44 = \frac{[1702\ \mu M][\text{AMP}]}{[\text{ADP}]^2}$$

Since $[\text{AMP}] = 298\ \mu M - [\text{ADP}]$,

$$0.44 = \frac{1702\ (298 - [\text{ADP}])}{[\text{ADP}]^2}$$

$$[\text{ADP}] = 278\ \mu M$$
$$[\text{AMP}] = 20\ \mu M$$

Thus, an 8% decrease in [ATP] results in a 20/5 or *fourfold increase in the concentration of AMP.*

Clearly, the activity of phosphofructokinase depends on both ATP and AMP levels and is a function of the cellular energy status. Phosphofructokinase activity is increased when the energy status falls and is decreased when the energy status is high. The rate of glycolysis activity thus decreases when ATP is plentiful and increases when more ATP is needed.

Glycolysis and the citric acid cycle (to be discussed in Chapter 19) are coupled via phosphofructokinase, because *citrate,* an intermediate in the citric acid cycle, is an allosteric inhibitor of phosphofructokinase. When the citric acid cycle reaches saturation, glycolysis (which "feeds" the citric acid cycle under aerobic conditions) slows down. The citric acid cycle directs electrons into the electron transport chain (for the purpose of ATP synthesis in oxidative phosphorylation) and also provides precursor molecules for biosynthetic pathways. Inhibition of glycolysis by citrate ensures that glucose will not be committed to these activities if the citric acid cycle is already saturated.

Phosphofructokinase is also regulated by β-D-fructose-2,6-bisphosphate, a potent allosteric activator that increases the affinity of phosphofructokinase for the substrate fructose-6-phosphate (Figure 18.9). Stimulation of phosphofructokinase is also achieved by decreasing the inhibitory effects of ATP (Figure 18.10). Fructose-2,6-bisphosphate increases the net flow of glucose through glycolysis by stimulating phosphofructokinase and, as we shall see in Chapter 22, by inhibiting fructose-1,6-bisphosphatase, the enzyme that catalyzes this reaction in the opposite direction.

Reaction 4: Cleavage by Fructose Bisphosphate Aldolase Creates Two 3-Carbon Intermediates

Fructose bisphosphate aldolase cleaves fructose-1,6-bisphosphate between the C-3 and C-4 carbons to yield two triose phosphates. The products are dihydroxyacetone phosphate (DHAP) and glyceraldehyde-3-phosphate. The reaction (Figure 18.11) has an equilibrium constant of approximately $10^{-4}\ M$, and a corresponding $\Delta G^{\circ\prime}$ of $+23.9$ kJ/mol. These values might imply that the reaction does not proceed effectively from left to right as written. However, the reaction makes two molecules (glyceraldehyde-3-P and dihydroxyacetone-P) from one molecule (fructose-1,6-bisphosphate), and the equilibrium is thus greatly influenced by concentration. The value of ΔG in erythrocytes is actually -0.23 kJ/mol (see Table 18.1). At physiological concentrations, the reaction is essentially at equilibrium.

Two classes of aldolase enzymes are found in nature. Animal tissues produce a Class I aldolase, characterized by the formation of a covalent Schiff base intermediate between an active-site lysine and the carbonyl group of the substrate. Class I aldolases do not require a divalent metal ion (and thus are not inhibited by EDTA) but are inhibited by sodium borohydride, $NaBH_4$, in the presence of

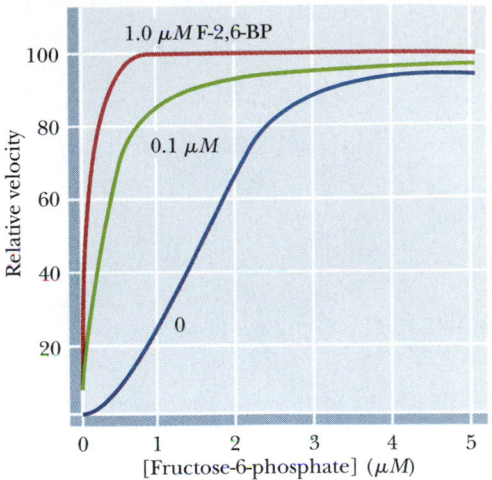

FIGURE 18.9 Fructose-2,6-bisphosphate activates phosphofructokinase, increasing the affinity of the enzyme for fructose-6-phosphate and restoring the hyperbolic dependence of enzyme activity on substrate.

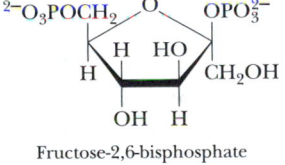

Fructose-2,6-bisphosphate

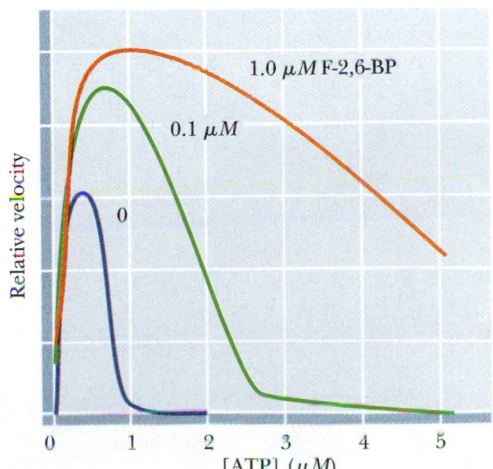

FIGURE 18.10 Fructose-2,6-bisphosphate decreases the inhibition of phosphofructokinase due to ATP.

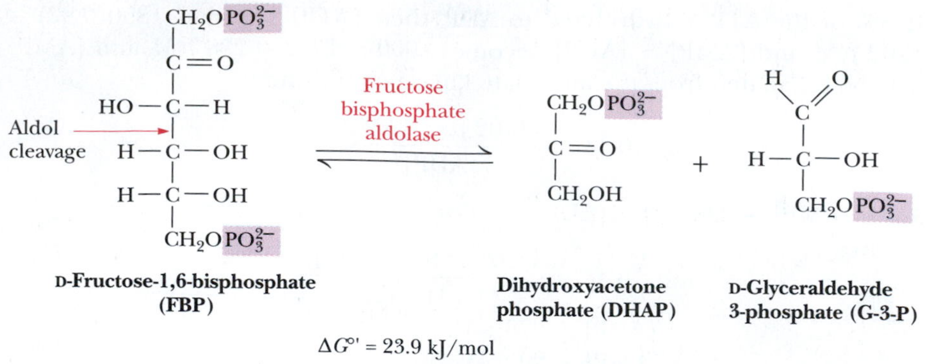

FIGURE 18.11 The fructose-1,6-bisphosphate aldolase reaction.

$\Delta G^{\circ\prime} = 23.9$ kJ/mol

R′ = H (aldehyde)
R′ = alkyl, etc. (ketone)

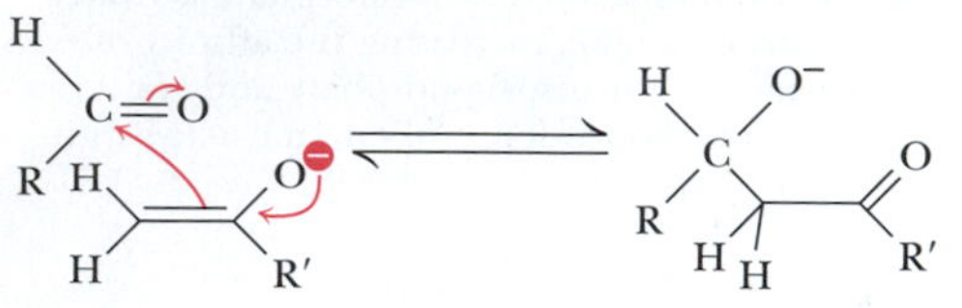

FIGURE 18.12 An aldol condensation reaction.

substrate (see A Deeper Look box, page 590). Class II aldolases are produced mainly in bacteria and fungi and are not inhibited by borohydride, but they do contain an active-site metal (normally zinc, Zn^{2+}) and are inhibited by EDTA. Cyanobacteria and some other simple organisms possess both classes of aldolase.

The aldolase reaction is merely the reverse of the **aldol condensation** well known to organic chemists. The latter reaction involves an attack by a nucleophilic enolate anion of an aldehyde or ketone on the carbonyl carbon of an aldehyde (Figure 18.12). The opposite reaction, aldol cleavage, begins with removal of a proton from the β-hydroxyl group, which is followed by the elimination of the enolate anion. A mechanism for the aldol cleavage reaction of fructose-1,6-bisphosphate in the Class I–type aldolases is shown in Figure 18.13a.

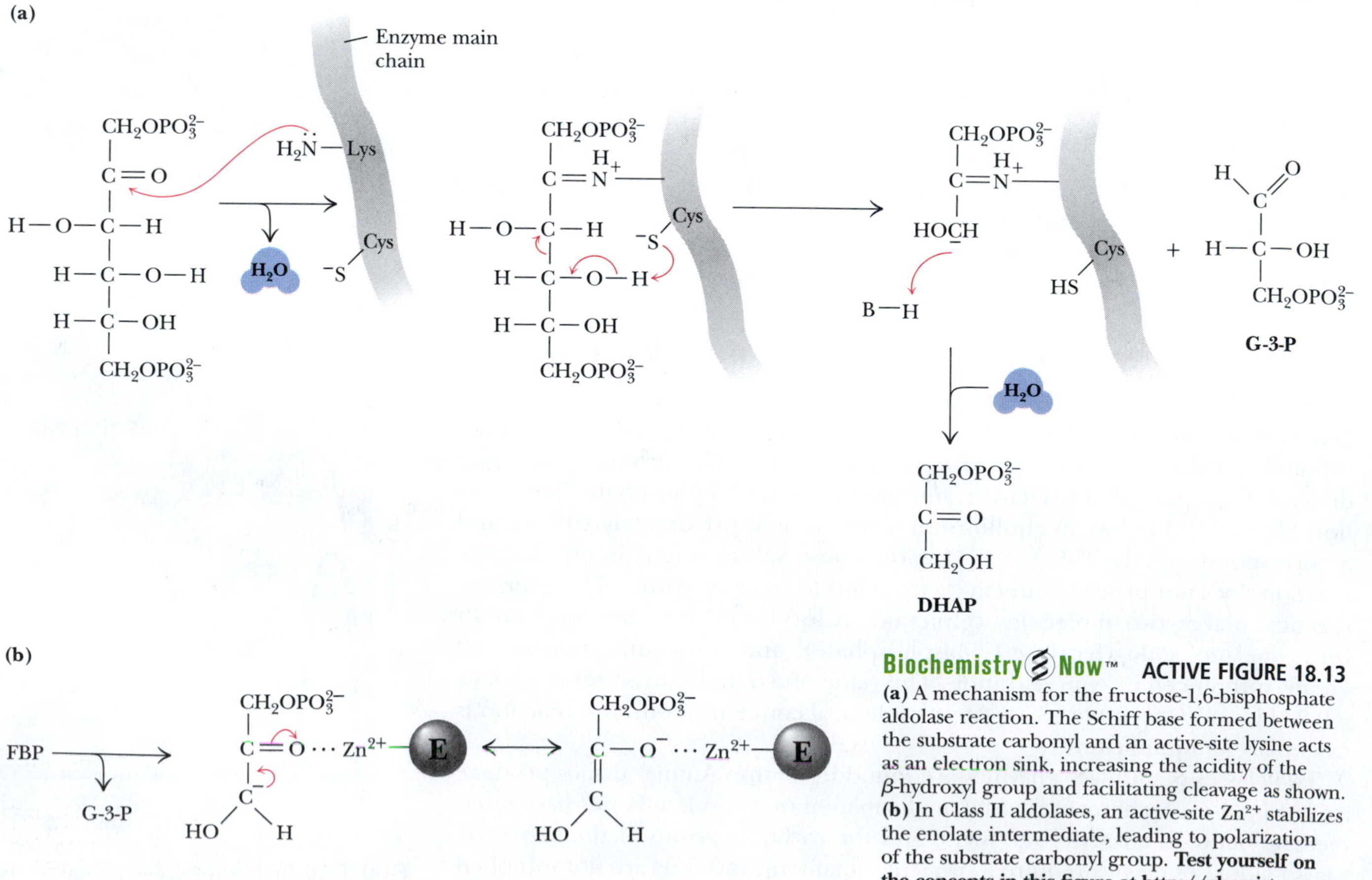

(a)

(b)

Biochemistry ⊜ Now™   ACTIVE FIGURE 18.13
(a) A mechanism for the fructose-1,6-bisphosphate aldolase reaction. The Schiff base formed between the substrate carbonyl and an active-site lysine acts as an electron sink, increasing the acidity of the β-hydroxyl group and facilitating cleavage as shown. **(b)** In Class II aldolases, an active-site Zn^{2+} stabilizes the enolate intermediate, leading to polarization of the substrate carbonyl group. **Test yourself on the concepts in this figure at http://chemistry. brookscole.com/ggb3**

In Class II aldolases, an active-site metal such as Zn^{2+} behaves as an electrophile, polarizing the carbonyl group of the substrate and stabilizing the enolate intermediate (Figure 18.13b).

Reaction 5: Triose Phosphate Isomerase Completes the First Phase of Glycolysis

Of the two products of the aldolase reaction, only glyceraldehyde-3-phosphate goes directly into the second phase of glycolysis. The other triose phosphate, dihydroxyacetone phosphate, must be converted to glyceraldehyde-3-phosphate by the enzyme **triose phosphate isomerase** (Figure 18.14). This reaction thus permits both products of the aldolase reaction to continue in the glycolytic pathway and in essence makes the C-1, C-2, and C-3 carbons of the starting glucose molecule equivalent to the C-6, C-5, and C-4 carbons, respectively. The reaction mechanism involves an enediol intermediate that can donate either of its hydroxyl protons to a basic residue on the enzyme and thereby become either dihydroxyacetone phosphate or glyceraldehyde-3-phosphate (Figure 18.15). Triose phosphate isomerase is one of the enzymes that have evolved to a state of "catalytic perfection," with a turnover number near the diffusion limit (see Chapter 13, Table 13.5).

The triose phosphate isomerase reaction completes the first phase of glycolysis, each glucose that passes through being converted to two molecules of glyceraldehyde-3-phosphate. Although the last two steps of the pathway are energetically unfavorable, the overall five-step reaction sequence has a net $\Delta G^{\circ\prime}$ of $+2.2$ kJ/mol ($K_{eq} \approx 0.43$). It is the free energy of hydrolysis from the two priming molecules of ATP that brings the overall equilibrium constant close to 1 under standard-state conditions. The net ΔG under cellular conditions is quite negative (-53.4 kJ/mol in erythrocytes).

18.4 | What Are the Chemical Principles and Features of the Second Phase of Glycolysis?

The second half of the glycolytic pathway involves the reactions that convert the metabolic energy in the glucose molecule into ATP. Altogether, four new ATP molecules are produced. If two are considered to offset the two ATPs consumed in phase 1, a net yield of two ATPs per glucose is realized. Phase 2 starts with the oxidation of glyceraldehyde-3-phosphate, a reaction with a large enough energy "kick" to produce a high-energy phosphate, namely, 1,3-bisphosphoglycerate

FIGURE 18.14 The triose phosphate isomerase reaction.

DHAP → G-3-P

$\Delta G^{\circ\prime} = +7.56$ kJ/mol

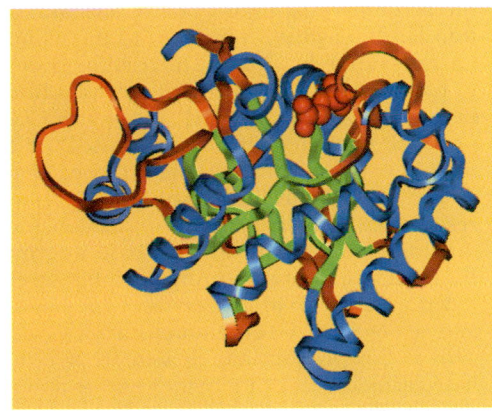

Triose phosphate isomerase with substrate analog 2-phosphoglycerate shown in red.

DHAP

Enediol intermediate

Glyceraldehyde-3-P

Biochemistry⑧Now™ ACTIVE FIGURE 18.15
A reaction mechanism for triose phosphate isomerase. **Test yourself on the concepts in this figure at http://chemistry.brookscole.com/ggb3**

The Chemical Evidence for the Schiff Base Intermediate in Class I Aldolases

Fructose bisphosphate aldolase of animal muscle is a Class I aldolase, which forms a Schiff base or imine intermediate between the substrate (fructose-1,6-bisP or dihydroxyacetone-P) and a lysine amino group at the enzyme active site. The chemical evidence for this intermediate comes from studies with the aldolase and the reducing agent sodium borohydride, $NaBH_4$. Incubation of fructose bisphosphate aldolase with dihydroxyacetone-P and $NaBH_4$ inactivates the enzyme. Interestingly, no inactivation is observed if $NaBH_4$ is added to the enzyme in the absence of substrate.

These observations are explained by the mechanism shown in the accompanying figure. $NaBH_4$ inactivates Class I aldolases by transfer of a hydride ion ($H:^-$) to the imine carbon atom of the enzyme–substrate adduct. The resulting secondary amine is stable to hydrolysis, and the active-site lysine is thus permanently modified and inactivated. $NaBH_4$ inactivates Class I aldolases in the presence of either dihydroxyacetone-P or fructose-1,6-bisP, but inhibition doesn't occur in the presence of glyceraldehyde-3-P.

Definitive identification of lysine as the modified active-site residue has come from radioisotope-labeling studies. $NaBH_4$ reduction of the aldolase Schiff base intermediate formed from ^{14}C-labeled dihydroxyacetone-P yields an enzyme covalently labeled with [14]C. Acid hydrolysis of the inactivated enzyme liberates a novel ^{14}C-labeled amino acid, *N⁶-dihydroxypropyl-L-lysine*. This is the product anticipated from reduction of the Schiff base formed between a lysine residue and the ^{14}C-labeled dihydroxyacetone-P. (The phosphate group is lost during acid hydrolysis of the inactivated enzyme.) The use of ^{14}C labeling in a case such as this facilitates the separation and identification of the telltale amino acid.

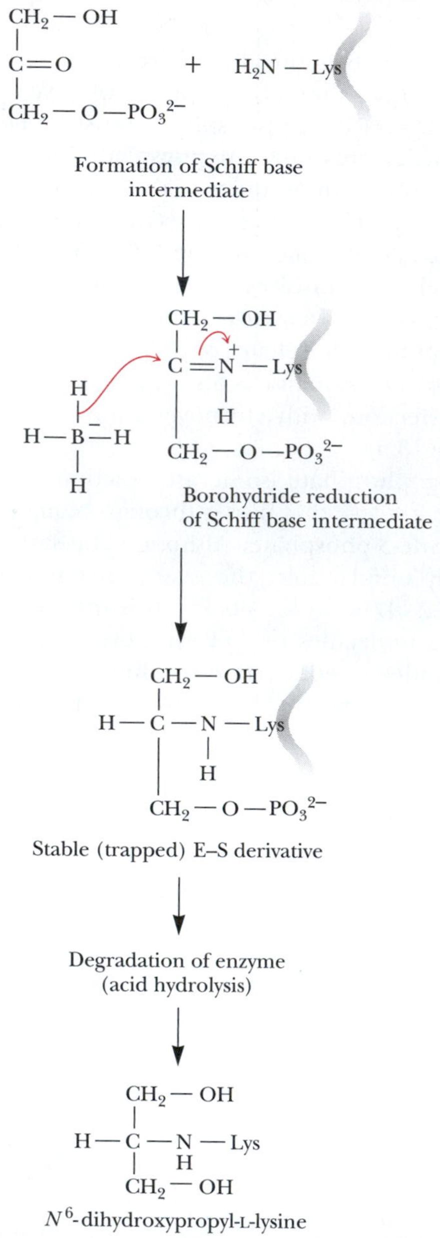

(Figure 18.16). Phosphoryl transfer from 1,3-BPG to ADP to make ATP is highly favorable. The product, 3-phosphoglycerate, is converted via several steps to phosphoenolpyruvate (PEP), another high-energy phosphate. PEP readily transfers its phosphoryl group to ADP in the pyruvate kinase reaction to make another ATP.

Reaction 6: Glyceraldehyde-3-Phosphate Dehydrogenase Creates a High-Energy Intermediate

In the first glycolytic reaction to involve oxidation–reduction, glyceraldehyde-3-phosphate is oxidized to 1,3-bisphosphoglycerate by **glyceraldehyde-3-phosphate dehydrogenase.** Although the oxidation of an aldehyde to a carboxylic acid is a

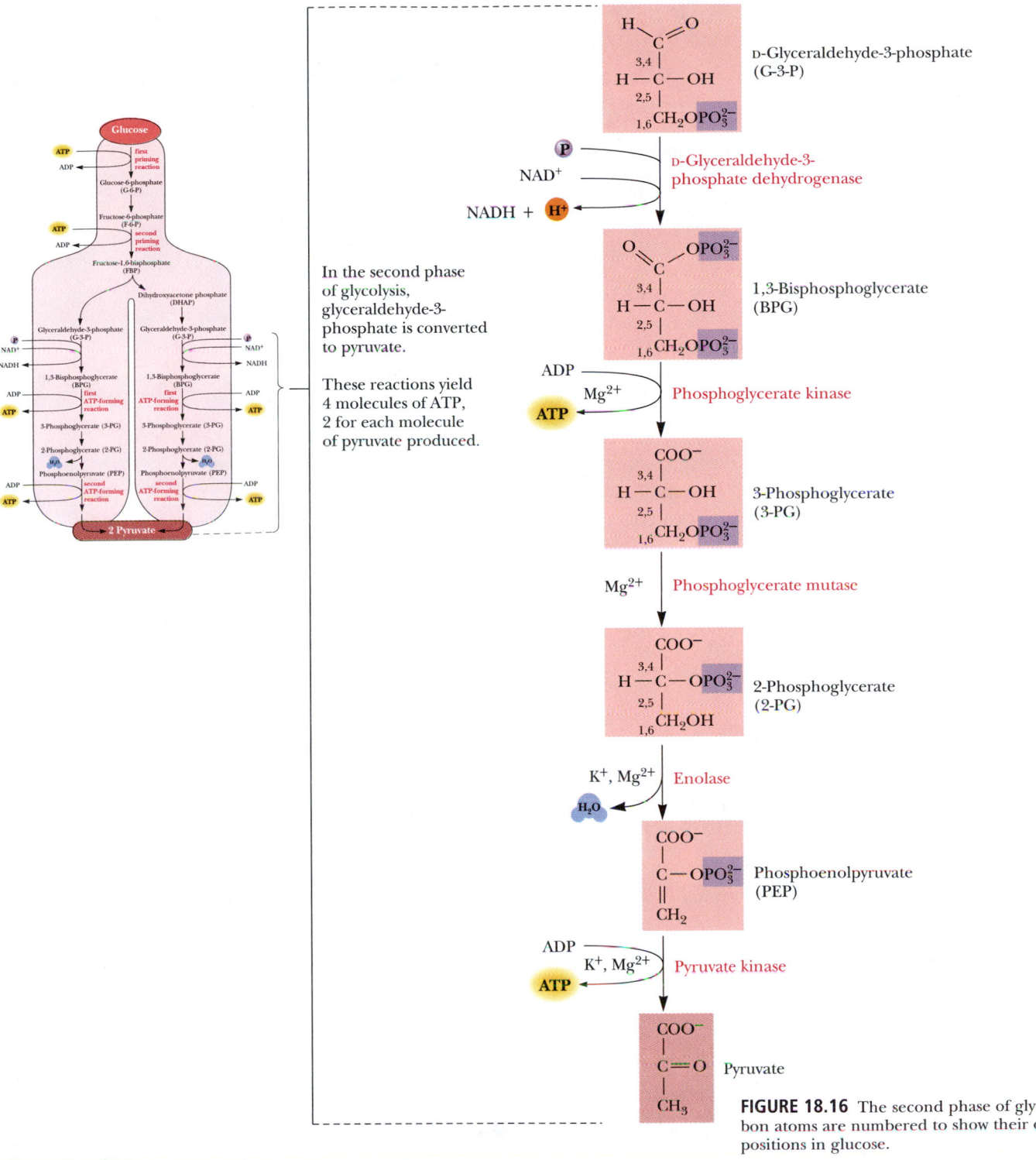

In the second phase of glycolysis, glyceraldehyde-3-phosphate is converted to pyruvate.

These reactions yield 4 molecules of ATP, 2 for each molecule of pyruvate produced.

FIGURE 18.16 The second phase of glycolysis. Carbon atoms are numbered to show their original positions in glucose.

highly exergonic reaction, the overall reaction (Figure 18.17) involves both formation of a carboxylic–phosphoric anhydride and the reduction of NAD⁺ to NADH and is therefore slightly endergonic at standard state, with a $\Delta G^{\circ\prime}$ of +6.30 kJ/mol. The free energy that might otherwise be released as heat in this reaction is directed into the formation of a high-energy phosphate compound, 1,3-bisphosphoglycerate, and the reduction of NAD⁺. The reaction mechanism involves nucleophilic attack by a cysteine —SH group on the carbonyl carbon of glyceraldehyde-3-phosphate to form a hemithioacetal (Figure 18.18). The hemithioacetal intermediate decomposes by hydride (H:⁻) transfer to NAD⁺ to form a high-energy thioester. Nucleophilic attack by phosphate displaces the product,

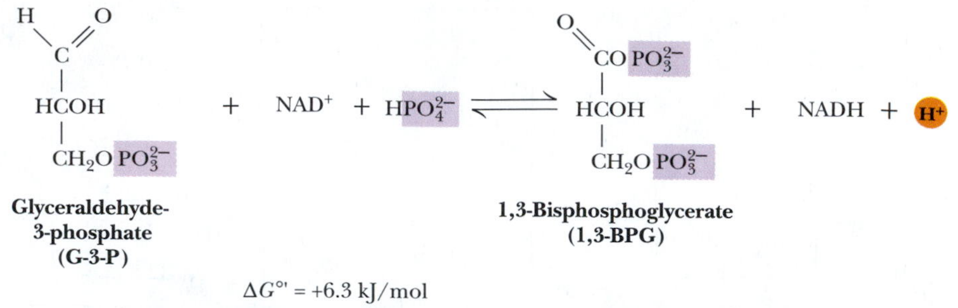

FIGURE 18.17 The glyceraldehyde-3-phosphate dehydrogenase reaction.

$\Delta G^{\circ\prime} = +6.3$ kJ/mol

1,3-bisphosphoglycerate, from the enzyme. The enzyme can be inactivated by reaction with iodoacetate, which reacts with and blocks the essential cysteine sulfhydryl.

The glyceraldehyde-3-phosphate dehydrogenase reaction is the site of action of *arsenate* (AsO_4^{3-}), an anion analogous to phosphate. Arsenate is an effective substrate in this reaction, forming *1-arseno-3-phosphoglycerate* (Figure 18.19), but acyl arsenates are quite unstable and are rapidly hydrolyzed. 1-Arseno-3-phosphoglycerate breaks down to yield *3-phosphoglycerate*, the product of the seventh reaction of glycolysis. The result is that glycolysis continues in the presence of arsenate, but the molecule of ATP formed in reaction 7 (phosphoglycerate kinase) is not made because this step has been bypassed. The lability of 1-arseno-3-phosphoglycerate effectively uncouples the oxidation and phosphorylation events, which are normally tightly coupled in the glyceraldehyde-3-phosphate dehydrogenase reaction.

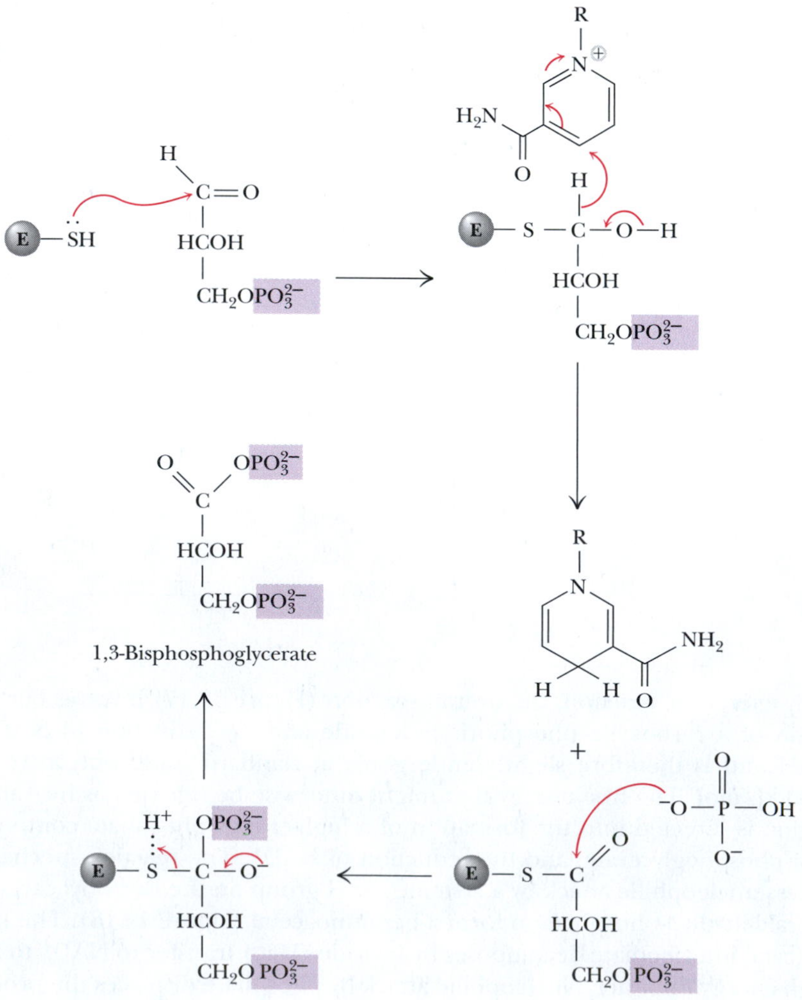

Biochemistry Now™   ACTIVE FIGURE 18.18
A mechanism for the glyceraldehyde-3-phosphate dehydrogenase reaction. Reaction of an enzyme sulfhydryl with the carbonyl carbon of glyceraldehyde-3-P forms a thiohemiacetal, which loses a hydride to NAD⁺ to become a thioester. Phosphorolysis of this thioester releases 1,3-bisphosphoglycerate. **Test yourself on the concepts in this figure at http://chemistry. brookscole.com/ggb3**

Reaction 7: Phosphoglycerate Kinase Is the Break-Even Reaction

The glycolytic pathway breaks even in terms of ATPs consumed and produced with this reaction. The enzyme **phosphoglycerate kinase** transfers a phosphoryl group from 1,3-bisphosphoglycerate to ADP to form an ATP (Figure 18.20). Because each glucose molecule sends two molecules of glyceraldehyde-3-phosphate into the second phase of glycolysis and because two ATPs were consumed per glucose in the first-phase reactions, the phosphoglycerate kinase reaction "pays off" the ATP debt created by the priming reactions. As might be expected for a phosphoryl transfer enzyme, Mg^{2+} ion is required for activity and the true nucleotide substrate for the reaction is $MgADP^-$. It is appropriate to view the sixth and seventh reactions of glycolysis as a coupled pair, with 1,3-bisphosphoglycerate as an intermediate. The phosphoglycerate kinase reaction is sufficiently exergonic at standard state to pull the G-3-P dehydrogenase reaction along. (In fact, the aldolase and triose phosphate isomerase are also pulled forward by phosphoglycerate kinase.) The net result of these coupled reactions is

Glyceraldehyde-3-phosphate $+$ ADP $+$ P_i $+$ NAD^+ $\longrightarrow$
$\quad$ 3-phosphoglycerate $+$ ATP $+$ NADH $+$ H^+
$$\Delta G^{\circ\prime} = -12.6 \text{ kJ/mol} \quad (18.9)$$

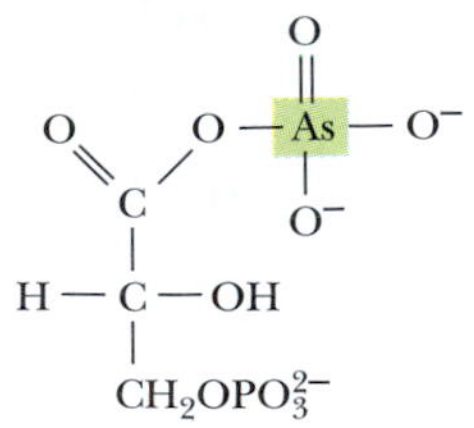

Another reflection of the coupling between these reactions lies in their values of ΔG under cellular conditions (Table 18.1). Despite its strongly negative $\Delta G^{\circ\prime}$, the phosphoglycerate kinase reaction operates at equilibrium in the erythrocyte ($\Delta G = 0.1$ kJ/mol). In essence, the free energy available in the phosphoglycerate kinase reaction is used to bring the three previous reactions closer to equilibrium. Viewed in this context, it is clear that ADP has been phosphorylated to form ATP at the expense of a substrate, namely, glyceraldehyde-3-phosphate. This is an example of **substrate-level phosphorylation,** a concept that will be encountered again. (The other kind of phosphorylation, *oxidative phosphorylation,* is driven energetically by the transport of electrons from appropriate coenzymes and substrates to oxygen. Oxidative phosphorylation will be covered in detail in Chapter 20.) Even though the coupled reactions exhibit a very favorable $\Delta G^{\circ\prime}$, there are conditions (i.e., high ATP and 3-phosphoglycerate levels) under which Equation 18.9 can be reversed so that 3-phosphoglycerate is phosphorylated from ATP.

An important regulatory molecule, 2,3-bisphosphoglycerate, is synthesized and metabolized by a pair of reactions that make a detour around the phosphoglycerate kinase reaction. 2,3-BPG, which stabilizes the deoxy form of hemoglobin and is primarily responsible for the cooperative nature of oxygen binding by hemoglobin (see Chapter 15), is formed from 1,3-bisphosphoglycerate by **bisphosphoglycerate mutase** (Figure 18.21). Interestingly, 3-phosphoglycerate is required for this reaction, which involves phosphoryl transfer from the C-1 position of 1,3-bisphosphoglycerate to the C-2 position of 3-phosphoglycerate (Figure 18.22). Hydrolysis of 2,3-BPG is carried out by 2,*3-bisphosphoglycerate phosphatase.* Although other cells contain only a trace of 2,3-BPG, erythrocytes typically contain 4 to 5 mM 2,3-BPG.

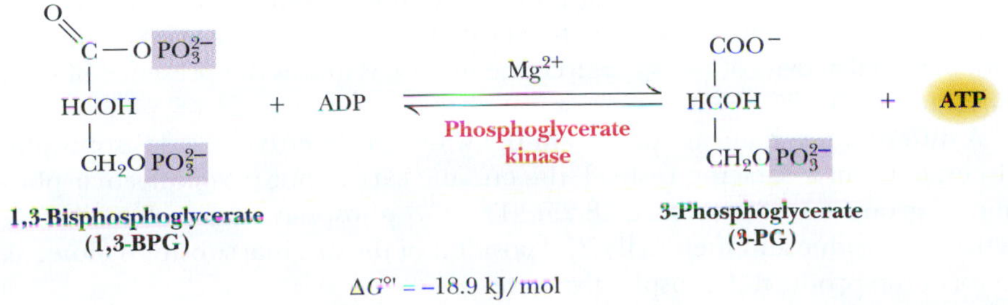

1-Arseno-3-phosphoglycerate

FIGURE 18.19

1,3-Bisphosphoglycerate
(1,3-BPG)

$$\Delta G^{\circ\prime} = -18.9 \text{ kJ/mol}$$

3-Phosphoglycerate
(3-PG)

FIGURE 18.20 The phosphoglycerate kinase reaction.

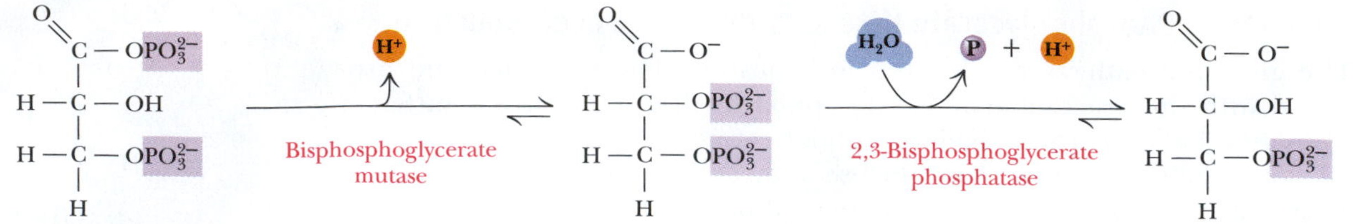

1,3-Bisphosphoglycerate (1,3-BPG)   **2,3-Bisphosphoglycerate (2,3-BPG)**   **3-Phosphoglycerate**

FIGURE 18.21 Formation and decomposition of 2,3-bisphosphoglycerate.

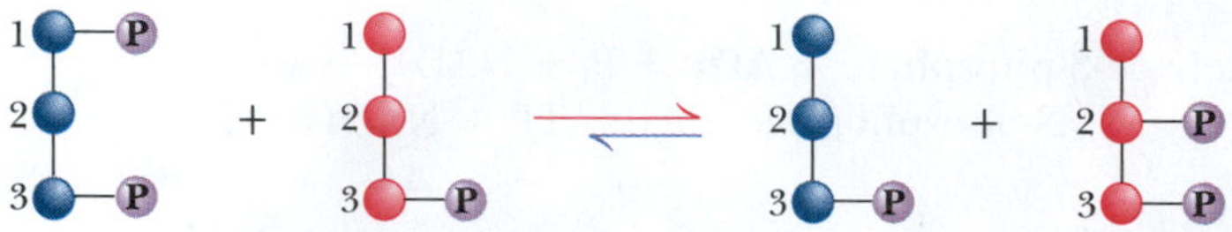

FIGURE 18.22 The mutase that forms 2,3-BPG from 1,3-BPG requires 3-phosphoglycerate. The reaction is actually an intermolecular phosphoryl transfer from C-1 of 1,3-BPG to C-2 of 3-PG.

Reaction 8: Phosphoglycerate Mutase Catalyzes a Phosphoryl Transfer

The remaining steps in the glycolytic pathway prepare for synthesis of the second ATP equivalent. This begins with the **phosphoglycerate mutase** reaction (Figure 18.23), in which the phosphoryl group of 3-phosphoglycerate is moved from C-3 to C-2. (The term *mutase* is applied to enzymes that catalyze migration of a functional group within a substrate molecule.) The free energy change for this reaction is very small under cellular conditions ($\Delta G = 0.83$ kJ/mol in erythrocytes). Phosphoglycerate mutase enzymes isolated from different sources exhibit different reaction mechanisms. As shown in Figure 18.24, the enzymes isolated from yeast and from rabbit muscle form *phosphoenzyme* intermediates, use *2,3-bisphosphoglycerate* as a cofactor, and undergo *inter*molecular phosphoryl group transfers (in which the phosphate of the product 2-phosphoglycerate is not that from the 3-phosphoglycerate substrate). The prevalent form of phosphoglycerate mutase is a *phosphoenzyme,* with a phosphoryl group covalently bound to a histidine residue at the active site. This phosphoryl group is transferred to the C-2 position of the substrate to form a transient, enzyme-bound 2,3-bisphosphoglycerate, which then decomposes by a second phosphoryl transfer from the C-3 position of the intermediate to the histidine residue on the enzyme. About once in every 100 enzyme turnovers, the intermediate, 2,3-bisphosphoglycerate, dissociates from the active site, leaving an inactive, unphosphorylated enzyme. The unphosphorylated enzyme can be reactivated by binding 2,3-BPG. For this reason, maximal activity of phosphoglycerate mutase requires the presence of small amounts of 2,3-BPG.

A different mechanism operates in the wheat germ enzyme. 2,3-Bisphosphoglycerate is not a cofactor. Instead, the enzyme carries out *intra*molecular phosphoryl group transfer (Figure 18.25). The C-3 phosphate is transferred to an active-site residue and then to the C-2 position of the original substrate molecule to form the product, 2-phosphoglycerate.

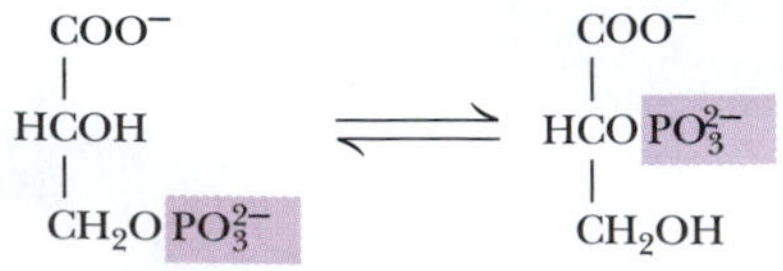

3-Phosphoglycerate (3-PG)   **2-Phosphoglycerate (2-PG)**

$\Delta G^{\circ\prime} = +4.4$ kJ/mol

FIGURE 18.23 The phosphoglycerate mutase reaction.

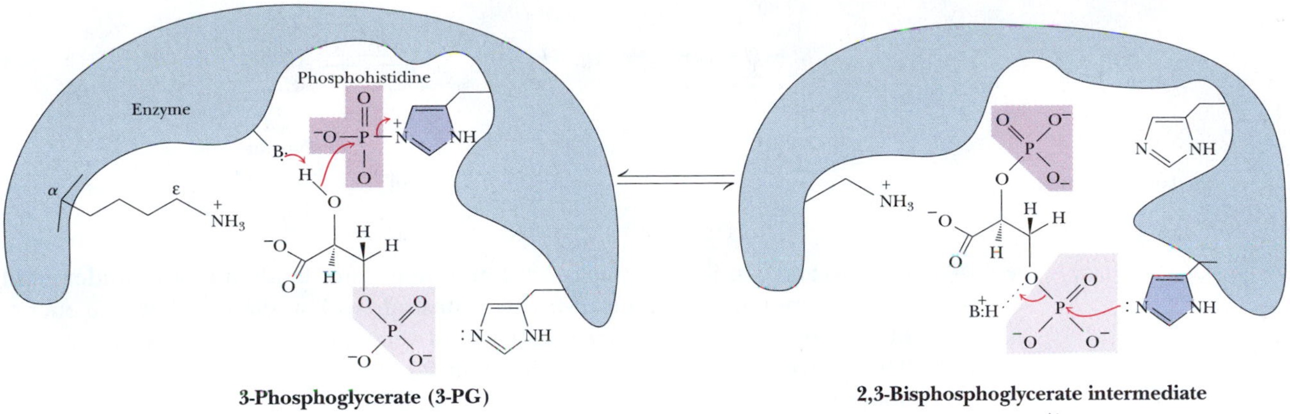

3-Phosphoglycerate (3-PG)

2,3-Bisphosphoglycerate intermediate

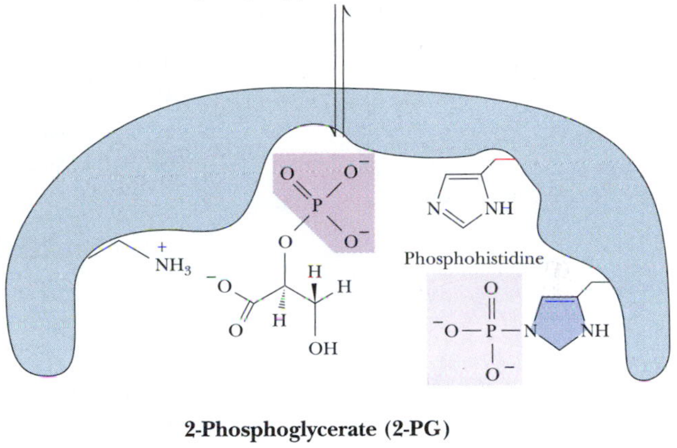

2-Phosphoglycerate (2-PG)

FIGURE 18.24 A mechanism for the phosphoglycerate mutase reaction in rabbit muscle and in yeast. Zelda Rose of the Institute for Cancer Research in Philadelphia showed that the enzyme requires a small amount of 2,3-BPG to phosphorylate the histidine residue before the mechanism can proceed. Prior to her work, the role of the phosphohistidine in this mechanism was not understood.

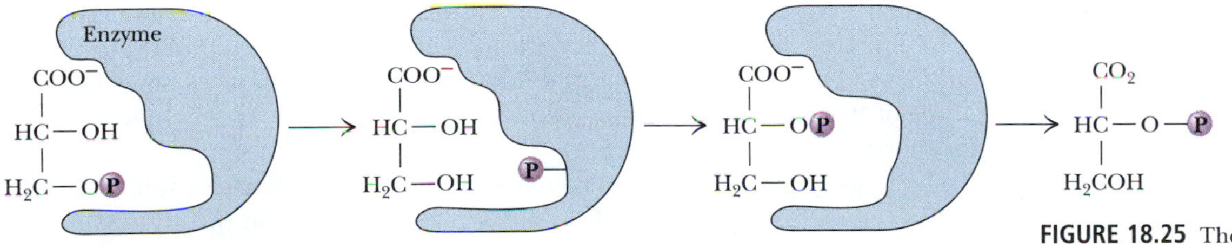

FIGURE 18.25 The phosphoglycerate mutase of wheat germ catalyzes an intramolecular phosphoryl transfer.

Reaction 9: Dehydration by Enolase Creates PEP

Recall that prior to synthesizing ATP in the phosphoglycerate kinase reaction, it was necessary to first make a substrate having a high-energy phosphate. Reaction 9 of glycolysis similarly makes a high-energy phosphate in preparation for ATP synthesis. **Enolase** catalyzes the formation of *phosphoenolpyruvate* from 2-phosphoglycerate (Figure 18.26). The reaction involves the removal of a water molecule to form the enol structure of PEP. The $\Delta G^{\circ\prime}$ for this reaction is relatively small at 1.8 kJ/mol ($K_{eq} = 0.5$); and, under cellular conditions, ΔG is very

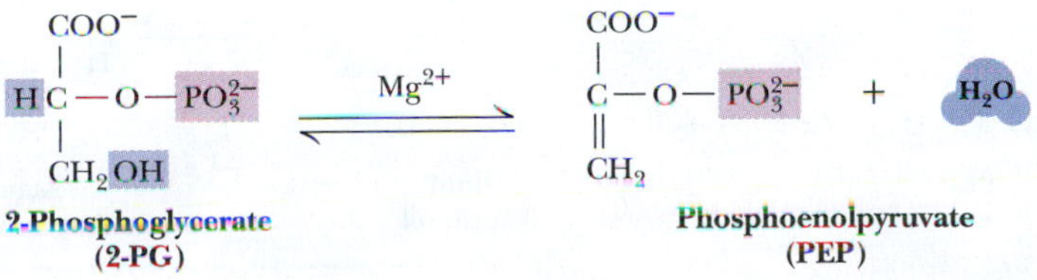

2-Phosphoglycerate (2-PG)

Phosphoenolpyruvate (PEP)

$\Delta G^{\circ\prime} = +1.8$ kJ/mol

FIGURE 18.26 The enolase reaction.

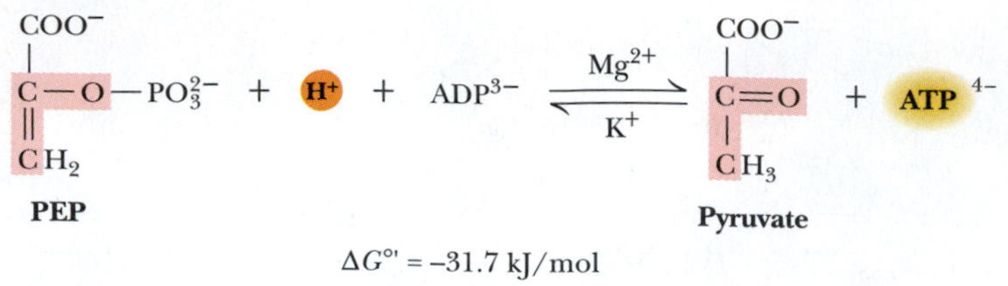

FIGURE 18.27 The pyruvate kinase reaction.

$$\Delta G^{\circ\prime} = -31.7 \text{ kJ/mol}$$

close to zero. In light of this condition, it may be difficult at first to understand how the enolase reaction transforms a substrate with a relatively low free energy of hydrolysis into a product (PEP) with a very high free energy of hydrolysis. This puzzle is clarified by realizing that 2-phosphoglycerate and PEP contain about the same amount of *potential* metabolic energy, with respect to decomposition to P_i, CO_2, and H_2O. What the enolase reaction does is rearrange the substrate into a form from which more of this potential energy can be released upon hydrolysis. The enzyme is strongly inhibited by fluoride ion in the presence of phosphate. Inhibition arises from the formation of *fluorophosphate* (FPO_3^{2-}), which forms a complex with Mg^{2+} at the active site of the enzyme.

Reaction 10: Pyruvate Kinase Yields More ATP

The second ATP-synthesizing reaction of glycolysis is catalyzed by **pyruvate kinase,** which brings the pathway at last to its pyruvate branch point. Pyruvate kinase mediates the transfer of a phosphoryl group from phosphoenolpyruvate to ADP to make ATP and pyruvate (Figure 18.27). The reaction requires Mg^{2+} ion and is stimulated by K^+ and certain other monovalent cations.

The corresponding K_{eq} at 25°C is 3.63×10^5, and it is clear that the pyruvate kinase reaction equilibrium lies very far to the right. Concentration effects reduce the magnitude of the free energy change somewhat in the cellular environment, but the ΔG in erythrocytes is still quite favorable at -23.0 kJ/mol. The high free energy change for the conversion of PEP to pyruvate is due largely to the highly favorable and spontaneous conversion of the enol tautomer of pyruvate to the more stable keto form (Figure 18.28) following the phosphoryl group transfer step.

The large negative ΔG of this reaction makes pyruvate kinase a suitable target site for regulation of glycolysis. For each glucose molecule in the glycolysis pathway, two ATPs are made at the pyruvate kinase stage (because two triose molecules were produced per glucose in the aldolase reaction). Because the pathway broke even in terms of ATP at the phosphoglycerate kinase reaction (two ATPs consumed and two ATPs produced), the two ATPs produced by pyruvate kinase represent the "payoff" of glycolysis—a net yield of two ATP molecules.

Pyruvate kinase possesses allosteric sites for numerous effectors. It is activated by AMP and fructose-1,6-bisphosphate and inhibited by ATP, acetyl-CoA, and alanine. (Note that alanine is the α-amino acid counterpart of the α-keto acid, pyruvate.) Furthermore, liver pyruvate kinase is regulated by covalent modification. Hormones such as *glucagon* activate a cAMP-dependent protein kinase, which transfers a phosphoryl group from ATP to the enzyme. The phos-

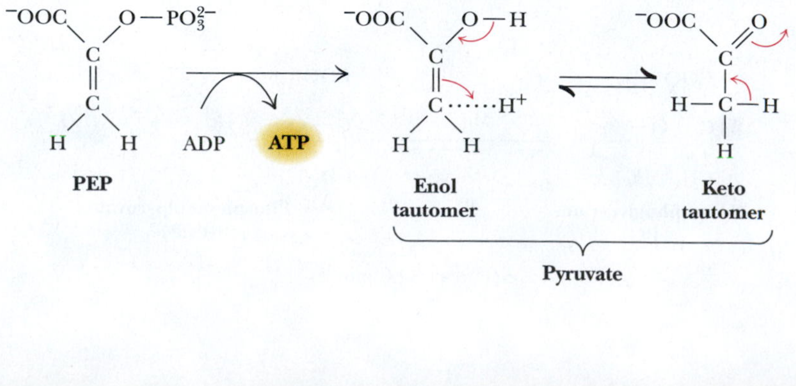

FIGURE 18.28 The conversion of phosphoenolpyruvate (PEP) to pyruvate may be viewed as involving two steps: phosphoryl transfer followed by an enol–keto tautomerization. The tautomerization is spontaneous ($\Delta G^{\circ\prime} \approx -35$–$40$ kJ/mol) and accounts for much of the free energy change for PEP hydrolysis.

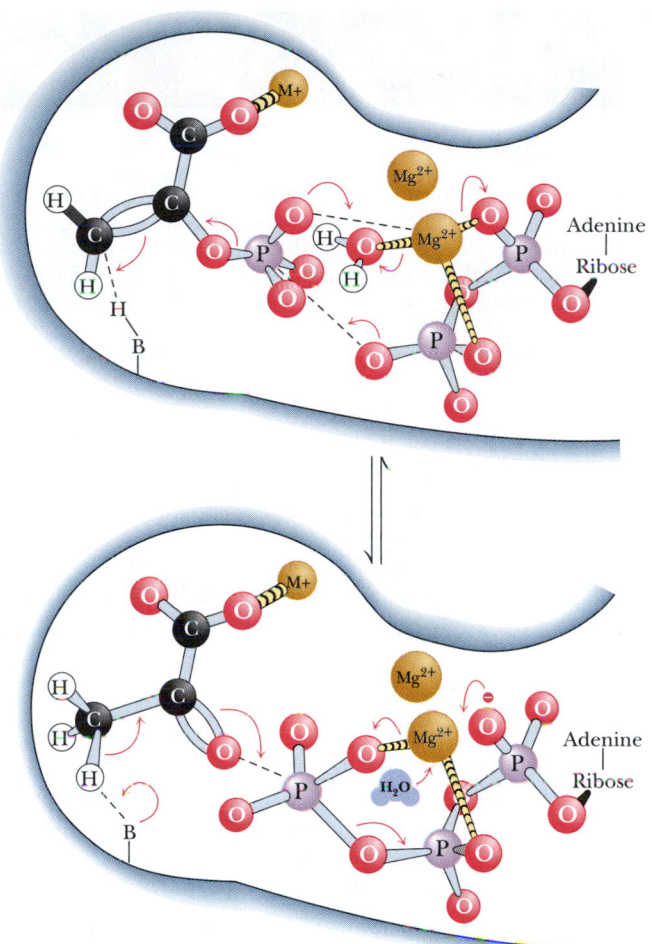

FIGURE 18.29 A mechanism for the pyruvate kinase reaction, based on NMR and EPR studies by Albert Mildvan and colleagues. Phosphoryl transfer from phosphoenolpyruvate (PEP) to ADP occurs in four steps: (1) A water on the Mg^{2+} ion coordinated to ADP is replaced by the phosphoryl group of PEP, (2) Mg^{2+} dissociates from the α-P of ADP, (3) the phosphoryl group is transferred, and (4) the enolate of pyruvate is protonated. (*Adapted from Mildvan, A., 1979. The role of metals in enzyme-catalyzed substitutions at each of the phosphorus atoms of ATP. Advances in Enzymology 49:103–126.*)

phorylated form of pyruvate kinase is more strongly inhibited by ATP and alanine and has a higher K_m for PEP, so in the presence of physiological levels of PEP, the enzyme is inactive. Then PEP is used as a substrate for glucose synthesis in the *gluconeogenesis* pathway (to be described in Chapter 22), instead of going on through glycolysis and the citric acid cycle (or fermentation routes). A suggested active-site geometry for pyruvate kinase, based on NMR and EPR studies by Albert Mildvan and colleagues, is presented in Figure 18.29. The carbonyl oxygen of pyruvate and the γ-phosphorus of ATP lie within 0.3 nm of each other at the active site, consistent with direct transfer of the phosphoryl group without formation of a phosphoenzyme intermediate.

18.5 | What Are the Metabolic Fates of NADH and Pyruvate Produced in Glycolysis?

In addition to ATP, the products of glycolysis are NADH and pyruvate. Their processing depends upon other cellular pathways. NADH must be recycled to NAD^+, lest NAD^+ become limiting in glycolysis. NADH can be recycled by both aerobic and anaerobic paths, either of which results in further metabolism of pyruvate. What a given cell does with the pyruvate produced in glycolysis depends in part on the availability of oxygen. Under aerobic conditions, pyruvate can be sent into the citric acid cycle (also known as the tricarboxylic acid cycle; see Chapter 19), where it is oxidized to CO_2 with the production of additional NADH (and $FADH_2$). Under aerobic conditions, the NADH produced in glycolysis and the citric acid cycle is reoxidized to NAD^+ in the mitochondrial electron transport chain (see Chapter 20).

Human Biochemistry

Pyruvate Kinase Deficiencies and Hemolytic Anemia

Erythrocytes, or **red blood cells,** do not have nuclei or intracellular organelles such as mitochondria. As such, they have restricted metabolic capabilities, and their ability to adapt to changing environments and conditions is limited. At the same time, they depend upon a constant supply of energy to maintain their structural integrity. Energy is required to maintain gradients of Na^+ and K^+ across the erythrocyte membrane and also to generate and preserve membrane lipids and proteins. If the erythrocyte's energy requirements are not met, **hemolysis** (rupture of the erythrocyte membrane) can occur, and the resulting red blood cell loss is termed **hemolytic anemia.**

Glycolysis is the primary source of ATP energy for the erythrocyte, with additional energy supplied by the pentose monophosphate pathway (to be covered in Chapter 22). For this reason, deficiencies of one or more of the glycolytic enzymes are likely to result in substantial hemolysis. The most common form

of hemolytic anemia results from a deficiency of pyruvate kinase. Individuals with one defective pyruvate kinase gene *(heterozygous carriers)* exhibit erythrocyte pyruvate kinase activities that are 40% to 60% of normal subjects. Those with two defective genes (and thus *homozygous* for the condition) exhibit pyruvate kinase activities that are 5% to 25% of normal.

In addition to the obvious reduction of glucose flux through glycolysis, other changes occur in cases of pyruvate kinase deficiency. Absence of pyruvate kinase activity causes glycolytic intermediates such as 3-phosphoglycerate to accumulate in affected cells. Ironically, 2,3-bisphosphoglycerate levels also rise, shifting hemoglobin's oxygen binding curve (see Figure 15.34) to the right and releasing more oxygen to affected tissues and compensating to some extent for the attendant anemia. However, high levels of 2,3-bisphosphoglycerate also inhibit hexokinase and phosphofructokinase, further inhibiting glycolysis.

Anaerobic Metabolism of Pyruvate Leads to Lactate or Ethanol

Under anaerobic conditions, the pyruvate produced in glycolysis is processed differently. In yeast, it is reduced to ethanol; in other microorganisms and in animals, it is reduced to lactate. These processes are examples of **fermentation**—the production of ATP energy by reaction pathways in which organic molecules function as donors and acceptors of electrons. In either case, reduction of pyruvate provides a means of reoxidizing the NADH produced in the glyceraldehyde-3-phosphate dehydrogenase reaction of glycolysis (Figure 18.30). In yeast, **alcoholic fermentation** is a two-step process. Pyruvate is decarboxylated to acetaldehyde by **pyruvate decarboxylase** in an essentially irreversible reaction. Thiamine pyrophosphate is a required cofactor for this enzyme. The second step, the reduction of acetaldehyde to ethanol by NADH, is catalyzed by **alcohol dehydrogenase** (Figure 18.30). At pH 7, the reaction equilibrium strongly favors ethanol. The end products of alcoholic fermentation are thus ethanol and carbon diox-

FIGURE 18.30 (a) Pyruvate reduction to ethanol in yeast provides a means for regenerating NAD^+ consumed in the glyceraldehyde-3-P dehydrogenase reaction. (b) In oxygen-depleted muscle, NAD^+ is regenerated in the lactate dehydrogenase reaction.

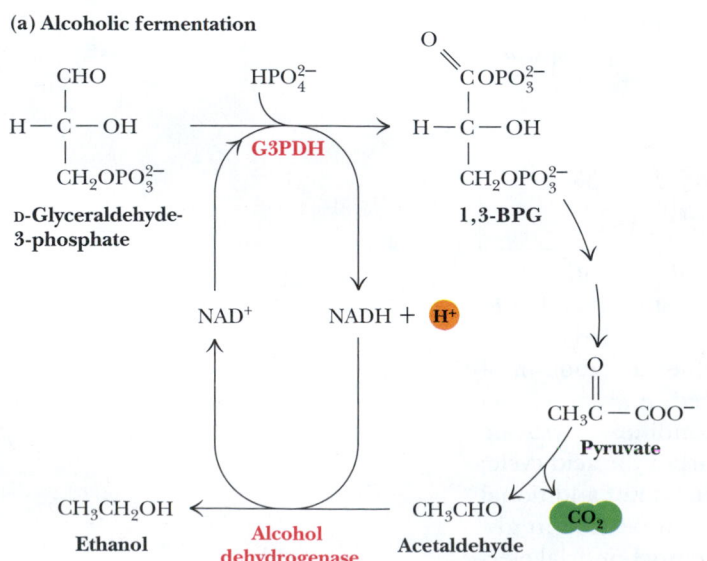

ide. Alcoholic fermentations are the basis for the brewing of beers and the fermentation of grape sugar in wine making. Lactate produced by anaerobic microorganisms during **lactic acid fermentation** is responsible for the taste of sour milk and for the characteristic taste and fragrance of sauerkraut, which in reality is fermented cabbage.

Lactate Accumulates Under Anaerobic Conditions in Animal Tissues

In animal tissues experiencing anaerobic conditions, pyruvate is reduced to lactate. Pyruvate reduction occurs in tissues that normally experience minimal access to blood flow (e.g., the cornea of the eye) and also in rapidly contracting skeletal muscle. When skeletal muscles are exercised strenuously, the available tissue oxygen is consumed and the pyruvate generated by glycolysis can no longer be oxidized in the TCA cycle. Instead, excess pyruvate is reduced to lactate by **lactate dehydrogenase** (Figure 18.30). In anaerobic muscle tissue, lactate represents the end of glycolysis. Anyone who exercises to the point of depleting available muscle oxygen stores knows the cramps and muscle fatigue associated with the buildup of lactic acid in the muscle. Most of this lactate must be carried out of the muscle by the blood and transported to the liver, where it can be resynthesized into glucose in gluconeogenesis. Moreover, because glycolysis generates only a fraction of the total energy available from the breakdown of glucose (the rest is generated by the TCA cycle and oxidative phosphorylation), the onset of anaerobic conditions in skeletal muscle also means a reduction in the energy available from the breakdown of glucose.

18.6	How Do Cells Regulate Glycolysis?

The elegance of nature's design for the glycolytic pathway may be appreciated through an examination of Figure 18.31. The standard-state free energy changes for the ten reactions of glycolysis and the lactate dehydrogenase reaction (Figure 18.31a) are variously positive and negative and, taken together, offer little insight into the coupling that occurs in the cellular milieu. On the other hand, the values of ΔG under cellular conditions (Figure 18.31b) fall into two distinct classes. For reactions 2 and 4 through 9, ΔG is very close to zero, meaning these reactions operate essentially at equilibrium. Small changes in the concentrations of reactants and products could "push" any of these reactions either forward or backward. By contrast, the hexokinase, phosphofructokinase, and pyruvate kinase reactions all exhibit large negative ΔG values under cellular conditions. These reactions are thus the sites of glycolytic regulation. When these three enzymes are active, glycolysis proceeds and glucose is readily metabolized to pyruvate or lactate. Inhibition of the three key enzymes by allosteric effectors brings glycolysis to a halt. When we consider **gluconeogenesis**—the biosynthesis of glucose—in Chapter 22, we will see that different enzymes are used to carry out reactions 1, 3, and 10 in reverse, effecting the net synthesis of glucose. The maintenance of reactions 2 and 4 through 9 at or near equilibrium permits these reactions (and their respective enzymes!) to operate effectively in *either* the forward or reverse direction.

18.7	Are Substrates Other Than Glucose Used in Glycolysis?

The glycolytic pathway described in this chapter begins with the breakdown of glucose, but other sugars, both simple and complex, can enter the cycle if they can be converted by appropriate enzymes to one of the intermediates of glycolysis. Figure 18.32 shows the mechanisms by which several simple metabolites can enter the glycolytic pathway. **Fructose,** for example, which is

(a) ΔG at standard state ($\Delta G^{\circ\prime}$)

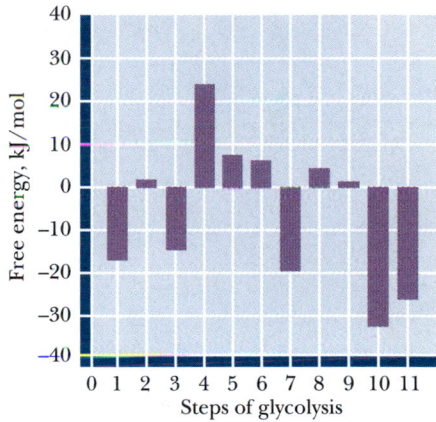

(b) ΔG in erythrocytes (ΔG)

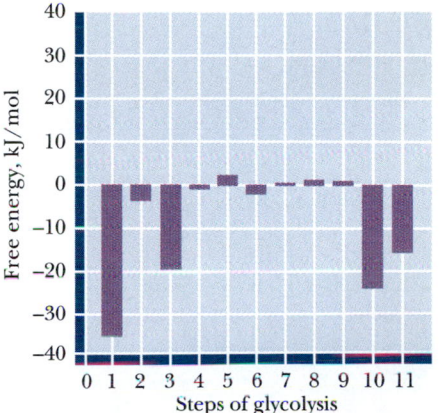

FIGURE 18.31 A comparison of free energy changes for the reactions of glycolysis (step 1 = hexokinase) under **(a)** standard-state conditions and **(b)** actual intracellular conditions in erythrocytes. The values of $\Delta G^{\circ\prime}$ provide little insight into the actual free energy changes that occur in glycolysis. On the other hand, under intracellular conditions, seven of the glycolytic reactions operate near equilibrium (with ΔG near zero). The driving force for glycolysis lies in the hexokinase (1), phosphofructokinase (3), and pyruvate kinase (10) reactions. The lactate dehydrogenase (step 11) reaction also exhibits a large negative ΔG under cellular conditions.

Human Biochemistry

Tumor Diagnosis Using Positron Emission Tomography (PET)

More than 70 years ago, Otto Warburg at the Kaiser Wilhelm Institute of Biology in Germany demonstrated that most animal and human tumors displayed a very high rate of glycolysis compared to that of normal tissue. This observation from long ago is the basis of a very modern diagnostic method for tumor detection called **positron emission tomography,** or **PET.** PET uses molecular probes that contain a neutron-deficient, radioactive element such as carbon-11 or fluorine-18. An example is 2-[18F]fluoro-2-deoxy-glucose (FDG), a molecular mimic of glucose. The 18F nucleus is unstable and spontaneously decays by emission of a positron (an antimatter particle) from a proton, thus converting a proton to a neutron and transforming the 18F to 18O. The emitted positron typically travels a short distance (less than a millimeter) and collides with an electron, annihilating both particles and creating a pair of high-energy photons—gamma rays. Detection of the gamma rays with special cameras can be used to construct three-dimensional models of the location of the radiolabeled molecular probe in the tissue of interest.

FDG is taken up by human cells and converted by hexokinase to 2-[18F]fluoro-2-deoxy-glucose-6-phosphate in the first step of glycolysis. Cells of a human brain, for example, accumulate FDG in direct proportion to the amount of glycolysis occurring in those cells. Tumors can be identified in PET scans as sites of unusually high FDG accumulation.

(a)

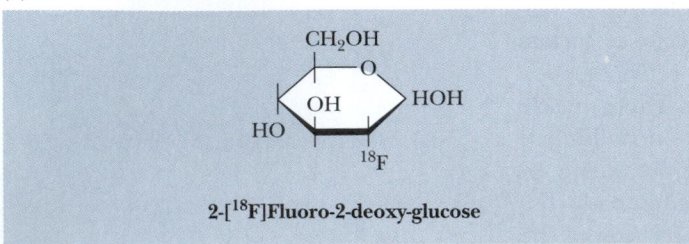

2-[18F]Fluoro-2-deoxy-glucose

(b)

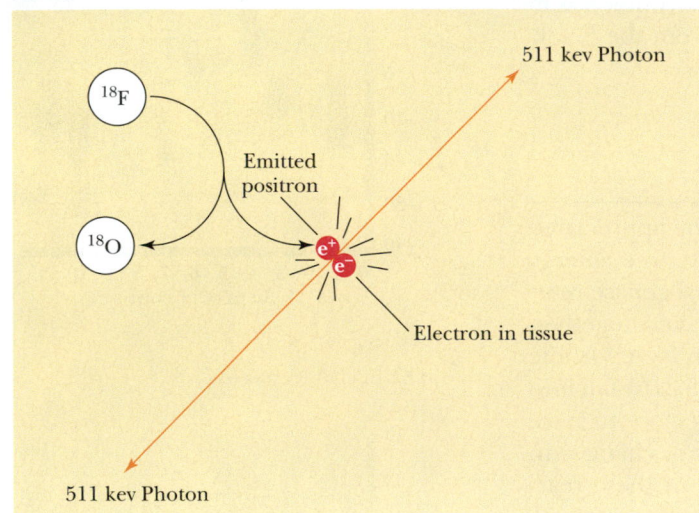

(c)   PET image of human brain following administration of 18FDG. Red area indicates a large malignant tumor

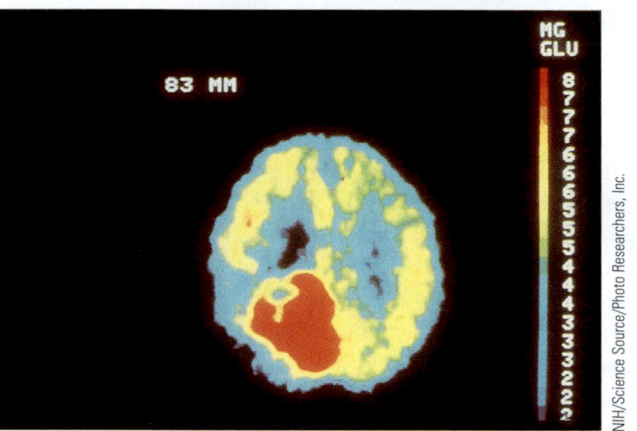

produced by breakdown of sucrose, may participate in glycolysis by at least two different routes. In the liver, fructose is phosphorylated at C-1 by the enzyme **fructokinase:**

$$\text{D-Fructose} + \text{ATP}^{4-} \longrightarrow \text{D-fructose-1-phosphate}^{2-} + \text{ADP}^{3-} + \text{H}^+ \qquad (18.10)$$

Subsequent action by **fructose-1-phosphate aldolase** cleaves fructose-1-P in a manner like the fructose bisphosphate aldolase reaction to produce dihydroxyacetone phosphate and D-glyceraldehyde:

$$\text{D-Fructose-1-P}^{2-} \longrightarrow \text{D-glyceraldehyde} + \text{dihydroxyacetone phosphate}^{2-} \qquad (18.11)$$

Dihydroxyacetone phosphate is of course an intermediate in glycolysis. D-Glyceraldehyde can be phosphorylated by **triose kinase** in the presence of ATP to form D-glyceraldehyde-3-phosphate, another glycolytic intermediate.

In the kidney and in muscle tissues, fructose is readily phosphorylated by hexokinase, which, as pointed out previously, can utilize several different hexose substrates. The free energy of hydrolysis of ATP drives the reaction forward:

$$\text{D-Fructose} + \text{ATP}^{4-} \longrightarrow \text{D-fructose-6-phosphate}^{2-} + \text{ADP}^{3-} + \text{H}^+ \qquad (18.12)$$

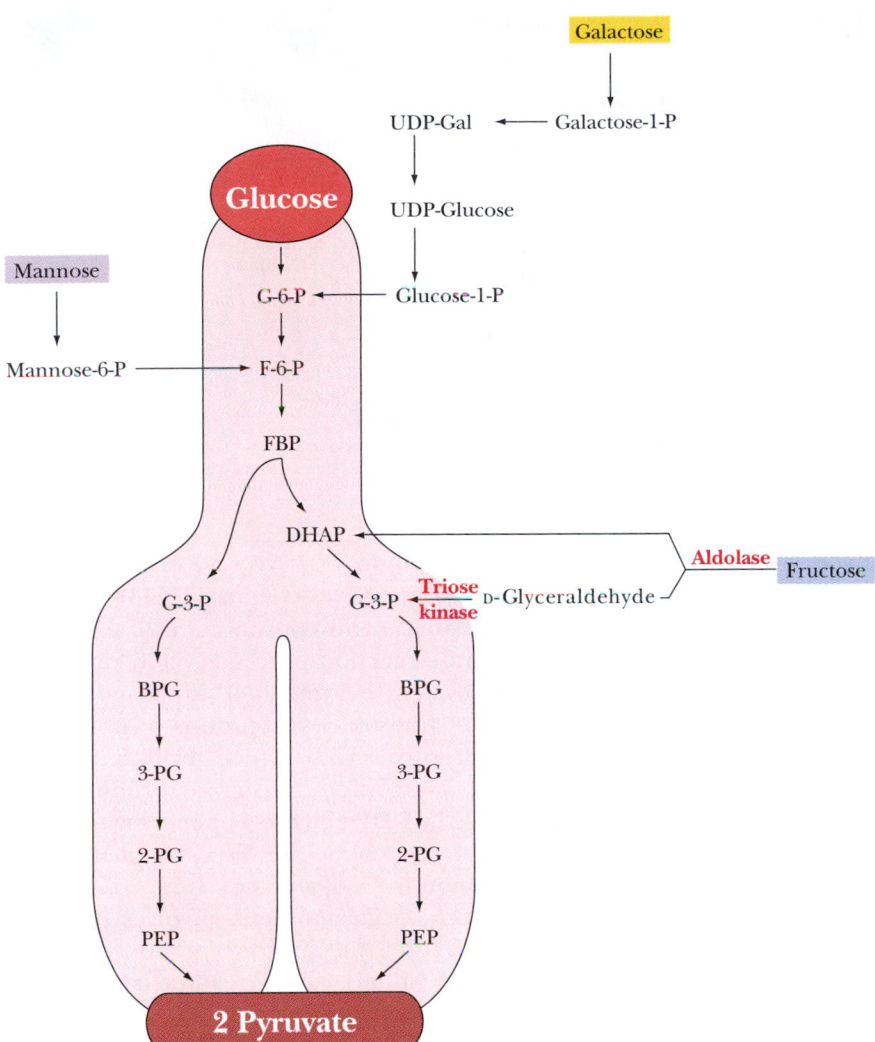

FIGURE 18.32 Mannose, galactose, fructose, and other simple metabolites can enter the glycolytic pathway.

Fructose-6-phosphate generated in this way enters the glycolytic pathway directly in step 3, the second priming reaction. This is the principal means for channeling fructose into glycolysis in adipose tissue, which contains high levels of fructose.

Mannose Enters Glycolysis in Two Steps

Another simple sugar that enters glycolysis at the same point as fructose is **mannose,** which occurs in many glycoproteins, glycolipids, and polysaccharides (see Chapter 7). Mannose is also phosphorylated from ATP by hexokinase, and the mannose-6-phosphate thus produced is converted to fructose-6-phosphate by **phosphomannoisomerase.**

$$\text{D-Mannose} + \text{ATP}^{4-} \longrightarrow \text{D-mannose-6-phosphate}^{2-} + \text{ADP}^{3-} + \text{H}^+ \qquad (18.13)$$

$$\text{D-Mannose-6-phosphate}^{2-} \longrightarrow \text{D-fructose-6-phosphate}^{2-} \qquad (18.14)$$

Galactose Enters Glycolysis Via the Leloir Pathway

A somewhat more complicated route into glycolysis is followed by **galactose,** another simple hexose sugar. The process, called the **Leloir pathway** after Luis Leloir, its discoverer, begins with phosphorylation from ATP at the C-1 position by **galactokinase:**

$$\text{D-Galactose} + \text{ATP}^{4-} \longrightarrow \text{D-galactose-1-phosphate}^{2-} + \text{ADP}^{3-} + \text{H}^+ \qquad (18.15)$$

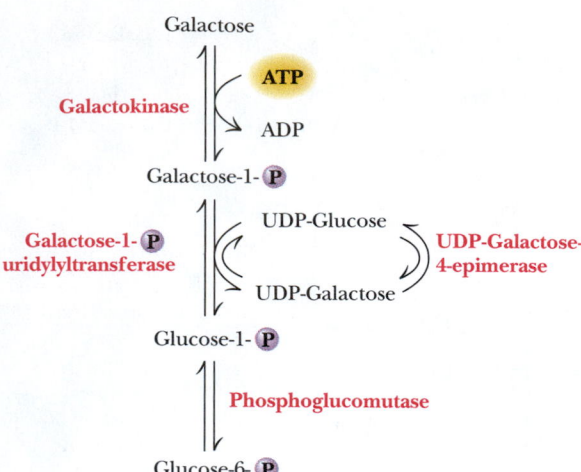

FIGURE 18.33 Galactose metabolism via the Leloir pathway.

Galactose-1-phosphate is then converted into *UDP-galactose* (a sugar nucleotide) by **galactose-1-phosphate uridylyltransferase** (Figure 18.33), with concurrent production of glucose-1-phosphate and consumption of a molecule of UDP-glucose. The uridylyltransferase reaction proceeds via a "ping-pong" mechanism (Figure 18.34) with a covalent enzyme-UMP intermediate. The glucose-1-phosphate produced by the transferase reaction is a substrate for the **phosphoglucomutase** reaction (Figure 18.33), which produces glucose-6-phosphate, a glycolytic substrate. The other transferase product, UDP-galactose, is converted to UDP-glucose by **UDP-glucose-4-epimerase.** The combined action of the uridylyltransferase and epimerase thus produces glucose-1-P from galactose-1-P, with regeneration of UDP-glucose.

A rare hereditary condition known as **galactosemia** involves defects in galactose-1-P uridylyltransferase that render the enzyme inactive. Toxic levels of galactose accumulate in afflicted individuals, causing cataracts and permanent neurological disorders. These problems can be prevented by removing galactose and lactose from the diet. In adults, the toxicity of galactose appears

FIGURE 18.34 The galactose-1-phosphate uridylyltransferase reaction involves a "ping-pong" kinetic mechanism.

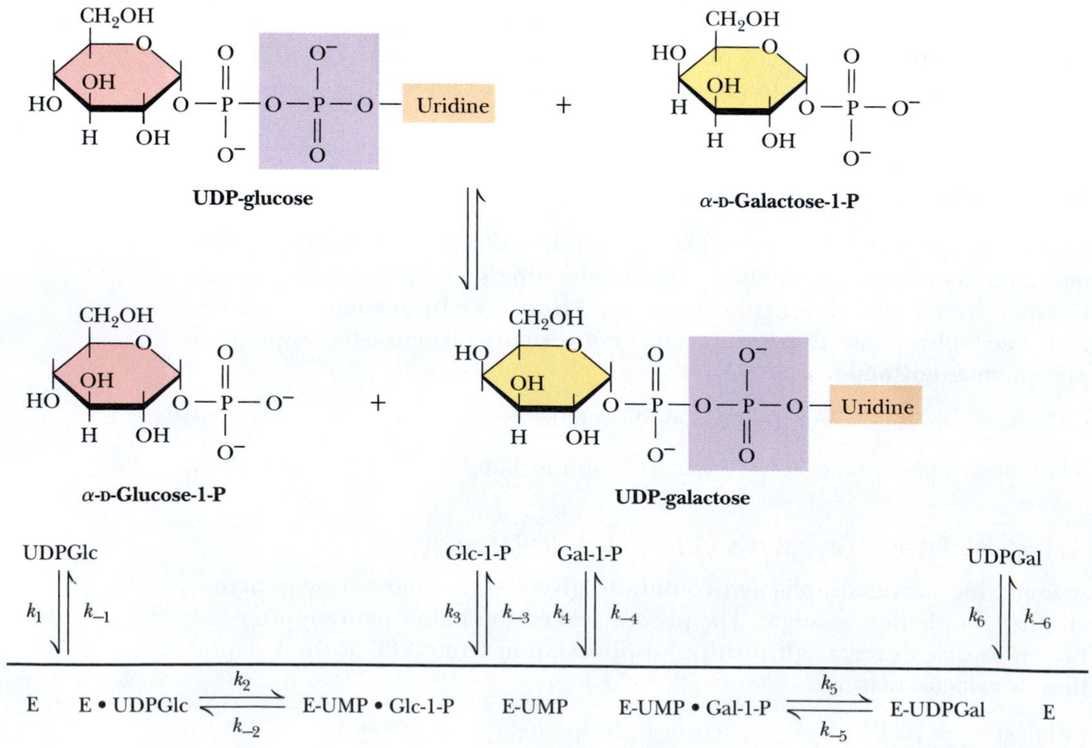

to be less severe, due in part to the metabolism of galactose-1-P by **UDP-glucose pyrophosphorylase,** which apparently can accept galactose-1-P in place of glucose-1-P (Figure 18.35). The levels of this enzyme may increase in galactosemic individuals in order to accommodate the metabolism of galactose.

An Enzyme Deficiency Causes Lactose Intolerance

A much more common metabolic disorder, **lactose intolerance,** occurs commonly in most parts of the world (notable exceptions being some parts of Africa and northern Europe). Lactose intolerance is an inability to digest

---

Human Biochemistry

Lactose—From Mother's Milk to Yogurt—and Lactose Intolerance

Lactose is an interesting sugar in many ways. In placental mammals, it is synthesized only in the mammary gland, and then only during late pregnancy and lactation. The synthesis is carried out by **lactose synthase,** a dimeric complex of two proteins: galactosyl transferase and α-lactalbumin. Galactosyl transferase is present in all human cells, and it is normally involved in incorporation of galactose into glycoproteins. In late pregnancy, the pituitary gland in the brain releases a protein hormone, prolactin, which triggers production of α-lactalbumin by certain cells in the breast. α-Lactalbumin, a 123-residue protein, associates with galactosyl transferase to form lactose synthase, which catalyzes the reaction:

$$UDP\text{-galactose} + glucose \longrightarrow lactose + UDP$$

Lactose breakdown by **lactase** in the small intestine provides newborn mammals with essential galactose for many purposes, including the synthesis of gangliosides in the developing brain. Lactase is a β-**galactosidase** that cleaves lactose to yield galactose and glucose—in fact the only human enzyme that can cleave a β-glycosidic linkage:

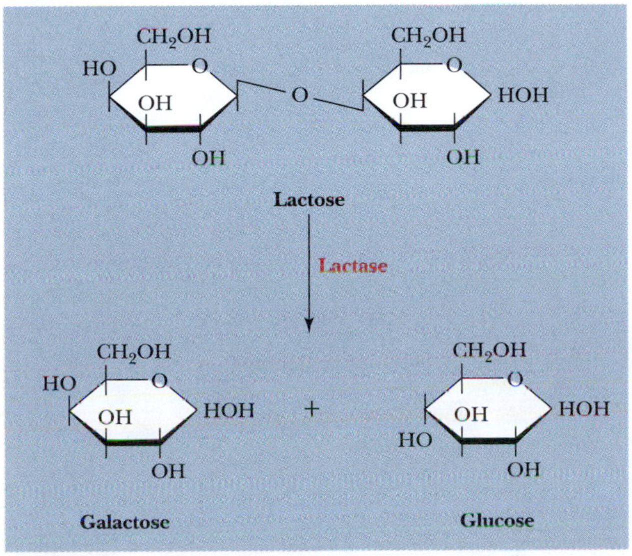

Lactase is an inducible enzyme in mammals, and it appears in the fetus only during the late stages of gestation. Lactase activity peaks shortly after birth, but by the age of 3 to 5 years, it declines to a low level in nearly all human children. Low levels of lactase make many adults **lactose intolerant.** Lactose intolerance occurs commonly in most parts of the world (with the notable exception of some parts of Africa and northern Europe; see table). The symptoms of lactose intolerance, including diarrhea and general discomfort, can be relieved by eliminating milk from the diet. Alternatively, products containing β-galactosidase are available commercially.

Certain bacteria, including several species of *Lactobacillus,* thrive on the lactose in milk and carry out lactic acid fermentation, converting lactose to lactate via glycolysis. This is the basis of production of yogurt, which is now popular in the Western world but of Turkish origin. Other cultures also produce yogurtlike foods. Nomadic Tatars in Siberia and Mongolia used camel milk to make *koumiss,* which was used for medicinal purposes. In the Caucasus, *kefir* is made much like yogurt, except that the starter culture contains (in addition to *Lactobacillus*) *Streptococcus lactis* and yeast, which convert some of the glucose to ethanol and CO_2, producing an effervescent and slightly intoxicating brew.

Breakdown of lactose to galactose and glucose by lactase.

Percentage of Population with Lactase Persistence	
Country	**Lactase Persistence (%)**
Sweden	99
Denmark	97
United Kingdom (Scotland)	95
Germany	88
Switzerland	84
Australia	82
United States (Iowa)	81
Bedouin tribes (North Africa)	75
Spain	72
France	58
Italy	49
India	36
Japan	10
China (Shanghai)	8
China (Singapore)	0

Portions adapted from Hill, R., and Brew, K., 1975. Lactose synthetase. *Advances in Enzymology* **43**:411–485; and Bloch, K., 1994. *Blondes in Venetian Paintings, the Nine-Banded Armadillo, and Other Essays in Biochemistry.* New Haven, CT: Yale University Press.

Adapted from Bloch, K., 1994. *Blondes in Venetian Paintings, the Nine-Banded Armadillo, and Other Essays in Biochemistry.* New Haven, CT: Yale University Press.

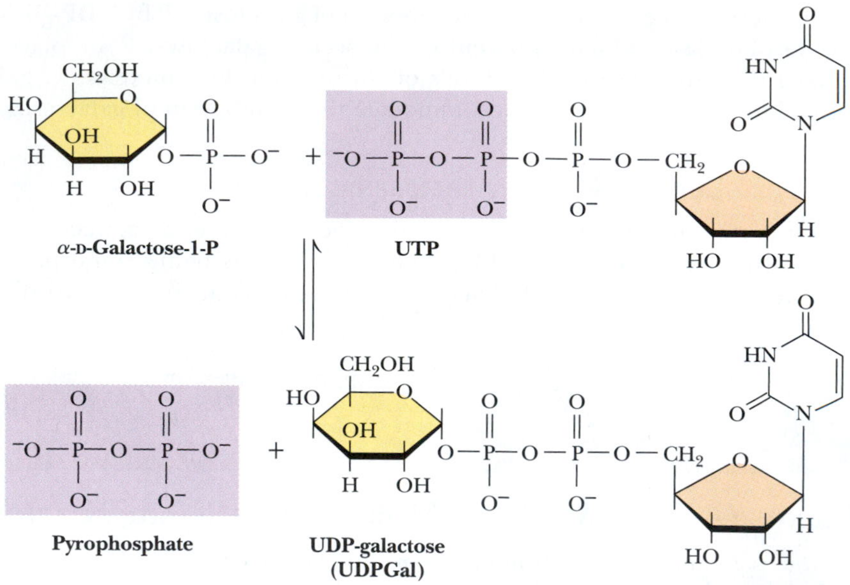

FIGURE 18.35 The UDP-glucose pyrophosphorylase reaction.

lactose because of the absence of the enzyme **lactase** in the intestines of adults. The symptoms of this disorder, which include diarrhea and general discomfort, can be relieved by eliminating milk from the diet.

Glycerol Can Also Enter Glycolysis

Glycerol is the last important simple substance whose ability to enter the glycolytic pathway must be considered. This metabolite, which is produced in substantial amounts by the decomposition of triacylglycerols (see Chapter 23), can be converted to glycerol-3-phosphate by the action of **glycerol kinase** and then oxidized to dihydroxyacetone phosphate by the action of **glycerol phosphate dehydrogenase,** with NAD^+ as the required coenzyme (Figure 18.36). The dihydroxyacetone phosphate thereby produced enters the glycolytic pathway as a substrate for triose phosphate isomerase.

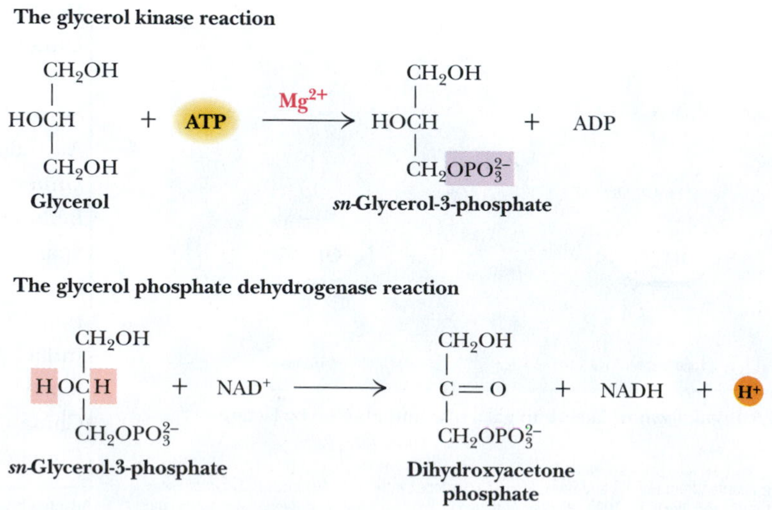

FIGURE 18.36 The glycerol kinase and glycerol phosphate dehydrogenase reactions.

Summary

Nearly every living cell carries out a catabolic process known as glycolysis—the stepwise degradation of glucose (and other simple sugars). Glycolysis is a paradigm of metabolic pathways. Carried out in the cytosol of cells, it is basically an anaerobic process; its principal steps occur with no requirement for oxygen.

18.1 What Are the Essential Features of Glycolysis?
Glycolysis consists of two phases. In the first phase, a series of five reactions, glucose is broken down to two molecules of glyceraldehyde-3-phosphate. In the second phase, five subsequent reactions convert these two molecules of glyceraldehyde-3-phosphate into two molecules of pyruvate. Phase 1 consumes two molecules of ATP (Figure 18.2). The later stages of glycolysis result in the production of four molecules of ATP. The net is $4 - 2 = 2$ molecules of ATP produced per molecule of glucose.

18.2 Why Are Coupled Reactions Important in Glycolysis?
Coupled reactions permit the energy of glycolysis to be used for generation of ATP. Conversion of one molecule of glucose to pyruvate in glycolysis drives the production of two molecules of ATP.

18.3 What Are the Chemical Principles and Features of the First Phase of Glycolysis?
In the first phase of glycolysis, glucose is converted into two molecules of glyceraldehyde-3-phosphate. Glucose is phosphorylated to glucose-6-P, which is isomerized to fructose-6-P. Another phosphorylation and then cleavage yields two 3-carbon intermediates. One of these is glyceraldehyde-3-P, and the other, dihydroxyacetone-P, is converted to glyceraldehyde-3-P. Energy released from this high-energy molecule in the second phase of glycolysis is then used to synthesize ATP.

18.4 What Are the Chemical Principles and Features of the Second Phase of Glycolysis?
The second half of the glycolytic pathway involves the reactions that convert the metabolic energy in the glucose molecule into ATP. Phase 2 starts with the oxidation of glyceraldehyde-3-phosphate, a reaction with a large enough energy "kick" to produce a high-energy phosphate, namely, 1,3-bisphosphoglycerate. Phosphoryl transfer from 1,3-BPG to ADP to make ATP is highly favorable. The product, 3-phosphoglycerate, is converted via several steps to phosphoenolpyruvate (PEP), another high-energy phosphate. PEP readily transfers its phosphoryl group to ADP in the pyruvate kinase reaction to make another ATP.

18.5 What Are the Metabolic Fates of NADH and Pyruvate Produced in Glycolysis?
In addition to ATP, the products of glycolysis are NADH and pyruvate. Their processing depends upon other cellular pathways. NADH must be recycled to NAD^+, lest NAD^+ become limiting in glycolysis. NADH can be recycled by both aerobic and anaerobic paths, either of which results in further metabolism of pyruvate. What a given cell does with the pyruvate produced in glycolysis depends in part on the availability of oxygen. Under aerobic conditions, pyruvate can be sent into the citric acid cycle, where it is oxidized to CO_2 with the production of additional NADH (and $FADH_2$). Under aerobic conditions, the NADH produced in glycolysis and the citric acid cycle is reoxidized to NAD^+ in the mitochondrial electron transport chain.

Under anaerobic conditions, the pyruvate produced in glycolysis is not sent to the citric acid cycle. Instead, it is reduced to ethanol in yeast; in other microorganisms and in animals, it is reduced to lactate. These processes are examples of fermentation—the production of ATP energy by reaction pathways in which organic molecules function as donors and acceptors of electrons. In either case, reduction of pyruvate provides a means of reoxidizing the NADH produced in the glyceraldehyde-3-phosphate dehydrogenase reaction of glycolysis.

18.6 How Do Cells Regulate Glycolysis?
The standard-state free energy changes for the ten reactions of glycolysis are variously positive and negative and, taken together, offer little insight into the coupling that occurs in the cellular milieu. On the other hand, the values of ΔG under cellular conditions fall into two distinct classes. For reactions 2 and 4 through 9, ΔG is very close to zero, meaning these reactions operate essentially at equilibrium. Small changes in the concentrations of reactants and products could "push" any of these reactions either forward or backward. By contrast, the hexokinase, phosphofructokinase, and pyruvate kinase reactions all exhibit large negative ΔG values under cellular conditions. These reactions are thus the sites of glycolytic regulation.

18.7 Are Substrates Other Than Glucose Used in Glycolysis?
Fructose enters glycolysis by either of two routes. Mannose, galactose, and glycerol enter via reactions that are linked to the glycolytic pathway, as shown in Figures 18.33 through 18.36.

Problems

1. List the reactions of glycolysis that
   a. are energy consuming (under standard-state conditions).
   b. are energy yielding (under standard-state conditions).
   c. consume ATP.
   d. yield ATP.
   e. are strongly influenced by changes in concentration of substrate and product because of their molecularity.
   f. are at or near equilibrium in the erythrocyte (see Table 18.2).

2. Determine the anticipated location in pyruvate of labeled carbons if glucose molecules labeled (in separate experiments) with ^{14}C at each position of the carbon skeleton proceed through the glycolytic pathway.

3. In an erythrocyte undergoing glycolysis, what would be the effect of a sudden increase in the concentration of
   a. ATP?   b. AMP?
   c. fructose-1,6-bisphosphate?   d. fructose-2,6-bisphosphate?
   e. citrate?   f. glucose-6-phosphate?

4. Discuss the cycling of NADH and NAD^+ in glycolysis and the related fermentation reactions.

5. For each of the following reactions, name the enzyme that carries out this reaction in glycolysis and write a suitable mechanism for the reaction.

6. Write the reactions that permit galactose to be utilized in glycolysis. Write a suitable mechanism for one of these reactions.

7. (Integrates with Chapters 4 and 14.) How might iodoacetic acid affect the glyceraldehyde-3-phosphate dehydrogenase reaction in glycolysis? Justify your answer.

8. If ^{32}P-labeled inorganic phosphate were introduced to erythrocytes undergoing glycolysis, would you expect to detect ^{32}P in glycolytic intermediates? If so, describe the relevant reactions and the ^{32}P incorporation you would observe.

9. Sucrose can enter glycolysis by either of two routes:

   Sucrose phosphorylase:
   Sucrose + $P_i \rightleftharpoons$ fructose + glucose-1-phosphate

   Invertase:
   Sucrose + $H_2O \rightleftharpoons$ fructose + glucose

   Would either of these reactions offer an advantage over the other in the preparation of hexoses for entry into glycolysis?

10. What would be the consequences of a Mg^{2+} ion deficiency for the reactions of glycolysis?

11. (Integrates with Chapter 3.) Triose phosphate isomerase catalyzes the conversion of dihydroxyacetone-P to glyceraldehyde-3-P. The standard free energy change, $\Delta G°'$, for this reaction is $+7.6$ kJ/mol. However, the observed free energy change (ΔG) for this reaction in erythrocytes is $+2.4$ kJ/mol.
    a. Calculate the ratio of [dihydroxyacetone-P]/[glyceraldehyde-3-P] in erythrocytes from ΔG.
    b. If [dihydroxyacetone-P] = 0.2 mM, what is [glyceraldehyde-3-P]?

12. (Integrates with Chapter 3.) Enolase catalyzes the conversion of 2-phosphoglycerate to phosphoenolpyruvate + H_2O. The standard free energy change, $\Delta G°'$, for this reaction is $+1.8$ kJ/mol. If the concentration of 2-phosphoglycerate is 0.045 mM and the concentration of phosphoenolpyruvate is 0.034 mM, what is ΔG, the free energy change for the enolase reaction, under these conditions?

13. (Integrates with Chapter 3.) The standard free energy change ($\Delta G°'$) for hydrolysis of phosphoenolpyruvate (PEP) is -61.9 kJ/mol. The standard free energy change ($\Delta G°'$) for ATP hydrolysis is -30.5 kJ/mol.
    a. What is the standard free energy change for the pyruvate kinase reaction:

    $$ADP + phosphoenolpyruvate \longrightarrow ATP + pyruvate$$

    b. What is the equilibrium constant for this reaction?
    c. Assuming the intracellular concentrations of [ATP] and [ADP] remain fixed at 8 mM and 1 mM, respectively, what will be the ratio of [pyruvate]/[phosphoenolpyruvate] when the pyruvate kinase reaction reaches equilibrium?

14. (Integrates with Chapter 3.) The standard free energy change ($\Delta G°'$) for hydrolysis of fructose-1,6-bisphosphate (FBP) to fructose-6-phosphate (F-6-P) and P_i is -16.7 kJ/mol:

    $$FBP + H_2O \longrightarrow fructose-6-P + P_i$$

The standard free energy change ($\Delta G°'$) for ATP hydrolysis is -30.5 kJ/mol:

$$ATP + H_2O \longrightarrow ADP + P_i$$

   a. What is the standard free energy change for the phosphofructokinase reaction:

   $$ATP + fructose-6-P \longrightarrow ADP + FBP$$

   b. What is the equilibrium constant for this reaction?
   c. Assuming the intracellular concentrations of [ATP] and [ADP] are maintained constant at 4 mM and 1.6 mM, respectively, in a rat liver cell, what will be the ratio of [FBP]/[fructose-6-P] when the phosphofructokinase reaction reaches equilibrium?

15. (Integrates with Chapter 3.) The standard free energy change ($\Delta G°'$) for hydrolysis of 1,3-bisphosphoglycerate (1,3-BPG) to 3-phosphoglycerate (3-PG) and P_i is -49.6 kJ/mol:

    $$1,3\text{-BPG} + H_2O \longrightarrow 3\text{-PG} + P_i$$

The standard free energy change ($\Delta G°'$) for ATP hydrolysis is -30.5 kJ/mol:

$$ATP + H_2O \longrightarrow ADP + P_i$$

   a. What is the standard free energy change for the phosphoglycerate kinase reaction:

   $$ADP + 1,3\text{-BPG} \longrightarrow ATP + 3\text{-PG}$$

   b. What is the equilibrium constant for this reaction?
   c. If the steady-state concentrations of [1,3-BPG] and [3-PG] in an erythrocyte are 1 μM and 120 μM, respectively, what will be the ratio of [ATP]/[ADP], assuming the phosphoglycerate kinase reaction is at equilibrium?

Preparing for the MCAT Exam

16. Regarding phosphofructokinase, which of the following statements is true:
    a. Low ATP stimulates the enzyme, but fructose-2,6-bisphosphate inhibits.
    b. High ATP stimulates the enzyme, but fructose-2,6-bisphosphate inhibits.
    c. High ATP stimulates the enzyme, but fructose-2,6-bisphosphate inhibits.
    d. The enzyme is more active at low ATP than at high, and fructose-2,6-bisphosphate activates the enzyme.
    e. ATP and fructose-2,6-bisphosphate both inhibit the enzyme.

17. Based on your reading of this chapter, what would you expect to be the most immediate effect on glycolysis if the steady-state concentration of glucose-6-P were 8.3 mM instead of 0.083 mM?

Biochemistry⊜Now™   Preparing for an exam? Test yourself on key questions at http://chemistry.brookscole.com/ggb3

Further Reading

General

Arkin, A., Shen, P., and Ross, J., 1997. A test case of correlation metric construction of a reaction pathway from measurements. *Science* **277:**1275–1279.

Beitner, R., 1985. *Regulation of Carbohydrate Metabolism.* Boca Raton, FL: CRC Press.

Bendjelid, K., Canet, E., Rayan, E., Casali, C., Revel, D., and Janier, M., 2003. Role of glycolysis in energy production for the non-mechanical myocardial work in isolated pig hearts. *Current Medical Research Opinions* **19:**51–58.

Bioteux, A., and Hess, A., 1981. Design of glycolysis. *Philosophical Transactions, Royal Society of London B* **293:**5–22.

Bodner, G. M., 1986. Metabolism: Part I, Glycolysis. *Journal of Chemical Education* **63:**566–570.

Braun, L., Puskas, F., Csala, M., et al., 1997. Ascorbate as a substrate for glycolysis or gluconeogenesis: Evidence for an interorgan ascorbate cycle. *Free Radical Biology and Medicine* **23:**804–808.

Fothergill-Gilmore, L., 1986. The evolution of the glycolytic pathway. *Trends in Biochemical Sciences* **11:**47–51.

Lakhdar-Ghazal, F., Blonski, C., Willson, M., Michels, P., and Perie, J., 2002. Glycolysis and proteases as targets for the design of new anti-trypanosome drugs. *Current Topics in Medicinal Chemistry* **2**:439–456.

Sparks, S., 1997. The purpose of glycolysis. *Science* **277**:459–460.

Waddell, T. G., et al., 1997. Optimization of glycolysis: A new look at the efficiency of energy coupling. *Biochemical Education* **25**:204–205.

Enzymes of Glycolysis

Bosca, L., and Corredor, C., 1984. Is phosphofructokinase the rate-limiting step of glycolysis? *Trends in Biochemical Sciences* **9**:372–373.

Boyer, P. D., 1972. *The Enzymes*, 3rd ed., vols. 5–9. New York: Academic Press.

Knowles, J., and Albery, W., 1977. Perfection in enzyme catalysis: The energetics of triose phosphate isomerase. *Accounts of Chemical Research* **10**:105–111.

Saier, M., Jr., 1987. *Enzymes in Metabolic Pathways*. New York: Harper and Row.

Vertessy, B. G., Orosz, F., Kovacs, J., and Ovadi, J., 1997. Alternative binding of two sequential glycolytic enzymes to microtubules. Molecular studies in the phosphofructokinase/aldolase/microtubule system. *Journal of Biological Chemistry* **272**:25542–25546.

Wilson, J. E., 2003. Isozymes of mammalian hexokinase: Structure, subcellular localization and metabolic function. *Journal of Experimental Biology* **206**:2049–2057.

Hormones and Signaling in Glycolysis

Goncalves, P. M., Giffioen, G., Bebelman, J. P., and Planta, R. J., 1997. Signalling pathways leading to transcriptional regulation of genes involved in the activation of glycolysis in yeast. *Molecular Microbiology* **25**:483–493.

Jiang, G., and Zhang, B. B., 2003. Glucagon and regulation of glucose metabolism. *American Journal of Physiology, Endocrinology and Metabolism* **284**:E671–E678.

Newsholme, E., Challiss, R., and Crabtree, B., 1984. Substrate cycles: Their role in improving sensitivity in metabolic control. *Trends in Biochemical Sciences* **9**:277–280.

Pilkus, S., and El-Maghrabi, M., 1988. Hormonal regulation of hepatic gluconeogenesis and glycolysis. *Annual Review of Biochemistry* **57**:755–783.

Muscle Biochemistry

Conley, K. E., Blei, M. L., Richards, T. L., et al., 1997. Activation of glycolysis in human muscle in vivo. *American Journal of Physiology* **273**:C306–C315.

Green, H. J., 1997. Mechanisms of muscle fatigue in intense exercise. *Journal of Sports Sciences* **15**:247–256.

Jucker, B. M., Rennings, A. J., Cline, G. W., et al., 1997. In vivo NMR investigation of intramuscular glucose metabolism in conscious rats. *American Journal of Physiology* **273**:E139–E148.

Wackerhage, H., Mueller, K., Hoffmann, U., et al., 1996. Glycolytic ATP production estimated from ^{31}P magnetic resonance spectroscopy measurements during ischemic exercise in vivo. *Magma* **4**:151–155.

CHAPTER 19

The Tricarboxylic Acid Cycle

A time-lapse photograph of a ferris wheel at night. Aerobic cells use a metabolic wheel—the tricarboxylic acid cycle—to generate energy by acetyl-CoA oxidation.

Thus times do shift, each thing his turn does hold;
New things succeed, as former things grow old.

Robert Herrick (*Hesperides* [1648], "Ceremonies for Christmas Eve")

Key Questions

Biochemistry ⊜ Now™ Test yourself on these Key Questions at BiochemistryNow at http://chemistry.brookscole.com/ggb3

© Richard Cummins/CORBIS

Essential Question

The glycolytic pathway converts glucose to pyruvate and produces two molecules of ATP per glucose—only a small fraction of the potential energy available from glucose. Under anaerobic conditions, pyruvate is reduced to lactate in animals and to ethanol in yeast, and much of the potential energy of the glucose molecule remains untapped. In the presence of oxygen, however, a much more interesting and thermodynamically complete story unfolds. *How is pyruvate oxidized under aerobic conditions, and what is the chemical logic that dictates how this process occurs?*

Under aerobic conditions, pyruvate from glycolysis is converted to acetyl-coenzyme A and oxidized to CO_2 in the **tricarboxylic acid (TCA) cycle** (also called the **citric acid cycle**). The electrons liberated by this oxidative process are passed via NADH and $FADH_2$ through an elaborate, membrane-associated **electron-transport pathway** to O_2, the final electron acceptor. Electron transfer is coupled to creation of a proton gradient across the membrane. Such a gradient represents an energized state, and the energy stored in this gradient is used to drive the synthesis of many equivalents of ATP.

ATP synthesis as a consequence of electron transport is termed **oxidative phosphorylation;** the complete process is diagrammed in Figure 19.1. Aerobic pathways permit the production of 30 to 38 molecules of ATP per glucose oxidized. Although two molecules of ATP come from glycolysis and two more directly out of the TCA cycle, most of the ATP arises from oxidative phosphorylation. Specifically, reducing equivalents released in the oxidative reactions of glycolysis, pyruvate decarboxylation, and the TCA cycle are captured in the form of NADH and enzyme-bound $FADH_2$, and these reduced coenzymes fuel the electron-transport pathway and oxidative phosphorylation. The path to oxidative phosphorylation winds through the TCA cycle, and we will examine this cycle in detail in this chapter.

19.1 How Did Hans Krebs Elucidate the TCA Cycle?

Within the orderly and logical confines of a textbook, it is difficult to appreciate the tortuous path of the research scientist through the labyrinth of scientific discovery, the patient sifting and comparing of hypotheses, and the often plodding progress toward new information. The elucidation of the TCA cycle in the first part of the 20th century is a typical case, and one worth recounting. Armed with accumulated small contributions—pieces of the puzzle—from many researchers over many years, Hans Krebs, in a single, seminal inspiration, put the pieces together and finally deciphered the cyclic nature of pyruvate oxidation. In his honor, the TCA cycle is often referred to as the **Krebs cycle.**

In 1932 Krebs was studying the rates of oxidation of small organic acids by kidney and liver tissue. Only a few substances were active in these experiments—notably succinate, fumarate, acetate, malate, and citrate (Figure 19.2). Later it was found that oxaloacetate could be made from pyruvate in such tissues and that it could be further oxidized like the other dicarboxylic acids.

In 1935 in Hungary, a crucial discovery was made by Albert Szent-Györgyi, who was studying the oxidation of similar organic substrates by pigeon breast muscle, an active flight muscle with very high rates of oxidation and metabolism. Carefully measuring the amount of oxygen consumed, he observed that addition of any of three four-carbon dicarboxylic acids—fumarate, succinate, or malate—caused the consumption of much more oxygen than was required for the oxidation of the added substance itself. He concluded that these substances were limiting in the cell and, when provided, stimulated oxidation of endogenous glucose and other carbohydrates in the tissues. He also found that **malonate,** a competitive inhibitor of succinate dehydrogenase (see Chapter 13), inhibited these oxidative processes; this finding

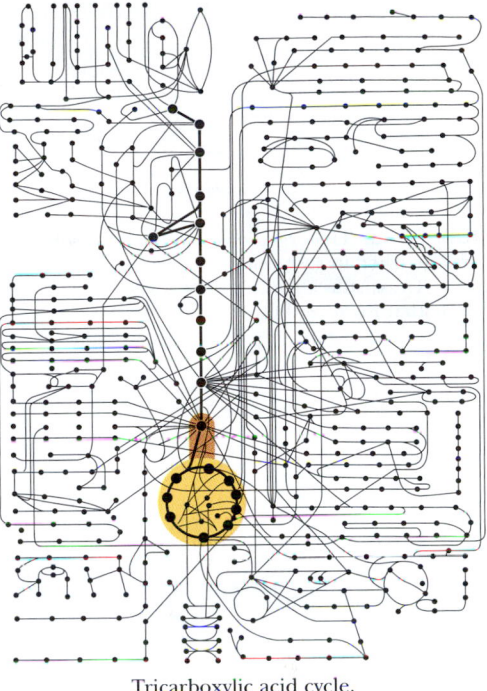

Tricarboxylic acid cycle.

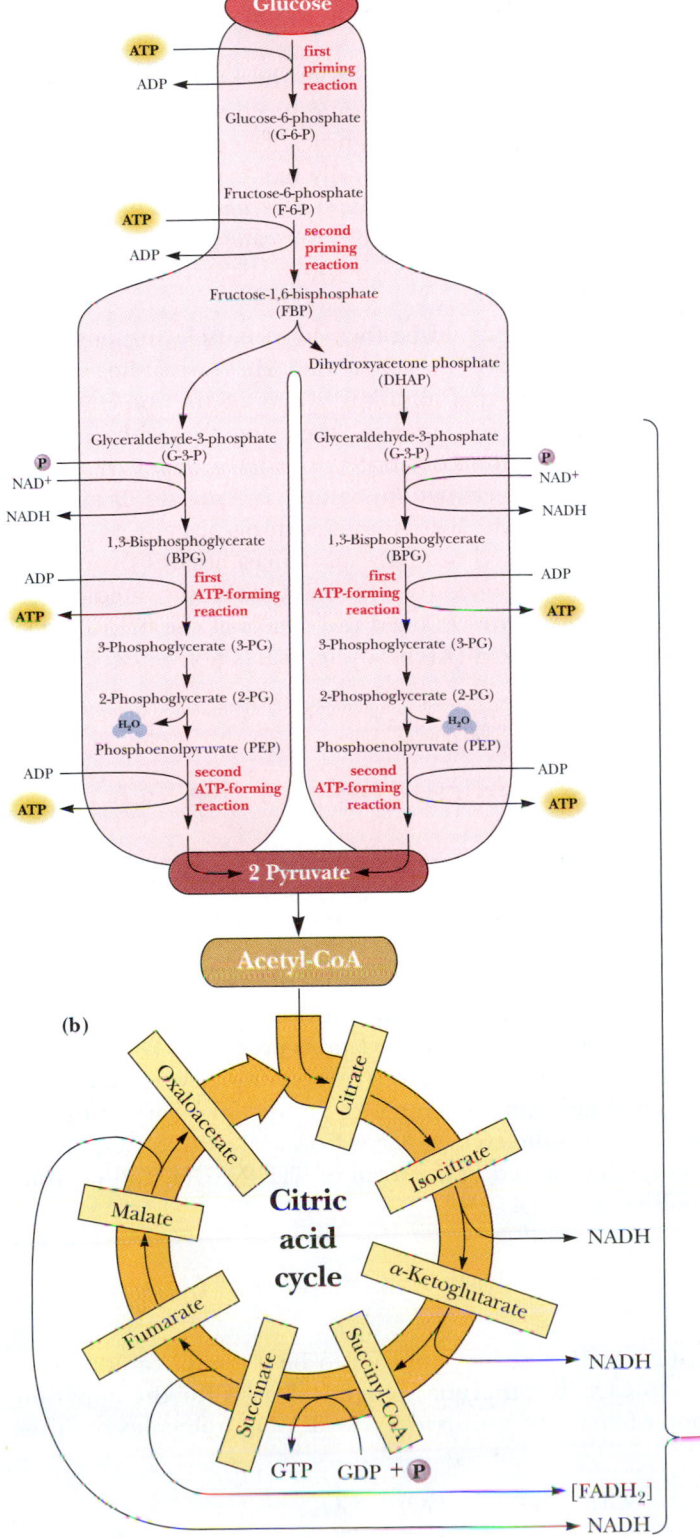

(a) Glycolysis

FIGURE 19.1 **(a)** Pyruvate produced in glycolysis is oxidized in **(b)** the tricarboxylic acid (TCA) cycle. **(c)** Electrons liberated in this oxidation flow through the electron-transport chain and drive the synthesis of ATP in oxidative phosphorylation. In eukaryotic cells, this overall process occurs in mitochondria.

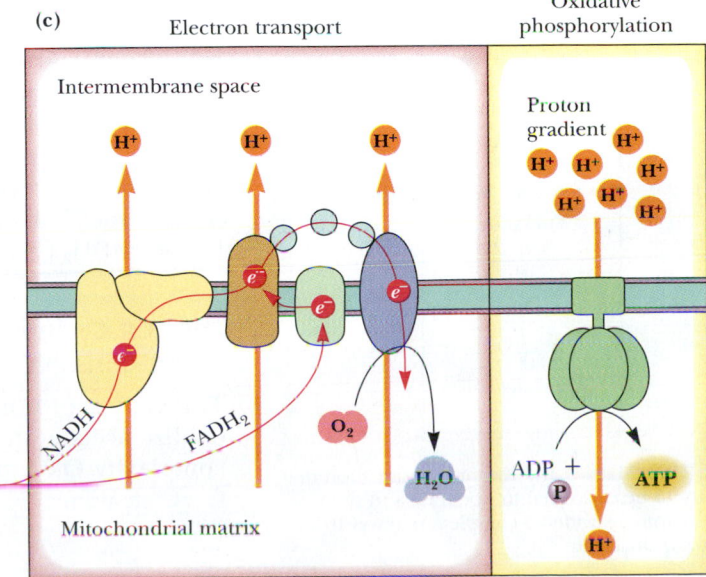

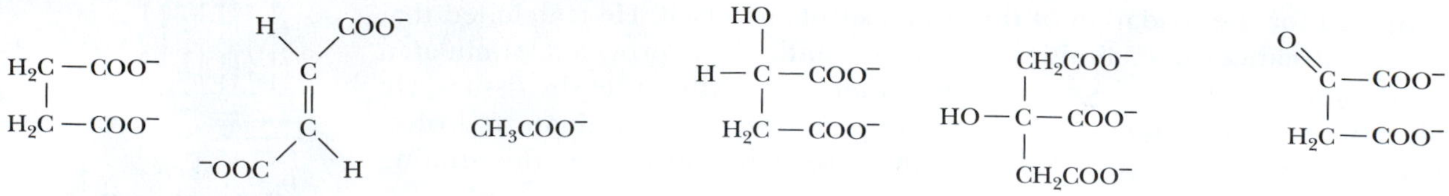

$$H_2C-COO^-$$
$$H_2C-COO^-$$

Succinate

(structure with C=C, H, COO⁻, ⁻OOC, H)

Fumarate

$$CH_3COO^-$$

Acetate

$$H-C-COO^-$$ (HO above)
$$H_2C-COO^-$$

Malate

$$CH_2COO^-$$
$$HO-C-COO^-$$
$$CH_2COO^-$$

Citrate

$$C-COO^-$$ (O above)
$$H_2C-COO^-$$

Oxaloacetate

FIGURE 19.2 The organic acids observed by Krebs to be oxidized in suspensions of liver and kidney tissue. These substances were the pieces in the TCA puzzle that Krebs and others eventually solved.

suggested that succinate oxidation is a crucial step. Szent-Györgyi hypothesized that these dicarboxylic acids were linked by an enzymatic pathway that was important for aerobic metabolism.

Another important piece of the puzzle came from the work of Carl Martius and Franz Knoop, who showed that citric acid could be converted to isocitrate and then to α-ketoglutarate. This finding was significant because it was already known that α-ketoglutarate could be enzymatically oxidized to succinate. At this juncture, the pathway from citrate to oxaloacetate seemed to be as shown in Figure 19.3. Whereas the pathway made sense, the catalytic effect of succinate and the other dicarboxylic acids from Szent-Györgyi's studies remained a puzzle.

In 1937 Krebs found that citrate could be formed in muscle suspensions if oxaloacetate and either pyruvate or acetate were added. He saw that he now had a cycle, not a simple pathway, and that addition of any of the intermediates could generate all of the others. The existence of a cycle, together with the entry of pyruvate into the cycle in the synthesis of citrate, provided a clear explanation for the accelerating properties of succinate, fumarate, and malate. If all these intermediates led to oxaloacetate, which combined with pyruvate from glycolysis, they could stimulate the oxidation of many substances besides themselves. (Krebs' conceptual leap to a cycle was not his first. Together with medical student Kurt Henseleit, he had already elucidated the details of the urea cycle in 1932.) The complete tricarboxylic acid (Krebs) cycle, as it is now understood, is shown in Figure 19.4.

19.2 | What Is the Chemical Logic of the TCA Cycle?

The entry of new carbon units into the cycle is through acetyl-CoA. This entry metabolite can be formed either from pyruvate (from glycolysis) or from oxidation of fatty acids (discussed in Chapter 23). Transfer of the two-carbon acetyl group from acetyl-CoA to the four-carbon oxaloacetate to yield six-carbon citrate is catalyzed by citrate synthase. A dehydration–rehydration rearrangement of citrate yields isocitrate. Two successive decarboxylations produce α-ketoglutarate and then succinyl-CoA, a CoA conjugate of a four-carbon unit. Several steps later, oxaloacetate is regenerated and can combine with another two-carbon unit of acetyl-CoA. Thus, carbon enters the cycle as acetyl-CoA and exits as CO_2. In the process, metabolic energy is captured in the form of ATP, NADH, and enzyme-bound $FADH_2$ (symbolized as $[FADH_2]$).

The TCA Cycle Provides a Chemically Feasible Way of Cleaving a Two-Carbon Compound

The cycle shown in Figure 19.4 at first appears to be a complicated way to oxidize acetate units to CO_2, but there is a chemical basis for the apparent complexity. Oxidation of an acetyl group to a pair of CO_2 molecules requires C—C cleavage:

$$CH_3COO^- \longrightarrow CO_2 + CO_2$$

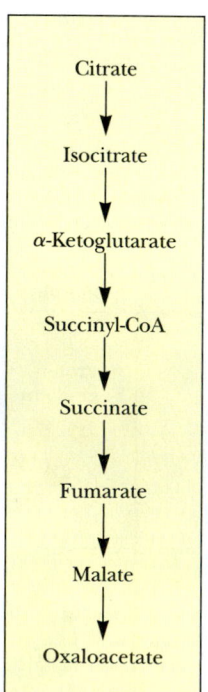

Citrate

↓

Isocitrate

↓

α-Ketoglutarate

↓

Succinyl-CoA

↓

Succinate

↓

Fumarate

↓

Malate

↓

Oxaloacetate

FIGURE 19.3 Martius and Knoop's observation that citrate could be converted to isocitrate and then α-ketoglutarate provided a complete pathway from citrate to oxaloacetate.

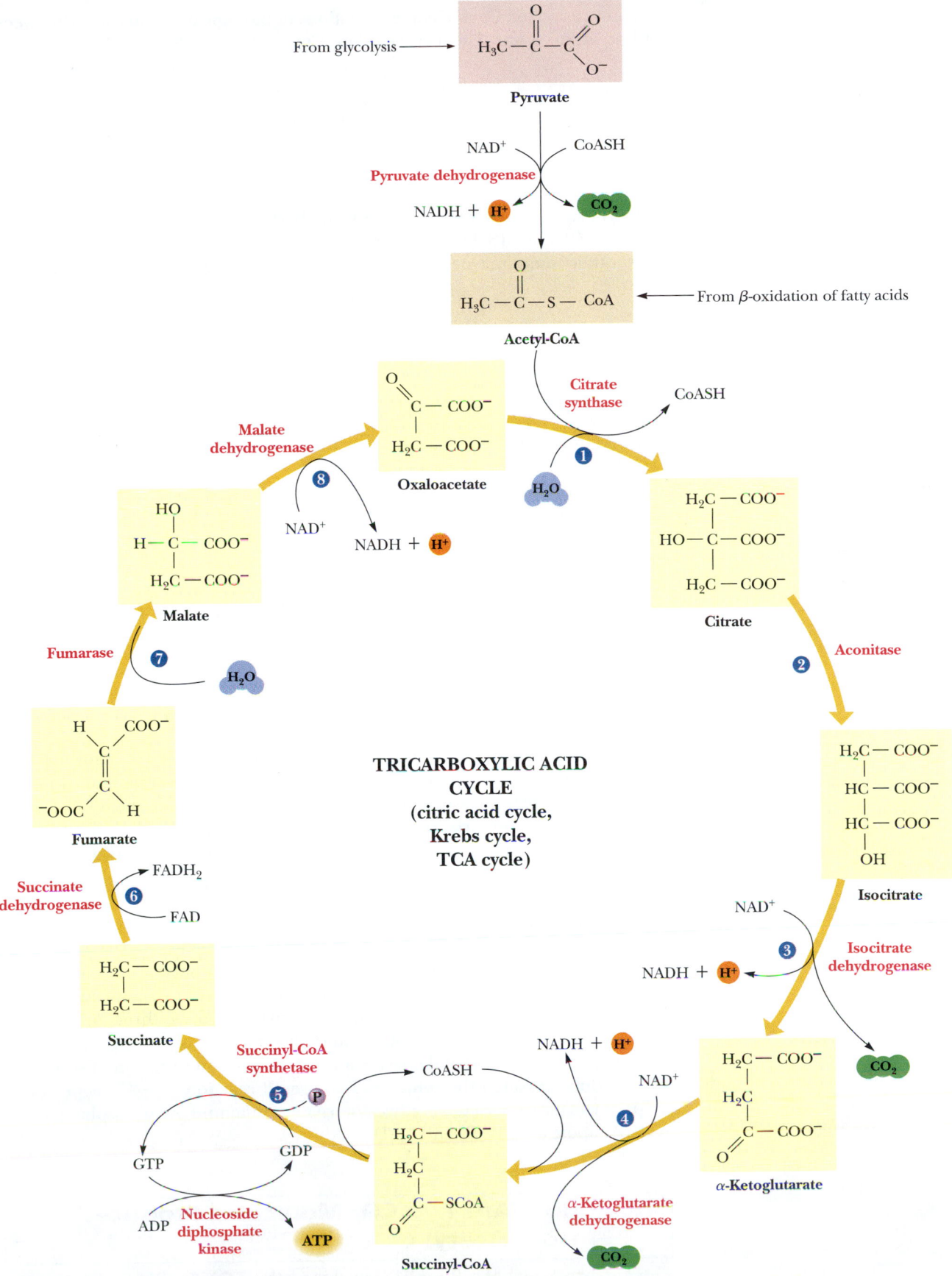

TRICARBOXYLIC ACID
CYCLE
(citric acid cycle,
Krebs cycle,
TCA cycle)

Biochemistry🔵Now™   **ACTIVE FIGURE 19.4** The tricarboxylic acid cycle. Test yourself on the concepts in this figure at http://chemistry.brookscole.com/ggb3

In many instances, C—C cleavage reactions in biological systems occur between carbon atoms α and β to a carbonyl group:

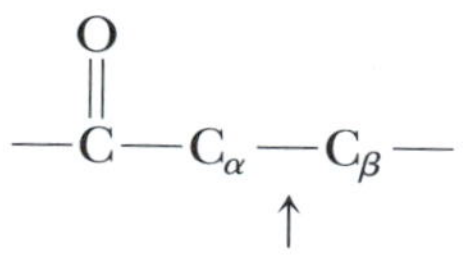

$$-\overset{\overset{\displaystyle O}{\|}}{C}-C_{\alpha}-C_{\beta}-$$

$\uparrow$
Cleavage

A good example of such a cleavage is the fructose bisphosphate aldolase reaction (see Chapter 18, Figure 18.13a).

Another common type of C—C cleavage is α-cleavage of an α-hydroxyketone:

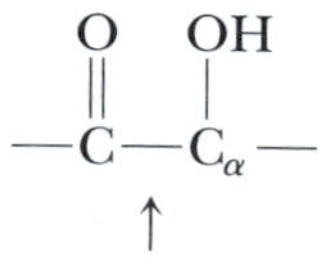

$$-\overset{\overset{\displaystyle O}{\|}}{C}-\overset{\overset{\displaystyle OH}{|}}{C_{\alpha}}-$$

$\uparrow$
Cleavage

(We see this type of cleavage in the transketolase reaction described in Chapter 22.)

Neither of these cleavage strategies is suitable for acetate. It has no β-carbon, and the second method would require hydroxylation—not a favorable reaction for acetate. Instead, living things have evolved the clever chemistry of condensing acetate with oxaloacetate and then carrying out a β-cleavage. The TCA cycle combines this β-cleavage reaction with oxidation to form CO_2, regenerate oxaloacetate, and capture the liberated metabolic energy in NADH and ATP.

19.3 How Is Pyruvate Oxidatively Decarboxylated to Acetyl-CoA?

Pyruvate produced by glycolysis is a significant source of acetyl-CoA for the TCA cycle. Because, in eukaryotic cells, glycolysis occurs in the cytoplasm, whereas the TCA cycle reactions and all subsequent steps of aerobic metabolism take place in the mitochondria, pyruvate must first enter the mitochondria to enter the TCA cycle. The oxidative decarboxylation of pyruvate to acetyl-CoA,

$$\text{Pyruvate} + \text{CoA} + \text{NAD}^+ \longrightarrow \text{acetyl-CoA} + CO_2 + \text{NADH} + \text{H}^+$$

is the connecting link between glycolysis and the TCA cycle. The reaction is catalyzed by pyruvate dehydrogenase, a multienzyme complex.

The **pyruvate dehydrogenase complex (PDC)** is a noncovalent assembly of three different enzymes operating in concert to catalyze successive steps in the conversion of pyruvate to acetyl-CoA. The active sites of all three enzymes are not far removed from one another, and the product of the first enzyme is passed directly to the second enzyme and so on, without diffusion of substrates and products through the solution. The overall reaction (see A Deeper Look on page 614) involves a total of five coenzymes: thiamine pyrophosphate, coenzyme A, lipoic acid, FAD, and NAD^+.

19.4 How Are Two CO_2 Molecules Produced from Acetyl-CoA?

The Citrate Synthase Reaction Initiates the TCA Cycle

The first reaction within the TCA cycle, the one by which carbon atoms are introduced, is the **citrate synthase reaction** (Figure 19.5). Here acetyl-CoA reacts with oxaloacetate in a **Perkin condensation** (a carbon–carbon condensation between a

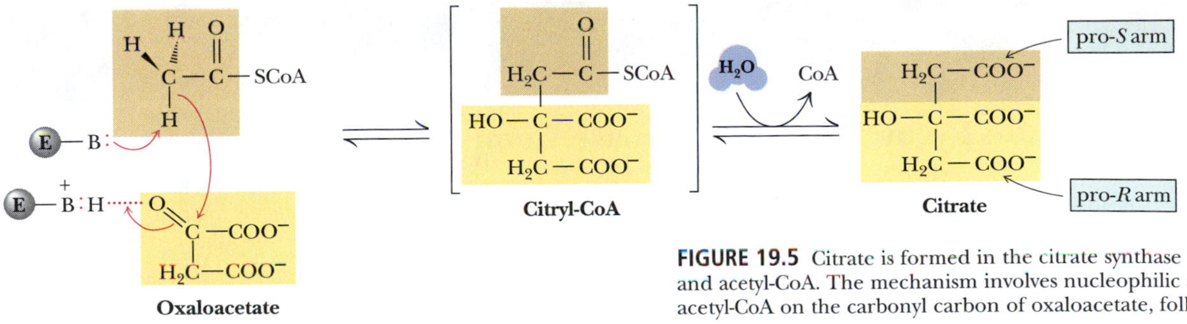

FIGURE 19.5 Citrate is formed in the citrate synthase reaction from oxaloacetate and acetyl-CoA. The mechanism involves nucleophilic attack by the carbanion of acetyl-CoA on the carbonyl carbon of oxaloacetate, followed by thioester hydrolysis.

ketone or aldehyde and an ester). The acyl group is activated in two ways in an acyl-CoA molecule: The carbonyl carbon is activated for attack by nucleophiles, and the C$_\alpha$ carbon is more acidic and can be deprotonated to form a carbanion. The citrate synthase reaction depends upon the latter mode of activation. As shown in Figure 19.5, a general base on the enzyme accepts a proton from the methyl group of acetyl-CoA, producing a stabilized α-carbanion of acetyl-CoA. This strong nucleophile attacks the α-carbonyl of oxaloacetate, yielding citryl-CoA. This part of the reaction has an equilibrium constant near 1, but the overall reaction is driven to completion by the subsequent hydrolysis of the high-energy thioester to citrate and free CoA. The overall $\Delta G^{\circ\prime}$ is -31.4 kJ/mol, and under standard conditions the reaction is essentially irreversible. Although the mitochondrial concentration of oxaloacetate is very low (much less than 1 μM—see example in Section 19.5), the strong, negative $\Delta G^{\circ\prime}$ drives the reaction forward.

Biochemistry Now™ Go to BiochemistryNow and click BiochemistryInteractive to explore the citrate synthase reaction.

Citrate Synthase Is a Dimer Citrate synthase in mammals is a dimer of 49-kD subunits (Table 19.1). On each subunit, oxaloacetate and acetyl-CoA bind to the active site, which lies in a cleft between two domains and is surrounded mainly by α-helical segments (Figure 19.6). Binding of oxaloacetate induces a conformational change that facilitates the binding of acetyl-CoA and closes the active site so that the reactive carbanion of acetyl-CoA is protected from protonation by water.

NADH Is an Allosteric Inhibitor of Citrate Synthase Citrate synthase is the first step in this metabolic pathway, and as stated the reaction has a large negative $\Delta G^{\circ\prime}$. As might be expected, it is a highly regulated enzyme. NADH, a product of the TCA cycle, is an allosteric inhibitor of citrate synthase, as is succinyl-CoA, the product of the fifth step in the cycle (and an acetyl-CoA analog).

Citrate Is Isomerized by Aconitase to Form Isocitrate

Citrate itself poses a problem: It is a poor candidate for further oxidation because it contains a tertiary alcohol, which could be oxidized only by breaking a carbon–carbon bond. An obvious solution to this problem is to isomerize the tertiary alcohol to a secondary alcohol, which the cycle proceeds to do in the next step.

Citrate is isomerized to isocitrate by **aconitase** in a two-step process involving aconitate as an intermediate (Figure 19.7). In this reaction, the elements of water are first abstracted from citrate to yield aconitate, which is then rehydrated with H— and HO— adding back in opposite positions to produce isocitrate. The net effect is the conversion of a tertiary alcohol (citrate) to a secondary alcohol (isocitrate). Oxidation of the secondary alcohol of isocitrate involves breakage of a C—H bond, a simpler matter than the C—C cleavage required for the direct oxidation of citrate.

Inspection of the citrate structure shows a total of four chemically equivalent hydrogens, but only one of these—the pro-R H atom of the pro-R arm of citrate—is abstracted by aconitase, which is quite stereospecific. Formation of the double

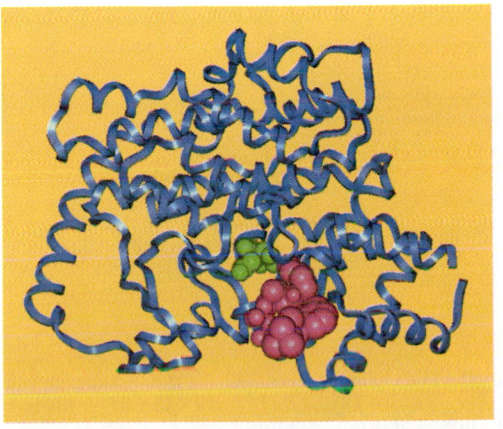

FIGURE 19.6 Citrate synthase. In the monomer shown here, citrate is shown in green, and CoA is pink.

Reaction Mechanism of the Pyruvate Dehydrogenase Complex

The pyruvate dehydrogenase reaction is a tour de force of mechanistic chemistry, involving as it does a total of three enzymes (part a of the accompanying figure) and five different coenzymes—thiamine pyrophosphate, lipoic acid, coenzyme A, FAD, and NAD$^+$ (part b of the figure).

The first step of this reaction, decarboxylation of pyruvate and transfer of the acetyl group to lipoic acid, depends on accumulation of negative charge on the transferred two-carbon fragment. This is facilitated by the quaternary nitrogen on the thiazolium group of thiamine pyrophosphate. As shown in part (c) of the figure, this cationic imine nitrogen plays two distinct and important roles in TPP-catalyzed reactions:

1. It provides electrostatic stabilization of the carbanion formed upon removal of the C-2 proton. (The sp^2 hybridization and the availability of vacant d orbitals on the adjacent sulfur probably also facilitate proton removal at C-2.)

2. TPP attack on pyruvate leads to decarboxylation. The TPP cationic imine nitrogen can act as an effective electron sink to stabilize the negative charge that must develop on the carbon that has been attacked. This stabilization takes place by resonance interaction through the double bond to the nitrogen atom.

(a)

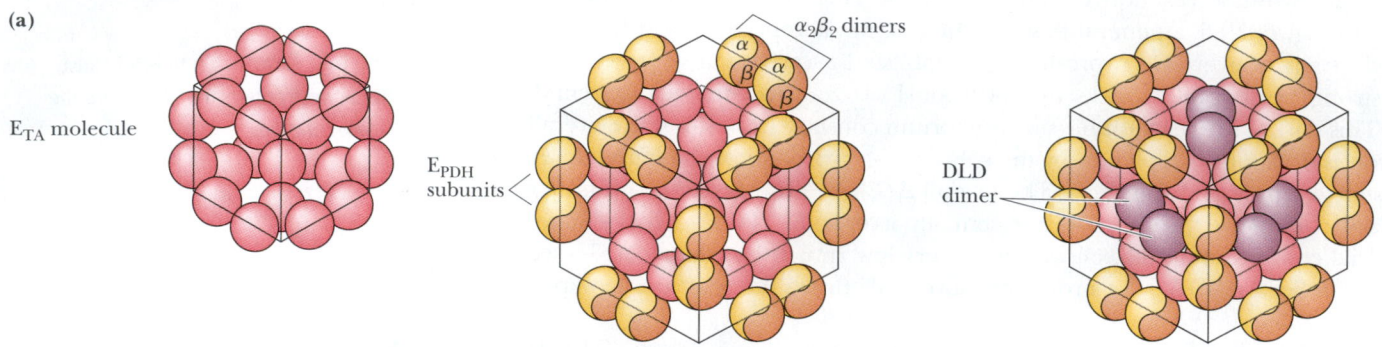

(a) The structure of the pyruvate dehydrogenase complex. This complex consists of three enzymes: pyruvate dehydrogenase (PDH), dihydrolipoyl transacetylase (TA), and dihydrolipoyl dehydrogenase (DLD). (i) Twenty-four dihydrolipoyl transacetylase subunits form a cubic core structure. (ii) Twenty-four $\alpha\beta$-dimers of pyruvate dehydrogenase are added to the cube (two per edge). (iii) Addition of 12 dihydrolipoyl dehydrogenase subunits (two per face) completes the complex.

(b)

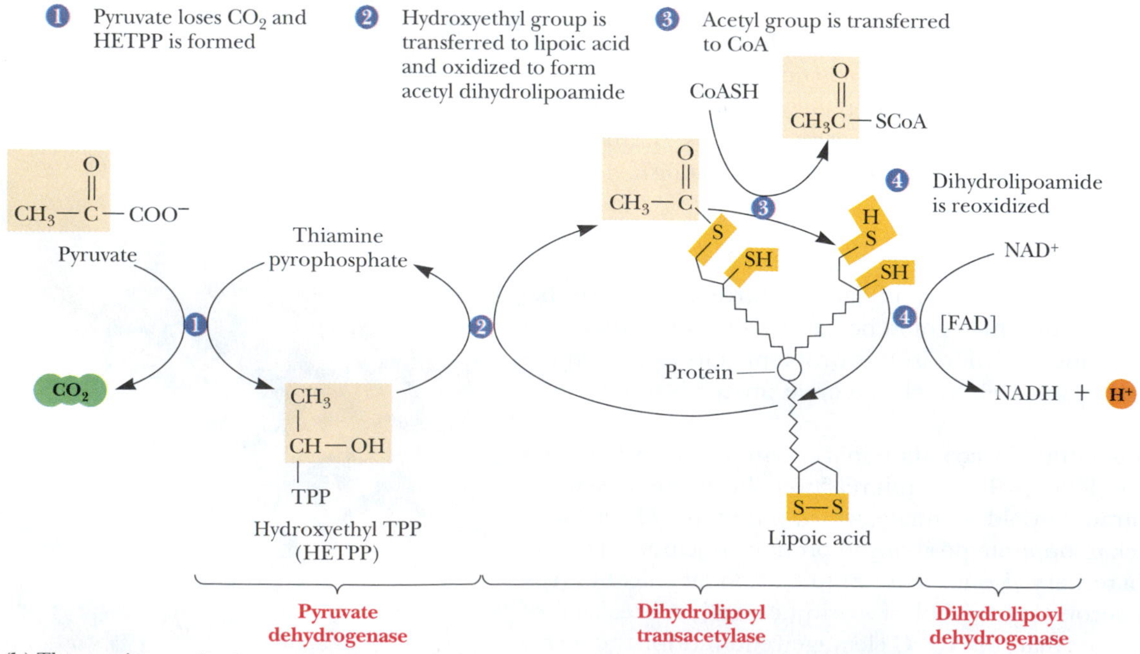

① Pyruvate loses CO$_2$ and HETPP is formed

② Hydroxyethyl group is transferred to lipoic acid and oxidized to form acetyl dihydrolipoamide

③ Acetyl group is transferred to CoA

④ Dihydrolipoamide is reoxidized

(b) The reaction mechanism of the pyruvate dehydrogenase complex. Decarboxylation of pyruvate occurs with formation of hydroxyethyl-TPP (Step 1). Transfer of the two-carbon unit to lipoic acid in Step 2 is followed by formation of acetyl-CoA in Step 3. Lipoic acid is reoxidized in Step 4 of the reaction.

This resonance-stabilized intermediate can be protonated to give **hydroxyethyl-TPP.** This well-characterized intermediate was once thought to be so unstable that it could not be synthesized or isolated. However, its synthesis and isolation are actually routine. (In fact, a substantial amount of the thiamine pyrophosphate in living things exists as the hydroxyethyl form.)

The reaction of hydroxyethyl-TPP with the oxidized form of lipoic acid yields the energy-rich thiol ester of reduced lipoic acid and results in oxidation of the hydroxyl-carbon of the two-carbon substrate unit (c). This is followed by nucleophilic attack by coenzyme A on the carbonyl-carbon (a characteristic feature of CoA chemistry). The result is transfer of the acetyl group from lipoic acid to CoA. The subsequent oxidation of lipoic acid is catalyzed by the FAD-dependent dihydrolipoyl dehydrogenase, and NAD^+ is reduced.

(c)

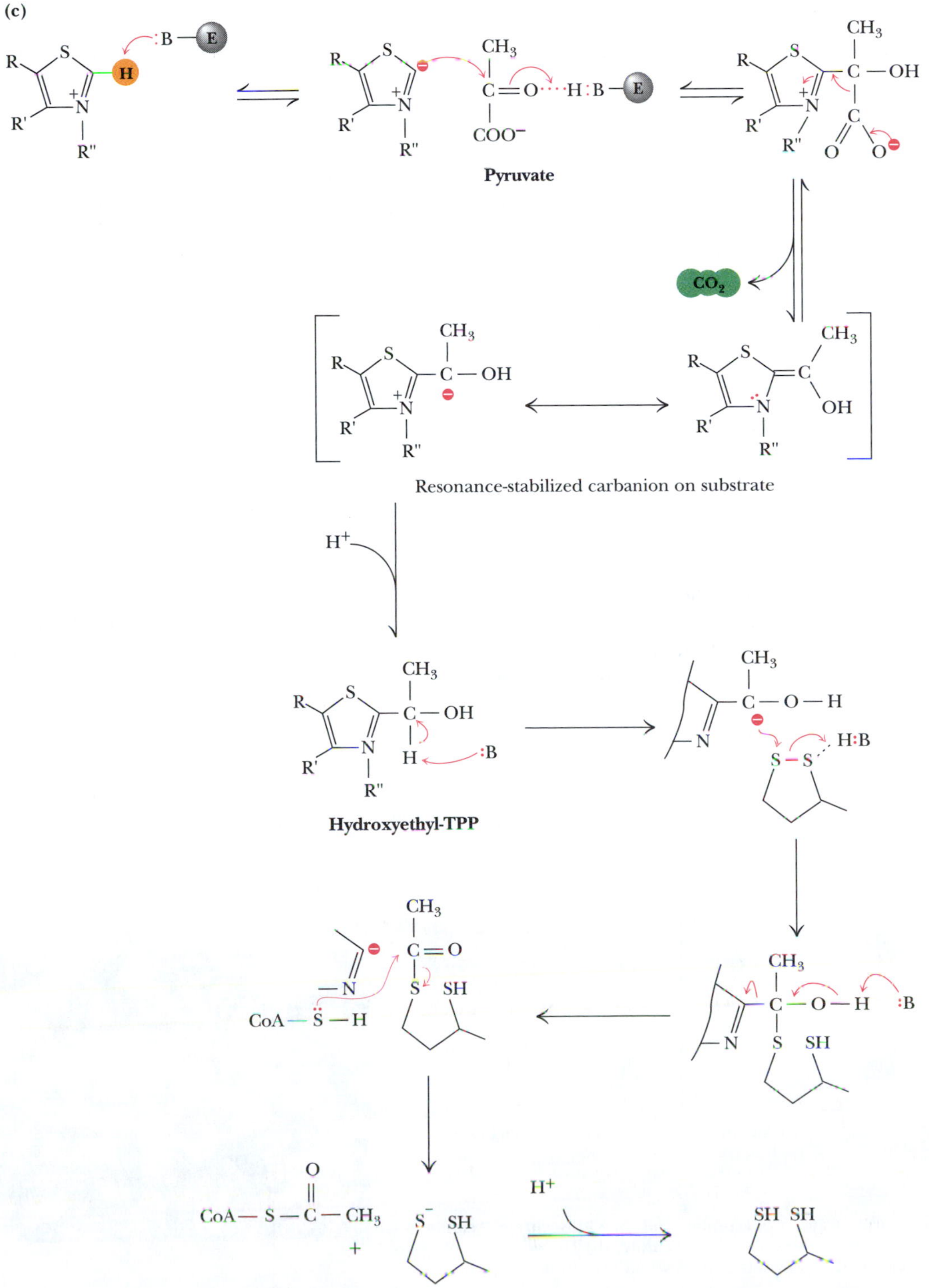

Pyruvate

CO_2

Resonance-stabilized carbanion on substrate

Hydroxyethyl-TPP

(c) The mechanistic details of the first three steps of the pyruvate dehydrogenase complex reaction.

Table 19.1

The Enzymes and Reactions of the TCA Cycle

Reaction	Enzyme
1. Acetyl-CoA + oxaloacetate + H_2O $\rightleftharpoons$ CoASH + citrate	Citrate synthase
2. Citrate $\rightleftharpoons$ isocitrate	Aconitase
3. Isocitrate + NAD^+ $\rightleftharpoons$ α-ketoglutarate + NADH + CO_2 + H^+	Isocitrate dehydrogenase
4. α-Ketoglutarate + CoASH + NAD^+ $\rightleftharpoons$ succinyl-CoA + NADH + CO_2 + H^+	α-Ketoglutarate dehydrogenase complex
5. Succinyl-CoA + GDP + P_i $\rightleftharpoons$ succinate + GTP + CoASH	Succinyl-CoA synthetase
6. Succinate + [FAD] $\rightleftharpoons$ fumarate + [$FADH_2$]	Succinate dehydrogenase
7. Fumarate + H_2O $\rightleftharpoons$ L-malate	Fumarase
8. L-Malate + NAD^+ $\rightleftharpoons$ oxaloacetate + NADH + H^+	Malate dehydrogenase

Net for reactions 1 – 8:

Acetyl-CoA + 3 NAD^+ + [FAD] + GDP + P_i + 2 H_2O $\rightleftharpoons$ CoASH + 3 NADH + [$FADH_2$] + GTP + 2 CO_2 + 3 H^+

Simple combustion of acetate: Acetate + 2 O_2 + H^+ $\rightleftharpoons$ 2 CO_2 + 2 H_2O

bond of aconitate following proton abstraction requires departure of hydroxide ion from the C-3 position. Hydroxide is a relatively poor leaving group, and its departure is facilitated in the aconitase reaction by coordination with an iron atom in an iron–sulfur cluster.

Aconitase Utilizes an Iron–Sulfur Cluster   Aconitase contains an **iron–sulfur cluster** consisting of three iron atoms and four sulfur atoms in a near-cubic arrangement (Figure 19.8). This cluster is bound to the enzyme via three cysteine groups from the protein. One corner of the cube is vacant and binds Fe^{2+}, which activates aconitase. The iron atom in this position can coordinate the C-3 carboxyl and hydroxyl groups of citrate. This iron atom thus acts as a Lewis acid, accepting an unshared pair of electrons from the hydroxyl, making it a better leaving group. The equilibrium for the aconitase reaction favors citrate, and an equilibrium mixture typically contains about 90% citrate, 4% *cis*-aconitate, and 6% isocitrate. The $\Delta G°'$ is +6.7 kJ/mol.

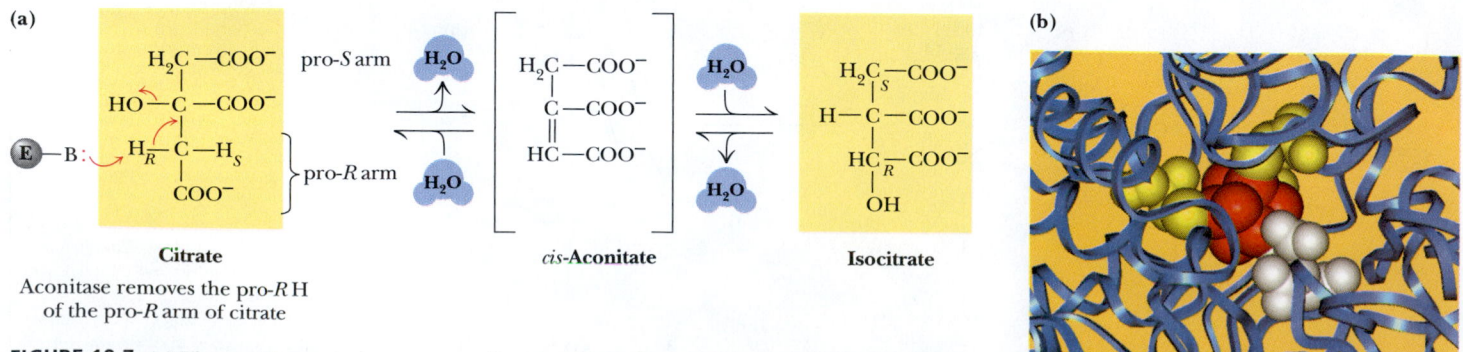

FIGURE 19.7 (a) The aconitase reaction converts citrate to *cis*-aconitate and then to isocitrate. Aconitase is stereospecific and removes the pro-R hydrogen from the pro-R arm of citrate. **(b)** The active site of aconitase. The iron–sulfur cluster (red) is coordinated by cysteines (yellow) and isocitrate (white).

Subunit M_r	Oligomeric Composition	$\Delta G°'$ (kJ/mol)	K_{eq} at 25°C	ΔG (kJ/mol)
49,000*	Dimer	−31.4	3.2×10^5	−53.9
44,500	Dimer	+6.7	0.067	+0.8
	$\alpha_2\beta\gamma$	−8.4	29.7	−17.5
E_1 96,000	Dimer			
E_2 70,000	24-mer	−30	1.8×10^5	−43.9
E_3 56,000	Dimer			
α 34,500	$\alpha\beta$	−3.3	3.8	≈0
β 42,500				
α 70,000	$\alpha\beta$	+0.4	0.85	≠0
β 27,000				
48,500	Tetramer	−3.8	4.6	≈0
35,000	Dimer	+29.7	6.2×10^{-6}	≈0
		−40		≈(−115)
		−849		

*CS in mammals, A in pig heart, αKDC in *E. coli*, S-CoA S in pig heart, SD in bovine heart, F in pig heart, MD in pig heart. ΔG values from Newsholme, E. A., and Leech, A. R., 1983. *Biochemistry for the Medical Sciences.* New York: Wiley.

Fluoroacetate Blocks the TCA Cycle   Fluoroacetate is an extremely poisonous agent that blocks the TCA cycle in vivo, although it has no apparent effect on any of the isolated enzymes. Its LD_{50}, the lethal dose for 50% of animals consuming it, is 0.2 mg per kilogram of body weight; it has been used as a rodent poison. The action of fluoroacetate has been traced to aconitase, which is inhibited in vivo by fluorocitrate, which is formed from fluoroacetate in two steps (Figure 19.9). Fluoroacetate readily crosses both the cellular and mitochondrial membranes, and in mitochondria it is converted to fluoroacetyl-CoA by acetyl-CoA synthetase. Fluoroacetyl-CoA is a substrate for citrate synthase, which condenses it with oxaloacetate to form fluorocitrate. Fluoroacetate may

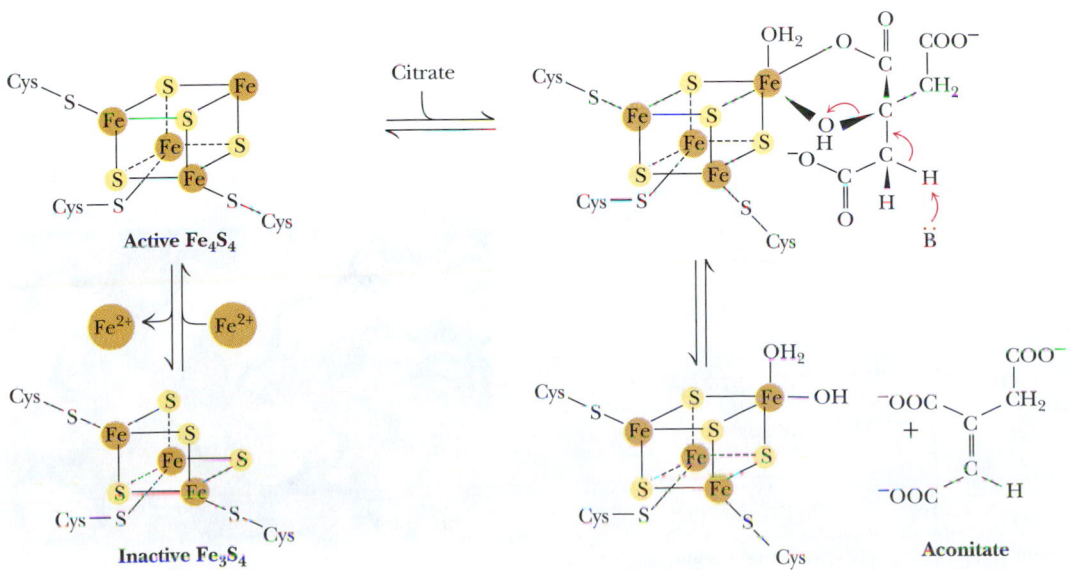

Biochemistry Now™ ACTIVE FIGURE 19.8 The iron–sulfur cluster of aconitase. Binding of Fe^{2+} to the vacant position of the cluster activates aconitase. The added iron atom coordinates the C-3 carboxyl and hydroxyl groups of citrate and acts as a Lewis acid, accepting an electron pair from the hydroxyl group and making it a better leaving group. **Test yourself on the concepts in this figure at http://chemistry.brookscole.com/ggb3**

FIGURE 19.9 The conversion of fluoroacetate to fluorocitrate.

$$\text{Fluoroacetate} \xrightarrow[\text{synthetase}]{\text{Acetyl-CoA}} \text{Fluoroacetyl-CoA} \xrightarrow[\text{synthase}]{\text{Citrate}} (2R, 3S)\text{-Fluorocitrate}$$

thus be viewed as a **trojan horse inhibitor.** Analogous to the giant Trojan Horse of legend—which the soldiers of Troy took into their city, not knowing that Greek soldiers were hidden inside it and waiting to attack—fluoroacetate enters the TCA cycle innocently enough, in the citrate synthase reaction. Citrate synthase converts fluoroacetate to inhibitory fluorocitrate for its TCA cycle partner, aconitase, blocking the cycle.

Isocitrate Dehydrogenase Catalyzes the First Oxidative Decarboxylation in the Cycle

In the next step of the TCA cycle, isocitrate is oxidatively decarboxylated to yield α-ketoglutarate, with concomitant reduction of NAD^+ to NADH in the isocitrate dehydrogenase reaction (Figure 19.10). The reaction has a net $\Delta G°'$ of -8.4 kJ/mol, and it is sufficiently exergonic to pull the aconitase reaction forward. This two-step reaction involves (1) oxidation of the C-2 alcohol of isocitrate to form oxalosuccinate, followed by (2) a β-decarboxylation reaction that expels the central carboxyl group as CO_2, leaving the product α-ketoglutarate. Oxalosuccinate, the β-keto acid produced by the initial dehydrogenation reaction, is unstable and thus is readily decarboxylated.

Isocitrate Dehydrogenase Links the TCA Cycle and Electron Transport Isocitrate dehydrogenase provides the first connection between the TCA cycle and the electron-transport pathway and oxidative phosphorylation, via its production of NADH. As a connecting point between two metabolic pathways, isocitrate dehydrogenase is a regulated reaction. NADH and ATP are allosteric inhibitors, whereas ADP acts as an allosteric activator, lowering the K_m for isocitrate by a factor of 10. The enzyme is virtually inactive in the absence of ADP. Also, the product, α-ketoglutarate, is a crucial α-keto acid for aminotransferase reactions (see Chapters 13 and 25), connecting the TCA cycle (that is, carbon metabolism) with nitrogen metabolism.

Biochemistry Now™ ANIMATED FIGURE 19.10
(a) The isocitrate dehydrogenase reaction. **(b)** The active site of isocitrate dehydrogenase. Isocitrate is shown in green, $NADP^+$ is shown in gold, with Ca^{2+} in red. **See this figure animated at http://chemistry. brookscole.com/ggb3**

(a)

Isocitrate

NAD^+ → Isocitrate dehydrogenase → NADH + **H⁺**

Oxalosuccinate

→ CO_2 → α-Ketoglutarate

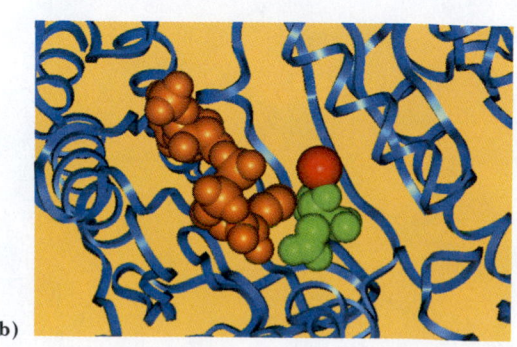

(b)

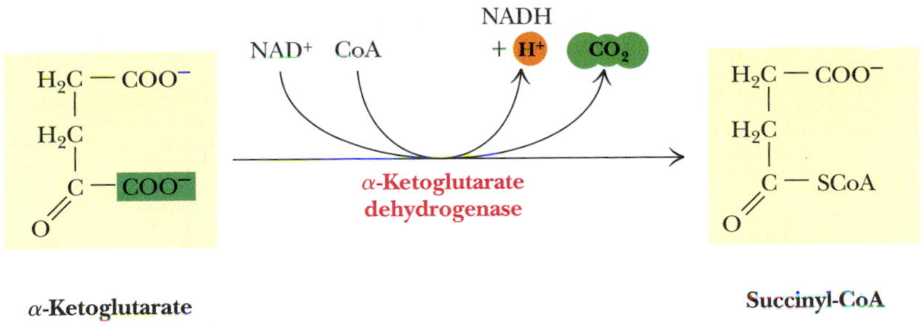

α-Ketoglutarate   Succinyl-CoA

FIGURE 19.11 The α-ketoglutarate dehydrogenase reaction.

α-Ketoglutarate Dehydrogenase Catalyzes the Second Oxidative Decarboxylation of the TCA Cycle

A second oxidative decarboxylation occurs in the α-ketoglutarate dehydrogenase reaction (Figure 19.11). Like the pyruvate dehydrogenase complex, α-ketoglutarate dehydrogenase is a multienzyme complex—consisting of *α-ketoglutarate dehydrogenase, dihydrolipoyl transsuccinylase,* and *dihydrolipoyl dehydrogenase*—that employs five different coenzymes (Table 19.2). The dihydrolipoyl dehydrogenase in this reaction is identical to that in the pyruvate dehydrogenase reaction. The mechanism is analogous to that of pyruvate dehydrogenase, and the free energy changes for these reactions are -29 to -33.5 kJ/mol. As with the pyruvate dehydrogenase reaction, this reaction produces NADH and a thioester product—in this case, succinyl-CoA. Succinyl-CoA and NADH products are energy-rich species that are important sources of metabolic energy in subsequent cellular processes.

19.5 How Is Oxaloacetate Regenerated to Complete the TCA Cycle?

Succinyl-CoA Synthetase Catalyzes Substrate-Level Phosphorylation

The NADH produced in the foregoing steps can be routed through the electron-transport pathway to make high-energy phosphates via oxidative phosphorylation. However, succinyl-CoA is itself a high-energy intermediate and is utilized in the next step of the TCA cycle to drive the phosphorylation of GDP to GTP (in mammals) or ADP to ATP (in plants and bacteria). The reaction (Figure 19.12) is catalyzed by **succinyl-CoA synthetase,** sometimes called **succinate thiokinase.** The free energies of hydrolysis of succinyl-CoA and GTP or ATP are similar, and the net reaction has a $\Delta G°'$ of -3.3 kJ/mol. Succinyl-CoA synthetase provides another example of a **substrate-level phosphorylation** (see Chapter 18), in which a substrate, rather than an electron-transport chain or proton gradient, provides the energy for phosphorylation. It is the only such reaction in the TCA cycle. The GTP produced by mammals in this reaction can exchange its terminal phosphoryl group with ADP via the **nucleoside diphosphate kinase reaction:**

$$\text{GTP} + \text{ADP} \underset{\text{kinase}}{\overset{\text{Nucleoside diphosphate}}{\rightleftharpoons}} \text{ATP} + \text{GDP}$$

Table 19.2

Composition of the α-Ketoglutarate Dehydrogenase Complex from *Escherichia coli*

Enzyme	Coenzyme	Enzyme M_r	Number of Subunits	Subunit M_r	Number of Subunits per Complex
α-Ketoglutarate dehydrogenase	Thiamine pyrophosphate	192,000	2	96,000	24
Dihydrolipoyl transsuccinylase	Lipoic acid, CoASH	1,700,000	24	70,000	24
Dihydrolipoyl dehydrogenase	FAD, NAD$^+$	112,000	2	56,000	12

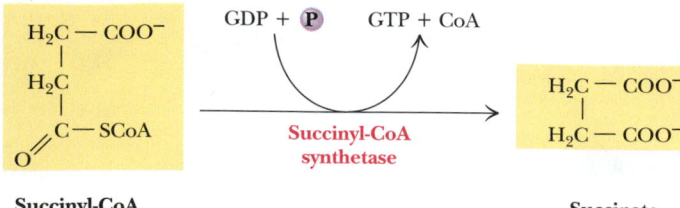

FIGURE 19.12 The succinyl-CoA synthetase reaction.

Succinyl-CoA

Succinate

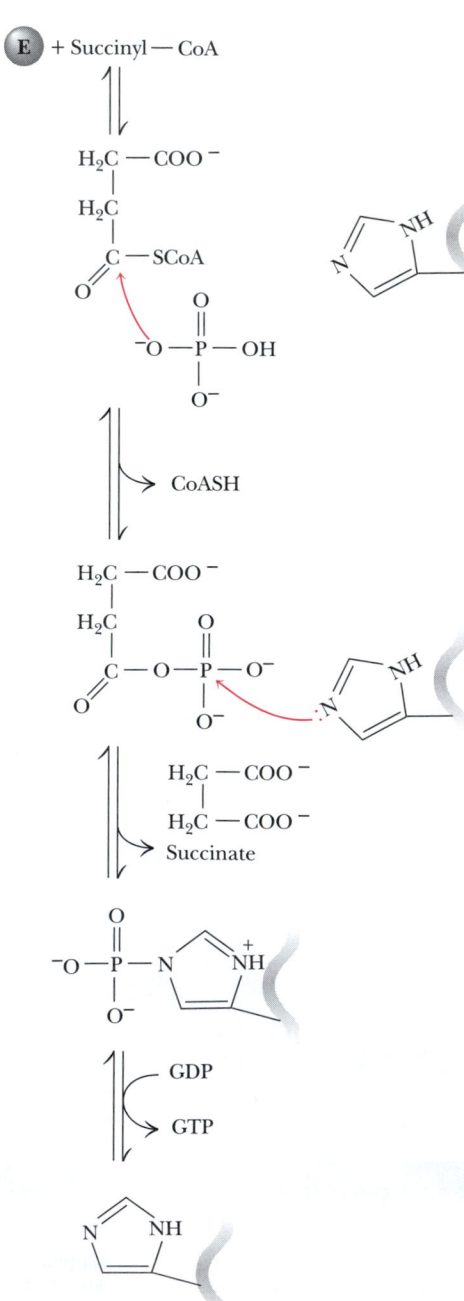

The Mechanism of Succinyl-CoA Synthetase Involves a Phosphohistidine   The
mechanism of succinyl-CoA synthetase is postulated to involve displacement of
CoA by phosphate, forming succinyl phosphate at the active site, followed by
transfer of the phosphoryl group to an active-site histidine (making a phosphohistidine intermediate) and release of succinate. The phosphoryl moiety is
then transferred to GDP to form GTP (Figure 19.13). This sequence of steps
"preserves" the energy of the thioester bond of succinyl-CoA in a series of high-
energy intermediates that lead to a molecule of ATP:

$$\text{Thioester} \longrightarrow [\text{succinyl-P}] \longrightarrow [\text{phosphohistidine}] \longrightarrow \text{GTP} \longrightarrow \text{ATP}$$

The First Five Steps of the TCA Cycle Produce NADH, CO_2, GTP (ATP), and Succinate
This is a good point to pause in our trip through the TCA cycle and see what
has happened. A two-carbon acetyl group has been introduced as acetyl-CoA
and linked to oxaloacetate, and two CO_2 molecules have been liberated. The
cycle has produced two molecules of NADH and one of GTP or ATP and has
left a molecule of succinate.

The TCA cycle can now be completed by converting succinate to oxaloacetate. This latter process represents a net oxidation. The TCA cycle breaks it
down into (consecutively) an oxidation step, a hydration reaction, and a second oxidation step. The oxidation steps are accompanied by the reduction of
an [FAD] and an NAD^+. The reduced coenzymes, [$FADH_2$] and NADH, subsequently provide reducing power in the electron-transport chain. (It will be seen
in Chapter 23 that virtually the same chemical strategy is used in β-oxidation of
fatty acids.)

Succinate Dehydrogenase Is FAD-Dependent
The oxidation of succinate to fumarate (Figure 19.14) is carried out by **succinate dehydrogenase**, a membrane-bound enzyme that is actually part of the
electron-transport chain. As will be seen in Chapter 20, succinate dehydrogenase is part of the succinate–coenzyme Q reductase of the electron-transport
chain. In contrast with all of the other enzymes of the TCA cycle, which are
soluble proteins found in the mitochondrial matrix, succinate dehydrogenase
is an integral membrane protein tightly associated with the inner mitochondrial membrane. Succinate oxidation involves removal of H atoms across
a C—C bond, rather than a C—O or C—N bond, and produces the *trans*-
unsaturated fumarate. This reaction (the oxidation of an alkane to an alkene)
is not sufficiently exergonic to reduce NAD^+, but it does yield enough energy
to reduce [FAD]. (By contrast, oxidations of alcohols to ketones or aldehydes
are more energetically favorable and provide sufficient energy to reduce
NAD^+.) This important point is illustrated and clarified in an example in
Chapter 20.

Succinate dehydrogenase is a dimeric protein, with subunits of molecular
masses 70 and 27 kD (see Table 19.1). FAD is covalently bound to the larger
subunit; the bond involves a methylene group of C-8a of FAD and N-3 of a
histidine on the protein (Figure 19.15). Succinate dehydrogenase also contains three different iron–sulfur clusters (Figure 19.16). Viewed from either
end of the succinate molecule, the reaction involves dehydrogenation α,β to

a carbonyl (actually, a carboxyl) group. The dehydrogenation is stereospe-cific (Figure 19.14), with the pro-*S* hydrogen removed from one carbon atom and the pro-*R* hydrogen removed from the other. The electrons cap-tured by [FAD] in this reaction are passed directly into the iron–sulfur clus-ters of the enzyme and on to coenzyme Q (UQ). The covalently bound FAD is first reduced to [FADH$_2$] and then reoxidized to form [FAD] and the re-duced form of coenzyme Q, UQH$_2$. Electrons captured by UQH$_2$ then flow through the rest of the electron-transport chain in a series of events that will be discussed in detail in Chapter 20.

Note that flavin coenzymes can carry out either one-electron or two-electron transfers. The succinate dehydrogenase reaction represents a net two-electron reduction of FAD.

Fumarase Catalyzes the *Trans*-Hydration of Fumarate to Form L-Malate

Fumarate is hydrated in a stereospecific reaction by fumarase to give L-malate (Figure 19.17). The reaction involves *trans*-addition of the elements of water across the double bond. Recall that aconitase carries out a similar reaction and that *trans*-addition of —H and —OH occurs across the double bond of *cis*-aconitate. Although the exact mechanism is uncertain, it may involve proton-ation of the double bond to form an intermediate carbonium ion (Figure 19.18) or possibly attack by water or OH$^-$ anion to produce a carbanion, followed by protonation.

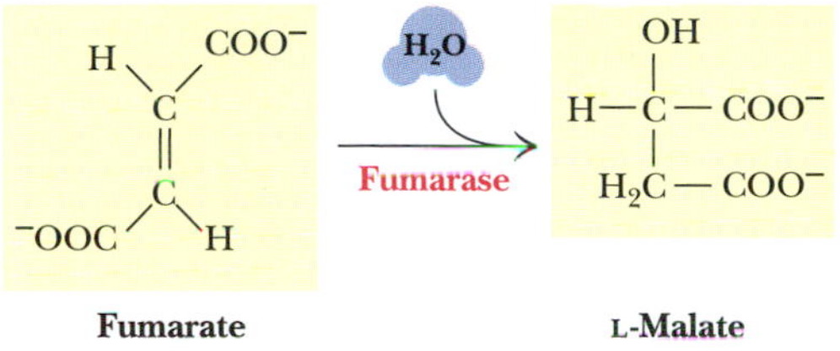

Malate Dehydrogenase Completes the Cycle by Oxidizing Malate to Oxaloacetate

In the last step of the TCA cycle, L-malate is oxidized to oxaloacetate by malate dehydrogenase (Figure 19.19). This reaction is very endergonic, with a $\Delta G^{\circ\prime}$ of +30 kJ/mol. Consequently, the concentration of oxaloacetate in the mitochondrial matrix is usually quite low (see the following example). The reaction, however, is pulled forward by the favorable citrate synthase reaction. Oxidation of malate is coupled to reduction of yet another mole-cule of NAD$^+$, the third one of the cycle. Counting the [FAD] reduced by suc-cinate dehydrogenase, this makes the fourth coenzyme reduced through oxi-dation of a single acetate unit.

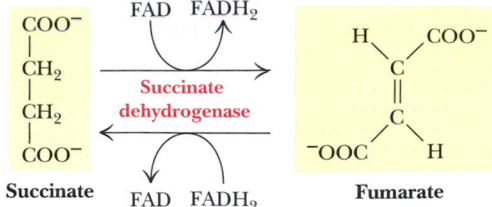

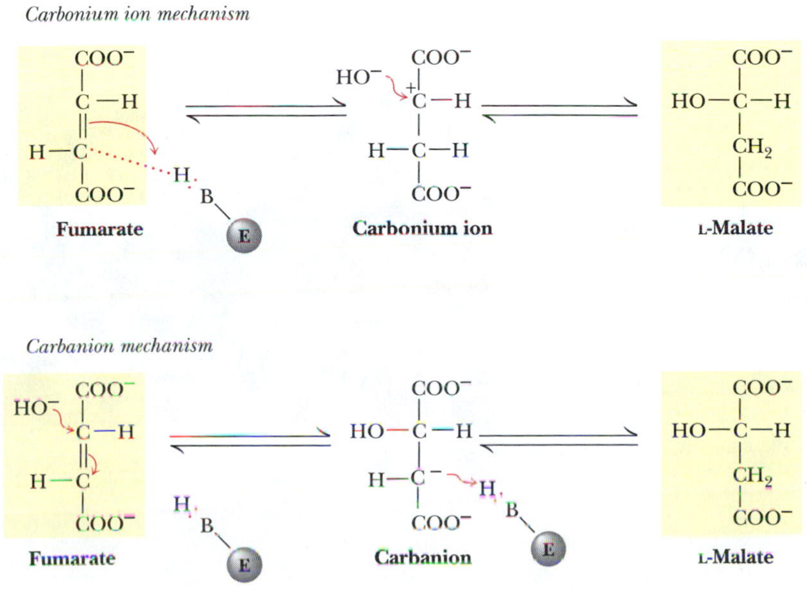

FIGURE 19.14 The succinate dehydrogenase reac-tion. Oxidation of succinate occurs with reduction of [FAD]. Reoxidation of [FADH$_2$] transfers electrons to coenzyme Q.

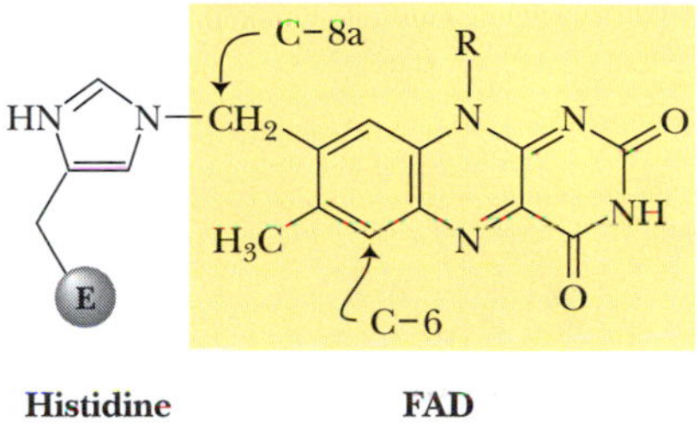

FIGURE 19.15 The covalent bond between FAD and succinate dehydrogenase involves the C-8a methy-lene group of FAD and the N-3 of a histidine residue on the enzyme.

FIGURE 19.16 The Fe$_2$S$_2$ cluster of succinate dehydrogenase.

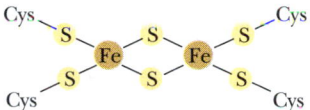

FIGURE 19.17 The fumarase reaction.

Carbonium ion mechanism

Fumarate   **Carbonium ion**   L-**Malate**

Carbanion mechanism

Fumarate   **Carbanion**   L-**Malate**

FIGURE 19.18 Two possible mechanisms for the fumarase reaction.

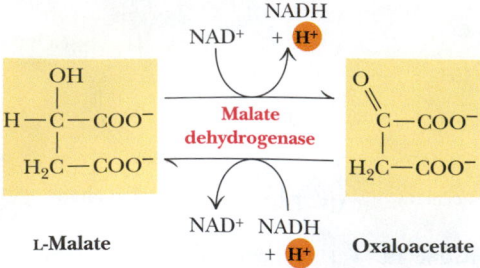

FIGURE 19.19 The malate dehydrogenase reaction.

Biochemistry ⒺNow™ Go to BiochemistryNow and click BiochemistryInteractive to understand the structure and function of malate dehydrogenase.

EXAMPLE

A typical intramitochondrial concentration of malate is 0.22 mM. If the [NAD^+]/[NADH] ratio in mitochondria is 20 and if the malate dehydrogenase reaction is at equilibrium, calculate the intramitochondrial concentration of oxaloacetate at 25°C.

Answer

For the malate dehydrogenase reaction,

$$Malate + NAD^+ \rightleftharpoons oxaloacetate + NADH + H^+$$

with the value of $\Delta G^{\circ\prime}$ being +30 kJ/mol. Then

$$\Delta G^{\circ\prime} = -RT \ln K_{eq}$$

$$= -(8.314 \text{ J/mol} \cdot \text{K})(298)\ln\left(\frac{[1]x}{[20][2.2 \times 10^{-4}]}\right)$$

$$\frac{-30,000 \text{ J/mol}}{2478 \text{ J/mol}} = \ln(x/4.4 \times 10^{-3})$$

$$-12.1 = \ln(x/4.4 \times 10^{-3})$$

$$x = (5.6 \times 10^{-6})(4.4 \times 10^{-3})$$

$$x = [oxaloacetate] = 0.024 \ \mu M$$

Malate dehydrogenase is structurally and functionally similar to other dehydrogenases, notably lactate dehydrogenase (Figure 19.20). Both consist of alternating β-sheet and α-helical segments. Binding of NAD^+ causes a conformational change in the 20-residue segment that connects the D and E strands of the β-sheet. The change is triggered by an interaction between the adenosine phosphate moiety of NAD^+ and an arginine residue in this loop region. Such a conformational change is consistent with an ordered single-displacement mechanism for NAD^+-dependent dehydrogenases (see Chapter 13).

19.6 | What Are the Energetic Consequences of the TCA Cycle?

The net reaction accomplished by the TCA cycle, as follows, shows two molecules of CO_2, one ATP, and four reduced coenzymes produced per acetate group oxidized. The cycle is exergonic, with a net $\Delta G^{\circ\prime}$ for one pass around the

(a)

(b)

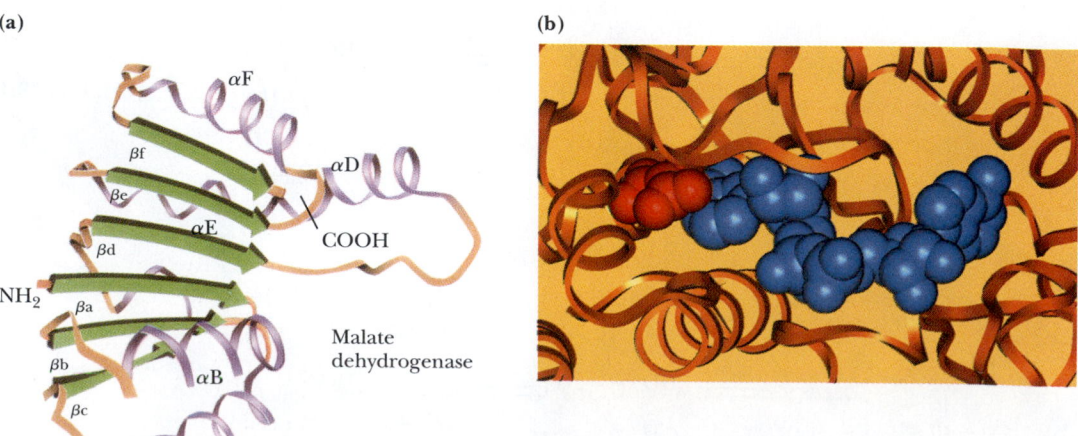

FIGURE 19.20 (a) The structure of malate dehydrogenase. (b) The active site of malate dehydrogenase. Malate is shown in red; NAD^+ is blue.

cycle of approximately -40 kJ/mol. Table 19.1 compares the $\Delta G^{\circ\prime}$ values for the individual reactions with the overall $\Delta G^{\circ\prime}$ for the net reaction.

$$\text{Acetyl-CoA} + 3\text{ NAD}^+ + [\text{FAD}] + \text{ADP} + \text{P}_i + 2\text{ H}_2\text{O} \longrightarrow$$
$$2\text{ CO}_2 + 3\text{ NADH} + 3\text{ H}^+ + [\text{FADH}_2] + \text{ATP} + \text{CoASH}$$
$$\Delta G^{\circ\prime} = -40\text{ kJ/mol}$$

Glucose metabolized via glycolysis produces two molecules of pyruvate and thus two molecules of acetyl-CoA, which can enter the TCA cycle. Combining glycolysis and the TCA cycle gives the net reaction shown:

$$\text{Glucose} + 2\text{ H}_2\text{O} + 10\text{ NAD}^+ + 2\text{ [FAD]} + 4\text{ ADP} + 4\text{ P}_i \longrightarrow$$
$$6\text{ CO}_2 + 10\text{ NADH} + 10\text{ H}^+ + 2\text{ [FADH}_2] + 4\text{ ATP}$$

All six carbons of glucose are liberated as CO_2, and a total of four molecules of ATP are formed thus far in substrate-level phosphorylations. The 12 reduced coenzymes produced up to this point can eventually produce a maximum of 34 molecules of ATP in the electron-transport and oxidative phosphorylation pathways. A stoichiometric relationship for these subsequent processes is

$$\text{NADH} + \text{H}^+ + \tfrac{1}{2}\text{O}_2 + 3\text{ ADP} + 3\text{ P}_i \longrightarrow \text{NAD}^+ + 3\text{ ATP} + 4\text{ H}_2\text{O}$$
$$[\text{FADH}_2] + \tfrac{1}{2}\text{O}_2 + 2\text{ ADP} + 2\text{ P}_i \longrightarrow [\text{FAD}] + 2\text{ ATP} + 2\text{ H}_2\text{O}$$

Thus, a total of 3 ATP per NADH and 2 ATP per $FADH_2$ may be produced through the processes of electron-transport and oxidative phosphorylation.

The Carbon Atoms of Acetyl-CoA Have Different Fates in the TCA Cycle

It is instructive to consider how the carbon atoms of a given acetate group are routed through several turns of the TCA cycle. As shown in Figure 19.21, neither of the carbon atoms of a labeled acetate unit is lost as CO_2 in the first turn of the cycle. The CO_2 evolved in any turn of the cycle derives from the carboxyl groups of the oxaloacetate acceptor (from the previous turn), not from incoming acetyl-CoA. On the other hand, succinate labeled on one end from the original labeled acetate forms two different labeled oxaloacetates. The carbonyl carbon of acetyl-CoA is evenly distributed between the two carboxyl carbons of oxaloacetate, and the labeled methyl carbon of incoming acetyl-CoA ends up evenly distributed between the methylene and carbonyl carbons of oxaloacetate.

When these labeled oxaloacetates enter a second turn of the cycle, both of the carboxyl carbons are lost as CO_2, but the methylene and carbonyl carbons survive through the second turn. Thus, the methyl carbon of a labeled acetyl-CoA survives two full turns of the cycle. In the third turn of the cycle, one-half of the carbon from the original methyl group of acetyl-CoA has become one of the carboxyl carbons of oxaloacetate and is thus lost as CO_2. In the fourth turn of the cycle, further "scrambling" results in loss of half of the remaining labeled carbon (one-fourth of the original methyl carbon label of acetyl-CoA), and so on.

It can be seen that the carbonyl and methyl carbons of labeled acetyl-CoA have very different fates in the TCA cycle. The carbonyl carbon survives the first turn intact but is completely lost in the second turn. The methyl carbon survives two full turns, then undergoes a 50% loss through each succeeding turn of the cycle.

It is worth noting that the carbon–carbon bond cleaved in the TCA pathway entered as an acetate unit in the previous turn of the cycle. Thus, the oxidative decarboxylations that cleave this bond are just a cleverly disguised acetate C—C cleavage and oxidation.

A Deeper Look

Steric Preferences in NAD⁺-Dependent Dehydrogenases

As noted in Chapter 17, the enzymes that require nicotinamide coenzymes are stereospecific and transfer hydride to either the pro-R or the pro-S positions selectively. The table (facing page) lists the preferences of several dehydrogenases.

What accounts for this stereospecificity? It arises from the fact that the enzymes (and especially the active sites of enzymes) are inherently asymmetric structures. The nicotinamide coenzyme (and the substrate) fit the active site in only one way. Malate dehydrogenase, the citric acid cycle enzyme, transfers hydride to the H_R position of NADH, but glyceraldehyde-3-P dehydrogenase in the glycolytic pathway transfers hydride to the H_S position, as shown in the accompanying table. Dehydrogenases are stereospecific with respect to the substrates as well. Note that alcohol dehydrogenase removes hydrogen from the pro-R position of ethanol and transfers it to the pro-R position of NADH.

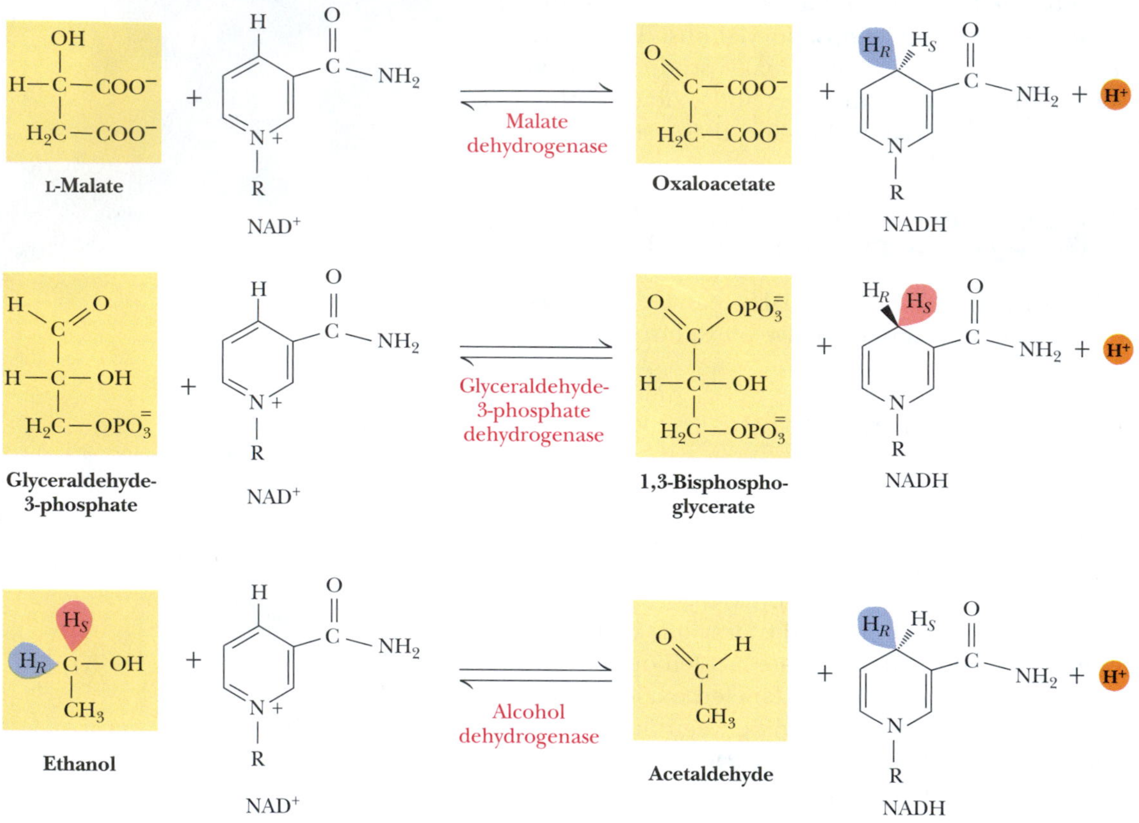

NAD(P)⁺-dependent enzymes are stereospecific. Malate dehydrogenase, for example, transfers a hydride to the pro-R position of NADH, whereas glyceraldehyde-3-phosphate dehydrogenase transfers a hydride to the pro-S position of the nicotinamide. Alcohol dehydrogenase removes a hydride from the pro-R position of ethanol and transfers it to the pro-R position of NADH.

Adapted from Kaplan, N. O., 1960. In *The Enzymes,* vol. 3, p. 115, Boyer, P. D., Lardy, H. A., and Myrbäck, K., eds. New York: Academic Press.

19.7 | Can the TCA Cycle Provide Intermediates for Biosynthesis?

Until now we have viewed the TCA cycle as a catabolic process because it oxidizes acetate units to CO_2 and converts the liberated energy to ATP and reduced coenzymes. The TCA cycle is, after all, the end point for breakdown of food materials, at least in terms of carbon turnover. However, as shown in Figure 19.22,

Steric Specificity for NAD of Various Pyridine Nucleotide-Linked Enzymes		
Dehydrogenase	**Source**	**Steric Specificity**
Alcohol (with ethanol)	Yeast, *Pseudomonas*, liver, wheat germ	
Alcohol (with isopropyl alcohol)	Yeast	
Acetaldehyde	Liver	
L-Lactate	Heart muscle, *Lactobacillus*	H_R
L-Malate	Pig heart, wheat germ	
D-Glycerate	Spinach	
Dihydroorotate	*Zymobacterium oroticum*	
α-Glycerophosphate	Muscle	
Glyceraldehyde-3-P	Yeast, muscle	
L-Glutamate	Liver	
D-Glucose	Liver	
β-Hydroxysteroid	*Pseudomonas*	H_S
NADH cytochrome *c* reductase	Rat liver mitochondria, pig heart	
NADPH transhydrogenase	*Pseudomonas*	
NADH diaphorase	Pig heart	
L-β-Hydroxybutyryl-CoA	Heart muscle	

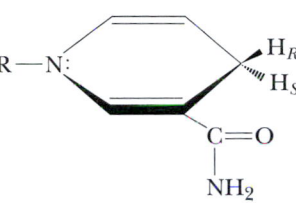

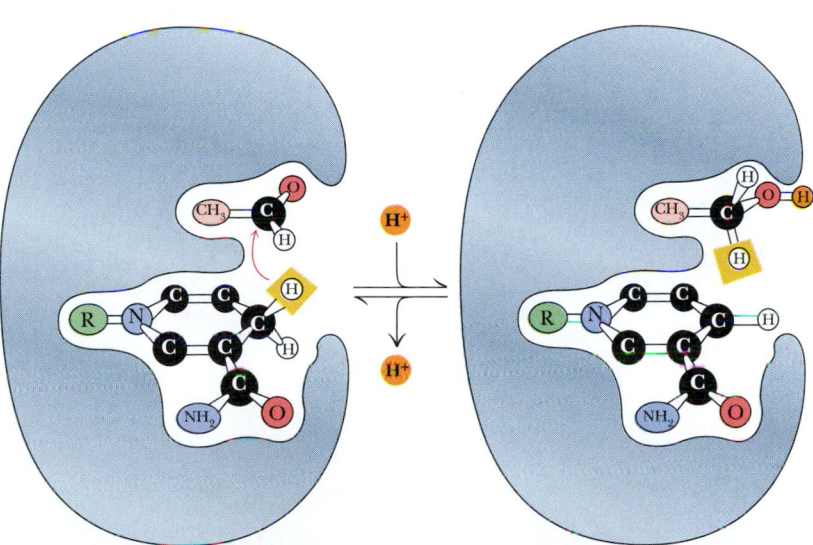

The stereospecificity of hydride transfer in dehydrogenases is a consequence of the asymmetric nature of the active site.

four-, five-, and six-carbon species produced in the TCA cycle also fuel a variety of **biosynthetic processes.** α-Ketoglutarate, succinyl-CoA, fumarate, and oxaloacetate are all precursors of important cellular species. (In order to participate in eukaryotic biosynthetic processes, however, they must first be transported out of the mitochondria.) A transamination reaction converts α-ketoglutarate directly to glutamate, which can then serve as a versatile precursor for proline, arginine, and glutamine (as described in Chapter 25). Succinyl-CoA provides most of the carbon atoms of the porphyrins. Oxaloacetate can be transaminated to produce aspartate. Aspartic acid itself is a precursor of the pyrimidine nucleotides and, in addition, is a key precursor for the synthesis of asparagine, methionine, lysine, threonine, and isoleucine. Oxaloacetate can also be decarboxylated to yield PEP, which is a key element of several pathways, namely

(a) Fate of the carboxyl carbon of acetate unit

(b) Fate of methyl carbon of acetate unit

All labeled carboxyl carbon removed by these two steps

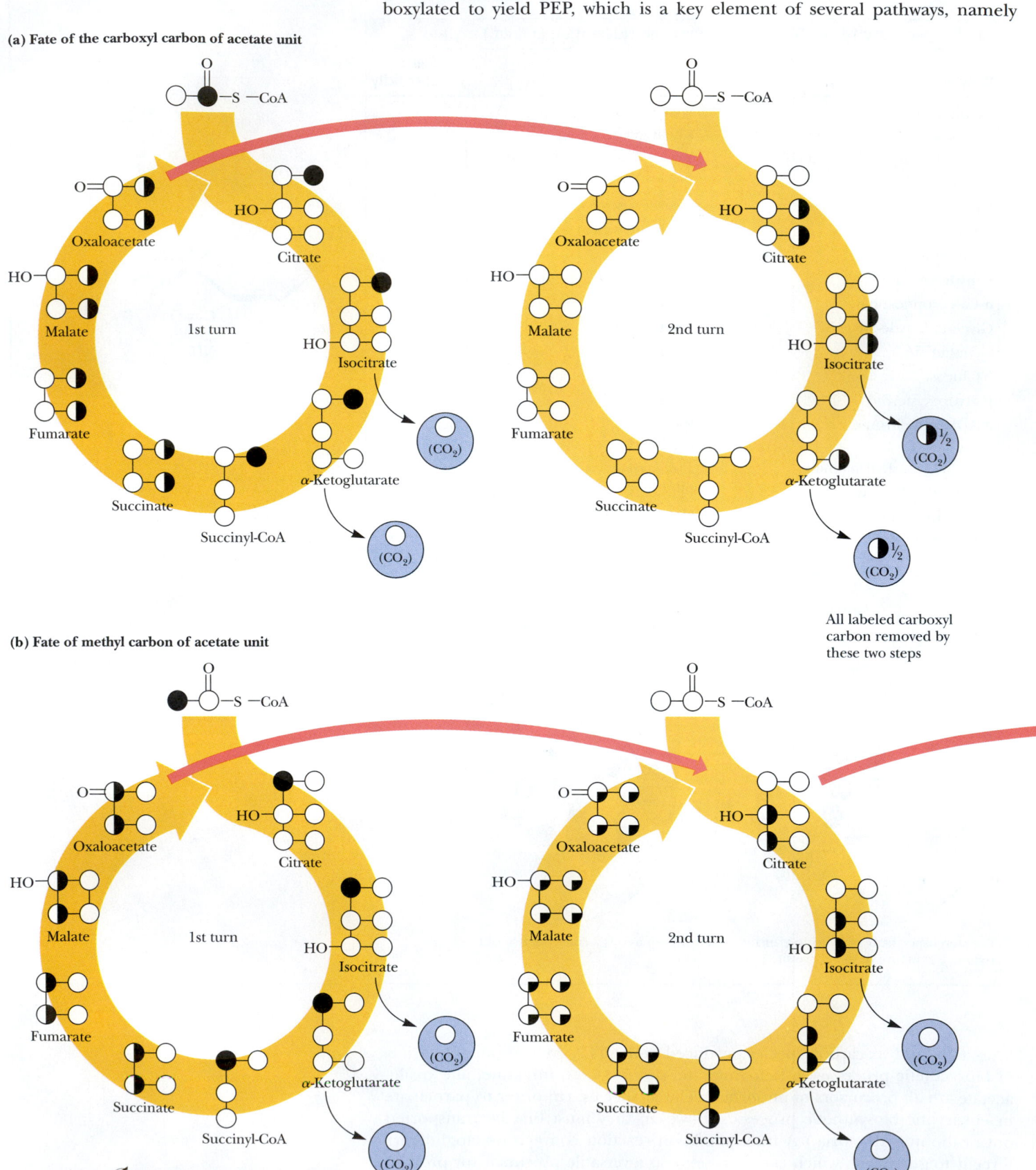

Biochemistry⊛Now™   ACTIVE FIGURE 19.21

continued

Human Biochemistry

Mitochondrial Diseases Are Rare

Diseases arising from defects in mitochondrial enzymes are quite rare, because major defects in the TCA cycle (and the respiratory chain) are incompatible with life and affected embryos rarely survive to birth. Even so, about 150 different hereditary mitochondrial diseases have been reported. Even though mitochondria carry their own DNA, many of the reported diseases map to the nuclear genome, because most of the mitochondrial proteins are imported from the cytosol.

An interesting disease linked to mitochondrial DNA mutations is that of Leber's hereditary optic neuropathy (LHON), in which the genetic defects are located primarily in the mitochondrial DNA coding for the subunits of NADH–CoQ reductase, also known as Complex I of the electron-transport chain (see Chapter 20). Leber's disease is the most common form of blindness in otherwise healthy young men and occurs less often in women.

(1) synthesis (in plants and microorganisms) of the aromatic amino acids phenylalanine, tyrosine, and tryptophan; (2) formation of 3-phosphoglycerate and conversion to the amino acids serine, glycine, and cysteine; and (3) gluconeogenesis, which, as we will see in Chapter 22, is the pathway that synthesizes new glucose and many other carbohydrates.

Finally, citrate can be exported from the mitochondria and then broken down by **ATP–citrate lyase** to yield oxaloacetate and acetyl-CoA, a precursor of fatty acids (Figure 19.23). Oxaloacetate produced in this reaction is rapidly reduced to

Biochemistry ⊘ Now™ **ACTIVE FIGURE 19.21** (*here and on the facing page*) The fate of the carbon atoms of acetate in successive TCA cycles. (a) The carbonyl carbon of acetyl-CoA is fully retained through one turn of the cycle but is lost completely in a second turn of the cycle. (b) The methyl carbon of a labeled acetyl-CoA survives two full turns of the cycle but becomes equally distributed among the four carbons of oxaloacetate by the end of the second turn. In each subsequent turn of the cycle, one-half of this carbon (the original labeled methyl group) is lost. **Test yourself on the concepts in this figure at http://chemistry.brookscole.com/ggb3**

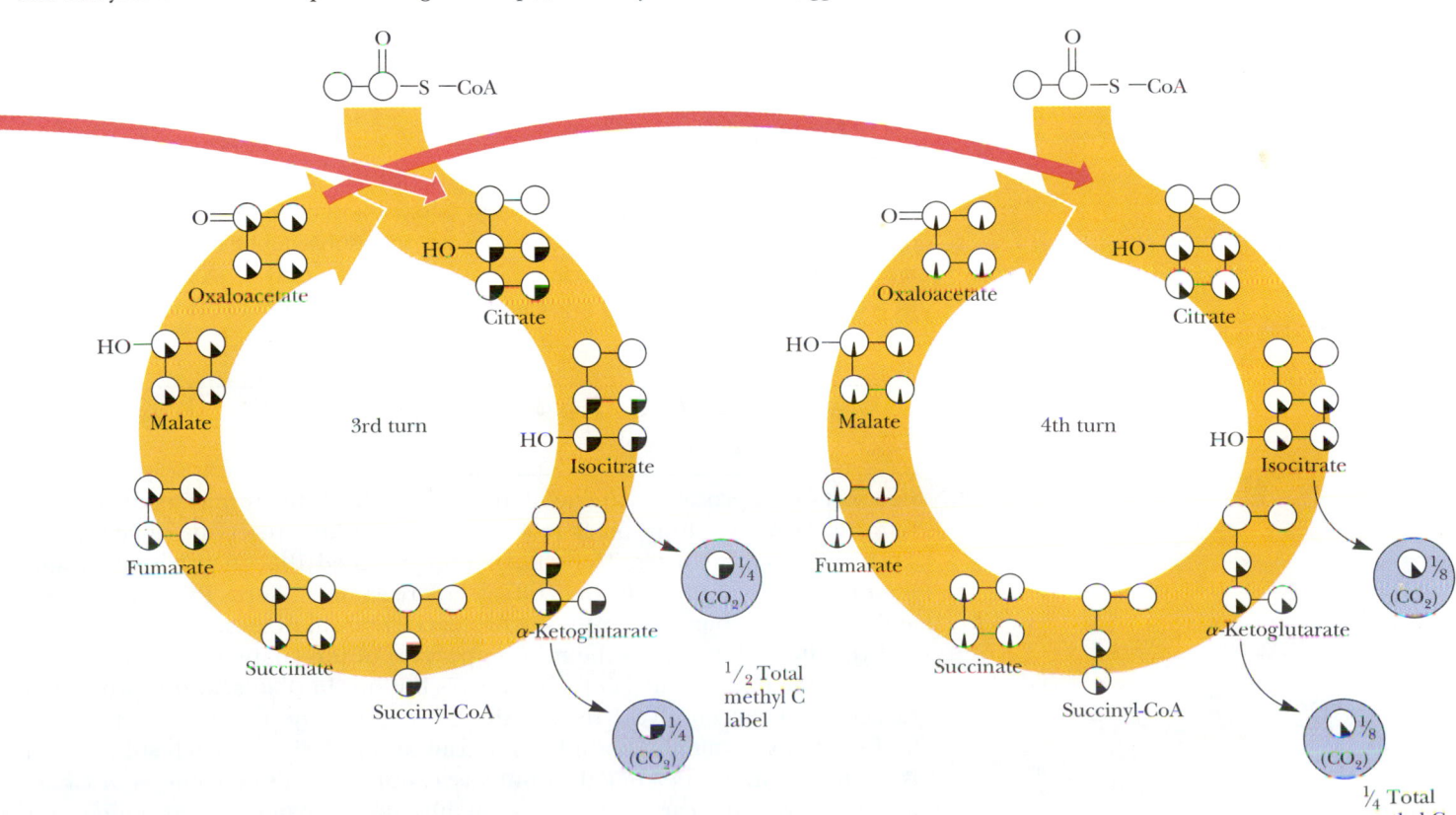

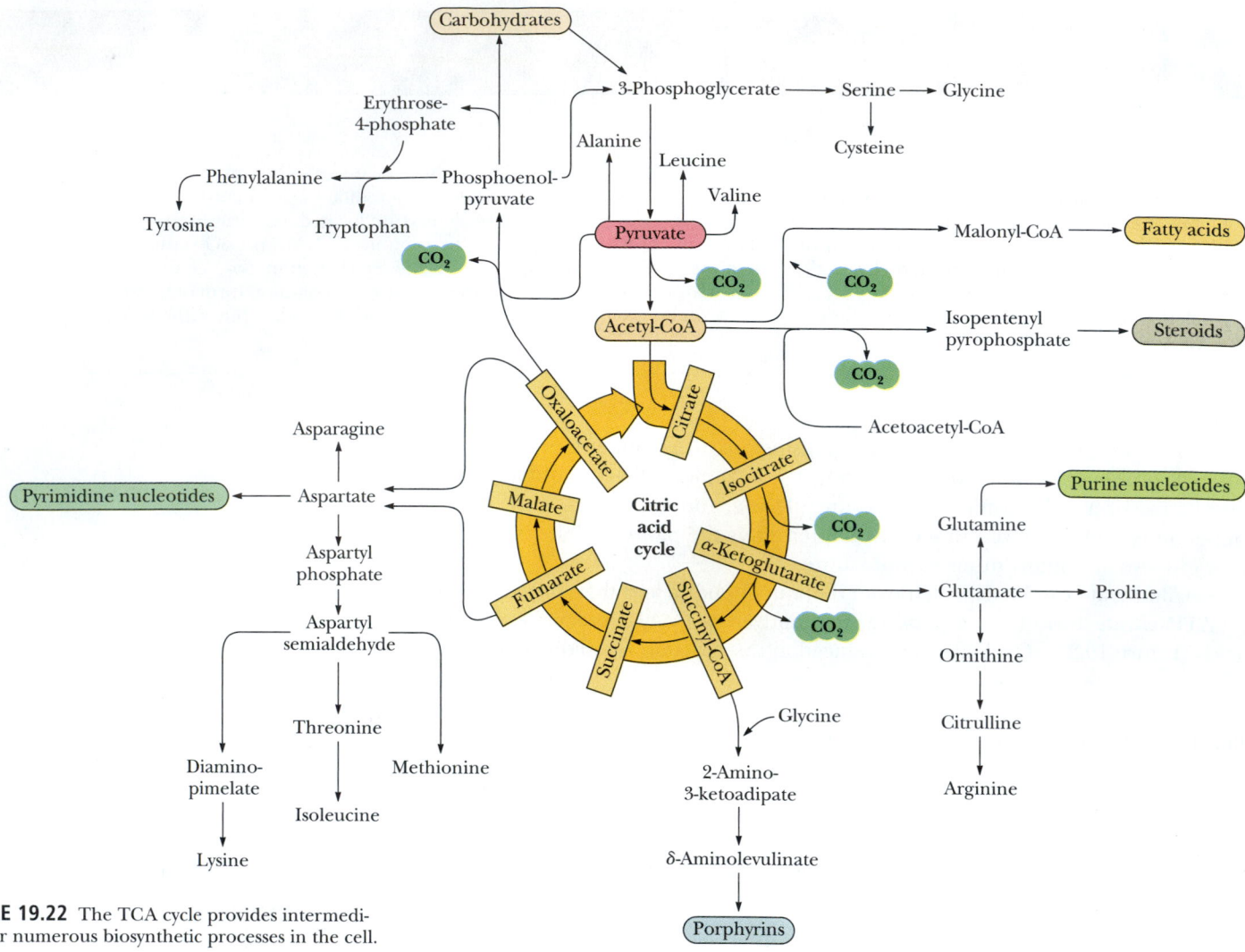

FIGURE 19.22 The TCA cycle provides intermediates for numerous biosynthetic processes in the cell.

malate, which can then be processed in either of two ways: It may be transported into mitochondria, where it is reoxidized to oxaloacetate, or it may be oxidatively decarboxylated to pyruvate by **malic enzyme**, with subsequent mitochondrial uptake of pyruvate. This cycle permits citrate to provide acetyl-CoA for biosynthetic processes, with return of the malate and pyruvate by-products to the mitochondria.

19.8 What Are the Anaplerotic, or "Filling Up," Reactions?

In a sort of reciprocal arrangement, the cell also feeds many intermediates back into the TCA cycle from other reactions. Because such reactions replenish the TCA cycle intermediates, Hans Kornberg proposed that they be called **anaplerotic reactions** (literally, the "filling up" reactions). Thus, **PEP carboxylase** and **pyruvate carboxylase** synthesize oxaloacetate from pyruvate (Figure 19.24).

Pyruvate carboxylase is the most important of the anaplerotic reactions. It exists in the mitochondria of animal cells but not in plants, and it provides a direct link between glycolysis and the TCA cycle. The enzyme is tetrameric and contains covalently bound biotin and an Mg^{2+} site on each subunit. (It is examined in greater detail in our discussion of gluconeogenesis in Chapter 22.) Pyruvate carboxylase has an absolute allosteric requirement for acetyl-CoA. Thus, when acetyl-CoA levels exceed the oxaloacetate supply,

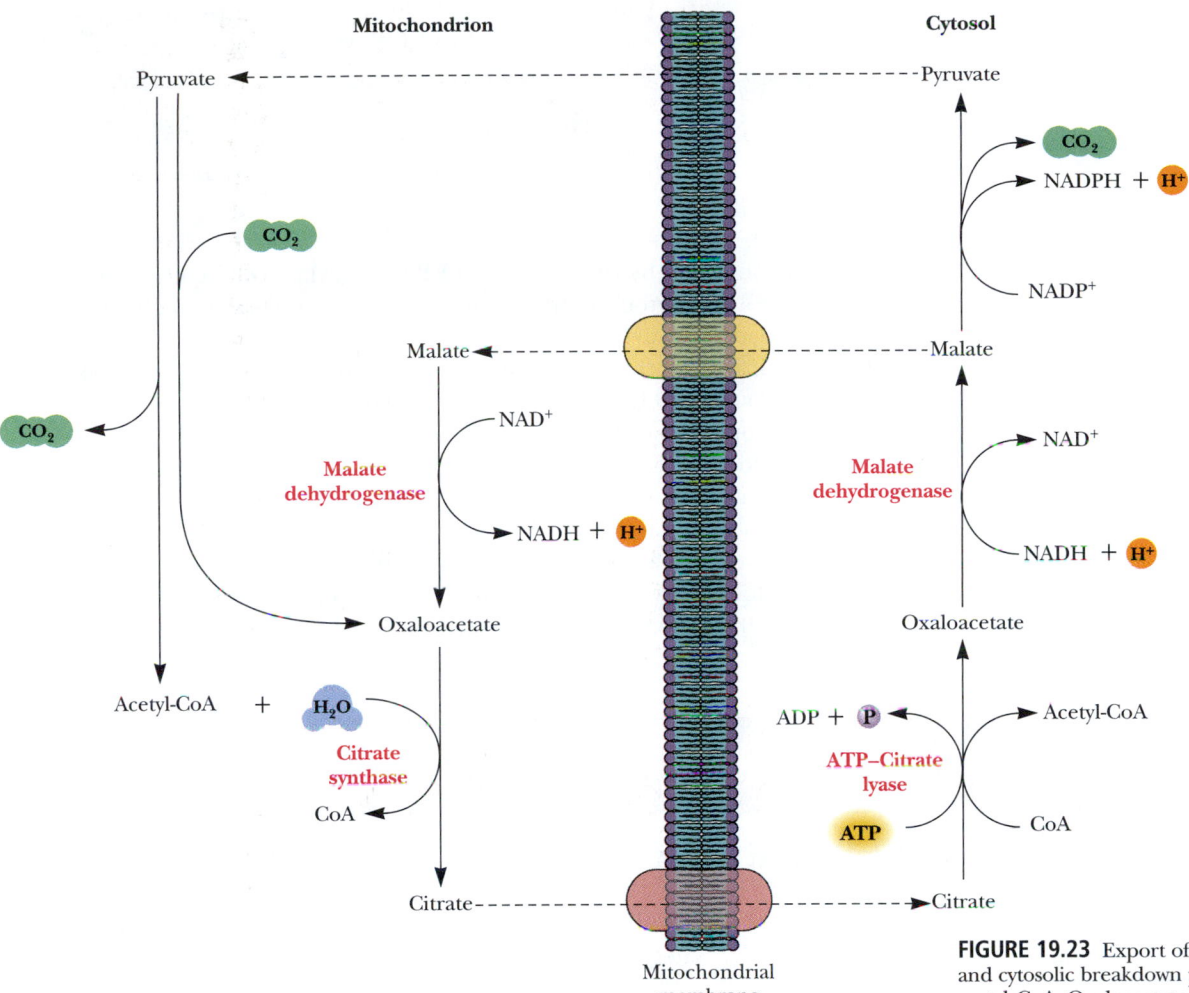

FIGURE 19.23 Export of citrate from mitochondria and cytosolic breakdown produces oxaloacetate and acetyl-CoA. Oxaloacetate is recycled to malate or pyruvate, which reenters the mitochondria. This cycle provides acetyl-CoA for fatty acid synthesis in the cytosol.

allosteric activation of pyruvate carboxylase by acetyl-CoA raises oxaloacetate levels, so the excess acetyl-CoA can enter the TCA cycle.

PEP carboxylase occurs in yeast, bacteria, and higher plants, but not in animals. The enzyme is specifically inhibited by aspartate, which is produced by transamination of oxaloacetate. Thus, organisms utilizing this enzyme control

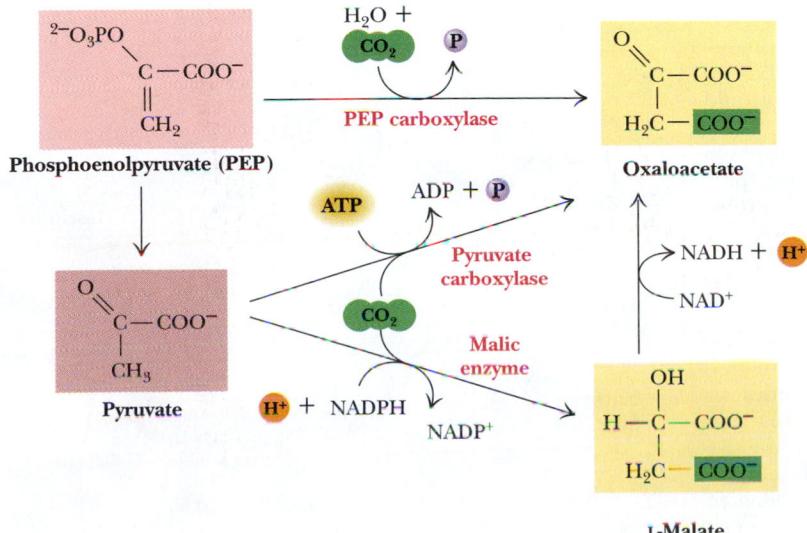

FIGURE 19.24 Phosphoenolpyruvate (PEP) carboxylase, pyruvate carboxylase, and malic enzyme catalyze anaplerotic reactions, replenishing TCA cycle intermediates.

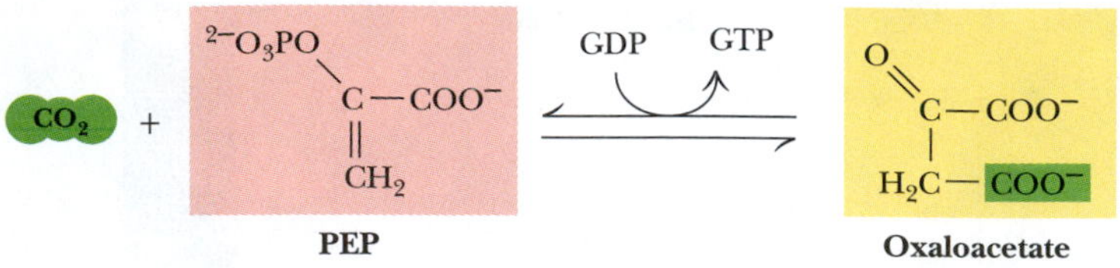

FIGURE 19.25 The phosphoenolpyruvate carboxy-kinase reaction.

aspartate production by regulation of PEP carboxylase. Malic enzyme is found in the cytosol or mitochondria of many animal and plant cells and is an NADPH-dependent enzyme.

It is worth noting that the reaction catalyzed by **PEP carboxykinase** (Figure 19.25) could also function as an anaplerotic reaction, were it not for the particular properties of the enzyme. CO_2 binds weakly to PEP carboxykinase, whereas oxaloacetate binds very tightly ($K_D = 2 \times 10^{-6}$ M), and, as a result, the enzyme favors formation of PEP from oxaloacetate.

The catabolism of amino acids provides pyruvate, acetyl-CoA, oxaloacetate, fumarate, α-ketoglutarate, and succinate, all of which may be oxidized by the TCA cycle. In this way, proteins may serve as excellent sources of nutrient energy, as seen in Chapter 25.

A Deeper Look

Fool's Gold and the Reductive Citric Acid Cycle—The First Metabolic Pathway?

How did life arise on the planet earth? It was once supposed that a reducing atmosphere, together with random synthesis of organic compounds, gave rise to a prebiotic "soup," in which the first living things appeared. However, certain key compounds, such as arginine, lysine, and histidine; the straight-chain fatty acids; porphyrins; and essential coenzymes, have not been convincingly synthesized under simulated prebiotic conditions. This and other problems have led researchers to consider other models for the evolution of life.

One of these alternative models, postulated by Günter Wächtershäuser, involves an archaic version of the TCA cycle running in the reverse (reductive) direction. Reversal of the TCA cycle results in assimilation of CO_2 and fixation of carbon as shown. For each turn of the reversed cycle, two carbons are fixed in the formation of isocitrate and two more are fixed in the reductive transformation of acetyl-CoA to oxaloacetate. Thus, for every succinate that enters the reversed cycle, two succinates are returned, making the cycle highly autocatalytic. Because TCA cycle intermediates are involved in many biosynthetic pathways (see Section 19.7), a reversed TCA cycle would be a bountiful and broad source of metabolic substrates.

A reversed, reductive TCA cycle would require energy input to drive it. What might have been the thermodynamic driving force for such a cycle? Wächtershäuser hypothesizes that the anaerobic reaction of FeS and H_2S to form insoluble FeS_2 (pyrite, also known as fool's gold) in the prebiotic milieu could have been the driving reaction:

$$FeS + H_2S \longrightarrow FeS_2 \text{ (pyrite)} \downarrow + H_2$$

This reaction is highly exergonic, with a standard-state free energy change ($\Delta G^{\circ\prime}$) of -38 kJ/mol. Under the conditions that might have existed in a prebiotic world, this reaction would have been sufficiently exergonic to drive the reductive steps of a reversed TCA cycle. In addition, in an H_2S-rich prebiotic environment, organic compounds would have been in equilibrium with their thio-organic counterparts. High-energy thioesters formed in this way may have played key roles in the energetics of early metabolic pathways.

Wächtershäuser has also suggested that early metabolic processes first occurred on the surface of pyrite and other related mineral materials. The iron–sulfur chemistry that prevailed on these mineral surfaces may have influenced the evolution of the iron–sulfur proteins that control and catalyze many reactions in modern pathways (including the succinate dehydrogenase and aconitase reactions of the TCA cycle).

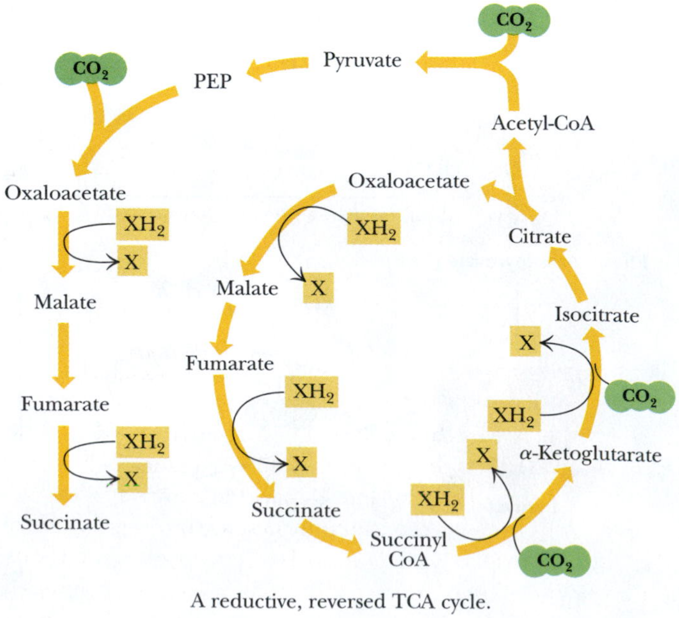

A reductive, reversed TCA cycle.

19.9 | **How Is the TCA Cycle Regulated?**

Situated as it is between glycolysis and the electron-transport chain, the TCA cycle must be carefully controlled. If the cycle were permitted to run unchecked, large amounts of metabolic energy could be wasted in overproduction of reduced coenzymes and ATP; conversely, if it ran too slowly, ATP would not be produced rapidly enough to satisfy the needs of the cell. Also, as just seen, the TCA cycle is an important source of precursors for biosynthetic processes and must be able to provide them as needed.

What are the sites of regulation in the TCA cycle? Based on our experience with glycolysis (see Figure 18.31), we might anticipate that some of the reactions of the TCA cycle would operate near equilibrium under cellular conditions (with $\Delta G < 0$), whereas others—the sites of regulation—would be characterized by large negative ΔG values. Estimates for the values of ΔG in mitochondria, based on mitochondrial concentrations of metabolites, are summarized in Table 19.1. Three reactions of the cycle—citrate synthase, isocitrate dehydrogenase, and α-ketoglutarate dehydrogenase—operate with large negative ΔG values under mitochondrial conditions and are thus the primary sites of regulation in the cycle.

The regulatory actions that control the TCA cycle are shown in Figure 19.26. As one might expect, the principal regulatory "signals" are the concentrations of acetyl-CoA, ATP, NAD$^+$, and NADH, with additional effects provided by several other metabolites. The main sites of regulation are pyruvate dehydrogenase, citrate synthase, isocitrate dehydrogenase, and α-ketoglutarate dehydrogenase. All of these enzymes are inhibited by NADH, so when the cell has produced all the NADH that can conveniently be turned into ATP, the cycle shuts down. For similar reasons, ATP is an inhibitor of pyruvate dehydrogenase and isocitrate dehydrogenase. The TCA cycle is turned on, however, when either the ADP/ATP or NAD$^+$/NADH ratio is high, an indication that the cell has run low on ATP or NADH. Regulation of the TCA cycle by NADH, NAD$^+$, ATP, and ADP thus reflects the energy status of the cell. On the other hand, succinyl-CoA is an intracycle regulator, inhibiting citrate synthase and α-ketoglutarate dehydrogenase. Acetyl-CoA acts as a signal to the TCA cycle that glycolysis or fatty acid breakdown is producing two-carbon units. Acetyl-CoA activates pyruvate carboxylase, the anaplerotic reaction that provides oxaloacetate, the acceptor for increased flux of acetyl-CoA into the TCA cycle.

Pyruvate Dehydrogenase Is Regulated by Phosphorylation/Dephosphorylation

As we shall see in Chapter 22, most organisms can synthesize sugars such as glucose from pyruvate. However, animals cannot synthesize glucose from acetyl-CoA. For this reason, the pyruvate dehydrogenase complex, which converts pyruvate to acetyl-CoA, plays a pivotal role in metabolism. Conversion to acetyl-CoA commits nutrient carbon atoms either to oxidation in the TCA cycle or to fatty acid synthesis (see Chapter 24). Because this choice is so crucial to the organism, pyruvate dehydrogenase is a carefully regulated enzyme. It is subject to product inhibition and is further regulated by nucleotides. Finally, activity of pyruvate dehydrogenase is regulated by phosphorylation and dephosphorylation of the enzyme complex itself.

High levels of either product, acetyl-CoA or NADH, allosterically inhibit the pyruvate dehydrogenase complex. Acetyl-CoA specifically blocks dihydrolipoyl transacetylase, and NADH acts on dihydrolipoyl dehydrogenase. The mammalian pyruvate dehydrogenase is also regulated by covalent modifications. As shown in Figure 19.27, a Mg^{2+}-dependent **pyruvate dehydrogenase kinase** is associated with the enzyme in mammals. This kinase is allosterically activated by NADH and acetyl-CoA, and when levels of these metabolites rise in the mitochondrion, they stimulate phosphorylation of a serine residue on the pyruvate dehydrogenase subunit, blocking the first step of the pyruvate dehydrogenase

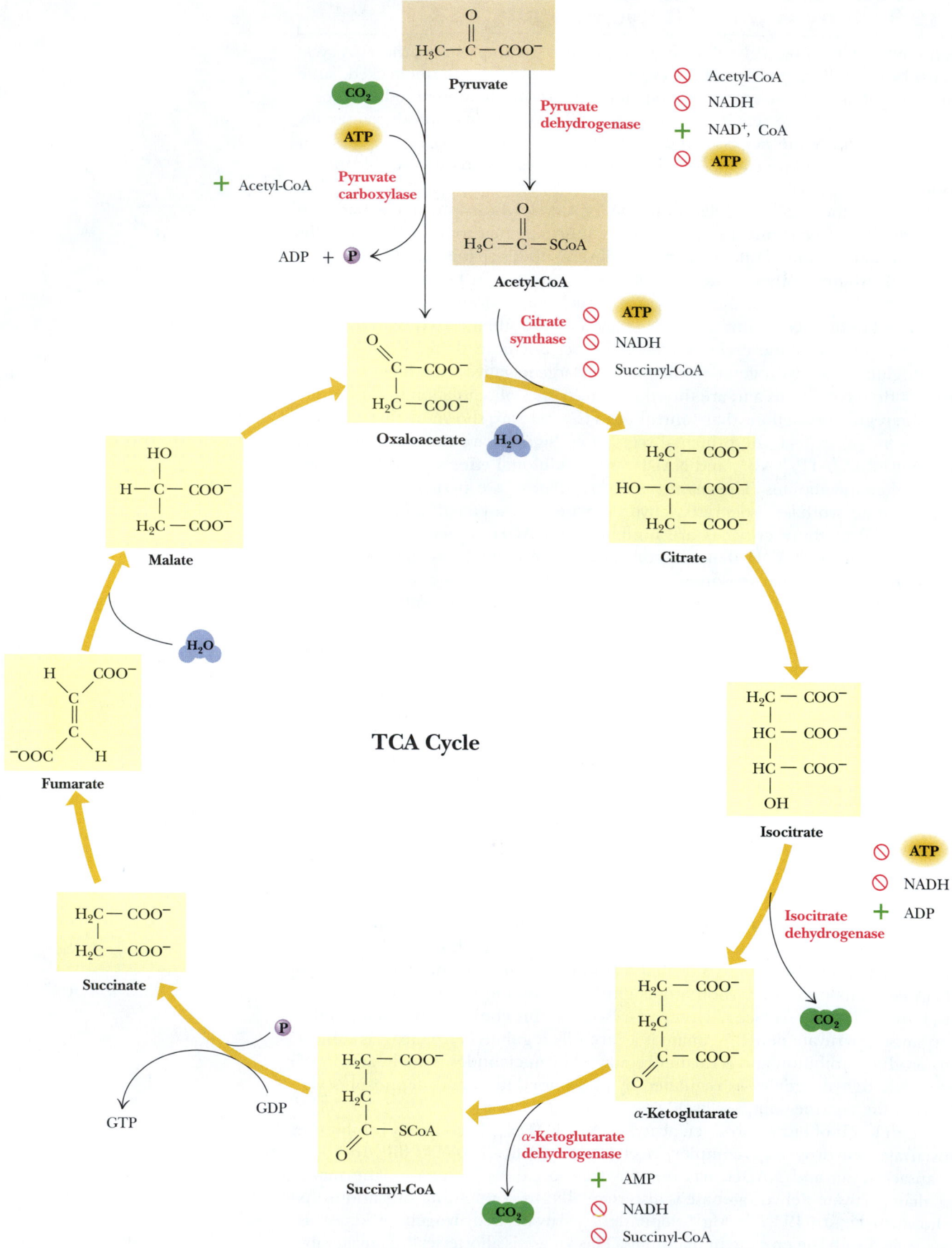

FIGURE 19.26 Regulation of the TCA cycle.

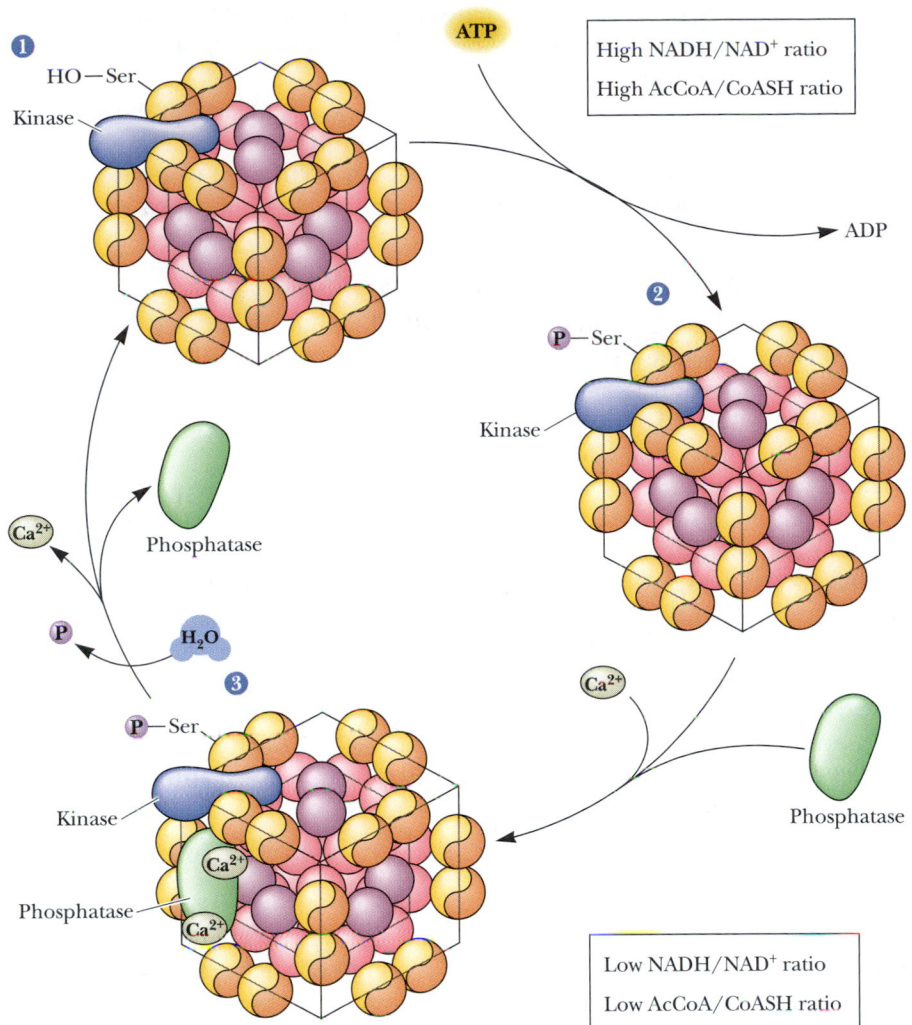

High NADH/NAD$^+$ ratio

High AcCoA/CoASH ratio

Low NADH/NAD$^+$ ratio

Low AcCoA/CoASH ratio

FIGURE 19.27 Regulation of the pyruvate dehydrogenase reaction.

reaction, the decarboxylation of pyruvate. Inhibition of the dehydrogenase in this manner eventually lowers the levels of NADH and acetyl-CoA in the matrix of the mitochondrion. Reactivation of the enzyme is carried out by **pyruvate dehydrogenase phosphatase,** a Ca^{2+}-activated enzyme that binds to the dehydrogenase complex and hydrolyzes the phosphoserine moiety on the dehydrogenase subunit. At low ratios of NADH to NAD$^+$ and low acetyl-CoA levels, the phosphatase maintains the dehydrogenase in an activated state, but a high level

Human Biochemistry

Therapy for Heart Attacks by Alterations of Heart Muscle Metabolism?

Ischemia, the state of reduced blood flow to a tissue, occurs in heart muscle during a heart attack. One strategy for minimizing tissue damage during and immediately following a heart attack involves interventions to alter heart muscle metabolism. One drug being studied for this purpose is dichloroacetate, which specifically inhibits pyruvate dehydrogenase kinase, thus activating pyruvate dehydrogenase. The primary energy source for heart muscle is normally the oxidation of long-chain fatty acids. By contrast, carbohydrate metabolism is less important,

except during periods of high workload. However, treatment with dichloroacetate activates pyruvate dehydrogenase in the heart muscle and increases the flow of carbon from pyruvate (and thus glucose) into the TCA cycle. This is advantageous for the ischemic heart (for which oxygen is limited), because the ATP yield per oxygen consumed is higher for glucose oxidation than for fatty acid oxidation (see Chapter 24). Research to date indicates that heart function is improved with intravenous dichloroacetate in animal models.

of acetyl-CoA or NADH once again activates the kinase and leads to the inhibition of the dehydrogenase. Insulin and Ca^{2+} ions activate dephosphorylation, and pyruvate inhibits the phosphorylation reaction.

Pyruvate dehydrogenase is also sensitive to the energy status of the cell. AMP activates pyruvate dehydrogenase, whereas GTP inhibits it. High levels of AMP are a sign that the cell may become energy-poor. Activation of pyruvate dehydrogenase under such conditions commits pyruvate to energy production.

Isocitrate Dehydrogenase Is Strongly Regulated

The mechanism of regulation of isocitrate dehydrogenase is in some respects the reverse of pyruvate dehydrogenase. The mammalian isocitrate dehydrogenase is subject only to allosteric activation by ADP and NAD^+ and to inhibition by ATP and NADH. Thus, high NAD^+/NADH and ADP/ATP ratios stimulate isocitrate dehydrogenase and TCA cycle activity.

It may seem surprising that isocitrate dehydrogenase is strongly regulated, because it is not an apparent branch point within the TCA cycle. However, the citrate/isocitrate ratio controls the rate of production of cytosolic acetyl-CoA, because acetyl-CoA in the cytosol is derived from citrate exported from the mitochondrion. (Breakdown of cytosolic citrate produces oxaloacetate and acetyl-CoA, which can be used in a variety of biosynthetic processes.) Thus, isocitrate dehydrogenase activity in the mitochondrion favors catabolic TCA cycle activity over anabolic utilization of acetyl-CoA in the cytosol.

Interestingly, the *Escherichia coli* isocitrate dehydrogenase is regulated by covalent modification. Serine residues on each subunit of the dimeric enzyme are phosphorylated by a protein kinase, causing inhibition of the isocitrate dehydrogenase activity. Activity is restored by the action of a specific phosphatase. When TCA cycle and glycolytic intermediates—such as isocitrate, 3-phosphoglycerate, pyruvate, PEP, and oxaloacetate—are high, the kinase is inhibited, the phosphatase is activated, and the TCA cycle operates normally. When levels of these intermediates fall, the kinase is activated, isocitrate dehydrogenase is inhibited, and isocitrate is diverted to the glyoxylate pathway, as explained in the next section.

19.10 | Can Any Organisms Use Acetate as Their Sole Carbon Source?

Plants (particularly seedlings, which cannot yet accomplish efficient photosynthesis), as well as some bacteria and algae, can use acetate as the only source of carbon for all the carbon compounds they produce. Although we saw that the TCA cycle can supply intermediates for some biosynthetic processes, the cycle gives off 2 CO_2 for every two-carbon acetate group that enters and cannot effect the net synthesis of TCA cycle intermediates. Thus, it would not be possible for the cycle to produce the massive amounts of biosynthetic intermediates needed for acetate-based growth unless alternative reactions were available. In essence, the TCA cycle is geared primarily to energy production, and it "wastes" carbon units by giving off CO_2. Modification of the cycle to support acetate-based growth would require eliminating the CO_2-producing reactions and enhancing the net production of four-carbon units (i.e., oxaloacetate). Plants and bacteria employ a modification of the TCA cycle called the **glyoxylate cycle** to produce four-carbon dicarboxylic acids (and eventually even sugars) from two-carbon acetate units. The glyoxylate cycle bypasses the two oxidative decarboxylations of the TCA cycle and instead routes isocitrate through the **isocitrate lyase** and **malate synthase** reactions (Figure 19.28). Glyoxylate produced by isocitrate lyase reacts with a second molecule of acetyl-CoA to form L-malate. The net effect is to conserve carbon units, using two acetyl-CoA molecules per cycle to generate oxaloacetate. Some of this is converted to PEP and then to glucose by pathways discussed in Chapter 22.

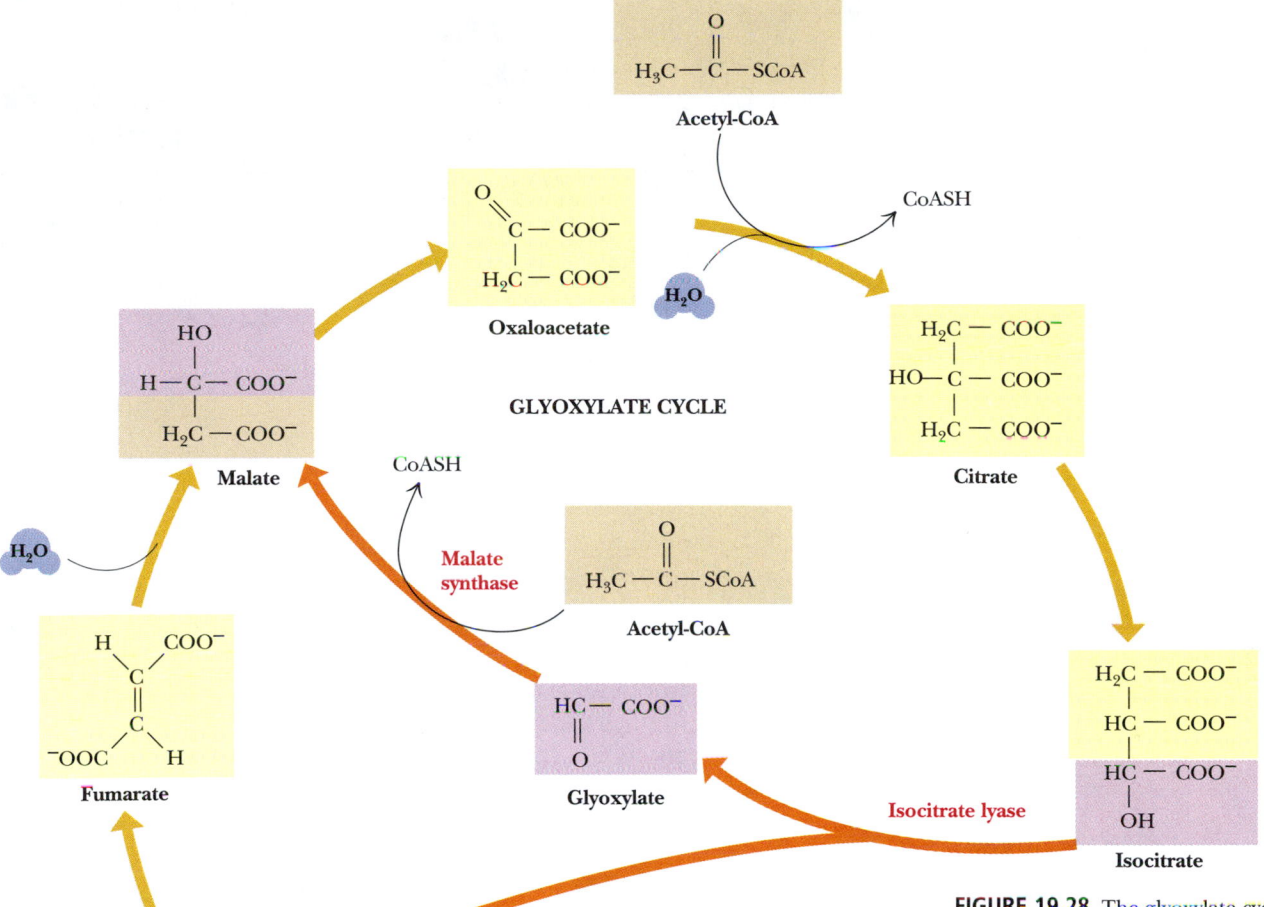

FIGURE 19.28 The glyoxylate cycle. The first two steps are identical to TCA cycle reactions. The third step bypasses the CO_2-evolving steps of the TCA cycle to produce succinate and glyoxylate. The malate synthase reaction forms malate from glyoxylate and another acetyl-CoA. The result is that one turn of the cycle consumes one oxaloacetate and two acetyl-CoA molecules but produces two molecules of oxaloacetate. The net for this cycle is one oxaloacetate from two acetyl-CoA molecules.

The Glyoxylate Cycle Operates in Specialized Organelles

The enzymes of the glyoxylate cycle in plants are contained in **glyoxysomes,** organelles devoted to this cycle. Yeast and algae carry out the glyoxylate cycle in the cytoplasm. The enzymes common to both the TCA and glyoxylate pathways exist as isozymes, with spatially and functionally distinct enzymes operating independently in the two cycles.

Isocitrate Lyase Short-Circuits the TCA Cycle by Producing Glyoxylate and Succinate

The **isocitrate lyase** reaction (Figure 19.29) produces succinate, a four-carbon product of the cycle, as well as glyoxylate, which can then combine with a second molecule of acetyl-CoA. Isocitrate lyase catalyzes an aldol cleavage and is

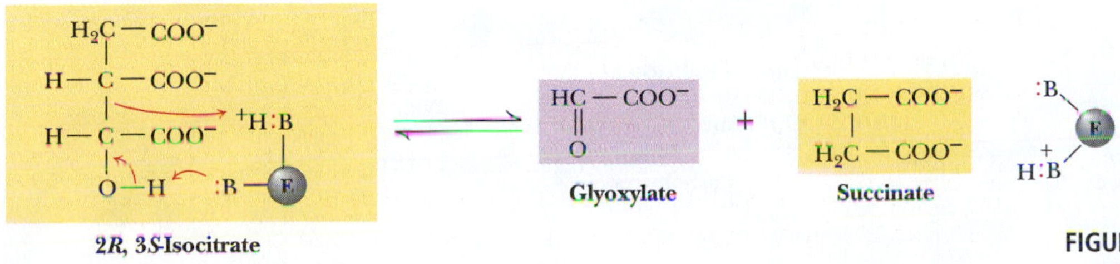

FIGURE 19.29 The isocitrate lyase reaction.

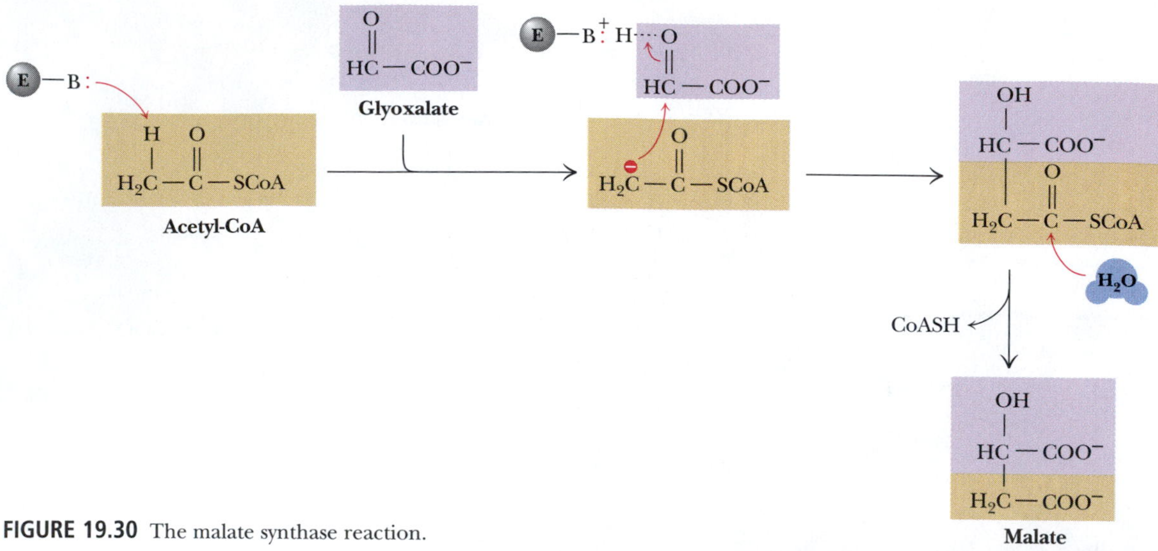

FIGURE 19.30 The malate synthase reaction.

similar to the reaction mediated by aldolase in glycolysis. The **malate synthase** reaction (Figure 19.30), a Claisen condensation of acetyl-CoA with the aldehyde of glyoxylate to yield malate, is quite similar to the citrate synthase reaction. Compared with the TCA cycle, the glyoxylate cycle (1) contains only five steps (as opposed to eight), (2) lacks the CO_2-liberating reactions, (3) consumes two molecules of acetyl-CoA per cycle, and (4) produces four-carbon units (oxaloacetate) as opposed to one-carbon units.

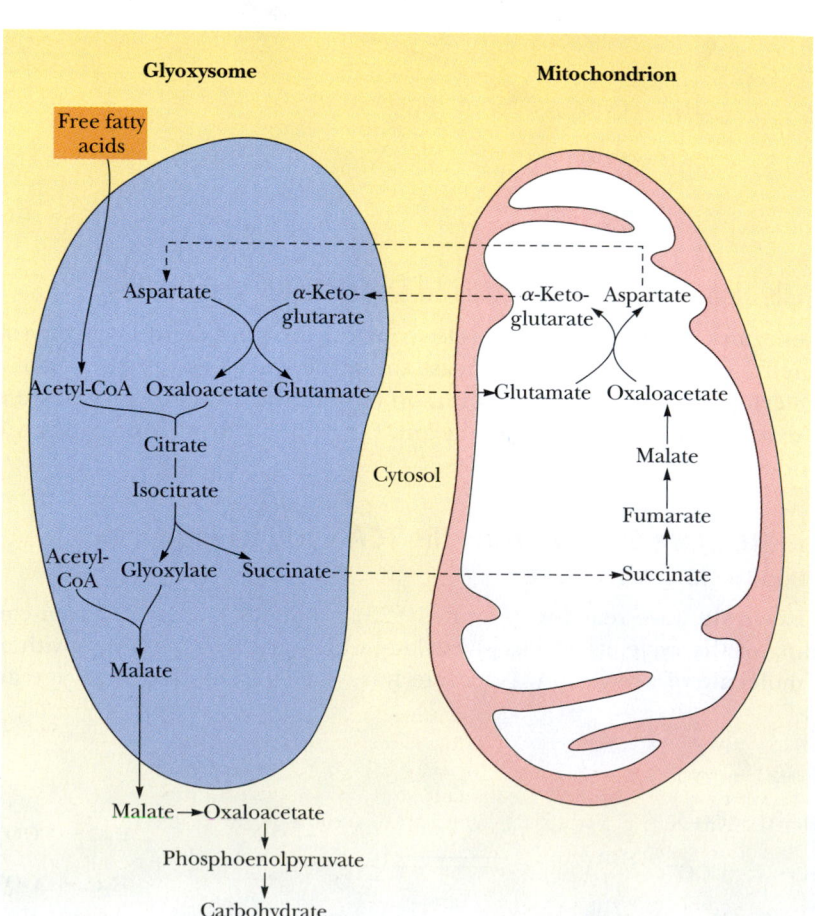

FIGURE 19.31 Glyoxysomes lack three of the enzymes needed to run the glyoxylate cycle. Succinate dehydrogenase, fumarase, and malate dehydrogenase are all "borrowed" from the mitochondria in a shuttle in which succinate and glutamate are passed to the mitochondria and α-ketoglutarate and aspartate are passed to the glyoxysome.

The Glyoxylate Cycle Helps Plants Grow in the Dark

The existence of the glyoxylate cycle explains how certain seeds grow underground (or in the dark), where photosynthesis is impossible. Many seeds (peanuts, soybeans, and castor beans, for example) are rich in lipids, and as we will see in Chapter 23, most organisms degrade the fatty acids of lipids to acetyl-CoA. Glyoxysomes form in seeds as germination begins, and the glyoxylate cycle uses the acetyl-CoA produced in fatty acid oxidation to provide large amounts of oxaloacetate and other intermediates for carbohydrate synthesis. Once the growing plant begins photosynthesis and can fix CO_2 to produce carbohydrates (see Chapter 21), the glyoxysomes disappear.

Glyoxysomes Must Borrow Three Reactions from Mitochondria

Glyoxysomes do not contain all the enzymes needed to run the glyoxylate cycle: Succinate dehydrogenase, fumarase, and malate dehydrogenase are absent. Consequently, glyoxysomes must cooperate with mitochondria to run their cycle (Figure 19.31). Succinate travels from the glyoxysomes to the mitochondria, where it is converted to oxaloacetate. Transamination to aspartate follows because oxaloacetate cannot be transported out of the mitochondria. Aspartate formed in this way then moves from the mitochondria back to the glyoxysomes, where a reverse transamination with α-ketoglutarate forms oxaloacetate, completing the shuttle. Finally, to balance the transaminations, glutamate shuttles from glyoxysomes to mitochondria.

Summary

The glycolytic pathway converts glucose to pyruvate and produces two molecules of ATP per glucose—only a small fraction of the potential energy available from glucose. In the presence of oxygen, pyruvate is oxidized to CO_2, releasing the rest of the energy available from glucose via the TCA cycle.

19.1 How Did Hans Krebs Elucidate the TCA Cycle?
In 1937, Krebs found that citrate could be formed in muscle suspensions if oxaloacetate and either pyruvate or acetate were added and realized that pyruvate oxidation must involve a cyclic series of reactions.

19.2 What Is the Chemical Logic of the TCA Cycle?
The entry of new carbon units into the cycle is through acetyl-CoA. Transfer of the two-carbon acetyl group from acetyl-CoA to the four-carbon oxaloacetate to yield six-carbon citrate is catalyzed by citrate synthase. A dehydration–rehydration rearrangement of citrate yields isocitrate. Two successive decarboxylations produce α-ketoglutarate and then succinyl-CoA, a CoA conjugate of a four-carbon unit. Several steps later, oxaloacetate is regenerated and can combine with another two-carbon unit of acetyl-CoA.

19.3 How Is Pyruvate Oxidatively Decarboxylated to Acetyl-CoA?
The pyruvate dehydrogenase complex (PDC) is a noncovalent assembly of three different enzymes operating in concert to catalyze successive steps in the conversion of pyruvate to acetyl-CoA.

19.4 How Are Two CO_2 Molecules Produced from Acetyl-CoA?
Citrate synthase combines acetyl-CoA with oxaloacetate in a Perkin condensation (a carbon–carbon condensation between a ketone or aldehyde and an ester). A general base on the enzyme accepts a proton from the methyl group of acetyl-CoA, producing a stabilized α-carbanion of acetyl-CoA. This strong nucleophile attacks the α-carbonyl of oxaloacetate, yielding citryl-CoA.

Citrate is isomerized to isocitrate by aconitase in a two-step process involving aconitate as an intermediate. The elements of water are first abstracted from citrate to yield aconitate, which is then rehydrated with H— and HO— adding back in opposite positions to produce isocitrate. The net effect is the conversion of a tertiary alcohol (citrate) to a secondary alcohol (isocitrate).

The two-step isocitrate dehydrogenase reaction involves (1) oxidation of the C-2 alcohol of isocitrate to form oxalosuccinate, followed by (2) a β-decarboxylation reaction that expels the central carboxyl group as CO_2, leaving the product α-ketoglutarate. Oxalosuccinate, the β-keto acid produced by the initial dehydrogenation reaction, is unstable and thus is readily decarboxylated.

α-Ketoglutarate dehydrogenase is a multienzyme complex—consisting of α-ketoglutarate dehydrogenase, dihydrolipoyl transsuccinylase, and dihydrolipoyl dehydrogenase—that employs five different coenzymes. The dihydrolipoyl dehydrogenase in this reaction is identical to that in the pyruvate dehydrogenase reaction. The mechanism is an oxidative phosphorylation analogous to that of pyruvate dehydrogenase. Succinyl-CoA is the product.

19.5 How Is Oxaloacetate Regenerated to Complete the Cycle?
Succinyl-CoA synthetase catalyzes a substrate-level phosphorylation: Succinyl-CoA is a high-energy intermediate and is used to drive the phosphorylation of GDP to GTP (in mammals) or ADP to ATP (in plants and bacteria).

Succinate dehydrogenase (succinate–coenzyme Q reductase of the electron-transport chain) catalyzes removal of H atoms across a C—C bond and produces the *trans*-unsaturated fumarate.

Fumarate is hydrated in a stereospecific reaction by fumarase to give L-malate. The reaction involves *trans*-addition of the elements of water across the double bond.

Malate dehydrogenase completes the TCA cycle. This reaction is very endergonic, with a $\Delta G^{\circ\prime}$ of +30 kJ/mol. Consequently, the concentration of oxaloacetate in the mitochondrial matrix is usually quite low. Oxidation of malate to oxaloacetate is coupled to reduction of yet another molecule of NAD^+, the third one of the cycle.

19.6 What Are the Energetic Consequences of the TCA Cycle?

The cycle is exergonic, with a net $\Delta G^{\circ\prime}$ for one pass around the cycle of approximately -40 kJ/mol. Three NADH, one [FADH$_2$], and one ATP equivalent are produced in each turn of the cycle.

19.7 Can the TCA Cycle Provide Intermediates for Biosynthesis?

α-Ketoglutarate, succinyl-CoA, fumarate, and oxaloacetate are all precursors of important cellular species. A transamination reaction converts α-ketoglutarate directly to glutamate, which can then serve as a precursor for proline, arginine, and glutamine. Succinyl-CoA provides most of the carbon atoms of the porphyrins. Oxaloacetate can be transaminated to produce aspartate. Aspartic acid itself is a precursor of the pyrimidine nucleotides and, in addition, is a key precursor for the synthesis of asparagine, methionine, lysine, threonine, and isoleucine. Oxaloacetate can also be decarboxylated to yield PEP, which is a key element of several pathways.

19.8 What Are the Anaplerotic, or "Filling Up," Reactions?

Anaplerotic reactions replenish the TCA cycle intermediates. Examples include PEP carboxylase and pyruvate carboxylase, both of which synthesize oxaloacetate from pyruvate.

19.9 How Is the TCA Cycle Regulated?

The main sites of regulation are pyruvate dehydrogenase, citrate synthase, isocitrate dehydrogenase, and α-ketoglutarate dehydrogenase. All of these enzymes are inhibited by NADH. ATP is an inhibitor of pyruvate dehydrogenase and isocitrate dehydrogenase. The TCA cycle is turned on, however, when either the ADP/ATP or NAD$^+$/NADH ratio is high. Regulation of the TCA cycle by NADH, NAD$^+$, ATP, and ADP thus reflects the energy status of the cell. Succinyl-CoA is an intracycle regulator, inhibiting citrate synthase and α-ketoglutarate dehydrogenase. Acetyl-CoA activates pyruvate carboxylase, the anaplerotic reaction that provides oxaloacetate, the acceptor for acetyl-CoA entry into the TCA cycle.

19.10 Can Any Organisms Use Acetate as Their Sole Carbon Source?

Plants and bacteria employ a modification of the TCA cycle called the glyoxylate cycle to produce four-carbon dicarboxylic acids (and eventually even sugars) from two-carbon acetate units. The glyoxylate cycle bypasses the two oxidative decarboxylations of the TCA cycle and instead routes isocitrate through the isocitrate lyase and malate synthase reactions. Glyoxylate produced by isocitrate lyase reacts with a second molecule of acetyl-CoA to form L-malate. The net effect is to conserve carbon units, using two acetyl-CoA molecules per cycle to generate oxaloacetate.

Problems

1. Describe the labeling pattern that would result from the introduction into the TCA cycle of glutamate labeled at C$_\gamma$ with ^{14}C.

2. Describe the effect on the TCA cycle of (a) increasing the concentration of NAD$^+$, (b) reducing the concentration of ATP, and (c) increasing the concentration of isocitrate.

3. (Integrates with Chapter 15.) The serine residue of isocitrate dehydrogenase that is phosphorylated by protein kinase lies within the active site of the enzyme. This situation contrasts with most other examples of covalent modification by protein phosphorylation, where the phosphorylation occurs at a site remote from the active site. What direct effect do you think such active-site phosphorylation might have on the catalytic activity of isocitrate dehydrogenase? (See Barford, D., 1991. Molecular mechanisms for the control of enzymic activity by protein phosphorylation. *Biochimica et Biophysica Acta* **1133**:55–62.)

4. The first step of the α-ketoglutarate dehydrogenase reaction involves decarboxylation of the substrate and leaves a covalent TPP intermediate. Write a reasonable mechanism for this reaction.

5. In a tissue where the TCA cycle has been inhibited by fluoroacetate, what difference in the concentration of each TCA cycle metabolite would you expect, compared with a normal, uninhibited tissue?

6. (Integrates with Chapter 17.) On the basis of the description in Chapter 17 of the physical properties of FAD and FADH$_2$, suggest a method for the measurement of the enzyme activity of succinate dehydrogenase.

7. Starting with citrate, isocitrate, α-ketoglutarate, and succinate, state which of the individual carbons of the molecule undergo oxidation in the next step of the TCA cycle. Which molecules undergo a net oxidation?

8. In addition to fluoroacetate, consider whether other analogs of TCA cycle metabolites or intermediates might be introduced to inhibit other, specific reactions of the cycle. Explain your reasoning.

9. (Integrates with Chapter 17.) Based on the action of thiamine pyrophosphate in catalysis of the pyruvate dehydrogenase reaction, suggest a suitable chemical mechanism for the pyruvate decarboxylase reaction in yeast:

$$\text{Pyruvate} \longrightarrow \text{acetaldehyde} + CO_2$$

10. (Integrates with Chapter 3.) Aconitase catalyzes the citric acid cycle reaction:

$$\text{Citrate} \rightleftharpoons \text{isocitrate}$$

The standard free energy change, $\Delta G^{\circ\prime}$, for this reaction is $+6.7$ kJ/mol. However, the observed free energy change (ΔG) for this reaction in pig heart mitochondria is $+0.8$ kJ/mol. What is the ratio of [isocitrate]/[citrate] in these mitochondria? If [isocitrate] $= 0.03$ mM, what is [citrate]?

11. Describe the labeling pattern that would result if ^{14}CO$_2$ were incorporated into the TCA cycle via the pyruvate carboxylase reaction.

12. Describe the labeling pattern that would result if the reductive, reversed TCA cycle (see A Deeper Look on page 630) operated with ^{14}CO$_2$.

13. Describe the labeling pattern that would result in the glyoxylate cycle if a plant were fed acetyl-CoA labeled at the —CH$_3$ carbon.

Preparing for the MCAT Exam

14. Complete oxidation of a 16-carbon fatty acid can yield 129 molecules of ATP. Study Figure 19.4 and determine how many ATP molecules would be generated if a 16-carbon fatty acid were metabolized solely by the TCA cycle, in the form of 8 acetyl-CoA molecules.

15. Study Figure 19.26 and decide which of the following statements is false?

    a. Pyruvate dehydrogenase is inhibited by NADH.
    b. Pyruvate dehydrogenase is inhibited by ATP.
    c. Citrate synthase is inhibited by NADH.
    d. Succinyl-CoA activates citrate synthase.
    e. Acetyl-CoA activates pyruvate carboxylase.

Biochemistry🇪Now™ Preparing for an exam? Test yourself on key questions at http://chemistry.brookscole.com/ggb3

Further Reading

General

Bodner, G. M., 1986. The tricarboxylic acid (TCA), citric acid or Krebs cycle. *Journal of Chemical Education* **63:**673–677.

Gibble, G. W., 1973. Fluoroacetate toxicity. *Journal of Chemical Education* **50:**460–462.

Hansford, R. G., 1980. Control of mitochondrial substrate oxidation. In *Current Topics in Bioenergetics,* vol. 10, pp. 217–278. New York: Academic Press.

Hawkins, R. A., and Mans, A. M., 1983. Intermediary metabolism of carbohydrates and other fuels. In *Handbook of Neurochemistry,* 2nd ed., Lajtha, A., ed., pp. 259–294. New York: Plenum Press.

Kelly, R. M., and Adams, M. W., 1994. Metabolism in hyperthermophilic microorganisms. *Antonie van Leeuwenhoek* **66:**247–270.

Krebs, H. A., 1970. The history of the tricarboxylic acid cycle. *Perspectives in Biology and Medicine* **14:**154–170.

Krebs, H. A., 1981. *Reminiscences and Reflections.* Oxford, England: Oxford University Press.

Lowenstein, J. M., 1967. The tricarboxylic acid cycle. In *Metabolic Pathways,* 3rd ed., Greenberg, D., ed., vol. 1, pp. 146–270. New York: Academic Press.

Lowenstein, J. M., ed., 1969. *Citric Acid Cycle: Control and Compartmentation.* New York: Marcel Dekker.

Maden, B. E., 1995. No soup for starters? Autotrophy and the origins of metabolism. *Trends in Biochemical Sciences* **20:**337–341.

Newsholme, E. A., and Leech, A. R., 1983. *Biochemistry for the Medical Sciences.* New York: John Wiley & Sons.

Enzymes

Akiyama, S. K., and Hammes, G. G., 1980. Elementary steps in the reaction mechanism of the pyruvate dehydrogenase multienzyme complex from *Escherichia coli:* Kinetics of acetylation and deacetylation. *Biochemistry* **19:**4208–4213.

Akiyama, S. K., and Hammes, G. G., 1981. Elementary steps in the reaction mechanism of the pyruvate dehydrogenase multienzyme complex from *Escherichia coli:* Kinetics of flavin reduction. *Biochemistry* **20:**1491–1497.

Frey, P. A., 1982. Mechanism of coupled electron and group transfer in *Escherichia coli* pyruvate dehydrogenase. *Annals of the New York Academy of Sciences* **378:**250–264.

Srere, P. A., 1975. The enzymology of the formation and breakdown of citrate. *Advances in Enzymology* **43:**57–101.

Srere, P. A., 1987. Complexes of sequential metabolic enzymes. *Annual Review of Biochemistry* **56:**89–124.

Walsh, C., 1979. *Enzymatic Reaction Mechanisms.* San Francisco: W. H. Freeman.

Wiegand, G., and Remington, S. J., 1986. Citrate synthase: Structure, control and mechanism. *Annual Review of Biophysics and Biophysical Chemistry* **15:**97–117.

Regulation

Atkinson, D. E., 1977. *Cellular Energy Metabolism and Its Regulation.* New York: Academic Press.

Gibson, D., and Harris, R., 2001. *Metabolic Regulation in Mammals.* New York: Taylor and Francis.

Williamson, J. R., 1980. Mitochondrial metabolism and cell regulation. In *Mitochondria: Bioenergetics, Biogenesis and Membrane Structure,* Packer, L., and Gomez-Puyou, A., eds. New York: Academic Press.

Electron Transport and Oxidative Phosphorylation

Wall Piece #IV (1985), a kinetic sculpture by George Rhoads. This complex mechanical art form can be viewed as a metaphor for the molecular apparatus underlying electron transport and ATP synthesis by oxidative phosphorylation.

George Rhoads/Rock Stream Studios

In all things of nature there is something of the marvelous.
Aristotle (384–322 B.C.)

Key Questions

Biochemistry⊘Now™ Test yourself on these Key Questions at BiochemistryNow at http://chemistry.brookscole.com/ggb3

Essential Question

Living cells save up metabolic energy predominantly in the form of fats and carbohydrates, and they "spend" this energy for biosynthesis, membrane transport, and movement. In both directions, energy is exchanged and transferred in the form of ATP. In Chapters 18 and 19 we saw that glycolysis and the TCA cycle convert some of the energy available from stored and dietary sugars directly to ATP. However, most of the metabolic energy that is obtainable from substrates entering glycolysis and the TCA cycle is funneled via oxidation–reduction reactions into NADH and reduced flavoproteins, the latter symbolized by [FADH$_2$]. *How do cells oxidize NADH and [FADH$_2$] and convert their reducing potential into the chemical energy of ATP?*

Whereas ATP made in glycolysis and the TCA cycle is the result of substrate-level phosphorylation, NADH-dependent ATP synthesis is the result of **oxidative phosphorylation.** Electrons stored in the form of the reduced coenzymes, NADH or [FADH$_2$], are passed through an elaborate and highly organized chain of proteins and coenzymes, the so-called **electron-transport chain,** finally reaching O$_2$ (molecular oxygen), the terminal electron acceptor. Each component of the chain can exist in (at least) two oxidation states, and each component is successively reduced and reoxidized as electrons move through the chain from NADH (or [FADH$_2$]) to O$_2$. In the course of electron transport, a proton gradient is established across the inner mitochondrial membrane. It is the energy of this proton gradient that drives ATP synthesis.

20.1 Where in the Cell Are Electron Transport and Oxidative Phosphorylation Carried Out?

The processes of electron transport and oxidative phosphorylation are **membrane associated.** Bacteria are the simplest life form, and bacterial cells typically consist of a single cellular compartment surrounded by a plasma membrane and a more rigid cell wall. In such a system, the conversion of energy from NADH and [FADH$_2$] to the energy of ATP via electron transport and oxidative phosphorylation is carried out at (and across) the plasma membrane.

In eukaryotic cells, electron transport and oxidative phosphorylation are localized in mitochondria, which are also the sites of TCA cycle activity and (as we shall see in Chapter 23) fatty acid oxidation. Mammalian cells contain 800 to 2500 mitochondria; other types of cells may have as few as one or two or as many as half a million mitochondria. Human erythrocytes, whose purpose is simply to transport oxygen to tissues, contain no mitochondria at all. The typical mitochondrion is about 0.5 ± 0.3 micron in diameter and from 0.5 micron to several microns long; its overall shape is sensitive to metabolic conditions in the cell.

Mitochondrial Functions Are Localized in Specific Compartments

Mitochondria are surrounded by a simple **outer membrane** and a more complex **inner membrane** (Figure 20.1). The space between the inner and outer membranes is referred to as the **intermembrane space.** Several enzymes that utilize ATP (such as creatine kinase and adenylate kinase) are found in the intermembrane space. The smooth outer membrane is about 30% to 40% lipid and 60% to 70% protein and has a relatively high concentration of phosphatidylinositol. The outer membrane contains significant amounts of **porin—**

a transmembrane protein, rich in β-sheets, that forms large channels across the membrane, permitting free diffusion of molecules with molecular weights of about 10,000 or less. Apparently, the outer membrane functions mainly to maintain the shape of the mitochondrion. The inner membrane is richly packed with proteins, which account for nearly 80% of its weight; thus, its density is higher than that of the outer membrane. The fatty acids of inner membrane lipids are highly unsaturated. Cardiolipin and diphosphatidylglycerol (see Chapter 8) are abundant. The inner membrane lacks cholesterol and is quite impermeable to molecules and ions. Species that must cross the mitochondrial inner membrane—ions, substrates, fatty acids for oxidation, and so on—are carried by specific transport proteins in the membrane. Notably, the inner membrane is extensively folded (Figure 20.1). The folds, known as **cristae,** provide the inner membrane with a large surface area in a small volume. During periods of active respiration, the inner membrane appears to shrink significantly, leaving a comparatively large intermembrane space.

The Mitochondrial Matrix Contains the Enzymes of the TCA Cycle

The space inside the inner mitochondrial membrane is called the **matrix,** and it contains most of the enzymes of the TCA cycle and fatty acid oxidation. (An important exception, succinate dehydrogenase of the TCA cycle, is located in the inner membrane itself.) In addition, mitochondria contain circular DNA molecules, along with ribosomes and the enzymes required to synthesize proteins coded within the mitochondrial genome. Although some of the mitochondrial proteins are made this way, most are encoded by nuclear DNA and synthesized by cytosolic ribosomes.

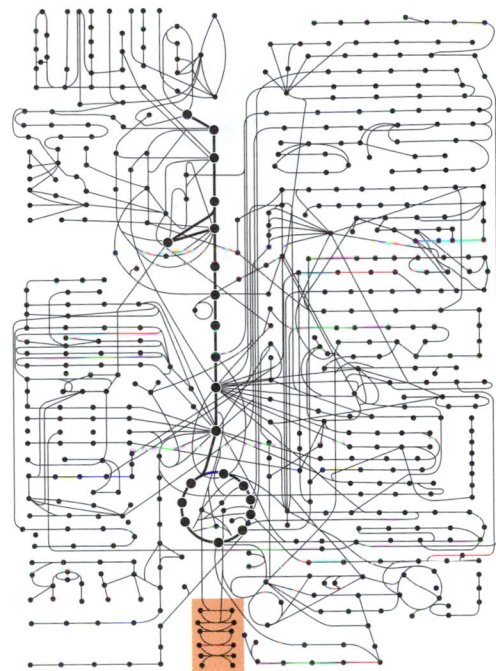

Electron transport and oxidative phosphorylation.

20.2	**What Are Reduction Potentials, and How Are They Used to Account for Free Energy Changes in Redox Reactions?**

On numerous occasions in earlier chapters, we have stressed that NADH and reduced flavoproteins ([FADH$_2$]) are forms of metabolic energy. These reduced coenzymes have a strong tendency to be oxidized—that is, to transfer electrons to other species. The electron-transport chain converts the energy of electron transfer into the energy of phosphoryl transfer stored in the phosphoric anhydride bonds of ATP. Just as the group transfer potential was used in Chapter 3 to

FIGURE 20.1 (a) A drawing of a mitochondrion with components labeled. (b) Tomography of a rat liver mitochondrion. The tubular structures in red, yellow, green, purple, and aqua represent individual cristae formed from the inner mitochondrial membrane. (b, Frey, T. G., and Mannella, C. A., 2000. *The internal structure of mitochondria.* Trends in Biochemical Sciences **25**:319–324.)

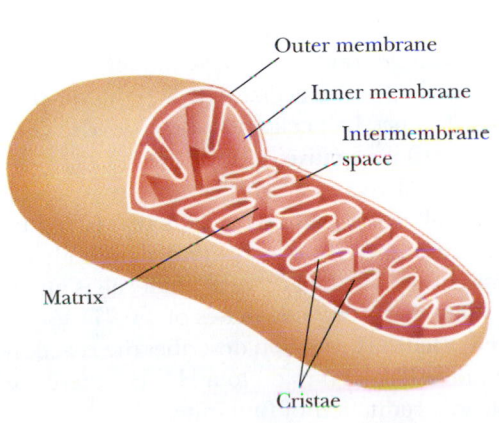

Outer membrane
Inner membrane
Intermembrane space
Matrix
Cristae

(a)

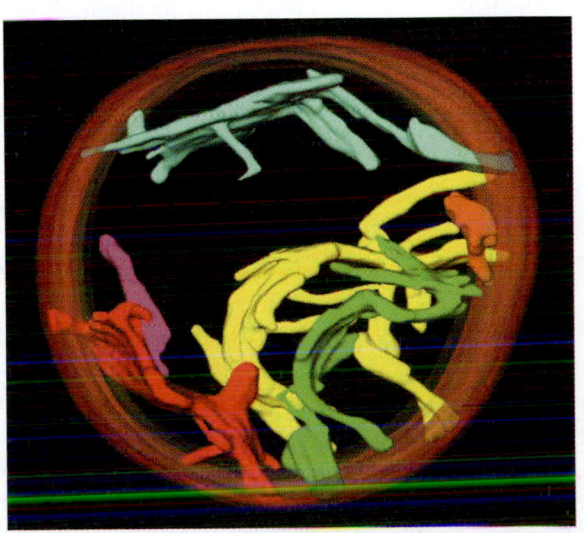

(b)

(a) Ethanol ⟶ acetaldehyde
−0.197 V

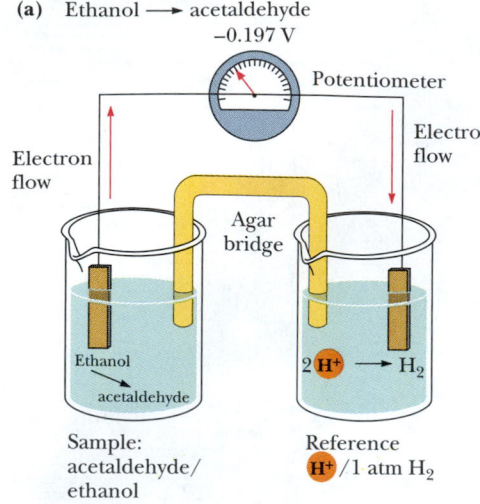

Sample:
acetaldehyde/
ethanol

Reference
H⁺/1 atm H₂

(b) Fumarate ⟶ succinate
+0.031 V

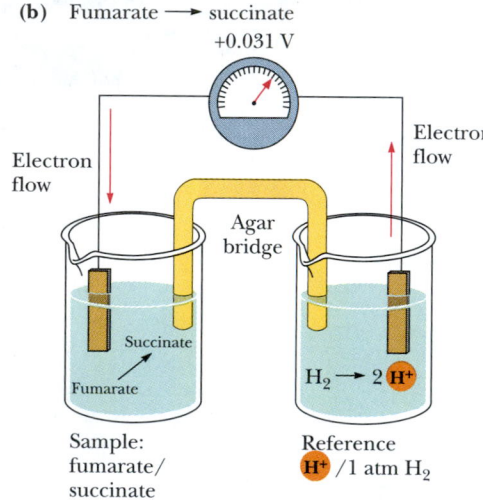

Sample:
fumarate/
succinate

Reference
H⁺/1 atm H₂

(c) Fe^{3+} ⟶ Fe^{2+}
+0.771 V

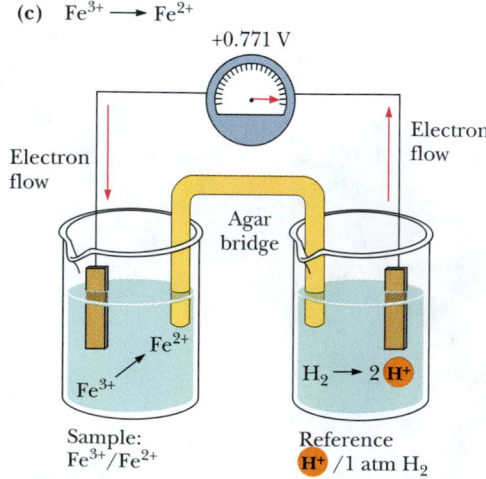

Sample:
Fe^{3+}/Fe^{2+}

Reference
H⁺/1 atm H₂

Biochemistry ⓔNow™ ACTIVE FIGURE 20.2
Experimental apparatus used to measure the standard reduction potential of the indicated redox couples: **(a)** the acetaldehyde/ethanol couple, **(b)** the fumarate/succinate couple, **(c)** the Fe^{3+}/Fe^{2+} couple. **Test yourself on the concepts in this figure at http://chemistry.brookscole.com/ggb3**

quantitate the energy of phosphoryl transfer, the **standard reduction potential,** denoted by $\mathscr{E}_o'$, quantitates the tendency of chemical species to be reduced or oxidized. The standard reduction potential difference describing electron transfer between two species,

Reduced donor ⟶ Oxidized acceptor
ne^-
Oxidized donor ⟵ Reduced acceptor

is related to the free energy change for the process by

$$\Delta G°' = -n\mathscr{F}\Delta\mathscr{E}_o' \qquad (20.2)$$

where n represents the number of electrons transferred; $\mathscr{F}$ is Faraday's constant, 96,485 J/V · mol; and $\Delta\mathscr{E}_o'$ is the difference in reduction potentials between the donor and acceptor. This relationship is straightforward, but it depends on a standard of reference by which reduction potentials are defined.

Standard Reduction Potentials Are Measured in Reaction Half-Cells

Standard reduction potentials are determined by measuring the voltages generated in **reaction half-cells** (Figure 20.2). A half-cell consists of a solution containing 1 M concentrations of both the oxidized and reduced forms of the substance whose reduction potential is being measured and a simple electrode. (Together, the oxidized and reduced forms of the substance are referred to as a **redox couple.**) Such a **sample half-cell** is connected to a **reference half-cell** and electrode via a conductive bridge (usually a salt-containing agar gel). A sensitive potentiometer (voltmeter) connects the two electrodes so that the electrical potential (voltage) between them can be measured. The reference half-cell normally contains 1 M H⁺ in equilibrium with H₂ gas at a pressure of 1 atm. The H⁺/H₂ reference half-cell is arbitrarily assigned a standard reduction potential of 0.0 V. The standard reduction potentials of all other redox couples are defined relative to the H⁺/H₂ reference half-cell on the basis of the sign and magnitude of the voltage (electromotive force, emf) registered on the potentiometer (Figure 20.2).

If electron flow between the electrodes is toward the sample half-cell, reduction occurs spontaneously in the sample half-cell and the reduction potential is said to be positive. If electron flow between the electrodes is away from the sample half-cell and toward the reference cell, the reduction potential is said to be negative because electron loss (oxidation) is occurring in the sample half-cell. Strictly speaking, the standard reduction potential, $\mathscr{E}_o'$, is the electromotive force generated at 25°C and pH 7.0 by a sample half-cell (containing 1 M concentrations of the oxidized and reduced species) with respect to a reference half-cell. (Note that the reduction potential of the hydrogen half-cell is pH-dependent. The standard reduction potential, 0.0 V, assumes 1 M H⁺. The hydrogen half-cell measured at pH 7.0 has an $\mathscr{E}_o'$ of −0.421 V.)

Several Examples Figure 20.2a shows a sample/reference half-cell pair for measurement of the standard reduction potential of the acetaldehyde/ethanol couple. Because electrons flow toward the reference half-cell and away from the sample half-cell, the standard reduction potential is negative, specifically −0.197 V. In contrast, the fumarate/succinate couple and the Fe^{3+}/Fe^{2+} couple both cause electrons to flow from the reference half-cell to the sample half-cell; that is, reduction occurs spontaneously in each system, and the reduction potentials of both are thus positive. The standard reduction potential for the Fe^{3+}/Fe^{2+} half-cell is much larger than that for the fumarate/succinate half-cell, with values of +0.771 V and +0.031 V, respectively. For each half-cell, a **half-cell reaction** describes the reaction taking place. For the fumarate/succinate half-cell coupled to a H⁺/H₂ reference half-cell, the reaction occurring is indeed a reduction of fumarate.

Fumarate + 2 H⁺ + 2 e^- ⟶ succinate $\mathscr{E}_o' = +0.031$ V   (20.3)

Similarly, for the Fe^{3+}/Fe^{2+} half-cell,

$$Fe^{3+} + e^- \longrightarrow Fe^{2+} \qquad \mathcal{E}_o' = +0.771 \text{ V} \qquad (20.4)$$

However, the reaction occurring in the acetaldehyde/ethanol half-cell is the oxidation of ethanol:

$$\text{Ethanol} \longrightarrow \text{acetaldehyde} + 2 \text{ H}^+ + 2 \text{ } e^- \qquad \mathcal{E}_o' = -0.197 \text{ V} \qquad (20.5)$$

$\mathcal{E}_o'$ Values Can Be Used to Predict the Direction of Redox Reactions

Some typical half-cell reactions and their respective standard reduction potentials are listed in Table 20.1. Whenever reactions of this type are tabulated, they are uniformly written as reduction reactions, regardless of what occurs in the given

Table 20.1

Standard Reduction Potentials for Several Biological Reduction Half-Reactions

Reduction Half-Reaction	$\mathcal{E}_o'$ (V)
$\frac{1}{2}O_2 + 2 \text{ H}^+ + 2 \text{ } e^- \longrightarrow H_2O$	0.816
$Fe^{3+} + e^- \longrightarrow Fe^{2+}$	0.771
Photosystem P700	0.430
$NO_3^- + 2 \text{ H}^+ + 2 \text{ } e^- \longrightarrow NO_2^- + H_2O$	0.421
Cytochrome f $(Fe^{3+}) + e^- \longrightarrow$ cytochrome f (Fe^{2+})	0.365
Cytochrome a_3 $(Fe^{3+}) + e^- \longrightarrow$ cytochrome a_3 (Fe^{2+})	0.350
Cytochrome a $(Fe^{3+}) + e^- \longrightarrow$ cytochrome a (Fe^{2+})	0.290
Rieske Fe-S $(Fe^{3+}) + e^- \longrightarrow$ Rieske Fe-S (Fe^{2+})	0.280
Cytochrome c $(Fe^{3+}) + e^- \longrightarrow$ cytochrome c (Fe^{2+})	0.254
Cytochrome c_1 $(Fe^{3+}) + e^- \longrightarrow$ cytochrome c_1 (Fe^{2+})	0.220
$UQH \cdot + \text{H}^+ + e^- \longrightarrow UQH_2$ (UQ = coenzyme Q)	0.190
$UQ + 2 \text{ H}^+ + 2 \text{ } e^- \longrightarrow UQH_2$	0.060
Cytochrome b_H $(Fe^{3+}) + e^- \longrightarrow$ cytochrome b_H (Fe^{2+})	0.050
Fumarate $+ 2 \text{ H}^+ + 2 \text{ } e^- \longrightarrow$ succinate	0.031
$UQ + \text{H}^+ + e^- \longrightarrow UQH \cdot$	0.030
Cytochrome b_5 $(Fe^{3+}) + e^- \longrightarrow$ cytochrome b_5 (Fe^{2+})	0.020
$[FAD] + 2 \text{ H}^+ + 2 \text{ } e^- \longrightarrow [FADH_2]$	0.003–0.091*
Cytochrome b_L $(Fe^{3+}) + e^- \longrightarrow$ cytochrome b_L (Fe^{2+})	−0.100
Oxaloacetate $+ 2 \text{ H}^+ + 2 \text{ } e^- \longrightarrow$ malate	−0.166
Pyruvate $+ 2 \text{ H}^+ + 2 \text{ } e^- \longrightarrow$ lactate	−0.185
Acetaldehyde $+ 2 \text{ H}^+ + 2 \text{ } e^- \longrightarrow$ ethanol	−0.197
$FMN + 2 \text{ H}^+ + 2 \text{ } e^- \longrightarrow FMNH_2$	−0.219
$FAD + 2 \text{ H}^+ + 2 \text{ } e^- \longrightarrow FADH_2$	−0.219
Glutathione (oxidized) $+ 2 \text{ H}^+ + 2 \text{ } e^- \longrightarrow 2$ glutathione (reduced)	−0.230
Lipoic acid $+ 2 \text{ H}^+ + 2 \text{ } e^- \longrightarrow$ dihydrolipoic acid	−0.290
1,3-Bisphosphoglycerate $+ 2 \text{ H}^+ + 2 \text{ } e^- \longrightarrow$ glyceraldehyde-3-phosphate $+ P_i$	−0.290
$NAD^+ + 2 \text{ H}^+ + 2 \text{ } e^- \longrightarrow NADH + \text{H}^+$	−0.320
$NADP^+ + 2 \text{ H}^+ + 2 \text{ } e^- \longrightarrow NADPH + \text{H}^+$	−0.320
Lipoyl dehydrogenase $[FAD] + 2 \text{ H}^+ + 2 \text{ } e^- \longrightarrow$ lipoyl dehydrogenase $[FADH_2]$	−0.340
α-Ketoglutarate $+ CO_2 + 2 \text{ H}^+ + 2 \text{ } e^- \longrightarrow$ isocitrate	−0.380
$2 \text{ H}^+ + 2 \text{ } e^- \longrightarrow H_2$	−0.421
Ferredoxin (spinach) $(Fe^{3+}) + e^- \longrightarrow$ ferredoxin (spinach) (Fe^{2+})	−0.430
Succinate $+ CO_2 + 2 \text{ H}^+ + 2 \text{ } e^- \longrightarrow \alpha$-ketoglutarate $+ H_2O$	−0.670

*Typical values for reduction of bound FAD in flavoproteins such as succinate dehydrogenase (see Bonomi, F., Pagani, S., Cerletti, P., and Giori, C., 1983. Modification of the thermodynamic properties of the electron-transferring groups in mitochondrial succinate dehydrogenase upon binding of succinate. *European Journal of Biochemistry* **134**:439–445).

half-cell. The sign of the standard reduction potential indicates which reaction really occurs when the given half-cell is combined with the reference hydrogen half-cell. Redox couples that have large positive reduction potentials have a strong tendency to accept electrons, and the oxidized form of such a couple (O_2, for example) is a strong oxidizing agent. Redox couples with large negative reduction potentials have a strong tendency to undergo oxidation (that is, donate electrons), and the reduced form of such a couple (NADPH, for example) is a strong reducing agent.

$\mathscr{E}_o'$ Values Can Be Used to Analyze Energy Changes of Redox Reactions

The half-reactions and reduction potentials in Table 20.1 can be used to analyze energy changes in redox reactions. The oxidation of NADH to NAD^+ can be coupled with the reduction of α-ketoglutarate to isocitrate:

$$NAD^+ + \text{isocitrate} \longrightarrow NADH + H^+ + \alpha\text{-ketoglutarate} + CO_2 \quad (20.6)$$

This is the isocitrate dehydrogenase reaction of the TCA cycle. Writing the two half-cell reactions, we have

$$NAD^+ + 2\,H^+ + 2\,e^- \longrightarrow NADH + H^+$$
$$\mathscr{E}_o' = -0.32\text{ V} \quad (20.7)$$

$$\alpha\text{-Ketoglutarate} + CO_2 + 2\,H^+ + 2\,e^- \longrightarrow \text{isocitrate}$$
$$\mathscr{E}_o' = -0.38\text{ V} \quad (20.8)$$

In a spontaneous reaction, electrons are donated by (flow away from) the half-reaction with the more negative reduction potential and are accepted by (flow toward) the half-reaction with the more positive reduction potential. Thus, in the present case, isocitrate donates electrons and NAD^+ accepts electrons. The convention defines $\Delta\mathscr{E}_o'$ as

$$\Delta\mathscr{E}_o' = \mathscr{E}_o'\,(\text{acceptor}) - \mathscr{E}_o'\,(\text{donor}) \quad (20.9)$$

In the present case, isocitrate is the donor and NAD^+ the acceptor, so we write

$$\Delta\mathscr{E}_o' = -0.32\text{ V} - (-0.38\text{ V}) = +0.06\text{ V} \quad (20.10)$$

From Equation 20.2, we can now calculate $\Delta G^{\circ\prime}$ as

$$\Delta G^{\circ\prime} = -\,(2)\,(96.485\text{ kJ/V} \cdot \text{mol})\,(0.06\text{ V}) \quad (20.11)$$

$$\Delta G^{\circ\prime} = -11.58\text{ kJ/mol}$$

Note that a reaction with a net positive $\Delta\mathscr{E}_o'$ yields a negative $\Delta G^{\circ\prime}$, indicating a spontaneous reaction.

The Reduction Potential Depends on Concentration

We have already noted that the standard free energy change for a reaction, $\Delta G^{\circ\prime}$, does not reflect the actual conditions in a cell, where reactants and products are not at standard-state concentrations (1 M). Equation 3.12 was introduced to permit calculations of actual free energy changes under non–standard-state conditions. Similarly, standard reduction potentials for redox couples must be modified to account for the actual concentrations of the oxidized and reduced species. For any redox couple,

$$\text{ox} + n e^- \rightleftharpoons \text{red} \quad (20.12)$$

the actual reduction potential is given by

$$\mathscr{E} = \mathscr{E}_o' + (RT/n\mathscr{F}) \ln \frac{[\text{ox}]}{[\text{red}]} \quad (20.13)$$

Reduction potentials can also be quite sensitive to molecular environment. The influence of environment is especially important for flavins, such as $FAD/FADH_2$ and $FMN/FMNH_2$. These species are normally bound to their respective flavoproteins; the reduction potential of bound FAD, for example, can be very different from the value shown in Table 20.1 for the free $FAD/FADH_2$ couple of -0.219 V. Problem 7 at the end of the chapter addresses this case.

<h2>20.3 | How Is the Electron-Transport Chain Organized?</h2>

As we have seen, the metabolic energy from oxidation of food materials—sugars, fats, and amino acids—is funneled into formation of reduced coenzymes (NADH) and reduced flavoproteins ($[FADH_2]$). The electron-transport chain reoxidizes the coenzymes and channels the free energy obtained from these reactions into the synthesis of ATP. This reoxidation process involves the removal of both protons and electrons from the coenzymes. Electrons move from NADH and $[FADH_2]$ to molecular oxygen, O_2, which is the terminal acceptor of electrons in the chain. The reoxidation of NADH,

$$\text{NADH (reductant)} + H^+ + O_2 \text{ (oxidant)} \longrightarrow NAD^+ + H_2O \quad (20.14)$$

involves the following half-reactions:

$$NAD^+ + 2\,H^+ + 2\,e^- \longrightarrow NADH + H^+ \qquad \mathscr{E}_o' = -0.32 \text{ V} \quad (20.15)$$

$$\tfrac{1}{2}\,O_2 + 2\,H^+ + 2\,e^- \longrightarrow H_2O \qquad\qquad \mathscr{E}_o' = +0.816 \text{ V} \quad (20.16)$$

Here, half-reaction 20.16 is the electron acceptor and half-reaction 20.15 is the electron donor. Then

$$\Delta\mathscr{E}_o' = 0.816 - (-0.32) = 1.136 \text{ V}$$

and, according to Equation 20.2, the standard-state free energy change, $\Delta G^{\circ\prime}$, is -219 kJ/mol. Molecules along the electron-transport chain have reduction potentials between the values for the $NAD^+/NADH$ couple and the oxygen/H_2O couple, so electrons move down the energy scale toward progressively more positive reduction potentials (Figure 20.3).

Although electrons move from more negative to more positive reduction potentials in the electron-transport chain, it should be emphasized that the electron carriers do not operate in a simple linear sequence. This will become evident when the individual components of the electron-transport chain are discussed in the following paragraphs.

The Electron-Transport Chain Can Be Isolated in Four Complexes

The electron-transport chain involves several different molecular species, including:

1. **Flavoproteins,** which contain tightly bound FMN or FAD as prosthetic groups and which (as noted in Chapter 17) may participate in one- or two-electron transfer events.
2. **Coenzyme Q,** also called **ubiquinone** (and abbreviated **CoQ** or **UQ**) (Figure 8.18), which can function in either one- or two-electron transfer reactions.
3. Several **cytochromes** (proteins containing heme prosthetic groups [see Chapter 5], which function by carrying or transferring electrons), including cytochromes b, c, c_1, a, and a_3. Cytochromes are one-electron transfer agents in which the heme iron is converted from Fe^{2+} to Fe^{3+} and back.
4. A number of **iron–sulfur proteins,** which participate in one-electron transfers involving the Fe^{2+} and Fe^{3+} states.
5. Protein-bound **copper,** a one-electron transfer site that converts between Cu^+ and Cu^{2+}.

All these intermediates except for cytochrome c are membrane associated (either in the mitochondrial inner membrane of eukaryotes or in the plasma membrane of prokaryotes). All three types of proteins involved in this

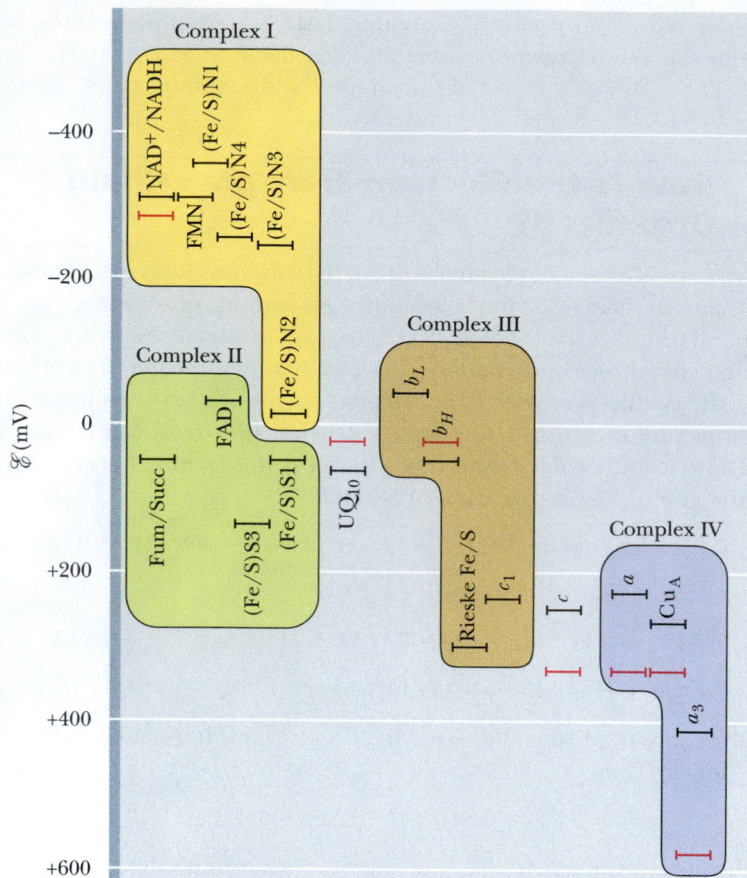

FIGURE 20.3 $\mathscr{E}_o'$ and $\mathscr{E}$ values for the components of the mitochondrial electron-transport chain. Values indicated are consensus values for animal mitochondria. Black bars represent $\mathscr{E}_o'$; red bars, $\mathscr{E}$.

chain—flavoproteins, cytochromes, and iron–sulfur proteins—possess electron-transferring **prosthetic groups.**

The components of the electron-transport chain can be purified from the mitochondrial inner membrane. Solubilization of the membranes containing the electron-transport chain results in the isolation of four distinct protein complexes, and the complete chain can thus be considered to be composed of four parts: (I) **NADH–coenzyme Q reductase,** (II) **succinate–coenzyme Q reductase,** (III) **coenzyme Q–cytochrome c reductase,** and (IV) **cytochrome c oxidase** (Figure 20.4). Complex I accepts electrons from NADH, serving as a link between glycolysis, the TCA cycle, fatty acid oxidation, and the electron-transport chain. Complex II includes succinate dehydrogenase and thus forms a direct link between the TCA cycle and electron transport. Complexes I and II produce a common product, reduced coenzyme Q (UQH_2), which is the substrate for coenzyme Q–cytochrome c reductase (Complex III). As shown in Figure 20.4, there are two other ways to feed electrons to UQ: the **electron-transferring flavoprotein,** which transfers electrons from the flavoprotein-linked step of fatty acyl-CoA dehydrogenase, and **sn-glycerophosphate dehydrogenase.** Complex III oxidizes UQH_2 while reducing cytochrome c, which in turn is the substrate for Complex IV, cytochrome c oxidase. Complex IV is responsible for reducing molecular oxygen. Each of the complexes shown in Figure 20.4 is a large multisubunit complex embedded within the inner mitochondrial membrane.

Complex I Oxidizes NADH and Reduces Coenzyme Q

As its name implies, this complex transfers a pair of electrons from NADH to coenzyme Q, a small, hydrophobic, yellow compound. Another common name for this enzyme complex is *NADH dehydrogenase.* The complex (with an estimated mass of 850 kD) involves more than 30 polypeptide chains, 1 molecule of flavin mononucleotide (FMN), and as many as seven Fe-S clusters, together containing

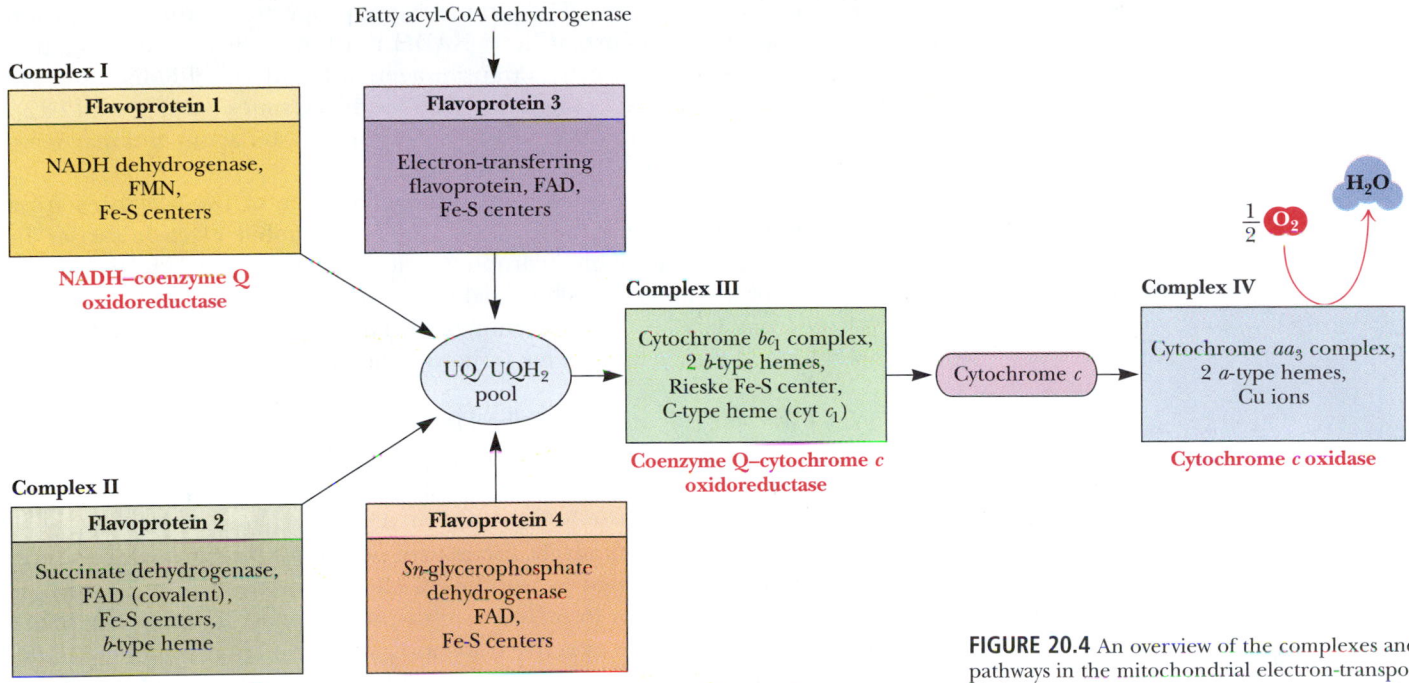

FIGURE 20.4 An overview of the complexes and pathways in the mitochondrial electron-transport chain. *(Adapted from Nicholls, D. G., and Ferguson, S. J., 2002. Bioenergetics 3. London: Academic Press.)*

a total of 20 to 26 iron atoms (Table 20.2). By virtue of its dependence on FMN, NADH–UQ reductase is a flavoprotein.

Although the precise mechanism of the NADH–UQ reductase is unknown, the first step involves binding of NADH to the enzyme on the matrix side of the inner mitochondrial membrane and transfer of electrons from NADH to tightly bound FMN:

$$\text{NADH} + [\text{FMN}] + \text{H}^+ \longrightarrow [\text{FMNH}_2] + \text{NAD}^+ \qquad (20.17)$$

The second step involves the transfer of electrons from the reduced $[\text{FMNH}_2]$ to a series of Fe-S proteins, including both 2Fe-2S and 4Fe-4S clusters

Table 20.2

Protein Complexes of the Mitochondrial Electron-Transport Chain

Complex	Mass (kD)	Subunits	Prosthetic Group	Binding Site for:
NADH–UQ reductase	850	>30	FMN Fe-S	NADH (matrix side) UQ (lipid core)
Succinate–UQ reductase	140	4	FAD Fe-S	Succinate (matrix side) UQ (lipid core)
UQ–Cyt c reductase	250	9–10	Heme b_L Heme b_H Heme c_1 Fe-S	Cyt c (intermembrane space side)
Cytochrome c	13	1	Heme c	Cyt c_1 Cyt a
Cytochrome c oxidase	162	>10	Heme a Heme a_3 Cu_A Cu_B	Cyt c (intermembrane space side)

Adapted from Hatefi, Y., 1985. The mitochondrial electron transport chain and oxidative phosphorylation system. *Annual Review of Biochemistry* **54**:1015–1069; and DePierre, J., and Ernster, L., 1977. Enzyme topology of intracellular membranes. *Annual Review of Biochemistry* **46**:201–262.

(see Figures 20.8 and 20.16). The unique redox properties of the flavin group of FMN are probably important here. NADH is a two-electron donor, whereas the Fe-S proteins are one-electron transfer agents. The flavin of FMN has three redox states—the oxidized, semiquinone, and reduced states (see Figure 17.22). It can act as either a one-electron or a two-electron transfer agent and may serve as a critical link between NADH and the Fe-S proteins.

The final step of the reaction involves the transfer of two electrons from iron–sulfur clusters to coenzyme Q. Coenzyme Q is a **mobile electron carrier.** Its isoprenoid tail makes it highly hydrophobic, and it diffuses freely in the hydrophobic core of the inner mitochondrial membrane. As a result, it shuttles electrons from Complexes I and II to Complex III. The redox cycle of UQ is shown in Figure 20.5, and the overall scheme is shown schematically in Figure 20.6.

Complex I Transports Protons from the Matrix to the Cytosol The oxidation of one NADH and the reduction of one UQ by NADH–UQ reductase results in the net transport of protons from the matrix side to the cytosolic side of the inner membrane. The cytosolic side, where H^+ accumulates, is referred to as the **P** (for positive) face; similarly, the matrix side is the **N** (for negative) face. Some of the energy liberated by the flow of electrons through this complex is used in a coupled process to drive the transport of protons across the membrane. (This is an example of active transport, a phenomenon examined in detail in Chapter 9.) Available experimental evidence suggests a stoichiometry of four H^+ transported per two electrons passed from NADH to UQ.

Complex II Oxidizes Succinate and Reduces Coenzyme Q

Complex II is perhaps better known by its other name—succinate dehydrogenase, the only TCA cycle enzyme that is an integral membrane protein in the inner mitochondrial membrane. This enzyme has a mass of approxi-

Human Biochemistry

Solving a Medical Mystery Revolutionized Our Treatment of Parkinson's Disease

A tragedy among illegal drug users was the impetus for a revolutionary treatment of Parkinson's disease. In 1982, several mysterious cases of paralysis came to light in southern California. The victims, some of them teenagers, were frozen like living statues, unable to talk or move. The case was baffling at first, but it was soon traced to a batch of synthetic heroin that contained MPTP (1-methyl-4-phenyl-1,2,3,6-tetrahydropyridine) as a contaminant. MPTP is rapidly converted in the brain to MPP$^+$ (1-methyl-4-phenylpyridine) by the enzyme monoamine oxidase B. MPP$^+$ is a potent inhibitor of mitochondrial Complex I (NADH–UQ reductase), and it acts preferentially in the *substantia nigra,* an area of the brain that is essential to movement and also the region of the brain that deteriorates slowly in Parkinson's disease.

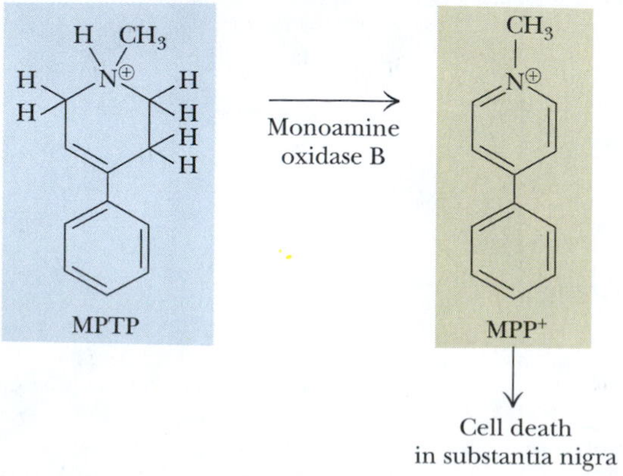

Parkinson's disease results from the inability of the brain to produce sufficient quantities of dopamine, a neurotransmitter. Neurologist J. William Langston, asked to consult on the treatment of some of these patients, recognized that the symptoms of this drug-induced disorder were in fact similar to those of parkinsonism. He began treatment of the patients with L-dopa, which is decarboxylated in the brain to produce dopamine. The treated patients immediately regained movement. Langston then took a bold step. He implanted fetal brain tissue into the brains of several of the affected patients, prompting substantial recovery from the Parkinson-like symptoms. Langston's innovation sparked a revolution in the use of tissue implantation for the treatment of neurodegenerative diseases.

Other toxins may cause similar effects in neural tissue. Timothy Greenmyre at Emory University has shown that rats exposed to the pesticide rotenone (see Figure 20.29) over a period of weeks experience a gradual loss of function in dopaminergic neurons and then develop symptoms of parkinsonism, including limb tremors and rigidity. This finding supports earlier research that links long-term pesticide exposure to Parkinson's disease.

MPTP

Monoamine oxidase B

MPP$^+$

Cell death in substantia nigra

(a)

Coenzyme Q, oxidized form
(Q, ubiquinone)

e^- + H^+

Semiquinone
intermediate
(QH·)

e^- + H^+

Coenzyme Q,
reduced form
(QH₂, ubiquinol)

(b)

FIGURE 20.5 **(a)** The three oxidation states of coenzyme Q. **(b)** A space-filling model of coenzyme Q.

mately 100 to 140 kD and is composed of four subunits: two Fe-S proteins of masses 70 and 27 kD, and two other peptides of masses 15 and 13 kD. Also known as flavoprotein 2 (FP₂), it contains an FAD covalently bound to a histidine residue (see Figure 20.15), and three Fe-S centers: a 4Fe-4S cluster, a 3Fe-4S cluster, and a 2Fe-2S cluster. When succinate is converted to fumarate in the TCA cycle, concomitant reduction of bound FAD to FADH₂ occurs in succinate dehydrogenase. This FADH₂ transfers its electrons

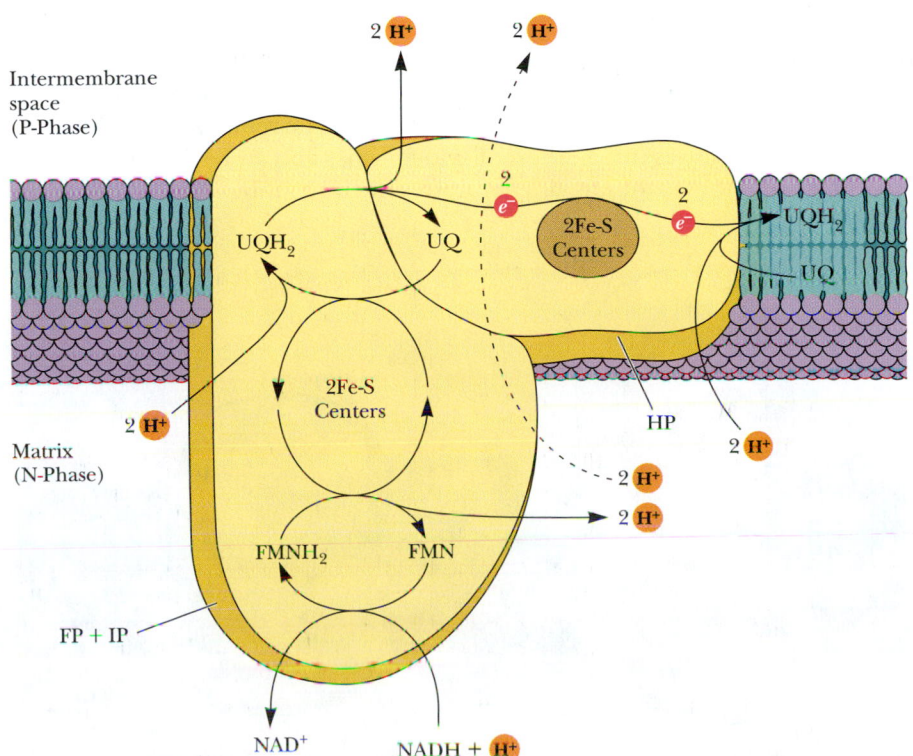

Biochemistry Now™ ACTIVE FIGURE 20.6
Proposed structure and electron-transport pathway for Complex I. Three protein complexes have been isolated, including the **flavoprotein (FP), iron–sulfur protein (IP),** and **hydrophobic protein (HP).** FP contains three peptides (of masses 51, 24, and 10 kD) and bound FMN and has 2 Fe-S centers (a 2Fe-2S center and a 4Fe-4S center). IP contains six peptides and at least three Fe-S centers. HP contains at least seven peptides and one Fe-S center. Note: Although the L-shape of Complex I shown here is purely schematic, there is evidence from structural analysis of Complex I from *E. coli* that the complex is in fact L-shaped. *(Sazanov, L., Carroll, J., Holt, P., Toime, L., and Fearnley, I., 2003. A role for native lipids in the stabilization and two-dimensional crystallization of the* Escherichia coli *NADH–ubiquinone oxidoreductase (Complex I).* Journal of Biological Chemistry *278:19483–19491.)* **Test yourself on the concepts in this figure at http://chemistry.brookscole. com/ggb3**

FIGURE 20.7 The fatty acyl-CoA dehydrogenase reaction, emphasizing that the reaction involves reduction of enzyme-bound FAD (indicated by brackets).

immediately to Fe-S centers, which pass them on to UQ. Electron flow from succinate to UQ,

$$\text{Succinate} \longrightarrow \text{fumarate} + 2\,H^+ + 2\,e^- \tag{20.18}$$

$$UQ + 2\,H^+ + 2\,e^- \longrightarrow UQH_2 \tag{20.19}$$

Net rxn: Succinate + UQ $\longrightarrow$ fumarate + UQH$_2$ $\Delta\mathscr{E}_o' = 0.029\ \text{V}$ (20.20)

yields a net reduction potential of 0.029 V. (Note that the first half-reaction is written in the direction of the e^- flow. As always, $\Delta\mathscr{E}_o'$ is calculated according to Equation 20.9.) The small free energy change of this reaction is not sufficient to drive the transport of protons across the inner mitochondrial membrane.

This is a crucial point because (as we will see) proton transport is coupled with ATP synthesis. Oxidation of one FADH$_2$ in the electron-transport chain results in synthesis of approximately two molecules of ATP, compared with the approximately three ATPs produced by the oxidation of one NADH. Other enzymes can also supply electrons to UQ, including mitochondrial *sn*-glycerophosphate dehydrogenase, an inner membrane-bound shuttle enzyme, and the fatty acyl-CoA dehydrogenases, three soluble matrix enzymes involved in fatty acid oxidation (Figure 20.7; also see Chapter 23). The path of electrons from succinate to UQ is shown in Figure 20.8.

Complex III Mediates Electron Transport from Coenzyme Q to Cytochrome *c*

In the third complex of the electron-transport chain, reduced coenzyme Q (UQH$_2$) passes its electrons to cytochrome *c* via a unique redox pathway known as the **Q cycle**. UQ–cytochrome *c* reductase (UQ–cyt *c* reductase), as this complex is known, involves three different cytochromes and an Fe-S protein. In the cytochromes of these and similar complexes, the iron atom at the center of the

Biochemistry❋Now™ ACTIVE FIGURE 20.8
A probable scheme for electron flow in Complex II. Oxidation of succinate occurs with reduction of [FAD]. Electrons are then passed to Fe-S centers and then to coenzyme Q (UQ). Proton transport does not occur in this complex. **Test yourself on the concepts in this figure at http://chemistry.brookscole. com/ggb3**

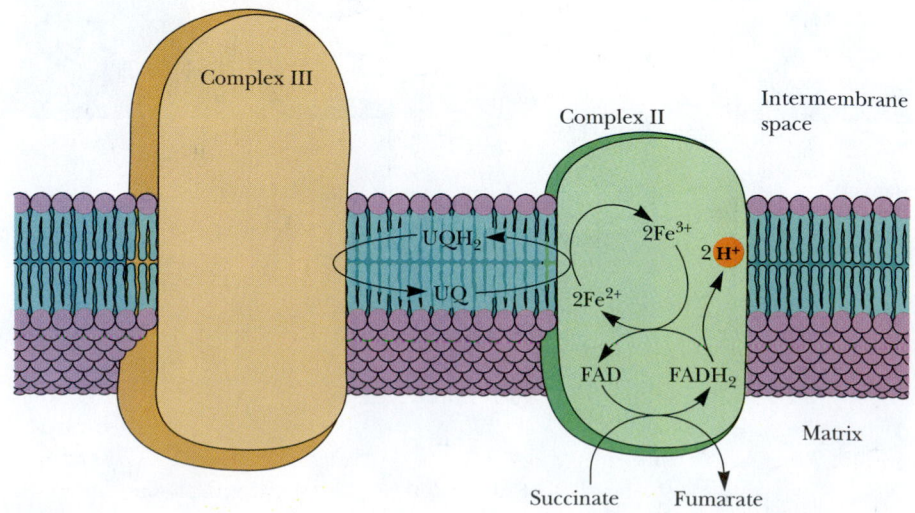

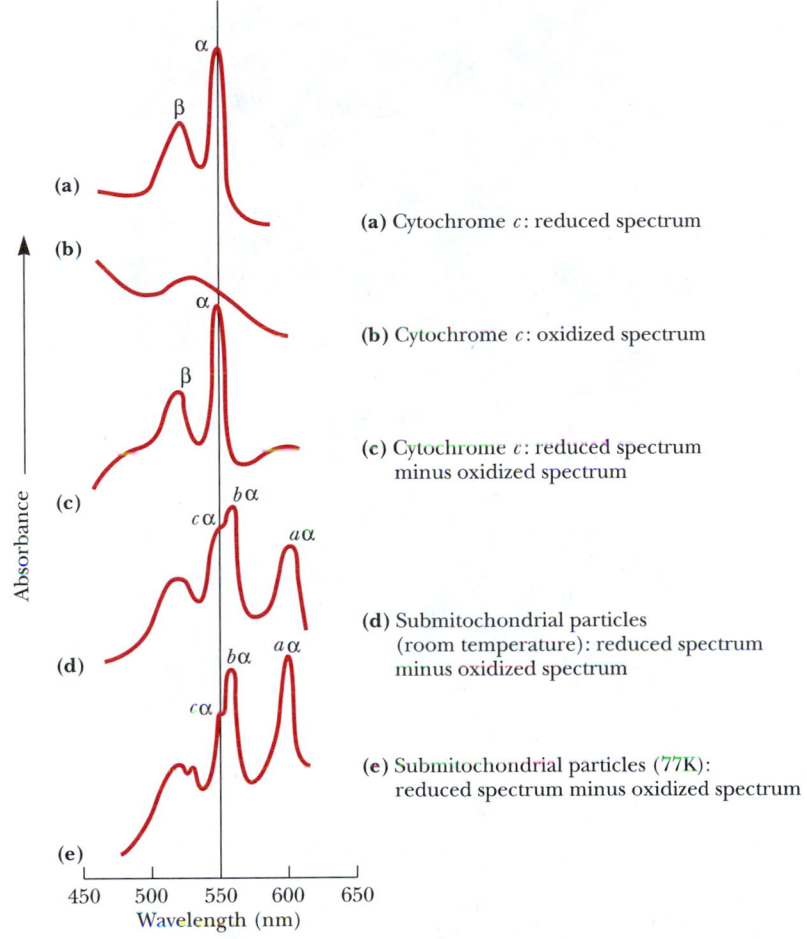

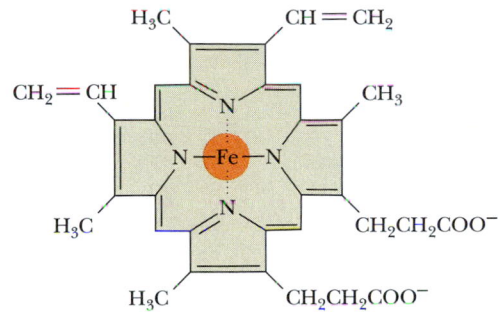

Iron protoporphyrin IX
(found in cytochrome b,
myoglobin, and hemoglobin)

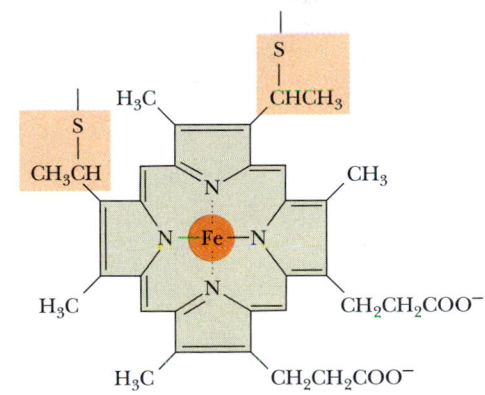

Heme c
(found in cytochrome c)

FIGURE 20.9 Typical visible absorption spectra of cytochromes. **(a)** Cytochrome c, reduced spectrum; **(b)** cytochrome c, oxidized spectrum; **(c)** the difference spectrum: (a) minus (b); **(d)** beef heart mitochondrial particles: room temperature difference (reduced minus oxidized) spectrum; **(e)** beef heart submitochondrial particles: same as (d) but at 77 K. α- and β-bands are labeled, and in (d) and (e) the bands for cytochromes a, b, and c are indicated.

porphyrin ring cycles between the reduced Fe^{2+} (ferrous) and oxidized Fe^{3+} (ferric) states.

Cytochromes were first named and classified on the basis of their absorption spectra (Figure 20.9), which depend upon the structure and environment of their heme groups. The **b cytochromes** contain *iron protoporphyrin IX* (Figure 20.10), the same heme found in hemoglobin and myoglobin. The **c cytochromes** contain *heme c*, derived from iron protoporphyrin IX by the covalent attachment of cysteine residues from the associated protein. UQ–cyt c reductase contains a b-type cytochrome, of 30 to 40 kD, with two different heme sites (Figure 20.11) and one c-type cytochrome. (One other variation, heme a, contains a 15-carbon isoprenoid chain on a modified vinyl group and a formyl group in place of one of the methyls [see Figure 20.10]. **Cytochrome a** is found in two forms in Complex IV of the electron-transport chain, as we shall see.) The two hemes on the b cytochrome polypeptide in UQ–cyt c reductase are distinguished by their reduction potentials and the wavelength (λ_{max}) of the so-called α-band (see Figure 20.9). One of these hemes, known as b_L or b_{566}, has a standard reduction potential, $\mathscr{E}_o'$, of -0.100 V and a wavelength of maximal absorbance (λ_{max}) of 566 nm. The other, known as b_H or b_{562}, has a standard reduction potential of $+0.050$ V and a λ_{max} of 562 nm. (H and L here refer to high and low reduction potentials.)

The structure of the UQ–cyt c reductase, also known as the **cytochrome bc_1 complex,** has been determined by Johann Deisenhofer and his colleagues. (Deisenhofer was a co-recipient of the Nobel Prize in Chemistry for his work

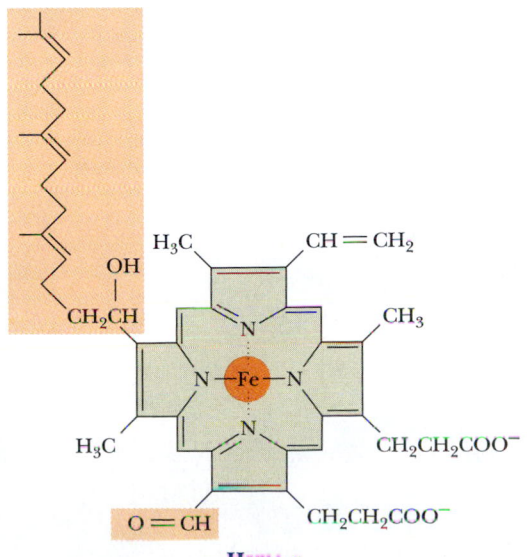

Heme a
(found in cytochrome a)

FIGURE 20.10 The structures of iron protoporphyrin IX, heme c, and heme a.

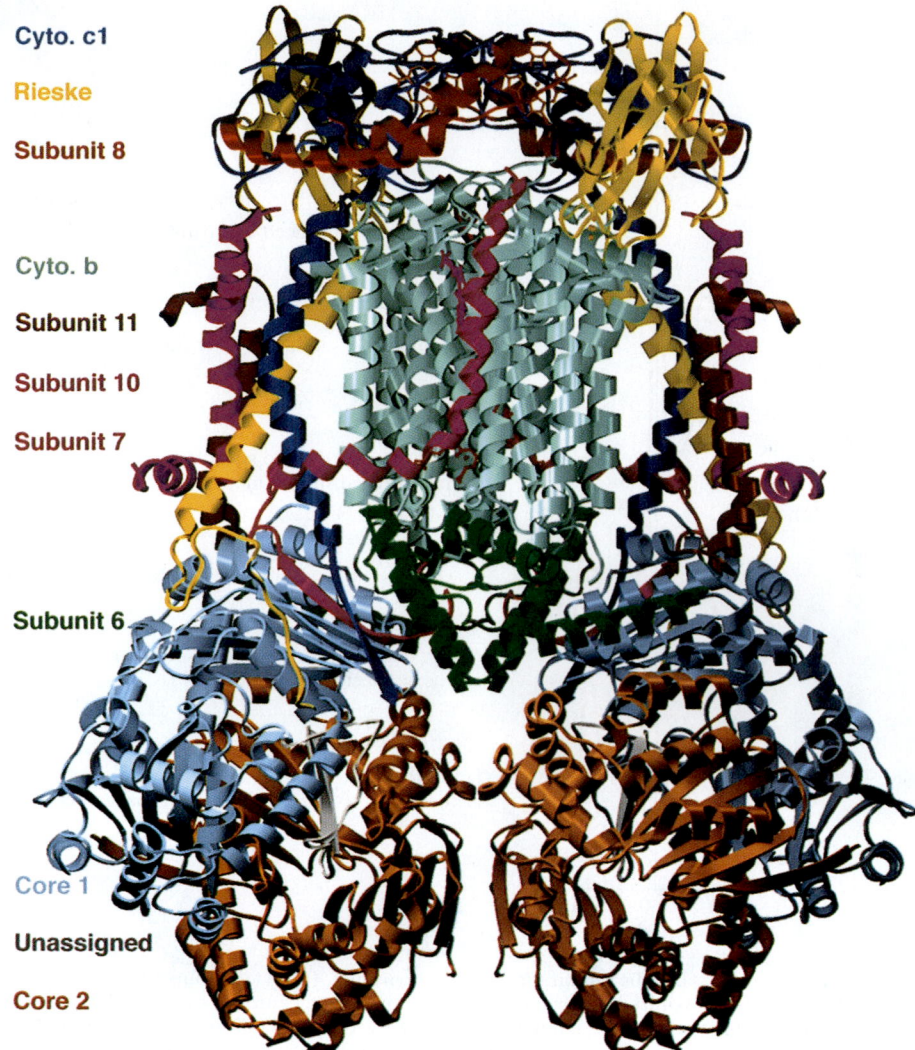

FIGURE 20.11 The structure of UQ–cyt c reductase, also known as the cytochrome bc_1 complex. The α-helices of cytochrome b (pale green) define the transmembrane domain of the protein. The bottom of the structure as shown extends approximately 75 Å into the mitochondrial matrix, and the top of the structure as shown extends about 38 Å into the intermembrane space. *(Photograph kindly provided by Di Xia and Johann Deisenhofer [From Xia, D., Yu, C-A., Kim, H., Xia, J-Z., Kachurin, A. M., Zhang, L., Yu, L., and Deisenhofer, J., 1997. The crystal structure of the cytochrome* bc₁ *complex from bovine heart mitochondria.* Science **277**:60–66.])

Cyto. c1

Rieske

Subunit 8

Cyto. b

Subunit 11

Subunit 10

Subunit 7

Subunit 6

Core 1

Unassigned

Core 2

on the structure of a photosynthetic reaction center [see Chapter 21]). The complex is a dimer, with each monomer consisting of 11 protein subunits and 2165 amino acid residues (monomer mass, 248 kD). The dimeric structure is pear-shaped and consists of a large domain that extends 75 Å into the mitochondrial matrix, a transmembrane domain consisting of 13 transmembrane α-helices in each monomer and a small domain that extends 38 Å into the intermembrane space (Figure 20.11). Most of the **Rieske protein** (an Fe-S protein named for its discoverer) is mobile in the crystal (only 62 of 196 residues are shown in the structure in Figure 20.11), and Deisenhofer has postulated that mobility of this subunit could be required for electron transfer in the function of this complex.

Complex III Drives Proton Transport As with Complex I, passage of electrons through the Q cycle of Complex III is accompanied by proton transport across the inner mitochondrial membrane. The postulated pathway for electrons in this system is shown in Figure 20.12. A large pool of UQ and UQH_2 exists in the inner mitochondrial membrane. The Q cycle is initiated when a molecule of UQH_2 from this pool diffuses to a site (called Q_p) on Complex III near the cytosolic face of the membrane.

Oxidation of this UQH_2 occurs in two steps. First, an electron from UQH_2 is transferred to the Rieske protein and then to cytochrome c_1. This releases two H^+ to the cytosol and leaves $UQ \cdot {}^-$, a semiquinone anion form of UQ, at the Q_p

(a) First half of Q cycle

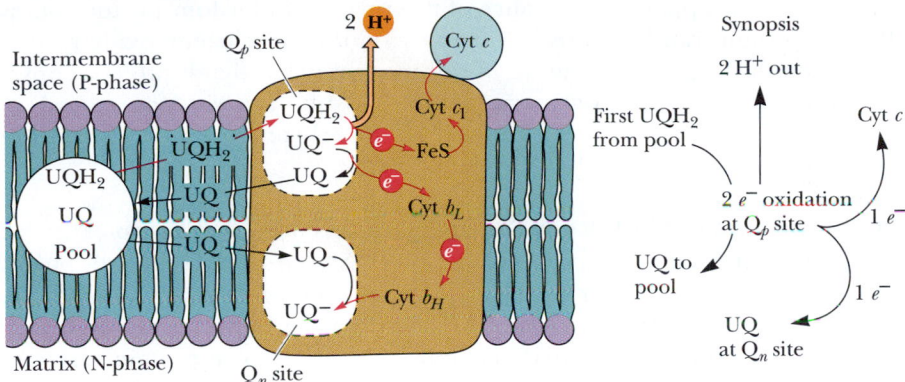

(b) Second half of Q cycle

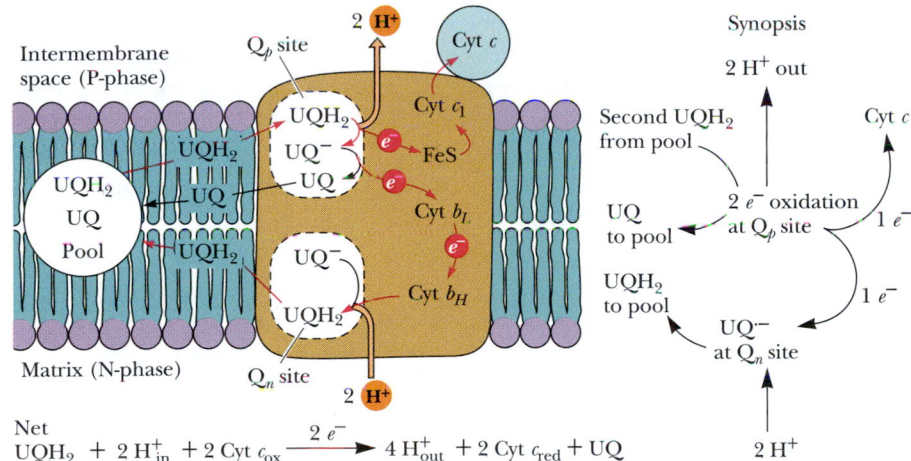

Net
$$UQH_2 + 2 H^+_{in} + 2\ Cyt\ c_{ox} \xrightarrow{2\ e^-} 4\ H^+_{out} + 2\ Cyt\ c_{red} + UQ$$

Biochemistry **Now**™   **ACTIVE FIGURE 20.12**
The Q cycle in mitochondria. **(a)** The electron-transfer pathway following oxidation of the first UQH_2 at the Q_p site near the cytosolic face of the membrane. **(b)** The pathway following oxidation of a second UQH_2. **Test yourself on the concepts in this figure at http://chemistry.brookscole.com/ggb3**

site. The second electron is then transferred to the b_L heme, converting $UQ\cdot^-$ to UQ. The Rieske protein and cytochrome c_1 are similar in structure; each has a globular domain and is anchored to the inner mitochondrial membrane by a hydrophobic segment. However, the hydrophobic segment is N-terminal in the Rieske protein and C-terminal in cytochrome c_1.

The electron on the b_L heme facing the cytosolic side of the membrane is now passed to the b_H heme on the matrix side of the membrane. This electron transfer occurs against a membrane potential of 0.15 V and is driven by the loss of redox potential as the electron moves from b_L ($\mathscr{E}_o' = -0.100$ V) to b_H ($\mathscr{E}_o' = +0.050$ V). The electron is then passed from b_H to a molecule of UQ at a second quinone-binding site, $\mathbf{Q}_n$, converting this UQ to $UQ\cdot^-$. The resulting $UQ\cdot^-$ remains firmly bound to the Q_n site. This completes the first half of the Q cycle (Figure 20.12a).

The second half of the cycle (Figure 20.12b) is similar to the first half, with a second molecule of UQH_2 oxidized at the Q_p site, one electron being passed to cytochrome c_1 and the other transferred to heme b_L and then to heme b_H. In this latter half of the Q cycle, however, the b_H electron is transferred to the semiquinone anion, $UQ\cdot^-$, at the Q_n site. With the addition of two H^+ from the mitochondrial matrix, this produces a molecule of UQH_2, which is released from the Q_n site and returns to the coenzyme Q pool, completing the Q cycle.

The Q Cycle Is an Unbalanced Proton Pump   Why has nature chosen this rather convoluted path for electrons in Complex III? First of all, Complex III takes up two protons on the matrix side of the inner membrane and releases four

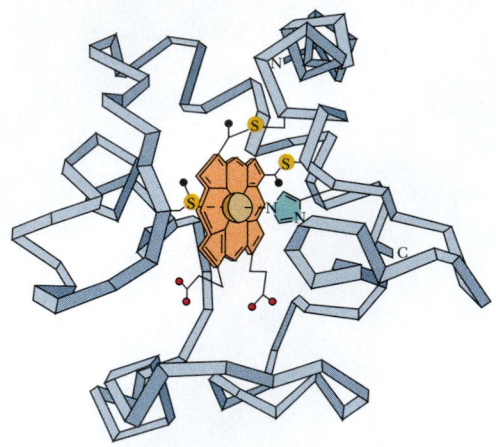

FIGURE 20.13 The structure of mitochondrial cyto-chrome *c*. The heme is shown at the center of the structure, covalently linked to the protein via its two sulfur atoms (yellow). A third sulfur from a methionine residue coordinates the iron.

protons on the cytoplasmic side for each pair of electrons that passes through the Q cycle. The apparent imbalance of two protons in for four protons out is offset by proton translocations in Complex IV, the cytochrome oxidase complex. The other significant feature of this mechanism is that it offers a convenient way for a two-electron carrier, UQH_2, to interact with the b_L and b_H hemes, the Rieske protein Fe-S cluster, and cytochrome c_1, all of which are one-electron carriers.

Cytochrome *c* Is a Mobile Electron Carrier Electrons traversing Complex III are passed through cytochrome c_1 to cytochrome *c*. Cytochrome *c* is the only one of the cytochromes that is water soluble. Its structure, determined by X-ray crystallography (Figure 20.13), is globular; the planar heme group lies near the center of the protein, surrounded predominantly by hydrophobic protein residues. The iron in the porphyrin ring is coordinated both to a histidine nitrogen and to the sulfur atom of a methionine residue. Coordination with ligands in this manner on both sides of the porphyrin plane precludes the binding of oxygen and other ligands, a feature that distinguishes the cytochromes from hemoglobin (see Chapter 15).

Cytochrome *c*, like UQ, is a mobile electron carrier. It associates loosely with the inner mitochondrial membrane (in the intermembrane space on the cytosolic side of the inner membrane) to acquire electrons from the Fe-S–cyt c_1 aggregate of Complex III, and then it migrates along the membrane surface in the reduced state, carrying electrons to cytochrome *c* oxidase, the fourth complex of the electron-transport chain.

Complex IV Transfers Electrons from Cytochrome *c* to Reduce Oxygen on the Matrix Side

Complex IV is called cytochrome *c* oxidase because it accepts electrons from cytochrome *c* and directs them to the four-electron reduction of O_2 to form H_2O:

$$4 \text{ cyt } c \text{ (Fe}^{2+}) + 4 \text{ H}^+ + O_2 \longrightarrow 4 \text{ cyt } c \text{ (Fe}^{3+}) + 2 \text{ H}_2O \qquad (20.21)$$

Thus, O_2 and cytochrome *c* oxidase are the final destination for the electrons derived from the oxidation of food materials. In concert with this process, cytochrome *c* oxidase also drives transport of protons across the inner mitochondrial membrane. These important functions are carried out by a transmembrane protein complex consisting of more than ten subunits (Table 20.2).

An electrophoresis gel of the bovine heart complex is shown in Figure 20.14. The total mass of the protein in the complex, composed of 13 subunits, is 204 kD. Subunits I through III, the largest ones, are encoded by mitochondrial DNA, synthesized in the mitochondrion, and inserted into the inner membrane from the matrix side. The smaller subunits are coded by nuclear DNA and synthesized in the cytosol.

The structure of cytochrome *c* oxidase has been solved. The essential Fe and Cu sites are contained entirely within the structures of subunits I, II, and III. None of the ten nuclear DNA–derived subunits directly impinges on the essential metal sites. The implication is that subunits I to III actively participate in the events of electron transfer but that the other ten subunits play regulatory roles in this process. Subunit I is cylindrical in shape and consists of 12 transmembrane helices, without any significant extramembrane parts (Figure 20.15). Hemes *a* and a_3, which lie perpendicular to the membrane plane, are cradled by the helices of subunit I. Subunits II and III lie on opposite sides of subunit I and do not contact each other. Subunit II has an extramembrane domain on the outer face of the inner mitochondrial membrane. This domain consists of a ten-strand β-barrel that holds Cu_A 7 Å from the nearest surface atom of the subunit. Subunit III consists of seven transmembrane helices with no significant extramembrane domains. Figure 20.16 presents a molecular graphic image of cytochrome *c* oxidase.

I

II
III

IV

Vab

VIa
VIb
VIc

VIIa
VIIbc
VIII

Photo kindly provided by Professor Roderick Capaldi

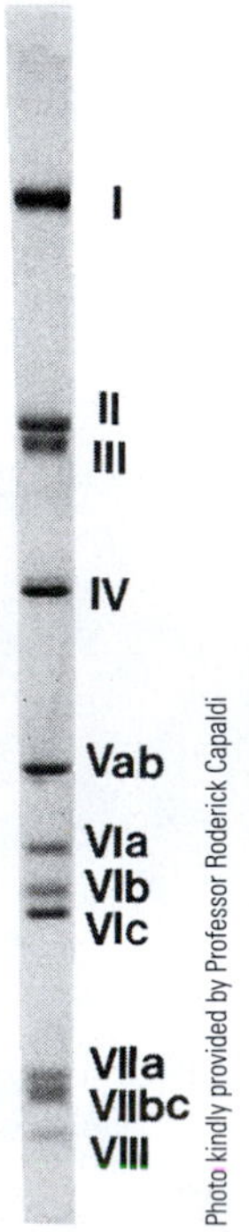

FIGURE 20.14 An electrophoresis gel showing the complex subunit structure of bovine heart cytochrome *c* oxidase. The three largest subunits, I, II, and III, are coded for by mitochondrial DNA. The others are encoded by nuclear DNA.

Electron Transfer in Complex IV Involves Two Hemes and Two Copper Sites   Cytochrome c oxidase contains two heme centers (cytochromes a and a_3) as well as two copper atoms (Figure 20.17). The copper sites, Cu_A and Cu_B, are associated with cytochromes a and a_3, respectively. The copper sites participate in electron transfer by cycling between the reduced (cuprous) Cu^+ state and the oxidized (cupric) Cu^{2+} state. (Remember, the cytochromes and copper sites are one-electron transfer agents.) Reduction of one oxygen molecule requires passage of four electrons through these carriers—one at a time (Figure 20.17).

Electrons from cytochrome c are transferred to Cu_A sites and then passed to the heme iron of cytochrome a. Cu_A is liganded by two cysteines and two histidines (Figure 20.18). The heme of cytochrome a is liganded by imidazole rings of histidine residues (Figure 20.18). The Cu_A and the Fe of cytochrome a are within 1.5 nm of each other.

Cu_B and the iron atom of cytochrome a_3 are also situated close to each other and are thought to share a ligand, which may be a cysteine sulfur (Figure 20.19). This closely associated pair of metal ions is referred to as a **binuclear center.**

As shown in Figure 20.20, the electron pathway through Complex IV continues as Cu_B accepts a single electron from cytochrome a (state O→state H). A second electron then reduces the iron center to Fe^{2+} (H→R), leading to the binding of O_2 (R→A) and the formation of a peroxy bridge between heme a_3 and Cu_B (A→P). *This amounts to the transfer of two electrons from the binuclear center to the bound O_2.* The next step involves uptake of two H^+ and a third electron (P→F), which leads to cleavage of the O—O bond and generation of an unusual Fe^{4+} state at the heme. Uptake of a fourth e^- facilitates formation of ferric hydroxide at the heme center (F→O'). In the final step of the cycle (O'→O), protons from the mitochondrial matrix are accepted by the coordinated hydroxyl groups, and the resulting water molecules dissociate from the binuclear center.

Complex IV Also Transports Protons Across the Inner Mitochondrial Membrane   The reduction of oxygen in Complex IV is accompanied by transport of protons across the inner mitochondrial membrane. Transfer of four electrons through this complex drives the transport of approximately four protons. The mechanism of proton transport is unknown but is thought to involve the steps from state P to state O (Figure 20.20). Four protons are taken up on the matrix side for every two protons transported to the cytoplasm (see Figure 20.17).

The Four Electron-Transport Complexes Are Independent

It should be emphasized here that the four major complexes of the electron-transport chain operate quite independently in the inner mitochondrial membrane. Each is a multiprotein aggregate maintained by numerous strong associations between peptides of the complex, but there is no evidence that the complexes associate with one another in the membrane. Measurements of the lateral diffusion rates of the four complexes, of coenzyme Q, and of cytochrome c in the inner mitochondrial membrane show that the rates differ considerably, indicating that these complexes do not move together in the membrane. Kinetic studies with reconstituted systems show that electron transport does not operate by means of connected sets of the four complexes.

The Model of Electron Transport Is a Dynamic One   The model that emerges for electron transport is shown in Figure 20.21. The four complexes are independently mobile in the membrane. Coenzyme Q collects electrons from NADH–UQ reductase and succinate–UQ reductase and delivers them (by diffusion through the membrane core) to UQ–cyt c reductase. Cytochrome c is water soluble and moves freely in the intermembrane space, carrying electrons from UQ–cyt c reductase to cytochrome c oxidase. In the process of these electron transfers, protons are driven across the inner membrane (from the matrix side to the intermembrane space). The proton gradient generated by electron

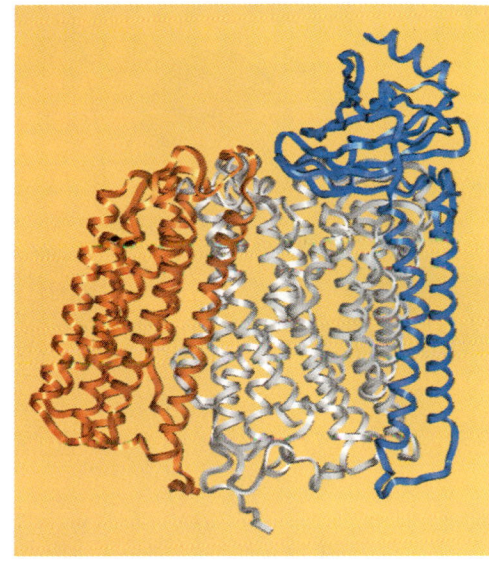

FIGURE 20.15 Molecular graphic image of subunits I, II, and III of cytochrome c oxidase.

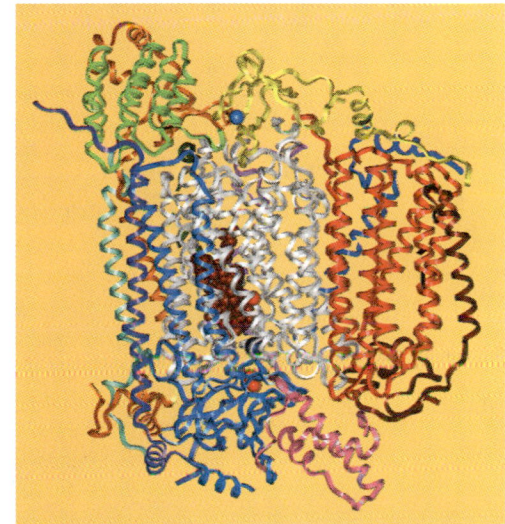

FIGURE 20.16 Molecular graphic image of cytochrome c oxidase. Seven of the ten nuclear DNA–derived subunits (IV, VIa, VIc, VIIa, VIIb, VIIc, and VIII) possess transmembrane segments. Three (Va, Vb, and VIb) do not. Subunits IV and VIc are transmembrane and dumbbell-shaped. Subunit Va is globular and bound to the matrix side of the complex, whereas VIb is a globular subunit on the cytosolic side of the membrane complex. Vb is globular and matrix side associated as well, but it has an N-terminal extended domain. VIa has a transmembrane helix and a small globular domain. Subunit VIIa consists of a tilted transmembrane helix, with another short helical segment on the matrix side of the membrane. Subunits VIIa, VIIb, and VIII consist of transmembrane segments with short extended regions outside the membrane.

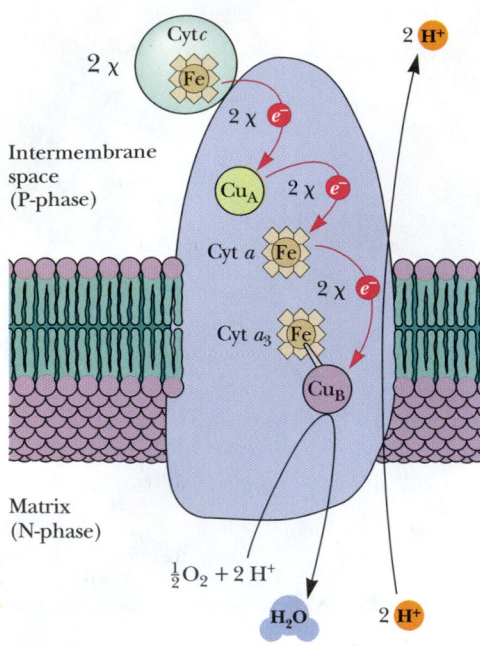

Biochemistry⊛Now™ **ACTIVE FIGURE 20.17**
The electron-transfer pathway for cytochrome oxidase. Cytochrome *c* binds on the cytosolic side, transferring electrons through the copper and heme centers to reduce O_2 on the matrix side of the membrane. **Test yourself on the concepts in this figure at http:// chemistry.brookscole.com/ggb3**

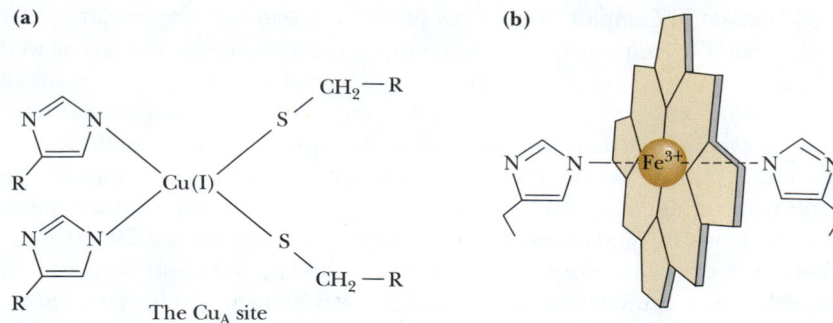

FIGURE 20.18 **(a)** The Cu_A site of cytochrome oxidase. Copper ligands include two histidine imidazole groups and two cysteine side chains from the protein. **(b)** The coordination of histidine imidazole ligands to the iron atom in the heme *a* center of cytochrome oxidase.

transport represents an enormous source of potential energy. As seen in the next section, this potential energy is used to synthesize ATP as protons flow back into the matrix.

The H⁺/2e⁻ Ratio for Electron Transport Is Uncertain

In 1961, Peter Mitchell, a British biochemist, proposed that the energy stored in a proton gradient across the inner mitochondrial membrane by electron transport drives the synthesis of ATP in cells. The proposal became known as **Mitchell's chemiosmotic hypothesis.** The ratio of protons transported per pair of electrons passed through the chain—the so-called **H⁺/2e⁻** ratio—has been an object of great interest for many years. Nevertheless, the ratio has remained extremely difficult to determine. The consensus estimate for the electron-transport pathway from succinate to O_2 is $6H^+/2e^-$. The ratio for Complex I by itself remains uncertain, but recent best estimates place it as high as $4H^+/2e^-$. On the basis of this value, the stoichiometry of transport for the pathway from NADH to O_2 is $10H^+/2e^-$. Although this is the value assumed in Figure 20.21, it is important to realize that this represents a consensus drawn from many experiments.

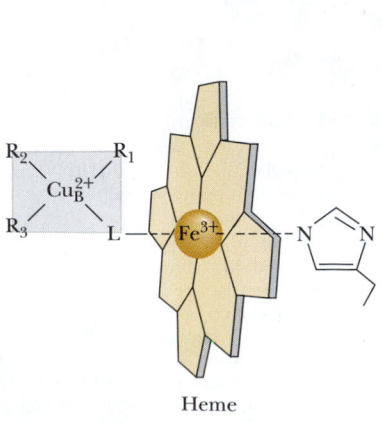

FIGURE 20.19 The binuclear center of cytochrome oxidase. A ligand, L (probably a cysteine S), is shown bridging the Cu_B and Fe of heme a_3 metal sites.

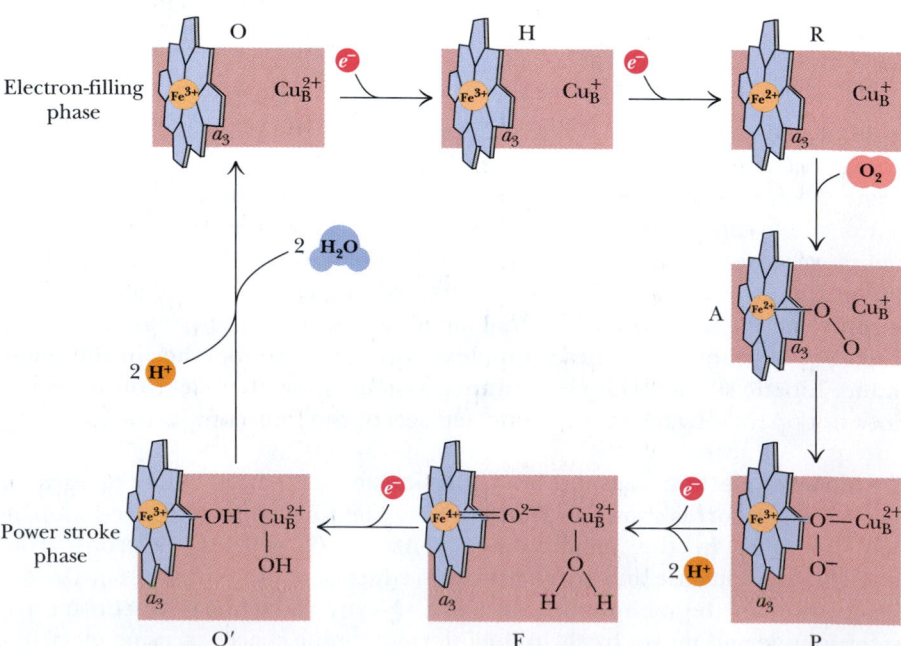

FIGURE 20.20 A model for the mechanism of O_2 reduction by cytochrome oxidase. *(Adapted from Nicholls, D. G., and Ferguson, S. J., 1992. Bioenergetics 2. London: Academic Press; and Babcock, G. T., and Wikström, M., 1992. Oxygen activation and the conservation of energy in cell respiration. Nature 356:301–309.)*

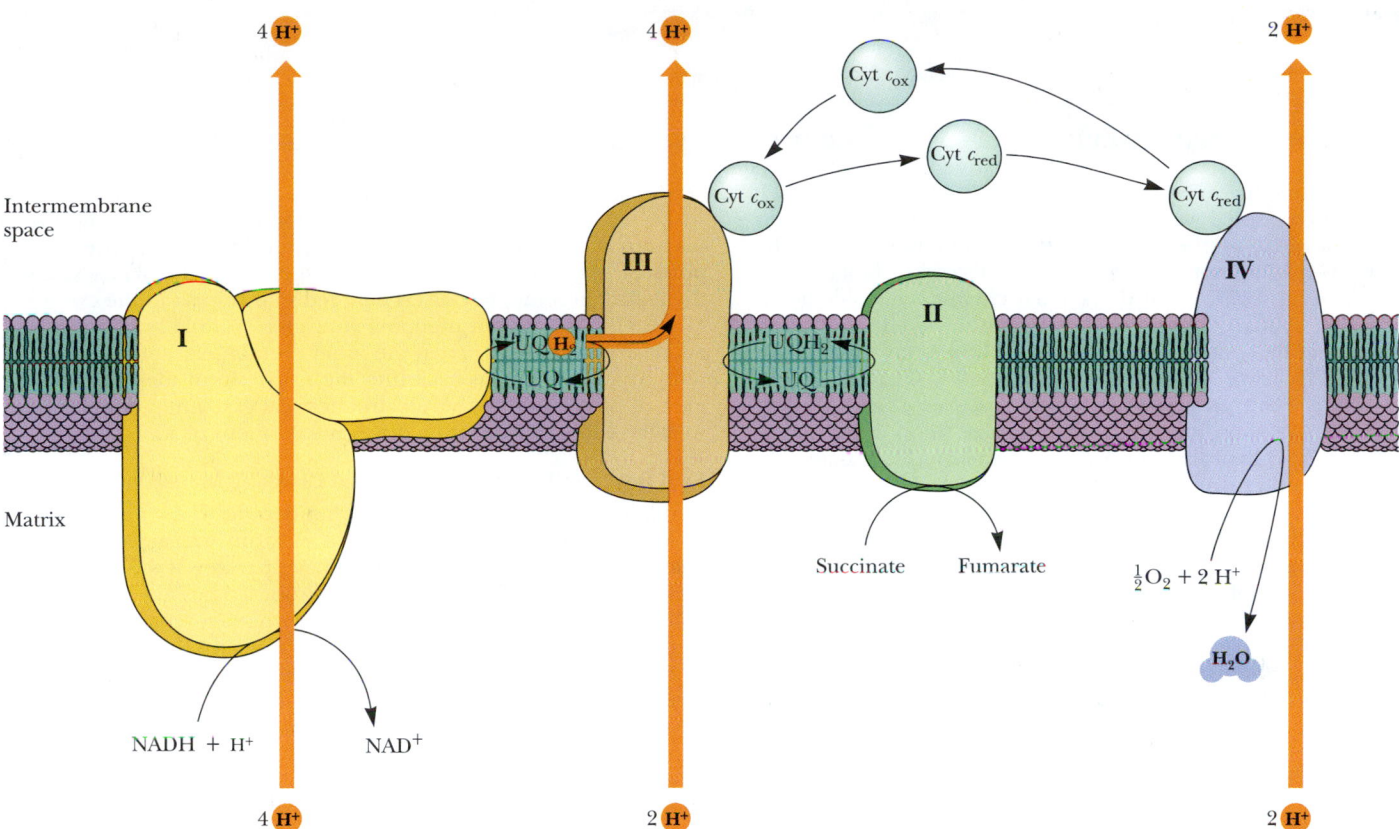

FIGURE 20.21 A model for the electron-transport pathway in the mitochondrial inner membrane. UQ/UQH_2 and cytochrome c are mobile electron carriers and function by transferring electrons between the complexes. The proton transport driven by Complexes I, III, and IV is indicated.

20.4 What Are the Thermodynamic Implications of Chemiosmotic Coupling?

Peter Mitchell's chemiosmotic hypothesis revolutionized our thinking about the energy coupling that drives ATP synthesis by means of an electrochemical gradient. How much energy is stored in this electrochemical gradient? For the transmembrane flow of protons across the inner membrane (from inside [matrix] to outside), we could write

$$H^+_{in} \longrightarrow H^+_{out} \qquad (20.22)$$

The free energy difference for protons across the inner mitochondrial membrane includes a term for the concentration difference and a term for the electrical potential. This is expressed as

$$\Delta G = RT \ln \frac{[c_2]}{[c_1]} + Z\mathscr{F}\Delta\psi \qquad (20.23)$$

where c_1 and c_2 are the proton concentrations on the two sides of the membrane, Z is the charge on a proton, $\mathscr{F}$ is Faraday's constant, and $\Delta\psi$ is the potential difference across the membrane. For the case at hand, this equation becomes

$$\Delta G = RT \ln \frac{[H^+_{out}]}{[H^+_{in}]} + Z\mathscr{F}\Delta\psi \qquad (20.24)$$

In terms of the matrix and cytoplasm pH values, the free energy difference is

$$\Delta G = -2.303 \, RT(\text{pH}_{out} - \text{pH}_{in}) + \mathscr{F}\Delta\psi \qquad (20.25)$$

Reported values for $\Delta\psi$ and ΔpH vary, but the membrane potential is always found to be positive outside and negative inside, and the pH is always more acidic outside and more basic inside. Taking typical values of $\Delta\psi = 0.18$ V and

Critical Developments in Biochemistry

Oxidative Phosphorylation—The Clash of Ideas and Energetic Personalities

For many years, the means by which electron transport and ATP synthesis are coupled was unknown. It is no exaggeration to say that the search for the coupling mechanism was one of the largest, longest, most bitter fights in the history of biochemical research. Since 1777, when the French chemist Lavoisier determined that foods undergo oxidative combustion in the body, chemists and biochemists have wondered how energy from food is captured by living things. A piece of the puzzle fell into place in 1929, when Fiske and Subbarow first discovered and studied adenosine 5′-triphosphate in muscle extracts. Soon it was understood that ATP hydrolysis provides the energy for muscle contraction and other processes.

Engelhardt's experiments in 1930 led to the notion that ATP is synthesized as the result of electron transport, and by 1940, Severo Ochoa had carried out a measurement of the P/O ratio, the number of molecules of ATP generated per atom of oxygen consumed in the electron-transport chain. Because two electrons are transferred down the chain per oxygen atom reduced, the P/O ratio also reflects the ratio of ATPs synthesized per pair of electrons consumed. After many tedious and careful measurements, scientists decided that the P/O ratio was 3 for NADH oxidation and 2 for succinate (that is, [FADH$_2$]) oxidation. Electron flow and ATP synthesis are very tightly coupled in the sense that, in normal mitochondria, neither occurs without the other.

A High-Energy Chemical Intermediate Coupling Oxidation and Phosphorylation Proved Elusive

Many models were proposed to account for the coupling of electron transport and ATP synthesis. A persuasive model, advanced by E. C. Slater in 1953, proposed that energy derived from electron transport was stored in a high-energy intermediate (symbolized as X ~ P). This chemical species—in essence an activated form of phosphate—functioned according to certain relations according to Equations 20.26–20.29 (see following) to drive ATP synthesis.

This hypothesis was based on the model of substrate-level phosphorylation in which a high-energy substrate intermediate is a precursor to ATP. A good example is the 3-phosphoglycerate kinase reaction of glycolysis, where 1,3-bisphosphoglycerate serves as a high-energy intermediate leading to ATP. Literally hundreds of attempts were made to isolate the high-energy intermediate, X ~ P. Among the scientists involved in the research, rumors that one group or another had isolated X ~ P circulated frequently, but none was substantiated. Eventually it became clear that the intermediate could not be isolated because it did not exist.

Peter Mitchell's Chemiosmotic Hypothesis

In 1961, Peter Mitchell proposed a novel coupling mechanism involving a proton gradient across the inner mitochondrial membrane. In Mitchell's chemiosmotic hypothesis, protons are driven across the membrane from the matrix to the intermembrane space and cytosol by the events of electron transport. This mechanism stores the energy of electron transport in an electrochemical potential. As protons are driven out of the matrix, the pH rises and the matrix becomes negatively charged with respect to the cytosol (Figure 20.22). Proton pumping thus creates a pH gradient and an electrical gradient across the inner membrane, both of which tend to attract protons back into the matrix from the cytoplasm. Flow of protons down this electrochemical gradient, an energetically favorable process, then drives the synthesis of ATP.

Paul Boyer and the Conformational Coupling Model

Another popular model invoked what became known as conformational coupling. If the energy of electron transport was not stored in some high-energy intermediate, perhaps it was stored in a high-energy protein conformation. Proposed by Paul Boyer, this model suggested that reversible conformation changes transferred energy from proteins of the electron-transport chain to the enzymes involved in ATP synthesis. This model was consistent with some of the observations made by others, and it eventually evolved into the binding change mechanism (the basis for the model in Figure 20.27). Boyer's model is supported by a variety of binding experiments and is essentially consistent with Mitchell's chemiosmotic hypothesis.

Electron transport drives H$^+$ out and creates an electrochemical gradient

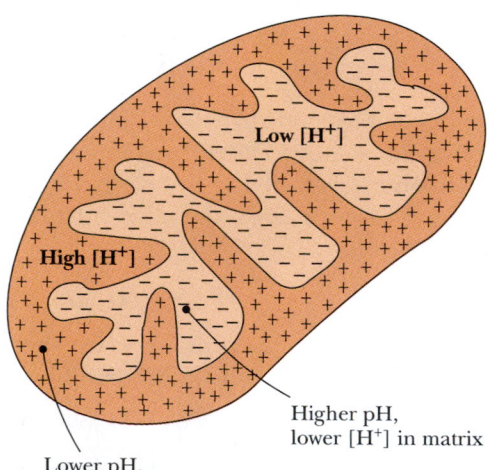

Low [H$^+$]

High [H$^+$]

Higher pH, lower [H$^+$] in matrix

Lower pH, higher [H$^+$] in intermembrane space

FIGURE 20.22 The proton and electrochemical gradients existing across the inner mitochondrial membrane. The electrochemical gradient is generated by the transport of protons across the membrane.

$$NADH + H^+ + FMN + X \longrightarrow NAD^+ \!-\! X + FMNH_2 \qquad (20.26)$$
$$NAD^+ \!-\! X + P_i \longrightarrow NAD^+ + X \!\sim\! P \qquad (20.27)$$
$$X \!\sim\! P + ADP \longrightarrow X + ATP + H_2O \qquad (20.28)$$

Net reaction:
$$NADH + H^+ + FMN + ADP + P_i \longrightarrow NAD^+ + FMNH_2 + ATP + H_2O \quad (20.29)$$

$\Delta pH = 1$ unit, the free energy change associated with the movement of one mole of protons from inside to outside is

$$\Delta G = 2.3\ RT + \mathscr{F}(0.18\ \text{V}) \tag{20.30}$$

With $\mathscr{F} = 96.485$ kJ/V · mol, the value of ΔG at 37°C is

$$\Delta G = 5.9\ \text{kJ} + 17.4\ \text{kJ} = 23.3\ \text{kJ} \tag{20.31}$$

which is the free energy change for movement of a mole of protons across a typical inner membrane. Note that the free energy terms for both the pH difference and the potential difference are unfavorable for the outward transport of protons, with the latter term making the greater contribution. On the other hand, the ΔG for inward flow of protons is -23.3 kJ/mol. It is this energy that drives the synthesis of ATP, in accord with Mitchell's model. Peter Mitchell was awarded the Nobel Prize in Chemistry in 1978.

20.5 | How Does a Proton Gradient Drive the Synthesis of ATP?

The mitochondrial complex that carries out ATP synthesis is called **ATP synthase,** or sometimes **F_1F_0–ATPase** (for the reverse reaction it catalyzes). ATP synthase was observed in early electron micrographs of submitochondrial particles (prepared by sonication of inner membrane preparations) as round, 8.5-nm-diameter projections or particles on the inner membrane (Figure 20.23). In micrographs of native mitochondria, the projections appear on the matrix-facing surface of the inner membrane. The purified particles catalyze ATP hydrolysis, the reverse reaction of the ATP synthase. Stripped of these particles, the membranes can still carry out electron transfer but cannot synthesize ATP. In one of the first reconstitution experiments with membrane proteins, Efraim Racker showed that adding the particles back to stripped membranes restored electron transfer-dependent ATP synthesis.

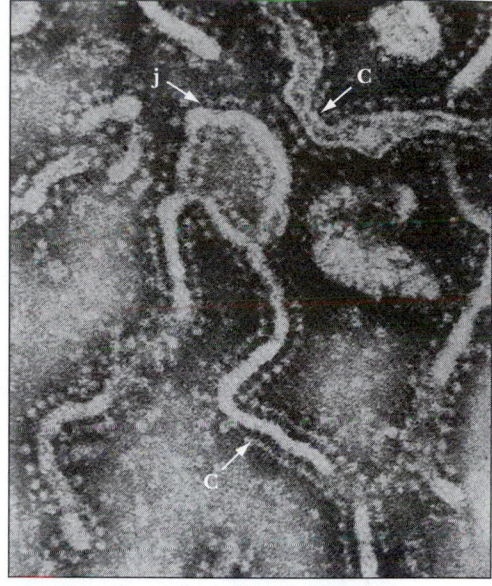

FIGURE 20.23 Electron micrograph of submitochondrial particles showing the 8.5-nm projections or particles on the inner membrane, eventually shown to be F_1–ATP synthase. *(From Parsons, D. F., 1963. Mitochondrial structure: Two types of subunits on negatively stained mitochondrial membranes.* Science **140:**985. *Reprinted with permission of the AAAS.)*

ATP Synthase Consists of Two Complexes—F_1 and F_0

ATP synthase actually consists of two principal complexes. The spheres observed in electron micrographs make up the **F_1 unit,** which catalyzes ATP synthesis. These F_1 spheres are attached to an integral membrane protein aggregate called the **F_0 unit.** F_1 consists of five polypeptide chains named α, β, γ, δ, and ϵ, with a subunit stoichiometry $\alpha_3\beta_3\gamma\delta\epsilon$ (Table 20.3). F_0 consists of three hydrophobic subunits denoted by a, b, and c, with an apparent stoichiometry of $a_1b_2c_{9-12}$. F_0 forms the transmembrane pore or channel through which protons move to drive ATP synthesis. The α-, β-, γ-, δ-, and ϵ-subunits of F_1 contain 510, 482, 272,

Biochemistry Now™ Go to BiochemistryNow and click BiochemistryInteractive to learn more about the mitochondrial complex that carries out ATP synthesis.

Table 20.3

Escherichia coli F_1F_0–ATP Synthase Subunit Organization

Complex	Protein Subunit	Mass (kD)	Stoichiometry
F_1	α	55	3
	β	52	3
	γ	30	1
	δ	15	1
	ϵ	5.6	1
F_0	a	30	1
	b	17	2
	c	8	9–12

(a)

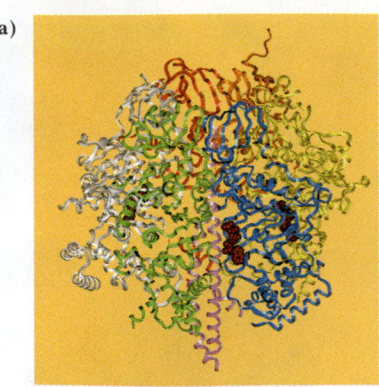

(b)

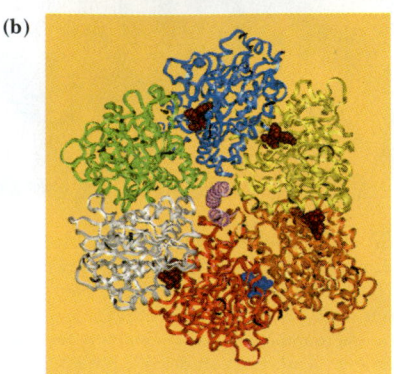

FIGURE 20.24 Molecular graphic images: **(a)** Side view and **(b)** top view of the F₁–ATP synthase showing the individual component peptides. The γ-subunit is the pink structure visible in the center of view (b).

146, and 50 amino acids, respectively, with a total molecular mass for F_1 of 371 kD. The α- and β-subunits are homologous, and each of these subunits binds a single ATP. The catalytic sites are in the β-subunits; the function of the ATP sites in the α-subunits is unknown (deletion of the sites does not affect activity).

John Walker and his colleagues have determined the structure of the F_1 complex (Figure 20.24). The F_1–ATPase is an inherently asymmetric structure, with the three β-subunits having three different conformations. In the structure solved by Walker, one of the β-subunit ATP sites contains AMP-PNP (a nonhydrolyzable analog of ATP), and another contains ADP, with the third site being empty. This state is consistent with the **binding change mechanism** for ATP synthesis proposed by Paul Boyer, in which three reaction sites cycle concertedly through the three intermediate states of ATP synthesis (take a look at Figure 20.27 on page 661).

How might such cycling occur? Important clues have emerged from several experiments that show that the γ-subunit rotates with respect to the $\alpha\beta$ complex. How such rotation might be linked to transmembrane proton flow and ATP synthesis is shown in Figure 20.25. In this model, the c-subunits of F_0 are arranged in a ring. Several lines of evidence suggest that each c-subunit consists of a pair of antiparallel transmembrane helices with a short hairpin loop on the cytosolic side of the membrane. A ring of c-subunits could form a **rotor** that turns with respect to the a-subunit, a **stator** consisting of five transmembrane α-helices with proton access channels on either side of the membrane. The γ-subunit is postulated to be the link between F_1 and F_0. Several experiments have shown that γ rotates relative to the $(\alpha\beta)_3$ complex during ATP synthesis. If γ is anchored to the c-subunit rotor, then the c rotor–γ complex can rotate together relative to the $(\alpha\beta)_3$ complex. Subunit b possesses a single transmembrane segment and a long hydrophilic head domain, and the complete stator may consist of the b-subunits anchored at one end to the a-subunit and linked at the other end to the $(\alpha\beta)_3$ complex via the δ-subunit, as shown in Figure 20.25.

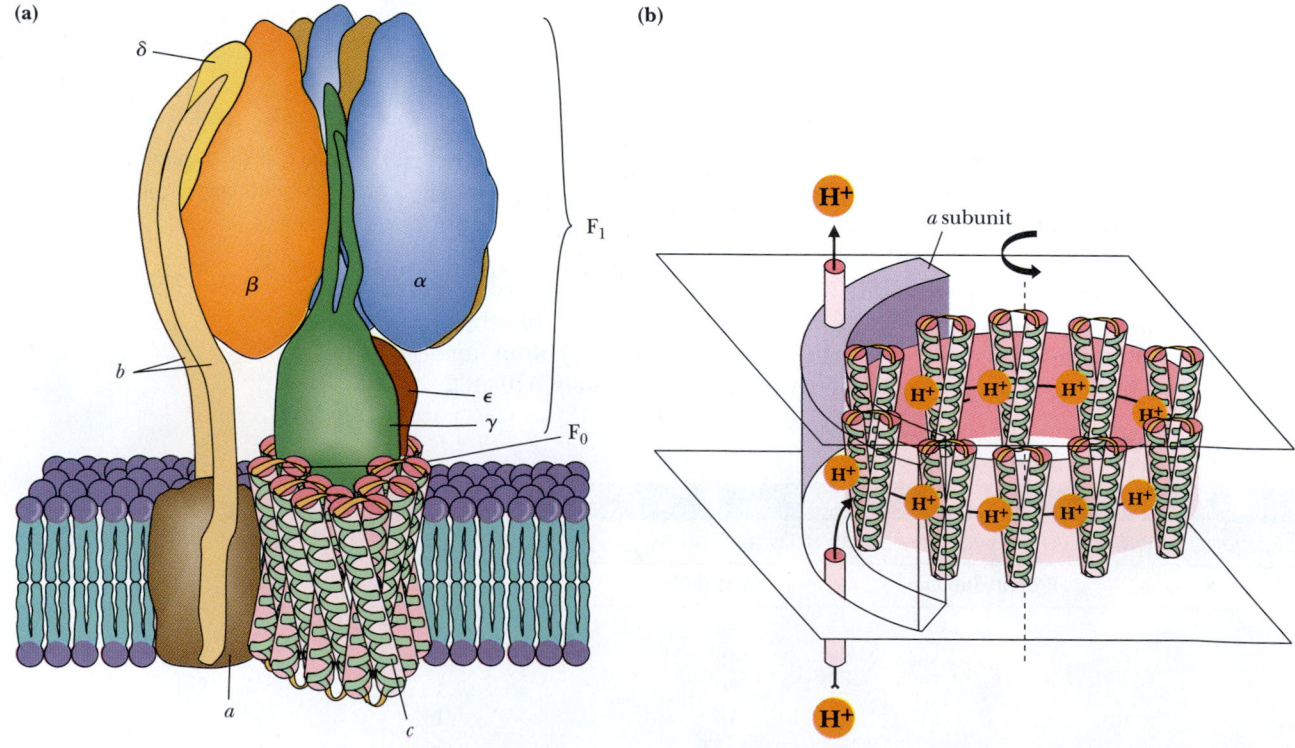

Biochemistry ⒺNow™ **ANIMATED FIGURE 20.25** A model of the F_1 and F_0 components of the ATP synthase, a rotating molecular motor. The a-, b-, α-, β-, and δ-subunits constitute the stator of the motor, and the c-, γ-, and ϵ-subunits form the rotor. Flow of protons through the structure turns the rotor and drives the cycle of conformational changes in α and β that synthesize ATP. **See this figure animated at http://chemistry.brookscole.com/ggb3**

In the presence of a proton gradient:

$$\text{ADP} + \text{P} \longrightarrow \text{ATP} \quad \text{is released}$$

In the absence of a proton gradient:

$$\text{ADP} + \text{P} \rightleftharpoons [\text{ATP}] \rightleftharpoons \text{ADP} + {}^{18}\text{O}-\overset{\displaystyle O}{\underset{\displaystyle O^-}{\overset{\|}{\underset{|}{P}}}}-\text{OH}$$

FIGURE 20.26 ATP production in the presence of a proton gradient and ATP/ADP exchange in the absence of a proton gradient. Exchange leads to incorporation of ^{18}O in phosphate as shown.

What, then, is the mechanism for ATP synthesis? The c rotor subunits each carry an essential residue, Asp[61]. (Changing this residue to Asn abolishes ATP synthase activity.) Rotation of the c rotor relative to the stator may depend upon neutralization of the negative charge on each c-subunit Asp[61] as the rotor turns. Protons taken up from the cytosol by one of the proton access channels in a protonate an Asp[61] and then ride the rotor until they reach the other proton access channel on a, from which they would be released into the matrix. Such rotation would cause the γ-subunit to turn relative to the three β-subunit nucleotide sites of F_1, changing the conformation of each in sequence, so ADP is first bound, then phosphorylated, then released, according to Boyer's binding change mechanism. Paul Boyer and John Walker shared in the 1997 Nobel Prize for Chemistry for their work on the structure and mechanism of ATP synthase.

Boyer's ^{18}O Exchange Experiment Identified the Energy-Requiring Step

The elegant studies by Paul Boyer of ^{18}O exchange in ATP synthase have provided other important insights into the mechanism of the enzyme. Boyer and his colleagues studied the ability of the synthase to incorporate labeled oxygen from $H_2{}^{18}\text{O}$ into P_i. This reaction (Figure 20.26) occurs via synthesis of ATP from ADP and P_i, followed by hydrolysis of ATP with incorporation of oxygen atoms from the solvent. Although net production of ATP requires coupling with a proton gradient, Boyer observed that this exchange reaction occurs readily, even in the absence of a proton gradient. His finding indicated that the formation of enzyme-bound ATP does not require energy. Indeed, movement of protons through the F_0 channel causes the release of newly synthesized ATP from the enzyme. Thus, the energy provided by electron transport creates a proton gradient that drives enzyme conformational changes resulting in the binding of substrates on ATP synthase, ATP synthesis, and the release of products. The mechanism involves catalytic cooperativity between three interacting sites (Figure 20.27).

Racker and Stoeckenius Confirmed the Mitchell Model in a Reconstitution Experiment

When Mitchell first described his chemiosmotic hypothesis in 1961, little evidence existed to support it and it was met with considerable skepticism by the scientific community. Eventually, however, considerable evidence accumulated

▶ **Biochemistry**⊜**Now**™ **ANIMATED FIGURE 20.27** The binding change mechanism for ATP synthesis by ATP synthase. This model assumes that F_1 has three interacting and conformationally distinct active sites. The open (O) conformation is inactive and has a low affinity for ligands; the L conformation (with "loose" affinity for ligands) is also inactive; the tight (T) conformation is active and has a high affinity for ligands. Synthesis of ATP is initiated (step 1) by binding of ADP and P_i to an L site. In the second step, an energy-driven conformational change converts the L site to a T conformation and also converts T to O and O to L. In the third step, ATP is synthesized at the T site and released from the O site. Two additional passes through this cycle produce two more ATPs and return the enzyme to its original state. **See this figure animated at http://chemistry.brookscole.com/ggb3**

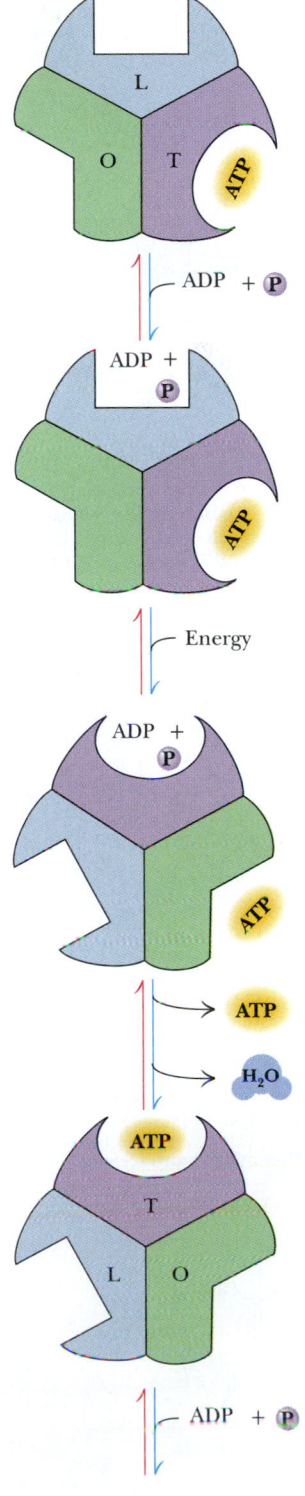

Cycle repeats

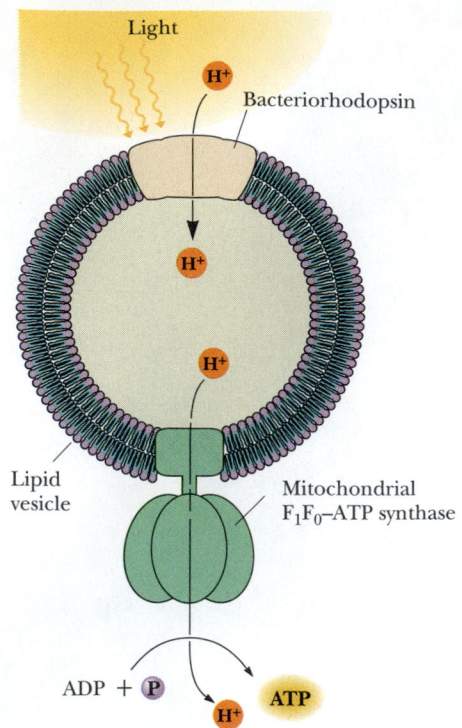

The reconstituted vesicles containing ATP synthase and bacteriorhodopsin used by Stoeckenius and Racker to confirm the Mitchell chemiosmotic hypothesis. **See this figure animated at http://chemistry.brookscole. com/ggb3**

to support this model. It is now clear that the electron-transport chain generates a proton gradient, and careful measurements have shown that ATP is synthesized when a pH gradient is applied to mitochondria that cannot carry out electron transport. Even more relevant is a simple but crucial experiment reported in 1974 by Efraim Racker and Walther Stoeckenius, which provided specific confirmation of the Mitchell hypothesis. In this experiment, the bovine mitochondrial ATP synthase was reconstituted in simple lipid vesicles with **bacteriorhodopsin,** a light-driven proton pump from *Halobacterium halobium*. As shown in Figure 20.28, upon illumination, bacteriorhodopsin pumped protons into these vesicles, and the resulting proton gradient was sufficient to drive ATP synthesis by the ATP synthase. Because the only two kinds of proteins present were one that produced a proton gradient and one that used such a gradient to make ATP, this experiment essentially verified Mitchell's chemiosmotic hypothesis.

Inhibitors of Oxidative Phosphorylation Reveal Insights About the Mechanism

The unique properties and actions of an inhibitory substance can often help identify aspects of an enzyme mechanism. Many details of electron transport and oxidative phosphorylation mechanisms have been gained from studying the effects of particular inhibitors. Figure 20.29 presents the structures of some electron transport and oxidative phosphorylation inhibitors. The sites of inhibition by these agents are indicated in Figure 20.30.

Inhibitors of Complexes I, II, and III Block Electron Transport Rotenone is a common insecticide that strongly inhibits the NADH–UQ reductase. Rotenone is obtained from the roots of several species of plants. Tribes in certain parts of the world have made a practice of beating the roots of trees along riverbanks to release rotenone into the water, where it paralyzes fish and makes them easy

FIGURE 20.29 The structures of several inhibitors of electron transport and oxidative phosphorylation.

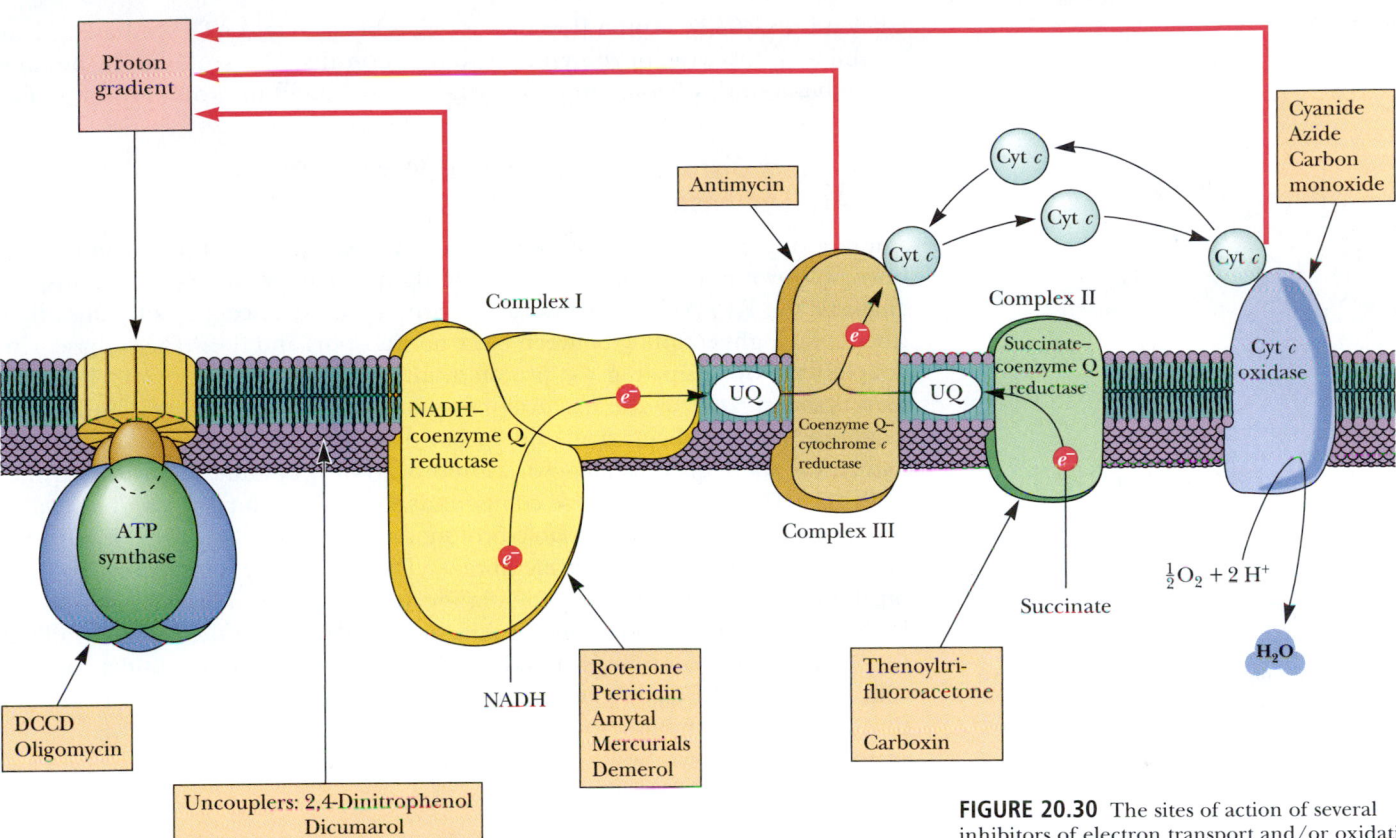

FIGURE 20.30 The sites of action of several inhibitors of electron transport and/or oxidative phosphorylation.

prey. Ptericidin, Amytal, and other barbiturates; mercurial agents; and the widely prescribed painkiller Demerol also exert inhibitory actions on this enzyme complex. All these substances appear to inhibit reduction of coenzyme Q and the oxidation of the Fe-S clusters of NADH–UQ reductase.

2-Thenoyltrifluoroacetone and carboxin and its derivatives specifically block Complex II, the succinate–UQ reductase. Antimycin, an antibiotic produced by *Streptomyces griseus*, inhibits the UQ–cytochrome *c* reductase by blocking electron transfer between b_H and coenzyme Q in the Q_n site. Myxothiazol inhibits the same complex by acting at the Q_p site.

Cyanide, Azide, and Carbon Monoxide Inhibit Complex IV   Complex IV, the cytochrome *c* oxidase, is specifically inhibited by cyanide (CN^-), azide (N_3^-), and carbon monoxide (CO). Cyanide and azide bind tightly to the ferric form of cytochrome a_3, whereas carbon monoxide binds only to the ferrous form. The inhibitory actions of cyanide and azide at this site are very potent, whereas the principal toxicity of carbon monoxide arises from its affinity for the iron of hemoglobin. Herein lies an important distinction between the poisonous effects of cyanide and carbon monoxide. Because animals (including humans) carry many, many hemoglobin molecules, they must inhale a large quantity of carbon monoxide to die from it. These same organisms, however, possess comparatively few molecules of cytochrome a_3. Consequently, a limited exposure to cyanide can be lethal. The sudden action of cyanide attests to the organism's constant and immediate need for the energy supplied by electron transport.

Oligomycin and DCCD Are ATP Synthase Inhibitors   Inhibitors of ATP synthase include dicyclohexylcarbodiimide (DCCD) and oligomycin (Figure 20.29). DCCD bonds covalently to carboxyl groups in hydrophobic domains of proteins in general and to a glutamic acid residue of the *c*-subunit of F_0, the proteolipid forming the proton channel of the ATP synthase, in particular. If the *c*-subunit is

Dinitrophenol

Dicumarol

Carbonyl cyanide-*p*-trifluoro-methoxyphenyl hydrazone
—best known as **FCCP**; for **F**luoro **C**arbonyl **C**yanide **P**henylhydrazone

$F_3C - O - \bigcirc - N - N = C \Big\langle{}^{C \equiv N}_{C \equiv N}$

FIGURE 20.31 Structures of several uncouplers, molecules that dissipate the proton gradient across the inner mitochondrial membrane and thereby destroy the tight coupling between electron transport and the ATP synthase reaction.

labeled with DCCD, proton flow through F_0 is blocked and ATP synthase activity is inhibited. Likewise, oligomycin acts directly on the ATP synthase. By binding to a subunit of F_0, oligomycin also blocks the movement of protons through F_0.

Uncouplers Disrupt the Coupling of Electron Transport and ATP Synthase

Another important class of reagents affects ATP synthesis, but in a manner that does not involve direct binding to any of the proteins of the electron-transport chain or the F_1F_0–ATPase. These agents are known as **uncouplers** because they disrupt the tight coupling between electron transport and the ATP synthase. Uncouplers act by dissipating the proton gradient across the inner mitochondrial membrane created by the electron-transport system. Typical examples include 2,4-dinitrophenol, dicumarol, and carbonyl cyanide-*p*-trifluoro-methoxyphenyl hydrazone (perhaps better known as fluorocarbonyl cyanide phenylhydrazone, or FCCP) (Figure 20.31). These compounds share two common features: hydrophobic character and a dissociable proton. As uncouplers, they function by carrying protons across the inner membrane. Their tendency is to acquire protons on the cytosolic surface of the membrane (where the proton concentration is high) and carry them to the matrix side, thereby destroying the proton gradient that couples electron transport and the ATP synthase. In mitochondria treated with uncouplers, electron transport continues and protons are driven out through the inner membrane. However, they leak back in so rapidly via the uncouplers that ATP synthesis does not occur. Instead, the energy released in electron transport is dissipated as heat.

Human Biochemistry

Endogenous Uncouplers Enable Organisms to Generate Heat

Certain cold-adapted animals, hibernating animals, and newborn animals generate large amounts of heat by uncoupling oxidative phosphorylation. These organisms have a type of fat known as brown adipose tissue, so called for the color imparted by the many mitochondria this adipose tissue contains. The inner membrane of brown adipose tissue mitochondria contains large amounts of an endogenous protein called **thermogenin** (literally, "heat maker"), or **uncoupling protein 1 (UCP1)**. UCP1 creates a passive proton channel through which protons flow from the cytosol to the matrix. Mice that lack UCP1 cannot maintain their body temperature in cold conditions, whereas normal animals produce larger amounts of UCP1 when they are cold-adapted. Two other mitochondrial proteins, designated UCP2 and UCP3, have sequences similar to UCP1.

Because the function of UCP1 is so closely linked to energy utilization, there has been great interest in the possible roles of UCP1, UCP2, and UCP3 as metabolic regulators and as factors in obesity. Under fasting conditions, expression of UCP1 mRNA is decreased, but expression of UCP2 and UCP3 is increased. There is no indication, however, that UCP2 and UCP3 actually function as uncouplers. There has also been interest in the possible roles of UCP2 and UCP3 in the development of obesity, especially because the genes for these proteins lie on chromosome 7 of the mouse, close to other genes linked to obesity.

Certain plants use the heat of uncoupled proton transport for a special purpose. Skunk cabbage and related plants contain floral spikes that are maintained as much as 20° above ambient temperature in this way. The warmth of the spikes serves to vaporize odiferous molecules, which attract insects that fertilize the flowers. Red tomatoes have very small mitochondrial membrane proton gradients compared with green tomatoes—evidence that uncouplers are more active in red tomatoes, whose energy needs are less.

Endogenous Uncouplers Enable Organisms to Generate Heat   Ironically, certain cold-adapted animals, hibernating animals, and newborn animals generate large amounts of heat by uncoupling oxidative phosphorylation. Adipose tissue in these organisms contains so many mitochondria that it is called *brown adipose tissue* for the color imparted by the mitochondria. The inner membrane of brown adipose tissue mitochondria contains an endogenous protein called **thermogenin** (literally, "heat maker"), or uncoupling protein, that creates a passive proton channel through which protons flow from the cytosol to the matrix. Certain plants also use the heat of uncoupled proton transport for a special purpose. Skunk cabbage and related plants contain floral spikes that are maintained as much as 20° above ambient temperature in this way. The warmth of the spikes serves to vaporize odiferous molecules, which attract insects that fertilize the flowers.

ATP–ADP Translocase Mediates the Movement of ATP and ADP Across the Mitochondrial Membrane

ATP, the cellular energy currency, must exit the mitochondria to carry energy throughout the cell, and ADP must be brought into the mitochondria for reprocessing. Neither of these processes occurs spontaneously because the highly charged ATP and ADP molecules do not readily cross biological membranes. Instead, these processes are mediated by a single transport system, the **ATP–ADP translocase.** This protein tightly couples the exit of ATP with the entry of ADP so that the mitochondrial nucleotide levels remain approximately constant. For each ATP transported out, one ADP is transported into the matrix. The translocase, which accounts for approximately 14% of the total mitochondrial membrane protein, is a homodimer of 30-kD subunits. Transport occurs via a single nucleotide-binding site, which alternately faces the matrix and the intermembrane space (Figure 20.32). It binds ATP on the matrix side, reorients to face the outside, and exchanges ATP for ADP, with subsequent movement back to the matrix face of the inner membrane.

Outward Movement of ATP Is Favored over Outward ADP Movement   The charge on ATP at pH 7.2 or so is about -4, and the charge on ADP at the same pH is about -3. Thus, net exchange of an ATP (out) for an ADP (in) results in the net movement of one negative charge from the matrix to the cytosol. (This process is equivalent to the movement of a proton from the cytosol to the matrix.) Recall that the inner membrane is positive outside (see Figure 20.22), and it becomes clear that outward movement of ATP is favored over outward ADP transport, ensuring that ATP will be transported out (Figure 20.32). Inward movement of ADP is favored over inward movement of ATP for the same reason. Thus, the membrane electrochemical potential itself controls the specificity of the ATP–ADP translocase. However, the electrochemical potential is diminished by the ATP–ADP translocase cycle and therefore operates with an energy

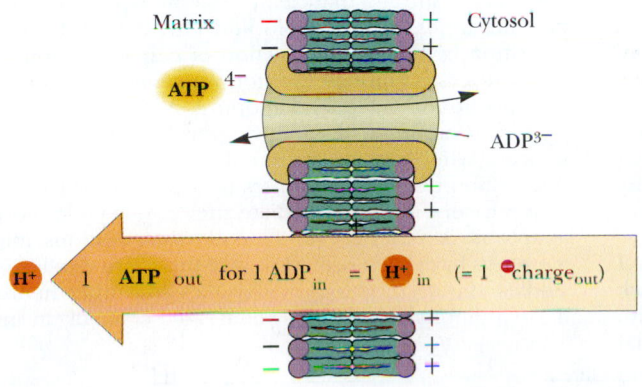

FIGURE 20.32 Outward transport of ATP (via the ATP/ADP translocase) is favored by the membrane electrochemical potential.

cost to the cell. The cell must compensate by passing yet more electrons down the electron-transport chain.

What is the cost of ATP–ADP exchange relative to the energy cost of ATP synthesis itself? We already noted that moving one ATP out and one ADP in is the equivalent of one proton moving from the cytosol to the matrix. Synthesis of an ATP results from the movement of approximately three protons from the cytosol into the matrix through F_0. Altogether this means that approximately four protons are transported into the matrix per ATP synthesized. Thus, approximately one-fourth of the energy derived from the respiratory chain (electron transport and oxidative phosphorylation) is expended as the electrochemical energy devoted to mitochondrial ATP–ADP transport.

| 20.6 | **What Is the P/O Ratio for Mitochondrial Electron Transport and Oxidative Phosphorylation?** |

$$\left(\frac{1 \text{ ATP}}{4 \text{ H}^+}\right)\left(\frac{10 \text{ H}^+}{2 \, e^- \, [\text{NADH}\rightarrow\frac{1}{2}\text{O}_2]}\right) = \frac{10}{4} = \frac{\text{P}}{\text{O}}$$

The **P/O ratio** is the number of molecules of ATP formed in oxidative phosphorylation per two electrons flowing through a defined segment of the electron-transport chain. In spite of intense study of this ratio, its actual value remains a matter of contention. If we accept the value of 10 H^+ transported out of the matrix per 2 e^- passed from NADH to O_2 through the electron-transport chain, and also agree (as previously) that 4 H^+ are transported into the matrix per ATP synthesized (and translocated), then the mitochondrial P/O ratio is 10/4, or 2.5, for the case of electrons entering the electron-transport chain as NADH. This is somewhat lower than earlier estimates, which placed the P/O ratio at 3 for mitochondrial oxidation of NADH. For the portion of the chain from succinate to O_2, the $H^+/2e^-$ ratio is 6 (as noted previously), and the P/O ratio in this case would be 6/4, or 1.5; earlier estimates placed this number at 2. The consensus of more recent experimental measurements of P/O ratios for these two cases has

Human Biochemistry

Mitochondria Play a Central Role in Apoptosis

Apoptosis (the second "p" is silent in this word) is the programmed death of cells—a mechanism through which certain cells are eliminated from higher organisms. It is central to the development and homeostasis of multicellular organisms, and it is the route by which unwanted or harmful cells are eliminated from the organism. Under normal circumstances, apoptosis is suppressed, as a result of the careful compartmentation of the involved activators and enzymes. Mitochondria play a major role in this subcellular partitioning of the apoptotic activator molecules. One such activator is cytochrome c, which normally resides in the intermembrane space. A variety of triggering agents, including Ca^{2+}, reactive oxygen species (ROS), certain lipid molecules, and certain protein kinases, can induce a mitochondrial membrane permeabilization (MMP). Permeabilization events, which occur at points where outer and inner mitochondrial membranes are in contact, involve association of the ATP–ADP translocase in the inner membrane and the **voltage-dependent anion channel (VDAC)** in the outer membrane. This interaction leads to the opening of protein-permeable pores, which release several proteins, including cytochrome c, Smac/Diablo, AIF, heat-shock protein 60, HtrA2/Omi, and endonuclease G, to the cytoplasm. Membrane permeabilization also dissipates the mitochondrial

transmembrane potential, $\Delta\Psi$. The pore formation is carefully regulated by the **BCL-2** family of proteins. This family of related proteins includes both pro-apoptotic members, including proteins known as Bax and Bid and Bad, as well as anti-apoptotic members such as BCL-2 itself, and also BCL-X_L and BCL-W.

Each of the released proteins plays a role in the apoptotic process. Cytochrome c activates **caspases** (where "c" is for cysteine and "asp" is for aspartic acid), a family of proteases that have Cys at the active site and that cleave after an Asp residues in their peptide substrates. Smac/Diablo and HtrA2/Omi facilitate caspase activation by blocking the action of caspase inhibitors. AIF and endonuclease G induce apoptotic changes in the nucleus.

Mitochondria-mediated apoptosis is the mode of cell death of many neurons in the brain during strokes and other brain-trauma injuries. When a stroke occurs, the neurons at the site of oxygen deprivation die within minutes by a process called necrosis, but cells adjacent to the immediate site of injury die more slowly by apoptosis. A variety of therapeutic interventions that suppress apoptosis have been proved to save these latter cells in laboratory studies, raising the hope that strokes and other neurodegenerative conditions may someday be treated clinically in similar ways.

been closer to the values of 2.5 and 1.5. Many chemists and biochemists, accustomed to the integral stoichiometries of chemical and metabolic reactions, were once reluctant to accept the notion of nonintegral P/O ratios. At some point, as we learn more about these complex coupled processes, it may be necessary to reassess the numbers.

<table>
<tr><td>**20.7**</td><td>**How Are the Electrons of Cytosolic NADH Fed into Electron Transport?**</td></tr>
</table>

Most of the NADH used in electron transport is produced in the mitochondrial matrix, an appropriate site because NADH is oxidized by Complex I on the matrix side of the inner membrane. Furthermore, the inner mitochondrial membrane is impermeable to NADH. Recall, however, that NADH is produced in glycolysis by glyceraldehyde-3-P dehydrogenase in the cytosol. If this NADH were not oxidized to regenerate NAD^+, the glycolytic pathway would cease to function due to NAD^+ limitation. Eukaryotic cells have a number of **shuttle systems** that harvest the electrons of cytosolic NADH for delivery to mitochondria without actually transporting NADH across the inner membrane (Figures 20.33 and 20.34).

The Glycerophosphate Shuttle Ensures Efficient Use of Cytosolic NADH

In the **glycerophosphate shuttle,** two different **glycerophosphate dehydrogenases,** one in the cytoplasm and one on the outer face of the mitochondrial inner membrane, work together to carry electrons into the mitochondrial matrix (Figure 20.33). NADH produced in the cytosol transfers its electrons to dihydroxyacetone phosphate, thus reducing it to glycerol-3-phosphate. This metabolite is reoxidized by the FAD^+-dependent mitochondrial membrane enzyme to reform dihydroxyacetone phosphate and enzyme-bound $FADH_2$. The two electrons of $[FADH_2]$ are passed directly to UQ, forming UQH_2. Thus, via this shuttle, cytosolic NADH can be used to produce mitochondrial $[FADH_2]$ and, subsequently, UQH_2. As a result, cytosolic NADH oxidized via this shuttle route yields only 1.5 molecules of ATP. The cell "pays" with a potential ATP molecule for the convenience of getting cytosolic NADH into the mitochondria. Although this may seem wasteful, there is an important payoff. The glycerophosphate shuttle is essentially irreversible, and even when NADH levels are very low relative to NAD^+, the cycle operates effectively.

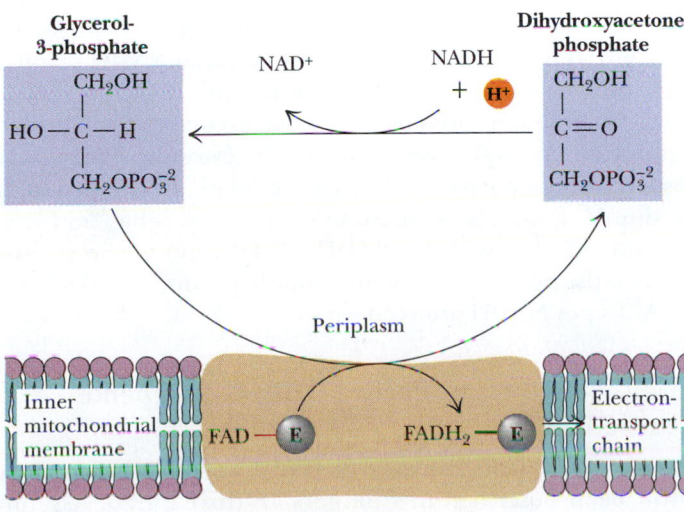

FIGURE 20.33 The glycerophosphate shuttle (also known as the glycerol phosphate shuttle) couples the cytosolic oxidation of NADH with mitochondrial reduction of [FAD].

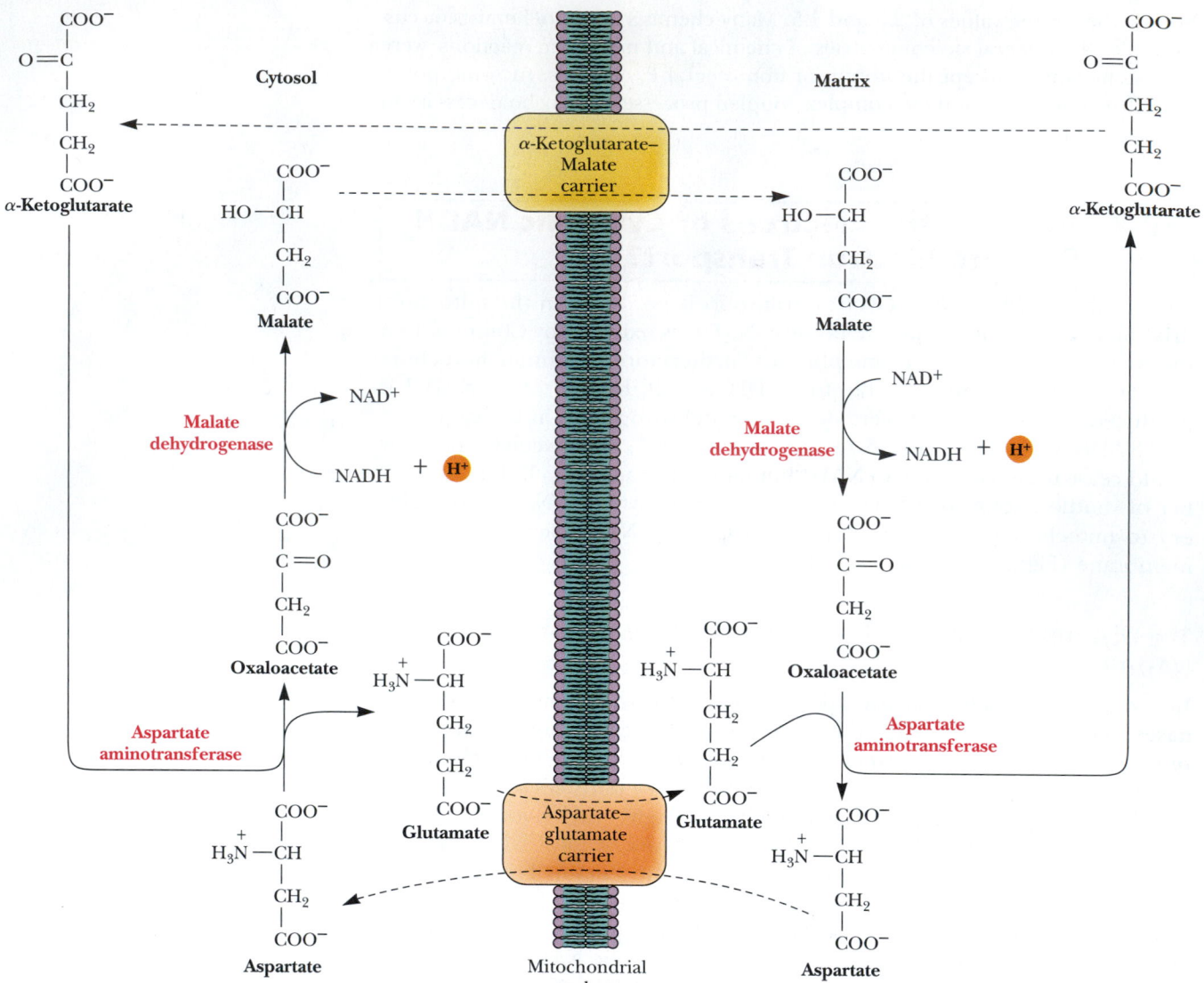

FIGURE 20.34 The malate (oxaloacetate)–aspartate shuttle, which operates across the inner mitochondrial membrane.

The Malate–Aspartate Shuttle Is Reversible

The second electron shuttle system, called the **malate–aspartate shuttle,** is shown in Figure 20.34. Oxaloacetate is reduced in the cytosol, acquiring the electrons of NADH (which is oxidized to NAD$^+$). Malate is transported across the inner membrane, where it is reoxidized by malate dehydrogenase, converting NAD$^+$ to NADH in the matrix. This mitochondrial NADH readily enters the electron-transport chain. The oxaloacetate produced in this reaction cannot cross the inner membrane and must be transaminated to form aspartate, which can be transported across the membrane to the cytosolic side. Transamination in the cytosol recycles aspartate back to oxaloacetate. In contrast to the glycerol phosphate shuttle, the malate–aspartate cycle is reversible, and it operates as shown in Figure 20.34 only if the NADH/NAD$^+$ ratio in the cytosol is higher than the ratio in the matrix. Because this shuttle produces NADH in the matrix, the full 2.5 ATPs per NADH are recovered.

The Net Yield of ATP from Glucose Oxidation Depends on the Shuttle Used

The complete route for the conversion of the metabolic energy of glucose to ATP has now been described in Chapters 18 through 20. Assuming appropriate P/O ratios, the number of ATP molecules produced by the complete

Table 20.4

Yield of ATP from Glucose Oxidation

Pathway	Glycerol–Phosphate Shuttle	Malate–Aspartate Shuttle
Glycolysis: glucose to pyruvate (cytosol)		
Phosphorylation of glucose	−1	−1
Phosphorylation of fructose-6-phosphate	−1	−1
Dephosphorylation of 2 molecules of 1,3-BPG	+2	+2
Dephosphorylation of 2 molecules of PEP	+2	+2
Oxidation of 2 molecules of glyceraldehyde-3-phosphate yields 2 NADH		
Pyruvate conversion to acetyl-CoA (mitochondria)		
2 NADH		
Citric acid cycle (mitochondria)		
2 molecules of GTP from 2 molecules of succinyl-CoA	+2	+2
Oxidation of 2 molecules each of isocitrate, α-ketoglutarate, and malate yields 6 NADH		
Oxidation of 2 molecules of succinate yields 2 [FADH$_2$]		
Oxidative phosphorylation (mitochondria)		
2 NADH from glycolysis yield 1.5 ATPs each if NADH is oxidized by glycerol–phosphate shuttle; 2.5 ATP by malate–aspartate shuttle	+3	+5
Oxidative decarboxylation of 2 pyruvate to 2 acetyl-CoA: 2 NADH produce 2.5 ATPs each	+5	+5
2 [FADH$_2$] from each citric acid cycle produce 1.5 ATPs each	+3	+3
6 NADH from citric acid cycle produce 2.5 ATPs each	+15	+15
Net Yield	30	32

Note: These P/O ratios of 2.5 and 1.5 for mitochondrial oxidation of NADH and [FADH$_2$] are "consensus values." Because they may not reflect actual values and because these ratios may change depending on metabolic conditions, these estimates of ATP yield from glucose oxidation are approximate.

oxidation of a molecule of glucose can be estimated. Keeping in mind that P/O ratios must be viewed as approximate, for all the reasons previously cited, we will assume the values of 2.5 and 1.5 for the mitochondrial oxidation of NADH and succinate, respectively. In eukaryotic cells, the combined pathways of glycolysis, the TCA cycle, electron transport, and oxidative phosphorylation then yield a net of approximately 30 to 32 molecules of ATP per molecule of glucose oxidized, depending on the shuttle route used (Table 20.4).

The net stoichiometric equation for the oxidation of glucose, using the glycerol phosphate shuttle, is

$$\text{Glucose} + 6\ O_2 + {\sim}30\ \text{ADP} + {\sim}30\ P_i \longrightarrow$$
$$6\ CO_2 + {\sim}30\ \text{ATP} + {\sim}36\ H_2O \qquad (20.32)$$

Because the 2 NADH formed in glycolysis are "transported" by the glycerol phosphate shuttle in this case, they each yield only 1.5 ATP, as already described. On the other hand, if these 2 NADH take part in the malate–aspartate shuttle, each yields 2.5 ATP, giving a total (in this case) of 32 ATP formed per glucose oxidized. Most of the ATP—26 out of 30 or 28 out of 32—is produced by oxidative phosphorylation; only 4 ATP molecules result from direct synthesis during glycolysis and the TCA cycle.

The situation in bacteria is somewhat different. Prokaryotic cells need not carry out ATP/ADP exchange. Thus, bacteria have the potential to produce approximately 38 ATP per glucose.

3.5 Billion Years of Evolution Have Resulted in a Very Efficient System

Hypothetically speaking, how much energy does a eukaryotic cell extract from the glucose molecule? Taking a value of 50 kJ/mol for the hydrolysis of ATP under cellular conditions (see Chapter 3), the production of 32 ATPs per glucose oxidized yields 1600 kJ/mol of glucose. The cellular oxidation (combustion) of glucose yields $\Delta G = -2937$ kJ/mol. We can calculate an efficiency for the pathways of glycolysis, the TCA cycle, electron transport, and oxidative phosphorylation of $1600/2937 \times 100\% = 54\%$. This is the result of approximately 3.5 billion years of evolution.

Summary

20.1 Where in the Cell Are Electron Transport and Oxidative Phosphorylation Carried Out? The processes of electron transport and oxidative phosphorylation are membrane associated. In bacteria, the conversion of energy from NADH and [FADH$_2$] to the energy of ATP via electron transport and oxidative phosphorylation is carried out at (and across) the plasma membrane. In eukaryotic cells, electron transport and oxidative phosphorylation are localized in mitochondria. Mitochondria are surrounded by a simple outer membrane and a more complex inner membrane (Figure 20.1). The space between the inner and outer membranes is referred to as the intermembrane space.

20.2 What Are Reduction Potentials, and How Are They Used to Account for Free Energy Changes in Redox Reactions? Just as the group transfer potential is used to quantitate the energy of phosphoryl transfer, the standard reduction potential, denoted by $\mathscr{E}_o'$, quantitates the tendency of chemical species to be reduced or oxidized. Standard reduction potentials are determined by measuring the voltages generated in reaction half-cells. A half-cell consists of a solution containing 1 M concentrations of both the oxidized and reduced forms of the substance whose reduction potential is being measured and a simple electrode.

20.3 How Is the Electron-Transport Chain Organized? The components of the electron-transport chain can be purified from the mitochondrial inner membrane as four distinct protein complexes: (I) NADH–coenzyme Q reductase, (II) succinate–coenzyme Q reductase, (III) coenzyme Q–cytochrome c reductase, and (IV) cytochrome c oxidase.

Complex I (NADH dehydrogenase) involves more than 30 polypeptide chains, 1 molecule of flavin mononucleotide (FMN), and as many as seven Fe-S clusters, together containing a total of 20 to 26 iron atoms. The complex transfers electrons from NADH to FMN, then to a series of FeS proteins, and finally to coenzyme Q.

Complex II (succinate dehydrogenase) oxidizes succinate to fumarate, with concomitant reduction of bound FAD to FADH$_2$. This FADH$_2$ transfers its electrons immediately to Fe-S centers, which pass them on to UQ. Electrons flow from succinate to UQ.

Complex III drives electron transport from coenzyme Q to cytochrome c via a unique redox pathway known as the Q cycle. UQ–cytochrome c reductase (UQ–cyt c reductase), as this complex is known, involves three different cytochromes and an Fe-S protein. In the cytochromes of these and similar complexes, the iron atom at the center of the porphyrin ring cycles between the reduced Fe^{2+} (ferrous) and oxidized Fe^{3+} (ferric) states.

Complex IV transfers electrons from cytochrome c to reduce oxygen on the matrix side. Complex IV (cytochrome c oxidase) accepts electrons from cytochrome c and directs them to the four-electron reduction of O$_2$ to form H$_2$O via Cu$_A$ sites, the heme iron of cytochrome a, Cu$_B$, and the heme iron of a_3.

20.4 What Are the Thermodynamic Implications of Chemiosmotic Coupling? Peter Mitchell's chemiosmotic hypothesis revolutionized our thinking about the energy coupling that drives ATP synthesis by means of an electrochemical gradient. The free energy difference for protons across the inner mitochondrial membrane includes a term for the concentration difference and a term for the electrical potential. It is this energy that drives the synthesis of ATP, in accord with Mitchell's model.

20.5 How Does a Proton Gradient Drive the Synthesis of ATP? The mitochondrial complex that carries out ATP synthesis is ATP synthase (F$_1$F$_0$–ATPase). ATP synthase consists of two principal complexes, designated F$_1$ and F$_0$. Protons taken up from the cytosol by one of the proton access channels in the a-subunit of F$_0$ ride the rotor of c-subunits until they reach the other proton access channel on a, from which they are released into the matrix. Such rotation causes the γ-subunit of F$_1$ to turn relative to the three β-subunit nucleotide sites of F$_1$, changing the conformation of each in sequence, so ADP is first bound, then phosphorylated, then released, according to Boyer's binding change mechanism.

The inhibitors of oxidative phosphorylation include rotenone, a common insecticide that strongly inhibits the NADH–UQ reductase. 2-Thenoyltrifluoroacetone and carboxin and its derivatives specifically block Complex II. Antimycin, an antibiotic produced by *Streptomyces griseus*, inhibits the UQ–cytochrome c reductase by blocking electron transfer between b_H and coenzyme Q in the Q$_n$ site. Myxothiazol inhibits the same complex by acting at the Q$_p$ site. Complex IV is specifically inhibited by cyanide (CN$^-$), azide (N$_3^-$), and carbon monoxide (CO). Cyanide and azide bind tightly to the ferric form of cytochrome a_3, whereas carbon monoxide binds only to the ferrous form.

Uncouplers disrupt the coupling of electron transport and ATP synthase. Uncouplers share two common features: hydrophobic character and a dissociable proton. They function by carrying protons across the inner membrane, acquiring protons on the outer surface of the membrane (where the proton concentration is high) and carrying them to the matrix side. Uncouplers destroy the proton gradient that couples electron transport and the ATP synthase.

ATP–ADP translocase mediates the movement of ATP and ADP across the mitochondrial membrane. The ATP–ADP translocase is an inner membrane protein that tightly couples the exit of ATP with the entry of ADP so that the mitochondrial nucleotide levels remain approximately constant. For each ATP transported out, one ADP is transported into the matrix. ATP–ADP translocase binds ATP on the matrix side, reorients to face the intermembrane space, and exchanges ATP for ADP, with subsequent movement back to the matrix face of the inner membrane.

20.6 What Is the P/O Ratio for Mitochondrial Electron Transport and Oxidative Phosphorylation?
The P/O ratio is the number of molecules of ATP formed in oxidative phosphorylation per two electrons flowing through a defined segment of the electron-transport chain. The consensus value for the mitochondrial P/O ratio is 10/4, or 2.5, for the case of electrons entering the electron-transport

chain as NADH. For succinate to O_2, the P/O ratio in this case would be 6/4, or 1.5.

20.7 How Are the Electrons of Cytosolic NADH Fed into Electron Transport?
Eukaryotic cells have a number of shuttle systems that harvest the electrons of cytosolic NADH for delivery to mitochondria without actually transporting NADH across the inner membrane. In the glycerophosphate shuttle, two different glycerophosphate dehydrogenases, one in the cytoplasm and one on the outer face of the mitochondrial inner membrane, work together to carry electrons into the mitochondrial matrix. In the malate–aspartate shuttle, oxaloacetate is reduced in the cytosol, acquiring the electrons of NADH (which is oxidized to NAD^+). Malate is transported across the inner membrane, where it is reoxidized by malate dehydrogenase, converting NAD^+ to NADH in the matrix.

Problems

1. For the following reaction,

$$[FAD] + 2 \text{ cyt } c \text{ (Fe}^{2+}) + 2 \text{ H}^+ \longrightarrow [FADH_2] + 2 \text{ cyt } c \text{ (Fe}^{3+})$$

determine which of the redox couples is the electron acceptor and which is the electron donor under standard-state conditions, calculate the value of $\Delta \mathscr{E}_o'$, and determine the free energy change for the reaction.

2. Calculate the value of $\Delta G^{o\prime}$ for the glyceraldehyde-3-phosphate dehydrogenase reaction, and calculate the free energy change for the reaction under standard-state conditions.

3. For the following redox reaction,

$$NAD^+ + 2 \text{ H}^+ + 2 \text{ } e^- \longrightarrow NADH + H^+$$

suggest an equation (analogous to Equation 20.13) that predicts the pH dependence of this reaction, and calculate the reduction potential for this reaction at pH 8.

4. Sodium nitrite ($NaNO_2$) is used by emergency medical personnel as an antidote for cyanide poisoning (for this purpose, it must be administered immediately). Based on the discussion of cyanide poisoning in Section 20.5, suggest a mechanism for the lifesaving effect of sodium nitrite.

5. A wealthy investor has come to you for advice. She has been approached by a biochemist who seeks financial backing for a company that would market dinitrophenol and dicumarol as weight-loss medications. The biochemist has explained to her that these agents are uncouplers and that they would dissipate metabolic energy as heat. The investor wants to know if you think she should invest in the biochemist's company. How do you respond?

6. Assuming that 3 H^+ are transported per ATP synthesized in the mitochondrial matrix, the membrane potential difference is 0.18 V (negative inside), and the pH difference is 1 unit (acid outside, basic inside), calculate the largest ratio of [ATP]/[ADP][P_i] under which synthesis of ATP can occur.

7. Of the dehydrogenase reactions in glycolysis and the TCA cycle, all but one use NAD^+ as the electron acceptor. The lone exception is the succinate dehydrogenase reaction, which uses covalently bound FAD of a flavoprotein as the electron acceptor. The standard reduction potential for this bound FAD is in the range of 0.003 to 0.091 V (Table 20.1). Compared with the other dehydrogenase reactions of glycolysis and the TCA cycle, what is unique about succinate dehydrogenase? Why is bound FAD a more suitable electron acceptor in this case?

8. **a.** What is the standard free energy change ($\Delta G^{o\prime}$) for the reduction of coenzyme Q by NADH as carried out by Complex I (NADH–coenzyme Q reductase) of the electron-transport pathway if $\mathscr{E}_o'$ ($NAD^+/NADH$) = −0.320 V and $\mathscr{E}_o'$ ($CoQ/CoQH_2$) = +0.060 V.
   b. What is the equilibrium constant (K_{eq}) for this reaction?

   c. Assume that (1) the actual free energy release accompanying the NADH–coenzyme Q reductase reaction is equal to the amount released under standard conditions (as calculated in part a), (2) this energy can be converted into the synthesis of ATP with an efficiency = 0.75 (that is, 75% of the energy released upon NADH oxidation is captured in ATP synthesis), and (3) the oxidation of 1 equivalent of NADH by coenzyme Q leads to the phosphorylation of 1 equivalent of ATP.
   Under these conditions, what is the maximum ratio of [ATP]/[ADP] attainable by oxidative phosphorylation when [P_i] = 1 mM? (Assume $\Delta G^{o\prime}$ for ATP synthesis = +30.5 kJ/mol.)

9. Consider the oxidation of succinate by molecular oxygen as carried out via the electron-transport pathway

$$\text{Succinate} + \tfrac{1}{2}O_2 \longrightarrow \text{fumarate} + H_2O$$

   a. What is the standard free energy change ($\Delta G^{o\prime}$) for this reaction if $\mathscr{E}_o'$ (Fum/Succ) = +0.031 V and $\mathscr{E}_o'$ ($\tfrac{1}{2}O_2/H_2O$) = +0.816 V.
   b. What is the equilibrium constant (K_{eq}) for this reaction?
   c. Assume that (1) the actual free energy release accompanying succinate oxidation by the electron-transport pathway is equal to the amount released under standard conditions (as calculated in part a), (2) this energy can be converted into the synthesis of ATP with an efficiency = 0.7 (that is, 70% of the energy released upon succinate oxidation is captured in ATP synthesis), and (3) the oxidation of 1 succinate leads to the phosphorylation of 2 equivalents of ATP.
   Under these conditions, what is the maximum ratio of [ATP]/[ADP] attainable by oxidative phosphorylation when [P_i] = 1 mM? (Assume $\Delta G^{o\prime}$ for ATP synthesis = +30.5 kJ/mol.)

10. Consider the oxidation of NADH by molecular oxygen as carried out via the electron-transport pathway

$$NADH + H^+ + \tfrac{1}{2}O_2 \longrightarrow NAD^+ + H_2O$$

   a. What is the standard free energy change ($\Delta G^{o\prime}$) for this reaction if $\mathscr{E}_o'$ ($NAD^+/NADH$) = −0.320 V and $\mathscr{E}_o'$ (O_2/H_2O) = +0.816 V.
   b. What is the equilibrium constant (K_{eq}) for this reaction?
   c. Assume that (1) the actual free energy release accompanying NADH oxidation by the electron-transport pathway is equal to the amount released under standard conditions (as calculated in part a), (2) this energy can be converted into the synthesis of ATP with an efficiency = 0.75 (that is, 75% of the energy released upon NADH oxidation is captured in ATP synthesis), and (3) the oxidation of 1 NADH leads to the phosphorylation of 3 equivalents of ATP.
   Under these conditions, what is the maximum ratio of [ATP]/[ADP] attainable by oxidative phosphorylation when [P_i] = 2 mM? (Assume $\Delta G^{o\prime}$ for ATP synthesis = +30.5 kJ/mol.)

11. Write a balanced equation for the reduction of molecular oxygen by reduced cytochrome c as carried out by Complex IV (cytochrome oxidase) of the electron-transport pathway.
    a. What is the standard free energy change ($\Delta G°'$) for this reaction if

    $$\Delta \mathscr{E}_o' \text{ cyt } c(Fe^{3+})/\text{cyt } c(Fe^{2+}) = +0.254 \text{ volts and}$$
    $$\mathscr{E}_o' \left(\tfrac{1}{2}O_2/H_2O\right) = 0.816 \text{ volts}$$

    b. What is the equilibrium constant (K_{eq}) for this reaction?
    c. Assume that (1) the actual free energy release accompanying cytochrome c oxidation by the electron-transport pathway is equal to the amount released under standard conditions (as calculated in part a), (2) this energy can be converted into the synthesis of ATP with an efficiency = 0.6 (that is, 60% of the energy released upon cytochrome c oxidation is captured in ATP synthesis), and (3) the reduction of 1 molecule of O_2 by reduced cytochrome c leads to the phosphorylation of 2 equivalents of ATP.
    Under these conditions, what is the maximum ratio of [ATP]/[ADP] attainable by oxidative phosphorylation when [P_i] = 3 mM? (Assume $\Delta G°'$ for ATP synthesis = +30.5 kJ/mol.)

12. The standard reduction potential for (NAD^+/NADH) is -0.320 V, and the standard reduction potential for (pyruvate/lactate) is -0.185 V.
    a. What is the standard free energy change ($\Delta G°'$) for the lactate dehydrogenase reaction:

    $$NADH + H^+ + \text{pyruvate} \longrightarrow \text{lactate} + NAD^+$$

    b. What is the equilibrium constant (K_{eq}) for this reaction?
    c. If [pyruvate] = 0.05 mM and [lactate] = 2.9 mM and ΔG for the lactate dehydrogenase reaction = -15 kJ/mol in erythrocytes, what is the [NAD^+]/[NADH] ratio under these conditions?

13. Assume that the free energy change (ΔG) associated with the movement of 1 mole of protons from the outside to the inside of a bacterial cell is -23 kJ/mol and 3 H$^+$ must cross the bacterial plasma membrane per ATP formed by the bacterial F_1F_0–ATP synthase. ATP synthesis thus takes place by the coupled process:

    $$3 \text{ H}^+_{out} + ADP + P_i \rightleftharpoons 3 \text{ H}^+_{in} + ATP + H_2O$$

    a. If the overall free energy change ($\Delta G_{overall}$) associated with ATP synthesis in these cells by the coupled process is -21 kJ/mol, what is the equilibrium constant (K_{eq}) for the process?
    b. What is $\Delta G_{synthesis}$, the free energy change for ATP synthesis, in these bacteria under these conditions?
    c. The standard free energy change for ATP hydrolysis ($\Delta G°'_{hydrolysis}$) is -30.5 kJ/mol. If [P_i] = 2 mM in these bacterial cells, what is the [ATP]/[ADP] ratio in these cells?

14. Describe in your own words the path of electrons through the Q cycle of Complex III.

15. Describe in your own words the path of electrons through the copper and iron centers of Complex IV.

Preparing for the MCAT Exam

16. Based on your reading on the F_1F_0–ATPase, what would you conclude about the mechanism of ATP synthesis:
    a. The reaction proceeds by nucleophilic substitution via the S_N2 mechanism.
    b. The reaction proceeds by nucleophilic substitution via the S_N1 mechanism.
    c. The reaction proceeds by electrophilic substitution via the E1 mechanism.
    d. The reaction proceeds by electrophilic substitution via the E2 mechanism.

17. Imagine that you are working with isolated mitochondria and you manage to double the ratio of protons outside to protons inside. In order to maintain the overall ΔG at its original value (whatever it is), how would you have to change the mitochondria membrane potential?

Biochemistry ⓔNow™   Preparing for an exam? Test yourself on key questions at http://chemistry.brookscole.com/ggb3

Further Reading

Bioenergetics

Babcock, G. T., and Wikström, M., 1992. Oxygen activation and the conservation of energy in cell respiration. *Nature* **356:**301–309.

Mitchell, P., 1979. Keilin's respiratory chain concept and its chemiosmotic consequences. *Science* **206:**1148–1159.

Nicholls, D. G., and Ferguson, S. J., 2002. *Bioenergetics 3*. London: Academic Press.

Nicholls, D. G., and Rial, E., 1984. Brown fat mitochondria. *Trends in Biochemical Sciences* **9:**489–491.

F_1-ATP synthase

Abraham, J. P., Leslie, A. G. W., Lutter, R., and Walker, J. E., 1994. Structure at 2.8 Å resolution of F_1-ATPase from bovine heart mitochondria. *Nature* **370:**621–628.

Boyer, P. D., 1989. A perspective of the binding change mechanism for ATP synthesis. *The FASEB Journal* **3:**2164–2178.

Boyer, P. D., 1997. The ATP synthase—a splendid molecular machine. *Annual Review of Biochemistry* **66:**717–750.

Cross, R. L., 1994. Our primary source of ATP. *Nature* **370:**594–595.

Junge, W., Lill, H., and Engelbrecht, S., 1997. ATP synthase: An electrochemical transducer with rotatory mechanics. *Trends in Biochemical Sciences* **22:**420–423.

Mitchell, P., and Moyle, J., 1965. Stoichiometry of proton translocation through the respiratory chain and adenosine triphosphatase systems of rat mitochondria. *Nature* **208:**147–151.

Noji, H., Yasuda, R., Yoshida, M., and Kinosita, K., 1997. Direct observation of the rotation of F_1-ATPase. *Nature* **386:**299–302.

Pedersen, P., and Carafoli, E., 1987. Ion-motive ATPases. I. Ubiquity, properties and significance to cell function. *Trends in Biochemical Sciences* **12:**146–150.

Sabbert, D., Engelbrecht, S., and Junge, W., 1996. Intersubunit rotation in active F_1-ATPase. *Nature* **381:**623–625.

Wilkens, S., Dunn, S. D., Chandler, J., et al., 1997. Solution structure of the N-terminal domain of the δ subunit of the *E. coli* ATP synthase. *Nature Structural Biology* **4:**198–201.

Cytochrome c Oxidase

Ferguson-Miller, S., 1996. Mammalian cytochrome c oxidase, a molecular monster subdued. *Science* **272:**1125.

Iwata, S., Ostermeier, C., Ludwig, B., and Michel, H., 1995. Structure at 2.8 Å resolution of cytochrome c oxidase from *Paracoccus denitrificans*. *Nature* **376:**660–669.

Tsukihara, T., Aoyama, H., Yamashita, E., et al., 1996. The whole structure of the 13-subunit oxidized cytochrome c oxidase at 2.8 Å. *Science* **272:**1136–1144.

Apoptosis

Kroemer, G., 2003. Mitochondrial control of apoptosis: An introduction. *Biochemical and Biophysical Research Communications* **304**: 433–435.

Mattson, M. P., and Kroemer, G., 2003. Mitochondria in cell death: novel targets for neuroprotection and cardioprotection. *Trends in Molecular Medicine* **9**:196–205.

Van Gurp, M., Festjens, N., van Loo, G., Saelens, X., and Verdenebeele, P., 2003. Mitochondrial intermembrane proteins in cell death. *Biochemical and Biophysical Research Communications* **304**:487–497.

Electron Transfer

Moser, C. C., et al. 1992. Nature of biological electron transfer. *Nature* **355**:796–802.

Naqui, A., Chance, B., and Cadenas, E., 1986. Reactive oxygen intermediates in biochemistry. *Annual Review of Biochemistry* **55**:137.

Slater, E. C., 1983. The Q cycle: An ubiquitous mechanism of electron transfer. *Trends in Biochemical Sciences* **8**:239–242.

Trumpower, B. L., 1990. Cytochrome bc_1 complexes of microorganisms. *Microbiological Reviews* **54**:101–129.

Trumpower, B. L., 1990. The protonmotive Q cycle—energy transduction by coupling of proton translocation to electron transfer by the cytochrome bc_1 complex. *Journal of Biological Chemistry* **265**:11409–11412.

Walker, J. E., 1992. The NADH:ubiquinone oxidoreductase (Complex I) of respiratory chains. *Quarterly Reviews of Biophysics* **25**:253–324.

Weiss, H., Friedrich, T., Hofhaus, G., and Preis, D., 1991. The respiratory-chain NADH dehydrogenase (Complex I) of mitochondria. *European Journal of Biochemistry* **197**:563–576.

Xia, D., Yu, C-A., Kim, H., et al., 1997. The crystal structure of the cytochrome bc_1 complex from bovine heart mitochondria. *Science* **277**:60–66.

CHAPTER 21

Photosynthesis

Essential Question

Photosynthesis is the primary source of energy for all life forms (except chemolithotrophic bacteria). Much of the energy of photosynthesis is used to drive the synthesis of organic molecules from atmospheric CO_2. *How is solar energy captured and transformed into metabolically useful chemical energy? How is the chemical energy produced by photosynthesis used to create organic molecules from carbon dioxide?*

The vast majority of energy consumed by living organisms stems from solar energy captured by the process of photosynthesis. Only chemolithotrophic bacteria (see Chapter 17) are independent of this energy source. Of the 1.5×10^{22} kJ of energy reaching the earth each day from the sun, 1% is absorbed by photosynthetic organisms and transduced into chemical energy.[1] This energy, in the form of biomolecules, becomes available to other members of the biosphere through food chains. The transduction of solar, or light, energy into chemical energy is often expressed in terms of **carbon dioxide fixation,** in which hexose is formed from carbon dioxide and oxygen is evolved:

$$6\ CO_2 + 6\ H_2O \xrightarrow{\text{Light}} C_6H_{12}O_6 + 6\ O_2 \qquad (21.1)$$

Estimates indicate that 10^{11} tons of carbon dioxide are fixed globally per year, of which one-third is fixed in the oceans, primarily by photosynthetic marine microorganisms.

Although photosynthesis is traditionally equated with CO_2 fixation, light energy (or rather the chemical energy derived from it) can be used to drive virtually any cellular process. The assimilation of inorganic forms of nitrogen and sulfur into organic molecules (see Chapter 25) represents two other metabolic conversions driven by light energy in green plants. Our previous considerations of aerobic metabolism (Chapters 18 through 20) treated cellular respiration (precisely the reverse of Equation 21.1) as the central energy-releasing process in life. It necessarily follows that the formation of hexose from carbon dioxide and water, the products of cellular respiration, must be endergonic. The necessary energy comes from light. Note that in the carbon dioxide fixation reaction described, light is used to drive a chemical reaction against its thermodynamic potential.

21.1 What Are the General Properties of Photosynthesis?

Photosynthesis Occurs in Membranes

Organisms capable of photosynthesis are very diverse, ranging from simple prokaryotic forms to the largest organisms of all, *Sequoia gigantea*, the giant redwood trees of California. Despite this diversity, we find certain generalities regarding photosynthesis. An important one is that *photosynthesis occurs in membranes*. In photosynthetic prokaryotes, the photosynthetic membranes fill up the cell interior; in photosynthetic eukaryotes, the photosynthetic membranes are localized in large organelles known as **chloroplasts** (Figures 21.1 and 21.2). Chloroplasts are one member in a family of related plant-specific organelles known as **plastids.** Chloroplasts themselves show a range of diversity, from the single, spiral chloroplast that gives *Spirogyra* its name to the multitude of ellipsoidal plastids typical of higher plant cells (Figure 21.3).

[1]Of the remaining 99%, two-thirds is absorbed by the earth and oceans, thereby heating the planet; the remaining one-third is lost as light reflected back into space.

Field of goldenrod.

In a sun-flecked lane,
Beside a path where cattle trod,
Blown by wind and rain,
Drawing substance from air and sod;

In ruggedness, it stands aloof,
The ragged grass and puerile leaves,
Lending a hand to fill the woof
In the pattern that beauty makes.

What mystery this, hath been wrought;
Beauty from sunshine, air, and sod!
Could we thus gain the ends we sought-
Tell us thy secret, Goldenrod.

Rosa Staubus, Oklahoma pioneer (1886–1966)

Key Questions

Biochemistry Now™ Test yourself on these Key Questions at BiochemistryNow at http://chemistry.brookscole.com/ggb3

Characteristic of all chloroplasts, however, is the organization of the inner membrane system, the so-called **thylakoid membrane.** The thylakoid membrane is organized into paired folds that extend throughout the organelle, as in Figure 21.1. These paired folds, or **lamellae,** give rise to flattened sacs or discs, **thylakoid vesicles** (from the Greek *thylakos,* meaning "sack"), which occur in stacks called **grana.** A single stack, or **granum,** may contain dozens of thylakoid vesicles, and different grana are joined by lamellae that run through the soluble portion, or **stroma,** of the organelle. Chloroplasts thus possess three membrane-bound aqueous compartments: the intermembrane space, the stroma, and the interior of the thylakoid vesicles, the so-called **thylakoid space** (also known as the **thylakoid lumen**). As we shall see, this third compartment serves an important function in the transduction of light energy into ATP formation. The thylakoid membrane has a highly characteristic lipid composition and, like the inner membrane of the mitochondrion, is impermeable to most ions and molecules. Chloroplasts, like their mitochondrial counterparts, possess DNA, RNA, and ribosomes and consequently display a considerable amount of autonomy. However, many critical chloroplast components are encoded by nuclear genes, so autonomy is far from absolute.

Photosynthesis Consists of Both Light Reactions and Dark Reactions

If a chloroplast suspension is illuminated in the absence of carbon dioxide, oxygen is evolved. Furthermore, if the illuminated chloroplasts are now placed in the dark and supplied with CO_2, net hexose synthesis can be observed (Figure 21.4). Thus, the evolution of oxygen can be temporally separated from CO_2 fixation and also has a light dependency that CO_2 fixation lacks. The **light reactions** of photosynthesis, of which O_2 evolution is only one part, are associated with the thylakoid membranes. In contrast, the light-independent reactions, or so-called **dark reactions,** notably CO_2 fixation, are located in the stroma. A concise summary of the photosynthetic process is that radiant electromagnetic energy (light) is transformed by a specific photochemical system located in the thylakoids to yield chemical energy in the form of reducing potential (NADPH) and high-energy phosphate (ATP). NADPH and ATP can then be used to drive the endergonic process of hexose formation from CO_2 by a series of enzymatic reactions found in the stroma (see Equation 21.3, which follows).

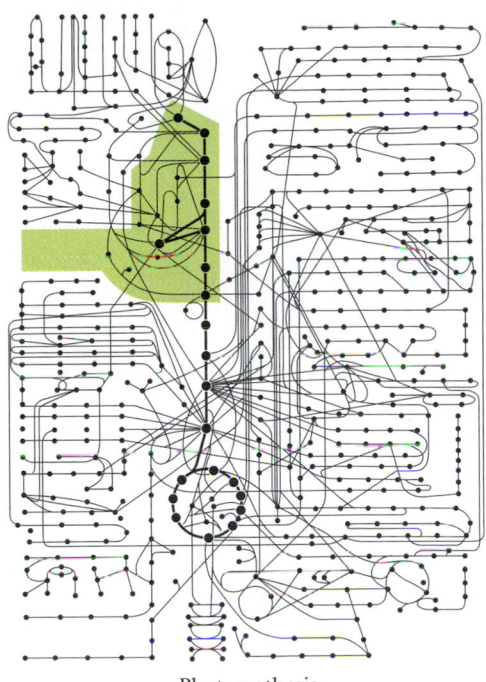

Photosynthesis

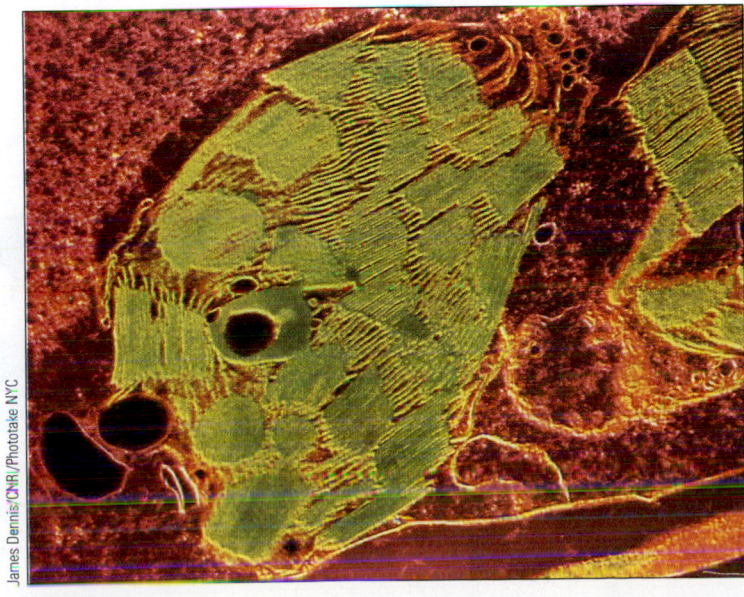

James Dennis/CNRI/Phototake NYC

FIGURE 21.1 Electron micrograph of a representative chloroplast.

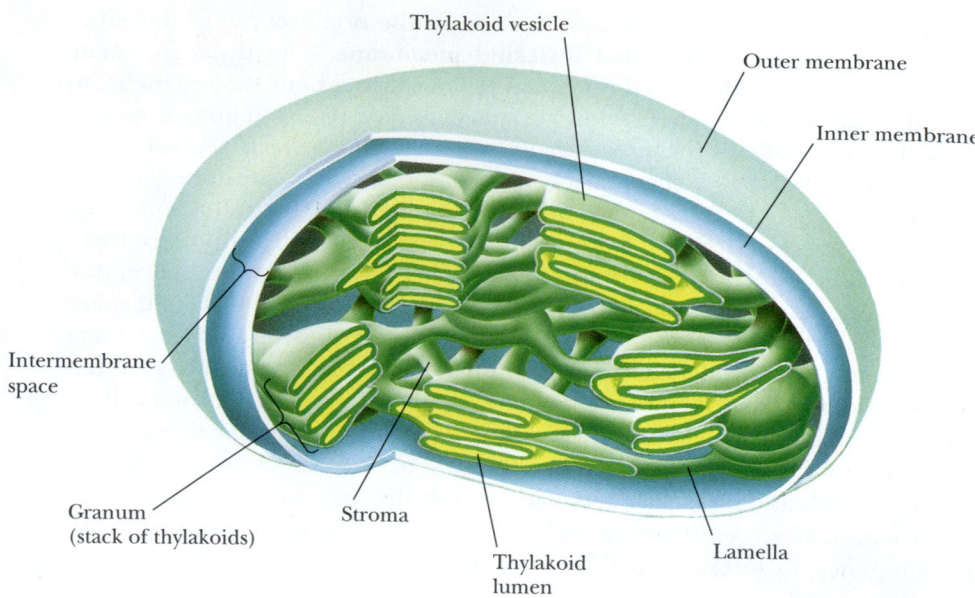

FIGURE 21.2 Schematic diagram of an idealized chloroplast.

Water Is the Ultimate e^- Donor for Photosynthetic NADP$^+$ Reduction

In green plants, water serves as the ultimate electron donor for the photosynthetic generation of reducing equivalents. The reaction sequence

$$2 \, H_2O + 2 \, NADP^+ + x \, ADP + x \, P_i \xrightarrow{nh\nu}$$
$$O_2 + 2 \, NADPH + 2 \, H^+ + x \, ATP + x \, H_2O \quad (21.2)$$

describes the process, where $nh\nu$ symbolizes light energy (n is some number of photons of energy $h\nu$, where h is Planck's constant and ν is the frequency of the light). Light energy is necessary to make the unfavorable reduction of NADP$^+$ by H_2O ($\Delta \mathscr{E}_o' = -1.136$ V; $\Delta G^{\circ\prime} = +219$ kJ/mol NADP$^+$) thermodynamically favorable. Thus, the light energy input, $nh\nu$, must exceed 219 kJ/mol NADP$^+$. The stoichiometry of ATP formation depends on the pattern of photophosphorylation operating in the cell at the time and on the ATP yield in terms of the chemiosmotic ratio, ATP/H$^+$, as we will see later. Nevertheless, the stoichiometry of the metabolic pathway of CO_2 fixation is certain:

$$12 \, NADPH + 12 \, H^+ + 18 \, ATP + 6 \, CO_2 + 12 \, H_2O \longrightarrow$$
$$C_6H_{12}O_6 + 12 \, NADP^+ + 18 \, ADP + 18 \, P_i \quad (21.3)$$

FIGURE 21.3 **(a)** *Spirogyra*—a freshwater green alga. **(b)** A higher plant cell.

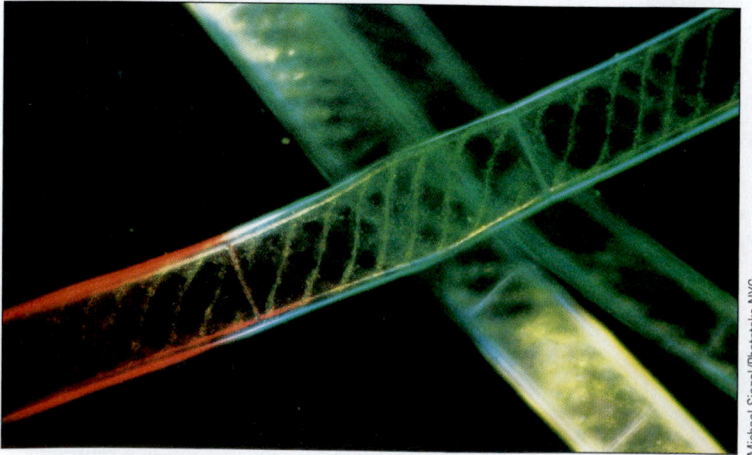

(a)

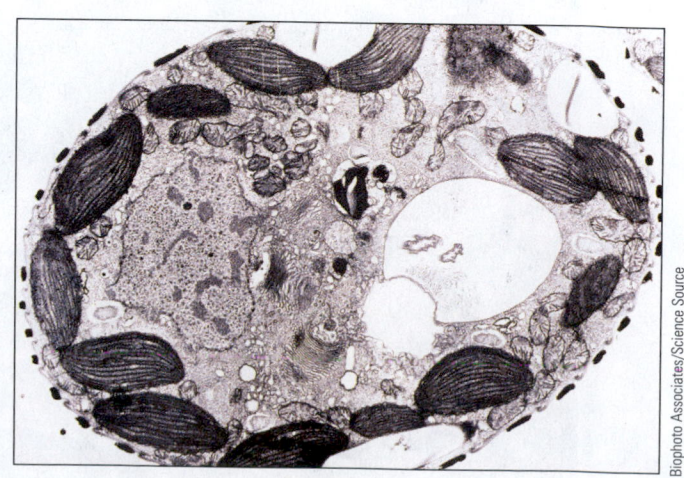

(b)

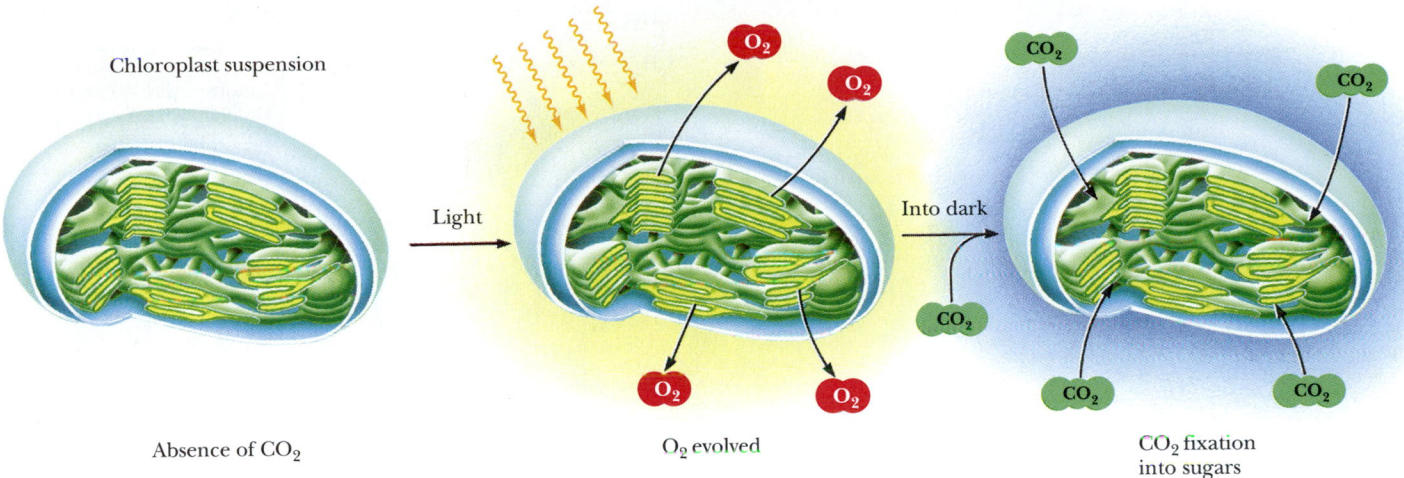

Chloroplast suspension

Light

Into dark

Absence of CO_2

O_2 evolved

CO_2 fixation
into sugars

Biochemistry ⊜Now™ **ANIMATED FIGURE 21.4**
The light-dependent and light-independent reactions
of photosynthesis. Light reactions are associated with
the thylakoid membranes, and light-independent reactions are associated with the stroma. **See this figure
animated at http://chemistry.brookscole.com/ggb3**

A More Generalized Equation for Photosynthesis   In 1931, comparative study of photosynthesis in bacteria led van Niel to a more general formulation of the overall reaction:

$$CO_2 \; + \; 2\,H_2A \; \xrightarrow{\text{Light}} \; (CH_2O) \; + \; 2A \; + \; H_2O \qquad (21.4)$$

| **Hydrogen acceptor** | **Hydrogen donor** | **Reduced acceptor** | **Oxidized donor** |

In photosynthetic bacteria, H_2A is variously H_2S (photosynthetic green and purple sulfur bacteria), isopropanol, or some similar oxidizable substrate. [(CH_2O) symbolizes a carbohydrate unit.]

$$CO_2 + 2\,H_2S \longrightarrow (CH_2O) + H_2O + 2\,S$$

$$CO_2 + 2\,CH_3-CHOH-CH_3 \longrightarrow (CH_2O) + H_2O + 2\,CH_3-\overset{\overset{\displaystyle O}{\displaystyle \|}}{C}-CH_3$$

In cyanobacteria and the eukaryotic photosynthetic cells of algae and higher plants, H_2A is H_2O, as implied earlier, and 2 A is O_2. The accumulation of O_2 to constitute 20% of the earth's atmosphere is the direct result of eons of global oxygenic photosynthesis.

21.2 How Is Solar Energy Captured by Chlorophyll?

Photosynthesis depends on the photoreactivity of **chlorophyll.** Chlorophylls are magnesium-containing substituted tetrapyrroles whose basic structure is reminiscent of heme, the iron-containing porphyrin (see Chapters 5 and 20). Chlorophylls differ from heme in a number of properties: Magnesium instead of iron is coordinated in the center of the planar conjugated ring structure; a long-chain alcohol, **phytol,** is esterified to a pyrrole ring substituent; and the methine bridge linking pyrroles III and IV is substituted and crosslinked to ring III, leading to the formation of a fifth five-membered ring. The structures of chlorophyll *a* and *b* are shown in Figure 21.5.

Chlorophylls are excellent light absorbers because of their aromaticity. That is, they possess delocalized π electrons above and below the planar ring structure. The energy differences between electronic states in these π orbitals correspond to the energies of visible light photons. When light energy is absorbed, an electron is promoted to a higher orbital, enhancing the potential for transfer of this electron to a suitable acceptor. Loss of such a photoexcited electron to an acceptor is an oxidation–reduction reaction. The net result is the transduction of light energy into the chemical energy of a redox reaction.

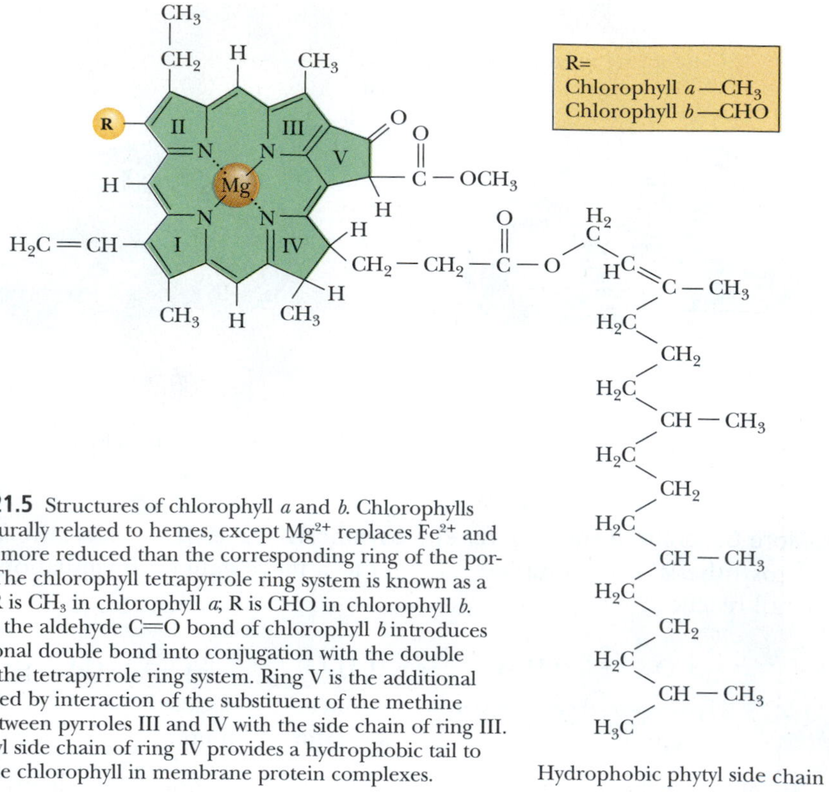

R=
Chlorophyll a —CH_3
Chlorophyll b —CHO

FIGURE 21.5 Structures of chlorophyll a and b. Chlorophylls are structurally related to hemes, except Mg^{2+} replaces Fe^{2+} and ring IV is more reduced than the corresponding ring of the porphyrins. The chlorophyll tetrapyrrole ring system is known as a chlorin. R is CH_3 in chlorophyll a; R is CHO in chlorophyll b. Note that the aldehyde C=O bond of chlorophyll b introduces an additional double bond into conjugation with the double bonds of the tetrapyrrole ring system. Ring V is the additional ring created by interaction of the substituent of the methine bridge between pyrroles III and IV with the side chain of ring III. The phytyl side chain of ring IV provides a hydrophobic tail to anchor the chlorophyll in membrane protein complexes.

Hydrophobic phytyl side chain

Chlorophylls and Accessory Light-Harvesting Pigments Absorb Light of Different Wavelengths

The absorption spectra of chlorophylls a and b (Figure 21.6) differ somewhat. Plants that possess both chlorophylls can harvest a wider spectrum of incident energy. Other pigments in photosynthetic organisms, so-called **accessory light-harvesting pigments** (Figure 21.7), increase the possibility for absorption of incident light of wavelengths not absorbed by the chlorophylls. Carotenoids and phycocyanobilins, like chlorophyll, possess many conjugated double bonds and thus absorb visible light. Carotenoids have two primary roles in photosynthesis—light harvesting and photoprotection through destruction of reactive oxygen species that arise as by-products of photoexcitation.

The Light Energy Absorbed by Photosynthetic Pigments Has Several Possible Fates

Each photon represents a quantum of light energy. A quantum of light energy absorbed by a photosynthetic pigment has four possible fates (Figure 21.8):

1. **Loss as heat.** The energy can be dissipated as heat through redistribution into atomic vibrations within the pigment molecule.
2. **Loss of light.** Energy of excitation reappears as **fluorescence** (light emission); a photon of fluorescence is emitted as the e^- returns to a lower orbital. This fate is common only in saturating light intensities. For thermodynamic reasons, the photon of fluorescence has a longer wavelength and hence lower energy than the quantum of excitation.
3. **Resonance energy transfer.** The excitation energy can be transferred by resonance energy transfer to a neighboring molecule if the energy level difference between the two corresponds to the quantum of excitation energy. In this process, the energy transferred raises an electron in the receptor molecule to a higher energy state as the photoexcited e^- in the original absorbing

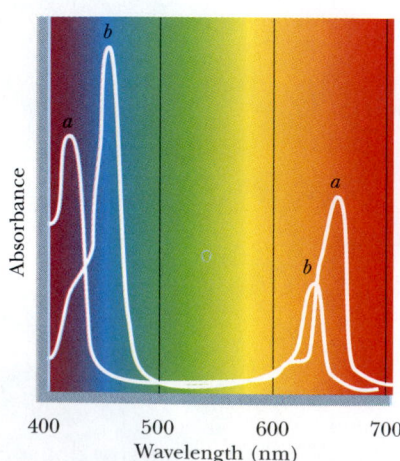

FIGURE 21.6 Absorption spectra of chlorophylls a and b.

(a)

β-Carotene

(b)

Phycocyanobilin

FIGURE 21.7 Structures of representative accessory light-harvesting pigments in photosynthetic cells. **(a)** β-Carotene, an accessory light-harvesting pigment in leaves. Note the many conjugated double bonds. **(b)** Phycocyanobilin, a blue pigment found in cyanobacteria. It is a linear or open pyrrole.

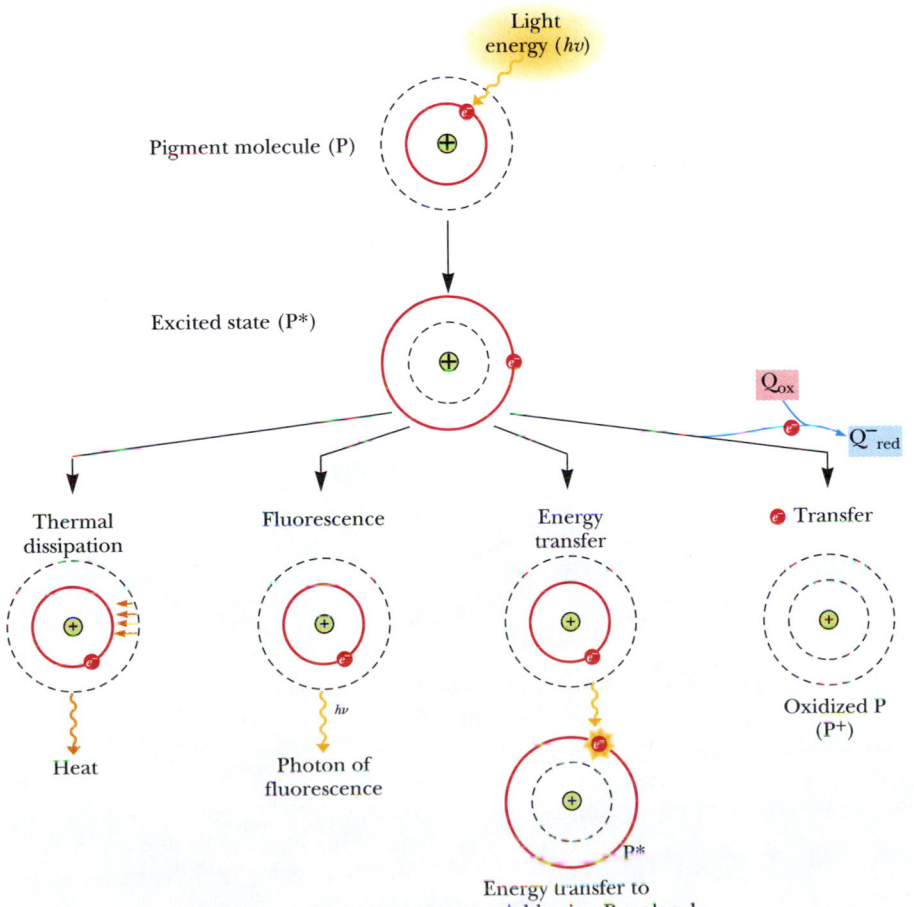

Light energy ($h\nu$)

Pigment molecule (P)

Excited state (P*)

Q_{ox}

Q^-_{red}

Thermal dissipation

Fluorescence

Energy transfer

Transfer

Oxidized P (P+)

Heat

Photon of fluorescence

$h\nu$

P*

Energy transfer to neighboring P molecule

Biochemistry Now™ ANIMATED FIGURE 21.8
Possible fates of the quantum of light energy absorbed by photosynthetic pigments. **See this figure animated at http://chemistry.brookscole.com/ggb3**

molecule returns to ground state. This so-called *Förster resonance energy transfer* is the mechanism whereby quanta of light falling anywhere within an array of pigment molecules can be transferred ultimately to specific photochemically reactive sites.

4. **Energy transduction.** The energy of excitation, in raising an electron to a higher energy orbital, dramatically changes the standard reduction potential, $\mathscr{E}_o{}'$, of the pigment such that it becomes a much more effective electron donor. That is, the excited-state species, by virtue of having an electron at a higher energy level through light absorption, has become a potent electron donor. Reaction of this excited-state electron donor with an electron acceptor situated in its vicinity leads to the transformation, or **transduction,** of light energy (photons) to chemical energy (reducing power, the potential for electron-transfer reactions). *Transduction of light energy into chemical energy, the photochemical event, is the essence of photosynthesis.*

The Transduction of Light Energy into Chemical Energy Involves Oxidation–Reduction

The diagram presented in Figure 21.9 illustrates the fundamental transduction of light energy into chemical energy (an oxidation–reduction reaction) that is the basis of photosynthesis. Chlorophyll (Chl) resides in a membrane in close association with molecules competent in e^- transfer, symbolized here as A and B. Chl absorbs a photon of light, becoming activated to Chl* in the process. Electron transfer from Chl* to A leads to oxidized Chl (Chl·$^+$, a **cationic free radical**) and reduced A (A$^-$ in the diagram). Subsequent oxidation of A$^-$ eventually culminates in reduction of NADP$^+$ to NADPH. The **electron "hole"** in oxidized Chl (Chl·$^+$) is filled by transfer of an electron from B to Chl·$^+$, restoring Chl and creating B$^+$. B$^+$ is restored to B by an e^- donated by water. O$_2$ is the product of water oxidation. Note that the system is restored to its original state once NADPH is formed and H$_2$O is oxidized.

Photosynthetic Units Consist of Many Chlorophyll Molecules but Only a Single Reaction Center

In the early 1930s, Emerson and Arnold investigated the relationship between the amount of incident light energy, the amount of chlorophyll present, and the amount of oxygen evolved by illuminated algal cells. Emerson and Arnold were seeking to determine the **quantum yield of photosynthesis:** the number of electrons transferred per photon of light. Their studies gave an unexpected result: When algae were illuminated with very brief light flashes that could excite every

FIGURE 21.9 Model for light absorption by chlorophyll and transduction of light energy into an oxidation–reduction reaction. Chlorophyll is represented by Chl; A and B represent electron-transfer molecules adjacent to Chl in the membrane. **I:** Photoexcitation of Chl creates Chl*. **II:** Electron transfer from Chl* to A yields oxidized Chl (Chl$^+$) and reduced A (A$^-$) **III:** An electron-transfer pathway from A$^-$ to NADP$^+$ leads to NADPH formation and restoration of oxidized A (A). **IV:** Chl$^+$ accepts an electron from B, restoring Chl and generating oxidized B (B$^+$). **V:** B$^+$ is reduced back to B by an electron originating in H$_2$O. Water oxidation is the source of O$_2$ formation. Not shown here are the H$^+$ translocations that accompany these light-driven electron-transport reactions; such proton translocations establish a chemiosmotic gradient across the photosynethetic membrane that can drive ATP synthesis.

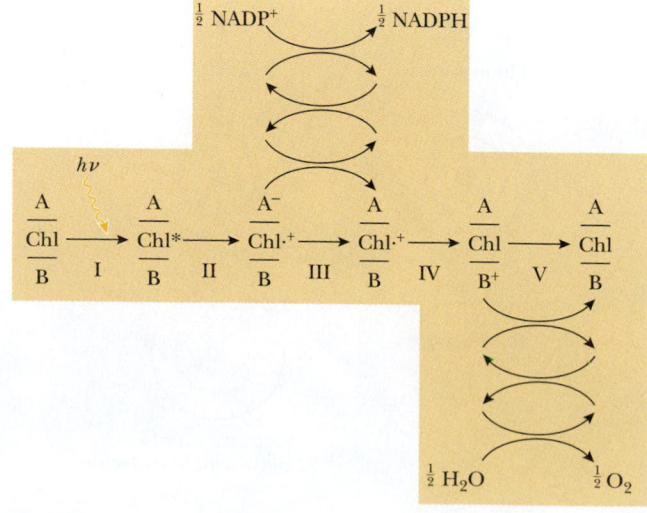

chlorophyll molecule at least once, only one molecule of O_2 was evolved per 2400 chlorophyll molecules. This result implied that not all chlorophyll molecules are photochemically reactive, and it led to the concept that photosynthesis occurs in functionally discrete units.

Chlorophyll serves two roles in photosynthesis. It is involved in light harvesting and the transfer of light energy to photoreactive sites by exciton transfer, and it participates directly in the photochemical events whereby light energy becomes chemical energy. A **photosynthetic unit** can be envisioned as an antenna of several hundred light-harvesting chlorophyll molecules plus a special pair of photochemically reactive chlorophyll *a* molecules called the **reaction center.** The purpose of the vast majority of chlorophyll in a photosynthetic unit is to harvest light incident within the unit and funnel it, via resonance energy transfer, to special reaction center chlorophyll molecules that are photochemically active. Most chlorophyll thus acts as a large light-collecting antenna, and it is at the reaction centers that the photochemical event occurs (Figure 21.10). Oxidation of chlorophyll leaves a **cationic free radical,** $Chl^{.+}$, whose properties as an electron acceptor have important consequences for photosynthesis. Note that the Mg^{2+} ion does not change in valence during these redox reactions.

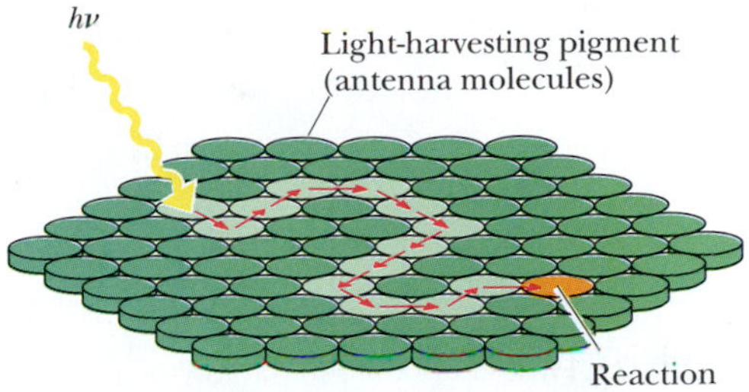

hv

Light-harvesting pigment (antenna molecules)

Reaction center

Biochemistry ⊗ Now™ ANIMATED FIGURE 21.10 Schematic diagram of a photosynthetic unit. The light-harvesting pigments, or antenna molecules (green), absorb and transfer light energy to the specialized chlorophyll dimer that constitutes the reaction center (orange). **See this figure animated at http://chemistry.brookscole.com/ggb3**

21.3 | What Kinds of Photosystems Are Used to Capture Light Energy?

All photosynthetic cells contain some form of photosystem. Photosynthetic bacteria have only one photosystem; furthermore, they lack the ability to use light energy to split H_2O and release O_2. Cyanobacteria, green algae, and higher plants are *oxygenic phototrophs* because they can generate O_2 from water. Oxygenic phototrophs have two distinct photosystems: **Photosystem I (PSI)** and **Photosystem II (PSII).** Type I photosytems use ferredoxins as terminal electron acceptors; type II photosystems use quinones as terminal electron acceptors. PSI is defined by reaction center chlorophylls with maximal red light absorption at 700 nm; PSII uses reaction centers that exhibit maximal red light absorption at 680 nm. The reaction center Chl of PSI is referred to as **P700** because it absorbs light of 700-nm wavelength; the reaction center Chl of PSII is called **P680** for analogous reasons. Both P700 and P680 are chlorophyll *a* dimers situated within specialized protein complexes. A distinct property of PSII is its role in light-driven O_2 evolution. Interestingly, the photosystems of photosynthetic bacteria are type II photosystems that resemble eukaryotic PSII more than PSI, even though these bacteria lack O_2-evolving capacity.

Biochemistry ⊗ Now™ Go to BiochemistryNow and click BiochemistryInteractive to examine the structure of *Synechococcus* PSI and see its similarities to the *R. viridis* reaction center.

Chlorophyll Exists in Plant Membranes in Association with Proteins

Detergent treatment of a suspension of thylakoids dissolves the membranes, releasing complexes containing both chlorophyll and protein. These chlorophyll–protein complexes represent integral components of the thylakoid membrane, and their organization reflects their roles as either **light-harvesting complexes (LHC), PSI complexes,** or **PSII complexes.** All chlorophyll is apparently localized within these three macromolecular assemblies.

PSI and PSII Participate in the Overall Process of Photosynthesis

What are the roles of the two photosystems, and what is their relationship to each other? PSI provides reducing power in the form of NADPH. PSII splits water, producing O_2, and feeds the electrons released into an electron-transport chain that couples PSII to PSI. Electron transfer between PSII and PSI pumps protons for chemiosmotic ATP synthesis. As summarized by Equation 21.2, photosynthesis involves the reduction of $NADP^+$, using electrons derived from water and activated by light, *hv*. ATP is generated in the process. The standard reduction potential for

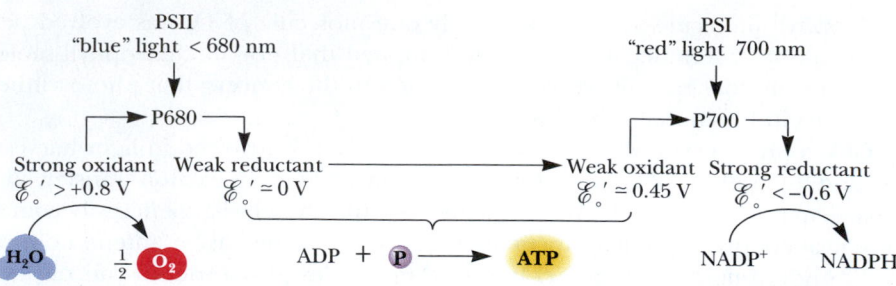

FIGURE 21.11 Roles of the two photosystems, PSI and PSII.

the $NADP^+/NADPH$ couple is -0.32 V. Thus, a strong reductant with an $\mathscr{E}_o{}'$ more negative than -0.32 V is required to reduce $NADP^+$ under standard conditions. By similar reasoning, a very strong oxidant will be required to oxidize water to oxygen because $\mathscr{E}_o{}'(\frac{1}{2}O_2/H_2O)$ is $+0.82$ V. Separation of the oxidizing and reducing aspects of Equation 21.2 is accomplished in nature by devoting PSI to $NADP^+$ reduction and PSII to water oxidation. PSI and PSII are linked via an electron-transport chain so that the weak reductant generated by PSII can provide an electron to reduce the weak oxidant side of P700 (Figure 21.11). Thus, *electrons flow from H_2O to $NADP^+$*, driven by light energy absorbed at the reaction centers. Oxygen is a by-product of the **photolysis,** literally "light-splitting," of water. Accompanying electron flow is production of a proton gradient and ATP synthesis (see Section 21.6). This light-driven phosphorylation is termed *photophosphorylation.*

The Pathway of Photosynthetic Electron Transfer Is Called the *Z* Scheme

Photosystems I and II contain unique complements of electron carriers, and these carriers mediate the stepwise transfer of electrons from water to $NADP^+$. When the individual redox components of PSI and PSII are arranged as an e^- transport chain according to their standard reduction potentials, the zigzag result resembles the letter *Z* laid sideways (Figure 21.12). The various electron carriers are indicated as follows: "Mn complex" symbolizes the manganese-containing oxygen-evolving complex; D is its e^- acceptor and the immediate e^- donor to $P680^+$; Q_A and Q_B represent special plastoquinone molecules (see Figure 21.14) and PQ the plastoquinone pool; Fe-S stands for the Rieske iron–sulfur center, and cyt *f,* cytochrome *f.* PC is the abbreviation for plastocyanin, the immediate e^- donor to $P700^+$; and F_A, F_B, and F_X represent the membrane-associated ferredoxins downstream from A_0 (a specialized Chl *a*) and A_1 (a specialized PSI quinone). Fd is the soluble ferredoxin pool that serves as the e^- donor to the flavoprotein (Fp), called **ferredoxin–NADP$^+$ reductase,** which catalyzes reduction of $NADP^+$ to NADPH. $Cyt(b_6)_n,(b_6)_p$ symbolizes the cytochrome b_6 moieties functioning to transfer e^- from F_A/F_B back to $P700^+$ during cyclic photophosphorylation (the pathway symbolized by the dashed arrow).

▶ **Biochemistry ⑤ Now™**   **ACTIVE FIGURE 21.12** The *Z* scheme of photosynthesis. **(a)** The *Z* scheme is a diagrammatic representation of photosynthetic electron flow from H_2O to $NADP^+$. The energy relationships can be derived from the $\mathscr{E}_o{}'$ scale beside the *Z* diagram, with lower standard potentials and hence greater energy as you go from bottom to top. Energy input as light is indicated by two broad arrows, one photon appearing in P680 and the other in P700. P680* and P700* represent photoexcited states. Electron loss from P680* and P700* creates P680$^+$ and P700$^+$. The representative components of the three supramolecular complexes (PSI, PSII, and the cytochrome b_6/cytochrome *f* complex) are in shaded boxes enclosed by solid black lines. Proton translocations that establish the proton-motive force driving ATP synthesis are illustrated as well. **(b)** Figure showing the functional relationships among PSII, the cytochrome *b*/cytochrome *f* complex, PSI, and the photosynthetic CF_1CF_0–ATP synthase within the thylakoid membrane. Note that e^- acceptors Q_A (for PSII) and A_1 (for PSI) are at the stromal side of the thylakoid membrane, whereas the e^- donors to P680$^+$ and P700$^+$ are situated at the lumenal side of the membrane. The consequence is charge separation ($-_{\text{stroma}}$, $+_{\text{lumen}}$) across the membrane. Also note that protons are translocated into the thylakoid lumen, giving rise to a chemiosmotic gradient that is the driving force for ATP synthesis by CF_1CF_0–ATP synthase. **Test yourself on the concepts in this figure at** http://chemistry.brookscole.com/ggb3

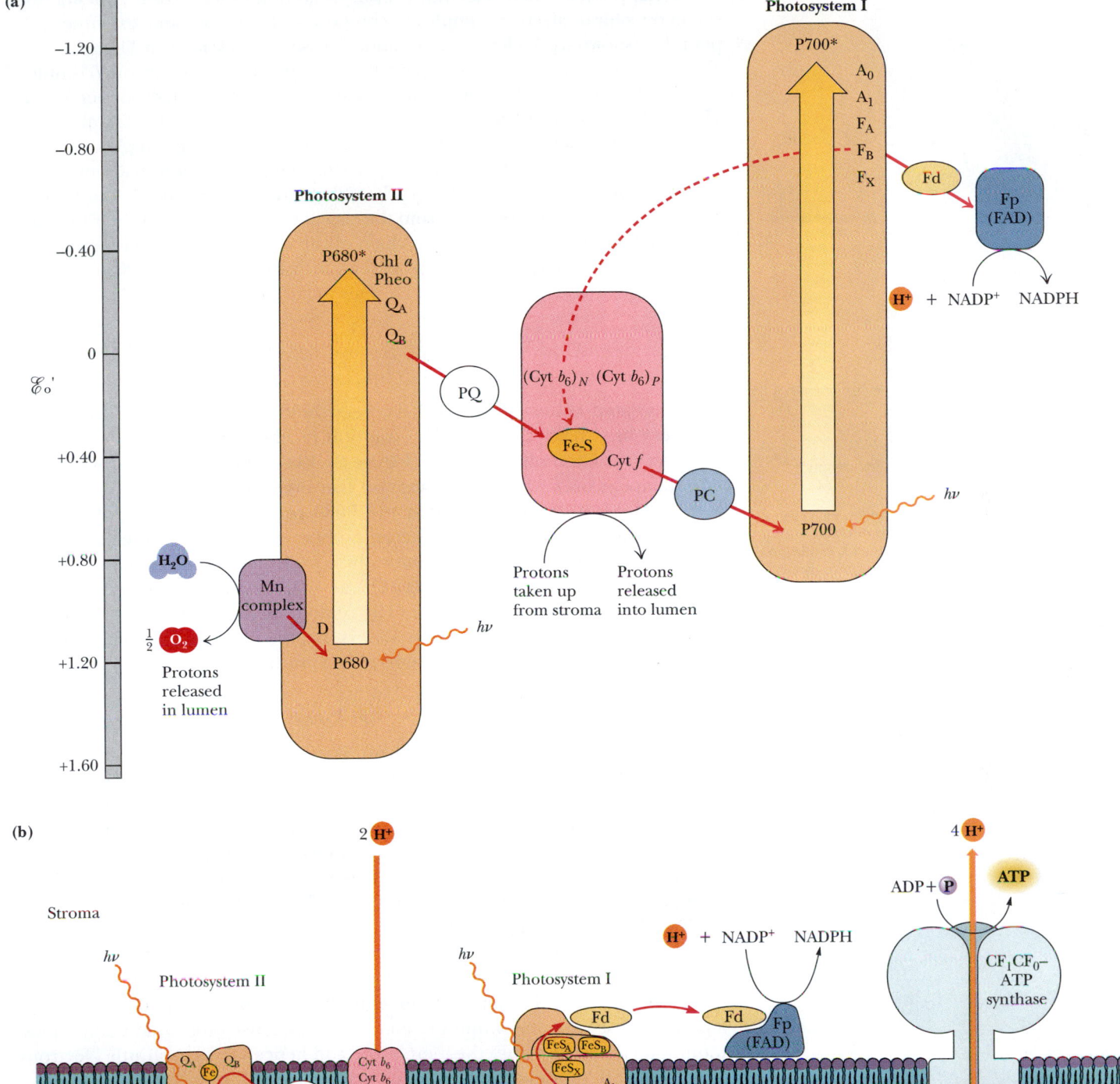

(a)

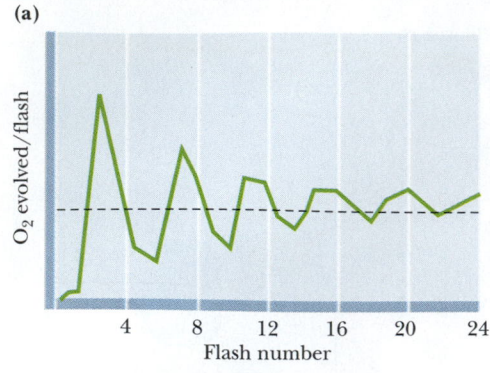

(b)

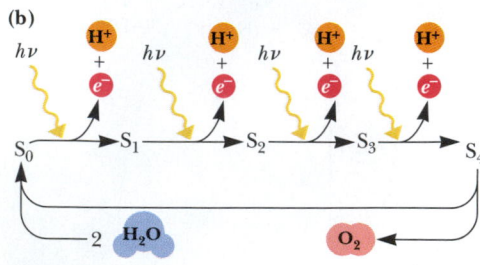

FIGURE 21.13 Oxygen evolution requires the accumulation of four oxidizing equivalents in PSII. (a) Dark-adapted chloroplasts show little O_2 evolution after two brief light flashes. Oxygen evolution then shows a peak on the third flash and every fourth flash thereafter. The oscillation in O_2 evolution is dampened by repeated flashes and converges to an average value after 20 or so flashes. (b) The oscillation in O_2 evolution per light flash is due to the cycling of the PSII reaction center through five different oxidation states, S_0 to S_4. When S_4 is reached, O_2 is released. One e^- is removed photochemically at each light flash, moving the reaction center successively through S_1, S_2, S_3, and S_4. S_4 decays spontaneously to S_0 by oxidizing $2 H_2O$ to O_2. The peak of O_2 evolution at flash 3 in part (a) is due to the fact that the isolated chloroplast suspension is already at the S_1 stage.

Overall photosynthetic electron transfer is accomplished by three **membrane-spanning supramolecular complexes** composed of intrinsic and extrinsic polypeptides (shown as shaded boxes bounded by solid black lines in Figure 22.12). These complexes are the PSII complex, the cytochrome b_6/cytochrome f complex, and the PSI complex. The PSII complex is aptly described as a light-driven **water: plastoquinone oxidoreductase;** it is the enzyme system responsible for photolysis of water, and as such, it is also referred to as the **oxygen-evolving complex, or OEC.** PSII possesses a metal cluster containing $4 Mn^{2+}$ atoms that coordinates two water molecules. As P680 undergoes four cycles of light-induced oxidation, four protons and four electrons are removed from the two water molecules and their O atoms are joined to form O_2. A tyrosyl side chain of the PSII complex (see following discussion) mediates electron transfer between the Mn^{2+} cluster and P680. The O_2-evolving reaction requires Ca^{2+} and Cl^- ions in addition to the $(Mn^{2+})_4$ cluster.

Oxygen Evolution Requires the Accumulation of Four Oxidizing Equivalents in PSII

When isolated chloroplasts that have been held in the dark are illuminated with very brief flashes of light, O_2 evolution reaches a peak on the third flash and every fourth flash thereafter (Figure 21.13a). The oscillation in O_2 evolution dampens over repeated flashes and converges to an average value. These data are interpreted to mean that the P680 reaction center complex cycles through five different oxidation states, numbered S_0 to S_4. One electron and one proton are removed photochemically in each step. When S_4 is attained, an O_2 molecule is released (Figure 21.13b) as PSII returns to oxidation state S_0 and two new water molecules bind. (The reason the first pulse of O_2 release occurred on the third flash [Figure 21.13a] is that the PSII reaction centers in the isolated chloroplasts were already poised at S_1 reduction level.)

Electrons Are Taken from H_2O to Replace Electrons Lost from P680

The events intervening between H_2O and P680 involve D, the name assigned to a specific protein tyrosine residue that mediates e^- transfer from H_2O via the Mn complex to $P680^+$ (Figure 21.12). The oxidized form of D is a tyrosyl free radical species, $D^{\cdot+}$. To begin the cycle, an exciton of energy excites P680 to P680*, whereupon P680* transfers an electron to a nearby Chl a molecule, which is the direct electron acceptor from P680*. This Chl a then reduces a molecule of **pheophytin,** symbolized by "Pheo" in Figure 21.12. Pheophytin is like chlorophyll a, except $2 H^+$ replace the centrally coordinated Mg^{2+} ion. This special pheophytin is the direct electron acceptor from P680*. Loss of an electron from P680* creates $P680^+$, the electron acceptor for D. Electrons flow from Pheo via specialized molecules of **plastoquinone,** represented by "Q" in Figure 21.12, to a pool of plastoquinone within the membrane. Because of its lipid nature, plastoquinone is mobile within the membrane and hence serves to shuttle electrons from the PSII supramolecular complex to the cytochrome b_6/cytochrome f complex. Alternate oxidation–reduction of plastoquinone to its hydroquinone form involves the uptake of protons (Figure 21.14). The asymmetry of the thylakoid membrane is designed to exploit this proton uptake and release so that protons (H^+) accumulate within the lumen of thylakoid vesicles, establishing an electrochemical gradient. Note that plastoquinone is an analog of coenzyme Q, the mitochondrial electron carrier (see Chapter 20).

Electrons from PSII Are Transferred to PSI Via the Cytochrome b_6/Cytochrome f Complex

The cytochrome b_6/cytochrome f or **plastoquinol:plastocyanin oxidoreductase** is a large (210 kD) multimeric protein possessing 22 to 24 transmembrane α-helices. It includes the two heme-containing electron transfer proteins for which it is named, as well as *iron–sulfur clusters* (see Chapter 20), which also participate in electron transport. The purpose of this complex is to mediate the transfer of electrons from PSII to PSI and to pump protons across the thylakoid membrane via a plastoquinone-mediated Q cycle, analogous to that found in mitochondrial e^- transport (see Chapter 20). Cytochrome f (f from the Latin *folium*, meaning "foliage") is a c-type cytochrome, with an α-absorbance band at 553 nm and a reduction potential of +0.365 V. Cytochrome b_6 in two forms (low- and high-potential) participates in the oxidation of plastoquinol and the Q cycle of the b_6/f complex. This cytochrome, whose absorbance band lies at 559 nm and whose $\mathscr{E}_0{}'$ is −0.06 V, can also serve in an alternative **cyclic electron transfer pathway.** Under certain conditions, electrons derived from P700* are not passed on to NADP+ but instead cycle down an alternative path via ferredoxins in the PSI complex to cytochrome b_6, plastoquinone, and ultimately back to P700+. This cyclic flow yields no O_2 evolution or NADP+ reduction but can lead to ATP synthesis via so-called cyclic photophosphorylation, discussed later.

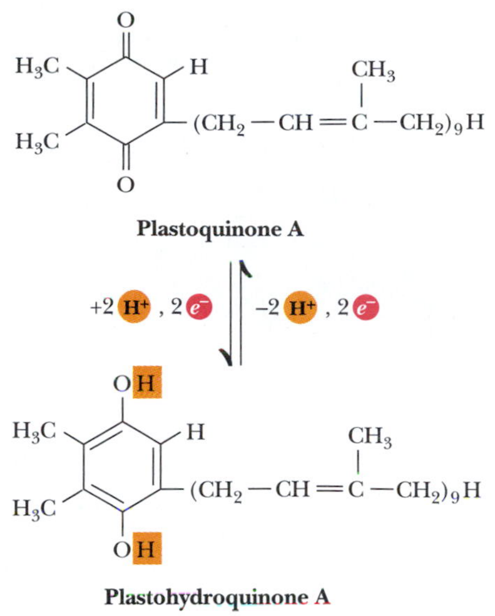

Plastocyanin Transfers Electrons from the Cytochrome b_6/Cytochrome f Complex to PSI

Plastocyanin ("**PC**" in Figure 21.12) is an electron carrier capable of diffusion along the inside of the thylakoid and migration in and out of the membrane, aptly suited to its role in shuttling electrons between the cytochrome b_6/cytochrome f complex and PSI. Plastocyanin is a low-molecular-weight (10.4 kD) protein containing a single copper atom. PC functions as a single-electron carrier ($\mathscr{E}_0{}' = $ +0.32 V) as its copper atom undergoes alternate oxidation–reduction between the cuprous (Cu+) and cupric (Cu²⁺) states. PSI is a light-driven **plastocyanin: ferredoxin oxidoreductase.** When P700, the specialized chlorophyll a dimer of PSI, is excited by light and oxidized by transferring its e^- to an adjacent chlorophyll a molecule that serves as its immediate e^- acceptor, P700+ is formed. (The standard reduction potential for the P700+/P700 couple lies near +0.45 V.) P700+ readily gains an electron from plastocyanin.

The immediate electron acceptor for P700* is a special molecule of chlorophyll. This unique Chl a (A_0) rapidly passes the electron to a specialized quinone (A_1), which in turn passes the e^- to the first in a series of *membrane-bound ferredoxins* (Fd; see Chapter 20). This Fd series ends with a soluble form of ferredoxin, Fd_s, which serves as the immediate electron donor to the flavoprotein (Fp) that catalyzes NADP+ reduction, namely, **ferredoxin:NADP+ reductase.**

The Initial Events in Photosynthesis Are Very Rapid Electron-Transfer Reactions

Electron transfer from P680 to Q and from P700 to Fd occurs on a picosecond-to-microsecond time scale. The necessity for such rapid reaction becomes obvious when one realizes that light-induced Chl excitation followed by electron transfer leads to separation of opposite charges in close proximity, as in P700+: A_0^-. Accordingly, subsequent electron-transfer reactions occur rapidly in order to shuttle the electron away quickly, before the wasteful back reaction of charge recombination (and dissipation of excitation energy), as in return to P700: A_0, can happen.

Plastoquinone A

Plastohydroquinone A

FIGURE 21.14 The structures of plastoquinone and its reduced form, plastohydroquinone (or plastoquinol). The oxidation of the hydroquinone releases 2 H+ as well as 2 e^-. The form shown (plastoquinone A) has nine isoprene units and is the most abundant plastoquinone in plants and algae. Other plastoquinones have different numbers of isoprene units and may vary in the substitutions on the quinone ring.

21.4 What Is the Molecular Architecture of Photosynthetic Reaction Centers?

What molecular architecture couples the absorption of light energy to rapid electron-transfer events, in turn coupling these e^- transfers to proton translocations so that ATP synthesis is possible? Part of the answer to this question lies in the membrane-associated nature of the photosystems. Membrane proteins have been difficult to study because they are not soluble in the aqueous solvents usually employed in protein biochemistry. A major breakthrough occurred in

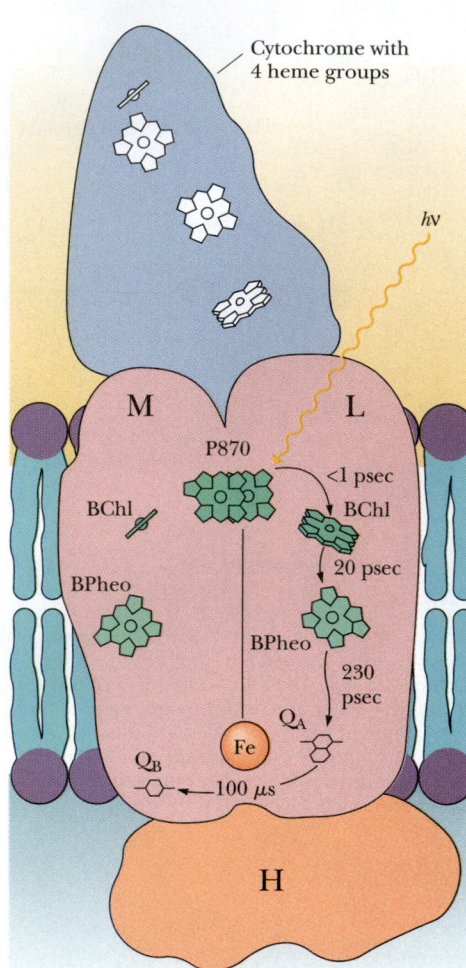

Note: The cytochrome subunit is membrane associated via a diacylglycerol moiety on its N-terminal Cys residue:

Membrane anchor

◄ **FIGURE 21.15** Model of the structure and activity of the *R. viridis* reaction center. Four polypeptides (designated cytochrome, *M*, *L*, and *H*) make up the reaction center, an integral membrane complex. The cytochrome maintains its association with the membrane via a diacylglyceryl group linked to its N-terminal Cys residue by a thioether bond. *M* and *L* both consist of five membrane-spanning α-helices; *H* has a single membrane-spanning α-helix. The prosthetic groups are spatially situated so that rapid e^- transfer from P870* to Q_B is facilitated. Photoexcitation of P870 leads in less than 1 picosecond (psec) to reduction of the *L*-branch BChl only. P870$^+$ is re-reduced via an e^- provided through the heme groups of the cytochrome.

1984, when Johann Deisenhofer, Hartmut Michel, and Robert Huber reported the first X-ray crystallographic analysis of a membrane protein. To the great benefit of photosynthesis research, this protein was the reaction center from the photosynthetic purple bacterium *Rhodopseudomonas viridis*. This research earned these three scientists the 1984 Nobel Prize in Chemistry.

The *R. viridis* Photosynthetic Reaction Center Is an Integral Membrane Protein

R. viridis is a photosynthetic prokaryote with a single type II photosystem. The reaction center (145 kD) of the *R. viridis* photosystem is localized in the plasma membrane of these photosynthetic bacteria and is composed of four different polypeptides, designated *L* (273 amino acid residues), *M* (323 residues), *H* (258 residues), and *cytochrome* (333 amino acid residues). *L* and *M* each consist of five membrane-spanning α-helical segments; *H* has one such helix, the majority of the protein forming a globular domain in the cytoplasm (Figure 21.15). The cytochrome subunit contains four heme groups; the N-terminal amino acid of this protein is cysteine. This cytochrome is anchored to the periplasmic face of the membrane via the hydrophobic chains of two fatty acid groups that are esterified to a glyceryl moiety joined via a thioether bond to the Cys (Figure 21.15). *L* and *M* each bear two *bacteriochlorophyll* molecules (the bacterial version of Chl) and one *bacteriopheophytin*. *L* also has a bound quinone molecule, Q_A. Together, *L* and *M* coordinate an Fe atom. The photochemically active species of the *R. viridis* reaction center, **P870,** is composed of two bacteriochlorophylls, one contributed by *L* and the other by *M*.

Photosynthetic Electron Transfer in the *R. viridis* Reaction Center Begins at P870

The prosthetic groups of the *R. viridis* reaction center (P870, BChl, BPheo, and the bound quinones) are fixed in a spatial relationship to one another that favors photosynthetic e^- transfer (Figure 21.16). Photoexcitation of P870 (creation of P870*) leads to e^- loss (P870$^+$) via electron transfer to the nearby bacteriochlorophyll (BChl). The e^- is then transferred via the *L* bacteriopheophytin (BPheo) to Q_A, which is also an *L* prosthetic group. The corresponding site on *M* is occupied by a loosely bound quinone, Q_B, and electron transfer from Q_A to Q_B takes place. An interesting aspect of the system is that *no* electron transfer occurs through *M*, even though it has components apparently symmetric to and identical to the *L* e^- transfer pathway.

The reduced quinone formed at the Q_B site is free to diffuse to a neighboring *cytochrome b/cytochrome c_1* membrane complex, where its oxidation is coupled to H$^+$ translocation (and, hence, ultimately to ATP synthesis) (Figure 21.16). *Cytochrome c_2*, a periplasmic protein, serves to cycle electrons back to P870$^+$ via the four hemes of the reaction center cytochrome subunit. A specific tyrosine residue of *L* (Try[162]) is situated between P870 and the closest cytochrome heme. This Tyr is the immediate e^- donor to P870$^+$ and completes the light-driven electron transfer cycle. The structure of the *R. viridis* reaction center (derived from X-ray crystallographic data) is modeled in Figure 21.17.

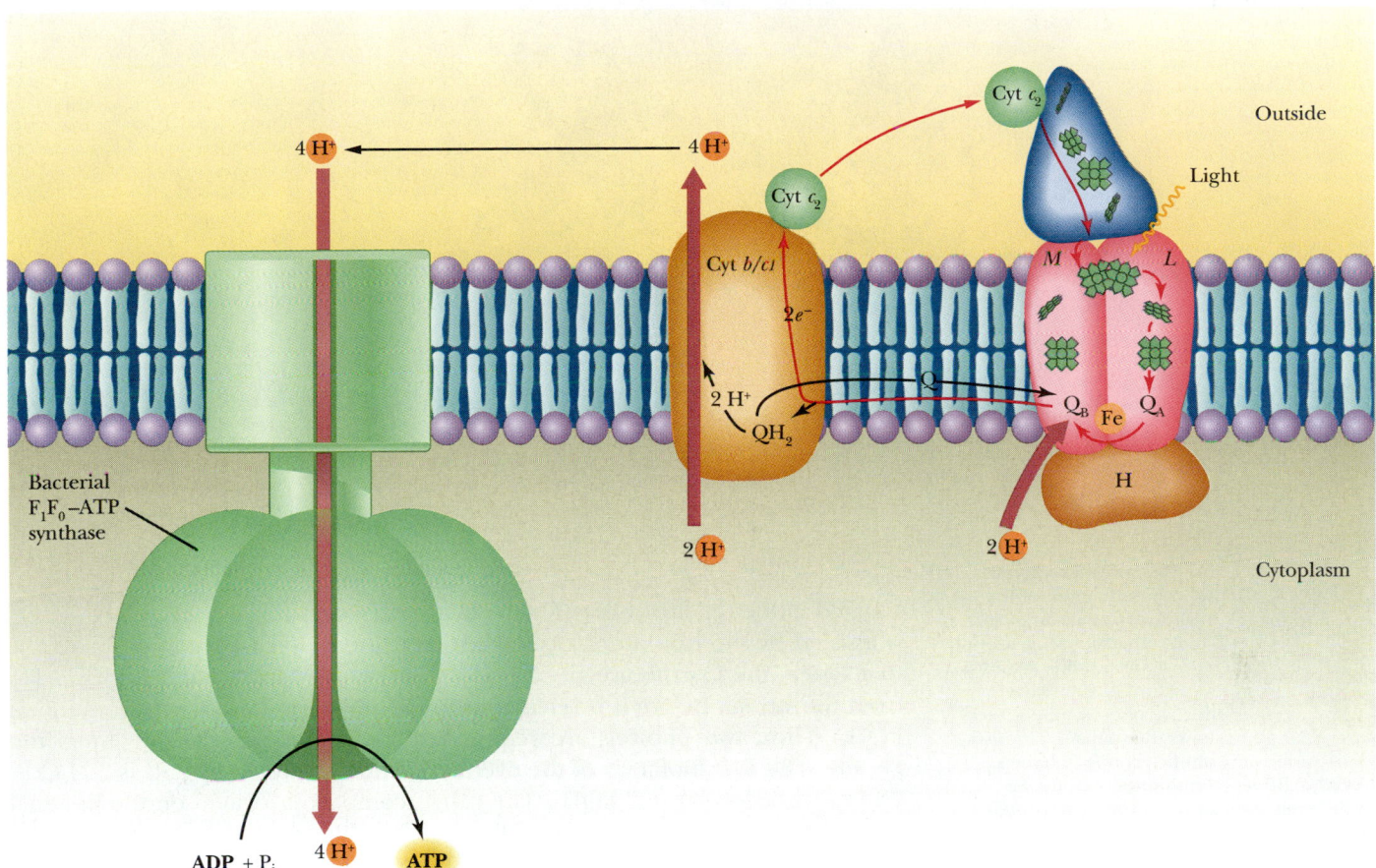

Biochemistry ⊜ Now™ ACTIVE FIGURE 21.16
The *R. viridis* reaction center is coupled to the cytochrome b/c_1 complex through the quinone pool (Q). Quinone molecules are photoreduced at the reaction center Q_B site (2 e^- [2 $h\nu$] per Q reduced) and then diffuse to the cytochrome b/c_1 complex, where they are reoxidized. Note that e^- flow from cytochrome b/c_1 back to the reaction center occurs via the periplasmic protein cytochrome c_2. Note also that a total of 4 H^+ are translocated into the periplasmic space for each Q molecule oxidized at cytochrome b/c_1. The resultant proton-motive force drives ATP synthesis by the bacterial F_1F_0–ATP synthase. (*Adapted from Deisenhofer, J., and Michel, H., 1989. The photosynthetic reaction center from the purple bacterium* Rhodopseudomonas viridis. Science *245:1463.*) **Test yourself on the concepts in this figure at http://chemistry.brookscole.com/ggb3**

The Molecular Architecture of PSII Resembles the *R. viridis* Reaction Center Architecture

Type II photosystems of higher plants, green algae, and cyanobacteria contain more than 20 subunits and are considerably more complex than the *R. viridis* reaction center. The structure of PSII from the thermophilic cyanobacterium *Synechococcus elongatus* has been revealed by X-ray crystallography, providing insight into PSII structures in general. Interestingly, both type II and type I photosystems show significant similarity to the *R. viridis* reaction center, thus establishing a strong evolutionary connection between reaction centers.

S. elongatus PSII is a homodimeric structure. Each "monomer" has a mass of almost 350 kD and 23 different protein subunits, the 4 largest being the reaction center pair of subunits (**D1** and **D2**) and two chlorophyll-containing inner antenna subunits (**CP43** and **CP47**) that bracket D1 and D2. Together, CP43 and CP47 have a total of 26 Chl *a* molecules, and exciton energy is collected and transferred from them to P680. Collectively, the protein subunits in a PSII "monomer" have at least 34 transmembrane α-helical segments, 22 of which are found in the D1-D2-CP43-CP47 "core" structure. D1 and D2 each have five membrane-spanning α-helices. Structurally and functionally, these two subunits are a direct counterpart of the *L* and *M* subunits of the *R. viridis* reaction center. P680 consists of a pair of Chl *a* molecules, with D1 and D2 each contributing one. D1 and D2 each have two other Chl *a* molecules, one near each P680 (**Chl$_{D1}$** and **Chl$_{D2}$**, respectively) and another that interacts with CP43/CP47 (Chl$_{Z-D1}$ and Chl$_{Z-D2}$, respectively) (Figure 21.18). Two equivalents of pheophytin (Pheo) are located on D1 and D2. The tyrosine species D is Tyr[161] in the D1 amino acid sequence. Complexed to D2 is a tightly bound plastoquinone molecule, Q_A. Electrons flow from P680* to Chl$_{D1}$ and on to Pheo$_{D1}$. Pheo$_{D1}$ then transfers the electron to Q_A on D2, where it then moves to a second plastoquinone

Biochemistry ⊜ Now™ Go to BiochemistryNow and click BiochemistryInteractive to explore the *R. viridis* reaction center, a complex scaffold for transduction of light energy.

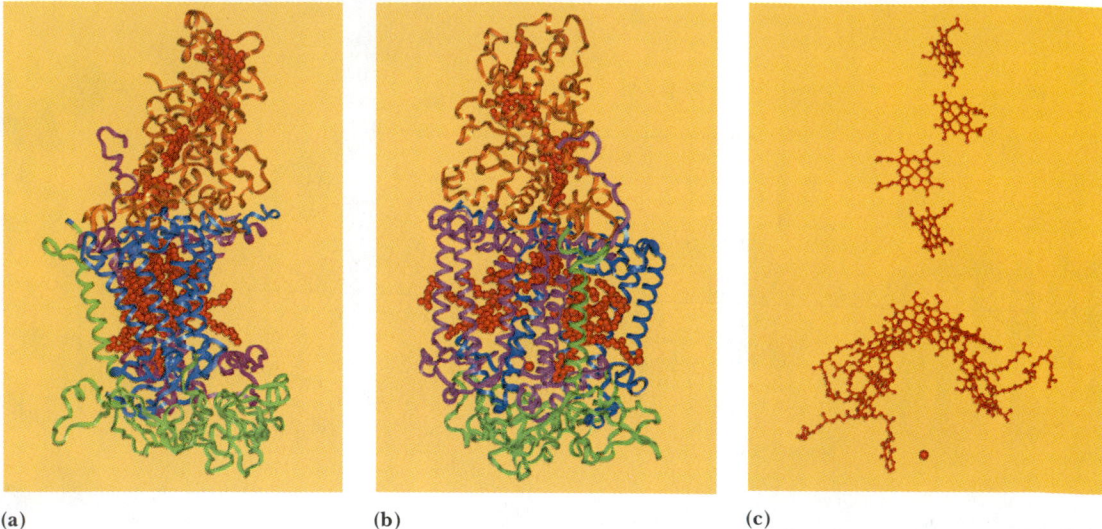

(a)   (b)   (c)

FIGURE 21.17 Model of the *R. viridis* reaction center. **(a, b)** Two views of the ribbon diagram of the reaction center. *M* and *L* subunits appear in purple and blue, respectively. Cytochrome subunit is brown; *H* subunit is green. These proteins provide a scaffold upon which the prosthetic groups of the reaction center are situated for effective photosynthetic electron transfer. Panel **(c)** shows the spatial relationship between the various prosthetic groups (4 hemes, P870, 2 BChl, 2 BPheo, 2 quinones, and the Fe atom) in the same view as in **(b)**, but with protein chains deleted.

situated in the Q_B site on D1 (Figure 21.18). Electron transfer from Q_A and Q_B is assisted by the iron atom located between them. Each plastoquinone (PQ) that enters the Q_B site accepts two electrons derived from water and two H^+ from the stroma before it is released into the membrane as the hydroquinone PQH_2. Thus, two photons are required to reduce each PQ that enters the Q_B site. The stoichiometry of the overall reaction catalyzed by PSII is $2\ H_2O + 2\ PQ + 4\ h\nu \longrightarrow O_2 + 2\ PQH_2$. The $(Mn)_4$ complex is located on the lumenal side of the thylakoid membrane. Thus, protons liberated from H_2O molecules at the Mn site are deposited directly into the lumen.

The Molecular Architecture of PSI Resembles the *R. viridis* Reaction Center and PSII Architecture

The structure of PSI from the cyanobacterium *Synechococcus elongatus* also has been solved by X-ray crystallography, completing our view of reaction center structure and confirming the fundamental similarities in organization that exist in these energy-transducing integral membrane proteins. Because of direct correlations with information about eukaryotic PSI, this cyanobacterial PSI provides a general model for all P700-dependent photosystems.

S. elongatus PSI exists as a cloverleaf-shaped trimeric structure. Each "monomer" (356 kD) consists of 12 different protein subunits and 127 cofactors: 96 chlorophyll *a* molecules, 2 phylloquinones, 3 Fe_4S_4 clusters, 22 carotenoids, and 4 lipids that are an intrinsic part of the protein complex. All of the electron-transferring prosthetic groups essential to PSI function are localized to just three polypeptides: **PsaA, PsaB,** and **PsaC.** PsaA and PsaB (83 kD each) compose the reaction center heterodimer, a structural pattern now seen as universal in photo-

FIGURE 21.18 Molecular architecture of the *Synecheococcus elongatus* PSII. The core of the PSII complex consists of the two polypeptides (D1 and D2) that bind P680, Chl *a*, pheophytin (Pheo), and the quinones (Q_A and Q_B). CP43 and CP47 are the two inner antenna subunits of PSII that bracket D1 and D2 and funnel light energy into P680. The arrow shows the path of electron transfer from P680* to Chl_{D1} to $Pheo_{D1}$ to Q_A on D2 and then, via the Fe atom, to Q_B on D1. The Tyr[161] residue of D1, symbolized by Y_Z, is situated between P680 and the $(Mn)_4$ cluster. Electrons flow from H_2O bound at the $(Mn)_4$ cluster to Y_Z and on to P680+ to fill the electron hole created when P680* loses an electron to Chl_{D1} to become P680+. *(Adapted from Barber, J., 2003. Photosystem II: The engine of life.* Quarterly Review of Biophysics *36:71–89.)*

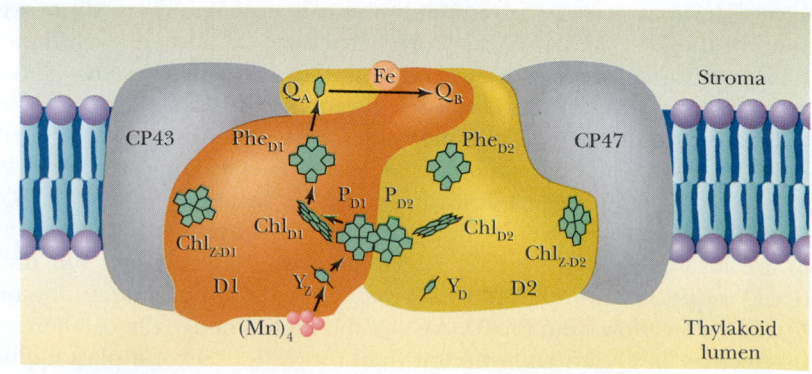

synthetic reaction centers (Figure 21.19). PsaA and PsaB each have 11 transmembrane α-helices, with the 5 most C-terminal α-helices of each serving as the scaffold for the reaction center photosynthetic electron-transfer apparatus. PsaC interacts with the stromal face of the PsaA–PsaB heterodimer. PsaC carries the two Fe_4S_4 clusters, F_A and F_B, and interacts with **PsaD.** Together they provide a docking site for ferredoxin. The electron-transfer system of PSI consists of three pairs of chlorophyll molecules: **P700** (a heterodimer of Chl a and an epimeric form, Chl a') and two additional Chl a pairs (symbolized by A_0) that mediate e^- transfer to the quinone acceptor. The *S. elongatus* quinone acceptor (A_1) is **phylloquinone** (also known as vitamin K_1; see Chapter 17). The **Fx** Fe_4S_4 cluster bridges PsaA and PsaB; two of its four cysteine ligands come from PsaA, the other two from PsaB. Photochemistry begins with exciton absorption at P700, almost instantaneous electron transfer and charge separation ($P700^+ : A_0^-$), followed by transfer of the electron from A_0 to A_1 and on to F_X and then F_A and F_B, where it goes on to reduce a ferredoxin molecule at the "stromal" side of the membrane. The positive charge at $P700^+$ and the e^- at F_A/F_B represent a charge separation across the membrane, an energized condition created by light. Plastocyanin (or in cyanobacteria, a lumenal cytochrome designated *cytochrome c_6*) delivers an electron to fill the electron hole in $P700^+$.

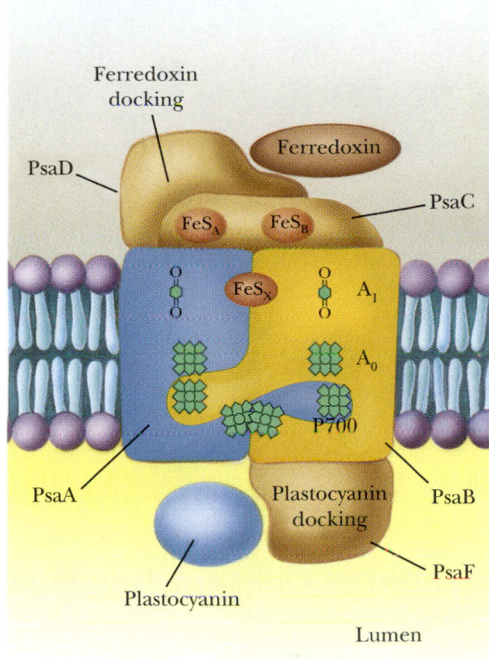

FIGURE 21.19 The molecular architecture of PSI. PsaA and PsaB constitute the reaction center dimer, an integral membrane complex; P700 is located at the lumenal side of this dimer. PsaC, which bears Fe-S centers F_A and F_B, and PsaD, the interaction site for ferredoxin, are on the stromal side of the thylakoid membrane. In *S. elongatus* PSI, electrons are delivered to $P700^+$ by cytochrome c_6; in higher plants, plastocyanin fills this role (as shown here). PsaF, which provides the plastocyanin interaction site, is on the lumenal side. (In *S. elongatus* PSI, the PsaA–PsaB heterodimer, not PsaF, provides the cytochrome c_6 docking site.) *(Adapted from Golbeck, J. H., 1992. Structure and function of photosystem I. Annual Review of Plant Physiology and Plant Molecular Biology **43:**293–324; and Fromme, P., Jordan, P., and Krausse, N., 2001. Structure of photosystem I. Biochimica Biophysica Acta **1507:**5–31.)*

21.5 | What Is the Quantum Yield of Photosynthesis?

The quantum yield of photosynthesis can be defined as the amount of product formed per equivalent of light input. In terms of exciton delivery to reaction center Chl dimers and subsequent e^- transfer, the quantum yield of photosynthesis typically approaches the theoretical limit of 1. The quantum yield of photosynthesis can also be expressed as the ratio of CO_2 fixed or O_2 evolved per photon absorbed. Interestingly, an overall stoichiometry of three H^+ translocated into the thylakoid vesicle has been observed for each electron passing from H_2O to $NADP^+$. Two photons per center would allow a pair of electrons to flow from H_2O to $NADP^+$ (Figure 21.12), resulting in the formation of 1 NADPH and $\frac{1}{2}O_2$. More appropriately, 4 $h\upsilon$ per center (8 quanta total) would drive the evolution of 1 O_2, the reduction of 2 $NADP^+$, and the translocation of 12 H^+. Current estimates suggest that 3 ATPs are formed for every 14 H^+ translocated, so $(12/14)3 = 2.57$ ATP would be synthesized from an input of 8 quanta.

The energy of a photon depends on its wavelength, according to the equation $E = h\upsilon = hc/\lambda$, where E is energy, c is the speed of light, and λ is its wavelength. Expressed in molar terms, an Einstein is the amount of energy in Avogadro's number of photons: $E = Nhc/\lambda$. Light of 700-nm wavelength is the longest-wavelength and the lowest-energy light acting in the eukaryotic photosystems. An Einstein of 700-nm light is equivalent in energy to approximately 170 kJ. Eight Einsteins of this light, 1360 kJ, theoretically generate 2 moles of NADPH, 2.57 moles of ATP, and 1 mole of O_2.

Calculation of the Photosynthetic Energy Requirements for Hexose Synthesis Depends on $H^+/h\upsilon$ and ATP/H^+ Ratios

The fixation of carbon dioxide to form hexose, the dark reactions of photosynthesis, requires considerable energy. The overall stoichiometry of this process (see Equation 21.3) involves 12 NADPH and 18 ATPs. To generate 12 equivalents of NADPH necessitates the consumption of 48 Einsteins of light, minimally 170 kJ each. However, if the preceding ratio of 1.29 ATPs per NADPH is correct, only 15.5 or so ATPs would be produced for CO_2 fixation. To make up the deficit of 2.5 ATPs would require 35 H^+ or about 12 more e^- transferred from H_2O to $NADP^+$ (an additional 24 Einsteins). From 72 Einsteins, or 12,240 kJ, 1 mole of hexose would be synthesized. The standard free energy change, $\Delta G^{\circ\prime}$, for hexose formation from carbon dioxide and water (the exact reverse of cellular respiration) is $+2870$ kJ/mol. Note that many assumptions underlie these calculations, including assumptions about the ATP/H$^+$ ratio, the H$^+$/e^- ratio, and ultimately, the relationship between quantum input and overall yields of NADPH and ATP. Also, cyclic photophosphorylation (see later section titled Cyclic Photophosphorylation Generates ATP but Not NADPH or O$_2$) leads to ATP synthesis and may aid in making up the ATP deficit just mentioned.

21.6 | How Does Light Drive the Synthesis of ATP?

Light-driven ATP synthesis, termed **photophosphorylation,** is a fundamental part of the photosynthetic process. The conversion of light energy to chemical energy results in electron-transfer reactions, which lead to the generation of reducing power (NADPH). Coupled with these electron transfers, protons are driven across the thylakoid membranes from the stromal side to the lumenal side. These proton translocations occur in a manner analogous to the proton translocations accompanying mitochondrial electron transport that provide the driving force for oxidative phosphorylation (see Chapter 20). Figure 21.12 indicates that proton translocations can occur at a number of sites. For example, protons are produced in the thylakoid lumen upon photolysis of water by PSII. The oxidation–reduction events as electrons pass through the plastoquinone pool and the Q cycle are another source of proton translocations. The proton transfer accompanying NADP$^+$ reduction also can be envisioned as protons being taken from the stromal side of the thylakoid vesicle. The current view is that three protons are translocated for each electron that flows from H$_2$O to NADP$^+$. Because this electron transfer requires two photons, one falling at PSII and one at PSI, the overall yield is 1.5 protons per quantum of light.

The Mechanism of Photophosphorylation Is Chemiosmotic

The thylakoid membrane is asymmetrically organized, or "sided," like the mitochondrial membrane. It also shares the property of being a barrier to the passive diffusion of H$^+$ ions. Photosynthetic electron transport thus establishes an electrochemical gradient, or proton-motive force, across the thylakoid membrane with the interior, or lumen, side accumulating H$^+$ ions relative to the stroma of the chloroplast. Like oxidative phosphorylation, the mechanism of photophosphorylation is chemiosmotic.

   A proton-motive force of approximately -250 mV is needed to achieve ATP synthesis. This proton-motive force, Δp, is composed of a membrane potential, $\Delta \psi$, and a pH gradient, ΔpH (see Chapter 20). The proton-motive force is defined as the free energy difference, ΔG, divided by $\mathscr{F}$, Faraday's constant:

$$\Delta p = \Delta G / \mathscr{F} = \Delta \psi - (2.3\ RT/\mathscr{F})\Delta \text{pH} \tag{21.5}$$

In chloroplasts, the value of $\Delta \psi$ is typically -50 to -100 mV and the pH gradient is equivalent to about 3 pH units, so $-(2.3\ RT/\mathscr{F})\Delta$pH $= -200$ mV. This situation contrasts with the mitochondrial proton-motive force, where the membrane potential contributes relatively more to Δp than does the pH gradient.

CF$_1$CF$_0$–ATP Synthase Is the Chloroplast Equivalent of the Mitochondrial F$_1$F$_0$–ATP Synthase

The transduction of the electrochemical gradient into the chemical energy represented by ATP is carried out by the chloroplast ATP synthase, which is highly analogous to the mitochondrial F$_1$F$_0$–ATP synthase. The chloroplast enzyme complex is called **CF$_1$CF$_0$–ATP synthase,** "C" symbolizing chloroplast. Like the mitochondrial complex, CF$_1$CF$_0$–ATP synthase is a heteromultimer of α-, β-, γ-, δ-, ϵ-, a-, b-, and c-subunits (see Chapter 20), consisting of a knoblike structure some 9 nm in diameter (CF$_1$) attached to a stalked base (CF$_0$) embedded in the

Critical Developments in Biochemistry

Biochemistry ❁ Now™

Experiments with Isolated Chloroplasts Provided the First Direct Evidence for the Chemiosmotic Hypothesis

Experimental proof that the mechanism of photophosphorylation is chemiosmotic was provided in an elegant experiment by Andre Jagendorf and Ernesto Uribe in 1966 (see accompanying figure). Jagendorf and Uribe reasoned that if photophosphorylation were indeed driven by an electrochemical gradient established by photosynthetic electron-transfer reactions, they might artificially generate such a gradient by first incubating chloroplasts in an acid bath in the dark and then quickly raising the pH of the external medium. The resulting inequality in hydrogen ion electrochemical activity across the membrane should mimic the pH gradient normally found upon illumination of chloroplasts and should provide the energized condition necessary to drive ATP formation. To test this idea, Jagendorf and Uribe bathed isolated chloroplasts in a weakly acid (pH 4) medium for 60 seconds, allowing the pH inside the chloroplasts to equilibrate with the external medium. The pH of the solution was then quickly raised to

slightly alkaline pH (pH 8), artificially creating a pH gradient across the thylakoid membrane. When ADP and radioactive $^{32}P_i$ were added, synthesis of radioactive ATP was observed as the pH gradient collapsed. This experiment was the first real proof of Mitchell's chemiosmotic hypothesis and directed the scientific community to a greater acceptance of Mitchell's interpretations. Mitchell's chemiosmotic hypothesis for ATP synthesis now occupies the position of dogma, due to the weight of evidence that has accumulated in its favor. Photophosphorylation then can be concisely summarized by noting that thylakoid vesicles accumulate H^+ upon illumination and that the electrochemical gradient thus created represents an energized state that can be tapped to drive ATP synthesis. Collapse of the gradient—that is, equilibration of the ion concentration difference across the membrane—is the energy-transducing mechanism: The chemical potential of a concentration difference is transduced into the synthesis of ATP.

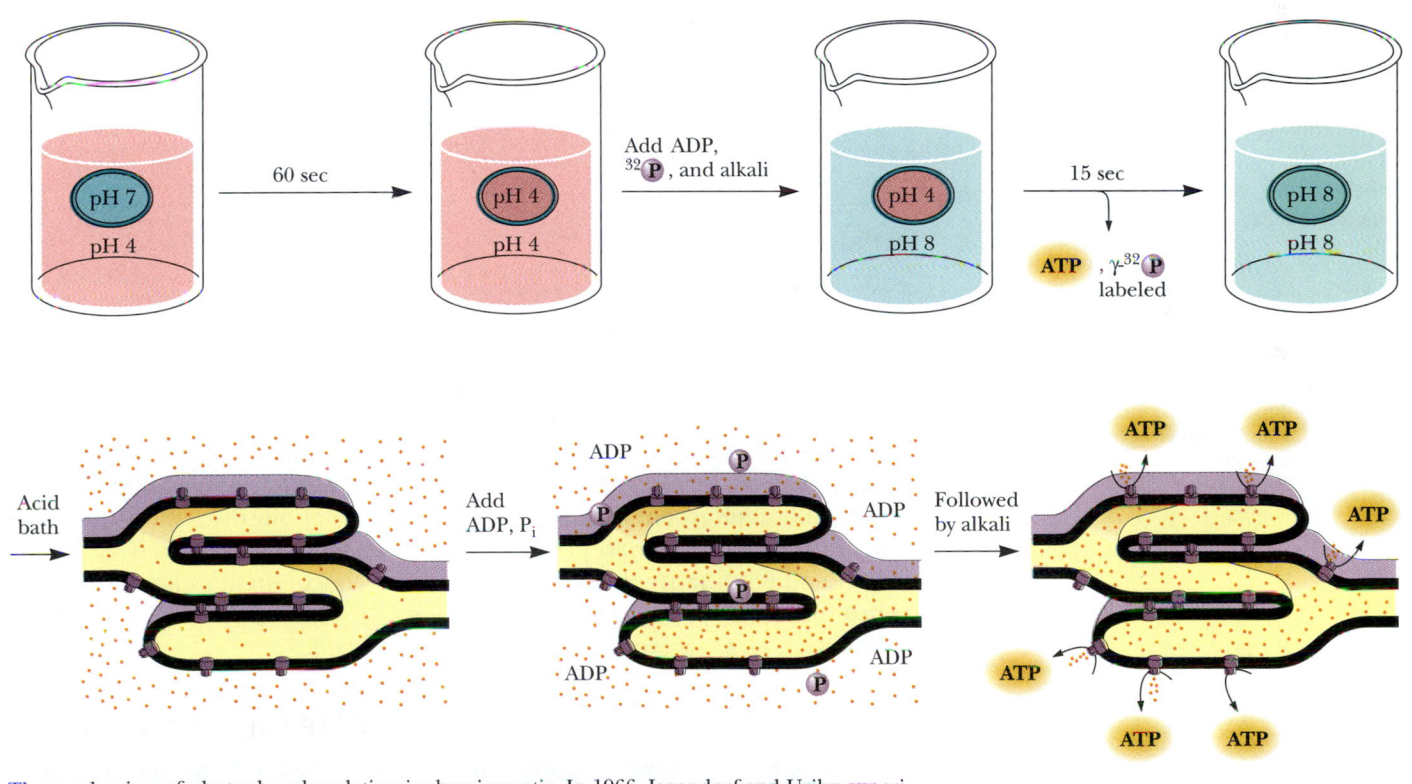

The mechanism of photophosphorylation is chemiosmotic. In 1966, Jagendorf and Uribe experimentally demonstrated for the first time that establishment of an electrochemical gradient across the membrane of an energy-transducing organelle could lead to ATP synthesis. They equilibrated isolated chloroplasts for 60 seconds in a pH 4 bath, adjusted the pH to 8 in the presence of ADP and P_i, and allowed phosphophosphorylation to proceed for 15 seconds. The entire experiment was carried out in the dark.

thylakoid membrane. The mechanism of action of CF_1CF_0–ATP synthase in coupling ATP synthesis to the collapse of the pH gradient is similar to that of the mitochondrial ATP synthase described in Chapter 20. However, higher plant CF_1CF_0–ATP synthase is believed to have 14 c-subunits in its F_0 rotor, implying that one turn of F_0 would require 14 H^+ and lead to synthesis of 3 ATPs. The mechanism of photophosphorylation is summarized schematically in Figure 21.20.

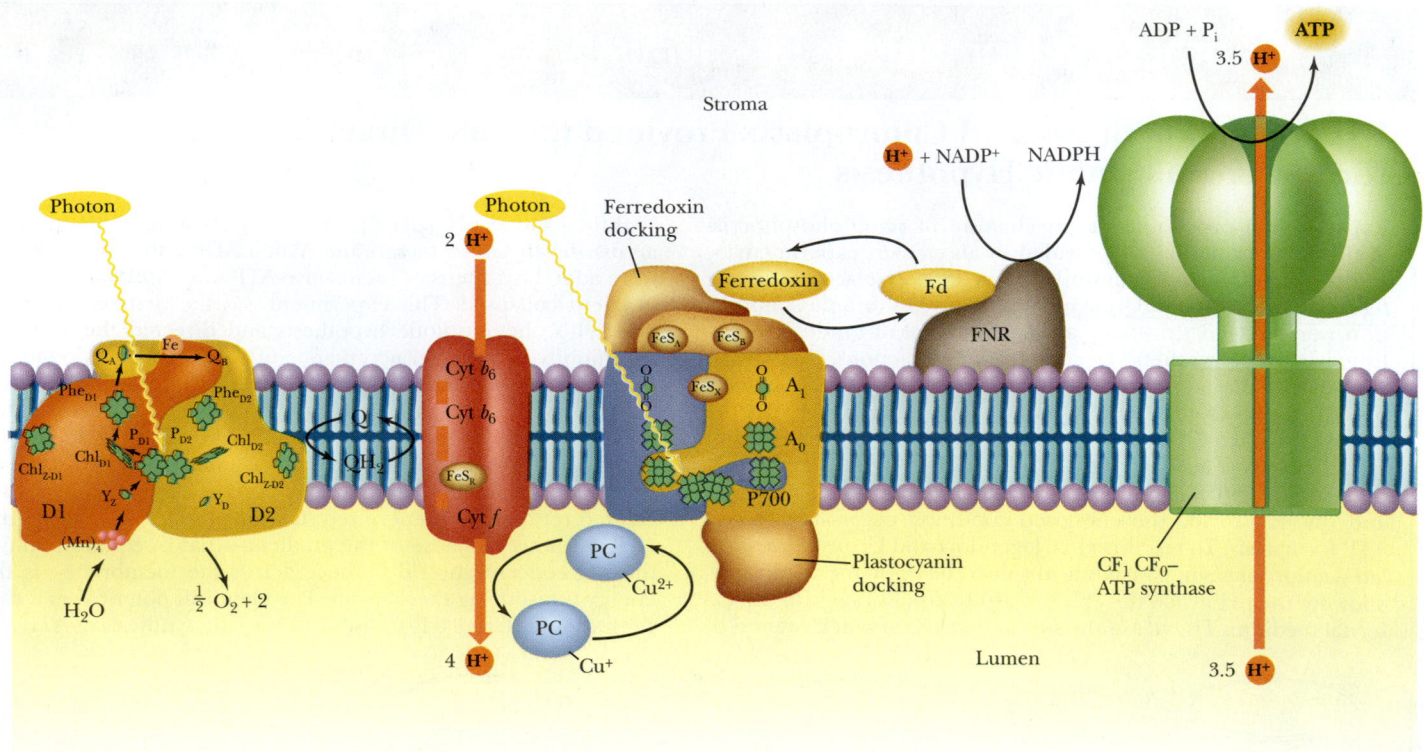

The mechanism of photophosphorylation. Photosynthetic electron transport establishes a proton gradient that is tapped by the CF_1CF_0–ATP synthase to drive ATP synthesis. Critical to this mechanism is the fact that the membrane-bound components of light-induced electron transport and ATP synthesis are asymmetric with respect to the thylakoid membrane so that vectorial discharge and uptake of H^+ ensue, generating the proton-motive force. **Test yourself on the concepts in this figure at http://chemistry. brookscole.com/ggb3**

Photophosphorylation Can Occur in Either a Noncyclic or a Cyclic Mode

Photosynthetic electron transport, which pumps H^+ into the thylakoid lumen, can occur in two modes, both of which lead to the establishment of a transmembrane proton-motive force. Thus, both modes are coupled to ATP synthesis and are considered alternative mechanisms of photophosphorylation, even though they are distinguished by differences in their electron transfer pathways. The two modes are cyclic and noncyclic photophosphorylation. **Noncyclic photophosphorylation** has been the focus of our discussion and is represented by the scheme in Figure 21.20, where electrons activated by quanta at PSII and PSI flow from H_2O to $NADP^+$, with concomitant establishment of the proton-motive force driving ATP synthesis. Note that in noncyclic photophosphorylation, O_2 is evolved and $NADP^+$ is reduced.

Cyclic Photophosphorylation Generates ATP but Not NADPH or O₂

In cyclic photophosphorylation, the "electron hole" in $P700^+$ created by electron loss from P700 is filled *not* by an electron derived from H_2O via PSII but by a cyclic pathway in which the photoexcited electron returns ultimately to $P700^+$. This pathway is schematically represented in Figure 21.12 by the dashed line connecting F_B and cytochrome b_6. Thus, one function of cytochrome b_6 (b_{563}) is to couple the bound ferredoxin carriers of the PSI complex with the cytochrome b_6/cytochrome f complex via the plastoquinone pool. This pathway diverts the activated e^- from $NADP^+$ reduction back through plastocyanin to re-reduce $P700^+$ (Figure 21.21).

Proton translocations accompany these cyclic electron transfer events so that ATP synthesis can be achieved. In cyclic photophosphorylation, ATP is the sole product of energy conversion. No NADPH is generated, and because PSII is not involved, no oxygen is evolved. Cyclic photophosphorylation theoretically yields 2 H^+ per e^- (2 $H^+/h\nu$) from the operation of the cytochrome b_6/cytochrome f

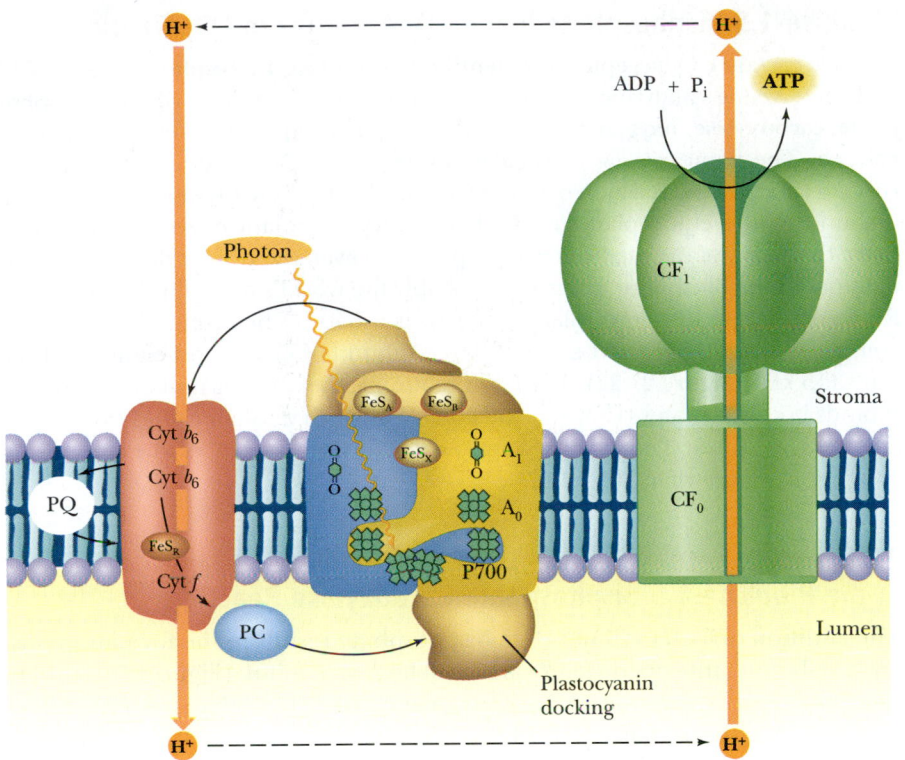

Biochemistry⊜Now™ **ACTIVE FIGURE 21.21**
The pathway of cyclic photophosphorylation by PSI.
(Adapted from Arnon, D. I., 1984. The discovery of photosynthetic phosphorylation. Trends in Biochemical Sciences 9:258–262.)
Test yourself on the concepts in this figure at
http://chemistry.brookscole.com/ggb3

complex. Thus, cyclic photophosphorylation provides a possible mechanism for overcoming the ATP deficit for CO_2 fixation (see the previous section titled Calculation of the Photosynthetic Energy Requirements for Hexose Synthesis Depends on H^+/hv and ATP/H^+ Ratios). The maximal rate of cyclic photophosphorylation is less than 5% of the rate of noncyclic photophosphorylation. Cyclic photophosphorylation depends only on PSI.

21.7 | How Is Carbon Dioxide Used to Make Organic Molecules?

As we began this chapter, we saw that photosynthesis traditionally is equated with the process of CO_2 fixation, that is, the net synthesis of carbohydrate from CO_2. Indeed, the capacity to perform net accumulation of carbohydrate from CO_2 distinguishes the phototrophic (and autotrophic) organisms from heterotrophs. Although animals possess enzymes capable of linking CO_2 to organic acceptors, they cannot achieve a net accumulation of organic material by these reactions. For example, fatty acid biosynthesis is primed by covalent attachment of CO_2 to acetyl-CoA to form malonyl-CoA (see Chapter 24). Nevertheless, this "fixed CO_2" is liberated in the very next reaction, so no net CO_2 incorporation occurs.

Elucidation of the pathway of CO_2 fixation represents one of the earliest applications of radioisotope tracers to the study of biology. In 1945, Melvin Calvin and his colleagues at the University of California, Berkeley, were investigating photosynthetic CO_2 fixation in *Chlorella*. Using $^{14}CO_2$, they traced the incorporation of radioactive ^{14}C into organic products and found that the earliest labeled product was **3-phosphoglycerate** (see Figure 17.13). Although this result suggested that the CO_2 acceptor was a two-carbon compound, further investigation revealed that, in reality, 2 equivalents of 3-phosphoglycerate were formed following addition of CO_2 to a five-carbon (pentose) sugar:

$$CO_2 + \text{5-carbon acceptor} \longrightarrow \text{[6-carbon intermediate]} \longrightarrow$$
$$\text{two 3-phosphoglycerates}$$

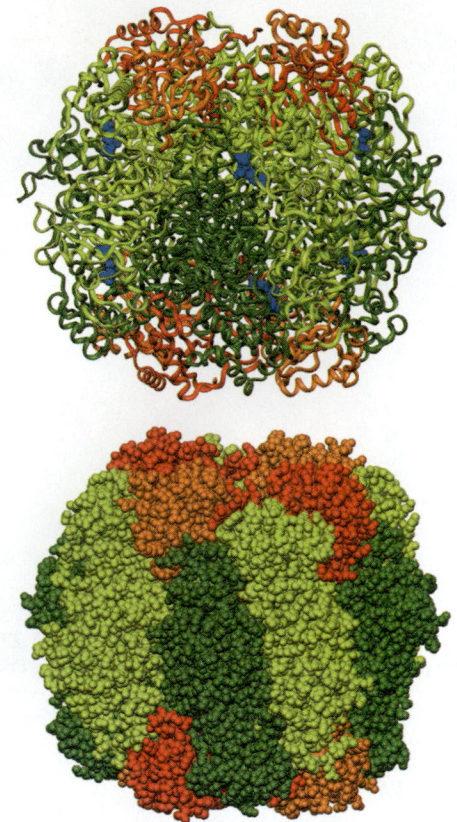

Ribulose-1,5-Bisphosphate Is the CO_2 Acceptor in CO_2 Fixation

The five-carbon CO_2 acceptor was identified as **ribulose-1,5-bisphosphate (RuBP)**, and the enzyme catalyzing this key reaction of CO_2 fixation is **ribulose bisphosphate carboxylase/oxygenase,** or, in the jargon used by workers in this field, **rubisco.** The name *ribulose bisphosphate carboxylase/oxygenase* reflects the fact that rubisco catalyzes the reaction of either CO_2 or, alternatively, O_2 with RuBP. Rubisco is found in the chloroplast stroma. It is a very abundant enzyme, constituting more than 15% of the total chloroplast protein. Given the preponderance of plant material in the biosphere, rubisco is probably the world's most abundant protein. Rubisco is large: In higher plants, rubisco is a 550-kD heteromultimeric ($\alpha_8\beta_8$) complex consisting of eight identical large subunits (55 kD) and eight small subunits (15 kD) (Figure 21.22). The large subunit is the catalytic unit of the enzyme. It binds both substrates (CO_2 and RuBP) and Mg^{2+} (a divalent cation essential for enzymatic activity). The small subunit modulates the activity of the enzyme, increasing k_{cat} more than 100-fold.[2]

2-Carboxy-3-Keto-Arabinitol Is an Intermediate in the Ribulose-1,5-Bisphosphate Carboxylase Reaction

The addition of CO_2 to ribulose-1,5-bisphosphate results in the formation of an enzyme-bound intermediate, **2-carboxy-3-keto-arabinitol** (Figure 21.23). This intermediate arises when CO_2 adds to the enediol intermediate generated from ribulose-1,5-bisphosphate. Hydrolysis of the C_2—C_3 bond of the intermediate generates two molecules of 3-phosphoglycerate. The CO_2 ends up as the carboxyl group of one of the two molecules.

Ribulose-1,5-Bisphosphate Carboxylase Exists in Inactive and Active Forms

Rubisco exists in three forms: an inactive form, designated $E;$ a carbamylated, but inactive, form, designated $EC;$ and an active form, ECM, which is carbamylated and has Mg^{2+} at its active sites as well. Carbamylation of rubisco takes place by addition of CO_2 to its Lys^{201} ϵ-NH_2 groups (to give ϵ—NH—COO^- derivatives). The CO_2 molecules used to carbamylate Lys residues do not become substrates. The carbamylation reaction occurs spontaneously at slightly alkaline pH (pH 8). Carbamylation of rubisco completes the formation of a binding site for the Mg^{2+} that participates in the catalytic reaction. Once Mg^{2+} binds to EC, rubisco achieves its active ECM form. Activated rubisco displays a K_m for CO_2 of 10 to 20 μM.[3]

Substrate RuBP binds much more tightly to the inactive E form of rubisco ($K_D = 20$ nM) than to the active ECM form (K_m for RuBP = 20 μM). Thus, RuBP is also a potent inhibitor of rubisco activity. Release of RuBP from the active site of rubisco is mediated by **rubisco activase.** Rubisco activase is a *regulatory protein;*

[2]The rubisco large subunit is encoded by a gene within the chloroplast DNA, whereas the small subunit is encoded by a multigene family in the nuclear DNA. Assembly of active rubisco heteromultimers occurs within chloroplasts following transit of the small subunit polypeptide across the chloroplast membrane.

[3]The relative abundance of CO_2 in the atmosphere is low, about 0.03%. The concentration of CO_2 dissolved in aqueous solutions equilibrated with air is about 10 μM.

▲ **FIGURE 21.22** Ribbon diagram (top) and space-filling model of ribulose bisphosphate carboxylase. The enzyme consists of 8 equivalents each of two types of subunits, large *L* (55 kD) and small *S* (15 kD). Clusters of four small subunits (orange and red) are located at each end of the symmetric octamer formed by the *L* subunits (light and dark green). The active sites are revealed in the ribbon diagram by bound 2-carboxyarabinitol bisphosphate (blue). *(From Andersson, I., 1996. Large structures at high resolution: The 1.6Å crystal structure of spinach ribulose-1,5-bisphosphate carboxylase/oxygenase complexed with 2-carboxyarabinitol bisphosphate.* Journal of Molecular Biology *259:160–174; image courtesy of Vijay Chandrasekaran and Robert J. Spreitzer, University of Nebraska.)*

▼ **FIGURE 21.23** The ribulose bisphosphate carboxylase reaction. Enzymatic abstraction of the C-3 proton of RuBP yields a 2,3-enediol intermediate (**I**), which is stereospecifically carboxylated at C-2 to create the six-carbon β-keto acid intermediate (**II**) known as 2-carboxy-3-keto-arabinitol. Intermediate II is rapidly hydrated to give the gem-diol form (**III**). Deprotonation of the C-3 hydroxyl and cleavage yield two 3-phosphoglycerates. Mg^{2+} at the active site aids in stabilizing the 2,3-enediol transition state for CO_2 addition and in facilitating the carbon–carbon bond cleavage that leads to product formation. Note that CO_2, not HCO_3^- (its hydrated form), is the true substrate.

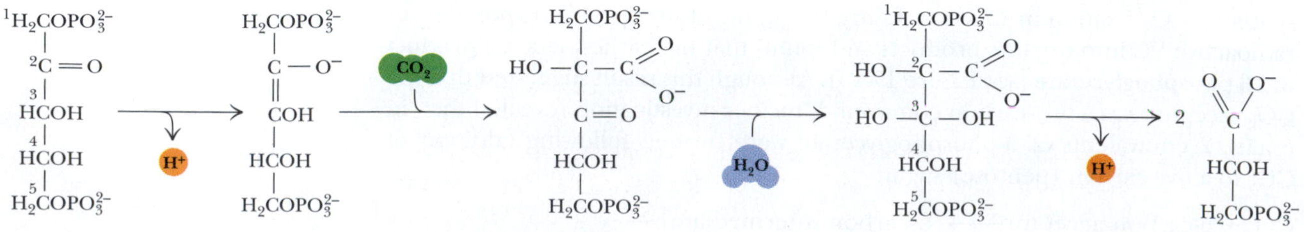

$^1H_2COPO_3^{2-}$
$^2C=O$
^3HCOH
^4HCOH
$^5H_2COPO_3^{2-}$

→ (H+)

$H_2COPO_3^{2-}$
$C—O^-$
COH
$HCOH$
$H_2COPO_3^{2-}$

I

→ (CO₂)

$H_2COPO_3^{2-}$
$HO—C—C$ (O, O$^-$)
$C=O$
$HCOH$
$H_2COPO_3^{2-}$

II

→ (H₂O)

$^1H_2COPO_3^{2-}$
$HO—^2C—C$ (O, O$^-$)
$HO—^3C—OH$
^4HCOH
$^5H_2COPO_3^{2-}$

→ (H+)

2
O O$^-$
C
$HCOH$
$H_2COPO_3^{2-}$

III

it binds to *E*-form rubisco and, in an ATP-dependent reaction, promotes the release of RuBP. Rubisco then becomes activated by carbamylation and Mg^{2+} binding. Rubisco activase itself is activated in an indirect manner by light. Thus, light is the ultimate activator of rubisco.

Biochemistry Now™ Go to BiochemistryNow and click BiochemistryInteractive to examine the structure of Rubisco, the principal CO_2-fixing enzyme in nature.

CO_2 Fixation into Carbohydrate Proceeds Via the Calvin–Benson Cycle

The immediate product of CO_2 fixation, 3-phosphoglycerate, must undergo a series of transformations before the net synthesis of carbohydrate is realized. Among carbohydrates, hexoses (particularly glucose) occupy center stage. Glucose is the building block for both cellulose and starch synthesis. These plant polymers constitute the most abundant organic material in the living world, and thus, the central focus on glucose as the ultimate end product of CO_2 fixation is amply justified. Also, sucrose (α-D-glucopyranosyl-(1→2)-β-D-fructofuranoside) is the major carbon form translocated out of leaves to other plant tissues. In nonphotosynthetic tissues, sucrose is metabolized via glycolysis and the TCA cycle to produce ATP.

The set of reactions that transforms 3-phosphoglycerate into hexose is named the **Calvin–Benson cycle** (often referred to simply as the Calvin cycle) for its discoverers. The reaction series is indeed cyclic because not only must carbohydrate appear as an end product, but the five-carbon acceptor, RuBP, must be regenerated to provide for continual CO_2 fixation. Balanced equations that schematically represent this situation are

$$6(1) + 6(5) \longrightarrow 12(3)$$
$$12(3) \longrightarrow 1(6) + 6(5)$$
$$\textit{Net: } 6(1) \longrightarrow 1(6)$$

Each number in parentheses represents the number of carbon atoms in a compound, and the number preceding the parentheses indicates the stoichiometry of the reaction. Thus, 6(1), or 6 CO_2, condense with 6(5) or 6 RuBP to give 12 3-phosphoglycerates. These 12(3)s are then rearranged in the Calvin cycle to form one hexose, 1(6), and regenerate the six 5-carbon (RuBP) acceptors.

The Enzymes of the Calvin Cycle Serve Three Metabolic Purposes

The Calvin cycle enzymes serve three important ends:

1. They constitute the predominant CO_2 fixation pathway in nature.
2. They accomplish the reduction of 3-phosphoglycerate, the primary product of CO_2 fixation, to glyceraldehyde-3-phosphate so that carbohydrate synthesis becomes feasible.
3. They catalyze reactions that transform three-carbon compounds into four-, five-, six-, and seven-carbon compounds.

Most of the enzymes mediating the reactions of the Calvin cycle also participate in either glycolysis (see Chapter 18) or the pentose phosphate pathway (see Chapter 22). The aim of the Calvin scheme is to account for hexose formation from 3-phosphoglycerate. In the course of this metabolic sequence, the NADPH and ATP produced in the light reactions are consumed, as indicated earlier in Equation 21.3.

The Calvin cycle of reactions starts with *ribulose bisphosphate carboxylase* catalyzing formation of 3-phosphoglycerate from CO_2 and RuBP and concludes with **ribulose-5-phosphate kinase** (also called *phosphoribulose kinase*), which forms RuBP (Figure 21.24 and Table 21.1). The carbon balance is given at the right side of Table 21.1. Several features of the reactions in this table merit discussion. Note that the 18 equivalents of ATP consumed in hexose formation are expended in reactions 2 and 15: 12 to form 12 equivalents of 1,3-bisphosphoglycerate from 3-phosphoglycerate by a reversal of the normal glycolytic reaction catalyzed by

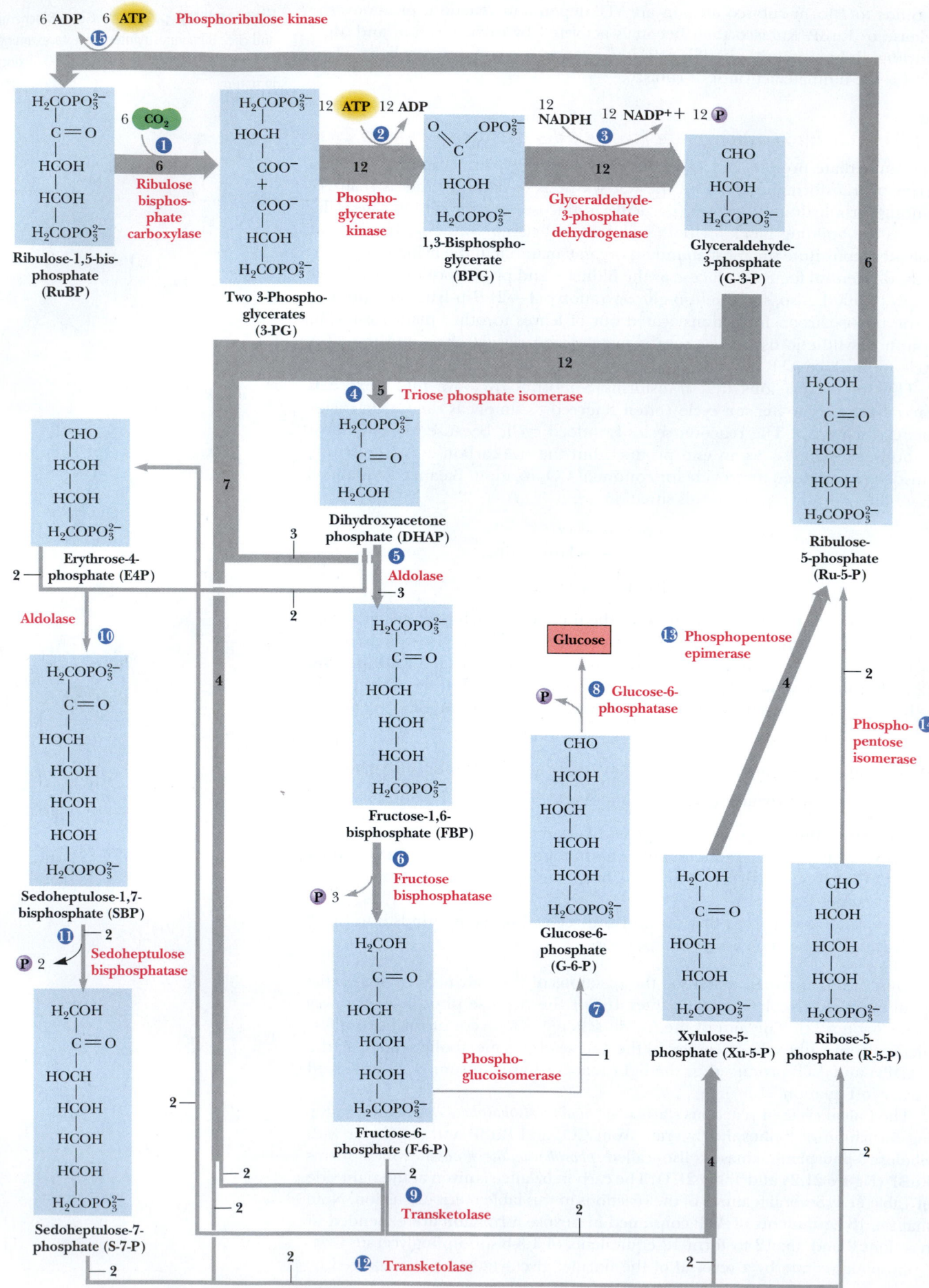

◀ **Biochemistry** ⓔ **Now**™   **ACTIVE FIGURE 21.24** The Calvin–Benson cycle of reactions. The number associated with the arrow at each step indicates the number of molecules reacting in a turn of the cycle that produces one molecule of glucose. Reactions are numbered as in Table 21.1. **Test yourself on the concepts in this figure at http://chemistry.brookscole.com/ggb3**

3-phosphoglycerate kinase and six to phosphorylate Ru-5-P to regenerate 6 RuBP. All 12 NADPH equivalents are used in reaction 3. Plants possess an **NADPH-specific glyceraldehyde-3-phosphate dehydrogenase,** which contrasts with its glycolytic counterpart in its specificity for NADP over NAD and in the direction in which the reaction normally proceeds.

The Calvin Cycle Reactions Can Account for Net Hexose Synthesis

When carbon rearrangements are balanced to account for net hexose synthesis, five of the glyceraldehyde-3-phosphate molecules are converted to dihydroxyacetone phosphate (DHAP). Three of these DHAPs then condense with three glyceraldehyde-3-P via the aldolase reaction to yield three hexoses in the form of fructose bisphosphate (Figure 21.24). (Recall that the $\Delta G°'$ for the aldolase reaction in the glycolytic direction is +23.9 kJ/mol. Thus, the aldolase reaction running "in reverse" in the Calvin cycle would be thermodynamically favored under standard-state conditions.) Taking one FBP to glucose, the desired product of this scheme, leaves 30 carbons, distributed as 2 fructose-6-phosphates, 4 glyceraldehyde-3-phosphates, and 2 DHAP. These 30 Cs are reorganized into 6 RuBP by reactions 9 through 15. Step 9 and steps 12 through 14 involve carbohydrate rearrangements like those in the pentose phosphate pathway (see Chapter 22). Reaction 11 is mediated by **sedoheptulose-1,7-bisphosphatase.** This phosphatase is unique to plants; it generates sedoheptulose-7-P, the seven-carbon sugar serving as the transketolase substrate. Likewise, **phosphoribulose kinase** carries out the unique plant function of providing RuBP from Ru-5-P

Table 21.1

The Calvin Cycle Series of Reactions

Reactions 1 through 15 constitute the cycle that leads to the formation of one equivalent of glucose. The enzyme catalyzing each step, a concise reaction, and the overall carbon balance are given. Numbers in parentheses show the numbers of carbon atoms in the substrate and product molecules. Prefix numbers indicate in a stoichiometric fashion how many times each step is carried out in order to provide a balanced net reaction.

Reaction	Carbon balance
1. Ribulose bisphosphate carboxylase: $6\ CO_2 + 6\ H_2O + 6\ RuBP \longrightarrow 12\ $3-PG	$6(1) + 6(5) \longrightarrow 12(3)$
2. 3-Phosphoglycerate kinase: $12\ $3-PG$ + 12\ ATP \longrightarrow 12\ $1,3-BPG$ + 12\ ADP$	$12(3) \longrightarrow 12(3)$
3. NADP⁺-glyceraldehyde-3-P dehydrogenase: $12\ $1,3-BPG$ + 12\ NADPH \longrightarrow 12\ NADP^+ + 12\ $G-3-P$ + 12\ P_i$	$12(3) \longrightarrow 12(3)$
4. Triose-P isomerase: $5\ $G-3-P$ \longrightarrow 5\ $DHAP	$5(3) \longrightarrow 5(3)$
5. Aldolase: $3\ $G-3-P$ + 3\ $DHAP$ \longrightarrow 3\ $FBP	$3(3) + 3(3) \longrightarrow 3(6)$
6. Fructose bisphosphatase: $3\ FBP + 3\ H_2O \longrightarrow 3\ $F6P$ + 3\ P_i$	$3(6) \longrightarrow 3(6)$
7. Phosphoglucoisomerase: $1\ $F-6-P$ \longrightarrow 1\ $G-6-P	$1(6) \longrightarrow 1(6)$
8. Glucose phosphatase: $1\ $G-6-P$ + 1\ H_2O \longrightarrow 1\ $GLUCOSE$ + 1\ P_i$	$1(6) \longrightarrow 1(6)$
The remainder of the pathway involves regenerating six RuBP acceptors ($= 30$ C) from the leftover two F-6-P (12 C), four G-3-P (12 C), and two DHAP (6 C).	
9. Transketolase: $2\ $F-6-P$ + 2\ $G-3-P$ \longrightarrow 2\ $Xu-5-P$ + 2\ $E-4-P	$2(6) + 2(3) \longrightarrow 2(5) + 2(4)$
10. Aldolase: $2\ $E-4-P$ + 2\ $DHAP$ \longrightarrow 2\ $sedoheptulose-1,7-bisphosphate (SBP)	$2(4) + 2(3) \longrightarrow 2(7)$
11. Sedoheptulose bisphosphatase: $2\ SBP + 2\ H_2O \longrightarrow 2\ $S-7-P$ + 2\ P_i$	$2(7) \longrightarrow 2(7)$
12. Transketolase: $2\ $S-7-P$ + 2\ $G-3-P$ \longrightarrow 2\ $Xu-5-P$ + 2\ $R-5-P	$2(7) + 2(3) \longrightarrow 4(5)$
13. Phosphopentose epimerase: $4\ $Xu-5-P$ \longrightarrow 4\ $Ru-5-P	$4(5) \longrightarrow 4(5)$
14. Phosphopentose isomerase: $2\ $R-5-P$ \longrightarrow 2\ $Ru-5-P	$2(5) \longrightarrow 2(5)$
15. Phosphoribulose kinase: $6\ $Ru-5-P$ + 6\ ATP \longrightarrow 6\ $RuBP$ + 6\ ADP$	$6(5) \longrightarrow 6(5)$
Net: $6\ CO_2 + 18\ ATP + 12\ NADPH + 12\ H^+ + 12\ H_2O \longrightarrow$ glucose $+ 18\ ADP + 18\ P_i + 12\ NADP^+$	$6(1) \longrightarrow 1(6)$

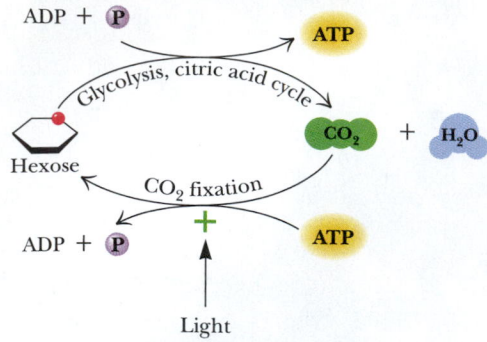

FIGURE 21.25 Light regulation of CO_2 fixation prevents a substrate cycle between cellular respiration and hexose synthesis by CO_2 fixation. Because plants possess mitochondria and are capable of deriving energy from hexose catabolism (glycolysis and the citric acid cycle), regulation of photosynthetic CO_2 fixation by light activation controls the net flux of carbon between these opposing routes.

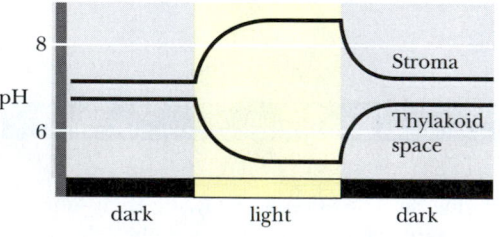

Biochemistry Now™ ANIMATED FIGURE 21.26
Light-induced pH changes in chloroplast compartments. Illumination of chloroplasts leads to proton pumping and pH changes in the chloroplast, such that the pH within the thylakoid space falls and the pH of the stroma rises. These pH changes modulate the activity of key Calvin cycle enzymes. **See this figure animated at http://chemistry.brookscole.com/ggb3**

(reaction 15). The net conversion accounts for the fixation of 6 equivalents of carbon dioxide into 1 hexose at the expense of 18 ATP and 12 NADPH.

The Carbon Dioxide Fixation Pathway Is Indirectly Activated by Light

Plant cells contain mitochondria and can carry out cellular respiration (glycolysis, the citric acid cycle, and oxidative phosphorylation) to provide energy in the dark. Futile cycling of carbohydrate to CO_2 by glycolysis and the citric acid cycle in one direction, and CO_2 to carbohydrate by the CO_2 fixation pathway in the opposite direction, is thwarted through regulation of the Calvin cycle (Figure 21.25). In this regulation, the activities of key Calvin cycle enzymes are coordinated with the output of photosynthesis. In effect, these enzymes respond indirectly to **light activation.** Thus, when light energy is available to generate ATP and NADPH for CO_2 fixation, the Calvin cycle proceeds. In the dark, when ATP and NADPH cannot be produced by photosynthesis, fixation of CO_2 ceases. The light-induced changes in the chloroplast which regulate key Calvin cycle enzymes include (1) *changes in stromal pH,* (2) *generation of reducing power,* and (3) *Mg^{2+} efflux from the thylakoid lumen.*

Light Induces pH Changes in Chloroplast Compartments As discussed in Section 21.7, illumination of chloroplasts leads to light-driven pumping of protons into the thylakoid lumen, which causes pH changes in both the stroma and the thylakoid lumen (Figure 21.26). The stromal pH rises, typically to pH 8. Because rubisco and rubisco activase are more active at pH 8, CO_2 fixation is activated as stromal pH rises. *Fructose-1,6-bisphosphatase, ribulose-5-phosphate kinase,* and *glyceraldehyde-3-phosphate dehydrogenase* all have alkaline pH optima. Thus, their activities increase as a result of the light-induced pH increase in the stroma.

Light Energy Generates Reducing Power Illumination of chloroplasts initiates photosynthetic electron transport, which generates reducing power in the form of reduced ferredoxin and NADPH. Several enzymes of CO_2 fixation, notably *fructose-1,6-bisphosphatase, sedoheptulose-1,7-bisphosphatase,* and *ribulose-5-phosphate kinase,* are activated upon reduction of specific Cys-Cys disulfide bonds to cysteine sulfhydryls. The reduced form of thioredoxin mediates this reaction. Thioredoxin is a small (12 kD) protein possessing in its reduced state a pair of sulfhydryls (—SH HS—), which upon oxidation form a disulfide bridge (—S—S—). Thioredoxin serves as the hydrogen carrier between NADPH or Fd_{red} and enzymes regulated by light (Figure 21.27).

Light Induces Movement of Mg^{2+} Ions from the Thylakoid Vesicles into the Stroma When light-driven proton pumping across the thylakoid membrane occurs, a concomitant efflux of Mg^{2+} ions from vesicles into the stroma is observed. This efflux of Mg^{2+} somewhat counteracts the charge accumulation due to H^+ influx and is one reason why the membrane potential change in response to proton pumping is less in chloroplasts than in mitochondria (Equation 21.5). Both *ribulose bisphosphate carboxylase* and *fructose-1,6-bisphosphatase* are Mg^{2+}-activated enzymes, and Mg^{2+} flux into the stroma as a result of light-driven proton pumping stimulates

FIGURE 21.27 The pathway for light-regulated reduction of Calvin cycle enzymes. Light-generated reducing power (Fd_{red} = reduced ferredoxin) provides e^- for reduction of thioredoxin (T) by FTR (ferredoxin–thioredoxin reductase). Several Calvin cycle enzymes have pairs of Cys residues that are involved in the disulfide–sulfhydryl transition between an inactive (—S—S—) form and an active (—SH HS—) form, as shown here. These enzymes include *fructose-1,6-bisphosphatase* (residues Cys[174] and Cys[179]), *NADP+-malate dehydrogenase* (residues Cys[10] and Cys[15]), and *ribulose-5-P kinase* (residues Cys[16] and Cys[55]).

the CO$_2$ fixation pathway at these key steps. Activity measurements have indicated that fructose bisphosphatase may be the rate-limiting step in the Calvin cycle. The recurring theme of fructose bisphosphatase as the target of the light-induced changes in the chloroplasts implicates this enzyme as a key point of control in the Calvin cycle.

21.8 How Does Photorespiration Limit CO$_2$ Fixation?

As indicated, ribulose bisphosphate carboxylase/oxygenase catalyzes an alternative reaction in which O$_2$ replaces CO$_2$ as the substrate added to RuBP (Figure 21.28a). The *ribulose-1,5-bisphosphate oxygenase* reaction diminishes plant productivity because it leads to loss of RuBP, the essential CO$_2$ acceptor. The K_m for O$_2$ in this oxygenase reaction is about 200 μM. Given the relative abundance of CO$_2$ and O$_2$ in the atmosphere and their relative K_m values in these rubisco-mediated reactions, the ratio of carboxylase to oxygenase activity in vivo is about 3 or 4 to 1.

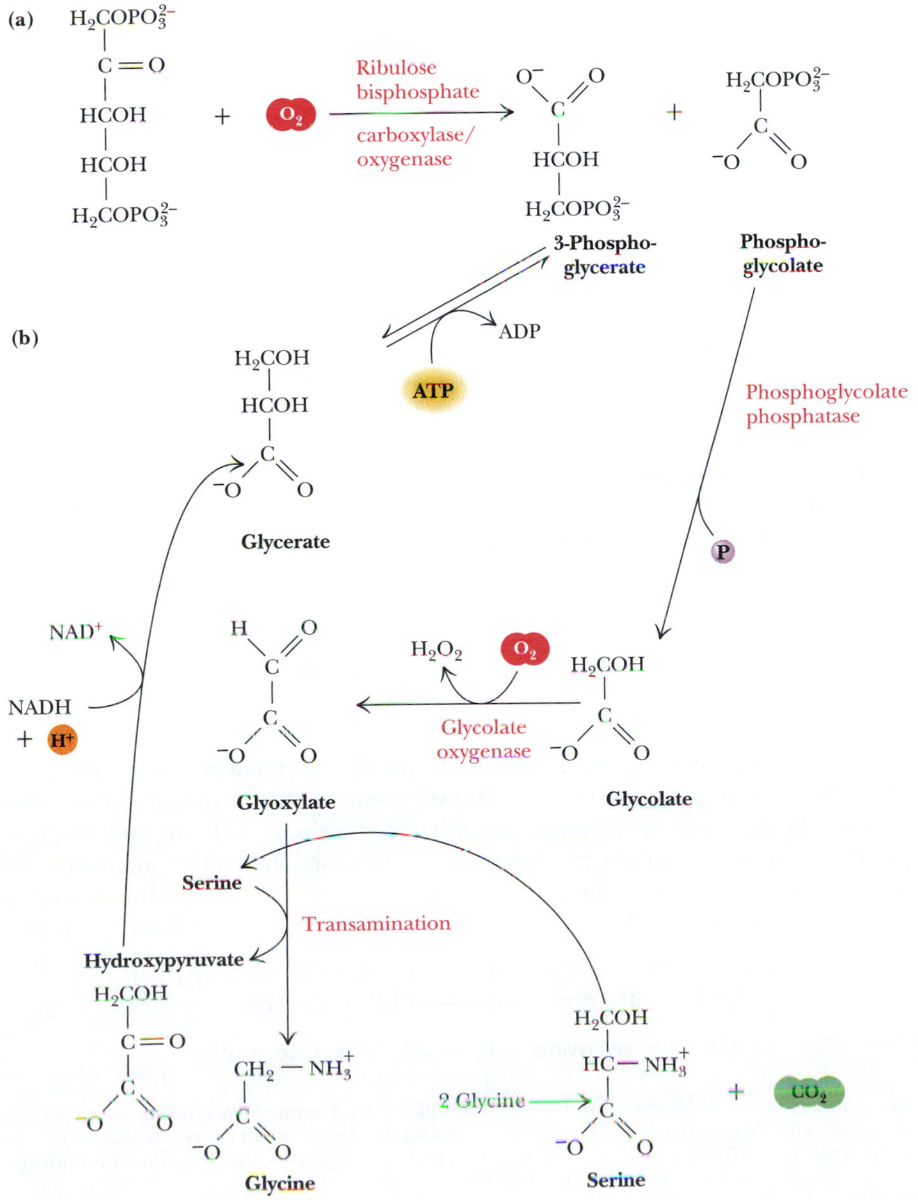

FIGURE 21.28 The oxygenase reaction of rubisco. (a) The reaction of ribulose bisphosphate carboxylase with O$_2$ in the presence of ribulose bisphosphate leads to wasteful cleavage of RuBP to yield 3-phosphoglycerate and phosphoglycolate. (b) Conversion of phosphoglycolate to glycine. In mitochondria, two glycines from photorespiration are converted into one serine plus CO$_2$. This step is the source of the CO$_2$ evolved in photorespiration. Transamination of glyoxylate to glycine by the product serine yields hydroxypyruvate; reduction of hydroxypyruvate yields glycerate, which can be phosphorylated to 3-phosphoglycerate. 3-Phosphoglycerate can fuel resynthesis of ribulose bisphosphate by the Calvin cycle (Figure 21.24).

The products of ribulose bisphosphate oxygenase activity are *3-phosphoglycerate* and *phosphoglycolate*. Dephosphorylation and oxidation convert phosphoglycolate to **glyoxylate,** the α-keto acid of glycine (Figure 21.28b). Transamination yields glycine. Other fates of phosphoglycolate are also possible, including oxidation to CO_2, with the released energy being dissipated as heat. Obviously, agricultural productivity is dramatically lowered by this phenomenon, which, because it is a light-related uptake of O_2 *and* release of CO_2, is termed **photorespiration.** As we shall see, certain plants, particularly tropical grasses, have evolved means to circumvent photorespiration. These plants are more efficient users of light for carbohydrate synthesis.

Tropical Grasses Use the Hatch–Slack Pathway to Capture Carbon Dioxide for CO_2 Fixation

Tropical grasses are less susceptible to the effects of photorespiration, as noted earlier. Studies using $^{14}CO_2$ as a tracer indicated that the first organic intermediate labeled in these plants was not a three-carbon compound but a four-carbon compound. Hatch and Slack, two Australian biochemists, first discovered this C-4 product of CO_2 fixation, and the C-4 pathway of CO_2 incorporation is named the **Hatch–Slack pathway** after them. The C-4 pathway is not an alternative to the Calvin cycle series of reactions or even a net CO_2 fixation scheme. Instead, it functions as a CO_2 delivery system, carrying CO_2 from the relatively oxygen-rich surface of the leaf to interior cells where oxygen is lower in concentration and hence less effective in competing with CO_2 in the rubisco reaction. Thus, the C-4 pathway is a means of avoiding photorespiration by sheltering the rubisco reaction in a cellular compartment away from high $[O_2]$. The C-4 compounds serving as CO_2 transporters are malate or *aspartate*.

Compartmentation of these reactions to prevent photorespiration involves the interaction of two cell types: *mesophyll cells* and *bundle sheath cells*. The mesophyll cells take up CO_2 at the leaf surface, where O_2 is abundant, and use it to carboxylate phosphoenolpyruvate to yield OAA in a reaction catalyzed by **PEP carboxylase** (Figure 21.29). This four-carbon dicarboxylic acid is then either reduced to malate by an **NADPH-specific malate dehydrogenase** or transaminated to give *aspartate* in the mesophyll cells.[4] The 4-C CO_2 carrier (malate or aspartate) is then transported to the bundle sheath cells, where it is decarboxylated to yield CO_2 and a 3-C product. The CO_2 is then fixed into organic carbon by the Calvin cycle localized within the bundle sheath cells, and the 3-C product is returned to the mesophyll cells, where it is reconverted to PEP in preparation to accept another CO_2 (Figure 21.29). Plants that use the C-4 pathway are termed **C4 plants,** in contrast to those plants with the conventional pathway of CO_2 uptake (**C3 plants**).

Intercellular Transport of Each CO_2 Via a C-4 Intermediate Costs 2 ATPs The transport of each CO_2 requires the expenditure of two high-energy phosphate bonds. The energy of these bonds is expended in the phosphorylation of pyruvate to PEP (phosphoenolpyruvate) by the plant enzyme **pyruvate-P_i dikinase;** the products are PEP, AMP, and pyrophosphate (PP_i). This represents a unique phosphotransferase reaction in that both the β- and γ-phosphates of a single ATP are used to phosphorylate the two substrates, pyruvate and P_i. The reaction mechanism involves an enzyme phosphohistidine intermediate. The γ-phosphate of ATP is transferred to P_i, whereas formation of E-His-P occurs by addition of the β-phosphate from ATP:

$$E\text{—}His + AMP_\alpha\text{—}P_\beta\text{—}P_\gamma + P_i \longrightarrow E\text{—}His\text{—}P_\beta + AMP_\alpha + P_\gamma P_i$$
$$E\text{—}His\text{—}P_\beta + pyruvate \longrightarrow PEP + E\text{—}His$$

$$Net:\ ATP + pyruvate + P_i \longrightarrow AMP + PEP + PP_i$$

[4]A number of different biochemical subtypes of C4 plants are known. They differ in whether OAA or malate is the CO_2 carrier to the bundle sheath cell and in the nature of the reaction by which the CO_2 carrier is decarboxylated to regenerate a 3-C product. In all cases, the 3-C product is returned to the mesophyll cell and reconverted to PEP.

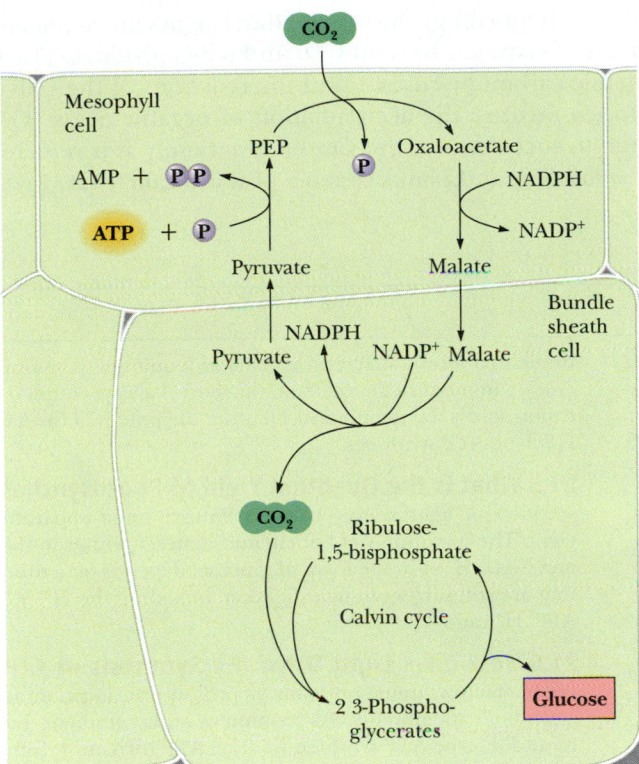

FIGURE 21.29 Essential features of the compartmentation and biochemistry of the Hatch–Slack pathway of carbon dioxide uptake in C4 plants. Carbon dioxide is fixed into organic linkage by PEP carboxylase of mesophyll cells, forming OAA. Either malate (the reduced form of OAA) or aspartate (the aminated form) serves as the carrier transporting CO_2 to the bundle sheath cells. Within the bundle sheath cells, CO_2 is liberated by decarboxylation of malate or aspartate; the C-3 product is returned to the mesophyll cell. Formation of PEP by pyruvate : P_i dikinase reinitiates the cycle. The CO_2 liberated in the bundle sheath cell is used to synthesize hexose by the conventional rubisco–Calvin cycle series of reactions.

Pyruvate-P_i dikinase is regulated by reversible phosphorylation of a threonine residue, the nonphosphorylated form being active. Interestingly, ADP is the phosphate donor in this interconvertible regulation. Despite the added metabolic expense of two phosphodiester bonds for each equivalent of carbon dioxide taken up, CO_2 fixation is more efficient in C4 plants, provided that light intensities and temperatures are both high. (As temperature rises, photorespiration in C3 plants rises and efficiency of CO_2 fixation falls.) Tropical grasses that are C4 plants include sugarcane, maize, and crabgrass. In terms of photosynthetic efficiency, cultivated fields of sugarcane represent the pinnacle of light-harvesting efficiency. Approximately 8% of the incident light energy on a sugarcane field appears as chemical energy in the form of CO_2 fixed into carbohydrate. This efficiency compares dramatically with the estimated photosynthetic efficiency of 0.2% for uncultivated plant areas. Research on photorespiration is actively pursued in hopes of enhancing the efficiency of agriculture by controlling this wasteful process. Only 1% of the 230,000 different plant species known are C4 plants; most are in hot climates.

Cacti and Other Desert Plants Capture CO_2 at Night

In contrast to C4 plants, which have separated CO_2 uptake and fixation into distinct cells in order to minimize photorespiration, succulent plants native to semiarid and tropical environments separate CO_2 uptake and fixation in time. Carbon dioxide (as well as O_2) enters the leaf through microscopic pores known as **stomata,** and water vapor escapes from plants via these same openings. In nonsucculent plants, the stomata are open during the day, when light can drive photosynthetic CO_2 fixation, and closed at night. Succulent plants, such as the *Cactaceae* (cacti) and *Crassulaceae*, cannot open their stomata during the heat of day because any loss of precious H_2O in their arid habitats would doom them. Instead, these plants open their stomata to take up CO_2 only at night, when temperatures are lower and water loss is less likely. This carbon dioxide is immediately incorporated into PEP to form OAA by PEP carboxylase; OAA is then reduced to malate by malate dehydrogenase and stored within vacuoles until morning. During the day, the malate is released from the vacuoles and decarboxylated to yield CO_2 and a 3-C product. The CO_2 is then fixed into organic carbon by rubisco and the reactions of the Calvin cycle. Because this process involves the accumulation of organic acids (OAA, malate) and is common to succulents of the *Crassulaceae* family, it is referred to as *crassulacean acid metabolism,* and plants capable of it are called *CAM plants.*

Summary

21.1 What Are the General Properties of Photosynthesis?
Photosynthesis takes place in membranes. In photosynthetic eukaryotes, the photosynthetic membranes form an inner membrane system within chloroplasts that is called the thylakoid membrane system. Photosynthesis is traditionally broken down into two sets of reactions: the light reactions, whereby light energy is used to generate NADPH and ATP concomitant with O_2 evolution, and the dark reactions in which NADPH and ATP provide the chemical energy for fixation of CO_2 into glucose. Water is the ultimate e^- donor for $NADP^+$ reduction.

21.2 How Is Solar Energy Captured by Chlorophyll?
Chlorophyll and various accessory light-harvesting pigments absorb light throughout the visible spectrum and use the light energy to initiate electron-transfer reactions. The absorption of a photon of light by a pigment molecule promotes an electron of the pigment molecule to a higher orbital (and higher energy level). As a result, the pigment molecule is a much better electron donor. Photosynthetic units consist of arrays of hundreds of chlorophyll molecules and accessory light-harvesting pigments, but only a single reaction center. The reaction center is formed from a pair of Chl molecules.

21.3 What Kinds of Photosystems Are Used to Capture Light Energy?
Photosynthetic bacteria have a single photosystem, but eukaryotic phototrophs have two distinct photosystems. Type I photosystems use proteins with Fe_4S_4 clusters as terminal e^- acceptors; type II photosystems reduce quinones, such as plastoquinone or phylloquinone. In oxygenic phototrophs (cyanobacteria, green algae, and higher plants), photosystem II (PSII) generates a strong oxidant that functions in O_2 evolution through the photolysis of water and a weak reductant that reduces plastoquinone to plastohydroquinone (PQH_2). Photosystem I (PSI) generates a weak oxidant that accepts electrons from plastohydroquinone via the cytochrome b_6/cytochrome f complex and a strong reductant capable of reducing $NADP^+$ to NADPH. Overall photosynthetic electron transfer is accomplished by three supramolecular membrane-spanning complexes: PSII, the cytochrome b_6/cytochrome f complex, and PSI. Oxygen evolution requires the accumulation of four oxidizing equivalents in PSII. The electrons withdrawn from water are used to re-reduce $P680^+$ back to P680, restoring its ability to absorb another photon, become P680*, and transfer an e^- once again. Electrons from P680* traverse PSII and reduce plastoquinone. Plastohydroquinone is oxidized via the cytochrome b_6/cytochrome f complex, with plastocyanin serving as e^- acceptor. The cytochrome b_6/cytochrome f complex catalyzes a Q cycle: It translocates 4 H^+ from the stroma to the thylakoid lumen for each molecule of PQH_2 that it oxidizes. PSI is a light-driven plastocyanin:ferredoxin oxidoreductase having P700 as its reaction center Chl dimer. Electrons from P700* are transferred to the Fe_4S_4 cluster of ferredoxin. Reduced ferredoxin reduces $NADP^+$ via the ferredoxin:$NADP^+$ reductase flavoprotein. The electron "hole" in $P700^+$ is filled by reduced plastocyanin.

21.4 What Is the Molecular Architecture of Photosynthetic Reaction Centers?
All known photosynthetic reaction centers have a universal molecular architecture. The "core" structure is a pair of protein subunits having (at least) five transmembrane α-helical segments that provide a scaffold upon which the reaction center Chl pair and its associated chain of electron transfer cofactors are arrayed in a characteristic spatial pattern that facilitates rapid removal of an electron from the photoactivated RC and efficient transfer of the e^- across the membrane to a terminal acceptor (such as a quinone or a ferredoxin molecule). Photosynthetic electron transport is always coupled to H^+ translocation across the membrane, creating the potential for ATP synthesis by F_1F_0-type ATP synthases.

21.5 What Is the Quantum Yield of Photosynthesis?
The absorption of light energy by the photosynthetic apparatus is very efficient. The quantum yield of chemical energy, either in the form of ATP and NADPH, or in the form of glucose, depends on a number of factors that are still subject to investigation, including the H^+/e^- ratio and the ATP/H^+ ratio.

21.6 How Does Light Drive the Synthesis of ATP?
Photosynthetic electron transport leads to proton translocation across the photosynthetic membrane and creation of an H^+ gradient that can be used by an F_1F_0-type ATP synthase to drive ATP formation from ADP and P_i. Photophosphorylation occurs by either of two modes: noncyclic and cyclic. Noncyclic photophosphorylation depends on both PSI and PSII and leads to O_2 evolution, $NADP^+$ reduction, and ATP synthesis. In cyclic photophosphorylation, only PSI is used, no $NADP^+$ is reduced, and no O_2 is evolved. However, the electron-transfer events of cyclic photophosphorylation lead to H^+ translocation and ATP synthesis.

21.7 How Is Carbon Dioxide Used to Make Organic Molecules?
Ribulose-1,5-bisphosphate is the CO_2 acceptor in the key reaction for conversion of carbon dioxide into organic compounds. The reaction is catalyzed by rubisco (ribulose bisphosphate carboxylase/oxygenase); the products of CO_2 fixation by the rubisco reaction are 2 equivalents of 3-phosphoglycerate.

The Calvin–Benson cycle is a series of reactions that converts the 3-phosphoglycerates formed by rubisco into carbohydrates such as glyceraldehyde-3-P, dihydroxyacetone-P, and glucose. CO_2 fixation is activated by light through a variety of mechanisms, including changes in stromal pH, generation of reducing power in the form of ferredoxin by photosynthetic electron transport, and increased Mg^{2+} efflux from the thylakoid lumen to the stroma.

21.8 How Does Photorespiration Limit CO₂ Fixation?
When O_2 replaces CO_2 in the rubisco, or ribulose bisphosphate carboxylase/oxygenase, reaction, ribulose-1,5-bisP is destroyed through conversion into 3-phosphoglyerate and phosphoglycolate. Phosphoglycolate is oxidized to form CO_2, with loss of organic substance from the cell. Because O_2 is taken up and CO_2 is released in these reactions, the process is called *photorespiration.*

Tropical grasses carry out the Calvin–Benson cycle of reactions in cells shielded from high O_2 levels. CO_2 is first incorporated into PEP by PEP carboxylase to form oxaloacetate (OAA) in the mesophyll cells on the leaf surface. OAA is then reduced to malate and transported to bundle sheath cells. There, CO_2 is released and taken up by the rubisco reaction to form 3-phosphoglyerate, initiating the Calvin–Benson cycle. Plants capable of doing this are called C4 plants.

Succulent plants of semiarid and tropical regions such as *Cactaceae* and *Crassulaceae* exchange gases through their stomata only at night in order to avoid precious water loss. CO_2 taken up at night is added to PEP by PEP carboxylase to form OAA, which is then reduced to malate in the dark. During the day, malate is decarboxylated to yield CO_2, which then enters the Calvin–Benson cycle. This metabolic variation is referred to as *crassulacean acid metabolism.*

Problems

1. In Photosystem I, P700 in its ground state has an $\mathscr{E}_o' = -0.4$ V. Excitation of P700 by a photon of 700-nm light alters the $\mathscr{E}_o'$ of P700* to -0.6 V. What is the efficiency of energy capture in this light reaction of P700?

2. What is the $\mathscr{E}_o'$ for the light-generated primary oxidant of Photosystem II if the light-induced oxidation of water (which leads to O_2 evolution) proceeds with a $\Delta G^{\circ}{}'$ of -25 kJ/mol?

3. (Integrates with Chapters 3 and 20.) Assuming that the concentrations of ATP, ADP, and P_i in chloroplasts are 3, 0.1, and 10 mM, respectively, what is the ΔG for ATP synthesis under these conditions? Photosynthetic electron transport establishes the proton-motive force driving photophosphorylation. What redox potential difference is necessary to achieve ATP synthesis under the foregoing conditions, assuming 1.3 ATP equivalents are synthesized for each electron pair transferred?

4. (Integrates with Chapter 20.) Write a balanced equation for the Q cycle as catalyzed by the cytochrome b_6/cytochrome f complex of chloroplasts.

5. If noncyclic photosynthetic electron transport leads to the translocation of 3 H^+/e^- and cyclic photosynthetic electron transport leads to the translocation of 2 H^+/e^-, what is the relative photosynthetic efficiency of ATP synthesis (expressed as the number of photons absorbed per ATP synthesized) for noncyclic versus cyclic photophosphorylation? (Assume that the CF_1CF_0–ATP synthase yields 3 ATP/14 H^+.)

6. (Integrates with Chapter 20.) In mitochondria, the membrane potential ($\Delta\psi$) contributes relatively more to Δp (proton-motive force) than does the pH gradient (ΔpH). The reverse is true in chloroplasts. Why do you suppose that the proton-motive force in chloroplasts can depend more on ΔpH than mitochondria can? Why is ($\Delta\psi$) less in chloroplasts than in mitochondria?

7. Predict the consequences of a Y161F mutation in the amino acid sequence of the D1 and the D2 subunits of PSII.

8. Why was the Jagendorf–Uribe experiment performed in the dark?

9. (Integrates with Chapter 20.) Calculate (in Einsteins and in kJ/mol) how many photons would be required by the *Rhodopseudomonas viridis* photophosphorylation system to synthesize 3 ATPs? (Assume that the *R. viridis* F_1F_0–ATP synthase c-subunit rotor contains 12 c-subunits and that the *R. viridis* cytochrome b/c_1 complex translocates 2 H^+/e^-.)

10. (Integrates with Chapters 18 and 20.) Calculate $\Delta G^{\circ}{}'$ for the $NADP^+$-specific glyceraldehyde-3-P dehydrogenase reaction of the Calvin–Benson cycle.

11. Write a balanced equation for the synthesis of a glucose molecule from ribulose-1,5-bisphosphate and CO_2 that involves the first three reactions of the Calvin cycle and subsequent conversion of the two glyceraldehyde-3-P molecules into glucose.

12. ^{14}C-labeled carbon dioxide is administered to a green plant, and shortly thereafter the following compounds are isolated from the plant: 3-phosphoglycerate, glucose, erythrose-4-phosphate, sedoheptulose-1,7-bisphosphate, and ribose-5-phosphate. In which carbon atoms will radioactivity be found?

13. The photosynthetic CO_2 fixation pathway is regulated in response to specific effects induced in chloroplasts by light. What is the nature of these effects, and how do they regulate this metabolic pathway?

14. Write a balanced equation for the conversion of phosphoglycolate to glycerate-3-P by the reactions of photorespiration. Does this balanced equation demonstrate that photorespiration is a wasteful process?

15. The overall equation for photosynthetic CO_2 fixation is

$$6\ CO_2 + 6\ H_2O \longrightarrow C_6H_{12}O_6 + 6\ O_2$$

All the O atoms evolved as O_2 come from water; *none* comes from carbon dioxide. But 12 O atoms are evolved as 6 O_2, and only 6 O atoms appear as 6 H_2O in the equation. Also, 6 CO_2 have 12 O atoms, yet there are only 6 O atoms in $C_6H_{12}O_6$. How can you account for these discrepancies? (*Hint:* Consider the partial reactions of photosynthesis: ATP synthesis, $NADP^+$ reduction, photolysis of water, and the overall reaction for hexose synthesis in the Calvin–Benson cycle.)

Preparing for the MCAT Exam

16. From Figure 21.6, predict the spectral properties of accessory light-harvesting pigments found in plants.

17. Draw a figure analogous to Figure 21.26, plotting [Mg^{2+}] in the stroma and thylakoid lumen on the *y*-axis and *dark-light-dark* on the *x*-axis.

Biochemistry⟨S⟩Now™ Preparing for an exam? Test yourself on key questions at http://chemistry.brookscole.com/ggb3

Further Reading

General References

Blankenship, R. E., 2002. *Molecular Mechanisms of Photosynthesis*. Malden, MA: Blackwell Science.

Buchanan, B. B., Gruissem, W., and Jones, R. I., 2000. *Biochemistry and Molecular Biology of Plants*. Rockville, MD: American Society of Plant Physiologists.

Cramer, W. A., and Knaff, D. B., 1990. *Energy Transduction in Biological Membranes—A Textbook of Bioenergetics*. New York: Springer-Verlag. A textbook on bioenergetics by two prominent workers in photosynthesis.

Harold, F. M., 1987. *The Vital Force: A Study of Bioenergetics*. Chapter 8: Harvesting the Light. San Francisco: Freeman & Company.

Heathcote, P., Fyfe, P. K., and Jones, M. R., 2002. Reaction centers: The structure and evolution of biological solar power. *Trends in Biochemical Sciences* **27**:79–87.

Photosynthetic Pigments

Glazer, A. N., 1983. Comparative biochemistry of photosynthetic light-harvesting pigments. *Annual Review of Biochemistry* **52**:125–157.

Green, B. R., and Durnford, D. G., 1996. The chlorophyll-carotenoid proteins of oxygenic photosynthesis. *Annual Review of Plant Physiology and Plant Molecular Biology* **47**:685–714.

Hoffman, E., et al., 1996. Structural basis of light harvesting by carotenoids: Peridinin-chlorophyll protein from *Amphidinium carterae*. *Science* **272**:1788–1791.

Properties of the Thylakoid Membranes

Anderson, J. M., 1986. Photoregulation of the composition, function and structure of the thylakoid membrane. *Annual Review of Plant Physiology* **37**:93–136.

Anderson, J. M., and Andersson, B., 1988. The dynamic photosynthetic membrane and regulation of solar energy conversion. *Trends in Biochemical Sciences* **13**:351–355.

Photosynthetic Reaction Centers of Photosynthetic Bacteria

Deisenhofer, J., and Michel, H., 1989. The photosynthetic reaction center from the purple bacterium *Rhodopseudomonas viridis*. *Science* **245**:1463–

1473. Published version of the Nobel laureate address by the researchers who first elucidated the molecular structure of a photosynthetic reaction center.

Deisenhofer, J., Michel, H., and Huber, R., 1985. The structural basis of light reactions in bacteria. *Trends in Biochemical Sciences* **10**:243–248.

Deisenhofer, J., et al., 1985. Structure of the protein subunits in the photosynthetic reaction center of *Rhodopseudomonas viridis* at 3 Å resolution. *Nature* **318**:618–624; also *The Journal of Molecular Biology* (1984) **180**:385–398. These papers are the original reports of the crystal structure of a photosynthetic reaction center.

Photosystem I

Barber, J., 2001. The structure of photosystem I. *Nature Structural Biology* **8**:577–579.

Fromme, P., Jordan, P., and Krausse, N., 2001. Structure of photosystem I. *Biochimica et Biophysica Acta* **1507**:5–31.

Jordan, P., et al., 2001. Three-dimensional structure of cyanobacterial photosystem I at 2.5 Å resolution. *Nature* **411**:909–917.

Photosystem II and Oxygenic Photosynthesis

Barber, J., 2003. Photosystem II: The engine of life. *Quarterly Review of Biophysics* **36**:71–89.

Blankenship, R. E., and Hartman, H., 1998. The origin and evolution of oxygenic photosynthesis. *Trends in Biochemical Sciences* **23**:94–97.

Cramer, W. A., et al., 1996. Some new structural aspects and old controversies concerning the cytochrome b_6f complex of oxygenic photosynthesis. *Annual Review of Plant Physiology and Plant Molecular Biology* **47**:477–508.

Hankamer, B., Barber, J., and Boekema, E. J., 1997. Structure and membrane organization of photosystem II in green plants. *Annual Review of Plant Physiology and Plant Molecular Biology* **48**:641–671.

Hoganson, C. W., and Babcock, G. T., 1997. A metalloradical mechanism for the generation of oxygen in photosynthesis. *Science* **277**:1953–1956.

Rögner, M., Boekema, E. J., and Barber, J., 1996. How does photosystem 2 split water? The structural basis of energy conversion. *Trends in Biochemical Sciences* **21**:44–49.

Vrettos, J. S., Limburg, J., and Brudvig, G. W., 2001. Mechanism of photosynthetic water oxidation: Combining biophysical studies of photosystem II with inorganic model chemistry. *Biochimica Biophysica Acta* **1503**:229–245.

Yommos, C., and Babcock, G. T., 2000. Proton and hydrogen currents in photosynthetic water oxidation. *Biochimica Biophysica Acta* **1458**:199–219.

Zouni, A., et al., 2001. Crystal structure of photosystem II from *Synechococcus elongates* at 3.8 Å resolution. *Nature* **409**:739–743.

Photophosphorylation

Allen, J. F., 2002. Photosynthesis of ATP—electrons, proton pumps, rotors, and poise. *Cell* **110**:273–276. This article provides an up-to-date and interesting discussion of the quantum yield of photosynthesis.

Arnon, D. I., 1984. The discovery of photosynthetic phosphorylation. *Trends in Biochemical Sciences* **9**:258–262. A historical account of photophosphorylation by its discoverer.

Bendall, D. S., and Manasse, R. S., 1995. Cyclic photophosphorylation and electron transport. *Biochimica et Biophysica Acta* **1229**:23–38.

Berry, S., and Rumberg, B., 1996. H^+/ATP coupling ratio at the unmodulated CF_1CF_0-ATP synthase determined by proton flux measurements. *Biochimica et Biophysica Acta* **1276**:51–56.

Jagendorf, A. T., and Uribe, E., 1966. ATP formation caused by acid-base transition of spinach chloroplasts. *Proceedings of the National Academy of Sciences, U.S.A.* **55**:170–177. The classic paper providing the first experimental verification of Mitchell's chemiosmotic hypothesis.

Remy, A., and Gerwert, K., 2003. Coupling of light-induced electron transfer to proton uptake in photosynthesis. *Nature Structural Biology* **10**:637–644.

Carbon Dioxide Fixation

Andersson, I., 1996. Large structures at high resolution. The 1.64 Å crystal structure of spinach ribulose1,5-bisphosphate carboxylase/oxygenase complexed with carboxyarabinitol bisphosphate. *Journal of Molecular Biology* **259**:160–174.

Burnell, J. N., and Hatch, M. D., 1985. Light-dark modulation of leaf pyruvate, P_i dikinase. *Trends in Biochemical Sciences* **10**:288–291. Regulation of a key enzyme in C4 CO_2 fixation.

Chapman, M. S., et al., 1988. Tertiary structure of plant rubisco: Domains and their contacts. *Science* **241**:71–74. Structural details of rubisco.

Cushman, J. C., and Bohnert, H. J., 1999. Crassulacean acid metabolism: Molecular genetics. *Annual Review of Plant Physiology and Plant Molecular Biology* **50**:305–332.

Hatch, M. D., 1987. C_4 photosynthesis: A unique blend of modified biochemistry, anatomy, and ultrastructure. *Biochimica Biophysica Acta* **895**:81–106. A review of C_4 biochemistry by its discoverer.

Kaplan, A., and Reinhold, L., 1999. CO_2-concentrating mechanisms in photosynthetic organisms. *Annual Review of Plant Physiology and Plant Molecular Biology* **50**:539–570.

Knaff, D. B., 1989. The regulatory role of thioredoxin in chloroplasts. *Trends in Biochemical Sciences* **14**:433–434.

Miziorko, H. M., and Lorimer, G. H., 1983. Ribulose-1,5-bisphosphate carboxylase/oxygenase. *Annual Review of Biochemistry* **52**:507–535. An early review of the enzymological properties of rubisco.

Ogren, W. L., 1984. Photorespiration: Pathways, regulation and modification. *Annual Review of Plant Physiology* **35**:415–442.

Portis, A. R., Jr., 1992. Regulation of ribulose 1,5-bisphosphate carboxylase/oxygenase activity. *Annual Review of Plant Physiology and Plant Molecular Biology* **43**:415–437.

Spreitzer, R. J., and Salvucci, M. E., 2002. Rubisco: Structure, regulatory interactions, and possibilities for a better enzyme. *Annual Review of Plant Biology* **53**:449–475.

Ting, I. P., 1985. Crassulacean acid metabolism. *Annual Review of Plant Physiology* **36**:595–622.

Wingler, A., et al., 2000. Photorespiration: Metabolic pathways and their role in stress protection. *Philosophical Transactions of the Royal Society of London B* **355**:1517–1529.

Gluconeogenesis, Glycogen Metabolism, and the Pentose Phosphate Pathway

A basket of fresh bread. Carbohydrates such as these provide a significant portion of human caloric intake.

Con pan y vino se anda el camino.
(With bread and wine you can walk
  your road.)
Spanish proverb

Essential Question

As shown in Chapters 18 and 19, the metabolism of sugars is an important source of energy for cells. Animals, including humans, typically obtain significant amounts of glucose and other sugars from the breakdown of starch and glycogen in their diets. Glucose can also be supplied via breakdown of cellular reserves of glycogen (in animals) or starch (in plants). *What is the nature of gluconeogenesis, the pathway that synthesizes glucose from noncarbohydrate precursors, how is glycogen synthesized from glucose, and how are electrons from glucose used in biosynthesis?*

22.1 What Is Gluconeogenesis, and How Does It Operate?

The ability to synthesize glucose from common metabolites is very important to most organisms. Human metabolism, for example, consumes about 160 ± 20 grams of glucose per day, about 75% of this in the brain. Body fluids carry only about 20 grams of free glucose, and glycogen stores normally can provide only about 180 to 200 grams of free glucose. Thus, the body carries only a little more than a 1-day supply of glucose. If glucose is not obtained in the diet, the body must produce new glucose from noncarbohydrate precursors. The term for this activity is **gluconeogenesis,** which means the generation *(genesis)* of new *(neo)* glucose.

Furthermore, muscles consume large amounts of glucose via glycolysis, producing large amounts of pyruvate. In vigorous exercise, muscle cells become anaerobic and pyruvate is converted to lactate. Gluconeogenesis salvages this pyruvate and lactate and reconverts it to glucose.

Another pathway of glucose catabolism, the *pentose phosphate pathway*, is the primary source of NADPH, the reduced coenzyme essential to most reductive biosynthetic processes. For example, NADPH is crucial to the biosynthesis of fatty acids (see Chapter 24) and amino acids (see Chapter 25). The pentose phosphate pathway also results in the production of ribose-5-phosphate, an essential component of ATP, NAD^+, FAD, coenzyme A, and particularly DNA and RNA. This important pathway will also be considered in this chapter.

The Substrates of Gluconeogenesis Include Pyruvate, Lactate, and Amino Acids

In addition to pyruvate and lactate, other noncarbohydrate precursors can be used as substrates for gluconeogenesis in animals. These include most of the amino acids, as well as glycerol and all the TCA cycle intermediates. On the other hand, fatty acids are not substrates for gluconeogenesis in animals, because most fatty acids yield only acetyl-CoA upon degradation, and animals cannot carry out net synthesis of sugars from acetyl-CoA. Lysine and leucine are the only amino acids that are not substrates for gluconeogenesis. These amino acids produce only acetyl-CoA upon degradation. Note also that acetyl-CoA is a substrate for gluconeogenesis when the glyoxylate cycle is operating (see Chapter 19).

Key Questions

22.1 What Is Gluconeogenesis, and How Does It Operate?

22.2 How Is Gluconeogenesis Regulated?

22.3 How Are Glycogen and Starch Catabolized in Animals?

22.4 How Is Glycogen Synthesized?

22.5 How Is Glycogen Metabolism Controlled?

22.6 Can Glucose Provide Electrons for Biosynthesis?

Biochemistry ⊜ Now™ Test yourself on these Key Questions at BiochemistryNow at **http://chemistry.brookscole.com/ggb3**

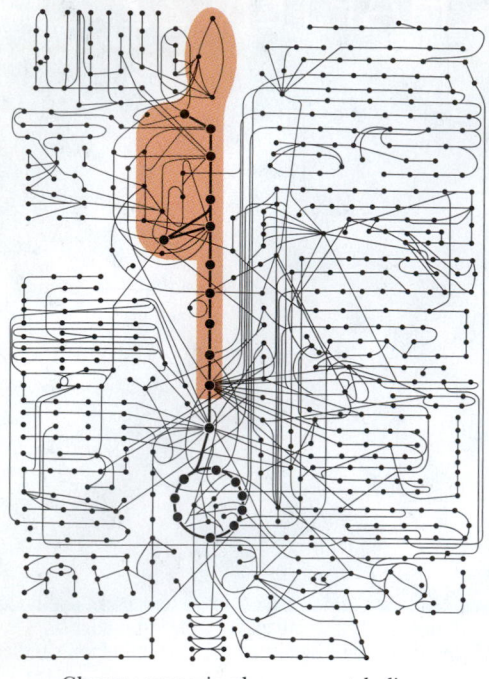

Gluconeogenesis, glycogen metabolism, and the pentose phosphate pathway.

Nearly All Gluconeogenesis Occurs in the Liver and Kidneys in Animals

Interestingly, the mammalian organs that consume the most glucose, namely, brain and muscle, carry out very little glucose synthesis. The major sites of gluconeogenesis are the liver and kidneys, which account for about 90% and 10% of the body's gluconeogenic activity, respectively. Glucose produced by gluconeogenesis in the liver and kidney is released into the blood and is subsequently absorbed by brain, heart, muscle, and red blood cells to meet their metabolic needs. In turn, pyruvate and lactate produced in these tissues are returned to the liver and kidney to be used as gluconeogenic substrates.

Gluconeogenesis Is Not Merely the Reverse of Glycolysis

In some ways, gluconeogenesis is the reverse, or antithesis, of glycolysis. Glucose is synthesized, not catabolized; ATP is consumed, not produced; and NADH is oxidized to NAD^+, rather than the other way around. However, gluconeogenesis cannot be merely the reversal of glycolysis, for two reasons. First, glycolysis is exergonic, with a $\Delta G°'$ of approximately -74 kJ/mol. If gluconeogenesis were merely the reverse, it would be a strongly endergonic process and could not occur spontaneously. Somehow the energetics of the process must be augmented so that gluconeogenesis can proceed spontaneously. Second, the processes of glycolysis and gluconeogenesis must be regulated in a reciprocal fashion so that when glycolysis is active, gluconeogenesis is inhibited, and when gluconeogenesis is proceeding, glycolysis is turned off. Both of these limitations are overcome by having unique reactions within the routes of glycolysis and gluconeogenesis, rather than a completely shared pathway.

Human Biochemistry

The Chemistry of Glucose Monitoring Devices

Individuals with diabetes must measure their serum glucose concentration frequently, often several times a day. The advent of computerized, automated devices for glucose monitoring has made this necessary chore easier, far more accurate, and convenient than it once was. These devices all use a simple chemical scheme for glucose measurement that involves oxidation of glucose to gluconic acid by glucose oxidase. This reaction produces two molecules of hydrogen peroxide per molecule of glucose oxidized. The H_2O_2 is then used to oxidize a dye, such as *o*-dianisidine, to a colored product that can be measured:

Glucose + 2 H_2O + $O_2 \longrightarrow$ gluconic acid + 2 H_2O_2
o-dianisidine (colorless) + $H_2O_2 \longrightarrow$ oxidized *o*-anisidine
(colored) + H_2O

The amount of colored dye produced is directly proportional to the amount of glucose in the sample.

The patient typically applies a drop of blood (from a finger-prick*) to a plastic test strip that is then inserted into the glucose monitor. Within half a minute, a digital readout indicates the blood glucose value. Modern glucose monitors store several days of glucose measurements, and the data can be easily transferred to a computer for analysis and graphing.

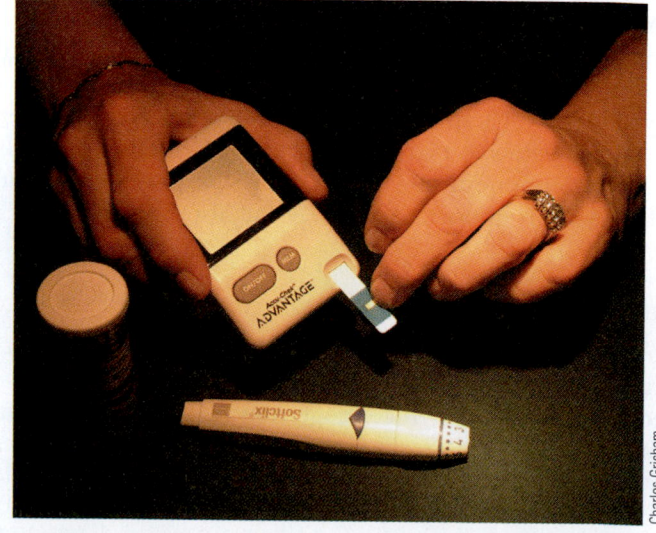

Charles Grisham

*How does the monitor deal with getting just the right amount of blood? The blood flows up an absorbent "wick" by capillary action. It is impossible to overfill this device, but the monitor will give an error signal if not enough blood flows up the strip.

Gluconeogenesis—Something Borrowed, Something New

The complete route of gluconeogenesis is shown in Figure 22.1, side by side with the glycolytic pathway. Gluconeogenesis employs three different reactions, catalyzed by three different enzymes, for the three steps of glycolysis that are highly exergonic (and highly regulated). In essence, seven of the ten steps of glycolysis are merely reversed in gluconeogenesis. The six reactions between fructose-1,6-bisphosphate and PEP are shared by the two pathways, as is the isomerization of glucose-6-P to fructose-6-P. The three exergonic regulated reactions—the hexokinase (glucokinase), phosphofructokinase, and pyruvate kinase reactions—are replaced by alternative reactions in the gluconeogenic pathway.

The conversion of pyruvate to PEP that initiates gluconeogenesis is accomplished by two unique reactions. **Pyruvate carboxylase** catalyzes the first,

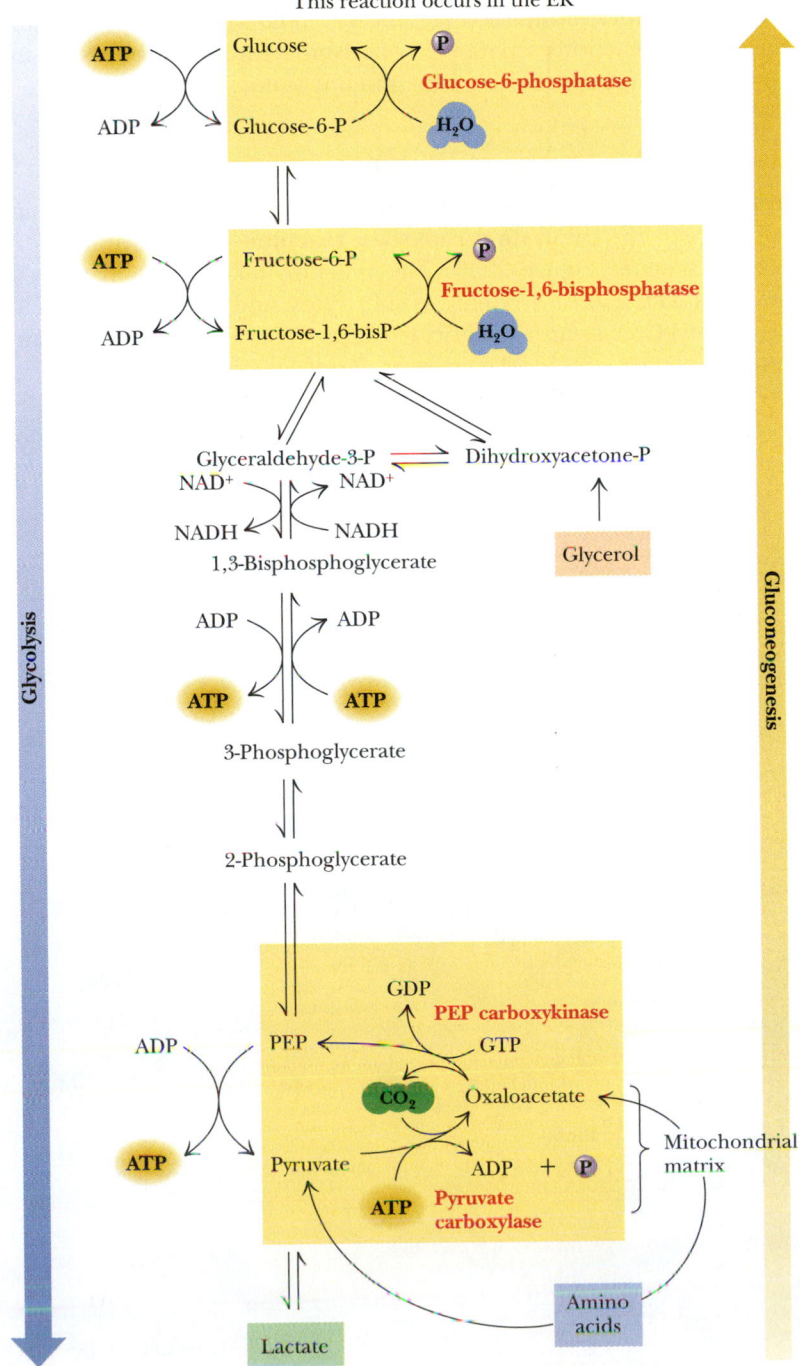

FIGURE 22.1 The pathways of gluconeogenesis and glycolysis. Species in blue, green, and peach-colored shaded boxes indicate other entry points for gluconeogenesis (in addition to pyruvate).

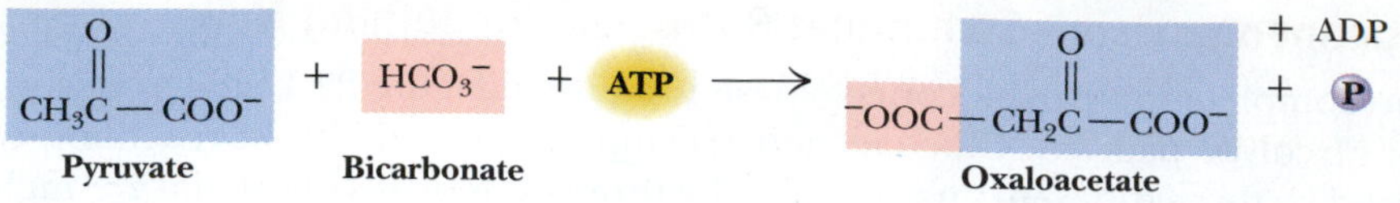

FIGURE 22.2 The pyruvate carboxylase reaction.

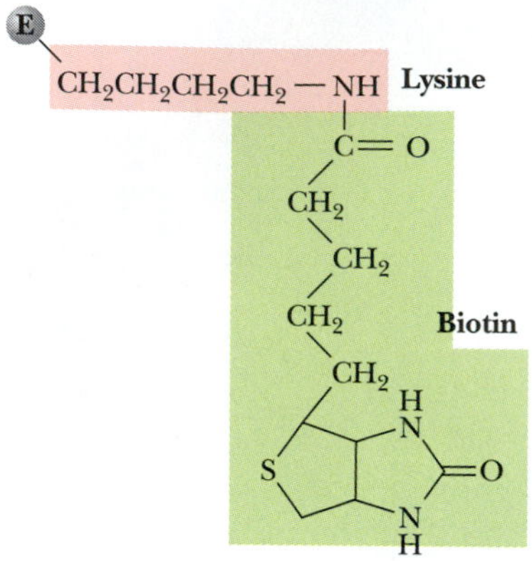

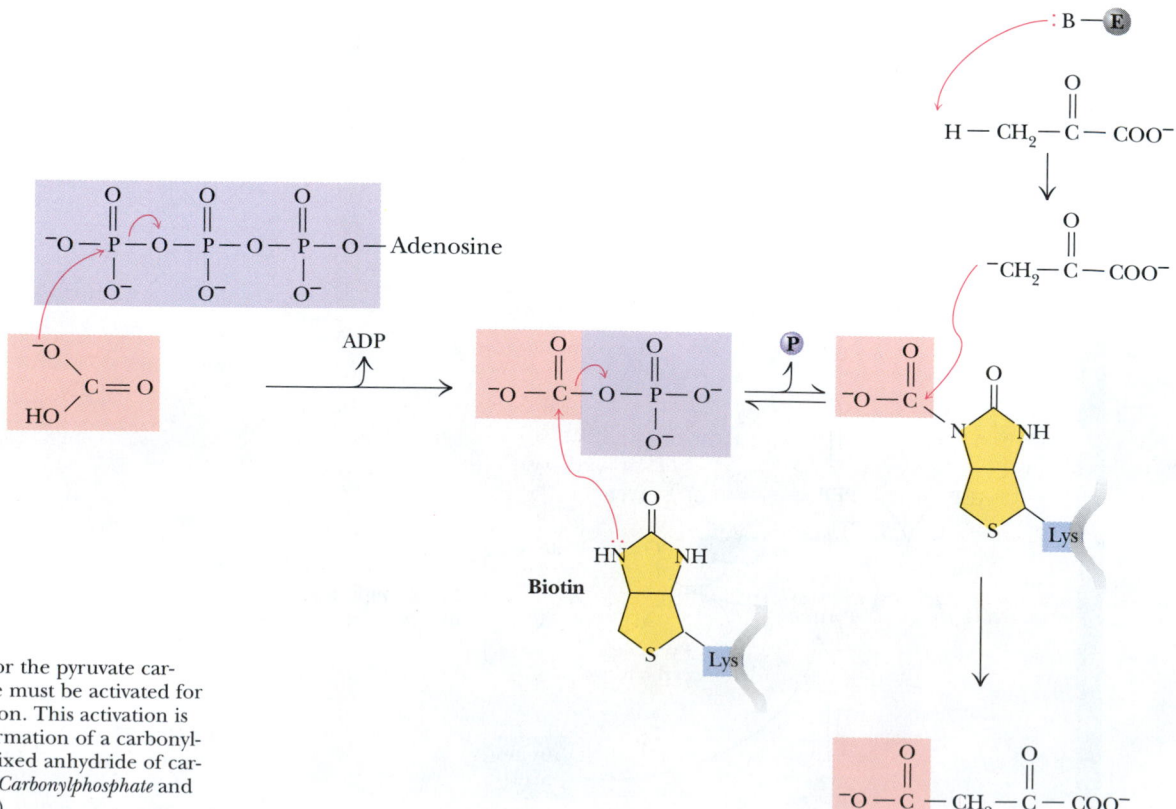

FIGURE 22.3 Covalent linkage of biotin to an active-site lysine in pyruvate carboxylase.

converting pyruvate to oxaloacetate. Then, **PEP carboxykinase** catalyzes the conversion of oxaloacetate to PEP. Conversion of fructose-1,6-bisphosphate to fructose-6-phosphate is catalyzed by a specific phosphatase, **fructose-1,6-bisphosphatase.** The final step to produce glucose, hydrolysis of glucose-6-phosphate, is mediated by **glucose-6-phosphatase.** Each of these steps is considered in detail in the following paragraphs. The overall conversion of pyruvate to PEP by pyruvate carboxylase and PEP carboxykinase has a $\Delta G^{\circ\prime}$ close to zero but is pulled along by subsequent reactions. The conversion of fructose-1,6-bisphosphate to glucose in the last three steps of gluconeogenesis is strongly exergonic, with a $\Delta G^{\circ\prime}$ of about -30.5 kJ/mol. This sequence of two phosphatase reactions separated by an isomerization accounts for most of the free energy release that makes the gluconeogenesis pathway spontaneous.

Four Reactions Are Unique to Gluconeogenesis

1. Pyruvate Carboxylase—A Biotin-Dependent Enzyme
Initiation of gluconeogenesis occurs in the **pyruvate carboxylase reaction**—the conversion of pyruvate to oxaloacetate (Figure 22.2). The reaction takes place in two discrete steps, involves ATP and bicarbonate as substrates, and utilizes biotin as a coenzyme and acetyl-CoA as an allosteric activator. Pyruvate carboxylase is a tetrameric enzyme (with a molecular mass of about 500 kD). Each monomer possesses a biotin covalently linked to the ϵ-amino group of a lysine residue at the active site (Figure 22.3). The first step of the reaction involves nucleophilic attack of a bicarbonate oxygen at the γ-P of ATP to form **carbonylphosphate,**

FIGURE 22.4 A mechanism for the pyruvate carboxylase reaction. Bicarbonate must be activated for attack by the pyruvate carbanion. This activation is driven by ATP and involves formation of a carbonylphosphate intermediate—a mixed anhydride of carbonic and phosphoric acids. (*Carbonylphosphate* and *carboxyphosphate* are synonyms.)

an activated form of CO_2, and ADP (Figure 22.4). Reaction of carbonylphosphate with biotin occurs rapidly to form N-carboxybiotin, liberating inorganic phosphate. The third step involves abstraction of a proton from the C-3 of pyruvate, forming a carbanion that can attack the carbon of N-carboxybiotin to form oxaloacetate.

PYRUVATE CARBOXYLASE IS ALLOSTERICALLY ACTIVATED BY ACYL-COENZYME A   Two particularly interesting aspects of the pyruvate carboxylase reaction are (1) allosteric activation of the enzyme by acyl-CoA derivatives and (2) compartmentation of the reaction in the mitochondrial matrix. The carboxylation of biotin requires the presence (at an allosteric site) of acetyl-CoA or other acylated CoA derivatives. The second half of the carboxylase reaction—the attack by pyruvate to form oxaloacetate—is not affected by CoA derivatives.

Activation of pyruvate carboxylase by acetyl-CoA provides an important physiological regulation. Acetyl-CoA is the primary substrate for the TCA cycle, and oxaloacetate (formed by pyruvate carboxylase) is an important intermediate in both the TCA cycle and the gluconeogenesis pathway. If levels of ATP and/or acetyl-CoA (or other acyl-CoAs) are low, pyruvate is directed primarily into the TCA cycle, which eventually promotes the synthesis of ATP. If ATP and acetyl-CoA levels are high, pyruvate is converted to oxaloacetate and consumed in gluconeogenesis. Clearly, high levels of ATP and CoA derivatives are signs that energy is abundant and that metabolites will be converted to glucose (and perhaps even glycogen). If the energy status of the cell is low (in terms of ATP and CoA derivatives), pyruvate is consumed in the TCA cycle. Also, as noted in Chapter 19, pyruvate carboxylase is an important anaplerotic enzyme. Its activation by acetyl-CoA leads to oxaloacetate formation, replenishing the level of TCA cycle intermediates.

COMPARTMENTALIZED PYRUVATE CARBOXYLASE DEPENDS ON METABOLITE CONVERSION AND TRANSPORT   The second interesting feature of pyruvate carboxylase is that it is found only in the matrix of the mitochondria. By contrast, the next enzyme in the gluconeogenic pathway, PEP carboxykinase, may be localized in the cytosol, in the mitochondria, or both. For example, rabbit liver PEP carboxykinase is predominantly mitochondrial, whereas the rat liver enzyme is strictly cytosolic. In human liver, PEP carboxykinase is found both in the cytosol and in the mitochondria. Pyruvate is transported into the mitochondrial matrix (Figure 22.5), where it can be converted to acetyl-CoA (for use in the TCA cycle) and then to citrate (for fatty acid synthesis; see Figure 24.1). Alternatively, it may be converted directly to OAA by pyruvate carboxylase and used in gluconeogenesis. In tissues where PEP carboxykinase is found only in the mitochondria, oxaloacetate is converted to PEP, which is then transported to the cytosol for gluconeogenesis (Figure 22.6). However, in tissues that must convert some oxaloacetate to PEP in the cytosol, a problem arises. Oxaloacetate cannot be transported directly across the mitochondrial membrane. Instead, it must first be transformed into malate or aspartate for transport across the mitochondrial inner membrane (Figure 22.5). Cytosolic malate and aspartate must be reconverted to oxaloacetate before continuing along the gluconeogenic route.

2. PEP Carboxykinase   The second reaction in the gluconeogenic pyruvate–PEP bypass is the conversion of oxaloacetate to PEP. Production of a high-energy metabolite such as PEP requires energy. The energetic requirements are handled

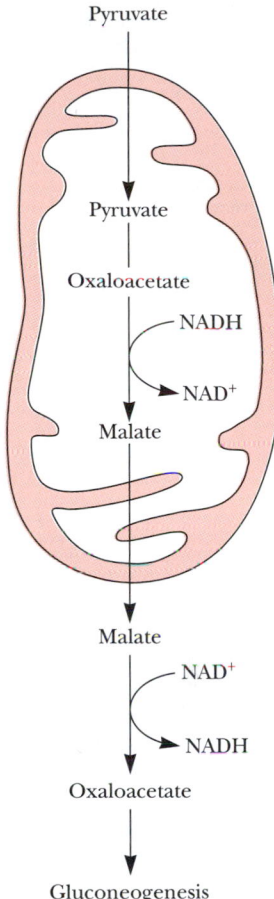

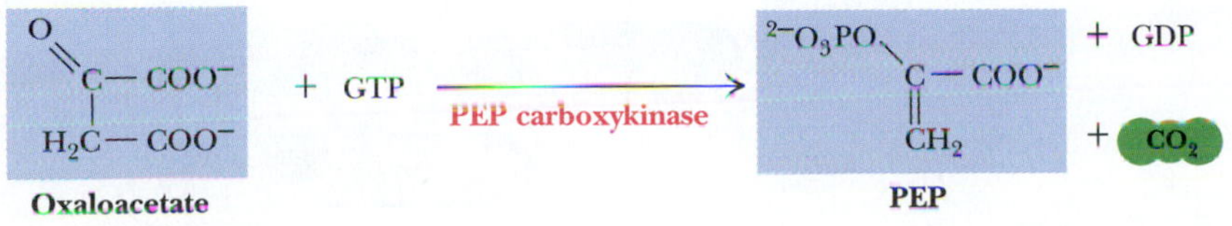

FIGURE 22.5 Pyruvate carboxylase is a compartmentalized reaction. Pyruvate is converted to oxaloacetate in the mitochondria. Because oxaloacetate cannot be transported across the mitochondrial membrane, it must be reduced to malate, transported to the cytosol, and then oxidized back to oxaloacetate before gluconeogenesis can continue.

Biochemistry Now™   Go to BiochemistryNow and click BiochemistryInteractive to learn more about the pyruvate carboxylase reaction.

FIGURE 22.6 The PEP carboxykinase reaction. GTP formed in this reaction can be converted to ATP by nucleoside diphosphate kinase, although liver cells in some species may not contain this enzyme.

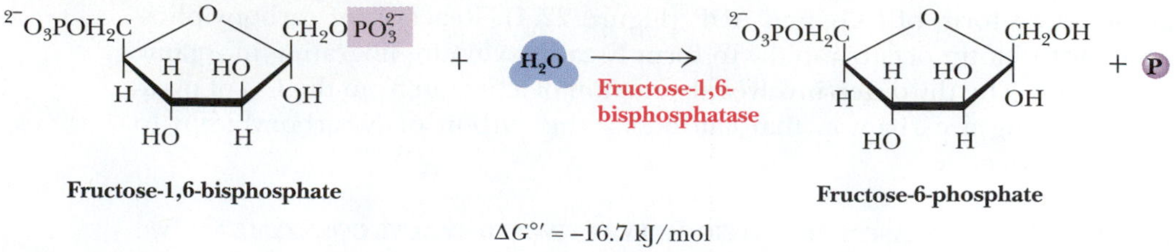

FIGURE 22.7 The fructose-1,6-bisphosphatase reaction.

$$\Delta G^{\circ\prime} = -16.7 \text{ kJ/mol}$$

Fructose-1,6-bisphosphate + H_2O $\xrightarrow{\text{Fructose-1,6-bisphosphatase}}$ Fructose-6-phosphate + P

in two ways here. First, the CO_2 added to pyruvate in the pyruvate carboxylase step is removed in the PEP carboxykinase reaction. Decarboxylation is a favorable process and helps drive the formation of the very high-energy enol phosphate in PEP. This decarboxylation drives a reaction that would otherwise be highly endergonic. Note the inherent metabolic logic in this pair of reactions: Pyruvate carboxylase consumed an ATP to drive a carboxylation so that the PEP carboxykinase could use the decarboxylation to facilitate formation of PEP. Second, as shown in Figure 22.6, another high-energy phosphate is consumed by the carboxykinase. Mammals and several other species use GTP in this reaction, rather than ATP. The use of GTP here is equivalent to the consumption of an ATP, due to the activity of the **nucleoside diphosphate kinase** (see Figure 19.4). The substantial free energy of hydrolysis of GTP is crucial to the synthesis of PEP in this step. The overall ΔG for the pyruvate carboxylase and PEP carboxykinase reactions under physiological conditions in the liver is -22.6 kJ/mol. Once PEP is formed in this way, the phosphoglycerate mutase, phosphoglycerate kinase, glyceraldehyde-3-P dehydrogenase, aldolase, and triose phosphate isomerase reactions act to eventually form fructose-1,6-bisphosphate, as shown in Figure 22.1.

3. Fructose-1,6-Bisphosphatase The hydrolysis of fructose-1,6-bisphosphate to fructose-6-phosphate (Figure 22.7), like all phosphate ester hydrolyses, is a thermodynamically favorable (exergonic) reaction under standard-state conditions ($\Delta G^{\circ\prime} = -16.7$ kJ/mol). Under physiological conditions in the liver, the reaction is also exergonic ($\Delta G = -8.6$ kJ/mol). **Fructose-1,6-bisphosphatase** is an allosterically regulated enzyme. Citrate stimulates bisphosphatase activity, but *fructose-2,6-bisphosphate* is a potent allosteric inhibitor. AMP also inhibits the bisphosphatase; the inhibition by AMP is enhanced by fructose-2,6-bisphosphate.

4. Glucose-6-Phosphatase The final step in the gluconeogenesis pathway is the conversion of glucose-6-phosphate to glucose by the action of **glucose-6-phosphatase.** This enzyme is present in the membranes of the endoplasmic reticulum of liver and kidney cells but is absent in muscle and brain. For this reason, gluconeogenesis is not carried out in muscle and brain. Its membrane association is important to its function because the substrate is hydrolyzed as it passes into the endoplasmic reticulum itself (Figure 22.8). Vesicles form from the endoplasmic reticulum membrane and diffuse to the plasma membrane and fuse with it, releasing their glucose contents into the bloodstream. The glucose-6-phosphatase reaction involves a phosphorylated enzyme inter-

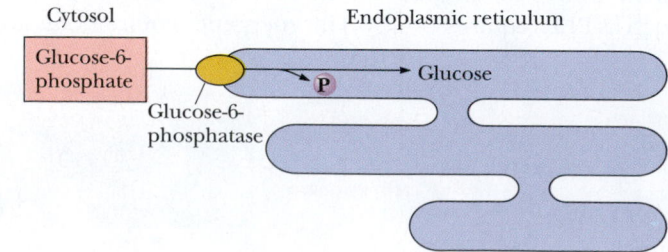

FIGURE 22.8 Glucose-6-phosphatase is localized in the endoplasmic reticulum membrane. Conversion of glucose-6-phosphate to glucose occurs during transport into the endoplasmic reticulum.

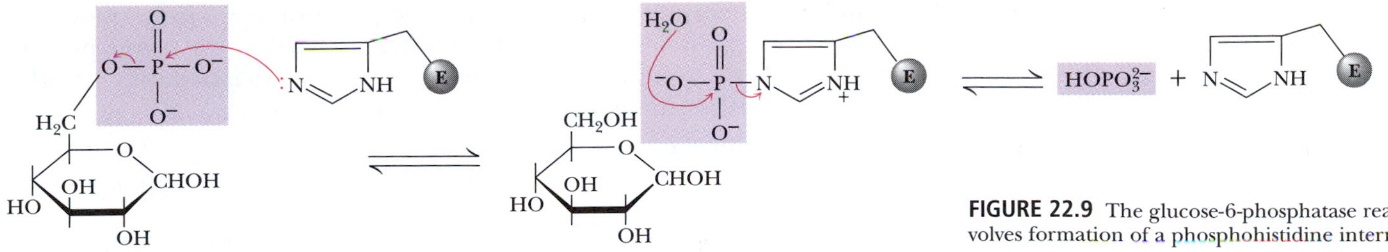

FIGURE 22.9 The glucose-6-phosphatase reaction involves formation of a phosphohistidine intermediate.

mediate, which may be a phosphohistidine (Figure 22.9). The ΔG for the glucose-6-phosphatase reaction in liver is -5.1 kJ/mol.

COUPLING WITH HYDROLYSIS OF ATP AND GTP DRIVES GLUCONEOGENESIS The net reaction for the conversion of pyruvate to glucose in gluconeogenesis is

$$2\ \text{Pyruvate} + 4\ \text{ATP} + 2\ \text{GTP} + 2\ \text{NADH} + 2\ \text{H}^+ + 6\ \text{H}_2\text{O} \longrightarrow$$
$$\text{glucose} + 4\ \text{ADP} + 2\ \text{GDP} + 6\ \text{P}_i + 2\ \text{NAD}^+$$

Human Biochemistry

Gluconeogenesis Inhibitors and Other Diabetes Therapy Strategies

Diabetes, the inability to assimilate and metabolize blood glucose, afflicts millions of people. People with type 1 diabetes are unable to synthesize and secrete insulin. On the other hand, people with type 2 diabetes make sufficient insulin, but the molecular pathways that respond to insulin are defective. Many type 2 diabetic people exhibit a condition termed **insulin resistance** even before the onset of diabetes. **Metformin** (see accompanying figure) is a drug that improves sensitivity to insulin, primarily by stimulating glucose uptake by glucose transporters in peripheral tissues. It also increases binding of insulin to insulin receptors, stimulates tyrosine kinase activity (see Chapter 32) of the insulin receptor, and inhibits gluconeogenesis in the liver.

Gluconeogenesis inhibitors may be the next wave in diabetes therapy. Drugs that block gluconeogenesis without affecting glycolysis would need to target one of the enzymes unique to gluconeogenesis. 3-Mercaptopicolinate and hydrazine specifically inhibit PEP carboxykinase, and **chlorogenic acid,** a natural product found in the skin of peaches, inhibits the transport activity of the glucose-6-phosphatase system (but not the glucose-6-phosphatase enzyme activity). The drug S-3483, a derivative of chlorogenic acid, also inhibits the glucose-6-phosphatase transport activity and binds a thousand times more tightly to the transporter than chlorogenic acid. Drugs of this type may be useful in the treatment of type 2 diabetes.

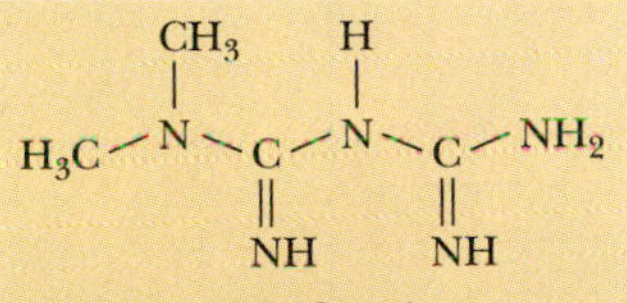

Metformin

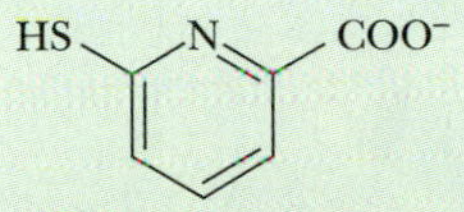

3-Mercaptopicolinate

Hydrazine

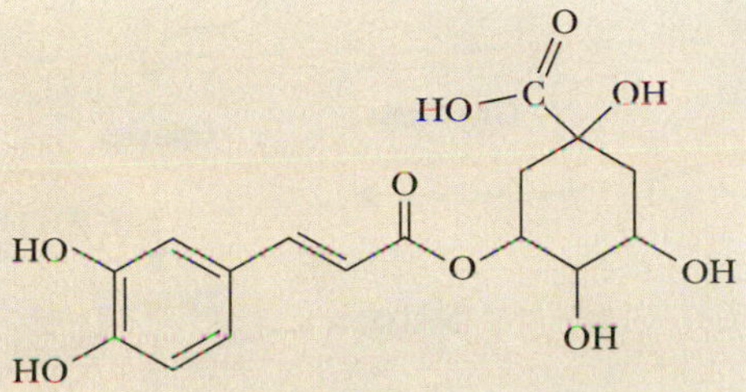

Chlorogenic acid

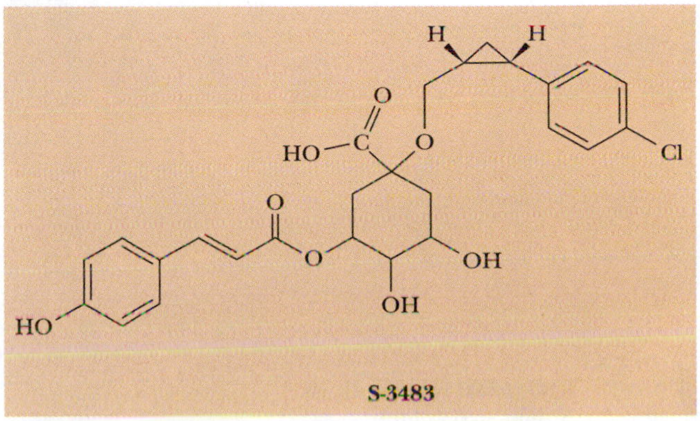

S-3483

The net free energy change, $\Delta G^{\circ\prime}$, for this conversion is -37.7 kJ/mol. The consumption of a total of six nucleoside triphosphates drives this process forward. If glycolysis were merely reversed to achieve the net synthesis of glucose from pyruvate, the net reaction would be

$$2 \text{ Pyruvate} + 2 \text{ ATP} + 2 \text{ NADH} + 2 \text{ H}^+ + 2 \text{ H}_2\text{O} \longrightarrow$$
$$\text{glucose} + 2 \text{ ADP} + 2 \text{ P}_i + 2 \text{ NAD}^+$$

and the overall $\Delta G^{\circ\prime}$ would be about $+74$ kJ/mol. Such a process would be highly endergonic and therefore thermodynamically unfeasible. Hydrolysis of four additional high-energy phosphate bonds makes gluconeogenesis thermodynamically favorable. Under physiological conditions, however, gluconeogenesis is somewhat less favorable than at standard state, with an overall ΔG of -15.6 kJ/mol for the conversion of pyruvate to glucose.

LACTATE FORMED IN MUSCLES IS RECYCLED TO GLUCOSE IN THE LIVER   A final point on the redistribution of lactate and glucose in the body serves to emphasize the metabolic interactions between organs. Vigorous exercise can lead to oxygen shortage (anaerobic conditions), and energy requirements must be met by increased levels of glycolysis. Under such conditions, glycolysis converts NAD$^+$ to NADH, yet O$_2$ is unavailable for regeneration of NAD$^+$ via cellular respiration. Instead, large amounts of NADH are reoxidized by the reduction of pyruvate to lactate. The lactate thus produced can be transported from muscle to the liver, where it is reoxidized by liver lactate dehydrogenase to yield pyruvate, which is converted eventually to glucose. In this way, the liver shares in the metabolic stress created by vigorous exercise. It exports glucose to muscle, which produces lactate, and lactate from muscle can be processed by the liver into new glucose. This is referred to as the Cori cycle (Figure 22.10). Liver, with a typically high NAD$^+$/NADH ratio (about 700), readily produces more glucose than it can use. Muscle that is vigorously exercising will enter anaerobiosis and show a decreasing NAD$^+$/NADH ratio, which favors reduction of pyruvate to lactate.

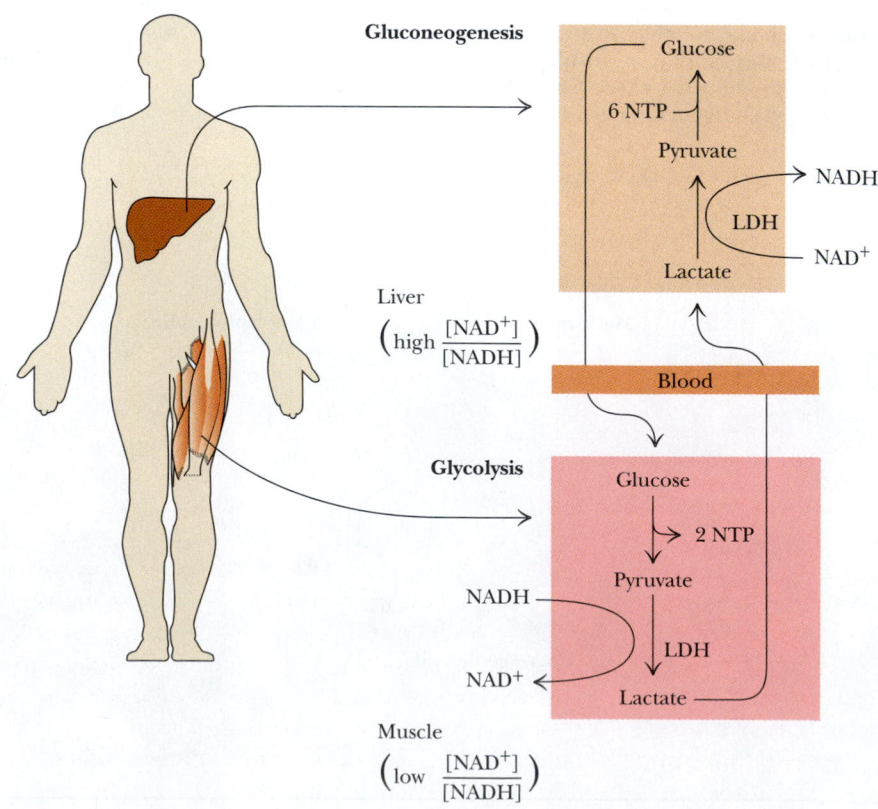

Biochemistry 𝒩ow™   ANIMATED FIGURE 22.10
The Cori cycle. See this figure animated at http://chemistry.brookscole.com/ggb3

Critical Developments in Biochemistry

The Pioneering Studies of Carl and Gerty Cori

The Cori cycle is named for Carl and Gerty Cori, who received the Nobel Prize in Physiology or Medicine in 1947 for their studies of glycogen metabolism and blood glucose regulation. Carl Ferdinand Cori and Gerty Theresa Radnitz were both born in Prague (then in Austria). They earned medical degrees from the German University of Prague in 1920 and were married later that year. They joined the faculty of the Washington University School of Medicine in St. Louis in 1931. Their remarkable collaboration resulted in many fundamental advances in carbohydrate and glycogen metabolism. They were credited with the discovery of glucose-1-phosphate, also known at the time as the "Cori ester." They also showed that glucose-6-phosphate was produced from glucose-1-P by the action of phosphoglucomutase. They isolated and crystallized glycogen phosphorylase and elucidated the pathway of glycogen breakdown. In 1952, they showed that absence of glucose-6-phosphatase in the liver was the enzymatic defect in von Gierke's disease, an inherited glycogen-storage disease. Six eventual Nobel laureates received training in their laboratory. Gerty Cori was the first American woman to receive a Nobel Prize. Carl Cori said of their remarkable collaboration: "Our efforts have been largely complementary and one without the other would not have gone so far…"

22.2 | How Is Gluconeogenesis Regulated?

Nearly all of the reactions of glycolysis and gluconeogenesis take place in the cytosol. If metabolic control were not exerted over these reactions, glycolytic degradation of glucose and gluconeogenic synthesis of glucose could operate simultaneously, with no net benefit to the cell and with considerable consumption of ATP. This is prevented by a sophisticated system of **reciprocal control,** which inhibits glycolysis when gluconeogenesis is active, and vice versa. Reciprocal regulation of these two pathways depends largely on the energy status of the cell. When the energy status of the cell is low, glucose is rapidly degraded to produce needed energy. When the energy status is high, pyruvate and other metabolites are utilized for synthesis (and storage) of glucose.

In glycolysis, the three regulated enzymes are those catalyzing the strongly exergonic reactions: hexokinase (glucokinase), phosphofructokinase, and pyruvate kinase. As noted, the gluconeogenic pathway replaces these three reactions with corresponding reactions that are exergonic in the direction of glucose synthesis: glucose-6-phosphatase, fructose-1,6-bisphosphatase, and the pyruvate carboxylase–PEP carboxykinase pair, respectively. These are the three most appropriate sites of regulation in gluconeogenesis.

Gluconeogenesis Is Regulated by Allosteric and Substrate-Level Control Mechanisms

The mechanisms of regulation of gluconeogenesis are shown in Figure 22.11. Control is exerted at all of the predicted sites, but in different ways. Glucose-6-phosphatase is not under allosteric control. However, the K_m for the substrate, glucose-6-phosphate, is considerably higher than the normal range of substrate concentrations. As a result, glucose-6-phosphatase displays a near-linear dependence of activity on substrate concentrations and is thus said to be under **substrate-level control** by glucose-6-phosphate.

Acetyl-CoA is a potent allosteric effector of glycolysis and gluconeogenesis. It allosterically inhibits pyruvate kinase (as noted in Chapter 18) and activates pyruvate carboxylase. Because it also allosterically inhibits pyruvate dehydrogenase (the enzymatic link between glycolysis and the TCA cycle), the cellular fate of pyruvate is strongly dependent on acetyl-CoA levels. A rise in [acetyl-CoA] indicates that cellular energy levels are high and that carbon metabolites can be directed to glucose synthesis and storage. When acetyl-CoA levels drop, the activities of pyruvate kinase and pyruvate dehydrogenase increase and flux through the TCA cycle increases, providing needed energy for the cell.

Fructose-1,6-bisphosphatase is another important site of gluconeogenic regulation. This enzyme is inhibited by AMP and activated by citrate. These effects

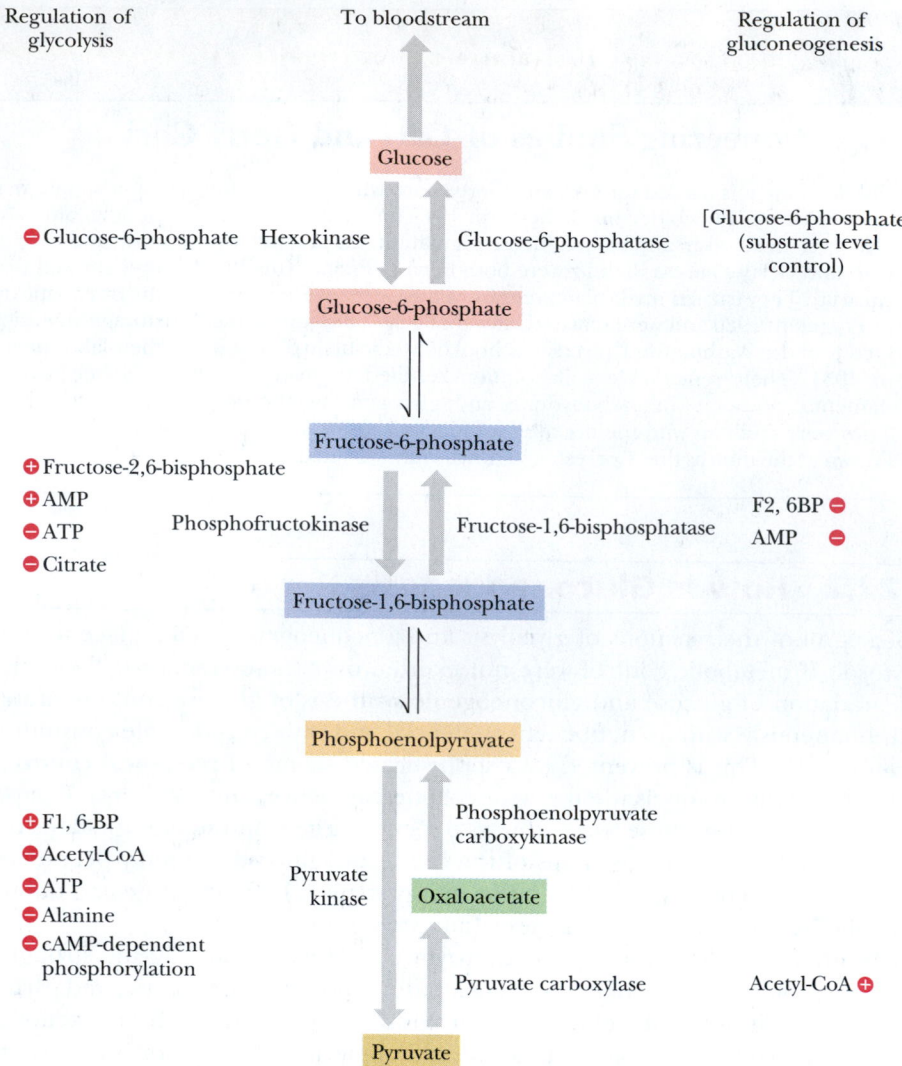

Biochemistry Now™　**ACTIVE FIGURE 22.11**
The principal regulatory mechanisms in glycolysis and gluconeogenesis. Activators are indicated by plus signs and inhibitors by minus signs. **Test yourself on the concepts in this figure at http://chemistry. brookscole.com/ggb3**

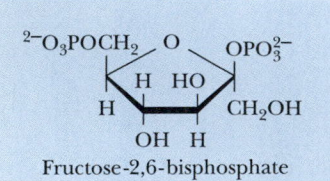

Fructose-2,6-bisphosphate

by AMP and citrate are the opposites of those exerted on phosphofructokinase in glycolysis, providing another example of reciprocal regulatory effects. When AMP levels increase, gluconeogenic activity is diminished and glycolysis is stimulated. An increase in citrate concentration signals that TCA cycle activity can be curtailed and that pyruvate should be directed to sugar synthesis instead.

Fructose-2,6-Bisphosphate—Allosteric Regulator of Gluconeogenesis　As described in Chapter 18, Emile Van Schaftingen and Henri-Géry Hers demonstrated in 1980 that fructose-2,6-bisphosphate is a potent stimulator of phosphofructokinase. Cognizant of the reciprocal nature of regulation in glycolysis and gluconeogenesis, Van Schaftingen and Hers also considered the possibility of an opposite effect— inhibition—for fructose-1,6-bisphosphatase. In 1981, they reported that fructose-2,6-bisphosphate was indeed a powerful inhibitor of fructose-1,6-bisphosphatase (Figure 22.12). Inhibition occurs in either the presence or absence of AMP, and the effects of AMP and fructose-2,6-bisphosphate are synergistic.

Cellular levels of fructose-2,6-bisphosphate are controlled by **phosphofructokinase-2 (PFK-2)**, an enzyme distinct from the phosphofructokinase of the glycolytic pathway, and by **fructose-2,6-bisphosphatase (F-2,6-BPase)**. Remarkably, these two enzymatic activities are both found in the same protein molecule, which is an example of a **bifunctional**, or **tandem, enzyme** (Figure 22.13). The opposing activities of this bifunctional enzyme are themselves regulated in two ways. First, fructose-6-phosphate, the substrate of phosphofructokinase and the

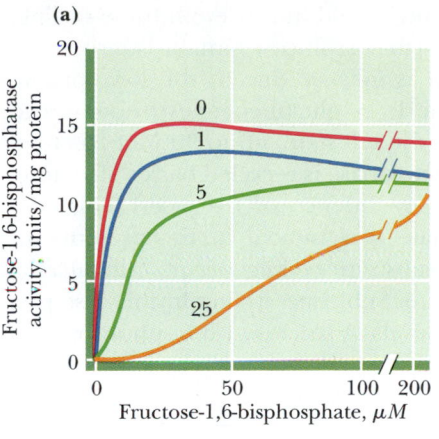

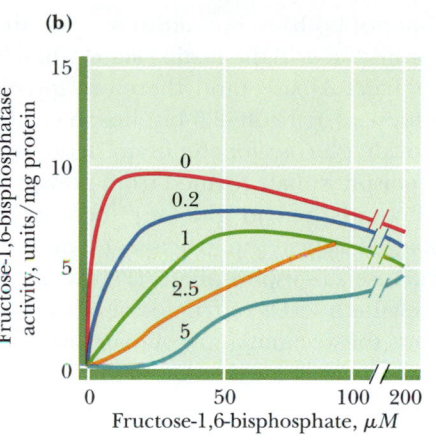

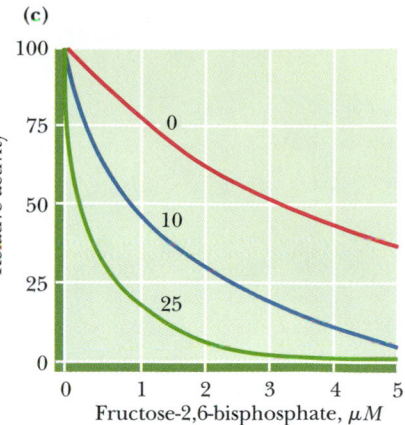

FIGURE 22.12 Inhibition of fructose-1,6-bisphospha-tase by fructose-2,6-bisphosphate in the **(a)** absence and **(b)** presence of 25 mM AMP. In (a) and (b), enzyme activity is plotted against substrate (fructose-1,6-bisphosphate) concentration. Concentrations of fructose-2,6-bisphosphate (in mM) are indicated above each curve. **(c)** The effect of AMP (0, 10, and 25 mM) on the inhibition of fructose-1,6-bisphos-phatase by fructose-2,6-bisphosphate. Activity was measured in the presence of 10 mM fructose-1,6-bisphosphate. (*Adapted from Van Schaftingen, E., and Hers, H-G., 1981. Inhibition of fructose-1,6-bisphosphatase by fructose-2,6-bisphosphate.* Proceedings of the National Academy of Science, U.S.A. **78:**2861–2863.)

product of fructose-1,6-bisphosphatase, allosterically activates PFK-2 and inhibits F-2,6-BPase. Second, the phosphorylation by **cAMP-dependent protein kinase** of a single Ser residue on the 49-kD subunit of this dimeric enzyme exerts reciprocal control of the PFK-2 and F-2,6-BPase activities. Phosphorylation inhibits PFK-2 activity (by increasing the K_m for fructose-6-phosphate) and stimulates F-2,6-BPase activity.

Substrate Cycles Provide Metabolic Control Mechanisms

If fructose-1,6-bisphosphatase and phosphofructokinase acted simultaneously, they would constitute a **substrate cycle** in which fructose-1,6-bisphosphate and fructose-6-phosphate became interconverted with net consumption of ATP:

$$\text{Fructose-1,6-bisP} + H_2O \longrightarrow \text{fructose-6-P} + P_i$$
$$\text{Fructose-6-P} + \text{ATP} \longrightarrow \text{fructose-1,6-bisP} + \text{ADP}$$

Net: $$\text{ATP} + H_2O \longrightarrow \text{ADP} + P_i$$

Because substrate cycles such as this appear to operate with no net benefit to the cell, they were once regarded as metabolic quirks and were referred to as futile cycles. More recently, substrate cycles have been recognized as important devices for controlling metabolite concentrations.

The three steps in glycolysis and gluconeogenesis that differ constitute three such substrate cycles, each with its own particular metabolic raison d'être. Consider, for example, the regulation of the fructose-1,6-bisP-fructose-6-P cycle by fructose-2,6-bisphosphate. As already noted, fructose-1,6-bisphosphatase is subject to allosteric inhibition by fructose-2,6-bisphosphate, whereas phosphofructokinase is allosterically activated by fructose-2,6-bisP. The combination of these effects should permit either phosphofructokinase or fructose-1,6-bisphosphatase

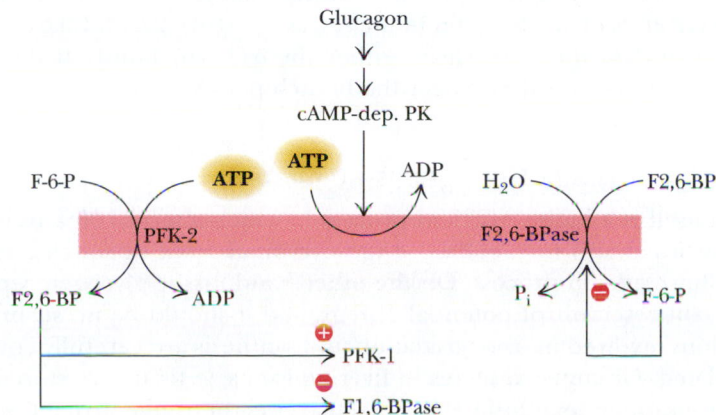

Biochemistry ⊜Now™ ACTIVE FIGURE 22.13
Synthesis and degradation of fructose-2,6-bisphosphate are catalyzed by the same bifunctional enzyme. **Test yourself on the concepts in this figure at http://chemistry.brookscole.com/ggb3**

(but not both) to operate at any one time and should thus prevent futile cycling. For instance, in the **fasting state,** when food (that is, glucose) intake is zero, phosphofructokinase (and therefore glycolysis) is inactive due to the low concentration of fructose-2,6-bisphosphate. In the liver, gluconeogenesis operates to provide glucose for the brain. However, in the fed state, up to 30% of fructose-1,6-bisphosphate formed from phosphofructokinase is recycled back to fructose-6-P (and then to glucose). Because the dependence of fructose-1,6-bisphosphatase activity on fructose-1,6-bisphosphate is sigmoidal in the presence of fructose-2,6-bisphosphate (Figure 22.12), substrate cycling occurs only at relatively high levels of fructose-1,6-bisphosphate. Substrate cycling in this case prevents the accumulation of excessively high levels of fructose-1,6-bisphosphate.

22.3 | How Are Glycogen and Starch Catabolized in Animals?

Dietary Glycogen and Starch Breakdown Provide Metabolic Energy

As noted earlier, well-fed adult human beings normally metabolize about 160 grams of carbohydrates each day. A balanced diet easily provides this amount, mostly in the form of starch, with smaller amounts of glycogen. If too little carbohydrate is supplied by the diet, glycogen reserves in liver and muscle tissue can also be mobilized. The reactions by which ingested starch and glycogen are digested are shown in Figure 22.14. The enzyme known as **α-amylase** is an important component of saliva and pancreatic juice. (**β-Amylase** is found in plants. The α- and β-designations for these enzymes serve only to distinguish the two and do not refer to glycosidic linkage nomenclature.) α-Amylase is an endoglycosidase that hydrolyzes $\alpha(1\rightarrow4)$ linkages of amylopectin and glycogen at random positions, eventually producing a mixture of maltose, maltotriose [with three $\alpha(1\rightarrow4)$-linked glucose residues], and other small oligosaccharides. α-Amylase can cleave on either side of a glycogen or amylopectin branch point, but activity is reduced in highly branched regions of the polysaccharide and stops four residues from any branch point.

The highly branched polysaccharides that are left after extensive exposure to α-amylase are called **limit dextrins.** These structures can be further degraded by the action of a **debranching enzyme,** which carries out two distinct reactions. The first of these, known as **oligo(α1,4→α1,4) glucanotransferase** activity, removes a trisaccharide unit and transfers this group to the end of another, nearby branch (Figure 22.15). This leaves a single glucose residue in $\alpha(1\rightarrow6)$ linkage to the main chain. The **$\alpha(1\rightarrow6)$ glucosidase** activity of the debranching enzyme then cleaves this residue from the chain, leaving a polysaccharide chain with one branch fewer. Repetition of this sequence of events leads to complete degradation of the polysaccharide.

β-Amylase is an exoglycosidase that cleaves maltose units from the free, nonreducing ends of amylopectin branches, as in Figure 22.14. Like α-amylase, however, β-amylase does not cleave either the $\alpha(1\rightarrow6)$ bonds at the branch points or the $\alpha(1\rightarrow4)$ linkages near the branch points.

Metabolism of Tissue Glycogen Is Regulated

Digestion itself is a highly efficient process in which almost 100% of ingested food is absorbed and metabolized. Digestive breakdown of starch and glycogen is an unregulated process. On the other hand, tissue glycogen represents an important reservoir of potential energy, and it should be no surprise that the reactions involved in its degradation and synthesis are carefully controlled and regulated. Glycogen reserves in liver and muscle tissue are stored in the cytosol as granules exhibiting a molecular weight range from 6×10^6 to

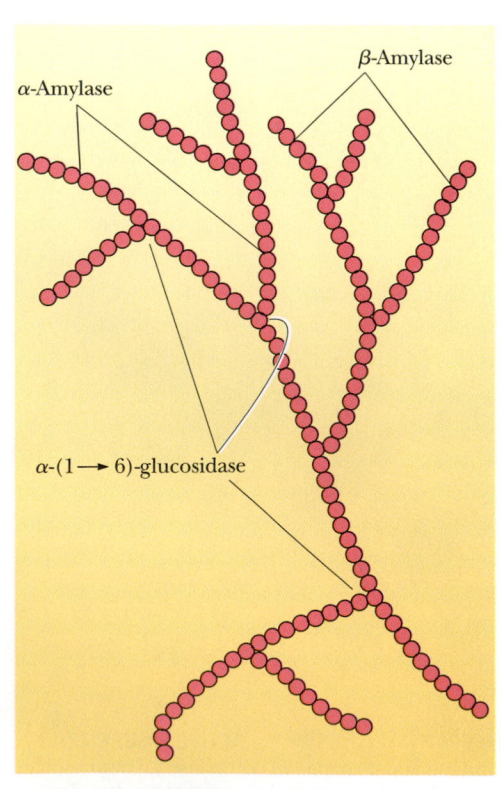

FIGURE 22.14 Hydrolysis of glycogen and starch by α- and β-amylase.

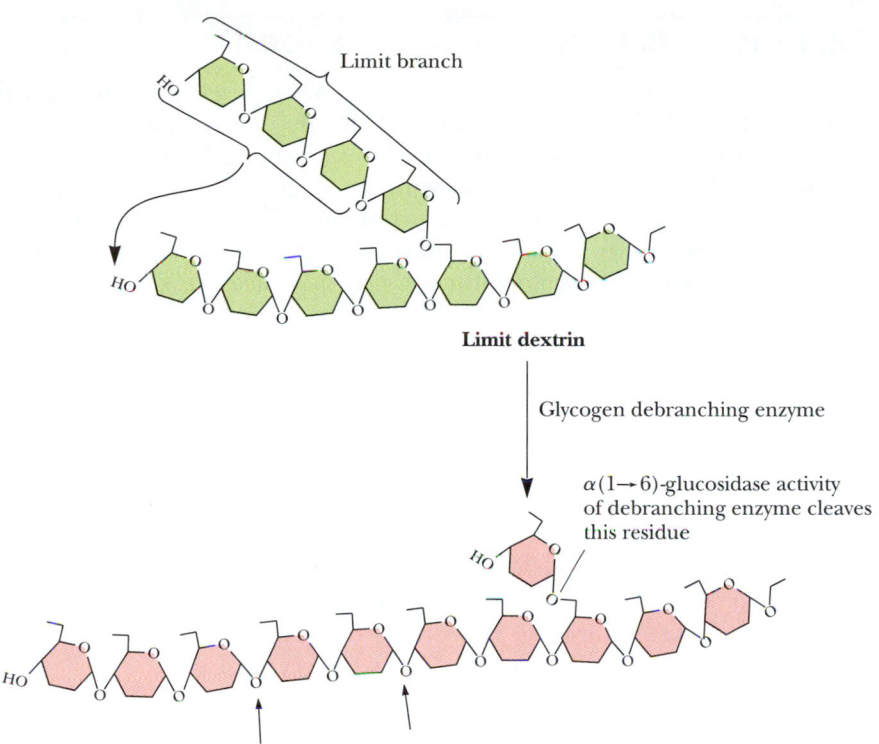

Limit branch

Limit dextrin

Glycogen debranching enzyme

$\alpha(1\rightarrow6)$-glucosidase activity
of debranching enzyme cleaves
this residue

Further cleavage by α-amylase

FIGURE 22.15 The reactions of glycogen debranching enzyme. Transfer of a group of three $\alpha(1\rightarrow4)$-linked glucose residues from a limit branch to another branch is followed by cleavage of the $\alpha(1\rightarrow6)$ bond of the residue that remains at the branch point.

1600×10^6. These granular aggregates contain the enzymes required to synthesize and catabolize the glycogen, as well as all the enzymes of glycolysis.

The principal enzyme of glycogen catabolism is **glycogen phosphorylase,** a highly regulated enzyme that was discussed extensively in Chapter 15. The glycogen phosphorylase reaction (Figure 22.16) involves phosphorolysis at a nonreducing end of a glycogen polymer. The standard-state free energy change for this reaction is +3.1 kJ/mol, but the intracellular ratio of $[P_i]$ to [glucose-1-P] approaches 100, and thus the actual ΔG in vivo is approximately -6 kJ/mol. There is an energetic advantage to the cell in this phosphorolysis reaction. If glycogen breakdown were hydrolytic and yielded glucose as a product, it would be necessary to phosphorylate the product glucose (with the expenditure of a molecule of ATP) to initiate its glycolytic degradation.

The glycogen phosphorylase reaction degrades glycogen to produce limit dextrins, which are further degraded by debranching enzyme, as already described.

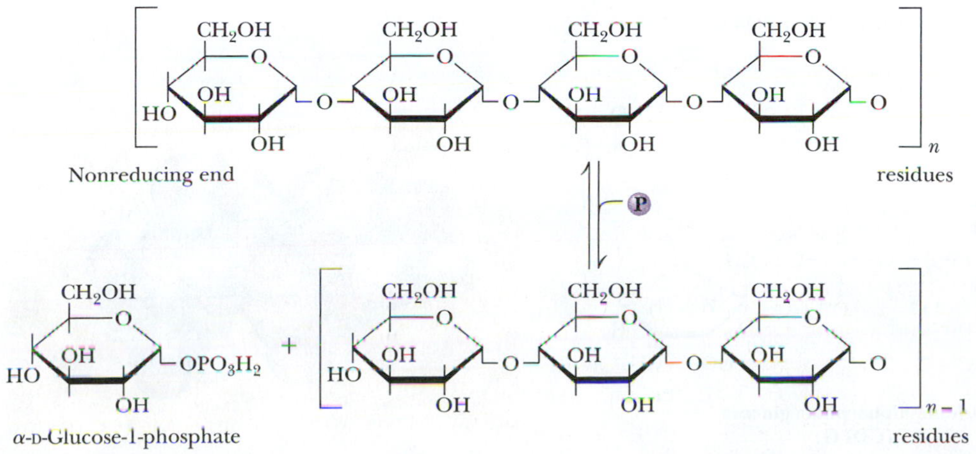

Nonreducing end

n residues

α-D-Glucose-1-phosphate

$n-1$ residues

FIGURE 22.16 The glycogen phosphorylase reaction.

22.4 | How Is Glycogen Synthesized?

Animals synthesize and store glycogen when glucose levels are high, but the synthetic pathway is not merely a reversal of the glycogen phosphorylase reaction. High levels of phosphate in the cell favor glycogen breakdown and prevent the phosphorylase reaction from synthesizing glycogen in vivo, despite the fact that $\Delta G°'$ for the phosphorylase reaction actually favors glycogen synthesis. Hence, another reaction pathway must be employed in the cell for the net synthesis of glycogen. In essence, this pathway must activate glucose units for transfer to glycogen chains.

Glucose Units Are Activated for Transfer by Formation of Sugar Nucleotides

We are familiar with several examples of chemical activation as a strategy for group transfer reactions. Acetyl-CoA is an activated form of acetate, biotin and tetrahydrofolate activate one-carbon groups for transfer, and ATP is an activated form of phosphate. Luis Leloir, a biochemist in Argentina, showed in the 1950s that glycogen synthesis depended upon **sugar nucleotides,** which may be thought of as activated forms of sugar units (Figure 22.17). For example, formation of an ester linkage between the C-1 hydroxyl group and the β-phosphate of UDP activates the glucose moiety of **UDP–glucose.**

UDP–Glucose Synthesis Is Driven by Pyrophosphate Hydrolysis

Sugar nucleotides are formed from sugar-1-phosphates and nucleoside triphosphates by specific **pyrophosphorylase** enzymes (Figure 22.18). For example, **UDP–glucose pyrophosphorylase** catalyzes the formation of UDP–glucose from glucose-1-phosphate and uridine 5'-triphosphate:

$$\text{Glucose-1-P} + \text{UTP} \longrightarrow \text{UDP–glucose} + \text{pyrophosphate}$$

The reaction proceeds via attack by a phosphate oxygen of glucose-1-phosphate on the α-phosphorus of UTP, with departure of the pyrophosphate anion. The reaction is a reversible one, but—as is the case for many biosynthetic reactions—it is driven forward by subsequent hydrolysis of pyrophosphate:

$$\text{Pyrophosphate} + \text{H}_2\text{O} \longrightarrow 2\,\text{P}_\text{i}$$

The net reaction for sugar nucleotide formation (combining the preceding two equations) is thus

$$\text{Glucose-1-P} + \text{UTP} + \text{H}_2\text{O} \longrightarrow \text{UDP–glucose} + 2\,\text{P}_\text{i}$$

Sugar nucleotides of this type act as donors of sugar units in the biosynthesis of oligosaccharides and polysaccharides. In animals, UDP–glucose is the donor of glucose units for glycogen synthesis, but ADP–glucose is the glucose source for starch synthesis in plants.

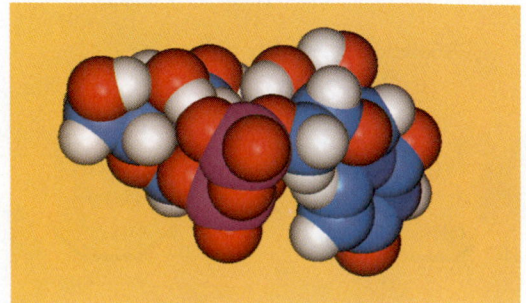

FIGURE 22.17 The structure of UDP–glucose, a sugar nucleotide.

Uridine diphosphate glucose
(UDPG)

Glycogen Synthase Catalyzes Formation of α(1→4) Glycosidic Bonds in Glycogen

The very large glycogen polymer is built around a tiny protein core. The first glucose residue is covalently joined to the protein **glycogenin** via an acetal linkage to a tyrosine–OH group on the protein. Sugar units are added to the glycogen polymer by the action of **glycogen synthase.** The reaction involves transfer of a glucosyl unit from UDP–glucose to the C-4 hydroxyl group at a nonreducing end of a glycogen strand. The mechanism proceeds by cleavage of the C—O bond between the glucose moiety and the β-phosphate of UDP–glucose, leaving an oxonium ion intermediate, which is rapidly attacked by the C-4 hydroxyl oxygen of a terminal glucose unit on glycogen (Figure 22.19). The reaction is exergonic and has a $\Delta G°'$ of -13.3 kJ/mol.

Glycogen Branching Occurs by Transfer of Terminal Chain Segments

Glycogen is a branched polymer of glucose units. The branches arise from α(1→6) linkages, which occur every 8 to 12 residues. As noted in Chapter 7, the branches provide multiple sites for rapid degradation or elongation of the polymer and also increase its solubility. Glycogen branches are formed by **amylo-(1,4→1,6)-transglycosylase,** also known as *branching enzyme*. The reaction involves the transfer of a 6- or 7-residue segment from the nonreducing end of a linear chain at least 11 residues in length to the C-6 hydroxyl of a glucose residue of the same chain or another chain (Figure 22.20). For each branching reaction, the resulting polymer has gained a new terminus at which growth can occur.

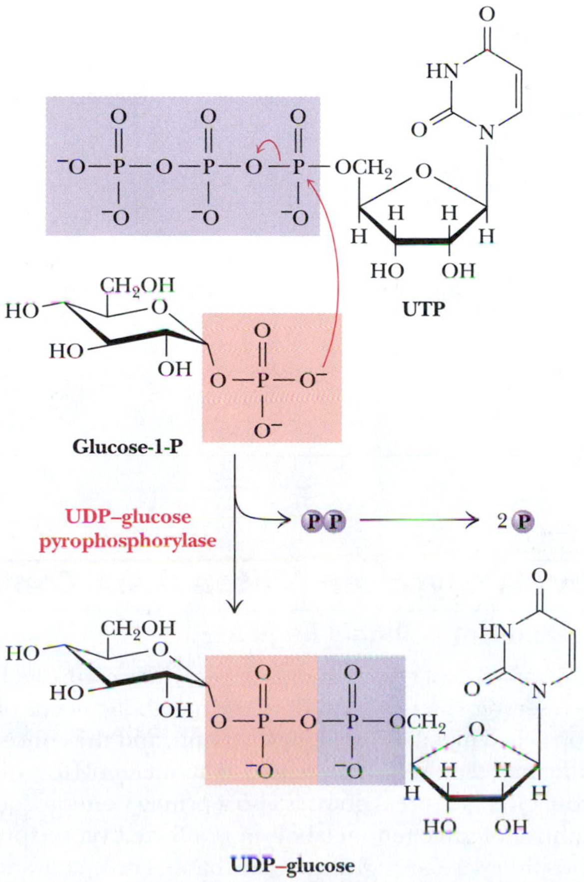

Biochemistry⟳Now™ **ANIMATED FIGURE 22.18** The UDP–glucose pyrophosphorylase reaction is a phosphoanhydride exchange, with a phosphoryl oxygen of glucose-1-P attacking the α-phosphorus of UTP to form UDP–glucose and pyrophosphate. **See this figure animated at http://chemistry.brookscole.com/ggb3**

Human Biochemistry

Advanced Glycation End Products—A Serious Complication of Diabetes

Covalent linkage of sugars to proteins to form glycoproteins normally occurs through the action of enzymes that use sugar nucleotides as substrates. However, sugars may also react nonenzymatically with proteins. The C-1 carbonyl group of glucose forms Schiff base linkages with lysine side chains of proteins. These Schiff base adducts undergo Amadori rearrangements and subsequent oxidations to form irreversible "glycation" products, including carboxymethyllysine and pentosidine derivatives (see accompanying figure). These **advanced glycation end products (AGEs)** can alter the function of the protein. Such AGE-dependent changes are thought to contribute to circulation, joint, and vision problems in people with diabetes.

Nonenzymatic glycation of hemoglobin is a useful diagnostic yardstick of long-term serum glucose levels. Red blood cells have an average life expectancy of about 4 months. By measuring the concentration of "glycated hemoglobin" in a patient, it is possible to determine the average glucose concentration in the blood over the past several months.

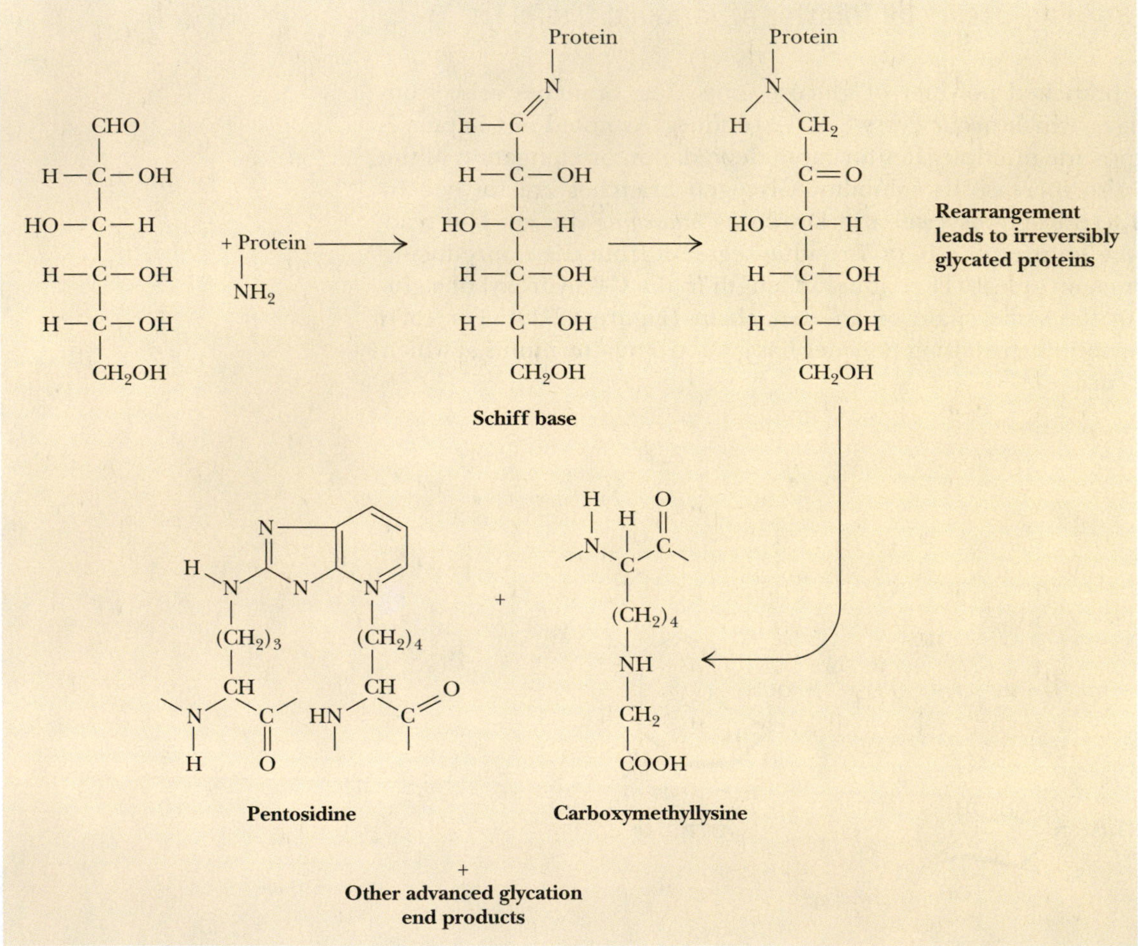

22.5 | How Is Glycogen Metabolism Controlled?

Glycogen Metabolism Is Highly Regulated

Synthesis and degradation of glycogen must be carefully controlled so that this important energy reservoir can properly serve the metabolic needs of the organism. Glucose is the principal metabolic fuel for the brain, and the concentration of glucose in circulating blood must be maintained at about 5 mM for this purpose. Glucose derived from glycogen breakdown is also a primary energy source for muscle contraction. Control of glycogen metabolism is effected via reciprocal regulation of glycogen phosphorylase and glycogen synthase. Thus, activation of glycogen

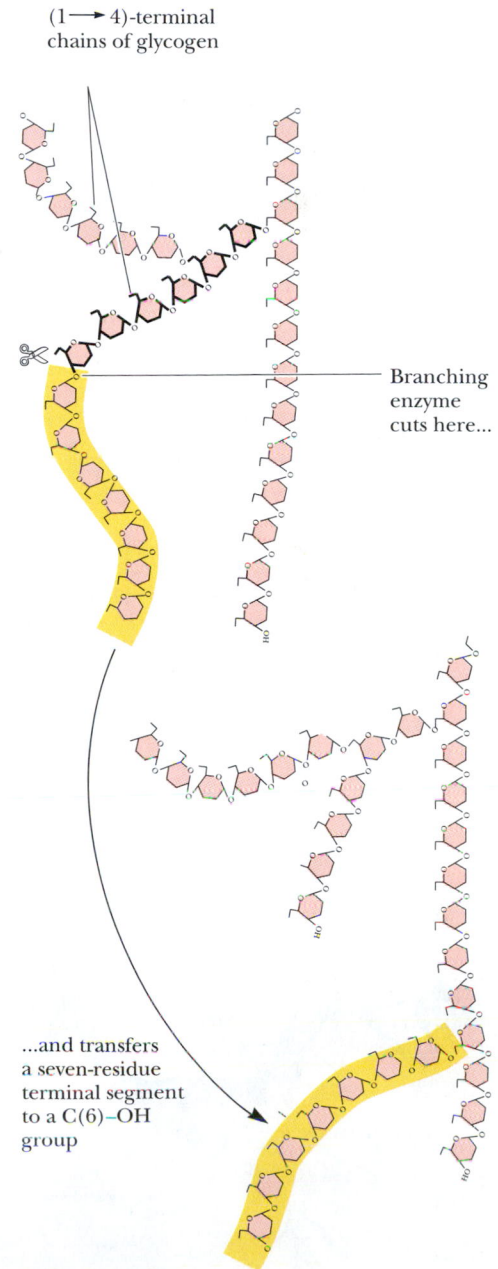

Oxonium ion intermediate

UDP–glucose

UDP

Glycogen (n residues)

H⁺

Glycogen (n + 1 residues)

Biochemistry Now™   ANIMATED FIGURE 22.19 The glycogen synthase reaction. Cleavage of the C—O bond of UDP–glucose yields an oxonium intermediate. Attack by the hydroxyl oxygen of the terminal residue of a glycogen molecule completes the reaction. **See this figure animated at http://chemistry.brookscole.com/ggb3**

phosphorylase is tightly linked to inhibition of glycogen synthase, and vice versa. Regulation involves both allosteric control and covalent modification, with the latter being under hormonal control. The regulation of glycogen phosphorylase is discussed in detail in Chapter 15.

Glycogen Synthase Is Regulated by Covalent Modification

Glycogen synthase also exists in two distinct forms that can be interconverted by the action of specific enzymes: active, dephosphorylated **glycogen synthase I** (glucose-6-P–independent) and less active, phosphorylated **glycogen synthase D** (glucose-6-P–dependent). The nature of phosphorylation is more complex with glycogen synthase. As many as nine serine residues on the enzyme appear to be subject to phosphorylation, each site's phosphorylation having some effect on enzyme activity.

Dephosphorylation of both glycogen phosphorylase and glycogen synthase is carried out by **phosphoprotein phosphatase-1.** The action of phosphoprotein phosphatase-1 inactivates glycogen phosphorylase and activates glycogen synthase.

Hormones Regulate Glycogen Synthesis and Degradation

Storage and utilization of tissue glycogen, maintenance of blood glucose concentration, and other aspects of carbohydrate metabolism are meticulously regulated by hormones, including *insulin, glucagon, epinephrine,* and the *glucocorticoids.*

Insulin Is a Response to Increased Blood Glucose   The primary hormone responsible for conversion of glucose to glycogen is **insulin** (see Figures 5.13 and 6.35). Insulin is secreted by special cells in the pancreas called the **islets of Langerhans.**

(1 → 4)-terminal chains of glycogen

Branching enzyme cuts here...

...and transfers a seven-residue terminal segment to a C(6)–OH group

FIGURE 22.20 Formation of glycogen branches by the branching enzyme. Six- or seven-residue segments of a growing glycogen chain are transferred to the C-6 hydroxyl group of a glucose residue on the same or a nearby chain.

A Deeper Look

Carbohydrate Utilization in Exercise

Animals have a remarkable ability to "shift gears" metabolically during periods of strenuous exercise or activity. Metabolic adaptations allow the body to draw on different sources of energy (all of which produce ATP) for different types of activity. During periods of short-term, high-intensity exercise (such as a 100-m dash), most of the required energy is supplied directly by existing stores of ATP and creatine phosphate (see figure, part a). Long-term, low-intensity exercise (such as a 10-km run or a 42.2-km marathon) is fueled almost entirely by aerobic metabolism. Between these extremes is a variety of activities (an 800-m run, for example) that rely on anaerobic glycolysis—conversion of glucose to lactate in the muscles and utilization of the Cori cycle.

For all these activities, breakdown of muscle glycogen provides much of the needed glucose. The rate of glycogen consumption depends on the intensity of the exercise (see figure, part b). By contrast, glucose derived from gluconeogenesis makes only small contributions to total glucose consumed during exercise. During prolonged mild exercise, gluconeogenesis accounts for only about 8% of the total glucose consumed. During heavy exercise, this percentage becomes even lower.

Choice of diet has a dramatic effect on glycogen recovery following exhaustive exercise. A diet consisting mainly of protein and fat results in very little recovery of muscle glycogen, even after 5 days (see figure, part c). On the other hand, a high-carbohydrate diet provides faster restoration of muscle glycogen. Even in this case, however, complete recovery of glycogen stores takes about 2 days.

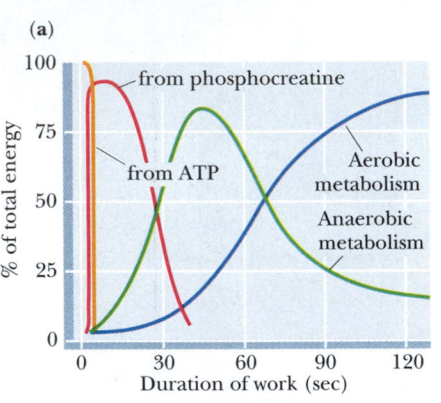
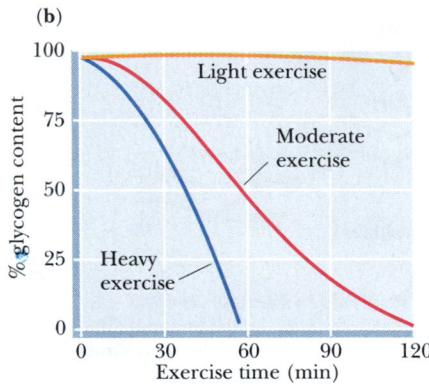
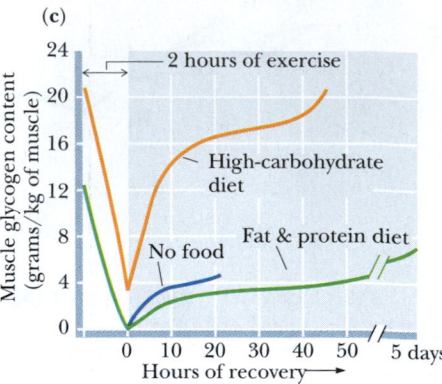

(a) Contributions of the various energy sources to muscle activity during mild exercise. **(b)** Consumption of glycogen stores in fast-twitch muscles during light, moderate, and heavy exercise. **(c)** Rate of glycogen replenishment following exhaustive exercise. *(a and c adapted from Rhodes, R., and Pflanzer, R. G., 1992. Human Physiology. Philadelphia: Saunders College Publishing; b adapted from Horton, E. S., and Terjung, R. L., 1988. Exercise, Nutrition and Energy Metabolism. New York: Macmillan.)*

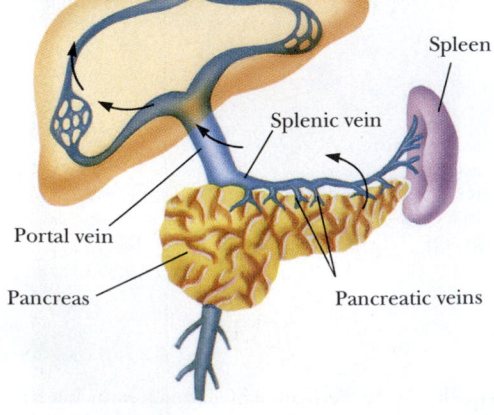

FIGURE 22.21 The portal vein system carries pancreatic secretions such as insulin and glucagon to the liver and then into the rest of the circulatory system.

Secretion of insulin is a response to increased glucose in the blood. When blood glucose levels rise (after a meal, for example), insulin is secreted from the pancreas into the *pancreatic vein,* which empties into the **portal vein system** (Figure 22.21), so insulin traverses the liver before it enters the systemic blood supply. Insulin acts to rapidly lower blood glucose concentration in several ways. Insulin stimulates glycogen synthesis and inhibits glycogen breakdown in liver and muscle.

Several other physiological effects of insulin also serve to lower blood and tissue glucose levels (Figure 22.22). Insulin stimulates the active transport of glucose (and amino acids) across the plasma membranes of muscle and adipose tissue. Insulin also increases cellular utilization of glucose by inducing the synthesis of several important glycolytic enzymes, namely, glucokinase, phosphofructokinase, and pyruvate kinase. In addition, insulin acts to inhibit several enzymes of gluconeogenesis. These various actions enable the organism to respond quickly to increases in blood glucose levels.

Glucagon and Epinephrine Stimulate Glycogen Breakdown Catabolism of tissue glycogen is triggered by the actions of the hormones epinephrine and glucagon (Figure 22.23). *In response to decreased blood glucose,* glucagon is released from the α-cells in pancreatic islets of Langerhans. This peptide hormone travels through the blood to specific receptors on liver cell membranes. (Glucagon is active in

Human Biochemistry

von Gierke Disease—A Glycogen-Storage Disease

In 1929, the physician Edgar von Gierke treated a patient with a very enlarged abdomen. The patient's liver and kidneys were severely enlarged due to massive accumulations of glycogen, and von Gierke appropriately called the condition "hepato-nephromegalia glycogenica." Now termed **von Gierke's disease,** or Type Ia glycogen storage disease, this condition results from the absence of glucose-6-phosphatase activity in the affected organs. This simple genetic defect causes a host of difficult complications, including a striking elevation of serum triglycerides, excess adipose tissue in the cheeks, thin extremities, short stature, excessive curvature of the lumbar spine, and delay of puberty.

The absence of glucose-6-phosphatase activity in the liver blocks the last steps of glycogenolysis and gluconeogenesis, inter-

rupting the recycling of glucose and causing affected individuals to be hypoglycemic. The accumulation of glucose-6-phosphate in the liver leads to greatly increased glycolytic activity, with consequent elevation of lactic acid, a condition known more commonly as **lactic acidosis.** Large amounts of uric acid and lipids are produced, and the high rates of glycolysis produce excess NADH.

The treatment of von Gierke's disease consists of trying to maintain normal levels of glucose in the patient's serum. This often requires oral administration of large amounts of glucose, in its various forms, including, for example, uncooked cornstarch, which acts as a slow-release form of glucose.

liver and adipose tissue but not in other tissues.) Similarly, signals from the central nervous system cause release of *epinephrine* (Figure 22.24)—also known as adrenaline—from the adrenal glands into the bloodstream. Epinephrine acts on liver and muscles. When either hormone binds to its receptor on the outside surface of the cell membrane, a cascade is initiated that activates glycogen phosphorylase and inhibits glycogen synthase. The result of these actions is *tightly coordinated stimulation of glycogen breakdown and inhibition of glycogen synthesis.*

The Phosphorylase Cascade Amplifies the Hormonal Signal   Stimulation of glycogen breakdown involves consumption of molecules of ATP at three different steps in the hormone-sensitive adenylyl cyclase cascade (see Figure 15.18). Note that the cascade mechanism is a means of chemical amplification, because the binding of just a few molecules of epinephrine or glucagon results in the synthesis of many molecules of cyclic AMP, which, through the action of cAMP-dependent protein kinase, can activate many more molecules of phosphorylase kinase and even more molecules of phosphorylase. For example, an extracellular level of 10^{-10} to 10^{-8} M epinephrine prompts the formation of 10^{-6} M cyclic

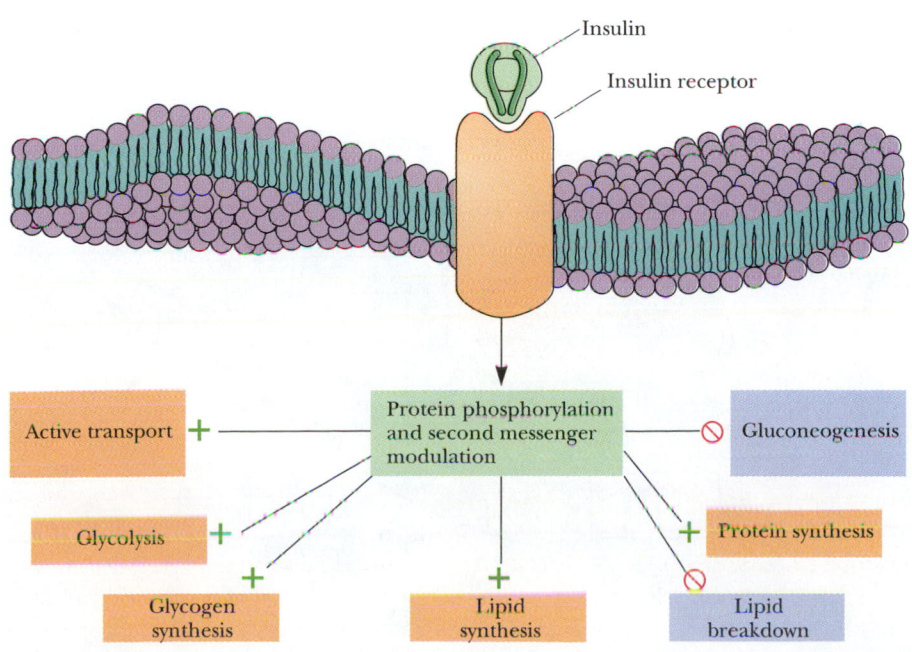

FIGURE 22.22 The metabolic effects of insulin. As described in Chapter 32, binding of insulin to membrane receptors stimulates the protein kinase activity of the receptor. Subsequent phosphorylation of target proteins modulates the effects indicated.

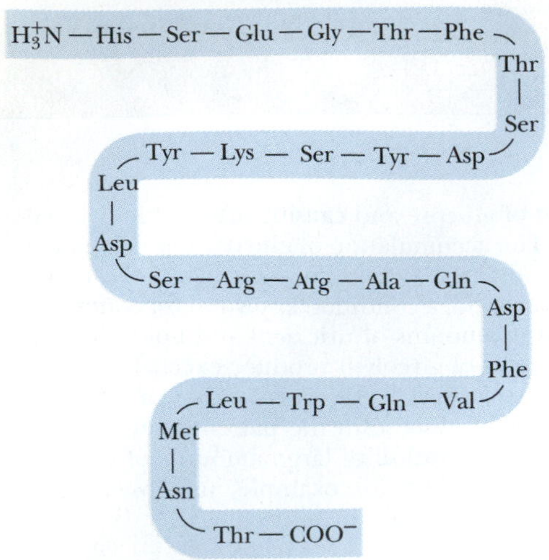

H₃⁺N—His—Ser—Glu—Gly—Thr—Phe
Thr
Ser
Tyr—Lys—Ser—Tyr—Asp
Leu
Asp
Ser—Arg—Arg—Ala—Gln
Asp
Phe
Leu—Trp—Gln—Val
Met
Asn
Thr—COO⁻

FIGURE 22.23 The amino acid sequence of glucagon.

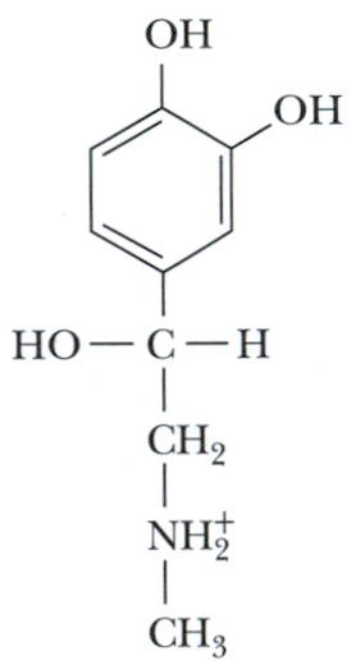

FIGURE 22.24 Epinephrine.

AMP, and for each protein kinase activated by cyclic AMP, approximately 30 phosphorylase kinase molecules are activated; these in turn activate some 800 molecules of phosphorylase. Each of these catalyzes the formation of many molecules of glucose-1-P.

The Difference Between Epinephrine and Glucagon Although both epinephrine and glucagon exert glycogenolytic effects, they do so for quite different reasons. Epinephrine is secreted as a response to anger or fear and may be viewed as an alarm or danger signal for the organism. Called the "fight or flight" hormone, it prepares the organism for mobilization of large amounts of energy. Among the many physiological changes elicited by epinephrine, one is the initiation of the enzyme cascade, as in Figure 15.18, which leads to rapid breakdown of glycogen, inhibition of glycogen synthesis, stimulation of glycolysis, and production of energy. The burst of energy produced is the result of a 2000-fold amplification of the rate of glycolysis. Because a fear or anger response must include generation of energy (in the form of glucose)—both immediately in localized sites (the muscles) and eventually throughout the organism (as supplied by the liver)—epinephrine must be able to activate glycogenolysis in both liver and muscles.

Glucagon is involved in the long-term maintenance of steady-state levels of glucose in the blood and other tissues. It performs this function by stimulating the liver to release glucose from glycogen stores into the bloodstream. To further elevate glucose levels, glucagon also stimulates liver gluconeogenesis by activating F-2,6-BPase activity (Figure 22.13). It is important to note, however, that stabilization of blood glucose levels is managed almost entirely by the liver. Glucagon does not activate the phosphorylase cascade in muscle (muscle membranes do not contain glucagon receptors). Muscle glycogen breakdown occurs only in response to epinephrine release, and muscle tissue does not participate in maintenance of steady-state glucose levels in the blood.

Cortisol and Glucocorticoid Effects on Glycogen Metabolism Glucocorticoids are a class of steroid hormones that exert distinct effects on liver, skeletal muscle, and adipose tissue. The effects of cortisol, a typical glucocorticoid, are best described as catabolic because cortisol promotes protein breakdown and decreases protein synthesis in skeletal muscle. In the liver, however, it stimulates gluconeogenesis and increases glycogen synthesis. Cortisol-induced gluconeogenesis results primarily from increased conversion of amino acids into glucose (Figure 22.25). Specific effects of cortisol in the liver include increased expression of several genes encoding enzymes of the gluconeogenic pathway, activation of enzymes

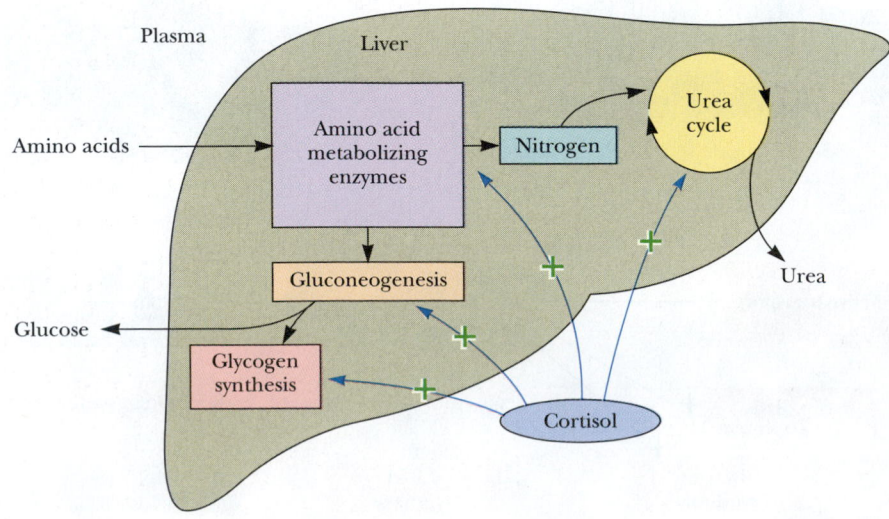

FIGURE 22.25 The effects of cortisol on carbohydrate and protein metabolism in the liver.

involved in amino acid metabolism, and stimulation of the urea cycle, which disposes of nitrogen liberated during amino acid catabolism (see Chapter 25).

| 22.6 | Can Glucose Provide Electrons for Biosynthesis? |

Cells require a constant supply of NADPH for reductive reactions vital to biosynthetic purposes. Much of this requirement is met by a glucose-based metabolic sequence variously called the **pentose phosphate pathway,** the **hexose monophosphate shunt,** or the **phosphogluconate pathway.** In addition to providing NADPH for biosynthetic processes, this pathway produces *ribose-5-phosphate,* which is essential for nucleic acid synthesis. Several metabolites of the pentose phosphate pathway can also be shuttled into glycolysis.

The Pentose Phosphate Pathway Operates Mainly in Liver and Adipose Cells

The pentose phosphate pathway begins with glucose-6-phosphate, a six-carbon sugar, and produces three-, four-, five-, six-, and seven-carbon sugars (Figure 22.26). As we will see, two successive oxidations lead to the reduction of NADP$^+$ to NADPH and the release of CO_2. Five subsequent nonoxidative steps produce a variety of carbohydrates, some of which may enter the glycolytic pathway. The enzymes of the pentose phosphate pathway are particularly abundant in the cytoplasm of liver and adipose cells. These enzymes are largely absent in muscle, where glucose-6-phosphate is utilized primarily for energy production via glycolysis and the TCA cycle. These pentose phosphate pathway enzymes are located in the cytosol, which is the site of fatty acid synthesis, a pathway heavily dependent on NADPH for reductive reactions.

The Pentose Phosphate Pathway Begins with Two Oxidative Steps

1. Glucose-6-Phosphate Dehydrogenase The pentose phosphate pathway begins with the oxidation of glucose-6-phosphate. The products of the reaction are a cyclic ester (the lactone of phosphogluconic acid) and NADPH (Figure 22.27). **Glucose-6-phosphate dehydrogenase,** which catalyzes this reaction, is highly specific for NADP$^+$. As the first step of a major pathway, the reaction is irreversible and highly regulated. Glucose-6-phosphate dehydrogenase is strongly inhibited by the product coenzyme, NADPH, and also by fatty acid esters of coenzyme A (which are intermediates of fatty acid biosynthesis). Inhibition due to NADPH depends upon the cytosolic NADP$^+$/NADPH ratio, which in the liver is about 0.015 (compared to about 725 for the NAD$^+$/NADH ratio in the cytosol).

2. Gluconolactonase The gluconolactone produced in step 1 is hydrolytically unstable and readily undergoes a spontaneous ring-opening hydrolysis, although an enzyme, gluconolactonase, accelerates this reaction (Figure 22.28). The linear product, the sugar acid 6-phospho-D-gluconate, is further oxidized in step 3.

3. 6-Phosphogluconate Dehydrogenase The oxidative decarboxylation of 6-phosphogluconate by **6-phosphogluconate dehydrogenase** yields D-ribulose-5-phosphate and another equivalent of NADPH. There are two distinct steps in this reaction (Figure 22.29): The initial NADP$^+$-dependent dehydrogenation yields a β-keto acid, 3-keto-6-phosphogluconate, which is very susceptible to decarboxylation (the second step). The resulting product, D-ribulose-5-P, is the substrate for the nonoxidative reactions composing the rest of this pathway.

Biochemistry$\textcircled{s}$**Now**™ Go to BiochemistryNow and click BiochemistryInteractive to interact with the reaction mechanism for 6-phosphogluconate.

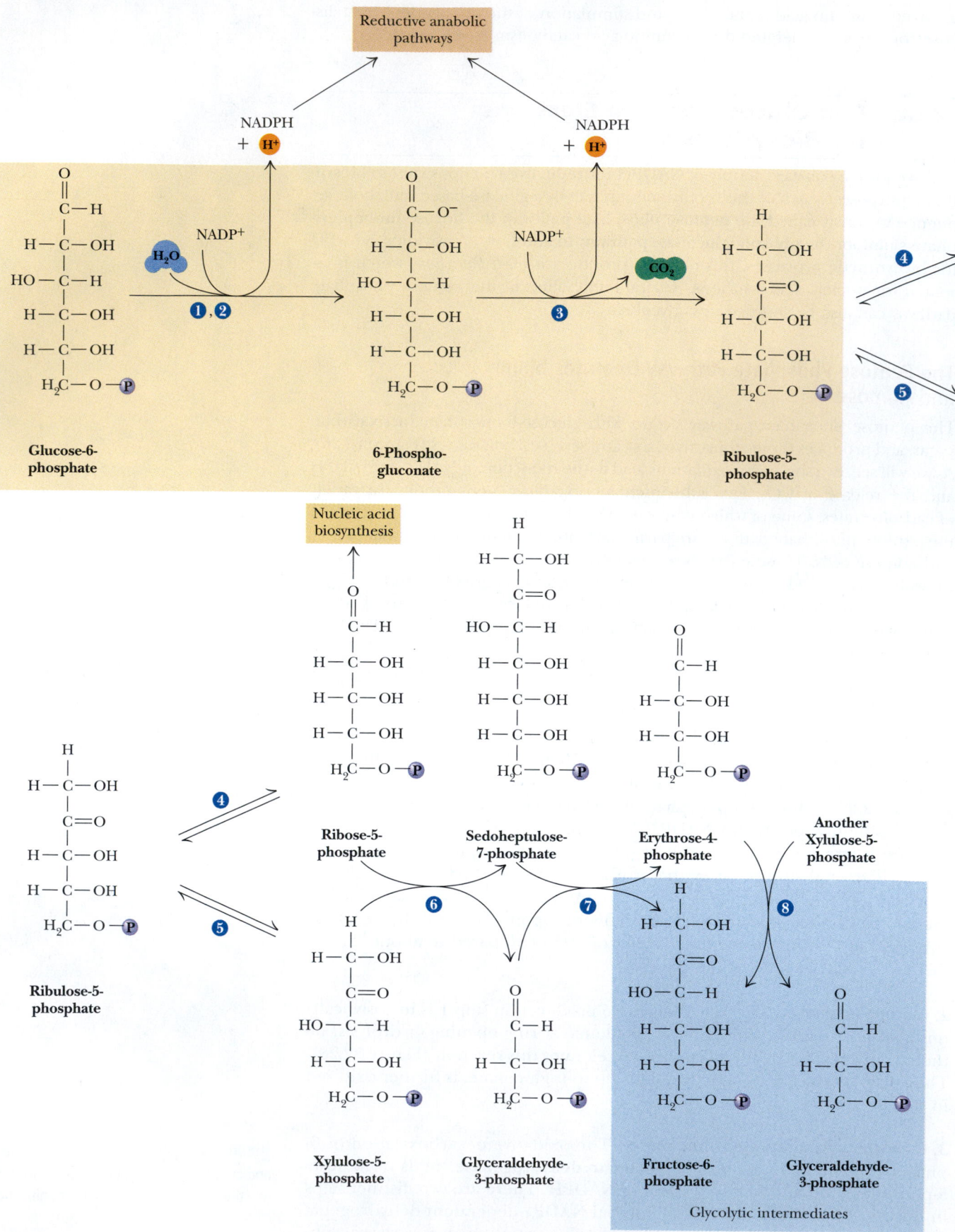

Biochemistry Now™ **ACTIVE FIGURE 22.26** The pentose phosphate pathway. The numerals in the blue circles indicate the steps discussed in the text. **Test yourself on the concepts in this figure** at **http://chemistry.brookscole.com/ggb3**

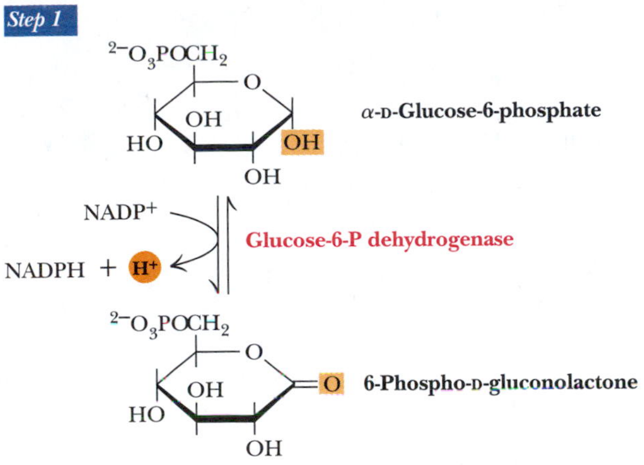

Step 1

α-D-**Glucose-6-phosphate**

NADP$^+$

Glucose-6-P dehydrogenase

NADPH + H$^+$

6-Phospho-D-gluconolactone

FIGURE 22.27 The glucose-6-phosphate dehydrogenase reaction is the committed step in the pentose phosphate pathway.

There Are Four Nonoxidative Reactions in the Pentose Phosphate Pathway

This portion of the pathway begins with an isomerization and an epimerization, and it leads to the formation of either D-ribose-5-phosphate or D-xylulose-5-phosphate. These intermediates can then be converted into glycolytic intermediates or directed to biosynthetic processes.

4. Phosphopentose Isomerase   This enzyme interconverts ribulose-5-P and ribose-5-P via an enediol intermediate (Figure 22.30). The reaction (and mechanism) is quite similar to the phosphoglucoisomerase reaction of glycolysis, which interconverts glucose-6-P and fructose-6-P. The ribose-5-P produced in this reaction is utilized in the biosynthesis of coenzymes (including NADH, NADPH, FAD, and B$_{12}$), nucleotides, and nucleic acids (DNA and RNA). The net reaction for the first four steps of the pentose phosphate pathway is

$$\text{Glucose-6-P} + 2\text{ NADP}^+ + \text{H}_2\text{O} \longrightarrow \text{ribose-5-P} + 2\text{ NADPH} + 2\text{ H}^+ + \text{CO}_2$$

5. Phosphopentose Epimerase   This reaction converts ribulose-5-P to another ketose, namely, xylulose-5-P. This reaction also proceeds by an enediol intermediate but involves an inversion at C-3 (Figure 22.31). In the reaction, an acidic proton located α- to a carbonyl carbon is removed to generate the enediolate, but the proton is added back to the same carbon from the opposite side. Note the distinction in nomenclature here. Interchange of groups on a single carbon is an epimerization, and interchange of groups between carbons is an isomerization.

To this point, the pathway has generated a pool of pentose phosphates. The $\Delta G°'$ for each of the last two reactions is small, and the three pentose-5-phosphates coexist at equilibrium. The pathway has also produced two molecules of NADPH for each glucose-6-P converted to pentose-5-phosphate. The

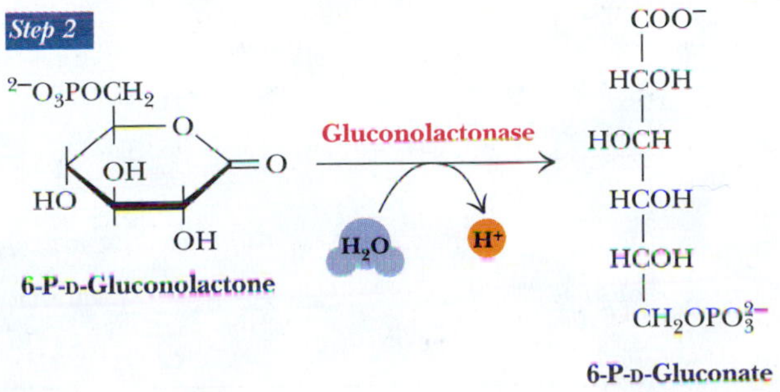

Step 2

6-P-D-Gluconolactone

Gluconolactonase

H$_2$O   H$^+$

COO$^-$
|
HCOH
|
HOCH
|
HCOH
|
HCOH
|
CH$_2$OPO$_3^{2-}$

6-P-D-Gluconate

FIGURE 22.28 The gluconolactonase reaction.

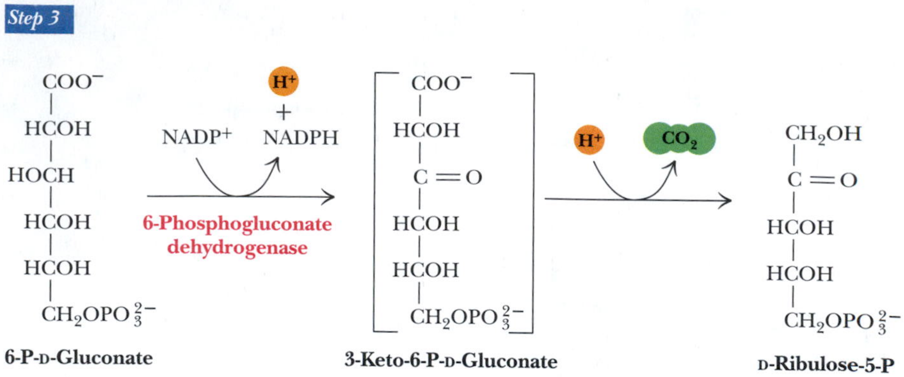

Step 3

FIGURE 22.29 The 6-phosphogluconate dehydrogenase reaction.

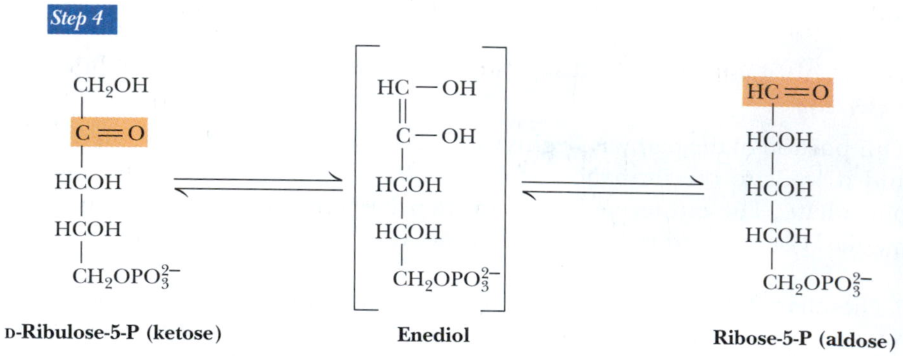

Step 4

FIGURE 22.30 The phosphopentose isomerase reaction involves an enediol intermediate.

Human Biochemistry

Aldose Reductase and Diabetic Cataract Formation

The complications of diabetes include a high propensity for cataract formation in later life, both in type 1 and type 2 diabetics. Hyperglycemia is the suspected cause, but by what mechanism? Several lines of evidence point to the **polyol pathway,** in which glucose and other simple sugars are reduced in NADPH-dependent reactions. Glucose, for example, is reduced by *aldose reductase* to sorbitol (see accompanying figure), which accumulates in lens fiber cells, increasing the intracellular osmotic pressure and eventually rupturing the cells. The involvement of aldose reductase in this process is supported by the fact that animals that have high levels of this enzyme in their lenses (such as rats and dogs) are prone to develop diabetic cataracts, whereas mice that have low levels of lens aldose reductase activity are not. Moreover, aldose reductase inhibitors such as **tolrestat** and **epalrestat** suppress cataract formation. These drugs or derivatives from them may represent an effective preventive therapy against cataract formation in people with diabetes.

(a)

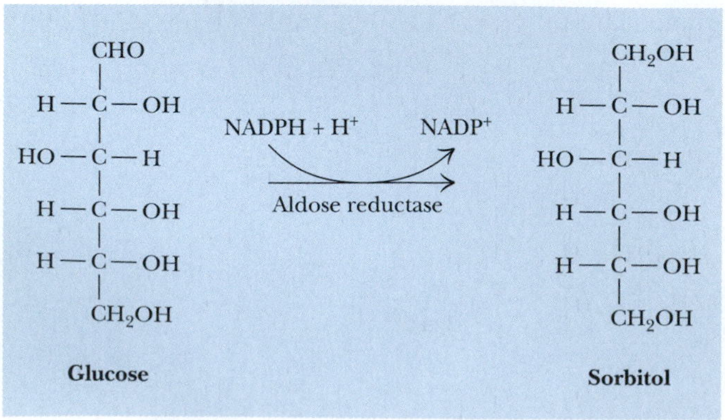

(b)

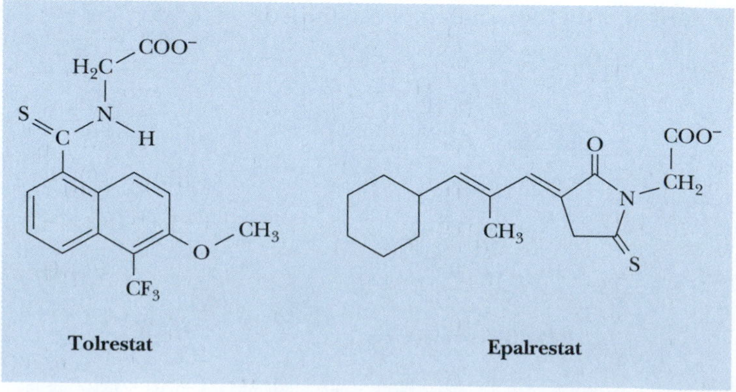

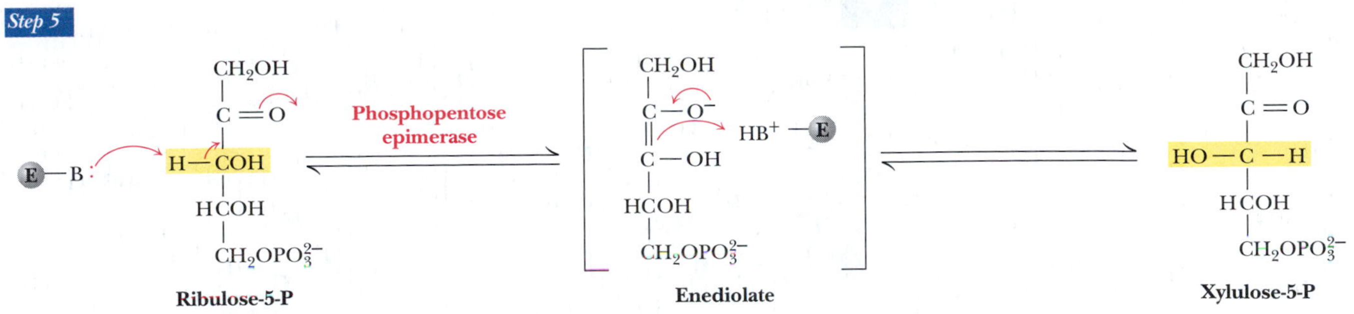

Step 5

Ribulose-5-P   **Enediolate**   **Xylulose-5-P**

FIGURE 22.31 The phosphopentose epimerase reaction interconverts ribulose-5-P and xylulose-5-phosphate. The mechanism involves an enediol intermediate and occurs with inversion at C-3.

next three steps rearrange the five-carbon skeletons of the pentoses to produce three-, four-, six-, and seven-carbon units, which can be used for various metabolic purposes. Why should the cell do this? Very often, the cellular need for NADPH is considerably greater than the need for ribose-5-phosphate. The next three steps thus return some of the five-carbon units to glyceraldehyde-3-phosphate and fructose-6-phosphate, which can enter the glycolytic pathway. The advantage of this is that the cell has met its needs for NADPH and ribose-5-phosphate in a single pathway, yet at the same time it can return the excess carbon metabolites to glycolysis.

6 and 8. Transketolase   The transketolase enzyme acts at both steps 6 and 8 of the pentose phosphate pathway. In both cases, the enzyme catalyzes the transfer of two-carbon units. In these reactions (and also in step 7, the transaldolase reaction, which transfers three-carbon units), the donor molecule is a ketose and the recipient is an aldose. In step 6, xylulose-5-phosphate transfers a two-carbon unit to ribose-5-phosphate to form glyceraldehyde-3-phosphate and sedoheptulose-7-phosphate (Figure 22.32). Step 8 involves a two-carbon transfer from xylulose-5-phosphate to erythrose-4-phosphate to produce another glyceraldehyde-3-phosphate and a fructose-6-phosphate (Figure 22.33). Three of these products enter directly into the glycolytic pathway. (The sedoheptulose-7-phosphate is taken care of in step 7, as we shall see.) Transketolase is a thiamine pyrophosphate–dependent enzyme, and the mechanism (Figure 22.34) involves abstraction of the acidic thiazole proton of TPP, attack by the resulting carbanion at the carbonyl carbon of the ketose phosphate substrate, expulsion of the glyceraldehyde-3-phosphate product, and transfer of the two-carbon unit. Transketolase can process a variety of 2-keto sugar phosphates in a similar manner. It is specific for ketose substrates with the configuration shown but can accept a variety of aldose phosphate substrates.

7. Transaldolase   The transaldolase functions primarily to make a useful glycolytic substrate from the sedoheptulose-7-phosphate produced by the first transketolase reaction. This reaction (Figure 22.35) is quite similar to the

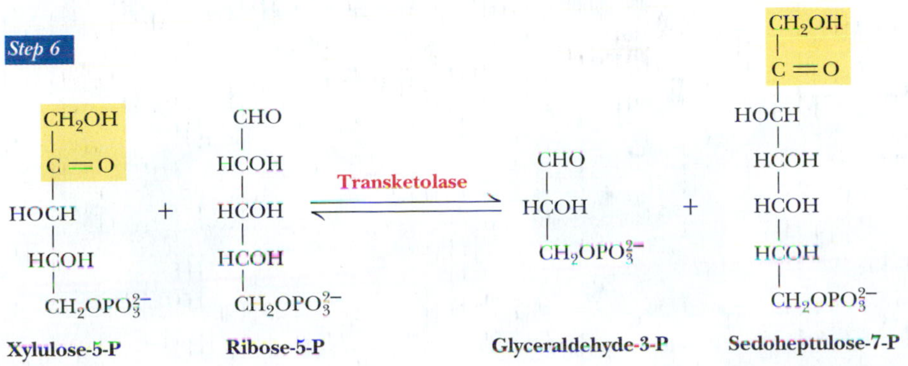

Biochemistry ⧖ Now™   Go to BiochemistryNow and click BiochemistryInteractive to learn more about the reaction of the transketolase enzyme.

Step 6

Xylulose-5-P   **Ribose-5-P**   **Glyceraldehyde-3-P**   **Sedoheptulose-7-P**

FIGURE 22.32 The transketolase reaction of step 6 in the pentose phosphate pathway.

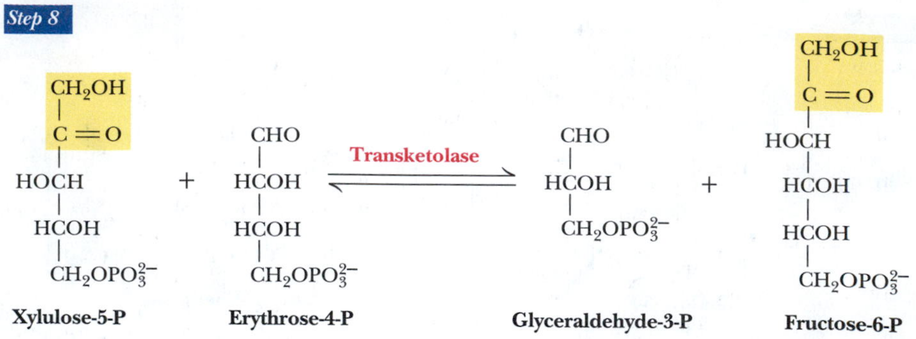

FIGURE 22.33 The transketolase reaction of step 8 in the pentose phosphate pathway.

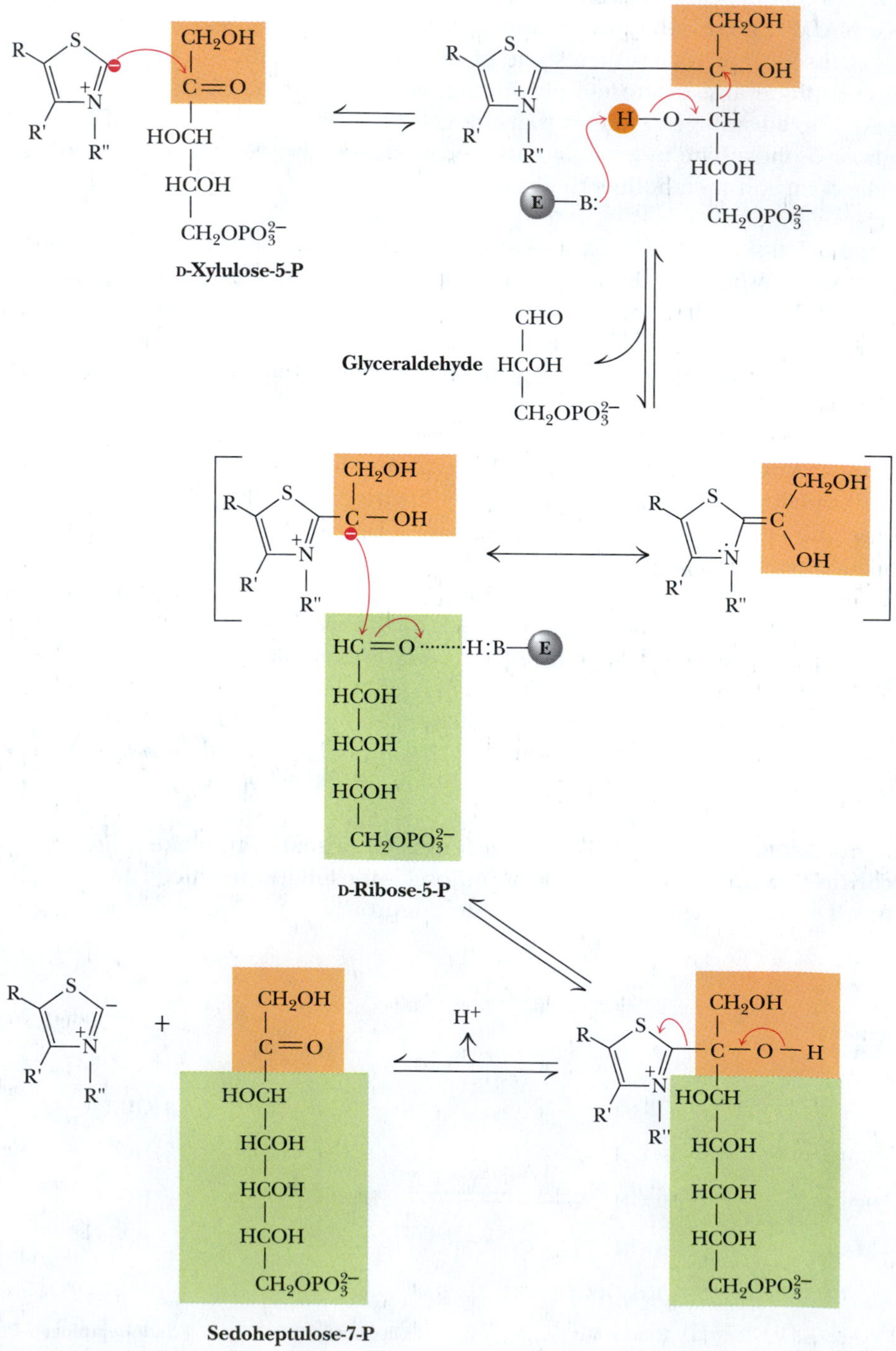

FIGURE 22.34 The mechanism of the TPP-dependent transketolase reaction. Ironically, the group transferred in the transketolase reaction might best be described as an aldol, whereas the transferred group in the transaldolase reaction is actually a ketol. Despite the irony, these names persist for historical reasons.

Step 7

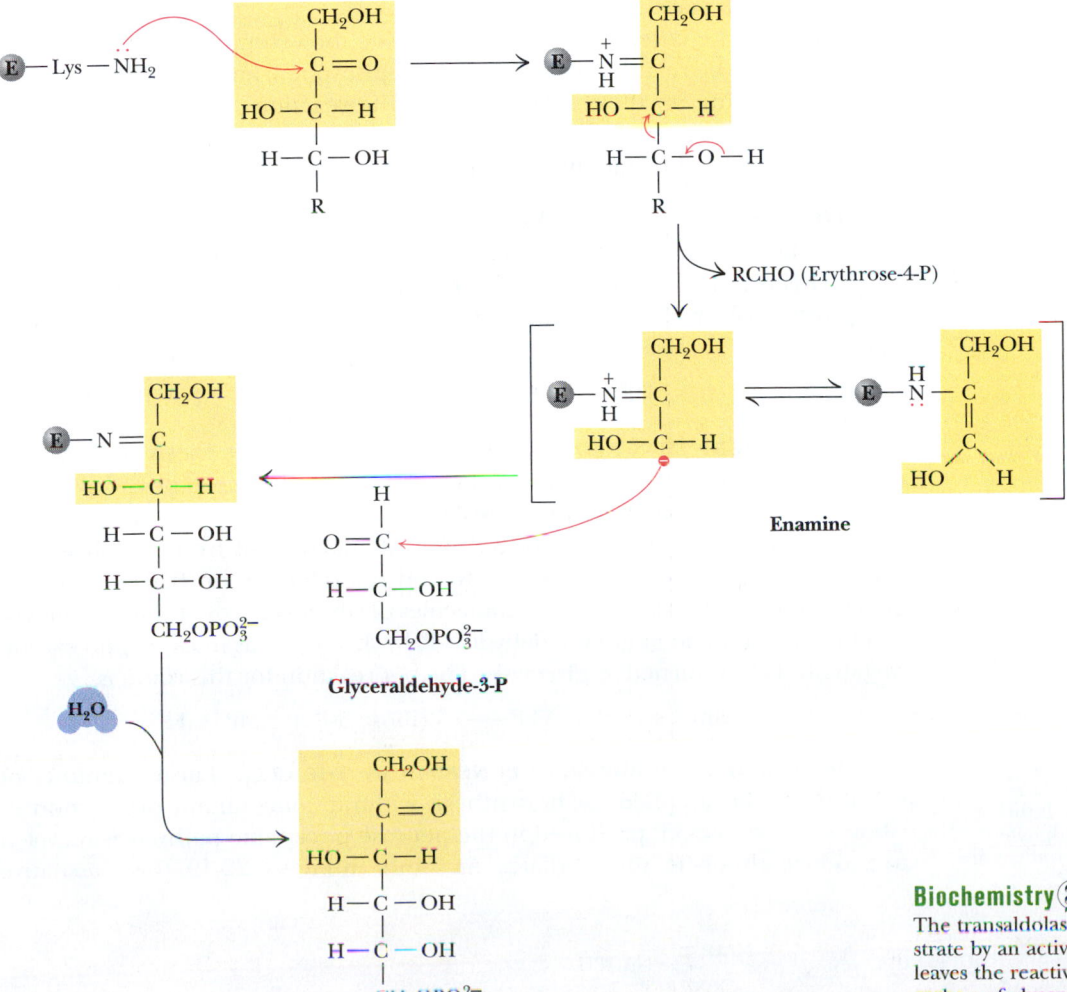

Sedoheptulose-7-P   Glyceraldehyde-3-P   Erythrose-4-P   Fructose-6-P   **FIGURE 22.35** The transaldolase reaction.

aldolase reaction of glycolysis, involving formation of a Schiff base intermediate between the sedoheptulose-7-phosphate and an active-site lysine residue
(Figure 22.36). Elimination of the erythrose-4-phosphate product leaves an
enamine of dihydroxyacetone, which remains stable at the active site (without
imine hydrolysis) until the other substrate comes into position. Attack of the
enamine carbanion at the carbonyl carbon of glyceraldehyde-3-phosphate is
followed by hydrolysis of the Schiff base (imine) to yield the product fructose-
6-phosphate.

Biochemistry Now™   ANIMATED FIGURE 22.36
The transaldolase mechanism involves attack on the substrate by an active-site lysine. Departure of erythrose-4-P
leaves the reactive enamine, which attacks the aldehyde
carbon of glyceraldehyde-3-P. Schiff base hydrolysis yields
the second product, fructose-6-P. **See this figure animated
at http://chemistry.brookscole.com/ggb3**

Utilization of Glucose-6-P Depends on the Cell's Need for ATP, NADPH, and Ribose-5-P

It is clear that glucose-6-phosphate can be used as a substrate either for glycolysis or for the pentose phosphate pathway. The cell makes this choice on the basis of its relative needs for biosynthesis and for energy from metabolism. ATP can be produced in abundance if glucose-6-phosphate is channeled into glycolysis. On the other hand, if NADPH or ribose-5-phosphate is needed, glucose-6-phosphate can be directed to the pentose phosphate pathway. The molecular basis for this regulatory decision depends on the enzymes that metabolize glucose-6-phosphate in glycolysis and the pentose phosphate pathway. In glycolysis, phosphoglucoisomerase converts glucose-6-phosphate to fructose-6-phosphate, which is utilized by phosphofructokinase (a highly regulated enzyme) to produce fructose-1,6-bisphosphate. In the pentose phosphate pathway, glucose-6-phosphate dehydrogenase (also highly regulated) produces gluconolactone from glucose-6-phosphate. Thus, the fate of glucose-6-phosphate is determined to a large extent by the relative activities of phosphofructokinase and glucose-6-P dehydrogenase. Recall from Chapter 18 that PFK is inhibited when the ATP/AMP ratio increases and that it is inhibited by citrate but activated by fructose-2,6-bisphosphate. Thus, when the energy charge is high, glycolytic flux decreases. Glucose-6-P dehydrogenase, on the other hand, is inhibited by high levels of NADPH and also by the intermediates of fatty acid biosynthesis. Both of these are indicators that biosynthetic demands have been satisfied. If that is the case, glucose-6-phosphate dehydrogenase and the pentose phosphate pathway are inhibited. If NADPH levels drop, the pentose phosphate pathway turns on and NADPH and ribose-5-phosphate are made for biosynthetic purposes.

Even when the latter choice has been made, however, the cell must still be "cognizant" of the relative needs for ribose-5-phosphate and NADPH (as well as ATP). Depending on these relative needs, the reactions of glycolysis and the pentose phosphate pathway can be combined in novel ways to emphasize the synthesis of needed metabolites. There are four principal possibilities.

1. BOTH RIBOSE-5-P AND NADPH ARE NEEDED BY THE CELL   In this case, the first four reactions of the pentose phosphate pathway predominate (Figure 22.37). NADPH is produced by the oxidative reactions of the pathway, and ribose-5-P is the principal product of carbon metabolism. As stated earlier, the net reaction for these processes is

$$\text{Glucose-6-P} + 2\ \text{NADP}^+ + H_2O \longrightarrow \text{ribose-5-P} + CO_2 + 2\ \text{NADPH} + 2\ H^+$$

2. MORE RIBOSE-5-P THAN NADPH IS NEEDED BY THE CELL   Synthesis of ribose-5-P can be accomplished without production of NADPH if the oxidative steps of the pentose phosphate pathway are bypassed. The key to this route is the extraction of fructose-6-P and glyceraldehyde-3-P, but not glucose-6-P, from glycolysis (Figure 22.38). The action of transketolase and transaldolase on fructose-6-P and glyceraldehyde-3-P produces three molecules of ribose-5-P from two molecules of fructose-6-P and one of glyceraldehyde-3-P. In this route, as in case 1, no carbon metabolites are returned to glycolysis. The net reaction for this route is

$$5\ \text{Glucose-6-P} + \text{ATP} \longrightarrow 6\ \text{ribose-5-P} + \text{ADP} + H^+$$

3. MORE NADPH THAN RIBOSE-5-P IS NEEDED BY THE CELL   Large amounts of NADPH can be supplied for biosynthesis without concomitant production of ribose-5-P if ribose-5-P produced in the pentose phosphate pathway is recycled to produce glycolytic intermediates. As shown in Figure 22.39, this alternative

FIGURE 22.37 When biosynthetic demands dictate, the first four reactions of the pentose phosphate pathway predominate and the principal products are ribose-5-P and NADPH.

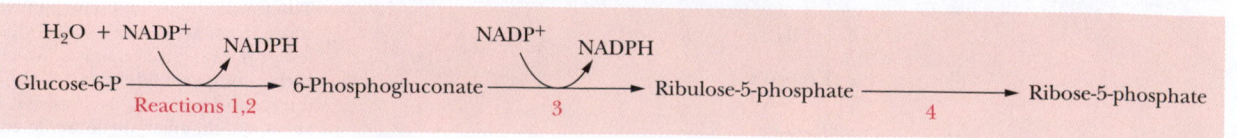

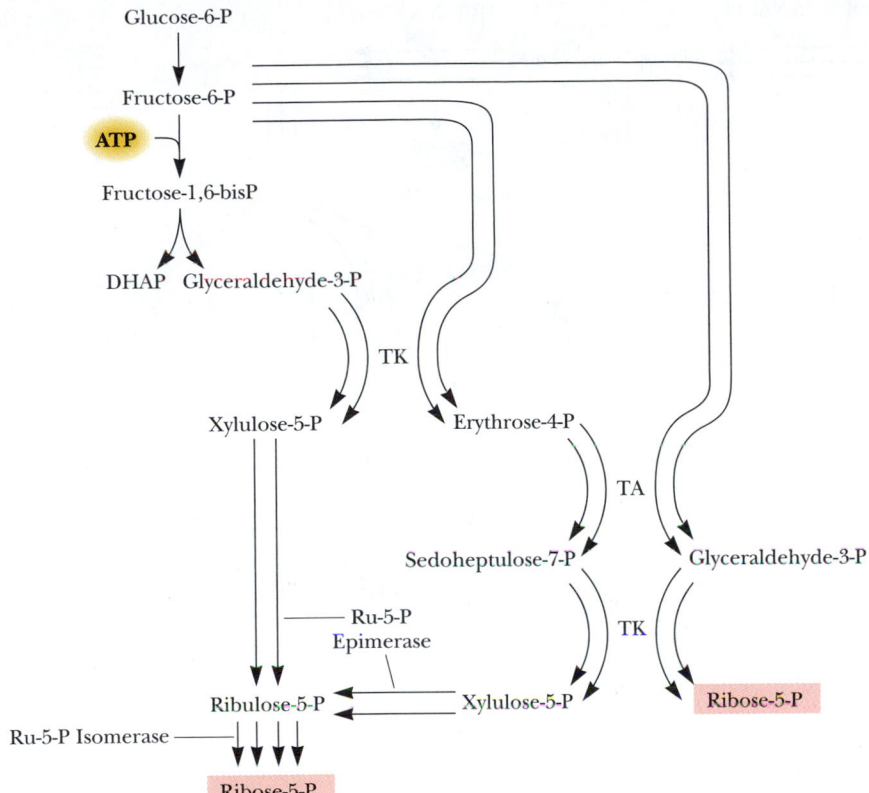

Glucose-6-P

Fructose-6-P

ATP

Fructose-1,6-bisP

DHAP   Glyceraldehyde-3-P

TK

Xylulose-5-P   Erythrose-4-P

TA

Sedoheptulose-7-P   Glyceraldehyde-3-P

Ru-5-P
Epimerase   TK

Ribulose-5-P ◄── Xylulose-5-P   Ribose-5-P

Ru-5-P Isomerase ───

Ribose-5-P

FIGURE 22.38 The oxidative steps of the pentose phosphate pathway can be bypassed if the primary need is for ribose-5-P.

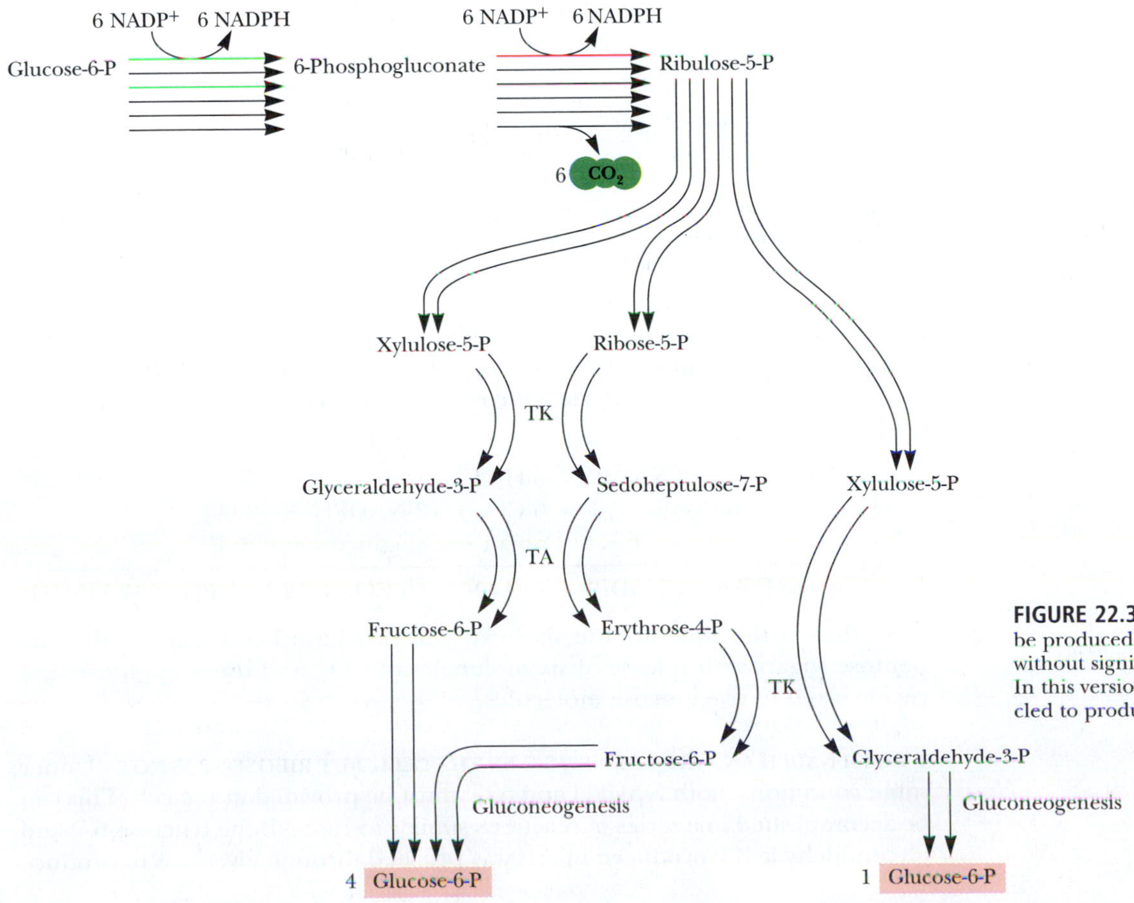

6 NADP$^+$ 6 NADPH   6 NADP$^+$ 6 NADPH

Glucose-6-P ──► 6-Phosphogluconate ── Ribulose-5-P

6 **CO_2**

Xylulose-5-P   Ribose-5-P

TK

Glyceraldehyde-3-P   Sedoheptulose-7-P   Xylulose-5-P

TA

Fructose-6-P   Erythrose-4-P

TK

Fructose-6-P   Glyceraldehyde-3-P

Gluconeogenesis   Gluconeogenesis

4 Glucose-6-P   1 Glucose-6-P

FIGURE 22.39 Large amounts of NADPH can be produced by the pentose phosphate pathway without significant net production of ribose-5-P. In this version of the pathway, ribose-5-P is recycled to produce glycolytic intermediates.

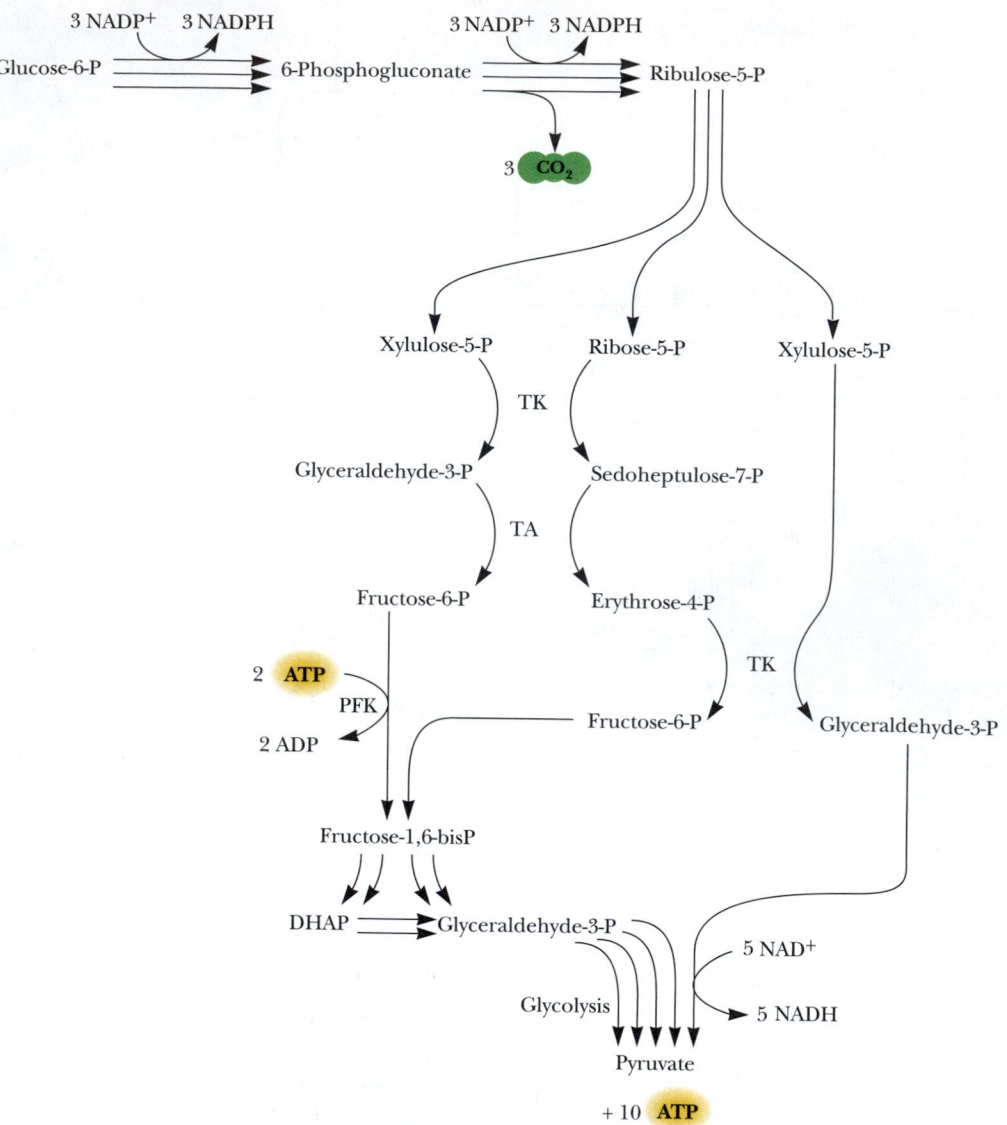

FIGURE 22.40 Both ATP and NADPH (as well as NADH) can be produced by this version of the pentose phosphate and glycolytic pathways.

involves a complex interplay between the transketolase and transaldolase reactions to convert ribulose-5-P to fructose-6-P and glyceraldehyde-3-P, which can be recycled to glucose-6-P via gluconeogenesis. The net reaction for this process is

6 Glucose-6-P + 12 NADP$^+$ + 6 H$_2$O$\longrightarrow$
6 ribulose-5-P + 6 CO$_2$ + 12 NADPH + 12 H$^+$
6 Ribulose-5-P$\longrightarrow$5-glucose-6-P + P$_i$

Net: Glucose-6-P + 12 NADP$^+$ + 6 H$_2$O$\longrightarrow$6 CO$_2$ + 12 NADPH + 12 H$^+$ + P$_i$

Note that in this scheme, the six hexose sugars have been converted to six pentose sugars with release of six molecules of CO$_2$, and the six pentoses are reconverted to five glucose molecules.

4. BOTH NADPH AND ATP ARE NEEDED BY THE CELL, BUT RIBOSE-5-P IS NOT Under some conditions, both NADPH and ATP must be provided in the cell. This can be accomplished in a series of reactions similar to case 3 if the fructose-6-P and glyceraldehyde-3-P produced in this way proceed through glycolysis to produce

ATP and pyruvate, which itself can yield even more ATP by continuing on to the TCA cycle (Figure 22.40). The net reaction for this alternative is

$$3 \text{ Glucose-6-P} + 5 \text{ NAD}^+ + 6 \text{ NADP}^+ + 8 \text{ ADP} + 5 \text{ P}_i \longrightarrow$$
$$5 \text{ pyruvate} + 3 \text{ CO}_2 + 5 \text{ NADH} + 6 \text{ NADPH} + 8 \text{ ATP} + 2 \text{ H}_2\text{O} + 8 \text{ H}^+$$

Note that, except for the three molecules of CO_2, all the other carbon from glucose-6-P is recovered in pyruvate.

Summary

22.1 What Is Gluconeogenesis, and How Does It Operate?
Gluconeogenesis is the generation (*genesis*) of new (*neo*) glucose. In addition to pyruvate and lactate, other noncarbohydrate precursors can be used as substrates for gluconeogenesis in animals, including most of the amino acids, as well as glycerol and all the TCA cycle intermediates. On the other hand, fatty acids are not substrates for gluconeogenesis in animals. Lysine and leucine are the only amino acids that are not substrates for gluconeogenesis. These amino acids produce only acetyl-CoA upon degradation. Acetyl-CoA is a substrate for gluconeogenesis when the glyoxylate cycle is operating. The major sites of gluconeogenesis are the liver and kidneys, which account for about 90% and 10% of the body's gluconeogenic activity, respectively.

22.2 How Is Gluconeogenesis Regulated?
Glycolysis and gluconeogenesis are under reciprocal control, so glycolysis is inhibited when gluconeogenesis is active, and vice versa. When the energy status of the cell is low, glucose is rapidly degraded to produce needed energy. When the energy status is high, pyruvate and other metabolites are utilized for synthesis (and storage) of glucose. The three sites of regulation in the gluconeogenic pathway are glucose-6-phosphatase, fructose-1,6-bisphosphatase, and the pyruvate carboxylase–PEP carboxykinase pair, respectively. These are the three most appropriate sites of regulation in gluconeogenesis. Glucose-6-phosphatase is under substrate-level control by glucose-6-phosphate. Acetyl-CoA allosterically activates pyruvate carboxylase. Fructose-1,6-bisphosphatase is inhibited by AMP and activated by citrate. Fructose-2,6-bisphosphate is a powerful inhibitor of fructose-1,6-bisphosphatase.

22.3 How Are Glycogen and Starch Catabolized in Animals?
Almost 100% of digestible food is absorbed and metabolized. Digestive breakdown of starch and glycogen is an unregulated process. On the other hand, tissue glycogen represents an important reservoir of potential energy, and the reactions involved in its degradation and synthesis are carefully controlled and regulated. Glycogen reserves in liver and muscle tissue are stored in the cytosol as granules exhibiting a molecular weight range from 6×10^6 to 1600×10^6. These granular aggregates contain the enzymes required to synthesize and catabolize the glycogen, as well as all the enzymes of glycolysis. The principal enzyme of glycogen catabolism is glycogen phosphorylase, a highly regulated enzyme. The glycogen phosphorylase reaction involves phosphorolysis at a nonreducing end of a glycogen polymer.

22.4 How Is Glycogen Synthesized?
Luis Leloir, a biochemist in Argentina, showed in the 1950s that glycogen synthesis depended upon sugar nucleotides. The glycogen polymer is built around a tiny protein core. The first glucose residue is covalently joined to the protein glycogenin via an acetal linkage to a tyrosine–OH group on the protein. Sugar units are added to the glycogen polymer by the action of glycogen synthase. The reaction involves transfer of a glucosyl unit from UDP–glucose to the C-4 hydroxyl group at a nonreducing end of a glycogen strand. The mechanism proceeds by cleavage of the C—O bond between the glucose moiety and the β-phosphate of UDP–glucose, leaving an oxonium ion intermediate, which is rapidly attacked by the C-4 hydroxyl oxygen of a terminal glucose unit on glycogen.

22.5 How Is Glycogen Metabolism Controlled?
Activation of glycogen phosphorylase is tightly linked to inhibition of glycogen synthase, and vice versa. Regulation involves both allosteric control and covalent modification, with the latter being under hormonal control. Glycogen synthase is regulated by covalent modification. Storage and utilization of tissue glycogen are regulated by hormones, including insulin, glucagon, epinephrine, and the glucocorticoids. Insulin stimulates glycogen synthesis and inhibits glycogen breakdown in liver and muscle, whereas glucagons and epinephrine stimulate glycogen breakdown.

22.6 Can Glucose Provide Electrons for Biosynthesis?
The pentose phosphate pathway is a collection of eight reactions that provide NADPH for biosynthetic processes and ribose-5-phosphate for nucleic acid synthesis. Several metabolites of the pentose phosphate pathway can also be shuttled into glycolysis. Utilization of glucose-6-P in the pentose phosphate pathway depends on the cell's need for ATP, NADPH, and ribose-5-P.

Problems

1. Consider the balanced equation for gluconeogenesis in Section 22.1. Account for each of the components of this equation and the indicated stoichiometry.

2. (Integrates with Chapters 3 and 18.) Calculate $\Delta G^{\circ\prime}$ and ΔG for gluconeogenesis in the erythrocyte, using data in Table 18.2 (assume $\text{NAD}^+/\text{NADH} = 20$, $[\text{GTP}] = [\text{ATP}]$, and $[\text{GDP}] = [\text{ADP}]$). See how closely your values match those in Section 22.1.

3. Use the data of Figure 22.12 to calculate the percent inhibition of fructose-1,6-bisphosphatase by 25 mM fructose-2,6-bisphosphate when fructose-1,6-bisphosphate is (a) 25 mM and (b) 100 mM.

4. (Integrates with Chapter 3.) Suggest an explanation for the exergonic nature of the glycogen synthase reaction ($\Delta G^{\circ\prime} = -13.3$ kJ/mol).

Consult Chapter 3 to review the energetics of high-energy phosphate compounds if necessary.

5. Using the values in Table 23.1 for body glycogen content and the data in part b of the illustration for A Deeper Look (page 722), calculate the rate of energy consumption by muscles in heavy exercise (in J/sec). Use the data for fast-twitch muscle.

6. What would be the distribution of carbon from positions 1, 3, and 6 of glucose after one pass through the pentose phosphate pathway if the primary need of the organism is for ribose-5-P and the oxidative steps are bypassed (Figure 22.38)?

7. What is the fate of carbon from positions 2 and 4 of glucose-6-P after one pass through the scheme shown in Figure 22.40?

8. Which reactions of the pentose phosphate pathway would be inhibited by $NaBH_4$? Why?

9. (Integrates with Chapter 7.) Imagine a glycogen molecule with 8000 glucose residues. If branches occur every eight residues, how many reducing ends does the molecule have? If branches occur every 12 residues, how many reducing ends does it have? How many nonreducing ends does it have in each of these cases?

10. Explain the effects of each of the following on the rates of gluconeogenesis and glycogen metabolism:
   a. Increasing the concentration of tissue fructose-1,6-bisphosphate
   b. Increasing the concentration of blood glucose
   c. Increasing the concentration of blood insulin
   d. Increasing the amount of blood glucagon
   e. Decreasing levels of tissue ATP
   f. Increasing the concentration of tissue AMP
   g. Decreasing the concentration of fructose-6-phosphate

11. (Integrates with Chapters 3 and 15.) The free energy change of the glycogen phosphorylase reaction is $\Delta G°' = +3.1$ kJ/mol. If $[P_i] = 1$ mM, what is the concentration of glucose-1-P when this reaction is at equilibrium?

12. Based on the mechanism for pyruvate carboxylase (Figure 22.4), write reasonable mechanisms for the reactions that follow:

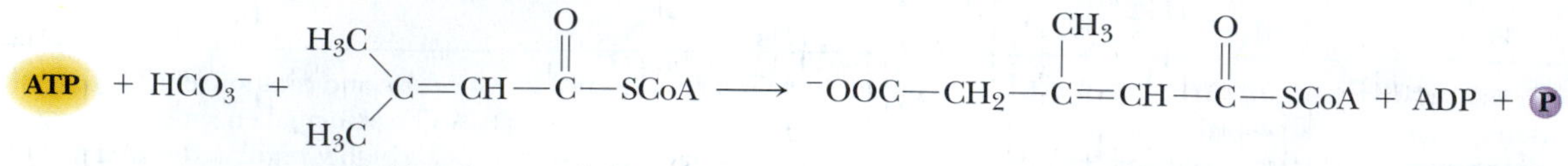

β-Methylcrotonyl-CoA → β-Methylglutaconyl-CoA

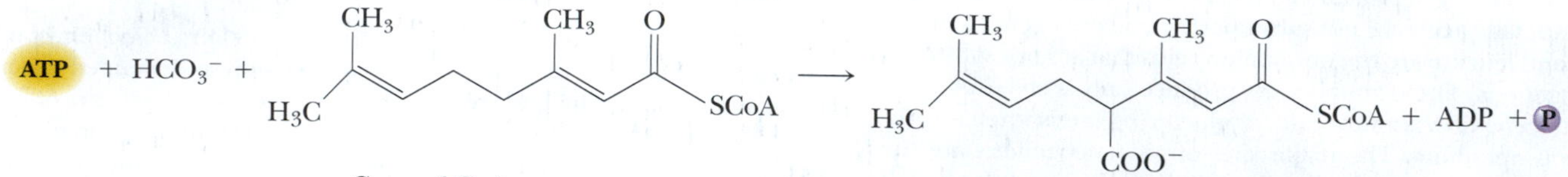

Geranyl-CoA → γ-Carboxygeranyl-CoA

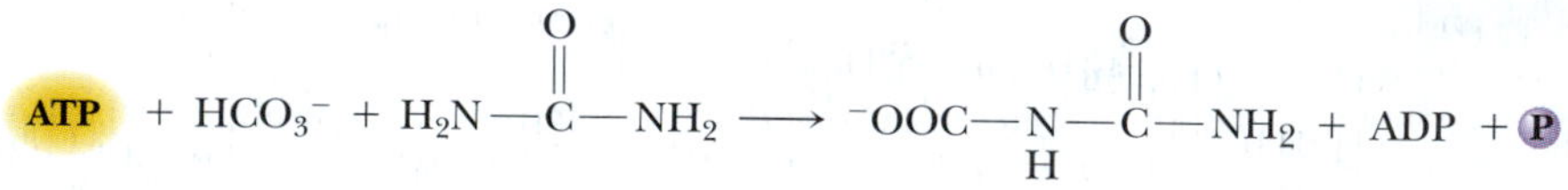

Urea → N-Carboxyurea

Transcarboxylase

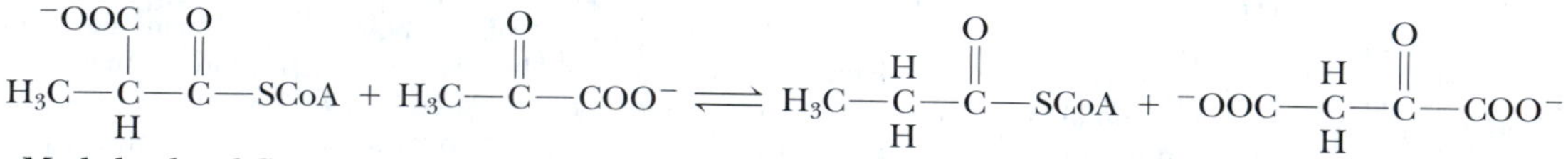

Methylmalonyl-CoA + Pyruvate ⇌ Propionyl-CoA + Oxaloacetate

13. The mechanistic chemistry of the acetolactate synthase and phosphoketolase reactions (shown here) is similar to that of the transketolase reaction (Figure 22.34). Write suitable mechanisms for these reactions.

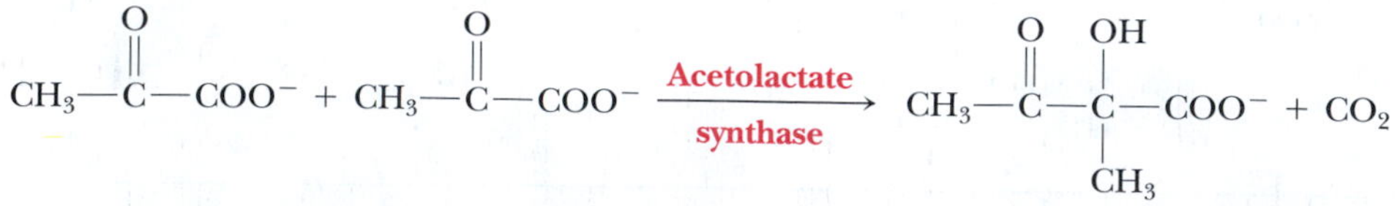

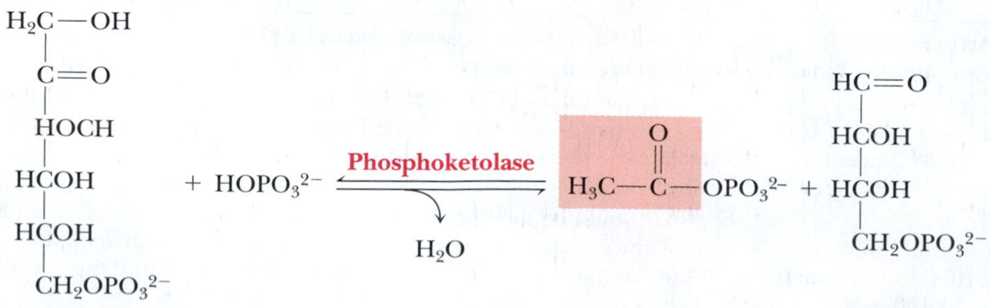

Fructose-6-P + $HOPO_3^{2-}$ → (Phosphoketolase) Acetyl-P + Erythrose-4-P + H_2O

14. Metaglip is a prescribed preparation (from Bristol-Myers Squibb) for treatment of type 2 diabetes. It consists of metformin (see Human Biochemistry, page 711) together with glipizide. The actions of metformin and glipizide are said to be complementary. Suggest a mechanism for the action of glipizide.

15. Study the structures of tolrestat and epalrestat in the Human Biochemistry box on page 728 and suggest a mechanism of action for these inhibitors of aldose reductase.

Preparing for the MCAT Exam

16. Study the graphs in the Deeper Look box (page 722) and explain the timing of the provision of energy from different metabolic sources during periods of heavy exercise.

17. (Integrates with Chapters 3 and 14.) What is the structure of creatine phosphate? Write reactions to indicate how it stores and provides energy for exercise.

Biochemistry ⊜ Now™   Preparing for an exam? Test yourself on key questions at http://chemistry.brookscole.com/ggb3

Further Reading

Gluconeogenesis

Boden, G., 2003. Effects of free fatty acids on gluconeogenesis and glycogenolysis. *Life Science* **72**:977–988.

Gerich, J. E., Meyer, C., Woerle, H. J., and Stumvoll, M., 2001. Renal gluconeogenesis: its importance in human glucose homeostasis. *Diabetes Care* **24**:382–391.

Hers, H-G., and Hue, L., 1983. Gluconeogenesis and related aspects of glycolysis. *Annual Review of Biochemistry* **52**:617–653.

Regulation of Gluconeogenesis

Hanson, R. W., and Reshef, L., 1997. Regulation of phosphoenolpyruvate carboxykinase (GTP) gene expression. *Annual Review of Biochemistry* **66**:581–611.

Moller, D. E., 2001. New drug targets for type 2 diabetes and the metabolic syndrome. *Nature* **414**:821–827.

Newsholme, E. A., Chaliss, R. A. J., and Crabtree, B., 1984. Substrate cycles: Their role in improving sensitivity in metabolic control. *Trends in Biochemical Sciences* **9**:277–280.

Newsholme, E. A., and Leech, A. R., 1983. *Biochemistry for the Medical Sciences.* New York: Wiley.

Pilkis, S. J., El-Maghrabi, M. R., and Claus, T. H., 1988. Hormonal regulation of hepatic gluconeogenesis and glycolysis. *Annual Review of Biochemistry* **57**:755–783.

Rolfe, D. F., and Brown, G. C., 1997. Cellular energy utilization and molecular origin of standard metabolic rate in mammals. *Physiological Reviews* **77**:731–758.

Sies, H., ed., 1982. *Metabolic Compartmentation.* London: Academic Press.

Sukalski, K. A., and Nordlie, R. C., 1989. Glucose-6-phosphatase: Two concepts of membrane-function relationship. *Advances in Enzymology* **62**:93–117.

Taylor, S. S., et al., 1993. A template for the protein kinase family. *Trends in Biochemical Sciences* **18**:84–89.

Van Schaftingen, E., and Hers, H-G., 1981. Inhibition of fructose-1,6-bisphosphatase by fructose-2,6-bisphosphate. *Proceedings of the National Academy of Sciences, U.S.A.* **78**:2861–2863.

Williamson, D. H., Lund, P., and Krebs, H. A., 1967. The redox state of free nicotinamide-adenine dinucleotide in the cytoplasm and mitochondria of rat liver. *Biochemical Journal* **103**:514–527.

Exercise Physiology

Akermark, C., Jacobs, I., Rasmusson, M., and Karlsson, J., 1996. Diet and muscle glycogen concentration in relation to physical performance in Swedish elite ice hockey players. *International Journal of Sport Nutrition* **6**:272–284.

Fox, E. L., 1984. *Sports Physiology,* 2nd ed. Philadelphia: Saunders College Publishing.

Hargreaves, M., 1997. Interactions between muscle glycogen and blood glucose during exercise. *Exercise and Sport Sciences Reviews* **25**:21–39.

Horton, E. S., and Terjung, R. L., eds. 1988. *Exercise, Nutrition and Energy Metabolism.* New York: Macmillan.

Rhoades, R., and Pflanzer, R., 1992. *Human Physiology.* Philadelphia: Saunders College Publishing.

Shulman, R. G., and Rothman, D. L., 1996. Nuclear magnetic resonance studies of muscle and applications to exercise and diabetes. *Diabetes* **45**:S93–S98.

Tarnopolsky, M., ed., 1999. *Gender Differences in Metabolism.* Boca Raton, FL: CRC Press.

Glycogen Metabolism

Browner, M. F., and Fletterick, R. J., 1992. Phosphorylase: A biological transducer. *Trends in Biochemical Sciences* **17**:66–71.

Huang, D., Wilson, W. A., and Roach, P. J., 1997. Glucose-6-P control of glycogen synthase phosphorylation in yeast. *Journal of Biological Chemistry* **272**:22495–22501.

Johnson, L. N., 1992. Glycogen phosphorylase: Control by phosphorylation and allosteric effectors. *The FASEB Journal* **6**:2274–2282.

Larner, J., 1990. Insulin and the stimulation of glycogen synthesis: The road from glycogen structure to glycogen synthase to cyclic AMP-dependent protein kinase to insulin mediators. *Advances in Enzymology* **63**:173–231.

Rybicka, K. K., 1996. Glycosomes—The organelles of glycogen metabolism. *Tissue and Cell* **28**:253–265.

Stalmans, W., Cadefau, J., Wera, S., and Bollen, M., 1997. New insight into the regulation of liver glycogen metabolism by glucose. *Biochemical Society Transactions* **25**:19–25.

Woodget, J. R., 1991. A common denominator linking glycogen metabolism, nuclear oncogenes, and development. *Trends in Biochemical Sciences* **16**:177–181.

Fatty Acid Catabolism

The hummingbird's tremendous capacity to store and use fatty acids enables it to make migratory journeys of remarkable distances.

The fat is in the fire.
Proverbs, **John Heywood** (1497–1580)

Key Questions

23.1 How Are Fats Mobilized from Dietary Intake and Adipose Tissue?

23.2 How Are Fatty Acids Broken Down?

23.3 How Are Odd-Carbon Fatty Acids Oxidized?

23.4 How Are Unsaturated Fatty Acids Oxidized?

23.5 Are There Other Ways to Oxidize Fatty Acids?

23.6 What Are Ketone Bodies, and What Role Do They Play in Metabolism?

Biochemistry ⊖ Now™ Test yourself on these Key Questions at BiochemistryNow at
http://chemistry.brookscole.com/ggb3

Essential Question

Fatty acids represent the principal form of stored energy for many organisms. There are two important advantages to storing energy in the form of fatty acids. (1) The carbon in fatty acids (mostly —CH_2— groups) is almost completely reduced compared to the carbon in other simple biomolecules (sugars, amino acids). Therefore, oxidation of fatty acids will yield more energy (in the form of ATP) than any other form of carbon. (2) Fatty acids are not generally hydrated as monosaccharides and polysaccharides are, and thus they can pack more closely in storage tissues. *How are fatty acids catabolized, and how is their inherent energy captured by organisms?*

23.1 How Are Fats Mobilized from Dietary Intake and Adipose Tissue?

Modern Diets Are Often High in Fat

Fatty acids are acquired readily in the diet and can also be made from carbohydrates and the carbon skeletons of amino acids. Fatty acids provide 30% to 60% of the calories in the diets of most Americans. For our caveman and cavewoman ancestors, the figure was probably closer to 20%. Dairy products were apparently not part of their diet, and the meat they consumed (from fast-moving animals) was low in fat. In contrast, modern domesticated cows and pigs are actually bred for high fat content (and better taste). However, woolly mammoth burgers and saber-toothed tiger steaks are hard to find these days—even in the gourmet sections of grocery stores—and so, by default, we consume (and metabolize) large quantities of fatty acids.

Triacylglycerols Are a Major Form of Stored Energy in Animals

Although some of the fat in our diets is in the form of phospholipids, triacylglycerols are a major source of fatty acids. Triacylglycerols are also our principal stored energy reserve. As shown in Table 23.1, the energy available in stores of fat in the average person far exceeds the energy available from protein, glycogen, and glucose. Overall, fat accounts for approximately 83% of available energy, partly because more fat is stored than protein and carbohydrate and partly because of the substantially higher energy yield per gram for fat compared with protein and carbohydrate. Complete combustion of fat yields about 37 kJ/g, compared with about 16 to 17 kJ/g for sugars, glycogen, and amino acids. In animals, fat is stored mainly as triacylglycerols in specialized cells called **adipocytes** or **adipose cells.** As shown in Figure 23.1, triacylglycerols, aggregated to form large globules, occupy most of the volume of adipose cells. Much smaller amounts of triacylglycerols are stored as small, aggregated globules in muscle tissue.

Hormones Trigger the Release of Fatty Acids from Adipose Tissue

The pathways for liberation of fatty acids from triacylglycerols, either from adipose cells or from the diet, are shown in Figures 23.2 and 23.3. Fatty acids are mobilized from adipocytes in response to hormone messengers such as adrenaline, glucagon, and adrenocorticotropic hormone (ACTH). These signal molecules bind to receptors on the plasma membrane of adipose cells and lead to the activation of adenylyl cyclase, which forms cyclic AMP from ATP. (Second messengers and hormonal signaling are discussed in Chapter 32.) In adipose cells, cAMP activates protein kinase A, which phosphorylates and activates a **triacylglycerol lipase** (also termed **hormone-sensitive lipase**) that hydrolyzes a fatty acid from C-1 or C-3 of triacylglycerols. Subsequent actions of **diacylglycerol lipase** and **monoacylglycerol lipase** yield fatty acids and glycerol. The cell then releases the fatty acids into the blood, where they are carried (in complexes with *serum albumin*) to sites of utilization.

Degradation of Dietary Fatty Acids Occurs Primarily in the Duodenum

Dietary triacylglycerols are degraded to a small extent (via fatty acid release) by lipases in the low-pH environment of the stomach, but mostly they pass untouched into the duodenum. Alkaline pancreatic juice secreted into the duodenum (Figure 23.3a) raises the pH of the digestive mixture, allowing hydrolysis of the triacylglycerols by pancreatic lipase and by nonspecific esterases, which hydrolyze the fatty acid ester linkages. Pancreatic lipase cleaves fatty acids from the C-1 and C-3 positions of triacylglycerols, and other lipases and esterases attack the C-2 position (Figure 23.3b). These processes depend upon the presence of **bile salts,** a family of carboxylic acid salts with steroid backbones (see also Chapter 24). These agents act as detergents to emulsify the triacylglycerols and facilitate the hydrolytic activity of the lipases and esterases. Short-chain fatty acids (ten or fewer carbons) released in this way are absorbed directly into the villi of the intestinal mucosa, whereas long-chain fatty acids, which are less soluble, form mixed micelles with bile salts and are carried in this fashion to the surfaces of the epithelial cells that cover the villi (Figure 23.4). The fatty acids pass into the epithelial cells, where they are condensed with glycerol to form new triacylglycerols. These triacylglycerols aggregate with lipoproteins to form particles called **chylomicrons,** which are then transported into the lymphatic system and on to the bloodstream, where they circulate to the liver, lungs, heart, muscles, and other organs (see Chapter 24). At these sites, the triacylglycerols are hydrolyzed to release fatty acids, which can then be oxidized in a highly exergonic metabolic pathway known as *β-oxidation*.

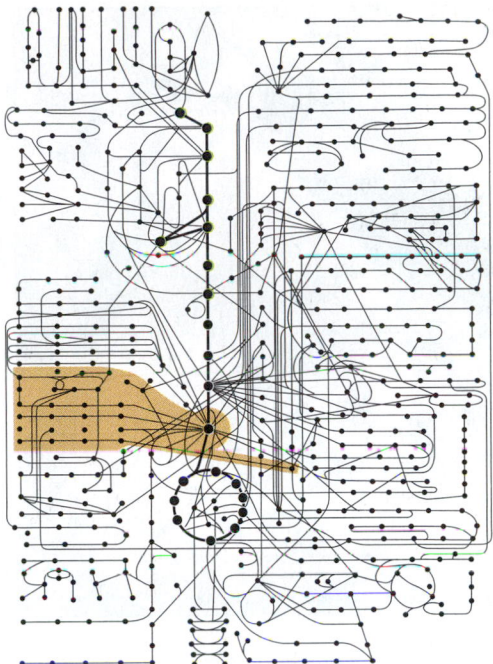

Fatty acid catabolism.

23.2 | How Are Fatty Acids Broken Down?

Franz Knoop Elucidated the Essential Feature of β-Oxidation

The earliest clue to the secret of fatty acid oxidation and breakdown came in the early 1900s, when Franz Knoop carried out experiments in which he fed dogs fatty acids in which the terminal methyl group had been replaced with a phenyl ring (Figure 23.5). Knoop discovered that fatty acids containing an even number of carbon atoms were broken down to yield phenyl acetate as the

Table 23.1			
Stored Metabolic Fuel in a 70-kg Person			
Constituent	**Energy (kJ/g dry weight)**	**Dry Weight (g)**	**Available Energy (kJ)**
Fat (adipose tissue)	37	15,000	555,000
Protein (muscle)	17	6,000	102,000
Glycogen (muscle)	16	120	1,920
Glycogen (liver)	16	70	1,120
Glucose (extracellular fluid)	16	20	320
Total			660,360

Sources: Owen, O. E., and Reichard, G. A., Jr., 1971. Fuels consumed by man: The interplay between carbohydrates and fatty acids. *Progress in Biochemistry and Pharmacology* **6**:177; and Newsholme, E. A., and Leech, A. R., 1983. *Biochemistry for the Medical Sciences.* New York: Wiley.

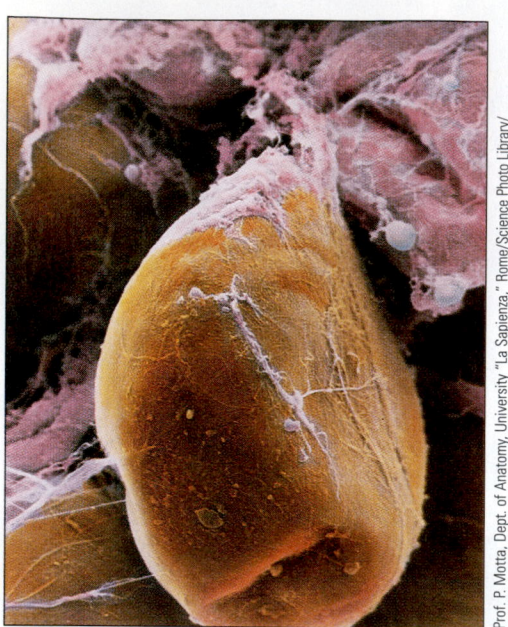

Prof. P. Motta, Dept. of Anatomy, University "La Sapienza," Rome/Science Photo Library/ Photo Researchers, Inc.

FIGURE 23.1 Scanning electron micrograph of an adipose cell (fat cell). Globules of triacylglycerols occupy most of the volume of such cells.

final product, whereas fatty acids with an odd number of carbon atoms yielded benzoate as the final product (Figure 23.5). From these experiments, Knoop concluded that the fatty acids must be degraded by *oxidation at the β-carbon* (Figure 23.6), followed by cleavage of the C_α—C_β bond. Repetition of this process yielded two-carbon units, which Knoop assumed must be acetate. Much later, Albert Lehninger showed that this degradative process took place in the mitochondria, and F. Lynen and E. Reichart showed that the two-carbon unit released is *acetyl-CoA,* not free acetate. Because the entire process begins with oxidation of the carbon that is "β" to the carboxyl carbon, the process has come to be known as **β-oxidation.**

Coenzyme A Activates Fatty Acids for Degradation

The process of β-oxidation begins with the formation of a thiol ester bond between the fatty acid and the thiol group of coenzyme A. This reaction, shown in Figure 23.7, is catalyzed by **acyl-CoA synthetase,** which is also called **acyl-CoA**

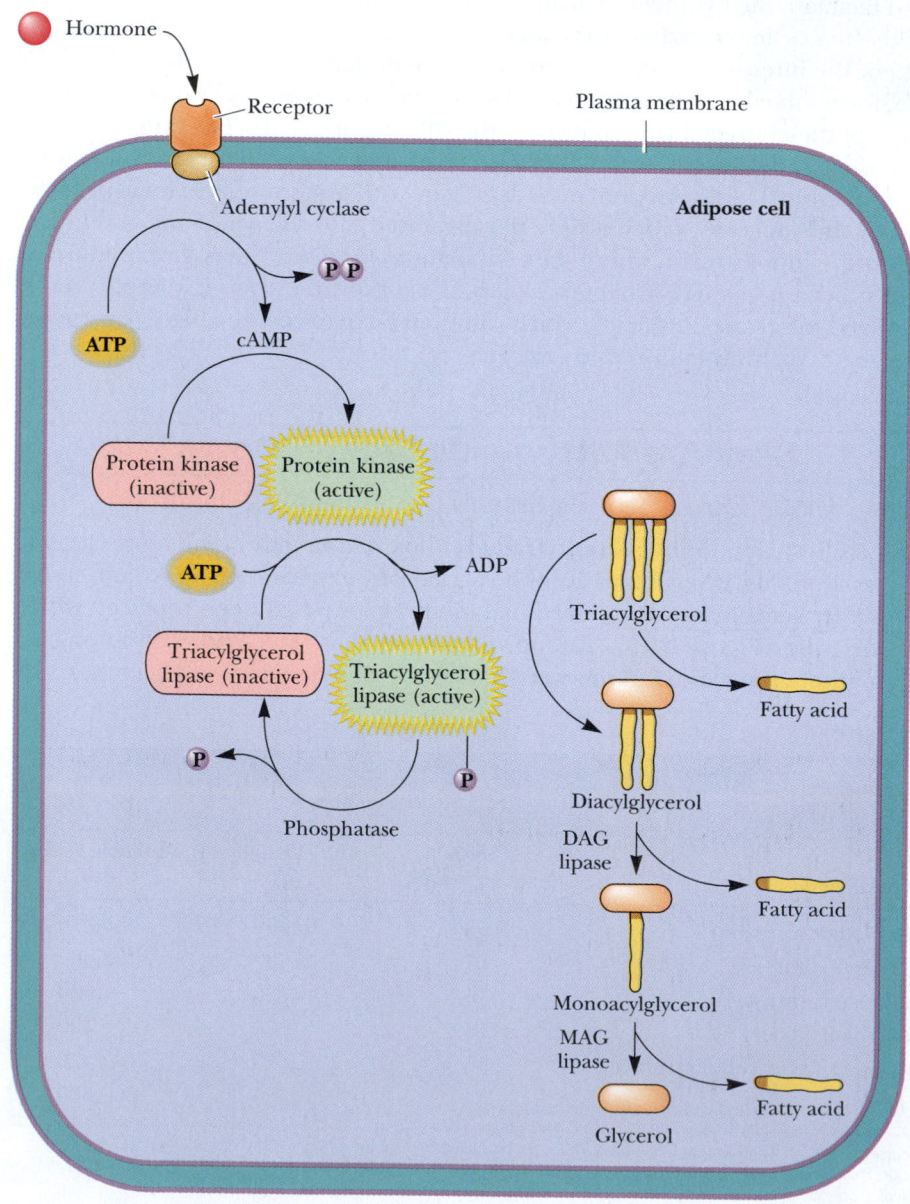

Biochemistry ⊜ Now™ ANIMATED FIGURE 23.2
Liberation of fatty acids from triacylglycerols in adipose tissue is hormone-dependent. **See this figure animated at** http://chemistry.brookscole.com/ggb3

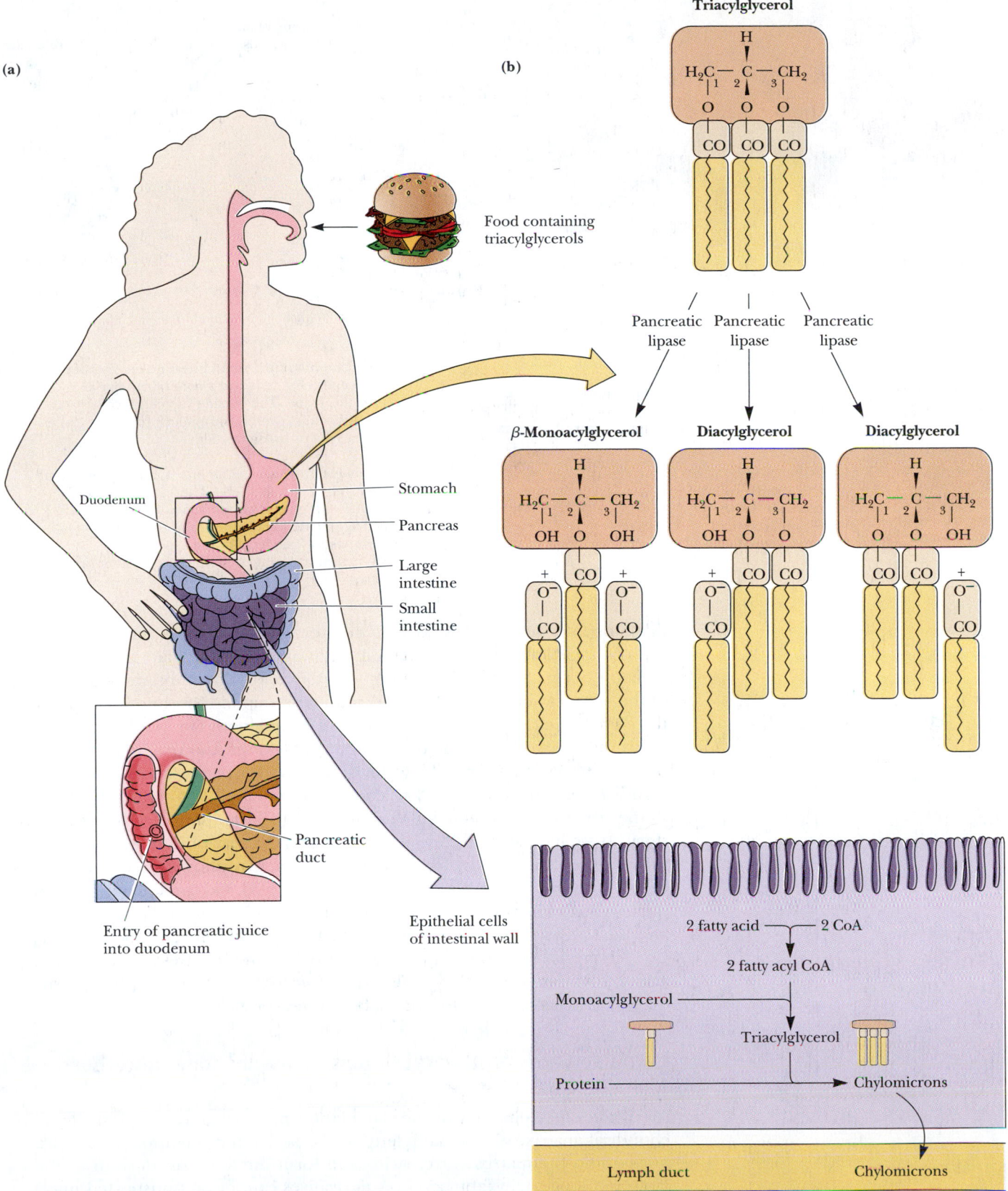

FIGURE 23.3 **(a)** A duct at the junction of the pancreas and duodenum secretes pancreatic juice into the duodenum, the first portion of the small intestine. **(b)** Hydrolysis of triacylglycerols by pancreatic and intestinal lipases. Pancreatic lipases cleave fatty acids at the C-1 and C-3 positions. Resulting monoacylglycerols with fatty acids at C-2 are hydrolyzed by intestinal lipases. Fatty acids and monoacylglycerols are absorbed through the intestinal wall and assembled into lipoprotein aggregates termed chylomicrons (discussed in Chapter 24).

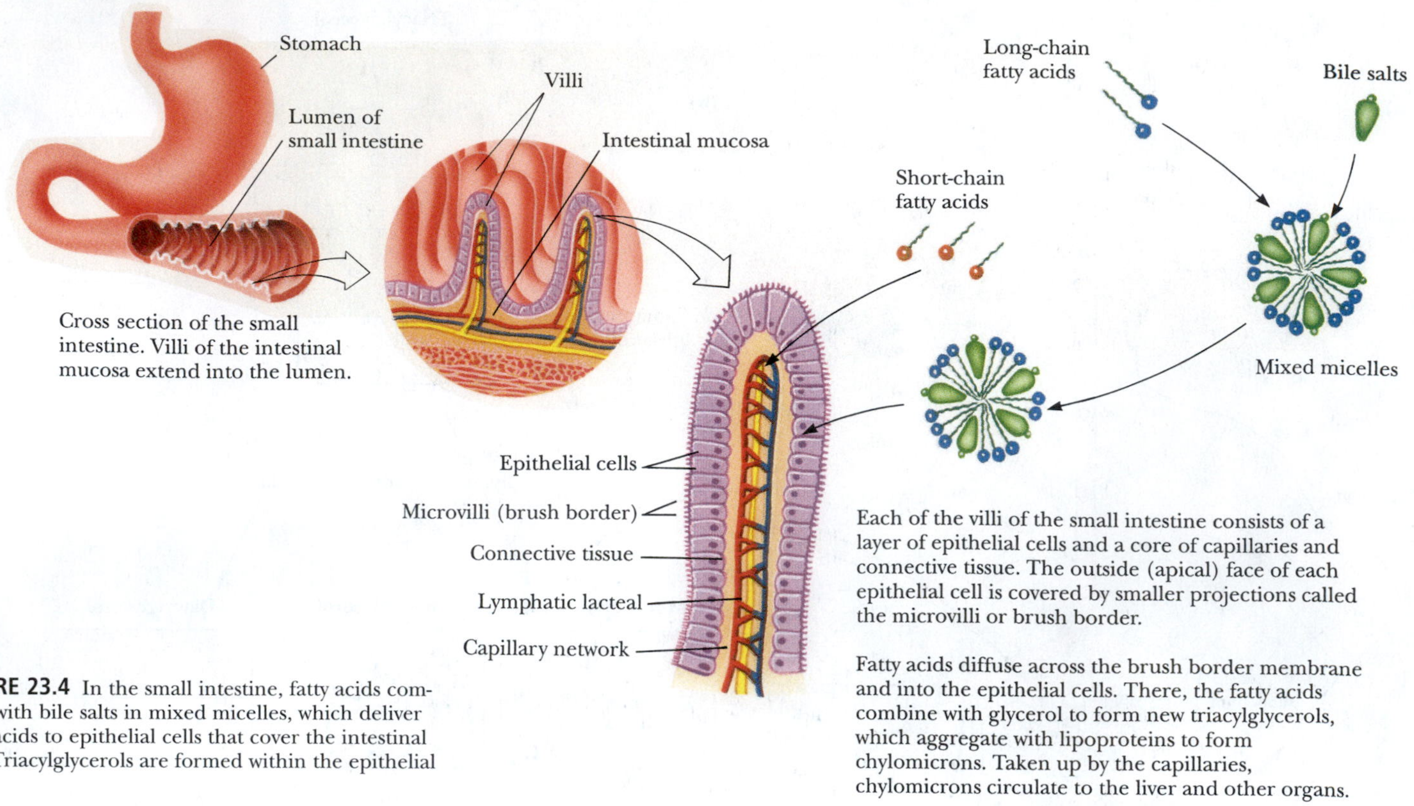

FIGURE 23.4 In the small intestine, fatty acids combine with bile salts in mixed micelles, which deliver fatty acids to epithelial cells that cover the intestinal villi. Triacylglycerols are formed within the epithelial cells.

Each of the villi of the small intestine consists of a layer of epithelial cells and a core of capillaries and connective tissue. The outside (apical) face of each epithelial cell is covered by smaller projections called the microvilli or brush border.

Fatty acids diffuse across the brush border membrane and into the epithelial cells. There, the fatty acids combine with glycerol to form new triacylglycerols, which aggregate with lipoproteins to form chylomicrons. Taken up by the capillaries, chylomicrons circulate to the liver and other organs.

ligase or **fatty acid thiokinase.** This condensation with CoA activates the fatty acid for reaction in the β-oxidation pathway. For long-chain fatty acids, this reaction normally occurs at the outer mitochondrial membrane, before entry of the fatty acid into the mitochondrion, but it may also occur at the surface of the endoplasmic reticulum. Short- and medium-length fatty acids undergo this activating reaction in the mitochondria. In all cases, the reaction is accompanied by the hydrolysis of ATP to form AMP and pyrophosphate. As shown in Figure 23.7, the two combined reactions have a net $\Delta G^{\circ\prime}$ of about -0.8 kJ/mol, so the reaction is favorable but easily reversible. However, there is more to the story. As we have seen in several similar cases, the pyrophosphate produced in this reaction is rapidly hydrolyzed by inorganic pyrophosphatase to two molecules of phosphate, with a net $\Delta G^{\circ\prime}$ of about -33.6 kJ/mol. Thus, pyrophosphate is maintained at a low concentration in the cell (usually less than 10 μM), and the synthetase reaction is strongly promoted. The mechanism of the acyl-CoA synthetase reaction is shown in Figure 23.8 and involves attack of the fatty acid carboxylate on ATP to form an *acyladenylate intermediate,* which is subsequently attacked by CoA, forming a fatty acyl-CoA thioester.

Carnitine Carries Fatty Acyl Groups Across the Inner Mitochondrial Membrane

All of the other enzymes of the β-oxidation pathway are located in the mitochondrial matrix. Short-chain fatty acids, as already mentioned, are transported into the matrix as free acids and form the acyl-CoA derivatives there. However, long-chain fatty acyl-CoA derivatives cannot be transported into the matrix directly. These long-chain derivatives must first be converted to *acylcarnitine* derivatives, as shown in Figure 23.9. **Carnitine acyltransferase I,** associated with the outer mitochondrial membrane, catalyzes the formation of the *O*-acylcarnitine, which is then transported across the inner membrane by a **translocase.** At this point, the acylcarnitine is passed to **carnitine acyltrans-**

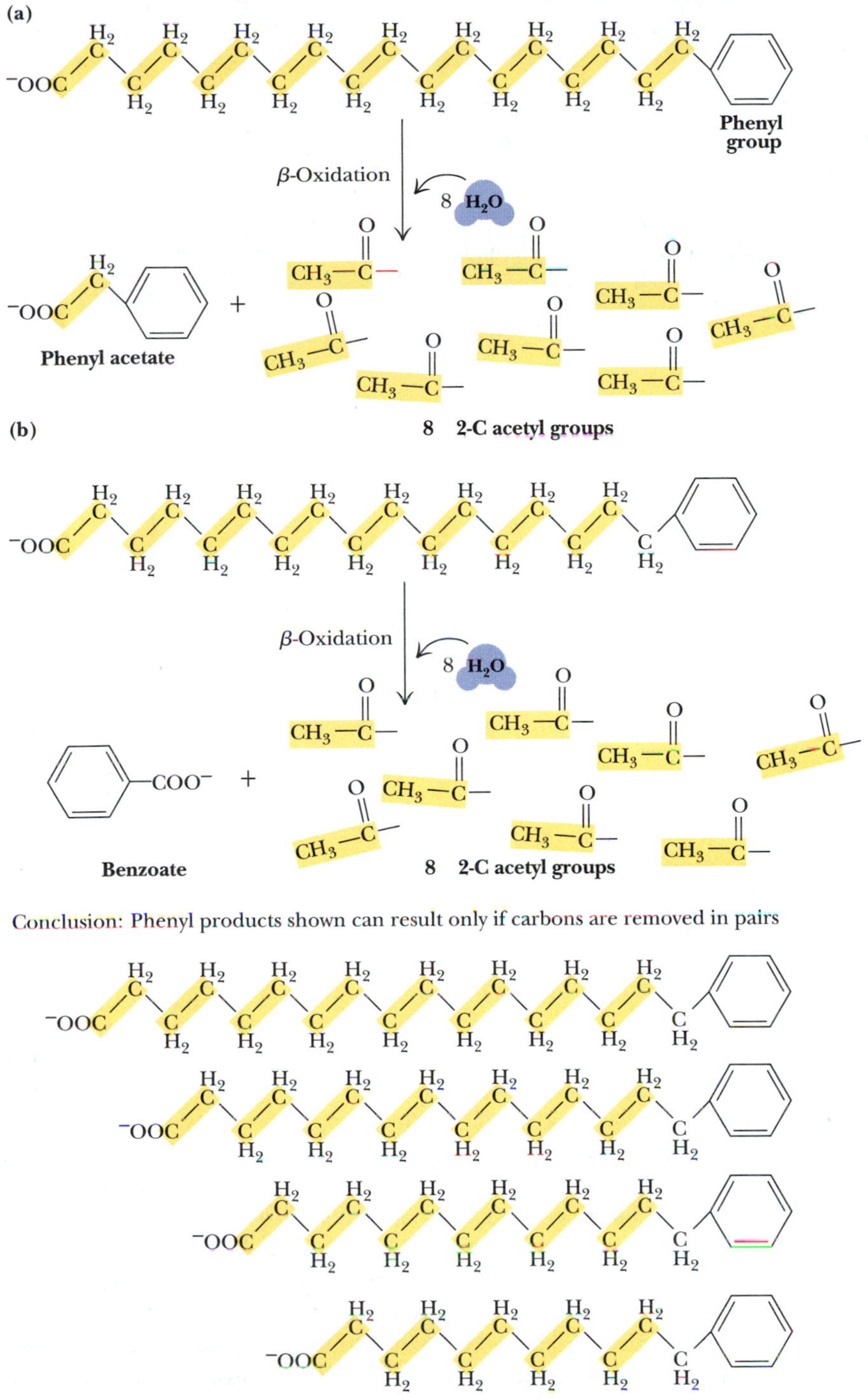

FIGURE 23.5 The oxidative breakdown of phenyl fatty acids observed by Franz Knoop. He observed that fatty acid analogs with even numbers of carbon atoms yielded phenyl acetate, whereas compounds with odd numbers of carbon atoms produced only benzoate.

ferase II on the matrix side of the inner membrane, which transfers the fatty acyl group back to CoA to re-form the fatty acyl-CoA, leaving free carnitine, which can return across the membrane via the translocase.

Several additional points should be made. First, although oxygen esters usually have lower group-transfer potentials than thiol esters, the O—acyl bonds in acylcarnitines have high group-transfer potentials, and the transesterification reactions mediated by the acyltransferases have equilibrium constants close to 1. Second, note that eukaryotic cells maintain separate pools of CoA in the mitochondria and in the cytosol. The cytosolic pool is utilized principally in fatty acid biosynthesis (see Chapter 24), and the mitochondrial pool is important in the oxidation of fatty acids and pyruvate, as well as some amino acids.

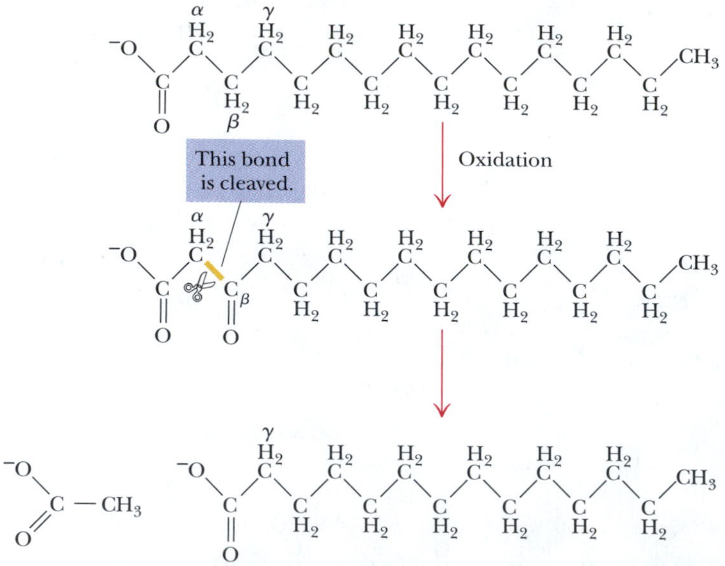

FIGURE 23.6 Fatty acids are degraded by repeated cycles of oxidation at the β-carbon and cleavage of the C_α—C_β bond to yield acetate units.

β-Oxidation Involves a Repeated Sequence of Four Reactions

For saturated fatty acids, the process of β-oxidation involves a recurring cycle of four steps, as shown in Figure 23.10. The overall strategy in the first three steps is to create a carbonyl group on the β-carbon by oxidizing the C_α—C_β bond to form an olefin, with subsequent hydration and oxidation. In essence, this cycle is directly analogous to the sequence of reactions converting succinate to oxaloacetate in the TCA cycle. The fourth reaction of the cycle cleaves the β-keto ester in a reverse Claisen condensation, producing an acetate unit and leaving a fatty acid chain that is two carbons shorter than it began. (Recall from Chapter 19 that Claisen condensations involve attack by a nucleophilic agent on a carbonyl carbon to yield a β-keto acid.)

Acyl-CoA Dehydrogenase—The First Reaction of β-Oxidation The first reaction, the oxidation of the C_α—C_β bond, is catalyzed by **acyl-CoA dehydrogenases,** a family of three soluble matrix enzymes (with molecular weights of 170 to 180 kD) that differ in their specificity for either long-, medium-, or short-chain acyl-CoAs. They carry noncovalently (but tightly) bound FAD, which is reduced during the oxidation of the fatty acid. As shown in Figure 23.11, $FADH_2$ transfers its electrons to an **electron transfer flavoprotein (ETF).** Reduced ETF is reoxidized by a specific oxidoreductase (an iron–sulfur protein), which in turn sends the electrons on to the electron-transport chain at the level of coenzyme Q. Recall from Chapter 20 that mitochondrial oxidation of FAD in this way eventually results in the net formation of about 1.5 ATPs. The mechanism of the acyl-CoA dehydrogenase (Figure 23.12) involves deprotonation of the fatty acid chain at the

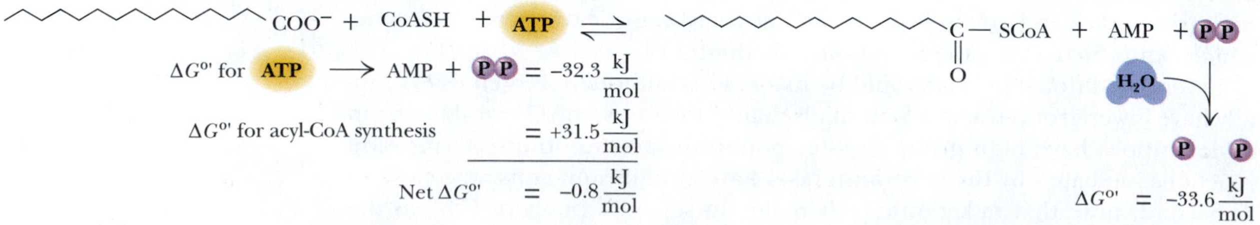

FIGURE 23.7 The acyl-CoA synthetase reaction activates fatty acids for β-oxidation. The reaction is driven by hydrolysis of ATP to AMP and pyrophosphate and by the subsequent hydrolysis of pyrophosphate.

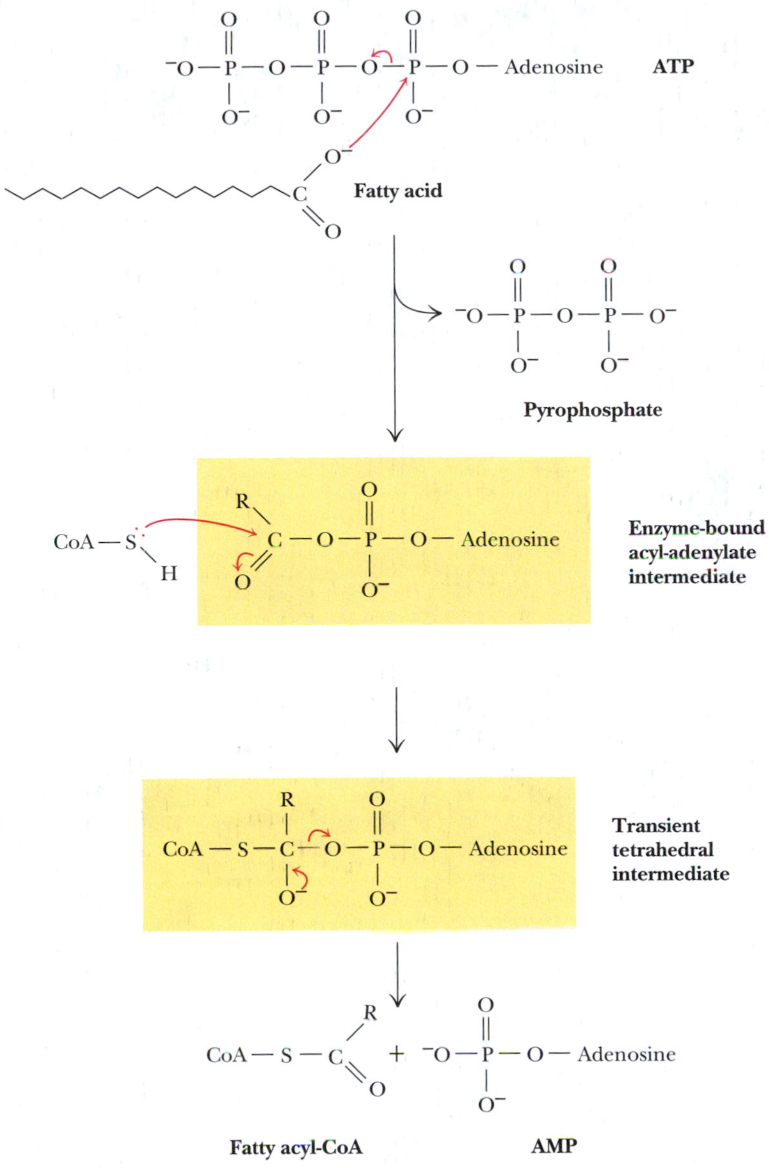

Biochemistry ⓔNow™   **ANIMATED FIGURE 23.8**
The mechanism of the acyl-CoA synthetase reaction
involves fatty acid carboxylate attack on ATP to form
an acyl-adenylate intermediate. The fatty acyl CoA
thioester product is formed by CoA attack on this
intermediate. **See this figure animated at http://
chemistry.brookscole.com/ggb3**

α-carbon, followed by hydride transfer from the β-carbon to FAD. The structure
of the medium-chain dehydrogenase from pig liver places an FAD molecule
in an extended conformation between a bundle of α-helices and a distorted
β-barrel (Figure 23.13).

A Metabolite of Hypoglycin from Akee Fruit Inhibits Acyl-CoA Dehydrogenase

The unripened fruit of the **akee tree** contains **hypoglycin,** a rare amino acid
(Figure 23.14). Metabolism of hypoglycin yields *methylenecyclopropylacetyl-CoA*
(MCPA-CoA). Acyl-CoA dehydrogenase will accept MCPA-CoA as a substrate,
removing a proton from the α-carbon to yield an intermediate that irreversibly inactivates acyl-CoA dehydrogenase by reacting covalently with FAD
on the enzyme. For this reason, consumption of unripened akee fruit can
lead to vomiting and, in severe cases, convulsions, coma, and death. The condition is most severe in individuals with low levels of acyl-CoA dehydrogenase.

Enoyl-CoA Hydratase Adds Water Across the Double Bond   The next step in β-
oxidation is the addition of the elements of H_2O across the new double bond
in a stereospecific manner, yielding the corresponding hydroxyacyl-CoA

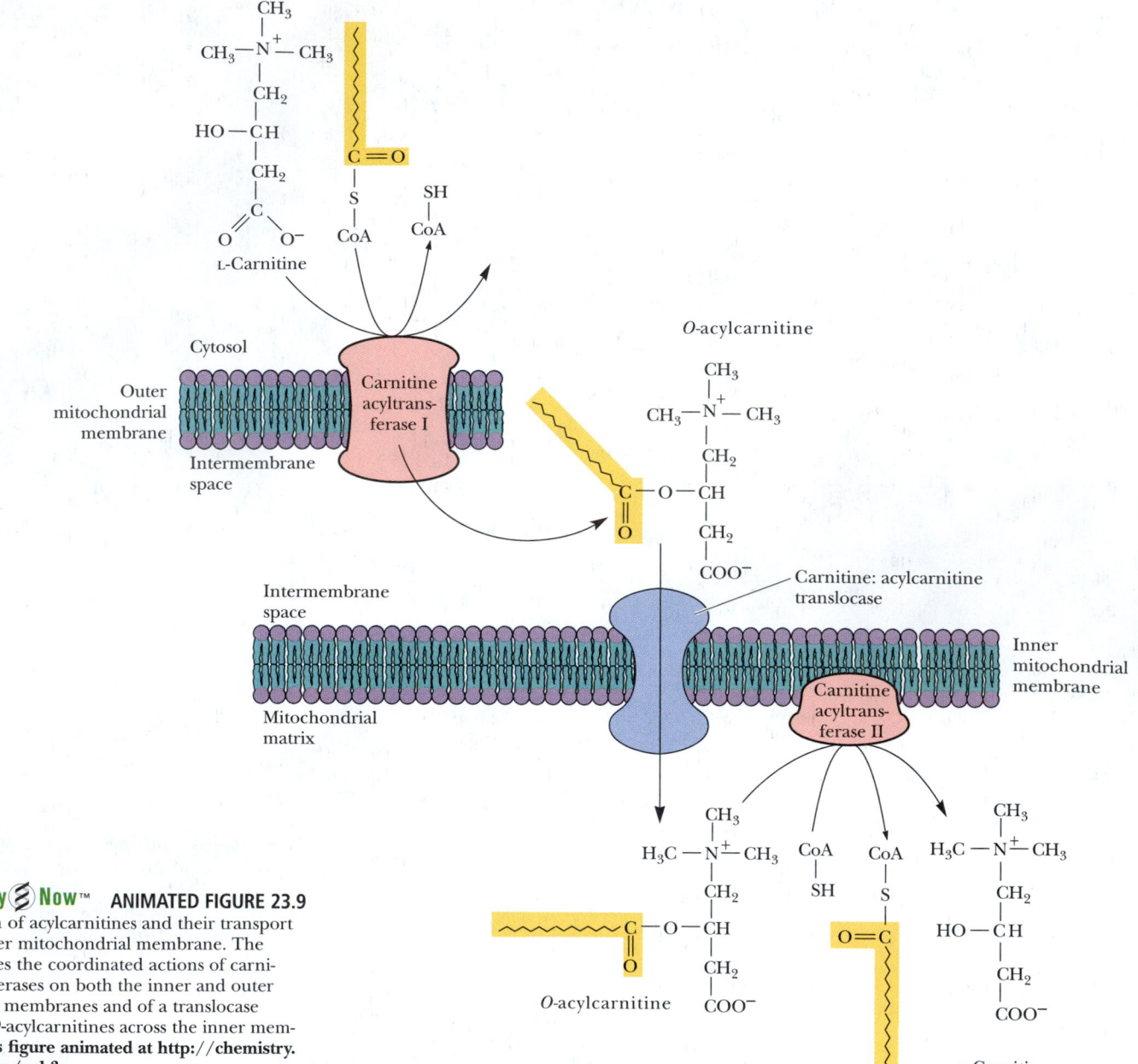

Biochemistry⬧Now™ ANIMATED FIGURE 23.9
The formation of acylcarnitines and their transport
across the inner mitochondrial membrane. The
process involves the coordinated actions of carnitine acyltransferases on both the inner and outer
mitochondrial membranes and of a translocase
that shuttles *O*-acylcarnitines across the inner membrane. **See this figure animated at http://chemistry.
brookscole.com/ggb3**

(Figure 23.15). The reaction is catalyzed by **enoyl-CoA hydratase.** A number
of different enoyl-CoA hydratase activities have been detected in various tissues. Also called **crotonases,** these enzymes specifically convert *trans*-enoyl-
CoA derivatives to L-β-hydroxyacyl-CoA. As shown in Figure 23.15, these enzymes will also metabolize *cis*-enoyl-CoA (at slower rates) to give specifically
D-β-hydroxyacyl-CoA. In addition, there is a novel enoyl-CoA hydratase that
converts *trans*-enoyl-CoA to D-β-hydroxyacyl-CoA, as shown in Figure 23.15.

L-Hydroxyacyl-CoA Dehydrogenase Oxidizes the β-Hydroxyl Group The third
reaction of this cycle is the oxidation of the hydroxyl group at the β-position
to produce a β-ketoacyl-CoA derivative. This second oxidation reaction is catalyzed by **L-hydroxyacyl-CoA dehydrogenase,** an enzyme that requires NAD$^+$
as a coenzyme. NADH produced in this reaction represents metabolic energy.
Each NADH produced in mitochondria by this reaction drives the synthesis of
2.5 molecules of ATP in the electron-transport pathway. L-Hydroxyacyl-CoA
dehydrogenase shows absolute specificity for the L-hydroxyacyl isomer of the

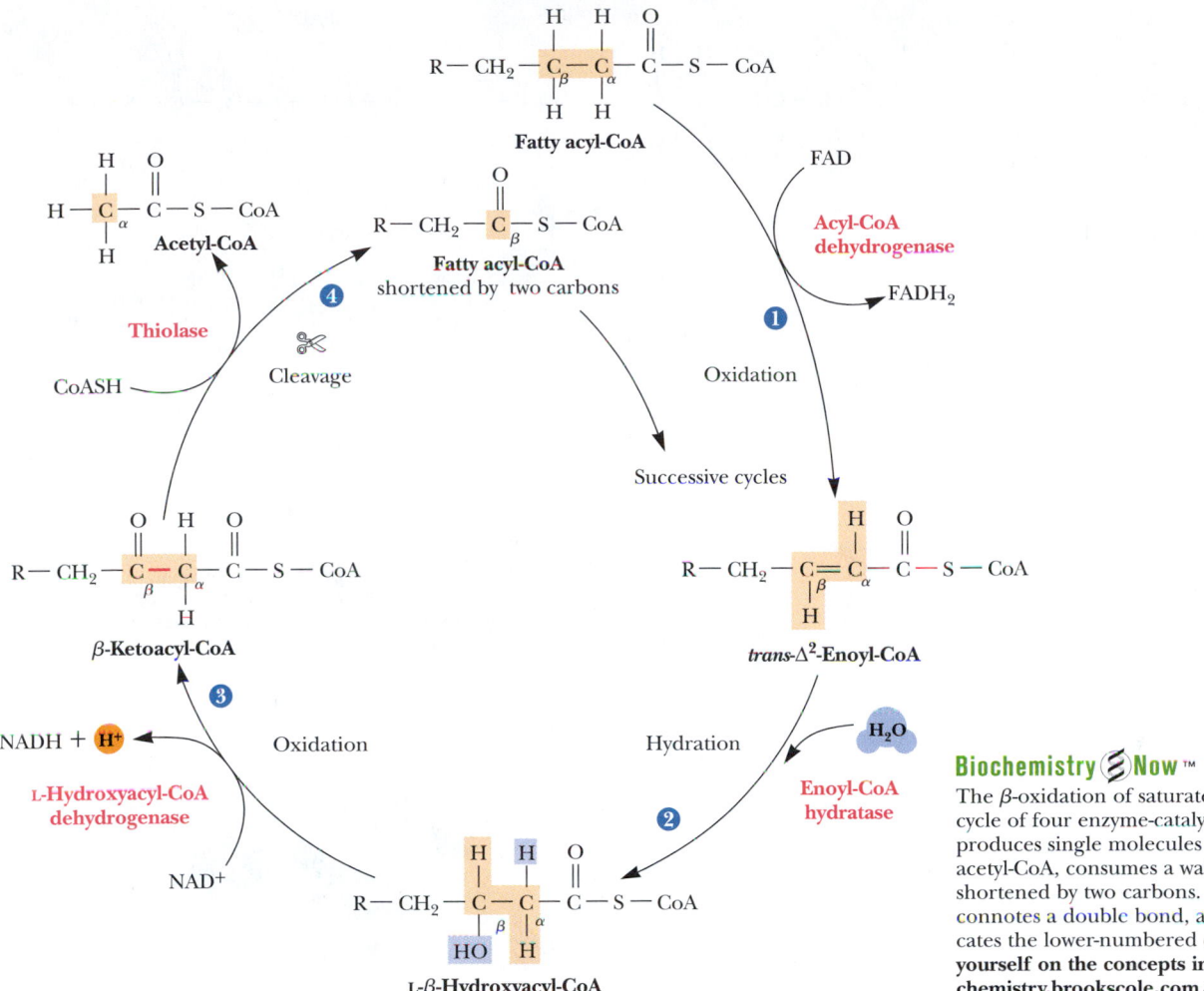

Biochemistry ⒺNow™ ACTIVE FIGURE 23.10
The β-oxidation of saturated fatty acids involves a cycle of four enzyme-catalyzed reactions. Each cycle produces single molecules of FADH₂, NADH, and acetyl-CoA, consumes a water, and yields a fatty acid shortened by two carbons. (The delta [Δ] symbol connotes a double bond, and its superscript indicates the lower-numbered carbon involved.) **Test yourself on the concepts in this figure at http://chemistry.brookscole.com/ggb3**

substrate (Figure 23.16). (D-Hydroxyacyl isomers, which arise mainly from oxidation of unsaturated fatty acids, are handled differently.)

β-Ketoacyl-CoA Intermediates Are Cleaved in the Thiolase Reaction

The final step in the β-oxidation cycle is the cleavage of the β-ketoacyl-CoA. This reaction, catalyzed by **thiolase** (also known as **β-ketothiolase**), involves the attack of a cysteine thiolate from the enzyme on the β-carbonyl carbon, followed by

Biochemistry ⒺNow™ ACTIVE FIGURE 23.11
The acyl-CoA dehydrogenase reaction. The two electrons removed in this oxidation reaction are delivered to the electron-transport chain in the form of reduced coenzyme Q (UQH₂). **Test yourself on the concepts in this figure at http://chemistry.brookscole.com/ggb3**

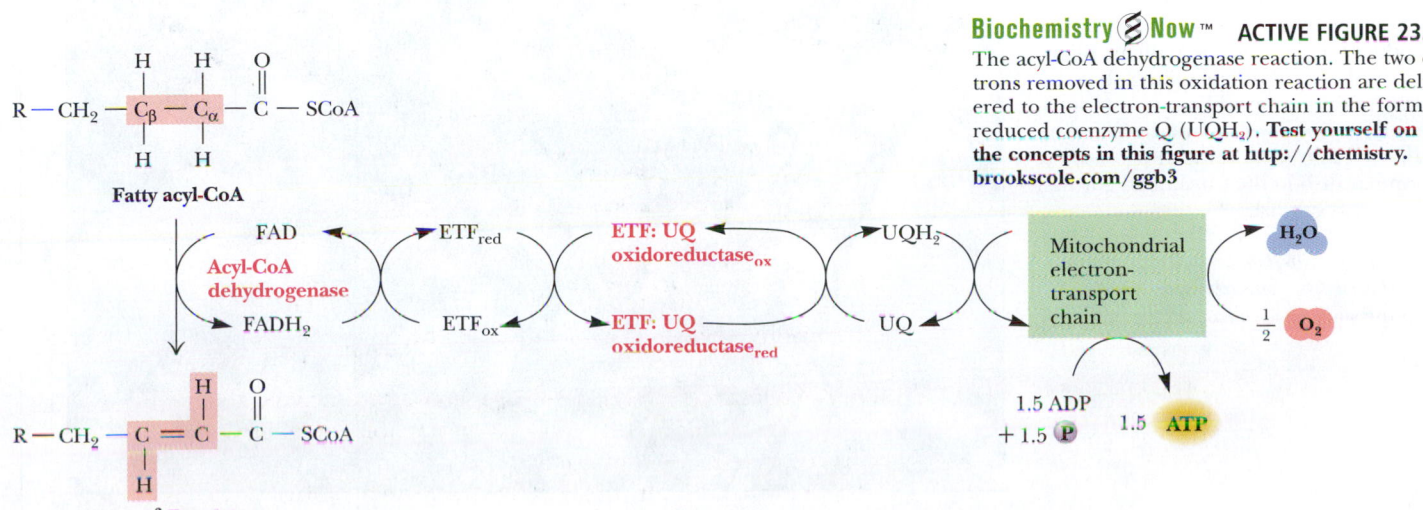

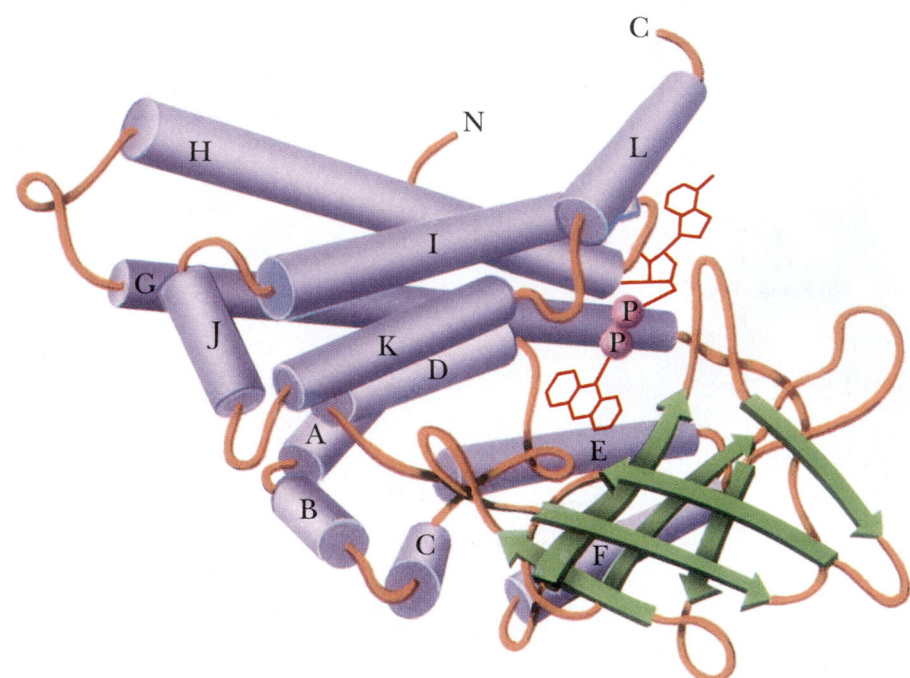

FIGURE 23.12 The mechanism of acyl-CoA dehydrogenase. Removal of a proton from the α-C is followed by hydride transfer from the β-carbon to FAD.

FIGURE 23.13 The subunit structure of medium-chain acyl-CoA dehydrogenase from pig liver mitochondria. Note the location of the bound FAD (red). *(Adapted from Kim, J-T., and Wu, J., 1988. Structure of the medium-chain acyl-CoA dehydrogenase from pig liver mitochondria at 3-Å resolution.* Proceedings of the National Academy of Sciences, U.S.A. *85:6671–6681.)*

A Deeper Look

The Akee Tree

The akee (also spelled *ackee*) tree is native to West Africa and was brought to the Caribbean by African slaves. It was introduced to science by William Bligh, captain of the infamous sailing ship the *Bounty,* and its botanical name is (appropriately) *Blighia sapida* (the latter name from the Latin *sapidus* meaning "tasty"). A popular dish in the Caribbean consists of akee and salt fish.

"Akee, rice, salt fish are nice,
And the rum is fine any time of year."
From the song *Jamaica Farewell*

R. R. Head/Earth Sciences/Animals, Animals

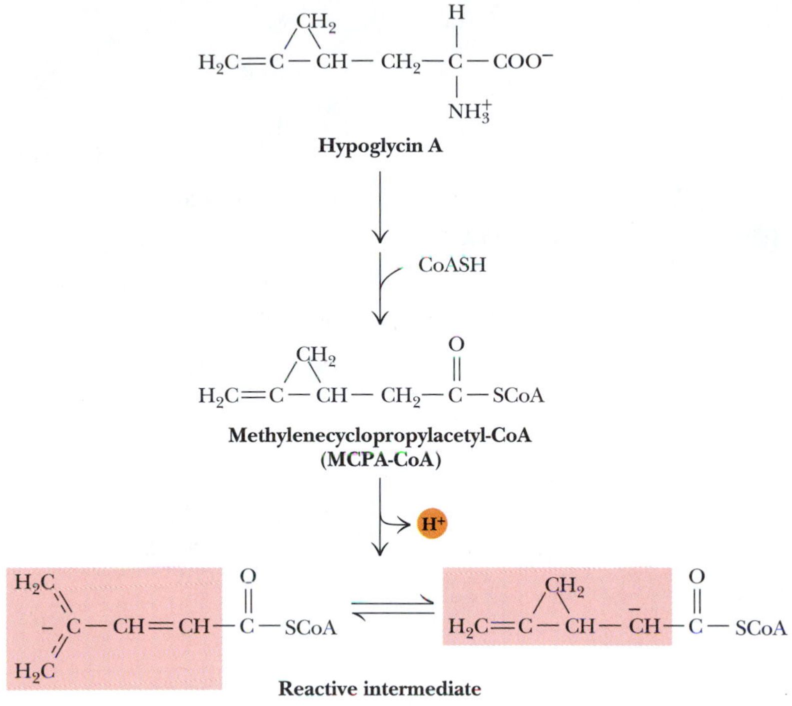

Hypoglycin A

CoASH

**Methylenecyclopropylacetyl-CoA
(MCPA-CoA)**

H⁺

Reactive intermediate

FIGURE 23.14 The conversion of hypoglycin from akee fruit to a form that inhibits acyl-CoA dehydrogenase.

trans-Enoyl-CoA

H₂O

Crotonase

L-β-Hydroxyacyl-CoA

cis-Enoyl-CoA

H₂O

Crotonase

D-β-Hydroxyacyl-CoA

trans-Enoyl-CoA

H₂O

D-β-Hydroxyacyl-CoA

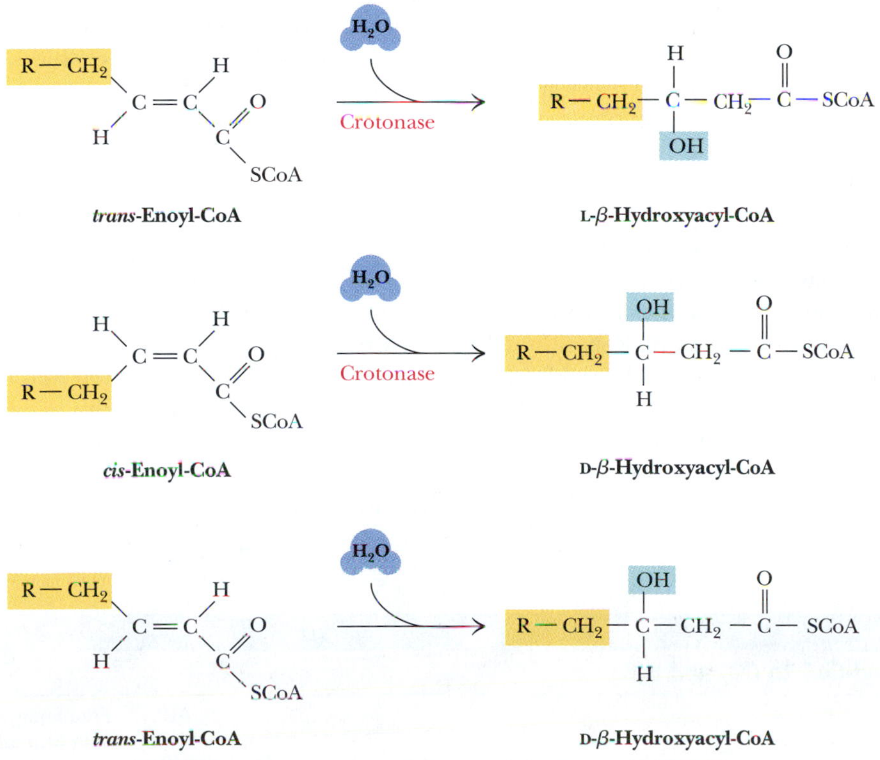

FIGURE 23.15 The conversion of *trans*- and *cis*-enoyl CoA derivatives to L- and D-β-hydroxyacyl CoA, respectively. These reactions are catalyzed by enoyl-CoA hydratases (also called crotonases), enzymes that vary in their acyl-chain length specificity. There is also an enzyme that converts *trans*-enoyl-CoA directly to D-β-hydroxyacyl-CoA.

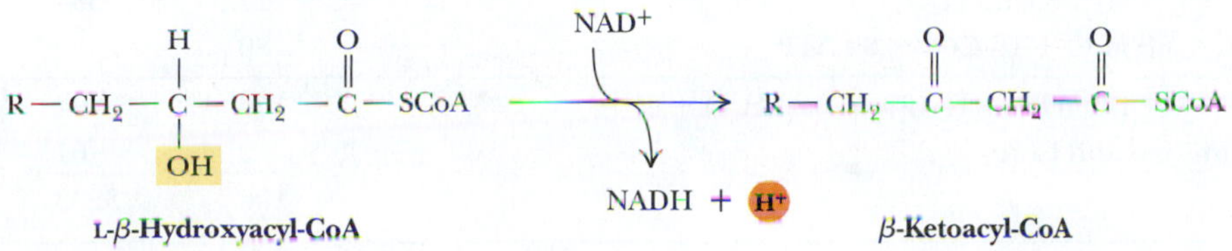

NAD⁺

NADH + H⁺

L-β-Hydroxyacyl-CoA

β-Ketoacyl-CoA

FIGURE 23.16 The L-β-hydroxyacyl-CoA dehydrogenase reaction.

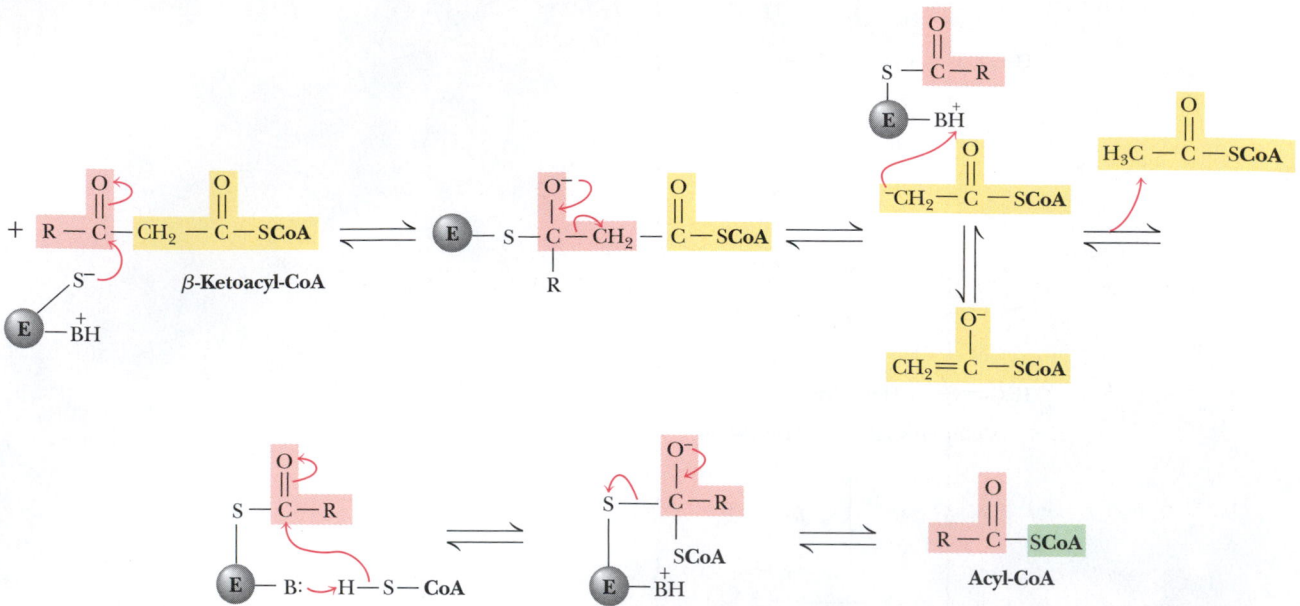

FIGURE 23.17 The mechanism of the thiolase reaction. Attack by an enzyme cysteine thiolate group at the β-carbonyl carbon produces a tetrahedral intermediate, which decomposes with departure of acetyl-CoA, leaving an enzyme thioester intermediate. Attack by the thiol group of a second CoA yields a new (shortened) acyl-CoA.

cleavage to give the enolate of acetyl-CoA and an enzyme-thioester intermediate (Figure 23.17). Subsequent attack by the thiol group of a second CoA and departure of the cysteine thiolate yields a new (shorter) acyl-CoA. If the reaction in Figure 23.17 is read in reverse, it is easy to see that it is a Claisen condensation—an attack of the enolate anion of acetyl-CoA on a thioester. Despite the formation of a second thioester, this reaction has a very favorable K_{eq}, and it drives the three previous reactions of β-oxidation.

Repetition of the β-Oxidation Cycle Yields a Succession of Acetate Units

Biochemistry Now™   Go to BiochemistryNow and click BiochemistryInteractive to discover the main functions of coenzyme A.

In essence, this series of four reactions has yielded a fatty acid (as a CoA ester) that has been shortened by two carbons and one molecule of acetyl-CoA. The shortened fatty acyl-CoA can now go through another β-oxidation cycle, as shown in Figure 23.10. Repetition of this cycle with a fatty acid with an even number of carbons eventually yields two molecules of acetyl-CoA in the final step. As noted in the first reaction in Table 23.2, complete β-oxidation of palmitic acid yields eight molecules of acetyl-CoA as well as seven molecules of $FADH_2$ and seven molecules of NADH. The acetyl-CoA can be further metabolized in the TCA cycle (as we have already seen). Alternatively, acetyl-CoA can also be used as

Table 23.2

Equations for the Complete Oxidation of Palmitoyl-CoA to CO_2 and H_2O

Equation	ATP Yield	Free Energy Yield (kJ/mol)
$CH_3(CH_2)_{14}CO\text{-}CoA + 7\ [FAD] + 7\ H_2O + 7\ NAD^+ + 7\ CoA \longrightarrow 8\ CH_3CO\text{-}CoA + 7\ [FADH_2] + 7\ NADH + 7\ H^+$		
$7\ [FADH_2] + 10.5\ P_i + 10.5\ ADP + 3.5\ O_2 \longrightarrow 7\ [FAD] + 17.5\ H_2O + 10.5\ ATP$	10.5	320
$7\ NADH + 7\ H^+ + 17.5\ P_i + 17.5\ ADP + 3.5\ O_2 \longrightarrow 7\ NAD^+ + 24.5\ H_2O + 17.5\ ATP$	17.5	534
$8\ Acetyl\text{-}CoA + 16\ O_2 + 80\ ADP + 80\ P_i \longrightarrow 8\ CoA + 88\ H_2O + 16\ CO_2 + 80\ ATP$	80	2440
$CH_3\text{—}(CH_2)_{14}CO\text{-}CoA + 108\ P_i + 108\ ADP + 23\ O_2 \longrightarrow 108\ ATP + 16\ CO_2 + 123\ H_2O + CoA$	108	3294
Energetic "cost" of forming palmitoyl-CoA from palmitate and CoA	−2	−61
Total	106	3233

a substrate in amino acid biosynthesis (see Chapter 25). As noted in Chapter 22, however, acetyl-CoA cannot be used as a substrate for gluconeogenesis.

Complete β-Oxidation of One Palmitic Acid Yields 106 Molecules of ATP

If the acetyl-CoA is directed entirely to the TCA cycle in mitochondria, it can eventually generate approximately ten high-energy phosphate bonds—that is, ten molecules of ATP synthesized from ADP (Table 23.2). Including the ATP formed from $FADH_2$ and NADH, complete β-oxidation of a molecule of palmitoyl-CoA in mitochondria yields 108 molecules of ATP. Subtracting the two high-energy bonds needed to form palmitoyl-CoA, the substrate for β-oxidation, one concludes that β-oxidation of a molecule of palmitic acid yields 106 molecules of ATP. The $\Delta G°'$ for complete combustion of palmitate to CO_2 is -9790 kJ/mol. The hydrolytic energy embodied in 106 ATPs is 106×30.5 kJ/mol $= 3233$ kJ/mol, so the overall efficiency of β-oxidation under standard-state conditions is approximately 33%. The large energy yield from fatty acid oxidation is a reflection of the highly reduced state of the carbon in fatty acids. Sugars, in which the carbon is already partially oxidized, produce much less energy, carbon for carbon, than do fatty acids. The breakdown of fatty acids is regulated by a variety of metabolites and hormones. Details of this regulation are described in Chapter 24, following a discussion of fatty acid synthesis.

Migratory Birds Travel Long Distances on Energy from Fatty Acid Oxidation

Because they represent the most highly concentrated form of stored biological energy, fatty acids are the metabolic fuel of choice for sustaining the incredibly long flights of many migratory birds. Although some birds migrate over landmasses and eat frequently, other species fly long distances without stopping to eat. The American golden plover flies directly from Alaska to Hawaii, a 3300-km flight requiring 35 hours (at an average speed of nearly 60 miles/hr) and more than 250,000 wing beats! The ruby-throated hummingbird, which winters in Central America and nests in southern Canada, often flies nonstop across the Gulf of Mexico. These and similar birds accomplish these prodigious feats by storing large amounts of fatty acids (as triacylglycerols) in the days before their migratory flights. The percentage of dry-weight body fat in these birds may be as high as 70% when migration begins (compared with values of 30% and less for nonmigratory birds).

Fatty Acid Oxidation Is an Important Source of Metabolic Water for Some Animals

Large amounts of metabolic water are generated by β-oxidation (123 H_2O per palmitoyl-CoA). For certain animals—including desert animals (such as gerbils) and killer whales (which do not drink seawater)—the oxidation of fatty acids can be a significant source of dietary water. A striking example is the camel (Figure 23.18), whose hump is essentially a large deposit of fat. Metabolism of fatty acids from this store provides needed water (as well as metabolic energy) during periods when drinking water is not available. It might well be said that "the ship of the desert" sails on its own metabolic water!

23.3 | How Are Odd-Carbon Fatty Acids Oxidized?

β-Oxidation of Odd-Carbon Fatty Acids Yields Propionyl-CoA

Fatty acids with odd numbers of carbon atoms are rare in mammals but fairly common in plants and marine organisms. Humans and animals whose diets include these food sources metabolize odd-carbon fatty acids via the β-oxidation pathway.

The final product of β-oxidation in this case is the three-carbon propionyl-CoA instead of acetyl-CoA. Three specialized enzymes then carry out the reactions that convert propionyl-CoA to succinyl-CoA, a TCA cycle intermediate. (Because propionyl-CoA is a degradation product of methionine, valine, and isoleucine, this sequence of reactions is also important in amino acid catabolism, as we shall see in Chapter 25.) The pathway involves an initial carboxylation at the α-carbon of propionyl-CoA to produce D-methylmalonyl-CoA (Figure 23.19). The reaction is catalyzed by a biotin-dependent enzyme, propionyl-CoA carboxylase. The mechanism involves ATP-driven carboxylation of biotin at N_1, followed by nucleophilic attack by the α-carbanion of propionyl-CoA in a stereospecific manner.

D-Methylmalonyl-CoA, the product of this reaction, is converted to the L-isomer by methylmalonyl-CoA epimerase (Figure 23.19). (This enzyme has often and incorrectly been called "methylmalonyl-CoA racemase." It is not a racemase because the CoA moiety contains five other asymmetric centers.) The epimerase reaction involves a carbanion at the α-position formed via a reversible dissociation of the acidic α-proton (Figure 23.20). The L-isomer is the substrate for methylmalonyl-CoA mutase. Methylmalonyl-CoA epimerase is an impressive catalyst. The pK_a for the proton that must dissociate to initiate this reaction is approximately 21! If binding of a proton to the α-anion is diffusion limited, with $k_{on} = 10^9\ M^{-1}\ sec^{-1}$, then the initial proton dissociation must be rate limiting and the rate constant must be

$$k_{off} = K_a \cdot k_{on} = (10^{-21}\ M) \cdot (10^9\ M^{-1}\ sec^{-1}) = 10^{-12}\ sec^{-1}$$

The turnover number of methylmalonyl-CoA epimerase is 100 sec^{-1}, and thus the enzyme enhances the reaction rate by a factor of 10^{14}.

A B$_{12}$-Catalyzed Rearrangement Yields Succinyl-CoA from L-Methylmalonyl-CoA

The third reaction, catalyzed by **methylmalonyl-CoA mutase,** is quite unusual because it involves a migration of the carbonyl-CoA group from one carbon to its neighbor (Figure 23.21). The mutase reaction is vitamin B$_{12}$–dependent and begins with **homolytic cleavage** of the Co^{3+}—C bond in cobalamin, reducing the cobalt to Co^{2+}. Transfer of a hydrogen atom from the substrate to the deoxyadenosyl group produces a methylmalonyl-CoA radical, which then can undergo a classic B$_{12}$-catalyzed rearrangement to yield a succinyl-CoA radical. Hydrogen transfer from the deoxyadenosyl group yields succinyl-CoA and regenerates the B$_{12}$ coenzyme.

(a) Gerbil

(b) Ruby-throated hummingbird

(c) Golden plover

(d) Orca

(e) Camels

FIGURE 23.18 Animals whose existence is strongly dependent on fatty acid oxidation: **(a)** gerbil, **(b)** ruby-throated hummingbird, **(c)** golden plover, **(d)** orca (killer whale), and **(e)** camels.

Human Biochemistry

Metabolic Therapy for the Treatment of Heart Disease

Myocardial ischemia is a condition of reduced oxygen in heart muscle—as might occur during and after a heart attack. During ischemia, ATP production in mitochondria decreases and the heart responds with accelerated anaerobic glycolysis, lactate accumulation, and cell acidosis. Classic drug therapy for ischemia typically involves the use of long-acting nitrates or Ca^{2+}-channel antagonists to increase oxygen delivery, or β-blockers and Ca^{2+}-channel antagonists to reduce blood pressure and heart rate. However, animal studies have shown that fatty acids are the primary mitochondrial substrate during moderately severe ischemia and that they inhibit carbohydrate oxidation and drive pyruvate to lactate. Drugs that inhibit myocardial fatty acid oxidation increase carbohydrate oxidation, resulting in reduced lactate production and a higher cell pH during ischemia. The first

drug to be used for this purpose was trimetazidine (1-[2,3,4-trimethoxybenzyl]-piperazine). Trimetazidine selectively inhibits β-ketothiolase in the β-oxidation pathway. It reduces symptoms of angina pectoris (chest pain) and aids in the recovery from ischemic stress.

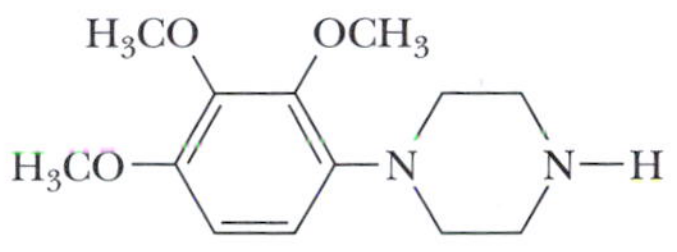

Trimetazidine

A Deeper Look

The Activation of Vitamin B_{12}

Conversion of inactive vitamin B_{12} to active *5'-deoxyadenosylcobalamin* involves three steps (see accompanying figure). Two flavoprotein reductases sequentially convert Co^{3+} in cyanocobalamin to the Co^{2+} state and then to the Co^{+} state. Co^{+} is an extremely powerful nucleophile. It attacks the C-5' carbon of ATP as shown, expelling the triphosphate anion to form 5'-deoxyadenosylcobalamin.

Because two electrons from Co^{+} are donated to the Co—carbon bond, the oxidation state of cobalt reverts to Co^{3+} in the active coenzyme. This is one of only two known adenosyl transfers (that is, nucleophilic attack on the ribose 5'-carbon of ATP) in biological systems. (The other is the formation of *S*-adenosylmethionine; see Chapter 25.)

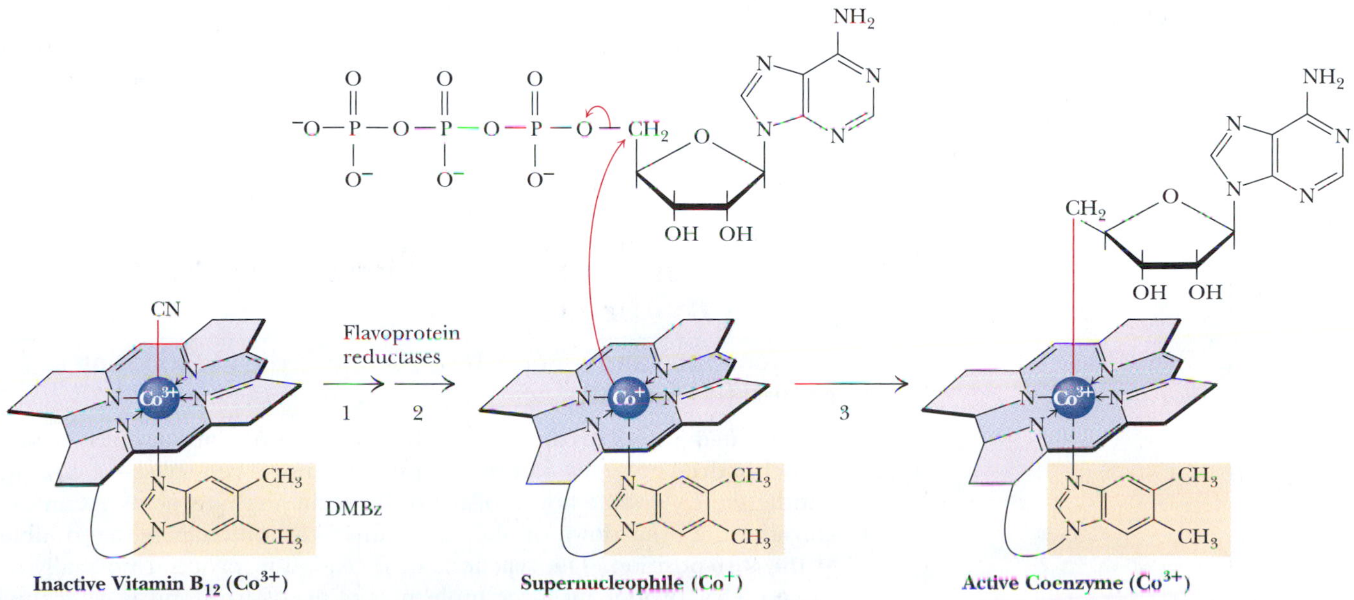

Formation of the active coenzyme 5'-deoxyadenosylcobalamin from inactive vitamin B_{12} is initiated by the action of flavoprotein reductases. The resulting Co^{+} species, dubbed a supernucleophile, attacks the 5'-carbon of ATP in an unusual adenosyl transfer.

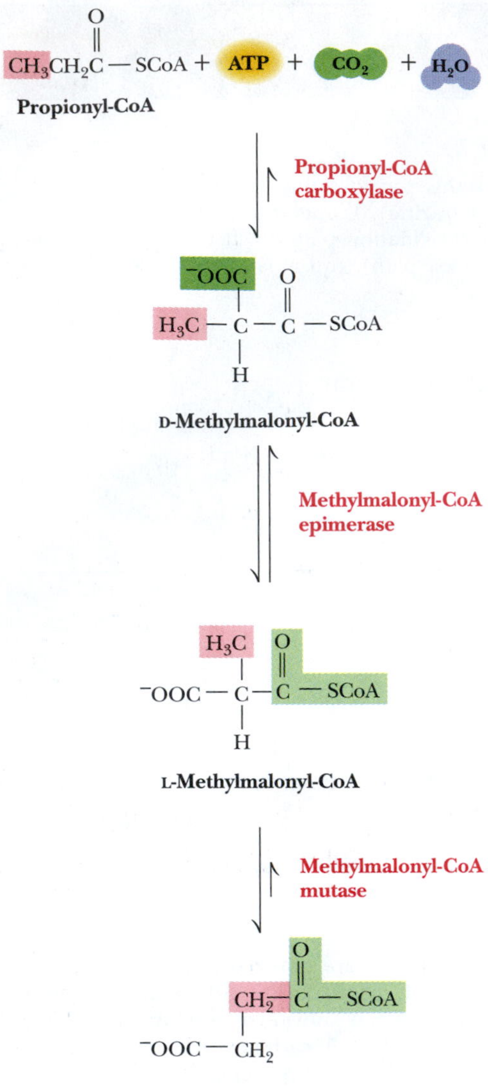

Biochemistry🌐Now™ ACTIVE FIGURE 23.19
The conversion of propionyl-CoA (formed from β-oxidation of odd-carbon fatty acids) to succinyl-CoA is carried out by a trio of enzymes, as shown. Succinyl-CoA can enter the TCA cycle. **Test yourself on the concepts in this figure at http://chemistry. brookscole.com/ggb3**

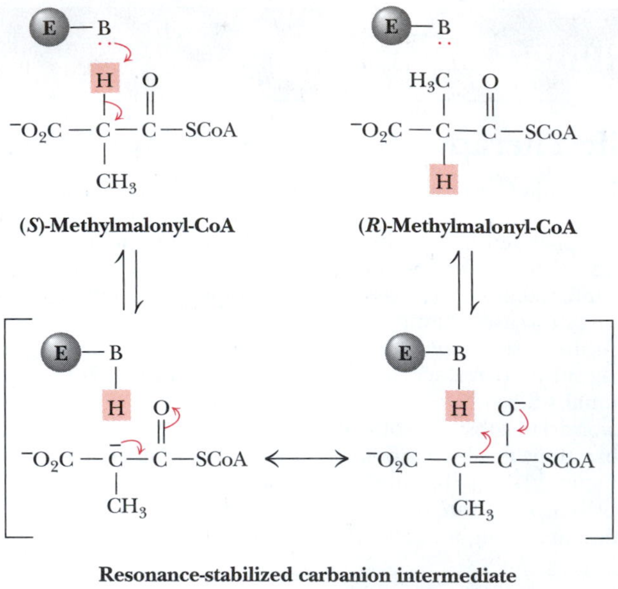

FIGURE 23.20 The methylmalonyl-CoA epimerase mechanism involves a resonance-stabilized carbanion at the α-position.

Net Oxidation of Succinyl-CoA Requires Conversion to Acetyl-CoA

Succinyl-CoA derived from propionyl-CoA can enter the TCA cycle. Oxidation of succinate to oxaloacetate provides a substrate for glucose synthesis. Thus, although the acetate units produced in β-oxidation cannot be utilized in gluconeogenesis by animals, the occasional propionate produced from oxidation of odd-carbon fatty acids *can* be used for sugar synthesis. Alternatively, succinate introduced to the TCA cycle from odd-carbon fatty acid oxidation may be oxidized to CO_2. However, all of the four-carbon intermediates in the TCA cycle are regenerated in the cycle and thus should be viewed as catalytic species. Net consumption of succinyl-CoA thus does not occur directly in the TCA cycle. Rather, the succinyl-CoA generated from β-oxidation of odd-carbon fatty acids must be converted to pyruvate and then to acetyl-CoA (which is completely oxidized in the TCA cycle). To follow this latter route, succinyl-CoA entering the TCA cycle must be first converted to malate in the usual way and then transported from the mitochondrial matrix to the cytosol, where it is oxidatively decarboxylated to pyruvate and CO_2 by **malic enzyme**, as shown in Figure 23.22. Pyruvate can then be transported back to the mitochondrial matrix, where it enters the TCA cycle via pyruvate dehydrogenase. Note that malic enzyme plays an important role in fatty acid synthesis (see Figure 24.1).

23.4 How Are Unsaturated Fatty Acids Oxidized?

An Isomerase and a Reductase Facilitate the β-Oxidation of Unsaturated Fatty Acids

Unsaturated fatty acids are also catabolized by β-oxidation, but two additional mitochondrial enzymes—an isomerase and a novel reductase—are required to handle the *cis* double bonds of naturally occurring fatty acids. As an example, consider the breakdown of oleic acid, an 18-carbon chain with a double bond at the 9,10-position. The reactions of β-oxidation proceed normally through three cycles, producing three molecules of acetyl-CoA and leaving the degradation product *cis*-Δ³-dodecenoyl-CoA, shown in Figure 23.23. This intermediate is not a substrate for acyl-CoA dehydrogenase. With a double bond at the 3,4-position, it is not possible to form another double bond at the 2,3- (or β-)

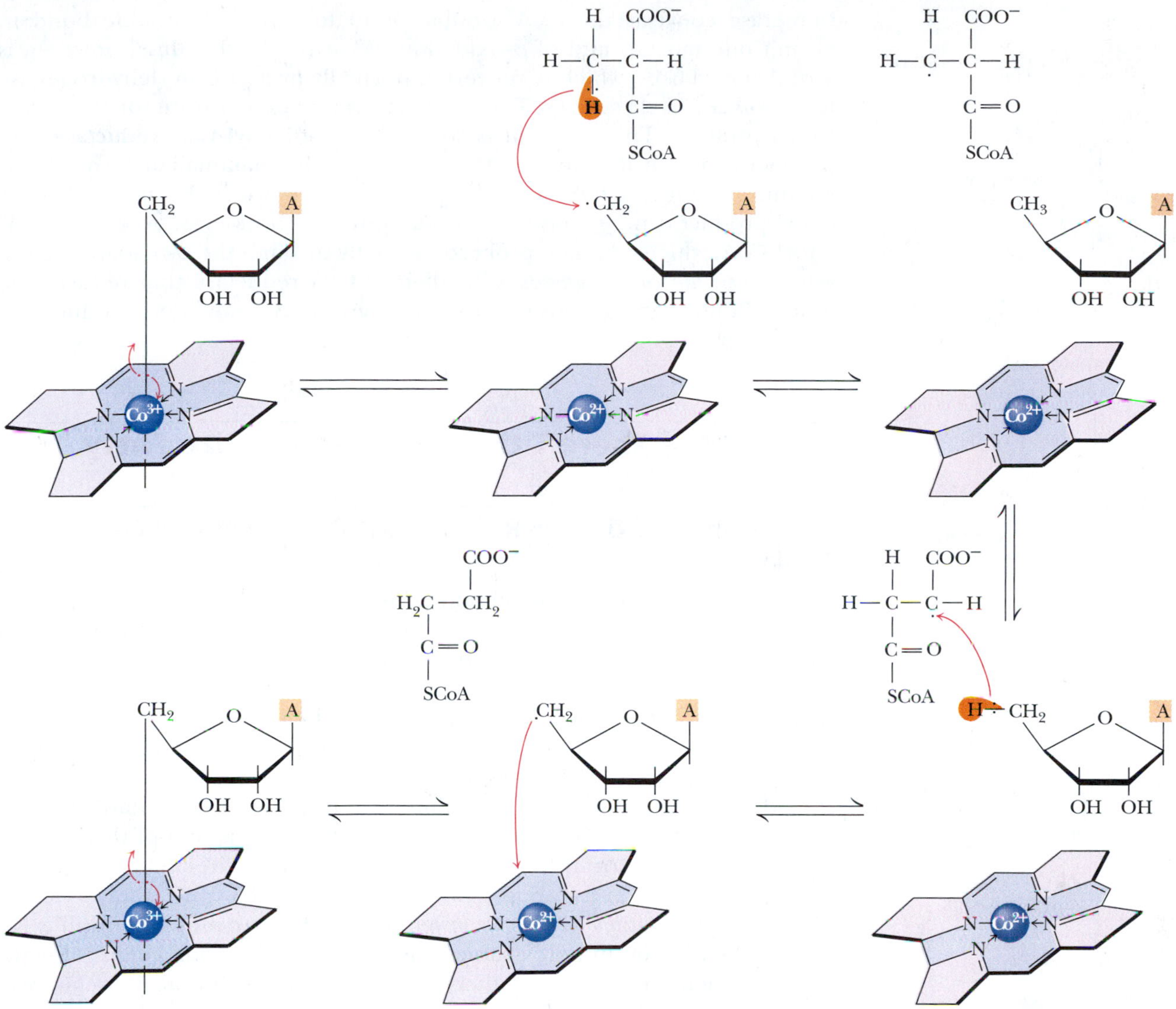

FIGURE 23.21 A mechanism for the methylmalonyl-CoA mutase reaction. In the first step, Co³⁺ is reduced to Co²⁺ due to homolytic cleavage of the Co³⁺—C bond in cobalamin. Hydrogen atom transfer from methylmalonyl-CoA yields a methylmalonyl-CoA radical that can undergo rearrangement to form a succinyl-CoA radical. Transfer of an H atom regenerates the coenzyme and yields succinyl-CoA.

position. As shown in Figure 23.23, this problem is solved by **enoyl-CoA isomerase,** an enzyme that rearranges this *cis-Δ³* double bond to a *trans-Δ²* double bond. This latter species can proceed through the normal route of β-oxidation.

Degradation of Polyunsaturated Fatty Acids Requires 2,4-Dienoyl-CoA Reductase

Polyunsaturated fatty acids pose a slightly more complicated situation for the cell. Consider, for example, the case of linoleic acid shown in Figure 23.24. As with oleic acid, β-oxidation proceeds through three cycles, and enoyl-CoA

Biochemistry⊘Now™ Go to BiochemistryNow and click BiochemistryInteractive to interact with a mechanism for the methylmalonyl mutase reaction.

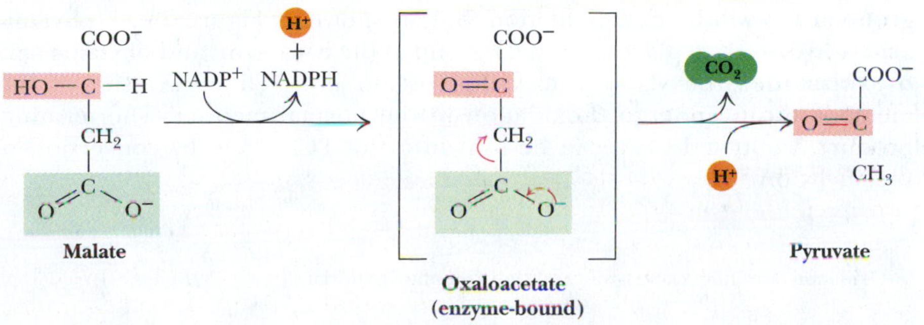

FIGURE 23.22 The malic enzyme reaction proceeds by oxidation of malate to oxaloacetate, followed by decarboxylation to yield pyruvate.

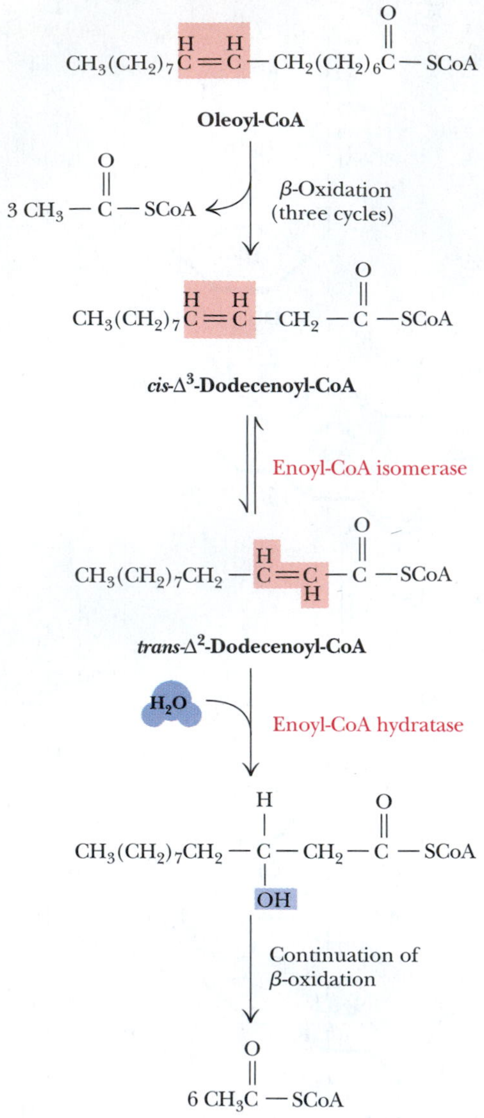

$$CH_3(CH_2)_7 \overset{H}{C} = \overset{H}{C} - CH_2(CH_2)_6 \overset{O}{\overset{\|}{C}} - SCoA$$

Oleoyl-CoA

$$3\ CH_3 - \overset{O}{\overset{\|}{C}} - SCoA \longleftarrow \begin{array}{c} \beta\text{-Oxidation} \\ \text{(three cycles)} \end{array}$$

$$CH_3(CH_2)_7 \overset{H}{C} = \overset{H}{C} - CH_2 - \overset{O}{\overset{\|}{C}} - SCoA$$

cis-Δ^3-**Dodecenoyl-CoA**

Enoyl-CoA isomerase

$$CH_3(CH_2)_7CH_2 - \overset{H}{\underset{H}{\overset{\|}{C}} = C} - \overset{O}{\overset{\|}{C}} - SCoA$$

trans-Δ^2-**Dodecenoyl-CoA**

H_2O

Enoyl-CoA hydratase

$$CH_3(CH_2)_7CH_2 - \overset{H}{\underset{OH}{\overset{|}{C}}} - CH_2 - \overset{O}{\overset{\|}{C}} - SCoA$$

Continuation of
β-oxidation

$$6\ CH_3 \overset{O}{\overset{\|}{C}} - SCoA$$

FIGURE 23.23 β-Oxidation of unsaturated fatty acids. In the case of oleoyl-CoA, three β-oxidation cycles produce three molecules of acetyl-CoA and leave *cis*-Δ^3-dodecenoyl-CoA. Rearrangement of enoyl-CoA isomerase gives the *trans*-Δ^2 species, which then proceeds normally through the β-oxidation pathway.

isomerase converts the *cis*-Δ^3 double bond to a *trans*-Δ^2 double bond to permit one more round of β-oxidation. What results this time, however, is a *cis*-Δ^4 enoyl-CoA, which is converted normally by acyl-CoA dehydrogenase to a *trans*-Δ^2, *cis*-Δ^4 species. This, however, is a poor substrate for the enoyl-CoA hydratase. This problem is solved by **2,4-dienoyl-CoA reductase,** the product of which depends on the organism. The mammalian form of this enzyme produces a *trans*-Δ^3 enoyl product, as shown in Figure 23.24; this enoyl product can be converted by an enoyl-CoA isomerase to the *trans*-Δ^2 enoyl-CoA, which can then proceed normally through the β-oxidation pathway. *Escherichia coli* possesses a 2,4-dienoyl-CoA reductase that reduces the double bond at the 4,5-position to yield the *trans*-Δ^2 enoyl-CoA product in a single step.

23.5 Are There Other Ways to Oxidize Fatty Acids?

Peroxisomal β-Oxidation Requires FAD-Dependent Acyl-CoA Oxidase

Although β-oxidation in mitochondria[1] is the principal pathway of fatty acid catabolism, several other minor pathways play important roles in fat catabolism. For example, organelles other than mitochondria, including *peroxisomes* and *glyoxysomes,* carry out β-oxidation processes. **Peroxisomes** are so named because they carry out a variety of flavin-dependent oxidation reactions, regenerating oxidized flavins by reaction with oxygen to produce hydrogen peroxide, H_2O_2. Peroxisomal β-oxidation is similar to mitochondrial β-oxidation, except that the initial double bond formation is catalyzed by an FAD-dependent acyl-CoA oxidase (Figure 23.25). The action of this enzyme in the peroxisomes transfers the liberated electrons directly to oxygen instead of the electron-transport chain. As a result, each two-carbon unit oxidized in peroxisomes produces fewer ATPs. The enzymes responsible for fatty acid oxidation in peroxisomes are inactive with carbon chains of eight or fewer. Such short-chain products must be transferred to the mitochondria for further breakdown. Similar β-oxidation enzymes are also found in **glyoxysomes**—peroxisomes in plants that also carry out the reactions of the glyoxylate pathway.

Branched-Chain Fatty Acids Are Degraded Via α-Oxidation

Although β-oxidation is universally important, there are some instances in which it cannot operate effectively. For example, branched-chain fatty acids with alkyl branches at odd-numbered carbons are not effective substrates for β-oxidation. For such species, α-**oxidation** is a useful alternative. Consider **phytol,** a breakdown product of chlorophyll that occurs in the fat of ruminant animals such as sheep and cows and also in dairy products. Ruminants oxidize phytol to phytanic acid, and digestion of phytanic acid in dairy products is thus an important dietary consideration for humans. The methyl group at C-3 will block β-oxidation, but, as shown in Figure 23.26, **phytanic acid α-hydroxylase** places an —OH group at the α-carbon, and **phytanic acid α-oxidase** decarboxylates it to yield *pristanic acid.* The CoA ester of this metabolite can undergo β-oxidation in the normal manner. The terminal product, isobutyryl-CoA, can be sent into the TCA cycle by conversion to succinyl-CoA.

---

[1]β-Oxidation does not occur significantly in plant mitochondria.

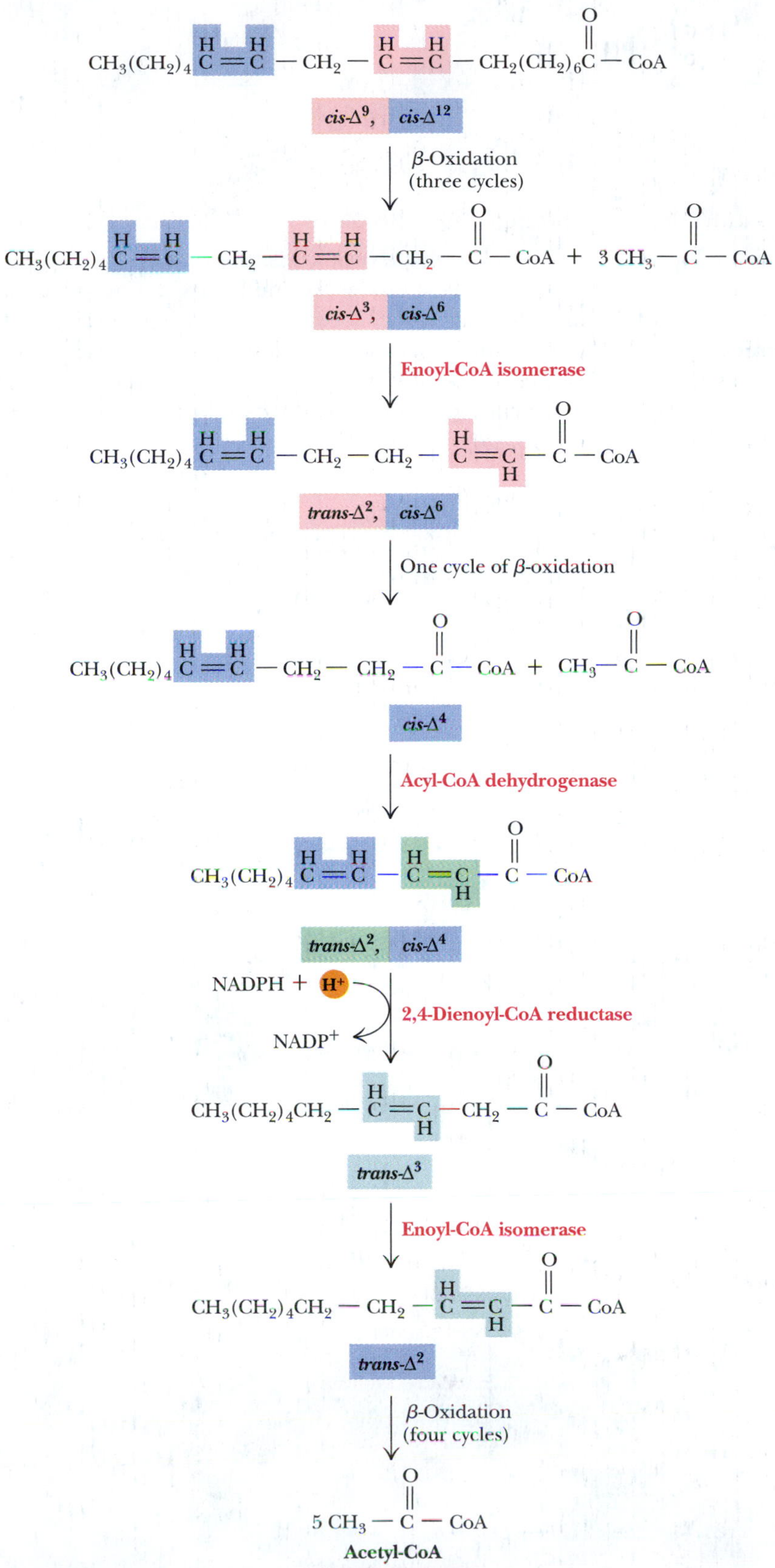

FIGURE 23.24 The oxidation pathway for polyunsaturated fatty acids, illustrated for linoleic acid. Three cycles of β-oxidation on linoleoyl-CoA yield the *cis*-Δ³, *cis*-Δ⁶ intermediate, which is converted to a *trans*-Δ², *cis*-Δ⁶ intermediate. An additional round of β-oxidation gives *cis*-Δ⁴ enoyl-CoA, which is oxidized to the *trans*-Δ², *cis*-Δ⁴ species by acyl-CoA dehydrogenase. The subsequent action of 2,4-dienoyl-CoA reductase yields the *trans*-Δ³ product, which is converted by enoyl-CoA isomerase to the *trans*-Δ² form. Normal β-oxidation then produces five molecules of acetyl-CoA.

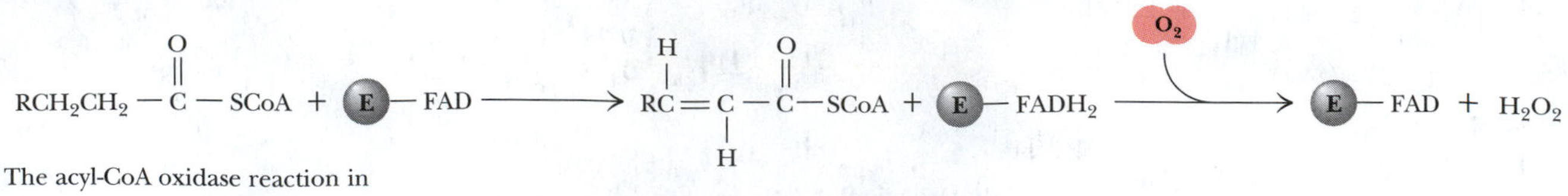

$$RCH_2CH_2 - \overset{\overset{\displaystyle O}{\|}}{C} - SCoA + \enspace \textcircled{E} - FAD \longrightarrow RC = \overset{\overset{\displaystyle H}{|}}{\underset{\underset{\displaystyle H}{|}}{C}} - \overset{\overset{\displaystyle O}{\|}}{C} - SCoA + \enspace \textcircled{E} - FADH_2 \quad \overset{\textcircled{O_2}}{\longrightarrow} \quad \textcircled{E} - FAD + H_2O_2$$

FIGURE 23.25 The acyl-CoA oxidase reaction in peroxisomes.

ω-Oxidation of Fatty Acids Yields Small Amounts of Dicarboxylic Acids

In the endoplasmic reticulum of eukaryotic cells, the oxidation of the terminal carbon of a normal fatty acid—a process termed ω-oxidation—can lead to the synthesis of small amounts of dicarboxylic acids (Figure 23.27). **Cytochrome P-450,** a monooxygenase enzyme that requires NADPH as a coenzyme and uses O_2 as a substrate, places a hydroxyl group at the terminal carbon. Subsequent oxidation to a carboxyl group produces a dicarboxylic acid. Either end can form an ester linkage to CoA and be subjected to β-oxidation, producing a variety of smaller dicarboxylic acids. (Cytochrome P-450–dependent monooxygenases also play an important role as agents of **detoxication,** the degradation and metabolism of toxic hydrocarbon agents.)

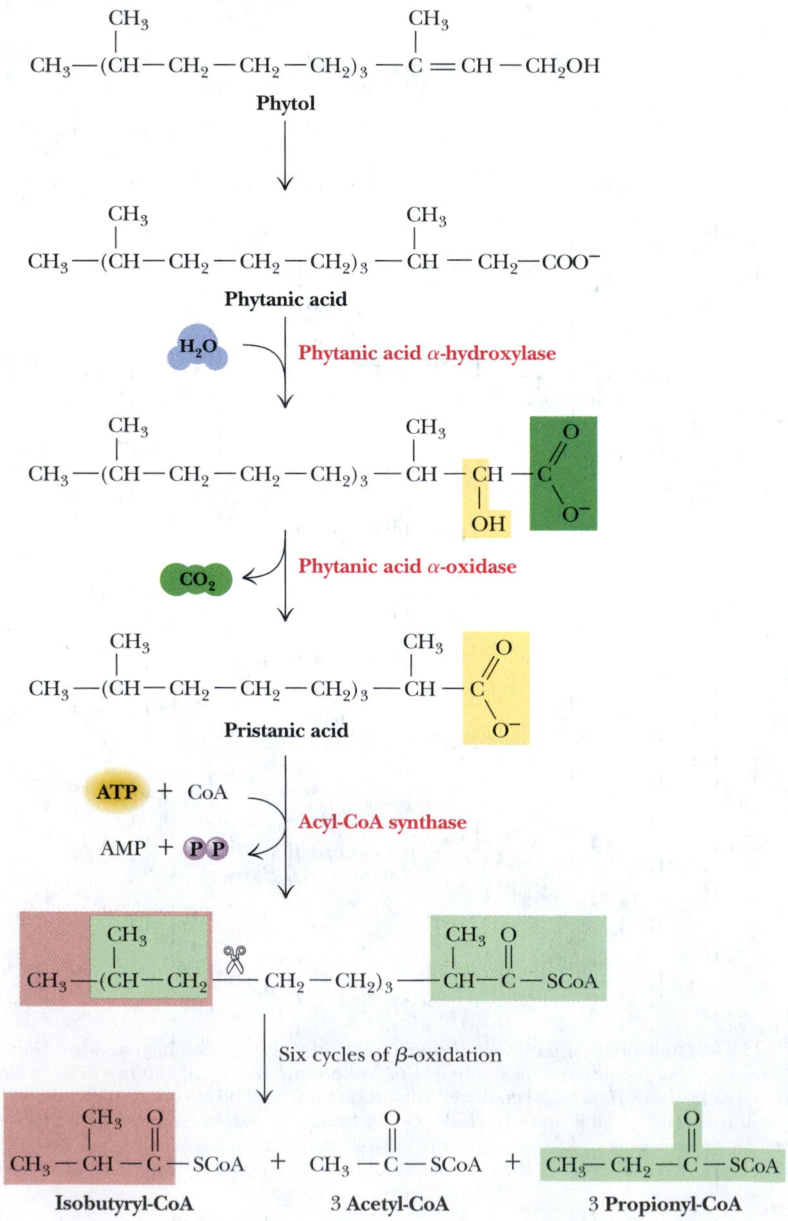

FIGURE 23.26 Branched-chain fatty acids are oxidized by α-oxidation, as shown for phytanic acid. The product of the phytanic acid oxidase, pristanic acid, is a suitable substrate for normal β-oxidation. Isobutyryl-CoA and propionyl-CoA can both be converted to succinyl-CoA, which can enter the TCA cycle.

Human Biochemistry

Refsum's Disease Is a Result of Defects in α-Oxidation

The α-oxidation pathway is defective in **Refsum's disease,** an inherited metabolic disorder that results in defective night vision, tremors, and other neurologic abnormalities. These symptoms are caused by accumulation of phytanic acid in the body. Treatment of Refsum's disease requires a diet free of chlorophyll, the precursor of phytanic acid. This regimen is difficult to implement because all green vegetables and even meat from plant-eating animals, such as cows, pigs, and poultry, must be excluded from the diet.

23.6 | What Are Ketone Bodies, and What Role Do They Play in Metabolism?

Ketone Bodies Are a Significant Source of Fuel and Energy for Certain Tissues

Most of the acetyl-CoA produced by the oxidation of fatty acids in liver mitochondria undergoes further oxidation in the TCA cycle, as stated earlier. However, some of this acetyl-CoA is converted to three important metabolites: acetone, acetoacetate, and β-hydroxybutyrate. The process is known as **ketogenesis,** and these three metabolites are traditionally known as **ketone bodies,** despite the fact that β-hydroxybutyrate does not contain a ketone function. These three metabolites are synthesized primarily in the liver but are important sources of fuel and energy for many tissues, including brain, heart, and skeletal muscle. The brain, for example, normally uses glucose as its source of metabolic energy. However, during periods of starvation, ketone bodies may be the major energy source for the brain. Acetoacetate and 3-hydroxybutyrate are the preferred and normal substrates for kidney cortex and for heart muscle.

Ketone body synthesis occurs only in the mitochondrial matrix. The reactions responsible for the formation of ketone bodies are shown in Figure 23.28. The first reaction—the condensation of two molecules of acetyl-CoA to form acetoacetyl-CoA—is catalyzed by thiolase, which is also known as **acetoacetyl-CoA thiolase** or **acetyl-CoA acetyltransferase.** This is the same enzyme that carries out the thiolase reaction in β-oxidation, but here it runs in reverse. The

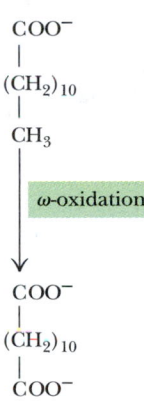

FIGURE 23.27 Dicarboxylic acids can be formed by oxidation of the methyl group of fatty acids in a cytochrome P-450–dependent reaction.

Human Biochemistry

Large Amounts of Ketone Bodies Are Produced in Diabetes Mellitus

Diabetes mellitus is the most common endocrine disease and the third leading cause of death in the United States, with approximately 6 million diagnosed cases and an estimated 4 million more borderline but undiagnosed cases. Diabetes is characterized by an abnormally high level of glucose in the blood. In **type 1 diabetes** (representing 10% or less of all cases), elevated blood glucose results from inadequate secretion of insulin by the islets of Langerhans in the pancreas. **Type 2 diabetes** (at least 90% of all cases) results from an insensitivity to insulin. Type 2 diabetics produce normal or even elevated levels of insulin, but their cells are not responsive to insulin, often due to a shortage of insulin receptors (see Chapter 32). In both cases, transport of glucose into muscle, liver, and adipose tissue is significantly reduced, and despite abundant glucose in the blood, the cells are metabolically starved. They respond by turning to increased gluconeogenesis and catabolism of fat and protein. In type 1 diabetes, increased gluconeogenesis consumes most of the available oxaloacetate, but breakdown of fat (and, to a lesser extent, protein) produces large amounts of acetyl-CoA. This increased acetyl-CoA would normally be directed into the TCA cycle, but with oxaloacetate in short supply, it is used instead for production of unusually large amounts of ketone bodies. Acetone can often be detected on the breath of type 1 diabetics, an indication of high plasma levels of ketone bodies.

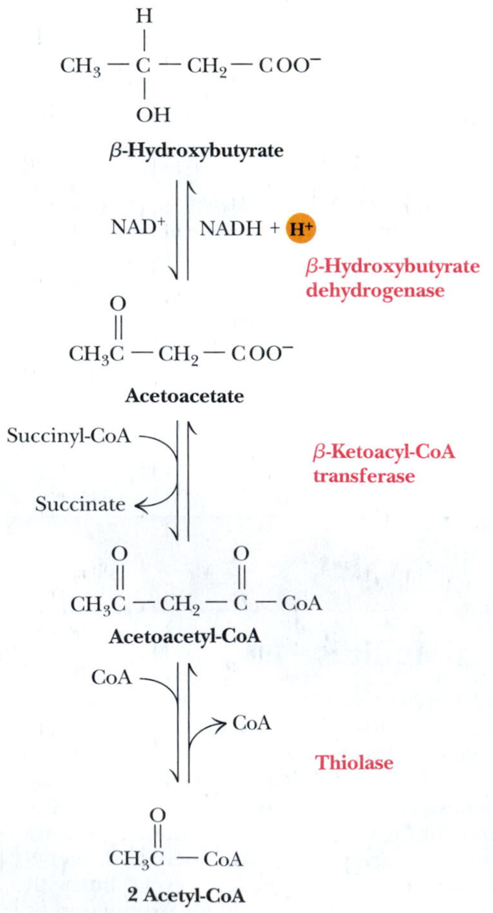

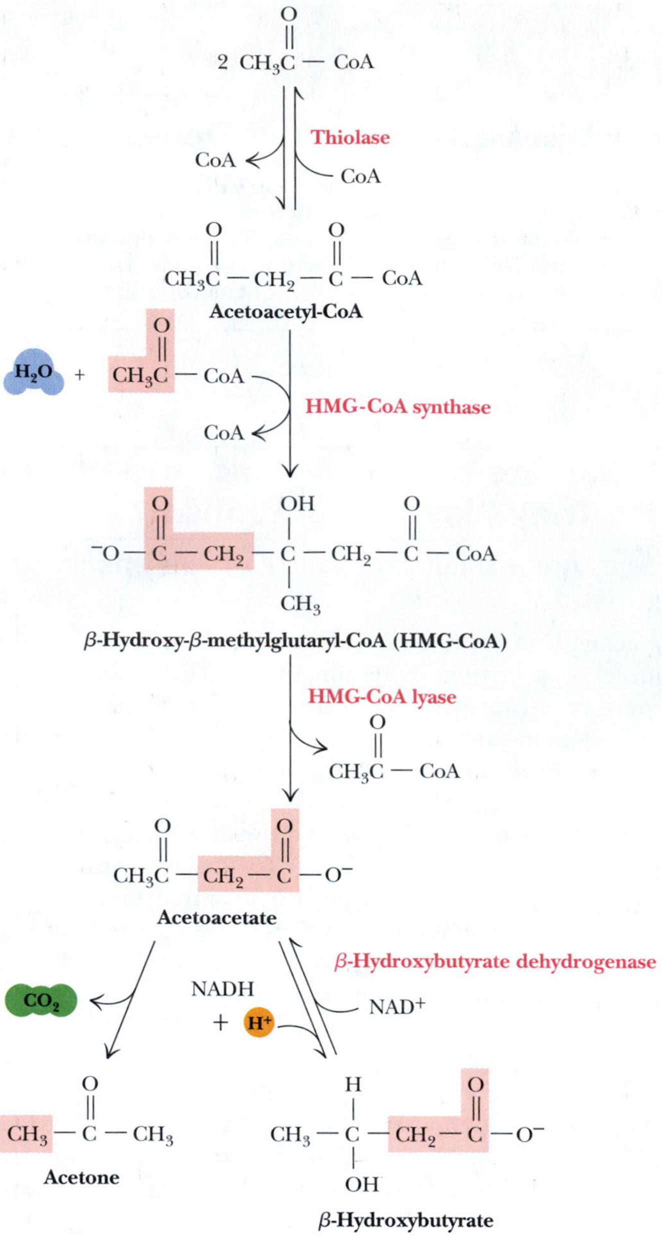

FIGURE 23.28 The formation of ketone bodies, synthesized primarily in liver mitochondria.

FIGURE 23.29 Reconversion of ketone bodies to acetyl-CoA in the mitochondria of many tissues (other than liver) provides significant metabolic energy.

second reaction adds another molecule of acetyl-CoA to give *β-hydroxy-β-methylglutaryl-CoA,* commonly abbreviated HMG-CoA. These two mitochondrial matrix reactions are analogous to the first two steps in cholesterol biosynthesis, a cytosolic process, as we shall see in Chapter 24. HMG-CoA is converted to acetoacetate and acetyl-CoA by the action of **HMG-CoA lyase** in a mixed aldol-Claisen ester cleavage reaction. This reaction is mechanistically similar to the reverse of the citrate synthase reaction in the TCA cycle. A membrane-bound enzyme, **β-hydroxybutyrate dehydrogenase,** then can reduce acetoacetate to β-hydroxybutyrate.

Acetoacetate and β-hydroxybutyrate are transported through the blood from liver to target organs and tissues, where they are converted to acetyl-CoA (Figure 23.29). *Ketone bodies are easily transportable forms of fatty acids that move through the circulatory system without the need for complexation with serum albumin and other fatty acid-binding proteins.*

Summary

23.1 How Are Fats Mobilized from Dietary Intake and Adipose Tissue?
Triacylglycerols are a major source of fatty acids in the diet, and they are also our principal stored energy reserve in adipose tissue. Hormone messengers such as adrenaline, glucagon, and ACTH bind to receptors on the plasma membrane of adipose cells and lead to the activation of a triacylglycerol lipase that hydrolyzes a fatty acid from C-1 or C-3 of triacylglycerols. Subsequent actions of diacylglycerol lipase and monoacylglycerol lipase yield fatty acids and glycerol. The cell then releases the fatty acids into the blood, where they are carried to sites of utilization. Dietary triacylglycerols are degraded by lipases and esterases in the stomach and duodenum. Pancreatic lipase cleaves fatty acids from the C-1 and C-3 positions of triacylglycerols, and other lipases and esterases attack the C-2 position. Bile salts act as detergents to emulsify the triacylglycerols and facilitate the hydrolytic activity of the lipases and esterases.

23.2 How Are Fatty Acids Broken Down?
The process of β-oxidation begins with the formation of a thiol ester bond between the fatty acid and the thiol group of coenzyme A, catalyzed by acyl-CoA synthetase. All of the other enzymes of the β-oxidation pathway are located in the mitochondrial matrix. Short-chain fatty acids are transported into the matrix as free acids and form the acyl-CoA derivatives there. However, long-chain fatty acyl-CoA derivatives must first be converted to acylcarnitine derivatives, which are transported across the inner membrane by a translocase. On the matrix side of the inner membrane, a second acyl carnitine transferase reforms the fatty acyl-CoA. The process of β-oxidation involves a recurring cycle of four steps. A double bond is formed, water is added across the double bond, and the resulting alcohol is oxidized to a carbonyl group. The fourth reaction of the cycle cleaves the resulting β-keto ester, producing an acetate unit and leaving a fatty acid chain that is two carbons shorter.

23.3 How Are Odd-Carbon Fatty Acids Oxidized?
Humans and animals metabolize odd-carbon fatty acids via the β-oxidation pathway, with the final product being propionyl-CoA. Three specialized enzymes then convert propionyl-CoA to succinyl-CoA, a TCA cycle intermediate. The pathway involves an initial carboxylation (by propionyl-CoA carboxylase) at the α-carbon of propionyl-CoA to produce D-methylmalonyl-CoA, which is converted to the L-isomer by methylmalonyl-CoA epimerase. The L-isomer is the substrate for methylmalonyl-CoA mutase,
which catalyzes a migration of a carbonyl-CoA group from one carbon to its neighbor, yielding succinyl-CoA.

23.4 How Are Unsaturated Fatty Acids Oxidized?
Two additional mitochondrial enzymes—an isomerase and a novel reductase—are required to handle the *cis*-double bonds of naturally occurring fatty acids. Consider the breakdown of oleic acid. The reactions of β-oxidation proceed normally through three cycles, producing three molecules of acetyl-CoA and leaving the product *cis*-Δ^3-dodecenoyl-CoA. This intermediate is not a substrate for acyl-CoA dehydrogenase. Instead, enoyl-CoA isomerase rearranges the *cis*-Δ^3 double bond to a *trans*-Δ^2 double bond, which can proceed through the normal route of β-oxidation.

23.5 Are There Other Ways to Oxidize Fatty Acids?
Organelles other than mitochondria, including peroxisomes and glyoxysomes, carry out β-oxidation processes. Peroxisomal β-oxidation is similar to mitochondrial β-oxidation, except that the initial double bond formation is catalyzed by an FAD-dependent acyl-CoA oxidase, which transfers the liberated electrons directly to oxygen instead of the electron-transport chain. Short-chain products must be transferred to the mitochondria for further breakdown. Similar β-oxidation enzymes are also found in glyoxysomes.

Branched-chain fatty acids with alkyl branches at odd-numbered carbons are not effective substrates for β-oxidation. For such species, α-oxidation is a useful alternative. Ruminants oxidize phytol to phytanic acid. The methyl group at C-3 will block β-oxidation, but phytanic acid α-hydroxylase places an —OH group at the α-carbon, and phytanic acid α-oxidase decarboxylates it to yield pristanic acid. The CoA ester of this metabolite can undergo β-oxidation in the normal manner. The terminal product, isobutyryl-CoA, can be sent into the TCA cycle by conversion to succinyl-CoA.

23.6 What Are Ketone Bodies, and What Role Do They Play in Metabolism?
Acetone, acetoacetate, and β-hydroxybutyrate are known as ketone bodies. These three metabolites are synthesized primarily in the liver but are important sources of fuel and energy for many peripheral tissues, including brain, heart, and skeletal muscle. During periods of starvation, ketone bodies may be the major energy source for the brain. Acetoacetate and 3-hydroxybutyrate are the preferred and normal substrates for kidney cortex and for heart muscle.

Problems

1. Calculate the volume of metabolic water available to a camel through fatty acid oxidation if it carries 30 pounds of triacylglycerol in its hump.

2. Calculate the approximate number of ATP molecules that can be obtained from the oxidation of *cis*-11-heptadecenoic acid to CO_2 and water.

3. Phytanic acid, the product of chlorophyll that causes problems for individuals with Refsum's disease, is 3,7,11,15-tetramethyl hexadecanoic acid. Suggest a route for its oxidation that is consistent with what you have learned in this chapter. (*Hint:* The methyl group at C-3 effectively blocks hydroxylation and normal β-oxidation. You may wish to initiate breakdown in some other way.)

4. Even though acetate units, such as those obtained from fatty acid oxidation, cannot be used for net synthesis of carbohydrate in animals, labeled carbon from ^{14}C-labeled acetate can be found in newly synthesized glucose (for example, in liver glycogen) in animal tracer studies. Explain how this can be. Which carbons of glucose would you expect to be the first to be labeled by ^{14}C-labeled acetate?

5. What would you expect to be the systemic metabolic effects of consuming unripened akee fruit?

6. Overweight individuals who diet to lose weight often view fat in negative ways because adipose tissue is the repository of excess caloric intake. However, the "weighty" consequences might be even worse if excess calories were stored in other forms. Consider a person who is 10 pounds "overweight," and estimate how much more he or she would weigh if excess energy were stored in the form of carbohydrate instead of fat.

7. What would be the consequences of a deficiency in vitamin B_{12} for fatty acid oxidation? What metabolic intermediates might accumulate?

8. Write properly balanced chemical equations for the oxidation to CO_2 and water of (a) myristic acid, (b) stearic acid, (c) α-linolenic acid, and (d) arachidonic acid.

9. How many tritium atoms are incorporated into acetate if a molecule of palmitic acid is oxidized in 100% tritiated water?

10. What would be the consequences of a carnitine deficiency for fatty acid oxidation?

11. Based on the mechanism for the methylmalonyl-CoA mutase (Figure 23.21), write reasonable mechanisms for the following reactions shown.

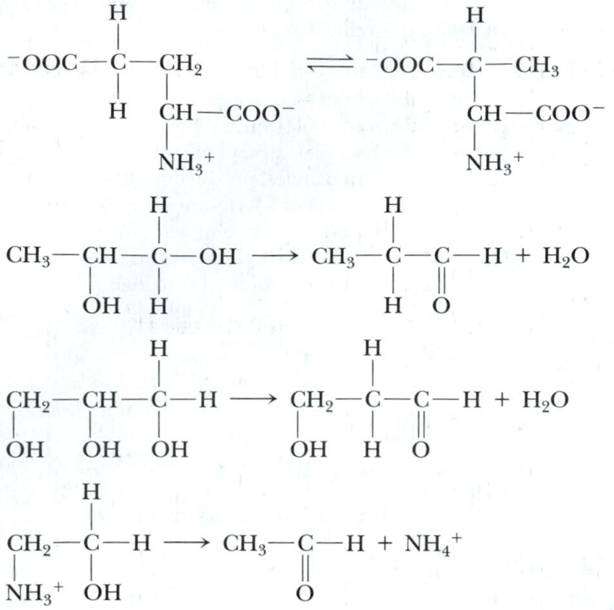

12. The ruby-throated hummingbird flies 500 miles nonstop across the Gulf of Mexico. The flight takes 10 hours at 50 mph. The hummingbird weighs about 4 grams at the start of the flight and about 2.7 grams at the end. Assuming that all the lost weight is fat burned for the flight, calculate the total energy required by the hummingbird in this prodigious flight. Does anything about the results of this calculation strike you as unusual?

13. Energy production in animals is related to oxygen consumption. The ruby-throated hummingbird consumes about 250 mL of oxygen per hour during its migration across the Gulf of Mexico. Use this number and the data in problem 12 to determine a conversion factor for energy expended per liter of oxygen consumed. If a human being consumes 12.7 kcal/min while running 8-minute miles, how long could a human run on the energy that the hummingbird consumes in its trans-Gulf flight? How many 8-minute miles would a person have to run to lose 1 pound of body fat?

14. Write a reasonable mechanism for the HMG-CoA synthase reaction shown in Figure 23.28.

15. Discuss the changes of the oxidation state of cobalt in the course of the methylmalonyl-CoA mutase reaction. Why do they occur as shown in Figure 23.21?

Preparing for the MCAT Exam

16. Study Figure 23.11 and comment on why nature uses $FAD/FADH_2$ as a cofacator in the acyl-CoA dehydrogenase reaction rather than $NAD^+/NADH$.

17. Study Figure 23.10. Where else in metabolism have you seen the chemical strategy and logic of the β-oxidation pathway? Why is it that these two pathways are carrying out the same chemistry?

Biochemistry ✑Now™   Preparing for an exam? Test yourself on key questions at http://chemistry.brookscole.com/ggb3

Further Reading

β-Oxidation

Bennett, M. J., 1994. The enzymes of mitochondrial fatty acid oxidation. *Clinica Chimica Acta* **226**:211–224.

Eder, M., Krautle, F., Dong, Y., et al., 1997. Characterization of human and pig kidney long-chain-acyl-CoA dehydrogenases and their role in beta-oxidation. *European Journal of Biochemistry* **245**:600–607.

Hiltunen, J. K., Palosaari, P., and Kunau, W-H., 1989. Epimerization of 3-hydroxyacyl-CoA esters in rat liver. *Journal of Biological Chemistry* **264**:13535–13540.

Pollitt, R. J., 1995. Disorders of mitochondrial long-chain fatty acid oxidation. *Journal of Inherited Metabolic Disease* **18**:473–490.

Srere, P. A., and Sumegi, B., 1994. Processivity and fatty acid oxidation. *Biochemical Society Transactions* **22**:446–450.

General

Bieber, L. L., 1988. Carnitine. *Annual Review of Biochemistry* **88**:261–283.

Boyer, P. D., ed., 1983. *The Enzymes*, 3rd ed., vol. 16. New York: Academic Press.

Halpern, J., 1985. Mechanisms of coenzyme B_{12}-dependent rearrangements. *Science* **227**:869–875.

Khanolkar, A. D., and Makriyannis, A., 1999. Structure-activity relationships of anandamide, an endogenous cannabinoid ligand. *Life Science* **65**:607–616.

Newsholme, E. A., and Leech, A. R., 1983. *Biochemistry for the Medical Sciences*. New York: Wiley.

Schulz, H., 1987. Inhibitors of fatty acid oxidation. *Life Sciences* **40**:1443–1449.

Scriver, C. R., et al., 1995. *The Metabolic and Molecular Bases of Inherited Disease*, 7th ed. New York: McGraw-Hill.

Romijn, J. A., Coyle, E. F., Sidossis, L. S., et al., 1996. Relationship between fatty acid delivery and fatty acid oxidation during strenuous exercise. *Journal of Applied Physiology* **79**:1939–1945.

Tolbert, N. E., 1981. Metabolic pathways in peroxisomes and glyoxysomes. *Annual Review of Biochemistry* **50**:133–157.

Vance, D. E., and Vance, J. E., eds., 1985. *Biochemistry of Lipids and Membranes*. Menlo Park, CA: Benjamin/Cummings.

Yagoob, P., Newsholme, E. A., and Calder, P. C., 1994. Fatty acid oxidation by lymphocytes. *Biochemical Society Transactions* **22**:116S.

Regulation of Fatty Acid Oxidation

Macfarlane S., and Macfarlane, G. T., 2003. Regulation of short-chain fatty acid production. *Proceedings of the Nutrition Society* **62**:67–72.

McGarry, J. D., and Foster, D. W., 1980. Regulation of hepatic fatty acid oxidation and ketone body production. *Annual Review of Biochemistry* **49**:395–420.

Sherratt, H. S., 1994. Introduction: The regulation of fatty acid oxidation in cells. *Biochemical Society Transactions* **22**:421–422.

Sherratt, H. S., and Spurway, T. D., 1994. Regulation of fatty acid oxidation in cells. *Biochemical Society Transactions* **22**:423–427.

Branched-Chain and Unsaturated Fatty Acid Oxidation

Graham, I. A., and Eastmond, P. J., 2002. Pathways of straight and branched chain fatty acid catabolism in higher plants. *Progress in Lipid Research* **41**:156–181.

Schulz, H., and Kunau, W-H., 1987. β-Oxidation of unsaturated fatty acids: A revised pathway. *Trends in Biochemical Sciences* **12**:403–406.

Trimetazidine

Stanley, W. C., and Marzilli, M., 2003. Metabolic therapy in the treatment of ischaemic heart disease: the pharmacology of trimetazidine. *Fundamentals of Clinical Pharmacology* **17**:133–145.

Phytanic Acid Oxidation

VanVeldhoven, P. P., Mannaerts, G. P., Casteels, M., and Cross, K., 1999. Hepatic alpha-oxidation of phytanic acid. A revised pathway. *Advances in Experimental Medicine and Biology* **466**:273–281.

Wanders, R. J., Jansen, G. A., and Skjeldal, O. H., 2001. Refsum disease, peroxisomes and phytanic acid oxidation: a review. *Journal of Neuropathology and Experimental Neurology* **60**:1021–1031.

Lipid Biosynthesis

CHAPTER 24

Essential Question

We turn now to the biosynthesis of lipid structures. We begin with a discussion of the biosynthesis of fatty acids, stressing the basic pathways, additional means of elongation, mechanisms for the introduction of double bonds, and regulation of fatty acid synthesis. Sections then follow on the biosynthesis of glycerophospholipids, sphingolipids, eicosanoids, and cholesterol. The transport of lipids through the body in lipoprotein complexes is described, and the chapter closes with discussions of the biosynthesis of bile salts and steroid hormones. *What are the pathways of lipid synthesis in biological systems?*

Walruses basking on the beach.

To everything there is a season, and a time for every purpose under heaven…A time to break down and a time to build up.
Ecclesiastes 3:1–3.

We have already seen several cases in which the *synthesis* of a class of biomolecules is conducted differently from degradation (glycolysis versus gluconeogenesis and glycogen or starch breakdown versus polysaccharide synthesis, for example). Likewise, the synthesis of fatty acids and other lipid components is different from their degradation. Fatty acid synthesis involves a set of reactions that follow a strategy different in several ways from the corresponding degradative process:

1. Intermediates in fatty acid synthesis are linked covalently to the sulfhydryl groups of special proteins, the **acyl carrier proteins.** In contrast, fatty acid breakdown intermediates are bound to the —SH group of coenzyme A.
2. Fatty acid synthesis occurs in the cytosol, whereas fatty acid degradation takes place in mitochondria.
3. In animals, the enzymes of fatty acid synthesis are components of one long polypeptide chain, the **fatty acid synthase,** whereas no similar association exists for the degradative enzymes. (Plants and bacteria employ separate enzymes to carry out the biosynthetic reactions.)
4. The coenzyme for the oxidation–reduction reactions of fatty acid synthesis is $NADP^+/NADPH$, whereas degradation involves the $NAD^+/NADH$ couple.

Key Questions

24.1 | How Are Fatty Acids Synthesized?

Formation of Malonyl-CoA Activates Acetate Units for Fatty Acid Synthesis

The design strategy for fatty acid synthesis is this:

1. Fatty acid chains are constructed by the addition of two-carbon units derived from *acetyl-CoA*.
2. The acetate units are activated by formation of *malonyl-CoA* (at the expense of ATP).
3. The addition of two-carbon units to the growing chain is driven by decarboxylation of malonyl-CoA.
4. The elongation reactions are repeated until the growing chain reaches 16 carbons in length (palmitic acid).
5. Other enzymes then add double bonds and additional carbon units to the chain.

Fatty Acid Biosynthesis Depends on the Reductive Power of NADPH

The net reaction for the formation of palmitate from acetyl-CoA is

$$\text{Acetyl-CoA} + 7 \text{ malonyl-CoA}^- + 14 \text{ NADPH} + 13 \text{ H}^+ + \text{H}_2\text{O} \longrightarrow$$
$$\text{palmitate} + 7 \text{ HCO}_3^- + 8 \text{ CoASH} + 14 \text{ NADP}^+ \qquad (24.1)$$

Biochemistry❀Now™ Test yourself on these Key Questions at BiochemistryNow at **http://chemistry.brookscole.com/ggb3**

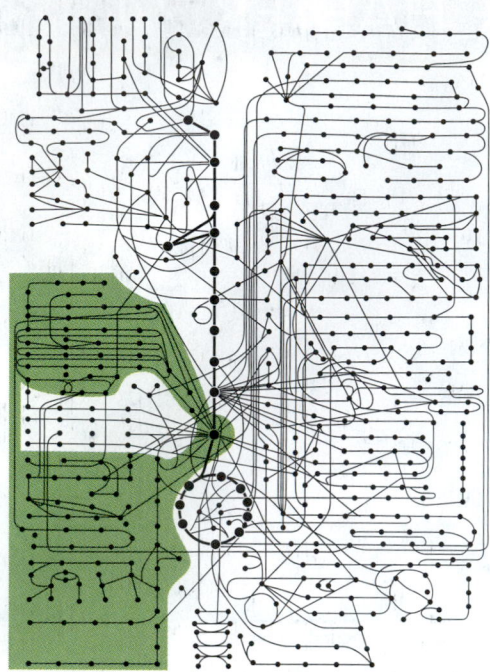

Lipid biosynthesis.

(Levels of free fatty acids are very low in the typical cell. The palmitate made in this process is rapidly converted to CoA esters in preparation for the formation of triacylglycerols and phospholipids.)

Cells Must Provide Cytosolic Acetyl-CoA and Reducing Power for Fatty Acid Synthesis

Eukaryotic cells face a dilemma in providing suitable amounts of substrate for fatty acid synthesis. Sufficient quantities of acetyl-CoA, malonyl-CoA, and NADPH must be generated *in the cytosol* for fatty acid synthesis. Malonyl-CoA is made by carboxylation of acetyl-CoA, so the problem reduces to generating sufficient acetyl-CoA and NADPH.

There are three principal sources of acetyl-CoA (Figure 24.1):

1. Amino acid degradation produces cytosolic acetyl-CoA.
2. Fatty acid oxidation produces mitochondrial acetyl-CoA.
3. Glycolysis yields cytosolic pyruvate, which (after transport into the mitochondria) is converted to acetyl-CoA by pyruvate dehydrogenase.

The acetyl-CoA derived from amino acid degradation is normally insufficient for fatty acid biosynthesis, and the acetyl-CoA produced by pyruvate dehydrogenase and by fatty acid oxidation cannot cross the mitochondrial membrane to participate directly in fatty acid synthesis. Instead, acetyl-CoA is linked with oxaloacetate to form citrate, which is transported from the mitochondrial matrix to the cytosol (Figure 24.1). Here it can be converted back into acetyl-CoA and oxaloacetate by **ATP–citrate lyase.** In this manner, mitochondrial acetyl-CoA becomes the substrate for cytosolic fatty acid synthesis. (Oxaloacetate returns to the mitochondria in the form of either pyruvate or malate, which is then reconverted to acetyl-CoA and oxaloacetate, respectively.)

NADPH can be produced in the pentose phosphate pathway as well as by malic enzyme (Figure 24.1). Reducing equivalents (electrons) derived from glycolysis in the form of NADH can be transformed into NADPH by the combined action of malate dehydrogenase and malic enzyme:

$$\text{Oxaloacetate} + \text{NADH} + \text{H}^+ \longrightarrow \text{malate} + \text{NAD}^+$$
$$\text{Malate} + \text{NADP}^+ \longrightarrow \text{pyruvate} + \text{CO}_2 + \text{NADPH} + \text{H}^+$$

How many of the 14 NADPH needed to form one palmitate (Equation 24.1) can be made in this way? The answer depends on the status of malate. Every citrate entering the cytosol produces one acetyl-CoA and one malate (Figure 24.1). Every malate oxidized by malic enzyme produces one NADPH, at the expense of a decarboxylation to pyruvate. Thus, when malate is oxidized, one NADPH is produced for every acetyl-CoA. Conversion of 8 acetyl-CoA units to one palmitate would then be accompanied by production of 8 NADPH. (The other 6 NADPH required [Equation 24.1] would be provided by the pentose phosphate pathway.) On the other hand, for every malate returned to the mitochondria, one NADPH fewer is produced.

Acetate Units Are Committed to Fatty Acid Synthesis by Formation of Malonyl-CoA

Rittenberg and Bloch showed in the late 1940s that acetate units are the building blocks of fatty acids. Their work, together with the discovery by Salih Wakil that bicarbonate is required for fatty acid biosynthesis, eventually made clear that this pathway involves synthesis of *malonyl-CoA*. The carboxylation of acetyl-CoA to form malonyl-CoA is essentially irreversible and is the **committed step** in the synthesis of fatty acids (Figure 24.2). The reaction is catalyzed by **acetyl-CoA carboxylase,** which contains a biotin prosthetic group. This carboxylase is the only enzyme of fatty acid synthesis in animals that is not part of the multienzyme complex called fatty acid synthase.

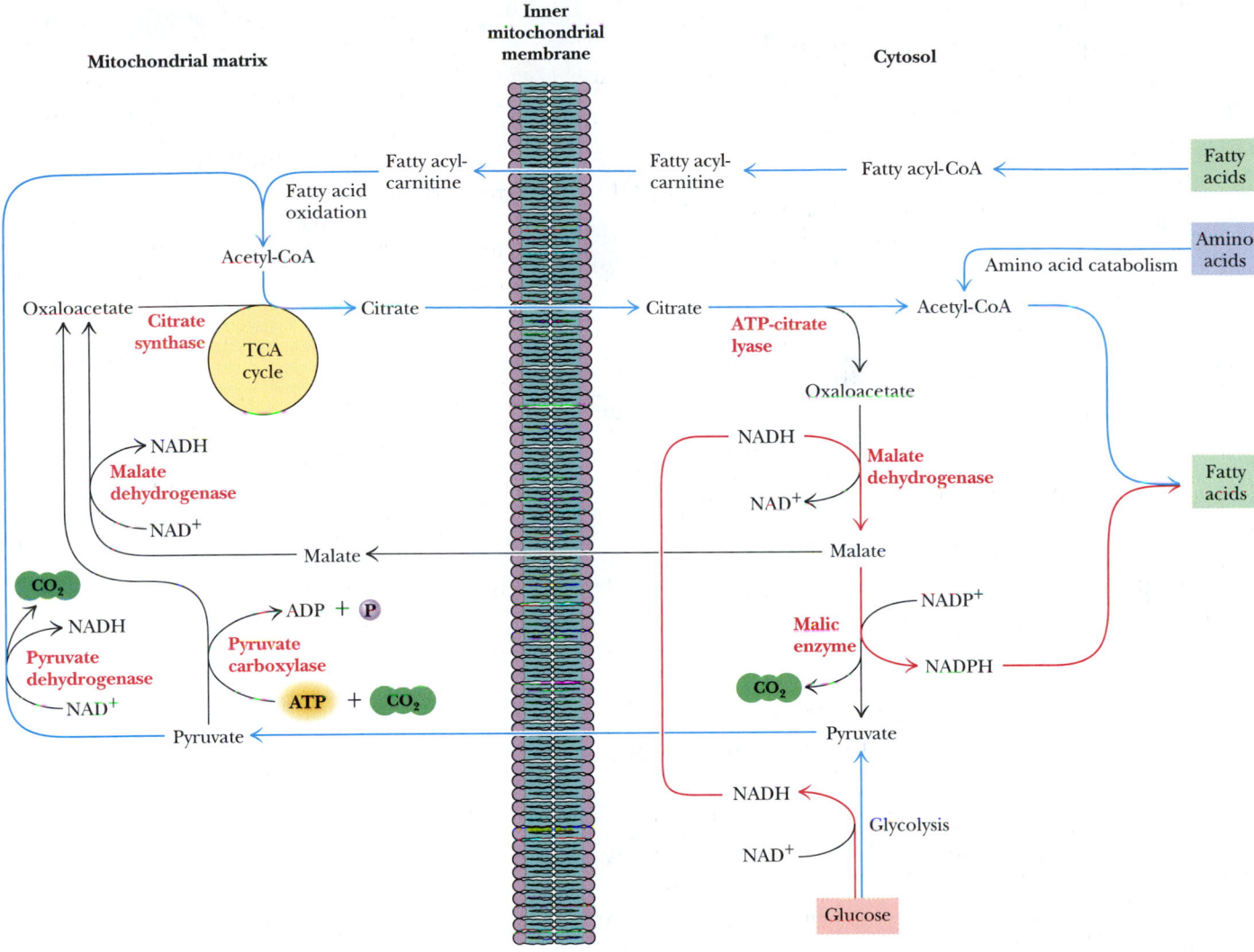

FIGURE 24.1 The citrate–malate–pyruvate shuttle provides cytosolic acetate units and reducing equivalents (electrons) for fatty acid synthesis. The shuttle collects carbon substrates, primarily from glycolysis but also from fatty acid oxidation and amino acid catabolism. Most of the reducing equivalents are glycolytic in origin. Pathways that provide carbon for fatty acid synthesis are shown in blue; pathways that supply electrons for fatty acid synthesis are shown in red.

Acetyl-CoA Carboxylase Is Biotin-Dependent and Displays Ping-Pong Kinetics

The biotin prosthetic group of acetyl-CoA carboxylase is covalently linked to the ε-amino group of an active-site lysine in a manner similar to pyruvate carboxylase (Figure 23.3). The reaction mechanism is also analogous to that of pyruvate carboxylase (Figure 23.4): ATP-driven carboxylation of biotin is followed by transfer of the activated CO_2 to acetyl-CoA to form malonyl-CoA. The enzyme from *Escherichia coli* has three subunits: (1) a **biotin carboxyl carrier protein** (a dimer of 22.5-kD subunits); (2) **biotin carboxylase** (a dimer of 51-kD subunits), which adds CO_2 to the prosthetic group; and (3) **transcarboxylase** (an $\alpha_2\beta_2$ tetramer with 30- and 35-kD subunits), which transfers the activated CO_2 unit to acetyl-CoA. The long, flexible biotin–lysine chain (biocytin) enables the activated carboxyl group to be carried between the biotin carboxylase and the transcarboxylase (Figure 24.3).

Acetyl-CoA Carboxylase in Animals Is a Multifunctional Protein

In animals, acetyl-CoA carboxylase (ACC) is a filamentous polymer (4 to 8 × 10^6 D) composed of 230-kD protomers. Each of these subunits contains the biotin carboxyl carrier moiety, biotin carboxylase, and transcarboxylase activities, as well as allosteric regulatory sites. Animal ACC is thus a multifunctional protein. The polymeric form is active, but the 230-kD protomers are inactive. The activity of

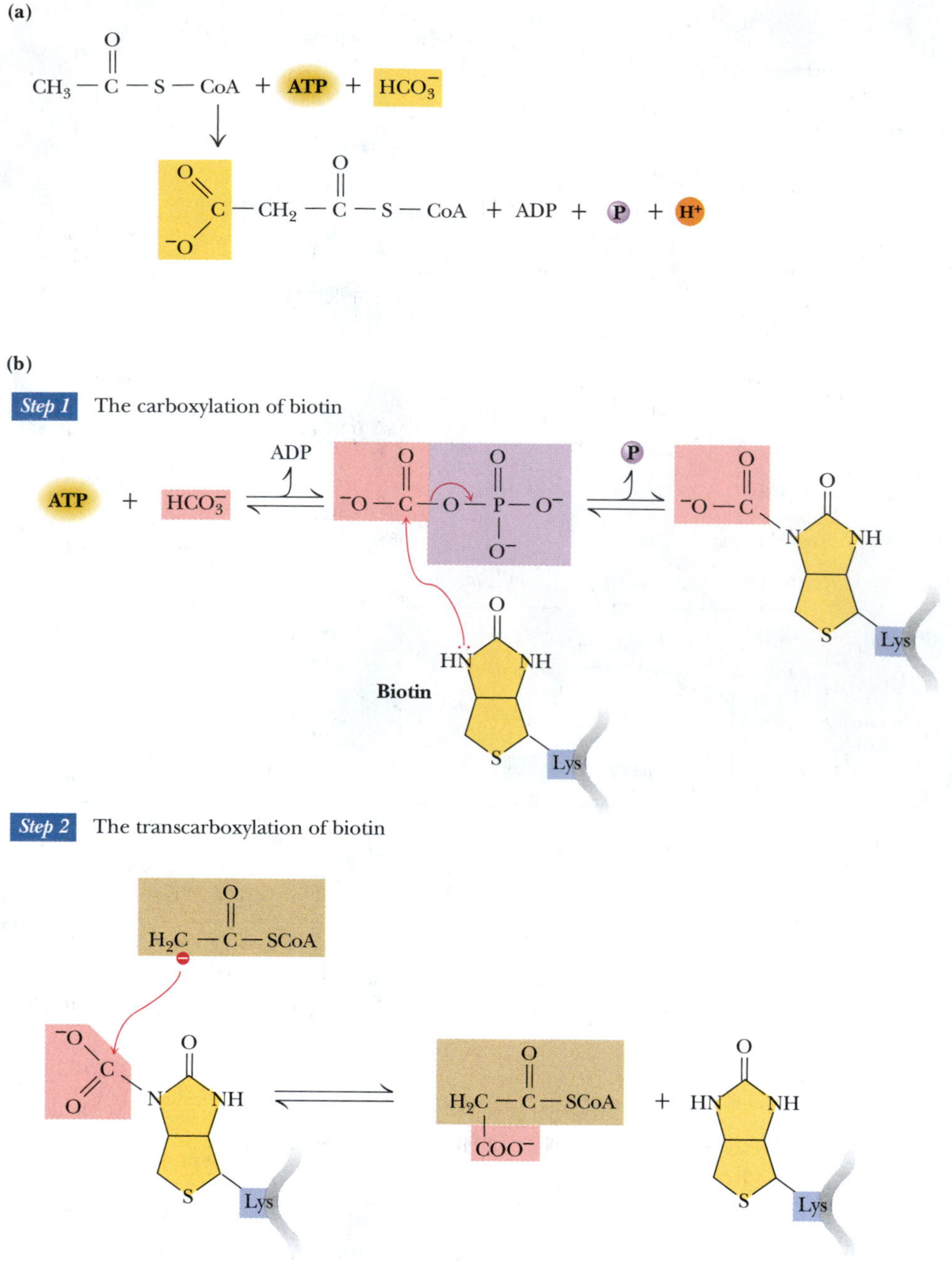

(a)

(b)

Step 1   The carboxylation of biotin

Step 2   The transcarboxylation of biotin

Biochemistry ⒺNow™   ACTIVE FIGURE 24.2
(a) The acetyl-CoA carboxylase reaction produces malonyl-CoA for fatty acid synthesis. **(b)** A mechanism for the acetyl-CoA carboxylase reaction. Bicarbonate is activated for carboxylation reactions by formation of N-carboxybiotin. ATP drives the reaction forward, with transient formation of a carbonyl-phosphate intermediate **(Step 1)**. In a typical biotin-dependent reaction, nucleophilic attack by the acetyl-CoA carbanion on the carboxyl carbon of N-carboxybiotin—a transcarboxylation—yields the carboxylated product **(Step 2)**. **Test yourself on the concepts in this figure at http://chemistry. brookscole.com/ggb3**

ACC is thus dependent upon the position of the equilibrium between these two forms:

$$\text{Inactive protomers} \rightleftharpoons \text{active polymer}$$

Because this enzyme catalyzes the committed step in fatty acid biosynthesis, it is carefully regulated. *Palmitoyl-CoA,* the final product of fatty acid biosynthesis, shifts the equilibrium toward the inactive protomers, whereas *citrate,* an important allosteric activator of this enzyme, shifts the equilibrium toward the active polymeric form of the enzyme. Acetyl-CoA carboxylase shows the kinetic behavior of a Monod–Wyman–Changeux V-system allosteric enzyme (see Chapter 15).

Phosphorylation of ACC Modulates Activation by Citrate and Inhibition by Palmitoyl-CoA

The regulatory effects of citrate and palmitoyl-CoA are dependent on the phosphorylation state of acetyl-CoA carboxylase. The animal enzyme is phosphorylated at eight to ten sites on each enzyme subunit (Figure 24.4). Some of these sites are regulatory,

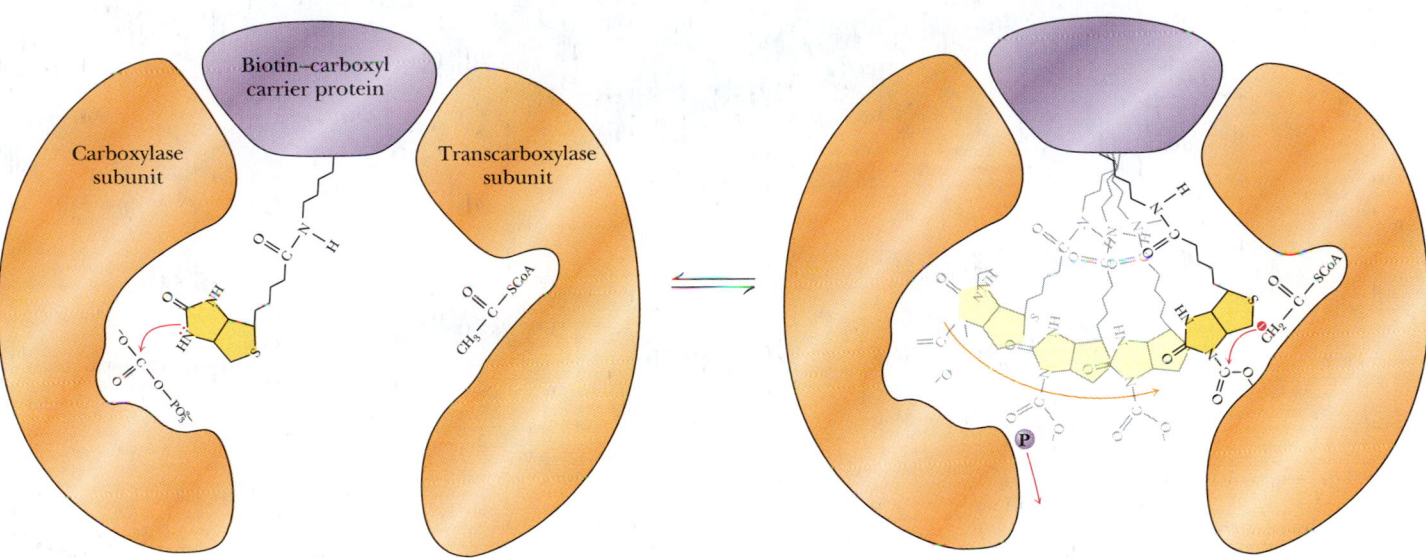

FIGURE 24.3 In the acetyl-CoA carboxylase reaction, the biotin ring, on its flexible tether, acquires carboxyl groups from carbonylphosphate on the carboxylase subunit and transfers them to acyl-CoA molecules on the transcarboxylase subunits.

whereas others are "silent" and have no effect on enzyme activity. Unphosphory-lated acetyl-CoA carboxylase binds citrate with high affinity and thus is active at very low citrate concentrations (Figure 24.5). Phosphorylation of the regulatory sites decreases the affinity of the enzyme for citrate, and in this case, high levels of citrate are required to activate the carboxylase. The inhibition by fatty acyl-CoAs operates in a similar but opposite manner. Thus, low levels of fatty acyl-CoA inhibit the phosphorylated carboxylase, but the dephosphoenzyme is inhibited

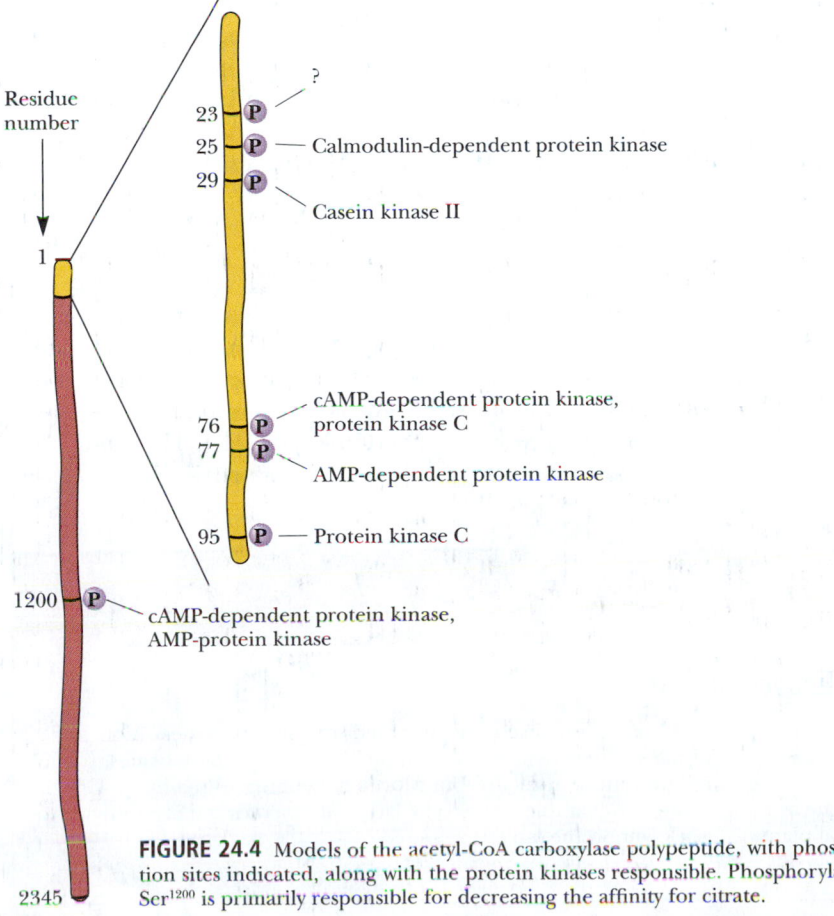

FIGURE 24.4 Models of the acetyl-CoA carboxylase polypeptide, with phosphorylation sites indicated, along with the protein kinases responsible. Phosphorylation at Ser^{1200} is primarily responsible for decreasing the affinity for citrate.

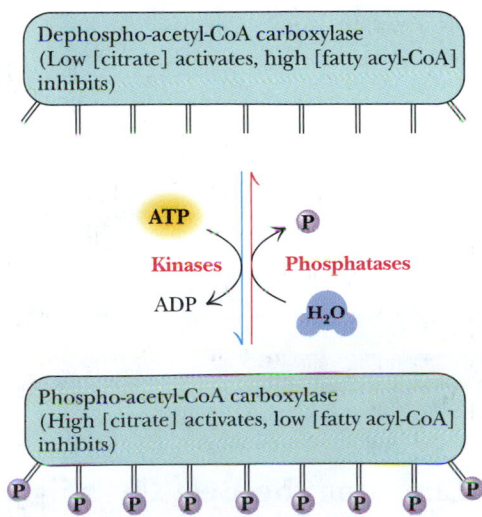

FIGURE 24.5 The activity of acetyl-CoA carboxylase is modulated by phosphorylation and dephosphorylation. The dephospho form of the enzyme is activated by low [citrate] and inhibited only by high levels of fatty acyl-CoA. In contrast, the phosphorylated form of the enzyme is activated only by high levels of citrate but is very sensitive to inhibition by fatty acyl-CoA.

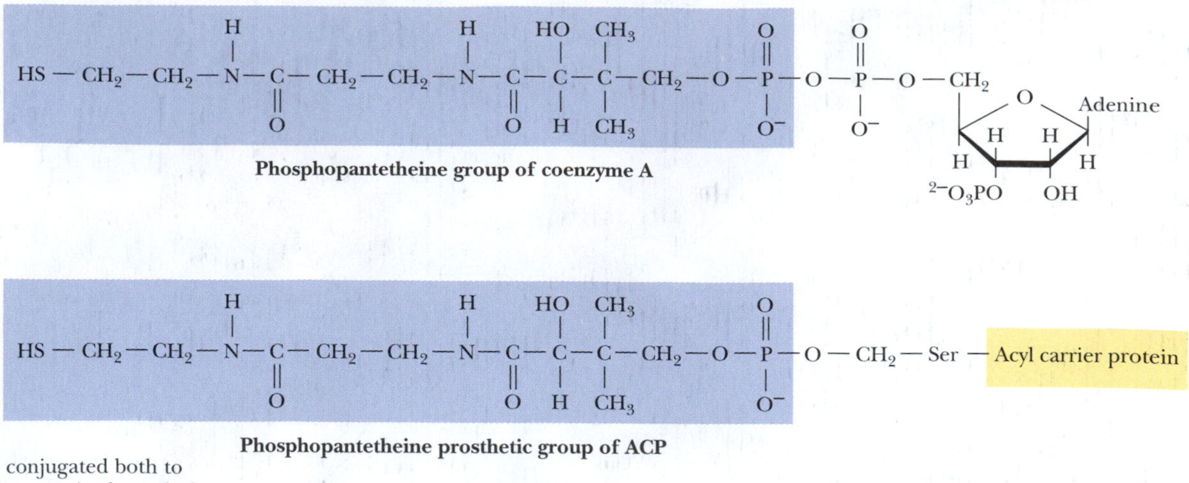

Phosphopantetheine group of coenzyme A

Phosphopantetheine prosthetic group of ACP

FIGURE 24.6 Fatty acids are conjugated both to coenzyme A and to acyl carrier protein through the sulfhydryl of phosphopantetheine prosthetic groups.

only by high levels of fatty acyl-CoA. Specific phosphatases act to dephosphory-late ACC, thereby increasing the sensitivity to citrate.

Acyl Carrier Proteins Carry the Intermediates in Fatty Acid Synthesis

The basic building blocks of fatty acid synthesis are acetyl and malonyl groups, but they are not transferred directly from CoA to the growing fatty acid chain. Rather, they are first passed to **acyl carrier protein** (or simply ACP), discovered by P. Roy Vagelos. This protein consists (in *E. coli*) of a single polypeptide chain of 77 residues to which is attached (on a serine residue) a **phosphopantetheine group,** the same group that forms the "business end" of coenzyme A. Thus, acyl carrier protein is a somewhat larger version of coenzyme A, specialized for use in fatty acid biosynthesis (Figure 24.6).

The enzymes that catalyze formation of acetyl-ACP and malonyl-ACP and the subsequent reactions of fatty acid synthesis are organized quite differently in different organisms. We first discuss fatty acid biosynthesis in bacteria and plants, where the various reactions are catalyzed by separate, independent proteins. Then we discuss the animal version of fatty acid biosynthesis, which involves a single multienzyme complex called **fatty acid synthase.**

Fatty Acid Synthesis Was Elucidated First in Bacteria and Plants

The individual steps in the elongation of the fatty acid chain are quite similar in bacteria, fungi, plants, and animals. The ease of purification of the separate enzymes from bacteria and plants made it possible in the beginning to sort out each step in the pathway, and then by extension to see the pattern of biosynthesis in animals. The reactions are summarized in Figure 24.7. The elongation reactions begin with the formation of acetyl-ACP and malonyl-ACP, which are

A Deeper Look

Choosing the Best Organism for the Experiment

The selection of a suitable and relevant organism is an important part of any biochemical investigation. The studies that revealed the secrets of fatty acid synthesis are a good case in point.

The paradigm for fatty acid synthesis in plants has been the avocado, which has one of the highest fatty acid contents in the plant kingdom. Early animal studies centered primarily on pigeons, which are easily bred and handled and which possess high levels of fats in their tissues. Other animals, richer in fatty tissues, might be even more attractive but more challenging to maintain. Grizzly bears, for example, carry very large fat reserves but are difficult to work with in the lab!

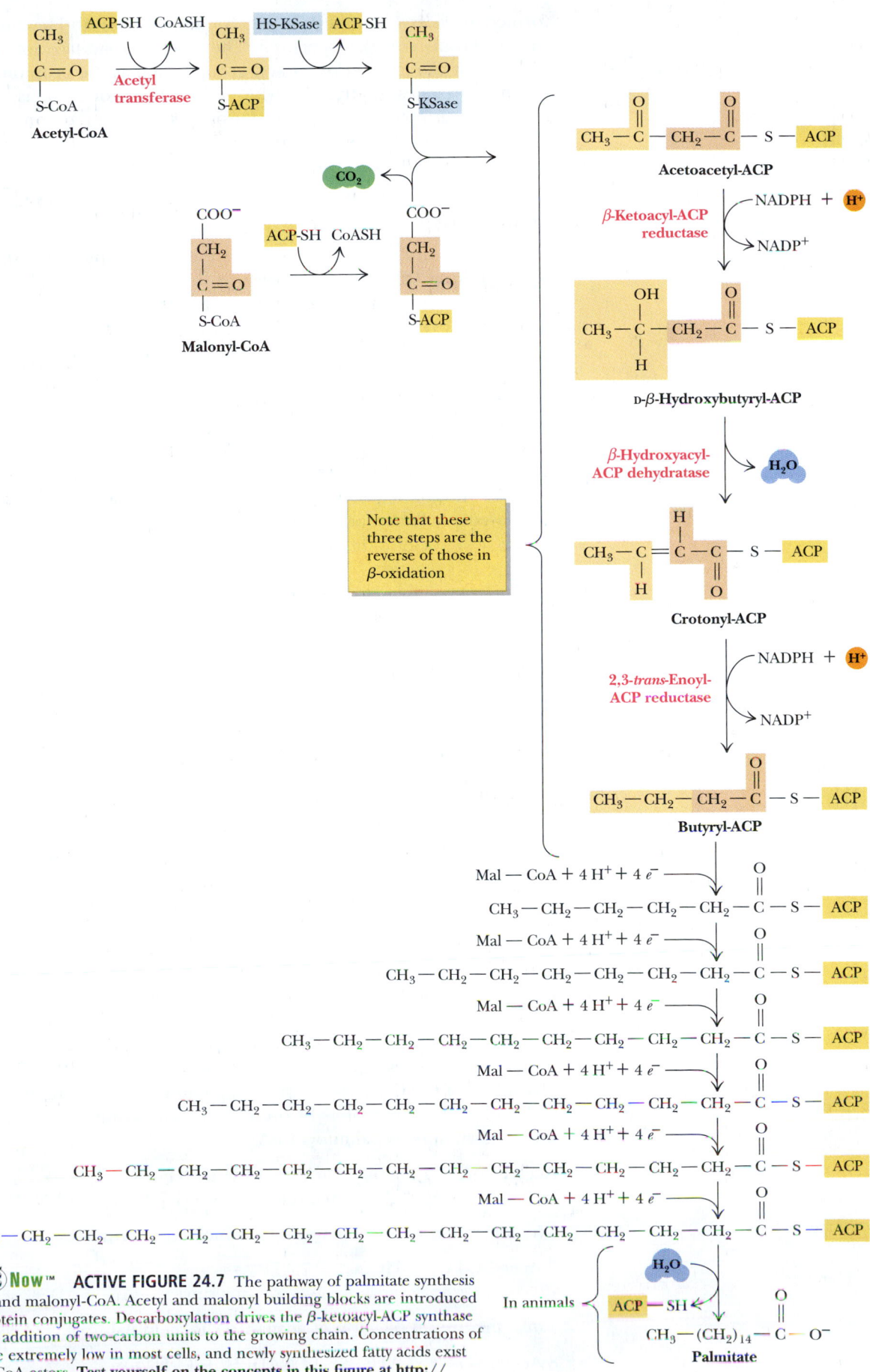

Biochemistry Now™   ACTIVE FIGURE 24.7 The pathway of palmitate synthesis from acetyl-CoA and malonyl-CoA. Acetyl and malonyl building blocks are introduced as acyl carrier protein conjugates. Decarboxylation drives the β-ketoacyl-ACP synthase and results in the addition of two-carbon units to the growing chain. Concentrations of free fatty acids are extremely low in most cells, and newly synthesized fatty acids exist primarily as acyl-CoA esters. **Test yourself on the concepts in this figure at http://chemistry.brookscole.com/ggb3**

formed by **acetyl transacylase (acetyl transferase)** and **malonyl transacylase (malonyl transferase)**, respectively. The acetyl transacylase enzyme is not highly specific—it can transfer other acyl groups, such as the propionyl group, but at much lower rates. (Fatty acids with odd numbers of carbons are made beginning with a propionyl group transfer by this enzyme.) Malonyl transacylase, on the other hand, is highly specific.

Decarboxylation Drives the Condensation of Acetyl-CoA and Malonyl-CoA

Another transacylase reaction transfers the acetyl group from ACP to β-**ketoacyl-ACP synthase (KSase)**, also known as **acyl-malonyl-ACP condensing enzyme.** The first actual elongation reaction involves the condensation of acetyl-ACP and malonyl-ACP by the β-ketoacyl-ACP synthase to form acetoacetyl-ACP (Figure 24.7). *One might ask at this point: Why is the three-carbon malonyl group used here as a two-carbon donor?* The answer is that this is yet another example of a decarboxylation driving a desired but otherwise thermodynamically unfavorable reaction. The decarboxylation that accompanies the reaction with malonyl-ACP drives the synthesis of acetoacetyl-ACP. Note that hydrolysis of ATP drove the carboxylation of acetyl-CoA to form malonyl-ACP, so, indirectly, ATP is responsible for the condensation reaction to form acetoacetyl-ACP. Malonyl-CoA can be viewed as a form of stored energy for driving fatty acid synthesis.

It is also worth noting that the carbon of the carboxyl group that was added to drive this reaction is the one removed by the condensing enzyme. Thus, all the carbons of acetoacetyl-ACP (*and* of the fatty acids to be made) are derived from acetate units of acetyl-CoA.

Reduction of the β-Carbonyl Group Follows a Now-Familiar Route

The next three steps—reduction of the β-carbonyl group to form a β-alcohol, followed by dehydration and reduction to saturate the chain (Figure 24.7)—look very similar to the fatty acid degradation pathway in reverse. However, there are two crucial differences between fatty acid biosynthesis and fatty acid oxidation (besides the fact that different enzymes are involved): First, the alcohol formed in the first step has the D-configuration rather than the L-form seen in catabolism; second, the reducing coenzyme is NADPH, whereas NAD^+ and FAD are the oxidants in the catabolic pathway.

The net result of this biosynthetic cycle is the synthesis of a four-carbon unit, a butyryl group, from two smaller building blocks. In the next cycle of the process, this butyryl-ACP condenses with another malonyl-ACP to make a six-carbon β-ketoacyl-ACP and CO_2. Subsequent reduction to a β-alcohol, dehydration, and another reduction yield a six-carbon saturated acyl-ACP. This cycle continues with the net addition of a two-carbon unit in each turn until the chain is 16 carbons long (Figure 24.7). The β-ketoacyl-ACP synthase cannot accommodate larger substrates, so the reaction cycle ends with a 16-carbon chain. Hydrolysis of the C_{16}-acyl-ACP yields a palmitic acid and the free ACP.

In the end, seven malonyl-CoA molecules and one acetyl-CoA yield a palmitate (shown here as palmitoyl-CoA):

$$\text{Acetyl-CoA} + 7 \text{ malonyl-CoA}^- + 14 \text{ NADPH} + 14 \text{ H}^+ \longrightarrow$$
$$\text{palmitoyl-CoA} + 7 \text{ HCO}_3^- + 14 \text{ NADP}^+ + 7 \text{ CoASH}$$

The formation of seven malonyl-CoA molecules requires

$$7 \text{ Acetyl-CoA} + 7 \text{ HCO}_3^- + 7 \text{ ATP}^{4-} \longrightarrow$$
$$7 \text{ malonyl-CoA}^- + 7 \text{ ADP}^{3-} + 7 \text{ P}_i^{2-} + 7 \text{ H}^+$$

Thus, the overall reaction of acetyl-CoA to yield palmitic acid is

$$8 \text{ Acetyl-CoA} + 7 \text{ ATP}^{4-} + 14 \text{ NADPH} + 14 \text{ H}^+ \longrightarrow$$
$$\text{palmitoyl-CoA} + 14 \text{ NADP}^+ + 7 \text{ CoASH} + 7 \text{ ADP}^{3-} + 7 \text{ P}_i^{2-}$$

Note: These equations are stoichiometric and are charge balanced. See problem 1 at the end of the chapter for practice in balancing these equations.

Fatty Acid Synthesis in Eukaryotes Occurs on a Multienzyme Complex

In contrast to bacterial and plant systems, the reactions of fatty acid synthesis beyond the acetyl-CoA carboxylase in animal systems are carried out by a special multienzyme complex called **fatty acid synthase (FAS)**. In yeast, this 2.4×10^6 D complex contains two different peptide chains, an α-subunit of 213 kD and a β-subunit of 203 kD, arranged in an $\alpha_6\beta_6$ dodecamer. The separate enzyme activities associated with each chain are shown in Figure 24.8. In animal systems, FAS is a dimer of identical 250-kD *multifunctional polypeptides*. Studies of the action of proteolytic enzymes on this polypeptide have led to a model involving three separate domains joined by flexible connecting sequences (Figure 24.9). The first domain is responsible for the binding of acetyl and malonyl building blocks and for the condensation of these units. This domain includes the acetyl transferase, the malonyl transferase, and the acyl-malonyl-ACP condensing enzyme (the β-ketoacyl synthase). The second domain is primarily responsible for the reduction of the intermediate synthesized in domain 1 and contains the acyl carrier protein, the β-ketoacyl reductase, the dehydratase, and the enoyl-ACP reductase. The third domain contains the thioesterase that liberates the product palmitate when the growing acyl chain reaches its limit length of 16 carbons. The close association of activities in this complex permits efficient exposure of intermediates to one active site and then the next. The presence of all these activities on a single polypeptide ensures that the cell will simultaneously synthesize all the enzymes needed for fatty acid synthesis.

The Mechanism of Fatty Acid Synthase Involves Condensation of Malonyl-CoA Units

The first domain of one subunit of the fatty acid synthase interacts with the second and third domains of the other subunit; that is, the subunits are arranged in a head-to-tail fashion (Figure 24.9). The first step in the fatty acid synthase reaction is the formation of an acetyl-*O*-enzyme intermediate between the acetyl group of an acetyl-CoA and an active-site serine of the acetyl transferase (Figure 24.10). In a similar manner, a malonyl-*O*-enzyme intermediate is formed between malonyl-CoA and a serine residue of the malonyl transferase. The acetyl group on the acetyl transferase is then transferred to the —SH group of the acyl carrier protein, as shown in Figure 24.11. The next step is the transfer of the acetyl group to the β-ketoacyl-ACP synthase, or condensing enzyme. This frees the acyl carrier protein to acquire the malonyl group from the malonyl transferase. The next step is the condensation reaction, in which decarboxylation facilitates the concerted attack of the remaining two-carbon unit of the acyl carrier protein at the carbonyl carbon of the acetate group on the condensing enzyme. Note that decarboxylation forms a transient, highly nucleophilic carbanion that can attack the acetate group.

The next three steps—reduction of the carbonyl to an alcohol, dehydration to yield a *trans*-α,β double bond, and reduction to yield a saturated chain—are identical to those occurring in bacteria and plants (Figure 24.7) and resemble the reverse of the reactions of fatty acid oxidation (and the conversion of succinate to oxaloacetate in the TCA cycle). This synthetic cycle now repeats until the growing chain is 16 carbons long. It is then released by the thioesterase domain on the synthase. The amino acid sequence of the thioesterase domain is homologous with serine proteases; the enzyme has an active-site serine that carries out nucleophilic attack on the carbonyl carbon of the fatty acyl thioester to be cleaved.

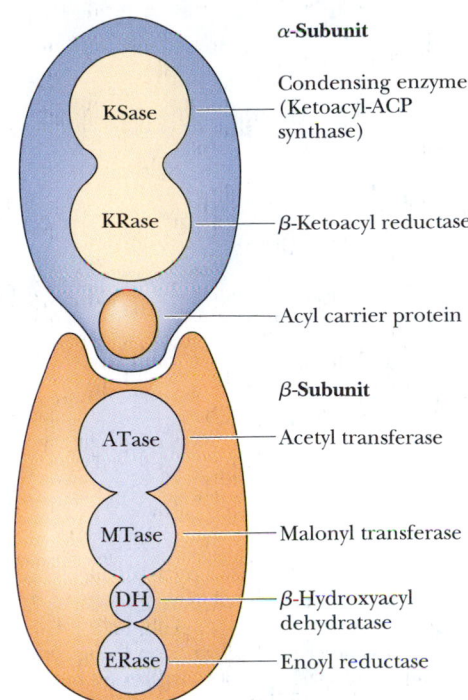

FIGURE 24.8 In yeast, the functional groups and enzyme activities required for fatty acid synthesis are distributed between α- and β-subunits.

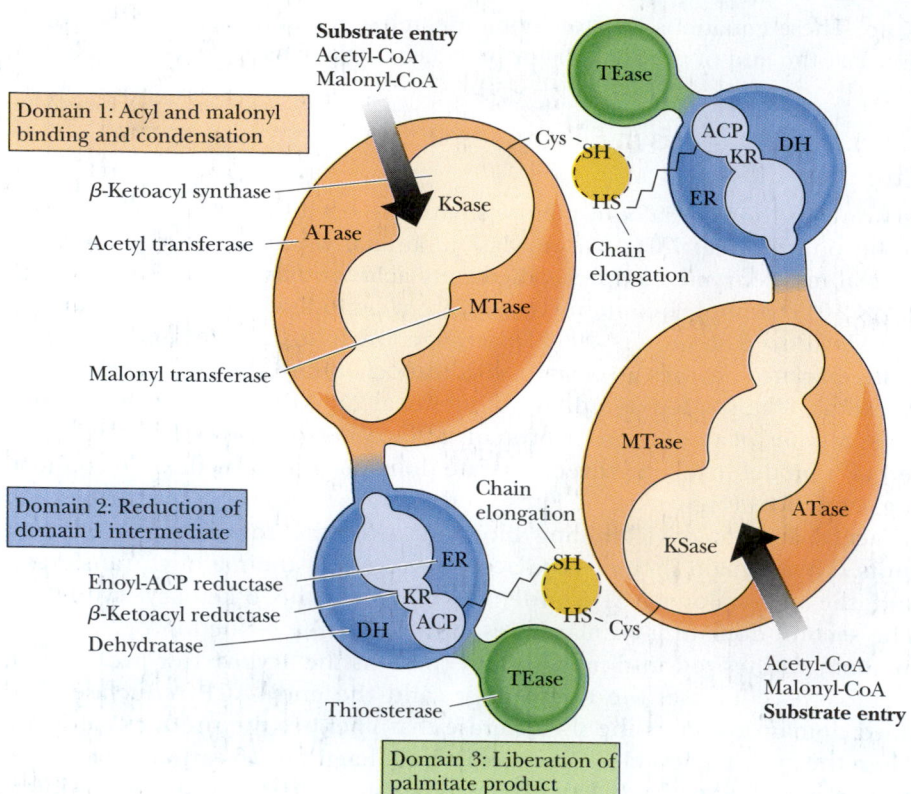

ANIMATED FIGURE 24.9
Fatty acid synthase in animals contains all the functional groups and enzyme activities on a single multifunctional subunit. The active enzyme is a head-to-tail dimer of identical subunits. *(Adapted from Wakil, S. J., Stoops, J. K., and Joshi, V. C., 1983. Fatty acid synthesis and its regulation. Annual Review of Biochemistry 52:537–579.)*
See this figure animated at http://chemistry. brookscole.com/ggb3

C₁₆ Fatty Acids May Undergo Elongation and Unsaturation

Additional Elongation   As seen already, palmitate is the primary product of the fatty acid synthase. Cells synthesize many other fatty acids. Shorter chains are easily made if the chain is released before reaching 16 carbons in length. Longer chains are made through special elongation reactions, which occur both in the mitochondria and at the surface of the endoplasmic reticulum (ER). The ER reactions are actually quite similar to those we have just discussed: addition of two-carbon units at the carboxyl end of the chain by means of oxidative decarboxylations involving malonyl-CoA. As was the case for the fatty acid synthase, this decarboxylation provides the thermodynamic driving force for the condensation reaction. The mitochondrial reactions involve addition (and subsequent reduction) of acetyl units. These reactions (Figure 24.12) are essentially a reversal of fatty acid oxidation, with the exception that NADPH is utilized in the saturation of the double bond, instead of FADH₂.

Introduction of a Single *cis* Double Bond   Both prokaryotes and eukaryotes are capable of introducing a single *cis* double bond in a newly synthesized fatty acid. Bacteria such as *E. coli* carry out this process in an O₂-independent pathway, whereas eukaryotes have adopted an O₂-dependent pathway. There is a fundamental chemical difference between the two. The O₂-dependent reaction

FIGURE 24.10 Acetyl units are covalently linked to a serine residue at the active site of the acetyl transferase in eukaryotes. A similar reaction links malonyl units to the malonyl transferase.

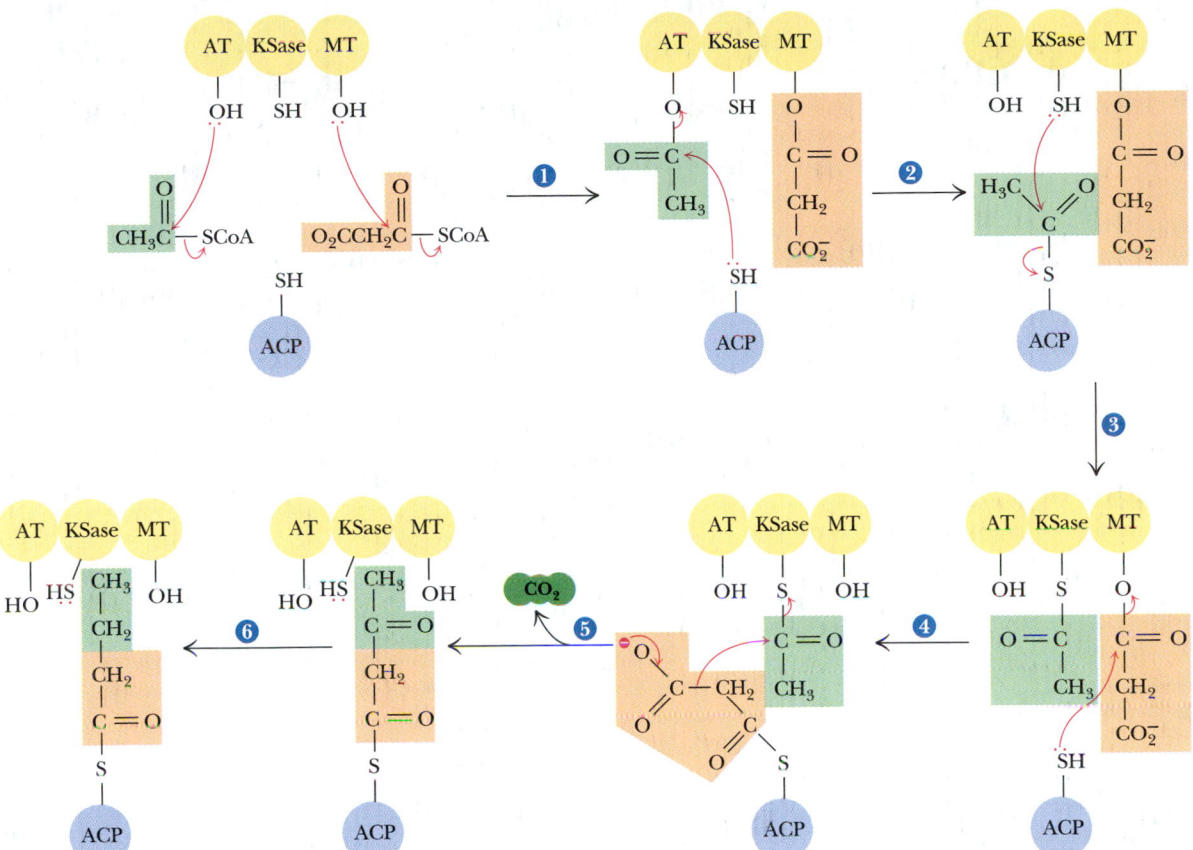

Biochemistry ✶Now™ ACTIVE FIGURE 24.11
The mechanism of the fatty acyl synthase reaction in eukaryotes. (**1**) Acetyl and malonyl groups are loaded onto acetyl transferase and malonyl transferase, respectively. (**2**) The acetate unit that forms the base of the nascent chain is transferred first to the acyl carrier protein domain and (**3**) then to the β-ketoacyl synthase. (**4**) Attack by ACP on the carbonyl carbon of a malonyl unit on malonyl transferase forms malonyl-ACP. (**5**) Decarboxylation leaves a reactive, transient carbanion that can attack the carbonyl carbon of the acetyl group on the β-ketoacyl synthase. (**6**) Reduction of the keto group, dehydration, and saturation of the resulting double bond follow, leaving an acyl group on ACP, and steps 3 through 6 repeat to lengthen the nascent chain. **Test yourself on the concepts in this figure at** http://chemistry.brookscole.com/ggb3

can occur anywhere in the fatty acid chain, with no (additional) need to activate the desired bond toward dehydrogenation. However, in the absence of O_2, some other means must be found to activate the bond in question. Thus, in the bacterial reaction, dehydrogenation occurs while the bond of interest is still near the β-carbonyl or β-hydroxy group and the thioester group at the end of the chain.

In *E. coli*, the biosynthesis of a monounsaturated fatty acid begins with four normal cycles of elongation to form a ten-carbon intermediate, β-hydroxydecanoyl-ACP (Figure 24.13). At this point, **β-hydroxydecanoyl thioester dehydrase** forms a double bond β,γ to the thioester and in the *cis* configuration. This is followed by three rounds of the normal elongation reactions to form *palmitoleoyl-ACP*. Elongation may terminate at this point or may be followed by additional biosynthetic events. The principal unsaturated fatty acid in *E. coli*, *cis-vaccenic acid*, is formed by an additional elongation step, using palmitoleoyl-ACP as a substrate.

Unsaturation Reactions Occur in Eukaryotes in the Middle of an Aliphatic Chain

The addition of double bonds to fatty acids in eukaryotes does not occur until the fatty acyl chain has reached its full length (usually 16 to 18 carbons). Dehydrogenation of stearoyl-CoA occurs in the middle of the chain, despite the absence of any useful functional group on the chain to facilitate activation:

$$CH_3{-}(CH_2)_{16}CO{-}SCoA \longrightarrow CH_3{-}(CH_2)_7CH{=}CH(CH_2)_7CO{-}SCoA$$

This impressive reaction is catalyzed by **stearoyl-CoA desaturase,** a 53-kD enzyme containing a nonheme iron center. NADH and O_2 are required, as are two other proteins: **cytochrome b_5 reductase** (a 43-kD flavoprotein) and **cytochrome b_5** (16.7 kD). All three proteins are associated with the ER membrane. Cytochrome b_5 reductase transfers a pair of electrons from NADH through FAD to

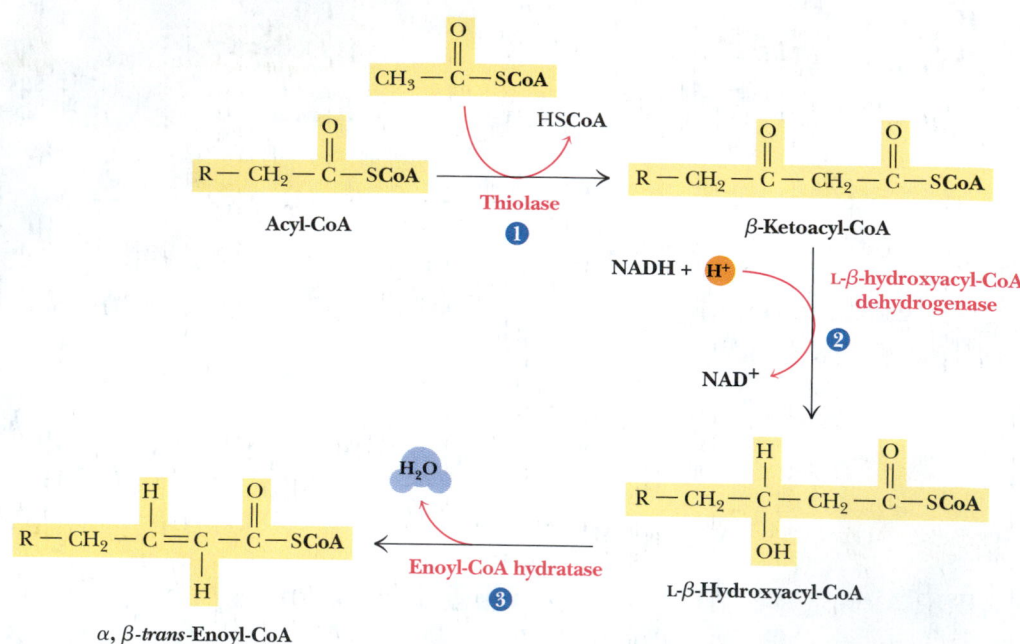

FIGURE 24.12 (1) Elongation of fatty acids in mitochondria is initiated by the thiolase reaction. (2) The β-ketoacyl intermediate thus formed undergoes the same three reactions (in reverse order) that are the basis of β-oxidation of fatty acids. (3) Reduction of the β-keto group is followed by dehydration to form a double bond. (4) Reduction of the double bond yields a fatty acyl-CoA that is elongated by two carbons. Note that the reducing coenzyme for the second step is NADH, whereas the reductant for the fourth step is NADPH.

cytochrome b_5 (Figure 24.14). Oxidation of reduced cytochrome b_5 is coupled to reduction of nonheme Fe^{3+} to Fe^{2+} in the desaturase. The Fe^{3+} accepts a pair of electrons (one at a time in a cycle) from cytochrome b_5 and creates a *cis* double bond at the 9,10-position of the stearoyl-CoA substrate. O_2 is the terminal electron acceptor in this fatty acyl desaturation cycle. Note that two water molecules are made, which means that four electrons are transferred overall. Two of these come through the reaction sequence from NADH, and two come from the fatty acyl substrate that is being dehydrogenated.

The Unsaturation Reaction May Be Followed by Chain Elongation

Additional chain elongation can occur following this single desaturation reaction. The oleoyl-CoA produced can be elongated by two carbons to form a 20:1 *cis*-Δ^{11} fatty acyl-CoA. If the starting fatty acid is palmitate, reactions similar to the preceding scheme yield palmitoleoyl-CoA (16:1 *cis*-Δ^9), which subsequently can be elongated to yield *cis*-vaccenic acid (18:1 *cis*-Δ^{11}). Similarly, C_{16} and C_{18} fatty acids can be elongated to yield C_{22} and C_{24} fatty acids, such as are often found in sphingolipids.

Mammals Cannot Synthesize Most Polyunsaturated Fatty Acids

Organisms differ with respect to formation, processing, and utilization of polyunsaturated fatty acids. *E. coli*, for example, does not have any polyunsaturated fatty acids. Eukaryotes *do* synthesize a variety of polyunsaturated fatty acids, certain organisms more than others. For example, plants manufacture double bonds between the Δ^9 and the methyl end of the chain, but mammals cannot. Plants readily desaturate oleic acid at the 12-position (to give linoleic acid) or at both the 12- and 15-positions (producing linolenic acid). Mammals require polyunsaturated fatty acids but must acquire them in their diet. As such, these fatty acids are referred to as **essential fatty acids.** On the other hand, mammals *can* introduce double bonds between the double bond at the 8- or 9-position and the carboxyl group. Enzyme complexes in the ER desaturate the 5-position, provided a double bond exists at the 8-position, and form a double bond at the 6-position if one already exists at the 9-position. Thus, oleate can be unsaturated at the 6,7-position to give an $18:2$ *cis-*Δ^6,Δ^9 fatty acid.

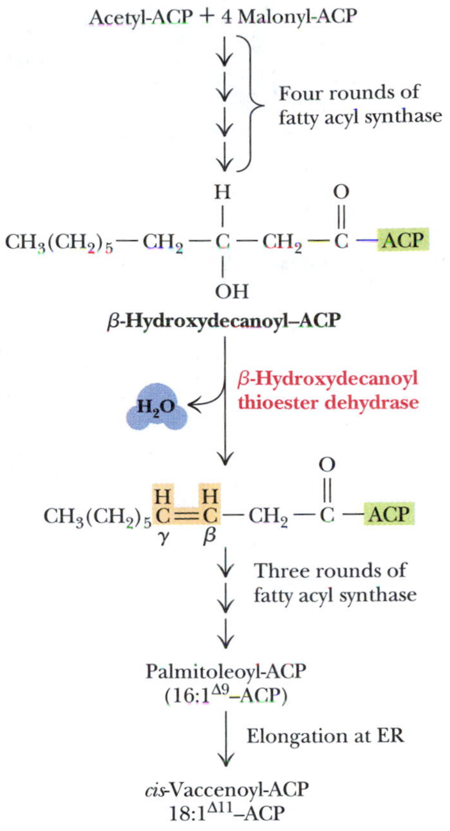

Arachidonic Acid Is Synthesized from Linoleic Acid by Mammals

Mammals can add additional double bonds to unsaturated fatty acids in their diets. Their ability to make arachidonic acid from linoleic acid is one example (Figure 24.15). This fatty acid is the precursor for prostaglandins and other biologically active derivatives such as leukotrienes. Synthesis involves formation of a linoleoyl ester of CoA from dietary linoleic acid, followed by introduction of a double bond at the 6-position. The triply unsaturated product is then elongated (by malonyl-CoA with a decarboxylation step) to yield a 20-carbon fatty acid with double bonds at the 8-, 11-, and 14-positions. A second desaturation reaction at the 5-position, followed by a reverse **acyl-CoA synthetase** reaction (see Chapter 23), liberates the product, a 20-carbon fatty acid with double bonds at the 5-, 8-, 11-, and 14-positions.

Regulatory Control of Fatty Acid Metabolism Is an Interplay of Allosteric Modifiers and Phosphorylation–Dephosphorylation Cycles

The control and regulation of fatty acid synthesis is intimately related to regulation of fatty acid breakdown, glycolysis, and the TCA cycle. Acetyl-CoA is an important intermediate metabolite in all these processes. In these terms, it is easy to appreciate the interlocking relationships in Figure 24.16. Malonyl-CoA can act to prevent fatty acyl-CoA derivatives from entering the mitochondria by inhibiting the carnitine acyltransferase of the outer mitochondrial membrane that initiates this transport. In this way, when fatty acid synthesis is turned on (as signaled by higher levels of malonyl-CoA), β-oxidation is inhibited. As we pointed out earlier, citrate is an important allosteric activator of acetyl-CoA carboxylase, and fatty acyl-CoAs are inhibitors. The degree of inhibition is proportional to the chain length of the fatty acyl-CoA; longer chains show a higher affinity for the allosteric inhibition site on acetyl-CoA carboxylase. Palmitoyl-CoA, stearoyl-CoA, and arachidyl-CoA are the most potent inhibitors of the carboxylase.

FIGURE 24.13 Double bonds are introduced into the growing fatty acid chain in *E. coli* by specific dehydrases. Palmitoleoyl-ACP is synthesized by a sequence of reactions involving four rounds of chain elongation, followed by double bond insertion by β-hydroxydecanoyl thioester dehydrase and three additional elongation steps. Another elongation cycle produces *cis*-vaccenic acid.

Biochemistry Now™ ANIMATED FIGURE 24.14 The conversion of stearoyl-CoA to oleoyl-CoA in eukaryotes is catalyzed by stearoyl-CoA desaturase in a reaction sequence that also involves cytochrome b_5 and cytochrome b_5 reductase. Two electrons are passed from NADH through the chain of reactions as shown, and two electrons are also derived from the fatty acyl substrate. **See this figure animated at http://chemistry.brookscole.com/ggb3**

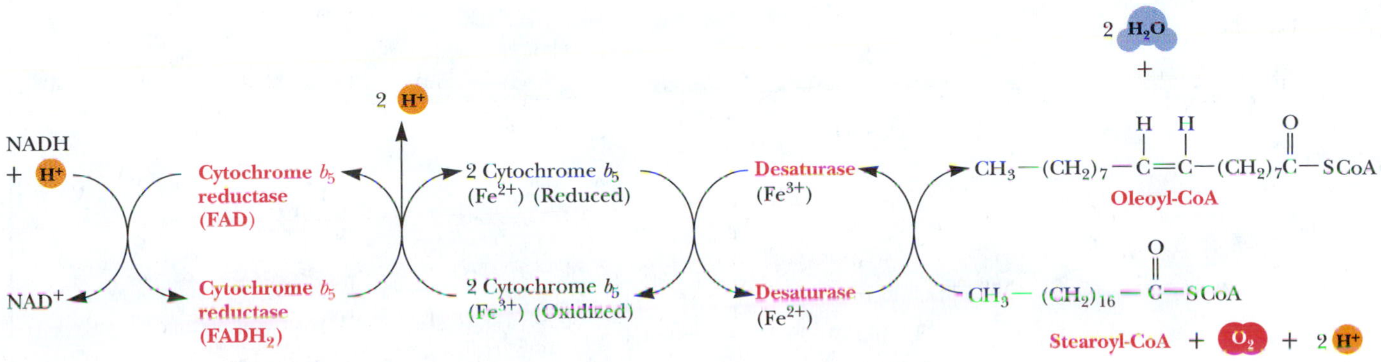

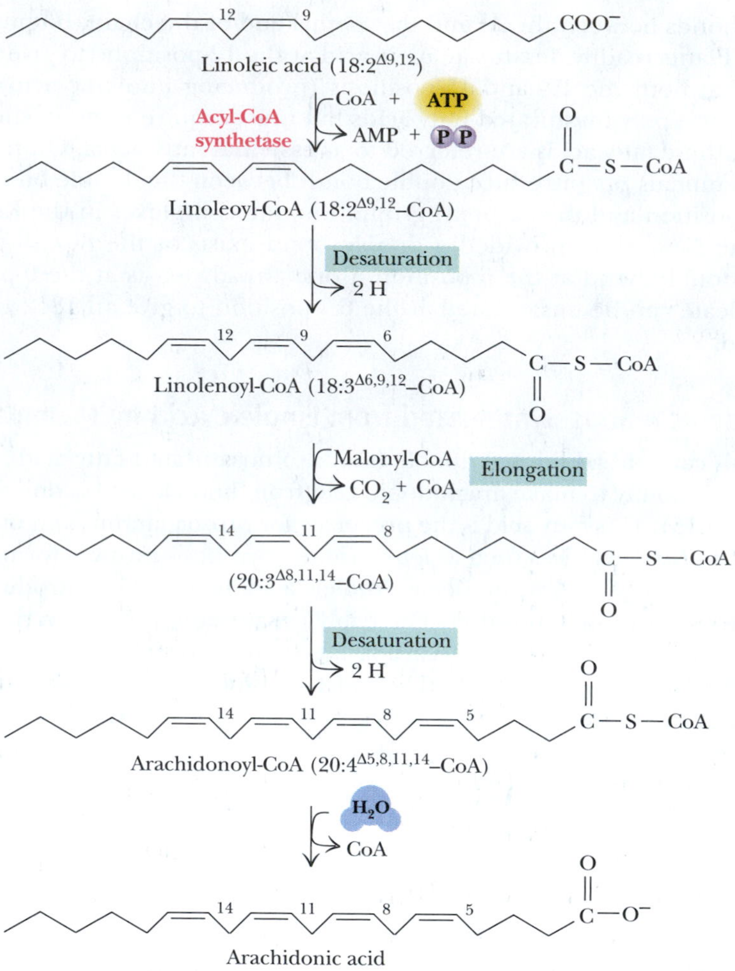

FIGURE 24.15 Arachidonic acid is synthesized from linoleic acid in eukaryotes. This is the only means by which animals can synthesize fatty acids with double bonds at positions beyond C-9.

Linoleic acid $(18:2^{\Delta 9,12})$

Linoleoyl-CoA $(18:2^{\Delta 9,12}$–CoA)

Linolenoyl-CoA $(18:3^{\Delta 6,9,12}$–CoA)

$(20:3^{\Delta 8,11,14}$–CoA)

Arachidonoyl-CoA $(20:4^{\Delta 5,8,11,14}$–CoA)

Arachidonic acid

Human Biochemistry

Docosahexaenoic Acid—A Major Polyunsaturated Fatty Acid in Retina and Brain

Two long-chain, polyunsaturated fatty acids, **docosahexaenoic acid (DHA)** and **eicosapentaenoic acid,** are known as "omega-3 fatty acids," because, counting from the end (omega) of the chain, the first double bond is at the third position (see accompanying figure). These fatty acids have beneficial effects in a variety of organs and biological processes, including growth regulation, modulation of inflammation, platelet activation, and lipoprotein metabolism. Interestingly, especially high levels of DHA have been found in rod cell membranes in animal retina and in neural tissue. DHA is approximately 22% of total fatty acids in animal retina and 35% to 40% of the fatty acids in retinal phosphatidylethanolamine. DHA supports neural and visual development, in part because it is a precursor for eicosanoids that regulate numerous cell and organ functions. Infants can synthesize DHA and other polyunsaturated fatty acids, but the rates of synthesis are low. Strong evidence exists for the importance of these fatty acids in infant nutrition.

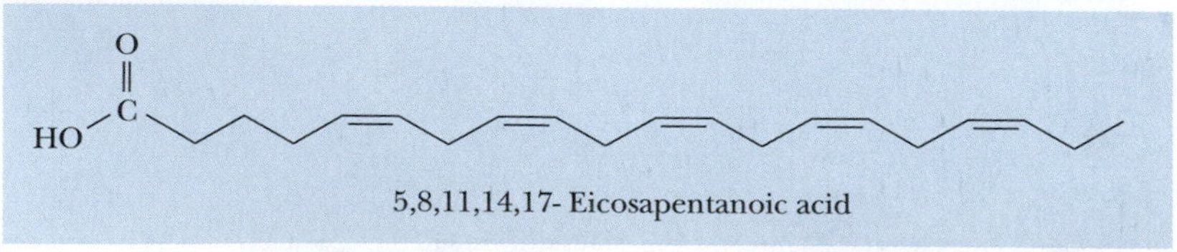

5,8,11,14,17- Eicosapentanoic acid

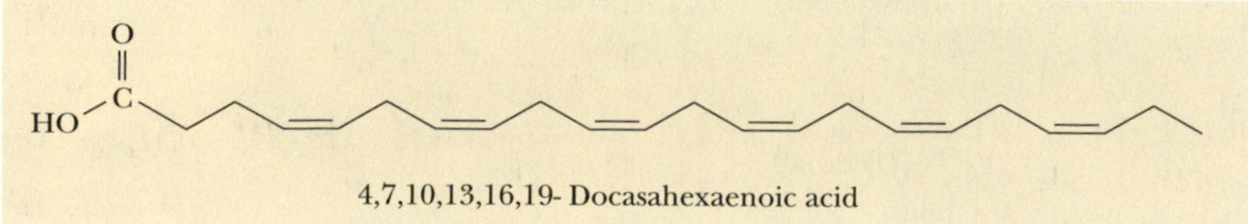

4,7,10,13,16,19- Docasahexaenoic acid

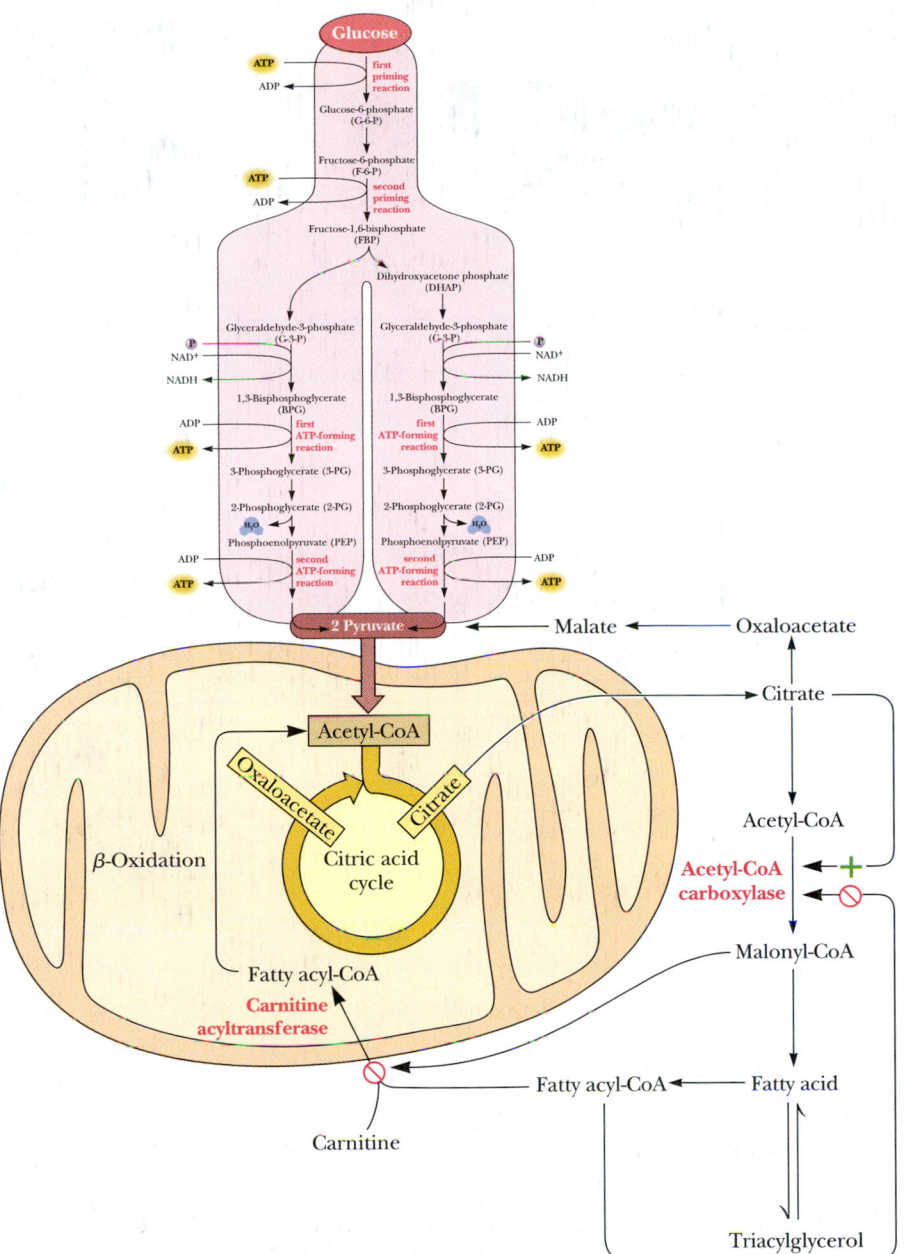

FIGURE 24.16 Regulation of fatty acid synthesis and fatty acid oxidation are coupled as shown. Malonyl-CoA, produced during fatty acid synthesis, inhibits the uptake of fatty acylcarnitine (and thus fatty acid oxidation) by mitochondria. When fatty acyl-CoA levels rise, fatty acid synthesis is inhibited and fatty acid oxidation activity increases. Rising citrate levels (which reflect an abundance of acetyl-CoA) similarly signal the initiation of fatty acid synthesis.

Hormonal Signals Regulate ACC and Fatty Acid Biosynthesis

As described earlier, citrate activation and palmitoyl-CoA inhibition of acetyl-CoA carboxylase are strongly dependent on the phosphorylation state of the enzyme. This provides a crucial connection to hormonal regulation. Many of the enzymes that act to phosphorylate acetyl-CoA carboxylase (Figure 24.4) are controlled by hormonal signals. Glucagon is a good example (Figure 24.17). As noted in Chapter 22, glucagon binding to membrane receptors activates an intracellular cascade involving activation of adenylyl cyclase. Cyclic AMP produced by the cyclase activates a protein kinase, which then phosphorylates acetyl-CoA carboxylase. Unless citrate levels are high, phosphorylation causes inhibition of fatty acid biosynthesis. The carboxylase (and fatty acid synthesis) can be reactivated by a specific phosphatase, which dephosphorylates the carboxylase. Also indicated in Figure 24.17 is the simultaneous activation by glucagon of triacylglycerol lipases, which hydrolyze triacylglycerols, releasing fatty acids for β-oxidation. Both the inactivation of acetyl-CoA carboxylase and the activation of triacylglycerol lipase are counteracted by insulin, whose receptor acts to stimulate a phosphodiesterase that converts cAMP to AMP.

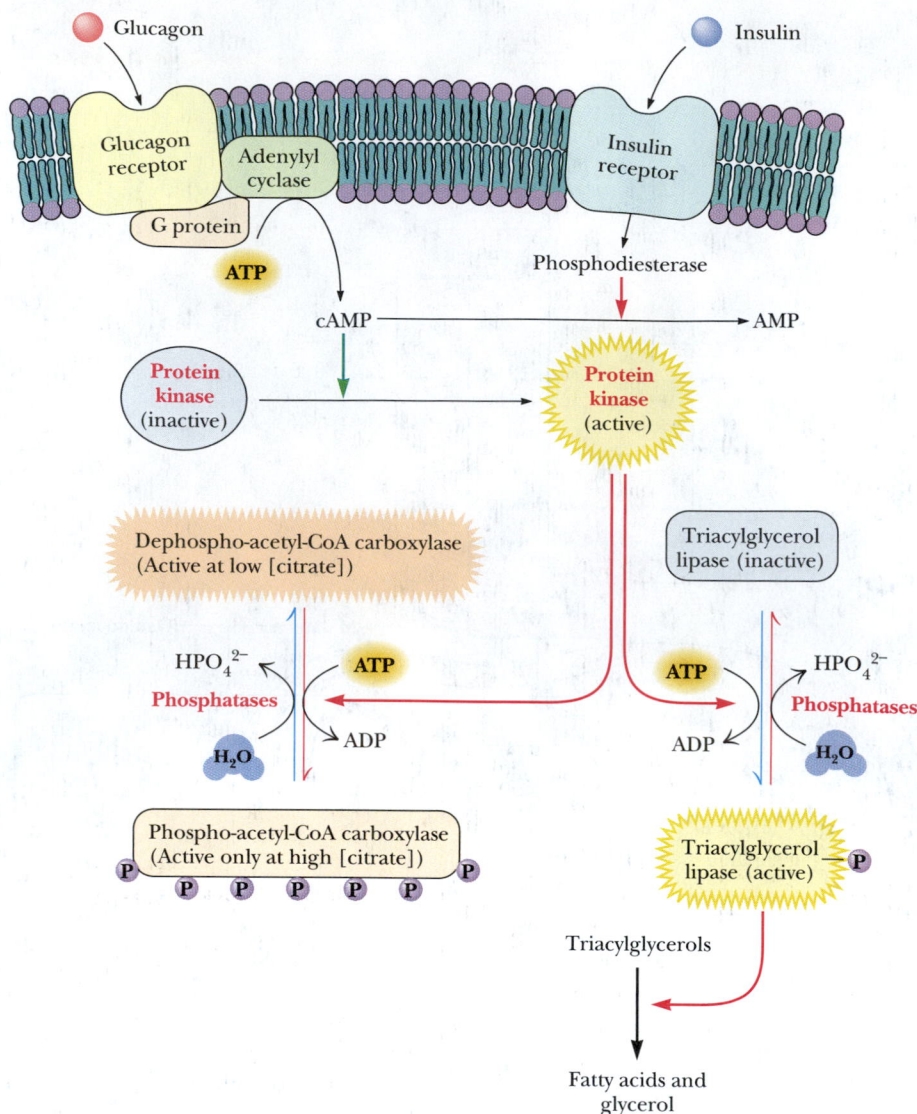

Biochemistry⑤Now™ ANIMATED FIGURE 24.17
Hormonal signals regulate fatty acid synthesis, primarily through actions on acetyl-CoA carboxylase. Availability of fatty acids also depends upon hormonal activation of triacylglycerol lipase. **See this figure animated at http://chemistry.brookscole.com/ggb3**

24.2 | How Are Complex Lipids Synthesized?

Complex lipids consist of backbone structures to which fatty acids are covalently bound. Principal classes include the **glycerolipids,** for which glycerol is the backbone, and **sphingolipids,** which are built on a sphingosine backbone. The two major classes of glycerolipids are **glycerophospholipids** and **triacylglycerols.** The **phospholipids,** which include both glycerophospholipids and sphingomyelins, are crucial components of membrane structure. They are also precursors of hormones such as the *eicosanoids* (for example, *prostaglandins*) and signal molecules (such as the breakdown products of *phosphatidylinositol*).

Different organisms possess greatly different complements of lipids and therefore invoke somewhat different lipid biosynthetic pathways. For example, sphingolipids and triacylglycerols are produced only in eukaryotes. In contrast, bacteria usually have rather simple lipid compositions. Phosphatidylethanolamine accounts for at least 75% of the phospholipids in *E. coli,* with phosphatidylglycerol and cardiolipin accounting for most of the rest. *E. coli* membranes possess no phosphatidylcholine, phosphatidylinositol, sphingolipids, or cholesterol. On the other hand, some bacteria (such as *Pseudomonas*) can synthesize phosphatidylcholine, for example. In this section and the one following, we consider some of the pathways for the synthesis of glycerolipids, sphingolipids, and the eicosanoids, which are derived from phospholipids.

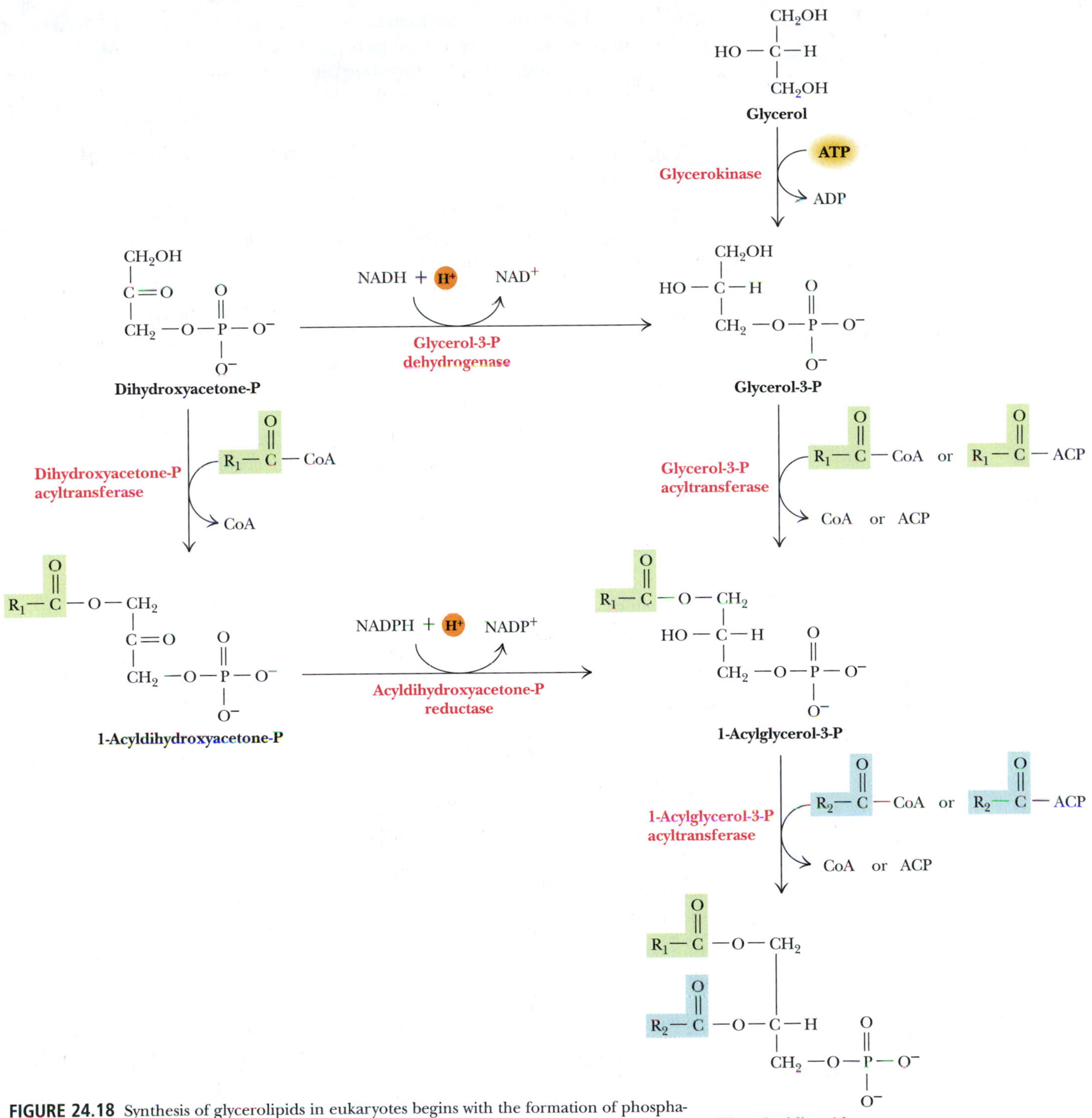

FIGURE 24.18 Synthesis of glycerolipids in eukaryotes begins with the formation of phosphatidic acid, which may be formed from dihydroxyacetone phosphate or glycerol as shown.

Glycerolipids Are Synthesized by Phosphorylation and Acylation of Glycerol

A common pathway operates in nearly all organisms for the synthesis of **phosphatidic acid,** the precursor to other glycerolipids. **Glycerokinase** catalyzes the phosphorylation of glycerol to form glycerol-3-phosphate, which is then acylated at both the 1- and 2-positions to yield phosphatidic acid (Figure 24.18). The first acylation, at position 1, is catalyzed by **glycerol-3-phosphate acyltransferase,** an enzyme that in most organisms is specific for saturated fatty acyl groups. Eukaryotic systems can also utilize **dihydroxyacetone phosphate** as a starting point for synthesis of phosphatidic acid (Figure 24.18). Again a specific acyltransferase adds the first acyl chain, followed by reduction of the backbone keto group by **acyldihydroxyacetone phosphate reductase,** using NADPH as the reductant. Alternatively, dihydroxyacetone phosphate can be reduced to glycerol-3-phosphate by **glycerol-3-phosphate dehydrogenase.**

Eukaryotes Synthesize Glycerolipids from CDP-Diacylglycerol or Diacylglycerol

In eukaryotes, phosphatidic acid is converted directly either to diacylglycerol or to *cytidine diphosphodiacylglycerol* (or simply *CDP-diacylglycerol;* Figure 24.19). From these two precursors, all other glycerophospholipids in eukaryotes are derived. Diacylglycerol is a precursor for synthesis of triacylglycerol, phosphatidylethanolamine, and phosphatidylcholine. Triacylglycerol is synthesized mainly in adipose tissue, liver, and intestines and serves as the principal energy storage molecule in eukaryotes. Triacylglycerol biosynthesis in liver and adipose tissue occurs via **diacylglycerol acyltransferase,** an enzyme bound to the cytoplasmic face of the ER. A different route is used, however, in intestines. Recall (see Figure 23.3) that triacylglycerols from the diet are broken down to 2-monoacylglycerols by specific lipases. Acyltransferases then acylate 2-monoacylglycerol to produce new triacylglycerols (Figure 24.20).

Phosphatidylethanolamine Is Synthesized from Diacylglycerol and CDP-Ethanolamine

Phosphatidylethanolamine synthesis begins with phosphorylation of ethanolamine to form phosphoethanolamine (Figure 24.19). The next reaction involves transfer of a cytidylyl group from CTP to form CDP-ethanolamine and pyrophosphate. As always, PP_i hydrolysis drives this reaction forward. A specific **phosphoethanolamine transferase** then links phosphoethanolamine to the diacylglycerol backbone. Biosynthesis of phosphatidylcholine is entirely analogous because animals synthesize it directly. All of the choline utilized in this pathway must be acquired from the diet. Yeast, certain bacteria, and animal livers, however, can convert phosphatidylethanolamine to phosphatidylcholine by methylation reactions involving *S*-adenosylmethionine (see Chapter 25).

Exchange of Ethanolamine for Serine Converts Phosphatidylethanolamine to Phosphatidylserine

Mammals synthesize phosphatidylserine (PS) in a calcium ion–dependent reaction involving aminoalcohol exchange (Figure 24.21). The enzyme catalyzing this reaction is associated with the ER and will accept phosphatidylethanolamine (PE) and other phospholipid substrates. A mitochondrial **PS decarboxylase** can subsequently convert PS to PE. No other pathway converting serine to ethanolamine has been found.

Eukaryotes Synthesize Other Phospholipids Via CDP-Diacylglycerol

Eukaryotes also use CDP-diacylglycerol, derived from phosphatidic acid, as a precursor for several other important phospholipids, including phosphatidylinositol (PI), phosphatidylglycerol (PG), and cardiolipin (Figure 24.22). PI accounts for only about 2% to 8% of the lipids in most animal membranes, but breakdown products of PI, including inositol-1,4,5-trisphosphate and diacylglycerol, are second messengers in a vast array of cellular signaling processes.

Dihydroxyacetone Phosphate Is a Precursor to the Plasmalogens

Certain glycerophospholipids possess alkyl or alkenyl ether groups at the 1-position in place of an acyl ester group. These glyceroether phospholipids are synthesized from dihydroxyacetone phosphate (Figure 24.23). Acylation of dihy-

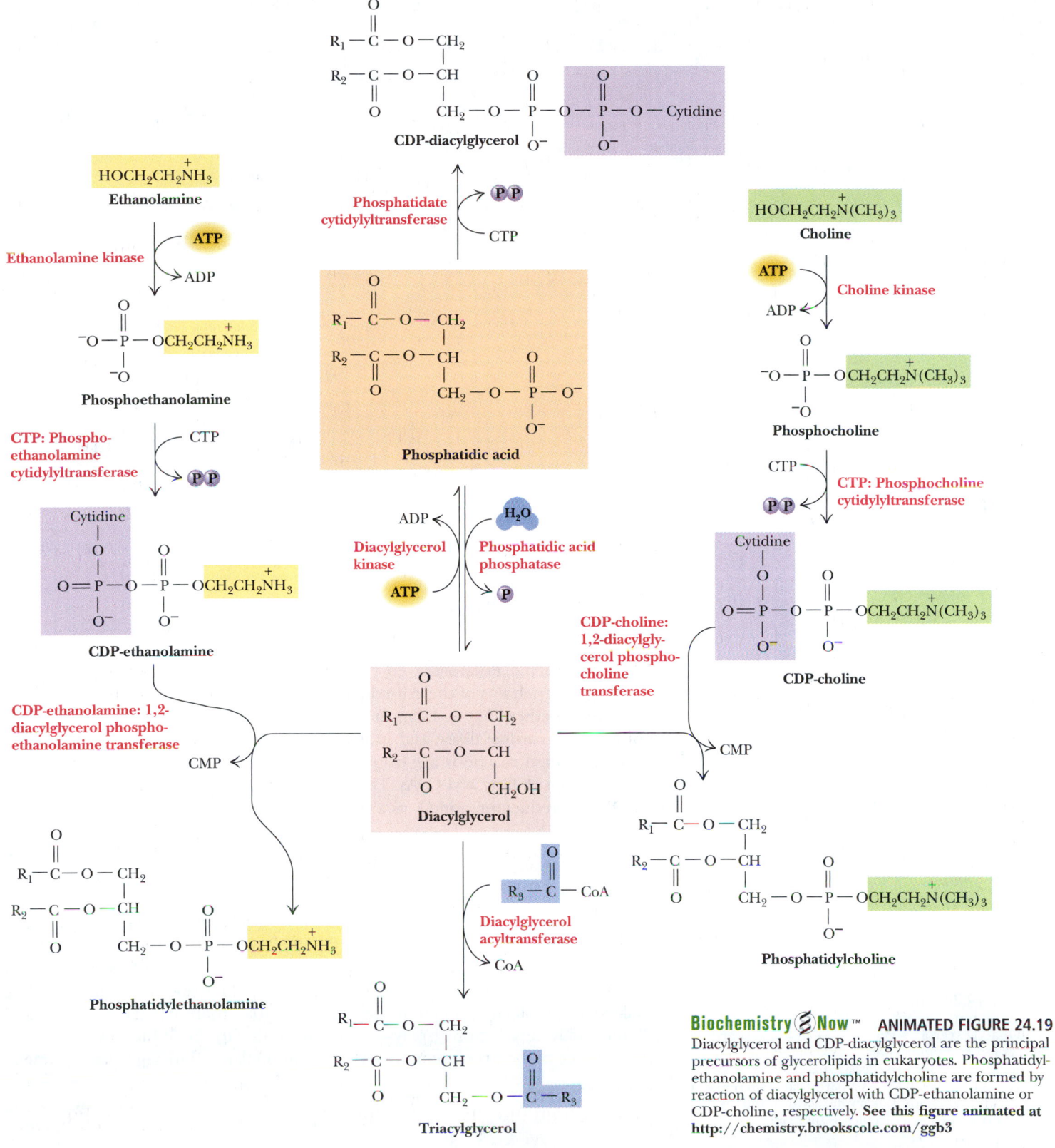

Biochemistry⊗Now™ ANIMATED FIGURE 24.19
Diacylglycerol and CDP-diacylglycerol are the principal precursors of glycerolipids in eukaryotes. Phosphatidylethanolamine and phosphatidylcholine are formed by reaction of diacylglycerol with CDP-ethanolamine or CDP-choline, respectively. **See this figure animated at http://chemistry.brookscole.com/ggb3**

droxyacetone phosphate (DHAP) is followed by an exchange reaction, in which the acyl group is removed as a carboxylic acid and a long-chain alcohol adds to the 1-position. This long-chain alcohol is derived from the corresponding acyl-CoA by means of an **acyl-CoA reductase** reaction involving oxidation of two molecules of NADH. The *2-keto* group of the DHAP backbone is then reduced to an alcohol, followed by acylation. The resulting 1-alkyl-2-acylglycero-3-phosphate

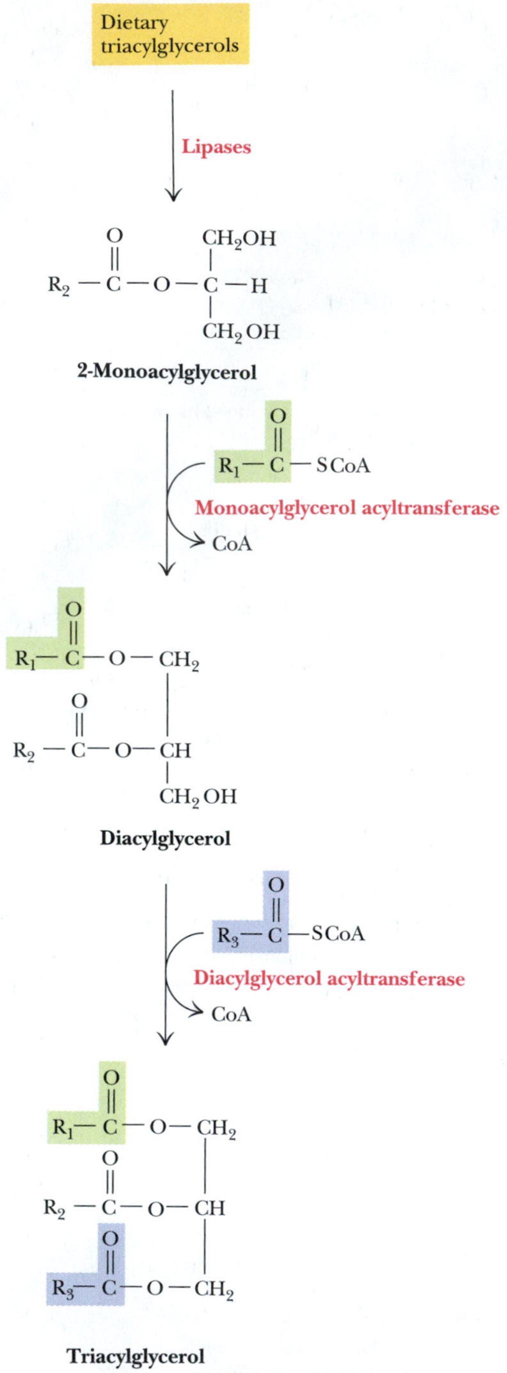

FIGURE 24.20 Triacylglycerols are formed primarily by the action of acyltransferases on monoglycerol and diacylglycerol. Acyltransferase in *E. coli* is an integral membrane protein (83 kD) and can utilize either fatty acyl-CoAs or acylated acyl carrier proteins as substrates. It shows a particular preference for palmitoyl groups. Eukaryotic acyltransferases use only fatty acyl-CoA molecules as substrates.

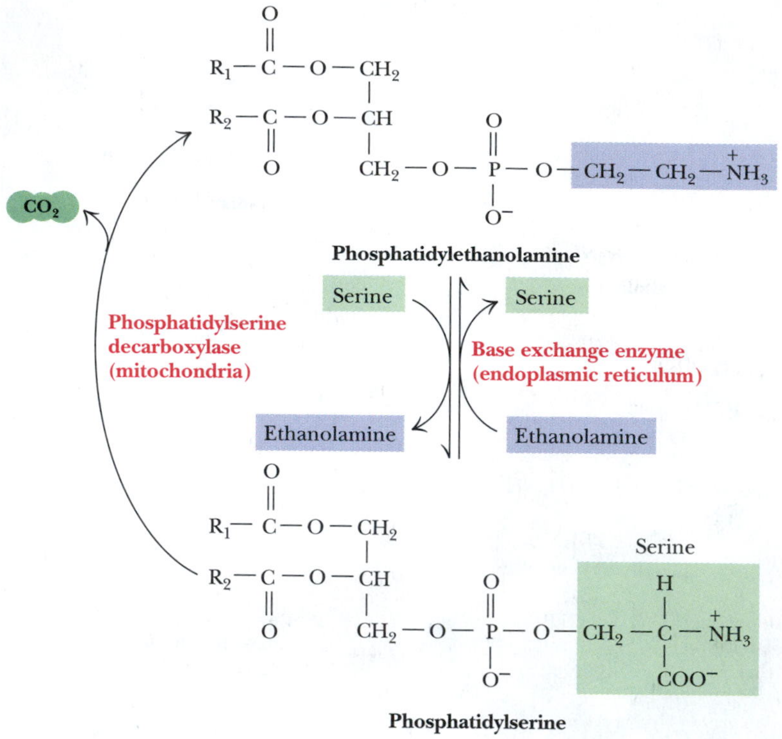

FIGURE 24.21 The interconversion of phosphatidylethanolamine and phosphatidylserine in mammals.

can react in a manner similar to phosphatidic acid to produce ether analogs of phosphatidylcholine, phosphatidylethanolamine, and so forth (Figure 24.23). In addition, specific **desaturase** enzymes associated with the ER can desaturate the alkyl ether chains of these lipids as shown. The products, which contain α,β-unsaturated ether-linked chains at the C-1 position, are **plasmalogens;** they are abundant in cardiac tissue and in the central nervous system. The desaturases catalyzing these reactions are distinct from but similar to those that introduce unsaturations in fatty acyl-CoAs. These enzymes use cytochrome b_5 as a cofactor, NADH as a reductant, and O_2 as a terminal electron acceptor.

Platelet-Activating Factor Is Formed by Acetylation of 1-Alkyl-2-Lysophosphatidylcholine

A particularly interesting ether phospholipid with unusual physiological properties has recently been characterized. As shown in Figure 24.24, **1-alkyl-2-acetylglycerophosphocholine,** also known as **platelet-activating factor,** possesses an alkyl ether at C-1 and an acetyl group at C-2. The very short chain at C-2 makes this molecule much more water soluble than typical glycerolipids. Platelet-activating factor displays a dramatic ability to dilate blood vessels (and thus reduce blood pressure in hypertensive animals) and to aggregate platelets.

Sphingolipid Biosynthesis Begins with Condensation of Serine and Palmitoyl-CoA

Sphingolipids, ubiquitous components of eukaryotic cell membranes, are present at high levels in neural tissues. The myelin sheath that insulates nerve axons is particularly rich in sphingomyelin and other related lipids. Prokaryotic organisms normally do not contain sphingolipids. Sphingolipids are built upon sphingosine backbones rather than glycerol. The initial reaction, which involves condensation of serine and palmitoyl-CoA with release of bicarbonate, is catalyzed by

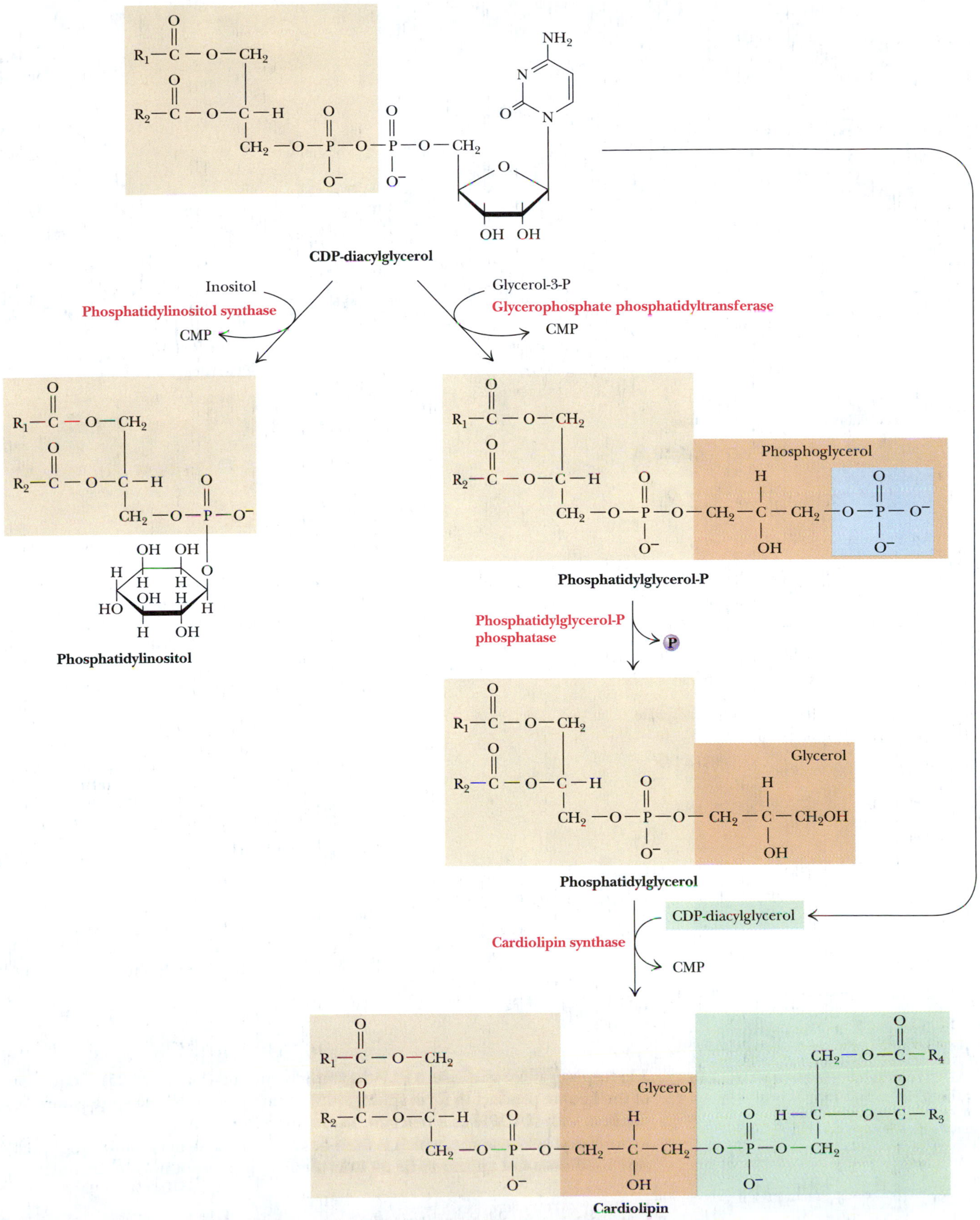

FIGURE 24.22 CDP-diacylglycerol is a precursor of phosphatidylinositol, phosphatidylglycerol, and cardiolipin in eukaryotes.

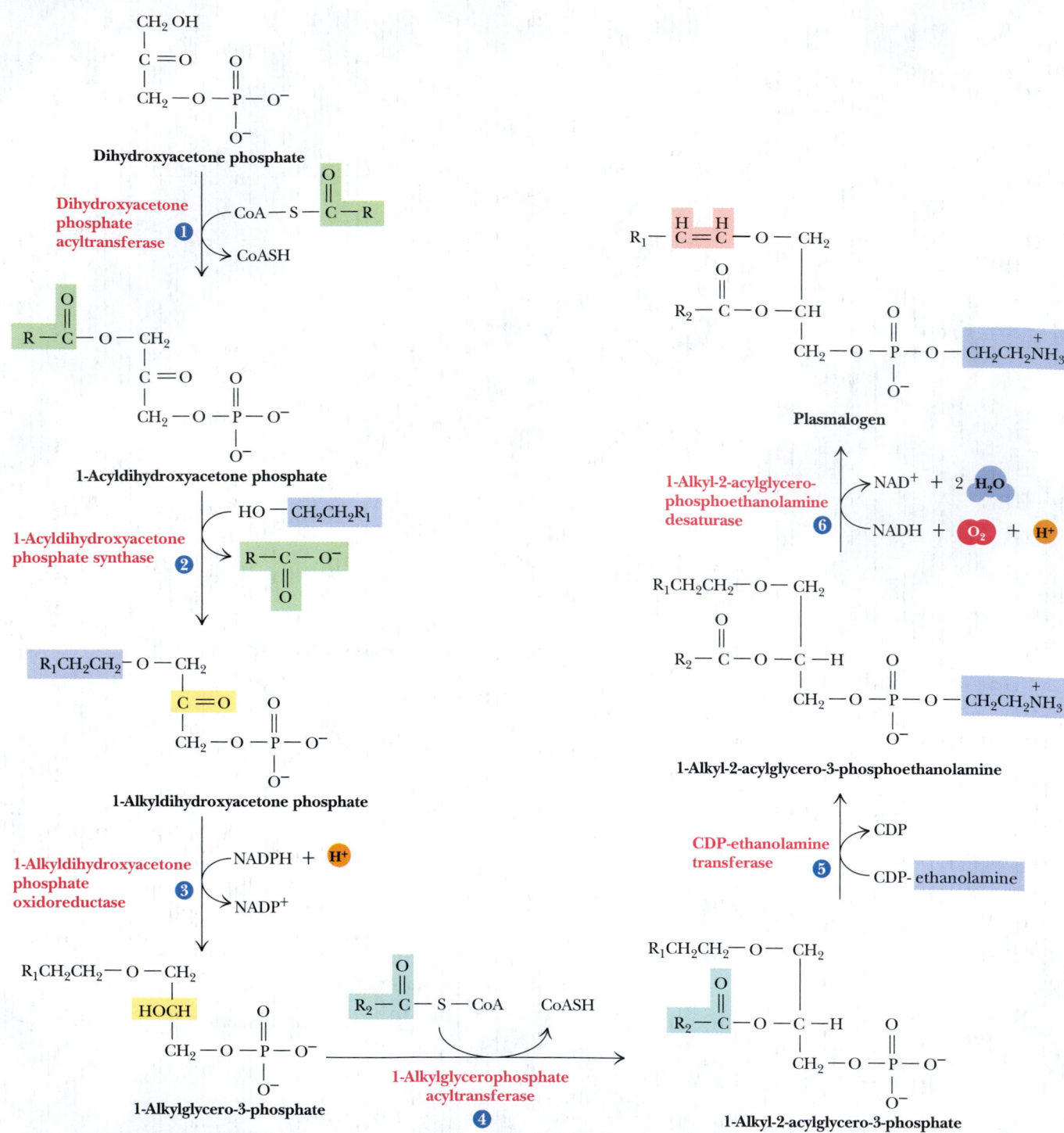

FIGURE 24.23 Biosynthesis of plasmalogens in animals. **(1)** Acylation at C-1 is followed by **(2)** exchange of the acyl group for a long-chain alcohol. **(3)** Reduction of the keto group at C-2 is followed by **(4 and 5)** transferase reactions, which add an acyl group at C-2 and a polar head-group moiety, and a **(6)** desaturase reaction that forms a double bond in the alkyl chain. The first two enzymes are of cytoplasmic origin, and the last transferase is located at the endoplasmic reticulum.

3-ketosphinganine synthase, a PLP-dependent enzyme (Figure 24.25). Reduction of the ketone product to form sphinganine is catalyzed by **3-keto-sphinganine reductase,** with NADPH as a reactant. In the next step, sphinganine is acylated to form *N*-acyl sphinganine, which is then desaturated to form ceramide. Sphingosine itself does not appear to be an intermediate in this pathway in mammals.

Ceramide Is the Precursor for Other Sphingolipids and Cerebrosides

Ceramide is the building block for all other sphingolipids. Sphingomyelin, for example, is produced by transfer of phosphocholine from phosphatidylcholine (Figure 24.26). Glycosylation of ceramide by sugar nucleotides yields

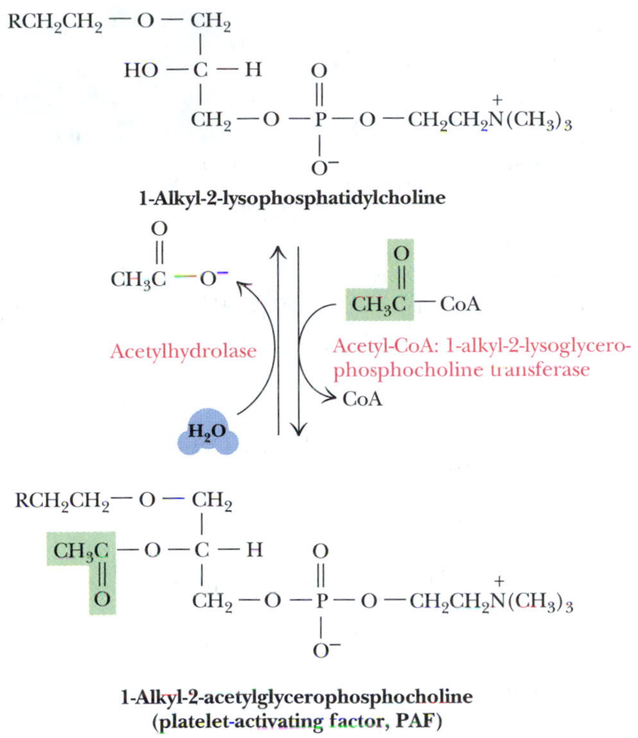

$$RCH_2CH_2 - O - CH_2$$

$$HO - C - H$$

$$CH_2 - O - P - O - CH_2CH_2N(CH_3)_3$$

$$O^-$$

1-Alkyl-2-lysophosphatidylcholine

$$CH_3C - O^-$$

Acetylhydrolase

$$CH_3C - CoA$$

Acetyl-CoA: 1-alkyl-2-lysoglycero-phosphocholine transferase

CoA

H₂O

$$RCH_2CH_2 - O - CH_2$$

$$CH_3C - O - C - H$$

$$O$$

$$CH_2 - O - P - O - CH_2CH_2N(CH_3)_3$$

$$O^-$$

1-Alkyl-2-acetylglycerophosphocholine
(platelet-activating factor, PAF)

FIGURE 24.24 Platelet-activating factor, formed from 1-alkyl-2-lysophosphatidylcholine by acetylation at C-2, is degraded by the action of acetylhydrolase.

cerebrosides, such as galactosylceramide, which makes up about 15% of the lipids of myelin sheath structures. Cerebrosides that contain one or more sialic acid (*N*-acetylneuraminic acid) moieties are called **gangliosides.** Several dozen gangliosides have been characterized, and the general form of the biosynthetic pathway is illustrated for the case of ganglioside GM$_2$ (Figure 24.26). Sugar units are added to the developing ganglioside from nucleotide derivatives, including UDP–*N*-acetylglucosamine, UDP–galactose, and UDP–glucose.

24.3 How Are Eicosanoids Synthesized, and What Are Their Functions?

Eicosanoids, so named because they are all derived from 20-carbon fatty acids, are ubiquitous breakdown products of phospholipids. In response to appropriate stimuli, cells activate the breakdown of selected phospholipids (Figure 24.27). Phospholipase A$_2$ (see Chapter 8) selectively cleaves fatty acids from the C-2 position of phospholipids. Often these are unsaturated fatty acids, among which is arachidonic acid. Arachidonic acid may also be released from phospholipids by the combined actions of phospholipase C (which yields diacylglycerols) and diacylglycerol lipase (which releases fatty acids).

Eicosanoids Are Local Hormones

Animal cells can modify arachidonic acid and other polyunsaturated fatty acids, in processes often involving cyclization and oxygenation, to produce so-called local hormones that (1) exert their effects at very low concentrations and (2) usually act near their sites of synthesis. These substances include the prostaglandins (PG) (Figure 24.27) as well as **thromboxanes** (Tx), **leukotrienes,** and other **hydroxy-eicosanoic acids.** Thromboxanes, discovered in blood platelets (*thrombocytes*), are cyclic ethers (TxB$_2$ is actually a hemiacetal; see Figure 24.27) with a hydroxyl group at C-15.

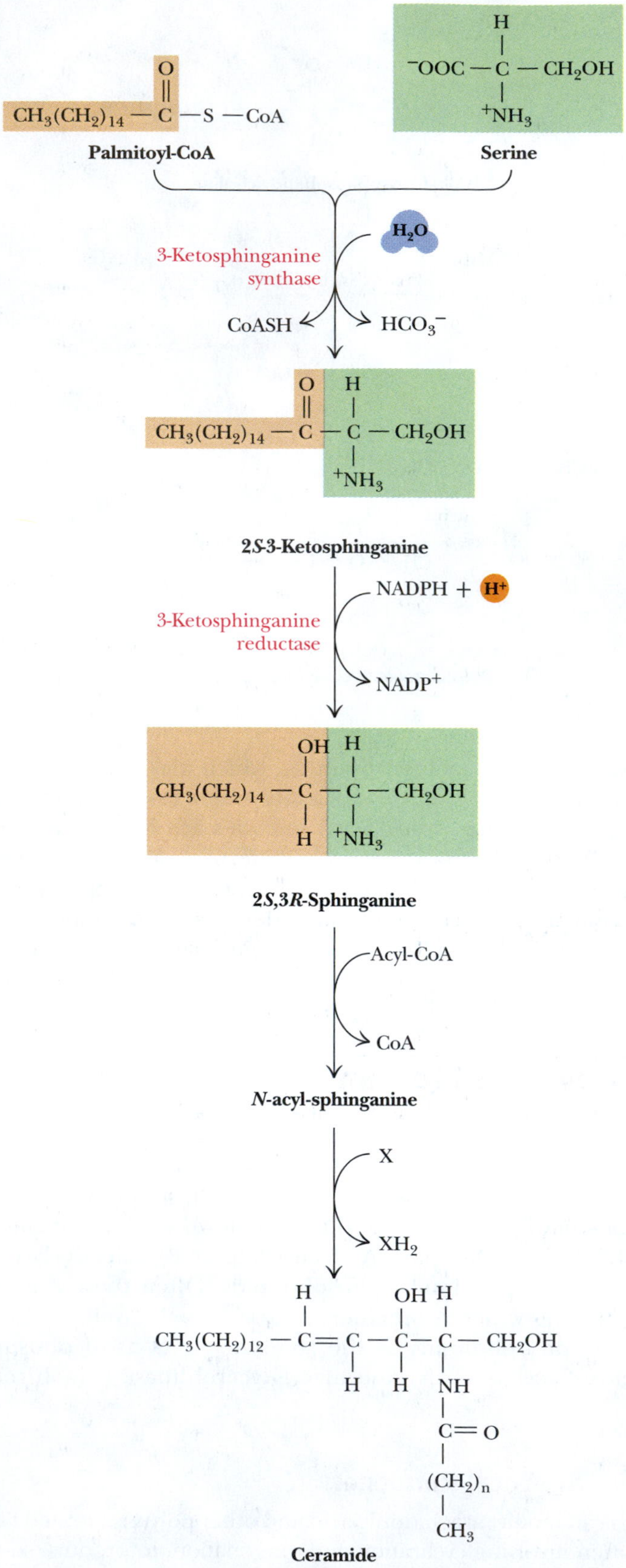

FIGURE 24.25 Biosynthesis of sphingolipids in animals begins with the 3-ketosphinganine synthase reaction, a PLP-dependent condensation of palmitoyl-CoA and serine. Subsequent reduction of the keto group, acylation, and desaturation (via reduction of an electron acceptor, X) form ceramide, the precursor of other sphingolipids.

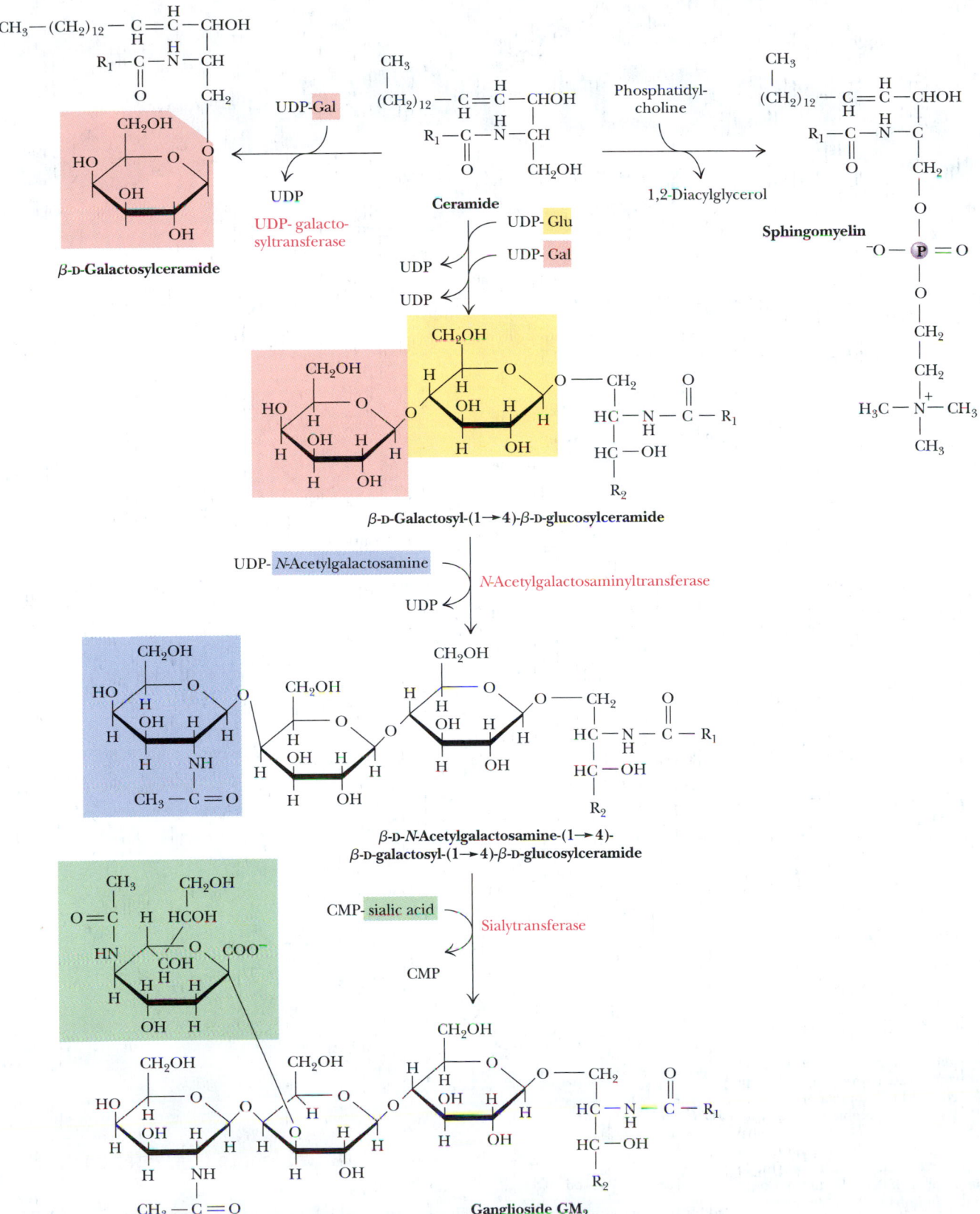

FIGURE 24.26 Glycosylceramides (such as galactosylceramide), gangliosides, and sphingomyelins are synthesized from ceramide in animals.

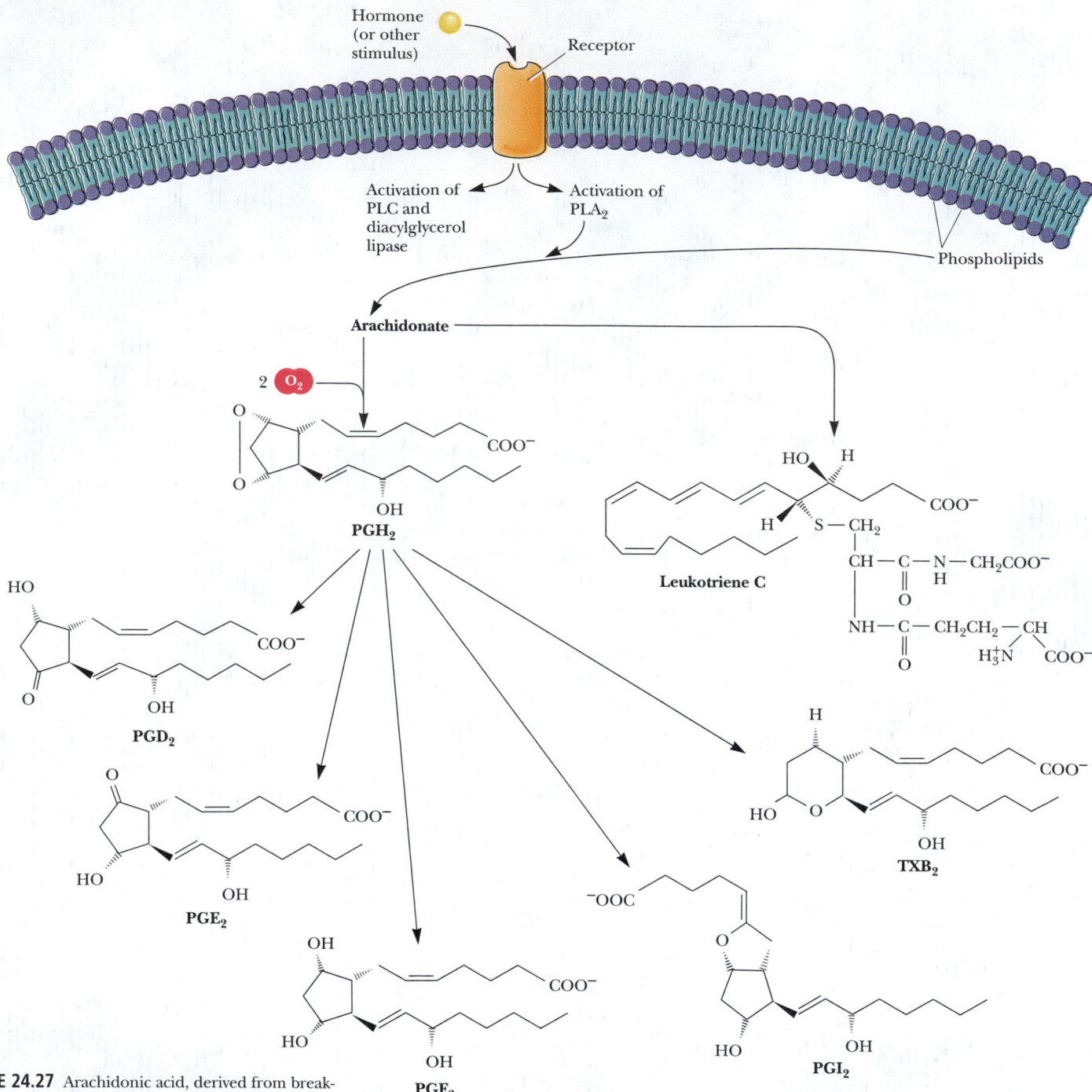

FIGURE 24.27 Arachidonic acid, derived from breakdown of phospholipids (PL), is the precursor of prostaglandins, thromboxanes, and leukotrienes. The letters used to name the prostaglandins are assigned on the basis of similarities in structure and physical properties. The class denoted PGE, for example, consists of β-hydroxyketones that are soluble in ether, whereas PGF denotes 1,3-diols that are soluble in phosphate buffer. PGA denotes prostaglandins possessing α,β-unsaturated ketones. The number following the letters refers to the number of carbon–carbon double bonds. Thus, PGE$_2$ contains two double bonds.

Prostaglandins Are Formed from Arachidonate by Oxidation and Cyclization

All prostaglandins are cyclopentanoic acids derived from arachidonic acid. The biosynthesis of prostaglandins is initiated by an enzyme associated with the ER, called **prostaglandin endoperoxide synthase**, also known as **cyclooxygenase.** The enzyme catalyzes simultaneous oxidation and cyclization of arachidonic acid. The enzyme is viewed as having two distinct activities, cyclooxygenase and peroxidase, as shown in Figure 24.28.

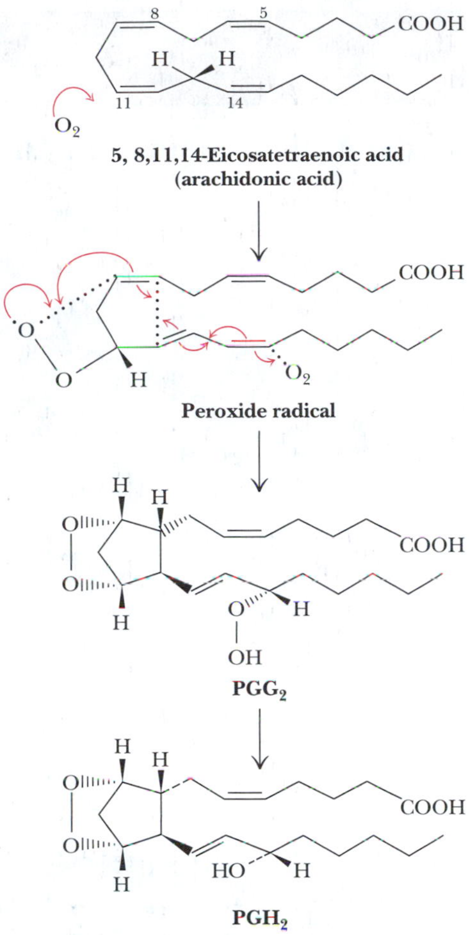

5, 8,11,14-Eicosatetraenoic acid
(arachidonic acid)

Peroxide radical

PGG$_2$

PGH$_2$

FIGURE 24.28 Prostaglandin endoperoxide synthase, the enzyme that converts arachidonic acid to prostaglandin PGH$_2$, possesses two distinct activities: cyclooxygenase (steps 1 and 2) and glutathione (GSSG)–dependent hydroperoxidase (step 3). Cyclooxygenase is the site of action of aspirin and many other analgesic agents.

A Variety of Stimuli Trigger Arachidonate Release and Eicosanoid Synthesis

The release of arachidonate and the synthesis or interconversion of eicosanoids can be initiated by a variety of stimuli, including histamine, hormones such as epinephrine and bradykinin, proteases such as thrombin, and even serum albumin. An important mechanism of arachidonate release and eicosanoid synthesis involves tissue injury and inflammation. When tissue damage or injury occurs,

---

A Deeper Look

The Discovery of Prostaglandins

The name *prostaglandin* was given to this class of compounds by Ulf von Euler, their discoverer, in Sweden in the 1930s. He extracted fluids containing these components from human semen. Because he thought they originated in the prostate gland, he named them prostaglandins. Actually, they were synthesized in the seminal vesicles, and it is now known that similar substances are synthesized in most animal tissues (both male and female). Von Euler observed that injection of these substances into animals caused smooth muscle contraction and dramatic lowering of blood pressure.

Von Euler (and others) soon found that it is difficult to analyze and characterize these obviously interesting compounds because they are present at extremely low levels. Prostaglandin E$_2\alpha$, or

PGE$_2\alpha$, is present in human serum at a level of less than $10^{-14} M!$ In addition, they often have half-lives of only 30 seconds to a few minutes, not lasting long enough to be easily identified. Moreover, most animal tissues upon dissection and homogenization rapidly synthesize and degrade a variety of these substances, so the amounts obtained in isolation procedures are extremely sensitive to the methods used and highly variable even when procedures are carefully controlled. Sune Bergstrom and his colleagues described the first structural determinations of prostaglandins in the late 1950s. In the early 1960s, dramatic advances in laboratory techniques, such as NMR spectroscopy and mass spectrometry, made further characterization possible.

A Deeper Look

The Molecular Basis for the Action of Nonsteroidal Anti-inflammatory Drugs

Prostaglandins are potent mediators of inflammation. The first and committed step in the production of prostaglandins from arachidonic acid is the bis-oxygenation of arachidonate to prostaglandin PGG_2. This is followed by reduction to PGH_2 in a peroxidase reaction. Both these reactions are catalyzed by prostaglandin endoperoxide synthase, also known as PGH_2 synthase or cyclooxygenase, thus abbreviated COX. This enzyme is inhibited by the family of drugs known as nonsteroidal anti-inflammatory drugs, or NSAIDs. Aspirin, ibuprofen, flurbiprofen, and acetaminophen (trade name Tylenol) are all NSAIDs.

There are two isoforms of COX in animals: COX-1 (figure a), which carries out normal, physiological production of prostaglandins, and COX-2 (figure b), which is induced by cytokines, mitogens, and endotoxins in inflammatory cells and is responsible for the production of prostaglandins in inflammation.

The enzyme structure shown here is that of residues 33 to 583 of COX-1 from sheep, inactivated by bromoaspirin. These 551 residues comprise three distinct domains. The first of these, residues 33 to 72 (purple), form a small, compact module that is similar to epidermal growth factor. The second domain, composed of residues 73 to 116 (yellow), forms a right-handed spiral of four α-helical segments along the base of the protein. These α-helical segments form a membrane-binding motif. The helical segments are amphipathic, with most of the hydrophobic residues (shown in green) facing away from the protein, where they can interact with a lipid bilayer. The third domain of the COX enzyme, the catalytic domain (in blue), is a globular structure that contains both the cyclooxygenase and peroxidase active sites.

The cyclooxygenase active site lies at the end of a long, narrow, hydrophobic tunnel or channel. Three of the α-helices of the membrane-binding domain lie at the entrance to this tunnel. The walls of the tunnel are defined by four α-helices, formed by residues 106 to 123, 325 to 353, 379 to 384, and 520 to 535 (shown in orange).

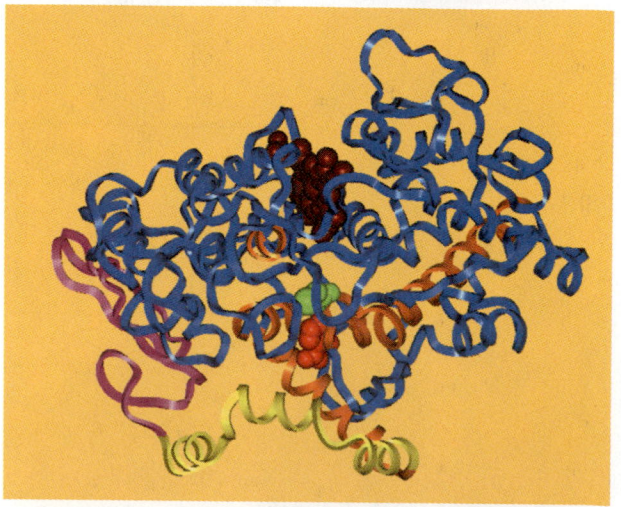

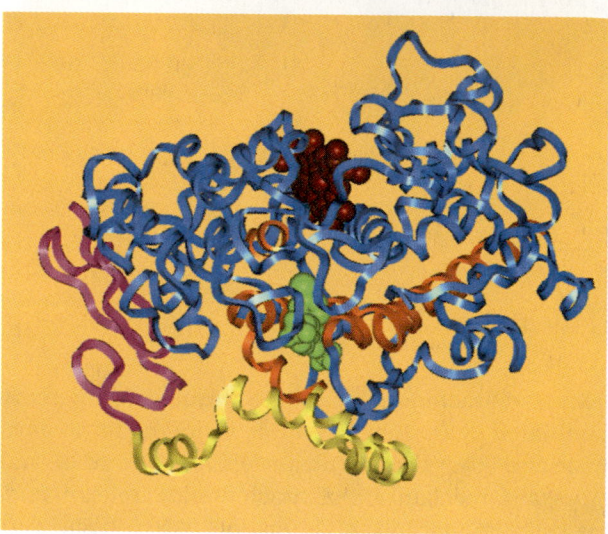

▲ **(a)** COX-1. **(b)** COX-2.

special inflammatory cells, **monocytes** and **neutrophils,** invade the injured tissue and interact with the resident cells (such as smooth muscle cells and fibroblasts). *This interaction typically leads to arachidonate release and eicosanoid synthesis.* Examples of tissue injury in which eicosanoid synthesis has been characterized include heart attack (myocardial infarction), rheumatoid arthritis, and ulcerative colitis.

"Take Two Aspirin and…" Inhibit Your Prostaglandin Synthesis

In 1971, biochemist John Vane was the first to show that **aspirin** (acetylsalicylate; Figure 24.29) exerts most of its effects by inhibiting the biosynthesis of prostaglandins. Its site of action is prostaglandin endoperoxide synthase. Cyclooxygenase activity is destroyed when aspirin O-acetylates Ser^{530} on the enzyme. From this you may begin to infer something about how prostaglandins (and aspirin) function. Prostaglandins are known to enhance inflammation in animal tissues. Aspirin exerts its powerful anti-inflammatory effect by inhibiting this first step in their synthesis. Aspirin does not have any measurable effect on the

In this bromoaspirin-inactivated structure, Ser[530], which lies along the wall of the tunnel, is bromoacetylated, and a molecule of salicylate is also bound in the tunnel. Deep in the tunnel, at the far end, lies Tyr[385], a catalytically important residue. Heme-dependent peroxidase activity is implicated in the formation of a proposed Tyr[385] radical, which is required for cyclooxygenase activity. Aspirin and other NSAIDs block the synthesis of prostaglandins by filling and blocking the tunnel, preventing the migration of arachidonic acid to Tyr[385] in the active site at the back of the tunnel.

There are thought to be at least four different mechanisms of action for NSAIDs. Aspirin (and also bromoaspirin) covalently modifies a residue in the tunnel, thus irreversibly inactivating both COX-1 and COX-2. Ibuprofen acts instead by competing in a reversible fashion for the substrate-binding site in the tunnel.

Flurbiprofen and indomethacin, which comprise the third class of inhibitors, cause a slow, time-dependent inhibition of COX-1 and COX-2, apparently via formation of a salt bridge between a carboxylate on the drug and Arg[120], which lies in the tunnel.

The new drugs Vioxx (from Merck) and Celebrex (from Pfizer) act by a fourth mechanism, specifically inhibition of COX-2. Selective COX-2 inhibitors will likely be the drugs of the future, because they selectively block the inflammation mediated by COX-2, without the potential for stomach lesions and renal toxicity that arise from COX-1 inhibition.

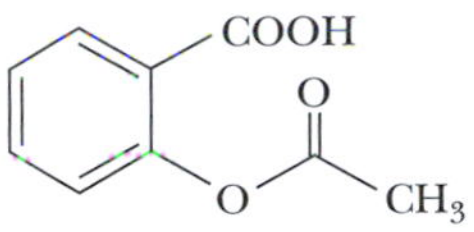

Aspirin

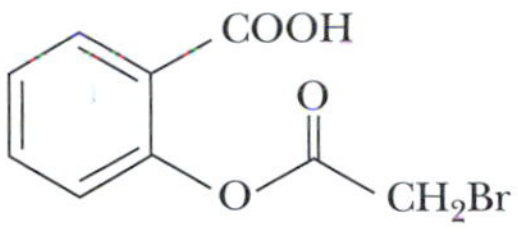

Bromoaspirin

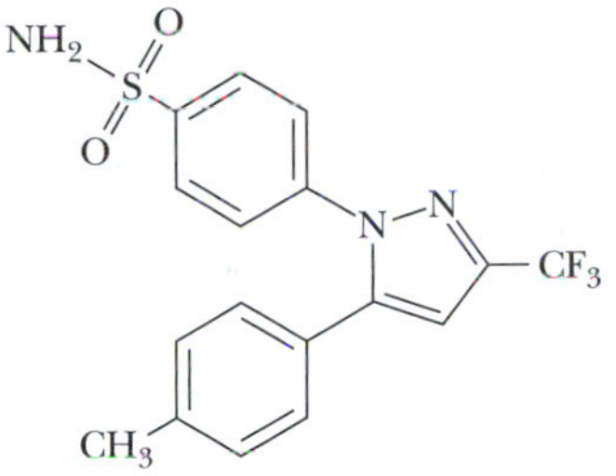

Celebrex

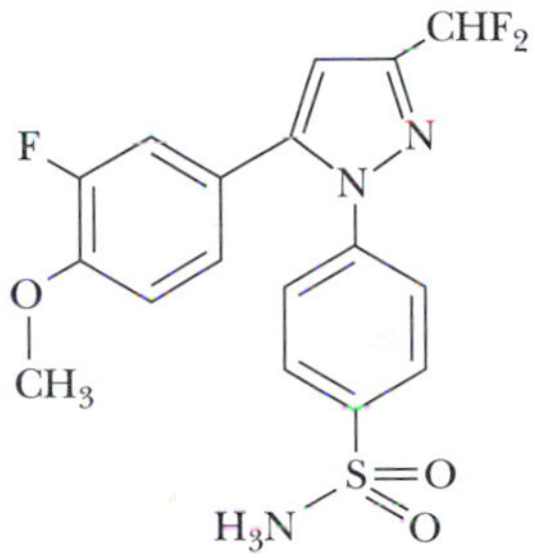

Deramaxx for dogs*

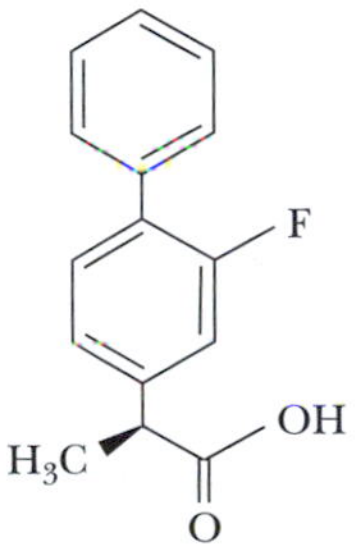

Flurbiprofen

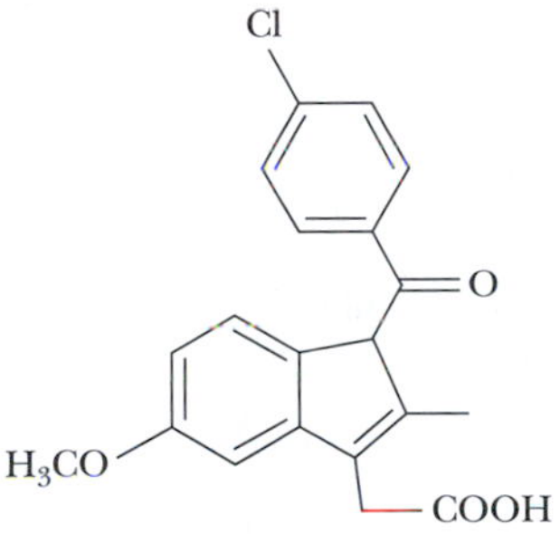

Indomethacin

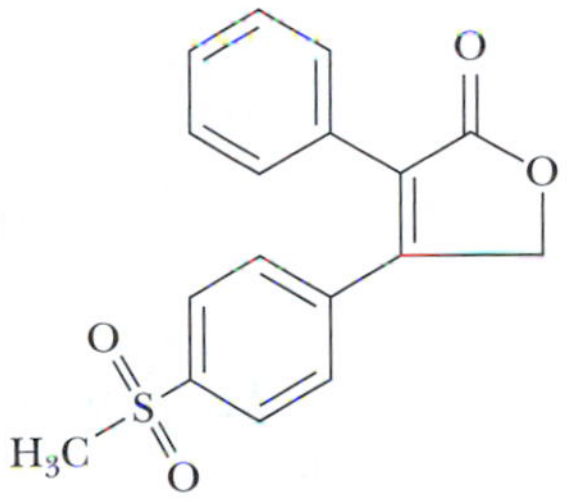

Vioxx

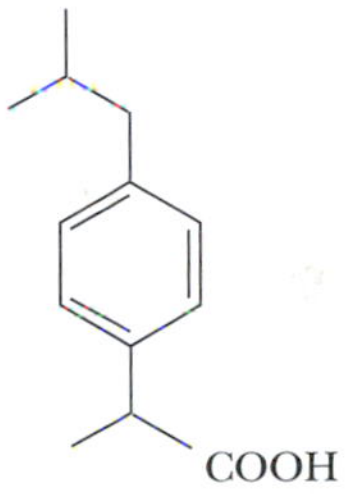

Ibuprofen

* Clancy Garrett takes this.

(a)

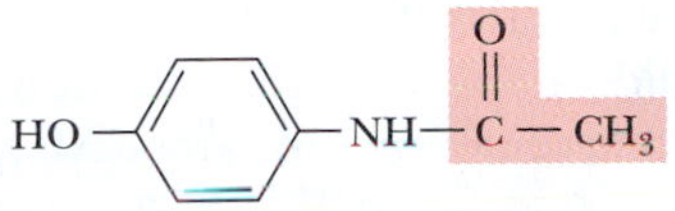

Acetaminophen

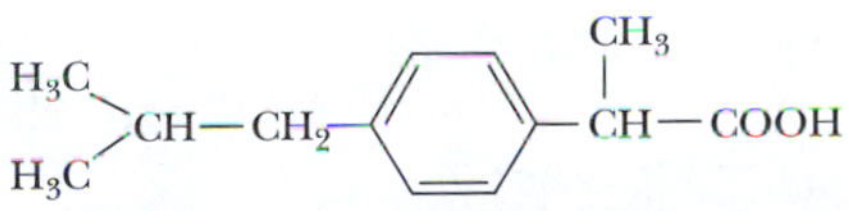

Ibuprofen

(b)

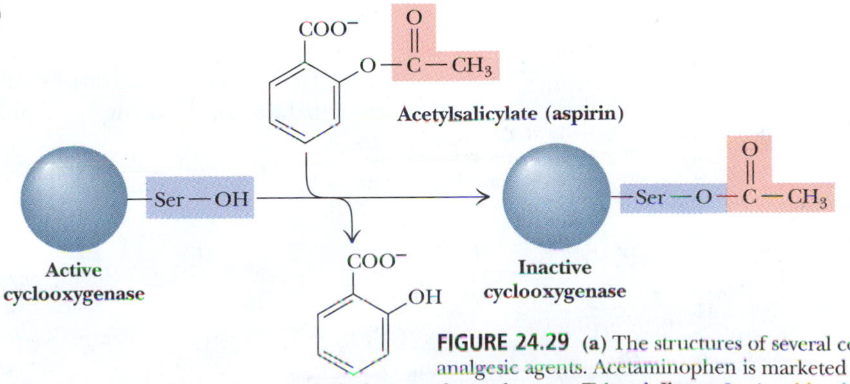

Acetylsalicylate (aspirin)

Active cyclooxygenase

Salicylate

Inactive cyclooxygenase

FIGURE 24.29 (a) The structures of several common analgesic agents. Acetaminophen is marketed under the trade name Tylenol. Ibuprofen is sold as Motrin, Nuprin, and Advil. (b) Acetylsalicylate (aspirin) inhibits the cyclooxygenase activity of endoperoxide synthase via acetylation (covalent modification) of Ser[530].

peroxidase activity of the synthase. Other nonsteroidal anti-inflammatory agents, such as **ibuprofen** (Figure 24.29) and phenylbutazone, inhibit the cyclo-oxygenase by competing at the active site with arachidonate or with the peroxy-acid intermediate (PGG$_2$, Figure 24.28). See A Deeper Look on page 790.

24.4 | How Is Cholesterol Synthesized?

The most prevalent steroid in animal cells is **cholesterol** (Figure 24.30). Plants contain no cholesterol, but they *do* contain other steroids very similar to cholesterol in structure (see page 263). Cholesterol serves as a crucial component of cell membranes and as a precursor to bile acids (such as cholate, glycocholate, taurocholate) and steroid hormones (such as testosterone, estradiol, progesterone). Also, vitamin D$_3$ is derived from *7-dehydrocholesterol,* the immediate precursor of cholesterol. Liver is the primary site of cholesterol biosynthesis.

Mevalonate Is Synthesized from Acetyl-CoA Via HMG-CoA Synthase

The cholesterol biosynthetic pathway begins in the cytosol with the synthesis of mevalonate from acetyl-CoA (Figure 24.31). The first step is the **β-ketothiolase–**catalyzed Claisen condensation of two molecules of acetyl-CoA to form acetoacetyl-CoA. In the next reaction, acetyl-CoA and acetoacetyl-CoA join to form *3-hydroxy-3-methylglutaryl-CoA,* which is abbreviated *HMG-CoA.* The reaction, a second Claisen condensation, is catalyzed by **HMG-CoA synthase.** The third step in the pathway is the rate-limiting step in cholesterol biosynthesis. Here, HMG-CoA undergoes two NADPH-dependent reductions to produce *3R-mevalonate* (Figure 24.32). The reaction is catalyzed by **HMG-CoA reductase,** a 97-kD glycoprotein that spans the ER membrane with its active site facing the cytosol. As the rate-limiting step, HMG-CoA reductase is the principal site of regulation in cholesterol synthesis.

Three different regulatory mechanisms are involved:

1. Phosphorylation by cAMP-dependent protein kinases inactivates the reductase. This inactivation can be reversed by two specific phosphatases (Figure 24.33).
2. Degradation of HMG-CoA reductase. This enzyme has a half-life of only 3 hours, and the half-life itself depends on cholesterol levels: High [cholesterol] means a short half-life for HMG-CoA reductase.
3. Gene expression. Cholesterol levels control the amount of mRNA. If [cholesterol] is high, levels of mRNA coding for the reductase are reduced. If [cholesterol] is low, more mRNA is made. (Regulation of gene expression is discussed in Chapter 29.)

A Thiolase Brainteaser Asks Why Thiolase Can't Be Used in Fatty Acid Synthesis

If acetate units can be condensed by the thiolase reaction to yield acetoacetate in the first step of cholesterol synthesis, why couldn't this same reaction also be used in fatty acid synthesis, avoiding all the complexity of the fatty acyl syn-

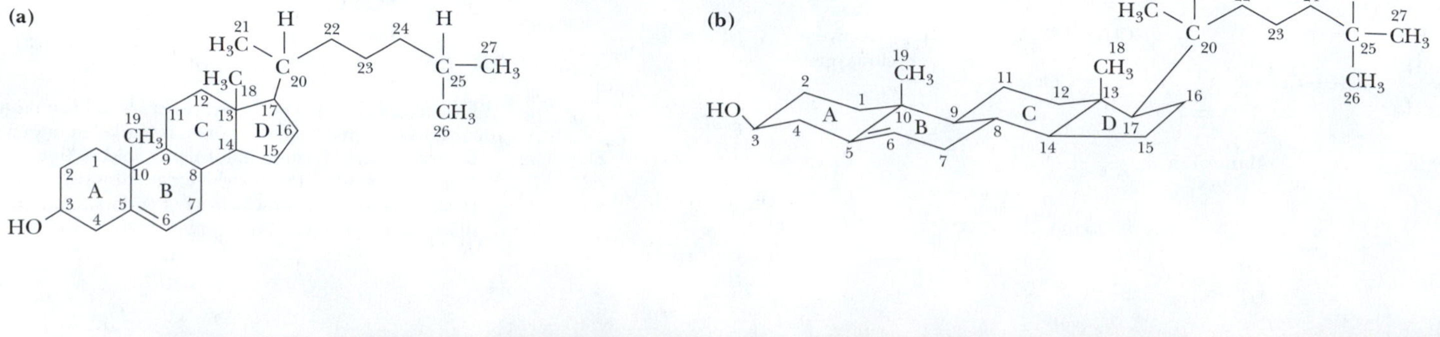

FIGURE 24.30 The structure of cholesterol, drawn **(a)** in the traditional planar motif and **(b)** in a form that more accurately describes the conformation of the ring system.

(a)

(b)

thase? The answer is that the thiolase reaction is more or less reversible but slightly favors the cleavage reaction. In the cholesterol synthesis pathway, subsequent reactions, including HMG-CoA reductase and the following kinase reactions, pull the thiolase-catalyzed condensation forward. However, in the case of fatty acid synthesis, a succession of eight thiolase condensations would be distinctly unfavorable from an energetic perspective. Given the necessity of repeated reactions in fatty acid synthesis, it makes better energetic sense to use a reaction that is favorable in the desired direction.

Squalene Is Synthesized from Mevalonate

The biosynthesis of squalene involves conversion of mevalonate to two key 5-carbon intermediates, *isopentenyl pyrophosphate* and *dimethylallyl pyrophosphate,* which join to yield *farnesyl pyrophosphate* and then squalene. A series of four reactions converts mevalonate to isopentenyl pyrophosphate and then to dimethylallyl pyrophosphate (Figure 24.34). The first three steps each consume an ATP, two for the purpose of forming a pyrophosphate at the 5-position and the third to drive the decarboxylation and double bond formation in the third step. **Pyrophospho-mevalonate decarboxylase** phosphorylates the 3-hydroxyl group, and this is followed by *trans* elimination of the phosphate and carboxyl groups to form the double bond in isopentenyl pyrophosphate. Isomerization of the double bond yields the dimethylallyl pyrophosphate. Condensation of these two 5-carbon intermediates produces *geranyl pyrophosphate;* addition of another 5-carbon isopentenyl

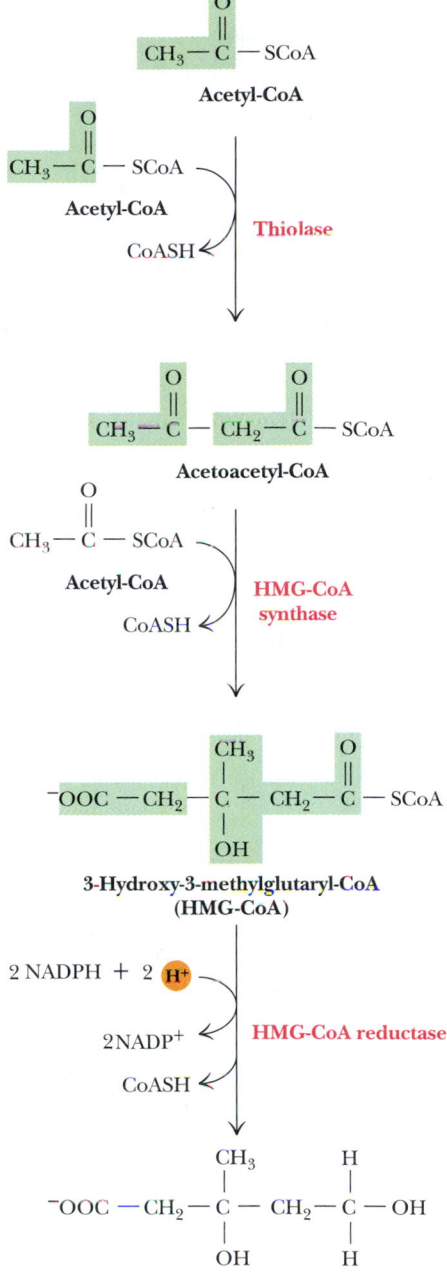

FIGURE 24.31 The biosynthesis of 3R-mevalonate from acetyl-CoA.

Biochemistry⊗**Now™ ANIMATED FIGURE 24.32**
A reaction mechanism for HMG-CoA reductase. Two successive NADPH-dependent reductions convert the thioester, HMG-CoA, to a primary alcohol. **See this figure animated at http://chemistry.brookscole. com/ggb3**

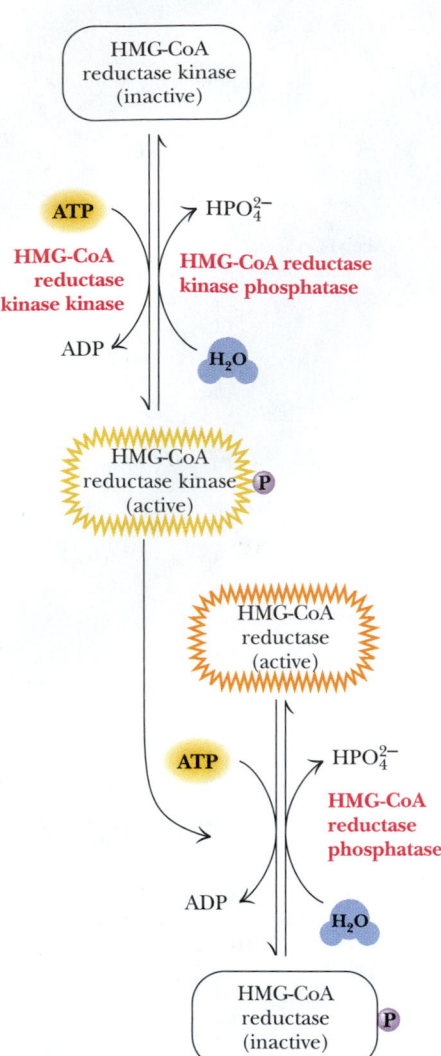

FIGURE 24.33 HMG-CoA reductase activity is modulated by a cycle of phosphorylation and dephosphorylation.

Mevalonate

Mevalonate kinase — ATP → ADP

Phosphomevalonate kinase — ATP → ADP

5-Pyrophosphomevalonate

Pyrophosphomevalonate decarboxylase — ATP → ADP + P + CO_2

Isopentenyl pyrophosphate

Isopentenyl pyrophosphate isomerase

Dimethylallyl pyrophosphate

Isopentenyl pyrophosphate → P P

Isopentenyl pyrophosphate → P P

Farnesyl pyrophosphate

NADPH + H⁺ → NADP⁺ + 2 P P

Squalene

Biochemistry Now™ **ACTIVE FIGURE 24.34** The conversion of mevalonate to squalene. Test yourself on the concepts in this figure at **http://chemistry.brookscole.com/ggb3**

group gives *farnesyl pyrophosphate*. Both steps in the production of farnesyl pyrophosphate occur with release of pyrophosphate, hydrolysis of which drives these reactions forward. Note too that the linkage of isoprene units to form farnesyl pyrophosphate occurs in a head-to-tail fashion. This is the general rule in biosynthesis of molecules involving isoprene linkages. The next step—the joining of two farnesyl pyrophosphates to produce squalene—is a "tail-to-tail" condensation and represents an important exception to the general rule.

Squalene monooxygenase, an enzyme bound to the ER, converts squalene to *squalene-2,3-epoxide* (Figure 24.35). This reaction employs FAD and NADPH as coenzymes and requires O_2 as well as a cytosolic protein called **soluble protein activator.** A second ER membrane enzyme, **2,3-oxidosqualene lanosterol cyclase,** catalyzes the second reaction, which involves a succession of 1,2 shifts of hydride ions and methyl groups.

Critical Developments in Biochemistry

The Long Search for the Route of Cholesterol Biosynthesis

Heilbron, Kamm, and Owens suggested as early as 1926 that squalene is a precursor of cholesterol. That same year, H. J. Channon demonstrated that animals fed squalene from shark oil produced more cholesterol in their tissues. Bloch and Rittenberg showed in the 1940s that a significant amount of the carbon in the tetracyclic moiety and in the aliphatic side chain of cholesterol was derived from acetate. In 1934, Sir Robert Robinson suggested a scheme for the cyclization of squalene to form cholesterol before the biosynthetic link between acetate and squalene was understood. Squalene is actually a polymer of isoprene units, and Bonner and Arreguin suggested in 1949 that three acetate units could join to form five-carbon *isoprene* units (see figure, part a).

In 1952, Konrad Bloch and Robert Langdon showed conclusively that labeled squalene is synthesized rapidly from labeled acetate and also that cholesterol is derived from squalene. Langdon, a graduate student of Bloch's, performed the critical experiments in Bloch's laboratory at the University of Chicago while Bloch spent the summer in Bermuda attempting to demonstrate that radioactively labeled squalene would be converted to cholesterol in shark livers. As Bloch himself admitted, "All I was able to learn was that sharks of manageable length are very difficult to catch and their oily livers impossible to slice" (Bloch, 1987).

In 1953, Bloch, together with the eminent organic chemist R. B. Woodward, proposed a new scheme (see figure, part b) for the cyclization of squalene. (Together with Fyodor Lynen, Bloch received the Nobel Prize in Medicine or Physiology in 1964 for his work.) The picture was nearly complete, but one crucial question remained: How could isoprene be the intermediate in the transformation of acetate into squalene? In 1956, Karl Folkers

and his colleagues at Merck Sharpe & Dohme isolated mevalonic acid and also showed that mevalonate was the precursor of isoprene units. The search for the remaining details (described in the text) made the biosynthesis of cholesterol one of the most enduring and challenging bioorganic problems of the 1940s, 1950s, and 1960s. Even today, several of the enzyme mechanisms remain poorly understood.

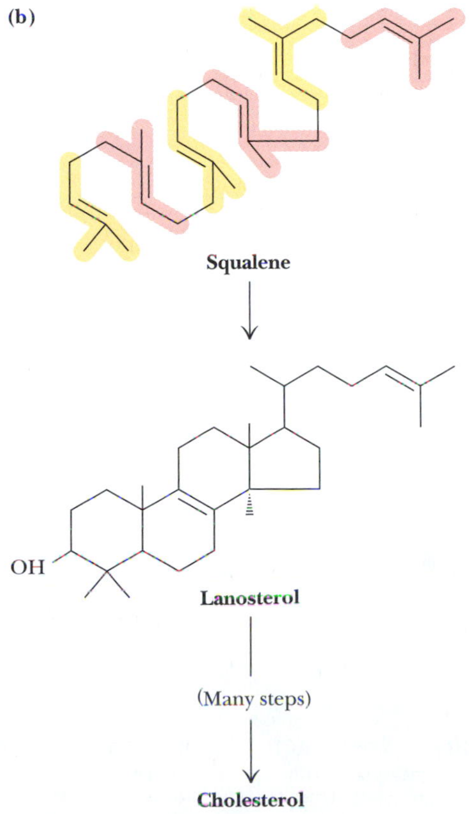

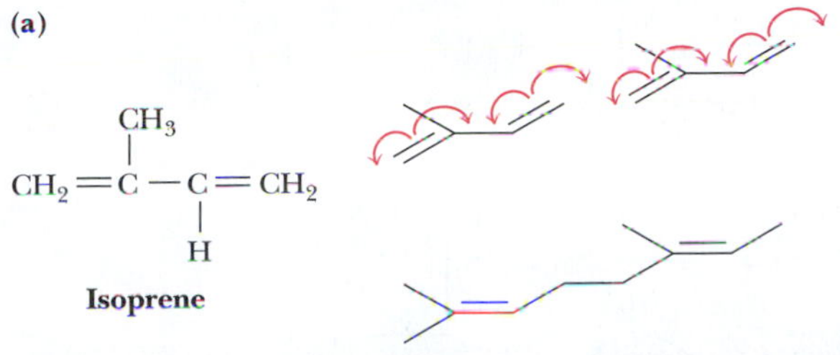

(a) An isoprene unit and a scheme for head-to-tail linking of isoprene units. **(b)** The cyclization of squalene to form lanosterol, as proposed by Bloch and Woodward.

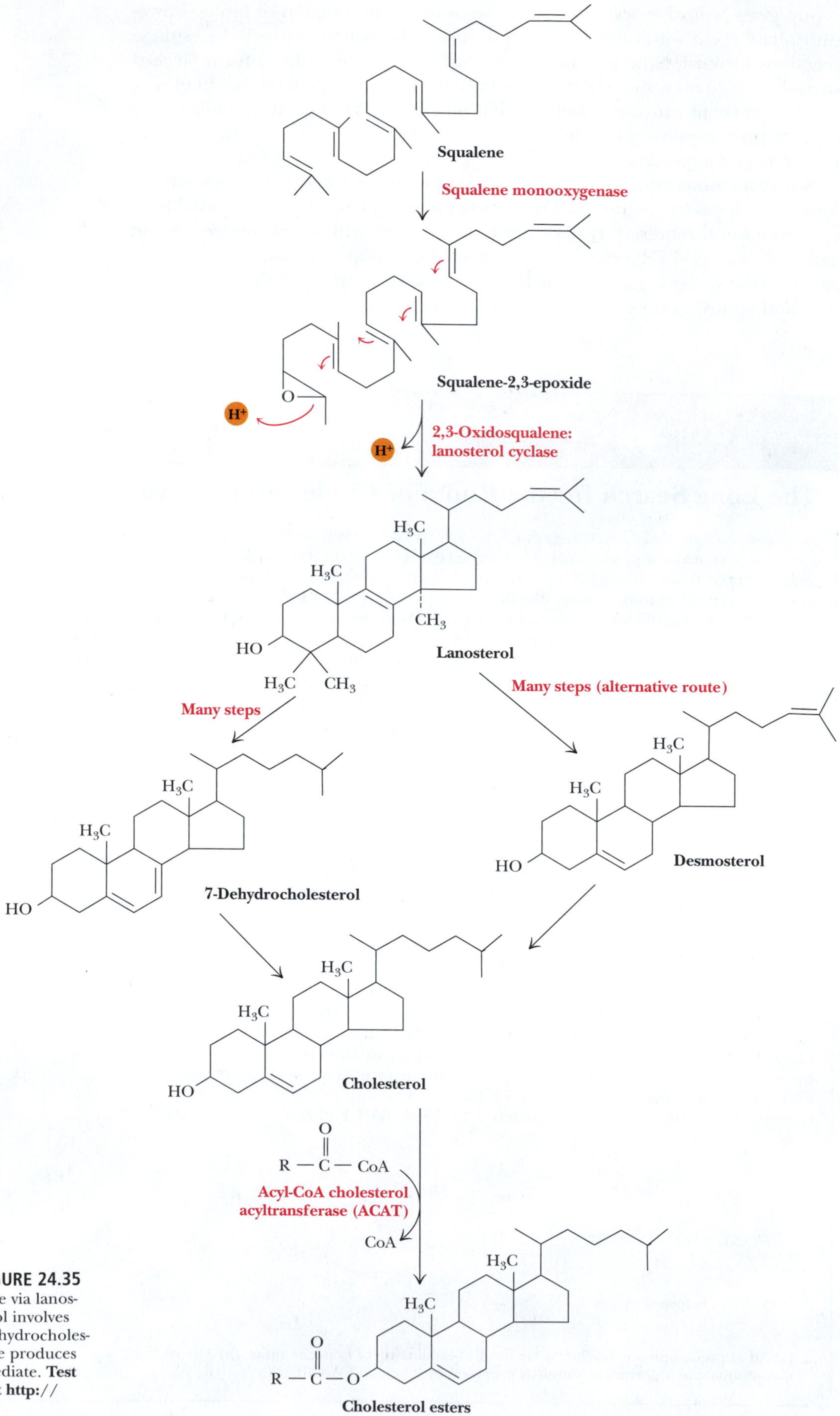

Squalene

Squalene monooxygenase

Squalene-2,3-epoxide

H⁺

H⁺

2,3-Oxidosqualene: lanosterol cyclase

H₃C

H₃C

CH₃

HO

H₃C   CH₃

Lanosterol

Many steps

Many steps (alternative route)

H₃C

H₃C

HO

7-Dehydrocholesterol

H₃C

H₃C

HO

Desmosterol

H₃C

H₃C

HO

Cholesterol

$$R - \overset{O}{\underset{}{C}} - CoA$$

Acyl-CoA cholesterol acyltransferase (ACAT)

CoA

H₃C

H₃C

$$R - \overset{O}{\underset{}{C}} - O$$

Cholesterol esters

Biochemistry ⓈNow™   ACTIVE FIGURE 24.35
Cholesterol is synthesized from squalene via lanosterol. The primary route from lanosterol involves
20 steps, the last of which converts 7-dehydrocholes-
terol to cholesterol. An alternative route produces
desmosterol as the penultimate intermediate. **Test
yourself on the concepts in this figure at http://
chemistry.brookscole.com/ggb3**

Human Biochemistry

Lovastatin Lowers Serum Cholesterol Levels

Chemists and biochemists have long sought a means of reducing serum cholesterol levels to reduce the risk of heart attack and cardiovascular disease. Because HMG-CoA reductase is the rate-limiting step in cholesterol biosynthesis, this enzyme is a likely drug target. **Mevinolin**, also known as **lovastatin** (see accompanying figure), was isolated from a strain of *Aspergillus terreus* and developed at Merck Sharpe & Dohme for this purpose. It is now a widely prescribed cholesterol-lowering drug. Dramatic reductions of serum cholesterol are observed at dosages of 20 to 80 mg per day.

Lovastatin is administered as an inactive lactone. After oral ingestion, it is hydrolyzed to the active **mevinolinic acid,** a competitive inhibitor of the reductase with a K_I of 0.6 nM. Mevinolinic acid is thought to behave as a transition-state analog (see Chapter 14) of the tetrahedral intermediate formed in the HMG-CoA reductase reaction (see figure).

Derivatives of lovastatin have been found to be even more potent in cholesterol-lowering trials. **Synvinolin** lowers serum cholesterol levels at much lower dosages than lovastatin.

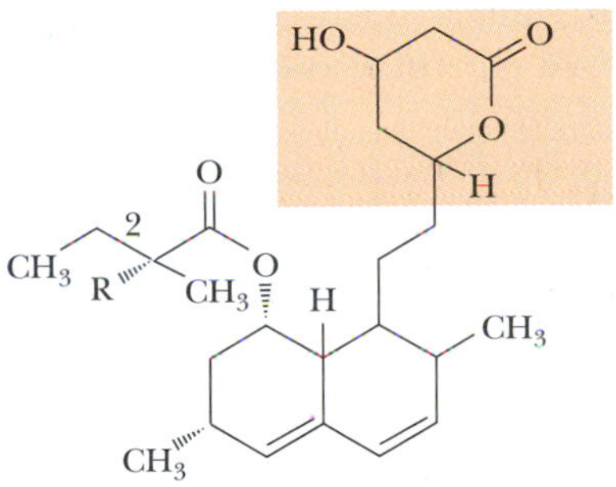

1 R=H Mevinolin (Lovastatin, MEVACOR®)
2 R=CH₃ Synvinolin (Simnastatin, ZOCOR®)

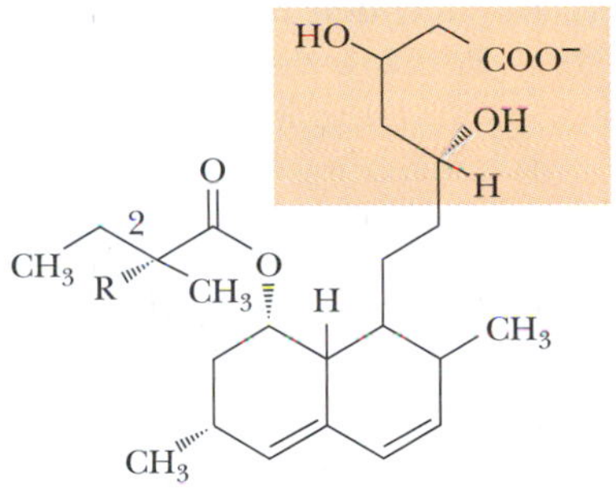

Mevinolinic acid

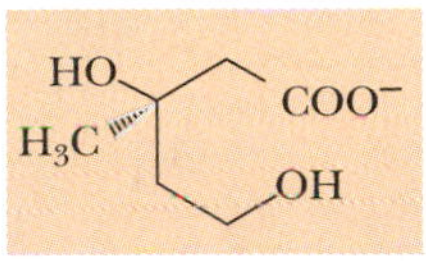

Mevalonate

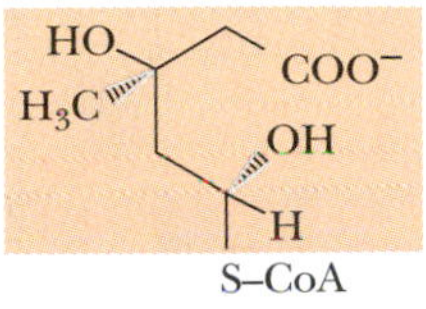

Tetrahedral intermediate in HMG-CoA reductase mechanism

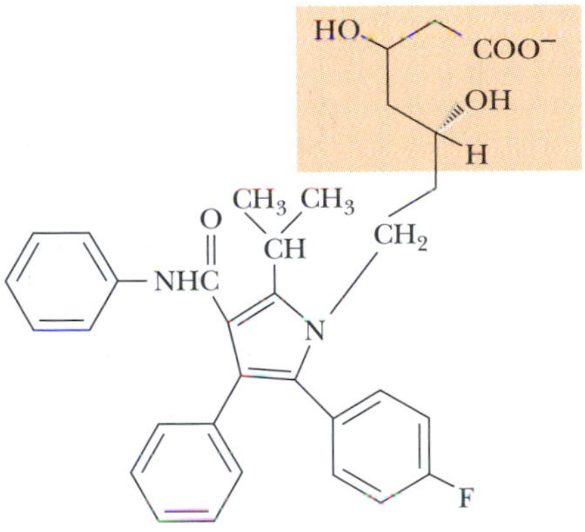

Lipitor®
(Atorvastatin)

▲ The structures of (inactive) lovastatin, (active) mevinolinic acid, mevalonate, and synvinolin

Conversion of Lanosterol to Cholesterol Requires 20 Additional Steps

Although lanosterol may appear similar to cholesterol in structure, another 20 steps are required to convert lanosterol to cholesterol (Figure 24.35). The enzymes responsible for this are all associated with the ER. The primary pathway involves *7-dehydrocholesterol* as the penultimate intermediate. An alternative pathway, also composed of many steps, produces the intermediate *desmosterol*. Reduction of the double bond at C-24 yields cholesterol. Cholesterol esters—a principal form of circulating cholesterol—are synthesized by **acyl-CoA:cholesterol acyltransferases (ACAT)** on the cytoplasmic face of the ER.

24.5 How Are Lipids Transported Throughout the Body?

When most lipids circulate in the body, they do so in the form of **lipoprotein complexes.** Simple, unesterified fatty acids are merely bound to serum albumin and other proteins in blood plasma, but phospholipids, triacylglycerols, cholesterol, and cholesterol esters are all transported in the form of lipoproteins. At various sites in the body, lipoproteins interact with specific receptors and enzymes that transfer or modify their lipid cargoes. It is now customary to classify lipoproteins according to their densities (Table 24.1). The densities are related to the relative amounts of lipid and protein in the complexes. Because most proteins have densities of about 1.3 to 1.4 g/mL, and lipid aggregates usually possess densities of about 0.8 g/mL, the more protein and the less lipid in a complex, the denser the lipoprotein. Thus, there are **high-density lipoproteins** (HDLs), **low-density lipoproteins** (LDLs), **intermediate-density lipoproteins** (IDLs), **very-low-density lipoproteins** (VLDLs), and also **chylomicrons.** Chylomicrons have the lowest protein-to-lipid ratio and thus are the lowest-density lipoproteins. They are also the largest.

Lipoprotein Complexes Transport Triacylglycerols and Cholesterol Esters

HDL and VLDL are assembled primarily in the ER of the liver (with smaller amounts produced in the intestine), whereas chylomicrons form in the intestine. LDL is not synthesized directly but rather is made from VLDL. LDL appears to be the major circulatory complex for cholesterol and cholesterol esters. The primary task of chylomicrons is to transport triacylglycerols. Despite all this, it is extremely important to note that each of these lipoprotein classes contains some of each type of lipid. The relative amounts of HDL and LDL are important in the disposition of cholesterol in the body and in the development of arterial plaques (Figure 24.36). The structures of the various lipoproteins are approximately similar, and they consist of a core of mobile triacylglycerols or cholesterol esters surrounded by a single layer of phospholipid, into which is inserted a mixture of cholesterol and proteins (Figure 24.37). Note that the phospholipids are oriented with their polar head groups facing outward to interact with solvent water and that the phospholipids thus shield the hydrophobic lipids inside from the solvent water outside. The proteins also function as recognition sites for the various lipoprotein receptors throughout the body. A number of different apoproteins have been identified in lipoproteins (Table 24.2), and others may exist as well. The apoproteins are abundant in hydrophobic amino acid residues, as is appropriate for interactions with lipids. A **cholesterol ester transfer protein** also associates with lipoproteins.

Table 24.1						
Composition and Properties of Human Lipoproteins						
				Composition (% dry weight)		
Lipoprotein Class	Density (g/mL)	Diameter (nm)	Protein	Cholesterol	Phospholipid	Triacylglycerol
HDL	1.063–1.21	5–15	33	30	29	8
LDL	1.019–1.063	18–28	25	50	21	4
IDL	1.006–1.019	25–50	18	29	22	31
VLDL	0.95–1.006	30–80	10	22	18	50
Chylomicrons	<0.95	100–500	1–2	8	7	84

Adapted from Brown, M., and Goldstein, J., 1987. In Braunwald, E., et al., eds., *Harrison's Principles of Internal Medicine,* 11th ed. New York: McGraw-Hill; and Vance, D., and Vance, J., eds., 1985. *Biochemistry of Lipids and Membranes.* Menlo Park, CA: Benjamin/Cummings.

Lipoproteins in Circulation Are Progressively Degraded by Lipoprotein Lipase

The livers and intestines of animals are the primary sources of circulating lipids. Chylomicrons carry triacylglycerol and cholesterol esters from the intestines to other tissues, and VLDLs carry lipid from liver, as shown in Figure 24.38. At various target sites, particularly in the capillaries of muscle and adipose cells, these particles are degraded by **lipoprotein lipase,** which hydrolyzes triacylglycerols. Lipase action causes progressive loss of triacylglycerol (and apoprotein) and makes the lipoproteins smaller. This process gradually converts VLDL particles to IDL and then LDL particles, which are either returned to the liver for reprocessing or redirected to adipose tissues and adrenal glands. Every 24 hours, nearly half of all circulating LDL is removed from circulation in this way. The LDL binds to specific LDL receptors, which cluster in domains of the plasma membrane known as **coated pits** (discussed in subsequent paragraphs). These domains eventually invaginate to form **coated vesicles** (Figure 24.39). Within the cell, these vesicles fuse with lysosomes, and the LDLs are degraded by **lysosomal acid lipases.**

HDLs have much longer life spans in the body (5 to 6 days) than other lipoproteins. Newly formed HDL contains virtually no cholesterol ester. However, over time, cholesterol esters are accumulated through the action of **lecithin:cholesterol acyltransferase** (LCAT), a 59-kD glycoprotein associated with HDLs. Another associated protein, **cholesterol ester transfer protein,** transfers some of these esters to VLDL and LDL. Alternatively, HDLs function to return cholesterol and cholesterol esters to the liver. This latter process apparently explains the correlation between high HDL levels and reduced risk of cardiovascular disease. (High LDL levels, on the other hand, are correlated with an *increased* risk of coronary artery and cardiovascular disease.)

The Structure of the LDL Receptor Involves Five Domains

The LDL receptor in plasma membranes (Figure 24.40; see also Figure 7.40) consists of 839 amino acid residues and is composed of five domains. These domains include an LDL-binding domain of 292 residues, a segment of about 350 to 400 residues containing N-linked oligosaccharides, a 58-residue segment of O-linked oligosaccharides, a 22-residue membrane-spanning segment, and a 50-residue segment extending into the cytosol. The clustering of

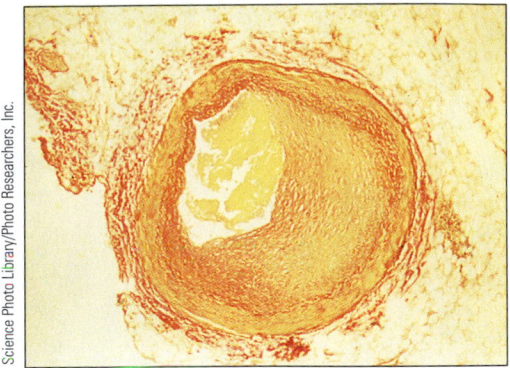

FIGURE 24.36 Photograph of an arterial plaque.

Biochemistry⊛Now™ Go to BiochemistryNow and click BiochemistryInteractive to learn about the function of apolipoprotein.

(a)   **(b)**

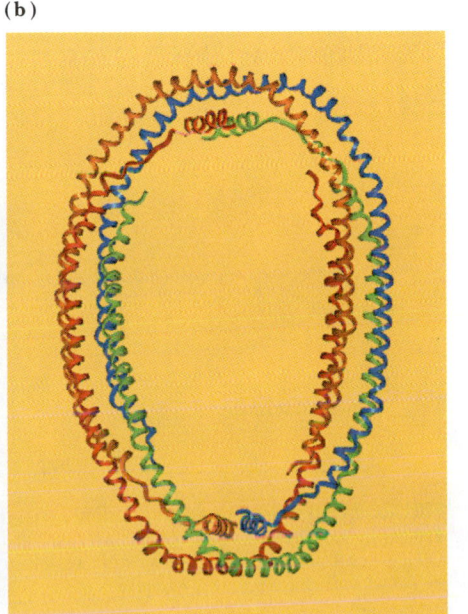

FIGURE 24.37 A model for the structure of a typical lipoprotein. **(a)** A core of cholesterol and cholesteryl esters is surrounded by a phospholipid (monolayer) membrane. Apolipoprotein A-I is modeled here as a long amphipathic α-helix, with the nonpolar face of the helix embedded in the hydrophobic core of the lipid particle and the polar face of the helix exposed to solvent. **(b)** A ribbon diagram of apolipoprotein A-I. *(Adapted from Borhani, D. W., Rogers, D. P., Engler, J. A., and Brouillette, C. G., 1997. Crystal structure of truncated human apolipoprotein A-I suggests a lipid-bound conformation.* Proceedings of the National Academy of Sciences **94:**12291–12296.)

Table 24.2

Apoproteins of Human Lipoproteins

Apoprotein	M_r	Concentration in Plasma (mg/100 mL)	Distribution
A-1	28,300	90–120	Principal protein in HDL
A-2	8,700	30–50	Occurs as dimer mainly in HDL
B-48	240,000	<5	Found only in chylomicrons
B100	500,000	80–100	Principal protein in LDL
C-1	7,000	4–7	Found in chylomicrons, VLDL, HDL
C-2	8,800	3–8	Found in chylomicrons, VLDL, HDL
C-3	8,800	8–15	Found in chylomicrons, VLDL, IDL, HDL
D	32,500	8–10	Found in HDL
E	34,100	3–6	Found in chylomicrons, VLDL, IDL, HDL

Adapted from Brown, M., and Goldstein, J., 1987. In Braunwald, E., et al., eds., *Harrison's Principles of Internal Medicine,* 11th ed. New York: McGraw-Hill; and Vance, D., and Vance, J., eds., 1985. *Biochemistry of Lipids and Membranes,* Menlo Park, CA: Benjamin/Cummings.

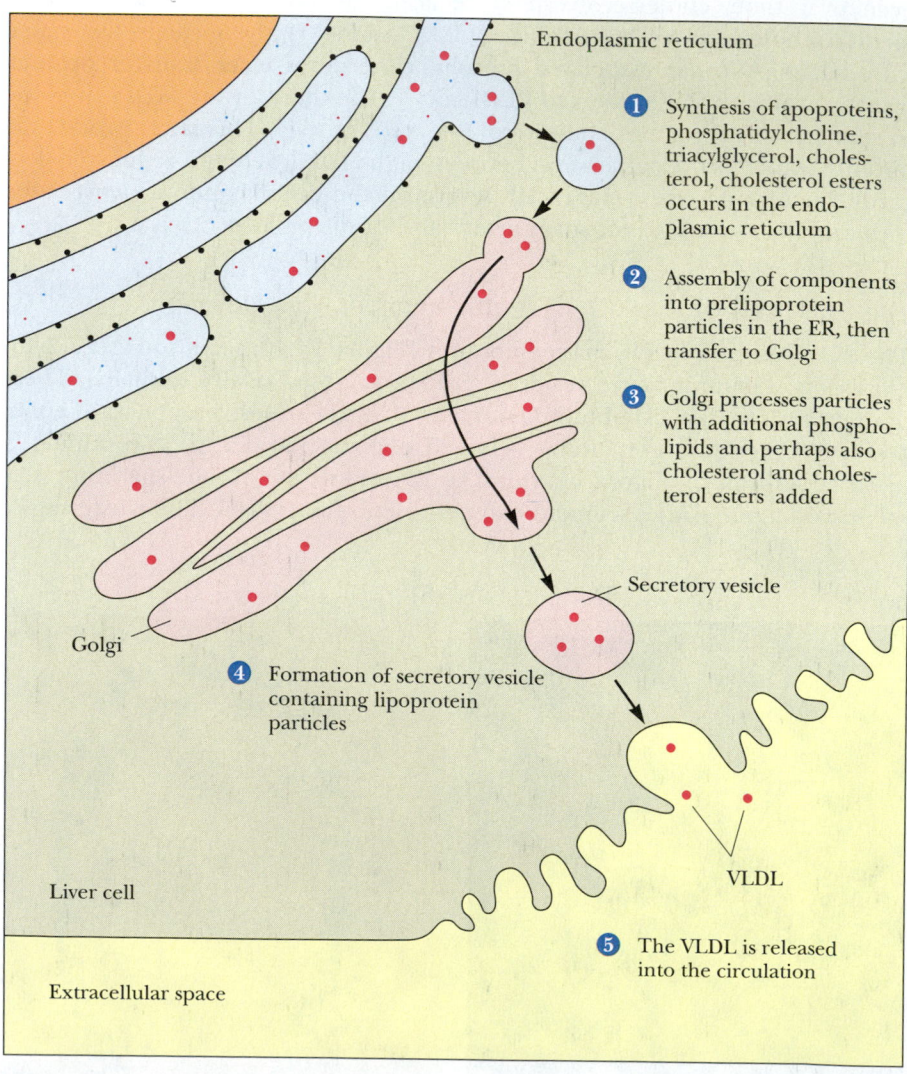

FIGURE 24.38 Lipoprotein components are synthesized predominantly in the ER of liver cells. Following assembly of lipoprotein particles (red dots) in the ER and processing in the Golgi, lipoproteins are packaged in secretory vesicles for export from the cell (via exocytosis) and released into the circulatory system.

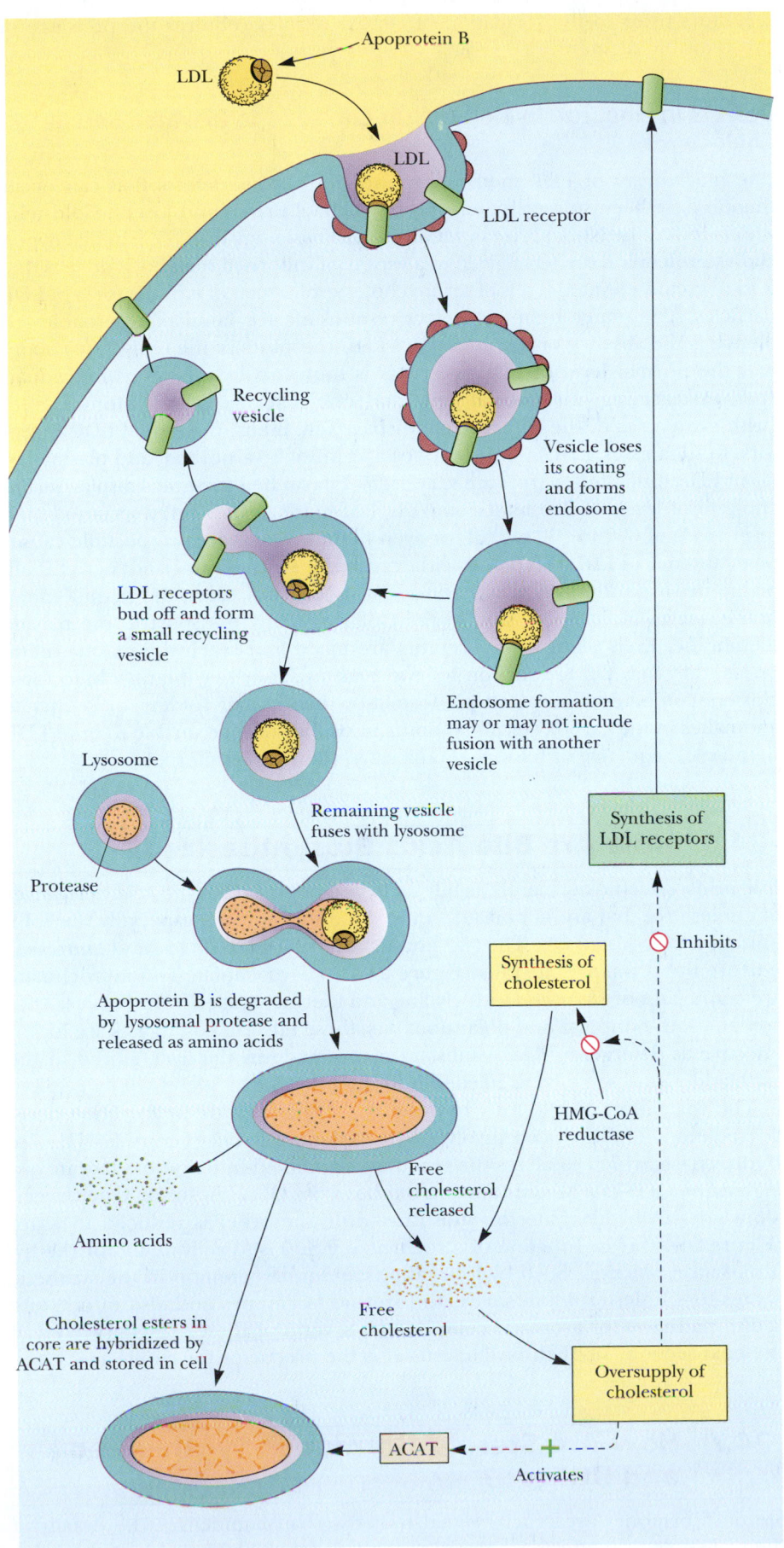

FIGURE 24.39 Endocytosis and degradation of lipoprotein particles. (ACAT is acyl-CoA cholesterol acyltransferase.)

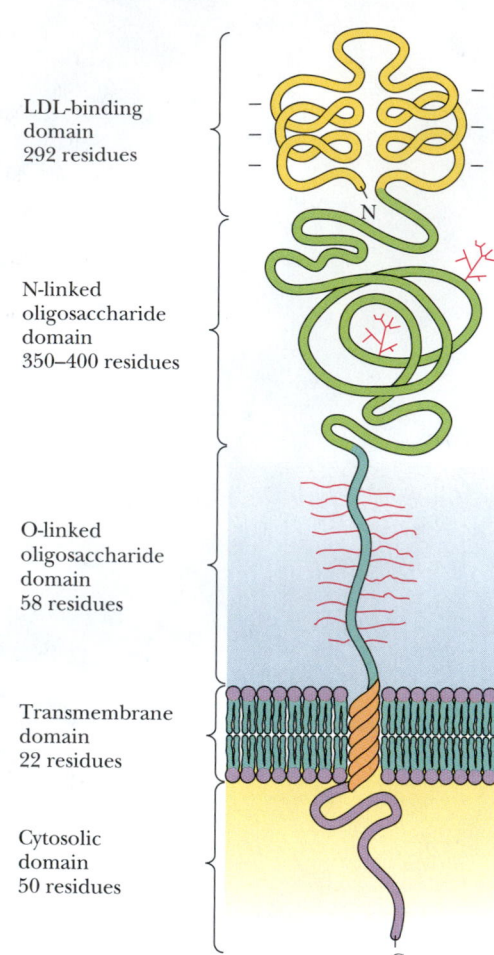

LDL-binding
domain
292 residues

N-linked
oligosaccharide
domain
350–400 residues

O-linked
oligosaccharide
domain
58 residues

Transmembrane
domain
22 residues

Cytosolic
domain
50 residues

FIGURE 24.40 The structure of the LDL receptor. The amino-terminal binding domain is responsible for recognition and binding of LDL apoprotein. The O-linked oligosaccharide-rich domain may act as a molecular spacer, raising the binding domain above the glycocalyx. The cytosolic domain is required for aggregation of LDL receptors during endocytosis.

receptors prior to the formation of coated vesicles requires the presence of this cytosolic segment.

Defects in Lipoprotein Metabolism Can Lead to Elevated Serum Cholesterol

The mechanism of LDL metabolism and the various defects that can occur therein have been studied extensively by Michael Brown and Joseph Goldstein, who received the Nobel Prize in Medicine or Physiology in 1985. **Familial hyper-cholesterolemia** is the term given to a variety of inherited metabolic defects that lead to greatly elevated levels of serum cholesterol, much of it in the form of LDL particles. The general genetic defect responsible for familial hypercholester-olemia is the absence or dysfunction of LDL receptors in the body. Only about half the normal level of LDL receptors is found in heterozygous individuals (persons carrying one normal gene and one defective gene). Homozygotes (with two copies of the defective gene) have few, if any, functional LDL recep-tors. In such cases, LDLs (and cholesterol) cannot be absorbed, and plasma lev-els of LDL (and cholesterol) are very high. Typical heterozygotes display serum cholesterol levels of 300 to 400 mg/dL, but homozygotes carry serum choles-terol levels of 600 to 800 mg/dL or even higher. There are two possible causes of an absence of LDL receptors—either receptor synthesis does not occur at all, or the newly synthesized protein does not successfully reach the plasma mem-brane due to faulty processing in the Golgi or faulty transport to the plasma membrane. Even when LDL receptors are made and reach the plasma mem-brane, they may fail to function for two reasons. They may be unable to form clusters competent in coated pit formation because of folding or sequence anomalies in the carboxy-terminal domain, or they may be unable to bind LDL because of sequence or folding anomalies in the LDL-binding domain.

24.6 | How Are Bile Acids Biosynthesized?

Bile acids, which exist mainly as **bile salts,** are polar carboxylic acid derivatives of cholesterol that are important in the digestion of food, especially the solu-bilization of ingested fats. The Na^+ and K^+ salts of *glycocholic acid* and *taurocholic acid* are the principal bile salts (Figure 24.41). Glycocholate and taurocholate are conjugates of *cholic acid* with glycine and taurine, respectively. Because they contain both nonpolar and polar domains, these bile salt conjugates are highly effective as detergents. These substances are made in the liver, stored in the gallbladder, and secreted as needed into the intestines.

The formation of bile salts represents the major pathway for cholesterol degradation. The first step involves hydroxylation at C-7 (Figure 24.41). **7α-Hydroxylase,** which catalyzes the reaction, is a mixed-function oxidase involv-ing *cytochrome P-450*. **Mixed-function oxidases** use O_2 as substrate. One oxygen atom goes to hydroxylate the substrate while the other is reduced to water (Figure 24.42). The function of cytochrome P-450 is to activate O_2 for the hy-droxylation reaction. Such hydroxylations are quite common in the synthetic routes for cholesterol, bile acids, and steroid hormones and also in detoxifi-cation pathways for aromatic compounds. Several of these are considered in the next section. 7α-Hydroxycholesterol is the precursor for cholic acid.

24.7 | How Are Steroid Hormones Synthesized and Utilized?

Steroid hormones are crucial signal molecules in mammals. (The details of their physiological effects are described in Chapter 32.) Their biosynthesis be-gins with the **desmolase** reaction, which converts cholesterol to pregnenolone

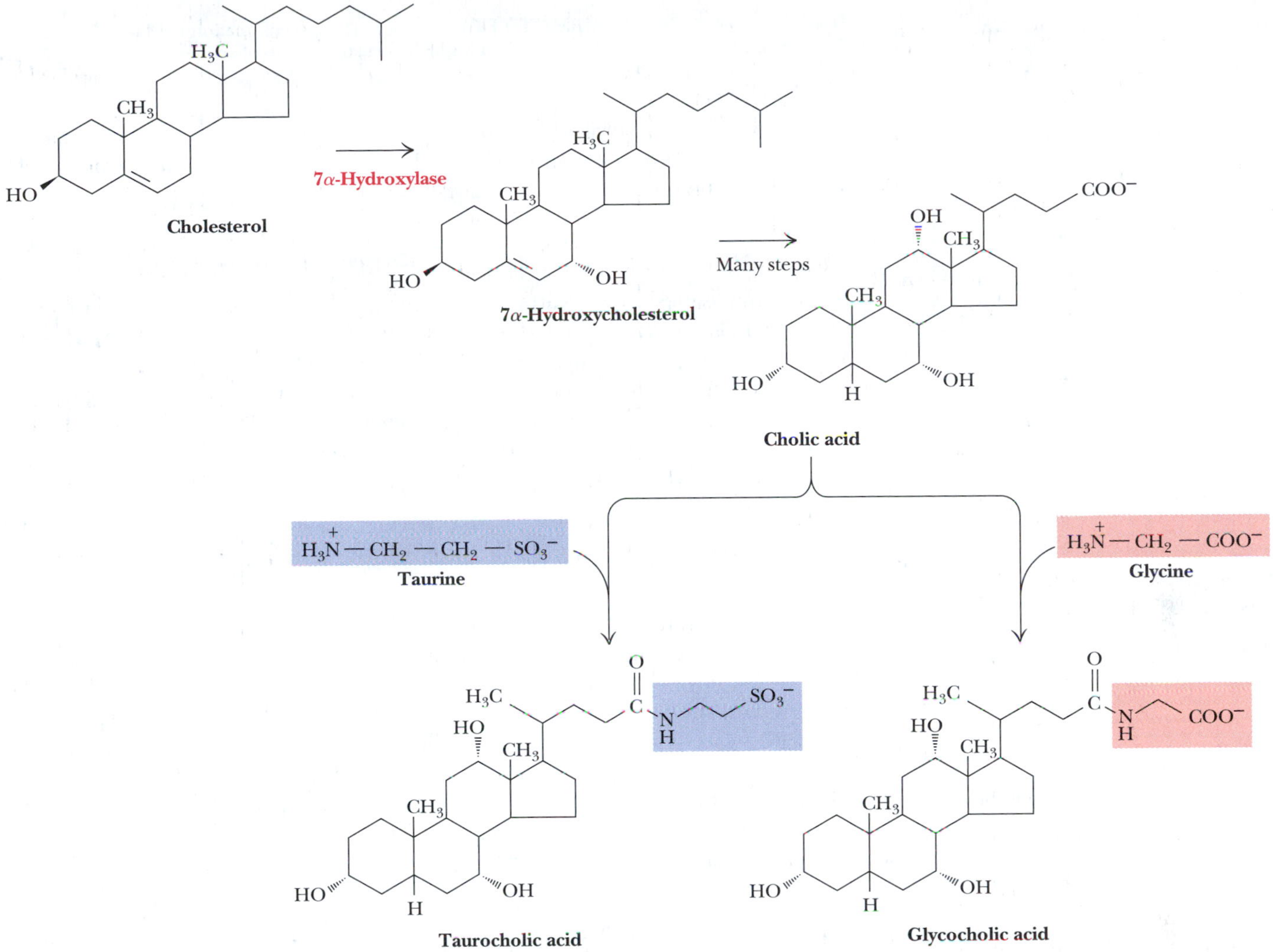

FIGURE 24.41 Cholic acid, a bile salt, is synthesized from cholesterol via 7α-hydroxycholesterol. Conjugation with taurine or glycine produces taurocholic acid and glycocholic acid, respectively. Taurocholate and glycocholate are freely water soluble and are highly effective detergents.

(Figure 24.43). Desmolase is found in the mitochondria of tissues that synthesize steroids (mainly the adrenal glands and gonads). Desmolase activity includes two hydroxylases and utilizes cytochrome P-450.

Pregnenolone and Progesterone Are the Precursors of All Other Steroid Hormones

Pregnenolone is transported from the mitochondria to the ER, where a hydroxyl oxidation and migration of the double bond yield progesterone. Pregnenolone synthesis in the adrenal cortex is activated by **adrenocorticotropic hormone** (ACTH), a peptide of 39 amino acid residues secreted by the anterior pituitary gland.

Progesterone is secreted from the corpus luteum during the latter half of the menstrual cycle and prepares the lining of the uterus for attachment of a fertilized ovum. If an ovum attaches, progesterone secretion continues to ensure the successful maintenance of a pregnancy. Progesterone is also the precursor for synthesis of the other **sex hormone steroids** and the **corticosteroids.** Male sex hormone steroids are called **androgens,** and female hormones, **estrogens.** Testosterone is an androgen synthesized in males primarily in the testes (and in much smaller amounts in the adrenal cortex). Androgens are necessary for sperm maturation. Even nonreproductive tissue (liver, brain, and skeletal muscle) is susceptible to the effects of androgens.

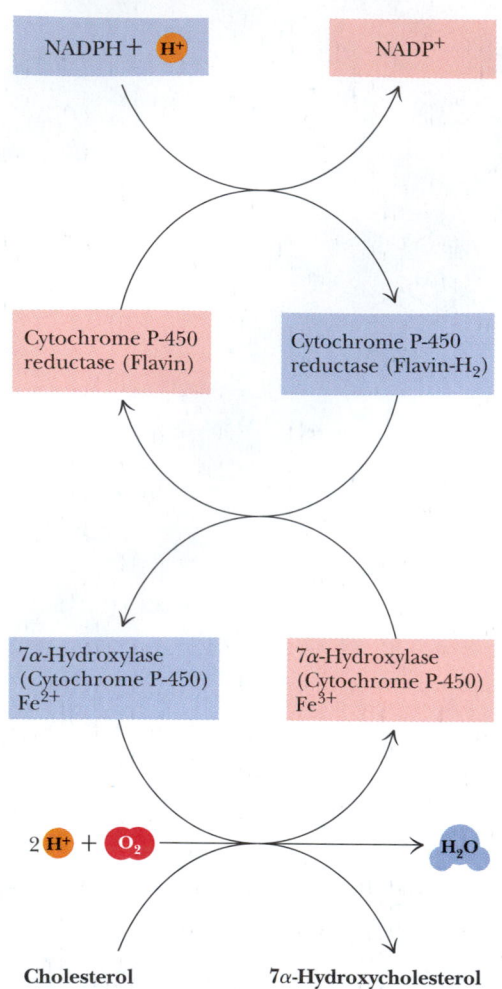

Cholesterol   **7α-Hydroxycholesterol**

Biochemistry ❸ Now™   **ANIMATED FIGURE 24.42** The mixed-function oxidase activity of 7α-hydroxylase. **See this figure animated at http://chemistry.brookscole.com/ggb3**

Testosterone is also produced primarily in the ovaries (and in much smaller amounts in the adrenal glands) of females as a precursor for the estrogens. β-Estradiol is the most important estrogen (Figure 24.43).

Steroid Hormones Modulate Transcription in the Nucleus

Steroid hormones act in a different manner from most hormones we have considered. In many cases, they do not bind to plasma membrane receptors but rather pass easily across the plasma membrane. Steroids may bind directly to receptors in the nucleus or may bind to cytosolic steroid hormone receptors, which then enter the nucleus. In the nucleus, the hormone-receptor complex binds directly to specific nucleotide sequences in DNA, increasing transcription of DNA to RNA (Chapters 29 and 32).

Cortisol and Other Corticosteroids Regulate a Variety of Body Processes

Corticosteroids, including the *glucocorticoids* and *mineralocorticoids,* are made by the cortex of the adrenal glands on top of the kidneys. **Cortisol** (Figure 24.43) is representative of the **glucocorticoids,** a class of compounds that (1) stimulate gluconeogenesis and glycogen synthesis in liver (by signaling the synthesis of PEP carboxykinase, fructose-1,6-bisphosphatase, glucose-6-phosphatase, and glycogen synthase); (2) inhibit protein synthesis and stimulate protein degradation in peripheral tissues such as muscle; (3) inhibit allergic and inflammatory responses; (4) exert an immunosuppressive effect, inhibiting DNA replication and mitosis and repressing the formation of antibodies and lym-

Human Biochemistry

Steroid 5α-Reductase—A Factor in Male Baldness, Prostatic Hyperplasia, and Prostate Cancer

An enzyme that metabolizes testosterone may be involved in the benign conditions of male-pattern baldness (also known as *androgenic alopecia*) and benign prostatic hyperplasia (prostate gland enlargement), as well as potentially fatal prostate cancers. Steroid 5α-reductases are membrane-bound enzymes that catalyze the NADPH-dependent reduction of testosterone to dihydrotestosterone (DHT) (see accompanying figure). Two isoforms of 5α-reductase have been identified. In humans; the type I enzyme

predominates in the sebaceous glands of skin and liver, whereas type II is most abundant in the prostate, seminal vesicles, liver, and epididymis. DHT is a contributory factor in male baldness and prostatic hyperplasia, and it has also been shown to act as a mitogen (a stimulator of cell division). For these reasons, 5α-reductase inhibitors are potential candidates for treatment of these human conditions.

Finasteride (see figure) is a specific inhibitor of type II 5α-reductase. It has been used clinically for treatment of benign prostatic hyperplasia, and it is also marketed under the trade name Propecia by Merck as a treatment for male baldness. Type II 5α-reductase inhibitors may also be potential therapeutic agents for treatment of prostate cancer. Somatic mutations occur in the gene for type II 5α-reductase during prostate cancer progression.

Because type I 5α-reductase is the predominant form of the enzyme in human scalp, the mechanism of finasteride's promotion of hair growth in men with androgenic alopecia has been uncertain. However, scientists at Merck have shown that whereas type I 5α-reductase predominates in sebaceous ducts of the skin, type II 5α-reductase is the only form of the enzyme present in hair follicles. Thus, finasteride's therapeutic effects may arise from inhibition of the type II enzyme in the hair follicle itself.

Dihydrotestosterone   Finasteride

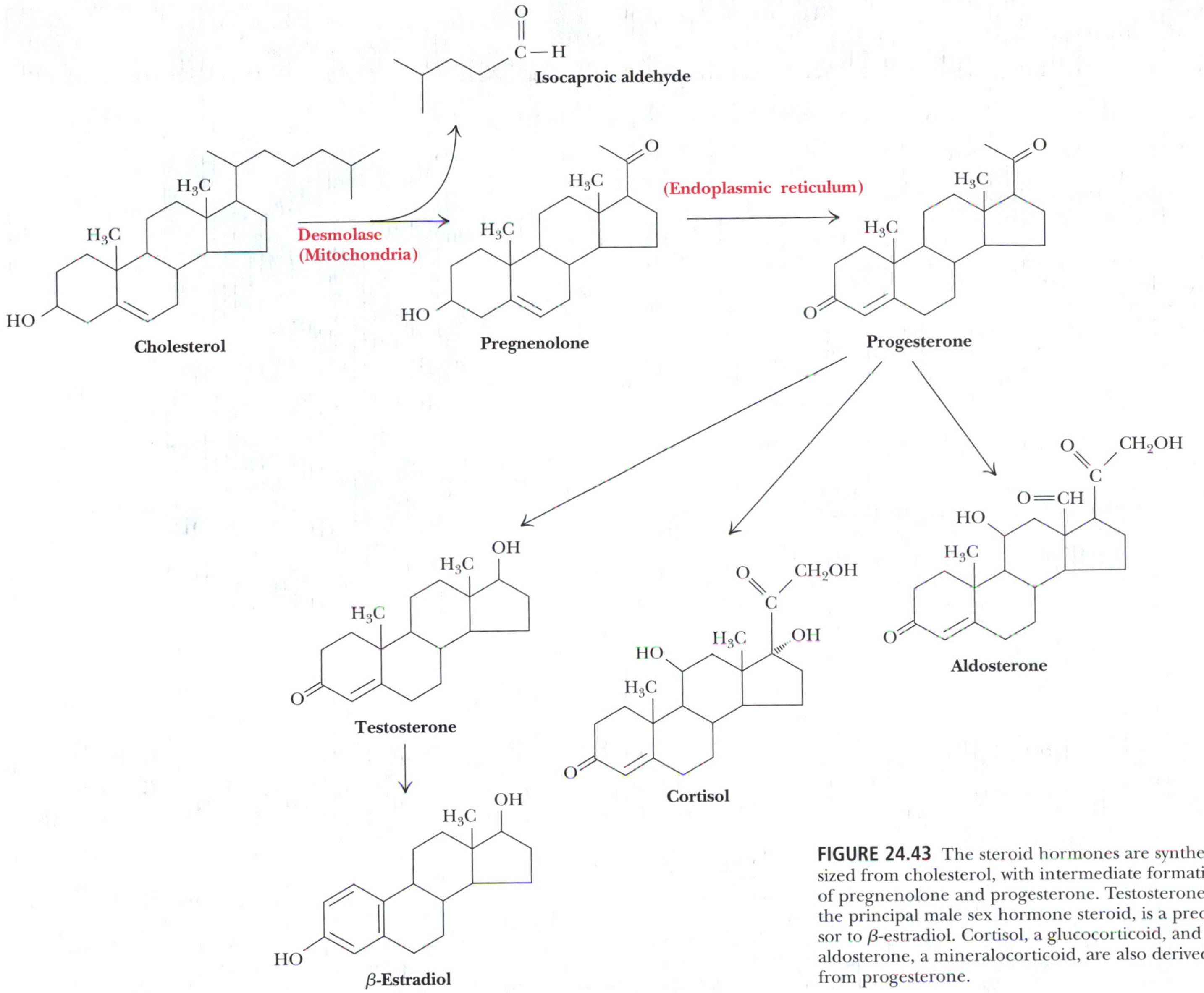

FIGURE 24.43 The steroid hormones are synthesized from cholesterol, with intermediate formation of pregnenolone and progesterone. Testosterone, the principal male sex hormone steroid, is a precursor to β-estradiol. Cortisol, a glucocorticoid, and aldosterone, a mineralocorticoid, are also derived from progesterone.

phocytes; and (5) inhibit formation of fibroblasts involved in healing wounds and slow the healing of broken bones.

Aldosterone, the most potent of the **mineralocorticoids** (Figure 24.43), is involved in the regulation of sodium and potassium balances in tissues. Aldosterone increases the kidney's capacity to absorb Na^+, Cl^-, and H_2O from the glomerular filtrate in the kidney tubules.

Anabolic Steroids Have Been Used Illegally to Enhance Athletic Performance

The dramatic effects of androgens on protein biosynthesis have led many athletes to the use of *synthetic androgens*, which go by the blanket term **anabolic steroids.** Despite numerous warnings from the medical community about side effects, which include kidney and liver disorders, sterility, and heart disease, abuse of such substances is epidemic. **Stanozolol** (Figure 24.44) was one of the agents found in the blood and urine of Ben Johnson following his record-setting performance in the 100-meter dash in the 1988 Olympic Games. Because use of such substances is disallowed, Johnson lost his gold medal and Carl Lewis was declared the official winner.

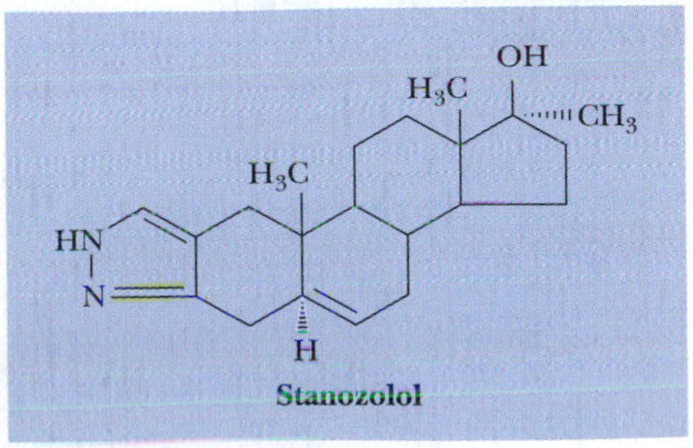

FIGURE 24.44 The structure of stanozolol, an anabolic steroid.

Human Biochemistry

Salt and Water Balances and Deaths in Marathoners

Marathon running puts unusual demands on the human body. Dehydration is an obvious problem, and runners are often encouraged to drink substantial amounts of water during a race. However, as Allen Arieff of the University of California at San Francisco has shown, drinking too much water during a race can lead to **hyponatremia**—a shortage of salt in the blood. Ordinarily, aldosterone carefully regulates salt balances in the body. However, exercise leads the body to release antidiuretic hormone (arginine vasopressin), which blocks normal elimination of excess fluid as urine. At the same time, during a race, so much blood is diverted to the muscles that the intestines are unable to absorb any water that the runner drinks. Once the race is over, however, the water floods into the bloodstream, disrupting the body's salt balance. This can cause the brain to swell, which in turn can lead to release of water into the lungs. On the occasions when a runner collapses during or after a run, the observation of fluid in the lungs and the circumstances of the collapse often lead emergency room doctors to suspect heart trouble and to order treatments that make the problem worse.

Allen Arieff has suggested that the amount of water a runner should drink during a race will vary from person to person and that the amount of water consumed should probably not exceed the amount of water actually lost during the race.

▲ Runners at Nice, France, Marathon, 1986.

Human Biochemistry

Androstenedione—A Steroid of Uncertain Effects

In 1998, Mark McGwire and Sammy Sosa electrified the baseball world by hitting unprecedented numbers of home runs for the St. Louis Cardinals and Chicago Cubs, respectively. The previous records for home runs in one season in the Major Leagues had been 60 in a 154-game season (held by Babe Ruth) and 61 in a 162-game season (held by Roger Maris). In 1998, McGwire hit an astounding and unprecedented 70 home runs, and Sosa followed close behind with 68. In the aftermath of this historic performance, it was reported that McGwire had used androstenedione as a dietary supplement related to his weight-lifting regimen during this record-setting season. "Andro," as it is called by athletes, is a precursor to testosterone, the male sex steroid hormone. Its use is permitted by Major League Baseball but banned by the National Football League, the National Collegiate Athletic Association, and the International Olympic Committee. In spite of its precursor relationship to testosterone, little actual research on the conversion of orally administered andro to testosterone has been reported, and its efficacy for bodybuilding and the enhancement of performance in sports is uncertain at best. In Mark McGwire's case, the issue is certainly moot. McGwire's use of andro was confined to the 1998 season, but his home run production both before and after this time have been prodigious. McGwire hit an unprecedented total of 245 home runs over the four seasons of 1996 to 1999.

4-Androstene-3,17-dione

Summary

24.1 How Are Fatty Acids Synthesized?

The synthesis of fatty acids and other lipid components is different from their degradation. Fatty acid synthesis involves a set of reactions that follow a strategy different in several ways from the corresponding degradative process:

1. Intermediates in fatty acid synthesis are linked covalently to the sulfhydryl groups of the acyl carrier proteins. In contrast, fatty acid breakdown intermediates are bound to the —SH group of coenzyme A.
2. Fatty acid synthesis occurs in the cytosol, whereas fatty acid degradation takes place in mitochondria.
3. In animals, the enzymes of fatty acid synthesis are components of one long polypeptide chain, the fatty acid synthase, whereas no similar association exists for the degradative enzymes.
4. The coenzyme for the oxidation–reduction reactions of fatty acid synthesis is $NADP^+/NADPH$, whereas degradation involves the $NAD^+/NADH$ couple.

24.2 How Are Complex Lipids Synthesized?

A common pathway operates in nearly all organisms for the synthesis of phosphatidic acid, the precursor to other glycerolipids. Glycerokinase catalyzes the phosphorylation of glycerol to form glycerol-3-phosphate, which is then acylated at both the 1- and 2-positions to yield phosphatidic acid. In eukaryotes, phosphatidic acid is converted directly either to diacylglycerol or to cytidine diphosphodiacylglycerol (or simply CDP-diacylglycerol). From these two precursors, all other glycerophospholipids in eukaryotes are derived. Phosphatidylethanolamine synthesis begins with phosphorylation of ethanolamine to form phosphoethanolamine. The next reaction involves transfer of a cytidylyl group from CTP to form CDP-ethanolamine and pyrophosphate. Eukaryotes also use CDP-diacylglycerol, derived from phosphatidic acid, as a precursor for several other important phospholipids, including phosphatidylinositol (PI), phosphatidylglycerol (PG), and cardiolipin.

24.3 How Are Eicosanoids Synthesized, and What Are Their Functions?

Eicosanoids are ubiquitous breakdown products of phospholipids. In response to appropriate stimuli, cells activate the breakdown of selected phospholipids. Phospholipase A_2 selectively cleaves fatty acids from the C-2 position of phospholipids. Often these are unsaturated fatty acids, among which is arachidonic acid. Arachidonic acid may also be released from phospholipids by the combined actions of phospholipase C (which yields diacylglycerols) and diacylglycerol lipase (which releases fatty acids). Animal cells can modify arachidonic acid and other polyunsaturated fatty acids to produce so-called local hormones. These substances include the prostaglandins (PG) as well as thromboxanes (Tx), leukotrienes, and other hydroxyeicosanoic acids.

24.4 How Is Cholesterol Synthesized?

The cholesterol biosynthetic pathway begins in the cytosol with the synthesis of mevalonate from acetyl-CoA. The first step is the condensation of two molecules of acetyl-CoA to form acetoacetyl-CoA. In the next reaction, acetyl-CoA and acetoacetyl-CoA join to form 3-hydroxy-3-methylglutaryl-CoA, which is abbreviated HMG-CoA, in a reaction catalyzed by HMG-CoA synthase. The third step in the pathway is the rate-limiting step in cholesterol biosynthesis; HMG-CoA undergoes two NADPH-dependent reductions to produce 3R-mevalonate. Biosynthesis of squalene involves conversion of mevalonate to isopentenyl pyrophosphate and dimethylallyl pyrophosphate, which join to yield farnesyl pyrophosphate and then squalene. A series of four reactions converts mevalonate to isopentenyl pyrophosphate and then to dimethylallyl pyrophosphate. Condensation of these two 5-carbon intermediates produces geranyl pyrophosphate; addition of another 5-carbon isopentenyl group gives farnesyl pyrophosphate. Both steps in the production of farnesyl pyrophosphate occur with release of pyrophosphate, hydrolysis of which drives these reactions forward. Two farnesyl pyrophosphates join to produce squalene. Squalene monooxygenase converts squalene to squalene-2,3-epoxide. A second ER membrane enzyme produces lanosterol, and another 20 steps are required to convert lanosterol to cholesterol.

24.5 How Are Lipids Transported Throughout the Body?

Most lipids circulate in the body in the form of lipoprotein complexes. Simple, unesterified fatty acids are merely bound to serum albumin and other proteins in blood plasma, but phospholipids, triacylglycerols, cholesterol, and cholesterol esters are all transported in the form of lipoproteins. At various sites in the body, lipoproteins interact with specific receptors and enzymes that transfer or modify their lipid cargoes.

24.6 How Are Bile Acids Biosynthesized?

The formation of bile salts represents the major pathway for cholesterol degradation. The first step involves hydroxylation at C-7. 7α-Hydroxylase is a mixed-function oxidase involving cytochrome P-450. Mixed-function oxidases use O_2 as substrate. One oxygen atom goes to hydroxylate the substrate while the other is reduced to water. The function of cytochrome P-450 is to activate O_2 for the hydroxylation reaction. Such hydroxylations are quite common in the synthetic routes for cholesterol, bile acids, and steroid hormones and also in detoxification pathways for aromatic compounds.

24.7 How Are Steroid Hormones Synthesized and Utilized?

Biosynthesis of steroid hormones begins with the desmolase reaction, which converts cholesterol to pregnenolone. Desmolase activity includes two hydroxylases and utilizes cytochrome P-450. Pregnenolone is transported from the mitochondria to the ER, where a hydroxyl oxidation and migration of the double bond yield progesterone. Progesterone is also the precursor for synthesis of the sex hormone steroids and the corticosteroids. Testosterone is an androgen synthesized in males primarily in the testes. β-Estradiol is the most important estrogen. Aldosterone, the most potent of the mineralocorticoids, is involved in the regulation of sodium and potassium balances in tissues.

Problems

1. Carefully count and account for each of the atoms and charges in the equations for the synthesis of palmitoyl-CoA, the synthesis of malonyl-CoA, and the overall reaction for the synthesis of palmitoyl-CoA from acetyl-CoA.

2. (Integrates with Chapters 18 and 19.) Use the relationships shown in Figure 24.1 to determine which carbons of glucose will be incorporated into palmitic acid. Consider the cases of both citrate that is immediately exported to the cytosol following its synthesis and citrate that enters the TCA cycle.

3. Based on the information presented in the text and in Figures 24.4 and 24.5, suggest a model for the regulation of acetyl-CoA carboxylase. Consider the possible roles of subunit interactions, phosphorylation, and conformation changes in your model.

4. (Integrates with Chapter 17.) Consider the role of the pantothenic acid groups in animal fatty acyl synthase and the size of the pantothenic acid group itself, and estimate a maximal separation between the malonyl transferase and the ketoacyl-ACP synthase active sites.

5. Carefully study the reaction mechanism for the stearoyl-CoA desaturase in Figure 24.14, and account for all of the electrons flowing through the reactions shown. Also account for all of the hydrogen and oxygen atoms involved in this reaction, and convince yourself that the stoichiometry is correct as shown.

6. Write a balanced, stoichiometric reaction for the synthesis of phosphatidylethanolamine from glycerol, fatty acyl-CoA, and ethanolamine. Make an estimate of the $\Delta G^{\circ\prime}$ for the overall process.

7. Write a balanced, stoichiometric reaction for the synthesis of cholesterol from acetyl-CoA.

8. Trace each of the carbon atoms of mevalonate through the synthesis of cholesterol, and determine the source (that is, the position in the mevalonate structure) of each carbon in the final structure.

9. Suggest a structural or functional role for the *O*-linked saccharide domain in the LDL receptor (Figure 24.40).

10. Identify the lipid synthetic pathways that would be affected by abnormally low levels of CTP.

11. Determine the number of ATP equivalents needed to form palmitic acid from acetyl-CoA. (Assume for this calculation that each NADPH is worth 3.5 ATPs.)

12. Write a reasonable mechanism for the 3-ketosphinganine synthase reaction, remembering that it is a pyridoxal phosphate–dependent reaction.

13. Why is the involvement of FAD important in the conversion of stearic acid to oleic acid?

14. Write a suitable mechanism for the HMG-CoA synthase reaction. What is the chemistry that drives this condensation reaction.

15. Write a suitable reaction mechanism for the β-ketoacyl ACP synthase, showing how the involvement of malonyl-CoA drives this reaction forward.

Preparing for the MCAT Exam

16. Consider the synthesis of linoleic acid from palmitic acid and identify a series of three consecutive reactions that embody chemistry similar to three reactions in the tricarboxylic acid cycle.

17. Rewrite the equation in Section 24.1 to describe the synthesis of behenic acid (see Table 8.1).

Biochemistry ⊜ Now™   Preparing for an exam? Test yourself on key questions at http://chemistry.brookscole.com/ggb3

Further Reading

General

Borhani, D. W., Rogers, D. P., Engler, J. A., and Brouillette, C. G., 1997. Crystal structure of truncated human apolipoprotein A-I suggests a lipid-bound conformation. *Proceedings of the National Academy of Sciences* **94:**12291–12296.

Boyer, P. D., ed., 1983. *The Enzymes,* 3rd ed., Vol. 16. New York: Academic Press.

Friedman, J. M., 1997. The alphabet of weight control. *Nature* **385:**119–120.

Lardy, H., and Shrago, E., 1990. Biochemical aspects of obesity. *Annual Review of Biochemistry* **59:**689–710.

Roberts, L. M., Ray, M. J., Shih, T. W., et al., 1997. Structural analysis of apolipoprotein A-I: Limited proteolysis of methionine-reduced and -oxidized lipid-free and lipid-bound human Apo A-I. *Biochemistry* **36:**7615–7624.

Schewe, T., and Kuhn, H., 1991. Do 15-lipoxygenases have a common biological role? *Trends in Biochemical Sciences* **16:**369–373.

Smith, W. L., Garavito, R. M., and DeWitt, D. L., 1996. Prostaglandin endoperoxide H synthases (cyclooxygenases)-1 and -2. *Journal of Biological Chemistry* **271:**33157–33160.

Tataranni, P. A., and Ravussin, E., 1997. Effect of fat intake on energy balance. *Annals of the New York Academy of Sciences* **819:**37–43.

Vance, D. E., and Vance, J. E., eds., 1985. *Biochemistry of Lipids and Membranes.* Menlo Park, CA: Benjamin/Cummings.

Fatty Acid Synthesis

Athappilly, F. K., and Hendrickson, W. A., 1995. Structure of the biotinyl domain of acetyl-CoA carboxylase determined by mad phasing. *Structure* **3:**1407.

Chang, S. I., and Hammes, G. G., 1990. Structure and mechanism of action of a multifunctional enzyme: Fatty acid synthase. *Accounts of Chemical Research* **23:**363–369.

Hansen, H. S., 1985. The essential nature of linoleic acid in mammals. *Trends in Biochemical Sciences* **11:**263–265.

Jeffcoat, R., 1979. The biosynthesis of unsaturated fatty acids and its control in mammalian liver. *Essays in Biochemistry* **15:**1–36.

Jump, D. B., 2002. The biochemistry of n-3 polyunsaturated fatty acids. *Journal of Biological Chemistry* **277:**8755–8758.

Kim, K-H., et al., 1989. Role of reversible phosphorylation of acetyl-CoA carboxylase in long-chain fatty acid synthesis. *The FASEB Journal* **3:**2250–2256.

Smith, W. L., Garavito, R. M., and DeWitt, D. L., 1996. Prostaglandin endoperoxide H synthases (cyclooxygenases)-1 and -2. *Journal of Biological Chemistry* **271:**33157–33160.

Sohlencamp, C., Lopez-Lara, I. M., and Geiger, O., 2003. Biosynthesis of phosphatidylcholine in bacteria. *Progress in Lipid Research* **42:**115–162.

Wakil, S., 1989. Fatty acid synthase, a proficient multifunctional enzyme. *Biochemistry* **28:**4523–4530.

Wakil, S., Stoops, J. K., and Joshi, V. C., 1983. Fatty acid synthesis and its regulation. *Annual Review of Biochemistry* **52:**537–579.

Phospholipid and Triacylglycerol Synthesis

Carman, G. M., and Henry, S. A., 1989. Phospholipid biosynthesis in yeast. *Annual Review of Biochemistry* **58:**635–669.

Carman, G. M., and Zeimetz, G. M., 1996. Regulation of phospholipid biosynthesis in the yeast *Saccharomyces cerevisiae. Journal of Biological Chemistry* **271:**13292–13296.

Dunne, S. J., Cornell, R. B., Johnson, J. E., et al., 1996. Structure of the membrane-binding domain of CTP phosphocholine cytidylyltransferase. *Biochemistry* **35:**11975–11984.

Jackowski, S., 1996. Cell cycle regulation of membrane phospholipid metabolism. *Journal of Biological Chemistry* **271:**20219–20222.

Kent, C., 1995. Eukaryotic phospholipid biosynthesis. *Annual Review of Biochemistry* **64:**315–343.

Sohlencamp, C., Lopez-Lara, I. M., and Geiger, O., 2003. Biosynthesis of phosphatidylcholine in bacteria. *Progress in Lipid Research* **42:**115–162.

Sorger, D., and Daum, G., 2003. Triacylglycerol biosynthesis in yeast. *Applied Microbiology and Biotechnology* **61:**289–299.

Cholesterol Synthesis

Bloch, K., 1965. The biological synthesis of cholesterol. *Science* **150:**19–28.

Bloch, K., 1987. Summing up. *Annual Review of Biochemistry* **56:**1–19.

Dietschy, J. M., and Turley, S. D., 2002. Control of cholesterol turnover in the mouse. *Journal of Biological Chemistry* **277:**3801–3804.

Liscum, L., and Underwood, K. W., 1995. Intracellular cholesterol transport and compartmentation. *Journal of Biological Chemistry* **270:**15443–15446.

Nitrogen Acquisition and Amino Acid Metabolism

CHAPTER 25

Essential Question

Nitrogen is an essential nutrient for all cells. Amino acids provide nitrogen for the synthesis of other nitrogen-containing biomolecules. Excess amino acids in the diet can be converted into α-keto acids and used for energy production. *What are the biochemical pathways that form ammonium from inorganic nitrogen compounds prevalent in the inanimate environment? How is ammonium incorporated into organic compounds? How are amino acids synthesized and degraded?*

Nitrogen is a vital macronutrient for all life, and in this chapter we begin our consideration of the pathways of nitrogen metabolism. We start with a presentation of the two principal routes for nitrogen acquisition from the inanimate environment: nitrate assimilation and nitrogen fixation. The reactions of ammonium assimilation follow. Glutamine synthetase merits particular attention because it conveys several important lessons in metabolic regulation. The pathways of amino acid biosynthesis and degradation are described; those involving the sulfur-containing amino acids provide an opportunity to introduce aspects of sulfur metabolism.

Soybeans. Only plants and certain microorganisms are able to transform the oxidized, inorganic forms of nitrogen available in the inanimate environment into reduced, biologically useful forms. Soybean plants can meet their nitrogen requirements both by assimilating nitrate and, in symbiosis with bacteria, fixing N_2.

25.1	**Which Metabolic Pathways Allow Organisms to Live on Inorganic Forms of Nitrogen?**

Nitrogen Is Cycled Between Organisms and the Inanimate Environment

Nitrogen exists predominantly in an oxidized state in the inanimate environment, occurring principally as N_2 in the atmosphere or as nitrate ion (NO_3^-) in the soils and oceans. Its acquisition by biological systems is accompanied by its reduction to ammonium ion (NH_4^+) and the incorporation of NH_4^+ into organic linkage as amino or amido groups (Figure 25.1). The reduction of NO_3^- to NH_4^+ occurs in green plants, various fungi, and certain bacteria in a two-step metabolic pathway known as **nitrate assimilation.** The formation of NH_4^+ from N_2 gas is termed **nitrogen fixation.** N_2 fixation is an exclusively prokaryotic process, although bacteria in symbiotic association with certain green plants also carry out nitrogen fixation. No animals are capable of either nitrogen fixation or nitrate assimilation, so they are totally dependent on plants and microorganisms for the synthesis of organic nitrogenous compounds, such as amino acids and proteins, to satisfy their requirements for this essential element.

Animals release excess nitrogen in a reduced form, either as NH_4^+ or as organic nitrogenous compounds such as urea. The release of N occurs both during life and as a consequence of microbial decomposition following death. Various bacteria return the reduced forms of nitrogen back to the environment by oxidizing them. The oxidation of NH_4^+ to NO_3^- by **nitrifying bacteria,** a group of chemoautotrophs, provides the sole source of chemical energy for the life of these microbes. Nitrate nitrogen also returns to the atmosphere as N_2 as a result of the metabolic activity of **denitrifying bacteria.** These bacteria are capable of using NO_3^- and similar oxidized inorganic forms of nitrogen as electron acceptors in place of O_2 in energy-producing pathways. The NO_3^- is reduced ultimately to *dinitrogen* (N_2). These bacteria thus deplete the levels of *combined*

© Royalty-Free/CORBIS

I was determined to know beans.
Henry David Thoreau (1817–1862), *The Writings of Henry David Thoreau,* vol. 2, p. 178, Houghton Mifflin (1906).

Key Questions

25.1   Which Metabolic Pathways Allow Organisms to Live on Inorganic Forms of Nitrogen?

25.2   What Is The Metabolic Fate of Ammonium?

25.3   What Regulatory Mechanisms Act on *Escherichia coli* Glutamine Synthetase?

25.4   How Do Organisms Synthesize Amino Acids?

25.5   How Does Amino Acid Catabolism Lead into Pathways of Energy Production?

Biochemistry⑂Now™ Test yourself on these Key Questions at BiochemistryNow at **http://chemistry.brookscole.com/ggb3**

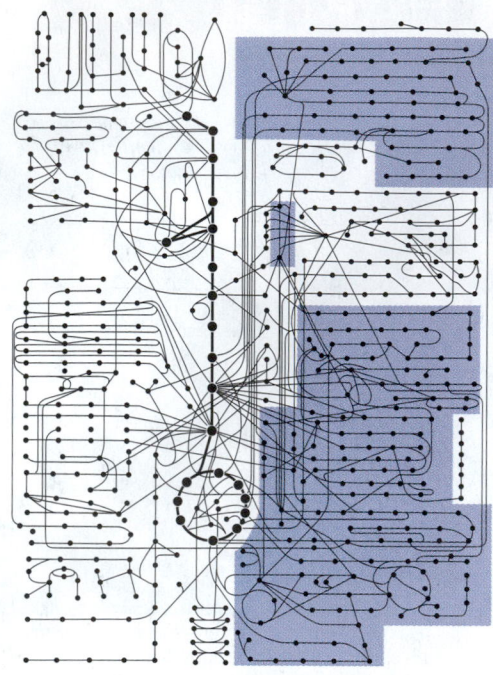

Nitrogen acquisition and amino acid metabolism.

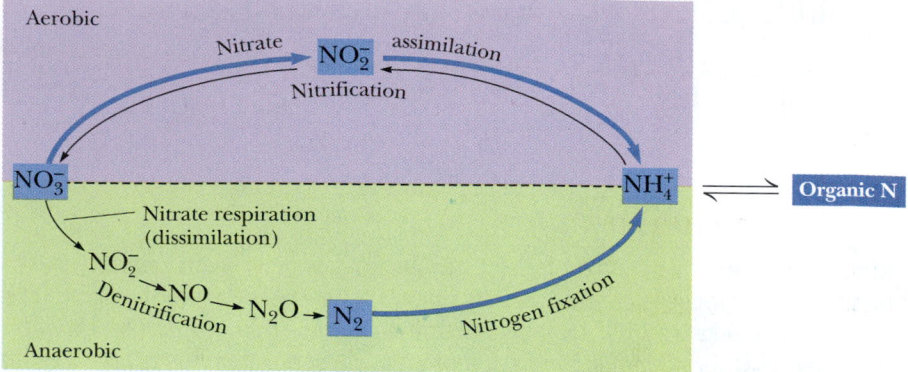

FIGURE 25.1 The nitrogen cycle. Organic nitrogenous compounds are formed by the incorporation of NH_4^+ into carbon skeletons. Ammonium can be formed from oxidized inorganic precursors by reductive reactions: Nitrogen fixation reduces N_2 to NH_4^+; nitrate assimilation reduces NO_3^- to NH_4^+. Nitrifying bacteria can oxidize NH_4^+ back to NO_3^- and obtain energy for growth in the process of nitrification. Denitrification is a form of bacterial respiration whereby nitrogen oxides serve as electron acceptors in the place of O_2 under anaerobic conditions.

nitrogen,[1] important as a natural fertilizer, that might otherwise be available. However, such bacterial activity is being exploited in water treatment plants to reduce the load of combined nitrogen that might otherwise enter lakes, streams, and bays.

Nitrate Assimilation Is the Principal Pathway for Ammonium Biosynthesis

Nitrate assimilation occurs in two steps: the two-electron reduction of nitrate to nitrite, catalyzed by **nitrate reductase** (Equation 25.1), followed by the six-electron reduction of nitrite to ammonium, catalyzed by **nitrite reductase** (Equation 25.2).

$$(1) \quad NO_3^- + 2\,H^+ + 2\,e^- \longrightarrow NO_2^- + H_2O \tag{25.1}$$

$$(2) \quad NO_2^- + 8\,H^+ + 6\,e^- \longrightarrow NH_4^+ + 2\,H_2O \tag{25.2}$$

Nitrate assimilation is the predominant means by which green plants, algae, and many microorganisms acquire nitrogen. The pathway of nitrate assimilation accounts for more than 99% of the inorganic nitrogen (nitrate or N_2) assimilated into organisms.

Nitrate Reductase Contains Cytochrome b_{557} and Molybdenum Cofactor

$$
\begin{array}{ccc}
\text{NADH} & & NO_3^- \\
& [-SH \rightarrow FAD \rightarrow \text{cytochrome } b_{557} \rightarrow MoCo] & \\
\text{NADH}^+ & & NO_2^-
\end{array}
$$

A pair of electrons is transferred from NADH via enzyme-associated sulfhydryl groups, FAD, cytochrome b_{557}, and **MoCo** (an essential molybdenum-containing cofactor) to nitrate, reducing it to nitrite. The brackets [ ] denote the protein-bound prosthetic groups that constitute an e^- transport chain between NADH and nitrate. Nitrate reductases typically are cytosolic 220-kD dimeric proteins. The structure of the molybdenum cofactor (MoCo) is shown in Figure 25.2a. Molybdenum cofactor is necessary for both nitrate reductase activity and the assembly of nitrate reductase subunits into the active holoenzyme dimer form. Molybdenum cofactor is also an essential cofactor for a variety of enzymes that catalyze hydroxylase-type reactions, including xanthine dehydrogenase, aldehyde oxidase, and sulfite oxidase.

[1]N joined with other elements in chemical compounds.

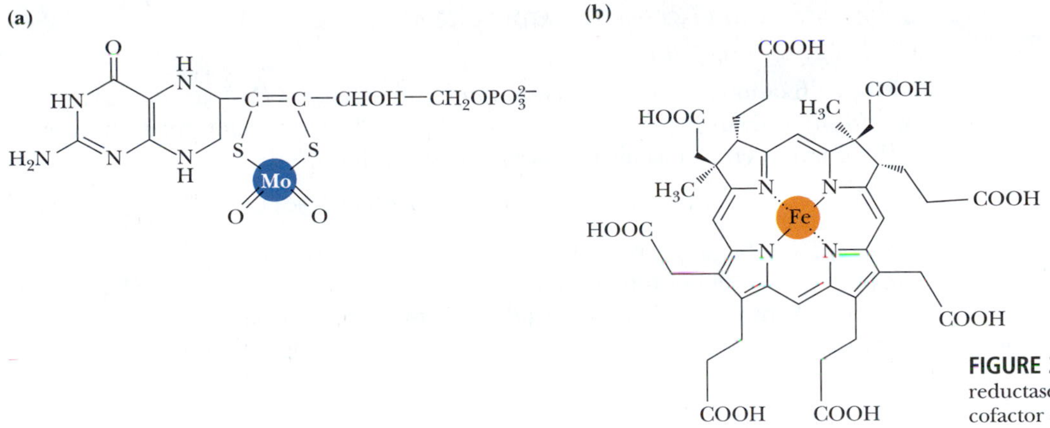

(a)

(b)

FIGURE 25.2 The novel prosthetic groups of nitrate reductase and nitrite reductase. **(a)** The molybdenum cofactor of nitrate reductase. The molybdenum-free version of this compound is a pterin derivative called **molybdopterin**. **(b)** Siroheme, a uroporphyrin derivative, is a member of the **isobacteriochlorin class** of hemes, a group of porphyrins in which adjacent pyrrole rings are reduced. Siroheme is novel in having eight carboxylate-containing side chains. These carboxylate groups may act as H^+ donors during the reduction of NO_2^- to NH_4^+.

Nitrite Reductase Contains Siroheme Six electrons are required to reduce NO_2^- to NH_4^+. Nitrite reductases in photosynthetic organisms obtain these electrons from six molecules of photosynthetically reduced ferredoxin (Fd_{red}).

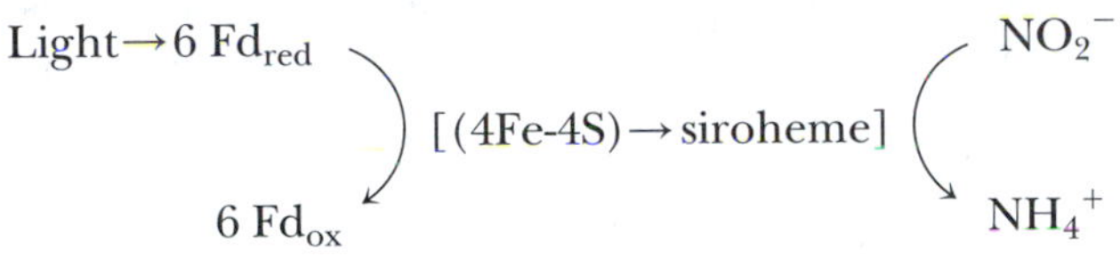

Photosynthetic nitrite reductases are 63-kD monomeric proteins having a tetranuclear iron–sulfur cluster and a novel heme, termed **siroheme,** as prosthetic groups. The [4Fe-4S] cluster and the siroheme act as a coupled e^- transfer center. Nitrite binds directly to siroheme, providing the sixth ligand, much as O_2 binds to the heme of hemoglobin. Nitrite is reduced to ammonium while liganded to siroheme. The structure of siroheme is shown in Figure 25.2b.

In higher plants, nitrite reductase is found in chloroplasts, where it has ready access to its primary reductant, photosynthetically reduced ferredoxin. Microbial nitrite reductases are larger and more complex than plant nitrite reductases. Indeed, microbial nitrite reductases closely resemble nitrate reductases in having essential —SH groups and FAD prosthetic groups to couple enzyme-mediated NADPH oxidation to nitrite reduction (Figure 25.3).

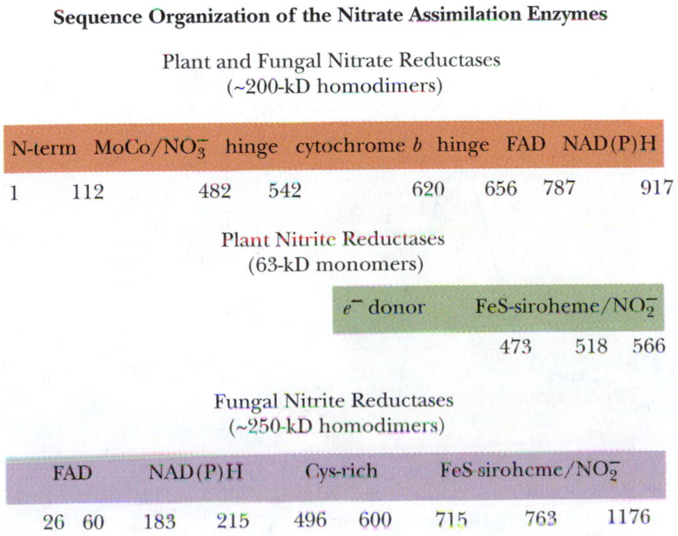

Sequence Organization of the Nitrate Assimilation Enzymes

Plant and Fungal Nitrate Reductases
(~200-kD homodimers)

N-term	MoCo/NO_3^-	hinge	cytochrome b	hinge	FAD	NAD(P)H
1 112	482	542	620	656	787	917

Plant Nitrite Reductases
(63-kD monomers)

e^- donor	FeS-siroheme/NO_2^-
473	518 566

Fungal Nitrite Reductases
(~250-kD homodimers)

FAD	NAD(P)H	Cys-rich	FeS siroheme/NO_2^-
26 60	183 215	496 600	715 763 1176

FIGURE 25.3 Domain organization within the enzymes of nitrate assimilation. The numbers denote residue number along the amino acid sequence of the proteins. The numbering for nitrate reductase is that from the green plant *Arabidopsis thaliana;* the plant nitrite reductase sequence shown here is spinach; the fungal nitrite reductase is *Neurospora crassa. (Adapted in part from Campbell, W. H., and Kinghorn, K. R., 1990. Functional domains of assimilatory nitrate reductases and nitrite reductases.* Trends in Biochemical Sciences *15:315–319.)*

Organisms Gain Access to Atmospheric N_2 Via the Pathway of Nitrogen Fixation

Nitrogen fixation involves the reduction of nitrogen gas (N_2) via an enzyme system found only in prokaryotic cells. The heart of the nitrogen fixation process is the enzyme known as **nitrogenase,** which catalyzes the reaction

$$N_2 + 10\ H^+ + 8\ e^- \longrightarrow 2\ NH_4^+ + H_2 \qquad (25.3)$$

Note that an obligatory reduction of two protons to hydrogen gas accompanies the biological reduction of N_2 to ammonia. Less than 1% of the inorganic N incorporated into organic compounds by organisms can be attributed to nitrogen fixation; however, this process provides the only direct biological access to the enormous reservoir of N_2 in the atmosphere.

Although nitrogen fixation is exclusively prokaryotic, N_2-fixing bacteria may be either free-living or living as symbionts with higher plants. For example, *Rhizobia* are bacteria that fix nitrogen in symbiotic association with leguminous plants. Because nitrogen in a metabolically useful form is often the limiting nutrient for plant growth, such symbiotic associations can be an important factor in plant growth and agriculture.

Despite the wide diversity of bacteria in which nitrogen fixation takes place, all N_2-fixing systems are nearly identical and all have four fundamental requirements: (1) the enzyme *nitrogenase;* (2) a strong reductant, such as reduced ferredoxin; (3) ATP; and (4) O_2-free conditions. In addition, several modes of regulation act to control nitrogen fixation.

The Nitrogenase Complex Is Composed of Two Metalloproteins   Two metalloproteins constitute the nitrogenase complex: the **Fe-protein** or nitrogenase reductase and the **MoFe-protein,** which is another name for nitrogenase. Nitrogenase reductase is a 60-kD homodimer possessing a single [4Fe-4S] cluster as a prosthetic group. **Nitrogenase reductase** is extremely O_2 sensitive. Nitrogenase reductase binds MgATP and hydrolyzes two ATPs per electron transferred during nitrogen fixation. Because reduction of N_2 to $2\ NH_4^+ + H_2$ requires 8 electrons, 16 ATPs are consumed per N_2 reduced.

This ATP requirement seems paradoxical because the reaction is thermodynamically favorable: The $\mathcal{E}_o{}'$ for the half-reaction ($N_2 + 8\ e^- + 10\ H^+ \rightarrow 2\ NH_4^+ + H_2$) is -0.314 V, and ferredoxin, the most common e^- donor for nitrogen fixation, has an $\mathcal{E}_o{}'$ that is more negative (see Table 20.1). The solution to the paradox is found in the very strong bonding between the two N atoms in N_2 (Figure 25.4). Substantial energy input is needed to overcome this large activation energy and break the $N{\equiv}N$ triple bond. In this biological system, the energy is provided by ATP.

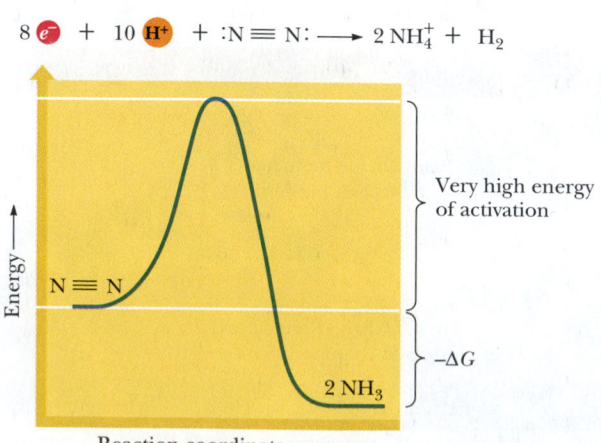

FIGURE 25.4 The triple bond in N_2 must be broken during nitrogen fixation. A substantial energy input is needed to overcome this thermodynamic barrier, even though the overall free energy change ($\Delta G^{\circ\prime}$) for biological N_2 reduction is negative.

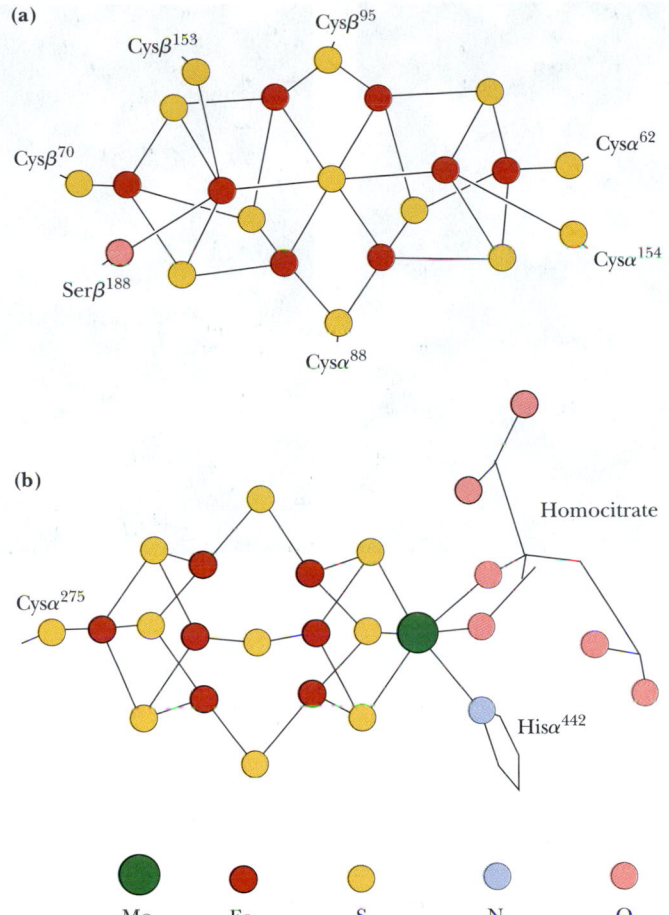

FIGURE 25.5 Structures of the two types of metal clusters found in nitrogenase. **(a)** The P-cluster. Two Fe_4S_3 clusters share a fourth S and are bridged by two thiol ligands from the protein ($Cys\alpha^{88}$ and $Cys\beta^{95}$). **(b)** The FeMo-cofactor. This novel molybdenum-containing Fe-S complex contains 1 Mo, 7 Fe, and 9 S atoms; it is liganded to the protein via a $Cys\alpha^{275}$—S linkage to an Fe atom and a $His\alpha^{442}$—N linkage to the Mo atom. Homocitrate provides two oxo ligands to the Mo atom. *(Adapted from Leigh, G. J., 1995. The mechanism of dinitrogen reduction by molybdenum nitrogenases.* European Journal of Biochemistry **229**:14–20.)

Nitrogenase, the MoFe-protein, is a 240-kD $\alpha_2\beta_2$-type heterotetramer. An $\alpha\beta$-dimer serves as the functional unit, and each $\alpha\beta$-dimer contains two types of metal centers: an unusual 8Fe-7S center known as the **P-cluster** (Figure 25.5a) and the novel 7Fe-1Mo-9S cluster known as the **FeMo-cofactor** (Figure 25.5b). Nitrogenase under unusual circumstances may contain an **iron:vanadium cofactor** instead of the molybdenum-containing one. Like nitrogenase reductase, nitrogenase is very oxygen labile.

The Nitrogenase Reaction In the nitrogenase reaction (Figure 25.6), electrons from reduced ferredoxin pass to nitrogenase reductase, which serves as electron donor to nitrogenase, the enzyme that actually catalyzes N_2 fixation. Electron transfer from nitrogenase reductase to nitrogenase takes place through docking of nitrogenase reductase with an $\alpha\beta$-subunit pair of nitrogenase (Figure 25.7) and transfer of an electron according to the following sequence: Fe-protein $\rightarrow$ P-cluster $\rightarrow$ FeMo-cofactor $\rightarrow N_2$. ATP hydrolysis is coupled to the transfer of an electron from the Fe-protein to the P-cluster. ATP hydrolysis leads to conformational change in the nitrogenase reductase, so it no longer binds to nitrogenase. The ADP:oxidized nitrogenase reductase complex dissociates,

FIGURE 25.6 The nitrogenase reaction. Depending on the bacterium, electrons for N_2 reduction may come from light, NADH, hydrogen gas, or pyruvate. The primary e^- donor for the nitrogenase system is reduced ferredoxin. Reduced ferredoxin passes electrons directly to nitrogenase reductase. A total of six electrons are required to reduce N_2 to 2 NH_4^+, and another two electrons are consumed in the obligatory reduction of 2 H^+ to H_2. Nitrogenase reductase transfers e^- to nitrogenase one electron at a time. N_2 is bound at the critical FeMoCo prosthetic group of nitrogenase until all electrons and protons are added; no free intermediates such as HN=NH or H_2N—NH_2 are detectable.

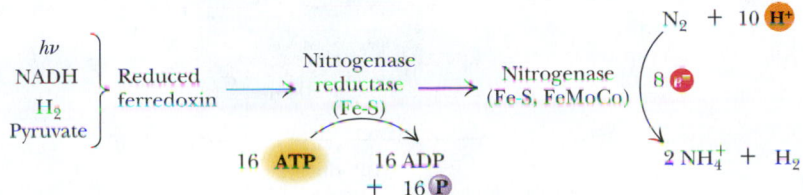

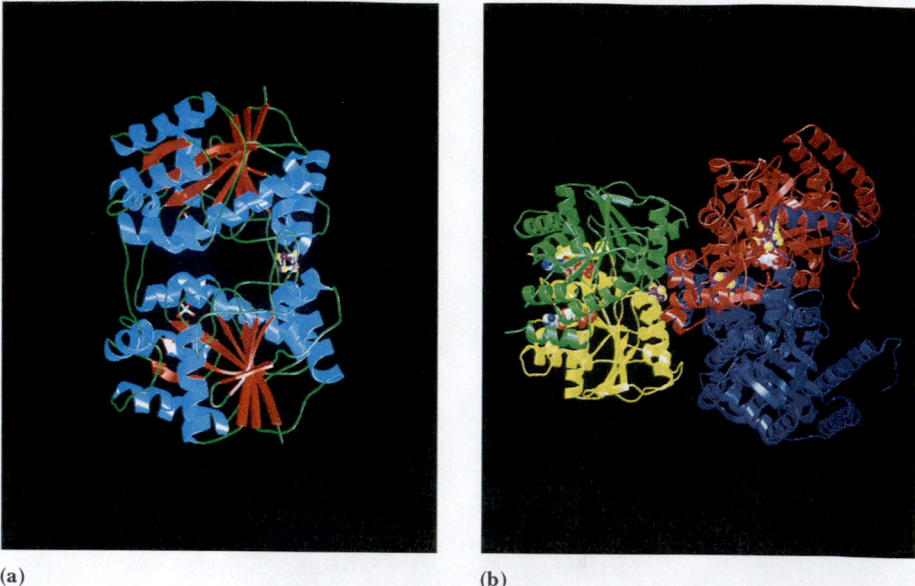

(a) (b)

FIGURE 25.7 (a) Ribbon diagram of nitrogenase reductase (the Fe-protein). The ATP-binding site (here occupied by ADP, green) is at the left in this orientation, and the Fe_4S_4 cluster (brown) is at the right. (b) Model of the complex formed between nitrogenase reductase and a nitrogenase $\alpha\beta$-dimer. Within the nitrogenase reductase (shown in purple and light blue here), the Fe_4S_4 center is closest to the nitrogenase $\alpha\beta$-dimer. The nitrogenase $\alpha\beta$-dimer (gold and blue) is to the right of the nitrogenase reductase, with its FeMo-cofactor and P-cluster as space-filling models (red). *(Adapted from Kim, J., and Rees, D. 1994. Nitrogenase and biological nitrogen fixation. Biochemistry 33:389–396.)*

Biochemistry⊛Now™ Go to BiochemistryNow and click BiochemistryInteractive to learn more about nitrogenase reductase and explore the novel MoFe prosthetic group in nitrogenase.

making way for another ATP:reduced nitrogenase reductase complex to bind to nitrogenase. Nitrogenase is a rather slow enzyme: Its optimal rate of e^- transfer is about 12 e^- pairs per second per enzyme molecule; that is, it reduces only three molecules of nitrogen gas per second. Because its activity is so weak, nitrogen-fixing cells maintain large amounts of nitrogenase so that their requirements for reduced N can be met. As much as 5% of the cellular protein may be nitrogenase. As indicated earlier, nitrogenase catalyzes the concomitant reduction of protons to hydrogen gas, its so-called **hydrogenase activity.** This activity accompanies N_2 reduction in vivo and is energy depleting. One equivalent

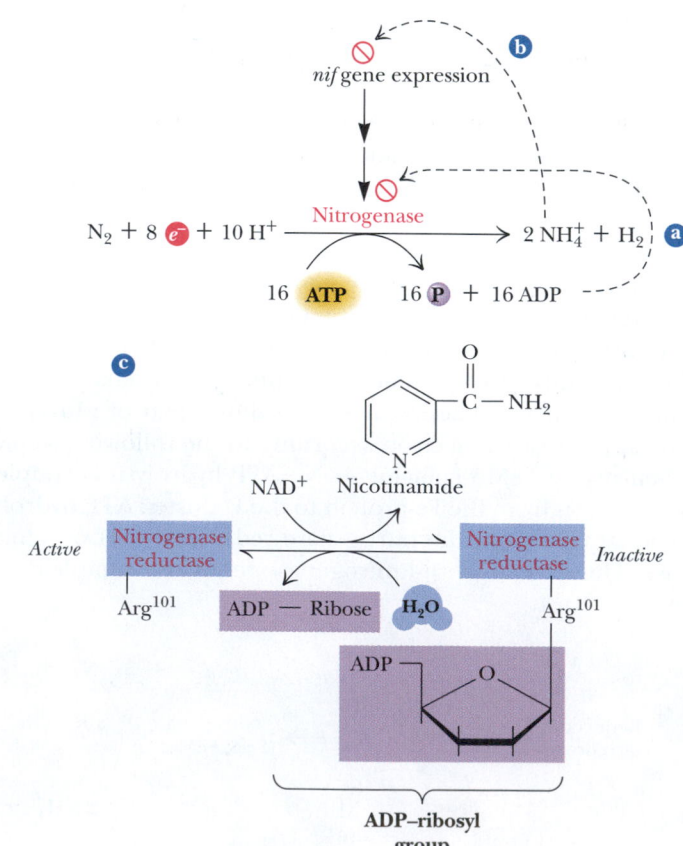

FIGURE 25.8 Regulation of nitrogen fixation. (a) ADP inhibits nitrogenase activity. (b) NH_4^+ represses *nif* gene expression. (c) In some organisms, the nitrogenase complex is regulated by covalent modification. ADP–ribosylation of nitrogenase reductase leads to its inactivation. Nitrogenase reductase is a distant relative of the signal-transducing G-protein superfamily.

of H_2 is formed for every 2 NH_3 (this unavoidable reaction leads to the stoichiometry given in Equation 25.3).

The Regulation of Nitrogen Fixation   To a first approximation, two regulatory controls are paramount (Figure 25.8): (1) ADP inhibits the activity of nitrogenase; thus, as the ATP/ADP ratio drops, nitrogen fixation is blocked. (2) NH_4^+ represses the expression of the *nif* genes, the genes that encode the proteins of the nitrogen-fixing system. To date, some 20 *nif* genes have been identified with the nitrogen fixation process. Repression of *nif* gene expression by ammonium, the primary product of nitrogen fixation, is an efficient and effective way of shutting down N_2 fixation when its end product is not needed.

25.2 | What Is the Metabolic Fate of Ammonium?

Ammonium enters organic linkage via three major reactions that are found in all cells. The enzymes mediating these reactions are (1) *carbamoyl-phosphate synthetase I*, (2) *glutamate dehydrogenase*, and (3) *glutamine synthetase*.

   Carbamoyl-phosphate synthetase I catalyzes one of the steps in the *urea cycle*. Two ATPs are consumed, one in the activation of HCO_3^- for reaction with ammonium and the other in the phosphorylation of the carbamate formed:

$$NH_4^+ + HCO_3^- + 2\,ATP \rightarrow H_2N-\overset{\overset{\displaystyle O}{\|}}{C}-O-PO_3^{2-} + 2\,ADP + P_i + 2\,H^+$$

N-acetylglutamate is an essential allosteric activator for this enzyme.

   Glutamate dehydrogenase (GDH) catalyzes the reductive amination of α-ketoglutarate to yield glutamate. Reduced pyridine nucleotides (NADH or NADPH) provide the reducing power:

$$NH_4^+ + \alpha\text{-ketoglutarate} + NADPH + H^+ \longrightarrow glutamate + NADP^+ + H_2O$$

This reaction provides an important interface between nitrogen metabolism and cellular pathways of carbon and energy metabolism because α-ketoglutarate is a citric acid cycle intermediate. In vertebrates, GDH is an α_6-type multimeric enzyme localized in the mitochondrial matrix that uses NADPH as electron donor when operating in the biosynthetic direction (the direction of glutamate synthesis) (Figure 25.9). In contrast, when GDH acts in the catabolic direction to generate α-ketoglutarate from glutamate, NAD^+, not $NADP^+$, is usually the electron acceptor. The catabolic activity is allosterically activated by ADP and inhibited by GTP. Some organisms (the fungus *Neurospora crassa* is one example) have two GDH isozymes: an $NADP^+$-specific cytosolic enzyme that functions in the direction of glutamate synthesis and an NAD^+-specific mitochondrial enzyme acting in the catabolic direction to convert excess glutamate into α-ketoglutarate for energy metabolism.

   Glutamine synthetase (GS) catalyzes the ATP-dependent amidation of the γ-carboxyl group of glutamate to form glutamine. The reaction proceeds via a γ-glutamyl-phosphate intermediate, and GS activity depends on the presence

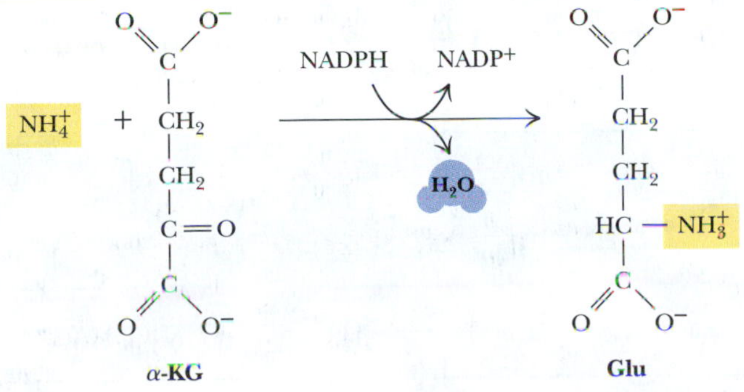

FIGURE 25.9 The glutamate dehydrogenase reaction.

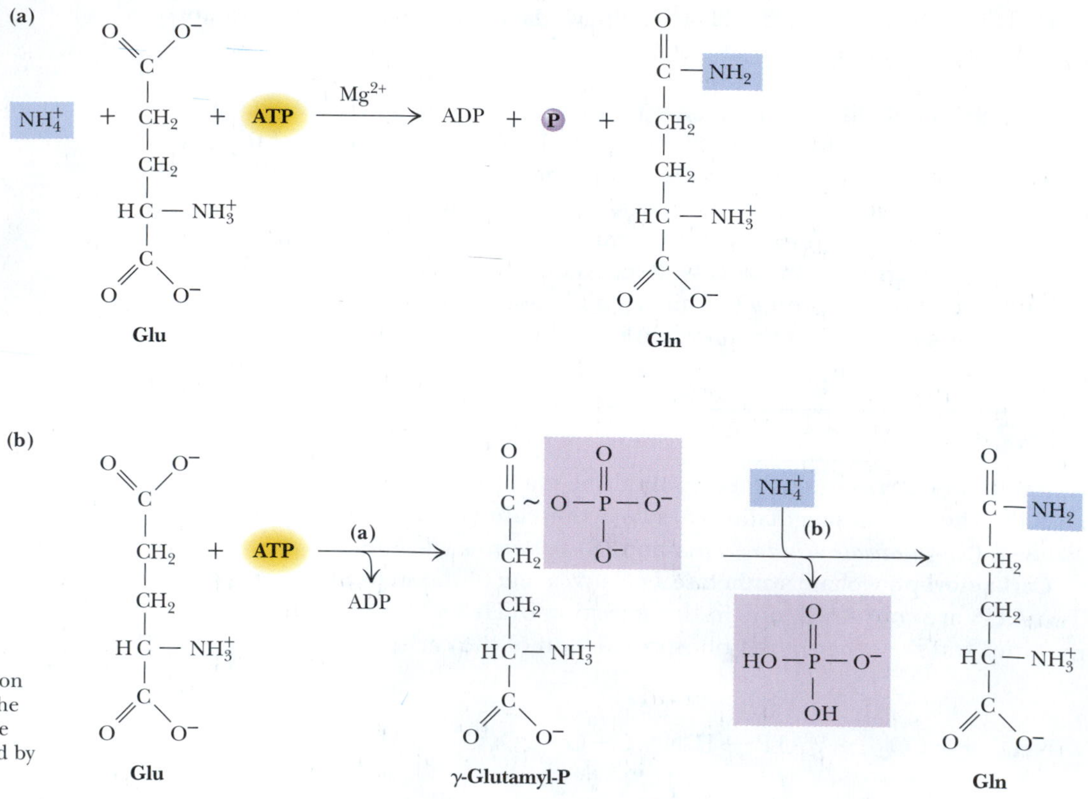

(a)

(b)

FIGURE 25.10 (a) The enzymatic reaction catalyzed by glutamine synthetase. (b) The reaction proceeds by (a) activation of the γ-carboxyl group of Glu by ATP, followed by (b) amidation by NH_4^+.

of divalent cations such as Mg^{2+} (Figure 25.10). **Glutamine** is a major N donor in the biosynthesis of many organic N compounds such as purines, pyrimidines, and other amino acids, and GS activity is tightly regulated, as we shall soon see. The amide-N of glutamine provides the nitrogen atom in these biosyntheses. In quantitative terms, GDH and GS are responsible for most of the ammonium assimilated into organic compounds.

The Major Pathways of Ammonium Assimilation Lead to Glutamine Synthesis

In organisms that enjoy environments rich in nitrogen, GDH and GS acting in sequence furnish the principal route of NH_4^+ incorporation (Figure 25.11). However, GDH has a significantly higher K_m for NH_4^+ than does GS. Consequently, in organisms such as green plants that grow under conditions where little NH_4^+ is available, GDH is not effective and GS is the only NH_4^+-assimilative reaction. Such a situation creates the need for an alternative mode of glutamate synthesis to replenish the glutamate consumed by the GS reaction. This need is filled by **glutamate synthase** (also known as *GOGAT*, the acronym for the other name of this enzyme—glutamate:oxo-glutarate amino-transferase). Glutamate synthase catalyzes the reductive amination of α-ketoglutarate using the amide-N of glutamine as the N donor:

$$\text{Reductant} + \alpha\text{-KG} + \text{Gln} \longrightarrow 2 \text{ Glu} + \text{oxidized reductant}$$

FIGURE 25.11 The GDH/GS pathway of ammonium assimilation. The sum of these reactions is the conversion of 1 α-ketoglutarate to 1 glutamine at the expense of 2 NH_4^+, 1 ATP, and 1 NADPH.

(a) $NH_4^+ + \alpha$-ketoglutarate + NADPH $\xrightarrow{\text{GDH}}$ glutamate + $NADP^+$ + H_2O

(b) Glutamate + NH_4^+ + ATP $\xrightarrow{\text{GS}}$ glutamine + ADP + P_i

SUM: 2 $NH_4^+ + \alpha$-ketoglutarate + NADPH + ATP $\longrightarrow$ glutamine + $NADP^+$ + ADP + P_i + H_2O

$$\left.\begin{array}{l} \text{NADH (yeast, } N.\ crassa\text{)} + \text{H}^+ \\ \text{NADPH } (E.\ coli) + \text{H}^+ \text{ or} \\ 2\,\text{H}^+ + 2\text{ reduced ferredoxin (plants)} \end{array}\right\} + \alpha\text{-KG} + \text{Gln} \longrightarrow 2\,\text{Glu} + \begin{array}{l} \text{NAD}^+ \\ \text{NADP}^+ \text{ or} \\ 2\text{ oxidized ferredoxin} \end{array}$$

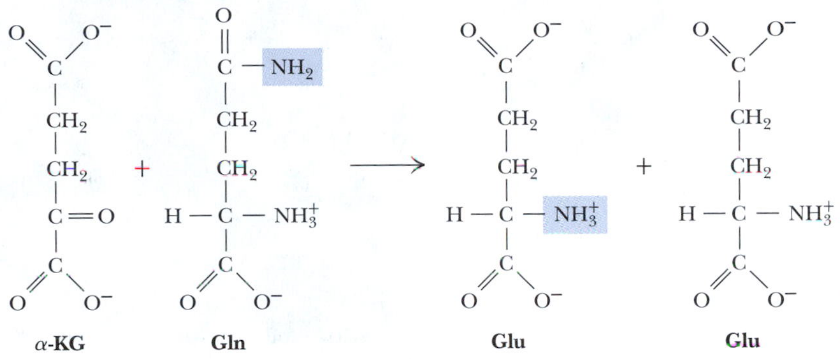

FIGURE 25.12 The glutamate synthase reaction, showing the reductants exploited by different organisms in this reductive amination reaction.

Two glutamates are formed—one from amination of α-ketoglutarate and the other from deamidation of Gln (Figure 25.12). These glutamates can now serve as ammonium acceptors for glutamine synthesis by GS. Organisms variously use NADH, NADPH, or reduced ferredoxin as reductant. Glutamate synthases are typically large, complex proteins; in *Escherichia coli*, GOGAT is an 800-kD flavoprotein containing both FMN and FAD, as well as [4Fe-4S] clusters.

Together, GS and GOGAT constitute a second pathway of ammonium assimilation, in which GS is the only NH_4^+-fixing step; the role of GOGAT is to regenerate glutamate (Figure 25.13). Note that this pathway consumes 2 equivalents of ATP and 1 NADPH (or similar reductant) per pair of N atoms introduced into Gln, in contrast to the GDH/GS pathway, in which only 1 ATP and 1 NADPH are used up per pair of NH_4^+ fixed. Clearly, coping with a nitrogen-limited environment has its cost.

25.3 What Regulatory Mechanisms Act on *Escherichia coli* Glutamine Synthetase?

As indicated earlier, glutamine plays a pivotal role in nitrogen metabolism by donating its amide nitrogen to the biosynthesis of many important organic N compounds. Consistent with its metabolic importance, in enteric bacteria such as *E. coli*, GS is regulated at three different levels:

1. Its activity is regulated allosterically by *feedback inhibition*.
2. GS is interconverted between active and inactive forms by *covalent modification*.
3. Cellular amounts of GS are carefully controlled at the level of *gene expression* and *protein synthesis*.

Eukaryotic versions of glutamine synthetase show none of these regulatory features.

E. coli GS is a 600-kD dodecamer (α_{12}-type subunit organization) of identical 52-kD monomers (each monomer contains 468 amino acid residues).

(a) $2\,NH_4^+ + 2\,ATP + 2\text{ glutamate} \xrightarrow{\text{GS}} 2\text{ glutamine} + 2\,ADP + 2\,P_i$

(b) $NADPH + \alpha\text{-ketoglutarate} + \text{glutamine} \xrightarrow{\text{GOGAT}} 2\text{ glutamate} + NADP^+$

SUM: $2\,NH_4^+ + \alpha\text{-ketoglutarate} + NADPH + 2\,ATP \longrightarrow \text{glutamine} + NADP^+ + 2\,ADP + 2\,P_i$

FIGURE 25.13 The GS/GOGAT pathway of ammonium assimilation. The sum of these reactions results in the conversion of 1 α-ketoglutarate to 1 glutamine at the expense of 2 ATP and 1 NADPH.

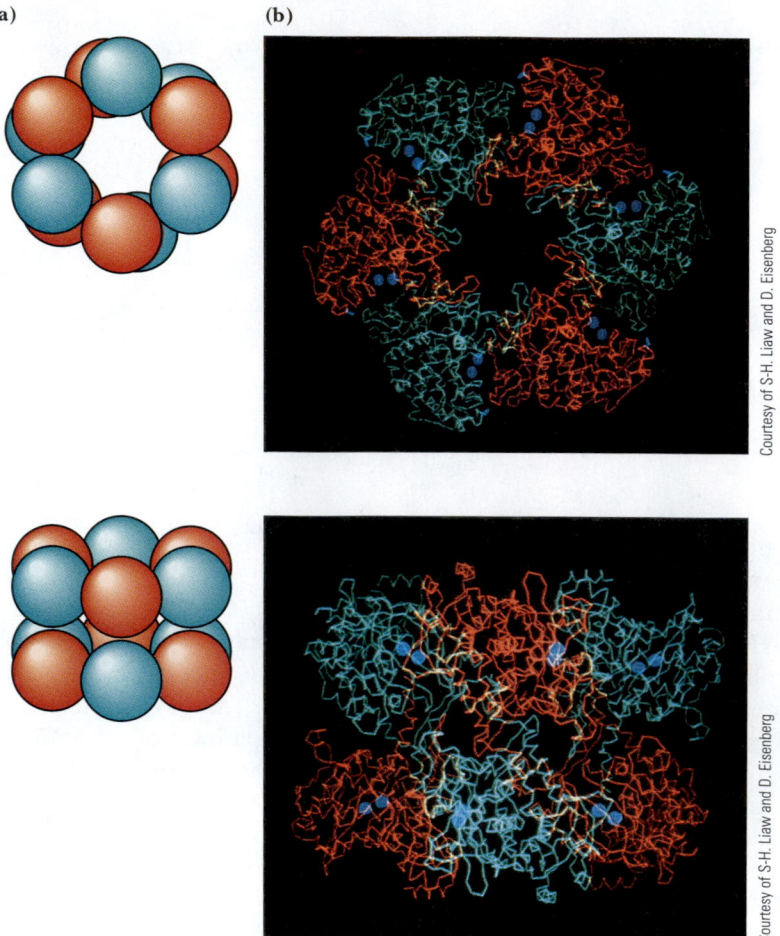

(a)   (b)

Courtesy of S-H. Liaw and D. Eisenberg

Courtesy of S-H. Liaw and D. Eisenberg

FIGURE 25.14   The subunit organization of bacterial glutamine synthetase. **(a)** Diagram showing its dodecameric structure as a stack of two hexagons. **(b)** Molecular structure of glutamine synthetase from *Salmonella typhimurium* (a close relative of *E. coli*), as revealed by X-ray crystallographic analysis. *(From Almassy, R. J., Janson, C. A., Hamlin, R., Xuong, N-H., and Eisenberg, D., 1986. Novel subunit-subunit interactions in the structure of glutamine synthetase.* Nature *323:304.)*

These monomers are arranged as a stack of two hexagons (Figure 25.14). The active sites are located at subunit interfaces within the hexagons; these active sites are recognizable in the X-ray crystallographic structure by the pair of divalent cations that occupy them. Adjacent subunits contribute to each active site, thus accounting for the fact that GS monomers are catalytically inactive.

Biochemistry⊜Now™   Go to BiochemistryNow and click BiochemistryInteractive to explore the structure of dodecameric glutamine synthetase and discover why it is inactive as a monomer.

Glutamine Synthetase Is Allosterically Regulated

Nine distinct feedback inhibitors (Gly, Ala, Ser, His, Trp, CTP, AMP, carbamoyl-P, and glucosamine-6-P) act on GS. Gly, Ala, and Ser are key indicators of amino acid metabolism in the cell; each of the other six compounds represents an end product of a biosynthetic pathway dependent on Gln (Figure 25.15). AMP competes with ATP for binding at the ATP substrate site. Gly, Ala, and Ser compete with Glu for binding at the active site. Carbamoyl-P binds at a site that overlaps both the Glu site and the site occupied by the γ-PO$_4$ of ATP.

Glutamine Synthetase Is Regulated by Covalent Modification

Each GS subunit can be adenylylated at a specific tyrosine residue (Tyr[397]) in an ATP-dependent reaction (Figure 25.16). Adenylylation inactivates GS. If we define n as the average number of adenylyl groups per GS molecule, GS activity is inversely proportional to n. The number n varies from 0 (no adenylyl groups) to 12 (every subunit in each GS molecule is adenylylated). Adenylylation of GS is catalyzed by the *converter enzyme* **ATP:GS:adenylyl transferase,** or simply *adenylyl transferase* **(AT).** However, whether or not this covalent modification occurs is determined by a highly regulated cycle

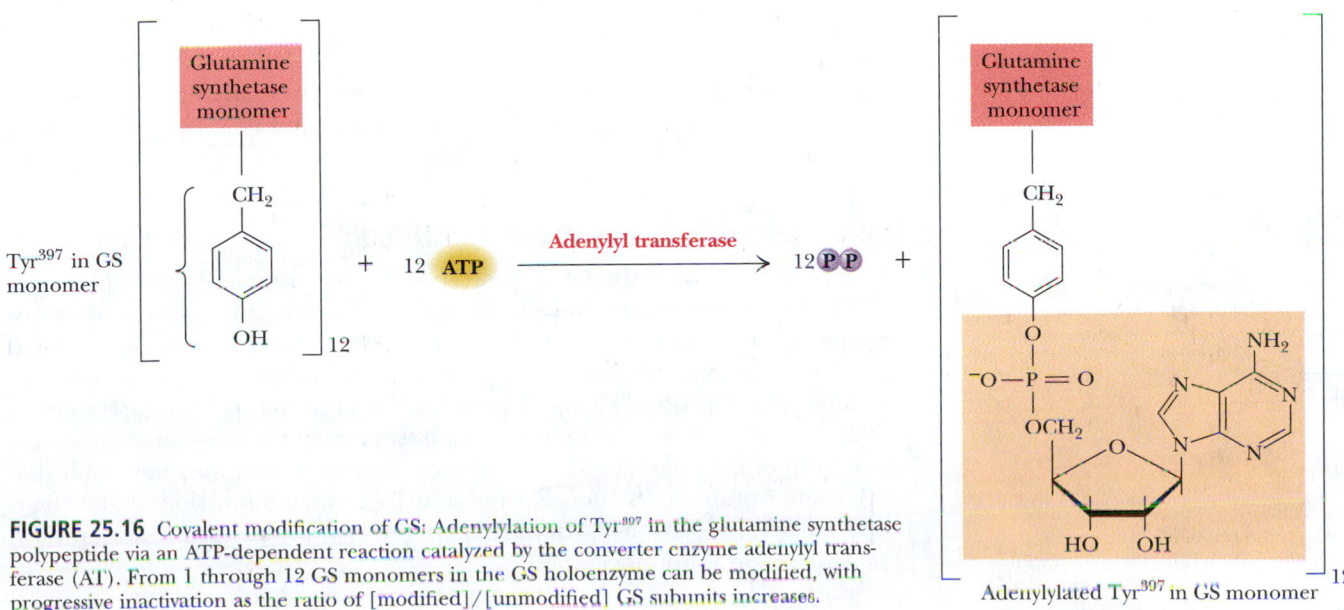

Glutamate

NH_4^+ + **ATP** — ⊘ Glycine $^+H_3N - \overset{H}{\underset{H}{C}} - C \overset{O}{\underset{O^-}{}}$

Glutamine synthetase Mg^{2+}

ADP + Ⓟ — ⊘ Alanine $^+H_3N - \overset{H}{\underset{CH_3}{C}} - C \overset{O}{\underset{O^-}{}}$

Serine

$^+H_3N - \overset{OH}{\underset{H}{\underset{|}{C}}} \overset{CH_2}{\underset{}{|}} - C \overset{O}{\underset{O^-}{}}$

Serine

Histidine

$HN \diagdown N$ $CH_2 - CH - C \overset{O}{\underset{O^-}{}}$ $\underset{^+NH_3}{|}$

Histidine

Tryptophan

$CH_2 - \overset{^+NH_3}{\underset{H}{\underset{|}{C}}} - C \overset{O}{\underset{O^-}{}}$

Tryptophan

Glutamine

Carbamoyl-P $^+H_3N - C \overset{O}{} - O - \overset{O}{\underset{O^-}{P}} - O^-$

Glucosamine-6-P Ⓟ $- CH_2$ ... NH_2

Glucosamine-6-P

AMP ... NH_2

CTP ⓅⓅⓅ

BiochemistryⒺNow™ ACTIVE FIGURE 25.15 The allosteric regulation of glutamine synthetase activity by feedback inhibition. **Test yourself on the concepts in this figure at http://chemistry.brookscole.com/ggb3**

FIGURE 25.16 Covalent modification of GS: Adenylylation of Tyr[397] in the glutamine synthetase polypeptide via an ATP-dependent reaction catalyzed by the converter enzyme adenylyl transferase (AT). From 1 through 12 GS monomers in the GS holoenzyme can be modified, with progressive inactivation as the ratio of [modified]/[unmodified] GS subunits increases.

Glutamine synthetase monomer

Tyr^{397} in GS monomer CH_2 ... OH + 12 **ATP** — **Adenylyl transferase** → 12 ⓅⓅ +

Adenylylated Tyr^{397} in GS monomer

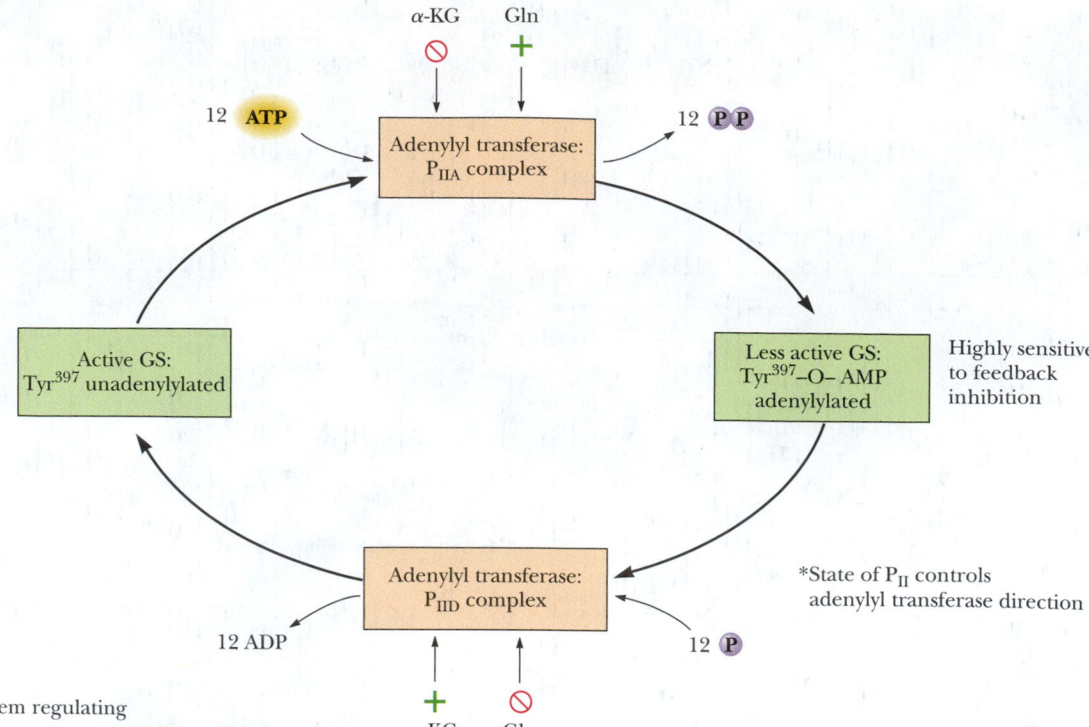

FIGURE 25.17 The cyclic cascade system regulating the covalent modification of GS.

(Figure 26.17). AT not only catalyzes adenylylation of GS, it also catalyzes **deadenylylation**—the phosphorolytic removal of the Tyr-linked adenylyl groups as ADP. The direction in which AT operates depends on the nature of a regulatory protein P_{II} associated with it. **P_{II}** is a 44-kD protein (tetramer of 11-kD subunits): The state of P_{II} controls the direction in which AT acts. If P_{II} is in its so-called P_{IIA} form, the AT:P_{IIA} complex acts to adenylylate GS. When P_{II} is in its so-called P_{IID} form, the AT:P_{IID} complex catalyzes the de-adenylylation of GS. The active sites of AT:P_{IIA} and AT:P_{IID} are different, consistent with the difference in their catalytic roles. In addition, the AT:P_{IIA} and AT:P_{IID} complexes are allosterically regulated in a reciprocal fashion by the effectors α-KG and Gln. Gln activates AT:P_{IIA} activity and inhibits AT:P_{IID} activity; the effect of α-KG on the activities of these two complexes is diametrically opposite (Figure 25.17).

Clearly, the determining factor regarding the degree of adenylylation, n, and hence the relative activity of GS, is the [Gln]/[α-KG] ratio. A high [Gln] level signals cellular nitrogen sufficiency, and GS becomes adenylylated and inactivated. In contrast, a high [α-KG] level is an indication of nitrogen limitation and a need for ammonium fixation by GS.

Glutamine Synthetase Is Regulated Through Gene Expression

The gene that encodes the GS subunit in *E. coli* is designated *GlnA*. The *GlnA* gene is actively transcribed to yield GS mRNA for translation and synthesis of GS protein only if a *specific transcriptional enhancer*, **NR_I**, is in its phosphorylated form, NR_I-P. In turn, NR_I is phosphorylated in an ATP-dependent reaction catalyzed by **NR_{II}**, a protein kinase (Figure 25.18). However, if NR_{II} is complexed with P_{IIA}, it acts not as a kinase but as a phosphatase, and the transcriptionally active form of NR_I, namely NR_I-P, is converted back to NR_I with the result that *GlnA* transcription halts. Recall from the foregoing discussion that a high [Gln]/[α-KG] ratio favors P_{IIA} at the expense of P_{IID}. Under such conditions, GS gene expression is not necessary.

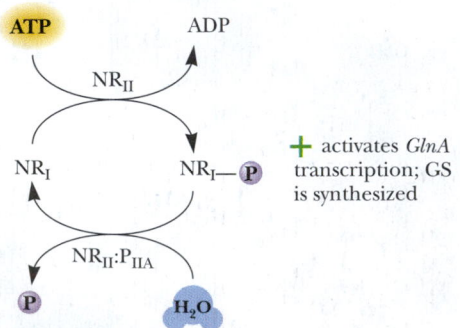

FIGURE 25.18 Transcriptional regulation of *GlnA* expression through the reversible phosphorylation of NR_I, as controlled by NR_{II} and its association with P_{IIA}.

How Do Organisms Synthesize Amino Acids?

Organisms show substantial differences in their capacity to synthesize the 20 amino acids common to proteins. Typically, plants and microorganisms can form all of their nitrogenous metabolites, including all of the amino acids, from inorganic forms of N such as NH_4^+ and NO_3^-. In these organisms, the α-amino group for all amino acids is derived from glutamate, usually via transamination of the corresponding α-keto acid analog of the amino acid (Figure 25.19). In many cases, amino acid biosynthesis is thus a matter of synthesizing the appropriate α-keto acid carbon skeleton, followed by transamination with Glu. The amino acids can be classified according to the source of intermediates for the α-keto acid biosynthesis (Table 25.1). For example, the amino acids Glu, Gln, Pro, and Arg (and, in some instances, Lys) are all members of the α-ketoglutarate family because they are all derived from the citric acid cycle intermediate α-ketoglutarate. We return to this classification scheme later when we discuss the individual biosynthetic pathways.

Amino Acids Are Formed from α-Keto Acids by Transamination

Transamination involves transfer of an α-amino group from an amino acid to the α-keto position of an α-keto acid (Figure 25.19). In the process, the amino donor becomes an α-keto acid while the α-keto acid acceptor becomes an α-amino acid:

$$\text{Amino acid}_1 + \alpha\text{-keto acid}_2 \longrightarrow \alpha\text{-keto acid}_1 + \text{amino acid}_2$$

The predominant amino acid/α-keto acid pair in these reactions is glutamate/α-ketoglutarate, with the net effect that glutamate is the primary amino donor for the synthesis of amino acids. Transamination reactions are catalyzed by **aminotransferases** (the preferred name for enzymes formerly termed *transaminases*). Aminotransferases are named according to their amino acid substrates, as in **glutamate–aspartate aminotransferase.** Aminotransferases are prime examples of enzymes that catalyze double displacement (ping-pong)–type bisubstrate reactions (see Figure 13.23).

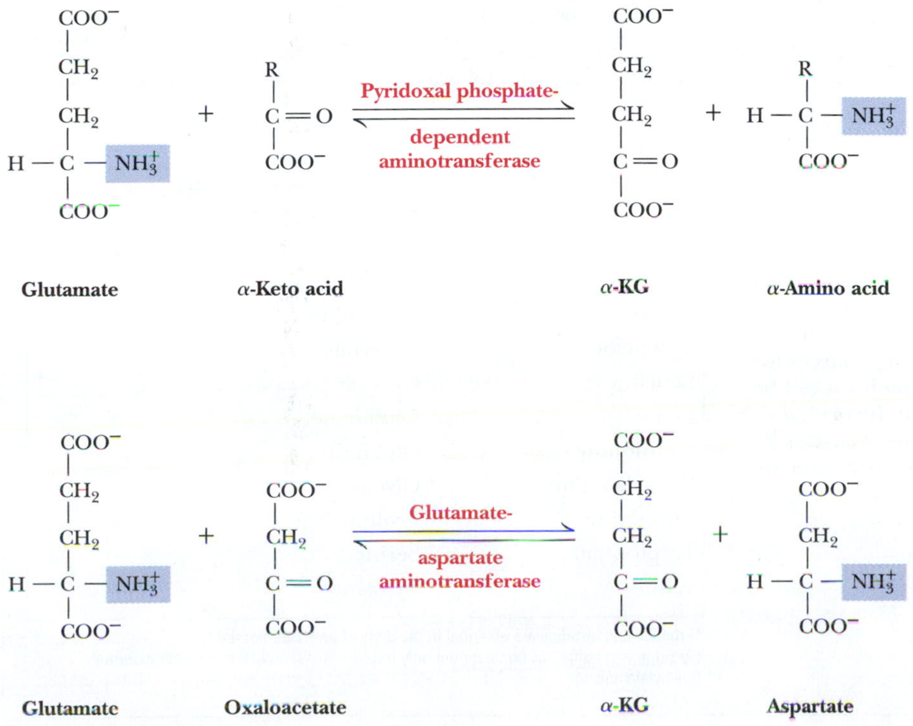

Biochemistry ⊗Now™ ACTIVE FIGURE 25.19

Glutamate-dependent transamination of α-keto acid carbon skeletons is a primary mechanism for amino acid synthesis. The generic transamination–aminotransferase reaction involves the transfer of the α-amino group of glutamate to an α-keto acid acceptor (see Figure 13.23). The transamination of oxaloacetate by glutamate to yield aspartate and α-ketoglutarate is a prime example. **Test yourself on the concepts in this figure at http://chemistry. brookscole.com/ggb3**

Table 25.1

The Grouping of Amino Acids into Families According to the Metabolic Intermediates That Serve as Their Progenitors

α-Ketoglutarate Family	Aspartate Family
Glutamate	Aspartate
Glutamine	Asparagine
Proline	Methionine
Arginine	Threonine
Lysine*	Isoleucine
	Lysine*

Pyruvate Family	3-Phosphoglycerate Family
Alanine	Serine
Valine	Glycine
Leucine	Cysteine

Phosphoenolpyruvate and Erythrose-4-P Family
The aromatic amino acids
Phenylalanine
Tyrosine
Tryptophan
The remaining amino acid, *histidine,* is derived from PRPP (5-phosphoribosyl-1-pyrophosphate) and ATP.

*Different organisms use different precursors to synthesize lysine.

Human Biochemistry

Human Dietary Requirements for Amino Acids

Humans can synthesize only 10 of the 20 common amino acids (see table); the others must be obtained in the diet. Those that can be synthesized are classified as **nonessential,** meaning it is not essential that these amino acids be part of the diet. In effect, humans can synthesize the α-keto acid analogs of nonessential amino acids and form the amino acids by transamination. In contrast, humans are incapable of constructing the carbon skeletons of **essential** amino acids, so they must rely on dietary sources for these essential metabolites. Excess dietary amino acids cannot be stored for future use, nor are they excreted unused. Instead, they are converted to common metabolic intermediates that can be either oxidized by the citric acid cycle or used to form glucose (see Section 25.5).

Essential and Nonessential Amino Acids in Humans

Essential	Nonessential
Arginine*	Alanine
Histidine*	Asparagine
Isoleucine	Aspartate
Leucine	Cysteine
Lysine	Glutamate
Methionine	Glutamine
Phenylalanine	Glycine
Threonine	Proline
Tryptophan	Serine
Valine	Tyrosine†

*Arginine and histidine are essential in the diets of juveniles, not adults.
†Tyrosine is classified as nonessential only because it is readily formed from essential phenylalanine.

A Deeper Look

The Mechanism of the Aminotransferase (Transamination) Reaction

The aminotransferase (transamination) reaction is a workhorse in biological systems. It provides a general means for exchange of nitrogen between amino acids and α-keto acids. This vital reaction is catalyzed by pyridoxal phosphate (see Figures 17.26 and 17.27). The mechanism involves loss of the C_α proton, followed by an aldimine–ketimine tautomerization—literally a "flip-flop" of the Schiff base double bond from the pyridoxal aldehyde carbon to the α-carbon of the amino acid substrate. This is followed by hydrolysis of the ketimine intermediate to yield the product α-keto acid. Left in the active site is a pyridoxamine phosphate intermediate, which combines with another (substrate) α-keto acid to form a second ketimine. Transaldiminization with a lysine at the active site completes the reaction.

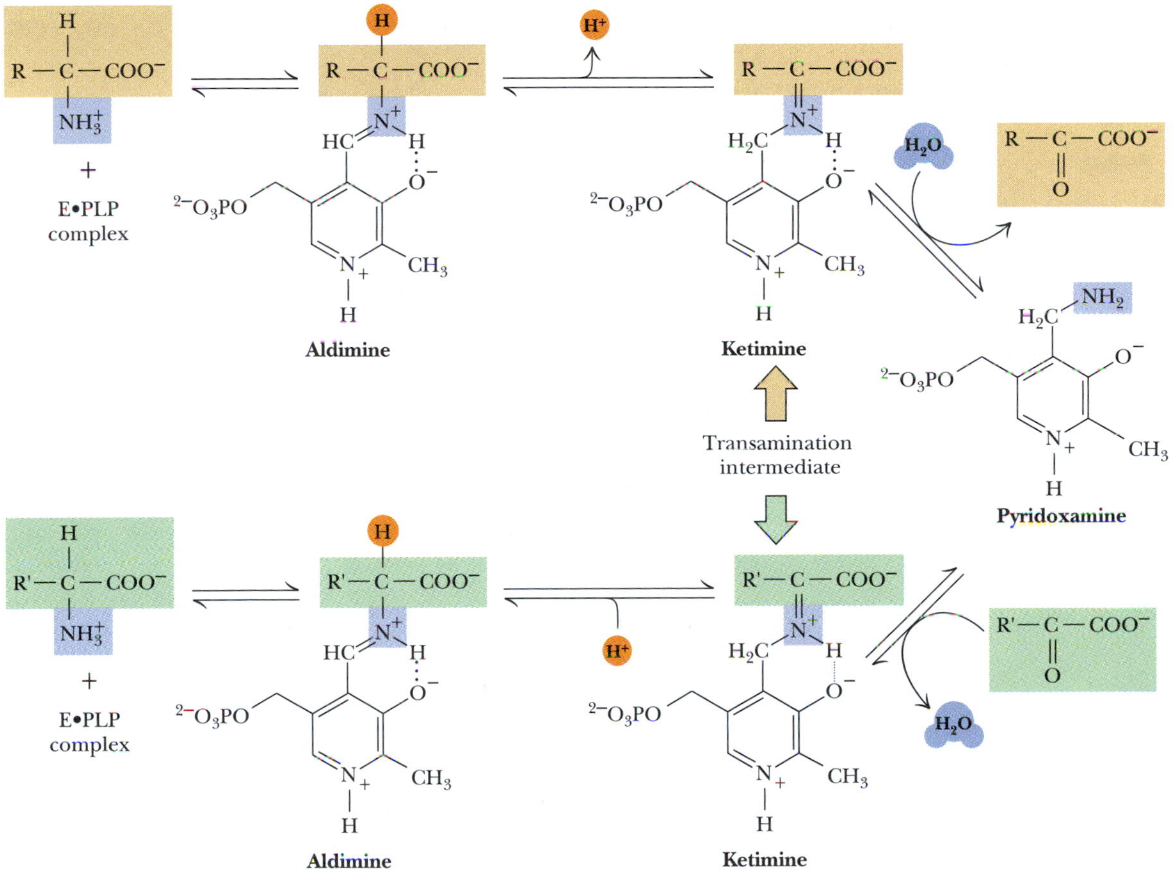

The mechanism of PLP-catalyzed transamination reactions.

The Pathways of Amino Acid Biosynthesis Can Be Organized into Families

As indicated in Table 25.1, the amino acids can be grouped into families on the basis of the metabolic intermediates that serve as their precursors.

The α-Ketoglutarate Family of Amino Acids Includes Glu, Gln, Pro, Arg, and Lys

Amino acids derived from α-ketoglutarate include glutamate (Glu), glutamine (Gln), proline (Pro), arginine (Arg), and in fungi and protoctists such as *Euglena*, lysine (Lys). We discussed the routes for Glu and Gln synthesis when we considered pathways of ammonium assimilation.

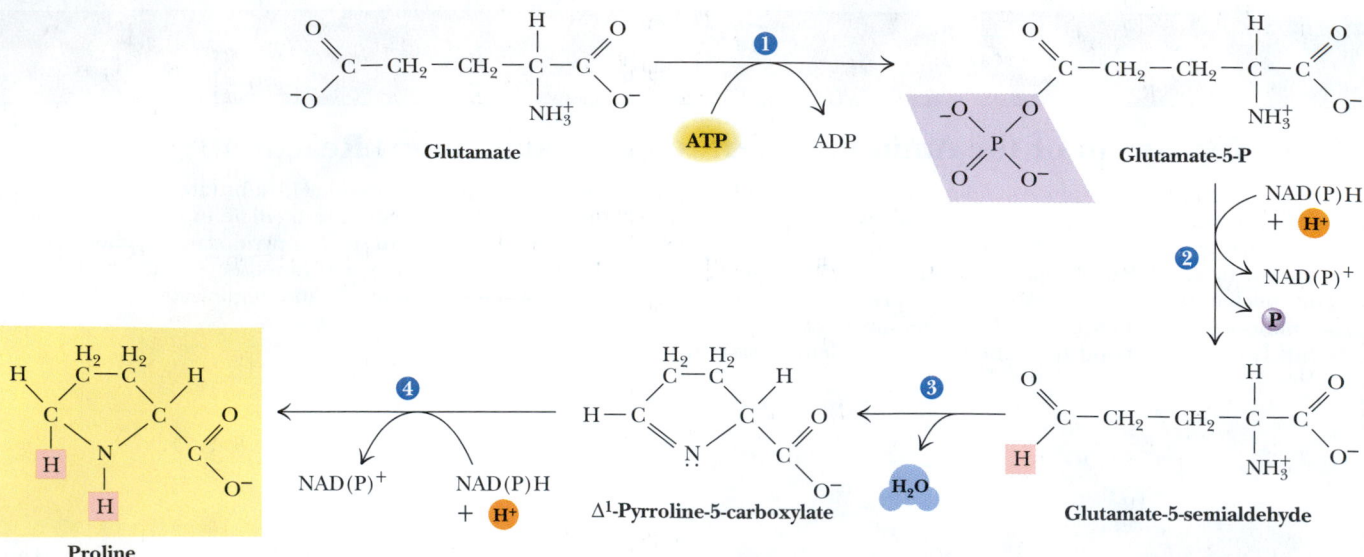

FIGURE 25.20 The pathway of proline biosynthesis from glutamate. The enzymes are (1) γ-glutamyl kinase, (2) glutamate-5-semialdehyde dehydrogenase, and (4) Δ¹-pyrroline-5-carboxylate reductase; reaction (3) occurs nonenzymatically.

Proline is derived from glutamate via a series of four reactions involving activation, then reduction, of the γ-carboxyl group to an aldehyde *(glutamate-5-semialdehyde)*, which spontaneously cyclizes to yield the internal Schiff base, *Δ¹-pyrroline-5-carboxylate* (Figure 25.20). NADPH-dependent reduction of the pyrroline double bond gives proline.

Arginine biosynthesis involves enzymatic steps that are also part of the **urea cycle,** a metabolic pathway that functions in N excretion in certain animals. Net synthesis of arginine depends on the formation of **ornithine.** Interestingly, ornithine is derived from glutamate via a reaction pathway reminiscent of the proline biosynthetic pathway (Figure 25.21). Glutamate is first *N*-acetylated in an acetyl-CoA–dependent reaction to yield *N-acetylglutamate* (Figure 25.21). An ATP-dependent phosphorylation of *N*-acetylglutamate to give *N-acetylglutamate-5-phosphate* primes this substrate for a reduced pyridine nucleotide-dependent reduction to the semialdehyde. *N*-acetylglutamate-5-semialdehyde then is aminated by a glutamate-dependent aminotransferase, giving *N-acetylornithine,* which is deacylated to ornithine.

Ornithine has three metabolic roles: (1) to serve as a precursor to arginine, (2) to function as an intermediate in the urea cycle, and (3) to act as an intermediate in Arg degradation. In any case, the δ-NH₃⁺ of ornithine is carbamoylated in a reaction catalyzed by **ornithine transcarbamoylase.** The carbamoyl group is derived from carbamoyl-P synthesized by **carbamoyl-phosphate synthetase I** *(CPS-I).* CPS-I is the mitochondrial CPS isozyme; it uses two ATPs in catalyzing the formation of carbamoyl-P from NH₃ and HCO₃⁻ (Figure 25.22). CPS-I represents the committed step in the urea cycle, and CPS-I is allosterically activated by *N*-acetylglutamate. Because *N*-acetylglutamate is both a precursor to ornithine synthesis and essential to the operation of the urea cycle, it serves to coordinate these related pathways.

The product of the ornithine transcarbamoylase reaction is *citrulline* (Figure 25.23). Ornithine and citrulline are two α-amino acids of metabolic importance that nevertheless are *not* among the 20 α-amino acids commonly found in proteins. Like CPS-I, ornithine transcarbamoylase is a mitochondrial enzyme. The reactions of ornithine synthesis and the rest of the urea cycle enzymes occur in the cytosol.

The pertinent feature of the citrulline side chain is the **ureido group.** In a complex reaction catalyzed by **argininosuccinate synthetase,** this ureido group is first activated by ATP to yield a citrullyl-AMP derivative, followed by displacement of AMP by aspartate to give *argininosuccinate* (Figure 25.23). The formation of arginine is then accomplished by **argininosuccinase,** which catalyzes the

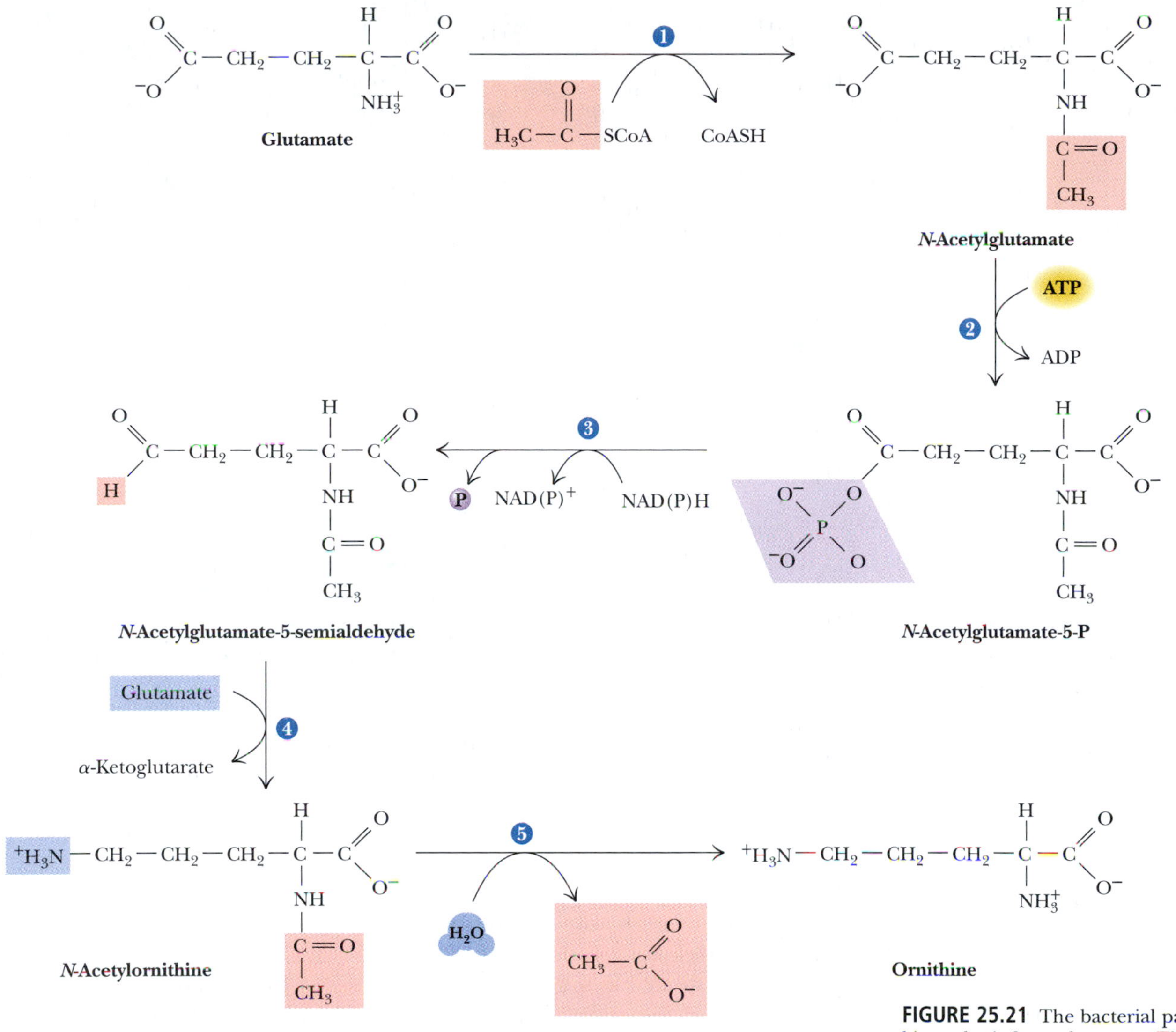

FIGURE 25.21 The bacterial pathway of ornithine biosynthesis from glutamate. The enzymes are (1) *N*-acetylglutamate synthase, (2) *N*-acetylglutamate kinase, (3) *N*-acetylglutamate-5-semialdehyde dehydrogenase, (4) *N*-acetylornithine δ-aminotransferase, and (5) *N*-acetylornithine deacetylase. In mammals, ornithine is synthesized directly from glutamate-5-semialdehyde by a pathway that does not involve an *N*-acetyl block.

nonhydrolytic elimination of fumarate from argininosuccinate. This reaction completes the biosynthesis of Arg.

The Urea Cycle Acts to Excrete Excess N Through Arg Breakdown

The carbon skeleton of arginine is derived principally from α-ketoglutarate, but the N and C atoms composing the **guanidino group** (Figure 26.23) of the Arg side chain come from NH_4^+, HCO_3^- (as carbamoyl-P), and the α-NH_2 groups of glutamate and aspartate. The circle of the urea cycle is closed when ornithine is regenerated from Arg by the arginase-catalyzed hydrolysis of arginine. Urea is the other product of this reaction and lends its name to the cycle. In terrestrial vertebrates, urea synthesis is required to excrete excess nitrogen generated by increased amino acid catabolism—for example, following dietary consumption of more than adequate amounts of protein. Urea formation is basically confined to the liver.

A healthy adult human male eating a typical American diet will consume about 100 g of protein per day. Because such an individual will be in *nitrogen balance* (neither increasing or decreasing his net protein levels), his body must dispose of about 1 mole of excess N derived from the amino acids in this dietary protein. Glutamate is key to this process because of the position of glutamate dehydrogenase at the interface of amino acid and carbohydrate metabolism and the importance of glutamate to the urea cycle.

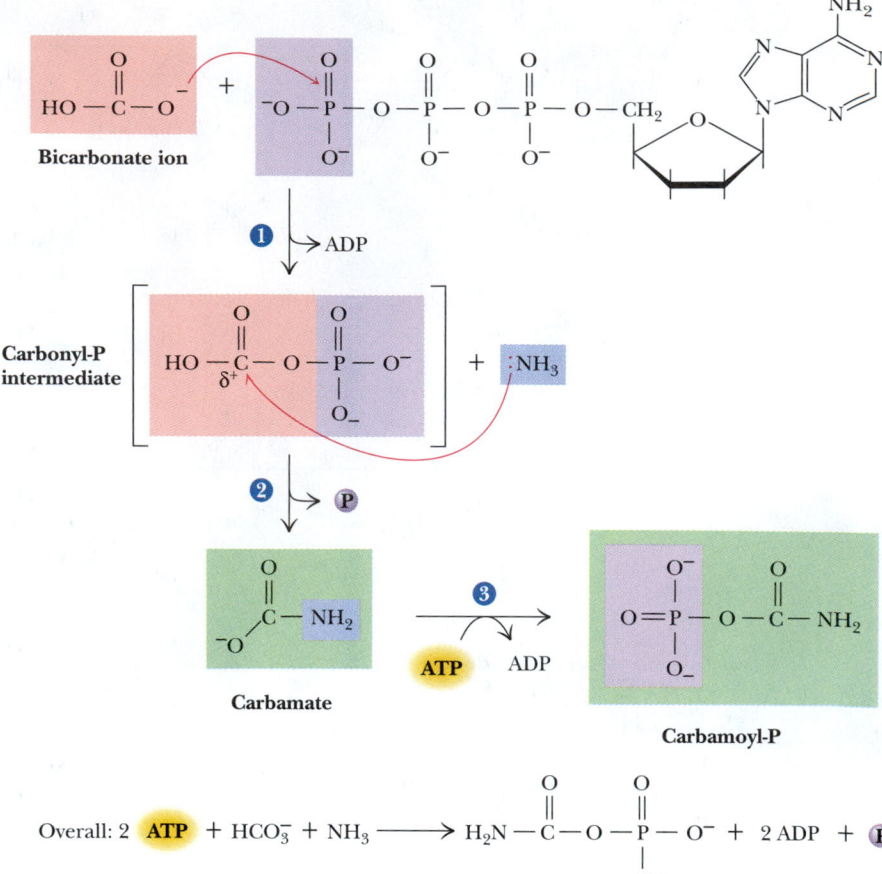

ANIMATED FIGURE 25.22
The mechanism of action of CPS-I, the NH_3-dependent mitochondrial CPS isozyme. (1) HCO_3^- is activated via an ATP-dependent phosphorylation. (2) Ammonia attacks the carbonyl carbon of carbonyl-P, displacing P_i to form carbamate. (3) Carbamate is phosphorylated via a second ATP to give carbamoyl-P. **See this figure animated at http://chemistry.brookscole. com/ggb3**

Increases in amino acid catabolism lead to elevated glutamate levels and a rise in *N*-acetylglutamate, the allosteric activator of CPS-I. Stimulation of CPS-I raises overall urea cycle activity because activities of the remaining enzymes of the cycle simply respond to increased substrate availability. Removal of potentially toxic NH_4^+ by CPS-I is another important aspect of this regulation. The urea cycle is linked to the citric acid cycle through *fumarate,* a by-product of the action of *argininosuccinase* (Figure 25.23, reaction 3).

Lysine biosynthesis in some fungi and in the protoctist *Euglena* also stems from α-ketoglutarate, making lysine a member of the α-ketoglutarate family of amino acids in these organisms. (As we shall see, the other organisms capable of lysine synthesis—namely, bacteria, other fungi, algae, and green plants— use aspartate as a precursor.) To make lysine from α-ketoglutarate requires a lengthening of the carbon skeleton by one CH_2 unit to yield *α-ketoadipate* (Figure 25.24). This addition is accomplished by a series of reactions reminiscent of the initial stages of the citric acid cycle. First, a two-carbon acetyl-CoA unit is added to the α-carbon of α-ketoglutarate to form *homocitrate.* Then, in a reaction

▶ **Biochemistry ⊜ Now™ ACTIVE FIGURE 25.23** The urea cycle series of reactions: Transfer of the carbamoyl group of carbamoyl-P to ornithine by *ornithine transcarbamoylase* (OTCase, **reaction 1**) yields citrulline. The citrulline ureido group is then activated by reaction with ATP to give a citrullyl–AMP intermediate **(reaction 2a)**; AMP is then displaced by aspartate, which is linked to the carbon framework of citrulline via its α-amino group **(reaction 2b).** The course of reaction 2 was verified using ^{18}O-labeled citrulline. The [18]O label (indicated by the asterisk, *) was recovered in AMP. Citrulline and AMP are joined via the ureido *O atom. The product of this reaction is argininosuccinate; the enzyme catalyzing the two steps of reaction 2 is *argininosuccinate synthetase.* The next step **(reaction 3)** is carried out by *argininosuccinase,* which catalyzes the nonhydrolytic removal of fumarate from argininosuccinate to give arginine. Hydrolysis of Arg by *arginase* **(reaction 4)** yields urea and ornithine, completing the urea cycle. **Test yourself on the concepts in this figure at http://chemistry.brookscole.com/ggb3**

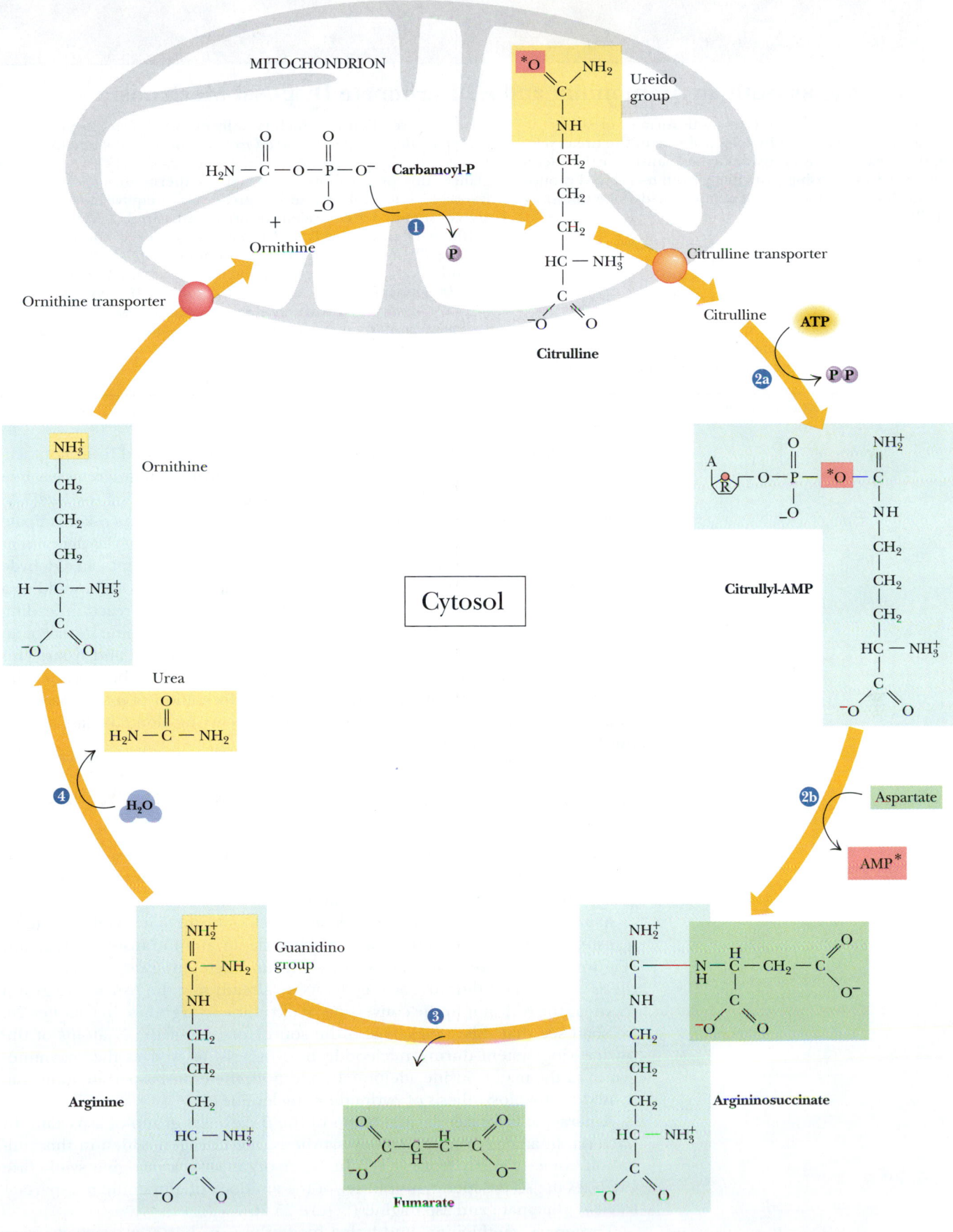

A Deeper Look

The Urea Cycle as Both an Ammonium and a Bicarbonate Disposal Mechanism

Excretion of excess NH_4^+ in the innocuous form of urea has traditionally been viewed as the physiological role of the urea cycle. However, the urea cycle also provides a mechanism for the excretion of excess HCO_3^- arising principally from α-carboxyl groups generated during the catabolism of α-amino acids. The following equations illustrate this property:

(1) $HCO_3^- + 2\, NH_4^+ \longrightarrow H_2NCONH_2 + 2\, H_2O + H^+$
(2) $HCO_3^- + H^+ \longrightarrow H_2O + CO_2$
Sum:
$2\, HCO_3^- + 2\, NH_4^+ \longrightarrow H_2NCONH_2 + CO_2 + 3\, H_2O$

That is, *2* moles of HCO_3^- are eliminated in the synthesis of each mole of urea: One is incorporated into the product, urea (reaction 1), and the second is simply protonated and dehydrated to form CO_2 (reaction 2), which is easily excreted. *One interpretation of the preceding is that these coupled reactions allow a weak acid (NH_4^+) to protonate the conjugate base of a stronger acid (HCO_3^-). At first glance, this protonation would appear thermodynamically un-* favorable, but recall that in the urea cycle, 4 equivalents of ATP are consumed per equivalent of urea synthesized: 2 ATPs in the synthesis of carbamoyl-P, and 2 more as 1 ATP is converted to AMP + PP$_i$ in the synthesis of argininosuccinate from citrulline (Figure 25.23). If this interpretation is correct, *the urea cycle may be considered an ATP-driven proton pump that transfers H^+ ions from NH_4^+ to HCO_3^- against a thermodynamic barrier. In the process, the potentially toxic waste products, ammonium and bicarbonate, are rendered innocuous and excreted.*

---

sequence like that catalyzed by aconitase, *homoisocitrate* is formed from homocitrate. Oxidative decarboxylation (as in isocitrate dehydrogenase) removes one carbon (the original α-carboxyl group of α-ketoglutarate), leaving *α-ketoadipate*. A glutamate-dependent aminotransferase enzyme then aminates α-ketoadipate to give *α-aminoadipate*. Next, the δ-COO^- group is activated in an ATP-dependent adenylylation reaction, priming this δ-COO^- group for reduction to an aldehyde by NADPH. *α-Aminoadipic-6-semialdehyde* is then reductively aminated by addition of glutamate to its aldehydic carbon in an NADPH-dependent reaction leading to the formation of *saccharopine*. Oxidative cleavage of saccharopine by way of an NAD^+-dependent dehydrogenase activity yields α-ketoglutarate and *lysine*. This pathway is known as the α-aminoadipic acid pathway of lysine biosynthesis. Interestingly, lysine degradation in animals leads to formation of α-aminoadipate by a reverse series of reactions identical to those occurring along the last steps of this biosynthetic pathway.

The Aspartate Family of Amino Acids Includes Asp, Asn, Lys, Met, Thr, and Ile

The members of the aspartate family of amino acids include aspartate (Asp), asparagine (Asn), lysine (via the diaminopimelic acid pathway), methionine (Met), threonine (Thr), and isoleucine (Ile).

Aspartate is formed from the citric acid cycle intermediate oxaloacetate by transfer of an amino group from glutamate via an aminotransferase reaction (Figure 25.25). Like glutamate synthesis from α-ketoglutarate, aspartate synthesis is a drain on the citric acid cycle. As we already saw, the Asp amino group serves as the N donor in the conversion of citrulline to arginine. In Chapter 26, we shall see that this —NH_2 is also the source of one of the N atoms of the purine ring system during nucleotide biosynthesis, as well as the six-amino-group of the major purine adenine. In addition, the entire aspartate molecule is used in the biosynthesis of pyrimidine nucleotides.

Asparagine is formed by amidation of the β-carboxyl group of aspartate. In bacteria, in analogy with glutamine synthesis, the nitrogen added in this amidation comes directly from NH_4^+. In other organisms, **asparagine synthetase** catalyzes the ATP-dependent transfer of the amido-N of glutamine to aspartate to yield glutamate and asparagine (Figure 25.26).

Threonine, methionine, and **lysine** biosynthesis in bacteria proceeds from the common precursor aspartate, which is converted first to *aspartyl-β-phosphate* and then to *β-aspartyl-semialdehyde*. The first reaction is an ATP-dependent

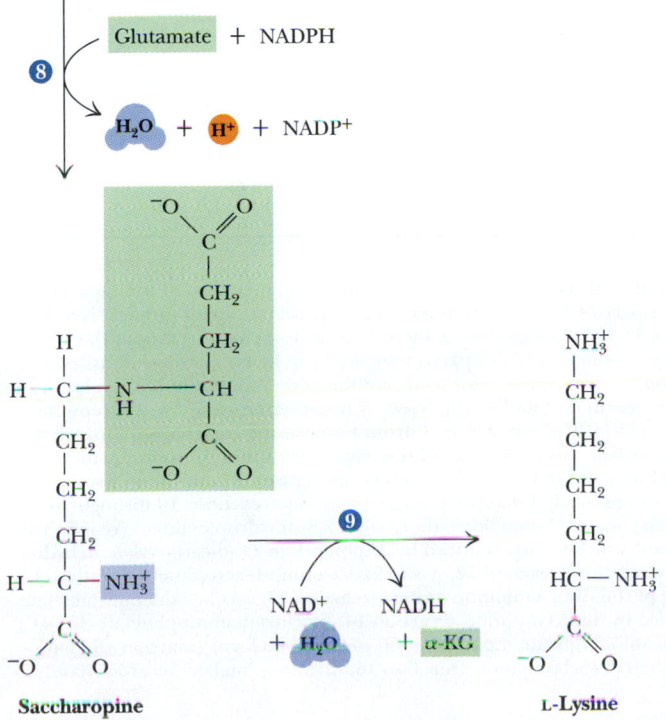

FIGURE 25.24 Lysine biosynthesis in certain fungi and *Euglena:* the α-aminoadipic acid pathway. **Reactions 1 through 4** are reminiscent of the first four reactions in the citric acid cycle, except that the product α-ketoadipate has an additional —CH₂— unit. **Reaction 5** is catalyzed by a glutamate-dependent aminotransferase; **reaction 6** is the adenylylation of the δ-carboxyl of α-aminoadipate to give the 6-adenylyl derivative. Reductive deadenylylation by an NADPH-dependent dehydrogenase in **reaction 7** gives α-aminoadipic-6-semialdehyde, which in **reaction 8** is coupled with glutamate via its amino group by a second NADPH-dependent dehydrogenase. Oxidative removal of the α-ketoglutarate moiety by NAD⁺-dependent saccharopine dehydrogenase in **reaction 9** leaves this amino group as the ε-NH₃⁺ of lysine.

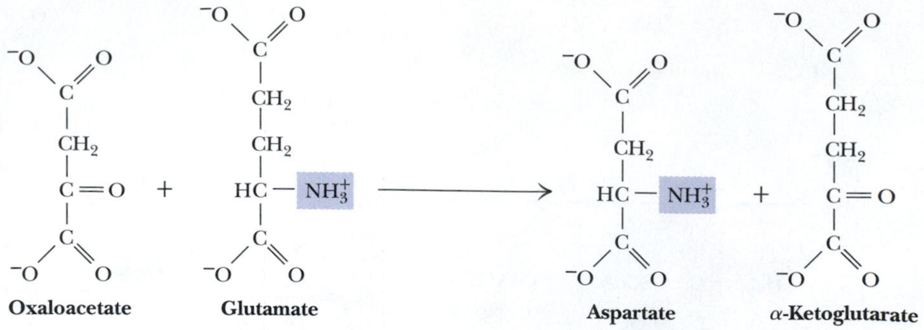

FIGURE 25.25 Aspartate biosynthesis via transamination of oxaloacetate by glutamate. The enzyme responsible is PLP-dependent glutamate:aspartate aminotransferase.

phosphorylation catalyzed by **aspartokinase** (Figure 25.27). In *E. coli*, there are three isozymes of aspartokinase, designated **aspartokinases I, II,** and **III.** Each of these isozymes is uniquely controlled by one of the three end-product amino acids (Table 25.2). Thus, the biosynthesis of each of the three amino acids may be independently regulated through controls exerted on the formation or activity of a particular aspartokinase isozyme.

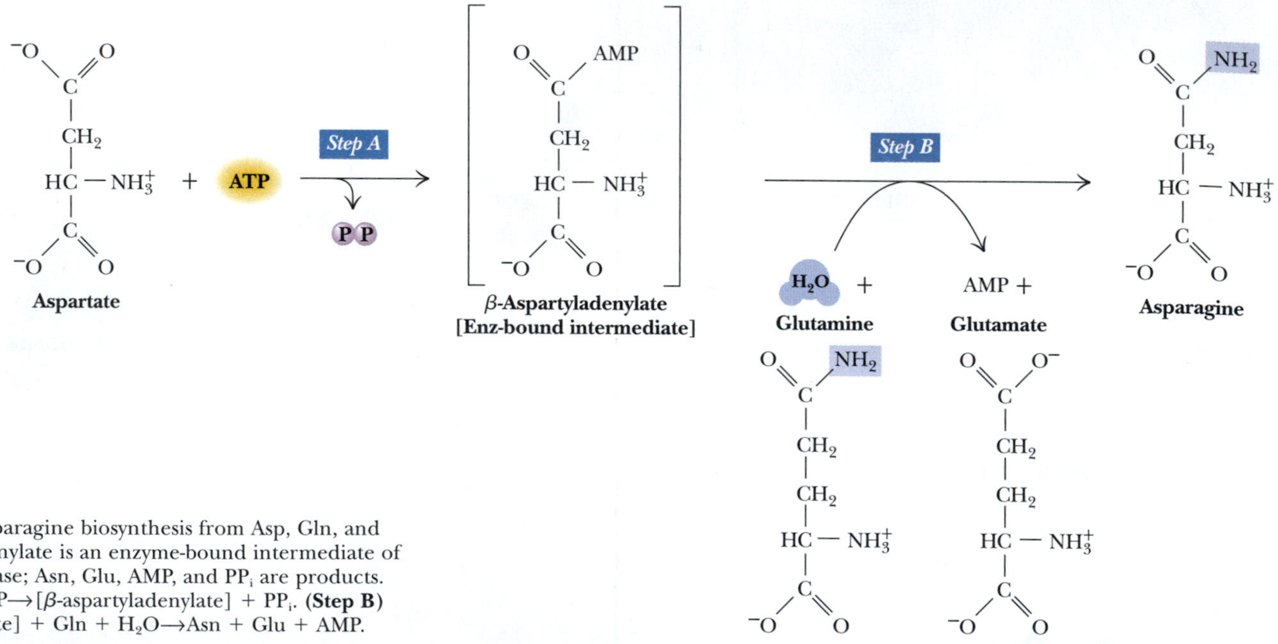

FIGURE 25.26 Asparagine biosynthesis from Asp, Gln, and ATP. β-Aspartyladenylate is an enzyme-bound intermediate of asparagine synthetase; Asn, Glu, AMP, and PPᵢ are products. **(Step A)** Asp + ATP→[β-aspartyladenylate] + PPᵢ. **(Step B)** [β-Aspartyladenylate] + Gln + H₂O→Asn + Glu + AMP.

▶ **FIGURE 25.27** Biosynthesis of threonine, methionine, and lysine, members of the aspartate family of amino acids. β-Aspartyl-semialdehyde is a common precursor to all three. It is formed by aspartokinase (**reaction 1**) and β-aspartyl-semialdehyde dehydrogenase (**reaction 2**). From here, the pathways diverge. Reduction of β-aspartyl-semialdehyde by homoserine dehydrogenase (**reaction 3**) gives homoserine, a precursor to threonine and methionine but not lysine. The branch designated by **reactions 4 and 5** (catalyzed by homoserine kinase and threonine synthase) gives rise to threonine. The other branch from homoserine (**reactions 6 through 9**) leads to methionine (the enzymes are, in order, homoserine acyltransferase, cystathionine synthase, cystathionine-β-lyase, and homocysteine methyltransferase). The route to lysine from β-aspartyl-semialdehyde is the so-called diaminopimelate pathway (**reactions 10 through 16**). Pyruvate is condensed with β-aspartyl-semialdehyde to yield 2,3-dihydropicolinate (**reaction 10,** dihydropicolinate synthase), which is then reduced by Δ^1-piperidine-2,6-dicarboxylate dehydrogenase (**reaction 11**). Succinylation (**reaction 12,** N-succinyl-2-amino-6-ketopimelate synthase) is accompanied by opening of the ring; amination ensues (**reaction 13,** succinyl-diaminopimelate aminotransferase), followed by desuccinylation (**reaction 14,** succinyl-diaminopimelate desuccinylase) to give L-L-α,ε-diaminopimelate. Epimerization to the meso form (**reaction 15,** diaminopimelate epimerase), then decarboxylation (**reaction 16,** diaminopimelate decarboxylase), yields lysine.

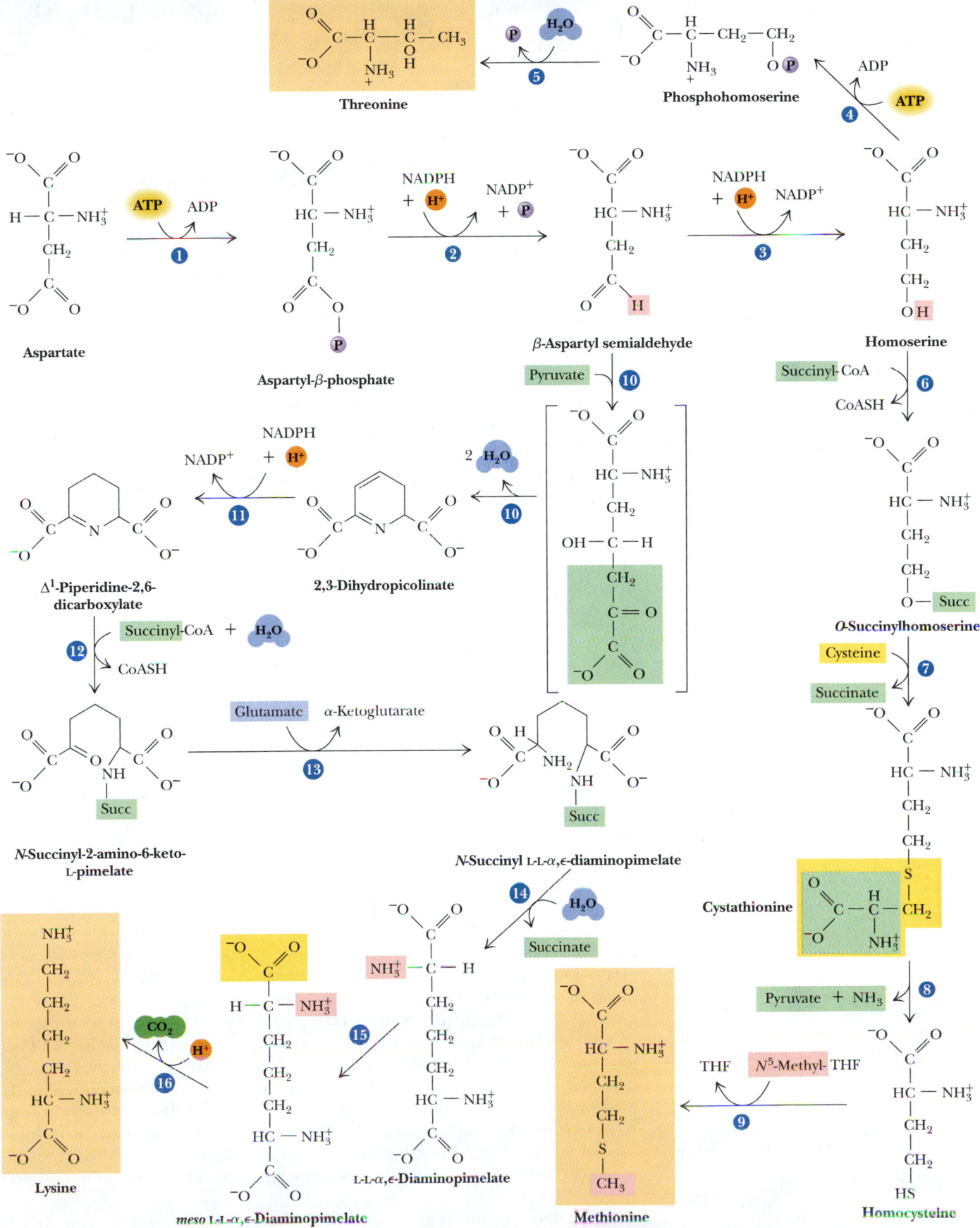

Table 25.2

Regulation of the Three Aspartokinase Isozymes of _E. coli_

Enzyme	Feedback Inhibitor	Co-repressor*
Aspartokinase I	Threonine	Threonine and isoleucine
Aspartokinase II	None	Methionine
Aspartokinase III	Lysine	Lysine

*_Co-repressor_ is the term given to metabolites that can act in repressing expression of specific genes.

β-Aspartyl-semialdehyde is formed via NADPH-dependent reduction of aspartyl-β-phosphate in a reaction catalyzed by **β-aspartyl-semialdehyde dehydrogenase** (Figure 25.27). From here, the pathway of lysine synthesis diverges. The methyl carbon of pyruvate is condensed with β-aspartyl-semialdehyde, and H_2O is eliminated to yield the cyclic compound _2,3-dihydropicolinate_ (Figure 25.27). Thus, lysine synthesized by this pathway must be considered a member of both the aspartate and the pyruvate families of amino acids. Lysine is a feedback inhibitor of this branch-point enzyme. _Dihydropicolinate_ is then reduced in an NADPH-dependent reaction to Δ^1-_piperidine-2,6-dicarboxylate_. A series of reactions, including a hydrolytic opening of the piperidine ring, a succinylation, a glutamate-dependent amination, and the hydrolytic removal of succinate, results in the formation of the symmetric _L,L-α,ϵ-diaminopimelate_. Epimerization of this intermediate to the _meso_ form, followed by decarboxylation, yields the end product _lysine_. Because this pathway proceeds through the symmetric _L,L-α,ϵ-diaminopimelate_, one-half of the CO_2 evolved in the terminal decarboxylase step is derived from the carboxyl group of pyruvate and one-half from the α-carboxyl of Asp.

The other metabolic branch diverging from β-aspartyl-semialdehyde leads to _threonine and methionine via homoserine_, an analog of serine that is formed by the NADPH-dependent reduction of β-aspartyl-semialdehyde (Figure 25.27) catalyzed by **homoserine dehydrogenase.** From homoserine, the biosynthetic pathways leading to methionine and threonine separate. To form **methionine,** the —OH group of homoserine is first succinylated by **homoserine acyltransferase.** Methionine is a feedback inhibitor of this enzyme. The succinyl group of _O-succinylhomoserine_ is then displaced by cysteine to yield _cystathionine_ (Figure 25.27). The sulfur atom in methionine is contributed by a cysteine sulfhydryl. _Cystathionine_ is then split to give pyruvate, NH_4^+, and _homocysteine_, a nonprotein amino acid whose side chain is one —CH_2— group longer than that of Cys. Methylation of the homocysteine —SH via methyl transfer from the methyl donor, N^5-methyl-THF (see Chapters 17 and 26) gives methionine.

In passing, it is important to note the role of methionine itself in methylation reactions. The enzyme **S-adenosylmethionine synthase** catalyzes the reaction of methionine with ATP to form _S-adenosylmethionine_, or SAM (Figure 25.28). SAM serves as a methyl group donor in many methylation reactions, such as the formation of phosphatidylcholine from phosphatidylethanolamine (see Figure 8.6).

The remaining amino acids of the aspartate family are threonine and isoleucine. **Threonine,** like methionine, is synthesized from homoserine. Indeed, homoserine is the primary alcohol analog of the secondary alcohol Thr. To move this —OH from C-4 to C-3 requires activation of the hydroxyl through ATP-dependent phosphorylation by **homoserine kinase.** As the first reaction unique to Thr biosynthesis, homoserine kinase is feedback inhibited by threonine. The last step is catalyzed by **threonine synthase,** a PLP-dependent enzyme (Figure 25.27).

Isoleucine is included in the aspartate family of amino acids because four of its six carbons derive from Asp (via threonine) and only two come from pyruvate. Nevertheless, four of the five enzymes necessary for isoleucine synthesis are

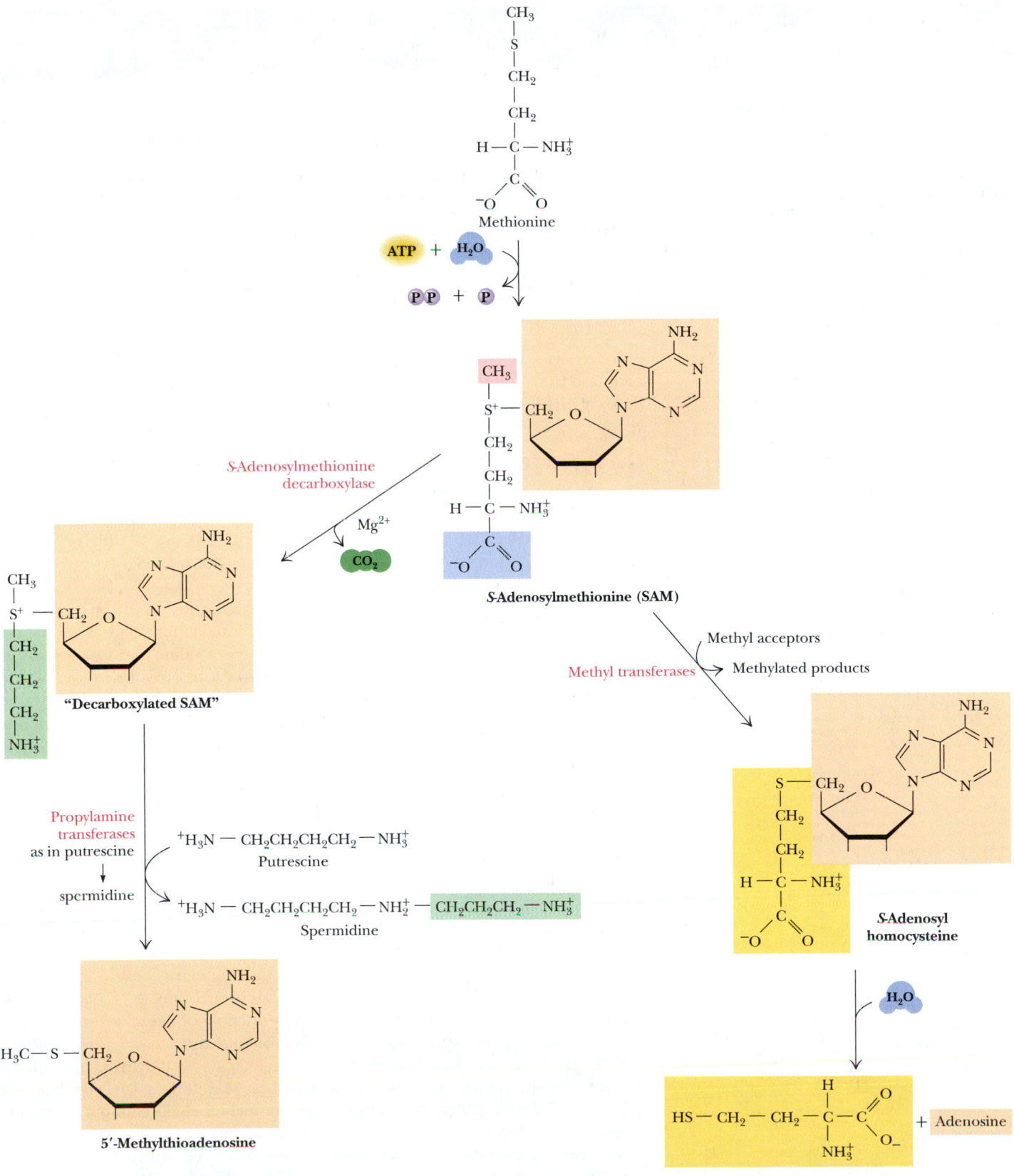

FIGURE 25.28 The synthesis of *S*-adenosylmethionine (SAM) from methionine plus ATP, and the role of SAM as a substrate of methyltransferases in methyl donor reactions and in propylamine transfer reactions, as in the synthesis of polyamines.

Human Biochemistry

Homocysteine and Heart Attacks

A rare inherited disease known as *homocystinuria* results in very high levels of homocysteine in the bloodstream. Children born with this disease seldom survive to be teenagers and die from such cardiovascular problems as stroke and arteriosclerosis (hardening of the arteries), diseases usually associated with old age.

Damage to blood vessels by homocysteine is the basis of the disease. Furthermore, studies indicate that adults with elevated levels of homocysteine (hyperhomocysteinemia) in their blood are at higher risk for heart attack and stroke. As early as 1969, Dr. Kilmer

McCully, a physician interested in homocystinuria, suggested that homocysteine might cause heart disease and that many people may have high plasma levels of homocysteine because their diets are deficient in folic acid, but his work went unheeded for 25 years. Fortunately, supplementing the amounts of folic acid (a B vitamin) in the diet reduces blood concentrations of homocysteine to a safe level, presumably by enhancing the conversion of homocysteine to methionine by the tetrahydrofolate (THF)–dependent enzyme homocysteine methyltransferase (reaction 9 of Figure 25.27).

common to the pathway for biosynthesis of valine, so discussion of isoleucine synthesis is presented under the biosynthesis of the pyruvate family of amino acids.

The Pyruvate Family of Amino Acids Includes Ala, Val, and Leu

The pyruvate family of amino acids includes alanine (Ala), valine (Val), and leucine (Leu). Transamination of pyruvate, with glutamate as amino donor, gives **alanine.** Because these transamination reactions are readily reversible, alanine degradation occurs via the reverse route, with α-ketoglutarate serving as amino acceptor.

Transamination of pyruvate to alanine is a reaction found in virtually all organisms, but valine, leucine, and isoleucine are essential amino acids, and as such, they are not synthesized in mammals. The pathways of **valine** and **isoleucine** synthesis can be considered together because one set of four enzymes is common to the last four steps of both pathways (Figure 25.29). Both pathways begin with an α-keto acid. Isoleucine can be considered a structural analog of valine that has one extra —CH_2— unit, and its α-keto acid precursor, namely, *α-ketobutyrate,* is one carbon longer than the valine precursor, pyruvate. Interestingly, α-ketobutyrate is formed from threonine by the action of **threonine deaminase,** an enzyme that both deaminates and dehydrates Thr. Threonine deaminase, a PLP-dependent enzyme, is feedback sensitive to isoleucine (this enzyme is also known as *threonine dehydratase* and *serine dehydratase*). So, part of the carbon skeleton for Ile comes from Asp by way of Thr. From here on, the Val and Ile pathways employ the same set of enzymes. The first reaction involves the generation of hydroxyethyl-thiamine pyrophosphate from pyruvate in a reaction analogous to those catalyzed by transketolase and the pyruvate dehydrogenase complex. The two-carbon hydroxyethyl group is transferred from TPP to the respective keto acid acceptor by **acetohydroxy acid synthase** (acetolactate synthase) to give *α-acetolactate* or *α-aceto-α-hydroxybutyrate.* NAD(P)H-dependent reduction of these α-keto hydroxy acids yields the dihydroxy acids *α,β-dihydroxy-isovalerate* and *α,β-dihydroxy-β-methylvalerate.* Dehydration of each of these dihydroxy acids by **dihydroxy acid dehydratase** gives the appropriate α-keto acid carbon skeletons *α-ketoisovalerate* and *α-keto-β-methylvalerate.* Transamination by the **branched-chain amino acid aminotransferase** yields Val or Ile, respectively (Figure 25.29).

Leucine synthesis depends on these reactions as well, because α-ketoisovalerate is a precursor common to both Val and Leu (Figure 25.30). Although Val and Leu differ by only a single —CH_2— in their respective side chains, the carboxyl group of α-ketoisovalerate first picks up *two* carbons from acetyl-CoA to give *α-isopropylmalate* in a reaction catalyzed by **isopropylmalate synthase;** the enzyme is sensitive to feedback inhibition by Leu (Figure 25.30). **Isopropylmalate dehydratase** converts the α-isomer to the β-form, which undergoes an NAD⁺-dependent oxidative decarboxylation by **isopropylmalate dehydrogenase,** so the carboxyl group of

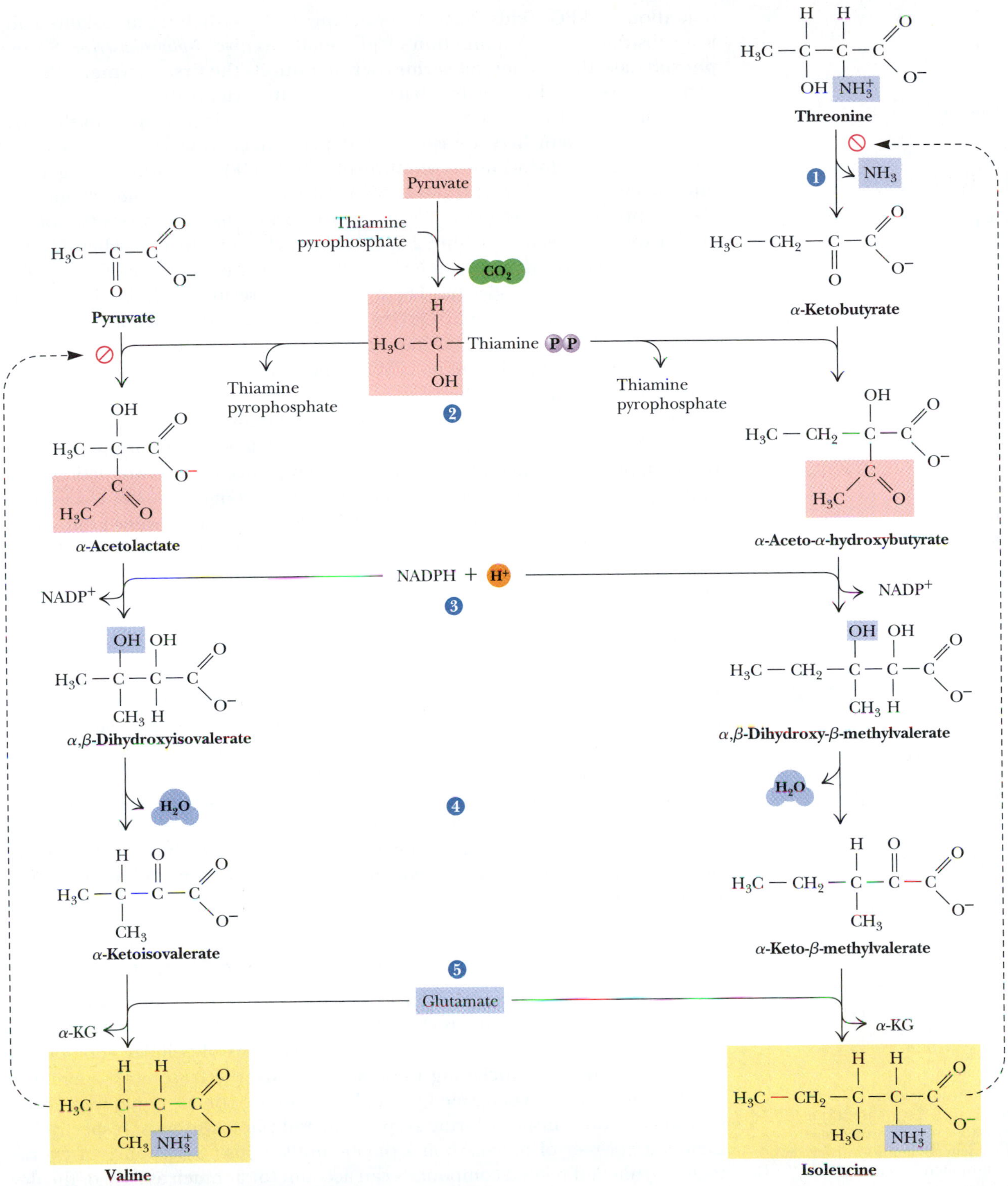

FIGURE 25.29 Biosynthesis of valine and isoleucine. The enzymes are (**1**) threonine deaminase, (**2**) aceto-hydroxy acid synthase, (**3**) acetohydroxy acid isom-eroreductase, (**4**) dihydroxy acid dehydratase, and (**5**) glutamate-dependent aminotransferase. Feedback inhibition regulates this pathway: Enzyme 1 is isoleucine sensitive, and enzyme 2 is valine sensitive.

α-ketoisovalerate is lost as CO_2. Amination of α-ketoisocaproate by **leucine aminotransferase** gives Leu.

The 3-Phosphoglycerate Family of Amino Acids Includes Ser, Gly, and Cys

Serine, glycine, and cysteine are derived from the glycolytic intermediate 3-phosphoglycerate. The diversion of 3-PG from glycolysis is achieved via **3-phosphoglycerate dehydrogenase** (Figure 25.31). This NAD^+-dependent

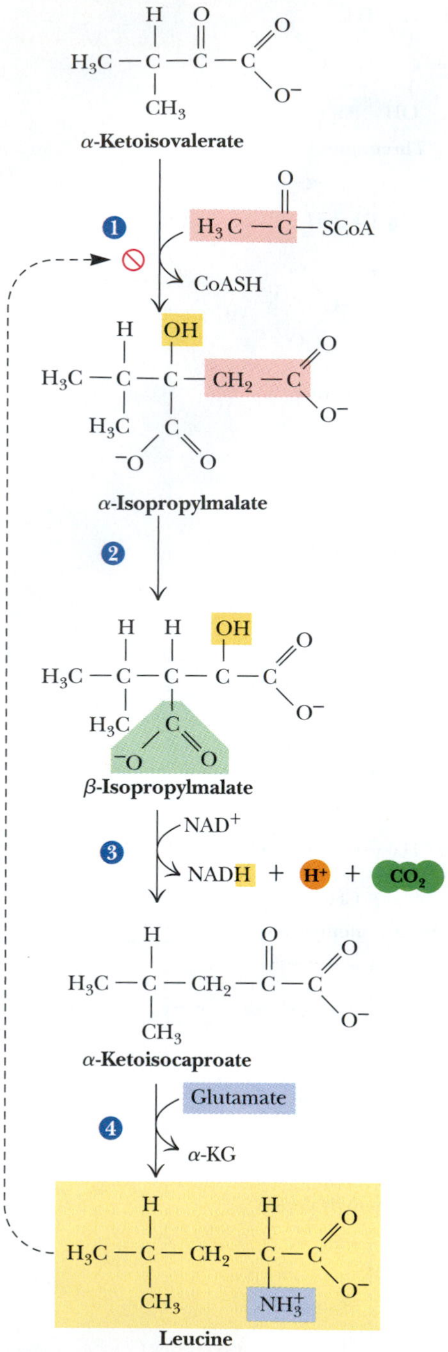

FIGURE 25.30 Biosynthesis of leucine. The enzymes are **(1)** α-isopropylmalate synthase, **(2)** α-isopropylmalate dehydratase, **(3)** isopropylmalate dehydrogenase, and **(4)** leucine aminotransferase. Enzyme 1 is feedback-inhibited by leucine.

oxidation of 3-PG yields *3-phosphohydroxypyruvate*—which, as an α-keto acid, is a substrate for transamination by glutamate to give *3-phosphoserine*. **Serine phosphatase** then generates **serine.** Serine inhibits the first enzyme, 3-PG dehydrogenase, and thereby feedback-regulates its own synthesis.

Glycine is made from serine via two related enzymatic processes. In the first, **serine hydroxymethyltransferase,** a PLP-dependent enzyme, catalyzes the transfer of the serine β-carbon to tetrahydrofolate (THF), the principal agent of one-carbon metabolism (Figure 25.32a). Glycine and N^5, N^{10}-methylene-THF are the products. In addition, glycine can be synthesized by a reversal of the **glycine oxidase** reaction (Figure 25.32b). Here, glycine is formed when N^5, N^{10}-methylene-THF condenses with NH_4^+ and CO_2. Via this route, the β-carbon of serine becomes part of glycine. The conversion of serine to glycine is a prominent means of generating one-carbon derivatives of THF, which are so important for the biosynthesis of purines and the C-5 methyl group of thymine (a pyrimidine), as well as the amino acid methionine. Glycine itself contributes to both purine and heme synthesis.

Cysteine synthesis is accomplished by sulfhydryl transfer to serine. In some bacteria, H_2S condenses directly with serine via a PLP-dependent enzyme-catalyzed reaction, but in most microorganisms and green plants, the sulfhydrylation reaction requires an activated form of serine, *O-acetylserine* (Figure 25.33). *O*-acetylserine is made by **serine acetyltransferase,** with the transfer of an acetyl group from acetyl-CoA to the —OH of Ser. This enzyme is inhibited by Cys. *O*-Acetylserine then undergoes sulfhydrylation by H_2S with elimination of acetate; the enzyme is ***O*-acetylserine sulfhydrylase.**

Sulfide Synthesis from Sulfate Involves S-Containing ATP Derivatives Given the prevailing oxidative nature of our environment and the reactivity and toxicity of H_2S, the source of sulfide for Cys synthesis merits discussion. In microorganisms and plants, sulfide is the product of sulfate assimilation. Sulfate is the common inorganic form of combined sulfur, and its assimilation involves several interesting ATP derivatives (Figure 25.34). **3'-Phosphoadenosine-5'-phosphosulfate** (PAPS) is not only an intermediate in sulfate assimilation; it also serves as the substrate for synthesis of sulfate esters, such as the sulfated polysaccharides found in the glycocalyx of animal cells. The "activated sulfate" of PAPS is reduced to sulfite (SO_3^{2-}) in a thioredoxin-dependent reaction, and sulfite is then reduced to sulfide (S^{2-}).

The Aromatic Amino Acids Are Synthesized from Chorismate

The aromatic amino acids, phenylalanine, tyrosine, and tryptophan, are derived from a shared pathway that has **chorismic acid** (Figure 25.35) as a key intermediate. Indeed, chorismate is common to the synthesis of cellular compounds having benzene rings, including these amino acids, the fat-soluble vitamins E and K, folic acid, and coenzyme Q and plastoquinone (the two quinones necessary to electron transport during respiration and photosynthesis, respectively). **Lignin,** a polymer of nine-carbon aromatic units, is also a derivative of chorismate. Lignin and related compounds can account for as much as 35% of the dry weight of higher plants; clearly, enormous amounts of carbon pass through the chorismate biosynthetic pathway.

CHORISMATE IS SYNTHESIZED FROM PEP AND ERYTHROSE-4-P. Chorismate biosynthesis occurs via the **shikimate pathway** (Figure 25.36). The precursors for this pathway are the common metabolic intermediates *phosphoenolpyruvate* and *erythrose-4-phosphate*. These intermediates are linked to form *3-deoxy-D-arabino-heptulosonate-7-phosphate* (DAHP) by **DAHP synthase.** Although this reaction is remote from the ultimate aromatic amino acid end products, it is an important point for regulation of aromatic amino acid biosynthesis, as we shall see. In the next step on the way to chorismate, DAHP is cyclized to form a six-membered saturated ring compound, *3-dehydroquinate.* A sequence of reactions ensues that introduces unsaturations into the ring, yielding *shikimate,* then *chorismate.* Note that the side chain of chorismate is derived from a second equivalent of phosphoenolpyruvate.

PHENYLALANINE AND TYROSINE. At chorismate, the pathway separates into three branches, each leading specifically to one of the aromatic amino acids. The branches leading to phenylalanine and tyrosine both pass through *prephenate* (Figure 25.37). In some organisms, such as *E. coli,* the branches are truly distinct because prephenate does not occur as a free intermediate but rather remains bound to the bifunctional enzyme that catalyzes the first two reactions after chorismate. In any case, **chorismate mutase** is the first reaction leading to Phe or Tyr. In the Phe branch, the —OH group *para* to the prephenate carboxyl is removed by a **dehydratase;** in the Tyr branch, this —OH is retained and becomes the phenolic —OH of Tyr. Glutamate-dependent aminotransferases introduce the amino groups into the two α-keto acids, *phenylpyruvate* and *4-hydroxy-phenylpyruvate,* to give Phe and Tyr, respectively. Some mammals can synthesize Tyr from Phe obtained in the diet via **phenylalanine-4-monooxygenase,** using O_2 and *tetrahydrobiopterin,* an analog of tetrahydrofolic acid, as co-substrates (Figure 25.38).

TRYPTOPHAN. The pathway of tryptophan synthesis is perhaps the most thoroughly studied of any biosynthetic sequence, particularly in terms of its genetic organization and expression. The principal stalwart of this research is Charles Yanofsky of Stanford University, and his many original insights represent themes of general significance in metabolic regulation. Synthesis of Trp from chorismate requires six steps (Figure 25.37). In most microorganisms, the first enzyme, **anthranilate synthase,** is an $\alpha_2\beta_2$-type protein, with the β-subunit acting in a glutamine–amidotransferase role to provide the —NH_2 group of anthranilate. Or, given high levels of NH_4^+, the α-subunit can carry out the formation of anthranilate directly by a process in which the activity of the β-subunit is unnecessary. Furthermore, in certain enteric bacteria, such as *E. coli* and *Salmonella typhimurium,* the second reaction of the pathway, the **phosphoribosyl-anthranilate transferase** reaction, is an activity catalyzed by the α-subunit of anthranilate synthase. *PRPP (5-phosphoribosyl-1-pyrophosphate),*

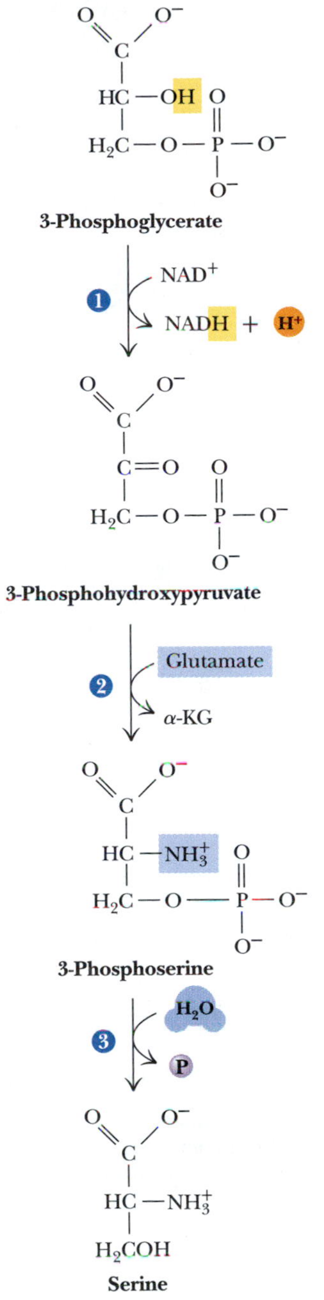

FIGURE 25.31 Biosynthesis of serine from 3-phosphoglycerate. The enzymes are **(1)** 3-phosphoglycerate dehydrogenase, **(2)** 3-phosphoserine aminotransferase, and **(3)** phosphoserine phosphatase.

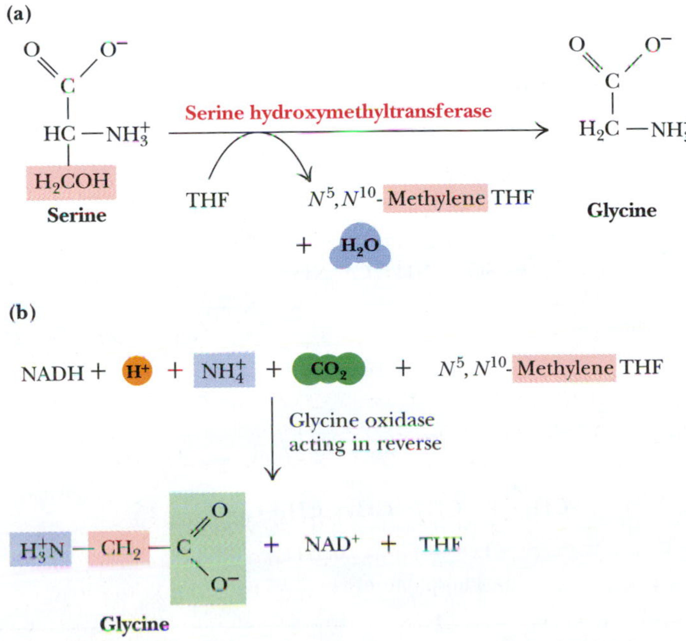

FIGURE 25.32 Biosynthesis of glycine from serine **(a)** via serine hydroxymethyltransferase and **(b)** via glycine oxidase.

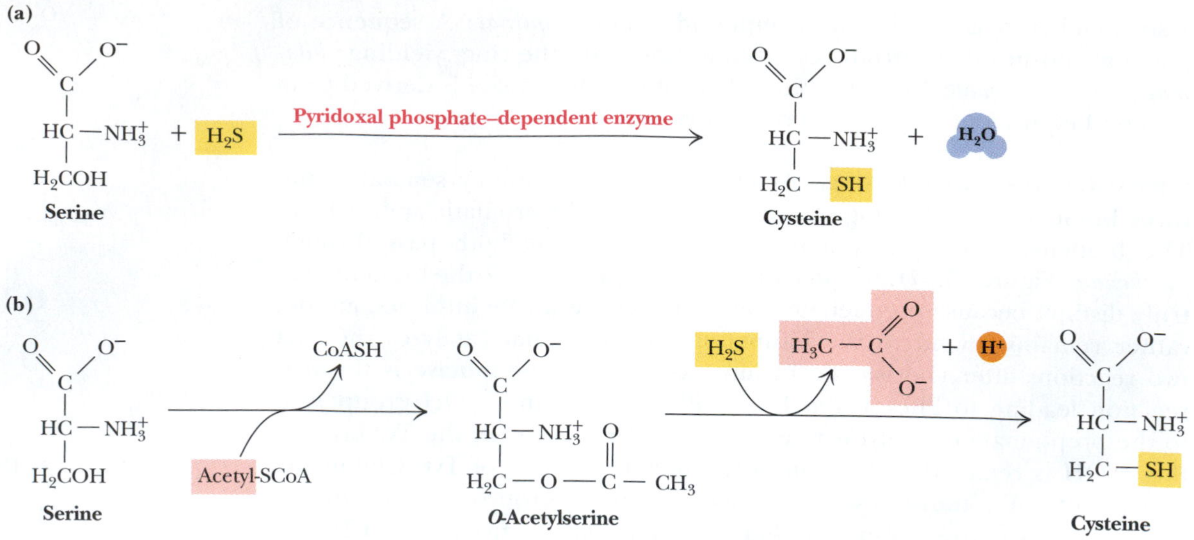

(a)

(b)

FIGURE 25.33 Cysteine biosynthesis. **(a)** Direct sulfhydrylation of serine by H_2S. **(b)** H_2S-dependent sulfhydrylation of O-acetylserine.

the substrate of this reaction, is also a precursor for purine biosynthesis (see Chapter 26). *Phosphoribosyl-anthranilate* then undergoes a rearrangement wherein the ribose moiety is isomerized to the ribulosyl form in *enol-1-(o-carboxyphenylamino)-1-deoxyribulose-5-phosphate* (reaction 8). Decarboxylation and ring closure ensue to yield the indole nucleus as *indole-3-glycerol phosphate* (**indole-3-glycerol phosphate synthase,** reaction 9). In the last reaction, serine displaces glyceraldehyde-3-phosphate to give Trp. The enzyme **tryptophan synthase** is also an $\alpha_2\beta_2$-type protein. The α-subunit cleaves indoleglycerol-3-phosphate to form indole and

A Deeper Look

Amino Acid Biosynthesis Inhibitors as Herbicides

Unlike animals, plants can synthesize all 20 of the common amino acids. Inhibitors acting specifically on the plant enzymes that are capable of carrying out the biosynthesis of the "essential" amino acids (that is, enzymes that animals lack) have been developed. These substances appear to be ideal for use as herbicides because they should show no effect on animals. **Glyphosate,** sold commercially as *RoundUp,* is a PEP analog that acts as an uncompetitive inhibitor of 3-enolpyruvylshikimate-5-P synthase (Figure 25.36). **Sulfmeturon methyl,** a sulfonylurea herbicide that inhibits *acetohy-droxy acid synthase,* an enzyme common to Val, Leu, and Ile biosynthesis (Figure 25.29), is the active ingredient in *Oust.* **Aminotriazole,** sold as *Amitrole,* blocks His biosynthesis by inhibiting *imidazole glycerol-P dehydratase* (Figure 25.40). **PPT (phosphinothricin)** is a potent inhibitor of *glutamine synthetase.* Although Gln is a nonessential amino acid and glutamine synthetase is a ubiquitous enzyme, PPT is relatively safe for animals because it does not cross the blood–brain barrier and is rapidly cleared by the kidneys.

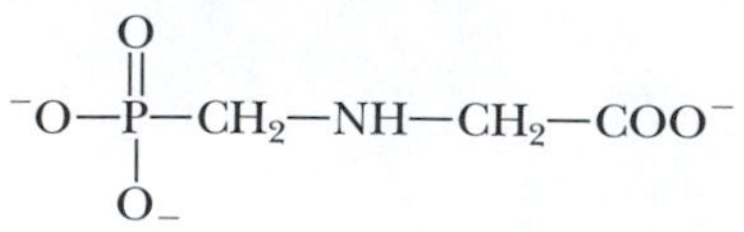

Glyphosate

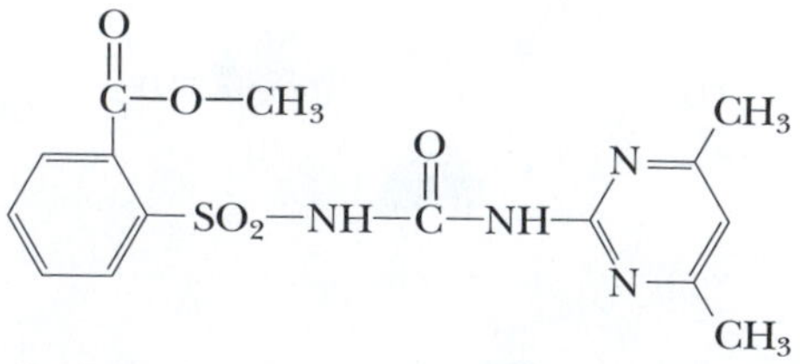

Sulfmeturon methyl

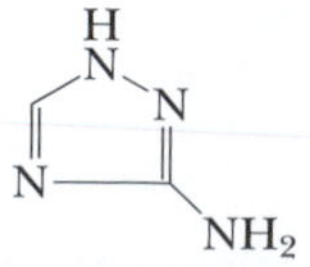

Aminotriazole

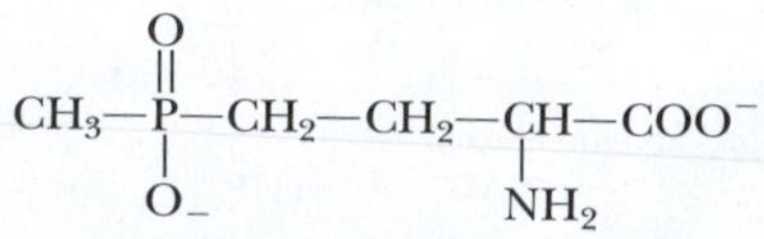

DL-Phosphinothricin (PPT)

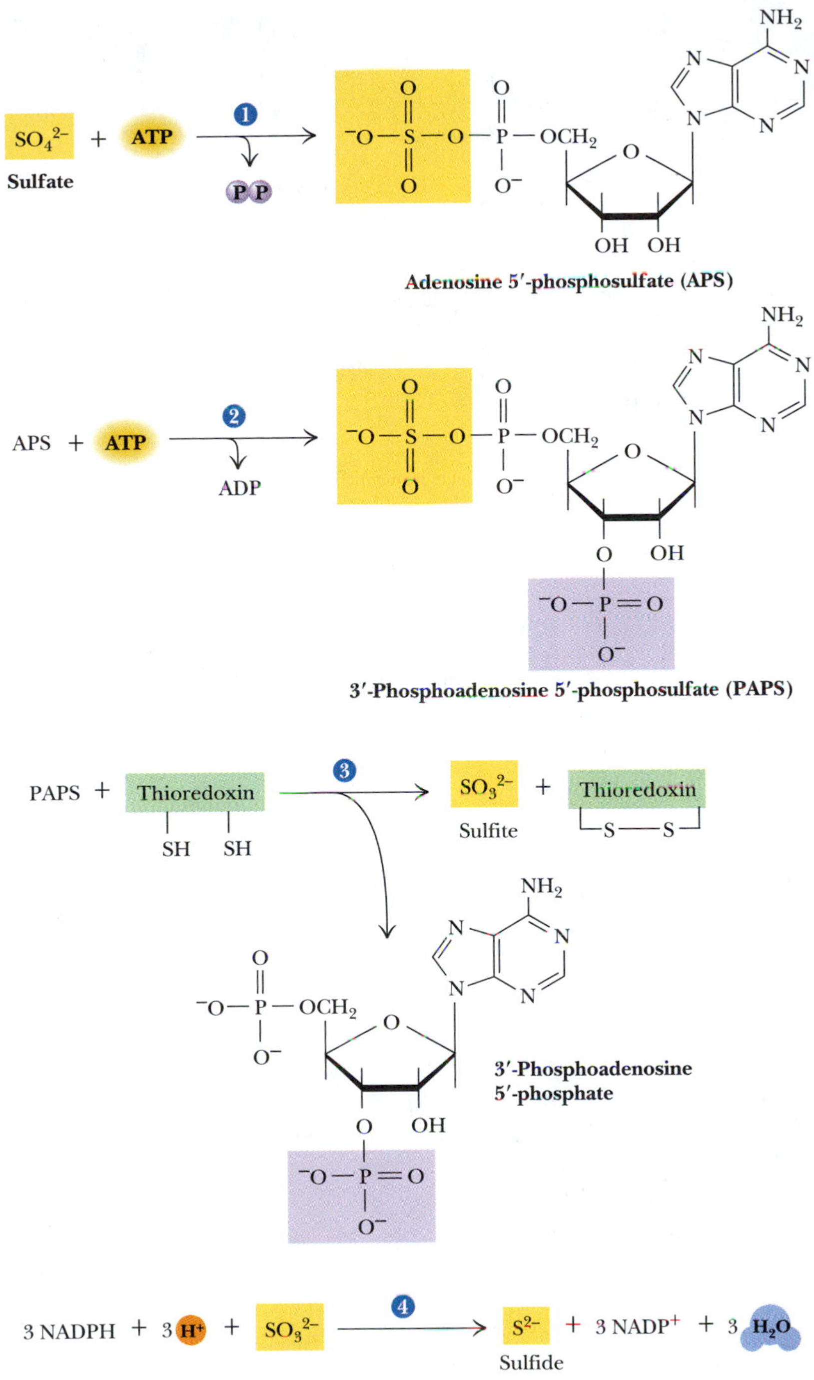

FIGURE 25.34 Sulfate assimilation and the generation of sulfide for synthesis of organic S compounds. In **reaction 1,** ATP sulfurylase catalyzes the formation of adenosine-5′-phosphosulfate (APS) + PP$_i$. In **reaction 2,** adenosine-5′-phosphosulfate 3′-phosphokinase catalyzes the reaction of adenosine 5′-phosphosulfate with a second ATP to form 3′-phosphoadenosine-5′-phosphosulfate (PAPS) + ADP. Both enzymes are Mg^{2+}-dependent. In **reaction 3,** PAPS is reduced to sulfite (SO$_3^{2-}$) in a *thioredoxin*-dependent reaction. Thioredoxin is a small (12-kD) protein that functions in a number of biological reductions (see Chapter 26). In **reaction 4,** *sulfite reductase* catalyzes the six-electron reduction of sulfite to sulfide. NADPH is the electron donor. Sulfite reductase possesses siroheme as a prosthetic group, the same heme found in nitrite reductase (Figure 25.2), which also catalyzes a six-electron transfer reaction.

3-glycerol phosphate. The indole is then channeled directly to the β-subunit, which adds serine in a PLP-dependent reaction.

X-ray crystallographic analysis of the structure of tryptophan synthase from *S. typhimurium* reveals that the active sites of the α- and β-subunits of the enzyme, although separated from each other by 2.5 nm, are connected by a hydrophobic tunnel wide enough to accommodate the bound indole intermediate (Figure 25.39). Thus, indole can be transferred directly from one active site to the other without being lost from the enzyme complex and diluted in the surrounding milieu. This phenomenon of direct transfer of enzyme-bound metabolic intermediates, or **channeling,** increases the efficiency of the overall pathway by preventing loss and dilution of the intermediate. Channeling is a widespread mechanism for substrate transfer in metabolism, particularly among the enzymes of higher organisms.

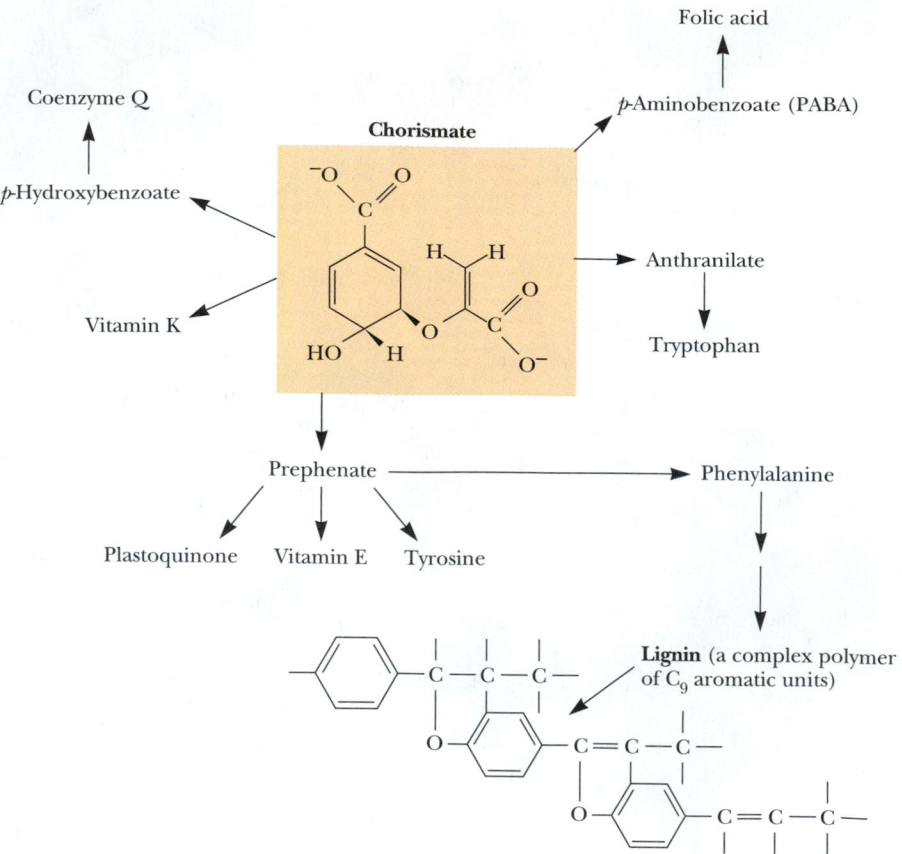

FIGURE 25.35 Some of the aromatic compounds derived from chorismate.

Histidine Biosynthesis and Purine Biosynthesis Are Connected by Common Intermediates

Like aromatic amino acid biosynthesis, **histidine** biosynthesis shares metabolic intermediates with the pathway of purine nucleotide synthesis. The pathway involves ten separate steps, the first being an unusual reaction that links ATP and PRPP (Figure 25.40). Five carbon atoms from PRPP and one from ATP end up in histidine. Step 5 involves some novel chemistry: The substrate, *phosphoribulosylformimino-5-aminoimidazole-4-carboxamide ribonucleotide,* picks up an amino group (from the amide of glutamine) in a reaction accompanied by cleavage and ring closure to yield two imidazole compounds— the histidine precursor, *imidazole glycerol phosphate,* and a purine nucleotide precursor, *5-aminoimidazole-4-carboxamide ribonucleotide* (AICAR). Note that AICAR as a purine nucleotide precursor can ultimately replenish the ATP consumed in reaction 1. Nine enzymes act in histidine's ten synthetic steps. Reactions 9 and 10, the successive NAD⁺-dependent oxidations of an alcohol to an aldehyde and then to a carboxylic acid, are catalyzed by the same dehydrogenase.

25.5	How Does Amino Acid Catabolism Lead into Pathways of Energy Production?

In normal human adults, close to 90% of the energy requirement is met by oxidation of carbohydrates and fats; the remainder comes from oxidation of the carbon skeletons of amino acids. The primary physiological purpose of amino acids is to serve as the building blocks for protein biosynthesis. The dietary amount of free amino acids is trivial under most circumstances. However, if ex-

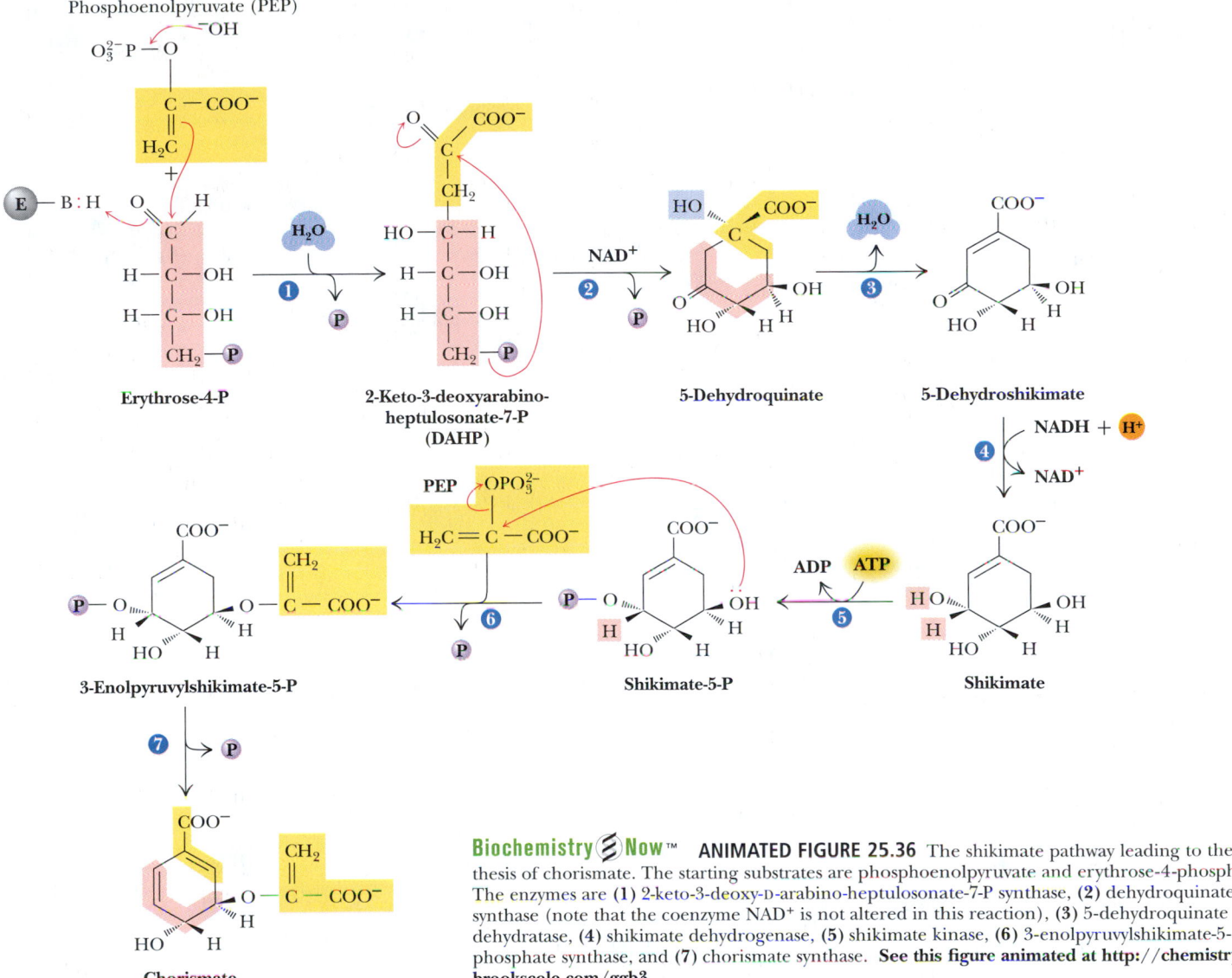

Biochemistry⊛Now™   ANIMATED FIGURE 25.36 The shikimate pathway leading to the synthesis of chorismate. The starting substrates are phosphoenolpyruvate and erythrose-4-phosphate. The enzymes are (**1**) 2-keto-3-deoxy-D-arabino-heptulosonate-7-P synthase, (**2**) dehydroquinate synthase (note that the coenzyme NAD⁺ is not altered in this reaction), (**3**) 5-dehydroquinate dehydratase, (**4**) shikimate dehydrogenase, (**5**) shikimate kinase, (**6**) 3-enolpyruvylshikimate-5-phosphate synthase, and (**7**) chorismate synthase. **See this figure animated at http://chemistry. brookscole.com/ggb3**

cess protein is consumed in the diet or if the amount of amino acids released during normal turnover of cellular proteins exceeds the requirements for new protein synthesis, the amino acid surplus must be catabolized. Also, if carbohydrate intake is insufficient (as during fasting or starvation) or if carbohydrates cannot be appropriately metabolized due to disease (as in *diabetes mellitus*), body protein becomes an important fuel for metabolic energy.

The 20 Common Amino Acids Are Degraded by 20 Different Pathways That Converge to Just 7 Metabolic Intermediates

Because the 20 common amino acids of proteins are distinctive in terms of their carbon skeletons, each amino acid requires its own unique degradative pathway. Because amino acid degradation normally supplies only 10% of the body's energy, then, on average, degradation of any given amino acid will satisfy less than 1% of energy needs. Therefore, we will not discuss these pathways in detail. It so happens, however, that degradation of the carbon skeletons of the 20 common α-amino acids converges to just 7 metabolic intermediates: *acetyl-CoA, succinyl-CoA, pyruvate, α-ketoglutarate, fumarate, oxaloacetate,* and *acetoacetate*. Because succinyl-CoA, pyruvate, α-ketoglutarate, fumarate,

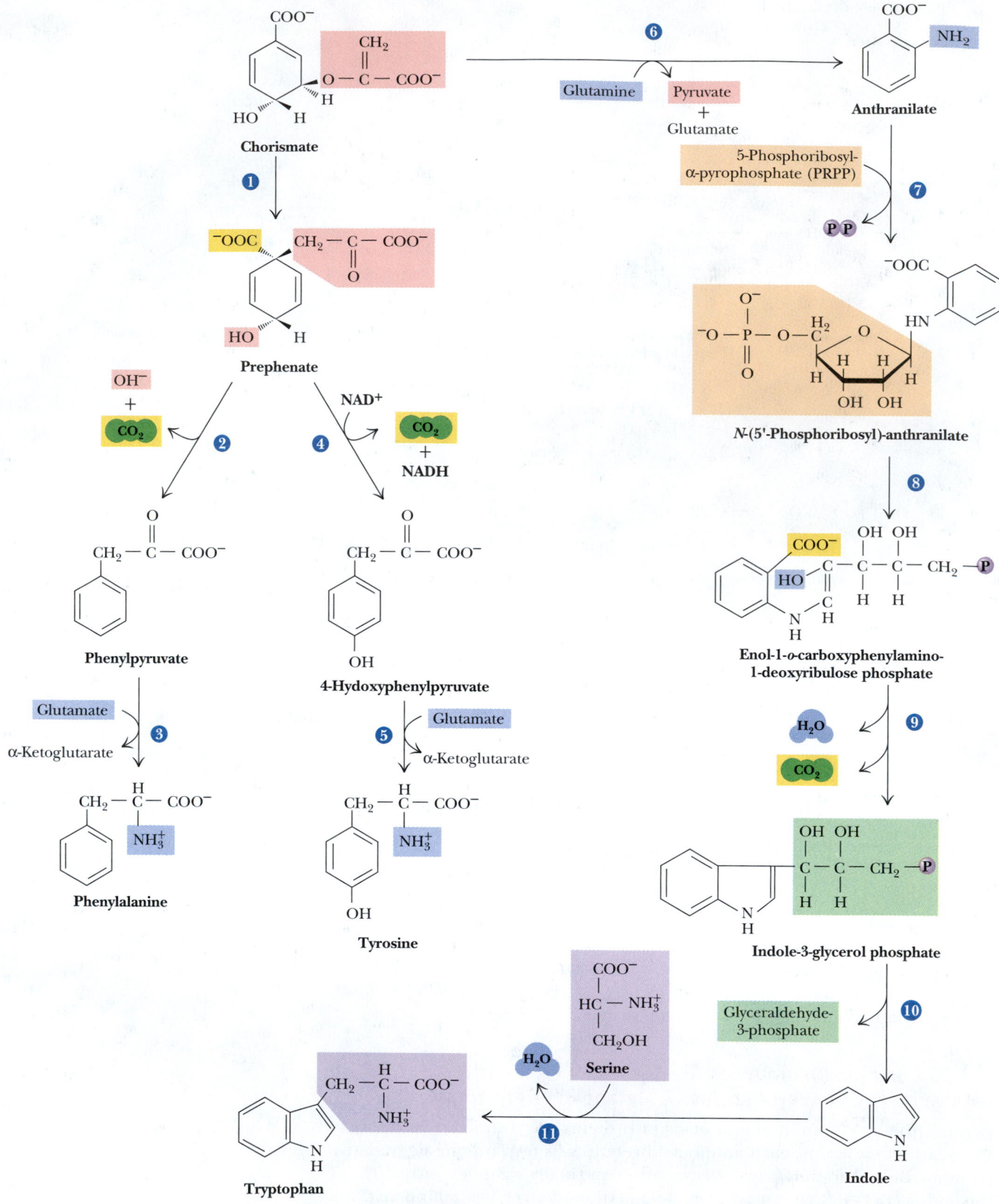

Biochemistry☰Now™ **ANIMATED FIGURE 25.37** The biosynthesis of phenylalanine, tyrosine, and tryptophan from chorismate. The enzymes are (1) chorismate mutase, (2) prephenate dehydratase, (3) phenylalanine aminotransferase, (4) prephenate dehydrogenase, (5) tyrosine aminotransferase, (6) anthranilate synthase, (7) anthranilate-phosphoribosyl transferase, (8) N-(5'-phosphoribosyl)-anthranilate isomerase, (9) indole-3-glycerol phosphate synthase, (10) tryptophan synthase (α-subunit), and (11) tryptophan synthase (β-subunit). See this figure animated at http://chemistry.brookscole.com/ggb3

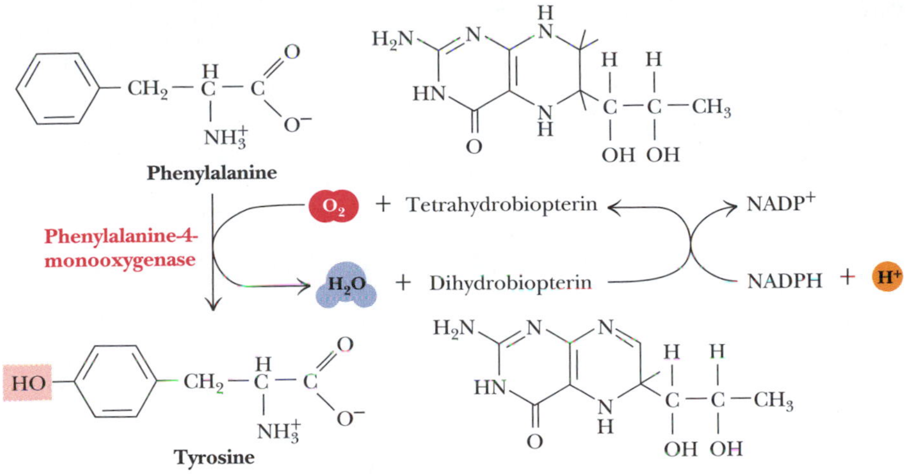

Phenylalanine

Phenylalanine-4-monooxygenase

O_2 + Tetrahydrobiopterin ⇌ NADP⁺

H_2O + Dihydrobiopterin ⇌ NADPH + H⁺

Tyrosine

FIGURE 25.38 The formation of tyrosine from phenylalanine. This reaction is normally the first step in phenylalanine degradation in most organisms; in mammals, however, it provides a route for the biosynthesis of Tyr from Phe. (Phenylalanine-4-monooxygenase is also known as phenylalanine hydroxylase.)

and oxaloacetate can serve as precursors for glucose synthesis, amino acids giving rise to these intermediates are termed **glucogenic.** Those degraded to yield acetyl-CoA or acetoacetate are termed **ketogenic,** because these substances can be used to synthesize fatty acids or ketone bodies. Some amino acids are both glucogenic and ketogenic (Figure 25.41).

The C-3 Family of Amino Acids: Alanine, Serine, and Cysteine

The carbon skeletons of alanine, serine, and cysteine all converge to *pyruvate* (Figure 25.42). Transamination of alanine yields pyruvate:

$$\text{Alanine} + \alpha\text{-ketoglutarate} \rightleftharpoons \text{pyruvate} + \text{glutamate}$$

Deamination of serine by **serine dehydratase** also yields pyruvate. Cysteine is converted to pyruvate via a number of paths.

The carbon skeletons of three other amino acids also become pyruvate. *Glycine* is convertible to serine and thus to pyruvate. The three carbon atoms of *tryptophan* that are not part of its indole ring appear as alanine (and, hence, pyruvate) upon Trp degradation. *Threonine* by one of its degradation routes is cleaved to glycine and acetaldehyde. The glycine is then converted to pyruvate via serine; the acetaldehyde is oxidized to acetyl-CoA (Figure 26.42).

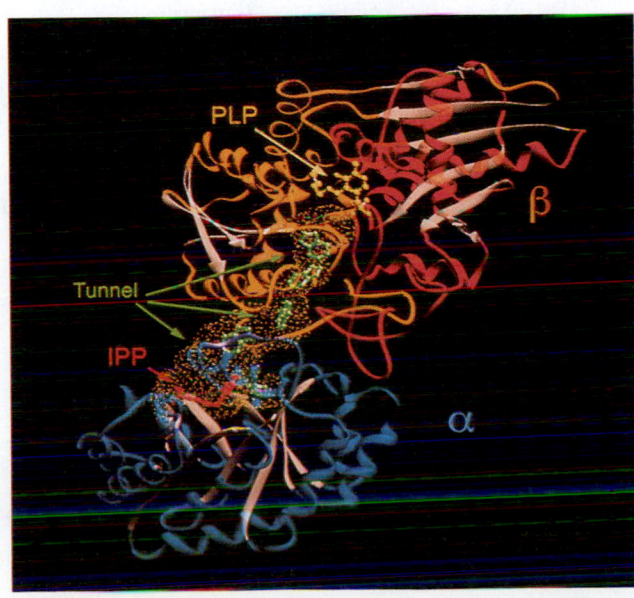

FIGURE 25.39 Tryptophan synthase is an example of a "channeling" multienzyme complex in which indole, the product of the α-reaction catalyzed by the α-subunit, passes intramolecularly to the β-subunit. In the β-subunit, the hydroxyl of the substrate L-serine is replaced with indole via a complicated pyridoxal phosphate–catalyzed reaction to produce the final product, L-tryptophan. The schematic figure shown here is a ribbon diagram of one α-subunit (blue) and neighboring β-subunit (the N-terminal domain of the β-subunit is in orange, C-terminal domain in red). The tunnel is outlined by the yellow dot surface and is shown with several indole molecules (green) packed in head-to-tail fashion. The labels "IPP" and "PLP" point to the active sites of the α- and β-subunits, respectively, in which a competitive inhibitor (indole propanol phosphate, IPP) and the coenzyme PLP are bound. (*Adapted from Hyde, C. C., et al., 1988. Three-dimensional structure of the tryptophan synthase multienzyme complex from* Salmonella typhimurium. *Journal of Biological Chemistry* **263:**17857–17871.)

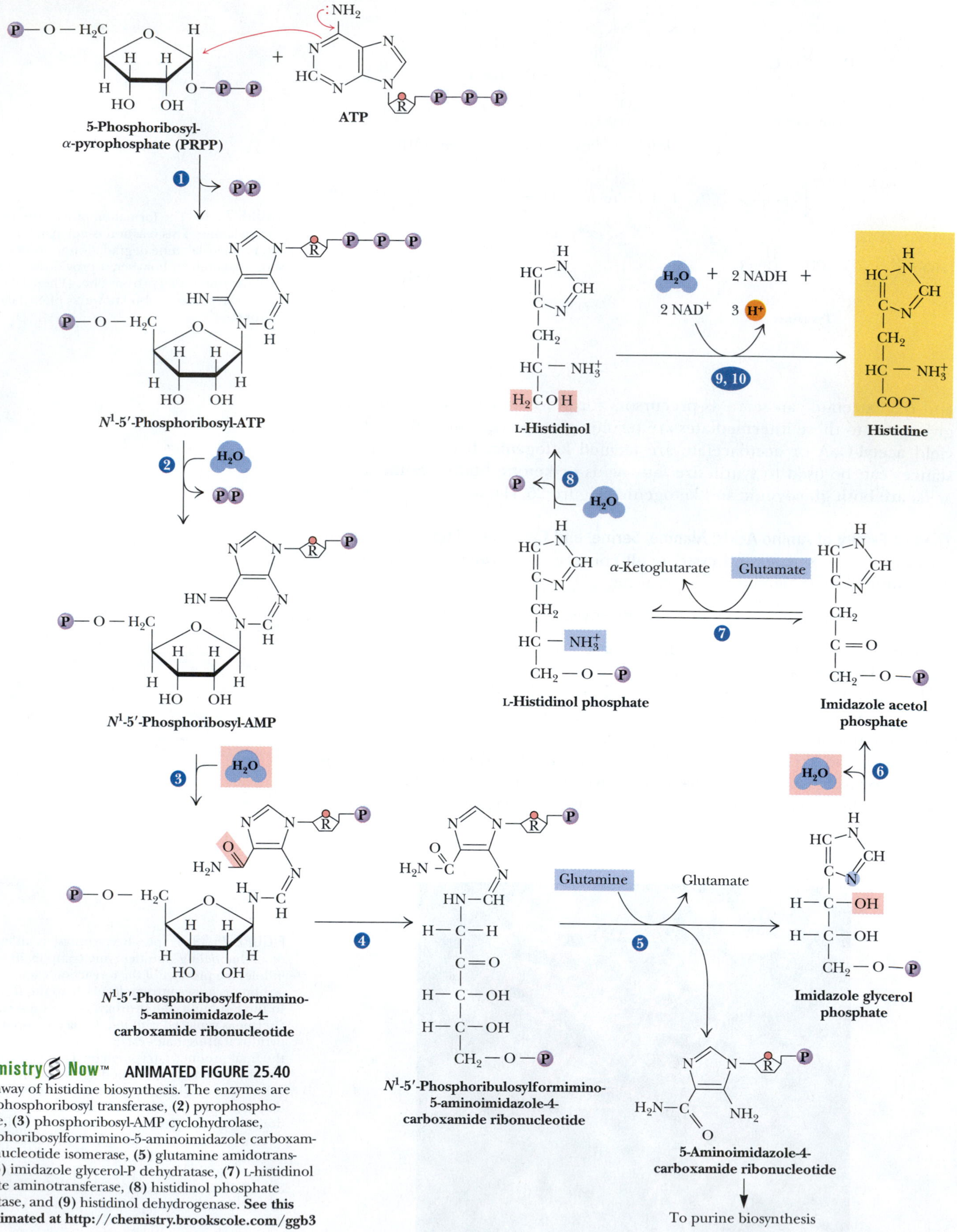

Biochemistry 🔵 Now™ ANIMATED FIGURE 25.40
The pathway of histidine biosynthesis. The enzymes are
(1) ATP-phosphoribosyl transferase, (2) pyrophospho-
hydrolase, (3) phosphoribosyl-AMP cyclohydrolase,
(4) phosphoribosylformimino-5-aminoimidazole carboxamide ribonucleotide isomerase, (5) glutamine amidotransferase, (6) imidazole glycerol-P dehydratase, (7) L-histidinol
phosphate aminotransferase, (8) histidinol phosphate
phosphatase, and (9) histidinol dehydrogenase. **See this
figure animated at http://chemistry.brookscole.com/ggb3**

Biochemistry Now™ ACTIVE FIGURE 25.41 Metabolic degradation of the common amino acids. The 20 common amino acids can be classified according to their degradation products. Those that give rise to precursors for glucose synthesis, such as α-ketoglutarate, succinyl-CoA, fumarate, oxaloacetate, and pyruvate, are termed *glucogenic* (shown in pink). Those degraded to acetyl-CoA or acetoacetate are called *ketogenic* (shown in blue) because they can be converted to fatty acids or ketone bodies. Some amino acids are both glucogenic and ketogenic. **Test yourself on the concepts in this figure at http://chemistry.brookscole.com/ggb3**

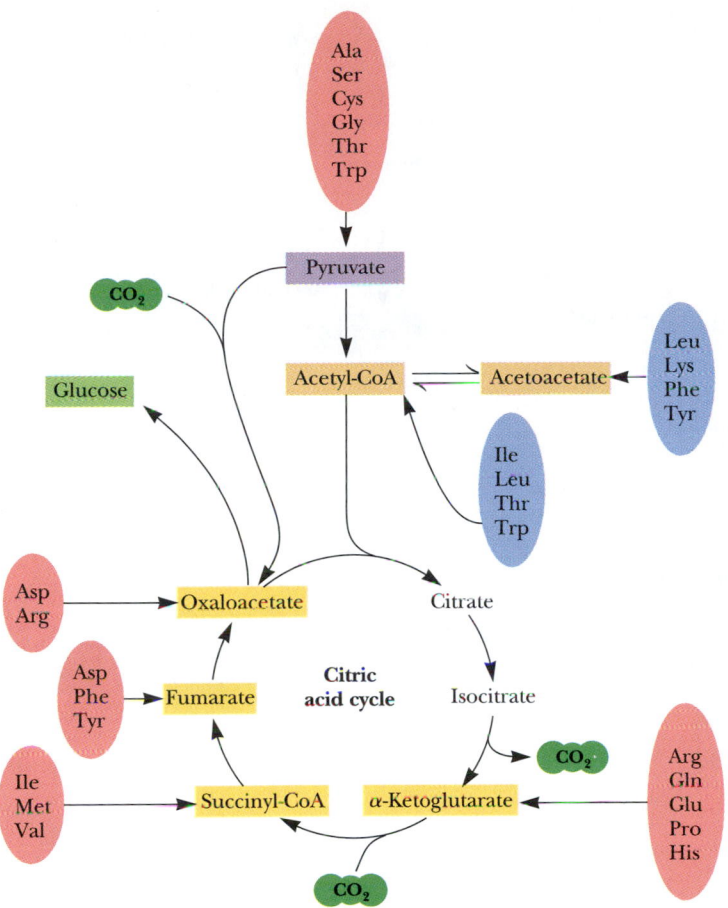

The C-4 Family of Amino Acids: Aspartate and Asparagine Transamination of aspartate gives *oxaloacetate:*

$$\text{Aspartate} + \alpha\text{-ketoglutarate} \rightleftharpoons \text{oxaloacetate} + \text{glutamate}$$

Hydrolysis of asparagine by **asparaginase** yields aspartate and NH_4^+. Alternatively, aspartate degradation via the urea cycle leads to a different citric acid cycle intermediate, namely, *fumarate* (Figure 25.23).

The C-5 Family of Amino Acids Are Converted to α-Ketoglutarate Via Glutamate The five-carbon citric acid cycle intermediate α-ketoglutarate is always a product of transamination reactions involving *glutamate.* Thus, glutamate *and* any amino acid convertible to glutamate are classified within the C-5 family. These amino acids include *glutamine, proline, arginine,* and *histidine* (Figure 25.43).

A Deeper Look

Biochemistry Now™

Histidine—A Clue to Understanding Early Evolution?

Histidine residues in the active sites of enzymes often act directly in the enzyme's catalytic mechanism. Catalytic participation by the imidazole group of His and the presence of imidazole as part of the purine ring system support a current speculation that life before the full evolution of protein molecules must have been RNA based. This notion correlates with recent revelations that RNA molecules can have catalytic activity, an idea captured in the term ribozyme (see Chapter 13).

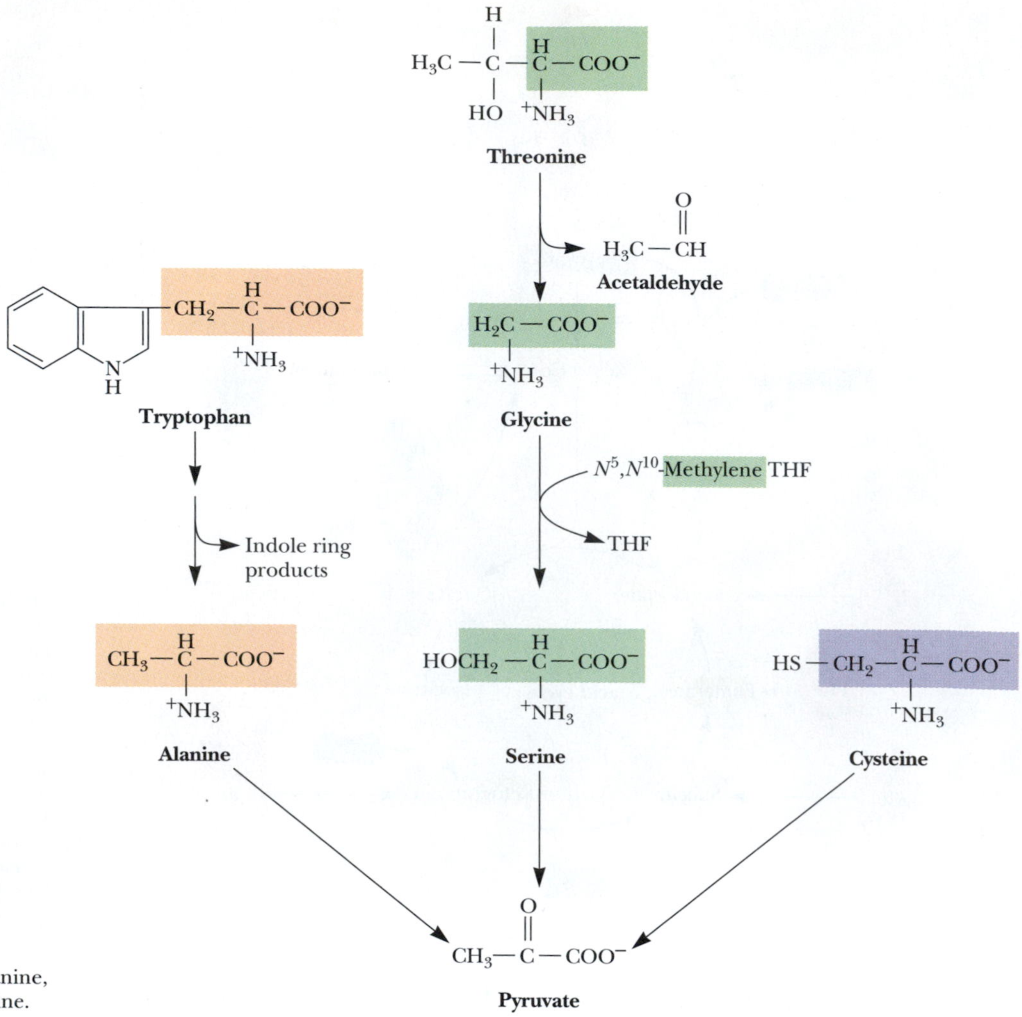

FIGURE 25.42 Formation of pyruvate from alanine, serine, cysteine, glycine, tryptophan, or threonine.

Degradation of Valine, Isoleucine, and Methionine Leads to Succinyl-CoA The breakdown of valine, isoleucine, and methionine converges at *propionyl-CoA* (Figure 25.44). Propionyl-CoA is subsequently converted to methylmalonyl-CoA and thence to *succinyl-CoA* via the same reactions mediating the oxidation of fatty acids that have odd numbers of carbon atoms (see Chapter 23).

Leucine Is Degraded to Acetyl-CoA and Acetoacetate **Leucine** is one of only two purely ketogenic amino acids; the other is lysine. Deamination of leucine via a transamination reaction yields α-ketoisocaproate, which is oxidatively decarboxylated to *isovaleryl-CoA* (Figure 25.45). Subsequent reactions, one of which is a biotin-dependent carboxylation, give β-hydroxy-β-methylglutaryl-CoA, which is then cleaved to yield *acetyl-CoA* and *acetoacetate*. Neither of these products is convertible to glucose.

The initial steps in valine, leucine, and isoleucine degradation are identical. All three are first deaminated to α-keto acids, which are then oxidatively decarboxylated to form CoA derivatives. **Maple syrup urine disease** is a hereditary defect in the oxidative decarboxylation of these branched-chain α-keto acids. The metabolic block created by this defect leads to elevated levels of valine, leucine, and isoleucine (and their corresponding branched-chain α-keto acids) in the blood and urine. The urine of individuals with this disease smells like maple syrup. The defect is fatal unless dietary intake of these amino acids is greatly restricted early in life.

Lysine Degradation Lysine degradation proceeds by several pathways, but the *saccharopine pathway* found in liver predominates (Figure 25.46). This degradative

A Deeper Look

The Serine Dehydratase Reaction—A β-Elimination

The degradation of serine to pyruvate is an example of a pyridoxal phosphate–catalyzed β-elimination reaction. β-Eliminations mediated by PLP yield products that have undergone a two-electron oxidation at C_α. Serine is thus oxidized to pyruvate, with release of ammonium ion (see accompanying figure). At first, this looks like a transaminase half-reaction, but there is an important difference. In each transaminase half-reaction, PLP undergoes a net two-electron reduction or oxidation (depending on the direction), whereas β-eliminations occur with no net oxidation or reduction of PLP. Note too that the aminoacrylate released from PLP is unstable in aqueous solution. It rapidly tautomerizes to the preferred imine form, which is spontaneously hydrolyzed to yield the α-keto acid product—pyruvate in this case.

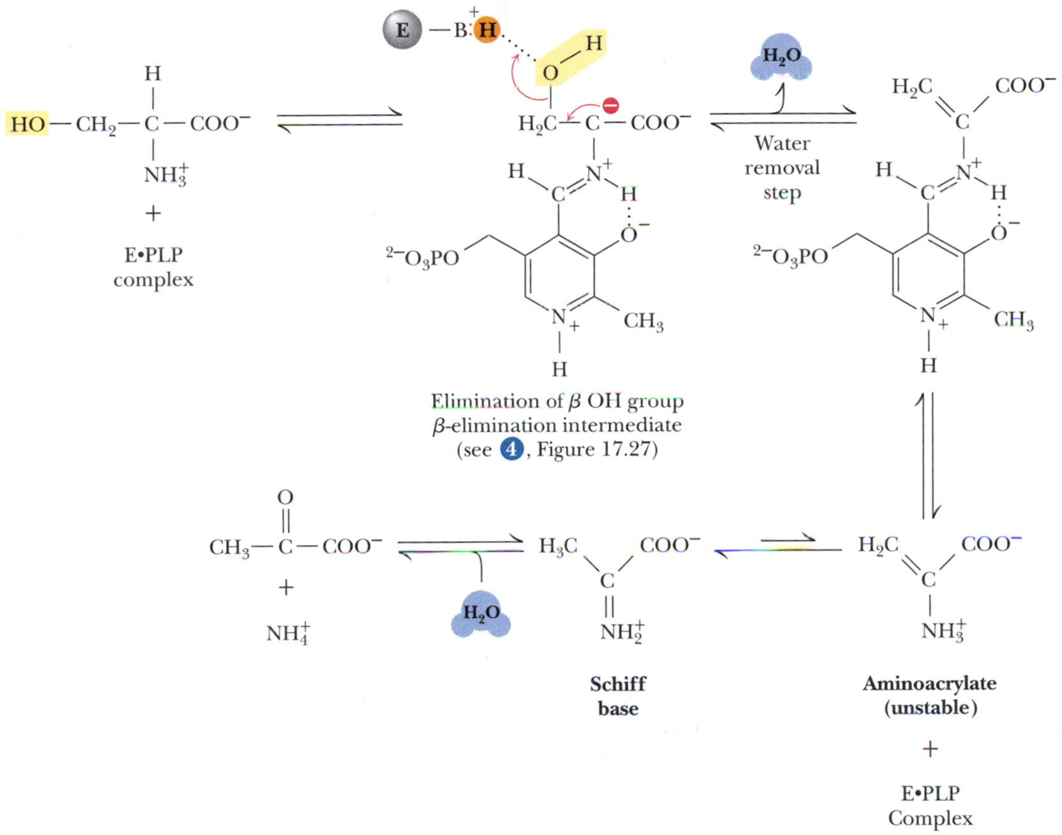

Elimination of β OH group
β-elimination intermediate
(see ❹, Figure 17.27)

Schiff
base

Aminoacrylate
(unstable)

The serine dehydratase reaction mechanism—an example of a PLP-dependent β-elimination reaction.

route proceeds backward along the lysine biosynthetic pathway through saccharopine and α-aminoadipate to α-ketoadipate (Figure 25.24). Next, α-ketoadipate undergoes oxidative decarboxylation to *glutaryl-CoA,* which is then transformed into *acetoacetyl-CoA* and ultimately into *acetoacetate,* a ketone body (see Chapter 23).

As indicated earlier, degradation of the nonindole carbons of tryptophan yields pyruvate. The *indole ring* of Trp is converted by a series of reactions to α-ketoadipate and ultimately *acetoacetate* by these same reactions of Lys degradation.

Phenylalanine and Tyrosine Are Degraded to Acetoacetate and Fumarate The first reaction in phenylalanine degradation is the hydroxylation reaction of *tyrosine* biosynthesis (Figure 25.38). Both these amino acids thus share a common degradative pathway. Transamination of Tyr yields the α-keto acid *p-hydroxyphenylpyruvate* (Figure 25.47). ***p*-Hydroxyphenylpyruvate dioxygenase,** a vitamin C–dependent enzyme, then carries out a ring hydroxylation–oxidative decarboxylation to yield homogentisate. Ring opening and isomerization give *4-fumaryl-acetoacetate,* which is hydrolyzed to *acetoacetate* and *fumarate.*

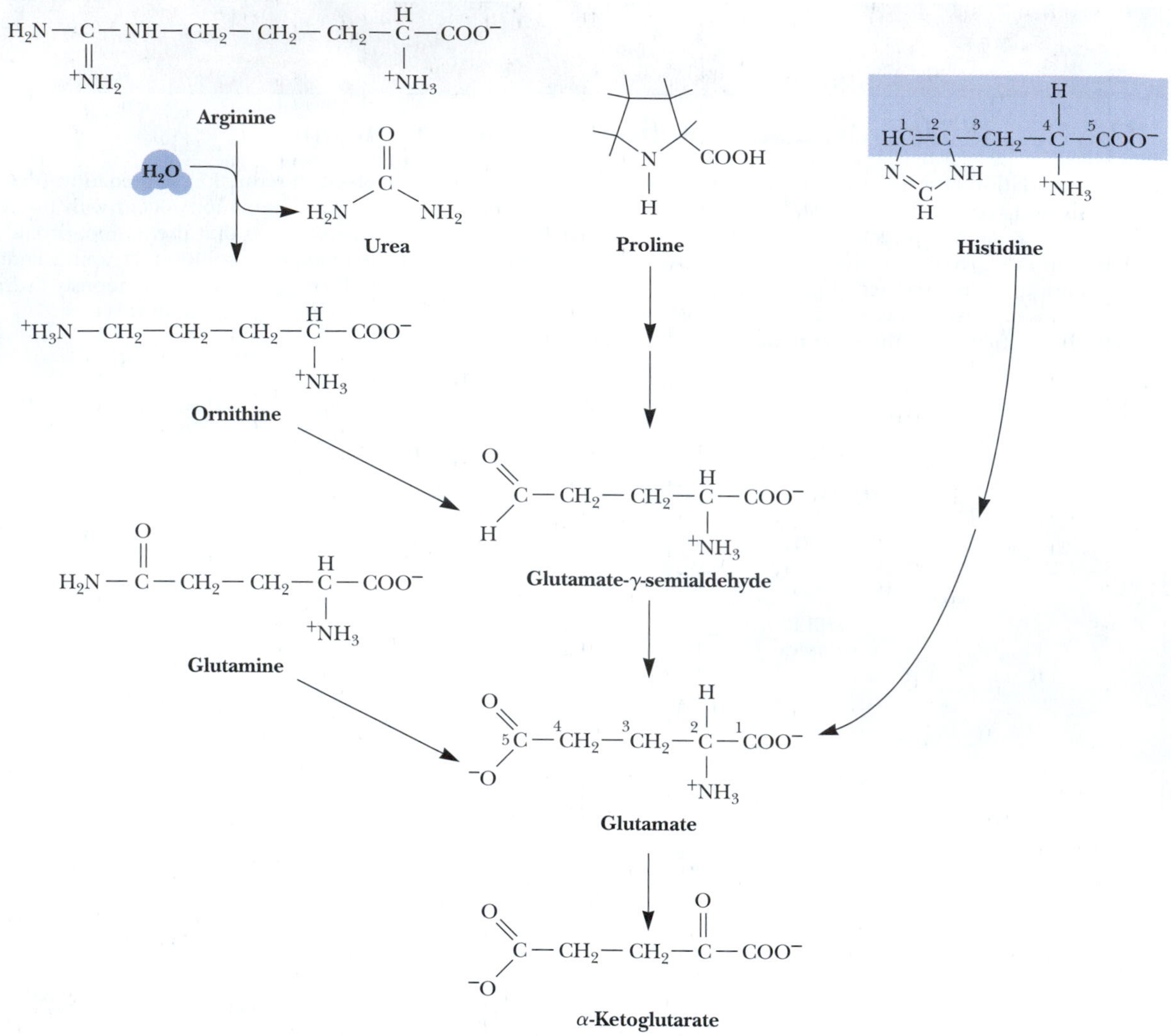

FIGURE 25.43 The degradation of the C-5 family of amino acids leads to α-ketoglutarate via glutamate. The histidine carbons, numbered 1 through 5, become carbons 1 through 5 of glutamate, as indicated.

Human Biochemistry

Hereditary Defects in Phe Catabolism Underlie Alkaptonuria and Phenylketonuria

Alkaptonuria and phenylketonuria are two human genetic diseases arising from specific enzyme defects in phenylalanine degradation. **Alkaptonuria** is characterized by urinary excretion of large amounts of homogentisate and results from a deficiency in **homogentisate dioxygenase** (Figure 25.47). Air oxidation of homogentisate causes urine to turn dark on standing, but the only malady suffered by carriers of this disease is a tendency toward arthritis later in life.

In contrast, **phenylketonurics,** whose urine contains excessive *phenylpyruvate* (see accompanying figure), suffer severe mental retardation if the defect is not recognized immediately after birth and treated by putting the victim on a diet low in phenylalanine. These individuals are deficient in phenylalanine hydroxylase (Figure 25.38), and the excess Phe that accumulates is transaminated to phenylpyruvate and excreted.

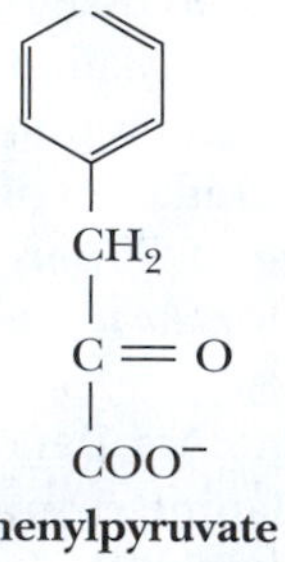

Phenylpyruvate

The structure of phenylpyruvate.

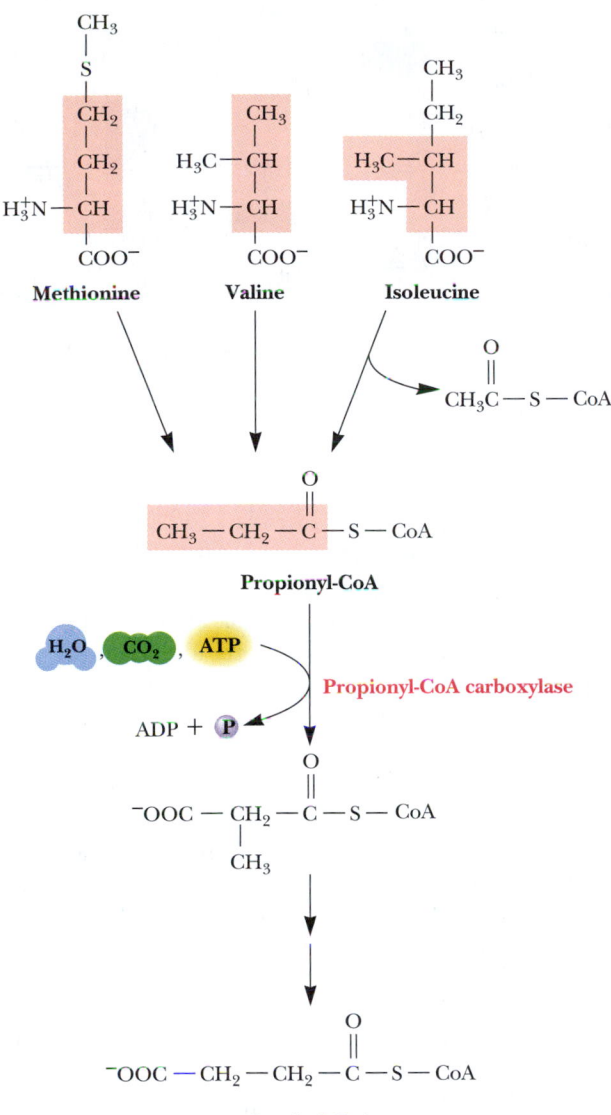

FIGURE 25.44 Valine, isoleucine, and methionine are converted via propionyl-CoA to succinyl-CoA for entry into the citric acid cycle. The shaded carbon atoms of the three amino acids give rise to propionyl-CoA. All three amino acids lose their α-carboxyl group as CO_2. Methionine first becomes S-adenosylmethionine, then homocysteine (see Figure 25.28). The terminal two carbons of isoleucine become acetyl-CoA.

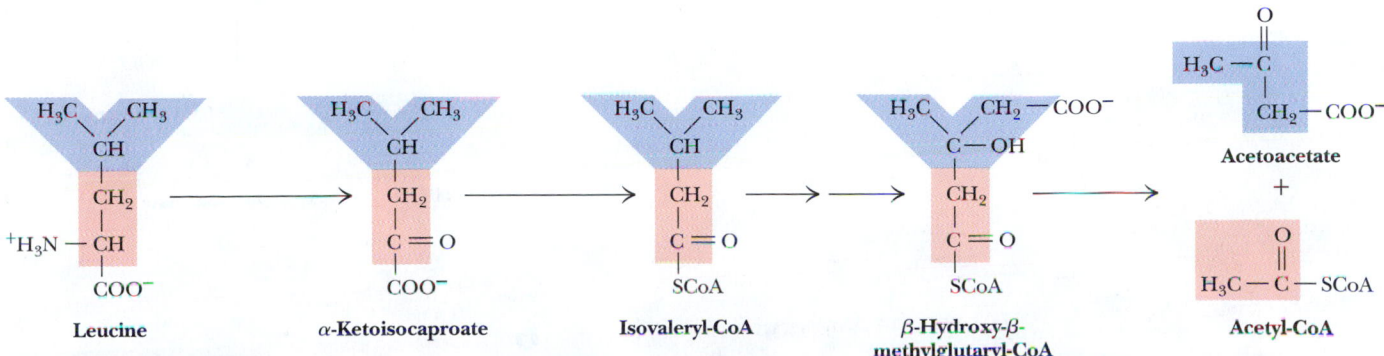

FIGURE 25.45 Leucine is degraded to acetyl-CoA and acetoacetate via β-hydroxy-β-methylglutaryl-CoA, which is also the intermediate in ketone body formation from fatty acids (see Chapter 23).

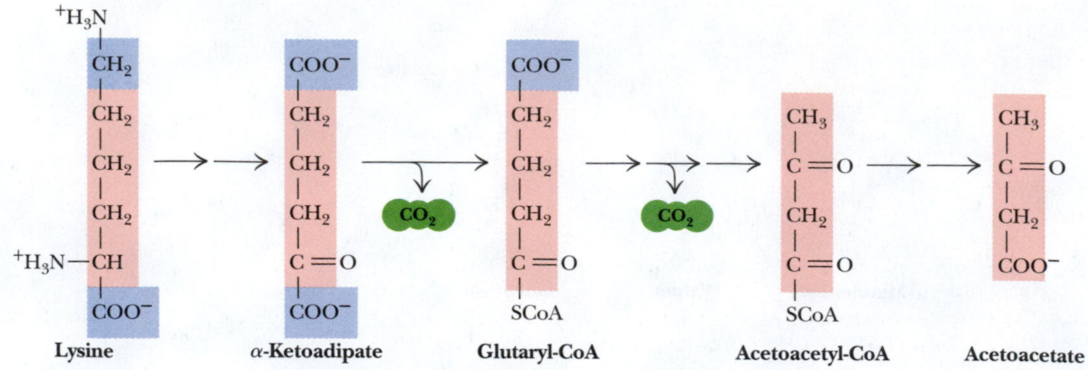

FIGURE 25.46 Lysine degradation via the saccharopine, α-ketoadipate pathway culminates in the formation of acetoacetyl-CoA.

Animals Differ in the Form of Nitrogen That They Excrete

Animals often enjoy a dietary surplus of nitrogen. Excess nitrogen liberated upon metabolic degradation of amino acids is excreted by animals in three different ways, in accord with the availability of water. Aquatic animals simply release free ammonia to the surrounding water; such animals are termed **ammonotelic** (from the Greek *telos,* meaning "end"). On the other hand, terrestrial and aerial species employ mechanisms that convert ammonium to less toxic waste compounds that require little H_2O for excretion. Many terrestrial vertebrates, including humans, are **ureotelic,** meaning that they excrete excess N as **urea,** a highly water-soluble nonionic substance. Urea is formed by ureoteles via the urea cycle (see earlier). The **uricotelic** organisms are those animals using the third means of N excretion, conversion to **uric acid,** a rather insoluble purine analog. Birds and reptiles are uricoteles. Uric acid metabolism is discussed in the next chapter. Some animals can switch from ammonotelic to ureotelic to uricotelic metabolism, depending on water availability.

FIGURE 25.47 Phenylalanine and tyrosine degradation. **(1)** Transamination of Tyr gives *p*-hydroxyphenylpyruvate, which **(2)** is oxidized to homogentisate by *p*-hydroxyphenylpyruvate dioxygenase in an ascorbic acid (vitamin C)–dependent reaction. **(3)** The ring opening of homogentisate by homogentisate dioxygenase gives 4-maleylacetoacetate. **(4)** 4-Maleylacetoacetate isomerase gives 4-fumarylacetoacetate, which **(5)** is hydrolyzed by fumarylacetoacetase.

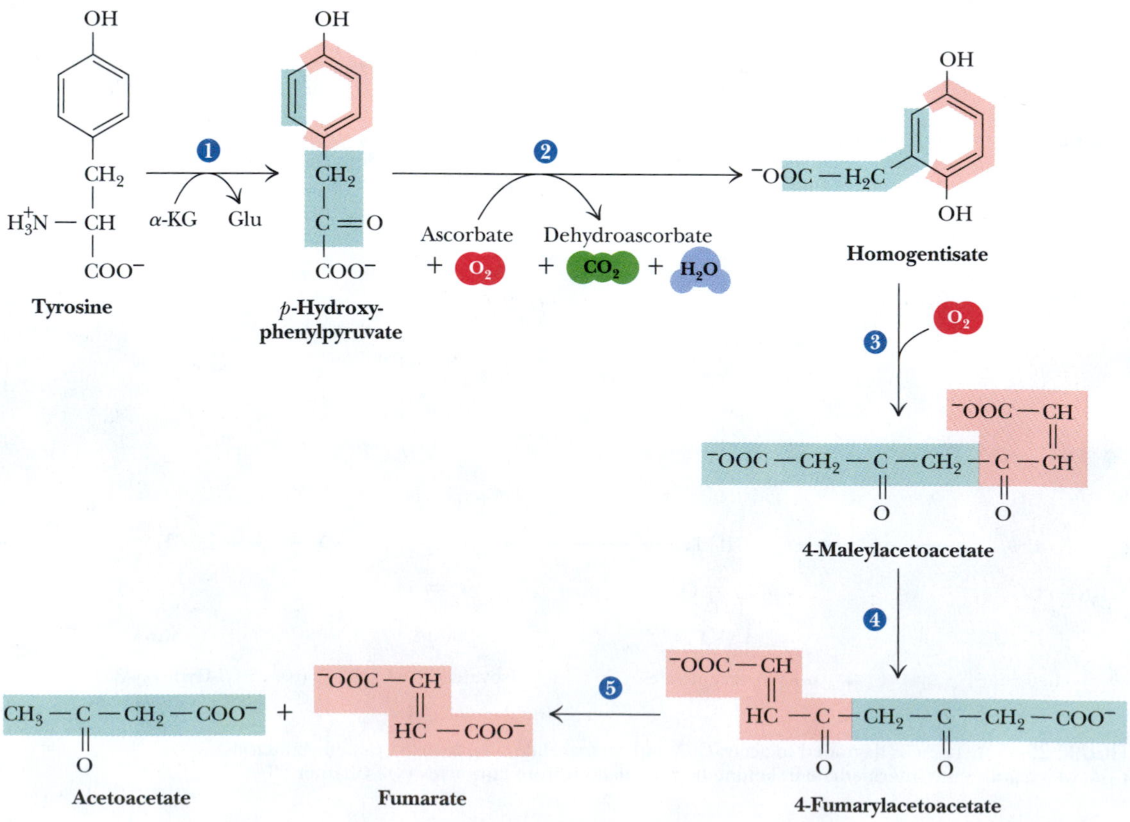

Summary

25.1 Which Metabolic Pathways Allow Organisms to Live on Inorganic Forms of Nitrogen?
Nitrogen, an element essential to life, occurs in the environment principally as atmospheric N_2 and as NO_3^- ions in solution in soils and water. The metabolic pathways of nitrogen fixation and nitrate assimilation reduce these oxidized forms of nitrogen into the metabolically useful form, ammonium. Nitrate assimilation is a two-enzyme pathway: nitrate reductase and nitrite reductase. Nitrate reductase is a molybdenum cofactor-dependent flavohemoprotein. Nitrite reductase catalyzes the six-electron reduction of NO_2^- to NH_4^+ via a siroheme-dependent reaction. Nitrogen fixation is carried out by the nitrogenase system; biological reduction of N_2 to $2 NH_4^+$ is an ATP-dependent eight-electron transfer reaction, with H_2 as an obligatory by-product. Nitrogenase is a metal-rich enzyme having an 8Fe-7S cluster as well as a 7Fe-1Mo-9S cluster known as the FeMo-cofactor. Nitrogenase is regulated in two ways: Its activity is inhibited by ADP, and its synthesis is repressed by NH_4^+.

25.2 What Is the Metabolic Fate of Ammonium?
Despite the great diversity of organic nitrogenous compounds found in cells, only a limited set of reactions incorporate ammonium ions into organic linkage: (1) carbamoyl-P synthetase, (2) glutamate dehydrogenase (GDH), and (3) glutamine synthetase (GS). Of these, the latter two are quantitatively more important. Glutamate dehydrogenase, by adding NH_4^+ to the citric acid cycle intermediate α-ketoglutarate, sits at the interface between nitrogen metabolism and carbohydrate (and energy) metabolism. Glutamine synthetase (α_{12}-subunit organization) catalyzes the ATP-dependent amidation of the γ-carboxyl group of Glu. Glutamine is the major donor of $—NH_2$ groups for the synthesis of many nitrogen-containing organic compounds, including purines, pyrimidines, and other amino acids. As such, its activity is tightly regulated. Ammonium assimilation into organic linkage proceeds by one of two routes, depending on NH_4^+ availability: the GDH–GS route when ammonium is abundant and the glutamate synthase (GOGAT)–GS route when $[NH_4^+]$ is limiting.

25.3 What Regulatory Mechanisms Act on *Escherichia coli* Glutamine Synthetase?
Glutamine synthetase is a paradigm of enzyme regulation, because its activity can be modulated at three different levels: (1) allosteric regulation by feedback inhibition, (2) covalent modification through adenylylation of Tyr^{397} in each of the 12 GS polypeptide chains, and (3) regulation of gene expression by the phosphorylated form of the transcriptional enhancer NR$_1$. Allosteric inhibitors of GS include five amino acids (Gly, Ser, Ala, His, and Trp), two nucleotides (one purine [AMP] and one pyrimidine [CTP]), one aminosugar (glucosamine-6-P), and carbamoyl-P. Adenylylation of GS converts it from a more active, allosterically unresponsive form to a less active, allosterically sensitive form. The ratio of adenylylated GS to deadenylylated GS is ultimately determined by the $[Gln]/[\alpha\text{-}KG]$ ratio, with a low ratio favoring the deadenylylated state and thus greater synthesis of glutamine.

25.4 How Do Organisms Synthesize Amino Acids?
In many cases, amino acid biosynthesis is a matter of synthesizing the appropriate α-keto acid carbon skeleton for the amino acid and then transaminating this α-keto acid using Glu as amino donor by action of an aminotransferase reaction. Mammals have retained the ability to synthesize the α-keto acid analog for 10 of the 20 common amino acids (the so-called nonessential amino acids), but the ability to make the other 10 α-keto acid analogs has been lost over evolutionary time, rendering these 10 amino acids as essential in the diet. The common amino acids can be grouped into families on the basis of the metabolic progenitor that serves as their precursor: The α-ketoglutarate family includes Glu, Gln, Pro, Arg, and (sometimes) Lys; the pyruvate family includes Ala, Val, and Leu; the aspartate family includes Asp, Asn, Met, Thr, Ile, and (sometimes) Lys; the 3-phosphoglycerate family includes Ser, Gly, and Cys; and the PEP and erythrose-4-P family includes the aromatic amino acids Phe, Tyr, and Trp. Histidine is a special case—it is formed from PRPP and ATP, with ATP providing its imidazole ring to His.

25.5 How Does Amino Acid Catabolism Lead into Pathways of Energy Production?
The 20 common amino acids are degraded by 20 different pathways that converge to just 7 metabolic intermediates: pyruvate, acetyl-CoA, acetoacetate, oxaloacetate, α-ketoglutarate, succinyl-CoA, and fumarate. All seven of these compounds are intermediates in or readily feed into the pathways of energy production (citric acid cycle and oxidative phosphorylation).

Problems

1. What is the oxidation number of N in nitrate, nitrite, NO, N_2O, and N_2?

2. How many ATP equivalents are consumed per N atom of ammonium formed by (a) the nitrate assimilation pathway and (b) the nitrogen fixation pathway? (Assume NADH, NADPH, and reduced ferredoxin are each worth 3 ATPs.)

3. Suppose at certain specific metabolite concentrations in vivo the cyclic cascade regulating *E. coli* glutamine synthetase has reached a dynamic equilibrium where the average state of GS adenylylation is poised at $n = 6$. Predict what change in n will occur if:
   a. [ATP] increases.
   b. P_{IID}/P_{IIA} increases.
   c. $[\alpha\text{-}KG]/[Gln]$ increases.
   d. $[P_i]$ decreases.

4. How many ATP equivalents are consumed in the production of 1 equivalent of urea by the urea cycle?

5. Why are persons on a high-protein diet (such as the Atkins diet) advised to drink lots of water?

6. How many ATP equivalents are consumed in the biosynthesis of lysine from aspartate by the pathway shown in Figure 25.27?

7. If PEP labeled with ^{14}C in the 2-position serves as the precursor to chorismate synthesis, which C atom in chorismate is radioactive?

8. (Integrates with Chapter 22.) Write a balanced equation for the synthesis of glucose (by gluconeogenesis) from aspartate.

9. For each of the 20 common amino acids, give the name of the enzyme that catalyzes the reaction providing its α-amino group.

10. (Integrates with Chapter 17.) Which vitamin is central in amino acid metabolism? Why?

11. (Integrates with Chapter 17.) Vitamins B_6, B_{12}, and folate may be recommended for individuals with high blood serum levels of homocysteine (a condition called *hyperhomocysteinemia*). How might these vitamins ameliorate homocysteinemia?

12. (Integrates with Chapters 17 and 19.) On the basis of the following information, predict a reaction mechanism for the mammalian branched-chain α-keto acid dehydrogenase complex (the BCKAD complex). This complex carries out the oxidative decarboxylation of the α-keto acids derived from methionine, valine, and isoleucine (see Figure 25.45).
   a. One form of maple syrup urine disease responds well to administration of thiamine.
   b. Lipoic acid is an essential coenzyme.
   c. The enzyme complex contains a flavoprotein.

13. People with phenylketonuria must avoid foods containing the low-calorie sweetener *Aspartame*, also known as *NutraSweet*. Find the

structure of the low-calorie sweetener *Aspartame* in the Merck Index (or other scientific source) and state why these people must avoid this substance.

14. Glyphosate (otherwise known as *RoundUp*) is an analog of PEP. It acts as a noncompetitive inhibitor of 3-enolpyruvylshikimate-5-P synthase; it has the following structure in its fully protonated state:

$$HOOC-CH_2-NH-CH_2-PO_3H_2$$

Consult Figures 25.35 and 25.36 and construct a list of the diverse metabolic consequences that might be experienced by a plant cell exposed to glyphosate.

15. (Integrates with Chapter 18.) When cells convert glucose to glycine, which carbon atoms of glucose are represented in glycine?

Preparing for the MCAT Exam

16. From the dodecameric (α_{12}) structure of glutamine synthetase shown in Figure 25.14, predict the relative enzymatic activity of GS monomers (isolated α-subunits).

17. Consider the synthesis and degradation of tyrosine as shown in Figures 25.37, 25.38, and 25.47 to determine where the carbon atoms in PEP and erythrose-4-P would end up in acetoacetate and fumarate.

Biochemistry ⬡ Now™ Preparing for an exam? Test yourself on key questions at http://chemistry.brookscole.com/ggb3

Further Reading

Nitrate Assimilation and Nitrogen Fixation

Brewin, A. J., and Legocki, A. B., 1996. Biological nitrogen fixation for sustainable agriculture. *Trends in Microbiology* **4**:476–477.

Burris, R. H., 1991. Nitrogenases. *The Journal of Biological Chemistry* **266**:9339–9342.

Kim, J., and Rees, D. C., 1994. Nitrogenase and biological nitrogen fixation. *Biochemistry* **33**:389–396.

Leigh, G. J., 1995. The mechanism of dinitrogen reduction by molybdenum nitrogenases. *European Journal of Biochemistry* **229**:14–20.

Lin, J. T., and Stewart, V., 1998. Nitrate assimilation in bacteria. *Advances in Microbial Physiology* **39**:1–30.

Mortenson, L. E., Seefeldt, L. C., Morgan, T. V., and Bolin, J. T., 1993. The role of metal clusters and MgATP in nitrogenase catalysis. *Advances in Enzymology* **67**:299–374.

Peters, J. W., Fisher, K., and Dean, D. R., 1995. Nitrogenase structure and function: A biochemical-genetic perspective. *Annual Review of Microbiology* **49**:335–366.

Rhee, C., and Stadtman, E. R., 1989. Regulation of *E. coli* glutamine synthetase. *Advances in Enzymology* **62**:37–92.

Stacey, G., Burris, R. H., and Evans, H. J., 1992. *Biological Nitrogen Fixation.* New York: Chapman & Hall.

Wray, J. L., and Kinghorn, J. R., 1989. *Molecular and Genetic Aspects of Nitrate Assimilation.* New York: Oxford Science.

Glutamate Dehydrogenase and Glutamine Synthetase

Brosnan, J. T., 2000. Glutamate, at the interface between amino acid and carbohydrate metabolism. *Journal of Nutrition* **130**(4S Suppl):988S–990S.

Hudson, R. C., and Daniel, R. M., 1993. L-Glutamate dehydrogenases: Distribution, properties, and mechanism. *Comparative Biochemistry* **106B**:767–792.

Liaw, S-H., and Eisenberg, D. S., 1995. Discovery of the ammonium substrate site on glutamine synthetase, a third cation binding site. *Protein Science* **4**:2358–2365.

Liaw, S-H., Pan, C., and Eisenberg, D. S., 1993. Feedback inhibition of fully unadenylylated glutamine synthetase from *Salmonella typhimurium* by glycine, alanine, and serine. *Proceedings of the National Academy of Sciences, U.S.A.* **90**:4996–5000.

Morris, S. M., Jr., 2002. Regulation of enzymes of the urea cycle and arginine metabolism. *Annual Review of Nutrition* **22**:87–105.

Mutalik, V. K., Shah, P., and Venkatesh, K.V., 2003. Allosteric interactions and bifunctionality make the response of glutamine synthetase cascade system of *Escherichia coli* robust and ultrasensitive. *Journal of Biological Chemistry* **278**: 26327–26332.

Stadtman, E. R., 2001. The story of glutamine synthetase regulation. *Journal of Biological Chemistry* **276**:44357–44364.

The Urea Cycle

Atkinson, D. E., and Camien, M. N., 1982. The role of urea synthesis in the removal of metabolic bicarbonate and the regulation of blood pH. *Current Topics in Cellular Regulation* **21**:261–302. Describes the reasoning behind the proposal that the urea cycle eliminates bicarbonate as well as ammonium when urea is formed.

Amino Acid Metabolism

Bender, D. A., 1985. *Amino Acid Metabolism.* New York: Wiley. A general review of amino acid metabolism.

Srere, P. A., 1987. Complexes of sequential metabolic enzymes. *Annual Review of Biochemistry* **56**:89–124. A review of the evidence that enzymes in a pathway sequence are often physically associated with one another in vivo, particularly in eukaryotic cells.

Wagenmakers, A. J., 1998. Protein and amino acid metabolism in human muscle. *Advances in Experimental Medicine and Biology* **441**:307–319.

Clinical Disorders in Amino Acid Metabolism

Boushey, C. J., et al., 1995. A quantitative assessment of plasma homocysteine as a risk factor for vascular disease. *Journal of the American Medical Association* **274**:1049–1057.

Fernandez-Canon, J. M., et al., 1996. The molecular basis of alkaptonuria. *Nature Genetics* **14**:19–24.

Scriver, C. R., et al., 1995. *The Metabolic and Molecular Bases of Inherited Disease,* 7th ed. New York: McGraw-Hill. A three-volume treatise on the biochemistry and genetics of inherited metabolic disorder, including disorders of amino acid metabolism.

Amino Acid Biosynthesis Inhibitors as Herbicides

Kishore, G. M., and Shah, D. M., 1988. Amino acid biosynthesis inhibitors as herbicides. *Annual Review of Biochemistry* **57**:627–663.

The Synthesis and Degradation of Nucleotides

Pigeon drinking at Gaia Fountain, Siena, Italy. The basic features of purine biosynthesis were elucidated initially from metabolic studies of nitrogen metabolism in pigeons. Pigeons excrete excess N as uric acid, a purine analog.

Guano, a substance found on some coasts frequented by sea birds, is composed chiefly of the birds' partially decomposed excrement.... The name for the purine guanine derives from the abundance of this base in guano.

J. C. Nesbit, *On Agricultural Chemistry and the Nature and Properties of Peruvian Guano* (1850)

Essential Question

Virtually all cells are capable of synthesizing purine and pyrimidine nucleotides. These compounds then serve as essential intermediates in metabolism and as the building blocks for DNA and RNA synthesis. ***How do cells synthesize purines and pyrimidines?***

Nucleotides are ubiquitous constituents of life, actively participating in the majority of biochemical reactions. Recall that ATP is the "energy currency" of the cell, that uridine nucleotide derivatives of carbohydrates are common intermediates in cellular transformations of carbohydrates (see Chapter 22), and that biosynthesis of phospholipids proceeds via cytosine nucleotide derivatives (see Chapter 24). In Chapter 30, we will see that GTP serves as the immediate energy source driving the endergonic reactions of protein synthesis. Many of the coenzymes (such as coenzyme A, NAD, NADP, and FAD) are derivatives of nucleotides. Nucleotides also act in metabolic regulation, as in the response of key enzymes of intermediary metabolism to the relative concentrations of AMP, ADP, and ATP (PFK is a prime example here; see also Chapter 18). Furthermore, cyclic derivatives of purine nucleotides such as cAMP and cGMP have no other role in metabolism than regulation. Last but not least, nucleotides are the monomeric units of nucleic acids. Deoxynucleoside triphosphates (dNTPs) and nucleoside triphosphates (NTPs) serve as the immediate substrates for the biosynthesis of DNA and RNA, respectively (see Part 4).

26.1 | Can Cells Synthesize Nucleotides?

Nearly all organisms can make the purine and pyrimidine nucleotides via so-called de novo biosynthetic pathways. (*De novo* means "anew"; a less literal but more apt translation might be "from scratch" because de novo pathways are metabolic sequences that form complex end products from rather simple precursors.) Many organisms also have salvage pathways to recover purine and pyrimidine compounds obtained in the diet or released during nucleic acid turnover and degradation. Whereas the ribose of nucleotides can be catabolized to generate energy, the nitrogenous bases do *not* serve as energy sources; their catabolism does not lead to products used by pathways of energy conservation. Compared to slowly dividing cells, rapidly proliferating cells synthesize larger amounts of DNA and RNA per unit time. To meet the increased demand for nucleic acid synthesis, substantially greater quantities of nucleotides must be produced. The pathways of nucleotide biosynthesis thus become attractive targets for the clinical control of rapidly dividing cells such as cancers or infectious bacteria. Many antibiotics and anticancer drugs are inhibitors of purine or pyrimidine nucleotide biosynthesis.

26.2 | How Do Cells Synthesize Purines?

Substantial insight into the de novo pathway for purine biosynthesis was provided in 1948 by John Buchanan, who cleverly exploited the fact that birds excrete excess nitrogen principally in the form of uric acid, a water-insoluble purine analog. Buchanan fed isotopically labeled compounds to pigeons and then examined the distribution of the labeled atoms in *uric acid* (Figure 26.1). By tracing the metabolic source of the various atoms in this end product, he

Key Questions

26.1 Can Cells Synthesize Nucleotides?
26.2 How Do Cells Synthesize Purines?
26.3 Can Cells Salvage Purines?
26.4 How Are Purines Degraded?
26.5 How Do Cells Synthesize Pyrimidines?
26.6 How Are Pyrimidines Degraded?
26.7 How Do Cells Form the Deoxyribonucleotides That Are Necessary for DNA Synthesis?
26.8 How Are Thymine Nucleotides Synthesized?

© Adam Woolfitt/CORBIS

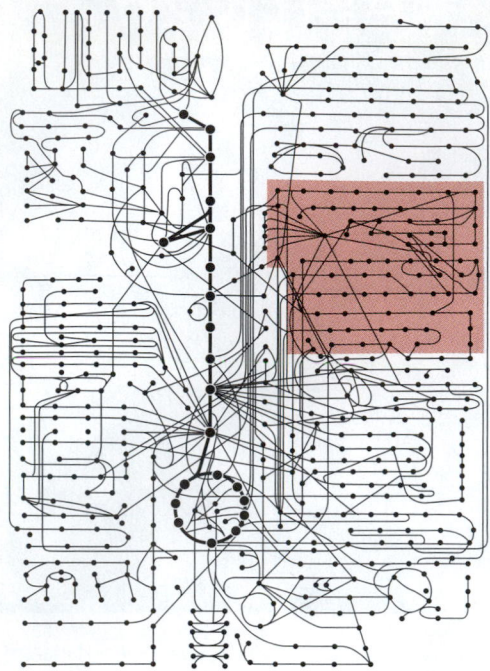

The synthesis and degradation of nucleotides

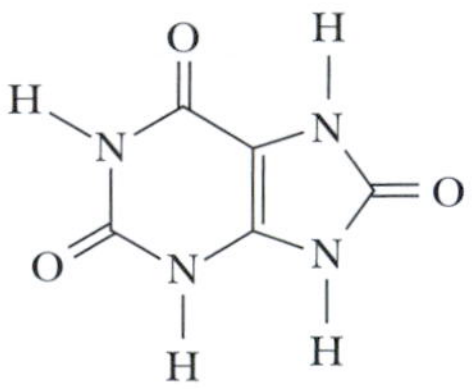

Uric acid

FIGURE 26.1 Nitrogen waste is excreted by birds principally as the purine analog, uric acid.

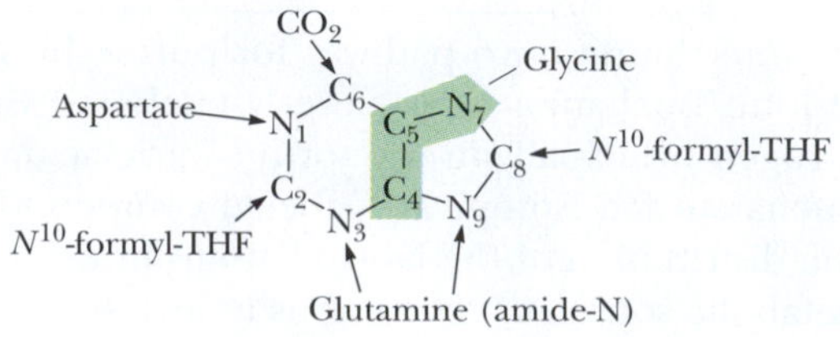

FIGURE 26.2 The metabolic origin of the nine atoms in the purine ring system.

showed that the nine atoms of the purine ring system (Figure 26.2) are contributed by aspartic acid (N-1), glutamine (N-3 and N-9), glycine (C-4, C-5, and N-7), CO_2 (C-6), and THF one-carbon derivatives (C-2 and C-8). The coenzyme THF and its role in one-carbon metabolism were introduced in Chapter 17.

Inosinic Acid (IMP) Is the Immediate Precursor to GMP and AMP

The de novo synthesis of purines occurs in an interesting manner: The atoms forming the purine ring are successively added to *ribose-5-phosphate*; thus, purines are directly synthesized as nucleotide derivatives by assembling the atoms that comprise the purine ring system directly on the ribose. In step 1, ribose-5-phosphate is activated via the direct transfer of a pyrophosphoryl group from ATP to C-1 of the ribose, yielding *5-phosphoribosyl-α-pyrophosphate (PRPP)* (Figure 26.3). The enzyme is **ribose-5-phosphate pyrophosphokinase.** PRPP is the limiting substance in purine biosynthesis. The two major purine nucleoside diphosphates, ADP and GDP, are negative effectors of ribose-5-phosphate pyrophosphokinase. However, because PRPP serves additional metabolic needs, the next reaction is actually the committed step in the pathway.

Step 2 (Figure 26.3) is catalyzed by **glutamine phosphoribosyl pyrophosphate amidotransferase.** The anomeric carbon atom of the substrate PRPP is in the α-configuration; the product is a β-glycoside (recall that all the biologically important nucleotides are β-glycosides). The N atom of this *N*-glycoside becomes N-9 of the nine-membered purine ring; it is the first atom added in the construction of this ring. Glutamine phosphoribosyl pyrophosphate amidotransferase is subject to feedback inhibition by GMP, GDP, and GTP as well as AMP, ADP, and ATP. The G series of nucleotides interacts at a guanine-specific allosteric site on the enzyme, whereas the adenine nucleotides act at an A-specific site. The pattern of inhibition by these nucleotides is competitive, thus ensuring that residual enzyme activity is expressed until sufficient amounts of both adenine and guanine nucleotides are synthesized. Glutamine phosphoribosyl pyrophosphate amidotransferase is also sensitive to inhibition by the glutamine analog **azaserine** (Figure 26.4). Azaserine has been used as an antitumor agent because it causes inactivation of glutamine-dependent enzymes in the purine biosynthetic pathway.

Step 3 is carried out by **glycinamide ribonucleotide synthetase** *(GAR synthetase)* via its ATP-dependent condensation of the glycine carboxyl group with the amine of *5-phosphoribosyl-β-amine* (Figure 26.3). The reaction proceeds in two stages. First, the glycine carboxyl group is activated via ATP-dependent phosphorylation. Next, an amide bond is formed between the activated carboxyl group of glycine and the β-amine. Glycine contributes C-4, C-5, and N-7 of the purine.

Step 4 is the first of two THF-dependent reactions in the purine pathway. **GAR transformylase** transfers the N^{10}-formyl group of N^{10}-formyl-THF to the free amino group of GAR to yield *α-N-formylglycinamide ribonucleotide (FGAR)*. Thus, C-8 of the purine is "formyl-ly" introduced. Although all of the atoms of the imidazole portion of the purine ring are now present, the ring is not closed until Reaction 6.

Step 5 is catalyzed by **FGAR amidotransferase** (also known as *FGAM synthetase*). ATP-dependent transfer of the glutamine amido group to the C-4-carbonyl of FGAR yields *formylglycinamidine ribonucleotide (FGAM)*. As a glutamine-dependent

A Deeper Look

Tetrahydrofolate (THF) and One-Carbon Units

An elaborate ensemble of enzymatic reactions serves to introduce one-carbon units into THF and to interconvert the various oxidation states (see accompanying figure and Table 17.6). N^5-methyltetrahydrofolate can be oxidized directly to N^5,N^{10}-methylenetetrahydrofolate, which can be further oxidized to N^5,N^{10}-methenyltetrahydrofolate. The N^5-formimino-, N^5-formyl-, and N^{10}-formyltetrahydrofolates can be formed from N^5,N^{10}-methenyltetrahydrofolate (all of these being at the same oxidation level), or they can be formed by one-carbon addition reactions from tetrahydrofolate itself. The principal

pathway for incorporation of one-carbon units into tetrahydrofolate is the **serine hydroxymethyltransferase** reaction (see Figure 25.32), which converts serine to glycine and forms N^5,N^{10}-methylenetetrahydrofolate.

The biosynthetic pathways for methionine, purines (Figure 26.3), and the pyrimidine thymine (see Figure 26.27) all rely on the incorporation of one-carbon units from tetrahydrofolate derivatives. N^{10}-formyl-THF is the source of carbons 2 and 8 of the purine ring and N^5,N^{10}-methylene-THF, the 5-CH_3 group of the pyrimidine thymine.

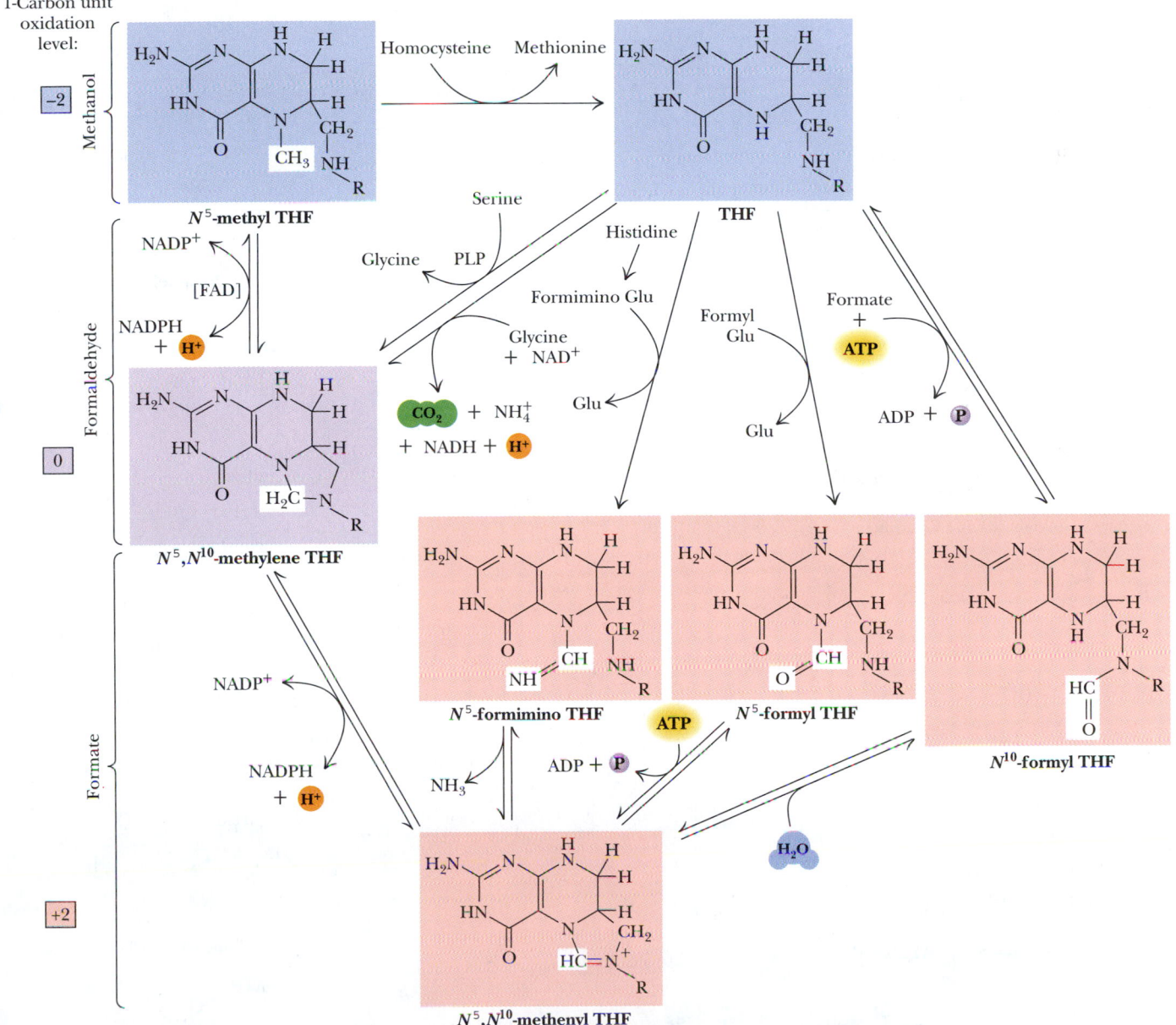

▲ The reactions that introduce one-carbon units into tetrahydrofolate (THF) link seven different folate intermediates that carry one-carbon units in three different oxidation states (−2, 0, and +2). (Adapted from Brody, T., et al., in Machlin, L. J., 1984. Handbook of Vitamins. New York: Marcel Dekker.)

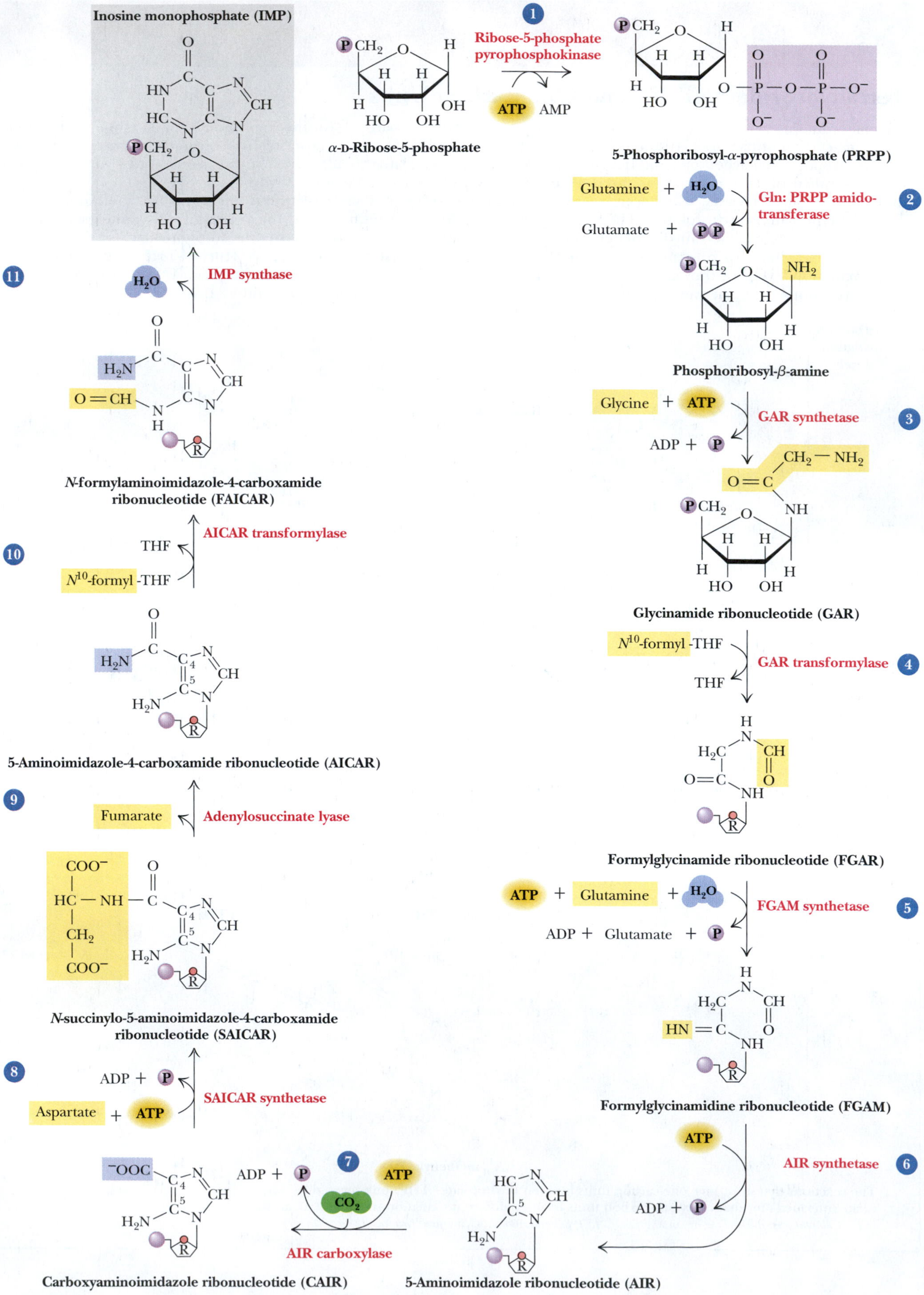

Inosine monophosphate (IMP)

1 Ribose-5-phosphate pyrophosphokinase

α-D-Ribose-5-phosphate

5-Phosphoribosyl-α-pyrophosphate (PRPP)

Glutamine + H_2O

2 Gln: PRPP amido-transferase

Glutamate + P P

Phosphoribosyl-β-amine

Glycine + **ATP**

3 GAR synthetase

ADP + P

11 IMP synthase

H_2O

N-formylaminoimidazole-4-carboxamide ribonucleotide (FAICAR)

10 AICAR transformylase

THF

N^{10}-formyl -THF

Glycinamide ribonucleotide (GAR)

N^{10}-formyl -THF

4 GAR transformylase

THF

5-Aminoimidazole-4-carboxamide ribonucleotide (AICAR)

9 Adenylosuccinate lyase

Fumarate

Formylglycinamide ribonucleotide (FGAR)

ATP + Glutamine + H_2O

5 FGAM synthetase

ADP + Glutamate + P

N-succinylo-5-aminoimidazole-4-carboxamide ribonucleotide (SAICAR)

8 SAICAR synthetase

ADP + P

Aspartate + **ATP**

Formylglycinamidine ribonucleotide (FGAM)

ATP

6 AIR synthetase

ADP + P

7

ADP + P

CO_2 **ATP**

AIR carboxylase

Carboxyaminoimidazole ribonucleotide (CAIR)

5-Aminoimidazole ribonucleotide (AIR)

◄ **Biochemistry ⑤ Now™ ACTIVE FIGURE 26.3** The de novo pathway for purine synthesis. The first purine product of this pathway, IMP (inosinic acid or inosine monophosphate), serves as a precursor to AMP and GMP. **Step 1:** PRPP synthesis from ribose-5-phosphate and ATP by ribose-5-phosphate pyrophosphokinase. **Step 2:** 5-Phosphoribosyl-β-1-amine synthesis from α-PRPP, glutamine, and H_2O by glutamine phosphoribosyl pyrophosphate amidotransferase. **Step 3:** Glycinamide ribonucleotide (GAR) synthesis from glycine, ATP, and 5-phosphoribosyl-β-amine by glycinamide ribonucleotide synthetase. **Step 4:** Formylglycinamide ribonucleotide synthesis from N^{10}-formyl-THF and GAR by GAR transformylase. **Step 5:** Formylglycinamidine ribonucleotide (FGAM) synthesis from FGAR, ATP, glutamine, and H_2O by FGAM synthetase (FGAR amidotransferase). The other products are ADP, P_i, and glutamate. **Step 6:** 5-Aminoimidazole ribonucleotide (AIR) synthesis is achieved via the ATP-dependent closure of the imidazole ring, as catalyzed by FGAM cyclase (AIR synthetase). (Note that the ring closure changes the numbering system.) **Step 7:** Carboxyaminoimidazole ribonucleotide (CAIR) synthesis from CO_2, ATP, and AIR by AIR carboxylase. **Step 8:** N-succinylo-5-aminoimidazole-4-carboxamide ribonucleotide (SAICAR) synthesis from aspartate, CAIR, and ATP by SAICAR synthetase. **Step 9:** 5-Aminoimidazole carboxamide ribonucleotide (AICAR) formation by the nonhydrolytic removal of fumarate from SAICAR. The enzyme is adenylosuccinase. **Step 10:** 5-Formylaminoimidazole carboxamide ribonucleotide (FAICAR) formation from AICAR and N^{10}-formyl-THF by AICAR transformylase. **Step 11:** Dehydration/ring closure yields the authentic purine ribonucleotide IMP. The enzyme is IMP synthase. **Test yourself on the concepts in this figure at** http://chemistry.brookscole.com/ggb3

enzyme, FGAR amidotransferase is, like glutamine phosphoribosyl pyrophosphate amidotransferase (Reaction 2), irreversibly inactivated by azaserine. The imino-N becomes N-3 of the purine.

Step 6 is an ATP-dependent dehydration that leads to formation of the imidazole ring. ATP is used to phosphorylate the oxygen atom of the formyl group, activating it for the ring closure step that follows. Because the product is *5-aminoimidazole ribonucleotide*, or *AIR*, this enzyme is called **AIR synthetase.** In avian liver, the enzymatic activities for steps 3, 4, and 6 (GAR synthetase, GAR transformylase, and AIR synthetase) reside on a single, 110-kD multifunctional polypeptide.

In step 7, carbon dioxide is added at the C-4 position of the imidazole ring by **AIR carboxylase** in an ATP-dependent reaction; the carbon of CO_2 will become C-6 of the purine ring. The product is *carboxyaminoimidazole ribonucleotide (CAIR)*.

In step 8, the amino-N of aspartate provides N-1 through linkage to the C-6 carboxyl function of CAIR. ATP hydrolysis drives the condensation of Asp with CAIR. The product is *N-succinylo-5-aminoimidazole-4-carboxamide ribonucleotide (SAICAR)*. **SAICAR synthetase** catalyzes the reaction. The enzymatic activities for steps 7 and 8 reside on a single, bifunctional polypeptide in avian liver.

Step 9 removes the four carbons of Asp as fumarate in a nonhydrolytic cleavage. The product is *5-aminoimidazole-4-carboxamide ribonucleotide (AICAR);* the enzyme is **adenylosuccinase** *(adenylosuccinate lyase)*. Adenylosuccinase acts again in that part of the purine pathway leading from IMP to AMP and takes its name from this latter reaction (see following). AICAR is also an intermediate in the histidine biosynthetic pathway (see Chapter 25), but because ATP is the precursor to AICAR in that pathway, no net purine synthesis is achieved.

Step 10 adds the formyl carbon of N^{10}-formyl-THF as the ninth and last atom necessary for forming the purine nucleus. The enzyme is called **AICAR transformylase;** the products are THF and *N-formylaminoimidazole-4-carboxamide ribonucleotide (FAICAR)*.

Step 11 involves dehydration and ring closure and completes the initial phase of purine biosynthesis. The enzyme is **IMP cyclohydrolase** (also known as *IMP synthase* and *inosinicase*). Unlike step 6, this ring closure does not require ATP. In avian liver, the enzymatic activities catalyzing steps 10 and 11 (AICAR transformylase and inosinicase) activities reside on 67-kD bifunctional polypeptides organized into 135-kD dimers.

Note that 6 ATPs are required in the purine biosynthetic pathway from ribose-5-phosphate to IMP: one each at steps 1, 3, 5, 6, 7, and 8. However, 7 high-energy phosphate bonds (equal to 7 ATP equivalents) are consumed because α-PRPP formation in Reaction 1 followed by PP_i release in Reaction 2 represents the loss of 2 ATP equivalents.

Azaserine

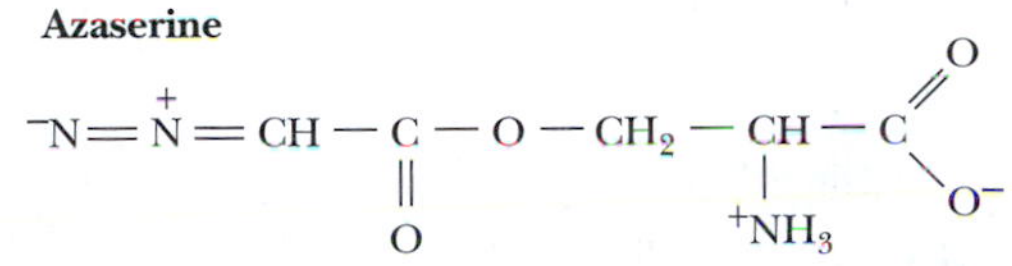

Glutamine

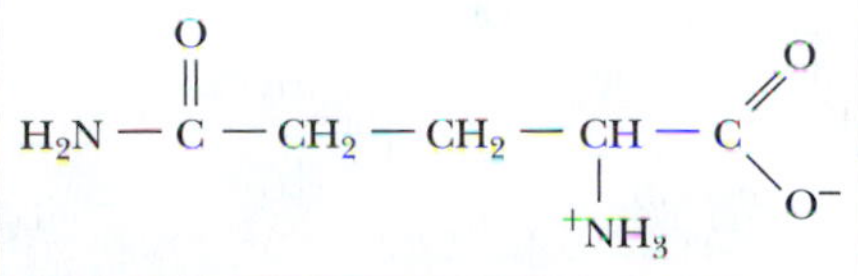

FIGURE 26.4 The structure of azaserine. Azaserine acts as an irreversible inhibitor of glutamine-dependent enzymes by covalently attaching to nucleophilic groups in the glutamine-binding site.

Human Biochemistry

Folate Analogs as Anticancer and Antimicrobial Agents

The dependence of de novo purine biosynthesis on folic acid compounds at steps 4 and 10 means that antagonists of folic acid metabolism (for example, methotrexate; see Figure 26.28) indirectly inhibit purine formation and, in turn, nucleic acid synthesis, cell growth, and cell division. Clearly, rapidly dividing cells such as malignancies or infective bacteria are more susceptible to these antagonists than slower-growing normal cells. Also among the folic acid antagonists are *sulfonamides* (see accompanying figure). Folic acid is a vitamin for animals and is obtained in the diet. In contrast, bacteria synthesize folic acid from precursors, including *p-aminobenzoic acid (PABA)*, and thus are more susceptible to sulfonamides than are animal cells.

Sulfonamides have the generic structure:

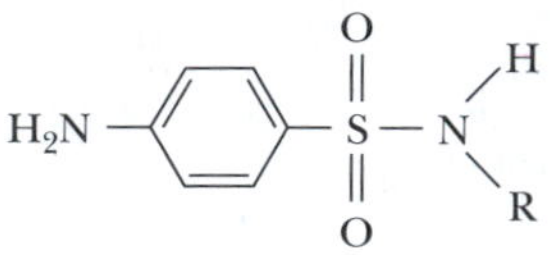

PABA (*p*-aminobenzoic acid)

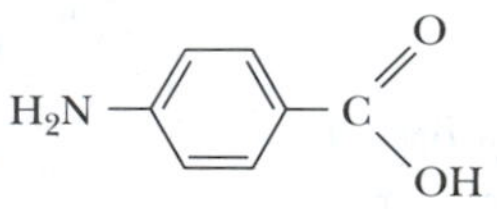

THF (tetrahydrofolate)

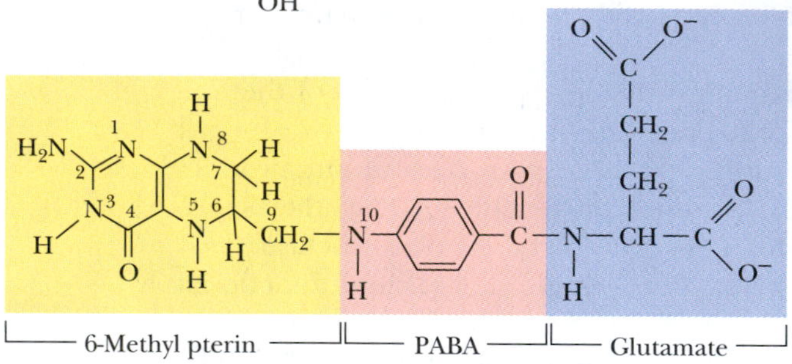

Additional γ-glutamyl residues (up to a maximum of seven) may add here

6-Methyl pterin ⎯⎯⎯⎯ PABA ⎯⎯⎯⎯ Glutamate

▲ Sulfa drugs, or sulfonamides, owe their antibiotic properties to their similarity to *p*-aminobenzoate (PABA), an important precursor in folic acid synthesis. Sulfonamides block folic acid formation by competing with PABA.

AMP and GMP Are Synthesized from IMP

IMP is the precursor to both AMP and GMP. These major purine nucleotides are formed via distinct two-step metabolic pathways that diverge from IMP. The branch leading to AMP (adenosine 5′-monophosphate) involves the displacement of the 6-O group of inosine with aspartate (Figure 26.5) in a GTP-dependent reaction, followed by the nonhydrolytic removal of the four-carbon skeleton of Asp as fumarate; the Asp amino group remains as the 6-amino group of AMP. **Adenylosuccinate synthetase** and **adenylosuccinase** are the two enzymes. Recall that adenylosuccinase also acted at step 9 in the pathway from ribose-5-phosphate to IMP. Fumarate production provides a connection between purine synthesis and the citric acid cycle.

The formation of GMP from IMP requires oxidation at C-2 of the purine ring, followed by a glutamine-dependent amidotransferase reaction that replaces the oxygen on C-2 with an amino group to yield *2-amino,6-oxy purine nucleoside monophosphate,* or as this compound is commonly known, *guanosine monophosphate.* The enzymes in the GMP branch are **IMP dehydrogenase** and **GMP synthetase.** Note that, starting from ribose-5-phosphate, 8 ATP equivalents are consumed in the synthesis of AMP and 9 in the synthesis of GMP.

The Purine Biosynthetic Pathway Is Regulated at Several Steps

The regulatory network that controls purine synthesis is schematically represented in Figure 26.6. To recapitulate, the purine biosynthetic pathway from ribose-5-phosphate to IMP is allosterically regulated at the first two steps. Ribose-

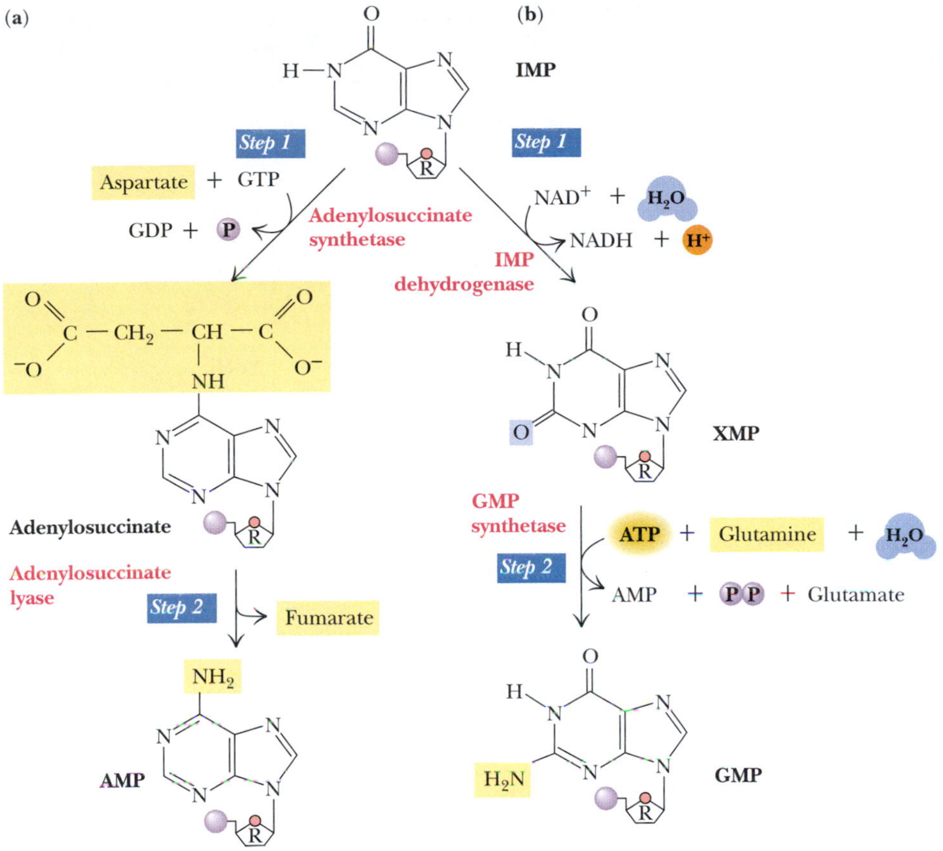

(a)

(b)

Biochemistry ⓔNow™ **ANIMATED FIGURE 26.5**
The synthesis of AMP and GMP from IMP. **(a)** AMP synthesis: The two reactions of AMP synthesis mimic steps 8 and 9 in the purine pathway leading to IMP. In **step 1,** the 6-*O* of inosine is displaced by aspartate to yield adenylosuccinate. The energy required to drive this reaction is derived from GTP hydrolysis. The enzyme is adenylosuccinate synthetase. AMP is a competitive inhibitor (with respect to the substrate IMP) of adenylosuccinate synthetase. In **step 2,** adenylosuccinase (also known as adenylosuccinate lyase, the same enzyme catalyzing step 9 in the purine pathway) carries out the nonhydrolytic removal of fumarate from adenylosuccinate, leaving AMP. **(b)** GMP synthesis: The two reactions of GMP synthesis are an NAD⁺-dependent oxidation followed by an amidotransferase reaction. In **step 1,** IMP dehydrogenase employs the substrates NAD⁺ and H_2O in catalyzing oxidation of IMP at C-2. The products are xanthylic acid (XMP or xanthosine monophosphate), NADH, and H⁺. GMP is a competitive inhibitor (with respect to IMP) of IMP dehydrogenase. In **step 2,** transfer of the amido-N of glutamine to the C-2 position of XMP yields GMP. This ATP-dependent reaction is catalyzed by GMP synthetase. Besides GMP, the products are glutamate, AMP, and PP_i. Hydrolysis of PP_i to two P_i by ubiquitous pyrophosphatases pulls this reaction to completion. **See this figure animated at http://chemistry.brookscole.com/ggb3**

5-phosphate pyrophosphokinase, although not the committed step in purine synthesis, is subject to feedback inhibition by ADP and GDP. The enzyme catalyzing the next step, glutamine phosphoribosyl pyrophosphate amidotransferase, has two allosteric sites, one where the "A" series of nucleoside phosphates (AMP, ADP, and ATP) binds and feedback-inhibits, and another where the corresponding "G" series binds and inhibits. Furthermore, PRPP is a "feed-forward" activator of this enzyme. Thus, the rate of IMP formation by this pathway is governed by the levels of the final end products, the adenine and guanine nucleotides.

The purine pathway splits at IMP. The first enzyme in the AMP branch, adenylosuccinate synthetase, is competitively inhibited by AMP. Its counterpart in the GMP branch, IMP dehydrogenase, is inhibited in a similar fashion by GMP. Thus, the fate of IMP is determined by the relative levels of AMP and GMP, so any deficiency in the amount of either of the principal purine nucleotides is self-correcting. This reciprocity of regulation is an effective mechanism for balancing the formation of AMP and GMP to satisfy cellular needs. Note also that reciprocity is even manifested at the level of energy input: GTP provides the energy to drive AMP synthesis, whereas ATP serves this role in GMP synthesis (Figure 26.6).

ATP-Dependent Kinases Form Nucleoside Diphosphates and Triphosphates from the Nucleoside Monophosphates

The products of de novo purine biosynthesis are the nucleoside monophosphates AMP and GMP. These nucleotides are converted by successive phosphorylation reactions into their metabolically prominent triphosphate forms, ATP and GTP. The first phosphorylation, to give the nucleoside diphosphate forms, is carried out by two base-specific, ATP-dependent kinases, **adenylate kinase** and **guanylate kinase.**

Adenylate kinase: AMP + ATP $\longrightarrow$ 2 ADP
Guanylate kinase: GMP + ATP $\longrightarrow$ GDP + ADP

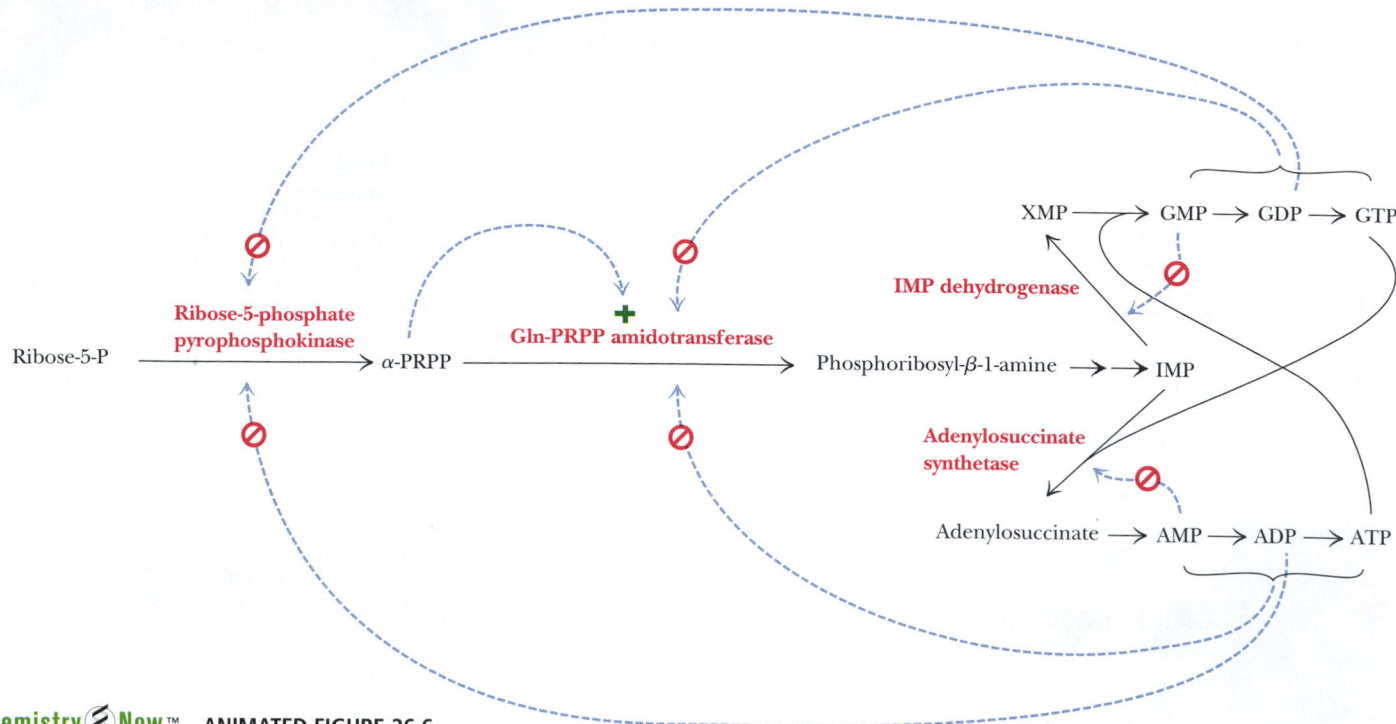

ANIMATED FIGURE 26.6
The regulatory circuit controlling purine biosynthesis. ADP and GDP are feedback inhibitors of ribose-
5-phosphate pyrophosphokinase, the first reaction
in the pathway. The second enzyme, glutamine
phosphoribosyl pyrophosphate amidotransferase,
has two distinct feedback inhibition sites, one for
A nucleotides and one for G nucleotides. Also, this
enzyme is allosterically activated by PRPP. In the
branch leading from IMP to AMP, the first enzyme
is feedback-inhibited by AMP, while the corresponding enzyme in the branch from IMP to GMP is
feedback-inhibited by GMP. Last, ATP is the energy
source for GMP synthesis, whereas GTP is the
energy source for AMP synthesis. **See this figure animated at http://chemistry.brookscole.com/ggb3**

These nucleoside monophosphate kinases also act on deoxynucleoside monophosphates to give dADP or dGDP.

Oxidative phosphorylation (see Chapter 20) is primarily responsible for the
conversion of ADP into ATP. ATP then serves as the phosphoryl donor for synthesis of the other nucleoside triphosphates from their corresponding NDPs in
a reaction catalyzed by **nucleoside diphosphate kinase,** a nonspecific enzyme.
For example,

$$GDP + ATP \rightleftharpoons GTP + ADP$$

Because this enzymatic reaction is readily reversible and nonspecific with respect to both phosphoryl acceptor and donor, in effect any NDP can be phosphorylated by any NTP, and vice versa. The preponderance of ATP over all
other nucleoside triphosphates means that, in quantitative terms, it is the principal nucleoside diphosphate kinase substrate. The enzyme does not discriminate between the ribose moieties of nucleotides and thus functions in phosphoryl transfers involving deoxy-NDPs and deoxy-NTPs as well.

26.3 | Can Cells Salvage Purines?

Nucleic acid turnover (synthesis and degradation) is an ongoing metabolic
process in most cells. Messenger RNA in particular is actively synthesized and
degraded. These degradative processes can lead to the release of free purines
in the form of adenine, guanine, and hypoxanthine (the base in IMP). These
substances represent a metabolic investment by cells. So-called salvage pathways exist to recover them in useful form. Salvage reactions involve resynthesis
of nucleotides from bases via **phosphoribosyltransferases.**

$$Base + PRPP \rightleftharpoons nucleoside-5'-phosphate + PP_i$$

The subsequent hydrolysis of PP_i to inorganic phosphate by pyrophosphatases
renders the phosphoribosyltransferase reaction effectively irreversible.

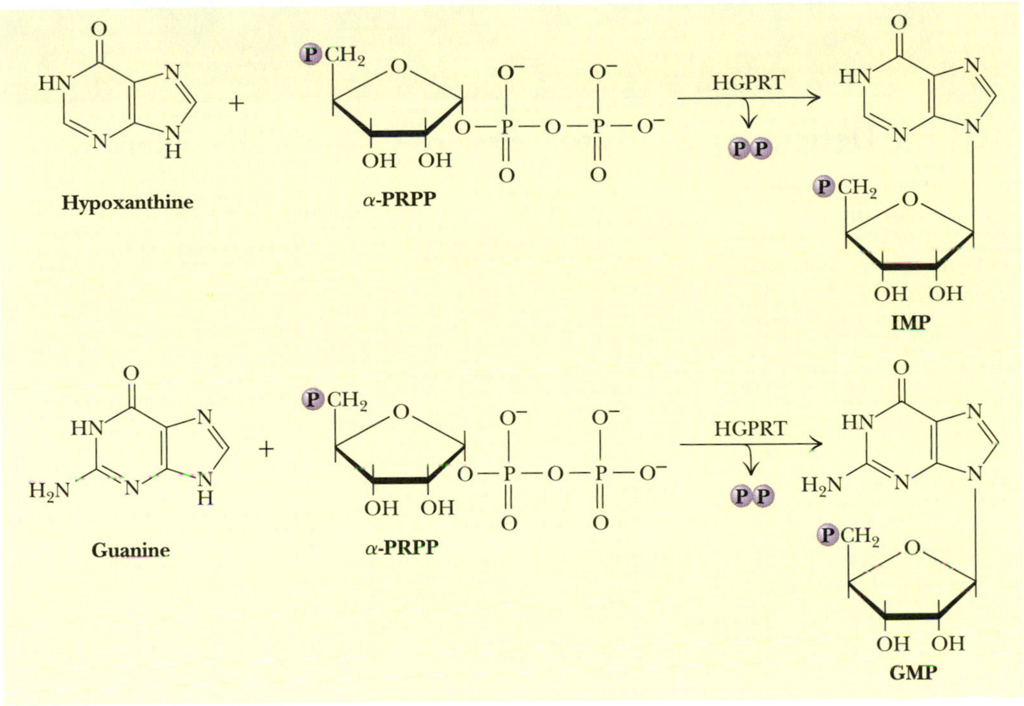

FIGURE 26.7 Purine salvage by the HGPRT reaction.

The purine phosphoribosyltransferases are **adenine phosphoribosyltransferase (APRT),** which mediates AMP formation, and **hypoxanthine-guanine phosphoribosyltransferase (HGPRT),** which can act on either hypoxanthine to form IMP or guanine to form GMP (Figure 26.7).

26.4 | How Are Purines Degraded?

Because nucleic acids are ubiquitous in cellular material, significant amounts are ingested in the diet. Nucleic acids are degraded in the digestive tract to nucleotides by various nucleases and phosphodiesterases. Nucleotides are then converted to nucleosides by base-specific nucleotidases and nonspecific phosphatases.

$$NMP + H_2O \longrightarrow nucleoside + P_i$$

Nucleosides are hydrolyzed by nucleosidases or nucleoside phosphorylases to release the purine base:

$$Nucleoside + H_2O \xrightarrow{\textit{nucleosidase}} base + ribose$$

$$Nucleoside + P_i \xrightarrow{\textit{nucleoside phosphorylase}} base + ribose\text{-}1\text{-}P$$

The pentoses liberated in these reactions provide the only source of metabolic energy available from purine nucleotide degradation.

Feeding experiments using radioactively labeled nucleic acids as metabolic tracers have demonstrated that little of the nucleotide ingested in the diet is incorporated into cellular nucleic acids. Dietary purines are converted to uric acid (see following discussion) in the gut and excreted, and pyrimidine nucleosides are inefficiently absorbed into the bloodstream. These findings confirm the de novo pathways of nucleotide biosynthesis as the primary source of nucleic acid precursors. Ingested bases are, for the most part, excreted. Nevertheless, cellular nucleic acids do undergo degradation in the course of the continuous recycling of cellular constituents.

Human Biochemistry

Lesch-Nyhan Syndrome: HGPRT Deficiency Leads to a Severe Clinical Disorder

The symptoms of **Lesch-Nyhan syndrome** are tragic: a crippling gouty arthritis due to excessive uric acid accumulation (uric acid is a purine degradation product, discussed in the next section) and, worse, severe malfunctions in the nervous system that lead to mental retardation, spasticity, aggressive behavior, and self-mutilation. Lesch-Nyhan syndrome results from a complete deficiency in HGPRT activity. The structural gene for HGPRT is located on the X chromosome, and the disease is a congenital, recessive, sex-linked trait manifested only in males. The severe consequences of HGPRT deficiency argue that purine salvage has greater metabolic importance than simply the energy-saving recovery of bases. Although HGPRT might seem to play a minor role in purine metabolism, its absence has profound consequences: De novo purine biosynthesis is dramatically increased, and uric acid levels in the blood are elevated. Presumably, these changes ensue because lack of consumption of PRPP by HGPRT elevates its availability for glutamine-PRPP amidotransferase, enhancing overall de novo purine synthesis and, ultimately, uric acid production (see accompanying figure). Despite these explanations, it remains unclear why deficiency in this single enzyme leads to the particular neurological aberrations characteristic of the syndrome. Fortunately, deficiencies in HGPRT activity in fetal cells can be detected following amniocentesis.

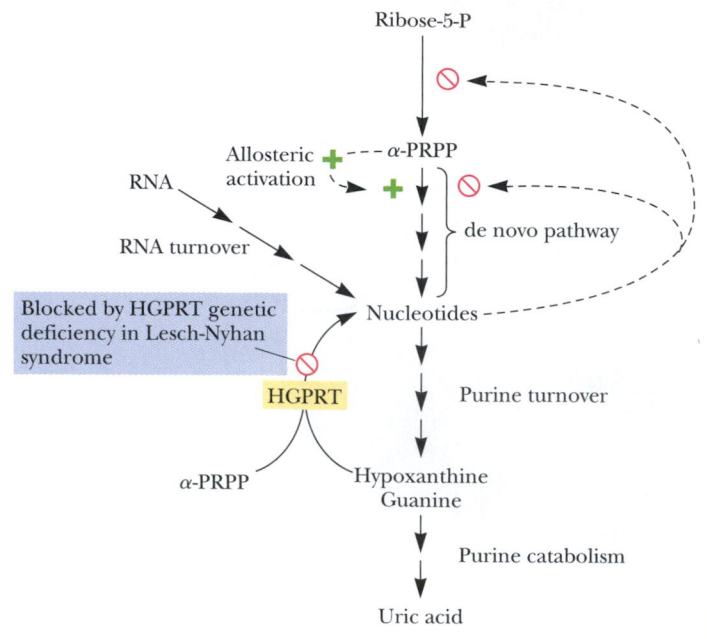

Human Biochemistry

Severe Combined Immunodeficiency Syndrome—A Lack of Adenosine Deaminase Is One Cause of This Inherited Disease

Severe combined immunodeficiency syndrome, or **SCID,** is a group of related inherited disorders characterized by the lack of an immune response to infectious disease. This immunological insufficiency is attributable to the inability of B and T lymphocytes to proliferate and produce antibodies in response to an antigenic challenge. About 30% of SCID patients suffer from a deficiency in the enzyme *adenosine deaminase (ADA)*. ADA deficiency is also implicated in a variety of other diseases, including AIDS, anemia, and various lymphomas and leukemias. **Gene therapy,** *the repair of a genetic deficiency through introduction of a functional recombinant version of the gene,* has been attempted on individuals with

SCID due to a defective ADA gene (see Section 12.4). ADA is a Zn^{2+}-dependent enzyme, and Zn^{2+} deficiency can also lead to reduced immune function.

In the absence of ADA, deoxyadenosine is not degraded but instead is converted into dAMP and then into dATP. dATP is a potent feedback inhibitor of deoxynucleotide biosynthesis (discussed later in this chapter). Without deoxyribonucleotides, DNA cannot be replicated and cells cannot divide (see accompanying figure). Rapidly proliferating cell types such as lymphocytes are particularly susceptible if DNA synthesis is impaired.

▶ The effect of elevated levels of deoxyadenosine on purine metabolism. If ADA is deficient or absent, deoxyadenosine is not converted into deoxyinosine as normal (see Figure 26.8). Instead, it is salvaged by a nucleoside kinase, which converts it to dAMP, leading to accumulation of dATP and inhibition of deoxynucleotide synthesis (see Figure 26.24). Thus, DNA replication is stalled.

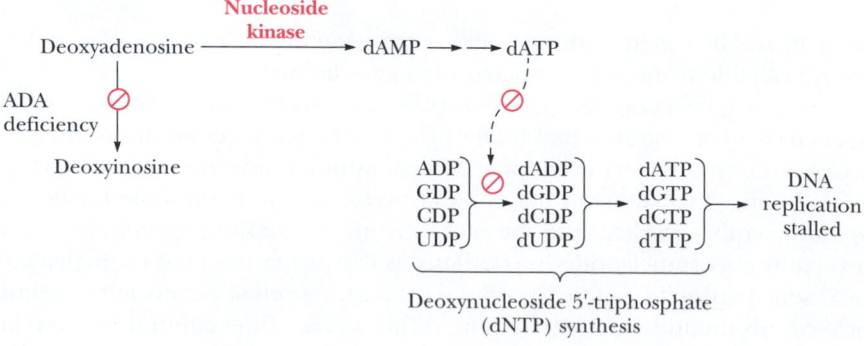

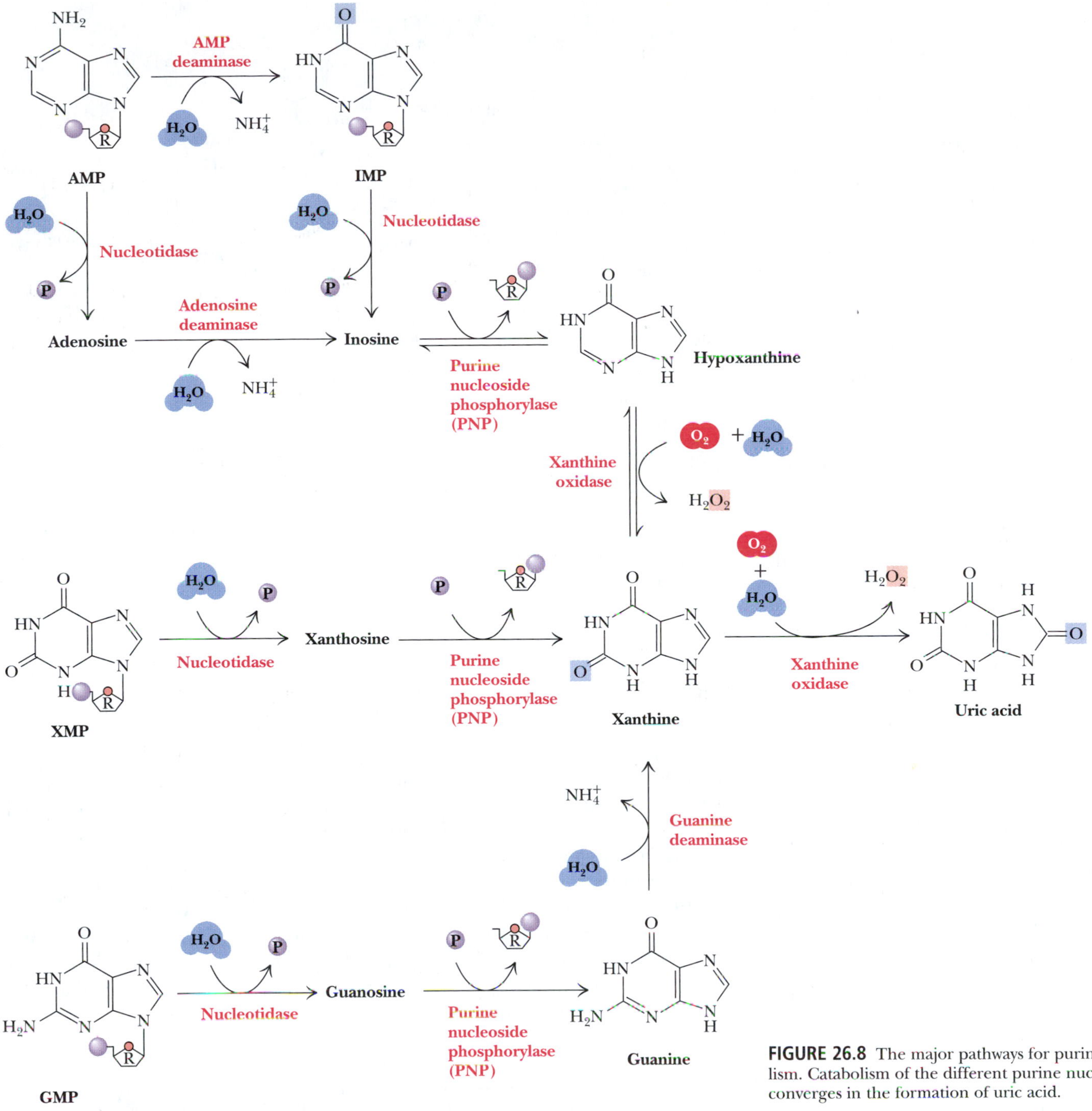

FIGURE 26.8 The major pathways for purine catabolism. Catabolism of the different purine nucleotides converges in the formation of uric acid.

The Major Pathways of Purine Catabolism Lead to Uric Acid

The major pathways of purine catabolism in animals are outlined in Figure 26.8. The various nucleotides are first converted to nucleosides by **intracellular nucleotidases.** These nucleotidases are under strict metabolic regulation to ensure that their substrates, which act as intermediates in many vital processes, are not depleted below critical levels. Nucleosides are then degraded by the enzyme **purine nucleoside phosphorylase (PNP)** to release the purine base and ribose-l-P. Note that neither adenosine nor deoxyadenosine is a substrate for PNP. Instead, these nucleosides are first converted to inosine by **adenosine deaminase.** The PNP products are merged into *xanthine* by **guanine deaminase** and **xanthine oxidase,** and xanthine is then oxidized to uric acid by this latter enzyme.

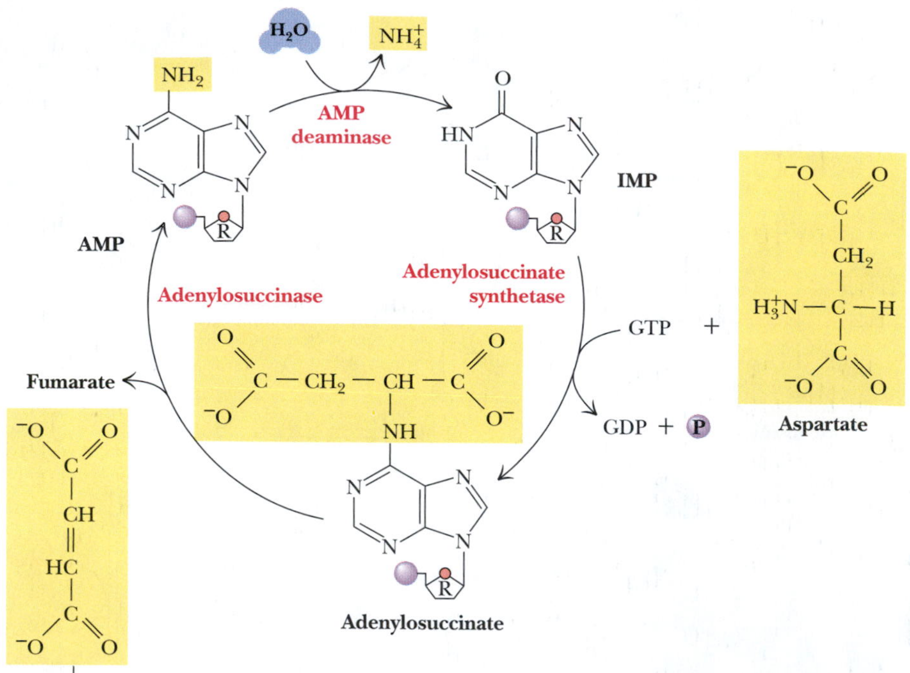

FIGURE 26.9 The purine nucleoside cycle for anaplerotic replenishment of citric acid cycle intermediates in skeletal muscle.

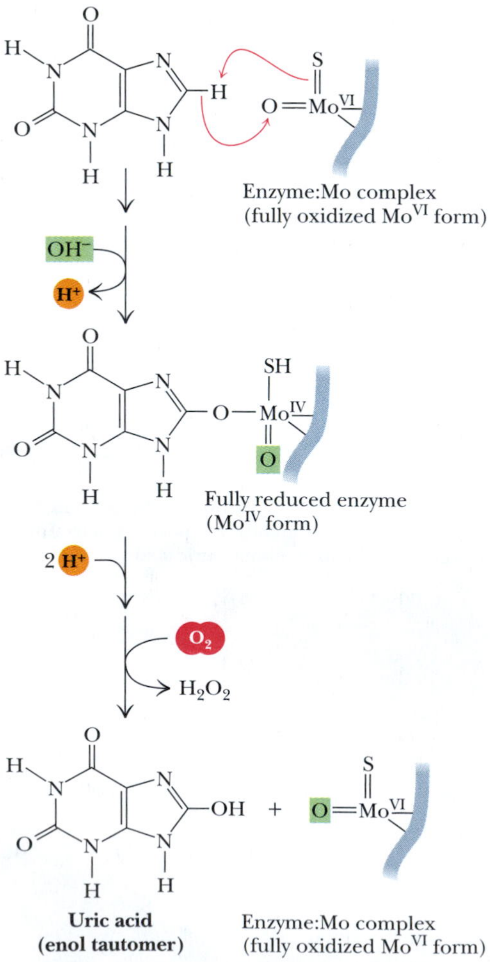

FIGURE 26.10 Xanthine oxidase catalyzes a hydroxylase-type reaction.

The Purine Nucleoside Cycle in Skeletal Muscle Serves as an Anaplerotic Pathway

Deamination of AMP to IMP by **AMP deaminase** (Figure 26.8) followed by resynthesis of AMP from IMP by the de novo purine pathway enzymes, *adenylosuccinate synthetase* and *adenylosuccinate lyase,* constitutes a purine nucleoside cycle (Figure 26.9). This cycle has the net effect of converting aspartate to fumarate plus NH_4^+. Although this cycle might seem like senseless energy consumption, it plays an important role in energy metabolism in skeletal muscle: The fumarate that it generates replenishes the levels of citric acid cycle intermediates lost in amphibolic side reactions (see Chapter 19). Skeletal muscle lacks the usual complement of anaplerotic enzymes and relies on enhanced levels of AMP deaminase, adenylosuccinate synthetase, and adenylosuccinate lyase to compensate.

Xanthine Oxidase

Xanthine oxidase (Figure 26.8) is present in large amounts in liver, intestinal mucosa, and milk. It oxidizes hypoxanthine to xanthine and xanthine to uric acid. Xanthine oxidase is a rather indiscriminate enzyme, using molecular oxygen to oxidize a wide variety of purines, pteridines, and aldehydes, producing H_2O_2 as a product. Xanthine oxidase possesses FAD, nonheme Fe-S centers, and *molybdenum cofactor* (a molybdenum-containing pterin complex) as electron-transferring prosthetic groups. Its mechanism of action is diagrammed in Figure 26.10. In humans and other primates, uric acid is the end product of purine catabolism and is excreted in the urine. Birds, terrestrial reptiles, and many insects also excrete uric acid, but in these organisms, uric acid represents the major nitrogen excretory compound, because, unlike mammals, they do not also produce urea (see Chapter 25). Instead, the catabolism of all nitrogenous compounds, including amino acids, is channeled into uric acid. This route of nitrogen catabolism allows these animals to conserve water by excreting crystals of uric acid in pastelike solid form.

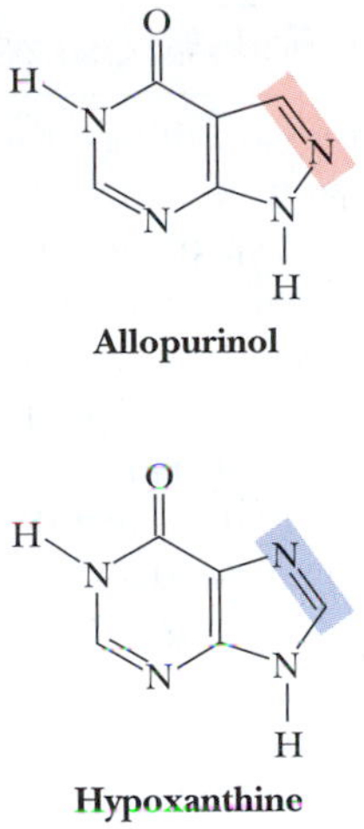

Allopurinol

Hypoxanthine

FIGURE 26.11 Allopurinol, an analog of hypoxanthine, is a potent inhibitor of xanthine oxidase.

Gout Is a Disease Caused by an Excess of Uric Acid

Gout is the clinical term describing the physiological consequences accompanying excessive uric acid accumulation in body fluids. Uric acid and urate salts are rather insoluble in water and tend to precipitate from solution if produced in excess. The most common symptom of gout is arthritic pain in the joints as a result of urate deposition in cartilaginous tissue. The joint of the big toe is particularly susceptible. Urate crystals may also appear as kidney stones and lead to painful obstruction of the urinary tract. **Hyperuricemia,** chronic elevation of blood uric acid levels, occurs in about 3% of the population as a consequence of impaired excretion of uric acid or overproduction of purines. Purine-rich foods such as caviar (fish eggs rich in nucleic acids) may exacerbate the condition. The biochemical causes of gout are varied. However, a common treatment is *allopurinol* (Figure 26.11). This hypoxanthine analog binds tightly to xanthine oxidase, thereby inhibiting its activity and preventing uric acid formation. Hypoxanthine and xanthine do not accumulate to harmful concentrations because they are more soluble and thus more easily excreted.

Animals Other Than Humans Oxidize Uric Acid to Form Excretory Products

The subsequent metabolism of uric acid in organisms that don't excrete it is shown in Figure 26.12. In molluscs and in mammals other than primates, uric acid is oxidized by **urate oxidase** to *allantoin* and excreted. In bony fishes (teleosts), uric acid degradation proceeds through yet another step wherein allantoin is hydrolyzed to *allantoic acid* by **allantoinase** before excretion. Cartilaginous fish (sharks and rays) and amphibians further degrade allantoic acid via the enzyme **allantoicase** to liberate glyoxylic acid and 2 equivalents of *urea.* Even simpler animals, such as most marine invertebrates (crustacea and so forth), use **urease** to hydrolyze urea to CO_2 and ammonia. In contrast to animals that must rid themselves of potentially harmful nitrogen waste products, microorganisms often are limited in growth by nitrogen availability. Many possess an identical pathway of uric acid degradation, using it instead to liberate NH_3 from uric acid so that it can be assimilated into organic-N compounds essential to their survival.

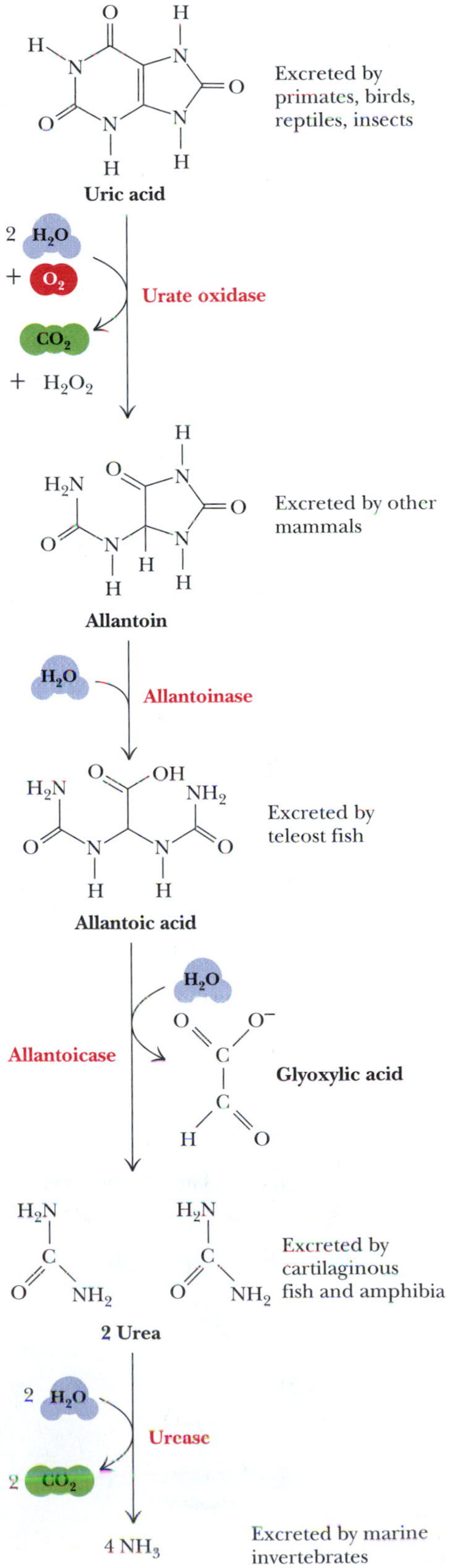

▶ **FIGURE 26.12** The catabolism of uric acid to allantoin, allantoic acid, urea, or ammonia in various animals.

Uric acid — Excreted by primates, birds, reptiles, insects

$2\ H_2O$
$+\ O_2$ **Urate oxidase**
CO_2
$+\ H_2O_2$

Allantoin — Excreted by other mammals

H_2O **Allantoinase**

Allantoic acid — Excreted by teleost fish

H_2O
Allantoicase — **Glyoxylic acid**

2 Urea — Excreted by cartilaginous fish and amphibia

$2\ H_2O$ **Urease**
$2\ CO_2$

$4\ NH_3$ — Excreted by marine invertebrates

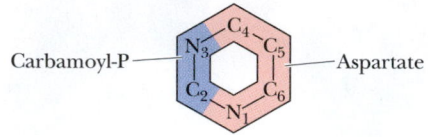

FIGURE 26.13 The metabolic origin of the six atoms of the pyrimidine ring.

26.5 | How Do Cells Synthesize Pyrimidines?

In contrast to purines, pyrimidines are not synthesized as nucleotide derivatives. Instead, the pyrimidine ring system is completed before a ribose-5-P moiety is attached. Also, only two precursors, carbamoyl-P and aspartate, contribute atoms to the six-membered pyrimidine ring (Figure 26.13), compared to seven precursors for the nine purine atoms.

Mammals have two enzymes for carbamoyl phosphate synthesis. Carbamoyl phosphate for pyrimidine biosynthesis is formed by **carbamoyl phosphate synthetase II (CPS-II),** a cytosolic enzyme. Recall that carbamoyl phosphate synthetase I is a mitochondrial enzyme dedicated to the urea cycle and arginine biosynthesis (see Figures 25.22 and 25.23). The substrates of carbamoyl phosphate synthetase II are HCO_3^-, H_2O, glutamine, and 2 ATPs (Figure 26.14). Because carbamoyl phosphate made by CPS-II in mammals has no fate other than incorporation into pyrimidines, mammalian CPS-II can be viewed as the committed step in the pyrimidine de novo pathway. Bacteria have but one CPS, and its carbamoyl phosphate product is incorporated into arginine as well as pyrimidines. Thus, the committed step in bacterial pyrimidine synthesis is the next reaction, which is mediated by **aspartate transcarbamoylase (ATCase).**

ATCase catalyzes the condensation of carbamoyl phosphate with aspartate to form carbamoyl-aspartate (Figure 26.15). No ATP input is required at this step because carbamoyl phosphate represents an "activated" carbamoyl group.

Step 3 of pyrimidine synthesis involves ring closure and dehydration via linkage of the —NH_2 group introduced by carbamoyl-P with the former β-COO^- of aspartate; this reaction is mediated by the enzyme **dihydroorotase.** The product of the reaction is *dihydroorotate,* a six-membered ring compound. Dihydroorotate

Biochemistry Now™ ANIMATED FIGURE 26.14
The reaction catalyzed by carbamoyl phosphate synthetase II (CPS-II). Note that, in contrast to carbamoyl phosphate synthetase I, CPS-II uses the amide of glutamine, not NH_4^+, to form carbamoyl-P. **Step 1:** The first ATP consumed in carbamoyl phosphate synthesis is used in forming carboxy-phosphate, an activated form of CO_2. **Step 2:** Carboxy-phosphate (also called carbonylphosphate) then reacts with the glutamine amide to yield carbamate and glutamate. **Step 3:** Carbamate is phosphorylated by the second ATP to give ADP and carbamoyl phosphate. **See this figure animated at** http://chemistry.brookscole.com/ggb3

$$HCO_3^- + \text{Glutamine} + 2\ \text{ATP} + H_2O \xrightarrow{\text{Carbamoyl phosphate synthetase II (CPS-II)}} \text{Carbamoyl-P} + \text{Glutamate} + 2\ \text{ADP} + P$$

VIA:

Bicarbonate $+$ ATP $\xrightarrow{\text{Step 1}}$ Carboxy-phosphate $+$ ADP $+$ H^+

Carboxy-phosphate $+$ Gln $\xrightarrow{\text{Step 2}}$ P $+$ Carbamate $+$ Glu $+$ H^+

Carbamate $+$ ATP $\xrightarrow{\text{Step 3}}$ Carbamoyl-P (CP) $+$ ADP

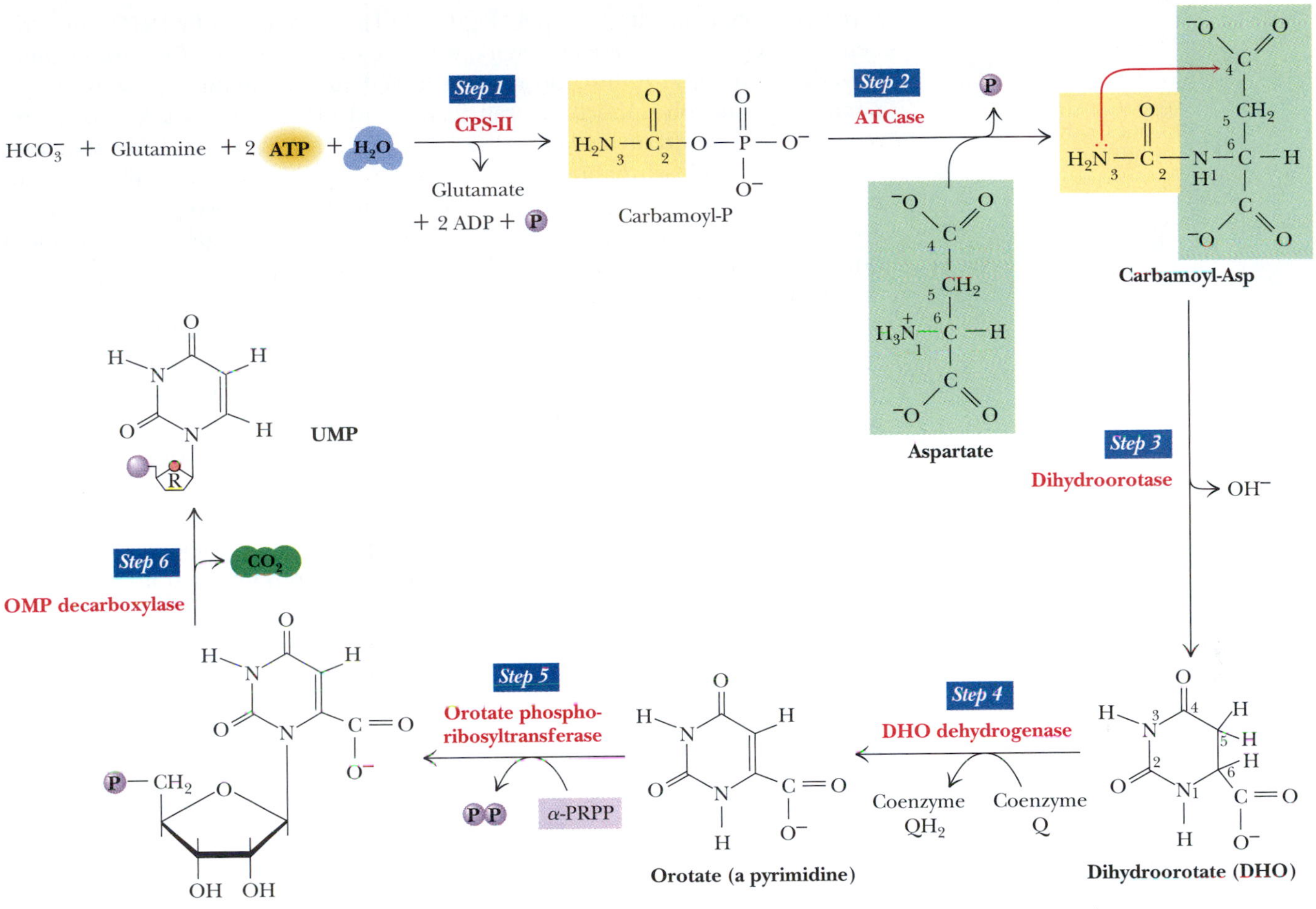

Biochemistry⊗Now™ ACTIVE FIGURE 26.15
The de novo pyrimidine biosynthetic pathway.
Step 1: Carbamoyl-P synthesis. **Step 2:** Condensation of carbamoyl phosphate and aspartate to yield carbamoyl-aspartate is catalyzed by aspartate transcarbamoylase (ATCase). **Step 3:** An intramolecular condensation catalyzed by dihydroorotase gives the six-membered heterocyclic ring characteristic of pyrimidines. The product is dihydroorotate (DHO). **Step 4:** The oxidation of DHO by dihydroorotate dehydrogenase gives orotate. (In bacteria, NAD$^+$ is the electron acceptor from DHO.) **Step 5:** PRPP provides the ribose-5-P moiety that transforms orotate into orotidine-5′-monophosphate, a pyrimidine nucleotide. Note that orotate phosphoribosyltransferase joins N-1 of the pyrimidine to the ribosyl group in appropriate β-configuration. PP$_i$ hydrolysis renders this reaction thermodynamically favorable. **Step 6:** Decarboxylation of OMP by OMP decarboxylase yields UMP. **Test yourself on the concepts in this figure at http://chemistry.brookscole.com/ggb3**

is not a true pyrimidine, but its oxidation yields *orotate,* which is. This oxidation (step 4) is catalyzed by **dihydroorotate dehydrogenase.** Bacterial dihydroorotate dehydrogenases are NAD$^+$-linked flavoproteins, which are somewhat unusual in possessing both FAD and FMN; these enzymes also have nonheme Fe-S centers as additional redox prosthetic groups. The eukaryotic version of dihydroorotate dehydrogenase is a protein component of the inner mitochondrial membrane; its immediate e^- acceptor is a quinone, and oxidation of the reduced quinone by the mitochondrial e^- transport chain can drive ATP synthesis via oxidative phosphorylation. At this stage, ribose-5-phosphate is joined to N-1 of orotate, giving the pyrimidine nucleotide *orotidine-5′-monophosphate,* or *OMP* (step 5, Figure 26.15). The ribose phosphate donor is PRPP; the enzyme is **orotate phosphoribosyltransferase.** The next reaction is catalyzed by **OMP decarboxylase.** Decarboxylation of OMP gives *UMP* (*uridine-5′-monophosphate,* or *uridylic acid*), one of the two common pyrimidine ribonucleotides.

Pyrimidine Biosynthesis in Mammals Is Another Example of "Metabolic Channeling"

In bacteria, the six enzymes of de novo pyrimidine biosynthesis exist as distinct proteins, each independently catalyzing its specific step in the overall pathway. In contrast, in mammals, the six enzymatic activities are distributed among only three proteins, two of which are **multifunctional polypeptides:** single polypeptide chains having two or more enzymic centers. The first three steps of pyrimidine synthesis, CPS-II, aspartate transcarbamoylase, and dihydroorotase, are all localized on a single 210-kD cytosolic polypeptide. This multifunctional enzyme is the product of a solitary gene, yet it is equipped with the active sites for all three enzymatic activities. Step 4 (Figure 26.15) is catalyzed by DHO dehydrogenase, a separate enzyme associated with the outer surface of the inner mitochondrial membrane, but the enzymatic activities mediating steps 5 and 6, namely, orotate phosphoribosyltransferase and OMP decarboxylase in mammals, are also found on a single cytosolic polypeptide known as **UMP synthase.**

The purine biosynthetic pathway of avian liver also provides examples of metabolic channeling. Recall that steps 3, 4, and 6 of de novo purine synthesis are catalyzed by three enzymatic activities localized on a single multifunctional polypeptide, and steps 7 and 8 and steps 10 and 11 by respective bifunctional polypeptides (see Figure 26.3).

Such multifunctional enzymes confer an advantage: The product of one reaction in a pathway is the substrate for the next. In multifunctional enzymes, such products remain bound and are channeled directly to the next active site, rather than dissociated into the surrounding medium for diffusion to the next enzyme. This **metabolic channeling** is more efficient for a variety of reasons: Transit time for movement from one active site to the next is shortened, substrates are not diluted into the solvent phase, chemically reactive intermediates are protected from decomposition in the aqueous milieu, no pools of intermediates accumulate, and intermediates are shielded from interactions with other enzymes that might metabolize them.

UMP Synthesis Leads to Formation of the Two Most Prominent Ribonucleotides—UTP and CTP

The two prominent pyrimidine ribonucleotide products are derived from UMP via the same unbranched pathway. First, UDP is formed from UMP via an ATP-dependent *nucleoside monophosphate kinase.*

$$UMP + ATP \rightleftharpoons UDP + ADP$$

Then, UTP is formed by *nucleoside diphosphate kinase.*

$$UDP + ATP \rightleftharpoons UTP + ADP$$

Amination of UTP at the 6-position gives CTP. The enzyme **CTP synthetase** is a glutamine amidotransferase (Figure 26.16). ATP hydrolysis provides the energy to drive the reaction.

Pyrimidine Biosynthesis Is Regulated at ATCase in Bacteria and at CPS-II In Animals

Pyrimidine biosynthesis in bacteria is allosterically regulated at aspartate transcarbamoylase (ATCase). *Escherichia coli* ATCase is feedback-inhibited by the end product, CTP. ATP, which can be viewed as a signal of both energy availability and purine sufficiency, is an allosteric activator of ATCase. CTP and ATP compete for a common allosteric site on the enzyme. In many bacteria, UTP, not CTP, acts as the ATCase feedback inhibitor.

In animals, CPS-II catalyzes the committed step in pyrimidine synthesis and serves as the focal point for allosteric regulation. UDP and UTP are feedback

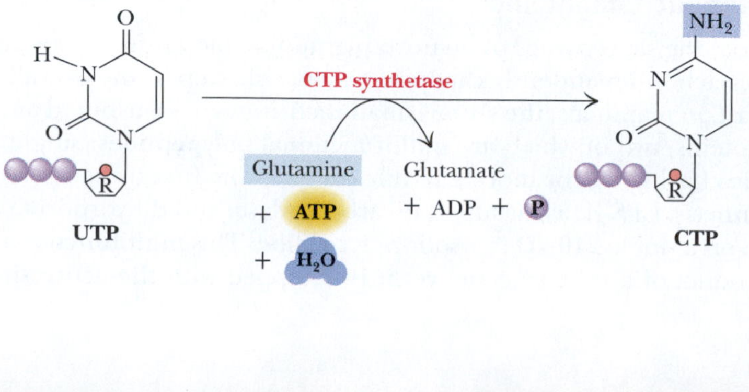

FIGURE 26.16 CTP synthesis from UTP. CTP synthetase catalyzes amination of the 4-position of the UTP pyrimidine ring, yielding CTP. In eukaryotes, this NH_2 comes from the amide-N of glutamine; in bacteria, NH_4^+ serves this role.

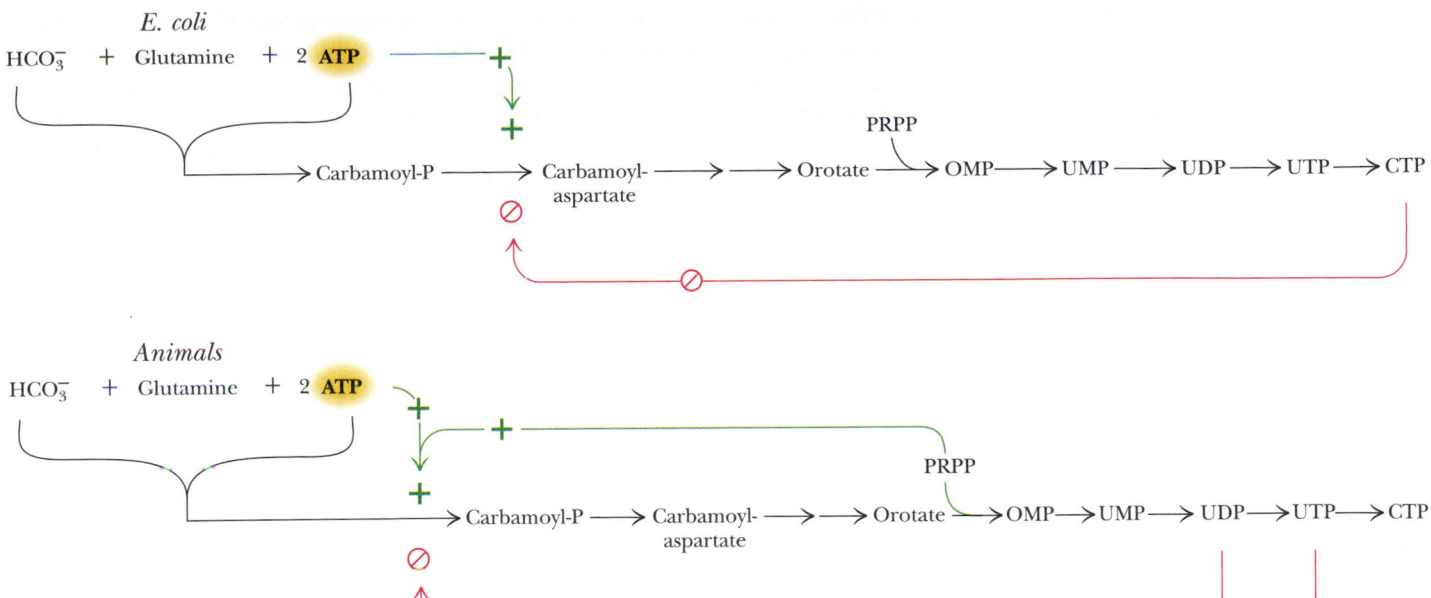

Biochemistry ⊘ Now™ ANIMATED FIGURE 26.17
A comparison of the regulatory circuits that control pyrimidine synthesis in *E. coli* and animals. **See this figure animated at http://chemistry.brookscole. com/ggb3**

inhibitors of CPS-II, whereas PRPP and ATP are allosteric activators. With the exception of ATP, none of these compounds are substrates of CPS-II or of either of the two other enzymic activities residing with it on the trifunctional polypeptide. Figure 26.17 compares the regulatory circuits governing pyrimidine synthesis in bacteria and animals.

26.6 | How Are Pyrimidines Degraded?

In some organisms, free pyrimidines, like purines, are salvaged and recycled to form nucleotides via phosphoribosyltransferase reactions similar to those discussed earlier. In humans, however, pyrimidines are recycled from nucleosides, but free pyrimidine bases are not salvaged. Pyrimidine catabolism results in degradation of the pyrimidine ring to products reminiscent of the original substrates, aspartate, CO_2, and ammonia (Figure 26.18). β-Alanine can be recycled into the synthesis of coenzyme A. Catabolism of the pyrimidine base, thymine (5-methyluracil), yields β-aminoisobutyric acid instead of β-alanine.

Human Biochemistry

Mammalian CPS-II Is Activated In Vitro by MAP Kinase and In Vivo by Epidermal Growth Factor

The rate-limiting step in mammalian de novo pyrimidine synthesis is catalyzed by CPS-II, and proliferating cells require lots of pyrimidine nucleotides for growth and cell division. Normally, CPS-II is feedback-inhibited by UTP, but in vitro phosphorylation of CPS-II by **MAP kinase** (*mitogen-activated protein kinase*) creates a covalently modified (phosphorylated) CPS-II that is no longer sensitive to UTP inhibition. Furthermore, phosphorylated CPS-II is more responsive to PRPP activation. Both of these responses favor enhanced pyrimidine biosynthesis. This regulation occurs in vivo when **epidermal growth factor (EGF)**, a **mitogen**, initiates an intracellular cascade of reactions that culminates in MAP kinase activation. By this action, pyrimidine nucleotides are made available for RNA and DNA synthesis, processes that are central to the cell proliferation that follows EGF activation.

A mitogen is a hormone that stimulates mitosis (cell division).

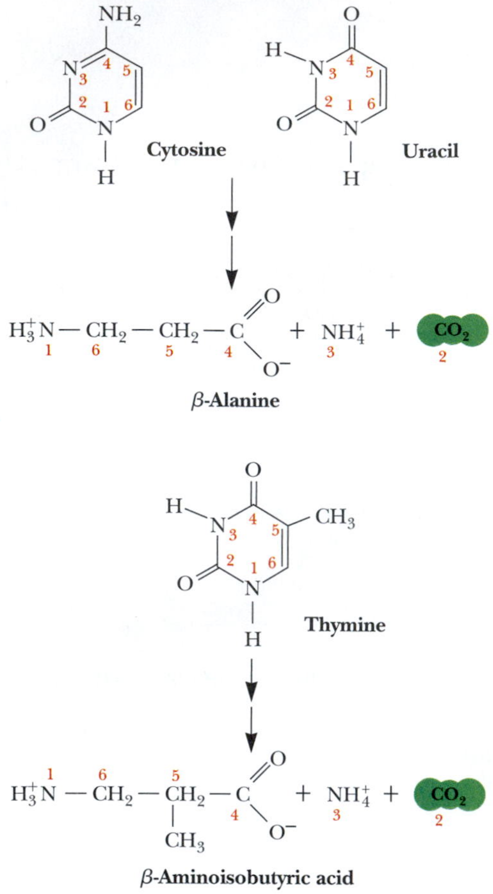

FIGURE 26.18 Pyrimidine degradation. Carbons 4, 5, and 6 plus N-1 are released as β-alanine, N-3 as NH_4^+, and C-2 as CO_2. (The pyrimidine thymine yields β-aminoisobutyric acid.) Recall that aspartate was the source of N-1 and C-4, C-5, and C-6, whereas C-2 came from CO_2 and N-3 from NH_4^+ via glutamine.

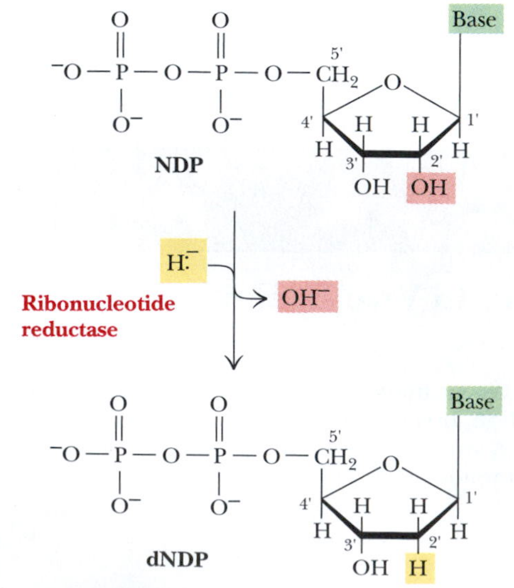

FIGURE 26.19 Deoxyribonucleotide synthesis involves reduction at the 2′-position of the ribose ring of nucleoside diphosphates.

Pathways presented thus far in this chapter account for the synthesis of the four principal ribonucleotides: ATP, GTP, UTP, and CTP. These compounds serve important coenzymic functions in metabolism and are the immediate precursors for ribonucleic acid (RNA) synthesis. Roughly 90% of the total nucleic acid in cells is RNA, with the remainder being deoxyribonucleic acid (DNA). DNA differs from RNA in being a polymer of deoxyribonucleotides, one of which is deoxythymidylic acid. We now turn to the synthesis of these compounds.

26.7 How Do Cells Form the Deoxyribonucleotides That Are Necessary for DNA Synthesis?

The deoxyribonucleotides have only one metabolic purpose: to serve as precursors for DNA synthesis. In most organisms, ribonucleoside diphosphates (NDPs) are the substrates for deoxyribonucleotide formation. Reduction at the 2′-position of the ribose ring in NDPs produces 2′-deoxy forms of these nucleotides (Figure 26.19). This reaction involves replacement of the 2′-OH by a hydride ion ($H:^-$) and is catalyzed by an enzyme known as **ribonucleotide reductase.** Enzymatic ribonucleotide reduction involves a free radical mechanism, and three classes of ribonucleotide reductases are known, differing from each other in their mechanisms of free radical generation. Class I enzymes, found in *E. coli* and virtually all eukaryotes, are Fe-dependent and generate the required free radical on a specific tyrosyl side chain.

E. coli Ribonucleotide Reductase Has Three Different Nucleotide-Binding Sites

The enzyme system for dNDP formation consists of four proteins, two of which constitute the ribonucleotide reductase proper, an enzyme of the $\alpha_2\beta_2$ type. The other two proteins, **thioredoxin** and **thioredoxin reductase,** function in the delivery of reducing equivalents, as we shall see shortly. The two proteins of ribonucleotide reductase are designated **R1** (86 kD) and **R2** (43.5 kD), and each is a homodimer in the holoenzyme (Figure 26.20). The R1 homodimer carries two types of regulatory sites in addition to the **catalytic site** (the active site). Substrates (ADP, CDP, GDP, and UDP) bind at the catalytic site. One regulatory site—the **substrate specificity site**—binds ATP, dATP, dGTP, or dTTP, and which of these nucleotides is bound there determines which nucleoside diphosphate is bound at the catalytic site. The other regulatory site, the **overall activity site,** binds either the activator ATP or the negative effector dATP; the nucleotide bound here determines whether the enzyme is active or inactive. Activity depends also on residues Cys^{439}, Cys^{225}, and Cys^{462} in R1. The 2 Fe atoms within the single active site formed by the R2 homodimer generate the free radical required for ribonucleotide reduction on a specific R2 residue, Tyr^{122}, which in turn generates a thiyl free radical (Cys-S·) on Cys^{439}. Cys^{439}-S· initiates ribonucleotide reduction by abstracting the 3′-H from the ribose ring of the nucleoside diphosphate substrate (Figure 26.21) and forming a free radical on C-3′. Subsequent dehydration forms the deoxyribonucleotide product.

Thioredoxin Provides the Reducing Power for Ribonucleotide Reductase

NADPH is the ultimate source of reducing equivalents for ribonucleotide reduction, but the immediate source is reduced **thioredoxin,** a small (12-kD) protein with reactive Cys-sulfhydryl groups situated next to one another in the sequence Cys-Gly-Pro-Cys. These Cys residues are able to undergo reversible

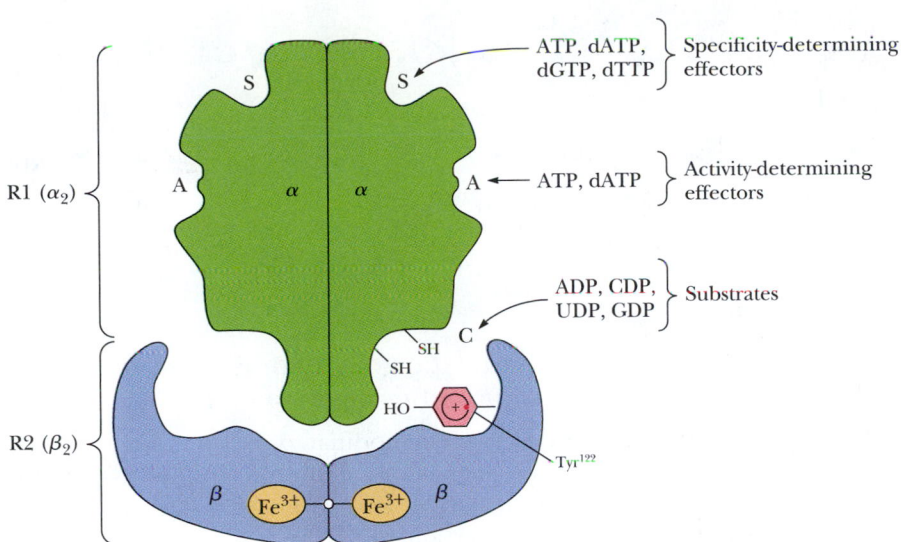

FIGURE 26.20 *E. coli* ribonucleotide reductase: its binding sites and subunit organization. Two proteins, R1 and R2 (each a dimer of identical subunits), combine to form the holoenzyme. The holoenzyme has three classes of nucleotide binding sites: S, the specificity-determining sites; A, the activity-determining sites; and C, the catalytic or active site. These various sites bind different nucleotide ligands. Note that the holoenzyme apparently possesses only one active site formed by interaction between Fe^{3+} atoms in each R2 subunit.

oxidation–reduction between (—S—S—) and (—SH HS—) and, in their reduced form, serve as primary electron donors to regenerate the reactive —SH pair of the ribonucleotide reductase active site (Figure 26.21). In turn, the sulfhydryls of thioredoxin must be restored to the (—SH HS—) state for another catalytic cycle. **Thioredoxin reductase,** an α_2-type enzyme composed of 58-kD flavoprotein subunits, mediates the NADPH-dependent reduction of thioredoxin (Figure 26.22). Thioredoxin functions in a number of metabolic roles besides deoxyribonucleotide synthesis, the common denominator of which is reversible sulfide:sulfhydryl transitions. Another sulfhydryl protein similar to thioredoxin, called **glutaredoxin,** can also function in ribonucleotide reduction. Oxidized glutaredoxin is re-reduced by 2 equivalents of **glutathione** (γ-glutamylcysteinylglycine; Figure 26.23), which in turn is re-reduced by glutathione reductase, another NADPH-dependent flavoenzyme.

The substrates for ribonucleotide reductase are CDP, UDP, GDP, and ADP, and the corresponding products are dCDP, dUDP, dGDP, and dADP. Because CDP is not an intermediate in pyrimidine nucleotide synthesis, it must arise by dephosphorylation of CTP, for instance, via nucleoside diphosphate kinase action. Although uridine nucleotides do not occur in DNA, UDP is a substrate. The formation of dUDP is justified because it is a precursor to dTTP, a necessary substrate for DNA synthesis (see following discussion).

Biochemistry Now™   ACTIVE FIGURE 26.21
The free radical mechanism of ribonucleotide reduction. H_a designates the C-3′ hydrogen and H_b the C-2′ hydrogen atom. Formation of a thiyl radical on Cys^{439} **(a)** of the *E. coli* ribonucleotide reductase R1 homodimer through reaction with a Tyr^{122} free radical on R2 leads to removal of the H_a hydrogen and creation of a C-3′ radical **(b)**. Dehydration via removal of H_b together with the C-2′—OH group and restoration of H_a to C-3′ forms the dNDP product, accompanied by oxidation of R1 Cys^{225} and Cys^{462} —SH groups to form a disulfide **(c)**. (*Adapted from Reichard, P., 1997. The evolution of ribonucleotide reduction.* Trends in Biochemical Sciences *22:81–85. This free radical mechanism of ribonucleotide reduction was originally proposed by JoAnn Stubbe of MIT.*) **Test yourself on the concepts in this figure at http://chemistry. brookscole.com/ggb3**

(a)   Cys$_{439}$
S
•

P P   O
H$_a$ H$_b$
OH OH
SH SH
Cys^{225} Cys^{462}

→

(b)   Cys$_{439}$
SH$_a$

P P   O   N
•   H$_b$
OH OH
SH SH
Cys^{225} Cys^{462}

→   H$_2$O

(c)   Cys$_{439}$
S
•

P P   O   N
H$_a$ H$_b$
OH H
S — S
Cys^{225} Cys^{462}

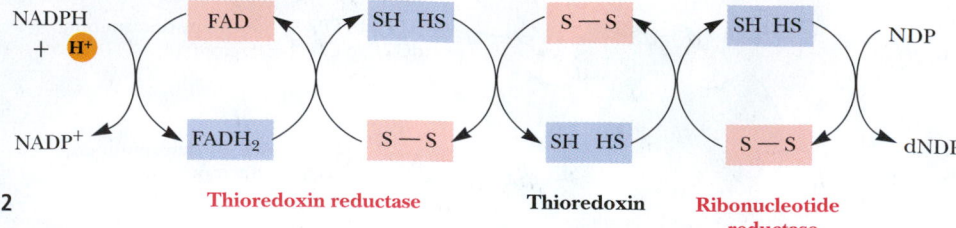

Biochemistry ⓔ Now™ **ANIMATED FIGURE 26.22**
The (—S—S—)/(—SH HS—) oxidation–reduction
cycle involving ribonucleotide reductase, thioredoxin,
thioredoxin reductase, and NADPH. **See this figure
animated at http://chemistry.brookscole.com/ggb3**

Both the Specificity and the Catalytic Activity of Ribonucleotide Reductase Are Regulated by Nucleotide Binding

Ribonucleotide reductase activity must be modulated in two ways in order to maintain an appropriate balance of the four deoxynucleotides essential to DNA synthesis, namely, dATP, dGTP, dCTP, and dTTP. First, the overall activity of the enzyme must be turned on and off in response to the need for dNTPs. Second, the relative amounts of each NDP substrate transformed into dNDP must be controlled so that the right balance of dATP:dGTP:dCTP:dTTP is produced. The two different effector-binding sites on ribonucleotide reductase, *discrete from the substrate-binding catalytic site,* are designed to serve these purposes. As noted previously, these two regulatory sites are designated the *overall activity site* and the *substrate specificity site.* Only ATP and dATP are able to bind at the *overall activity site.* ATP is an allosteric activator and dATP is an allosteric inhibitor, and they compete for the same site. If ATP is bound, the enzyme is active, whereas if its deoxy counterpart, dATP, occupies this site, the enzyme is inactive. That is, ATP is a positive effector and dATP is a negative effector with respect to enzyme activity, and they compete for the same site.

The second regulatory site, the *substrate specificity site,* can bind either ATP, dTTP, dGTP, or dATP, and the substrate specificity of the enzyme is determined by which of these nucleotides occupies this site. If ATP is in the *sub-*

Reduced glutathione (γ-glutamylcysteinylglycine, GSH)

Oxidized glutathione (GSSG) or glutathione disulfide

FIGURE 26.23 The structure of glutathione.

Energy status of cell is robust; [ATP] is high. Make DNA:

❶ ATP occupies activity site A: ribonucleotide reductase *ON*

❷ ATP in specificity site S favors CDP or UDP in catalytic site C ⟶ [dCDP], [dUDP]↑

❸ dCDP ⎫
         ⎬ ⟶ ⟶ dUMP ⟶ dTMP ⟶ ⟶ dTTP
   dUDP ⎭

❹ dTTP occupies specificity site S, favoring GDP or ADP in catalytic site C
   GDP ⟶ dGDP ⟶ dGTP

❺ dGTP occupies specificity site S, favoring ADP in catalytic site C ⟶ [dADP]↑

❻ dATP replaces ATP in activity site A: ribonucleotide reductase *OFF*

FIGURE 26.24 Regulation of deoxynucleotide biosynthesis: the rationale for the various affinities displayed by the two nucleotide-binding regulatory sites on ribonucleotide reductase.

strate specificity site, ribonucleotide reductase preferentially binds pyrimidine nucleotides (UDP or CDP) at its active site and reduces them to dUDP and dCDP. With dTTP in the specificity-determining site, GDP is the preferred substrate. When dGTP binds to the specificity site, ADP becomes the favored substrate for reduction. The rationale for these varying affinities is as follows (Figure 26.24): High [ATP] is consistent with cell growth and division and, consequently, the need for DNA synthesis. Thus, ATP binds in the *overall activity site* of ribonucleotide reductase, turning it on and promoting production of dNTPs for DNA synthesis. Under these conditions, ATP is also likely to occupy the *substrate specificity site,* so UDP and CDP are bound at the *catalytic site* and reduced to dUDP and dCDP. Both of these pyrimidine deoxynucleoside diphosphates are precursors to dTTP. Thus, elevation of dUDP and dCDP levels leads to an increase in [dTTP]. High dTTP levels increase the likelihood that it will occupy the *substrate specificity site,* in which case GDP becomes the preferred substrate and dGTP levels rise. Upon dGTP association with the *substrate specificity site,* ADP is the favored substrate, leading to ADP reduction and the eventual accumulation of dATP. Binding of dATP to the *overall activity site* then shuts the enzyme down. In summary, the relative affinities of the three classes of nucleotide binding sites in ribonucleotide reductase for the various substrates, activators, and inhibitors are such that the formation of dNDPs proceeds in an orderly and balanced fashion. As these dNDPs are formed in amounts consistent with cellular needs, their phosphorylation by nucleoside diphosphate kinases produces dNTPs, the actual substrates of DNA synthesis.

26.8 How Are Thymine Nucleotides Synthesized?

The synthesis of thymine nucleotides proceeds from other pyrimidine deoxyribonucleotides. Cells have no requirement for free thymine ribonucleotides and do not synthesize them. Small amounts of thymine ribonucleotides do occur in tRNA (tRNA is notable for having unusual nucleotides), but these Ts arise via methylation of U residues already incorporated into the tRNA. Both dUDP and dCDP can lead to formation of dUMP, the immediate precursor for dTMP synthesis (Figure 26.25). Interestingly, formation of

From dUDP: dUDP ⟶ dUTP ⟶ dUMP ⟶ dTMP
From dCDP: dCDP ⟶ dCMP ⟶ dUMP ⟶ dTMP

FIGURE 26.25 Pathways of dTMP synthesis. dTMP production is dependent on dUMP formation from dCDP and dUDP synthesis. If the dCDP pathway is traced from the common pyrimidine precursor, UMP, it will proceed as follows:
UMP → UDP → UTP → CTP → CDP → dCDP → dCMP → dUMP → dTMP

FIGURE 26.26 The dCMP deaminase reaction.

dUMP from dUDP passes through dUTP, which is then cleaved by **dUTPase,** a pyrophosphatase that removes PP$_i$ from dUTP. The action of dUTPase prevents dUTP from serving as a substrate in DNA synthesis. An alternative route to dUMP formation starts with dCDP, which is dephosphorylated to dCMP and then deaminated by **dCMP deaminase** (Figure 26.26), leaving dUMP. dCMP deaminase provides a second point for allosteric regulation of dNTP synthesis; it is allosterically activated by dCTP and feedback-inhibited by dTTP. Of the four dNTPs, only dCTP does not interact with either of the

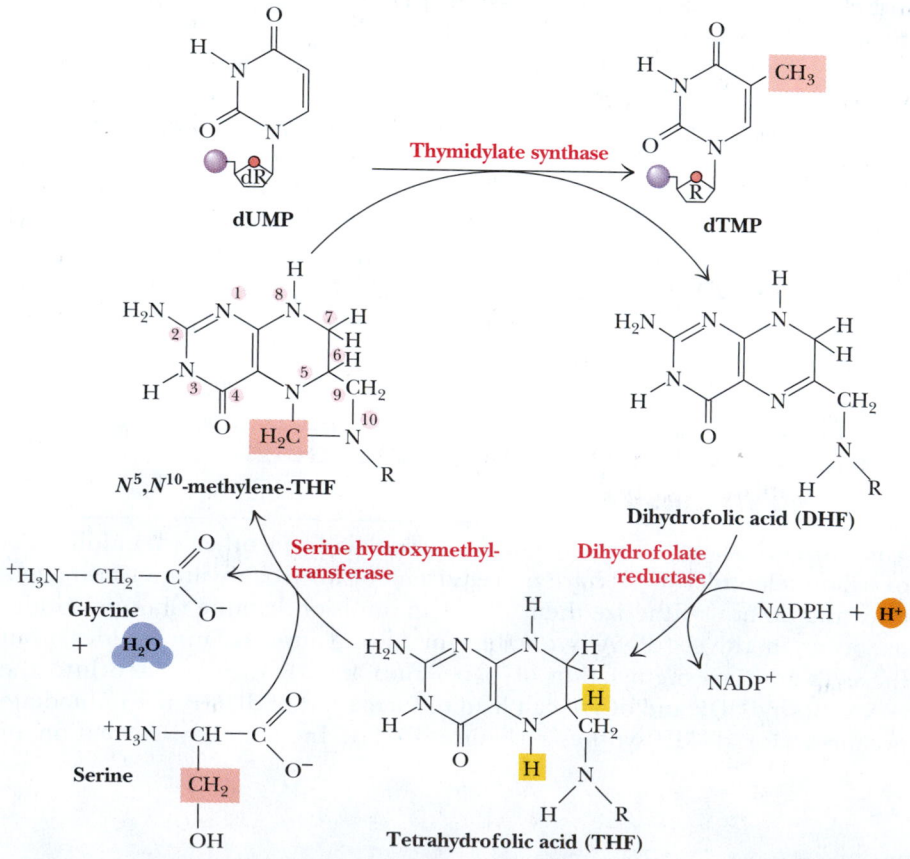

Biochemistry ⊜ Now™ ACTIVE FIGURE 26.27 The thymidylate synthase reaction. The 5-CH$_3$ group is ultimately derived from the β-carbon of serine. **Test yourself on the concepts in this figure at http://chemistry.brookscole.com/ggb3**

A Deeper Look

Fluoro-Substituted Analogs as Therapeutic Agents

Carbon–fluorine bonds are exceedingly rare in nature, and fluorine is an uncommon constituent of biological molecules. F has three properties attractive to drug designers: (1) It is the smallest replacement for an H atom in organic synthesis, (2) fluorine is the most electronegative element, and (3) the F—C bond is relatively unreactive. This steric compactness and potential for strong inductive effects through its electronegativity renders F a useful substituent in the construction of inhibitory analogs of enzyme substrates. One interesting strategy is to devise fluorinated precursors that are taken up and processed by normal metabolic pathways to generate a potent antimetabolite. A classic example is *fluoroacetate*. FCH_2COO^- is exceptionally toxic because it is readily converted to fluorocitrate by citrate synthase of the citric acid cycle (see Chapter 19). In turn, fluorocitrate is a powerful inhibitor of aconitase. The metabolic transformation of an otherwise innocuous compound into a poisonous derivative is termed **lethal synthesis.** *5-Fluorouracil* and *5-fluorocytosine* are also examples of this strategy (see text).

Unlike hydrogen, which is often abstracted from substrates as H^+, electronegative fluorine cannot be readily eliminated as the corresponding F^+. Thus, enzyme inhibitors can be fashioned in which F replaces H at positions where catalysis involves H removal as H^+. Thymidylate synthase catalyzes removal of H from dUMP as H^+ through a covalent catalysis mechanism. A thiol group on this enzyme normally attacks the 6-position of the uracil moiety of 2'-deoxyuridylic acid so that C-5 can act as a carbanion in attack on the methylene carbon of N^5,N^{10}-methylene-THF (see accompanying figure). Regeneration of free enzyme then occurs through loss of the C-5 H atom as H^+ and dissociation of product dTMP. If F replaces H at C-5 as in 2'-deoxy-5-fluorouridylate (dFUMP), the enzyme is immobilized in a very stable ternary [enzyme:dFUMP: methylene-THF] complex and effectively inactivated. Enzyme inhibitors like dFUMP whose adverse properties are elicited only through direct participation in the catalytic cycle are variously called **mechanism-based inhibitors, suicide substrates,** or **Trojan horse substrates.**

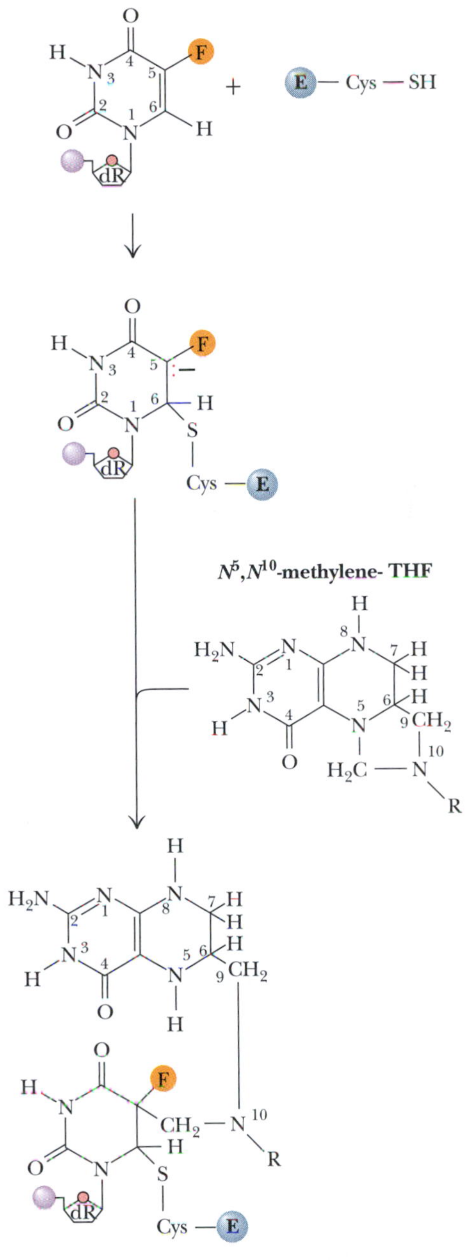

▶ The effect of the 5-fluoro substitution on the mechanism of action of thymidylate synthase. An enzyme thiol group (from a Cys side chain) ordinarily attacks the 6-position of dUMP so that C-5 can react as a carbanion with N^5,N^{10}-methylene-THF. Normally, free enzyme is regenerated following release of the hydrogen at C-5 as a proton. Because release of fluorine as F^+ cannot occur, the ternary (three-part) complex of [enzyme:fluorouridylate:methylene-THF] is stable and persists, preventing enzyme turnover. (The N^5,N^{10}-methylene-THF structure is given in abbreviated form.)

N^5,N^{10}-methylene- THF

Ternary complex

regulatory sites on ribonucleotide reductase (Figure 26.20). Instead, it acts upon dCMP deaminase.

Synthesis of dTMP from dUMP is catalyzed by **thymidylate synthase** (Figure 26.27). This enzyme methylates dUMP at the 5-position to create dTMP; the methyl donor is the one-carbon folic acid derivative N^5,N^{10}-methylene-THF. The reaction is actually a reductive methylation in which the one-carbon unit is transferred at the methylene level of reduction and then reduced to the methyl level. The THF cofactor is oxidized at the expense of

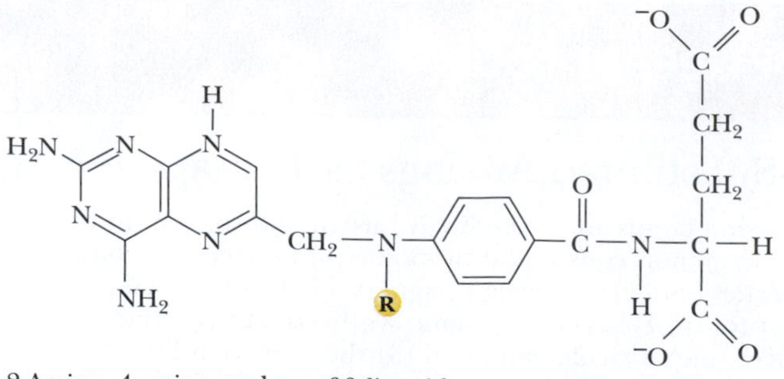

2-Amino, 4-amino analogs of folic acid

R = H — **Aminopterin**

R = CH$_3$ — **Amethopterin (methotrexate)**

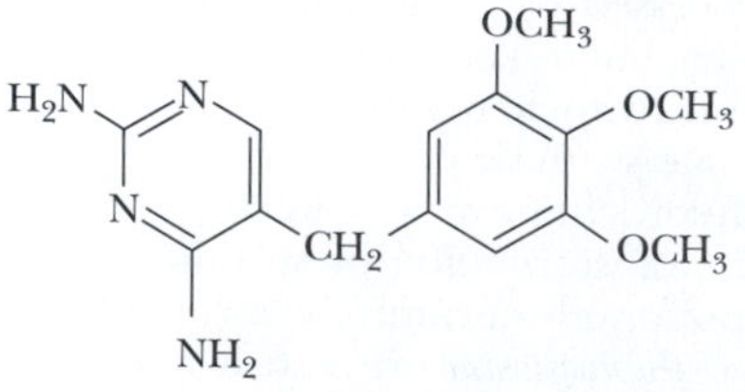

Trimethoprim

FIGURE 26.28 Precursors and analogs of folic acid employed as antimetabolites include sulfonamides (see Human Biochemistry box on page 858), as well as methotrexate, aminopterin, and trimethoprim, whose structures are shown here. The compounds shown here bind to dihydrofolate reductase with about 1000-fold greater affinity than DHF and thus act as virtually irreversible inhibitors.

methylene reduction to yield dihydrofolate, or DHF. Dihydrofolate reductase then reduces DHF back to THF for service again as a one-carbon vehicle (see Figure 17.35). Thymidylate synthase sits at a junction connecting dNTP synthesis with folate metabolism. It has become a preferred target for inhibitors designed to disrupt DNA synthesis. An indirect approach is to employ folic acid precursors or analogs as antimetabolites of dTMP synthesis (Figure 26.28). Purine synthesis is affected as well because it is also dependent on THF (Figure 26.3).

Human Biochemistry

Fluoro-Substituted Pyrimidines in Cancer Chemotherapy, Fungal Infections, and Malaria

5-Fluorouracil (*5-FU;* see accompanying figure) is a thymine analog. It is converted in vivo to *5′-fluorouridylate* by a PRPP-dependent phosphoribosyltransferase and passes through the reactions of dNTP synthesis, culminating ultimately as *2′-deoxy-5-fluorouridylic acid,* a potent inhibitor of dTMP synthase (see the A Deeper Look box on page 875). 5-FU is used as a chemotherapeutic agent in the treatment of human cancers. Similarly, *5-fluorocytosine* (see figure) is used as an antifungal drug because fungi, unlike mammals, can convert it to 2′-deoxy-5-fluorouridylate. Furthermore, malarial parasites can use exogenous orotate to make pyrimidines for nucleic acid synthesis, whereas mammals cannot. Thus, *5-fluoroorotate* (see figure) is an effective antimalarial drug because it is selectively toxic to these parasites.

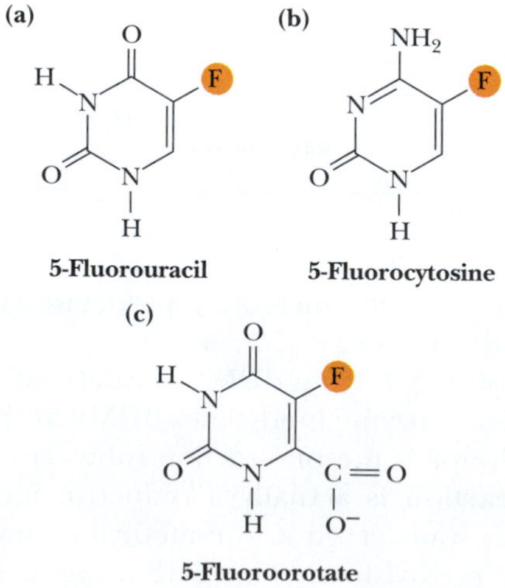

(a) **5-Fluorouracil** (b) **5-Fluorocytosine**

(c) **5-Fluoroorotate**

▶ The structures of 5-fluorouracil (5-FU), 5-fluorocytosine, and 5-fluoroorotate.

Summary

26.1 Can Cells Synthesize Nucleotides?
Nucleotides are ubiquitous constituents of life and nearly all cells are capable of synthesizing them "from scratch" via de novo pathways. Rapidly proliferating cells must make lots of purine and pyrimidine nucleotides to satisfy demands for DNA and RNA synthesis. Nucleotide biosynthetic pathways are attractive targets for the clinical control of rapidly dividing cells such as cancers or infectious bacteria. Many antibiotics and anticancer drugs are inhibitors of purine or pyrimidine nucleotide biosynthesis.

26.2 How Do Cells Synthesize Purines?
The nine atoms of the purine ring system are derived from aspartate (N-1), glutamine (N-3 and N-9), glycine (C-4, C-5, and N-7), CO_2 (C-6), and THF one-carbon derivatives (C-2 and C-8). The atoms of the purine ring are successively added to *ribose-5-phosphate*, so purines begin as nucleotide derivatives through assembly of the purine ring system directly on the ribose. Because purine biosynthesis depends on folic acid derivatives, it is sensitive to inhibition by folate analogs. Distinct, two-step metabolic pathways diverge from IMP, one leading to AMP and the other to GMP. Purine biosynthesis is regulated at several stages: Reaction 1 (ribose-5-phosphate pyrophosphokinase) is feedback-inhibited by ADP and GDP; the enzyme catalyzing reaction 2 (glutamine phosphoribosyl pyrophosphate amidotransferase) has two inhibitory allosteric sites, one where adenine nucleotides bind and another where guanine nucleotides bind. PRPP is a "feed-forward" activator of this enzyme. The first reaction in the conversion of IMP to AMP involves adenylosuccinate synthetase, which is inhibited by AMP; the first step in the conversion of IMP to GMP is catalyzed by IMP dehydrogenase and is inhibited by GMP. ATP-dependent kinases form nucleoside diphosphates and triphosphates from AMP and GMP.

26.3 Can Cells Salvage Purines?
Purine ring systems represent a metabolic investment by cells, and salvage pathways exist to recover them when degradation of nucleic releases free purines in the form of adenine, guanine, and hypoxanthine (the base in IMP). Hypoxanthine-guanine phosphoribosyltransferase (HGPRT) acts on either hypoxanthine to form IMP or guanine to form GMP; an absence of HGPRT is the basis of Lesch-Nyhan syndrome.

26.4 How Are Purines Degraded?
Dietary nucleic acids are digested to nucleotides by various nucleases and phosphodiesterases, the nucleotides are converted to nucleosides by base-specific nucleotidases and nonspecific phosphatases, and then nucleosides are hydrolyzed to release the purine base. Only the pentoses of nucleotides serve as sources of metabolic energy. In humans, the purine ring is oxidized to uric acid by xanthine oxidase and excreted. Gout occurs when bodily fluids accumulate an excess of uric acid. Skeletal muscle operates a purine nucleoside cycle as an anaplerotic pathway.

26.5 How Do Cells Synthesize Pyrimidines?
In contrast to formation of the purine ring system, the pyrimidine ring system is completed before a ribose-5-P moiety is attached. Only two precursors, carbamoyl-P and aspartate, contribute atoms to the six-membered pyrimidine ring. The first step in humans is catalyzed by CPS-II. ATCase then links carbamoyl-P with aspartate. Subsequent reactions close the ring and oxidize it, before adding ribose-5-P, using α-PRPP as donor. Decarboxylation gives UMP. In mammals, the six enzymatic activities of pyrimidine biosynthesis are distributed among only three proteins, two of which are multifunctional polypeptides. Purine and pyrimidine synthesis in mammals are two prominent examples of metabolic channeling. UMP leads to UTP, the substrate for formation of CTP via CTP synthetase. Regulation of pyrimidine synthesis in animals occurs at CPS-II. UDP and UTP are feedback inhibitors, whereas PRPP and ATP are allosteric activators. In bacteria, regulation acts at ATCase through feedback inhibition by CTP (or UTP) and activation by ATP.

26.6 How Are Pyrimidines Degraded?
Degradation of the pyrimidine ring generates β-alanine, CO_2, and ammonia. In humans, pyrimidines are recycled from nucleosides, but free pyrimidine bases are not salvaged.

26.7 How Do Cells Form the Deoxyribonucleotides That Are Necessary for DNA Synthesis?
2'-Deoxyribonucleotides are formed from ribonucleotides through reduction at the 2'-position of the ribose ring in NDPs. The reaction, catalyzed by ribonucleotide reductase, involves a free radical mechanism that replaces the 2'-OH by a hydride ion (H:$^-$). Thioredoxin provides the reducing power for ribonucleotide reduction. *E. coli* ribonucleotide reductase has three different nucleotide-binding sites: the catalytic site (or active site), which binds substrates (ADP, CDP, GDP, and UDP); the substrate specificity site, which can bind ATP, dATP, dGTP, or dTTP; and the overall activity site, which binds either the activator ATP or the negative effector dATP. The relative affinities of the three classes of nucleotide binding sites in ribonucleotide reductase for the various substrates, activators, and inhibitors are such that the various dNDPs are formed in amounts consistent with cellular needs.

26.8 How Are Thymine Nucleotides Synthesized?
Both dUDP and dCDP can lead to formation of dUMP, the immediate precursor for dTMP synthesis. Formation of dTMP from dUMP is catalyzed by thymidylate synthase through reductive methylation of dUMP at the 5-position. The methyl donor is the one-carbon folic acid derivative N^5, N^{10}-methylene-THF. Fluoro-substituted pyrimidine analogs such as 5-fluorouracil (5-FU), 5-fluorocytosine, and 5-fluoroorotate inhibit thymidylate synthase. These fluoro compounds have found a range of therapeutic uses in treating diseases from cancer to malaria.

Problems

1. Draw the purine and pyrimidine ring structures, indicating the metabolic source of each atom in the rings.

2. Starting from glutamine, aspartate, glycine, CO_2 and N^{10}-formyl-THF, how many ATP equivalents are expended in the synthesis of (a) ATP, (b) GTP, (c) UTP, and (d) CTP?

3. Illustrate the key points of regulation in (a) the biosynthesis of IMP, AMP, and GMP; (b) *E. coli* pyrimidine biosynthesis; and (c) mammalian pyrimidine biosynthesis.

4. Indicate which reactions of purine or pyrimidine metabolism are affected by the inhibitors (a) azaserine, (b) methotrexate, (c) sulfonamides, (d) allopurinol, and (e) 5-fluorouracil.

5. Since dUTP is not a normal component of DNA, why do you suppose ribonucleotide reductase has the capacity to convert UDP to dUDP?

6. Describe the underlying rationale for the regulatory effects exerted on ribonucleotide reductase by ATP, dATP, dTTP, and dGTP.

7. (Integrates with Chapters 18–20 and 22.) By what pathway(s) does the ribose released upon nucleotide degradation enter intermediary metabolism and become converted to cellular energy? How many ATP equivalents can be recovered from one equivalent of ribose?

8. (Integrates with Chapter 25.) At which steps does the purine biosynthetic pathway resemble the pathway for biosynthesis of the amino acid histidine?

9. Write reasonable chemical mechanisms for steps 6, 8, and 9 in purine biosynthesis (Figure 26.3).

10. Write a balanced equation for the conversion of aspartate to fumarate by the purine nucleoside cycle in skeletal muscle.

11. Write a balanced equation for the oxidation of uric acid to glyoxylic acid, CO_2, and NH_3, showing each step in the process and naming all the enzymes involved.

12. (Integrates with Chapter 15.) *E. coli* aspartate transcarbamoylase (ATCase) displays classic allosteric behavior. This $\alpha_6\beta_6$ enzyme is

activated by ATP and feedback-inhibited by CTP. In analogy with the behavior of glycogen phosphorylase shown in Figure 15.15, illustrate the allosteric v versus [aspartate] curves for ATCase (a) in the absence of effectors, (b) in the presence of CTP, and (c) in the presence of ATP.

*13. (Integrates with Chapter 15.) Unlike its allosteric counterpart glycogen phosphorylase (an α_2 enzyme), *E. coli* ATCase has a heteromeric ($\alpha_6\beta_6$) organization. The α-subunits bind aspartate and are considered catalytic subunits, whereas the β-subunits bind CTP or ATP and are considered regulatory subunits. How would you describe the subunit organization of ATCase from a functional point of view?

14. (Integrates with Chapter 20.) Starting from HCO_3^-, glutamine, aspartate, and ribose-5-P, how many ATP equivalents are consumed in the synthesis of dTTP in a eukaryotic cell, assuming dihydroorotate oxidation is coupled to oxidative phosphorylation? How does this result compare with the ATP costs of purine nucleotide biosynthesis calculated in problem 2?

15. (Integrates with Chapter 17.) Write a *balanced* equation for the synthesis of dTMP from UMP and N^5,N^{10}-methylene-THF. Thymidylate synthase has four active-site arginine residues (Arg[23], Arg[178'], Arg[179'], and Arg[218]) involved in substrate binding. Postulate a role for the side chains of these Arg residues.

Preparing for the MCAT Exam

16. Examine Figure 26.6 and predict the relative rates of the regulated reactions in the purine biosynthetic pathway from ribose-5-P to GMP and AMP under conditions in which GMP levels are very high.

17. Decide from Figures 18.2, 18.16, 25.31, 26.27, and the Deeper Look box on page 855 which carbon atom(s) in glucose would be most likely to end up as the 5-CH$_3$ carbon in dTMP.

Biochemistry ℰ Now™ Preparing for an exam? Test yourself on key questions at http://chemistry.brookscole.com/ggb3

Further Reading

Purine Metabolism

Caroline Kisker, K., Schindelin, H., and, Rees, D. C., 1997. Molybdenum-containing enzymes: Structure and mechanism. *Annual Review of Biochemistry* **66:**233–267. Discusses xanthine oxidase.

Mueller, E. J., et al., 1994. N^5-carboxyaminoimidazole ribonucleotide: Evidence for a new intermediate and two new enzymatic activities in the de novo purine biosynthetic pathway of *Escherichia coli*. *Biochemistry* **33:**2269–2278.

Watts, R. W. E., 1983. Some regulatory and integrative aspects of purine nucleotide synthesis and its control: An overview. *Advances in Enzyme Regulation* **21:**33–51.

Wilson, D. K., Rudolph, F. B., and Quiocho, F. A., 1991. Atomic structure of adenosine deaminase complexed with a transition-state analog: Understanding catalysis and immunodeficient mutations. *Science* **252:**1279–1284.

Pyrimidine Metabolism

Connolly, G. P., and Duley, J. A., 1999. Uridine and its nucleotides: Biological actions, therapeutic potentials. *Trends in Pharmacological Sciences* **20:**218–225.

Graves, L. M., et al., 2000. Regulation of carbamoyl phosphate synthetase by MAP kinase. *Nature* **403:**328–331.

Jones, M. E., 1980. Pyrimidine nucleotide biosynthesis in animals: Genes, enzymes and regulation of UMP biosynthesis. *Annual Review of Biochemistry* **49:**253–279.

Metabolic Disorders of Purine and Pyrimidine Metabolism

Scriver, C. R., et al., 1995. *The Metabolic and Molecular Bases of Inherited Disease,* 7th ed. New York: McGraw-Hill. A three-volume treatise on the biochemistry and genetics of inherited metabolic disorders, including disorders of purine and pyrimidine metabolism.

Metabolic Channeling

Benkovic, S. J., 1984. The transformylase enzymes in de novo purine biosynthesis. *Trends in Biochemical Sciences* **9:**320–322. These enzymes provide an instance of metabolic channeling in one-carbon metabolism.

Henikoff, S., 1987. Multifunctional polypeptides for purine de novo synthesis. *BioEssays* **6:**8–13.

Huang, X., Holden, H. M., and Raushel, F. M., 2001. Channeling of substrates and intermediates in enzyme-catalyzed reactions. *Annual Review of Biochemistry* **70:**149–180.

Srere, P. A., 1987. Complexes of sequential metabolic enzymes. *Annual Review of Biochemistry* **56:**89–124. A discussion of how the enzymes acting sequentially in a metabolic pathway are often organized into multienzyme complexes, or even synthesized as multifunctional proteins, especially in eukaryotes.

Deoxyribonucleotide Biosynthesis

Carreras, C. W., and Santi, D. V., 1995. The catalytic mechanism and structure of thymidylate synthase. *Annual Review of Biochemistry* **64:**721–762.

Frey, P. A., 2001. Radical mechanisms of enzymatic catalysis. *Annual Review of Biochemistry* **70:**121–148. A review of enzymatic mechanisms proceeding via carbon-based free radicals, the most prominent examples of which are ribonucleotide reductases.

Holmgren, A., 1989. Thioredoxin and glutaredoxin systems. *The Journal of Biological Chemistry* **264:**13963–13966.

Jordan, A., and Reichard, P., 1998. Ribonucleotide reductases. *Annual Review of Biochemistry* **67:**71–98.

Licht, S., Gerfen, G. J., and Stubbe, J., 1996. Thiyl radicals in ribonucleotide reductases. *Science* **271:**477–481.

Marsh, E. N. G., 1995. A radical approach to enzyme catalysis. *BioEssays* **17:**431–441.

Nordlund, P., and Eklund, H., 1993. Structure and function of the *Escherichia coli* ribonucleotide reductase protein R2. *Journal of Molecular Biology* **232:**123–164.

Reichard, P., 1988. Interactions between deoxyribonucleotide and DNA synthesis. *Annual Review of Biochemistry* **57:**349–374. A review of the regulation of ribonucleotide reductase by the scientist who discovered these phenomena.

Reichard, P., 1997. The evolution of ribonucleotide reduction. *Trends in Biochemical Sciences* **22:**81–85.

Stubbe, J., Ge, J., and Yee, C. S., 2001. The evolution of ribonucleotide reduction revisited. *Trends in Biochemical Sciences* **26:**93–99.

Uhlin, U., and Eklund, H., 1996. The ten-stranded β/α barrel in ribonucleotide reductase protein R1. *Journal of Molecular Biology* **262:**358–369.

Inhibitors of Purine, Pyrimidine, and Deoxyribonucleotide Biosynthesis as Therapeutic Agents

Abeles, R. H., and Alston, T. A., 1990. Enzyme inhibition by fluoro compounds. *The Journal of Biological Chemistry* **265:**16705–16708. A brief review of the usefulness of fluoro derivatives in probing reaction mechanisms.

Galmarini, C. M., Mackey, J. R., and Dumontet, C., 2002. Nucleoside analogues and nucleobases in cancer treatment. *The Lancet Oncology* **3:**415–424.

Hitchings, G. H., 1992. Antagonists of nucleic acid derivatives as medicinal agents. *Annual Review of Pharmacology and Toxicology* **32:**1–9.

Park, B. K., Kitteringham, N. R., and O'Neill, P. M., 2001. Metabolism of fluorine-containing drugs. *Annual Review of Pharmacology and Toxicology* **41:**443–470.

Metabolic Integration and Organ Specialization

Colorful gear wheels are a metaphor for metabolic integration, achieved through the highly regulated coordination of thousands of enzymatic reactions.

© Royalty-Free/CORBIS

Essential Question

Cells are systems in a dynamic steady state, maintained by a constant flux of nutrients that serve as energy sources or as raw material for the maintenance of cellular structures. Catabolism and anabolism are ongoing, concomitant processes. *What principles underlie the integration of catabolism and energy production with anabolism and energy consumption? How is metabolism integrated in complex organisms with multiple organ systems?*

In the preceding chapters in this section (Part 3), we have explored the major metabolic pathways—glycolysis, the citric acid cycle, electron transport and oxidative phosphorylation, photosynthesis, gluconeogenesis, fatty acid oxidation, lipid biosynthesis, amino acid metabolism, and nucleotide metabolism. Several of these pathways are catabolic and serve to generate chemical energy useful to the cell; others are anabolic and use this energy to drive the synthesis of essential biomolecules. Despite their opposing purposes, these reactions typically occur at the same time as food molecules are broken down to provide the building blocks and energy for ongoing biosynthesis. Cells maintain a dynamic steady state through processes that involve considerable metabolic flux. The metabolism that takes place in just a single cell is so complex that it defies meaningful quantitative description. However, an appreciation of overall relationships can be achieved by stepping back and considering intermediary metabolism at a systems level of organization.[1]

Study of an enzyme, a reaction, or a sequence can be biologically relevant only if its position in the hierarchy of function is kept in mind.

Daniel E. Atkinson, *Cellular Energy Metabolism and Its Regulation* (1977)

Key Questions

27.1 Can Systems Analysis Simplify the Complexity of Metabolism?

27.2 What Underlying Principle Relates ATP Coupling to the Thermodynamics of Metabolism?

27.3 Can Cellular Energy Status Be Quantified?

27.4 How Is Metabolism Integrated in a Multicellular Organism?

27.1	Can Systems Analysis Simplify the Complexity of Metabolism?

The metabolism of a typical heterotrophic cell can be portrayed by a schematic diagram consisting of just three interconnected functional blocks: (1) catabolism, (2) anabolism, and (3) macromolecular synthesis and growth (Figure 27.1).

1. CATABOLISM. Foods are oxidized to CO_2 and H_2O in catabolism, and most of the electrons liberated are passed to oxygen via an electron-transport pathway coupled to oxidative phosphorylation, resulting in the formation of ATP. Some electrons go to reduce $NADP^+$ to NADPH, the source of reducing power for anabolism. Glycolysis, the citric acid cycle, electron transport and oxidative phosphorylation, and the pentose phosphate pathway are the principal pathways within this block. The metabolic intermediates in these pathways also serve as substrates for processes within the anabolic block.

2. ANABOLISM. The biosynthetic reactions that form the many cellular molecules collectively comprise anabolism. For thermodynamic reasons, the chemistry of anabolism is more complex than that of catabolism (that is, it takes more energy [and often more steps] to synthesize a molecule than can be produced from its degradation). Metabolic intermediates derived from glycolysis and the citric acid cycle are the precursors for this synthesis, with NADPH supplying the reducing power and ATP the coupling energy.

[1] Many of the ideas presented in this chapter originated in an insightful book by Daniel E. Atkinson of the University of California, Los Angeles, titled *Cellular Energy Metabolism and Its Regulation* (New York: Academic Press, 1977).

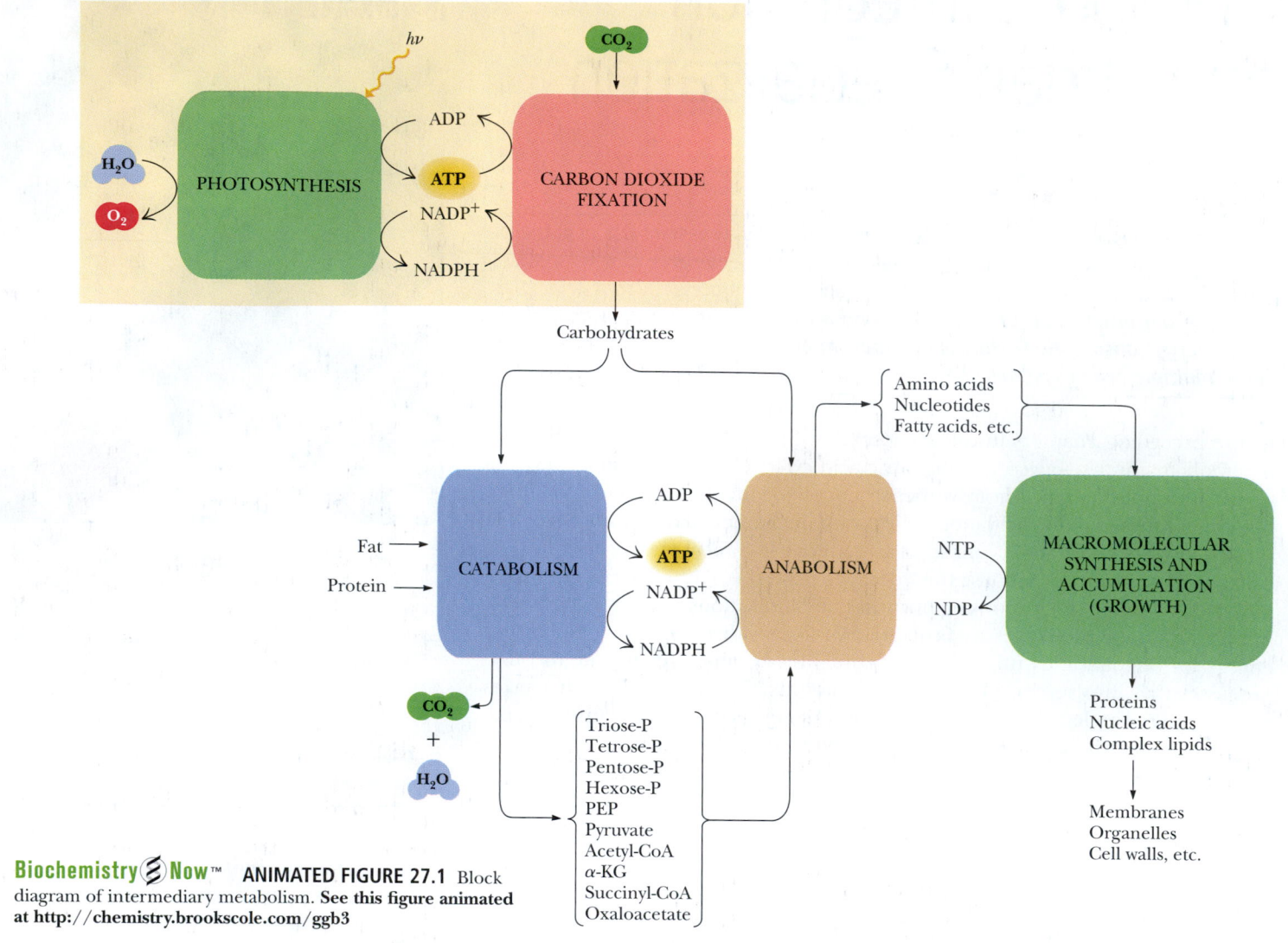

ANIMATED FIGURE 27.1 Block diagram of intermediary metabolism. **See this figure animated at http://chemistry.brookscole.com/ggb3**

3. MACROMOLECULAR SYNTHESIS AND GROWTH. The organic molecules produced in anabolism are the building blocks for creation of macromolecules. Like anabolism, macromolecular synthesis is driven by energy from ATP, although indirectly in some cases: GTP is the principal energy source for protein synthesis, CTP for phospholipid synthesis, and UTP for polysaccharide synthesis. However, keep in mind that ATP is the ultimate phosphorylating agent for formation of GTP, CTP, and UTP from GDP, CDP, and UDP, respectively. Macromolecules are the agents of biological function and information—proteins, nucleic acids, lipids that self-assemble into membranes, and so on. Growth can be represented as cellular accumulation of macromolecules and the partitioning of these materials of function and information into daughter cells in the process of cell division.

Only a Few Intermediates Interconnect the Major Metabolic Systems

Despite the complexity of processes taking place within each block, the connections between blocks involve only a limited number of substances. Just ten or so kinds of catabolic intermediates from glycolysis, the pentose phosphate pathway, and the citric acid cycle serve as the raw material for most of anabolism: four kinds of sugar phosphates (triose-P, tetrose-P, pentose-P, and hexose-P), three α-keto acids (pyruvate, oxaloacetate, and α-ketoglutarate), two coenzyme A derivatives (acetyl-CoA and succinyl-CoA), and PEP (phosphoenolpyruvate).

ATP and NADPH Couple Anabolism and Catabolism

Metabolic intermediates are consumed by anabolic reactions and must be continuously replaced by catabolic processes. In contrast, the energy-rich compounds ATP and NADPH are recycled rather than replaced. When these substances are used in biosynthesis, the products are ADP and $NADP^+$, and ATP and NADPH are regenerated by oxidative reactions that occur in catabolism. ATP and NADPH are unique in that they are the only compounds whose purpose is to couple the energy-yielding processes of catabolism to the energy-consuming reactions of anabolism. Certainly, other coupling agents serve essential roles in metabolism. For example, NADH and $[FADH_2]$ participate in the transfer of electrons from substrates to O_2 during oxidative phosphorylation. However, these reactions are solely catabolic, and the functions of NADH and $[FADH_2]$ are fulfilled within the block called catabolism.

Phototrophs Have an Additional Metabolic System— The Photochemical Apparatus

The systems in Figure 27.1 reviewed thus far are representative only of metabolism as it exists in aerobic heterotrophs. The photosynthetic production of ATP and NADPH in photoautotrophic organisms entails a fourth block, the photochemical system (Figure 27.1). This block consumes H_2O and releases O_2. When this fourth block operates, energy production within the catabolic block can be largely eliminated. Yet another block, one to account for the fixation of carbon dioxide into carbohydrates, is also required for photoautotrophs. The inputs to this fifth block are the products of the photochemical system (ATP and NADPH) and CO_2 derived from the environment. The carbohydrate products of this block may enter catabolism, but not primarily for energy production. In photoautotrophs, carbohydrates are fed into catabolism to generate the metabolic intermediates needed to supply the block of anabolism. Although these diagrams are oversimplifications of the total metabolic processes in heterotrophic or phototrophic cells, they are useful illustrations of functional relationships between the major metabolic subdivisions. This general pattern provides an overall perspective on metabolism, making its purpose easier to understand.

27.2	**What Underlying Principle Relates ATP Coupling to the Thermodynamics of Metabolism?**

Virtually every metabolic pathway either consumes or produces ATP. The amount of ATP involved—that is, the stoichiometry of ATP synthesis or hydrolysis—lies at the heart of metabolic relationships. The overall thermodynamic efficiency of any metabolic sequence, be it catabolic or anabolic, is determined by ATP coupling. In every case, *the overall reaction mediated by any metabolic pathway is energetically favorable because of its particular ATP stoichiometry.* In the highly exergonic reactions of catabolism, much of the energy released is captured in ATP synthesis. In turn, the thermodynamically unfavorable reactions of anabolism are driven by energy released upon ATP hydrolysis.

To illustrate this principle, we must first consider the three types of stoichiometries. The first two are fixed by the laws of chemistry, but the third is unique to living systems and reveals a fundamental difference between the inanimate world of chemistry and physics and the world of functional design—that is, the world of living organisms. The fundamental difference is the stoichiometry of ATP coupling.

Stoichiometry is the measurement of the amounts of chemical elements and molecules involved in chemical reactions (from the Greek *stoicheion*, meaning "element," and *metria*, meaning "measure").

1. Reaction Stoichiometry   This is simple chemical stoichiometry—the number of each kind of atom in any chemical reaction remains the same, and thus equal numbers must be present on both sides of the equation. This requirement holds even for a process as complex as cellular respiration:

$$C_6H_{12}O_6 + 6\ O_2 \longrightarrow 6\ CO_2 + 6\ H_2O$$

The six carbons in glucose appear as 6 CO_2, the 12 H of glucose appear as the 12 H in six molecules of water, and the 18 oxygens are distributed between CO_2 and H_2O.

2. Obligate Coupling Stoichiometry   Cellular respiration is an oxidation–reduction process, and the oxidation of glucose is coupled to the reduction of NAD^+ and [FAD]. (Brackets here denote that the relevant FAD is covalently linked to succinate dehydrogenase [see Chapter 20]). The NADH and [$FADH_2$] thus formed are oxidized in the electron-transport pathway:

(a) $C_6H_{12}O_6 + 10\ NAD^+ + 2\ [FAD] + 6\ H_2O \longrightarrow$
$$6\ CO_2 + 10\ NADH + 10\ H^+ + 2\ [FADH_2]$$

(b) $10\ NADH + 10\ H^+ + 2\ [FADH_2] + 6\ O_2 \longrightarrow$
$$12\ H_2O + 10\ NAD^+ + 2\ [FAD]$$

Sequence (a) accounts for the oxidation of glucose via glycolysis and the citric acid cycle. Sequence (b) is the overall equation for electron transport per glucose. The stoichiometry of coupling by the biological e^- carriers, NAD^+ and FAD, is fixed by the chemistry of electron transfer; each of the coenzymes serves as an e^- pair acceptor. Reduction of each O atom takes an e^- pair. Metabolism must obey these facts of chemistry: Biological oxidation of glucose releases 12 e^- pairs, creating a requirement for 12 equivalents of e^- pair acceptors, which transfer the electrons to 12 O atoms.

3. Evolved Coupling Stoichiometries   The participation of ATP is fundamentally different from the role played by pyridine nucleotides and flavins. The stoichiometry of adenine nucleotides in metabolic sequences is not fixed by chemical necessity. Instead, the "stoichiometries" we observe are the consequences of evolutionary design. The overall equation for cellular respiration,[2] including the coupled formation of ATP by oxidative phosphorylation, is

$$C_6H_{12}O_6 + 6\ O_2 + 38\ ADP + 38\ P_i \longrightarrow 6\ CO_2 + 38\ ATP + 44\ H_2O$$

The "stoichiometry" of ATP formation, 38 ADP + 38 $P_i \to$ 38 ATP + 38 H_2O, cannot be predicted from any chemical considerations. The value of 38 ATP is an end result of biological adaptation. It is a trait that evolved through interactions between chemistry, heredity, and the environment over the course of evolution. Like any evolved character, ATP stoichiometry is the result of compromise. The final trait is one particularly suited to the fitness of the organism.

The number 38 is not magical. Recall that in eukaryotes, the net yield of ATP per glucose is 30 to 32, not 38 (Table 20.4). Also, the value of 38 was established a long time ago in evolution, when the prevailing atmospheric conditions and the competitive situation were undoubtedly very different from those today. The significance of this number is that it provides a high yield of ATP for each glucose molecule, yet the yield is still low enough that essentially all of the glucose is metabolized.

---

[2]This overall equation for cellular respiration is for the reaction within an uncompartmentalized (bacterial) cell. In eukaryotes, where much of the cellular respiration is compartmentalized within mitochondria, mitochondrial ADP/ATP exchange imposes a metabolic cost on the proton gradient of 1 H^+ per ATP, so the overall yield of ATP per glucose is 32, not 38.

ATP Coupling Stoichiometry Determines the K_{eq} for Metabolic Sequences

The fundamental biological purpose of ATP as an energy-coupling agent is to drive thermodynamically unfavorable reactions. (As a corollary, metabolic sequences composed of thermodynamically favorable reactions are exploited to drive the phosphorylation of ADP to make ATP.) Nature has devised enzymatic mechanisms that couple unfavorable reactions with ATP hydrolysis. In effect, the energy release accompanying ATP hydrolysis is transmitted to the unfavorable reaction so that the overall free energy change for the coupled process is negative (that is, favorable). The involvement of ATP serves to alter the free energy change for a reaction; or to put it another way, the role of ATP is to change the equilibrium ratio of [reactants] to [products] for a reaction. (See the A Deeper Look box on page 65.)

Another way of viewing these relationships is to note that, at equilibrium, the concentrations of ADP and P_i will be vastly greater than that of ATP because $\Delta G°'$ for ATP hydrolysis is a large negative number.[3] However, the cell where this reaction is at equilibrium is a dead cell. The living cell metabolizes food molecules to generate ATP. These catabolic reactions proceed with a very large overall decrease in free energy. Kinetic controls over the rates of the catabolic pathways are designed to ensure that the $[ATP]/([ADP][P_i])$ ratio is maintained very high. *The cell, by employing kinetic controls over the rates of metabolic pathways, maintains a very high $[ATP]/([ADP][P_i])$ ratio so that ATP hydrolysis can serve as the driving force for virtually all biochemical events.*

ATP Has Two Metabolic Roles

The role of ATP in metabolism is twofold:

1. It serves in a stoichiometric role to establish large equilibrium constants for metabolic conversions and to render metabolic sequences thermodynamically favorable. This is the role referred to when we call ATP the *energy currency* of the cell.
2. ATP also serves as an important allosteric effector in the kinetic regulation of metabolism. Its concentration (relative to those of ADP and AMP) is an index of the energy status of the cell and determines the rates of regulatory enzymes situated at key points in metabolism, such as PFK in glycolysis and FBPase in gluconeogenesis.

27.3 | Can Cellular Energy Status Be Quantified?

Is there a convenient measure of the energy status of the cell? Energy transduction and energy storage in the *adenylate system*—ATP, ADP, and AMP—lie at the very heart of metabolism. The amount of ATP a cell uses per minute is roughly equivalent to the steady-state amount of ATP it contains. Thus, the metabolic lifetime of an ATP molecule is brief. ATP, ADP, and AMP are all important effectors in exerting kinetic control on regulatory enzymes situated at key points in metabolism, so uncontrolled changes in their concentrations could have drastic consequences. The regulation of metabolism by adenylates in turn requires close control of the relative concentrations of ATP, ADP, and AMP. Some ATP-consuming reactions produce ADP; PFK and hexokinase are examples. Others lead to the formation of AMP, as in fatty acid activation by acyl-CoA synthetases:

Fatty acid + ATP + coenzyme A $\longrightarrow$ AMP + PP_i + fatty acyl-CoA

[3]Since $\Delta G°' = -30.5$ kJ/mol, ln $K_{eq} = 12.3$. So $K_{eq} = 2.2 \times 10^5$. Choosing starting conditions of [ATP] = 8 mM, [ADP] = 8 mM, and [P_i] = 1 mM, we can assume that, at equilibrium, [ATP] has fallen to some insignificant value x, [ADP] = approximately 16 mM, and [P_i] = approximately 9 mM. The concentration of ATP at equilibrium, x, then calculates to be about 1 nM.

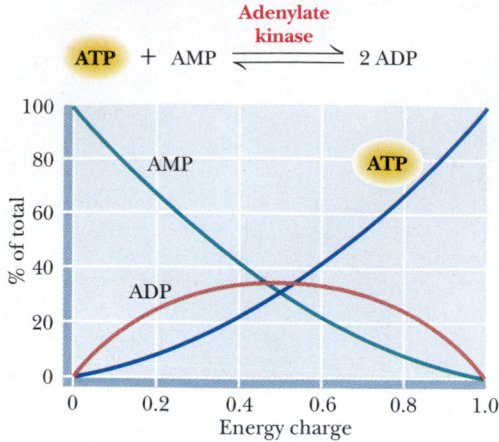

FIGURE 27.2 Relative concentrations of AMP, ADP, and ATP as a function of energy charge. (This graph was constructed assuming that the adenylate kinase reaction is at equilibrium and that $\Delta G^{\circ\prime}$ for the reaction is -473 J/mol; $K_{eq} = 1.2$.)

Adenylate Kinase Interconverts ATP, ADP, and AMP

Adenylate kinase (see Chapter 18), by catalyzing the reversible phosphorylation of AMP by ATP, provides a direct connection among all three members of the adenylate pool:

$$\text{ATP} + \text{AMP} \rightleftharpoons 2\,\text{ADP}$$

The free energy of hydrolysis of a phosphoanhydride bond is essentially the same in ADP and ATP (see Chapter 3), and the standard free energy change for this reaction is close to zero ($K_{eq} = 2.27$).

Energy Charge Relates the ATP Levels to the Total Adenine Nucleotide Pool

The role of the adenylate system is to provide phosphoryl groups at high group-transfer potential in order to drive thermodynamically unfavorable reactions. The capacity of the adenylate system to fulfill this role depends on how fully charged it is with phosphoric anhydrides. Energy charge is an index of this capacity:

$$\text{Energy charge} = \frac{1}{2}\left(\frac{2\,[\text{ATP}] + [\text{ADP}]}{[\text{ATP}] + [\text{ADP}] + [\text{AMP}]}\right)$$

The denominator represents the total adenylate pool ($[\text{ATP}] + [\text{ADP}] + [\text{AMP}]$); the numerator is the number of phosphoric anhydride bonds in the pool, two for each ATP and one for each ADP. The factor $\frac{1}{2}$ normalizes the equation so that energy charge, or **E.C.**, has the range 0 to 1.0. If all the adenylate is in the form of ATP, E.C. = 1.0, and the potential for phosphoryl transfer is maximal. At the other extreme, if AMP is the only adenylate form present, E.C. = 0. It is reasonable to assume that the adenylate kinase reaction is never far from equilibrium in the cell. Then the relative amounts of the three adenine nucleotides are fixed by the energy charge. Figure 27.2 shows the relative changes in the concentrations of the adenylates as energy charge varies from 0 to 1.0.

Key Enzymes Are Regulated by Energy Charge

Regulatory enzymes typically respond in reciprocal fashion to adenine nucleotides. For example, PFK is stimulated by AMP and inhibited by ATP. If the activities of various regulatory enzymes are examined in vitro as a function of energy charge, an interesting relationship appears. Regulatory enzymes in energy-producing catabolic pathways show greater activity at low energy charge, but the activity falls off abruptly as E.C. approaches 1.0. In contrast, regulatory enzymes of anabolic sequences are not very active at low energy charge, but their activities increase exponentially as E.C. nears 1.0 (Figure 27.3). These contrasting responses are termed **R,** for ATP-regenerating, and **U,** for ATP-utilizing. Regulatory enzymes such as PFK and pyruvate kinase in glycolysis follow the **R** response curve as E.C. is varied. Note that PFK itself is an ATP-utilizing enzyme, using ATP to phosphorylate fructose-6-phosphate to yield fructose-1,6-bisphosphate. Nevertheless, because PFK acts physiologically as the valve controlling the flux of carbohydrate down the catabolic pathways of cellular respiration that lead to ATP regeneration, it responds as an "**R**" enzyme to energy charge. Regulatory enzymes in anabolic pathways, such as acetyl-CoA carboxylase, which initiates fatty acid biosynthesis, respond as "**U**" enzymes.

The overall purposes of the **R** and **U** pathways are diametrically opposite in terms of ATP involvement. Note in Figure 27.3 that the **R** and **U** curves intersect at a rather high E.C. value. As E.C. increases past this point, **R** activities decline precipitously and **U** activities rise. That is, when E.C. is very high, biosynthesis is accelerated while catabolism diminishes. The consequence of these effects is that ATP is used up faster than it is regenerated, and so E.C. begins to fall. As E.C. drops below the point of intersection, **R** processes are favored over **U**. Then, ATP is generated faster than it is consumed, and E.C. rises again. The

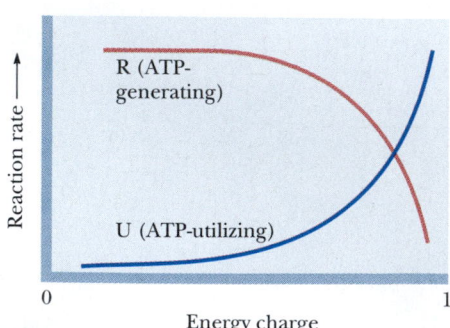

FIGURE 27.3 Responses of regulatory enzymes to variation in energy charge. Enzymes in catabolic pathways have as their ultimate metabolic purpose the regeneration of ATP from ADP. Such enzymes show an **R** pattern of response to energy charge. Enzymes in biosynthetic pathways utilize ATP to drive anabolic reactions; these enzymes follow the **U** curve in response to energy charge.

net result is that the value of energy charge oscillates about a point of **steady state** (Figure 27.4). The experimental results obtained from careful measurement of the relative amounts of AMP, ADP, and ATP in living cells reveals that normal cells have an energy charge in the neighborhood of 0.85 to 0.88. Maintenance of this steady-state value is one criterion of cell health and normalcy.

Phosphorylation Potential Is a Measure of Relative ATP Levels

Because energy charge is maintained at a relatively constant value in normal cells, it is not an informative index of cellular capacity to carry out phosphorylation reactions. The relative concentrations of ATP, ADP, and P_i do provide such information, and a function called **phosphorylation potential** has been defined in terms of these concentrations:

$$ADP + P_i \rightleftharpoons ATP + H_2O$$

Phosphorylation potential, Γ, is equal to $[ATP]/([ADP][P_i])$.

Note that this expression includes a term for the concentration of inorganic phosphate. $[P_i]$ has substantial influence on the thermodynamics of ATP hydrolysis. In contrast with energy charge, phosphorylation potential varies over a significant range as the actual proportions of ATP, ADP, and P_i in cells vary in response to metabolic state. Γ ranges from 200 to 800 M^{-1}, higher levels signifying more ATP and correspondingly greater phosphorylation potential.

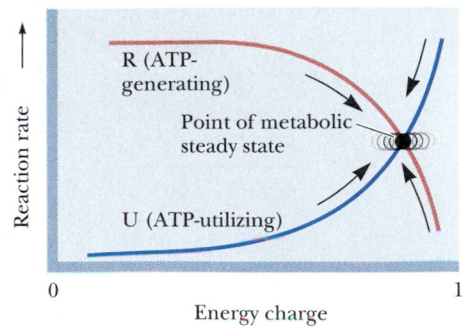

Biochemistry ⊛ **Now**™   **ACTIVE FIGURE 27.4**
The oscillation of energy charge (E.C.) about a steady-state value as a consequence of the offsetting influences of **R** and **U** processes on the production and consumption of ATP. As E.C. increases, the rates of **R** reactions decline, but **U** reactions go faster. ATP is consumed, and E.C. drops. Below the point of intersection, **R** processes are more active and **U** processes are slower, so E.C. recovers. Energy charge oscillates about a steady-state value determined by the intersection point of the **R** and **U** curves. **Test yourself on the concepts in this figure at http://chemistry.brookscole.com/ggb3**

27.4 | How Is Metabolism Integrated in a Multicellular Organism?

In complex multicellular organisms, organ systems have arisen to carry out specific physiological functions. Each organ expresses a repertoire of metabolic pathways that is consistent with its physiological purpose. Such specialization depends on coordination of metabolic responsibilities among organs so that the organism as a whole may thrive. Essentially all cells in animals have the set of enzymes common to the central pathways of intermediary metabolism, especially the enzymes involved in the formation of ATP and the synthesis of glycogen and lipid reserves. Nevertheless, organs differ in the metabolic fuels they prefer as substrates for energy production. Important differences also occur in the ways ATP is used to fulfill the organs' specialized metabolic functions. To illustrate these relationships, we will consider the metabolic interactions among the major organ systems found in humans: brain, skeletal muscle, heart, adipose tissue, and liver. In particular, the focus will be on energy metabolism in these organs (Figure 27.5). The major fuel depots in animals are *glycogen* in liver and muscle; *triacylglycerols* (fats) stored in adipose tissue; and *protein,* most of which is in skeletal muscle. In general, the order of preference for the use of these fuels is the order given: glycogen > triacylglycerol > protein. Nevertheless, the tissues of the body work together to maintain **caloric homeostasis,** defined as *a constant availability of fuels in the blood.*

The Major Organ Systems Have Specialized Metabolic Roles

Table 27.1 summarizes the energy metabolism of the major human organs.

BRAIN. The brain has two remarkable metabolic features. First, it has a very high respiratory metabolism. In resting adult humans, 20% of the oxygen consumed is used by the brain, even though it constitutes only 2% or so of body mass. Interestingly, this level of oxygen consumption is independent of mental activity, continuing even during sleep. Second, the brain is an organ with no significant fuel reserves—no glycogen, usable protein, or fat (even in "fatheads"!). Normally, the brain uses only glucose as a fuel and is totally dependent on the blood for a continuous incoming supply. Interruption of glucose supply for

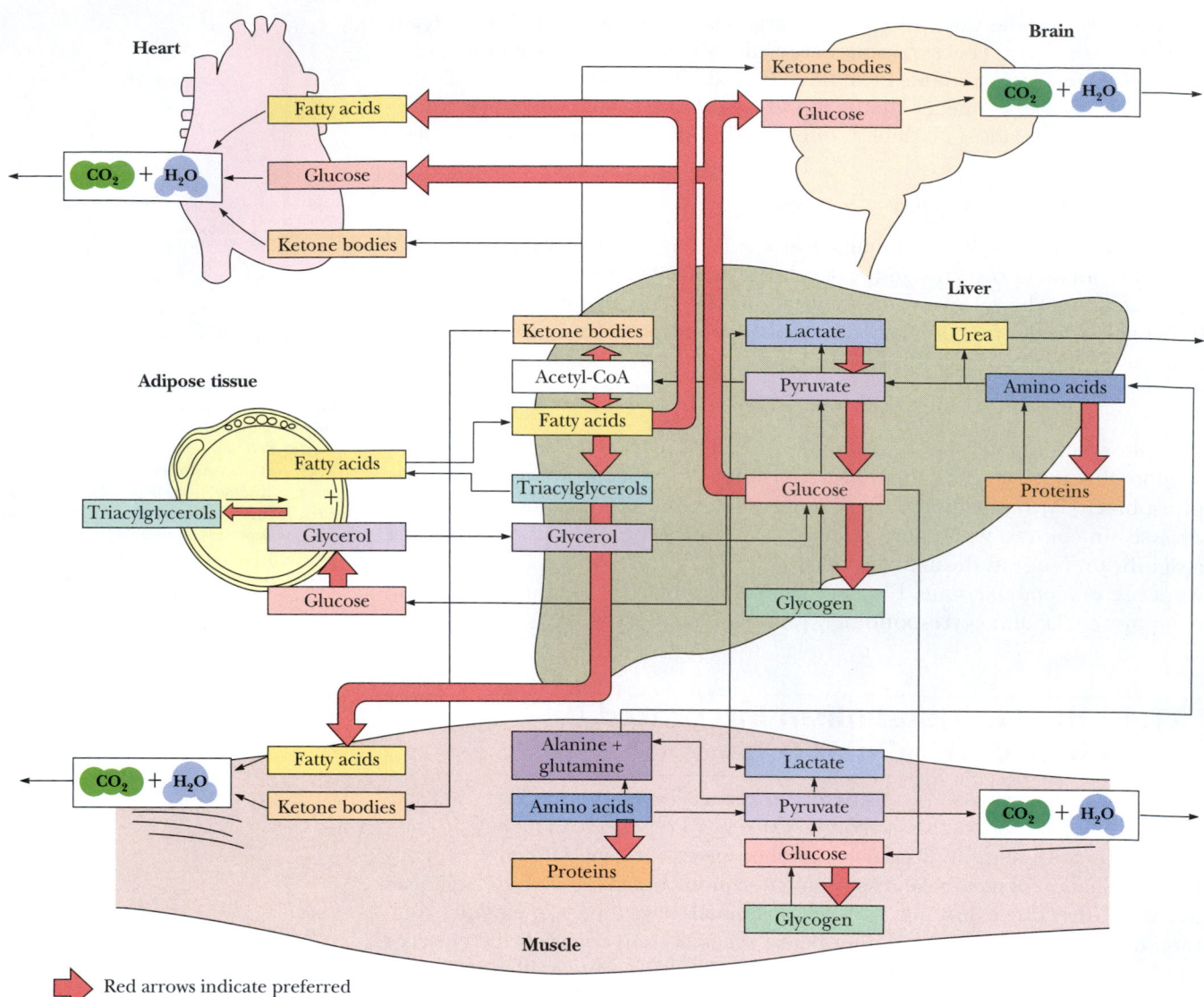

Red arrows indicate preferred routes in the well-fed state

Biochemistry⊛Now™ ANIMATED FIGURE 27.5
Metabolic relationships among the major human organs: brain, muscle, heart, adipose tissue, and liver. **See this figure animated at http://chemistry. brookscole.com/ggb3**

Table 27.1			
Energy Metabolism in Major Vertebrate Organs			
Organ	Energy Reservoir	Preferred Substrate	Energy Sources Exported
Brain	None	Glucose (ketone bodies during starvation)	None
Skeletal muscle (resting)	Glycogen	Fatty acids	None
Skeletal muscle (prolonged exercise)	None	Glucose	Lactate
Heart muscle	Glycogen	Fatty acids	None
Adipose tissue	Triacylglycerol	Fatty acids	Fatty acids, glycerol
Liver	Glycogen, triacylglycerol	Amino acids, glucose, fatty acids	Fatty acids, glucose, ketone bodies

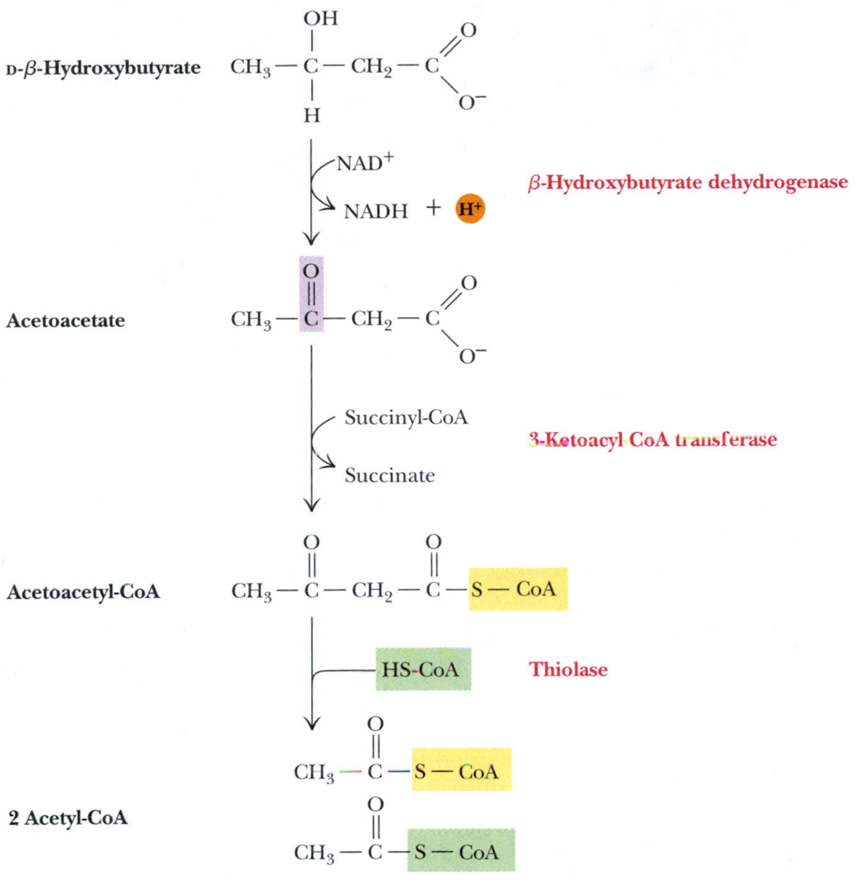

FIGURE 27.6 The structure of β-hydroxybutyrate and its conversion to acetyl-CoA for combustion in the citric acid cycle.

even brief periods of time (as in a stroke) can lead to irreversible losses in brain function. The brain uses glucose to carry out ATP synthesis via cellular respiration. High rates of ATP production are necessary to power the plasma membrane Na^+,K^+-ATPase so that the membrane potential essential for transmission of nerve impulses is maintained.

During prolonged fasting or starvation, the body's glycogen reserves are depleted. Under such conditions, the brain adapts to use β-hydroxybutyrate (Figure 27.6) as a source of fuel, converting it to acetyl-CoA for energy production via the citric acid cycle. β-Hydroxybutyrate (see Chapter 23) is formed from fatty acids in the liver. Although the brain cannot use free fatty acids or lipids directly from the blood as fuel, the conversion of these substances to β-hydroxybutyrate in the liver allows the brain to use body fat as a source of energy. The brain's other potential source of fuel during starvation is glucose obtained from gluconeogenesis in the liver (see Chapter 22), using the carbon skeletons of amino acids derived from muscle protein breakdown. The adaptation of the brain to use β-hydroxybutyrate from fat spares protein from degradation until lipid reserves are exhausted.

MUSCLE.   Skeletal muscle is responsible for about 30% of the O_2 consumed by the human body at rest. During periods of maximal exertion, skeletal muscle can account for more than 90% of the total metabolism. Muscle metabolism is primarily dedicated to the production of ATP as the source of energy for contraction and relaxation. Muscle contraction occurs when a motor nerve impulse causes Ca^{2+} release from specialized endomembrane compartments (the transverse tubules and sarcoplasmic reticulum). Ca^{2+} floods the *sarcoplasm* (the term denoting the cytosolic compartment of muscle cells), where it binds to **troponin C,** a regulatory

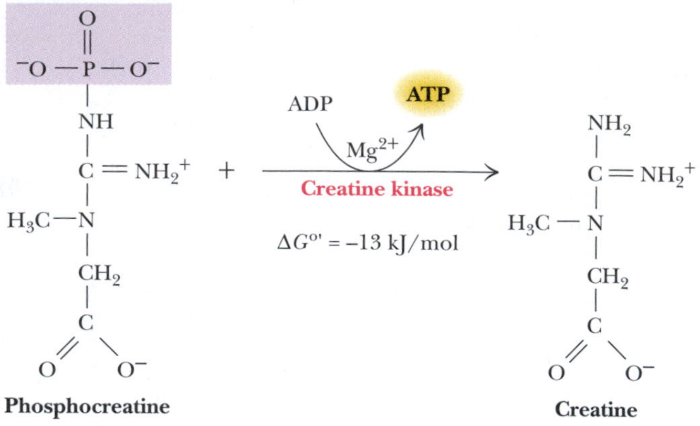

Biochemistry ⓔ Now™   ANIMATED FIGURE 27.7
Phosphocreatine serves as a reservoir of ATP-synthesizing potential. When ADP accumulates as a consequence of ATP hydrolysis, creatine kinase catalyzes the formation of ATP at the expense of phosphocreatine. During periods of rest, when ATP levels are restored by oxidative phosphorylation, creatine kinase acts in reverse to restore the phosphocreatine supply. **See this figure animated at** http://chemistry.brookscole.com/ggb3

protein, initiating a series of events that culminate in the sliding of myosin thick filaments along actin thin filaments. This mechanical movement is driven by energy released upon hydrolysis of ATP (see Chapter 16). The net result is that the muscle shortens. Relaxation occurs when the Ca^{2+} ions are pumped back into the sarcoplasmic reticulum by the action of a Ca^{2+}-transporting membrane ATPase. Two Ca^{2+} are translocated per ATP hydrolyzed. The amount of ATP used during relaxation is almost as much as that consumed during contraction.

Because muscle contraction is an intermittent process that occurs upon demand, muscle metabolism is designed for a demand response. Muscle at rest uses free fatty acids, glucose, or ketone bodies as fuel and produces ATP via oxidative phosphorylation. Resting muscle also contains about 2% glycogen by weight and an amount of phosphocreatine (Figure 27.7) capable of providing enough ATP to power about 4 seconds of exertion. During strenuous exertion, such as a 100-meter sprint, once the phosphocreatine is depleted, muscle relies solely on its glycogen reserves, making the ATP for contraction via glycolysis. In contrast with the citric acid cycle and oxidative phosphorylation pathways, glycolysis is capable of explosive bursts of activity, and the flux of glucose-6-phosphate through this pathway can increase 2000-fold almost instantaneously. The triggers for this activation are Ca^{2+} and the "fight or flight" hormone *epinephrine* (see Chapters 22 and 32). Little interorgan cooperation occurs during strenuous (anaerobic) exercise.

Muscle fatigue is the inability of a muscle to maintain power output. During maximum exertion, the onset of fatigue takes only 20 seconds or so. Fatigue is not the result of exhaustion of the glycogen reserves, nor is it a consequence of

---

Human Biochemistry

Biochemistry ⓔ Now™

Athletic Performance Enhancement with Creatine Supplements?

The creatine pool in a 70-kg (154-lb) human body is about 120 grams. This pool includes dietary creatine (from meat) and creatine synthesized by the human body from its precursors (arginine, glycine, and methionine). Of this creatine, 95% is stored in the skeletal and smooth muscles, about 70% of which is in the form of phosphocreatine. Supplementing the diet with 20 to 30 grams of creatine per day for 4 to 21 days can increase the muscle creatine pool by as much as 50% in someone with a previously low creatine level. Thereafter, supplements of 2 grams per day will maintain elevated creatine stores. Studies indicate that creatine supplementation gives some improvement in athletic performance during high-intensity, short-duration events (such as weight lifting), but no benefit in endurance events (such as distance running). The distinction makes sense in light of phosphocreatine's role as the substrate that creatine kinase uses to regenerate ATP from ADP. Intense muscular activity quickly (less than 2 seconds) exhausts ATP supplies; [phosphocreatine]$_{muscle}$ is sufficient to restore ATP levels for a few extra seconds, but no more. The U.S. Food and Drug Administration advises consumers to consult with their doctors before using creatine as a dietary supplement.

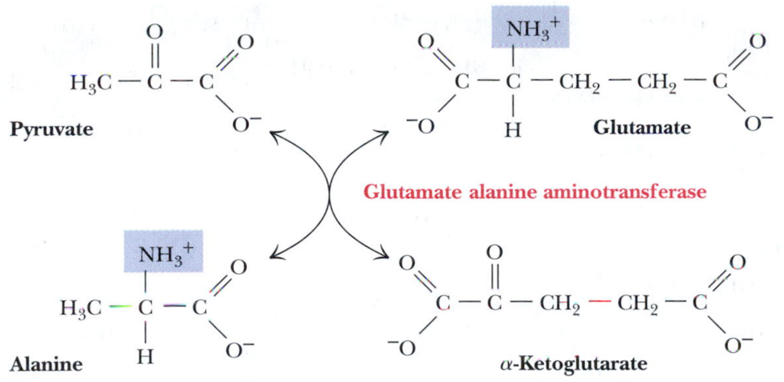

FIGURE 27.8 The transamination of pyruvate to alanine by glutamate : alanine aminotransferase.

lactate accumulation in the muscle. Instead, it is caused by a decline in intramuscular pH as protons are generated during glycolysis. (The overall conversion of glucose to 2 lactate in glycolysis is accompanied by the release of 2 H^+.) The pH may fall as low as 6.4. It is likely that the decline in PFK activity at low pH leads to a lowered flux of hexose through glycolysis and inadequate ATP levels, causing a feeling of fatigue. One benefit of PFK inhibition is that the ATP remaining is not consumed in the PFK reaction, thereby sparing the cell from the more serious consequences of losing all of its ATP.

During fasting or excessive activity, skeletal muscle protein is degraded to amino acids so that their carbon skeletons can be used as fuel. Many of the skeletons are converted to pyruvate, which can be transaminated back into alanine for export via the circulation (Figure 27.8). Alanine is carried to the liver, which in turn transaminates it back into pyruvate so that it can serve as a substrate for gluconeogenesis. Although muscle protein can be mobilized as an energy source, it is not efficient for an organism to consume its muscle and lower its overall fitness for survival. Muscle protein represents a fuel of last resort.

HEART.   In contrast with the intermittent work of skeletal muscle, the activity of heart muscle is constant and rhythmic. The range of activity in heart is also much less than that in muscle. Consequently, the heart functions as a completely aerobic organ and, as such, is very rich in mitochondria. Roughly half the cytoplasmic volume of heart muscle cells is occupied by mitochondria. Under normal working conditions, the heart prefers fatty acids as fuel, oxidizing acetyl-CoA units via the citric acid cycle and producing ATP for contraction via oxidative phosphorylation. Heart tissue has minimal energy reserves: a small amount of phosphocreatine and limited quantities of glycogen. As a result, the heart must be continually nourished with oxygen and free fatty acids, glucose, or ketone bodies as fuel.

ADIPOSE TISSUE.   Adipose tissue is an amorphous tissue that is widely distributed about the body—around blood vessels, in the abdominal cavity and mammary glands, and most prevalently, as deposits under the skin. It consists principally of cells known as adipocytes that no longer replicate. However, adipocytes can increase in number as adipocyte precursor cells divide, and obese individuals tend to have more of them. As much as 65% of the weight of adipose tissue is triacylglycerol that is stored in adipocytes, essentially as oil droplets. The average 70-kg man has enough caloric reserve stored as fat to sustain a 6000 kJ/day rate of energy production for 3 months, which is adequate for survival, assuming no serious metabolic aberrations (such as nitrogen, mineral, or vitamin deficiencies). Despite their role as energy storage depots, adipocytes have a high rate of metabolic activity, synthesizing and breaking

Human Biochemistry

Fat-Free Mice: A Model for One Form of Diabetes

Scientists at the National Institutes of Health have created transgenic mice that lack white adipose tissue throughout their lifetimes. These mice were created by blocking the normal differentiation of stem cells into adipocytes so that essentially no white adipose tissue can be formed in these animals. These "fat-free" mice have double the food intake and five times the water intake of normal mice. Fat-free mice also show decreased physical activity and must be kept warm on little heating pads to survive, because they lack insulating fat. They are also diabetic, with three times normal blood glucose and triacylglycerol levels and only 5% of normal leptin levels; they die prematurely. Like type 2 diabetic patients, fat-free mice have markedly elevated insulin levels (50–400 times normal) but are unresponsive to insulin. These mice serve as an excellent model for the disease *lipoatrophic diabetes,* an inherited disease characterized by the absence of adipose tissue and severe diabetic symptoms. Indeed, transplantation of adipose tissue into these fat-free mice cured their diabetes. As the major organ for triacylglycerol storage, white adipose tissue helps control energy homeostasis (food intake and energy expenditure) via the release of leptin and other hormonelike substances (see box on page 892). Clearly, absence of adipose tissue has widespread, harmful consequences for metabolism.

down triacylglycerol so that the average turnover time for a triacylglycerol molecule is just a few days. Adipocytes actively carry out cellular respiration, transforming glucose to energy via glycolysis, the citric acid cycle, and oxidative phosphorylation. If glucose levels in the diet are high, glucose is converted to acetyl-CoA for fatty acid synthesis. However, under most conditions, free fatty acids for triacylglycerol synthesis are obtained from the liver. Because adipocytes lack glycerol kinase, they cannot recycle the glycerol of triacylglycerol but rather depend on glycolytic conversion of glucose to dihydroxyacetone-3-phosphate (DHAP) and the reduction of DHAP to glycerol-3-phosphate for triacylglycerol biosynthesis. Adipocytes also require glucose to feed the pentose phosphate pathway for NADPH production.

Glucose plays a pivotal role for adipocytes. If glucose levels are adequate, glycerol-3-phosphate is formed in glycolysis and the free fatty acids liberated in triacylglycerol breakdown are re-esterified to glycerol to re-form triacylglycerols. However, if glucose levels are low, [glycerol-3-phosphate] falls and free fatty acids are released to the bloodstream (see Chapter 23).

"Brown Fat."   A specialized type of adipose tissue, so-called **brown fat,** is found in newborns and hibernating animals. The abundance of mitochondria, which are rich in cytochromes, is responsible for the brown color of this fat. As usual, these mitochondria are very active in electron transport–driven proton translocation, but these particular mitochondria contain in their inner membranes a protein, **thermogenin,** also known as *uncoupling protein-1* (see Chapter 20), that creates a passive proton channel, permitting the H^+ ions to reenter the mitochondrial matrix without generating ATP. Instead, the energy of oxidation is dissipated as heat. Indeed, brown fat is specialized to oxidize fatty acids for heat production rather than ATP synthesis.

LIVER.   The liver serves as the major metabolic processing center in vertebrates. Except for dietary triacylglycerols, which are metabolized principally by adipose tissue, most of the incoming nutrients that pass through the intestinal tract are routed via the portal vein to the liver for processing and distribution. Much of the liver's activity centers around conversions involving glucose-6-phosphate (Figure 27.9). Glucose-6-phosphate can be converted to glycogen, released as blood glucose, used to generate NADPH and pentoses via the pentose phosphate cycle, or catabolized to acetyl-CoA for fatty acid synthesis or for energy production via oxidative phosphorylation. Most of the liver glucose-6-phosphate arises from dietary carbohydrate, from degradation of glycogen reserves, or from muscle lactate that enters the gluconeogenic pathway.

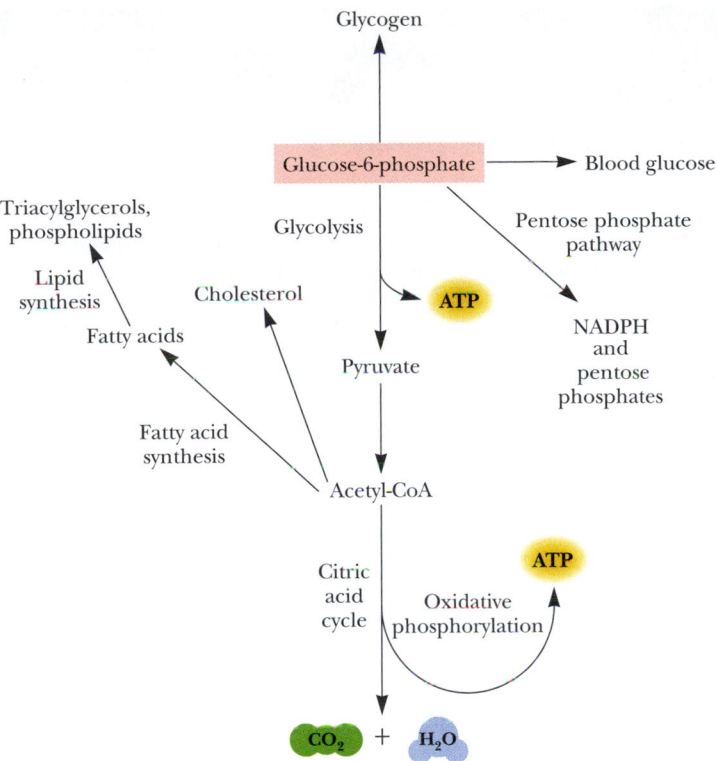

FIGURE 27.9 Metabolic conversions of glucose-6-phosphate in the liver.

The liver plays an important regulatory role in metabolism by buffering the level of blood glucose. Liver has two enzymes for glucose phosphorylation: hexokinase and glucokinase. Unlike hexokinase, glucokinase has a low affinity for glucose. Its K_m for glucose is high, on the order of 10 mM. When blood glucose levels are high, glucokinase activity augments hexokinase in phosphorylating glucose as an initial step leading to its storage in glycogen. The major metabolic hormones—epinephrine, glucagon, and insulin—all influence glucose metabolism in the liver to keep blood glucose levels relatively constant (see Chapters 22 and 32).

The liver is a major center for fatty acid turnover. When the demand for metabolic energy is high, triacylglycerols are broken down and fatty acids are degraded in the liver to acetyl-CoA to form ketone bodies, which are exported to the heart, brain, and other tissues. If energy demands are low, fatty acids are incorporated into triacylglycerols that are carried to adipose tissue for deposition as fat. Cholesterol is also synthesized in the liver from two-carbon units derived from acetyl-CoA.

In addition to these central functions in carbohydrate and fat-based energy metabolism, the liver serves other purposes. For example, the liver can use amino acids as metabolic fuels. Amino acids are first converted to their corresponding α-keto acids by aminotransferases. The amino group is excreted after incorporation into urea in the urea cycle. The carbon skeletons of gluconeogenic amino acids can be used for glucose synthesis, whereas those of ketogenic amino acids appear in ketone bodies (see Figure 25.41). The liver is also the principal detoxification organ in the body. The endoplasmic reticulum of liver cells is rich in enzymes that convert biologically active substances such as hormones, poisons, and drugs into less harmful by-products.

Liver disease leads to serious metabolic derangements, particularly in amino acid metabolism. In cirrhosis, the liver becomes defective in converting NH_4^+ to urea for excretion, and blood levels of NH_4^+ rise. Ammonia is toxic to the central nervous system, and coma ensues.

Human Biochemistry

Are You Hungry? The Hormones That Control Eating Behavior

Approximately 65% of Americans are overweight, and one in three Americans is clinically obese (overweight by 20% or more). Obesity is the single most important cause of type 2 (adult-onset insulin-independent) diabetes. Research into the regulatory controls that govern our feeding behavior has become a medical urgency with great financial incentives, given the epidemic proportions of obesity and widespread preoccupation with dieting and weight loss. Appetite and weight regulation is determined by a complex neuro-endocrine system that involves hormones produced in the stomach, liver, pancreas, fat tissue, and the digestive tract. These hormones move to the brain and act on neurons within the arcuate nucleus region of the hypothalamus, an anatomically distinct brain area that functions in homeostasis of body weight, body temperature, blood pressure, and other vital functions (see accompanying figure). The neurons respond to these signals by activating, or not, pathways involved in eating (food intake) and energy expenditure. Hormones that regulate eating behavior can be divided into short-term regulators that determine individual meals and long-term regulators that act to stabilize the levels of body fat deposits. Two subsets of neurons are involved: (1) the **NPY/AgRP**-producing neurons that release *NPY* (**neuropeptide Y),** the protein that stimulates the neurons that initiate eating behavior, and (2) the **melanocortin**-producing neurons, whose products inhibit the neurons initiating eating behavior. *AgRP* is **agouti-related peptide,** a protein that blocks the activity of melanocortin-producing neurons. *Melanocortins* are a group of peptide hormones that includes **α-melanocyte–stimulating hormone (α-MSH).** Melanocortins act on melanocortin receptors (MCRs), which are members of the 7-TMS G-protein–coupled receptor (GPCR) family of membrane receptors; MCRs trigger cellular responses through adenylyl cyclase activation (see Chapters 15 and 32).

Short-term regulators of eating include **ghrelin** and **cholecystokinin.** Ghrelin is an appetite-stimulating peptide hormone produced in the stomach. Production of ghrelin is maximal when the stomach is empty, but ghrelin levels fall quickly once food is consumed. Cholecystokinin is a peptide hormone released from the gastrointestinal tract during eating. In contrast to ghrelin, cholecystokinin signals satiety (the sense of fullness) and tends to curtail further eating. Together, ghrelin and cholecystokinin constitute a meal-to-meal control system that regulates the onset and end of eating behavior. The activity of this control system is also modulated by the long-term regulators.

Long-term regulators include **insulin** and **leptin,** both of which inhibit eating and promote energy expenditure. Insulin is produced in the β-cells of the pancreas when blood glucose levels rise. A major role of insulin is to stimulate glucose uptake from the blood into muscle, fat, and other tissues. Blood insulin levels correlate with body fat amounts. Insulin also stimulates fat cells to make leptin. Leptin (from the Greek word *lepto,* meaning "thin") is a 16-kD, 146–amino acid residue protein produced principally in adipocytes (fat cells). Leptin has a four-helix bundle tertiary structure similar to that of cytokines (protein hormones involved in cell–cell communication). Normally, as fat deposits accumulate in adipocytes, more and more leptin is produced in these cells and spewed into the bloodstream. Leptin levels in the blood communicate the status of triacylglycerol levels in the adipocytes to the central nervous system so that appropriate changes in appetite take place. If leptin levels are low ("starvation"), appetite increases; if leptin levels are high ("overfeeding"), appetite is suppressed. Leptin also regulates fat metabolism in adipocytes, inhibiting fatty acid biosynthesis and stimulating fat metabolism. In the latter case, leptin induces synthesis of the enzymes in the fatty acid oxidation pathway and increases expression of *uncoupling protein 2 (UCP2),* a mitochondrial protein that uncouples oxidation from phosphorylation so that the energy of oxidation is lost as heat (thermogenesis). Leptin binding to leptin receptors in the hypothalamus inhibits release of NPY. Because NPY is a potent *orexic* (appetite-stimulating) peptide hormone, leptin is therefore an *anorexic* (appetite-suppressing) agent. Functional leptin receptors are also essential for pituitary function, growth hormone secretion, and normal puberty. When body fat stores decline, the circulating levels of leptin and insulin also decline. Hypothalamic neurons sense this decline and act to increase appetite to restore body fat levels.

Intermediate regulation of eating behavior is accomplished by the gut hormone PYY_{3-36}. PYY_{3-36} is produced in endocrine cells found in distal regions of the small intestine and in the large intestine, areas that receive ingested food some time after a meal is eaten. PYY_{3-36} inhibits eating for many hours after a meal by acting on the NPY/AgRP-producing neurons in the arcuate nucleus. Clearly, the regulatory controls that govern eating are complex and layered. Some believe that defects in these controls are common and biased in favor of overeating, an advantageous evolutionary strategy that may have unforeseen consequences in these bountiful times.

Human Biochemistry

The Metabolic Effects of Alcohol Consumption

Ethanol metabolism alters the $NAD^+/NADH$ ratio. Ethanol is metabolized to acetate in the liver by alcohol dehydrogenase and aldehyde dehydrogenase:

$$CH_3CH_2OH + NAD^+ \rightleftharpoons CH_3CHO + NADH + H^+$$
$$CH_3CHO + NAD^+ \rightleftharpoons CH_3COO^- + NADH + H^+$$

The excess NADH produced inhibits NAD^+-requiring reactions, such as gluconeogenesis and fatty acid oxidation. Inhibition of fatty acid oxidation causes elevated triacylglycerol levels in the liver. Over

time, these triacylglycerols accumulate as fatty deposits, which ultimately contribute to cirrhosis of the liver. Inhibition of gluconeogenesis leads to buildup of this pathway's substrate, lactate. Lactic acid accumulation in the blood causes acidosis. A further consequence is that acetaldehyde can form adducts with protein $-NH_2$ groups, which may impair protein function. Because gluconeogenesis is impaired, alcohol consumption can cause *hypoglycemia* (low blood sugar) in someone who is undernourished. In turn, hypoglycemia can cause irreversible damage to the central nervous system.

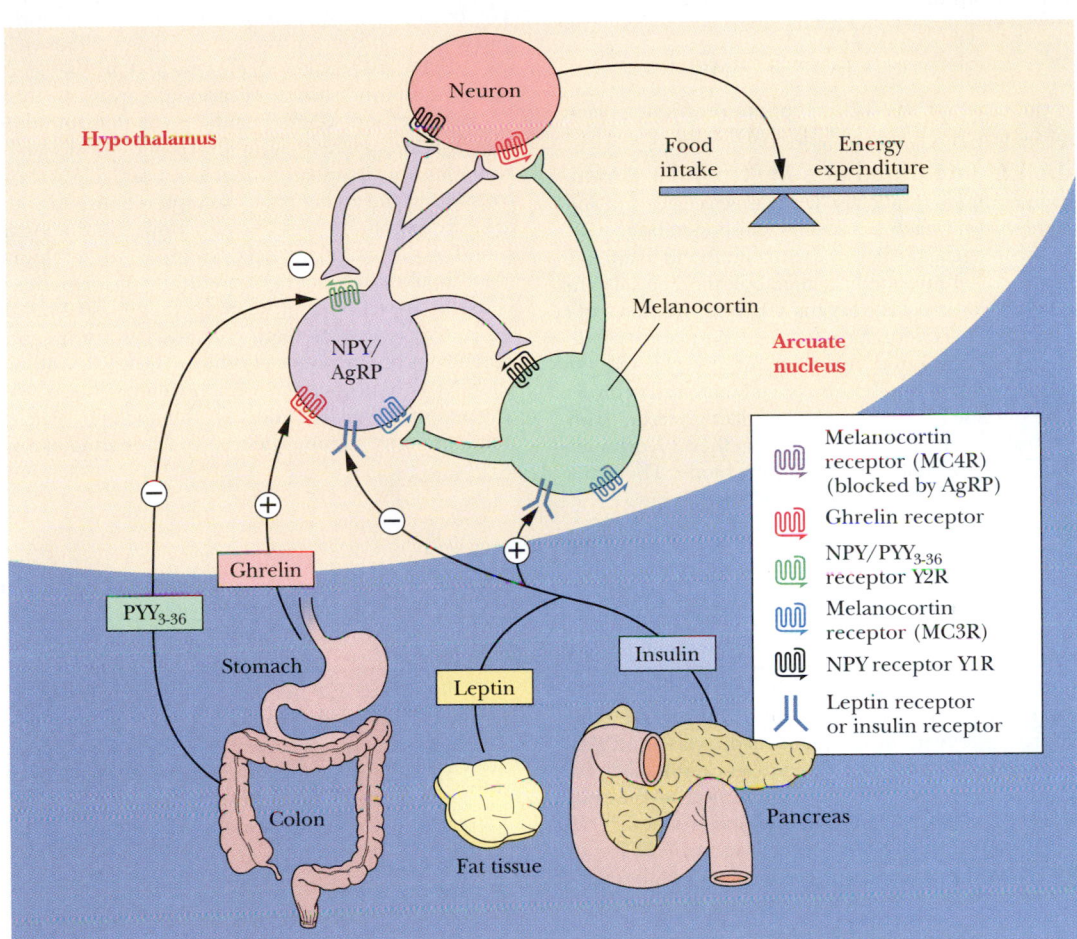

▲ The regulatory pathways that control eating. The long-term regulators leptin and insulin suppress appetite by inhibiting NPY/AgRP-producing neurons in the arcuate nucleus of the hypothalamus. NPY/AgRP-producing neurons release NPY and AgRP, which stimulate eating through actions on other neurons that initiate feeding. At the same time, leptin and insulin are stimulating the melanocortin-producing neurons. Melanocortins inhibit eating through effects on other neurons. Ghrelin is a short-term regulator that also activates the NPY/AgRP-expressing neurons. PYY_{3-36}, released from the lower intestinal tract, inhibits the NPY/AgRP-releasing neurons, thereby decreasing appetite. *(Adapted from Figure 1 in Schwartz, M. W., and Morton, G. J., 2002. Obesity: Keeping hunger at bay. Nature **418**:595–597.)*

Summary

27.1 Can Systems Analysis Simplify the Complexity of Metabolism?
Cells are in a dynamic steady state maintained by considerable metabolic flux. The metabolism going on in even a single cell is so complex that it defies meaningful quantitative description. Nevertheless, overall relationships become more obvious by a systems analysis approach to intermediary metabolism. The metabolism of a typical heterotrophic cell can be summarized in three interconnected functional blocks: (1) catabolism, (2) anabolism, and (3) macromolecular synthesis and growth. Phototrophic cells require a fourth block: photosynthesis. Only a few metabolic intermediates connect these systems, and ATP and NADPH serve as the carriers of chemical energy and reducing power, respectively, between these various blocks.

27.2 What Underlying Principle Relates ATP Coupling to the Thermodynamics of Metabolism?
ATP coupling determines the thermodynamics of metabolic sequences. The ATP coupling stoichiometry cannot be predicted from chemical considerations; instead it is a quantity selected by evolution and the fundamental need for metabolic sequences to be emphatically favorable from a thermodynamic perspective. Catabolic sequences generate ATP with an overall favorable K_{eq} (and hence, a negative ΔG) and anabolic sequences consume this energy with an overall favorable K_{eq}, even though such sequences may span the same starting and end points (as in fatty acid oxidation of palmitoyl-CoA to 8 acetyl-CoA versus synthesis of palmitoyl-CoA from 8 acetyl-CoA). ATP has two metabolic roles: a stoichiometric role in

rendering metabolic sequences thermodynamically favorable and a regulatory role as an allosteric effector.

27.3 Can Cellular Energy Status Be Quantified?

The level of phosphoric anhydride bonds in the adenylate system of ATP, ADP, and AMP can be expressed in terms of the energy charge equation:

$$\text{Energy charge} = \frac{1}{2}\left(\frac{2\,[\text{ATP}] + [\text{ADP}]}{[\text{ATP}] + [\text{ADP}] + [\text{AMP}]}\right)$$

More revealing of the potential for an ATP-dependent reaction to occur is Γ, the phosphorylation potential: $\Gamma = [\text{ATP}]/([\text{ADP}][\text{P}_i])$.

27.4 How Is Metabolism Integrated in a Multicellular Organism?

Organ systems in complex multicellular organisms carry out specific physiological functions, with each expressing those metabolic pathways appropriate to its physiological purpose. Essentially all cells in animals carry out the central pathways of intermediary metabolism, especially the reactions of ATP synthesis. Nevertheless, organs differ in the metabolic fuels they prefer as substrates for energy production. The major fuel depots in animals are glycogen in liver and muscle; triacylglycerols (fats) in adipose tissue; and protein, most of which is in skeletal muscle. The order of preference for the use of these fuels is glycogen > triacylglycerol > protein. Nevertheless, the tissues of the body work together to maintain caloric homeostasis: Constant availability of fuels in the blood. The major organ systems have specialized metabolic roles within the organism.

The brain has a strong reliance on glucose as fuel. Muscle at rest primarily relies on fatty acids, but under conditions of strenuous contraction when O_2 is limiting, muscle shifts to glycogen as its primary fuel. The heart is a completely aerobic organ, rich in mitochondria, with a preference for fatty acids as fuel under normal operating conditions. The liver is the body's metabolic processing center, taking in nutrients and sending out products such as glucose, fatty acids, and ketone bodies. Adipose tissue takes up glucose and, to a lesser extent, fatty acids for the synthesis and storage of triacylglycerols.

Appetite and weight regulation is determined by a complex neuroendocrine system involving hormones produced in the stomach, liver, pancreas, fat tissue, and intestinal tract. These hormones act on neurons within the arcuate nucleus region of the hypothalamus, an anatomically distinct brain area that functions in homeostasis. The neurons respond to these signals by activating, or not, pathways involved in eating (food intake) and energy expenditure. Hormones that regulate eating behavior are either short-term regulators, such as ghrelin and cholecystokinin, or long-term regulators, such as insulin and leptin. Two subsets of neurons are involved: (1) the NPY/AgRP-producing neurons that release NPY (neuropeptide Y), a hormone that triggers neurons controlling eating behavior, and (2) the melanocortin-producing neurons, whose products inhibit the neurons initiating eating behavior.

Problems

1. (Integrates with Chapters 3, 18, and 22.) The conversion of PEP to pyruvate by pyruvate kinase (glycolysis) and the reverse reaction to form PEP from pyruvate by pyruvate carboxylase and PEP carboxykinase (gluconeogenesis) represent a so-called substrate cycle. The direction of net conversion is determined by the relative concentrations of allosteric regulators that exert kinetic control over pyruvate kinase, pyruvate carboxylase, and PEP carboxykinase. Recall that the last step in glycolysis is catalyzed by pyruvate kinase:

$$\text{PEP} + \text{ADP} \rightleftharpoons \text{pyruvate} + \text{ATP}$$

The standard free energy change is -31.7 kJ/mol.
   a. Calculate the equilibrium constant for this reaction.
   b. If $[\text{ATP}] = [\text{ADP}]$, by what factor must [pyruvate] exceed [PEP] for this reaction to proceed in the reverse direction?

The reversal of this reaction in eukaryotic cells is essential to gluconeogenesis and proceeds in two steps, each requiring an equivalent of nucleoside triphosphate energy:

Pyruvate carboxylase
$$\text{Pyruvate} + \text{CO}_2 + \text{ATP} \longrightarrow \text{oxaloacetate} + \text{ADP} + \text{P}_i$$

PEP carboxykinase
$$\underline{\text{Oxaloacetate} + \text{GTP} \longrightarrow \text{PEP} + \text{CO}_2 + \text{GDP}}$$

Net: $\text{Pyruvate} + \text{ATP} + \text{GTP} \longrightarrow \text{PEP} + \text{ADP} + \text{GDP} + \text{P}_i$

   c. The $\Delta G^{\circ\prime}$ for the overall reaction is $+0.8$ kJ/mol. What is the value of K_{eq}?
   d. Assuming $[\text{ATP}] = [\text{ADP}]$, $[\text{GTP}] = [\text{GDP}]$, and $\text{P}_i = 1$ mM when this reaction reaches equilibrium, what is the ratio of [PEP]/[pyruvate]?
   e. Are both directions in the substrate cycle likely to be strongly favored under physiological conditions?

2. (Integrates with Chapter 3.) Assume the following intracellular concentrations in muscle tissue: ATP = 8 mM, ADP = 0.9 mM, AMP = 0.04 mM, P$_i$ = 8 mM. What is the *energy charge* in muscle? What is the *phosphorylation potential*?

3. Strenuous muscle exertion (as in the 100-meter dash) rapidly depletes ATP levels. How long will 8 mM ATP last if 1 gram of muscle consumes 300 μmol of ATP per minute? (Assume muscle is 70% water.) Muscle contains phosphocreatine as a reserve of phosphorylation potential. Assuming [phosphocreatine] = 40 mM, [creatine] = 4 mM, and $\Delta G^{\circ\prime}$ (phosphocreatine + $H_2O \rightleftharpoons$ creatine + P$_i$) = -43.3 kJ/mol, how low must [ATP] become before it can be replenished by the reaction: phosphocreatine + ADP $\rightleftharpoons$ ATP + creatine? [Remember, $\Delta G^{\circ\prime}$ (ATP hydrolysis) = -30.5 kJ/mol.]

4. (Integrates with Chapter 20.) The standard reduction potentials for the (NAD^+/NADH) and ($\text{NADP}^+/\text{NADPH}$) couples are identical, namely, -320 mV. Assuming the in vivo concentration ratios $\text{NAD}^+/\text{NADH} = 20$ and $\text{NADP}^+/\text{NADPH} = 0.1$, what is ΔG for the following reaction?

$$\text{NADPH} + \text{NAD}^+ \rightleftharpoons \text{NADP}^+ + \text{NADH}$$

Assuming standard state conditions for the reaction, ADP + P$_i \longrightarrow$ ATP + H_2O, calculate how many ATP equivalents can be formed from ADP + P$_i$ by the energy released in this reaction.

5. (Integrates with Chapter 3.) Assume the total intracellular pool of adenylates (ATP + ADP + AMP) = 8 mM, 90% of which is ATP. What are [ADP] and [AMP] if the adenylate kinase reaction is at equilibrium? Suppose [ATP] drops suddenly by 10%. What are the concentrations now for ADP and AMP, assuming that the adenylate kinase reaction is at equilibrium? By what factor has the AMP concentration changed?

6. (Integrates with Chapters 18 and 22.) The reactions catalyzed by PFK and FBPase constitute another substrate cycle. PFK is AMP activated; FBPase is AMP inhibited. In muscle, the maximal activity of PFK (mmol of substrate transformed per minute) is ten times greater than FBPase activity. If the increase in [AMP] described in problem 5 raised PFK activity from 10% to 90% of its maximal value but lowered FBPase activity from 90% to 10% of its maximal value, by what factor is the flux of fructose-6-P through the glycolytic pathway changed? (*Hint:* Let PFK maximal activity = 10, FBPase maximal activity = 1; calculate the relative activities of the two enzymes at low [AMP] and at high [AMP]; let *J*, the flux of F-6-P through the substrate cycle under any condition, equal the velocity of the PFK reaction *minus* the velocity of the FBPase reaction.)

7. (Integrates with Chapters 23 and 24.) Leptin not only induces synthesis of fatty acid oxidation enzymes and uncoupling protein 2 in adipocytes, but it also causes inhibition of acetyl-CoA carboxylase, resulting in a decline in fatty acid biosynthesis. This effect on acetyl-CoA carboxylase, as an additional consequence, enhances fatty acid oxidation. Explain how leptin-induced inhibition of acetyl-CoA carboxylase might promote fatty acid oxidation.

8. (Integrates with Chapters 19 and 20.) Acetate produced in ethanol metabolism can be transformed into acetyl-CoA by the acetyl thiokinase reaction:

$$Acetate + ATP + CoASH \longrightarrow acetyl\text{-}CoA + AMP + PP_i$$

Acetyl-CoA then can enter the citric acid cycle and undergo oxidation to 2 CO_2. How many ATP equivalents can be generated in a liver cell from the oxidation of one molecule of ethanol to 2 CO_2 by this route, assuming oxidative phosphorylation is part of the process? (Assume all reactions prior to acetyl-CoA entering the citric acid cycle occur outside the mitochondrion.) Per carbon atom, which is a better metabolic fuel, ethanol or glucose? That is, how many ATP equivalents per carbon atom are generated by combustion of glucose versus ethanol to CO_2?

9. (Integrates with Chapter 23.) Assuming each NADH is worth 3 ATP, each FADH$_2$ is worth 2 ATP, and each NADPH is worth 4 ATP: How many ATP equivalents are produced when one molecule of palmitoyl-CoA is oxidized to 8 molecules of acetyl-CoA by the fatty acid β-oxidation pathway? How many ATP equivalents are consumed when 8 molecules of acetyl-CoA are transformed into one molecule of palmitoyl-CoA by the fatty acid biosynthetic pathway? Can both of these metabolic sequences be metabolically favorable at the same time if ΔG for ATP synthesis is +50 kJ/mol?

10. (Integrates with Chapters 18–21.) If each NADH is worth 3 ATP, each FADH$_2$ is worth 2 ATP, and each NADPH is worth 4 ATP, calculate the equilibrium constant for cellular respiration, assuming synthesis of each ATP costs 50 kJ/mol of energy. Calculate the equilibrium constant for CO_2 fixation under the same conditions, except here ATP will be hydrolyzed to ADP + P_i with the release of 50 kJ/mol. Comment on whether these reactions are thermodynamically favorable under such conditions.

11. (Integrates with Chapter 22.) In type 2 diabetics, glucose production in the liver is not appropriately regulated, so glucose is overproduced. One strategy to treat this disease focuses on the development of drugs targeted against regulated steps in glycogenolysis and gluconeogenesis, the pathways by which liver produces glucose for release into the blood. Which enzymes would you select for as potential targets for such drugs?

12. As chief scientist for drug development at PhatFarmaceuticals, Inc., you want to create a series of new diet drugs. You have a grand plan to design drugs that might limit production of some hormones or promote the production of others. Which hormones are on your "limit production" list and which are on your "raise levels" list?

13. The existence of leptin was revealed when the *ob/ob* genetically obese strain of mice was discovered. These mice have a defective leptin gene. Predict the effects of daily leptin injections into *ob/ob* mice on food intake, fatty acid oxidation, and body weight. Similar clinical trials have been conducted on humans, with limited success. Suggest a reason why this therapy might not be a miracle cure for overweight individuals.

14. Would it be appropriate to call neuropeptide Y (NPY) the obesity-promoting hormone? What would be the phenotype of a mouse whose melanocortin-producing neurons failed to produce melanocortin? What would be the phenotype of a mouse lacking a functional MC3R gene? What would be the phenotype of a mouse lacking a functional leptin receptor gene?

15. The Human Biochemistry box The Metabolic Effects of Alcohol Consumption, points out that ethanol is metabolized to acetate in the liver by alcohol dehydrogenase and aldehyde dehydrogenase:

$$CH_3CH_2OH + NAD^+ \rightleftharpoons CH_3CHO + NADH + H^+$$
$$CH_3CHO + NAD^+ \rightleftharpoons CH_3COO^- + NADH + H^+$$

These reactions alter the NAD$^+$/NADH ratio in liver cells. From your knowledge of glycolysis, gluconeogenesis, and fatty acid oxidation, what might be the effect of an altered NAD$^+$/NADH ratio on these pathways? What is the basis of this effect?

Preparing for the MCAT Exam

16. Consult Figure 27.5 and answer the following questions: Which organs use both fatty acids and glucose as a fuel in the well-fed state, which rely mostly on glucose, which rely most mostly on fatty acids, which one never uses fatty acids, and which one produces lactate.

17. Figure 27.3 illustrates the response of R (ATP-regenerating) and U (ATP-utilizing) enzymes to energy charge.
    a. Would hexokinase be an R enzyme or a U enzyme? Would glutamine:PRPP amidotransferase, the second enzyme in purine biosynthesis, be an R enzyme or a U enzyme?
    b. If energy charge = 0.5: Is the activity of hexokinase high or low? Is ribose-5-P pyrophosphokinase activity high or low?
    c. If energy charge = 0.95: Is the activity of hexokinase high or low? Is ribose-5-P pyrophosphokinase activity high or low?

Biochemistry ⒺNow™ Preparing for an exam? Test yourself on key questions at http://chemistry.brookscole.com/ggb3

Further Reading

Systems Analysis of Metabolism
Brand, M. D., and Curtis, R. K., 2002. Simplifying metabolic complexity. *Biochemical Society Transactions* 30:25–30.

ATP Coupling and the Thermodynamics of Metabolism
Atkinson, D. E., 1977. *Cellular Energy Metabolism and Its Regulation.* New York: Academic Press. A very readable book on the design and purpose of cellular energy metabolism. Its emphasis is the evolutionary design of metabolism within the constraints of chemical thermodynamics. The book is filled with novel insights regarding why metabolism is organized as it is and why ATP occupies a central position in biological energy transformations.

Newsholme, E. A., Challiss, R. A. J., and Crabtree, B., 1984. Substrate cycles: Their role in improving sensitivity in metabolic control. *Trends in Biochemical Sciences* 9:277–280. A review suggesting that substrate cycles provide a mechanism for greater responsiveness to regulatory signals. (See end-of-chapter problems 1 and 6.)

Newsholme, E. A., and Leech, A. R., 1983. *Biochemistry for the Medical Sciences.* New York: John Wiley & Sons.

Metabolic Relationships Between Organ Systems
Harris, R., and Crabb, D. W., 1997. Metabolic interrelationships. In *Textbook of Biochemistry with Clinical Correlations,* 4th ed., Devlin, T. M., ed. New York: Wiley-Liss. A synopsis of the interdependence of metabolic processes in the major tissues of the human body—brain, liver, muscle, kidney, gut, and adipose tissue. Metabolic aberrations that occur in certain disease states are also discussed.

Sugden, M. C., Holness, M. J., and Palmer, T. N., 1989. Fuel selection and carbon flux during the starved-to-fed transition. *Biochemical Journal* **263**:313–323. Changes in lipid and carbohydrate metabolism of rats upon carbohydrate feeding after prolonged starvation.

Creatine as a Nutritional Supplement

Ekblom, B., 1999. Effects of creatine supplementation on performance. *American Journal of Sports Medicine* **24**:S-38.

Kreider, R., 1998. Creatine supplementation: Analysis of ergogenic value, medical safety, and concerns. *Journal of Exercise Physiology* **1,** an international online journal available at http://www.css.edu/users/tboone2/asep/jan3.htm

Fat-Free Mice

Gavrilova, O., et al., 2000. Surgical implantation of adipose tissue reverses diabetes in lipoatrophic mice. *Journal of Clinical Investigation* **105**:271–278.

Moitra, J., et al., 1998. Life without white fat: a transgenic mouse. *Genes and Development* **12**:3168–3181.

Leptin and Hormonal Regulation of Eating Behavior

Barinaga, M., 1995. "Obese" protein slims mice. *Science* **269**:475–476, and references therein.

Clement, K., et al., 1998. A mutation in the human leptin receptor gene causes obesity and pituitary dysfunction. *Nature* **392**:398–401.

Erickson, J. C., et al., 1996. Attenuation of the obesity syndrome of *ob/ob* mice by the loss of neuropeptide Y. *Science* **274**:1704–1707.

Flier, J. S., 1997. Leptin expression and action: New experimental paradigms. *Proceedings of the National Academy of Science, U.S.A.* **94**:4242–4245.

Gura, T., 1997. Obesity sheds its secrets. *Science* **275**:751–753.

Heymsfield, S. B., et al., 1999. Recombinant leptin for weight loss in obese and lean adults: A randomized, controlled, dose-escalation trial. *Journal of the American Medical Association* **282**:1568–1575.

Lonnqvist, F., et al., 1999. Leptin and its potential role in human obesity. *Journal of Internal Medicine* **245**:643–652.

Saper, C. B., Chou. T. C., and Elmquist, J. K., 2002. The need to feed: Homeostatic and hedonic control of eating. *Neuron* **36**:199–211.

Schwartz, M. W., and Morton, G. J., 2002. Obesity: Keeping hunger at bay. *Nature* **418**:595–597.

Tartaglia, L. A., 1997. The leptin receptor. *Journal of Biological Chemistry* **272**:6093–6096.

Vaisse, C., et al., 2000. Melanocortin-4 receptor mutations are a frequent and heterogeneous cause of morbid obesity. *Journal of Clinical Investigation* **106**:253–262.

Zhou, Y-T., et al., 1997. Induction by leptin of uncoupling protein-2 and enzymes of fatty acid oxidation. *Proceedings of the National Academy of Sciences, U.S.A.* **94**:6386–6390.

How Do Cells Coordinate Their Activities?

An Essay by David L. Brautigan, University of Virginia

Despite great diversity among eukaryotic organisms, from single cell microbes like yeast to complex organisms like human beings, the biochemistry is fundamentally the same. This poses a question for continuing research: *What is common, and what is distinctive about divergent organisms?* Part 4 offers us a view of the basis for both unity and diversity.

The relative number of genes, and the conservation of what those genes encode, delivers a message of the fundamental unity among all eukaryotes. Sequence analysis of genomes reveals not a large difference in the number of genes between metazoan organisms, especially those with complex embryology and specialized tissues (see Table 1.5). Many human genes are recognizable as common to worms or fruit flies, implying that they serve the same basic life functions. Indeed, human genes can even replace the corresponding yeast genes, showing that the encoded proteins are functionally equivalent. Consider further that all organisms use the same amino acids, produce proteins and enzymes that fold into the same conserved motifs and three-dimensional conformations, and catalyze the same basic set of chemical reactions. Most of our metabolic machinery is the same (e.g., TCA cycle), yeast to human, from common molecular blueprints in our genomes. Indeed, yeast and other simple organisms can act as models for understanding human diseases and for developing treatments for defects in these conserved processes. On the other hand, although the number of human genes is only about 30,000, we still do not know the function of more than 40% of them! A lot of biochemistry remains to be done.

Animal cells have to perform together like a marching band; in contrast, a yeast cell is more like a solo musician.

Diversity among eukaryotic organisms appears on this background of biochemical unity. Worms and fruit flies and humans have many different types of cells. Furthermore, humans have five senses and cognitive abilities not shared by simpler eukaryotes. Without accumulating many new genes, the protein repertoire has been expanded in higher eukaryotes. For example, single gene transcripts are spliced into multiple messages, each encoding a variant protein, and single proteins are cleaved into multiple products, each tailored for a specific biological function. Diversity arises in part from subtle variations.

Specialization of cell function allows for more complex organisms, but the activities of the different cells must be coordinated with one another. Animal cells have to perform together like a marching band; in contrast, a yeast cell is more like a solo musician. To coordinate different biochemical activities, cells communicate with neighboring cells and distant tissues. Over the past 50 years, biochemists have discovered hormones and growth factors that function as chemical messengers between cells. These messengers bind to membrane receptors and trigger a series of intracellular reactions collectively referred to as hormone action, signal transduction, or cell signaling. This has been my area of research specialization. Over the past 20 years, we have deciphered many different linear reaction sequences for signal transduction; however, we are just beginning to learn how these pathways are networked together, governed by feedback loops and multiple inputs, and localized to certain sites within a cell. In my laboratory, we look for sequences conserved between species and among related proteins as keys to the common function, and test distinctive sequences for encoded specificity. Even on a molecular basis, we seek and study unity and diversity.

CHAPTER 28

DNA Metabolism: Replication, Recombination, and Repair

Julie Newdoll's painting "Dawn of the Double Helix" composes the DNA duplex as human figures. Her theme in this painting is "Life Forms: The basic structures that make our existence possible." Thomas Hardy's poem captures the immortality of DNA as "the eternal thing in man, That heeds no call to die," or as Richard Dawkins wryly comments, "the individual is just DNA's way of making more DNA!"

"Dawn of the Double Helix," by Julie Newdoll

Heredity

I am the family face; Flesh perishes, I live on,
Projecting trait and trace Through time to
*   times anon,*
And leaping from place to place Over
*   oblivion.*
The years-heired feature that can In curve
*   and voice and eye*
Despise the human span Of durance—that
*   is I;*
The eternal thing in man, That heeds no
*   call to die.*

Thomas Hardy (First published in *Moments of Vision and Miscellaneous Verses*, Macmillan, 1917)

Key Questions

Biochemistry ⊜ Now™ Test yourself on these Key Questions at BiochemistryNow at **http://chemistry.brookscole.com/ggb3**

Essential Question

DNA is the physical repository of genetic information in the cell and the material of heredity that is passed on to progeny. *How is this genetic information in the form of DNA replicated, how is the information rearranged, and how is its integrity maintained in the face of damage?*

Heredity, which we can define generally as the tendency of an organism to possess the characteristics of its parent(s), is clearly evident throughout nature and since the dawn of history has served to justify the classification of organisms according to shared similarities. The basis of heredity, however, was a mystery. Early in the 20th century, geneticists demonstrated that **genes,** the elements or units carrying and transferring inherited characteristics from parent to offspring, are contained within the nuclei of cells in association with the chromosomes. Yet the chemical identity of genes remained unknown, and genetics was an abstract science. Even the realization that chromosomes are composed of proteins and nucleic acids did little to define the molecular nature of the gene because, at the time, no one understood either of these substances.

The material of heredity must have certain properties. It must be very stable so that genetic information can be stored in it and transmitted countless times to subsequent generations. It must be capable of precise copying or replication so that its information is not lost or altered. And, although stable, it must also be subject to change in order to account, in the short term, for the appearance of mutant forms and, in the long term, for evolution. DNA is the material of heredity.

28.1 | How Is DNA Replicated?

Transfer of genetic information from generation to generation requires the faithful reproduction of the parental DNA. DNA reproduction produces two identical copies of the original DNA in a process termed **DNA replication.**

The publication of Watson and Crick's famous paper titled *Molecular Structure of Nucleic Acids: A Structure for Deoxyribose Nucleic Acid* (Figure 28.1) marked the dawn of a new scientific epoch, the age of molecular biology. As these authors drew to a close their brief but far-reaching description of the DNA double helix, they pointedly commented, "It has not escaped our notice that the specific [base] pairing we have postulated immediately suggests a possible copying mechanism for the genetic material." The mechanism for DNA replication that Watson and Crick viewed as intuitively obvious is *strand separation followed by the copying of each strand.* In the process, each separated strand acts as a **template** for the synthesis of a new complementary strand whose nucleotide sequence is fixed by the base-pairing rules Watson and Crick proposed. Strand separation is achieved by untwisting the double helix (Figure 28.2). Base pairing then dictates an accurate replication of the original DNA double helix.

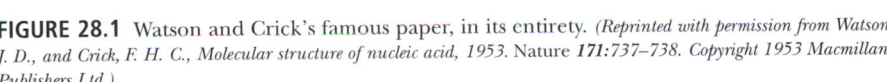

FIGURE 28.1 Watson and Crick's famous paper, in its entirety. *(Reprinted with permission from Watson, J. D., and Crick, F. H. C., Molecular structure of nucleic acid, 1953. Nature **171**:737–738. Copyright 1953 Macmillan Publishers Ltd.)*

A *template* is something whose edge is shaped in a particular way so that it can serve as a guide in making a similar object with a corresponding contour.

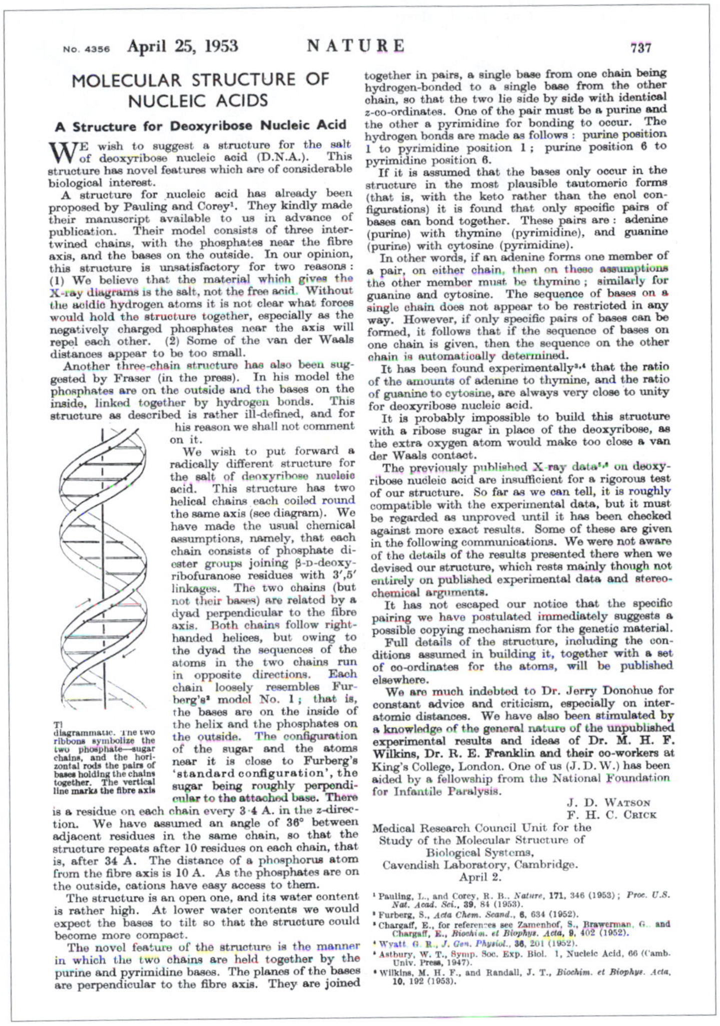

DNA Replication Is Semiconservative

According to Watson and Crick's base-pairing rules, the nucleotide sequence in one strand dictates the sequence of nucleotides in the other. Given these rules, three basic models for DNA replication are possible. These three models—conservative, semiconservative, and dispersive—are diagrammed in Figure 28.3. In 1958, Matthew Meselson and Franklin Stahl provided experimental proof for the **semiconservative model** of DNA replication. *Escherichia coli* cells were grown for many generations in medium containing $^{15}NH_4Cl$ as the sole nitrogen source. Thus, the nitrogen atoms in the purine and pyrimidine bases of the DNA in these cells were mostly ^{15}N, the stable heavy isotope of nitrogen. Then, a tenfold excess of ordinary $^{14}NH_4Cl$ was added to the growing culture, and at appropriate intervals, cells were collected from the culture and lysed. The DNA they contained was analyzed by CsCl density gradient ultracentrifugation (see the Appendix to Chapter 11, Isopycnic Centrifugation and Buoyant Density of DNA box). This technique can resolve macromolecules differing in density by less than 0.01 g/mL.

DNA isolated from cells grown on $^{15}NH_4^+$ (the "0" generation cells) banded in the ultracentrifuge at a density corresponding to 1.724 g/mL, whereas DNA from cells grown for 4.1 generations on ^{14}N had a density of 1.710 g/mL (Figure 28.4).

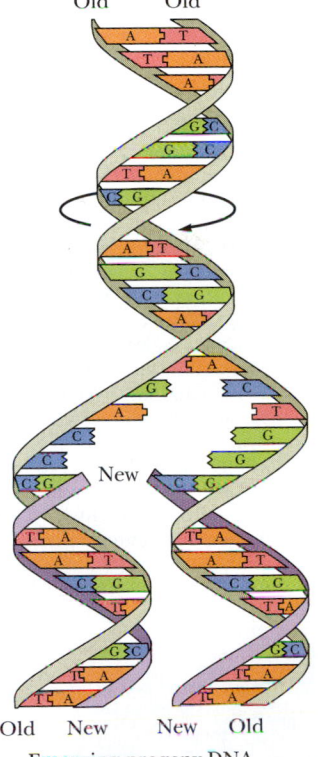

Emerging progeny DNA

FIGURE 28.2 Untwisting of DNA strands exposes their bases for hydrogen bonding. Base pairing ensures that appropriate nucleotides are inserted in the correct positions as the new complementary strands are synthesized. By this mechanism, the nucleotide sequence of one strand dictates a complementary sequence in its daughter strand. The original strands untwist by rotating about the axis of the unreplicated DNA double helix.

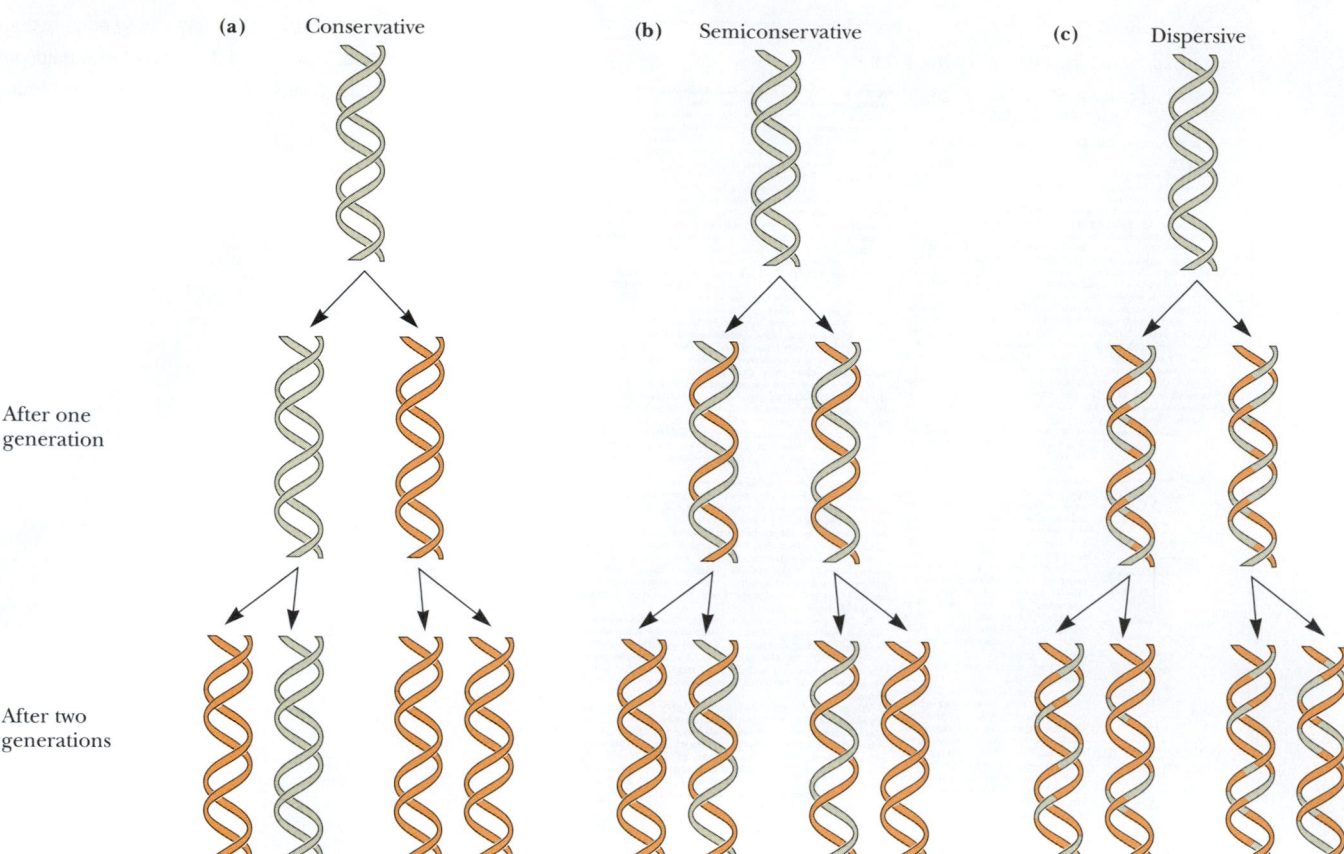

(a) Conservative   **(b)** Semiconservative   **(c)** Dispersive

After one
generation

After two
generations

Biochemistry Now™   ANIMATED FIGURE 28.3
Three models of DNA replication prompted by
Watson and Crick's double helix structure for DNA.
(a) *Conservative:* Each strand of the DNA duplex is
replicated, and the two newly synthesized strands
join to form one DNA double helix while the two
parental strands remain associated with each other.
The products are one completely new DNA duplex
and the original DNA duplex. **(b)** *Semiconservative:*
The two strands separate, and each strand is copied
to generate a complementary strand. Each parental
strand remains associated with its newly synthesized
complement, so each DNA duplex contains one
parental strand and one new strand. **(c)** *Dispersive:*
This model predicts that each of the four strands in
the two daughter DNA duplexes contains both newly
synthesized segments and segments derived from the
parental strands. **See this figure animated at http://
chemistry.brookscole.com/ggb3**

These bands represent "heavy" and "light" DNA, respectively. Significantly, DNA
isolated from cells grown for just one generation on ^{14}N yielded just a single band
corresponding to a density of 1.717 g/mL, halfway between heavy and light DNA,
indicating that *each* DNA duplex molecule contained equal amounts of ^{15}N and
^{14}N. This result is consistent with the semiconservative model for DNA replication,
at the same time ruling out the conservative model. After approximately two generations on ^{14}N, cells yielded DNA that gave two essentially equal bands upon
ultracentrifugation, one at the intermediate density of 1.717 g/mL and the other
at the light position, 1.710 g/mL, also in accord with semiconservative replication.

It remained conceivable that DNA replication might follow a dispersive
model, so both strands of DNA in the chromosomes of cells after one generation on ^{14}N might be intermediate in density. Meselson and Stahl eliminated
this possibility with the following experiment: DNA isolated from ^{15}N-labeled
cells kept one generation on ^{14}N was heated at 100°C so that the DNA duplexes
were denatured into their component single strands. When this heat-denatured
DNA was analyzed by CsCl density gradient ultracentrifugation, two distinct
bands were observed, showing that one strand was ^{15}N-labeled and the other
strand was ^{14}N-labeled. A dispersive mode of replication would have yielded a
single band of DNA of intermediate density.

DNA Replication Is Bidirectional

Replication of DNA molecules begins at one or more specific regions called the
origin(s) of replication (discussed in Section 28.3) and, excepting certain bacteriophage chromosomes and plasmids, proceeds in both directions from this
origin (Figure 28.5). For example, replication of *E. coli* DNA begins at *oriC*, a
unique 245-bp chromosomal site. From this site, replication advances in both
directions around the circular chromosome. That is, bidirectional replication
involves two **replication forks** that move in opposite directions.

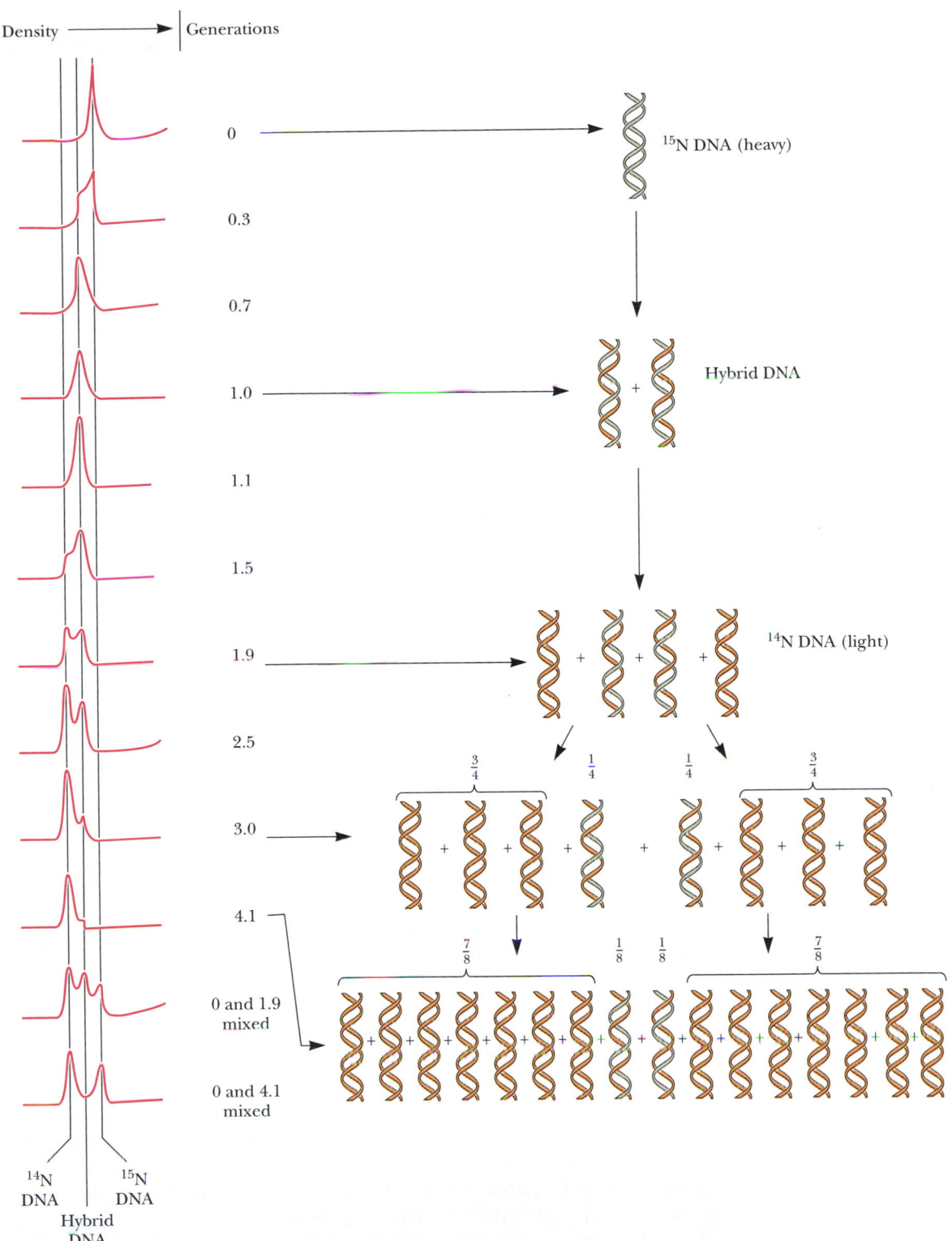

Density ——→ | Generations

0 — ¹⁵N DNA (heavy)

0.3

0.7

1.0 — Hybrid DNA

1.1

1.5

1.9 — ¹⁴N DNA (light)

2.5

3.0 — $\frac{3}{4}$ $\frac{1}{4}$ $\frac{1}{4}$ $\frac{3}{4}$

4.1 — $\frac{7}{8}$ $\frac{1}{8}$ $\frac{1}{8}$ $\frac{7}{8}$

0 and 1.9 mixed

0 and 4.1 mixed

¹⁴N DNA   ¹⁵N DNA

Hybrid DNA

FIGURE 28.4 The Meselson and Stahl experiment demonstrating that DNA replication is semi-conservative. On the left are densitometric traces made of UV absorption photographs taken of the ultracentrifugation cells containing DNA isolated from *E. coli* grown for various generation times after ¹⁵N-labeling. The photographs were taken once the migration of the DNA in the density gradient had reached equilibrium. Density increases from left to right. The peaks reveal the positions of the banded DNA with respect to the density of the solution. The number of generations that the *E. coli* cells were grown (following 14 generations of ¹⁵N density-labeling) is shown down the middle. A schematic representation interpreting the pattern expected of semiconservative replication is shown on the right side of this figure. (*Adapted from Meselson, M., and Stahl, F. W., 1958. The replication of DNA.* Proceedings of the National Academy of Sciences, U.S.A. **44**:671–682.)

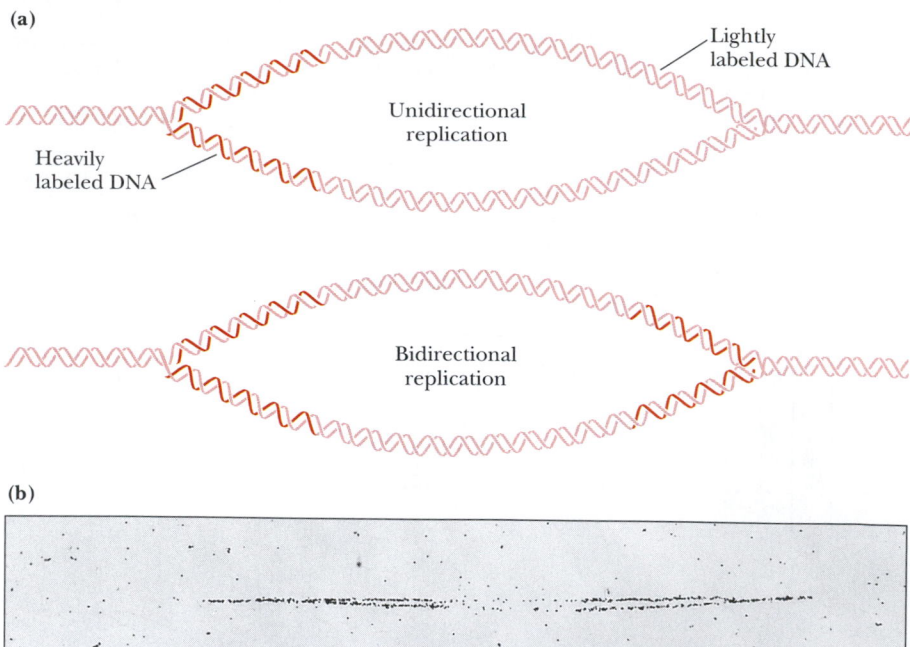

(a)

Lightly labeled DNA

Unidirectional replication

Heavily labeled DNA

Bidirectional replication

(b)

FIGURE 28.5 Bidirectional replication of the *E. coli* chromosome. **(a)** If replication is bidirectional, autoradiograms of radioactively labeled replicating chromosomes should show two replication forks heavily labeled with radioactive thymidine. **(b)** An autoradiogram of the chromosome from a dividing *E. coli* cell shows bidirectional replication. *(Photo courtesy of David M. Prescott, University of Colorado.)*

Replication Requires Unwinding of the DNA Helix

Semiconservative replication depends on unwinding the DNA double helix to expose single-stranded templates to polymerase action. For a double helix to unwind, it must either rotate about its axis (while the end of its strands are held fixed), or positive supercoils must be introduced, one for each turn of the helix unwound (see Chapter 11). If the chromosome is circular, as in *E. coli,* only the latter alternative is possible. Because DNA replication in *E. coli* proceeds at a rate approaching 1000 nucleotides per second and there are about 10 bp per helical turn, the chromosome would accumulate 100 positive supercoils per second! In effect, the DNA would become too tightly supercoiled to allow unwinding of the strands.

DNA gyrase, a Type II topoisomerase, acts to overcome the torsional stress imposed upon unwinding; DNA gyrase introduces negative supercoils at the expense of ATP hydrolysis. The unwinding reaction is driven by **helicases** (see also Chapter 16), a class of proteins that catalyze the ATP-dependent unwinding of DNA double helices. Unlike topoisomerases that alter the linking number of dsDNA through phosphodiester bond breakage and reunion (see Chapter 11), helicases simply disrupt the hydrogen bonds that hold the two strands of duplex DNA together. A helicase molecule requires a single-stranded region for binding. It then moves along the single strand, its translocation coupled to ATP hydrolysis and to unwinding the double-stranded DNA it encounters. **SSB (single-stranded DNA-binding protein)** binds to the unwound strands, preventing their re-annealing. At least ten distinct DNA helicases involved in different aspects of DNA and RNA metabolism have been found in *E. coli* alone. **DnaB** is the DNA helicase acting in *E. coli* DNA replication. DnaB helicase assembles as a hexameric (α_6) "doughnut"-shaped protein ring, with DNA passing through its hole.

DNA Replication Is Semidiscontinuous

Incorporation of radioactively labeled thymine nucleotides into DNA during replication, followed by autoradiography of the replicating DNA (Figure 28.5), reveals that the two strands of duplex DNA are both replicated at each advancing replication fork by **DNA polymerase.** DNA polymerase uses single-stranded DNA (ssDNA) as a template and makes a complementary strand by polymerizing de-

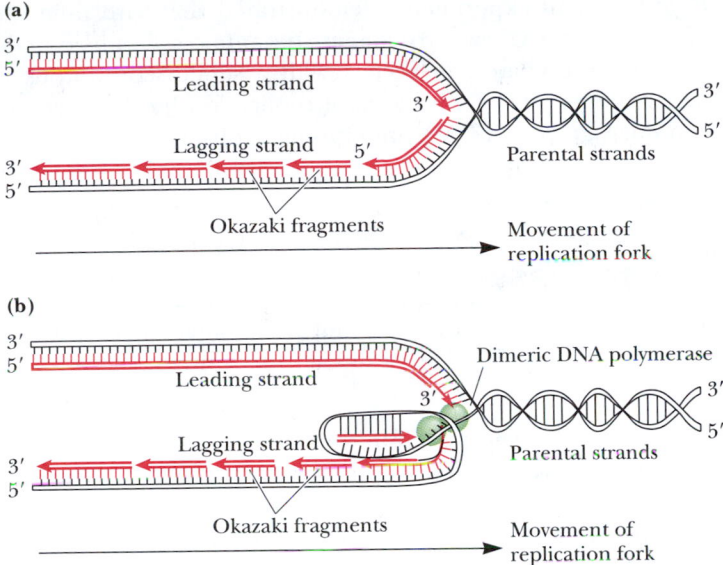

Biochemistry ⒺNow™ **ANIMATED FIGURE 28.6** The semidiscontinuous model for DNA replication. Newly synthesized DNA is shown as red. Because DNA polymerases only polymerize nucleotides 5'→3', both strands must be synthesized in the 5'→3' direction. Thus, the copy of the parental 3'→5' strand is synthesized continuously; this newly made strand is designated the leading strand. **(a)** As the helix unwinds, the other parental strand (the 5'→3' strand) is copied in a discontinuous fashion through synthesis of a series of fragments 1000 to 2000 nucleotides in length, called the Okazaki fragments; the strand constructed from the Okazaki fragments is called the lagging strand. **(b)** Because both strands are synthesized in concert by a dimeric DNA polymerase situated at the replication fork, the 5'→3' parental strand must wrap around in *trombone fashion* so that the unit of the dimeric DNA polymerase replicating it can move along it in the 3'→5' direction. This parental strand is copied in a discontinuous fashion because the DNA polymerase must occasionally dissociate from this strand and rejoin it further along. The Okazaki fragments are then covalently joined by DNA ligase to form an uninterrupted DNA strand. **See this figure animated at http://chemistry.brookscole.com/ggb3**

oxynucleotides in the order specified by their base pairing with bases in the template. DNA polymerases synthesize DNA only in a 5'→3' direction, reading the antiparallel template strand in a 3'→5' sense. A dilemma arises: How does DNA polymerase copy the parent strand that runs in the 5'→3' direction at the replication fork? It turns out that *replication is semidiscontinuous* (Figure 28.6): As the DNA helix is unwound during its replication, the 3'→5' strand (as defined by the direction that the replication fork is moving) can be copied continuously by DNA polymerase proceeding in the 5'→3' direction behind the replication fork. The other parental strand is copied only when a sufficient stretch of its sequence has been exposed for DNA polymerase to move along it in the 5'→3' mode. Thus, one parental strand is copied continuously to give a newly synthesized copy, called the **leading strand,** at each replication fork. The other parental strand is copied in an intermittent, or discontinuous, mode to yield a set of fragments. These fragments are then joined to form an intact **lagging strand.** Overall, each of the two DNA duplexes produced in DNA replication contain one "old" and one "new" DNA strand, and half of the new strand was formed by leading strand synthesis and the other half by lagging strand synthesis.

The Lagging Strand Is Formed from Okazaki Fragments

In 1968, Tuneko and Reiji Okazaki provided biochemical verification of the semidiscontinuous pattern of DNA replication just described. The Okazakis exposed a rapidly dividing *E. coli* culture to ³H-labeled thymidine for 30 seconds, quickly collected the cells, and found that half of the label incorporated into nucleic acid appeared in short ssDNA chains just 1000 to 2000 nucleotides in length. (The other half of the radioactivity was recovered in very large DNA molecules.) Subsequent experiments demonstrated that with time, the newly synthesized short ssDNA **Okazaki fragments** became covalently joined to form longer polynucleotide chains, in accord with a semidiscontinuous mode of replication. The generality of this mode of replication has been corroborated with electron micrographs of DNA undergoing replication in eukaryotic cells.

28.2	What Are the Properties of DNA Polymerases?

The enzymes that replicate DNA are called DNA polymerases. All DNA polymerases, whether from prokaryotic or eukaryotic sources, share the following properties:

1. The incoming base is selected within the DNA polymerase active site, as determined by Watson–Crick geometric interactions with the corresponding base in the template strand.
2. Chain growth is in the $5'→3'$ direction and is antiparallel to the template strand.
3. DNA polymerases cannot initiate DNA synthesis de novo—all require a primer oligonucleotide with a free 3'-OH to build upon.

E. coli Cells Have Several Different DNA Polymerases

Table 28.1 compares the properties of the principal DNA polymerases in *E. coli*. These enzymes are numbered **I** through **V** in order of their discovery. DNA polymerases I, II, and V function principally in DNA repair; **DNA polymerase III** is the chief DNA-replicating enzyme of *E. coli*. Only 10 to 20 copies of this enzyme are present per cell.

The First DNA Polymerase Discovered Was *E. coli* DNA Polymerase I

In 1957, Arthur Kornberg and his colleagues discovered the first DNA polymerase, **DNA polymerase I.** DNA polymerase I catalyzed the synthesis of DNA in vitro if provided with all four deoxynucleoside-5'-triphosphates (dATP, dTTP, dCTP, dGTP), a template DNA strand to copy, *and* a **primer.** A primer is essential because DNA polymerases can elongate only preexisting chains; they cannot join two deoxyribonucleoside-5'-phosphates together to make the initial phosphodiester bond. The primer base pairs with the template DNA, forming a short, double-stranded region. This primer must possess a free 3'-OH end to which an incoming deoxynucleoside monophosphate is added. One of the four dNTPs is selected as substrate, pyrophosphate (PP_i) is released, and the

Table 28.1			
Properties of the DNA Polymerases of *E. coli*			
Property	**Pol I**	**Pol II**	**Pol III (core)***
Mass (kD)	103	90	$130(\alpha), 27.5(\epsilon), 8.6(\theta)$
Molecules/cell	400	?	40
Turnover number[†]	600	30	1200
Polymerization $5'→3'$	Yes	Yes	Yes
Exonuclease $3'→5'$	Yes	Yes	Yes
Exonuclease $5'→3'$	Yes	No	No

*α-, ϵ-, and θ-subunits.
[†]Nucleotides polymerized at 37°C/minute/molecule of enzyme.
Source: Adapted from Kornberg, A., and Baker, T. A., 1991. *DNA Replication,* 2nd ed. New York: W. H. Freeman; and Kelman, Z., and O'Donnell, M., 1995. DNA polymerase III holoenzyme: Structure and function of a chromosomal replicating machine. *Annual Review of Biochemistry* **64**:171–200.

dNMP is linked to the 3'-OH of the primer chain through formation of a phosphoester bond (Figure 28.7). The deoxynucleoside monophosphate selected as substrate is chosen through its geometric fit with the template base to form a Watson–Crick base pair. As DNA polymerase I catalyzes the successive addition of deoxynucleotide units to the 3'-end of the primer, the chain is elongated in the 5'→3' direction, forming a polynucleotide sequence that is antiparallel and complementary to the template. DNA polymerase I can proceed along the template strand, synthesizing a complementary strand of about 20 bases before it "falls off" (dissociates from) the template. The degree to which the enzyme remains associated with the template through successive cycles of nucleotide addition is referred to as its **processivity.** As DNA polymerases go, DNA polymerase I is a modestly processive enzyme. Arthur Kornberg was awarded the Nobel Prize in Physiology or Medicine in 1959 for his discovery of this DNA polymerase. DNA polymerase I is the best characterized of these enzymes.

E. coli DNA Polymerase I Has Three Active Sites on Its Single Polypeptide Chain

In addition to its 5'→3' polymerase activity, *E. coli* DNA polymerase I has two other catalytic functions: a *3'→5' exonuclease (3'-exonuclease)* activity and a *5'→3' exonuclease (5'-exonuclease)* activity. The three distinct catalytic activities of DNA polymerase I reside in separate active sites in the enzyme.

E. coli DNA Polymerase I Is Its Own Proofreader and Editor

The exonuclease activities of *E. coli* DNA polymerase I are functions that enhance the accuracy of DNA replication. The 3'-exonuclease activity removes nucleotides from the 3'-end of the growing chain (Figure 28.8), an action that negates the action of the polymerase activity. Its purpose, however, is to remove incorrect (mismatched) bases. Although the 3'-exonuclease works slowly compared to the polymerase, the polymerase cannot elongate an improperly base-paired primer terminus. Thus, the relatively slow 3'-exonuclease has time to act and remove the mispaired nucleotide. Therefore, the polymerase active site is a proofreader, and the 3'-exonuclease activity is an editor. This check on the accuracy of base pairing enhances the overall precision of the process.

The 5'-exonuclease of DNA polymerase I acts upon duplex DNA, degrading it from the 5'-end by releasing mononucleotides and oligonucleotides. It can

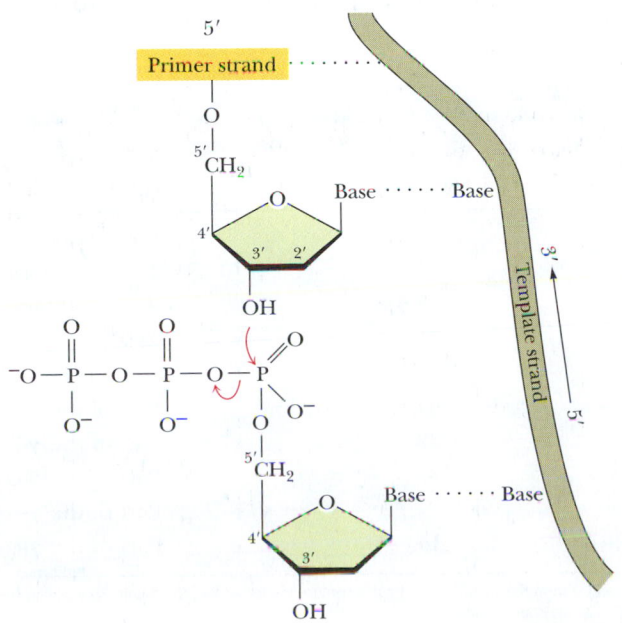

FIGURE 28.7 The chain elongation reaction catalyzed by DNA polymerase. DNA polymerase I joins deoxynucleoside monophosphate units to the 3'-OH end of the primer, employing dNTPs as substrates. The 3'-OH carries out a nucleophilic attack on the α-phosphoryl group of the incoming dNTP to form a phosphoester bond, and PP_i is released. The subsequent hydrolysis of PP_i by inorganic pyrophosphatase renders the reaction effectively irreversible.

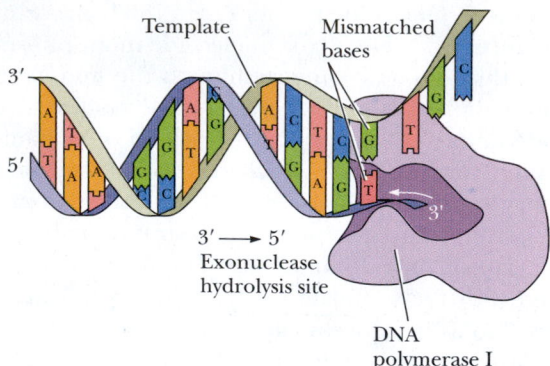

Template   Mismatched bases

3′ → 5′
Exonuclease
hydrolysis site

DNA polymerase I

FIGURE 28.8 The 3′→5′ exonuclease activity of DNA polymerase I removes nucleotides from the 3′-end of the growing DNA chain.

remove distorted (mispaired) segments lying in the path of the advancing polymerase. Its biological roles depend on the ability of DNA polymerase I to bind at nicks (single-stranded breaks) in dsDNA and move in the 5′→3′ direction, removing successive nucleotides with its 5′-exonucleolytic activity. (This overall process is known as "nick translation," because the nick is translated [that is, moved] down the DNA.) This 5′-exonuclease activity plays an important role in primer removal during DNA replication, as we shall soon see. DNA polymerase I is also involved in DNA repair processes (see Section 28.7).

E. coli DNA Polymerase III Holoenzyme Replicates the *E. coli* Chromosome

In its holoenzyme form, DNA polymerase III is the enzyme responsible for replication of the *E. coli* chromosome. The simplest form of DNA polymerase III showing any DNA-synthesizing activity in vitro, "core" DNA polymerase III, is 165 kD in size and consists of three polypeptides: α (130 kD), ϵ (27.5 kD), and θ (10 kD). In vivo, core DNA polymerase III functions as part of a multisubunit complex, the **DNA polymerase III holoenzyme,** which is composed of ten different kinds of subunits (Table 28.2). The various auxiliary subunits increase both the polymerase activity of the core enzyme and its processivity. DNA polymerase III holoenzyme synthesizes DNA strands at a speed of nearly 1 kb/sec. DNA polymerase III holoenzyme is organized in the following way: Two core ($\alpha\epsilon\theta$) DNA polymerase III units and one γ-complex are held together by a dimer of τ-subunits in a structure known as **DNA polymerase III*.** In turn, each core polymerase within DNA polymerase III* binds to a β-subunit dimer

Table 28.2		
Subunits of *E. coli* DNA Polymerase III Holoenzyme		
Subunit	**Mass (kD)**	**Function**
α	130	Polymerase
ϵ	27.5	3′-Exonuclease
θ	8.6	α, ϵ assembly?
τ	71	Assembly, processivity switch
β	41	Sliding clamp, processivity
γ	47.5	Part of the γ-complex*
δ	39	Part of the γ-complex*
δ'	37	Part of the γ-complex*
χ	17	Part of the γ-complex*
ψ	15	Part of the γ-complex*

*Subunits γ, δ, δ', χ, and ψ form the so-called γ-complex responsible for adding β-subunits (the sliding clamp) to DNA. The γ-complex is referred to as the clamp loader.

▶ **FIGURE 28.9** (a) Ribbon diagram of the β-subunit dimer of the DNA polymerase III holoenzyme on B-DNA, viewed down the axis of the DNA. One monomer of the β-subunit dimer is colored red and the other yellow. The centrally located DNA is mostly blue. (b) Space-filling model of the β-subunit dimer of the DNA polymerase III holoenzyme on B-DNA. One monomer is shown in red, the other in yellow. The B-DNA has one strand colored white and the other blue. The hole formed by the β-subunits (diameter $\approx$ 3.5 nm) is large enough to easily accommodate DNA (diameter $\approx$ 2.5 nm) with no steric repulsion. The rest of polymerase III holoenzyme ("core" polymerase + γ-complex) associates with this sliding clamp to form the replicative polymerase (not shown). *(Adapted from Kong, X-P., et al., 1992. Three-dimensional structure of the beta subunit of* E. coli *DNA polymerase III holoenzyme: a sliding DNA clamp.* Cell **69**:425–437; *photos courtesy of John Kuriyan of the Rockefeller University.)*

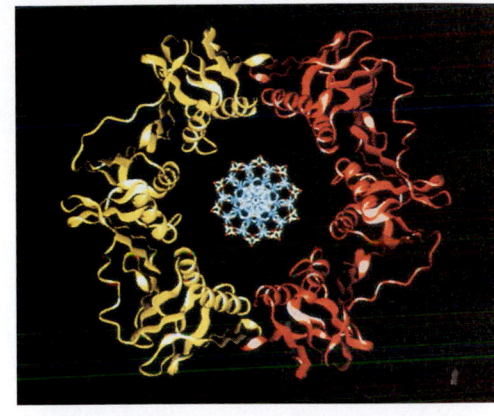

(a)

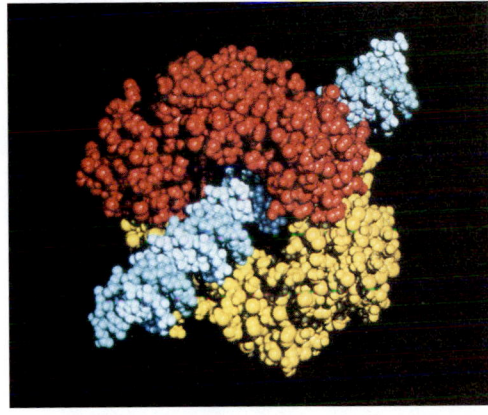

(b)

to create **DNA polymerase III holoenzyme.** The γ-complex is responsible for assembly of the DNA polymerase III holoenzyme complex onto DNA. The γ-complex of the holoenzyme acts as a **clamp loader** by catalyzing the ATP-dependent transfer of a pair of β-subunits to each strand of the DNA template. Each β-subunit dimer forms a closed ring around a DNA strand and acts as a tight clamp that can slide along the DNA (Figure 28.9). Each β_2-**sliding clamp** tethers a core polymerase to the template, accounting for the great processivity of the DNA polymerase holoenzyme. This complex can replicate an entire strand of the *E. coli* genome (more than 4.6 megabases) without dissociating. Compare this to the processivity of DNA polymerase I, which is only 20!

The core polymerase synthesizing the lagging strand must release from the DNA template when synthesis of an Okazaki fragment is completed and rejoin the template at the next RNA primer to begin synthesis of the next Okazaki fragment. The τ-subunit serves as a "processivity switch" that accomplishes this purpose. The τ-subunit is usually "off" and is turned "on" only on the lagging strand *and* only when synthesis of an Okazaki fragment is completed. When activated, τ ejects the β_2-sliding clamp bound to the lagging strand core polymerase. Almost immediately, the lagging strand core polymerase is reloaded onto a new β_2-sliding clamp at the 3'-end of next RNA primer, and synthesis of the next Okazaki fragment commences.

A DNA Polymerase III Holoenzyme Sits at Each Replication Fork

We now can present a snapshot of the enzymatic apparatus assembled at a replication fork (Figure 28.10 and Table 28.3). DNA gyrase (topoisomerase) and DnaB helicase unwind the DNA double helix, and the unwound, single-stranded regions of DNA are maintained through interaction with SSB. **Primase (DnaG)** synthesizes an RNA primer on the lagging strand; the leading strand, which needs priming only once, was primed when replication was initiated. The lagging strand template is looped around, and each replicative DNA polymerase moves 5'→3' relative to its strand, copying template and synthesizing a new DNA strand.

▼ **Biochemistry ⊜ Now™   ACTIVE FIGURE 28.10** General features of a replication fork. The DNA duplex is unwound by the action of DNA gyrase and helicase, and the single strands are coated with SSB (ssDNA-binding protein). Primase periodically primes synthesis on the lagging strand. Each half of the dimeric replicative polymerase is a "core" polymerase bound to its template strand by a β-subunit sliding clamp. DNA polymerase I and DNA ligase act downstream on the lagging strand to remove RNA primers, replace them with DNA, and ligate the Okazaki fragments. **Test yourself on the concepts in this figure at http://chemistry.brookscole.com/ggb3**

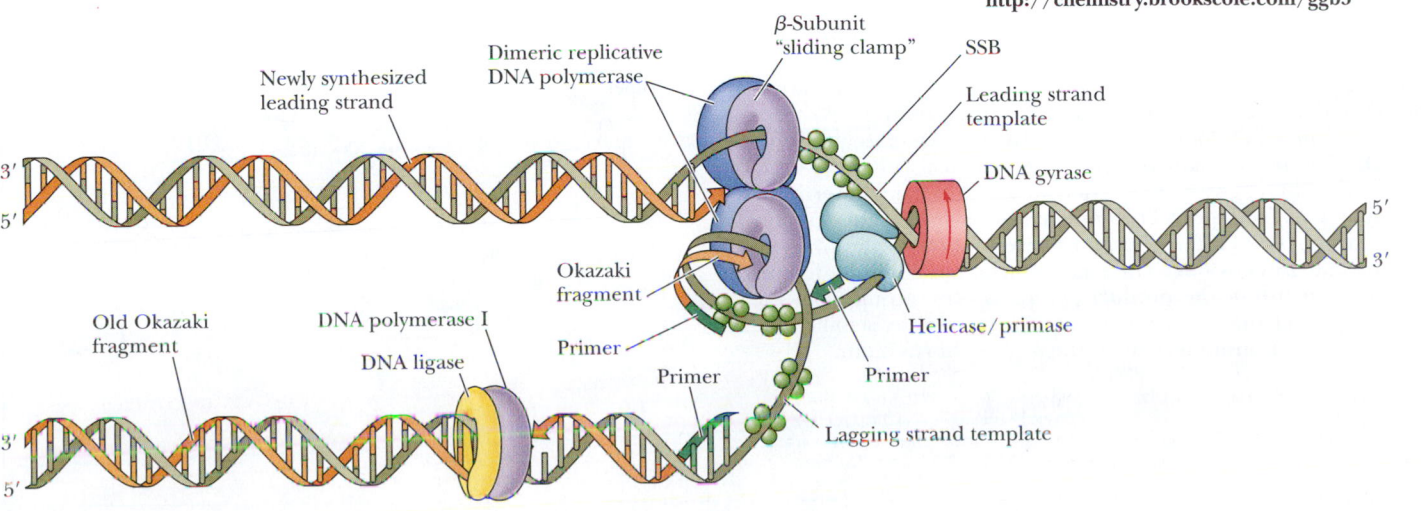

Table 28.3

Proteins Involved in DNA Replication in *E. coli*

Protein	Function
DNA gyrase	Unwinding DNA
SSB	Single-stranded DNA binding
DnaA	Initiation factor; origin-binding protein
DnaB	$5' \rightarrow 3'$ helicase (DNA unwinding)
DnaC	DnaB chaperone; loading DnaB on DNA
DnaT	Assists DnaC in delivery of DnaB
Primase (DnaG)	Synthesis of RNA primer
DNA polymerase III holoenzyme	Elongation (DNA synthesis)
DNA polymerase I	Excises RNA primer, fills in with DNA
DNA ligase	Covalently links Okazaki fragments
Tus	Termination

Each replicative polymerase is tethered to the DNA by its β-subunit sliding clamp. The DNA polymerase III γ-complex periodically unclamps and then reclamps β-subunits on the lagging strand as the primer for each new Okazaki fragment is encountered. Downstream on the lagging strand, DNA polymerase I excises the RNA primer and replaces it with DNA, and DNA ligase seals the remaining nick.

DNA Ligase Seals the Nicks Between Okazaki Fragments

DNA ligase (see Chapter 12) seals nicks in double-stranded DNA where a 3'-OH and a 5'-phosphate are juxtaposed. This enzyme is responsible for joining Okazaki fragments together to make the lagging strand a covalently contiguous polynucleotide chain.

A Deeper Look

A Mechanism for All Polymerases

Thomas A. Steitz of Yale University has suggested that biosynthesis of nucleic acids proceeds by an enzymatic mechanism that is universal among polymerases. His suggestion is based on structural studies indicating that DNA polymerases use a "two-metal-ion" mechanism to catalyze nucleotide addition during elongation of a growing polynucleotide chain (see accompanying figure). The incoming nucleotide has two Mg^{2+} ions coordinated to its phosphate groups, and these metal ions interact with two aspartate residues that are highly conserved in DNA (and RNA) polymerases. These residues in phage T7 DNA polymerase are D705 and D882. One metal ion, designated A, interacts with the O atom of the free 3'-OH group on the polynucleotide chain, lowering its affinity for its hydrogen. This interaction promotes nucleophilic attack of the 3'-O on the phosphorus atom in the α-phosphate of the incoming nucleotide. The second metal ion (B in the figure) assists departure of the product pyrophosphate group from the incoming nucleotide. Together, the two metal ions stabilize the pentacovalent transition state on the α-phosphorus atom.

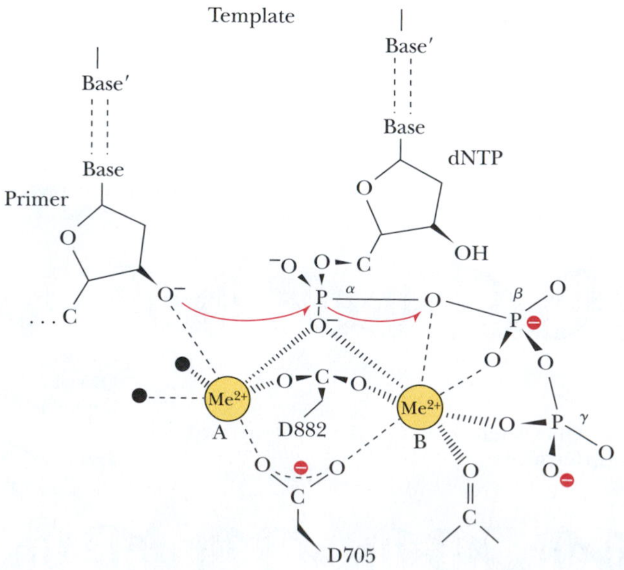

Adapted from Steitz, T., 1998. A mechanism for all polymerases. *Nature* **391**:231–232. (See also Doublié, S., et al., 1998. Crystal structure of bacteriophage T7 DNA replication complex at 2.2 Å resolution. *Nature* **391**:251–258; and Kiefer, J. R., et al., 1998. Visualizing DNA replication in a catalytically active *Bacillus* DNA polymerase crystal. *Nature* **391**:304–307.)

DNA Replication Terminates at the *Ter* Region

Located diametrically opposite from *oriC* on the *E. coli* circular map is a terminus region, the *Ter,* or *t,* locus. The oppositely moving replication forks meet here, and replication is terminated. The *Ter* region contains a number of short DNA sequences, with a consensus core element 5′-GTGTGTTGT. These *Ter* sequences act as terminators; clusters of three or four *Ter* sequences are organized into two sets inversely oriented with respect to one another. One set blocks the clockwise-moving replication fork, and its inverted counterpart blocks the counterclockwise-moving replication fork. Termination requires that a specific replication termination protein, **Tus protein,** be bound to *Ter.* Tus protein is a **contrahelicase.** That is, Tus protein prevents the DNA duplex from unwinding by blocking progression of the replication fork and inhibiting the ATP-dependent DnaB helicase activity. Final synthesis of both duplexes is completed.

DNA Polymerases Are Immobilized in Replication Factories

Most drawings of DNA replication (such as Figure 28.10) suggest that the DNA polymerases are tracking along the DNA, like locomotives along train tracks, synthesizing DNA as they go. Recent evidence, however, favors the view that the DNA polymerases are immobilized, either via attachment to the cell membrane in prokaryotic cells or to the nuclear matrix in eukaryotic cells. All the associated proteins of DNA replication, as well as proteins necessary to hold DNA polymerase at its fixed location, constitute **replication factories.** The DNA is then fed through the DNA polymerases within the replication factory, much like tape is fed past the heads of a tape player, with all four strands of newly replicated DNA looping out from this fixed structure (Figure 28.11).

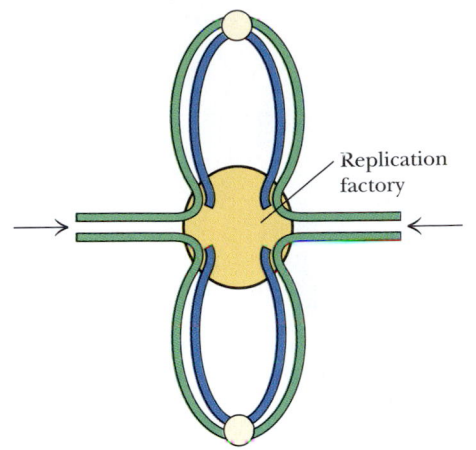

Replication factory

Biochemistry ✺ Now™   ANIMATED FIGURE 28.11
A replication factory "fixed" to a cellular substructure extrudes loops of newly synthesized DNA as parental DNA duplex is fed in from the sides. Parental DNA strands are blue; newly synthesized strands are green; small circles indicate origins of replication. *(Adapted from Cook, P. R., 1999. The organization of replication and transcription. Science 284:1790–1795.)* **See this figure animated at http://chemistry.brookscole.com/ggb3**

28.3 | How Is DNA Replicated in Eukaryotic Cells?

DNA replication in eukaryotic cells shows strong parallels with prokaryotic DNA replication, but it is vastly more complex. First, eukaryotic DNA is organized into chromosomes, which are compartmentalized within the nucleus. Furthermore, these chromosomes must be duplicated with high fidelity once (and only once!) each cell cycle. For example, in a dividing human cell, a carefully choreographed replication of 6 billion bp of DNA distributed among 46 chromosomes occurs. The events associated with cell growth and division in eukaryotic cells fall into a general sequence having four distinct phases: M, G_1, S, and G_2 (Figure 28.12). Eukaryotic cells have solved the problem of replicating their enormous genomes in the few hours allotted to the S phase by initiating DNA replication at multiple origins of replication distributed along each chromosome. Depending on the organism and cell type, replication origins, also called **replicators,** are DNA regions 0.5 to 2 kbp in size that occur every 3 to 300 kbp (for example, an average human chromosome has several hundred replication origins). Since eukaryotic DNA replication proceeds concomitantly throughout the genome, each eukaryotic chromosome must contain many units of replication, called **replicons.**

The Cell Cycle Controls the Timing of DNA Replication

Checkpoints, Cyclins, and CDKs   Progression through the cell cycle is regulated through a series of **checkpoints** that control whether the cell continues into the next phase. These checkpoints are situated to ensure that *all* the necessary steps in each phase of the cycle have been satisfactorily completed before the next phase is entered. If conditions for advancement to the next phase are not met, the cycle is arrested until they are. Checkpoints depend on **cyclins** and **cyclin-dependent protein kinases (CDKs).** *Cyclin* is the name given to a class of proteins synthesized at one phase of the cell cycle and degraded at another. Thus, cyclins appear and then disappear at specific times during the cell cycle. Cyclins bind to CDKs and are essential for the protein kinase activity displayed by these

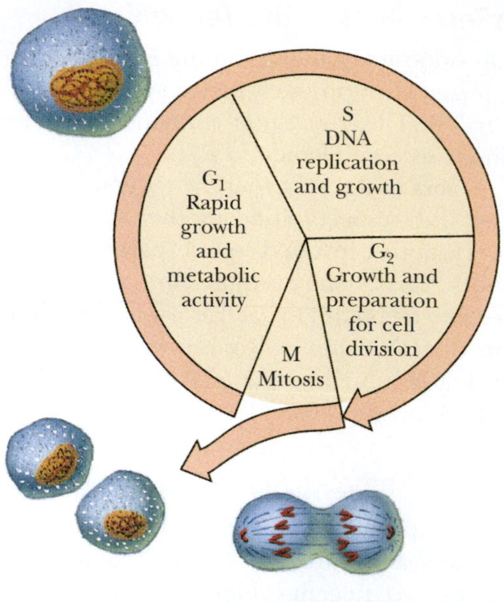

FIGURE 28.12 The eukaryotic cell cycle. The stages of mitosis and cell division define the M phase (*M* for *mitosis*). G₁ (*G* for *gap*, not growth) is typically the longest part of the cell cycle; G₁ is characterized by rapid growth and metabolic activity. Cells that are quiescent, that is, not growing and dividing (such as neurons), are said to be in G₀. The S phase is the time of DNA synthesis. S is followed by G₂, a relatively short period of growth when the cell prepares for cell division. Cell cycle times vary from less than 24 hours (rapidly dividing cells such as the epithelial cells lining the mouth and gut) to hundreds of days.

CDKs. In turn, these CDKs control events at each phase of the cycle by targeting specific proteins for phosphorylation. Destruction of the phase-specific cyclin at the end of the phase inactivates the CDK.

Initiation of Replication   Initiation of replication involves *replicators* and the **origin recognition complex,** or **ORC,** a protein complex that binds to replicators. Yeast, a simple eukaryote, provides an informative model for initiation of eukaryotic DNA replication. Early in G₁ (just after M), ORC serves as a "landing pad" for proteins essential to replication control. Proteins binding to ORC establish a **pre-replication complex (pre-RC),** but the pre-RC can be formed only during a window of opportunity during G₁. One of the principal proteins in assembly of the pre-RC in yeast is **Cdc6p,** the **replication activator protein** (Figure 28.13). Once Cdc6p binds to ORC, **MCM proteins** then bind to the chromosomes. MCM proteins are essential replication initiation factors; they are also known as **replication licensing factors (RLFs)** because they "license," or permit, DNA replication to occur. The MCM proteins render the chromosomes competent for DNA replication. The pre-RC therefore consists of ORC, Cdc6p, the MCM complex, and other proteins.

At this point, two protein kinases act upon the pre-RC to directly trigger DNA replication. One of these protein kinases is the **cyclin B–CDK** complex. B-Cyclin is a cyclin that accumulates to high levels just before S phase. Cyclin B–CDK can phosphorylate sites in ORC, Cdc6p, and several MCM subunits. Phosphorylation of Cdc6p causes it to dissociate from ORC, whereupon it is degraded. Cyclin B–CDK also phosphorylates other protein kinases essential to activation of DNA replication (Figure 28.13). The MCM complex is one phosphorylation target of these CDKs, and some of the MCM proteins dissociate from the chromosomes when they are phosphorylated. The consequence of these actions brings the cell into S phase.

These phosphorylation events serve as a **replication switch** because once proteins in the pre-RC are phosphorylated (and perhaps destroyed, as Cdc6p is), the **post-RC** state is achieved. The post-RC state is incapable of reinitiating DNA replication. This transformation ensures that eukaryotic DNA replication occurs once, and only once, per cell cycle.

Eukaryotic Cells Contain a Number of Different DNA Polymerases

At least 19 different DNA polymerases have been described in eukaryotic cells thus far. These various polymerases have been assigned Greek letters in the order of their discovery (Table 28.4 lists the principal ones). Three of these DNA poly-

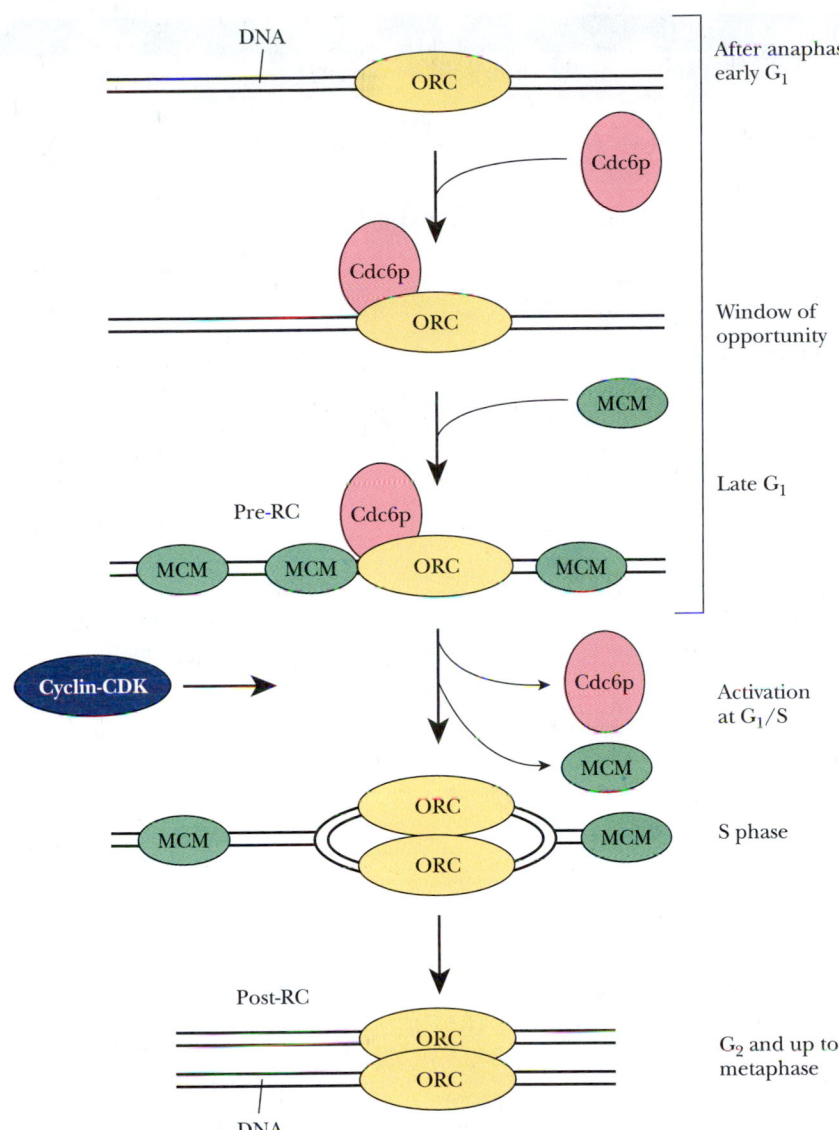

DNA

ORC

After anaphase
early G₁

Cdc6p

Cdc6p
ORC

Window of
opportunity

MCM

Pre-RC   Cdc6p

Late G₁

MCM   MCM   ORC   MCM

Cyclin-CDK →   Cdc6p

Activation
at G₁/S

MCM

ORC

MCM   ORC   MCM

S phase

Post-RC

ORC

G₂ and up to
metaphase

ORC

DNA

FIGURE 28.13 Model for initiation of the DNA
replication cycle in eukaryotes. ORC is present at
the replicators throughout the cell cycle. The pre-
replication complex (pre-RC) is assembled through
the sequential addition of Cdc6p and MCM proteins
during a window of opportunity defined by the state
of the cyclin-CDKs. After initiation, a post-RC state is
established. (*Adapted from Figure 2 in Stillman, B., 1996. Cell
cycle control of DNA replication.* Science **274:**1659–1663.)

merases—α, δ, and ϵ—share the major burden of genome replication. **DNA
polymerase α** has an associated primase subunit and functions in the initiation
of nuclear DNA replication. Given a template, it not only synthesizes an RNA
primer of about 10 nucleotides, but it then adds 20 to 30 deoxynucleotides to
extend the chain in the $5'\rightarrow3'$ direction. **DNA polymerase δ,** a heterotetrameric
enzyme, is the principal DNA polymerase in eukaryotic DNA replication. It interacts with **PCNA** protein (PCNA stands for proliferating cell nuclear antigen).
Through its association with PCNA, DNA polymerase δ carries out highly processive DNA synthesis. PCNA is the eukaryotic counterpart of the *E. coli* β_2-sliding
clamp; it clamps DNA polymerase δ to the DNA. Like β_2, PCNA encircles the
double helix, but in contrast to the prokaryotic β_2-sliding clamp, PCNA is a homotrimer, not a homodimer (Figure 28.14). **DNA polymerase ϵ** also plays a major
role in DNA replication. Its precise role is unclear, but it has an acidic C-terminal
extension lacking in other DNA polymerases and evidence suggests that this
domain is a sensor for DNA damage checkpoint control, halting DNA replication
until the damage is repaired. **DNA polymerase γ** is the DNA-replicating enzyme
of mitochondria; **DNA polymerase β** functions in DNA repair. The more recently
discovered eukaryotic DNA polymerases (including ζ, η, ι, κ, and Rev1) are novel
in that they are more error-prone, resulting in lower fidelity of DNA replication.
Nevertheless, they have the important ability to function in DNA replication and
repair when damaged regions of DNA are encountered.

Biochemistry Now™   Go to BiochemistryNow
and click BiochemistryInteractive to discover how
PCNA is a trimeric analog of the prokaryotic beta subunit dimer sliding clamp.

Table 28.4

Biochemical Properties of The Principal Human DNA Polymerases

Polymerase Localization and Function	Subunits (mass in kD)	Subunit Function
DNA polymerase α		
Nuclear; initiation of nuclear DNA replication	180	Catalytic subunit
	68	Protein–protein interactions
	55	Primase
	48	Primase
DNA polymerase δ		
Nuclear; principal polymerase in leading and lagging strand synthesis; highly processive	125	Catalytic subunit
	66	Structural
	50	Interaction with PCNA
	12	Protein–protein interactions
DNA polymerase ϵ		
Nuclear; leading and lagging strand synthesis, sensor of DNA damage checkpoint control	261	Catalytic subunit
	59	Multimerization
	17	Protein–protein interactions
	12	Protein–protein interactions
DNA polymerase γ		
Mitochondria; mitochondrial DNA replication	140	Catalytic subunit
	55	Processivity
DNA polymerase β		
DNA repair	39	

(a)

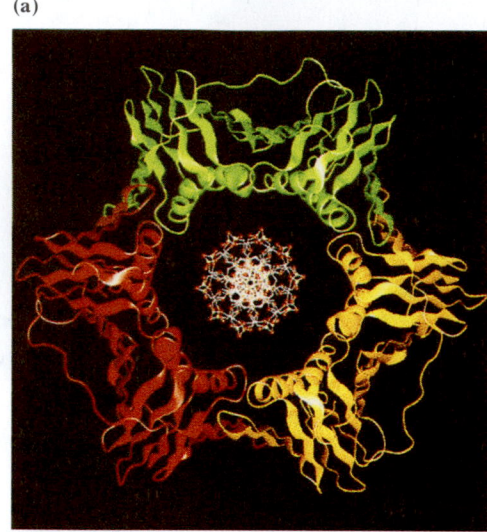

(b)

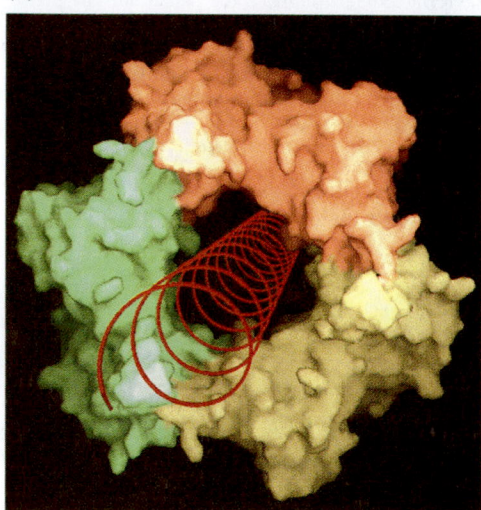

FIGURE 28.14 Structure of the PCNA homotrimer. Note that the trimeric PCNA ring of eukaryotes is remarkably similar to its prokaryotic counterpart, the dimeric β_2-sliding clamp (Figure 28.9). **(a)** Ribbon representation of the PCNA trimer with an axial view of a B-form DNA duplex in its center. The molecular mass of each PCNA monomer is 37 kD. **(b)** Molecular surface of the PCNA trimer with each monomer colored differently. The red spiral represents the sugar–phosphate backbone of a strand of B-form DNA. *(Adapted from Figure 3 in Krishna, T. S., et al., 1994. Crystal structure of the eukaryotic DNA polymerase processivity factor PCNA. Cell **79:**1233–1243. Photos courtesy of John Kuriyan of the Rockefeller University.)*

Other proteins involved in eukaryotic DNA replication include **replication protein A (RPA)**, a ssDNA-binding protein that is the eukaryotic counterpart of SSB, and **replication factor C (RFC)**, the eukaryotic counterpart of the prokaryotic γ-complex.

28.4 How Are the Ends of Chromosomes Replicated?

Telomeres are the structures at the ends of eukaryotic chromosomes. Telomeres are short (5 to 8 bp), tandemly repeated, G-rich nucleotide sequences that form protective caps 1 to 12 kbp long on the chromosome ends (see Chapter 11). Vertebrate telomeres have a TTAGGG consensus sequence. Telomeres are necessary for chromosome maintenance and stability. DNA polymerases cannot replicate the extreme 5'-ends of chromosomes because these enzymes require a template and a primer and replicate only in the 5'→3' direction. Thus, lagging strand synthesis at the 3'-ends of chromosomes is primed by RNA primase to form Okazaki fragments, but these RNA primers are subsequently removed, resulting in gaps in the progeny 5'-terminal strands at each end of the chromosome after each round of replication ("primer gap"; see Figure 28.15).

Telomerase is an RNA-dependent DNA polymerase. Telomerase maintains telomere length by restoring telomeres at the 3'-ends of chromosomes. The RNA upon which telomerase activity depends is actually part of the enzyme's structure. That is, telomerase is a ribonucleoprotein, and its RNA component contains a 9- to 30-nucleotide-long region that serves as a template for the synthesis of telomeric repeats at DNA ends. The human telomerase RNA component is 450 nucleotides long; its template sequence is CUAACCCUAAC. Telomerase uses the 3'-end of the DNA as a primer and adds successive

A Deeper Look

Protein Rings in DNA Metabolism

Many of the proteins involved in DNA metabolism (DNA replication, recombination, and repair) adopt a ring-shaped (toroidal) quaternary structure (see table). An obvious advantage of this quaternary structure is that the DNA is enclosed within the "hole" of the ring and therefore remains stably associated with and contained within the protein that is operating on it.

Protein	Examples	Action on DNA
Sliding clamps	*E. coli* β; human PCNA	Protein ring encircles DNA, tethers DNA polymerase, and increases processivity
Hexameric helicases	*E. coli* DnaB and RuvB; SV40 T antigen; MCM proteins	Hexameric protein rings unwind dsDNA to form ssDNA substrates for replication, recombination, or repair
Circular topoisomerases	*E. coli* and human topoisomerases	Change the superhelicity of DNA
DNA recombination proteins	Human RAD51, 52	Facilitates homologous DNA recombination

Adapted from Hingorani, M. M., and O'Donnell, M., 2000. A tale of toroids in DNA metabolism. *Nature Reviews Molecular Cell Biology* **1:**22–30.

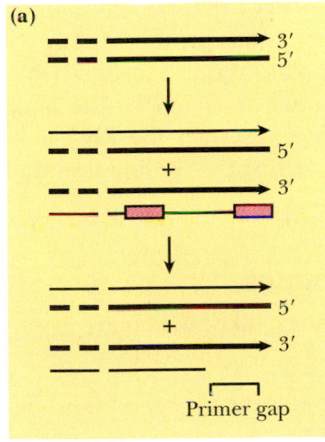

(a)

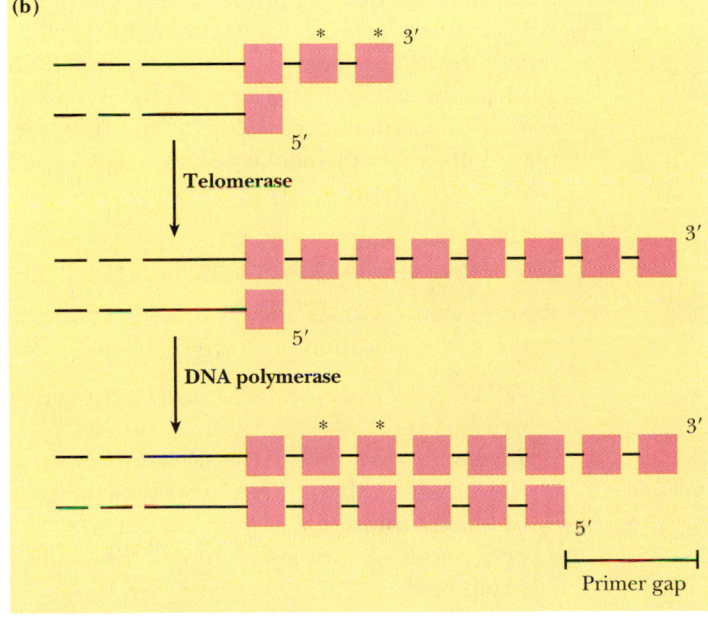

(b)

Biochemistry ✌Now™
ANIMATED FIGURE 28.15 Telomere replication. **(a)** In replication of the lagging strand, short RNA primers are added (pink) and extended by DNA polymerase. When the RNA primer at the 5′-end of each strand is removed, there is no nucleotide sequence to read in the next round of DNA replication. The result is a gap (primer gap) at the 5′-end of each strand (only one end of a chromosome is shown in this figure). **(b)** Asterisks indicate sequences at the 3′-end that cannot be copied by conventional DNA replication. Synthesis of telomeric DNA by telomerase extends the 5′-ends of DNA strands, allowing the strands to be copied by normal DNA replication. **See this figure animated at http://chemistry.brookscole.com/ggb3**

Human Biochemistry

Telomeres—A Timely End to Chromosomes?

Mammalian cells in culture undergo only 50 or so cell divisions before they die. Somatic cells are known to lack telomerase activity, and thus, they inevitably lose bits of their telomeres with each cell division. Telomerase activity is missing because the telomerase–reverse transcriptase gene (the *TRT* gene) is switched off. This fact has led to a telomere theory of cell aging, which suggests that cells senesce and die when their telomeres are gone. In support of this notion, a team of biologists headed by Calvin B. Harley at Geron Corporation used recombinant DNA techniques to express the catalytic subunit of human telomerase in skin cells in culture and observed that such cells divide 40 times more after cells lacking this treatment have become senescent. These results, although controversial, may have relevance to the aging process.

TTAGGG repeats to it, employing its RNA as template over and over again (Figure 28.15, see also figure in the Chapter 11 Human Biochemistry box Telomeres and Tumors).

28.5 | How Are RNA Genomes Replicated?

Many viruses have genomes composed of RNA, not DNA. How then is the information in these RNA genomes replicated? In 1964, Howard Temin noted that inhibitors of DNA synthesis prevented infection of cells in culture by RNA tumor viruses such as avian sarcoma virus. On the basis of this observation, Temin made the bold proposal that DNA is an intermediate in the replication of such viruses; that is, *an RNA tumor virus can use viral RNA as the template for DNA synthesis.*

RNA viral chromosome⟶DNA intermediate⟶RNA viral chromosome

In 1970, Temin and David Baltimore independently discovered a viral enzyme capable of mediating such a process, namely, an **RNA-directed DNA polymerase** or, as it is usually called, **reverse transcriptase.** All RNA tumor viruses contain such an enzyme within their virions (viral particles), so they are now classified as **retroviruses.**

Like other DNA and RNA polymerases, reverse transcriptase synthesizes polynucleotides in the 5′→3′ direction, and like all DNA polymerases, reverse transcriptase requires a primer. Interestingly, the primer is a specific tRNA molecule captured by the virion from the host cell in which it was produced. The 3′-end of the tRNA is base-paired with the viral RNA template at the site where DNA synthesis initiates, and its free 3′-OH accepts the initial deoxynucleotide once transcription commences. Reverse transcriptase then transcribes the RNA template into a complementary DNA (cDNA) strand to form a double-stranded DNA:RNA hybrid.

The Enzymatic Activities of Reverse Transcriptases

Reverse transcriptases possess three enzymatic activities, all of which are essential to viral replication:

1. *RNA-directed DNA polymerase activity,* for which the enzyme is named (see Figure 12.14).
2. *RNase H activity.* Recall that RNase H is a nuclease that specifically degrades RNA chains in DNA:RNA hybrids (see Figure 12.14). The RNase H function of reverse transcriptase is an exonuclease activity that degrades the template genomic RNA and also removes the priming tRNA after DNA synthesis is completed.

A Deeper Look

RNA as Genetic Material

Whereas the genetic material of cells is dsDNA, virtually all plant viruses, several bacteriophages, and many animal viruses have genomes consisting of RNA. In most cases, this RNA is single stranded. Viruses with single-stranded genomes use the single strand as a template for synthesis of a complementary strand, which can then serve as template in replicating the original strand. **Retroviruses** are an interesting group of eukaryotic viruses with single-stranded RNA genomes that replicate through a dsDNA intermediate. Furthermore, the life cycle of retroviruses includes an obligatory step in which the dsDNA is inserted into the host cell genome in a transposition event. Retroviruses are responsible for many diseases, including tumors and other disorders. **HIV-1,** the **human immunodeficiency virus** that causes **AIDS,** is a retrovirus. **Tobacco mosaic virus (TMV),** a rodlike RNA virus that infects plants, consists of an RNA genome packaged in a protein coat made of 2130 identical protein chains of 18 kD each (see Figure 1.24).

3. *DNA-directed DNA polymerase activity.* This activity replicates the ssDNA remaining after RNase H degradation of the viral genome, yielding a DNA duplex. This DNA duplex directs the remainder of the viral infection process or becomes integrated into the host chromosome, where it can lie dormant for many years as a **provirus.** Activation of the provirus restores the infectious state.

HIV reverse transcriptase is of great clinical interest because it is the enzyme for replication of the AIDS virus. DNA synthesis by HIV reverse transcriptase is blocked by nucleotide analogs such as AZT (Figure 28.16). HIV reverse transcriptase is error-prone: It incorporates the wrong base at a frequency of 1 per 2000 to 4000 nucleotides polymerized. This high error rate during replication of the HIV genome means that the virus is ever changing, a feature that makes it difficult to devise an effective vaccine.

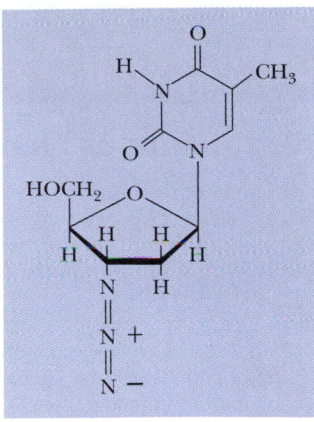

FIGURE 28.16 The structures of AZT (3′-azido-2′,3′-dideoxythymidine). This nucleoside was the first approved drug for treatment of AIDS. AZT is phosphorylated in vivo to give AZTTP (AZT 5′-triphosphate), a substrate analog that binds to HIV reverse transcriptase. HIV reverse transcriptase incorporates AZTTP into growing DNA chains in place of dTTP. Incorporated AZTMP blocks further chain elongation because its 3′-azido group cannot form a phosphodiester bond with an incoming nucleotide. Host cell DNA polymerases have little affinity for AZTTP.

28.6 How Is the Genetic Information Shuffled by Genetic Recombination?

Genetic recombination is the natural process by which genetic information is rearranged to form new associations. For example, compared to their parents, progeny may have new combinations of traits because of genetic recombination. At the molecular level, genetic recombination is the exchange (or incorporation) of one DNA sequence with (or into) another. When recombination involves reaction between very similar sequences (homologous sequences) of DNA, the process is called **homologous recombination.** When very different nucleotide sequences recombine, it's **nonhomologous recombination. Transposition**—the enzymatic insertion of a *transposon* (a mobile segment of DNA, see page 924) into a new location in the genome—and **nonhomologous recombination** (incorporation of a DNA segment whose sequence differs greatly from the DNA at the point of insertion) are two types of recombination that play a significant evolutionary role. Nonhomologous recombination occurs at a low frequency in all cells and serves as a powerful genetic force that reshapes the genomes of all organisms. Homologous recombination involves an exchange of DNA sequences between homologous chromosomes, resulting in the arrangement of genes into new combinations. Homologous recombination is generally used to fix the DNA so that information is not lost. For example, large lesions in DNA are repaired via recombination of the damaged chromosome with a homologous chromosome.

The process underlying homologous recombination is termed **general recombination** because the enzymatic machinery that mediates the exchange can use essentially any pair of homologous DNA sequences as substrates. Homologous recombination occurs in all organisms and is particularly prevalent during the production of gametes (meiosis) in diploid organisms. In higher animals—that is, those with immune systems—recombination also occurs in the DNA of somatic cells responsible for expressing proteins of the immune response, such as the immunoglobulins. This **somatic recombination** rearranges the immunoglobulin genes, dramatically increasing the potential diversity of immunoglobulins available from a fixed amount of genetic information. Homologous recombination can also occur in bacteria. Indeed, even viral chromosomes undergo recombination. For example, if two mutant viral particles simultaneously infect a host cell, a recombination event between the two viral genomes can lead to the formation of a virus chromosome that is wild-type.

General Recombination Requires Breakage and Reunion of DNA Strands

Recombination occurs by the breakage and reunion of DNA strands so that a physical exchange of parts takes place. Matthew Meselson and J. J. Weigle demonstrated this in 1961 by coinfecting *E. coli* with two genetically distinct

Human Biochemistry

Biochemistry Now™

Prions: Proteins as Genetic Agents?

DNA is the genetic material in organisms, although some viruses have RNA genomes. The idea that proteins could carry genetic information was considered early in the history of molecular biology and dismissed for lack of evidence. *Prions* may be an exception to this rule.

Prion is an acronym derived from the words *proteinaceous infectious particle*. The term *prion* was coined to distinguish such particles, which are pathogenic and thus capable of causing disease, from nucleic acid–containing infectious particles such as viruses and virions. Prions are transmissible agents (genetic material?) that are apparently composed only of a protein that has adopted an abnormal conformation. They produce fatal degenerative diseases of the central nervous system in mammals and are believed to be the agents responsible for the human diseases kuru, Creutzfeldt-Jakob disease, Gerstmann-Straussler-Sheinker syndrome, and fatal familial insomnia. Prions also cause diseases in animals, including scrapie (in sheep), "mad cow disease" (bovine spongiform encephalopathy), and chronic wasting disease (in elk and mule deer). All attempts to show that the infectivity of these diseases is due to a nucleic

acid–carrying agent have been unsuccessful. Prion diseases are novel in that they are genetic and infectious; their occurrence may be sporadic, dominantly inherited, or acquired by infection. Their inheritability questions the principle that nucleic acids are the sole genetic agents.

PrP, the prion protein, comes in various forms, such as **Prp^c**, the normal cellular prion protein, and **PrP^sc,** the scrapie form of PrP, a conformational variant of PrP^c that is protease resistant. These two forms are thought to differ only in terms of their secondary and tertiary structure. One model suggests that PrP^c is dominated by α-helical elements (see figure, part a), whereas PrP^sc has both α-helices and β-strands (see figure, part b). It has been hypothesized that the presence of PrP^sc can cause PrP^c to adopt the PrP^sc conformation. The various diseases are a consequence of the accumulation of the abnormal PrP^sc form, which accumulates as amyloid plaques (amyloid = starchlike), that cause destruction of tissues in the central nervous system. The 1997 Nobel Prize in Physiology or Medicine was awarded to Stanley B. Prusiner for his discovery of prions.

(a)   **(b)**

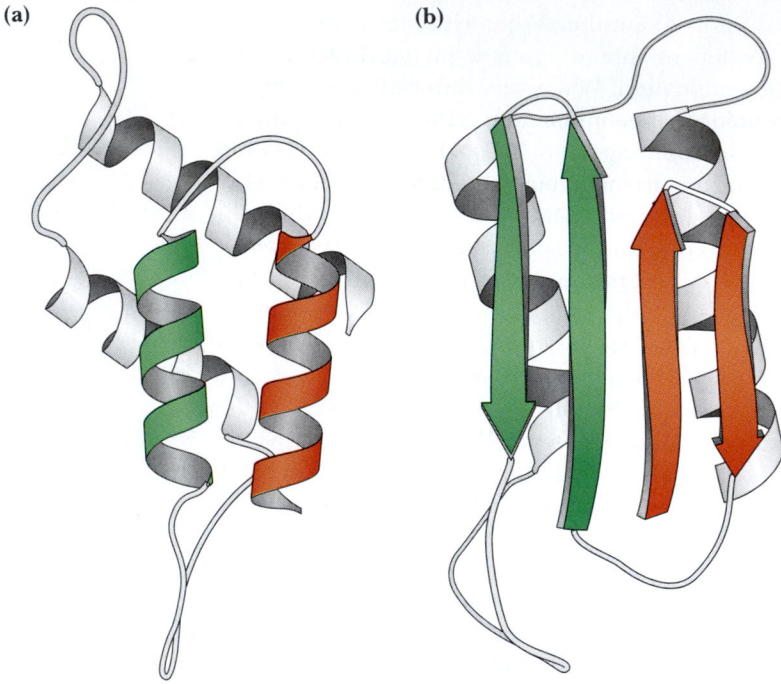

Adapted from Figure 1 in Prusiner, S. B., 1996. Molecular biology and the pathogenesis of prion diseases. *Trends in Biochemical Sciences* **21:**482–487.

bacteriophage λ strains, one of which had been density-labeled by growth in ^{13}C- and ^{15}N-containing media (Figure 28.17). The phage progeny were recovered and separated by CsCl density gradient centrifugation. Phage particles that displayed recombinant genotypes were distributed throughout the gradient while parental (nonrecombinant) genotypes were found within discrete "heavy" and "light" bands in the density gradient. The results showed that recombinant phage contained DNA derived in varying proportions from both parents. The obvious explanation is that these recombinant DNAs arose via the breakage and rejoining of DNA molecules.

▶ **Biochemistry** 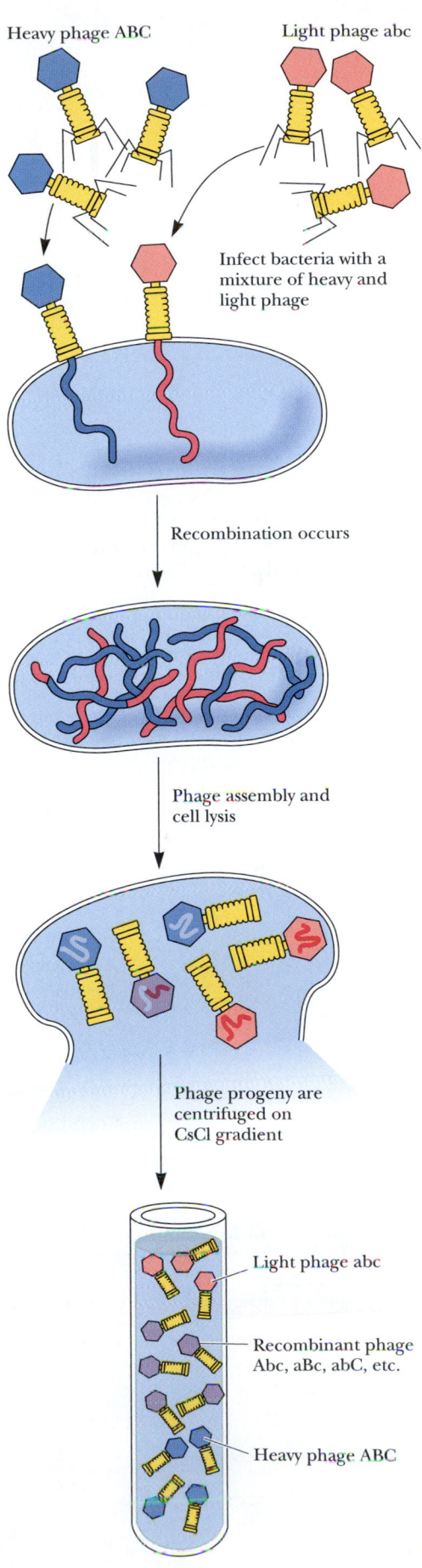 **Now™** **ANIMATED FIGURE 28.17** Meselson and Weigle's experiment demonstrated that a physical exchange of chromosome parts actually occurs during recombination. Density-labeled, "heavy" phage, symbolized as ABC phage in the diagram, was used to coinfect bacteria along with "light" phage, the abc phage. The progeny from the infection were collected and subjected to CsCl density gradient centrifugation. Parental-type ABC and abc phage were well separated in the gradient, but recombinant phage (ABc, Abc, aBc, aBC, and so on) were distributed diffusely between the two parental bands because they contained chromosomes constituted from fragments of both "heavy" and "light' DNA. These recombinant chromosomes formed by breakage and reunion of parental "heavy" and "light" chromosomes. **See this figure animated at http://chemistry.brookscole.com/ggb3**

A second important observation made during this type of experiment was that some of the plaques formed by the phage progeny contained phage of two different genotypes, even though each plaque was caused by a single phage infecting one bacterium. Therefore, some infecting phage chromosomes must have contained a region of **heteroduplex DNA,** duplex DNA in which a part of each strand is contributed by a different parent (Figure 28.18).

Homologous Recombination Proceeds According to the Holliday Model

In 1964, Robin Holliday proposed a model for homologous recombination that has proved to be correct in its essential features (Figure 28.19). The two homologous DNA duplexes are first juxtaposed so that their sequences are aligned. This process of **chromosome pairing** is called **synapsis** (Figure 28.19a). Holliday suggested that recombination begins with the introduction of single-stranded nicks in the DNA at homologous sites on the two paired chromosomes (Figure 28.19b). The two duplexes partially unwind, and the free, single-stranded end of one duplex begins to base-pair with its nearly complementary, single-stranded region along the intact strand in the other duplex, and vice versa (Figure 28.19c). This **strand invasion** is followed by ligation of the free ends from different duplexes to create a cross-stranded intermediate known as a **Holliday junction** (Figure 28.19d). The cross-stranded junction can now migrate in either direction (**branch migration**) by unwinding and rewinding of the two duplexes (Figure 28.19e). Branch migration results in **strand exchange;** heteroduplex regions of varying length are possible. In order for the joint molecule formed by strand exchange to be resolved into two DNA duplex molecules, another pair of nicks must be introduced. Resolution can be represented best if the duplexes are drawn with the chromosome arms bent "up" or "down" to give a planar representation (Figure 28.19f). Nicks then take place, either at E and W, that is, in the − strands that were

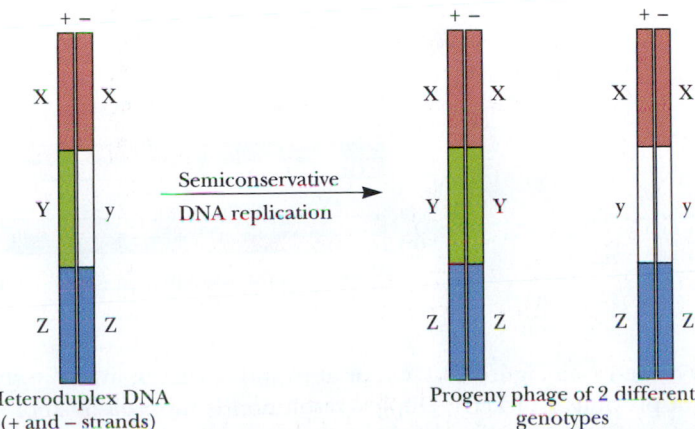

FIGURE 28.18 The generation of progeny bacteriophage of two different genotypes from a single phage particle carrying a heteroduplex DNA region within its chromosome. The heteroduplex DNA is composed of one strand that is genotypically XYZ (the + strand), and the other strand that is genotypically XyZ (the − strand). That is, the genotype of the two parental strands for gene Y is different (one is Y, the other y).

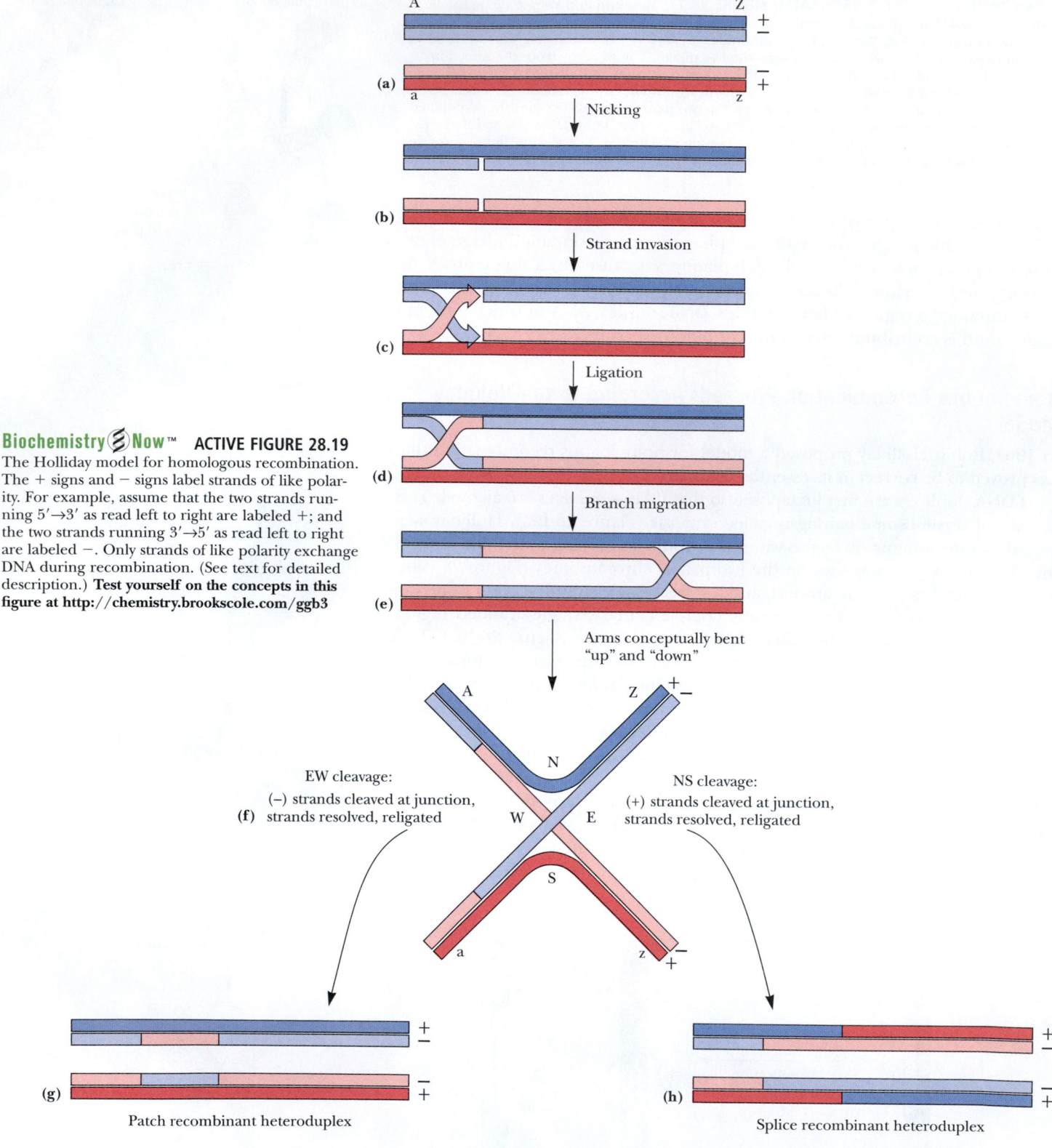

Biochemistry ⓔNow™   ACTIVE FIGURE 28.19
The Holliday model for homologous recombination. The + signs and − signs label strands of like polarity. For example, assume that the two strands running 5′→3′ as read left to right are labeled +; and the two strands running 3′→5′ as read left to right are labeled −. Only strands of like polarity exchange DNA during recombination. (See text for detailed description.) **Test yourself on the concepts in this figure at http://chemistry.brookscole.com/ggb3**

originally nicked (see Figure 28.19b), or at N and S, that is, in the + strands (the strands not previously nicked). Duplex resolution is most easily kept straight by remembering that + strands are complementary to − strands and any resultant duplex must have one of each. Nicks made in the strands originally nicked lead to DNA duplexes in which one strand of each remains intact. Although these duplexes contain heteroduplex regions, they are not recombinant for the markers (AZ, az) that flank the heteroduplex region; such heteroduplexes are called **patch**

recombinants (Figure 28.19g). Nicks introduced into the two strands not previously nicked yield DNA molecules that are both heteroduplex and recombinant for the markers A/a and Z/z; these heteroduplexes are termed **splice recombinants** (Figure 28.19h). Although this Holliday model explains the outcome of recombination, it provides no mechanistic explanation for the strand exchange reactions and other molecular details of the process.

The Enzymes of General Recombination Include RecA, RecBCD, RuvA, RuvB, and RuvC

To illustrate recombination mechanisms, we focus on general recombination as it occurs in *E. coli*. The principal players in the process are the **RecBCD** enzyme complex, which initiates recombination; the **RecA** protein, which binds single-stranded DNA, forming a nucleoprotein filament capable of strand invasion and homologous pairing; and the **RuvA, RuvB,** and **RuvC** proteins, which drive branch migration and process the Holliday junction into recombinant products. Eukaryotic homologs of these prokaryotic recombination proteins have been identified, indicating that the fundamental process of general recombination is conserved across all organisms.

The RecBCD Enzyme Complex Unwinds dsDNA and Cleaves Its Single Strands

The RecBCD complex is composed of the proteins **RecB** (140 kD; 1180 amino acids), **RecC** (130 kD; 1122 amino acids), and **RecD** (67 kD; 608 amino acids). This multifunctional enzyme complex has both helicase and nuclease activity and initiates recombination by attaching to the end of a DNA duplex (or at a double-stranded break in the DNA) and using its ATP-dependent helicase function to unwind the dsDNA (Figure 28.20a). As RecBCD progresses along unwinding the duplex, its nuclease activity cleaves both of the newly formed single strands (although the strand that provided the 3′-end at the RecBCD entry site is cut more frequently than the 5′-terminal strand [Figure 28.20b]). SSB (and some RecA protein) readily binds to the emerging single strands. Sooner or later, RecBCD encounters a particular nucleotide sequence, a so-called **Chi** (or χ) site, characterized by the sequence **5′-GCTGGTGG-3′.** These χ sites are recombinational "hot spots"; 1009 χ sites have been identified in the *E. coli* genome (on average, about one every 4.5 kb of DNA). When a χ sequence is encountered by a RecBCD complex approaching its 3′-side (the ..G-3′-side), RecBCD cleaves the χ-bearing DNA strand four to six bases to the 3′ side of χ (Figure 28.20c). The D-subunit of RecBCD becomes irreversibly altered such that the RecBCD complex no longer expresses nuclease activity against the 3′-terminal strand, but nuclease activity against the 5′-terminal strand increases (Figure 28.20).

Resuming its helicase function, RecBCD unwinds the dsDNA, and collectively these processes generate an ssDNA tail bearing a χ site at its 3′-terminal end. This ssDNA may reach several kilobases in length. RecA protein now binds to the 3′-terminal strand to form a **nucleoprotein filament** (Figure 28.20e). This filament is active in pairing and strand invasion with a homologous region in another dsDNA molecule.

The RecA Protein Can Bind ssDNA and Then Interact with Duplex DNA

The **RecA** protein, or **recombinase,** is a multifunctional protein that acts in general recombination to catalyze the ATP-dependent **DNA strand exchange reaction,** leading to formation of a Holliday junction (Figure 28.19b–f). RecA protein (Figure 28.21a) crystallizes in the absence of DNA to form a helical filament having six monomers per turn (Figure 28.21b). This filament has a deep spiral groove large enough to accommodate three strands of DNA. The nucleoprotein

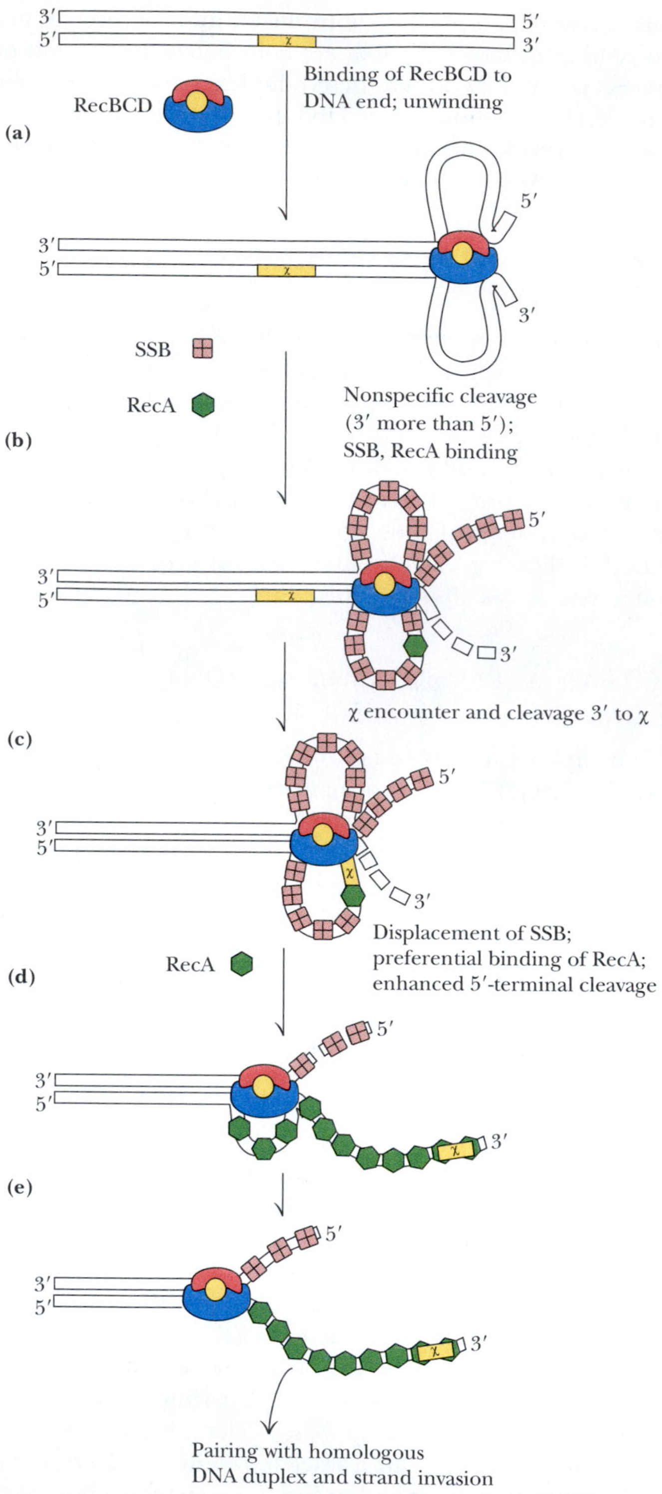

FIGURE 28.20 Model of RecBCD-dependent initiation of recombination. **(a)** RecBCD binds to a duplex DNA end, and its helicase activity begins to unwind the DNA double helix. "Rabbit ears" of ssDNA loop out from RecBCD because the rate of DNA unwinding exceeds the rate of ssDNA release by RecBCD. **(b)** As it unwinds the DNA, SSB (and some RecA) bind to the single-stranded regions; the RecBCD endonuclease activity randomly cleaves the ssDNA, showing a greater tendency to cut the 3'-terminal strand rather than the 5'-terminal strand. **(c)** When RecBCD encounters a properly oriented χ site, the 3'-terminal strand is cleaved just below the 3'-end of χ. **(d)** RecBCD now directs the binding of RecA to the 3'-terminal strand, as RecBCD endonuclease activity now acts more often on the 5'-terminal strand. **(e)** A nucleoprotein filament consisting of RecA-coated 3'-strand ssDNA is formed. This nucleoprotein filament is capable of homologous pairing with a dsDNA and strand invasion. *(Adapted from Figure 2 in Eggleston, A. K., and West, S. C., 1996. Exchanging partners: recombination in E. coli. Trends in Genetics **12**:20–25; and Figure 3 in Eggleston, A. K., and West, S. C., 1997. Recombination initiation: Easy as A, B, C, D ....χ? Current Biology **7**:R745–R749.)*

filament formed by binding of RecA protein to the 3'-terminal ssDNA has affinity for other DNA molecules. In fact, binding of multiple DNA strands is the hallmark of RecA function. In recombination, RecA uses its high-affinity, primary DNA-binding site to bind ssDNA. This complex then interacts with other DNA molecules through a secondary DNA-binding site within RecA.

Procession of strand separation of dsDNA and the re-pairing into hybrid strands along the DNA duplex initiates **branch migration** (Figure 28.22b). Branch migration drives the displacement of the homologous DNA strand from the DNA duplex and its replacement with the ssDNA strand, a process known as

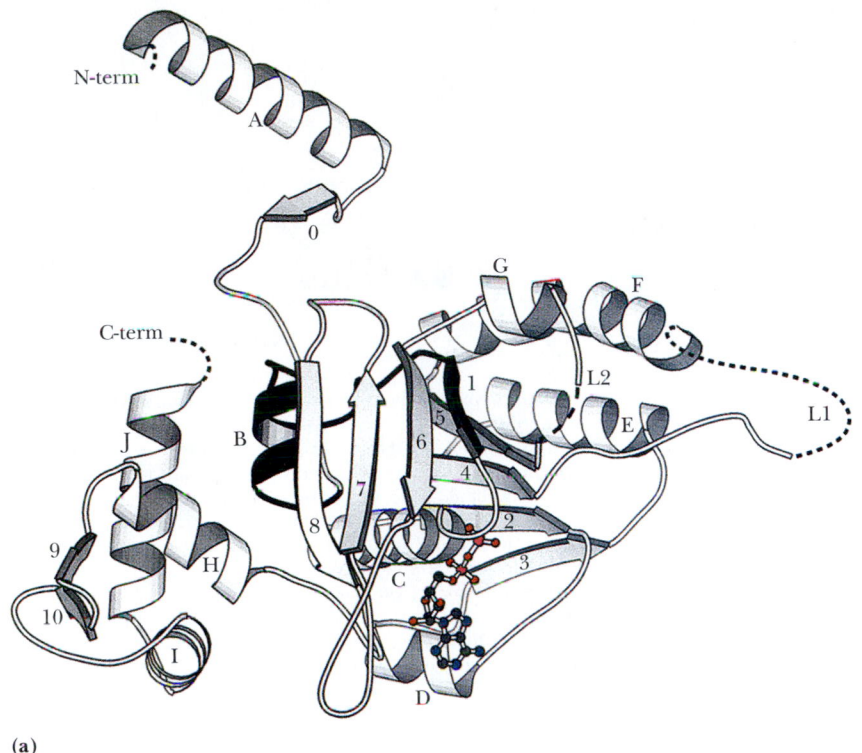

(a)

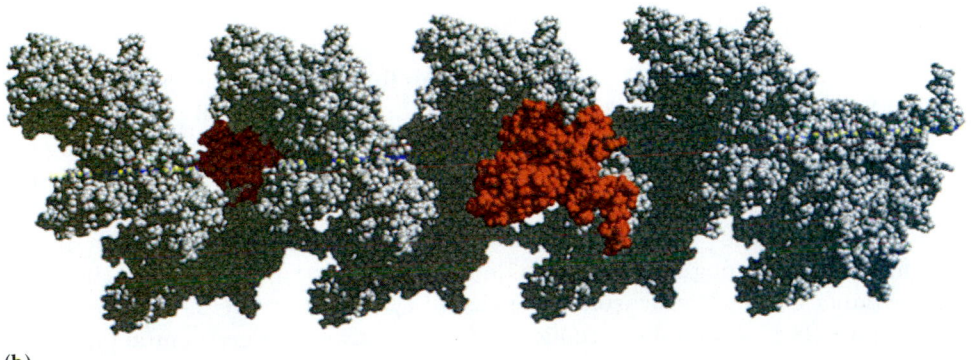

(b)

◀ **FIGURE 28.21** The structure of RecA, a 352-residue, 38-kD protein. **(a)** Ribbon diagram of the RecA monomer. Note the ADP bound at the site near helices C and D. **(b)** RecA filament. Four turns of a helical filament that has six RecA monomers per turn. A RecA monomer is highlighted in red. *(Adapted from Figures 2 and 3 in Roca, A. I., and Cox, M. M., 1997. RecA protein: Structure, function, and role in recombinational DNA repair.* Progress in Nucleic Acid Research and Molecular Biology **56:**127–223. *Photos courtesy of Michael M. Cox, University of Wisconsin.)*

single-strand assimilation (or **single-strand uptake**). Strand assimilation does not occur if there is no sequence homology between the ssDNA and the invaded DNA duplex. The DNA strand displaced by the invading 3′-terminal ssDNA is free to anneal with the 5′-terminal strand in the original DNA, a step that is also mediated by RecA protein and SSB (Figure 28.22c). The result is a Holliday junction, the classic intermediate in genetic recombination.

RuvA, RuvB, and RuvC Proteins Resolve the Holliday Junction to Form the Recombination Products

The Holliday junction is processed into recombination products by **RuvA** (203 amino acids), **RuvB** (336 amino acids), and **RuvC** (173 amino acids). Specifically, RuvA and RuvB work together as a Holliday junction–specific helicase complex that dissociates the RecA filament and catalyzes branch migration. An RuvA

▶ **Biochemistry Now™** **ACTIVE FIGURE 28.22** Model for homologous recombination as promoted by RecA enzyme. **(a)** RecA protein (and SSB) aid strand invasion of the 3′-ssDNA into a homologous DNA duplex, **(b)** forming a D-loop. **(c)** The D-loop strand that has been displaced by strand invasion pairs with its complementary strand in the original duplex to form a Holliday junction as strand invasion continues. **Test yourself on the concepts in this figure at http:// chemistry.brookscole.com/ggb3**

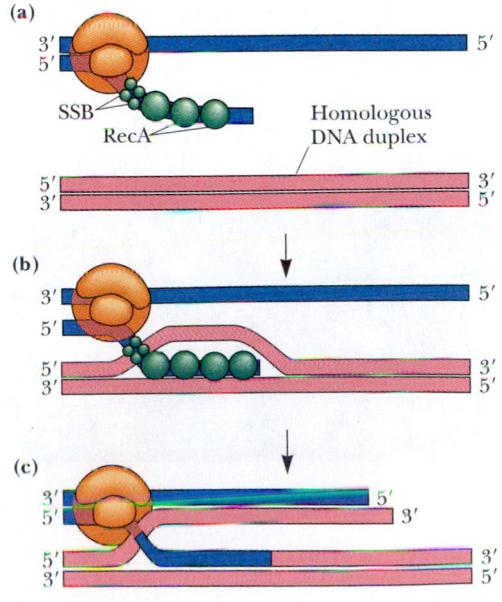

(a)

3′
5′

SSB

RecA

Homologous DNA duplex

5′
3′

3′
5′

(b)

3′

5′

5′
3′

5′

3′
5′

(c)

3′
5′

5′

3′

5′
3′

3′
5′

(a)

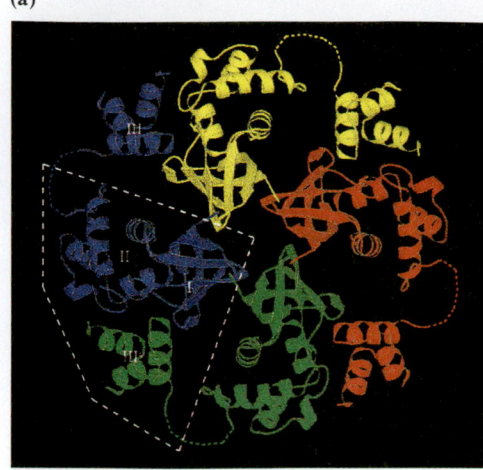

(b)

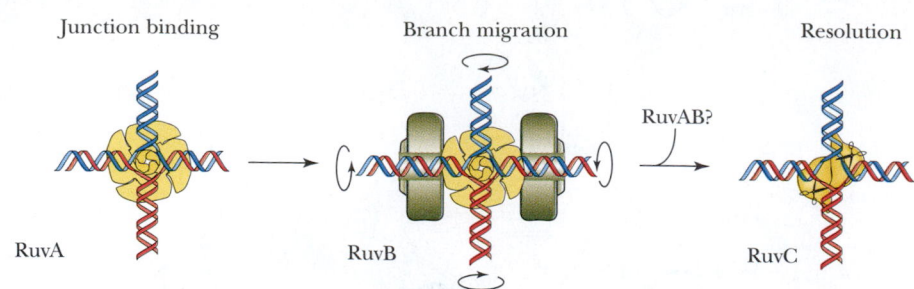

Junction binding	Branch migration	Resolution
RuvA	RuvB	RuvC

(c)

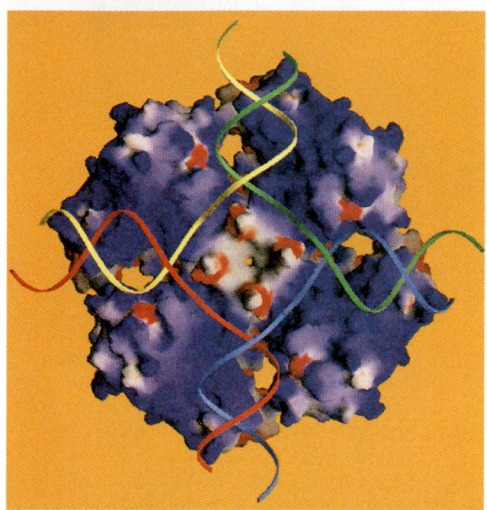

FIGURE 28.23 Model for the resolution of a Holliday junction in *E. coli* by the RuvA, RuvB, and RuvC proteins. **(a)** Ribbon diagram of the RuvA tetramer. RuvA monomers have an overall L shape (one of them is outlined by the dashed white line); four of them form a tetramer with fourfold rotational symmetry, a structure reminiscent of a four-petaled flower. **(b)** Model for RuvA/RuvB action (first suggested by Parsons, C. A., et al., 1995. Structure of a multisubunit complex that promotes DNA branch migration. *Nature* **374:**375–378.) (left) The RuvA tetramer fits snugly within the Holliday junction point. (center) Oppositely facing RuvB hexameric rings assemble on the heteroduplexes, with the DNA passing through their centers. These RuvB hexamers act as motors to promote branch migration by driving the passage of the DNA duplexes through themselves. (right) Binding of RuvC at the Holliday junction and strand scission by its nuclease activity. The locations of the RuvC active sites are indicated by the scissors. **(c)** Charge distribution on the concave surface of an RuvA tetramer. Blue indicates positive charge and red, negative charge. Note the overall positive charge on this surface of (RuvA)$_4$, with the exception of the four red (negatively charged) pins at its center. **(d)** Structural model for the interaction of (RuvA)$_4$ with the hypothesized square-planar Holliday junction center. *(Adapted from Figures 1, 2, and 3 in Rafferty, J. B., et al., 1996. Crystal structure of DNA recombination protein RuvA and a model for its binding to the Holliday junction. Science **274:**415–421.)*

(d)

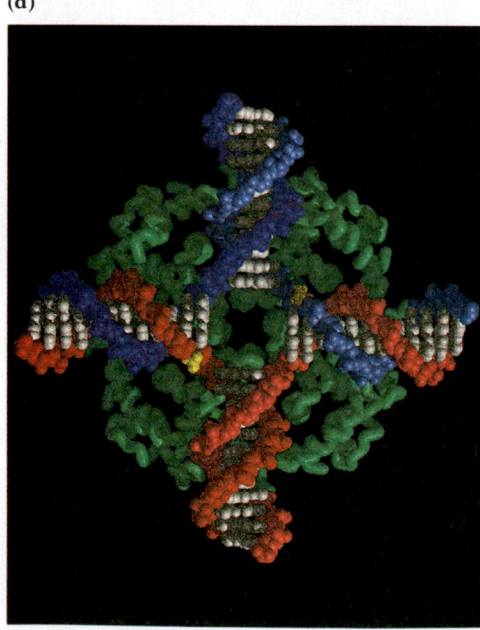

tetramer (Figure 28.23a) fits precisely within the junction point (Figure 28.23b), which has a square-planar geometry, and this RuvA tetramer targets the assembly of RuvB around opposite arms of the DNA junction. The RuvB protein binds to form two oppositely oriented, hexameric [(RuvB)$_6$] ring structures encircling the dsDNAs, one on each side of the Holliday junction. Rotation of the dsDNAs by the RuvB hexameric rings pulls the dsDNAs through (RuvB)$_6$ and unwinds the DNA strands across the "spool" of RuvA, which threads the separated single strands into newly forming hybrid (recombinant) duplexes (Figure 28.23b). The RuvA tetramer is a disclike structure, one face of which has an overall positive charge (Figure 28.23c), with the exception of four negatively charged central pins, each contributed by an RuvA monomer. These four pins fit neatly into the hole at the center of the Holliday junction. The negatively charged sugar–phosphate backbones of the four DNA duplexes of the Holliday junction are threaded along grooves in the positively charged RuvA face, with the negatively charged central pins appropriately situated to transiently separate the dsDNA molecules into their component single strands through repulsive electrostatic interactions with the phosphate backbones of the DNA. The separated strands of each parental duplex are then channeled into grooves in the RuvA face, where they are led into hydrogen-bonding interactions with bases contributed by strands of the other parental DNA to form the two daughter hybrid duplexes flowing out from the RuvAB complex (Figure 28.23b). Figure 28.23d illustrates a model for the RuvA tetramer with the square-planar Holliday junction.

Depending on how the strands in the Holliday junction are cleaved and resolved, patch or splice recombinant duplexes result (Figure 28.19g and h). RuvC is an endonuclease that resolves Holliday junctions into heteroduplex recombinant products (**RuvC resolvase**). An RuvC dimer binds at the Holliday junction and cuts pairs of DNA strands of similar polarity (Figure 28.23b); whether a patch or a splice recombinant results depends on which DNA pair is cleaved.

Hexameric ring helicases such as RuvB are DNA-driving molecular motors; similar motors act during DNA replication to propel strand separation and initiate DNA synthesis. Thus, the RuvABC system for processing Holliday junctions may represent a general paradigm for DNA manipulation in all cells.

A Deeper Look

The Three R's of Genomic Manipulation: Replication, Recombination, and Repair

DNA replication, recombination, and repair have traditionally been treated as separate aspects of DNA metabolism. In recent years, however, scientists have come to realize that DNA replication is an essential component of both DNA recombination and DNA repair processes. Furthermore, recombination mechanisms play an absolutely vital role in restarting replication forks that become halted at breaks or other lesions in the DNA strands. If a double-stranded break (DSB) or a nick in just one of the DNA strands (called a *single-stranded gap*, or *SSG*) is present in the DNA undergoing replication, the replication fork stalls and the replication complex dissociates *(replication fork "collapse")*. Significantly,

the whole process of homologous recombination can initiate only at SSGs or DSBs, and establishment of homologous recombination at such sites can rescue DNA replication. This **recombination-dependent replication (RDR)** has the interesting property of initiating DNA replication at sites other than the *oriC* site, and thus RDR is an important mechanism for restarting DNA replication if the replication fork is disrupted for any reason. As might be expected from the close relationships between replication, recombination, and repair, many of the same proteins are involved in all three, and all three must be viewed as essential processes in the perpetuation of the genome.

Recombination-Dependent Replication Restarts DNA Replication at Stalled Replication Forks

It is likely that most replication forks that begin at the *E. coli oriC* initiation sites (or analogous initiation sites in eukaryotes) are derailed by nicks or more extensive DNA damage lying downstream. However, DNA replication can be reinitiated (and genome replication can be completed) following **replication fork restart.** Recombinational repair of stalled replication forks requires the action

A Deeper Look

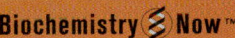

"Knockout" Mice: A Method to Investigate the Essentiality of a Gene

Homologous recombination can be used to replace a gene with an inactivated equivalent of itself. Inactivation is accomplished by inserting a foreign gene, such as *neo,* a gene encoding resistance to the drug *G418,* within one of the exons of a copy of the gene of interest. Homologous recombination between the *neo*-bearing transgene and DNA in wild-type mouse embryonic stem (ES) cells replaces the target gene with the inactive transgene (see accompanying figure). ES cells in which homologous recombination has occurred will be resistant to G418, and such cells can be se-

lected. These recombinant ES cells can then be injected into early-stage mouse embryos, where they have a chance of becoming the germline cells of the newborn mouse. If they do, an inactivated target gene is then present in the gametes of this mouse. Mating between male and female mice with inactive target genes yields a generation of homozygous "knockout" mice—mice lacking a functional copy of the targeted gene. Characterization of these knockout mice reveals which physiological functions the gene directs.

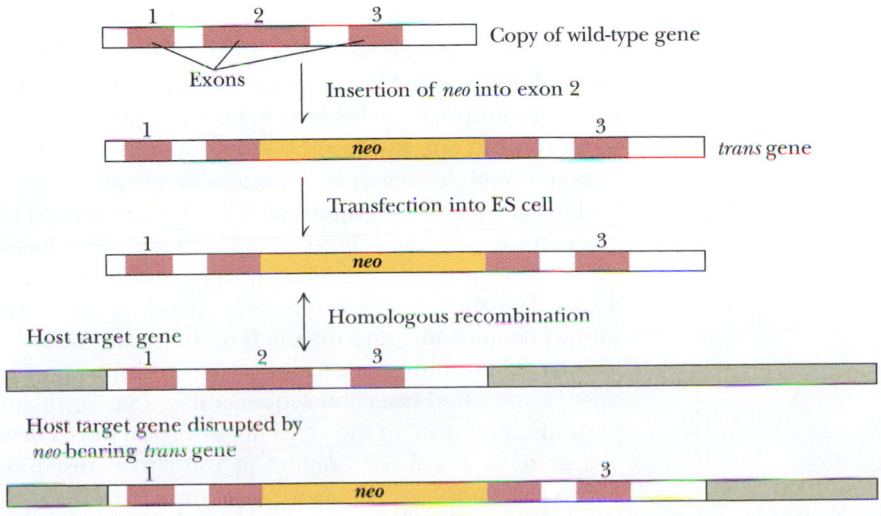

Human Biochemistry

The Breast Cancer Susceptibility Genes BRCA1 and BRCA2 Are Involved in DNA Damage Control and DNA Repair

Mutations in the BRCA1 and BRCA2 genes cause increased likelihood of breast, ovarian, and other cancers. The BRCA1 protein functions in regulation of the cell cycle in response to DNA damage control. Phosphorylation of BRCA1 by DNA damage-response proteins controls the expression, phosphorylation, and cellular localization of specific cyclin-CDKs involved in the cell cycle G_2/M checkpoint and the onset of mitosis. Activation of these cyclin-CDKs leads to arrest of the cell cycle in G_2 until the damage to DNA is repaired. Mutations in BRCA1 that impair its function allow the cell cycle to enter mitosis and DNA damage to accumulate, raising the risk of cancer.

The BRCA2 protein participates in the pathway for DNA repair by homologous recombination. The BRCA2 protein (3418 amino acids) is a very large protein with 8 conserved sequence motifs of about 30 amino acids each, known as the *BRC repeats*. These repeats act as binding sites for RAD51, the eukaryotic analog of RecA. BRCA2 transfers RAD51 to a ssDNA strand coated with RPA (the eukaryotic counterpart of SSB), allowing formation of the RAD51–ssDNA nucleoprotein filament that is an essential intermediate in eukaryotic homologous recombination (see Figure 28.20). Mutations that impair BRCA2 function prevent DNA repair by homologous recombination, leading to accumulation of DNA damage and a greater likelihood of cancer.

Source: Yarden, R. I., et al., 2003. BRCA1 regulates the G_2/M checkpoint by activating Chk1 kinase upon DNA damage. *Nature Genetics* **30**:285–289; Simon, N., Powell, S. M., Willers, H., and Xia, F., 2003. BRCA2 keeps Rad51 in line: High-fidelity homologous recombination prevents breast and ovarian cancer? *Molecular Cell* **10**:1262–1263; and Pelligrini, L., et al., 2002. Insights into DNA recombination from the structure of a RAD51-BRCA2 complex. *Nature* **420**:287–293.

of enzymes from every aspect of DNA metabolism: replication, recombination, and repair. The initial steps in restoration of a replication fork depend on the recombination proteins RecA and RecBCD and the formation of a D-loop (Figure 28.22). The *E. coli* protein **PriA** recognizes and binds with high affinity to D-loops. Once bound, PriA coordinates resumption of DNA replication by recruiting DnaB helicase to the D-loop and reestablishing a replication fork complete with two copies of the replicative DNA polymerase.

Transposons Are DNA Sequences That Can Move from Place to Place in the Genome

In 1950, Barbara McClintock reported the results of her studies on an **activator gene** in maize (*Zea mays,* or as it's usually called, corn) that was recognizable principally by its ability to cause mutations in a second gene. Activator genes were thus an internal source of mutation. A most puzzling property was their ability to move relatively freely about the genome. Scientists had labored to establish that chromosomes consisted of genes arrayed in a fixed order, so most geneticists viewed as incredible this idea of genes moving around. The recognition that McClintock so richly deserved for her explanation of this novel phenomenon had to await verification by molecular biologists. In 1983, Barbara McClintock was finally awarded the Nobel Prize in Physiology or Medicine. By this time, it was appreciated that many organisms, from bacteria to humans, possessed similar "jumping genes" able to move from one site to another in the genome. This mobility led to their designation as **mobile elements, transposable elements,** or, simply, **transposons.**

Transposons are segments of DNA that are moved enzymatically from place to place in the genome (Figure 28.24). That is, their location within the DNA is unstable. Transposons range in size from several hundred base pairs to more than 8 kbp. Transposons contain a gene encoding an enzyme necessary for insertion into a chromosome and for the remobilization of the transposon to different locations. These movements are termed **transposition events.** The smallest transposons are called **insertion sequences,** or **ISs,** signifying their ability to insert apparently at random in the genome. Insertion into a new site can cause a mutation if a gene or regulatory region at the site is disrupted. Because transposition events can move genes to new places or lead to the duplication of existing genes, transposition is a major force in evolution.

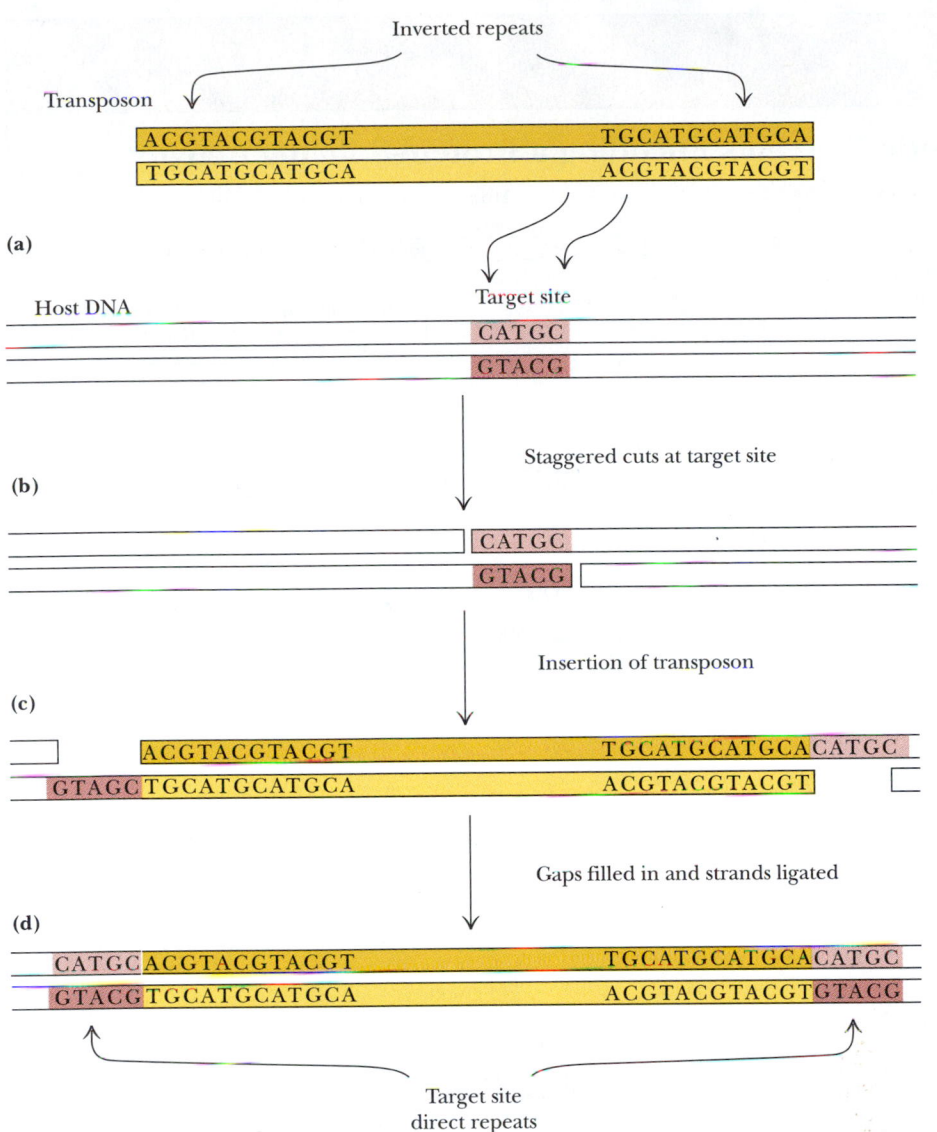

Inverted repeats

Transposon

| ACGTACGTACGT | TGCATGCATGCA |
| TGCATGCATGCA | ACGTACGTACGT |

(a)

Host DNA

Target site

CATGC
GTACG

Staggered cuts at target site

(b)

CATGC
GTACG

Insertion of transposon

(c)

| ACGTACGTACGT | TGCATGCATGCACATGC |
| GTAGCTGCATGCATGCA | ACGTACGTACGT |

Gaps filled in and strands ligated

(d)

| CATGCACGTACGTACGT | TGCATGCATGCACATGC |
| GTACGTGCATGCATGCA | ACGTACGTACGTGTACG |

Target site
direct repeats

Biochemistry ⊘ Now™   ACTIVE FIGURE 28.24
The typical transposon has inverted nucleotide-sequence repeats at its termini, represented here as the 12-bp sequence ACGTACGTACGT (a). It acts at a target sequence (shown here as the sequence CATGC) within host DNA by creating a staggered cut (b) whose protruding single-stranded ends are then ligated to the transposon (c). The gaps at the target site are then filled in, and the filled-in strands are ligated (d). Transposon insertion thus generates direct repeats of the target site in the host DNA, and these direct repeats flank the inserted transposon. **Test yourself on the concepts in this figure at http://chemistry.brookscole.com/ggb3**

28.7 | Can DNA Be Repaired?

Biological macromolecules are susceptible to chemical alterations that arise from environmental damage or errors during synthesis. For RNAs, proteins, or other cellular molecules, most consequences of such damage are avoided by replacement of these molecules through normal turnover (synthesis and degradation). However, the integrity of DNA is vital to cell survival and reproduction. Its information content must be protected over the life span of the cell and preserved from generation to generation. Safeguards include (1) high-fidelity replication systems and (2) repair systems that correct DNA damage that might alter its information content. DNA is the only molecule that, if damaged, is repaired by the cell. Such repair is possible because the information content of duplex DNA is inherently redundant. The most common forms of damage are (1) replication errors resulting in a missing or incorrect base; (2) bulges due to deletions or insertions; (3) UV-induced base alterations, such as pyrimidine dimers (Figure 28.25); (4) strand breaks at phosphodiester bonds or within deoxyribose rings; and (5) covalent crosslinking of strands. Cells have extraordinarily diverse and effective DNA repair systems to deal with these problems, some of which are also involved in DNA replication and recombination. For example, RecBCD can catalyze a complex reaction that repairs double-stranded breaks in DNA by

A Deeper Look

Inteins—Bizarre Parasitic Genetic Elements Encoding a Protein-Splicing Activity

Inteins are parasitic genetic elements found within protein-coding regions of genes. These selfish DNA elements are transcribed and translated along with the flanking host gene sequences. The typical intein consists of two domains: One domain is capable of self-catalyzed **protein splicing;** the other is an endonuclease that mediates the insertion of the intein nucleotide sequence into host genes. After the full protein is synthesized, the intein catalyzes excision of itself from the host protein *and* ligation of adjacent host polypeptide regions to form the functional protein that the host gene encodes. These adjacent polypeptides are termed **exteins ("external proteins")** to distinguish them from the intein ("internal protein"). Inteins have been found across all domains of life—archaea, bacteria, and eukaryotes—although thus far only in unicellular organisms. Inteins vary in size from about 130 to 600 amino acid residues. The protein splicing function of inteins is found in its N-terminal and its C-terminal regions; the endonuclease function that carries out parasitic insertion of the intein sequence into host genes is found in the central part of the intein. Splicing of the protein is an intramolecular process that liberates the intein sequence and ligates the host protein sequences (see accompanying figure).

Inteins are usually found as inserts in highly conserved host genes that have essential functions, such as genes encoding DNA or RNA polymerases, proton-translocating ATPases, or other vital metabolic enzymes. Their location in such genes means that removal of the intein via deletion or genetic rearrangement is more difficult. The endonuclease activity of the intein recognizes a 14– to 40–base-pair sequence in a potential host gene and cleaves the DNA there. During repair of the double-stranded DNA break, the intein gene is copied into the cleavage site, thereby establishing the parasitic genetic element in the host gene.

▶ Transcription and translation of the combined intein-host gene flanking sequences leads to synthesis of a fused intein–extein protein. The intein splices itself out when (1) the C-terminal residue of the N-extein is shifted to the O (or S) atom of a neighboring intein Ser (or Cys) residue, (2) the N-extein C-terminal carbonyl undergoes nucleophilic attack by the O (or S) atom of a Ser (or Cys) residue at the end of the C-extein in a transesterification reaction that creates a branched protein intermediate, (3) cyclization of the intein C-terminal asparagine residue excises the intein, and (4) the two exteins are properly united via a peptide bond when the N-extein C-terminus spontaneously shifts to the C-extein N-terminus to form an intact host protein.

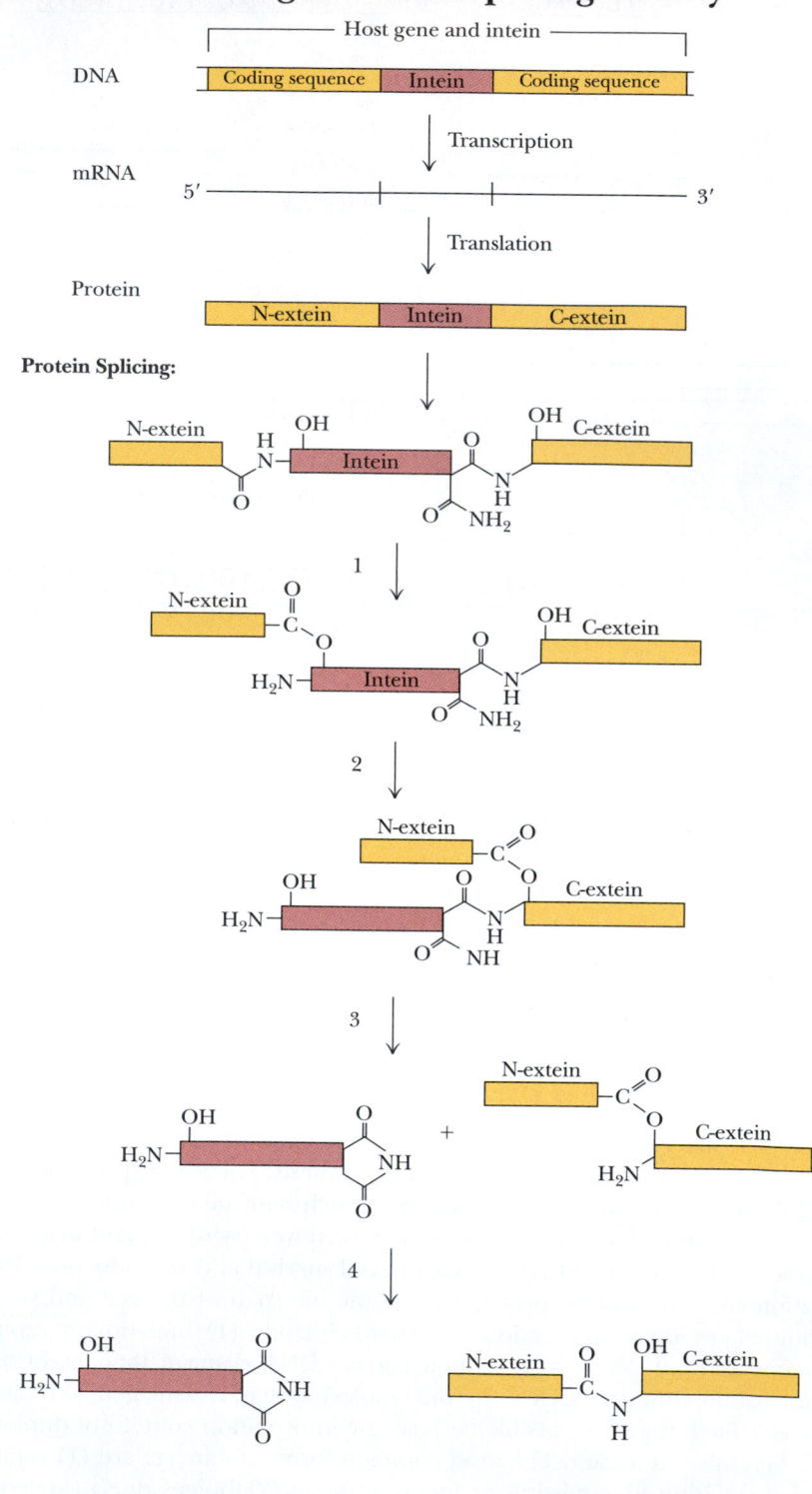

Adapted from Paulus, H., 2000. Protein splicing and related forms of protein autoprocessing. *Annual Review of Biochemistry* **69:**447–496; and Gogarten, J. P., et al., 2002. Inteins: Structure, function, and evolution. *Annual Review of Microbiology* **56:**263–287.

A Deeper Look

Transgenic Animals Are Animals Carrying Foreign Genes

Experimental advances in gene transfer techniques have made it possible to introduce genes into animals by **transfection.** Transfection is defined as the uptake or injection of plasmid DNA into recipient cells. Animals that have acquired new genetic information as a consequence of the introduction of foreign genes are termed **transgenic.** Plasmids carrying the gene of interest are injected into the nucleus of an oocyte or fertilized egg, and the egg is then implanted into a receptive female. The technique has been perfected for mice (see figure, part a). In a small number of cases—10% or so—the mice that develop from the injected eggs carry the transfected gene integrated into a single chromosomal site. The gene is subsequently inherited by the progeny of the transfected animal as if it were a normal gene. Expression of the donor gene in the transgenic animals is variable because the gene is randomly integrated into the host genome and gene expression is often influenced by chromosomal location. Nevertheless, transfection of animals has produced some startling results, as in the case of the transfection of mice with the gene encoding the **rat growth hormone (rGH).** The transgenic mice grew to nearly twice the normal size (see figure, part b). Growth hormone levels in these animals were several hundred times greater than normal. Similar results were obtained in transgenic mice transfected with the **human growth hormone (hGH)** gene. The biotechnology of transfection has been extended to farm animals, and transgenic chickens, cows, pigs, rabbits, sheep, and even fish have been produced.

The first animal cloned from an adult cell, a sheep named Dolly, represented a milestone in cloning technology. Subsequent accomplishments include incorporation of the human gene encoding blood coagulation factor IX into sheep. Fetal sheep fibroblast cells were transfected with the human factor IX gene, nuclei from the transfected cells were transferred into sheep oocytes lacking nuclei, and these transgenic oocytes were placed in the uterus of receptive female sheep, which subsequently gave birth to transgenic lambs. The introduced factor IX transgene was specifically designed so that factor IX protein, a medically useful product for the treatment of hemophiliacs, would be expressed in the milk of the transgenic sheep. Similar successes in cows, which produce much more milk, has brought the potential for commercial production of virtually any protein into the realm of reality.

Transfection technology also holds promise as a mechanism for "gene therapy" by replacing defective genes in animals with functional genes (see Chapter 12). Problems concerning delivery, integration and regulation of the transfected gene, including its appropriate expression in the right cells at the proper time during development and growth of the organism, must be brought under control before gene therapy becomes commonplace in humans.

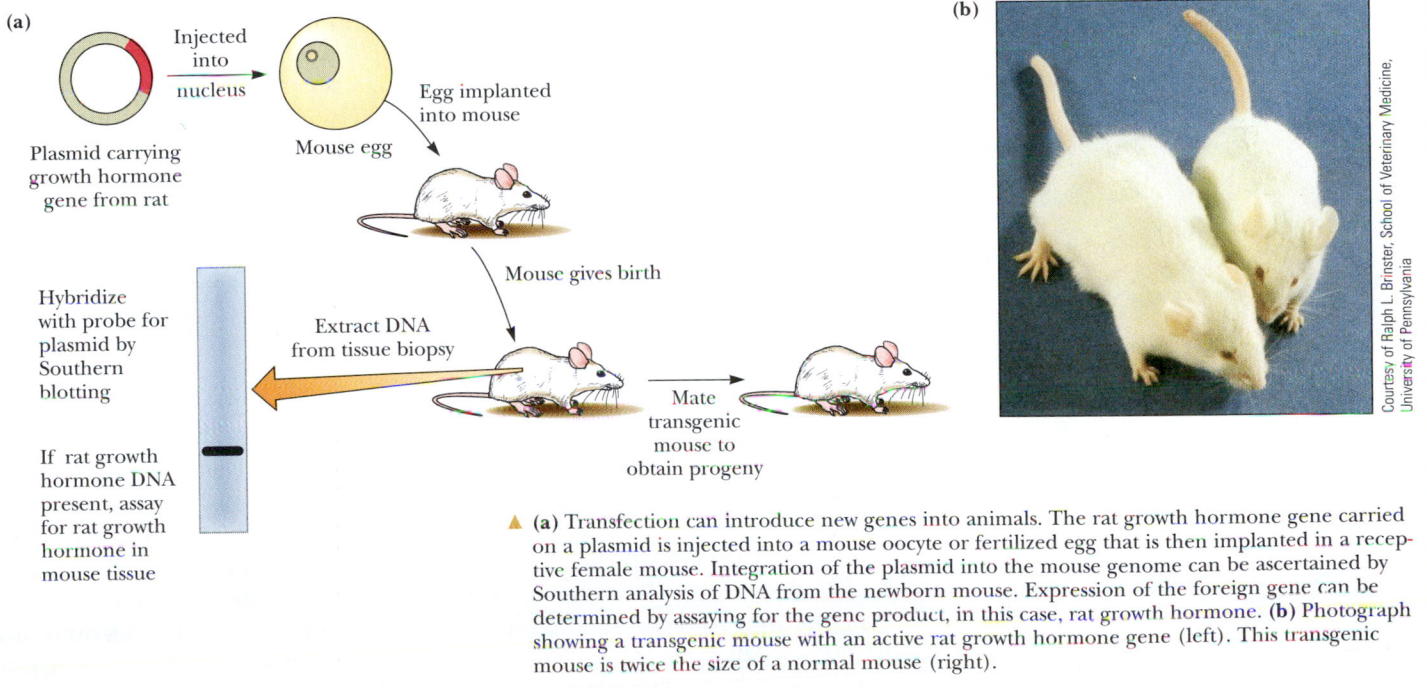

(a) Transfection can introduce new genes into animals. The rat growth hormone gene carried on a plasmid is injected into a mouse oocyte or fertilized egg that is then implanted in a receptive female mouse. Integration of the plasmid into the mouse genome can be ascertained by Southern analysis of DNA from the newborn mouse. Expression of the foreign gene can be determined by assaying for the gene product, in this case, rat growth hormone. **(b)** Photograph showing a transgenic mouse with an active rat growth hormone gene (left). This transgenic mouse is twice the size of a normal mouse (right).

homologous recombination. When repair systems fail, the genome may still be preserved if an "error-prone" mode of replication allows the lesion to be bypassed.

Human DNA replication has an error rate of about three base-pair mistakes during copying the 6 billion base pairs in the diploid human genome. The low error rate is due to the DNA repair systems that review and edit the newly replicated DNA. Furthermore, about 10^4 bases (mostly purines) are lost per cell per

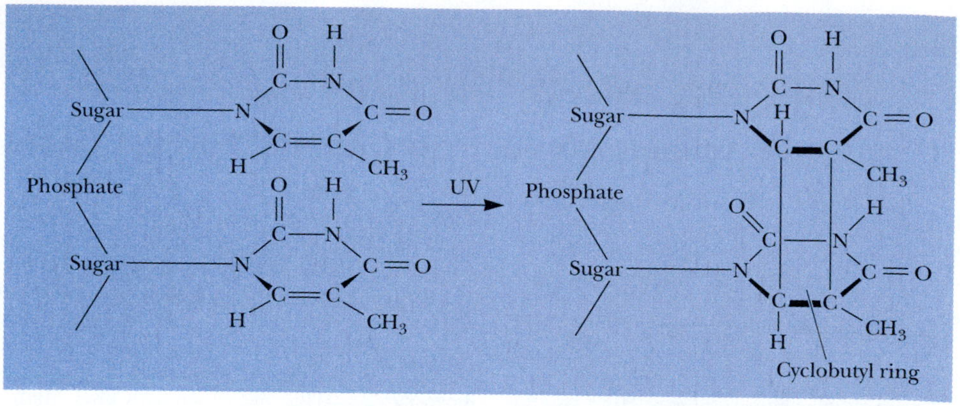

FIGURE 28.25 UV irradiation causes dimerization of adjacent thymine bases. A cyclobutyl ring is formed between carbons 5 and 6 of the pyrimidine rings. Normal base pairing is disrupted by the presence of such dimers.

day from spontaneous breakdown in human DNA; the repair systems must replace these bases to maintain the fidelity of the encoded information.

Usually, the complementary structure of duplex DNA ensures that information lost through damage to one strand can be recovered from the other. However, even errors involving both strands can be corrected. For example, deletions or insertions can be repaired by replacing the missing or superfluous region through recombination (see Section 28.6). Double-stranded breaks, potentially the most serious lesions, can be repaired by DNA ligases or recombination events.

Molecular Mechanisms of DNA Repair Include Mismatch Repair and Excision Repair

Several fundamental types of molecular mechanisms for DNA repair can be distinguished: mechanisms that excise and replace damaged regions by recombination-dependent replication (discussed earlier), **mismatch repair** (discussed in the following section), and mechanisms such as **excision repair** systems that reverse damaging chemical changes in DNA (also discussed in the following section).

Mismatch Repair Corrects Errors Introduced During DNA Replication

The *mismatch repair system* corrects errors introduced when DNA is replicated. It scans newly synthesized DNA for mispaired bases, excises the mismatched region, and then replaces it by DNA polymerase–mediated local replication. The key to such replacement is to know which base of the mismatched pair is correct.

The *E. coli* **methyl-directed pathway** of mismatch repair relies on methylation patterns in the DNA to determine which strand is the newly synthesized one and which one was the parental (template) strand. DNA methylation, often an identifying and characteristic feature of a prokaryote's DNA, occurs just after DNA replication. During methylation, methyl groups are added to certain bases along the new DNA strand. However, a window of opportunity exists between the start of methylation and the end of replication, when only the parental strand of a dsDNA is methylated. This window in time provides an opportunity for the mismatch repair system to review the dsDNA for mismatched bases that arose as a consequence of replication errors. By definition, the newly synthesized strand is the one containing the error, and the methylated strand is the one having the correct nucleotide sequence. When the methyl-directed mismatch repair system encounters a mismatched base pair, it searches along the DNA—through thousands of base pairs if necessary—until it finds a methylated base.

The system identifies the strand bearing the methylated base as parental, assumes its sequence is the correct one, and replaces the entire stretch of nucleotides within the new strand from this recognition point to and including the mismatched base. Mismatch repair does this by using an endonuclease to cut the new, unmethylated strand and an exonuclease to remove the mis-matched bases, creating a gap in the newly synthesized strand. DNA polymerase III holoenzyme then fills in the gap, using the methylated strand as template. Finally, DNA ligase reseals the strand.

Damage to DNA by UV Light or Chemical Modification Can Also Be Repaired

Repair of Pyrimidine Dimers Formed by UV Light   UV irradiation promotes the formation of covalent bonds between adjacent thymine residues in a DNA strand, creating a cyclobutyl ring (Figure 28.25). Because the C—C bonds in this ring are shorter than the normal 0.34-nm base stacking in B-DNA, the DNA is distorted at this spot and is no longer a proper template for either replication or transcription. **Photolyase** (also called **photoreactivating enzyme**), a flavin- and pterin-dependent enzyme, binds at the dimer and uses the energy of visible light to break the cyclobutyl ring, restoring the pyrimidines to their original form.

Excision Repair   Replacement of chemically damaged or modified bases occurs via **excision repair systems.** There are two fundamental excision repair systems—**base excision** and **nucleotide excision.** *Base excision repair* acts on single bases that have been damaged through oxidation or other chemical modifications during normal cellular processes. The damaged base is removed by **DNA glycosylase,** which cleaves the glycosidic bond, creating an apurinic acid (AP) site where the sugar–phosphate backbone is intact but a purine *(apurinic site)* or a pyrimidine *(apyrimidinic site)* is missing. An **AP endonuclease** then cleaves the backbone, an exonuclease removes the deoxyribose-P and a number of additional residues, and the gap is repaired by DNA polymerase and DNA ligase (Figure 28.26). The information of the complementary strand is used to dictate which bases are added in refilling the gap. The DNA polymerase I binds at the gap and moves in the 5′→3′ direction, removing nucleotides with its 5′-exonuclease activity. The 5′→3′ DNA polymerase activity of DNA polymerase I fills in the sequence behind the 5′-exonuclease action. No net synthesis of DNA results, but this action of DNA polymerase I "edits out" sections of damaged DNA. Excision is coordinated with 5′→3′ polymerase-catalyzed replacement of the damaged nucleotides so that DNA of the right sequence is restored.

Nucleotide excision repair recognizes and repairs larger regions of damaged DNA than base excision repair. The nucleotide excision repair system cuts the sugar–phosphate backbone of a DNA strand in two places, one on each side of the lesion, and removes the region. The region removed in prokaryotic nucleotide excision repair spans 12 or 13 nucleotides; in eukaryotic excision repair, an oligonucleotide stretch 27 to 29 units long is removed. The resultant gap is then filled in using DNA polymerase (DNA polymerase I in prokaryotes or DNA polymerase δ or ε and PCNA plus RFC in eukaryotes), and the sugar–phosphate backbone is covalently closed by DNA ligase.

In mammalian cells, nucleotide excision repair is the main pathway for removal of carcinogenic (cancer-causing) lesions caused by sunlight or other mutagenic agents. Such lesions are recognized by **XPA** protein, named for *xeroderma pigmentosum,* an inherited human syndrome whose victims suffer serious skin lesions if exposed to sunlight. At sites recognized by XPA, a multiprotein endonuclease is assembled and the damaged strand is cleaved and repaired.

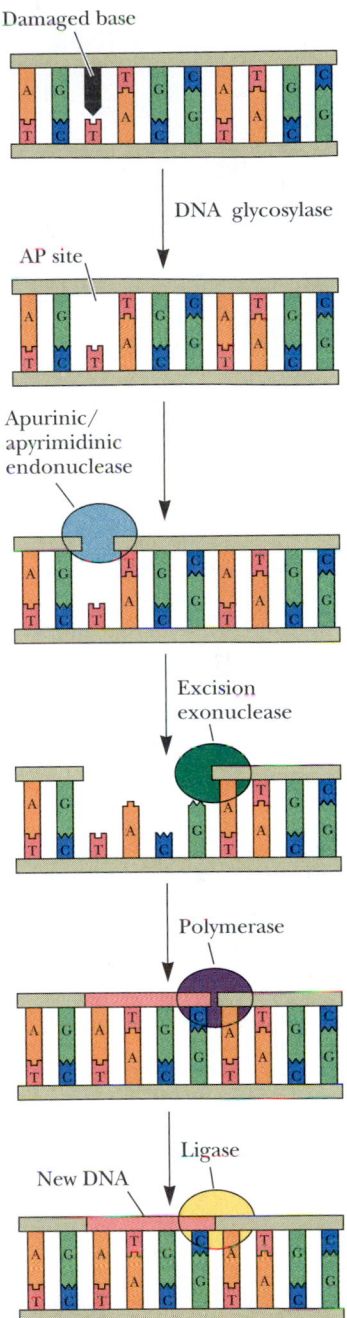

FIGURE 28.26 Base excision repair. A damaged base (■) is excised from the sugar-phosphate backbone by DNA glycosylase, creating an AP site. Then, an apurinic/apyrimidinic endonuclease severs the DNA strand, and an excision nuclease removes the AP site and several nucleotides. DNA polymerase I and DNA ligase then repair the gap.

28.8 | What Is the Molecular Basis of Mutation?

Genes are normally transmitted unchanged from generation to generation, owing to the great precision and fidelity with which genes are copied during chromosome duplication. However, on rare occasions, genetically heritable changes **(mutations)** occur and result in altered forms. Most mutated genes function less effectively than the unaltered, wild-type allele, but occasionally mutations arise that give the organism a selective advantage. When this occurs, they are propagated to many offspring. Together with recombination, mutation provides for genetic variability within species and, ultimately, the evolution of new species.

Mutations change the sequence of bases in DNA, either by the substitution of one base pair for another (so-called **point mutations**) or by the insertion or deletion of one or more base pairs (**insertions** and **deletions**).

Point Mutations Arise by Inappropriate Base-Pairing

Point mutations arise when a base pairs with an inappropriate partner. The two possible kinds of point mutations are **transitions,** in which one purine (or pyrimidine) is replaced by another, as in A→G (or T→C), and **transversions,** in which a purine is substituted for a pyrimidine, or vice versa.

Point mutations arise by the pairing of bases with inappropriate partners during DNA replication, by the introduction of base analogs into DNA, or by chemical mutagens. Bases may rarely mispair (Figure 28.27), either because of their tautomeric properties (see Chapter 10) or because of other influences. Even in mispairing, the C_1'–C_1' distances between bases must still be close to that of a Watson–Crick base pair (11 nm or so; see Figure 11.7) to maintain the mismatched base pair in the double helix. In tautomerization, for example, an amino group (—NH_2), usually an H-bond donor, can tautomerize to an imino form (=NH) and become an H-bond acceptor. Or a keto group (C=O), normally an H-bond acceptor, can tautomerize to an enol C—OH, an H-bond donor. Proofreading mechanisms operating during DNA replication catch most mispairings. The frequency of spontaneous mutation in both *E. coli* and fruit flies *(Drosophila melanogaster)* is about 10^{-10} per base pair per replication.

Mutations Can Be Induced by Base Analogs

Base analogs that become incorporated into DNA can induce mutations through changes in base-pairing possibilities. Two examples are **5-bromouracil (5-BU)** and **2-aminopurine (2-AP).** 5-Bromouracil is a thymine analog and becomes inserted

FIGURE 28.27 Point mutations due to base mispairings. **(a)** An example based on tautomeric properties. The rare imino tautomer of adenine base pairs with cytosine rather than thymine. **(1)** The normal A-T base pair. **(2)** The A*–C base pair is possible for the adenine tautomer in which a proton has been transferred from the 6-NH_2 of adenine to N-1. **(3)** Pairing of C with the imino tautomer of A (A*) leads to a transition mutation (A–T to G–C) appearing in the next generation. **(b)** A in the syn conformation pairing with G (G is in the usual anti conformation). **(c)** T and C form a base pair by H-bonding interactions mediated by a water molecule.

(a)

(1)

T · · · A

(2)

Cytosine · · · Rare imino tautomer of adenine

(3)

A—T
A—T → A—T
A—T → A*—C → A—T
                  G—C

(b)

A · · · G

(c)

T · · · C

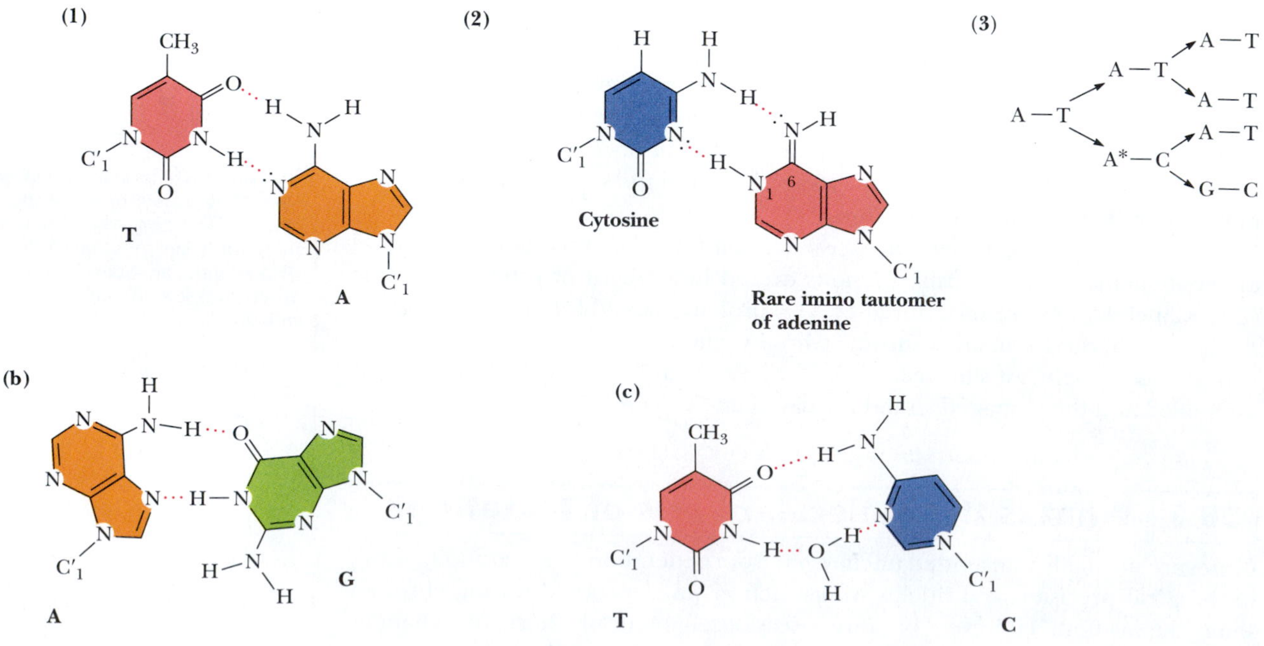

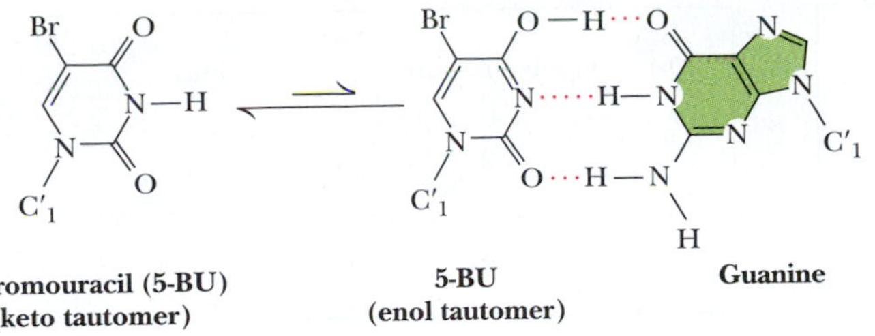

FIGURE 28.28 5-Bromouracil usually favors the keto tautomer that mimics the base-pairing properties of thymine, but it frequently shifts to the enol form, whereupon it can base-pair with guanine, causing a T–A to C–G transition.

into DNA at sites normally occupied by T; its 5-Br group sterically resembles thymine's 5-methyl group. However, because 5-BU frequently assumes the enol tautomeric form and pairs with G instead of A, a point mutation of the transition type may be induced (Figure 28.28). Less often, 5-BU is inserted into DNA at cytosine sites, not T sites. Then, if it base-pairs in its keto form, mimicking T, a C–G to T–A transition ensues. The adenine analog, 2-aminopurine (recall that adenine is 6-aminopurine) normally behaves like A and base-pairs with T. However, 2-AP can form a single H bond of sufficient stability with cytosine (Figure 28.29) that occasionally C replaces T in DNA replicating in the presence of 2-AP. Hypoxanthine (Figure 28.30) is an adenine analog that arises in situ in DNA through oxidative deamination of A. Hypoxanthine base-pairs with cytosine, creating an A–T to G–C transition.

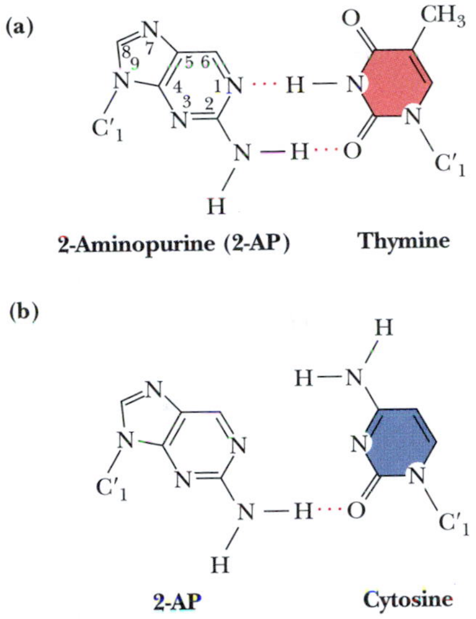

Chemical Mutagens React with the Bases in DNA

Chemical mutagens are agents that chemically modify bases so that their base-pairing characteristics are altered. For instance, *nitrous acid* (HNO_2) causes the oxidative deamination of primary amine groups in adenine and cytosine. Oxidative deamination of cytosine yields uracil, which base-pairs the way T does and gives a C–G to T–A transition (Figure 28.31a). *Hydroxylamine* specifically causes C–G to T–A transitions because it reacts specifically with cytosine, converting it to a derivative that base-pairs with adenine instead of guanine. **Alkylating agents** are also chemical mutagens. Alkylation of reactive sites on the bases to add methyl or ethyl groups alters their H bonding and hence base pairing. For example, methylation of O^6 on guanine (giving O^6-methylguanine) causes this G to mispair with thymine, resulting in a G–C to A–T transition (Figure 28.31d). Alkylating agents can also induce point mutations of the transversion type. Alkylation of N^7 of guanine labilizes its *N*-glycosidic bond, which leads to elimination of the purine ring, creating a gap in the base sequence. An enzyme, AP endonuclease, then cleaves the sugar–phosphate backbone of the DNA on the 5'-side, and the gap can be repaired by enzymatic removal of the 5'-sugar–phosphate and insertion of a new nucleotide. A transversion results if a pyrimidine nucleotide is inserted in place of the purine during enzymatic repair of this gap.

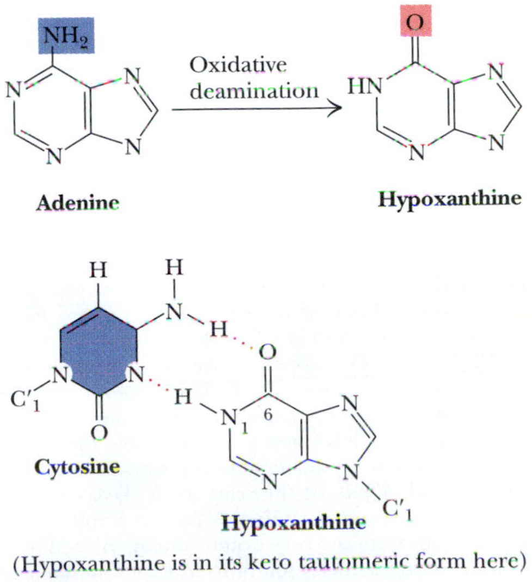

Insertions and Deletions

The addition or removal of one or more base pairs leads to *insertion* or *deletion* mutations, respectively. Either shifts the triplet reading frame of codons, causing **frameshift mutations** (misincorporation of all subsequent amino acids) in the protein encoded by the gene. Such mutations can arise if flat aromatic molecules such as *acridine orange* insert themselves between successive bases in one or both strands of the double helix. This insertion or, more aptly, **intercalation,** doubles the distance between the bases as measured along the helix axis. This distortion of the DNA results in inappropriate insertion or deletion of bases when the DNA is replicated. Disruptions that arise from the insertion of a transposon within a gene also fall in this category of mutation.

FIGURE 28.29 (a) 2-Aminopurine normally base-pairs with T but (b) may also pair with cytosine through a single hydrogen bond.

(Hypoxanthine is in its keto tautomeric form here)

FIGURE 28.30 Oxidative deamination of adenine in DNA yields hypoxanthine, which base-pairs with cytosine, resulting in an A–T to G–C transition.

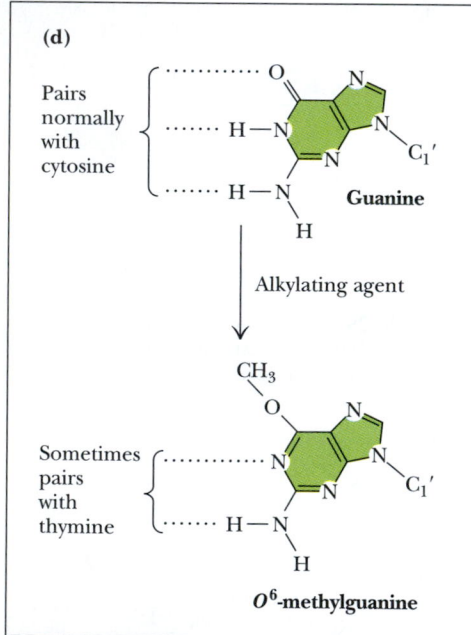

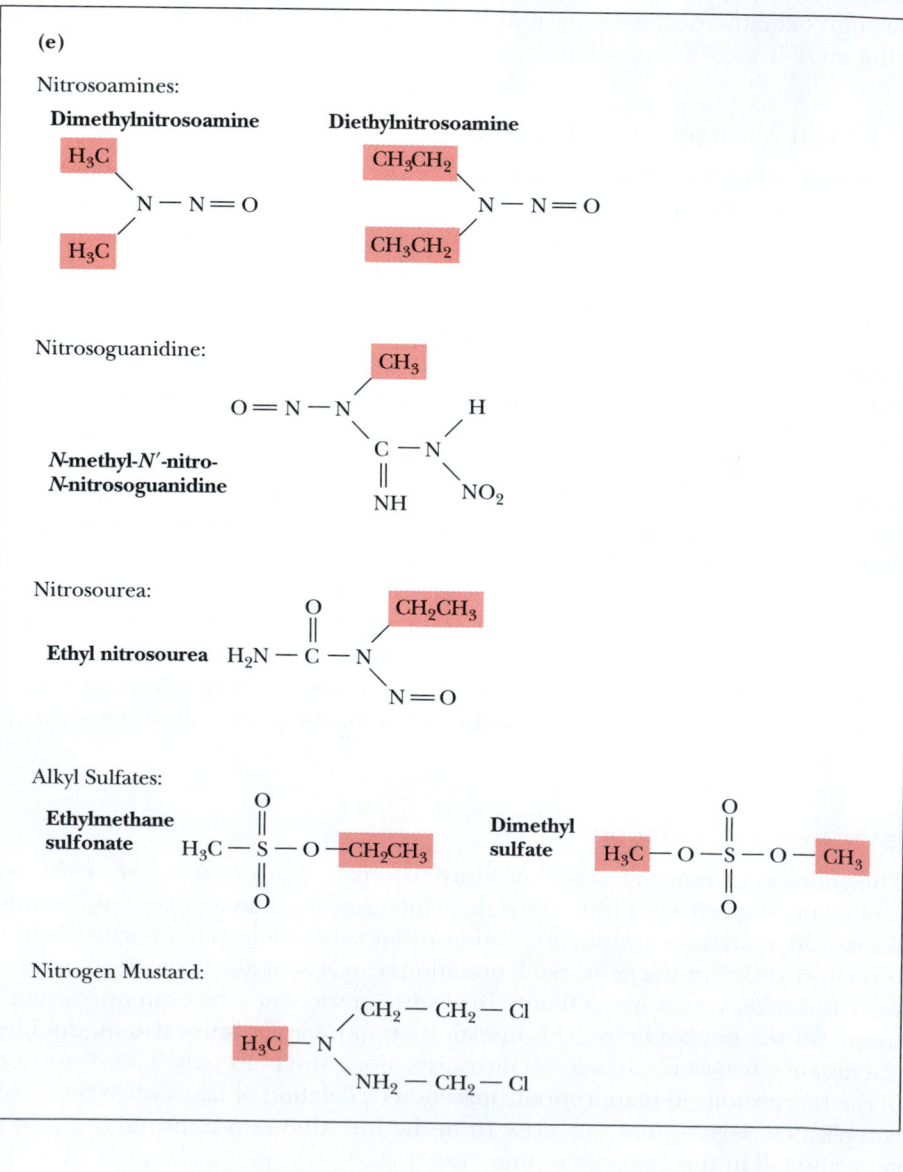

FIGURE 28.31 Chemical mutagens. **(a)** HNO_2 (nitrous acid) converts cytosine to uracil and adenine to hypoxanthine. **(b)** Nitrosoamines, organic compounds that react to form nitrous acid, also lead to the oxidative deamination of A and C. **(c)** Hydroxylamine (NH_2OH) reacts with cytosine, converting it to a derivative that base-pairs with adenine instead of guanine. The result is a C–G to T–A transition. **(d)** Alkylation of G residues to give O^6-methylguanine, which base-pairs with T. **(e)** Alkylating agents include nitrosoamines, nitrosoguanidines, nitrosoureas, alkyl sulfates, and nitrogen mustards. Note that nitrosoamines are mutagenic in two ways: They can react to yield HNO_2, or they can act as alkylating agents. The nitrosoguanidine, *N*-methyl-*N'*-nitro-*N*-nitrosoguanidine, is a very potent mutagen used in laboratories to induce mutations in experimental organisms such as *Drosophila melanogaster*. Ethylmethane sulfonate (EMS) and dimethyl sulfate are also favorite mutagens among geneticists.

Special Focus: Gene Rearrangements and Immunology—Is It Possible to Generate Protein Diversity Using Genetic Recombination?

Animals have evolved a way to exploit genetic recombination in order to generate protein diversity. This development was crucial to the evolution of the immune system. For example, the immunoglobulin genes are a highly evolved system for maximizing protein diversity from a finite amount of genetic information. This diversity is essential for gaining immunity to the great variety of infectious organisms and foreign substances that cause disease.

Cells Active in the Immune Response Are Capable of Gene Rearrangement

Only vertebrates show an immune response. If a foreign substance, called an **antigen,** gains entry to the bloodstream of a vertebrate, the animal responds via a protective system called the *immune response.* The immune response involves production of proteins capable of recognizing and destroying the antigen. This response is mounted by certain white blood cells—the **B-** and **T-cell lymphocytes** and the **macrophages.** B cells are so named because they mature in the bone marrow; T cells mature in the thymus gland. Each of these cell types is capable of gene rearrangement as a mechanism for producing proteins essential to the immune response. **Antibodies,** which can recognize and bind antigens, are immunoglobulin proteins secreted from B cells. Because antigens can be almost anything, the immune response must have an incredible repertoire of structural recognition. Thus, vertebrates must have the potential to produce immunoglobulins of great diversity in order to recognize virtually any antigen.

Immunoglobulin G Molecules Constitute the Major Class of Circulating Antibodies

Immunoglobulin G (**IgG** or **γ-globulin**) is the major class of antibody molecules found circulating in the bloodstream. IgG is a very abundant protein, amounting to 12 mg per mL of serum. It is a 150-kD $\alpha_2\beta_2$-type tetramer. The α or *H* (for *heavy*) chain is 50 kD; the β or *L* (for *light*) chain is 25 kD. A preparation of IgG from serum is heterogeneous in terms of the amino acid sequences represented in its L and H chains. However, the IgG L and H chains produced from any given B lymphocyte are homogeneous in amino acid sequence. L chains consist of 214 amino acid residues and are organized into two roughly equal segments: the V_L and C_L regions. The V_L designation reflects the fact that L chains isolated from serum IgG show variations in amino acid sequence over the first 108 residues, V_L symbolizing this "variable" region of the L polypeptide. The amino acid sequence for residues 109 to 214 of the L polypeptide is constant, as represented by its designation as the "constant light," or C_L, region. The heavy, or H, chains consist of 446 amino acid residues. Like L chains, the amino acid sequence for the first 108 residues of H polypeptides is variable, ergo its designation as the V_H region, while residues 109 to 446 are constant in amino acid sequence. This "constant heavy" region consists of three quite equivalent domains of homology designated C_H1, C_H2, and C_H3. Each L chain has two intrachain disulfide bonds: one in the V_L region and the other in the C_L region. The C-terminal amino acid in L chains is cysteine, and it forms an interchain disulfide bond to a neighboring H chain. Each H chain has four intrachain disulfide bonds, one in each of the four regions. Figure 28.32 presents a diagram of IgG organization. Within the variable regions of the L and H chains, certain positions are **hypervariable** with regard to amino acid composition. These hypervariable residues occur at positions 24 to 34, 50 to 55, and 89 to 96 in the L chains and at positions 31 to 35, 50 to 65, 81 to 85, and 91 to

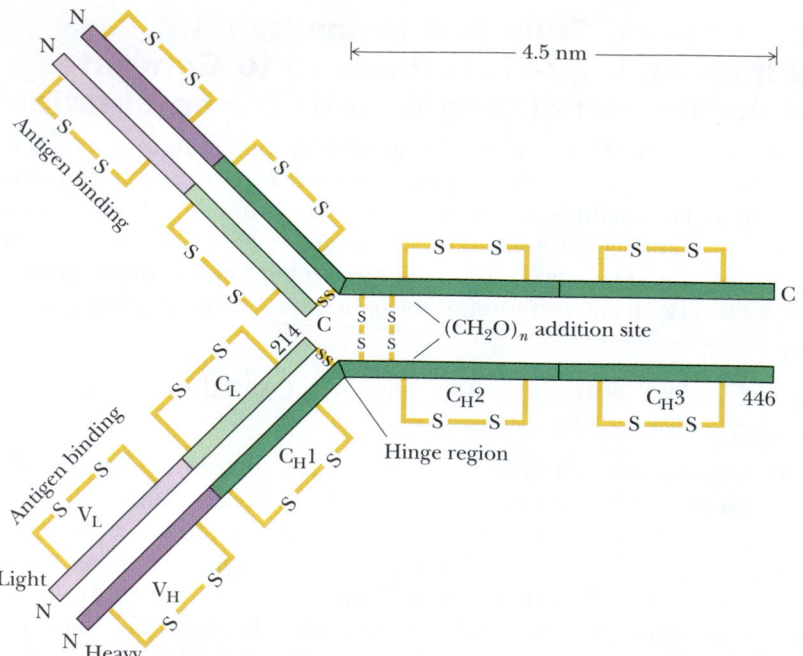

FIGURE 28.32 Diagram of the organization of the IgG molecule. Two identical L chains are joined with two identical H chains. Each L chain is held to an H chain via an interchain disulfide bond. The variable regions of the four polypeptides lie at the ends of the arms of the Y-shaped molecule. These regions are responsible for the antigen recognition function of the antibody molecules. The actual antigen-binding site is constituted from hypervariable residues within the V_L and V_H regions. For purposes of illustration, some features are shown on only one or the other L chain or H chain, but all features are common to both chains.

102 in the H chains. The hypervariable regions are also called **complementarity-determining regions,** or **CDRs,** because it is these regions that form the structural site that is complementary to some part of an antigen's structure, providing the basis for antibody : antigen recognition.

In the immunoglobulin genes, the arrangement of exons correlates with protein structure. In terms of its tertiary structure, the IgG molecule is composed of 12 discrete *collapsed β-barrel domains,* each domain having a *Greek key* motif (see Figure 6.33). The characteristic structure of this domain is referred to as the **immunoglobulin fold** (Figure 28.33). Each of IgG's two heavy chains contributes four of these domains and each of its light chains contributes two. The four *variable-region* domains (one on each chain) are encoded by multiple exons, but the eight constant-region domains are each the product of a single exon. All of these *constant-region* exons are derived from a single ancestral exon encoding an immunoglobulin fold. The major variable-region exon probably derives from this ancestral exon also. Contemporary immunoglobulin genes are a consequence of multiple duplications of the ancestral exon.

The discovery of variability in amino acid sequence in otherwise identical polypeptide chains was surprising and almost heretical to protein chemists. For geneticists, it presented a genuine enigma. They noted that mammals, which can make millions of different antibodies, don't have millions of different antibody genes. How can the mammalian genome encode the diversity seen in L and H chains?

The Immunoglobulin Genes Undergo Gene Rearrangement

The answer to the enigma of immunoglobulin sequence diversity is found in the organization of the immunoglobulin genes. The genetic information for an immunoglobulin polypeptide chain is scattered among multiple gene segments along a chromosome in germline cells (sperm and eggs). During vertebrate development and the formation of B lymphocytes, these segments are brought together and assembled by **DNA rearrangement** (that is, genetic recombination) into complete genes. DNA rearrangement, or **gene reorganization,** provides a mechanism for generating a variety of protein isoforms from a limited number of genes. DNA rearrangement occurs in only a few genes, namely, those encoding the antigen-binding proteins of the immune response—the immunoglobulins

(a)

N

Immunoglobulin
V_L domain

(b)

Immunoglobulin
C_L domain
($\approx C_H$ domains)

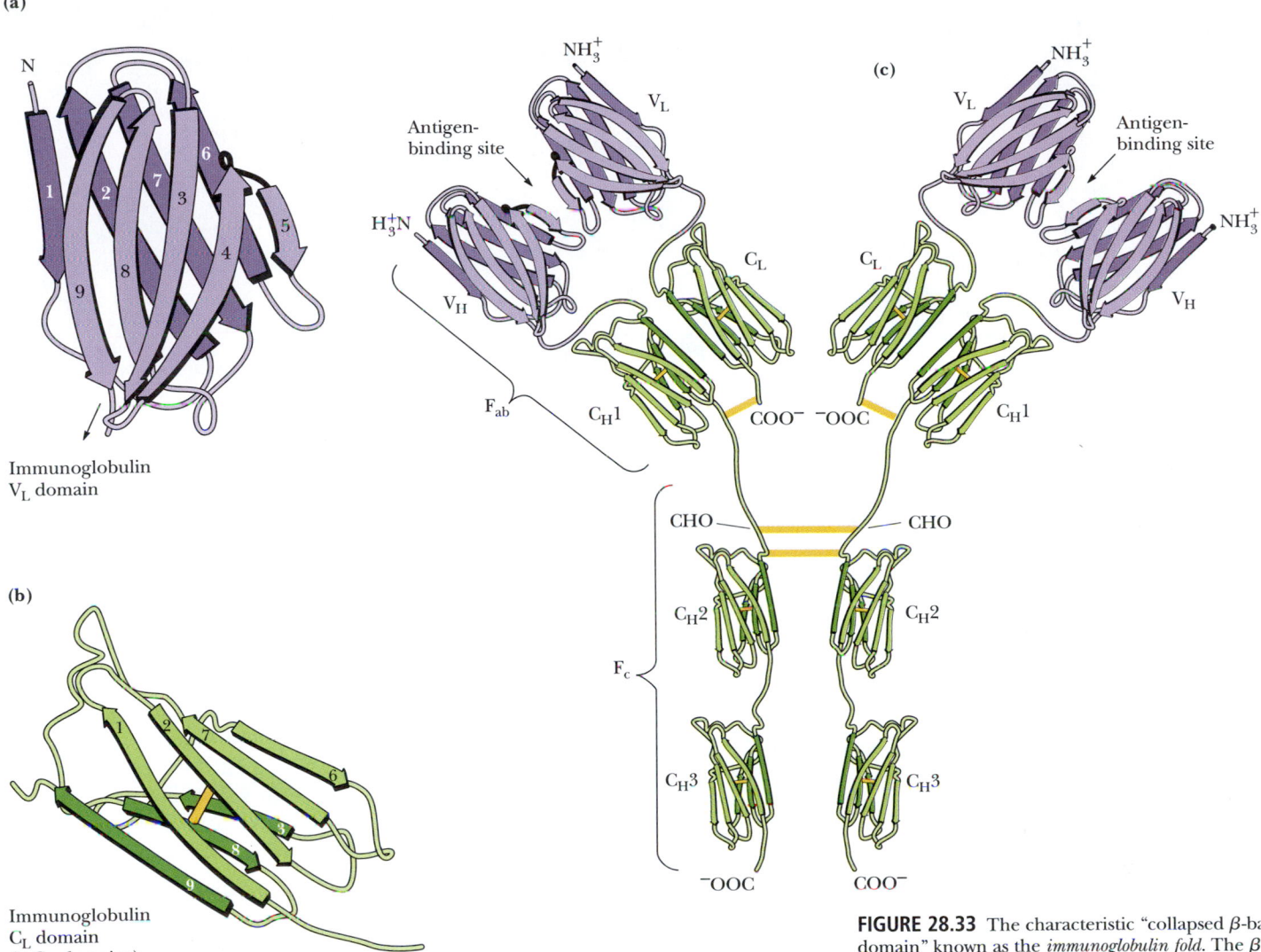

FIGURE 28.33 The characteristic "collapsed β-barrel domain" known as the *immunoglobulin fold*. The β-barrel structures for both **(a)** *variable* and **(b)** *constant regions* are shown. **(c)** A schematic diagram of the 12 collapsed β-barrel domains that make up an IgG molecule. CHO indicates the carbohydrate addition site; F_{ab} denotes one of the two antigen-binding fragments of IgG, and F_c, the proteolytic fragment consisting of the pairs of C_H2 and C_H3 domains.

and the T-cell receptors. The gene segments encoding the amino-terminal portion of the immunoglobulin polypeptides are also unusually susceptible to mutation events. The result is a population of B cells whose antibody-encoding genes collectively show great sequence diversity even though a given cell can make only a limited set of immunoglobulin chains. Hence, at least one cell among the B-cell population will likely be capable of producing an antibody that will specifically recognize a particular antigen.

DNA Rearrangements Assemble an L-Chain Gene by Combining Three Separate Genes

The organization of various immunoglobulin gene segments in the human genome is shown in Figure 28.34. L-chain variable-region genes are assembled from two kinds of **germline genes**: V_L and J_L (*J* stands for *joining*). In mammals, there are two different families of **L-chain genes**: the **κ, or kappa, gene family** and the **λ, or lambda, gene family**; each family has V and J members. These families are on different chromosomes. Humans have 40 functional V genes and 5 functional J genes for the κ light chains and 31 V genes and 4 J genes for the λ light chain. The V and J genes lie upstream from the single $C_κ$ **gene** that encodes the L-chain constant region. Each $V_κ$ gene has its own $L_κ$ segment for encoding the L-chain leader peptide that targets the L chain to the endoplasmic reticulum for

(a)

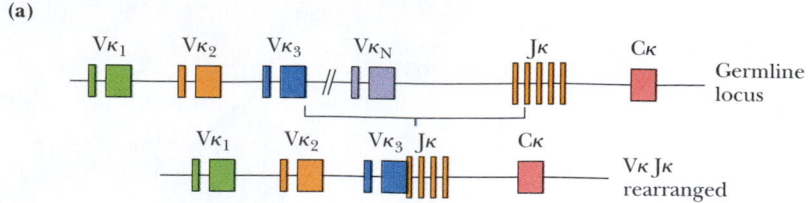

FIGURE 28.34 Organization of human immunoglobulin gene segments. Green, orange, blue, or purple colors indicate the exons of a particular V_L or V_H gene. **(a)** L-chain gene assembly: During B-lymphocyte maturation in the bone marrow, one of the 40 V genes combines with one of the 5 J genes and is joined with a C gene. During the recombination process, the intervening DNA between the gene segments is deleted (see Figure 28.36). These rearrangements occur by a mostly random process, giving rise to many possible light-chain sequences from each gene family. **(b)** H-chain gene assembly: H chains are encoded by V, D, J, and C genes. In H-chain gene rearrangements, a D gene joins with a J gene and then one of the V genes adds to the DJ assembly. *(Adapted from Figure 2b and c in Nossal, G. J. V., 2003. The double helix and immunology.* Nature *421:440–444.)*

(b)

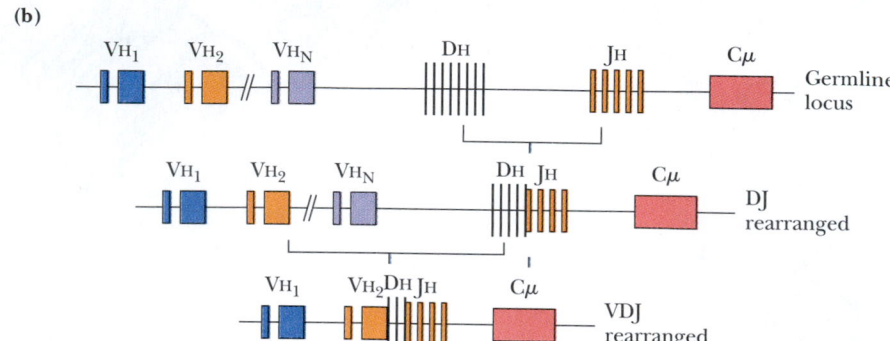

IgG assembly and secretion. (This leader peptide is cleaved once the L chain reaches the ER lumen.) The λ family of L-chain genes is organized similarly (Figure 28.34). In different mature B-lymphocyte cells, V_κ and J_κ genes have joined in different combinations, and along with the C–V_κ gene, form complete L–V_κ chains with a variety of V_κ regions. However, any given B lymphocyte expresses only one V_κ–J_κ combination. Construction of the mature B-lymphocyte L-chain gene has occurred by DNA rearrangements that combine three genes (L–$V_{\kappa,\lambda}$, $J_{\kappa,\lambda}$, $C_{\kappa,\lambda}$) to make one polypeptide!

DNA Rearrangements Assemble an H-Chain Gene by Combining Four Separate Genes

The first 98 amino acids of the 108-residue, H-chain variable region are encoded by a **V_H gene.** Each V_H gene has an accompanying L_H gene that encodes its essential leader peptide. It is estimated that there are 200 to 1000 V_H genes and that they can be subdivided into eight distinct families based on nucleotide sequence homology. The members of a particular V_H family are grouped together on the chromosome, separated from one another by 10 to 20 bp. In assembling a mature H-chain gene, a V_H gene is joined to a **D gene** (D for *diversity*), which encodes amino acids 99 to 113 of the H chain. These amino acids comprise the core of the third CDR in the variable region of H chains. The V_H–D gene assemblage is linked in turn to a **J_H gene,** which encodes the remaining part of the variable region of the H chain. The V_H, D, and J_H genes are grouped in three separate clusters on the same chromosome. The four J_H genes lie 7 kb upstream of the eight C genes, the closest of which is C_μ. Any of four **C genes** may encode the constant region of IgG H chains: $C_{\gamma1}$, $C_{\gamma2a}$, $C_{\gamma2b}$, and $C_{\gamma3}$. Each C gene is composed of multiple exons (only C_μ is shown in Figure 28.34, none of the other C genes). Ten to twenty D genes are found 1 to 80 kb farther upstream. The V_H genes lie even farther upstream. In B lymphocytes, the variable region of an H-chain gene is composed of one each of the L_H–V_H genes, a D gene, and a J_H gene joined head to tail. Because the H-chain variable region is encoded in three genes and the joinings can occur in various combinations, the H chains have a greater potential for diversity than the L-chain variable re-

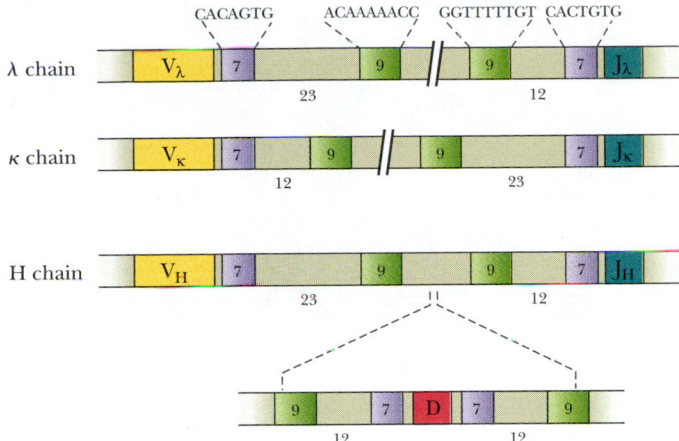

FIGURE 28.35 Consensus elements are located above and below germline variable-region genes that recombine to form genes encoding immunoglobulin chains. These consensus elements are complementary and are arranged in a heptamer-nonamer, 12- to 23-bp spacer pattern. *(Adapted from Tonewaga, S., 1983. Somatic generation of antibody diversity.* Nature **302**:575.)

gions that are assembled from just two genes (for example, L_κ–V_κ and J_κ). In making H-chain genes, four genes have been brought together and reorganized by DNA rearrangement to produce a single polypeptide!

V–J and V–D–J Joining in Light- and Heavy-Chain Gene Assembly Is Mediated by the RAG proteins

Specific nucleotide sequences adjacent to the various variable-region genes suggest a mechanism in which these sequences act as joining signals. All germline V and D genes are followed by a consensus CACAGTG heptamer separated from a consensus ACAAAAACC nonamer by a short, nonconserved 23-bp spacer. Likewise, all germline D and J genes are immediately preceded by a consensus GGTTTTTGT nonamer separated from a consensus CACTGTG heptamer by a short nonconserved 12-bp spacer (Figure 28.35). Note that the consensus elements downstream of a gene are complementary to those upstream from the gene with which it recombines. Indeed, it is these complementary consensus sequences that serve as **recombination recognition signals** (**RSSs**) and determine the site of recombination between variable-region genes. Functionally meaningful recombination happens only where one has a 12-bp spacer and the other has a 23-bp spacer (Figure 28.35). Lymphoid cell-specific **recombination-activating gene** proteins 1 and 2 (**RAG1** and **RAG2**) recognize and bind at these RSSs, presumably through looping out of the 12- and 23-bp spacers and alignment of the homologous heptamer and nonamer regions (Figure 28.36). RAG1 and RAG2 together function as the **V(D)J recombinase.** The similarity between the organization of flanking repeats in immunoglobulin genes and the reaction catalyzed by RAG1/RAG2 proteins suggests that these genes and the RAG recombinase may have evolved from an ancestral transposon.

Imprecise Joining of Immunoglobulin Genes Creates New Coding Arrangements

Joining of the ends of the immunoglobulin-coding regions during gene reorganization is somewhat imprecise. This imprecision actually leads to even greater antibody diversity because new coding arrangements result. Position 96 in κ chains is typically encoded by the first triplet in the J_κ element. Most κ chains have one of four amino acids here, depending on which J_κ gene was recruited in gene assembly. However, occasionally only the second and third bases or just the third base of the codon for position 96 is contributed by the J_κ gene, with the other one or two nucleotides supplied by the V_κ segment (Figure 28.37). So, the precise

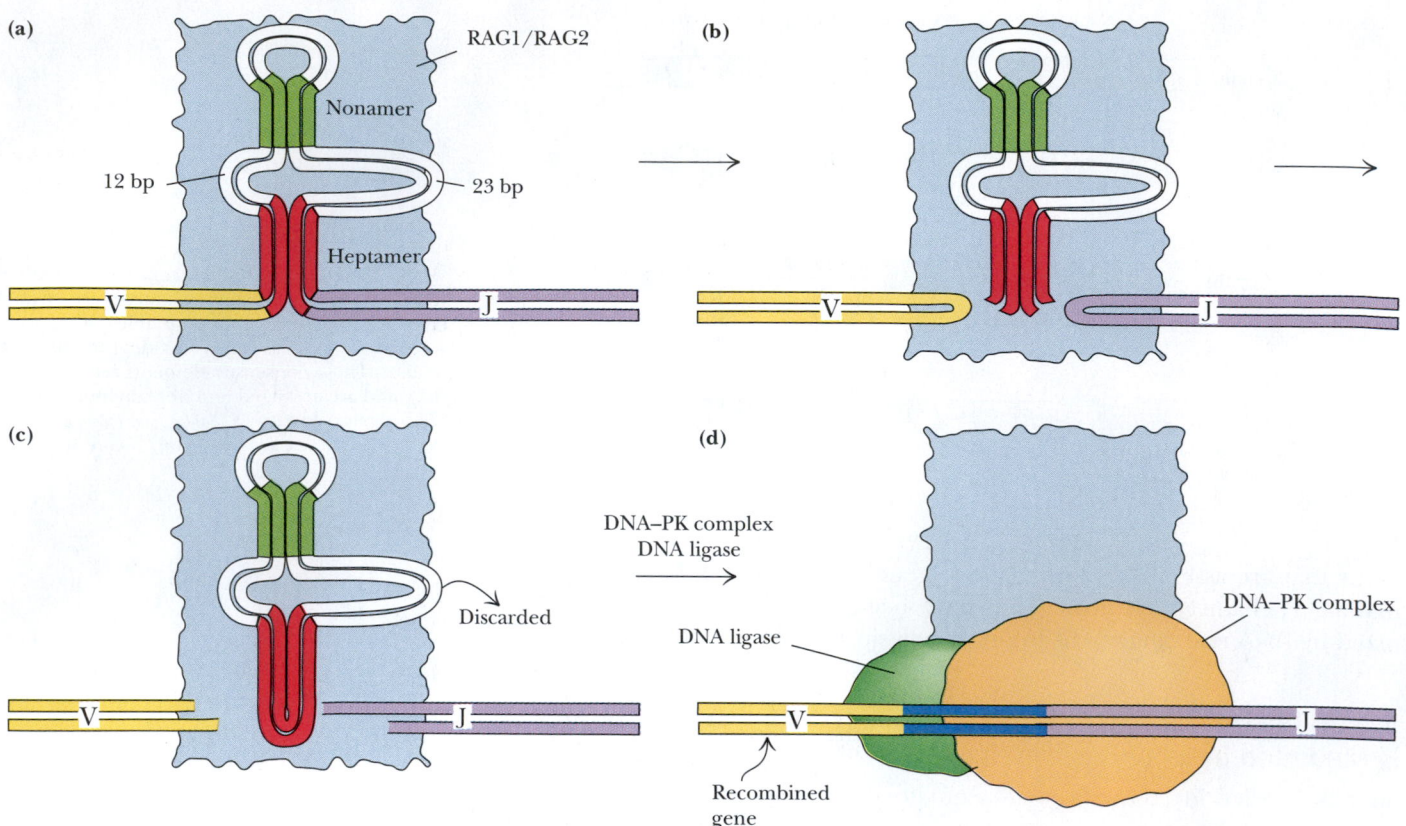

BiochemistryⒺNow™ **ACTIVE FIGURE 28.36** Model for V(D)J recombination. A RAG1 : RAG2 complex is assembled on DNA in the region of recombination signal sequences **(a),** and this complex introduces double-stranded breaks in the DNA at the borders of protein-coding sequences and the recombination signal sequences **(b).** The products of RAG1 : RAG2 DNA cleavage are novel: The DNA bearing the recombination signal sequences has blunt ends, whereas the coding DNA has hairpin ends. That is, the two strands of the V and J coding DNA segments are covalently joined as a result of transesterification reactions catalyzed by RAG1 : RAG2. To complete the recombination process, the two RSS ends are precisely joined to make a covalently closed circular dsDNA, but the V and J coding ends undergo further processing before they are joined **(c).** Coding-end processing involves opening of the V and J hairpins and the addition or removal of nucleotides from the strands. This processing means that joining of the V and J coding ends is imprecise, providing an additional means for introducing antibody diversity. Finally, the V and J coding segments are then joined to create a recombinant immunoglobulin-encoding gene **(d).** The processing and joining reactions require RAG1 : RAG2, DNA-dependent protein kinase (DNA-PK, which consists of three subunits—Ku70, Ku80, and DNA-PK$_{CS}$), and DNA ligase. *(Adapted from Figure 1 in Weaver, D. T., and Alt, F. W., 1997. From RAGs to stitches. Nature **388:**428–429.)* **Test yourself on the concepts in this figure at http://chemistry.brookscole.com/ggb3**

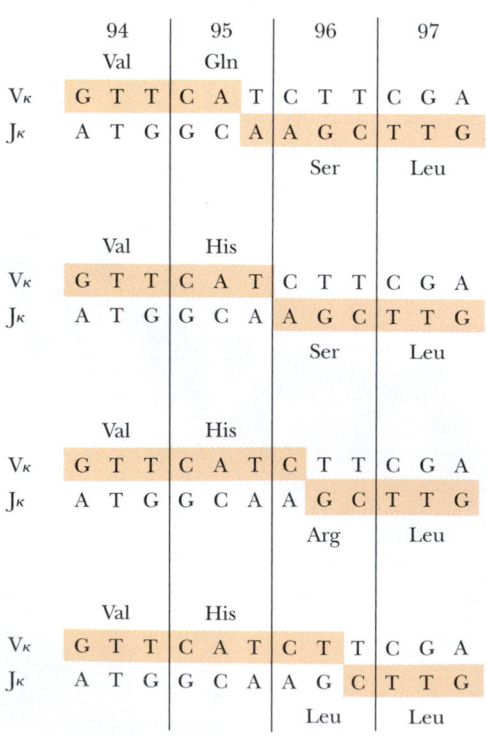

FIGURE 28.37 Recombination between the V$_\kappa$ and J$_\kappa$ genes can vary by several nucleotides, giving rise to variations in amino acid sequence and hence diversity in immunoglobulin L chains.

point where recombination occurs during gene reorganization can vary over several nucleotides, creating even more diversity.

Antibody Diversity Is Due to Immunoglobulin Gene Rearrangements

Taking as an example the mouse with perhaps 300 V$_\kappa$ genes, 4 J$_\kappa$ genes, 200 V$_H$ genes, 12 D genes, and 4 J$_H$ genes, the number of possible combinations is given by $300 \times 4 \times 200 \times 12 \times 4$. Thus, more than 10^7 different antibody molecules can be created from roughly 500 or so different mouse variable-region genes. Including the possibility for V$_\kappa$–J$_\kappa$ joinings occurring within codons adds to this diversity, as does the high rate of somatic mutation associated with the variable-region genes. (Somatic mutations are mutations that arise in diploid cells and are transmitted to the progeny of these cells within the organism, but not to the offspring of the organism.) Clearly, gene rearrangement is a powerful mechanism for dramatically enhancing the protein-coding potential of genetic information.

Summary

28.1 How Is DNA Replicated?

DNA replication is accomplished through strand separation and the copying of each strand. Strand separation is achieved by untwisting the double helix. Each separated strand acts as a template for the synthesis of a new complementary strand whose nucleotide sequence is fixed by Watson–Crick base-pairing rules. Base pairing then dictates an accurate replication of the original DNA double helix. DNA replication follows a semiconservative mechanism where each original strand is copied to yield a complete complementary strand and these paired strands, one old and one new, remain together as a duplex DNA molecule. Replication begins at specific regions called origins of replication and proceeds in both directions. Bidirectional replication involves two replication forks, which move in opposite directions. Helicases unwind the double helix, and DNA gyrases act to overcome torsional stress by introducing negative supercoils at the expense of ATP hydrolysis. Because DNA polymerases synthesize DNA only in a $5' \rightarrow 3'$ direction, replication is semidiscontinuous: The $3' \rightarrow 5'$ strand can be copied continuously by DNA polymerase proceeding in the $5' \rightarrow 3'$ direction. The other parental strand is copied only when a sufficient stretch of its sequence has been exposed for DNA polymerase to move along it in the $5' \rightarrow 3'$ mode. Thus, one parental strand is copied continuously to form the leading strand, while the other parental strand is copied in an intermittent, or discontinuous, mode to yield a set of Okazaki fragments that are joined later to give the lagging strand.

28.2 What Are the Properties of DNA Polymerases?

All DNA polymerases share the following properties: (1) The incoming base is selected within the DNA polymerase active site through base-pairing with the corresponding base in the template strand, (2) chain growth is in the $5' \rightarrow 3'$ direction antiparallel to the template strand, and (3) DNA polymerases cannot initiate DNA synthesis de novo—all require a primer with a free 3'-OH to build upon. DNA polymerase III holoenzyme, the enzyme that replicates of the *E. coli* chromosome, is composed of ten different kinds of subunits. DNA polymerases are immobilized in replication factories.

28.3 How Is DNA Replicated in Eukaryotic Cells?

Eukaryotic DNA is organized into chromosomes within the nucleus. These chromosomes must be replicated once (and only once!) each cell cycle. Progression through the cell cycle is regulated through checkpoints that control whether the cell continues into the next phase. Cyclins and CDKs maintain these checkpoints. Replication licensing factors (MCM proteins) interact with origins of replication and render chromosomes competent for replication. Three DNA polymerases—α, δ, and ϵ—carry out genome replication. DNA polymerase α initiates replication through synthesis of an RNA. DNA polymerase δ is the principal DNA polymerase in eukaryotic DNA replication.

28.4 How Are the Ends of Chromosomes Replicated?

Telomeres are short, tandemly repeated, G-rich nucleotide sequences that form protective caps on the chromosome ends. DNA polymerases cannot replicate the extreme 5'-ends of chromosomes, but a special polymerase called telomerase maintains telomere length. Telomerase is a ribonucleoprotein, and its RNA component serves as template for telomere synthesis.

28.5 How Are RNA Genomes Replicated?

Many viruses have genomes composed of RNA, not DNA. DNA may be an intermediate in the replication of such viruses; that is, viral RNA serves as the template for DNA synthesis. This reaction is catalyzed by reverse transcriptase, an RNA-dependent DNA polymerase.

28.6 How Is the Genetic Information Shuffled by Genetic Recombination?

Genetic recombination is the exchange (or incorporation) of one DNA sequence with (or into) another. Recombination between very similar DNA sequences is called homologous recombination. Homologous recombination proceeds according to the Holliday model. The RecBCD enzyme complex unwinds dsDNA and cleaves its single strands. RecA protein acts in recombination to catalyze the ATP-dependent DNA strand exchange reaction. Procession of strand separation and repairing into hybrid strands along the DNA duplex initiates branch migration, displacing the homologous DNA strand from the DNA duplex and replacing it with the ssDNA strand. RuvA, RuvB, and RuvC resolve the Holliday junction to form the recombination products. DNA replication is an essential component of both DNA recombination and DNA repair processes. Furthermore, recombination mechanisms can restart replication forks that have halted at breaks or other lesions in the DNA strands.

Transposons are mobile DNA segments ranging in size from several hundred base pairs to more than 8 kbp that move enzymatically from place to place in the genome.

28.7 Can DNA Be Repaired?

Repair systems correct damage to DNA in order to maintain its information content. The most common forms of damage are (1) replication errors, (2) deletions or insertions, (3) UV-induced alterations, (4) DNA strand breaks, and (5) covalent crosslinking of strands. Cells have extraordinarily diverse and effective DNA repair systems to deal with these problems, some of which are also involved in DNA replication and recombination. When repair systems fail, the genome may still be preserved if an "error-prone" mode of replication allows the lesion to be bypassed.

28.8 What Is the Molecular Basis of Mutation?

Mutations change the sequence of bases in DNA, either by the substitution of one base pair for another (point mutations) or by the insertion or deletion of one or more base pairs. Point mutations arise by the pairing of bases with inappropriate partners during DNA replication, by the introduction of base analogs into DNA, or by chemical mutagens. Chemical mutagens are agents that chemically modify bases so that their base-pairing characteristics are altered.

Special Focus: Gene Rearrangements and Immunology—Is It Possible to Generate Protein Diversity Using Genetic Recombination?

Animals have evolved a way to exploit genetic recombination in order to generate protein diversity. The immunoglobulin genes are a highly evolved system for maximizing protein diversity from a finite amount of genetic information. Cells active in the immune response are capable of gene rearrangements. The antibody diversity found in IgG molecules are a prime example of proteins produced via gene rearrangements. IgG L-chain genes are created by combining three separate genes, and H-chain genes by combining four. V–J and V–D–J joining in L- and H-chain gene assembly is mediated by RAG proteins.

Problems

1. If ^{15}N-labeled *E. coli* DNA has a density of 1.724 g/mL, ^{14}N-labeled DNA has a density of 1.710 g/mL, and *E. coli* cells grown for many generations on $^{14}NH_4^+$ as a nitrogen source are transferred to media containing $^{15}NH_4^+$ as the sole N-source, (a) what will be the density of the DNA after one generation, assuming replication is semiconservative? (b) Supposing replication took place by a dispersive mechanism, what would be the density of DNA after one generation? (c) Design an experiment to distinguish between semiconservative and dispersive modes of replication.

2. (a) What are the respective roles of the 5'-exonuclease and 3'-exonuclease activities of DNA polymerase I? (b) What might be a feature of an *E. coli* strain that lacked DNA polymerase I 3'-exonuclease activity?

3. Assuming DNA replication proceeds at a rate of 750 base pairs per second, calculate how long it will take to replicate the entire *E. coli* genome. Under optimal conditions, *E. coli* cells divide every 20 minutes. What is the minimal number of replication forks per *E. coli* chromosome in order to sustain such a rate of cell division?

4. On the basis of Figure 28.5, draw a simple diagram illustrating replication of the circular *E. coli* chromosome (a) at an early stage, (b) when one-third completed, (c) when two-thirds completed, and (d) when almost finished, assuming the initiation of replication at *oriC* has occurred only once. Then, draw a diagram showing the *E. coli* chromosome in problem 3 where the *E. coli* cell is dividing every 20 minutes.

5. It is estimated that there are ten molecules of DNA polymerase III per *E. coli* cell. Is it likely that the growth rate of *E. coli* is limited by DNA polymerase III availability?

6. Approximately how many Okazaki fragments are synthesized in the course of replicating an *E. coli* chromosome? How many in replicating an "average" human chromosome?

7. How do DNA gyrases and helicases differ in their respective functions and modes of action?

8. Assuming DNA replication proceeds at a rate of 100 base pairs per second in human cells and origins of replication occur every 300 kbp, how long would it take to replicate the entire diploid human genome? How many molecules of DNA polymerase does each cell need to carry out this task?

9. From the information in Figure 28.18, diagram the recombinational event leading to the formation of a heteroduplex DNA region within a bacteriophage chromosome.

10. Homologous recombination in *E. coli* leads to the formation of regions of heteroduplex DNA. By definition, such regions contain mismatched bases. Why doesn't the mismatch repair system of *E. coli* eliminate these mismatches?

11. If RecA protein unwinds duplex DNA so that there are about 18.6 bp per turn, what is the change in $\Delta\phi$, the helical twist of DNA, compared to its value in B-DNA?

12. Diagram a Holliday junction between two duplex DNA molecules and show how the action of resolvase might give rise to either patch or splice recombinant DNA molecules.

13. Show the nucleotide sequence changes that might arise in a dsDNA (coding strand segment GCTA) upon mutagenesis with (a) HNO_2, (b) bromouracil, and (c) 2-aminopurine.

14. Transposons are mutagenic agents. Why?

15. Give a plausible explanation for the genetic and infectious properties of PrP[sc].

Preparing for the MCAT Exam

16. Figure 28.12 depicts the eukaryotic cell cycle. Many cell types "exit" the cell cycle and don't divide for prolonged periods, a state termed G_0; some, for example, neurons, never divide again.
    a. What stage of the cell cycle do you suppose a cell might be in when it exits the cell cycle and enters G_0?
    b. The cell cycle is controlled by checkpoints, cyclins, and CDKs. Describe how biochemical events involving cyclins and CDKs might control passage of a dividing cell through the cell cycle.

17. Figure 28.37 gives some examples of recombination in IgG codons 95 and 96, as specified by the V_κ and J_κ genes. List the codon possibilities and the amino acids encoded if recombination occurred in codon 97. Which of these possibilities is least desirable?

Biochemistry⊗Now™ Preparing for an exam? Test yourself on key questions at http://chemistry.brookscole.com/ggb3

Further Reading

General

Holliday, R., 1964. A mechanism for gene conversion in fungi. *Genetic Research* **5**:282–304. The classic model for the mechanism of DNA strand exchange during homologous recombination.

Kornberg, A., and Baker, T. A., 1992. *DNA Replication*, 2nd ed. New York: W. H. Freeman. A comprehensive detailed account of the enzymology of DNA metabolism, including replication, recombination, repair, and more.

Lewin, B., 2004. *Genes VIII*. New York: Prentice Hall. A contemporary genetics text that seeks to explain heredity in terms of molecular structures.

Meselson, M., and Stahl, F. W., 1958. The replication of DNA in *Escherichia coli*. *Proceedings of the National Academy of Sciences, U.S.A.* **44**:671–682. The classic paper showing that DNA replication is semiconservative.

Meselson, M., and Weigle, J. J., 1961. Chromosome breakage accompanying genetic recombination in bacteriophage. *Proceedings of the National Academy of Sciences, U.S.A.* **47**:857–869. The experiments demonstrating that physical exchange of DNA occurs during recombination.

Ogawa, T., and Okazaki T., 1980. Discontinuous DNA replication. *Annual Review of Biochemistry* **49**:421–457. Okazaki fragments and their implications for the mechanism of DNA replication.

Palmiter, R. D., et al., 1982. Dramatic growth of mice that develop from eggs microinjected with metallothionein-growth hormone fusion genes. *Nature* **300**:611–615.

DNA Replication

Baker, T. A., and Bell, S. P., 1998. Polymerases and the replisome: Machines within machines. *Cell* **92**:295–305.

Bell, S. P., and Dutta, A., 2002. DNA replication in eukaryotic cells. *Annual Review of Biochemistry* **71**:333–374.

Boehmer, P. E., and Lehman, I. R., 1997. Herpes simplex virus DNA replication. *Annual Review of Biochemistry* **66**:347–384.

Botchan, M., 1996. Coordinating DNA replication with cell division: current status of the licensing concept. *Proceedings of the National Academy of Sciences, U.S.A.* **93**:9997–10000.

Cook, P. R., 1999. The organization of replication and transcription. *Science* **284**:1790–1795.

Frick, D. N., and Richardson, C. C., 2001. DNA primases. *Annual Review of Biochemistry* **70**:39–80.

Goodman, M. F., 2002. Error-prone repair DNA polymerases in prokaryotes and eukaryotes. *Annual Review of Biochemistry* **71**:17–50.

Hübscher, U., Maga, G., and Spadari, S., 2002. Eukaryotic DNA polymerases. *Annual Review of Biochemistry* **71**:133–163.

Hübscher, U., et al., 2000. Eukaryotic DNA polymerases, a growing family. *Trends in Biochemical Sciences* **25**:143–147.

Jallepalli, P. V., and Kelly, T. J., 1997. Cyclin-dependent kinase and initiation at eukaryotic origins: A replication switch? *Current Opinion in Cell Biology* **9**:358–363.

Keck, J. L., 2000. Structure of the RNA polymerase domain of the *E. coli* primase. *Science* **287**:2482–2486.

Kool, E. T., 2002. Active site tightness and substrate fit in DNA replication. *Annual Review of Biochemistry* **71**:191–219.

Leu, F. P., Georgescu, R., and O'Donnell, M., 2003. Mechanism of the *E. coli* τ processivity switch during lagging-strand synthesis. *Molecular Cell* **11**:315–327.

Page, A. M., and Hieter, P., 1999. The anaphase-promoting complex: New subunits and regulators. *Annual Review of Biochemistry* **68:**583–609.

Russell, P., 1998. Checkpoints on the road to mitosis. *Trends in Biochemical Sciences* **23:**399–402.

Steitz, T. A., 1998. A mechanism for all polymerases. *Nature* **391:**231–232.

Tye, B. K., 1999. MCM proteins in DNA replication. *Annual Review of Biochemistry* **68:**649–686.

Waga, S., and Stillman, B., 1998. The DNA replication fork in eukaryotic cells. *Annual Review of Biochemistry* **67:**721–751.

Wold, M. S., 1997. Replication protein A: A heterotrimeric, single-stranded DNA-binding protein required for eukaryotic DNA metabolism. *Annual Review of Biochemistry* **66:**61–92.

Protein Rings in DNA Metabolism

Hingorani, M. M., and O'Donnell, M., 2000. A tale of toroids in DNA metabolism. *Nature Reviews Molecular Cell Biology* **1:**22–30.

Wyman, C., and Botchan, M., 1995. DNA replication: a familiar ring to DNA polymerase processivity. *Current Biology* **5:**334–337.

Telomerase

Blackburn, E. H., 1992. Telomerases. *Annual Review of Biochemistry* **61:**113–129.

Collins, K., 1999. Ciliate telomerase biochemistry. *Annual Review of Biochemistry* **68:**187–218.

Kim, N. W., 1994. Specific association of human telomerase activity with immortal cells and cancer. *Science* **266:**2011–2015.

Nakamura, T. M., et al., 1997. Telomerase catalytic subunit homologs from fission yeast and human. *Science* **277:**955–959.

Prions

Cohen, F. E., and Prusiner, S. B., 1998. Pathological conformations of prion proteins. *Annual Review of Biochemistry* **67:**793–819.

Prusiner, S. B., 1996. Molecular biology and pathogenesis of prion diseases. *Trends in Biochemical Sciences* **21:**482–487.

Prusiner, S. B., 1997. Prion diseases and the BSE crisis. *Science* **278:**245–251.

Recombination

Alberts, B., 2003. DNA replication and recombination. *Nature* **421:**431–435.

Anderson, D. G., and Kowalczykowski, S. C., 1997. The translocating RecBCD enzyme stimulates recombination by directing RecA protein onto ssDNA in a χ-regulated manner. *Cell* **90:**77–86.

Baumann, P., and West, S. C., 1998. Role of the human RAD51 protein in homologous recombination and double-stranded-break repair. *Trends in Biochemical Sciences* **23:**247–252.

Beernink, H. T. H., and Morrical, S. W., 1999. RMPs: Recombination/replication proteins. *Trends in Biochemical Sciences* **24:**385–389.

Cox, M. M., 1999. Recombinational DNA repair in bacteria and the RecA protein. *Progress in Nucleic Acid Research and Molecular Biology* **63:**311–366.

Haber, J. E., 1999. DNA recombination: The replication connection. *Trends in Biochemical Sciences* **24:**271–275.

Kowalczykowski, S. C., 2000. Initiation of genetic recombination and recombination-dependent replication. *Trends in Biochemical Sciences* **25:**156–165.

Lovett, S. T., 2003. Connecting replication and recombination. *Molecular Cell* **11:**554–556.

Lusetti, S. L., and Cox, M. M., 2002. The bacterial RecA protein and the recombinational DNA repair of replication forks. *Annual Review of Biochemistry* **71:**71–100.

Rafferty, J. B., et al., 1996. Crystal structure of DNA recombination protein RuvA and a model for its binding to the Holliday junction. *Science* **274:**415–421.

Roca, A. I., and Cox, M. M., 1997. RecA protein: Structure, function, and role in recombinational DNA repair. *Progress in Nucleic Acid Research and Molecular Biology* **56:**127–223.

Taylor, A. F., and Smith, G. R., 2003. RecBCD enzyme is a DNA helicase with fast and slow motors of opposite polarity. *Nature* **423:**889–893. See also Dillingham, M. S., Spies, M., and Kowalczykowski, S. C., 2003. RecBCD is a bipolar DNA helicase. *Nature* **423:**893–897.

Transposons

Lambowitz, A. M., and Belfort, M., 1993. Introns as mobile genetic elements. *Annual Review of Biochemistry* **62:**587–622.

Stellwagen, A. E., and Craig, N. L., 1998. Mobile DNA elements: Controlling transposition with ATP-dependent molecular switches. *Trends in Biochemical Sciences* **23:**486–490.

V(D)J Recombination and the Immunoglobulin Genes

Gellert, M., 2002. V(D)J recombination: RAG proteins, repair factors, and regulation. *Annual Review of Biochemistry* **71:**101–132.

Hiom, K., and Gellert, M., 1997. A stable RAG1-RAG2-DNA complex that is active in V(D)J cleavage. *Cell* **88:**65–72.

Lewis, S. M., and Wu, G. E., 1997. The origins of V(D)J recombination. *Cell* **88:**159–162.

Nossal, G. J. V., 2003. The double helix and immunology. *Nature* **421:**440–444.

Transgenic Animals

Morgan, R. A., and Anderson, W.F., 1993. Human gene therapy. *Annual Review of Biochemistry* **62:**192–217.

Schnieke, A. E., et al., 1997. Human factor IX transgenic sheep produced by transfer of nuclei from transfected fetal fibroblasts. *Science* **278:**2130–2133.

Wilmut, I., et al., 1997. Viable offspring derived from fetal and adult mammalian cells. *Nature* **385:**810–818. See also Campbell, K. H. S., et al., 1996. Sheep cloned by nuclear transfer from a cultured cell line. *Nature* **380:**64–66.

Repair

Bartek, J., and Lukas, J., 2003. Damage alert. *Nature* **421:**486–488.

Friedberg, E. C., 2003. DNA damage and repair. *Nature* **421:**436–440.

Friedberg, E. C., Walker, G. C., and Siede, W., 1995. *DNA Repair and Mutagenesis.* Washington, DC: ASM Press.

Marians, K. J., 2000. PriA-directed replication fork restart in *Escherichia coli. Trends in Biochemical Sciences* **25:**185–189.

McCollough, A. K., et al., 1999. Initiation of base excision repair: Glycosylase mechanisms and structures. *Annual Review of Biochemistry* **68:**255–285.

Michel, B., 2000. Replication fork arrest and DNA recombination. *Trends in Biochemical Sciences* **25:**173–178.

Modrich, P., and Lahue, R., 1996. Mismatch repair in replication fidelity, genetic recombination, and cancer biology. *Annual Review of Biochemistry* **65:**101–133.

Mol, C. D., et al., 1999. DNA repair mechanisms for the recognition and removal of damaged DNA bases. *Annual Review of Biophysics and Biomolecular Structure* **28:**101–128.

Morgan, A. R., 1993. Base mismatches and mutagenesis: How important is tautomerism? *Trends in Biochemical Sciences* **18:**160–163.

Parikh, S. S., et al., 1999. Envisioning the molecular choreography of DNA base excision repair. *Current Opinion in Structural Biology* **9:**37–47.

Sancar, A., 1994. Mechanisms of DNA excision repair. *Science* **266:**1954–1956. (*Science* named the extended family of DNA repair enzymes its "Molecules of the Year" in 1994. See the 23 December 1994 issue of *Science* for additional readings.)

CHAPTER 29

Transcription and the Regulation of Gene Expression

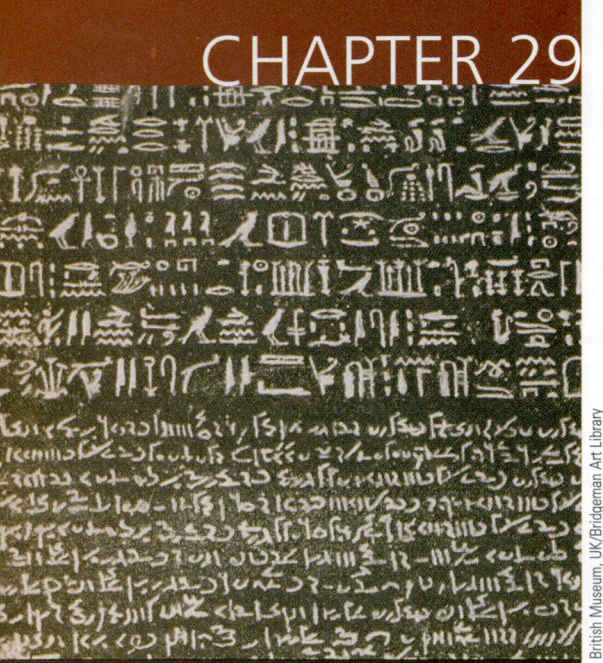

The Rosetta stone, inscribed in 196 B.C. The writing on the Rosetta stone is in three forms: hieroglyphs, Demotic (the conventional Egyptian script of the time), and Greek (the Greeks ruled Egypt in 196 B.C.). The Rosetta stone represents the transcription of hieroglyphic symbols into two living languages. Shown here is an expanded view of part of the interface where hieroglyphs and Demotic meet.

Now that we have all this useful information, it would be nice to do something with it.

From the *Unix Programmer's Manual*

Key Questions

Essential Question

Expression of the information encoded in DNA depends on transcription of that information into RNA. *How are the genes of prokaryotes and eukaryotes transcribed to form RNA products that can be translated into proteins?*

In 1958, Francis Crick enunciated the "central dogma of molecular biology" (Figure 29.1). This scheme outlined the residue-by-residue transfer of biological information as encoded in the primary structure of the informational biopolymers, nucleic acids and proteins. The predominant path of information transfer, DNA→RNA→protein, postulated that RNA was an information carrier between DNA and proteins, the agents of biological function. In 1961, François Jacob and Jacques Monod extended this hypothesis to predict that the RNA intermediate, which they dubbed **messenger RNA,** or **mRNA,** would have the following properties:

1. Its base composition would reflect the base composition of DNA (a property consistent with genes as protein-encoding units).
2. It would be very heterogeneous with respect to molecular mass, yet the average molecular mass would be several hundred kilodaltons. (A 200-kD RNA contains roughly 750 nucleotides, which could encode a protein of about 250 amino acids—approximately 30 kD—a reasonable estimate for the average size of polypeptides.)
3. It would be able to associate with ribosomes because ribosomes are the site of protein synthesis.
4. It would have a high rate of turnover. (That is, mRNA would be rapidly degraded. Turnover of mRNA would allow the rate of mRNA synthesis to control the rate of protein synthesis.)

Since Jacob and Monod's 1961 hypothesis, it has been realized that cells contain three major classes of RNA—mRNA, ribosomal RNA (rRNA), and transfer RNA (tRNA)—all of which participate in protein synthesis (see Chapters 10 and 30). All of these RNAs are synthesized from DNA templates by **DNA-dependent RNA polymerases** in the process known as **transcription.** However, only mRNAs direct the synthesis of proteins. Thus, not all genes encode proteins; some encode rRNAs or tRNAs. Protein synthesis occurs via the process of **translation,** wherein the instructions encoded in the sequence of bases in mRNA are translated into a specific amino acid sequence by ribosomes, the "workbenches" of polypeptide synthesis (see Chapter 30).

Transcription is tightly regulated in all cells. In prokaryotes, only 3% or so of the genes are undergoing transcription at any given time. The metabolic conditions and the growth status of the cell dictate which gene products are needed at any moment. Similarly, differentiated eukaryotic cells express only a small percentage of their genes in fulfilling their biological functions, not the full genetic potential encoded in their chromosomes.

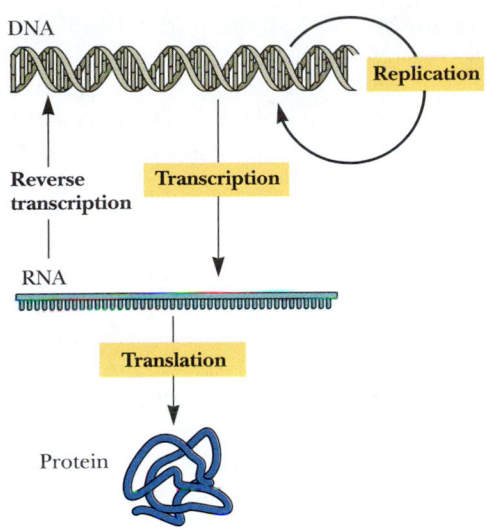

FIGURE 29.1 Crick's 1958 view of the "central dogma of molecular biology": Directional flow of detailed sequence information includes DNA→DNA (replication), DNA→RNA (transcription), RNA→protein (translation), RNA→DNA (reverse transcription). Note that no pathway exists for the flow of information from proteins to nucleic acids, that is, protein→RNA or DNA. A possible path from DNA to protein has since been discounted. Interestingly, in 1958, mRNA had not yet been discovered.

29.1 | How Are Genes Transcribed in Prokaryotes?

In prokaryotes, virtually all RNA is synthesized by a single species of DNA-dependent RNA polymerase. (The only exception is the short RNA primers formed by primase during DNA replication.) Like DNA polymerases, RNA polymerase links ribonucleoside 5′-triphosphates (ATP, GTP, CTP, and UTP, represented generically as NTPs) in an order specified by base pairing with a DNA template:

$$n\,\text{NTP} \longrightarrow (\text{NMP})_n + n\,\text{PP}_i$$

A Deeper Look

Conventions Used in Expressing the Sequences of Nucleic Acids and Proteins

Certain conventions are useful in tracing the course of information transfer from DNA to protein. The strand of duplex DNA that is read by RNA polymerase is termed the **template strand.** Thus, the strand that is not read is the **nontemplate strand.** Because the template strand is read by the RNA polymerase moving 3′→5′ along it, the RNA product, called the **transcript,** grows in the 5′→3′ direction (see accompanying figure). Note that the nontemplate strand has a nucleotide sequence and direction identical to those of the RNA transcript, except that the transcript has U residues in place of T. The RNA transcript will eventually be translated into the amino acid sequence of a protein (see Chapter 30) by a process in which successive triplets of bases (termed **codons**), read 5′→3′, specify a particular amino acid. Polypeptide chains are synthesized in the N→C direction, and the 5′-end of mRNA encodes the N-terminus of the protein.

By convention, when the order of nucleotides in DNA is specified, it is the 5′→3′ sequence of nucleotides in the nontemplate strand that is presented. Consequently, if convention is followed, DNA sequences are written in terms that correspond directly to mRNA sequences, which correspond in turn to the amino acid sequences of proteins as read beginning with the N-terminus.

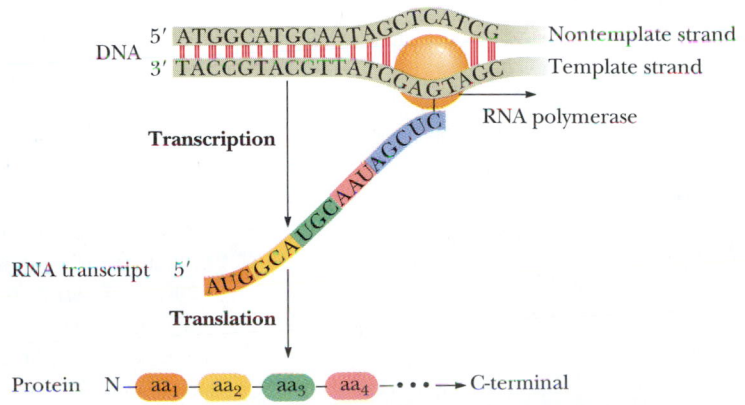

The enzyme moves along a DNA strand in the $3' \rightarrow 5'$ direction, joining the 5'-phosphate of an incoming ribonucleotide to the 3'-OH of the previous residue. Thus, the RNA chain grows $5' \rightarrow 3'$ during transcription, just as DNA chains do during replication. The reaction is driven by subsequent hydrolysis of PP_i to inorganic phosphate by the pyrophosphatases present in all cells.

Escherichia coli RNA Polymerase Is a Complex Multimeric Protein

The RNA polymerase of *E. coli* that carries out transcription, termed **RNA polymerase holoenzyme,** is a complex multimeric protein (450 kD) large enough to be visible in the electron microscope. Its subunit composition is $\alpha_2\beta\beta'\sigma$. The largest subunit, β' (155 kD), functions in DNA binding; β (151 kD) binds the nucleoside triphosphate substrates and interacts with the σ (sigma)-subunit. A number of related proteins, **the sigma (σ) factors,** can serve as the σ-subunit. The different σ factors allow the RNA polymerase holoenzyme to recognize different DNA sequences that act as **promoters.** Promoters are nucleotide sequences that identify the location of *transcription start sites,* where transcription begins. Both β and β' contribute to formation of the catalytic site for RNA synthesis. The two α-subunits (36.5 kD each) are essential for assembly of the enzyme and activation by some regulatory proteins. Dissociation of the σ-subunit from the holoenzyme leaves the **core polymerase** $(\alpha_2\beta\beta')$, which can transcribe DNA into RNA but is unable to recognize promoters and initiate transcription.

The Process of Transcription Has Four Stages

Transcription can be divided into four stages: (1) binding of RNA polymerase holoenzyme at promoter sites, (2) initiation of polymerization, (3) chain elongation, and (4) chain termination.

Binding of RNA Polymerase to Template DNA   The process of transcription begins when the σ-subunit of RNA polymerase recognizes a promoter sequence (Figure 29.2), and RNA polymerase holoenzyme and the promoter form a **closed promoter complex** (Figure 29.2, Step 2). This stage in RNA polymerase:DNA interaction is referred to as the *closed* promoter complex because the dsDNA has not yet been "opened" (unwound) so that the RNA polymerase can read the base sequence of the DNA template strand and transcribe it into a complementary RNA sequence.

Once the closed promoter complex is established, the RNA polymerase holoenzyme unwinds about 14 base pairs of DNA (base pairs located at positions -12 to $+2$, relative to the transcription start site; see later discussion), forming the very stable **open promoter complex** (Figure 29.2, Step 3). Promoter sequences can be identified in vitro by **DNA footprinting:** RNA polymerase holoenzyme is bound to a putative promoter sequence in a DNA duplex, and the DNA:protein complex is treated with DNase I. DNase I cleaves the DNA at sites not protected by bound protein, and the set of DNA fragments left after DNase I digestion reveals the promoter (by definition, the promoter is the RNA polymerase holoenzyme binding site[1]).

RNA polymerase binding typically protects a nucleotide sequence spanning the region from -70 to $+20$, where the $+1$ position is defined as the **transcription start site:** that base in DNA that specifies the first base in the RNA transcript. The next base, $+2$, specifies the second base in the transcript. Bases in the 5', or "minus," direction from the **transcript start site** are numbered -1, -2, and so on. (Note that there is no zero.) Nucleotides in the "minus" direction are said to lie **upstream** of the transcription start site, whereas nucleotides

---

[1]Promoters can also be defined genetically in terms of mutations (nucleotide changes) in this region that block gene expression because they are no longer recognizable by the σ-subunit.

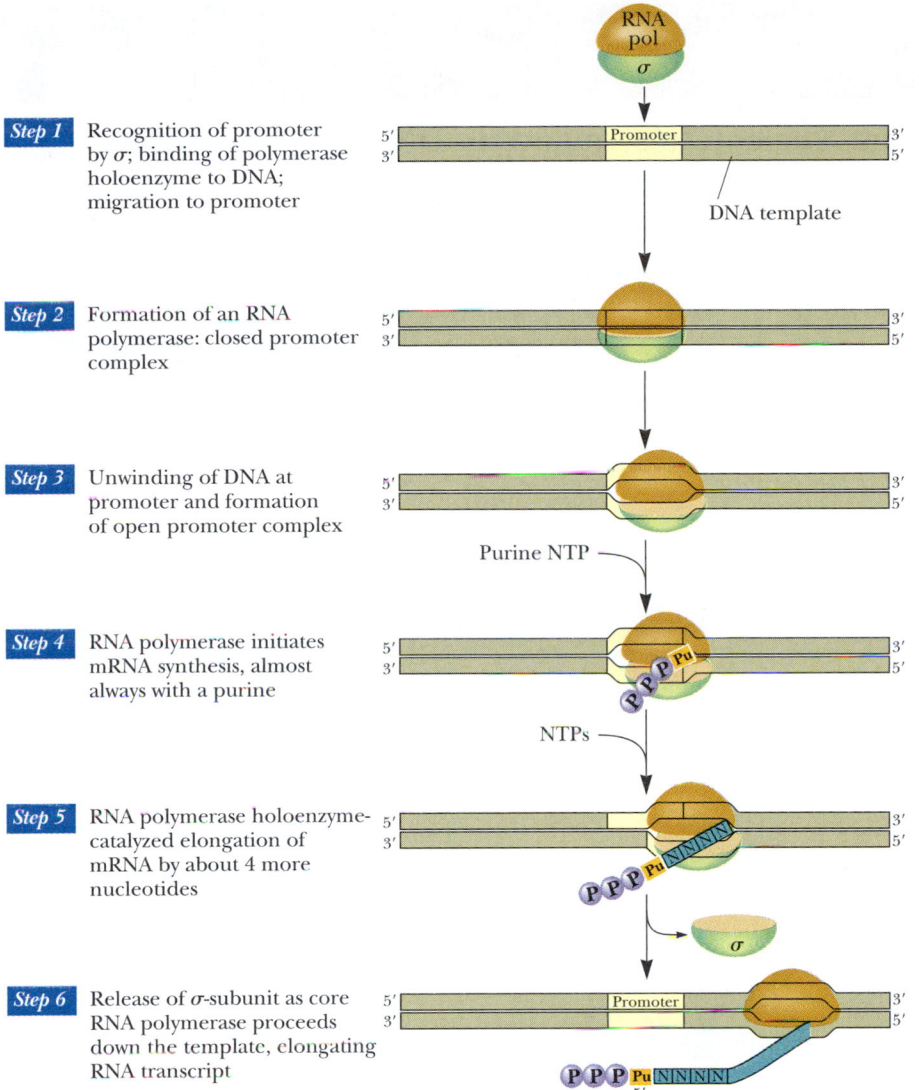

Step 1 Recognition of promoter by σ; binding of polymerase holoenzyme to DNA; migration to promoter

DNA template

Step 2 Formation of an RNA polymerase: closed promoter complex

Step 3 Unwinding of DNA at promoter and formation of open promoter complex

Purine NTP

Step 4 RNA polymerase initiates mRNA synthesis, almost always with a purine

NTPs

Step 5 RNA polymerase holoenzyme-catalyzed elongation of mRNA by about 4 more nucleotides

Step 6 Release of σ-subunit as core RNA polymerase proceeds down the template, elongating RNA transcript

Biochemistry Now™   ACTIVE FIGURE 29.2
Sequence of events in the initiation and elongation phases of transcription as it occurs in prokaryotes. Nucleotides in this region are numbered with reference to the base at the transcription start site, which is designated +1. **Test yourself on the concepts in this figure at http://chemistry.brookscole.com/ggb3**

in the 3′, or "plus," direction are **downstream** of the transcription start site. The transcript start site on the template strand is almost always a pyrimidine, so almost all transcripts begin with a purine. RNA polymerase binding protects 90 bp of DNA, equivalent to a distance of 30 nm along B-DNA. Because RNA polymerase is only 16 nm in its longest dimension, the DNA must be wrapped around the enzyme.

PROPERTIES OF PROKARYOTIC PROMOTERS.   Promoters recognized by the principal σ factor, σ^{70}, serve as the paradigm for prokaryotic promoters. These promoters vary in size from 20 to 200 bp but typically consist of a 40-bp region located on the 5′-side of the transcription start site. Within the promoter are two **consensus sequence elements.** These two elements are the **Pribnow box**[2] near −10, whose consensus sequence is the hexameric TATAAT, and a sequence in the **−35 region** containing the hexameric consensus TTGACA (Figure 29.3). The Pribnow box and the −35 region are separated by about 17 bp of nonconserved sequence. RNA polymerase holoenzyme uses its σ-subunit to

A *consensus sequence* can be defined as the bases that appear with highest frequency at each position when a series of sequences believed to have common function is compared.

[2]Named for David Pribnow, who, along with David Hogness, first recognized the importance of this sequence element in transcription.

A Deeper Look

DNA Footprinting—Identifying the Nucleotide Sequence in DNA Where a Protein Binds

DNA footprinting is a widely used technique to identify the nucleotide sequence within DNA where a specific protein binds, such as the **promoter** sequence(s) bound by RNA polymerase holoenzyme. In this technique, the protein is incubated with a labeled (*) DNA fragment containing the nucleotide sequence where the protein is believed to bind. (The DNA fragment is labeled at only one end.) Then, a DNA cleaving agent, such as DNase I, is added to the solution containing the DNA:protein complex. DNase I cleaves the DNA backbone in exposed regions—that is, wherever the presence of the DNA-binding protein does not prevent DNase I from binding. A control solution containing naked DNA (a sample of the same labeled DNA fragment with no DNA-binding protein added) is also treated with DNase I. When these DNase I digests are analyzed by gel electrophoresis, a difference is found between the set of labeled fragments from the DNA:protein complex and the set from naked DNA. The absence of certain fragments in the digest of the DNA:protein complex reveals the location of the protein-binding site on the DNA (see accompanying figure).

Adapted from Rhodes, D., and Fairall, L., 1997. Analysis of sequence-specific DNA-binding proteins. In *Protein Function: A Practical Approach*, Creighton, T. E., ed., Oxford: IRL Press at Oxford University Press.

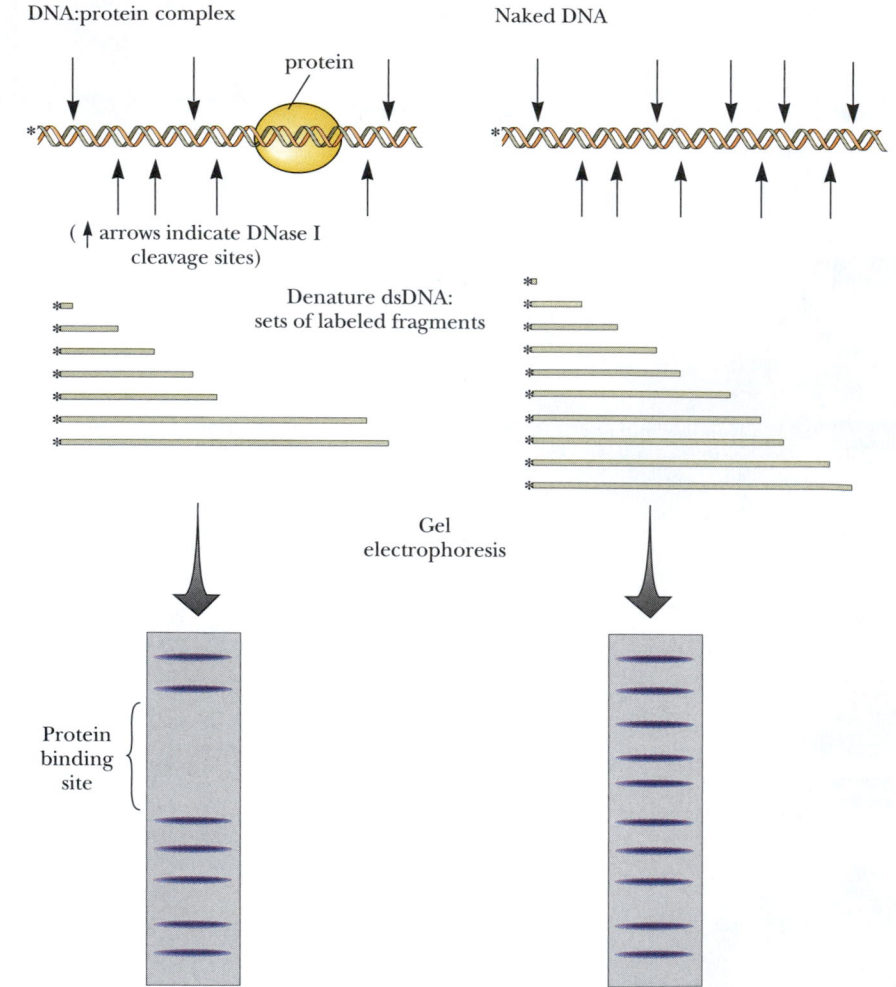

bind to the conserved sequences, and the more closely the -35 region sequence corresponds to its consensus sequence, the greater is the efficiency of transcription of the gene. The highly expressed *rrn* genes in *E. coli* that encode ribosomal RNA (rRNA) have a third sequence element in their promoters, the **upstream element** (**UP** element), located about 20 bp immediately upstream of the -35 region. (Transcription from the *rrn* genes accounts for more than 60% of total RNA synthesis in rapidly growing *E. coli* cells.) Whereas the σ-subunit recognizes the -10 and -35 elements, the C-terminal domains (CTD) of the α-subunits of RNA polymerase recognize and bind the UP element.

In order for transcription to begin, the DNA duplex must be "opened" so that RNA polymerase has access to single-stranded template. The efficiency of initiation is inversely proportional to the melting temperature, T_m, in the Pribnow box, suggesting that the A:T-rich nature of this region is aptly suited for easy "melting" of the DNA duplex and creation of the open promoter complex (see Figure 29.2). Negative supercoiling facilitates transcription initiation by favoring DNA unwinding.

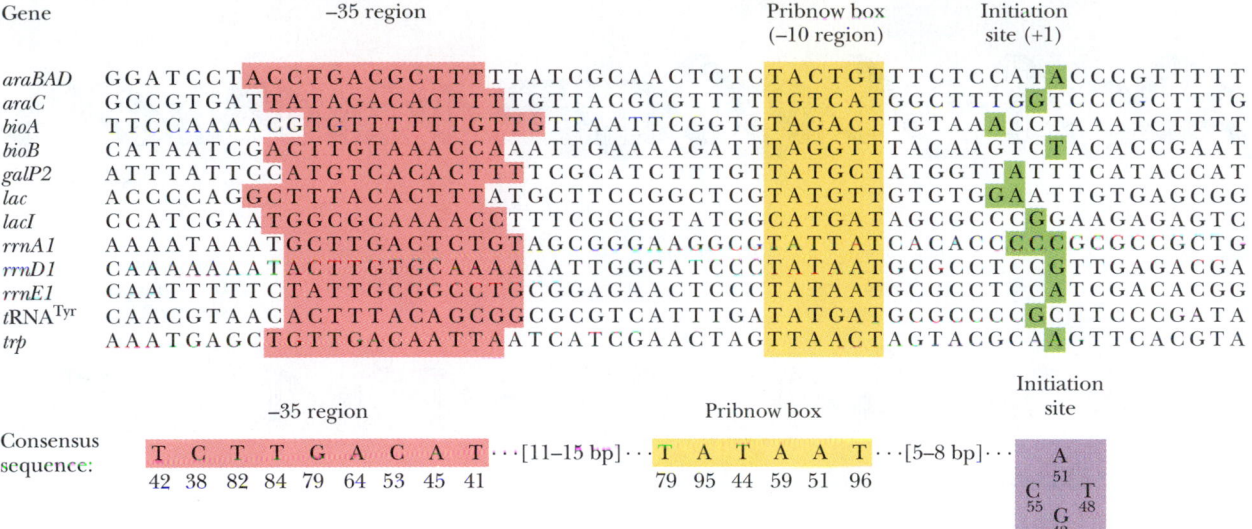

FIGURE 29.3 The nucleotide sequences of representative *E. coli* promoters. (In accordance with convention, these sequences are those of the nontemplate strand where RNA polymerase binds.) Consensus sequences for the −35 region, the Pribnow box, and the initiation site are shown at the bottom. The numbers represent the percent occurrence of the indicated base. (Note: The −35 region is only roughly 35 nucleotides from the transcription start site; the Pribnow box [the −10 region] likewise is located at approximately position −10.) In this figure, sequences are aligned relative to the Pribnow box.

The RNA polymerase σ-subunit is directly involved in melting the dsDNA. Interaction of the σ-subunit with the nontemplate strand maintains the open complex formed between RNA polymerase and promoter DNA, with the σ-subunit acting as a sequence-specific single-stranded DNA-binding protein. Association of the σ-subunit with the nontemplate strand stabilizes the open promoter complex and leaves the bases along the template strand available to the catalytic site of the RNA polymerase.

Initiation of Polymerization   RNA polymerase has two binding sites for NTPs: the initiation site and the elongation site. The **initiation site** binds the purine nucleotides ATP and GTP preferentially; most RNAs begin with a purine at the 5′-end. The first nucleotide binds at the initiation site, base-pairing with the +1 base exposed within the *open promoter complex* (Figure 29.2, Step 4). The second incoming nucleotide binds at the elongation site, base-pairing with the +2 base. The ribonucleotides are then united when the 3′-O of the first nucleotide makes a phosphoester bond with the α-phosphorus atom of the second nucleotide, and PP_i is eliminated. Note that the 5′-end of the transcript starts out with a triphosphate attached to it. Movement of RNA polymerase along the template strand *(translocation)* to the next base prepares the RNA polymerase to add the next nucleotide (Figure 29.2, Step 5). Once an oligonucleotide 9 to 12 residues long has been formed, the σ-subunit dissociates from RNA polymerase, signaling the completion of initiation (Figure 29.2, Step 6). The core RNA polymerase is highly processive and goes on to synthesize the remainder of the mRNA. As the core RNA polymerase progresses, advancing the 3′-end of the RNA chain, the DNA duplex is unwound just ahead of it. About 12 base pairs of the growing RNA remain base-paired to the DNA template at any time, with the RNA strand becoming displaced as the DNA duplex rewinds behind the advancing RNA polymerase.

Chain Elongation   Elongation of the RNA transcript is catalyzed by the *core polymerase,* because once a short oligonucleotide chain has been synthesized, the σ-subunit dissociates. The accuracy of transcription is such that about once every 10^4 nucleotides, an error is made and the wrong base is inserted. Because many transcripts are made per gene and most transcripts are smaller than 10 kb, this error rate is acceptable.

Two possibilities can be envisioned for the course of the new RNA chain. In one, the RNA chain is wrapped around the DNA as the RNA polymerase follows the template strand around the axis of the DNA duplex, but this possibility seems unlikely due to its potential for tangling the nucleic acid strands (Figure 29.4a). In reality, transcription involves supercoiling of the DNA, so positive

Biochemistry ℰ Now™   Go to BiochemistryNow and click BiochemistryInteractive to explore RNA polymerase II as the machine of transcription.

(a)

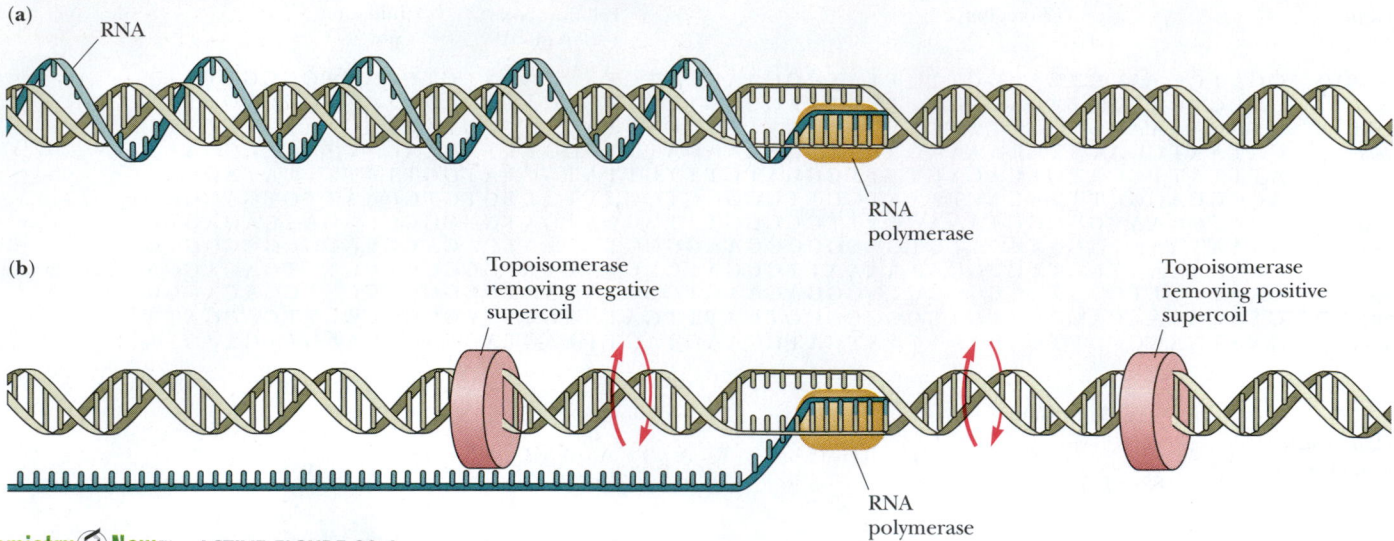

(b)

Topoisomerase removing negative supercoil

Topoisomerase removing positive supercoil

RNA polymerase

RNA polymerase

Biochemistry ⊜ Now™ **ACTIVE FIGURE 29.4**
Supercoiling versus transcription. **(a)** If the RNA polymerase followed the template strand around the axis of the DNA duplex, no supercoiling of the DNA would occur but the RNA chain would be wrapped around the double helix once every 10 bp. This possibility seems unlikely because it would be difficult to disentangle the transcript from the DNA duplex. **(b)** Alternatively, topoisomerases could remove the supercoils. A topoisomerase capable of relaxing positive supercoils situated ahead of the advancing transcription bubble would "relax" the DNA. A second topoisomerase behind the bubble would remove the negative supercoils. *(Adapted from Futcher, B., 1988. Supercoiling and transcription, or vice versa? Trends in Genetics 4:271–272.)* **Test yourself on the concepts in this figure at http://chemistry.brookscole.com/ggb3**

supercoils are created ahead of the transcription bubble and negative supercoils are created behind it (Figure 29.4b). To prevent torsional stress from inhibiting transcription, topoisomerases act to remove these supercoils from the DNA segment undergoing transcription (Figure 29.4b).

Chain Termination   Two types of transcription termination mechanisms operate in bacteria: one that is dependent on a specific protein called **rho termination factor** (for the Greek symbol, **ρ**) and another, **intrinsic termination**, that is not. In intrinsic termination, termination is determined by specific sequences in the DNA called **termination sites.** These sites are not indicated by particular bases showing where transcription halts. Instead, these sites consist of three structural features whose base-pairing possibilities lead to termination:

1. Inverted repeats, which are typically G:C-rich, so a stable **stem-loop structure** can form in the transcript via intrachain base-pairing (Figure 29.5)
2. A nonrepeating segment that punctuates the inverted repeats
3. A run of 6 to 8 As in the DNA template, coding for Us in the transcript

Termination then occurs as follows: A G:C-rich, stem-loop structure, or "hairpin," forms in the transcript. The hairpin apparently causes the RNA polymerase to pause, whereupon the A:U base pairs between the transcript and the DNA template strand are displaced through formation of somewhat more stable A:T base pairs between the template and nontemplate strands of the DNA. The result is spontaneous dissociation of the nascent transcript from DNA.

The alternative mechanism of termination—factor-dependent termination—is less common and mechanistically more complex. Rho factor is an ATP-

FIGURE 29.5 The termination site for the *E. coli trp* operon (the *trp* operon encodes the enzymes of tryptophan biosynthesis). The inverted repeats give rise to a stem-loop, or "hairpin," structure ending in a series of U residues.

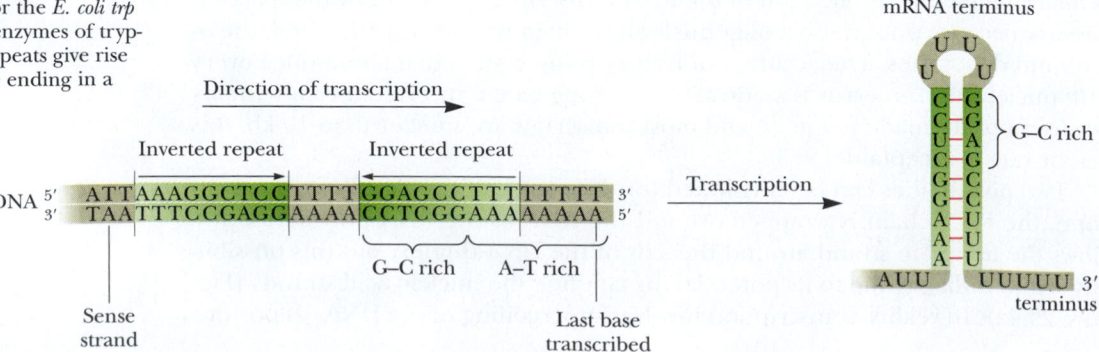

(a)

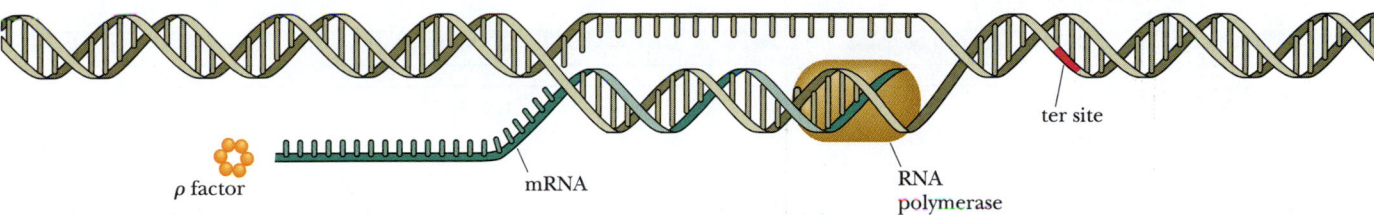

ρ factor   mRNA   RNA
polymerase

ter site

(b)

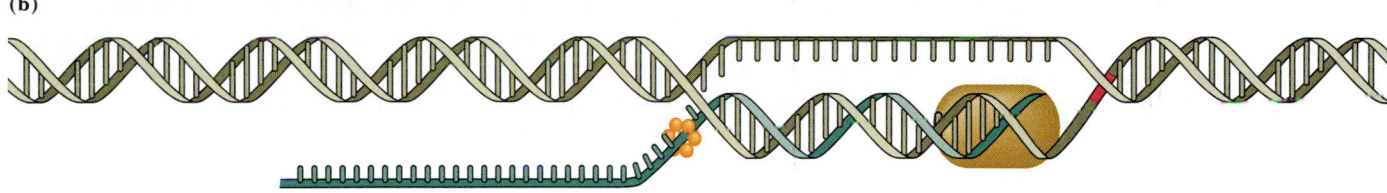

(c)

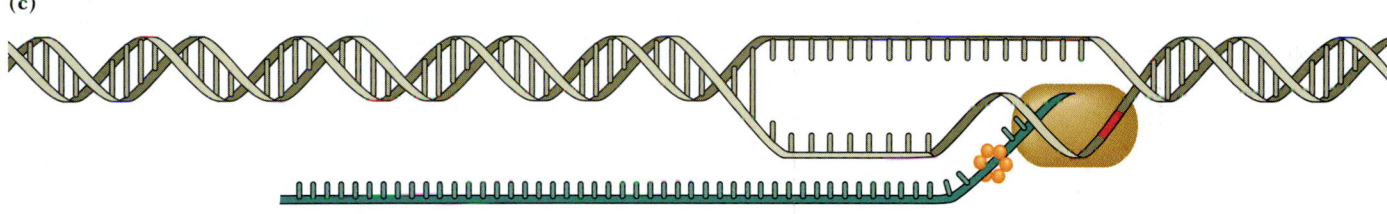

(d)

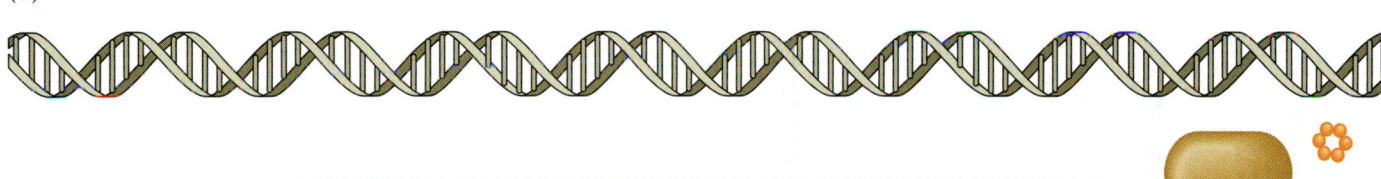

mRNA

Biochemistry � Now™   **ANIMATED FIGURE 29.6**
The rho factor mechanism of transcription termination. Rho factor **(a)** attaches to a recognition site on
mRNA and **(b)** moves along it behind RNA polymerase. **(c)** When RNA polymerase pauses at the
termination site, rho factor unwinds the DNA:RNA
hybrid in the transcription bubble, **(d)** releasing the
nascent mRNA. **See this figure animated at http://
chemistry.brookscole.com/ggb3**

dependent helicase (hexamer of 50-kD subunits) that catalyzes the unwinding
of RNA:DNA hybrid duplexes (or RNA:RNA duplexes). The rho factor recognizes and binds to C-rich regions in the RNA transcript. These regions must be
unoccupied by translating ribosomes for rho factor to bind. Once bound, rho
factor advances in the 5′→3′ direction until it reaches the transcription bubble
(Figure 29.6). There it catalyzes the unwinding of the transcript and template,
releasing the nascent RNA chain. It is likely that the RNA polymerase stalls in a
G:C-rich termination region, allowing rho factor to overtake it.

29.2 | How Is Transcription Regulated in Prokaryotes?

In bacteria, genes encoding the enzymes of a particular metabolic pathway are
often grouped adjacent to one another in a cluster on the chromosome. This
pattern of organization allows all of the genes in the group to be expressed in
a coordinated fashion through transcription into a **single polycistronic mRNA**
encoding all the enzymes of the metabolic pathway.[3] A regulatory sequence lying

[3]A **polycistronic mRNA** is a single RNA transcript that encodes more than one polypeptide. *Cistron*
is a genetic term for a DNA region representing a protein: *Cistron* and *gene* are essentially equivalent terms.

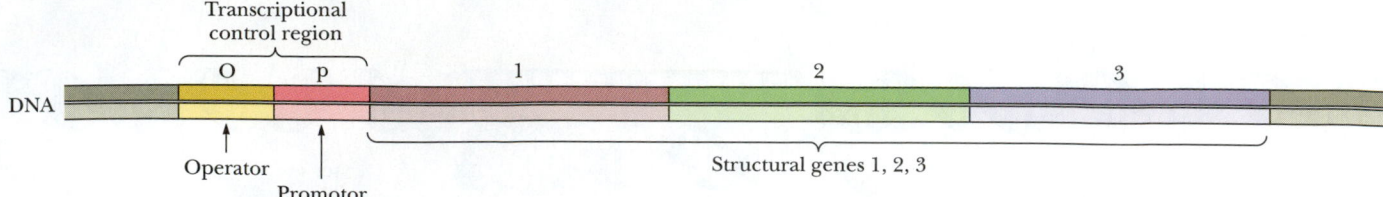

FIGURE 29.7 The general organization of operons. Operons consist of transcriptional control regions and a set of related structural genes, all organized in a contiguous linear array along the chromosome. The transcriptional control regions are the promoter and the operator, which lie next to, or overlap, each other, upstream from the structural genes they control. Operators may lie at various positions relative to the promoter, either upstream or downstream. Expression of the operon is determined by access of RNA polymerase to the promoter, and occupancy of the operator by regulatory proteins influences this access. Induction activates transcription from the promoter; repression prevents it.

adjacent to the DNA being transcribed determines whether transcription takes place. This sequence is termed the **operator** (Figure 29.7). The operator is located next to a promoter. Interaction of a **regulatory protein** with the operator controls transcription of the gene cluster by controlling access of RNA polymerase to the promoter.[4] Such co-expressed gene clusters, together with the operator and promoter sequences that control their transcription, are called **operons.**

Transcription of Operons Is Controlled by Induction and Repression

In prokaryotes, gene expression is often responsive to small molecules serving as signals of the nutritional or environmental conditions confronting the cell. Increased synthesis of enzymes in response to the presence of a particular substrate is termed **induction.** For example, lactose (Figure 29.8) can serve as both carbon and energy source for *E. coli.* Metabolism of lactose depends on hydrolysis into its component sugars, glucose and galactose, by the enzyme **β-galactosidase.** In the absence of lactose, *E. coli* cells contain very little β-galactosidase (less than 5 molecules per cell). However, lactose availability *induces* the synthesis of β-galactosidase by activating transcription of the *lac* **operon.** One of the genes in the *lac* operon, *lacZ*, is the structural gene for β-galactosidase. When its synthesis is fully induced, β-galactosidase can amount to almost 10% of the total soluble protein in *E. coli.* When lactose is removed from the culture, synthesis of β-galactosidase halts.

The alternative to induction—namely *decreased* synthesis of enzymes in response to a specific metabolite—is termed **repression.** For example, the enzymes of tryptophan biosynthesis in *E. coli* are encoded in the *trp* **operon.** If sufficient Trp is available to the growing bacterial culture, the *trp* operon is not transcribed, so the Trp biosynthetic enzymes are not made; that is, their synthesis is *repressed.* Repression of the *trp* operon in the presence of Trp is an eminently logical control mechanism: If the end product of the pathway is present, why waste cellular resources making unneeded enzymes?

Induction and repression are two faces of the same phenomenon. In induction, a substrate activates enzyme synthesis. Substrates capable of activating synthesis of the enzymes that metabolize them are called **co-inducers,** or, often, simply **inducers.** Some substrate analogs can induce enzyme synthesis even though the enzymes are incapable of metabolizing them. These analogs are called **gratuitous inducers.** A number of thiogalactosides, such as **IPTG** (isopropyl β-thiogalactoside, Figure 29.9), are excellent gratuitous inducers of β-galactosidase activity in *E. coli.* In repression, a metabolite, typically an end product, depresses synthesis of its own biosynthetic enzymes. Such metabolites are called **co-repressors.**

Lactose
(O-β-D-galactopyranosyl (1 → 4) β-D-glucopyranose)

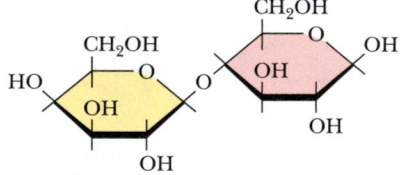

FIGURE 29.8 The structure of lactose, a β-galactoside.

The *lac* Operon Serves as a Paradigm of Operons

In 1961, François Jacob and Jacques Monod proposed the **operon hypothesis** to account for the coordinate regulation of related metabolic enzymes. The operon was considered to be the unit of gene expression, consisting of two classes of genes: the **structural genes** for the enzymes and **regulatory genes** that

Isopropyl β-thiogalactoside (IPTG)

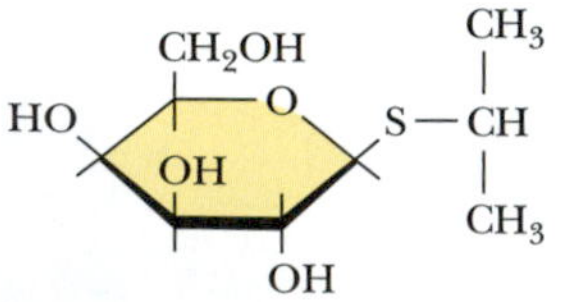

FIGURE 29.9 The structure of IPTG (isopropyl β-thiogalactoside).

[4]Although this is the paradigm for prokaryotic gene regulation, it must be emphasized that many regulated prokaryotic genes do not contain operators and are regulated in ways that do not involve protein:operator interactions.

		lacI	p_{lac} O	lacZ	lacY	lacA
DNA	p					

bp		1080	82	3069	1251	609
mRNA						

Polypeptide	Amino acids	360		1023	417	203
	kD	38.6		116.4	46.5	22.7
Protein	Structure	Tetramer		Tetramer	Membrane protein	Dimer
	kD	154.4		465	46.5	45.4
Function		Repressor		β-Galactosidase	Permease	Trans-acetylase

FIGURE 29.10 The *lac* operon. The operon consists of two transcription units. In one unit, there are three structural genes, *lacZ*, *lacY*, and *lacA*, under control of the promoter, p_{lac}, and the operator *O*. In the other unit, there is a regulator gene, *lacI*, with its own promoter, p_{lacI}. *lacI* encodes a 360-residue, 38.6-kD polypeptide that forms a tetrameric *lac* repressor protein. *lacZ* encodes β-galactosidase, a tetrameric enzyme of 116-kD subunits. *lacY* is the β-galactoside permease structural gene, a 46.5-kD integral membrane protein active in β-galactoside transport into the cell. The remaining structural gene encodes a 22.7-kD polypeptide that forms a dimer displaying thiogalactoside transacetylase activity in vitro, transferring an acetyl group from acetyl-CoA to the C-6 OH of thiogalactosides, but the metabolic role of this protein in vivo remains uncertain. *lacA* mutants show no identifiable metabolic deficiency. Perhaps the lacA protein acts to detoxify toxic analogs of lactose through acetylation.

control expression of the structural genes. The two kinds of genes could be distinguished by mutation. Mutations in a structural gene would abolish one particular enzymatic activity, but mutations in a regulatory gene would affect all of the different enzymes under its control. Mutations of both kinds were known in *E. coli* for lactose metabolism. Bacteria with mutations in either the *lacZ* gene or the *lacY* gene (Figure 29.10) could no longer metabolize lactose—the *lacZ* mutants (*lacZ⁻* strains) because β-galactosidase activity was absent, the *lacY* mutants because lactose was no longer transported into the cell. Other mutations defined another gene, the *lacI* gene. *lacI* mutants were different because they both expressed β-galactosidase activity and immediately transported lactose, *without prior exposure to an inducer*. That is, a single mutation led to the expression of lactose metabolic functions independently of inducer. Expression of genes independently of regulation is termed **constitutive expression.** Thus, *lacI* had the properties of a regulatory gene. The *lac* operon includes the regulatory gene *lacI*; its promoter *p*; and three structural genes, *lacZ*, *lacY*, and *lacA*, with their own promoter p_{lac} and operator *O* (Figure 29.10).

lac Repressor Is a Negative Regulator of the *lac* Operon

The structural genes of the *lac* operon are controlled by **negative regulation.** That is, they are transcribed to give an mRNA unless turned off by the *lacI* gene product. This gene product is the *lac* **repressor,** a tetrameric protein (Figure 29.11). (Note that the language can be misleading: *Inducers and corepressors* are small molecules/metabolites; *repressors* are proteins.) The *lac* repressor has an N-terminal DNA-binding domain; the rest of the protein functions in inducer binding and tetramer formation. In the absence of an inducer, *lac* repressor blocks *lac* gene expression by binding to the operator DNA site upstream from the *lac* structural genes. The *lac* operator is a palindromic DNA sequence (Figure 29.12). **Palindromes,** or "inverted repeats" (see Chapter 11), provide a twofold, or dyad, symmetry, a structural feature common at sites in DNA where proteins specifically bind. Despite the presence of *lac* repressor, RNA polymerase can still initiate transcription at the promoter (p_{lac}), but *lac* repressor blocks elongation of transcription, so initiation is aborted. In *lacI* mutants, the *lac* repressor is absent or defective in binding to operator DNA, *lac* gene transcription is not blocked, and the *lac* operon is constitutively expressed in these mutants. Note that *lacI* is normally expressed constitutively from its promoter, so *lac* repressor protein is always available to fill its regulatory role. About ten molecules of *lac* repressor are present in an *E. coli* cell.

Derepression of the *lac* operon occurs when appropriate β-galactosides occupy the inducer site on *lac* repressor, causing a conformational change in the protein that lowers the repressor's affinity for operator DNA. As a tetramer, *lac* repressor has four inducer binding sites, and its response to inducer shows cooperative allosteric effects. Thus, as a consequence of the "inducer"-induced

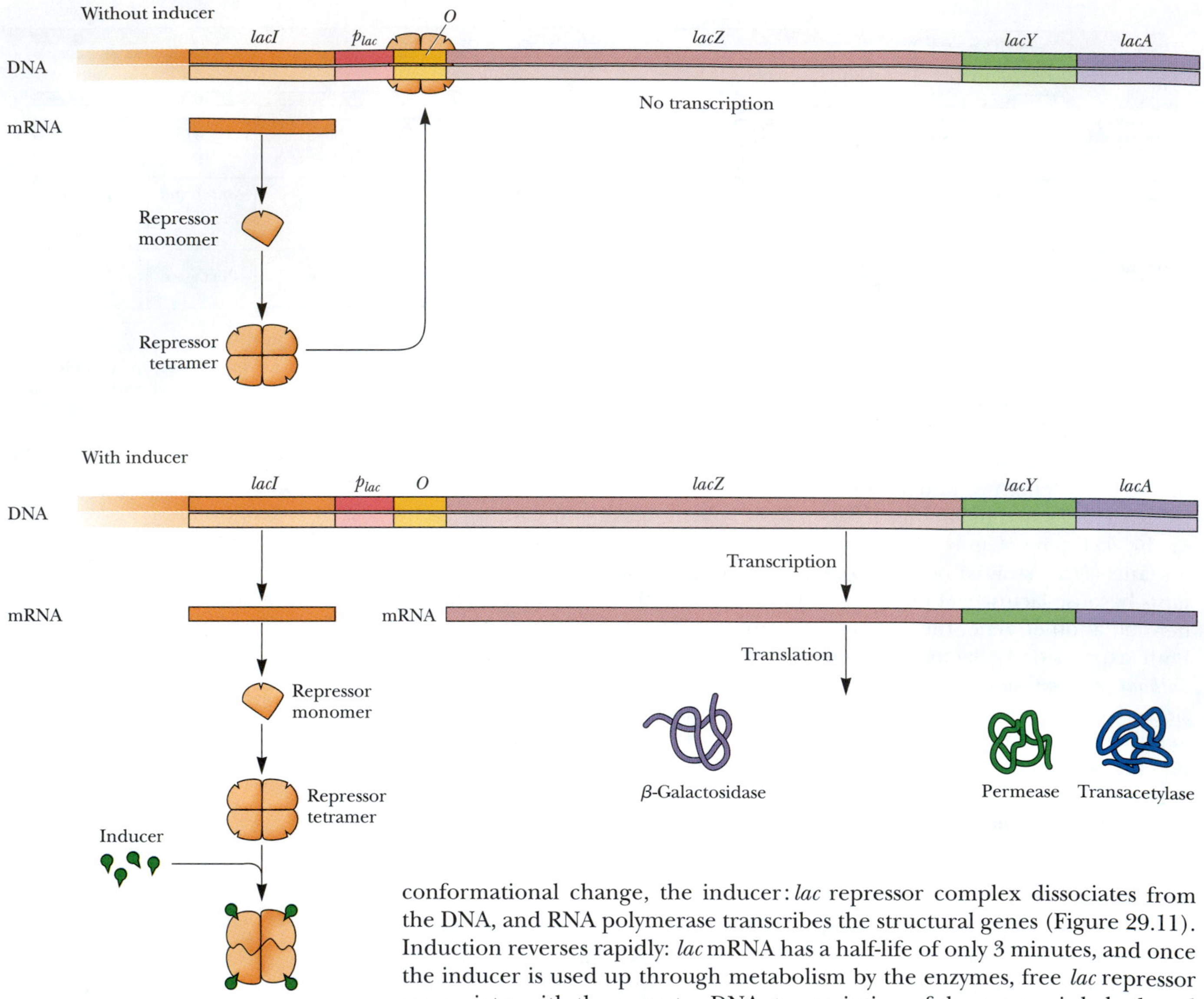

Without inducer

lacI p_{lac} O   lacZ   lacY   lacA

DNA

No transcription

mRNA

Repressor monomer

Repressor tetramer

With inducer

lacI p_{lac} O   lacZ   lacY   lacA

DNA

Transcription

mRNA

mRNA

Translation

β-Galactosidase   Permease   Transacetylase

Repressor monomer

Repressor tetramer

Inducer

conformational change, the inducer: *lac* repressor complex dissociates from the DNA, and RNA polymerase transcribes the structural genes (Figure 29.11). Induction reverses rapidly: *lac* mRNA has a half-life of only 3 minutes, and once the inducer is used up through metabolism by the enzymes, free *lac* repressor reassociates with the operator DNA, transcription of the operon is halted, and any residual *lac* mRNA is degraded.

In the absence of inducer, *lac* repressor binds nonspecifically to duplex DNA with an association constant, K_A, of $2 \times 10^6 \ M^{-1}$ (Table 29.1) and to the *lac* operator DNA sequence with much higher affinity, $K_A = 2 \times 10^{13} \ M^{-1}$. Thus, *lac* repressor binds 10^7 times better to *lac* operator DNA than to any random DNA sequence. IPTG binds to *lac* repressor with an association constant of about $10^6 \ M^{-1}$. The IPTG: *lac* repressor complex binds to operator DNA with an association constant, $K_A = 2 \times 10^{10} \ M^{-1}$. Although this affinity is high, it is 3 orders of magnitude *less* than the affinity of inducer-free repressor for *lac* operator. There is no difference in the affinity of free *lac* repressor and *lac*

FIGURE 29.12 The nucleotide sequence of the *lac* operator. This sequence comprises 36 bp showing nearly palindromic symmetry. The inverted repeats that constitute this approximate twofold symmetry are shaded in rose. The bases are numbered relative to the +1 start site for transcription. The G:C base pair at position +11 represents the axis of symmetry. In vitro studies show that bound *lac* repressor protects a 26-bp region from −5 to +21 against nuclease digestion. Bases that interact with bound *lac* repressor are indicated below the operator. Note the symmetry of protection at +1 through +4 TTAA to +18 through +21 AATT.

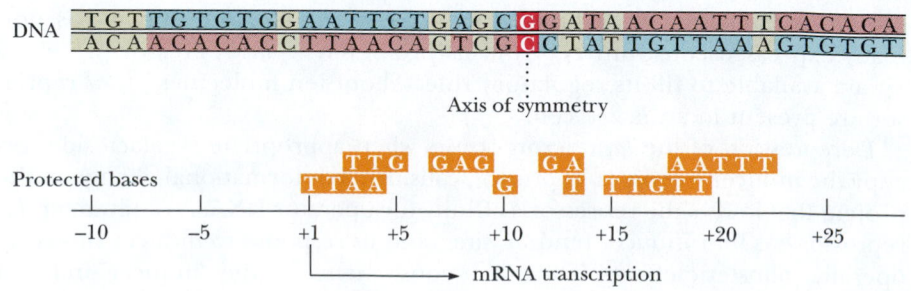

repressor with IPTG bound for nonoperator DNA. The *lac* repressor apparently acts by binding to DNA and sliding along it, testing sequences in a one-dimensional search until it finds the *lac* operator. The *lac* repressor then binds there with high affinity until inducer causes this affinity to drop by 3 orders of magnitude (Table 29.1).

CAP Is a Positive Regulator of the *lac* Operon

Transcription by RNA polymerase from some promoters proceeds with low efficiency unless assisted by an accessory protein that acts as a *positive regulator.* One such protein is **CAP,** or **catabolite activator protein.** Its name derives from the phenomenon of catabolite repression in *E. coli.* Catabolite repression is a global control that coordinates gene expression with the total physiological state of the cell: As long as glucose is available, *E. coli* catabolizes it in preference to any other energy source, such as lactose or galactose. Catabolite repression ensures that the operons necessary for metabolism of these alternative energy sources, that is, the *lac* and *gal* operons, remain repressed until the supply of glucose is exhausted. Catabolite repression overrides the influence of any inducers that might be present.

Catabolite repression is maintained until the *E. coli* cells become starved of glucose. Glucose starvation leads to activation of adenylyl cyclase, and the cells begin to make cAMP. (In contrast, glucose uptake is accompanied by deactivation of adenylyl cyclase.) The action of CAP as a positive regulator is cAMP-dependent. cAMP is a small-molecule inducer for CAP, and cAMP binding enhances CAP's affinity for DNA. CAP, also referred to as **CRP** (for **cAMP receptor protein**), is a dimer of identical 22.5-kD polypeptides. The N-terminal domains bind cAMP; the C-terminal domains constitute the DNA-binding site. Two molecules of cAMP are bound per dimer. The CAP–(cAMP)$_2$ complex binds to specific target sites near the promoters of operons (Figure 29.13). Binding of CAP–(cAMP)$_2$ to DNA causes the DNA to bend more than 80° (Figure 29.14). This CAP-induced DNA bending near the promoter assists RNA polymerase holoenzyme binding and closed promoter complex formation. Contacts made between the CAP–(cAMP)$_2$ complex and the α-subunit of RNA polymerase holoenzyme activate transcription.

Table 29.1

The Affinity of *lac* Repressor for DNA*

DNA	Repressor	Repressor + Inducer
lac operator	$2 \times 10^{13}\ M^{-1}$	$2 \times 10^{10}\ M^{-1}$
All other DNA	$2 \times 10^6\ M^{-1}$	$2 \times 10^6\ M^{-1}$
Specificity†	10^7	10^4

*Values for repressor:DNA binding are given as association constants, K_A, for the formation of DNA:repressor complex from DNA and repressor.
†Specificity is defined as the ratio (K_A for repressor binding to operator DNA)/(K_A for repressor binding to random DNA).

A Deeper Look

Quantitative Evaluation of *lac* Repressor:DNA Interactions

The affinity of *lac* repressor for random DNA ensures that essentially all repressor is DNA bound. Assume that *E. coli* DNA has a single specific *lac* operator site for repressor binding and 4.64×10^6 base pairs and any nucleotide sequence even one base out of phase with the operator constitutes a nonspecific binding site. Thus, there are 4.64×10^6 nonspecific sites for repressor binding.

The binding of repressor to DNA is given by the association constant, K_A:

$$K_A = \frac{[\text{repressor:DNA}]}{[\text{repressor}][\text{DNA}]}$$

where [repressor:DNA] is the concentration of repressor:DNA complex, [repressor] is the concentration of free repressor, and [DNA] is the concentration of nonspecific binding sites. Rearranging gives the following:

$$\frac{[\text{repressor}]}{[\text{repressor:DNA}]} = \frac{1}{K_A[\text{DNA}]}$$

If the number of nonspecific binding sites is 4.64×10^6, there are $(4.64 \times 10^6)/(6.023 \times 10^{23}) = 0.77 \times 10^{-17}$ "moles of binding sites contained in the volume of a bacterial cell (roughly 10^{-15} liters). Therefore, [DNA] $= (0.77 \times 10^{-17})/(10^{-15}) = 0.77 \times 10^{-2}$ M. Since $K_A = 2 \times 10^6\ M^{-1}$ (Table 29.1),

$$\frac{[\text{repressor}]}{[\text{repressor:DNA}]} = \frac{1}{(2 \times 10^6)\ (0.77 \times 10^{-2})} = \frac{1}{(1.54 \times 10^4)}$$

So, the ratio of free repressor to DNA-bound repressor is 6.5×10^{-5}. *Less than 0.01% of repressor is not bound to DNA!* The behavior of *lac* repressor is characteristic of DNA-binding proteins. These proteins bind with low affinity to random DNA sequences, but with much higher affinity to their unique target sites (Table 29.1).

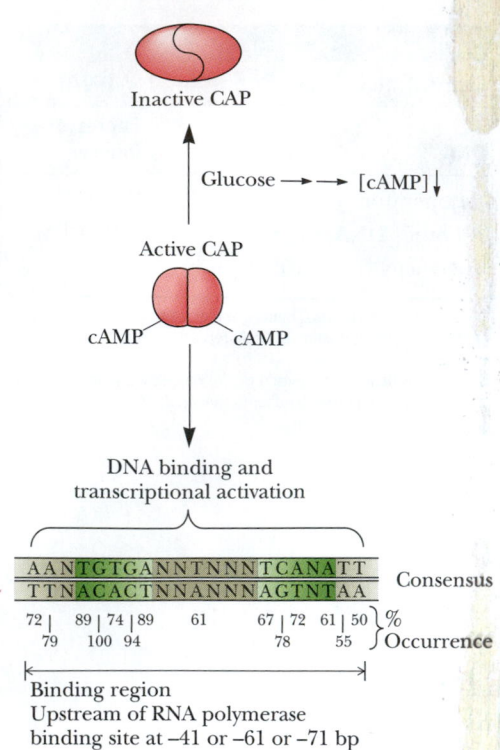

Inactive CAP

Glucose ⟶ [cAMP]↓

Active CAP

cAMP ⟍ ⟋ cAMP

DNA binding and
transcriptional activation

```
AANTGTGANNTNNNTCANATT
TTNACACTNNANNNAGTNTAA    Consensus
```

72 | 89 | 74 | 89   61   67 | 72 61 | 50 ⎱ %
79   100 94   78   55 ⎰ Occurrence

Binding region
Upstream of RNA polymerase
binding site at –41 or –61 or –71 bp

FIGURE 29.13 The mechanism of catabolite repression and CAP action. Glucose instigates catabolite repression by lowering cAMP levels. cAMP is necessary for CAP binding near promoters of operons whose gene products are involved in the metabolism of alternative energy sources such as lactose, galactose, and arabinose. The binding sites for the CAP–(cAMP)₂ complex are consensus DNA sequences containing the conserved pentamer TGTGA and a less well conserved inverted repeat, TCANA (where N is any nucleotide).

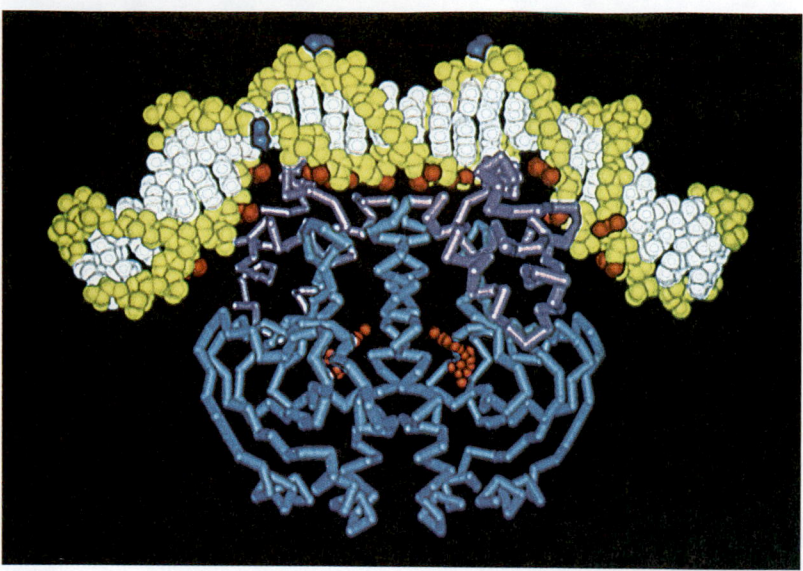

FIGURE 29.14 Binding of CAP–(cAMP)₂ induces a severe bend in DNA about the center of dyad symmetry at the CAP-binding site. The CAP dimer with two molecules of cAMP bound interacts with 27 to 30 base pairs of duplex DNA. Two α-helices of the CAP dimer insert into the major groove of the DNA at the dyad-symmetric CAP-binding site. The cAMP-binding domain of CAP protein is shown in blue and the DNA-binding domain in purple. The two cAMP molecules bound by the CAP dimer are indicated in red. For DNA, the bases are shown in white and the sugar–phosphate backbone in yellow. DNA phosphates that interact with CAP are highlighted in red. Binding of CAP–(cAMP)₂ to its specific DNA site involves H bonding and ionic interactions between protein functional groups and DNA phosphates, as well as H-bonding interactions in the DNA major groove between amino acid side chains of CAP and DNA base pairs. *(Adapted from Schultz, S. C., Shields, G. C., and Steitz, T. A., 1991. Crystal structure of a CAP-DNA complex: The DNA is bent by 90°. Science 253:1001–1007. Photograph courtesy of Professor Thomas A. Steitz of Yale University.)*

Negative and Positive Control Systems Are Fundamentally Different

Negative and positive control systems operate in fundamentally different ways (although in some instances both govern the expression of the same gene). Genes under negative control are transcribed unless they are turned off by the presence of a repressor protein. Often, transcription activation is merely the release from negative control. In contrast, genes under positive control are expressed only if an active regulator protein is present. The *lac* operon illustrates these differences. The action of *lac* repressor is negative. It binds to operator DNA and blocks transcription; expression of the operon occurs only when this negative control is lifted through the release of the repressor. In contrast, regulation of the *lac* operon by CAP is positive: Transcription of the operon by RNA polymerase is stimulated by CAP's action as a positive regulator.

Operons can also be classified as **inducible, repressible,** or both, depending on how they respond to the small molecules that mediate their expression. Repressible operons are expressed only in the absence of their co-repressors. Inducible operons are transcribed only in the presence of small-molecule co-inducers (Figure 29.15).

The *araBAD* Operon Is Both Positively and Negatively Controlled by *AraC*

E. coli can use the plant pentose L-arabinose as sole source of carbon and energy. Arabinose is metabolized via conversion to D-xylulose-5-P (a pentose phosphate pathway intermediate and transketolase substrate [see Chapter 22]) by three enzymes encoded in the **araBAD operon**. Transcription of this operon is regulated by both catabolite repression and arabinose-mediated induction. CAP functions in catabolite repression; arabinose induction is achieved via the product of the *araC* gene, which lies next to the *araBAD* operon on the *E. coli* chro-

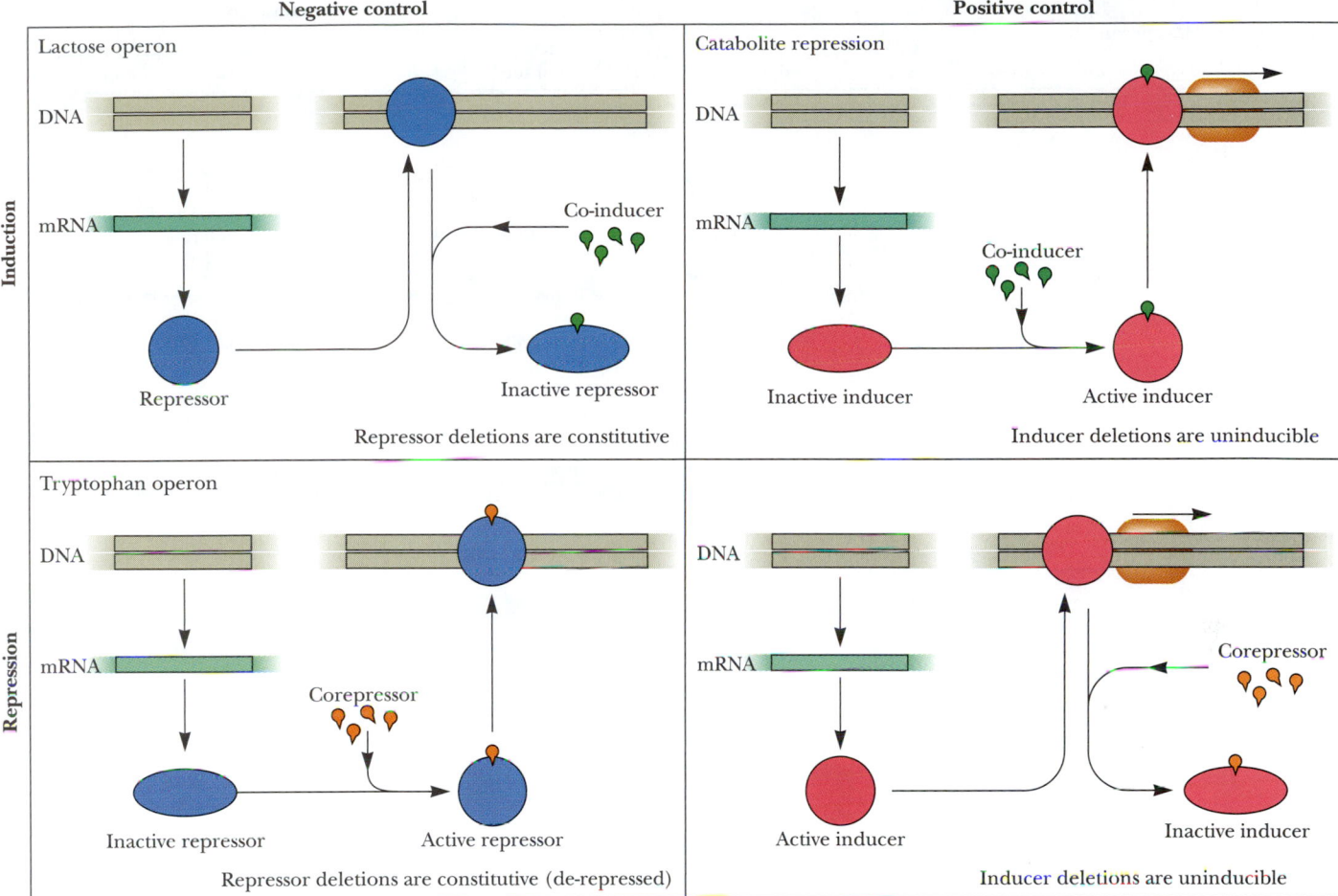

| Negative control | Positive control |

Induction

Lactose operon

Catabolite repression

Repressor deletions are constitutive

Inducer deletions are uninducible

Repression

Tryptophan operon

Repressor deletions are constitutive (de-repressed)

Inducer deletions are uninducible

Biochemistry ⊗ Now™   **ANIMATED FIGURE 29.15**
Control circuits governing the expression of genes.
These circuits can be either negative or positive,
inducible or repressible. **See this figure animated at**
http://chemistry.brookscole.com/ggb3

mosome. The *araC* gene product, the protein **AraC**,[5] is a 292-residue protein
consisting of an N-terminal domain (residues 1 to 170) that binds arabinose
and acts as a dimerization motif and a C-terminal (residues 178 to 292) DNA-
binding domain. Regulation of *araBAD* by AraC is novel in that it acts both negatively and positively. The *ara* operon has three binding sites for AraC: *araO₁*,
located at nucleotides 2106 to 2144 relative to the *araBAD* transcription start
site; *araO₂* (spanning positions 2265 to 2294); and *araI*, the *araBAD* promoter.
The *araI* site consists of two "half-sites"; *araI₁* (nucleotides 256 to 278) and *araI₂*
(235 to 251). (The *araO₁* site contributes minimally to *ara* operon regulation.)

The details of *araBAD* regulation are as follows: When AraC protein levels are
low, the *araC* gene is transcribed from its promoter p_c (adjacent to *araO₁*) by RNA
polymerase (Figure 29.16). *araC* is transcribed in the direction away from *araBAD*.
When cAMP levels are low and arabinose is absent, an AraC protein dimer binds
to two sites, *araO₂* and the *araI₁* half-site, forming a DNA loop between them and
restricting transcription of *araBAD* (Figure 29.16). In the presence of L-arabinose,
the monomer of AraC bound to the *araO₂* site is released from that site; it then associates with the unoccupied *araI* half-site, *araI₂*. L-Arabinose thus behaves as an
allosteric effector that alters the conformation of AraC. In the arabinose-liganded
conformation, the AraC dimer interacts with CAP–(cAMP)₂ to activate transcription by RNA polymerase. Thus, AraC protein is both a repressor and an activator.

Positive control of the *araBAD* operon occurs in the presence of L-arabinose
and cAMP. Arabinose binding by AraC protein causes the release of *araO₂*,
opening of the DNA loop, and association of AraC with *araI₂*, CAP–(cAMP)₂
binds at a site between *araO₁* and *araI*, and together the AraC–(arabinose)₂ and

[5]Proteins are often named for the genes encoding them. By convention, the name of the protein
is capitalized but not italicized.

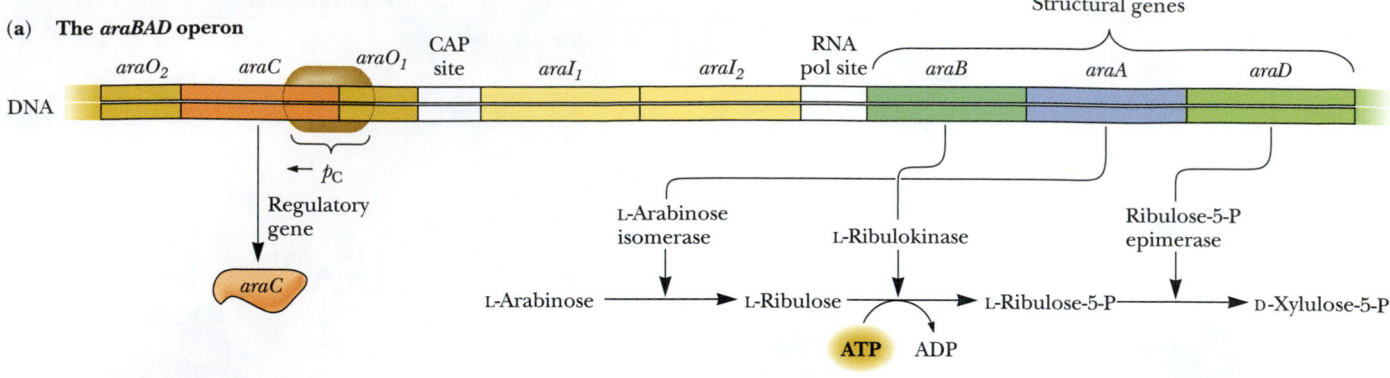

(a) The *araBAD* operon

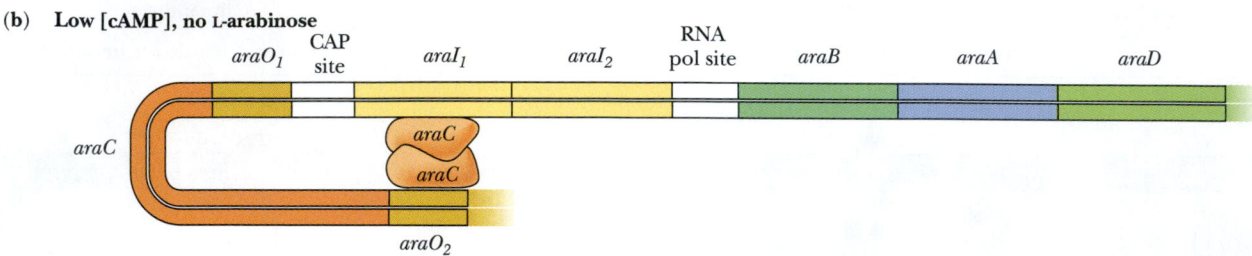

(b) Low [cAMP], no L-arabinose

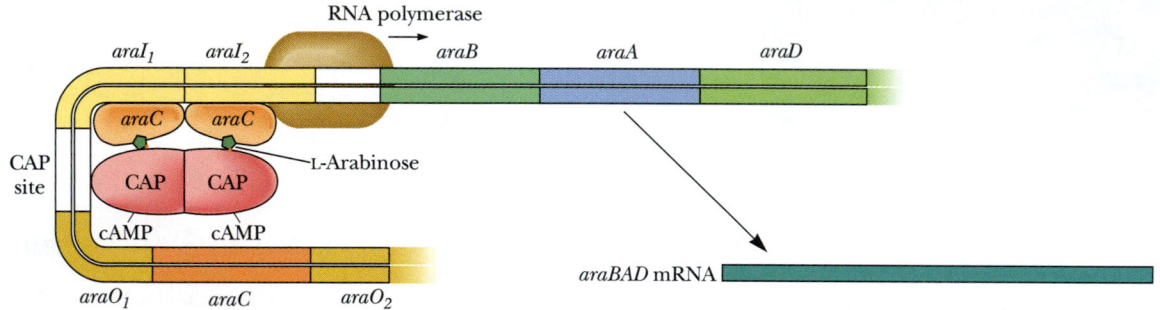

(c) High [cAMP], L-arabinose present

Biochemistry🔵Now™ ANIMATED FIGURE 29.16
Regulation of the *araBAD* operon by the combined action of CAP and AraC protein. **See this figure animated at http://chemistry.brookscole.com/ggb3**

CAP–(cAMP)$_2$ complexes influence RNA polymerase, through protein:protein interactions, to create an active transcription initiation complex. Supercoiling-induced DNA looping may promote protein:protein interactions between DNA-binding proteins by bringing them into juxtaposition.

The *trp* Operon Is Regulated Through a Co-Repressor–Mediated Negative Control Circuit

The *trp* operon of *E. coli* (and *S. typhimurium*) encodes the five polypeptides, *trpE* through *trpA* (Figure 29.17), that assemble into the three enzymes catalyzing tryptophan synthesis from chorismate (see Chapter 25). Expression of the *trp* operon is under the control of **Trp repressor,** a dimer of 108-residue polypeptide chains. When tryptophan is plentiful, Trp repressor binds two molecules of tryptophan and associates with the *trp* operator that is located within the *trp* promoter. Trp repressor binding excludes RNA polymerase from the promoter, preventing transcription of the *trp* operon. When Trp becomes limiting, repression is lifted because Trp repressor lacking bound Trp (Trp apo-repressor) has a lowered affinity for the *trp* promoter. Thus, the behavior of Trp repressor corresponds to a co-repressor–mediated, negative control circuit (Figure 29.15). Trp repressor not only is encoded by the *trpR* operon but also regulates expression of the *trpR* operon. This is an example of **autogenous regulation (autoregulation):** regulation of gene expression by the product of the gene.

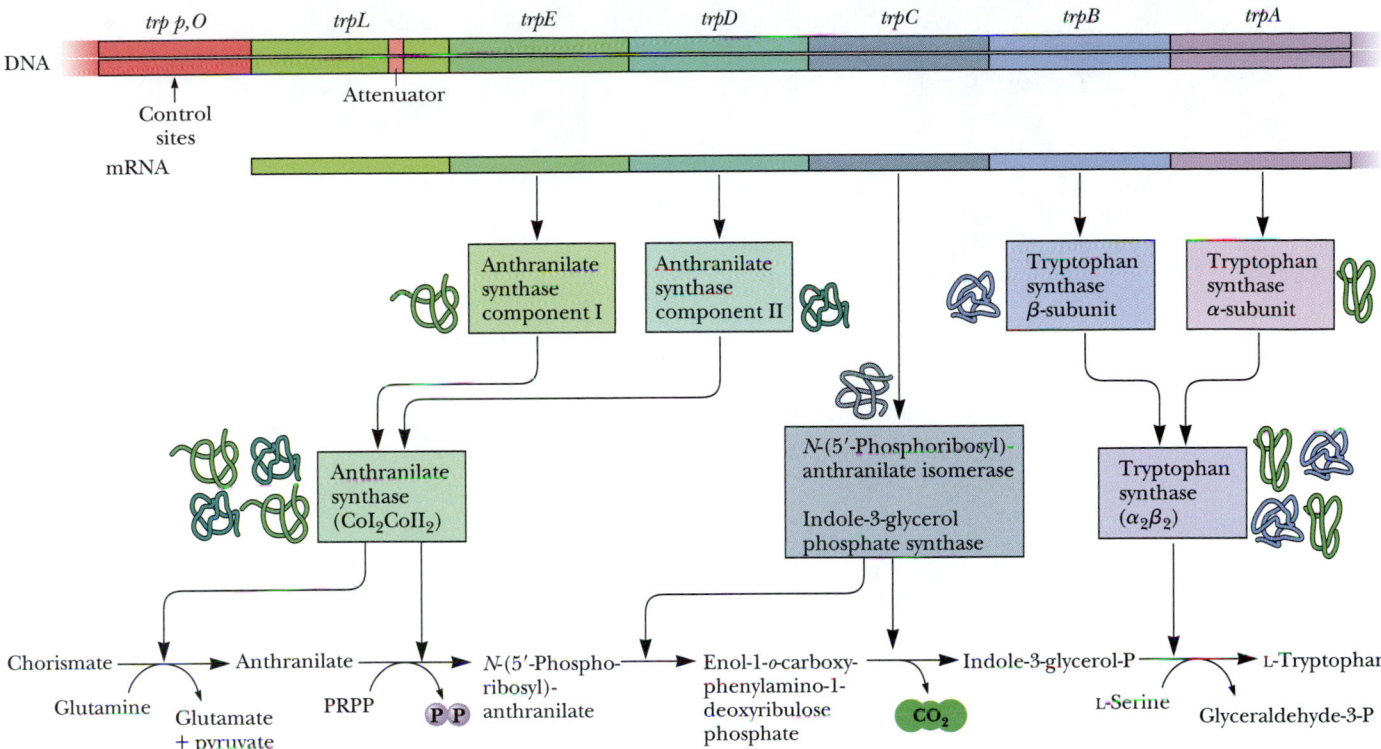

Biochemistry (E) Now™ **ANIMATED FIGURE 29.17** The *trp* operon of *E. coli*. **See this figure animated at** http://chemistry.brookscole.com/ggb3

Attenuation Is a Prokaryotic Mechanism for Post-Transcriptional Regulation of Gene Expression

In addition to repression, expression of the *trp* operon is controlled by **transcription attenuation.** Unlike the mechanisms discussed thus far, attenuation regulates transcription after it has begun. Charles Yanofsky, the discoverer of this phenomenon, has defined attenuation as *any regulatory mechanism that manipulates transcription termination or transcription pausing to regulate gene transcription downstream.* In prokaryotes, transcription and translation (see Chapters 10 and 30) are coupled, and the translating ribosome is affected by the formation and persistence of secondary structures in the mRNA. In many operons encoding enzymes of amino acid biosynthesis, a transcribed 150- to 300-bp leader region is positioned between the promoter and the first major structural gene. These regions encode a short leader peptide containing **multiple codons** for the pertinent amino acid. For example, the leader peptide of the *leu* operon has four leucine codons, the *trp* operon has two tandem tryptophan codons, and so forth (Figure 29.18). Translation of these codons depends on an adequate supply of the relevant aminoacyl-tRNA, which in turn rests on the availability of the amino acid. When tryptophan is scarce, the entire *trp* operon from *trpL* to *trpA* is transcribed to give a polycistronic mRNA. But, as [Trp] increases, more and more of the *trp* transcripts consist of only a 140-nucleotide fragment corresponding to the

FIGURE 29.18 Amino acid sequences of leader peptides in various amino acid biosynthetic operons regulated by attenuation. Color indicates amino acids synthesized in the pathway catalyzed by the operon's gene products. (The *ilv* operon encodes enzymes of isoleucine, leucine, and valine biosynthesis.)

Operon	Amino acid Sequence
his	Met—Thr—Arg—Val—Gln—Phe—Lys—His—His—His—His—His—His—His—Pro—Asp
ilv	Met—Thr—Ala—Leu—Leu—Arg—Val—Ile—Ser—Leu—Val—Val—Ile—Ser—Val—Val—Val—Ile—Ile—Ile—Pro—Pro—Cys—Gly—Ala—Ala—Leu—Gly—Arg—Gly—Lys—Ala
leu	Met—Ser—His—Ile—Val—Arg—Phe—Thr—Gly—Leu—Leu—Leu—Leu—Asn—Ala—Phe—Ile—Val—Arg—Gly—Arg—Pro—Val—Gly—Gly—Ile—Gln—His
pheA	Met—Lys—His—Ile—Pro—Phe—Phe—Phe—Ala—Phe—Phe—Phe—Thr—Phe—Pro
thr	Met—Lys—Arg—Ile—Ser—Thr—Thr—Ile—Thr—Thr—Thr—Ile—Thr—Ile—Thr—Thr—Gln—Asn—Gly—Ala—Gly
trp	Met—Lys—Ala—Ile—Phe—Val—Leu—Lys—Gly—Trp—Trp—Arg—Thr—Ser

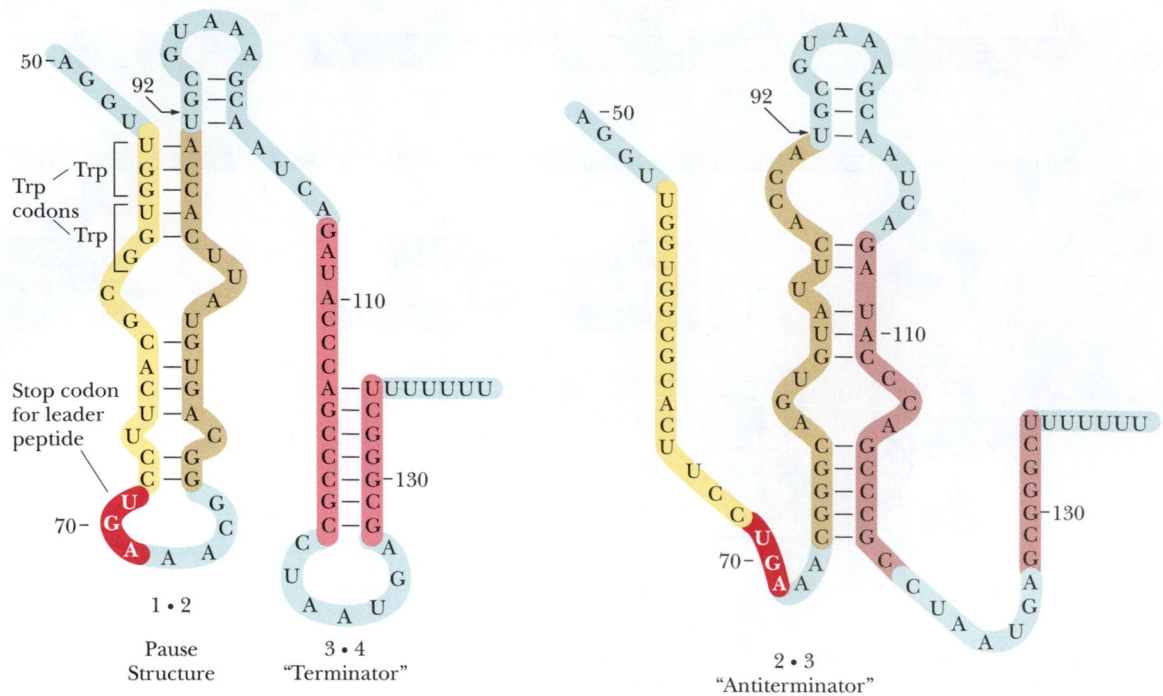

FIGURE 29.19 Alternative secondary structures for the leader region (*trpL* mRNA) of the *trp* operon transcript.

5'-end of *trpL*. Tryptophan availability is causing premature termination of *trp* transcription, that is, transcription attenuation. Although attenuation occurs when tryptophan is abundant, attenuation is blocked when levels of tryptophan are low and little tryptophanyl-tRNA is available. The secondary structure of the 160-bp leader region transcript is the principal control element in transcription attenuation (Figure 29.19). This RNA segment includes the coding region for the 14-residue leader peptide. Three critical base-paired hairpins can form in this RNA: the **1:2 pause** structure, the **3:4 terminator,** and the **2:3 antiterminator.** Obviously, the 1:2 pause, 3:4 terminator, and the 2:3 antiterminator represent mutually exclusive alternatives. A significant feature of this coding region is the tandem UGG tryptophan codons.

Transcription of the *trp* operon by RNA polymerase begins and progresses until position 92 is reached, whereupon the 1:2 hairpin is formed, causing RNA polymerase to pause in its elongation cycle. While RNA polymerase is paused, a ribosome begins to translate the leader region of the transcript. Translation by the ribosome releases the paused RNA polymerase and transcription continues, with RNA polymerase and the ribosome moving in unison. As long as tryptophan is plentiful enough that tryptophanyl-tRNA^{Trp} is not limiting, the ribosome is not delayed at the two tryptophan codons and follows closely behind RNA polymerase, translating the message soon after it is transcribed. The presence of the ribosome atop segment 2 blocks formation of the 2:3 antiterminator hairpin, allowing the alternative 3:4 terminator hairpin to form (Figure 29.20). Stable hairpin structures followed by a run of Us are features typical of *rho*-independent transcription termination signals, so the RNA polymerase perceives this hairpin as a transcription stop signal and transcription is terminated at this point. On the other hand, a paucity of tryptophan and hence low availability of tryptophanyl-tRNA^{Trp} causes the ribosome to stall on segment 1. This leaves segment 2 free to pair with segment 3 and to form the 2:3 antiterminator hairpin in the transcript. Because this hairpin precludes formation of the 3:4 terminator, termination is prevented and the entire operon is transcribed. Thus, transcription attenuation is determined by the availability of tyrptophanyl-tRNA^{Trp} and its transitory influence over the formation of alternative secondary structures in the mRNA.

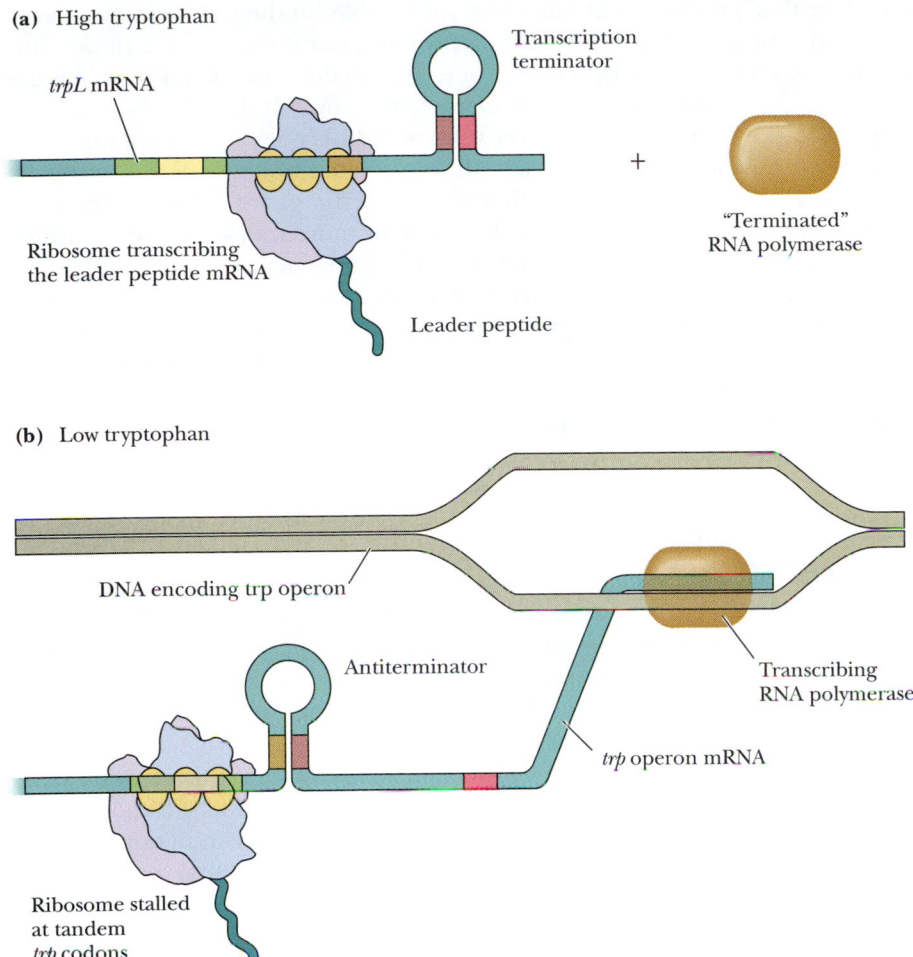

(a) High tryptophan

trpL mRNA

Transcription terminator

+

"Terminated" RNA polymerase

Ribosome transcribing the leader peptide mRNA

Leader peptide

(b) Low tryptophan

DNA encoding trp operon

Antiterminator

Transcribing RNA polymerase

trp operon mRNA

Ribosome stalled at tandem *trp* codons

FIGURE 29.20 The mechanism of attenuation in the *trp* operon.

DNA:Protein Interactions and Protein:Protein Interactions Are Essential to Transcription Regulation

Quite a variety of control mechanisms regulate transcription in prokaryotes. Several organizing principles materialize. First, **DNA:protein interactions** are a central feature in transcriptional control, and the DNA sites where regulatory proteins bind commonly display at least partial dyad symmetry or inverted repeats. Furthermore, DNA-binding proteins themselves are generally even-numbered oligomers (for example, dimers, tetramers) that have an innate twofold rotational symmetry. Second, **protein:protein interactions** are an essential component of transcriptional activation. We see this latter feature in the activation of RNA polymerase by CAP–(cAMP)$_2$, for example. Third, the regulator proteins receive cues that signal the status of the environment (for example, Trp, lactose, cAMP) and act to communicate this information to the genome, typically via the medium of conformational changes and DNA:protein interactions.

Proteins That Activate Transcription Work Through Protein:Protein Contacts with RNA Polymerase

Although transcriptional control is governed by a variety of mechanisms, an underlying principle of transcriptional activation has emerged. Transcriptional activation can take place when a **transcriptional activator** protein [such as CAP–(cAMP)$_2$] bound to DNA makes protein:protein contacts with RNA polymerase, and the degree of transcriptional activation is proportional to the strength of the protein:protein interaction. Generally speaking, a nucleotide sequence that provides a binding site for a DNA-binding protein can serve as an **activator site** if the DNA-binding protein bound there can interact with promoter-bound RNA polymerase. These interactions can involve either the α-, β-, β'-, or σ-subunits of RNA polymerase. Moreover, if the DNA-bound transcriptional activator makes contacts with two different components of RNA polymerase, a synergistic effect takes place such that transcription is markedly elevated. Thus, transcriptional activation at specific genes relies on the presence of one or more activator sites where one or more transcriptional activator proteins can bind and make contacts with RNA polymerase bound at the promoter of the gene. Indeed, transcriptional activators may facilitate the recruitment and binding of RNA polymerase to the promoter. This general principle applies to transcriptional activation in both prokaryotic and eukaryotic cells. In eukaryotes, transcriptional activators typically have discrete domains of protein structure dedicated to DNA binding (DB domains) and transcriptional activation (TA domains).

DNA Looping Allows Multiple DNA-Binding Proteins to Interact with One Another

Because transcription must respond to a variety of regulatory signals, multiple proteins are essential for appropriate regulation of gene expression. These regulatory proteins are the **sensors** of cellular circumstances, and they communicate this information to the genome by binding at specific nucleotide sequences. However, DNA is virtually a one-dimensional polymer, and there is little space for a lot of proteins to bind at (or even near) a transcription initiation site. DNA looping permits additional proteins to convene at the initiation site and to exert their influence on creating and activating an RNA polymerase initiation complex (Figure 29.21). The number of participants in transcriptional regulation is greatly expanded by DNA looping.

| 29.3 | **How Are Genes Transcribed in Eukaryotes?** |

Although the mechanism of transcription in prokaryotes and eukaryotes is fundamentally similar, transcription is substantially more complicated in eukaryotes. The significant difference is that the DNA of eukaryotes is wrapped around

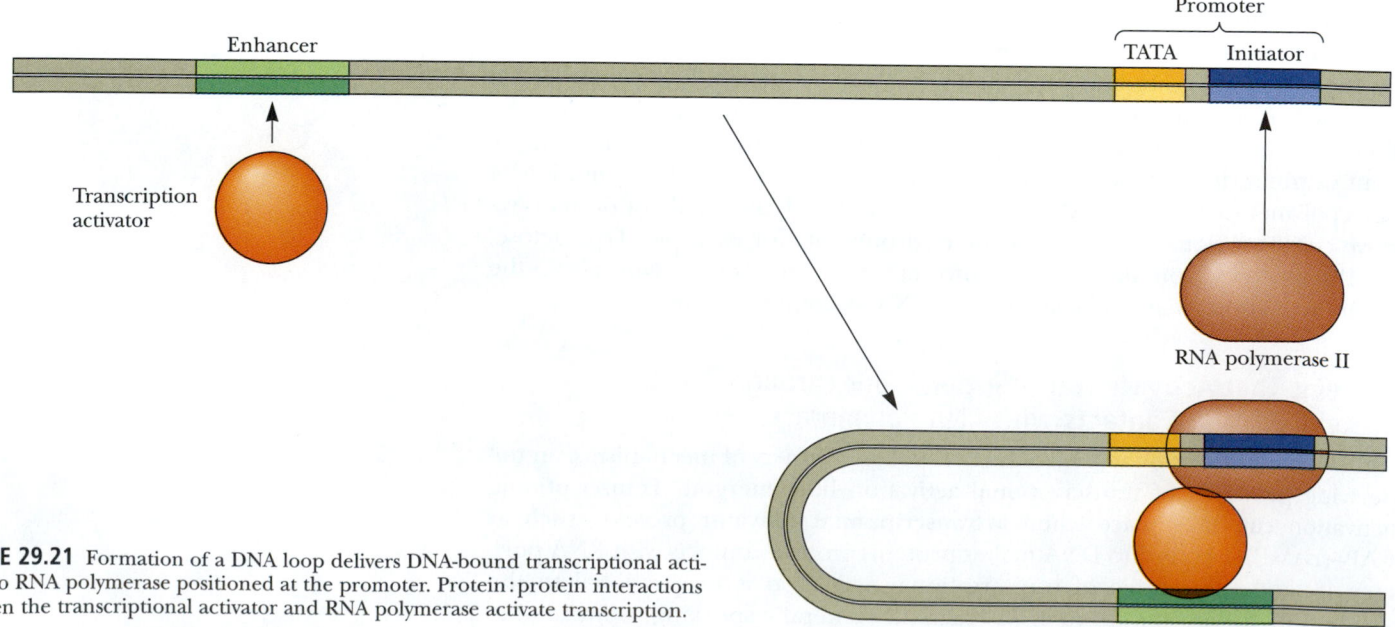

FIGURE 29.21 Formation of a DNA loop delivers DNA-bound transcriptional activator to RNA polymerase positioned at the promoter. Protein:protein interactions between the transcriptional activator and RNA polymerase activate transcription.

histones to form nucleosomes, and the nucleosomes are further organized into chromatin (see Chapter 11). *Nucleosomes repress gene expression.* Nucleosomes control gene expression by controlling access of the transcriptional apparatus to genes. Two classes of transcriptional co-regulators are necessary to overcome nucleosome repression: (1) enzymes that covalently modify the nucleosome histone proteins and thereby loosen histone:DNA interactions and (2) ATP-dependent chromatin-remodeling complexes. However, gene activation depends not only on relief from nucleosome repression but also on interaction of RNA polymerase with the promoter. Only those genes activated by specific positive regulatory mechanisms are transcribed. A general understanding of transcription in eukaryotes rests on the following topics:

- The three classes of RNA polymerase in eukaryotes: RNA polymerases I, II, and III
- The structure and function of RNA polymerase II, the mRNA-synthesizing RNA polymerase
- Transcription regulation in eukaryotes, including:
  - General features of gene regulatory sequences: promoters, enhancers, and response elements
  - Transcription initiation by RNA polymerase II
    The general transcription factors (GTFs)
    Alleviating the repression due to nucleosomes
      Histone acetyl transferases (HATs)
      Chromatin-remodeling complexes
- A general model for eukaryotic gene activation, based on the preceding

We turn now to a review of these various features of eukaryotic transcription.

Eukaryotes Have Three Classes of RNA Polymerases

Eukaryotic cells have three classes of RNA polymerase, each of which synthesizes a different class of RNA. All three enzymes are found in the nucleus. **RNA polymerase I** is localized to the nucleolus and transcribes the major ribosomal RNA genes. **RNA polymerase II** transcribes protein-encoding genes, and thus it is responsible for the synthesis of mRNA. **RNA polymerase III** transcribes tRNA genes, the ribosomal RNA genes encoding 5S rRNA, and a variety of other small RNAs, including several involved in mRNA processing and protein transport.

All three RNA polymerase types are large, complex multimeric proteins (500 to 700 kD), consisting of ten or more types of subunits. Although the three differ in overall subunit composition, they have several smaller subunits in common. Furthermore, all possess two large subunits (each 140 kD or greater) having sequence similarity to the large β- and β'-subunits of *E. coli* RNA polymerase, indicating that the fundamental catalytic site of RNA polymerase is conserved among its various forms.

In addition to their different functions, the three classes of RNA polymerase can be distinguished by their sensitivity to α-**amanitin** (Figure 29.22), a bicyclic octapeptide produced by the poisonous mushroom *Amanita phalloides* (the "destroying angel" mushroom). α-Amanitin blocks RNA chain elongation. Although RNA polymerase I is resistant to this compound, RNA polymerase II is very sensitive and RNA polymerase III is less sensitive.

The existence of three classes of RNA polymerases acting on three distinct sets of genes implies that at least three categories of promoters exist to maintain this specificity. Eukaryotic promoters are very different from prokaryotic promoters. All three eukaryotic RNA polymerases interact with their promoters via so-called **transcription factors**—DNA-binding proteins that recognize and accurately initiate transcription at specific promoter sequences. For RNA polymerase I, its templates are the rRNA genes. Ribosomal RNA genes are present in multiple copies. Optimal expression of these genes requires the first 150 nucleotides in the immediate 5′-upstream region.

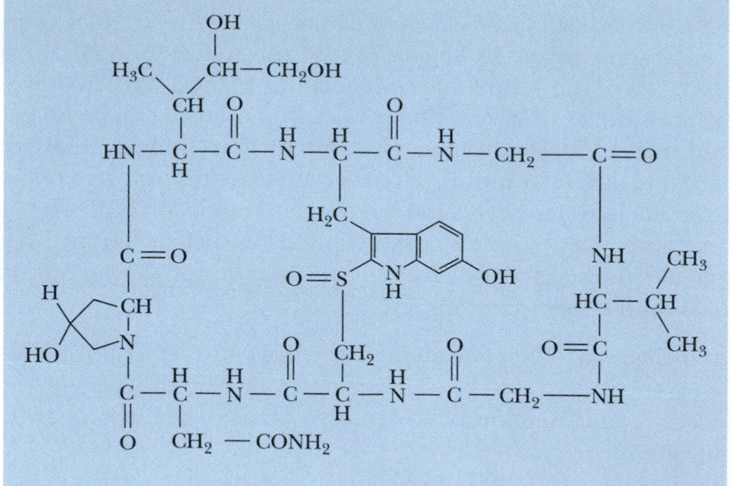

FIGURE 29.22 The structure of α-amanitin, one of a series of toxic compounds known as amatoxins that are found in the mushroom *Amanita phalloides*.

RNA polymerase III interacts with transcription factors **TFIIIA, TFIIIB,** and **TFIIIC.** Interestingly, TFIIIA and/or TFIIIC bind to specific recognition sequences that in some instances are located *within* the coding regions of the genes, not in the 5′-untranscribed region upstream from the transcription start site. TFIIIB associates with TFIIIA or TFIIIC already bound to the DNA. RNA polymerase III then binds to TFIIIB to establish an initiation complex.

RNA Polymerase II Transcribes Protein-Coding Genes

As the enzyme responsible for the regulated synthesis of mRNA, RNA polymerase II has aroused greater interest than RNA polymerases I and III. RNA polymerase II must be capable of transcribing a great diversity of genes, yet it must carry out its function at any moment only on those genes whose products are appropriate to the needs of the cell in its ever-changing metabolism and growth. The RNA polymerase II from yeast *(Saccharomyces cerevisiae)* has been extensively characterized, and its structure has been solved by x-ray crystallography (Figure 29.23). Strong homology between yeast and human RNA polymerase II subunits suggests that the yeast RNA polymerase II is an excellent model for human RNA polymerase II. The yeast RNA polymerase II consists of 12 different polypeptides, designated RPB1 through RPB12 and ranging in size from 192 to 8 kD (Table 29.2). RPB3, RPB4, and RPB7 are unique to RNA polymerase II, whereas RPB5, RPB6, RPB8, and RPB10 are common to all three eukaryotic RNA polymerases.

The RPB1 subunit has an unusual structural feature not found in prokaryotes: Its **C-terminal domain (CTD)** contains 27 repeats of the amino acid sequence YSPTSPS. (The analogous subunit in RNA polymerase II enzymes of other eukaryotes has this heptapeptide tandemly repeated as many as 52 times.) Note that the side chains of 5 of the 7 residues in this repeat have —OH groups, endowing the CTD with considerable hydrophilicity *and* multiple sites for phosphorylation. A number of CTD kinases have been described, targeting different residues at different stages of the transcription process. The CTD domain may project more than 50 nm from the surface of RNA polymerase II.

The CTD is essential to RNA polymerase II function. Only RNA polymerase II whose CTD is not phosphorylated can initiate transcription. However, transcription elongation proceeds only after protein phosphorylation within the CTD, suggesting that phosphorylation triggers the conversion of an initiation complex into an elongation complex. Such a mechanism would allow protein phosphorylation to regulate gene expression. Following termination of transcription, a phosphatase recycles RNA polymerase II to its unphosphorylated

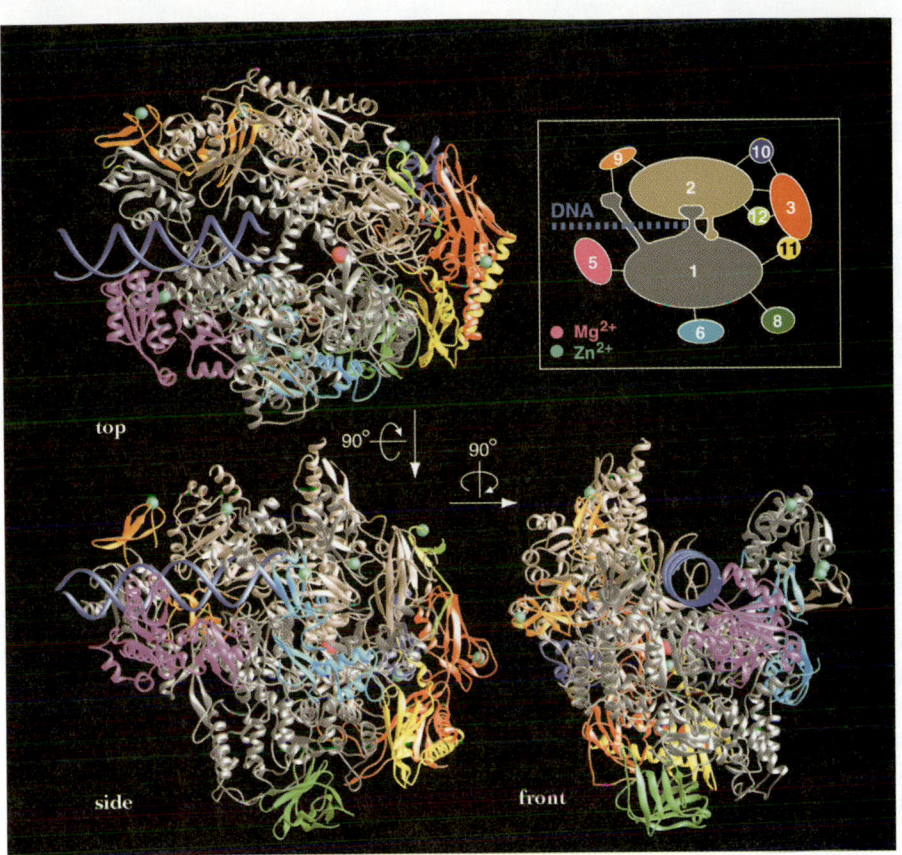

FIGURE 29.23 Crystal structure of a ten-subunit version of the yeast RNA polymerase II. (Subunits RPB4 and RPB7 are absent from this structure.) Protein is shown in the form of ribbon diagrams; a length of DNA corresponding to 20 base pairs is shown as a blue double-helical coil. The inset shows the color code and location of principal RPB subunits. The active-site Mg^{2+} atom (pink), as well as eight Zn^{2+} atoms (green) associated with this enzyme preparation, is also highlighted. (*Adapted from Figure 3 in Cramer, P., et al., 2000. Architecture of RNA polymerase II and implications for the transcription mechanism.* Science **288**:640–649. *Figure courtesy of Patrick Cramer and Roger D. Kornberg of Stanford University.*)

form. The CTD also plays a prominent role in orchestrating subsequent events in the transcription process. A multitude of additional proteins are essential to the formation of a translatable mRNA from the primary RNA polymerase II transcript; these proteins (described later in this chapter) include 5'-capping enzymes, splicing factors, and 3'-polyadenylylation complexes. All of these proteins are recruited to the transcript through interactions with the CTD as it is choreographed through various phosphorylated states.

Table 29.2			
Yeast* RNA Polymerase II Subunits			
Subunit	**Side (kD)**	**Features**	**Prokaryotic Homolog**
RPB1[†]	192	YSPTSPS CTD	β'
RPB2	139	NTP binding	β
RPB3	35	Core assembly	α
RPB4	25	Promoter recognition	σ
RPB5	25	In polymerases I, II, and III	
RPB6	18	In polymerases I, II, and III	
RPB7	19	Unique to polymerase II	
RPB8	17	In polymerases I, II, and III	
RPB9	14	Nonessential	
RPB10	8	In polymerases I, II, and III	
RPB11	14		
RPB12	8	In polymerases I, II, and III	

*A very similar RNA polymerase II can be isolated from human cells.
[†]RPB stands for RNA polymerase B; RNA polymerases I, II, and III are sometimes called RPA, RPB, and RPC.
Adapted from Myer, V. E., and Young, R. A., 1998. RNA polymerase II holoenzymes and subcomplexes. *The Journal of Biological Chemistry* **273**:27757–27760.

Transcription Regulation Is Much More Complex in Eukaryotes

Not only metabolic activity and cell division but also complex patterns of embryonic development and cell differentiation must be coordinated through transcriptional regulation. All this coordinated regulation takes place in cells where the relative quantity (and diversity) of DNA is very great: A typical mammalian cell has 1500 times as much DNA as an *E. coli* cell. The structural genes of eukaryotes are rarely organized in clusters akin to operons. Each eukaryotic gene typically possesses a discrete set of regulatory sequences appropriate to the requirements for its expression. Certain of these sequences provide sites of interaction for general transcription factors, whereas others endow the gene with great specificity in expression by providing targets for specific transcription factors.

Gene Regulatory Sequences in Eukaryotes Include Promoters, Enhancers, and Response Elements

RNA polymerase II promoters commonly consist of two separate sequence features: the **core** element, near the transcription start site, where **general transcription factors (GTFs)** bind, and more distantly located **regulatory elements,** known variously as **enhancers** (or **silencers**). These regulatory sequences are recognized by specific DNA-binding proteins that activate transcription above basal levels (*enhancers* bind *transcriptional activators*) or repress transcription (*silencers* bind *repressors*). The core region often consists of a **TATA box** (a TATAAA consensus element) indicating the transcription start site; the TATA motif is usually located at position −25 (Figure 29.24). Genes that lack a TATA often have an **initiator element,** or *Inr,* where transcription is initiated. The *Inr* sequence encompasses the transcription start site, but this sequence is not highly conserved among eukaryotic genes. Other regulatory elements include short nucleotide sequences (sometimes called **response elements**) found near the promoter (the *promoter-proximal region*) that can bind certain specific transcription factors, such as proteins that trigger expression of a related set of genes in response to some physiological signal (hormone) or challenge (temperature shock).

Promoters The promoters of eukaryotic genes encoding proteins can be quite complex and variable, but they typically contain modules of short conserved sequences, such as the TATA box, the CAAT box, and the GC box. Sets of such modules embedded in the upstream region collectively define the promoter. The presence of a CAAT box, usually located around −80 relative to the transcription start site, signifies a strong promoter. One or more copies of the sequence GGGCGG or its complement (referred to as the GC box) have been found upstream from the transcription start sites of "housekeeping genes." Housekeeping genes encode proteins commonly present in all cells and essential to normal function; such genes are typically transcribed at more or less steady levels. Figure 29.25 depicts the promoter regions of several representative eukaryotic genes. Table 29.3 lists

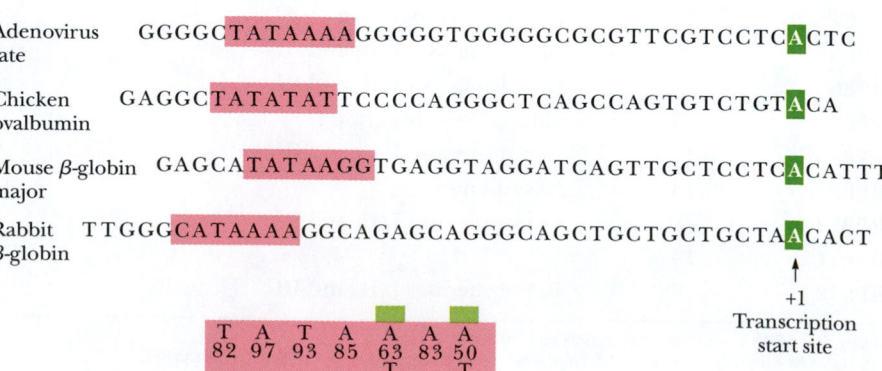

FIGURE 29.24 The TATA box in selected eukaryotic genes. The consensus sequence of a number of such promoters is presented in the lower part of the figure, the numbers giving the percent occurrence of various bases at the positions indicated.

(a)

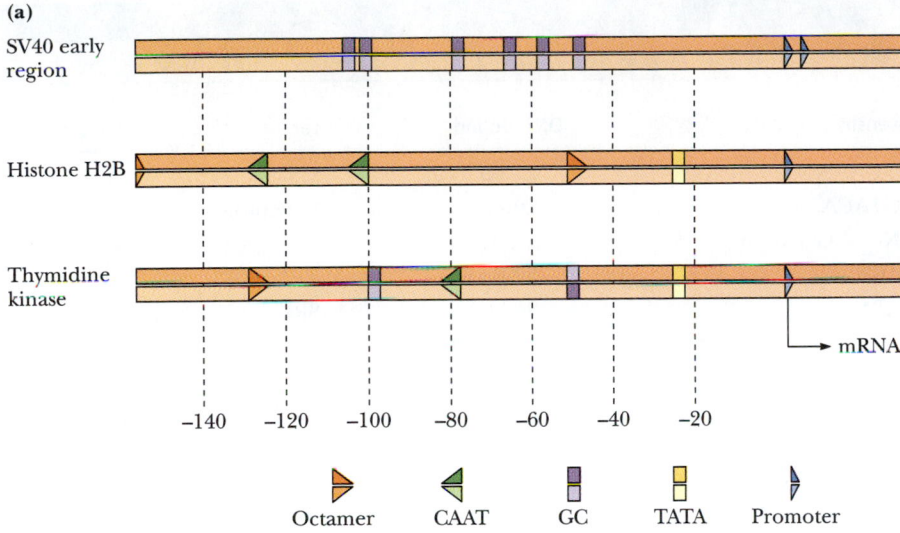

(b)

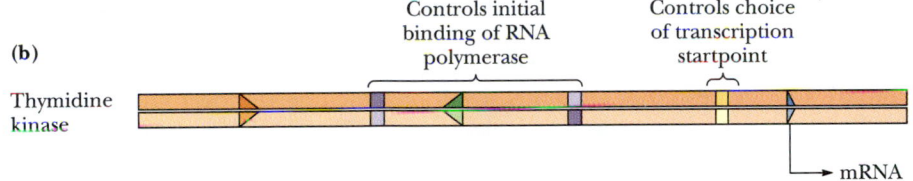

FIGURE 29.25 Promoter regions of several representative eukaryotic genes. (a) The SV40 early genes, the histone H2B gene, and the thymidine kinase gene. Note that these promoters contain different combinations of the various modules. In (b), the function of the modules within the thymidine kinase gene is shown.

transcription factors that bind to respective modules. These transcription factors typically behave as positive regulatory proteins essential to transcriptional activation by RNA polymerase II at these promoters.

Enhancers   Eukaryotic genes have, in addition to promoters, regulatory sequences known as **enhancers.** Enhancers (also called **upstream activation sequences,** or **UASs**) assist initiation. Enhancers differ from promoters in two fundamental ways. First, the location of enhancers relative to the transcription start site is not fixed. Enhancers may be several thousand nucleotides away from the promoter, and they act to enhance transcription initiation even if positioned *downstream* from the gene. Second, enhancer sequences are *bidirectional* in that they function in either orientation. That is, enhancers can be removed and then

Table 29.3					
A Selection of Consensus Sequences That Define Various RNA Polymerase II Promoter Modules and the Transcription Factors That Bind to Them					
Sequence Module	Consensus Sequence	DNA Bound	Factor	Size (kD)	Abundance (molecules/cell)
TATA box	TATAAAA	~10 bp	TBP	27	?
CAAT box	GGCCAATCT	~22 bp	CTF/NF1	60	300,000
GC box	GGGCGG	~20 bp	SP1	105	60,000
Octamer	ATTTGCAT	~20 bp	Oct-1	76	?
"	"	23 bp	Oct-2	52	?
κB	GGGACTTTCC	~10 bp	NFκB	44	?
"	"	~10 bp	H2-TF1	?	?
ATF	GTGACGT	~20 bp	ATF	?	?

Adapted from Lewin, B., 1994. *Genes V.* Cambridge, MA: Cell Press.

Table 29.4

Response Elements That Identify Genes Coordinately Regulated in Response to Particular Physiological Challenges

Physiological Challenge	Response Element	Consensus Sequence	DNA Bound	Factor	Size (kD)
Heat shock	HSE	CNNGAANNTCCNNG	27 bp	HSTF	93
Glucocorticoid	GRE	TGGTACAAATGTTCT	20 bp	Receptor	94
Cadmium	MRE	CGNCCCGGNCNC			
Phorbol ester	TRE	TGACTCA	22 bp	AP1	39
Serum	SRE	CCATATTAGG	20 bp	SRF	52

Adapted from Lewin, B., 1994. *Genes V.* Cambridge, MA: Cell Press.

reinserted in the reverse sequence orientation without impairing their function. Like promoters, enhancers represent modules of consensus sequence. Enhancers are "promiscuous," because they stimulate transcription from any promoter that happens to be in their vicinity. Nevertheless, *enhancer function is dependent on recognition by a specific transcription factor.* A specific transcription factor bound at an enhancer element stimulates transcription by interacting with RNA polymerase II at a nearby promoter.

Response Elements   Promoter modules in genes responsive to common regulation are termed **response elements.** Examples include the **heat shock element (HSE),** the **glucocorticoid response element (GRE),** and the **metal response element (MRE).** These various elements are found in the promoter regions of genes whose transcription is activated in response to a sudden increase in temperature (heat shock), glucocorticoid hormones, or toxic heavy metals, respectively (Table 29.4). HSE sequences are recognized by a specific transcription factor, **HSTF** (for **heat shock transcription factor**). HSEs are located about 15 bp upstream from the transcription start site of a variety of genes whose expression is dramatically enhanced in response to elevated temperature. Similarly, the response to steroid hormones depends on the presence of a GRE positioned 250 bp upstream of the transcription start point. Activation of the **steroid receptor** (a specific transcription factor) at a GRE occurs when certain steroids bind to the steroid receptor.

Many genes are subject to multiple regulatory influences. Regulation of such genes is achieved through the presence of an array of different regulatory elements. The **metallothionein** gene is a good example (Figure 29.26). Metallothionein is a metal-binding protein that protects cells against metal toxicity by binding excess amounts of heavy metals and removing them from the cell. This protein is always present at low levels, but its concentration increases in response to heavy metal ions such as cadmium or in response to glucocorticoid hormones. The metallothionein gene promoter consists of two general promoter elements, namely, a TATA box and a GC box, two basal-level enhancers, four MREs, and one GRE. These elements function independently of one another; any one is able to activate transcription of the gene.

Biochemistry Now™   ANIMATED FIGURE 29.26
The metallothionein gene possesses several constitutive elements in its promoter (the TATA and GC boxes) as well as specific response elements such as MREs and a GRE. The BLEs are elements involved in basal level expression (constitutive expression). TRE is a tumor response element activated in the presence of tumor-promoting phorbol esters such as TPA (tetradecanoyl phorbol acetate). **See this figure animated at http://chemistry.brookscole.com/ggb3**

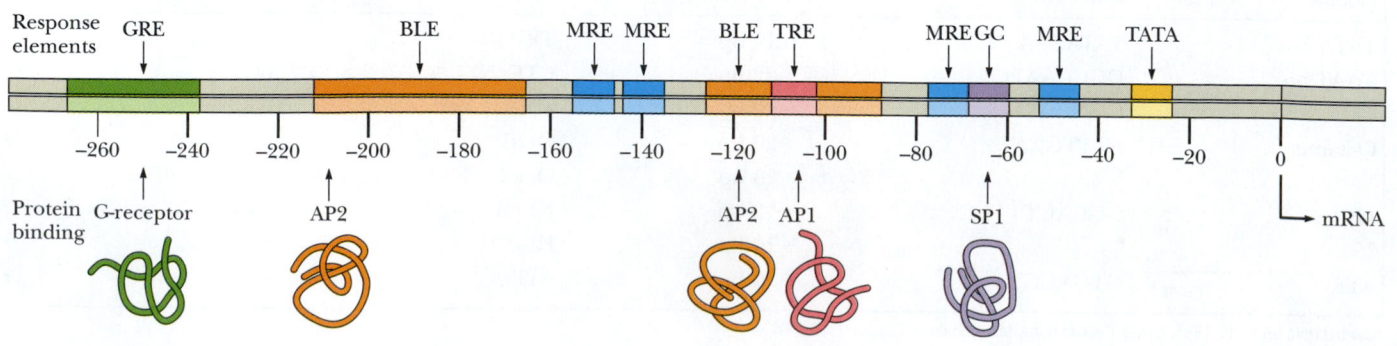

Transcription Initiation by RNA Polymerase II Requires TBP and the GTFs

A eukaryotic transcription initiation complex consists of RNA polymerase II, five **general transcription factors (GTFs),** and an 18-subunit complex called **Mediator** (or **Srb/Med**). The CTD of RNA polymerase II anchors Mediator to the polymerase. Mediator allows RNA polymerase II to communicate with transcriptional activators bound at sites distal from the promoter. There are six GTFs (Table 29.5), five of which are required for transcription: **TFIIB, TFIID, TFIIE, TFIIF,** and **TFIIH.** The sixth, **TFIIA,** stimulates transcription by stabilizing the interaction of TFIID with the TATA box. TFIID consists of **TBP** (TATA-**b**inding **p**rotein), which directly recognizes the TATA box within the core promoter, and a set of **T**BP-**a**ssociated **f**actors (**TAFs** or **TAF$_{II}$s**).[6] The TBP–TAF$_{II}$ complexes serve as a bridge between the promoter and RNA polymerase II. Some are capable of recognizing core promoters lacking a TATA box. TBP binds to the core promoter through contacts made with the minor groove of the DNA, distorting and bending the DNA so that DNA sequences upstream and downstream of the TATA box come into closer proximity (Figure 29.27a). Once TBP/TFIID is bound at the core promoter, a complex containing RNA polymerase II and the remaining GTFs convenes at this site, establishing a competent transcription *preinitiation complex* (Figure 29.27b). An *open complex* then forms, and transcription begins.

Chromatin-Remodeling Complexes and HATs Alleviate the Repression Due to Nucleosomes

The central structural unit of nucleosomes, the histone "core octamer" (Figure 11.27), is constructed from the eight *histone-fold protein domains* of the eight various histone monomers comprising the octamer. Successive histone octamers are linked via histone H1, which is not part of the octamer. Each histone monomer in the core octamer has an unstructured N-terminal tail that extends outside the core octamer. Interactions between histone tails contributed by core histones in adjacent nucleosomes are an important influence in establishing the higher orders of chromatin organization. Activation of eukaryotic transcription is dependent on *two* sets of circumstances: (1) relief from the repression imposed by chromatin structure and (2) interaction of

---

[6]For many genes, another transcription factor called SAGA (which also contains TAF$_{II}$s) can serve instead of TFIID to initiate transcription.

Table 29.5		
General Transcription Initiation Factors from Human Cells		
Factor	**Number of Subunits**	**Function**
TFIID		
TBP	1	Core promoter recognition (TATA); TFIIB recruitment
TAFs	14	Core promoter recognition (non-TATA elements); positive and negative regulatory functions; HAT (histone acetyltransferase) activity
TFIIA	3	Stabilization of TBP binding; stabilization of TAF–DNA interactions
TFIIB	1	RNA polymerase II-TFIIF recruitment; start-site selection by RNA polymerase II
TFIIF	2	Promoter targeting of polymerase II; destabilization of nonspecific RNA polymerase II–DNA interactions
RNA pol II	12	Enzymatic synthesis of RNA; TFIIE recruitment
TFIIE	2	TFIIH recruitment; modulation of TFIIH helicase, ATPase, and kinase activities; promoter melting
TFIIH	9	Promoter melting using helicase activity; promoter clearance via CTD phosphorylation (2 subunits of TFHII are a cyclin : CDK pair)

Adapted from Table 1 in Roeder, R. G., 1996. The role of general initiation factors in transcription by RNA polymerase II. *Trends in Biochemical Sciences* **21**:327–335; and Reese, J. C., 2003. Basal transcription factors. *Current Opinion in Genetics and Development* **13**:114–118.

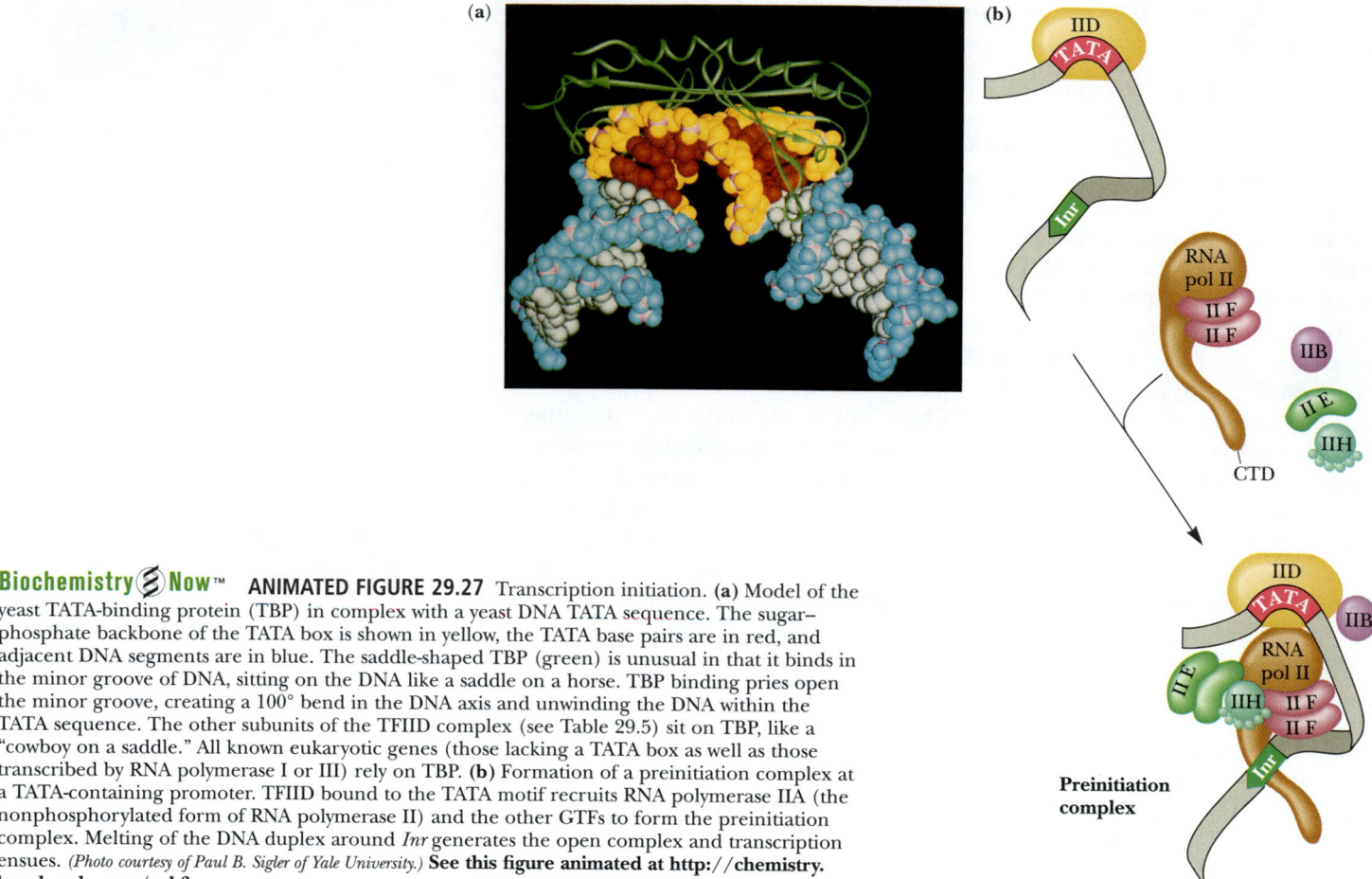

Biochemistry Now™   ANIMATED FIGURE 29.27 Transcription initiation. **(a)** Model of the yeast TATA-binding protein (TBP) in complex with a yeast DNA TATA sequence. The sugar–phosphate backbone of the TATA box is shown in yellow, the TATA base pairs are in red, and adjacent DNA segments are in blue. The saddle-shaped TBP (green) is unusual in that it binds in the minor groove of DNA, sitting on the DNA like a saddle on a horse. TBP binding pries open the minor groove, creating a 100° bend in the DNA axis and unwinding the DNA within the TATA sequence. The other subunits of the TFIID complex (see Table 29.5) sit on TBP, like a "cowboy on a saddle." All known eukaryotic genes (those lacking a TATA box as well as those transcribed by RNA polymerase I or III) rely on TBP. **(b)** Formation of a preinitiation complex at a TATA-containing promoter. TFIID bound to the TATA motif recruits RNA polymerase IIA (the nonphosphorylated form of RNA polymerase II) and the other GTFs to form the preinitiation complex. Melting of the DNA duplex around *Inr* generates the open complex and transcription ensues. *(Photo courtesy of Paul B. Sigler of Yale University.)* **See this figure animated at http://chemistry. brookscole.com/ggb3**

RNA polymerase II with the promoter and transcription regulatory proteins. Relief from repression requires factors that can reorganize the chromatin and then alter the nucleosomes so that promoters become accessible to the transcriptional machinery. Two such factors are important: **chromatin-remodeling complexes** that mediate ATP-dependent conformational (noncovalent) changes in nucleosome structure and **HATs** that covalently modify histones. (*HATs* is the acronym for histone acetyltransferases, enzymes that acetylate the ϵ-NH_3^+ groups of lysine residues in the histone tails).

HATs   Initial events in transcriptional activation include acetyl-CoA–dependent acetylation of histone tails by HATs (Figure 29.28). The histone transacetylases responsible are essential components of several megadalton-size complexes known to be required for transcription co-activation (*co-activation* in the sense that they are required along with RNA polymerase II and other components of the transcriptional apparatus). Examples of such complexes include the **TFIID** (some of whose TAF_{II}s have HAT activity), the **SAGA complex** (which also contains TAF_{II}s), and the **ADA complex.** *N*-Acetylation suppresses the positive charge in histone tails, diminishing their interaction with the negatively charged DNA. Phosphorylation of Ser residues and methylation of Lys residues in histone tails also contribute to transcription regulation (Figure 29.28). Along with lysine acetylations, these modifications create binding sites for proteins that modulate chromatin structure (such as the chromatin-remodeling complexes). Such proteins have domains that recognize covalently modified histone sites, such as **bromodomains** that interact specifically with acetylated lysine residues and **chromodomains** that bind to methylated lysine residues. A "histone code" hypothesis has emerged, suggesting that a code based on various covalent modifications of histone tails deter-

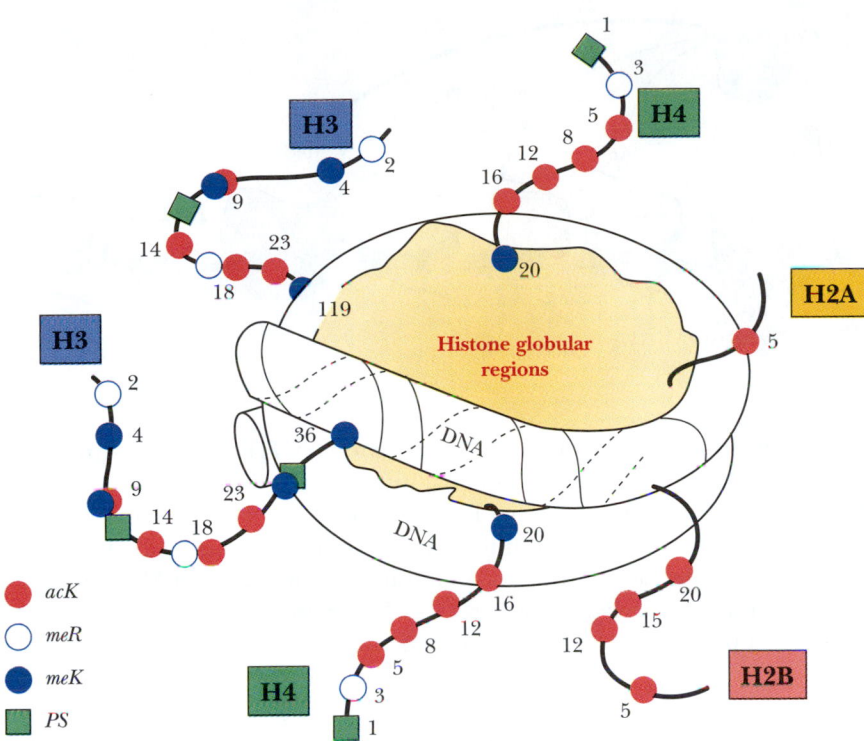

FIGURE 29.28 A schematic diagram of the nucleosome illustrating the various covalent modifications on histone tails. *acK* = acetylated lysine residue; *meK* = methylated lysine residue; *PS* = phosphorylated serine residue. The numbers indicate the positions of the amino acids in the amino acid sequences. Note the prevalence of modifiable sites, particularly acetylatable lysines, on the N-terminal tails of histones H2B, H3, and H4. (*Adapted from Figure 1 in Turner, B. M., 2002. Cellular memory and the histone code.* Cell **111**:285–291.)

mines gene expression through protein recruitment. Such a histone code could be quite versatile in that different combinations of modifications could recruit different sets of regulatory proteins. Deacetylation of histones is a biologically relevant matter, and enzyme complexes that carry out such reactions have been characterized. Known as **histone deacetylase complexes,** or **HDACs,** they catalyze the removal of acetyl groups from lysine residues along the histone tails, restoring the chromatin to a repressed state.

Chromatin-Remodeling Complexes Although acetylation of histone tails disrupts chromatin structure, it does not expose the DNA within nucleosome "core" particles for transcription. Such exposure requires *chromatin-remodeling complexes,* huge (1 to 2 megadalton) multisubunit entities that mediate ATP-dependent conformational changes that peel about 50 bp of DNA from the edge of nucleosome core particle and create a "bulge" in the DNA:core nucleosome association. These changes allow DNA-binding proteins such as RNA polymerase II, GTFs, and other transcription factors to gain access to the DNA. Prominent among the various chromatin-remodeling complexes are **SWI/SNF**, **RSC,** and **ISWI.**

Nucleosome Alteration and Interaction of RNA Polymerase II with the Promoter Are Two Essential Features in Eukaryotic Gene Activation

Gene activation (the initiation of transcription) can thus be viewed as a process requiring two principal steps: (1) alterations in nucleosomes (and thus, chromatin) that relieve the general repressed state imposed by chromatin structure, followed by (2) the interaction of RNA polymerase II and the GTFs with the promoter. **Transcription activators** (proteins that bind to enhancers and response elements) initiate the process by recruiting chromatin-altering proteins (the *chromatin-remodeling complexes* and *histone acetyltransferases* described previously). Once these alterations have occurred, promoter DNA is accessible to TBP:TFIID, the other GTFs, and RNA polymerase II. Transcription activation, however, requires communication between RNA polymerase II and the *transcription activator* for transcription to take place. Mediator (or Srb/Med) fulfills this function. Mediator

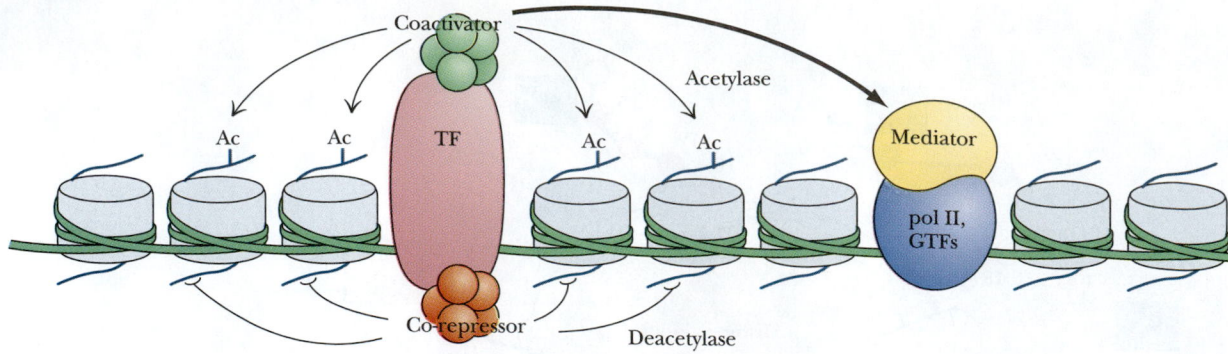

Biochemistry ⑤ Now™   ACTIVE FIGURE 29.29
A model for the transcriptional regulation of eukaryotic genes. The DNA is a dark blue ribbon wrapped
around disc-like nucleosomes. A specific transcription
factor (TF, blue) is bound to a regulatory element
(either an enhancer or silencer). RNA polymerase II
and its associated GTFs (also blue) are bound at the
promoter. The N-terminal tails of histones are shown
as wavy lines (gray) emanating from the nucleosome
discs. A specific transcription factor that is a transcription activator stimulates transcription through interaction with a co-activator whose HAT activity renders
the DNA more accessible *and* through interactions
with the Mediator complex associated with RNA polymerase II. A specific transcription factor that is a repressor interacts with a co-repressor that has HDAC
activity that deacetylates histones, restructuring the
nucleosomes into a repressed state. *(From Figure 1 in
Kornberg, R. D., 1999. Eukaryotic transcription control.* Trends in
Biochemical Sciences *24:M46–M49.)* **Test yourself on the
concepts in this figure at http://chemistry.brookscole.
com/ggb3**

interacts with *both* the transcription activator *and* the CTD of RNA polymerase II.
This Mediator bridge provides an essential interface for communication between
enhancers and promoters, triggering RNA polymerase II to begin transcription. A
general model for transcription initiation is shown in Figure 29.29. Once transcription begins, Mediator is replaced by another complex called **Elongator.** *Elongator* has HAT subunits whose activity remodels downstream nucleosomes as RNA
polymerase II progresses along the chromatin-associated DNA.

29.4 | How Do Gene Regulatory Proteins Recognize Specific DNA Sequences?

Proteins that recognize nucleic acids do so by the basic rule of macromolecular
recognition. That is, such proteins present a three-dimensional shape or contour
that is structurally and chemically complementary to the surface of a DNA sequence. When the two molecules come into close contact, the numerous atomic
interactions that underlie recognition and binding can take place. Nucleotide
sequence–specific recognition by the protein involves a set of atomic contacts with
the bases and the sugar–phosphate backbone. Hydrogen bonding is critical for
recognition, with amino acid side chains providing most of the critical contacts
with DNA. Protein contacts with the bases of DNA usually occur within the major
groove (but not always). Protein contacts with the DNA backbone involve both
H bonds and salt bridges with electronegative oxygen atoms of the phosphodiester linkages. Structural studies on regulatory proteins that bind to specific

Human Biochemistry

Storage of Long-Term Memory Depends on Gene Expression Activated by CREB-Type Transcription Factors

Learning can be defined as the process whereby new information is
acquired and *memory* as the process by which this information is retained. Short-term memory (which lasts minutes or hours) requires
only the covalent modification of preexisting proteins, but long-
term memory (which lasts days, weeks, or a lifetime) depends on
gene expression, protein synthesis, and the establishment of new
neuronal connections.

The macromolecular synthesis underlying long-term memory
storage requires **cAMP-response element-binding (CREB)** protein–
related transcription factors and the activation of cAMP-dependent
gene expression. Serotonin (5-hydroxytryptamine, or 5-HT, a hormone implicated in learning and memory) acting on neurons promotes cAMP synthesis, which in turn stimulates protein kinase A to

phosphorylate CREB protein–related transcription factors that
activate transcription of cAMP-inducible genes. These genes are
characterized by the presence of **CRE** (**c**AMP **r**esponse **e**lement)
consensus sequences containing the 8-bp TGACGTCA palindrome.
CREB transcription factors are *bZIP*-type proteins (see later discussion). These exciting findings opened a new arena in molecular
biology, the molecular biology of **cognition.** Eric Kandel was
awarded the 2000 Nobel Prize in Medicine for, among other things,
his discovery of the role of CREB-type transcription factors in long-
term memory storage.

Cognition is the act or process of knowing; the acquisition of knowledge.

DNA sequences have revealed that roughly 80% of such proteins can be assigned to one of three principal classes based on their possession of one of three kinds of small, distinctive structural motifs: the **helix-turn-helix** (or **HTH**), the **zinc finger** (or **Zn-finger**), and the **leucine zipper-basic region** (or *bZIP*). The latter two motifs are found only in DNA-binding proteins from eukaryotic organisms.

In addition to their DNA-binding domains, these proteins commonly possess other structural domains that function in protein:protein recognitions essential to oligomerization (for example, dimer formation), DNA looping, transcriptional activation, and signal reception (for example, effector binding).

α-Helices Fit Snugly into the Major Groove of B-DNA

A recurring structural feature in DNA-binding proteins is the presence of α-helical segments that fit directly into the major groove of B-form DNA. The diameter of an α-helix (including its side chains) is about 1.2 nm. The dimensions of the major groove in B-DNA are 1.2 nm wide by 0.6 to 0.8 nm deep. Thus, one side of an α-helix can fit snugly into the major groove. Although examples of β-sheet DNA recognition elements in proteins are known, the α-helix and B-form DNA are the predominant structures involved in protein:DNA interactions. Significantly, proteins can recognize specific sites in "normal" B-DNA; the DNA need not assume any unusual, alternative conformation (such as Z-DNA).

Proteins with the Helix-Turn-Helix Motif Use One Helix to Recognize DNA

The HTH motif is a protein structural domain consisting of two successive α-helices separated by a sharp β-turn (Figure 29.30). Within this domain, the α-helix situated more toward the C-terminal end of the protein, the so-called **helix 3,** is the DNA recognition helix; it fits nicely into the major groove, with several of its side chains touching DNA base pairs. **Helix 2,** the helix at the beginning of the HTH motif, creates a stable structural domain through hydrophobic interactions with helix 3 that locks helix 3 into its DNA interface. Proteins with HTH motifs bind to DNA as dimers. In the dimer, the two helix 3 cylinders are antiparallel to each other, such that their N→C orientations match the inverted relationship of nucleotide sequence in the dyad-symmetric DNA-binding site. An example is **Antp**. *Antp* is a member of a family of eukaryotic proteins involved in the regulation of early embryonic development that have in common an amino acid sequence element known as the **homeobox**[6] **domain.** The homeobox is a DNA motif that encodes a related 60–amino acid sequence (the homeobox domain) found among proteins of virtually every eukaryote, from yeast to man. Embedded within the homeobox domain is an HTH motif (Figure 29.31). Homeobox domain proteins act as **sequence-specific transcription factors.** Typically, the homeobox portion comprises only 10% or so of the protein's mass, with the remainder of the protein serving in protein:protein interactions essential to transcription regulation.

How Does the Recognition Helix Recognize Its Specific DNA-Binding Site? The edges of base pairs in dsDNA present a pattern of hydrogen-bond donor and acceptor groups within the major and minor grooves, but only the pattern displayed on the major-groove side is distinctive for each of the four base pairs A:T, T:A, C:G, and G:C. (You can get an idea of this by inspecting the structures of the base pairs in Figure 11.7.) Thus, the base-pair edges in the major groove act as a **recognition matrix** identifiable through H bonding with a specific protein, so it is not necessary to melt the base pairs to read the base sequence. Although formation of such H bonds is very important in DNA:protein recognition, other interactions also play a significant role. For example,

[6]*Homeo* derives from homeotic genes, a set of genes originally discovered in the fruit fly *Drosophila melanogaster* through their involvement in the specification of body parts during development.

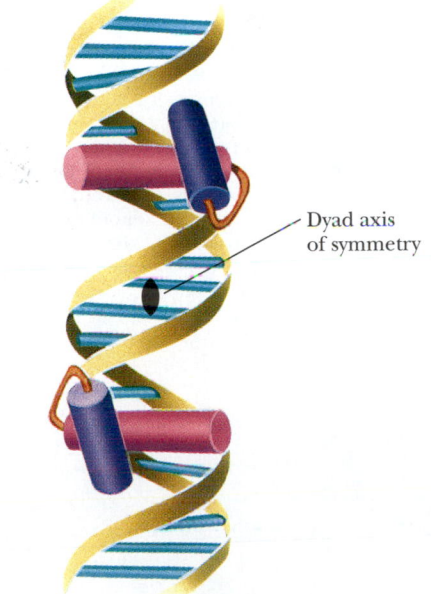

Dyad axis of symmetry

FIGURE 29.30 Schematic representation of the helix-turn-helix (HTH) motif. A dimeric HTH protein bound to a dyad-symmetric DNA site. (Note dyad axis of symmetry.) The recognition helix (helix 3, red) is represented as a dark barrel situated in the major groove of each half of the dyad site. Helix 2 sits above helix 3 and aids in locking it into place. (*Adapted from Figure 1 in Johnson, P. F., and McKnight, S. L., 1989. Eukaryotic transcriptional regulatory proteins.* Annual Review of Biochemistry **58**:799.)

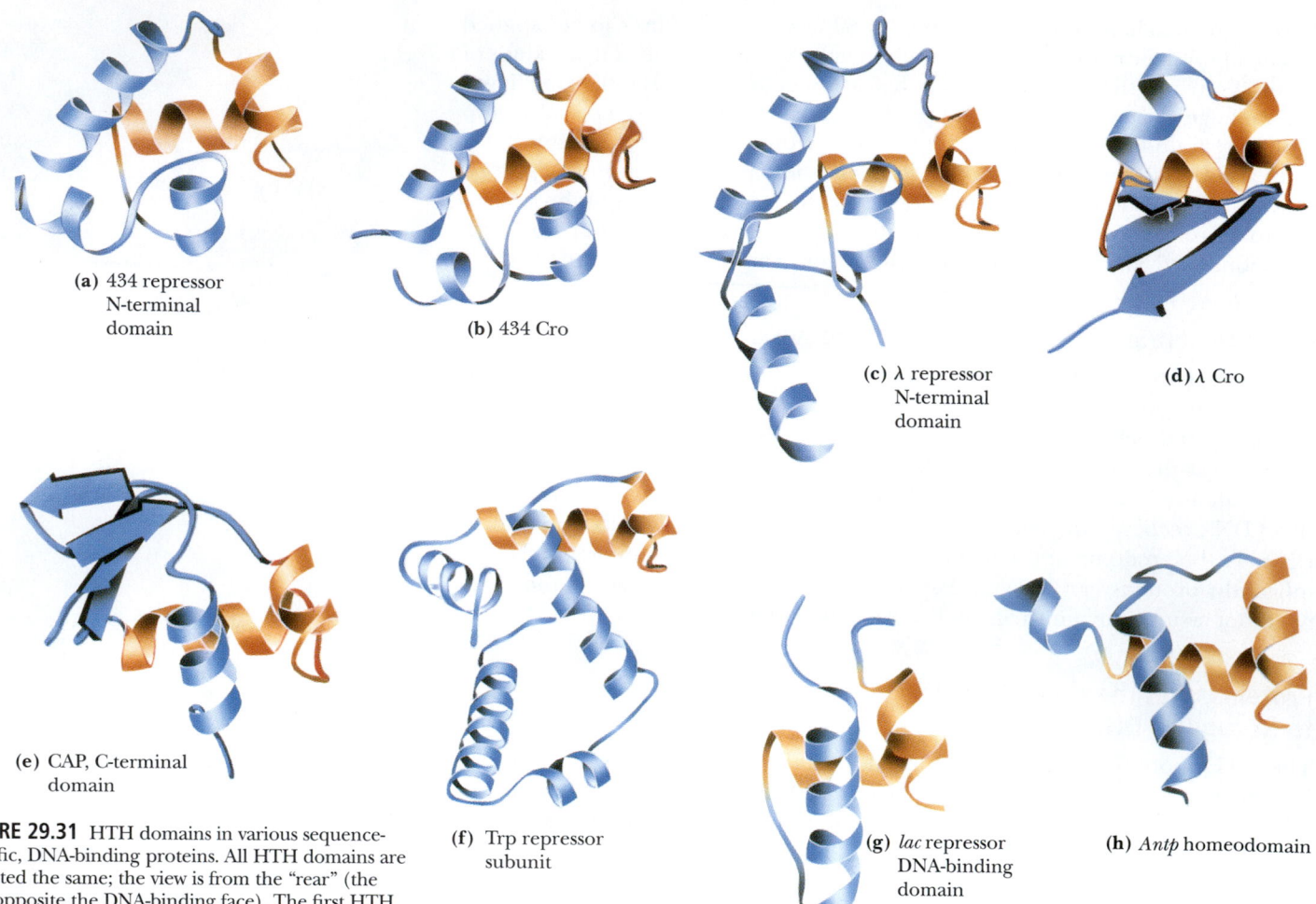

(a) 434 repressor N-terminal domain

(b) 434 Cro

(c) λ repressor N-terminal domain

(d) λ Cro

(e) CAP, C-terminal domain

(f) Trp repressor subunit

(g) *lac* repressor DNA-binding domain

(h) *Antp* homeodomain

FIGURE 29.31 HTH domains in various sequence-specific, DNA-binding proteins. All HTH domains are oriented the same; the view is from the "rear" (the side opposite the DNA-binding face). The first HTH helix runs vertically down from the upper right; the turn is right center, and the "recognition" helix runs across the center from right to left "behind" the domain in this perspective. **(a)** 434 repressor N-terminal domain; **(b)** 434 Cro; **(c)** λ repressor N-terminal domain; **(d)** λ Cro; **(e)** CAP, C-terminal domain; **(f)** Trp repressor subunit; **(g)** *lac* repressor DNA-binding domain; **(h)** *Antp* homeodomain. Domains (a) through (d) are in bacteriophage proteins, (e) through (g) are bacterial representatives, and (h) is a motif found in a eukaryotic DNA-binding protein. *(Adapted from Figure 2 in Harrison, S. C., and Aggarwal, A. K., 1990. DNA recognition by proteins with the helix-turn-helix motif. Annual Review of Biochemistry 59:933; and Pabo, C. O., and Sauer, R. T., 1984. Protein-DNA recognition. Annual Review of Biochemistry 53:293.)*

the C-5-methyl groups unique to thymine residues are nonpolar "knobs" projecting into the major groove.

Proteins Also Recognize DNA via "Indirect Readout"   **Indirect readout** is the term for the ability of a protein to indirectly recognize a particular nucleotide sequence by recognizing local conformational variations resulting from the effects that base sequence has on DNA structure. Superficially, the B-form structure of DNA appears to be a uniform cylinder. Nevertheless, the conformation of DNA over a short distance along its circumference varies subtly according to local base sequence. That is, base sequences generate unique contours that proteins can recognize. Because these contours arise from the base sequence, the DNA-binding protein "indirectly reads out" the base sequence through interactions with the DNA backbone. In the *E. coli* Trp repressor:*trp* operator DNA complex, the Trp repressor engages in 30 specific hydrogen bonds to the DNA: 28 involve phosphate groups in the backbone; only 2 are to bases. Thus, some sequence-specific DNA-binding proteins are able to recognize an overall DNA conformation caused by the specific DNA sequence.

Some Proteins Bind to DNA via Zn-Finger Motifs

There are many classes of Zn-finger motifs. The prototype Zn-finger is a structural feature formed by a pair of Cys residues separated by 2 residues, then a run of 12 amino acids, and finally a pair of His residues separated by 3 residues (Cys-x_2-Cys-x_{12}-His-x_3-His). This motif may be repeated as many as 13 times over the primary structure of a Zn-finger protein. Each repeat coordinates a zinc ion via its 2 Cys and 2 His residues (Figure 29.32). The 12 or so residues separating the Cys and His coordination sites are looped out and form a distinct DNA interaction module, the so-called Zn-finger. When Zn-finger proteins associate with DNA, each Zn-finger binds in the major groove and interacts with about five nucleotides, adjacent fingers interacting with contiguous stretches of DNA. Many DNA-binding proteins with this motif have been identified. In all cases, the finger motif is repeated at least two times, with at least a 7– to 8–amino acid linker between Cys/Cys and His/His sites. Proteins with this general pattern are assigned to the **C₂H₂ class** of Zn-finger proteins to distinguish them from proteins bearing another kind of Zn-finger, the **Cₓ type,** which includes the C_4 and C_6 Zn-finger proteins. The C_x proteins have a variable number of Cys residues available for Zn **chelation.** For example, the vertebrate steroid receptors have two sets of Cys residues, one with four conserved cysteines (C_4) and the other with five (C_5).

Chelation is from the Greek word *chele*, meaning "claw"; it refers to the binding of a metal ion to two or more nonmetallic atoms in the same molecule.

Some DNA-Binding Proteins Use a Basic Region-Leucine Zipper (bZIP) Motif

bZIP is a structural motif characterizing the third major class of sequence-specific, DNA-binding proteins. This motif was first recognized by Steve McKnight in **C/EBP,** a heat-stable, DNA-binding protein isolated from rat liver nuclei that binds to both CCAAT promoter elements and certain enhancer core elements.[7] The DNA-binding domain of C/EBP was localized to the C-terminal region of the protein. This region shows a notable absence of Pro residues, suggesting it might be arrayed in an α-helix. Within this region are two clusters of basic residues: A and B. Further along is a 28-residue sequence. When this latter region is displayed end-to-end down the axis of a hypothetical α-helix, beginning at Leu[315], an amphipathic cylinder is generated, similar to the one shown in Figure 6.24. One side of this amphipathic helix consists principally of hydrophobic residues (particularly leucines), whereas the other side has an array of negatively and positively charged side chains (Asp, Glu, Arg, and Lys), as well as many uncharged polar side chains (glutamines, threonines, and serines).

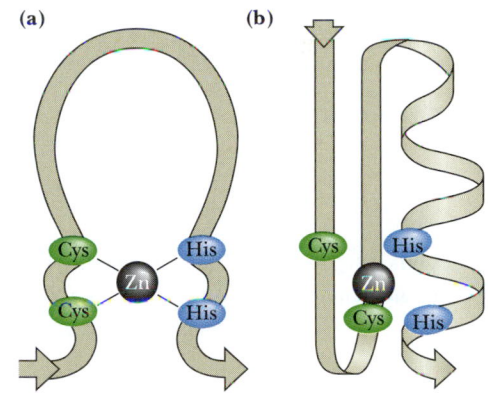

FIGURE 29.32 The Zn-finger motif of the C_2H_2 type showing **(a)** the coordination of Cys and His residues to Zn and **(b)** the secondary structure. (*Adapted from Figure 1 in Evans, R. M., and Hollenberg, S. M., 1988. Zinc fingers: gilt by association. Cell* **52:**1.)

The Zipper Motif of bZIP Proteins Operates Through Intersubunit Interaction of Leucine Side Chains

The leucine zipper motif arises from the periodic repetition of leucine residues within this helical region. The periodicity causes the Leu side chains to protrude from the same side of the helical cylinder, where they can enter into hydrophobic interactions with a similar set of Leu side chains extending from a matching helix in a second polypeptide. These hydrophobic interactions establish a stable noncovalent linkage, fostering dimerization of the two polypeptides (as shown in Figure 29.33). The leucine zipper is not a DNA-binding domain. Instead, it functions in protein dimerization. Leucine zippers have been found in other mammalian transcriptional regulatory proteins, including *Myc, Fos,* and *Jun.*

The Basic Region of bZIP Proteins Provides the DNA-Binding Motif

The actual DNA contact surface of *bZIP* proteins is contributed by a 16-residue segment that ends exactly **7** residues before the first Leu residue of the Leu zipper. This DNA contact region is rich in basic residues and hence is referred to as the **basic region.** Two *bZIP* polypeptides join via a Leu zipper to form a Y-shaped molecule in which the stem of the Y corresponds to a coiled pair of α-helices held by the leucine zipper. The arms of the Y are the respective basic regions of each polypeptide; they act as a linked set of DNA contact surfaces (Figure 29.33). The dimer interacts with a DNA target site by situating the fork of the Y at the center of the dyad-symmetric DNA sequence. The two arms of the Y can then track along the major groove of the DNA in opposite

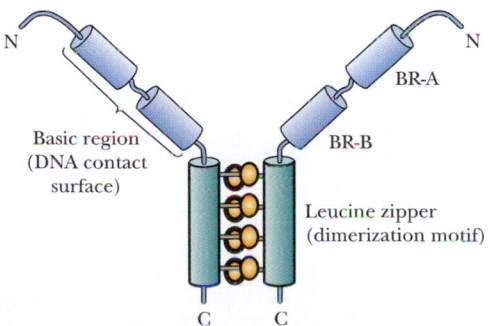

FIGURE 29.33 Model for a dimeric *bZIP* protein. Two *bZIP* polypeptides dimerize to form a Y-shaped molecule. The stem of the Y is the Leu zipper, and it holds the two polypeptides together. Each arm of the Y is the basic region from one polypeptide. Each arm is composed of two α-helical segments: BR-A and BR-B (basic regions A and B).

[7]The acronym *C/EBP* designates this protein as a "CCAAT and enhancer-binding protein."

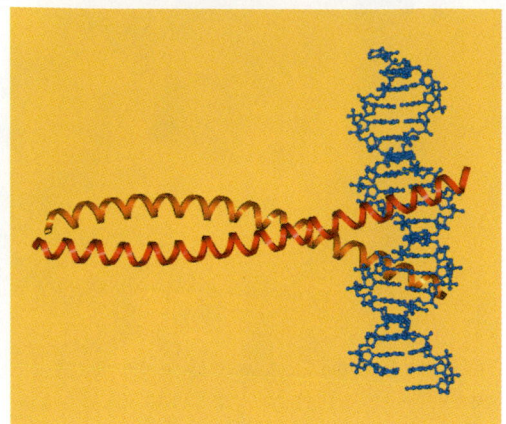

FIGURE 29.34 Model for the heterodimeric *bZIP* transcription factor *c-Fos : c-Jun* bound to a DNA oligomer containing the AP-1 consensus target sequence TGACTCA. *(Adapted from Glover, J. N. M., and Harrison, S. C., 1995. Crystal structure of the heterodimeric bZIP transcription factor c-Fos : c-Jun bound to DNA. Nature 373:257–261.)*

directions, reading the specific recognition sequence (Figure 29.34). An interesting aspect of *bZIP* proteins is that the two polypeptides need not be identical (Figure 29.34). Heterodimers can form, provided both polypeptides possess a leucine zipper region. An important consequence of heterodimer formation is that the DNA target site need not be a palindromic sequence. The respective basic regions of the two different *bZIP* polypeptides (for example, *Fos* and *Jun*) can track along the major groove reading two different base sequences. Heterodimer formation expands enormously the DNA recognition and regulatory possibilities of this set of proteins.

| 29.5 | **How Are Eukaryotic Transcripts Processed and Delivered to the Ribosomes for Translation?** |

Transcription and translation are concomitant processes in prokaryotes, but in eukaryotes, the two processes are spatially separated (see Chapter 10). *Transcription occurs on DNA in the nucleus, and translation occurs on ribosomes in the cytoplasm.* Consequently, transcripts must be transported from the nucleus to the cytosol to be translated. On the way, these transcripts undergo **processing:** alterations that convert the newly synthesized RNAs, or *primary transcripts,* into mature messenger RNAs. Also, unlike prokaryotes, in which many mRNAs encode more than one polypeptide (that is, they are polycistronic), eukaryotic mRNAs encode only one polypeptide (that is, they are exclusively monocistronic).

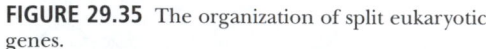

Biochemistry Now™ Go to BiochemistryNow and click BiochemistryInteractive to learn more about the Fos/Jun complex as a hetrodimeric leucine zipper.

Eukaryotic Genes Are Split Genes

Most genes in higher eukaryotes are split into coding regions, called **exons,**[8] and noncoding regions, called **introns** (Figure 29.35; see also Figure 10.24). Introns are the intervening nucleotide sequences that are removed from the primary transcript when it is processed into a mature RNA. Gene expression in eukaryotes entails not only transcription but also the *processing of primary transcripts* to yield the mature RNA molecules we classify as mRNAs, tRNAs, rRNAs, and so forth.

[8]Although the term *exon* is commonly used to refer to the protein-coding regions of an interrupted or split gene, a more precise definition would specify exons as sequences that are represented in mature RNA molecules. This definition encompasses not only protein-coding genes but also the genes for various RNAs (such as tRNAs or rRNAs) from which intervening sequences must be excised in order to generate the mature gene product.

FIGURE 29.35 The organization of split eukaryotic genes.

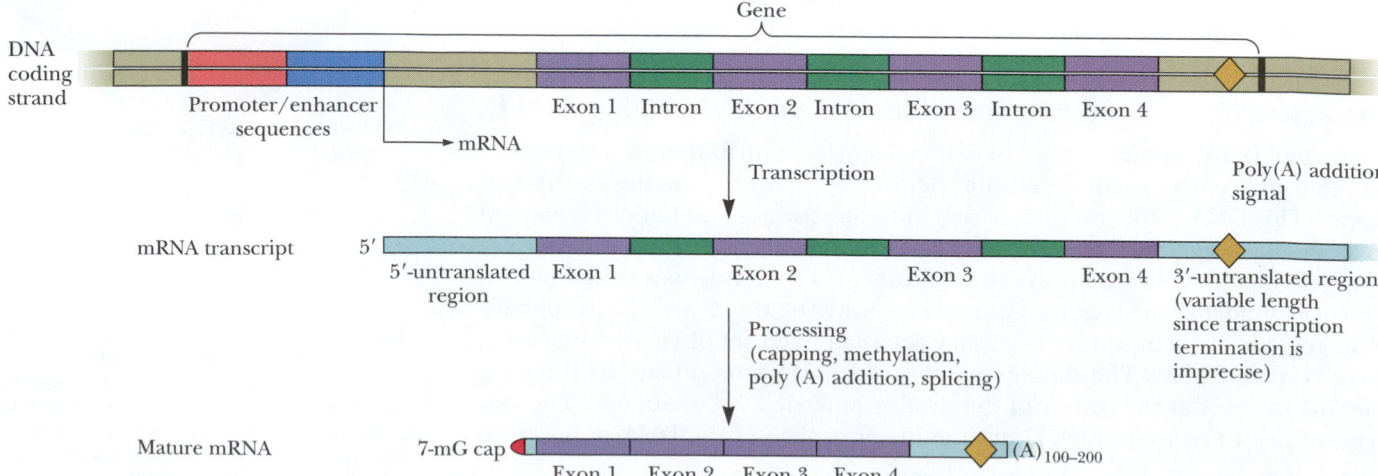

The Organization of Exons and Introns in Split Genes Is Both Diverse and Conserved

Split genes occur in an incredible variety of interruptions and sizes. The yeast **actin gene** is a simple example, having only a single 309-bp intron that separates the nucleotides encoding the first 3 amino acids from those encoding the remaining 350 or so amino acids in the protein. The chicken **ovalbumin gene** is composed of 8 exons and 7 introns. The two **vitellogenin genes** of the African clawed toad *Xenopus laevis* are both spread over more than 21 kbp of DNA; their primary transcripts consist of just 6 kb of message that is punctuated by 33 introns. The chicken **pro α-2 collagen gene** has a length of about 40 kbp; the coding regions constitute only 5 kb distributed over 51 exons within the primary transcript. The exons are quite small, ranging from 45 to 249 bases in size.

Clearly, the mechanism by which introns are removed and multiple exons are spliced together to generate a continuous, translatable mRNA must be both precise and complex. If one base too many or too few is excised during splicing, the coding sequence in the mRNA will be disrupted. The mammalian **DHFR (dihydrofolate reductase) gene** is split into 6 exons spread over more than 31 kbp of DNA. The 6 exons are spliced together to give a 6-kb mRNA (Figure 29.36). Note that, in three different mammalian species, the size and position of the exons are essentially the same but that the lengths of the corresponding introns vary considerably. Indeed, the lengths of introns in vertebrate genes range from a minimum of about 60 bases to more than 10,000 bp. Many introns have nonsense codons in all three reading frames and thus are untranslatable. Introns are found in the genes of mitochondria and chloroplasts as well as in nuclear genes. Although introns have been observed in Archaebacteria and even bacteriophage T4, none are known in the genomes of eubacteria.

Post-Transcriptional Processing of Messenger RNA Precursors Involves Capping, Methylation, Polyadenylylation, and Splicing

Capping and Methylation of Eukaryotic mRNAs The protein-coding genes of eukaryotes are transcribed by RNA polymerase II to form primary transcripts or **pre-mRNAs** that serve as precursors to mRNA. As a population, these RNA molecules are very large and their nucleotide sequences are very heterogeneous because they represent the transcripts of many different genes, hence the designation **heterogeneous nuclear RNA,** or **hnRNA.** Shortly after transcription of hnRNA is initiated, the 5′-end of the growing transcript is capped by addition of a guanylyl residue. This reaction is catalyzed by the nuclear enzyme **guanylyl transferase** using GTP as substrate (Figure 29.37). The **cap structure** is methylated at the 7-position of the G residue. Additional methylations may occur at the 2′-O positions of the two nucleosides following the 7-methyl-G cap and at the 6-amino group of a first base adenine (Figure 29.38).

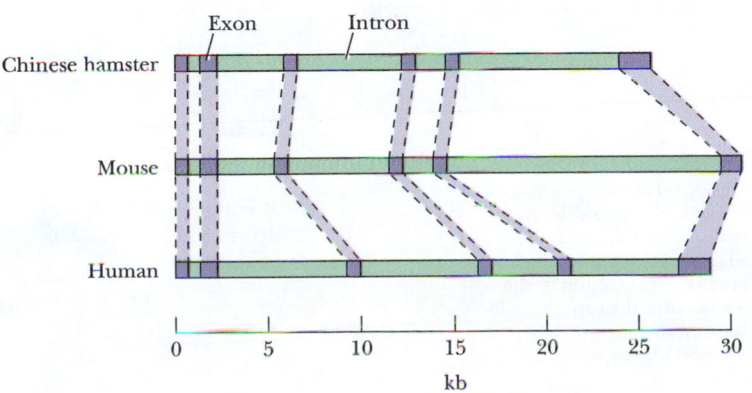

FIGURE 29.36 The organization of the mammalian DHFR gene in three representative species. Note that the exons are much shorter than the introns. Note also that the exon pattern is more highly conserved than the intron pattern.

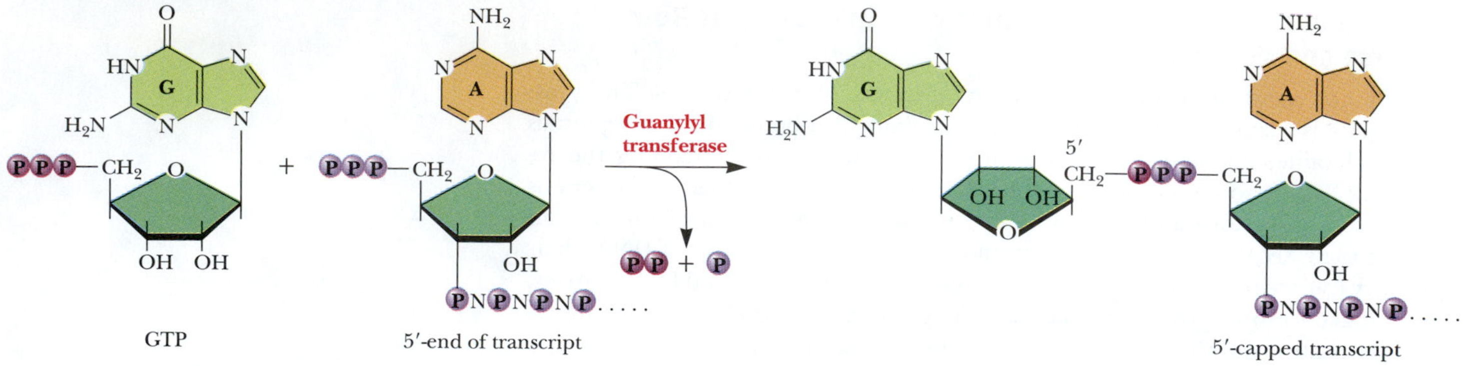

FIGURE 29.37 The capping of eukaryotic pre-mRNAs. Guanylyl transferase catalyzes the addition of a guanylyl residue (G_p) derived from GTP to the 5'-end of the growing transcript, which has a 5'-triphosphate group already there. In the process, pyrophosphate (pp) is liberated from GTP and the terminal phosphate (p) is removed from the transcript. $Gppp + pppApNpNpNp... \rightarrow GpppApNpNpNp... + pp + p$ (A is often the initial nucleotide in the primary transcript).

3'-Polyadenylation of Eukaryotic mRNAs Transcription by RNA polymerase II typically continues past the 3'-end of the mature messenger RNA. Primary transcripts show heterogeneity in sequence at their 3'-ends, indicating that the precise point where termination occurs is nonspecific. However, termination does not normally occur until RNA polymerase II has transcribed past a consensus AAUAAA sequence known as the **polyadenylylation signal.**

Most eukaryotic mRNAs have 100 to 200 adenine residues attached at their 3'-end, the **poly(A) tail.** [Histone mRNAs are the only common mRNAs that lack poly(A) tails.] These A residues are not encoded in the DNA but are added post-transcriptionally by the enzyme **poly(A) polymerase,** using ATP as a substrate. The consensus AAUAAA is not itself the poly(A) addition site; instead it defines the position where poly(A) addition occurs (Figure 29.39). The consensus AAUAAA is found 10 to 35 nucleotides upstream from where the nascent primary transcript is cleaved by an endonuclease to generate a new 3'-OH end. This end is where the poly(A) tail is added. The processing events of mRNA capping, poly(A) addition, and splicing of the primary transcript create the mature mRNA. Interestingly, both the guanylyl transferase that adds the 5'-cap structure and the enzymes that process the 3'-end of the transcript and add the poly (A) tail are anchored to RNA polymerase II via interactions with its RPB1 CTD.

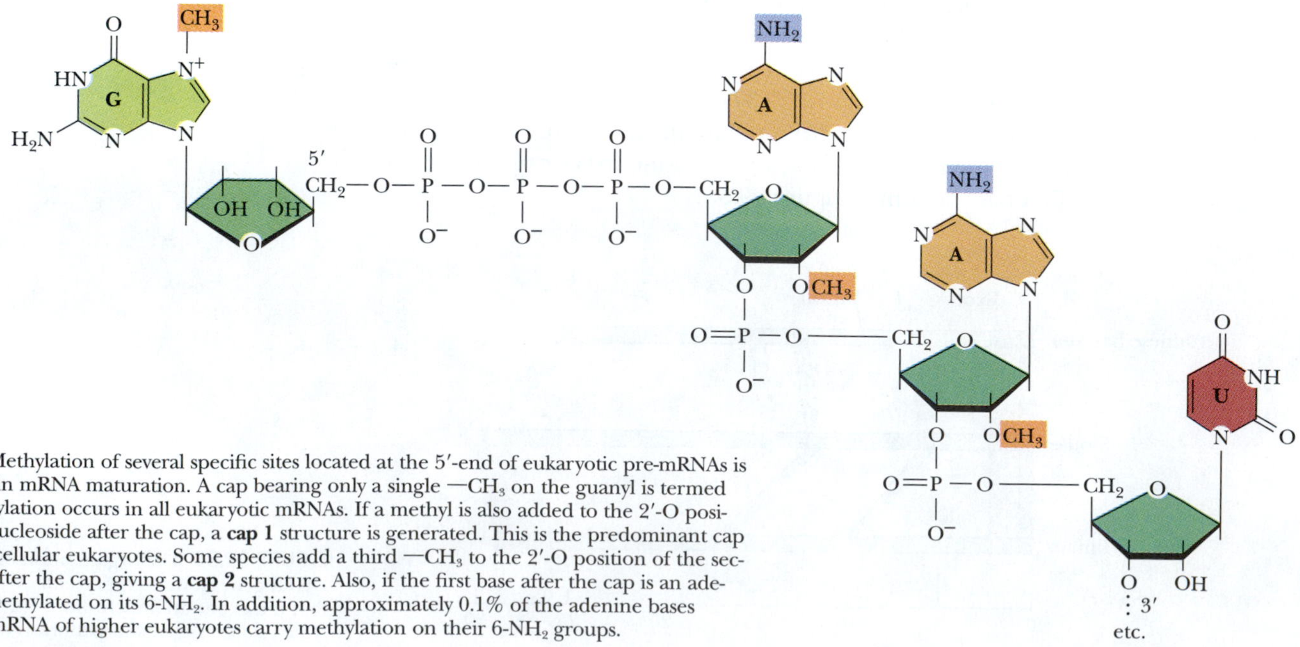

FIGURE 29.38 Methylation of several specific sites located at the 5'-end of eukaryotic pre-mRNAs is an essential step in mRNA maturation. A cap bearing only a single —CH₃ on the guanyl is termed **cap 0.** This methylation occurs in all eukaryotic mRNAs. If a methyl is also added to the 2'-O position of the first nucleoside after the cap, a **cap 1** structure is generated. This is the predominant cap form in all multicellular eukaryotes. Some species add a third —CH₃ to the 2'-O position of the second nucleoside after the cap, giving a **cap 2** structure. Also, if the first base after the cap is an adenine, it may be methylated on its 6-NH₂. In addition, approximately 0.1% of the adenine bases throughout the mRNA of higher eukaryotes carry methylation on their 6-NH₂ groups.

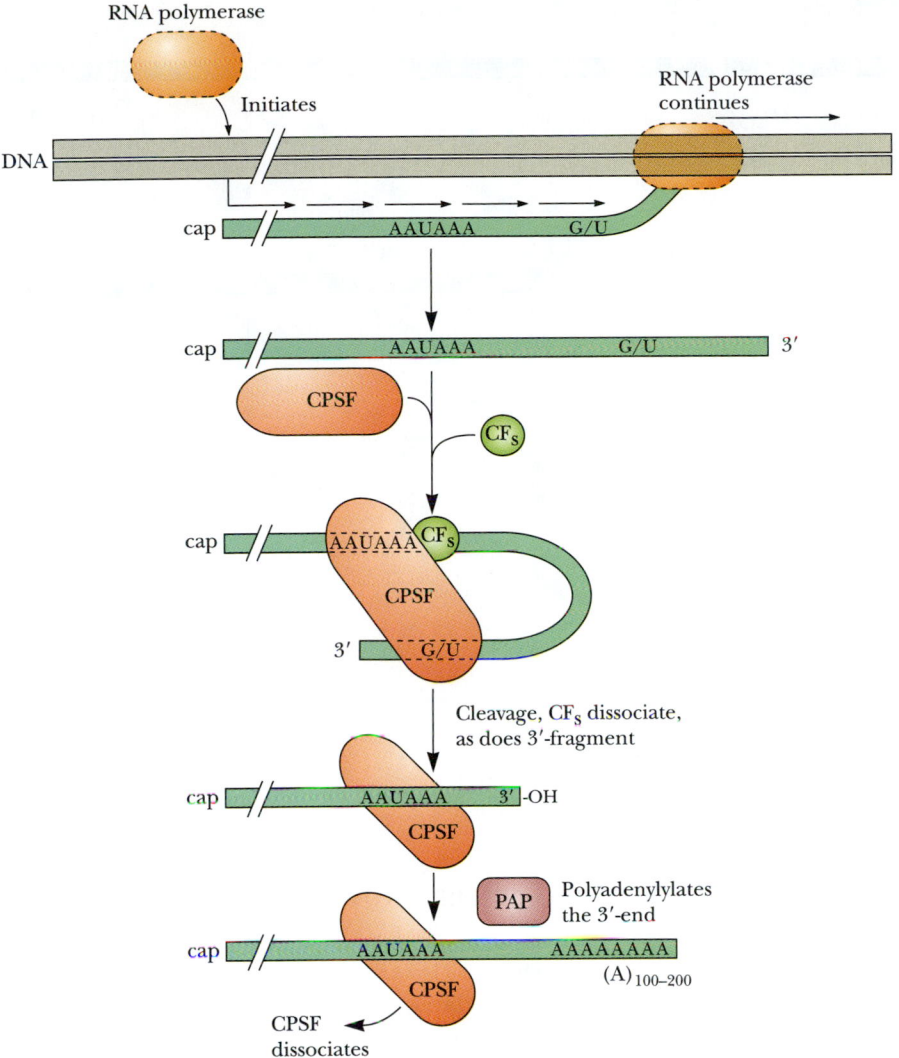

FIGURE 29.39 Poly(A) addition to the 3′-ends of transcripts occurs 10 to 35 nucleotides downstream from a consensus AAUAAA sequence, defined as the *polyadenylylation signal*. CPSF *(cleavage and polyadenylylation specificity factor)* binds to this signal sequence and mediates looping of the 3′-end of the transcript through interactions with a G/U-rich sequence even further downstream. Cleavage factors *(CFs)* then bind and bring about the endonucleolytic cleavage of the transcript to create a new 3′-end 10 to 35 nucleotides downstream from the polyadenylylation signal. Poly(A) polymerase (PAP) then successively adds 200 to 250 adenylyl residues to the new 3′-end. (RNA polymerase II is also a significant part of the polyadenylylation complex at the 3′-end of the transcript, but for simplicity in illustration, its presence is not shown in the lower part of the figure.)

Nuclear Pre-mRNA Splicing

Within the nucleus, hnRNA forms **ribonucleoprotein particles (RNPs)** through association with a characteristic set of nuclear proteins. These proteins interact with the nascent RNA chain as it is synthesized, maintaining the hnRNA in an untangled, accessible conformation. The substrate for splicing, that is, intron excision and exon ligation, is the capped primary transcript emerging from the RNA polymerase II transcriptional apparatus, in the form of an RNP complex. Splicing occurs exclusively in the nucleus. The mature mRNA that results is then exported to the cytoplasm to be translated. Splicing requires precise cleavage at the 5′- and 3′-ends of introns and the accurate joining of the two ends. Consensus sequences define the exon/intron junctions in eukaryotic mRNA precursors, as indicated from an analysis of the splice sites in vertebrate genes (Figure 29.40). Note that the sequences GU and AG are found at the 5′- and 3′-ends, respectively, of introns in pre-mRNAs from higher eukaryotes. In addition to the splice junctions, a conserved sequence within the intron, the **branch site,** is also essential to pre-mRNA splicing. The site lies 18 to 40 nucleotides upstream from the 3′-splice site and is represented in higher eukaryotes by the consensus sequence YNYRAY, where Y is any pyrimidine, R is any purine, and N is any nucleotide.

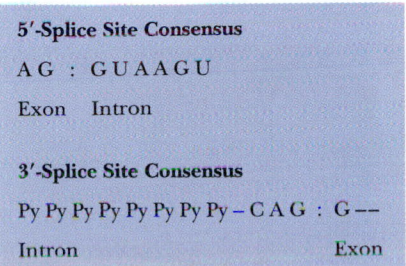

5′-Splice Site Consensus

A G : G U A A G U

Exon   Intron

3′-Splice Site Consensus

Py Py Py Py Py Py Py Py – C A G : G ––

Intron   Exon

FIGURE 29.40 Consensus sequences at the splice sites in vertebrate genes.

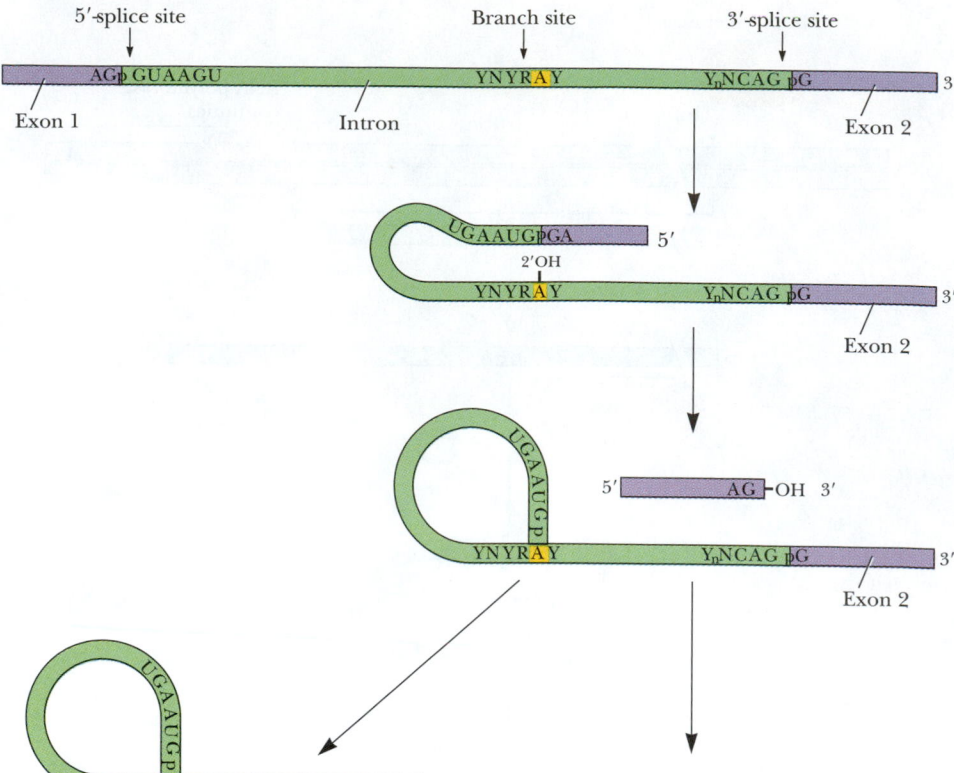

FIGURE 29.41 Splicing of mRNA precursors. A representative precursor mRNA is depicted. Exon 1 and Exon 2 indicate two exons separated by an intervening sequence (or intron) with consensus 5′, 3′, and branch sites. The fate of the phosphates at the 5′- and 3′-splice sites can be followed by tracing the fate of the respective *p*s. The products of the splicing reaction, the lariat form of the excised intron and the united exons, are shown at the bottom of the figure. The lariat intermediate is generated when the invariant G at the 5′-end of the intron attaches via its 5′-phosphate to the 2′-OH of the invariant A within the branch site. The consensus guanosine residue at the 3′-end of Exon 1 (the 5′-splice site) then reacts with the 5′-phosphate at the 3′-splice site (the 5′-end of Exon 2), ligating the two exons and releasing the lariat structure. Although the reaction is shown here in a stepwise fashion, 5′-cleavage, lariat formation, and exon ligation/lariat excision are believed to occur in a concerted fashion. (*Adapted from Figure 1 in Sharp, P. A., 1987. Splicing of messenger RNA precursors.* Science **235**:766.)

The Splicing Reaction Proceeds via Formation of a Lariat Intermediate

The mechanism for splicing nuclear mRNA precursors is shown in Figure 29.41. A covalently closed loop of RNA, the **lariat,** is formed by attachment of the 5′-phosphate group of the intron's invariant 5′-G to the 2′-OH at the invariant branch site A to form a 2′-5′ phosphodiester bond. Note that lariat formation creates an unusual branched nucleic acid. The lariat structure is excised when the 3′-OH of the consensus G at the 3′-end of the 5′ exon (Exon 1, Figure 29.41) covalently joins with the 5′-phosphate at the 5′-end of the 3′ exon (Exon 2). The reactions that occur are transesterification reactions where an OH group reacts with a phosphoester bond, displacing an —OH to form a new phosphoester link. Because the reactions lead to no net change in the number of phosphodiester linkages, no energy input (for example, as ATP) is needed. The lariat product is unstable; the 2′-5′ phosphodiester branch is quickly cleaved to give a linear excised intron that is rapidly degraded in the nucleus.

Splicing Depends on snRNPs

The hnRNA (pre-mRNA) substrate is not the only RNP complex involved in the splicing process. Splicing also depends on a unique set of small nuclear ribonucleoprotein particles, so-called **snRNPs** (pronounced "snurps"). In higher eukaryotes, each snRNP consists of a small RNA molecule 100 to 200 nucleotides long and a set of about 10 different proteins. Some of the different proteins form a "core" set common to all snRNPs, whereas others are unique to a specific snRNP. The major snRNP species are very abundant, present at greater than 100,000 copies per nucleus. The RNAs of snRNPs are typically rich in uridine, hence the classification of particular snRNPs as U1, U2, and so on. The prominent snRNPs are given in Table 29.6). U1 snRNA folds into a secondary structure that leaves the 11 nucleotides at its 5′-end single-stranded. The 5′-end of U1

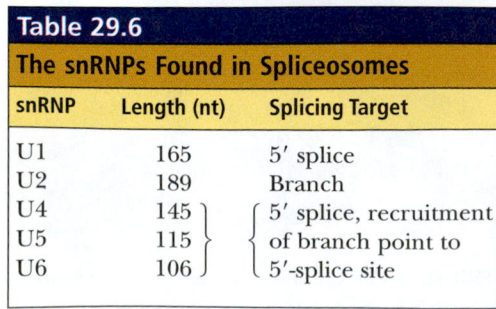

Table 29.6		
The snRNPs Found in Spliceosomes		
snRNP	**Length (nt)**	**Splicing Target**
U1	165	5′ splice
U2	189	Branch
U4	145	5′ splice, recruitment
U5	115	of branch point to
U6	106	5′-splice site

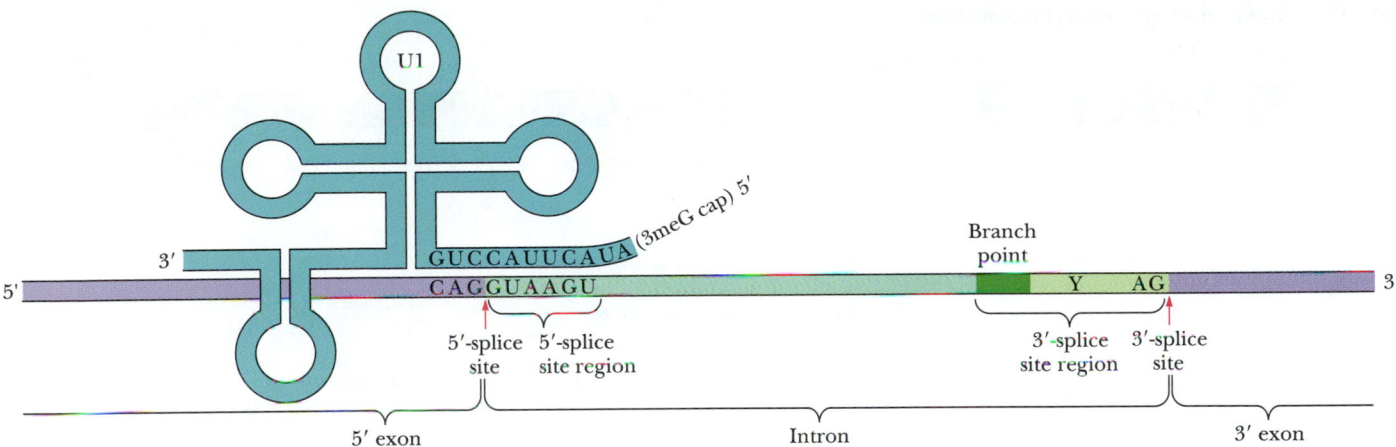

FIGURE 29.42 Mammalian U1 snRNA can be arranged in a secondary structure where its 5′-end is single-stranded and can base-pair with the consensus 5′-splice site of the intron. (*Adapted from Figure 1 in Rosbash, M., and Seraphin, B., 1991. Who's on first? The U1 snRNP-5′ splice site interaction and splicing.* Trends in Biochemical Sciences **16**:187.)

snRNA is complementary to the consensus sequence at the 5′-splice junction of the pre-mRNA (Figure 29.42), as is a region at the 5′-end of U6 snRNA. U2 snRNA is complementary to the consensus branch site sequence.

snRNPs Form the Spliceosome

Splicing occurs when the various snRNPs come together with the pre-mRNA to form a multicomponent complex called the **spliceosome.** The spliceosome is a large complex, roughly equivalent to a ribosome in size, and its assembly requires ATP. Assembly of the spliceosome begins with the binding of U1 snRNP at the 5′-splice site of the pre-mRNA (Figure 29.43). The branch-point sequence (UACUAAC in yeast) binds U2 snRNP, and then the triple snRNP complex of U4/U6·U5 replaces U1 at the 5′-splice site. The substitution of base-pairing interactions between U1 and the pre-mRNA 5′-splice site by base-pairing between U6 and the 5′-splice site is just one of the many RNA rearrangements that accompany the splicing reaction. Base-pairings between U6 and U2 RNA bring the 5′-splice site and the branch point RNA sequences into proximity. Interactions between U2 and U6 lead to release of U4 snRNP. The spliceosome is now activated for catalysis: A transesterification reaction involving the 2′-O of the invariant A residue in the branch-point sequence displaces the 5′-exon from the intron, creating the lariat intermediate. The free 3′-O of the 5′-exon now triggers a second transesterification reaction through attack on the P atom at the 3′-exon splice site. This second reaction joins the two exons and releases the intron as a lariat structure. In addition to the snRNPs, a number of proteins with RNA-annealing functions as well as proteins with ATP-dependent RNA-unwinding activity participate in spliceosome function. The spliceosome is thus a dynamic structure that uses the pre-mRNA as a template for assembly, carries out its transesterification reactions, and then disassembles when the splicing reaction is over. Thomas Cech's discovery of catalytic RNA (see Chapter 13) raises the possibility that the transesterification reactions are catalyzed not by snRNP proteins but by the snRNAs themselves.

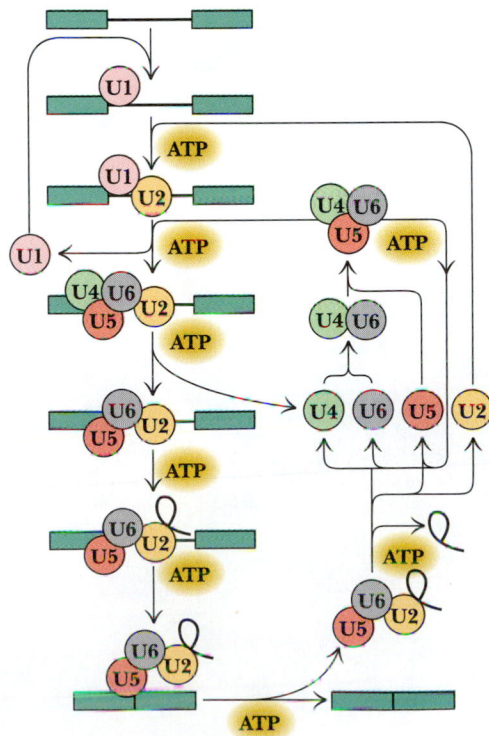

▶ **FIGURE 29.43** Events in spliceosome assembly. U1 snRNP binds at the 5′-splice site, followed by the association of U2 snRNP with the UACUAA*C branch-point sequence. The triple U4/U6-U5 snRNP complex replaces U1 at the 5′-splice site and directs the juxtaposition of the branch-point sequence with the 5′-splice site, whereupon U4 snRNP is released. Lariat formation occurs, freeing the 3′-end of the 5′-exon to join with the 5′-end of the 3′-exon, followed by exon ligation. U2, U5, and U6 snRNPs dissociate from the lariat following exon ligation. Spliceosome assembly, rearrangement, and disassembly require ATP as well as various RNA-binding proteins (not shown). (*Adapted from Figure 2 in Staley, J. P., and Guthrie, C., 1998. Mechanical devices of the spliceosome: Motors, clocks, springs, and things.* Cell **92**:315–326.)

Fast skeletal troponin T gene and spliced mRNAs

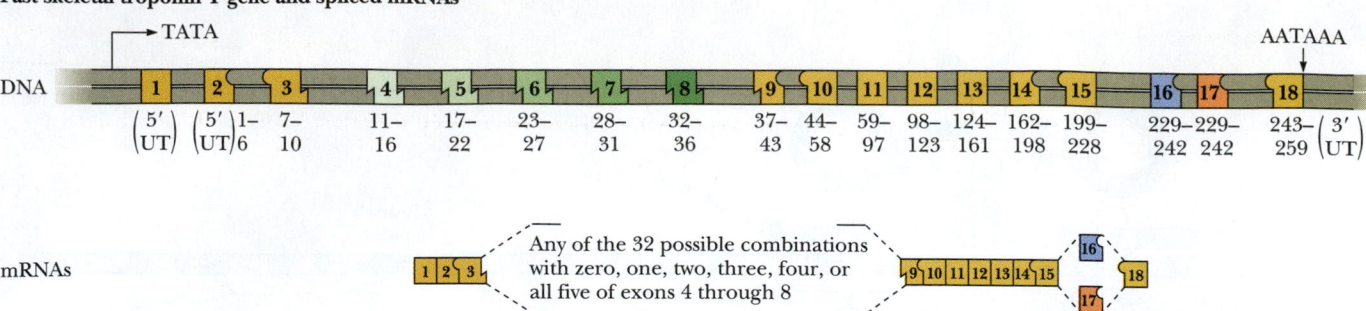

FIGURE 29.44 Organization of the fast skeletal muscle troponin T gene and the 64 possible mRNAs that can be generated from it. Exons are constitutive (yellow), combinatorial (green), or mutually exclusive (blue or orange). Exon 1 is composed of 5′-untranslated (UT) sequences, and Exon 18 includes the polyadenylation site (AATAAA) and 3′-UT sequences. The TATA box indicates the transcription start site. The amino acid residues encoded by each exon are indicated below. Many exon:intron junctions fall between codons. The "sawtooth" exon boundaries indicate that the splice site falls between the first and second nucleotides of a codon, the "concave/convex" exon boundaries indicate that the splice site falls between the second and third nucleotides of a codon, and flush boundaries between codons signify that the splice site falls between intact codons. Each mRNA includes all constitutive exons, 1 of the 32 possible combinations of Exons 4 to 8, and either Exon 16 or 17. (*Adapted from Figure 2 in Breitbart, R. E., Andreadis, A., and Nadal-Ginard, B., 1987. Alternative splicing: a ubiquitous mechanism for the generation of multiple protein isoforms from single genes. Annual Review of Biochemistry* **56:**467.)

Alternative RNA Splicing Creates Protein Isoforms

In one mode of splicing, every intron is removed and every exon is incorporated into the mature RNA without exception. This type of splicing, termed **constitutive splicing,** results in a single form of mature mRNA from the primary transcript. However, many eukaryotic genes can give rise to multiple forms of mature RNA transcripts. The mechanisms for production of multiple transcripts from a single gene include use of different promoters, selection of different polyadenylation sites, **alternative splicing** of the primary transcript, or even a combination of the three.

Different transcripts from a single gene make possible a set of related polypeptides, or **protein isoforms,** each with slightly altered functional capability. Such variation serves as a useful mechanism for increasing the apparent coding capacity of the genome. Furthermore, alternative splicing offers another level at which regulation of gene expression can operate. For example, mRNAs unique to particular cells, tissues, or developmental stages could be formed from a single gene by choosing different 5′- or 3′-splice sites or by omitting entire exons. Translation of these mature mRNAs produces cell-specific protein isoforms that display properties tailored to the needs of the particular cell. Such regulated expression of distinct protein isoforms is a fundamental characteristic of eukaryotic cell differentiation and development.

Fast Skeletal Muscle Troponin T Isoforms Are an Example of Alternative Splicing

In addition to many other instances, alternative splicing is a prevalent mechanism for generating protein isoforms from the genes encoding muscle proteins (see Chapter 16), allowing distinctive isoforms aptly suited to the function of each muscle. An impressive manifestation of alternative splicing is seen in the expression possibilities for the rat **fast skeletal muscle troponin T gene** (Figure 29.44). This gene consists of 18 exons, 11 of which are found in all mature mRNAs (exons 1 through 3, 9 through 15, and 18) and thus are **constitutive.** Five exons, those numbered 4 through 8, are **combinatorial** in that they may be individually included or excluded, in any combination, in the mature

A Deeper Look

RNA Editing: Another Mechanism That Increases the Diversity of Genomic Information

RNA editing is a process that changes one or more nucleotides in an RNA transcript by deaminating a base, either A→I (adenine to inosine, through deamination at the 6-position in a purine ring) or C→U (cytosine to uracil, through deamination at the 4-position in a pyrimidine ring). These changes alter the coding possibilities in a transcript, because I will pair with G (not U as A does) and U will pair with A (not G as C does). RNA editing has the potential to increase protein diversity by (1) altering amino acid coding possibilities, (2) introducing premature stop codons, or (3) changing splice sites in a transcript. If RNA splicing is cutting-and-pasting, then these single-base changes are aptly termed RNA editing.

A-to-I editing is carried out by adenosine deaminases that act on RNA (the ADAR family of RNA-editing enzymes). ADARs act only on double-stranded regions of RNA. Typically, such regions form when an exon region containing an A to be edited base-pairs with a complementary base sequence in an intron known as the *editing site complementary sequence,* or *ECS.* ADARs are abundant in the nervous system of animals. A prominent example of RNA editing occurs in transcripts encoding mammalian glutamate re-

ceptors (GluRs; see Chapter 32). Deamination of the GluR-B gene transcript changes a glutamine codon CAG to CIG, which is read by the translational machinery as an arginine codon (CGG), dramatically altering the conductance properties of the membrane receptor produced from the edited transcript, as compared to the receptor produced from the unedited transcript.

C-to-I editing is carried out on single-stranded regions of transcripts by an *editosome* core structure consisting of a cytosine deaminase and an adapter protein that brings the deaminase and the transcript together. A prominent example of C-to-I editing targets a single C residue in a 14-kb transcript encoding the 4536-residue apolipoprotein B100 protein (see Chapter 24). ApoB RNA editing changes codon 2153 (a CAA [glutamine] codon) to a UAA stop codon, which leads to a shortened protein product, ApoB48, consisting of the N-terminal 48% of apoB100. In humans, apoB100 is made in the liver and found in liver-derived VLDL serum lipoprotein complexes. In contrast, apoB48 is made in intestinal cells and found in intestinal-derived lipid complexes.

Sources: Samuel, C. E., 2003. RNA editing minireview series. *The Journal of Biological Chemistry* **278:**1389–1390; Maas, S., Rich, A., and Nishikura, K., 2003. A-to-I editing: Recent news and residual mysteries. *The Journal of Biological Chemistry* **278:**1391–1394; and Blanc, V., and Davidson, N. O., 2003. C-to-U RNA editing: mechanisms leading to genetic diversity. *The Journal of Biological Chemistry* **278:**1395–1398.

---

mRNA. Two exons, 16 and 17, are mutually exclusive: One or the other is always present, but never both. Sixty-four different mature mRNAs can be generated from the primary transcript of this gene by alternative splicing. Because each exon represents a cassette of genetic information encoding a segment of protein, alternative splicing is a versatile way to introduce functional variation within a common protein theme.

29.6	**Can We Propose a Unified Theory of Gene Expression?**

The stages of eukaryotic gene expression—from transcriptional activation, transcription, transcript processing, nuclear export of mRNA, to translation—have traditionally been presented as a linear series of events, that is, as a pathway of discrete, independent steps. However, it now is clear that each stage is part of a continuous process, with physical and functional connections between the various transcriptional and processing machineries. This realization led George Orphanides and Danny Reinberg to propose a "unified theory of gene expression." The principal tenet of this theory is that eukaryotic gene expression is a continuous process, from transcription through processing and protein synthesis: DNA→RNA→protein (Figure 29.45). Furthermore, regulation occurs at multiple levels in this continuous process, in a coordinated fashion. Tom Maniatis and Robin Reed provide additional support for this theory by pointing out that eukaryotic gene expression depends on an interacting network of multicomponent protein machines—nucleosomes, HATs, and the remodeling apparatus; RNA polymerase II and its associated factors, which include capping, splicing, and polyadenylylation enzymes; and the proteins

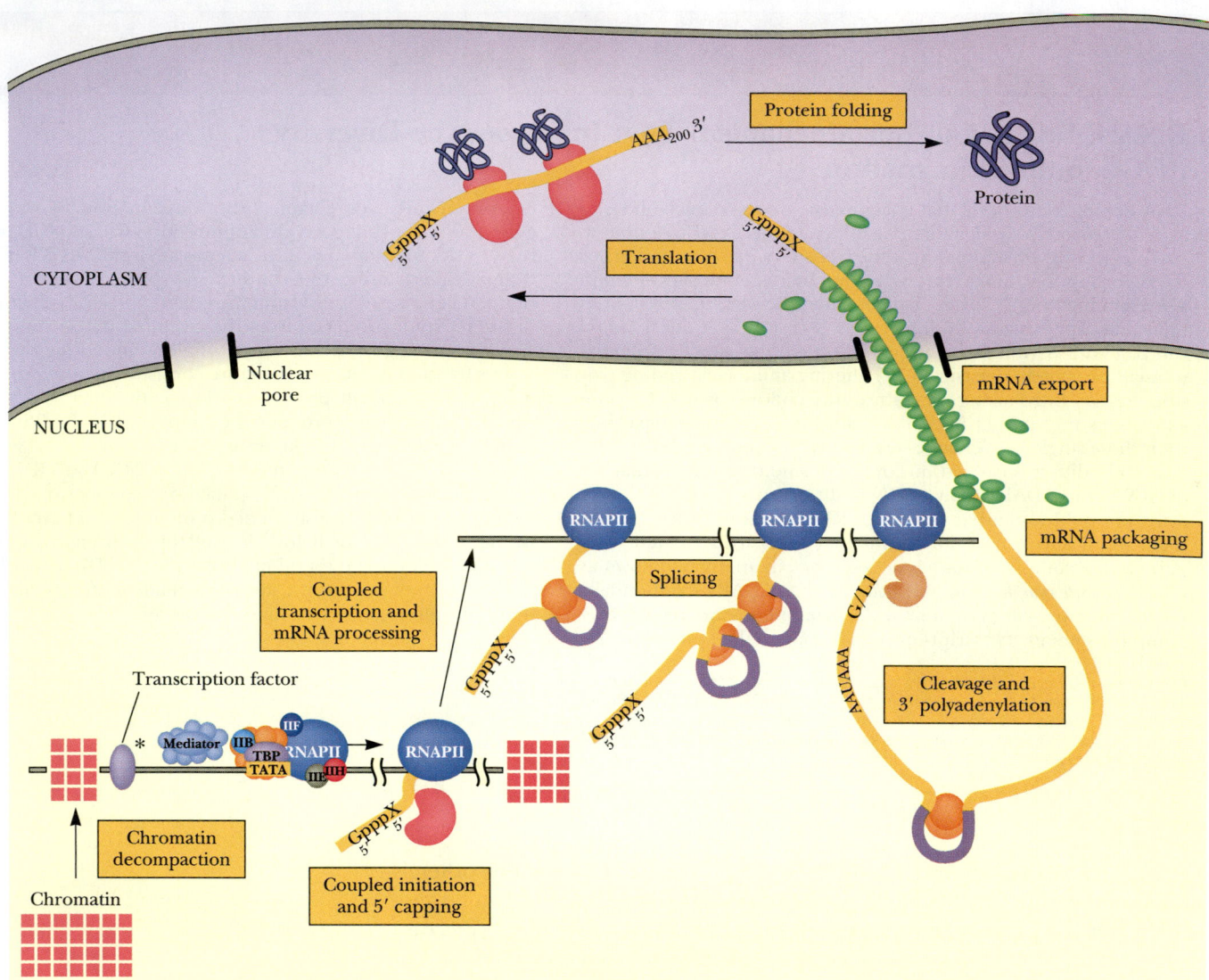

A unified theory of gene expression. Each step in gene expression, from transcription to translation, is but a stage within a continuous process. Each stage is physically and functionally connected to the next, ensuring that all steps proceed in an appropriate fashion and overall regulation of gene expression is tightly integrated. *(From Figure 2 in Orphanides, G., and Reinberg, D., 2002. A unified theory of gene expression. Cell 108:439–451. With permission from Elsevier.)* **Test yourself on the concepts in this figure at http://chemistry. brookscole.com/ggb3**

involved in mRNA export to the cytoplasm for translation on ribosomes, the topic of the next chapter. In closing, it should be mentioned that eukaryotic cells have elaborate systems for mRNA surveillance; these systems destroy any messages containing errors, such as the **nonsense mediated decay (NMD)** system. NMD degrades any message that has premature nonsense, or "stop," codons. Furthermore, the regulation of mRNA levels through selective destruction provides a mechanism for the post-transcriptional regulation of gene expression.

Summary

29.1 How Are Genes Transcribed in Prokaryotes?
In prokaryotes, virtually all RNA synthesis is carried out by a single species of DNA-dependent RNA polymerase. RNA polymerase links ribonucleoside 5′-triphosphates in an order specified by base pairing with a DNA template. The enzyme moves along a DNA strand in the 3′→5′ direction, joining the 5′-phosphate of an incoming ribonucleotide to the 3′-OH of the previous residue, so the RNA chain grows 5′→3′ during transcription. Transcription begins when the σ-subunit of RNA polymerase recognizes a promoter and forms a complex with

it. Next, the RNA polymerase holoenzyme unwinds about 14 base pairs of DNA, and transcription commences. Once an oligonucleotide 9 to 12 residues long has been formed, the σ-subunit dissociates. The core RNA polymerase is highly processive and goes on to synthesize the remainder of the mRNA. Prokaryotes have two types of transcription termination mechanisms: one that is dependent on ρ termination factor protein and another that depends on dissociation of the mRNA through reestablishment of DNA base pairs.

29.2 How Is Transcription Regulated in Prokaryotes?

Bacterial genes encoding a common metabolic pathway are often grouped adjacent to one another in an operon, allowing all of the genes to be expressed in a coordinated fashion. The operator, a regulatory sequence adjacent to the structural genes, determines whether transcription takes place. The operator is located next to a promoter. Interaction of a regulatory protein with the operator controls transcription. Small molecules act as signals of the nutritional or environmental conditions. These small molecules interact with operator-binding regulatory proteins and determine whether transcription occurs. Induction is the increased synthesis of enzymes in response to a small molecule called a co-inducer. Repression is the decreased transcription in response to a specific metabolite termed a co-repressor. The *lac* operon provides an example of induction, and the *trp* operon, repression. Operon regulation depends on the interaction of sequence-specific DNA-binding proteins with regulatory sequences along the DNA. DNA looping increases the regulatory input available to a specific gene.

29.3 How Are Genes Transcribed in Eukaryotes?

Transcription is more complicated in eukaryotes because eukaryotic DNA is wrapped around histones to form nucleosomes, and nucleosomes repress gene expression by limiting access of the transcriptional apparatus to genes. Two classes of transcriptional co-regulators are necessary to overcome nucleosome repression: (1) HATs that acetylate lysine ϵ-NH_3^+ groups and (2) ATP-dependent chromatin-remodeling complexes. Gene activation also requires interaction of RNA polymerase with the promoter. RNA polymerase II consists of 12 different polypeptides. The largest, RPB1, has a C-terminal domain (CTD) containing multiple repeats of the heptapeptide sequence YSPTSPS; 5 of these 7 residues can be phosphorylated by protein kinases. The CTD orchestrates events in the transcription process. RNA polymerase II promoters commonly consist of the core element, near the transcription start site, where general transcription factors bind, and more distantly located regulatory elements, known as enhancers or silencers. These regulatory sequences are recognized by specific DNA-binding proteins that activate transcription above basal levels. A eukaryotic transcription initiation complex consists of RNA polymerase II, five general transcription factors, and Mediator. The CTD of RNA polymerase II anchors Mediator to the polymerase and allows RNA polymerase II to communicate with transcriptional activators bound at sites distal from the promoter.

29.4 How Do Gene Regulatory Proteins Recognize Specific DNA Sequences?

Proteins that recognize nucleic acids present a three-dimensional shape or contour that is structurally and chemically complementary to the surface of a DNA sequence. Nucleotide sequence–specific recognition by the protein involves a set of atomic contacts with the bases and the sugar–phosphate backbone. Protein contacts with the bases of DNA usually occur within the major groove (but not always). Roughly 80% of DNA-binding regulatory proteins fall into one of three principal classes based on distinctive structural motifs: the helix-turn-helix (or HTH), the zinc finger (or Zn-finger), and the leucine zipper-basic region (or bZIP). In addition to their DNA-binding domains, these proteins also have protein:protein recognition domains essential to oligomerization, DNA looping, transcriptional activation, and signal reception (effector binding).

29.5 How Are Eukaryotic Transcripts Processed and Delivered to the Ribosomes for Translation?

In eukaryotes, primary transcripts must be processed to form mature messenger RNAs and exported from the nucleus to the cytosol for translation. Shortly after transcription initiation, the 5′-end of the growing transcript is capped with a guanylyl residue that is then methylated at the 7-position. Additional methylations may occur at the 2′-O positions of the next two nucleosides and at the 6-amino group of a first adenine. Transcription termination does not normally occur until RNA polymerase II has transcribed past the polyadenylation signal. Most eukaryotic mRNAs have a poly(A) tails consisting of 100 to 200 adenine residues at their 3′-end, added posttranscriptionally by poly(A) polymerase. Most eukaryotic genes are split genes, subdivided into coding regions, called exons, and noncoding regions, called introns. Intron excision and exon ligation, a process called splicing, also occurs in the nucleus. Splicing is mediated by the spliceosome, which is assembled from a set of small nuclear ribonucleoprotein particles called snRNPs. Splicing requires precise cleavage at the 5′- and 3′-ends of introns and the accurate joining of the two exons. Exon/intron junctions are defined by consensus sequences recognized by the spliceosome. In addition, a conserved sequence within the intron, the branch site, is also essential to splicing. The splicing reaction involves formation of a lariat intermediate through attachment of the 5′-phosphate group of the intron's invariant 5′-G to the 2′-OH at the invariant branch site A to form a 2′-5′ phosphodiester bond. The lariat structure is excised when the exons are ligated. In constitutive splicing, every intron is removed and every exon is incorporated into the mature RNA without exception. However, alternative splicing can give rise to different transcripts from a single gene, making possible a set of protein isoforms, each with slightly altered functional capability. Fast skeletal muscle troponin T isoforms are an example of alternative splicing.

29.6 Can We Propose a Unified Theory of Gene Expression?

Each stage in eukaryotic transcription is part of a continuous process, with physical and functional connections between the various transcriptional and processing machineries. These multicomponent protein machines are organized into an interacting network, and regulation occurs in a coordinated fashion at multiple levels in the continuous process. Eukaryotic cells also have elaborate systems for mRNA surveillance.

Problems

1. The 5′-end of an mRNA has the sequence

   ...AGAUCCGUAUGGCGAUCUCGACGAAGACUC-
   CUAGGGAAUCC...

   What is the nucleotide sequence of the DNA template strand from which it was transcribed? If this mRNA is translated beginning with the first AUG codon in its sequence, what is the N-terminal amino acid sequence of the protein it encodes? (See Table 30.1 for the genetic code.)

2. Describe the sequence of events involved in the initiation of transcription by *E. coli* RNA polymerase. Include in your description those features a gene must have for proper recognition and transcription by RNA polymerase.

3. RNA polymerase has two binding sites for ribonucleoside triphosphates: the initiation site and the elongation site. The initiation site has a greater K_m for NTPs than the elongation site. Suggest what possible significance this fact might have for the control of transcription in cells.

4. Make a list of the ways that transcription in eukaryotes differs from transcription in prokaryotes.

5. DNA-binding proteins may recognize specific DNA regions either by reading the base sequence or by "indirect readout." How do these two modes of protein:DNA recognition differ?

6. (Integrates with Chapter 11.) The metallothionein promoter is illustrated in Figure 29.26. How long is this promoter, in nm? How many turns of B-DNA are found in this length of DNA? How many nucleosomes (approximately) would be bound to this much DNA? (Consult Chapter 11 to review the properties of nucleosomes.)

7. Describe why the ability of *bZIP* proteins to form heterodimers increases the repertoire of genes whose transcription might be responsive to regulation by these proteins.

8. Suppose exon 17 were deleted from the fast skeletal muscle troponin T gene (Figure 29.44). How many different mRNAs could now be generated by alternative splicing? Suppose that exon 7 in a wild-type troponin T gene were duplicated. How many different mRNAs might be generated from a transcript of this new gene by alternative splicing?

9. Figure 29.25 illustrates the various covalent modifications that occur on histone tails. How might each of these modifications influence DNA : histone interactions?

10. (Integrates with Chapter 15.) Predict from Figure 29.11 whether the interaction of *lac* repressor with inducer might be cooperative. Would it be advantageous for inducer to show cooperative binding to *lac* repressor? Why?

11. What might be the advantages of capping, methylation, and polyadenylylation of eukaryotic mRNAs?

12. (Integrates with Chapter 28.) Although Figure 29.23 shows only one Mg^{2+} ion in the RNA polymerase II active site, more recent studies reveal the presence of two. Why is the presence of two Mg^{2+} ions significant?

13. (Integrates with Chapter 11.) The SWI/SNF chromatin-remodeling complex peels about 50 bp from the nucleosome. Assuming B-form DNA, how long is this DNA segment? In forming nucleosomes, DNA is wrapped in turns about the histone core octamer. What fraction of a DNA turn around the core octamer does 50 bp of DNA comprise? How does 50 bp of DNA compare to the typical size of eukaryotic promoter modules and response elements?

14. Draw the structures that comprise the lariat branch point formed during mRNA splicing: the invariant A, its 5′-R neighbor, its 3′-Y neighbor, and its 2′-G neighbor.

15. (Integrates with Chapters 6 and 11.) The α-helices in HTH (helix-turn-helix motif) DNA-binding proteins are formed from 7– or 8–amino acid residues. What is the overall length of these α-helices? How does their length compare with the diameter of B-form DNA?

Preparing for the MCAT Exam

16. Figure 24.14 highlights in red the DNA phosphates that interact with catabolite activator protein (CAP). What kind of interactions do you suppose predominate and what kinds of CAP amino acid side chains might be involved in these interactions?

17. Chromatin decompaction is a preliminary step in gene expression (Figure 29.45). How is chromatin decompacted?

Biochemistry ⓔNow™   Preparing for an exam? Test yourself on key questions at http://chemistry.brookscole.com/ggb3

Further Reading

Transcription in Prokaryotes

Busby, S., and Ebright, R. H., 1994. Promoter structure, promoter recognition, and transcription activation in prokaryotes. *Cell* **79:**743–746.

Chan, C., Lonetto, M. A., and Gross, C. A., 1996. Sigma domain structure: One down, one to go. *Current Biology* **4:**1235–1238.

Roberts, C. W., and Roberts, J. W., 1996. Base-specific recognition of the nontemplate strand of promoter DNA by *E. coli* RNA polymerase. *Cell* **86:**495–501.

Vassylyev, D. G., et al., 2002. Crystal structure of a bacterial RNA polymerase at 2.6 Å resolution. *Nature* **417:**712–719.

Regulation of Transcription in Prokaryotes

Berg, O. G., and von Hippel, P. H., 1988. Selection of DNA binding sites by regulatory proteins. *Trends in Biochemical Sciences* **13:**207–211. A discussion of the quantitative binding aspects of DNA : protein interactions.

de Crombrugghe, B., Busby, S., and Buc, H., 1985. Cyclic AMP receptor protein: Role in transcription activation. *Science* **224:**831–838. The role of CAP in activating expression of prokaryotic genes.

Dover, S. L., et al., 1997. Activation of prokaryotic transcription through arbitrary protein–protein contacts. *Nature* **386:**627–630.

Jacob, F., and Monod, J., 1961. Genetic regulatory mechanisms in the synthesis of proteins. *Journal of Molecular Biology* **3:**318–356. The classic paper presenting the operon hypothesis.

Matthews, K. S., 1992. DNA looping. *Microbiological Reviews* **56:**123–136. A review of DNA looping as a general mechanism in the regulation of gene expression.

Platt, T., 1998. RNA structure in transcription elongation, termination, and antitermination. Pages 541–574 in Simons, R. W., and Grunberg-Monago, M., eds. *RNA Structure and Function*. Cold Spring Harbor, NY: Cold Spring Harbor Press.

Schleif, R., 1992. DNA looping. *Annual Review of Biochemistry* **61:**199–223. An excellent review of DNA looping as a regulatory mechanism in gene expression.

Transcription in Eukaryotes

Burley, S., 1998. X-ray crystallographic studies of eukaryotic transcription factors. *Cold Spring Harbor Symposium on Quantitative Biology* **LXIII:**33–40.

Burley, S. K., and Roeder, R. G., 1996. Biochemistry and structural biology of transcription factor IID (TFIID). *Annual Review of Biochemistry* **65:**769–799.

Conaway, R. C., and Conaway, J. W., 1999. Transcription elongation and human disease. *Annual Review of Biochemistry* **68:**301–319.

Cramer, D., et al., 2000. Architecture of RNA polymerase II and implications for the transcription mechanism. *Science* **288:**640–649.

Kornberg, R. D., 1998. Mechanism and regulation of yeast RNA polymerase II transcription. *Cold Spring Harbor Symposium on Quantitative Biology* **LXIII:**229–233.

Reinberg, D., et al., 1998. The RNA polymerase II general transcription factors: Past, present, and future. *Cold Spring Harbor Symposium on Quantitative Biology* **LXIII:**83–103.

Thomas, M. J., et al., 1998. Transcription fidelity and proofreading by RNA polymerase II. *Cell* **93:**627–637.

Regulation of Transcription in Eukaryotes

Bailey, C. H., Bartsch, D., and Kandel, E. R., 1996. Toward a molecular definition of long-term memory storage. *Proceedings of the National Academy of Sciences, U.S.A.* **93:**13445–13452.

Bjorklund, S., et al., 1999. Global transcription regulators of eukaryotes. *Cell* **96:**759–767.

Carey, M., and Smale, S. T., 2000. *Transcriptional Regulation in Eukaryotes: Concepts, Strategies, and Techniques*. New York: Cold Spring Harbor Laboratory Press.

Lucas, P. C., and Granner, D. K., 1992. Hormone response domains in gene transcription. *Annual Review of Biochemistry* **61:**1131–1173.

Moore, M. J., 2002. Nuclear RNA turnover. *Cell* **108:**431–434.

Struhl, K., 1999. Fundamentally different logic of gene regulation in prokaryotes and eukaryotes. *Cell* **98:**1–4.

Utley, R. T., et al., 1998. Transcriptional activators direct histone acetyl-transferase complexes to promoters. *Nature* **394**:498–502.

Tully, T., 1997. Regulation of gene expression and its role in long-term memory and synaptic plasticity. *Proceedings of the National Academy of Sciences, U.S.A.* **94**:4239–4241.

Winston, F., and Sudarsanam, P., 1998. The SAGA of Spt proteins and transcriptional analysis in yeast: Past, present, and future. *Cold Spring Harbor Symposium on Quantitative Biology* **LXIII**:553–560.

Nucleosome Structure and Gene Expression

Adelman, K., and Lis, J. T., 2002. How does pol II overcome the nucleosome barrier? *Molecular Cell* **9**:451–452.

Bell, A., et al., 1998. The establishment of active chromatin domains. *Cold Spring Harbor Symposium on Quantitative Biology* **LXIII**:509–514.

Boeger, H., et al., 2003. Nucleosomes unfold completely at a transcriptionally active promoter. *Molecular Cell* **11**:1587–1598.

Brown, C. E., et al., 2000. The many HATs of transcriptional coactivators. *Trends in Biochemical Sciences* **25**:15–19.

Fan, H. Y., et al., 2003. Distinct strategies to make nucleosomal DNA accessible. *Molecular Cell* **11**:1311–1322.

Felsenfeld, G., and Groudine, M., 2003. Controlling the double helix. *Nature* **421**:448–453. A review of the dynamic properties of chromatin structure.

Kassabov, S. R., et al., 2003. SWI/SNF unwraps, slides, and rewraps the nucleosome. *Molecular Cell* **11**:391–403.

Kornberg, R. D., and Lorch, Y., 1999. Twenty-five years of the nucleosome, fundamental particle of the eukaryotic chromosome. *Cell* **98**:285–294.

Ng, H. H., and Bird, A., 2000. Histone deacylases: Silencers for hire. *Trends in Biochemical Sciences* **25**:121–126.

Scxhnitzler, G. R., et al., 1998. A model for chromatin remodeling by the SW1/SNF family. *Cold Spring Harbor Symposium on Quantitative Biology* **LXIII**:535–542.

Struhl, K., 1998. Histone acetylation and transcriptional regulatory mechanisms. *Genes and Development* **12**:599–606.

Suka, N., et al., 1998. The regulation of gene activity by histones and the histone deacylase RPD3. *Cold Spring Harbor Symposium on Quantitative Biology* **LXIII**:391–399.

Workman, J. L., ed., 2003. *Protein Complexes that Modify Chromatin.* New York: Springer.

Wu, C., et al., 1998. ATP-dependent remodeling of chromatin. *Cold Spring Harbor Symposium on Quantitative Biology* **LXIII**:525–534.

Zaman, Z., et al., 1998. Gene transcription by recruitment. *Cold Spring Harbor Symposium on Quantitative Biology* **LXIII**:167–171.

Zlatanova, J., et al., 2000. Linker histone binding and displacement: Versatile mechanism for transcriptional regulation. *The FASEB Journal* **14**:1697–1704.

DNA-Binding Gene Regulatory Proteins

Berg, J. M., and Shi, Y., 1996. The galvanization of biology: A growing appreciation for the roles of zinc. *Science* **271**:1081–1085.

Edmondson, D. G., and Olson, E. N., 1993. Helix-loop-helix proteins as regulators of muscle-specific transcription. *Journal of Biological Chemistry* **268**:755–758.

Glover, J. N. M., and Harrison, S. C., 1995. Crystal structure of the heterodimeric *bZIP* transcription factor *c-Fos-c-Jun* bound to DNA. *Nature* **373**:257–261.

Johnson, P. F., and McKnight, S. L., 1989. Eukaryotic transcriptional regulatory proteins. *Annual Review of Biochemistry* **58**:799–839. A review of the structure and function of eukaryotic DNA-binding proteins that activate transcription.

Landschulz, W. H., Johnson, P. F., and McKnight, S. L., 1988. The leucine zipper: A hypothetical structure common to a new class of DNA-binding proteins. *Science* **240**:1759–1764.

Pabo, C. O., and Sauer, R. T., 1992. Transcription factors: Structural families and principles of DNA recognition. *Annual Review of Biochemistry* **61**:1053–1095.

Patikoglou, G., and Burley, S. K., 1997. Eukaryotic transcription factor-DNA complexes. *Annual Review of Biophysics and Biomolecular Structure* **26**:289–325.

Vinson, C. R., Sigler, P. B., and McKnight, S. L., 1989. Scissors-grip model for DNA recognition by a family of leucine zipper proteins. *Science* **246**:911–916. A model for the interaction of a *bZIP* protein with its DNA site.

Processing of Eukaryotic Transcripts

Breitbart, R. E., Andreadis, A., and Nadal-Ginard, B., 1987. Alternative splicing: A ubiquitous mechanism for the generation of multiple protein isoforms from single genes. *Annual Review of Biochemistry* **56**:467–495.

Kramer, A., 1996. The structure and function of proteins involved in mammalian pre-mRNA splicing. *Annual Review of Biochemistry* **65**:367–409.

Leff, S. E., Rosenfeld, M. G., and Evans, R. M., 1986. Complex transcriptional units: Diversity in gene expression by alternative RNA processing. *Annual Review of Biochemistry* **55**:1091–1117.

Sachs, A., and Wahle, E., 1993. Poly (A) tail metabolism and function in eucaryotes. *Journal of Biological Chemistry* **268**:22955–22958.

Sharp, P. A., 1987. Splicing of messenger RNA precursors. *Science* **235**:766–771.

Staley, J. P., and Guthrie, C., 1998. Mechanical devices of the spliceosome: Motors, clocks, springs, and things. *Cell* **92**:315–326.

A Unified Theory of Gene Expression

Maniatis, T., and Reed, R., 2002. An extensive network of coupling among gene expression machines. *Nature* **416**:499–506.

Narliker, G. J., Fan, H-Y., and Kingston, R. E., 2002. Cooperation between complexes that regulate chromatin structure and transcription. *Cell* **108**:475–487.

Orphanides, G., and Reinberg, D., 2002. A unified theory of gene expression. *Cell* **108**:439–451.

Schreiber, S. L., and Bernstein, B. E., 2002. Signaling network model of chromatin. *Cell* **111**:771–778. These authors argue that similar principles underlie chromatin remodeling and signal transduction cascades (see Chapter 32).

Woychik, N. A., and Hampsey, M., 2002. The RNA polymerase II machinery: Structure illuminates function. *Cell* **108**:439–451.

CHAPTER 30

Protein Synthesis

The Maya encoded their history in hieroglyphs carved on stelae and temples like these ruins in Tikal, Guatemala.

We are a spectacular, splendid manifestation of life. We have language and can build metaphors as skillfully and precisely as ribosomes make proteins. We have affection. We have genes for usefulness, and usefulness is about as close to a "common goal" of nature as I can guess at. And finally, and perhaps best of all, we have music.

Lewis Thomas (1913–1994), "The Youngest and Brightest Thing Around" in *The Medusa and the Snail* (1979)

Key Questions

30.1   What Is the Genetic Code?

30.2   How Is an Amino Acid Matched with Its Proper tRNA?

30.3   What Are the Rules in Codon–Anticodon Pairing?

30.4   What Is the Structure of Ribosomes, and How Are They Assembled?

30.5   What Are the Mechanics of mRNA Translation?

30.6   How Are Proteins Synthesized in Eukaryotic Cells?

Biochemistry⊘Now™   Test yourself on these Key Questions at BiochemistryNow at **http://chemistry.brookscole.com/ggb3**

Essential Question

Ribosomes synthesize proteins by reading the nucleotide sequence of mRNAs and polymerizing amino acids in an N→C direction. *How is the nucleotide sequence of an mRNA molecule translated into the amino acid sequence of a protein molecule?*

We turn now to the problem of how the sequence of nucleotides in an mRNA molecule is translated into the specific amino acid sequence of a protein. The problem raises both informational and mechanical questions. First, what is the **genetic code** that allows the information specified in a sequence of bases to be translated into the amino acid sequence of a polypeptide? That is, how is the 4-letter language of nucleic acids translated into the 20-letter language of proteins? Implicit in this question is a mechanistic problem: It is easy to see how base pairing establishes a one-to-one correspondence that allows the template-directed synthesis of polynucleotide chains in the processes of replication and transcription. However, there is no obvious chemical affinity between the purine and pyrimidine bases and the 20 different amino acids. Nor is there any obvious structural or stereochemical connection between polynucleotides and amino acids that might guide the translation of information.

Francis Crick reasoned that **adapter molecules** must bridge this information gap. These adapter molecules must interact specifically with both nucleic acids (mRNAs) and amino acids. At least 20 different adapter molecules would be needed, at least one for each amino acid. The various adapter molecules would be able to read the genetic code in an mRNA template and align the amino acids according to the template's directions so that they could be polymerized into a unique polypeptide. Transfer RNAs (tRNAs; Figure 30.1) are the adapter molecules (see Chapter 10). Amino acids are attached to the 3′-OH at the 3′-CCA end of tRNAs as aminoacyl esters. The formation of these aminoacyl-tRNAs, so-called charged tRNAs, is catalyzed by specific **aminoacyl-tRNA synthetases.** There is one of these enzymes for each of the 20 amino acids and each aminoacyl-tRNA synthetase loads its amino acid only onto tRNAs designed to carry it. In turn, these tRNAs specifically recognize unique sequences of bases in the mRNA through complementary base pairing.

30.1 | What Is the Genetic Code?

Once it was realized that the sequence of bases in a gene specified the sequence of amino acids in a protein, various possibilities for such a genetic code were considered. How many bases were necessary to specify each amino acid? Is the code overlapping or nonoverlapping (Figure 30.2)? Was the code punctuated or continuous? Mathematical considerations favored a triplet of bases as the minimal code word, or **codon,** for each amino acid: A doublet code based on pairs of the four possible bases, A, C, G, and U, has $4^2 = 16$ unique arrangements, an insufficient number to encode the 20 amino acids. A triplet code of four bases has $4^3 = 64$ possible code words, more than enough for the task.

The Genetic Code Is a Triplet Code

The genetic code is a triplet code read continuously from a fixed starting point in each mRNA. Specifically, it is defined by the following:

1. A group of three bases codes for one amino acid.
2. The code is not overlapping.

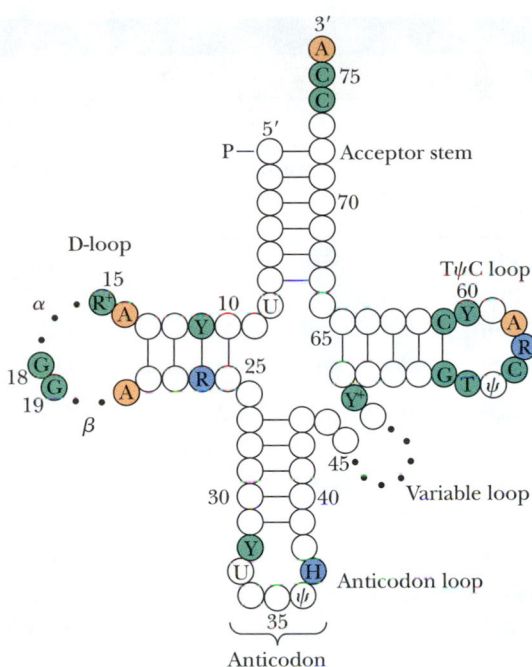

FIGURE 30.1 The general structure of tRNA molecules. Circles represent nucleotides in the tRNA sequence. The numbers given indicate the standardized numbering system for tRNAs (which differ in total number of nucleotides). Dots indicate places where the number of nucleotides may vary in different tRNA species. All tRNAs have the invariant 3-base sequence CCA at their 3′-ends. Recall from Chapter 10 that tRNA molecules often have modified or unusual bases.

3. The base sequence is read from a fixed starting point without punctuation. That is, the mRNA sequences contain no "commas" signifying appropriate groupings of triplets. If the reading frame is displaced by one base, it remains shifted throughout the subsequent message; no "commas" are present to restore the "correct" frame.

4. The code is **degenerate,** meaning that, in most cases, each amino acid can be coded by any of several triplets. Recall that a triplet code yields 64 codons for 20 amino acids. Most codons (61 of 64) code for some amino acid.

Codons Specify Amino Acids

The complete translation of the genetic code is presented in Table 30.1. Codons, like other nucleotide sequences, are read 5′→3′. Codons represent triplets of bases in mRNA or, replacing U with T, triplets along the nontranscribed (nontemplate) strand of DNA.

Several noteworthy features characterize the genetic code:

1. *All the codons have meaning.* Of the 64 codons, 61 specify particular amino acids. The remaining 3—UAA, UAG, and UGA—specify no amino acid and thus they are **nonsense codons.** Nonsense codons serve as **termination codons;** they are "stop" signals indicating that the end of the protein has been reached.
2. *The genetic code is unambiguous.* Each of the 61 "sense" codons encodes only one amino acid.
3. *The genetic code is degenerate.* With the exception of Met and Trp, every amino acid is coded by more than one codon. Several—Arg, Leu, and Ser—are represented by six different codons. Codons coding for the same amino acid are called **synonymous codons.**
4. *Codons representing the same amino acid or chemically similar amino acids tend to be similar in sequence.* Often the third base in a codon is irrelevant, so, for example, all four codons in the GGX family specify Gly, and the UCX family specifies Ser (Table 30.1). This feature is known as **third-base degeneracy.** Note also that codons with a pyrimidine as second base likely encode amino acids with hydrophobic side chains, and codons with a purine in the second-base position typically specify polar or charged amino acids.

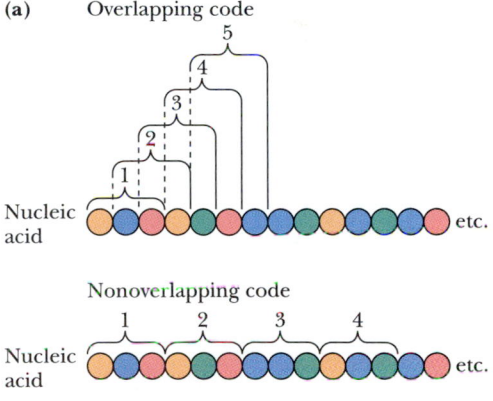

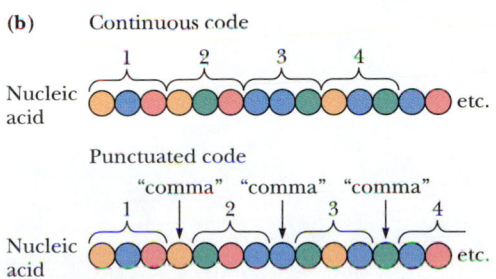

FIGURE 30.2 **(a)** An overlapping versus a nonoverlapping code. **(b)** A continuous versus a punctuated code.

Table 30.1

The Genetic Code

First Position (5'-end)	Second Position				Third Position (3'-end)
	U	**C**	**A**	**G**	
U	UUU Phe	UCU Ser	UAU Tyr	UGU Cys	U
	UUC Phe	UCC Ser	UAC Tyr	UGC Cys	C
	UUA Leu	UCA Ser	UAA Stop	UGA Stop	A
	UUG Leu	UCG Ser	UAG Stop	UGG Trp	G
C	CUU Leu	CCU Pro	CAU His	CGU Arg	U
	CUC Leu	CCC Pro	CAC His	CGC Arg	C
	CUA Leu	CCA Pro	CAA Gln	CGA Arg	A
	CUG Leu	CCG Pro	CAG Gln	CGG Arg	G
A	AUU Ile	ACU Thr	AAU Asn	AGU Ser	U
	AUC Ile	ACC Thr	AAC Asn	AGC Ser	C
	AUA Ile	ACA Thr	AAA Lys	AGA Arg	A
	AUG Met*	ACG Thr	AAG Lys	AGG Arg	G
G	GUU Val	GCU Ala	GAU Asp	GGU Gly	U
	GUC Val	GCC Ala	GAC Asp	GGC Gly	C
	GUA Val	GCA Ala	GAA Glu	GGA Gly	A
	GUG Val	GCG Ala	GAG Glu	GGG Gly	G

*AUG signals translation initiation as well as coding for Met residues.

Third-Base Degeneracy Is Color-Coded

	Third-Base Relationship	Third Bases with Same Meaning	Number of Codons
	Third base irrelevant	U, C, A, G	32 (8 families)
	Purines	A or G	12 (6 pairs)
	Pyrimidines	U or C	14 (7 pairs)
	Three out of four	U, C, A	3 (AUX = Ile)
	Unique definitions	G only	2 (AUG = Met) (UGG = Trp)
	Unique definition	A only	1 (UGA = Stop)

The two negatively charged amino acids, Asp and Glu, are encoded by GAX codons; GA–pyrimidine gives Asp and GA–purine specifies Glu. The consequence of these similarities is that mutations are less likely to be harmful because single base changes in a codon will result either in no change or in a substitution with an amino acid similar to the original amino acid. The degeneracy of the code is evolution's buffer against mutational disruption.

5. *The genetic code is "universal."* Although certain minor exceptions in codon usage occur (see A Deeper Look box on page 989), the more striking feature of the code is its universality: Codon assignments are virtually the same throughout all organisms—archaea, eubacteria, and eukaryotes. This conformity means that all extant organisms use the same genetic code, providing strong evidence that they all evolved from a common primordial ancestor.

A Deeper Look

Natural Variations in the Standard Genetic Code

The genomes of some lower eukaryotes, prokaryotes, and mitochondria show some exceptions to the standard genetic code (Table 30.1) in codon assignments. The phenomenon is more common in mitochondria. For example, the termination codon UGA codes for tryptophan in mitochondria from various animals, protozoans, and fungi. AUA, normally an Ile codon, codes for methionine in some animal and fungal mitochondrial genomes, and AGA (an Arg codon) is a termination codon in vertebrate mitochondria but is a Ser codon in fruit fly mitochondria. Mitochondria in several species of yeast use the CUX codons to specify Thr instead of Leu. Some yeast and algal mitochondria use CGG, normally an Arg codon, as a stop codon.

Less common are genomic codon variations within the genomes of prokaryotic and eukaryotic cells. Among the lower eukaryotes, certain ciliated protozoans (*Tetrahymena* and *Paramecium*) use UAA and UGA as glutamine codons rather than stop codons. Instances in prokaryotes include use of the stop codon UGA to specify Trp by *Mycoplasma*. Perhaps most interesting is

the use of some UGA codons by both prokaryotes and eukaryotes (including humans) to specify **selenocysteine,** an analog of cysteine in which the sulfur atom is replaced by a selenium atom. Indeed, the identification of selenocysteine residues in proteins from bacteria, archaea, and eukaryotes has led some people to nominate selenocysteine as the 21st amino acid! Selenocysteine formation requires a novel selenocysteine-specific tRNA known as tRNA^{Sec}. This tRNA^{Sec} is loaded with a Ser residue by seryl-tRNA synthetase, the aminoacyl-tRNA synthetase for serine. Then, in an ATP-dependent process, the Ser-O is replaced by Se. Translation of UGA codons by selenocysteinyl-tRNA^{Sec} depends on the presence of specific stem-loop secondary structures in the mRNA called **SECIS elements** that recode the UGA codon from "stop" to "Sec." SECIS elements are recognized by specific proteins that recruit selenocysteinyl-tRNA^{Sec} to the UGA codon during protein synthesis.

Adapted from Fox, T. D., 1987. Natural variation in the genetic code. *Annual Review of Genetics* **21:**67–91; Knight, R. D., et al., 1999. Selection, history, and chemistry: Three faces of the genetic code. *Trends in Biochemical Sciences* **24:**241–247; Low, S. C., and Berry, M. J., 1996. Knowing when not to stop. *Trends in Biochemical Sciences* **21:**203–208; and Zavacki, A. M., et al., 2003. Coupled tRNA^{Sec}-dependent assembly of the selenocysteine decoding apparatus. *Molecular Cell* **11:**773–781.

$$H-Se-CH_2-\overset{\overset{\displaystyle H}{|}}{\underset{\underset{\displaystyle NH_3^+}{|}}{C}}-COO^-$$

Selenocysteine

30.2 | How Is an Amino Acid Matched with Its Proper tRNA?

Codon recognition is achieved by aminoacyl-tRNAs. In order for accurate translation to occur, the appropriate aminoacyl-tRNA must "read" the codon through base pairing via its **anticodon loop** (see Chapter 11). Once an aminoacyl-tRNA has been synthesized, the amino acid part makes no contribution to accurate translation of the mRNA. That is, the amino acid is passively chauffeured by its tRNA and becomes inserted into a growing peptide chain following codon–anticodon recognition between the mRNA and tRNA.

Aminoacyl-tRNA Synthetases Interpret the Second Genetic Code

A **second genetic code** must exist, the code by which each aminoacyl-tRNA synthetase matches up its amino acid with tRNAs that can interact with codons specifying its amino acid. To interpret this second genetic code, an aminoacyl-tRNA synthetase must discriminate between the 20 amino acids and the many tRNAs and uniquely picks out its proper substrates—one specific amino acid and the tRNA(s) appropriate to it—from among the more than 400 possible combinations. The appropriate tRNAs are those having anticodons that can base-pair with the codons specifying the particular amino acid. It is imperative that the proper amino acids be loaded onto the various tRNAs so that the mRNA can be translated with fidelity. Although the primary genetic code is key to understanding the central dogma of molecular biology on how DNA encodes proteins, the second genetic code is just as crucial to the fidelity of information transfer.

Cells have 20 different aminoacyl-tRNA synthetases, one for each amino acid. Each of these enzymes catalyzes ATP-dependent attachment of its specific

Cognate kindred; in this sense, cognate refers to those tRNAs having anticodons that can read one or more of the codons that specify one particular amino acid.

amino acid to the 3′-end of its **cognate tRNA molecules** (Figure 30.3). The aminoacyl-tRNA synthetase reaction serves two purposes:

1. It activates the amino acid so that it will readily react to form a peptide bond.
2. It bridges the information gap between amino acids and codons.

The underlying mechanisms of molecular recognition used by each aminoacyl-tRNA synthetase to bring the proper amino acid to its cognate tRNA are the embodiment of the second genetic code.

Evolution Has Provided Two Distinct Classes of Aminoacyl-tRNA Synthetases

Despite their common enzymatic function, aminoacyl-tRNA synthetases are a diverse group of proteins in terms of size, amino acid sequence, and oligomeric structure. In higher eukaryotes, some aminoacyl-tRNA synthetases are assembled into large multiprotein complexes. The aminoacyl-tRNA synthetases fall into two fundamental classes on the basis of similar amino acid sequence motifs, oligomeric state, and acylation function (Table 30.2): class I and class II. Class I aminoacyl-tRNA synthetases first add the amino acid to the 2′-OH of the terminal adenylate residue of tRNA before shifting it to the 3′-OH; class II enzymes add it directly to the 3′-OH (Figure 30.3). The catalytic domains of these enzymes evolved from two different ancestral predecessors. Aminoacyl-tRNA synthetases are ranked among the oldest proteins because the different classes of these enzymes were present very early in evolution. Class I and class II aminoacyl-tRNA synthetases interact with the tRNA 3′-terminal CCA and acceptor stem in a mirror-symmetric fashion with respect to each other (Figure 30.4). Class I enzymes bind to the tRNA acceptor stem helix from the minor-groove side, whereas class II enzymes bind it from the major-groove side.

Both class I and class II aminoacyl-tRNA synthetases can be approximated as two-domain structures, as can their L-shaped tRNA substrates, which have the acceptor stem/CCA-3′-OH at one end and the anticodon stem-loop at the other (see Figures 11.34 and 30.5). This L-shaped tertiary structure of tRNAs separates the 3′-CCA acceptor end from the anticodon loop by a distance of 7.6 nm. The two domains of tRNAs have distinct functions: The 3′-CCA end is the site of aminoacylation, and the anticodon-containing domain interacts with the mRNA template. The two domains of tRNAs interact with the separate domains in the synthetases. One of the two major aminoacyl-tRNA synthetase domains is the catalytic domain (which defines the difference between class I and class II enzymes); this domain interacts with the tRNA 3′-CCA end. The other major domain in aminoacyl-tRNA synthetases is highly variable and interacts with parts of the tRNA beyond the acceptor-TΨC stem-loop domain, including, in some cases, the anticodon.

Aminoacyl-tRNA Synthetases Can Discriminate Between the Various tRNAs

Aside from the need to uniquely recognize their cognate amino acids, aminoacyl-tRNA synthetases must be able to discriminate between the various tRNAs. The structural features that permit the synthetases to recognize and aminoacylate their cognate tRNA(s) are *not* universal. That is, a common set of rules does not govern tRNA recognition by these enzymes. Most surprising is the fact that the recognition features are not limited to the anticodon and, in some instances, do not even include the anticodon. For most tRNAs, a set of sequence elements is recognized by its specific aminoacyl-tRNA synthetase, rather than a single distinctive nucleotide or base pair. These elements include one or more of the following: (1) at least one base in the anticodon; (2) one or more of the three base pairs in the acceptor stem; and (3) the base at canonical position 73 (the unpaired base preceding the CCA end), re-

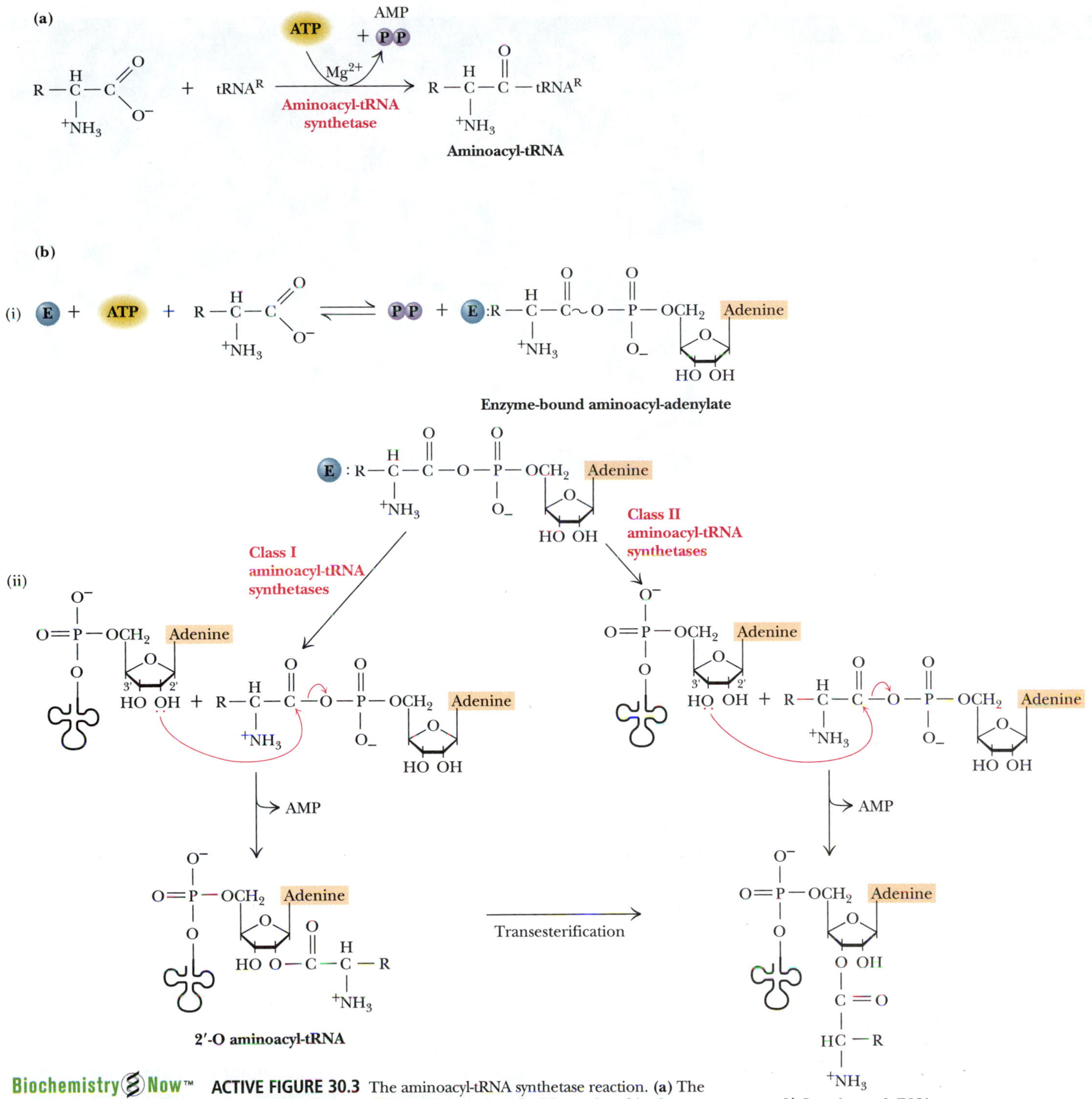

(a)

Aminoacyl-tRNA

(b)

(i)

Enzyme-bound aminoacyl-adenylate

(ii)

Class I aminoacyl-tRNA synthetases

Class II aminoacyl-tRNA synthetases

AMP

Transesterification

AMP

2′-O aminoacyl-tRNA

3′-O aminoacyl-tRNA

Biochemistry⊗Now™   ACTIVE FIGURE 30.3 The aminoacyl-tRNA synthetase reaction. **(a)** The overall reaction. Ever-present pyrophosphatases in cells quickly hydrolyze the PP$_i$ produced in the aminoacyl-tRNA synthetase reaction, rendering aminoacyl-tRNA synthesis thermodynamically favorable and essentially irreversible. **(b)** The overall reaction commonly proceeds in two steps: (i) formation of an aminoacyl-adenylate and (ii) transfer of the activated amino acid moiety of the mixed anhydride to either the 2′-OH (class I aminoacyl-tRNA synthetases) or 3′-OH (class II aminoacyl-tRNA synthetases) of the ribose on the terminal adenylic acid at the 3′-CCA terminus common to all tRNAs. Those aminoacyl-tRNAs formed as 2′-aminoacyl esters undergo a transesterification that moves the aminoacyl function to the 3′-O of tRNA. Only the 3′-esters are substrates for protein synthesis. **Test yourself on the concepts in this figure at http://chemistry.brookscole.com/ggb3**

ferred to as the **discriminator base** because this base is invariant in the tRNAs for a particular amino acid. Figure 30.5 presents a ribbon diagram of a tRNA molecule showing the location of nucleotides that contribute to specific recognition by the respective aminoacyl-tRNA synthetases for each of the 20 amino acids. Interestingly, the same set of tRNA features that serves as

Table 30.2

The Two Classes of Aminoacyl-tRNA Synthetases

Class I	Class II
Arg	Ala
Cys	Asn
Gln	Asp
Glu	Gly
Ile	His
Leu	Lys
Met	Phe
Trp	Pro
Tyr	Ser
Val	Thr

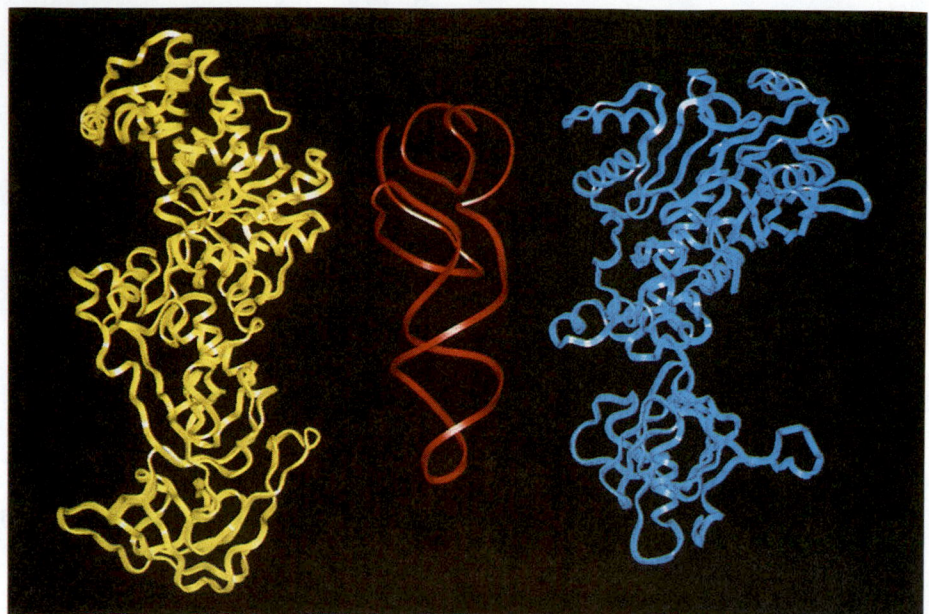

FIGURE 30.4 Mirror-symmetric interactions of class I versus class II aminoacyl-tRNA synthetases with their tRNA substrates. The two different classes of aminoacyl-tRNA synthetases bind to opposite faces of tRNA molecules. On the left is a space-filling model of the class I glutaminyl-tRNA^{Gln} synthetase. Class I synthetases bind to the side of their tRNA substrates shown as closest in this figure (the model tRNA structure is tRNA^{Phe} for purposes of illustration). On the right is a space-filling model of the class II aspartyl-tRNA^{Asp} synthetase; this class of synthetase binds to the side of tRNA closest to it here. *(Adapted from Figure 5 in Arnez, J. G., and Moras, D., 1997. Structural and functional considerations of the aminoacylation reaction.* Trends in Biochemical Sciences **22:**211–216.)

positive determinants for binding and aminoacylation of the tRNA by its cognate aminoacyl-tRNA synthetase may act as negative determinants that prohibit binding and aminoacylation by other (noncognate) aminoacyl-tRNA synthetases. Because no common set of rules exists, the second genetic code is an **operational code** based on aminoacyl-tRNA synthetase recognition of varying sequence and structural features in the different tRNA molecules during the operation of aminoacyl-tRNA synthesis. Some examples of this code are given in Figure 30.6.

Escherichia coli Glutaminyl-tRNA^{Gln} Synthetase Recognizes Specific Sites on tRNA^{Gln}

E. coli glutaminyl-tRNA^{Gln} synthetase, a class I enzyme, provides a good illustration of aminoacyl-tRNA synthetase:cognate tRNA interactions. This glutaminyl-tRNA^{Gln} synthetase shares a continuous interaction with its cognate tRNA that extends from the anticodon to the acceptor stem along the entire inside of the L-shaped tRNA (Figure 30.7). Specific recognition elements include enzyme contacts with the discriminator base, acceptor stem, and anticodon, particularly the central U in the CUG anticodon. The carboxylate group of Asp^{235} makes sequence-specific H bonds in the tRNA minor groove with the 2-NH$_2$ group of G3 in the base pair G3:C70 of the acceptor stem. A mutant glutaminyl-tRNA^{Gln} synthetase with Asn substituted for Asp at position 235 shows relaxed specificity; that is, it now will acylate noncognate tRNAs with Gln.

The Identity Elements Recognized by Some Aminoacyl-tRNA Synthetases Reside in the Anticodon

Alteration of the anticodons of either tRNA^{Trp} or tRNA^{Val} to CAU, the anticodon for the methionine codon AUG, transforms each of the tRNAs into a substrate for methionyl-tRNA synthetase, and they are loaded with methio-

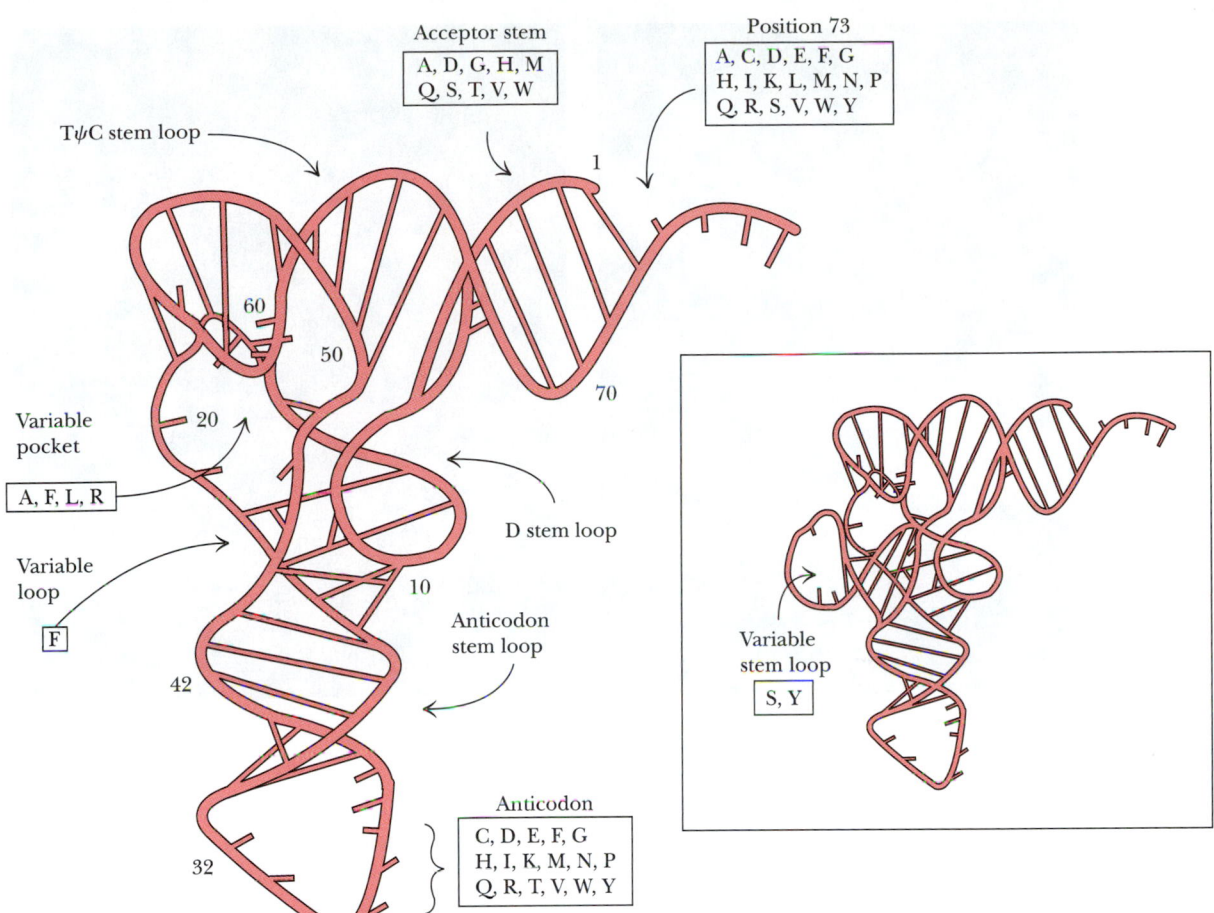

FIGURE 30.5 Ribbon diagram of tRNA tertiary structure. Numbers represent the consensus nucleotide sequence (see Figure 30.1). The locations of nucleotides recognized by the various aminoacyl-tRNA synthetases are indicated; shown within the boxes are one-letter designations of the amino acids whose respective aminoacyl-tRNA synthetases interact at the discriminator base (position 73), acceptor stem, variable pocket and/or loop, or anticodon. The inset shows additional recognition sites in those tRNAs having a variable loop that forms a stem-loop structure. *(Adapted from Figure 2 in Saks, M. E., Sampson, J. R., and Abelson, J. N., 1994. The transfer RNA problem: A search for rules. Science 263:191–197.)*

nine. Similarly, reversing the methionine CAU anticodon of tRNA^Met to UAC transforms it into a substrate for valyl-tRNA^Val synthetase. Clearly, methionyl-tRNA synthetase and valyl-tRNA synthetase rely on the anticodon in selecting tRNAs for loading.

A Single G:U Base Pair Defines tRNA^Ala s

The noncanonical base pair, G3:U70, is the singular feature by which alanyl-tRNA^Ala synthetase recognizes tRNAs as its substrates. All tRNA^Ala representatives, from Archaebacteria to eukaryotes, possess this G3:U70 acceptor stem base pair. Altering this unusual G3:U70 base pair of tRNA^Ala to G:C, A:U, or even U:G abolishes its ability to be aminoacylated with alanine. On the other hand, provided the G3:U70 base pair is present, alanyl-tRNA^Ala synthetase aminoacylates a 24-nucleotide stem-loop analog of tRNA^Ala (Figure 30.8). The key feature of the G3:U70 base pair is the 2-NH$_2$ group of G3. In the RNA A-form double-helical structure adopted by the tRNA acceptor stem, the G3 2-HN$_2$ group is exposed in the minor groove of the helix, and if the G3 pairing partner is a U, this 2-NH$_2$ group lacks an H-bonding partner (Figure 30.8). Thus, an unpaired G 2-amino group at the right place in a tRNA acceptor stem marks a tRNA for aminoacylation by alanyl-tRNA^Ala synthetase.

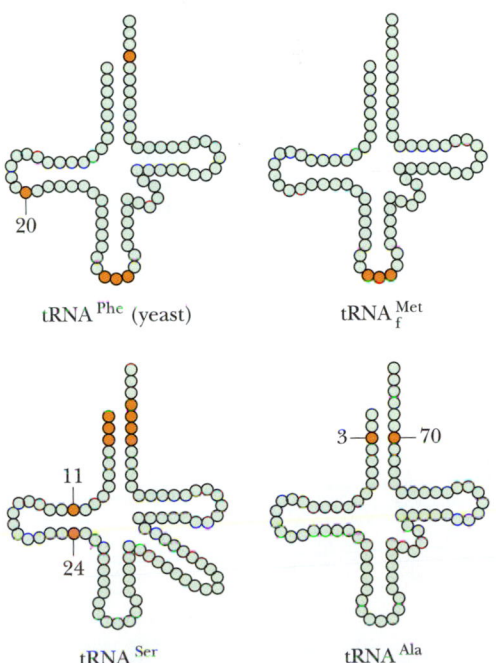

FIGURE 30.6 Major identity elements in four tRNA species. Each base in the tRNA is represented by a circle. Numbered filled circles indicate positions of identity elements within the tRNA that are recognized by its specific aminoacyl-tRNA synthetase. *(Adapted from Schulman, L. H., and Abelson, J., 1988. Recent excitement in understanding transfer RNA identity. Science 240:1591–1592.)*

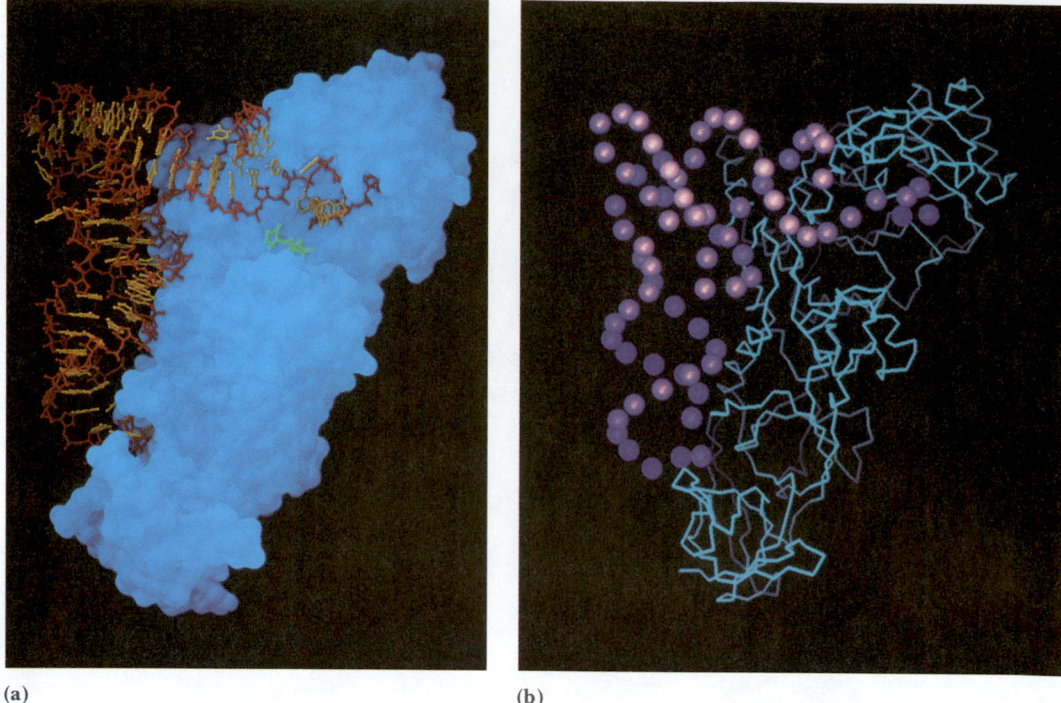

(a)   (b)

FIGURE 30.7 **(a)** A solvent-accessible representation of *E. coli* glutaminyl-tRNA^{Gln} synthetase complexed with tRNA^{Gln} and ATP, derived from analysis of the crystal structure of the complex. The protein is colored blue. The sugar-phosphate backbone of the tRNA is red; its bases are yellow. The protein:tRNA contact region extends along one side of the entire length of this extended protein. The acceptor stem of the tRNA and the ATP *(green)* fit into a cleft at the top of the protein in this view. The enzyme also interacts extensively with the anticodon (lower tip of tRNA^{Gln}). **(b)** Diagram showing the structure of tRNA^{Gln}, as represented by its phosphorus atoms *(purple spheres),* in complex with *E. coli* glutaminyl-tRNA^{Gln} synthetase, as represented in the terms of its Cα atoms *(blue).* [*(a) adapted from Rould, M. A., et al., 1989. Structure of* E. coli *glutaminyl-tRNA synthetase complexed with* tRNA^{Gln} *and ATP at 2.8 Å resolution.* Science **246**:1135; *photo courtesy of Thomas A. Steitz of Yale University.*]

30.3 What Are the Rules in Codon–Anticodon Pairing?

Protein synthesis depends on the codon-directed binding of the proper aminoacyl-tRNAs so that the right amino acids are sequentially aligned according to the specifications of the mRNA undergoing translation. This alignment is achieved via codon–anticodon pairing in antiparallel orientation (Figure 30.9). However, considerable degeneracy exists in the genetic code at the third position. Conceivably, this degeneracy could be handled in either of two ways: (1) Codon–anticodon recognition could be highly specific so that a complementary anticodon is required for each codon, or (2) fewer than 61 anticodons could be used for the "sense" codons if certain allowances were made in the base-pairing rules. Then, some anticodons could recognize more than one codon. As early as 1965, it was known that poly(U) bound *all* Phe-tRNA^{Phe} molecules even though UUC is also a Phe codon. The phenylalanine-specific tRNAs could recognize either UUU or UUC. Also, one particular yeast tRNA^{Ala} was able to bind to three codons: GCU, GCC, and GCA.

Francis Crick Proposed the "Wobble" Hypothesis for Codon:Anticodon Pairing

Francis Crick considered these results and tested alternative base-pairing possibilities by model building. He hypothesized that the first two bases of the codon and the last two bases of the anticodon form canonical Watson–Crick A:U or G:C base pairs, but pairing between the third base of the codon and the first base of the anticodon follows less stringent rules. That is, a certain amount of play, or **wobble,** might be allowed in base pairing at this position. The third base of the codon is sometimes referred to as the **wobble position.**

Crick's investigations suggested a set of rules for pairing between the third base of the codon and the first base of the anticodon (Table 30.3). The wobble rules indicate that a first-base anticodon U could recognize either an A or G in the codon third-base position; first-base anticodon G might recognize either U

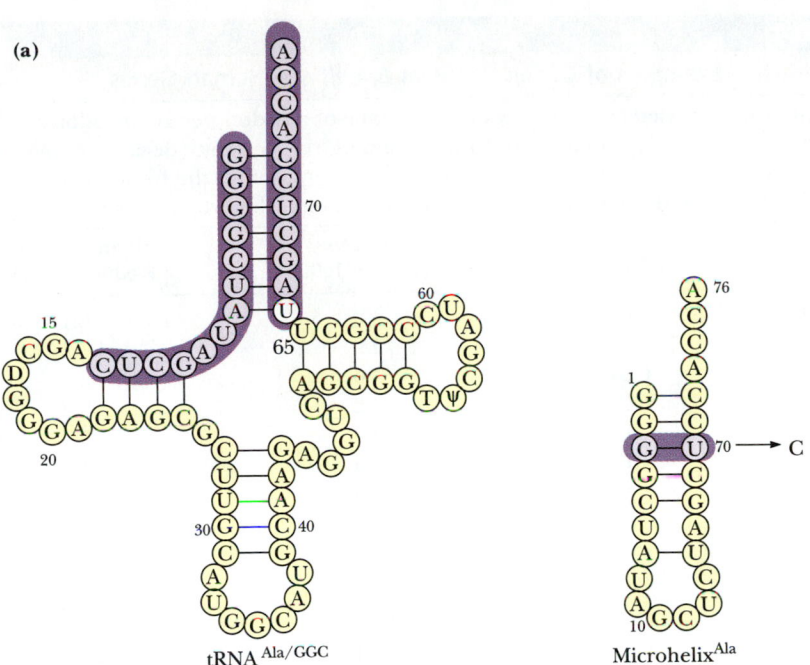

tRNA$^{Ala/GGC}$ Microhelix^{Ala}

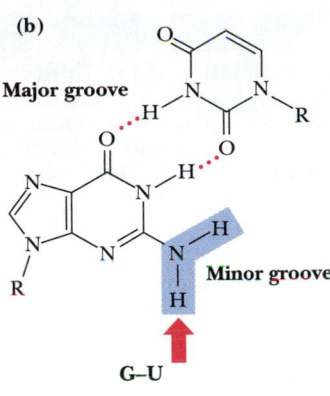

FIGURE 30.8 A **(a)** A microhelix analog of tRNA^{Ala} is aminoacylated by alanyl-tRNA^{Ala} synthetase, provided it has the characteristic tRNA^{Ala} G3:U70 acceptor stem base pair. The microhelix^{Ala} consists of nucleotides 1 through 13 of tRNA$^{Ala/GGC}$ connected directly to 66 through 76 to re-create the tRNA^{Ala} 7-bp acceptor stem. **(b)** The unbonded 2-HN$_2$ group of the G3 in the G3:U70 base pair would lie within the minor groove of an A-form RNA double helix. [(a) *Adapted from Schimmel, P., 1989. Parameters for molecular recognition of transfer RNAs.* Biochemistry **28**:2747–2759.]

or C in the third-base position of the codon; and first-base anticodon I[1] might interact with U, C, or A in the codon third position.[2]

The wobble rules also predict that four-codon families (like Pro or Thr), where any of the four bases may be in the third position, require at least two different tRNAs. However, all members of the set of tRNAs specific for a particular amino acid—termed **isoacceptor tRNAs**—are served by one aminoacyl-tRNA synthetase.

Some Codons Are Used More Than Others

Because more than one codon exists for most amino acids, the possibility for variation in codon usage arises. Indeed, variation in codon usage accommodates the fact that the DNA of different organisms varies in relative A:T/G:C content. Nevertheless, even in organisms of average base composition, codon usage may be biased. Table 30.4 gives some examples from *E. coli* and humans reflecting the nonrandom usage of codons. Of more than 109,000 Leu codons tabulated in human genes, CUG was used in excess of 48,000 times, CUC more than 23,000 times, but UUA just 6000 times.

The occurrence of codons in *E. coli* mRNAs correlates well with the relative abundance of the tRNAs that read them. Preferred codons are represented by the most abundant isoacceptor tRNAs. Furthermore, mRNAs for proteins that are synthesized in abundance tend to employ preferred codons. Rare tRNAs correspond to rarely used codons, and messages containing such codons might experience delays in translation.

Nonsense Suppression Occurs When Suppressor tRNAs Read Nonsense Codons

Mutations that alter a sense codon to one of the three nonsense codons—UAA, UAG, or UGA—result in premature termination of protein synthesis and the release of truncated (incomplete) polypeptides. Geneticists found that second

---

[1]I is inosine (6-OH purine).
[2]Thus, the first base of the anticodon indicates whether the tRNA can read one, two, or three different codons: Anticodons beginning with A or C read only one codon, those beginning with G or U read two, and anticodons beginning with I can read three codons.

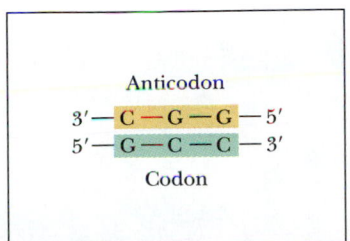

Biochemistry ⊗ **Now**™ **ANIMATED FIGURE 30.9**
Codon–anticodon pairing. Complementary trinucleotide sequence elements align in antiparallel fashion. **See this figure animated at http://chemistry. brookscole.com/ggb3**

Table 30.3
Base-Pairing Possibilities at the Third Position of the Codon

Base on the Anticodon	Bases Recognized on the Codon
U	A, G
C	G
A	U
G	U, C
I	U, C, A

Adapted from Crick, F. H. C., 1966. Codon–anticodon pairing: The wobble hypothesis. *Journal of Molecular Biology* **19**:548–555.

Table 30.4
Representative Examples of Codon Usage in *E. coli* and Human Genes
The results are expressed as frequency of occurrence of a codon per 1000 codons tabulated in 1562 *E. coli* genes and 2681 human genes, respectively. (Because *E. coli* and human proteins differ somewhat in amino acid composition, the frequencies for a particular amino acid do not correspond exactly between the two species.)

Amino Acid	Codon	*E. coli* Gene Frequency/1000	Human Gene Frequency/1000
Leu	CUA	3.2	6.1
	CUC	9.9	20.1
	CUG	54.6	42.1
	CUU	10.2	10.8
	UUA	10.9	5.4
	UUG	11.5	11.1
Pro	CCA	8.2	15.4
	CCC	4.3	20.6
	CCG	23.8	6.8
	CCU	6.6	16.1
Ala	GCA	15.6	14.4
	GCC	34.4	29.7
	GCG	32.9	7.2
	GCU	13.4	18.9
Lys	AAA	36.5	21.9
	AAG	12.0	35.2
Glu	GAA	43.5	26.4
	GAG	19.2	41.6

Adapted from Wada, K., et al., 1992. Codon usage tabulated from Genbank genetic sequence data. *Nucleic Acids Research* **20**:2111–2118.

mutations elsewhere in the genome were able to *suppress* the effects of nonsense mutations so that the organism survived, a phenomenon termed **nonsense suppression.** The molecular basis for nonsense suppression was a mystery until it was realized that **suppressors** were mutations in tRNA genes that altered the anticodon so that the mutant tRNA could now read a particular "stop" codon and insert an amino acid. For example, alteration of the anticodon of a tRNA^{Tyr} from GUA to CUA allows this tRNA to read the *amber* stop codon, UAG, and insert Tyr. (The nonsense codons are named *amber* [UAG], *ochre* [UAA], and *opal* [UGA]). **Suppressor tRNAs** are typically generated from minor tRNA species within a set of isoacceptor tRNAs, so their recruitment to a new role via mutation does not involve loss of an essential tRNA; that is, the mutation is not particularly deleterious to the organism. Suppressor tRNAs don't necessarily carry the same amino acid as the tRNA produced by the unmutated tRNA gene.

30.4	**What Is the Structure of Ribosomes, and How Are They Assembled?**

Protein biosynthesis is achieved by the process of **translation.** Translation converts the language of genetic information embodied in the base sequence of a messenger RNA molecule into the amino acid sequence of a polypeptide chain. During translation, proteins are synthesized on ribosomes by linking amino acids together in the specific linear order stipulated by the sequence of codons in an mRNA. Ribosomes are the agents of protein synthesis.

Ribosomes are compact ribonucleoprotein particles found in the cytosol of all cells, as well as in the matrix of mitochondria and the stroma of chloroplasts. The general structure of ribosomes is described in Chapter 10; here we consider

Table 30.5

Structural Organization of *E. coli* Ribosomes

	Ribosome	Small Subunit	Large Subunit
Sedimentation coefficient	70S	30S	50S
Mass (kD)	2520	930	1590
Major RNAs		16S = 1542 bases	23S = 2904 bases
Minor RNAs			5S = 120 bases
RNA mass (kD)	1664	560	1104
RNA proportion	66%	60%	70%
Protein number		21 polypeptides	31 polypeptides
Protein mass (kD)	857	370	487
Protein proportion	34%	40%	30%

their structure in light of their function in synthesizing proteins. Ribosomes are mechanochemical systems that move along mRNA templates, orchestrating the interactions between successive codons and the corresponding anticodons presented by aminoacyl-tRNAs. As they align successive amino acids via codon–anticodon recognition, ribosomes also catalyze the formation of peptide bonds between adjacent amino acid residues.

Prokaryotic Ribosomes Are Composed of 30S and 50S Subunits

E. coli ribosomes are representative of the structural organization of the prokaryotic versions of these supramolecular protein-synthesizing machines (Table 30.5, see also Figure 10.25). The *E. coli* ribosome is a roughly globular particle with a diameter of 25 nm, a sedimentation coefficient of 70S, and a mass of about 2520 kD. It consists of two unequal subunits that dissociate from each other at Mg^{2+} concentrations below 1 mM. The smaller, or **30S,** subunit is composed of 21 different proteins and a single rRNA, **16S ribosomal RNA (rRNA).** The larger **50S** subunit consists of 31 different proteins and two rRNAs: **23S rRNA** and **5S rRNA.** Ribosomes are roughly two-thirds RNA and one-third protein by mass. An *E. coli* cell contains around 20,000 ribosomes, constituting about 20% of the dry cell mass.

Prokaryotic Ribosomes Are Made from 50 Different Proteins and Three Different RNAs

Ribosomal Proteins   There is one copy of each ribosomal protein per 70S ribosome, excepting protein L7/L12 (L7 and L12 have identical amino acid sequences and differ only in the degree of N-terminal acetylation). Only one protein is common to both the small and large subunit: S20 = L26. The largest ribosomal protein is S1 (557 residues, 61.2 kD); the smallest is L34 (46 residues, 5.4 kD). The sequences of ribosomal proteins share little similarity. These proteins are typically rich in the cationic amino acids Lys and Arg and have few aromatic amino acid residues, properties appropriate to proteins intended to interact strongly with polyanionic RNAs.

rRNAs   The three *E. coli* rRNA molecules—23S, 16S, and 5S—are derived from a single 30S rRNA precursor transcript that also includes several tRNAs (Figure 30.10). Ribosomal RNAs show extensive potential for intrachain hydrogen bonding and assume secondary structures reminiscent of tRNAs, although substantially more complex (see Figures 11.35 and 11.37). About two-thirds of rRNA is double-helical. Double-helical regions are punctuated by short, single-stranded stretches, generating hairpin conformations that dominate the molecule; four

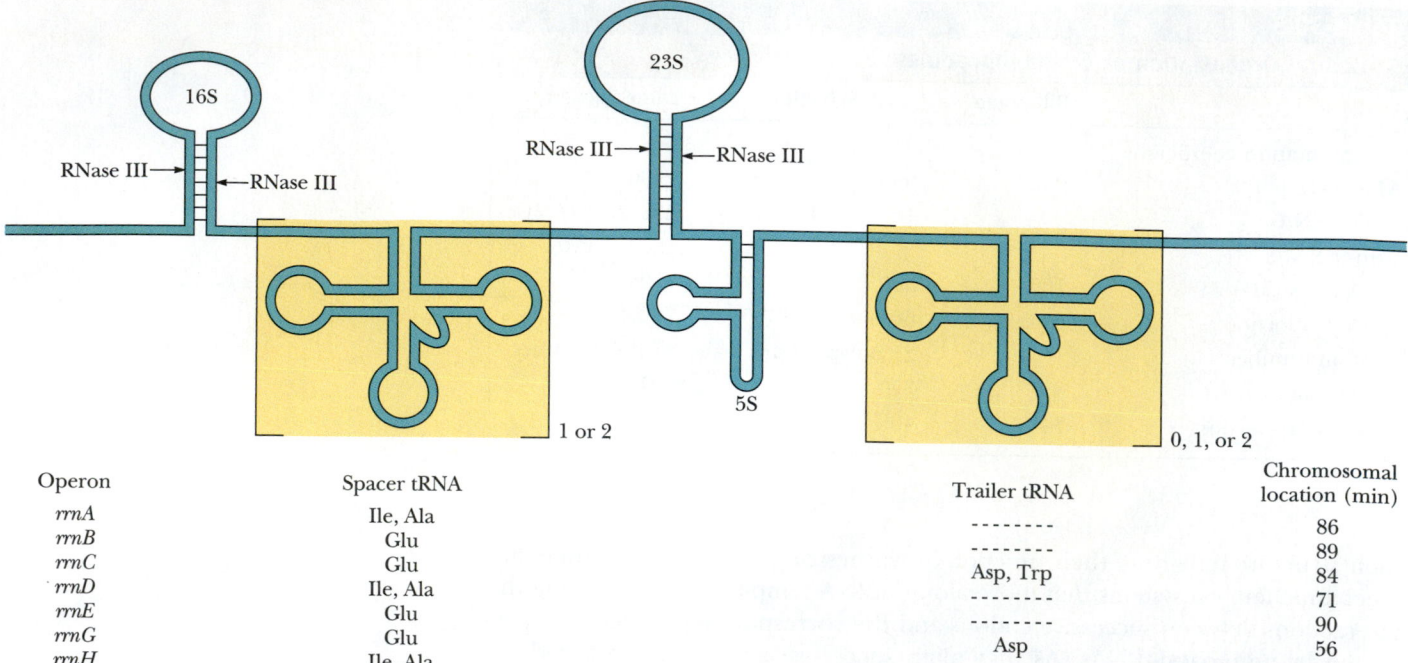

Operon	Spacer tRNA	Trailer tRNA	Chromosomal location (min)
rrnA	Ile, Ala	--------	86
rrnB	Glu	--------	89
rrnC	Glu	Asp, Trp	84
rrnD	Ile, Ala	--------	71
rrnE	Glu	--------	90
rrnG	Glu	Asp	56
rrnH	Ile, Ala		5

FIGURE 30.10 The seven ribosomal RNA operons in *E. coli*. These operons, or gene clusters, are transcribed to give a precursor RNA that is subsequently cleaved by RNase III and other nucleases, at the sites indicated, to generate 23S, 16S, and 5S rRNA molecules, as well as several tRNAs that are unique to each operon. Numerals to the right of the brackets indicate the number of species of tRNA encoded by each transcript.

distinct domains (I through IV) can be discerned in the secondary structure. The three-dimensional structures of both the 30S and 50S ribosomal subunits show that the general shapes of the ribosomal subunits are determined by the conformation of the rRNA molecules within them. Figure 30.11 illustrates the three-dimensional structure of 16S rRNA within the 30S subunit, revealing that the overall form of the 30S structure is essentially that of the rRNA. Ribosomal proteins serve a largely structural role in ribosomes; their primary function is to brace and stabilize the rRNA conformations within the ribosomal subunits.

Ribosomes Spontaneously Self-Assemble In Vitro

Ribosomal subunit self-assembly is one of the paradigms for the spontaneous formation of supramolecular complexes from their macromolecular components. If the individual proteins and rRNAs composing ribosomal subunits are mixed together in vitro under appropriate conditions of pH and ionic strength, spontaneous self-assembly into functionally competent subunits takes place without the intervention of any additional factors or chaperones. The rRNA acts as a scaffold upon which the various ribosomal proteins convene. Ribosomal proteins bind in a specified order.

Ribosomes Have a Characteristic Anatomy

Ribosomal subunits have a characteristic three-dimensional architecture that has been revealed by image reconstructions from cryoelectron microscopy, X-ray crystallography, and X-ray and neutron solution scattering. Such analyses provide images as depicted in Figure 30.12. The 30S, or small, subunit features a "head" and a "base," or "body," from which a "platform" projects. A cleft is defined by the spatial relationship between the head, base, and platform (Figure 30.12a). The mRNA passes across this cleft. The platform represents the central domain of the 30S subunit; it contains one-third of the 16S rRNA. This central domain binds mRNA and the anticodon stem-loop end of aminoacyl-tRNAs, providing the framework for decoding the genetic information in mRNA by mediating codon—anticodon recognition. As such, this central domain of the 30S subunit serves as the **decoding center**. This center is composed only of 16S rRNA; no ribosomal proteins are involved in decoding the message.

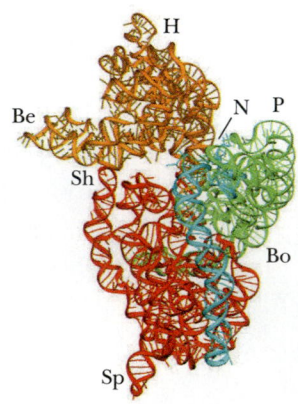

FIGURE 30.11 Tertiary structure of the 16S rRNA within the *Thermus thermophilus* 30S ribosomal subunit. This view is of the face that interacts with the 50S subunit (see Figures 30.12 and 30.14). Red: 5′ domain; green: central domain; orange: 3′ major domain; cyan: 3′ minor domain. H, head; Be, beak; N, neck; P, platform; Sh, shoulder; Sp, spur; Bo, body. *(Adapted from Figure 2 in Wimberly, B. T, et al., 2000. Structure of the 30S ribosomal subunit.* Nature **407**:*327–339. Image courtesy of V. Ramakrishnan, University of Cambridge, U.K.)*

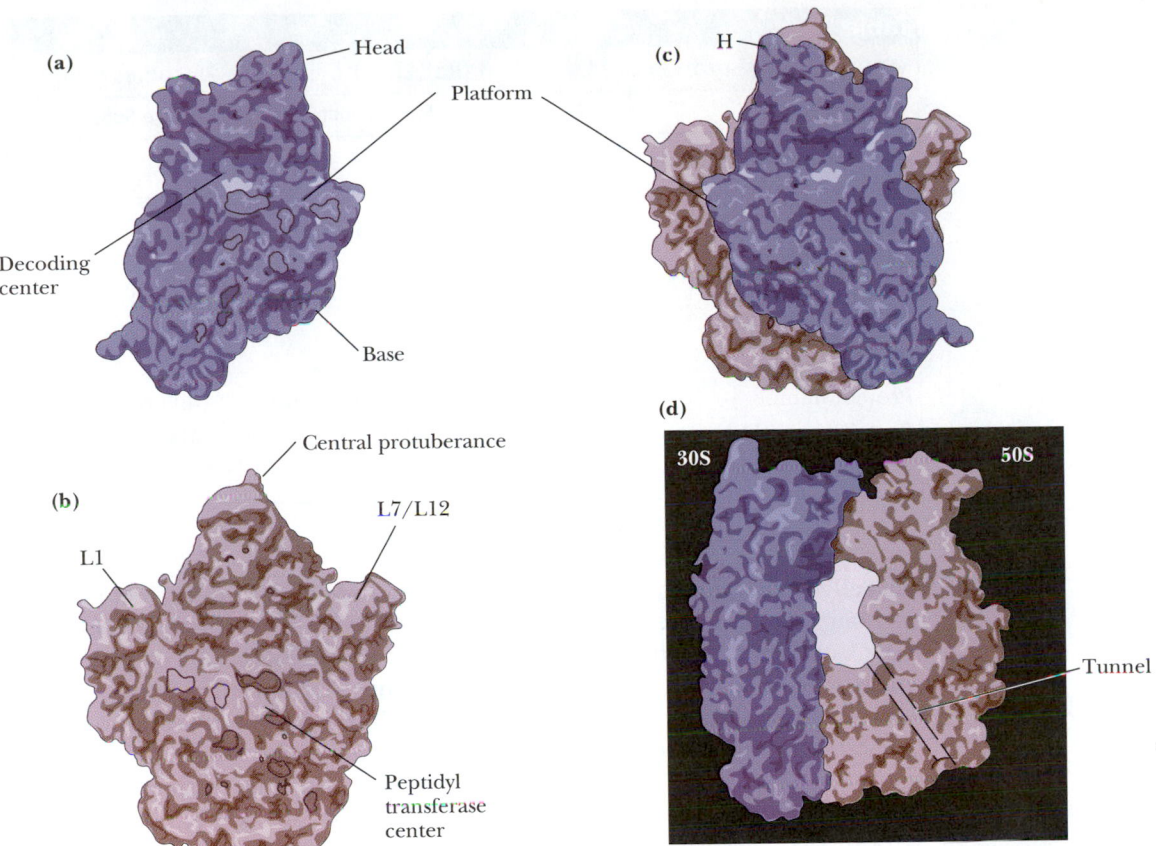

FIGURE 30.12 Structure of the *E. coli* ribosomal subunits and 70S ribosome, as deduced by X-ray crystallography. Prominent structural features are labeled. A and B present views of the 30S **(a)** and 50S **(b)** subunits. These views show the sides of these two that form the interface between them when they come together to form a 70S subunit **(c)**. **(d)** is a side view of the 70S ribosome; the white area represents the region where mRNA and tRNAs are bound and peptide bond formation occurs. The tunnel through the 50S subunit that the growing peptide chain transits is shown as a dashed line. The approximate dimensions of the 30S subunit are $5.5 \times 22 \times 22$ nm; the 50S subunit dimensions are $15 \times 20 \times 20$ nm. *(Adapted from Figures 2 and 3 in Cate, J. H., et al., 1999. X-ray crystal structures of 70S ribosomal functional complexes.* Science **285**:2095–2104.*)*

The 50S, or large, subunit is a mitt-like globular structure with three distinctive projections: a "central protuberance," the "stalk" containing protein L1, and a winglike ridge known as the "L7/L12 region" (Figure 30.12b). The large subunit binds the aminoacyl-acceptor ends of the tRNAs and is responsible for catalyzing formation of the peptide bond formed between successive amino acids in the polypeptide chain. This catalytic center, the **peptidyl transferase,** is located at the bottom of a deep cleft. From it, a 10-nm-long tunnel passes outward through the back of the large subunit.

The small and large subunits associate with each other in the manner shown in Figure 30.12c and d. The contacts between the 30S and 50S subunits are rather limited, and the subunit interface contains mostly rRNA, with relatively little contribution from ribosomal proteins. The decoding center in the 30S subunit is aligned somewhat with the peptidyl transferase and the tunnel in the large subunit, and the growing peptidyl chain is threaded through this tunnel as protein synthesis proceeds. Even though the ribosomal proteins are arranged peripherally around the rRNAs in ribosomes, rRNA occupies 30% to 40% of the ribosomal subunit surface areas.

Both subunits are involved in **translocation,** the process by which the mRNA moves through the ribosome, one codon at a time. (Although it is physically more likely for the mRNA to move through the ribosome, the descriptions of the events in protein synthesis that follow infer that the ribosome moves along the mRNA.)

The Cytosolic Ribosomes of Eukaryotes Are Larger Than Prokaryotic Ribosomes

Eukaryotic cells have ribosomes in their mitochondria (and chloroplasts) as well as in the cytosol. The mitochondrial and chloroplastic ribosomes resemble prokaryotic ribosomes in size, overall organization, structure, and function, a

Table 30.6

Structural Organization of Mammalian (Rat Liver) Cytosolic Ribosomes

	Ribosome	Small Subunit	Large Subunit
Sedimentation coefficient	80S	40S	60S
Mass (kD)	4220	1400	2820
Major RNAs		18S = 1874 bases	28S = 4718 bases
Minor RNAs			5.8S = 160 bases
			5S = 120 bases
RNA mass (kD)	2520	700	1820
RNA proportion	60%	50%	65%
Protein number		33 polypeptides	49 polypeptides
Protein mass (kD)	1700	700	1000
Protein proportion	40%	50%	35%

fact reflecting the prokaryotic origins of these organelles. Although eukaryotic cytosolic ribosomes are larger and considerably more complex, they retain the "core" structural and functional properties of their prokaryotic counterparts, confirming that the fundamental ribosome organization and operation has been conserved across evolutionary time. Higher eukaryotes have more complex ribosomes than lower eukaryotes. For example, the yeast cytosolic ribosomes have major rRNAs of 3392 (large subunit) and 1799 nucleotides (small subunit); the major rRNAs of mammalian cytosolic ribosomes are 4718 and 1874 nucleotides, respectively. Table 30.6 lists the properties of cytosolic ribosomes in a mammal, the rat. Comparison of base sequences and secondary structures of rRNAs from different organisms suggests that evolution has worked to conserve the secondary structure of these molecules, although not necessarily the nucleotide sequences creating such structure. That is, the retention of a base pair at a particular location seems more important than whether the base pair is G:C or A:U.

30.5 | What Are the Mechanics of mRNA Translation?

In translating an mRNA, a ribosome must move along it in the 5′→3′ direction, recruiting aminoacyl-tRNAs whose anticodons match up with successive codons and joining amino acids in peptide bonds in a polymerization process that forms a particular protein. Like chemical polymerization processes, protein biosynthesis in all cells is characterized by three distinct phases: initiation, elongation, and termination. At each stage, the energy driving the assembly process is provided by GTP hydrolysis, and specific soluble protein factors participate in the events. These soluble proteins are often **G-protein family** members that use the energy released upon hydrolysis of bound GTP to fuel switchlike conformational changes. Such conformational changes are at the heart of the mechanical steps necessary move a ribosome along an mRNA and to deliver an aminoacyl-tRNA into appropriate register with a codon.

Initiation involves binding of mRNA by the small ribosomal subunit, followed by association of a particular **initiator aminoacyl-tRNA** that recognizes the first codon. This codon often lies within the first 30 nucleotides or so of mRNA spanned by the small subunit. The large ribosomal subunit then joins the initiation complex, preparing it for the elongation stage.

Elongation includes the synthesis of all peptide bonds from the first to the last. The ribosome remains associated with the mRNA throughout elongation, moving along it and translating its message into an amino acid sequence. This is accomplished via a repetitive cycle of events in which successive aminoacyl-tRNAs are added to the ribosome:mRNA complex as directed by codon binding, the 50S subunit catalyzes peptide bond formation, and the polypeptide chain grows by one amino acid at a time.

Three tRNA molecules may be associated with the ribosome:mRNA complex at any moment. Each lies in a distinct site (Figure 30.13). The **A,** or **acceptor, site** is the attachment site for an incoming aminoacyl-tRNA. The **P,** or **peptidyl, site** is occupied by peptidyl-tRNA, the tRNA carrying the growing polypeptide chain. The elongation reaction transfers the peptide chain from the peptidyl-tRNA in the P site to the aminoacyl-tRNA in the A site. This transfer occurs through covalent attachment of the α-amino group of the aminoacyl-tRNA to the α-carboxyl group of the peptidyl-tRNA, forming a new peptide bond. The new, longer peptidyl-tRNA now moves from the A site into the P site as the ribosome moves one codon further along the mRNA. The A site, left vacant by this translocation, can accept the next incoming aminoacyl-tRNA. The **E,** or **exit, site,** is transiently occupied by the "unloaded," or deacylated, tRNA, which has lost its peptidyl chain through the peptidyl transferase reaction. These events are summarized in Figure 30.13. The contributions made to each of the three tRNA-binding sites by each ribosomal subunit are shown in Figure 30.14.

Termination is triggered when the ribosome reaches a "stop" codon on the mRNA. At this point, the polypeptide chain is released and the ribosomal subunits dissociate from the mRNA.

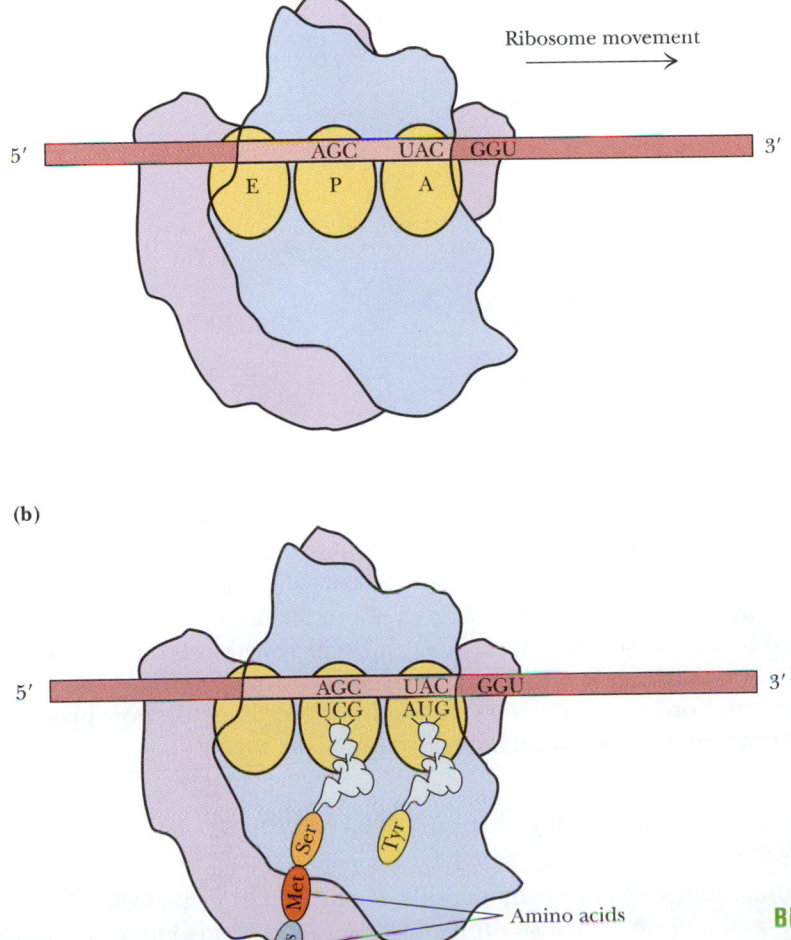

Biochemistry ⊜**Now**™ **ACTIVE FIGURE 30.13** The basic steps in protein synthesis. The ribosome has three distinct binding sites for tRNA: the A, or acceptor, site; the P, or peptidyl, site; and the E, or exit, site. **Test yourself on the concepts in this figure at** http://chemistry.brookscole.com/ggb3   (continued)

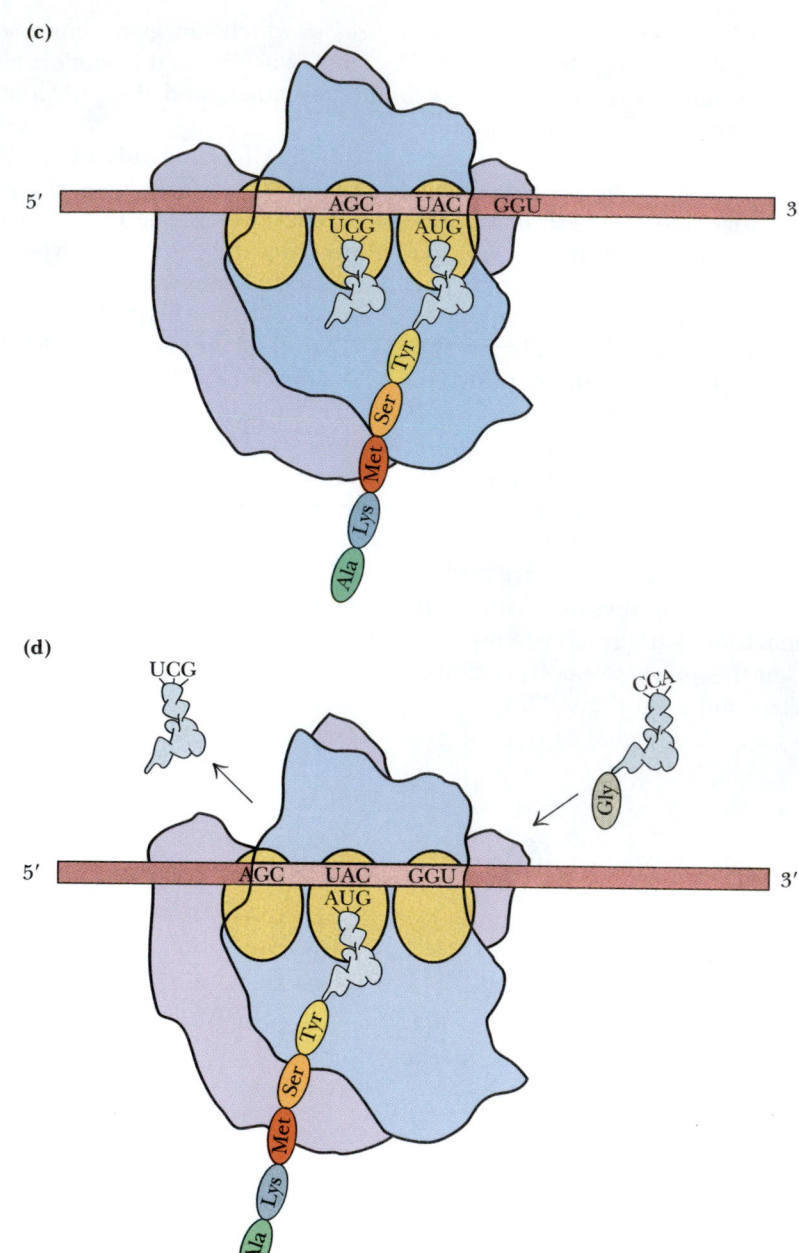

(c)

(d)

Biochemistry ⑧ Now™   **ACTIVE FIGURE 30.13 continued**

Protein synthesis proceeds rapidly. In vigorously growing bacteria, about 20 amino acid residues are added to a growing polypeptide chain each second. So an average protein molecule of about 300 amino acid residues is synthesized in only 15 seconds. Eukaryotic protein synthesis is only about 10% as fast. Protein synthesis is also highly accurate: An inappropriate amino acid is incorporated only once in every 10^4 codons. We focus first on protein synthesis in *E. coli*, the system for which we know the most.

Peptide Chain Initiation in Prokaryotes Requires a G-Protein Family Member

The components required for peptide chain initiation include (1) mRNA; (2) 30S and 50S ribosomal subunits; (3) a set of proteins known as **initiation factors;** (4) GTP; and (5) a specific charged tRNA, **f-Met-tRNA$_i^{fMet}$**. A discussion of the properties of these components and their interaction follows.

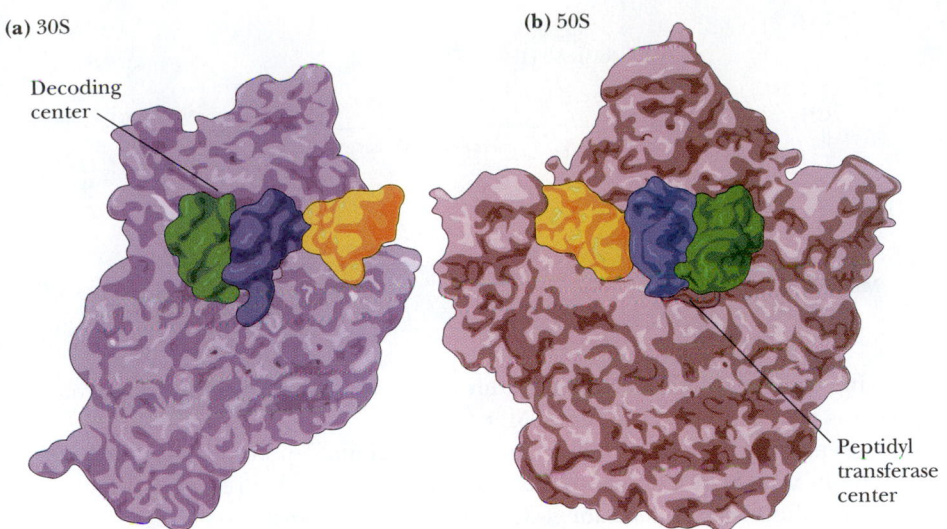

(a) 30S

Decoding center

(b) 50S

Peptidyl transferase center

FIGURE 30.14 The three tRNA-binding sites on ribosomes. The view shows the ribosomal surfaces that form the interface between the 30S **(a)** and 50S **(b)** subunits in a 70S ribosome (as if a 70S ribosome has been "opened" like a book to expose facing "pages"). The A (green), P (blue), and E (yellow) sites are occupied by tRNAs. The decoding center on the 30S subunit (b) lies behind the top of the tRNAs in the A and P sites (which is where the anticodon ends of the tRNAs are located). The peptidyl transferase center on the 50S subunit (b) lies at the lower tips (acceptor ends) of the A- and P-site tRNAs. *(Adapted from Figure 5 in Cate, J. H., et al., 1999. X-ray crystal structures of 70S ribosome functional complexes. Science **285**:2095–2104.)*

Initiator tRNA   tRNA$_i^{fMet}$ is a particular tRNA for reading an AUG (or GUG, or even UUG) codon that signals the start site, or N-terminus, of a polypeptide chain; the $_i$ signifies "initiation." This tRNA$_i^{fMet}$ does not read internal AUG codons, so it does not participate in chain elongation. Instead, that role is filled by another methionine-specific tRNA, referred to as tRNA^{Met}, which cannot replace tRNA$_i^{fMet}$ in peptide chain initiation. (However, both of these tRNAs are loaded with Met by the same methionyl-tRNA synthetase.) The structure of *E. coli* tRNA$_i^{fMet}$ has several distinguishing features (Figure 30.15). Collectively, these features identify this tRNA as essential to initiation and inappropriate for chain elongation.

The synthesis of all *E. coli* polypeptides begins with the incorporation of a modified methionine residue, *N*-formyl-Met, as N-terminal amino acid. However, in about half of the *E. coli* proteins, this Met residue is removed once the

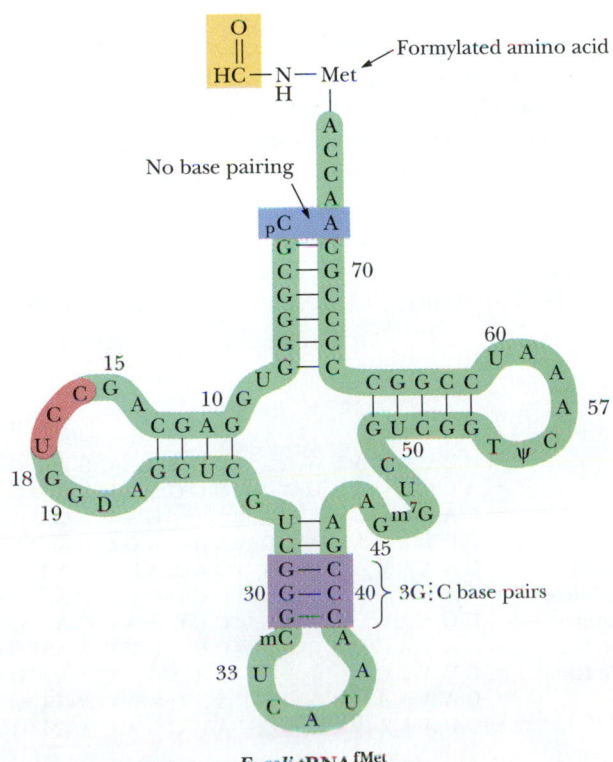

E. coli tRNA$_i^{fMet}$

FIGURE 30.15 The structure of *E. coli* N-formyl-methionyl-tRNA$_i^{fMet}$. The features distinguishing it from noninitiator tRNAs are highlighted.

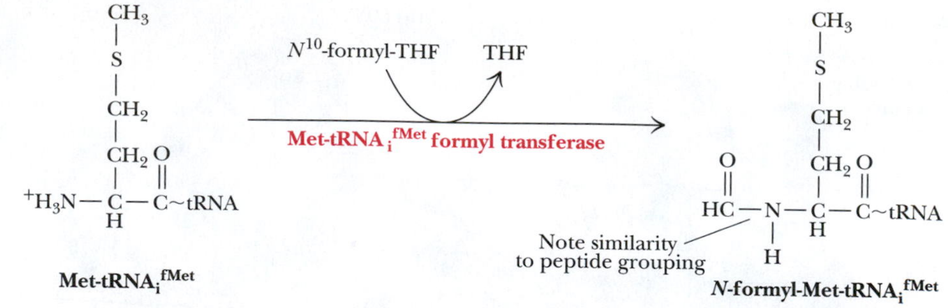

CH_3
S
CH_2
CH_2 O
$^+H_3N—C—C\sim tRNA$
H
Met-tRNA$_i^{fMet}$
N^{10}-formyl-THF
THF
Met-tRNA$_i^{fMet}$ formyl transferase
CH_3
S
CH_2
O
CH_2 O
$HC—N—C—C\sim tRNA$
H
H
Note similarity to peptide grouping
N-formyl-Met-tRNA$_i^{fMet}$

Biochemistry Now™ ANIMATED FIGURE 30.16 Methionyl-tRNA$_i^{fMet}$ formyl transferase catalyzes the transformylation of methionyl-tRNA$_i^{fMet}$ using N^{10}-formyl-THF as formyl donor. The tRNA for reading Met codons within a protein (tRNA^{Met}) is not a substrate for this transformylase. **See this figure animated at http://chemistry.brookscole.com/ggb3**

growing polypeptide is ten or so residues long; as a consequence, many mature proteins in *E. coli* lack N-terminal Met.

The methionine contributed in peptide chain initiation by tRNA$_i^{fMet}$ is unique in that its amino group has been formylated. This reaction is catalyzed by a specific enzyme, **methionyl-tRNA$_i^{fMet}$ formyl transferase** (Figure 30.16). Note that the addition of the formyl group to the α-amino group of Met creates an N-terminal block resembling a peptidyl grouping. That is, the initiating Met is transformed into a minimal analog of a peptidyl chain.

mRNA Recognition and Alignment In order for the mRNA to be translated accurately, its sequence of codons must be brought into proper register with the translational apparatus. Recognition of translation initiation sequences on mRNAs involves the 16S rRNA component of the 30S ribosomal subunit. Base pairing between a pyrimidine-rich sequence at the 3′-end of 16S rRNA and complementary purine-rich tracts at the 5′-end of prokaryotic mRNAs positions the 30S ribosomal subunit in proper alignment with an initiation codon on the mRNA. The purine-rich mRNA sequence, the **ribosome-binding site,** is often called the **Shine–Dalgarno sequence** in honor of its discoverers. Figure 30.17 shows various Shine–Dalgarno sequences found in prokaryotic mRNAs, along with the complementary 3′-tract on *E. coli* 16S rRNA. The 3′-end of 16S rRNA resides in the "head" region of the 30S small subunit.

Initiation Factors Initiation involves interaction of the **initiation factors (IFs)** with GTP, *N*-formyl-Met-tRNA$_i^{fMet}$, mRNA, and the 30S subunit to give a **30S initiation complex** to which the 50S subunit then adds to form a **70S initiation complex.** The initiation factors are soluble proteins required for assembly of proper initiation complexes. Their properties are summarized in Table 30.7.

Events in Initiation Initiation begins when a 30S subunit: (IF-3 : IF-1) complex binds mRNA and a complex of IF-2, GTP, and f-Met-tRNA$_i^{fMet}$. The sequence of events is summarized in Figure 30.18. Although IF-3 is absolutely essential for

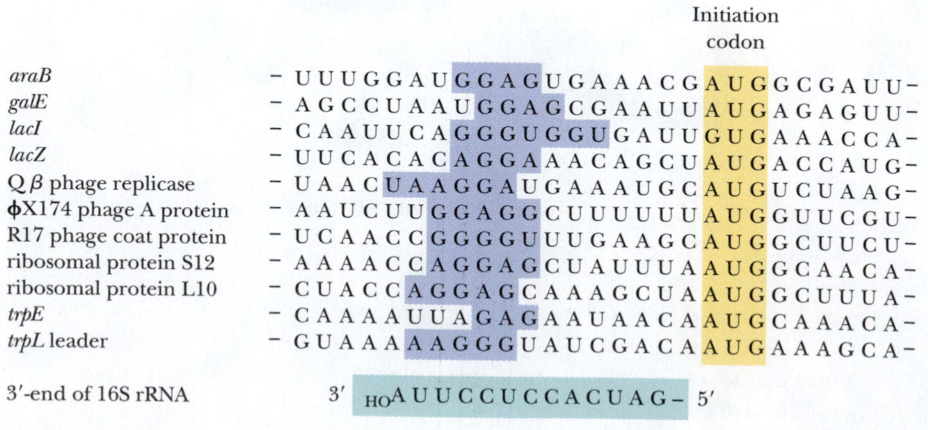

Initiation codon
araB – U U U G G A U G G A G U G A A A C G A U G G C G A U U –
galE – A G C C U A A U G G A G C G A A U U A U G A G A G U U –
lacI – C A A U U C A G G G U G G U G A U U G U G A A A C C A –
lacZ – U U C A C A C A G G A A A C A G C U A U G A C C A U G –
Qβ phage replicase – U A A C U A A G G A U G A A A U G C A U G U C U A A G –
φX174 phage A protein – A A U C U U G G A G G C U U U U U U A U G G U U C G U –
R17 phage coat protein – U C A A C C G G G G U U U G A A G C A U G G C U U C U –
ribosomal protein S12 – A A A A C C A G G A G C U A U U U A A U G G C A A C A –
ribosomal protein L10 – C U A C C A G G A G C A A A G C U A A U G G C U U U A –
trpE – C A A A A U U A G A G A A U A A C A A U G C A A A C A –
trpL leader – G U A A A A A G G G U A U C G A C A A U G A A A G C A –
3′-end of 16S rRNA 3′ HOA U U C C U C C A C U A G – 5′

FIGURE 30.17 Various Shine–Dalgarno sequences recognized by *E. coli* ribosomes. These sequences lie about ten nucleotides upstream from their respective AUG initiation codon and are complementary to the UCCU core sequence element of *E. coli* 16S rRNA. G:U as well as canonical G:C and A:U base pairs are involved here.

Table 30.7			
Properties of *E. coli* Initiation Factors			
Factor	**Mass (kD)**	**Molecules/ Ribosome**	**Function**
IF-1	9	0.15	Binds to 30S A site and prevents tRNA binding
IF-2	97		G-protein that binds fMet-tRNA$_i^{fMet}$; interacts with IF-1
IF-3	23	0.25	Binds to 30S E site; prevents 50S binding

mRNA binding by the 30S subunit, it is not involved in locating the proper translation initiation site on the message. The presence of IF-3 on 30S subunits also prevents them from reassociating with 50S subunits. IF-3 must dissociate before the 50S subunit will associate with the mRNA:30S subunit complex.

IF-2 delivers the initiator f-Met-tRNA$_i^{fMet}$ in a GTP-dependent process. Apparently, the 30S subunit is aligned with the mRNA such that the initiation codon is situated within the "30S part" of the P site. Upon binding, f-Met-tRNA$_i^{fMet}$ enters this 30S portion of the P site. GTP hydrolysis is necessary to form an active 70S ribosome. GTP hydrolysis is triggered when the 50S subunit joins and is accompanied by IF-1 and IF-2 release. The A site of the *70S initiation complex* is ready to accept an incoming aminoacyl-tRNA; the 70S ribosome is poised to begin chain elongation.

Peptide Chain Elongation Requires Two G-Protein Family Members

The requirements for peptide chain elongation are (1) an mRNA:70S ribosome:peptidyl-tRNA complex (peptidyl-tRNA in the P site), (2) aminoacyl-tRNAs, (3) a set of proteins known as **elongation factors,** and (4) GTP. Chain elongation can be divided into three principal steps:

1. Codon-directed binding of the incoming aminoacyl-tRNA at the A site. Decoding center regions of 16S rRNA makes sure the proper aminoacyl-tRNA is in the A site by direct surveillance of codon–anticodon base pairing geometry.
2. Peptide bond formation: transfer of the peptidyl chain from the tRNA bearing it to the —NH$_2$ group of the new amino acid.
3. Translocation of the "one-residue-longer" peptidyl-tRNA to the P site to make room for the next aminoacyl-tRNA at the A site. These shifts are coupled with movement of the ribosome one codon further along the mRNA.

The Elongation Cycle

The properties of the soluble proteins essential to peptide chain elongation are summarized in Table 30.8. These proteins are present in large quantities, reflecting the great importance of protein synthesis to cell vitality. For example, **elongation factor Tu (EF-Tu)** is the most abundant protein in *E. coli,* accounting for 5% of total cellular protein.

Aminoacyl-tRNA Binding

EF-Tu binds aminoacyl-tRNA and GTP. There is only one EF-Tu species serving all the different aminoacyl-tRNAs, and aminoacyl-tRNAs are accessible to the A site of active 70S ribosomes only in the form of aminoacyl-tRNA:EF-Tu:GTP complexes. Once correct base pairing between codon and anticodon has been established within the A site, the GTP is hydrolyzed to GDP and P$_i$, the aminoacyl end of the tRNA is properly oriented in the peptidyl transferase site of the 50S subunit, and the EF-Tu molecule is released as a EF-Tu:GDP complex (Figure 30.19).

Elongation factor Ts (EF-Ts) is a **guanine-nucleotide exchange factor (GEF)** that catalyzes the recycling of EF-Tu by mediating the displacement of GDP and

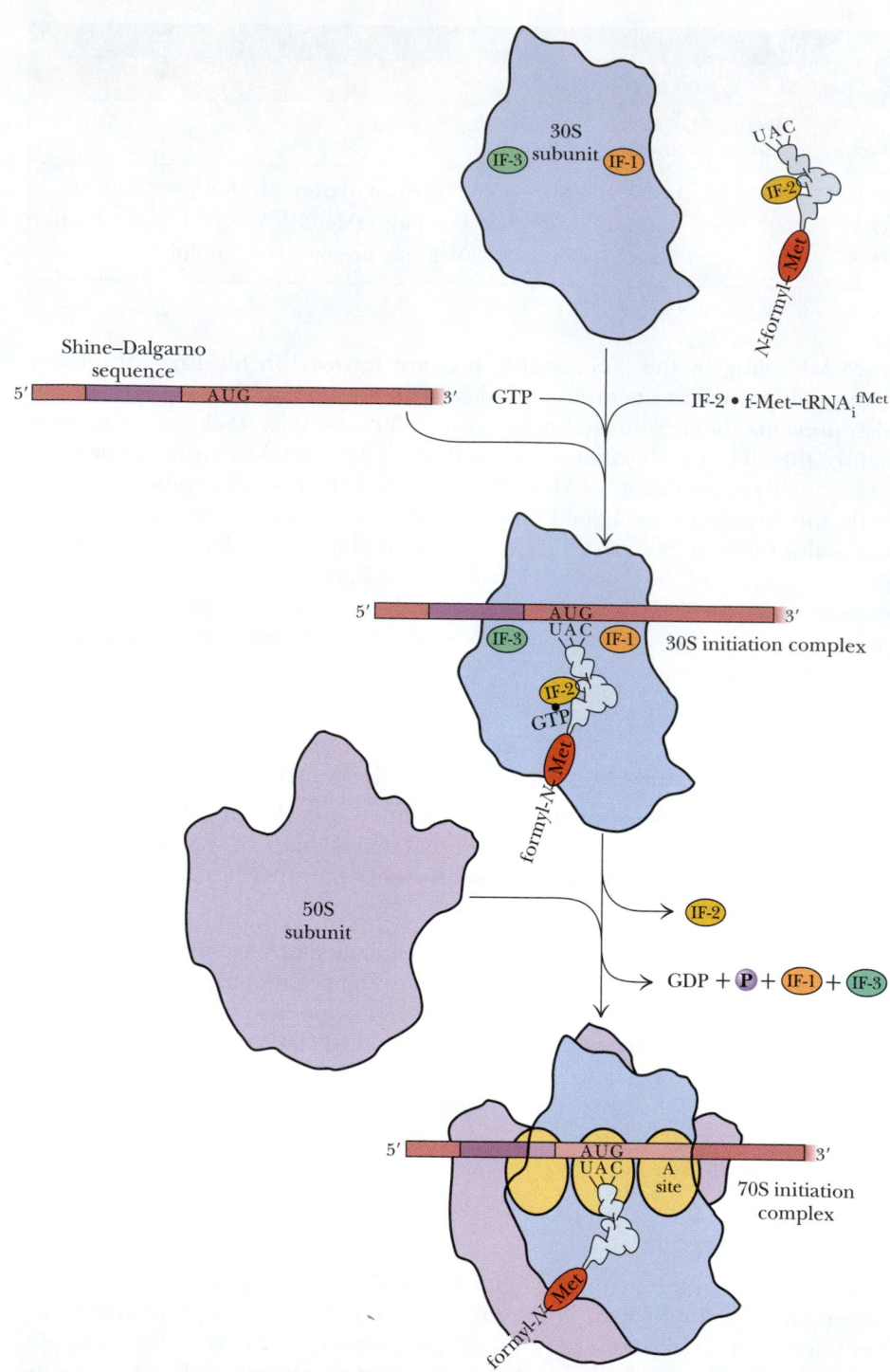

Biochemistry⊗Now™ ACTIVE FIGURE 30.18
The sequence of events in peptide chain initiation.
**Test yourself on the concepts in this figure at http://
chemistry.brookscole.com/ggb3**

Table 30.8			
Properties of *E. coli* Elongation Factors			
Factor	**Mass (kD)**	**Molecules/ Cell**	**Function**
EF-Tu	43	70,000	G protein that binds aminoacyl-tRNA and delivers it to the A site
EF-Ts	74	10,000	Guanine-nucleotide exchange factor (GEF) that replaces GDP on EF-Tu with GTP
EF-G	77	20,000	G protein that promotes translocation of mRNA

its replacement by GTP. EF-Ts forms a transient complex with EF-Tu by displacing GDP, whereupon GTP displaces EF-Ts from EF-Tu (Figure 30.19).

The Decoding Center: A 16S rRNA Function Analysis of the structures of the 70S ribosome:tRNA complexes and isolated 30S subunits has revealed the decoding center in the 30S subunit. This decoding center, where anticodon loops of the A- and P-site tRNAs and the codons of the mRNA are matched up, is a property of 16S rRNA. No ribosomal proteins segments are found in this region; only the 16S rRNA conformation determines how codon–anticodon interactions occur (Figure 30.20).

Peptidyl Transfer Peptidyl transfer, or **transpeptidation,** is the central reaction of protein synthesis, the actual peptide bond–forming step. No energy input (for example, in the form of ATP) is needed; the ester bond linking the peptidyl

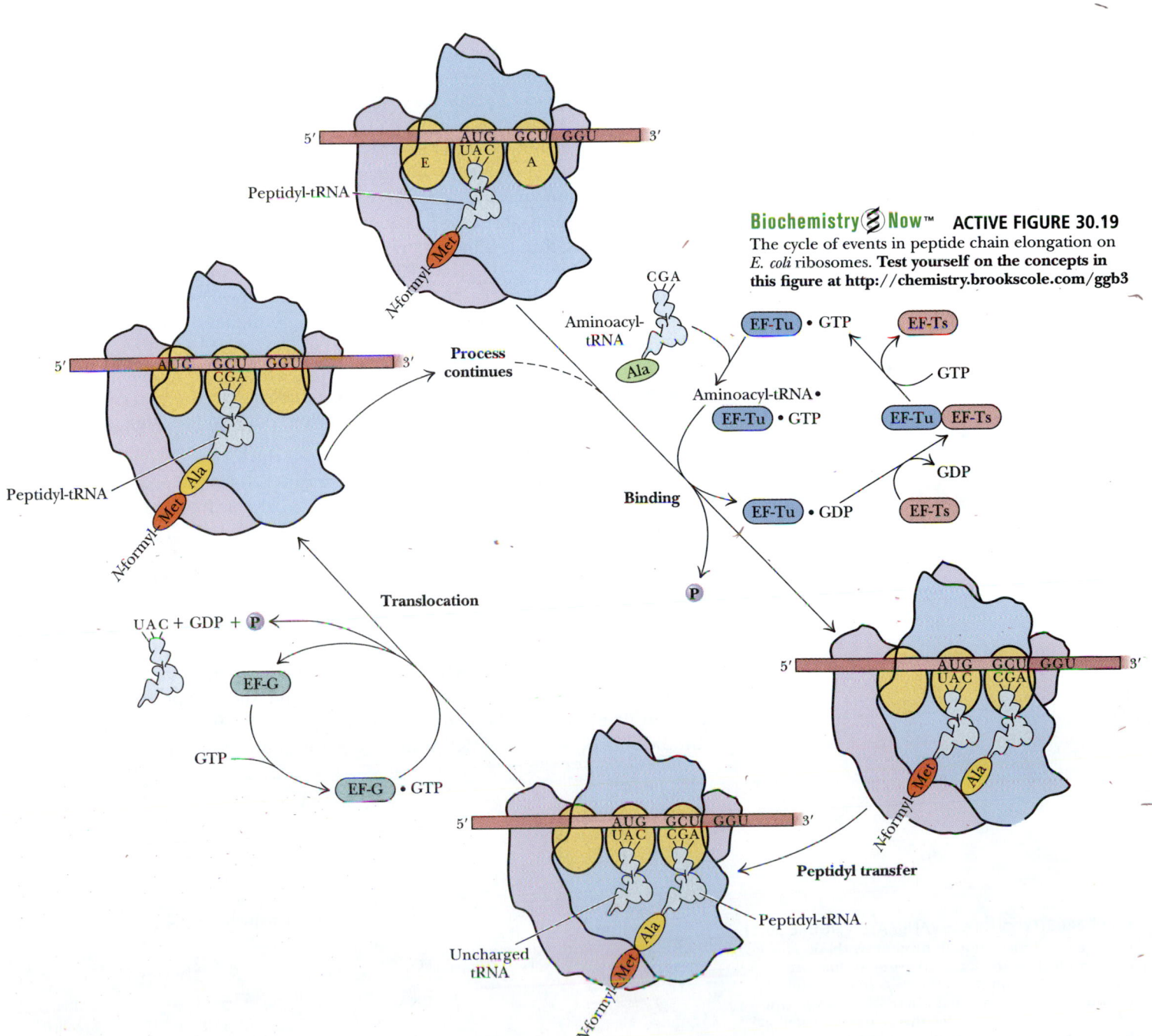

Biochemistry⊗Now™ ACTIVE FIGURE 30.19
The cycle of events in peptide chain elongation on *E. coli* ribosomes. **Test yourself on the concepts in this figure at http://chemistry.brookscole.com/ggb3**

FIGURE 30.20 The decoding center of the 30S ribosomal subunit is composed only of 16S rRNA. **(a)** The 30S subunit, as viewed from the 50S subunit. The circle (*cyan*) shows the latch structure of the 30S subunit that encircles and encloses the mRNA. The mRNA enters along the path indicated by the arrow and follows a groove along this face of the subunit. The location of the decoding center is indicated by the *red* circle. **(b)** Enlarged view of the decoding center. Helix H44 of the 16S rRNA is at the bottom, in the *olive* shaded region. Two codons of the mRNA (the *blue* shaded region) are immediately above H44. Three anticodon loops are shown above the mRNA codons: The A-site tRNA anticodon loop is in *green,* the P-site tRNA anticodon loop is in *magenta,* and the E-site tRNA anticodon loop is colored *gray.* (*Adapted from Figure 3 in Schluenzen, F., et al., 2000. Structure of the functionally activated small ribosomal subunit at 3.3 Å resolution.* Cell **102**:615–623.)

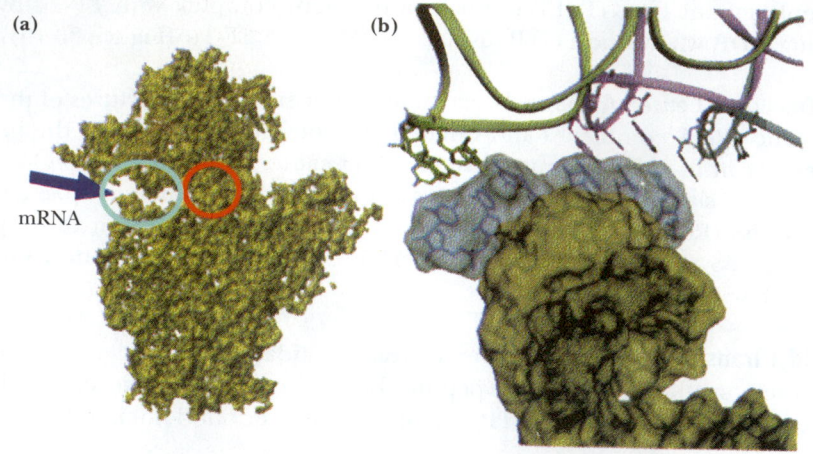

moiety to tRNA is intrinsically reactive. As noted earlier, *peptidyl transferase,* the activity catalyzing peptide bond formation, is associated with the 50S ribosomal subunit. Indeed, this reaction is a property of the 23S rRNA in the 50S subunit.

23S rRNA Is the Peptidyl Transferase Enzyme   A possible mechanism for the reaction catalyzed by the **peptidyl transferase center** of 23S rRNA is depicted in Figure 30.21. The lone electron pair on N-3 of the purine A^{2451} of 23S rRNA serves as a general base. It abstracts a proton from the α-amino group on the A-site aminoacyl-tRNA, thus facilitating nucleophilic attack by this amino group on the carbonyl-C of the P-site peptidyl-tRNA and peptide bond formation via transfer of the peptidyl chain to the aminoacyl-tRNA. The peptidyl transferase region of 23S rRNA is shown in Figure 30.22. Nucleotide sequences in this region of 23S rRNA are among the most highly conserved in all biology.

Translocation   Three things remain to be accomplished in order to return the active 70S ribosome : mRNA complex to the starting point in the elongation cycle:

1. The deacylated tRNA must be removed from the P site.
2. The peptidyl-tRNA must be moved (translocated) from the A site to the P site.
3. The ribosome must move one codon down the mRNA so that the next codon is positioned in the A site.

The precise events in translocation are still being resolved, but several distinct steps are clear. The acceptor ends (the aminoacylated ends) of both A- and P-site tRNAs interact with the **peptidyl transferase** center of the 50S subunit. Because the growing peptidyl chain doesn't move during peptidyl transfer, the acceptor end of the A-site aminoacyl-tRNA must move into the P site as its aminoacyl function picks up the peptidyl chain. At the same time, the acceptor end of the deacylated P-site tRNA is shunted into the E site (Figure 30.14). Then, the mRNA and the anticodon ends of tRNAs move together with respect to the 30S subunit

Biochemistry⊜Now™   ANIMATED FIGURE 30.21
Peptide bond formation in protein synthesis. Nucleophilic attack by the α-amino group of the A-site aminoacyl-tRNA on the carbonyl-C of the P-site peptidyl-tRNA is facilitated when 32S rRNA purine A^{2451} abstracts a proton. **See this figure animated at** http://chemistry.brookscole.com/ggb3

so that the mRNA is passively dragged one codon further through the ribosome. With this movement, the anticodon end of the now one-residue-longer peptidyl-tRNA goes from the A site of the 30S subunit to the P site. Concomitantly, the anticodon end of the deacylated tRNA is moved into the E site. These ratchetlike movements of the 30S subunit relative to the 50S subunit are catalyzed by the translocation protein **elongation factor G (EF-G),** which apparently couples the energy of GTP hydrolysis to movement. Note that translocation of the mRNA relative to the 30S subunit will deliver the next codon to the 30S A site.

EF-G binds to the ribosome as an EF-G:GTP complex. GTP hydrolysis is essential not only for translocation but also for subsequent EF-G dissociation. Because EF-G and EF-Tu compete for a common binding site on the ribosome (the **factor-binding center**) adjacent to the A site, EF-G release is a prerequisite for return of the 70S ribosome:mRNA to the beginning point in the elongation cycle.

In this simple model of peptidyl transfer and translocation, the ends of both tRNAs move relative to the two ribosomal subunits in two discrete steps, the acceptor ends moving first and then the anticodon ends. Furthermore, the readjustments needed to reposition the ribosomal subunits relative to the mRNA and to one another imply that the 30S and 50S subunits must move relative to one another in ratchetlike fashion. *This model provides a convincing explanation for why ribosomes are universally organized into a two-subunit structure: The small and large subunits* must *move relative to each other, as opposed to moving as a unit, in order to carry out the process of translation.*

GTP Hydrolysis Fuels the Conformational Changes That Drive Ribosomal Functions

Two GTPs are hydrolyzed for each amino acid residue incorporated into peptide during chain elongation: one upon EF-Tu–mediated binding of aa-tRNA and one more in translocation. The role of GTP (with EF-Tu as well as EF-G) is mechanical, in analogy with the role of ATP in driving muscle contraction (see Chapter 16). GTP binding induces conformational changes in ribosomal components that actively engage these components in the mechanics of protein synthesis; subsequent GTP hydrolysis followed by GDP and Pi release relax the system back to the initial conformational state so that another turn in the cycle can take place. The energy expenditure for protein synthesis is at least four high-energy phosphoric anhydride bonds per amino acid. In addition to the two provided by GTP, two from ATP are expended in amino acid activation via aminoacyl-tRNA synthesis (Figure 30.3).

Peptide Chain Termination Requires a G-Protein Family Member

The elongation cycle of polypeptide synthesis continues until the 70S ribosome encounters a "stop" codon. At this point, polypeptidyl-tRNA occupies the P site and the arrival of a "stop" or nonsense codon in the A site signals that the end of the polypeptide chain has been reached (Figure 30.23) These nonsense codons are not "read" by any "terminator tRNAs" but instead are recognized by specific proteins known as **release factors,** so named because they promote polypeptide release from the ribosome. The release factors bind at the A site. **RF-1** recognizes UAA and UAG, whereas **RF-2** recognizes UAA and UGA. RF-1 and RF-2 are members of the guanine nucleotide exchange factor (GEF) family of proteins, which includes EF-Ts. Like EF-G, RF-1 and RF-2 interact well with the ribosomal A-site structure. Furthermore, these release factors "read" the nonsense codons through specific tripeptide sequences that serve as the RF protein equivalent of the tRNA anticodon loop. There is about one molecule each of RF-1 and RF-2 per 50 ribosomes. Ribosomal binding of RF-1 or RF-2 is competitive with EF-G. RF-1 or RF-2 recruit a third release factor, **RF-3,** complexed with GTP. RF-3 is the fourth G-protein family member (the other three are IF-2, EF-Tu, and EF-G) involved in protein synthesis. All share the same ribosomal

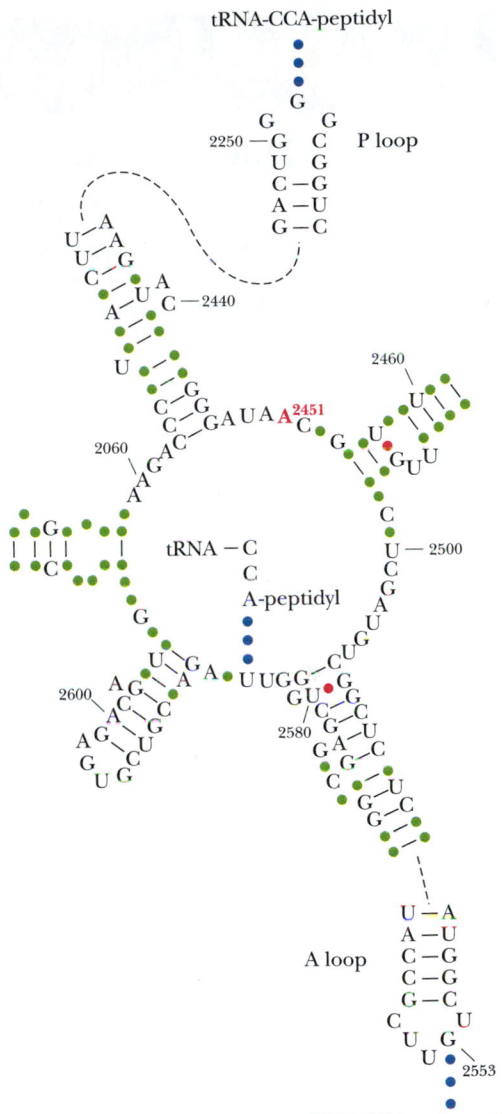

FIGURE 30.22 The peptidyl transferase center of 23S rRNA. 23S rRNA has a highly conserved secondary structure reminiscent of that of 16S rRNA (see Figures 11.36 and 11.37). Only sequences implicated in the peptidyl transferase function are presented here. This region corresponds to region V of 23S rRNA. Numbers indicate base positions in the 23S rRNA nucleotide sequence. Green dots symbolize bases in the central region that are not conserved in 23S rRNAs from different sources. Purine A2451, the catalytic purine in the peptidyl transferase reaction, is highlighted. 23S rRNA sites involved in interactions with peptidyl-tRNA and aminoacyl-tRNA, the substrates of protein synthesis, are indicated. Residue G2252 forms close contacts with the CCA-end of the P-site (peptidyl) tRNA, and residue G2553 forms close contacts with the CCA-end of the A-site (aminoacyl) tRNA. The loops in which these residues are found are labeled P loop and A loop, respectively. The results indicate that the tertiary structure of 23S rRNA brings the P loop–bound peptidyl-tRNA into an interaction with base U2585 in the central region. (*Adapted from Pace, N. R., 1992. New horizons for RNA catalysis. Science* **256:**1402–1403; *Porse, B., et al., 1997. The donor substrate site within the peptidyl transferase loop of 23S rRNA and its putative interactions with the CCA-end of N-blocked aminoacyl-tRNA. Journal of Molecular Biology* **266:**472–483; *and Green, R., et al., 1998. Ribosome-catalyzed peptide-bond formation with an A-site substrate covalently linked to 23S ribosomal RNA. Science* **280:**286–289.)

A Deeper Look

Biochemistry Now™

Molecular Mimicry—The Structures of EF-Tu:Aminoacyl-tRNA and EF-G

EF-Tu and EF-G compete for binding to ribosomes. EF-Tu has the unique capacity to recognize and bind any aminoacyl-tRNA and deliver it to the ribosome in a GTP-dependent reaction. EF-G catalyzes GTP-dependent translocation. The ternary structure of the *Thermus aquaticus* EF-Tu:Phe-tRNA^Phe complexed with GMPPNP (a nonhydrolyzable analog of GTP) is remarkably similar to the structure of EF-G:GDP (see accompanying figure). EF-Tu is a three-domain protein (shown in red, green, and light blue here); a space-filling representation of its bound tRNA is shown in magenta. EF-G is a six-domain protein, five of which correspond to the overall EF-Tu:tRNA structure. Domains 1 and 2 in EF-G correspond to the EF-Tu protein domains colored red

and green; domains 3, 4, and 5 of EF-G show a striking structural resemblance to the tRNA component in EF-Tu:tRNA and are colored magenta. (EF-G has an extra domain, colored dark blue here.) Thus, parts of the EF-G structure mimic the structure of a tRNA molecule.

One view of early evolution suggests that RNA was the primordial macromolecule, fulfilling all biological functions, including those of catalysis and information storage that are now assumed for the most part by proteins and DNA. The mimicry of EF-Tu:tRNA by EF-G may represent a fossil of early macromolecular evolution when the proteins first began to take over some functions of RNA by mimicking shapes known to work as RNAs.

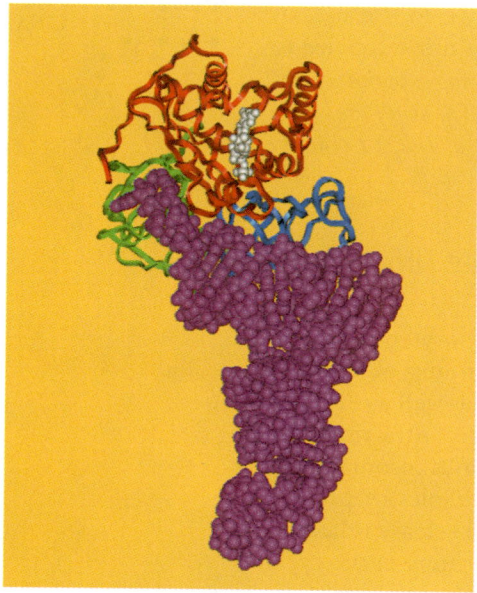 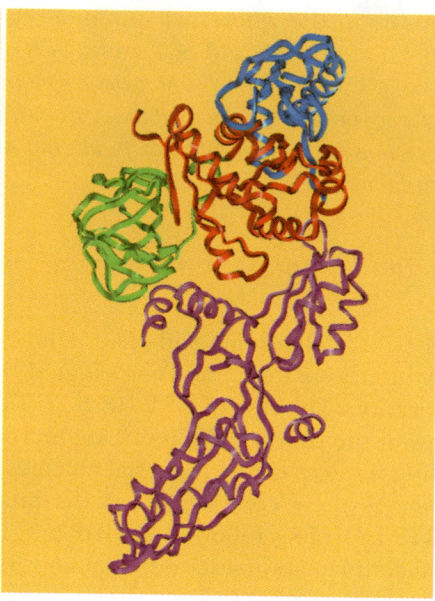

Adapted from Nyborg, J., et al., 1996. Structure of the ternary complex of EF-Tu: Macromolecular mimicry in translation. *Trends in Biochemical Sciences* **21**:81–82.

binding site. The state of the peptidyl-tRNA in the P site determines which is bound and, importantly, the progression of protein synthesis through initiation, elongation, and termination.

The presence of release factors with a nonsense codon in the A site creates a **70S ribosome:RF-1** (or **RF-2):RF-3-GTP:termination signal** complex that transforms the ribosomal peptidyl transferase into a hydrolase. That is, instead of catalyzing the transfer of the polypeptidyl chain from a polypeptidyl-tRNA to an acceptor aminoacyl-tRNA, the peptidyl transferase hydrolyzes the ester bond linking the polypeptidyl chain to its tRNA carrier. In actuality, peptidyl transferase transfers the polypeptidyl chain to a water molecule instead of an aminoacyl-tRNA. Peptide release is followed by expulsion of RF-1 (or RF-2) from the ribosome (Figure 30.23). This leaves a ribosome:mRNA:P-site tRNA complex that must be disassembled by a protein **ribosome recycling factor (RRF)** with the help of EF-G. The structure of RRF shows that it is an excellent molecular mimic of a tRNA.

We can now recount the central role played by GTP in protein synthesis. IF-2, EF-Tu, EF-G, and RF-3 are all GTP-binding proteins, and all are part of the

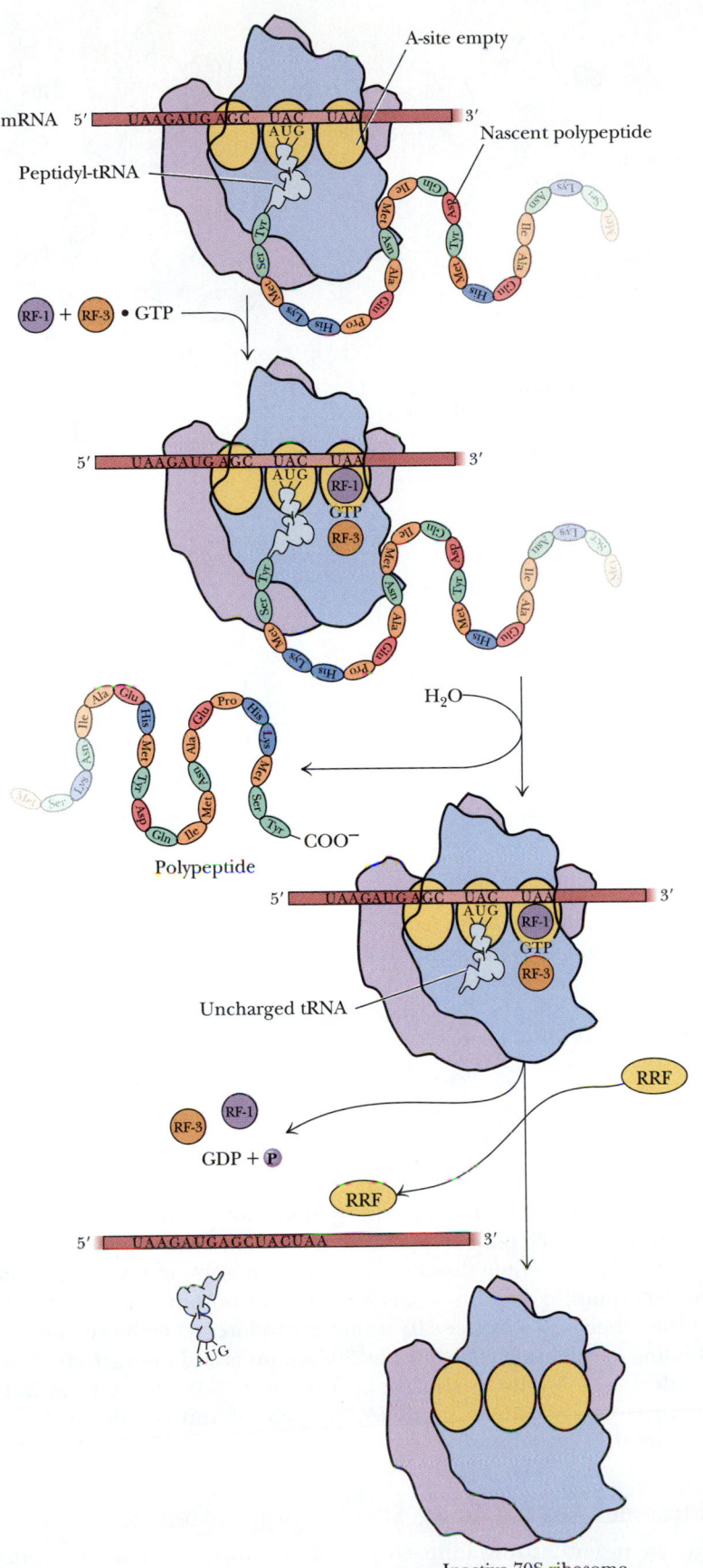

A-site empty

mRNA 5′

Peptidyl-tRNA

Nascent polypeptide

RF-1 + RF-3 • GTP

3′

5′

RF-1
GTP
RF-3

3′

H₂O

Polypeptide

COO⁻

5′

RF-1
GTP
RF-3

3′

Uncharged tRNA

RRF

RF-3   RF-1

GDP + P

RRF

5′ UAAGAUGAGCUACUAA 3′

AUG

Inactive 70S ribosome

Biochemistry ⒺNow™ ACTIVE FIGURE 30.23
The events in peptide chain termination. **Test yourself on the concepts in this figure at http:// chemistry.brookscole.com/ggb3**

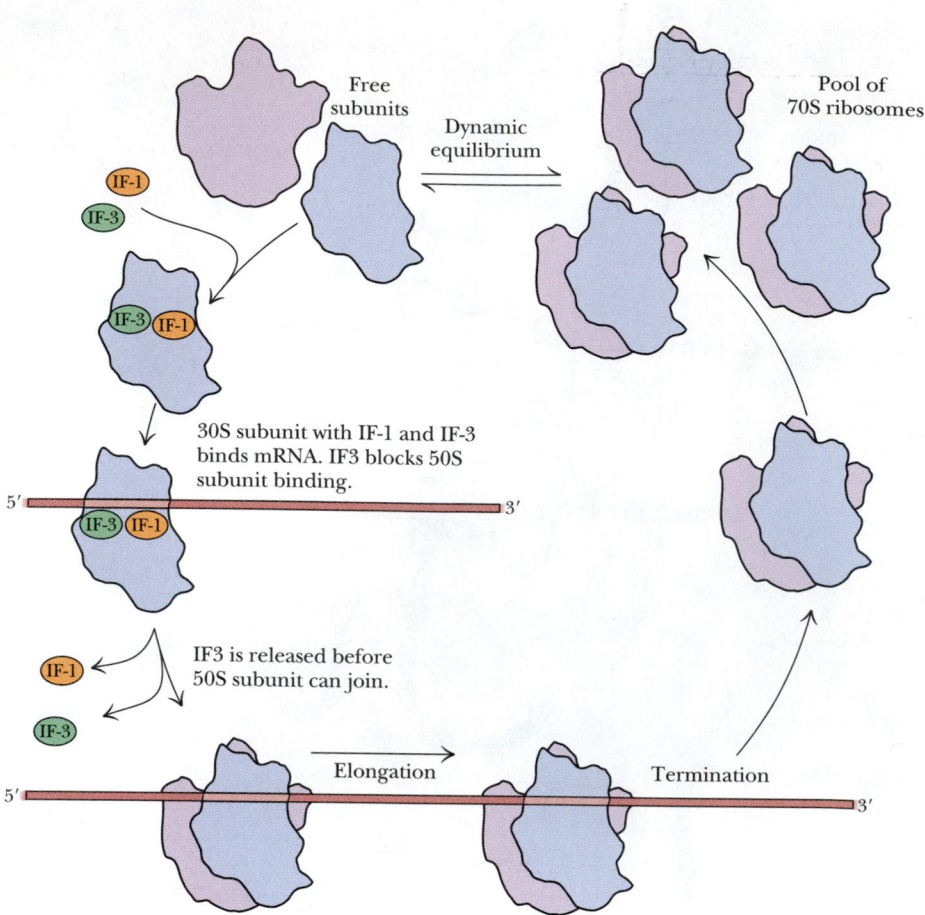

FIGURE 30.24 The ribosome life cycle. Note that IF-3 is released prior to 50S addition.

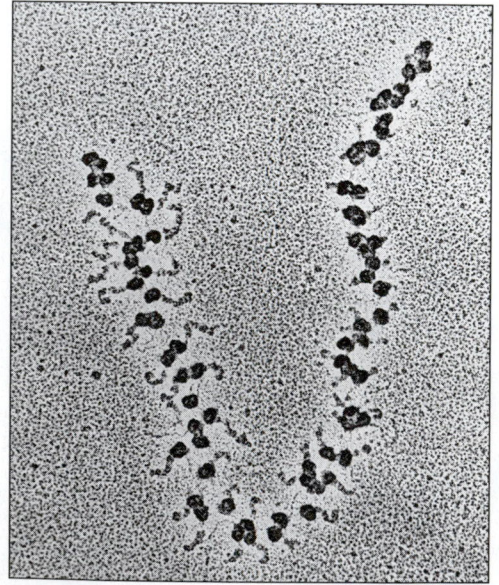

FIGURE 30.25 Electron micrograph of polysomes: multiple ribosomes translating the same mRNA. *(From Francke, C., et al., 1982. Electron microscopic visualization of a discrete class of giant translation units in salivary gland cells of* Chironomus tentans. *The EMBO Journal 1:59–62. Photo courtesy of Oscar L. Miller, Jr., University of Virginia.)*

G-protein superfamily (whose name is derived from the heterotrimeric G proteins that function in transmembrane signaling pathways, as in Figure 15.20). IF-2, EF-Tu, EF-G, and RF-3 interact with the same site on the 50S subunit, the *factor-binding center,* in the 50S cleft. This factor-binding center activates the GTPase activity of these factors, once they become bound.

The Ribosomal Subunits Cycle Between 70S Complexes and a Pool of Free Subunits

Ribosomal subunits cycle rapidly through protein synthesis. In actively growing bacteria, 80% of the ribosomes are engaged in protein synthesis at any instant. Once a polypeptide chain is synthesized and the nascent polypeptide chain is released, the 70S ribosome dissociates from the mRNA and separates into free 30S and 50S subunits (Figure 30.24). Intact 70S ribosomes are inactive in protein synthesis because only free 30S subunits can interact with the initiation factors. Binding of initiation factor IF-3 by 30S subunits and interaction of 30S subunits with 50S subunits are mutually exclusive. 30S subunits with bound initiation factors associate with mRNA, but 50S subunit addition requires IF-3 release from the 30S subunit.

Polyribosomes Are the Active Structures of Protein Synthesis

Active protein-synthesizing units consist of an mRNA with several ribosomes attached to it. Such structures are **polyribosomes,** or, simply, **polysomes** (Figure 30.25). All protein synthesis occurs on polysomes. In the polysome, each ribosome is traversing the mRNA and independently translating it into polypeptide. The further a ribosome has moved along the mRNA, the greater the length of its associated polypeptide product. In prokaryotes, as many as ten ribosomes may be found in a polysome. Ultimately, as many as 300 ribosomes may translate an mRNA, so as many as 300 enzyme molecules may be produced from a single transcript. Eukaryotic polysomes typically contain fewer than 10 ribosomes.

30.6   How Are Proteins Synthesized in Eukaryotic Cells?

Eukaryotic mRNAs are characterized by two post-transcriptional modifications: the 5′-terminal **⁷methyl-GTP cap** and the 3′-terminal **poly(A) tail** (Figure 30.26). The ⁷methyl-GTP cap is essential for mRNA binding by eukaryotic ribosomes and also enhances the stability of these mRNAs by preventing their degradation by 5′-exonucleases. The poly(A) tail enhances both the stability and translational efficiency of eukaryotic mRNAs. The Shine–Dalgarno sequences found at the 5′-end of prokaryotic mRNAs are absent in eukaryotic mRNAs.

Peptide Chain Initiation in Eukaryotes

The events in eukaryotic peptide chain initiation are summarized in Figure 30.27, and the properties of **eukaryotic initiation factors,** symbolized **eIFs,** are presented in Table 30.9. As might be expected, eukaryotic protein synthesis is considerably more complex than prokaryotic protein synthesis. The eukaryotic initiator tRNA is a unique tRNA functioning only in initiation. Like the prokaryotic initiator tRNA, the eukaryotic version carries only Met. However, unlike prokaryotic f-Met-tRNA$_i^{fMet}$, the Met on this tRNA is not formylated. The eukaryotic initiator tRNA is usually designated **tRNA$_i^{Met}$,** with the "i" indicating "initiation."

Eukaryotic initiation can be divided into three fundamental stages:

Stage 1: Formation of the **43S preinitiation complex** (Figure 30.27, stage 1). Initiation factors eIF-1A and eIF-3 bind to a 40S ribosomal subunit. Then, Met-tRNA$_i^{Met}$ (in the form of an eIF-2:GTP:Met-tRNA$_i^{Met}$ ternary complex) is delivered to the eIF-1A:eIF-3:40S subunit complex. (Unlike in prokaryotes, binding of Met-tRNA$_i^{Met}$ by eukaryotic ribosomes occurs in the absence of mRNA, so Met-tRNA$_i^{Met}$ binding is not codon-directed.)

Stage 2: Formation of the **48S initiation complex** (Figure 30.27, stage 2). This stage involves binding of the 43S preinitiation complex to mRNA and migration of the 40S ribosomal subunit to the correct AUG initiation codon. Binding of mRNA by the 43S preinitiation complex requires a set of proteins termed the **eIF-4 group.** Collectively, these proteins recognize the 5′-terminal cap and 3′-terminal poly(A)

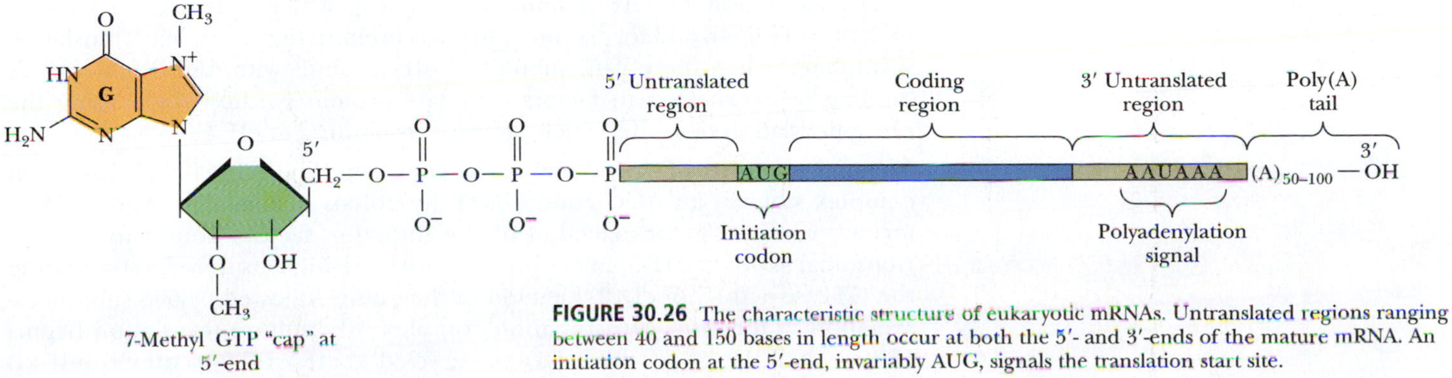

FIGURE 30.26 The characteristic structure of eukaryotic mRNAs. Untranslated regions ranging between 40 and 150 bases in length occur at both the 5′- and 3′-ends of the mature mRNA. An initiation codon at the 5′-end, invariably AUG, signals the translation start site.

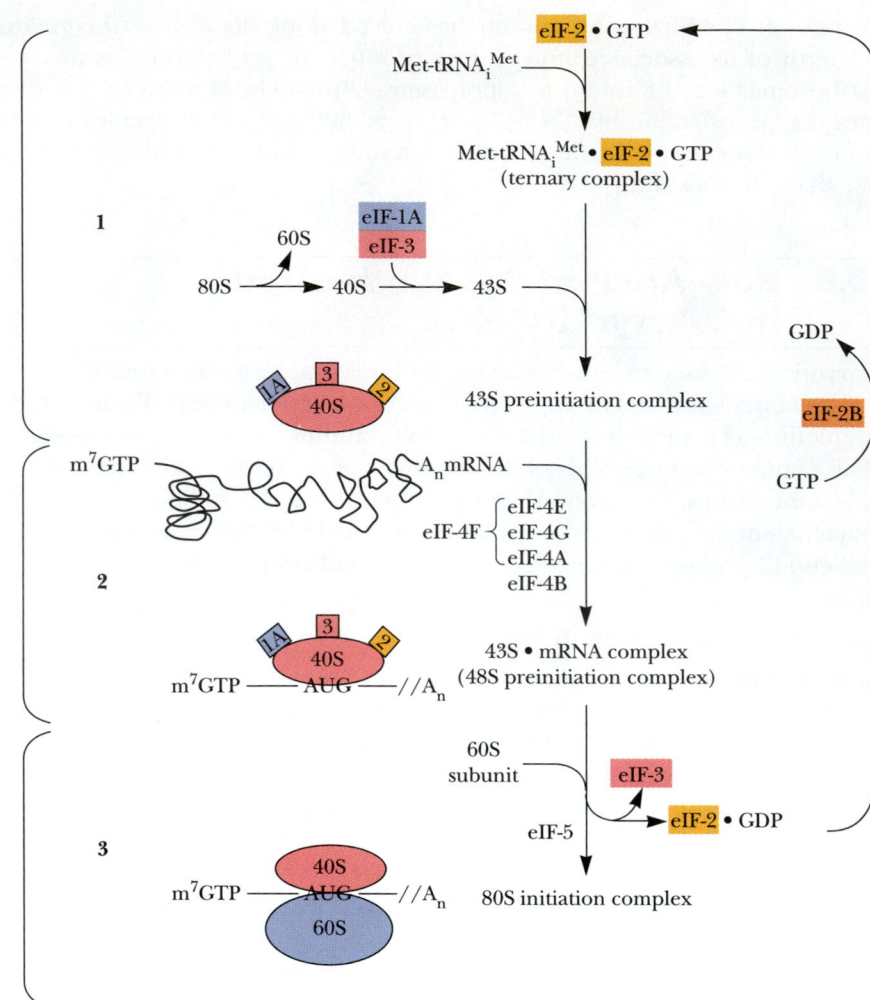

FIGURE 30.27 The three stages in the initiation of translation in eukaryotic cells. See Table 30.9 for a description of the functions of the eukaryotic initiation factors (eIFs). *(Adapted from Figure 1 in Pain, V. M., 1996. Initiation of protein synthesis in eukaryotic cells. European Journal of Biochemistry **236**:747–771; and Figure 1 in Gingras, A-C., et al., 1999. EIF4 initiation factors: Effectors of mRNA recruitment to ribosomes and regulators of translation. Annual Review of Biochemistry **68**:913–963.)*

tail of an mRNA, unwind any secondary structure in the mRNA, and transfer the mRNA to the 43S preinitiation complex. The eIF-4 group includes eIF-4B and eIF-4F. eIF-4F is a trimeric complex consisting of eIF-4A (an ATP-dependent RNA helicase), eIF-4E (which binds the 5′-terminal ⁷methyl-GTP of mRNAs), and eIF-4G. Because eIF-4G interacts with **Pab1p,** the **p**oly(**A**)-**b**inding **p**rotein that binds to the poly(A) tract on mRNAs (Figure 30.28), eIF-4G serves as the bridge between the cap-binding eIF-4E, the poly(A) tail of the mRNA, and the 40S subunit (through interaction with eIF-3). These interactions between the 5′-terminal ⁷methyl-GTP cap and the 3′-poly(A) tail initiate scanning of the 40S subunit in search of an AUG codon.

eIF-4E, the mRNA cap-binding protein, represents a key regulatory element in eukaryotic translation. eIF-4F binding to the cap structure is necessary for association of eIF-4B and formation of the 48S preinitiation complex. Translation is inhibited when the eIF-4E subunit of eIF-4F binds with **4E-BP** (the eIF-4E binding protein). Growth factors stimulate protein synthesis by causing the phosphorylation of 4E-BP, which prevents its binding to eIF-4E.

Stage 3: Formation of the **80S initiation complex.** When the 48S preinitiation complex stops at an AUG codon, GTP hydrolysis in the eIF-2:Met-tRNA$_i^{Met}$ ternary complex causes ejection of the initiation factors bound to the 40S ribosomal subunit. EIF-5, in conjunction with eIF-5B, acts here by stimulating the GTPase activity of eIF-2. Ejection of the eIFs is followed by 60S subunit association to form the 80S initiation complex, whereupon translation begins (Figure 30.27, stage 3). eIF-2:GDP is recycled to eIF-2:GTP by eIF-2B (eIF-2B is a *guanine nucleotide exchange factor*).

Table 30.9

Properties of Eukaryotic Translation Initiation Factors

Factor	Subunit	Size (kD)	Function
eIF-1		15	Enhances initiation complex formation
eIF-1A		17	Stabilizes Met-tRNA$_i$ binding to 40S ribosomes
eIF-2		125	GTP-dependent Met-tRNA$_i$ binding to 40S ribosomes
	α	36	Regulated by phosphorylation
	β	50	Binds Met-tRNA$_i$
	γ	55	Binds GTP, Met-tRNA$_i$
eIF-2B		270	Promotes guanine nucleotide exchange on eIF-2
eIF-2C		94	Stabilizes ternary complex in presence of RNA
eIF-3		550	Promotes Met-tRNA$_i$ and mRNA binding
eIF-4F		243	Binds to mRNA caps and poly(A) tails; consists of eIF-4A, eIF-4E, and eIF-4G; RNA helicase activity unwinds mRNA 2° structure
eIF-4A		46	Binds RNA; ATP-dependent RNA helicase; promotes mRNA binding to 40S ribosomes
eIF-4E		24	Binds to 5′-terminal ^{7}methyl-GTP cap on mRNA
eIF-4G		173	Binds to Pab1p
eIF-4B		80	Binds mRNA; promotes RNA helicase activity and mRNA binding to 40S ribosomes
eIF-5		49	Promotes GTPase of eIF-2, ejection of eIF-2 and eIF-3
eIF-5B		175	Ribosome-dependent GTPase activity; mediates 40S and 60S joining
eIF-6			Dissociates 80S; binds to 60S

Adapted from Clark, B. F. C., et al., eds. 1996. Prokaryotic and eukaryotic translation factors. *Biochimie* **78**:1119–1122; Dever, T.. E., 1999. Translation initiation: Adept at adapting. *Trends in Biochemical Sciences* **24**:398–403; and Gingras, A-C., et al., 1999. eIF-4 initiation factors: Effectors of mRNA recruitment to ribosomes and regulators of translation. *Annual Review of Biochemistry* **68**:913-963.

Control of Eukaryotic Peptide Chain Initiation Is One Mechanism for Post-Transcriptional Regulation of Gene Expression

Regulation of gene expression can be exerted post-transcriptionally through control of mRNA translation. Phosphorylation/dephosphorylation of translational components is a dominant mechanism for control of protein synthesis. Thus far, seven proteins—eIF-2α; eIF-2B; eIF-4E; eIF-4G; 40S ribosomal protein S6; and the two eukaryotic elongation factors, eEF-1 and eEF-2 (see following)—have been identified as targets of regulatory controls. Modification of some factors affects the rate of mRNA translation; modification of others affects which mRNAs are selected for translation. Peptide chain initiation, the initial phase of the synthetic process, is the optimal place for such control. Phosphorylation of S6 facilitates initiation of protein

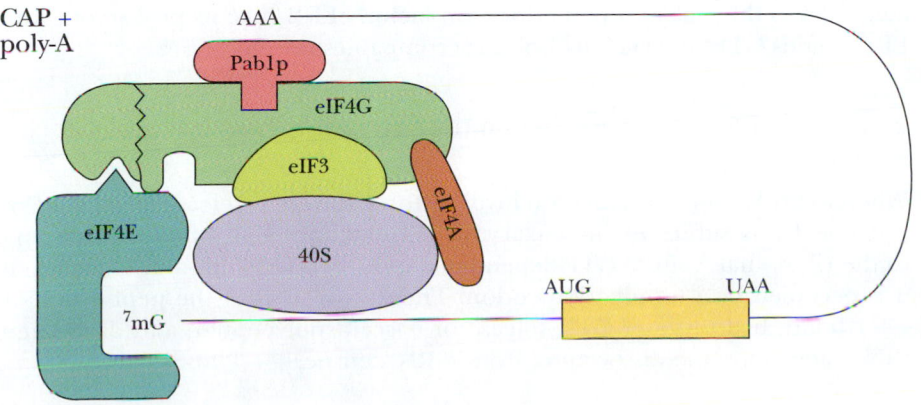

Biochemistry ⊜Now™ ANIMATED FIGURE 30.28
Initiation factor eIF-4G serves as a multipurpose adapter to engage the ^{7}methyl-G cap : eIF-4E complex, the Pab1p : poly(A) tract, and the 40S ribosomal subunit in eukaryotic translation initiation. (*Adapted from Hentze, M. W., 1997. eIF4G: A multipurpose ribosome adapter?* Science *275:500–501.*) **See this figure animated at** http://chemistry.brookscole.com/ggb3

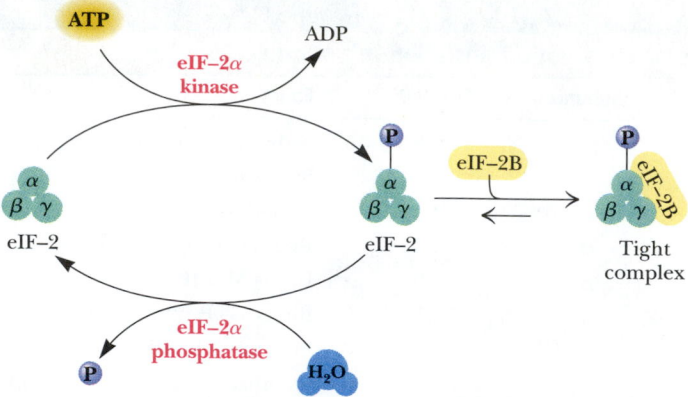

FIGURE 30.29 Control of eIF-2 functions through reversible phosphorylation of a Ser residue on its α-subunit. The phosphorylated form of eIF-2 (eIF-2–P) enters a tight complex with eIF-2B and is unavailable for initiation.

synthesis, resulting in a shift of the ribosomal population from inactive ribosomes to actively translating polysomes. S6 phosphorylation is stimulated by serum growth factors (see Chapter 32). The action of eIF-4F, the mRNA cap-binding complex, is promoted by phosphorylation. On the other hand, the phosphorylation of other translational components inhibits protein synthesis. For example, the α-subunit of eIF-2 can be reversibly phosphorylated at a specific Ser residue by an eIF-2α kinase/phosphatase system (Figure 30.29). Phosphorylation of eIF-2α inhibits peptide chain initiation. Reversible phosphorylation of eIF-2 is an important control governing globin synthesis in reticulocytes. If heme for hemoglobin synthesis becomes limiting in these cells, eIF-2α is phosphorylated, so globin mRNA is not translated and chains are not synthesized. Availability of heme reverses the inhibition through phosphatase-mediated removal of the phosphate group from the Ser residue.

Peptide Chain Elongation in Eukaryotes Resembles the Prokaryotic Process

Eukaryotic peptide elongation occurs in very similar fashion to the process in prokaryotes. An incoming aminoacyl-tRNA enters the ribosomal A site while peptidyl-tRNA occupies the P site. Peptidyl transfer then occurs, followed by translocation of the ribosome one codon further along the mRNA. Two elongation factors, eEF-1 and eEF-2, mediate the elongation steps. eEF-1 consists of two components: eEF-1A and eEF-1B. eEF-1A is the eukaryotic counterpart of EF-Tu; it serves as the aminoacyl-tRNA binding factor and requires GTP. eEF-1B is the eukaryotic equivalent of prokaryotic EF-Ts; it catalyzes the exchange of bound GDP on eEF-1 : GDP for GTP so that active eEF-1 : GTP can be regenerated. EF-2 is the eukaryotic translocation factor. eEF-2 (like its prokaryotic kin, EF-G) binds GTP, and GTP hydrolysis accompanies translocation.

Eukaryotic Peptide Chain Termination Requires Just One Release Factor

Whereas prokaryotic termination involves three different release factors (RFs), just one RF is sufficient for eukaryotic termination. Eukaryotic RF binding to the ribosomal A site is GTP-dependent, and RF : GTP binds at this site when it is occupied by a termination codon. Then, hydrolysis of the peptidyl-tRNA ester bond, hydrolysis of GTP, release of nascent polypeptide and deacylated tRNA, and ribosome dissociation from mRNA ensue.

Human Biochemistry

Diphtheria Toxin ADP-Ribosylates eEF-2

Diphtheria arises from infection by *Corynebacterium diphtheriae* bacteria carrying bacteriophage *corynephage β*. Diphtheria toxin is a phage-encoded enzyme secreted by these bacteria that is capable of inactivating a number of GTP-dependent enzymes through covalent attachment of an ADP-ribosyl moiety derived from NAD^+. That is, diphtheria toxin is an NAD^+-dependent ADP-ribosylase. One target of diphtheria toxin is the eukaryotic translocation factor, eEF-2. This protein has a modified His residue known as diphthamide. Diphthamide is generated post-translationally on eEF-2; its biological function is unknown. (EF-G of prokaryotes lacks this unusual modification and is not susceptible to diphtheria toxin.) Diphtheria toxin specifically ADP-ribosylates an imidazole-N within the diphthamide moiety of eEF-2 (see accompanying figure). ADP-ribosylated eEF-2 retains the ability to bind GTP but is unable to function in protein synthesis. Because diphtheria toxin is an enzyme and can act catalytically to modify many molecules of its target protein, just a few micrograms suffice to cause death.

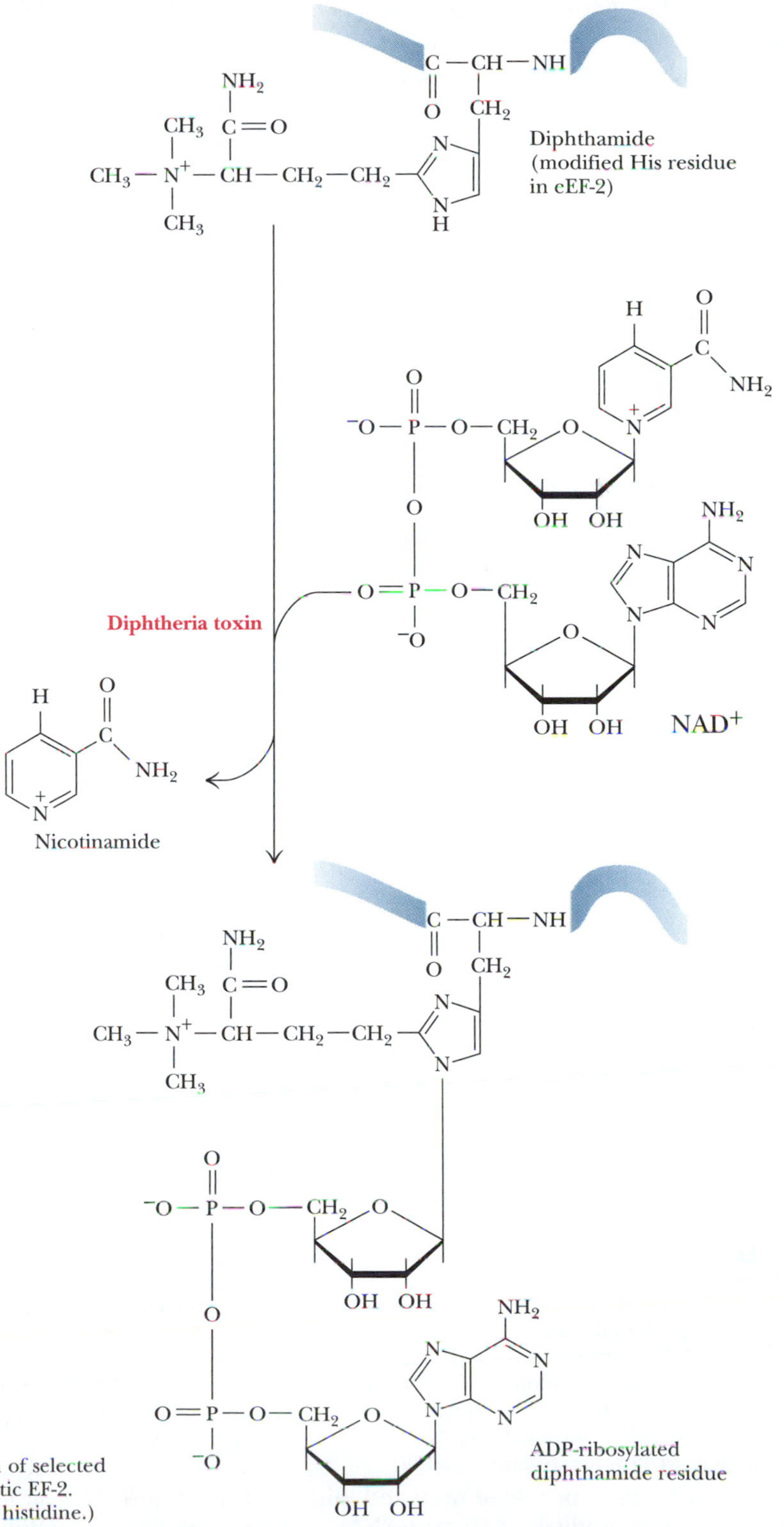

▶ Diphtheria toxin catalyzes the NAD^+-dependent ADP-ribosylation of selected proteins. ADP-ribosylation of the diphthamide moiety of eukaryotic EF-2. (Diphthamide = 2-[3-carboxamido-3-(trimethylammonio)propyl]histidine.)

Table 30.10		
Some Protein Synthesis Inhibitors		
Inhibitor	**Cells Inhibited**	**Mode of Action**
Initiation		
Aurintricarboxylic acid	Prokaryotic	Prevents IF binding to 30S subunit
Kasugamycin	Prokaryotic	Inhibits f Met-tRNA$_i^{fMet}$ binding
Streptomycin	Prokaryotic	Prevents formation of initiation complexes
Elongation: Aminoacyl-tRNA Binding		
Tetracycline	Prokaryotic	Inhibits aminoacyl-tRNA binding at A site
Streptomycin	Prokaryotic	Codon misreading, insertion of improper amino acid
Kirromycin	Prokaryotic	Binds to EF-Tu, preventing conformational switch from EF-Tu:GTP to EF-Tu:GDP
Elongation: Peptide Bond Formation		
Sparsomycin	Prokaryotic	Peptidyl transferase inhibitor
Chloramphenicol	Prokaryotic	Binds to 50S subunit, blocks the A site and inhibits peptidyl transferase activity
Clindamycin	Prokaryotic	Binds to 50S subunit, overlapping the A and P sites and blocking peptidyl transferase activity
Erythromycin	Prokaryotic	Blocks the 50S subunit tunnel, causing premature peptidyl-tRNA dissociation
Elongation: Translocation		
Fusidic acid	Both	Inhibits EF-G:GDP dissociation from ribosome
Thiostrepton	Prokaryotic	Inhibits ribosome-dependent EF-Tu and EF-G GTPase activity
Diphtheria toxin	Eukaryotic	Inactivates eEF-2 through ADP-ribosylation
Cycloheximide	Eukaryotic	Inhibits translocation of peptidyl-tRNA
Premature Termination		
Puromycin	Both	Aminoacyl-tRNA analog, binds at A site and acts as peptidyl acceptor, aborting peptide elongation
Ribosome Inactivation		
Ricin	Eukaryotic	Catalytic inactivation of 28S rRNA via *N*-glycosidase action on A^{4256}

Inhibitors of Protein Synthesis

Protein synthesis inhibitors have served two major, and perhaps complementary, purposes. First, they have been very useful scientifically in elucidating the biochemical mechanisms of protein synthesis. Second, some of these inhibitors affect prokaryotic but not eukaryotic protein synthesis and thus are medically important antibiotics. Table 30.10 is a partial list of these inhibitors and their mode of action. The structures of some of these compounds are given in Figure 30.30.

Summary

30.1 What Is the Genetic Code?
The genetic code is the code of bases that specifies the sequence of amino acids in a protein. The genetic code is a triplet code. Given the four RNA bases—A, C, G, and U—a total of $4^3 = 64$ three-letter codons are available to specify the 20 amino acids found in proteins. Of these 64 codons, 61 are used for amino acids, and the remaining 3 are nonsense, or "stop," codons. The genetic code is unambiguous, degenerate, and universal.

30.2 How Is an Amino Acid Matched with Its Proper tRNA?
During protein synthesis, aminoacyl-tRNAs recognize the codons through base pairing using their anticodon loops. A second genetic code exists, the code by which each aminoacyl-tRNA synthetase adds its amino acid to tRNAs that can interact with the codons that specify its amino acid. A common set of rules does not govern tRNA recognition by aminoacyl-tRNA synthetases. The tRNA features recognized are not limited to the anticodon and in some instances do not even include the anticodon. Usually, an aminoacyl-tRNA synthetase recognizes a set of sequence elements in its cognate tRNAs.

30.3 What Are the Rules In Codon–Anticodon Pairing?
Anticodon are paired with codons in antiparallel orientation. There are more codons than there are amino acids, and considerable degeneracy exists in the genetic code at the third base position. The first two bases of the codon and the last two bases of the anticodon form canonical Watson–Crick base pairs, but pairing between the third base of the codon and the first base of the anticodon follows less stringent rules, allowing some anticodons to recognize more than one codon, in accordance with Crick's wobble hypothesis. Some codons for a particular amino acid are used more than the others. Nonsense suppression occurs when suppressor tRNAs read nonsense codons.

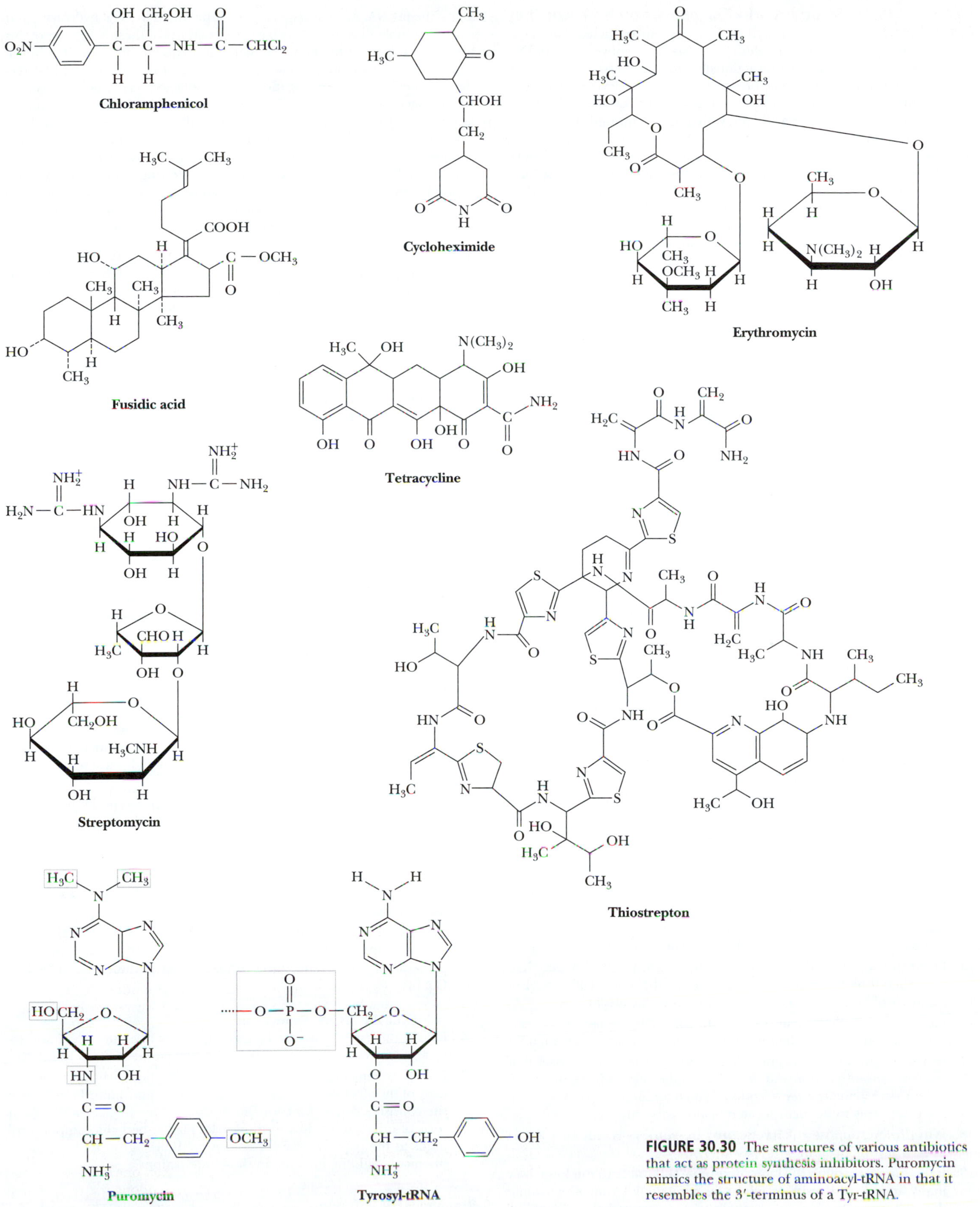

Chloramphenicol

Cycloheximide

Erythromycin

Fusidic acid

Tetracycline

Streptomycin

Thiostrepton

Puromycin

Tyrosyl-tRNA

FIGURE 30.30 The structures of various antibiotics that act as protein synthesis inhibitors. Puromycin mimics the structure of aminoacyl-tRNA in that it resembles the 3'-terminus of a Tyr-tRNA.

30.4 What Is the Structure of Ribosomes, and How Are They Assembled?

Ribosomes are ribonucleoprotein particles that act as mechanochemical systems in protein synthesis. They move along mRNA templates, orchestrating the interactions between successive codons and the corresponding anticodons presented by aminoacyl-tRNAs. Ribosomes catalyze the formation of peptide bonds. Prokaryotic ribosomes consist of two subunits, 30S and 50S, which are composed of 50 different proteins and 3 rRNAs—16S, 23S, and 5S. The general shapes of the ribosomal subunits are determined by their rRNA molecules; ribosomal proteins serve a largely structural role in ribosomes. Ribosomes spontaneously self-assemble in vitro. The 30S subunit provides the decoding center that matches up the tRNA anticodons with the mRNA codons. The 50S subunit has the peptidyl transferase center that catalyzes peptide bond formation. This center consists solely of 23S rRNA; the ribosome is a ribozyme. Eukaryotic cytosolic ribosomes are larger than prokaryotic ribosomes.

30.5 What Are the Mechanics of mRNA Translation?

Ribosomes move along the mRNA in the 5′→3′ direction, recruiting aminoacyl-tRNAs whose anticodons match up with successive codons and joining amino acids in peptide bonds in a polymerization process that forms a particular protein. Protein synthesis proceeds in three distinct phases: initiation, elongation, and termination. Elongation involves two steps: peptide bond formation and translocation of the ribosome one codon further along the mRNA. At each stage, energy is provided by GTP hydrolysis, and specific soluble protein factors participate. Many of these soluble proteins are G-protein family members. Initiation involves binding of mRNA by the small ribosomal subunit, followed by binding of fMet-tRNA$_i^{fMet}$ that recognizes the first codon. Elongation is accomplished via a repetitive cycle in which successive aminoacyl-tRNAs add to the ribosome:mRNA complex as directed by codon binding, the 50S subunit catalyzes peptide bond formation, and the polypeptide chain grows by one amino acid at a time. Ribosomes have three tRNA-binding sites: the A site, where incoming aminoacyl-tRNAs bind; the P site, where the growing peptidyl-tRNA chain is bound; and the E site, where deacylated tRNAs exit the ribosome. Termination occurs when the ribosome encounters a stop codon in the mRNA. Polysomes are the active structures in protein synthesis.

30.6 How Are Proteins Synthesized in Eukaryotic Cells?

The process of protein synthesis in eukaryotes strongly resembles that in prokaryotes, but the events are more complicated. Eukaryotic mRNAs have 5′-terminal ^{7}methyl G caps and 3′-polyadenylylated tails. Initiation of eukaryotic protein synthesis involves three stages and multiple proteins. This complexity offers many opportunities for regulation, and eukaryotic cells employ a variety of mechanisms for post-transcriptional regulation of gene expression. Many antibiotics are specific inhibitors of prokaryotic protein synthesis, making them particularly useful for the treatment of bacterial infections and diseases.

Problems

1. (Integrates with Chapter 12.) The following sequence represents part of the nucleotide sequence of a cloned cDNA:

   ...CAATACGAAGCAATCCCGCGACTAGACCTTAAC...

   Can you reach an unambiguous conclusion from these data about the partial amino acid sequence of the protein encoded by this cDNA?

2. A random (AG) copolymer was synthesized using a mixture of 5 parts adenine nucleotide to 1 part guanine nucleotide as substrate. If this random copolymer is used as an mRNA in a cell-free protein synthesis system, which amino acids will be incorporated into the polypeptide product? What will be the relative abundances of these amino acids in the product?

3. Review the evidence establishing that aminoacyl-tRNA synthetases bridge the information gap between amino acids and codons. Indicate the various levels of specificity possessed by aminoacyl-tRNA synthetases that are essential for high-fidelity translation of messenger RNA molecules.

4. (Integrates with Chapter 11.) Draw base-pair structures for (a) a G:C base pair, (b) a C:G base pair, (c) a G:U base pair, and (d) a U:G base pair. Note how these various base pairs differ in the potential hydrogen-bonding patterns they present within the major groove and minor groove of a double-helical nucleic acid.

5. Point out why Crick's wobble hypothesis would allow fewer than 61 anticodons to be used to translate the 61 sense codons. How might "wobble" tend to accelerate the rate of translation?

6. How many codons can mutate to become nonsense codons through a single base change? Which amino acids do they encode?

7. Nonsense suppression occurs when a suppressor mutant arises that reads a nonsense codon and inserts an amino acid, as if the nonsense codon were actually a sense codon. Which amino acids do you think are most likely to be incorporated by nonsense suppressor mutants?

8. Why do you suppose eukaryotic protein synthesis is only 10% as fast as prokaryotic protein synthesis?

9. If the tunnel through the large ribosomal subunit is 10 nm long, how many amino acid residues might be contained within it? (Assume that the growing polypeptide chain is in an extended β-sheet–like conformation.)

10. Eukaryotic ribosomes are larger and more complex than prokaryotic ribosomes. What advantages and disadvantages might this greater ribosomal complexity bring to a eukaryotic cell?

11. What ideas can you suggest to explain why ribosomes invariably exist as two-subunit structures, instead of a larger, single-subunit entity?

12. How do prokaryotic cells determine whether a particular methionyl-tRNA^{Met} is intended to initiate protein synthesis or to deliver a Met residue for internal incorporation into a polypeptide chain? How do the Met codons for these two different purposes differ? How do eukaryotic cells handle these problems?

13. What is the Shine–Dalgarno sequence? What does it do? The efficiency of protein synthesis initiation may vary by as much as 100-fold for different mRNAs. How might the Shine–Dalgarno sequence be responsible for this difference?

14. In the protein synthesis elongation events described under the section on translocation, which of the following seems the most apt account of the peptidyl transfer reaction: (a) The peptidyl-tRNA delivers its peptide chain to the newly arrived aminoacyl-tRNA situated in the A site, or (b) the aminoacyl end of the aminoacyl-tRNA moves toward the P site to accept the peptidyl chain? Which of these two scenarios makes more sense to you? Why?

15. (Integrates with Chapter 15.) Why might you suspect that the elongation factors EF-Tu and EF-Ts are evolutionarily related to the G proteins of membrane signal transduction pathways described in Chapter 15?

Preparing for the MCAT Exam

16. Review the list of Shine–Dalgarno sequences in Figure 30.17 and select the one that will interact best with the 3′-end of *E. coli* 16S rRNA.

17. Chloramphenicol (Figure 30.30) inhibits the peptidyl transferase activity of the 50S ribosomal subunit. The 50S peptidyl transferase active site consists solely of functionalities provided by the 23S rRNA. What sorts of interactions do you think take place when chloramphenicol binds to the peptidyl transferase center? Which groups on chloramphenicol might be involved in these interactions?

Biochemistry ⒺNow™ Preparing for an exam? Test yourself on key questions at http://chemistry.brookscole.com/ggb3

Further Reading

General

Gesteland, R. F., Cech, T. R., and Atkins, J. F., eds. 1999. *The RNA World*, 2nd ed. Cold Spring Harbor, NY: Cold Spring Harbor Laboratory Press.

The Genetic Code

Cedergren, R., and Miramontes, P., 1996. The puzzling origin of the genetic code. *Trends in Biochemical Sciences* **21**:199–200.

Huttenhofer, A., and Bock, A., 1998. RNA structures involved in selenoprotein synthesis, in *RNA Structure and Function*, Simons, R. W., and Grunberg-Monago, M., eds., pp. 603–639. Cold Spring Harbor, NY: Cold Spring Harbor Laboratory Press.

Khorana, H. G., et al., 1966. Polynucleotide synthesis and the genetic code. *Cold Spring Harbor Symposium on Quantitative Biology* **31**:39–49. The use of synthetic polyribonucleotides in elucidating the genetic code.

Knight, R. D., et al., 1999. Selection, history, and chemistry: Three faces of the genetic code. *Trends in Biochemical Sciences* **24**:241–247.

Low, S. C., and Berry, M. J., 1996. Knowing when not to stop: Selenocysteine incorporation in eukaryotes. *Trends in Biochemical Sciences* **21**:203–208.

Nirenberg, M. W., and Leder, P., 1964. RNA codewords and protein synthesis. *Science* **145**:1399–1407. The use of simple trinucleotides and a ribosome-binding assay to decipher the genetic code.

Nirenberg, M. W., and Matthaei, J. H., 1961. The dependence of cell-free protein synthesis in *E. coli* upon naturally occurring or synthetic polyribonucleotides. *Proceedings of the National Academy of Sciences, U.S.A.* **47**:1588–1602.

Speyer, J. F., et al., 1963. Synthetic polynucleotides and the amino acid code. *Cold Spring Harbor Symposium on Quantitative Biology* **28**:559–567.

Aminoacylation of tRNAs and the Second Genetic Code

Arnez, J. G., and Moras, D., 1997. Structural and functional considerations of the aminoacylation reaction. *Trends in Biochemical Sciences* **22**:211–216.

Burbaum, J. J., and Schimmel, P., 1991. Structural relationships and the classification of aminoacyl-tRNA synthetases. *Journal of Biological Chemistry* **266**:16965–16968.

Carter, C. W., Jr., 1993. Cognition, mechanism, and evolutionary relationships in aminoacyl-tRNA synthetases. *Annual Review of Biochemistry* **62**:715–748.

Eriani, G., et al., 1990. Partition of tRNA synthetases into two classes based on mutually exclusive sets of sequence motifs. *Nature* **347**:203–206.

Francklyn, C., Shi, J-P., and Schimmel, P., 1992. Overlapping nucleotide determinants for specific aminoacylation of RNA microhelices. *Science* **255**:1121–1125. One of a series of papers from Schimmel's laboratory elucidating tRNA features that serve as aminoacyl-tRNA synthetase recognition elements.

Hale, S. P., et al., 1997. Discrete determinants in transfer RNA for editing and aminoacylation. *Science* **276**:1250–1252.

Ibba, M., Curnow, A. W., and Söll, D., 1997. Aminoacyl-tRNA synthesis: Divergent routes to a common goal. *Trends in Biochemical Sciences* **22**:39–42.

Ibba, M., Hong, W.-W., and Söll, D., 1996. Glutaminyl-tRNA synthetase: From genetics to molecular recognition. *Genes to Cells* **1**:421–427.

Musier-Forsyth, K., and Schimmel, P., 1993. Aminoacylation of RNA oligonucleotides: Minimalist structures and origins of specificity. *The FASEB Journal* **7**:282–289.

Normanly, J., and Abelson, J., 1989. tRNA identity. *Annual Review of Biochemistry* **58**:1029–1049. Review of the structural features of tRNA that are recognized by aminoacyl-tRNA synthetases.

Rould, M. A., et al., 1989. Structure of *E. coli* glutaminyl-tRNA synthetase complexed with tRNA^Gln and ATP at 2.8Å resolution. *Science* **246**:1135–1142. One of the first high-resolution, three-dimensional structures of an aminoacyl-tRNA synthetase complexed with its cognate tRNA provides insights into the features employed by these enzymes in recognizing unique tRNAs and translating the genetic code.

Saks, M. E., Sampson, J. R., and Abelson, J. N., 1994. The transfer RNA problem: A search for rules. *Science* **263**:191–197.

Schimmel, P., 1987. Aminoacyl-tRNA synthetases: General scheme of structure-function relationships in the polypeptides and recognition of transfer RNAs. *Annual Review of Biochemistry* **56**:125–158.

Schimmel, P., 1995. An operational RNA code for amino acids and variations in critical nucleotide sequences in evolution. *Journal of Molecular Evolution* **40**:531–536.

Schimmel, P., and Ribas de Pouplana, L., 1995. Transfer RNA: From minihelix to genetic code. *Cell* **81**:983–986.

Schimmel, P., and Schmidt, E., 1995. Making connections: RNA-dependent amino acid recognition. *Trends in Biochemical Sciences* **20**:1–2.

Codon–Anticodon Recognition

Crick, F. H. C., 1966. Codon–anticodon pairing: The wobble hypothesis. *Journal of Molecular Biology* **19**:548–555. Crick's original paper on wobble interactions between tRNAs and mRNA.

Crick, F. H. C., et al., 1961. General nature of the genetic code for proteins. *Nature* **192**:1227–1232. An insightful paper on insertion/deletion mutants providing convincing genetic arguments that the genetic code was a triplet code, read continuously from a fixed starting point. This genetic study foresaw the nature of the genetic code, as later substantiated by biochemical results.

Ribosome Structure and Function

Ban, N., et al., 2000. The complete atomic structure of the large ribosomal subunit at 2.4 Å resolution. *Science* **289**:905–920.

Carter, A. P., et al., 2000. Functional insights from the structure of the 30S ribosomal subunit and its interactions with antibiotics. *Nature* **407**:340–348.

Cate, J. H., et al., 1999. X-ray crystal structure of 70S functional ribosomal complexes. *Science* **285**:2095–2104.

Clemons, W. M., Jr., 1999. Structure of a bacterial 30S ribosomal subunit at 5.5 Å resolution. *Nature* **400**:833–840.

Doudna, J. A., and Rath, V. L., 2002. Structure and function of the eukaryotic ribosome: The next frontier. *Cell* **109**:153–156.

Draper, D. E., 1995. Protein-RNA recognition. *Annual Review of Biochemistry* **64**:593–620.

Frank, J., 1997. The ribosome at higher resolution—The donut takes shape. *Current Opinion in Structural Biology* **7**:266–272.

Gabashvili, I. S., et al., 2000. Solution structure of the *E. coli* 70S ribosome at 11.5Å resolution. *Cell* **100**:537–549.

Garrett, R., 1999. Mechanics of the ribosome. *Nature* **400**:811–812.

Green, R., and Noller, H. F., 1997. Ribosomes and translation. *Annual Review of Biochemistry* **66**:679–716.

Moore, P. B., 1997. The conformation of ribosomes and rRNA. *Current Opinion in Structural Biology* **7**:343–347.

Moore, P. B., and Steitz, T. A., 2002. The involvement of RNA in ribosome function. *Nature* **418**:229–235.

Mueller, F., and Brimacombe, R., 1997. A new model for the three-dimensional folding of *Escherichia coli* 16S ribosomal RNA: I. Fitting the RNA to a 3D electron microscopic map at 20 Å. II. The RNA–protein interaction. III. The topography of the functional center (with van Heel, H. S., and Rinke-Appel, J.). *Journal of Molecular Biology* **271**:524–587.

Ogle, J. M., Carter, A. P., and Ramakrishnan, V., 2003. Insights into the decoding mechanism from recent ribosome structures. *Trends in Biochemical Sciences* **28**:259–266.

Porse, B., Thi-Ngoc, H. P., and Garrett, R. A., 1997. The donor substrate site within the peptidyl transferase loop of 23S rRNA and its putative interactions with the CCA-end of N-blocked aminoacyl-tRNA. *Journal of Molecular Biology* **266**:472–483.

Ramakrishnan, V., 2002. Ribosome structure and the mechanism of translation. *Cell* **108**:557–572.

Rodnina, M. V., and Wintermeyer, W., 2001. Fidelity of aminoacyl-tRNA selection on the ribosome: Kinetic and structural mechanisms. *Annual Review of Biochemistry* **70**:415–435.

Simonson, A. B., and Lake, J. A., 2002. The transorientation hypothesis for codon recognition during protein synthesis. *Nature* **416**:281–284.

Spahn, C. M. T., et al., 2001. Structure of the 80S ribosome from *Saccharomyces cerevisiae*–tRNA-ribosome and subunit–subunit interactions. *Cell* **107**:373–386.

Stark, H., et al., 2002. Ribosome interactions of aminoacyl-tRNA and elongation factor Tu in the codon-recognition complex. *Nature Structural Biology* **9**:849–854.

Svergun, D. I., et al., 1997. Solution scattering analysis of the 70S *Escherichia coli* ribosome by contrast variation. I. Invariants and validation of electron microscopy models. II. A model of the ribosome and its RNA at 3.5 nm resolution. *Journal of Molecular Biology* **271**:588–618.

Tenson, T., and Ehrenberg, M., 2002. Regulatory nascent peptides in the ribosomal tunnel. *Cell* **108**:591–594.

Valle, M., *et al.*, 2002. Locking and unlocking of ribosomal motions. *Cell* **114**:123–134.

Wilson, K., and Noller, H. F., 1998. Molecular movement inside the translational engine. *Cell* **92**:337–349.

Wimberly, B. T., et al., 2000. Structure of the 30S ribosomal subunit. *Nature* **407**:327–339.

The Ribosome Is a Ribozyme

Cech, T. R., The ribosome is a ribozyme. *Science* **289**:878–879.

Green, R., Samaha, R. R., and Noller, H. F., 1997. Mutations at nucleotides G2251 and U2585 of 23 S rRNA perturb the peptidyl transferase center of the ribosome. *Journal of Molecular Biology* **266**:40–50.

Green, R., Switzer, C., and Noller, H. F., 1998. Ribosome-catalyzed peptide-bond formation with an A-site substrate covalently linked to 23S ribosomal RNA. *Science* **280**:286–289.

Muth, G. W., et al., 2000. A single adenosine with a neutral pK_a in the ribosomal peptidyl transferase center. *Science* **289**:947–950.

Nissen, P., et al., 2000. The structural basis of ribosome activity in peptide bond synthesis. *Science* **289**:920–930.

Noller, H. F., Hoffarth, V., and Zimniak, L., 1992. Unusual resistance of peptidyl transferase to protein extraction procedures. *Science* **256**:1416–1419. Research paper presenting evidence that the peptide bond–forming step in protein synthesis—the peptidyl transferase reaction—is catalyzed by 23S rRNA.

Protein Synthesis: Initiation, Elongation, and Termination Factors

Clark, B. F. C., and Nyborg, J., 1997. The ternary complex of EF-Tu and its role in protein synthesis. *Current Opinion in Structural Biology* **7**:110–116.

Clark, B. F. C., et al., eds., 1996. Prokaryotic and eukaryotic translation factors. *Biochimie* **78**:1119–1122.

Dever, T. E., 1999. Translation initiation: Adept at adapting. *Trends in Biochemical Sciences* **24**:398–403.

Ehrenberg, M., and Tenson, T., 2002. A new beginning to the end of translation. *Nature Structural Biology* **9**:85–87.

Ibba, M., and Söll, D., 1999. Quality control mechanisms during translation. *Science* **286**:1893–1897.

Nakamura, Y., et al., 2000. Mimicry grasps reality in transcription termination. *Cell* **101**:349–352.

Nissen, P., et al., 1995. Crystal structure of the ternary complex of Phe-tRNA^{Phe}, EF-Tu, and a GTP analog. *Science* **270**:1464–1472.

Nyborg, J., et al., 1996. Structure of the ternary complex of EF-Tu: Macromolecular mimicry in translation. *Trends in Biochemical Sciences* **21**:81–82.

Weijland, A., and Parmeggiani, A., 1993. Toward a model for the interaction between elongation factor Tu and the ribosome. *Science* **259**:1311–1314.

Wilson, K., and Noller, H. F., 1998. Mapping the position of translational elongation factor EF-G in the ribosome by directed hydroxyl radical probing. *Cell* **92**:131–139.

Zavialov, A. V., and Ehrenberg, M., 2003. Peptidyl-tRNA regulates the GTPase activity of translation factors. *Cell* **114**:113–122.

Eukaryotic Protein Synthesis

Gingras, A.-C., et al., 1999. eIF4 initiation factors: Effectors of mRNA recruitment to ribosomes and regulators of translation. *Annual Review of Biochemistry* **68**:913–963.

Hentze, M. W., 1997. eIF4G: A multipurpose ribosome adapter? *Science* **275**:500–501.

Hershey, J. W. B., Mathews, M. B., and Sonenberg, N., eds., 1996. *Translational Control*. Cold Spring Harbor, NY: Cold Spring Harbor Laboratory Press.

Marcotrigiano, J., et al., 1997. Cocrystal structure of the messenger RNA 5′ cap-binding protein (eIF4E) bound to 7-methyl-GTP. *Cell* **89**:951–961.

Matsuo, H., et al., 1997. Structure of translation factor eIF4E bound to 7mGDP and interaction with 4E-binding protein. *Nature Structural Biology* **4**:717–724.

Pain, V. M., 1996. Initiation of protein synthesis in eukaryotic cells. *European Journal of Biochemistry* **236**:747–771.

Pestova, T. V., et al., 2000. The joining of ribosomal subunits requires eIF5B. *Nature* **403**:332–335.

Rhoads, R. E., 1999. Signal transduction pathways that regulate eukaryotic protein synthesis. *Journal of Biological Chemistry* **274**:30337–30340.

Sachs, A. B., and Varani, G., 2000. Eukaryotic translation initiation: There are two sides (at least) to every story. *Nature Structural Biology* **7**:356–361.

Samuel, C. E., 1993. The eIF-2α protein kinases, regulators of translation in eukaryotes from yeast to humans. *Journal of Biological Chemistry* **268**:7603–7606.

Tarun, S. Z., Jr., et al., 1997. Translation factor eIF4G mediates in vitro poly(A) tail-dependent translation. *Proceedings of the National Academy of Sciences, U.S.A.* **94**:9046–9051.

Protein Synthesis Inhibitors

Caskey, C. T., 1973. *Inhibitors of protein synthesis*. In *Metabolic Inhibitors*, Volume 4, Hochster, R. M., Kates, M., and Quastel, J. H., eds. New York: Academic Press.

Endo, Y., et al., 1987. The mechanism of action of ricin and related toxic lectins on eukaryotic ribosomes. The site and the characteristics of the modification in 28S ribosomal RNA caused by the toxins. *Journal of Biological Chemistry* **262**:5098–5912.

Schlünzen, F., et al., 2000. Structural basis for the interaction of antibiotics with the peptidyl transferase center in eubacteria. *Nature* **413**:814–821.

Completing the Protein Life Cycle: Folding, Processing, and Degradation

Vong's Flying Crane. *Origami*—the Asian art of paper folding—arose in China almost 2000 years ago when paper was rare and expensive and the folded shape added special meaning. Protein folding, like origami, takes a functionless form and creates a structure with unique identity and purpose.

Life is a process of becoming, a combination of states we have to go through.
Anais Nin (1903–1977)

Essential Question

Proteins are the agents of biological function. Protein turnover (synthesis and decay) is a fundamental aspect of each protein's natural history. *How are newly synthesized polypeptide chains transformed into mature, active proteins? How are undesired proteins removed from cells?*

The human genome apparently contains less than 30,000 genes, but some estimates suggest that the total number of proteins in the human proteome may approach 1 million. What processes introduce such dramatically increased variation into the products of protein-encoding genes? We've reviewed (or will soon cover) many of these processes; a partial list (with examples) includes:

1. Gene rearrangements (immunoglobulin G)
2. Alternative splicing (fast skeletal muscle troponin T)
3. RNA editing (apolipoprotein B)
4. Proteolytic processing (chymotrypsinogen or prepro-opiomelanocortin; see Chapter 32)
5. Isozymes (lactate dehydrogenase)
6. Protein sharing (the glycolytic enzyme enolase is identical to τ-crystallin in the eye)
7. Protein–protein interactions at many levels (oligomerization, supramolecular complexes, assembly of signaling pathway protein complexes upon scaffold proteins)
8. Covalent modifications of many kinds (phosphorylation or glycosylation, with multisite phosphorylation or variable degrees of glycosylation, to name just two of the dozens of possibilities)

Thus, the nascent polypeptide emerging from a ribosome is not yet the agent of biological function that is its destiny. First, the polypeptide must fold into its native tertiary structure. Even then, seldom is the nascent, folded protein in its final functional state. Proteins often undergo various proteolytic processing reactions and covalent modifications as steps in their maturation to functional molecules. Finally, at the end of their usefulness, damaged by chemical reactions or denatured due to partial unfolding, they are degraded. In addition, some proteins are targeted for early destruction as part of regulatory programs that carefully control available amounts of particular proteins. Damaged or misfolded proteins are a serious hazard; accumulation of protein aggregates can be a cause of human disease, including the prion diseases (see Chapter 28) and diseases of amyloid accumulation, such as Alzheimer's, Parkinson's, or Huntington's disease.

Key Questions

31.1 How Do Newly Synthesized Proteins Fold?

31.2 How Are Proteins Processed Following Translation?

31.3 How Do Proteins Find Their Proper Place in the Cell?

31.4 How Does Protein Degradation Regulate Cellular Levels of Specific Proteins?

Nascent means "undergoing the process of being born," or in the molecular sense, "newly synthesized."

31.1 | How Do Newly Synthesized Proteins Fold?

As Christian Anfinsen pointed out 40 years ago, the information for folding each protein into its unique three-dimensional architecture resides within its amino acid sequence or primary structure (see Chapter 6). Proteins begin to fold even before their synthesis by ribosomes is completed (Figure 31.1a). However, the

Human Biochemistry

Alzheimer's, Parkinson's, and Huntington's Disease Are Late-Onset Neurodegenerative Disorders Caused by the Accumulation of Protein Deposits

As noted in Chapter 6, protein misfolding problems can cause disease by a variety of mechanisms. For example, protein aggregates can impair cell function. **Amyloid plaques** (so named because they resemble the intracellular starch, or amyloid, deposits found in plant cells) and **neurofibrillary tangles (NFTs)** are proteinaceous deposits found in the brains of individuals suffering from any of several neurogenerative diseases. In each case, the protein is different. In Alzheimer's, disease is caused both by extracellular amyloid deposits composed of proteolytic fragments of the amyloid precursor protein (APP) termed **amyloid-β (Aβ)** and intracellular NFTs composed of the **microtubule-binding protein tau (τ).** Aβ is a peptide 39 to 43 amino acids long that polymerizes to form long, highly ordered, insoluble fibrils consisting of a hydrogen-bonded parallel β-sheet structure in which identical residues on adjacent chains are aligned directly, in register (see accompanying figure). Why Aβ aggregates in some people but not others is not clear. In Parkinson's, the culprit is NFTs composed of polymeric τ; no amyloid plaques are evident. In Huntington's disease, the protein deposits occur as nuclear inclusions composed of polyglutamine (polyQ) aggregates. PolyQ aggregates arise from mutant forms of **huntingtin,** a protein that characteristically has a stretch of glutamine residues close to its N-terminus. Huntingtin is a 3144-residue protein encoded by the *IT15* gene, which has 67 exons. Exon 1 encodes the polyglutamine region. Individuals whose huntingtin gene has fewer that 35 CAG (glutamine codon) repeats never develop the disease; those with 40 or more always develop the disease within a normal lifetime. The nuclear inclusions in Huntington's disease are huntingtin-derived polyglutamine fragments that have aggregated to form β-sheet–containing amyloid fibrils.

Impairment of cellular function by proteinaceous deposits may be a general phenomenon. In vitro experiments have demonstrated that aggregates of proteins not associated with disease can be cytotoxic, and the ability to form amyloid deposits is a general property of proteins. The evolution of chaperones to assist protein folding and proteasomes to destroy improperly folded proteins may have been driven by the necessity to prevent protein aggregation.

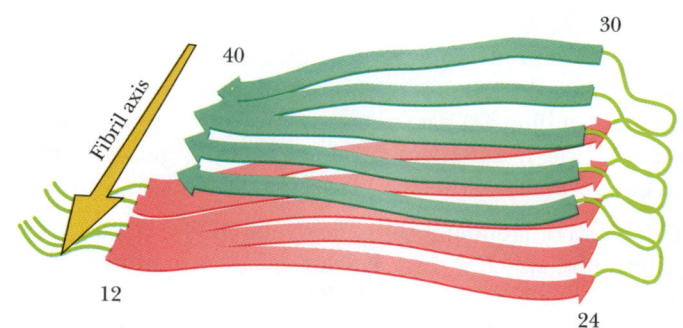

▲ A model for the Aβ$_{1-40}$ structural unit in β-amyloid fibrils. Fibrils contain β-strands perpendicular to the fibril axis, with interstrand hydrogen bonding parallel to the fiber axis. The top face of the β-sheet is hydrophobic and presumably interacts with neighboring Aβ molecules in fibril formation. (*Figure adapted from Figure 1 in Thompson, L. K., 2003. Unraveling the secrets of Alzheimer's β-amyloid fibrils.* Proceedings of the National Academy of Sciences, U.S.A. *100:383–385.*)

cytosolic environment is a very crowded place, with effective protein concentrations as high as 0.3 grams/mL. Macromolecular crowding enhances the likelihood of nonspecific protein association and aggregation. The primary driving force for protein folding is the burial of hydrophobic side chains away from the aqueous solvent and reduction in solvent-accessible surface area (see Chapter 6). The folded protein typically has a buried hydrophobic core and a hydrophilic surface. Protein aggregation is typically driven by hydrophobic interactions, so burial of hydrophobic regions through folding is a crucial factor in preventing aggregation. To evade such problems, nascent proteins are often assisted in folding by a family of helper proteins known as **molecular chaperones** (see Chapter 6), because, like the chaperones at a prom, their purpose is to prevent inappropriate liaisons. Chaperones also serve to shepherd proteins to their ultimate cellular destinations. Also, mature proteins that have become partially unfolded may be rescued by chaperone-assisted refolding.

Chaperones Help Some Proteins Fold

A number of chaperone systems are found in all cells. Many of the proteins in these systems are designated by the acronym **Hsp** (for **heat shock protein**) and a number indicating their relative mass in kilodaltons (as in Hsp60). Hsps were originally observed as abundant proteins in cells given brief exposure to high temperature (42° C or so). The principal Hsp chaperones are **Hsp70, Hsp60** (the **chaperonins**), and **Hsp90.** In general, proteins whose folding is chaperone-dependent pass down a pathway in which Hsp70 acts first on the newly synthesized protein and then passes the partially folded intermediate to a chaperonin for completion of folding.

Nascent polypeptide chains exiting the large ribosomal subunits are met by ribosome-associated chaperones (**TF,** or **trigger factor,** in *Escherichia coli;* **NAC [nascent chain-associated complex]** in eukaryotes). In *E. coli*, the 50S ribosomal protein L23, which is situated at the peptide exit tunnel, serves as the docking site for TF, directly linking protein synthesis with chaperone-assisted protein folding. TF and NAC mediate transfer of the emerging nascent polypeptide chain to the Hsp70 class of chaperones, although many proteins do not require this step for proper folding.

Hsp70 Chaperones Bind to Hydrophobic Regions of Extended Polypeptides

In Hsp70-assisted folding, proteins of the Hsp70 class bind to nascent polypeptide chains while they are still on ribosomes (Figure 31.1b). Hsp70 (known as **DnaK** in *E. coli*) recognizes exposed, extended regions of polypeptides that are rich in hydrophobic residues. By interacting with these regions, Hsp70 prevents nonproductive associations and keeps the polypeptide in an unfolded (or partially folded) state until productive folding interactions can occur. Completion of folding requires release of the protein from Hsp70; release is energy-dependent and is driven by ATP hydrolysis.

Hsp70 proteins such as DnaK consist of two domains: a 44-kD N-terminal ATP-binding domain and an 18-kD central domain that binds polypeptides with exposed hydrophobic regions (Figure 31.2a). The DnaK:ATP complex receives an unfolded (or partially folded) polypeptide chain from **DnaJ** (Figure 31.2b). DnaJ is an **Hsp40** family member. Interaction of DnaK with DnaJ triggers the ATPase activity of DnaK; the DnaK:ADP complex forms a stable complex with the unfolded polypeptide, preventing its aggregation with other proteins. A third protein, **GrpE,** catalyzes nucleotide exchange on DnaK, replacing ADP with ATP, which converts DnaK back to a conformational form having low affinity for its polypeptide substrate. Release of the polypeptide gives it the opportunity to fold. Multiple cycles of interaction with DnaK (or Hsp70) give rise to partially folded intermediates or, in some cases, completely folded proteins. The partially folded intermediates may be passed along to the Hsp60/chaperonin system for completion of folding (Figure 31.1c).

The GroES–GroEL Complex of *E. coli* is an Hsp60 Chaperonin

The Hsp60 class of chaperones, also known as **chaperonins,** assists some partially folded proteins to complete folding after their release from ribosomes. Chaperonins sequester partially folded molecules from one another (and from extraneous interactions), allowing folding to proceed in a protected environment. This protected environment is sometimes referred to as an **"Anfinsen cage"** because it provides an enclosed space where proteins fold spontaneously, free from the possibility of aggregation with other proteins. Chaperonins are large, cylindrical protein complexes formed from two stacked rings of subunits. The chaperonins have been organized into two groups, I and II, on the basis of their source and structure. Group I chaperonins are found in eubacteria, group II in archaea and eukaryotes. The group I chaperonin in *E. coli* is the **GroES–GroEL complex** (Figure 31.1c). GroEL is made of two stacked seven-membered rings of 60-kD subunits that form a cylindrical α_{14} oligomer 15 nm high and 14 nm wide (Figure 31.3). Each GroEL ring has a 5-nm central cavity where folding can take place. This cavity can accommodate proteins up to 60 kD in size. GroES, sometimes referred to as a **co-chaperonin,** consists of a single seven-membered ring of 10-kD subunits that sits like a dome on one end of GroEL (Figure 31.3). The end of GroEL where GroES is sitting is referred to as the *apical end.* Each GroEL subunit has two structural domains: an equatorial domain that binds ATP and

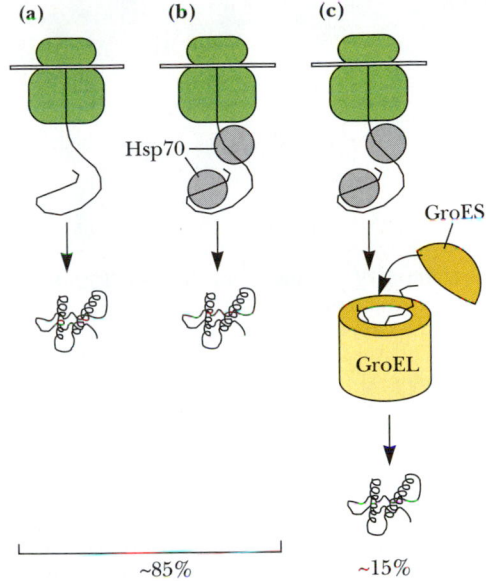

(a)   (b)   (c)

Hsp70

GroES

GroEL

~85%   ~15%

Biochemistry 🔵 Now™ ANIMATED FIGURE 31.1
Protein folding pathways. **(a)** Chaperone-independent folding. The protein folds as it is synthesized on the ribosome *(green)* (or shortly thereafter). **(b)** Hsp70-assisted protein folding. Hsp70 *(gray)* binds to nascent polypeptide chains as they are synthesized and assists their folding. **(c)** Folding assisted by Hsp70 and chaperonin complexes. The chaperonin complex in *E. coli* is GroES–GroEL. The chaperonin complex in eukaryotic cells is known as TRiC (for TCP-1 ring complex) or CCT (cytosolic chaperonin containing TCP-1). The majority of proteins fold by pathways (a) or (b). *(Adapted from Figure 2 in Netzer, W. J., and Hartl, F. U., 1998. Protein folding in the cytosol: Chaperonin-dependent and -independent mechanisms.* Trends in Biochemical Sciences **23:**68–73; *and Figure 2 in Hartl, F. U., and Hayer-Hartl, M., 2002. Molecular chaperones in the cytosol: From nascent chain to folded protein,* Science **295:**1852–1858.) **See this figure animated at http:// chemistry.brookscole.com/ggb3**

(a)   Domain organization and structure of the Hsp70 family member, DnaK

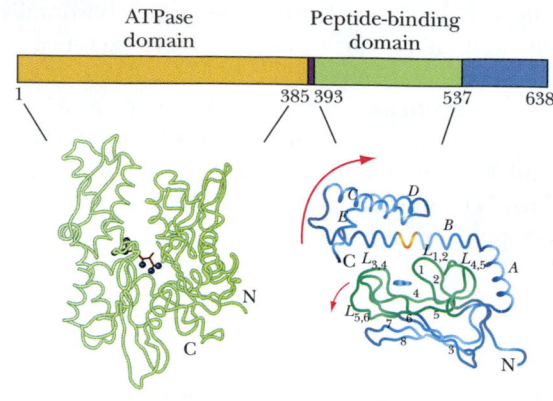

Biochemistry ⊜ Now™   ANIMATED FIGURE 31.2

Structure and function of DnaK: **(a)** Domain organization and structure of the Hsp70 family member, DnaK. The N-terminal domain (residues 1–385) of DnaK is the ATP-binding domain; the polypeptide-binding domain encompasses residues 393 to 537 of the 638-residue protein. The ribbon diagram on the lower left is the ATP-binding domain of the DnaK analog, bovine Hsc70; bound ADP is shown as a stick diagram (*purple*). The ribbon diagram on the lower right is the polypeptide-binding domain of DnaK. The small blue ovals highlight the position of the polypeptide substrate; the protein regions that bind the polypeptide substrate are blue-green. **(b)** DnaK mechanism of action: DnaJ binds an unfolded protein (U) or partially folded intermediate (I) and delivers it to the DnaK:ATP complex. DnaJ:DnaK interaction stimulates the ATPase activity of DnaK, converting bound ATP to ADP, which stabilizes the DnaK:unfolded polypeptide association. The nucleotide exchange protein GrpE replaces ADP with ATP on DnaK and the partially folded intermediate ([I]) is released. I has several possible fates: It may fold into the native state, N; it may undergo another cycle of interaction with DnaJ and DnaK; or it may be become a substrate for folding by the GroEL chaperonin system. (*Adapted from Figures 1a and 2a in Frydman, J., 2001. Folding of newly translated proteins in vivo: The role of molecular chaperones.* Annual Review of Biochemistry **70**:603–647.) **See this figure animated at http://chemistry.brookscole.com/ggb3**

(b)   DnaK mechanism of action

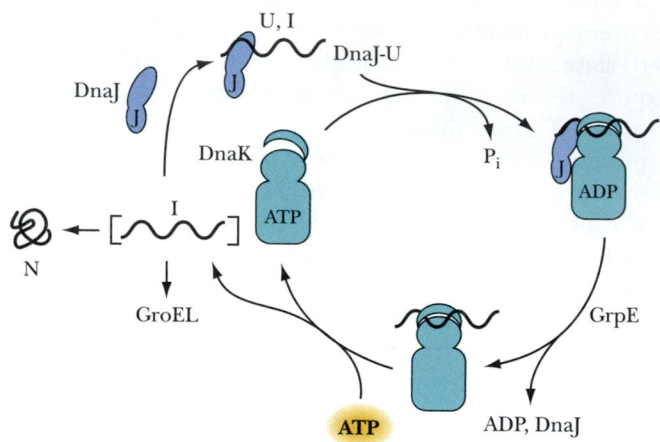

interacts with neighbors in the other α_7 ring and a apical domain that has with hydrophobic residues that can interact with hydrophobic regions on partially folded proteins. The apical domain hydrophobic patches face the interior of the central cavity. An unfolded (or partially folded) protein binds to the apical patches and is delivered to the central cavity of the upper α_7 ring (Figure 31.3c). ATP binding to the subunits of the upper α_7 ring has two consequences: (1) It recruits GroES to GroEL, and (2) it causes the α-subunits to undergo a conformational change that buries their hydrophobic patches. The α-subunits now present a hydrophilic surface to the central cavity. This change displaces the bound partially folded polypeptide into the sheltered hydrophilic environment of the central cavity, where it can fold, free from danger of aggregation with other proteins. GroEs also promotes ATP hydrolysis (Figure 31.3c). The GroEL:ADP:GroES complex dissociates when ATP binds to the subunits of the other (lower) α_7 ring. Dissociation of GroES allows the partially folded (or folded) protein to escape from GroEL. If the protein has achieved its native conformation, its hydrophobic residues will be buried in its core and the hydrophobic patches on the α_7 rings will have no affinity for it. On the other hand, if the protein is only partially folded, it may be bound again, gaining access to the Anfinsen cage of the α_7 ring and another cycle of folding. The folding of rhodanese, a 33-kD protein, requires the hydrolysis of about 130 equivalents of ATP.

The group II chaperonin and eukaryotic analog of GroEL, **CCT** (also called **TriC**) is also a double-ring structure, but each ring consists of eight subunits and the subunits are not identical; they vary in size from 50 to 60 kD. Furthermore, group II chaperonins lack a GroES counterpart. **Prefoldin** (also known

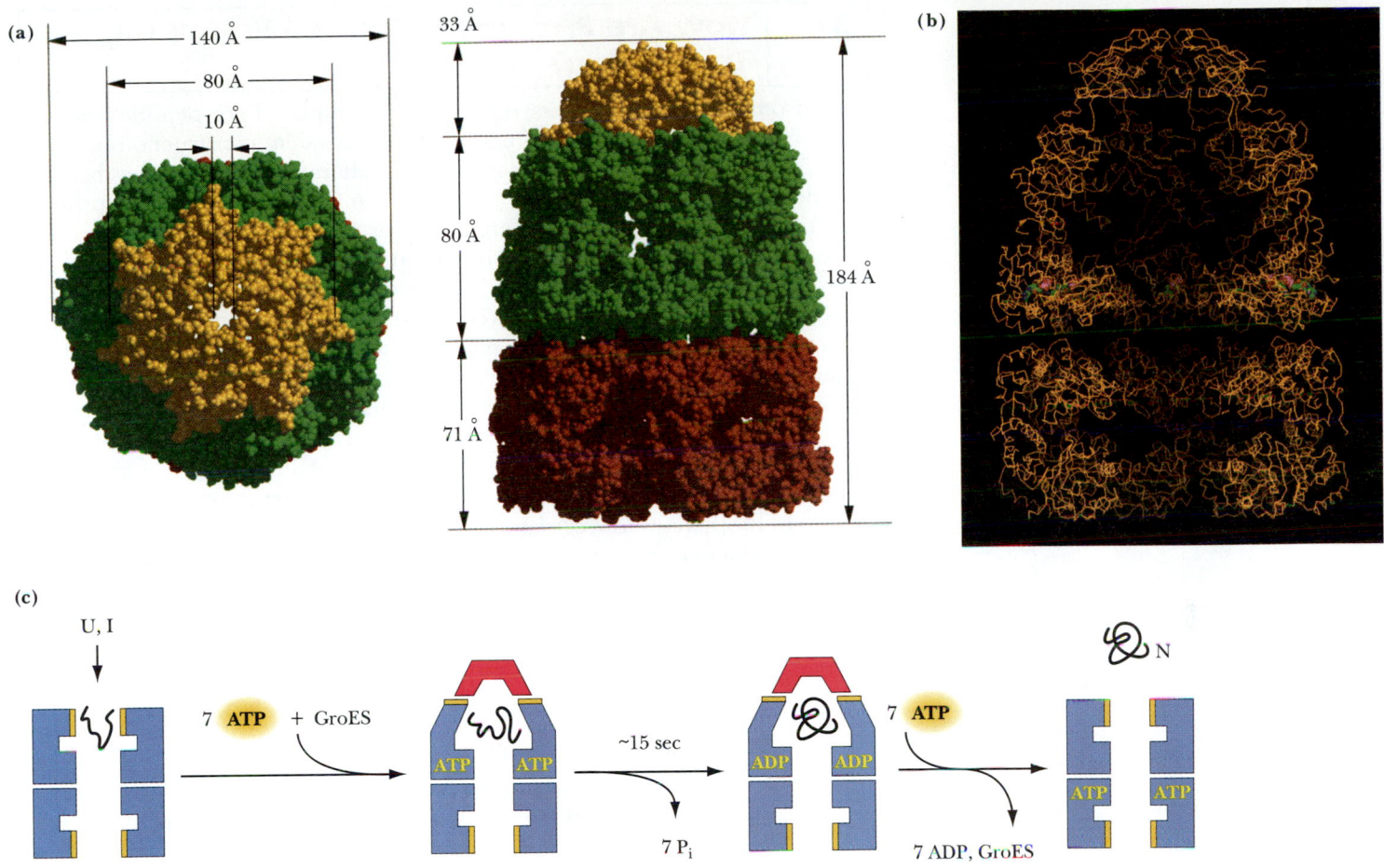

Biochemistry ⊜ Now™ ACTIVE FIGURE 31.3

Structure and function of the GroEL–GroES complex. **(a)** Space-filling representation and overall dimensions of GroEL–GroES (top view, left; side view, right). GroES is *gold;* the top, or apical, GroEL ring is *green,* and the bottom GroEL ring is *red.* **(b)** Section through the center of the complex to reveal the central cavity. The GroEL–GroES structure is shown as a C_α carbon trace. ADP molecules bound to GroEL are shown as space-filling models. **(c)** Model of the GroEL cylinder *(blue)* in action. An unfolded (U) or partially folded (I) polypeptide binds to hydrophobic patches on the apical ring of α_7-subunits, followed by ATP binding and GroES *(red)* association. ATP binding triggers a conformational change that buries the α_7-subunit hydrophobic patches *(yellow)*, releasing the polypeptide into the central cavity ("Anfinsen cage"). After about 15 seconds, ATP hydrolysis takes place, followed by binding of ATP to the lower α_7-subunit ring, which causes dissociation of GroEL:ADP:GroES. Loss of the GroES cap allows the folded protein to escape from GroEL into the cytosol. If it has not folded completely, successive cycles of protein binding, GroES recruitment and ATP-dependent release of polypeptide into the central cavity, ATP hydrolysis, and complex dissociation will take place until the protein achieves its fully folded form. *(Figure parts [a] and [b] adapted from Figure 1 in Xu, Z., Horwich, A. L., and Sigler, P. B., 1997. The crystal structure of the asymmetric GroEL-GroES-(ADP)7 chaperonin complex.* Nature **388**:741–750. *Molecular graphics courtesy of Paul B. Sigler, Yale University.)* **Test yourself on the concepts in this figure at http://chemistry.brookscole.com/ggb3**

as **GimC**), a hexameric protein composed of six subunits from two related classes (two α and four β), can serve as a co-chaperone for CCT, much as GroES does for GroEL. However, prefoldin also acts like an Hsp70 protein, because it binds unfolded polypeptide chains emerging from ribosomes and delivers them to CCT. Prefoldin resembles a jellyfish, with six tentacle-like coiled coils extending from a barrel-shaped body. The ends of the tentacles have hydrophobic patches for binding unfolded proteins.

The Eukaryotic Hsp90 Chaperone System Acts on Proteins of Signal Transduction Pathways

Hsp 90 constitutes 1% to 2% of the total cytosolic proteins of eukaryotes, its abundance reflecting its importance. Like other Hsp chaperones, its action depends on cyclic binding and hydrolysis of ATP. Conformational regulation of signal transduction molecules seems to be a major purpose of Hsp90. Receptor tyrosine kinases, soluble tyrosine kinases, and steroid hormone receptors are some of the signal transduction molecules (see Chapter 32) that must associate with Hsp90 in order to become fully competent; proteins fitting this description are called Hsp90 "client proteins." The maturation of Hsp70 client proteins requires other proteins as well, and together with Hsp90, these proteins come together to form an assembly that has been called a **foldosome.** CFTR (cystic fibrosis transmembrane regulator), telomerase, and nitric oxide synthase are also Hsp90-dependent.

Association of nascent polypeptide chains with proteins of the various chaperone systems commits them to a folding pathway, redirecting them away from degradation pathways that would otherwise eliminate them from the cell. However, if these protein chains fail to fold, they are recognized as non-native and targeted for destruction.

31.2 How Are Proteins Processed Following Translation?

Aside from these folding events, release of the completed polypeptide from the ribosome is not necessarily the final step in the covalent construction of a protein. Many proteins must undergo covalent alterations before they become functional. In the course of these **post-translational modifications,** the primary structure of a protein may be altered, and/or novel derivations may be introduced into its amino acid side chains. Hundreds of different amino acid variations have been described in proteins, virtually all arising post-translationally. The list of such modifications is very large; some are rather commonplace, whereas others are peculiar to a single protein. The diphthamide moiety in elongation factor eEF-2 is one example of an amino acid modification (see the Human Biochemistry box on page 1017 in Chapter 30); the fluorescent group of green fluorescent protein (GFP; see Chapter 4) is another. In addition, common chemical groups such as carbohydrates and lipids may be covalently attached to a protein during its maturation. Phosphorylation of proteins is a prevalent mechanism for regulating protein function. (A survey of some of the more prominent chemical groups conjugated to proteins is given in Chapter 5.) To put a number on the significance of post-translational modifications, we have seen that the number of human proteins is estimated to exceed the number of human genes (30,000 or so) by more than an order of magnitude.

Proteolytic Cleavage Is the Most Common Form of Post-Translational Processing

Proteolytic cleavage, as the most prevalent form of protein post-translational modification, merits special attention. The very occurrence of proteolysis as a processing mechanism seems strange: Why join a number of amino acids in sequence and then eliminate some of them? Three reasons can be cited. First, diversity can be introduced where none exists. For example, a simple form of proteolysis, enzymatic removal of N-terminal Met residues, occurs in many proteins. **Met-aminopeptidase,** by removing the invariant Met initiating all polypeptide chains, introduces diversity at N-termini. Second, proteolysis serves as an activation mechanism so that expression of the biological activity of a protein can be delayed until appropriate. A number of metabolically active proteins, including digestive enzymes and hormones, are synthesized as larger inactive precursors termed **pro-proteins** that are activated through proteolysis (see **zymogens,** Chapter 15). The N-terminal prosequence on such proteins may act as an intramolecular chaperone to ensure correct folding of the active site. Third, proteolysis is involved in the targeting of proteins to their proper destinations in the cell, a process known as **protein translocation.**

31.3 How Do Proteins Find Their Proper Place in the Cell?

Proteins are targeted to their proper cellular locations by **signal sequences:** Proteins destined for service in membranous organelles or for export from the cell are synthesized in precursor form carrying an N-terminal stretch of amino acid residues, or **leader peptide,** that serves as a *signal sequence.* In effect, signal sequences serve as "zip codes" for sorting and dispatching proteins to their proper compartments. Thus, the information specifying the correct cellular localization of a protein is found within its structural gene. Once the protein is routed to its destination, the signal sequence is often, but not always, proteolytically clipped from the protein by a signal sequence-specific endopeptidase called a **signal peptidase.**

Proteins Are Delivered to the Proper Cellular Compartment by Translocation

Protein translocation is the name given to the process whereby proteins are inserted into membranes or delivered across membranes. Protein translocation occurs in all cells. Newly synthesized chains of membrane proteins or secretory proteins are targeted to the plasma membrane (in prokaryotes) or the endoplasmic reticulum (in eukaryotes) by their signal sequences. In addition to the ER, a number of eukaryotic membrane systems are competent in protein translocation, including the membranes of the nucleus, mitochondria, chloroplasts, and peroxisomes. Several common features characterize protein translocation systems:

1. Proteins to be translocated are made as preproteins containing contiguous blocks of amino acid sequence that act as organelle-specific sorting signals.
2. **Signal recognition particles (SRPs)** recognize the sorting signals as they emerge from the ribosome and, together with **signal receptors (SRs),** deliver the nascent chain to specific membrane protein complexes called *translocons* that mediate protein integration into the membrane or protein translocation across the membrane.
3. **Translocons** are selectively permeable protein-conducting channels that catalyze movement of the proteins across the membrane, and metabolic energy in the form of ATP, GTP, or a membrane potential is essential. In eukaryotes, ATP-dependent chaperone proteins within the membrane compartment usually associate with the entering polypeptide and provide the energy for translocation. Proteins destined for membrane integration contain amino acid sequences that act as **stop-transfer signals,** allowing diffusion of transmembrane segments into the bilayer.
4. Preproteins are maintained in a loosely folded, translocation-competent conformation through interaction with molecular chaperones.

Prokaryotic Proteins Destined for Translocation Are Synthesized as Preproteins

Gram-negative bacteria typically have four compartments: cytoplasm, plasma (or inner) membrane, periplasmic space (or periplasm), and outer membrane. Most proteins destined for any location other than the cytoplasm are synthesized with amino-terminal leader sequences 16 to 26 amino acid residues long. These leader sequences, or *signal sequences,* consist of a basic N-terminal region, a central domain of 7 to 13 hydrophobic residues, and a nonhelical C-terminal region (Figure 31.4). The conserved features of the last part of the leader, the C-terminal region, include a helix-breaking Gly or Pro residue and amino acids with small side chains located one and three residues before the proteolytic cleavage site. Unlike the basic N-terminal and nonpolar central regions, the C-terminal features are not essential for translocation but instead serve as recognition signals for the **leader peptidase,** which removes the leader sequence. The exact amino acid sequence of the leader peptide is unimportant. Nonpolar residues in the center and a few Lys residues at the amino terminus are sufficient for successful translocation. The functions of leader peptides are to retard the folding of the preprotein so that molecular chaperones have a chance to interact with it and to provide recognition signals for the translocation machinery and leader peptidase.

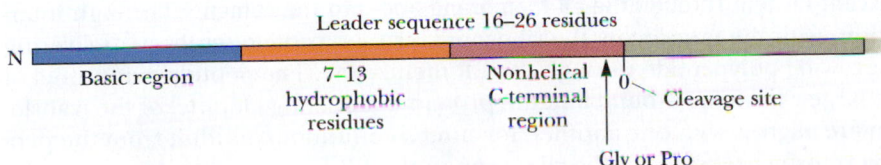

FIGURE 31.4 General features of the N-terminal signal sequences on *E. coli* proteins destined for translocation: a basic N-terminal region, a central apolar domain, and a nonhelical C-terminal region.

Eukaryotic Proteins Are Routed to Their Proper Destinations by Protein Sorting and Translocation

Eukaryotic cells are characterized by many membrane-bounded compartments. In general, signal sequences targeting proteins to their appropriate compartments are located at the N-terminus as *cleavable presequences,* although many proteins have N-terminal localization signals that are not cleaved and others have internal targeting sequences that may or may not be cleaved. Proteolytic removal of the leader sequences is also catalyzed by specialized proteases, but removal is not essential to translocation. No sequence similarity is found among the targeting signals for each compartment. Thus, the targeting information resides in more generalized features of the leader sequences such as charge distribution, relative polarity, and secondary structure. For example, proteins destined for secretion enter the lumen of the endoplasmic reticulum (ER) and reach the plasma membrane via a series of vesicles that traverse the endomembrane system. Recognition by the ER depends on an N-terminal amino acid sequence that contains one or more basic amino acids followed by a run of 6 to 12 hydrophobic amino acids. An example is serum albumin, which is synthesized in precursor form (**preproalbumin**) having a M*K*WVT**FLLLLFISGSAFS**R N-terminal signal sequence. The italicized K highlights the basic residue in the sequence, and the bold residues denote a continuous stretch of (mostly) hydrophobic residues. A signal peptidase in the ER removes the signal sequence by cleaving the preproprotein between the S and R.

The Synthesis of Secretory Proteins and Many Membrane Proteins Is Coupled to Translocation Across the ER Membrane   The signals recognized by the ER translocation system are virtually indistinguishable from bacterial signal sequences; indeed, the two are interchangeable in vitro. In addition, the translocon systems in prokaryotes and eukaryotes are highly analogous. In higher eukaryotes, translation and translocation of many proteins destined for processing via the ER are tightly coupled. Translocation across the ER occurs co-translationally (that is, as the protein is being translated on the ribosome). As the N-terminal signal sequence of a preprotein undergoing synthesis emerges from the ribosome, it is detected by a **signal recognition particle** (**SRP;** Figure 31.5). SRP is a 325-kD nucleoprotein assembly that contains six polypeptides and a 300-nucleotide **7S RNA. SRP54,** a 54-kD subunit of SRP and a G-protein family member, recognizes the nascent protein's signal sequence, and SRP binding of the signal sequence causes the ribosome to cease translation. This arrest prevents release of the growing protein into the cytosol before it reaches the ER and its intended translocation. The SRP–ribosome complex is referred to as the **RNC–SRP (ribosome nascent chain:SRP complex).**

Interaction Between the RNC–SRP and the SR Delivers the RNC to the Membrane   The RNC–SRP is then directed to the cytosolic face of the ER, where it binds to the signal receptor (SR), an $\alpha\beta$ heterodimeric protein. The 70-kD α-subunit is anchored to the membrane by the transmembrane β-subunit; both subunits are G-protein family members, and both have bound GTP. When SRP54 docks with SRα, the RNC–SRP becomes membrane associated (Figure 31.5). When they interact with each other, SRP54 and SRα function as reciprocal GTPase-activating proteins. GTP hydrolysis causes the dissociation of SRP from SR and transfer of the RNC to the translocon.

The Ribosome and the Translocon Form a Common Conduit for Transfer of the Nascent Protein Through the ER Membrane and into the Lumen   Through interactions with the translocon, the ribosome resumes protein synthesis, delivering its growing polypeptide through the ER membrane. The peptide exit tunnel of the large ribosomal subunit and the protein-conducting channel of the translocon are aligned with one another, forming a continuous conduit from the peptidyl transferase center of the ribosome to the ER lumen.

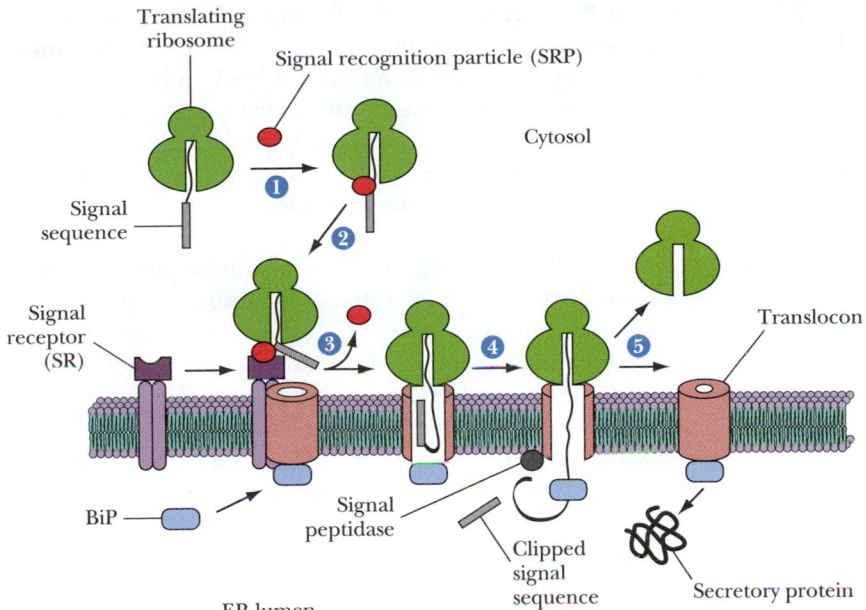

Biochemistry ⧖ Now™ ACTIVE FIGURE 31.5
Synthesis of a eukaryotic secretory protein and its
translocation into the endoplasmic reticulum. **(1)** The
signal recognition particle (SRP, *red*) recognizes the
signal sequence emerging from a translating ribosome (ribosome nascent complex [RNC], *green*).
(2) The RNS-SRP interacts with the signal receptor
(SR, *purple*) and is transferred to the translocon (*pink*).
(3) Release of the SRP and alignment of the peptide
exit tunnel of the RNC with the protein-conducting
channel of the translocon stimulates the ribosome to
resume translation. **(4)** The membrane-associated signal peptidase (*black*) clips off the N-terminal signal
sequence, and BiP (the ER lumen Hsp70 chaperone,
blue) binds the nascent chain mediating its folding
into its native conformation. **(5)** Following dissociation of the ribosome, BiP plugs the translocon channel. Not shown are subsequent secretory protein maturation events, such as glycosylation. (*Adapted from
Figures 1a and 2a in Frydman, J., 2001. Folding of newly translated proteins in vivo: The role of molecular chaperones. Annual
Review of Biochemistry **70**:603–647.*) **Test yourself on the
concepts in this figure at http://chemistry.brookscole.
com/ggb3**

The mammalian translocon is a complex, multifunctional entity that has as
its core the **Sec61 complex,** a heterotrimeric complex of membrane proteins,
and a unique fourth subunit, **TRAM,** that is required for insertion of nascent
integral membrane proteins into the membrane. The 53-kD α-subunit of
Sec61p has ten membrane-spanning segments, whereas the β- and γ-subunits
are single TMS proteins. Sec61α forms the transmembrane protein-conducting
channel through which the nascent polypeptide is transported into the ER lumen (Figure 31.5). The pore size of Sec61p is very dynamic, ranging from
about 0.6 to 6 nm in diameter. Thus, a great variety of protein structures could
be easily accommodated within the translocon. This flexibility allows the
Sec61p translocon complex to function in post-translational translocation
(translocation of completely formed proteins) as well as co-translational
translocation.

As the protein is threaded through the Sec61p channel into the lumen, an
Hsp70 chaperone family member called **BiP** binds to it and mediates proper
folding. BiP function, like that of other Hsp70 proteins, is ATP-dependent, and
ATP-dependent protein folding provides the driving force for translocation of
the polypeptide into the lumen. When the ribosome dissociates from the
translocon, BiP serves as a plug to block the protein-conducting channel, preventing ions and other substances from moving between the ER lumen and the
cytosol.

A Signal Peptidase Within the ER Lumen Clips Off the Signal Peptide   Soon after
it enters the ER lumen, the signal peptide is clipped off by membrane-bound
signal peptidase (also called *leader peptidase*), which is a complex of five proteins. Other modifying enzymes within the lumen introduce additional post-
translational alterations into the polypeptide, such as glycosylation with specific
carbohydrate residues. ER-processed proteins destined for secretion from the
cell or inclusion in vesicles such as lysosomes end up contained within the soluble phase of the ER lumen. On the other hand, polypeptides destined to become membrane proteins carry **stop-transfer** sequences within their mature
domains. These stop-transfer sequences are typically a 20-residue stretch of hydrophobic amino acids that arrests the passage across the ER membrane. Proteins with stop-transfer sequences remain embedded in the ER membrane with
their C-termini on the cytosolic face of the ER. Such membrane proteins arrive
at their intended destinations via subsequent processing of the ER.

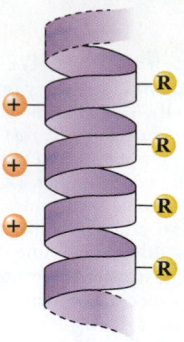

FIGURE 31.6 Structure of an amphipathic α-helix having basic (+) residues on one side and uncharged and hydrophobic (R) residues on the other.

Retrograde Translocation Prevents Secretion of Damaged Proteins and Recycles Old ER Proteins To prevent secretion of inappropriate proteins, fragmented or misfolded secretory proteins are passed from the ER back into the cytosol via Sec61p. Thus, Sec61p also serves as a channel for aberrant secretory proteins to be returned to the cytosol so that they can be destroyed by the proteasome degradation apparatus (see Section 31.4). Among these proteins are ER membrane proteins that are damaged or no longer needed.

Mitochondrial Protein Import Most mitochondrial proteins are encoded by the nuclear genome and synthesized on cytosolic ribosomes. Mitochondria consist of four principal subcompartments: the outer membrane, the intermembrane space, the inner membrane, and the matrix. Thus, not only must mitochondrial proteins find mitochondria, they must gain access to the proper subcompartment; and once there, they must attain a functionally active conformation. As a consequence, mitochondria possess multiple preprotein translocons and chaperones. Similar considerations apply to protein import to chloroplasts, organelles with five principal subcompartments (outer membrane, intermembrane space, inner/thylakoid membrane, stroma, and thylakoid lumen; see Chapter 21).

Signal sequences on nuclear-encoded proteins destined for the mitochondria are N-terminal cleavable presequences 10 to 70 residues long. These mitochondrial presequences lack contiguous hydrophobic regions. Instead, they have positively charged and hydroxy amino acid residues spread along their entire length. These sequences form **amphipathic α-helices** (Figure 31.6) with basic residues on one side of the helix and uncharged and hydrophobic residues on the other; that is, mitochondrial presequences are positively charged amphiphatic sequences. In general, mitochondrial targeting sequences share no sequence homology. Once synthesized, mitochondrial preproteins are retained in an unfolded state with their target sequences exposed, through association with Hsp70 molecular chaperones. Import involves binding of a preprotein to the **mitochondrial outer membrane translocon (TOM)** (Figure 31.7). If the protein is destined to be an outer mitochondrial membrane protein, it is transferred from the TOM to the **sorting and assembly complex (SAM)** and inserted in the outer membrane. If it is an integral protein of the inner mitochondrial membrane, it traverses the TOM complex, enters the intermembrane space, and is taken up by the **inner mitochondrial membrane translocon (TIM22)** and inserted into the inner membrane. On the other hand, if it is destined to be a mitochondrial matrix protein, a different TIM complex, **TIM23,** binds the preprotein and threads it across the inner mi-

Biochemistry⑤Now™ **ANIMATED**
FIGURE 31.7 Translocation of mitochondrial preproteins involves distinct translocons. All mitochondrial proteins must interact with the outer mitochondrial membrane (TOM). From there, depending on their destiny, they are (1) passed to the SAM complex if they are integral proteins of the outer mitochondrial membrane or (2) traverse the TOM and enter the intermembrane space, where they are taken up by either TIM22 or TIM23, depending on whether they are integral membrane proteins of the inner mitochondrial membrane (TIM22) or mitochondrial matrix proteins (TIM23). *(Adapted from Figure 1 in Mihara, K., 2003. Moving inside membranes.* Nature *424:505–506.)* **See this figure animated at http://chemistry.brookscole. com/ggb3**

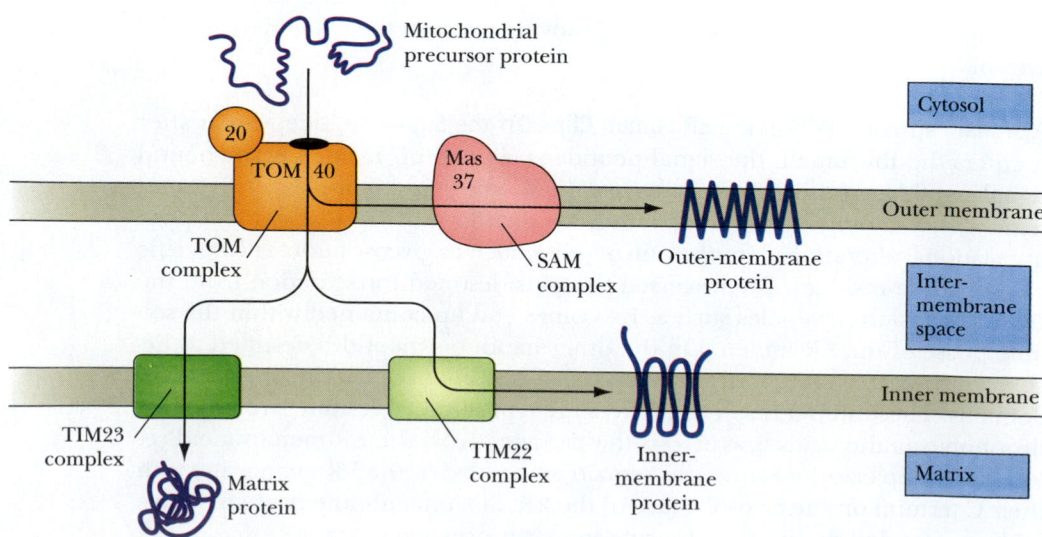

tochondrial membrane into the matrix. Chloroplasts have **TOCs** (translocon outer chloroplast membrane) and **TICs** (translocon inner chloroplast membrane) for these purposes.

<table>
<tr><td>**31.4**</td><td>**How Does Protein Degradation Regulate Cellular Levels of Specific Proteins?**</td></tr>
</table>

Cellular proteins are in a dynamic state of turnover, with the relative rates of protein synthesis and protein degradation ultimately determining the amount of protein present at any point in time. In many instances, transcriptional regulation determines the concentrations of specific proteins expressed within cells, with protein degradation playing a minor role. In other instances, the amounts of key enzymes and regulatory proteins, such as cyclins and transcription factors, are controlled via selective protein degradation. In addition, abnormal proteins arising from biosynthetic errors or postsynthetic damage must be destroyed to prevent the deleterious consequences of their buildup. The elimination of proteins typically follows first-order kinetics, with half-lives ($t_{1/2}$) of different proteins ranging from several minutes to many days. A single, random proteolytic break introduced into the polypeptide backbone of a protein is believed sufficient to trigger its rapid disappearance because no partially degraded proteins are normally observed in cells.

Protein degradation poses a real hazard to cellular processes. To control this hazard, protein degradation is compartmentalized, either in macromolecular structures known as **proteasomes** or in degradative organelles such as lysosomes. Protein degradation within lysosomes is largely nonselective; selection occurs during lysosomal uptake. Proteasomes are found in eukaryotic as well as prokaryotic cells. The proteasome is a functionally and structurally sophisticated counterpart to the ribosome. Regulation of protein levels via degradation is an essential cellular mechanism. Regulation by degradation is both rapid and irreversible.

Eukaryotic Proteins Are Targeted for Proteasome Destruction by the Ubiquitin Pathway

Ubiquitination is the most common mechanism to label a protein for proteasome degradation in eukaryotes. **Ubiquitin** is a highly conserved, 76-residue (8.5-kD) polypeptide widespread in eukaryotes. Proteins are condemned to degradation through ligation to ubiquitin. Three proteins in addition to ubiquitin are involved in the ligation process: E_1, E_2, and E_3 (Figure 31.8). E_1 is the **ubiquitin-activating enzyme** (105-kD dimer). It becomes attached via a thioester bond to the C-terminal Gly residue of ubiquitin through ATP-driven formation of an activated ubiquitin-adenylate intermediate. Ubiquitin is then transferred from E_1 to an SH group on E_2, the **ubiquitin-carrier protein.** (E_2 is actually a family of at least seven different small proteins, several of which are heat shock proteins; there is also a variety of E_3 proteins) In protein degradation, E_2-S ~ ubiquitin transfers ubiquitin to free amino groups on proteins selected by E_3 (180 kD), the **ubiquitin-protein ligase.** Upon binding a protein substrate, E_3 catalyzes the transfer of ubiquitin from E_2-S ~ ubiquitin to free amino groups (usually Lys ϵ-NH$_2$) on the protein. More than one ubiquitin may be attached to a protein substrate, and tandemly linked chains of ubiquitin also occur via *isopeptide bonds* between the C-terminal glycine residue of one ubiquitin and the ϵ-amino of Lys residues in another. Ubiquitin has seven lysine residues, at positions 6, 11, 27, 29, 33, 48, and 63. Only isopeptide linkages to K^{11}, K^{29}, K^{48}, and K^{63} have been found, with the K^{48}-type being most common as a degradation signal.

E_3 plays a central role in recognizing and selecting proteins for degradation. E_3 selects proteins by the nature of the N-terminal amino acid. Proteins must have a free α-amino terminus to be susceptible. Proteins having either

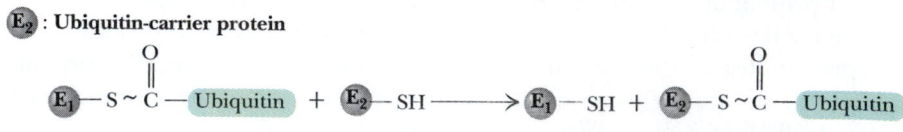

E₁ : Ubiquitin-activating enzyme

① Ubiquitin—C(=O)—O⁻ + **ATP** $\xrightarrow{E_1}$ **P P** + Ubiquitin—C(=O)~AMP

(C-term. Gly)

Ubiquitinyl-acyladenylate

② Ubiquitin—C(=O)~AMP + E₁—SH ⟶ AMP + E₁—S~C(=O)—Ubiquitin

(Thioester)

E₂ : Ubiquitin-carrier protein

E₁—S~C(=O)—Ubiquitin + E₂—SH ⟶ E₁—SH + E₂—S~C(=O)—Ubiquitin

Biochemistry⊗Now™ ANIMATED FIGURE 31.8
Enzymatic reactions in the ligation of ubiquitin to proteins. Ubiquitin is attached to selected proteins via isopeptide bonds formed between the ubiquitin carboxy-terminus and free amino groups (α-NH₂ terminus, Lys ϵ-NH₂ side chains) on the protein. **See this figure animated at http://chemistry.brookscole. com/ggb3**

E₃ : Ligase

E₃ + Protein ⟶ E₃ : Protein

(substrate)

E₃ : Protein + E₂~S~C(=O)—Ubiquitin ⟶ E₂—SH + E₃ + Protein—Ubiquitin

Met, Ser, Ala, Thr, Val, Gly, or Cys at the amino terminus are resistant to the ubiquitin-mediated degradation pathway. However, proteins having Arg, Lys, His, Phe, Tyr, Trp, Leu, Asn, Gln, Asp, or Glu N-termini have half-lives of only 2 to 30 minutes.

Interestingly, proteins with acidic N-termini (Asp or Glu) show a tRNA requirement for degradation (Figure 31.9). Transfer of Arg from Arg-tRNA to the N-terminus of these proteins alters their N-terminus from acidic to basic, rendering the protein susceptible to E₃. It is also interesting that Met is less likely to be cleaved from the N-terminus if the next amino acid in the chain is one particularly susceptible to ubiquitin-mediated degradation.

Most proteins with susceptible N-terminal residues are *not* normal intracellular proteins but tend to be secreted proteins in which the susceptible residue has been exposed by action of a signal peptidase. Perhaps part of the function of the N-terminal recognition system is to recognize and remove from the cytosol any invading "foreign" or secreted proteins.

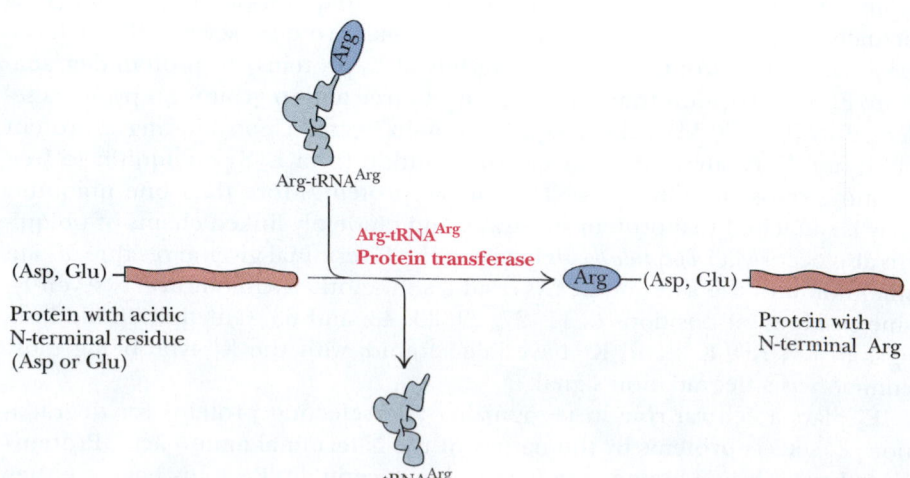

FIGURE 31.9 Proteins with acidic N-termini show a tRNA requirement for degradation. Arginyl-tRNA^{Arg}: protein transferase catalyzes the transfer of Arg to the free α-NH₂ of proteins with Asp or Glu N-terminal residues. Arg-tRNA^{Arg}:protein transferase serves as part of the protein degradation recognition system.

Other proteins targeted for ubiquitin ligation and proteasome degradation contain **PEST sequences**—short, highly conserved sequence elements rich in proline (P), glutamate (E), serine (S), and threonine (T) residues.

Proteins Targeted for Destruction Are Degraded by Proteasomes

Proteasomes are large oligomeric structures enclosing a central cavity where proteolysis takes place. The 20S proteasome from the archaebacterium *Thermoplasma acidophilum* is a 700-kD barrel-shaped structure composed of two different kinds of polypeptide chains, α and β, arranged to form four stacked rings of $\alpha_7\beta_7\beta_7\alpha_7$-subunit organization. The barrel is about 15 nm in height and 11 nm in diameter, and it contains a three-part central cavity (Figure 31.10a). The proteolytic sites of the 20S proteasome are found within this cavity. Access to the cavity is controlled through a 1.3-nm opening formed by the outer α_7 rings. These rings are believed to unfold proteins destined for degradation and transport them into the central cavity. The β-subunits possess the proteolytic activity. Proteolysis occurs when the β-subunit N-terminal threonine side-chain O atom makes nucleophilic attack on the carbonyl-C of a peptide bond in the target protein. The products of proteasome degradation are oligopeptides seven to nine residues long.

Eukaryotic cells contain two forms of proteasomes: the **20S proteasome,** and its larger counterpart, the **26S proteasome.** The eukaryotic 26S proteasome is a 45-nm-long structure composed of a 20S proteasome plus two additional substructures known as **19S regulators** (also called **19S caps** or **PA700** [for **protea**some **a**ctivator-**700** kD]) (Figure 31.10b). Overall, the 26S proteasome (approximately 2.5 megadaltons) has 2 copies each of 32 to 34 distinct subunits, 14 in the 20S core and 18 to 20 in the cap structures. Unlike the archaeal 20S proteasome, the eukaryotic 20S core structure contains seven different kinds of α-subunits and seven different kinds of β-subunits. Interestingly, only three of the seven different β-subunits have protease active sites. The 26S proteasome forms when the 19S regulators dock to the two outer α_7 rings of the 20S proteasome cylinder. Many of the 19S regulator subunits have ATPase activity. Replacement of certain 19S regulator subunits with others changes the specificity of the proteasome. The 19S regulators cause the proteolytic function of the 20S proteasome to become ATP-dependent and specific for ubiquitinylated proteins as

Biochemistry ⓢ**Now**™   Go to BiochemistryNow and click BiochemistryInteractive to discover why proteasomes are the protein degradation counterparts to ribosomes.

FIGURE 31.10 Model for the structure of the 26S proteasome. **(a)** The *Thermoplasma acidophilum* 20S proteasome core structure. **(b)** Composite model of the 26S proteasome. The 20S proteasome core is shown in yellow; the 19S regulator (19S cap) structures are in blue. *(Adapted from Figures 3 and 5 in Voges, D., Zwickl, P., and Baumeister, W., 1999. The 26S proteasome: A molecular machine designed for controlled proteolysis. Annual Review of Biochemistry **68**:1015–1068.)*

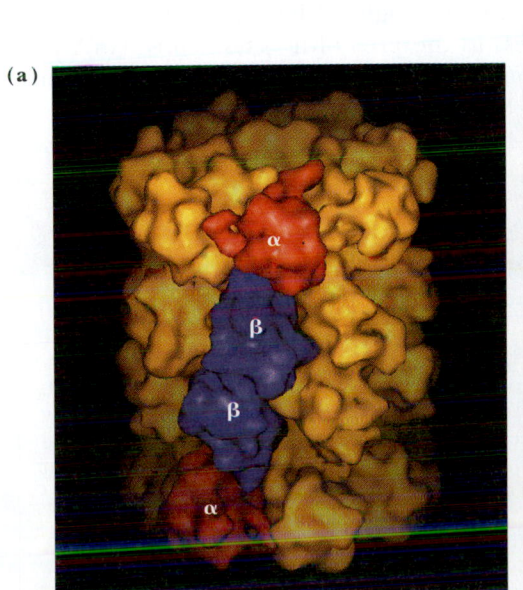

(a)

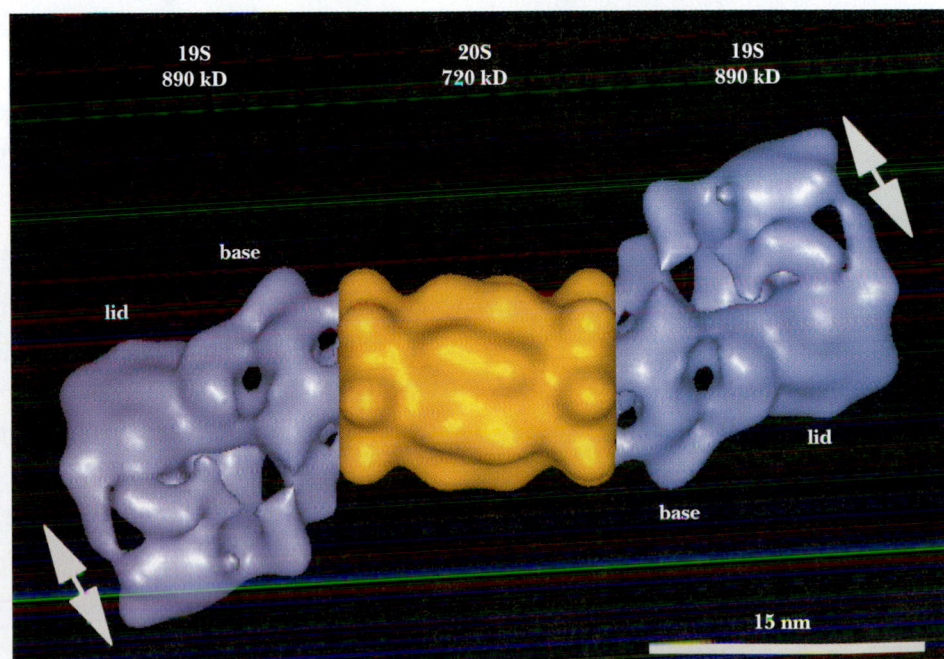

(b)

19S
890 kD

20S
720 kD

19S
890 kD

base

lid

lid

base

15 nm

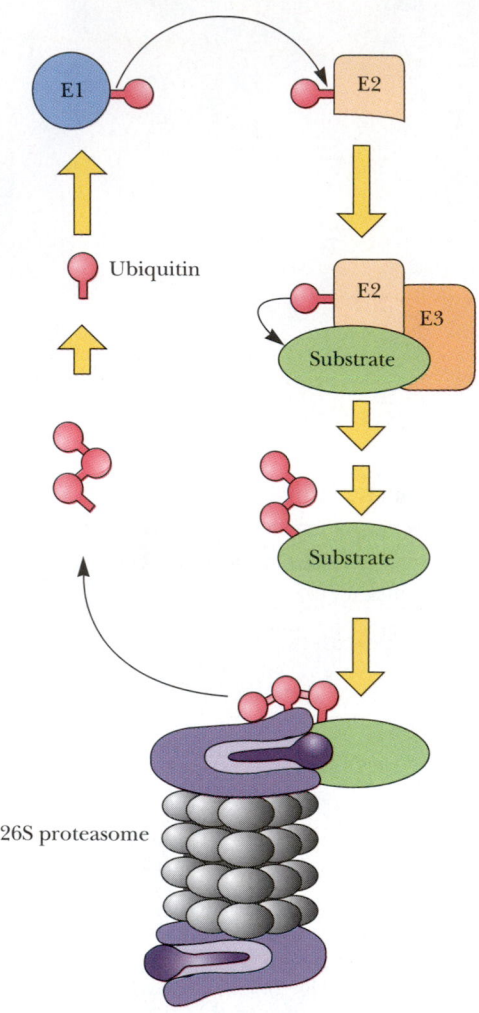

Biochemistry ◉ **Now**™   **ACTIVE FIGURE 31.11**
Diagram of the ubiquitin-proteasome degradation
pathway. Pink "lollipop" structures symbolize ubiquitin molecules. *(Adapted from Figure 1 in Hilt, W., and Wolf,
D. H., 1996. Proteasomes: Destruction as a program.* Trends in
Biochemical Sciences *21:96–102.)* **Test yourself on the
concepts in this figure at http://chemistry.
brookscole.com/ggb3**

substrates. That is, these 19S caps act as regulatory complexes for the recognition and selection of ubiquitinylated proteins for degradation by the 20S proteasome core (Figure 31.11). The 26S proteasome shows a preference for proteins having four or more ubiquitin molecules attached to them. The 19S
regulators also carry out the unfolding and transport of ubiquitinylated protein
substrates into the proteolytic central cavity.

The 19S regulators consist of two parts: the base and the lid. The base subcomplex connects to the 20S proteasome and contains the six ATPase subunits
that unfold proteasome substrates. These subunits are members of the **AAA family of ATPases** (**A**TPases **a**ssociated with various cellular **a**ctivities); AAA-ATPases
are an evolutionarily ancient family of proteins involved in a variety of cellular
functions requiring energy-dependent unfolding, disassembly, and remodeling
of proteins. The lid subcomplex acts as a cap on the base subcomplex and one
of its subunits functions in recognition and ubiquitin-chain processing of proteasome protein substrates.

HtrA Proteases Also Function in Protein Quality Control

The discussion thus far has stressed the importance of protein quality control to
cellular health. **HtrA proteases** are a class of proteins involved in quality control
that combine the dual functions of chaperones and proteasomes. (The acronym
Htr comes from "**h**igh **t**emperature **r**equirement" because *E. coli* strains bearing
mutations in HtrA genes do not grow at elevated temperatures.) In addition to
their novel ability to be either chaperones or proteases, HtrA proteases are the
only known protein quality control factor that is not ATP-dependent. Prokaryotic HtrA proteases act as chaperones at low temperatures (20°C) where they
have negligible protease activity. However, as the temperature increases, these
proteins switch from a chaperone function to a protease function to remove misfolded or unfolded proteins from the cell. With this functional duality, HtrA
proteases have the potential to mediate quality control through protein triage
(see A Deeper Look box on page 1037).

The *E. coli* HtrA protein **DegP** is the best characterized HtrA protease. DegP
is localized in the *E. coli* periplasmic space, where it oversees quality control of
proteins involved in the cell envelope. It is a 448-residue protein containing a
central protease domain with a classic Ser protease Asp-His-Ser catalytic triad (see
Chapter 14) and two C-terminal **PDZ domains.** PDZ domains are structural modules involved in protein–protein interactions (see Chapter 32); PDZ domains recognize and bind selectively to the C-terminal three or four residues of target proteins. Like other quality control systems, HtrA proteases have a central cavity
where proteolysis occurs (Figure 31.12); the height of this cavity is 1.5 nm, which
excludes folded proteins from access to the proteolytic sites. Thus, HtrA proteases can act only on misfolded proteins. As we have seen, limited access to pro-

---

Human Biochemistry

Proteasome Inhibitors in Cancer Chemotherapy

Proteasome inhibition offers a promising approach to treating cancer. The counterintuitive rationale goes like this: The proteasome is
responsible for the regulated destruction of proteins involved in cell
cycle progression and the control of apoptosis (programmed cell
death). Inhibition of proteasome function leads to cell cycle arrest
and apoptosis. In clinical trials, proteasome inhibitors have retarded

cancer progression by interfering with the programmed degradation of regulatory proteins, causing cancer cells to self-destruct.
Bortezomib, a small-molecule proteasome inhibitor developed by
Millenium Pharmaceuticals, Inc., has received FDA approval for the
treatment of multiple myeloma, a cancer of plasma cells that
accounts for 10% of all cancers of the blood.

Source: Adams, J., 2003. The proteasome: Structure, function, and role in the cell. *Cancer Treatment
Reviews Supplement* **1:**3–9.

(a)

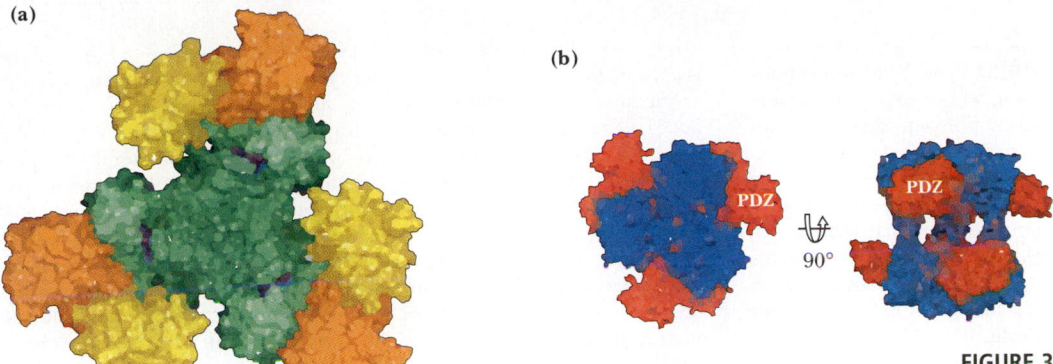

(b)

teolytic sites is an important regulatory feature of quality control proteases; only proteins targeted for destruction have access to such sites. The PDZ domains of the HtrA proteases act as gatekeepers, determining access of protein substrates to the proteolytic centers. Human HtrA proteases are implicated in stress response pathways. Human HtrA1 is expressed at higher levels in osteoarthritis and aging and lower levels in ovarian cancer and melanoma. Secreted human Htr1 may be involved in degradation of extracellular matrix proteins involved in arthritis as well as tumor progression and invasion.

FIGURE 31.12 The HtrA protease structure. **(a)** A trimer of DegP subunits represents the HtrA functional unit. The different domains are color-coded: The protease domain is green, PDZ domain 1 (PDZ1) is yellow, and PDZ domain 2 (PDZ2) is orange. Protease active sites are highlighted in blue. The trimer has somewhat of a funnel shape, with the protease in the center and the PDZ domains on the rim. **(b)** Two HtrA trimers come together to form a hexameric structure in which the two protease domains form a rigid molecular cage *(blue)* and the six PDZ domains are like tentacles *(red)* that both bind protein substrate targets and control lateral access into the protease cavity. *(Adapted from Figure 3 in Clausen, T., Southan, C., and Ehrmann, M., 2002. The HtrA family of proteases: Implications for protein composition and cell fate. Molecular Cell **10**:443–455.)*

A Deeper Look

Protein Triage—A Model for Quality Control

Triage is a medical term for the sorting of patients according to their need for (and their likelihood to benefit from) medical treatment. Sue Wickner, Michael Maurizi, and Susan Gottesman have pointed out that cells control the quality of their proteins through a system of triage based on the chaperones and the ubiquitination-proteasome degradation pathway. These systems recognize non-native proteins (proteins that are only partially folded, misfolded, incorrectly modified, damaged, or in an inappropriate compartment). Depending on the severity of its damage, a non-native protein is directed to chaperones for refolding or targeted for destruction by a proteasome (see accompanying figure).

Adapted from Figures 2 and 3 in Wickner, S., Maurizi, M., and Gottesman, S., 1999. Posttranslational quality control: Folding, refolding, and degrading proteins. *Science* **286**:1888–1893.

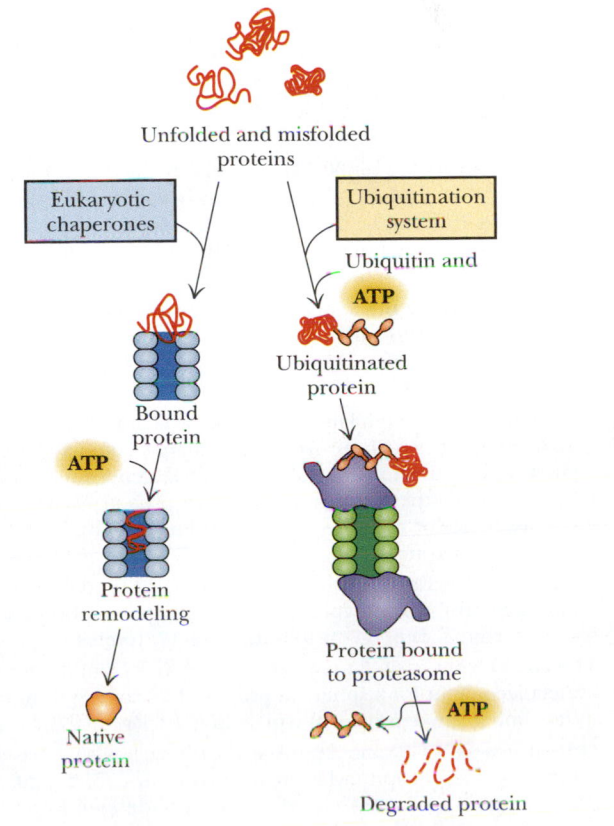

Summary

31.1 How Do Newly Synthesized Proteins Fold?
Most proteins fold spontaneously, as anticipated by Anfinsen, whose experiments suggested that all of the information necessary for a polypeptide chain to assume its active, folded conformation resides in its amino acid sequence. However, some proteins depend on molecular chaperones to achieve their folded conformation within the crowded intracellular environment. Hsp70 chaperones are ATP-dependent proteins that bind to exposed hydrophobic regions of polypeptides, preventing nonproductive associations with other proteins and keeping the protein in an unfolded state until productive folding steps can take place. Hsp60 chaperones such as GroEL–GroES are ATP-dependent cylindrical chaperonins that provide a protected central cavity or "Anfinsen cage," where partially folded proteins can spontaneously fold, free from the danger of nonspecific hydrophobic interactions with other unfolded protein chains. Hsp90 chaperones assist a subset of "client proteins" involved in signal transduction pathways in assuming their active conformations.

31.2 How Are Proteins Processed Following Translation?
Nascent proteins are seldom produced in their final, functional form. Maturation typically involves proteolytic cleavage and may require post-translational modification, such as phosphorylation, glycosylation, or other covalent substitutions. Removal of nascent N-terminal methionine residues is a common form of protein processing by proteolysis. The number of human proteins is believed to exceed the number of human genes (30,000 or so) by an order of magnitude or more. The great number of proteins available from a fixed set of genes is attributed to a variety of processes, including alternative splicing of mRNAs and post-translational modification of proteins.

31.3 How Do Proteins Find Their Proper Place in the Cell?
Most proteins destined for compartments other than the cytosol are synthesized with N-terminal signal sequences that target them to their proper destinations. These signal sequences are recognized by signal recognition particles as they emerge from translating ribosomes. The signal recognition particle interacts with a membrane-bound signal receptor, delivering the translating ribosome to a translocon, a multimeric integral membrane protein structure having at its core the Sec61p complex. The translocon transfers the growing protein chain across the membrane (or in the case of nascent integral membrane proteins, inserts the protein into the membrane). Signal peptidases within the membrane compartment clip off the signal sequence. Other post-translational modifications may follow. Mitochondrial protein import and membrane insertion are mediated by specific translocon complexes in the outer mitochondrial membrane called TOM and SAM. Proteins destined for the inner mitochondrial membrane or mitochondrial matrix must interact with inner mitochondrial translocons (either TIM23 or TIM22) as well as TOM.

31.4 How Does Protein Degradation Regulate Cellular Levels of Specific Proteins?
Protein degradation is potentially hazardous to cells, because cell function depends on active proteins. Therefore, protein degradation is compartmentalized in lysosomes or in proteasomes. Proteins targeted for destruction are selected by ubiquitination. A set of three enzymes (E_1, E_2, and E_3) mediate transfer of ubiquitin to free $-NH_2$ groups in targeted proteins. Proteins with charged or hydrophobic residues at their N-termini are particularly susceptible to ubiquitination and destruction. The ubiquitin moieties are recognized by 19S cap structures found at either end of 26S proteasomes. Protein degradation occurs when, in an ATP-dependent process, the ubiquitinated protein is unfolded and threaded into the central cavity of the cylindrical $\alpha_7\beta_7\beta_7\alpha_7$ 20S part of the 26S proteasome. The β-subunits possess protease active sites that chop the protein substrate into short (seven- to nine-residue) fragments; the ubiquitin moieties are recycled. HtrA proteases also function in protein quality control. HtrA proteins are novel in two aspects: Unlike other chaperones and proteasomes, they are ATP-independent; and unlike the others, they have dual chaperone and protease activities and switch between these two functions in response to stress conditions.

Problems

1. (Integrates with Chapter 30.) Human rhodanese (33 kD) consists of 296 amino acid residues. Approximately how many ATP equivalents are consumed in the synthesis of the rhodanese polypeptide chain from its constituent amino acids *and* the folding of this chain into an active tertiary structure?

2. A single proteolytic break in a polypeptide chain of a native protein is often sufficient to initiate its total degradation. What does this fact suggest to you regarding the structural consequences of proteolytic nicks in proteins?

3. Protein molecules, like all molecules, can be characterized in terms of general properties such as size, shape, charge, solubility/hydrophobicity. Consider the influence of each of these general features on the likelihood of whether folding of a particular protein will require chaperone assistance or not. Be specific regarding just Hsp70 chaperones or Hsp70 chaperones *and* Hsp60 chaperonins.

4. Many multidomain proteins apparently do not require chaperones to attain the fully folded conformations. Suggest a rational scenario for chaperone-independent folding of such proteins.

5. The GroEL ring has a 5-nm central cavity. Calculate the maximum molecular weight for a spherical protein that can just fit in this cavity, assuming the density of the protein is 1.25 g/mL.

6. (Integrates with Chapter 24.) Acetyl-CoA carboxylase has at least seven possible phosphorylation sites (residues 23, 25, 29, 76, 77, 95, and 1200) in its 2345-residue polypeptide (see Figure 24.4). How many different covalently modified forms of acetyl-CoA carboxylase protein are possible if there are seven phosphorylation sites?

7. (Integrates with Chapter 30.) In what ways are the mechanisms of action of EF-Tu/EF-Ts and DnaK/GrpE similar? What mechanistic functions do the ribosome A-site and DnaJ have in common?

8. The amino acid sequence deduced from the nucleotide sequence of a newly discovered human gene begins: MRSLLILVLCFLPLAALGK.... Is this a signal sequence? If so, where does the signal peptidase act on it? What can you surmise about the intended destination of this protein?

9. Not only is the Sec61p translocon complex essential for translocation of proteins into the ER lumen, it also mediates the incorporation of integral membrane proteins into the ER membrane. The mechanism for integration is triggered by stop-transfer signals that cause a pause in translocation. Figure 31.5 show the translocon as a closed cylinder spanning the membrane. Suggest a mechanism for lateral transfer of an integral membrane protein from the protein-conducting channel of the translocon into the hydrophobic phase of the ER membrane.

10. The Sec61p core complex of the translocon has a highly dynamic pore whose internal diameter varies from 0.6 to 6 nm. In post-translational translocation, folded proteins can move across the ER membrane through this pore. What is the molecular weight of

a spherical protein that would just fit through a 6-nm pore? (Adopt the same assumptions used in problem 5.)

11. (Integrates with Chapters 6, 9, and 30.) During co-translational translocation, the peptide tunnel running from the peptidyl transferase center of the large ribosomal subunit and the protein-conducting channel are aligned. If the tunnel through the ribosomal subunit is 10 nm and the translocon channel has the same length as the thickness of a phospholipid bilayer, what is the minimum number of amino acid residues sequestered in this common conduit?

12. Draw the structure of the isopeptide bond formed between Gly[76] of one ubiquitin molecule and Lys[48] of another ubiquitin molecule.

13. Assign the 20 amino acids to either of two groups based on their susceptibility to ubiquitin ligation by E_3 ubiquitin protein ligase. Can you discern any common attributes among the amino acids in the less susceptible versus the more susceptible group?

14. Lactacystin is a *Streptomyces* natural product that acts as an irreversible inhibitor of 26S proteasome β-subunit catalytic activity by covalent attachment to N-terminal threonine —OH groups. Predict the effects of lactacystin on cell cycle progression.

15. HtrA proteases are dual-function chaperone-protease protein quality control systems. The protease activity of HtrA proteases depends on a proper spatial relationship between the Asp-His-Ser catalytic triad. Propose a mechanism for the temperature-induced switch of HtrA proteases from chaperone function to protease function.

Preparing for the MCAT Exam

16. A common post-translational modification is removal of the universal N-terminal methionine in many proteins by Met-aminopeptidase. How might Met removal affect the half-life of the protein?

17. Figure 31.6 shows the generalized amphipathic α-helix structure found as an N-terminal presequence on a nuclear-encoded mitochondrial protein. Write out a 20-residue-long amino acid sequence that would give rise to such an amphipathic α-helical secondary structure.

Biochemistry ⊗ Now™   Preparing for an exam? Test yourself on key questions at http://chemistry.brookscole.com/ggb3

Further Reading

General
Wickner, S., Mauzis, M. R., and Gottesman, S., 1999. Posttranslational quality control: Folding, refolding, and degrading proteins. *Science* **286:**1888–1893.

Protein-Folding Diseases
Bates, G., 2003. Huntingtin aggregation and toxicity in Huntington's disease. *The Lancet* **361:**1642–1644.

Bucciantini, M., et al., 2002. Inherent toxicity of aggregates implies a common mechanism for protein misfolding diseases. *Nature* **416:**507–511.

Gamblin, T. C., et al., 2003. Caspase cleavage of tau: Linking amyloid and neurofibrillary tangles in Alzheimer's disease. *Proceedings of the National Academy of Sciences, U.S.A.* **100:**10032–10037.

Chaperone-Assisted Protein Folding
Bukau, B., and Horwich, A. L., 1998. The Hsp70 and Hsp60 chaperone machines. *Cell* **92:**351–366.

Bukau, B., et al., 2000. Getting newly synthesized proteins into shape. *Cell* **101:**119–122.

Ellis, J. R., 2001. Molecular chaperones: Inside and outside the Anfinsen cage. *Current Biology* **11:**R1038–R1040.

Frydman, J., 2001. Folding of newly translated proteins in vivo: The role of molecular chaperones. *Annual Review of Biochemistry* **70:**603–647.

Hartl, F. U., and Hayer-Hartl, M., 2002. Molecular chaperones in the cytosol: From nascent chain to folded protein. *Science* **295:**1852–1858.

Kramer, G., et al., 2002. L23 protein functions as a chaperone docking site on the ribosome. *Nature* **419:**171–174.

Netzer, W. J., and Hartl, F. U., 1998. Protein folding in the cytosol: Chaperonin-dependent and -independent mechanisms. *Trends in Biochemical Sciences* **23:**68–73.

Xu, Z., Horwich, A. L., and Sigler, P. B., 1997. The crystal structure of the asymmetric GroEL-GroES-(ADP)₇ chaperonin complex. *Nature* **388:**741–750.

Protein Translocation
Matlack, K. E. S., Mothes, W., and Rapoport, T. A., 1998. Protein translocation: Tunnel vision. *Cell* **92:**381–390.

Mihara, K., 2003. Moving inside membranes. *Nature* **424:**505–506. A description of the sorting and assembly pathways for mitochondrial proteins.

Neuport, W., 1997. Protein import into mitochondria. *Annual Review of Biochemistry* **66:**863–917.

Rapoport, T. A., Jungnickel, B., and Kytay, U., 1996. Protein transport across the eukaryotic endoplasmic reticulum and bacterial inner membranes. *Annual Review of Biochemistry* **65:**271–303.

Schnell, D. J., and Hebert, D. N., 2003. Protein translocons: Multifunctional mediators of protein translocation across membranes. *Cell* **112:**491–505.

Schwartz, S., and Blobel, G., 2003. Structural basis for the function of the β subunit of the eukaryotic signal recognition particle. *Cell* **112:**793–803.

Wirth, A., et al., 2003. The Sec61p complex is a dynamic precursor activated channel. *Molecular Cell* **12:**261–268.

Ubiquitin Selection of Proteins for Degradation
Haas, A. L., 1997. Introduction: Evolving roles for ubiquitin in cellular regulation. *The FASEB Journal* **11:**1053–1054.

Haas, A. L., and Siepman, T. J., 1997. Pathways of ubiquitin conjugation. *The FASEB Journal* **11:**1257–1268.

Hershko, A., 1996. Lessons from the discovery of ubiquitin system. *Trends in Biochemical Sciences* **21:**445–449.

Hochstrasser, M., 1996. Ubiquitin-dependent protein degradation. *Annual Review of Genetics* **30:**405–439.

Pagano, M., 1997. Cell cycle regulation by the ubiquitin pathway. *The FASEB Journal* **11:**1067–1075.

Rechsteiner, M., and Rogers, W. S., 1996. PEST sequences and regulation by proteolysis. *Trends in Biochemical Sciences* **21:**267–271.

Varshavsky, A., 1997. The ubiquitin system. *Trends in Biochemical Sciences* **22:**383–387.

Wilkinson, K. D., 1997. Regulation of ubiquitin-dependent processes by deubiquitinating enzymes. *The FASEB Journal* **11:**1245–1256.

Proteasome-Mediated Protein Degradation
Baumeister, W., et al., 1998. The proteasome: Paradigm of a self-compartmentalizing protease. *Cell* **92:**367–380.

Coux, O., Tanaka, K., and Goldberg, A. L., 1996. Structure and functions of the 20S and 26S proteasomes. *Annual Review of Biochemistry* **65**:801–847.

Ferrel, K., et al., 2000. Regulatory subunit interactions of the 26S proteasome, a complex problem. *Trends in Biochemical Sciences* **25**:83–88.

Groll, M., et al., 1997. Structure of 20S proteasome from yeast at 2.4Å resolution. *Nature* **386**:463–471.

Hartmann-Petersen, R., Seeger, M., and Gordon, C., 2003. Transferring substrates to the 26S proteasome. *Trends in Biochemical Sciences* **28**:26–31.

Hilt, W., and Wolf, D. H., 1996. Proteasomes: Destruction as program. *Trends in Biochemical Sciences* **21**:96–101.

Lowe, J., et al., 1995. Crystal structure of the 20S proteasome from the archaeon *T. acidophilum* at 3.4 Å resolution. *Science* **268**:533–539.

Pickart, C. M., 1997. Targeting of substrates to the 26S proteasome. *The FASEB Journal* **11**:1055–1066.

Schmidt, M., and Kloetzel, P-M., 1997. Biogenesis of 20S proteasomes: The complex maturation pathway of a complex enzyme. *The FASEB Journal* **11**:1235–1243.

Voges, D., et al., 1999. The 26S proteasome: A molecular machine for controlled proteolysis. *Annual Review of Biochemistry* **68**:1015–1068.

HtrA Proteases

Clausen, T., Southan, C., and Ehrmann, M., 2002. The HtrA family of proteases: Implications for protein composition and cell fate. *Molecular Cell* **10**:443–455.

CHAPTER 32

The Reception and Transmission of Extracellular Information

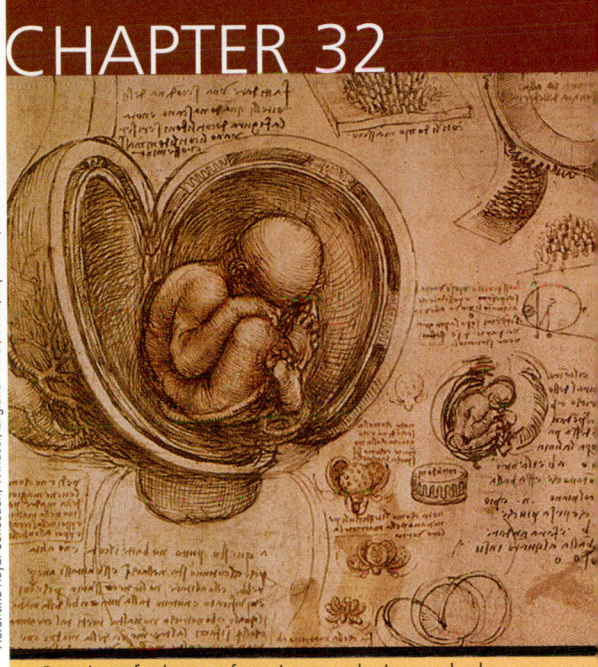

Florentine Royal Collection, Windsor, England A.K.G. Berlin/Superstock, International

Drawing of a human fetus in utero, by Leonardo da Vinci. Human sexuality and embryonic development represent two hormonally regulated processes of universal interest.

Essential Question

Higher life forms must have molecular mechanisms for detecting environmental information as well as mechanisms that allow for communication at the cell and tissue levels. Sensory systems detect and integrate physical and chemical information from the environment and pass this information along by the process of **neurotransmission.** Control and coordination of processes at the cell and tissue levels are achieved not only by neurotransmission but also by chemical signals in the form of **hormones** that are secreted by one set of cells to direct the activity of other cells. *What are these mechanisms of information transfer that mediate the molecular basis of hormone action and that use excitable membranes to transduce the signals of neurotransmission and sensory systems?*

Hormones are secreted by certain cells, usually located in glands, and travel, either by simple diffusion or circulation in the bloodstream, to specific target cells. As we shall see, some hormones bind to specialized receptors on the plasma membrane and induce responses within the cell without themselves entering the target cell (Figure 32.1). Other hormones actually enter the target cell and interact with specific receptors there. By these mechanisms, hormones regulate the metabolic processes of various organs and tissues; facilitate and control growth, differentiation, reproductive activities, and learning and memory; and help the organism cope with changing conditions and stresses in its environment.

"How little we know, how much to discover,
What chemical forces flow from lover to
*   lover.*
How little we understand what touches off
*   that tingle,*
That sudden explosion when two tingles
*   intermingle.*
Who cares to define what chemistry this is?
Who cares with your lips on mine
How ignorant bliss is,
So long as you kiss me and the world
*   around us shatters?*
How little it matters how little we know"
"How Little We Know"
by **P. Springer** and **C. Leigh,** as recorded by
Frank Sinatra, April 5, 1956. Capitol Records, Inc.

32.1 | What Are Hormones?

Many different chemical species act as hormones. **Steroid hormones,** all derived from cholesterol, regulate metabolism, salt and water balances, inflammatory processes, and sexual function. Several hormones are **amino acid derivatives.** Among these are *epinephrine* and *norepinephrine* (which regulate smooth muscle contraction and relaxation, blood pressure, cardiac rate, and the processes of lipolysis and glycogenolysis) and the *thyroid hormones* (which stimulate metabolism). **Peptide hormones** are a large group of hormones that regulate processes in all body tissues, including the release of yet other hormones.

Hormones and other signal molecules in biological systems bind with very high affinities to their receptors, displaying K_D values in the range of 10^{-12} to 10^{-6} M. The hormones are produced at concentrations equivalent to or slightly above these K_D values. Once hormonal effects have been induced, the hormone is usually rapidly metabolized.

Steroid Hormones Act in Two Ways

The steroid hormones include the glucocorticoids (cortisol and corticosterone), the mineralocorticoids (aldosterone), vitamin D, and the sex hormones (progesterone and testosterone, for example) (Figure 32.2; see Chapter 24 for the details of their synthesis). The steroid hormones exert their effects in two ways: First, by entering cells and migrating to the nucleus, steroid hormones act as transcription regulators, modulating gene expression. These effects of the

Key Questions

32.1 What Are Hormones?

32.2 What Are Signal Transduction Pathways?

32.3 How Do Signal-Transducing Receptors Respond to the Hormonal Message?

32.4 How Are Receptor Signals Transduced?

32.5 How Do Effectors Convert the Signals to Actions in the Cell?

32.6 What Is the Role of Protein Modules in Signal Transduction?

32.7 How Do Neurotransmission Pathways Control the Function of Sensory Systems?

Biochemistry Now™ Test yourself on these Key Questions at BiochemistryNow at **http://chemistry.brookscole.com/ggb3**

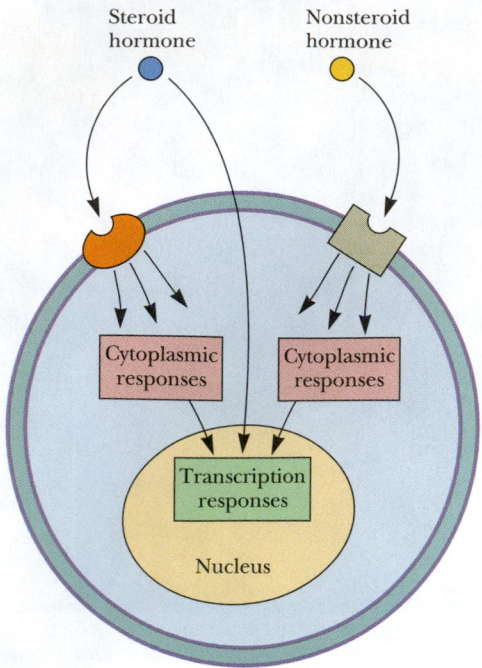

Biochemistry ⊗ Now™   ANIMATED FIGURE 32.1
Nonsteroid hormones bind exclusively to plasma membrane receptors, which mediate the cellular responses to the hormone. Steroid hormones exert their effects either by binding to plasma membrane receptors or by diffusing to the nucleus, where they modulate transcriptional events. **See this figure animated at http://chemistry.brookscole.com/ggb3**

steroid hormones occur on time scales of hours and involve synthesis of new proteins. Steroids can also act at the cell membrane, directly regulating ligand-gated ion channels and perhaps other processes. These latter processes take place very rapidly, on time scales of seconds and minutes.

Polypeptide Hormones Share Similarities of Synthesis and Processing

The largest class of hormones in vertebrate organisms is that of the **polypeptide hormones** (Table 32.1). One of the first polypeptide hormones to be discovered, **insulin,** was described by Banting and Best in 1921. Insulin, a secretion of the pancreas, controls glucose utilization and promotes the synthesis of proteins, fatty acids, and glycogen. Insulin, which is typical of the **secreted polypeptide hormones,** is discussed in detail in Chapters 5, 15, and 22.

Many other polypeptide hormones are produced and processed in a manner similar to that of insulin. Three unifying features of their synthesis and cellular processing should be noted. First, all secreted polypeptide hormones are originally synthesized with a signal sequence, which facilitates their eventual direction to secretory granules, and thence to the extracellular milieu. Second, peptide hormones are usually synthesized from mRNA as inactive precursors, termed **preprohormones,** which become activated by proteolysis. Third, a single polypeptide precursor or preprohormone may produce several different peptide hormones by suitable proteolytic processing.

An impressive example of the production of many hormone products from a single precursor is the case of **prepro-opiomelanocortin,** a 250-residue precursor peptide synthesized in the pituitary gland. A cascade of proteolytic steps produces, as the name implies, a natural *opi*ate substance (**endorphin**) and several other hormones (Figure 32.3). Endorphins and other opiatelike hormones are produced by the body in response to systemic stress. These substances probably contribute to the "runner's high" that marathon runners describe.

32.2 | What Are Signal Transduction Pathways?

Hormonal regulation depends on the transduction of the hormonal signal across the plasma membrane to specific intracellular sites, particularly the nucleus. Signal transduction pathways consist of a stepwise progression of

FIGURE 32.2 Structures of some steroid hormones.

A Deeper Look

The Acrosome Reaction

Steroid hormones affect ion channels in the **acrosome reaction,** which must occur before human sperm can fertilize an egg. The **acrosome** is an organelle that surrounds the head of a sperm (see accompanying figure) and lies just inside and juxtaposed with the plasma membrane. The acrosome itself is essentially a large vesicle of hydrolytic and proteolytic enzymes. In the acrosome reaction, influx of Ca^{2+} ions causes the outer acrosomal membrane to fuse with the plasma membrane. These fused membrane segments separate from each other and diffuse away, freeing the acrosomal enzymes to attack the egg, and exposing

binding sites on the inner acrosomal membrane that interact with the egg in the fertilization process.

This acrosome reaction is induced by **progesterone,** a female hormone secreted by the cumulus oophorus, a collection of ovarian follicle cells surrounding the egg! Intracellular Ca^{2+} levels increase within seconds of treating human sperm with progesterone. These effects must occur via binding of the steroid to the sperm plasma membrane. A far longer time would be required for progesterone to act through an enhancement of transcription.

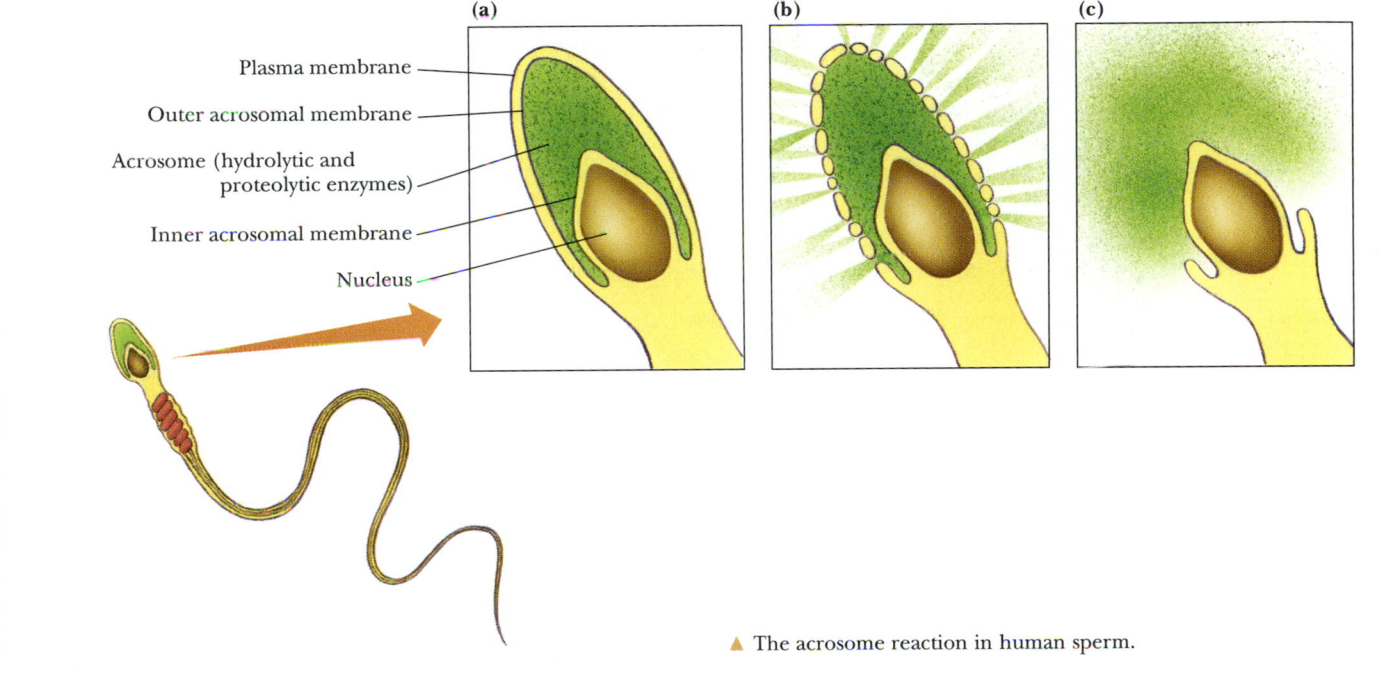

(a)   (b)   (c)

Plasma membrane
Outer acrosomal membrane
Acrosome (hydrolytic and proteolytic enzymes)
Inner acrosomal membrane
Nucleus

▲ The acrosome reaction in human sperm.

signaling stages: receptor→transducer→effector. The receptor perceives the signal, transducers relay the signal, and the effectors convert the signal into an intracellular response. Often, effector action involves a series of steps, each of which is mediated by an enzyme, and each of these enzymes can be considered as an amplifier in a pathway connecting the hormonal signal to its intracellular targets. Such enzymes are typically protein kinases (and protein phosphatases), and many steps in these signaling pathways involve phosphorylation of serine, threonine, and tyrosine residues on target proteins. The complexity of signal transduction is thus manifested in the estimates that the human genome contains about a thousand protein kinase and protein phosphatase genes. Although many of these kinases—and other signaling proteins and molecules—have not been identified, the principles that govern these pathways are now becoming apparent (see Chapter 15).

Many protein kinases and phosphatases have relatively broad specificities, but they interact with their substrates via specialized recognition domains on the involved proteins (see box on page 476). There are also cellular strategies for localization of these signaling molecules. As we shall see later, modular proteins having one or more protein–protein or lipid–protein recognition domains recruit kinases and other proteins into heteromultimeric scaffolds that mediate those intracellular events the hormone is intended to elicit.

Table 32.1

Polypeptide Hormones

Hormone	Amino Acid Residues	Source	Target Cells	Function
Adrenocorticotropic hormone (ACTH)	39	Anterior pituitary	Adrenal cortex	Promotes adrenal steroid production
Bradykinin	9	Kidney, other tissues	Blood vessels	Causes vasodilation
Calcitonin	33	Thyroid gland	Bone	Regulates plasma Ca^{2+} and phosphate
Chorionic gonadotropin	α, 96 β, 147	Placenta	Various reproductive tissues	Maintains pregnancy
Follicle-stimulating hormone (FSH)	α, 96 β, 120	Anterior pituitary	Gonads	Stimulates growth and development
Gastrin	17	Gastrointestinal tract	GI tract, gallbladder, pancreas	Regulates digestion
Glucagon	29	Pancreas	Primarily liver	Regulates metabolism and blood glucose
Growth hormone (GH)	191	Anterior pituitary	Many: bone, fat, liver	Stimulates skeleton and muscle growth
Insulin	A, 21 B, 30	Pancreas	Primarily liver, muscle, and fat	Regulates metabolism and blood glucose
Luteinizing hormone (LH)	α, 96 β, 121	Anterior pituitary	Gonads, ovarian follicle cells	Triggers ovulation
Prolactin	197	Anterior pituitary	Breast	Stimulates milk production
Somatostatin	14	Hypothalamus	Anterior pituitary	Inhibits growth hormone secretion

Adapted from Rhoades, R., and Pflanzer, R., 1992. *Human Physiology*, 2nd ed. Philadelphia: Saunders College Publishing.

Many Signaling Pathways Involve Enzyme Cascades

Signaling pathways must operate with speed and precision, facilitating the accurate relay of intracellular signals to specific targets. But how does this happen? Enzyme cascades are one answer to this question. Enzymes can produce (or modify) a large number of molecules rapidly and specifically. The hormonal activation of glycogen phosphorylase is mediated by cAMP and involves a cascade of three enzymes: adenylyl cyclase, cAMP-dependent kinase (PKA), and phosphorylase kinase (see Figure 15.18). As indicated in the previous section, the cascade acts like a series of amplifiers, dramatically increasing the magnitude of the intracellular response available from a very small amount of hormone.

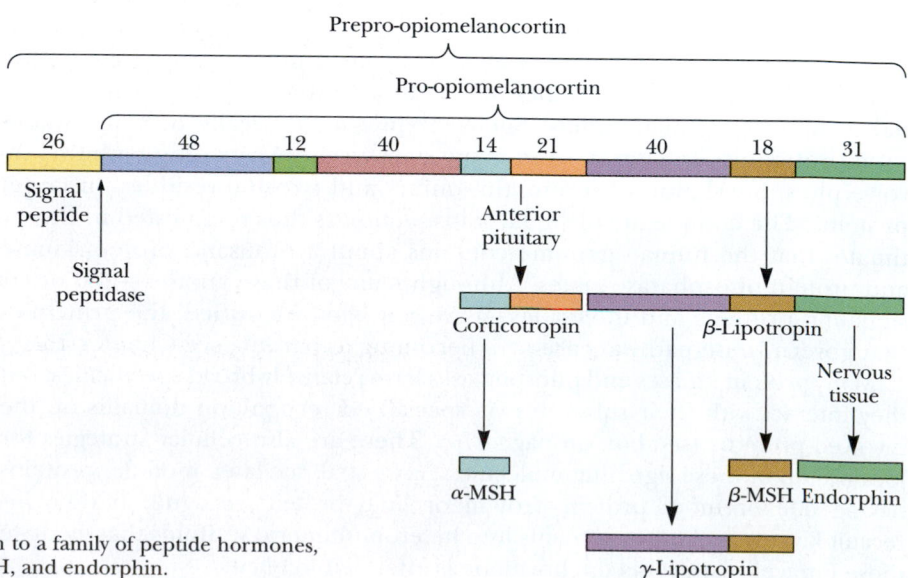

FIGURE 32.3 The conversion of prepro-opiomelanocortin to a family of peptide hormones, including corticotropin, β- and γ-lipotropin, α- and β-MSH, and endorphin.

Signaling Pathways Connect Membrane Interactions with Events in the Nucleus

The complete pathway from hormone binding at the plasma membrane to modulation of transcription in the nucleus is understood for a few signaling pathways. Figure 32.4 shows a complete signal transduction pathway that connects receptor tyrosine kinases, the Ras GTPase, cytoplasmic Raf, and two other protein kinases with transcription factors that alter gene expression in the nucleus. This pathway represents just one component of a complex signaling network that involves many other proteins and signaling factors. The existence of nearly 4000 human genes devoted to signal transduction portends a complex and interwoven network of signaling interactions in nearly all human cells.[1]

32.3	How Do Signal-Transducing Receptors Respond to the Hormonal Message?

Very often in life, *the message is more important than the messenger,* and this is certainly true for hormones. The structure and chemical properties of a hormone are important only for specific binding of the hormone to its appropriate receptor. Of much greater interest and importance is the metabolic information carried by the hormonal signal. The information implicit in the hormonal signal is interpreted by the cell, and an intricate pattern of cellular responses ensues.

Steroid hormones may either bind to plasma membrane receptors or exert their effects within target cells, entering the cell and migrating to their sites of action via specific cytoplasmic receptor proteins. The nonsteroid hormones, which act by binding to outward-facing plasma membrane receptors, activate various **signal transduction pathways** that mobilize various **second messengers**—cyclic nucleotides, Ca^{2+} ions, and other substances that activate or inhibit enzymes or cascades of enzymes in very specific ways. These hormonally activated processes are the focus of this chapter.

All receptors that mediate transmembrane signaling processes fit into one of three **receptor superfamilies:**

1. The **G-protein–coupled receptors** (see Section 32.4) are integral membrane proteins with an extracellular recognition site for ligands and an intracellular recognition site for a **GTP-binding protein** (see following discussion).
2. The **single-transmembrane segment (1-TMS) catalytic receptors** are proteins with only a single transmembrane segment and substantial globular domains on both the extracellular and intracellular faces of the membrane. The extracellular domain in the ligand recognition site and the intracellular catalytic domain is either a **tyrosine kinase** or a **guanylyl cyclase.**
3. **Oligomeric ion channels** consist of associations of protein subunits, each of which contains several transmembrane segments. These oligomeric structures are **ligand-gated ion channels.** Binding of the specific ligand typically opens the ion channel. The ligands for these ion channels are neurotransmitters.

The G-Protein–Coupled Receptors Are 7-TMS Integral Membrane Proteins

The G-protein–coupled receptors (GPCRs) have primary and secondary structure similar to that of bacteriorhodopsin (see Chapter 9), with seven apparent transmembrane α-helical segments; they are thus known as *7-transmembrane segment* (7-TMS) proteins. Rhodopsin and the β-adrenergic receptor, for which epinephrine is a ligand, are good examples (Figure 32.5). The site for binding

---

[1]The American Association for the Advancement of Science oversees a consortium of researchers who have established an outstanding Web site—the Signal Transduction Knowledge Environment (STKE). This site is an up-to-date and ongoing compilation of information about cell signaling and signal transduction. The URL is *http://www.stke.org.*

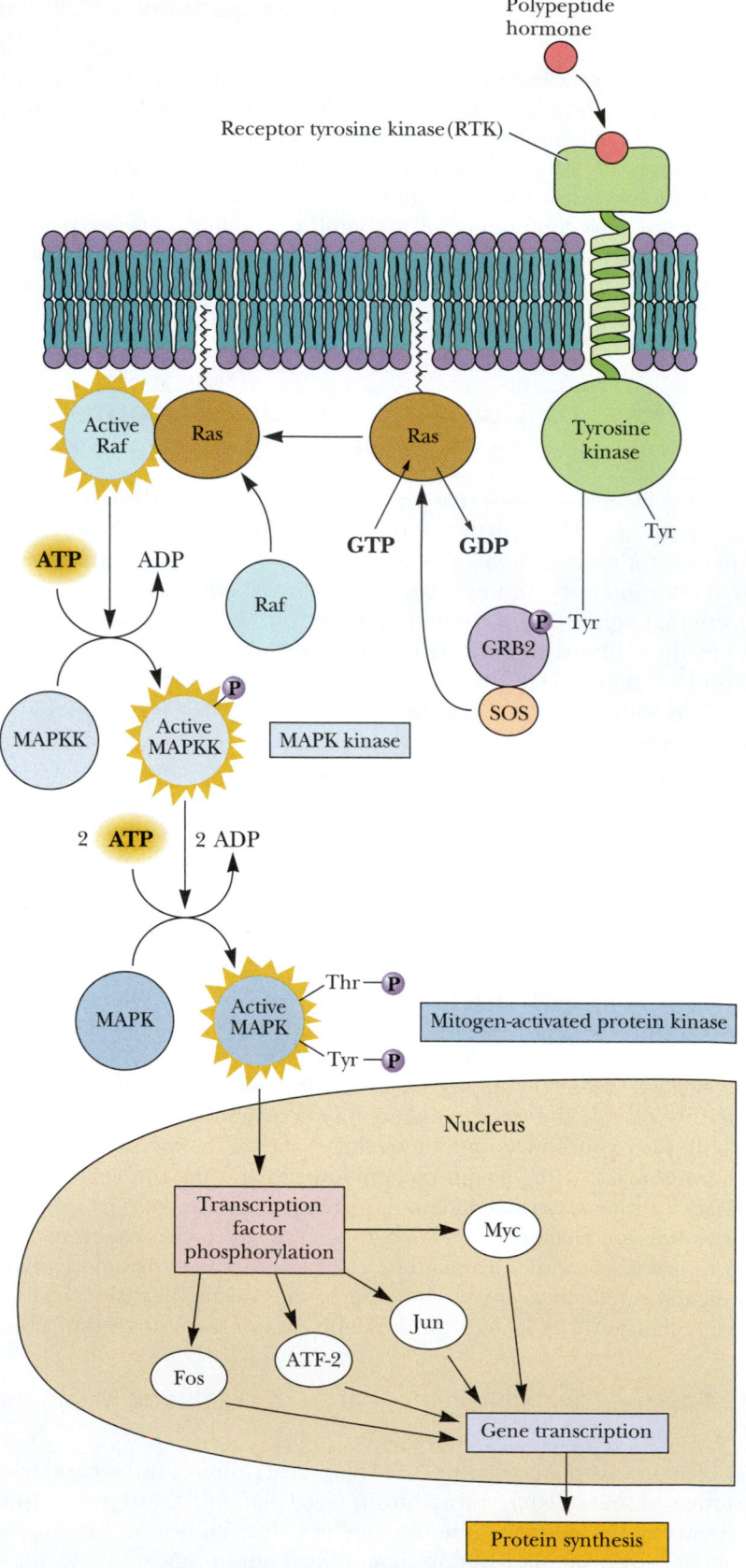

Biochemistry📖Now™   **ACTIVE FIGURE 32.4** A complete signal transduction pathway that connects a hormone receptor with transcription events in the nucleus. A number of similar pathways have been characterized. **Test yourself on the concepts in this figure at http://chemistry. brookscole.com/ggb3**

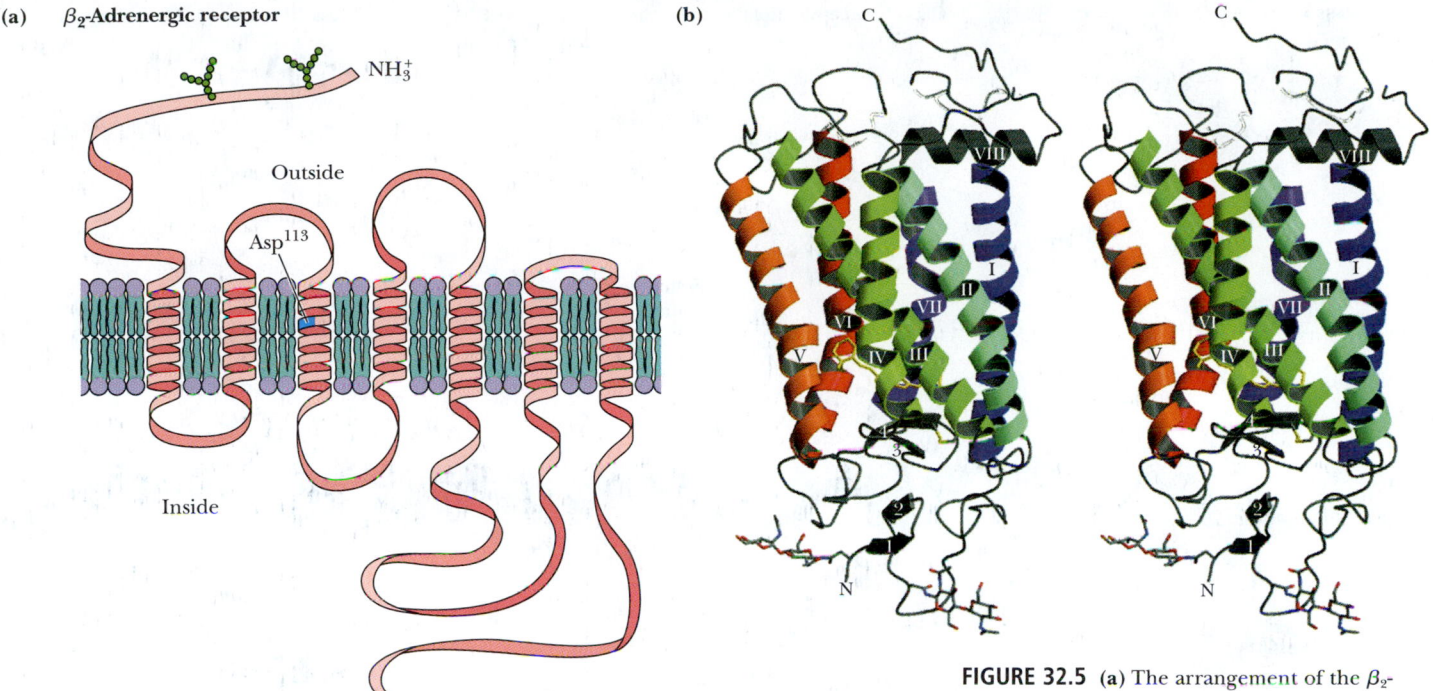

(a) β_2-Adrenergic receptor

(b)

FIGURE 32.5   (a) The arrangement of the β_2-adrenergic receptor in the membrane. Substitution of Asp[113] in the third hydrophobic domain of the β-adrenergic receptor with an Asn or Gln by site-directed mutagenesis results in a dramatic decrease in affinity of the receptor for both agonists and antagonists. Significantly, this Asp residue is conserved in all other G-protein–coupled receptors that bind biogenic amines but is absent in receptors whose ligands are not amines. Asp[113] appears to be the counterion of the amine moiety of adrenergic ligands. (b) Stereo image of rhodopsin, viewed parallel to the plane of the membrane. The seven-transmembrane α-helices are indicated by roman numerals. *(From Figure 2 from Palczewski, R., et al., 2000. Crystal structure of rhodopsin: A G-protein-coupled receptor. Science **289**:730–745. Reprinted with permission of the AAAS.)*

of cationic catecholamines to the adrenergic receptors is located within the hydrophobic core of the receptor.

Binding of hormone to a GPCR induces a conformation change that activates a GTP-binding protein, also known as a G protein (discussed in Section 32.4). Activated G proteins trigger a variety of cellular effects, including activation of adenylyl and guanylyl cyclases (which produce cAMP and cGMP from ATP), activation of phospholipases (which produce second messengers from phospholipids) and activation of Ca^{2+} and K^+ channels (which leads to elevation of cell $[Ca^{2+}]$ and $[K^+]$). (All of these effects are described in Section 32.4.)

The Single TMS Receptors Are Guanylyl Cyclases or Tyrosine Kinases

Receptor proteins that span the plasma membrane with a single helical transmembrane segment possess an external ligand recognition site and an internal domain with enzyme activity—either **receptor tyrosine kinase** or **receptor guanylyl cyclase.** Each of these enzyme activities is manifested in two different cellular forms. Thus, guanylyl cyclase is found in both membrane-bound receptors and in soluble, cytoplasmic proteins. Tyrosine kinase activity, on the other hand, is exhibited by two different types of membrane proteins: The receptor tyrosine kinases are integral transmembrane proteins, whereas the nonreceptor tyrosine kinases are peripheral, lipid-anchored proteins.

Receptor Tyrosine Kinases Are Membrane-Associated Allosteric Enzymes

The binding of polypeptide hormones and growth factors to receptor tyrosine kinases activates the tyrosine kinase activity of these proteins. These catalytic receptors are composed of three domains (Figure 32.6): a glycosylated extracellular receptor-binding domain, a transmembrane domain consisting

Biochemistry ⊜ Now™   Go to BiochemistryNow and click BiochemistryInteractive to interact with β_2 adrenergic receptor, a classic 7-TMS receptor protein.

Class I receptor (EGF receptor)   **Class II receptor (insulin receptor)**   **Class III receptor (FGF receptor)**

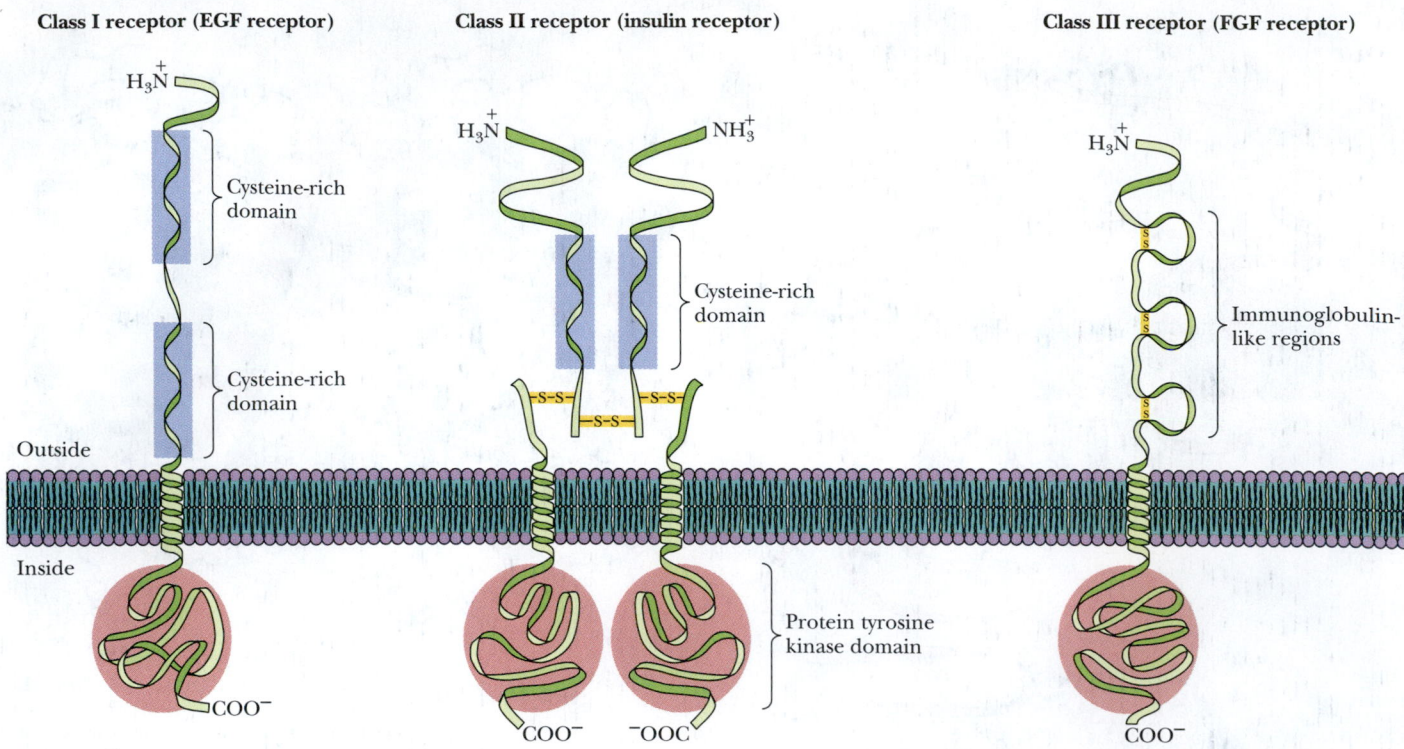

Biochemistry⊗Now™   ANIMATED FIGURE 32.6
The three classes of receptor tyrosine kinases. Class I
receptors are monomeric and contain a pair of Cys-
rich repeat sequences. The insulin receptor, a typical
class II receptor, is a glycoprotein composed of two
kinds of subunits in an $\alpha_2\beta_2$ tetramer. The α- and
β-subunits are synthesized as a single peptide chain,
together with an N-terminal signal sequence. Subsequent proteolytic processing yields the separate α-
and β-subunits. The β-subunits of 620 residues each
are integral transmembrane proteins, with only a
single transmembrane α-helix and with the amino
terminus outside the cell and the carboxyl terminus
inside. The α-subunits of 735 residues each are extracellular proteins that are linked to the β-subunits and
to each other by disulfide bonds. The insulin-binding
domain is located in a cysteine-rich region on the
α-subunits. Class III receptors contain multiple
immunoglobulin-like domains. Shown here is fibroblast growth factor (FGF) receptor, which has three
immunoglobulin-like domains. *(Adapted from Ullrich A.,
and Schlessinger, J., 1990. Signal transduction by receptors with
tyrosine kinase activity. Cell* **61:**203–212.*)* **See this figure animated at http://chemistry.brookscole.com/ggb3**

of a single transmembrane α-helix, and an intracellular domain that includes
a tyrosine kinase domain that mediates the biological response to the hormone or growth factor via its catalytic activity and a regulatory domain that
contains multiple phosphorylation sites.

There are three classes of receptor tyrosine kinases (Figure 32.6). Class I, exemplified by the **epidermal growth factor (EGF) receptor,** has an extracellular
domain containing two Cys-rich repeat sequences. Class II, typified by the **insulin
receptor,** has an $\alpha_2\beta_2$ tetrameric structure with transmembrane β-subunits and a
Cys-rich domain in the extracellular α-subunit. Class III receptors, such as the
platelet-derived growth factor (PDGF) receptor, have five (or sometimes three)
immunoglobulin-like extracellular domains.

Given that the extracellular and intracellular domains of receptor tyrosine
kinases are joined by only a single transmembrane helical segment, how does
extracellular hormone binding activate intracellular tyrosine kinase activity? How
is the signal transduced? As shown in Figure 32.7, signal transduction occurs by
hormone-induced oligomeric association of receptors. Hormone binding triggers
a conformational change in the *extracellular* domain, which induces oligomeric
association. Oligomeric association allows adjacent *cytoplasmic* domains to interact,
leading to phosphorylation of the cytoplasmic domains and stimulation of cytoplasmic tyrosine kinase activity. By virtue of these ligand-induced conformation
changes and oligomeric interactions, receptor tyrosine kinases are **membrane-
associated allosteric enzymes.**

Receptor Tyrosine Kinases Phosphorylate a Variety of Cellular Target Proteins

Receptor tyrosine kinases catalyze the phosphorylation of numerous cellular target proteins, producing coordinated changes in cell behavior, including alterations in membrane transport of ions and amino acids, the transcription of
genes, and the synthesis of proteins. Many individual phosphorylation targets
have been characterized, including the γ-isozymes of phospholipase C and
phosphatidylinositol-3-kinase.

Membrane-Bound Guanylyl Cyclases Are Single-TMS Receptors

Another cellular second messenger, **guanosine 3′,5′-cyclic monophosphate (cGMP),** is formed from GTP by **guanylyl cyclase,** an enzyme found in several different forms in different cellular locations. **Membrane-bound guanylyl cyclases** constitute the second class of single-TMS receptors that have an extracellular hormone-binding domain; a single, α-helical transmembrane segment; and an intracellular catalytic domain (Figure 32.8). A variety of peptides act to stimulate the membrane-bound guanylyl cyclases, including **atrial natriuretic peptide (ANP),** which regulates body fluid homeostasis and cardiovascular function; the **heat-stable enterotoxins** from *Escherichia coli;* and a series of peptides secreted by mammalian ova (eggs), which stimulate sperm motility and act as sperm chemoattractant signals. Binding of these peptides to an extracellular site on the guanylyl cyclase in the sperm plasma membrane induces a conformational change that activates the intracellular catalytic site for cyclase activity. Activation involves oligomerization of receptors in the membrane, as for the RTKs discussed previously.

Nonreceptor Tyrosine Kinases Are Typified by pp60src

The first tyrosine kinases to be discovered were associated with **viral transforming proteins.** These proteins, produced by **oncogenic viruses,** enable the virus to *transform* animal cells, that is, to convert them to the cancerous state. A prime example is the tyrosine kinase expressed by the **src gene** of **Rous** or **avian sarcoma virus.** The protein product of this gene is **pp60^{v-src}** (the abbreviation refers to phosphoprotein, 60 kD, viral origin, sarcoma-causing). The v-src gene was derived from the avian proto-oncogenic gene c-src during the original formation of the virus. The cellular proto-oncogene homolog of pp60^{v-src} is referred to as pp60^{c-src}. pp60^{v-src} is a 526-residue peripheral membrane protein. It undergoes two post-translational modifications: First, the amino group of the NH$_2$-terminal glycine is modified by the covalent attachment of a **myristyl** group (this modification is required for membrane association of the kinase; see Figure 32.9a). Then Ser^{17} and Tyr^{416} are phosphorylated. The phosphorylation at Tyr^{416}, which increases kinase activity twofold to threefold, appears to be an autophosphorylation. On the other hand, phosphorylation at Tyr^{527} is inhibitory and is catalyzed by another kinase known as CSK. The significance of nonreceptor tyrosine kinase activity to cell growth and transformation is only partially understood, but 1% of all cellular proteins (many of which are also kinases) are phosphorylated by these kinases.

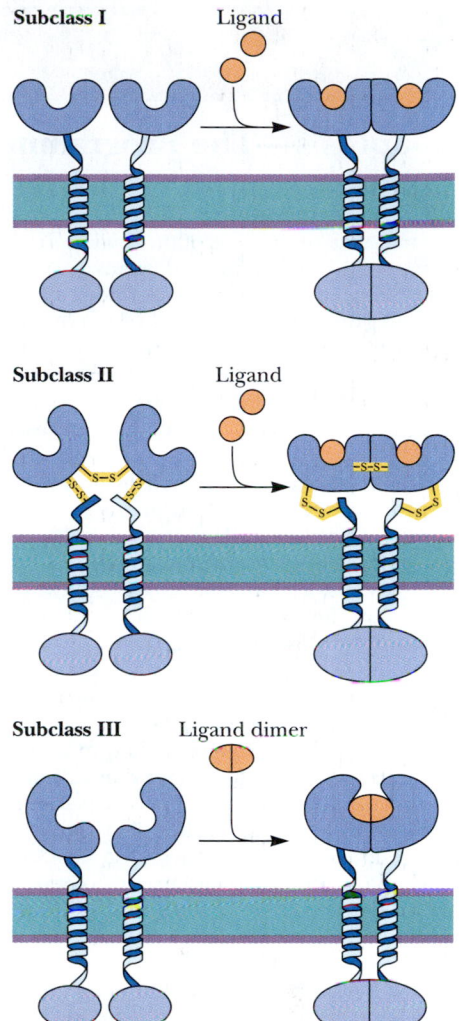

Biochemistry ⊜ Now™ ANIMATED FIGURE 32.7 Ligand (hormone)-stimulated oligomeric association of receptor tyrosine kinases. **See this figure animated at http://chemistry.brookscole.com/ggb3**

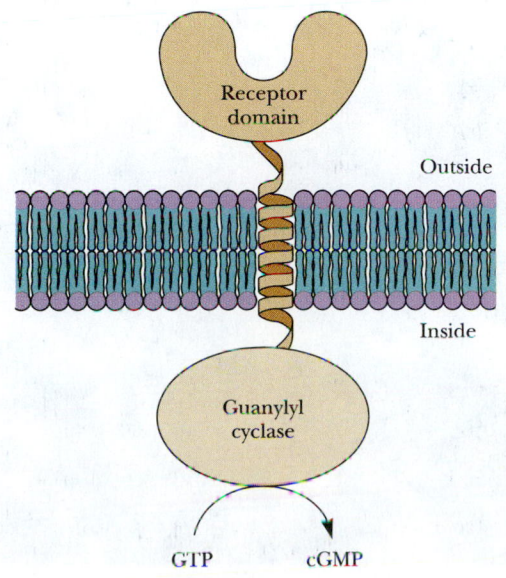

FIGURE 32.8 The structure of membrane-bound guanylyl cyclases.

A Deeper Look

Apoptosis—The Programmed Suicide of Cells

According to a Japanese saying, "Once we are in the land of the living, we will eventually die." True as this may be for human beings, it is also true of the cells of our bodies. In the ongoing cycle of cell division and differentiation, many surplus and/or harmful cells are produced. These cells must be removed or killed to maintain the integrity and homeostasis of the organism. The magnitude of such cell death is often surprising: More than 95% of thymocytes die during maturation of the thymus! The cell death that occurs during embryogenesis, metamorphosis, and normal cell turnover is termed *programmed cell death,* or **apoptosis** (where the second "p" is silent).

Apoptosis can be initiated in a variety of ways. One of these involves "death factors"—proteins such as **Fas ligand** and **tumor necrosis factor,** or **TNF,** that are members of the **cytokine** family of proteins. These factors bind to specific plasma membrane receptors, **Fas** and **TNF-receptor,** respectively, prompting trimeric associations of the receptor proteins in the membrane. These receptor proteins (see accompanying figure) possess intracellular *death domains,* polypeptide motifs that adopt an unusual structure consisting of six antiparallel, amphipathic α-helices. These receptor death domains form oligomeric associations with other, similar death domains on *adaptor proteins* (known as **FADD** or **MORT1**). This prompts special **death-effector domains** on FADD/MORT1 to bind to cysteine proteases known as **caspases** (so-named with a "c" for cysteine and "asp" indicating that these proteases cleave after an Asp residue in their protein substrates). Sequential activation in a cascade of caspases, together with other events, triggers apoptosis (cell death).

Inappropriate apoptosis underlies the cause of certain human diseases, including neurodegenerative disorders and cancer. An unusual caspase cleavage appears to play a role in Huntington's disease (HD), the cause of the death of folk-singer Woody Guthrie. The mutation responsible for HD is an expansion of a CAG trinucleotide repeat at the 5′-end of the gene *Hdh,* which encodes an essential protein (**huntingtin**) of unknown function. The CAG extension results in an extended polyglutamine domain on the N-terminus of the protein. HD is manifested if this Gln repeat region exceeds 35 residues. Downstream of the poly-Gln region is a cluster of five DXXD motifs that are cleaved specifically by *caspase 3.* The longer the poly-Gln region, the greater the activity of caspase 3 in cleaving the DXXD motifs. The released poly-Gln domains are cytotoxic (see Chapter 31), provoking death of affected neurons, progressive loss of control of voluntary movements, general loss of neural function, and eventual death.

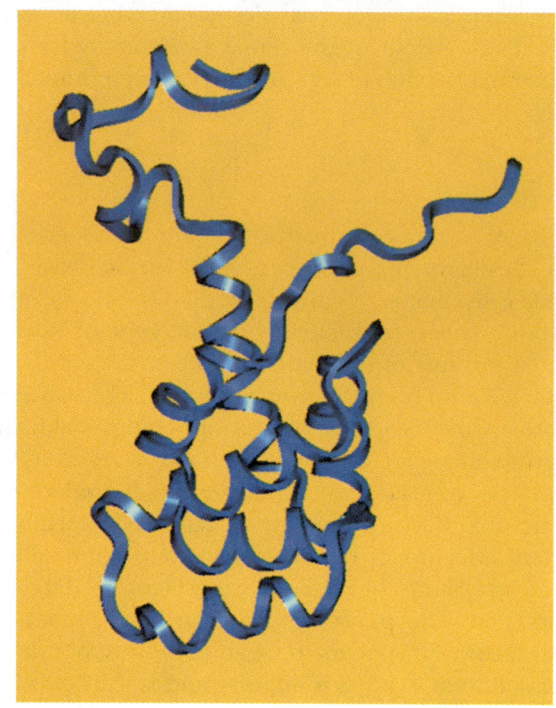

▲ The Fas "death domain."

FIGURE 32.9 **(a)** The soluble tyrosine kinase pp60[v-src] is anchored to the plasma membrane via an N-terminal myristyl group. **(b)** The structure of protein tyrosine kinase pp60[c-src], showing AMP–PNP in the active site (ball-and-stick), Tyr[416] (red), and Tyr[527] (yellow). Tyr[527] is phosphorylated (purple).

(a)

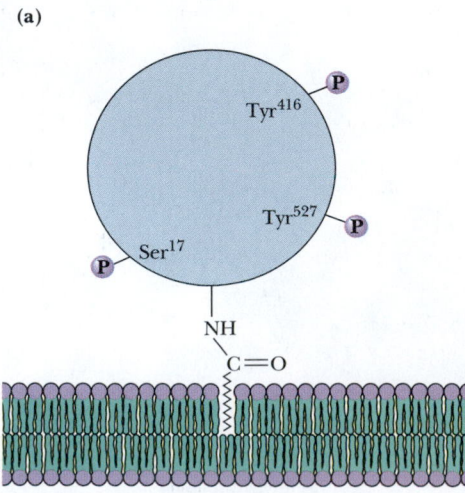

(b)

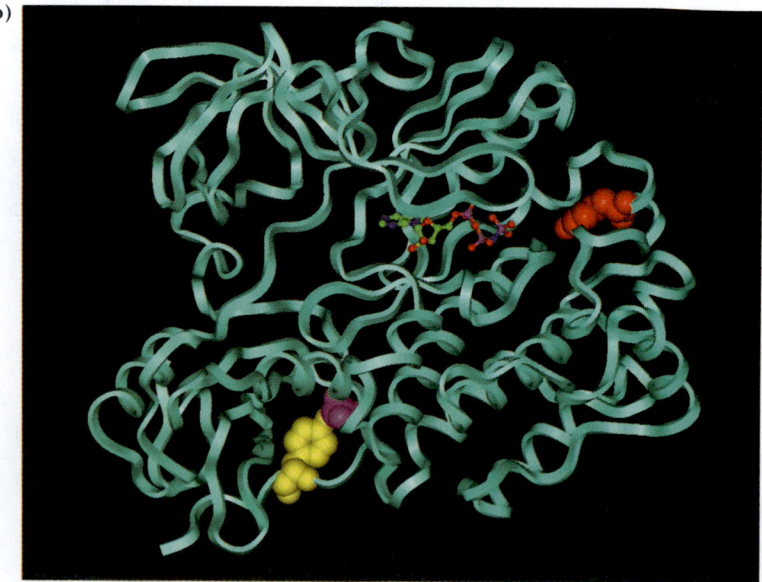

A Deeper Look

Nitric Oxide, Nitroglycerin, and Alfred Nobel

NO· is the active agent released by **nitroglycerin** (see accompanying figure), a powerful drug that ameliorates the symptoms of heart attacks and **angina pectoris** (chest pain due to coronary artery disease) by causing the dilation of coronary arteries. Nitroglycerin is also the active agent in dynamite. Ironically, Alfred Nobel, the inventor of dynamite who also endowed the Nobel prizes, himself suffered from angina pectoris. In a letter to a friend in 1885, Nobel wrote, "It sounds like the irony of fate that I should be ordered by my doctor to take nitroglycerin internally."

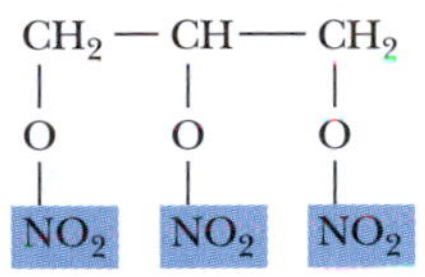

$$CH_2 - CH - CH_2$$

▲ The structure of nitroglycerin, a potent vasodilator.

Soluble Guanylyl Cyclases Are Receptors for Nitric Oxide

Nitric oxide, or **NO·,** a reactive free radical, acts as a neurotransmitter and as a second messenger, activating soluble guanylyl cyclase more than 400-fold. The cGMP thus produced also acts as a second messenger, inducing relaxation of vascular smooth muscle and mediating penile erection. As a dissolved gas, NO· is capable of rapid diffusion across membranes in the absence of any apparent carrier mechanism. This property makes NO· a particularly attractive second messenger because NO· generated in one cell can exert its effects quickly in many neighboring cells. NO· has a very short cellular half-life (1 to 5 seconds) and is rapidly degraded by nonenzymatic pathways.

32.4 | How Are Receptor Signals Transduced?

Receptor signals are *transduced* in one of three ways to initiate actions inside the cell:

1. Exchange of GDP for GTP by GTP-binding proteins (G proteins), which leads to generation of *second messengers,* including cAMP, phospholipid breakdown products, and Ca^{2+}.
2. Receptor-mediated activation of phosphorylation cascades that in turn trigger activation of various enzymes. This is the action of the receptor tyrosine kinases described in Section 32.3. Protein kinases and protein phosphatases acting as effectors will be discussed in Section 32.5.
3. Conformation changes that open ion channels or recruit proteins into nuclear transcription complexes. Ion channels are discussed in Section 32.7, and the formation of nuclear transcription complexes was described in Chapter 29.

GPCR Signals Are Transduced by G Proteins

The signals of G-protein–coupled receptors (GPCRs) are transduced by GTP-binding proteins, known more commonly as G proteins. The large G proteins are heterotrimers consisting of α- (45 to 47 kD), β- (35 kD), and γ- (7 to 9 kD) subunits. The α-subunit binds GDP or GTP and has an intrinsic, slow GTPase activity. The $G_{\alpha\beta\gamma}$ complex in the unactivated state has GDP at the nucleotide site (Figure 32.10). Binding of hormone to receptor stimulates a rapid exchange of GTP for GDP on G_α. The binding of GTP causes G_α to dissociate from $G_{\beta\gamma}$ and to associate with an effector protein such as adenylyl cyclase. *Binding of $G_\alpha(GTP)$ activates adenylyl cyclase.* The adenylyl cyclase actively synthesizes cAMP as long as $G_\alpha(GTP)$ remains bound to it. However, the intrinsic GTPase activity of G_α eventually hydrolyzes GTP to GDP, leading to dissociation of $G_\alpha(GDP)$ from adenylyl cyclase and reassociation with the $G_{\beta\gamma}$ dimer, regenerating the inactive heterotrimeric $G_{\alpha\beta\gamma}$ complex.

Biochemistry ⊗ Now™ Go to BiochemistryNow and click BiochemistryInteractive to learn more about the heterotrimeric G-protein complex.

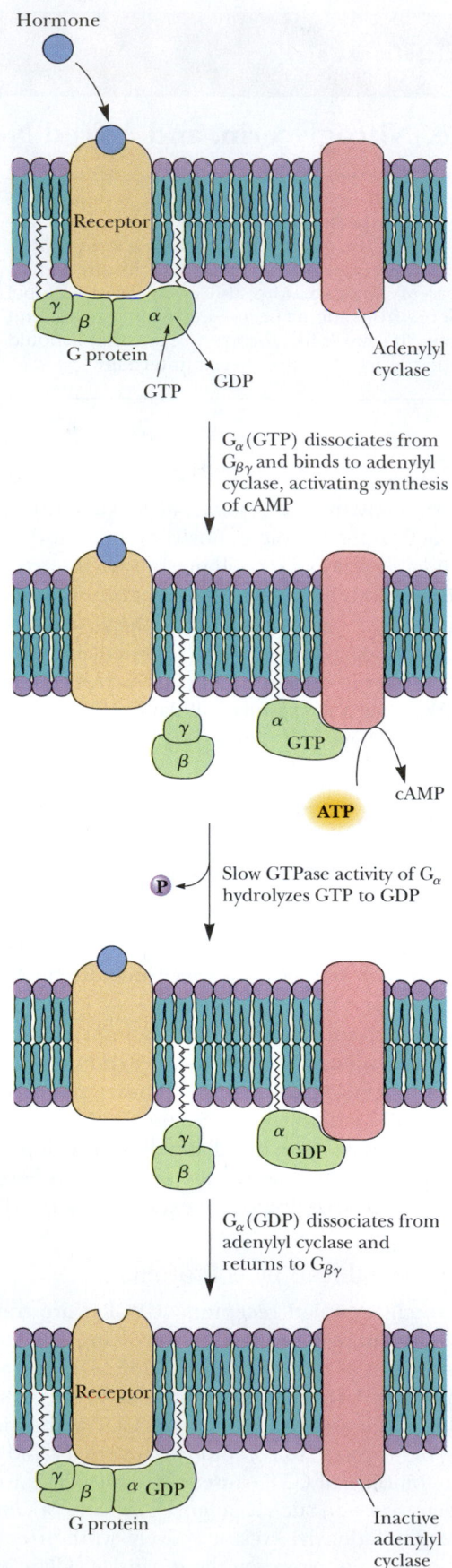

FIGURE 32.10 Activation of adenylyl cyclase by heterotrimeric G proteins. Binding of hormone to the receptor causes a conformational change that induces the receptor to catalyze a replacement of GDP by GTP on G_α. The G_α(GTP) complex dissociates from $G_{\beta\gamma}$ and binds to adenylyl cyclase, stimulating synthesis of cAMP. Bound GTP is slowly hydrolyzed to GDP by the intrinsic GTPase activity of G_α. G_α(GDP) dissociates from adenylyl cyclase and reassociates with $G_{\beta\gamma}$. G_α and G_γ are lipid-anchored proteins. Adenylyl cyclase is an integral membrane protein consisting of 12 transmembrane α-helical segments.

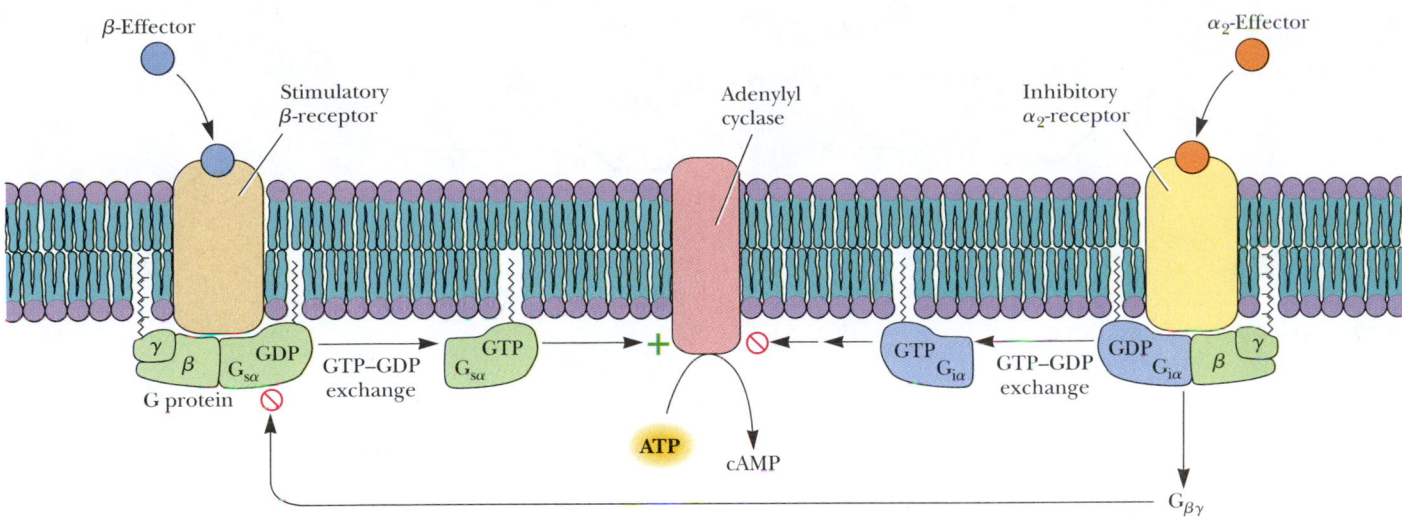

Biochemistry ⟨Ξ⟩ Now™ **ACTIVE FIGURE 32.11**
Adenylyl cyclase activity is modulated by the interplay of stimulatory (G_s) and inhibitory (G_i) G proteins. Binding of hormones to β_1- and β_2-adrenergic receptors activates adenylyl cyclase via G_s, whereas hormone binding to α_2-receptors leads to the inhibition of adenylyl cyclase. Inhibition may occur by direct inhibition of cyclase activity by $G_{i\alpha}$ or by binding of $G_{i\beta\gamma}$ to $G_{s\alpha}$. **Test yourself on the concepts in this figure at http://chemistry.brookscole.com/ggb3**

Two stages of amplification occur in the G-protein–mediated hormone response. First, a single hormone-receptor complex can activate many G proteins before the hormone dissociates from the receptor. Second, and more obvious, the G_α-activated adenylyl cyclase synthesizes many cAMP molecules. Thus, the binding of hormone to a very small number of membrane receptors stimulates a large increase in concentration of cAMP within the cell. The hormone receptor, G protein, and cyclase constitute a complete hormone **signal transduction unit** (Figure 32.11).

Hormone-receptor–mediated processes regulated by G proteins may be stimulatory or inhibitory. Each hormone receptor interacts specifically with either a stimulatory G protein, denoted G_s, or an inhibitory G protein, denoted G_i.

Cyclic AMP Is a Second Messenger

Cyclic AMP (denoted *cAMP*) was identified in 1956 by Earl Sutherland, who termed cAMP a **second messenger,** because it is the intracellular response provoked by binding of hormone (the first messenger) to its receptor. Since Sutherland's discovery of cAMP, many other second messengers have been identified (Table 32.2). The concentrations of second messengers in cells are carefully regulated. Synthesis or release of a second messenger is followed quickly by degradation or removal from the cytosol. Following its synthesis

Table 32.2		
Intracellular Second Messengers*		
Messenger	**Source**	**Effect**
cAMP	Adenylyl cyclase	Activates protein kinases
cGMP	Guanylyl cyclase	Activates protein kinases, regulates ion channels, regulates phosphodiesterases
Ca^{2+}	Ion channels in ER and plasma membrane	Activates protein kinases, activates Ca^{2+}-modulated proteins
IP_3	PLC action on PI	Activates Ca^{2+} channels
DAG	PLC action on PI	Activates protein kinase C
Phosphatidic acid	Membrane component and product of PLD	Activates Ca^{2+} channels, inhibits adenylyl cyclase
Ceramide	PLC action on sphingomyelin	Activates protein kinases
Nitric oxide (NO)	NO synthase	Activates guanylyl cyclase, relaxes smooth muscle
Cyclic ADP-ribose	cADP-ribose synthase	Activates Ca^{2+} channels

*IP_3 is inositol-1,4,5-trisphosphate; PLC is phospholipase C; PLD is phospholipase D; PI is phosphatidylinositol; DAG is diacylglycerol.

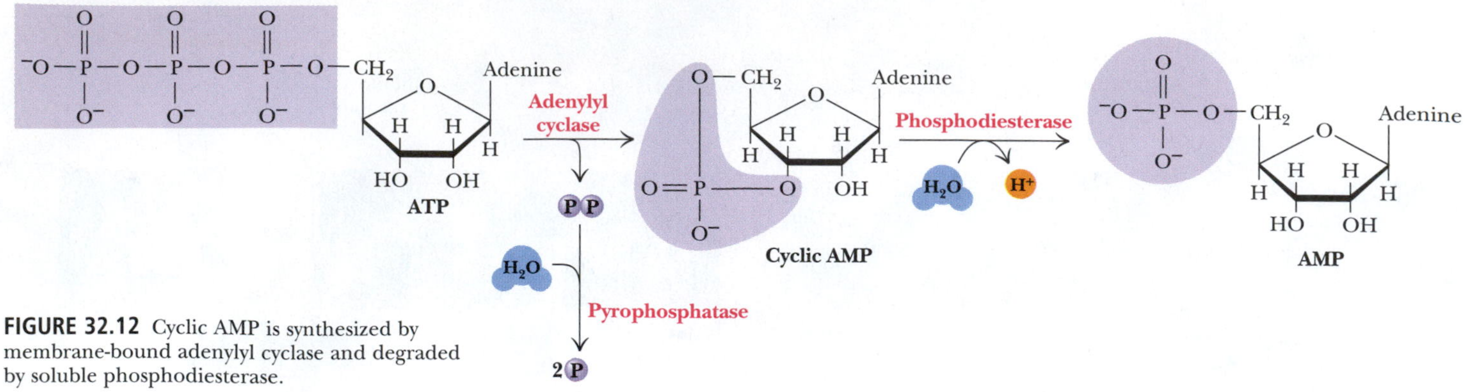

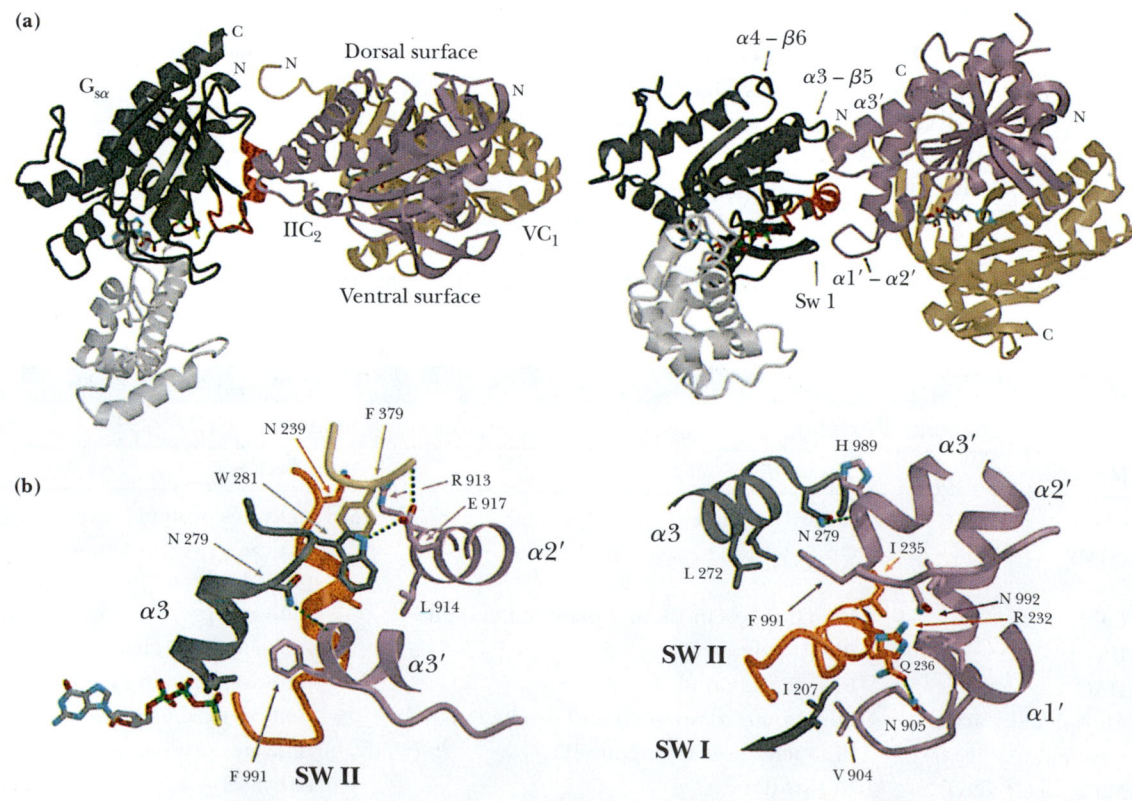

FIGURE 32.12 Cyclic AMP is synthesized by membrane-bound adenylyl cyclase and degraded by soluble phosphodiesterase.

Biochemistry ⓔ Now™ Go to BiochemistryNow and click BiochemistryInteractive to explore the structure and function of adenylyl cyclase.

by adenylyl cyclase, cAMP is broken down to 5′-AMP by phosphodiesterase (Figure 32.12).

Adenylyl cyclase (AC) is an integral membrane enzyme. Its catalytic domain, on the cytoplasmic face of the plasma membrane, includes two subdomains, denoted VC_1 and IIC_2. Binding of the α-subunit of G_s (denoted $G_{s\alpha}$) activates the AC catalytic domain.

Alfred Gilman, Stephen Sprang, and co-workers have determined the structure of a complex of $G_{s\alpha}$ (with bound GTP) with the cytoplasmic domains (VC_1 and IIC_2) of adenylyl cyclase (Figure 32.13). The $G_{s\alpha}$ complex binds to a cleft at one corner of the C_2 domain, and the surface of $G_{s\alpha}$-GTP that contacts adenylyl cyclase is the same surface that binds the $G_{\beta\gamma}$ dimer. The catalytic site, where ATP is converted to cyclic AMP, is far removed from the bound G protein.

FIGURE 32.13 (a) Two views of the complex of the VC_1–IIC_2 catalytic domain of adenylyl cyclase and $G_{s\alpha}$. (b) Details of the $G_{s\alpha}$ complex in the same orientation as the structures in (a). SW-1 and SW-2 are "switch regions," whose conformations differ greatly depending on whether GTP or GDP is bound. (*Courtesy of Alfred Gilman, University of Texas Southwestern Medical Center.*)

cAMP Activates Protein Kinase A

All second messengers exert their cellular effects by binding to one or more target molecules. cAMP produced by adenylyl cyclase activates a protein kinase, which is thus known as *cAMP-dependent protein kinase;* protein kinase A, as this enzyme is also known, activates many other cellular proteins by phosphorylation. The activation of protein kinase A by cAMP and regulation of the enzyme by intrasteric control was described in detail in Chapter 15. The structure of protein kinase A has served as a paradigm for understanding many related protein kinases (Figure 32.14).

Ras and the Small GTP-Binding Proteins Are Often Proto-Oncogene Products

GTP-binding proteins are implicated in growth control mechanisms in higher organisms. Certain tumor virus genomes contain genes encoding 21-kD proteins that bind GTP and show regions of homology with other G proteins. The first of these genes to be identified was found in *rat* *s*arcoma virus and was dubbed the *ras* **gene.** Genes implicated in tumor formation are known as **oncogenes;** they are often mutated versions of normal, noncancerous genes involved in growth regulation, so-called **proto-oncogenes.** The normal, cellular Ras protein is a GTP-binding protein that functions in a manner similar to that of other G proteins described previously, activating metabolic processes when GTP is bound and becoming inactive when GTP is hydrolyzed to GDP. The GTPase activity of the normal Ras p21 is very low, as is appropriate for a G protein that regulates long-term effects like growth and differentiation. A specific **GTPase-activating protein (GAP)** increases the GTPase activity of the Ras protein. Mutant (oncogenic) Ras proteins have severely impaired GTPase activity, which apparently causes serious alterations of cellular growth and metabolism in tumor cells. The conformations of Ras proteins (Figure 32.15) in complexes with GDP are different from the corresponding complexes with GTP analogs such as GMP–PNP (a nonhydrolyzable analog of GTP in which the β-P and γ-P are linked by N rather than by O). Two regions of the Ras structure change conformation upon GTP hydrolysis. These conformation changes mediate the interactions of Ras with other proteins, termed **effectors.**

G Proteins Are Universal Signal Transducers

A given G protein can be activated by several different hormone-receptor complexes. For example, either glucagon or epinephrine, binding to their distinctive receptor proteins, can activate the same species of G protein in liver cells. The effects are additive, and combined stimulation by glucagon

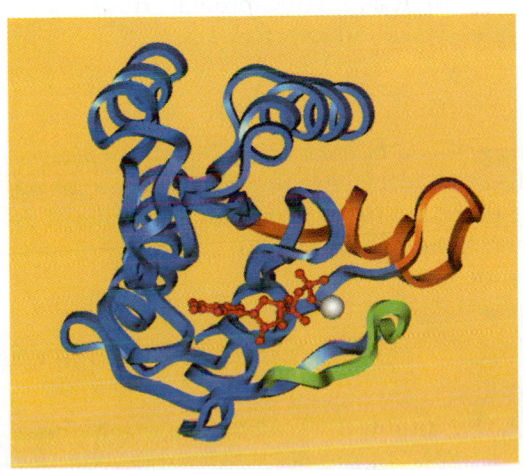

FIGURE 32.14 Cyclic AMP–dependent protein kinase is shown complexed with a pseudosubstrate peptide (red). This complex also includes ATP (yellow) and two Mn^{2+} ions (violet) bound at the active site.

(a)

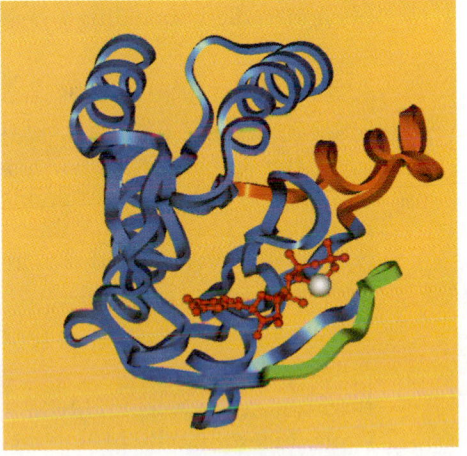

(b)

FIGURE 32.15 The structure of Ras complexed with **(a)** GDP and **(b)** GMP–PNP. The Ras p21–GMP–PNP complex is the active conformation of this protein.

A Deeper Look

RGSs and GAPs—Switches That Turn Off G Proteins

Nature has made Ras p21 and $G_{s\alpha}$ very poor enzymes by design. For example, Ras p21 hydrolyzes GTP with a rate constant of only 0.02 min^{-1}. These G proteins are active only in the GTP-bound state, and downstream targets will dissociate upon GTP hydrolysis. If Ras p21 and $G_{s\alpha}$ were efficient enzymes, the GTP-bound state would be short lived, and G-protein–mediated signaling would be ineffective.

But how can G proteins be switched off if they are inherently poor GTPases? The answer is provided by **GTPase-activating proteins (GAPs)** and **regulators of G-protein signaling (RGS),** which cause dramatic increases in GTPase activity when bound to G pro-

teins. The figure on the left shows Ras p21 (white) with a fragment of a GAP (blue) bound to it. GAPs increase the GTPase activity of Ras p21 by a factor of 10^5. The figure on the right shows RGS (blue) bound to $G_{i\alpha}$ (yellow). RGS proteins accelerate $G_{s\alpha}$-catalyzed GTP hydrolysis by nearly 100-fold. In both Ras p21 and $G_{s\alpha}$, GTPase activity and the conversion from GTP-bound to GDP-bound forms of the protein involve conformation changes in the *switch regions,* portions of the G-protein structure that surround the GTP/GDP binding site. GAPs and RGS increase GTPase activities of their respective G proteins by binding near the active site and stabilizing the transition state of the GTP hydrolysis reaction.

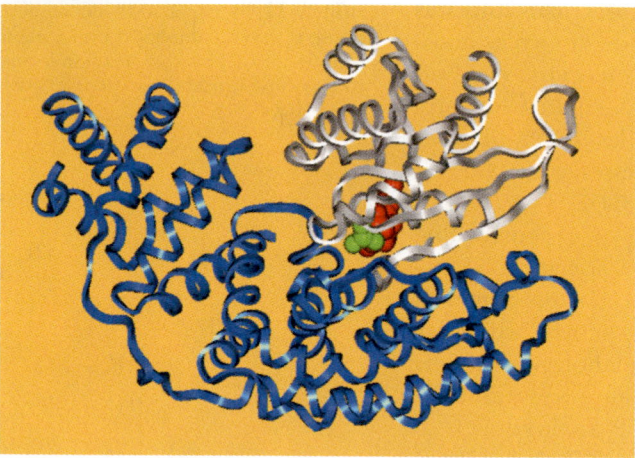

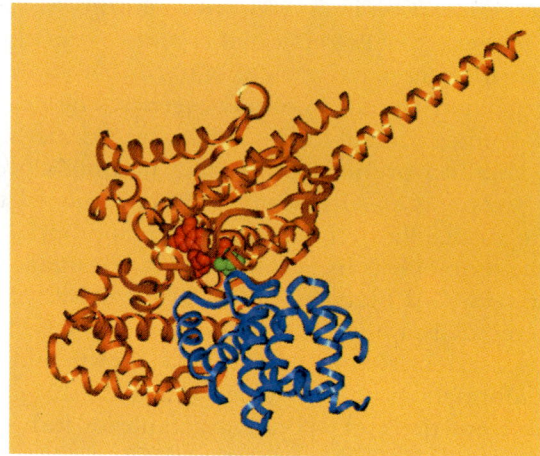

and epinephrine leads to higher cytoplasmic concentrations of cAMP than activation by either hormone alone.

G proteins are a universal means of signal transduction in higher organisms, activating many hormone-receptor–initiated cellular processes in addition to adenylyl cyclase. Such processes include, but are not limited to, activation of phospholipases C and A_2 and the opening or closing of transmembrane channels for K^+, Na^+, and Ca^{2+} in brain, muscle, heart, and other organs (Table 32.3). G proteins are integral components of sensory pathways such as vision and olfaction. More than 100 different G-protein–coupled receptors and at least 21 distinct G proteins are known. At least a dozen different G-protein effectors have been identified, including a variety of enzymes and ion channels.

Specific Phospholipases Release Second Messengers

A diverse array of second messengers are generated by breakdown of membrane phospholipids. Binding of certain hormones and growth factors to their respective receptors triggers a sequence of events that can lead to the activation of **specific phospholipases.** The action of these phospholipases on membrane lipids produces the second messengers shown in Figure 32.16.

Inositol Phospholipid Breakdown Yields Inositol-1,4,5-Trisphosphate and Diacylglycerol

Breakdown of **phosphatidylinositol (PI)** and its derivatives by **phospholipase C** produces a family of second messengers. In the best-understood pathway, successive phosphorylations of PI produce **phosphatidylinositol-4-P (PIP)**

Table 32.3

G Proteins and Their Physiological Effects

G Protein	Location	Stimulus	Effector	Effect
G_s	Liver	Epinephrine, glucagon	Adenylyl cyclase	Glycogen breakdown
G_s	Adipose tissue	Epinephrine, glucagon	Adenylyl cyclase	Fat breakdown
G_s	Kidney	Antidiuretic hormone	Adenylyl cyclase	Conservation of water
G_s	Ovarian follicle	Luteinizing hormone	Adenylyl cyclase	Increased estrogen and progesterone synthesis
G_i	Heart muscle	Acetylcholine	Potassium channel	Decreased heart rate and pumping force
G_i/G_o	Brain neurons	Enkephalins, endorphins, opioids	Adenylyl cyclase, potassium channels, calcium channels	Changes in neuron electrical activity
G_q	Smooth muscle cells in blood vessels	Angiotensin	Phospholipase C	Muscle contraction, blood pressure elevation
G_{olf}	Neuroepithelial cells in the nose	Odorant molecules	Adenylyl cyclase	Odorant detection
Transducin (G_t)	Retinal rod and cone cells	Light	cGMP phosphodiesterase	Light detection
GPA1	Baker's yeast	Pheromones	Unknown	Mating

Adapted from Hepler, J., and Gilman, A., 1992. G proteins. *Trends in Biochemical Sciences* **17**:383–387.

and **phosphatidylinositol-4,5-bisphosphate (PIP$_2$)**. Four isozymes of phospholipase C (denoted α, β, γ, and δ) hydrolyze PI, PIP, and PIP$_2$. Hydrolysis of PIP$_2$ by phospholipase C yields the second messenger **inositol-1,4,5-trisphosphate (IP$_3$)**, as well as another second messenger, **diacylglycerol (DAG)** (Figure 32.17). IP$_3$ is water soluble and diffuses to intracellular organelles where release of Ca^{2+} is activated. DAG, on the other hand, is lipophilic and remains in the plasma membrane, where it activates a Ca^{2+}-dependent protein kinase known as **protein kinase C** (see following discussion).

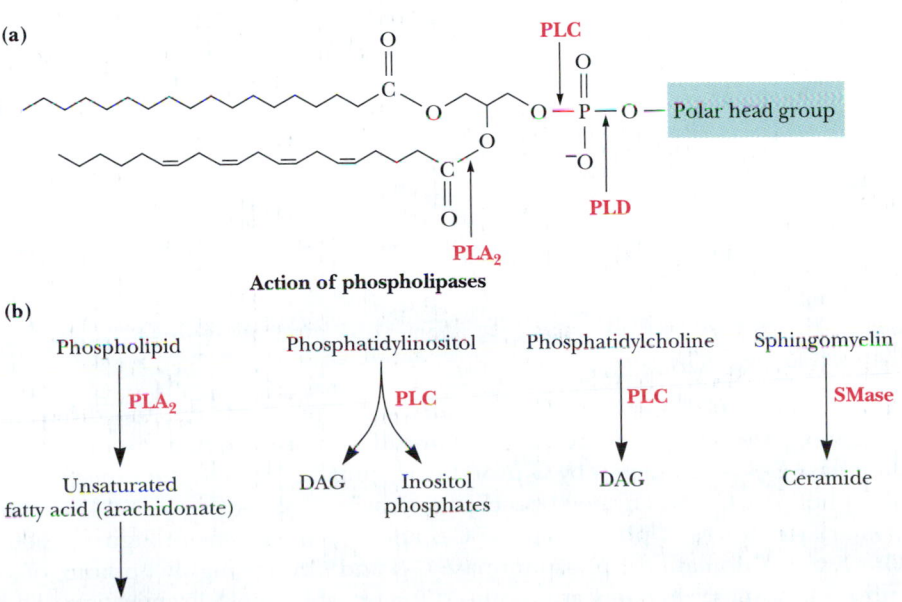

FIGURE 32.16 (a) The general action of phospholipase A$_2$ (PLA$_2$), phospholipase C (PLC), and phospholipase D (PLD). **(b)** The synthesis of second messengers from phospholipids by the action of phospholipases and sphingomyelinase.

Human Biochemistry

Cancer, Oncogenes, and Tumor Suppressor Genes

The disease state known as **cancer** is the uncontrolled growth and proliferation of one or more cell types in the body. Control of cell growth and division is an incredibly complex process, involving the signal-transducing proteins (and small molecules) described in this chapter and many others like them. The genes that give rise to these growth-controlling proteins are of two distinct types:

1. **Oncogenes:** These genes code for proteins that are capable of stimulating cell growth and division. In normal tissues and organisms, such growth-stimulating proteins are regulated so that growth is appropriately limited. However, mutations in these genes may result in loss of growth regulation, leading to uncontrolled cell proliferation and tumor development. These mutant genes are known *as oncogenes* because they induce the oncogenic state—cancer. The normal versions of these genes are termed **proto-oncogenes;** proto-oncogenes are essential for normal cell growth and differentiation. Oncogenes are *dominant,* because mutation of only one of the cell's two copies of the gene can lead to tumor formation. Table A lists a few of the known oncogenes (more than 60 are now known).

2. **Tumor suppressor genes:** These genes code for proteins whose normal function is to *turn off* cell growth. A mutation in one of these growth-limiting genes may result in a protein product that has lost its growth-limiting ability. Since the normal products suppress tumor growth, the genes are known as *tumor suppressor genes.* Because both cellular copies of a tumor suppressor gene must be mutated to foil its growth-limiting action, these genes are *recessive* in nature. Table B presents several recognized tumor suppressor genes.

Careful molecular analysis of cancerous tissue has shown that tumor development may result from mutations in several proto-oncogenes or tumor suppressor genes. The implication is that *there is redundancy in cellular growth regulation.* Many (if not all) tumors are either the result of interactions of two or more oncogene products or arise from simultaneous mutations in a proto-oncogene and both copies of a tumor suppressor gene. Cells have thus evolved with overlapping growth-control mechanisms. When one is compromised by mutation, others take over.

Table A

A Representative List of Proto-Oncogenes Implicated in Human Tumors

Proto-Oncogene	Neoplasm(s)
Abl	Chronic myelogenous leukemia
ErbB-1	Squamous cell carcinoma; astrocytoma
ErbB-2 (Neu)	Adenocarcinoma of breast, ovary, and stomach
Myc	Burkitt's lymphoma; carcinoma of lung, breast, and cervix
H-Ras	Carcinoma of colon, lung, and pancreas; melanoma
N-Ras	Carcinoma of genitourinary tract and thyroid; melanoma
Ros	Astrocytoma
Src	Carcinoma of colon
Jun / Fos	Several

Adapted from Bishop, J. M., 1991. Molecular themes in oncogenesis. *Cell* **64:**235–248.

Table B

Representative Tumor Suppressor Genes Implicated in Human Tumors

Tumor Suppressor Gene	Neoplasm(s)
RB1	Retinoblastoma; osteosarcoma; carcinoma of breast, bladder, and lung
p53	Astrocytoma; carcinoma of breast, colon, and lung; osteosarcoma
WT1	Wilms' tumor
DCC	Carcinoma of colon
NF1	Neurofibromatosis type 1
FAP	Carcinoma of colon
MEN-1	Tumors of parathyroid, pancreas, pituitary, and adrenal cortex

Adapted from Bishop, J. M., 1991. Molecular themes in oncogenesis. *Cell* **64:**235–248.

Activation of Phospholipase C Is Mediated by G Proteins or by Tyrosine Kinases

Phospholipase C-β, C-γ, and C-δ are all Ca^{2+}-dependent, but the different phospholipase C isozymes are activated by different intracellular events. Phospholipase C-β is stimulated by G proteins (Figure 32.18). On the other hand, phospholipase C-γ is activated by **receptor tyrosine kinases** (Figure 32.19). The primary structures of phospholipase C-β and C-γ are shown in Figure 32.20. The X and Y domains of phospholipase C-β and C-γ are highly homologous, and both of these domains are required for phospholipase C activation. The other domains of these isozymes confer specificity for G-protein activation or tyrosine kinase activation.

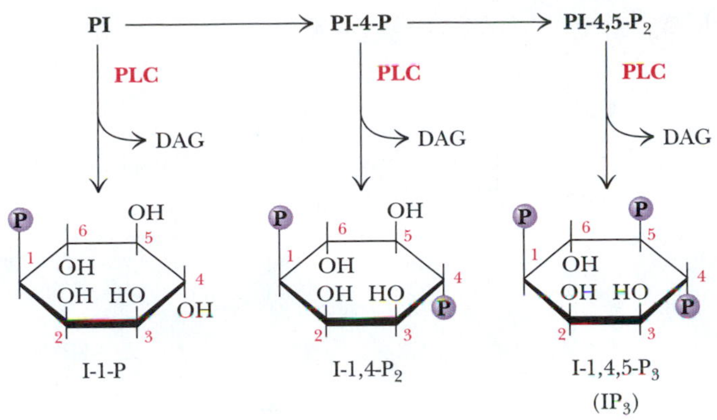

PI $\longrightarrow$ PI-4-P $\longrightarrow$ PI-4,5-P$_2$

PLC   PLC   PLC

DAG   DAG   DAG

I-1-P   I-1,4-P$_2$ I-1,4,5-P$_3$
(IP$_3$)

FIGURE 32.17 The family of second messengers produced by phosphorylation and breakdown of phosphatidylinositol. PLC action instigates a bifurcating pathway culminating in two distinct and independent second messengers: DAG and IP$_3$.

Phosphatidylcholine, Sphingomyelin, and Glycosphingolipids Also Generate Second Messengers

In addition to PI, other phospholipids serve as sources of second messengers. Breakdown of phosphatidylcholine by phospholipases yields a variety of second messengers, including DAG, phosphatidic acid, and prostaglandins. The action of **sphingomyelinase** on sphingomyelin produces **ceramide,** which stimulates **ceramide-activated protein kinase.** Similarly, gangliosides (such as ganglioside G$_{M3}$; see Chapter 8) and their breakdown products modulate the activity of protein kinases and G-protein–coupled receptors.

Calcium Is a Second Messenger

Calcium ion is an important intracellular signal. Binding of certain hormones and signal molecules to plasma membrane receptors can cause transient increases in cytoplasmic Ca^{2+} levels, which in turn can activate a wide variety of enzymatic processes, including smooth muscle contraction, exocytosis, and glycogen metabolism. (Most of these activation processes depend on special Ca^{2+}-binding proteins discussed in the following section.) Cytoplasmic [Ca^{2+}] can be increased in two ways (Figure 32.21). As mentioned briefly earlier, cAMP can activate the opening of plasma membrane Ca^{2+} channels, allowing extracellular Ca^{2+} to stream in. On the other hand, cells also contain intracellular reservoirs of Ca^{2+}, within the endoplasmic reticulum and **calciosomes,** small membrane vesicles that are similar in some ways to muscle sarcoplasmic reticulum. These special intracellular Ca^{2+} stores are *not* released by cAMP. They respond to IP$_3$, a second messenger derived from PI.

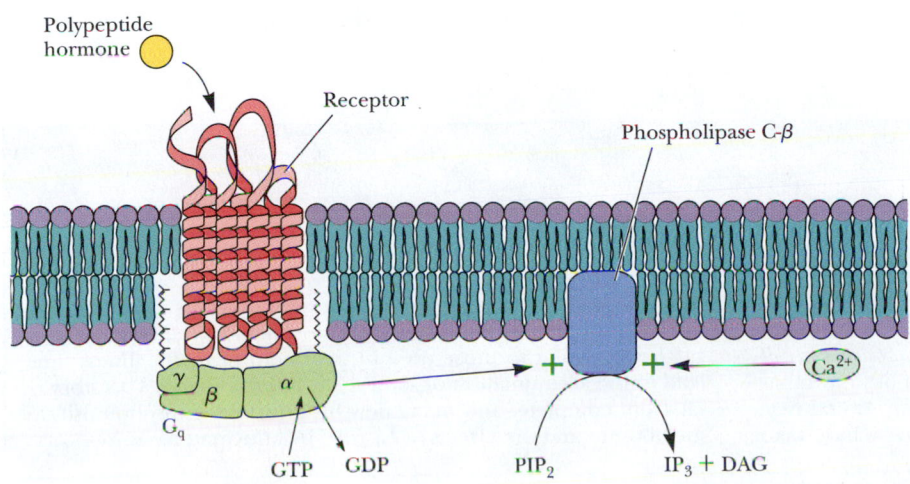

FIGURE 32.18 Phospholipase C-β is activated specifically by G$_q$, a GTP-binding protein, and also by Ca^{2+}.

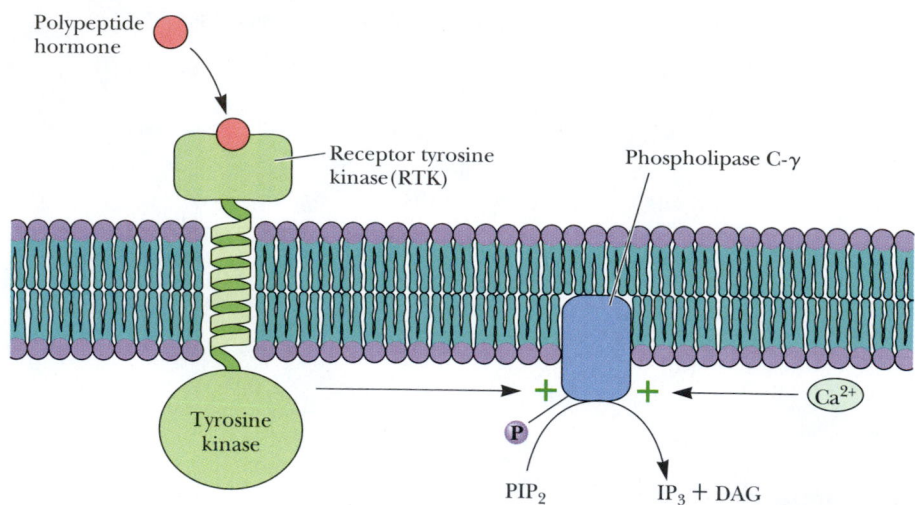

FIGURE 32.19 Phospholipase C-γ is activated by receptor tyrosine kinases and by Ca²⁺.

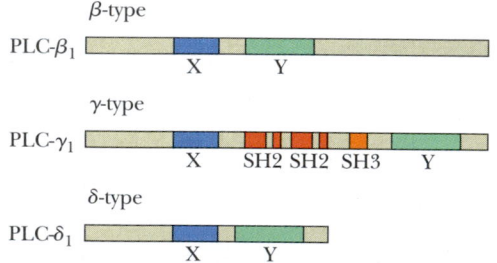

FIGURE 32.20 The amino acid sequences of phospholipase C isozymes β, γ, and δ share two homologous domains, denoted X and Y. The sequence of γ-isozyme contains src homology domains, denoted SH2 and SH3. SH2 domains (approximately 100 residues in length) interact with phosphotyrosine-containing proteins (such as RTKs), whereas SH3 domains mediate interactions with cytoskeletal proteins. *(Adapted from Dennis, E., Rhee, S., Gillah, M., and Hannun, E., 1991. Role of phospholipases in generating lipid second messengers in signal transduction. The FASEB Journal 5:2068–2077.)*

Intracellular Calcium-Binding Proteins Mediate the Calcium Signal

Given the central importance of Ca²⁺ as an intracellular messenger, it should not be surprising that complex mechanisms exist in cells to manage and control Ca²⁺. When Ca²⁺ signals are generated by cAMP, IP₃, and other agents, these signals are translated into the desired intracellular responses by **calcium-binding proteins,** which in turn regulate many cellular processes (Figure 32.22). One of these, protein kinase C, is described in Section 32.5. The other important Ca²⁺-binding proteins can, for the most part, be divided into two groups on the basis of structure and function: (1) the **calcium-modulated proteins,** including **calmodulin, parvalbumin, troponin C,** and many others, all of which have in common a structural feature called the **EF hand** (Figure 5.26), and (2) the **annexin proteins,** a family of homologous proteins that interact with membranes and phospholipids in a Ca²⁺-dependent manner.

More than 170 calcium-modulated proteins are known (Table 32.4). All possess a characteristic peptide domain consisting of a short α-helix, a loop of 12 amino acids, and a second α-helix. Robert Kretsinger at the University of Virginia initially discovered this pattern in parvalbumin, a protein first identified in the carp fish and later in neurons possessing a high firing rate and a high oxidative metabolism. Kretsinger lettered the six helices of parvalbumin A through F. He noticed that the E and F helices, joined by a loop, resembled the thumb and forefinger of a right hand (see Figure 6.26) and named this structure the *EF hand*, a name in common use today to identify the helix-loop-helix motif in calcium-binding proteins. In the EF hand, Ca²⁺ is coordinated by six carboxyl oxygens contributed by a glutamate and three aspartates, by a carbonyl oxygen from a peptide bond, and by the oxygen of

Human Biochemistry

PI Metabolism and the Pharmacology of Li⁺

An intriguing aspect of the phosphoinositide story is the specific action of lithium ion, Li⁺, on several steps of PI metabolism. Lithium salts have been used in the treatment of *manic-depressive illnesses* for more than 30 years, but the mechanism of lithium's therapeutic effects had been unclear. Recently, however, several reactions in the phosphatidylinositol degradation pathway have been shown to be sensitive to Li⁺ ion. For example, Li⁺ is an uncompetitive inhibitor of *myo*-inositol monophosphatase (see Chapter 13). Li⁺ levels similar to those used in treatment of manic illness thus lead to the accumulation of several key intermediates. This story is far from complete, and many new insights into phosphoinositide metabolism and the effects of Li⁺ can be anticipated.

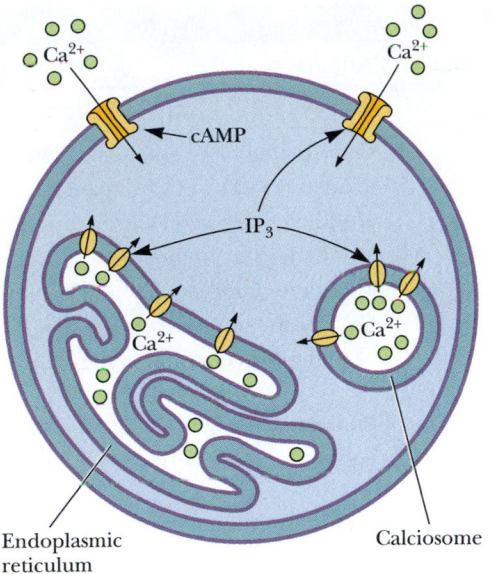

Endoplasmic reticulum

Calciosome

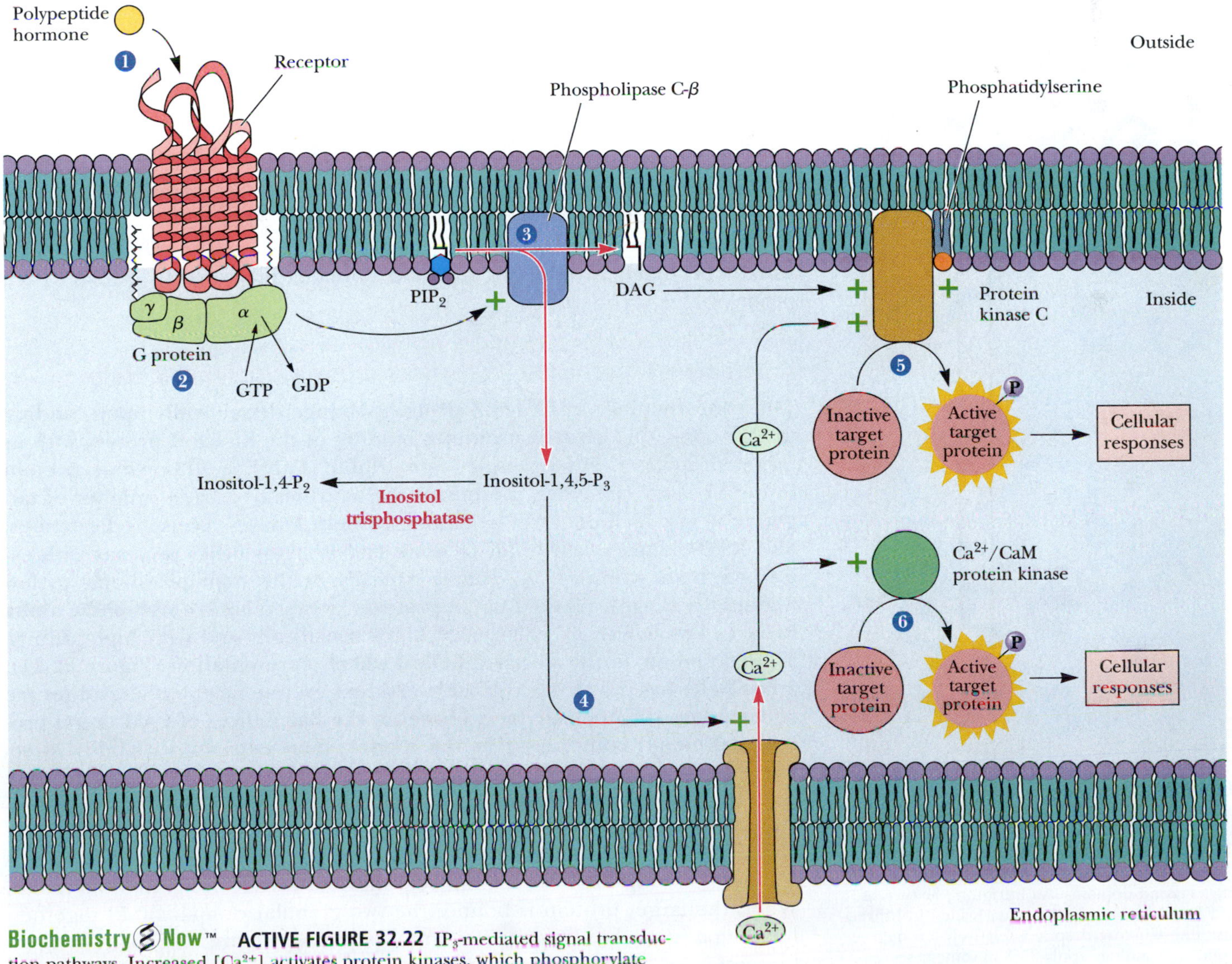

Polypeptide hormone ①

Receptor

Phospholipase C-β

Phosphatidylserine

Outside

G protein
γ β α
② GTP GDP

PIP₂ ③ +

DAG

Protein kinase C

Inside

⑤

Inactive target protein → Active target protein P → Cellular responses

Ca²⁺

Inositol-1,4-P₂ ← Inositol-1,4,5-P₃

Inositol trisphosphatase

Ca²⁺/CaM protein kinase

⑥

Inactive target protein → Active target protein P → Cellular responses

④

Ca²⁺ +

Ca²⁺

Endoplasmic reticulum

(a)

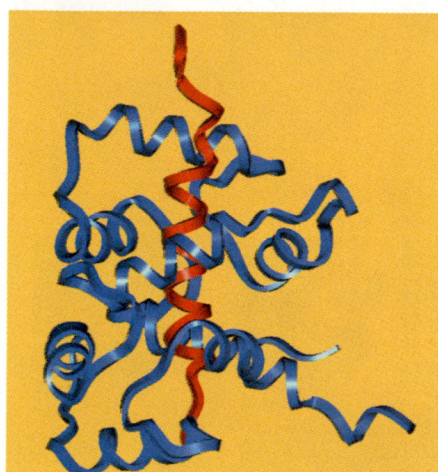

(b)

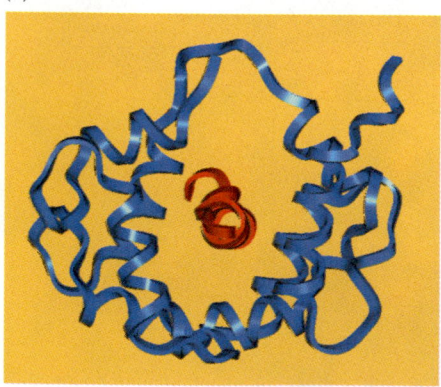

(c)

FIGURE 32.23 **(a)** Structure of uncomplexed cal- modulin (gold). Calmodulin, with four Ca²⁺-binding domains, forms a dumbbell-shaped structure with two globular domains joined by an extended, central helix. Each globular domain juxtaposes two Ca²⁺-binding EF-hand domains. An intriguing feature of these EF-hand domains is their nearly identical three-dimensional structure despite a relatively low degree of sequence homology (only 25% in some cases). **(b, c)** Complex of calmodulin (gold) with a peptide from myosin light chain kinase (white); (b) side view; (c) top view.

Table 32.4

Some Calcium-Modulated Proteins

Protein	Function
α-Actinin	Crosslinking of cytoskeletal F-actin
Calcineurin B	Protein Ser/Thr phosphatase
Calmodulin	Modulates activity of Ca²⁺-dependent proteins
Calretinin	Modulates Ca²⁺-dependent neural processes
Caltractin	Modulates Ca²⁺-sensitive contractile fibers
β- and γ-Crystallins	Ca²⁺-modulated processes in eye lens
Flagellar Ca²⁺-binding protein	Flagellar function and cell motility
Frequinin	Phototransduction in retinal cone cells
Inositol phospholipid-specific phospholipase C	Second messenger release and cell signaling
Myeloperoxidase	Inflammatory action of neutrophils
Parvalbumin	Acceleration of muscle relaxation, Ca²⁺ sequestration
S-100	Cell cycle progression, cell differentiation, cytoskeleton-membrane interactions
Thioredoxin reductase	Electron transfer processes in keratinocytes
Troponin C	Activation of muscle contraction

Adapted from Heizmann, C. W., ed., 1991. *Novel Calcium Binding Proteins—Fundamentals and Clinical Implications.* New York: Springer-Verlag.

a coordinated water molecule. The EF hand was subsequently identified in calmodulin, troponin C, and **calbindin-9K** (Figure 32.23). Most of the known EF-hand proteins possess two or more (as many as eight) EF-hand domains, usually arranged so that two EF-hand domains may directly contact each other.

Calmodulin Target Proteins Possess a Basic Amphiphilic Helix

The conformations of EF-hand proteins change dramatically upon binding of Ca²⁺ ions. This change promotes binding of the EF-hand protein with its target protein(s). For example, calmodulin (CaM), a 148-residue protein found in many cell types, modulates the activities of a large number of target proteins, including Ca²⁺-ATPases, protein kinases, phosphodiesterases, and NAD⁺ kinase. CaM binds to these and to many other proteins with extremely high affinities (K_D values typically in the high picomolar to low nanomolar range). All CaM target proteins possess a *basic amphiphilic alpha helix* (a **Baa helix**), to which CaM binds specifically and with high affinity. Viewed end-on, in the so-called **helical wheel** representation (Figure 32.24), a Baa helix has mostly hydrophobic residues on one face; basic residues are collected on the opposite face. However, the Baa helices of CaM target proteins, although conforming to the model, show extreme variability in sequence. How does CaM, itself a highly conserved protein, accommodate such variety of sequence and structure? Each globular domain consists of a large hydrophobic surface flanked by regions of highly negative electrostatic potential—a surface suitable for interacting with a Baa helix. The long central helix joining the two globular regions behaves as a long, flexible tether. When the target protein is bound, the two globular domains fold together, forming a single binding site for target peptides (Figure 32.23b). The flexible nature of the tethering helix allows the two globular domains to adjust their orientation synergistically for maximal binding of the target protein or peptide.

(a)

(b)

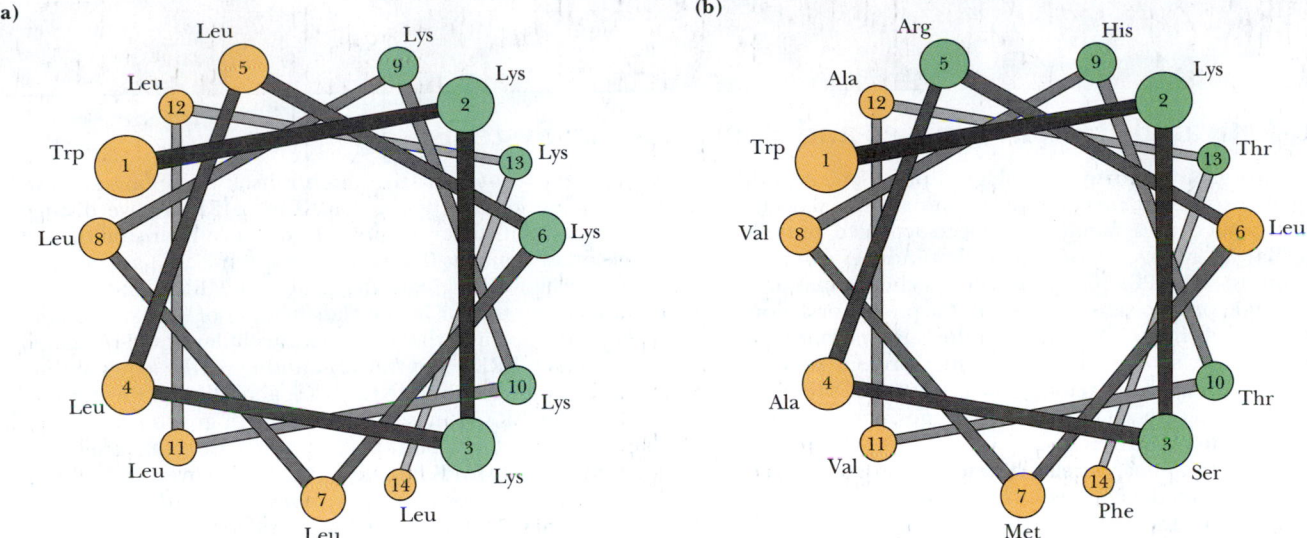

FIGURE 32.24 Helical wheel representations of **(a)** a model peptide, Ac-WKKLLKLLKKLLKL-CONH₂, and **(b)** the calmodulin-binding domain of spectrin. Positively charged and polar residues are indicated in green, and hydrophobic residues are orange. *(Adapted from O'Neil, K., and DeGrado, W., 1990. How calmodulin binds its targets: Sequence independent recognition of amphiphilic α-helices. Trends in Biochemical Sciences 15:59–64.)*

32.5 How Do Effectors Convert the Signals to Actions in the Cell?

Transduction of the hormonal signal leads to activation of **effectors**—usually protein kinases and protein phosphatases—that elicit a variety of actions that regulate discrete cellular functions. Of the thousands of mammalian kinases and phosphatases, the structures and functions of a few are representative.

Protein Kinase A Is a Paradigm of Kinases

Most protein kinases share a common catalytic core first characterized in protein kinase A (PKA), the enzyme that phosphorylates phosphorylase kinase (see A Deeper Look box on page 476 and Figure 15.18). The active site of the catalytic subunit of PKA in a ternary complex with MnAMP–PNP and a pseudo-substrate inhibitor peptide (Figure 32.25) includes a glycine-rich β-strand that acts as a flap over the triphosphate moiety of the bound nucleotide. A conserved

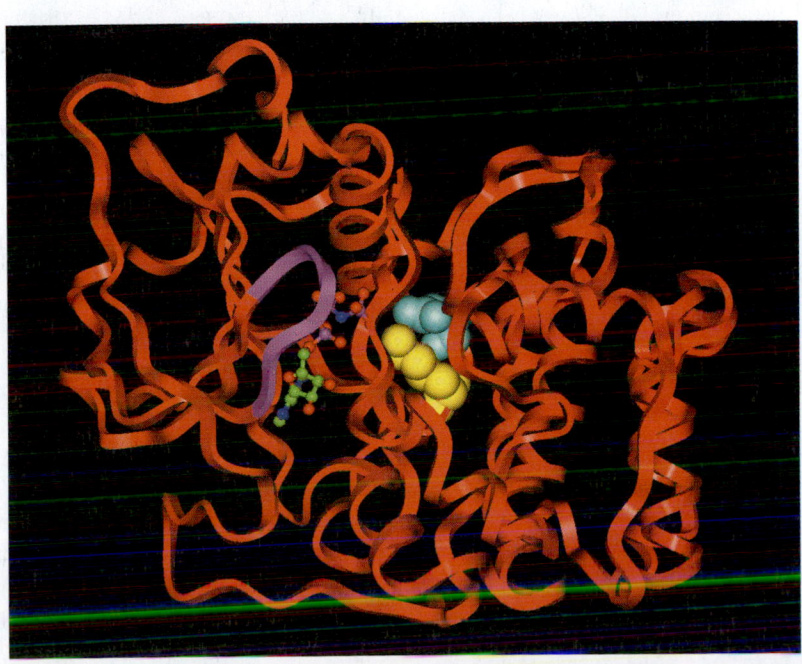

FIGURE 32.25 The structure of the catalytic subunit of PKA in a ternary complex with MnAMP–PNP and a pseudosubstrate inhibitor peptide. A glycine-rich β-strand acts as a flap over the triphosphate moiety of the bound nucleotide. The glycine-rich flap that covers the ATP-binding site is shown in magenta, AMP–PNP is bound in the ATP site, Asp[166] is shown in blue, and Lys[168] is shown in yellow.

A Deeper Look

Mitogen-Activated Protein Kinases and Phosphorelay Systems

In multicellular organisms, many physiological processes, including mitosis, gene expression, metabolism, and programmed death of cells, are regulated by a family of **mitogen-activated protein kinases (MAPKs).** (A mitogen is any agent that induces cell division, that is, mitosis.) MAPKs phosphorylate specific serines and threonines of target protein substrates, and these phosphorylation events function as switches to turn on or off the activity of the substrate proteins. These "substrates" may be other protein kinases, phospholipases, transcription factors, and cytoskeletal proteins. Protein phosphatases reverse the process, removing the phosphates that were added by MAPKs.

MAPKs are part of a **phosphorelay system** composed of three kinases that are activated in sequence (see accompanying figure). In such systems, the MAPK itself is phosphorylated by a

MAPK kinase (denoted MKK), which is itself phosphorylated by a MAPK kinase kinase (denoted MKKK). MKKKs have distinct domains and motifs that recognize different cellular stimuli, and they have other domains that recognize specific MKKs. The same kinds of specificities regulate the action of MKKs. These specificities are accounted for in the classification of four subfamilies of MAPKs: One group is that of the extracellular signal-regulated kinases, notably ERK1 and ERK2; another is the c-Jun-amino-terminal kinases, including JNK, JNK1, and JNK2; a third group depends on the ERK5 kinase; and the fourth group involves the p38 kinases, including p38α, p38β, p38γ, and p38δ. There are undoubtedly other MAPK families yet to be discovered. As shown in the figure, these phosphorelay systems link a variety of stimuli to substrates that affect many cellular functions.

▶ MAPK phosphorelay systems. The left column is a general model. The four columns to the right show the four known MAPK phosphorelay families.

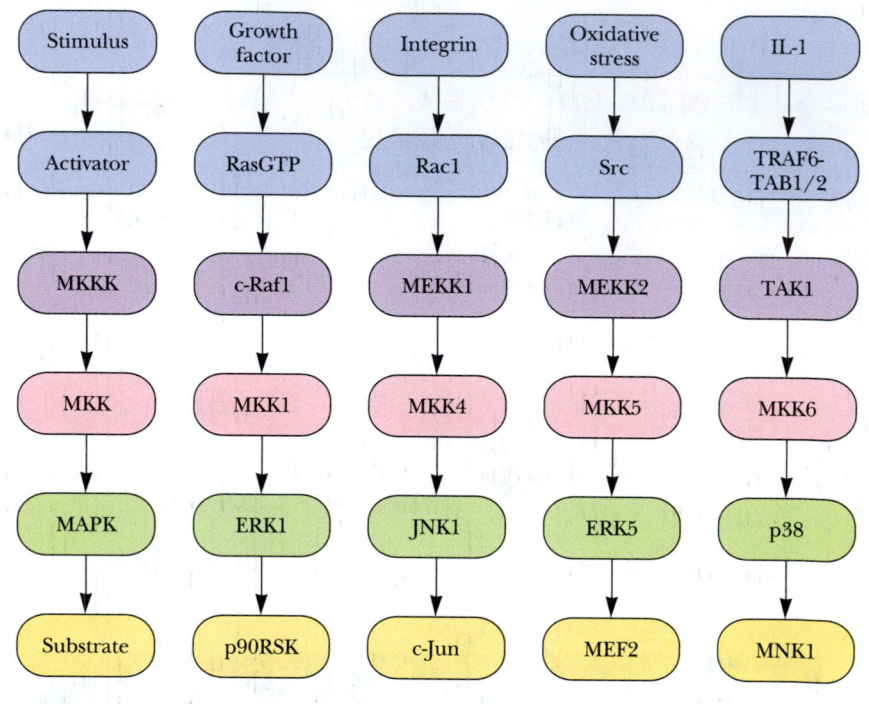

From Johnson, G. L., and Lapadat, R., 2002. Mitogen-activated protein kinase pathways mediated by ERK, JNK, and p38 protein kinases. *Science* **298:**1911–1912.

residue, Asp[166], is the catalytic base that deprotonates the Ser/Thr-OH during phosphorylation, and Lys[168] stabilizes the transition state of the reaction. Three Glu residues on the enzyme are involved in recognition of the pseudosubstrate inhibitor peptide.

Biochemistry⑤Now™ Go to BiochemistryNow and click BiochemistryInteractive to explore the structure and function of protein kinase C.

Protein Kinase C Is a Family of Isozymes

The enzymes called *protein kinase C* are actually a family of similar enzymes—isozymes—that encompass three subclasses. The "conventional PKCs," α, βI, βII, and γ, are regulated by Ca^{2+}, diacylglycerol (DAG), and phosphatidylserine (PS). Because Ca^{2+} levels increase in the cell in response to IP_3, the activation of conventional PKCs depends on both of the second messengers released by the hydrolysis of PIP_2. The "novel PKCs," δ, ε, θ, and η, are Ca^{2+}-independent but are regulated by DAG and PS. PKCs ζ, ι, and λ are termed "atypical" and are

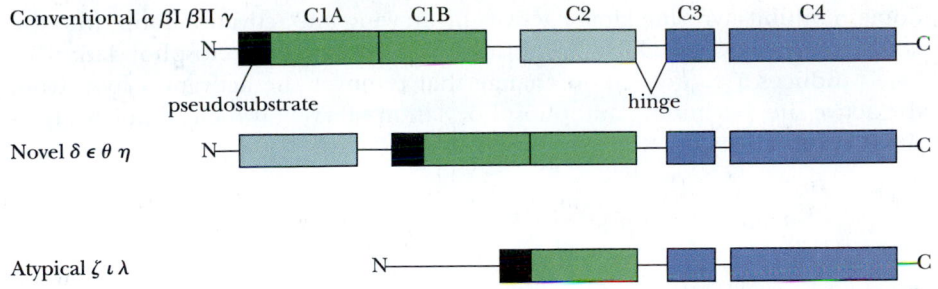

FIGURE 32.26. The primary structures of the PKC isozymes. Conserved domains C1–C4 are indicated. Variable regions are shown as simple lines.

activated by PS alone. These various cofactor requirements are imparted by subdomains represented in the conventional PKC polypeptide sequence. Conventional PKCs are comprised (Figure 32.26) of four conserved domains (C1–C4) and five variable regions (V1–V5). Domain CI is a *pseudosubstrate sequence* that regulates the kinase by *intrasteric control* (see A Deeper Look box on page 476), C2 is a Ca^{2+}-binding domain, C3 is the ATP-binding domain, and C4 binds peptide substrates.

PKC phosphorylates serine and threonine residues on a wide range of protein substrates. A role for protein kinase C in cellular growth and division is demonstrated by its strong activation by **phorbol esters** (Figure 32.27). These compounds, from the seeds of *Croton tiglium*, are **tumor promoters**—agents that do not themselves cause tumorigenesis but that potentiate the effects of carcinogens. The phorbol esters mimic DAG, bind to the regulatory pseudosubstrate domain of the enzyme, and activate protein kinase C.

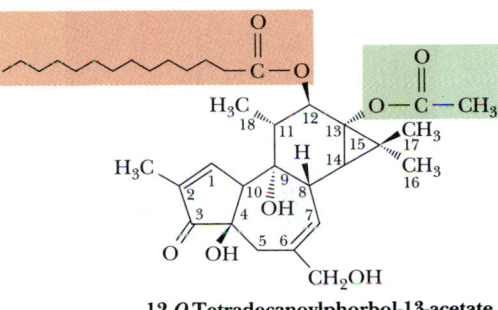

12-*O*-Tetradecanoylphorbol-13-acetate

FIGURE 32.27 The structure of a phorbol ester. Long-chain fatty acids predominate at the 12-position, whereas acetate is usually found at the 13-position.

Protein Tyrosine Kinase pp60[c-src] Is Regulated by Phosphorylation/Dephosphorylation

The structure of protein tyrosine kinase pp60[c-src] (see also Figure 32.9b) consists of an N-terminal "unique domain," an SH2 domain, an SH3 domain, and a kinase domain that includes a small lobe comprised mainly of a twisted β-sheet and a large lobe that is predominantly α-helical (Figure 32.28). (SH2 and SH3 domains are discussed in Section 32.6.) Phosphorylation of Tyr[527] in the SH2

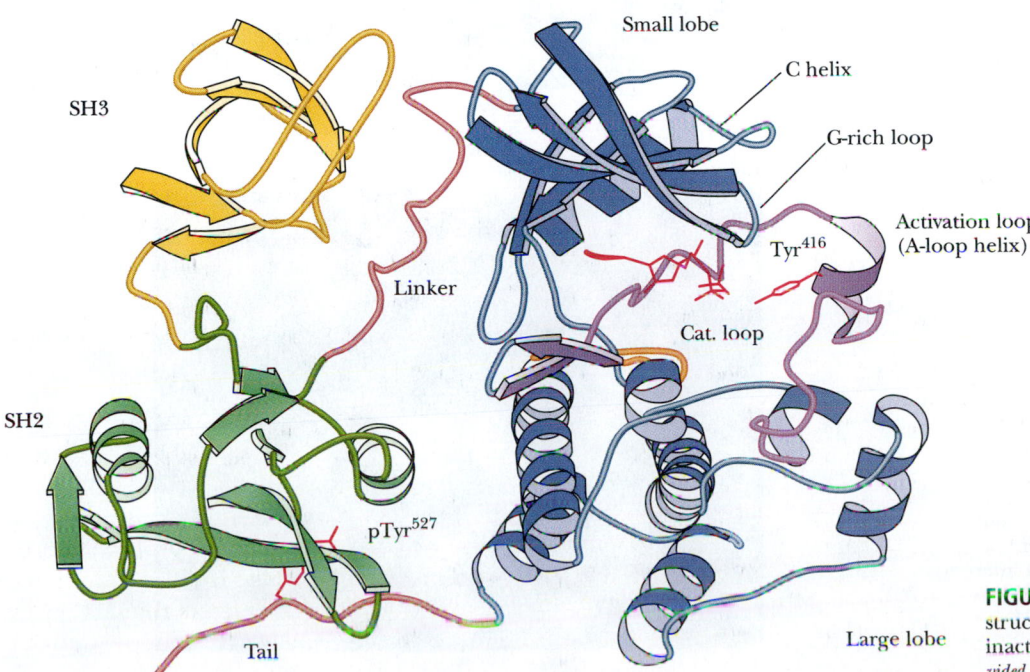

FIGURE 32.28 A ribbon diagram showing the structure of protein tyrosine kinase pp60[c-src] in its inactive state bound to AMP–PNP. (*Image kindly provided by Stephen C. Harrison.*)

domain inhibits tyrosine kinase activity by drawing an "activation loop" into the active site, blocking ATP and/or substrate binding. Dephosphorylation of Tyr[527] induces a conformation change that removes the activation loop from the active site, permitting autophosphorylation of Tyr[416], which stimulates tyrosine kinase activity.

Protein Tyrosine Phosphatase SHP-2 Is a Nonreceptor Tyrosine Phosphatase

The human phosphatase SHP-2 is a cytosolic nonreceptor tyrosine phosphatase. It comprises two SH2 domains, a catalytic phosphatase domain, and a C-terminal tail domain. The SH2 domains enable the enzyme to bind to its target substrates, and they also regulate the phosphatase activity. The catalytic domain of SHP-2 consists of nine α-helices and a ten-stranded mixed β-sheet that wraps around one of the helices (Figure 32.29) The other eight helices pack together on the opposite side of the β-sheet.

The N-terminal SH2 domain regulates phosphatase activity by binding to the phosphatase domain and directly blocking the active site. When a target peptide containing a phosphotyrosine group binds to the SH2 domain, a conformation change causes this domain to dissociate from the catalytic domain, exposing the active site and allowing peptide substrate to bind. Binding of phosphotyrosine-containing peptide to the second SH2 domain provides additional activation. Target peptides with two phosphotyrosines (one to bind to each SH2 domain) provide maximal activation of the phosphatase activity.

Biochemistry Now™ Go to BiochemistryNow and click BiochemistryInteractive to learn more about the human phosphatase SHP-2, a cystolic nonreceptor tyrosine phosphate.

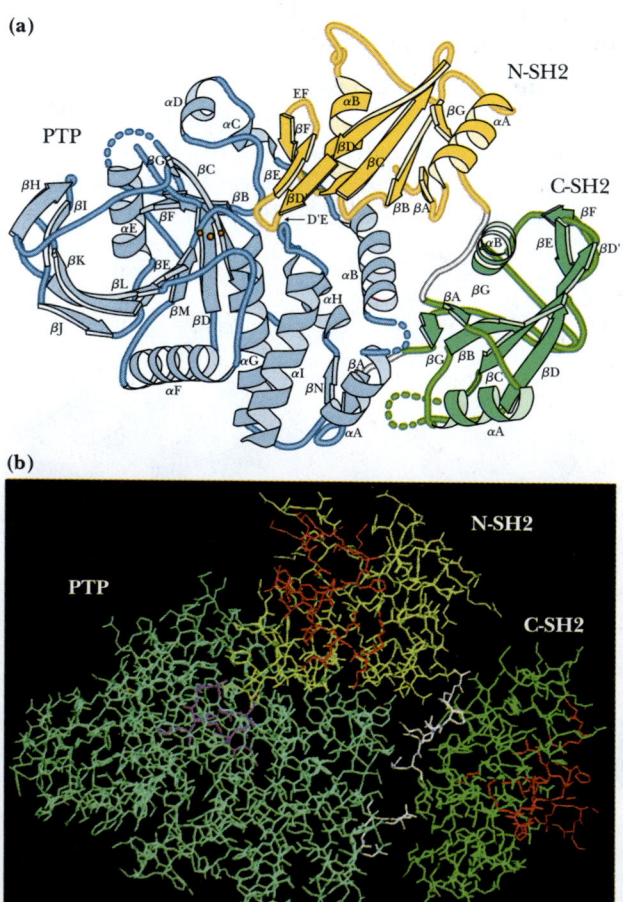

FIGURE 32.29 **(a)** A ribbon diagram showing the structure of protein tyrosine phosphatase SHP-2 in its autoinhibited, closed conformation. The N- and C-terminal SH2 domains are yellow and green, respectively. The catalytic domain is blue, and inter-domain linkers (residues 104–111 and 217–220) are white. **(b)** A wireframe model of SHP-2. *(Image kindly provided by Stephen C. Harrison.)*

| 32.6 | **What Is the Role of Protein Modules in Signal Transduction?** |

Signal transduction within cells occurs via protein–protein and protein–phospholipid interactions based on protein modules. Proteins with two (or more) such modules associate simultaneously with two (or more) binding partners, leading to assembly of functional complexes, either at an activated cell surface receptor or free in the cytoplasm. The SH2 domain, which binds with high affinity to peptide motifs containing a phosphotyrosine, is a good example (Figure 32.30). The binding of a particular SH2 domain to a particular phosphotyrosine motif depends on the particular sequence of residues that are C-terminal to the phosphotyrosine. The SH2 domain itself is a module of about 100 residues that consists of a small β-sheet flanked by α-helices.

SH2 domains are the prototype for a growing number of protein modules that play a role in cell signaling. **PTB** modules also recognize and bind to phosphotyrosine motifs, but in a manner different from that of SH2 domains. **SH3** and **WW** modules bind proline-rich target sequences, and **PDZ** modules bind to the terminal four or five residues of a target protein. The **PH** (pleckstrin homology) module, with a fold similar to the PTB module, functions quite differently, associating with specific phosphoinositides and directing target proteins to the plasma membrane. Figure 32.30 illustrates structures of these modules, together with examples of proteins that contain them.

Biochemistry Now™   Go to BiochemistryNow and click BiochemistryInteractive to explore and interact with several protein modules of signaling proteins.

A Deeper Look

Whimsical Names for Proteins and Genes

The study of cell signaling and the identification of hundreds of new signaling proteins provided an unprecedented creative opportunity for cell biologists and geneticists in the naming of these proteins. In the early days of molecular biology, such names were typically arcane abbreviations and acronyms. One such case is the family of **14-3-3 proteins,** named for the migration patterns of these proteins on DEAE-cellulose chromatography and starch-gel electrophoresis. In the 1970s, a few creative scientists suggested whimsical names for newly discovered genes, such as *sevenless,* named in reference to R7, one of the eight photoreceptor cells in the compound eye of *Drosophila,* the common fruit fly. What began as a trickle became a torrent of whimsical names for proteins and genes. *Sevenless* was followed by *bride of sevenless* (*boss,* a ligand of sevenless), and *son of sevenless* (*sos,* first isolated in a genetic screen of the sevenless receptor tyrosine kinase pathway in *Drosophila*). The *hedgehog* (*hh*) genes, including *sonic hedgehog (Shh),* play critical roles in the development and patterning of vertebrate embryonic tissues but were named for a popular video game.

The accompanying table lists a few notable examples of whimsically named genes and gene products, many of which were first identified in *Drosophila.*

Name	Role or Function
Armadillo	Plakoglobin $= \beta$-catenin
Bag of marbles	Novel protein involved in oogenesis and spermatogenesis
Bullwinkle	Oocyte protein
Cactus	Signaling protein—IkB homolog
Cheap date	Alcohol sensitivity
Chickadee	Profilin homolog—regulation of actin cytoskeleton
Corkscrew	A protein tyrosine phosphatase
Dachshund	Novel nuclear protein of unknown function
Dishevelled	Novel cytoplasmic protein in the wingless pathway
Dunce	A cAMP phosphodiesterase
Hopscotch	A Janus family tyrosine kinase
Naked	A segment polarity gene
Reaper	A death-domain protein functioning in apoptosis
Rutabaga	A Ca²⁺/calmodulin-dependent protein kinase
Shark	A tyrosine kinase SH2-nonreceptor
Yak	Literally "yet another kinase"

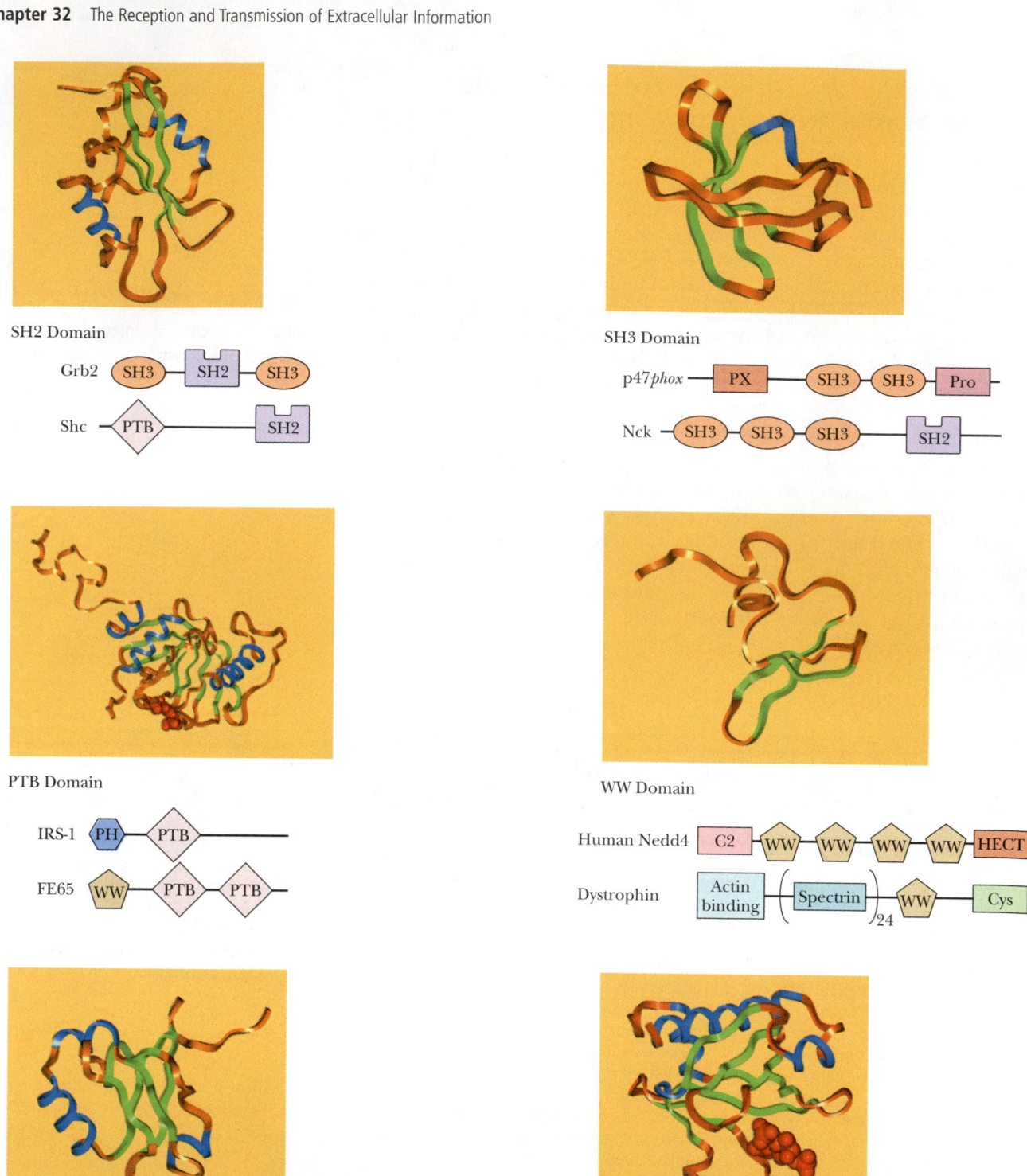

FIGURE 32.30 Six of the protein modules that are found in cell-signaling proteins. Shown for each are a molecular graphic image of the module, together with primary structures of several proteins in which they are found. (*WW domain coordinates kindly provided by Harmut Oschkinat, Forschungsinstitut für Molekulare Pharmakologie, and Marius Sudol, Mount Sinai School of Medicine.*)

Protein Scaffolds Localize Signaling Molecules

Membrane-bound receptors can amplify their signaling by means of adaptor proteins that provide docking sites for signaling modules on other proteins. Such docking proteins typically possess an N-terminal sequence that targets the docking protein to the membrane (for example, a PH domain or a myristoylation site) and a PTB domain that enables the docking protein to bind to a phosphorylated tyrosine on a receptor, as well as additional modules and phosphorylation sites that facilitate the binding of target proteins. A typical case is **IRS-1** (insulin receptor substrate-1), a substrate of the insulin receptor (see Figure 5.35). IRS-1 has an N-terminal PH domain followed by a PTB domain and 18 potential tyrosine phosphorylation sites. The PH and PTB domains direct IRS-1 to the membrane, facilitating tyrosine phosphorylation of IRS-1 by the insulin receptor tyrosine kinase and subsequent mediation of additional cell signaling events. Scaffolding proteins can thus assemble sequential components of a signaling pathway.

32.7 How Do Neurotransmission Pathways Control the Function of Sensory Systems?

The survival of higher organisms is predicated on the ability to respond rapidly to sensory input from physical signals (sights, sounds) and chemical cues (smells). The responses to such stimuli may include muscle movements and many forms of intercellular communication. Hormones (as described earlier in this chapter) can move through an organism only at speeds determined by the circulatory system. In most higher organisms, a faster means of communication is crucial. Nerve impulses, which can be propagated at speeds up to 100 m/sec, provide a means of intercellular signaling that is fast enough to encompass sensory recognition, movement, and other physiological functions and behaviors in higher animals. The generation and transmission of nerve impulses in vertebrates is mediated by an incredibly complicated neural network that connects every part of the organism with the brain—itself an interconnected array of as many as 10^{12} cells.

Despite their complexity and diversity, the nervous systems of animals all possess common features and common mechanisms. Physical or chemical stimuli are recognized by specialized **receptor proteins** in the membranes of **excitable cells.** Conformational changes in the receptor protein result in a change in enzyme activity or a change in the permeability of the membrane. These changes are then propagated throughout the cell or from cell to cell in specific and reversible ways to carry information through the organism. This section describes the characteristics of excitable cells and the mechanisms by which these cells carry information at high speeds through an organism.

Nerve Impulses Are Carried by Neurons

Neurons and **neuroglia** (or **glial cells**) are cell types unique to nervous systems. The reception and transmission of nerve impulses are carried out by neurons (Figure 32.31), whereas glial cells serve protective and supportive functions. (*Neuroglia* could be translated as "nerve glue.") Glial cells differ from neurons in several ways. Glial cells do not possess axons or synapses, and they retain the ability to divide throughout their life spans. Glial cells outnumber neurons by at least 10 to 1 in most animals.

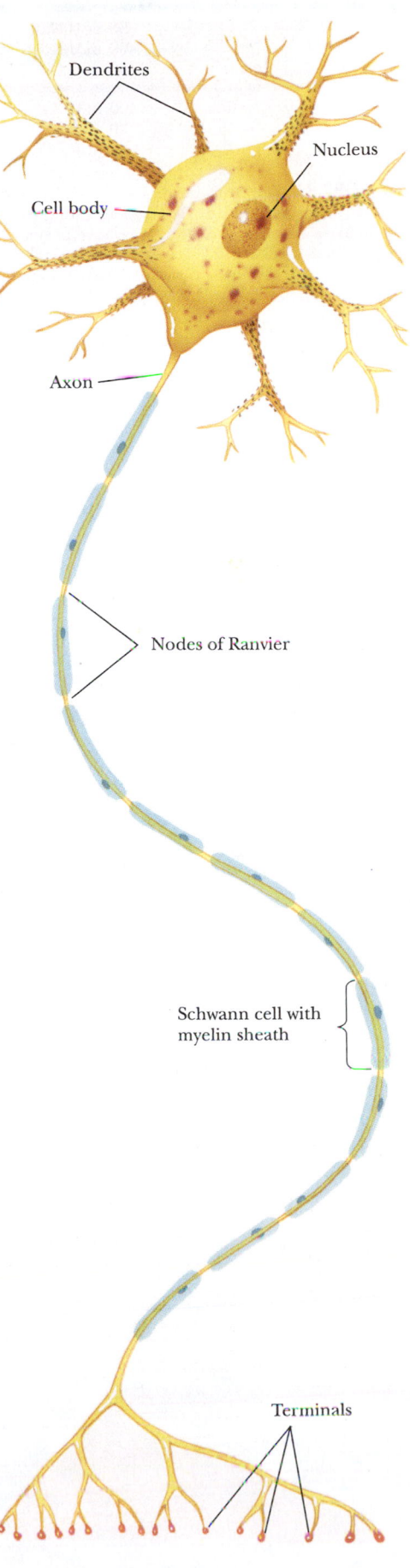

▶ **FIGURE 32.31** The structure of a mammalian motor neuron. The nucleus and most other organelles are contained in the cell body. One long axon and many shorter dendrites project from the body. The dendrites receive signals from other neurons and conduct them to the cell body. The axon transmits signals from this cell to other cells via the synaptic knobs. Glial cells called Schwann cells envelop the axon in layers of an insulating myelin membrane. Although glial cells lie in proximity to neurons in most cases, no specific connections (such as gap junctions, for example) connect glial cells and neurons. However, gap junctions can exist between adjacent glial cells.

Axon		50 mM Na⁺
	Inside	400 mM K⁺
		60 mM Cl⁻
	Outside	400 mM Na⁺
		20 mM K⁺
		560 mM Cl⁻

FIGURE 32.32 The concentrations of Na^+, K^+, and Cl^- ions inside and outside of a typical resting mammalian axon are shown. Assuming relative permeabilities for K^+, Na^+, and Cl^- are 1, 0.04, and 0.45, respectively, the Goldman equation yields a membrane potential of -60 mV. (See problem 14, page 1085.)

Neurons are distinguished from other cell types by their long cytoplasmic extensions or projections, called **processes.** Most neurons consist of three distinct regions (see Figure 32.31): the **cell body** (called the **soma**), the **axon,** and the **dendrites.** The axon ends in small structures called **synaptic terminals, synaptic knobs,** or **synaptic bulbs.** Dendrites are short, highly branched structures emanating from the cell body that receive neural impulses and transmit them to the cell body. The space between a synaptic knob on one neuron and a dendrite ending of an adjacent neuron is the **synapse** or **synaptic cleft.**

Three kinds of neurons are found in higher organisms: sensory neurons, interneurons, and motor neurons. **Sensory neurons** acquire sensory signals, either directly or from specific receptor cells, and pass this information along to either **interneurons** or **motor neurons.** Interneurons simply pass signals from one neuron to another, whereas motor neurons pass signals from other neurons to muscle cells, thereby inducing muscle movement (motor activity).

Ion Gradients Are the Source of Electrical Potentials in Neurons

The impulses that are carried along axons, as signals pass from neuron to neuron, are electrical in nature. *These electrical signals occur as transient changes in the electrical potential differences (voltages) across the membranes of neurons (and other cells). Such potentials are generated by* **ion gradients.** The cytoplasm of a neuron at rest is low in Na^+ and Cl^- and high in K^+, relative to the extracellular fluid (Figure 32.32). These gradients are generated by the Na^+,K^+-ATPase (see Chapter 9). A resting neuron exhibits a potential difference of approximately -60 mV (that is, negative inside).

Action Potentials Carry the Neural Message

Nerve impulses, also called **action potentials,** are transient changes in the membrane potential that move rapidly along nerve cells. Action potentials are created when the membrane is locally **depolarized** by approximately 20 mV—from the resting value of about -60 mV to a new value of approximately -40 mV. This small change is enough to have a dramatic effect on specific proteins in the axon membrane called **voltage-gated ion channels.** These proteins are ion channels that are specific either for Na^+ or K^+. These ion channels are normally closed at the resting potential of -60 mV. When the potential difference rises to -40 mV, the "gates" of the Na^+ channels are opened and Na^+ ions begin to flow into the cell. As Na^+ enters the cell, the membrane potential continues to increase and additional Na^+ channels are opened (Figure 32.33). The potential rises to more than $+30$ mV. At this point, Na^+ influx slows and stops. As the Na^+ channels close, K^+ channels begin to open and K^+ ions stream out of the cell, returning the membrane potential to negative values. The potential eventually overshoots its resting value a bit. At this point, K^+ channels close and the resting potential is eventually restored by action of the Na^+,K^+-ATPase and the other channels. Alan Hodgkin and Andrew Huxley originally observed these transient increases and decreases, first in Na^+ permeability and then in K^+ permeability. For this and related work, Hodgkin and Huxley, along with J. C. Eccles, won the Nobel Prize in Physiology or Medicine in 1963.

The Action Potential Is Mediated by the Flow of Na⁺ and K⁺ Ions

These changes in potential in one part of the axon are rapidly passed along the axonal membrane (Figure 32.34). The sodium ions that rush into the cell in one localized region actually diffuse farther along the axon, raising the Na^+ concentration *and* depolarizing the membrane, causing Na^+ gates to open in that adjacent region of the axon. In this way, the action potential moves down

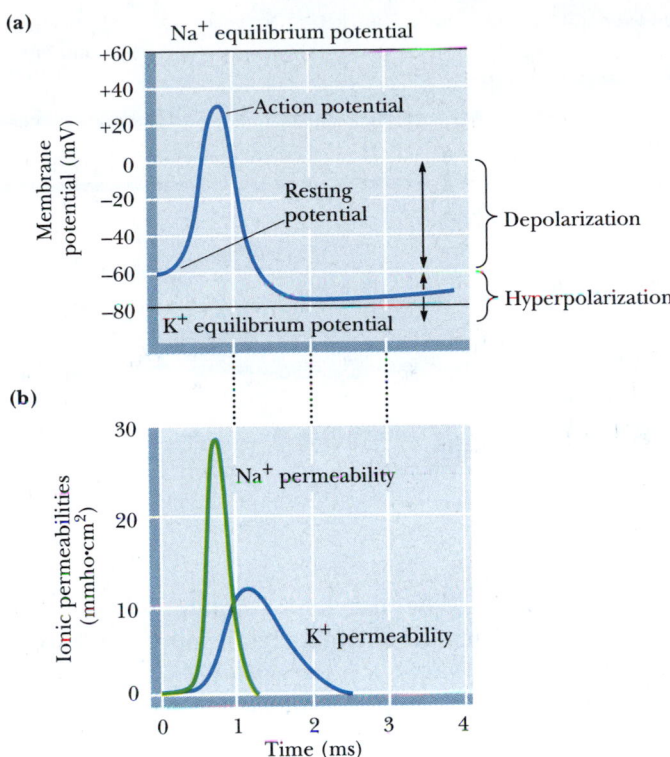

FIGURE 32.33 The time dependence of an action potential, compared with the ionic permeabilities of Na^+ and K^+. **(a)** The rapid rise in membrane potential from -60 mV to slightly more than $+30$ mV is referred to as a "depolarization." This depolarization is caused **(b)** by a sudden increase in the permeability of Na^+. As the Na^+ permeability decreases, K^+ permeability is increased and the membrane potential drops, eventually falling below the resting potential—a state of "hyperpolarization"—followed by a slow return to the resting potential. *(Adapted from Hodgkin, A., and Huxley, A., 1952. A quantitative description of membrane current and its application to conduction and excitation in nerve.* Journal of Physiology *117:500–544.)*

the axon in wavelike fashion. This simple process has several very dramatic properties:

1. Action potentials propagate very rapidly—up to and sometimes exceeding 100 m/sec.
2. The action potential is not attenuated (diminished in intensity) as a function of distance transmitted.

The input of energy all the way along an axon—in the form of ion gradients maintained by Na^+,K^+-ATPase—ensures that the shape and intensity of the action potential are maintained over long distances. The action potential has an all-or-none character. There are no gradations of amplitude; a given neuron is either at rest (with a polarized membrane) or is conducting a nerve impulse

Biochemistry⑧Now™ ACTIVE FIGURE 32.34
The propagation of action potentials along an axon. Figure 32.33 shows the time dependence of an action potential at a discrete point on the axon. This figure shows how the membrane potential varies along the axon as an action potential is propagated. (For this reason, the shape of the action potential is the apparent reverse of that shown in Figure 32.33.) At the leading edge of the action potential, membrane depolarization causes Na^+ channels to open briefly. As the potential moves along the axon, the Na^+ channels close and K^+ channels open, leading to a drop in potential and the onset of hyperpolarization. When the resting potential is restored, another action potential can be initiated. **Test yourself on the concepts in this figure at http://chemistry.brookscole.com/ggb3**

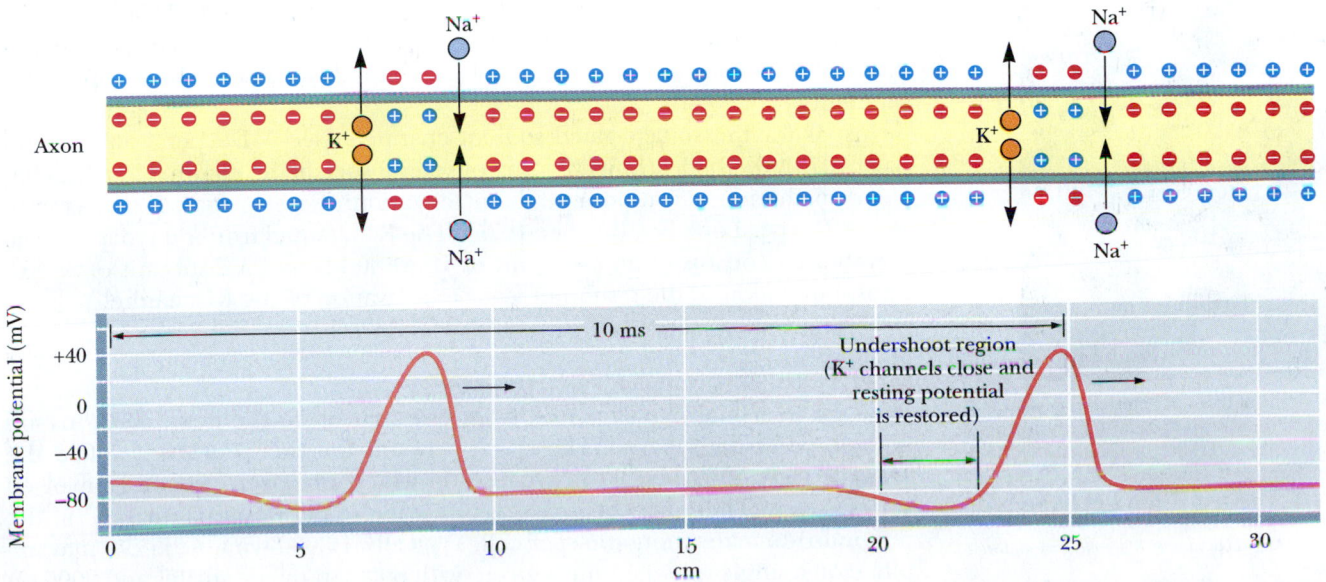

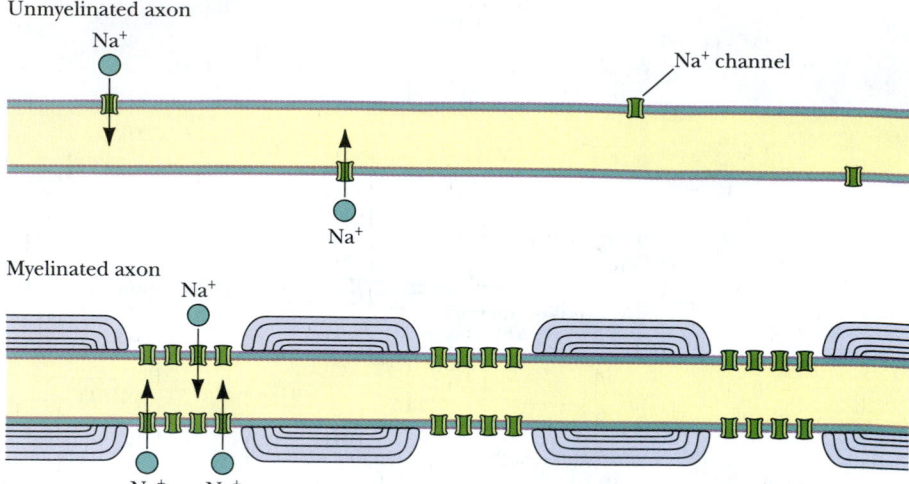

Unmyelinated axon

Na⁺

Na⁺ channel

Na⁺

Myelinated axon

Na⁺

Na⁺   Na⁺

FIGURE 32.35 Na⁺ channels are infrequently and randomly distributed in unmyelinated nerve. In myelinated axons, Na⁺ channels are clustered in large numbers in the nodes of Ranvier, between the regions surrounded by myelin sheath structures.

Turret   Selectivity filter

Outer helix

Inner helix

Turret   Selectivity filter

Outer helix

Inner helix

The potassium channel from *Streptomyces lividans* is a tetrameric integral membrane protein. Solution of this structure earned Roderick MacKinnon the 2003 Nobel Prize in Chemistry. *(Adapted from Figure 3 in D. A. Doyle, J. M. Cabral, R. A. Pfuetzner, A. Kuo, J. M. Gulbis, S. L. Cohen, B. T. Chait, and R. MacKinnon, 1998. The structure of the potassium channel: Molecular basis of K⁺ conduction and selectivity. Science **280**:69–77.)*

(with a reversed polarization). Because nerve impulses display no variation in amplitude, the size of the action potential is not important in processing signals in the nervous system. *Instead, it is the number of action potential firings and the frequency of firing that carry specific information.*

Sodium and Potassium Channels in Neurons Are Voltage Gated

The action potential is a delicately orchestrated interplay between the Na⁺,K⁺-ATPase and the voltage-gated Na⁺ and K⁺ channels that is initiated by a stimulus at the postsynaptic membrane. The density and distribution of Na⁺ channels along the axon are different for myelinated and unmyelinated axons (Figure 32.35). In unmyelinated axons, Na⁺ channels are uniformly distributed, although they are few in number—approximately 20 channels per μm^2. On the other hand, in myelinated axons, Na⁺ channels are **clustered** at the nodes of Ranvier. In these latter regions, they occur with a density of approximately 10,000 per μm^2. Elucidation of these distributions of Na⁺ channels was made possible by the use of several Na⁺-channel toxins (Figure 32.36).

The Na⁺ channel in mammalian brain is a heterotrimer consisting of α- (260 kD), β_1- (36 kD), and β_2- (33 kD) subunits (Figure 32.37). All three subunits are exposed to the extracellular surface and are heavily glycosylated. The α-subunit contains the binding site for toxins and has four domains of 300 to 400 amino acids each (Figure 32.38), with approximately 50% identity or conservation in their amino acid sequences. Each domain contains six regions of α-helical structure that are long enough to be membrane-spanning segments.

Just as for the voltage-gated sodium channels (see A Deeper Look box on page 1075), the high-affinity binding of several specific K⁺ channel blockers has aided in the identification and characterization of voltage-gated K⁺ channels (see A Deeper Look box on page 1075). The K⁺ channel from rat synaptosomal membranes consists of an α-subunit of 76 to 80 kD and a β-subunit of 38 kD. Phosphorylation of the α-subunit leads to activation of the K⁺ channel.

Neurons Communicate at the Synapse

How are neuronal signals passed from one neuron to the next? Neurons are juxtaposed at the synapse. The space between the two neurons is called the **synaptic cleft.** The number of synapses in which any given neuron is involved varies greatly. There may be as few as one synapse per postsynaptic cell (in the midbrain) to many thousands per cell. Typically, 10,000 synaptic knobs may impinge on a single spinal motor neuron, with 8000 on the dendrites and 2000 on

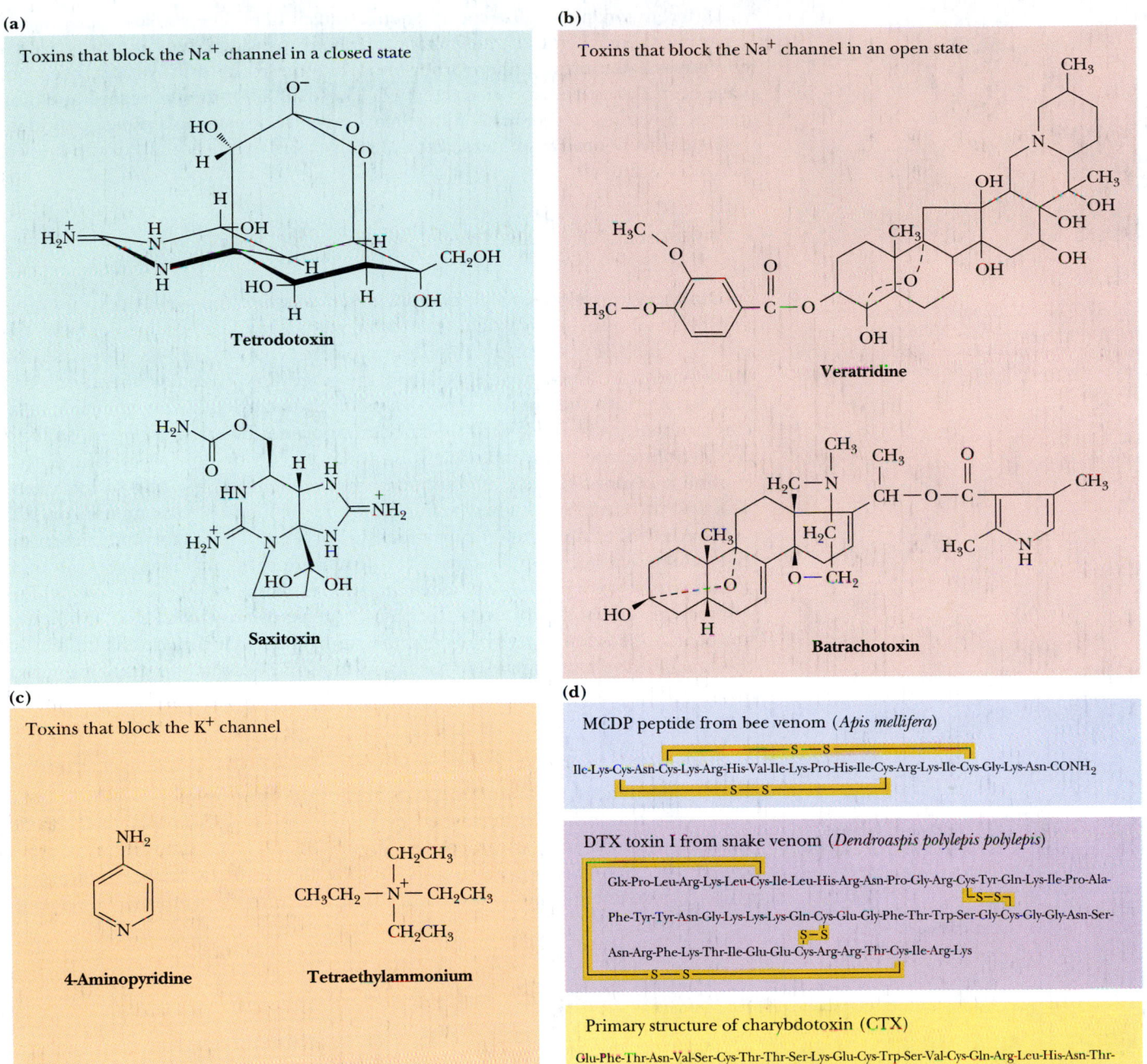

(a)

Toxins that block the Na⁺ channel in a closed state

Tetrodotoxin

Saxitoxin

(b)

Toxins that block the Na⁺ channel in an open state

Veratridine

Batrachotoxin

(c)

Toxins that block the K⁺ channel

4-Aminopyridine

Tetraethylammonium

(d)

MCDP peptide from bee venom (*Apis mellifera*)

Ile-Lys-Cys-Asn-Cys-Lys-Arg-His-Val-Ile-Lys-Pro-His-Ile-Cys-Arg-Lys-Ile-Cys-Gly-Lys-Asn-CONH₂

DTX toxin I from snake venom (*Dendroaspis polylepis polylepis*)

Glx-Pro-Leu-Arg-Lys-Leu-Cys-Ile-Leu-His-Arg-Asn-Pro-Gly-Arg-Cys-Tyr-Gln-Lys-Ile-Pro-Ala-
Phe-Tyr-Tyr-Asn-Gly-Lys-Lys-Lys-Gln-Cys-Glu-Gly-Phe-Thr-Trp-Ser-Gly-Cys-Gly-Gly-Asn-Ser-
Asn-Arg-Phe-Lys-Thr-Ile-Glu-Glu-Cys-Arg-Arg-Thr-Cys-Ile-Arg-Lys

Primary structure of charybdotoxin (CTX)

Glu-Phe-Thr-Asn-Val-Ser-Cys-Thr-Thr-Ser-Lys-Glu-Cys-Trp-Ser-Val-Cys-Gln-Arg-Leu-His-Asn-Thr-
Ser-Arg-Gly-Lys-Cys-Met-Asn-Lys-Lys-Cys-Arg-Cys-Tyr-Ser

FIGURE 32.36 Effectors of Na⁺ channels include **(a)** tetrodotoxin and saxitoxin, which block the Na⁺ channel in a closed state, and **(b)** veratridine and batrachotoxin, which block the Na⁺ channel in an open state. K⁺ channel blockers include **(c)** 4-aminopyridine, tetraethylammonium ion, mast cell degranulating peptide (MCDP), dendrotoxin (DTX), and charybdotoxin (CTX). (See both A Deeper Look boxes on page 1075.)

the soma or cell body. The ratio of synapses to neurons in the human forebrain is approximately 40,000 to 1!

Synapses are actually quite specialized structures, and several different types exist. *A minority of synapses in mammals, termed **electrical synapses**, are characterized by a very small gap—approximately 2 nm—between the **presynaptic cell** (which delivers the signal) and the **postsynaptic cell** (which receives the signal).* At electrical synapses, the arrival of an action potential on the presynaptic membrane leads directly to depolarization of the postsynaptic membrane, initiating a new action potential in the postsynaptic cell. However, most synaptic clefts are much wider—on the order of 20 to 50 nm. In these, an action potential in the presynaptic membrane causes secretion of a chemical substance—called a **neurotransmitter**—by the presynaptic cell. This substance binds to receptors on the postsynaptic cell, initiating a new action potential. Synapses of this type are thus **chemical synapses.**

Na⁺ channel

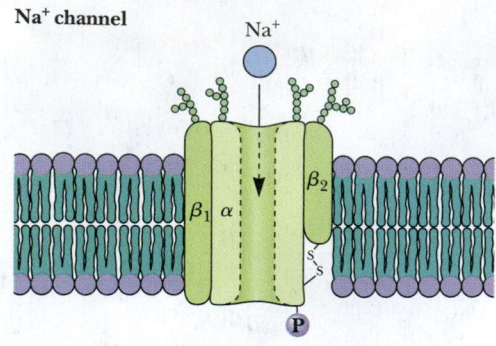

FIGURE 32.37 The Na⁺ channel comprises three subunits, denoted α, β₁, and β₂. A disulfide bridge links α and β₂ as shown. All three subunits are glycosylated, and the α-subunit can be phosphorylated on the cytoplasmic surface.

Different synapses utilize specific neurotransmitters. The **cholinergic synapse,** a paradigm for chemical transmission mechanisms at synapses, employs acetylcholine as a neurotransmitter. Other important neurotransmitters and receptors have been discovered and characterized. These all fall into one of several major classes, including **amino acids** (and their derivatives), **catecholamines, peptides,** and **gaseous neurotransmitters.** Table 32.5 lists some, but not all, of the known neurotransmitters.

Communication at Cholinergic Synapses Depends upon Acetylcholine

In **cholinergic synapses,** small **synaptic vesicles** inside the synaptic knobs contain large amounts of acetylcholine (approximately 10,000 molecules per vesicle; Figure 32.39). When the membrane of the synaptic knob is stimulated by an arriving action potential, special **voltage-gated Ca²⁺ channels** open and Ca²⁺ ions stream into the synaptic knob, causing the acetylcholine-containing vesicles to attach to and fuse with the knob membrane. The vesicles open, spilling acetylcholine into the synaptic cleft. Binding of acetylcholine to specific **acetylcholine receptors** in the postsynaptic membrane causes opening of ion channels and the creation of a new action potential in the postsynaptic neuron.

A variety of toxins can alter or affect this process. The anaerobic bacterium *Clostridium botulinum,* which causes botulism poisoning, produces several toxic proteins that strongly inhibit acetylcholine release. The black widow spider, *Lactrodectus mactans,* produces a venom protein, **α-latrotoxin,** that stimulates abnormal release of acetylcholine at the neuromuscular junction. The bite of the black widow causes pain, nausea, and mild paralysis of the diaphragm but is rarely fatal.

There Are Two Classes of Acetylcholine Receptors

Two different acetylcholine receptors are found in postsynaptic membranes. They were originally distinguished by their responses to **muscarine,** a toxic alkaloid in toadstools, and **nicotine** (Figure 32.40). The **nicotinic receptors** are cation channels in postsynaptic membranes, and the **muscarinic receptors** are transmembrane proteins that interact with G proteins. The recep-

FIGURE 32.38 Model for the arrangement of the Na⁺ channel α-subunit in the plasma membrane. The α-subunit consists of four domains (I through IV), each of which contains six transmembrane α-helices, designated S1 through S6. Phosphorylation sites (P) and location of positive charges on helix S4 are indicated.

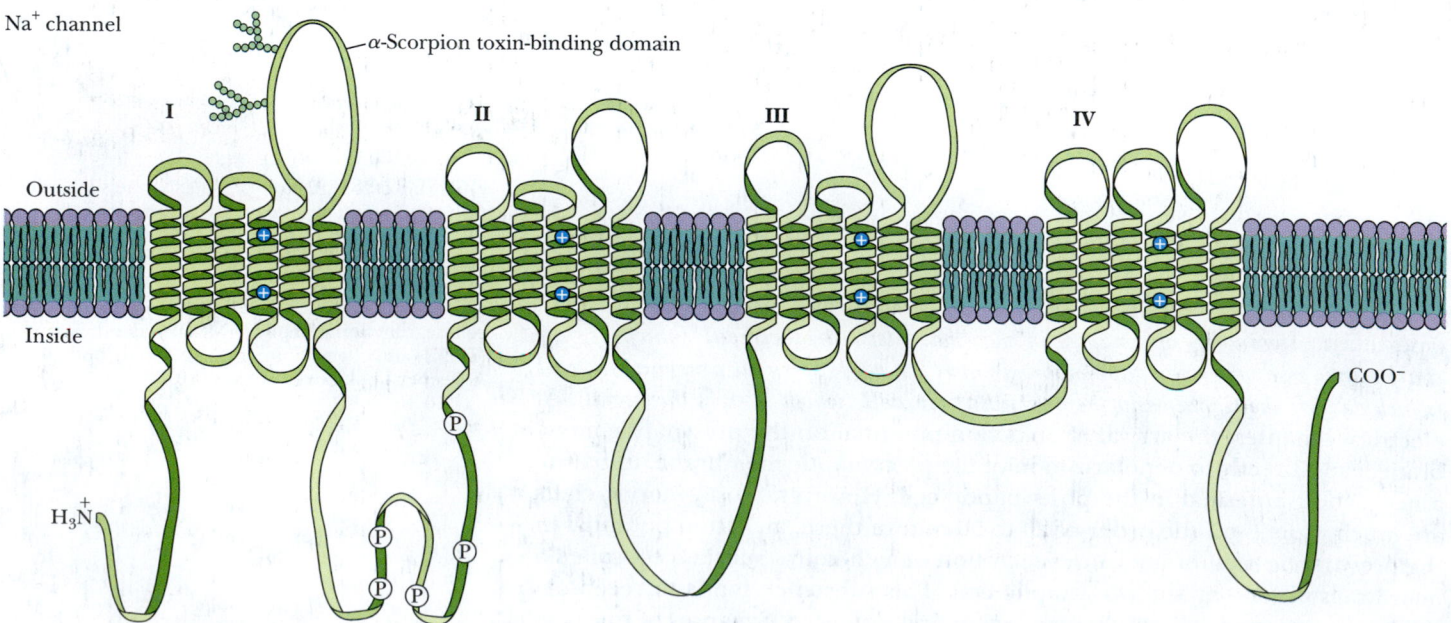

Na⁺ channel

A Deeper Look

Tetrodotoxin and Other Na⁺ Channel Toxins

Tetrodotoxin and **saxitoxin** are highly specific blockers of Na⁺ channels and bind with very high affinity ($K_D < 1$ nM). This unique specificity and affinity have made it possible to use radioactive forms of these toxins to purify Na⁺ channels and map their distribution on axons. Tetrodotoxin is found in the skin and several internal organs of puffer fish, also known as blowfish or swellfish, members of the family *Tetraodontidae*, which react to danger by inflating themselves with water or air to nearly spherical (and often comical) shapes (see accompanying figure). Although tetrodotoxin poisoning can easily be fatal, puffer fish are delicacies in Japan, where they are served in a dish called fugu. For this purpose, the puffer fish must be cleaned and prepared by specially trained chefs. Saxitoxin is made by *Gonyaulax catenella* and *G.*

tamarensis, two species of marine dinoflagellates or plankton that are responsible for "red tides" that cause massive fish kills. Saxitoxin is concentrated by certain species of mussels, scallops, and other shellfish that are exposed to red tides. Consumption of these shellfish by animals, including humans, can be fatal. In addition to these toxins, which prevent the Na⁺ channel from opening, there are equally poisonous agents that block the Na⁺ channel in an open state, permitting Na⁺ to stream into the cell without control, destroying the Na⁺ gradients. Included in this group of toxins (see Figure 32.36) are **veratridine** from *Schoenocaulon officinalis*, a member of the lily family, and **batrachotoxin,** a compound found in skin secretions of a Colombian frog, *Phyllobates aurotaenia*. These skin secretions have traditionally been used as arrow poisons.

◄ Tetrodotoxin is found in puffer fish, which are prepared and served in Japan as fugu. The puffer fish on the left is unexpanded; the one on the right is inflated.

tors in sympathetic ganglia and those in motor endplates of skeletal muscle are nicotinic receptors. Nicotine locks the ion channels of these receptors in their open conformation. Muscarinic receptors are found in smooth muscle and in glands. Muscarine mimics the effect of acetylcholine on these latter receptors.

The nicotinic acetylcholine receptor is a transmembrane glycoprotein with an approximate molecular mass of 270 kD, consisting of four different subunits, α (54 kD), β (56 kD), γ (58 kD), and δ (60 kD), with a quaternary structure of $\alpha_2\beta\gamma\delta$. Each α-subunit possesses a binding site for acetylcholine.

A Deeper Look

Potassium Channel Toxins

K⁺ channel blockers include **4-aminopyridine; tetraethylammonium ion;** and several peptide toxins, including the **dendrotoxins (DTX), mast cell degranulating peptide (MCDP),** and **charybdotoxin (CTX)** (see Figure 32.36). Dendrotoxin I is a 60-residue peptide from *Dendroaspis polylepis*, the dangerous black mamba snake of sub-Saharan Africa. MCDP, a bee venom toxin that has a degranulating action on mast cells, is a potent convulsant. It is a 22-residue peptide with two

disulfide bonds, one proline, and a C-terminal amide. Charybdotoxin is a minor component of the venom of the scorpion, *Leiurus quinquestriatus*. It is a 37-residue peptide with 8 positively charged residues (3 arginines, 4 lysines, and 1 histidine). All these agents bind with high affinity to membranes containing voltage-activated K⁺ channels.

Table 32.5
Families of Neurotransmitters
Cholinergic Agents
Acetylcholine
Catecholamines
Norepinephrine (noradrenaline)
Epinephrine (adrenaline)
L-Dopa
Dopamine
Octopamine
Amino Acids (and Derivatives)
γ-Aminobutyric acid (GABA)
Alanine
Aspartate
Cystathione
Glycine
Glutamate
Histamine
Proline
Serotonin
Taurine
Tyrosine
Peptide Neurotransmitters
Cholecystokinin
Enkephalins and endorphins
Gastrin
Gonadotropin
Neurotensin
Oxytocin
Secretin
Somatostatin
Substance P
Thyrotropin releasing factor
Vasopressin
Vasoactive intestinal peptide (VIP)
Gaseous Neurotransmitters
Carbon monoxide (CO)
Nitric oxide (NO)

The Nicotinic Acetylcholine Receptor Is a Ligand-Gated Ion Channel

The nicotinic acetylcholine receptor functions as a **ligand-gated ion channel,** and on the basis of its structure, it is also an **oligomeric ion channel.** When acetylcholine (the ligand) binds to this receptor, a conformational change opens the channel, which is equally permeable to Na^+ and K^+. Na^+ rushes in while K^+ streams out, but because the Na^+ gradient across this membrane is steeper than that of K^+, the Na^+ influx greatly exceeds the K^+ efflux. The influx of Na^+ depolarizes the postsynaptic membrane, initiating an action potential in the adjacent membrane. After a few milliseconds, the channel closes, even though acetylcholine remains bound to the receptor. At this point, the channel will remain closed until the concentration of acetylcholine in the synaptic cleft drops to about 10 nM.

Acetylcholinesterase Degrades Acetylcholine in the Synaptic Cleft

Following every synaptic signal transmission, the synapse must be readied for the arrival of another action potential. Several things must happen very quickly. First, the acetylcholine left in the synaptic cleft must be rapidly degraded to resensitize the acetylcholine receptor and to restore the excitability of the postsynaptic membrane. This reaction is catalyzed by **acetylcholinesterase** (Figure 32.41).

When [acetylcholine] has decreased to low levels, acetylcholine dissociates from the receptor, which thereby regains its ability to open in a ligand-dependent manner. Second, the synaptic vesicles must be reformed from the presynaptic membrane by endocytosis (Figure 32.42) and then must be restocked with acetylcholine. This occurs through the action of an ATP-driven H^+ pump and an **acetylcholine transport protein.** The H^+ pump in this case is a member of the family of **V-type ATPases.** It uses the free energy of ATP hydrolysis to create an H^+ gradient across the vesicle membrane. This gradient is used by the acetylcholine transport protein to drive acetylcholine into the vesicle, as shown in Figure 32.42.

Antagonists of the nicotinic acetylcholine receptor are particularly potent neurotoxins. These agents, which bind to the receptor and prevent opening of the ion channel, include **d-tubocurarine**, the active agent in the South American arrow poison **curare,** and several small proteins from poisonous snakes. These latter agents include **cobratoxin** from cobra venom, and **α-bungarotoxin,** *from Bungarus multicinctus,* a snake common in Taiwan (Figure 32.43).

Muscarinic Receptor Function Is Mediated by G Proteins

There are several different types of muscarinic acetylcholine receptors, with different structures and different apparent functions in synaptic transmission. However, certain structural and functional features are shared by this class of receptors. Muscarinic receptors are 70-kD glycoproteins and are members of the seven-transmembrane segment (7-TMS) family of receptors.

Activation of muscarinic receptors (by binding of acetylcholine) results in several effects, including the inhibition of **adenylyl cyclase,** the stimulation of **phospholipase C,** and the opening of K^+ channels. As shown in Figure 32.44, all of these effects of muscarinic receptors are mediated by G proteins. Many antagonists for muscarinic acetylcholine receptors are known, including **atropine** from *Atropa belladonna,* the deadly nightshade plant whose berries are sweet and tasty but highly poisonous (Figure 32.43).

Both the nicotinic and muscarinic acetylcholine receptors are sensitive to certain agents that inactivate acetylcholinesterase itself. Acetylcholinesterase is a serine esterase similar to trypsin and chymotrypsin (see Chapter 14). The reactive serine at the active site of such enzymes is a vulnerable target for organophosphorus inhibitors (Figure 32.45). **DIPF** and related agents form stable covalent complexes with the active-site serine, irreversibly blocking the

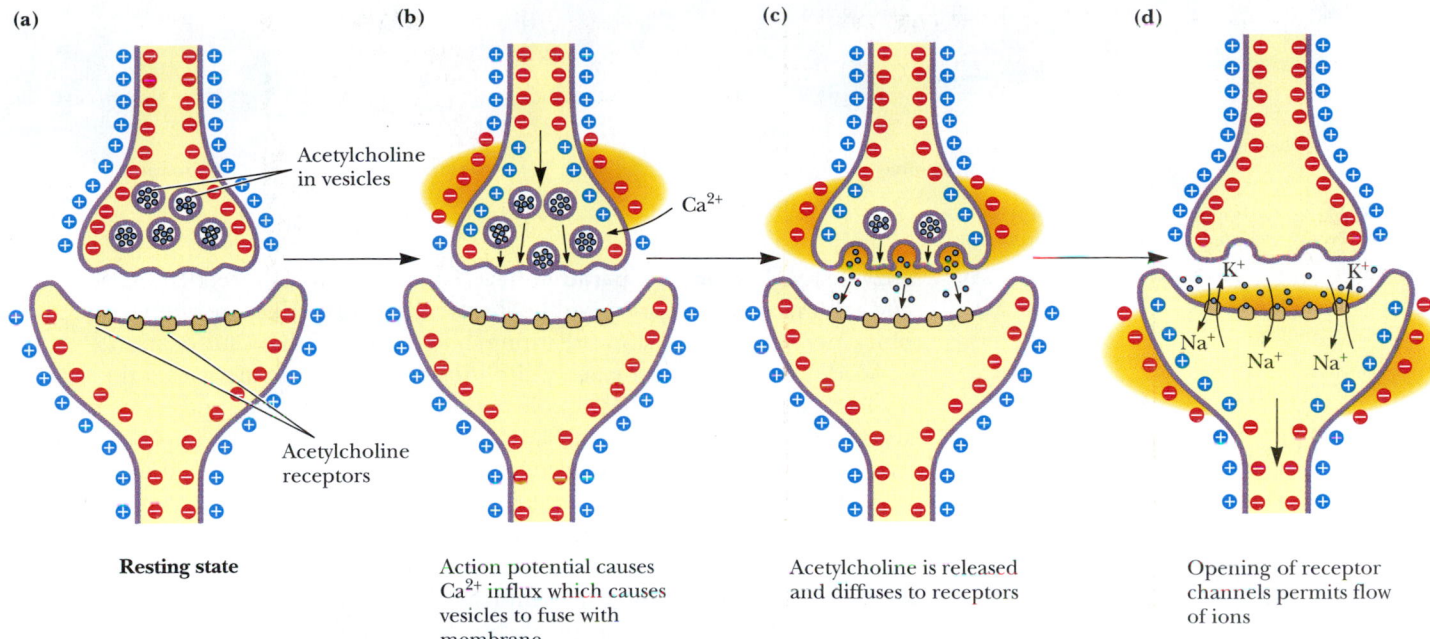

(a)

Acetylcholine
in vesicles

Acetylcholine
receptors

Resting state

(b)

Ca^{2+}

Action potential causes
Ca^{2+} influx which causes
vesicles to fuse with
membrane

(c)

Acetylcholine is released
and diffuses to receptors

(d)

K^+ K^+

Na^+ Na^+ Na^+

Opening of receptor
channels permits flow
of ions

Biochemistry⊜Now™   ANIMATED FIGURE 32.39 Cell–cell communication at the synapse
(a) is mediated by neurotransmitters such as acetylcholine, produced from choline by choline
acetyltransferase. The arrival of an action potential at the synaptic knob **(b)** opens Ca^{2+} channels
in the presynaptic membrane. Influx of Ca^{2+} induces the fusion of acetylcholine-containing vesicles with the plasma membrane and release of acetylcholine into the synaptic cleft **(c)**. Binding of
acetylcholine to receptors in the postsynaptic membrane opens Na^+ channels **(d)**. The influx of
Na^+ depolarizes the postsynaptic membrane, generating a new action potential. **See this figure
animated at http://chemistry.brookscole.com/ggb3**

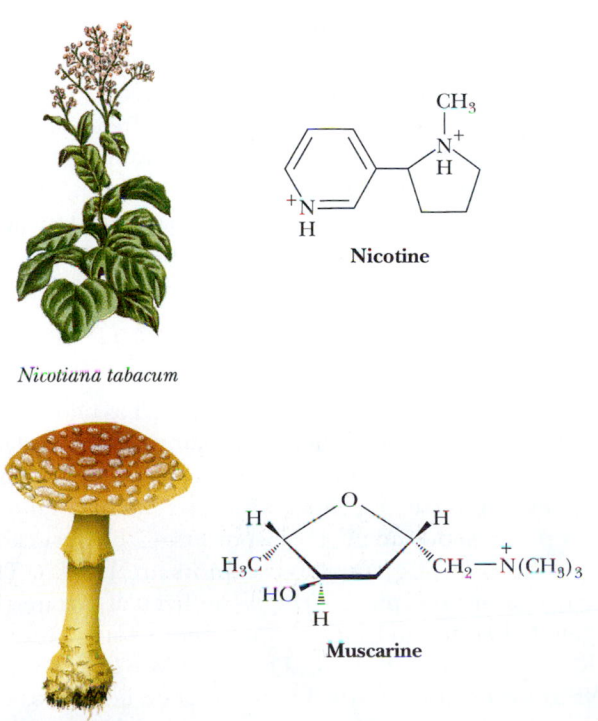

Nicotiana tabacum

Nicotine

Amanita muscaria

Muscarine

FIGURE 32.40 Two types of acetylcholine receptors are known. Nicotinic acetylcholine receptors
are locked in their open conformation by nicotine. Obtained from tobacco plants, nicotine is
named for Jean Nicot, French ambassador to Portugal, who sent tobacco seeds to France in 1550
for cultivation. Muscarinic acetylcholine receptors are stimulated by muscarine, obtained from
the intensely poisonous mushroom, *Amanita muscaria*.

FIGURE 32.41 Acetylcholine is degraded to acetate and choline by acetylcholinesterase, a serine protease.

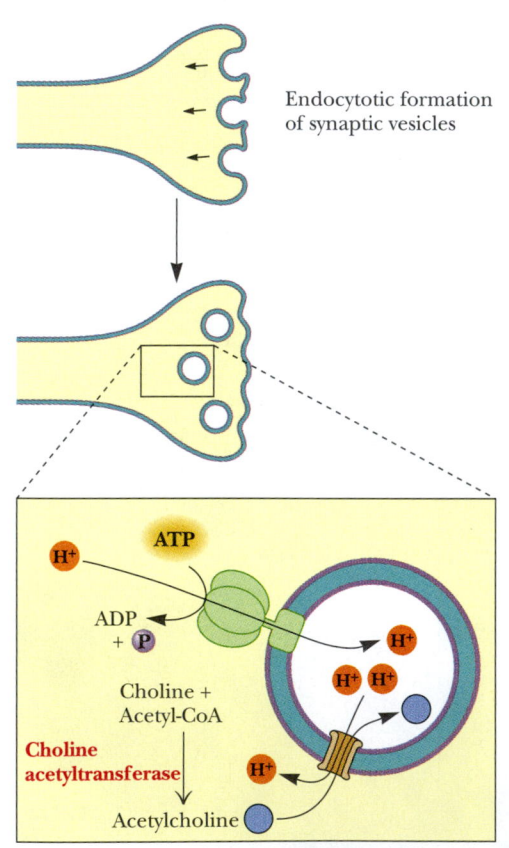

Biochemistry⊜Now™ **ANIMATED FIGURE 32.42**
Following a synaptic transmission event, acetylcholine is repackaged in vesicles in a multistep process. Synaptic vesicles are formed by endocytosis, and acetylcholine is synthesized by choline acetyltransferase. A proton gradient is established across the vesicle membrane by an H^+-transport ATPase, and a proton–acetylcholine transport protein transports acetylcholine into the synaptic vesicles, exchanging acetylcholine for protons in an electrically neutral antiport process. **See this figure animated at http://chemistry. brookscole.com/ggb3**

enzyme. **Malathion** and **parathion** are commonly used insecticides, and **sarin** and **tabun** are nerve gases used in chemical warfare. All these agents effectively block nerve impulses, stop breathing, and cause death by suffocation.

Milder inhibitors of the acetylcholinesterase reaction are useful therapeutic agents. **Physostigmine,** an alkaloid found in calabar beans, and **neostigmine,** a synthetic analog, contain carbamoyl ester groups (Figure 32.45). Reaction with the active-site serine of acetylcholinesterase leaves an intermediate that is hydrolyzed only very slowly, effectively inhibiting the enzyme. Physostigmine and neostigmine have been used to treat **myasthenia gravis,** a chronic disorder that causes muscle weakness and eventual paralysis. Myasthenia gravis is an autoimmune disease in which individuals produce antibodies that bind to their own acetylcholine receptors, blocking the response to acetylcholine. By blocking acetylcholinesterase (thus allowing acetylcholine levels in the synaptic cleft to remain high), physostigmine and neostigmine can suppress the symptoms of myasthenia gravis.

Other Neurotransmitters Can Act Within Synaptic Junctions

Synaptic junctions that use amino acids, catecholamines, and peptides (see Table 32.5) appear to operate much the way the cholinergic synapses do. Presynaptic vesicles release their contents into the synaptic cleft, where the neurotransmitter substance can bind to specific receptors on the postsynaptic membrane to induce a conformational change and elicit a particular response. Some of these neurotransmitters are **excitatory** in nature and stimulate postsynaptic neurons to transmit impulses, whereas others are **inhibitory** and prevent the postsynaptic neuron from carrying other signals. Just as acetylcholine acts on both nicotinic and muscarinic receptors, so most of the known neurotransmitters act on several (and in some cases, many) different kinds of receptors. Biochemists are just beginning to understand the sophistication and complexity of neuronal signal transmission.

Glutamate and Aspartate Are Excitatory Amino Acid Neurotransmitters

The common amino acids glutamate and aspartate act as neurotransmitters. Like acetylcholine, glutamate and aspartate are excitatory and stimulate receptors on the postsynaptic membrane to transmit a nerve impulse. No enzymes that degrade glutamate exist in the extracellular space, so glutamate must be cleared by the high-affinity presynaptic and glial transporters—a process called **reuptake.**

At least five subclasses of glutamate receptors are known. The best understood of these excitatory receptors is the N-methyl-D-aspartate (NMDA) receptor, a ligand-gated channel that, when open, allows Ca^{2+} and Na^+ to flow into the cell and K^+ to flow out of the cell. **Phencyclidine (PCP)** is a specific antagonist of the NMDA receptor (Figure 32.46). Phencyclidine was once used as an anesthetic agent, but legitimate human use was quickly discontinued when it was found to be responsible for bizarre psychotic reactions and behavior in its users. Since this time, PCP has been used illegally as a hallucinogenic drug, under the street name of **angel dust.** Sadly, it has caused many serious, long-term psychological problems in its users.

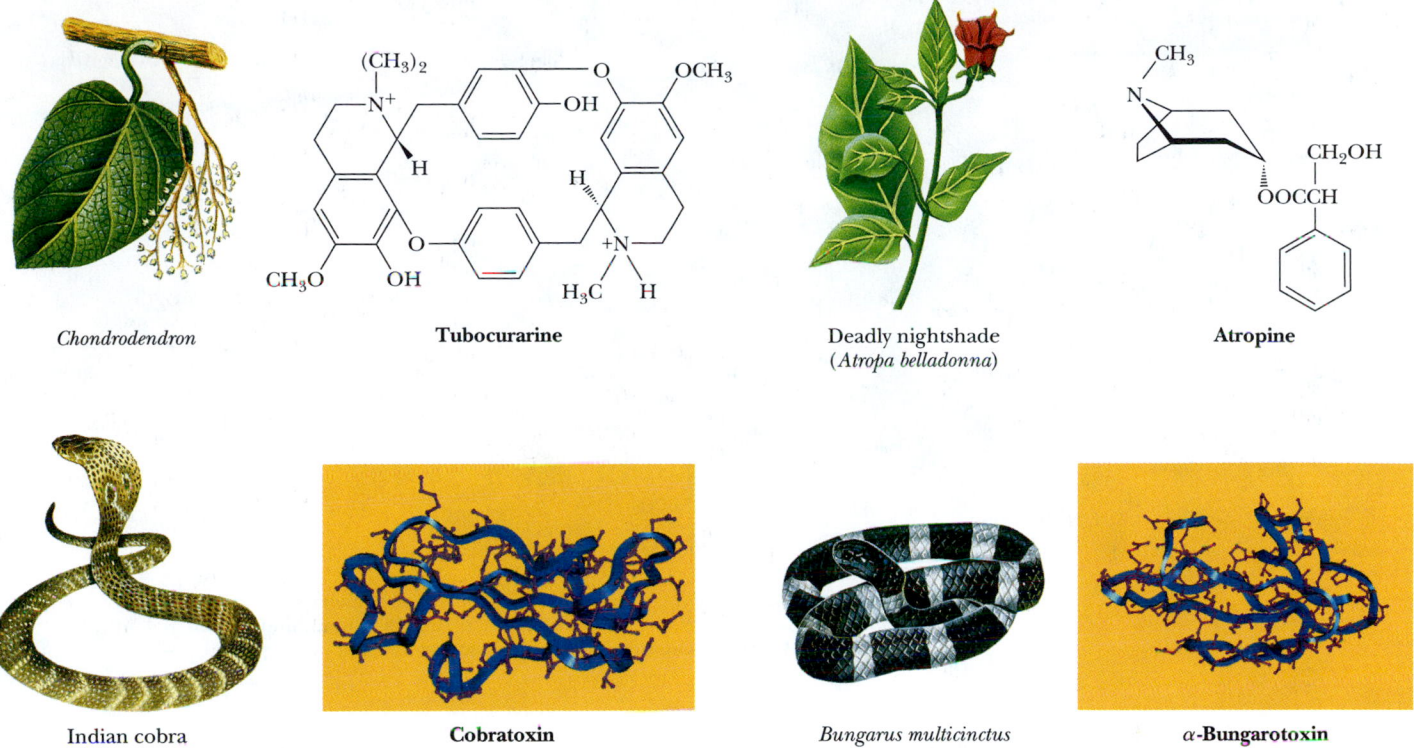

Chondrodendron   **Tubocurarine**   Deadly nightshade (*Atropa belladonna*)   **Atropine**

Indian cobra (*Naja naja*)   **Cobratoxin**   *Bungarus multicinctus* α-**Bungarotoxin**

FIGURE 32.43 Tubocurarine, obtained from the plant *Chondrodendron tomentosum*, is the active agent in "tube curare," named for the bamboo tubes in which it is kept by South American tribal hunters. Atropine is produced by *Atropa belladonna*, the poisonous deadly nightshade. The species name, which means "beautiful woman," derives from the use of atropine in years past by Italian women to dilate their pupils. Atropine is still used for pupil dilation in eye exams by ophthalmologists. Cobratoxin and α-bungarotoxin are produced by the cobra *(Naja naja)* and the banded krait snake *(Bungarus multicinctus),* respectively.

γ-Aminobutyric Acid and Glycine Are Inhibitory Neurotransmitters

Certain neurotransmitters, acting through their conjugate postsynaptic receptors, inhibit the postsynaptic neuron from propagating nerve impulses from other neurons. Two such inhibitory neurotransmitters are **γ-aminobutyric acid (GABA)** and **glycine.** These agents make postsynaptic membranes permeable to chloride ions and cause a net influx of Cl^-, which in turn causes **hyperpolarization** of the postsynaptic membrane (making the membrane potential more negative). Hyperpolarization of a neuron effectively raises the threshold for the onset of action potentials in that neuron, making the neuron resistant to stimulation by excitatory neurotransmitters. These effects are mediated by the GABA and glycine receptors, which are ligand-gated chloride channels (Figure 32.47). GABA is derived by a decarboxylation of glutamate (Figure 32.48) and appears to operate mainly in the brain, whereas glycine acts primarily in the spinal cord. The glycine receptor has a specific affinity for the convulsive alkaloid **strychnine** (Figure 32.49). The effects of ethanol on the brain arise in part from the opening of GABA receptor Cl^- channels.

The Catecholamine Neurotransmitters Are Derived from Tyrosine

Epinephrine, norepinephrine, dopamine, and **L-dopa** are collectively known as the **catecholamine** neurotransmitters. These compounds are synthesized from tyrosine (Figure 32.50), both in sympathetic neurons and in the adrenal glands. They function as neurotransmitters in the brain and as hormones in the circulatory system. However, these two pools operate independently, thanks to the

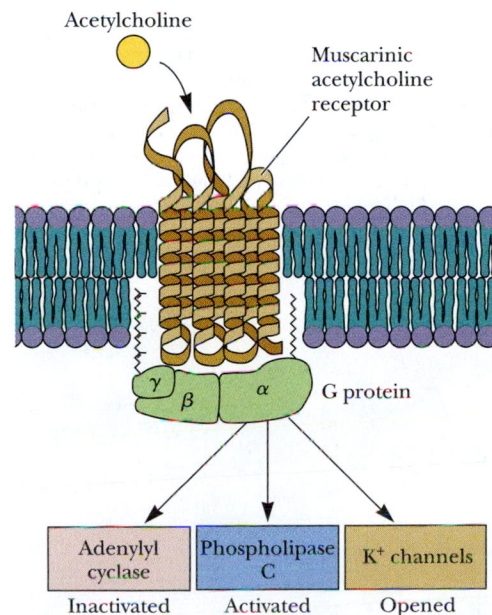

FIGURE 32.44 Muscarinic acetylcholine receptors are typical seven-transmembrane segment receptor proteins. Binding of acetylcholine to these receptors activates G proteins, which inactivate adenylyl cyclase, activate phospholipase C, and open K^+ channels.

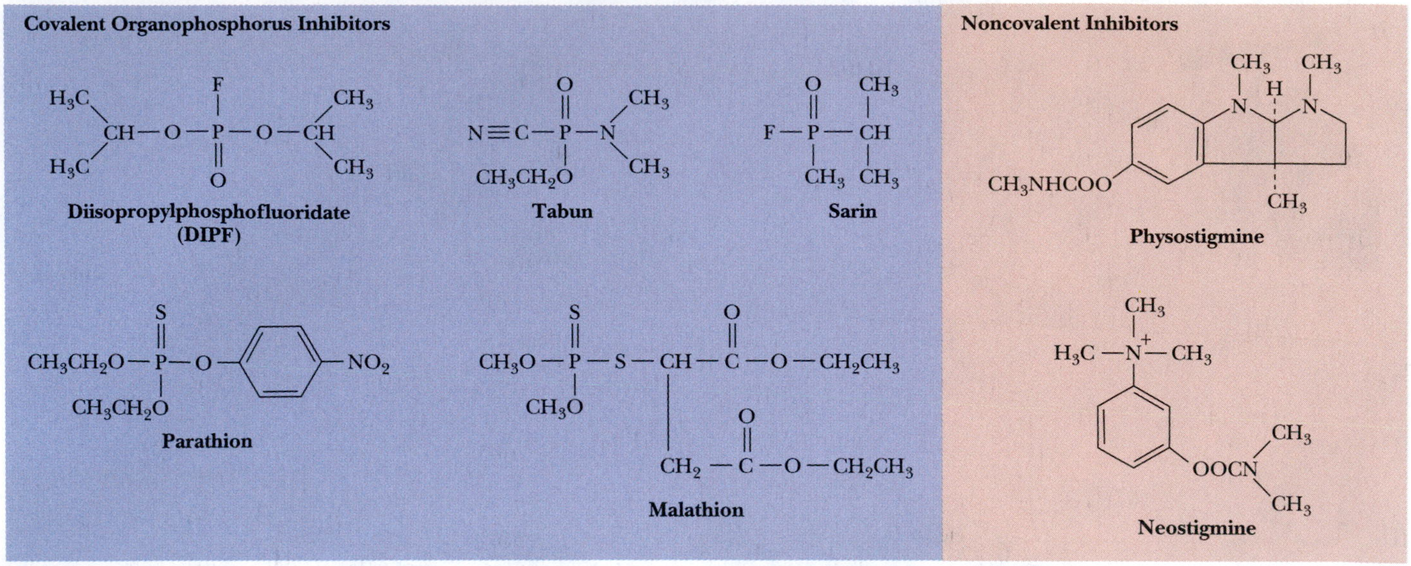

Covalent Organophosphorus Inhibitors

Diisopropylphosphofluoridate
(DIPF)

Tabun

Sarin

Parathion

Malathion

Noncovalent Inhibitors

Physostigmine

Neostigmine

FIGURE 32.45 Covalent inhibitors (blue) of acetylcholinesterase include DIFP, the nerve gases tabun and sarin, and the insecticides parathion and malathion. Milder, noncovalent (pink) inhibitors of acetylcholinesterase include physostigmine and neostigmine.

FIGURE 32.46 The NMDA receptor is an Na^+ and Ca^{2+} channel, which is regulated by Zn^{2+} and glycine, stimulated by *N*-methyl-D-aspartate, and inhibited by phencyclidine (PCP) and the anticonvulsant drug MK-801. *(Adapted from Young, A., and Fagg, G., 1990. Excitatory amino acid receptors in the brain. Membrane binding and receptor autoradiographic approaches.* Trends in Pharmacological Sciences *11:126–133.)*

N-Methyl-D-asparate

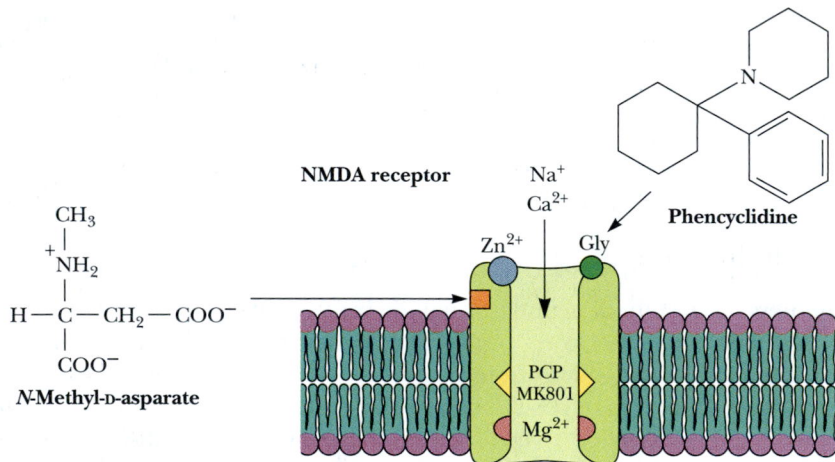

FIGURE 32.47 GABA (γ-aminobutyric acid) and glycine are inhibitory neurotransmitters that activate chloride channels. Influx of Cl^- causes a hyperpolarization of the postsynaptic membrane.

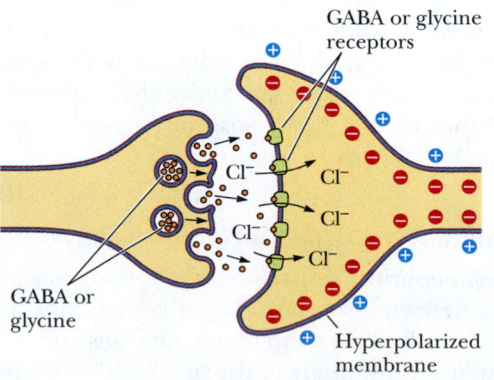

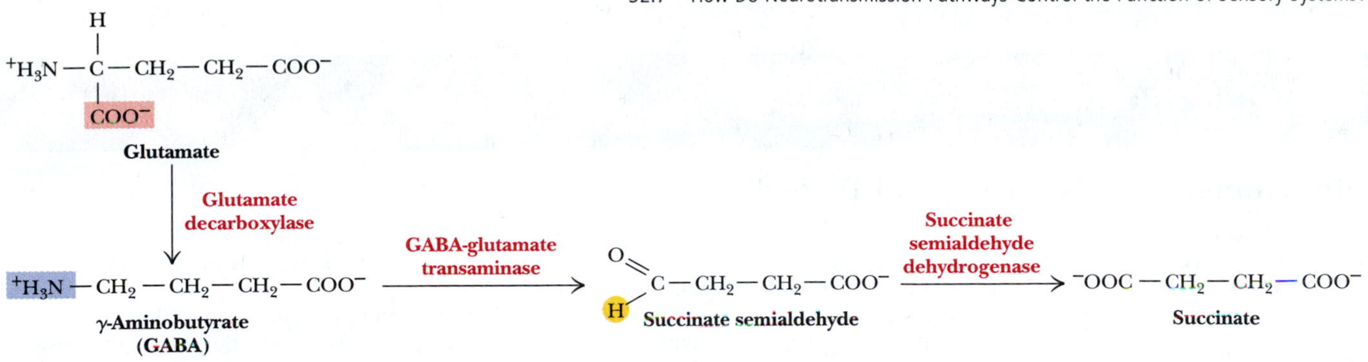

Glutamate

$$^+H_3N - \overset{\overset{\displaystyle H}{|}}{\underset{\underset{\displaystyle \boxed{COO^-}}{|}}{C}} - CH_2 - CH_2 - COO^-$$

Glutamate

Glutamate decarboxylase ↓

$$^+\boxed{H_3N} - CH_2 - CH_2 - CH_2 - COO^-$$

γ-Aminobutyrate (GABA)

GABA-glutamate transaminase →

$$\underset{\underset{\displaystyle \boxed{H}}{|}}{\overset{\overset{\displaystyle O}{\|}}{C}} - CH_2 - CH_2 - COO^-$$

Succinate semialdehyde

Succinate semialdehyde dehydrogenase →

$$^-OOC - CH_2 - CH_2 - COO^-$$

Succinate

FIGURE 32.48 Glutamate is converted to GABA by glutamate decarboxylase. GABA is degraded by the action of GABA–glutamate transaminase and succinate semialdehyde dehydrogenase to produce succinate.

blood–brain barrier, which permits only very hydrophobic species in the circulatory system to cross over into the brain. Hydroxylation of tyrosine (by **tyrosine hydroxylase**) to form **3,4-*dihydroxyphenylalanine*** (L-dopa) is the rate-limiting step in this pathway. Dopamine, a crucial catecholamine involved in several neurological diseases, is synthesized from L-dopa by a pyridoxal phosphate-dependent enzyme, **dopa decarboxylase.** Subsequent hydroxylation and methylation produce norepinephrine and epinephrine (Figure 32.50). The methyl group in the final reaction is supplied by *S*-adenosylmethionine.

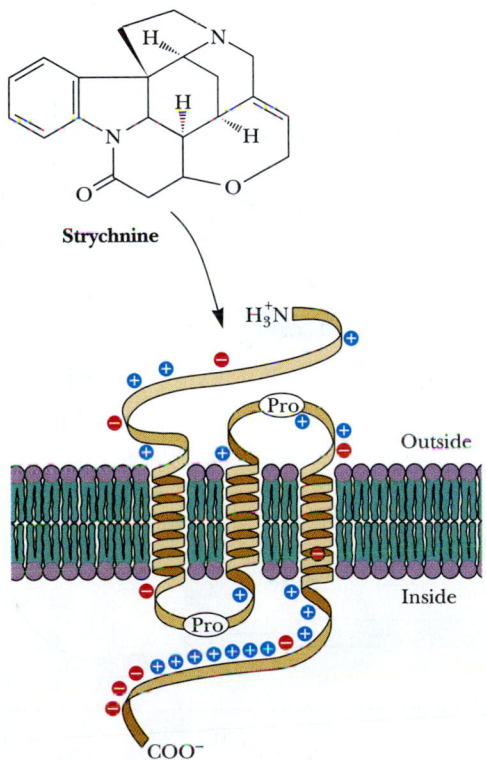

Strychnine

FIGURE 32.49 Glycine receptors are distinguished by their unique affinity for strychnine. A model for the arrangement of the 48-kD subunit of the glycine receptor in the postsynaptic membrane is shown.

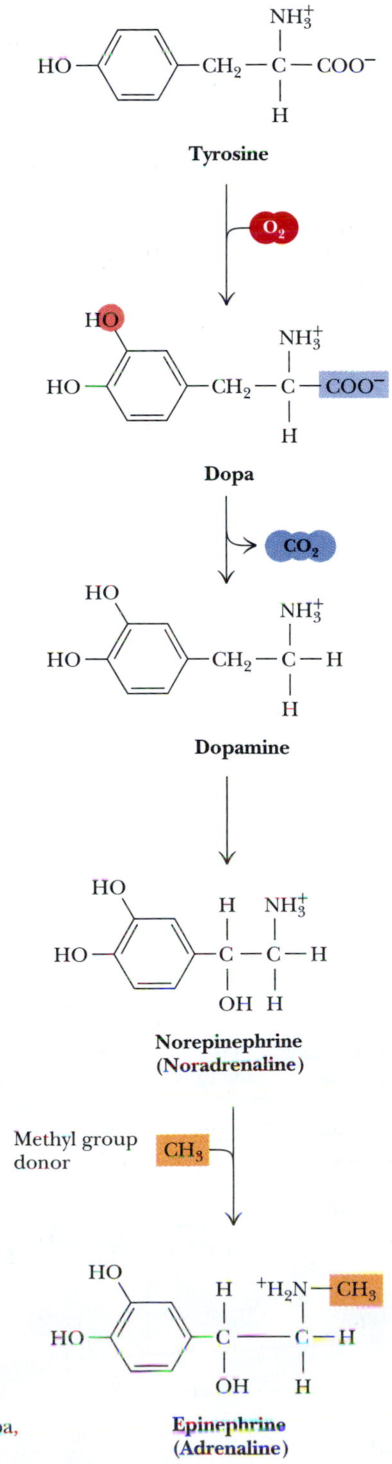

Tyrosine

$$HO - \langle\text{ring}\rangle - CH_2 - \overset{\overset{\displaystyle NH_3^+}{|}}{\underset{\underset{\displaystyle H}{|}}{C}} - COO^-$$

Tyrosine

↓ O_2

$$HO - \langle\text{ring (HO)}\rangle - CH_2 - \overset{\overset{\displaystyle NH_3^+}{|}}{\underset{\underset{\displaystyle H}{|}}{C}} - \boxed{COO^-}$$

Dopa

↓ CO_2

$$HO - \langle\text{ring (HO)}\rangle - CH_2 - \overset{\overset{\displaystyle NH_3^+}{|}}{\underset{\underset{\displaystyle H}{|}}{C}} - H$$

Dopamine

↓

$$HO - \langle\text{ring (HO)}\rangle - \overset{\overset{\displaystyle H}{|}}{\underset{\underset{\displaystyle OH}{|}}{C}} - \overset{\overset{\displaystyle NH_3^+}{|}}{\underset{\underset{\displaystyle H}{|}}{C}} - H$$

Norepinephrine (Noradrenaline)

Methyl group donor $\boxed{CH_3}$ ↓

$$HO - \langle\text{ring (HO)}\rangle - \overset{\overset{\displaystyle H}{|}}{\underset{\underset{\displaystyle OH}{|}}{C}} - \overset{\overset{\displaystyle ^+H_2N - \boxed{CH_3}}{|}}{\underset{\underset{\displaystyle H}{|}}{C}} - H$$

Epinephrine (Adrenaline)

▶ **FIGURE 32.50** The pathway for the synthesis of catecholamine neurotransmitters. Dopa, dopamine, noradrenaline, and adrenaline are synthesized sequentially from tyrosine.

Human Biochemistry

The Biochemistry of Neurological Disorders

Defects in catecholamine processing are responsible for the symptoms of many neurological disorders, including clinical depression (which involves norepinephrine [NE]) and parkinsonism (involving dopamine [DA]). Once these neurotransmitters have bound to and elicited responses from postsynaptic membranes, they must be efficiently cleared from the synaptic cleft (see accompanying figure, part a). Clearing can occur by several mechanisms. NE and DA transport or reuptake proteins exist both in the presynaptic membrane and in nearby glial cell membranes. On the other hand, catecholamine neurotransmitters can be metabolized and inactivated by two enzymes: **catechol-*O*-methyl-transferase** in the synaptic cleft and **monoamine oxidase** in the mitochondria (see figure, part b). Catecholamines transported back into the presynaptic neuron are accumulated in synaptic vesicles by the same H^+-ATPase/H^+-ligand exchange mechanism described for glutamate. Clinical depression has been treated by two different strategies. **Monoamine oxidase inhibitors** act as antidepressants by increasing levels of catecholamines in the brain. Another class of antidepressants, the **tricyclics,** such as desipramine (see figure, part c), act on several classes of neurotransmitter reuptake transporters and facilitate

more prolonged stimulation of postsynaptic receptors. **Prozac** is a more specific reuptake inhibitor and acts only on serotonin reuptake transporters.

Parkinsonism is characterized by degeneration of dopaminergic neurons, as well as consequent overproduction of postsynaptic dopamine receptors. In recent years, Parkinson's patients have been treated with dopamine agonists such as bromocriptine (see figure, part d) to counter the degeneration of dopamine neurons.

Catecholaminergic neurons are involved in many other interesting pharmacological phenomena. For example, **reserpine** (see figure, part e), an alkaloid from a climbing shrub of India, is a powerful sedative that depletes the level of brain monoamines by inhibiting the H^+–monoamine exchange protein in the membranes of synaptic vesicles.

Cocaine (see figure, part f), a highly addictive drug, binds with high affinity and specificity to reuptake transporters for the monoamine neurotransmitters in presynaptic membranes. Thus, at least one of the pharmacological effects of cocaine is to prolong the synaptic effects of these neurotransmitters.

(a)

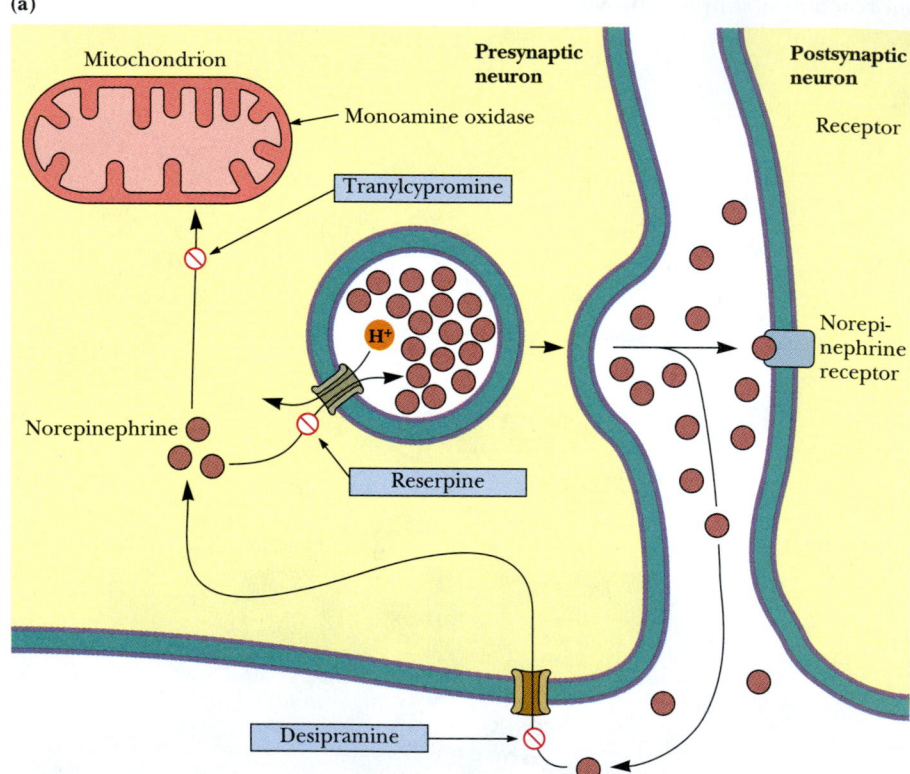

▶ **(a)** The pathway for reuptake and vesicular repackaging of the catecholamine neurotransmitters. The sites of action of desipramine, tranylcypromine, and reserpine are indicated.

Each of these catecholamine neurotransmitters is known to play a unique role in synaptic transmission. The neurotransmitter in junctions between sympathetic nerves and smooth muscle is norepinephrine. On the other hand, dopamine is involved in other processes. Either excessive brain production of dopamine or hypersensitivity of dopamine receptors is responsible for psychotic symptoms and schizophrenia, whereas lowered production of dopamine and the loss of dopamine neurons are important factors in Parkinson's disease.

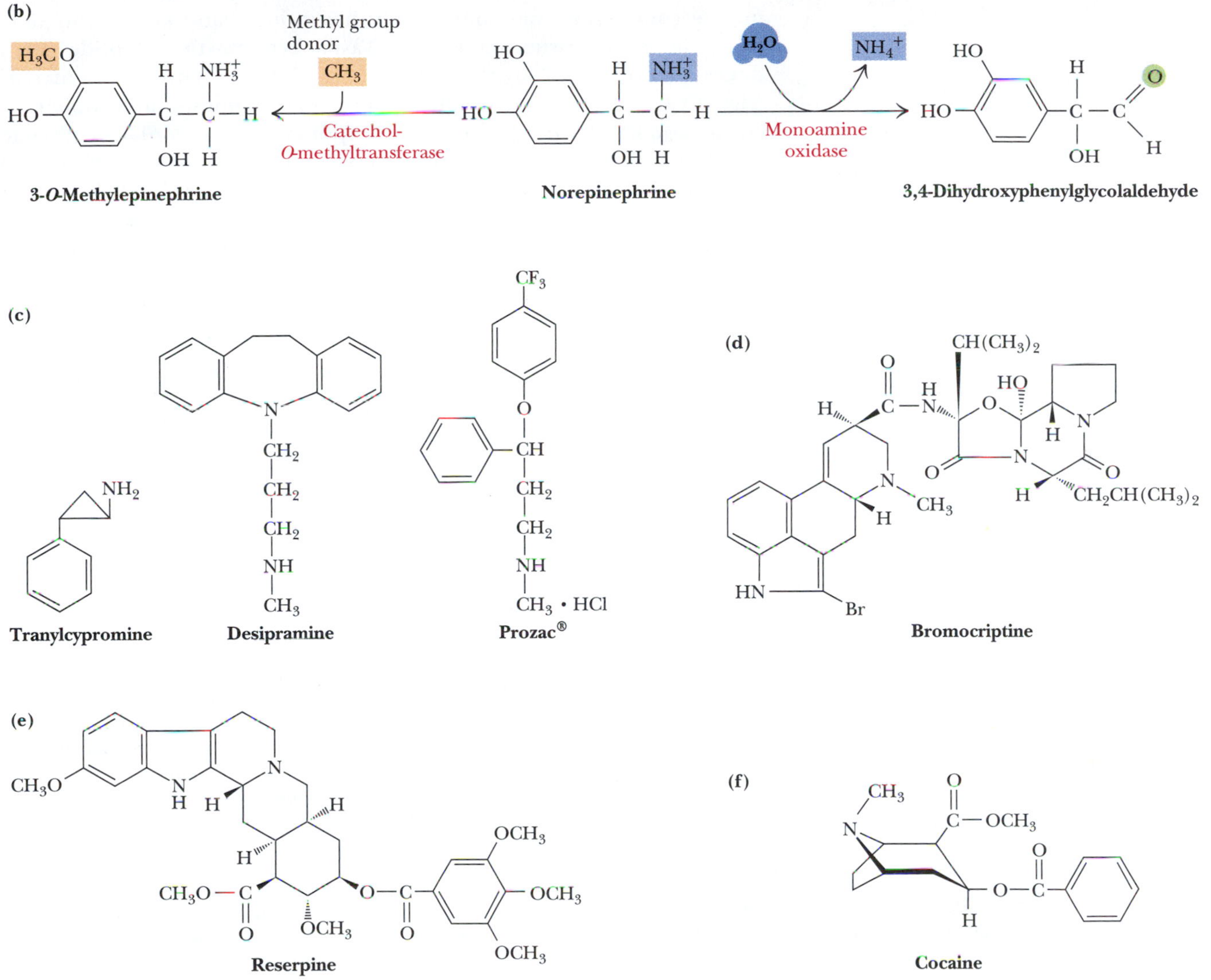

(b)

3-*O*-Methylepinephrine

Methyl group donor

Catechol-*O*-methyltransferase

Norepinephrine

Monoamine oxidase

3,4-Dihydroxyphenylglycolaldehyde

(c)

Tranylcypromine   Desipramine   Prozac®

(d)

Bromocriptine

(e)

Reserpine

(f)

Cocaine

▲ **(b)** Norepinephrine can be degraded in the synaptic cleft by catechol-*O*-methyltransferase or in the mitochondria of presynaptic neurons by monoamine oxidase. **(c)** The structures of tranylcypromine, desipramine, and Prozac. **(d)** The structure of bromocriptine. **(e)** The structure of reserpine. **(f)** The structure of cocaine.

Various Peptides Also Act as Neurotransmitters

Many relatively small peptides have been shown to possess neurotransmitter activity (see Table 32.5). One of the challenges of this field is that the known neuropeptides may represent a very small subset of the neuropeptides that exist. Another challenge arises from the small in vivo concentrations of these agents and the small number of receptors that are present in neural tissue.

Physiological roles for most of these peptides are complex. For example, the **endorphins** and **enkephalins** are natural opioid substances and potent pain relievers. The **endothelins** are a family of homologous regulatory peptides, synthesized by certain endothelial and epithelial cells that act on nearby smooth muscle and connective tissue cells. They induce or affect smooth muscle contraction; vasoconstriction; heart, lung, and kidney function; and mitogenesis and tissue remodeling. **Vasoactive intestinal peptide (VIP)** produces a G protein–adenylyl cyclase–mediated increase in cAMP, which in turn triggers a variety of protein phosphorylation cascades, one of which leads to conversion of phosphorylase b to phosphorylase a, stimulating glycogenolysis. Moreover, VIP has synergistic effects with other neurotransmitters, such as norepinephrine. In addition to increasing cAMP levels through β-adrenergic receptors, norepinephrine acting at α_1-adrenergic receptors markedly stimulates the increases in cAMP elicited by VIP. Many other effects have also been observed. For example, injection of VIP increases rapid eye movement (REM) sleep and decreases waking time in rats. VIP receptors exist in regions of the central nervous system involved in sleep modulation.

Summary

32.1 What Are Hormones?
Many different chemical species act as hormones. Steroid hormones, all derived from cholesterol, regulate metabolism, salt and water balances, inflammatory processes, and sexual function. Several hormones are amino acid derivatives. Among these are epinephrine and norepinephrine (which regulate smooth muscle contraction and relaxation, blood pressure, cardiac rate, and the processes of lipolysis and glycogenolysis) and the thyroid hormones (which stimulate metabolism). Peptide hormones are a large group of hormones that appear to regulate processes in all body tissues, including the release of yet other hormones. Hormones and other signal molecules in biological systems bind with very high affinities to their receptors, displaying K_D values in the range of 10^{-12} to 10^{-6} M. Hormones are produced at concentrations equivalent to or slightly above these K_D values. Once hormonal effects have been induced, the hormone is usually rapidly metabolized.

32.2 What Are Signal Transduction Pathways?
Hormonal regulation depends upon the transduction of the hormonal signal across the plasma membrane to specific intracellular sites, particularly the nucleus. Signal transduction pathways consist of a stepwise progression of signaling stages: receptor→transducer→effector. The receptor perceives the signal, transducers relay the signal, and the effectors convert the signal into an intracellular response. Often, effector action involves a series of steps, each of which is mediated by an enzyme, and each of these enzymes can be considered as an amplifier in a pathway connecting the hormonal signal to its intracellular targets.

32.3 How Do Signal-Transducing Receptors Respond to the Hormonal Message?
Steroid hormones may either bind to plasma membrane receptors or exert their effects within target cells, entering the cell and migrating to their sites of action via specific cytoplasmic receptor proteins. The nonsteroid hormones, which act by binding to outward-facing plasma membrane receptors, activate signal transduction pathways that mobilize various second messengers—cyclic nucleotides, Ca^{2+} ions, and other substances—that activate or inhibit enzymes or cascades of enzymes in very specific ways.

32.4 How Are Receptor Signals Transduced?
Receptor signals are transduced in one of three ways, to initiate actions inside the cell:
1. Exchange of GDP for GTP on GTP-binding proteins (G proteins), which in turn leads to generation of second messengers, including cAMP, phospholipid breakdown products, and Ca^{2+}.
2. Receptor-mediated activation of phosphorylation cascades that in turn trigger activation of various enzymes.
3. Conformation changes that open ion channels or recruit proteins into nuclear transcription complexes.

32.5 How Do Effectors Convert the Signals to Actions in the Cell?
Transduction of the hormonal signal leads to activation of effectors—usually protein kinases and protein phosphatases—that elicit a variety of actions that regulate discrete cellular functions. Of the thousands of mammalian kinases and phosphatases, the structures and functions of a few are representative, including protein kinase A (PKA), protein kinase C (PKC), and protein tyrosine phosphatase SHP-2.

32.6 What Is the Role of Protein Modules in Signal Transduction?
Signal transduction within cells occurs via protein–protein and protein–phospholipid interactions based on protein modules. Proteins with two (or more) such modules associate simultaneously with two (or more) binding partners, leading to assembly of functional complexes, either at an activated cell surface receptor or free in the cytoplasm. The SH2 domain, which binds with high affinity to peptide motifs containing a phosphotyrosine, is a good example.

32.7 How Do Neurotransmission Pathways Control the Function of Sensory Systems?
Nerve impulses, which can be propagated at speeds up to 100 m/sec, provide a means of intercellular signaling that is fast enough to encompass sensory recognition, movement, and other physiological functions and behaviors in higher animals. The generation and transmission of nerve impulses in vertebrates is mediated by an incredibly complicated neural network that connects every part of the organism with the brain—itself an interconnected array of as many as 10^{12} cells. Despite their complexity and diversity, the nervous systems of animals all possess common features and common mechanisms. Physical or chemical stimuli are recognized by specialized receptor proteins in the membranes of excitable cells. Conformational changes in the receptor protein result in a change in enzyme activity or a change in the permeability of the membrane. These changes are then propagated throughout the cell or from cell to cell in specific and reversible ways to carry information through the organism.

Problems

1. Compare and contrast the features and physiological advantages of each of the major classes of hormones, including the steroid hormones, polypeptide hormones, and the amino acid–derived hormones.
2. Compare and contrast the features and physiological advantages of each of the known classes of second messengers.
3. Nitric oxide may be merely the first of a new class of gaseous second messenger/neurotransmitter molecules. Based on your knowledge of the molecular action of nitric oxide, suggest another gaseous molecule that might act as a second messenger and propose a molecular function for it.
4. Herbimycin A is an antibiotic that inhibits tyrosine kinase activity by binding to SH groups of cysteine in the *src* gene tyrosine kinase and other similar tyrosine kinases. What effect might it have on normal rat kidney cells that have been transformed by Rous sarcoma virus? Can you think of other effects you might expect for this interesting antibiotic?
5. Monoclonal antibodies that recognize phosphotyrosine are commercially available. How could such an antibody be used in studies of cell signaling pathways and mechanisms?
6. Explain and comment on this statement: The main function of hormone receptors is that of signal amplification.
7. Synaptic vesicles are approximately 40 nm in outside diameter, and each vesicle contains about 10,000 acetylcholine molecules. Calculate the concentration of acetylcholine in a synaptic vesicle.
8. GTPγS is a nonhydrolyzable analog of GTP. Experiments with squid giant synapses reveal that injection of GTγS into the presynaptic end (terminal) of the neuron inhibits neurotransmitter release (slowly and irreversibly). The calcium signals produced by presynaptic action potentials and the number of synaptic vesicles docking on the presynaptic membrane are unchanged by GTPγS. Propose a model for neurotransmitter release that accounts for all of these observations.
9. A typical hormone binds to its receptor with an affinity (K_D) of approximately 1×10^{-9} M. Consider an in vitro (test-tube) system in which the total hormone concentration is approximately 1 nM and the total concentration of receptor sites is 0.1 nM. What fraction of the receptor sites is bound with hormone? If the concentration of receptors is decreased to 0.033 nM, what fraction of receptor is bound with the hormone?
10. (Integrates with Chapter 24.) All steroid hormones are synthesized in the human body from cholesterol. What is the consequence for steroid hormones and their action from taking a "statin" drug, such as Zocor, which blocks the synthesis of cholesterol in the body? Are steroid hormone functions compromised by statin action?
11. Given that β-strands provide a more genetically economical way for a polypeptide to cross a membrane, why has nature chosen α-helices as the membrane-spanning motif for G-protein–coupled receptors? That is, what other reasons can you imagine to justify the use of α-helices?
12. Write simple reaction mechanisms for the formation of cAMP from ATP by adenylyl cyclase and for the breakdown of cAMP to 5′-AMP by phosphodiesterases.
13. (Integrates with Chapter 9.) Consider the data in Figure 32.47. Recast Equation 9.2 to derive a form from which you could calculate the equilibrium electrochemical potential at which no net flow of potassium would occur. This is the Nernst equation. Calculate the equilibrium potential for K^+ and also for Na^+, assuming $t = 37°C$.
14. The calculation of the actual transmembrane potential difference for a neuron is accomplished with the Goldman equation:

$$\Delta\psi = \frac{RT}{\mathscr{F}} \ln\left(\frac{\Sigma P_C[C]_{outside} + \Sigma P_A[A]_{inside}}{\Sigma P_C[C]_{inside} + \Sigma P_A[A]_{outside}}\right)$$

where [C] and [A] are the cation and anion concentrations, respectively, and P_C and P_A are the respective permeability coefficients of cations and anions.

Assume relative permeabilities for K^+, Na^+, and Cl^- of 1, 0.04, and 0.45, respectively, and use this equation to calculate the actual transmembrane potential difference for the neuron whose ionic concentrations are those given in Figure 32.47.

15. Use the information in problems 13 and 14, together with Figure 32.34, to discuss the behavior of potassium, sodium, and chloride ions as an action potential propagates along an axon.

Preparing for the MCAT Exam

16. Malathion (Figure 32.45) is one of the secrets behind the near-complete eradication of the boll weevil from cotton fields in the United States. For most of the 20th century, boll weevils wreaked havoc on the economy of states from Texas to the Carolinas. When boll weevils attacked cotton fields in a farming community, the destruction of cotton plants meant loss of jobs for farm workers, bankruptcies for farm owners, and resulting hardship for the entire community. Relentless application of malathion to cotton crops and fields has turned the tide, however, and agriculture experts expect that boll weevils will be completely gone from cotton fields within a few years. Remarkably, malathion-resistant boll weevils have not emerged despite years of this pesticide's use. Consider the structure and chemistry of malathion and suggest what you would expect to be the ecological consequences of chronic malathion application to cotton fields.
17. Consult the excellent review article "Assembly of Cell Regulatory Systems Through Protein Interaction Domains" (*Science* **300**:445–452, 2003, by Pawson and Nash) and discuss the structural requirements for a regulatory protein operating in a signaling network.

Biochemistry Now™ Preparing for an exam? Test yourself on key questions at http://chemistry.brookscole.com/ggb3

Further Reading

Signal Transduction

Avruch, J., Zhang, X-F., and Kyriakis, J. M., 1994. Raf meets ras: Completing the framework of a signal transduction pathway. *Trends in Biochemical Sciences* **19**:279–283.

Bork, P., Schultz, J., and Ponting, C. P., 1997. Cytoplasmic signaling domains: The next generation. *Trends in Biochemical Sciences* **22**:296–298.

Casey, P. J., and Seabra, M. C., 1996. Protein prenyltransferases. *Journal of Biological Chemistry* **271**:5289–5292.

Dennis, E. A., 1997. The growing phospholipase A_2 superfamily of signal transduction enzymes. *Trends in Biochemical Sciences* **22**:1–2.

Feig, L. A., Urano, T., and Cantor, S., 1996. Evidence for a ras/ral signaling cascade. *Trends in Biochemical Sciences* **21**:438–441.

Graves, P. R., and Haystead. T. A., 2003. A functional proteomics approach to signal transduction. *Recent Progress in Hormone Research* **58**:1–24.

Kay, B. K., Williamson, M. P., and Sudol, M., 2000. The importance of being proline: the interaction of proline-rich motifs in signaling proteins with their cognate domains. *FASEB Journal* **14**:231–241.

Koch, C., et al., 1991. SH2 and SH3 domains: Elements that control interactions of cytoplasmic signaling proteins. *Science* **252**:668–674.

Li, G., and Qian, H., 2003. Sensitivity and specificity amplification in signal transduction. *Cell Biochemistry and Biophysics* **39**:45–60.

Marshall, M. S., 1993. The effector interactions of p21^ras. *Trends in Biochemical Sciences* **18**:250–254.

Milligan, G., Parenti, M., and Magee, A. I., 1995. The dynamic role of palmitoylation in signal transduction. *Trends in Biochemical Sciences* **20**:181–186.

Nicholson, D. W., and Thornberry, N. A., 1997. Caspases: Killer proteases. *Trends in Biochemical Sciences* **22**:299–306.

Omer, C. A., and Kohl, N. E., 1997. CA$_1$A$_2$X-competitive inhibitors of farnesyltransferase as anticancer agents. *Trends in Pharmacological Sciences* **18**:437–444.

Pawson, T., and Nash, P., 2003. Assembly of cell regulatory systems through protein interaction domains. *Science* **300**:445–452.

Pawson, T., and Scott, J. D., 1997. Signaling through scaffold, anchoring and adaptor proteins. *Science* **278**:2075–2080.

Salveson, G. S., and Dixit, V. M., 1997. Caspases: Intracellular signaling by proteolysis. *Cell* **91**:443–446.

Wittinghofer, A., and Pai, E., 1991. The structure of *ras* protein: A model for a universal molecular switch. *Trends in Biochemical Sciences* **16**:382–387.

G Proteins

Bourne, H. R., 1997. Pieces of the true grail: A G protein finds its target. *Science* **278**:1898–1899.

Cherfils, J., and Chabre, M., 2003. Activation of G-protein G$_\alpha$ subunits by receptors through G$_\alpha$-G$_\beta$ and G$_\alpha$-G$_\gamma$ interactions. *Trends in Biochemical Sciences* **28**:13–17.

Clapham, D. E., 1996. The G-protein nanomachine. *Nature* **379**:297–229.

Coleman, D. E., and Sprang, S. R., 1996. How G proteins work: A continuing story. *Trends in Biochemical Sciences* **21**:41–44.

Decaillot, F., Befort, K., Filliol, D., Yue, S., Walker, P., and Kieffer, B., 2003. Opioid receptor random mutagenesis reveals a mechanism for G protein-coupled receptor activation. *Nature Structural Biology* **10**:629–636.

Gudermann, T., Schönberg, T., and Schultz, G., 1997. Functional and structural complexity of signal transduction via G-protein-coupled receptors. *Annual Review of Neuroscience* **20**:399–427.

Sprang, S. R., 1997. G protein mechanisms: Insights from structural analysis. *Annual Review of Biochemistry* **66**:639–678.

Strader, C. D., Fong, T. M., Tota, M. R., and Underwood, D., 1994. Structure and function of G-protein-coupled receptors. *Annual Review of Biochemistry* **63**:101–132.

Traylor, T., and Sharma, V., 1992. Why NO? *Biochemistry* **31**:2847–2849.

Second Messengers

Asaoka, Y., et al., 1992. Protein kinase C, calcium and phospholipid degradation. *Trends in Biochemical Sciences* **17**:414–417.

Hannun, Y. A., and Obeid, L. M., 1995. Ceramide: An intracellular signal for apoptosis. *Trends in Biochemical Sciences* **20**:73–77.

Hofer, A. M., and Brown, E. M., 2003. Extracellular calcium sensing and signaling. *Nature Reviews of Molecular and Cellular Biology* **4**:530–538.

James, P., Vorherr, T., and Carafoli, E., 1995. Calmodulin-binding domains: Just two faced or multi-faceted? *Trends in Biochemical Sciences* **20**:38–42.

Nelson, E. J., Connolly, J., McArthur, P., 2003. Nitric oxide and S-nitrosylation: excitotoxic and cell signaling mechanism. *Biology of the Cell* **95**:3–8.

Pendaries, C., Tronchere, H., Plantavid, M., and Payrastre, B., 2003. Phosphoinositide signaling disorders in human disease. *FEBS Letters* **546**:25–31.

Plotkin, M., 1993. *Tales of a Shaman's Apprentice*. New York: Viking Penguin.

Putney, J. W., 1998. Calcium signaling: Up, down, up, down…what's the point? *Science* **279**:191–192.

Wickelgren, I., 1997. Biologists catch their first detailed look at NO enzyme. *Science* **278**:389.

Protein Kinases and Protein Phosphatases

Camps, M., Nichols, A., and Arkinstall, S., 2000. Dual specificity phosphatases: A gene family for control of MAP kinase function. *FASEB Journal* **14**:6–16.

Cohen, P., 1992. Signal integration at the level of protein kinases, protein phosphatases and their substrates. *Trends in Biochemical Sciences* **17**:408–413.

English, J., Pearson, G., Wilsbacher, J., Swantek, J., Karandikar, M., Xu, S., and Cobb, M. H., 1999. New insights into the control of MAP kinase pathways. *Experimental Cell Research* **253**:255–270.

Ferrell, J. E., 1997. How responses get more switch-like as you move down a protein kinase cascade. *Trends in Biochemical Sciences* **22**:288–289.

Johnson, G., and Lapadat, R., 2002. Mitogen-activated protein kinase pathways mediated by ERK, JNK, and p38 protein kinases. *Science* **298**:1911–1912.

Kemp, B., and Pearson, R., 1991. Intrasteric regulation of protein kinases and phosphatases. *Biochimica et Biophysica Acta* **1094**:67–76.

Keyse, S. M., 2000. Protein phosphatases and the regulation of mitogen-activated protein kinase signalling. *Current Opinions in Cell Biology* **12**:186–192.

Luttrell, L. M., Daaka, Y., and Lefkowitz, R. J., 1999. Regulation of tyrosine kinase cascades by G-protein-coupled receptors. *Current Opinions in Cell Biology* **11**:177–183.

Meskiene, I., and Hirt, H., 2000. MAP kinase pathways: Molecular plug-and-play chips for the cell. *Plant Molecular Biology* **42**:791–806.

Millward, T. A., Zolnierowicz, S., and Hemmings, B. A., 1999. Regulation of protein kinase cascades by protein phosphatase 2A. *Trends in Biochemical Sciences* **24**:186–191.

Neel, B. G., Gu, H., and Pao, L., 2003. The 'Shp'ing news: SH2 domain-containing tyrosine phosphatase in cell signaling. *Trends in Biochemical Sciences* **28**:284–293.

Virshup, D. M., 2000. Protein phosphatase 2A: A panoply of enzymes. *Current Opinions in Cell Biology* **12**:180–185.

Neurotransmission

Armstrong, C., 1998. The vision of the pore. *Science* **280**:56–57.

Bajjalieh, S. M., and Scheller, R. H., 1995. The biochemistry of neurotransmitter secretion. *The Journal of Biological Chemistry* **270**:1971–1974.

Clapham, D. E., 1997. Some like it hot: Spicing up ion channels. *Nature* **389**:783–784.

Doyle, D. A., Cabral, J. M., Pfuetzner, R. A., et al., 1998. The structure of the potassium channel: Molecular basis of K$^+$ conduction and selectivity. *Science* **280**:69–77.

For detailed answers to the end-of-chapter problems as well as additional problems to solve, see *The Student Solutions Manual, Study Guide and Problems Book* by David Jemiolo and Steven Theg that accompanies this textbook.

Chapter 1

1. Because bacteria (compared with humans) have simple nutritional requirements, their cells obviously contain enzyme systems that allow them to convert rudimentary precursors (even inorganic substances such as NH_4^+, NO_3^-, N_2, and CO_2) into complex biomolecules—proteins, nucleic acids, polysaccharides, and complex lipids. On the other hand, animals have an assortment of different cell types designed for specific physiological functions; these cells possess a correspondingly greater repertoire of complex biomolecules to accomplish their intricate physiology.

2. Consult Figures 1.21, 1.22, and 1.23 to confirm your answer.

3. a. Laid end to end, 250 *E. coli* cells would span the head of a pin.

   b. The volume of an *E. coli* cell is about 10^{-15} L.

   c. The surface area of an *E. coli* cell is about 6.3×10^{-12} m². Its surface-to-volume ratio is 6.3×10^6 m⁻¹.

   d. 600,000 molecules.

   e. 1.7 n*M*.

   f. Because we can calculate the volume of one ribosome to be 4.2×10^{-24} m³ (or 4.2×10^{-21} L), 15,000 ribosomes would occupy 6.3×10^{-17} L, or 6.3% of the total cell volume.

   g. Because the *E. coli* chromosome contains 4600 kilobase pairs (4.6×10^6 bp) of DNA, its total length would be 1.6 mm—approximately 800 times the length of an *E. coli* cell. This DNA would encode 4300 different proteins, each 360 amino acids long.

4. a. The volume of a single mitochondrion is about 4.2×10^{-16} L (about 40% the volume calculated for an *E. coli* cell in problem 3).

   b. A mitochondrion would contain on average fewer than eight molecules of oxaloacetate.

5. a. Laid end to end, 25 liver cells would span the head of a pin.

   b. The volume of a liver cell is about 8×10^{-12} L (8000 times the volume of an *E. coli* cell).

   c. The surface area of a liver cell is 2.4×10^{-9} m²; its surface-to-volume ratio is 3×10^5 m⁻¹, or about 0.05 (1/20) that of an *E. coli* cell. Cells with lower surface-to-volume ratios are limited in their exchange of materials with the environment.

   d. The number of base pairs in the DNA of a liver cell is 6×10^9 bp, which would amount to a total DNA length of 2 m (or 6 feet of DNA!) contained within a cell that is only 20 μm on a side. Maximal information content of liver-cell DNA = 3×10^9 bp, which, expressed in proteins 400 amino acids in length, could encode 2.5×10^6 proteins.

6. The amino acid side chains of proteins provide a range of shapes, polarity, and chemical features that allow a protein to be tailored to fit almost any possible molecular surface in a complementary way.

7. Biopolymers may be informational molecules because they are constructed of different monomeric units ("letters") joined head to tail in a particular order ("words, sentences"). Polysaccharides are often linear polymers composed of only one (or two repeating) monosaccharide unit(s) and thus display little information content. Polysaccharides with a variety of monosaccharide units may convey information through specific recognition by other biomolecules. Also, most monosaccharide units are typically capable of forming branched polysaccharide structures that are potentially very rich in information content (as in cell surface molecules that act as the unique labels displayed by different cell types in multicellular organisms).

8. Molecular recognition is based on structural complementarity. If complementary interactions involved covalent bonds (strong forces), stable structures would be formed that would be less responsive to the continually changing dynamic interactions that characterize living processes.

9. Two carbon atoms interacting through van der Waals forces are 0.34 nm apart; two carbon atoms joined in a covalent bond are 0.154 nm apart.

10. Slight changes in temperature, pH, ionic concentrations, and so forth may be sufficient to disrupt weak forces (H bonds, ionic bonds, van der Waals interactions, hydrophobic interactions).

11. Living systems are maintained by a continuous flow of matter and energy through them. Despite the ongoing transformations of matter and energy by these highly organized, dynamic systems, no overt changes seem to occur in them: They are in a *steady* state.

12. Increasing kinetic energy increases the motions of molecules and raises their average energy, which means that the difference between the energy to disrupt a weak force between two molecules and the energy of the weak force is smaller. Thus, increases in kinetic energy may break the weak forces between molecules.

13. Informational polymers must have "sense" or direction, and they must be composed of more than one kind of monomer unit.

Chapter 2

1. a. 3.3; b. 9.85; c. 5.7; d. 12.5; e. 4.4; f. 6.97.

2. a. 1.26 m*M*; b. 0.25 m*M*; c. 4×10^{-12} *M*; d. 2×10^{-4} *M*; e. 3.16×10^{-10} *M*; f. 1.26×10^{-7} *M* (0.126 μ*M*).

3. a. $[H^+] = 2.51 \times 10^{-5}$ *M*; b. $K_a = 3.13 \times 10^{-8}$ *M*; p$K_a = 7.5$.

4. a. pH = 2.38; b. pH = 4.23.

5. Combine 187 mL of 0.1 *M* acetic acid with 813 mL of 0.1 *M* sodium acetate.

6. $[HPO_4^{-2}]/[H_2PO_4^-] = 0.398$.

7. Combine 555.7 mL of 0.1 *M* Na_3PO_4 with 444.3 mL of 0.1 *M* H_3PO_4. Final concentrations of ions will be $[H_2PO_4^-] = 0.0333$ *M*; $[HPO_4^{2-}] = 0.0667$ *M*; $[Na^+] = 0.1667$ *M*; $[H^+] = 3.16 \times 10^{-8}$ *M*.

8. Add 432 mL of 0.1 N HCl to 1 L 0.05 *M* BICINE. $[BICINE]_{total} = 0.05$ *M*/1.432 L = 0.0349 *M* [protonated form] = 0.0302 *M*.

9. a. Fraction of H_3PO_4: @pH 0 = 0.993; @pH 2 = 0.58; @pH 4 = 0.01; negligible @pH 6.

   b. Fraction of $H_2PO_4^-$: @pH 0 = 0.007; @pH 2 = 0.41; @pH 4 = 0.986; @pH 6 = 0.94; @pH 8 = 0.14; negligible @pH 10.

   c. Fraction of HPO_4^{2-}: negligible @pH 0, 2, and 4; @pH 6 = 0.06; @pH 8 = 0.86; @pH 10 ≈ 1.0; @pH 12 = 0.72.

   d. Fraction of PO_4^{3-}: negligible at any pH <10; @pH 12 = 0.28.

10. At pH 5.2, $[H_3A] = 4.33 \times 10^{-5}$ M; $[H_2A^-] = 0.0051$ M; $[HA^{2-}] = 0.014$ M; $[A^{3-}] = 0.0009$ M.

11. a. pH = 7.02; $[H_2PO_4^-] = 0.0200$ M; $[HPO_4^{2-}] = 0.0133$ M.

    b. pH = 7.38; $[H_2PO_4^-] = 0.0133$ M; $[HPO_4^{2-}] = 0.0200$ M.

12. $[H_2CO_3] = 2.2$ μM; $[CO_{2(d)}] = 0.75$ mM. When $[HCO_3^-] = 15$ mM and $[CO_{2(d)}] = 3$ mM, pH = 6.8.

13. Titration of the fully protonated form of anserine will require the addition of three equivalents of OH^-. The pK_a values lie at 2.64 (COOH); 7.04 (imidazole-N$^+$H); and 9.49 (NH$_3^+$). Its isoelectric point lies midway between pK_2 and pK_3, so pH_I = 8.265. To prepare 1 L of 0.04 M anserine buffer, add 164 mL of 0.1 M HCl to 400 mL of 0.1 M anserine at its isoelectric point, and make up to 1 L final volume.

14. Add 410 mL of 0.1 M NaOH to 250 mL of 0.1 M HEPES in its fully protonated form and make up to 1 L final volume.

15. 166.7 g/mole.

16. A drop in blood pH would occur.

17. c.

Chapter 3

1. $K_{eq} = 613$ M; $\Delta G° = -15.9$ kJ/mol.

2. $\Delta G° = 1.69$ kJ/mol at 20°C; $\Delta G° = -5.80$ kJ/mol at 30°C. $\Delta S° = 0.75$ kJ/mol · K.

3. $\Delta G = -24.8$ kJ/mol.

4. State functions are quantities that depend on the state of the system and not on the path or process taken to reach that state. Volume, pressure, and temperature are state functions. Heat and all forms of work, such as mechanical work and electrical work, are not state functions.

5. $\Delta G°' = \Delta G° - 39.5$ n (in kJ/mol), where n is the number of H^+ produced in any process. So $\Delta G° = \Delta G°' + 39.5$ $n = -30.5$ kJ/mol + 39.5(1) kJ/mol.

   $\Delta G° = 9.0$ kJ/mol at 1 M $[H^+]$.

6. a. $K_{eq}(AC) = (0.02 \times 1000) = 20$.

   b. $\Delta G°(AB) = 10.1$ kJ/mol.

      $\Delta G°(BC) = -17.8$ kJ/mol.

      $\Delta G°(AC) = -7.7$ kJ/mol.

      $K_{eq} = 20$.

7. See *The Student Solutions Manual, Study Guide and Problems Book* for resonance structures.

8. $K_{eq} = [Cr][P_i]/[CrP][H_2O]$.

   $K_{eq} = 3.89 \times 10^7$.

9. CrP in the amount of 135.3 moles would be required per day to provide 5860 kJ energy. This corresponds to 17,730 g of CrP per day. With a body content of 20 g CrP, each molecule would recycle 886 times per day.

   Similarly, 637 moles of glycerol-3-P, or 108,300 g of glycerol-3-P, would be required. Each molecule would recycle 5410 times/day.

10. $\Delta G = -46.1$ kJ/mol.

11. The hexokinase reaction is a sum of the reactions for hydrolysis of ATP and phosphorylation of glucose:

$$ATP + H_2O \rightleftharpoons ADP + P_i$$
$$\underline{Glucose + P_i \rightleftharpoons G\text{-}6\text{-}P + H_2O}$$
$$Glucose + ATP \rightleftharpoons G\text{-}6\text{-}P + ADP$$

The free energy change for the hexokinase reaction can thus be obtained by summing the free energy changes for the first two reactions listed here.

$\Delta G°'$ for hexokinase = -30.5 kJ/mol + 13.9 kJ/mol = -16.6 kJ/mol

12. Comparing the acetyl group of acetoacetyl-CoA and the methyl group of acetyl-CoA, it is reasonable to suggest that the acetyl group is more electron-withdrawing in nature. For this reason, it tends to destabilize the thiol ester of acetoacetyl-CoA, and the free energy of hydrolysis of acetoacetyl-CoA should be somewhat larger than that of acetyl-CoA. In fact, $\Delta G°' = -43.9$ kJ/mol, compared with -31.5 kJ/mol for acetyl-CoA.

13. Carbamoyl phosphate should have a somewhat larger free energy of hydrolysis than acetyl phosphate, at least in part because of greater opportunities for resonance stabilization in the products. In fact, the free energy of hydrolysis of carbamoyl phosphate is -51.5 kJ/mol, compared with -43.3 kJ/mol for acetyl phosphate.

14. The denaturation of chymotrypsinogen is spontaneous at 58°C, because the $\Delta G°$ at this temperature is negative (at approximately -2.8 kJ/mol). The native and denatured forms are in equilibrium at approximately 56.6°C.

15. The positive values for ΔC_p for the protein denaturations described in Table 3.1 reflect an increase in motion of the peptide chain in the denatured state. This increased motion provides new ways to store heat energy.

Chapter 4

1. Structures for glycine, aspartate, leucine, isoleucine, methionine, and threonine are presented in Figure 4.3.

2. Asparagine = Asn = N.

   Arginine = Arg = R.

   Cysteine = Cys = C.

   Lysine = Lys = K.

   Proline = Pro = P.

   Tyrosine = Tyr = Y.

   Tryptophan = Trp = W.

3.

Alanine dissociation:

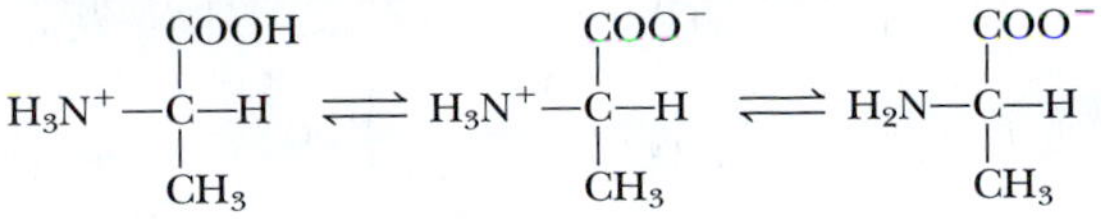

$$H_3N^+{-}\underset{\underset{CH_3}{|}}{\overset{\overset{COOH}{|}}{C}}{-}H \rightleftharpoons H_3N^+{-}\underset{\underset{CH_3}{|}}{\overset{\overset{COO^-}{|}}{C}}{-}H \rightleftharpoons H_2N{-}\underset{\underset{CH_3}{|}}{\overset{\overset{COO^-}{|}}{C}}{-}H$$

Glutamate dissociation:

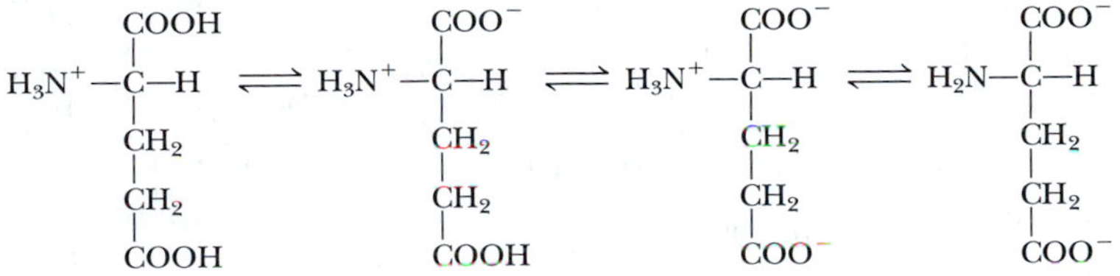

Histidine dissociation:

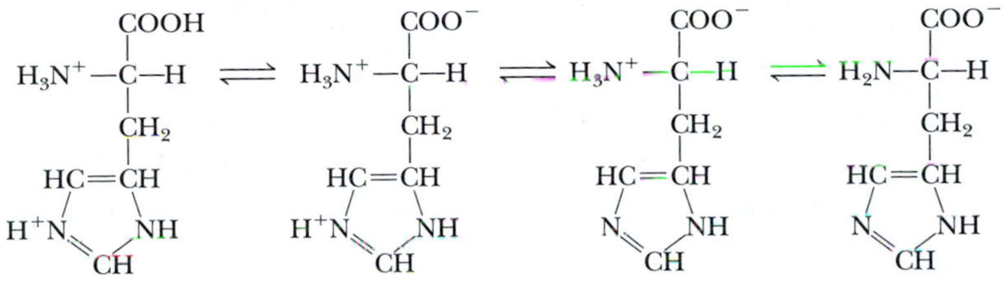

Lysine dissociation:

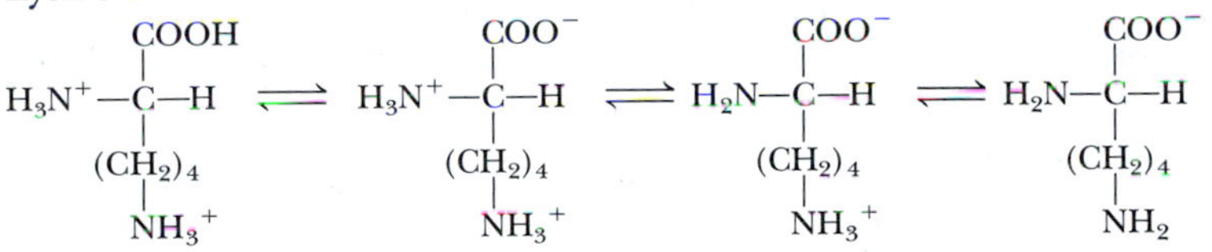

Phenylalanine dissociation:

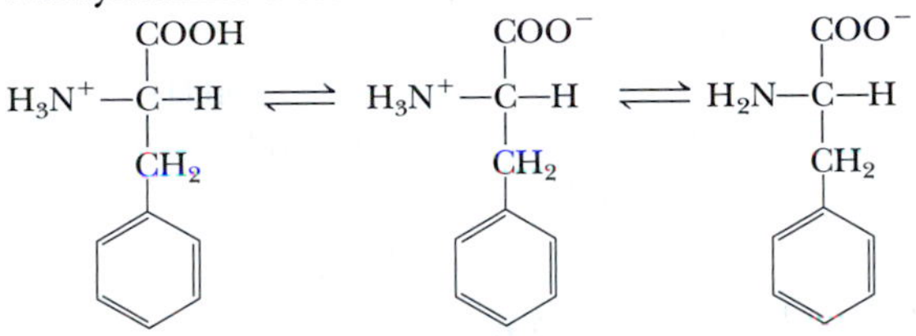

4. The proximity of the α-carboxyl group lowers the pK_a of the α-amino group.

5.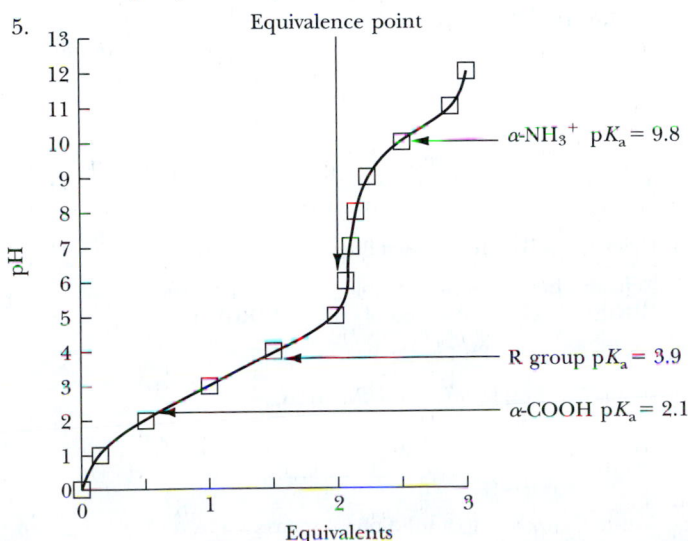

Equivalence point

α-NH$_3^+$ $pK_a = 9.8$

R group $pK_a = 3.9$

α-COOH $pK_a = 2.1$

Equivalents

6. Denoting the four histidine species as His^{2+}, His$^+$, His^0, and His$^-$, the concentrations are:

pH 2: $[\text{His}^{2+}] = 0.097\ M$, $[\text{His}^+] = 0.153\ M$, $[\text{His}^0] = 1.53 \times 10^{-5}\ M$, $[\text{His}^-] = 9.6 \times 10^{-13}\ M$.

pH 6.4: $[\text{His}^{2+}] = 1.78 \times 10^{-4}\ M$, $[\text{His}^+] = 0.071\ M$, $[\text{His}^0] = 0.179\ M$, $[\text{His}^-] = 2.8 \times 10^{-4}\ M$.

pH 9.3: $[\text{His}^{2+}] = 1.75 \times 10^{-12}\ M$, $[\text{His}^+] - 5.5 \times 10^{-5}\ M$, $[\text{His}^0] = 0.111\ M$, $[\text{His}^-] = 0.139\ M$.

7. $pH = pK_a + \log (2/1) = 4.3 + 0.3 = 4.6$.

The γ-carboxyl group of glutamic acid is 2/3 dissociated at pH = 4.6.

8. $pH = pK_a + \log (1/4) = 10.5 + (-0.6) = 9.9$.

9. a. The pH of a 0.3 M leucine hydrochloride solution is approximately 1.46.

 b. The pH of a 0.3 M sodium leucinate solution is approximately 11.5.

 c. The pH of a 0.3 M solution of isoelectric leucine is approximately 6.05.

10. $[\text{Arg}] = (35\ \text{degrees})/([\alpha]_D^{25} \times 1\ \text{dm})$.

Using $[\alpha]_D^{25}$ for L-arginine $= 12.5$ (from Table 4.2), then $[\text{Arg}] = 2.8\ \text{g/mL}$.

11. The sequence of reactions shown would demonstrate that L(−)-serine is related stereochemically to L(−)-glyceraldehyde:

$$
\begin{array}{ccc}
\text{COOH} & \text{COOH} & \text{COOH} \\
\text{H}_2\text{N}\!-\!|\!-\!\text{H} \;\longrightarrow\; & \text{H}_2\text{N}\!-\!|\!-\!\text{H} \;\longleftarrow\; & \text{H}\!-\!|\!-\!\text{Br} \\
\text{CH}_2\text{OH} & \text{CH}_3 & \text{CH}_3 \\
\text{L-}(-)\text{-Serine} & \text{L-}(-)\text{-Alanine} & \text{2-Bromopropanoic acid}
\end{array}
$$

$$
\begin{array}{ccc}
\text{CHO} & \text{COOH} & \text{COOH} \\
\text{HO}\!-\!|\!-\!\text{H} \;\longrightarrow\; & \text{HO}\!-\!|\!-\!\text{H} \;\longrightarrow\; & \text{HO}\!-\!|\!-\!\text{H} \\
\text{CH}_2\text{OH} & \text{CH}_2\text{OH} & \text{CH}_3 \\
\text{L-}(-)\text{-Glyceraldehyde} & \text{L-Glyceric acid} & \text{2-Hydroxypropanoic acid}
\end{array}
$$

Straight arrows indicate reactions that occur with retention of configuration. Looped arrows indicate inversion of configuration. (*From Kopple, K. D., 1966. Peptides and Amino Acids. New York: Benjamin Co.*)

12. Cystine (disulfide-linked cysteine) has two chiral carbons, the two α-carbons of the cysteine moieties. Each chiral center can exist in two forms, so there are four stereoisomers of cystine. However, it is impossible to distinguish the difference between L-cysteine/D-cysteine and D-cysteine/L-cysteine dimers. So three distinct isomers are formed:

$$
\begin{array}{cc}
\text{COO}^- \quad \text{COO}^- & \text{COO}^- \quad \text{COO}^- \\
\overset{+}{\text{H}_3\text{N}}\!-\!\text{C}\!-\!\text{H} \;\; \overset{+}{\text{H}_3\text{N}}\!-\!\text{C}\!-\!\text{H} & \text{H}\!-\!\text{C}\!-\!\overset{+}{\text{NH}_3} \;\; \text{H}\!-\!\text{C}\!-\!\overset{+}{\text{NH}_3} \\
\text{CH}_2\!-\!\text{S}\!-\!\text{S}\!-\!\text{CH}_2 & \text{CH}_2\!-\!\text{S}\!-\!\text{S}\!-\!\text{CH}_2 \\
\text{Mirror plane} & \\
\text{L-Cystine} & \text{D-Cystine}
\end{array}
$$

$$
\begin{array}{c}
\text{COO}^- \quad \text{COO}^- \\
\overset{+}{\text{H}_3\text{N}}\!-\!\text{C}\!-\!\text{H} \;\; \text{H}\!-\!\text{C}\!-\!\overset{+}{\text{NH}_3} \\
\text{CH}_2\!-\!\text{S}\!-\!\text{S}\!-\!\text{CH}_2 \\
\text{Mirror plane} \\
meso\text{-Cystine}
\end{array}
$$

13.

$$
\begin{array}{ccc}
& \text{O} & \text{NH}_2 \\
\text{I}\!-\!\text{CH}_2\!-\!\overset{\|}{\text{C}}\!-\!\text{NH}_2 & & \overset{|}{\text{C}}\!=\!\text{O} \\
\text{S}^- & \;\text{I}^- & \text{CH}_2 \\
\text{CH}_2 & \longrightarrow & \text{S} \\
\overset{+}{\text{H}_3\text{N}}\!-\!\text{C}\!-\!\text{COO}^- & & \text{CH}_2 \\
\text{H} & & \overset{+}{\text{H}_3\text{N}}\!-\!\text{C}\!-\!\text{COO}^- \\
& & \text{H}
\end{array}
$$

14. Basic amino acids should adhere strongly to Dowex-50, a sulfonated resin, whereas acidic amino acids should bind least strongly, with neutral and hydrophobic amino acids being intermediate in behavior. Therefore, the order of elution should be aspartate, valine, isoleucine, histidine, and finally arginine. (The pK_a values for valine are slightly lower than those for isoleucine—thus the elution time of valine should be somewhat slower than that of isoleucine.)

15. L-threonine is (2S, 3R)-threonine.

    D-threonine is (2R, 3S)-threonine.

    L-allothreonine is (2S, 3S)-threonine.

    D-allothreonine is (2R, 3R)-threonine.

Chapter 5

1. Nitrate reductase is a dimer (2 Mo/240,000 M_r).

2. Phe-Asp-Tyr-Met-Leu-Met-Lys.

3. Tyr-Asn-Trp-Met-(Glu-Leu)-Lys. Parentheses indicate that the relative positions of Glu and Leu cannot be assigned from the information provided.

4. Ser-Glu-Tyr-Arg-Lys-Lys-Phe-Met-Asn-Pro.

5. Ala-Arg-Met-Tyr-Asn-Ala-Val-Tyr or Asn-Ala-Val-Tyr-Ala-Arg-Met-Tyr sequences both fit the results. (That is, in one-letter code, either *ARMYNAVY* or *NAVYARMY*.)

6. Gly-Arg-Lys-Trp-Met-Tyr-Arg-Phe.

7. Actually, there are four possible sequences: NIGIRVIA, GINIRVIA, VIRNIGIA, and of course, VIRGINIA.

8. Gly-Trp-Arg-Met-Tyr-Lys-Gly-Pro.

9.   Leu-Met-Cys-Val-Tyr-Arg-Cys-Gly-Pro.

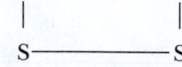

10. Alanine, attached to a solid-phase matrix via its α-carboxyl group, is reacted with diisopropylcarbodiimide-activated lysine. Both the α-amino and ε-amino groups of the lysine must be blocked with

9-fluorenyl-methoxycarbonyl (Fmoc) groups. To add leucine to Lys-Ala to form a linear tripeptide, precautions must be taken to prevent the incoming Leu α-carboxyl group from reacting inappropriately with the Lys ε-amino group instead of the Lys α-amino group.

11. The mass of the myoglobin chain is calculated to be $16{,}947 \pm 1$ daltons.

12. Unlike any amino acid side chain, the phosphate group (or more appropriately, the phosphoryl group) bears two equivalents of negative charge at physiological pH. Furthermore, replacing an H atom on an S, T, or Y side chain with a phosphoryl group introduces a very bulky substituent into the protein structure where none existed before.

13. Centromere protein F. It has a molecular weight of 367,594 and yields 283 tryptic peptides.

14. Nucleophilic attack by the hydroxyl O of the active-site serine on the carbonyl carbon of a peptide bond.

15. a. Amino acid changes in mutant hemoglobins that appear on the surface of the folded globin chains may affect quaternary structure.

    b. Amino acid substitutions on the surface on the quaternary hemoglobin structure that create hydrophobic patches might lead to polymerization. Such amino acids would include all of the hydrophobic amino acids.

Chapter 6

1. The central rod domain of keratin is composed of distorted α-helices, with 3.6 residues per turn, but a pitch of 0.51 nm, compared with 0.54 nm for a true α-helix.

   (0.51 nm/turn)(312 residues)/(3.6 residues/turn) =
   $$44.2 \text{ nm} = 442 \text{ Å}.$$

   For an α-helix, the length would be:

   (0.54 nm/turn)(312 residues)/(3.6 residues/turn) =
   $$46.8 \text{ nm} = 468 \text{ Å}.$$

   The distance between residues is 0.347 nm for antiparallel β-sheets and 0.325 nm for parallel β-sheets. So 312 residues of antiparallel β-sheet amount to 1083 Å and 312 residues of parallel β-sheet amount to 1014 Å.

2. The collagen helix has 3.3 residues per turn and 0.29 nm per residue, or 0.96 nm/turn. Then:

   (4 in/year)(2.54 cm/in)(10^7 nm/cm)/(0.96 nm/turn) =
   $$1.06 \times 10^8 \text{ turns/year}.$$

   (1.06×10^8 turns/year)(1 year/365 days)(1 day/24 hours)
   (1 hour/60 minutes) = 201 turns/minute.

3. **Asp:** The ionizable carboxyl can participate in ionic and hydrogen bonds. Hydrophobic and van der Waals interactions are negligible.

   Leu: The leucine side chain does not participate in hydrogen bonds or ionic bonds, but it will participate in hydrophobic and van der Waals interactions.

   Tyr: The phenolic hydroxyl of tyrosine, with a relatively high pK_a, will participate in ionic bonds only at high pH but can both donate and accept hydrogen bonds. Uncharged tyrosine is capable of hydrophobic interactions. The relatively large size of the tyrosine side chain will permit substantial van der Waals interactions.

   His: The imidazole side chain of histidine can act as both an acceptor and donor of hydrogen bonds and, when protonated, can participate in ionic bonds. Van der Waals interactions are expected, but hydrophobic interactions are less likely in most cases.

4. As an imido acid, proline has a secondary nitrogen with only one hydrogen. In a peptide bond, this nitrogen possesses no hydrogens and thus cannot function as a hydrogen-bond donor in α-helices. On the other hand, proline stabilizes the *cis*-configuration of a peptide bond and is thus well suited to β-turns, which require the *cis*-configuration.

5. For a right-handed crossover, moving in the N-terminal to C-terminal direction, the crossover moves in a clockwise direction when viewed from the C-terminal side toward the N-terminal side. The reverse is true for a left-handed crossover; that is, movement from N-terminus to C-terminus is accompanied by counterclockwise rotation.

6. The Ramachandran plot reveals allowable values of ϕ and ψ for α-helix and β-sheet formation. The plots consider steric hindrance and will be somewhat specific for individual amino acids. For example, peptide bonds containing glycine can adopt a much wider range of ϕ and ψ angles than can peptide bonds containing tryptophan.

7. The protein appears to be a tetramer of four 60-kD subunits. Each of the 60-kD subunits in turn is a heterodimer of two peptides, one of 34 kD and one of 26 kD, joined by at least one disulfide bond.

8. Hydrophobic interactions frequently play a major role in subunit–subunit interactions. The surfaces that participate in subunit–subunit interactions in the B_4 tetramer are likely to possess larger numbers of hydrophobic residues than the corresponding surfaces of protein A.

9. The length is given by (53 residues)×(0.15 nm run/residue) = 7.95 nm. The number of turns in the helix is given by (53 residues)/(3.6 residues/turn) = 14.7 turns. There are 49 hydrogen bonds in this helix.

10. Glycines are essential components of tight turns (β-turns) and thus are often essential for maintenance of protein structure.

11. Asp, Glu, Ser, Thr, His, and perhaps also Asn, Gln, Cys, Arg, Lys.

12. The ability of poly-Glu to form α-helices requires that the glutamate carboxyls be protonated. Deprotonation produces a polyanionic peptide that is not amenable to helix formation.

13. A coiled-coil formed from α-helices with 3.5 residues per turn would form a symmetrical seven-residue-repeating structure that would place the first and fourth residues of the seven-residue repeat at the same positions about the helix axis in every seven-residue repeat. This would allow the two helices of a coiled-coil structure to lie side by side with no twist about the coiled-coil axis. Such a structure would probably not be as stable as the twisted structure of coiled coils formed from α-helices with 3.6 residues per turn.

14. a. The third sequence would place hydrophobic residues on both sides of a β-strand and could thus be found in a parallel β-sheet.

    b. The second sequence would place hydrophobic residues on just one side of a β-strand and could thus be found in an anti-parallel β-sheet.

    c. The sixth sequence consists of GPX repeats (where X is any amino acid) and could thus be part of a tropocollagen molecule.

    d. The first sequence consists of seven-residue repeats, with first and fourth residues hydrophobic, and could thus be part of a coiled coil structure.

15. The solution to this exercise is to be completed by the student.

Chapter 7

1. See structures in *The Student Solutions Manual, Study Guide and Problems Book*.

2. See structures and titration curve in *The Student Solutions Manual, Study Guide and Problems Book*.

3. The systematic name for stachyose (Figure 7.19) is β-D-fructofuranosyl-*O*-α-D-galactopyranosyl-(1→6)-*O*-α-D-galactopyranosyl-(1→6)-α-D-glucopyranoside.

4. Glycated hemoglobin can be separated from ordinary hemoglobin on the basis of charge difference (by ion-exchange chromatography, high-performance liquid chromatography [HPLC] electrophoresis, and isoelectric focusing) or on the basis of structural difference (by affinity chromatography).

5. The systematic name for trehalose is α-D-glucopyranosyl-(1→1)-α-D-glucopyranoside.

   Trehalose is not a reducing sugar. Both anomeric carbons are occupied in the disaccharide linkage.

6. See structures in *The Student Solutions Manual, Study Guide and Problems Book*.

7. A sample that is 0.69 g α-D-glucose/mL and 0.31 g β-D-glucose/mL will produce a specific rotation of 83°.

9. A 0.2 g sample of amylopectin corresponds to 0.2 g/162 g/mole or 1.23×10^{-3} mole glucose residues; 50 μmole is 0.04 of the total sample or 4% of the residues. Methylation of such a sample should yield 1,2,3,6-tetramethylglucose for the glucose residues on the reducing ends of the sample. The amylopectin sample contains 1.2×10^{18} reducing ends.

10. There are several target sites for trypsin and chymotrypsin in the extracellular sequence of glycophorin, and it would be reasonable to expect that access to these sites by trypsin and chymotrypsin would be restricted by the presence of oligosaccharides in the extracellular domain of glycophorin.

11. Energy yield upon combustion (whether by metabolic pathways or other reactions) depends on the oxidation level. Carbohydrate and protein are at approximately the same oxidation level, and both of these are significantly less than that of fat.

12. This mechanism could involve either an S_N1 or S_N2 mechanism. In the former, protonation of the bridging oxygen would result in dissociation to produce a carbo-cation intermediate, which could be attacked by the phosphate nucleophile. The observation of retention of configuration at the anomeric carbon favors this mechanism. An S_N2 mechanism would presumably involve water attack at the anomeric carbon, with dissociation of the oxygen of the carbohydrate chain. This would be followed by S_N2 attack by phosphate. Two S_N2 attacks would result in retention of configuration at the anomeric carbon atom.

13. The enzyme in Beano is an α-galactosidase. This enzyme breaks down two of the three linkages in stachyose and similar oligosaccharides. Moreover, the third linkage requires α-glucosidase activity, and it is reasonable to expect that the α-galactosidase in Beano will have sufficient α-glucosidase activity to hydrolyze the third glycosidic bond in stachyose. The mechanism is shown in *The Student Solutions Manual, Study Guide and Problems Book*.

14. Laetrile contains a cyanide group. Breakdown of laetrile and release of this cyanide function in the body would be highly toxic.

15. Chondroitin and glucosamine are amino sugar components of cartilage and connective tissue. Dietary supplement with these substances could help replenish the carbohydrate groups of cartilage matrix proteoglycan, relieving pain and restoring the proper function of connective tissue structures.

16. Basic amino acid side chains (Arg, His, Lys) in antithrombin III present positively charged side chains for ionic interactions with sulfate functions on heparin; H-bond donating amino acid side chains could form H bonds with O atoms in —OH groups on the heparin carbohydrate residues; H-bond accepting amino acid side chains could form H bonds with H atoms in —OH groups of heparin.

17. Because these glycosaminoglycans are rich in hydroxyl groups, amine groups, and anionic functions (carboxylates, sulfates), they interact strongly with water. The heavily hydrated proteoglycans formed from these glycosaminoglycans are reversibly dehydrated in response to the pressure imposed on a joint during normal body movements. This dehydration has a cushioning effect on the joint; when the pressure is relieved, the glycosaminoglycans are spontaneously rehydrated due to their affinity for water.

Chapter 8

1. Because the question specifically asks for triacylglycerols that contain stearic acid *and* arachidonic acid, we can discount the triacylglycerols that contain only stearic or only arachidonic acid. In this case, there are six possibilities:

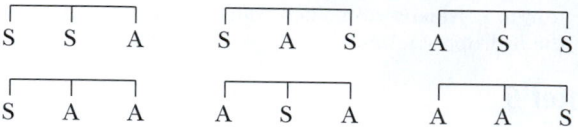

2. See *The Student Solutions Manual, Study Guide and Problems Book* for a discussion.

3. a. Phosphatidylethanolamine and phosphatidylserine have a net positive charge at low pH.

   b. Phosphatidic acid, phosphatidylglycerol, phosphatidylinositol, phosphatidylserine, and diphosphatidylglycerol normally carry a net negative charge.

   c. Phosphatidylethanolamine and phosphatidylcholine carry a net zero charge at neutral pH.

4. Diets high in cholesterol contribute to heart disease and stroke. On the other hand, plant sterols bind to cholesterol receptors in the intestines but are not taken up by the cells containing these receptors. There is substantial evidence that a diet that includes plant sterols can reduce serum cholesterol levels significantly.

5. Former Interior Secretary James Watt was well known during the Reagan administration for his comments on several occasions that trees caused and produced air pollution. As noted in this chapter, it is of course true that trees emit isoprenes that are the cause of the blue–gray haze that is common in still air in mid- to late-summer in the eastern United States. However, these isoprene compounds are not significantly toxic to living things, and it would be misleading to call them pollutants.

6. Louis L'Amour clearly knew his biochemistry. His protagonist knew that fat carries a higher energy content than protein or carbohydrate. If meals are going to be scarce (as for a person living in the wild and on the run), it is wise to consume fat rather than protein or carbohydrate. The same reasoning applies for migratory birds in the weeks preceding their long flights.

7. Benecol contains plant sterols (from pine trees, a plentiful source), which are not taken up by intestinal cholesterol receptors and which in fact bind to cholesterol receptors, blocking the absorption of cholesterol itself. See the Human Biochemistry box on page 263 for the structures of typical plant sterols.

8. The lethal dose for 50% of animals is referred to as the LD_{50}. The LD_{50} for dogs is approximately 3 mg/kg of body weight. Thus, for a 40-lb dog (18.2 kg), consumption of approximately 55 mg of warfarin would be lethal for 50% of animals.

9. See the *Student Solution Guide* and *http://chemistry.brookscole.com/ggb3* for the solutions to this problem.

10. Humans require approximately 2000 kcal per day. Seal blubber is predominantly triglycerides, which yield approximately 9 kcal per gram. This means that a typical human would need to consume 222 grams of seal blubber per day (about half a pound) in order to obtain all of his or her calories from this energy source.

11. Results for this problem will depend on the particular cookies chosen by the student.

12. The only structural differences between cholesterol and stigmasterol are the double bond of stigmasterol at C_{22}–C_{23}, and the ethyl group at C_{24}.

13. Androgens mediate the development of sexual characteristics and sexual function in animals. Glucocorticoids participate in the control of carbohydrate, protein, and lipid metabolism. Mineralocorticoids regulate salt (Na^+, K^+, and Cl^-) balances in tissues.

14. The answers to this question will depend on the household products chosen and the isoprene substances identified.

15. Gibberellic acid stimulates root formation, overcomes dormancy of seeds, promotes increased fruit production, stimulates premature flowering, and increases growth of plants.

16. Most obviously, a diet of triglycerides (from the blubber of seals, the polar bears' favorite food) provides high energy and material that can be reprocessed into other triglycerides and used as insulation under the skin. Less obvious is the need of the polar bear to stay warm and conserve water. (Polar bears cannot afford to eat snow or ice [they are too cold], and they cannot drink seawater [it is too salty].) The polar bear is adapted to conserve body water and stay warm, and thus it does not urinate for months at a time. (Urination would give up both water and heat.) To achieve this, it must consume little or no nitrogen, because a diet rich in nitrogen would require urination and defecation. Triglycerides contain no nitrogen and are thus ideal food for the adult polar bear. Juvenile polar bears, which have not yet reached their full adult body size, must consume protein in order to make their own proteins. Once the bear reaches its full size, it changes its diet and no longer consumes protein!

17. Snake venom phospholipase A_2 cleaves fatty acids from phospholipids at the C-2 position. The fatty acids behave as detergents and form micelles (see Chapter 9) that can remove lipids and proteins from the membrane and disrupt membrane structure, causing pores and eventually rupturing the cell itself.

Chapter 9

1. Glycerophospholipids with an unsaturated chain at the C-1 position and a saturated chain at the C-2 position are rare to nonexistent. Glycerophospholipids with two unsaturated chains, or with a saturated chain at C-1 and an unsaturated chain at C-2, are commonly found in biomembranes.

2. The PL/protein molar ratio in purple patches of *H. halobium* is 10.8.

3. See *The Student Solutions Manual, Study Guide and Problems Book* for plots of sucrose solution density versus percent by weight and by volume. The plot in terms of percent by weight exhibits a greater curve, because less water is required to form a solution that is, for example, 10% by weight (10 g sucrose/100 g total) than to form a solution that is 10% by volume (10 g sucrose/100 mL solution).

4. $r = (4Dt)^{1/2}$.

According to this equation, a phospholipid with $D = 1 \times 10^{-8}$ cm^2 will move approximately 200 nm in 10 milliseconds.

5. Fibronectin: For $t = 10$ msec, $r = 1.67 \times 10^{-7}$ cm = 1.67 nm.

Rhodopsin: $r = 110$ nm.

All else being equal, the value of D is roughly proportional to $(M_r)^{-1/3}$. Molecular weights of rhodopsin and fibronectin are 40,000 and 460,000, respectively. The ratio of diffusion coefficients is thus expected to be $(40,000)^{-1/3}/(460,000)^{-1/3} = 2.3$.

On the other hand, the values given for rhodopsin and fibronectin give an actual ratio of 4286. The explanation is that fibronectin is anchored in the membrane via interactions with cytoskeletal proteins, and its diffusion is severely restricted compared with that of rhodopsin.

6. a. Divalent cations increase T_m.

   b. Cholesterol broadens the phase transition without significantly changing T_m.

   c. Distearoylphosphatidylserine should increase T_m, due to increased chain length and also to the favorable interactions between the more negative PS head group and the more positive PC head groups.

   d. Dioleoylphosphatidylcholine, with unsaturated fatty acid chains, will increase T_m.

   e. Integral proteins will broaden the transition and could raise or lower T_m, depending on the nature of the protein.

7. $\Delta G = RT \ln([C_2]/[C_1])$.

   $\Delta G = +4.0$ kJ/mol.

8. $\Delta G = RT \ln([C_{out}]/[C_{in}]) + Z\mathcal{F}\Delta\psi$.

   $\Delta G = +4.08$ kJ/mol -2.89 kJ/mol.

   $\Delta G = +1.19$ kJ/mol.

   The unfavorable concentration gradient thus overcomes the favorable electrical potential and the outward movement of Na^+ is not thermodynamically favored.

9. One could solve this problem by going to the trouble of plotting the data in v vs. [S], $1/v$ vs. $1/[S]$, or $[S]/v$ vs. [S] plots, but it is simpler to examine the value of $[S]/v$ at each value of [S]. The Hanes–Woolf plot makes clear that $[S]/v$ should be constant for all [S] for the case of passive diffusion. In the present case, $[S]/v$ is a constant value of 0.0588 $(L/min)^{-1}$. It is thus easy to recognize that this problem describes a system that permits passive diffusion of histidine.

10. This is a two-part problem. First calculate the energy available from ATP hydrolysis under the stated conditions; then use the answer to calculate the maximal internal fructose concentration against which fructose can be transported by coupling to ATP hydrolysis. Using a value of -30.5 kJ/mol for the $\Delta G^{\circ\prime}$ of ATP and the indicated concentrations of ATP, ADP, and P_i, one finds that the ΔG for ATP hydrolysis under these conditions at 298 K is -52.0 kJ/mol. Putting the value of $+52.0$ kJ/mol into Equation 10.1 and solving for C_2 yields a value for the maximum possible internal fructose concentration of 1300 *M*! Thus, ATP hydrolysis could (theoretically) drive fructose transport against internal fructose concentrations up to this value. (In fact, this value is vastly in excess of the limit of fructose solubility.)

11. Each of the transport systems described can be inhibited (with varying degrees of specificity). Inhibition of the rhamnose transport system by one or more of these agents would be consistent with involvement of one of these transport systems with rhamnose transport. Thus, nonhydrolyzable ATP analogs

should inhibit ATP-dependent transport systems, ouabain should specifically block Na^+ (and K^+) transport, uncouplers should inhibit proton gradient–dependent systems, and fluoride should inhibit the PTS system (via inhibition of enolase).

12. *N*-myristoyl lipid anchors are found linked only to N-terminal Gly residues. Only the peptide in (e) of this problem contains an N-terminal Gly residue.

13. Only the peptide in (a) possesses a CAAX sequence (where C = Cys, A = aliphatic (Ala, Val, Leu, Ile), and X = any amino acid).

14. Point (b)—that proteins can be anchored to the membrane via covalent links to lipid molecules—was not part of the Singer–Nicolson fluid mosaic model.

15. Monensin is a mobile carrier ionophore, whereas cecropin is a channel-forming ionophore. As noted in this chapter, the transport rates for mobile carrier ionophores are quite sensitive to temperature (in the vicinity of the membrane phase transition), but the transport rates for channel-forming ionophores, by contrast, are relatively insensitive to temperature. The phase transition temperature for DPPC is 41.4°C (see Table 9.1). Therefore, the transport rates for monensin in DPPC should increase significantly between 35° and 50°C, but the rates observed for cecropin a in DPPC membranes at 50°C should be about the same as they are at 35°C.

Chapter 10

1.

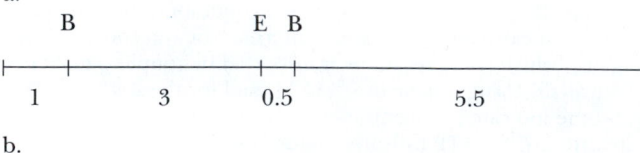

2. See Figure 10.17.

3. $f_A = 0.29$; $f_G = 0.22$; $f_C = 0.25$; $f_T = 0.28$.

4. The number of A and T residues = 7.86×10^8 each; the number of G and C residues = 6.69×10^8 each.

5. 5'-TAGTGACAGTTGCGAT-3'.

6. 5'-ATCGCAACTGTCACTA-3'.

7. 5'-TACGGTCTAAGCTGA-3'.

8. There are two possibilities, a and b. (E = *Eco*RI site; B = *Bam*HI site.) Note that b is the reverse of a.

a.

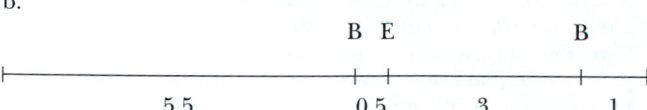

b.

9. a. GGATCCCGGGTCGACTGCAG;

b. GTCGACCCGGGATCCTGCAG.

*Sma*I products: a. GGATCCC and GGGTCGACTGCAG.

b. GTCGACCC and GGGATCCTGCAG.

10. Synthesis of a polynucleotide 100 residues long requires formation of 99 phosphodiester bonds. $\Delta G^{\circ\prime}$ for phosphodiester synthesis (assuming it is the same magnitude but opposite sign as that for phosphoric anhydride cleavage of an NTP to give NMP + PP_i) is +32.3 kJ/mol. $\Delta G^{\circ\prime}_{overall} = (99)(+32.3) = 3198$ kJ/mol.

11. a. Hydrogen bonding and van der Waals interactions between amino acid side chains and DNA, and ionic interactions of amino acid side chains with the nucleic acid backbone phosphate groups. Double-helical DNA does not present hydrophobic regions for interaction with proteins because of base-pair stacking.

b. Proteins can recognize specific base sequences if they can fit within the major or minor groove of DNA and "read" the H-bonding pattern presented by the edges of the bases in the groove. The dimensions of an α-helix are such that it fits snugly within the major groove of B-DNA; then, depending on the amino acid sequence of the protein, the side chains displayed on the circumference of the α-helix have the potential to form H bonds with H-bonding functions provided by the bases.

12. a. The restriction endonuclease must be able to recognize a specific nucleotide sequence in the DNA that has twofold rotational symmetry.

b. In order to read a base sequence within DNA, either the restriction endonuclease must interact with the bases by direct access, for example, by binding in the major groove, or the restriction endonuclease must be able to "read" the base sequence by indirect means, for example, if the base sequence imparts some local variation in the cylindrical surface of the DNA that the enzyme might recognize.

c. The restriction endonuclease must be able to cleave both DNA strands, often in a staggered fashion.

d. An obvious solution to the requirements listed in (a) and (c) would be a homodimeric subunit organization for restriction endonucleases.

13. a. The ribose group of nucleosides greatly increases the water solubility of the base.

b. Ribose has fewer hydroxyl groups and thus less likelihood to undergo unwanted side reactions.

c. The absence of the 2-OH in 2-deoxyribose leads to a polynucleotide sugar–phosphate backbone that is more stable because it is not susceptible to alkaline hydrolysis.

14. a. Phosphate groups bear a negative electrical charge at neutral pH.

b. Cleavage of phosphoric anhydride bonds is strongly exergonic and can provide the thermodynamic driving force for diverse metabolic reactions.

c. Cleavage of phosphoric anhydride bonds is strongly exergonic and can provide the thermodynamic driving force for phosphodiester bond formation in polynucleotide synthesis.

15. The use of bases as "information symbols" in metabolism allows the cell to allocate portions of its phosphorylation potential (as total NTP) to dedicated tasks, as in GTP for protein synthesis, CTP for phospholipid synthesis, UTP for carbohydrate synthesis, with ATP serving the central role. Enzymes in these pathways selectively bind the proper nucleotide through recognition of the particular base.

16. The most prominent structural feature of the DNA double helix in terms of structural complementarity is found in the canonical base-pairing of A with T and G with C. Base pairing is essential to DNA replication (preservation of the genetic information) and transcription (expression of genetic information).

Chapter 11

1. Sanger's dideoxy method:

A	G	C	T
			—
		—	
—			
	—		
	—		
—			
		—	
			—
—			

2. Original nucleotide: 5′-dAGACTTGACGCT.

3. a. 10.5 base pairs per turn; b. $\Delta\phi = 34.3°$; c (true repeat) = 6.72 nm.

4. 27.3 nm; 122 base pairs.

5. 4325 nm (4.33 μm).

6. $L_0 = 160$. If $W = -12$, $L = T + W = 160 + (-12) = 148$. $\sigma = \Delta L/L_0 = -12/160 = -0.075$.

7. For 1 turn of B-DNA (10 base pairs): $L_B = 1.0 + W_B$.

   For Z-DNA, 10 base pairs can only form 10/12 turn (0.833 turn), and $L_Z = 0.833 + W_Z$.

   For the transition B-DNA to Z-DNA, strands are not broken, so $L_B = L_Z$; that is, $1.0 + W_B = 0.833 = W_Z$, or $W_Z - W_B = +0.167$.

   (In going from B-DNA to Z-DNA, the change in W, the number of supercoils, is positive. This result means that, if B-DNA contains negative supercoils, their number will be reduced in Z-DNA. Thus, all else being equal, negative supercoils favor the B $\rightarrow$ Z transition.)

8. 6×10^9 bp/200 bp = 3×10^7 nucleosomes.

   The length of B-DNA 6×10^9 bp long = (0.34 nm)(6×10^9) = 2.04×10^9 nm (more than 2 meters!). The length of 3×10^7 nucleosomes = (6 nm)(3×10^7) = 18×10^7 nm (0.18 meter).

9. From Figure 11.30: Similarly shaded regions indicate complementary sequences joined via intrastrand hydrogen bonds:

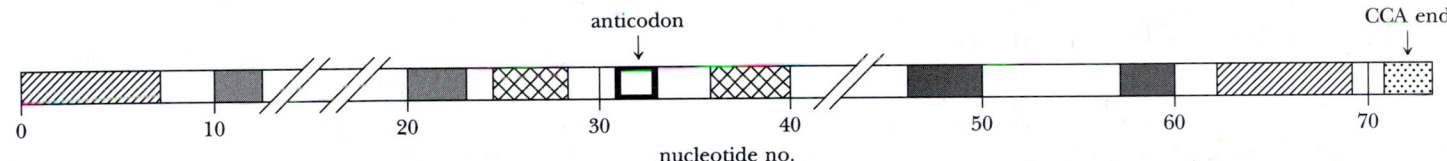

10. Increasing order of T_m: yeast < human < salmon < wheat < E. coli.

11. In 0.2 M Na⁺, $T_m(°C) = 69.3 + 0.41(\%G + C)$:

    Rats $(\%G + C) = 40\%$, $T_m = 69.3 + 0.41(40) = 85.7°C$

    Mice $(\%G + C) = 44\%$, $T_m = 69.3 + 0.41(44) = 87.3°C$

    Because mouse DNA differs in GC content from rat DNA, they could be separated by isopycnic centrifugation in a CsCl gradient.

12. GC content = 0.714 (from Table 10.3 and equations used in problem 3, Chapter 10). $\rho = 1.660 + 0.098(\text{GC}) = 1.730$ g/mL.

13. See Figure 11.34 and compare your structure with the base pair formed between G18 and ψ55 in yeast phenylalanine tRNA, which has a single H bond between the 2-amino group of G (2-NH$_2$···O=) and the O atom at position 4 in ψ.

14. Assuming the plasmid is in the B-DNA conformation, where each pair contributes 0.34 nm to the length of the molecule, the circumference of a perfect circle formed from pBR322 would be (0.34 nm/bp)(4353 bp) = 1480 nm = 1.48 μm.

    The E. coli K12 chromosome laid out as a perfect circle would have a circumference of (0.34 nm/bp)(4,639,000 bp) = 1,577,260 nm = 1.58 mm.

    The diameter of a B-DNA molecule is about 2.4 nm (see Table 11.1). For pBR322, the length/diameter ratio would be 1480/2.4 = 616.7. For the E. coli K12 chromosome, the length/diameter ratio would be 1,577,260/2.4 = 657,192.

15. The DNA in such a thermophilic organism would be subjected to very high temperatures that might denature the DNA. Because G:C base pairs are more heat stable than A:T base pairs (see Figure 11.15), one might expect DNA from a thermophilic organism to have a high G + C content.

16. DNA is the material of heredity; that is, genetic information. This information is encoded in the sequence of bases in DNA, so this is its most important structural feature. The double-stranded nature of DNA and the complementary base sequence of the two strands are also crucial structural features in the transmission of genetic information through DNA replication. Beyond these points, one might cite the structural features of DNA that impart stability to the double helix and structural aspects that render DNA less susceptible to degradation.

Chapter 12

1. Linear and circular DNA molecules consisting of one or more copies of just the genomic DNA fragment; linear and circular DNA molecules consisting of one or more copies of just the vector DNA; linear and circular DNA molecules containing one or more copies of both the genomic DNA fragment and plasmid DNA.

2.

-GAATTCCCGGGGATCCTCTAGAGTCGACCTGCAGGCATGC-

GAATTC	GGATCC	GTCGAC	GCATGC
EcoRI	*BamHI*	*SalI*	*SphI*

CCCGGG	TCTAGA	CTGCAG
SmaI	*XbaI*	*PstI*

3. a. AAGCTTGAGCTCGAGATCTAGATCGAT
      *Hind*III   *Xho*I   *Xba*I
              *Sac*I   *Bg*III   *Cla*I

   b.
   Vector: *Hind*III: 5'-A. . . . .-gap-. . . . . .CGAT-3': *Cla*I
                      3'-TTCGA. . . .-gap-. . . . . .TA-5'

   Fragment: *Hind*III: 5'-AGCTT(NNNN-etc-NNNN)AT-3'
                         3'-A(NNNN-etc-NNNN)TAGC-5'

4. N = 3480.

5. N = 10.4 million.

6. 5'-ATGCCGTAGTCGATCAT and 5'-ATGCTATCTGTCCTATG.

7. -Thr-Met-Ile-Thr-Asn-Ser-Pro-*Asp-Pro-Phe-Ile-His-Arg-Arg-Ala-Gly-Ile-Pro-Lys-Arg-Arg-Pro...*

   The junction between β-galactosidase and the insert amino acid sequence is between Pro and Asp, so the first amino acid encoded by the insert is Asp. (The polylinker itself codes for Asp just at the *Bam*HI site, but in constructing the fusion, this Asp and all of this downstream section of polylinker DNA is displaced to a position after the end of the insert.)

8. 5'(G)AATTCNGGNATGCAYCCNGGNAAR$_c$^{T}T$_N$^{Y}GCN**AGY**TGG-TTYGTNGGGAATTCN-

   (*Note:* The underlined triplet AGY represents the middle Ser residue. Ser codons are either AGY or TCN [where Y = pyrimidine and N = any base]; AGY was selected here so that the mutagenesis of this codon to a Cys codon [TGY] would involve only an A → T change in the nucleotide sequence.)

   Because the middle Ser residue lies nearer to the 3'-end of this *Eco*RI fragment, the mutant primer for PCR amplification should encompass this end. That is, it should be the primer for the 3 → 5' strand of the *Eco*RI fragment:

   5'-NNNGAATTCCCN$_c$ACR$_c$AACCA**R**$_c$**CA**N$_c$GC-3'

   where the mutated Ser → Cys triplet is underlined, NNN = several extra bases at the 5'-end of the primer to place the *Eco*RI site internal, N$_c$ = the nucleotide complementary to the nucleotide at this position in the 5'→3' strand, and R$_c$ = the pyrimidine complementary to the purine at this position in the 5'→3' strand.

9. The number of sequence possibilities for a polymer is given by x^y, where x is the number of different monomer types and y is the number of monomers in the oligomers. Thus, for RNA oligomers 15 nucleotides long: $x^y = 4^{15} = 1,073,741,824$. For pentapeptides, $x^y = 20^5 = 3,200,000$.

10. See A Deeper Look box on page 395. You would need a *GAL4*-deficient yeast strain expressing a fusion protein composed of the *GAL4* DB domain fused with a cytoskeletal protein (such as actin or tubulin) to serve as the bait. These *GAL4*⁻ cells would be transformed with a cDNA library of human epithelial proteins constructed so that the human proteins were expressed as *GAL4* TA fusion proteins; these fused proteins are the target. Interaction of a target protein with the cytoskeletal protein fused to the *GAL4* DB domain will lead to *GAL4*-driven expression of the *lacZ* gene, whose presence can be revealed by testing for β-galactosidase activity.

11. See Figure 12.13 for isolation of mRNA and Figure 12.14 for preparation of a cDNA library from mRNA. The mRNA would be isolated, and the cDNA libraries would be prepared from two batches of yeast cells, one grown aerobically and one grown anaerobically.

12. See Figure 12.15. Differently labeled single-stranded cDNA from separate libraries prepared from aerobically versus anaerobically grown yeast cells would be hybridized with a DNA microarray (gene chip) of yeast genes.

13. An antibody against hexokinase A could be used to screen the yeast cDNA library to identify a yeast colony expressing this protein (see discussion on page 393). Once a sample of the putative yeast hexokinase A protein was isolated, its identity could be confirmed by peptide mass fingerprinting using mass spectrometry (see page 127).

14. The experiment in problem 12 identified the cDNA clone for the protein, but this clone will not have the regulatory elements (promoter) necessary for the experiment at hand. Using the cDNA clone as a probe, the genomic clone for this *nox* gene could be identified in a yeast genomic library. Cloning and sequencing a genomic clone would identify putative promoter regions that could be fused to the coding region of green fluorescent protein (GFP) to create a reporter gene construct (see page 396) that would "light up" more and more as oxygen levels declined.

15. Go to the NCBI site map at *http://www.ncbi.nlm.nih.gov/Sitemap/index.html*. In the alphabetical index, click on Genomes and Maps to go to the Entrez Genome site at *http://www.ncbi.nlm.nih.gov/Sitemap/index.html#Genomes*.

    Under Organism Collections, Entrez Genomes section, click on the links for bacteria, archea, and eukaryotes to see:

    Number of bacterial genomes completed at *http://www.ncbi.nlm.nih.gov/PMGifs/Genomes/eub_g.html*

    Number of archeal genomes completed at

    http://www.ncbi.nlm.nih.gov/PMGifs/Genomes/a_g.html

    Completed eukaryotic genomes accessible at *http://www.ncbi.nlm.nih.gov/PMGifs/Genomes/euk_g.html*

16. Insertion of DNA at any of the following restriction sites would render cells harboring the recombinant plasmid ampicillin-sensitive: *Ssp*I, *Sca*I, *Pvu*I, *Pst*I, and *Ppa*I.

17. The respective codons are Trp = UGG; Cys = UGC and UGU. Thus, changing only the third base G in the Trp codon to a pyrimidine (either C or U) would yield a Cys codon. In this case, there are now 6 differences in codon possibilities, so $2^6 = 64$ different oligonucleotides must be synthesized.

Chapter 13

1. $v/V_{max} = 0.8$.

2. $v = 91$ μmol/mL · sec.

3. $K_s = 1.43 \times 10^{-5}$ M; $K_m = 3 \times 10^{-4}$ M. Because k_2 is 20 times greater than k_{-1}, the system behaves like a steady-state system.

4. If the data are graphed as double-reciprocal Lineweaver–Burk plots:

   a. $V_{max} = 51$ μmol/mL · sec and $K_m = 3.2$ mM.

   b. Inhibitor (2) shows competitive inhibition with a $K_I = 2.13$ mM. Inhibitor (3) shows noncompetitive inhibition with a $K_I = 4$ mM.

5.

(a)

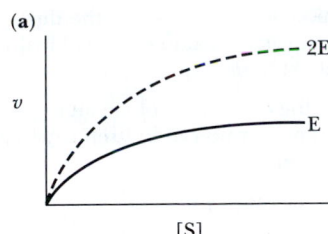

(b)

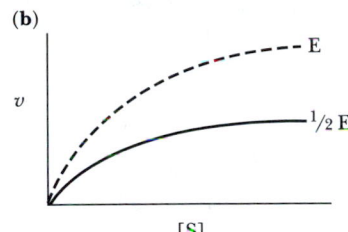

(c)

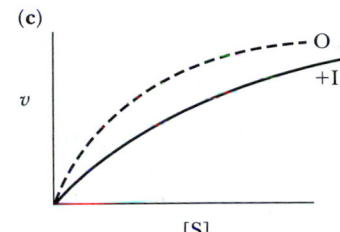

(d)

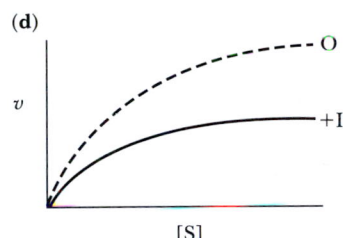

(e)

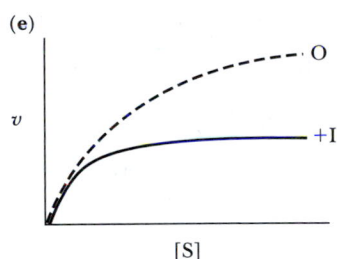

In (a), V_{max} doubles, but K_m is constant.

In (b), V_{max} halves, but K_m is constant.

In (c), V_{max} is constant, but the apparent K_m increases.

In (d), V_{max} decreases, but K_m is constant.

In (e), V_{max} decreases, K_m decreases, but the ratio K_m/V_{max} is constant.

6. a. The slope is given by $K_m^B/V_{max}(K_S^A/[A] + 1)$.

   b. y-intercept $= ((K_m^A/[A]) + 1)1/V_{max}$.

   c. The horizontal and vertical coordinates of the point of intersection are $1/[B] = -K_m^A/K_S^A K_m^B$ and $1/v = 1/V_{max}(1 - (K_m^A/K_S^A))$.

7. Top left: (1) Competitive inhibition (I competes with S for binding to E). (2) I binds to and forms a complex with S.

Top right: (1) Pure noncompetitive inhibition. (2) Random, single-displacement bisubstrate reaction, where A doesn't affect B binding, and vice versa. (Other possibilities include [3] Irreversible inhibition of E by I; [4] $1/v$ vs. $1/[S]$ plot at two different concentrations of enzyme, E.)

Bottom left: (1) Mixed noncompetitive inhibition. (2) Ordered single-displacement bisubstrate mechanism.

Bottom right: (1) Uncompetitive inhibition. (2) Double-displacement (ping-pong) bisubstrate mechanism.

8. Clancy must drink 694 mL of wine, or about one 750-mL bottle.

9. a. $K_S = 5\ \mu M$; b. $K_m = 30\ \mu M$; c. $k_{cat} = 5 \times 10^3\ sec^{-1}$; d. $k_{cat}/K_m = 1.67 \times 10^8\ M^{-1}sec^{-1}$; e. Yes, because k_{cat}/K_m approaches the limiting value of $10^9\ M^{-1}sec^{-1}$; f. $V_{max} = 10^{-5}\ mol/mL \cdot sec$; g. $[S] = 90\ \mu M$; h. V_{max} would equal $2 \times 10^{-5}\ mol/mL \cdot sec$, but $K_m = 30\ \mu M$, as before.

10. a. $V_{max} = 132\ \mu mol\ mL^{-1}\ sec^{-1}$; b. $k_{cat} = 44,000\ sec^{-1}$; c. $k_{cat}/K_m = 24.4 \times 10^8\ M^{-1}\ sec^{-1}$; d. Yes! k_{cat}/K_m actually exceeds the theoretical limit of $10^9\ M^{-1}\ sec^{-1}$ in this problem; e. The rate at which E encounters S; the ultimate limit is the rate of diffusion of S.

11. a. $V_{max} = 1.6\ \mu mol\ mL^{-1}\ sec^{-1}$; b. $v = 1.45\ \mu mol\ mL^{-1}\ sec^{-1}$; c. $k_{cat}/K_m = 1.6 \times 10^8\ M^{-1}\ sec^{-1}$; d. Yes.

12. a. $V_{max} = 6\ \mu mol\ mL^{-1}\ sec^{-1}$; b. $k_{cat} = 1.2 \times 10^6\ sec^{-1}$; c. $k_{cat}/K_m = 1 \times 10^8\ M^{-1}\ sec^{-1}$; d. Yes! k_{cat}/K_m approaches the theoretical limit of $10^9\ M^{-1}\ sec^{-1}$; e. The rate at which E encounters S; the ultimate limit is the rate of diffusion of S.

13. a. $V_{max} = 64\ \mu mol\ mL^{-1}\ sec^{-1}$; b. $k_{cat} = 1.28 \times 10^4\ sec^{-1}$; c. $k_{cat}/K_m = 1.4 \times 10^8\ M^{-1}\ sec^{-1}$; d. Yes! k_{cat}/K_m approaches the theoretical limit of $10^9\ M^{-1}\ sec^{-1}$.

14. a. $V_{max} = 120\ mmol\ mL^{-1}\ sec^{-1}$; b. $v = 104.7\ mmol\ mL^{-1}\ sec^{-1}$; c. $k_{cat}/K_m = 3.64 \times 10^8\ M^{-1}\ sec^{-1}$; d. Yes! k_{cat}/K_m approaches the theoretical limit of $10^9\ M^{-1}\ sec^{-1}$.

15. a. Starting from $V_{max}^f = k_2[E_T]$ and $V_{max}^r = k_{-1}[E_T]$ for the maximal rates of the forward and reverse reactions respectively, and the Michaelis constants $K_m^S = (k_{-1} + k_2)/k_1$ and $K_m^P = (k_{-1} + k_2)/k_{-2}$ for S and P, respectively,

$$v = (V_{max}^f\ [S]/K_m^S) - V_{max}^r[P]/K_m^P)/(1 + [S]/K_m^S + [P]/K_m^P)$$

   b. At equilibrium, $v = 0$ and $K_{eq} = [P]/[S] = V_{max}^f\ K_m^P/V_{max}^r K_m^S$.

16. a. S is the preferred substrate; its K_m is smaller than the K_m for T, so a lower $[S]$ will give $v = V_{max}/2$, compared with $[T]$.

   b. k_{cat}/K_m defines catalytic efficiency. k_{cat}/K_m for S $= 2 \times 10^7\ M^{-1}sec^{-1}$; k_{cat}/K_m for T $= 4 \times 10^7\ M^{-1}sec^{-1}$, so the enzyme is a more efficient catalysis with T as substrate.

17. a. Because the enzyme shows maximal activity at or below 40°C, it seems more like a mammalian enzyme than a plant enzyme, which would be expected to have a broader temperature optimum because plants experience a broader range of temperatures.

   b. An enzyme from a thermophilic bacterium growing at 80°C would show an activity versus temperature profile similar to this one but shifted much farther to the right.

Chapter 14

1. a. Nucleophilic attack by an imidazole nitrogen of His[57] on the —CH_2— carbon of the chloromethyl group of TPCK covalently inactivates chymotrypsin. (See *The Student Solutions Manual, Study Guide and Problems Book* for structures.)

b. TPCK is specific for chymotrypsin because the phenyl ring of the phenylalanine residue interacts effectively with the binding pocket of the chymotrypsin active site. This positions the chloromethyl group to react with His[57].

c. Replacement of the phenylalanine residue of TPCK with arginine or lysine produces reagents that are specific for trypsin.

2. a. The structure proposed by Craik et al., 1987 (*Science* **237**:905–907) is shown here. (If you look up this reference, note that the letters A and B of the figure legend for Figure 3 of this article actually refer to parts B and A, respectively. Reverse either the letters in the figure or the letters in the figure legend and it will make sense.)

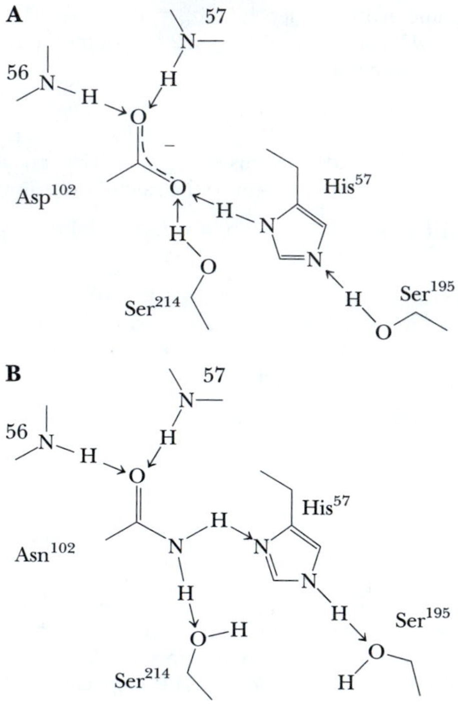

b. In the proposed model (shown for part a), Asn[102] of the mutant enzyme can serve only as a hydrogen-bond donor to His[57]. It is unable to act as a hydrogen-bond acceptor, as aspartate does in native trypsin. As a result, His[57] is unable to act as a general base in transferring a proton from Ser[195]. This presumably accounts for the diminished activity of the mutant trypsin.

3. a. The usual explanation for the inhibitory properties of pepstatin is that the central amino acid, statine, mimics the tetrahedral amide hydrate transition state of a good pepsin substrate with its unique hydroxyl group.

b. Pepsin and other aspartic proteases prefer to cleave peptide chains between a pair of hydrophobic residues, whereas HIV-1 protease preferentially cleaves a Tyr-Pro amide bond. Because pepstatin more closely fits the profile of a pepsin substrate, we would surmise that it is a better inhibitor of pepsin than of HIV-1 protease. In fact, pepstatin is a potent inhibitor of pepsin ($K_I < 1$ nm) but only a moderately good inhibitor of HIV-1 protease, with a K_I of about 1 μM.

4. This problem is solved best by using the equation derived in problem 6.

$$\frac{k_e}{k_u} = e^{(\Delta G_u^{\ddagger} - \Delta G_e^{\ddagger})/RT}$$

Using this equation, we can show that the difference in activation energies for the uncatalyzed and catalyzed hydrolysis reactions ($\Delta G_u - \Delta G_c$) is 92 kJ/mol.

5. Trypsin catalyzes the conversion of chymotrypsinogen to π-chymotrypsin, and chymotrypsin itself catalyzes the conversion of π-chymotrypsin to α-chymotrypsin.

6. The enzyme-catalyzed rate is given by:

$$v = k_e[ES] = k_e'[EX^{\ddagger}]$$

$$K_e^{\ddagger} = \frac{[EX^{\ddagger}]}{[ES]}$$

$$\Delta G_e^{\ddagger} = -RT \ln K_e^{\ddagger}$$

$$K_e^{\ddagger} = e^{-\Delta G_e^{\ddagger}/RT}$$

$$[EX^{\ddagger}] = K_e^{\ddagger}[ES] = e^{-\Delta G_e^{\ddagger}/RT}[ES]$$

So

$$k_e[ES] = k_e'e^{-\Delta G_e^{\ddagger}/RT}[ES]$$

or

$$k_e = k_e'e^{-\Delta G_e^{\ddagger}/RT}$$

Similarly, for the uncatalyzed reaction:

$$k_u = k_u'e^{-\Delta G_u^{\ddagger}/RT}$$

Assuming that $k_u' \cong k_e'$

Then

$$\frac{k_e}{k_u} = \frac{e^{-\Delta G_e^{\ddagger}/RT}}{e^{-\Delta G_u^{\ddagger}/RT}}$$

$$\frac{k_e}{k_u} = e^{(\Delta G_u^{\ddagger} - \Delta G_e^{\ddagger})/RT}$$

7. The mechanism suggested by Lipscomb is a general base pathway in which Glu[270] promotes the attack of water on the carbonyl carbon of the substrate:

Carboxypeptidase

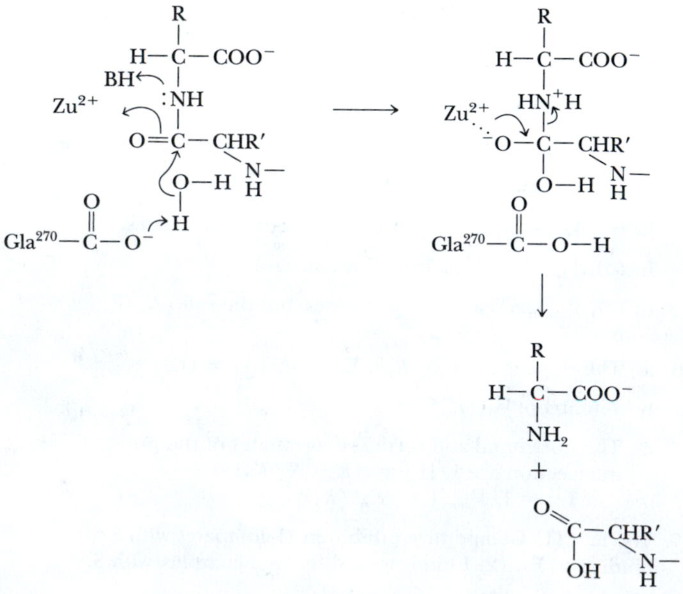

8. Using the equation derived in problem 6, it is possible to calculate the ratio k_e/k_u as 1.86×10^{12}.

9. The correct answer is c. The stomach is a very acidic environment, whereas the small intestine is slightly alkaline. The lower part of the figure shows that enzyme X has optimal activity near pH 2, whereas enzyme Y works best at a pH near 8.

10. The correct answer is a. The two enzymes have nonoverlapping pH ranges, so it is highly unlikely that they could operate in the same place at the same time.

11. The correct answer is b. Only enzyme A has a temperature range that encompasses human body temperature (37°C).

12. The correct answer is d. The activities of the two enzymes overlap between 40° and 50°C.

13. The correct answer is c. We have no information on the pH behavior of enzymes A and B, nor on the behavior of X and Y as a function of temperature. The only answer that is appropriate to the data shown is c.

14. The only possible answer is b, because a "leveling off" implies that all the enzyme is saturated with S.

15. The correct answer is c. In order to bring the substrate into the transition state, an enzyme must enjoy environmental conditions that favor catalysis. There is no activity apparent for enzyme Y below pH 5.5.

Chapter 15

1. a. As [P] rises, the rate of P formation shows an apparent decline, as enzyme-catalyzed conversion of $P \rightarrow S$ becomes more likely.

   b. Availability of substrates and cofactors.

   c. Changes in [enzyme] due to enzyme synthesis and degradation.

   d. Covalent modification.

   e. Allosteric regulation.

   f. Specialized controls, such as zymogen activation, isozyme variability, and modulator protein influences.

2. Proteolytic enzymes have the potential to degrade the proteins of the cell in which they are synthesized. Synthesis of these enzymes as zymogens is a way of delaying expression of their activity to the appropriate time and place.

3. Monod, Wyman, Changeux allosteric K system:

Lineweaver–Burk plot

Hanes–Woolf plot

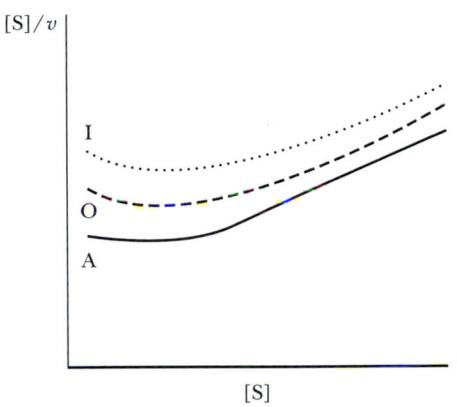

Monod, Wyman, Changeux allosteric V system:

Lineweaver–Burk plot

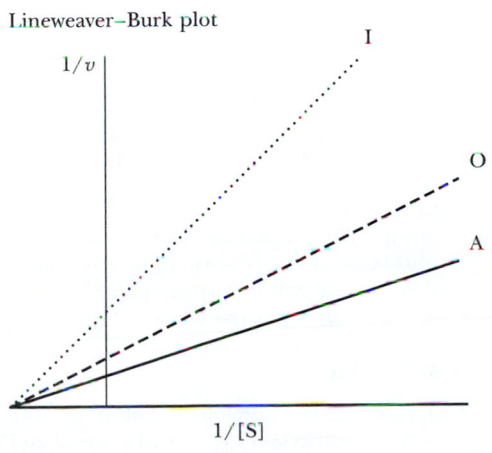

Hanes–Woolf plot

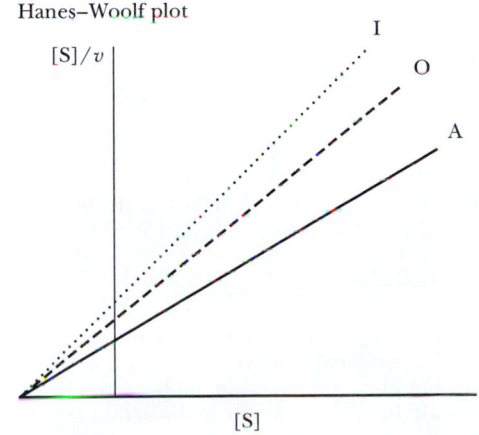

4. If L_0 is large, that is, $T_0 \gg R_0$, activator, A, will show positive homotropic binding. If L_0 is very small, that is, $R_0 \gg T_0$, inhibitor, I, will show positive homotropic binding.

5. Using the curves for negative cooperativity as shown in the figure in the box on page 485 as a guide:

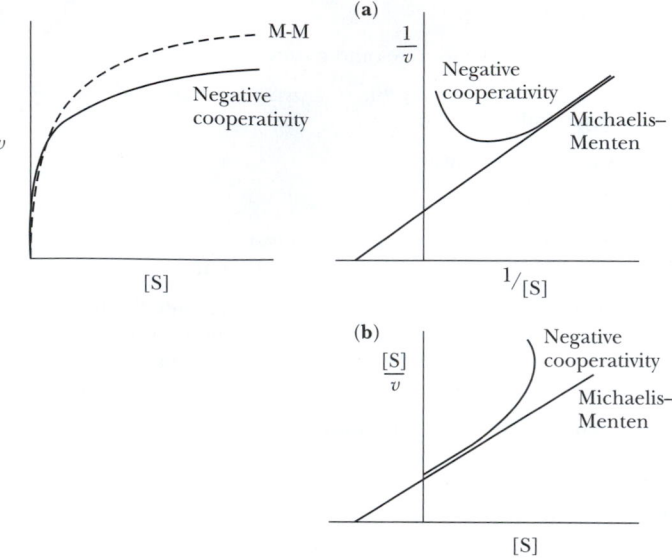

6. For $n = 2.8$, $Y_{lungs} = 0.98$ and $Y_{capillaries} = 0.77$.

   For $n = 1.0$, $Y_{lungs} = 0.79$ and $Y_{capillaries} = 0.61$.

   Thus, with an n of 2.8 and a P_{50} of 26 torr, hemoglobin becomes almost fully saturated with O_2 in the lungs and drops to 77% saturation in resting tissue, a change of 21%. If n of 1.0 and a P_{50} of 26 torr, hemoglobin would become only 79% saturated with O_2 in the lungs and would drop to 61% in resting tissue, a change of 18%. The difference in hemoglobin O_2 saturation conditions between the values for $n = 2.8$ and 1.0 (21% − 18%, or 3% saturation) seems small, but note that the potential for O_2 delivery (98% saturation versus 79% saturation) is large and becomes crucial when pO_2 in actively metabolizing tissue falls below 40 torr.

7. More glycogen phosphorylase will be in the glycogen phosphorylase a (more active) form, but caffeine promotes the less active T conformation of glycogen phosphorylase.

8. Over time, stored erythrocytes will metabolize 2,3-BPG via the pathway of glycolysis. If [BPG] drops, hemoglobin may bind O_2 with such great affinity that it will not be released to the tissues (see Figure 15.34). The patient receiving a transfusion of [BPG]-depleted blood may actually suffocate.

9. a. By definition, when $[P_i] = K_{0.5}$, $v = 0.5\ V_{max}$.

   b. In the presence of AMP, at $[P_i] = K_{0.5}$, $v = 0.85\ V_{max}$ (from Figure 15.16c).

   c. In the presence of ATP, at $[P_i] = K_{0.5}$, $v = 0.12\ V_{max}$ (from Figure 15.16b).

10. If G_α–GTPase activity is inactivated, the interaction between G_α and adenylyl cyclase will be persistent, adenylyl cyclase will be active, [cAMP] will rise, and glycogen levels will fall because glycogen phosphorylase will be predominantly in the active, phosphorylated a form.

11. An excess of a negatively cooperative allosteric inhibitor could never completely shut down the enzyme. Because the enzyme leads to several essential products, inhibition by one product might starve the cell for the others.

12. a. -R(R/K)X(S*/T*)-

    b. -KRKQIAVRGL-

13. Ligand binding is the basis of allosteric regulation, and allosteric effectors are common metabolites whose concentrations reflect prevailing cellular conditions. Through reversible binding of such ligands, enzymatic activity can be adjusted to the momentary needs of the cell. On the other hand, allosteric regulation is inevitably determined by the amounts of allosteric effectors at any moment, which can be disadvantageous. Covalent modification, like allosteric regulation, is also rapid and reversible, because the converter enzymes act catalytically. Furthermore, covalent modification allows cells to escape allosteric regulation by covalently locking the modified enzyme in an active (or inactive) state, regardless of effector concentrations. One disadvantage is that covalent modification systems are often elaborate cascades that require many participants.

14. Sickle-cell anemia is the consequence of Hb S polymerization through hydrophobic contacts between the side chain of Valβ6 and a pocket in the EF corner of β-subunits. Potential drugs might target this interaction directly or indirectly. For example, a drug might compete with Valβ6 side chain for binding in the EF corner. Alternatively, a useful drug might alter the conformation of Hb S such that the EF corner was no longer accessible or accommodating to the Valβ6 side chain. Another possibility might be to create drugs that deter the polymerization process in other ways through alterations in the surface properties between Hb S molecules.

15. Nitric oxide is covalently attached to Cys93β. Thus, the interaction is not reversible binding, as in allosteric regulation. On the other hand, the reaction of NO· with this cysteine residue is apparently spontaneous, and no converter enzyme is needed to add or remove it. Thus, the regulation of covalent modification that is afforded by converter enzyme involvement is obviated. The Hb:NO· interaction illustrates that nature does not always neatly fit the definitions that we create.

16. a. Probably not; c values of 0.1 or greater show hyperbolic v vs. [S] plots.

    b. An increase in cooperativity is defined as a shift of the saturation curve to the right, so an allosteric inhibitor increases the cooperativity of S binding.

17. When hormone disappears, the hormone:receptor complex dissociates, the GTPase activity of the G_α subunit cleaves bound GTP to GDP + P_i, and the affinity of G_α–GDP for adenylyl cyclase is low, so it dissociates and adenylyl cyclase is no longer activated. Residual cAMP will be converted to 5′-AMP by phosphodiesterase, and the catalytic subunits of protein kinase A will be bound again by the regulatory subunits, whose cAMP ligands have dissociated. When protein kinase A becomes inactive, phosphorylase kinase will revert to the unphosphorylated, inactive form through loss of phosphoryl groups. Phophoprotein phosphatase 1 will act on glycogen phosphorylase a, removing the phosphoryl group from Ser^{14} and thereby converting glycogen phosphorylase to the less active, allosterically regulated b form.

Chapter 16

1. The pronghorn antelope is truly a remarkable animal, with numerous specially evolved anatomical and molecular features. These include a large windpipe (to draw in more oxygen and exhale more carbon dioxide), lungs that are three times the size of those of comparable animals (such as goats), and lung alveoli with five times the surface area so that oxygen can diffuse more rapidly into the capillaries. The blood contains larger numbers of

red blood cells and thus more hemoglobin. The skeletal and heart muscles are likewise adapted for speed and endurance. The heart is three times the size of that of comparable animals and pumps a proportionally larger volume of blood per contraction. Significantly, the muscles contain much larger numbers of energy-producing mitochondria, and the muscle fibers themselves are shorter and thus designed for faster contractions. All these characteristics enable the pronghorn antelope to run at a speed nearly twice the top speed of a thoroughbred racehorse and to sustain such speed for up to 1 hour.

2. Refer to Figure 17.23. Step 5, in which the myosin head conformation change occurs, is the step that should be blocked by β,γ-methylene-ATP, because hydrolysis of ATP should occur in this step and β,γ-methylene-ATP is nonhydrolyzable.

3. Phosphocreatine is synthesized from creatine (via creatine kinase) primarily in muscle mitochondria (where ATP is readily generated) and then transported to the sarcoplasm, where it can act as an ATP buffer. The creatine kinase reaction in the sarcoplasm yields the product creatine, which is transported back into the mitochondria to complete the cycle. Like many mitochondrial proteins, the expression of mitochondrial creatine kinase is directed by mitochondrial DNA, whereas sarcoplasmic creatine kinase is derived from information encoded in nuclear DNA.

4. Note in step 4 of Figure 17.23 that it is ATP that stimulates dissociation of myosin heads from the actin filaments—the dissociation of the cross-bridge complex. When ATP levels decline (as happens rapidly after death), large numbers of myosin heads are unable to dissociate from actin and the muscle becomes stiff and unable to relax.

5. The skeletal muscles of the average adult male have a cross-sectional area of approximately 55,000 cm². The gluteus maximus muscles represent approximately 300 cm² of this total. Assuming 4 kg of maximal tension per square centimeter of cross-sectional area, one calculates a total tension for the gluteus maximus of 1200 kg (as stated in the problem). The same calculation shows that the total tension that could be developed by all the muscles in the body is 22,000 kg (or 24.2 tons)!

6. Taking 55,000 g/mol divided by 6.02×10^{23}/mol and by 1.3 g/mL, one obtains a volume of 7.03×10^{-20} mL. Assume a sphere and use the volume of a sphere ($V = (4/3)\pi r^3$) to obtain a radius of 25.6 Å. The diameter of the tubulin dimer (see Figure 16.2) is 8 nm, or 80 Å, and two times the radius we calculated here is approximately 51 Å, a reasonable value by comparison.

7. A liver cell is 20,000 nm long, which would correspond to 5000 tubulin monomers if using the value of Figure 16.2, and about 7800 tubulin monomers using the value of 25.6 Å calculated in problem 6.

8. 4 inches (length of the giant axon) = 10.16 cm. Movement at 2 to 5 μ/sec would correspond to a time of 5.6 to 14 hours to traverse this distance.

9. The 80 Å step of the kinesin motor is a 0.008 μm step. To cover 10.16 cm would require 12.7 million steps.

10. 14 nm is 140 Å, which would be approximately 93 residues of a coiled coil.

11. The correct answer is d, although each answer is reasonable. ATP is needed for several processes involved in muscle relaxation. Salt imbalances can also prevent normal muscle function, and interrupted blood flow could prevent efficient delivery of oxygen needed for ATP production during cellular respiration.

12. The correct answer is a. Understanding the inheritance pattern of sex-linked traits is essential. Males always inherit sex-linked traits (on the X chromosome) from their mothers. In addition,

sex-linked traits are much more commonly expressed in males because they have only one X chromosome.

13. Using Equation 9.1, one can calculate a ΔG of 17,100 J/mol for this calcium gradient.

14. Using Equation 3.12, one can calculate a cellular ΔG of $-48,600$ J/mol for ATP hydrolysis.

15. From the result of problem 13, 17,100 J are required to transport 1 mole of calcium ions. Two moles of calcium would require 34,200 J, and three would cost 51,300 J. Thus, the gradient of problem 14 would provide enough energy to drive the transport of two calcium ions per ATP hydrolyzed.

Chapter 17

1. 6.5×10^{12} (6.5 trillion) people.

2. Consult Table 18.2.

3. O_2, H_2O, and CO_2.

4. See Section 18.2.

5. Consult Figure 18.6.

6. See *Why Metabolic Pathways Have So Many Steps*, p. 552.

7. Consult *Corresponding Pathways of Catabolism and Anabolism Differ in Important Ways*, p. 576. See also Figure 18.8.

8. See *The ATP Cycle*, p. 577; *NAD+ Collects Electrons Released in Catabolism*, p. 577; and *NADPH Provides the Reducing Power for Anabolic Process*, p. 578.

9. In terms of quickness of response, the order is allosteric regulation > covalent modification > enzyme synthesis and degradation. See *The Student Solutions Manual, Study Guide and Problems Book* for further discussions.

10. See *Metabolic Pathways Are Compartmentalized within Cells*, p. 582, and discussions in *The Student Solutions Manual, Study Guide and Problems Book*.

11. A wide variety of mutations of mitochondrial branched-chain α-keto acid dehydrogenases (BCKD) have been identified in patients with MSUD. Some of these mutations produce BCKD with low enzyme activity, but others appear to shorten the biological half-life of the enzyme. Some mutant BCKD are degraded much more rapidly in the cell than are normal BCKD. The result is that although such mutant BCKD molecules show normal activity, each cell contains fewer enzyme molecules and displays a lower total enzyme activity. Studies by Louis Elsas at Emory University have led to a proposal that thiamine pyrophosphate stabilizes a conformation of BCKD that is more resistant to cellular degradation. For this reason, treatment of some MSUD patients with thiamine (100–500 mg/day) increases the biological half-life of the BCKD enzyme molecules and thus increases the total mitochondrial BCKD activity.

12. A mechanism for liver alcohol dehydrogenase:

13. A mechanism for tyrosine racemase.

Tyrosine racemase

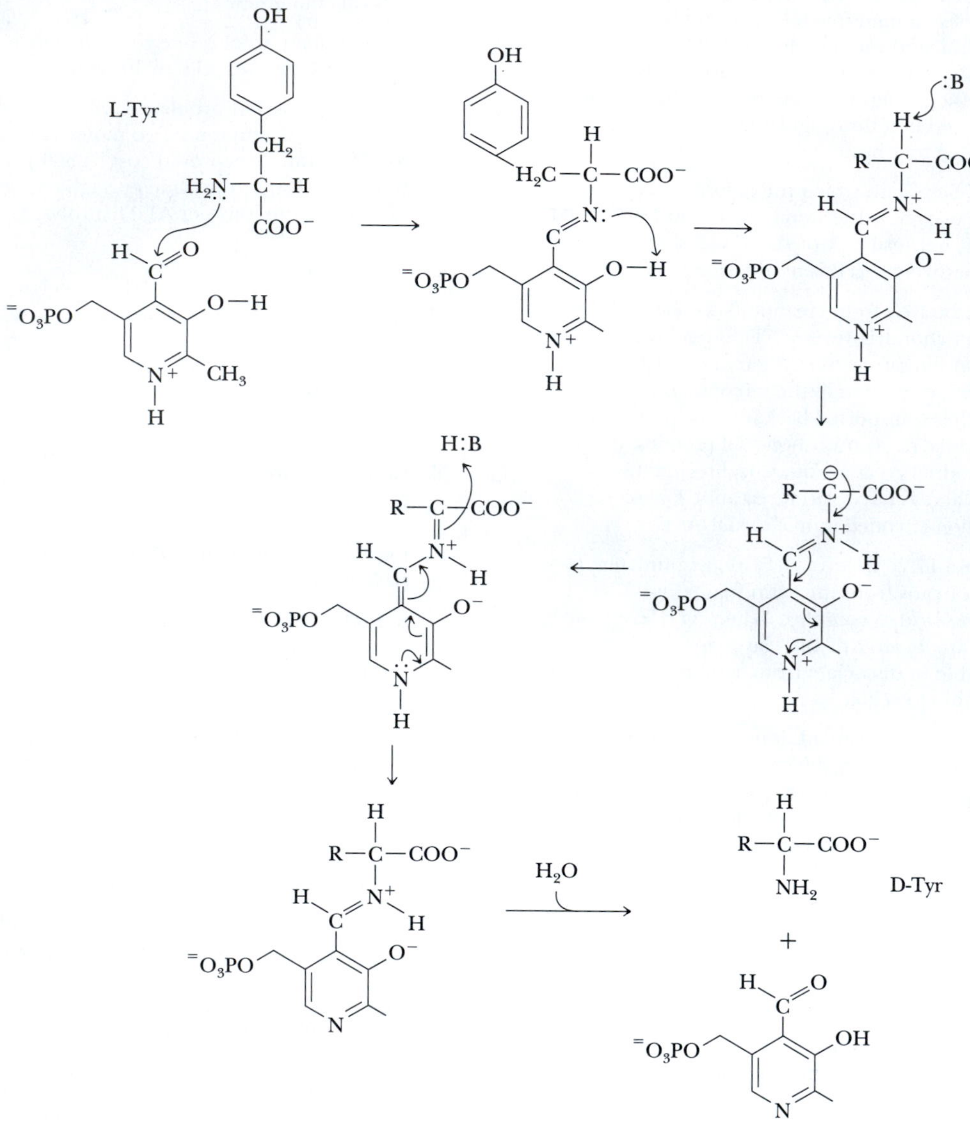

14. ^{32}P and ^{35}S both decay via beta particle emission. A beta particle is merely an electron emitted by a neutron in the nucleus. Thus, beta decay does not affect the atomic mass, but it does convert a neutron to a proton. Thus, beta emission changes ^{32}P to ^{32}S and converts ^{35}S to ^{35}Cl:

$$^{32}P \longrightarrow \beta^- + {}^{32}S, \; t_{1/2} = 14.3 \text{ days.}$$

$$^{35}S \longrightarrow \beta^- + {}^{35}Cl, \; t_{1/2} = 87.1 \text{ days.}$$

The decay equation is a first-order decay equation:

$$A/A_0 = e^{-0.693t/t(t1/2)}$$

Thus, after 100 days of decay, the fraction of ^{32}P remaining would be 0.786% and the fraction of ^{35}S remaining would be 45%.

15. The mechanism of pyruvate carboxylase is shown in Figure 22.4.

16. The correct answer is b. Obligate anaerobes can survive only in the absence of oxygen.

17. The correct answer is a. Fiber provides little or no nutrition. The foods containing fiber, however, have other nutritional substances that we are able to digest and absorb.

Chapter 18

1. a. Phosphoglucoisomerase, fructose bisphosphate aldolase, triose phosphate isomerase, glyceraldehyde-3-P dehydrogenase, phosphoglycerate mutase, and enolase.

    b. Hexokinase/glucokinase, phosphofructokinase, phosphoglycerate kinase, pyruvate kinase, and lactate dehydrogenase.

    c. Hexokinase and phosphofructokinase.

    d. Phosphoglycerate kinase and pyruvate kinase.

    e. According to Equation 3.12, reactions in which the number of reactant molecules differs from the number of product molecules exhibit a strong dependence on concentration. That is, such reactions are extremely sensitive to changes in concentration. Using this criterion, we can predict from Table 19.1 that the free energy changes of the fructose bisphosphate aldolase and glyceraldehyde-3-P dehydrogenase reactions will be strongly influenced by changes in concentration.

f. Reactions that occur with ΔG near zero operate at or near equilibrium. See Table 19.1 and Figure 19.31.

2. The carboxyl carbon of pyruvate derives from carbons 3 and 4 of glucose. The keto carbon of pyruvate derives from carbons 2 and 5 of glucose. The methyl carbon of pyruvate is obtained from carbons 1 and 6 of glucose.

3. Increased [ATP] or [citrate] inhibits glycolysis. Increased [AMP], [fructose-1,6-bisphosphate], [fructose-2,6-bisphosphate], or [glucose-6-P] stimulates glycolysis.

4. See *The Student Solutions Manual, Study Guide and Problems Book* for discussion.

5. The mechanisms for fructose bisphosphate aldolase and glyceraldehyde-3-P dehydrogenase are shown in Figures 19.13 and 19.18, respectively.

6. The relevant reactions of galactose metabolism are shown in Figure 19.33. See *The Student Solutions Manual, Study Guide and Problems Book* for mechanisms.

7. Iodoacetic acid would be expected to alkylate the reactive active-site cysteine that is vital to the glyceraldehyde-3-P dehydrogenase reaction. This alkylation would irreversibly inactivate the enzyme.

8. Ignoring the possibility that ^{32}P might be incorporated into ATP, the only reaction of glycolysis that utilizes P_i is glyceraldehyde-3-P dehydrogenase, which converts glyceraldehyde-3-P to 1,3-bisphosphoglycerate. $^{32}P_i$ would label the phosphate at carbon 1 of 1,3-bisphosphoglycerate. The label will be lost in the next reaction, and no other glycolytic intermediates will be directly labeled. (Once the label is incorporated into ATP, it will also show up in glucose-6-P, fructose-6-P, and fructose-1, 6-bisphosphate.)

9. The sucrose phosphorylase reaction produces glucose-6-P directly, bypassing the hexokinase reaction. Direct production offers the obvious advantage of saving a molecule of ATP.

10. All of the kinases involved in glycolysis, as well as enolase, are activated by Mg^{2+} ion. A Mg^{2+} deficiency could lead to reduced activity for some or all of these enzymes. However, other systemic effects of a Mg^{2+} deficiency might cause even more serious problems.

11. a. 0.133.

    b. 0.0266 mM.

12. +1.08 kJ/mol.

13. a. −13.8 kJ/mol.

    b. $K_{eq} = 194{,}850$.

    c. [Pyr]/[PEP] = 24,356.

14. a. −13.8 kJ/mol.

    b. $K_{eq} = 211$.

    c. [FBP]/[F-6-P] = 528.

15. a. −19.1 kJ/mol.

    b. $K_{eq} = 1651$.

    c. [ATP]/[ADP] = 13.8.

16. The best answer is d. PFK is more active at low ATP than at high ATP, and F-2,6-bisP activates the enzyme.

17. Hexokinase is inhibited by high concentrations of glucose-6-P, so glycolysis would probably stop at such a high level of glucose-6-P. Also, the increased concentration of glucose-6-P would change the cellular free energy of the hexokinase reaction to approximately −22,000 J/mol.

Chapter 19

1. Glutamate enters the TCA cycle via a transamination to form α-ketoglutarate. The γ-carbon of glutamate entering the cycle is equivalent to the methyl carbon of an entering acetate. Thus, no radioactivity from such a label would be lost in the first or second cycle, but in each subsequent cycle, 50% of the total label would be lost (see Figure 19.21).

2. NAD^+ activates pyruvate dehydrogenase and isocitrate dehydrogenase and thus would increase TCA cycle activity. If NAD^+ increases at the expense of NADH, the resulting decrease in NADH would likewise activate the cycle by stimulating citrate synthase and α-ketoglutarate dehydrogenase. ATP inhibits pyruvate dehydrogenase, citrate synthase, and isocitrate dehydrogenase; reducing the ATP concentration would thus activate the cycle. Isocitrate is not a regulator of the cycle, but increasing its concentration would mimic an increase in acetate flux through the cycle and increase overall cycle activity.

3. For most enzymes that are regulated by phosphorylation, the covalent binding of a phosphate group at a distant site induces a conformation change at the active site that either activates or inhibits the enzyme activity. On the other hand, X-ray crystallographic studies reveal that the phosphorylated and unphosphorylated forms of isocitrate dehydrogenase share identical structures with only small (and probably insignificant) conformation changes at Ser^{113}, the locus of phosphorylation. What phosphorylation *does* do is block isocitrate binding (with no effect on the binding affinity of $NADP^+$). As shown in Figure 2 of the paper cited in the problem (Barford, D., 1991. Molecular mechanisms for the control of enzymic activity by protein phosphorylation. *Biochimica et Biophysica Acta* **1133**:55–62), the γ-carboxyl group of bound isocitrate forms a hydrogen bond with the hydroxyl group of Ser^{113}. Phosphorylation apparently prevents isocitrate binding by a combination of a loss of the crucial H bond between substrate and enzyme and by repulsive electrostatic and steric effects.

4. A mechanism for the first step of the α-ketoglutarate dehydrogenase reaction:

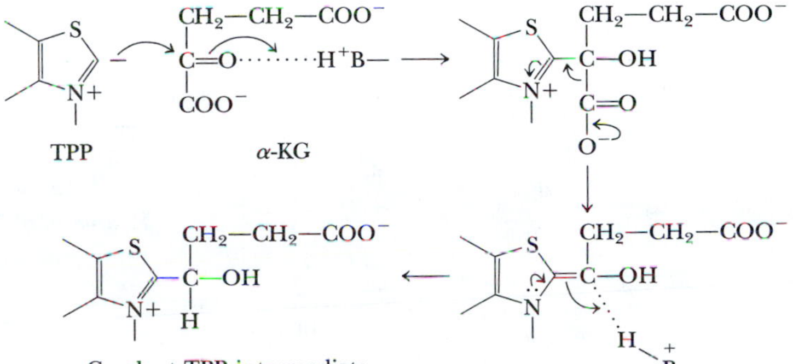

Covalent TPP intermediate

5. Aconitase is inhibited by fluorocitrate, the product of citrate synthase action on fluoroacetate. In a tissue where inhibition has occurred, all TCA cycle metabolites should be reduced in concentration. Fluorocitrate would replace citrate, and the concentrations of isocitrate and all subsequent metabolites would be reduced because of aconitase inhibition.

6. $FADH_2$ is colorless, but FAD is yellow, with a maximal absorbance at 450 nm. Succinate dehydrogenase could be conveniently assayed by measuring the decrease in absorbance in a solution of the flavoenzyme and succinate.

7. The central (C-3) carbon of citrate is reduced, and an adjacent carbon is oxidized in the aconitase reaction. The carbon bearing the hydroxyl group is obviously oxidized by isocitrate dehydrogenase. In the α-ketoglutarate dehydrogenase reaction, the departing carbon atom and the carbon adjacent to it are both

oxidized. Both of the —CH_2— carbons of succinate are oxidized in the succinate dehydrogenase reaction. Of these four molecules, all but citrate undergo a net oxidation in the TCA cycle.

8. Several TCA metabolite analogs are known, including malonate, an analog of succinate, and 3-nitro-2-S-hydroxypropionate, the anion of which is a transition-state analog for the fumarase reaction.

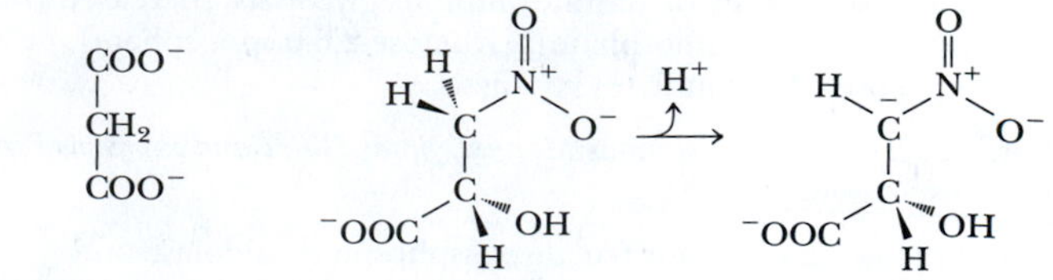

Malonate (succinate analog)   3-Nitro-2-S-hydroxypropionate   Transition-state analog for fumarase

9. A mechanism for pyruvate decarboxylase:

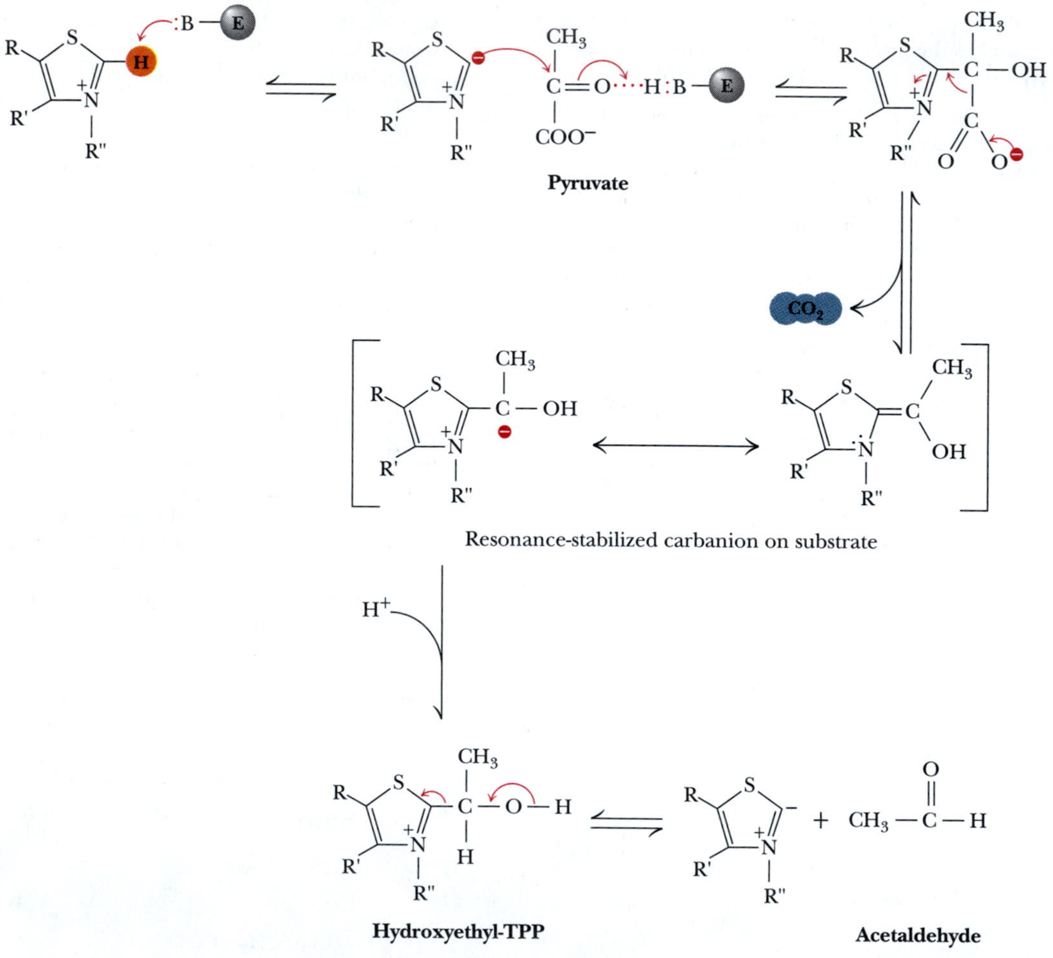

Pyruvate

Resonance-stabilized carbanion on substrate

Hydroxyethyl-TPP   **Acetaldehyde**

10. [isocitrate]/[citrate] = 0.1. When [isocitrate] = 0.03 mM, [citrate] = 0.3 mM.

11. $^{14}CO_2$ incorporation into TCA via the pyruvate carboxylase reaction would label the —CH_2—COOH carboxyl carbon in oxaloacetate. When this entered the TCA cycle, the labeled carbon would survive only to the α-ketoglutarate dehydrogenase reaction, where it would be eliminated as $^{14}CO_2$.

12. ^{14}C incorporated in a reversed TCA cycle would label the two carboxyl carbons of oxaloacetate in the first pass through the cycle. One of these would be eliminated in its second pass through the cycle. The other would persist for more than two cycles and would be eliminated slowly as methyl carbons in acetyl-CoA in the reversed citrate synthase reaction.

13. The constraints of the question appear to preclude the calculation of ATPs generated by the NADH and $FADH_2$ produced either in the breakdown of palmitic acid to eight acetate molecules or the NADH and $FADH_2$ generated in the TCA cycle itself. This leaves only the eight ATPs that would be generated in the succinyl-CoA synthetase (and nucleoside diphosphate kinase) reaction of the TCA cycle itself.

14. d is false. Succinyl-CoA is an inhibitor of citrate synthase.

15. The labeling pattern would be the same as if methyl-labeled acetyl-CoA was fed to the conventional TCA cycle (see Figure 19.21).

Chapter 20

1. The cytochrome couple is the acceptor, and the donor is the (bound) FAD/$FADH_2$ couple, because the cytochrome couple has a higher (more positive) reduction potential.

$$\Delta\mathscr{E}_o' = 0.254\ V - 0.02\ V^*$$

*This is a typical value for enzyme-bound [FAD].

$$\Delta G = -n\mathscr{F}\Delta\mathscr{E}_o' = -43.4\ kJ/mol.$$

2. $\Delta\mathscr{E}_o' = -0.03\ V; \Delta G = +5790\ J/mol.$

3. This situation is analogous to that described in Section 3.3. The net result of the reduction of NAD^+ is that a proton is consumed. The effect on the calculation of free energy change is similar to that described in Equation 3.22. Adding an appropriate term to Equation 20.13 yields:

$$\mathscr{E} = \mathscr{E}_o' - RT\ln\ [H^+] + (RT/n\mathscr{F}\ \ln\ ([ox]/[red])$$

4. Cyanide acts primarily via binding to cytochrome a_3, and the amount of cytochrome a_3 in the body is much lower than the amount of hemoglobin. Nitrite anion is an effective antidote for cyanide poisoning because of its unique ability to oxidize ferrohemoglobin to ferrihemoglobin, a form of hemoglobin that competes very effectively with cytochrome a_3 for cyanide. The amount of ferrohemoglobin needed to neutralize an otherwise lethal dose of cyanide is small compared with the total amount of hemoglobin in the body. Even though a small amount of hemoglobin is sacrificed by sequestering the cyanide in this manner, a "lethal dose" of cyanide can be neutralized in this manner without adversely affecting oxygen transport.

5. You should advise the wealthy investor that she should decline this request for financial backing. Uncouplers can indeed produce dramatic weight loss, but they can also cause death. Dinitrophenol was actually marketed for weight loss at one time, but sales were discontinued when the potentially fatal nature of such "therapy" was fully appreciated. Weight loss is best achieved by simply making sure that the number of calories of food eaten is less than the number of calories metabolized. A person who metabolizes about 2000 kJ (about 500 Cal) more than he or she consumes every day will lose about a pound in about 8 days.)

6. The calculation in Section 20.4 assumes the same conditions as in this problem (1 pH unit gradient across the membrane and an electrical potential difference of 0.18 V). The transport of 1 mole of H^+ across such a membrane generates 23.3 kJ of energy. For three moles of H^+, the energy yield is 69.9 kJ. Then, from Equation 3.12, we have:

$$69,900\ J/mol = 30,500\ J/mol + RT\ln([ATP]/[ADP][P_i])$$

$$39,400\ J/mol = RT\ln\ ([ATP]/[ADP][P_i])$$

$$[ATP]/[ADP][P_i] = 4.36 \times 10^6\ at\ 37°C.$$

In the absence of the proton gradient, this same ratio is 7.25×10^{-6} at equilibrium!

7. The succinate/fumarate redox couple has the highest (that is, most positive) reduction potential of any of the couples in glycolysis and the TCA cycle. Thus, oxidation of succinate by NAD^+ would be very unfavorable in the thermodynamic sense.

Oxidation of succinate by NAD^+:

$$\Delta\mathscr{E}_o' \approx -0.35\ V; \Delta G°' \approx 67,500\ J/mol.$$

On the other hand, oxidation of succinate is quite feasible using enzyme-bound FAD.

Oxidation of succinate by [FAD]:

$$\Delta\mathscr{E}_o' \approx 0, \Delta G°' \approx 0, \text{ depending on the exact reduction}$$
potential for bound FAD.

8. a. −73.3 kJ/mol.

   b. $K_{eq} = 7.1 \times 10^{12}$.

   c. [ATP]/[ADP] = 19.7.

9. a. −151.5 kJ/mol.

   b. $K_{eq} = 3.54 \times 10^{26}$.

   c. [ATP]/[ADP] = 8.8.

10. a. −219 kJ/mol.

    b. $K_{eq} = 2.63 \times 10^{38}$.

    c. [ATP]/[ADP] = 54.4.

11. a. −217 kJ/mol.

    b. $K_{eq} = 1.08 \times 10^{38}$.

    c. [ATP]/[ADP] = −3479.

12. a. −26 kJ/mol.

    b. $K_{eq} = 3.68 \times 10^4$.

    c. $[NAD^+]/[NADH] = 1.25$.

13. a. $K_{eq} = 4.8 \times 10^3$.

    b. +48 kJ/mol.

    c. [ATP]/[ADP] = 2.3.

14. This answer should be based on the information on pages 652–654 and should be in the student's own words.

15. This answer should be based on the information on pages 654–655 and should be in the student's own words.

16. The answer is a. The mitochondrial ATPase reaction proceeds via nucleophilic S_N2-type substitution.

17. At 298 K, doubling the proton gradient would change ΔG by 1717 J/mole. The mitochondrial membrane potential would have to change by approximately 18 mV. The student should decide whether this change should be positive or negative.

Chapter 21

1. Efficiency = $\Delta G^{\circ\prime}/(\text{light energy input}) = n\mathcal{F}\Delta\mathcal{E}_o{}'/(Nhc/\lambda) = 0.564$.

2. $\Delta G^{\circ\prime} = -n\mathcal{F}\Delta\mathcal{E}_o{}'$, so $\Delta G^{\circ\prime}/-n\mathcal{F} = \Delta\mathcal{E}_o{}'$. $n = 4$ for $2\ H_2O \longrightarrow 4\ e^- + 4\ H^+ + O_2$. $\Delta\mathcal{E}_o{}' = (\mathcal{E}_o{}'(\text{primary oxidant}) - \mathcal{E}_o{}'\ (\frac{1}{2}\ O_2/H_2O)) = +0.0648$ V. Thus, $\mathcal{E}_o{}'(\text{primary oxidant}) = 0.881$ V.

3. ΔG for ATP synthesis = $+50{,}336$ J/mol. $\Delta\mathcal{E}_o{}' = 0.26$ V.

4. Reduced plastoquinone = PQH_2; oxidized plastoquinone = PQ; Reduced plastocyanin = $PC(Cu^+)$; plastocyanin = $PC(Cu^{2+})$:

   $$PQH_2 + 2\ H^+{}_{\text{stroma}} + 2\ PC(Cu^{2+}) \longrightarrow$$
   $$PQ + 2\ PC(Cu^+) + 4\ H^+{}_{\text{thylakoid lumen}}$$

5. Noncyclic photophosphorylation: 2.57 ATP would be synthesized from 8 quanta, which equates to 3.11 $h\nu$/ATP. Cyclic photophosphorylation theoretically yields 2 $H^+/h\nu$, so if 14 H^+ yield 3 ATP, 7 $h\nu$ yield 3 ATP, which equates to 2.33 $h\nu$/ATP.

6. In mitochondria, H^+ translocation leads to a decline in intermembrane space pH and hence cytosolic pH, because the outer mitochondrial membrane is permeable to protons. Thus, the eukaryotic cytosol pH is at risk from mitochondrial proton translocation. So, mitochondria rely on a greater membrane potential ($\Delta\psi$) and a smaller ΔpH to achieve the same proton-motive force. Because proton translocation in eukaryotic photosynthesis deposits H^+ into the thylakoid lumen, the cytosol does not experience any pH change and ΔpH is not a problem. Moreover, the light-induced efflux of Mg^{2+} from the lumen diminishes $\Delta\psi$ across the thylakoid membrane, so a greater contribution of ΔpH to the proton-motive force is warranted.

7. Replacement of Tyr by Phe would greatly diminish the possibility of e^- transfer between water and P680$^+$, limiting the ability of P680$^+$ to regain an e^- and return to the P680 ground state.

8. These authors wished to demonstrate unambiguously that a pH gradient was the direct energy source for photosynthetic phosphorylation. ATP synthesis driven by an artificially imposed pH gradient in the absence of any light input proved their point.

9. Assuming 12 c-subunits means 12 H^+ are needed to drive one turn of the c-subunit rotor and the synthesis of 3 ATP by the CF$_1$ part of the ATP synthase. If the $R.\ viridis$ cytochrome bc_1 complex drives the translocation of 2 H^+/e^-, then 2 $h\nu$ gives 4 H^+, and 6 $h\nu$ gives 12 H^+ and thus 3 ATP (thus, 2 $h\nu$ yield 1 ATP).

10. $\Delta G^{\circ} = -n\mathcal{F}\Delta\mathcal{E}_o{}' = -2(96{,}485\ \text{J/volt}\cdot\text{mol})$ $(\mathcal{E}_o{}'(\text{1,3-BPG/Gal3P}) - \mathcal{E}_o{}'(\text{NADP}^+/\text{NADPH})) = -(192{,}970\ \text{J/volt}\cdot\text{mol})(-0.29\ \text{V} - (-0.32\ \text{V})) = -192{,}970$ (0.03) J/mol = $-5{,}789$ J/mol.

11. Use the first eight reactions in Table 21.1 to show that:

    $$CO_2 + \text{Ru-1,5-BP} + 3\ H_2O + 2\ \text{ATP} + 2\ \text{NADPH} + 2\ H^+ \longrightarrow$$
    $$\text{glucose} + 2\ \text{ADP} + 4\ P_i + 2\ \text{NADP}^+$$

12. Radioactivity will be found in C-1 of 3-phosphoglycerate; C-3 and C-4 of glucose; C-1 and C-2 of erythrose-4-P; C-3, C-4, and C-5 of sedoheptulose-1,7-bisP; and C-1, C-2, and C-3 of ribose-5-P.

13. Light induces three effects in chloroplasts: (1) pH increase in the stroma, (2) generation of reducing power (as ferredoxin),

and (3) Mg^{2+} efflux from the thylakoid lumen. Key enzymes in the Calvin–Benson CO_2 fixation pathway are activated by one or more of these effects. In addition, rubisco activase is activated indirectly by light, and, in turn, activates rubisco.

14. The following series of reactions accomplishes the conversion of 2-phosphoglycolate to 3-phosphoglycerate:

    1. 2 phosphoglycolate + 2 $H_2O \longrightarrow$ 2 glycolate + 2 P_i

    2. 2 glycolate + 2 $O_2 \longrightarrow$ 2 glyoxylate + 2 H_2O_2

    3. 2 glyoxylate + 2 serine $\longrightarrow$ 2 hydroxypyruvate + 2 glycine

    4. 2 glycine $\longrightarrow$ serine + CO_2 + NH_3

    5. hydroxypyruvate + glutamate $\longrightarrow$ serine + α-ketoglutarate (an aminotransferase reaction)

    6. α-ketoglutarate + NH_3 + NADH$^+$ + $H^+ \longrightarrow$ glutamate + H_2O + NAD$^+$ (the glutamate dehydrogenase reaction)

    7. hydroxypyruvate + NADH$^+$ + $H^+ \longrightarrow$ glycerate + NAD$^+$

    8. glycerate + ATP $\longrightarrow$ 3-phosphoglycerate + ADP

    Net: 2 phosphoglycolate + 2 NADH$^+$ + 2 H^+ + ATP + 2 O_2 + $H_2O \longrightarrow$ 3-phosphoglycerate + CO_2 + 2 P_i + ADP + 2 H_2O_2 + 2 NAD$^+$

15. Considering the reactions involving water separately:

    1. ATP synthesis: 18 ADP + 18 $P_i \longrightarrow$ 18 ATP + 18 H_2O

    2. NADP$^+$ reduction and the photolysis of water:

       12 H_2O + 12 NADP$^+ \longrightarrow$ 12 NADPH + 12 H^+ + 6 O_2

    3. Overall reaction for hexose synthesis:

       6 CO_2 + 12 NADPH + 12 H^+ + 18 ATP + 12 $H_2O \longrightarrow$ glucose + 12 NADP$^+$ + 6 O_2 + 18 ADP + 18 P_i

    Net: 6 CO_2 + 6 $H_2O \longrightarrow$ glucose + 6 O_2

    Of the 12 waters consumed in O_2 production in reaction 2 and the 12 waters consumed in the reactions of the Calvin–Benson cycle (reaction 3), 18 are restored by H_2O release in phosphoric anhydride bond formation in reaction 1.

16. Ideally, accessory light-harvesting pigments would absorb visible light of wavelengths that the chlorophylls do not absorb, that is, light in the 470–620 nm wavelength range.

17.

Chapter 22

1. The reactions that contribute to the equation on page 711 are:

   $$2\ \text{Pyruvate} + 2\ H^+ + 2\ H_2O \longrightarrow \text{glucose} + O_2$$
   $$2\ \text{NADH} + 2\ H^+ + O_2 \longrightarrow 2\ \text{NAD}^+ + 2\ H_2O$$
   $$4\ \text{ATP} + 4\ H_2O \longrightarrow 4\ \text{ADP} + 4\ P_i$$
   $$2\ \text{GTP} + 2\ H_2O \longrightarrow 2\ \text{GDP} + 2\ P_i$$

   Summing these four reactions produces the equation on page 711.

2. This problem essentially involves consideration of the three unique steps of gluconeogenesis. The conversion of PEP to pyruvate was shown in Chapter 18 to have a $\Delta G^{\circ\prime}$ of -31.7 kJ/mol. For the conversion of pyruvate to PEP, we need only add the conversion of a GTP to GDP (equivalent to ATP to ADP) to the reverse reaction. Thus:

Pyruvate $\longrightarrow$ PEP: $\Delta G^{\circ\prime}$ + +31.7 kJ/mol − 30.5 kJ/mol =
$$1.2 \text{ kJ/mol}$$

Then, using Equation 3.12:

$\Delta G = 1.2$ kJ/mol + $RT \ln$ ([PEP][ADP]^2[Pi][Py][ATP]^2)

$$\Delta G \text{ (in erythrocytes)} = -14.1 \text{ kJ/mol}$$

In the case of the fructose-1,6-bisphosphatase reaction, $\Delta G^{\circ\prime} = -16.3$ kJ/mol (see Equation 18.6).

$$\Delta G = -16.3 \text{ kJ/mol} + RT \ln \text{ ([F-6-P]/[F-1,6-P])}$$

ΔG (in erythrocytes) = -18.3 kJ/mol

For the glucose-6-phosphatase reaction, $\Delta G^{\circ\prime} = -13.8$ kJ/mol (see Table 3.3). $\Delta G = -13.8$ kJ/mol + $RT \ln$ ([Glu]/[G-6-P]) = -3.2 kJ/mol. From these ΔG values and those in Table 19.1, ΔG for gluconeogenesis = -35.6 kJ/mol.

3. Inhibition by 25 μM fructose-2,6-bisphosphate is approximately 94% at 25 μM fructose-1,6-bisphosphate and approximately 44% at 100 μM fructose-1,6-bisphosphate.

4. The hydrolysis of UDP–glucose to UDP and glucose is characterized by a $\Delta G^{\circ\prime}$ of -31.9 kJ/mol. This is more than sufficient to overcome the energetic cost of synthesizing a new glycosidic bond in a glycogen molecule. The net $\Delta G^{\circ\prime}$ for the glycogen synthase reaction is -13.3 kJ/mol.

5. According to Table 24.1, a 70-kg person possesses 1920 kJ of muscle glycogen. Without knowing how much of this is in fast-twitch muscle, we can simply use the fast-twitch data from A Deeper Look (page 722) to calculate a rate of energy consumption. The plot on page 722 shows that glycogen supplies are exhausted after 60 minutes of heavy exercise. Ignoring the curvature of the plot, 1920 kJ of energy consumed in 60 minutes corresponds to an energy consumption rate of 533 J/sec.

6. C-6 of glucose becomes C-5 of ribose for all five glucose molecules passing through the pentose phosphate pathway in Figure 22.38. C-3 of glucose becomes C-3 of ribose in one molecule, C-2 in two molecules of ribose, and C-1 in two molecules of ribose produced by one pass of the pathway in Figure 23.38. C-1 of glucose becomes C-5 of one molecule of ribose and C-1 of four molecules of ribose through one pass of the pathway.

7. Three molecules of glucose are required for one pass through the scheme of Figure 22.40. The C-2 of one glucose becomes the C-3 of pyruvate, and the C-4 of the same molecule of glucose becomes C-1 of a different molecule of pyruvate. For another glucose through the pathway, C-2 becomes C-3 of pyruvate and C-4 becomes C-1 of another pyruvate. For the third glucose through the pathway, both C-2 and C-4 carbons become C-1 of pyruvate molecules.

8. Although other inhibitory processes might also occur, enzymes with mechanisms involving formation of Schiff base intermediates with active-site lysine residues are likely to be inhibited by sodium borohydride (see A Deeper Look box, page 590). The transaldolase reaction of the pentose phosphate pathway involves this type of active-site intermediate and would be expected to be inhibited by sodium borohydride.

9. Glycogen molecules do not have any free reducing ends, regardless of the size of the molecule. If branching occurs every 8 residues and each arm of the branch has 8 residues (or 16 per branch point), a glycogen molecule with 8000 residues would have about 500 ends. If branching occurs every 12 residues, a glycogen molecule with 8000 residues would have about 334 ends.

10. a. Increased fructose-1,6-bisphosphate would activate pyruvate kinase, stimulating glycolysis.

   b. Increased blood glucose would decrease gluconeogenesis and increase glycogen synthesis.

   c. Increased blood insulin inhibits gluconeogenesis and stimulates glycogen synthesis.

   d. Increased blood glucagon inhibits glycogen synthesis and stimulates glycogen breakdown.

   e. Because ATP inhibits both phosphofructokinase and pyruvate kinase, and because its level reflects the energy status of the cell, a decrease in tissue ATP would have the effect of stimulating glycolysis.

   f. Increasing AMP would have the same effect as decreasing ATP—stimulation of glycolysis and inhibition of gluconeogenesis.

   g. Fructose-6-phosphate is not a regulatory molecule and decreases in its concentration would not markedly affect either glycolysis or gluconeogenesis (ignoring any effects due to decreased [G-6-P] as a consequence of decreased [F-6-P].

11. At 298 K, assuming roughly equal concentrations of glycogen molecules of different lengths, the glucose-1-P concentration would be about 3.5 mM.

12. All four of these reactions begin with the formation of N-carboxybiotin:

The carboxylation of β-methylcrotonyl-CoA occurs as follows:

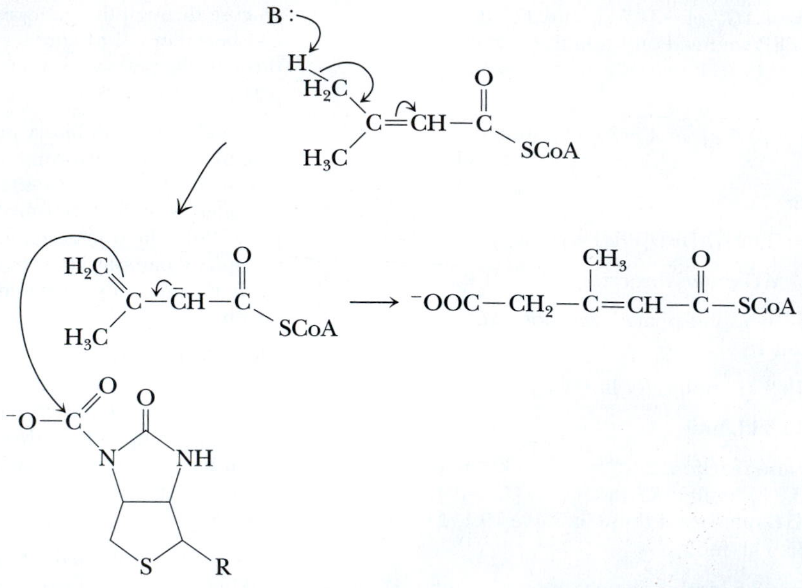

The carboxylations of geranyl-CoA and urea are shown below:

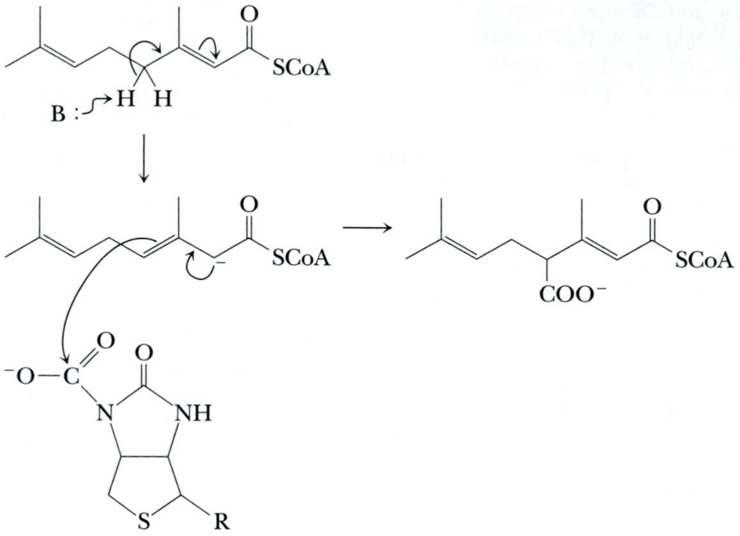

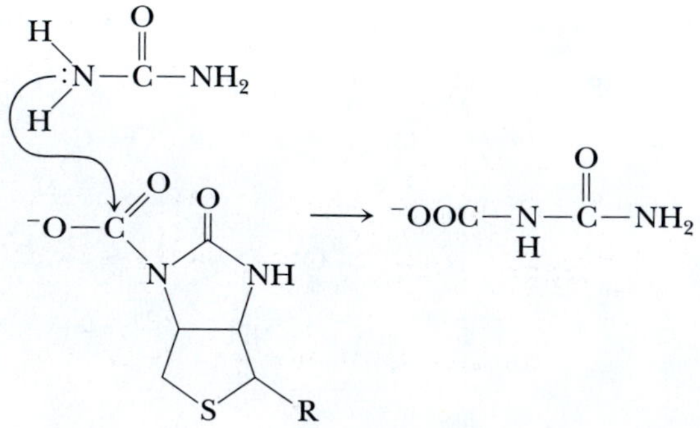

The mechanism of the transcarboxylase reaction is a combination of the pyruvate carboxylase mechanism (shown in Fig. 22.4) and a reverse propionyl-CoA carboxylase reaction.

13. a. The mechanism of the acetolactate synthase reaction.

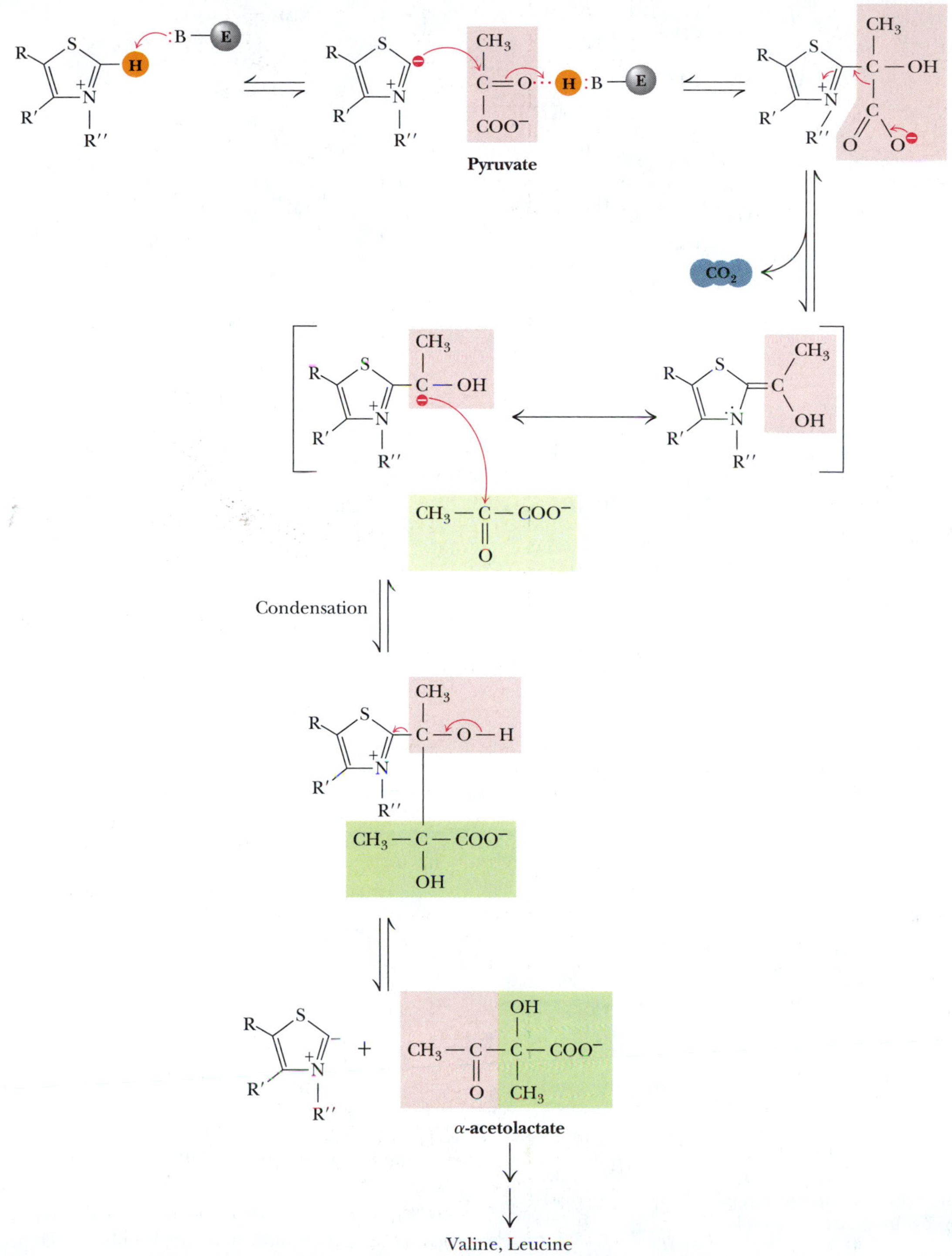

Pyruvate

CO_2

Condensation

$CH_3 - C - COO^-$
$\quad\quad \| $
$\quad\quad O$

α-acetolactate

Valine, Leucine

b. The phosphoketolase reaction.

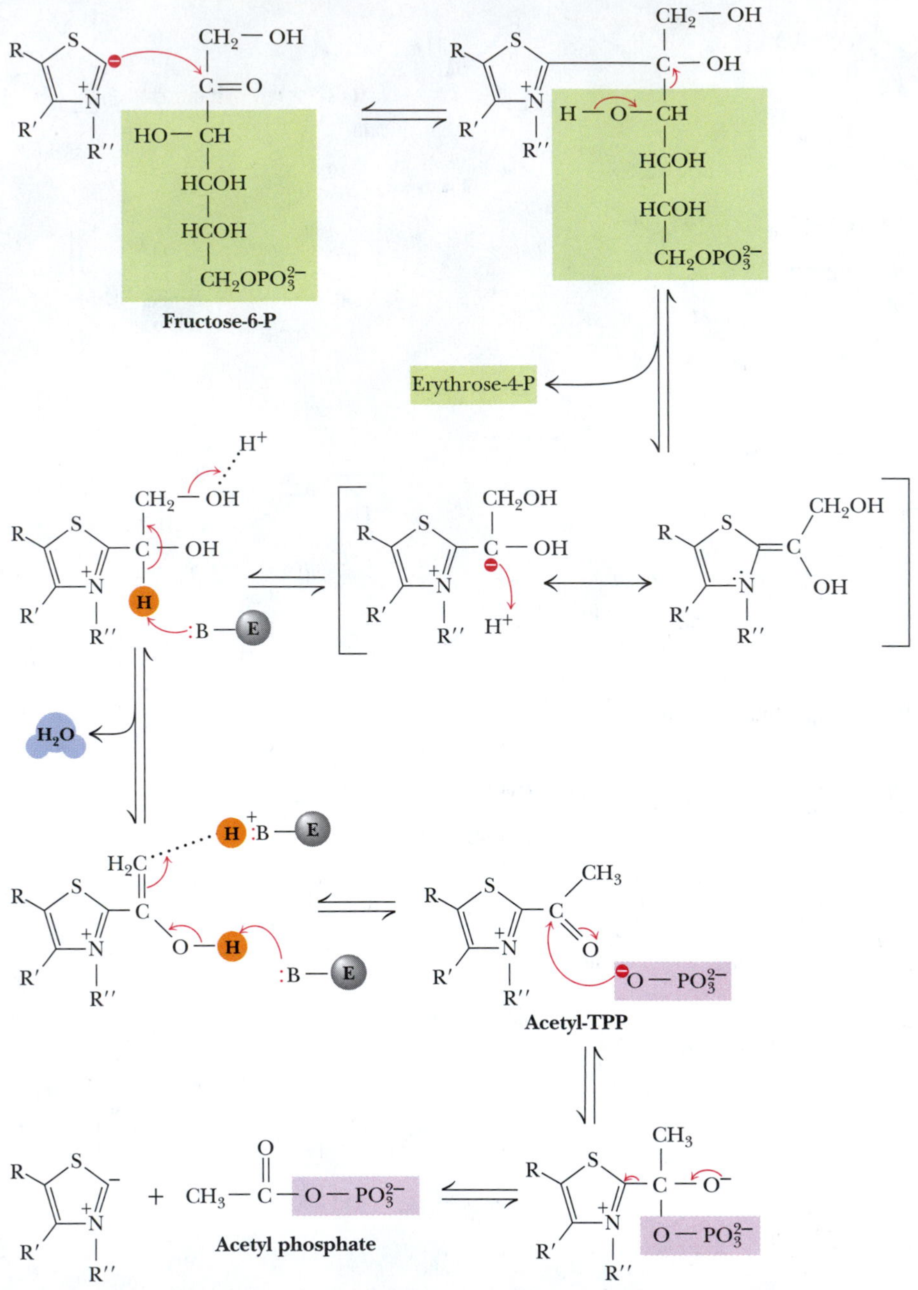

Fructose-6-P

Erythrose-4-P

Acetyl-TPP

Acetyl phosphate

14. Metformin both stimulates glucose uptake by peripheral tissues and enhances the binding of insulin to its receptors. Glipizide complements the actions of metformin by stimulating increased insulin secretion by the pancreas.

15. Epalrestat and tolrestat do not resemble the transition state for aldose reductase, and evidence from studies by Franklin Prendergast and others (Ehrig, T., Bohren, K., Prendergast, F., and Gabbay, K., 1994. Mechanism of aldose reductase inhibition: Binding of NADP+/NADPH and alrestatin-like inhibitors. *Biochemistry* **33**:7157–7165) show that these inhibitors probably

bind to the enzyme: NADP+ complex at a site other than the active site.

16. During the first few seconds of exercise, existing stores of ATP are consumed and creatine phosphate then provides additional ATP via the creatine kinase reaction. For the next 90 seconds or so, anaerobic metabolism (conversion of glucose to lactate via glycolysis) provides energy. At this point, aerobic metabolism begins in earnest, delivering significant energy resources to sustain long-term exercise.

17. Creatine phosphate provides ATP to muscle via the creatine kinase reaction:

The creatine kinase reaction

During periods of energy abundance, muscles store ATP equivalents in the form of creatine phosphate. When the muscles demand energy, creatine kinase runs in the reverse direction, converting creatine phosphate to ATP.

Chapter 23

1. Assuming that all fatty acid chains in the triacylglycerol are palmitic acid, the fatty acid content of the triacylglycerol is 95% of the total weight. On the basis of this assumption, one can calculate that 30 lb of triacylglycerol will yield 118.7 L of water.

2. 11-*cis*-Heptadecenoic acid is metabolized by means of seven cycles of β-oxidation, leaving a propionyl-CoA as the final product. However, the fifth cycle bypasses the acyl-CoA dehydrogenase reaction, because a *cis*-double bond is already present at the proper position. Thus, β-oxidation produces 7 NADH (= 17.5 ATP), 6 FADH$_2$ (= 9 ATP), and 7 acetyl-CoA (= 70 ATP), for a total of 96.5 ATP. Propionyl-CoA is converted to succinyl-CoA (with expenditure of 1 ATP), which can be converted to oxaloacetate in the TCA cycle (with production of 1 GTP, 1 FADH$_2$ and 1 NADH). Oxaloacetate can be converted to pyruvate (with no net ATP formed or consumed), and pyruvate can be metabolized in the TCA cycle (producing 1 GTP, 1 FADH$_2$, and 4 NADH). The net for these conversions of propionate is 16.5 ATP. Together with the results of β-oxidation, the total ATP yield for the oxidation of one molecule of 11-*cis*-heptadecenoic acid is 113 ATP.

3. Instead of invoking hydroxylation and β-oxidation, the best strategy for oxidation of phytanic acid is α-hydroxylation, which places a hydroxyl group at C-2. This facilitates oxidative α-decarboxylation, and the resulting acid can react with CoA to form a CoA ester. This product then undergoes six cycles of β-oxidation. In addition to CO$_2$, the products of this pathway are three molecules of acetyl-CoA, three molecules of propionyl-CoA, and one molecule of 2-propionyl-CoA.

4. Although acetate units cannot be used for net carbohydrate synthesis, oxaloacetate can enter the gluconeogenesis pathway in the PEP carboxykinase reaction. (For this purpose, it must be converted to malate for transport to the cytosol.) Acetate labeled at the carboxyl carbon will first label (equally) the C-3 and C-4 positions of newly formed glucose. Acetate labeled at the methyl carbon will label (equally) the C-1, C-2, C-5, and C-6 positions of newly formed glucose.

5. Hypoglycin in unripened akee fruit irreversibly inactivates acyl-CoA dehydrogenase, thereby blocking β-oxidation of fatty acids, a pathway that is a major source of energy. Victims of poisoning by unripened akee fruit are often found to be severely hypoglycemic because they have used up their carbohydrate reserves.

6. Fat is capable of storing more energy (37 kJ/g) than carbohydrate (16 kJ/g). Ten pounds of fat contains 10 × 454 × 37 = 167,980 kJ of energy. This same amount of energy would require 167,980/16 = 10,499 g, or 23 lb, of stored carbohydrate.

7. The enzyme methylmalonyl-CoA mutase, which catalyzes the third step in the conversion of propionyl-CoA to succinyl-CoA, is B$_{12}$-dependent. If a deficiency in this vitamin occurs, and if large amounts of odd-carbon fatty acids were ingested in the diet, L-methylmalonyl-CoA could accumulate.

8. a. Myristic acid:

$$CH_3(CH_2)_{12}CO\text{-}CoA + 94\ P_i + 94\ ADP + 20\ O_2 \longrightarrow$$
$$94\ ATP + 14\ CO_2 + 113\ H_2O + CoA$$

   b. Stearic acid:
$$CH_3(CH_2)_{16}CO\text{-}CoA + 122\ P_i + 122\ ADP + 26\ O_2 \longrightarrow$$
$$122\ ATP + 18\ CO_2 + 147\ H_2O + CoA$$

   c. α-Linolenic acid:
$$C_{17}H_{29}CO\text{-}CoA + 116.5\ P_i + 116.5\ ADP + 24.5\ O_2 \longrightarrow$$
$$116.5\ ATP + 18\ CO_2 + 138.5\ H_2O + CoA$$

   d. Arachidonic acid:
$$C_{19}H_{31}CO\text{-}CoA + 128\ P_i + 128\ ADP + 27\ O_2 \longrightarrow$$
$$128\ ATP + 20\ CO_2 + 152\ H_2O + CoA$$

9. During the hydration step, the elements of water are added across the double bond. Also, the proton transferred to the acetyl-CoA carbanion in the thiolase reaction is derived from the solvent, so each acetyl-CoA released by the enzyme would probably contain two tritiums. Seven tritiated acetyl-CoAs would thus derive from each molecule of palmitoyl-CoA metabolized, each with two tritiums at C-2.

10. A carnitine deficiency would presumably result in defective or limited transport of fatty acids into the mitochondrial matrix and reduced rates of fatty acid oxidation.

11. See the following mechanisms.

α-Methyleneglutarate mutase

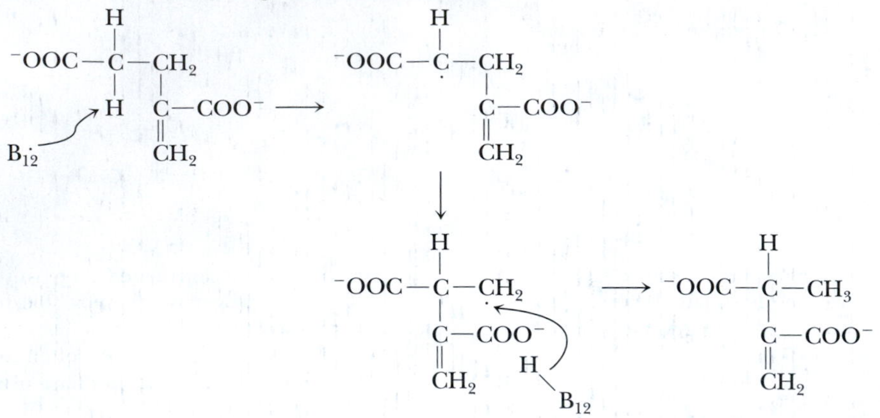

Diol dehydrase

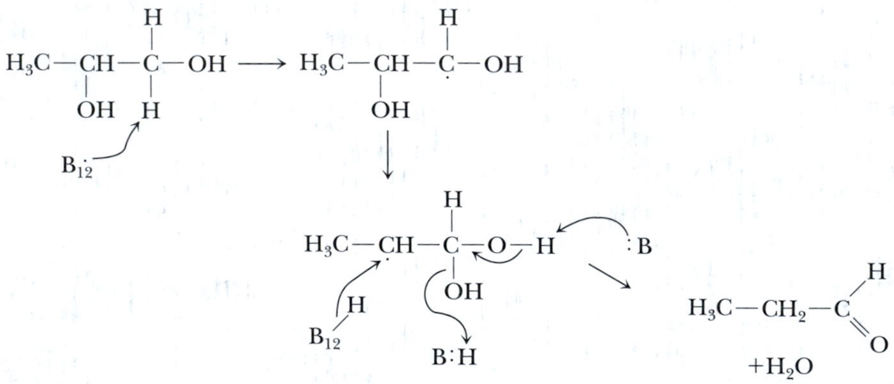

Glycerol dehydrase

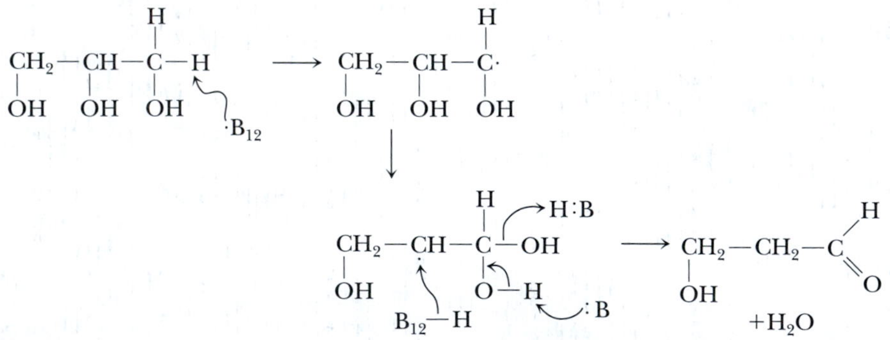

Ethanolamine ammonia-lyase

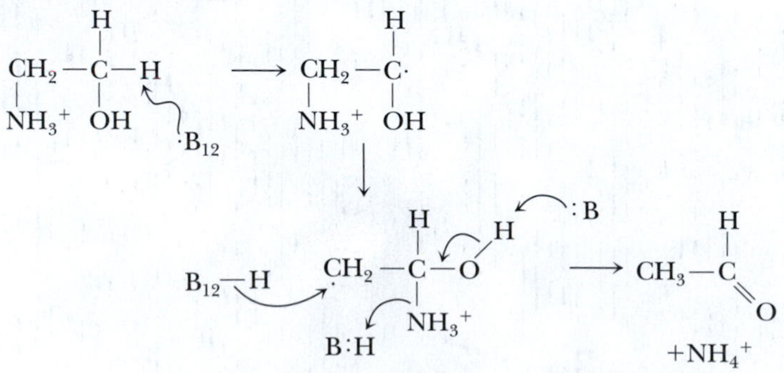

12. 1.3 grams × 37 kJ/gram = 48,000 J. This at first seems like a remarkably small amount of energy to sustain the hummingbird during a 500-mile flight at 50 mph. However, keep in mind that the hummingbird weighs only 3 to 4 grams, and see problem 13 following.

13. 48,000 J in 10 hours is 4800 J/hr. If the hummingbird consumes 250 mL per hour during migration, this means that it is consuming 4800/250 or 19.2 J/mL of oxygen consumed. If a human consumes 12.7 kcal/min while running, this is equivalent to 12,700 × 4.184 J/cal = 53.1 kJ/min; 48,000 J/53,100 J/min = 0.9 minute. So a human could run for only less than a minute (nowhere close to an 8-minute mile) on the energy consumed by the hummingbird (only 3 to 4 g) on a 500-mile flight. A typical person would have to run about 40 miles to lose 1 lb of fat.

14. A mechanism for the HMG-CoA synthase reaction:

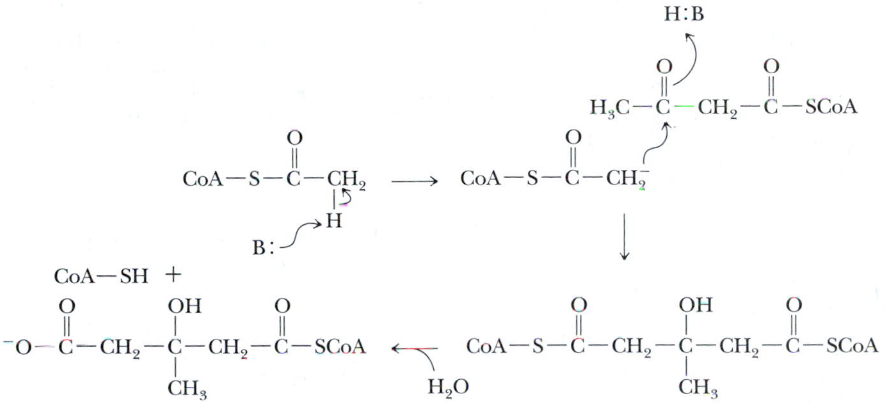

15. The changes in oxidation state of cobalt in the course of any B_{12}-dependent reaction arise from hemolytic cleavages of the Co^{3+}—C bonds involved. Moreover, in the Co^{3+}—C bonds shown in a typical B_{12}-dependent reaction, the two electrons of the bond are not ascribed to the Co^{3+} atom in the calculation of the Co oxidation state. Thus, when the Co—C bond is cleaved homolytically, one electron reverts to the Co, changing its oxidation state from 3+ to 2+.

16. It may be presumed that the oxidation of the acyl chain is accomplished via a two-electron transfer, whereas the steps involved in reoxidation of $FADH_2$ by ETF are one-electron transfers. $FAD/FADH_2$ can participate both in one-electron and two-electron transfers, whereas $NAD^+/NADH$ can participate only in two-electron transfers.

17. The sequence of reactions involving creation of a double bond, then hydration across it, followed by oxidation is the reverse of what happens to succinate in the TCA cycle. (This same sequence of reactions is also employed in fatty acid synthesis and in both the catabolism and anabolism of amino acids.) Clearly, this sequence of three reactions must represent an optimal mechanistic strategy for the chemistry achieved.

Chapter 24

1. The equations needed for this problem are found on page 770. See *The Student Solutions Manual, Study Guide and Problems Book* for details.

2. Carbons C-1 and C-6 of glucose become the methyl carbons of acetyl-CoA that is the substrate for fatty acid synthesis. Carbons C-2 and C-5 of glucose become the carboxyl carbon of acetyl-CoA for fatty acid synthesis. Only citrate that is immediately exported to the cytosol provides glucose carbons for fatty acid synthesis. Citrate that enters the TCA cycle does not immediately provide carbon for fatty acid synthesis.

3. A suitable model, based on the evidence presented in this chapter, would be that the fundamental regulatory mechanism in ACC is a polymerization-dependent conformation change in the protein. All other effectors—palmitoyl-CoA, citrate, and phosphorylation-dephosphorylation—may function primarily by shifting the inactive protomer-active polymer equilibrium. Polymerization may bring domains of the protomer (that is, bicarbonate-, acetyl-CoA-, and biotin-binding domains) closer together or may bring these domains on separate protomers close to each other. See *The Student Solutions Manual, Study Guide and Problems Book* for further details.

4. The pantothenic acid group may function, at least to some extent, as a flexible "arm" to carry acyl groups between the malonyl transferase and ketoacyl-ACP synthase active sites. The pantothenic acid moiety is approximately 1.9 nm in length, setting an absolute upper-limit distance between these active sites of 3.8 nm. However, on the basis of modeling considerations, it seems likely that the distance between these sites is smaller than this upper-limit value.

5. Two electrons pass through the chain from NADH to FAD to the two cytochromes of the cytochrome b_5 reductase and then to the desaturase. Together with two electrons from the fatty acyl substrate, these electrons reduce an O_2 to two molecules of water. The hydrogen for the waters that are formed in this way comes from the substrate (2H) and from two protons from solution.

6. Ethanolamine + glycerol + 2 fatty acyl-CoA + 2 ATP + CTP + $H_2O \longrightarrow$ phosphatidylethanolamine + 2 ADP + 2 CoA + CMP + PP_i + P_i

7. The conversion of acetyl-CoA to lanosterol can be written as:

18 Acetyl-CoA + 13 NADPH + 13 H^+ + 18 ATP + 0.5 $O_2 \longrightarrow$ lanosterol + 18 CoA + 13 $NADP^+$ + 18 ADP + 6 P_i + 6 PP_i + CO_2

The conversion of lanosterol to cholesterol is complicated; however, in terms of carbon counting, three carbons are lost in the conversion to cholesterol. This process might be viewed as 1.5 acetate groups for the purpose of completing the balanced equation.

8. The numbers 1–4 in the cholesterol structure indicate the carbon positions of mevalonate as shown (note that the numbering shown here is not based on the systematic numbering of mevalonate):

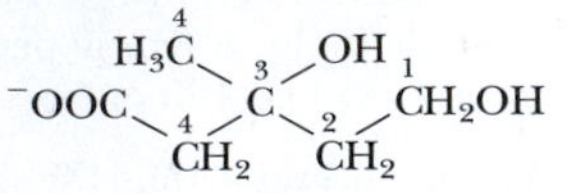

Mevalonate 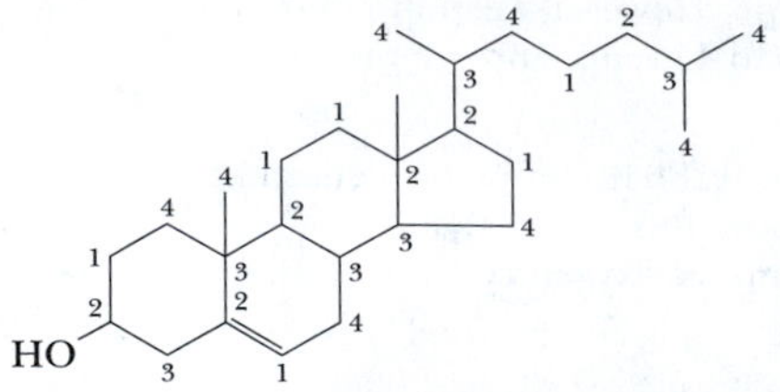   Cholesterol

9. The O-linked saccharide domain of the LDL receptor probably functions to extend the receptor domain away from the cell surface and above the glycocalyx coat so that the receptor can recognize circulating lipoproteins.

10. As shown in Figures 24.19 and 24.22, the syntheses (in eukaryotes) of phosphatidylcholine, phosphatidylethanolamine, phosphatidylinositol, and phosphatidylglycerol are dependent upon CTP. A CTP deficiency would be likely to affect all these synthetic pathways.

11. It "costs" 1 ATP to form a malonyl-CoA. Each cycle of the fatty acyl synthase consumes 2 NADPH molecules, each worth 3.5 ATP. Thus, each of the seven cycles required to form a palmitic acid consumes 8 ATPs. A total of 56 ATPs are consumed to synthesize one molecule of palmitic acid.

12. The mechanism of the 3-ketosphinganine synthase reaction: is shown in the following figure.

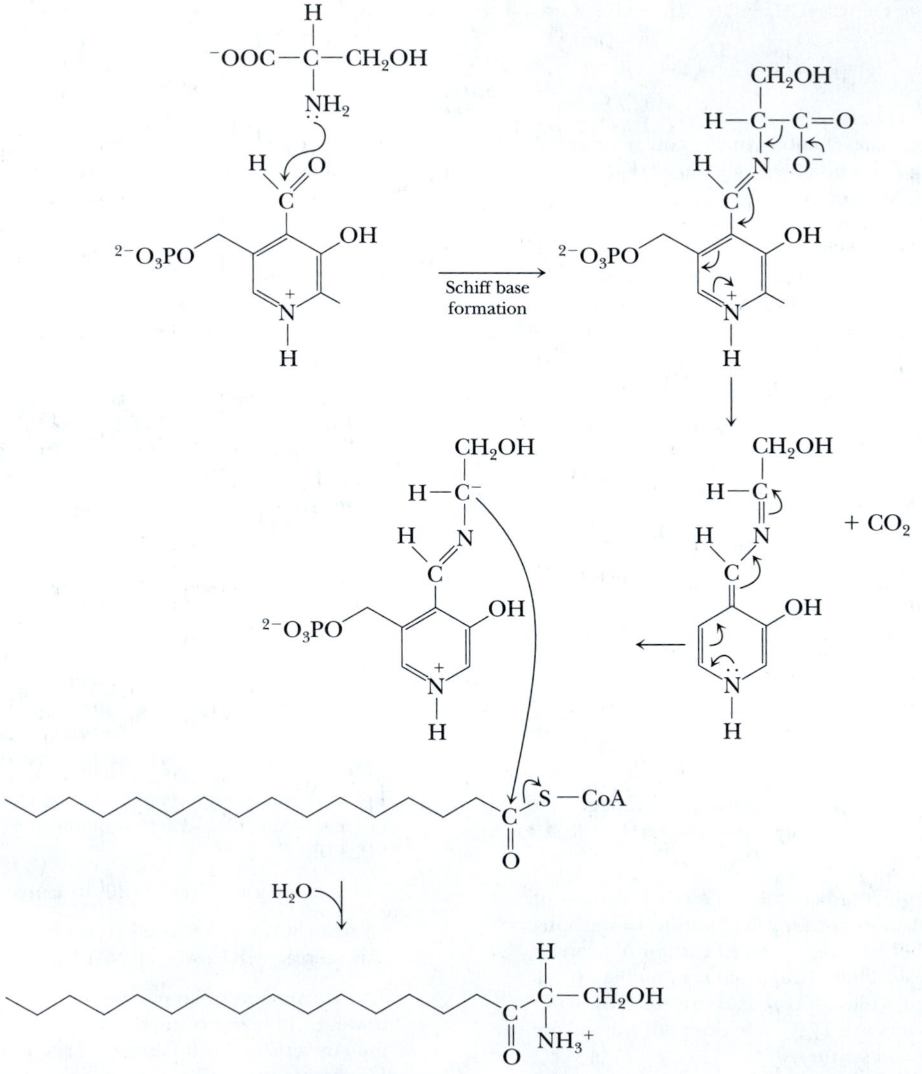

2 S-3-Ketosphinganine

13. FAD is required for the eukaryotic reaction that converts stearic acid to oleic acid because the oxidation of a single bond to a double bond in stearic acid involves NADH and thus requires a two-electron transfer, whereas the reoxidation of $FADH_2$ to FAD is accomplished by electron transfer to cytochrome b_5. Cytochromes are capable of one-electron transfers only, so $FADH_2$/FAD is required because it can participate both in one-electron and two-electron transfers.

14. See Chapter 23, problem 23.14 for a mechanism for the HMG-CoA synthase reaction.

15. A mechanism for the β-ketoacyl ACP synthase reaction:

β-Ketoacyl ACP synthase

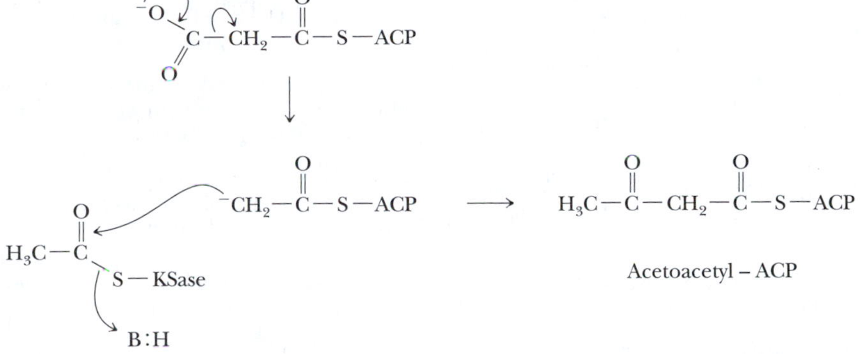

Acetoacetyl – ACP

16. Palmitic acid must be elongated and then unsaturated to form linoleic acid. The elongation process involves a thiolase reaction to add two carbons to palmitoyl-CoA and then reduction of a carbonyl to a hydroxyl, dehydration to form a double bond, and then reduction of the double bond to a single bond. These last three reactions are the reverse of what happens in TCA, in the conversion of succinate to fumarate, then malate, then oxaloacetate. These same three reactions occur in β-oxidation, in fatty acid synthesis, and in amino acid synthesis and degradation.

17. 11 Acetyl-CoA + 10 ATP^{4-} + 20 NADPH + 10 H^+ ⟶ behenoyl-CoA + 20 $NADP^+$ + 10 CoASH + 10 ADP^{3-} + 10 P_i^{2-}

Chapter 25

1. The oxidation number of N in nitrate is +5; in nitrite, +3; in NO, +2; in N_2O, +1; and in N_2, 0.

2. a. Assume that nitrate assimilation requires 4 NADPH equivalents per NO_3^- reduced to NH_4^+. Four NADPH have a metabolic value of 16 ATP.

   b. Nitrogen fixation requires 8 e^- (see Equation 25.3) and 16 ATPs (see Figure 25.6) per N_2 reduced. If 4 NADH provide the requisite 8 e^-, each NADH having a metabolic value of 3 ATPs, then 28 ATP equivalents are consumed per N_2 reduced in biological nitrogen fixation (or 14 ATP equivalents per NH_4^+ formed).

3. a. [ATP] increase will favor adenylylation; the value of n will be greater than 6 ($n>6$).

   b. An increase in P_{IIA}/P_{IID} will favor adenylylation; the value of n will be greater than 6 ($n>6$).

   c. An increase in the [αKG]/[Gln] ratio will favor deadenylylation; $n<6$.

   d. [P_i] decrease will favor adenylylation; $n>6$.

4. Two ATPs are consumed in the carbamoyl-P synthetase-I reaction, and 2 phosphoric anhydride bonds (equal to 2 ATP equivalents) are expended in the argininosuccinate synthetase reaction. Thus, 4 ATP equivalents are consumed in the urea cycle, as 1 urea and 1 fumarate are formed from 1 CO_2, 1 NH_3, and 1 aspartate.

5. Protein catabolism to generate carbon skeletons for energy production releases the amino groups of amino acids as excess nitrogen, which is excreted in the urine, principally as urea.

6. One ATP in reaction 1, one NADPH in reaction 2, one NADPH in reaction 11, and one succinyl-CoA in reaction 12 add up to 10 ATP equivalents, assuming each NADPH is worth 4 ATPs. (The succinyl-CoA synthetase reaction of the citric acid cycle [see Figure 20.12] fixes the metabolic value of succinyl-CoA versus succinate at 1 GTP [= 1 ATP].)

7. From Figure 25.36: ^{14}C-labeled carbon atoms derived from ^{14}C-2 of PEP are shaded yellow.

8. 1. 2 aspartate ⟶ *transamination* ⟶ 2 oxaloacetate.

   2. 2 oxaloacetate + 2 GTP ⟶ *PEP carboxykinase* ⟶ 2 PEP + 2 CO_2 + 2 GDP.

   3. 2 PEP + 2 H_2O ⟶ *enolase* ⟶ 2 2-PG.

   4. 2 2-PG ⟶ *phosphoglyceromutase* ⟶ 2 3-PG.

   5. 2 3-PG + 2 ATP ⟶ *3-P glycerate kinase* ⟶ 2 1,3-bisPG + 2 ADP.

   6. 2 1,3-bisPG + 2 NADH + 2 H^+ ⟶ *G-3-P dehydrogenase* ⟶ 2 G-3-P + 2 NAD^+ + 2 P_i.

   7. 1 G-3-P ⟶ *triose-P isomerase* ⟶ 1 DHAP.

   8. G-3-P + DHAP ⟶ *aldolase* ⟶ fructose-1,6-bisP.

   9. F-1,6-bisP + H_2O ⟶ *FBPase* ⟶ F-6-P + P_i.

   10. F-6-P ⟶ *phosphoglucoisomerase* ⟶ G-6-P.

   11. G-6-P + H_2O ⟶ *glucose phosphatase* ⟶ glucose + P_i.

   Net: 2 aspartate + 2 GTP + 2 ATP + 2 NADH + 6 H^+ + 4 H_2O ⟶ glucose + 2 CO_2 + 2 GDP + 2 ADP + 4 P_i + 2 NAD^+

   (Note that 4 of the 6 H^+ are necessary to balance the charge on the 4 carboxylate groups of the 2 OAA.)

   (As a consequence of reaction [1], 2 α-keto acids [for example, α-ketoglutarate] will receive amino groups to become 2 α-amino acids [for example, glutamate].)

9. Alanine: glutamate:pyruvate aminotransferase.

   Arginine: from glutamate via ornithine, so it's glutamate dehydrogenase.

   Aspartate: glutamate:oxaloacetate aminotransferase.

   Asparagine: from aspartate, so it's the glutamate:oxaloacetate aminotransferase.

Cysteine: cysteine is formed from serine, so it's glutamate via 3-phosphoserine aminotransferase.

Glutamate: glutamate dehydrogenase.

Glutamine: glutamine synthetase.

Glycine: glycine is formed from serine, so it's glutamate via 3-phosphoserine aminotransferase.

Histidine: from glutamate via L-histidinol phosphate aminotransferase.

Isoleucine: glutamate:α-keto-β-methylvalerate aminotransferase.

Leucine: glutamate:α-ketoisocaproate aminotransferase.

Lysine: from glutamate via saccharopine formation by a glutamate-dependent NADPH dehydrogenase; in bacteria, lysine is synthesized from aspartate, so it's glutamate:oxaloacetate aminotransferase.

Methionine: from aspartate, so it's glutamate:oxaloacetate aminotransferase.

Phenylalanine: glutamate:phenylpyruvate aminotransferase (= phenylalanine aminotransferase).

Proline: from glutamine, so it's glutamate dehydrogenase.

Serine: glutamate via 3-phosphoserine aminotransferase.

Threonine: from aspartate, so it's glutamate:oxaloacetate aminotransferase.

Tryptophan: from serine via tryptophan synthase, so its α-amino group comes from serine, which gets its amino group from glutamate via 3-phosphoserine aminotransferase.

Tyrosine: glutamate:4-hydroxyphenylpyruvate aminotransferase (= tyrosine aminotransferase).

Valine: glutamate:α-ketoisovalerate aminotransferase.

10. Pyridoxal (vitamin B_6), because it is the precursor to pyridoxal-P, the key coenzyme in aminotransferase reactions, as well as other aspects of amino acid metabolism.

11. The conversion of homocysteine to methionine is folate-dependent; dietary folate absorption is dependent on vitamin B_{12} for removal of methyl groups added to folate during digestion; finally, the α-amino group of homocysteine formed in the methione biosynthetic pathway comes from aspartate via the pyridoxal-P–dependent glutamate:oxaloacetate aminotransferase.

12. The BCKAD complex is structurally and functionally analogous to the pyruvate dehydrogenase complex and the α-ketoglutarate dehydrogenase complexes. See the A Deeper Look box in Chapter 19 that describes the reaction mechanism for the pyruvate dehydrogenase complex for the mechanism.

13. Aspartame is a N-α-L-aspartyl-L-phenylalanine-1-methyl ester that is broken down in the digestive tract and phenylalanine is released. Phenylalanine is the substance that phenylketonurics must avoid.

14. Glyphosate inhibits 3-enolpyruvylshikimate-5-P synthase, an essential enzyme in the biosynthesis of chorismate. Not only is chorismate the precursor for synthesis of the aromatic amino acids Phe, Tyr, and Trp, it is also the precursor for formation of lignin, a major structural component in plant cell walls, as well as other essential substances such as folate, coenzyme Q, plastoquinone, and vitamins E and K.

15. Glycine is formed from serine, which is formed from 3-phosphoglycerate. Carbons 3 and 4, 2 and 5, and 1 and 6 of glucose contribute carbons 1, 2, and 3 of 3-phosphoglycerate, respectively. The β-carbon of serine derives from the 3-C in 3-phosphoglycerate, the $C_α$-carbon atom in serine comes from 3-phosphoglycerate C-2, and the carboxyl-C in serine comes from C-1 of 3-PG. It is these latter two carbons of 3-PG that are found in glycine; that is, C-1 and C-2 of 3-PG; C-1 of 3-PG came from C-3 and C-4 in glucose, and C-2 of 3-PG came from C-2 and C-5 in glucose.

16. GS monomers are inactive because GS active sites require elements of protein structure contributed by adjacent subunits in the GS_{12} dodecamer.

17. Four of the six carbons in the ring of Tyr come from erythrose-4-P and the remaining two from PEP; the carboxyl-C, $C_α$, and $C_β$ of Tyr come from PEP. Degradation of Tyr to acetoacetate + fumarate yields acetoacetate composed from the $C_α$ and $C_β$ plus 2 C atoms from the ring which are either both from E-4-P or one each from E-4-P and PEP, depending on the orientation of the ring (rotation about the $C_β$-ring C bond). Fumarate will be composed from 2 C atoms from PEP and 2 from E-4-P *or* 3 C atoms from E-4-P and 1 from PEP.

Chapter 26

1. See Figure 26.2 (purines) and Figure 26.13 (pyrimidines).

2. Assume ribose-5-P is available.

   Purine synthesis: 2 ATP equivalents in the ribose-5-P pyrophosphokinase reaction, 1 in the GAR synthetase reaction, 1 in the FGAM synthetase reaction, 1 in the AIR carboxylase reaction, 1 in the CAIR synthetase reaction, and 1 in the SACAIR synthetase reaction yields IMP, the precursor common to ATP and GTP. Net: 7 ATP equivalents.

   a. *ATP:* 1 GTP (an ATP equivalent) is consumed in converting IMP to AMP; 2 more ATP equivalents are needed to convert AMP to ATP. Overall, ATP synthesis from ribose-5-P onward requires 10 ATP equivalents.

   b. *GTP:* 2 high-energy phosphoric anhydride bonds from ATP, but 1 NADH is produced in converting IMP to GMP; 2 more ATP equivalents are needed to convert GMP to GTP. Overall, GTP synthesis from ribose-5-P onward requires 8 ATP equivalents.

   Pyrimidine synthesis: Starting from HCO_3^- and Gln, 2 ATP equivalents are consumed by CPS-II, and an NADH equivalent is produced in forming orotate. OMP synthesis from ribose-5-P plus orotate requires conversion of ribose-5-P to PRPP at a cost of 2 ATP equivalents. Thus, the net ATP investment in UMP synthesis is just 1 ATP.

   c. *UTP:* Formation of UTP from UMP requires 2 ATP equivalents. Net ATP equivalents in UTP biosynthesis = 3.

   d. *CTP:* CTP biosynthesis from UTP by CTP synthetase consumes 1 ATP equivalent. Overall ATP investment in CTP synthesis = 4 ATP equivalents.

3. a. See Figure 26.6.

   b. See Figure 26.17.

   c. See Figure 26.17.

4. a. Azaserine inhibits glutamine-dependent enzymes, as in steps 2 and 5 of IMP synthesis (glutamine:PRPP amidotransferase

and FGAM synthetase), as well as GMP synthetase (step 2, Figure 26.5), and CTP synthetase (Figure 26.16).

b. Methotrexate, an analog of folic acid, antagonizes THF-dependent processes, such as steps 4 and 10 (GAR transformylase and AICAR transformylase) in purine biosynthesis (Figure 26.3), and the thymidylate synthase reaction (Figure 26.27) of pyrimidine metabolism.

c. Sulfonamides are analogs of *p*-aminobenzoic acid (PABA). Like methotrexate, sulfonamides antagonize THF formation. Thus, sulfonamides affect nucleotide biosynthesis at the same sites as methotrexate, but only in organisms such as prokaryotes that synthesize their THF from simple precursors such as PABA.

d. Allopurinol is an inhibitor of xanthine oxidase (Figure 26.10).

e. 5-Fluorouracil inhibits the thymidylate synthase reaction (Figure 26.27).

5. UDP, via conversion to dUDP (Figure 26.25), is ultimately a precursor to dTTP, which is essential to DNA synthesis.

6. See Figure 26.24.

7. Ribose, as ribose-5-P, is released during nucleotide catabolism (as in Figure 26.8). Ribose-5-P is catabolized via the pentose phosphate pathway and glycolysis to form pyruvate, which enters the citric acid cycle. From Figure 22.40, note that 3 ribose-5-P (rearranged to give 1 ribose-5-P and 2 xylulose-5-P) give a net consumption of 2 ATPs and a net production of 8 ATPs and 5 NADH (= 15 ATPs), when converted to 5 pyruvate. If each pyruvate is worth 15 ATP equivalents (as in a prokaryotic cell), the overall yield of ATP from 3 ribose-5-P is 75 ATP + 23 ATP − 2 ATP = 96 ATP. Net yield per ribose-5-P is thus 32 ATP equivalents.

8. Comparing Figures 26.3 and 25.40, note that AICAR (5-aminoimidazole-4-carboxamide ribonucleotide) is a common intermediate in both pathways. It is a product of step 5 of histidine biosynthesis (Figure 25.40) and step 9 of purine biosynthesis (Figure 26.3). Thus, formation of AICAR as a byproduct of histidine biosynthesis from PRPP and ATP bypasses the first nine steps in purine synthesis. However, cells require greater quantities of purine than of histidine, and these nine reactions of purine synthesis are essential in satisfying cellular needs for purines.

9. See the following figure.

STEP 6

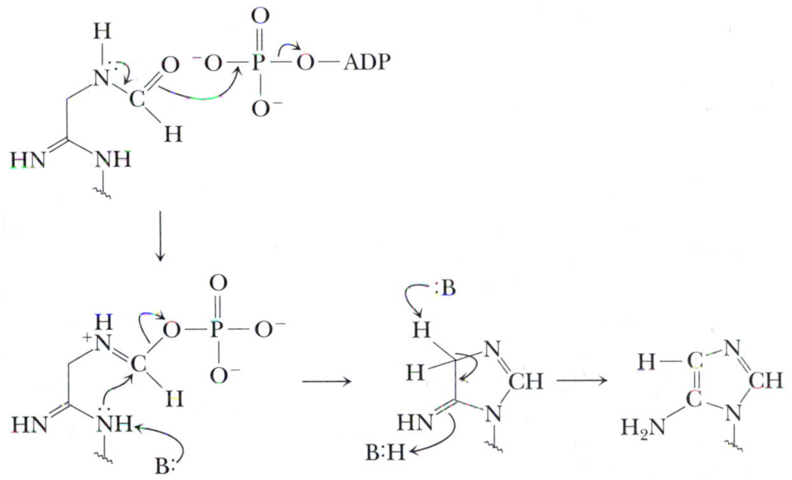

STEP 8

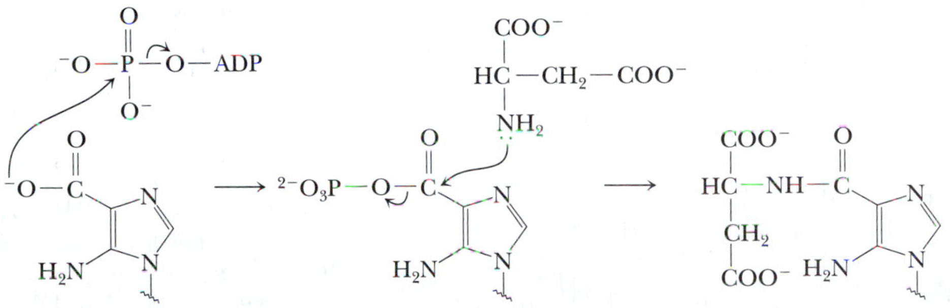

STEP 9

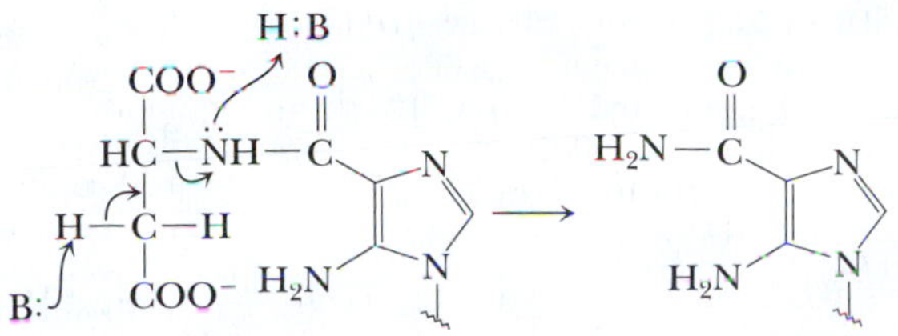

10. Aspartate + GTP^{4-} + $H_2O \longrightarrow$ fumarate + NH_4^+ + GDP^{3-} + P_i^{2-}

11. 1. Uric acid + 2 H_2O + $O_2 \longrightarrow$ *urate oxidase* $\longrightarrow$ allantoin + H_2O_2 + CO_2

    2. Allantoin + $H_2O \longrightarrow$ *allantoinase* $\longrightarrow$ allantoic acid + glyoxylic acid

    3. Allantoic acid + $H_2O \longrightarrow$ *allantoicase* $\longrightarrow$ 2 urea

    4. 2 Urea + 2 $H_2O \longrightarrow$ *urease* $\longrightarrow$ 2 CO_2 + 2 HN_3

12.

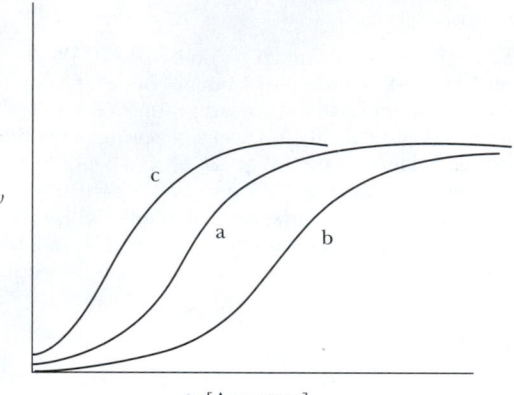

[Aspartate]

13. ATCase (subunit organization $\alpha_6\beta_6$) consists of 6 functional units, each composed of 1 α-subunit and 1 β-subunit, best described as $(\alpha\beta)_6$.

14. From these starting materials to UTP in a eukaryotic cell, where mitochondrial oxidation of dihydroorotate via a coenzyme Q–linked flavoprotein would yield 1.5 ATPs, has a net cost of 2.5 ATPs. From UTP to dTTP would require an ATP at the CTP synthetase step; recovery of an ATP in the nucleoside diphosphate kinase reaction ADP + CTP $\rightarrow$ CDP + ATP; an NADPH (= 4 ATP equivalents) in converting CDP $\rightarrow$ dCDP, recovery of an ATP equivalent in the deoxycytidylate kinase reaction dCDP + ADP $\rightarrow$ dCMP + ATP, deamination of dCMP to dUMP by dCMP deaminase; conversion of dUMP to dTMP by thymidylate synthase and then 2 ATP equivalents to form dTTP. Net: 7.5 ATP equivalents versus 10 ATP for ATP synthesis via the purine biosynthetic pathway.

15. 1. UMP + ATP $\longrightarrow$ UDP + ADP

    2. UDP + NADPH + $H_2O \longrightarrow$ dUDP + $NADP^+$ + OH^-

    3. dUDP + ATP $\longrightarrow$ dUTP + ADP

    4. dUTP + $H_2O \longrightarrow$ dUMP + PP_i

    5. PP_i + $H_2O \longrightarrow$ 2 P_i

    6. dUMP + N^5,N^{10}-methylene-THF $\longrightarrow$ dTMP + THF
    Net: UMP + N^5,N^{10}-methylene-THF + 2 ATP + NADPH + 3 $H_2O \longrightarrow$ dTMP + THF + 2 ADP + 2 P_i + OH^-

16. If GMP levels are high (but GDP or AMP is not), ribose-5-P pyrophosphokinase will not be inhibited, glutamine-PRPP amidotransferase will be roughly 50% inhibited due to GMP binding at the G nucleotide allosteric site, and IMP dehydrogenase will be blocked, so IMP is directed toward AMP synthesis to give a balanced amount of A versus G nucleotides.

17. Carbons 1 and 6.

Chapter 27

1. a. K_{eq} = 360,333.

    b. Assuming [ATP] = [ADP], [pyruvate] must be greater than 360,333 [PEP] for the reaction to proceed in reverse.

    c. K_{eq} = 0.724.

    d. [PEP]/[pyruvate] = 724.

    e. Yes. Both reactions will be favorable as long as [PEP]/[pyruvate] falls between 0.0000028 and 724.

2. Energy charge = 0.945; phosphorylation potential = 1111 M^{-1}.

3. 8 mM ATP will last 1.12 sec. Because the equilibrium constant for creatine-P + ADP $\rightarrow$ Cr + ATP is 175, [ATP] must be less than 1750 [ADP] for the reaction creatine-P + ADP $\rightarrow$ Cr + ATP to proceed to the right when [Cr-P] = 40 mM and [Cr] = 4 mM.

4. From Equation 20.13, $\mathscr{E}(NADP^+/NADPH)$ = −0.350 V; $\mathscr{E}(NAD^+/NADH)$ = −0.281V; thus, $\Delta\mathscr{E}$ = 0.069 V, and ΔG = −13,316 J/mol. If an ATP "costs" 50 kJ/mol, this reaction can produce about 0.27 ATP equivalent at these concentrations of NAD^+, NADH, $NADP^+$, and NADPH.

5. Assume that the K_{eq} for ATP + AMP $\rightarrow$ 2 ADP = 1.2 (see legend to Figure 27.2). When [ATP] = 7.2 mM, [ADP] = 0.737 mM and [AMP] = 0.063 mM. When [ATP] decreases by 10% to 6.48 mM, [ADP] + [AMP] = 1.52 mM and thus [ADP] = 1.30 mM and [AMP] = 0.22 mM. A 10% decrease in [ATP] has resulted in a 0.22/0.063 = 3.5-fold increase in [AMP].

6. J (the flux of F-6-P through the substrate cycle) at low [AMP] = 0.1; J at high [AMP] = 8.9. Therefore, the flux of F-6-P through the cycle has increased 89-fold.

7. Inhibition of acetyl-CoA carboxylase will lower the concentration of malonyl-CoA. Because malonyl-CoA inhibits uptake of fatty acids (see Figure 24.16), decreases in [malonyl-CoA] favor fatty acid synthesis.

8. Ethanol oxidation to acetate yields 2 NADH in the cytosol, each worth 1.5 ATPs. The acetyl thiokinase reaction consumes 2 ATP equivalents in converting acetate to acetyl-CoA (due to ATP $\rightarrow$ AMP + PP_i). Combustion of acetyl-CoA $\rightarrow$ 2 CO_2 in a liver cell yields a net of 10 ATPs. Therefore, the net yield of ATP from ethanol in a liver cell is 11 ATPs. Glucose $\rightarrow$ 6 CO_2 in a liver cell yields 30 ATPs (see Table 20.4) or 4 ATP/C atom. Ethanol $\rightarrow$ 2 CO_2 gives 11 ATPs or 5.5 ATP/C atom.

9. Palmitoyl-CoA $\rightarrow$ 8 acetyl-CoA yields 7 NADH and 7 [$FADH_2$] = 21 + 14 = 35. Eight acetyl-CoA $\rightarrow$ palmitoyl-CoA requires 14 NADPH + 7 ATP = −63 ATP (negative sign denotes ATP consumed). The palmitoyl-CoA $\rightleftharpoons$ 8 acetyl-CoA conversion is favorable in both directions provided the free energy release is more than 1750 kJ/mol in the catabolic (acetyl-CoA forming) direction (the energy necessary to produce 35 ATP equivalents at a cost of 50 kJ/mol each) and less than 3150 kJ/mol in the anabolic (palmitoyl-CoA forming) direction (the energy released from 63 ATP equivalents).

10. Cellular respiration releases 2870 kJ/mol of glucose under standard-state conditions (and about the same amount under cellular conditions). Assuming the cell is a bacterial cell where cellular respiration produces 38 ATPs, the total cost of ATP synthesis is 38(50) = 1900 kJ/mol. Thus, the overall free energy

change $= -2870 + 1900 = -970$ kJ/mol. Because $\Delta G°' = -RT \ln K_{eq}$, $K_{eq} = e^{391.4447} = 10^{168}$. This is a very large number! It will be even larger for a cell where ATP yields per glucose are less.

Carbon dioxide fixation leads to glucose synthesis at the cost of 12 NADPH and 18 ATP = 66 ATP equivalents. At 50 kJ/mol, the energy investment from ATP is -3300 kJ/mol and the value of a glucose is $+2870$ kJ/mol. Thus, $\Delta G°' = -430$ kJ/mol $= -RT \ln K_{eq}$. $K_{eq} = e^{173.52} = 10^{75}$, which is also a very large number!

11. Glycogen phosphorylase, phosphorylase kinase, glycogen synthase, PFK-1, PFK-2, FBPase, and glucose-6-P phosphatase are liver enzymes that act specifically in glycogenolysis or gluconeogenesis; these enzymes are potential targets to control blood glucose levels.

12. From the *Are You Hungry?* box on page 892: The "limit production" list would include NPY, AgRP, and ghrelin; the "raise levels" list would include cholecystokinin, leptin, insulin, melanocortins, and PYY$_{3-36}$.

13. Leptin injection in *ob/ob* mice decreases food intake, raises fatty acid oxidation levels, and lowers body weight. Obese humans are not usually defective in leptin production. Indeed, because leptin is produced in adipocytes, obese individuals may already have high levels of leptin, so leptin injection has limited success.

14. No, although NPY is an orexic agent, it is not necessarily an obesity-promoting hormone. Mice deficient in melanocortin production, melanocortin receptors, or leptin receptors would have an obese phenotype, because all these agents act in appetite-suppressing pathways.

15. Alcohol consumption lowers the NAD$^+$/NADH ratio, which limits glycolysis because glyceraldehyde-3-P dehydrogenase requires NAD$^+$, stimulates gluconeogenesis because NADH favors glyceraldehyde-3-P dehydrogenase working in the glucose synthesis direction, and limits fatty acid oxidation because β-oxidation requires NAD$^+$. NADH is also a negative regulator of several citric acid cycle enzymes.

16. The heart is the principal organ using both glucose and fatty acids in the well-fed state; the brain relies mostly on glucose (although adipose tissue also relies on glucose for glycerol production); muscle relies mostly on fatty acids; the brain never uses fatty acids; and muscle produces lactate, which is converted in the liver into glucose.

17. a. Hexokinase is an R enzyme because it is a glycolytic enzyme. Glutamine:PRPP amidotransferase is not an ATP-dependent enzyme, so it is neither an R or a U enzyme. It responds to the adenine nucleotide pool through feedback inhibition by AMP, ADP, and ATP.

    b. If E.C. = 0.5, hexokinase activity is high and ribose-5-P pyrophosphokinase activity is likely to be low because R-5-P pyrophosphokinase is probably a U enzyme. At E.C. = 0.95, the situation reverses and hexokinase activity is low and R-5-P pyrophosphokinase activity is high.

Chapter 28

1. Because DNA replication is semiconservative, the density of the DNA will be $(1.724 + 1.710)/2 = 1.717$ g/mL. If replication is dispersive, the density of the DNA will also be 1.717 g/mL. To distinguish between these possibilities, heat the dsDNA obtained after one generation in ^{15}N to 100°C to separate the strands; then examine the strands by density gradient ultracentrifugation. If replication is indeed semiconservative, two bands of ssDNA will be observed, the "heavy" one containing ^{15}N atoms and the "light" one containing ^{14}N atoms. If replication is dispersive, only a single band of intermediate density would be observed.

2. The 5'-exonuclease activity of DNA polymerase I removes mispaired segments of DNA sequence that lie in the path of the advancing polymerase. Its biological role is to remove mispaired bases during DNA repair. The 3'-exonuclease activity acts as a proofreader to see whether the base just added by the polymerase activity is properly base-paired with the template. If not (that is, if it is an improper base with respect to the template), the 3'-exonuclease removes it and the polymerase activity can try once more to insert the proper base. An *E. coli* strain lacking DNA polymerase I 3'-exonuclease activity would show a high rate of spontaneous mutation.

3. Assume that the polymerization rate achieved by each half of the DNA polymerase III homodimer is 750 nucleotides per sec. The entire *E. coli* genome consists of 4.64×10^6 bp. At a rate of 750 bp/sec per DNA polymerase III homodimer (one at each replication fork), DNA replication would take almost 3100 sec (51.7 min; 0.86 hr). When *E. coli* is dividing at a rate of once every 20 min, *E. coli* cells must be replicating DNA at the rate of 4.64×10^6 bp per 20 min (2.32×10^5 bp/min or 3867 bp/sec). To achieve this rate of replication would require initiation of DNA replication once every 20 min at *ori*, and a minimum of $3933/1500 = 2.57$ replication bubbles per *E. coli* chromosome, or 5.14 replication forks.

4.

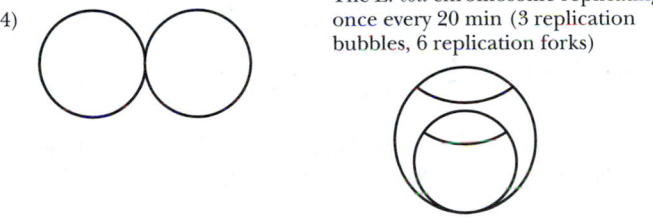

The *E. coli* chromosome replicating once every 20 min (3 replication bubbles, 6 replication forks)

5. If there are only 10 molecules of DNA polymerase III (as DNA polymerase III homodimers) per *E. coli* cell and *E. coli* growing at its maximum rate has about 5 replication forks per chromosome, it seems likely that DNA polymerase III availability is sufficient to sustain growth at this rate.

6. Okazaki fragments are 1000 to 2000 nucleotides in length. Because DNA replication in *E. coli* generates Okazaki fragments whose total length must be 4.64×10^6 nucleotides, a total of 2300 to 4700 Okazaki fragments must be synthesized. Consider in comparison a human cell carrying out DNA replication. The haploid human genome is 3×10^9 bp, but most cells are diploid (6×10^9 bp). Collectively, the Okazaki fragments must total 6×10^9 nucleotides in length, distributed over 3 to 6 million separate fragments.

7. DNA gyrases are ATP-dependent topoisomerases that introduce negative supercoils into DNA. DNA gyrases change the linking number, L, of double-helical DNA by breaking the sugar–phosphate backbone of its DNA strands and then religating them (see Figure 11.22). In contrast, helicases are ATP-dependent enzymes that disrupt the hydrogen bonds between base pairs that hold double-helical DNA together. Helicases move along dsDNA, leaving ssDNA in their wake.

8. A diploid human cell contains 6×10^9 bp of DNA. One origin of replication every 300 kbp gives 20,000 origins of replication in 6×10^9 bp. If DNA replication proceeds at a rate of 100 bp/sec at each of 40,000 replication forks (2 per origin), 6×10^9 bp could be replicated in 1500 sec (25 min). To provide 2 molecules of DNA polymerase per replication fork, a cell would require 80,000 molecules of this enzyme.

9. For purposes of illustration, consider how the heteroduplex bacteriophage chromosome in Figure 28.18 might have arisen:

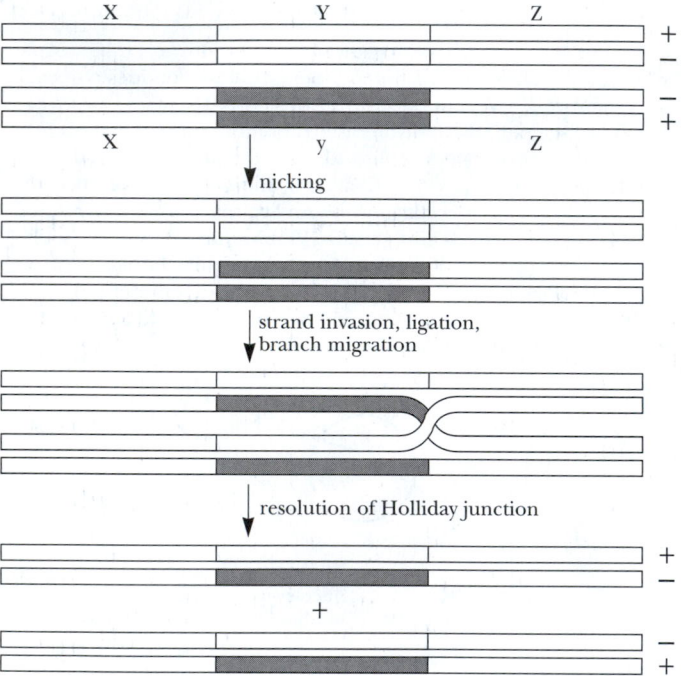

10. The mismatch repair system of *E. coli* relies on DNA methylation to distinguish which DNA strand is "correct" and which is "mismatched"; the unmethylated strand (the newly synthesized one that has not had sufficient time to become methylated is, by definition, the mismatched one). Homologous recombination involves DNA duplexes at similar stages of methylation, neither of which would be interpreted as "mismatched," and thus the mismatch repair system does not act.

11. B-DNA normally has a helical twist of about 10 bp/turn, so the rotation per residue (base pair), $\Delta\phi$, is 36°. If the DNA is unwound to 18.6 bp/turn, $\Delta\phi$ becomes 19.4°. Thus, the change in $\Delta\phi$ is 16.6°.

12. Consider the two DNA duplexes, *WC* and *wc*, respectively, displayed as "ladders" in the following figure. The gap at the right denotes the initial cleavage, in this case of the *W* and *w* strands. Cleavage by resolvase at the arrows labeled 1 and ligation of like strands (*W* with *w* and *C* with *c*) will yield patch

recombinants; cleavage at the arrows labeled 2 and ligation of like strands gives splice recombinants.

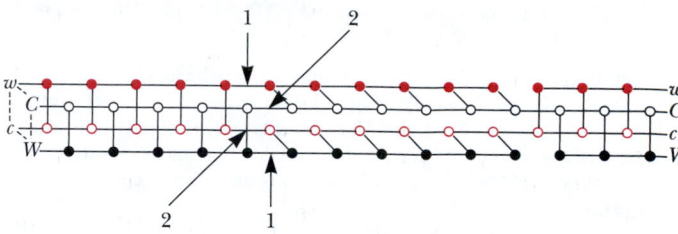

13. The DNA is: GCTA

    CGAT

   a. HNO$_2$ causes deamination of C and A. Deamination of C yields U, which pairs the way T does, giving:

   GTTA and ACTA and (rarely) ATTA

   CAAT   TGAT   TAAT

   Deamination of A yields I, which pairs as G does, giving:

   GCTG and GCCA and (rarely) GCCG

   CGAC   CGGT   CGGC

   b. Bromouracil usually replaces T, and pairs the way C does:

   GCCA and GCTG and (rarely) GCCG

   CGGT   CGAC   CGGC

   Less often, bromouracil replaces C, and mimics T in its base-pairing:

   GTTA and ACAT and (rarely) ATTA

   CAAT   TGTA   TAAT

   c. 2-Aminopurine replaces A and normally base-pairs with T but may also pair with C:

   GCTG and CGGT and (rarely) GCCG

   CGAC   GCCA   CGGC

14. Transposons are mobile genetic elements that can move from place to place in the genome. Insertion of a transposon within or near a gene may disrupt the gene or inactivate its expression.

15. The PrP^{SC} form of the prion protein accumulates in individuals with mutations in the gene for PrP^c and is the cause of inheritable (genetic) forms of Creutzfeldt-Jacob disease. Furthermore, prion-related diseases may be acquired by ingestion of PrP^{SC}, indicating the infectious properties of this form of the protein. Variant CJD (vCJD) is an example of an infectious version of a prion disease.

16. a. Cells exit the cell at G$_1$ and enter G$_0$, because this stage is prior to DNA replication. Once DNA replication takes place, cells move on to mitosis.

   b. Progression through the cell cycle is regulated at checkpoints. Checkpoints, which control whether the cell continues into the next phase, ensure that steps necessary for successful completion of each phase of the cycle are satisfactorily accomplished. If conditions for advancement to the next phase are not met, the cycle is arrested until they are. Checkpoints depend on cyclins and cyclin-dependent protein kinases (CDKs). Cyclins are synthesized at one phase of the cell cycle and degraded at another. CDK interaction with a specific cyclin is essential for the protein kinase activity of the CDK. In turn, these CDKs control events at each phase of the cycle by targeting specific proteins for activation (or inactivation) through phosphorylation. Destruction of the phase-specific cyclin at the end of the phase inactivates the CDK.

17. Recombination in codon 97 suggests that the V_κ gene provides codons 94–96, which are GTT CAT CTT and specify Val-Gln-Leu. The fourth codon is either TTG (Leu), CTG (Leu), or CGG (Arg). CGG (Arg) would be the change. However, the desirability of a change cannot be anticipated.

Chapter 29

1. The first AUG codon is underlined:

   5′-AGAUCCGU<u>AUG</u>GCGAUCUCGACGAAGACUCCUAGG GAAUCC...

   Reading from the AUG codon, the amino acid sequence of the protein is:

   (N-term) Met-Ala-Ile-Ser-Thr-Lys-Thr-Pro-Arg-Glu-Ser-...

2. See Figure 29.2 for a summary of the events in transcription initiation by *E. coli* RNA polymerase. A gene must have a promoter region for proper recognition and transcription by RNA polymerase. The promoter region (where RNA polymerase binds) typically consists of 40 bp to the 5′-side of the transcription start site. The promoter is characterized by two consensus sequence elements: a hexameric TTGACA consensus element in the −35 region and a Pribnow box (consensus sequence TATAAT) in the −10 region (see Figure 29.3).

3. The fact that the initiator site on RNA polymerase for nucleotide binding has a higher K_m for NTP than the elongation site ensures that initiation of mRNA synthesis will not begin unless the available concentration of NTP is sufficient for complete synthesis of an mRNA.

4. First, transcription is compartmentalized within the nucleus of eukaryotes. Furthermore, in contrast to prokaryotes (which have only a single kind of RNA polymerase), eukaryotes possess three distinct RNA polymerases—I, II, and III—acting on three distinct sets of genes (see Section 29.3). The subunit organization of these eukaryotic RNA polymerases is more complex than that of their prokaryotic counterparts. The three sets of genes are recognized by their respective RNA polymerases because they possess distinctive categories of promoters. In turn, all three polymerases interact with their promoters via particular transcription factors that recognize promoter elements within their respective genes. Protein-coding eukaryotic genes, in analogy with prokaryotic genes, have two consensus sequence elements within their promoters, a TATA box (consensus sequence TATAAA) located in the −25 region, and *Inr*, an initiator element that encompasses the transcription start site. In addition, eukaryotic promoters often contain additional short, conserved sequence modules such as enhancers and response elements for appropriate regulation of transcription.

   Transcription termination differs as well as prokaryotes and eukaryotes. In prokaryotes, two transcription basic termination mechanisms occur: *rho* protein-dependent termination and DNA-encoded termination sites. Transcription in eukaryotes tends to be imprecise, occurring downstream from consensus AAUAAA sequences known as poly(A) addition sites. An important aspect of eukaryotic transcription is post-transcriptional processing of mRNA (see Section 29.5).

5. When DNA-binding proteins "read" specific base sequences in DNA, they do so by recognizing a specific matrix of H-bond donors and acceptors displayed within the major groove of B-DNA by the edges of the bases. When DNA-binding proteins recognize a specific DNA region by "indirect readout," the protein is discerning local conformational variations in the cylindrical surface of the DNA double helix. These conformational variations are a consequence of unique sequence information contributed by the base pairs that make up this region of the DNA.

6. The metallothionine promoter is about 265 bp long. Because each bp of B-DNA contributes 0.34 nm to the length of DNA, this promoter is about 90 nm long, covering 26.5 turns of B-DNA (10 bp/turn). Each nucleosome spans about 146 bp, so about 2 nucleosomes would be bound to this promoter.

7. If cells contain a variety of proteins that share a leucine zipper dimerization motif but differ in their DNA recognition helices, heterodimeric *bZIP* proteins can form that will contain two different DNA contact basic regions (consider this possibility in light of Figures 29.33 and 29.34). These heterodimers are no longer restricted to binding sites on DNA that are dyad-symmetric; the repertoire of genes with which they can interact will thus be dramatically increased.

8. Exon 17 of the fast skeletal muscle troponin T gene is one of the mutually exclusive exons (see Figure 29.44). Deletion of this exon would cut by half the alternative splicing possibilities, so only 32 different mature mRNAs would be available from this gene. If exon 7, a combinatorial exon, were duplicated, it would likely occur as a tandem duplication. The different mature mRNAs could contain 0, 1, or 2 copies of exon 7. The 32 combinatorial possibilities shown in the center of Figure 29.44 represent those with 0 or 1 copy of exon 7. Those with 2 copies are 16 in number: 456778, 56778, 46778, 6778, 45778, 4778, 5778, 778, 4577, 577, 5677, 45677, 4677, 477, 677, and 77. Given the mutually exclusive exons 17 and 18, 96 different mature mRNAs would be possible.

9. Modifications include acetylation and methylation of Lys residues, thus suppressing their positive charge: methylation of Arg residues, again suppressing the positive charge, and phosphorylation of Ser residues, thus introducing negative charges. The positively charged Lys and Arg residues interact electrostatically with phosphate groups on the DNA backbone, so these interactions would be diminished through acetylation and methylation. Phosphorylation of Ser residues would lead to electrostatic repulsion between the DNA backbone and the histone tails. Collectively, these modifications weaken histone : DNA interactions.

10. Because the *lac* repressor is tetrameric and each subunit has an inducer-binding site, it seems likely that inducer binding is cooperative. Cooperative binding of inducer would be advantageous, because the slope of the *lac* operon expression versus [inducer] would be steeper over a narrower [inducer] range.

11. Capping of the 5′-end and polyadenylation of the 3′-end of mRNA protect the RNA from both 5′- and 3′-exonucleolytic degradation. Methylations at the 5′-end enhance the interaction between cap-binding protein eIF-4E (see Figure 30.28), thus increasing the likelihood that the mRNA will be translated. Note also in Figure 30.28 that polyadenylylation provides a protein-binding site for proper assembly of the 40S initiation complex essential to translation.

12. In the A Deeper Look box on page 908, the argument is made that polymerases have a common enzymatic mechanism for nucleotide addition to a growing polynucleotide chain based on

two metal ions coordinated to the incoming nucleotide. These metal ions interact with two aspartate residues that are highly conserved in both DNA and RNA polymerases. The discovery of a second Mg^{2+} ion in the RNA polymerase II active site confirms to the universality of this model.

13. 50 bp of DNA is $(50)(0.34 \text{ nm/bp}) = 17$ nm. Because 146 bp of DNA make 1.65 turns around the nucleosome (see Chapter 11) = 88.5 bp/turn of DNA, 50 bp = 0.565 turns, or slightly more than $\frac{1}{2}$ turn of DNA around the histone core octamer. Promoter and response modules are about 20 to 25 bp.

14.

15. α-Helices composed of 7 to 8 residues would have a length of 1.05 to 1.2 nm. The overall diameter of B-DNA is about 2.4 nm, but the diameter at the bottom of major groove is significantly less (as a reference, the C_1'–C_1' distance between base-paired nucleotides in separate chains is 1.1 nm). A distance along the B-DNA helix axis of 1.1 nm would correspond to about 3 base pairs, so if such an α-helix were laid along the major groove, it could not make contacts with more than 3 base pairs (see also Figures 29.30 and 29.34).

16. A protein such as CAP that interacts with phosphate groups in the DNA backbone does so via electrostatic interactions based on positively charged Arg and Lys side chains.

17. Chromatin is decompacted by HATs and chromatin remodeling complexes. Histone acetylation by HATs disrupts chromatin structure (see problem 9) and chromatin remodeling complexes then mediate ATP-dependent conformational changes in the chromatin that peels about 50 bp of DNA from the core octamer, exposing the DNA for access by the transcriptional machinery.

Chapter 30

1. The cDNA sequence as presented:

    CAATACGAAGCAATCCCGCGACTAGACCTTAAC...

represents six potential reading frames, the three inherent in the sequence as written and the three implicit in the complementary DNA strand, which is written $5' \rightarrow 3'$, is:

...GTTAAGGTCTAGTCGCGGGATTGCTTCGTATTG

Two of the three reading frames of the cDNA sequence given contain stop codons. The third reading frame is a so-called open reading frame (a stretch of coding sequence devoid of stop codons):

<u>CAA</u>TAC<u>GAA</u>GCA<u>ATC</u>CCG<u>CGA</u>CTA<u>GAC</u>CTT<u>AAC</u>...
(alternate codons underlined)

The amino acid sequence it encodes is:

Gln-Tyr-Glu-Ala-Ile-Pro-Arg-Leu-Asp-Leu-Asn....

Two of the three reading frames of the complementary DNA sequence also contain stop codons. The third may be an open reading frame:

...<u>GTT</u>AAG<u>GTC</u>TAG<u>TCG</u>CGG<u>GAT</u>TGC<u>TTC</u>GTA<u>TTG</u>
(alternate codons underlined)

An unambiguous conclusion about the partial amino acid sequence of this cDNA cannot be reached.

2. A random (AG) copolymer would contain varying amounts of the following codons:

AAA AAG AGA GAA AGG GAG GGA GGG

codons for Lys Lys Arg Glu Arg Glu Gly Gly, respectively.

Therefore, the random (AG) copolymer would direct the synthesis of a polypeptide consisting of Lys, Arg, Glu, and Gly. The relative frequencies of the various codons are a function of the probability that a base will occur in a codon. For example, if the A/G ratio is 5/1, the ratio of AAA/AAG is $(5 \times 5 \times 5)/(5 \times 5 \times 1) = 5/1$. If the 3A codon is assigned a value of 100, then the 2A1G codon has a frequency of 20. From this analysis, the relative abundances of these amino acids in the polypeptide should be:

Lys = 120; Arg = 24; Glu = 24; Gly = 4.8
(Normalized to Lys = 100: Arg = 20; Glu = 20; Gly = 4)

3. Review the information in Section 30.2, noting in particular the two levels of specificity exhibited by aminoacyl-tRNA synthetases (1: at the level at ATP $\rightleftharpoons$ PP$_i$ exchange and aminoacyl adenylate synthesis in the presence of amino acid and the absence of tRNA; and 2: at the level of loading the aminoacyl group on an acceptor tRNA).

4. Base pairs are drawn such that the B-DNA major groove is at the top (see following figure).

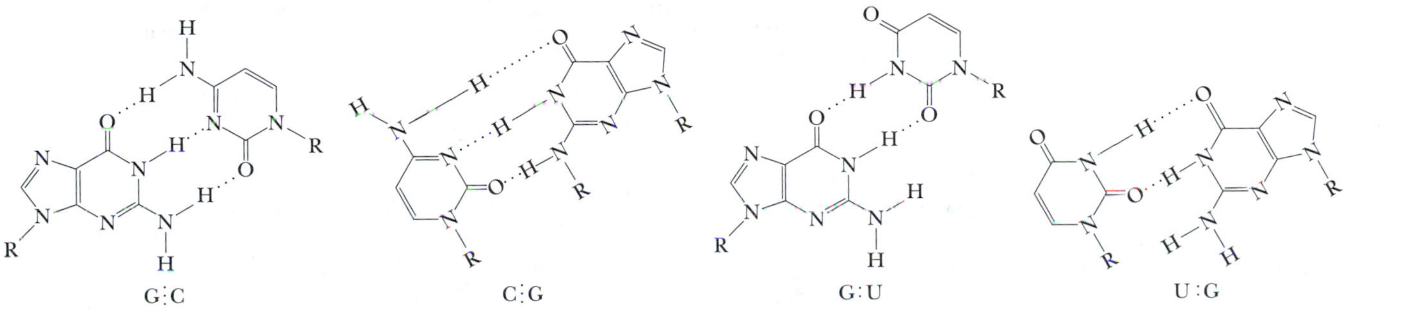

G:C   C:G   G:U   U:G

5. The wobble rules state that a first-base anticodon U can recognize either an A or a G in the codon third-base position; first-base anticodon G can recognize either U or C in the codon third-base position; and first-base anticodon I can recognize either U, C, or A in the codon third-base position. Thus, codons with third-base A or G, which are degenerate for a particular amino acid (Table 30.3 reveals that all codons with third-base purines are degenerate, except those for Met and Trp), could be served by single tRNA species with first-base anticodon U. More emphatically, codons with third-base pyrimidines (C or U) are always degenerate and could be served by single tRNA species with first-base anticodon G. Wobble involving first-base anticodon I further minimizes the number of tRNAs needed to translate the 61 sense codons. Wobble tends to accelerate the rate of translation because the noncanonical base pairs formed between bases in the third position of codons and bases occupying the first-base wobble position of anticodons are less stable. As a consequence, the codon:anticodon interaction is more transient.

6. The stop codons are UAA, UAG, and UGA.

Sense codons that are a single-base change from UAA include CAA (Pro), AAA (Lys), GAA (Glu), UUA (Leu), UCA (Ser), UAU (Tyr), and UAC (Tyr).

Sense codons that are a single-base change from UAG include CAG (Gln), AAG (Lys), GAG (Glu), UCG (Ser), UUG (Leu), UGG (Trp), UAU (Tyr), and UAC (Tyr).

Sense codons that are a single-base change from UGA include CGA (Arg), AGA (Arg), GGA (Gly), UUA (Leu), UCA (Ser), UGU (Cys), UGC (Cys), and UGG (Trp).

That is, 19 of the 61 sense codons are just a single base change from a nonsense codon.

7. The list of amino acids in problem 6 is a good place to start in considering the answer to this question. Amino acid codons in which the codon base (the wobble position) is but a single base change from a nonsense codon are the more likely among this list, because pairing is less stringent at this position. These include UAU (Tyr) and UAC (Tyr) for nonsense codons UAA and UAG, and UGU (Cys), UGC (Cys), and UGG (Trp) for nonsense codon UGA.

8. The more obvious answer to this question is that eukaryotic ribosomes are larger, more complex, and hence slower than prokaryotic ribosomes. In addition, initiation of translation requires a greater number of initiation factors in eukaryotes than in prokaryotes. It is also worth noting that eukaryotic cells, in contrast to prokaryotic cells, are typically under less selective pressure to multiply rapidly.

9. Each amino acid of a protein in an extended β-sheet–like conformation contributes about 0.35 nm to its length. 10 nm ÷ 0.35 nm per residue = 28.6 amino acids.

10. Larger, more complex ribosomes offer greater advantages in terms of their potential to respond to the input of regulatory influences, to have greater accuracy in translation, and to enter into interactions with subcellular structures. Their larger size may slow the rate of translation, which in some instances may be a disadvantage.

11. The universal organization of ribosomes as two-subunit structures in all cells—archaea, eubacteria, and eukaryotes—suggests that such an organization is fundamental to ribosome function. Translocation along mRNA, aminoacyl-tRNA binding, peptidyl transfer, and deacylated-tRNA release are processes that require repetitious uncoupling of physical interactions between the large and small ribosomal subunits.

12. Prokaryotic cells rely on N-formyl-Met-tRNA$_i^{fMet}$ to initiate protein synthesis. The tRNA$_i^{fMet}$ molecule has a number of distinctive features, not found in noninitiator tRNAs, that earmark it for its role in translation initiation (see Figure 30.15). Furthermore, N-formyl-Met-tRNA$_i^{fMet}$ interacts only with initiator codons (AUG or, less commonly, GUG). A second tRNA^{Met}, designated tRNA$_m^{Met}$, serves to deliver methionyl residues as directed by internal AUG codons. Both tRNA$_i^{fMet}$ and tRNA$_m^{Met}$ are loaded with methionine by the same methionyl-tRNA synthetase, and AUG is the Met codon, both in initiation and elongation. AUG initiation codons are distinctive in that they are situated about 10 nucleotides downstream from the Shine–Dalgarno sequence at the 5′-end of mRNAs; this sequence determines the translation start site (see Figure 30.17). Eukaryotic cells also have two tRNA^{Met} species, one of which is a unique tRNA$_i^{Met}$ that functions only in translation initiation. Eukaryotic mRNAs lack a counterpart to the

prokaryotic Shine–Dalgarno sequence; apparently the eukaryotic small ribosomal subunit binds to the 5′-end of a eukaryotic mRNA and scans along it until it encounters an AUG codon. This first AUG codon defines the eukaryotic translation start site.

13. The Shine–Dalgarno sequence is a purine-rich sequence element near the 5′-end of prokaryotic mRNAs (see Figure 30.17). It base-pairs with a complementary pyrimidine-rich region near the 3′-end of 16S rRNA, the rRNA component of the prokaryotic 30S ribosomal subunit. Base pairing between the Shine–Dalgarno sequence and 16S rRNA brings the translation start site of the mRNA into the P site on the prokaryotic ribosome. Because the nucleotide sequence of the Shine–Dalgarno element varies somewhat from mRNA to mRNA, whereas the pyrimidine-rich Shine–Dalgarno-binding sequence of 16S rRNA is invariant, different mRNAs vary in their affinity for binding to 30S ribosomal subunits. Those that bind with highest affinity are more likely to be translated.

14. The most apt account is b. Polypeptide chains typically contain hundreds of amino acid residues. Such chains, attached as a peptidyl group to a tRNA in the P site, would show significant inertia to movement, compared with an aminoacyl-tRNA in the A site.

15. Elongation factors EF-Tu and EF-Ts interact in a manner analogous to the GTP-binding G proteins of signal transduction pathways (see p. 491 and Figure 15.10). EF-Tu binds GTP and in the GTP-bound form delivers an aminoacyl-tRNA to the ribosome, whereupon the GTP is hydrolyzed to yield EF-Tu:GDP and P_i. EF-Ts mediates an exchange of the bound GDP on EF-Tu:GDP with free GTP, regenerating EF-Tu:GTP for another cycle of aminoacyl-tRNA delivery. The α-subunit of the heterotrimeric GTP-binding G proteins also binds GTP. G_α has an intrinsic GTPase activity and the GTP is eventually hydrolyzed to form G_α:GDP and P_i. Guanine nucleotide exchange factors facilitate the exchange of bound GDP on G_α for GTP, in analogy with EF-Ts for EF-Tu. The amino acid sequences of EF-Tu and G proteins reveal that they share a common ancestry.

16. The *lac*I gene shows the most matches (7) complementary to the Shine–Dalgarno sequence and thus would interact best with it.

17. Chloramphenicol consists of an aromatic ring and several polar functions (a nitro group, two —OH groups, and an amine-N). The aromatic ring might intercalate between bases in the 23S rRNA peptidyl transferase site, the polar functions might form H bonds with nitrogenous bases or sugar —OH groups in the rRNA. The —NO_2 group is also rather bulky in addition to being polar, so it could disrupt peptidyl transferase in a number of ways.

Chapter 31

1. Human rhodanese has 296 amino acid residues. Its synthesis would require the involvement of 4 ATP equivalents per residue or 1184 ATP equivalents. Folding of rhodanese by the Hsp60 α_{14} complex consumes another 130 ATP equivalents. The total number of ATP equivalents expended in the synthesis and folding of rhodanese is approximately 1314.

2. Because a single proteolytic nick in a protein can doom it to total degradation, nicked proteins are clearly not tolerated by cells and are quickly degraded. Cells are virtually devoid of partially degraded protein fragments, which would be the obvious intermediates in protein degradation. The absence of such intermediates and the rapid disappearance of nicked proteins from cells indicate that protein degradation is a rigorously selective, efficient cellular process.

3. *Hsp70:* Hsp70 binds to exposed hydrophobic regions of unfolded proteins. The Hsp70 domain involved in this binding is 18 kD in size and therefore would have a diameter of roughly 3 nm (see problem 5 for the math). Assuming the hydrophobic binding site of Hsp70 stretches across its diameter, it would interact with about 3 nm/0.35 nm = 8 or 9 amino acid residues. Multiple Hsp70 monomers could interact with longer hydrophobic stretches. So, for Hsp70 to interact with a polypeptide chain does not necessarily depend on the protein's size, but it does depend on its hydrophobicity (and absence of charged groups).

   Hsp70 and Hsp60 chaperonins: Proteins that interact with both of these classes of chaperones not only must fit the description for Hsp70 targets but must also be small enough to access the Hsp60 chaperonin chamber, which has a diameter of about 5 nm. This restriction means that Hsp60 cannot interact with proteins more than roughly 50 kD in mass (see problem 6).

4. Many eukaryotic proteins have a multidomain or modular organization, where each module is composed of a contiguous sequence of amino acid residues that folds independently into a discrete domain of structure (see Figure 6.37 for examples of this type of sequence and structure organization). Such proteins would be ideal for co-translational folding: As each newly synthesized contiguous sequence emerges from the ribosome tunnel, it begins folding into its characteristic domain structure. The final, fully folded state of the complete protein would be achieved when the various domains assumed their proper spatial relationships to one another through hinge motions occurring at intradomain regions of the protein.

5. The maximal diameter for a spherical protein would be 5 nm (radius = 2.5 nm). The volume of this protein is given by $V = \frac{4}{3}\pi r^3 = 4/3(3.14)(2.5 \times 10^{-9}\,m)^3 = 65.4 \times 10^{-21}\,mL$. If its density is 1.25 g/mL, its mass would be $(1.25\,g)(65.4 \times 10^{-21}\,mL) = 81.8 \times 10^{-21}$ g/molecule, so its molecular weight = $(6.023 \times 10^{23})(81.8 \times 10^{-21}\,g) = 492.5 \times 10^2$ g/mol = 49,250 g/mol, or about 50 kD.

6. There are 649 different phosphorylated forms for a protein having 7 separate phosphorylation sites.

7. GrpE catalyzes nucleotide exchange on DnaK, replacing ADP with ATP, which converts DnaK back to a conformational form having low affinity for its polypeptide substrate. This change leads to release of bound polypeptide, giving it the opportunity to fold, which is an important step in DnaK function. EF-Ts catalyzes nucleotide exchange on EF-Tu, converting it to the conformational form competent in aminoacyl-tRNA binding. Binding of the aminoacyl-tRNA:EF-Tu complex to the ribosome A site triggers the GTPase activity of EF-Tu as codon:anticodon recognition takes place and the aminoacyl-tRNA:EF-Tu complex conformationally adjusts to the A site. DnaJ delivers an unfolded polypeptide chain to DnaK, and its interaction with DnaK also triggers the ATPase activity of DnaK, whereupon the DnaK:ADP complex forms a stable complex with the unfolded polypeptide. In both instances, hydrolysis of bound nucleoside triphosphate is triggered and leads to conformational changes that stabilize the respective complexes.

8. Protein targeting information resides in more generalized features of the leader sequences such as charge distribution, relative polarity, and secondary structure, rather than amino acid sequence per se. Proteins destined for secretion have N-terminal amino acid sequence with one or more basic amino acids followed by a run of 6 to 12 hydrophobic amino acids. The sequence MRSLLILVLCFLPAALGK... has a basic residue (Arg) at position 2 and a run of 12 nonpolar residues (...LLILVLPLAALG...); thus, it appears to be a signal sequence

for a secretory protein. Cleavage by the signal peptidase would occur between the G and K.

9. Translocation proceeds until a stop transfer signal associated with a hydrophobic transmembrane protein segment is recognized. The stop-transfer signal induces a pause in translocation, and the translocon changes its conformation such that the wall of this closed cylindrical structure either opens or exposes a hydrophobic path, allowing the transmembrane segment to diffuse laterally into the hydrophobic phase of the membrane.

10. The maximal diameter for a spherical protein would be 6 nm (radius = 3 nm). The volume of this protein is given by $V = \frac{4}{3}\pi r^3 = 4/3(3.14)(3 \times 10^{-9}\,\text{m})^3 = 113 \times 10^{-21}\,\text{mL}$. If its density is 1.25 g/mL, its mass would be $(1.25\,\text{g})(113 \times 10^{-21}\,\text{mL}) = 141 \times 10^{-21}\,\text{g/molecule}$, so its molecular weight $= (6.023 \times 10^{23})(141 \times 10^{-21}\,\text{g}) = 851 \times 10^2\,\text{g/mol} = 85{,}100\,\text{g/mol}$, or about 85 kD.

11. The thickness of a phospholipid bilayer is 5 nm, so the overall channel formed by a 50S ribosome and the translocon channel is 15 nm; 15 nm ÷ 0.35 nm per amino acid = about 43 amino acid residues.

12.

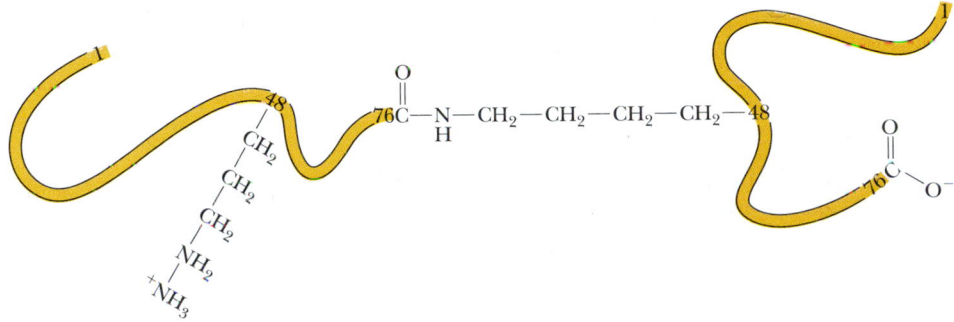

13. E₃ ubiquitin protein ligase selects proteins by the nature of the N-terminal amino acid. Proteins with Met, Ser, Ala, Thr, Val, Gly, or Cys at the amino terminus are resistant to its action. Proteins having Arg, Lys, His, Phe, Tyr, Trp, Leu, Ile, Asn, Gln, Asp, or Glu at their N-terminus are susceptible. N-terminal Pro residues lack a free α-amino group, so such proteins are not susceptible.

14. The cell cycle relies on the cyclic synthesis and destruction of cell cycle regulatory proteins. Lactacystin inhibits cell cycle progression by interfering with programmed destruction of proteins by proteasomes.

15. The temperature-induced switch could be based on a temperature-dependent conformational change in the HtrA protein that brings the His-Ser-Asp catalytic triad into the proper spatial relationship for protease function.

16. It depends on which amino acid is penultimate. If Arg, Lys, His, Phe, Tyr, Trp, Leu, Ile, Asn, Gln, Asp, or Glu follow immediately after Met, the protein will have a short half-life after Met removal. If Ser, Ala, Thr, Val, Gly, Cys, Pro, or another Met follow, the protein should have a long half-life after N-terminal Met removal.

17. To array positively charged residues on one side of an α-helix and hydrophobic residues on the other requires a pattern with positively charged residues positioned every 3.6 residues and similarly for hydrophobic residues. A sequence with Arg or Lys at positions 1, 4, 7, 11, 14, and 18 and an array of hydrophobic residues at 2, 3, 5, 6, 8, 9, 10, 12, 13, 15, 16, 17, 19, and 20 would fit this pattern.

Chapter 32

1. Polypeptide hormones constitute a larger and structurally more diverse group of hormones than either the steroid or amino acid–derived hormones, and it thus might be concluded that the specificity of polypeptide hormone-receptor interactions, at least in certain cases, should be extremely high. The steroid hormones may act either by binding to receptors in the plasma membrane or by entering the cell and acting directly with proteins controlling gene expression, whereas polypeptides and amino acid–derived hormones act exclusively at the membrane surface. Amino acid–derived hormones can be rapidly interconverted in enzyme-catalyzed reactions that provide rapid responses to changing environmental stresses and conditions. See *The Student Solutions Manual, Study Guide and Problems Book* for additional information.

2. The cyclic nucleotides are highly specific in their action, because cyclic nucleotides play no metabolic roles in animals. Ca²⁺ ion has an advantage over many second messengers because it can be very rapidly "produced" by simple diffusion processes, with no enzymatic activity required. IP₃ and DAG, both released by the metabolism of phosphatidylinositol, form a novel pair of effectors that can act either separately or synergistically to produce a variety of physiological effects. DAG and phosphatidic acid share the unique property that they can be prepared from several different lipid precursors. Nitric oxide, a gaseous second messenger, requires no transport or translocation mechanisms and can diffuse rapidly to its target sites.

3. Nitric oxide functions primarily by binding to the heme prosthetic group of soluble guanylyl cyclase, activating the enzyme. An agent that could bind in place of NO—but that does not activate guanylyl cyclase—could reverse the physiological effects of nitric oxide. Interestingly, carbon monoxide, which has long been known to bind effectively to heme groups, appears to function in this way. Solomon Snyder and his colleagues at Johns Hopkins University have shown that administration of CO to cells that have been stimulated with nitric oxide causes attenuation of the NO-induced effects.

4. Herbimycin, whose structure is shown in the following figure, reverses the transformation of cells by Rous sarcoma virus, presumably as a direct result of its inactivation of the viral tyrosine kinase. The manifestations of transformation (on rat kidney cells, for example) include rounded cell morphology, increased glucose uptake and glycolytic activity, and the ability to grow without being anchored to a physical support (termed *anchorage-independent growth*). Herbimycin reverses all these phenotypic changes. On the basis of these observations, one might predict that herbimycin might also inactivate tyrosine

kinases that bear homology to the viral pp60^{v-src} tyrosine kinase. This inactivation has in fact been observed, and herbimycin is used as a diagnostic tool for implicating tyrosine kinases in cell-signaling pathways.

5. The identification of phosphorylated tyrosine residues on cellular proteins is difficult. Quantities of phosphorylated proteins are generally extremely small, and to distinguish tyrosine phosphorylation from serine/threonine phosphorylation is tedious and laborious. On the other hand, monoclonal antibodies that recognize phosphotyrosine groups on protein provide a sensitive means of detecting and characterizing proteins with phosphorylated tyrosines, using, for example, Western blot methodology.

6. Hormones act at extremely low concentrations, but many of the metabolic consequences of hormonal activation (release of cyclic nucleotides, Ca^{2+} ions, DAG, etc., and the subsequent alterations of metabolic pathways) occur at and involve higher concentrations of the affected molecular species. As we have seen in this chapter, most of the known hormone receptors mediate hormonal signals by activating enzymes (adenylyl cyclase, phospholipases, protein kinases, and phosphatases). One activated enzyme can produce many thousands of product molecules before it is inactivated by cellular regulation pathways.

7. If side 1 is denoted as the side with 330 mM Na$^+$ and side 2 as the side with 70 mM Na$^+$, then the equilibrium potential $\Delta\psi = \psi_2 - \psi_1 = 40$ mV. Note that in any real situation there would have to be counteranions available, but this calculation considers only the sodium concentrations on either side of the membrane.

8. Use the permeability coefficients from Figure 32.47 and apply the Goldman equation to the concentrations as given. The calculated potential difference is -21 mV.

9. Vesicles with an outside diameter of 40 nm have an inside diameter of approximately 36 nm and an inside radius of 18 nm. The data correspond to a volume of 2.44×10^{-20} L. Then 10,000 molecules/6.02×10^{23} molecules/mole = 1.66×10^{-20} mole. The concentration of acetylcholine in the vesicle is thus $1.66/2.44$ M or 0.68 M.

10. Decamethonium and succinylcholine are acetylcholine analogs that bind to and activate the acetylcholine receptor. However, acetylcholinesterase shows no activity toward decamethonium, and succinylcholine is only slowly hydrolyzed by this enzyme. Thus, the effects of decamethonium are longer lasting than those of succinylcholine. Succinylcholine is used as a muscle relaxant in certain surgeries, because it blocks the transmission of nerve impulses to muscles. It is slowly hydrolyzed by acetylcholinesterase and also by other cholinesterases in liver and in the blood, so its effects diminish soon after drug treatment is stopped.

11. The evidence outlined in this problem points to a role for cAMP in fusion of synaptic vesicles with the presynaptic membrane and the release of neurotransmitters. GTPγS may activate an inhibitory G protein, releasing G$_{\alpha i}$(GTPγS), which inhibits adenylyl cyclase and prevents the formation of requisite cAMP.

12. The formation and breakdown of cAMP:

13. The point at which no ion flow occurs is the point of equilibrium that balances the chemical and electrical forces across a membrane. The Nernst equation is obtained by setting ΔG to zero in Equation 9.2. Rearrangement yields

$$RT \ln (C_2/C_1) = -Z\mathcal{F}\Delta\psi$$

Using the data in Figure 32.32, one can use this equation to yield an equilibrium potential for K$^+$ of -77 mV and an equilibrium potential for Na$^+$ of $+53.4$ mV.

14. Using the Goldman equation, one can calculate $\Delta\psi = -60$ mV, in agreement with values measured experimentally in neurons.

15. The resting potential in neurons is -60 mV. Local depolarization of the membrane causes the potential to rise approximately 20 mV to about -40 mV. This change causes the Na$^+$ channels to open and Na$^+$ ions begin to flow into the cell. This causes the potential to continue increasing to a value of about $+30$ mV. Because this value is close to the equilibrium potential for Na$^+$, the Na$^+$ channels begin to close and K$^+$ channels open. K$^+$ rushes out of the neuron, returning the membrane potential to a very negative value. There is a slight overshoot of the potential, and the K$^+$ channels close and the resting membrane potential is restored.

16. Malathion has been used for years in the eradication of boll weevils, but has not presented a serious toxicity problem in this program. Apparently, malathion is very toxic to insects, and relatively less toxic to humans, particularly in the low volume applications used for this purpose.

17. This exercise is left to the student, for obvious reasons.

B = box; D = definition; F = figure;
G = molecular graphic; S = structure;
T = table.

Common Abbreviations Used by Biochemists

A	adenine or the amino acid alanine
Ab	antibody
Ag	antigen
Ac-CoA	acetyl-coenzyme A
ACh	acetylcholine
ACP	acyl carrier protein
ADH	alcohol dehydrogenase
ADP	adenosine diphosphate
AIDS	acquired immunodeficiency syndrome
AMP	adenosine monophosphate
ALA	δ-aminolevulinic acid
ATCase	aspartate transcarbamoylase
atm	atmosphere
ATP	adenosine triphosphate
BChl	bacteriochlorophyll
bp	base pair
BPG	bisphosphoglycerate
BPheo	bacteriopheophytin
C	cytosine or the amino acid cysteine
cal	calorie
CaM	calmodulin
cAMP	cyclic $3',5'$-adenosine monophosphate
CAP	catabolite activator protein
cDNA	complementary DNA
CDP	cytidine diphosphate
CDR	complementarity-determining region
Chl	chlorophyll
CM	carboxymethyl
CMP	cytidine monophosphate
CoA or CoASH	coenzyme A
CoQ	coenzyme Q
cpm	counts per minute
CTP	cytidine triphosphate
D	dalton or the amino acid aspartate
d	deoxy
dd	dideoxy
DAG	diacylglycerol
DEAE	diethylaminoethyl
DHAP	dihydroxyacetone phosphate
DHF	dihydrofolate
DHFR	dihydrofolate reductase
DNP	dinitrophenol
Dopa	dihydroxyphenylalanine
DNA	deoxyribonucleic acid
$\mathscr{E}$	reduction potential
E4P	erythrose-4-phosphate
EF	elongation factor
EGF	epidermal growth factor
EPR	electron paramagnetic resonance
ER	endoplasmic reticulum
$\mathscr{F}$	Faraday's constant
F_{AB}	antibody molecule fragment that binds antigen
FAD	flavin adenine dinucleotide
$FADH_2$	reduced flavin adenine dinucleotide
FBP	fructose-1,6-bisphosphate
FBPase	fructose bisphosphatase
Fd	ferredoxin
fMet	N-formyl-methionine
FMN	flavin mononucleotide
F-1-P	fructose-1-phosphate

F-6-P	fructose-6-phosphate
G	guanine or Gibbs free energy or the amino acid glycine
GABA	γ-aminobutyric acid
Gal	galactose
GDP	guanosine diphosphate
GLC	gas-liquid chromatography
Glc	glucose
GMP	guanosine monophosphate
G-1-P	glucose-1-phosphate
G-3-P	glyceraldehyde-3-phosphate
G-6-P	glucose-6-phosphate
G6PD	glucose-6-phosphate dehydrogenase
GS	glutamine synthetase
GSH	glutathione (reduced glutathione)
GSSG	glutathione disulfide (oxidized glutathione)
GTP	guanosine triphosphate
h	hour
h	Planck's constant
Hb	hemoglobin
HDL	high-density lipoprotein
HGPRT	hypoxanthine-guanine phosphoribosyltransferase
HIV	human immunodeficiency virus
HMG-CoA	hydroxymethylglutaryl-coenzyme A
hnRNA	heterogeneous nuclear RNA
HPLC	high-pressure (or high-performance) liquid chromatography
HX	hypoxanthine
Hyl	hydroxylysine
Hyp	hydroxyproline
I	inosine or the amino acid isoleucine
IDL	intermediate density lipoprotein
IF	initiation factor
IgG	immunoglobulin G
IMP	inosine monophosphate
IP_3	inositol-1, 4, 5-trisphosphate
IPTG	isopropylthiogalactoside
IR	infrared
ITP	inosine triphosphate
J	joule
k	kilo (10^3)
K_m	Michaelis constant
kb	kilobases
kD	kilodaltons
L	liter
LDH	lactate dehydrogenase
LDL	low density lipoprotein
Lys	lysine
M	molar
m	milli (10^{-3})
mL	milliliter
mm	millimeter
Man	mannose
Mb	myoglobin
mol	mole
mRNA	messenger RNA
mV	millivolt
N	Avogadro's number
n	nano (10^{-9})
NAD^+	nicotinamide adenine dinucleotide